A Dictionary of Real Numbers

Jonathan Borwein
Dalhousie University

Peter Borwein
Dalhousie University

Wadsworth & Brooks/Cole Advanced Books & Software
Pacific Grove, California

Acknowledgments

The authors would like to thank Barbara Taylor
for her substantial contribution to the production of this volume.
All computations were made to high precision in MAPLE™ with
truncations then directly used to produce TEX output.

Brooks/Cole Publishing Company
A Division of Wadsworth, Inc.

Printed in the United States of America
10 9 8 7 6 5 4 3 2 1

Library of Congress Cataloging-in-Publication Data
Borwein, Jonathan M.
A dictionary of real numbers.

1. Numbers, Real—Tables. I. Borwein, Peter B. II.Title.
QA47.B625 1989 512′.7 89-22267 ISBN 0-534-12840-8

Sponsoring Editor: *John Kimmel*
Marketing Representative: *Gail Garber*
Editorial Assistant: *Mary Ann Zuzow*
Production Editor: *Linda Loba*
Permissions Editor: *Carline Haga*
Cover Design: *Vernon T. Boes*
Cover Printing: *Phoenix Color Corp.*
Printing and Binding: *BookCrafters*

Preface

How do we recognize that the number .93371663... is actually $2\log_{10}(e+\pi)/2$? Gauss observed that the number 1.85407467... is (essentially) a rational value of an elliptic integral—an observation that was critical in the development of nineteenth century analysis. How do we decide that such a number is actually a special value of a familiar function without the tools Gauss had at his disposal, which were, presumably, phenomenal insight and a prodigious memory? Part of the answer, we hope, lies in this volume.

This book is structured like a reverse telephone book, or more accurately, like a reverse handbook of special function values. It is a list of just over 100,000 eight-digit real numbers in the interval $[0,1)$ that arise as the first eight digits of special values of familiar functions. It is designed for people, like ourselves, who encounter various numbers computationally and want to know if these numbers have some simple form. This is not a particularly well-defined endeavor—every eight-digit number is rational and this is not interesting. However, the chances of an eight digit number agreeing with a small rational, say with numerator and denominator less than twenty-five, is small. Thus the list is comprised primarily of special function evaluations at various algebraic and simple transcendental values. The exact numbers included are described below.

Each entry consists of the first eight digits after the decimal point of the number in question. The values are truncated not rounded. The next part of the entry specifies the function, and the final part of the entry is the value at which the function is evaluated. So $-4.828313737\ldots$ is entered as "8283 1373 ln : 5^{-3}". The abbreviations are also described below. There are two exceptions to this format. One is to describe certain combinations of two functions, for instance "8064 9591 $\exp(\sqrt{3})+\exp(e)$" the meaning of which is self-evident. The other is for real roots of cubic polynomials, i.e. "2027 1481 rt : $(5,8,-8,-3)$" means that the polynomial $5x^3+8x^2-8x-3$ has a real root, the fractional part of which is .20271481.... Repeats have in general been excluded except for cases where two genuinely different numbers agree through at least eight digits. These latter coincidences account for the fewer than fifty repeat entries.

If the number you are checking is not in the list try checking one divided by the number and perhaps a few other variants such as one minus the number. Of course, finding the number in the list, in general, only indicates that this is a good candidate for the number. Proving agreement or checking to further accuracy is then appropriate.

Jonathan Borwein
Peter Borwein

Introduction

Numbers in the Standard Domain

The "standard domain" is the set of roughly four thousand numbers on which the elementary functions are calculated. These numbers can be grouped into the following seven categories.

C1) Rational numbers of the form a/b for $1 \le a, b \le 25$ in lowest terms.

C2) Rational multiples of certain constants, i.e. numbers of the form $a\alpha/b$ for $1 \le a, b \le 10$ and α equal to one of the following: π, π^2, $1/\pi$, $e\pi$, e, $e+\pi$, $\sqrt{2}$, $\sqrt{3}$, $\sqrt{5}$, $\sqrt[3]{2}$, $\sqrt[3]{3}$, $\sqrt[3]{5}$, $\varsigma(3)$, $\ln 2$.

C3) Simple algebraic numbers. For $1 \le n \le 25$ the following are included: $\sqrt{n}$, $\sqrt[3]{n}$, $\sqrt{n/2}$, $\sqrt[3]{n/2}$, $\sqrt{n/3}$, $\sqrt[3]{n/3}$, $\sqrt{n\pi}$, $1/\sqrt{n}$, $1/\sqrt[3]{n}$.

C4) Elementary functions evaluated at simple values. The functions are: e^x, $\sqrt{x}$, $\sqrt[3]{x}$, $\ln(x)$, x^2, x^3, e^{-x}, $1/\sqrt{x}$, $1/\sqrt[3]{x}$, $1/\ln x$, x^{-2}, x^{-3} for x equal to each of the following 63 constants: π, 2π, 3π, $\sqrt{2}\pi$, e, $2e$, $\ln 2$, $\ln 3$, $e+\pi$, 1, 2, 3, 4, 5, $1/\pi$, $1/e$, $\sqrt[3]{2}$, $\sqrt[3]{3}$, $\sqrt[3]{4}$, $\sqrt[3]{5}$, $\sqrt{2}$, $\sqrt{3}$, $\sqrt{5}$, $\pi^2/2$, $3\pi/2$, $\sqrt{2}\pi/2$, $e/2$, $\ln 2/2$, $\ln 3/2$, $(e+\pi)/2$, $1/2$, $3/2$, $5/2$, $1/2\pi$, $1/2e$, $\sqrt[3]{2}/2$, $\sqrt[3]{3}/2$, $\sqrt[3]{4}/2$, $\sqrt[3]{5}/2$, $1/\sqrt{2}$, $\sqrt{3}/2$, $\sqrt{5}/2$, $\pi/3$, $2\pi/3$, $\sqrt{2}\pi/3$, $e/3$, $2e/3$, $\ln 2/3$, $\ln 3/3$, $(e+\pi)/3$, $1/3$, $2/3$, $4/3$, $5/3$, $1/3\pi$, $1/3e$, $\sqrt[3]{2}/3$, $\sqrt[3]{3}/3$, $\sqrt[3]{4}/3$, $\sqrt[3]{5}/3$, $\sqrt{2}/3$, $\sqrt{3}/3$, $\sqrt{5}/3$.

C5) Sums and differences of squares: $\pm\sqrt{a} \pm \sqrt{b} \pm \sqrt{c}$ for $2 \le a < b < c \le 7$, a, b, c nonsquare.

C6) Rational combinations of constants: $a/b[\ ]+c/d[\ ]$ for $1 \le a, b, c, d \le 4$, a and b relatively prime, c and d relatively prime, with the following ordered pairs of constants $(1, \pi)$, $(1, e)$, $(1, \sqrt{2})$, $(\sqrt{2}, \sqrt{3})$, $(\sqrt{2}, \sqrt{5})$, $(\sqrt{3}, \sqrt{5})$, $(1, \sqrt{3})$, $(1, \sqrt{5})$, (e, π).

C7) Euler's constant (γ), Catalan's constant ($\beta(2)$), $\varsigma(5)$, $\varsigma(7)$, $\varsigma(9)$, $\varsigma(11)$ and 82 physical constants (labelled P_1 through P_{82}) which are normalized so that their mantissas are in the interval $[.1, 1)$; for example, P_{79}, the speed of light, is converted to .299792458 and P_{55}, Planck's constant, becomes .66260755. See the following table.

With duplicate entries removed, the standard domain contains 4258 numbers. Fifteen elementary functions are evaluated on this domain giving entries which comprise roughly half of this book.

Functions Evaluated on Standard Domain

abbreviation	*function*	*domain of definition*
as	$\arcsin(x)$	SD $\cap[-1,1]$
at	$\arctan(x)$	SD $\cap[-1,1]$
cr	$\sqrt[3]{x}$	SD $\cap(0,\infty)$
cu	x^3	SD
e^x	e^x	SD $\cap(-\infty,64]$
id	x	SD
ℓ_2	$\log_2(x)$	SD $\cap(0,\infty)$
ℓ_{10}	$\log_{10}(x)$	SD $\cap(0,\infty)$
ln	$\ln(x)$	SD $\cap(0,\infty)$
sπ	$\sin(\pi x)$	SD
sq	x^2	SD
sr	$\sqrt{x}$	SD $\cap(0,\infty)$
tπ	$\tan(\pi x)$	SD $\backslash\{\pm 1/2, \pm 3/2, \ldots\}$
2^x	2^x	SD $\cap(-\infty,64]$
10^x	10^x	SD $\cap(-\infty,25]$

Physical Constants*

(1) Atomic mass unit $u = 1.6605402 \times 10^{-27} kg$
(2) Avogadro constant $N_A, L = 6.0221367 \times 10^{23}/\text{mol}$
(3) Bohr radius $a_0 = 0.529177249 \times 10^{-10} m$
(4) Boltzmann constant $k = 1.380658 \times 10^{-23} J/K$
(5) Boltzmann constant in electron volts $8.617385 \times 10^{-5} eV/K$
(6) Classical electron radius $r_e = 2.81794092 \times 10^{-15} m$
(7) Deuteron mass $m_d = 3.3435860 \times 10^{-27} kg$
(8) Deuteron mass in mass units $2.013553214\ u$
(9) Deuteron mass in electron volts $1875.61339\ MeV$
(10) Deuteron magnetic moment $\mu_d = 0.43307375 \times 10^{-26} J/T$
(11) Electron mass $m_e = 9.1093897 \times 10^{-31} kg$
(12) Electron mass in mass units $5.48579903 \times 10^{-4} u$
(13) Electron mass in electron volts $0.51099906\ MeV$
(14) Electron charge to mass ratio $e/m_e = 1.75881962 \times 10^{11} C/kg$
(15) Electron Compton wavelength $\lambda_e = 2.42631058 \times 10^{-12} m$
(16) Electron g factor $g_e = 2.002319304386$
(17) Electron magnetic moment $\mu_e = 928.47701 \times 10^{-26} J/T$
(18) Electron magnetic moment in Bohr magnetons $\mu_e/\mu_B = 1.001159652193$
(19) Electron proton magnetic moment ratio $\mu_e/\mu_p = 658.2106881$
(20) Elementary charge $e = 1.60217733 \times 10^{-19} C$
(21) Elementary charge $e = 4.8032068 \times 10^{-10} esu$
(22) Faraday constant $F = 96486.309\ C/\text{mol}$
(23) Fine-structure constant $\alpha = 7.29735308 \times 10^{-3}$
(24) Fine-structure constant, inverse $\alpha^{-1} = 137.0359895$
(25) Gas constant, molar $R = 8.314510\ J/\text{mol} \cdot K$
(26) Gravitational constant $G = 6.67259 \times 10^{-11} m^3/kg \cdot s^2$
(27) Hall conductance $e^2/h = 3.87404614 \times 10^{-5} S$
(28) Hall resistance $R_H = 2.58128056 \times 10^4 \Omega$
(29) Hartree energy $E_h = 4.3597482 \times 10^{-18} J$
(30) Hartree energy in electron volts $27.2113961\ eV$
(31) Ideal gas, molar volume at s.t.p. $V_m = 22.41410 \times 10^{-3} m^3/\text{mol}$
(32) Josephson frequency-voltage ratio $2e/h = 4.8359767 \times 10^{14} Hz/V$
(33) Magnetic flux quantum $\phi_0 = 2.06783461 \times 10^{-15} Wb$
(34) Magneton, Bohr $\mu_B = 9.2740154 \times 10^{-24} J/T$
(35) Magneton, nuclear $\mu_N = 5.0507866 \times 10^{-27} J/T$
(36) Muon mass $m_\mu = 1.8835327 \times 10^{-28} kg$
(37) Muon mass in mass units $0.113428913\ u$
(38) Muon mass in electron volts $105.658389\ MeV$
(39) Muon electron mass ratio $m_\mu/m_e = 206.768262$
(40) Muon g factor $g_\mu = 2.002331846$
(41) Muon magnetic moment $\mu_\mu = 4.4904514$
(42) Muon magnetic moment in Bohr magnetons $\mu_\mu/\mu_B = 4.84197097 \times 10^{-3}$
(43) Muon magnetic moment in nuclear magnetons $\mu_\mu/\mu_N = 8.8905981$
(44) Neutron mass $m_n = 1.6749286 \times 10^{-27} kg$
(45) Neutron mass in mass units $1.008664904\ u$
(46) Neutron mass in electron volts $939.56563\ MeV$
(47) Neutron electron mass ratio $m_n/m_e = 1838.683662$
(48) Neutron proton mass ratio $m_n/m_p = 1.001378404$
(49) Neutron Compton wavelength $\lambda_n = 1.31959110 \times 10^{-15} m$
(50) Neutron magnetic moment $\mu_n = 0.96623707 \times 10^{-26} J/T$
(51) Neutron magnetic moment in Bohr magnetons $\mu_n/\mu_B = 1.04187563 \times 10^{-3}$
(52) Neutron magnetic moment in nuclear magnetons $\mu_n/\mu_N = 1.91304275$
(53) Permeability of vacuum $\mu_0 = 4\pi \times 10^{-7} N/A^2$

*Fischbeck and Fischbeck, *Formulas, Facts and Constants*, Springer-Verlag, 1987

(54) Permittivity of vacuum $\varepsilon_0 = 1/\mu_0 c^2 = 8{:}854187817\ldots \times 10^{-12} F/M$
(55) Planck constant $h = 6.6260755 \times 10^{-34} J \cdot s$
(56) Planck constant in electron volt sec $4.1356692 \times 10^{-15} eV \cdot s$
(57) Planck constant $/2\pi$ $h = 1.05457266 \times 10^{-34}$
(58) Planck constant$/2\pi$ in electron volt sec $6.5821220 \times 10^{-16} eV \cdot s$
(59) Planck constant, molar $N_A h = 3.99031323 \times 10^{-10} J \cdot s/\text{mol}$
(60) Proton mass $m_p = 1.6726231 \times 10^{-27} kg$
(61) Proton mass in mass units $1.007276470\ u$
(62) Proton mass in electron volts $938.27231\ MeV$
(63) Proton electron mass ratio $m_p/m_e = 1836.152701$
(64) Proton Compton wavelength $\lambda_p = 1.32141002 \times 10^{-15} m$
(65) Proton gyromagnetic ratio $\gamma_p = 2.67522128 \times 10^8 (s \cdot T)^{-1}$
(66) Proton gyromagnetic ratio in H_2O $\gamma'_p = 2.67515255 \times 10^8 (s \cdot T)^{-1}$
(67) Proton magnetic moment $\mu_p = 1.41060761 \times 10^{-26} J/T$
(68) Proton magnetic moment in Bohr magnetons $\mu_p/\mu_B = 1.521032202 \times 10^{-3}$
(69) Proton magnetic moment in nuclear magnetons $\mu_p/\mu_N = 2.792847386$
(70) Proton magnetic moment in H_2O (shielded) $\mu'_p = 1.41057138 \times 10^{-26} J/T$
(71) Proton magnetic moment in $H_2 0$ in Bohr magnetons $\mu'_p/\mu_B = 1.520993129 \times 10^{-3}$
(72) Proton magnetic moment in $H_2 0$ in nuclear magnetons $\mu'_p/\mu_N = 2.792775642$
(73) Quantum of circulation $h/m_e = 7.27389614 \times 10^{-4} m^2/s$
(74) First radiation constant $c_1 = 3.7417749 \times 10^{-16} W \cdot m^2$
(75) Second radiation constant $c_2 = 1.438769 \times 10^{-2} m \cdot K$
(76) Rydberg constant $R_\infty = 1.0973731534 \times 10^7 m^{-1}$
(77) Sackur-Tetrode constant ($p_0 = 101325$ Pa) $S_0/R = -1.164856$
(78) Sackur-Tetrode constant ($p_0 = 100000$ Pa) $S_0/R = -1.151693$
(79) Speed of light (vacuum) $c = 299792458\ m/s$
(80) Stefan-Boltzmann constant $\sigma = 5.67051 \times 10^{-8} W/m^2 K^4$
(81) Thomson cross section $\sigma_e = 0.66524616 \times 10^{-28} m^2$
(82) Wien displacement law constant $b = 2.897756 \times 10^{-3} m \cdot K$

Other Entries

There are three more collections of numbers which are contained in this book.

1) Combinations of values of elementary functions: i.e. numbers of the form $f(x) + g(y)$, $f(x) - g(y)$, $f(x)g(y)$ and $f(x)/g(y)$ for f and g equal to any of x, $\ln(x)$, e^x or $\sin(\pi x)$ and x and y equal to any of 1, 2, 3, 4, 5, 1/2, 1/3, 2/3, 1/4, 3/4, 1/5, 2/5, 3/5, 4/5, π, e, $\sqrt{2}$, $\sqrt{3}$ or $\sqrt{5}$. The functions are abbreviated as ln, exp and $s\pi$.

2) Real roots of cubic polynomials $ax^3 + bx^2 + cx + d$ for a, b, c and d integers, where $-9 \le b, c, d \le 9$, $0 \le a \le 9$. They are denoted in the form "rt : (a, b, c, d)", i.e. the entry "0900 5010 rt : $(3, 3, -5, -2)$" means that the polynomial $3x^3 + 3x^2 - 5x - 2$ has a real root, the fractional part of which is $.09005010\ldots.$

3) Special functions evaluated on subsets of the "restricted domain" (abbreviated RD) which is obtained from the standard domain by omitting **C6** and the physical constants. The subsets and functions are explained in the following tables.

Functions Evaluated on Restricted Domain

abbreviation	*function*	*domain of definition*
Γ	gamma function	$\mathrm{RD} \cap (-\infty, 25] \setminus \{0, -1\}$
ς	Riemann's zeta function	$\mathrm{RD} \setminus \{1\}$
θ_3	theta function	$\mathrm{RD} \cap (-1, 1)$
λ	modular function	$\mathrm{RD} \cap [0, .5] \cup \{e^{-p\pi/q} : 1 \leq p, q \leq 20\}$
Ψ	digamma function	$\mathrm{RD} \setminus \{0, -1\}$
erf	error function	$\mathrm{RD} \cap (-4, 4)$
E	complete elliptic integral of the second kind	$\mathrm{RD} \cap [-.95, .95]$
Ei	exponential integral	$\mathrm{RD} \cap (0, 20)$
J	Klein's absolute invariant	$\mathrm{RD} \cap (0, .5] \cup \{e^{-p\pi/q} : 1 \leq p, q \leq 20\}$
J_0, J_1, J_2	Bessel functions	$\mathrm{RD} \cap (-\infty, 50)$
K	complete elliptic integral of the first kind	$\mathrm{RD} \cap [-.95, .95]$
Li_2	dilogarithm	$\mathrm{RD} \cap [0, 1]$
sinh	hyperbolic sine	$\mathrm{RD} \cap (-\infty, 20]$
tanh	hyperbolic tangent	$\mathrm{RD} \cap [-9, 9]$

Formulae

$$\Gamma(x) := \int_0^\infty e^{-t} t^{x-1}\, dt$$

$$\varsigma(x) := \sum_{n=1}^{\infty} n^{-x}$$

$$\theta_3(x) := \sum_{n=-\infty}^{\infty} x^{n^2}$$

$$\lambda(x) := \left[\frac{\theta_2(x)}{\theta_3(x)}\right]^4, \quad \theta_2(x) := \sum_{n=-\infty}^{\infty} x^{(n+1/2)^2}$$

$$\Psi(x) := \Gamma'(x)/\Gamma(x)$$

$$erf(x) := \frac{2}{\sqrt{\pi}} \int_0^x e^{-t^2}\, dt$$

$$E(x) := \int_0^{\pi/2} \sqrt{1 - x^2 \sin^2\theta}\, d\theta$$

$$Ei(x) := \int_{-\infty}^x \frac{e^t}{t}\, dt$$

$$J(x) := \frac{4}{27} \frac{(1 - \lambda(x) + \lambda^2(x))^3}{\lambda^2(x)(1 - \lambda(x))^2}$$

$$J_\nu(x) := (x/2)^\nu \sum_{k=0}^{\infty} \frac{(-1/4x^2)^k}{k!\,\Gamma(\nu + k + 1)}$$

$$K(x) := \int_0^{\pi/2} \frac{d\theta}{\sqrt{1 - x^2 \sin^2\theta}}$$

$$Li_2(x) := \sum_{n=1}^{\infty} \frac{x^n}{n^2}$$

$$\sinh(x) := \frac{e^x - e^{-x}}{2}$$

$$\tanh(x) := \frac{e^x - e^{-x}}{e^x + e^{-x}}$$

0000 0000 id : 0
0000 0001 $10^x : 3 - 4e$
0000 0002 $10^x : 1/2 - 3e$
0000 0003 $10^x : 2 - 3\pi$
0000 0004 $10^x : 3e/4 - 3\pi$
0000 0006 $10^x : \sqrt{3} - 4\sqrt{5}$
0000 0007 $10^x : 1 - 3e$
0000 0008 $10^x : 4\sqrt{2}/3 - 4\sqrt{5}$
0000 0010 cu : $3/4 - \sqrt{5}/3$
0000 0011 $10^x : 2 - 4\sqrt{5}$
0000 0012 cu : $(e + \pi)^{-3}$
0000 0013 $10^x : 4 - 4e$
0000 0014 $10^x : -\sqrt{3} - \sqrt{6} - \sqrt{7}$
0000 0015 $10^x : 4/3 - 3e$
0000 0017 $J_2 : (3\pi)^{-3}$
0000 0019 $10^x : e - 3\pi$
0000 0020 $10^x : 1/4 - 4\sqrt{3}$
0000 0022 $10^x : 3/2 - 3e$
0000 0023 $10^x : 4\sqrt{3}/3 - 4\sqrt{5}$
0000 0024 $10^x : -\sqrt{3} - \sqrt{5} - \sqrt{7}$
0000 0025 $10^x : 1/3 - 4\sqrt{3}$
0000 0026 $10^x : \sqrt{2}/4 - 4\sqrt{3}$
0000 0030 $10^x : -\sqrt{2} - \sqrt{6} - \sqrt{7}$
0000 0034 $10^x : 1/4 - 3\sqrt{5}$
0000 0037 $10^x : 3 - 3\pi$
0000 0038 $10^x : -\sqrt{3} - \sqrt{5} - \sqrt{6}$
0000 0041 $10^x : 1/\ln(\sqrt[3]{5}/2)$
0000 0042 $10^x : 1/3 - 3\sqrt{5}$
0000 0043 $J_2 : \exp(-2\pi)$
0000 0044 $10^x : \sqrt{2}/4 - 3\sqrt{5}$
0000 0045 $10^x : 3\sqrt{3}/2 - 4\sqrt{5}$
0000 0050 $10^x : -\sqrt{2} - \sqrt{5} - \sqrt{7}$
0000 0051 cu : $(5)^{-3}$
0000 0053 $10^x : \sqrt{3}/4 - 3\sqrt{5}$
0000 0054 $10^x : 2/3 - 4\sqrt{3}$
0000 0057 $10^x : \sqrt{2}/3 - 3\sqrt{5}$
0000 0060 $10^x : \sqrt{2}/2 - 4\sqrt{3}$
0000 0061 $10^x : 1/2 - 3\sqrt{5}$
0000 0066 $10^x : 3/4 - 4\sqrt{3}$
0000 0070 $10^x : 2 - 3e$
0000 0072 cu : $\exp(-3\pi/2)$
0000 0073 $10^x : \sqrt{3}/3 - 3\sqrt{5}$
0000 0076 $10^x : 2\sqrt{2} - 4\sqrt{5}$
0000 0079 $10^x : -\sqrt{2} - \sqrt{5} - \sqrt{6}$
0000 0080 cu : $3/2 - 2\sqrt{5}/3$
0000 0084 $\lambda : \exp(-16\pi/3)$
0000 0087 cu : $(3\pi/2)^{-3}$
0000 0090 $10^x : 2/3 - 3\sqrt{5}$
0000 0092 $10^x : 1/4 - 2\pi$
0000 0099 $10^x : \sqrt{2}/2 - 3\sqrt{5}$
0000 0101 $J_2 : \exp(-(e + \pi))$
0000 0103 $10^x : 2\sqrt{2}/3 - 4\sqrt{3}$
0000 0110 $10^x : 3/4 - 3\sqrt{5}$
0000 0112 $10^x : 1/3 - 2\pi$
0000 0113 $10^x : 3 - 4\sqrt{5}$
0000 0117 $10^x : 1 - 4\sqrt{3}$
0000 0135 $10^x : 3\sqrt{2}/4 - 4\sqrt{3}$
0000 0142 sq : $(3\pi)^{-3}$
0000 0143 $10^x : \sqrt{3}/2 - 3\sqrt{5}$
0000 0145 $\Gamma : 1/\ln(e/3)$
0000 0148 cu : $(\sqrt{2}\pi)^{-3}$
0000 0158 $10^x : 4e/3 - 3\pi$
0000 0161 $10^x : -\sqrt{2} - \sqrt{3} - \sqrt{7}$
0000 0162 cu : $\exp(-\sqrt{2}\pi)$
0000 0164 $10^x : 1/2 - 2\pi$
0000 0171 $10^x : 2\sqrt{2}/3 - 3\sqrt{5}$
0000 0187 cu : $(\ln 2/3)^3$
0000 0195 $10^x : 1 - 3\sqrt{5}$
0000 0203 $J_2 : (2\pi)^{-3}$
0000 0214 cu : $2/3 - e/4$
0000 0225 $10^x : 3\sqrt{2}/4 - 3\sqrt{5}$
0000 0237 $J_2 : \exp(-2e)$
0000 0241 $10^x : 2/3 - 2\pi$
0000 0249 $10^x : e/4 - 2\pi$
0000 0253 $10^x : -\sqrt{2} - \sqrt{3} - \sqrt{6}$
0000 0254 $10^x : 4/3 - 4\sqrt{3}$
0000 0279 $10^x : 2\sqrt{3}/3 - 3\sqrt{5}$
0000 0292 $10^x : 3/4 - 2\pi$
0000 0306 $10^x : \sqrt{2} - 4\sqrt{3}$
0000 0308 $J_2 : (e + \pi)^{-3}$
0000 0331 $10^x : 2\sqrt{3} - 4\sqrt{5}$
0000 0340 cu : $(1/(3e))^2$
0000 0348 sq : $\exp(-2\pi)$
0000 0373 $10^x : 3/2 - 4\sqrt{3}$
0000 0376 $10^x : 4 - 3\pi$
0000 0381 cu : $(4)^{-3}$
0000 0389 $10^x : 3\sqrt{3}/4 - 3\sqrt{5}$
0000 0391 $10^x : 1/4 - 4\sqrt{2}$
0000 0414 $10^x : -\sqrt{2} - \sqrt{3} - \sqrt{5}$
0000 0419 $10^x : e/3 - 2\pi$
0000 0421 $10^x : 4/3 - 3\sqrt{5}$
0000 0447 $e^x : 1/4 - 4\pi$
0000 0449 $10^x : 3e/2 - 3\pi$
0000 0474 $10^x : 1/3 - 4\sqrt{2}$
0000 0484 $J_2 : (2e)^{-3}$
0000 0486 $e^x : 1/3 - 4\pi$
0000 0508 $10^x : \sqrt{2} - 3\sqrt{5}$
0000 0520 $10^x : 1 - 2\pi$
0000 0528 $\lambda : \exp(-19\pi/4)$
0000 0567 $J_2 : \exp(-5)$
0000 0574 $e^x : 1/2 - 4\pi$
0000 0614 cu : $\exp(-4)$
0000 0616 cu : $\sqrt{3}/3 - \sqrt{5}/4$
0000 0619 $10^x : 3/2 - 3\sqrt{5}$
0000 0641 cu : $3 - 4\sqrt{5}/3$
0000 0650 $10^x : 1/4 - 2e$
0000 0679 $e^x : 2/3 - 4\pi$
0000 0687 $\lambda : \exp(-14\pi/3)$
0000 0688 $e^x : e/4 - 4\pi$
0000 0696 $10^x : 1/2 - 4\sqrt{2}$
0000 0700 $10^x : 3 - 3e$
0000 0738 $e^x : 3/4 - 4\pi$
0000 0788 $10^x : 1/3 - 2e$
0000 0799 $J_2 : (5)^{-3}$
0000 0813 sq : $\exp(-(e + \pi))$
0000 0826 cu : $1/3 - \sqrt{2}/4$
0000 0862 $e^x : e/3 - 4\pi$
0000 0906 $10^x : 4\sqrt{2}/3 - 4\sqrt{3}$
0000 0947 $e^x : 1 - 4\pi$
0000 1008 $J_2 : \exp(-3\pi/2)$
0000 1022 $10^x : 2/3 - 4\sqrt{2}$
0000 1056 $10^x : \sqrt{3} - 3\sqrt{5}$
0000 1118 $\zeta : 5\pi^2/3$
0000 1122 $10^x : 4/3 - 2\pi$
0000 1132 $10^x : 1/4 - 3\sqrt{3}$
0000 1136 $10^x : 4 - 4\sqrt{5}$
0000 1141 $J_2 : (3\pi/2)^{-3}$
0000 1157 $10^x : 1/2 - 2e$
0000 1159 $\lambda : \exp(-9\pi/2)$
0000 1179 $10^x : 2 - 4\sqrt{3}$
0000 1191 $10^x : e/2 - 2\pi$
0000 1232 $\zeta : 6e$
0000 1239 $10^x : 3/4 - 4\sqrt{2}$
0000 1322 $e^x : 4/3 - 4\pi$
0000 1357 $e^x : e/2 - 4\pi$
0000 1371 $10^x : 1/3 - 3\sqrt{3}$
0000 1436 $10^x : \sqrt{2}/4 - 3\sqrt{3}$
0000 1504 $10^x : 4\sqrt{2}/3 - 3\sqrt{5}$
0000 1528 $\zeta : 16$
0000 1559 $10^x : 3\sqrt{2}/2 - 4\sqrt{3}$
0000 1562 $e^x : 3/2 - 4\pi$
0000 1584 $J_2 : (3\pi)^{-2}$
0000 1625 $J_2 : (\sqrt{2}\pi)^{-3}$
0000 1625 sq : $(2\pi)^{-3}$
0000 1647 $10^x : 3/2 - 2\pi$
0000 1698 $10^x : 2/3 - 2e$
0000 1718 cu : $4/3 - e/2$
0000 1729 $J_2 : \exp(-\sqrt{2}\pi)$
0000 1766 $\zeta : 8\pi^2/5$
0000 1871 $\zeta : 5\pi$
0000 1884 $10^x : \sqrt{2}/3 - 3\sqrt{3}$
0000 1896 sq : $\exp(-2e)$
0000 1901 $J_2 : (\ln 2/3)^3$
0000 1944 $\zeta : 7\sqrt{5}$
0000 1957 $10^x : 2 - 3\sqrt{5}$
0000 1958 $\lambda : \exp(-13\pi/3)$
0000 1980 $\zeta : 8(e + \pi)/3$
0000 1982 $\zeta : (5/2)^3$
0000 1987 $10^x : 3\sqrt{2} - 4\sqrt{5}$
0000 2013 $10^x : 1/2 - 3\sqrt{3}$
0000 2033 $\zeta : 9\sqrt{3}$
0000 2057 $10^x : 3/4 - 2e$
0000 2069 $e^x : 5\sqrt{3}$
0000 2135 $e^x : 2e/3 - 4\pi$
0000 2156 sq : $3/4 - \sqrt{5}/3$
0000 2203 $10^x : 1 - 4\sqrt{2}$
0000 2332 $\zeta : \sqrt{24}\pi$
0000 2333 $\zeta : 9\sqrt[3]{5}$
0000 2338 cu : $1/2 - \sqrt{2}/3$
0000 2363 $\zeta : 9e\pi/5$
0000 2434 $e^x : 1/4 - 4e$
0000 2469 sq : $(e + \pi)^{-3}$
0000 2543 $\lambda : \exp(-17\pi/4)$
0000 2576 $e^x : 2 - 4\pi$
0000 2588 $10^x : 3\sqrt{2}/2 - 3\sqrt{5}$
0000 2646 $e^x : 1/3 - 4e$
0000 2678 $e^x : 3e/4 - 4\pi$
0000 2748 $\zeta : \exp(e)$
0000 2826 $J_2 : (1/(3e))^2$
0000 2920 $\zeta : \sqrt{23}\pi$
0000 2954 $10^x : 2/3 - 3\sqrt{3}$
0000 3051 $J_2 : (4)^{-3}$
0000 3058 $\zeta : 15$
0000 3126 $e^x : 1/2 - 4e$
0000 3178 $\zeta : 7e\pi/4$
0000 3225 cu : $(1/\pi)/10$
0000 3243 $10^x : \sqrt{2}/2 - 3\sqrt{3}$
0000 3354 cu : $(\pi)^{-3}$
0000 3380 $10^x : 2e/3 - 2\pi$
0000 3404 $\theta_3 : (\sqrt{2}\pi/3)^{-2}$
0000 3448 $10^x : 1/4 - 3\pi/2$
0000 3503 $\zeta : 3\pi^2/2$
0000 3579 $10^x : 3/4 - 3\sqrt{3}$
0000 3659 $10^x : 1 - 2e$
0000 3675 $\zeta : \sqrt{22}\pi$
0000 3693 $e^x : 2/3 - 4e$
0000 3873 sq : $(2e)^{-3}$
0000 3876 $10^x : 3e - 4\pi$
0000 3900 $\zeta : 5(e + \pi)/2$
0000 3944 $e^x : 1/\ln(e/3)$
0000 3992 $10^x : 4\sqrt{3}/3 - 3\sqrt{5}$
0000 4014 $e^x : 3/4 - 4e$
0000 4020 $\theta_3 : (\sqrt{2}\pi/3)^{-1/2}$
0000 4033 cu : $4/3 - 3\sqrt{3}/4$
0000 4034 $J_1 : \exp(-3\pi)$
0000 4177 $10^x : 1/3 - 3\pi/2$
0000 4193 $J_2 : \exp(-4)$
0000 4424 cu : $(1/\pi)/9$
0000 4435 cu : $3/4 - \pi/4$
0000 4539 sq : $\exp(-5)$
0000 4567 $\zeta : 10\sqrt[3]{3}$
0000 4650 $\zeta : \sqrt{21}\pi$
0000 4698 $10^x : 1/\ln(\sqrt[3]{4}/2)$
0000 4747 $10^x : 4/3 - 4\sqrt{2}$
0000 4929 cu : $2\sqrt{3}/3 - \sqrt{5}/2$
0000 5080 cu : $(3)^{-3}$
0000 5154 $e^x : 1 - 4e$
0000 5209 $10^x : 2 - 2\pi$
0000 5210 $\zeta : 5e\pi/3$
0000 5284 $e^x : e - 4\pi$
0000 5424 $J : 1/23$
0000 5549 $\zeta : 10\sqrt{2}$
0000 5568 $\zeta : 9\pi/2$
0000 5579 $\lambda : \exp(-4\pi)$
0000 5580 $10^x : 2\sqrt{2}/3 - 3\sqrt{3}$
0000 5593 $2^x : \exp(-3\pi)$
0000 5595 cu : $\sqrt{2}/2 - \sqrt{5}/3$
0000 5658 cu : $\sqrt{2}/3 - \sqrt{3}/4$
0000 5695 $10^x : 3e/4 - 2\pi$
0000 5716 $\zeta : 10\pi^2/7$
0000 5801 cu : $2 - 3e/4$
0000 5917 $\zeta : \sqrt{20}\pi$
0000 5996 $10^x : 1/4 - 2\sqrt{5}$
0000 6124 $\zeta : 14$
0000 6132 $10^x : 1/2 - 3\pi/2$
0000 6256 $J_0 : 6\sqrt[3]{3}$
0000 6284 cu : $((e + \pi)/2)^{-3}$
0000 6299 cu : $(1/\pi)/8$
0000 6365 $10^x : 1 - 3\sqrt{3}$
0000 6400 sq : $(5)^{-3}$

0000 6613 cu : $2/3 - \sqrt{2}/2$
0000 6767 $\zeta : 8\sqrt{3}$
0000 6952 $\zeta : 7\pi^2/5$
0000 6968 $10^x : 3/2 - 4\sqrt{2}$
0000 7000 $10^x : 4 - 3e$
0000 7004 $e^x : 3 - 4\pi$
0000 7193 $e^x : 4/3 - 4e$
0000 7213 cu : $(\ln 2/2)^3$
0000 7233 cu : $1/24$
0000 7264 $10^x : 1/3 - 2\sqrt{5}$
0000 7319 $10^x : 3\sqrt{2}/4 - 3\sqrt{3}$
0000 7576 $\zeta : \sqrt{19}\pi$
0000 7610 $10^x : \sqrt{2}/4 - 2\sqrt{5}$
0000 7650 $\zeta : 8\sqrt[3]{5}$
0000 7686 $\zeta : 7(e + \pi)/3$
0000 7737 $\zeta : 8e\pi/5$
0000 7760 $10^x : 3\sqrt{3}/2 - 3\sqrt{5}$
0000 7884 $10^x : 4/3 - 2e$
0000 7891 cu : $3/4 - \sqrt{2}/2$
0000 7947 $10^x : 2\sqrt{2} - 4\sqrt{3}$
0000 8019 $J_2 : (2\pi)^{-2}$
0000 8069 id : $\exp(-3\pi)$
0000 8070 $e^x : \exp(-3\pi)$
0000 8135 $\zeta : 5e$
0000 8218 cu : $1/23$
0000 8497 $e^x : 3/2 - 4e$
0000 8626 sq : $3/2 - 2\sqrt{5}/3$
0000 8668 $\zeta : (\sqrt[3]{2}/3)^{-3}$
0000 9000 $10^x : 2/3 - 3\pi/2$
0000 9065 $\Psi : 19/13$
0000 9105 $erf : \exp(-3\pi)$
0000 9131 sq : $(3\pi/2)^{-3}$
0000 9138 $10^x : \sqrt{3}/4 - 2\sqrt{5}$
0000 9159 $Ei : 5\sqrt{5}/8$
0000 9187 $\zeta : 6\sqrt{5}$
0000 9272 $10^x : e/4 - 3\pi/2$
0000 9391 cu : $1/22$
0000 9402 cu : $(1/\pi)/7$
0000 9764 $\zeta : \sqrt{18}\pi$
0000 9808 cu : $\ln(\pi/3)$
0000 9983 $10^x : \sqrt{2}/3 - 2\sqrt{5}$
0001 0170 $10^x : 1/4 - 3\sqrt{2}$
0001 0275 $10^x : 2e - 3\pi$
0001 0362 $e^x : 1/4 - 3\pi$
0001 0458 $\lambda : \exp(-19\pi/5)$
0001 0513 cu : $1 - \pi/3$
0001 0600 $J_2 : (e + \pi)^{-2}$
0001 0662 $10^x : 1/2 - 2\sqrt{5}$
0001 0792 $\zeta : 9(e + \pi)/4$
0001 0797 cu : $1/21$
0001 0904 $10^x : 3/4 - 3\pi/2$
0001 0983 $\zeta : 4\pi^2/3$
0001 1262 $e^x : 1/3 - 3\pi$
0001 1513 $10^x : 1/4 - 4\pi/3$
0001 1572 $10^x : 3/2 - 2e$
0001 1797 $10^x : 3 - 4\sqrt{3}$
0001 1844 cu : $(\ln 3/3)^3$
0001 2237 $\lambda : \exp(-15\pi/4)$
0001 2271 $\zeta : 13$
0001 2322 $10^x : 1/3 - 3\sqrt{2}$
0001 2340 cu : $(e)^{-3}$
0001 2441 $\zeta : 9\sqrt[3]{3}$
0001 2500 cu : $1/20$
0001 2664 $J_2 : (1/\pi)/10$
0001 2674 sq : $(3\pi)^{-2}$
0001 2677 $\zeta : \sqrt{17}\pi$
0001 2741 $10^x : \sqrt{3}/3 - 2\sqrt{5}$
0001 3000 $J_2 : (\pi)^{-3}$
0001 3002 sq : $(\sqrt{2}\pi)^{-3}$
0001 3077 $e^x : 4e/3 - 4\pi$
0001 3189 $10^x : 2\sqrt{2} - 3\sqrt{5}$
0001 3305 $e^x : 1/2 - 3\pi$
0001 3714 $10^x : 4/3 - 3\sqrt{3}$
0001 3834 sq : $\exp(-\sqrt{2}\pi)$
0001 3949 $10^x : 1/3 - 4\pi/3$
0001 4008 $\zeta : 3e\pi/2$
0001 4010 $e^x : 2 - 4e$
0001 4307 $J_2 : (2e)^{-2}$
0001 4579 cu : $1/19$
0001 4827 $\zeta : 9\sqrt{2}$
0001 4879 $10^x : 1/4 - 3e/2$
0001 4931 cu : $(1/\pi)/6$
0001 5213 sq : $(\ln 2/3)^3$
0001 5227 cu : $\exp(-(e + \pi)/2)$
0001 5229 $\zeta : 9\pi^2/7$
0001 5620 $10^x : e/3 - 3\pi/2$
0001 5634 $J_2 : (1/\pi)/9$
0001 5650 $10^x : 2/3 - 2\sqrt{5}$
0001 5718 $e^x : 2/3 - 3\pi$
0001 5899 $\lambda : \exp(-11\pi/3)$
0001 5922 $e^x : e/4 - 3\pi$
0001 6139 $\theta_3 : \exp(-3\pi)$
0001 6216 $\zeta : 10\sqrt[3]{2}$
0001 6470 cr : $\zeta(11)$
0001 6521 $10^x : \sqrt{2} - 3\sqrt{3}$
0001 6591 $\zeta : 4\pi$
0001 6637 cu : $\sqrt{3} - 3\sqrt{5}/4$
0001 6650 sq : $2/3 - e/4$
0001 6754 $e^x : 1/4 - 4\sqrt{5}$
0001 7060 ln : $2\sqrt{3} - \sqrt{5}/3$
0001 7084 $e^x : 3/4 - 3\pi$
0001 7144 $J_2 : (3)^{-3}$
0001 7146 cu : $1/18$
0001 7177 $10^x : \sqrt{2}/2 - 2\sqrt{5}$
0001 7265 cu : $3e/4 - 2\pi/3$
0001 7375 $\zeta : 25/2$
0001 7859 $10^x : 3\sqrt{3} - 4\sqrt{5}$
0001 8026 $10^x : 1/3 - 3e/2$
0001 8086 $10^x : 1/2 - 3\sqrt{2}$
0001 8210 $e^x : 1/3 - 4\sqrt{5}$
0001 8582 $e^x : \sqrt{2}/4 - 4\sqrt{5}$
0001 8583 $10^x : \exp(-3\pi)$
0001 8706 cu : $1 - 2\sqrt{2}/3$
0001 8886 cu : $3\sqrt{2}/4 - \sqrt{5}/2$
0001 8898 $e^x : 2/3 + 2\sqrt{2}/3$
0001 8931 $e^x : 3\sqrt[3]{5}/2$
0001 8961 $10^x : 3/4 - 2\sqrt{5}$
0001 9040 $e^x : 4 - 4\pi$
0001 9391 $10^x : 1 - 3\pi/2$
0001 9462 $\zeta : 5\pi^2/4$
0001 9579 $10^x : 3 - 3\sqrt{5}$
0001 9606 $2^x : 1/4 - 4\pi$
0001 9756 $J_2 : ((e + \pi)/2)^{-3}$
0001 9786 $J_2 : (1/\pi)/8$
0001 9970 $e^x : e/3 - 3\pi$
0001 9997 $J_2 : 1/25$
0002 0119 $e^x : \sqrt{3}/4 - 4\sqrt{5}$
0002 0130 $10^x : 3/2 - 3\sqrt{3}$
0002 0354 cu : $1/17$
0002 0474 $10^x : 1/2 - 4\pi/3$
0002 0555 cu : $1/2 - \sqrt{5}/4$
0002 0572 $e^x : 3e/2 - 4\pi$
0002 0772 $2^x : 1/3 - 4\pi$
0002 0787 $\exp(2/5) \div s\pi(\sqrt{3})$
0002 0843 $10^x : \sqrt{2} - \sqrt{6} - \sqrt{7}$
0002 0906 $e^x : \sqrt{2}/3 - 4\sqrt{5}$
0002 0934 $\zeta : 9e/2$
0002 1414 $\ell_{10} : 3\sqrt{2}/4 + 4\sqrt{5}$
0002 1415 $\zeta : 10e\pi/7$
0002 1457 $\ell_{10} : \zeta(11)$
0002 1512 $e^x : 1/2 - 4\sqrt{5}$
0002 1658 $J_2 : (\ln 2/2)^3$
0002 1672 $\zeta : \exp(5/2)$
0002 1698 $J_2 : 1/24$
0002 1902 $\zeta : \sqrt{15}\pi$
0002 1936 $e^x : 1 - 3\pi$
0002 2036 $10^x : 2 - 4\sqrt{2}$
0002 2318 $t\pi : ((e + \pi)/3)^{1/3}$
0002 2320 cu : $1 - 3\sqrt{2}/4$
0002 2567 $\zeta : 7\sqrt{3}$
0002 2611 sq : $(1/(3e))^2$
0002 3242 $e^x : \sqrt{3}/3 - 4\sqrt{5}$
0002 3316 $2^x : 1/2 - 4\pi$
0002 3339 $J_2 : \exp(-\pi)$
0002 3625 $J_2 : 1/23$
0002 4258 $\zeta : 10\zeta(3)$
0002 4414 sq : $(4)^{-3}$
0002 4451 sr : $e + 2\pi$
0002 4608 $\zeta : 12$
0002 4689 cu : $3/2 + \pi$
0002 4706 sr : $\zeta(11)$
0002 4767 $10^x : \sqrt{3}/2 - 2\sqrt{5}$
0002 5131 $\zeta : 7\sqrt[3]{5}$
0002 5343 $J_2 : (3\pi/2)^{-2}$
0002 5352 $s\pi : \exp(-3\pi)$
0002 5381 $\zeta : 7e\pi/5$
0002 5414 $e^x : 2/3 - 4\sqrt{5}$
0002 5801 cu : $(1/\pi)/5$
0002 5821 $J_2 : 1/22$
0002 5842 $J_2 : (1/\pi)/7$
0002 6171 $2^x : 2/3 - 4\pi$
0002 6214 cu : $(5/2)^{-3}$
0002 6401 $Ei : ((e + \pi)/3)^{1/2}$
0002 6406 $2^x : e/4 - 4\pi$
0002 6459 $10^x : 1/2 - 3e/2$
0002 6463 $e^x : \sqrt{2}/2 - 4\sqrt{5}$
0002 6547 $10^x : 2/3 - 3\sqrt{2}$
0002 6580 $J_2 : \ln(\pi/3)$
0002 6837 $\lambda : \exp(-7\pi/2)$
0002 7233 $10^x : e - 2\pi$
0002 7442 $\zeta : 6\pi^2/5$
0002 7623 $e^x : 3/4 - 4\sqrt{5}$
0002 7727 $2^x : 3/4 - 4\pi$
0002 8339 $J_2 : 1/21$
0002 8733 cu : $\exp(-e)$
0002 9193 $\zeta : \sqrt{14}\pi$
0002 9557 $10^x : 2\sqrt{2}/3 - 2\sqrt{5}$
0002 9629 cu : $1/15$
0002 9825 $2^x : 9\pi^2/10$
0002 9914 $\zeta : 2(e + \pi)$
0003 0052 $10^x : 2/3 - 4\pi/3$
0003 0059 cu : $1/2 - \sqrt{3}/4$
0003 0141 $J_2 : (\ln 3/3)^3$
0003 0614 $e^x : 4/3 - 3\pi$
0003 0896 $2^x : e/3 - 4\pi$
0003 0958 $10^x : e/4 - 4\pi/3$
0003 0978 $J_2 : (e)^{-3}$
0003 1021 $e^x : \sqrt{3}/2 - 4\sqrt{5}$
0003 1243 $J_2 : 1/20$
0003 1415 $e^x : e/2 - 3\pi$
0003 2074 $J_2 : (\sqrt{2}\pi)^{-2}$
0003 2163 $10^x : 3/4 - 3\sqrt{2}$
0003 2974 $2^x : 1 - 4\pi$
0003 3302 cu : $\ln 2/10$
0003 3497 $e^x : 2\sqrt{2}/3 - 4\sqrt{5}$
0003 3546 sq : $\exp(-4)$
0003 3610 sq : $\sqrt{3}/3 - \sqrt{5}/4$
0003 3718 $10^x : 1 - 2\sqrt{5}$
0003 3954 $\zeta : 8\sqrt[3]{3}$
0003 4071 $10^x : \sqrt{2} - \sqrt{5} - \sqrt{7}$
0003 4506 sq : $3 - 4\sqrt{5}/3$
0003 4513 $\zeta : 7\pi^2/6$
0003 4618 $J_2 : 1/19$
0003 4865 $\zeta : 23/2$
0003 4935 cu : $3/4 - e/4$
0003 5172 $J_2 : (1/\pi)/6$
0003 5392 cu : $2(1/\pi)/9$
0003 5468 $e^x : 1 - 4\sqrt{5}$
0003 5483 cu : $3/2 - \pi/2$
0003 5614 $J_2 : (\ln 2/3)^2$
0003 5636 $J_2 : \exp(-(e + \pi)/2)$
0003 6167 $e^x : 3/2 - 3\pi$
0003 6409 $10^x : 3/4 - 4\pi/3$
0003 6443 cu : $1/14$
0003 6596 $10^x : 2 - 2e$
0003 6741 $\lambda : \exp(-17\pi/5)$
0003 6782 $\Gamma : -\sqrt{5} - \sqrt{6} - \sqrt{7}$
0003 6895 $e^x : 1/4 - 3e$
0003 6955 $J_0 : 2\zeta(3)$
0003 7686 $e^x : 3\sqrt{2}/4 - 4\sqrt{5}$
0003 7742 $\zeta : 4e\pi/3$
0003 8084 $e^x : 3 - 4e$
0003 8570 $J_2 : 1/18$
0003 8772 $10^x : 3\sqrt{2}/4 - 2\sqrt{5}$
0003 8836 $10^x : 2/3 - 3e/2$
0003 9000 $\zeta : 9\sqrt[3]{2}$
0003 9126 $10^x : \sqrt{2}/4 + 3\sqrt{5}/2$
0003 9331 $\zeta : \sqrt{13}\pi$
0003 9436 cu : $4\sqrt{3}/3 - \sqrt{5}$
0003 9573 $10^x : 1/\ln(\sqrt{5}/3)$

0003 9702 $\zeta : 8\sqrt{2}$
0004 0101 $e^x : 1/3 - 3e$
0004 0644 cu : $(\sqrt[3]{2}/3)^3$
0004 0659 $\zeta : 8\pi^2/7$
0004 0885 sq : $1/3 - \sqrt{2}/4$
0004 1403 $e^x : 2\sqrt{3}/3 - 4\sqrt{5}$
0004 1545 $2^x : 4/3 - 4\pi$
0004 1655 $Ei : 9\sqrt{5}/10$
0004 1777 $10^x : 4/3 - 3\pi/2$
0004 1951 at : $9\sqrt{3}/10$
0004 2230 $10^x : 1/4 - 4e/3$
0004 2294 $2^x : e/2 - 4\pi$
0004 2823 $e^x : \sqrt[3]{1/3}$
0004 3240 $J_2 : 1/17$
0004 3332 $10^x : \sqrt{3} - \sqrt{6} - \sqrt{7}$
0004 3422 $e^x : 3 + 2e/3$
0004 3573 $\zeta : 5\sqrt{5}$
0004 4335 $10^x : e/2 - 3\pi/2$
0004 4766 cu : $\sqrt{2} - 2\sqrt{5}/3$
0004 5269 cu : $2\sqrt{2}/3 - \sqrt{3}/2$
0004 5300 $\lambda : \exp(-10\pi/3)$
0004 5516 cu : $1/13$
0004 5682 cu : $\ln 2/9$
0004 5980 $\zeta : 9\pi^2/8$
0004 6279 cu : $1/2 - \sqrt{3}/3$
0004 6409 cu : $4 - 3e/2$
0004 6632 $2^x : 3/2 - 4\pi$
0004 6944 at : $1 + \sqrt{5}/4$
0004 7051 $10^x : 3/4 - 3e/2$
0004 7374 $e^x : 1/2 - 3e$
0004 7831 $e^x : 3\sqrt{3}/4 - 4\sqrt{5}$
0004 8146 $10^x : 2\sqrt{3}/3 - 2\sqrt{5}$
0004 8724 cu : $2/3 - \sqrt{5}/3$
0004 8812 $J_2 : 1/16$
0004 8904 id : $e/3 + 2\pi/3$
0004 8917 $10^x : 4\sqrt{2}/3 - 3\sqrt{3}$
0004 9406 ln : $\zeta(11)$
0004 9418 id : $\zeta(11)$
0004 9419 $e^x : 2e/3 - 3\pi$
0004 9500 $e^x : 4/3 - 4\sqrt{5}$
0004 9571 $\zeta : 7\pi/2$
0005 0125 $\zeta : 9e\pi/7$
0005 0168 cu : $\sqrt{2}/4 - \sqrt{3}/4$
0005 0393 cu : $(1/\pi)/4$
0005 0598 $\zeta : 10\pi^2/9$
0005 0643 $J_2 : (1/\pi)/5$
0005 0734 $\zeta : (\sqrt{2}\pi/2)^3$
0005 1163 $10^x : 1/3 - 4e/3$
0005 1182 $J_2 : (5/2)^{-3}$
0005 1200 cu : $2/25$
0005 1591 $10^x : 4\sqrt{2} - 4\sqrt{5}$
0005 2097 $10^x : 3 - 2\pi$
0005 2155 $10^x : e/3 - 4\pi/3$
0005 2389 sq : $5\sqrt[3]{3}/2$
0005 2908 cu : $4/3 - \sqrt{2}$
0005 3537 $10^x : \sqrt{2} - \sqrt{5} - \sqrt{6}$
0005 3633 $\zeta : \sqrt{12}\pi$
0005 3670 $e^x : \sqrt{2} - 4\sqrt{5}$
0005 3996 $\zeta : 4e$
0005 4410 $J_2 : \exp(-e)$
0005 4857 id : $\sqrt{2}/4 - 3\sqrt{5}/2$
0005 5258 t$\pi : 10(e + \pi)/9$
0005 5308 cu : $\exp(-5/2)$
0005 5534 $J_2 : 1/15$
0005 5886 $J_0 : 5\sqrt[3]{3}/3$
0005 5966 $e^x : 2/3 - 3e$
0005 6096 $\zeta : 9\zeta(3)$
0005 7002 $10^x : 2\sqrt{3} - 3\sqrt{5}$
0005 7103 cr : $5\zeta(3)/6$
0005 7195 $10^x : 1 - 3\sqrt{2}$
0005 7870 cu : $1/12$
0005 7898 $2^x : 2e/3 - 4\pi$
0005 7921 t$\pi : ((e + \pi)/3)^{1/2}$
0005 8138 at : $3/4 - 4\sqrt{3}/3$
0005 8478 $e^x : 3/2 - 4\sqrt{5}$
0005 8837 $\zeta : -\sqrt{2} - \sqrt{5} - \sqrt{6}$
0005 8853 $\lambda : \exp(-13\pi/4)$
0005 9629 $e^x : 2 - 3\pi$
0005 9725 $J_1 : (3\pi)^{-3}$
0006 0032 $J_2 : \ln 2/10$
0006 0332 cr : $3\sqrt{3}/4 + 3\sqrt{5}$
0006 0479 $2^x : 3\sqrt{2} + 3\sqrt{3}$
0006 0829 $e^x : 3/4 - 3e$
0006 1079 $10^x : 1/4 - 2\sqrt{3}$
0006 1321 $10^x : 3/2 - 3\pi/2$
0006 1982 $e^x : 3e/4 - 3\pi$
0006 2027 $\zeta : 5e\pi/4$
0006 2517 $J_2 : 2(1/\pi)/9$
0006 2622 t$\pi : 3\sqrt{2}/4 - 3\sqrt{5}$
0006 3132 cu : $3/2 - \sqrt{2}$
0006 3385 at : $9\ln 2/4$
0006 3404 $2^x : 1/4 - 4e$
0006 3657 $10^x : 2 - 3\sqrt{3}$
0006 3748 $J_2 : 1/14$
0006 3865 $\zeta : 1/\ln(\ln 3)$
0006 4162 sq : $(2\pi)^{-2}$
0006 4745 $10^x : 1 - 4\pi/3$
0006 5043 cu : $\ln 2/8$
0006 5471 $e^x : -\sqrt{5} - \sqrt{6} - \sqrt{7}$
0006 5529 $J_1 : 23/6$
0006 5593 $Ei : (\ln 3/3)^{-1/3}$
0006 5751 cu : $2/23$
0006 5948 $2^x : 2 - 4\pi$
0006 6454 $Ei : \sqrt{5}/6$
0006 6603 sq : $4/3 - e/2$
0006 6901 cr : $\zeta(9)$
0006 7127 $10^x : 3\sqrt{3}/4 - 2\sqrt{5}$
0006 7174 $2^x : 1/3 - 4e$
0006 7250 cu : $\sqrt{2}/3 - \sqrt{5}/4$
0006 7742 $2^x : 3e/4 - 4\pi$
0006 7781 $\zeta : 9(e + \pi)/5$
0006 8130 $Li_2 : 16/21$
0006 8520 $2^x : \zeta(11)$
0006 8555 $J_2 : (\sqrt[3]{2}/3)^3$
0006 8705 $10^x : 4(e + \pi)/7$
0006 8859 $\lambda : \exp(-16\pi/5)$
0007 0084 $\zeta : 21/2$
0007 0832 $10^x : \sqrt{3} - \sqrt{5} - \sqrt{7}$
0007 1251 cu : $2/3 - \sqrt{3}/3$
0007 1278 $\ell_2 : \zeta(11)$
0007 1471 $\zeta : 10\pi/3$
0007 1554 cu : $(\sqrt{5})^{-3}$
0007 2643 $10^x : 4/3 - 2\sqrt{5}$
0007 3143 id : $\sqrt{2}/3 - 2\sqrt{5}$
0007 3751 $e^x : \sqrt{3} - 4\sqrt{5}$
0007 3928 $J_2 : 1/13$
0007 3999 $10^x : 1/3 - 2\sqrt{3}$
0007 4107 $J_2 : \ln 2/9$
0007 4144 $\zeta : \sqrt{11}\pi$
0007 4378 $\ell_{10} : 5\zeta(3)/6$
0007 5097 $10^x : 1/2 - 4e/3$
0007 5131 cu : $1/11$
0007 5222 cu : $2(1/\pi)/7$
0007 5262 $\theta_3 : 7\sqrt[3]{2}/10$
0007 5401 $2^x : 1/2 - 4e$
0007 5567 $\zeta : 6\sqrt{3}$
0007 5907 cu : $(\sqrt{2}\pi/2)^{-3}$
0007 6459 $\lambda : \exp(-19\pi/6)$
0007 7243 $J_2 : 7\zeta(3)$
0007 7526 $10^x : \sqrt{2}/4 - 2\sqrt{3}$
0007 8107 $e^x : 1 - 3e$
0007 8686 $10^x : 1/4 - 3\sqrt{5}/2$
0007 9115 $J_2 : (1/\pi)/4$
0007 9957 $J_2 : 2/25$
0008 0087 $e^x : 2e - 4\pi$
0008 0788 sinh : $7\sqrt[3]{2}/10$
0008 1770 sq : $1/2 - \sqrt{2}/3$
0008 2809 cu : $1 - e/3$
0008 2830 $2^x : (3\pi)^{-3}$
0008 2907 $\zeta : 6\sqrt[3]{5}$
0008 3165 cu : $3\sqrt{2}/4 - 2\sqrt{3}/3$
0008 3185 cu : $\ln(\ln 3)$
0008 3202 $\zeta : 7(e + \pi)/4$
0008 3616 $\zeta : 6e\pi/5$
0008 3671 $10^x : 1 - 3e/2$
0008 3808 $J : \exp(-19\pi/3)$
0008 4110 cu : $2 - 2\pi/3$
0008 4172 $10^x : 3\sqrt{2}/2 - 3\sqrt{3}$
0008 4177 $J_2 : \exp(-5/2)$
0008 4619 $\Psi : 8\zeta(3)/3$
0008 4634 $2^x : 2/3 - 4e$
0008 4809 sq : $(e + \pi)^{-2}$
0008 5667 sr : $5\zeta(3)/6$
0008 5993 $e^x : 4\sqrt{2}/3 - 4\sqrt{5}$
0008 6383 cu : $2/21$
0008 6755 $J_2 : 1/12$
0008 7079 cu : $3(1/\pi)/10$
0008 7135 $\ell_{10} : \zeta(9)$
0008 7370 $10^x : 1/\ln(\sqrt[3]{3}/2)$
0008 7514 $10^x : \sqrt{2} - 2\sqrt{5}$
0008 8580 $2^x : 1/\ln(e/3)$
0008 9141 $e^x : \sqrt{2}/4 + \sqrt{5}/3$
0008 9667 $2^x : 3/4 - 4e$
0009 0105 $\ln(\sqrt{3}) \times \exp(3/5)$
0009 0122 sq : $\sqrt{23}\pi$
0009 2198 $J_1 : (1/(3e))^3$
0009 2986 $10^x : 3\sqrt{2}/2 - \sqrt{3}/2$
0009 3004 $\zeta : 7\sqrt[3]{3}$
0009 3372 $J_1 : \exp(-2\pi)$
0009 3779 $J_2 : \ln 2/8$
0009 4077 $\zeta : 8\sqrt[3]{2}$
0009 4435 $\exp(5) + s\pi(1/5)$
0009 4458 $J_2 : 2/23$
0009 4807 at : $1/2 + 3\sqrt{2}/4$
0009 5208 $\zeta : 9\sqrt{5}/2$
0009 5330 $10^x : 1/3 - 3\sqrt{5}/2$
0009 5660 $e^x : 1/\ln(\sqrt{3}/2)$
0009 5894 cu : $\ln(e/3)$
0009 6414 $e^x : 2 - 4\sqrt{5}$
0009 7091 cu : $\ln 2/7$
0009 7809 id : $2e/3 + 4\pi/3$
0009 8862 sq : $\zeta(11)$
0009 9041 cu : $1/3 - \sqrt{3}/4$
0009 9457 $\zeta : 10$
0009 9873 $10^x : \sqrt{2}/4 - 3\sqrt{5}/2$
0009 9933 $J_2 : (\sqrt{5})^{-3}$
0010 0000 cu : $1/10$
0010 0348 cu : P_{18}
0010 0369 sr : $\zeta(9)$
0010 0414 cu : P_{48}
0010 1321 sq : $(1/\pi)/10$
0010 1695 $10^x : \sqrt{2}/3 - 2\sqrt{3}$
0010 2068 $\zeta : 7e\pi/6$
0010 2198 cu : P_{61}
0010 2622 cu : P_{45}
0010 3234 $J_2 : 1/11$
0010 3317 $J_2 : 2(1/\pi)/7$
0010 3523 $e^x : 4 - 4e$
0010 3944 $J_2 : (\sqrt{2}\pi/2)^{-3}$
0010 4016 sq : $(\pi)^{-3}$
0010 4123 $\zeta : \sqrt{10}\pi$
0010 6626 $10^x : 3/2 - 2\sqrt{5}$
0010 6633 $2^x : 1 - 4e$
0010 6717 $\zeta : 7\sqrt{2}$
0010 8217 $J_1 : \sqrt{18}\pi$
0010 8378 $e^x : -\sqrt{3} - \sqrt{6} - \sqrt{7}$
0010 8500 $2^x : e - 4\pi$
0010 8617 $10^x : 1/2 - 2\sqrt{3}$
0010 8745 $10^x : \sqrt{2} - \sqrt{3} - \sqrt{7}$
0010 8850 $e^x : 3\sqrt{2}/2 - 4\sqrt{5}$
0010 8977 $\zeta : \pi^2$
0010 9007 $e^x : 4/3 - 3e$
0010 9425 cr : $3\sqrt{2}/2 - \sqrt{5}/2$
0010 9715 id : $\sqrt{2}/2 - 3\sqrt{5}$
0011 0227 $10^x : 2/3 - 4e/3$
0011 0480 $J_2 : \ln(\ln 3)$
0011 1043 cu : $1/4 - \sqrt{2}/4$
0011 1299 $10^x : \sqrt{3} - \sqrt{5} - \sqrt{6}$
0011 3096 cu : P_{51}
0011 3293 $J_2 : 2/21$
0011 3899 $J_2 : 3(1/\pi)/10$
0011 4472 sq : $(2e)^{-2}$
0011 4959 cu : $(\sqrt{2}/3)^3$
0011 5306 $\theta_3 : \ln(\sqrt{2}/3)$
0011 5819 cu : $3/2 + 4\sqrt{5}/3$
0011 6080 $\zeta : -\sqrt{5} - \sqrt{6} + \sqrt{7}$
0011 6635 cu : $2/19$
0011 7155 $\zeta : 5(e + \pi)/3$
0011 7281 cu : P_{57}
0011 7500 as : $16/19$

0011 7616 sq : $4/3-3\sqrt{3}/4$
0011 7712 ζ : $8e\pi/7$
0011 7953 cu : P_{38}
0011 7976 10^x : $4-4\sqrt{3}$
0011 8521 cu : $e/4-\pi/4$
0011 8918 cu : $\sqrt{2}/3-\sqrt{3}/3$
0011 9083 cu : $1/\ln(4/3)$
0011 9450 id : $(3\pi)^{-3}$
0011 9485 Li_2 : $(3\pi)^{-3}$
0011 9521 e^x : $(3\pi)^{-3}$
0011 9925 10^x : $\sqrt{3}/4-3\sqrt{5}/2$
0012 0226 ζ : $-\sqrt{2}-\sqrt{3}-\sqrt{7}$
0012 1456 J_2 : $\ln(e/3)$
0012 2085 cu : $\exp(-\sqrt{5})$
0012 2294 e^x : $e-3\pi$
0012 2464 J_2 : $\ln 2/7$
0012 3223 10^x : $4/3-3\sqrt{2}$
0012 3963 sq : $2e+3\pi/2$
0012 4098 2^x : $5e\pi$
0012 4749 cu : $2/3-\sqrt{5}/4$
0012 4895 J_2 : $1/10$
0012 5087 sq : $(1/\pi)/9$
0012 5302 sq : $3/4-\pi/4$
0012 5803 e^x : $1/4-4\sqrt{3}$
0012 5834 10^x : $2e/3-3\pi/2$
0012 7561 cu : $\exp(-\sqrt{2}\pi/2)$
0012 7895 2^x : $(1/(3e))^3$
0012 8215 J_2 : $(\pi)^{-2}$
0012 8353 10^x : $1/4-\pi$
0012 8776 e^x : $3/2-3e$
0012 8965 cu : $(2\pi/3)^{-3}$
0012 9035 λ : $\exp(-3\pi)$
0012 9525 2^x : $\exp(-2\pi)$
0013 0163 ζ : $8\zeta(3)$
0013 0199 10^x : $1/\ln(1/\sqrt{2})$
0013 0541 ℓ_2 : $3\sqrt{3}/4+3\sqrt{5}$
0013 1009 10^x : $\sqrt{2}/3-3\sqrt{5}/2$
0013 1012 ζ : $9e\pi/8$
0013 1374 e^x : $4\sqrt{3}/3-4\sqrt{5}$
0013 1897 2^x : $3-4\pi$
0013 2148 cu : P_{76}
0013 3098 cu : $2\sqrt{3}-3\sqrt{5}/2$
0013 3543 10^x : $3/4-4e/3$
0013 4162 e^x : $-\sqrt{3}-\sqrt{5}-\sqrt{7}$
0013 4349 2^x : $4/3-4e$
0013 4443 sq : $2\sqrt{3}/3-\sqrt{5}/2$
0013 4758 $1/5\div\exp(5)$
0013 4784 erf : $(3\pi)^{-3}$
0013 6503 cu : $6\zeta(3)/5$
0013 6736 e^x : $1/3-4\sqrt{3}$
0013 6772 ζ : $(\sqrt{2}/3)^{-3}$
0013 7048 J_2 : $(\sqrt{2}/3)^3$
0013 7174 sq : $(3)^{-3}$
0013 8125 cu : $3e/2-4\pi/3$
0013 8301 10^x : $\sqrt{5}-\sqrt{6}-\sqrt{7}$
0013 8376 J_2 : $2/19$
0013 8549 sinh : $15/17$
0013 8887 cu : $\ln(\sqrt{5}/2)$
0013 9242 at : $1/4-2e/3$
0013 9490 10^x : $4/3-4\pi/3$
0013 9529 e^x : $\sqrt{2}/4-4\sqrt{3}$
0013 9877 ζ : $7e/2$
0013 9925 10^x : $1/2-3\sqrt{5}/2$
0014 0591 J_2 : $(1/\pi)/3$
0014 1259 ζ : $19/2$
0014 1568 10^x : P_2
0014 2395 ζ : $10e\pi/9$
0014 2490 ℓ_{10} : $3\sqrt{2}/2-\sqrt{5}/2$
0014 2579 J_1 : $\exp(-(e+\pi))$
0014 2598 at : $3\sqrt{2}/2-\sqrt{5}/4$
0014 2650 J_2 : $\exp(-\sqrt{5})$
0014 2699 ln : $\sqrt{2}/3+4\sqrt{3}$
0014 3465 Γ : $-\sqrt{3}-\sqrt{5}-\sqrt{7}$
0014 3945 10^x : $1/4+2\pi/3$
0014 5422 Γ : $5\zeta(3)/3$
0014 5938 cu : P_{37}
0014 6286 id : $2\sqrt{2}/3-4\sqrt{5}$
0014 6300 sq : $\sqrt{2}/2-\sqrt{5}/3$
0014 6713 id : $e+2\pi$
0014 6880 J_2 : $\exp(-\sqrt{2}\pi/2)$
0014 7393 sq : $\sqrt{2}/3-\sqrt{3}/4$
0014 7954 J_2 : $(2\pi/3)^{-3}$
0014 8030 10^x : $e/2-4\pi/3$
0014 8313 at : $25/16$
0014 8329 cu : $\zeta(11)$
0014 8928 ζ : 3π
0014 8929 e^x : $-\sqrt{2}-\sqrt{6}-\sqrt{7}$
0014 9648 cu : $2-4\sqrt{2}/3$
0014 9857 sq : $2-3e/4$
0015 0352 $\ln(4/5)\div\exp(5)$
0015 0780 cr : $\sqrt{2}+\sqrt{5}-\sqrt{7}$
0015 0802 2^x : $3/2-4e$
0015 1088 cu : $3\sqrt{2}/2-\sqrt{5}$
0015 2760 cu : P_{78}
0015 2784 cu : $\sqrt{2}-3\sqrt{3}/4$
0015 3638 sπ : $e/3+2\pi/3$
0015 4146 ζ : $8(e+\pi)/5$
0015 4162 J_2 : $1/9$
0015 4178 cu : $\ln 2/6$
0015 5253 sπ : $\zeta(11)$
0015 5254 tπ : $\zeta(11)$
0015 5441 J_2 : $\ln(\sqrt{5}/2)$
0015 5503 10^x : $1/3-\pi$
0015 6192 cu : $3/4-\sqrt{3}/2$
0015 6256 ζ : $\exp(\sqrt{5})$
0015 6760 e^x : $1/4-3\sqrt{5}$
0015 6981 e^x : $\sqrt{2}/3-4\sqrt{3}$
0015 8058 cu : P_{77}
0015 8070 sq : $((e+\pi)/2)^{-3}$
0015 8314 sq : $(1/\pi)/8$
0015 9428 10^x : $2/3-2\sqrt{3}$
0016 0000 sq : $1/25$
0016 1535 e^x : $1/2-4\sqrt{3}$
0016 2089 e^x : $3-3\pi$
0016 2833 cu : $2/17$
0016 3255 e^x : $-\sqrt{3}-\sqrt{5}-\sqrt{6}$
0016 3540 sq : $2/3-\sqrt{2}/2$
0016 4182 sr : $3\sqrt{2}/2-\sqrt{5}/2$
0016 4445 cu : $1-\sqrt{5}/2$
0016 4896 Γ : $-\sqrt{2}-\sqrt{6}-\sqrt{7}$
0016 5589 Ψ : $4\pi^2/5$
0016 6452 10^x : $(\sqrt{2}/3)^{1/3}$
0016 6638 J_2 : $\ln 2/6$
0016 6758 cu : $1/\ln(2/3)$
0016 7204 10^x : $\sqrt{3}/3-3\sqrt{5}/2$
0016 7204 e^x : $4\sqrt{2}-3\sqrt{5}/2$
0016 7377 cu : $2/3-\pi/4$
0016 8448 $1/4\div\exp(5)$
0016 9016 e^x : $1/\ln(\sqrt[3]{5}/2)$
0016 9428 J_2 : $((e+\pi)/2)^{-2}$
0017 0076 cu : $3(1/\pi)/8$
0017 0384 e^x : $1/3-3\sqrt{5}$
0017 0645 Γ : $-\sqrt{3}-\sqrt{6}-\sqrt{7}$
0017 0872 10^x : $\sqrt{2}-\sqrt{3}-\sqrt{6}$
0017 0928 cu : $8\pi/5$
0017 1034 ζ : $-\sqrt{2}-\sqrt{5}-\sqrt{7}$
0017 1261 ln : $5\zeta(3)/6$
0017 1311 e^x : $\sqrt[3]{8/3}$
0017 1335 sr : $10\zeta(3)/3$
0017 1381 2^x : $3-\sqrt{2}$
0017 1408 id : $5\zeta(3)/6$
0017 1930 ζ : $\exp(\sqrt{2}\pi/2)$
0017 2340 tπ : $\sqrt{2}/4-3\sqrt{5}/2$
0017 2800 cu : $3/25$
0017 2810 J_2 : $2/17$
0017 2877 sq : $e/4-2\pi/3$
0017 3028 2^x : $1/4-3\pi$
0017 3289 sq : $(\ln 2/2)^3$
0017 3611 sq : $1/24$
0017 3690 cu : $\zeta(3)/10$
0017 3864 e^x : $\sqrt{2}/4-3\sqrt{5}$
0017 4574 J_2 : $3\pi^2/2$
0017 4986 10^x : $\sqrt{2}/2-2\sqrt{3}$
0017 5340 e^x : $3\sqrt{3}/2-4\sqrt{5}$
0017 5593 Ei : $(1/(3e))^{-1/3}$
0017 5708 cu : $\sqrt{3}/2-\sqrt{5}/3$
0017 5823 cu : $e/3-\pi/4$
0017 6032 ζ : $-\sqrt{3}+\sqrt{5}-\sqrt{6}$
0017 6041 ζ : $(2\pi/3)^3$
0017 7085 J_0 : $5\sqrt{3}$
0017 7769 ln : $3\pi^2/4$
0017 7892 J_2 : $3(1/\pi)/8$
0017 8416 Ei : $3\sqrt{3}/5$
0017 8566 cu : $2+3\sqrt{2}/2$
0017 9784 J_2 : $3/25$
0018 0123 J_2 : $(\ln 2/2)^2$
0018 0210 $\ln(3)\times\exp(3/5)$
0018 0264 10^x : $4/3-3e/2$
0018 0343 10^x : $\sqrt{3}/2+4\sqrt{5}/3$
0018 0400 J_2 : $\zeta(3)/10$
0018 0866 10^x : $3/2-3\sqrt{2}$
0018 0956 sinh : $13/9$
0018 1934 10^x : $\sqrt{3}-2\sqrt{5}$
0018 2548 sr : $4\sqrt{2}+3\sqrt{5}/2$
0018 2765 $\ln(2/5)-\exp(3)$
0018 3316 2^x : $1/3-3\pi$
0018 3982 sq : $3/4-\sqrt{2}/2$
0018 4360 e^x : $-\sqrt{2}-\sqrt{5}-\sqrt{7}$
0018 4396 id : $(1/(3e))^3$
0018 4481 Li_2 : $(1/(3e))^3$
0018 4566 e^x : $(1/(3e))^3$
0018 5599 10^x : $1/4-4\sqrt{5}/3$
0018 6744 id : $\exp(-2\pi)$
0018 6831 Li_2 : $\exp(-2\pi)$
0018 6888 10^x : $1/\ln(\ln 2)$
0018 6918 e^x : $\exp(-2\pi)$
0018 7402 10^x : $2e+3\pi/2$
0018 8243 e^x : $\sqrt{3}/4-3\sqrt{5}$
0018 8986 e^x : $4/3+4\sqrt{2}/3$
0018 9035 sq : $1/23$
0018 9318 J_2 : $\exp(-2\pi/3)$
0019 0831 e^x : $2/3-4\sqrt{3}$
0019 1460 J : $1/24$
0019 1680 e^x : $4\sqrt{3}/3+3\sqrt{5}$
0019 2394 ζ : $10e/3$
0019 3151 10^x : $3/4-2\sqrt{3}$
0019 3178 at : $4\sqrt{3}/3-\sqrt{5}/3$
0019 3529 J_2 : $3\sqrt[3]{5}$
0019 3838 $\ln(3/4)\div\exp(5)$
0019 3914 10^x : $2-3\pi/2$
0019 4353 ln : $9\sqrt{5}$
0019 5058 J_2 : $1/8$
0019 5312 cu : $1/8$
0019 5610 e^x : $\sqrt{2}/3-3\sqrt{5}$
0019 5792 10^x : $4-3\sqrt{5}$
0019 6301 ℓ_{10} : $\sqrt{2}+\sqrt{5}-\sqrt{7}$
0019 7517 cu : $3+2\pi$
0019 7853 2^x : $\exp(-(e+\pi))$
0019 8162 J_2 : $\sqrt[3]{2}/10$
0019 8440 cu : P_{53}
0019 8706 e^x : $\sqrt{2}/2-4\sqrt{3}$
0019 9881 $\sqrt{2}+s\pi(1/5)$
0020 0000 cu : $\sqrt[3]{2}/10$
0020 0058 cu : $\sqrt{3}/4-\sqrt{5}/4$
0020 0637 ln : $\zeta(9)$
0020 0839 id : $\zeta(9)$
0020 0874 Ψ : $1/\sqrt[3]{24}$
0020 1233 10^x : $3\sqrt{2}/2+\sqrt{5}/4$
0020 1263 J : $(\ln 2/2)^3$
0020 1284 e^x : $1/2-3\sqrt{5}$
0020 1571 J_1 : $(2\pi)^{-3}$
0020 2051 λ : $\exp(-20\pi/7)$
0020 2156 ℓ_{10} : $1/4+\sqrt{5}/3$
0020 2368 J_2 : $2(1/\pi)/5$
0020 2784 sq : $(3\pi/2)^{-2}$
0020 3325 2^x : $4e/3-4\pi$
0020 3788 Γ : $-\sqrt{3}-\sqrt{5}-\sqrt{6}$
0020 3994 e^x : $17/6$
0020 4620 cr : $2\sqrt{2}+3\sqrt{3}$
0020 4743 10^x : $3/2-4\pi/3$
0020 5383 10^x : $2/3-3\sqrt{5}/2$
0020 5532 Ψ : $1/\ln(e/3)$
0020 5541 e^x : $3(e+\pi)/8$
0020 5766 2^x : $1/2-3\pi$
0020 6171 ζ : $1/\ln(\sqrt{5}/2)$
0020 6270 10^x : $3\sqrt{2}-4\sqrt{3}$
0020 6409 cu : $2(1/\pi)/5$
0020 6611 sq : $1/22$
0020 6777 sq : $(1/\pi)/7$
0020 7415 e^x : $3/4-4\sqrt{3}$

0020 7937 cu : $2/3+3\sqrt{5}/2$
0020 8068 erf : $(1/(3e))^3$
0020 8889 ζ : $4\sqrt{5}$
0020 9557 sq : $5\sqrt[3]{3}$
0020 9589 J_1 : $8e\pi/3$
0021 0718 erf : $\exp(-2\pi)$
0021 0951 sq : $3+e/2$
0021 1993 10^x : $3e/4-3\pi/2$
0021 2316 e^x : $2-3e$
0021 2364 J_2 : $3/23$
0021 2683 sq : $\ln(\pi/3)$
0021 3266 2^x : $2-4e$
0021 7357 ζ : $1/\ln(\sqrt[3]{5}/2)$
0021 7472 e^x : $\sqrt{3}/3-3\sqrt{5}$
0021 7691 ζ : $\sqrt{8}\pi$
0021 7720 J_1 : $\exp(-2e)$
0021 7727 λ : $\exp(-17\pi/6)$
0021 8171 ζ : $9\pi^2/10$
0021 8467 cu : $\sqrt{2}/2-\sqrt{3}/3$
0021 9376 10^x : $4e/3-2\pi$
0021 9837 cu : $5(e+\pi)/4$
0022 0366 10^x : $3-4\sqrt{2}$
0022 0761 e^x : $2\sqrt{2}-4\sqrt{5}$
0022 1792 10^x : $1/\sqrt{11}$
0022 1807 J : $4/9$
0022 1893 J_2 : $2/15$
0022 1911 cu : $3/23$
0022 2653 J_2 : $\zeta(3)/9$
0022 2760 sq : $1-\pi/3$
0022 3127 ζ : $-\sqrt{2}-\sqrt{3}-\sqrt{6}$
0022 4338 e^x : $-\sqrt{2}-\sqrt{5}-\sqrt{6}$
0022 4466 J_2 : $(\ln 3/3)^2$
0022 4598 $1/3\div\exp(5)$
0022 4727 J_2 : $((e+\pi)/3)^{-3}$
0022 4858 10^x : $1/3-4\sqrt{5}/3$
0022 5426 10^x : $\sqrt{2}/2-3\sqrt{5}/2$
0022 5489 ζ : $1/\ln(\sqrt[3]{4}/2)$
0022 5750 Γ : $1/\ln(\sqrt[3]{5}/2)$
0022 6255 sr : $\sqrt{2}+\sqrt{5}-\sqrt{7}$
0022 6757 sq : $1/21$
0022 8053 at : $7\sqrt{5}/10$
0022 8123 ζ : $7\sqrt[3]{2}$
0022 8248 10^x : $1/2-\pi$
0022 8596 J_2 : $(e)^{-2}$
0022 9783 cu : P_{49}
0022 9787 tπ : $\sqrt{2}/3-2\sqrt{5}$
0023 0122 θ_3 : $\ln(\sqrt[3]{3}/3)$
0023 0723 tπ : $e/4+\pi/2$
0023 0724 $e+\exp(1/4)$
0023 0734 cu : P_{64}
0023 0936 10^x : $1/2+2\pi/3$
0023 0955 sinh : $\exp(1/e)$
0023 0964 2^x : $2/3-3\pi$
0023 2078 J_2 : $3/22$
0023 2264 J_2 : $3(1/\pi)/7$
0023 2947 ζ : $3(e+\pi)/2$
0023 3039 2^x : $e/4-3\pi$
0023 3139 tπ : $1/4-e$
0023 3519 $\ln(\sqrt{2})\div\exp(5)$
0023 4333 ζ : $\sqrt{3}/5$
0023 4562 sq : $4+\pi$
0023 5011 ζ : $-\sqrt{3}-\sqrt{5}-\sqrt{6}$
0023 5575 10^x : $\sqrt{2}/4-4\sqrt{5}/3$
0023 5698 ζ : $-\sqrt{3}+\sqrt{6}-\sqrt{7}$
0023 5733 ζ : $8\pi^2/9$
0023 7037 cu : $2/15$
0023 7426 10^x : $3(1/\pi)/2$
0023 7478 10^x : $1-4e/3$
0023 7763 2^x : $5\zeta(3)/6$
0023 7790 e^x : $2/3-3\sqrt{5}$
0023 8258 cu : $\zeta(3)/9$
0023 8900 θ_3 : $(3\pi)^{-3}$
0023 9737 $\ln(4/5)+\exp(4/5)$
0023 9784 e^x : $1/4-2\pi$
0023 9842 J_2 : $\ln 2/5$
0024 0473 cu : $1-\sqrt{3}/2$
0024 1178 sq : $(\ln 3/3)^3$
0024 1414 2^x : $1/4-4\sqrt{5}$
0024 1599 cu : $((e+\pi)/3)^{-3}$
0024 1735 λ : $\exp(-14\pi/5)$
0024 4569 J_2 : $\sqrt[3]{2}/9$
0024 4698 2^x : $3/4-3\pi$
0024 6283 ζ : $9\sqrt[3]{2}/8$
0024 7078 ℓ_2 : $5\zeta(3)/3$
0024 7603 e^x : $\sqrt{2}/2-3\sqrt{5}$
0024 7875 sq : $(e)^{-3}$
0024 8487 J_1 : $(e+\pi)^{-3}$
0024 8827 10^x : $3/4-3\sqrt{5}/2$
0024 9583 J_2 : $\sqrt{2}/10$
0024 9758 J_2 : $4(1/\pi)/9$
0025 0000 sq : $1/20$
0025 0394 ζ : $19/11$
0025 0679 J : $(3\pi/2)^{-2}$
0025 0768 J_0 : $12/5$
0025 0942 J_2 : $\exp(-(e+\pi)/3)$
0025 1522 e^x : $2\sqrt{2}/3-4\sqrt{3}$
0025 2125 Γ : $2/23$
0025 3568 cu : $3/22$
0025 3875 cu : $3(1/\pi)/7$
0025 4668 J_2 : $1/7$
0025 5278 ζ : $5\sqrt{3}$
0025 5769 2^x : $1/3-4\sqrt{5}$
0025 6500 ζ : $6\sqrt[3]{3}$
0025 6649 sq : $(\sqrt{2}\pi)^{-2}$
0025 7337 cu : P_{24}
0025 8182 J_2 : $\ln(\sqrt{3}/2)$
0025 8454 e^x : $3/4-3\sqrt{5}$
0025 8964 10^x : $3+3\sqrt{2}/4$
0025 9108 10^x : $4\sqrt{2}/3-2\sqrt{5}$
0025 9379 2^x : $\sqrt{2}/4-4\sqrt{5}$
0025 9560 J_2 : $\sqrt[3]{3}/10$
0025 9711 ζ : $7\pi^2/8$
0026 0622 e^x : $1/3-2\pi$
0026 2080 e^x : $3-4\sqrt{5}$
0026 3183 cu : P_4
0026 3214 cu : $1/3-\sqrt{2}/3$
0026 3794 2^x : $4-4\pi$
0026 3864 cr : $4\sqrt[3]{2}/5$
0026 3981 10^x : $1/4-2\sqrt{2}$
0026 4592 10^x : $3/2-3e/2$
0026 5032 $\ln(3/5)-\exp(2/5)$
0026 6326 e^x : $1-4\sqrt{3}$
0026 6419 cu : $\ln 2/5$
0026 9321 ζ : $((e+\pi)/2)^2$
0026 9517 $2/5\div\exp(5)$
0026 9576 ζ : $\ln(1/(3e))$
0026 9751 ℓ_{10} : $4\sqrt{5}/9$
0027 1230 2^x : $e/3+2\pi/3$
0027 1315 ζ : $\ln 2/2$
0027 1737 ζ : $-\sqrt{2}+\sqrt{3}-\sqrt{5}$
0027 2659 2^x : $e/3-3\pi$
0027 3200 $\ln(2/3)\div\exp(5)$
0027 3846 J_2 : $(\sqrt[3]{4}/3)^3$
0027 4066 2^x : $\sqrt{3}/4-4\sqrt{5}$
0027 4348 cu : $\sqrt[3]{2}/9$
0027 5422 10^x : $(3\pi)^{-3}$
0027 5709 J_0 : $\sqrt[3]{14}$
0027 6006 ζ : $5\sqrt[3]{5}$
0027 6173 $\ln(4)-\exp(2)$
0027 6762 J_1 : $\sqrt{5}\pi$
0027 7008 sq : $1/19$
0027 7306 ζ : $\sqrt{2}-\sqrt{6}-\sqrt{7}$
0027 7538 cr : $\zeta(7)$
0027 7610 10^x : $3\sqrt{2}+3\sqrt{5}/2$
0027 7995 ζ : $e\pi$
0027 8070 id : $(\ln 2)^{-3}$
0027 8337 2^x : $3e/2-4\pi$
0027 8439 at : $2-\sqrt{3}/4$
0027 8616 2^x : $\zeta(9)$
0027 8803 2^x : $10\sqrt[3]{3}/3$
0027 9003 ζ : $-\sqrt{2}-\sqrt{6}-\sqrt{7}$
0027 9315 10^x : $\sqrt{2}-\sqrt{3}-\sqrt{5}$
0027 9482 cu : $3/2-e/2$
0027 9829 2^x : $(2\pi)^{-3}$
0027 9841 J_2 : $7(e+\pi)/8$
0028 0663 cu : P_{70}
0028 0684 cu : P_{67}
0028 0723 J_2 : $3/20$
0028 0940 cu : $e/3-\pi/3$
0028 1121 J_0 : $7e\pi/4$
0028 1447 sq : $(1/\pi)/6$
0028 1457 2^x : $\sqrt{2}/3-4\sqrt{5}$
0028 1684 J_2 : $\zeta(3)/8$
0028 2794 λ : $\exp(-11\pi/4)$
0028 2842 cu : $\sqrt{2}/10$
0028 2871 10^x : $\sqrt{3}/4-4\sqrt{5}/3$
0028 2981 e^x : $3\sqrt{2}/4-4\sqrt{3}$
0028 3133 sq : $5\ln 2/2$
0028 3141 cu : $4(1/\pi)/9$
0028 3871 cu : $3-\pi$
0028 4293 ζ : $1/\ln(e/3)$
0028 4895 at : $\sqrt{2}-4\sqrt{5}/3$
0028 4981 sq : $(\ln 2/3)^2$
0028 5159 at : $\exp(-(e+\pi))$
0028 5160 id : $\exp(-(e+\pi))$
0028 5363 Li_2 : $\exp(-(e+\pi))$
0028 5567 e^x : $\exp(-(e+\pi))$
0028 5925 ζ : $17/2$
0028 6018 Ψ : $(\pi)^{1/3}$
0028 6773 10^x : $(2\pi/3)^{1/2}$
0028 7091 2^x : $1/2-4\sqrt{5}$
0028 8920 ζ : $6\sqrt{2}$
0028 9459 ℓ_2 : $\zeta(9)$
0029 0251 e^x : $\sqrt{3}/2-3\sqrt{5}$
0029 0997 2^x : $1-3\pi$
0029 1545 cu : $1/7$
0029 3273 sr : $4\sqrt{3}/3+3\sqrt{5}$
0029 3427 id : $2e+4\pi$
0029 3451 sq : $e/3+2\pi/3$
0029 4210 ζ : $6\pi^2/7$
0029 5274 J_2 : $2/13$
0029 5550 ζ : $1/\sqrt[3]{24}$
0029 5989 J_2 : $2\ln 2/9$
0029 7505 J_0 : $\exp(\sqrt[3]{5})$
0029 7610 cu : $\ln(\sqrt{3}/2)$
0029 7833 cu : P_{75}
0030 0000 cu : $\sqrt[3]{3}/10$
0030 1097 10^x : $2\sqrt{2}/3-2\sqrt{3}$
0030 1115 sinh : $9/2$
0030 1539 Ψ : $\ln 2/2$
0030 2281 2^x : $\exp(-2e)$
0030 2498 sq : $\sqrt{3}-3\sqrt{5}/4$
0030 2633 e^x : $4e/3-3\pi$
0030 2904 2^x : $\sqrt{3}/3-4\sqrt{5}$
0030 3248 e^x : $2+2\pi/3$
0030 3796 ζ : $7\zeta(3)$
0030 5182 e^x : $-\sqrt{2}-\sqrt{3}-\sqrt{7}$
0030 5813 cu : $3\sqrt{2}+4\sqrt{5}/3$
0030 6183 J_2 : $\ln(\sqrt[3]{5}/2)$
0030 7008 sr : $\sqrt{2}+3\sqrt{3}/2$
0030 7275 sπ : $2e/3+4\pi/3$
0030 7277 tπ : $2e/3+4\pi/3$
0030 7889 e^x : $1/2-2\pi$
0030 8007 J_2 : $\sqrt{2}/9$
0030 8641 sq : $1/18$
0030 9015 10^x : $\sqrt{2}/3-4\sqrt{5}/3$
0030 9028 cu : $3/2-2e$
0030 9398 J_2 : $\sqrt[3]{2}/8$
0031 0067 sq : $3e/4-2\pi/3$
0031 0591 Γ : $-\sqrt{2}-\sqrt{5}-\sqrt{7}$
0031 0987 J_2 : $3/19$
0031 1167 J_1 : $(2e)^{-3}$
0031 1791 Ei : $3\ln 2/2$
0031 1827 ζ : $8\pi/3$
0031 2171 10^x : $5\pi/3$
0031 3230 ζ : $10(e+\pi)/7$
0031 3415 e^x : $2\sqrt{2}/3-3\sqrt{5}$
0031 4043 ζ : $-\sqrt{2}-\sqrt{3}-\sqrt{5}$
0031 4078 cu : $1/2-\sqrt{2}/4$
0031 5960 J_2 : $(1/\pi)/2$
0031 6318 λ : $\exp(-19\pi/7)$
0031 8143 cr : $7\sqrt[3]{3}/10$
0031 8837 10^x : $25/3$
0031 9317 J_2 : $4/25$
0031 9820 10^x : $1/3-2\sqrt{2}$
0032 0294 e^x : $2/3+\sqrt{3}/4$
0032 0314 J_2 : $\sqrt[3]{3}/9$
0032 1764 ζ : $25/3$
0032 1767 erf : $\exp(-(e+\pi))$
0032 2249 2^x : $2/3-4\sqrt{5}$

0032 2369 $\Psi : (e+\pi)/4$
0032 2416 $2^x : 3/4 - \sqrt{5}/3$
0032 3295 $\zeta : -\sqrt{3} - \sqrt{5} - \sqrt{7}$
0032 3562 $\zeta : (\ln 2/2)^{-2}$
0032 4728 $10^x : 2 - 2\sqrt{3}/3$
0032 4769 $\mathrm{cu} : \sqrt{2}/2 - \sqrt{5}/4$
0032 5029 $10^x : \sqrt{3}/2 - 3\sqrt{5}/2$
0032 5153 $\mathrm{cu} : (\sqrt[3]{4}/3)^3$
0032 5678 $10^x : 8\sqrt[3]{5}/7$
0032 5963 $10^x : (\ln 3/2)^2$
0032 6699 $\zeta : \sqrt{7}\pi$
0032 6756 $\theta_3 : 12/17$
0032 7080 $\mathrm{sq} : 1 - 2\sqrt{2}/3$
0032 8096 $\ln : 3\sqrt{2}/2 - \sqrt{5}/2$
0032 8635 $\mathrm{id} : 3\sqrt{2}/2 - \sqrt{5}/2$
0032 9175 $\mathrm{sq} : 3\sqrt{2}/4 - \sqrt{5}/2$
0033 0047 $10^x : 1/2 - 4\sqrt{5}/3$
0033 1410 $2^x : \sqrt{2}/2 - 4\sqrt{5}$
0033 1862 $e^x : 1 - 3\sqrt{5}$
0033 2581 $J_2 : \exp(-2e/3)$
0033 3598 $\mathrm{cu} : ((e+\pi)/2)^3$
0033 5022 $10^x : 2/3 - \pi$
0033 6895 $J_1 : \exp(-5)$
0033 6897 $1/2 \div \exp(5)$
0033 7181 $10^x : 2 - 2\sqrt{5}$
0033 7279 $\mathrm{cr} : 5\sqrt{2}/7$
0033 7500 $\mathrm{cu} : 3/20$
0033 9238 $\mathrm{cu} : \zeta(3)/8$
0034 0187 $10^x : 1/4 - e$
0034 1411 $2^x : 3/4 - 4\sqrt{5}$
0034 1740 $10^x : 1/\ln(2/3)$
0034 1891 $\mathrm{cu} : \exp(\sqrt[3]{4})$
0034 2323 $10^x : 3\sqrt{2} - 3\sqrt{5}$
0034 2611 $J_2 : (\ln 3/2)^3$
0034 2763 $2^x : 4 - \sqrt{2}$
0034 2817 $\mathrm{id} : 5\zeta(3)/3$
0034 3111 $\mathrm{sq} : 5\zeta(3)/6$
0034 3331 $\ell_{10} : 8\sqrt[3]{2}$
0034 3477 $10^x : 1 - 2\sqrt{3}$
0034 4191 $\ln(3/5) \div \exp(5)$
0034 4681 $t\pi : \sqrt{2}/2 - 3\sqrt{5}$
0034 5071 $2^x : (e+\pi)^{-3}$
0034 5126 $10^x : e/4 - \pi$
0034 5546 $\mathrm{cr} : 4/3 + 3\sqrt{5}$
0034 5571 $\mathrm{cu} : \sqrt{2}/3 - \sqrt{3}$
0034 6020 $\mathrm{sq} : 1/17$
0034 6419 $J_2 : 1/6$
0034 6916 $\zeta : -\sqrt{2} + \sqrt{3} - \sqrt{6}$
0034 7532 $\ell_{10} : \sqrt{3}/4 + \sqrt{5}/4$
0034 7554 $\zeta : 5\pi^2/6$
0034 8224 $\mathrm{at} : 2/3 - \sqrt{5}$
0034 8300 $\mathrm{sq} : 1/2 - \sqrt{5}/4$
0034 9259 $\mathrm{cu} : 1/2 + 2\sqrt{2}/3$
0035 1240 $\exp(1/2) - \exp(\sqrt{3})$
0035 1869 $\mathrm{cu} : P_{71}$
0035 1896 $\mathrm{cu} : P_{68}$
0035 2102 $J_2 : (2e/3)^{-3}$
0035 2114 $e^x : 3\sqrt{2} + 2\sqrt{5}/3$
0035 2616 $e^x : 3\sqrt{2}/4 - 3\sqrt{5}$
0035 2736 $\zeta : 7(e+\pi)/5$
0035 3949 $2^x : 4 - 3\sqrt{5}/4$
0035 4687 $\ln : \sqrt{2}/2 + 3\sqrt{5}$
0035 5267 $\sinh : 5\sqrt{2}/8$
0035 8135 $\mathrm{cu} : 2\sqrt{2} - 4\sqrt{5}/3$
0036 1099 $\ell_{10} : \zeta(7)$
0036 1226 $\Gamma : 4\sqrt{5}/9$
0036 1514 $\ln(2/3) - \exp(4)$
0036 2156 $\mathrm{cu} : 4\sqrt{2}/3 - \sqrt{3}$
0036 3728 $e^x : 2/3 - 2\pi$
0036 4104 $\mathrm{as} : (5/3)^{-1/3}$
0036 4132 $\mathrm{cu} : 2/13$
0036 4612 $J_2 : \sqrt[3]{5}/10$
0036 5207 $\mathrm{id} : 4\sqrt{2}/3 + \sqrt{5}/2$
0036 5223 $\zeta : 3e$
0036 5459 $\mathrm{cu} : 2\ln 2/9$
0036 5962 $10^x : 3 - 2e$
0036 6049 $\mathrm{cu} : 2\sqrt{2} - \sqrt{5}/2$
0036 6312 $1/5 \div \exp(4)$
0036 6633 $2^x : 4/3 - 3\pi$
0036 7269 $\lambda : \exp(-8\pi/3)$
0036 7702 $J_2 : \zeta(3)/7$
0036 7965 $\mathrm{sq} : 3 - \sqrt{2}/4$
0036 8452 $e^x : e/4 - 2\pi$
0036 8793 $\theta_3 : (1/(3e))^3$
0036 9294 $\zeta : -\sqrt{3} + \sqrt{5} - \sqrt{7}$
0036 9547 $10^x : -\sqrt{5} + \sqrt{6} - \sqrt{7}$
0037 0003 $2^x : \sqrt{3}/2 - 4\sqrt{5}$
0037 0036 $J : 9/25$
0037 0119 $\ln(\sqrt{3}) \div \exp(5)$
0037 0122 $\ell_{10} : 3e/4 - \pi/3$
0037 0233 $\mathrm{cu} : 1 + 2\sqrt{3}/3$
0037 1359 $e^x : -\sqrt{2} - \sqrt{3} - \sqrt{6}$
0037 1688 $e^x : 4/3 - 4\sqrt{3}$
0037 3251 $2^x : e/2 - 3\pi$
0037 3318 $10^x : 4\sqrt{2} + \sqrt{3}$
0037 3488 $\theta_3 : \exp(-2\pi)$
0037 4063 $J_2 : \sqrt{3}/10$
0037 4239 $\zeta : \exp(2\pi/3)$
0037 4415 $J_2 : \ln 2/4$
0037 5262 $s\pi : (3\pi)^{-3}$
0037 5265 $t\pi : (3\pi)^{-3}$
0037 7119 $J_2 : 4/23$
0037 8662 $J : 1/22$
0037 9341 $\ell_{10} : 4e - \pi/4$
0038 0327 $\mathrm{cu} : 3/4 - e/3$
0038 3988 $t\pi : \sqrt{2}/2 - 4\sqrt{3}/3$
0038 4538 $\mathrm{cu} : \ln(\sqrt[3]{5}/2)$
0038 4570 $\mathrm{cr} : 4\sqrt{3} + \sqrt{5}/2$
0038 4722 $\Psi : (\sqrt[3]{3}/3)^{1/3}$
0038 4774 $J : (1/\pi)/7$
0038 5877 $\zeta : -\sqrt{5} - \sqrt{6} - \sqrt{7}$
0038 6448 $e^x : 3 + 4\sqrt{5}/3$
0038 6904 $\mathrm{cu} : 3 + 3e$
0038 7386 $e^x : 2\sqrt{3}/3 - 3\sqrt{5}$
0038 7857 $J_2 : (\sqrt[3]{2}/3)^2$
0038 7888 $10^x : 2\sqrt{2}/3 - 3\sqrt{5}/2$
0038 7987 $\mathrm{cu} : \sqrt{2}/9$
0038 8264 $J_2 : 3/17$
0038 8482 $\mathrm{at} : \pi/2$
0038 9608 $J_2 : \sqrt{2}/8$
0038 9796 $\mathrm{cu} : 4/3 - 2\sqrt{5}/3$
0038 9881 $J_2 : 5(1/\pi)/9$
0039 0229 $2^x : 2\sqrt{2}/3 - 4\sqrt{5}$
0039 0244 $J_2 : \exp(-\sqrt{3})$
0039 0277 $\zeta : -\sqrt{3} - \sqrt{6} - \sqrt{7}$
0039 0625 $\mathrm{sq} : 1/16$
0039 2293 $\Psi : 6\sqrt[3]{5}/7$
0039 3643 $\mathrm{cu} : 3/19$
0039 4390 $10^x : \sqrt{3}/3 - 4\sqrt{5}/3$
0039 4964 $10^x : 3\sqrt{2}/4 - 2\sqrt{3}$
0039 5334 $\ell_{10} : 2/3 + 3\pi$
0039 5337 $e^x : 3/4 - 2\pi$
0039 5997 $\mathrm{at} : 1/\ln(\sqrt[3]{4}/3)$
0039 6057 $\mathrm{sr} : 4\sqrt[3]{2}/5$
0039 9744 $e^x : 7\sqrt{2}/9$
0039 9996 $J_1 : (5)^{-3}$
0040 0641 $J_2 : 3e\pi/5$
0040 1351 $\mathrm{cu} : \sqrt{2}/2 - \sqrt{3}/2$
0040 2081 $\mathrm{sq} : \zeta(9)$
0040 2999 $e^x : \sqrt{2} - 4\sqrt{3}$
0040 3141 $\mathrm{at} : (2\pi)^{-3}$
0040 3144 $\mathrm{id} : (2\pi)^{-3}$
0040 3145 $\mathrm{as} : (2\pi)^{-3}$
0040 3330 $\mathrm{sq} : \sqrt{3}/2 - 2\sqrt{5}$
0040 3551 $Li_2 : (2\pi)^{-3}$
0040 3857 $J_2 : \sqrt[3]{2}/7$
0040 3957 $e^x : (2\pi)^{-3}$
0040 4276 $3/5 \div \exp(5)$
0040 4400 $\mathrm{at} : 10\sqrt{2}/9$
0040 5167 $J_2 : \sqrt[3]{3}/8$
0040 5284 $\mathrm{sq} : (1/\pi)/5$
0040 5352 $\Gamma : 7e\pi/8$
0040 5889 $10^x : 3/4 - \pi$
0040 6009 $2^x : 1 - 4\sqrt{5}$
0040 6710 $\mathrm{at} : 11/7$
0040 7735 $\zeta : 8$
0040 7811 $J_2 : \exp(-\sqrt[3]{5})$
0040 8701 $\ln(4/5) \div \exp(4)$
0040 9600 $\mathrm{sq} : (5/2)^{-3}$
0041 1274 $\mathrm{cu} : P_{20}$
0041 1522 $\mathrm{cu} : \sqrt[3]{3}/9$
0041 1532 $2^x : 3/2 - 3\pi$
0041 2085 $J_2 : 2/11$
0041 2088 $\zeta : 1/\ln(\sqrt{3}/2)$
0041 2146 $10^x : 1/3 - e$
0041 2416 $J_2 : 4(1/\pi)/7$
0041 3817 $\exp(1/5) - \exp(4/5)$
0041 3845 $\ell_{10} : 7\sqrt[3]{3}$
0041 6517 $3/4 + s\pi(\sqrt{3})$
0041 6596 $\mathrm{sr} : \zeta(7)$
0041 6602 $e^x : 11/10$
0041 6839 $\zeta : 1/\ln(\sqrt[3]{2}/2)$
0041 6861 $e^x : 2\sqrt{3} - 4\sqrt{5}$
0041 7009 $\mathrm{sr} : (\ln 2/3)^{-3}$
0041 7257 $2^x : 1/4 - 3e$
0041 8501 $J_1 : 5\pi^2/3$
0042 0143 $10^x : 2e/3 - 4\pi/3$
0042 2878 $10^x : 7(1/\pi)/2$
0042 3444 $2^x : 3\sqrt{2}/4 - 4\sqrt{5}$
0042 5491 $10^x : (1/(3e))^3$
0042 6532 $2^x : 3 - 4e$
0042 7257 $\ell_{10} : \ln 2/7$
0042 7445 $J_2 : (\sqrt[3]{5}/3)^3$
0042 7718 $\ell_{10} : 2\sqrt{3}/3 + 4\sqrt{5}$
0042 8289 $\mathrm{cr} : 1/3 + e/4$
0042 8819 $10^x : 2\sqrt{2} - 3\sqrt{3}$
0043 0920 $10^x : \exp(-2\pi)$
0043 2303 $2^x : (2e)^{-3}$
0043 3638 $\sinh : 1/\ln(\sqrt{5}/3)$
0043 5439 $\mathrm{at} : \exp(-2e)$
0043 5442 $\mathrm{id} : \exp(-2e)$
0043 5443 $\mathrm{as} : \exp(-2e)$
0043 5660 $10^x : 3/2 + e/2$
0043 5917 $Li_2 : \exp(-2e)$
0043 6391 $e^x : \exp(-2e)$
0043 8167 $J_2 : 3/16$
0043 8696 $\ell_{10} : 5\sqrt{2}/7$
0043 8912 $t\pi : 1/\ln(2\pi/3)$
0043 9097 $e^x : 3/2 - 4\sqrt{3}$
0043 9180 $\zeta : 4\pi^2/5$
0044 0604 $e^x : 4 - 3\pi$
0044 2069 $2^x : 1/3 - 3e$
0044 2484 $10^x : 1 - 3\sqrt{5}/2$
0044 2580 $\ell_2 : 2\sqrt{2} + 3\sqrt{3}$
0044 4444 $\mathrm{sq} : 1/15$
0044 4600 $J_2 : \exp(-5/3)$
0044 5081 $\mathrm{at} : 2/3 + e/3$
0044 5845 $10^x : 3\sqrt{2}/2 - 2\sqrt{5}$
0044 6718 $\mathrm{sq} : 3\zeta(3)$
0044 6890 $Ei : 7(1/\pi)/6$
0044 7537 $e^x : 3\sqrt{3}/4 - 3\sqrt{5}$
0044 7742 $\ell_{10} : 4\sqrt{3}/7$
0044 8572 $e^x : 1/4 - 4\sqrt{2}$
0044 8666 $10^x : 1/4 - 3\sqrt{3}/2$
0044 8729 $\mathrm{sq} : 1/2 - \sqrt{3}/4$
0044 9160 $J_1 : \exp(-3\pi/2)$
0044 9196 $2/3 \div \exp(5)$
0044 9881 $J_2 : \sqrt[3]{5}/9$
0045 0831 $\ell_{10} : e/4 + 3\pi$
0045 1965 $2^x : 2\sqrt{3}/3 - 4\sqrt{5}$
0045 1999 $\ln : \sqrt{2} + \sqrt{5} - \sqrt{7}$
0045 2145 $J_2 : 4/21$
0045 2431 $\zeta : 5\pi/2$
0045 2643 $\lambda : \exp(-13\pi/5)$
0045 3022 $\mathrm{id} : \sqrt{2} + \sqrt{5} - \sqrt{7}$
0045 4560 $J_2 : 3(1/\pi)/5$
0045 4897 $erf : (2\pi)^{-3}$
0045 5332 $\Psi : \sqrt{3}/5$
0045 5334 $\mathrm{cu} : (\ln 3/2)^3$
0045 5777 $\mathrm{at} : \sqrt{2}/2 + \sqrt{3}/2$
0045 5956 $\sinh : \exp(-1/(3e))$
0045 6104 $2^x : 3\sqrt{2}/2 - \sqrt{5}/2$
0045 6446 $\mathrm{id} : \ln(\ln 3/3)$
0045 7876 $\mathrm{cu} : P_1$
0045 7890 $1/4 \div \exp(4)$
0045 7918 $K : \ln(\sqrt{5})$
0045 8028 $\Gamma : 1/\ln(\sqrt{3}/2)$
0045 9576 $t\pi : 2\sqrt{2}/3 - 4\sqrt{5}$
0045 9708 $e^x : -\sqrt{2} - \sqrt{3} - \sqrt{5}$
0046 0912 $s\pi : e + 2\pi$

0046 0917 $\mathrm{t}\pi : e + 2\pi$
0046 1471 $\zeta : 7\sqrt{5}/2$
0046 1535 $J_2 : \sqrt{3}/9$
0046 1656 $10^x : e/9$
0046 2124 $e^x : e/3 - 2\pi$
0046 2962 $\mathrm{cu} : 1/6$
0046 3151 $e^x : 4/3 - 3\sqrt{5}$
0046 3679 $\mathrm{cu} : 3/2 + 2\sqrt{5}$
0046 4397 $\mathrm{at} : 3/4 - \sqrt{5}/3$
0046 4400 $\mathrm{id} : 3/4 - \sqrt{5}/3$
0046 4402 $\mathrm{as} : 3/4 - \sqrt{5}/3$
0046 5480 $e^x : 3/4 - \sqrt{5}/3$
0046 5482 $\ln : 1/4 + \sqrt{5}/3$
0046 5794 $\zeta : 4(e + \pi)/3$
0046 7038 $\ln(2) \div \exp(5)$
0046 7944 $\mathrm{cu} : P_{60}$
0046 8131 $2^x : \exp(-5)$
0046 8282 $J_1 : 7$
0046 9432 $10^x : 1/2 - 2\sqrt{2}$
0046 9882 $\mathrm{cu} : P_{44}$
0046 9987 $10^x : -\sqrt{2} + \sqrt{3} - \sqrt{7}$
0047 1895 $1/5 + \ln(\sqrt{5})$
0047 2130 $\zeta : 9\sqrt{3}/2$
0047 3010 $\sqrt{5} \div \exp(4/5)$
0047 3343 $\ell_2 : 3\sqrt{2} - \sqrt{5}$
0047 3650 $\zeta : \sqrt{2} - \sqrt{5} - \sqrt{7}$
0047 4211 $\mathrm{cu} : \sqrt{3}/3 - \sqrt{5}/3$
0047 4428 $\mathrm{cu} : (2e/3)^{-3}$
0047 6072 $e^x : 3e/2 - 3\pi$
0047 6899 $\mathrm{cu} : 1/4 - 3\sqrt{3}$
0047 7594 $\mathrm{sr} : 7\sqrt[3]{3}/10$
0047 7795 $J_1 : (3\pi/2)^{-3}$
0048 0453 $\mathrm{sq} : \ln 2/10$
0048 3972 $\mathrm{t}\pi : (\ln 3/3)^{-1/3}$
0048 4311 $J_0 : 7\pi^2/8$
0048 4443 $10^x : 2/3 - 4\sqrt{5}/3$
0048 5011 $10^x : 1/3 + 2e$
0048 6057 $\Psi : 22/15$
0048 7483 $J_1 : 6(e + \pi)/5$
0048 7555 $e^x : 1/3 - 4\sqrt{2}$
0048 8656 $e^x : \sqrt{2} + 3\sqrt{5}/4$
0048 8657 $J_2 : 2\ln 2/7$
0049 1340 $erf : \exp(-2e)$
0049 2221 $e^x : 3\sqrt[3]{5}$
0049 3102 $J_2 : 5(1/\pi)/8$
0049 3208 $\mathrm{id} : 3\sqrt{2}/4 + 4\sqrt{5}$
0049 4590 $\zeta : (\sqrt{2}\pi)^{-1/3}$
0049 4845 $10^x : \sqrt{3}/4 + 3\sqrt{5}$
0049 5047 $\lambda : \exp(-18\pi/7)$
0049 5431 $\mathrm{cu} : 9e\pi/10$
0049 5840 $e^x : \sqrt{23}$
0049 5853 $\mathrm{sq} : 4e + 3\pi$
0049 6032 $\mathrm{sq} : 3/4 - e/4$
0049 6205 $2^x : 1/2 - 3e$
0049 6276 $\Psi : \exp(-4/3)$
0049 6971 $\mathrm{at} : (e + \pi)^{-3}$
0049 6975 $\mathrm{id} : (e + \pi)^{-3}$
0049 6977 $\mathrm{as} : (e + \pi)^{-3}$
0049 7594 $Li_2 : (e + \pi)^{-3}$
0049 8212 $e^x : (e + \pi)^{-3}$
0049 8335 $J_2 : 1/5$
0049 9522 $2^x : 3\sqrt{3}/4 - 4\sqrt{5}$
0050 0000 $\mathrm{cu} : \sqrt[3]{5}/10$
0050 0039 $J_2 : \zeta(3)/6$
0050 0193 $\Psi : (\sqrt{2}/3)^{-1/2}$
0050 0351 $\mathrm{sq} : 2(1/\pi)/9$
0050 1211 $\mathrm{sq} : 3/2 - \pi/2$
0050 1626 $e^x : 16/23$
0050 2168 $e^x : \sqrt{2} - 3\sqrt{5}$
0050 5033 $\ln(5) + \exp(1/3)$
0050 5063 $\mathrm{cu} : 3 - 2\sqrt{2}$
0050 5253 $\ln : 1 + \sqrt{3}$
0050 5346 $3/4 \div \exp(5)$
0050 6345 $\mathrm{sr} : 5\sqrt{2}/7$
0050 6385 $\mathrm{cu} : \zeta(3)/7$
0050 6650 $\mathrm{at} : 5\sqrt[3]{2}/4$
0050 6686 $\zeta : \sqrt{6}\pi$
0050 6834 $\zeta : 9\sqrt[3]{5}/2$
0050 7511 $2^x : \sqrt[3]{4}$
0050 7623 $e^x : 1 - 2\pi$
0050 8411 $Ei : 5e/6$
0050 8470 $J_2 : \sqrt{2}/7$
0050 8813 $10^x : 3\sqrt{2}/4 - 3\sqrt{5}/2$
0051 0151 $\zeta : 9e\pi/10$
0051 0204 $\mathrm{sq} : 1/14$
0051 0954 $2^x : 2e/3 - 3\pi$
0051 1539 $2^x : 4/3 - 4\sqrt{5}$
0051 1544 $J_2 : (\sqrt{2}\pi/2)^{-2}$
0051 1631 $10^x : 4/3 - 4e/3$
0051 3091 $\sinh : (\ln 2)^{1/3}$
0051 3591 $\zeta : 7\pi^2/9$
0051 3732 $\mathrm{cr} : (\pi/3)^{1/3}$
0051 3990 $e^x : \sqrt[3]{9}$
0051 4163 $\Gamma : (\ln 2)^{-3}$
0051 4225 $\mathrm{id} : 5\zeta(3)/2$
0051 4633 $\mathrm{cu} : 3/4 - \sqrt{3}/3$
0051 5107 $\mathrm{cu} : 5\zeta(3)/6$
0051 7161 $\zeta : 23/3$
0051 8543 $\mathrm{sr} : 2/3 + 3\sqrt{5}/2$
0051 8878 $e^x : \sqrt{2} - \sqrt{6} + \sqrt{7}$
0051 9027 $J_2 : 1/\sqrt{24}$
0051 9028 $\exp(1/2) + \exp(\sqrt{5})$
0051 9615 $\mathrm{cu} : \sqrt{3}/10$
0052 0351 $\mathrm{cu} : \ln 2/4$
0052 0711 $J_2 : \exp(-\sqrt[3]{4})$
0052 0972 $10^x : 4 - 2\pi$
0052 4672 $J_0 : (\sqrt{5}/3)^{-3}$
0052 6009 $\sinh : (\sqrt[3]{3})^{-1/3}$
0052 6012 $\mathrm{cu} : 4/23$
0052 6908 $\ln(3/4) \div \exp(4)$
0052 7510 $Ei : \zeta(3)/7$
0052 8400 $\ln : 1/2 + 4\sqrt{3}$
0052 8758 $J_2 : \sqrt[3]{3}/7$
0053 0258 $e^x : 4\sqrt{2} + 4\sqrt{5}$
0053 1719 $10^x : \sqrt{2}/2 - 4\sqrt{5}/3$
0053 2110 $\zeta : 1/\ln(\sqrt{5}/3)$
0053 3250 $\Gamma : 9\sqrt{5}/10$
0053 4928 $e^x : \sqrt{2}/2 + 2\sqrt{5}/3$
0053 6034 $10^x : 3\sqrt{2} - 2\sqrt{3}$
0053 6166 $\Gamma : 7\pi/6$
0053 7774 $\mathrm{sq} : 4\sqrt{3}/3 - \sqrt{5}$
0053 8006 $\mathrm{cu} : 2\sqrt{2}/3 - \sqrt{5}/2$
0053 8231 $J_2 : \exp(-\pi/2)$
0053 8493 $\mathrm{s}\pi : 5\zeta(3)/6$
0053 8501 $\mathrm{t}\pi : 5\zeta(3)/6$
0053 8564 $J_2 : 3\ln 2/10$
0053 9035 $4/5 \div \exp(5)$
0054 0575 $J_2 : 5/24$
0054 0816 $2^x : 3e$
0054 1036 $2^x : \sqrt{2} - 4\sqrt{5}$
0054 1511 $J_2 : 1/\sqrt{23}$
0054 2215 $\ln(\sqrt{5}) \div \exp(5)$
0054 3572 $10^x : 1/3 - 3\sqrt{3}/2$
0054 4081 $\mathrm{cu} : P_{14}$
0054 4710 $\sinh : 25/4$
0054 5597 $\mathrm{t}\pi : \exp(\sqrt{2}/3)$
0054 5966 $\zeta : 8e\pi/9$
0054 6510 $2/5 + \ln(2/3)$
0054 7149 $e^x : 3/2 - 3\sqrt{5}$
0054 8696 $\mathrm{sq} : (\sqrt[3]{2}/3)^3$
0054 8857 $\Gamma : (1/(3e))^{-1/3}$
0054 9158 $J_2 : \sqrt[3]{2}/6$
0054 9562 $\mathrm{cu} : 3/17$
0055 1973 $J_2 : 4/19$
0055 2427 $\mathrm{cu} : \sqrt{2}/8$
0055 3009 $\mathrm{cu} : 5(1/\pi)/9$
0055 3783 $\mathrm{cu} : \exp(-\sqrt{3})$
0055 5002 $\mathrm{cu} : 3/2 - 3\sqrt{5}/4$
0055 5383 $\zeta : 8/23$
0055 6058 $2^x : (5)^{-3}$
0055 6615 $\ell_{10} : 4\sqrt{2} + 2\sqrt{5}$
0055 6819 $\ell_{10} : 1/3 + e/4$
0055 6972 $2^x : 2/3 - 3e$
0055 8354 $\zeta : 6\sqrt[3]{2}$
0055 8457 $3/4 \div s\pi(\sqrt{3})$
0055 9118 $e^x : 1/4 - 2e$
0056 0772 $erf : (e + \pi)^{-3}$
0056 0786 $J_2 : 2(1/\pi)/3$
0056 1613 $\ln : e + 3\pi/2$
0056 2886 $J_1 : (3\pi)^{-2}$
0056 4880 $\zeta : \ln(1/(3\pi))$
0056 5019 $J_0 : 7\sqrt[3]{5}/5$
0056 6032 $J_2 : 1/\sqrt{22}$
0056 6802 $\zeta : \sqrt{3} - \sqrt{6} - \sqrt{7}$
0056 8314 $\Psi : 18/23$
0056 8602 $\zeta : 9(e + \pi)/7$
0056 8925 $J_2 : \sqrt[3]{5}/8$
0056 9478 $10^x : \sqrt{2}/4 - 3\sqrt{3}/2$
0057 0011 $\mathrm{cu} : 4/3 - 2\sqrt{3}/3$
0057 0025 $\ell_{10} : \pi^2/10$
0057 0122 $J_1 : (\sqrt{2}\pi)^{-3}$
0057 0320 $\theta_3 : \exp(-(e + \pi))$
0057 1028 $10^x : \ln(1/(3\pi))$
0057 1786 $J_2 : 3/14$
0057 1951 $10^x : 2 - 3\sqrt{2}$
0057 2502 $e^x : \exp((e + \pi)/3)$
0057 4183 $2^x : 3/2 - 4\sqrt{5}$
0057 4570 $J_1 : 8\sqrt[3]{3}/3$
0057 4644 $\Gamma : 10\ln 2/7$
0057 5979 $e^x : 1/2 - 4\sqrt{2}$
0057 6938 $\mathrm{at} : 1 + \sqrt{3}/3$
0057 7132 $\mathrm{sr} : 2\sqrt{3} + \sqrt{5}/4$
0057 7137 $e^x : 3 - 3e$
0057 9034 $\mathrm{cu} : 1/2 - e/4$
0057 9295 $\mathrm{s}\pi : (1/(3e))^3$
0057 9305 $\mathrm{t}\pi : (1/(3e))^3$
0058 0935 $J_2 : (5/3)^{-3}$
0058 1435 $10^x : e/3 - \pi$
0058 1715 $\ell_{10} : \sqrt{3} - \sqrt{5}/3$
0058 1994 $2^x : 2 - 3\pi$
0058 2465 $\ln : 3\sqrt{3} + \sqrt{5}$
0058 2672 $\zeta : 15/2$
0058 3090 $\mathrm{cu} : \sqrt[3]{2}/7$
0058 3652 $J_2 : \sqrt{3}/8$
0058 5200 $\mathrm{sq} : \sqrt{2} - 2\sqrt{5}/3$
0058 5937 $\mathrm{cu} : \sqrt[3]{3}/8$
0058 6274 $\zeta : \ln(1/(2\pi))$
0058 6671 $\mathrm{s}\pi : \exp(-2\pi)$
0058 6681 $\mathrm{t}\pi : \exp(-2\pi)$
0058 6913 $Ei : e/4$
0058 6916 $10^x : 3/4 - 4\sqrt{5}/3$
0058 8088 $J_1 : \exp(-\sqrt{2}\pi)$
0058 8414 $J_2 : 5/23$
0058 9572 $\mathrm{sq} : 2\sqrt{2}/3 - \sqrt{3}/2$
0059 0091 $2^x : 3/4 - 3e$
0059 0214 $\Gamma : 7\sqrt{2}/10$
0059 1295 $J_1 : 9e\pi/2$
0059 1698 $\mathrm{cu} : \exp(-\sqrt[3]{5})$
0059 1706 $J_0 : 8e/9$
0059 1715 $\mathrm{sq} : 1/13$
0059 2879 $J_2 : 1/\sqrt{21}$
0059 3014 $\mathrm{cu} : 3\sqrt{3}/4 - \sqrt{5}/2$
0059 3151 $\mathrm{sq} : \ln 2/9$
0059 4368 $\zeta : 7e\pi/8$
0059 6390 $e - \ln(3/4)$
0059 6598 $2^x : 3\pi^2/8$
0059 6933 $Li_2 : 13/17$
0059 7822 $2^x : 3e/4 - 3\pi$
0059 8306 $\mathrm{sq} : 1/2 - \sqrt{3}/3$
0059 9428 $\mathrm{sq} : 4 - 3e/2$
0060 0454 $Li_2 : (\sqrt[3]{5})^{-1/2}$
0060 0988 $\zeta : (\ln 3/3)^{-2}$
0060 1051 $\mathrm{cu} : 2/11$
0060 1777 $\mathrm{cu} : 4(1/\pi)/7$
0060 1956 $\mathrm{sq} : 6e$
0060 2394 $\zeta : 10\sqrt{5}/3$
0060 2535 $\Gamma : 4\sqrt{3}/7$
0060 2857 $\zeta : ((e + \pi)/3)^3$
0060 3728 $\mathrm{cu} : \zeta(9)$
0060 4948 $10^x : 1/2 - e$
0060 7708 $e^x : 1/3 - 2e$
0060 9411 $\mathrm{at} : 4\sqrt{2}/3 - 2\sqrt{3}$
0060 9656 $e^x : \sqrt[3]{4/3}$
0061 0521 $1/3 \div \exp(4)$
0061 2976 $\mathrm{cu} : 1/4 - \sqrt{3}/4$
0061 4488 $\mathrm{id} : \sqrt{2}/2 + 3\sqrt{3}/4$
0061 4747 $J_2 : 2/9$
0061 5405 $\mathrm{cr} : 4 - 4\sqrt{5}/3$
0061 6073 $10^x : 3/4 + 2e$
0061 6700 $J_1 : (\ln 2/3)^3$

0061 7391 $\ln(2/5) \div \exp(5)$
0061 7657 cu : $3\sqrt{2} - 4\sqrt{5}/3$
0061 8028 $J_2 : 7(1/\pi)/10$
0061 9050 cu : P_{63}
0061 9200 $\lambda : \exp(-5\pi/2)$
0061 9201 sq : $2/3 - \sqrt{5}/3$
0061 9760 $J_2 : \exp(-3/2)$
0062 0949 $2^x : -\sqrt{5} - \sqrt{6} - \sqrt{7}$
0062 1125 ln : $4\sqrt{5}/9$
0062 1614 cu : P_{47}
0062 2091 sr : $9\sqrt{5}/5$
0062 2330 at : $(2e)^{-3}$
0062 2338 id : $(2e)^{-3}$
0062 2342 as : $(2e)^{-3}$
0062 2399 $J_2 : \sqrt{5}/10$
0062 2640 $10^x : 3e/2 - 2\pi$
0062 3309 $Li_2 : (2e)^{-3}$
0062 4064 sq : $2/3 + \pi/2$
0062 4278 $e^x : (2e)^{-3}$
0062 4616 $2^x : \exp(-3\pi/2)$
0062 4998 $\zeta : 3\pi^2/4$
0062 7668 id : $(\pi)^3$
0062 8020 $10^x : 21/22$
0062 9010 $2^x : \sqrt{2} + \sqrt{5} - \sqrt{7}$
0063 0295 $J : \ln(\pi/3)$
0063 0923 $\zeta : (e)^2$
0063 0951 $s\pi : \zeta(9)$
0063 0963 $t\pi : \zeta(9)$
0063 1378 sq : $\sqrt{2}/4 - \sqrt{3}/4$
0063 1827 $10^x : 2\sqrt{3}/3 - 3\sqrt{5}/2$
0063 2616 $J_1 : 9\sqrt[3]{5}/4$
0063 3257 sq : $(1/\pi)/4$
0063 4771 $\ln(\sqrt{2}) \div \exp(4)$
0063 5065 cu : $(\sqrt[3]{5}/3)^3$
0063 6572 $10^x : 3 - 3\sqrt{3}$
0063 7855 $\theta_3 : -\sqrt{2} - \sqrt{3} + \sqrt{6}$
0063 8058 $t\pi : P_2$
0064 0000 sq : $2/25$
0064 0886 ln : $2 + 2e$
0064 1071 cr : $3e/8$
0064 2122 $\ell_{10} : 2e + 3\pi/2$
0064 2886 $J_2 : 5/22$
0064 3119 cr : $4 + 3e/2$
0064 3121 sr : $1/3 + e/4$
0064 3192 sinh : $1/\ln(\sqrt{3})$
0064 3401 $J_2 : 5(1/\pi)/7$
0064 3710 $J_2 : \exp(-\sqrt{2}\pi/3)$
0064 4617 $\zeta : \sqrt{2} - \sqrt{5} - \sqrt{6}$
0064 5703 $e^x : 4\sqrt{2}/3 - 4\sqrt{3}$
0064 5872 $2^x : 3/2 - 2\sqrt{5}/3$
0064 6173 $\zeta : -\sqrt{2} + \sqrt{5} - \sqrt{7}$
0064 6835 $J_2 : (2\pi/3)^{-2}$
0064 7455 $10^x : 2 - 4\pi/3$
0064 8293 ln : $8\sqrt[3]{5}/5$
0064 8869 cr : $4\sqrt{2}/3 - \sqrt{3}/2$
0065 2097 $\ell_2 : \sqrt{2} + \sqrt{5} - \sqrt{7}$
0065 2184 ln : $1/2 + \sqrt{5}$
0065 4161 sq : $4/3 - \sqrt{2}$
0065 5013 $J_2 : 1/\sqrt{19}$
0065 6679 $\zeta : 22/3$
0065 7270 id : $3\sqrt{2} - \sqrt{5}$
0065 7393 $t\pi : 3\ln 2/4$
0065 7635 $\zeta : \sqrt{5} + \sqrt{6} + \sqrt{7}$
0065 7825 $e^x : 4/3 + 3\sqrt{2}$
0065 8073 $\zeta : 7\pi/3$
0065 8350 sq : $3\sqrt{2}/2 - \sqrt{5}/2$
0065 8765 $10^x : \exp(-(e + \pi))$
0065 9179 cu : $3/16$
0065 9826 cu : P_9
0066 0180 ln : $2\sqrt{2}/3 - \sqrt{3}/3$
0066 0229 $J_1 : ((e + \pi)/3)^2$
0066 0697 $\zeta : 5(e + \pi)/4$
0066 0787 $10^x : 1/2 + 4\sqrt{2}/3$
0066 1063 $\zeta : 1/\ln(\sqrt{3}/3)$
0066 2477 cu : $2 - 2e/3$
0066 2731 $J_2 : 3/13$
0066 3109 $\zeta : 6e\pi/7$
0066 4332 $J_2 : \ln 2/3$
0066 4570 $2^x : (3\pi/2)^{-3}$
0066 4740 $J_2 : (\sqrt[3]{3}/3)^2$
0066 4825 $\ln(2/5) \times \ln(3)$
0066 5323 cu : $3\sqrt{2}/2 - 4\sqrt{3}/3$
0066 7331 $\ell_{10} : 2 + 3e$
0066 7620 $\ell_{10} : (\pi/3)^{1/3}$
0066 8220 cu : P_{36}
0066 9249 $\zeta : 17/12$
0066 9509 cr : $2/3 + \sqrt{2}/4$
0067 0958 $e^x : 4 + \sqrt{5}/2$
0067 1549 $\ell_2 : 1/4 + \sqrt{5}/3$
0067 1770 $\theta_3 : 11/24$
0067 1803 $\exp(\pi) + s\pi(1/3)$
0067 2881 cu : $4 - 4\pi/3$
0067 3784 tanh : $\exp(-5)$
0067 3794 id : $\exp(-5)$
0067 3799 as : $\exp(-5)$
0067 4380 $2^x : \sqrt{3} - 4\sqrt{5}$
0067 4933 $Li_2 : \exp(-5)$
0067 5582 cr : $10e$
0067 6069 $e^x : \exp(-5)$
0067 8907 $t\pi : 2\sqrt{2} - \sqrt{3}/3$
0068 0438 $e^x : 2/3 - 4\sqrt{2}$
0068 1524 $\Gamma : 8\sqrt[3]{2}/5$
0068 4386 $J_0 : 11/2$
0068 5368 at : $\sqrt{5/2}$
0068 5421 $10^x : 1/\ln(\sqrt[3]{2}/2)$
0068 5634 id : $10\zeta(3)/3$
0068 5871 cu : $\sqrt[3]{5}/9$
0068 7487 $10^x : 4\sqrt{3}/3 - 2\sqrt{5}$
0068 7992 $\Psi : 16/11$
0068 8854 $J_2 : 4/17$
0068 9032 $10^x : 2/3 - 2\sqrt{2}$
0068 9495 $\zeta : -\sqrt{2} + \sqrt{3} - \sqrt{7}$
0069 0055 $e^x : \sqrt{3} - 3\sqrt{5}$
0069 1070 cu : $4/21$
0069 1235 $J_2 : \sqrt{2}/6$
0069 1510 sq : $e/2 - 4\pi/3$
0069 4171 $10^x : 25/24$
0069 4444 sq : $1/12$
0069 5287 $J_2 : \exp(-\sqrt[3]{3})$
0069 5337 $Ei : 10\sqrt[3]{2}/9$
0069 6601 cu : $3/4 - \sqrt{5}/4$
0069 6633 cu : $3(1/\pi)/5$
0069 7585 $10^x : 10\pi/3$
0069 7616 $t\pi : 7\pi/8$
0069 7847 $\zeta : 8e/3$
0069 7986 $\theta_3 : \ln(1/2)$
0069 8836 $2^x : 2/3 + 3\pi$
0070 0122 cu : P_{52}
0070 1741 $2^x : 1 - 3e$
0070 2224 $erf : (2e)^{-3}$
0070 3265 cu : $5\sqrt{3}/6$
0070 4188 cu : $3\sqrt{3}/4 - 2\sqrt{5}/3$
0070 5275 $J_2 : 5/21$
0070 7817 $10^x : 3e/4 - 4\pi/3$
0070 8445 $e^x : 4/3 - 2\pi$
0070 9037 $J_2 : 3(1/\pi)/4$
0071 1071 $e^x : 1/4 - 3\sqrt{3}$
0071 1158 $2^x : 2\sqrt{2} + 3\sqrt{3}/2$
0071 1521 $t\pi : (\sqrt[3]{2}/3)^{1/3}$
0071 1612 $Ei : 17/25$
0071 2222 ln : $3\sqrt{2}/4 + 3\sqrt{5}/4$
0071 2409 $e^x : 4 - 4\sqrt{5}$
0071 2540 as : $2 - 2\sqrt{3}/3$
0071 2778 cu : $\sqrt{3}/9$
0071 4026 $2^x : 2e - 4\pi$
0071 6373 $\zeta : 6\zeta(3)$
0071 6550 $J_2 : 6/25$
0071 6632 $10^x : \sqrt{3} + \sqrt{5} - \sqrt{7}$
0071 6773 cu : $3/4 - 2\sqrt{2}/3$
0071 6937 $\zeta : 5\sqrt[3]{3}$
0071 6945 cu : $10\pi^2/3$
0071 7922 $e^x : 1/2 - 2e$
0071 8779 $J_2 : \sqrt[3]{3}/6$
0071 8996 $J_2 : \zeta(3)/5$
0072 0746 $Ei : 7/5$
0072 1549 $10^x : -\sqrt{3} + \sqrt{5} - \sqrt{7}$
0072 1784 $10^x : 1 - \pi$
0072 3949 $e^x : 2 - 4\sqrt{3}$
0072 4203 id : $3\sqrt{3}/4 + 3\sqrt{5}$
0072 6611 $t\pi : 4\sqrt{3} - 3\sqrt{5}/4$
0072 6806 $e^x : \sqrt{2} + \sqrt{3} - \sqrt{6}$
0072 6966 $e^x : e/2 - 2\pi$
0072 9518 $\exp(\pi) \div s\pi(\sqrt{2})$
0073 1696 $J_2 : 1/\sqrt{17}$
0073 2625 $2/5 \div \exp(4)$
0073 5189 $J_2 : \exp(-\sqrt{2})$
0073 5931 sq : $3/2 - \sqrt{2}$
0073 5998 id : $2e + \pi/2$
0073 6172 $\Gamma : \exp(-\sqrt{2}\pi/3)$
0073 7328 cu : $3\sqrt{2}/4 - \sqrt{3}/2$
0073 8465 $\zeta : \sqrt{3} - \sqrt{5} - \sqrt{7}$
0073 8494 $10^x : -\sqrt{2} + \sqrt{3} - \sqrt{6}$
0073 9571 $e^x : 3/4 - 4\sqrt{2}$
0073 9999 $10^x : 4/3 - 2\sqrt{3}$
0074 0239 $\ln(3) \div \exp(5)$
0074 0880 cu : $1/\ln(\sqrt{5}/2)$
0074 1939 $\ell_{10} : 2\sqrt{3} + 3\sqrt{5}$
0074 2220 $J_2 : \sqrt[3]{5}/7$
0074 2635 $\ln(2/3) \div \exp(4)$
0074 4481 cu : $2/3 - \sqrt{2}/3$
0074 5151 ln : $\sqrt{15/2}$
0074 7013 $10^x : \sqrt{2}/3 - 3\sqrt{3}/2$
0074 7132 $\ell_2 : 4/3 + 3\sqrt{5}$
0074 8006 at : $19/12$
0075 0124 $2^x : 4\sqrt{2}/3 - 4\sqrt{5}$
0075 0707 sq : $\ln 2/8$
0075 0972 $10^x : 3/2 - 4e/3$
0075 1841 $J_1 : (1/(3e))^2$
0075 2494 $\ell_{10} : 3/4 + 3\pi$
0075 3251 $J_2 : 18$
0075 5209 $\zeta : 14/23$
0075 5565 $2^x : 9\sqrt{2}$
0075 6143 sq : $2/23$
0075 9955 $J_0 : 7\pi/4$
0076 0284 $erf : \exp(-5)$
0076 1408 $J_2 : \sqrt{3}/7$
0076 1805 $\pi + s\pi(1/3)$
0076 2257 $J_2 : 7(1/\pi)/9$
0076 6657 $10^x : \sqrt{3}/2 - 4\sqrt{5}/3$
0076 6919 sq : $4/3 - 4e$
0076 7594 sq : $\sqrt{2}/3 - \sqrt{5}/4$
0076 7616 $\zeta : 5e\pi/6$
0076 7643 $J_2 : \sqrt{5}/9$
0076 9399 $\ln(2/3) + \exp(5)$
0076 9684 $\Gamma : \pi^2/10$
0076 9824 cu : $2\sqrt{2}/3 - \sqrt{5}/3$
0077 1168 $e^x : (\ln 2/3)^{-1/2}$
0077 1310 $\exp(2/5) \times s\pi(\sqrt{5})$
0077 1312 $\ln(\pi) \div \exp(5)$
0077 1588 sr : $(\pi/3)^{1/3}$
0077 2866 $e^x : 1/3 - 3\sqrt{3}$
0077 4368 $\lambda : \exp(-17\pi/7)$
0077 6733 cu : $2\ln 2/7$
0077 7188 $J_2 : 1/4$
0078 1226 $J_1 : (4)^{-3}$
0078 3078 $10^x : 1/4 - 3\pi/4$
0078 3391 $2^x : (3\pi)^{-2}$
0078 7390 cu : $5(1/\pi)/8$
0078 8251 $10^x : 11/23$
0078 8653 $e^x : \sqrt{2}/4 - 3\sqrt{3}$
0078 9509 $J_2 : \sqrt[3]{2}/5$
0079 0550 ln : $4\sqrt[3]{2}/5$
0079 1639 cr : $2e/3 + 2\pi$
0079 2329 cu : $2/3 - \sqrt{3}/2$
0079 3158 $\zeta : 5\sqrt{2}$
0079 3502 $2^x : (\sqrt{2}\pi)^{-3}$
0079 3683 id : $4\sqrt[3]{2}/5$
0079 4582 $\zeta : 9\pi/4$
0079 6870 $10^x : \ln(1/(3e))$
0079 7741 sq : $2/3 - \sqrt{3}/3$
0079 7803 $\Psi : 7\sqrt[3]{2}/6$
0079 7854 $10^x : 1/2 - 3\sqrt{3}/2$
0079 8887 sinh : $(2\pi/3)^{1/2}$
0079 9345 $\ell_{10} : 4 - 4\sqrt{5}/3$
0079 9446 $\zeta : 1/\ln(\sqrt[3]{3}/2)$
0079 9982 at : $(5)^{-3}$
0080 0000 id : $(5)^{-3}$
0080 0008 as : $(5)^{-3}$
0080 0223 ln : $\sqrt{3}/4 + \sqrt{5}/4$
0080 1297 $\zeta : -\sqrt{5} + \sqrt{6} - \sqrt{7}$

0080 1605 $Li_2 : (5)^{-3}$
0080 2786 cu : P_{16}
0080 2801 cu : P_{40}
0080 3208 $e^x : (5)^{-3}$
0080 3433 id : $2\sqrt{3} - 2\sqrt{5}$
0080 4120 cu : $\zeta(3)/6$
0080 4231 $\zeta : \exp((e+\pi)/3)$
0080 4595 $e^x : 4\sqrt{2}/3 - 3\sqrt{5}$
0080 5476 $\zeta : 5\pi^2/7$
0080 6198 $J_2 : 4(1/\pi)/5$
0080 6288 $\theta_3 : (2\pi)^{-3}$
0080 7613 $2^x : 1/\ln(\sqrt{3}/2)$
0081 0268 $\Psi : (2\pi/3)^{-1/3}$
0081 0651 $\exp(\pi) \times s\pi(2/5)$
0081 1598 cu : $3/2 - 3\sqrt[3]{3}/4$
0081 2018 $2^x : 2 - 4\sqrt{5}$
0081 3352 cu : $3\sqrt{2}/4 - 4\sqrt{3}$
0081 5936 $\zeta : 6(e+\pi)/5$
0081 6374 cu : P_8
0081 7329 $e^x : 3\sqrt{2}/2 - 4\sqrt{3}$
0081 7879 at : $3 - \sqrt{2}$
0081 8610 $2^x : \exp(-\sqrt{2}\pi)$
0081 8811 $J_1 : 1/\ln(e/3)$
0082 0092 $\zeta : \sqrt{5}\pi$
0082 0323 $J_2 : \exp(-e/2)$
0082 2291 $J_0 : (e+\pi)^{1/2}$
0082 2974 $\exp(1/5) \div \exp(5)$
0082 4207 $\ell_{10} : \sqrt{2} - \sqrt{3}/4$
0082 4260 sr : $3\sqrt{2}/2 + 4\sqrt{3}$
0082 4614 cu : $\sqrt{2}/7$
0082 4767 $\Gamma : 7\sqrt[3]{3}/5$
0082 6446 sq : $1/11$
0082 6829 $\ln(4) \times \exp(e)$
0082 7072 tπ : $6/17$
0082 7111 sq : $2(1/\pi)/7$
0082 7522 $J_2 : (\pi/2)^{-3}$
0082 8629 $\zeta : 1/\ln(2/3)$
0082 8713 $J_2 : 1/\sqrt{15}$
0083 1427 $\ell_2 : 2\sqrt{3} + \sqrt{5}/4$
0083 1461 ln : $\zeta(7)$
0083 1823 $10^x : \sqrt{3}/2 + \sqrt{5}$
0083 2129 sq : $(\sqrt{2}\pi/2)^{-3}$
0083 2574 $\ell_{10} : 3e/8$
0083 2717 $e^x : 3\sqrt{2} - 3\sqrt{5}/4$
0083 4781 $10^x : 3/4 - 2\sqrt{2}$
0083 4927 id : $\zeta(7)$
0083 6714 $10^x : 2 - 3e/2$
0083 6929 $e^x : 3/2 - 2\pi$
0083 7679 $\zeta : \pi/9$
0083 8230 sq : $10\sqrt[3]{3}$
0083 8420 $\Gamma : 5(1/\pi)/7$
0083 9801 $J_2 : 3\ln 2/8$
0084 1167 $Ei : 25/24$
0084 2670 $\ell_{10} : 4\sqrt{2}/3 - \sqrt{3}/2$
0084 5847 $J_2 : 6/23$
0084 6100 $10^x : 3\sqrt{3}/2 + 3\sqrt{5}/4$
0084 6188 tπ : $5e/4$
0084 6781 $\lambda : \exp(-12\pi/5)$
0084 8125 $e^x : 2/3 - 2e$
0084 9823 $\zeta : 1/\ln(\sqrt[3]{5}/3)$
0085 0517 cu : $1/\sqrt{24}$
0085 1553 $J : 1/25$
0085 1947 $\zeta : \sqrt{2} - \sqrt{3} - \sqrt{7}$
0085 2016 $J_0 : 8\ln 2$
0085 2239 ln : $3e/4 - \pi/3$
0085 3065 $2^x : 4 - 4e$
0085 3795 $J_2 : ((e+\pi)/3)^{-2}$
0085 4675 cu : $\exp(-\sqrt[3]{4})$
0085 4768 ln : $((e+\pi)/3)^3$
0085 5922 $J_0 : \sqrt{14}\pi$
0085 6746 $\zeta : \sqrt{3} - \sqrt{5} - \sqrt{6}$
0085 7263 $e^x : 2\sqrt{3}/3 + 4\sqrt{5}$
0085 8609 $2^x : (\ln 2/3)^3$
0085 8789 cr : $3\sqrt[3]{5}/5$
0086 0666 $J_2 : 5/19$
0086 1514 $10^x : e/2 + 3\pi/4$
0086 3122 $\Psi : 25/17$
0086 3524 $J_2 : \exp(-4/3)$
0086 3784 at : $\sqrt[3]{4}$
0086 4278 $1/5 \div \exp(\pi)$
0086 5169 $\exp(1/4) \div \exp(5)$
0086 7370 cu : $\sqrt{2}/4 - \sqrt{5}/4$
0086 9176 ln : $10\sqrt{5}/3$
0086 9385 $\ell_{10} : 2/3 + \sqrt{2}/4$
0087 0269 $\exp(\sqrt{3}) + \exp(\sqrt{5})$
0087 0884 $\theta_3 : \exp(-2e)$
0087 2165 $10^x : 1/4 - 4\sqrt{3}/3$
0087 2652 as : $11/13$
0087 3573 sπ : $(\ln 2)^{-3}$
0087 3607 tπ : $(\ln 2)^{-3}$
0087 4378 $J_2 : 5(1/\pi)/6$
0087 4515 $Ei : 1/\sqrt[3]{19}$
0087 4635 cu : $\sqrt[3]{3}/7$
0087 4725 $10^x : \sqrt{3} + \sqrt{5} - \sqrt{6}$
0087 4845 $\sqrt{2} + \ln(2/3)$
0087 5516 $\Gamma : 10\sqrt{2}/7$
0087 6161 $2^x : 7e/6$
0087 7344 $\zeta : 10\ln 2$
0087 7908 $2^x : 2/3 + \pi$
0087 8846 $\theta_3 : 11/12$
0087 9421 $\zeta : 4\sqrt{3}$
0088 0113 $\ell_{10} : 7\sqrt[3]{2}/9$
0088 0603 $2^x : -\sqrt{3} - \sqrt{6} - \sqrt{7}$
0088 0683 $\zeta : \sqrt{2} - \sqrt{3} - \sqrt{5}$
0088 0919 $10^x : 3\sqrt{3}/4 - 3\sqrt{5}/2$
0088 1834 sq : $1 - e/3$
0088 2362 $1/3 + s\pi(\sqrt{5})$
0088 3256 $2^x : 3\sqrt{2}/2 - 4\sqrt{5}$
0088 3633 $J_2 : 4/15$
0088 3998 cu : P_{39}
0088 4138 $2^x : 4/3 - 3e$
0088 4193 cu : P_{33}
0088 4359 sq : $3\sqrt{2}/4 - 2\sqrt{3}/3$
0088 4499 sq : $\ln(\ln 3)$
0088 4816 $\zeta : 1/\ln(1/\sqrt{2})$
0088 5038 $e^x : 4\sqrt{2} + 2\sqrt{5}/3$
0088 5133 cr : $2e/3 - \pi/4$
0088 6646 $J_2 : 2\zeta(3)/9$
0088 7145 $\ell_{10} : 4/3 - \sqrt{2}/4$
0088 7295 $e^x : \sqrt{2}/3 - 3\sqrt{3}$
0088 7554 $J_2 : 1/\sqrt{14}$
0088 7942 $10^x : 2/3 - e$
0088 8347 cu : $1/2 - \sqrt{2}/2$
0089 1043 sq : $2 - 2\pi/3$
0089 1480 $10^x : \sqrt{2} - 2\sqrt{3}$
0089 1909 $\zeta : 7\pi^2/10$
0089 1956 cu : $\sqrt{2} - \sqrt{5} - \sqrt{6}$
0089 3453 $\zeta : \sqrt{5} - \sqrt{6} - \sqrt{7}$
0089 4025 $\Gamma : 7\sqrt{3}/6$
0089 5845 sπ : $\exp(-(e+\pi))$
0089 5881 tπ : $\exp(-(e+\pi))$
0089 6095 $\ell_2 : \sqrt{5}/9$
0089 7327 $Ei : 22/5$
0089 7411 $\theta_3 : 1/\ln(\ln 2/3)$
0089 8304 at : $\exp(-3\pi/2)$
0089 8329 id : $\exp(-3\pi/2)$
0089 8341 as : $\exp(-3\pi/2)$
0089 9166 cu : $3\ln 2/10$
0090 0354 $Li_2 : \exp(-3\pi/2)$
0090 2096 $e^x : 2 - 3\sqrt{5}$
0090 2376 $e^x : \exp(-3\pi/2)$
0090 2684 $erf : (5)^{-3}$
0090 3885 $\Gamma : (\pi/3)^{-1/3}$
0090 4224 cu : $5/24$
0090 4793 $e^x : 2/3 + \sqrt{3}$
0090 5445 sq : $9\sqrt[3]{2}/8$
0090 6584 cu : $1/\sqrt{23}$
0090 7029 sq : $2/21$
0090 7271 cu : $4\sqrt{2}/3 - 3\sqrt{5}/4$
0090 8045 $e^x : 3\sqrt{2} - 4\sqrt{5}$
0090 8098 sq : $\sqrt{2} + \sqrt{5} - \sqrt{7}$
0091 1890 sq : $3(1/\pi)/10$
0091 2417 $10^x : -\sqrt{5} - \sqrt{6} + \sqrt{7}$
0091 2892 ln : $(\ln 3/2)^2$
0091 3033 $e^x : 1/2 - 3\sqrt{3}$
0091 3037 $\zeta : \sqrt{2} - \sqrt{3} - \sqrt{6}$
0091 4924 $10^x : 2\sqrt{2}/3 - 4\sqrt{5}/3$
0091 4975 sq : $\ln(\ln 3/3)$
0091 5651 $\lambda : \exp(-19\pi/8)$
0091 5743 $J_1 : \exp(-4)$
0091 5781 $1/2 \div \exp(4)$
0091 5839 $\zeta : 1/\ln(\ln 2)$
0091 7772 $Li_2 : (\sqrt{2}\pi/2)^{-1/3}$
0091 7957 $J_2 : e/10$
0092 0174 sq : $4/3 + \sqrt{5}/2$
0092 1616 at : $\sqrt{2}/3 + \sqrt{5}/2$
0092 1815 sπ : $2e + 4\pi$
0092 1830 $e^x : 3/4 - 2e$
0092 1854 tπ : $2e + 4\pi$
0092 1918 $\ell_2 : (1/(3e))^{1/3}$
0092 3975 $\Psi : \zeta(3)/8$
0092 4002 $J_2 : 3/11$
0092 4527 sr : $4 - 4\sqrt{5}/3$
0092 4741 $J_2 : 6(1/\pi)/7$
0092 5672 tπ : $e/2 - 3\pi/4$
0092 5925 cu : $\sqrt[3]{2}/6$
0092 7738 tπ : $\sqrt{2}/2 - 3\sqrt{5}/2$
0092 7821 $\ell_2 : 4/3 + e/4$
0092 7925 $Ei : 19/11$
0092 8774 at : $3/2 - 2\sqrt{5}/3$
0092 8801 id : $3/2 - 2\sqrt{5}/3$
0092 8814 as : $3/2 - 2\sqrt{5}/3$
0093 1462 ln : $3/4 + 3\sqrt{5}$
0093 1806 $\ln(4) - \exp(1/3)$
0093 2205 $2^x : \sqrt{2}/3 + \sqrt{5}/2$
0093 2595 $10^x : (2\pi)^{-3}$
0093 3080 cu : $4/19$
0093 3128 $e^x : 3/2 - 2\sqrt{5}/3$
0093 4077 $\ln(4) \div \exp(5)$
0093 5609 $\ln(3/5) \div \exp(4)$
0093 7503 $\zeta : 4\sqrt[3]{5}$
0093 9806 $\zeta : 7(e+\pi)/6$
0094 0356 $\exp(1/3) \div \exp(5)$
0094 3036 $\zeta : 4e\pi/5$
0094 3791 $2/5 + \ln(5)$
0094 4238 cr : $3/2 - \sqrt{2}/3$
0094 6115 $\zeta : \sqrt{3} + \sqrt{6} + \sqrt{7}$
0094 8174 cu : $e/2 - \pi/2$
0094 8721 $10^x : 1/3 - 3\pi/4$
0094 9343 cu : $4 + 3e/2$
0094 9628 $e^x : 1 - 4\sqrt{2}$
0095 1350 cu : $2\sqrt{2}/3 - 2\sqrt{3}/3$
0095 1621 $2^x : 5\zeta(3)/3$
0095 2340 $\Gamma : 2/21$
0095 2342 $\Gamma : 5\zeta(3)/2$
0095 2668 cu : $1/\ln(\sqrt{3}/2)$
0095 2889 $\sqrt{2} \div \exp(5)$
0095 2915 ln : $7\sqrt[3]{3}/10$
0095 3235 $\ln(\pi) - \exp(e)$
0095 3303 $10^x : 4/3 - 3\sqrt{5}/2$
0095 4765 $J_2 : 2\ln 2/5$
0095 5389 $J_2 : 1/\sqrt{13}$
0095 5571 at : $(3\pi/2)^{-3}$
0095 5601 id : $(3\pi/2)^{-3}$
0095 5615 as : $(3\pi/2)^{-3}$
0095 6967 tπ : $5(1/\pi)/4$
0095 7469 id : $7\sqrt[3]{3}/10$
0095 7510 $2^x : e - 3\pi$
0095 7893 $Li_2 : (3\pi/2)^{-3}$
0095 8319 $J_2 : 5/18$
0095 9800 $10^x : 1 + \pi/4$
0096 0181 $e^x : (3\pi/2)^{-3}$
0096 0491 as : $2e - 2\pi$
0096 1603 $\Gamma : -\sqrt{2} - \sqrt{3} - \sqrt{6}$
0096 3146 sr : $3e/8$
0096 3422 $J_2 : 7(1/\pi)/8$
0096 3676 $10^x : 4\sqrt{3} - 4\sqrt{5}$
0096 4290 $\ln(4/5) \div \exp(\pi)$
0096 5454 sr : $2 + 3e/4$
0096 5590 at : $9\sqrt{2}/8$
0096 8044 $\zeta : 5e/2$
0096 9094 cu : $1/\sqrt{22}$
0096 9137 $Ei : 3/8$
0097 0220 $J_2 : \sqrt{5}/8$
0097 1126 $J : (1/\pi)/8$
0097 2438 sq : $\ln(e/3)$
0097 2686 $e^x : \sqrt{3}/2 + \sqrt{5}/3$
0097 3491 $J_2 : 2\sqrt[3]{2}/9$
0097 3612 $J_2 : 7/25$
0097 4881 sr : $4\sqrt{2}/3 - \sqrt{3}/2$

0097 6473 $2^x : 1/4 - 4\sqrt{3}$
0097 6562 cu : $\sqrt[3]{5}/8$
0097 7375 sr : $9\pi/7$
0098 0516 sq : $\ln 2/7$
0098 1421 at : $5(1/\pi)$
0098 3797 ln : $10 \ln 2/7$
0098 3965 cu : $3/14$
0098 8326 cu : $1 - \pi/4$
0098 9150 cu : $3\sqrt{2}/2 - \sqrt{5}/2$
0098 9194 $J : ((e+\pi)/2)^{-3}$
0098 9542 $\ln(\sqrt{2}) - \exp(\sqrt{5})$
0099 1075 ln : $2 + \sqrt{5}/3$
0099 2411 $2^x : 3/2 - 3e$
0099 3349 $J_2 : \sqrt{2}/5$
0099 3504 $2^x : 1/2 + 4\sqrt{3}/3$
0099 3597 sq : $1/3 - \sqrt{3}/4$
0099 3951 $\theta_3 : (e+\pi)^{-3}$
0099 4043 $J_2 : 8(1/\pi)/9$
0099 5741 $1/5 \div \exp(3)$
0099 5989 cr : $5\sqrt[3]{3}/7$
0099 7994 cu : $4/3 - \sqrt{5}/2$
0099 8823 $\theta_3 : 15/17$
0099 9175 $J_2 : \exp(-\sqrt[3]{2})$
0099 9392 $e^x : 2\pi/9$
0100 0000 sq : $1/10$
0100 0704 $\zeta : (\sqrt[3]{4}/3)^{-3}$
0100 1066 $\sqrt{5} + s\pi(e)$
0100 1094 cr : $6\zeta(3)/7$
0100 1430 $\ell_{10} : (\pi/3)^{1/2}$
0100 2320 sq : P_{18}
0100 2758 sq : P_{48}
0100 3937 cr : $\sqrt{2}/3 + \sqrt{5}/4$
0100 4804 $\ell_{10} : 4\sqrt[3]{5}/7$
0100 5183 $\exp(2/5) \div \exp(5)$
0100 5943 sr : $2/3 + \sqrt{2}/4$
0100 6089 $\ln(\sqrt{3}) \div \exp(4)$
0100 6247 $2^x : 4\sqrt{3}/3 - 4\sqrt{5}$
0100 7685 $10^x : \exp(-2e)$
0100 7769 cu : $(5/3)^{-3}$
0100 8429 $J_2 : \sqrt[3]{5}/6$
0100 9110 $\Gamma : 5/22$
0101 0135 ln : $5\sqrt{2}/7$
0101 0507 ln : $4 + 2\sqrt{3}$
0101 3484 $J_2 : 2/7$
0101 3628 $erf : \exp(-3\pi/2)$
0101 3661 $10^x : e - 3\pi/2$
0101 3943 sr : $10e/3$
0101 4605 sq : P_{61}
0101 4873 cu : $\sqrt{3}/8$
0101 4986 cr : $\sqrt{2} + 3\sqrt{5}$
0101 5254 id : $5\sqrt{2}/7$
0101 7404 sq : P_{45}
0101 8454 $e^x : 3\sqrt{2}/2 - 3\sqrt{5}$
0101 8878 $J_2 : 9(1/\pi)/10$
0102 1002 $2^x : -\sqrt{3} - \sqrt{5} - \sqrt{7}$
0102 1639 $J_1 : 5\pi^2/7$
0102 6293 $\ell_{10} : (e+\pi)/6$
0102 6598 sq : $(\pi)^{-2}$
0102 7337 $10^x : 3/2 - \sqrt{2}/4$
0102 7369 cu : $5/23$
0102 7395 $J_2 : \ln(4/3)$
0102 8451 id : $5\zeta(3)$
0102 9555 $Ei : \sqrt[3]{5}/10$
0103 0964 ln : $4\sqrt{3}/7$
0103 1565 $\zeta : 3\sqrt{5}$
0103 2420 $s\pi : 3\sqrt{2}/2 - \sqrt{5}/2$
0103 2475 $t\pi : 3\sqrt{2}/2 - \sqrt{5}/2$
0103 2599 $10^x : 1/4 - \sqrt{5}$
0103 2849 $J_2 : \sqrt[3]{3}/5$
0103 3626 sr : $7\sqrt{3}/3$
0103 4150 cu : $2/3 + \sqrt{2}$
0103 4451 $J_2 : \sqrt{3}/6$
0103 4537 $2^x : 1/3 - 4\sqrt{3}$
0103 6278 $J_0 : 7\sqrt{3}/5$
0103 6478 $\exp(4) - s\pi(1/5)$
0103 8077 cu : $7\sqrt{3}/3$
0103 8194 cu : $e + \pi/4$
0103 8431 id : $1/3 + 3\sqrt{5}/4$
0103 8797 cu : $3\sqrt{3} + 3\sqrt{5}$
0103 9132 cu : $1/\sqrt{21}$
0103 9158 $e^x : 25/18$
0104 0003 $\zeta : 8(e+\pi)/7$
0104 3039 $\lambda : \exp(-7\pi/3)$
0104 3700 $10^x : 1 - 4\sqrt{5}/3$
0104 4031 $\ell_{10} : 3\sqrt{3}/4 + 4\sqrt{5}$
0104 7550 sq : $1/2 + 4e/3$
0104 7753 $2^x : (1/(3e))^2$
0104 9139 $2^x : \sqrt{2}/4 - 4\sqrt{3}$
0104 9350 sr : $3\sqrt{2}/4 + 4\sqrt{5}/3$
0105 0454 cr : $(\ln 3)^{1/3}$
0105 1122 $\zeta : 7/20$
0105 1769 ln : $4e + 3\pi$
0105 1840 cu : $20/11$
0105 2704 $\sqrt{5} - \exp(4/5)$
0105 3207 sinh : $25/9$
0105 5149 $J_1 : 7e/5$
0105 5309 $2^x : 3\sqrt{3} + \sqrt{5}/4$
0105 5849 $J_2 : 7/24$
0105 6077 $\Gamma : -\sqrt{2} - \sqrt{3} - \sqrt{7}$
0105 6469 $10^x : 3/4 + 3\pi/4$
0105 6652 $10^x : 1/3 - 4\sqrt{3}/3$
0106 1862 at : $1/2 - 2\pi/3$
0106 2469 $\theta_3 : (\ln 3)^{-1/3}$
0106 2560 $\ell_{10} : 6e\pi/5$
0106 3207 $\zeta : 20/3$
0106 4789 cu : $4\sqrt{3} - 3\sqrt{5}$
0106 5658 sinh : $8/9$
0106 7288 $J_2 : 5\pi^2$
0106 9168 $\ell_{10} : 3e + 2\pi/3$
0107 0847 ln : $4/3 + \sqrt{2}$
0107 1915 $J_2 : \ln(\sqrt{5}/3)$
0107 2330 sq : $1/4 - \sqrt{2}/4$
0107 3540 $J_2 : 5/17$
0107 5059 $10^x : 3/4 - \sqrt{5}/3$
0107 5115 $10^x : 2\sqrt[3]{2}$
0107 5528 $\ln(\pi) + s\pi(1/3)$
0107 5766 $10^x : 3/4 - e$
0107 6610 $\Gamma : (\sqrt[3]{4}/3)^{-3}$
0107 6739 $\theta_3 : \sqrt{2}/2$
0107 6971 $s\pi : 5\zeta(3)/3$
0107 7033 $t\pi : 5\zeta(3)/3$
0107 8247 $erf : (3\pi/2)^{-3}$
0107 8622 $e^x : 2/3 - 3\sqrt{3}$
0107 8681 $J_2 : 8\pi/3$
0107 8761 sq : $1/4 + 4e/3$
0108 0347 $1/4 \div \exp(\pi)$
0108 1632 $2^x : 1 + 3e$
0108 2457 $\zeta : 7e\pi/9$
0108 2562 $1/2 + \ln(3/5)$
0108 4430 $\ln(5) \div \exp(5)$
0108 5324 cu : $1/4 - \sqrt{2}/3$
0108 5504 sq : P_{51}
0108 5599 $J_1 : \exp((e+\pi)/3)$
0108 5965 $Ei : ((e+\pi)/2)^2$
0108 6171 $10^x : 3/2 - 2\sqrt{3}$
0108 8323 $10^x : \exp(\sqrt{5})$
0108 8928 $2^x : (4)^{-3}$
0108 9387 $J_2 : (3/2)^{-3}$
0109 2470 $\zeta : (1/(3e))^{1/2}$
0109 2636 $\ell_{10} : 7(e+\pi)/4$
0109 4991 $J_2 : 3 \ln 2/7$
0109 5151 $J_0 : 7 \ln 2/2$
0109 5621 id : $4\sqrt{2} + 3\sqrt{5}/2$
0109 6013 $3 \div \exp(2/5)$
0109 7393 sq : $(\sqrt{2}/3)^3$
0109 7637 $2^x : -\sqrt{2} - \sqrt{6} - \sqrt{7}$
0109 8938 $3/5 \div \exp(4)$
0110 1716 cr : $(e/3)^{-1/3}$
0110 3311 $2^x : 4\sqrt[3]{2}/5$
0110 4877 $\zeta : \sqrt{3} + \sqrt{5} + \sqrt{7}$
0110 6227 cu : $7(1/\pi)/10$
0110 7011 $10^x : \sqrt{2}/4 - 4\sqrt{3}/3$
0110 7554 $2^x : \sqrt{3}/3 + 4\sqrt{5}$
0110 8033 sq : $2/19$
0110 8515 rt : (1,8,-9,1)
0111 0899 cu : $\exp(-3/2)$
0111 0966 $\ln(4/5) \div \exp(3)$
0111 2123 sq : P_{57}
0111 4125 $\ell_{10} : 6\sqrt[3]{5}$
0111 5838 cu : $2/3 + 4e/3$
0111 6369 sq : P_{38}
0111 6586 $J_2 : 3/10$
0111 7504 $e - s\pi(1/4)$
0111 8033 cu : $\sqrt{5}/10$
0111 8089 ln : $3 + 2\sqrt{5}$
0111 9849 ln : $7\pi/8$
0111 9950 sq : $e/4 - \pi/4$
0112 0388 $J_2 : \zeta(3)/4$
0112 0887 cu : $\sqrt{2}/4 - \sqrt{3}/3$
0112 2335 $\zeta : 9(e+\pi)/8$
0112 2450 sq : $\sqrt{2}/3 - \sqrt{3}/3$
0112 3135 $e^x : \sqrt{2}/2 - 3\sqrt{3}$
0112 4801 $J : (3\pi)^{-2}$
0112 5743 at : $(3\pi)^{-2}$
0112 5790 id : $(3\pi)^{-2}$
0112 5814 as : $(3\pi)^{-2}$
0112 6066 cu : P_{31}
0112 7779 $J_2 : 1/\sqrt{11}$
0112 8317 $10^x : 5\sqrt{5}/6$
0112 8975 $Li_2 : (3\pi)^{-2}$
0112 9456 $J_2 : (\ln 3/2)^2$
0112 9856 $t\pi : \exp(\sqrt[3]{5})$
0113 0913 $\exp(4) + \exp(5)$
0113 1642 $J_2 : e/9$
0113 2151 $e^x : (3\pi)^{-2}$
0113 2534 sq : $5 \ln 2$
0113 2711 $\zeta : 2\pi^2/3$
0113 3464 $10^x : 6\zeta(3)$
0113 3483 $J : 5/21$
0113 3776 $10^x : -\sqrt{3} + \sqrt{5} - \sqrt{6}$
0113 5620 sq : $4/3 + 2\pi$
0113 6619 $10^x : 5\sqrt[3]{2}/9$
0113 7331 $2^x : 1/4 - 3\sqrt{5}$
0113 8439 $2^x : \sqrt{2}/3 - 4\sqrt{3}$
0113 8558 $10^x : \zeta(11)$
0114 0214 at : $(\sqrt{2}\pi)^{-3}$
0114 0263 id : $(\sqrt{2}\pi)^{-3}$
0114 0288 as : $(\sqrt{2}\pi)^{-3}$
0114 0523 $\ell_2 : 4\sqrt[3]{2}/5$
0114 0708 at : $5\sqrt{5}/7$
0114 2289 sq : $\exp(-\sqrt{5})$
0114 2718 $e^x : ((e+\pi)/2)^{-1/3}$
0114 3531 $Li_2 : (\sqrt{2}\pi)^{-3}$
0114 3590 $e^x : 2e/3 - 2\pi$
0114 3836 $K : 4\sqrt{2}/7$
0114 3840 cu : $9\sqrt{2}/7$
0114 6789 $e^x : (\sqrt{2}\pi)^{-3}$
0114 7023 $\Gamma : (2e)^{-2}$
0114 7307 $s\pi : 4\sqrt{2}/3 + \sqrt{5}/2$
0114 7382 $t\pi : 4\sqrt{2}/3 + \sqrt{5}/2$
0114 7779 $e^x : 7\sqrt[3]{2}/8$
0114 7901 $J_2 : 6\pi^2/7$
0114 8151 $\ell_{10} : 2e/3 - \pi/4$
0114 8400 $2^x : 4\sqrt{2}/3 + 2\sqrt{5}$
0114 8448 $10^x : \sqrt{2} - 3\sqrt{5}/2$
0114 8933 $J_2 : 7/23$
0114 9477 cu : $1/3 - \sqrt{5}/4$
0115 0106 $J_2 : (2e/3)^{-2}$
0115 0901 $10^x : (e+\pi)^{-3}$
0115 1778 $e^x : 2/3 + \sqrt{2}$
0115 3477 $e^x : 1/4 - 3\pi/2$
0115 3904 sinh : $\sqrt{7}$
0115 4478 $\ell_2 : \sqrt{3}/4 + \sqrt{5}/4$
0115 8845 sq : $2/3 - \sqrt{5}/4$
0115 9611 sr : $(\pi/3)^{1/2}$
0116 0091 ln : $4/11$
0116 0291 sq : $1/2 + 2\pi$
0116 0811 $2^x : \zeta(7)$
0116 1229 $2^x : 1/2 - 4\sqrt{3}$
0116 3989 $2^x : 3 - 3\pi$
0116 5615 at : $1 - 3\sqrt{3}/2$
0116 6283 cr : $1/4 + \pi/4$
0116 7046 $\sqrt{3} \div \exp(5)$
0116 7775 $J_2 : \ln(e/2)$
0116 9788 $2^x : -\sqrt{3} - \sqrt{5} - \sqrt{6}$
0117 0624 $\zeta : (\sqrt{5}/2)^{1/2}$
0117 1090 $10^x : 2/3 - 3\sqrt{3}/2$
0117 1376 $Ei : (\pi/2)^3$
0117 2358 $e^x : 3/4 - 3\sqrt{3}$
0117 3055 $\Gamma : -\sqrt{2} - \sqrt{5} - \sqrt{6}$

0117 3804 sq : $2e/3 + 4\pi/3$
0117 3929 cu : $5/22$
0117 4122 J_2 : $4/13$
0117 5347 cu : $5(1/\pi)/7$
0117 5478 J_2 : $(\sqrt{2}\pi/3)^{-3}$
0117 6143 at : $\exp(-\sqrt{2}\pi)$
0117 6198 id : $\exp(-\sqrt{2}\pi)$
0117 6225 as : $\exp(-\sqrt{2}\pi)$
0117 6478 sr : $3 + \pi/3$
0117 6949 J_2 : $4\ln 2/9$
0117 9471 10^x : $-\sqrt{3} + \sqrt{6} - \sqrt{7}$
0117 9674 Li_2 : $\exp(-\sqrt{2}\pi)$
0118 0767 10^x : $9\sqrt[3]{5}$
0118 1309 Ψ : $5/19$
0118 3142 e^x : $\exp(-\sqrt{2}\pi)$
0118 3654 e^x : $1 - 2e$
0118 4808 sq : $(2\pi/3)^{-3}$
0118 5112 e^x : $9\sqrt[3]{5}/7$
0118 5857 10^x : $5e\pi/6$
0118 7233 10^x : $7/10$
0118 8596 $t\pi$: $3\sqrt{3} - 4\sqrt{5}$
0118 8632 sr : $e/3 + \pi$
0118 9193 ℓ_{10} : $4/3 + 4\sqrt{5}$
0118 9550 Γ : $\sqrt[3]{25/3}$
0119 1005 ζ : $10(e + \pi)/9$
0119 2326 ζ : $\sqrt{2} + \sqrt{6} + \sqrt{7}$
0119 3180 cu : $1/2 + 3\sqrt{2}/2$
0119 8248 2^x : $1/\ln(\sqrt[3]{5}/2)$
0119 8530 Γ : $7\sqrt[3]{2}/9$
0119 9545 ℓ_2 : $\zeta(7)$
0120 0151 10^x : $3\sqrt{2}/4 - 4\sqrt{5}/3$
0120 0589 ζ : $13/2$
0120 4227 sq : P_{76}
0120 4815 10^x : $4 + \sqrt{5}/3$
0120 4960 2^x : $1/3 - 3\sqrt{5}$
0120 5378 J_0 : $6\pi^2/5$
0120 7172 10^x : $-\sqrt{2} + \sqrt{3} - \sqrt{5}$
0120 7451 cu : $1/\sqrt{19}$
0120 7530 Ei : $17/5$
0120 8707 cu : $3\sqrt{2} - 2\sqrt{5}$
0120 9049 ℓ_{10} : $4\sqrt{3} + 3\sqrt{5}/2$
0120 9286 ζ : $9\sqrt[3]{3}/2$
0120 9992 sq : $2\sqrt{3} - 3\sqrt{5}/2$
0121 0331 λ : $\exp(-16\pi/7)$
0121 0799 J_2 : $5/16$
0121 2766 ℓ_{10} : $4 + 2\pi$
0121 3665 e^x : $3e - 4\pi$
0121 6076 cr : $\zeta(5)$
0121 8934 Ei : $(1/(3e))^{-1/2}$
0121 9330 cu : $3 + \sqrt{2}$
0121 9701 at : $8/5$
0121 9836 ln : $4\sqrt{2}/3 + \sqrt{3}/2$
0122 0130 Γ : $11/3$
0122 1042 $2/3 \div \exp(4)$
0122 1397 Γ : $\exp(1/\sqrt{2})$
0122 1474 $3/5 - s\pi(1/5)$
0122 1967 2^x : $\sqrt{2}/4 - 3\sqrt{5}$
0122 2277 cu : $2\sqrt{2} - 3\sqrt{3}/2$
0122 3585 J_2 : $\pi/10$
0122 4438 J_2 : $2\sqrt{2}/9$
0122 4461 ℓ_{10} : $3/2 - \sqrt{2}/3$
0122 6599 J_0 : $17/7$
0122 7733 $\exp(3/5) \div \exp(5)$
0122 8948 cu : $3/13$
0122 8977 id : $\sqrt{2} + 3\sqrt{3}/2$
0122 9149 2^x : $3\sqrt{3}/2 - 4\sqrt{5}$
0122 9204 e^x : $4\sqrt{3}/3 - 3\sqrt{5}$
0122 9521 ℓ_2 : $3e/4 - \pi/3$
0122 9935 J_2 : $\sqrt[3]{2}/4$
0123 0201 ln : $3\sqrt{2} - 2\sqrt{5}/3$
0123 1197 rt : (1,9,0,1)
0123 3362 at : $(\ln 2/3)^3$
0123 3424 id : $(\ln 2/3)^3$
0123 3455 as : $(\ln 2/3)^3$
0123 4567 sq : $1/9$
0123 4644 10^x : $3/2 + 4\sqrt{5}$
0123 6210 J_2 : $6/19$
0123 6359 $\sqrt{2} + \exp(4)$
0123 7119 cu : $8\sqrt{5}/3$
0123 7248 Li_2 : $(\ln 2/3)^3$
0123 8402 sq : $1/3 - 4\sqrt{5}/3$
0123 9161 Ei : $15/22$
0123 9615 J_2 : $1/\sqrt{10}$
0124 0271 sq : $3e/2 - 4\pi/3$
0124 0596 Ψ : $9\ln 2/8$
0124 1062 e^x : $(\ln 2/3)^3$
0124 3186 $\ln(3/4) \div \exp(\pi)$
0124 4676 $1/4 \div \exp(3)$
0124 4677 θ_3 : $(2e)^{-3}$
0124 4826 sq : $\ln(\sqrt{5}/2)$
0124 6117 id : $9\sqrt{5}/10$
0124 6501 rt : (1,9,-1,-8)
0124 7402 cr : $(\sqrt{5}/2)^{1/3}$
0124 9537 cu : $3/2 - \sqrt{3}$
0125 0470 Γ : $10e\pi/9$
0125 1023 10^x : $1/3 - \sqrt{5}$
0125 1227 J_2 : $10(e + \pi)/7$
0125 1341 e^x : $1/3 + 3e$
0125 1494 2^x : $4\sqrt{2}/3 + 4\sqrt{3}$
0125 2958 J_2 : $\sqrt{22}\pi$
0125 3719 e^x : $1/3 - 3\pi/2$
0125 4279 10^x : $16/13$
0125 4552 $t\pi$: $\sqrt[3]{2}/5$
0125 4853 J_2 : $7/22$
0125 5854 J_2 : $(1/\pi)$
0125 6001 K : $7\ln 2/6$
0125 6058 2^x : $9\sqrt{2}/8$
0125 8756 J_0 : $9(e + \pi)$
0125 9671 e^x : $3 + 3\sqrt{2}/4$
0125 9987 ℓ_{10} : $1/2 + \sqrt{2}/3$
0126 0272 $s\pi(1/3) \div s\pi(\pi)$
0126 1716 J_0 : $5\pi^2/9$
0126 1910 ζ : $\exp(-\pi/3)$
0126 4698 J_2 : $\sqrt{5}/7$
0126 6413 J_1 : $(2\pi)^{-2}$
0126 6480 $s\pi$: $(2\pi)^{-3}$
0126 6582 $t\pi$: $(2\pi)^{-3}$
0126 6958 $\pi - \exp(e)$
0126 7920 e^x : $\sqrt{2}/2 + 4\sqrt{3}$
0126 9112 J_2 : $8/25$
0126 9543 $\ln(2) \div \exp(4)$
0127 0062 $t\pi$: $\sqrt{3}/2 - \sqrt{5}/2$
0127 0265 erf : $(3\pi)^{-2}$
0127 0411 J_0 : $\sqrt{17/3}$
0127 2639 2^x : $-\sqrt{2} - \sqrt{5} - \sqrt{7}$
0127 3046 J_2 : $2\sqrt[3]{3}/9$
0127 5159 ζ : $\sqrt{3} + \sqrt{5} + \sqrt{6}$
0127 5614 cu : $2/3 - \sqrt{3}/4$
0127 7636 2^x : $\exp(-4)$
0127 7889 e^x : $\exp(2/3)$
0127 8874 2^x : $\sqrt{3}/3 - \sqrt{5}/4$
0128 1607 cu : $3/4 + 2\pi/3$
0128 2100 J_1 : $19/5$
0128 2124 ln : $1/3 + e/4$
0128 2142 id : $(1/(3e))^{-1/3}$
0128 2466 e^x : $25/4$
0128 2643 at : $\exp(\sqrt{2}/3)$
0128 4081 at : $\sqrt{2}/2 - 4\sqrt{3}/3$
0128 4601 $s\pi(2/5) + s\pi(\sqrt{2})$
0128 5377 10^x : $\sqrt{2}/2 - 3\sqrt{3}/2$
0128 6242 sr : $4/3 + e$
0128 6594 erf : $(\sqrt{2}\pi)^{-3}$
0128 6611 sq : P_{37}
0128 7175 ζ : $3e\pi/4$
0128 7566 K : $(\sqrt[3]{4}/3)^{1/3}$
0128 8044 ζ : $\ln(1/(2e))$
0128 9833 at : $10\sqrt[3]{3}/9$
0129 0307 at : $2/3 - e/4$
0129 0379 id : $1/3 + e/4$
0129 0414 as : $2/3 - e/4$
0129 0945 sr : $3\sqrt[3]{5}/5$
0129 1157 2^x : $\sqrt{3}/4 - 3\sqrt{5}$
0129 1238 ℓ_{10} : $5\sqrt[3]{3}/7$
0129 2442 $\sqrt{5} + \ln(4/5)$
0129 5917 2^x : $3 - 4\sqrt{5}/3$
0129 6176 e^x : $(\sqrt{5}/3)^{-1/3}$
0129 6643 at : $4\zeta(3)/3$
0129 7823 ℓ_{10} : $6\zeta(3)/7$
0130 1491 ℓ_{10} : $\sqrt{2}/3 + \sqrt{5}/4$
0130 2666 cu : $4/17$
0130 3435 2^x : $2/3 - 4\sqrt{3}$
0130 4650 ℓ_{10} : $7\ln 2/5$
0130 6872 cr : $3\ln 2/2$
0130 7858 J_2 : $(\sqrt[3]{5}/3)^2$
0130 8322 sq : $2 - 4\sqrt{2}/3$
0130 9457 cu : $\sqrt{2}/6$
0131 0646 10^x : $\sqrt{2}/4 - \sqrt{5}$
0131 0728 10^x : $20/17$
0131 1407 Ei : $8\sqrt[3]{2}/5$
0131 2372 $\exp(2/3) \div \exp(5)$
0131 2532 ln : $\pi^2/10$
0131 3271 $t\pi$: $e/2 - 3\pi/2$
0131 3444 10^x : $3 - 4\sqrt{2}/3$
0131 4603 J_2 : $(3\pi)^{-1/2}$
0131 5360 cu : $1/4 + \pi$
0131 5561 cu : $2 + \sqrt{5}$
0131 5910 $2/5 - \exp(5)$
0131 6701 sq : $3\sqrt{2}/2 - \sqrt{5}$
0131 9040 ln : $\zeta(3)/9$
0131 9278 e^x : $1/\ln(\sqrt[3]{4}/2)$
0131 9332 Γ : $11/25$
0131 9750 as : $3\sqrt{2}/5$
0131 9759 ln : $4\sqrt{3} + \sqrt{5}/4$
0131 9760 $1/5 \div \exp(e)$
0132 0637 cu : $e/3 + 2\pi/3$
0132 1042 cu : $\exp(-\sqrt[3]{3})$
0132 4025 $1/3 - \ln(\sqrt{2})$
0132 4114 J_2 : $\exp(-\sqrt{5}/2)$
0132 4285 J_2 : $\ln(\sqrt[3]{3}/2)$
0132 5283 id : $3e - \pi$
0132 5313 e^x : $4/3 - 4\sqrt{2}$
0132 5654 sr : 8π
0132 5753 θ_3 : $(3\pi/2)^{-1/2}$
0132 5977 2^x : $\sqrt{2}/3 - 3\sqrt{5}$
0132 6277 ζ : $9\sqrt{2}/2$
0132 6396 sq : P_{78}
0132 6538 sq : $\sqrt{2} - 3\sqrt{3}/4$
0132 6689 2^x : $2\sqrt{3} + 2\sqrt{5}/3$
0132 7136 erf : $\exp(-\sqrt{2}\pi)$
0133 0230 cr : $1/3 + \sqrt{2}/2$
0133 0633 sr : $2e/3 - \pi/4$
0133 1749 2^x : $7\sqrt[3]{3}/10$
0133 2829 $\sqrt{2} \div \exp(1/3)$
0133 4043 Ei : $24/23$
0133 4591 sq : $\ln 2/6$
0133 5580 J_0 : 18
0133 6411 10^x : $3\sqrt{3}/2 - 2\sqrt{5}$
0133 8666 $\exp(3/4) \times s\pi(2/5)$
0133 9022 cu : $4/3 - \pi/2$
0133 9449 ln : $\sqrt{3} - \sqrt{5}/3$
0133 9574 J_2 : $8\sqrt{5}$
0134 0489 2^x : $\sqrt{2}/2 - 4\sqrt{3}$
0134 2147 $t\pi$: $4\sqrt{2}/3 + 3\sqrt{5}/4$
0134 4867 10^x : $1/4 - 3\sqrt{2}/2$
0134 6189 sq : $3/4 - \sqrt{3}/2$
0134 7161 ζ : $7e/3$
0134 7589 θ_3 : $\exp(-5)$
0134 7589 $2 \div \exp(5)$
0134 9746 cu : $5/21$
0135 0710 $s\pi(\sqrt{2}) \div s\pi(2/5)$
0135 1613 sinh : $(\sqrt[3]{2})^{-1/2}$
0135 2022 Ψ : $\sqrt{21/2}$
0135 2067 2^x : $4\sqrt{3}/3 + 2\sqrt{5}$
0135 2522 2^x : $1/2 - 3\sqrt{5}$
0135 3065 λ : $\exp(-9\pi/4)$
0135 4559 cu : $3\sqrt{2}/4 - 3\sqrt{3}/4$
0135 6408 ζ : $19/3$
0135 6889 sq : P_{77}
0135 6959 sq : $((e + \pi)/2)^{-2}$
0136 0611 cu : $3(1/\pi)/4$
0136 1481 ℓ_{10} : $(\ln 3)^{1/3}$
0136 5234 cu : $\sqrt{2} + \sqrt{5} - \sqrt{7}$
0136 5517 rt : (1,7,-7,9)
0136 6438 e^x : $5\sqrt[3]{2}/9$
0136 7938 $s\pi$: $\exp(-2e)$
0136 8067 $t\pi$: $\exp(-2e)$
0136 8256 Γ : $(\pi/3)^{-1/2}$
0136 9338 ln : $(\ln 3/3)^3$
0137 0033 cr : $25/24$
0137 1742 cu : $13/9$

0137 2444 sq : $5\zeta(3)/3$
0137 2526 cu : $2/3 - e/3$
0137 2845 2^x : $5(1/\pi)$
0137 2991 Γ : $4\sqrt[3]{5}/7$
0137 3549 J_2 : $(\ln 2)^3$
0137 3672 $3/4 \div \exp(4)$
0137 4239 J_0 : $7e/8$
0137 4766 ℓ_2 : $7\sqrt[3]{3}/5$
0137 5270 e^x : $7/10$
0137 5598 cu : $\ln(\ln 3/3)$
0137 6073 J_2 : $1/3$
0137 6134 as : $8(1/\pi)/3$
0137 7839 10^x : $1/4 + \sqrt{2}/4$
0137 9863 e^x : $2 - 2\pi$
0138 0942 2^x : $3/4 - 4\sqrt{3}$
0138 2400 cu : $6/25$
0138 4083 sq : $2/17$
0138 4360 rt : (1,7,-8,1)
0138 5276 Γ : $7\pi^2/6$
0138 8888 cu : $\sqrt[3]{3}/6$
0138 9504 ℓ_2 : $2 + 3e/4$
0138 9520 cu : $\zeta(3)/5$
0139 0333 sπ : $7\pi/2$
0139 0412 ζ : $5\sqrt[3]{2}$
0139 0468 tπ : $7\pi/2$
0139 1700 erf : $(\ln 2/3)^3$
0139 2533 10^x : $1/2 - 3\pi/4$
0139 2981 sinh : $9\ln 2/7$
0139 3202 sq : $1 - \sqrt{5}/2$
0139 3220 at : $3/4 - 3\pi/4$
0139 4065 ζ : $\sqrt{2} + \sqrt{5} + \sqrt{7}$
0139 4402 e^x : $1/4 + \sqrt{5}$
0139 5960 $\ln(2) - s\pi(1/4)$
0139 8390 tπ : $2\sqrt{2}/3 + 4\sqrt{3}/3$
0139 9258 10^x : $3/2 - 3\sqrt{5}/2$
0140 1226 10^x : $3\sqrt{2} - \sqrt{3}/4$
0140 3175 Γ : $(e + \pi)/6$
0140 3482 2^x : $2 - 3e$
0140 3728 rt : (1,7,-9,-7)
0140 5613 Ψ : $(2\pi/3)^{1/2}$
0140 5664 cu : $\sqrt{3} - 2\sqrt{5}/3$
0140 6585 cu : $2e/3 - \pi/2$
0140 6620 ℓ_2 : $9\pi/7$
0140 7286 ζ : 2π
0140 8729 cu : $1/2 + \sqrt{2}$
0140 9716 sq : $2/3 - \pi/4$
0140 9929 Ei : $23/5$
0141 2405 2^x : $5\sqrt{2}/7$
0141 3697 ℓ_{10} : $e/3 + 3\pi$
0141 6081 Ei : $7\zeta(3)/6$
0141 7423 e^x : $2 + 2\sqrt{2}$
0141 8808 10^x : $3/4 - 3\sqrt{3}/2$
0141 9320 ℓ_2 : $5\ln 2/7$
0141 9580 J_0 : $\sqrt{5} - \sqrt{6} + \sqrt{7}$
0141 9696 sr : $3/2 - \sqrt{2}/3$
0142 1356 $2/5 - \sqrt{2}$
0142 1662 e^x : $2\sqrt{2}/3 - 3\sqrt{3}$
0142 2447 J_0 : 15
0142 2929 ζ : $1/\ln(\ln 3/2)$
0142 3165 sπ : $\sqrt{2} + \sqrt{5} - \sqrt{7}$
0142 3309 tπ : $\sqrt{2} + \sqrt{5} - \sqrt{7}$
0142 3657 J : $1/21$
0142 4829 sq : $3(1/\pi)/8$
0142 5124 10^x : $3(1/\pi)$
0142 5979 ζ : $1/\sqrt[3]{23}$
0142 6423 $\exp(3/4) \div \exp(5)$
0142 6680 cu : $1/\sqrt{17}$
0142 6701 cu : $3\sqrt{2}/2 + 3\sqrt{3}/4$
0142 7017 2^x : $\sqrt{3}/3 - 3\sqrt{5}$
0142 7549 tπ : $1/\ln(\ln 3/3)$
0142 7559 ℓ_{10} : $(e/3)^{1/3}$
0142 8087 e^x : $1/3 + 3\sqrt{3}$
0142 8364 cu : P_{15}
0142 8534 cu : $4 + 3\sqrt{2}$
0142 8764 cr : $24/23$
0142 9339 J_2 : $e/8$
0142 9552 J_0 : $\exp(\sqrt{3}/2)$
0143 0644 K : $17/21$
0143 0885 10^x : $1/4 - 2\pi/3$
0143 2284 $\ln(3/4) \div \exp(3)$
0143 4066 tπ : $\ln(\ln 3/3)$
0143 4327 e^x : $3e/4 - 2\pi$
0143 4935 sinh : $\sqrt{2} - \sqrt{5} - \sqrt{7}$
0143 6959 cu : $\exp(-\sqrt{2})$
0144 0000 sq : $3/25$
0144 0114 ζ : $4\sqrt[3]{2}/7$
0144 0463 $1/3 \div \exp(\pi)$
0144 1940 2^x : $2\sqrt{2} - 4\sqrt{5}$
0144 2039 ζ : $25/4$
0144 2719 sq : $(\ln 2/2)^2$
0144 3303 10^x : $(2e)^{-3}$
0144 3524 Γ : $-\sqrt{2} - \sqrt{3} - \sqrt{5}$
0144 4940 sq : $\zeta(3)/10$
0144 5081 as : $(\sqrt[3]{3}/2)^{1/2}$
0144 5422 10^x : $10e/3$
0144 7811 J_2 : $\sqrt[3]{5}/5$
0144 8870 tπ : $8\sqrt{5}/5$
0145 0853 cu : $\sqrt{3}/4 + \sqrt{5}$
0145 2123 10^x : $\sqrt{2}/3 - 4\sqrt{3}/3$
0145 2522 10^x : $\ln(1/(2\pi))$
0145 2980 cu : $1/3 - \sqrt{3}/3$
0145 4477 ζ : $9\ln 2$
0145 5952 J_1 : $(e + \pi)^{-2}$
0145 6110 sq : $\sqrt{3}/2 - \sqrt{5}/3$
0145 6279 Li_2 : $10/13$
0145 6747 sq : $e/3 - \pi/4$
0145 7317 ℓ_2 : $5\sqrt{2}/7$
0145 7725 cu : $\sqrt[3]{5}/7$
0145 8095 2^x : $-\sqrt{2} - \sqrt{5} - \sqrt{6}$
0145 8906 sπ : $1/4 + \sqrt{5}/3$
0145 9061 tπ : $1/4 + \sqrt{5}/3$
0145 9989 J_2 : $2\zeta(3)/7$
0146 3809 $\ln(3) - \exp(\sqrt{2})$
0146 5251 $4/5 \div \exp(4)$
0146 6521 $\sqrt{5} - \exp(1/5)$
0146 6728 e^x : $1/4 - 2\sqrt{5}$
0146 7407 J_1 : $4e\pi/9$
0146 8987 cr : $3\sqrt{3} + 4\sqrt{5}/3$
0147 1862 sq : $2 - 3\sqrt{2}/2$
0147 1996 id : $4e + \pi$
0147 2069 Ψ : $(2\pi)^{-2}$
0147 2480 $\ln(4/5) \div \exp(e)$
0147 3894 $\ln(\sqrt{5}) \div \exp(4)$
0147 4431 ℓ_{10} : $3\sqrt{2}/2 - 2\sqrt{3}/3$
0147 5417 at : $1/4 + e/2$
0147 5533 λ : $\exp(-20\pi/9)$
0147 7032 cu : $1/2 - \sqrt{5}/3$
0148 1094 e^x : $1/2 - 3\pi/2$
0148 1425 cu : $\sqrt{2}/4 - 3\sqrt{5}/2$
0148 3690 at : $\ln(5)$
0148 4081 ln : $\sqrt[3]{21}$
0148 4474 10^x : $1 - 2\sqrt{2}$
0148 4743 at : $2/3 + 2\sqrt{2}/3$
0148 5056 J_2 : $\sqrt{3}/5$
0148 5998 e^x : $3/4 + \sqrt{2}/4$
0148 6443 J_2 : $\ln 2/2$
0148 7352 J_2 : $1/\sqrt[3]{24}$
0148 7367 ℓ_2 : $\sqrt{3}/7$
0149 0302 ln : $2/15$
0149 0309 10^x : $2\sqrt{3}/3 - 4\sqrt{5}/3$
0149 1630 ℓ_{10} : P_{50}
0149 2757 cr : $23/22$
0149 4965 J_1 : $10(e + \pi)/3$
0149 5539 sq : $4e + 3\pi/4$
0149 6725 sq : $6e\pi/5$
0149 7098 J_2 : $8/23$
0149 7308 ζ : $18/25$
0149 7680 $\ln(\sqrt{2}) \div \exp(\pi)$
0149 7697 sr : $5\sqrt[3]{3}/7$
0149 9557 $\exp(4/5) \div \exp(5)$
0150 0040 10^x : $-\sqrt{2} + \sqrt{5} - \sqrt{7}$
0150 3452 sq : $4/3 + e/3$
0150 3612 at : $(1/(3e))^2$
0150 3725 id : $(1/(3e))^2$
0150 3782 as : $(1/(3e))^2$
0150 5093 10^x : $(e + \pi)/2$
0150 5338 e^x : $1 - 3\sqrt{3}$
0150 5393 sr : $6\zeta(3)/7$
0150 6650 $\sqrt{5} \div \exp(5)$
0150 7680 J_2 : $\pi/9$
0150 7818 tπ : $\sqrt{2}/3 - \sqrt{5}/2$
0150 9250 2^x : $e + 4\pi/3$
0150 9416 Li_2 : $(1/(3e))^2$
0150 9680 sr : $\sqrt{2}/3 + \sqrt{5}/4$
0150 9935 ℓ_2 : $3\sqrt{2}/4 + 4\sqrt{5}/3$
0151 0739 ℓ_{10} : $1/4 + \pi/4$
0151 0798 sr : $3 + 3\sqrt{2}/4$
0151 1605 cu : $\exp(1/e)$
0151 1894 10^x : $1/\ln(\sqrt{3}/3)$
0151 2566 at : $\sqrt{2} - \sqrt{6} + \sqrt{7}$
0151 4913 cu : $\sqrt{3}/7$
0151 5088 e^x : $(1/(3e))^2$
0151 5678 J_2 : $7/20$
0151 6461 sq : $\exp(-2\pi/3)$
0151 7227 J_2 : $(1/(3e))^{1/2}$
0151 7450 10^x : $3 + \pi/3$
0151 7459 cu : $7(1/\pi)/9$
0151 8154 2^x : $2/3 - 3\sqrt{5}$
0151 9099 sr : $((e + \pi)/2)^3$
0151 9232 Ei : $(\pi)^{-1/3}$
0152 1249 Ψ : $9\sqrt[3]{3}/4$
0152 3573 J_2 : $\exp(-\pi/3)$
0152 4088 J_0 : $5\pi^2$
0152 4409 sr : $\sqrt{2}/2 + 3\sqrt{5}/2$
0152 6969 2^x : $1/4 - 2\pi$
0152 8839 cu : $4e/3 + 3\pi$
0152 8909 2^x : $3\sqrt{2}/2 + 2\sqrt{3}$
0152 9626 ℓ_{10} : $\sqrt{2} + 4\sqrt{5}$
0152 9715 J_2 : $1/\sqrt[3]{23}$
0153 1223 ζ : $5\pi^2/8$
0153 2673 rt : (1,8,1,9)
0153 3654 cu : $\sqrt{5}/9$
0153 3768 10^x : $7/23$
0153 7253 ln : $(\pi/3)^{1/3}$
0153 7775 at : $\sqrt{3}/2 + \sqrt{5}/3$
0153 8208 sq : $4/3 + 2\sqrt{3}$
0153 8524 ℓ_{10} : $3e/2 + 2\pi$
0153 9328 J_1 : $\exp(4/3)$
0154 0992 J_2 : $6/17$
0154 2199 cu : $\sqrt{2}/3 + \sqrt{5}/2$
0154 3438 e^x : $(\ln 2)^{-2}$
0154 5414 cu : $24/19$
0154 6287 J_2 : $\sqrt{2}/4$
0154 7363 J_2 : $10(1/\pi)/9$
0154 9129 id : $(\pi/3)^{1/3}$
0154 9397 sπ : $3\sqrt{2}/4 + 4\sqrt{5}$
0154 9583 tπ : $3\sqrt{2}/4 + 4\sqrt{5}$
0155 0954 10^x : $1/2 - 4\sqrt{3}/3$
0155 3332 tπ : $1/3 + \sqrt{2}$
0155 3880 ℓ_{10} : P_{22}
0155 5036 10^x : $4/3 - \pi$
0155 5899 sq : $3e/4 - 2\pi$
0155 6437 rt : (1,8,0,1)
0155 7837 J_0 : $19/8$
0156 1035 cr : $4 + 4\pi/3$
0156 1231 sπ : $(e + \pi)^{-3}$
0156 1312 2^x : $\sqrt{2}/2 - 3\sqrt{5}$
0156 1421 tπ : $(e + \pi)^{-3}$
0156 1530 ln : $2/3 + 2\pi/3$
0156 2372 at : $(4)^{-3}$
0156 2450 2^x : $3\sqrt{3}/2 + 2\sqrt{5}/3$
0156 2500 id : $(4)^{-3}$
0156 2563 as : $(4)^{-3}$
0156 2752 cr : $22/21$
0156 3567 10^x : $\exp(-5)$
0156 4147 Ψ : 8
0156 4700 2^x : $3\sqrt[3]{5}$
0156 4907 Li_2 : $4\sqrt{3}/9$
0156 5673 e^x : $3/2 - 4\sqrt{2}$
0156 8193 Γ : $7\pi/2$
0156 8646 Li_2 : $(4)^{-3}$
0156 8821 e^x : $4 - 3e$
0156 9065 e^x : $2 - 3\sqrt{3}/4$
0156 9707 e^x : $2\sqrt{2}/3 + 2\sqrt{3}$
0157 3782 10^x : $\sqrt{3}/4 - \sqrt{5}$
0157 3918 θ_3 : $6/13$
0157 4770 e^x : $(4)^{-3}$
0157 4849 ℓ_{10} : $\zeta(5)$
0157 5237 J_2 : $1/\sqrt[3]{22}$
0157 7508 J_2 : $5/14$

0157 8399 2^x : $2\sqrt{2}/3 - 4\sqrt{3}$
0157 8835 10^x : $4/3 + e/4$
0157 9136 sq : P_{53}
0157 9546 $s\pi$: $7\sqrt[3]{5}/6$
0157 9743 $t\pi$: $7\sqrt[3]{5}/6$
0157 9812 sr : $(\ln 3)^{1/3}$
0158 0947 rt : (1,8,-1,-7)
0158 1378 λ : $\exp(-11\pi/5)$
0158 2524 ζ : $\exp(2e/3)$
0158 3813 J_1 : $9\pi/4$
0158 5731 sq : $3e/2 + 4\pi$
0158 5871 J_2 : $9(1/\pi)/8$
0158 6983 ln : $e - 3\pi/4$
0158 7367 id : $8\sqrt[3]{2}/5$
0158 7401 sq : $\sqrt[3]{2}/10$
0158 7708 sq : $\sqrt{3}/4 - \sqrt{5}/4$
0159 1347 J_1 : $(1/\pi)/10$
0159 1542 ζ : $9e/4$
0159 3667 sq : $4\sqrt[3]{2}/5$
0159 4192 e^x : $1/3 - 2\sqrt{5}$
0159 7657 Ei : $7/19$
0159 8529 as : $17/20$
0159 9441 cr : $\sqrt{2}/3 + \sqrt{3}/3$
0159 9478 e^x : $3\sqrt{2}/4 - 3\sqrt{3}$
0160 0000 θ_3 : $(5)^{-3}$
0160 0000 cu : $\sqrt[3]{2}/5$
0160 0464 cu : $\sqrt{3}/2 - \sqrt{5}/2$
0160 1959 2^x : $\sqrt{2}/2 + 2\sqrt{3}$
0160 2376 J_2 : $2\sqrt[3]{2}/7$
0160 2574 J_2 : $9/25$
0160 4495 Ψ : $(\sqrt{2}/3)^{1/3}$
0160 4681 Ei : $1/\sqrt[3]{20}$
0160 6433 10^x : $(2e/3)^{-2}$
0160 6867 id : $4\sqrt{3} - 4\sqrt{5}$
0160 7531 J_2 : $\sqrt[3]{3}/4$
0160 7622 $\sqrt{3} + \exp(1/4)$
0160 8013 J_2 : $3\zeta(3)/10$
0160 8429 2^x : $3/4 - 3\sqrt{5}$
0160 8832 ln : $3\sqrt{3}/2 - \sqrt{5}$
0160 9694 $t\pi$: $(2\pi)^{1/2}$
0161 0846 ζ : $5e\pi/7$
0161 0892 ζ : $\sqrt{2} + \sqrt{5} + \sqrt{6}$
0161 0950 sinh : $\ln(1/(3e))$
0161 1073 cr : $(\sqrt{3}/2)^{-1/3}$
0161 2367 J_1 : $(\pi)^{-3}$
0161 3323 sq : $4\sqrt{3} - \sqrt{5}$
0161 5166 ℓ_{10} : $(\sqrt{5}/2)^{1/3}$
0161 5417 $s\pi$: $5\zeta(3)/2$
0161 5628 $t\pi$: $5\zeta(3)/2$
0161 7767 2^x : $1/3 - 2\pi$
0162 1138 sq : $2(1/\pi)/5$
0162 1422 cr : $3 + 3\sqrt{3}$
0162 3577 10^x : $\sqrt[3]{21}$
0162 4036 2^x : $3 - 4\sqrt{5}$
0162 4311 J_2 : $1/\sqrt[3]{21}$
0162 4728 Ei : $9\sqrt[3]{2}/5$
0162 4906 J : $(1/(3e))^2$
0162 6629 10^x : $1/4 - 3e/4$
0162 6755 e^x : $\sqrt{2}/4 - 2\sqrt{5}$
0162 9344 10^x : $1/3 - 3\sqrt{2}/2$
0162 9713 10^x : $3/2 + 3\pi$
0162 9750 rt : (2,8,-9,6)
0163 0659 cu : $3e/4 + \pi$
0163 1098 Ψ : $(\sqrt{2}\pi/3)^3$
0163 4223 Li_2 : $10\ln 2/9$
0163 4753 J_2 : $4/11$
0163 6055 J_2 : $8(1/\pi)/7$
0163 7512 cr : $5\sqrt[3]{2}/6$
0163 9635 cr : $21/20$
0164 0567 e^x : $3\sqrt{3}/2 - 3\sqrt{5}$
0164 2226 2^x : $1 - 4\sqrt{3}$
0164 3491 2^x : $\sqrt{3}/3 + 3\sqrt{5}$
0164 8883 at : $21/13$
0164 9700 $1/4 \div \exp(e)$
0165 0244 10^x : $e/2 - \pi$
0165 0785 id : $\sqrt{2}/2 + 4\sqrt{3}/3$
0165 1202 cu : $1 - \sqrt{5}/3$
0165 1278 cu : $4(1/\pi)/5$
0165 1922 e^x : $4/3 - 2e$
0165 3156 J_1 : $\sqrt{15}$
0165 4456 Ψ : $\exp(1/e)$
0165 6283 ζ : $7\sqrt{3}/2$
0165 7118 sr : $(e/3)^{-1/3}$
0165 7638 e^x : $2\sqrt{2} - 4\sqrt{3}$
0165 7662 J_2 : $\ln(\sqrt[3]{3})$
0165 7693 J_1 : $5\sqrt{2}$
0165 9568 $1/3 \div \exp(3)$
0166 0428 J_2 : $\ln(\ln 2)$
0166 3497 cu : $4/3 + 3\sqrt{5}$
0166 3648 10^x : $1/\ln(\sqrt[3]{5}/3)$
0166 6630 at : $3/4 + \sqrt{3}/2$
0166 6870 10^x : $2e\pi/9$
0166 7744 sinh : $7\sqrt[3]{5}/3$
0166 9197 sq : $(\ln 2)^{-3}$
0167 0307 ζ : $-\sqrt{2} + \sqrt{5} - \sqrt{6}$
0167 1187 ℓ_{10} : $6\sqrt{3}$
0167 2221 J_2 : $9(e + \pi)$
0167 2386 J_0 : $2(e + \pi)$
0167 2685 at : $8\sqrt{2}/7$
0167 2692 J_2 : $\exp(-1)$
0167 6418 Ψ : $13/9$
0167 6826 sq : $\zeta(7)$
0167 7404 J_2 : $1/\sqrt[3]{20}$
0167 7565 J_2 : $7/19$
0167 8245 $\ln(2/5) \div \exp(4)$
0167 8419 J : $(3\pi)^{-3}$
0168 3675 sq : $\sqrt{2}/2 - \sqrt{3}/3$
0168 3793 10^x : $17/2$
0168 4311 cr : $7(e + \pi)/5$
0168 5901 10^x : $\exp(\sqrt{2}/3)$
0168 8084 10^x : $\sqrt[3]{19}$
0169 0900 $t\pi$: $e/3 - 3\pi$
0169 1448 J_1 : $(2e)^{-2}$
0169 1672 ℓ_{10} : $3\ln 2/2$
0169 2024 ζ : $(\ln 3/2)^{-3}$
0169 2327 ℓ_2 : $3 + \pi/3$
0169 4240 2^x : $4\zeta(3)$
0169 5109 cu : $\exp(-e/2)$
0169 6644 erf : $(1/(3e))^2$
0169 7675 cr : $7\zeta(3)/8$
0169 8498 ℓ_{10} : $4\zeta(3)/5$
0169 8863 Γ : $3e/4$
0169 9091 Ψ : $13/4$
0170 1245 J : $2/21$
0170 1323 sq : $3/23$
0170 1534 at : $\exp(\sqrt[3]{3}/3)$
0170 2861 e^x : $7\pi/10$
0170 3104 at : $7\ln 2/3$
0170 3828 10^x : $11/13$
0170 4141 J_2 : $7(1/\pi)/6$
0170 5084 ℓ_{10} : $2\sqrt[3]{3}/3$
0170 7357 2^x : $\sqrt{2}/2 + 3\sqrt{3}/4$
0170 9758 ℓ_2 : $e/3 + \pi$
0171 0738 $1/5 \times \exp(3)$
0171 2748 2^x : $3\sqrt{2}/4 - 4\sqrt{3}$
0171 3130 e^x : $9/2$
0171 4619 ℓ_{10} : $3 - 3e/4$
0171 6104 J_2 : $\sqrt{5}/6$
0171 7596 cu : $(\pi/2)^{-3}$
0171 9240 10^x : $\sqrt{2}/3 - \sqrt{5}$
0171 9909 cu : P_{28}
0172 0194 cr : $3/2 + 3\sqrt{5}$
0172 1140 ζ : $5\zeta(3)$
0172 1325 cu : $1/\sqrt{15}$
0172 1708 ℓ_{10} : $1/3 + \sqrt{2}/2$
0172 2196 at : $1/2 + \sqrt{5}/2$
0172 3074 Ψ : $7/9$
0172 4476 cr : $20/19$
0172 5488 $\ln(\sqrt{2}) \div \exp(3)$
0172 7318 ζ : $6/17$
0172 8044 sinh : $\sqrt{2} - \sqrt{5} + \sqrt{7}$
0172 8556 $2/5 \div \exp(\pi)$
0173 2049 θ_3 : $2\ln 2/3$
0173 3557 10^x : $1/3 - 2\pi/3$
0173 4154 cu : $\sqrt{2}/2 + 4\sqrt{5}$
0173 4306 ζ : 6
0173 5071 J_2 : $1/\sqrt[3]{19}$
0173 5176 $3/4 \times \exp(\sqrt{5})$
0173 7303 J_2 : $3/8$
0173 9984 $\ln(\sqrt{5}) - \exp(3/5)$
0174 1320 sq : P_{49}
0174 2485 sinh : $7\sqrt{5}/4$
0174 3126 2^x : $\sqrt{3}/2 - 3\sqrt{5}$
0174 6124 sq : P_{64}
0174 6350 rt : (1,9,9,9)
0174 7406 cu : $7(e + \pi)/10$
0174 7742 cu : $\sqrt{2} - 2\sqrt{3}/3$
0174 8743 Γ : $\sqrt{5} + \sqrt{6} - \sqrt{7}$
0174 9706 e^x : $2/3 - 3\pi/2$
0175 1550 ζ : $(2\pi/3)^{-1/3}$
0175 2173 $\ln(2/3) \div \exp(\pi)$
0175 3806 ζ : $7\sqrt[3]{5}/2$
0175 4422 λ : $\exp(-13\pi/6)$
0175 4516 sr : $1/4 + \pi/4$
0175 4667 $\exp(1/3) - \exp(5)$
0175 5993 cr : $(\sqrt[3]{5}/2)^{-1/3}$
0175 6184 cu : $3\ln 2/8$
0175 8700 cu : $5(e + \pi)/2$
0176 0500 id : $4\sqrt{3}/3 + 3\sqrt{5}$
0176 1290 e^x : $\sqrt{3}/4 - 2\sqrt{5}$
0176 2948 erf : $(4)^{-3}$
0176 3066 ζ : $7e\pi/10$
0176 4550 J_2 : $1/\sqrt{7}$
0176 4640 J_1 : $(\pi/2)^3$
0176 4659 J_2 : $3\sqrt[3]{2}/10$
0176 5688 $\ln(2) - s\pi(\sqrt{5})$
0176 7061 10^x : $1/3 + \pi/3$
0176 7579 at : $3\sqrt{2}/4 + \sqrt{5}/4$
0176 8111 J_1 : $(1/\pi)/9$
0177 1266 2^x : $(2\pi)^{-2}$
0177 2430 e^x : $e/4 - 3\pi/2$
0177 2876 ℓ_{10} : $24/25$
0177 3687 at : $9\sqrt[3]{2}/7$
0177 3894 rt : (2,9,-4,6)
0177 5293 cu : $6/23$
0177 7243 rt : (1,9,8,1)
0177 7755 ζ : $-\sqrt{2} + \sqrt{6} - \sqrt{7}$
0177 7777 sq : $2/15$
0178 1544 Ψ : $5\sqrt{3}/6$
0178 1546 e^x : $\sqrt{13/3}$
0178 2747 ζ : $8\sqrt{5}/3$
0178 3328 J_2 : $2\sqrt[3]{5}/9$
0178 3877 sq : $\zeta(3)/9$
0178 4621 ℓ_{10} : $\sqrt{11\pi}$
0178 6874 sq : $6\zeta(3)$
0178 7492 sinh : 3
0178 7640 Γ : $2/19$
0178 8186 ln : $-\sqrt{2} + \sqrt{3} + \sqrt{6}$
0179 2219 J_2 : $8/21$
0179 4340 2^x : $4e/3 - 3\pi$
0179 4919 sq : $1 - \sqrt{3}/2$
0179 5349 Γ : $\sqrt[3]{17/2}$
0179 6659 θ_3 : $\exp(-3\pi/2)$
0179 6869 2^x : $1/3 + e/4$
0179 7260 Γ : $7\ln 2/5$
0179 7977 J_2 : $1/\sqrt[3]{18}$
0179 8130 ζ : $(2e/3)^3$
0179 8426 sq : $(\ln 3/3)^2$
0179 8867 $t\pi$: $\sqrt{2}/4$
0179 9942 $t\pi$: $\sqrt[3]{16/3}$
0180 0489 cu : $10(e + \pi)/9$
0180 0518 sq : $((e + \pi)/3)^{-3}$
0180 1313 J_0 : $(4/3)^3$
0180 1707 J_2 : $6(1/\pi)/5$
0180 4168 J_0 : $9e$
0180 4801 2^x : $-\sqrt{2} - \sqrt{3} - \sqrt{7}$
0180 5710 rt : (2,9,-5,1)
0180 6681 ℓ_{10} : $1 + 3\pi$
0180 7392 $\sqrt{3} \times s\pi(1/5)$
0180 8589 J_0 : $2e\pi/7$
0180 8646 $s\pi(1/5) - s\pi(\pi)$
0180 9243 rt : (1,9,7,-7)
0181 0250 Ei : $\exp(-1)$
0181 2896 at : $1/2 - 3\sqrt{2}/2$
0181 5773 cu : $\sqrt{2} - 3\sqrt{5}/4$
0181 5882 2^x : $1/2 - 2\pi$
0181 8579 cr : $19/18$
0182 0550 sq : $1/3 - 3e/2$
0182 2423 cu : $5/19$
0182 6422 J_2 : $5/13$

0182 6499 $2^x : 3\sqrt{2} - \sqrt{5}$
0182 9095 $J_2 : 2\sqrt{3}/9$
0182 9649 sr : $\zeta(5)$
0182 9723 $10^x : -\sqrt{2} + \sqrt{3} + \sqrt{5}$
0182 9858 $\Gamma : \sqrt{17}$
0183 0224 $e^x : \sqrt{2}/3 - 2\sqrt{5}$
0183 0799 $J_2 : 5\ln 2/9$
0183 0892 $t\pi : 1/\ln(\sqrt{2}\pi/2)$
0183 1359 at : $\exp(-4)$
0183 1563 id : $\exp(-4)$
0183 1666 as : $\exp(-4)$
0183 2368 $\theta_3 : 17/24$
0183 3029 at : $\sqrt{3} - 3\sqrt{5}/2$
0183 3122 at : $\sqrt{3}/3 - \sqrt{5}/4$
0183 3327 id : $\sqrt{3}/3 - \sqrt{5}/4$
0183 3430 as : $\sqrt{3}/3 - \sqrt{5}/4$
0183 6250 $10^x : 1/2 - \sqrt{5}$
0183 7650 $J_2 : 6\sqrt{2}$
0183 7967 $\zeta : 3\pi^2/5$
0183 8413 $2^x : 2\sqrt{2}/3 - 3\sqrt{5}$
0183 8681 rt : (2,9,-6,-4)
0183 8772 $\ln(3) - \exp(3/4)$
0184 0019 $Li_2 : \exp(-4)$
0184 0560 ln : $4 - 4\sqrt{5}/3$
0184 4096 ln : $\sqrt{23/3}$
0184 4728 $J_1 : 6e\pi/5$
0184 5092 $e^x : 1/4 - 3\sqrt{2}$
0184 6235 at : $9\sqrt[3]{3}/8$
0184 8340 $\ell_{10} : 23/24$
0184 8439 $e^x : \exp(-4)$
0184 8546 $\Psi : 1/\ln(2)$
0184 8756 $10^x : 2e/9$
0184 9178 $\ell_{10} : 3\sqrt{5}/7$
0184 9714 $\ell_2 : 4/3 + e$
0185 0236 $e^x : \sqrt{3}/3 - \sqrt{5}/4$
0185 1504 $t\pi : e - 3\pi/2$
0185 1534 $J_1 : (3)^{-3}$
0185 3277 $e^x : 2e - 3\pi$
0185 4630 sinh : $2\sqrt{5}/5$
0185 4994 cr : $5\pi^2/6$
0185 5361 $e^x : 9\sqrt{3}/8$
0185 7389 at : $3 - 4\sqrt{5}/3$
0185 7507 $10^x : \sqrt[3]{23}$
0185 7603 id : $3 - 4\sqrt{5}/3$
0185 7709 as : $3 - 4\sqrt{5}/3$
0185 9138 $10^x : (5)^{-3}$
0185 9504 sq : $3/22$
0186 0113 $\sqrt{2} - \exp(1/3)$
0186 1001 sq : $3(1/\pi)/7$
0186 1387 $J_2 : e/7$
0186 3717 $e^x : (\ln 2/2)^{1/3}$
0186 6408 cu : $5(1/\pi)/6$
0186 6719 $J_2 : 7/18$
0186 6930 $J_2 : 1/\sqrt[3]{17}$
0186 8843 $Ei : 2\sqrt[3]{5}/5$
0186 8871 $\zeta : \sqrt{2}/4$
0186 9461 $Ei : 6\sqrt{5}/5$
0187 0878 $\Psi : 6\zeta(3)/5$
0187 1135 cr : $2 - 2\sqrt{2}/3$
0187 1770 $Ei : 23/22$
0187 2418 $\exp(\sqrt{5}) \times s\pi(\sqrt{2})$
0187 2869 rt : (2,9,-7,-9)
0187 4963 $e^x : 3 - 4\sqrt{5}/3$
0187 6926 sr : $(\sqrt{5}/2)^{1/3}$
0187 7886 sq : P_{24}
0187 8429 $\Gamma : (\ln 3)^{-1/3}$
0188 1667 sq : $2\sqrt{3}/3 + 3\sqrt{5}/4$
0188 3316 $e^x : 1/2 - 2\sqrt{5}$
0188 4377 $J_0 : ((e+\pi)/2)^2$
0188 7816 $\ell_{10} : 3/2 + 4\sqrt{5}$
0188 9390 $\lambda : \exp(-15\pi/7)$
0188 9682 $J_2 : 9/23$
0189 1340 $t\pi : \sqrt{2}/2 - 2\sqrt{3}/3$
0189 2422 $\Psi : \sqrt[3]{3}$
0189 3057 $\lambda : (3\pi)^{-3}$
0189 3583 $\ell_2 : \pi^2/10$
0189 6277 $\Psi : 5(e+\pi)/9$
0189 6296 cu : $4/15$
0189 6983 at : $2 - 4e/3$
0189 7635 $\zeta : 10(1/\pi)/9$
0189 7808 ln : $\sqrt{2} - \sqrt{3}/4$
0189 8357 $\ln(3/4) \div \exp(e)$
0189 8834 sinh : $17/19$
0190 1762 $e^x : 3/4 - 3\pi/2$
0190 2235 ln : $2e + 2\pi/3$
0190 2928 $J_2 : \ln(\sqrt{2}\pi/3)$
0190 3003 $J_2 : \pi/8$
0190 3960 $\ell_{10} : 1/4 + \sqrt{2}/2$
0190 6064 cu : $2\zeta(3)/9$
0190 6216 sq : P_4
0190 6365 sq : $1/3 - \sqrt{2}/3$
0190 7086 $J_2 : 8\sqrt[3]{3}$
0190 9008 cu : $1/\sqrt{14}$
0191 1203 $\theta_3 : (3\pi/2)^{-3}$
0191 2137 id : $7(e+\pi)$
0191 2175 $J_1 : 1/\ln(\sqrt{3}/2)$
0191 2422 $t\pi : 5\ln 2/6$
0191 2755 $2^x : 1 - 3\sqrt{5}$
0191 3014 $10^x : 1 - e$
0191 3612 $2^x : 3\sqrt{2}/4 + 2\sqrt{3}$
0191 4134 at : $13/8$
0191 4457 cu : P_{66}
0191 4604 cu : P_{65}
0191 4939 id : $7\sqrt[3]{3}/5$
0191 6335 $10^x : 4\sqrt{2}/3 + 3\sqrt{5}$
0191 7074 ln : $3e/8$
0192 1812 sq : $\ln 2/5$
0192 2794 $e^x : 4/3 + \sqrt{3}/2$
0192 3546 cr : $18/17$
0192 3788 cu : $2 - \sqrt{3}$
0192 4107 sq : $7\sqrt[3]{3}/10$
0192 4545 $\zeta : (e+\pi)$
0192 6292 sr : $3e/2$
0192 9754 cr : $1/2 + \sqrt{5}/4$
0193 0353 $s\pi : \sqrt{2}/2 + 3\sqrt{3}/4$
0193 0515 $\ell_{10} : 22/23$
0193 0712 $t\pi : \sqrt{2}/2 + 3\sqrt{3}/4$
0193 0746 id : $\sqrt{2}/3 - 2\sqrt{5}/3$
0193 0977 ln : $8\sqrt{3}/5$
0193 2416 $\ell_2 : \sqrt{3} - \sqrt{5}/3$
0193 2625 $Ei : 13/19$
0193 3598 $2^x : 15/2$
0193 5520 $J_2 : 4\ln 2/7$
0193 5568 id : $3e/8$
0193 7054 $10^x : 4 - \sqrt{5}/2$
0193 9148 $10^x : 3 - 3\pi/2$
0194 0321 ln : $4\sqrt{2}/3 - \sqrt{3}/2$
0194 0700 $2^x : \sqrt{2}/4 + \sqrt{5}$
0194 2654 sr : $3\sqrt{3}/5$
0194 2916 $J_2 : 1/\sqrt[3]{16}$
0194 3517 sr : $3\sqrt{3} - \sqrt{5}/2$
0194 4261 ln : $9(e+\pi)/7$
0194 4473 at : $\sqrt{2}/3 + 2\sqrt{3}/3$
0194 5154 $s\pi : 4\sqrt{5}/9$
0194 5522 $t\pi : 4\sqrt{5}/9$
0194 6244 $J_1 : 7\sqrt[3]{3}$
0194 7175 $e^x : 1/4 - 4\pi/3$
0195 1516 $e^x : 3/2 - 2e$
0195 2950 $J_2 : 5(1/\pi)/4$
0195 5009 $s\pi : (2e)^{-3}$
0195 5382 $t\pi : (2e)^{-3}$
0195 6723 $J_0 : 10\sqrt[3]{5}/7$
0195 6916 $J_2 : (e/2)^{-3}$
0195 9267 id : $4\sqrt{2}/3 - \sqrt{3}/2$
0195 9754 sq : $\sqrt[3]{2}/9$
0196 3122 $\ell_2 : \sqrt[3]{25/3}$
0196 3181 $J_2 : (2\pi)^{-1/2}$
0196 6699 sr : $3\ln 2/2$
0196 7899 $e^x : 3 - 4\sqrt{3}$
0196 9522 id : $\exp(\sqrt{2}\pi)$
0197 0706 $10^x : 1/3 - 3e/4$
0197 1749 $s\pi : (\pi)^3$
0197 2132 $t\pi : (\pi)^3$
0197 2339 $\ell_{10} : (3\pi/2)^3$
0197 3106 $\Gamma : (e/3)^{1/3}$
0197 3288 $\sqrt{3} - \ln(3/4)$
0197 3466 $J_2 : 2/5$
0197 6466 $e^x : 2\sqrt{2}/3 + \sqrt{5}$
0197 8144 ln : $4\ln 2$
0197 9185 $\Psi : (1/\pi)/4$
0198 0146 $J_2 : \zeta(3)/3$
0198 2445 cr : $3\sqrt{2}/4$
0198 3560 $\zeta : 7\sqrt[3]{3}/10$
0198 4128 sq : $3/2 - e/2$
0198 6436 at : $\sqrt{2} - \sqrt{5} + \sqrt{6}$
0198 7009 $2^x : 3/2 + 4\sqrt{3}/3$
0198 7509 $J_1 : ((e+\pi)/2)^{-3}$
0198 8908 rt : (1,7,1,8)
0198 9043 $J_1 : (1/\pi)/8$
0198 9310 $J_0 : 7\pi/9$
0198 9711 sq : P_{70}
0198 9813 sq : P_{67}
0199 1022 sq : $e/3 - \pi/3$
0199 1482 $2/5 \div \exp(3)$
0199 3785 $\Psi : 2e/7$
0199 3862 ln : $3\zeta(3)/10$
0199 4391 cr : $10(1/\pi)/3$
0199 4894 $2^x : 3\sqrt{2}/4 - 3\sqrt{5}$
0199 5172 $10^x : 12/25$
0199 5292 $\ln(\sqrt{2}) \times \exp(\pi)$
0199 7033 $t\pi : 10(1/\pi)/9$
0199 7319 $\lambda : \exp(-17\pi/8)$
0199 9600 $J_1 : 1/25$
0200 0000 sq : $\sqrt{2}/10$
0200 1406 sq : $4(1/\pi)/9$
0200 1834 ln : $2/3 + \sqrt{2}/4$
0200 1861 $2^x : 1/2 - \sqrt{2}/3$
0200 1900 cr : $4 + 3\sqrt{2}$
0200 1910 $e^x : 21/19$
0200 1966 sr : $1/3 + \sqrt{2}/2$
0200 2861 $\ell_{10} : \pi/3$
0200 4847 sq : $3 - \pi$
0200 5438 $e^x : 1/3 - 3\sqrt{2}$
0200 6139 cu : $2\sqrt{2}/3 - 3\sqrt{5}/2$
0200 8553 cu : $e/10$
0200 9026 ln : $\sqrt[3]{3}/4$
0200 9202 sr : $4\sqrt{2} + 2\sqrt{3}$
0201 0909 sq : $\exp(-(e+\pi)/3)$
0201 2178 $\ln(3) \div \exp(4)$
0201 3191 $J_2 : 2\sqrt{2}/7$
0201 3570 $J_2 : \exp(-e/3)$
0201 3944 $\ell_2 : \exp(1/\sqrt{2})$
0201 4654 $J_1 : 6\sqrt{5}$
0201 4895 cu : P_{30}
0201 8494 $\Psi : 1/\ln(e/2)$
0201 8691 $\ln(2/3) \div \exp(3)$
0202 0127 $J_1 : 7(e+\pi)/4$
0202 0338 $\ell_{10} : 21/22$
0202 1384 $3 \div \exp(5)$
0202 1730 at : $1/3 - \sqrt{2}/4$
0202 2005 id : $1/3 - \sqrt{2}/4$
0202 2143 as : $1/3 - \sqrt{2}/4$
0202 4402 $\zeta : \sqrt{2} + \sqrt{3} + \sqrt{7}$
0202 4756 cu : $\sqrt{3}/4 + 2\sqrt{5}$
0202 5236 $J_2 : (\pi/2)^{-2}$
0202 6536 ln : $7\sqrt[3]{2}/9$
0202 6547 $10^x : 4e - 4\pi$
0202 6995 $10^x : \ln(1/(2e))$
0202 7014 $J_2 : \ln(2/3)$
0202 7162 $J_2 : 1/\sqrt[3]{15}$
0202 7342 cu : $4/3 - 3\sqrt{2}/4$
0202 7704 $2^x : 4\sqrt{2}/3 + \sqrt{5}/2$
0202 7790 ln : $e/4 + 2\pi/3$
0202 8190 $J_2 : 21$
0202 8549 cu : $3/11$
0202 8764 $\exp(e) + s\pi(1/3)$
0202 9036 rt : (1,7,0,1)
0203 0508 id : $10\sqrt{2}/7$
0203 1000 cu : $6(1/\pi)/7$
0203 1314 id : $10e/9$
0203 4773 $e^x : \sqrt{3}/3 - 2\sqrt{5}$
0203 6688 $\zeta : 1/\ln(\sqrt[3]{4}/3)$
0203 8259 $2^x : 2/3 - 2\pi$
0203 9104 $2^x : (e+\pi)^{-2}$
0203 9512 $J_0 : 22/9$
0204 0816 sq : $1/7$
0204 1377 cr : $17/16$
0204 2032 at : $(\ln 2/3)^{-1/3}$
0204 2213 $t\pi : 7\pi/5$
0204 2222 $\ell_{10} : (\ln 3)^{1/2}$

0204 2274 cu : $4\sqrt{2}/3+\sqrt{5}$
0204 2728 ln : $4/3-\sqrt{2}/4$
0204 2936 J_2 : $3\sqrt{3}$
0204 3958 10^x : $2/3-3\pi/4$
0204 4739 Ei : $6\sqrt[3]{3}/5$
0204 8075 K : $13/16$
0204 8820 Ei : $\exp(4/3)$
0205 1580 e^x : $2\sqrt{2}+3\sqrt{3}/4$
0205 2399 Ψ : 21
0205 2577 ζ : $10\sqrt{3}/3$
0205 3767 cu : $3/4-3e$
0205 4548 J_2 : $1/\sqrt{6}$
0205 5266 cu : $\sqrt{13/3}$
0205 5989 cu : $\sqrt{2}/3-\sqrt{5}/3$
0205 6571 2^x : $e/4-2\pi$
0205 6903 id : $10\zeta(3)$
0205 9202 cu : $\sqrt{2}/2-\sqrt{3}/4$
0205 9495 ζ : $4\sqrt[3]{3}$
0206 0037 at : $\sqrt[3]{13/3}$
0206 1680 J_1 : $10e/7$
0206 2072 sr : $25/24$
0206 2919 J_2 : $9/22$
0206 4557 J_2 : $9(1/\pi)/7$
0206 4730 sπ : $3\sqrt{2}-\sqrt{5}$
0206 5171 tπ : $3\sqrt{2}-\sqrt{5}$
0206 5346 e^x : $4\pi/7$
0206 5543 e^x : $2\sqrt{2}-3\sqrt{5}$
0206 6467 erf : $\exp(-4)$
0206 6632 10^x : $4-\pi/3$
0206 7341 e^x : $1/\ln(e/2)$
0206 7395 ℓ_{10} : $\sqrt{2}/3+\sqrt{3}/3$
0206 7530 $\exp(2/3)\div s\pi(\sqrt{2})$
0206 7807 2^x : $-\sqrt{2}-\sqrt{3}-\sqrt{6}$
0206 9024 sq : $\ln(\sqrt{3}/2)$
0206 9075 2^x : $4/3-4\sqrt{3}$
0207 0056 sq : P_{75}
0207 0807 rt : (1,7,-1,-6)
0207 1165 $\sqrt{3}\div\ln(3/4)$
0207 2594 id : $7\sqrt{3}/6$
0207 3767 ζ : $\sqrt{2}$
0207 3812 J : $\sqrt[3]{3}/7$
0207 6001 $\sqrt{5}\div\ln(4/5)$
0207 6863 id : $1/3-3\sqrt{5}/2$
0207 7113 10^x : $6\sqrt[3]{2}/7$
0207 7656 at : $3e/5$
0207 7849 10^x : $3\sqrt{3}/4-4\sqrt{5}/3$
0207 8097 $3/5\div s\pi(1/5)$
0207 8889 2^x : $\sqrt{2}-\sqrt{3}+\sqrt{7}$
0207 9019 sq : $3/4+2\sqrt{5}/3$
0208 0083 sq : $\sqrt[3]{3}/10$
0208 0953 J_1 : $(\ln 2/2)^3$
0208 2312 ℓ_{10} : $(\sqrt{3}/2)^{1/3}$
0208 2881 J_1 : $1/24$
0208 5463 λ : $\exp(-19\pi/9)$
0208 7646 sq : $2/3-3\sqrt{5}/4$
0208 8940 ζ : $23/4$
0208 9115 ln : $7(e+\pi)/2$
0208 9590 J_2 : $7/17$
0209 0020 10^x : $\exp(-3\pi/2)$
0209 2659 J_2 : $2\sqrt[3]{3}/7$
0209 3068 sinh : $15/2$
0209 6645 $\ln(\pi)\div\exp(4)$
0209 9334 Ψ : $3/20$
0209 9713 ℓ_{10} : $2-\pi/3$
0210 0869 e^x : $4/3-3\sqrt{3}$
0210 2193 rt : (4,9,-7,7)
0210 2245 cr : $4\sqrt{2}+3\sqrt{3}/2$
0210 2977 J_2 : $(e+\pi)^{-1/2}$
0210 3442 2^x : $7e\pi$
0210 3723 tπ : $3e/4-\pi/4$
0210 4319 tπ : $(\pi/2)^{1/2}$
0210 5473 J_1 : $3\sqrt[3]{2}$
0210 5600 10^x : $e/4-3\pi/4$
0210 5712 cu : $2/3-2\sqrt{2}/3$
0210 6686 Ψ : $2\sqrt[3]{5}/7$
0210 9775 Γ : $19/3$
0211 1085 ζ : $10\zeta(3)/7$
0211 2884 J_2 : $(\sqrt{5}/3)^3$
0211 3212 cu : $\sqrt{11/3}$
0211 5947 at : $1/3+3\sqrt{3}/4$
0211 6208 ℓ_{10} : $5\sqrt[3]{2}/6$
0211 6393 e^x : $1/3-4\pi/3$
0211 6521 Ei : $1/\sqrt{7}$
0211 6630 sπ : $\exp(-5)$
0211 6681 Ei : $20/7$
0211 6788 $\pi\div\exp(5)$
0211 7104 tπ : $\exp(-5)$
0211 8929 ℓ_{10} : $2/21$
0211 9371 J : $5/22$
0211 9714 cu : $3-3\pi/2$
0212 1093 Ei : $3\sqrt[3]{2}/10$
0212 1206 J_2 : $1/\sqrt[3]{14}$
0212 3086 tπ : $2\sqrt{2}/3-3\sqrt{3}$
0212 3174 Ψ : $8\sqrt[3]{2}/7$
0212 4736 Li_2 : $17/22$
0212 7714 Ψ : $6e/5$
0212 9260 2^x : $2\sqrt{3}/3-3\sqrt{5}$
0213 1043 J_2 : $3\ln 2/5$
0213 1145 sinh : $5\sqrt[3]{2}/3$
0213 1357 cu : $2\ln 2/5$
0213 2908 at : $\sqrt{8/3}$
0213 3462 cu : $1/\sqrt{13}$
0213 4775 Ei : $\ln 2/10$
0213 6324 tπ : $2\sqrt[3]{2}$
0213 7558 $1/5\div\exp(\sqrt{5})$
0213 8423 $1/4\times\exp(3)$
0213 8912 J_2 : $5/12$
0214 0275 $1/5-\exp(1/5)$
0214 1338 ℓ_{10} : $(e/3)^{1/2}$
0214 2001 J_0 : $9e/10$
0214 2040 10^x : $1/\ln(\ln 3/2)$
0214 3347 cu : $5/18$
0214 4654 as : $\exp(-1/(2\pi))$
0214 4660 sq : $1/2-\sqrt{2}/4$
0214 5386 J_1 : $6\sqrt[3]{5}$
0214 6776 rt : (4,9,-8,4)
0214 7158 sq : $3e\pi/4$
0215 0783 sr : $24/23$
0215 1097 sq : $4\sqrt{3}/3-4\sqrt{5}$
0215 3751 10^x : $\sqrt{23}$
0215 3817 sπ : $10\zeta(3)/3$
0215 4317 tπ : $10\zeta(3)/3$
0215 4392 10^x : $10\ln 2$
0215 6143 cr : $4/3+4\sqrt{3}$
0215 7342 ℓ_2 : $\exp(2\pi/3)$
0215 8265 J_1 : $6e\pi$
0215 9121 2^x : $(\pi/3)^{1/3}$
0215 9460 2^x : $3/4-2\pi$
0216 0191 J_1 : $\exp(-\pi)$
0216 0600 cu : $7(1/\pi)/8$
0216 0695 $1/2\div\exp(\pi)$
0216 1355 cu : $7e\pi/9$
0216 1677 10^x : $3/2-2\sqrt{5}/3$
0216 2331 cu : $3/4-\sqrt{2}/3$
0216 3264 2^x : $2+3e$
0216 4568 ℓ_{10} : $3e+3\pi/4$
0216 5124 ln : $9/25$
0216 5266 $\ln(5)-s\pi(1/5)$
0216 8697 ζ : $10\sqrt[3]{5}/3$
0217 0614 sq : $(\pi/2)^3$
0217 1390 ln : $2\sqrt[3]{2}/7$
0217 1429 ℓ_2 : $3+3\sqrt{2}/4$
0217 1723 e^x : $1/4+2\sqrt{3}$
0217 2496 J_2 : $\sqrt[3]{2}/3$
0217 3399 J_1 : $1/23$
0217 4590 cr : $16/15$
0217 6563 e^x : $1/4-3e/2$
0217 8252 cu : P_{72}
0217 8419 cu : P_{69}
0217 9089 J_2 : $\exp(-\sqrt{3}/2)$
0217 9712 ζ : $2e\pi/3$
0218 3507 J_2 : $8/19$
0218 3660 cu : $\sqrt{5}/8$
0218 4827 cr : $3/2-\sqrt{3}/4$
0218 5674 sinh : $21/10$
0218 6093 sinh : $(\sqrt[3]{3}/2)^{1/3}$
0218 8384 2^x : $\sqrt{2}-4\sqrt{3}$
0219 0918 ℓ_2 : $\sqrt{2}+3\sqrt{5}$
0219 1918 J_2 : $(4/3)^{-3}$
0219 2576 sq : $4\sqrt{2}/3+\sqrt{5}/2$
0219 3058 sq : $\sqrt{2}/2-\sqrt{5}/4$
0219 3265 rt : (4,9,-9,1)
0219 3307 ℓ_{10} : $7\zeta(3)/8$
0219 4230 10^x : $\sqrt{3}/3-\sqrt{5}$
0219 4278 cr : $3e/2+4\pi/3$
0219 4787 sq : $(\sqrt[3]{4}/3)^3$
0219 5200 cu : $7/25$
0219 8954 $\exp(\pi)\div\exp(1/4)$
0219 9601 $1/3\div\exp(e)$
0220 0131 cr : $e\pi/8$
0220 0907 Ψ : $6\sqrt{3}/7$
0220 3665 10^x : $4-4\sqrt{2}$
0220 4413 at : $1/4-4\sqrt{2}/3$
0220 4456 e^x : $e/4+2\pi/3$
0220 6388 $\ln(\sqrt{2})+s\pi(\sqrt{5})$
0220 7477 $\ln(3/5)\div\exp(\pi)$
0220 7522 sr : $3\sqrt{3}/2+2\sqrt{5}/3$
0220 7532 2^x : $3\sqrt{2}-\sqrt{3}/4$
0220 8536 J_1 : $2\pi^2$
0221 1733 10^x : $2\sqrt{2}/3-3\sqrt{3}/2$
0221 1880 $1/5+\exp(3/5)$
0221 1881 rt : (1,8,-9,2)
0221 6439 J_2 : $3\sqrt{2}/10$
0221 7787 ℓ_2 : $(\pi/3)^{1/3}$
0221 7974 J_2 : $4(1/\pi)/3$
0221 8245 10^x : P_{21}
0221 8360 tπ : $(3\pi/2)^3$
0221 8459 ζ : $9\sqrt[3]{2}/2$
0222 3038 e^x : $e/3-3\pi/2$
0222 3200 10^x : $\sqrt{2}+\sqrt{5}/4$
0222 3419 ζ : $17/3$
0222 4565 id : $(\ln 2/2)^{-3}$
0222 4692 at : $18/11$
0222 4738 10^x : $(3\pi/2)^{-3}$
0222 4875 e^x : $2/3-2\sqrt{5}$
0222 6123 sinh : $8e\pi/9$
0222 6991 J_2 : $\exp(-\sqrt[3]{5}/2)$
0222 7013 J_2 : $1/\sqrt[3]{13}$
0222 7639 ℓ_{10} : $19/20$
0222 8250 ℓ_{10} : $5\sqrt[3]{5}/9$
0223 0876 2^x : $(1/\pi)/10$
0223 2473 J_2 : $17/2$
0223 2936 cr : $8\zeta(3)/9$
0223 3830 $3\div s\pi(\sqrt{3})$
0223 3985 rt : (7,8,-8,9)
0223 5859 cu : $3-e$
0223 7077 $\exp(1/5)\div\exp(4)$
0223 7667 cu : P_6
0223 9174 θ_3 : $\ln(\sqrt[3]{4}/3)$
0223 9842 ζ : $4\sqrt{2}$
0224 0290 2^x : $2\sqrt{3}-4\sqrt{5}$
0224 0588 cr : $5\sqrt[3]{5}/8$
0224 3184 ζ : $9\pi/5$
0224 3688 Γ : $(\sqrt{5}/2)^{-1/3}$
0224 7471 sr : $23/22$
0224 7524 cu : $2e/3-2\pi/3$
0224 7620 ζ : $\exp(\sqrt{3})$
0224 8585 10^x : $4/3-4\sqrt{5}/3$
0224 9797 J_2 : $\sqrt[3]{5}/4$
0225 0000 sq : $3/20$
0225 1011 J_1 : $(3\pi/2)^{-2}$
0225 1585 θ_3 : $(3\pi)^{-2}$
0225 3204 θ_3 : $14/17$
0225 4888 tπ : $(e/2)^{-3}$
0225 7719 sq : $\zeta(3)/8$
0225 8931 10^x : $4\sqrt{2}+4\sqrt{3}$
0225 9701 $\ln(5)+\exp(5)$
0226 0680 2^x : $(\pi)^{-3}$
0226 0977 J_2 : $3/7$
0226 2741 cu : $\sqrt{2}/5$
0226 4204 J_2 : $(2e)^{-1/2}$
0226 5128 cu : $8(1/\pi)/9$
0226 6160 Γ : $3\ln 2/7$
0226 7999 ℓ_{10} : $(\sqrt[3]{5}/2)^{1/3}$
0226 8735 ζ : $4\pi^2/7$
0226 9094 10^x : $9\ln 2/8$
0226 9437 at : $3\sqrt{2}/4+\sqrt{3}/3$
0227 1387 10^x : $2\sqrt{2}-2\sqrt{5}$
0227 2140 J_1 : $1/22$
0227 3054 J_1 : $(1/\pi)/7$

0227 4066 ln : $\sqrt{2}/3+4\sqrt{3}/3$
0227 4955 id : $6(e+\pi)/7$
0227 4957 sπ : $3\sqrt{3}/4+3\sqrt{5}$
0227 5546 tπ : $3\sqrt{3}/4+3\sqrt{5}$
0227 5781 ζ : $(e/3)^2$
0227 6489 10^x : $2/3-4\sqrt{3}/3$
0227 6791 as : $1/2+\sqrt{2}/4$
0227 7848 e^x : $\sqrt{2}-3\sqrt{3}$
0227 8769 2^x : $\sqrt{8\pi}$
0227 9815 ℓ_{10} : $e\pi/9$
0228 0531 θ_3 : $(\sqrt{2\pi})^{-3}$
0228 0852 ln : $6\sqrt[3]{2}$
0228 2481 10^x : $3/2-\pi$
0228 2809 cu : $\exp(-\sqrt[3]{2})$
0228 3244 ζ : $-\sqrt{3}-\sqrt{6}+\sqrt{7}$
0228 4134 rt : (7,8,-9,7)
0228 6088 J_2 : $25/3$
0228 6115 id : $1/3-3\pi/4$
0228 6971 $\ln(\sqrt{2})\div\exp(e)$
0229 4678 tπ : $\sqrt{2}/3-2\sqrt{3}$
0229 4744 e^x : $10e\pi/3$
0229 6696 J_0 : $\sqrt{6}$
0229 8515 10^x : $2-3\sqrt{3}/4$
0230 3800 J_2 : $3\sqrt[3]{3}/10$
0230 5051 cu : $4/3-3\sqrt{3}/2$
0230 5266 J_1 : $\ln(\pi/3)$
0230 5879 ln : $(\pi/3)^{1/2}$
0230 7342 J_2 : $\sqrt{3}/4$
0230 7726 sinh : $2\pi/7$
0230 8629 J_2 : $2(e+\pi)$
0230 9368 10^x : $3/2+2\pi/3$
0230 9428 rt : (1,8,8,8)
0230 9486 J_2 : $5\ln 2/8$
0230 9695 Γ : $6\sqrt[3]{5}/5$
0231 0348 10^x : $(\ln 2)^2$
0231 1790 sπ : $3/4+3\sqrt{2}$
0231 1860 id : $2\sqrt{3}+\sqrt{5}/4$
0231 2001 sπ : $2e+\pi/2$
0231 2408 tπ : $1/4-3\sqrt{2}$
0231 2619 tπ : $2e+\pi/2$
0231 3420 sq : P_{71}
0231 3538 sq : P_{68}
0231 3568 $1/3-\exp(\sqrt{5})$
0231 3648 ln : $4\sqrt[3]{5}/7$
0231 4098 10^x : $1/4-4\sqrt{2}/3$
0231 4355 $1/5+\ln(4/5)$
0231 4371 J : $\exp(-10\pi/7)$
0231 4814 cu : $\sqrt[3]{5}/6$
0231 5011 sinh : $\sqrt{10/3}$
0231 5883 θ_3 : $1/\sqrt[3]{10}$
0231 6081 ℓ_{10} : $9(e+\pi)/5$
0231 6693 e^x : $\sqrt{2}/2-2\sqrt{5}$
0231 6779 J_1 : $6(e+\pi)$
0231 6829 cu : $4(e+\pi)/5$
0231 7638 e^x : $12/5$
0231 9130 sq : $9\pi^2/7$
0232 2458 2^x : $3/2-4\sqrt{3}$
0232 4053 ζ : $11/18$
0232 4628 cu : $1/2-\pi/4$
0232 4995 Γ : $5\sqrt{3}/9$
0232 5944 J_2 : $10/23$
0232 6410 cr : $15/14$
0232 7979 2^x : $4-3\pi$
0233 1868 cu : $3e\pi/7$
0233 2361 cu : $2/7$
0233 2458 2^x : $7e\pi/10$
0233 2567 cr : $2+2\pi$
0233 2670 id : $(\pi/3)^{1/2}$
0233 4851 tπ : $4\sqrt{3}/3+4\sqrt{5}$
0233 7695 2^x : $4e+\pi$
0233 7797 rt : (7,9,-6,9)
0234 0803 sq : $2\sqrt{2}-4\sqrt{5}/3$
0234 0847 10^x : $1/3+\sqrt{5}$
0234 2699 cu : $4/3-\pi/3$
0234 5022 ζ : $\sqrt{2}+\sqrt{3}+\sqrt{6}$
0234 6061 Ei : $\pi/3$
0234 6116 sr : $2+2\pi/3$
0234 6684 at : $3-e/2$
0234 7130 J_2 : $1/\sqrt[3]{12}$
0234 8109 ℓ_{10} : $18/19$
0234 8900 2^x : $1+2e/3$
0235 0704 J : $1/\sqrt[3]{18}$
0235 1136 cu : $9(1/\pi)/10$
0235 1774 $\exp(1/4)\div\exp(4)$
0235 2399 θ_3 : $\exp(-\sqrt{2\pi})$
0235 3263 sr : $22/21$
0235 3308 2^x : $3\sqrt{3}/4-3\sqrt{5}$
0235 4642 J_2 : $7/16$
0235 4891 ζ : $5\sqrt{5}/2$
0235 6014 10^x : $\sqrt{3}/3+\sqrt{5}/3$
0235 6201 e^x : $3\sqrt{3}-4\sqrt{5}$
0235 7027 10^x : $-\sqrt{2}+\sqrt{5}-\sqrt{6}$
0235 7079 rt : (2,6,-6,9)
0235 7081 2^x : $1/4-4\sqrt{2}$
0235 7202 J_1 : $8\sqrt[3]{2}$
0235 7464 as : $e\pi/10$
0235 8290 sq : $4\sqrt{2}/3-\sqrt{3}$
0235 9478 $5\times\ln(\sqrt{5})$
0236 1354 Ψ : $23/16$
0236 3128 ln : $(e+\pi)/6$
0236 3628 rt : (1,8,7,1)
0236 4718 Γ : $4\zeta(3)/5$
0236 5147 $\sqrt{2}+\ln(5)$
0236 5613 e^x : $7\sqrt{5}$
0236 5716 e^x : $1/3-3e/2$
0236 6544 at : $3/2-\pi$
0236 6863 sq : $2/13$
0236 8023 J_0 : $5\sqrt{5}/2$
0236 9145 e^x : $1/2-3\sqrt{2}$
0236 9196 sπ : $7\sqrt[3]{5}/4$
0236 9261 $\exp(3/4)-\exp(\pi)$
0236 9862 tπ : $7\sqrt[3]{5}/4$
0237 0633 J_0 : $5e\pi/2$
0237 2607 sq : $2\ln 2/9$
0237 2897 2^x : $(2e)^{-2}$
0237 3767 $\ln(\sqrt{3})\div\exp(\pi)$
0237 4304 Γ : $2\sqrt[3]{3}/3$
0237 4784 10^x : $2-4e/3$
0237 7265 sr : $4e/3+4\pi$
0237 9054 sr : $(\ln 3)^{1/2}$
0238 0277 J_1 : $1/21$
0238 0884 cu : $\ln(4/3)$
0238 1192 J_2 : $11/25$
0238 1228 10^x : $5\pi^2/9$
0238 1417 tπ : $8e$
0238 4911 $\ln(4/5)\div\exp(\sqrt{5})$
0238 7530 10^x : $\sqrt{3}-3\sqrt{5}/2$
0238 8572 ln : P_{64}
0238 8866 cu : $\sqrt{21/2}$
0238 9754 2^x : $9\pi^2/4$
0239 1551 10^x : $1/2-3\sqrt{2}/2$
0239 2710 rt : (7,9,-7,7)
0239 3225 sq : $2-\sqrt{3}/3$
0239 3535 $\exp(1/5)\times\exp(\sqrt{2})$
0239 4268 $1/4+s\pi(e)$
0239 7430 at : $2/3-4\sqrt{3}/3$
0239 7488 2^x : $-\sqrt{2}-\sqrt{3}-\sqrt{5}$
0239 7948 sq : $\sqrt{2}/4+4\sqrt{3}$
0240 0000 cu : $\sqrt[3]{3}/5$
0240 0748 at : $23/14$
0240 5626 cu : $\sqrt{3}/6$
0240 5643 2^x : $4/3-3\sqrt{3}/4$
0240 6213 2^x : $e/3-2\pi$
0240 7537 e^x : $3\sqrt{2}/4+\sqrt{5}$
0240 8500 ζ : $-\sqrt{3}-\sqrt{5}+\sqrt{6}$
0240 8729 sr : $\sqrt{2}/3+\sqrt{3}/3$
0240 9920 2^x : $4/3-3\sqrt{5}$
0241 0300 $\exp(2)-\exp(5)$
0241 3385 rt : (2,6,-7,5)
0241 4329 tπ : $5\ln 2/7$
0241 5344 ℓ_{10} : $2-2\sqrt{2}/3$
0241 6119 $e\div s\pi(\sqrt{5})$
0241 8226 e^x : $3/4-2\sqrt{5}$
0242 0414 rt : (1,8,6,-6)
0242 2662 Ψ : $1/\ln(\sqrt[3]{4}/3)$
0242 2750 ℓ_{10} : $(\sqrt{5}/2)^{1/2}$
0242 3764 at : $2\sqrt{2}-2\sqrt{5}$
0242 6318 sr : $(\sqrt{3}/2)^{-1/3}$
0242 6376 at : $4-3\pi/4$
0242 8741 J_2 : $4/9$
0242 8947 sq : $\sqrt{2}-\sqrt{5}-\sqrt{7}$
0242 9719 cu : $2\sqrt{3}-2\sqrt{5}$
0242 9822 10^x : $(\sqrt[3]{2})^{1/3}$
0243 0246 at : $(\sqrt{2\pi})^{1/3}$
0243 3029 θ_3 : $20/21$
0243 3242 cu : P_{82}
0243 6021 ζ : $8\ln 2$
0243 6531 sq : $3/4-e/3$
0243 7057 sr : $3/2+3\sqrt{3}/2$
0243 8937 sinh : $5(e+\pi)/6$
0243 9083 ln : $2+\pi/4$
0243 9377 cr : $\ln((e+\pi)/2)$
0244 1542 J_2 : $7(1/\pi)/5$
0244 1675 J_1 : $\exp(e/2)$
0244 1911 e^x : $1-3\pi/2$
0244 2939 J_2 : $1/\ln(3\pi)$
0244 5005 Γ : $8(e+\pi)/9$
0244 6870 at : $3\sqrt{2}-3\sqrt{3}/2$
0244 8598 10^x : $\sqrt{2}/3+\sqrt{3}/4$
0244 8950 10^x : $\sqrt{2}+3\sqrt{3}/2$
0245 0213 rt : (7,9,-8,5)
0245 1496 sinh : $16/11$
0245 2024 10^x : $-\sqrt{2}+\sqrt{6}-\sqrt{7}$
0245 2152 e^x : $3-3\sqrt{5}$
0245 3508 J_1 : $8\sqrt{2}/3$
0245 4441 J_2 : $7\pi^2/6$
0245 4482 sq : $\ln(\sqrt[3]{5}/2)$
0245 4590 rt : (6,6,-8,9)
0245 4757 J_1 : $(\ln 3/3)^3$
0245 5350 cu : $3/2+3\sqrt{5}$
0245 5716 rt : (1,9,0,2)
0245 6324 2^x : $3e/2-3\pi$
0245 6845 at : $\pi^2/6$
0245 7954 id : $2\sqrt{2}+3\sqrt{3}$
0245 8592 J_2 : $\sqrt{5}/5$
0245 9263 $\pi-\exp(3/4)$
0245 9540 ℓ_{10} : $3\sqrt[3]{2}/4$
0246 0761 Ei : $22/21$
0246 1730 sq : $\sqrt{2}/2+3\sqrt{3}/4$
0246 4145 Ψ : $-\sqrt{2}+\sqrt{5}+\sqrt{6}$
0246 4911 2^x : $8\pi^2/5$
0246 5198 cu : $10\ln 2$
0246 5349 10^x : $(\ln 2/3)^{1/2}$
0246 5444 J_2 : $8(e+\pi)/9$
0246 6227 ζ : $\exp(\sqrt[3]{5})$
0246 6297 sr : $5\sqrt[3]{2}/6$
0246 6853 θ_3 : $(\ln 2/3)^3$
0246 9111 Ei : $\ln(\sqrt[3]{3})$
0246 9135 sq : $\sqrt{2}/9$
0246 9507 sr : $21/20$
0247 2231 cu : $3-4e$
0247 2416 rt : (2,6,-8,1)
0247 3181 Γ : $24/25$
0247 5160 ln : $\sqrt{3}/2+3\sqrt{5}$
0247 5760 J_2 : $\pi/7$
0247 6312 10^x : $3/4-3\pi/4$
0247 6804 sq : $4/3-2\sqrt{5}/3$
0248 0314 sq : $\sqrt[3]{2}/8$
0248 1056 Γ : $7/16$
0248 1192 cu : $7/24$
0248 1473 id : $\sqrt{5}\pi$
0248 1803 2^x : $(1/\pi)/9$
0248 1883 e^x : $3/2-3\sqrt{3}$
0248 2358 ℓ_{10} : $17/18$
0248 3610 $\ln(2/3)-s\pi(\pi)$
0248 3732 Li_2 : $(5/3)^{-1/2}$
0248 4938 J_2 : $1/\sqrt[3]{11}$
0248 5593 Ei : $\sqrt{3}$
0248 6074 rt : (1,9,-1,-7)
0248 8582 J_1 : $(e)^{-3}$
0248 8804 J_2 : $9/20$
0248 9353 $1/2\div\exp(3)$
0248 9564 Ψ : $(5/3)^{-1/2}$
0249 0292 ℓ_{10} : $1/2+\sqrt{5}/4$
0249 2235 id : $9\sqrt{5}/5$
0249 3074 sq : $3/19$
0249 3173 sπ : $4\sqrt[3]{2}/5$
0249 3948 tπ : $4\sqrt[3]{2}/5$
0249 4402 id : $\sqrt{2}/2-\sqrt{3}$
0249 5947 sr : $(e/3)^{-1/2}$

0249 6258 sq : $4/3+\pi$
0249 7198 J_2 : $3\zeta(3)/8$
0249 7241 2^x : $1/3-4\sqrt{2}$
0249 7633 at : $2-\sqrt{2}/4$
0249 8651 10^x : $\sqrt{2}/2-4\sqrt{3}/3$
0249 9218 J_1 : $1/20$
0250 0223 e^x : $1/2-4\pi/3$
0250 0792 J_2 : $(\ln 2/2)^{-2}$
0250 1029 cr : $14/13$
0250 1349 e^x : $11/5$
0250 1401 2^x : $6\pi^2$
0250 3683 $s\pi$: $\sqrt{3}/4+\sqrt{5}/4$
0250 4468 $t\pi$: $\sqrt{3}/4+\sqrt{5}/4$
0250 5720 J_0 : $5\sqrt{2}/3$
0250 6042 J_2 : $\ln(\pi/2)$
0250 7644 J_1 : $6\pi/5$
0250 7750 J : $(\ln 3/3)^3$
0251 0489 rt : (7,9,-9,3)
0251 1277 e^x : $4\sqrt{3}+2\sqrt{5}$
0251 2436 J_2 : $\exp(-\sqrt[3]{4}/2)$
0251 2626 cu : $1-\sqrt{2}/2$
0251 3009 $s\pi$: $(5)^{-3}$
0251 3464 $e-\ln(2)$
0251 3803 $t\pi$: $(5)^{-3}$
0251 4581 cr : $1/2+\sqrt{3}/3$
0251 5194 rt : (6,6,-9,7)
0251 7050 10^x : $\sqrt[3]{3}/3$
0251 9707 e^x : $\sqrt{2}-\sqrt{6}-\sqrt{7}$
0252 0457 ζ : $11/2$
0252 0545 $\sqrt{3}\div s\pi(\pi)$
0252 2041 J_2 : $e/6$
0252 3037 10^x : $1-3\sqrt{3}/2$
0252 4598 $t\pi$: $2\sqrt{3}-2\sqrt{5}$
0252 4672 ζ : $7\pi/4$
0252 4732 e^x : $4-e/2$
0252 5234 sq : $4-e/4$
0252 5512 sq : $\sqrt{2}/2-\sqrt{3}/2$
0252 5754 cu : $\zeta(7)$
0252 6317 ln : P_{49}
0253 1778 sr : $7(e+\pi)/10$
0253 1983 ln : $2/3+3\sqrt{2}/2$
0253 2217 J_1 : $(\sqrt{2}\pi)^{-2}$
0253 2320 Ei : $7\sqrt[3]{3}/5$
0253 2467 sq : $1/3+3\sqrt{2}/2$
0253 2487 tanh : $(2\pi)^{-2}$
0253 2488 at : $(2\pi)^{-2}$
0253 3029 id : $(2\pi)^{-2}$
0253 3284 Γ : $\exp(\sqrt[3]{3}/2)$
0253 3300 as : $(2\pi)^{-2}$
0253 4374 rt : (2,6,-9,-3)
0253 4434 e^x : $4\sqrt{3}+\sqrt{5}/2$
0253 6742 J_1 : $10\ln 2$
0253 8453 cu : $\ln(\sqrt{5}/3)$
0253 8463 J_2 : $5/11$
0253 9086 $\ln(4)\div\exp(4)$
0253 9641 cu : $4/3-4e/3$
0254 0472 J_2 : $10(1/\pi)/7$
0254 1905 at : $4/3-4\sqrt{5}/3$
0254 3251 $\ln(3/5)\div\exp(3)$
0254 4270 cu : $5/17$
0254 4486 Ψ : $\ln(\ln 2/2)$
0254 4514 10^x : $1/2-2\pi/3$
0254 7140 at : $8\sqrt[3]{3}/7$
0254 8884 2^x : $\sqrt{2}-3\sqrt{5}$
0254 9253 Li_2 : $(2\pi)^{-2}$
0254 9541 J_0 : $3\pi/4$
0255 1171 $t\pi$: $10\sqrt[3]{3}$
0255 2765 as : $\sqrt[3]{5}/2$
0255 2807 ζ : $5\pi^2/9$
0255 3729 sq : $4/3+3e$
0255 3738 2^x : $5\sqrt{5}/7$
0255 3850 J_2 : $(\sqrt{2}\pi/3)^{-2}$
0255 4092 $1/5-\exp(4/5)$
0255 5056 J_0 : $\sqrt{2}+\sqrt{3}+\sqrt{6}$
0255 5928 sinh : $\exp(-1/(3\pi))$
0255 6153 $\exp(1/3)\div\exp(4)$
0255 7035 cr : $1/3+\sqrt{5}/3$
0255 7290 sr : $7\zeta(3)/8$
0255 7626 ℓ_{10} : $2\sqrt{2}/3$
0255 7961 ln : $3/4+3e/4$
0255 8870 at : $(e)^{1/2}$
0256 0000 sq : $4/25$
0256 3322 e^x : $12/17$
0256 5368 ln : $3\sqrt[3]{5}/5$
0256 5383 e^x : $(2\pi)^{-2}$
0256 5536 e^x : $3\sqrt{2}-\sqrt{5}/3$
0256 5881 cr : $\sqrt{7}\pi$
0256 6972 sq : P_{20}
0256 7837 10^x : $2\zeta(3)/5$
0256 8004 sq : $\sqrt[3]{3}/9$
0256 8045 2^x : $1-2\pi$
0256 9752 sr : $1+3e$
0257 2887 ℓ_{10} : $3\pi/10$
0257 4103 2^x : $2\sqrt{3}/3-\sqrt{5}/2$
0257 5168 sq : $e\pi/6$
0257 8034 Ψ : $4/19$
0257 9581 cu : $\sqrt{3}+\sqrt{5}/4$
0258 0185 at : $4/3-e/2$
0258 0200 J_2 : $11/24$
0258 0489 rt : (6,7,-6,9)
0258 0758 id : $1/3-e/2$
0258 1044 as : $4/3-e/2$
0258 1358 e^x : $2-4\sqrt{2}$
0258 3649 Γ : $23/24$
0258 4878 Γ : $3\sqrt{5}/7$
0258 5832 sr : $3/4+3\sqrt{5}/2$
0258 7863 sq : $2\sqrt{2}/3+3\sqrt{3}/4$
0258 9646 10^x : $4+3\sqrt{2}/4$
0259 0222 $\sqrt{2}\div\exp(4)$
0259 0979 at : $7\sqrt{2}/6$
0259 1834 2^x : $4-4\sqrt{5}/3$
0259 2835 $3/5\div\exp(\pi)$
0259 2997 as : $\sqrt{2}-\sqrt{5}/4$
0259 4248 rt : (3,6,-7,7)
0259 4318 e^x : $4/3+3\pi/4$
0259 5201 ln : $3\sqrt{3}/4+2\sqrt{5}/3$
0259 6413 cr : $6\sqrt[3]{2}/7$
0259 7408 sq : $1/3+e/4$
0259 7835 sr : $20/19$
0259 8156 cu : $\sqrt{2}-\sqrt{5}/2$
0259 8222 2^x : $3\sqrt{3}+2\sqrt{5}/3$
0259 8411 10^x : $3+\sqrt{5}/2$
0259 8556 id : $3\sqrt[3]{5}/5$
0260 0448 2^x : $(3)^{-3}$
0260 1215 rt : (2,7,-2,9)
0260 1229 cu : $(3/2)^{-3}$
0260 4453 Ei : $(\ln 3)^{1/2}$
0260 5731 $4/5-s\pi(e)$
0260 5984 J_2 : $(3\pi/2)^{-1/2}$
0261 0000 $\exp(\sqrt{5})\times s\pi(\pi)$
0261 0412 J_1 : $9\sqrt{3}/4$
0261 2171 $\ln(\sqrt{5})+\exp(1/5)$
0261 3936 $\ln(2)\div s\pi(\sqrt{5})$
0261 4163 2^x : $9\pi^2/8$
0261 4735 J : $\sqrt[3]{3}/4$
0261 5768 J_2 : $6/13$
0261 7180 J_0 : $3\pi^2/2$
0261 9116 at : $\sqrt[3]{9/2}$
0262 1322 Ei : $(\sqrt{2}/3)^{1/2}$
0262 1476 cu : $3\ln 2/7$
0262 2002 J_2 : $2\ln 2/3$
0262 2702 $s\pi$: $\zeta(7)$
0262 2840 10^x : $3e+2\pi$
0262 3604 $t\pi$: $\zeta(7)$
0262 5038 cu : $3/4-\pi/3$
0262 6119 10^x : $(3\pi)^{-2}$
0263 0667 J_1 : $1/19$
0263 2893 ℓ_{10} : $16/17$
0263 3403 sq : $3\sqrt{2}-\sqrt{5}$
0263 4618 ζ : $\sqrt{3}\pi$
0263 5695 J_1 : $4\sqrt{3}$
0263 7104 e^x : $3\sqrt{3}/2-2\sqrt{5}/3$
0263 7502 e^x : $3+3\sqrt{5}$
0263 9468 10^x : $4\sqrt{2}/3-2\sqrt{3}$
0263 9521 $2/5\div\exp(e)$
0264 1059 sq : $e+2\pi$
0264 1821 ℓ_{10} : $3/2-\sqrt{5}/4$
0264 3591 sinh : $5\sqrt[3]{2}/7$
0264 3593 ζ : $1/\sqrt[3]{22}$
0264 3716 ln : $2e/3-\pi/4$
0264 4277 ζ : $2e$
0264 5017 J_2 : $1/\sqrt[3]{10}$
0264 5519 sr : $(\sqrt[3]{5}/2)^{-1/3}$
0264 6116 sinh : $9\sqrt{2}/4$
0264 7088 $3\times s\pi(\sqrt{5})$
0264 7277 cr : $3e/4+2\pi$
0264 7465 rt : (6,7,-7,7)
0264 8051 $1/3+\ln(2)$
0264 9990 $s\pi$: $\sqrt{2}+\sqrt{3}/3$
0265 0921 $t\pi$: $\sqrt{2}+\sqrt{3}/3$
0265 1649 J_1 : $(1/\pi)/6$
0265 1672 sinh : $9/10$
0265 1955 cr : $3\sqrt[3]{3}/4$
0265 2024 e^x : $3-\sqrt{3}/4$
0265 4824 id : $8\pi/5$
0265 5367 ℓ_2 : $4-4\sqrt{5}/3$
0265 6476 $\exp(\pi)\div s\pi(\sqrt{3})$
0265 7144 cr : $9\zeta(3)/10$
0266 0013 at : $(\ln 3/3)^{-1/2}$
0266 0326 10^x : $(\sqrt{2}\pi)^{-3}$
0266 1695 $t\pi$: $4\sqrt{3}-2\sqrt{5}/3$
0266 2320 rt : (3,6,-8,4)
0266 2778 at : $\sqrt{2}/4+3\sqrt{3}/4$
0266 3644 cu : $3e/4+\pi/2$
0266 5123 ℓ_{10} : $\ln(\ln 3)$
0266 5416 ζ : $12/7$
0266 5696 $s\pi$: $3e/4-\pi/3$
0266 6573 sq : $\exp(-2e/3)$
0266 6643 $t\pi$: $3e/4-\pi/3$
0266 8232 J_1 : $(\ln 2/3)^2$
0266 9067 J_1 : $\exp(-(e+\pi)/2)$
0266 9849 rt : (2,7,-3,5)
0267 1948 $1/4\div\exp(\sqrt{5})$
0267 3153 J_2 : $7/15$
0267 4112 cu : $1-3\sqrt{3}/4$
0267 5584 $\ln(2/3)\div\exp(e)$
0267 6075 $t\pi$: $3\sqrt{5}/2$
0267 6326 sq : $(\ln 2/2)^{-1/3}$
0267 8665 cr : $5\sqrt{3}/8$
0267 8972 id : $2e/3-\pi/4$
0267 9144 K : $3e/10$
0267 9367 rt : (1,6,1,7)
0268 3423 $2\div\exp(2/3)$
0268 4992 10^x : $1/\ln(\sqrt[3]{4}/3)$
0268 5644 $1/4+\ln(4/5)$
0268 6474 $\ln(4)-\exp(5)$
0269 3685 J_0 : $\sqrt{3}\pi$
0269 4400 cu : P_{79}
0269 4685 ln : $8\pi/9$
0269 5178 $4\div\exp(5)$
0269 5247 10^x : $2/3-\sqrt{5}$
0269 6842 2^x : $\sqrt{2}/3-\sqrt{3}/4$
0269 9544 ζ : $9\zeta(3)/2$
0269 9566 10^x : $\ln(e/2)$
0270 0000 cu : $3/10$
0270 1280 ln : $3\sqrt{2}/4+\sqrt{3}$
0270 1348 2^x : $3e/8$
0270 3701 Γ : $10\sqrt[3]{3}/7$
0270 4002 cr : $13/12$
0270 4423 ζ : $5/14$
0270 4425 Γ : $22/23$
0270 5044 2^x : $3/2-3\sqrt{5}$
0270 6705 $1/5\div\exp(2)$
0270 7287 ℓ_{10} : P_{46}
0270 8311 Ei : $4\zeta(3)/7$
0270 8878 10^x : $\sqrt{2}-4\sqrt{5}/3$
0270 8985 10^x : $2+\sqrt{3}/4$
0271 3907 cu : $\zeta(3)/4$
0271 5726 e^x : $\sqrt{3}/2-2\sqrt{5}$
0271 7433 J_2 : $8/17$
0271 7726 ℓ_{10} : $2-3\sqrt{2}/4$
0271 7935 rt : (6,7,-8,5)
0271 8553 ζ : $(\sqrt[3]{5}/3)^{-3}$
0271 9252 at : $1/2+2\sqrt{3}/3$
0271 9408 cu : $9\sqrt{2}/8$
0271 9903 rt : (6,8,-4,9)
0272 2033 $4/5\times\exp(1/4)$
0272 3910 rt : (5,4,-8,9)
0272 5473 sr : $4\sqrt{3}+\sqrt{5}$
0272 6693 J_2 : $\sqrt{2}/3$

0273 0639 $\ell_{10} : 1 - e/3$
0273 1384 $2^x : 17/4$
0273 2372 $\exp(2/5) \div \exp(4)$
0273 2554 $5 \times \ln(2/3)$
0273 3949 $t\pi : 7/16$
0273 4004 rt : (3,6,-9,1)
0273 4406 at : $2\sqrt{2}/3 - 3\sqrt{3}/2$
0273 4649 $2^x : 4\sqrt{2}/3 - \sqrt{3}/2$
0273 4834 $\ln(\sqrt{3}) \div \exp(3)$
0273 6762 $10^x : 2\sqrt{2}/3 + 2\sqrt{5}$
0273 7532 $e^x : 9\sqrt[3]{2}/4$
0273 7959 $\ell_2 : \sqrt{2} - \sqrt{3}/4$
0273 8229 $e^x : (e/2)^{1/3}$
0273 8807 $J_2 : 3\sqrt[3]{2}/8$
0273 8926 $e \times s\pi(\sqrt{3})$
0274 0066 id : $\sqrt[3]{25/3}$
0274 0233 sr : 19/18
0274 0388 $10^x : 1/4 - 2e/3$
0274 0995 $\exp(2) \times s\pi(2/5)$
0274 1012 cu : $1/\sqrt{11}$
0274 2154 rt : (2,7,-4,1)
0274 4225 $\exp(\pi) - \exp(\sqrt{2})$
0274 5303 $10^x : \exp(-\sqrt{2}\pi)$
0274 5925 $2^x : 1/4 - 2e$
0274 7177 sq : $(\ln 3/2)^3$
0274 7300 ln : $2/3 + 4\sqrt{3}$
0274 7768 $10^x : 4\sqrt[3]{3}$
0274 9575 $J_2 : (3\pi)^{-1/3}$
0275 2299 $Ei : 5\ln 2/2$
0275 2466 rt : (1,6,0,1)
0275 2469 $\Psi : (\sqrt{2}\pi/2)^{1/2}$
0275 2632 $J_2 : 9/19$
0275 5217 cu : $e/9$
0275 5233 $\zeta : \sqrt{2} + \sqrt{3} + \sqrt{5}$
0275 6302 $J_2 : 7\sqrt{5}/3$
0275 6505 at : $3/4 + e/3$
0275 7393 sq : P_1
0275 7738 $10^x : 10\sqrt{3}/3$
0275 8029 $10^x : 3/4 - 4\sqrt{3}/3$
0275 9765 rt : (1,7,-8,2)
0276 1086 $J_2 : (\sqrt{2}\pi)^{-1/2}$
0276 2689 $J_2 : \exp(-\sqrt{5}/3)$
0276 4571 cu : $2\sqrt{2}/3 - 2\sqrt{3}$
0276 5754 $\ell_2 : 3e$
0276 7110 $\ell_{10} : P_{62}$
0276 7890 cr : $4e/3 + 3\pi/2$
0276 9151 sq : $3 + \sqrt{5}/3$
0277 1486 $\exp(\sqrt{2}) \div \exp(5)$
0277 1955 ln : $3\sqrt{2} + 3\sqrt{5}/2$
0277 2198 $10^x : \sqrt{3}/2 - \sqrt{5}/4$
0277 4613 $10^x : (2\pi)^{1/3}$
0277 5105 $10^x : 4e/3 + 3\pi/2$
0277 6706 $J_1 : 1/18$
0277 6813 at : $4 - 4\sqrt{2}$
0277 7157 $\Psi : 2\sqrt{5}/3$
0277 7777 sq : 1/6
0277 7885 $\ln(2/5) \times \exp(4)$
0278 0227 $J_0 : \sqrt{23}\pi$
0278 0398 $s\pi : 7\pi$
0278 1283 $J_2 : 10/21$
0278 1474 $t\pi : 7\pi$
0278 4893 $J_2 : 1/\ln(1/(3e))$
0278 5363 $J_1 : 9\sqrt{5}/2$
0278 5840 as : $3/2 - 3\pi/4$
0278 6250 ln : $5\sqrt{5}/4$
0278 9964 $2^x : 6e/7$
0279 0367 $\ell_2 : 3\sqrt{3} - \sqrt{5}/2$
0279 1920 $\sqrt{2} - \ln(4)$
0279 2172 rt : (6,7,-9,3)
0279 2558 $Ei : \sqrt{7}\pi$
0279 2632 $\Gamma : \zeta(3)/9$
0279 3592 $J_2 : 23/2$
0279 4076 cu : $1 + 4\sqrt{2}/3$
0279 4141 $2^x : ((e+\pi)/2)^{-3}$
0279 4303 rt : (6,8,-5,7)
0279 4763 $e^x : 1/2 - 3e/2$
0279 5905 $J_2 : 3(1/\pi)/2$
0279 6328 $2^x : (1/\pi)/8$
0279 7279 id : $7e$
0279 7668 sq : P_{60}
0279 8123 rt : (1,7,-9,-6)
0279 8646 rt : (5,4,-9,7)
0279 8815 $e^x : 2/3 - 3\sqrt{2}$
0279 8972 $e^x : 3\sqrt{2} + 4\sqrt{3}$
0279 9292 $\ell_2 : 4\sqrt{2}/3 - \sqrt{3}/2$
0280 1408 $\Gamma : 2/17$
0280 2872 $\ell_{10} : 15/16$
0280 3058 $2^x : 1/2 - 4\sqrt{2}$
0280 3594 $10^x : 1/3 - 4\sqrt{2}/3$
0280 4342 $t\pi : \sqrt{3}/3 + 3\sqrt{5}/4$
0280 5057 $J_2 : 11/23$
0280 5385 sq : P_{44}
0280 6370 $J_0 : \exp(\sqrt[3]{5}/2)$
0280 6964 $2^x : 3 - 3e$
0280 6976 $J_0 : \sqrt[3]{13}$
0280 9208 $10^x : 3\sqrt{2} + \sqrt{5}/4$
0281 0785 $s\pi(\sqrt{3}) + s\pi(e)$
0281 1173 $\Gamma : 3(1/\pi)$
0281 1382 $2^x : 1/25$
0281 1498 id : $\exp(1/\sqrt{2})$
0281 1774 rt : (3,7,-4,7)
0281 3529 $\Gamma : 1/\sqrt[3]{12}$
0281 3703 cr : $\sqrt{2} + 4\sqrt{3}$
0281 5924 $\ell_{10} : 3/2 - \sqrt{3}/4$
0281 8372 cr : 25/23
0281 8431 rt : (2,7,-5,-3)
0281 8911 $\pi \times s\pi(\sqrt{2})$
0281 9100 cu : 7/23
0281 9425 ln : $3/2 - \sqrt{2}/3$
0281 9791 sr : $2 - 2\sqrt{2}/3$
0282 1809 $s\pi : \exp(-3\pi/2)$
0282 2313 $\zeta : \ln(\ln 2/3)$
0282 2592 sq : $\sqrt{3}/3 - \sqrt{5}/3$
0282 2830 $2^x : 2/3 + \sqrt{2}/4$
0282 2933 $t\pi : \exp(-3\pi/2)$
0282 3148 $\Psi : -\sqrt{2} - \sqrt{3} + \sqrt{6}$
0282 3453 sq : $(2e/3)^{-3}$
0282 5100 $J_2 : 12/25$
0282 6528 at : $\sqrt{3}/3 - \sqrt{5}$
0282 8559 sr : $(\sqrt{5}/2)^{1/2}$
0282 9602 cr : $2e/5$
0282 9626 rt : (1,6,-1,-5)
0282 9971 $e^x : e - 2\pi$
0283 0332 $J_2 : (\ln 2)^2$
0283 0938 $J_0 : 5\sqrt[3]{5}$
0283 1259 $1/4 \times \exp(\sqrt{2})$
0283 2894 $\Psi : \ln(\sqrt{2}\pi)$
0283 2905 $J_2 : (\ln 2/3)^{1/2}$
0283 3763 $J_2 : \sqrt[3]{3}/3$
0283 4606 $J_2 : 2\zeta(3)/5$
0283 5436 $\ell_{10} : e\pi/8$
0283 7019 $\Gamma : 21/22$
0284 0663 $Ei : 19/2$
0284 1070 $\Psi : ((e+\pi)/3)^{-2}$
0284 2169 $\ell_{10} : \sqrt{3} + 4\sqrt{5}$
0284 3223 $\ell_2 : \sqrt{5} + \sqrt{6} - \sqrt{7}$
0284 5125 $\exp(4) - s\pi(\pi)$
0285 1045 $\zeta : 5e\pi/8$
0285 3263 sinh : $5\sqrt[3]{3}/8$
0285 4292 $10^x : (\ln 3/3)^{1/2}$
0285 4582 sq : $3e/4 + \pi/2$
0285 6560 rt : (9,8,-9,-9)
0285 6561 $2^x : 5\zeta(3)/2$
0285 7606 $erf : (2\pi)^{-2}$
0285 8768 at : $1/2 - \sqrt{2}/3$
0285 9547 id : $1/2 - \sqrt{2}/3$
0285 9732 $\zeta : 16/3$
0285 9937 as : $1/2 - \sqrt{2}/3$
0286 0091 $J_0 : 2e$
0286 2878 cu : $4e/3 + 2\pi/3$
0286 7754 $\Psi : 9(1/\pi)/2$
0286 7881 $Ei : (\sqrt{3}/2)^{-1/3}$
0286 9415 $\Gamma : (\ln 3)^{-1/2}$
0287 2624 $J_2 : \sqrt{7}\pi$
0287 2792 rt : (6,8,-6,5)
0287 2873 sinh : $3\zeta(3)/4$
0287 5113 rt : (6,9,-2,9)
0287 7250 $\ell_{10} : 8\zeta(3)/9$
0287 7488 cu : $\sqrt{2} - \sqrt{3} - \sqrt{7}$
0287 9582 sinh : $(\sqrt[3]{4}/3)^{-3}$
0287 9733 $Li_2 : 2e/7$
0287 9843 rt : (5,5,-6,9)
0288 0299 $\ell_{10} : 3e/2 - \pi$
0288 0779 $10^x : (\ln 2/3)^3$
0288 0927 $2/3 \div \exp(\pi)$
0288 4889 $Ei : 11/16$
0288 5464 $J_2 : 7\ln 2/10$
0288 7001 $\ell_{10} : 5\sqrt[3]{5}/8$
0288 7640 sinh : $(\sqrt{2}/3)^{-1/2}$
0288 8036 $\ell_2 : 2/3 + \sqrt{2}/4$
0288 9284 cu : $\ln(e/2)$
0288 9530 $2^x : 1/3 + 3\sqrt{5}/4$
0289 1772 rt : (3,7,-5,4)
0289 2601 $10^x : 1/2 - 3e/4$
0289 3451 $J : \exp(-5/2)$
0289 3682 cu : $\sqrt{3}/2 - \sqrt{5}/4$
0289 5910 $K : \sqrt{2/3}$
0289 7159 $J_2 : \exp(-\sqrt[3]{3}/2)$
0289 7884 sq : $3/4 + e$
0289 9011 rt : (2,7,-6,-7)
0289 9065 $Ei : 2\sqrt[3]{5}/9$
0289 9151 sr : 18/17
0290 0000 cu : $7\sqrt[3]{3}/10$
0290 0825 $e^x : 1/2 - \sqrt{2}/3$
0290 1140 $e^x : 10\pi$
0290 1229 ln : $1/2 + \sqrt{2}/3$
0290 1242 $10^x : 3\sqrt{2}/4 - 3\sqrt{3}/2$
0290 1417 rt : (2,8,2,9)
0290 7259 $\lambda : (1/(3e))^3$
0290 7693 id : $7\pi^2/3$
0290 8551 sr : $1/2 + \sqrt{5}/4$
0290 8948 $e^x : 2\sqrt{2} - \sqrt{5}/3$
0290 9206 $2^x : 1/3 - 2e$
0291 1391 at : $(e+\pi)^{-2}$
0291 1734 $t\pi : \sqrt{7}$
0291 2130 $10^x : -\sqrt{3} - \sqrt{6} + \sqrt{7}$
0291 2214 id : $(e+\pi)^{-2}$
0291 2626 as : $(e+\pi)^{-2}$
0291 3063 cu : 4/13
0291 5428 $\ell_2 : \sqrt[3]{17/2}$
0291 5942 $\Psi : (2e/3)^2$
0291 7501 $s\pi : 1/2 + 2\sqrt{5}/3$
0291 8148 cu : $(\sqrt{2}\pi/3)^{-3}$
0291 8744 $t\pi : 1/2 + 2\sqrt{5}/3$
0292 1944 $\ell_{10} : 7\zeta(3)/9$
0292 3673 cu : $4\ln 2/9$
0292 3674 $\ell_2 : 7\sqrt[3]{2}/9$
0292 4017 sq : $\sqrt[3]{5}/10$
0292 4783 $J_2 : 2\sqrt[3]{5}/7$
0292 7470 $2^x : (\ln 2/2)^3$
0292 7582 ln : $3/2 + 3\sqrt{3}/4$
0292 7870 $Ei : 9\sqrt{2}$
0292 8399 cu : $4\sqrt{2} - \sqrt{5}$
0292 8535 $\zeta : 9(1/\pi)/8$
0292 8854 $\Gamma : (\sqrt{3}/2)^{1/3}$
0292 9285 $\zeta : \sqrt[3]{3}/2$
0292 9396 sr : $3 + \sqrt{5}/2$
0293 0223 $2^x : 1/24$
0293 2464 $e^x : 2\sqrt{2}/3 - 2\sqrt{5}$
0293 3696 $Li_2 : (e+\pi)^{-2}$
0293 9904 $J_1 : 1/17$
0294 0061 $J : \exp(-e/3)$
0294 0522 rt : (9,7,-8,-9)
0294 2281 $\Psi : 17/22$
0294 2849 cr : 12/11
0294 3068 rt : (8,9,-9,-9)
0294 3725 sq : $2 - 3\sqrt{2}$
0294 5559 $\zeta : \exp(5/3)$
0294 7033 $\ell_2 : 4/3 - \sqrt{2}/4$
0294 7788 $\ln(5) \div \exp(4)$
0294 8134 $2^x : \sqrt{3}/3 + \sqrt{5}$
0294 8858 sq : $\zeta(3)/7$
0295 0849 cu : $1/4 - \sqrt{5}/4$
0295 3664 $e^x : 2/3 - 4\pi/3$
0295 4111 $\ln(2/5) - \exp(\sqrt{2})$
0295 5034 $e^x : (e+\pi)^{-2}$
0295 5708 rt : (6,8,-7,3)
0295 6926 cr : $1/\ln(5/2)$
0295 6955 $t\pi : 3e/2 - \pi/2$
0295 8232 rt : (6,9,-3,7)

0295 8276 $10^x : \sqrt{2}/2 - \sqrt{5}$
0296 1330 $t\pi : \sqrt{5}/3$
0296 1866 $cu : 2 + 4\sqrt{3}/3$
0296 1941 $\ln : 5/14$
0296 3234 $J : 1/\sqrt{17}$
0296 3381 $rt : (5,5,-7,7)$
0296 3547 $Ei : 10\sqrt{2}/7$
0296 3855 $t\pi : 4(1/\pi)/5$
0296 8430 $cu : \sqrt{2}/2 + \sqrt{5}/4$
0296 9680 $as : 6/7$
0297 1663 $\ell_{10} : 4 + 3\sqrt{5}$
0297 2677 $at : 1/4 + \sqrt{2}$
0297 3185 $\ln : 5\sqrt[3]{3}/7$
0297 5979 $\Psi : 23/7$
0297 6373 $rt : (3,7,-6,1)$
0297 7856 $J_2 : \exp(-1/\sqrt{2})$
0297 8638 $t\pi : 5\pi/9$
0298 0039 $Ei : 10\pi/3$
0298 0792 $sq : 3/4 - \sqrt{3}/3$
0298 2078 $\zeta : 5\ln 2/3$
0298 2209 $e^x : 21/13$
0298 3250 $\Gamma : 20/21$
0298 4050 $J_1 : 2(e+\pi)/3$
0298 6024 $\exp(1/4) - s\pi(\sqrt{3})$
0298 6885 $rt : (2,8,1,5)$
0298 7224 $3/5 \div \exp(3)$
0298 8349 $\ln : 6\zeta(3)/7$
0298 8357 $sr : 3\sqrt{2}/4$
0298 9316 $as : (\sqrt[3]{4})^{-1/3}$
0299 2025 $e^x : e/4 - 4\pi/3$
0299 2191 $\zeta : 9(e+\pi)/10$
0299 2429 $t\pi : 1/2 + 3e/2$
0299 5360 $\ln(2) \div \exp(\pi)$
0299 6322 $\ell_{10} : 14/15$
0299 6795 $\ln : \sqrt{2}/3 + \sqrt{5}/4$
0299 8173 $cu : 3/4 - 3\sqrt{2}/4$
0299 9226 $J_2 : 2\sqrt{3}/7$
0300 0000 $sq : \sqrt{3}/10$
0300 0322 $10^x : (2\pi/3)^{1/3}$
0300 0380 $J_1 : 5e\pi/6$
0300 1658 $s\pi : (3\pi/2)^{-3}$
0300 1999 $J_2 : 5\ln 2/7$
0300 2831 $sq : \ln 2/4$
0300 3011 $t\pi : (3\pi/2)^{-3}$
0300 3048 $J_2 : 7\sqrt[3]{3}/2$
0300 4068 $\ln : 7\ln 2/5$
0300 6453 $sr : 10(1/\pi)/3$
0300 6715 $\zeta : 1/\ln(1/2)$
0300 7460 $\theta_3 : (1/(3e))^2$
0300 7526 $s\pi : 7\sqrt[3]{3}/10$
0300 7821 $sq : 1/4 + 2\sqrt{5}/3$
0300 8888 $t\pi : 7\sqrt[3]{3}/10$
0301 0353 $sr : 2 + 3\sqrt{2}/2$
0301 1244 $\ell_{10} : 1/2 + \sqrt{3}/4$
0301 3115 $\Psi : ((e+\pi)/2)^{1/3}$
0301 6588 $\Gamma : (e/3)^{1/2}$
0301 7770 $2^x : 3/4 - \sqrt{2}/2$
0301 7826 $id : 5\sqrt[3]{3}/7$
0301 9360 $sr : 4\sqrt{2}/3 + \sqrt{5}$
0301 9738 $\exp(1/2) \div \exp(4)$
0301 9919 $\ln(4/5) \div \exp(2)$
0302 2341 $J_2 : (1/(3e))^{1/3}$
0302 2519 $id : 3e/2 - \pi/3$
0302 3401 $J_2 : 2\sqrt{5}/9$
0302 3683 $e^x : 19/3$
0302 3847 $10^x : 4 + 3\sqrt{5}/4$
0302 4574 $sq : 4/23$
0302 5396 $J_2 : 5(e+\pi)/2$
0302 5988 $\ln(\sqrt{5}) + \exp(4/5)$
0302 6473 $\theta_3 : 4\sqrt[3]{3}/7$
0302 6874 $rt : (9,7,-9,-8)$
0302 9346 $t\pi : 3\sqrt{3}/4 + \sqrt{5}$
0302 9500 $10^x : -\sqrt{3} - \sqrt{5} + \sqrt{6}$
0302 9566 $rt : (9,6,-7,-9)$
0302 9716 $\theta_3 : 7/15$
0303 0121 $2/5 + s\pi(\pi)$
0303 1939 $\ln : 4/3 + 2\pi$
0303 2351 $rt : (8,8,-8,-9)$
0303 3448 $id : 6\zeta(3)/7$
0303 4105 $2^x : 4\sqrt{2}/3 - 4\sqrt{3}$
0303 4748 $\ln : \sqrt[3]{22}$
0303 5479 $cu : e/2 - \pi/3$
0303 7032 $cr : 2 - e/3$
0303 7682 $at : 5/3$
0303 9491 $2^x : 1/3 + 3\pi/2$
0304 0672 $2^x : \exp(-\pi)$
0304 0818 $J_1 : 9(e+\pi)/4$
0304 2043 $e^x : 3/4 - 3\sqrt{2}$
0304 2151 $id : \sqrt{2}/3 + \sqrt{5}/4$
0304 2519 $J_2 : 7e\pi/4$
0304 2622 $cu : 3/2 - 2e/3$
0304 3425 $rt : (6,8,-8,1)$
0304 6176 $rt : (6,9,-4,5)$
0304 7084 $cr : 10(e+\pi)/7$
0304 7169 $cu : \sqrt{3}/4 - \sqrt{5}/3$
0304 7239 $\zeta : 21/4$
0304 9445 $\sinh : (e/2)^{-1/3}$
0304 9606 $t\pi : 7(e+\pi)/4$
0305 0573 $\Psi : 1/\ln((e+\pi)/3)$
0305 1757 $cu : 5/16$
0305 1794 $rt : (5,5,-8,5)$
0305 1846 $Ei : (2e/3)^2$
0305 4318 $sr : \sqrt{17}$
0305 4576 $rt : (5,6,-4,9)$
0305 6787 $sq : 5\sqrt[3]{5}/6$
0305 7478 $t\pi : e/4 - 3\pi$
0305 9554 $2^x : 1/23$
0306 0256 $rt : (4,2,-8,9)$
0306 0402 $J_2 : 1/2$
0306 0409 $e^x : 17/24$
0306 0437 $id : \sqrt{2}/4 + 3\sqrt{5}/4$
0306 5978 $rt : (3,7,-7,-2)$
0306 6554 $J_2 : \sqrt{2} + \sqrt{3} - \sqrt{7}$
0306 7970 $\Psi : \sqrt[3]{10/3}$
0306 8812 $rt : (3,8,-1,7)$
0306 9128 $\zeta : 3\zeta(3)/5$
0306 9760 $cu : 2/3 - \sqrt{2}/4$
0307 0378 $sq : 2\sqrt{2}/3 - \sqrt{5}/2$
0307 4583 $sq : \sqrt{3}/2 + \sqrt{5}/4$
0307 4686 $\ln(3/4) \div \exp(\sqrt{5})$
0307 5053 $s\pi : 10\ln 2/7$
0307 5731 $10^x : 3\sqrt{3} - 3\sqrt{5}$
0307 6508 $t\pi : 10\ln 2/7$
0307 6790 $cu : 5\sqrt{2}/7$
0307 7463 $rt : (2,8,0,1)$
0307 7627 $\ln(\sqrt{5}) - s\pi(e)$
0307 7640 $sr : 17/16$
0307 8350 $e^x : 10(e+\pi)/7$
0307 8837 $cr : 23/21$
0308 0032 $\zeta : 5\pi/3$
0308 2158 $e^x : 1/3 + 4\pi/3$
0308 2173 $s\pi : 1/3 + 4\sqrt{2}$
0308 2239 $e^x : 3/4 + 3\sqrt{5}$
0308 3638 $t\pi : 1/3 + 4\sqrt{2}$
0308 5244 $1/5 \times \exp(e)$
0308 5592 $sr : 1/2 + 4e/3$
0308 5722 $10^x : 19/21$
0308 7999 $sq : 5\zeta(3)/2$
0308 8054 $\sinh : 6e/5$
0308 8585 $\Gamma : (3/2)^{-3}$
0309 0370 $as : (e/2)^{-1/2}$
0309 0812 $Ei : 5\sqrt[3]{2}/6$
0309 1543 $e^x : 1/3 + 3\sqrt{2}/4$
0309 1762 $10^x : 4/13$
0309 3446 $sq : P_{14}$
0309 4758 $J : (e)^{-3}$
0309 7239 $Li_2 : 7/9$
0309 8259 $cr : 8\pi/3$
0310 0627 $cu : \pi/10$
0310 0914 $rt : (1,8,0,2)$
0310 1384 $sr : (2\pi/3)^3$
0310 2691 $Ei : (2e/3)^{-3}$
0310 3652 $\sinh : 5e\pi/6$
0310 3898 $cu : 2\sqrt{2}/9$
0310 3933 $at : 1/\ln(\ln 3/2)$
0310 5063 $e^x : 1 - 2\sqrt{5}$
0310 8137 $J_2 : 2\sqrt[3]{2}/5$
0310 8204 $J_0 : e\pi$
0310 8718 $Ei : 21/20$
0310 8891 $id : 7\sqrt{3}/4$
0310 8930 $J_2 : \sqrt{14\pi}$
0310 9526 $\ell_{10} : \sqrt{2}/4 + \sqrt{3}/3$
0310 9660 $\Gamma : 7\pi$
0311 0916 $sq : (\sqrt[3]{2}/3)^2$
0311 1772 $J_2 : 9e$
0311 4186 $sq : 3/17$
0311 4533 $cr : 2\sqrt{2} - \sqrt{3}$
0311 5294 $id : 9\sqrt{5}/4$
0311 5810 $\ln : 3e + 4\pi$
0311 9162 $e^x : \sqrt{2} - \sqrt{5} - \sqrt{7}$
0311 9499 $e^x : 25/12$
0312 0291 $t\pi : 1/3 - 2e/3$
0312 0643 $Ei : 7\sqrt{3}/6$
0312 1218 $rt : (9,6,-8,-8)$
0312 2257 $sq : (\pi/3)^{1/3}$
0312 3449 $\Psi : \pi^2/3$
0312 3468 $\ell_{10} : (\sqrt{3}/2)^{1/2}$
0312 3474 $J_1 : 1/16$
0312 3893 $\zeta : 7\sqrt{5}/3$
0312 4168 $rt : (9,5,-6,-9)$
0312 4260 $rt : (8,8,-9,-8)$
0312 5000 $sq : \sqrt{2}/8$
0312 5011 $\theta_3 : (4)^{-3}$
0312 7197 $sq : 5(1/\pi)/9$
0312 7222 $rt : (8,7,-7,-9)$
0313 0111 $sq : \exp(-\sqrt{3})$
0313 0287 $rt : (7,9,-8,-9)$
0313 2190 $J_0 : \sqrt{11/2}$
0313 3203 $\ln : 7\zeta(3)/3$
0313 4650 $cu : 3\sqrt{2}/4 - \sqrt{5}/3$
0313 4705 $sq : 3/2 - 3\sqrt{5}/4$
0313 4927 $\ln : (\ln 3)^{1/3}$
0313 5045 $\ln : 4 + 4e/3$
0313 6361 $rt : (6,8,-9,-1)$
0313 9368 $rt : (6,9,-5,3)$
0314 0078 $\ell_{10} : \ln((e+\pi)/2)$
0314 0345 $t\pi : 2\sqrt{3} - 4\sqrt{5}$
0314 3313 $e^x : 8\ln 2/5$
0314 3858 $2^x : 5e$
0314 4180 $cu : 5(1/\pi)$
0314 4330 $\ln(2/5) + \exp(2/3)$
0314 4790 $\zeta : 8(e+\pi)/9$
0314 4894 $J_0 : \sqrt[3]{15}$
0314 5331 $\Gamma : 19/20$
0314 5512 $rt : (5,5,-9,3)$
0314 5960 $Ei : 8\pi^2/7$
0314 6245 $\Gamma : 5\sqrt[3]{5}/9$
0314 6326 $2^x : 2/3 - 4\sqrt{2}$
0314 6444 $\sinh : 6\zeta(3)$
0314 6645 $2^x : 3\sqrt{2}/4 + 4\sqrt{3}$
0314 6895 $t\pi : 4\sqrt{2} + 3\sqrt{3}/2$
0314 7757 $cu : 2/3 - 3\sqrt{5}/4$
0314 8554 $rt : (5,6,-5,7)$
0314 9147 $cu : 6/19$
0314 9363 $rt : (1,8,-1,-6)$
0314 9455 $\Psi : \ln(\ln 2/3)$
0314 9511 $J_0 : 1/\ln(2/3)$
0315 4391 $at : e - \pi/3$
0315 4770 $rt : (4,2,-9,7)$
0315 4788 $at : \sqrt[3]{14/3}$
0315 5982 $\zeta : ((e+\pi)/2)^{1/2}$
0315 6093 $\Psi : \exp(2\pi/3)$
0315 6935 $s\pi : 7\sqrt{2}/10$
0315 8509 $t\pi : 7\sqrt{2}/10$
0315 8699 $s\pi : 7\sqrt[3]{5}/3$
0316 0276 $t\pi : 7\sqrt[3]{5}/3$
0316 1035 $rt : (3,7,-8,-5)$
0316 1640 $sr : 2\sqrt{2} + 3\sqrt{3}/4$
0316 1648 $s\pi(1/4) - s\pi(\sqrt{5})$
0316 2277 $cu : 1/\sqrt{10}$
0316 2489 $e^x : 10/3$
0316 3796 $J_1 : 6e$
0316 4137 $rt : (3,8,-2,4)$
0316 7333 $\ell_2 : 3\sqrt{3} + 4\sqrt{5}/3$
0317 0580 $2^x : (3\pi/2)^{-2}$
0317 1768 $\ell_{10} : 1/4 + e/4$
0317 2361 $\sqrt{3} \div \exp(4)$
0317 2523 $cu : 7\sqrt{5}/3$
0317 2761 $J_2 : 8(1/\pi)/5$
0317 3603 $\ell_{10} : 4/3 + 3\pi$

0317 3616 rt : (2,8,-1,-3)
0317 5327 $\zeta : 3\sqrt{3}$
0317 5465 $10^x : (\sqrt{2}\pi/3)^{-3}$
0317 7083 $2^x : \sqrt{3} - 3\sqrt{5}$
0318 1486 $J_1 : (1/\pi)/5$
0318 2024 at : $(1/\pi)/10$
0318 3098 id : $(1/\pi)/10$
0318 3636 as : $(1/\pi)/10$
0318 4583 id : $(\ln 3)^{1/3}$
0318 4937 id : $6(e+\pi)/5$
0318 5118 cu : $3/4 - \sqrt{3}/4$
0318 8975 $s\pi : 5\sqrt{2}/7$
0318 9602 rt : (1,7,7,7)
0319 0578 $\Psi : -\sqrt{2} - \sqrt{3} + \sqrt{7}$
0319 0597 $t\pi : 5\sqrt{2}/7$
0319 0967 sq : $4/3 - 2\sqrt{3}/3$
0319 1432 $J_2 : \ln(5/3)$
0319 2797 $Li_2 : (\sqrt{2}/3)^{1/3}$
0319 3884 cr : $\sqrt{2}/4 + \sqrt{5}/3$
0319 5931 $\exp(\sqrt{5}) + s\pi(\sqrt{5})$
0319 8204 $10^x : 4/3 - 2\sqrt{2}$
0319 8361 $J_1 : (5/2)^{-3}$
0319 9062 $J_0 : 2(e+\pi)/5$
0320 0087 cu : $3e/4 - 3\pi/4$
0320 0827 $2^x : 1/22$
0320 2136 $2^x : (1/\pi)/7$
0320 4954 $J_0 : \pi^2/4$
0320 5732 $\Gamma : (\sqrt[3]{5}/2)^{1/3}$
0320 7453 $J_0 : 8\pi^2/9$
0320 7709 $\ell_{10} : 3/2 - 2\sqrt{5}/3$
0320 8794 $Li_2 : (1/\pi)/10$
0321 0350 $e^x : 3/4 - 4\pi/3$
0321 0808 cu : $\sqrt{2} - \sqrt{3}$
0321 2157 $10^x : 3\pi/2$
0321 3301 $\Psi : (\sqrt{5})^{1/2}$
0321 4231 $\zeta : 1/\ln(4)$
0321 5678 as : $4 - \pi$
0321 7506 $e^x : 2 - 2e$
0321 7971 $J_2 : 3\sqrt[3]{5}/10$
0321 7980 cr : $2/3 + \sqrt{3}/4$
0321 8394 rt : (9,6,-9,-7)
0321 8468 $\ell_{10} : 13/14$
0322 0578 sr : $1/4 + 4\sqrt{5}$
0322 1262 cu : 7/22
0322 1622 rt : (9,5,-7,-8)
0322 1673 $s\pi : 4\sqrt{3}/7$
0322 2884 $\ell_{10} : P_{17}$
0322 3179 $J_2 : \exp(-2/3)$
0322 3347 $t\pi : 4\sqrt{3}/7$
0322 3437 $\Gamma : e\pi/9$
0322 3982 at : $2\sqrt{2} - 2\sqrt{3}/3$
0322 4035 at : $(\pi)^{-3}$
0322 4554 sq : $1/2 - e/4$
0322 4863 rt : (9,4,-5,-9)
0322 4967 rt : (8,7,-8,-8)
0322 5153 id : $(\pi)^{-3}$
0322 5712 as : $(\pi)^{-3}$
0322 5800 $J_1 : 7\pi^2/10$
0322 6256 cr : $7\sqrt{2}/9$
0322 6646 sq : $2\sqrt{3} - 4\sqrt{5}$
0322 8011 cr : 11/10
0322 8222 rt : (8,6,-6,-9)
0322 8327 rt : (7,9,-9,-8)
0322 9629 cu : 19/15
0323 0413 $s\pi : 5\zeta(3)$
0323 1594 rt : (7,8,-7,-9)
0323 2100 $t\pi : 5\zeta(3)$
0323 4301 $e^x : (1/\pi)/10$
0323 5534 $2^x : 4/3 - 2\pi$
0323 5692 $\ell_{10} : 1/2 + \sqrt{3}/3$
0323 7744 at : $2(e+\pi)/7$
0323 8276 rt : (6,9,-6,1)
0323 9593 sq : $\sqrt[3]{2}/7$
0324 0700 rt : (2,8,-9,7)
0324 1043 $3/4 \div \exp(\pi)$
0324 1291 $J_1 : 7\sqrt{5}/4$
0324 1607 sinh : $\sqrt{15}$
0324 2915 sr : $2/3 + 2\sqrt{3}$
0324 3842 $2^x : 1/4 - 3\sqrt{3}$
0324 4701 $J_2 : 3\zeta(3)/7$
0324 4886 $\Gamma : 2e\pi/5$
0324 7285 $10^x : 3 - 2\sqrt{3}/3$
0324 8073 $2^x : 4 - 4\sqrt{5}$
0324 8103 cr : $\sqrt[3]{4/3}$
0324 8269 $2^x : \ln(\pi/3)$
0324 8347 rt : (5,6,-6,5)
0324 8542 $\Psi : 10/7$
0325 0130 sq : $\sqrt[3]{3}/8$
0325 1537 $Li_2 : (\pi)^{-3}$
0325 1585 sr : $4 + 3\sqrt{3}$
0325 1696 rt : (5,7,-2,9)
0325 5601 as : $5\zeta(3)/7$
0325 6207 $Ei : (e/3)^{-1/2}$
0325 8167 $10^x : 3e/4 + 3\pi$
0325 8411 sq : $\ln(3\pi)$
0325 8545 rt : (4,3,-6,9)
0325 9166 $e^x : 6e\pi/5$
0325 9574 cu : $\sqrt{5}/7$
0326 0052 $2^x : (\pi/3)^{1/2}$
0326 0054 $K : 9/11$
0326 0463 $e^x : 4e + \pi/2$
0326 1750 $s\pi : 1/3 + 3\sqrt{5}/4$
0326 2046 rt : (3,7,-9,-8)
0326 3487 $t\pi : 1/3 + 3\sqrt{5}/4$
0326 3833 $\Gamma : \sqrt{11}\pi$
0326 5329 $5 \div \exp(1/2)$
0326 5367 $10^x : 3/4 - \sqrt{5}$
0326 5451 rt : (3,8,-3,1)
0326 5474 $2^x : 1/2 - 2e$
0326 6314 $10^x : 4\ln 2/9$
0326 7325 $Ei : 8/21$
0326 8844 $\Psi : 2\zeta(3)/7$
0327 1400 sq : $\exp(-\sqrt[3]{5})$
0327 1737 $e^x : 3/4 + \pi/3$
0327 3218 $\ell_{10} : P_{34}$
0327 5865 rt : (2,8,-2,-7)
0327 6249 sq : $3\sqrt{3}/4 - \sqrt{5}/2$
0327 6800 cu : 8/25
0327 7661 $\ell_{10} : e/2 + 3\pi$
0327 7725 $e^x : (\pi)^{-3}$
0327 8354 ln : $\sqrt{2}/2 + 4\sqrt{3}$
0327 9328 rt : (2,9,6,9)
0327 9555 sr : 16/15
0328 1852 id : $e/4 - 3\pi/2$
0328 4452 $2^x : 2 - 4\sqrt{3}$
0328 5153 $erf : (e+\pi)^{-2}$
0328 5275 $J_2 : 4\sqrt[3]{2}$
0328 7076 ln : $(e/3)^{1/3}$
0328 8127 $10^x : \sqrt{2} + \sqrt{3} + \sqrt{6}$
0328 9326 cu : $2\sqrt{3}/3 + \sqrt{5}/4$
0328 9638 $\ell_{10} : 1/3 + \sqrt{5}/3$
0329 0013 cu : $1 - e/4$
0329 1020 sinh : $7(e+\pi)/4$
0329 1385 $J : 1/20$
0329 2181 cu : $2\sqrt[3]{3}/9$
0329 2424 $J_2 : e\pi$
0329 3442 rt : (1,7,6,1)
0329 3934 $2^x : e/2 - 2\pi$
0329 4119 $\Gamma : (\ln 3/2)^{-3}$
0329 5077 sr : $3/2 - \sqrt{3}/4$
0329 5711 $10^x : 1/4 - \sqrt{3}$
0329 5888 $\Psi : 7\sqrt[3]{5}/8$
0329 7132 ln : $1/2 + 4\sqrt{3}/3$
0329 7606 $J_1 : \exp(-e)$
0329 7880 at : $(3\pi/2)^{1/3}$
0329 9247 $e^x : 3\sqrt{2}/4 - 2\sqrt{5}$
0329 9401 $1/2 \div \exp(e)$
0329 9700 $J_2 : 3\sqrt{3}/10$
0330 0106 at : $e/4 - 3\pi/4$
0330 0429 sr : $3e/4 + 2\pi/3$
0330 0471 $10^x : 3/2 - 4\sqrt{5}/3$
0330 1622 $e^x : 2/3 - 3e/2$
0330 2744 $J_2 : 3\ln 2/4$
0330 3997 cr : $7\sqrt[3]{2}/8$
0330 4478 $J_2 : 13/25$
0330 4618 $e^x : 3/4 + \sqrt{3}/2$
0330 4738 $J_2 : (\sqrt[3]{3}/2)^2$
0330 5785 sq : 2/11
0330 8446 sq : $4(1/\pi)/7$
0331 0161 rt : (1,8,-9,3)
0331 0884 $\ln(\sqrt{3}) \times \exp(3)$
0331 1307 at : $3\sqrt{5}/4$
0331 5509 $t\pi : 1/3 - 4\sqrt{2}/3$
0331 8283 sr : $e\pi/8$
0331 8350 cr : $1/4 + 3e$
0331 8530 id : $1/4 - 2\pi$
0331 9137 $2/3 \div \exp(3)$
0331 9383 cr : $(\sqrt{5}/3)^{-1/3}$
0332 0056 $10^x : 1/3 - 2e/3$
0332 2934 $J_1 : 15/4$
0332 3034 $\Psi : \exp(-\sqrt[3]{3}/2)$
0332 3122 cu : $2/3 - 4\pi$
0332 5132 rt : (9,5,-8,-7)
0332 5979 $\Gamma : 18/19$
0332 6113 $J_2 : 12/23$
0332 6681 $\ell_2 : (\pi/3)^{1/2}$
0332 7096 $\exp(2/3) + \exp(3)$
0332 7203 $\Gamma : 9\sqrt[3]{5}/2$
0332 7959 $\ln(\sqrt{5}) \times \exp(1/4)$
0332 8677 $e^x : 1/\ln(\sqrt{5}/3)$
0332 8688 rt : (9,4,-6,-8)
0332 8806 rt : (8,7,-9,-7)
0332 9074 $\ln(2/5) - \exp(3/4)$
0333 1481 $J_1 : 1/15$
0333 2259 rt : (9,3,-4,-9)
0333 2373 id : $(\ln 3/2)^{-3}$
0333 2378 rt : (8,6,-7,-8)
0333 3417 $2^x : 3/4 - 4\sqrt{2}$
0333 5261 sinh : 19/21
0333 5965 rt : (8,5,-5,-9)
0333 6085 rt : (7,8,-8,-8)
0333 7326 $\exp(3/5) \div \exp(4)$
0333 7889 $\ell_2 : \sqrt[3]{5}/7$
0333 8818 $J_2 : 1/\sqrt[3]{7}$
0333 9046 cr : $3/4 + \sqrt{2}/4$
0333 9654 $\ell_{10} : 6\sqrt[3]{2}/7$
0333 9687 rt : (7,7,-6,-9)
0334 0778 $\zeta : 3\sqrt[3]{5}$
0334 0814 sq : $10e\pi/9$
0334 1697 id : $(e/3)^{-1/3}$
0334 3426 rt : (6,9,-7,-1)
0334 4777 $\ell_{10} : (\sqrt[3]{2})^{1/3}$
0334 5586 $10^x : \sqrt{2} + \sqrt{3} + \sqrt{5}$
0334 6216 $J : (3)^{-3}$
0334 7298 $\zeta : 7(e+\pi)/8$
0334 7594 $10^x : 3\sqrt{2} - \sqrt{3}$
0334 8694 sr : $3e + \pi/3$
0334 9320 $J_2 : \pi/6$
0334 9364 sq : $1/4 - \sqrt{3}/4$
0335 1690 $\ln(3/4) + s\pi(\sqrt{3})$
0335 1955 $J_2 : 11/21$
0335 4499 rt : (5,6,-7,3)
0335 4734 $Ei : -\sqrt{5} + \sqrt{6} + \sqrt{7}$
0335 5778 $2^x : 1/21$
0335 6442 $\zeta : 3e\pi/5$
0335 8181 rt : (5,7,-3,7)
0335 8854 as : $1/2 - e/2$
0335 9661 $e^x : 3e/2 + 4\pi$
0336 5032 $\ell_2 : 1 + \pi/3$
0336 5719 rt : (4,3,-7,7)
0336 6827 $4/5 \div s\pi(e)$
0336 8033 sr : $8\zeta(3)/9$
0336 8231 cu : $2 - 3\sqrt{5}/4$
0336 8973 $5 \div \exp(5)$
0337 0203 ln : $e/2 + 2\pi$
0337 0837 $\ln(3/5) \div \exp(e)$
0337 1109 $\zeta : 2\sqrt[3]{2}/7$
0337 1393 $J_2 : 5\pi/3$
0337 1456 sq : P_{63}
0337 1817 $10^x : 3 - 2\sqrt{5}$
0337 2291 $\Gamma : (\ln 2/3)^{-2}$
0337 3324 rt : (3,8,-4,-2)
0337 4642 sr : $2\sqrt{3}/3 + 4\sqrt{5}/3$
0337 6439 $\zeta : 9/25$
0337 7089 rt : (3,9,2,7)
0337 7171 sq : $4 + \pi/2$
0337 7318 at : $1 + e/4$
0337 8633 $2^x : (2e)^{1/2}$
0337 9638 sr : $5\sqrt[3]{5}/8$
0338 0757 sq : P_{47}

0338 0903 $J : 6(1/\pi)/5$
0338 2091 at : $(2e)^{-2}$
0338 3361 $J_2 : 10/19$
0338 3382 id : $(2e)^{-2}$
0338 3503 t$\pi : 3\sqrt{2}/2 - \sqrt{3}/2$
0338 4027 as : $(2e)^{-2}$
0338 4477 $10^x : e - 4\pi/3$
0338 5320 t$\pi : \sqrt{3}/3$
0338 5804 at : $4\sqrt[3]{2}/3$
0338 8481 $\pi \times \exp(1/4)$
0338 8614 rt : (2,9,5,5)
0339 2387 cr : 21/19
0339 4661 $\ln(4) \times s\pi(\sqrt{3})$
0339 5003 ln : $3\sqrt{2}/2 - 2\sqrt{3}/3$
0339 5351 cr : $7\zeta(3)$
0339 6278 ln : $1 + 2e/3$
0340 0291 $10^x : 4\sqrt{2}/3 - 3\sqrt{5}/2$
0340 1999 $\ell_{10} : (\sqrt[3]{5}/2)^{1/2}$
0340 2541 $1/4 - \exp(1/4)$
0340 4183 rt : (1,7,5,-5)
0340 5408 $\ln(2) \times \exp(2/5)$
0340 7962 $e^x : 4/3 - 3\pi/2$
0340 9273 $\ell_2 : (e+\pi)/6$
0341 0168 $\ell_{10} : 3\sqrt[3]{3}/4$
0341 2438 $Li_2 : (2e)^{-2}$
0341 4686 $e^x : 8\sqrt{2}/7$
0341 4821 $2^x : \sqrt{3}/4 + 4\sqrt{5}$
0341 6032 $10^x : 1/2 + e$
0341 6753 $\ell_{10} : 9\zeta(3)$
0341 8573 cu : $2/3 + \pi/3$
0342 1476 $2/5 \times \exp(3)$
0342 2345 $J_2 : 9/17$
0342 3580 $\ell_{10} : 4\ln 2/3$
0342 3948 $e^x : 1/4 - 4e/3$
0342 4745 $Ei : 8(1/\pi)/7$
0342 4760 $J_0 : 8\sqrt{5}$
0342 6689 $10^x : \ln(\ln 2/3)$
0342 8179 at : $4/3 - 3\sqrt{3}/4$
0342 8837 $e^x : 3\sqrt{3} - \sqrt{5}/2$
0342 9355 sq : $(\sqrt[3]{5}/3)^3$
0342 9522 id : $4/3 - 3\sqrt{3}/4$
0342 9536 $10^x : 4\sqrt{2} - \sqrt{5}$
0343 0195 as : $4/3 - 3\sqrt{3}/4$
0343 1739 cr : $4\sqrt{3} + 2\sqrt{5}/3$
0343 3187 $K : (\ln 3/2)^{1/3}$
0343 3948 $J_2 : 3\sqrt{2}/8$
0343 4606 ln : P_{50}
0343 4775 $10^x : 2 - 2\sqrt{3}$
0343 5257 rt : (9,5,-9,-6)
0343 6305 $J_2 : 5(1/\pi)/3$
0343 6731 $2^x : 1/3 - 3\sqrt{3}$
0343 8450 $\Gamma : (\sqrt{5}/2)^{-1/2}$
0343 9169 rt : (9,4,-7,-7)
0344 0020 ln : $\sqrt{3}/3 + \sqrt{5}$
0344 0497 at : $1/\ln(2e/3)$
0344 1269 $e^x : (2e)^{-2}$
0344 1317 s$\pi : 4\sqrt{2} + 3\sqrt{5}/2$
0344 3099 rt : (9,3,-5,-8)
0344 3235 rt : (8,6,-8,-7)
0344 3356 t$\pi : 4\sqrt{2} + 3\sqrt{5}/2$
0344 4064 $\ell_{10} : 5\sqrt{3}/8$
0344 5413 $\Gamma : 10\sqrt[3]{3}/3$
0344 7047 rt : (9,2,-3,-9)
0344 7183 rt : (8,5,-6,-8)
0344 7319 rt : (7,8,-9,-7)
0344 7831 $\zeta : \sqrt{5/3}$
0344 8535 sinh : $7\sqrt[3]{5}$
0345 0910 at : $3\sqrt{3}/4 - 4\sqrt{5}/3$
0345 0976 $\ln(2) \div \exp(3)$
0345 1149 rt : (8,4,-4,-9)
0345 1287 rt : (7,7,-7,-8)
0345 5272 rt : (7,6,-5,-9)
0345 5410 rt : (6,9,-8,-3)
0345 6155 sr : $(3\pi)^{-3}$
0345 7113 $4/5 \div \exp(\pi)$
0345 7140 at : $7\sqrt[3]{3}/6$
0345 7566 t$\pi : 2\sqrt{3}/3 + 2\sqrt{5}/3$
0345 7861 cr : $3\sqrt{3}/2 - 2\sqrt{5}/3$
0345 9414 rt : (6,8,-6,-9)
0346 0969 sq : $3/2 + 4\sqrt{3}$
0346 2404 $\ell_{10} : 4\sqrt{2}/3 + 4\sqrt{5}$
0346 2462 $e^x : \sqrt{3} - \sqrt{6} - \sqrt{7}$
0346 2643 $2^x : (\ln 3/3)^3$
0346 2869 $J_2 : \exp(-\sqrt[3]{2}/2)$
0346 3654 $J_1 : \ln 2/10$
0346 3799 at : $7\zeta(3)/5$
0346 7617 rt : (5,6,-8,1)
0346 8264 cr : $(e/2)^{1/3}$
0346 9936 $2^x : 9\sqrt{5}/10$
0347 1678 rt : (5,7,-4,5)
0347 2021 $J_2 : 8/15$
0347 5757 rt : (5,8,0,9)
0347 6210 $\ell_{10} : 12/13$
0347 6297 $\Psi : 7\sqrt{2}/3$
0347 6578 $\ln(2/5) + s\pi(2/5)$
0347 6885 rt : (1,6,-6,9)
0347 7343 cu : $4 - 2\sqrt{3}/3$
0347 7505 $\ln(\sqrt{5}) \div \exp(\pi)$
0347 8605 ln : $1/4 + \pi/4$
0347 9998 rt : (4,3,-8,5)
0348 2639 $Ei : 4/11$
0348 3649 $J_2 : 4\zeta(3)/9$
0348 3805 ln : $2\sqrt{2}/3 + 3\sqrt{5}$
0348 4116 rt : (4,4,-4,9)
0348 5237 $2^x : \sqrt{2}/4 - 3\sqrt{3}$
0348 7018 $\zeta : \sqrt[3]{5}$
0348 7147 $Li_2 : 9\ln 2/8$
0348 8398 rt : (3,8,-5,-5)
0348 9008 $e^x : 4/3 - 3\sqrt{3}/4$
0349 2556 rt : (3,9,1,4)
0349 2624 $\Gamma : 20/3$
0349 2739 $\zeta : 24/17$
0349 3164 $10^x : 3e/4 + 4\pi$
0349 4039 $\Gamma : 3\sqrt[3]{2}/4$
0349 4073 cu : $\exp(-\sqrt{5}/2)$
0349 4681 $\ell_2 : 3 + 3\sqrt{3}$
0349 4752 cu : $\ln(\sqrt[3]{3}/2)$
0349 6314 $\ell_{10} : 3/2 - \sqrt{3}/3$
0349 7058 $e^x : e/2 - 3\pi/2$
0349 7951 s$\pi : 3\sqrt{2}/4 + 4\sqrt{3}$
0350 0093 t$\pi : 3\sqrt{2}/4 + 4\sqrt{3}$
0350 1563 $\ln(4) + \exp(1/2)$
0350 5047 $Ei : 1/\sqrt[3]{18}$
0350 5297 rt : (2,9,4,1)
0350 6522 cr : $3/2 + 4\sqrt{3}$
0350 7643 t$\pi : 5(e+\pi)/9$
0350 7826 cu : $1/4 - \sqrt{3}/3$
0350 9033 sr : $1 + \pi$
0350 9448 $\zeta : \sqrt[3]{3}/4$
0350 9833 sr : 15/14
0350 9882 cr : $8\ln 2/5$
0351 0316 $10^x : 2/3 - 3\sqrt{2}/2$
0351 1213 $2^x : (e)^{-3}$
0351 1526 cu : $\sqrt{2}/3 - 2\sqrt{5}$
0351 2111 $J_2 : \ln(\sqrt[3]{5})$
0351 5625 sq : 3/16
0351 7925 sq : P_9
0352 2393 $\zeta : 3\zeta(3)/10$
0352 2526 rt : (1,4,-7,-9)
0352 3096 $10^x : (1/(3e))^2$
0352 4684 rt : (2,9,-4,7)
0352 6492 $2^x : 1/20$
0352 6898 sinh : $e/3$
0352 7339 sq : $2 - 2e/3$
0352 8568 $\Gamma : 17/18$
0353 1485 $10^x : 3 - 3\sqrt{5}/4$
0353 3922 $2^x : 4\sqrt{2}/3 - 3\sqrt{5}$
0353 4564 $J_1 : 2(1/\pi)/9$
0353 5302 at : $(1/\pi)/9$
0353 5678 $\sqrt{5} \div \ln(3)$
0353 6039 s$\pi : (3\pi)^{-2}$
0353 6776 id : $(1/\pi)/9$
0353 7436 sq : $3\sqrt{2}/2 - 4\sqrt{3}/3$
0353 7481 $J_2 : 7/13$
0353 7513 sinh : $(1/\pi)/9$
0353 7514 as : $(1/\pi)/9$
0353 7823 rt : (1,9,8,2)
0353 8251 t$\pi : (3\pi)^{-2}$
0353 8338 at : $3/4 - \pi/4$
0353 8424 $1/5 \div \exp(\sqrt{3})$
0353 9816 id : $1/4 + \pi/4$
0354 0376 $e^x : 3\sqrt{2}/2 + 3\sqrt{5}$
0354 0556 as : $3/4 - \pi/4$
0354 4344 cu : $4\sqrt{2} - 3\sqrt{5}/4$
0354 5856 $J_2 : 7\ln 2/9$
0354 7277 as : $3\sqrt{2}/2 - 4\sqrt{5}/3$
0354 7312 $10^x : e/3 - 3\pi/4$
0354 7695 sq : P_{36}
0355 1244 sq : $5\pi/7$
0355 3956 $\exp(\pi) \div \exp(1/2)$
0355 4736 $\theta_3 : 5\sqrt{3}/9$
0355 5007 $\ell_{10} : 4\sqrt{2} + 3\sqrt{3}$
0355 5706 $J_0 : 10\sqrt{3}/7$
0355 6794 $J_2 : 3\sqrt[3]{2}/7$
0355 6941 rt : (9,4,-8,-6)
0355 8662 $\zeta : 7\sqrt[3]{3}/2$
0355 9434 t$\pi : \sqrt{2}/4 - 3\sqrt{5}$
0356 0122 ln : $2 + 4\sqrt{2}$
0356 1279 rt : (9,3,-6,-7)
0356 1434 rt : (8,6,-9,-6)
0356 1552 $J_2 : 5\sqrt[3]{5}$
0356 2597 $1/3 \div \exp(\sqrt{5})$
0356 4174 sq : $4 - 4\pi/3$
0356 5637 rt : (9,2,-4,-8)
0356 5793 rt : (8,5,-7,-7)
0356 5923 $\Psi : 10\ln 2/9$
0356 6463 $2^x : 2\sqrt{3}$
0356 7399 sq : $\exp(-5/3)$
0356 8079 $J_2 : 3\sqrt[3]{3}/8$
0356 8182 at : $4/3 + \sqrt{2}/4$
0356 8550 $Li_2 : (1/\pi)/9$
0356 9151 $J_1 : 1/14$
0356 9761 $J_0 : \exp(e/3)$
0357 0016 rt : (9,1,-2,-9)
0357 0173 rt : (8,4,-5,-8)
0357 0329 rt : (7,7,-8,-7)
0357 0695 $2^x : (1/(3e))^{-1/3}$
0357 2597 $2^x : 3\sqrt{2}/2 - 4\sqrt{3}$
0357 2847 $\ell_{10} : 3\sqrt{3}/2 - 3\sqrt{5}/4$
0357 3790 cr : $3 + 2e$
0357 3906 $2^x : (\sqrt{2\pi})^{-2}$
0357 4416 cr : 10/9
0357 4486 $J_2 : (e/2)^{-2}$
0357 4574 rt : (8,3,-3,-9)
0357 4732 rt : (7,6,-6,-8)
0357 4890 rt : (6,9,-9,-5)
0357 4962 $\Gamma : 6\sqrt{3}/5$
0357 6138 $10^x : 4 - 3\pi/4$
0357 6494 id : $e/4 + 3\pi/4$
0357 6766 as : $(\pi/2)^{-1/3}$
0357 7942 ln : P_{22}
0357 8680 $J_2 : 13/24$
0357 9155 rt : (7,5,-4,-9)
0357 9314 rt : (6,8,-7,-8)
0357 9577 $J_2 : 9\sqrt{5}/4$
0358 1001 $\zeta : 3\sqrt{2}/5$
0358 1228 $\zeta : 4\sqrt[3]{2}$
0358 1478 s$\pi : (\sqrt{2\pi})^{-3}$
0358 1646 $J_0 : 7\sqrt{2}/4$
0358 2024 $J_2 : (2\pi)^{-1/3}$
0358 3761 rt : (6,7,-5,-9)
0358 3777 t$\pi : (\sqrt{2\pi})^{-3}$
0358 7067 rt : (2,9,-5,2)
0358 8389 rt : (5,9,-6,-9)
0358 8547 $e^x : 3/4 - 3e/2$
0359 0520 $10^x : (\sqrt[3]{3}/2)^{-1/3}$
0359 0529 $erf : (1/\pi)/10$
0359 2063 $J_1 : 4\pi^2/3$
0359 2881 rt : (5,7,-5,3)
0359 3738 $2^x : 4/3 + e/4$
0359 6922 sq : $2\sqrt{2} + 3\sqrt{3}/4$
0359 7394 rt : (5,8,-1,7)
0359 8565 $Ei : 7\zeta(3)/8$
0360 0000 $J_2 : 10e$
0360 0064 $e^x : (1/\pi)/9$
0360 0608 t$\pi : 1/3 - \sqrt{3}$
0360 0915 rt : (1,9,7,-6)
0360 1404 at : $2\sqrt{2}/3 + \sqrt{5}/3$
0360 2093 rt : (4,3,-9,3)
0360 3474 cr : $\sqrt{3} + 3\sqrt{5}$

0360 4367 $J_2 : e/5$
0360 4781 $\Psi : 3e$
0360 4900 $\zeta : 9\sqrt{5}/4$
0360 5067 $Ei : \pi^2/7$
0360 5108 at : $3\sqrt{2}/2 - \sqrt{3}/4$
0360 6653 rt : (4,4,-5,7)
0360 6797 $1/5 - \sqrt{5}$
0360 7513 $\Psi : 5\sqrt[3]{5}/6$
0360 8319 $10^x : 1/\ln(1/2)$
0360 9746 $1 + s\pi(\sqrt{2})$
0360 9898 sq : $\sqrt[3]{5}/9$
0361 0184 $J_2 : 1/\ln(2\pi)$
0361 1401 rt : (3,8,-6,-8)
0361 1435 sinh : $(\sqrt{5}/3)^{1/3}$
0361 4985 $2^x : \exp(\sqrt{2}/3)$
0361 6008 rt : (3,9,0,1)
0361 7746 $\zeta : 8\pi/5$
0362 1217 $\ell_{10} : 23/25$
0362 1782 sq : $9\sqrt[3]{2}/4$
0362 2960 $\Gamma : 3\ln 2$
0362 4566 $e^x : 2\sqrt{3}/3 - 2\sqrt{5}$
0362 4763 $\ln(\sqrt{3}) \div \exp(e)$
0362 6225 ln : $\zeta(5)$
0362 7654 $J_2 : 6/11$
0362 8117 sq : 4/21
0362 8381 $\zeta : 6(e+\pi)/7$
0362 9658 $\ell_2 : 7(e+\pi)/5$
0363 0138 rt : (2,9,3,-3)
0363 1765 $2^x : 3/2 - 2\pi$
0363 5002 $2^x : 3\sqrt[3]{5}/5$
0363 5447 $\ell_{10} : 4e$
0363 6781 at : $2/3 - 3\pi/4$
0363 7934 $erf : (\pi)^{-3}$
0364 2187 $10^x : 2e + \pi/2$
0364 2742 $\Gamma : 2\sqrt{2}/3$
0364 7450 sq : $3/4 - \sqrt{5}/4$
0364 7562 sq : $3(1/\pi)/5$
0364 8472 cu : $1/4 + \pi/2$
0364 8967 $J_2 : 9\pi^2/5$
0364 8997 $\Psi : 3/2$
0364 9206 tπ : $5e\pi/9$
0364 9666 $e^x : 4\sqrt{2}/3 - 3\sqrt{3}$
0365 1666 rt : (2,9,-6,-3)
0365 3979 $J_0 : 3(e+\pi)/2$
0365 4444 $\Gamma : \sqrt[3]{9}$
0365 4528 $\Psi : 4\sqrt{3}/9$
0365 4614 sr : $\exp(\sqrt{2}\pi/2)$
0365 7907 $J_0 : 21$
0365 8702 $Ei : 6(1/\pi)/5$
0365 9343 $\exp(3) + s\pi(2/5)$
0365 9622 $10^x : 4 - 2e$
0365 9634 $10^x : 2/3 + 4\sqrt{5}$
0365 9732 sq : P_{52}
0366 1464 sr : $(\sqrt{3}/2)^{-1/2}$
0366 1971 $\zeta : 2e/5$
0366 3127 $2 \div \exp(4)$
0366 3150 $\theta_3 : \exp(-4)$
0366 3292 $10^x : (4)^{-3}$
0366 4556 at : $4 - 4\sqrt{3}/3$
0366 5013 at : $2\sqrt{3}/3 - \sqrt{5}/2$
0366 5370 $2^x : 2/3 - 2e$
0366 5475 $J_1 : \sqrt{14}$
0366 5943 $\Gamma : 3\pi/10$
0366 6654 id : $2\sqrt{3}/3 - \sqrt{5}/2$
0366 6727 cr : $7(1/\pi)/2$
0366 7477 as : $2\sqrt{3}/3 - \sqrt{5}/2$
0366 8975 $2^x : 10\sqrt[3]{3}/9$
0366 9680 ln : $2\pi^2/7$
0367 0203 $\Gamma : (\ln 2/3)^{-1/2}$
0367 3652 rt : (1,9,0,3)
0367 3887 sq : $3\sqrt{3}/4 - 2\sqrt{5}/3$
0367 4049 $\ell_{10} : \sqrt{12\pi}$
0367 5948 cr : $3 - 4\sqrt{2}/3$
0367 7766 $J_2 : \ln(\sqrt{3})$
0368 0061 sinh : $7\zeta(3)/4$
0368 1290 sr : $\ln((e+\pi)/2)$
0368 2384 $J_1 : 5(e+\pi)$
0368 2719 rt : (9,4,-9,-5)
0368 3486 cu : $2 + \sqrt{2}/4$
0368 6827 $J_2 : 11/20$
0368 7520 rt : (9,3,-7,-6)
0368 7698 at : $e/3 + \pi/4$
0368 8192 ln : 3/23
0369 1025 $J_2 : 1/\sqrt[3]{6}$
0369 2346 rt : (9,2,-5,-7)
0369 2524 rt : (8,5,-8,-6)
0369 2775 id : $\zeta(5)$
0369 3423 cu : $(\ln 2)^3$
0369 4294 sπ : $\exp(-\sqrt{2}\pi)$
0369 5442 $\zeta : 1/\ln(\sqrt[3]{3}/3)$
0369 6817 tπ : $\exp(-\sqrt{2}\pi)$
0369 7196 rt : (9,1,-3,-8)
0369 7376 rt : (8,4,-6,-7)
0369 7556 rt : (7,7,-9,-6)
0369 9489 cr : $(\sqrt[3]{3}/2)^{-1/3}$
0370 0170 $e^x : 3(e+\pi)/4$
0370 1043 $\ell_2 : 3\sqrt[3]{5}/5$
0370 1164 $J_1 : (\sqrt[3]{2}/3)^3$
0370 1400 $Ei : 25/11$
0370 2011 at : $(3)^{-3}$
0370 2072 rt : (9,0,-1,9)
0370 2253 rt : (8,3,-4,-8)
0370 2434 rt : (7,6,-7,-7)
0370 3703 id : $(3)^{-3}$
0370 4106 $\ln(\sqrt{2}) \div \exp(\sqrt{5})$
0370 4278 tπ : $2e - 3\pi$
0370 4550 as : $(3)^{-3}$
0370 6657 $\ell_2 : 3/2 + 3\sqrt{5}$
0370 6669 $K : (2e/3)^{-1/3}$
0370 7156 rt : (8,2,-2,-9)
0370 7338 rt : (7,5,-5,-8)
0370 7520 rt : (6,8,-8,-7)
0370 8813 at : 22/13
0370 9436 cu : $3 + 4\sqrt{5}$
0370 9865 $10^x : 8\zeta(3)/3$
0371 2267 rt : (7,4,-3,-9)
0371 2451 rt : (6,7,-6,-8)
0371 3123 at : $1/3 + e/2$
0371 5504 $2^x : 1/19$
0371 7407 rt : (6,6,-4,-9)
0371 7532 sq : $3/4 - 2\sqrt{2}/3$
0371 7549 $\Gamma : (\ln 2)^{-2}$
0371 7592 rt : (5,9,-7,-8)
0371 8600 rt : (2,9,-7,-8)
0371 8819 rt : (1,9,-1,-6)
0371 9059 ln : $(\sqrt{5}/2)^{1/3}$
0372 0131 $2^x : 4\zeta(3)/3$
0372 1503 $e^x : 1/3 - 4e/3$
0372 1785 at : $3/4 + 2\sqrt{2}/3$
0372 2575 rt : (5,8,-5,-9)
0372 6222 $\exp(3/4) - \exp(e)$
0372 6890 cr : $1/4 + \sqrt{3}/2$
0372 7585 rt : (5,8,-2,5)
0373 0469 $10^x : 3\sqrt{2}/7$
0373 0531 at : $\ln(2e)$
0373 2154 $\Gamma : \sqrt{13/3}$
0373 2622 rt : (5,9,2,9)
0373 3017 at : $4e - 4\pi$
0373 4030 $3/4 \div \exp(3)$
0373 4705 $e^x : 2\sqrt{3}/3 - \sqrt{5}/2$
0373 4836 $10^x : 2/3 - 2\pi/3$
0373 5017 $e^x : 4\sqrt{2} - 4\sqrt{5}$
0373 7875 rt : (4,4,-6,5)
0373 7984 cu : P_7
0373 8542 $J_2 : 8\pi/5$
0373 8573 $Li_2 : (3)^{-3}$
0374 0666 $10^x : 1/4 - 3\sqrt{5}/4$
0374 1318 $\Psi : (\ln 2/2)^{-1/3}$
0374 1789 $\Psi : 5e\pi/2$
0374 2967 rt : (4,5,-2,9)
0374 4644 $e^x : (e)^{1/3}$
0374 4929 $1 \div s\pi(\sqrt{2})$
0374 5707 $2^x : (1/\pi)/6$
0374 6076 $J_2 : 4\ln 2/5$
0374 6460 tπ : $3\sqrt{3}/4 - 4\sqrt{5}$
0374 8123 $J_2 : (e+\pi)^{-1/3}$
0374 8279 rt : (3,9,-1,-2)
0374 8540 $2^x : 2e/3 - \pi/4$
0374 9712 sq : $4 - 4\sqrt{5}/3$
0375 0858 $e^x : 3 - 2\pi$
0375 2693 $e^x : e/3 - 4\pi/3$
0375 3428 rt : (3,1,-6,9)
0375 7340 $\Gamma : 16/17$
0375 8540 $e^x : 4/3 + 2\sqrt{3}/3$
0375 9399 cr : $6\pi^2/7$
0375 9747 $J_2 : 5/9$
0376 0381 $\Gamma : 10/23$
0376 0538 cu : $4\sqrt{2} + 2\sqrt{3}/3$
0376 2227 cu : $2 + 4\pi/3$
0376 4006 rt : (2,9,2,-7)
0376 4579 $J_0 : 7/3$
0376 8207 $1/4 + \ln(3/4)$
0376 8643 $s\pi(e) \div s\pi(\sqrt{3})$
0376 9241 rt : (2,4,-4,7)
0376 9586 $2^x : (\ln 2/3)^2$
0376 9968 sinh : $5e$
0377 0060 sπ : $3\sqrt{2} + \sqrt{5}/3$
0377 0789 $2^x : \exp(-(e+\pi)/2)$
0377 1136 $2^x : 3(e+\pi)/4$
0377 1238 $J_0 : 7e\pi/5$
0377 2742 tπ : $3\sqrt{2} + \sqrt{5}/3$
0377 3145 $e^x : (3)^{-3}$
0377 3177 $e^x : 10/9$
0377 4904 sr : 14/13
0377 6944 $\Psi : e\pi/6$
0377 7107 cr : 19/17
0377 7736 $\exp(1/2) + \exp(2)$
0377 8055 ln : $4\sqrt{3} + \sqrt{5}/3$
0377 8856 $\ell_{10} : 11/12$
0377 9373 $J_2 : 7(1/\pi)/4$
0377 9769 cu : $1/4 - e$
0378 1892 $2^x : \sqrt{2}/3 - 3\sqrt{3}$
0378 4440 tπ : $1/3 - 3e/2$
0378 5393 $J_1 : 5\pi/4$
0378 6104 cu : $1/4 + 3e/2$
0378 8269 sq : $3\sqrt{2}/4 - \sqrt{3}/2$
0378 8329 at : $\sqrt{3}/3 + \sqrt{5}/2$
0378 9081 id : $(\sqrt{5}/2)^{1/3}$
0379 0087 cu : 12/7
0379 0678 rt : (1,5,1,6)
0379 3624 $\Psi : 10/13$
0379 3693 cu : $2\sqrt{3}/3 - 2\sqrt{5}/3$
0379 5484 sr : $1/2 + \sqrt{3}/3$
0379 5537 $e^x : \sqrt{2} - \sqrt{5} - \sqrt{6}$
0379 6670 $\ell_{10} : \ln(5/2)$
0379 9952 sq : $3\sqrt{2}/2 + 4\sqrt{5}/3$
0380 0847 $\Gamma : 10e/9$
0380 1458 $J_0 : 9\zeta(3)/2$
0380 4929 $J_1 : (\sqrt[3]{2}/3)^{-3}$
0380 5512 $J_2 : \sqrt{5}/4$
0380 8445 $\exp(\sqrt{3}) \div \exp(5)$
0381 0660 cr : $1/3 + \pi/4$
0381 2083 $\ell_{10} : \beta(2)$
0381 2730 sq : $2/3 - \sqrt{2}/3$
0381 4076 $\ell_2 : 2e/3 - \pi/4$
0381 4282 $\Gamma : 25/12$
0381 4534 ln : $7\pi^2/9$
0381 6281 $erf : (2e)^{-2}$
0381 7036 ln : $4/3 + 2\sqrt{5}/3$
0381 8090 $J_2 : 4\sqrt[3]{2}/9$
0381 8556 $J_2 : 14/25$
0381 9310 ln : $3e/4 + \pi/4$
0382 0051 $e^x : 4e + 3\pi/2$
0382 2539 cu : $4e - 3\pi$
0382 2638 rt : (9,3,-8,-5)
0382 2694 $e^x : 3\sqrt{3} + 3\sqrt{5}/4$
0382 3057 at : $\sqrt{2}/2 - \sqrt{5}/3$
0382 4921 id : $\sqrt{2}/2 - \sqrt{5}/3$
0382 4967 $Ei : 20/19$
0382 5510 $2^x : 2 - 3\sqrt{5}$
0382 5854 as : $\sqrt{2}/2 - \sqrt{5}/3$
0382 7999 rt : (9,2,-6,-6)
0382 8204 rt : (8,5,-9,-5)
0383 0825 sr : $3 + 2\sqrt{3}/3$
0383 1454 at : $6\sqrt{2}/5$
0383 3389 rt : (9,1,-4,-7)
0383 3596 rt : (8,4,-7,-6)
0383 5290 $J_2 : 21/4$
0383 6953 $J_2 : \exp(-\sqrt{3}/3)$
0383 7297 at : $\sqrt{2}/3 - \sqrt{3}/4$

0383 7380 $Li_2 : (2\pi/3)^{-1/3}$
0383 8556 $\ln : 3/4 + 4\sqrt{3}$
0383 8809 rt : (9,0,-2,8)
0383 9018 rt : (8,3,-5,-7)
0383 9181 id : $\sqrt{2}/3 - \sqrt{3}/4$
0383 9227 rt : (7,6,-8,-6)
0383 9409 $\zeta : 7\sqrt{2}/2$
0383 9905 $J : \exp(-19\pi/20)$
0384 0125 as : $\sqrt{2}/3 - \sqrt{3}/4$
0384 0848 sq : $2/3 - 2\pi/3$
0384 1261 $\ell_{10} : 3/2 + 3\pi$
0384 1886 at : $\exp(\sqrt[3]{4}/3)$
0384 2668 $\Psi : 7\ln 2/10$
0384 2978 $2^x : 3\sqrt{2} - 4\sqrt{5}$
0384 3309 $J_1 : 1/13$
0384 4260 rt : (9,1,0,9)
0384 4470 rt : (8,2,-3,-8)
0384 4680 rt : (7,5,-6,-7)
0384 4890 rt : (6,8,-9,-6)
0384 5233 $10^x : 2\sqrt{2}/3 + \sqrt{5}/2$
0384 6927 $J_2 : \ln(\sqrt[3]{5}/3)$
0384 7470 $10^x : e/4 - 2\pi/3$
0384 7577 cu : $3 - \sqrt{3}$
0384 7712 cr : $8\sqrt[3]{2}/9$
0384 7910 $\ln(\sqrt{3}) - s\pi(1/5)$
0384 7963 $J_1 : \ln 2/9$
0384 8052 $\ln : 3\sqrt{3}/5$
0384 9952 rt : (8,1,-1,-9)
0385 0164 rt : (7,4,-4,-8)
0385 0376 rt : (6,7,-7,-7)
0385 1819 $J_2 : 9/16$
0385 2584 sr : $6e$
0385 2803 $\ln(4) + \exp(\sqrt{3})$
0385 4767 $J_0 : (2e)^{1/2}$
0385 4876 $\ell_{10} : 4 + 4\sqrt{3}$
0385 5678 rt : (7,3,-2,-9)
0385 5892 rt : (6,6,-5,-8)
0385 6105 rt : (5,9,-8,-7)
0385 7600 $2^x : 1/2 - 3\sqrt{3}$
0385 9508 cr : $4 + 2\sqrt{5}$
0385 9969 sr : $1/3 + \sqrt{5}/3$
0385 9987 $s\pi : \sqrt{2} + 3\sqrt{3}/2$
0386 1439 rt : (6,5,-3,-9)
0386 1492 id : $2\sqrt{2}/3 - 4\sqrt{5}/3$
0386 1654 rt : (5,8,-6,-8)
0386 2865 $t\pi : \sqrt{2} + 3\sqrt{3}/2$
0386 2876 $10^x : 2\sqrt{3} - 4\sqrt{5}/3$
0386 5912 $\Psi : \sqrt[3]{2}/6$
0386 7196 $2^x : 17/2$
0386 7234 rt : (5,8,-3,3)
0386 8637 as : P_5
0386 9205 at : $2 - 3e/4$
0386 9742 $J_2 : 6(e+\pi)/7$
0387 1137 id : $3e/4$
0387 2104 as : $2 - 3e/4$
0387 2336 at : $(\ln 2/2)^{-1/2}$
0387 2804 $s\pi(1/4) + s\pi(\sqrt{3})$
0387 2847 rt : (5,9,1,7)
0387 3065 rt : (4,9,-5,-9)
0387 3948 $s\pi : (\ln 2/3)^3$
0387 4374 $J_2 : (\pi)^{-1/2}$
0387 4894 $t\pi : 4\sqrt{3}/3 - 3\sqrt{5}$
0387 5105 $\Gamma : \exp(-4)$
0387 6858 $t\pi : (\ln 2/3)^3$
0387 7416 $\zeta : e\pi/5$
0387 7420 $\exp(3/4) \div \exp(4)$
0387 8711 rt : (4,4,-7,3)
0387 9465 $K : (e/3)^2$
0387 9607 at : $5e/8$
0388 2213 sinh : $(4/3)^{-1/3}$
0388 3172 cu : $2\sqrt{2} - \sqrt{5}/3$
0388 3325 $2^x : 3/4 - 2e$
0388 3910 $\zeta : 18/23$
0388 4217 $\zeta : \pi^2/2$
0388 4389 rt : (4,5,-3,7)
0388 5897 $2^x : \sqrt{3} - 3\sqrt{5}/4$
0388 8125 $J_2 : 13/23$
0388 9073 $\Psi : 5\zeta(3)/4$
0388 9573 $s\pi : 8\sqrt{5}/9$
0389 0324 rt : (3,9,-2,-5)
0389 0362 $\sqrt{3} - \ln(2)$
0389 2311 $J_2 : \exp(-\sqrt[3]{5}/3)$
0389 2518 $t\pi : 8\sqrt{5}/9$
0389 2845 $J_2 : 1/\ln(e+\pi)$
0389 3157 $Ei : (2\pi/3)^{-1/2}$
0389 3353 $\ln(3/4) \div \exp(2)$
0389 4394 $J_2 : 2\sqrt{2}/5$
0389 5218 $\ln : 3\ln 2/2$
0389 6070 rt : (3,1,-7,7)
0389 8002 $\ell_{10} : 2 - e/3$
0389 8770 sq : $2\sqrt{2}/3 - \sqrt{5}/3$
0390 0356 $e^x : 2\sqrt{3} - 3\sqrt{5}$
0390 2382 $t\pi : (e/3)^3$
0390 3932 cr : $7\sqrt[3]{3}/9$
0390 5921 id : $3\sqrt{3}/2 - \sqrt{5}/4$
0390 6061 $e^x : 1 - 3\sqrt{2}$
0390 6160 as : $4/3 - \sqrt{2}/3$
0390 7225 at : 17/10
0390 8601 sq : $3e/8$
0391 0937 $\ln : 4\zeta(3)/5$
0391 2325 id : $\sqrt{3}/4 - 2\sqrt{5}$
0391 3743 rt : (2,4,-5,4)
0391 3795 $s\pi : 9\sqrt{5}/10$
0391 3830 $e^x : \sqrt{2}/3 - \sqrt{3}/4$
0391 5940 sq : $2e + \pi/3$
0391 6630 $e^x : 4/3 + 4\sqrt{3}$
0391 6796 $t\pi : 9\sqrt{5}/10$
0391 8687 $\ell_{10} : 2 + 4\sqrt{5}$
0391 9055 id : $9\pi/7$
0391 9378 $J : (\sqrt{2}\pi)^{-1/2}$
0391 9792 sr : $6\sqrt[3]{2}/7$
0392 1304 cu : $1/3 + e/4$
0392 2065 sq : $2\ln 2/7$
0392 2956 cu : $e/8$
0392 3048 id : $3\sqrt{3}/5$
0392 4252 $\ln(2/5) \times \exp(4/5)$
0392 5545 cu : $2 + e$
0392 5922 $2^x : 1/18$
0392 6101 $\ln : 2\sqrt[3]{3}/3$
0392 7134 $e^x : 3\sqrt{2} + 4\sqrt{5}$
0392 8335 $\ln(5) - \exp(1/2)$
0393 1942 at : $3 - 3\sqrt{3}/4$
0393 3398 sr : $6\ln 2$
0393 5034 $J_2 : (2e)^{-1/3}$
0393 5761 $2^x : 1/2 + 2\pi/3$
0393 6681 $10^x : 4e/3 - \pi$
0393 6937 $\ln : 9\pi/10$
0393 7701 $J : (\sqrt{2}\pi)^{-2}$
0393 7732 rt : (1,5,0,1)
0393 9825 $e^x : \exp(\sqrt[3]{3}/3)$
0394 1778 $J : (2\pi)^{-3}$
0394 3684 $2^x : 7e/4$
0394 4761 $\ell_{10} : 3\sqrt{2} + 3\sqrt{5}$
0394 4944 $J_0 : 6e/7$
0394 7882 $\ln(4/5) \div \exp(\sqrt{3})$
0394 8057 $\ln : 3 - 3e/4$
0394 8441 $Ei : 1/\sqrt[3]{21}$
0395 0105 sq : $3\sqrt{2}/2 - 3\sqrt{5}$
0395 0854 $\ell_{10} : 21/23$
0395 1932 $J_1 : 23$
0395 2293 $J_2 : \sqrt[3]{5}/3$
0395 4627 $10^x : 5\zeta(3)/6$
0395 6055 rt : (1,7,1,9)
0395 6923 sq : $4\sqrt{2}/3 - \sqrt{3}/2$
0395 7858 sq : $5(1/\pi)/8$
0395 8738 sinh : 10/11
0395 9282 $3/5 \div \exp(e)$
0395 9536 cu : $4e/3 - 3\pi/4$
0395 9651 $\ln(2/5) \div \exp(\pi)$
0396 0065 $\zeta : 1/\sqrt[3]{21}$
0396 3453 cu : $9e\pi/5$
0396 4123 $2^x : 1 - 4\sqrt{2}$
0396 4381 $\ln : 1/3 + \sqrt{2}/2$
0396 4890 cr : $6\sqrt{2}$
0396 7555 rt : (9,3,-9,-4)
0396 8017 $J_2 : 5e\pi/2$
0396 8419 id : $4\sqrt[3]{2}$
0397 1694 $J_2 : 4/7$
0397 2077 id : $3\ln 2/2$
0397 3362 $\ell_{10} : \sqrt{2}/4 + \sqrt{5}/4$
0397 3530 rt : (9,2,-7,-5)
0397 3711 at : $((e+\pi)/2)^{-3}$
0397 3912 $\ln(5) - s\pi(\pi)$
0397 4390 sq : $2/3 - \sqrt{3}/2$
0397 4403 $s\pi : 3\sqrt{2}/2 + \sqrt{3}/2$
0397 5724 $J_1 : (1/\pi)/4$
0397 5804 id : $((e+\pi)/2)^{-3}$
0397 6371 $\zeta : (\ln 2/3)^{1/3}$
0397 6556 $\ln(\sqrt{5}) \div s\pi(e)$
0397 6775 at : $(1/\pi)/8$
0397 6852 as : $((e+\pi)/2)^{-3}$
0397 7546 $t\pi : 3\sqrt{2}/2 + \sqrt{3}/2$
0397 8873 id : $(1/\pi)/8$
0397 9541 rt : (9,1,-5,-6)
0397 9781 rt : (8,4,-8,-5)
0397 9923 sinh : $(1/\pi)/8$
0397 9924 as : $(1/\pi)/8$
0398 0640 id : $\sqrt{5} + \sqrt{6} - \sqrt{7}$
0398 0930 $\Psi : (\ln 3/2)^{-2}$
0398 2414 $J : \sqrt{2}/7$
0398 2965 $4/5 \div \exp(3)$
0398 5587 rt : (9,0,-3,7)
0398 5829 rt : (8,3,-6,-6)
0398 6070 rt : (7,6,-9,-5)
0398 6784 cu : $1 + \pi$
0398 7789 $J_2 : (\ln 2/2)^{-3}$
0398 8103 cr : $1/3 + 3e$
0398 9161 $erf : (1/\pi)/9$
0399 0462 $\ell_{10} : (\sqrt{2}\pi/2)^3$
0399 0585 $\ln(2) \times \exp(\pi)$
0399 0812 $10^x : (\ln 2/2)^{1/3}$
0399 1669 rt : (9,1,-1,8)
0399 1913 rt : (8,2,-4,-7)
0399 2156 rt : (7,5,-7,-6)
0399 2394 $J_2 : 9(1/\pi)/5$
0399 2845 $10^x : 1/3 - \sqrt{3}$
0399 3925 $\zeta : \sqrt{24}$
0399 4410 sr : $(\sqrt[3]{5}/2)^{-1/2}$
0399 5386 $t\pi : 1/\ln(3\pi/2)$
0399 5785 $\ell_2 : 5\pi^2/3$
0399 5964 $\ell_{10} : 2\sqrt{2} - \sqrt{3}$
0399 6485 sinh : $10(1/\pi)$
0399 6800 $J_1 : 2/25$
0399 7788 rt : (9,2,1,9)
0399 7868 at : 1/25
0399 8033 rt : (8,1,-2,-8)
0399 8279 rt : (7,4,-5,-7)
0399 8524 rt : (6,7,-8,-6)
0400 0000 id : 1/25
0400 1066 sinh : 1/25
0400 1067 as : 1/25
0400 3722 $2^x : 3/2 - \sqrt{2}/3$
0400 3932 sr : $4/3 + 2\sqrt{2}$
0400 4191 sr : $3\sqrt[3]{3}/4$
0400 4438 rt : (7,3,-3,-8)
0400 4685 rt : (6,6,-6,-7)
0400 4933 rt : (5,9,-9,-6)
0400 5723 $\ell_{10} : \pi^2/9$
0400 6459 $\ln(\sqrt{5}) \div \exp(3)$
0400 7558 $e^x : 4\pi/9$
0400 7585 $10^x : 3\sqrt{3} + 4\sqrt{5}/3$
0400 8260 $\lambda : \exp(-19\pi/10)$
0400 9282 sq : P_{16}
0400 9332 sq : P_{40}
0401 0635 rt : (7,2,-1,-9)
0401 0884 rt : (6,5,-4,-8)
0401 0911 $\theta_3 : \ln(\ln 3/2)$
0401 1134 rt : (5,8,-7,-7)
0401 2076 sr : $9\zeta(3)/10$
0401 2850 $\zeta : 9e/5$
0401 3724 sq : $\zeta(3)/6$
0401 6036 $Li_2 : ((e+\pi)/2)^{-3}$
0401 7121 rt : (6,4,-2,-9)
0401 7372 rt : (5,8,-4,1)
0401 7701 $\Gamma : 15/16$
0401 8749 $\zeta : \exp(\sqrt[3]{4})$
0401 9142 $e^x : 1/4 - 2\sqrt{3}$
0401 9168 $Li_2 : (1/\pi)/8$
0401 9684 $J_0 : 4\pi^2/7$
0402 0467 sq : $9\zeta(3)$

0402 3649 rt : (5,9,0,5)
0402 3902 rt : (4,9,-6,-8)
0402 4081 e^x : $\exp(1/(3\pi))$
0402 6031 e^x : $3/2-3\pi/2$
0402 6878 $s\pi$: $(1/(3e))^{-1/3}$
0402 7889 ℓ_2 : $\exp(\sqrt{2})$
0403 0147 $t\pi$: $(1/(3e))^{-1/3}$
0403 0189 10^x : $\sqrt{10}$
0403 0220 rt : (4,8,-4,-9)
0403 1862 id : $4\sqrt{2}/3+2\sqrt{3}/3$
0403 4981 rt : (1,7,0,2)
0403 5332 Li_2 : 18/23
0403 5495 sinh : 19/13
0403 6315 ℓ_2 : $\exp(\sqrt[3]{3}/2)$
0403 6576 rt : (4,5,-4,5)
0403 8568 sq : $3/2-3\sqrt{3}/4$
0403 9665 K : $(\sqrt{2}\pi/3)^{-1/2}$
0403 9671 J_0 : $\sqrt{22}\pi$
0404 0727 Li_2 : 1/25
0404 1809 at : $2/3-\sqrt{2}/2$
0404 2184 ζ : $1/\ln(\sqrt{2}/3)$
0404 2768 $\exp(\pi)\times s\pi(1/3)$
0404 2973 rt : (4,6,0,9)
0404 3215 ζ : $5(e+\pi)/6$
0404 3232 rt : (3,9,-3,-8)
0404 3797 2^x : $1-2\sqrt{2}/3$
0404 4011 id : $1/3+\sqrt{2}/2$
0404 4786 sr : $5\sqrt{3}/8$
0404 5114 as : $2/3-\sqrt{2}/2$
0404 5154 at : $1/3-3e/4$
0404 5981 J_2 : $2\sqrt[3]{3}/5$
0404 9670 rt : (3,1,-8,5)
0404 9884 Ei : $\sqrt{22}$
0405 0281 10^x : $10\sqrt[3]{3}/9$
0405 0288 J_2 : γ
0405 1071 ℓ_{10} : P_{11}
0405 1081 cu : $2\zeta(3)/7$
0405 2124 J_2 : $\sqrt{3}/3$
0405 2735 $s\pi$: $1/3+e/4$
0405 3879 ζ : $4e\pi/7$
0405 4085 J_0 : $\sqrt{2}-\sqrt{3}+\sqrt{7}$
0405 4396 sq : P_8
0405 5670 ln : $9\sqrt[3]{5}/2$
0405 5842 J_2 : $5\ln 2/6$
0405 5897 e^x : $((e+\pi)/2)^{-3}$
0405 6067 $t\pi$: $1/3+e/4$
0405 6150 rt : (3,2,-4,9)
0405 7912 Γ : 2/15
0405 8167 $\exp(3/4)\times s\pi(\sqrt{2})$
0405 8882 ℓ_{10} : $9e\pi/7$
0405 9091 e^x : $(1/\pi)/8$
0406 0742 cu : $4/3-3\sqrt{5}/4$
0406 0962 ln : $\sqrt{6}\pi$
0406 1084 J_1 : $9\pi^2/2$
0406 3248 ℓ_{10} : $1/3+\sqrt{3}/3$
0406 5932 J_0 : $10\sqrt{5}/9$
0406 6723 Ei : 1/6
0406 7571 ℓ_2 : $3-2\sqrt{2}/3$
0406 8051 cu : $6\sqrt{5}/7$
0406 8234 Ψ : $\sqrt{11}$
0406 8862 10^x : $7\sqrt{5}/3$
0406 9500 rt : (2,4,-6,1)
0407 3942 J_2 : 11/19
0407 6220 $\exp(4/5)\div\exp(4)$
0407 8693 $\sqrt{5}+\ln(\sqrt{5})$
0408 1077 e^x : 1/25
0408 1632 sq : 10/7
0408 2199 ln : 24/25
0408 2755 id : $\sqrt[3]{17/2}$
0408 3299 sr : 13/12
0408 4445 ℓ_{10} : $\ln(3)$
0408 4896 sq : $2/3+\sqrt{2}/4$
0408 6300 10^x : $3\zeta(3)$
0408 6489 ℓ_{10} : $\sqrt{2}+\sqrt{3}-\sqrt{5}$
0408 9353 at : $1+\sqrt{2}/2$
0408 9543 ln : $2\sqrt{3}/3+3\sqrt{5}/4$
0408 9635 $\exp(1/3)\times s\pi(\sqrt{3})$
0409 1367 Ei : $(\sqrt[3]{5}/2)^{-1/3}$
0409 1909 J_0 : $7\sqrt[3]{5}$
0409 1934 e^x : $2-3\sqrt{3}$
0409 5352 $s\pi$: $\pi^2/10$
0409 5501 $\sqrt{5}\div\exp(4)$
0409 6188 ℓ_{10} : $\sqrt{2}/4+\sqrt{5}/3$
0409 6459 rt : (1,5,-1,-4)
0409 8791 $t\pi$: $\pi^2/10$
0410 0794 J_1 : $\exp(-5/2)$
0410 1186 $\exp(\sqrt{2})-\exp(e)$
0410 3273 $t\pi$: $4\sqrt{2}+4\sqrt{5}$
0410 5002 cu : $2/3+2\sqrt{5}/3$
0410 6392 sq : $(\sqrt{2}\pi/2)^{-2}$
0410 8796 cu : 23/12
0411 0807 at : $e\pi/5$
0411 1299 sr : $4\sqrt{2}-2\sqrt{5}/3$
0411 3084 $\ln(\sqrt{3})+\exp(2/5)$
0411 3894 10^x : $7\ln 2/8$
0411 5114 10^x : $1/2-4\sqrt{2}/3$
0411 6135 ζ : $-\sqrt{3}-\sqrt{5}+\sqrt{7}$
0411 6951 $t\pi$: γ
0411 7064 cu : $3/2-2\sqrt{3}/3$
0411 7082 rt : (1,7,-1,-5)
0411 8265 sinh : $\sqrt{2}+\sqrt{3}-\sqrt{5}$
0412 0843 cu : $\sqrt{3}-\sqrt{6}-\sqrt{7}$
0412 0861 cu : $5\zeta(3)/3$
0412 1010 J_0 : $(\sqrt[3]{5}/3)^{-3}$
0412 1463 10^x : $4/3-e$
0412 1791 ℓ_{10} : $7\pi/2$
0412 2074 Ei : $4e\pi/9$
0412 2171 e^x : $1-4\pi/3$
0412 4145 sr : 25/6
0412 4335 Γ : $-\sqrt{5}+\sqrt{6}-\sqrt{7}$
0412 4448 sinh : $1/\ln(3)$
0412 5637 e^x : $3\sqrt{2}/2+2\sqrt{5}/3$
0412 6366 rt : (1,7,-8,3)
0412 6607 ℓ_{10} : $2/3+\sqrt{3}/4$
0412 8977 $\exp(2)+\exp(\sqrt{3})$
0413 0003 rt : (9,2,-8,-4)
0413 2706 J : $3(1/\pi)/10$
0413 4134 J_2 : 7/12
0413 4251 Ei : 24/17
0413 6623 cu : $3\sqrt{2}/4+4\sqrt{5}/3$
0413 6729 rt : (9,1,-6,-5)
0413 7008 rt : (8,4,-9,-4)
0413 7052 ℓ_{10} : $7\sqrt{2}/9$
0413 8904 10^x : $(e)^{-1/2}$
0413 9268 ℓ_{10} : 11
0414 2793 ζ : $7\ln 2$
0414 3498 rt : (9,0,-4,6)
0414 3779 rt : (8,3,-7,-5)
0414 3827 10^x : $\sqrt{2}+2\sqrt{3}$
0414 5106 J : $\exp(-\sqrt[3]{3})$
0414 5188 id : $7\sqrt{3}/3$
0414 5387 ln : 6/17
0414 6247 J_2 : $((e+\pi)/2)^{-1/2}$
0414 7654 λ : $\exp(-17\pi/9)$
0415 0310 rt : (9,1,-2,7)
0415 0594 rt : (8,2,-5,-6)
0415 0695 cu : $1/3-e/4$
0415 0877 rt : (7,5,-8,-5)
0415 2849 cr : $(\sqrt[3]{3})^{1/3}$
0415 3476 θ_3 : 8/17
0415 3726 id : $1/3+3\sqrt{5}$
0415 4400 J_2 : $1/\sqrt[3]{5}$
0415 5507 rt : (4,9,-7,8)
0415 6423 cr : $(\ln 2)^{-1/3}$
0415 6921 cu : $\sqrt{3}/5$
0415 7166 rt : (9,2,0,8)
0415 7167 J_2 : 15
0415 7452 rt : (8,1,-3,-7)
0415 7738 rt : (7,4,-6,-6)
0415 8024 rt : (6,7,-9,-5)
0416 0405 tanh : $(\ln 2/2)^3$
0416 0406 at : $(\ln 2/2)^3$
0416 1122 2^x : $3\sqrt{2}/2-3\sqrt{5}$
0416 1601 2^x : 1/17
0416 2297 $s\pi$: $3e-\pi$
0416 2353 sq : $3/2+\sqrt{5}/3$
0416 2563 at : $\sqrt[3]{5}$
0416 2808 id : $(\ln 2/2)^3$
0416 3050 J_1 : 1/12
0416 4010 sinh : $(\ln 2/2)^3$
0416 4011 as : $(\ln 2/2)^3$
0416 4067 rt : (9,3,2,9)
0416 4257 at : 1/24
0416 4355 rt : (8,0,-1,8)
0416 4509 sq : $1/3+3\sqrt{5}/4$
0416 4624 ℓ_{10} : $(4/3)^{-1/3}$
0416 4644 rt : (7,3,-4,-7)
0416 4932 rt : (6,6,-7,-6)
0416 5907 $t\pi$: $3e-\pi$
0416 6167 sr : $2e/3+3\pi/4$
0416 6666 id : 1/24
0416 7872 sinh : 1/24
0416 7873 as : 1/24
0417 0752 J_2 : $(e+\pi)/10$
0417 1303 rt : (8,1,1,9)
0417 1594 rt : (7,2,-2,-8)
0417 1885 rt : (6,5,-5,-7)
0417 2176 rt : (5,8,-8,-6)
0417 2239 cr : $\exp(1/(3e))$
0417 3193 at : $2\sqrt{2}-\sqrt{5}/2$
0417 7271 erf : $(3)^{-3}$
0417 8590 rt : (7,1,0,-9)
0417 8730 $s\pi$: $\sqrt{3}-\sqrt{5}/3$
0417 8884 rt : (6,4,-3,-8)
0417 9177 rt : (5,8,-5,-1)
0418 0237 sq : $\exp(-\sqrt[3]{4})$
0418 2383 $t\pi$: $\sqrt{3}-\sqrt{5}/3$
0418 3141 10^x : $\sqrt{2}/4-\sqrt{3}$
0418 3346 rt : (1,7,-9,-5)
0418 5151 as : $(\sqrt{5}/3)^{1/2}$
0418 5589 ℓ_2 : $1/2+\sqrt{2}/3$
0418 5929 rt : (6,3,-1,-9)
0418 6036 Li_2 : $(\sqrt[3]{3}/3)^{1/3}$
0418 6225 rt : (5,9,-1,3)
0418 6521 rt : (4,9,-7,-7)
0418 6918 2^x : $3+2\pi$
0418 7367 e^x : $3\sqrt{3}/4-2\sqrt{5}$
0419 0494 Γ : 23/11
0419 3319 rt : (5,5,-2,-9)
0419 3617 rt : (4,8,-5,-8)
0419 3925 J_0 : 17/2
0419 5573 ℓ_2 : $3+\sqrt{5}/2$
0419 7804 $\ln(\pi)\div\ln(3)$
0419 7927 ln : $9\sqrt[3]{2}/4$
0420 0146 cr : $4\sqrt{2}/5$
0420 0762 rt : (4,7,-3,-9)
0420 1377 Γ : $7\zeta(3)/9$
0420 1881 J_2 : 10/17
0420 6617 at : $((e+\pi)/2)^{1/2}$
0420 6951 Li_2 : $(\ln 2/2)^3$
0420 7954 rt : (4,6,-1,7)
0420 8034 $\ln(3)-\exp(\pi)$
0420 8103 cu : 8/23
0420 8258 rt : (3,9,-4,-9)
0420 8399 J_2 : $(\ln 2/2)^{1/2}$
0420 8414 id : $3\sqrt{2}/4+4\sqrt{5}/3$
0420 9556 ln : $3\sqrt{2}+2\sqrt{3}$
0421 0892 Li_2 : 1/24
0421 4093 J_2 : $\exp(-\sqrt[3]{4}/3)$
0421 5026 e^x : $\sqrt{3}/2+3\sqrt{5}/2$
0421 5502 rt : (3,1,-9,3)
0421 6439 cu : $2/3+2\sqrt{3}/3$
0421 8182 cr : $4e-3\pi/4$
0422 1529 sq : $\sqrt{2}/4-\sqrt{5}/4$
0422 2795 rt : (3,2,-5,7)
0422 3992 at : $3-3\pi/2$
0422 4537 cu : 25/12
0422 7658 2^x : $5\sqrt[3]{3}/7$
0422 8520 ln : $1+3\sqrt{5}$
0422 9559 cu : $1/2-2\sqrt{3}$
0423 2462 $t\pi$: $\pi^2/6$
0423 3510 Γ : $\sqrt{2}-\sqrt{3}-\sqrt{6}$
0423 4965 J_2 : $1/\ln(2e)$
0423 5138 ℓ_{10} : $7\sqrt[3]{2}/8$
0423 5364 sr : $\sqrt{2}/2+2\sqrt{3}$
0423 7846 rt : (2,4,-7,-2)
0423 9044 J_2 : 13/22
0424 0759 ζ : 4/11
0424 1445 e^x : 15/2
0424 1563 rt : (4,9,-8,5)

0424 1687 $\Psi : 7(e+\pi)/5$
0424 3119 $10^x : 4\pi/5$
0424 3203 $Ei : \exp(-1/e)$
0424 3906 as : $19/22$
0424 5069 sq : $\sqrt[3]{3}/7$
0424 5294 rt : (2,5,-1,7)
0424 5758 $10^x : 2/3 - 3e/4$
0424 8250 $\ln(\sqrt{2}) - \exp(2)$
0424 9774 $2^x : 6\zeta(3)/7$
0425 0667 $e^x : (\ln 2/2)^3$
0425 2845 $10^x : 3/4 - 3\sqrt{2}/2$
0425 3261 cu : $\pi/9$
0425 4541 $\ell_{10} : (\sqrt{5}/3)^{1/3}$
0425 4690 $e^x : 1/24$
0425 4831 $Ei : 7\ln 2$
0425 4925 $s\pi : 4\sqrt{3}/3 + 3\sqrt{5}/4$
0425 5961 ln : $23/24$
0425 7117 $e^x : 2/3 + 4e/3$
0425 7207 sr : $25/23$
0425 7758 at : $2\sqrt{3}/3 + \sqrt{5}/4$
0425 7890 ln : $3\sqrt{5}/7$
0425 8514 id : $4\sqrt{2}/3 - 4\sqrt{3}$
0425 8782 $t\pi : 4\sqrt{3}/3 + 3\sqrt{5}/4$
0425 9623 $t\pi : 2\pi/5$
0426 0360 cr : $17/15$
0426 1484 at : $2/3 + \pi/3$
0426 2096 $2^x : \sqrt{2}/3 + \sqrt{5}/4$
0426 2313 sq : $3 - \pi/2$
0426 5377 $10^x : \sqrt{3}/2 - \sqrt{5}$
0426 8278 rt : (1,5,-2,-9)
0427 2187 at : $12/7$
0427 2707 $e^x : 5/7$
0427 3902 $J : \exp(-19\pi/18)$
0427 4107 $Ei : 9/13$
0427 4288 sr : $2e/5$
0427 4947 sq : $3\sqrt{2} + 3\sqrt{5}/4$
0427 5117 $2/5 \div \exp(\sqrt{5})$
0427 5311 sq : P_{39}
0427 5762 $\zeta : 8(1/\pi)/7$
0427 5939 sq : P_{33}
0427 6846 $1/2 \times \exp(3)$
0427 7625 $J_1 : 5\sqrt{5}/3$
0427 8284 sinh : $\sqrt[3]{25/2}$
0427 8621 cr : $9\sqrt[3]{2}/10$
0427 9153 $e - s\pi(\sqrt{5})$
0427 9334 $\ell_{10} : 3/4 + \sqrt{2}/4$
0427 9552 $\zeta : \exp(\pi/2)$
0427 9576 $\ln(5) - \exp(\sqrt{3})$
0428 0020 cr : $2 - \sqrt{3}/2$
0428 2202 $\zeta : \sqrt{3/2}$
0428 2677 $\ell_{10} : e/3$
0428 2799 ln : $3 + 3\pi/2$
0428 3956 $J_2 : 6\ln 2/7$
0428 6204 $e^x : \sqrt{3} - \sqrt{5} - \sqrt{7}$
0428 6694 at : $3/4 - \sqrt{2}/2$
0428 7095 $\zeta : 4\zeta(3)$
0428 7435 $10^x : 10\sqrt{3}/9$
0428 7500 cu : $7/20$
0428 9321 id : $3/4 - \sqrt{2}/2$
0428 9400 $\ell_2 : 5\sqrt[3]{3}/7$
0428 9542 $\zeta : 10\sqrt[3]{3}/3$
0428 9677 $J_2 : \ln(2e/3)$
0429 0638 as : $3/4 - \sqrt{2}/2$
0429 4141 sr : $(1/(3e))^3$
0429 5973 sq : $1/3 - 2e$
0429 8036 $\pi - \ln(3)$
0429 8628 rt : (9,2,-9,-3)
0429 9401 $10^x : 2\sqrt{2}/3 - 4\sqrt{3}/3$
0430 4054 cu : $4\sqrt{2}/3 - \sqrt{5}$
0430 5570 $10^x : 3\sqrt{2}/4 - \sqrt{3}/3$
0430 6123 $\ell_{10} : 3 - 2\pi/3$
0430 6183 rt : (9,1,-7,-4)
0430 7524 $10^x : \exp(-4)$
0431 0660 $\ell_2 : 4 + 3\sqrt{2}$
0431 1277 $\ell_2 : 6\zeta(3)/7$
0431 1626 $10^x : 1/\ln(\sqrt[3]{3}/3)$
0431 1760 $10^x : \sqrt{3}/3 - \sqrt{5}/4$
0431 3790 rt : (9,0,-5,5)
0431 4119 rt : (8,3,-8,-4)
0431 4801 $\zeta : 24/5$
0431 6050 sq : $3(e+\pi)$
0431 6386 $\ell_2 : P_{15}$
0431 6564 $e^x : 1/2 + \sqrt{5}/2$
0431 6651 $\Gamma : 14/15$
0431 6767 $J_2 : (3\pi/2)^{-1/3}$
0431 8703 tanh : $\exp(-\pi)$
0431 8704 at : $\exp(-\pi)$
0432 0924 $t\pi : 4e/3 - \pi/3$
0432 1391 id : $\exp(-\pi)$
0432 1449 rt : (9,1,-3,6)
0432 1781 rt : (8,2,-6,-5)
0432 2113 rt : (7,5,-9,-4)
0432 2736 sinh : $\exp(-\pi)$
0432 2737 as : $\exp(-\pi)$
0432 3462 $\ell_2 : 4\sqrt{2}/3 + \sqrt{5}$
0432 3757 $t\pi : (2\pi)^{-1/2}$
0432 4077 sq : $3\ln 2/10$
0432 6528 $2^x : 3 + 3e$
0432 8105 $J_1 : \ln 2/8$
0432 8578 $\lambda : \exp(-15\pi/8)$
0432 8912 $\zeta : \sqrt{23}$
0432 9162 rt : (9,2,-1,7)
0432 9497 rt : (8,1,-4,-6)
0432 9832 rt : (7,4,-7,-5)
0433 0010 $2^x : 2/3 - 3\sqrt{3}$
0433 0249 $4 \times \ln(3/5)$
0433 1168 rt : (4,9,-9,2)
0433 3465 $e^x : 4/3 - 2\sqrt{5}$
0433 3526 $\ln(2/3) \div \exp(\sqrt{5})$
0433 3954 $\ell_2 : 7\ln 2/5$
0433 4298 $\zeta : 5\sqrt{2}/8$
0433 6928 rt : (9,3,1,8)
0433 7266 rt : (8,0,-2,7)
0433 7292 $e^x : 3/2 - \pi/4$
0433 7605 rt : (7,3,-5,-6)
0433 7689 $\zeta : 13/18$
0433 7943 rt : (6,6,-8,-5)
0433 9376 $J_2 : 24$
0434 0277 sq : $5/24$
0434 0419 $2^x : 2e/3 + 4\pi/3$
0434 0645 $10^x : 2\pi^2/9$
0434 3717 $J_1 : 2/23$
0434 4399 ln : $4e/3 - \pi/4$
0434 4750 rt : (9,4,3,9)
0434 5088 tanh : $1/23$
0434 5089 at : $1/23$
0434 5091 rt : (8,1,0,8)
0434 5432 rt : (7,2,-3,-7)
0434 5774 rt : (6,5,-6,-6)
0434 6115 rt : (5,8,-9,-5)
0434 6569 $\ell_{10} : 19/21$
0434 6695 at : $10\zeta(3)/7$
0434 7274 $J_1 : 10$
0434 7826 id : $1/23$
0434 9196 sinh : $1/23$
0434 9197 as : $1/23$
0435 0023 sq : $4\sqrt{2}/3 - 3\sqrt{5}/4$
0435 2155 $\exp(1/5) + \exp(3/5)$
0435 2971 rt : (8,2,2,9)
0435 3201 cr : $25/22$
0435 3312 cr : $\sqrt{3}/3 + \sqrt{5}/4$
0435 3315 rt : (7,1,-1,-8)
0435 3660 rt : (6,4,-4,-7)
0435 3805 $J_2 : \ln(\ln 3/2)$
0435 4005 rt : (5,8,-6,-3)
0435 6189 $10^x : 5(1/\pi)$
0436 1255 rt : (7,0,1,9)
0436 1602 rt : (6,3,-2,-8)
0436 1950 rt : (5,9,-2,1)
0436 2190 cu : $3e + 4\pi$
0436 2214 $t\pi : 3\sqrt{2}/2 - \sqrt{5}/4$
0436 2298 rt : (4,9,-8,-6)
0436 3002 sinh : $(\pi/3)^{-2}$
0436 3117 $\ell_{10} : \sqrt{2}/3 + \sqrt{3}/4$
0436 5296 $\Gamma : 2\pi/3$
0436 6509 $J_2 : 3/5$
0436 7169 sq : $3/4 + e/4$
0436 8021 $\Psi : 9\sqrt[3]{2}/8$
0436 8422 $e^x : 1/3 - 2\sqrt{3}$
0436 8996 $Li_2 : \exp(-\pi)$
0436 9602 rt : (6,2,0,-9)
0436 9953 rt : (5,5,-3,-8)
0437 0083 $10^x : 3 - 4\sqrt{5}/3$
0437 0304 rt : (4,8,-6,-7)
0437 1753 cu : $9\sqrt{3}/7$
0437 3142 $\ell_2 : \sqrt{17}$
0437 3177 cu : $22/7$
0437 3467 at : $1 - e$
0437 4494 $10^x : 3 - \sqrt{2}/4$
0437 4778 $t\pi : 3\sqrt{2}/4 + 3\sqrt{3}$
0437 5326 $10^x : 3\sqrt{2}/2 + 3\sqrt{5}$
0437 5476 $s\pi : 3/4 + \sqrt{5}$
0437 6895 $2^x : \sqrt{5} + \sqrt{6} + \sqrt{7}$
0437 8014 rt : (5,4,-1,-9)
0437 8368 rt : (4,7,-4,-8)
0437 9671 $t\pi : 1/4 - \sqrt{5}$
0438 1037 $J_2 : \zeta(3)/2$
0438 1499 $\Gamma : \ln 2/8$
0438 2642 $e^x : 3/4 - \sqrt{2}/2$
0438 2799 $10^x : e - \pi/2$
0438 3114 $e^x : 1/2 + 3\sqrt{3}/4$
0438 4031 ln : $1/4 + \sqrt{2}/2$
0438 6491 rt : (4,6,-2,5)
0438 6848 rt : (3,9,-5,-8)
0438 7312 cu : $\sqrt{5} + \sqrt{6} + \sqrt{7}$
0438 8556 $\theta_3 : \sqrt{2}/3$
0439 4673 rt : (4,7,2,9)
0439 5034 rt : (3,8,-3,-9)
0439 6021 $Li_2 : 1/23$
0439 6436 $e^x : 1/2 - 4e/3$
0439 6499 cu : $6/17$
0439 7466 $\ell_{10} : 3\sqrt{2}/2 + 4\sqrt{5}$
0439 9202 $2/3 \div \exp(e)$
0440 0263 cr : $e\pi$
0440 3280 rt : (3,2,-6,5)
0440 3414 rt : (1,8,-9,4)
0440 3576 $10^x : 1 - 3\pi/4$
0440 5443 cr : $2/3 + \sqrt{2}/3$
0440 6601 sq : $\sqrt{15}\pi$
0440 9447 sq : $\sqrt[3]{2}/6$
0441 1587 rt : (3,3,-2,9)
0441 6123 $e^x : \exp(-\pi)$
0441 7155 $J_1 : 7\pi^2/3$
0441 7578 $\ell_2 : 1/2 + 4e/3$
0441 8134 $t\pi : \zeta(3)/2$
0441 9417 cu : $\sqrt{2}/4$
0441 9929 $t\pi : \exp(-e/2)$
0442 0328 rt : (2,4,-8,-5)
0442 2060 $Ei : 2\sqrt{3}/5$
0442 2879 $J_1 : 3e\pi$
0442 3030 $1/4 \div \exp(\sqrt{3})$
0442 4023 $J_2 : 2e/9$
0442 4078 cu : $10(1/\pi)/9$
0442 4844 $10^x : 2 - 3\sqrt{5}/2$
0442 5205 $10^x : e\pi/5$
0442 5265 $Ei : 23/7$
0442 5333 $10^x : 1/4 + 2\sqrt{5}/3$
0442 5419 $2^x : 8\sqrt[3]{2}/5$
0442 6474 cu : $3\sqrt{2}/4 + 3\sqrt{5}/2$
0442 7378 $2^x : 1/16$
0442 8764 rt : (2,5,-2,4)
0442 9048 $\ell_{10} : 3\sqrt{3}/2 - 2\sqrt{5}/3$
0443 0880 $J_0 : \exp(\sqrt{3})$
0443 2132 sq : $4/19$
0443 3369 $\exp(1/2) + \exp(1/3)$
0443 5272 $J_0 : \sqrt[3]{25/2}$
0443 5593 $J_0 : 7\sqrt[3]{2}$
0443 8897 $2^x : 2\sqrt{2} + 4\sqrt{3}$
0443 9409 id : $3\sqrt{3}/4 + \sqrt{5}/3$
0443 9446 $J_2 : (\ln 3/3)^{1/2}$
0444 0020 cu : $3 - 3\sqrt{5}/2$
0444 2149 $\ell_{10} : (e/2)^{1/3}$
0444 2727 $K : 14/17$
0444 2974 $e^x : \sqrt{2}/2 + 3\sqrt{5}/2$
0444 3125 $\Psi : 17/12$
0444 3728 $e^x : 1/23$
0444 5176 ln : $22/23$
0444 6593 sr : $12/11$
0444 7917 at : $3e/2 - 3\pi/4$
0445 1323 $\exp(4) \div s\pi(1/3)$

0445 2243 ln : 21
0445 2874 J_2 : $3\sqrt{2}/7$
0445 3101 2^x : $\sqrt{2}/2 - 3\sqrt{3}$
0445 6415 2^x : $3/2 + 3\sqrt{5}/4$
0445 7651 e^x : $\sqrt{2}/4 - 2\sqrt{3}$
0445 8746 J_2 : $7\ln 2/8$
0445 9129 J_2 : $(e)^{-1/2}$
0445 9824 cr : $(\sqrt{2}\pi/3)^{1/3}$
0446 0090 λ : $\exp(-(e+\pi))$
0446 2813 ζ : $7e/4$
0446 2871 $1/5 \times \ln(4/5)$
0446 3546 rt : (1,3,-2,9)
0446 3856 2^x : $(\ln 3)^{1/3}$
0446 3902 cr : $2\sqrt[3]{5}/3$
0446 7665 J_1 : $(\sqrt{5})^{-3}$
0446 8017 sr : $1/\ln(5/2)$
0446 8042 cr : $\sqrt{2}/2 + \sqrt{3}/4$
0447 0088 id : $4\sqrt{3}/3 - 3\sqrt{5}/2$
0447 2397 cr : $4\sqrt{2}/3 - \sqrt{5}/3$
0447 5601 J_2 : $((e+\pi)/2)^2$
0447 8821 sq : $1 + 2\pi$
0447 9801 sq : $e/2 - \pi/2$
0448 1176 cr : $5\sqrt[3]{5}$
0448 1455 e^x : $4\sqrt{2} + 4\sqrt{3}/3$
0448 3852 erf : $((e+\pi)/2)^{-3}$
0448 4163 cr : $3\sqrt{3} + 3\sqrt{5}/2$
0448 4237 $\ln(3) \times s\pi(2/5)$
0448 4251 J_2 : $(\sqrt{2}\pi)^{-1/3}$
0448 6479 e^x : $1/4 - 3\sqrt{5}/2$
0448 7309 erf : $(1/\pi)/8$
0448 7396 ζ : 19/4
0448 7489 Ψ : $(\sqrt{2}\pi/2)^{-1/3}$
0448 9308 rt : (9,1,-8,-3)
0448 9800 sq : $2\sqrt{2}/3 - 2\sqrt{3}/3$
0449 0027 J_2 : 14/23
0449 0264 rt : (7,8,-9,8)
0449 4544 ℓ_{10} : $8\ln 2/5$
0449 6679 Γ : $\exp(\sqrt{3})$
0449 7891 rt : (9,0,-6,4)
0449 8279 rt : (8,3,-9,-3)
0450 0122 tanh : $(3\pi/2)^{-2}$
0450 0123 at : $(3\pi/2)^{-2}$
0450 1370 ℓ_{10} : $3\zeta(3)/4$
0450 2751 e^x : $8e\pi/9$
0450 3163 id : $(3\pi/2)^{-2}$
0450 3871 θ_3 : $5\sqrt{2}/8$
0450 4685 sinh : $(3\pi/2)^{-2}$
0450 4687 as : $(3\pi/2)^{-2}$
0450 5431 Γ : $5\ln 2/8$
0450 6538 rt : (9,1,-4,5)
0450 6930 rt : (8,2,-7,-4)
0450 7273 10^x : P_{32}
0450 7398 θ_3 : $1/\ln(1/(2e))$
0450 7553 ζ : $5e\pi/9$
0450 7956 ℓ_{10} : $5\sqrt[3]{3}/8$
0450 9030 e^x : $(\ln 3/3)^{1/3}$
0450 9160 2^x : $2e/3 - 2\pi$
0451 1110 erf : 1/25
0451 1176 $1/3 \div \exp(2)$
0451 1520 2^x : $(1/\pi)/5$
0451 1539 Ψ : $5\pi^2/6$
0451 3923 2^x : $3\sqrt{3} - 2\sqrt{5}/3$
0451 4667 Γ : $(\sqrt{3}/2)^{1/2}$
0451 4946 2^x : $1/2 + 2\sqrt{2}$
0451 5251 rt : (9,2,-2,6)
0451 5397 10^x : $10\sqrt[3]{2}$
0451 5646 rt : (8,1,-5,-5)
0451 6041 rt : (7,4,-8,-4)
0451 6378 Ei : $\ln 2$
0451 6473 ln : $\sqrt[3]{23}$
0451 7575 J_0 : $9\pi/5$
0451 8015 2^x : $4 + e/2$
0451 8344 10^x : $3/2 + 2\pi$
0451 9200 cu : $2 - 3\pi/4$
0452 0516 cu : $2\sqrt{2}/3 - 3\sqrt{3}/4$
0452 0603 10^x : $\sqrt{3} + 2\sqrt{5}$
0452 2003 10^x : $4/3 + 2e/3$
0452 2744 ℓ_2 : $(\ln 3)^{1/3}$
0452 3205 e^x : $((e+\pi)/3)^{-1/2}$
0452 4029 rt : (9,3,0,7)
0452 4308 2^x : $8\pi/7$
0452 4428 rt : (8,0,-3,6)
0452 4613 J_2 : 11/18
0452 4649 e^x : $5\sqrt{5}/8$
0452 4827 rt : (7,3,-6,-5)
0452 4857 10^x : $3/4 - 2\pi/3$
0452 5226 rt : (6,6,-9,-4)
0452 5606 ℓ_2 : $2\sqrt{2} + 3\sqrt{3}/4$
0452 6235 $\ln(4/5) - \exp(3/5)$
0452 9442 e^x : $(\sqrt{5}/2)^{-3}$
0452 9700 sinh : 21/23
0453 0139 sq : $10\ln 2$
0453 0737 ln : $4e - \pi$
0453 1921 10^x : $1/3 - 3\sqrt{5}/4$
0453 2875 rt : (9,4,2,8)
0453 2948 ζ : $(\sqrt{2}\pi/3)^{-1/2}$
0453 3277 rt : (8,1,-1,7)
0453 3680 rt : (7,2,-4,-6)
0453 4060 cu : $5e\pi/8$
0453 4083 rt : (6,5,-7,-5)
0453 5042 ln : $3/4 + 2\pi/3$
0453 6010 2^x : $(5/2)^{-3}$
0453 6146 2^x : $1/4 - 3\pi/2$
0454 0626 Γ : $1/\ln((e+\pi)/2)$
0454 0760 J_1 : 1/11
0454 0931 10^x : $1 + 3\sqrt{3}$
0454 0965 J_0 : $\exp(e)$
0454 1702 10^x : $3\sqrt{2}/2 - 2\sqrt{3}$
0454 1788 rt : (9,5,4,9)
0454 2194 rt : (8,2,1,8)
0454 2326 tanh : 1/22
0454 2327 at : 1/22
0454 2584 J_1 : $2(1/\pi)/7$
0454 2601 rt : (7,1,-2,-7)
0454 3008 rt : (6,4,-5,-6)
0454 3415 rt : (5,8,-7,-5)
0454 4152 tanh : $(1/\pi)/7$
0454 4153 at : $(1/\pi)/7$
0454 5226 ℓ_{10} : $9\pi^2/8$
0454 5454 id : 1/22
0454 6066 sq : $7\pi^2$
0454 7019 sinh : 1/22
0454 7021 as : 1/22
0454 7254 10^x : $4 - 4\sqrt{3}/3$
0454 7284 id : $(1/\pi)/7$
0454 8745 e^x : $((e+\pi)/3)^{1/2}$
0454 8851 sinh : $(1/\pi)/7$
0454 8852 as : $(1/\pi)/7$
0454 9618 $t\pi$: $8\sqrt{2}/9$
0455 1181 rt : (8,3,3,9)
0455 1591 cr : 8/7
0455 2002 rt : (6,3,-3,-7)
0455 2413 rt : (5,9,-3,-1)
0455 2825 rt : (4,9,-9,-5)
0455 4901 Li_2 : $(3\pi/2)^{-2}$
0455 5393 cu : 5/14
0455 6256 sq : $4(e+\pi)/3$
0455 6313 J_1 : $(\sqrt{2}\pi/2)^{-3}$
0455 7405 sinh : 17/4
0455 9150 ζ : $\pi^2/7$
0456 0279 ℓ_{10} : $4/3 - \sqrt{3}/4$
0456 0651 rt : (7,1,2,9)
0456 0702 J_2 : $(\ln 2/3)^{1/3}$
0456 1066 rt : (6,2,-1,-8)
0456 1481 rt : (5,5,-4,-7)
0456 1897 rt : (4,8,-7,-6)
0456 1942 $\ln(2/5) \div \exp(3)$
0456 3666 e^x : $\pi^2/2$
0456 5126 $1/3 + \ln(3/4)$
0456 6832 ln : $4 - 2\sqrt{3}/3$
0456 8777 sq : $\sqrt[3]{5}/8$
0457 0202 rt : (6,1,1,-9)
0457 0621 rt : (5,4,-2,-8)
0457 1040 rt : (4,7,-5,-7)
0457 1532 Γ : $2\sqrt[3]{5}$
0457 2059 Ei : $7(e+\pi)/4$
0457 2231 id : $1/3 + 3\pi/2$
0457 2560 λ : $\exp(-13\pi/7)$
0457 3942 $\ln(2) \div \exp(e)$
0457 5749 ℓ_{10} : 10/9
0457 8149 Ei : $\sqrt[3]{1/3}$
0457 8470 ℓ_{10} : $5\sqrt[3]{2}/7$
0457 9832 rt : (5,3,0,-9)
0458 0256 rt : (4,6,-3,3)
0458 0591 Li_2 : $\pi/4$
0458 0680 rt : (3,9,-6,-7)
0458 1060 K : $4\sqrt[3]{3}/7$
0458 2966 J_0 : $4\sqrt{2}$
0458 4021 e^x : $4/3 + 3\sqrt{3}$
0458 4780 as : $3\sqrt[3]{3}/5$
0458 6096 J_2 : 8/13
0458 7318 rt : (1,8,8,9)
0458 7382 at : $e/4 + \pi/3$
0458 7486 2^x : $3/4 - 3\sqrt{3}$
0458 7991 sr : $5(e+\pi)/7$
0458 9545 rt : (4,7,1,7)
0458 9964 sr : $2 - e/3$
0458 9973 rt : (3,8,-4,-8)
0459 0891 sinh : $(\ln 2)^{-3}$
0459 1615 sr : $1/3 + 4\sqrt{5}$
0459 1774 $t\pi$: $2\sqrt{2}/3 - 4\sqrt{3}$
0459 1836 sq : 3/14
0459 2064 cu : $9(1/\pi)/8$
0459 4185 Ψ : $5e/9$
0459 4215 10^x : $4/3 + \sqrt{5}/4$
0459 5301 cu : $1 - 3\sqrt{5}/2$
0459 5802 J_0 : $6\sqrt{2}$
0459 6870 J_2 : $8\ln 2/9$
0459 8178 Li_2 : 1/22
0459 8325 J_1 : $2\pi^2/5$
0459 9341 rt : (3,7,-2,-9)
0459 9581 $t\pi$: $(2e/3)^{-1/2}$
0460 0050 Li_2 : $(1/\pi)/7$
0460 0055 at : 19/11
0460 1413 at : $2/3 + 3\sqrt{2}/4$
0460 3081 Γ : $\sqrt{3}/4$
0460 3523 e^x : $(\ln 3/3)^{-1/3}$
0460 5394 sq : $1 - \pi/4$
0460 6095 e^x : $(3\pi/2)^{-2}$
0460 7785 e^x : $1 - 3e/2$
0460 8007 ℓ_{10} : $\exp(1/(3\pi))$
0460 8191 cr : $\sqrt[3]{3}/2$
0460 8493 tanh : $\ln(\pi/3)$
0460 8494 at : $\ln(\pi/3)$
0460 8668 cr : $\ln(\pi)$
0460 8785 rt : (3,3,-3,7)
0460 9222 rt : (2,9,-3,-9)
0461 1046 $t\pi$: P_{59}
0461 1759 id : $\ln(\pi/3)$
0461 3394 sinh : $\ln(\pi/3)$
0461 3396 as : $\ln(\pi/3)$
0461 5393 $t\pi$: $\sqrt[3]{23/2}$
0461 6128 ζ : $10\sqrt{2}/3$
0461 7419 sq : $5\pi/9$
0461 7451 e^x : $4 + 3\pi$
0461 8077 2^x : $1 - 2e$
0461 8747 rt : (2,4,-9,-8)
0461 9738 e^x : $3\sqrt{2}/2 - 3\sqrt{3}$
0462 0910 Ei : 19/18
0462 1737 J_2 : $9(e+\pi)/10$
0462 2154 ζ : $3\pi/2$
0462 2690 ℓ_{10} : $1/4 + 4e$
0462 2766 $s\pi$: $4e + \pi$
0462 3721 id : $4\sqrt{3} + \sqrt{5}/2$
0462 4494 sq : $4/3 - 3\pi/4$
0462 5457 J_2 : $3\sqrt[3]{3}/7$
0462 6754 sq : $1 + \sqrt{5}/3$
0462 7713 $t\pi$: $4e + \pi$
0462 8350 rt : (2,5,-3,1)
0462 8537 J_2 : $\exp(-\sqrt[3]{3}/3)$
0462 8601 2^x : $9\sqrt[3]{3}/5$
0463 2278 cu : $1 - e/2$
0463 3366 ℓ_2 : P_{42}
0463 3667 rt : (1,8,0,3)
0463 4632 $\ln(3) + \exp(2/3)$
0463 4693 Li_2 : $5\sqrt{2}/9$
0463 5204 10^x : $\zeta(9)$
0463 5380 sq : $4/3 - \sqrt{5}/2$
0463 5769 $t\pi$: $\sqrt{2}/2 - 3\sqrt{3}$
0463 9093 J_2 : 13/21

0464 1013 $\ell_2 : 1/3+\sqrt{3}$
0464 1164 sq : $1+4\pi$
0464 1902 $\Gamma : 5\sqrt[3]{2}/3$
0464 2552 $Li_2 : 11/14$
0464 6075 $10^x : 3-2\pi/3$
0464 7389 cr : 9π
0464 8584 $\Gamma : 21/10$
0465 0343 $e^x : 1/22$
0465 2001 ln : 21/22
0465 2258 $e^x : (1/\pi)/7$
0465 3623 sr : 23/21
0465 3764 $\Psi : 10/3$
0465 6453 $\ell_{10} : \exp(3\pi/2)$
0465 6511 $J_2 : 5$
0466 0347 $\Psi : \sqrt[3]{5}/5$
0466 0935 cr : $3/2-\sqrt{2}/4$
0466 3413 $\Gamma : 13/14$
0466 3553 $e^x : 9(1/\pi)/4$
0466 4375 ln : $1/4+3\sqrt{3}/2$
0466 4723 cu : $2\sqrt[3]{2}/7$
0466 5341 sr : $4\pi/3$
0466 5592 $2^x : 6\sqrt{5}$
0466 5600 sq : $(5/3)^{-3}$
0466 6049 $Li_2 : \ln(\pi/3)$
0466 8032 rt : (1,6,6,6)
0467 0549 $Ei : 5/13$
0467 0931 $3/4 \times \exp(1/3)$
0467 2324 $J_2 : 1/\ln(5)$
0467 7754 sq : $3\sqrt{2}/2+2\sqrt{5}/3$
0467 7788 $e^x : 7(1/\pi)/2$
0467 8477 $J_0 : 10e$
0468 0037 tπ : $7(1/\pi)/3$
0468 0160 $2^x : \exp(-e)$
0468 0251 $Ei : 3\zeta(3)/10$
0468 0380 $10^x : 1/\ln(\sqrt{2}/3)$
0468 0524 $s\pi(1/4) \div s\pi(\sqrt{5})$
0468 0923 $10^x : \sqrt{3}+\sqrt{5}+\sqrt{6}$
0468 3201 sr : $3+2\pi$
0468 3766 $10^x : 2e/3-\pi$
0468 4212 tπ : $3\sqrt{2}-\sqrt{3}$
0468 5952 at : $6\sqrt[3]{3}/5$
0468 6644 $2^x : (e/3)^{-1/3}$
0468 7500 sq : $\sqrt{3}/8$
0468 7718 rt : (9,1,-9,-2)
0468 9632 cu : $3\zeta(3)/10$
0469 0363 $\ln(\sqrt{2}) \div \exp(2)$
0469 0962 $\Psi : \sqrt{2}$
0469 1817 $\ell_{10} : 2\pi/7$
0469 2539 cr : $e-\pi/2$
0469 3262 sq : $3e+2\pi/3$
0469 3474 rt : (1,8,7,2)
0469 4321 $10^x : 3/2-2\sqrt{2}$
0469 4514 $erf : (\ln 2/2)^3$
0469 5167 $e^x : 1/\ln(\sqrt[3]{3}/2)$
0469 6125 $\theta_3 : 3\sqrt[3]{2}/8$
0469 7194 $J_1 : \ln(\ln 3)$
0469 7444 rt : (9,0,-7,3)
0469 7801 $2^x : 2+2e/3$
0469 8341 $10^x : 3-\sqrt{5}/4$
0469 8521 $e^x : \sqrt{2}-2\sqrt{5}$
0469 8587 rt : (7,9,-7,8)
0469 8860 $erf : 1/24$
0469 8926 $2^x : 3e-4\pi$
0470 1998 $Ei : \sqrt[3]{3}/4$
0470 2178 $10^x : 3/4+3e$
0470 2391 ln : $(\ln 3)^{1/2}$
0470 2911 $\zeta : \sqrt{22}$
0470 3405 $\ell_{10} : 3-4\sqrt{2}/3$
0470 5506 rt : (1,8,-1,-5)
0470 6351 cr : $2e+\pi$
0470 7249 rt : (9,1,-5,4)
0470 7713 rt : (8,2,-8,-3)
0470 7990 sr : $2\sqrt{2}-\sqrt{3}$
0470 8731 $J_2 : 9\ln 2/10$
0471 2256 $\zeta : 4(e+\pi)/5$
0471 3005 $J_2 : \exp(-\sqrt{2}/3)$
0471 3331 $J : 7(1/\pi)/10$
0471 3772 $e^x : 4\sqrt{2}+2\sqrt{5}$
0471 6255 at : $1-\pi/3$
0471 7134 rt : (9,2,-3,5)
0471 7603 rt : (8,1,-6,-4)
0471 8071 rt : (7,4,-9,-3)
0471 8956 $5 \times \ln(5)$
0471 9755 id : $\pi/3$
0472 0729 $Ei : (\ln 3/2)^3$
0472 1509 as : $1-\pi/3$
0472 2335 sπ : $(1/(3e))^2$
0472 2652 $\ell_2 : 3e/2+4\pi/3$
0472 5794 $J_2 : 5/8$
0472 5897 sq : 5/23
0472 7100 rt : (9,3,-1,6)
0472 7574 rt : (8,0,-4,5)
0472 7609 tπ : $(1/(3e))^2$
0472 8047 rt : (7,3,-7,-4)
0472 9412 $2^x : 1/15$
0473 0257 sinh : $(\pi)^{1/3}$
0473 0492 ln : $5\sqrt[3]{5}/3$
0473 1477 $J_0 : \sqrt{2}+\sqrt{3}+\sqrt{5}$
0473 1936 $e^x : (e)^{-1/3}$
0473 2985 $\ell_{10} : (\sqrt[3]{3}/2)^{1/3}$
0473 2990 sq : $7e\pi/3$
0473 4292 tπ : $1/2+3\sqrt{2}$
0473 4874 ln : $\sqrt{3}+\sqrt{5}/2$
0473 5724 $10^x : 22/23$
0473 7063 sπ : $7\sqrt[3]{5}/2$
0473 7149 rt : (9,4,1,7)
0473 7627 rt : (8,1,-2,6)
0473 8106 rt : (7,2,-5,-5)
0473 8584 rt : (6,5,-8,-4)
0474 0176 at : $5\ln 2/2$
0474 0546 $2^x : 4\sqrt{3}/3-3\sqrt{5}$
0474 1060 sinh : 11/6
0474 2248 $\ell_2 : (e/3)^{1/3}$
0474 2387 tπ : $7\sqrt[3]{5}/2$
0474 3494 sq : $3/2+2\sqrt{3}/3$
0474 4116 cu : $3\sqrt{3}/2-\sqrt{5}$
0474 6355 $\ell_{10} : 3+3e$
0474 7227 cu : $e-3\pi/4$
0474 7281 rt : (9,5,3,8)
0474 7534 $\ln(3) \div \exp(\pi)$
0474 7765 rt : (8,2,0,7)
0474 7909 $10^x : \sqrt{2}/4-3\sqrt{5}/4$
0474 8248 rt : (7,1,-3,-6)
0474 8732 rt : (6,4,-6,-5)
0474 9216 rt : (5,8,-8,-7)
0475 2690 $2^x : 1/2-\sqrt{3}/4$
0475 6069 $\Psi : 6\sqrt[3]{2}/5$
0475 6507 $J_1 : 2/21$
0475 7056 sq : $((e+\pi)/2)^{1/3}$
0475 7499 rt : (9,6,5,9)
0475 7987 rt : (8,3,2,8)
0475 8308 tanh : 1/21
0475 8310 at : 1/21
0475 8476 rt : (7,0,-1,7)
0475 8965 rt : (6,3,-4,-6)
0475 9454 rt : (5,9,-4,-3)
0476 0280 $10^x : -\sqrt{3}-\sqrt{5}+\sqrt{7}$
0476 0354 ln : $\sqrt{2}/3+\sqrt{3}/3$
0476 1180 $\ln(2/5) - s\pi(\sqrt{2})$
0476 1904 id : 1/21
0476 3704 sinh : 1/21
0476 3706 as : 1/21
0476 4091 $\Psi : 3/23$
0476 4787 $\Gamma : 3\sqrt[3]{3}/10$
0476 5972 $\exp(3/5) + \exp(4/5)$
0476 6480 $10^x : 4e+2\pi$
0476 7408 $\ell_{10} : 1/4+\sqrt{3}/2$
0476 8296 rt : (8,4,4,9)
0476 8387 $e^x : 3-4\sqrt{2}/3$
0476 8659 id : $e/3+\pi$
0476 8790 rt : (7,1,1,8)
0476 8931 $J_0 : 12$
0476 8955 cr : 23/20
0476 9207 $J_1 : 3(1/\pi)/10$
0476 9284 rt : (6,2,-2,-7)
0476 9779 rt : (5,5,-5,-6)
0477 0273 rt : (4,8,-8,-5)
0477 3345 $10^x : (2\pi/3)^{-1/3}$
0477 4443 $J_2 : \pi/5$
0477 7319 rt : (2,6,-7,6)
0477 7532 $10^x : 1/4-\pi/2$
0477 7688 $J_2 : 4\sqrt{2}/9$
0477 8137 sq : $3/4+4\sqrt{2}$
0477 9192 rt : (7,2,3,9)
0477 9302 $Ei : 2\sqrt{3}/9$
0477 9692 rt : (6,1,0,-8)
0478 0192 rt : (5,4,-3,-7)
0478 0692 rt : (4,7,-6,-6)
0478 4609 $\exp(1/3) + \exp(\sqrt{3})$
0478 5026 $\Psi : 6/23$
0478 6387 $J : \exp(-18\pi/17)$
0478 7349 id : $7\sqrt[3]{3}/2$
0478 9546 $3/5 \times \exp(5)$
0479 0189 rt : (6,0,2,9)
0479 0658 sπ : $(\pi/3)^{-1/3}$
0479 0694 rt : (5,3,-1,-8)
0479 1199 rt : (4,6,-4,1)
0479 1705 rt : (3,9,-7,-6)
0479 1922 $\zeta : 14/3$
0479 2232 $Ei : \exp(5/3)$
0479 4701 ln : $(\sqrt{3}/2)^{1/3}$
0479 6165 tπ : $(\pi/3)^{-1/3}$
0479 6876 $\exp(2/3) \div s\pi(2/5)$
0479 8597 $J_2 : \sqrt[3]{2}/2$
0480 1287 rt : (5,2,1,-9)
0480 1527 as : $5\ln 2/4$
0480 1798 rt : (4,7,0,5)
0480 2310 rt : (3,8,-5,-7)
0480 4584 rt : (1,8,6,-5)
0480 5628 cu : $3e-3\pi$
0480 5879 $2^x : 1/3-3\pi/2$
0480 7389 rt : (7,9,-8,6)
0480 8414 cu : 4/11
0480 8598 $\ell_{10} : 3\sqrt{2}+4\sqrt{3}$
0481 0348 $Ei : 7\sqrt{5}/9$
0481 0729 cr : $2\sqrt{2}-3\sqrt{5}/4$
0481 2080 $\ell_2 : P_{32}$
0481 2490 rt : (4,8,4,9)
0481 3007 rt : (3,7,-3,-8)
0481 4223 cu : $8(1/\pi)/7$
0481 4707 id : $(\ln 3)^{1/2}$
0481 6110 sinh : $(e+\pi)/4$
0481 7294 $2^x : 5\ln 2$
0481 7459 sr : $4\sqrt{2}/3+4\sqrt{3}/3$
0481 7611 $\exp(3/5) - s\pi(e)$
0481 9827 $Li_2 : 1/21$
0482 0009 at : $1/2-\sqrt{5}$
0482 2457 $J_2 : 12/19$
0482 3798 rt : (3,6,-1,-9)
0482 4321 rt : (2,9,-4,-8)
0482 7883 $10^x : 8\sqrt[3]{3}/5$
0482 7961 $\ell_2 : \sqrt{3}/3+2\sqrt{5}/3$
0482 8878 sr : $\sqrt{2}/4+\sqrt{5}/3$
0483 0467 $\ell_{10} : 17/19$
0483 0889 cr : $4\sqrt{2}/3+3\sqrt{5}$
0483 1978 $\Gamma : 7\zeta(3)/4$
0483 3406 rt : (2,8,-9,8)
0483 4307 tπ : $3\sqrt{2}-3\sqrt{3}/2$
0483 4685 rt : (3,4,0,9)
0483 4769 ln : $2-\pi/3$
0483 5214 rt : (2,8,-2,-9)
0483 5402 $J_2 : (5/2)^{-1/2}$
0483 8346 $2^x : \sqrt{2}+4\sqrt{5}/3$
0483 8377 $J_0 : 5/2$
0483 9969 sq : $3\sqrt{3}-4\sqrt{5}$
0484 1274 $e^x : 3-\pi/3$
0484 5500 $\ell_{10} : \sqrt{5}/2$
0484 6204 rt : (2,5,-4,-2)
0484 6247 $Ei : \sqrt{2}$
0484 8150 as : 13/15
0484 8618 $Ei : 5\ln 2/9$
0485 0113 $10^x : 4/3+2e$
0485 1803 sinh : $\ln(\ln 2/3)$
0485 2039 $\ln(3) \times \exp(5)$
0485 7256 $\zeta : \ln(\sqrt[3]{3})$
0485 7294 rt : (2,6,2,7)
0485 9810 $\exp(3/5) \times \exp(\sqrt{5})$
0486 2232 $\Gamma : (\sqrt[3]{2})^{-1/3}$
0486 2334 $1/5 \div \exp(\sqrt{2})$
0486 4813 sπ : $(\pi/3)^{1/3}$

0486 5598 sr : $2/3+\sqrt{3}/4$
0486 6780 Ei : $3\sqrt{5}$
0487 0580 $t\pi$: $(\pi/3)^{1/3}$
0487 2127 $2/5+\exp(1/2)$
0487 2586 ℓ_{10} : $1/3+\pi/4$
0487 2750 ln : $5\sqrt[3]{2}/6$
0487 3134 erf : $\exp(-\pi)$
0487 3176 10^x : $1/2-2e/3$
0487 3365 sq : $4+2e$
0487 3977 cu : $\sqrt{2}/3+3\sqrt{3}$
0487 5470 e^x : $4+\sqrt{5}/3$
0487 5478 id : $\sqrt{2}/3+\sqrt{3}/3$
0487 6372 e^x : $1/3-3\sqrt{5}/2$
0487 7104 e^x : $1/21$
0487 8209 sr : $7\sqrt{2}/9$
0487 9016 ln : $20/21$
0488 0330 rt : (1,4,-8,-9)
0488 0884 sr : $11/10$
0488 1071 cu : $2\sqrt{2}/3-\sqrt{3}/3$
0488 1165 ζ : $8(1/\pi)/3$
0488 1282 cu : $3\sqrt{2}+\sqrt{5}/2$
0488 3769 Γ : $2\ln 2/9$
0488 4215 cr : $4\sqrt[3]{3}/5$
0488 4388 $\ln(\sqrt{3})-\exp(4)$
0488 4973 cu : $2/3+2\pi/3$
0488 5096 rt : (1,9,0,4)
0488 5624 cr : $15/13$
0488 8836 $\sqrt{5}\times\ln(2/5)$
0489 0538 $\ln(2/5)\times\ln(\pi)$
0489 1733 rt : (2,6,-8,2)
0489 3306 J_2 : $7/11$
0489 4348 $1-s\pi(2/5)$
0489 5542 J_2 : $\ln(\sqrt[3]{4}/3)$
0489 6518 e^x : $10\sqrt[3]{2}/7$
0489 7038 at : $7\sqrt{5}/9$
0489 7111 J_2 : $2(1/\pi)$
0489 7954 ℓ_2 : $3\sqrt{2}/4-\sqrt{3}/3$
0489 9751 Ψ : $(\sqrt[3]{5})^{-1/2}$
0490 1996 sq : $1/4-\sqrt{2}/3$
0490 2906 erf : $1/23$
0490 3810 id : $1/4-3\sqrt{3}/4$
0490 3883 $\ln(\sqrt{2})-\exp(1/3)$
0490 4339 Ψ : $13/17$
0490 4547 10^x : $1-4\sqrt{3}/3$
0490 5202 J_0 : $17/3$
0490 6767 $s\pi$: $(4)^{-3}$
0490 6880 ℓ_{10} : $2/3-\sqrt{3}/3$
0490 7051 tanh : $(\ln 3/3)^3$
0490 7053 at : $(\ln 3/3)^3$
0491 0996 id : $(\ln 3/3)^3$
0491 1506 id : $(\sqrt{3}/2)^{-1/3}$
0491 1972 Ψ : $1/\ln(\sqrt[3]{3}/3)$
0491 2684 $t\pi$: $(4)^{-3}$
0491 2970 sinh : $(\ln 3/3)^3$
0491 2972 as : $(\ln 3/3)^3$
0491 4341 rt : (9,0,-8,2)
0491 5373 id : $4e/3+3\pi$
0491 7965 sq : $1/3+3e$
0491 8836 λ : $\exp(-11\pi/6)$
0491 9080 ℓ_{10} : $8\sqrt[3]{2}/9$
0492 1147 rt : (7,9,-9,4)
0492 1677 cr : $4\sqrt{3}+3\sqrt{5}/4$
0492 1818 2^x : $\ln 2/10$
0492 3355 sr : $\sqrt{2}/4+4\sqrt{5}$
0492 3431 cu : $\ln(\ln 2)$
0492 4623 J_1 : $\ln(e/3)$
0492 5509 rt : (9,1,-6,3)
0492 5801 Ei : $9/25$
0492 5924 cu : $e/3-3\pi/4$
0492 6062 rt : (8,2,-9,-2)
0492 7853 10^x : P_{42}
0492 8002 cr : $5\ln 2/3$
0493 0614 ln : $(e/3)^{1/2}$
0493 0700 J_2 : $2\sqrt{5}/7$
0493 2594 sr : $10\sqrt[3]{2}/3$
0493 4417 ζ : $(5/3)^3$
0493 4783 Ei : $2\sqrt[3]{2}/7$
0493 5475 at : $1/4+2\sqrt{5}/3$
0493 6777 rt : (9,2,-4,4)
0493 7336 rt : (8,1,-7,-3)
0493 8271 sq : $2/9$
0493 8728 rt : (6,6,-9,8)
0493 8948 Ψ : $24/17$
0493 9015 sr : $21/5$
0494 3945 $\exp(3)-s\pi(\sqrt{2})$
0494 4833 rt : (1,9,-1,-5)
0494 4985 J_1 : $\ln 2/7$
0494 5347 e^x : $\sqrt{2}/4+\sqrt{3}$
0494 6461 ℓ_{10} : $1/3+\sqrt{5}/4$
0494 6826 $\ln(\pi)\div\exp(\pi)$
0494 6847 ℓ_{10} : $1/3+4e$
0494 7459 J_2 : $16/25$
0494 8145 rt : (9,3,-2,5)
0494 8622 id : $\sqrt{3}/3+2\sqrt{5}$
0494 8710 rt : (8,0,-5,4)
0494 9014 ln : $1/2+3\pi/4$
0494 9102 $3/4\div\exp(e)$
0494 9276 rt : (7,3,-8,-3)
0495 2357 id : $3\sqrt{2}/2+4\sqrt{3}$
0495 2698 Γ : $(\sqrt[3]{5}/2)^{1/2}$
0495 2860 sinh : $beta(2)$
0495 3689 cr : $1/4+e/3$
0495 4034 $t\pi$: $3\sqrt{3}/7$
0495 5089 ℓ_2 : P_{50}
0495 9616 rt : (9,4,0,6)
0496 0188 rt : (8,1,-3,5)
0496 0760 rt : (7,2,-6,-4)
0496 1332 rt : (6,5,-9,-3)
0496 1619 $\ln(2)+\exp(\sqrt{5})$
0496 2394 J_2 : $4\sqrt[3]{3}/9$
0496 2946 id : $\sqrt{20}\pi$
0496 4737 sq : $7(1/\pi)/10$
0496 4746 e^x : $9/5$
0496 6000 J_2 : $3\sqrt[3]{5}/8$
0496 6880 $\ln(3)+s\pi(2/5)$
0496 7925 2^x : $1/4+\pi/4$
0496 8468 cu : $e/4-\pi/3$
0496 9580 θ_3 : $(3\pi)^{-1/3}$
0497 0725 $t\pi$: $(e+\pi)/4$
0497 1191 rt : (9,5,2,7)
0497 1742 id : $5\pi^2/7$
0497 1769 rt : (8,2,-1,6)
0497 2347 rt : (7,1,-4,-5)
0497 2644 Li_2 : $(\ln 3/3)^3$
0497 2926 rt : (6,4,-7,-4)
0497 3506 rt : (5,8,-9,-9)
0497 4597 tanh : $(e)^{-3}$
0497 4599 at : $(e)^{-3}$
0497 5006 10^x : $3/2+2e/3$
0497 5976 e^x : $\sqrt{2}/4-3\sqrt{5}/2$
0497 6665 10^x : $4/3+3\sqrt{3}/2$
0497 6730 10^x : $4e/3+3\pi$
0497 6941 ζ : $8\sqrt{3}/3$
0497 7963 $t\pi$: $(3\pi/2)^{-1/2}$
0497 8706 id : $(e)^{-3}$
0498 0763 sinh : $(e)^{-3}$
0498 0765 as : $(e)^{-3}$
0498 1420 sinh : $6\sqrt[3]{5}/7$
0498 2212 ln : $7/20$
0498 2871 rt : (9,6,4,8)
0498 3456 rt : (8,3,1,7)
0498 3499 Li_2 : $5\sqrt[3]{2}/8$
0498 4041 rt : (7,0,-2,6)
0498 4443 Ei : $3e\pi/5$
0498 4627 rt : (6,3,-5,-5)
0498 4796 $s\pi$: $8\sqrt[3]{2}/5$
0498 4984 cu : $3/4-\sqrt{5}/2$
0498 5213 rt : (5,9,-5,-5)
0498 6473 10^x : $2/3+3\pi/4$
0498 6880 Γ : $4\ln 2/3$
0498 8612 J_0 : $10\ln 2/3$
0498 8805 id : $\sqrt{2}-2\sqrt{3}$
0498 9594 ℓ_{10} : $7\sqrt[3]{3}/9$
0499 0193 J_2 : $9/14$
0499 1001 $t\pi$: $8\sqrt[3]{2}/5$
0499 3420 id : $5\sqrt[3]{2}/6$
0499 3752 J_1 : $1/10$
0499 4482 2^x : $4/3-4\sqrt{2}$
0499 4659 rt : (9,7,6,9)
0499 5251 rt : (8,4,3,8)
0499 5837 tanh : $1/20$
0499 5839 at : $1/20$
0499 5843 rt : (7,1,0,7)
0499 6436 rt : (6,2,-3,-6)
0499 6710 sr : $7\sqrt[3]{2}/8$
0499 7029 rt : (5,5,-6,-5)
0499 7229 sr : $4e-\pi/2$
0499 7622 rt : (4,8,-9,-4)
0499 9878 e^x : $(\sqrt[3]{3}/2)^{-1/3}$
0500 0000 id : $1/20$
0500 0240 $\exp(\sqrt{3})\div\exp(1/3)$
0500 0728 cu : $7/19$
0500 0742 10^x : $4(e+\pi)/3$
0500 1596 e^x : $4+3\pi/4$
0500 2083 sinh : $1/20$
0500 2085 as : $1/20$
0500 2798 J_0 : $9\sqrt[3]{2}/2$
0500 2926 2^x : $3/4-e/4$
0500 3006 ℓ_{10} : $9\ln 2/7$
0500 5796 $s\pi$: $\sqrt{3}/2+\sqrt{5}/2$
0500 7155 rt : (8,5,5,9)
0500 7755 rt : (7,2,2,8)
0500 8154 cr : $22/19$
0500 8354 rt : (6,1,-1,-7)
0500 8504 sq : $\sqrt{2}/4-\sqrt{3}/3$
0500 8954 rt : (5,4,-4,-6)
0500 9554 rt : (4,7,-7,-5)
0501 1637 rt : (2,6,-9,-2)
0501 2080 $t\pi$: $\sqrt{3}/2+\sqrt{5}/2$
0501 2396 ln : $3\sqrt{2}/4+3\sqrt{5}$
0501 5198 e^x : $\sqrt{2}/3-2\sqrt{3}$
0501 6119 Γ : $\exp(\sqrt{5}/3)$
0501 7166 ℓ_{10} : $(\sqrt[3]{2})^{1/2}$
0501 8566 ℓ_2 : $1+\pi$
0501 9778 rt : (7,3,4,9)
0502 0168 sr : $(\sqrt{5}/3)^{-1/3}$
0502 0384 rt : (6,0,1,8)
0502 0448 sinh : $5/2$
0502 0991 rt : (5,3,-2,-7)
0502 1344 id : $(2\pi)^3$
0502 1599 rt : (4,6,-5,-1)
0502 2207 rt : (3,9,-8,-5)
0502 3501 K : $19/23$
0502 3556 J_2 : $1/\ln(3\pi/2)$
0502 3918 sq : P_{31}
0502 5199 2^x : $2(1/\pi)/9$
0502 6532 $\ln(2/5)+s\pi(1/3)$
0502 6811 e^x : $8\sqrt[3]{3}/5$
0502 8144 as : $\ln(\sqrt[3]{2}/3)$
0502 9387 10^x : $(\sqrt{2}\pi)^{1/3}$
0502 9718 2^x : $2/3+3\sqrt{2}$
0503 2528 rt : (6,1,3,9)
0503 3143 rt : (5,2,0,-8)
0503 3584 e^x : $(\ln 3/3)^3$
0503 3757 rt : (4,7,-1,3)
0503 4372 rt : (3,8,-6,-6)
0503 6242 e^x : $\exp(e/2)$
0504 0007 sq : $4\sqrt{3}+3\sqrt{5}/4$
0504 2086 Li_2 : $(e)^{-3}$
0504 2618 sq : $5\pi^2/8$
0504 5410 rt : (5,1,2,-9)
0504 6032 rt : (4,8,3,7)
0504 6654 rt : (3,7,-4,-7)
0504 7233 θ_3 : $9/19$
0504 7236 J_0 : $4\sqrt{3}/3$
0504 8189 Γ : $(\sqrt{2}\pi)^{1/2}$
0504 9817 at : $5\pi/9$
0505 0149 sr : $3/4+\sqrt{2}/4$
0505 0278 ln : $7\zeta(3)/8$
0505 0478 at : $1+\sqrt{5}/3$
0505 1025 sq : $\sqrt{2}/2-\sqrt{3}$
0505 1738 ζ : $23/5$
0505 2120 ln : $3/2+e/2$
0505 2417 $t\pi$: $4\sqrt{3}-4\sqrt{5}$
0505 3244 ln : $-\sqrt{5}+\sqrt{6}+\sqrt{7}$
0505 3335 J_2 : $11/17$
0505 4191 id : $(e/3)^{-1/2}$
0505 4517 sinh : $11/12$
0505 5286 ℓ_2 : $4e/3-\pi$
0505 8425 rt : (4,3,1,-9)

0505 9055 rt : (3,6,-2,-8)
0505 9560 J_1 : $(\pi)^{-2}$
0505 9685 rt : (2,9,-5,-7)
0506 0671 ln : $2e/3+\pi/3$
0506 1729 tanh : $(\sqrt{2}\pi)^{-2}$
0506 1731 at : $(\sqrt{2}\pi)^{-2}$
0506 1963 sr : $3\sqrt{3}/2+3\sqrt{5}$
0506 3929 Li_2 : 1/20
0506 4906 ℓ_{10} : $2e/3+3\pi$
0506 4934 sq : $3+\sqrt{2}/3$
0506 6059 id : $(\sqrt{2}\pi)^{-2}$
0506 6141 θ_3 : $(2\pi)^{-2}$
0506 8226 sinh : $(\sqrt{2}\pi)^{-2}$
0506 8228 as : $(\sqrt{2}\pi)^{-2}$
0506 8338 J_2 : $(\sqrt[3]{2}/3)^{1/2}$
0506 9385 $3/5-\ln(\sqrt{3})$
0507 0126 2^x : $1/4+e/2$
0507 0390 Γ : 12/13
0507 1575 rt : (3,5,0,-9)
0507 2214 rt : (2,8,-3,-8)
0507 5663 2^x : 1/14
0507 6771 J_0 : $5\pi^2/2$
0507 7843 erf : $(3\pi/2)^{-2}$
0508 3037 $\exp(5)\div s\pi(2/5)$
0508 4865 rt : (2,7,-1,-9)
0508 6808 ℓ_2 : $\sqrt{3}-2\sqrt{5}/3$
0508 9470 10^x : $e/2-\pi/3$
0508 9705 $\ln(3/4)\div\exp(\sqrt{3})$
0509 0272 10^x : $2\sqrt{2}/3-\sqrt{5}$
0509 0470 J_2 : $3\sqrt{3}/8$
0509 0637 $\exp(e)\div\exp(2)$
0509 2170 Ψ : $7\sqrt{3}/8$
0509 3311 sq : $1/3-\sqrt{5}/4$
0509 5050 at : $\sqrt[3]{16/3}$
0509 7641 rt : (2,6,1,4)
0509 7743 J_2 : 13/20
0509 8296 rt : (1,9,-2,-9)
0510 3448 J_0 : $4e\pi$
0510 4038 10^x : $1/\ln(\sqrt{2})$
0510 4573 at : $1/3+\sqrt{2}$
0510 4727 e^x : $(e)^{-3}$
0510 6902 ℓ_{10} : P_{43}
0510 8374 J_2 : $6\pi^2/5$
0511 4353 J_2 : $(e+\pi)/9$
0511 4581 sr : $7\zeta(3)/2$
0511 5229 Ei : $(\sqrt{5}/2)^{1/2}$
0511 5252 ℓ_{10} : 8/9
0511 7379 rt : (3,6,-7,8)
0511 9384 e^x : $3/2-2\sqrt{5}$
0512 1424 cu : $7(1/\pi)/6$
0512 2149 ζ : $\sqrt{21}$
0512 2880 as : $1/4-\sqrt{5}/2$
0512 3752 Ψ : $\pi^2/7$
0512 4250 rt : (1,4,-9,-7)
0512 5466 erf : 1/22
0512 7109 e^x : 1/20
0512 7526 erf : $(1/\pi)/7$
0512 9329 ln : 19/20
0512 9429 ℓ_{10} : $4\sqrt{3}/3+4\sqrt{5}$
0513 0678 J_2 : 15/23
0513 0736 ln : $5\sqrt[3]{5}/9$
0513 1323 10^x : $3\sqrt{3}-2\sqrt{5}/3$
0513 1496 sr : 21/19
0513 1708 Li_2 : $(\sqrt{2}\pi)^{-2}$
0513 2215 $3/5\times\exp(3)$
0513 2364 ℓ_{10} : $3\sqrt{2}-3\sqrt{5}/2$
0513 2935 2^x : $\ln(5)$
0513 4968 id : $4\sqrt{2}-3\sqrt{5}$
0513 6091 2^x : $2-2\pi$
0513 7424 rt : (1,6,4,-4)
0514 0929 2^x : $2/3+2\sqrt{2}/3$
0514 2074 $1/3\times\exp(e)$
0514 3853 10^x : $3/4-3e/4$
0514 4232 J_0 : $8\sqrt[3]{3}/5$
0514 5007 sq : $(1/(3e))^{-1/3}$
0514 5598 cr : $(\pi/2)^{1/3}$
0514 5927 e^x : $3\sqrt{2}/4+\sqrt{5}/4$
0514 6222 $1\div s\pi(2/5)$
0514 8552 ζ : 8/13
0515 0766 rt : (9,0,-9,1)
0515 2356 λ : $\exp(-20\pi/11)$
0515 7388 e^x : $1/3+4\sqrt{3}/3$
0516 0681 e^x : $1/2-2\sqrt{3}$
0516 1516 id : $1/3+e$
0516 1880 ℓ_2 : P_{22}
0516 3329 e^x : $\sqrt{2}-\sqrt{3}-\sqrt{7}$
0516 3550 rt : (9,1,-7,2)
0516 4397 e^x : $\sqrt{2}+2\sqrt{3}/3$
0516 4681 cr : $7\pi^2/8$
0516 5021 at : 7/4
0516 5289 sq : 5/22
0516 5371 $t\pi$: $(\pi/2)^{-3}$
0516 7320 J_0 : $(2\pi)^{1/2}$
0516 9448 sq : $5(1/\pi)/7$
0516 9627 sr : $2\sqrt{3}+\sqrt{5}/3$
0517 1941 sq : $\exp(-\sqrt{2}\pi/3)$
0517 4474 at : $4\sqrt{3}/3-\sqrt{5}/4$
0517 5396 sq : $9(1/\pi)/2$
0517 6083 cu : $\sqrt{5}/6$
0517 6457 rt : (9,2,-5,3)
0517 6460 sinh : $\exp(\sqrt{5}/3)$
0517 7129 rt : (8,1,-8,-2)
0517 8166 sq : $4/3+e/4$
0517 9979 id : $7\zeta(3)/8$
0518 2327 Ψ : $4(e+\pi)/7$
0518 3482 cr : $4\sqrt{2}+4\sqrt{5}/3$
0518 3772 $s\pi$: $\sqrt{2}/2+4\sqrt{3}/3$
0518 5353 2^x : $\zeta(5)$
0518 5564 e^x : $7\sqrt[3]{5}/3$
0518 6057 id : $\exp((e+\pi)/3)$
0518 9490 rt : (9,3,-3,4)
0519 0170 rt : (8,0,-6,3)
0519 0750 $t\pi$: $\sqrt{2}/2+4\sqrt{3}/3$
0519 0850 rt : (7,3,-9,-2)
0519 1797 10^x : $(\sqrt[3]{3}/2)^{-1/2}$
0519 1863 rt : (6,7,-7,8)
0519 3775 Γ : $10\zeta(3)/3$
0519 3776 e^x : $2/3-4e/3$
0519 4717 cu : $7\pi^2/5$
0519 5904 cu : $1/\ln(\sqrt[3]{5}/2)$
0519 6567 ln : $3/2+2\pi$
0519 6578 e^x : $(\sqrt{2}\pi)^{-2}$
0519 7113 id : $6\sqrt[3]{5}/5$
0519 7153 sq : $(2\pi/3)^{-2}$
0520 0126 erf : $\ln(\pi/3)$
0520 2650 rt : (9,4,-1,5)
0520 3339 rt : (8,1,-4,4)
0520 3432 sinh : 22/15
0520 4028 rt : (7,2,-7,-3)
0520 4199 J_0 : $(\pi)^3$
0520 5106 J_2 : $\exp(-\sqrt[3]{2}/3)$
0521 4467 2^x : $4\sqrt{3}/3-\sqrt{5}$
0521 5242 ζ : $8\sqrt[3]{5}/3$
0521 5647 e^x : $\sqrt{3}-\sqrt{5}-\sqrt{6}$
0521 5941 rt : (9,5,1,6)
0521 6638 rt : (8,2,-2,5)
0521 7336 rt : (7,1,-5,-4)
0521 8034 rt : (6,4,-8,-3)
0521 8372 Γ : 19/9
0522 1841 cr : $3e/7$
0522 1861 cu : $1/3-\sqrt{2}/2$
0522 2262 ln : $(\sqrt[3]{5}/2)^{1/3}$
0522 3367 $2/5+\exp(\sqrt{3})$
0522 4606 J : 3/22
0522 4618 ζ : $7(e+\pi)/9$
0522 6595 $2-\exp(2/3)$
0522 7106 J_0 : $10\zeta(3)$
0522 7872 $\ln(4)\div s\pi(\sqrt{5})$
0522 8119 sq : $1/3-e/2$
0522 9169 ℓ_{10} : $8\pi^2/7$
0522 9365 rt : (9,6,3,7)
0523 0071 rt : (8,3,0,6)
0523 0646 J_1 : $(\sqrt{2}/3)^3$
0523 0777 rt : (7,0,-3,5)
0523 1374 sr : $3\sqrt{3}/2-2\sqrt{5}/3$
0523 1484 rt : (6,3,-6,-4)
0523 1538 ℓ_2 : $\zeta(5)$
0523 2192 rt : (5,9,-6,-7)
0523 2400 J : $\ln 2/5$
0523 4114 Γ : $5\sqrt[3]{3}$
0523 8767 Ei : 16/23
0523 8813 cu : P_{74}
0524 2002 $t\pi$: P_{28}
0524 2924 rt : (9,7,5,8)
0524 3437 2^x : $2\sqrt{2}/3-3\sqrt{3}$
0524 3639 rt : (8,4,2,7)
0524 4354 rt : (7,1,-1,6)
0524 5070 rt : (6,2,-4,-5)
0524 5787 rt : (5,5,-7,-4)
0524 6476 ζ : $2\sqrt{3}/3$
0524 7247 sr : $(e/2)^{1/3}$
0524 7921 rt : (3,6,-8,5)
0524 9469 ln : $\pi/9$
0525 0977 10^x : $4+e/2$
0525 3142 rt : (2,9,-4,8)
0525 5871 J_1 : 2/19
0525 6093 J_2 : $5\pi^2/6$
0525 6621 rt : (9,8,7,9)
0525 6724 10^x : $3e+\pi$
0525 7131 sq : $\sqrt[3]{16/3}$
0525 7345 rt : (8,5,4,8)
0525 7741 e^x : $9\sqrt[3]{2}/7$
0525 8069 rt : (7,2,1,7)
0525 8303 tanh : 1/19
0525 8306 at : 1/19
0525 8794 rt : (6,1,-2,-6)
0525 9520 rt : (5,4,-5,-5)
0526 0246 rt : (4,7,-8,-4)
0526 1958 θ_3 : $(\sqrt{2}\pi)^{-1/2}$
0526 2288 ζ : $\exp(-1)$
0526 2547 ζ : $-\sqrt{2}-\sqrt{6}+\sqrt{7}$
0526 3157 id : 1/19
0526 4638 sq : $3\sqrt[3]{5}/5$
0526 5588 sinh : 1/19
0526 5590 as : 1/19
0526 6807 sq : $2\sqrt{2}-3\sqrt{5}$
0526 8520 2^x : $(\sqrt[3]{2}/3)^3$
0526 8764 $\ln(2)+s\pi(\sqrt{3})$
0526 9682 e^x : $1/4+\sqrt{3}/2$
0527 1191 rt : (8,6,6,9)
0527 1300 J_1 : $7e\pi/6$
0527 1925 rt : (7,3,3,8)
0527 2659 cr : 7/6
0527 3021 2^x : $2e/3+\pi/4$
0527 3395 rt : (5,3,-3,-6)
0527 3437 cu : 3/8
0527 4130 rt : (4,6,-6,-3)
0527 4867 rt : (3,9,-9,-4)
0527 5772 2^x : $3e/4-2\pi$
0527 6682 rt : (2,7,-3,6)
0527 8159 $\exp(1/5)\div\exp(\pi)$
0527 9042 $4/5\div\exp(e)$
0527 9155 J_0 : $(\ln 2/2)^{-3}$
0528 0351 Γ : $(3\pi)^{1/3}$
0528 2144 rt : (1,9,8,3)
0528 2771 sr : $3/4+2\sqrt{3}$
0528 5126 ℓ_{10} : P_{54}
0528 5924 rt : (7,4,5,9)
0528 6668 rt : (6,1,2,8)
0528 7413 rt : (5,2,-1,-7)
0528 8158 rt : (4,7,-2,1)
0528 8904 rt : (3,8,-7,-5)
0528 8913 $t\pi$: $1/\sqrt{15}$
0529 2619 ln : $2+\sqrt{3}/2$
0529 4152 ℓ_{10} : $3e+\pi$
0529 4580 J_0 : $6\pi^2/7$
0529 5141 10^x : 24/23
0529 5411 J_2 : $\exp(5/3)$
0529 6102 $1/3-\ln(4)$
0529 7702 J_1 : $(1/\pi)/3$
0529 8119 $s\pi$: $2\sqrt{2}+2\sqrt{3}/3$
0529 9816 cu : $4-4e/3$
0530 0193 tanh : $(1/\pi)/6$
0530 0196 at : $(1/\pi)/6$
0530 0824 rt : (6,2,4,9)
0530 1347 ℓ_{10} : $(\sqrt[3]{3})^{1/3}$
0530 1555 Γ : 23/25
0530 1578 rt : (5,1,1,-8)
0530 1857 e^x : $3\sqrt{2}/2+\sqrt{5}$
0530 2333 rt : (4,8,2,5)

0530 2673 $\theta_3 : \exp(-\sqrt{5}/3)$
0530 3089 rt : (3,7,-5,-6)
0530 3911 cr : $6\sqrt[3]{3}$
0530 5164 id : $(1/\pi)/6$
0530 5570 tπ : $2\sqrt{2}+2\sqrt{3}/3$
0530 5817 ℓ_{10} : $(\ln 2)^{1/3}$
0530 6154 sq : $2\sqrt{2}-3\sqrt{3}/2$
0530 7653 sinh : $(1/\pi)/6$
0530 7656 as : $(1/\pi)/6$
0531 0182 $\ln(\sqrt{5}) \div \exp(e)$
0531 0500 cu : $1/2-2\pi/3$
0531 0753 sr : $8\ln 2/5$
0531 2805 $J_1 : 4\sqrt[3]{5}$
0531 3434 rt : (1,6,1,8)
0531 4571 cr : $1/2+3e$
0531 5328 ln : $3\sqrt{2}/2+\sqrt{5}/3$
0531 5893 rt : (5,0,3,9)
0531 6658 rt : (4,9,6,9)
0531 7424 rt : (3,6,-3,-7)
0531 8191 rt : (2,9,-6,-6)
0531 9485 ln : $2e+3\pi/4$
0532 0744 $e^x : 3+\pi/4$
0532 2369 $2^x : (\sqrt{5}/2)^{1/3}$
0532 2987 tπ : $1/4-2e/3$
0532 4649 rt : (6,7,-8,6)
0532 5443 sq : 3/13
0532 5600 $\ell_{10} : \exp(1/(3e))$
0532 9497 sπ : $3e/2-2\pi/3$
0533 0454 cr : $3+4\sqrt{2}$
0533 1136 rt : (4,2,2,-9)
0533 1912 rt : (3,5,-1,-8)
0533 2468 $\Psi : 1/\ln(\pi/3)$
0533 2689 rt : (2,8,-4,-7)
0533 3301 tanh : $(\ln 2/3)^2$
0533 3304 at : $(\ln 2/3)^2$
0533 3437 cr : $10e\pi/3$
0533 4079 $Li_2 : 1/19$
0533 4968 tanh : $\exp(-(e+\pi)/2)$
0533 4971 at : $\exp(-(e+\pi)/2)$
0533 6269 $J_1 : \exp(-\sqrt{5})$
0533 7082 tπ : $3e/2-2\pi/3$
0533 7818 $J_1 : 3\pi^2/8$
0533 8354 ln : $9\sqrt{3}/2$
0533 8366 id : $(\ln 2/3)^2$
0533 9186 $\zeta : 5e/3$
0534 0038 id : $\exp(-(e+\pi)/2)$
0534 0902 sinh : $(\ln 2/3)^2$
0534 0905 as : $(\ln 2/3)^2$
0534 1665 sq : $(\sqrt[3]{3}/3)^2$
0534 2577 sinh : $\exp(-(e+\pi)/2)$
0534 2580 as : $\exp(-(e+\pi)/2)$
0534 2640 $2/5-\ln(\sqrt{2})$
0534 3896 $1/2 \div \exp(\sqrt{5})$
0534 4733 $2^x : 7\sqrt[3]{3}/5$
0534 4912 rt : (2,9,-5,3)
0534 5442 $\Psi : \sqrt[3]{7/2}$
0534 6556 rt : (3,5,2,9)
0534 7343 rt : (2,7,-2,-8)
0534 9306 $e^x : 4-4\sqrt{3}$
0535 2388 $2^x : \sqrt{2}-\sqrt{6}+\sqrt{7}$
0535 2502 $10^x : 5/16$
0535 2540 sq : $1+\sqrt{3}/4$
0535 2595 $\zeta : 9\sqrt[3]{2}/10$
0535 2630 $J_2 : 2/3$
0535 3662 $10^x : 4\sqrt{2}-4\sqrt{3}$
0535 7330 cr : $5\sqrt{3}$
0535 7944 id : $4e/3-\pi/2$
0535 8095 $2^x : 1/4-2\sqrt{5}$
0536 0347 at : $3(e+\pi)/10$
0536 0498 $\ell_{10} : 8\sqrt{2}$
0536 1027 id : $(\sqrt[3]{5}/2)^{-1/3}$
0536 1344 $J_2 : 5e\pi$
0536 2157 rt : (2,6,0,1)
0536 2956 rt : (1,9,-3,-8)
0536 3873 $J_1 : 6\sqrt{3}$
0536 5468 $\ell_2 : (\sqrt{5}/2)^{1/3}$
0536 6847 $4 \div \exp(2/3)$
0536 8037 $\ln(\sqrt{2})+s\pi(1/4)$
0536 8456 $J_0 : (e/2)^3$
0536 9148 $10^x : 2(e+\pi)/9$
0536 9175 $erf : 1/21$
0536 9543 $\Gamma : 4\zeta(3)$
0537 0103 at : $(2e)^{1/3}$
0537 0307 $J_2 : 5\zeta(3)/9$
0537 1153 $10^x : 3e-3\pi$
0537 2062 $10^x : 4\sqrt{2}+4\sqrt{5}/3$
0537 5457 rt : (1,9,7,-5)
0537 6068 ln : $4\sqrt{3}/3+\sqrt{5}/4$
0537 6213 $\Psi : \sqrt{3}+\sqrt{5}-\sqrt{6}$
0537 6874 $\Psi : 3\sqrt{5}/2$
0537 7134 rt : (2,1,-2,9)
0537 7236 $Li_2 : (1/\pi)/6$
0537 7944 rt : (1,8,-1,-9)
0537 9718 $\Gamma : 7e/9$
0538 2057 $Li_2 : 15/19$
0538 3537 $10^x : 3/2+\pi/2$
0538 4165 tπ : $\sqrt{14}$
0538 4561 sr : $3/2+e$
0538 4726 $J_2 : (\sqrt{5})^{-1/2}$
0538 4757 sq : $3/2-\sqrt{3}$
0538 5092 rt : (3,6,-9,2)
0538 7497 $e^x : \sqrt{3}/4-3\sqrt{5}/2$
0538 9355 $\zeta : 1/\sqrt[3]{20}$
0539 2015 sq : $1/3+\sqrt{2}$
0539 3703 $\zeta : 7/19$
0539 4417 $2^x : 1/2-3\pi/2$
0539 6752 $J_2 : \ln((e+\pi)/3)$
0539 7640 $J : \exp(-16\pi/17)$
0539 9492 cu : $1/\sqrt{7}$
0540 0000 cu : $3\sqrt[3]{2}/10$
0540 1567 cu : $3\sqrt{3}/4-3\sqrt{5}/4$
0540 2204 $J_0 : 3(e+\pi)/7$
0540 2296 $10^x : 25/14$
0540 4124 $e^x : 1/19$
0540 6722 ln : 18/19
0540 8421 rt : (1,6,3,-9)
0540 9255 sr : 10/9
0541 1355 $Li_2 : (\ln 2/3)^2$
0541 2154 $\ell_{10} : \sqrt{13}\pi$
0541 2872 $J_2 : 1/\ln(\sqrt{2}\pi)$
0541 3073 $Li_2 : \exp(-(e+\pi)/2)$
0541 3411 $2/5 \div \exp(2)$
0541 4664 $J_1 : \exp(-\sqrt{2}\pi/2)$
0541 5398 $J_1 : 7(e+\pi)/6$
0541 6148 rt : (2,7,-4,2)
0541 6517 $4/5+s\pi(\sqrt{3})$
0541 7006 $J_2 : 3\sqrt{5}/10$
0541 8637 $\ln(2/3)-\exp(1/2)$
0541 8838 $J_2 : (\sqrt{2}\pi/2)^{-1/2}$
0542 0022 cr : $4-2\sqrt{2}$
0542 2384 $Ei : 4\pi^2/7$
0542 3918 rt : (1,4,2,9)
0542 3961 rt : (9,1,-8,1)
0542 4993 $\ln(4/5) \div \exp(\sqrt{2})$
0542 5344 $\Psi : (1/(3e))^2$
0542 8875 $\zeta : 17/10$
0542 9485 sr : $\sqrt{3}/2+3\sqrt{5}/2$
0542 9713 sq : $2e/3-\pi/4$
0543 0017 tπ : $3\sqrt{2}/4-\sqrt{3}/3$
0543 2078 cr : $(e+\pi)/5$
0543 2118 as : 20/23
0543 2196 $10^x : 7e/8$
0543 4052 $e^x : 4\sqrt[3]{2}/7$
0543 4390 $J_1 : (2\pi/3)^{-3}$
0543 5299 $J_2 : 4e\pi/3$
0543 5720 $10^x : 4/3-3\sqrt{3}/2$
0543 5766 $\ell_{10} : 15/17$
0543 6018 at : $1/3-2\pi/3$
0543 6657 sinh : $(\sqrt{2}\pi)^{1/2}$
0543 8823 rt : (9,2,-6,2)
0543 9636 rt : (8,1,-9,-1)
0543 9878 rt : (2,9,-6,-2)
0544 3321 $e^x : 18/25$
0544 6837 $\lambda : \exp(-9\pi/5)$
0544 8410 $e^x : (1/\pi)/6$
0544 9882 $\Gamma : 5/17$
0545 0096 sq : $e/2-3\pi$
0545 1347 $e^x : 4/3-3\sqrt{2}$
0545 3434 $J_1 : 7e\pi/3$
0545 3843 rt : (9,3,-4,3)
0545 4667 rt : (8,0,-7,2)
0545 4815 $\ell_{10} : 2-\sqrt{5}/2$
0545 5471 $2^x : 1-3\sqrt{3}$
0545 5862 $\ell_{10} : 7\sqrt[3]{2}/10$
0545 5882 rt : (1,6,0,2)
0545 6351 cu : $3-\sqrt{2}/2$
0545 8584 $\ell_{10} : 9\sqrt[3]{2}$
0545 9417 sq : $2/3-\sqrt{3}/4$
0545 9598 $\ln(3/5) \div \exp(\sqrt{5})$
0546 0332 $\ell_{10} : 2-\sqrt{3}/2$
0546 2768 $2^x : 21/4$
0546 3823 cr : $\sqrt{3}-\sqrt{5}/4$
0546 4106 rt : (6,7,-9,4)
0546 6413 $2^x : 2\sqrt{2}/3-\sqrt{3}/2$
0546 7368 $e^x : 3\sqrt{2}+4\sqrt{5}/3$
0546 9024 rt : (9,4,-2,4)
0546 9178 $\zeta : -\sqrt{2}-\sqrt{5}+\sqrt{6}$
0546 9668 $\ln(3) \div \exp(3)$
0546 9860 rt : (8,1,-5,3)
0547 0695 rt : (7,2,-8,-2)
0547 0751 $\zeta : 9/2$
0547 1895 $1/4+\ln(\sqrt{5})$
0547 1939 rt : (6,8,-5,8)
0547 2305 $J_2 : \sqrt{23}\pi$
0547 4449 $\ell_2 : 3+2\sqrt{3}/3$
0547 5706 id : $4\sqrt{3}/3+\sqrt{5}/3$
0547 6607 $2^x : 1/13$
0547 6973 $\sqrt{2} \times s\pi(\sqrt{3})$
0547 7197 ln : $3/4+3\sqrt{2}/2$
0547 8701 $2^x : 3\sqrt{3}/2$
0548 0276 $Ei : 25/6$
0548 1395 $10^x : 9e/8$
0548 2779 $e^x : 7\sqrt{5}/4$
0548 3427 $2^x : \ln 2/9$
0548 4309 rt : (1,7,-8,4)
0548 4370 rt : (9,5,0,5)
0548 4425 $e^x : 10\sqrt[3]{2}/9$
0548 5191 $e^x : \exp(-(e+\pi)/2)$
0548 5217 rt : (8,2,-3,4)
0548 6064 rt : (7,1,-6,-3)
0548 6912 rt : (6,4,-9,-2)
0548 6968 cu : $2\sqrt[3]{5}/9$
0548 7236 $10^x : \sqrt{2}/3-\sqrt{3}$
0548 7373 $\ln(2/3) \div \exp(2)$
0548 7936 $\exp(2/5) \times s\pi(1/4)$
0548 8175 rt : (5,4,-9,8)
0548 9776 sq : $10\zeta(3)/3$
0549 1702 rt : (1,8,-9,5)
0549 1724 cu : $4\sqrt{2}-4\sqrt{3}$
0549 4446 at : $\sqrt{3}-3\sqrt{5}/4$
0549 4691 $3 \div \exp(4)$
0549 6036 $J_0 : 4\pi/5$
0549 9884 rt : (9,6,2,6)
0549 9982 id : $\sqrt{3}-3\sqrt{5}/4$
0550 0247 $\Psi : 8\sqrt[3]{5}/9$
0550 0742 rt : (8,3,-1,5)
0550 1215 $e^x : 2e/3-3\pi/2$
0550 1602 rt : (7,0,-4,4)
0550 2461 rt : (6,3,-7,-3)
0550 2759 as : $\sqrt{3}-3\sqrt{5}/4$
0550 3322 rt : (5,9,-7,-9)
0550 4299 $\theta_3 : 19/23$
0550 4881 at : $7\sqrt[3]{2}/5$
0550 5916 at : $4-\sqrt{5}$
0550 6386 id : $3\sqrt{3}/4-3\sqrt{5}/2$
0551 0230 cu : $2/3-\pi/3$
0551 0638 $Ei : 18/17$
0551 3117 $2^x : 3\sqrt{3}/5$
0551 5569 rt : (9,7,4,7)
0551 5570 $\exp(3)-\exp(\pi)$
0551 6440 rt : (8,4,1,6)
0551 7311 rt : (7,1,-2,5)
0551 7359 $\ell_2 : \sqrt{7}\pi$
0551 8183 rt : (6,2,-5,-4)
0551 8229 $\zeta : (\sqrt[3]{3}/3)^{1/3}$
0551 9055 rt : (5,5,-8,-3)
0551 9996 $e^x : 7/5$
0552 3583 $\zeta : 10\pi/7$
0552 3700 at : $\sqrt{2}/3-\sqrt{5}$
0552 7957 sπ : $4\sqrt{3}/3+3\sqrt{5}$

0552 8560 cu : 8/21
0553 1429 rt : (9,8,6,8)
0553 1508 ℓ_2 : $(1/(3e))^2$
0553 2312 rt : (8,5,3,7)
0553 3195 rt : (7,2,0,6)
0553 4079 rt : (6,1,-3,-5)
0553 4965 rt : (5,4,-6,-4)
0553 5851 rt : (4,7,-9,-3)
0553 6111 at : $\sqrt[3]{11/2}$
0553 6332 sq : 4/17
0553 6423 $t\pi$: $4\sqrt{3}/3+3\sqrt{5}$
0553 7014 erf : $(\ln 3/3)^3$
0553 7601 rt : (3,7,-4,8)
0553 8206 rt : (2,9,-7,-7)
0553 8567 sinh : 23/25
0554 4263 2^x : $\sqrt{3}/2+\sqrt{5}/3$
0554 6986 J_1 : 1/9
0554 7468 rt : (9,9,8,9)
0554 8363 rt : (8,6,5,8)
0554 8776 e^x : $1/4-\pi$
0554 9259 rt : (7,3,2,7)
0554 9847 tanh : 1/18
0554 9850 at : 1/18
0555 0156 rt : (6,0,-1,6)
0555 0206 sr : $7(1/\pi)/2$
0555 0554 10^x : $3\sqrt{2}/5$
0555 1054 rt : (5,3,-4,-5)
0555 1522 ζ : exp(3/2)
0555 1565 ℓ_2 : $3\sqrt{3}/5$
0555 1732 ℓ_{10} : 22/25
0555 1871 ℓ_{10} : $\sqrt{3}/3+\sqrt{5}/4$
0555 1952 rt : (4,6,-7,-4)
0555 2672 10^x : $e/3$
0555 3721 J_2 : $e/4$
0555 3978 J_0 : 23/10
0555 4656 Γ : 11/12
0555 5555 id : 1/18
0555 8413 sinh : 1/18
0555 8417 as : 1/18
0555 8886 J_1 : $4e\pi/5$
0555 9554 rt : (1,7,-9,-4)
0555 9991 Ei : 17/12
0556 0470 J_2 : 17/25
0556 1535 ln : $2-2\sqrt{2}/3$
0556 2202 Ei : $\sqrt{2}+\sqrt{3}-\sqrt{6}$
0556 2628 at : $3e/4-2\pi/3$
0556 2685 θ_3 : 5/7
0556 2988 rt : (2,7,-5,-2)
0556 4289 sr : $3-4\sqrt{2}/3$
0556 4569 at : $4/3+\sqrt{3}/4$
0556 4597 rt : (8,7,7,9)
0556 4995 $\ln(3)-\exp(e)$
0556 5116 cu : $7\pi^2/8$
0556 5505 rt : (7,4,4,8)
0556 6415 rt : (6,1,1,7)
0556 6719 cr : 20/17
0556 7326 rt : (5,2,-2,-6)
0556 7415 ℓ_2 : $4/3+\sqrt{5}/3$
0556 8237 rt : (4,7,-3,-1)
0556 8373 id : $3e/4-2\pi/3$
0556 9150 rt : (3,8,-8,-4)
0556 9912 J_1 : $\ln(\sqrt{5}/2)$
0557 1254 as : $3e/4-2\pi/3$
0557 2276 $1/3-\exp(2)$
0557 2526 Γ : $\sqrt[3]{19/2}$
0557 2809 sq : $2-\sqrt{5}$
0557 3065 cu : $6(1/\pi)/5$
0557 3997 Ψ : $8\sqrt[3]{2}/3$
0557 7867 sr : $5\pi^2/3$
0557 8588 ln : $(\sqrt{5}/2)^{1/2}$
0558 1025 ln : $4(e+\pi)/3$
0558 1939 rt : (7,5,6,9)
0558 1996 $\exp(5)\times s\pi(\sqrt{2})$
0558 2862 rt : (6,2,3,8)
0558 2971 2^x : $3\ln 2/2$
0558 3300 e^x : $1/\ln(1/\sqrt{2})$
0558 3376 Γ : ln(5/2)
0558 3786 rt : (5,1,0,-7)
0558 4711 rt : (4,8,1,3)
0558 5636 rt : (3,7,-6,-5)
0558 7990 ℓ_{10} : $1/2+4e$
0558 8277 sq : $\exp(-\sqrt[3]{3})$
0558 9068 K : $(\ln 3)^{-2}$
0558 9081 J_2 : 15/22
0559 1649 sq : $3\sqrt{2}+\sqrt{5}/4$
0559 4180 ζ : $2\sqrt{5}$
0559 5001 2^x : $6\pi^2/7$
0559 7068 Ψ : 16/21
0559 8354 e^x : $\sqrt{2}/3-3\sqrt{5}/2$
0559 9037 at : $5\sqrt{2}/4$
0559 9499 rt : (6,3,5,9)
0560 0245 sr : $(\sqrt[3]{3}/2)^{-1/3}$
0560 0386 J_2 : $1/\ln(\ln 2/3)$
0560 0437 rt : (5,0,2,8)
0560 1375 rt : (4,9,5,7)
0560 1921 2^x : $2\sqrt{2}/3+\sqrt{5}$
0560 2315 rt : (3,6,-4,-6)
0560 3255 rt : (2,9,-7,-5)
0560 3338 ln : $9\sqrt{5}/7$
0560 4306 J_2 : $(\pi)^{-1/3}$
0560 5267 ln : 8/23
0560 6030 rt : (1,6,-1,-4)
0560 6117 2^x : $3/2-4\sqrt{2}$
0560 7792 at : $\ln(e+\pi)$
0560 8245 Γ : $\beta(2)$
0560 9655 J_0 : $9\sqrt{5}/8$
0561 0814 at : $\exp(\sqrt[3]{5}/3)$
0561 3230 erf : $(e)^{-3}$
0561 3928 2^x : $4-3e$
0561 5351 cr : $3\pi/8$
0561 6492 e^x : $3e/2+2\pi/3$
0561 6582 sq : $2+3\pi/2$
0561 6669 Ei : $\sqrt[3]{25/3}$
0561 6942 ℓ_{10} : $3-3\sqrt{2}/2$
0561 7284 rt : (5,1,4,9)
0561 8236 rt : (4,2,1,-8)
0561 8603 J : $(1/\pi)/9$
0561 9190 rt : (6,8,-6,6)
0561 9612 ℓ_2 : $3\ln 2$
0561 9832 $\pi\times\ln(5)$
0562 0144 rt : (2,8,-5,-6)
0562 3027 J_0 : 24
0562 3346 J_2 : $2\sqrt[3]{5}/5$
0562 3462 cr : $\sqrt{3}/4+\sqrt{5}/3$
0562 3543 Γ : $\sqrt{21}$
0562 4372 $\sqrt{5}\div\exp(3/4)$
0562 5182 e^x : $\sqrt{2}/3+\sqrt{3}$
0562 5259 e^x : $2+\pi/3$
0562 6121 cu : $5\sqrt{3}/3$
0562 6823 J_2 : 13/19
0562 7289 Ψ : $8e$
0562 7722 cr : $5\sqrt{2}/6$
0562 8501 sq : $4+4\pi/3$
0563 1535 10^x : $\sqrt{3}-4\sqrt{5}/3$
0563 2557 Ψ : $\sqrt{7}\pi$
0563 3506 cu : $3\sqrt{3}/2-4\sqrt{5}/3$
0563 4501 at : 23/13
0563 4682 Li_2 : 1/18
0563 4957 10^x : $7\ln 2/10$
0563 5126 e^x : $5(e+\pi)/6$
0563 5297 rt : (4,1,3,-9)
0563 5917 $s\pi$: $1/4+\sqrt{3}$
0563 6265 rt : (3,5,1,7)
0563 7197 erf : 1/20
0563 7234 rt : (2,7,-3,-7)
0563 8309 ℓ_{10} : $4e\pi/3$
0563 8867 sq : $4/3-\pi/2$
0563 9740 10^x : $3\sqrt{2}/4-4\sqrt{3}/3$
0564 1316 ζ : $5e/8$
0564 2103 sr : $1/4+\sqrt{3}/2$
0564 2290 ℓ_2 : $\zeta(3)/5$
0564 2689 10^x : $2/3-\sqrt{2}/4$
0564 4889 $t\pi$: $1/4+\sqrt{3}$
0564 5136 e^x : $3/4-4e/3$
0564 6711 Ei : $3\pi^2/7$
0564 8273 ζ : $8\ln 2/9$
0565 3116 cu : $2\sqrt{2}/3-\sqrt{5}/4$
0565 3545 rt : (3,3,2,-9)
0565 4042 e^x : $\sqrt{3}-3\sqrt{5}/4$
0565 4529 rt : (2,6,-1,-2)
0565 4840 10^x : $2\sqrt{3}/3+\sqrt{5}$
0565 5513 rt : (1,9,-4,-7)
0565 6051 $\sqrt{3}-s\pi(\sqrt{5})$
0565 6683 Γ : 2/13
0565 9356 cr : $1/2+e/4$
0566 0689 J_0 : $\sqrt{19/3}$
0566 1398 e^x : $3e+\pi/4$
0566 2518 $1/2\times\exp(\sqrt{2})$
0566 3301 ln : $3\sqrt[3]{2}/4$
0566 3827 at : $\sqrt{2}/3+3\sqrt{3}/4$
0566 4166 ℓ_2 : $\sqrt[3]{9}$
0566 4458 J_2 : $(\sqrt{2}/3)^{1/2}$
0566 4507 10^x : $2\pi/5$
0566 5516 sr : $3\pi^2/7$
0566 7677 10^x : $2\sqrt{3}-2\sqrt{5}/3$
0566 8934 sq : 5/21
0566 9008 $\exp(5)-\exp(\sqrt{5})$
0566 9222 J_2 : $4\zeta(3)/7$
0566 9570 2^x : $10\sqrt{2}/7$
0567 0851 2^x : $(1/\pi)/4$
0567 2034 rt : (2,7,5,7)
0567 3033 rt : (1,8,-2,-8)
0567 4530 id : $\exp(\sqrt[3]{3}/2)$
0567 6704 2^x : $1/3-2\sqrt{5}$
0567 6970 cu : $4-4\sqrt{5}/3$
0567 8554 Ψ : $\exp(\sqrt[3]{2}/3)$
0567 8901 J_2 : 11/16
0568 0976 as : $(\pi/3)^{-3}$
0568 2403 sq : $3\sqrt{2}/4-3\sqrt{3}/4$
0568 2553 2^x : $3+3e/2$
0568 4114 Ei : $9(1/\pi)/8$
0568 4787 ℓ_{10} : $(\sqrt{2}\pi/3)^{1/3}$
0568 5281 $3/4-\ln(2)$
0568 5503 2^x : $1/3+\sqrt{2}/2$
0568 6443 ℓ_2 : $(\ln 2/3)^{1/2}$
0568 7938 Γ : $9e\pi/7$
0568 9576 cu : 5/13
0568 9744 2^x : $3\sqrt{2}/4-3\sqrt{3}$
0568 9874 ℓ_{10} : $2\sqrt[3]{5}/3$
0569 0262 rt : (3,7,-5,5)
0569 0769 rt : (1,7,0,-9)
0569 1237 ℓ_2 : $3e/2+4\pi$
0569 1682 Ψ : $-\sqrt{3}+\sqrt{6}+\sqrt{7}$
0569 1777 ℓ_{10} : $4\sqrt{3}+2\sqrt{5}$
0569 5036 ℓ_{10} : $\sqrt{2}/2+\sqrt{3}/4$
0569 5158 J_1 : $\sqrt{3}+\sqrt{6}+\sqrt{7}$
0569 5842 ℓ_2 : $3-3e/4$
0569 6540 J : $\exp(-12\pi/5)$
0569 7023 e^x : $3\zeta(3)/5$
0569 7323 ζ : $(\ln 2/2)^{-1/2}$
0569 8336 $\ln(2/5)-\exp(\pi)$
0569 9274 $\ln(\pi)\div\exp(3)$
0569 9316 sq : $3(1/\pi)/4$
0569 9620 J_0 : $9\pi^2/5$
0570 0468 ℓ_{10} : $4\sqrt{2}/3-\sqrt{5}/3$
0570 1140 10^x : $\sqrt{3}/4-3\sqrt{5}/4$
0570 1804 2^x : 2/25
0570 2224 cu : $2\sqrt{3}/9$
0570 8664 cu : $3/4+4\sqrt{5}$
0570 8725 rt : (1,3,-7,-7)
0570 9757 rt : (9,1,-9,0)
0571 0299 cu : $5\ln 2/9$
0571 1549 erf : $(\sqrt{2}\pi)^{-2}$
0571 2431 at : $(\pi)^{1/2}$
0571 2774 e^x : 1/18
0571 2872 at : $1-2\sqrt{2}/3$
0571 5841 ln : 17/18
0571 7787 rt : (2,7,-6,-6)
0571 8761 J : 6/17
0571 8827 sr : 19/17
0571 9095 id : $1-2\sqrt{2}/3$
0571 9387 J_1 : $5e$
0571 9392 ℓ_2 : $2/3+\sqrt{2}$
0571 9516 10^x : $3-3\sqrt{2}$
0571 9892 $t\pi$: $(\ln 3/2)^{1/2}$
0572 0346 e^x : $1/\ln(4)$
0572 0744 10^x : $4/3+4\sqrt{2}$
0572 2218 as : $1-2\sqrt{2}/3$
0572 5751 cu : $1/3+\sqrt{5}/2$
0572 6427 cr : 13/11

0572 6907 rt : (1,4,1,5)
0572 6963 rt : (9,2,-7,1)
0572 7061 ζ : $\sqrt{2}\pi$
0572 7971 K : $(\sqrt[3]{5}/3)^{1/3}$
0573 1098 at : $3\sqrt{2}/4 - \sqrt{5}/2$
0573 1613 e^x : $\sqrt{5}-\sqrt{6}-\sqrt{7}$
0573 2073 e^x : $1/4+\sqrt{2}/3$
0573 2542 sq : $2/3 - e/3$
0573 2635 sq : $7e/6$
0573 4111 ln : $1/2+\sqrt{5}/4$
0573 4191 cu : $3/2 - 4\sqrt{2}/3$
0573 4354 J_2 : $(2\pi/3)^{-1/2}$
0573 5639 cr : $2+3\sqrt{5}$
0573 7126 id : $(\sqrt{5}/2)^{1/2}$
0573 7381 id : $3\sqrt{2}/4-\sqrt{5}/2$
0574 0534 as : $3\sqrt{2}/4-\sqrt{5}/2$
0574 4370 rt : (9,3,-5,2)
0574 4973 10^x : $1/4 - 2\sqrt{5}/3$
0574 5377 rt : (8,0,-8,1)
0574 8192 ln : $7\sqrt{5}/2$
0574 8675 tπ : $1/\ln(e/2)$
0575 0853 sπ : $\exp(-4)$
0575 1584 Γ : $3\sqrt{2}/2$
0575 2953 e^x : $4/3-4\pi/3$
0575 3274 ℓ_2 : $(\ln 2)^2$
0575 3641 $1/5\times\ln(3/4)$
0575 3681 J_2 : $\exp(-1/e)$
0575 4027 $\pi\div\exp(4)$
0575 5389 J_2 : 9/13
0575 5550 J_1 : $5\sqrt[3]{3}$
0575 6384 sπ : $\sqrt{3}/3-\sqrt{5}/4$
0575 6826 2^x : $\sqrt{2}/4 - 2\sqrt{5}$
0576 0000 sq : 6/25
0576 0386 tπ : $\exp(-4)$
0576 0753 e^x : $1/2-3\sqrt{5}/2$
0576 1982 rt : (9,4,-3,3)
0576 2035 cr : $3/4+\sqrt{3}/4$
0576 2858 Ψ : $7\sqrt[3]{3}/3$
0576 3004 rt : (8,1,-6,2)
0576 3571 J_2 : $2\sqrt{3}/5$
0576 4026 rt : (7,2,-9,-1)
0576 5125 e^x : 19/17
0576 5945 tπ : $\sqrt{3}/3-\sqrt{5}/4$
0576 6595 J_1 : $\ln 2/6$
0576 8591 cr : $4+3\pi/2$
0576 8790 J_2 : $\ln 2$
0576 9878 $\ln(2/3)-\exp(\sqrt{3})$
0576 9887 J_0 : $2e\pi/3$
0576 9928 id : $\sqrt{3}/4-2\sqrt{5}/3$
0577 0104 sr : $1/3+\pi/4$
0577 0410 J : 5/18
0577 2210 J_2 : $\sqrt[3]{1/3}$
0577 3438 10^x : $6\sqrt[3]{5}$
0577 3860 ℓ_2 : $\sqrt{13/3}$
0577 4208 rt : (6,8,-7,4)
0577 5076 θ_3 : 10/21
0577 6581 $\exp(1/5)\times s\pi(1/3)$
0577 6714 Ei : $9\sqrt[3]{2}/8$
0577 8010 sq : $\sqrt[3]{3}/6$
0577 9061 2^x : $4/3+3\sqrt{5}/4$
0577 9763 sq : $\zeta(3)/5$
0577 9804 rt : (9,5,-1,4)
0578 0755 10^x : $(\sqrt{2}\pi)^{-1/3}$
0578 0841 rt : (8,2,-4,3)
0578 1056 tπ : $\ln(\sqrt[3]{5}/3)$
0578 1879 rt : (7,1,-7,-2)
0578 3423 rt : (6,9,-3,8)
0578 4776 ℓ_{10} : $2+3\pi$
0578 6870 J_1 : $6\zeta(3)$
0578 7142 J_0 : $5(e+\pi)/2$
0578 7387 cr : $3\sqrt{2}+2\sqrt{5}$
0578 7831 ζ : $\sqrt{5}/2$
0578 7926 2^x : $7\sqrt{3}/6$
0578 8113 10^x : $1/3-\pi/2$
0578 8338 $\pi - \ln(2/5)$
0578 9542 Ψ : $\ln 2/6$
0579 0140 10^x : $-\sqrt{5}+\sqrt{6}+\sqrt{7}$
0579 0662 2^x : $3\sqrt{3}/2-3\sqrt{5}$
0579 2239 id : $\sqrt{2}-2\sqrt{5}$
0579 3017 ℓ_2 : P_{21}
0579 6810 $\exp(1/4)+s\pi(e)$
0579 7417 10^x : $\sqrt{6}\pi$
0579 7840 rt : (9,6,1,5)
0579 8893 rt : (8,3,-2,4)
0579 9194 ℓ_{10} : 7/8
0579 9947 rt : (7,0,-5,3)
0580 0441 e^x : $1/3+\sqrt{5}$
0580 0865 sq : $3/2-\sqrt{2}/3$
0580 1001 rt : (6,3,-8,-2)
0580 2571 rt : (5,5,-7,8)
0580 4140 $\ln(2/5)\div s\pi(1/3)$
0580 6303 Ψ : $(\ln 2/2)^{-2}$
0580 6705 id : $9e/8$
0580 6735 10^x : $1-\sqrt{5}$
0580 8860 J_2 : 16/23
0581 1430 cu : $4\sqrt{2}-\sqrt{3}/3$
0581 2838 cr : $4\sqrt{2}-2\sqrt{5}$
0581 4258 cu : P_{27}
0581 4262 J_2 : $7\pi^2/8$
0581 4555 J_1 : $((e+\pi)/2)^{-2}$
0581 5724 Li_2 : 19/24
0581 6096 rt : (9,7,3,6)
0581 7102 sr : $2+\sqrt{5}$
0581 7164 rt : (8,4,0,5)
0581 8234 rt : (7,1,-3,4)
0581 8413 2^x : $4/3-2e$
0581 8677 $\sqrt{5}+\exp(3/5)$
0581 9305 rt : (6,2,-6,-3)
0581 9808 J_0 : $2\sqrt[3]{2}$
0582 0378 rt : (5,5,-9,-2)
0582 4442 sq : $\sqrt{3}-2\sqrt{5}/3$
0582 4573 θ_3 : $(e+\pi)^{-2}$
0582 4669 J_2 : $7(e+\pi)/5$
0582 6736 sr : $8\sqrt[3]{2}/9$
0582 6854 J_2 : $\sqrt{2}+\sqrt{3}-\sqrt{6}$
0582 6988 sq : $2e/3-\pi/2$
0582 8999 λ : $\exp(-16\pi/9)$
0583 0322 sq : $1/\ln(\ln 3)$
0583 1243 e^x : $2\sqrt{3}/3+3\sqrt{5}$
0583 2361 2^x : $2\sqrt{2}-4\sqrt{3}$
0583 2519 sπ : $4\sqrt{5}/3$
0583 2951 Γ : 21/23
0583 3933 cu : $\sqrt{2}-\sqrt{3}+\sqrt{5}$
0583 4575 rt : (9,8,5,7)
0583 5660 rt : (8,5,2,6)
0583 6716 10^x : $e/4+3\pi$
0583 6747 rt : (7,2,-1,5)
0583 7835 rt : (6,1,-4,-4)
0583 8924 rt : (5,4,-7,-3)
0584 0007 cr : $\sqrt[3]{5/3}$
0584 0463 J_2 : $4e\pi$
0584 0686 at : 16/9
0584 2465 tπ : $4\sqrt{5}/3$
0584 3245 tπ : $(\sqrt{2}\pi)^{1/3}$
0584 3515 tπ : $2\sqrt[3]{3}/5$
0584 6669 sq : $4\sqrt{3}+\sqrt{5}/4$
0584 7270 ln : $4-\sqrt{5}/2$
0584 8642 J_2 : $2\pi/9$
0585 0000 $1/2\times\exp(3/4)$
0585 0842 10^x : $3\sqrt{2}/2-3\sqrt{5}/2$
0585 1284 rt : (3,7,-6,2)
0585 1357 2^x : $1/2+3e$
0585 1820 J : 1/9
0585 3284 rt : (9,9,7,8)
0585 4386 rt : (8,6,4,7)
0585 4675 2^x : $\exp(-5/2)$
0585 5490 rt : (7,3,1,6)
0585 5841 cu : $e/7$
0585 6595 rt : (6,0,-2,5)
0585 7701 rt : (5,3,-5,-4)
0585 8808 rt : (4,6,-8,-3)
0585 9895 Γ : $7\sqrt{3}/4$
0586 0100 J_2 : $((e+\pi)/2)^{-1/3}$
0586 0447 2^x : 25/24
0586 0695 10^x : $1/2-\sqrt{3}$
0586 3634 id : $1/\ln(\sqrt[3]{3}/2)$
0586 5382 id : $4\sqrt{2}/3-4\sqrt{5}$
0586 5596 $\ln(5)\times\exp(\sqrt{5})$
0586 6184 J_0 : $5e\pi$
0586 6987 tπ : $4\sqrt{2}-3\sqrt{3}$
0586 7465 10^x : $8(e+\pi)/9$
0586 8112 e^x : $8\sqrt[3]{5}/7$
0586 8571 at : $1/\ln(\sqrt[3]{5}/3)$
0586 9708 ℓ_{10} : $\sqrt[3]{2/3}$
0587 0302 ℓ_{10} : $\ln(\pi)$
0587 0870 $\ln(\sqrt{3})\div\exp(\sqrt{5})$
0587 1210 ℓ_2 : $2\sqrt{2}-\sqrt{5}/3$
0587 2181 J_1 : 2/17
0587 3347 rt : (8,7,6,8)
0587 4468 rt : (7,4,3,7)
0587 5217 cu : $1+3e/4$
0587 5577 tanh : 1/17
0587 5582 at : 1/17
0587 5591 rt : (6,1,0,6)
0587 6252 e^x : $(\sqrt[3]{4}/3)^{-3}$
0587 6714 rt : (5,2,-3,-5)
0587 7277 rt : (9,7,-9,-9)
0587 7803 id : $4\sqrt{2}-3\sqrt{3}/2$
0587 7840 rt : (4,7,-4,-3)
0587 7988 J_2 : $5\sqrt[3]{2}/9$
0587 8694 J_2 : 7/10
0587 8966 rt : (3,8,-9,-3)
0588 1344 cu : 7/18
0588 1898 ζ : $7\sqrt[3]{2}/2$
0588 2352 id : 1/17
0588 3458 id : $\exp(\sqrt{5}/2)$
0588 4289 Ei : $\exp(1/\sqrt{2})$
0588 5391 $\ln(\sqrt{3})\times\exp(2)$
0588 5745 sinh : 1/17
0588 5750 as : 1/17
0588 5798 e^x : $1-2\sqrt{2}/3$
0588 6983 sq : P_{15}
0588 7450 sq : $3-4\sqrt{2}$
0588 8412 $\ln(\sqrt{5})+s\pi(\sqrt{3})$
0588 9151 ln : $2\sqrt{2}/3$
0588 9368 ℓ_2 : 3/25
0589 1100 rt : (2,8,1,6)
0589 2548 rt : (8,8,8,9)
0589 2681 Ψ : $7\zeta(3)/6$
0589 3687 rt : (7,5,5,8)
0589 4395 J_0 : $8\sqrt[3]{3}$
0589 4828 rt : (6,2,2,7)
0589 4861 at : $1/2-\sqrt{5}/4$
0589 5516 $\ln(3)\times s\pi(\sqrt{2})$
0589 5589 cr : 19/16
0589 5970 rt : (5,1,-1,-6)
0589 7113 rt : (4,8,0,1)
0589 7373 $1/3\div\exp(\sqrt{3})$
0589 8258 rt : (3,7,-7,-4)
0589 8631 cu : $3\sqrt{2}/2-\sqrt{3}$
0589 9298 10^x : $5\ln 2/2$
0590 0369 e^x : 13/18
0590 1699 id : $1/2+\sqrt{5}/4$
0590 2491 sπ : $\sqrt{2}-\sqrt{3}/4$
0590 3355 e^x : $e/2-4\pi/3$
0590 4792 cu : $\sqrt{2}/3-2\sqrt{5}/3$
0590 4866 cr : $3-2e/3$
0590 5130 as : $1/2-\sqrt{5}/4$
0590 6525 Ψ : 25/3
0590 7190 2^x : $10\pi^2/7$
0591 0574 sq : $\exp(-\sqrt{2})$
0591 2684 sr : $7\sqrt[3]{3}/9$
0591 2800 tπ : $\sqrt{2}-\sqrt{3}/4$
0591 3153 rt : (7,6,7,9)
0591 3207 $\sqrt{3}\times\exp(5)$
0591 4313 rt : (6,3,4,8)
0591 5145 Li_2 : $8\ln 2/7$
0591 5473 rt : (5,0,1,7)
0591 6635 rt : (4,9,4,5)
0591 7762 J_2 : $(\ln 2/2)^{1/3}$
0591 7798 rt : (3,6,-5,-5)
0591 8940 2^x : $2\pi^2$
0591 8963 rt : (2,9,-8,-4)
0591 9823 cu : $3e/8$
0592 0582 $\pi\div s\pi(e)$
0592 2201 Γ : $(\pi/3)^{-2}$
0592 4291 ln : $3\pi/10$
0592 5509 at : $\exp(\sqrt{3}/3)$
0592 8172 ζ : 22/5
0593 1683 ℓ_{10} : $3e/4+3\pi$

0593 2225 $\zeta : \exp(\sqrt[3]{4}/3)$
0593 3358 $erf : 1/19$
0593 3699 $J : 1/\sqrt[3]{11}$
0593 4050 rt : (6,4,6,9)
0593 5127 $\ln : 2\sqrt[3]{3}$
0593 5230 rt : (5,1,3,8)
0593 5383 $\ell_{10} : 3/2 - \sqrt{2}/4$
0593 6411 rt : (4,2,0,-7)
0593 6640 $\zeta : 7\pi/5$
0593 7593 rt : (6,8,-8,2)
0593 8777 rt : (2,8,-6,-5)
0594 0107 $\mathrm{id} : 1/4 - 4\sqrt{3}/3$
0594 1719 $\zeta : \exp(\sqrt{2}\pi/3)$
0594 4394 $\sinh : 7\sqrt[3]{2}/6$
0594 6053 $\Gamma : 17/8$
0594 6309 $2^x : 1/12$
0594 7588 rt : (6,9,-4,6)
0594 8762 $\ell_2 : 2e/3 + 3\pi/4$
0594 9940 $\Gamma : 4e\pi/7$
0595 2575 $\zeta : 3(e+\pi)/4$
0595 2804 $\mathrm{at} : e/2 - \pi$
0595 3033 $Ei : 2\pi/9$
0595 4426 $\mathrm{sq} : 1/3 - \sqrt{3}/3$
0595 5247 rt : (5,2,5,9)
0595 5323 $10^x : 1/4 + \sqrt{2}/2$
0595 5855 $\mathrm{cu} : 4/3 - 2\sqrt{2}/3$
0595 6447 rt : (4,1,2,-8)
0595 7649 rt : (3,5,0,5)
0595 7686 $J_1 : 3(1/\pi)/8$
0595 8853 rt : (2,7,-4,-6)
0596 1788 $\mathrm{at} : \sqrt[3]{17/3}$
0596 6010 $\mathrm{id} : \ln(\ln 2/2)$
0596 7383 $\mathrm{sq} : \sqrt[3]{5}/7$
0596 8384 rt : (5,5,-8,6)
0596 9171 $\mathrm{cu} : 3/2 + e$
0596 9619 $J_0 : 8(e+\pi)$
0597 1197 $Li_2 : 1/17$
0597 3749 $J_2 : 12/17$
0597 3911 $\ln : 1 + 4\sqrt{2}/3$
0597 4720 $\ell_{10} : e - \pi/2$
0597 6714 $\mathrm{sr} : 3\sqrt{2}$
0597 6750 rt : (4,0,4,9)
0597 7973 rt : (3,6,4,9)
0597 9197 rt : (2,6,-2,-5)
0597 9385 $\mathrm{sq} : 4e/3 - \pi/4$
0597 9414 $J_1 : \sqrt{10}\pi$
0598 0422 rt : (1,9,-5,-6)
0598 0626 $erf : (1/\pi)/6$
0598 3983 cr : 25/21
0598 6423 sinh : 12/13
0598 9032 $e^x : -\sqrt{2} + \sqrt{5} + \sqrt{7}$
0598 9206 $J_1 : 3/25$
0598 9401 $J_0 : 10\sqrt[3]{5}/3$
0599 0721 $\ln(4) \div \exp(\pi)$
0599 1616 cu : 9/23
0599 3128 $\zeta : 4\pi^2/9$
0599 3618 $J_2 : \sqrt{2}/2$
0599 3717 $\mathrm{cu} : 4\sqrt{2}/3 - \sqrt{3}/2$
0599 4838 $J_1 : (\ln 2/2)^2$
0599 6670 $t\pi : \exp(-\sqrt[3]{3}/2)$
0599 8570 rt : (3,2,3,-9)
0599 9435 $J_1 : \zeta(3)/10$
0599 9815 rt : (2,7,4,4)
0600 1061 rt : (1,8,-3,-7)
0600 3544 $s\pi : 7(e+\pi)$
0600 4062 $\Gamma : 5e\pi/3$
0600 5963 $10^x : (2\pi)^{-2}$
0600 6591 $\mathrm{at} : 2/3 + \sqrt{5}/2$
0600 8585 $\ell_{10} : (\pi/3)^3$
0600 9565 $J_1 : \sqrt{11}\pi$
0601 0414 $\mathrm{cr} : 4\sqrt{3}/3 - \sqrt{5}/2$
0601 0747 $Ei : 3\sqrt{2}/4$
0601 2332 $s\pi : 7\sqrt[3]{3}/5$
0601 2841 $\zeta : 6\sqrt{2}/5$
0601 2992 $t\pi : 1/4 + 2\sqrt{5}/3$
0601 3176 $\ln : \sqrt{3}/5$
0601 3550 $J_2 : 17/24$
0601 4393 $t\pi : 7(e+\pi)$
0601 4533 $\mathrm{cu} : \sqrt{2}/4 - \sqrt{5}/3$
0601 7984 $erf : (\ln 2/3)^2$
0601 8511 rt : (1,7,0,3)
0601 9866 $erf : \exp(-(e+\pi)/2)$
0601 9956 $\mathrm{sq} : 1/2 - \sqrt{5}/3$
0602 0715 rt : (2,4,2,-9)
0602 1104 $J_0 : 7\sqrt[3]{3}/4$
0602 1349 rt : (3,7,-7,-1)
0602 1535 $2^x : 3 + \sqrt{2}/2$
0602 1983 rt : (1,7,-1,-8)
0602 3229 $t\pi : 7\sqrt[3]{3}/5$
0602 3252 $\mathrm{at} : 1 + \pi/4$
0603 0800 at : 25/14
0603 0988 $e^x : 1/3 - \pi$
0603 1912 rt : (3,8,-1,8)
0603 5215 $10^x : 8(1/\pi)/3$
0603 5652 $\mathrm{id} : 10\sqrt[3]{3}/7$
0604 0161 $10^x : 2/3 - 4\sqrt{2}/3$
0604 3194 rt : (1,6,1,-9)
0604 4486 rt : (9,2,-8,0)
0604 5038 $\mathrm{id} : 3e - 2\pi/3$
0604 6422 $\ln(2/5) \div \exp(e)$
0604 6539 $\ell_2 : \sqrt{2} + 4\sqrt{3}$
0604 8334 $\mathrm{sq} : \sqrt{3}/4 + 2\sqrt{5}$
0604 9481 $10^x : 3/2 - e$
0605 0645 $\Gamma : 1/\ln(3)$
0605 3166 $\mathrm{cr} : 1/4 + 2\sqrt{2}/3$
0605 3986 $\Gamma : \sqrt{2} + \sqrt{3} - \sqrt{5}$
0605 4077 $10^x : -\sqrt{2} - \sqrt{6} + \sqrt{7}$
0605 4897 rt : (9,6,-8,-9)
0605 5029 $2^x : 2/3 - 3\pi/2$
0605 5548 $\mathrm{cu} : \ln(\sqrt{2}\pi/3)$
0605 5669 $10^x : 3 + 3\sqrt{2}/2$
0605 5913 $\mathrm{cu} : \pi/8$
0605 8593 $\mathrm{at} : 1 - 3\sqrt{2}/4$
0605 8806 $e^x : 1/17$
0605 9983 $J_2 : (\pi)^3$
0606 0170 $e^x : e + 2\pi/3$
0606 2462 ln : 16/17
0606 2685 $\mathrm{at} : 2\sqrt{3} - 3\sqrt{5}/4$
0606 4702 rt : (2,8,0,2)
0606 4777 rt : (9,3,-6,1)
0606 6017 $\mathrm{id} : 3\sqrt{2}/4$
0606 6270 $Ei : 5/14$
0606 6668 $2^x : \exp(\pi/2)$
0606 9743 $\mathrm{as} : 1 - 3\sqrt{2}/4$
0606 9784 $\ell_{10} : 2/23$
0607 0396 $\Psi : 19/25$
0607 3058 $\Psi : 4\sqrt[3]{5}/9$
0607 3068 $t\pi : \sqrt{2}/3 - 2\sqrt{5}/3$
0607 5740 $\sinh : 6(e+\pi)/5$
0607 7021 $s\pi : 3e/8$
0607 7918 $1/4 \div \exp(\sqrt{2})$
0607 8536 $\zeta : 1/\ln(\sqrt[3]{2}/3)$
0607 8983 $2^x : 4/3 + 3\pi/2$
0607 9235 $t\pi : 3\pi/4$
0607 9784 $J_0 : 9\pi^2/10$
0608 1006 $\exp(1/5) \div \exp(3)$
0608 2788 $2^x : \sqrt{3}/4 - 2\sqrt{5}$
0608 3019 $\ln : 3/2 - \sqrt{5}/4$
0608 3642 $Ei : e/7$
0608 4303 $\mathrm{id} : 2\sqrt{2}/3 + \sqrt{5}/2$
0608 5005 $\mathrm{at} : 1/3 - 3\sqrt{2}/2$
0608 5332 rt : (9,4,-4,2)
0608 5635 $10^x : 1/2 + 4e/3$
0608 6592 rt : (8,1,-7,1)
0608 8186 $\ln(4/5) + \exp(1/4)$
0608 8273 $t\pi : 3e/8$
0609 0138 rt : (1,9,0,5)
0609 0722 $\sinh : 7e\pi/3$
0609 3601 $\exp(1/2) - s\pi(1/5)$
0609 3942 $\mathrm{id} : 10e/3$
0609 6624 $e^x : 2/3 - 2\sqrt{3}$
0609 6889 $e^x : 1/3 + \pi/4$
0609 8441 $\exp(2/3) + \exp(\sqrt{2})$
0609 9269 $t\pi : 3e/2 + 4\pi$
0609 9297 $\Psi : (3/2)^3$
0610 0199 sinh : 25/17
0610 0242 $s\pi : 4e/3 + 3\pi/4$
0610 2061 $\ln : 5\pi/2$
0610 2259 $\mathrm{at} : 1/4 - 3e/4$
0610 3295 $\mathrm{id} : 10(1/\pi)/3$
0610 5664 $\mathrm{at} : 4\sqrt{5}/5$
0610 6157 rt : (9,5,-2,3)
0610 7074 $2^x : 2\pi^2/7$
0610 7438 rt : (8,2,-5,2)
0610 8720 rt : (7,1,-8,-1)
0610 9429 $2^x : e/4 - 3\pi/2$
0611 0004 rt : (6,8,-9,0)
0611 0683 $J_2 : 5/7$
0611 1256 $\zeta : 7(1/\pi)/6$
0611 1314 $\Psi : (\sqrt{3})^{-1/2}$
0611 1370 $\sqrt{2} \div \exp(\pi)$
0611 1624 $t\pi : 4e/3 + 3\pi/4$
0611 2260 $Ei : 10(1/\pi)/3$
0611 3302 $e^x : 7e\pi$
0611 5995 $\mathrm{cu} : 2/3 - 3\sqrt{2}/4$
0611 9108 $2^x : 24/23$
0612 0866 rt : (6,9,-5,4)
0612 0874 $\mathrm{id} : \sqrt{2}/2 + 3\sqrt{5}/2$
0612 1722 $\ell_{10} : 2\sqrt{2} - 3\sqrt{5}/4$
0612 2448 $\mathrm{sq} : \sqrt{3}/7$
0612 4653 $\ell_{10} : 7\pi^2/6$
0612 5699 $J_0 : \sqrt[3]{12}$
0612 6609 $2^x : 3/2 - \sqrt{2}$
0612 6725 $\mathrm{sq} : 5\sqrt[3]{3}/7$
0612 7258 rt : (9,6,0,4)
0612 8251 $\zeta : \sqrt{19}$
0612 8560 rt : (8,3,-3,3)
0612 9306 $\mathrm{sq} : 7(1/\pi)/9$
0612 9628 $J_2 : (\ln 3/3)^{1/3}$
0612 9864 rt : (7,0,-6,2)
0613 0644 $\ln(\sqrt{5}) - s\pi(1/3)$
0613 0765 $J_2 : ((e+\pi)/3)^{-1/2}$
0613 1170 rt : (6,3,-9,-1)
0613 1265 $J_2 : (\sqrt{5}/2)^{-3}$
0613 1621 $\ln(\sqrt{2}) \div \exp(\sqrt{3})$
0613 4994 $J : \exp(-15\pi/16)$
0613 5946 $\Psi : 2\zeta(3)/5$
0613 6238 $t\pi : 2\sqrt{2}/3 - 3\sqrt{3}/4$
0613 8454 $\Psi : 7/5$
0613 8573 $10^x : \pi/10$
0613 9575 rt : (1,7,-1,-4)
0614 0054 $\ell_2 : 23/24$
0614 0373 $\Gamma : 10/11$
0614 2018 $J_2 : 9(1/\pi)/4$
0614 2124 $\mathrm{cr} : (\sqrt[3]{5})^{1/3}$
0614 2837 $\ell_2 : 3\sqrt{5}/7$
0614 3497 rt : (5,5,-9,4)
0614 5259 $\mathrm{cu} : \sqrt{2}/3 - \sqrt{3}/2$
0614 5571 $J_1 : \exp(-2\pi/3)$
0614 6664 $\theta_3 : 3(1/\pi)/2$
0614 7449 $\Psi : 10\sqrt[3]{2}/9$
0614 7501 $J_2 : (e)^{-1/3}$
0614 8642 rt : (9,7,2,5)
0614 9591 $\sinh : 4\ln 2/3$
0614 9966 rt : (8,4,-1,4)
0615 1292 rt : (7,1,-4,3)
0615 1335 $s\pi : 4\sqrt{2}/3 - \sqrt{3}/2$
0615 2348 $\exp(2/5) + s\pi(\pi)$
0615 2620 rt : (6,2,-7,-2)
0615 3504 $10^x : 1/\ln(e/2)$
0615 4153 $\theta_3 : \exp(-1/(3e))$
0615 4153 $\mathrm{sr} : 3e\pi$
0615 4595 rt : (5,6,-5,8)
0615 4927 rt : (1,8,0,4)
0615 5281 sr : 17/4
0615 5329 $\mathrm{sq} : e/4 - \pi$
0615 8228 $Ei : ((e+\pi)/2)^{-1/3}$
0615 8593 $\mathrm{cu} : 4\sqrt{2}/3 - 2\sqrt{5}/3$
0615 8598 $10^x : 14/23$
0615 8915 $\mathrm{sq} : 6\zeta(3)/7$
0615 9588 $J_1 : \exp(\pi)$
0616 0129 $J_0 : \sqrt{8}\pi$
0616 0529 $J : \exp(-13\pi/7)$
0616 0781 cu : 14/11
0616 2475 $\ln(3/4) - s\pi(e)$
0616 2802 rt : (4,9,-7,9)
0616 3006 $t\pi : 4\sqrt{2}/3 - \sqrt{3}/2$
0616 3239 $t\pi : \sqrt{2} - 2\sqrt{3}/3$
0616 4212 rt : (1,9,-1,-4)
0616 6288 $10^x : 7\sqrt[3]{2}/5$

0616 6747 cu : $2/3 + \pi/4$
0616 9519 J_1 : $\sqrt{2} - \sqrt{6} - \sqrt{7}$
0617 0315 Ei : $1/\sqrt[3]{22}$
0617 0315 rt : (9,8,4,6)
0617 0489 $2/5 \times \exp(e)$
0617 1663 rt : (8,5,1,5)
0617 2839 sq : $\sqrt{5}/9$
0617 3012 rt : (7,2,-2,4)
0617 4363 rt : (6,1,-5,-3)
0617 4418 Ψ : $\sqrt[3]{3}/3$
0617 5715 rt : (5,4,-8,-2)
0617 5732 id : $\sqrt{2}/4 + 3\sqrt{5}$
0617 6290 ζ : $8e/5$
0617 6730 cr : $7\sqrt[3]{5}/10$
0617 6849 sq : $\sqrt{2}/3 + \sqrt{5}/4$
0617 7727 rt : (4,2,-9,8)
0617 8465 $\ln(4) + s\pi(\sqrt{5})$
0618 1950 Γ : $(4/3)^{-1/3}$
0618 3189 ζ : $\ln(1/\pi)$
0618 3489 $s\pi$: $\exp(\sqrt{2}\pi)$
0618 3617 $t\pi$: $3\sqrt{2} + 4\sqrt{3}/3$
0618 8037 $\exp(1/2) + \exp(5)$
0618 8324 J : $1/19$
0618 9499 cu : $2/3 + \sqrt{2}/4$
0618 9667 2^x : $\ln 2/8$
0618 9899 10^x : $18/23$
0619 0988 10^x : $2\sqrt{2}/9$
0619 1147 Ψ : $\sqrt{7/3}$
0619 1751 id : $4e + 4\pi/3$
0619 2287 rt : (9,9,6,7)
0619 2337 cu : $9\sqrt{2}/10$
0619 3658 rt : (8,6,3,6)
0619 5030 rt : (7,3,0,5)
0619 5345 $t\pi$: $\exp(\sqrt{2}\pi)$
0619 6404 rt : (6,0,-3,4)
0619 7781 rt : (5,3,-6,-3)
0619 9158 rt : (4,6,-9,-2)
0619 9267 sr : $8(e + \pi)/5$
0620 0397 ζ : $3\sqrt[3]{5}/5$
0620 1208 rt : (3,7,-8,-4)
0620 1518 sinh : $\ln(2\pi)$
0620 1877 $s\pi$: $9\sqrt[3]{3}$
0620 3816 J_2 : $4\sqrt[3]{2}/7$
0620 4559 J_2 : $18/25$
0621 2533 2^x : $10\ln 2$
0621 2717 2^x : $2/23$
0621 2724 rt : (3,8,-2,5)
0621 3040 ℓ_{10} : $8\sqrt[3]{3}$
0621 3599 Ψ : $(\ln 2/3)^{1/2}$
0621 3839 $t\pi$: $9\sqrt[3]{3}$
0621 3871 cu : $4\ln 2/7$
0621 4790 ℓ_{10} : $13/15$
0621 5606 ln : $2\sqrt{3}/3 + 3\sqrt{5}$
0621 5640 Ψ : $10(e + \pi)$
0621 5958 rt : (8,7,5,7)
0621 6593 sinh : $(\sqrt[3]{5}/2)^{1/2}$
0621 7355 rt : (7,4,2,6)
0621 7782 id : $7\sqrt{3}/2$
0621 8753 rt : (6,1,-1,5)
0621 8788 id : $1/4 + 2e/3$
0621 9455 Li_2 : $\sqrt[3]{1/2}$
0621 9805 ℓ_{10} : $2/3 + 4e$
0622 0154 rt : (5,2,-4,-4)
0622 0768 J_1 : $7\sqrt[3]{5}/3$
0622 1556 rt : (4,7,-5,-5)
0622 3110 J_2 : $\sqrt[3]{3}/2$
0622 3523 sr : $\sqrt{3}/4 + 4\sqrt{5}$
0622 3596 sq : $3e/2 - 4\pi$
0622 4035 e^x : $\sqrt{3}/3 - 3\sqrt{5}/2$
0622 4915 J_2 : $3\zeta(3)/5$
0622 6452 ℓ_{10} : $\ln 2/8$
0622 6787 J_2 : $1/\ln(4)$
0622 7631 cu : $7(1/\pi)$
0622 7909 $1/3 - \exp(1/3)$
0622 8070 10^x : $\sqrt{2}/3 - 3\sqrt{5}/4$
0622 9416 ℓ_{10} : $3\sqrt{3}/2 + 4\sqrt{5}$
0623 0589 id : $9\sqrt{5}/2$
0623 0947 cu : $3/4 - \sqrt{2}/4$
0623 2166 rt : (9,6,-9,-8)
0623 3760 ln : P_{46}
0623 7800 J_1 : $1/8$
0623 8572 rt : (8,8,7,8)
0623 8697 $t\pi$: $2\sqrt{2} - 2\sqrt{5}$
0623 9994 rt : (7,5,4,7)
0624 1233 J_2 : $13/18$
0624 1373 sq : $4/3 + 3\sqrt{5}/4$
0624 1418 rt : (6,2,1,6)
0624 1874 tanh : $1/16$
0624 1880 at : $1/16$
0624 2843 rt : (5,1,-2,-5)
0624 3557 rt : (9,5,-7,-9)
0624 4270 rt : (4,8,-1,-1)
0624 4515 cr : $8\pi^2/9$
0624 5700 rt : (3,7,-8,-3)
0624 6807 λ : $(2\pi)^{-3}$
0624 6832 2^x : $\sqrt{2}/3 - 2\sqrt{5}$
0624 6924 cr : $1/3 + \sqrt{3}/2$
0624 6936 ℓ_{10} : $\sqrt{3}/2$
0624 7294 10^x : $1/4 + \sqrt{3}/2$
0624 8521 rt : (2,8,-1,-2)
0624 8683 as : $1/\ln(\pi)$
0624 9089 Ei : $\exp(\sqrt[3]{3})$
0624 9624 rt : (1,8,-1,-4)
0625 0000 id : $1/16$
0625 0135 2^x : $-\sqrt{2} + \sqrt{5} + \sqrt{7}$
0625 1136 as : $\sqrt[3]{2/3}$
0625 1482 $\ln(3) - s\pi(\sqrt{2})$
0625 3800 Ei : $9\sqrt[3]{2}$
0625 4069 sinh : $1/16$
0625 4076 as : $1/16$
0625 5749 rt : (8,7,-8,-9)
0625 6244 at : $4\pi/7$
0625 6458 ζ : $13/3$
0625 7312 Γ : $(1/\pi)/4$
0625 7796 ln : $2 - 3\sqrt{2}/4$
0626 1508 rt : (8,9,9,9)
0626 2265 $\exp(1/4) - \exp(1/5)$
0626 2329 erf : $1/18$
0626 2453 J_2 : $(3)^3$
0626 2956 rt : (7,6,6,8)
0626 4405 rt : (6,3,3,7)
0626 5856 cr : $6/5$
0626 7310 rt : (4,9,3,3)
0626 7421 ℓ_{10} : $5\ln 2/3$
0626 7727 J_2 : $6\sqrt[3]{3}$
0626 8037 rt : (7,9,-9,-9)
0626 8765 rt : (3,6,-6,-4)
0627 0136 e^x : $3\sqrt{3} - 2\sqrt{5}$
0627 0222 rt : (2,9,-9,-3)
0627 2746 ζ : $5\sqrt{3}/2$
0627 3027 J_1 : $8\pi^2/3$
0627 9852 e^x : $4 + 4\sqrt{3}$
0628 0833 ℓ_{10} : $3\sqrt[3]{3}/5$
0628 1561 Ψ : $3\ln 2/8$
0628 1963 2^x : $1/4 - 3\sqrt{2}$
0628 2972 e^x : $\sqrt{2} - \sqrt{3} - \sqrt{6}$
0628 3143 ζ : $1/\ln(\sqrt[3]{2})$
0628 6249 rt : (7,7,8,9)
0628 6626 2^x : $13/5$
0628 7113 J_1 : $\sqrt[3]{2}/10$
0628 7535 rt : (4,9,-8,6)
0628 7725 rt : (6,4,5,8)
0628 8261 J_0 : $5e\pi/8$
0628 8469 e^x : $5e\pi/6$
0628 9203 rt : (5,1,2,7)
0628 9224 cr : $\sqrt{2} + \sqrt{5} - \sqrt{6}$
0628 9568 cr : $4e - 2\pi/3$
0628 9957 ζ : $3\sqrt[3]{3}$
0629 0683 rt : (4,2,-1,-6)
0629 1425 10^x : $4\zeta(3)/3$
0629 2165 rt : (3,5,-4,-5)
0629 3483 $\sqrt{2} + \exp(1/2)$
0629 3507 sr : $(\sqrt[3]{3})^{1/3}$
0629 3649 rt : (2,8,-7,-4)
0629 7291 Ei : $7/18$
0629 8080 10^x : $-\sqrt{2} - \sqrt{5} + \sqrt{6}$
0629 8304 $s\pi$: $7\sqrt[3]{2}/9$
0629 8978 sr : $(\ln 2)^{-1/3}$
0629 9127 cu : $5(1/\pi)/4$
0629 9312 ℓ_{10} : $1/4 + e/3$
0629 9761 θ_3 : $(\ln 3/3)^{1/3}$
0630 0728 Γ : $\sqrt{2} - \sqrt{3} + \sqrt{6}$
0630 0882 e^x : $4 - 3\sqrt{3}/2$
0630 1266 2^x : $2e - 3\pi$
0630 2773 e^x : $2e - 3\pi/2$
0630 3608 at : $3/4 + \pi/3$
0630 3805 J_2 : $5\pi^2/2$
0630 3993 rt : (6,9,-6,2)
0630 4008 Ei : $1/15$
0630 4339 $\exp(\sqrt{5}) \div \exp(5)$
0630 5724 Ei : $1/\sqrt[3]{17}$
0630 8957 e^x : $4 - e/3$
0630 9540 rt : (1,7,7,8)
0631 0723 $s\pi$: $7\sqrt{2}/5$
0631 0834 $t\pi$: $7\sqrt[3]{2}/9$
0631 1385 rt : (6,5,7,9)
0631 2891 rt : (5,2,4,8)
0631 4398 rt : (4,1,1,-7)
0631 5908 rt : (3,5,-1,3)
0631 7394 cu : $4 + \sqrt{5}/2$
0631 7420 rt : (2,7,-5,-5)
0631 8581 cu : $(e/2)^{-3}$
0631 8667 ln : $2\sqrt{2}/3 + 4\sqrt{3}$
0631 9423 J : $((e + \pi)/2)^{-2}$
0632 1164 cr : $5\sqrt[3]{3}/6$
0632 1776 cu : $3\sqrt{2}/4 - 3\sqrt{5}/2$
0632 3190 sr : $\exp(1/(3e))$
0632 3327 $t\pi$: $7\sqrt{2}/5$
0632 3546 10^x : $8\pi/3$
0632 4659 sq : $2/3 + 2\sqrt{3}$
0632 4820 ℓ_2 : $1/4 + \sqrt{2}/2$
0632 4925 J_2 : $8/11$
0632 6538 cr : $\zeta(3)$
0632 6664 J_2 : $\exp(-1/\pi)$
0632 7035 J_0 : $16/7$
0632 7302 J_2 : $(\sqrt[3]{4}/3)^{1/2}$
0632 9780 Γ : $(\sqrt{5}/3)^{1/3}$
0633 1168 Ψ : $(\ln 2)^2$
0633 4323 ln : $3/2 - 2\sqrt{3}/3$
0633 6930 rt : (5,3,6,9)
0633 8466 rt : (4,0,3,8)
0633 8635 cu : $4/3 - \sqrt{3}$
0633 9903 Γ : $(\sqrt{2}/3)^3$
0634 0004 rt : (3,6,3,7)
0634 0734 $s\pi$: $4\sqrt{2} - 3\sqrt{5}/4$
0634 0774 rt : (5,6,-6,6)
0634 0962 10^x : $\exp(-\sqrt[3]{3}/2)$
0634 1544 rt : (2,6,-3,-6)
0634 1651 $t\pi$: $2\sqrt{3}/7$
0634 3087 rt : (1,9,-6,-5)
0634 3341 λ : $\exp(-7\pi/4)$
0634 4077 θ_3 : $((e + \pi)/3)^{-1/2}$
0634 7086 at : $1/2 + 3\sqrt{3}/4$
0634 8047 $s\pi$: $2/3 + \sqrt{2}/4$
0634 8225 e^x : $\sqrt{2}/2 - 2\sqrt{3}$
0634 9363 sr : $(2\pi)^{-3}$
0634 9604 sq : $\sqrt[3]{2}/5$
0635 0468 Li_2 : $1/16$
0635 0832 sq : $\sqrt{3}/2 - \sqrt{5}/2$
0635 2586 K : $6\ln 2/5$
0635 2886 Γ : $9e\pi/10$
0635 3305 J_1 : $2(1/\pi)/5$
0635 3519 $t\pi$: $4\sqrt{2} - 3\sqrt{5}/4$
0635 3616 cu : P_{59}
0635 7579 $\exp(\sqrt{5}) + s\pi(1/4)$
0635 7611 tanh : $(1/\pi)/5$
0635 7618 at : $(1/\pi)/5$
0635 9014 10^x : $4 - 2\sqrt{5}/3$
0636 0876 $t\pi$: $1/3 - \sqrt{2}/4$
0636 1336 e^x : $7\pi/6$
0636 2896 rt : (4,1,5,9)
0636 3580 θ_3 : $(\sqrt{5}/2)^{-3}$
0636 4463 rt : (3,2,2,-8)
0636 5720 10^x : $4 - 3\sqrt{3}$
0636 5917 sr : $4\sqrt{2}/5$
0636 6033 rt : (2,7,3,1)
0636 6197 id : $(1/\pi)/5$
0636 6403 θ_3 : $(1/\pi)/10$
0636 6631 e^x : $6\zeta(3)$
0636 6907 ℓ_{10} : $19/22$

0636 7122 at : $10\sqrt[3]{2}/7$
0636 7605 rt : (1,8,-4,-6)
0636 8377 e^x : $2+2\sqrt{3}$
0636 9782 at : $9/5$
0637 0498 sinh : $(1/\pi)/5$
0637 0505 as : $(1/\pi)/5$
0637 0700 Ψ : $(\ln 3/3)^{-1/3}$
0637 1506 ln : P_{62}
0637 1514 sq : $4-2e$
0637 1885 2^x : $1/2-2\sqrt{5}$
0637 3689 Γ : $3\pi^2/5$
0637 4669 sq : $8\sqrt[3]{2}/5$
0637 4706 sπ : $10\sqrt{2}/7$
0637 4872 10^x : $\sqrt{2}/4+3\sqrt{5}$
0637 4884 sq : $7e$
0637 6163 Γ : $e/3$
0637 6316 cr : $3(e+\pi)/2$
0637 7231 sπ : $10e/9$
0637 8940 10^x : $(\sqrt[3]{5}/2)^{-1/2}$
0637 9042 Ei : $(3\pi/2)^{-3}$
0637 9300 θ_3 : $11/23$
0638 1812 ℓ_{10} : $(\sqrt{5}/3)^{1/2}$
0638 4443 sq : $4\sqrt{3}/3-\sqrt{5}/4$
0638 4614 Ψ : $((e+\pi)/3)^{1/2}$
0638 5149 cu : $10\zeta(3)/7$
0638 6189 sπ : $9e\pi/7$
0638 6596 2^x : $2/3-\sqrt{3}/3$
0638 7070 J_2 : $(\sqrt[3]{5}/2)^2$
0638 7698 tπ : $10\sqrt{2}/7$
0638 8741 Ψ : $10(e+\pi)/7$
0638 9296 rt : (3,1,4,-9)
0639 0238 tπ : $10e/9$
0639 0736 Ψ : $5\sqrt{5}/8$
0639 0895 rt : (2,4,1,-8)
0639 1276 tanh : $(5/2)^{-3}$
0639 1283 at : $(5/2)^{-3}$
0639 1696 rt : (3,7,-9,-7)
0639 2476 tπ : $6\sqrt[3]{5}$
0639 2498 rt : (1,7,-2,-7)
0639 2786 $\exp(1/4)\div\exp(3)$
0639 3271 2^x : $21/13$
0639 3891 J_2 : $8(e+\pi)$
0639 4520 sinh : $(\sqrt[3]{2})^{-1/3}$
0639 5613 rt : (9,2,-9,-1)
0639 5911 2^x : $(\sqrt{5})^{-3}$
0639 9251 tπ : $9e\pi/7$
0639 9516 J_2 : $7\sqrt{2}/2$
0640 0000 id : $(5/2)^{-3}$
0640 1655 2^x : $23/22$
0640 2024 Ψ : $1/\ln(\ln 3/2)$
0640 2495 K : $\exp(-1/(2e))$
0640 4279 rt : (3,8,-3,2)
0640 4369 sinh : $(5/2)^{-3}$
0640 4377 as : $(5/2)^{-3}$
0640 8401 rt : (2,8,-9,9)
0640 9820 cu : $4(1/\pi)$
0641 0514 J_2 : $\ln(\sqrt[3]{3}/3)$
0641 1786 J_2 : $(e+\pi)/8$
0641 2675 $3/5\div\exp(\sqrt{5})$
0641 3033 ℓ_2 : $11/23$
0641 5066 sq : $e/4+\pi/2$
0641 5079 2^x : $3/4-3\pi/2$
0641 5269 $3/4\times\exp(3)$
0641 5598 10^x : $9(1/\pi)/2$
0641 6143 rt : (2,3,3,-9)
0641 6916 J_2 : $4\pi^2$
0641 7232 rt : (4,9,-9,3)
0641 7777 rt : (1,6,0,-8)
0641 7847 J_2 : $3\sqrt[3]{5}/7$
0641 8534 cu : $\sqrt{2}/2+\sqrt{5}/3$
0641 8764 2^x : $4+\pi/3$
0641 9413 rt : (9,3,-7,0)
0641 9439 $\ln(3/4)\times\ln(4/5)$
0642 5984 J_2 : $11/15$
0642 6730 J_0 : $16/3$
0643 0683 e^x : $4/3-3e/2$
0643 0689 $\exp(\sqrt{2})+s\pi(2/5)$
0643 1816 rt : (9,5,-8,-8)
0643 2116 tπ : $\sqrt{3}-2\sqrt{5}$
0643 2873 J_0 : $7\pi^2/6$
0643 2966 cu : $\zeta(3)/3$
0643 4759 ζ : $\sqrt{5}/6$
0643 6023 at : $5\sqrt[3]{3}/4$
0643 6268 tπ : $\sqrt[3]{2}$
0643 7115 cu : $1/4-4\sqrt{2}$
0643 7228 at : $(e+\pi)^{1/3}$
0644 0003 sπ : $8\sqrt{3}/7$
0644 0006 e^x : $3/2-3\sqrt{2}$
0644 1149 J_2 : $5\sqrt{3}$
0644 1747 at : $\sqrt{3}/4-\sqrt{5}$
0644 2395 tπ : $3\ln 2/8$
0644 2454 at : $3\zeta(3)/2$
0644 2612 θ_3 : $\ln(\sqrt[3]{5}/3)$
0644 3453 rt : (2,8,-2,-6)
0644 3554 rt : (9,4,-5,1)
0644 4313 rt : (9,4,-6,-9)
0644 4842 $\exp(\sqrt{3})-s\pi(1/5)$
0644 5105 cr : $(\sqrt[3]{5}/3)^{-1/3}$
0644 5122 rt : (8,9,-7,-8)
0644 6759 $\exp(2/5)\div\exp(\pi)$
0644 9445 e^x : $1/16$
0644 9510 id : $2\sqrt{2}+\sqrt{5}$
0645 0523 θ_3 : $(\pi)^{-3}$
0645 2860 ℓ_{10} : $4/3-\sqrt{2}/3$
0645 3292 sq : $4\sqrt{3}-4\sqrt{5}$
0645 3399 tπ : $8\sqrt{3}/7$
0645 3426 sq : $5\pi^2/9$
0645 3852 ln : $15/16$
0645 4639 $\exp(2)+s\pi(\sqrt{5})$
0645 6484 e^x : $\sqrt{3}-2\sqrt{5}$
0645 7455 sπ : $10\zeta(3)$
0645 7721 rt : (8,6,-7,-9)
0645 8129 sr : $17/15$
0645 8297 e^x : $5\sqrt[3]{3}/3$
0646 2450 ℓ_{10} : P_5
0646 3912 e^x : $8\sqrt[3]{2}/9$
0646 8045 rt : (9,5,-3,2)
0646 9035 Ψ : $8\pi/3$
0646 9641 rt : (8,2,-6,1)
0647 0492 Li_2 : $(1/\pi)/5$
0647 0583 sq : $(\ln 3)^{1/3}$
0647 0961 tπ : $10\zeta(3)$
0647 1240 rt : (7,8,-8,-9)
0647 2028 ℓ_2 : $\exp(2\pi)$
0647 3405 $\exp(2/3)+\exp(3/4)$
0647 4311 2^x : $2e/3+3\pi/2$
0647 4553 10^x : $3-4\pi/3$
0647 5031 tπ : $3\sqrt{3}/4-\sqrt{5}/4$
0647 5224 cr : $1/2+\sqrt{2}/2$
0647 5979 Ei : $\exp(-2e/3)$
0647 6386 ζ : $(\sqrt[3]{3}/2)^{1/2}$
0647 7127 Γ : $e\pi/4$
0647 8101 Ei : $5\sqrt[3]{2}/9$
0648 1299 at : $4/3+\sqrt{2}/3$
0648 1714 Γ : $19/21$
0648 1847 10^x : $e/3-2\pi/3$
0648 2761 tπ : $\sqrt{2}/3+4\sqrt{3}$
0648 3906 ln : $3/2-\sqrt{3}/4$
0648 4202 J_2 : $(5/2)^{-1/3}$
0648 4357 sq : $1-\sqrt{5}/3$
0648 4555 sq : $4(1/\pi)/5$
0648 4804 J_2 : $14/19$
0648 5541 $e+\ln(\sqrt{2})$
0648 6099 sr : $9\sqrt[3]{2}/10$
0648 8243 sr : $2-\sqrt{3}/2$
0648 9759 2^x : $7\sqrt{5}/4$
0649 0719 Ei : $7/10$
0649 2786 cu : $3-3\sqrt{3}/2$
0649 2895 rt : (9,6,-1,3)
0649 4299 e^x : $7\zeta(3)/6$
0649 4521 rt : (8,3,-4,2)
0649 6150 rt : (7,0,-7,1)
0649 7781 rt : (6,9,-7,0)
0649 9521 rt : (1,9,0,-9)
0650 4108 2^x : $1/11$
0650 4679 tπ : $10\pi/9$
0650 5161 cr : $4e/9$
0650 5422 Li_2 : $(5/2)^{-3}$
0650 6646 sπ : $7\sqrt{3}/6$
0650 6810 2^x : $2(1/\pi)/7$
0650 7644 cu : $7\zeta(3)/3$
0650 7879 J_1 : $3/23$
0650 7987 tπ : $\sqrt{3}+\sqrt{5}/3$
0650 8798 cu : $4\sqrt{2}/3-\sqrt{3}/4$
0651 0337 rt : (1,7,6,2)
0651 1311 at : $3\sqrt{2}/4+\sqrt{5}/3$
0651 1776 Ei : $17/16$
0651 2648 e^x : $1/4-4\sqrt{5}/3$
0651 7598 J_2 : $9\sqrt[3]{2}$
0651 8114 rt : (9,7,1,4)
0651 8236 2^x : $\sqrt{2}/4+3\sqrt{5}/2$
0651 9771 rt : (8,4,-2,3)
0652 0030 sπ : $2/3+3\sqrt{5}/2$
0652 0325 cu : $1/3+3\sqrt{3}$
0652 0464 tπ : $7\sqrt{3}/6$
0652 0712 Ψ : $\ln(\ln 3)$
0652 0876 2^x : $1/4-4\pi/3$
0652 1265 10^x : $3/4+2\sqrt{5}/3$
0652 1431 rt : (7,1,-5,2)
0652 2281 Ψ : $4\pi/9$
0652 3093 rt : (6,2,-8,-1)
0652 3287 J_2 : $17/23$
0652 5470 J_0 : $7\zeta(3)$
0652 5549 J_2 : $\ln(2\pi/3)$
0652 7145 2^x : $(\sqrt{2}\pi/2)^{-3}$
0652 8629 10^x : $\sqrt[3]{2}/4$
0652 8834 ln : $e\pi/8$
0652 9389 2^x : $3/4+\sqrt{3}/2$
0653 0948 2^x : $3/2-2e$
0653 2262 e^x : $1/\ln(\ln 2)$
0653 2299 ℓ_{10} : $3/4+4e$
0653 2557 tπ : $\exp(\sqrt{5})$
0653 3932 tπ : $1/3-3\sqrt{5}/2$
0653 5853 Ei : $20/3$
0653 7329 ℓ_{10} : $(\pi/2)^{1/3}$
0653 7980 rt : (5,6,-7,4)
0653 8414 id : $1/3+\sqrt{3}$
0653 8442 K : $(\ln 2)^{1/2}$
0653 9277 $\exp(5)+\exp(\sqrt{3})$
0654 3581 as : $7/8$
0654 3713 rt : (9,8,3,5)
0654 4293 ℓ_2 : $5(e+\pi)/7$
0654 5402 rt : (8,5,0,4)
0654 7093 rt : (7,2,-3,3)
0654 8432 10^x : $\sqrt{2}-3\sqrt{3}/2$
0654 8787 rt : (6,1,-6,-2)
0655 0484 rt : (5,4,-9,-1)
0655 1306 rt : (5,7,-3,8)
0655 2979 J_2 : $16/3$
0655 7363 J_2 : $(\ln 3/2)^{1/2}$
0655 9225 id : $3\sqrt{2}/2+4\sqrt{5}$
0655 9371 J_1 : $10\sqrt{5}$
0656 0569 ζ : $5\sqrt[3]{5}/2$
0656 2828 10^x : $8(e+\pi)/7$
0656 3704 id : $e/2-3\pi$
0656 3896 at : $4/3-\pi$
0656 4053 cr : $4\sqrt{2}/3+4\sqrt{3}$
0656 5327 ln : $2/3+\sqrt{5}$
0656 8905 2^x : $3-4\sqrt{3}$
0656 9253 K : $(\sqrt[3]{3})^{-1/2}$
0656 9701 rt : (9,9,5,6)
0657 0706 Ψ : $12/25$
0657 1423 rt : (8,6,2,5)
0657 3147 rt : (7,3,-1,4)
0657 3209 e^x : $(1/\pi)/5$
0657 4874 rt : (6,0,-4,3)
0657 5083 rt : (1,8,-9,6)
0657 5671 cr : $23/19$
0657 5853 2^x : $8\sqrt{2}/7$
0657 6604 rt : (5,3,-7,-2)
0657 6876 J_2 : $3\sqrt{3}/7$
0657 7362 at : $2\sqrt{2}/3+\sqrt{3}/2$
0657 9176 rt : (4,3,-7,8)
0657 9551 $5\times\exp(5)$
0658 0355 Ψ : $\exp(-2\pi)$
0658 1226 cu : $6\pi^2$
0658 3900 J_2 : $7(1/\pi)/3$
0658 4130 cu : $e\pi/3$
0658 5559 e^x : $7(e+\pi)/4$
0658 5967 J_2 : $(2e/3)^{-1/2}$

0658 7751 $\zeta : e\pi/2$
0658 9242 $\tanh : \exp(-e)$
0658 9250 at : $\exp(-e)$
0658 9291 $\Psi : (e)^{1/3}$
0658 9470 $e^x : 9\sqrt[3]{3}/8$
0659 0624 at : $1/2 - 4\sqrt{3}/3$
0659 6568 $\Gamma : 3(e+\pi)/4$
0659 6643 ln : $4\sqrt{2} + \sqrt{5}$
0659 6914 cu : $2\sqrt{2}/7$
0659 7847 rt : (8,7,4,6)
0659 8803 id : $\exp(-e)$
0659 9289 cu : $e/4 + 4\pi$
0659 9409 $\Gamma : (2e)^{-1/2}$
0659 9605 rt : (7,4,1,5)
0660 0358 sr : 25/22
0660 0528 sr : $\sqrt{3}/3 + \sqrt{5}/4$
0660 1366 rt : (6,1,-2,4)
0660 2540 $1/5 + s\pi(1/3)$
0660 3130 rt : (5,2,-5,-3)
0660 3308 $J : 4/17$
0660 3593 sinh : $\exp(-e)$
0660 3601 as : $\exp(-e)$
0660 3957 $J_2 : (e/3)^3$
0660 4332 $J_1 : 4$
0660 4896 rt : (4,7,-6,-7)
0660 7324 $\exp(1/3) \div s\pi(\sqrt{5})$
0660 7523 rt : (3,8,-4,-1)
0660 7874 $10^x : 3/2 + 4\sqrt{2}/3$
0660 7995 cr : $7\sqrt[3]{2}$
0660 9239 $e^x : (5/2)^{-3}$
0661 3374 ln : $4/3 + \pi/2$
0661 3924 sq : $4e - 2\pi/3$
0661 4159 $\Gamma : 5\sqrt[3]{5}/4$
0661 9671 $\ell_2 : 3 - e/3$
0662 0068 at : $3/4 + 3\sqrt{2}/4$
0662 0300 $\ell_{10} : 5\zeta(3)/7$
0662 1310 sq : $4\sqrt{2}/3 + \sqrt{3}/3$
0662 1418 rt : (3,9,2,8)
0662 1768 $\ln(3) \times \exp(3)$
0662 4117 cr : $2/3 + 3e$
0662 4269 cr : $6\sqrt{2}/7$
0662 4684 rt : (8,8,6,7)
0662 4971 rt : (7,8,-9,9)
0662 5113 ln : $8\zeta(3)/9$
0662 6445 $e^x : 3/4 - 2\sqrt{3}$
0662 6477 rt : (7,5,3,6)
0662 7711 $10^x : (\sqrt[3]{3}/2)^{1/2}$
0662 8273 rt : (6,2,0,5)
0662 8287 cu : $3/4 - 2\sqrt{3}/3$
0662 8472 $J_2 : \sqrt{5}/3$
0662 9876 $erf : 1/17$
0663 0072 rt : (5,1,-3,-4)
0663 0657 $\ell_{10} : 4 - \pi$
0663 0973 rt : (9,5,-9,-7)
0663 1622 ln : $4\pi^2/5$
0663 1672 cr : $7\sqrt{3}/10$
0663 1769 $\ell_{10} : 3e/7$
0663 1874 rt : (4,8,-2,-3)
0663 2134 ln : $3e/2 - \pi$
0663 2725 $Li_2 : 5(1/\pi)/2$
0663 3679 rt : (3,7,-9,-2)
0663 6567 sr : $e\pi/2$
0663 7061 id : $1/2 + 4\pi$
0663 7804 $e^x : 2 - 3\pi/2$
0663 9001 $J_0 : 4\sqrt[3]{5}/3$
0664 0192 $\Gamma : (e/2)^{-1/3}$
0664 0625 sq : 23/16
0664 0776 ln : $\sqrt{2} + 2\sqrt{5}/3$
0664 2575 $10^x : (\sqrt{5}/3)^{1/3}$
0664 4144 sinh : $5\sqrt[3]{5}$
0664 4614 rt : (9,4,-7,-8)
0664 4723 $\sqrt{2} + \exp(\sqrt{3})$
0664 6251 ln : $2\sqrt{3} - \sqrt{5}/4$
0664 7567 ln : $5\sqrt[3]{5}/8$
0664 8439 cr : $7\ln 2/4$
0665 1172 $2^x : \pi/3$
0665 1862 $J_1 : 2/15$
0665 1865 $J_1 : 5e/2$
0665 1947 rt : (8,9,8,8)
0665 1950 $10^x : 1/2 - 3\sqrt{5}/4$
0665 3362 $\ell_2 : \pi/3$
0665 3777 rt : (7,6,5,7)
0665 4579 $\Psi : \sqrt{23/2}$
0665 4908 id : $\sqrt{23}\pi$
0665 5508 $2^x : 1/3 - 3\sqrt{2}$
0665 5609 rt : (6,3,2,6)
0665 5751 at : $2e/3$
0665 5796 $Ei : 4\sqrt[3]{5}$
0665 5918 $\ln(5) \times \exp(1/4)$
0665 6807 tanh : 1/15
0665 6816 at : 1/15
0665 7033 sq : $(\pi/2)^{-3}$
0665 7444 rt : (5,0,-1,5)
0665 8136 $\ell_{10} : 3/2 - \sqrt{2}$
0665 8363 rt : (9,3,-5,-9)
0665 8551 $\ln(\sqrt{2}) - \exp(5)$
0665 9283 rt : (8,6,-8,-8)
0666 1124 rt : (3,6,-7,-3)
0666 3009 sq : P_{28}
0666 3213 $J_1 : \zeta(3)/9$
0666 3224 $\ell_{10} : (e/2)^{1/2}$
0666 4362 $\ell_{10} : 4e + \pi/4$
0666 4588 $e^x : 3e + 4\pi$
0666 4756 rt : (2,9,5,6)
0666 5641 $e^x : 4 - 3\sqrt{5}$
0666 5925 cu : $\ln(2/3)$
0666 6666 id : 1/15
0666 8278 $e^x : 5\sqrt[3]{3}/4$
0666 9524 $\Gamma : 9e$
0667 1606 sinh : 1/15
0667 1614 as : 1/15
0667 3153 rt : (8,5,-6,-9)
0667 4082 rt : (7,8,-9,-8)
0667 7166 $10^x : 3\sqrt{2}/4 - \sqrt{5}$
0667 8241 $\Gamma : 8(e+\pi)/3$
0668 0419 sr : $2/3 + \sqrt{2}/3$
0668 0989 ln : $2 + e/3$
0668 1516 rt : (7,7,7,8)
0668 1524 $J_2 : 5e\pi/8$
0668 2725 $10^x : 3\sqrt{2}/4 - \sqrt{5}/3$
0668 3312 $J_1 : 20$
0668 3386 rt : (6,4,4,7)
0668 3590 $s\pi(1/4) - s\pi(e)$
0668 5259 rt : (5,1,1,6)
0668 5884 cr : 17/14
0668 6654 cr : $3\sqrt{2}/2 + 3\sqrt{5}$
0668 7135 rt : (4,2,-2,-5)
0668 8074 rt : (7,7,-7,-9)
0668 8185 $J_2 : (\sqrt[3]{2}/3)^{1/3}$
0668 8296 $t\pi : 4\sqrt{2}/3 - 3\sqrt{5}/2$
0668 8736 at : $1/2 - \sqrt{3}/4$
0668 9014 rt : (3,5,-5,-4)
0668 9555 $\ell_{10} : (\sqrt[3]{4})^{1/3}$
0668 9615 $\zeta : 3\sqrt{2}/4$
0669 0209 $J_1 : (\ln 3/3)^2$
0669 0896 rt : (2,8,-8,-3)
0669 3970 $2^x : 3\sqrt[3]{3}$
0669 4081 $J_1 : ((e+\pi)/3)^{-3}$
0669 4678 $\ell_{10} : 6/7$
0669 5141 cr : $2 - \pi/4$
0669 5419 $\zeta : 17/4$
0669 6988 cu : $1/2 - e/3$
0669 8729 id : $1/2 - \sqrt{3}/4$
0669 9132 $10^x : 3 + 4\sqrt{2}$
0669 9356 $e^x : (e+\pi)^{1/3}$
0670 3130 rt : (6,9,-8,-2)
0670 3749 as : $1/2 - \sqrt{3}/4$
0670 5939 $e^x : 3/2 + 4\sqrt{5}$
0670 7399 $J_2 : 3/4$
0670 8898 $\Gamma : 3\pi^2/7$
0670 9708 rt : (7,8,9,9)
0671 0980 $Li_2 : \exp(-e)$
0671 1419 $\ell_2 : 21/22$
0671 1556 $2^x : 22/21$
0671 1617 rt : (6,5,6,8)
0671 3529 rt : (5,2,3,7)
0671 5444 rt : (4,1,0,-6)
0671 7363 rt : (3,5,-2,1)
0671 9284 rt : (2,7,-6,-4)
0671 9812 $\Psi : 23/15$
0672 2838 $2^x : \sqrt{3}/3 - 2\sqrt{5}$
0672 4008 rt : (1,7,5,-4)
0672 4251 $t\pi : \sqrt{3} + \sqrt{5} - \sqrt{6}$
0672 5493 $K : 5/6$
0672 5584 $2^x : 1 - e/3$
0672 7277 sq : $3\pi^2/5$
0672 8025 ln : $7\zeta(3)/9$
0672 9318 $J_2 : 5\zeta(3)/8$
0672 9732 $J : (1/\pi)/6$
0673 0076 $\lambda : \exp(-2e)$
0673 4700 sq : $2\sqrt{2} - \sqrt{3}/3$
0673 5210 $10^x : 4 - \sqrt{5}$
0673 5796 $\zeta : 3\sqrt{2}$
0673 6072 $2^x : \ln(\ln 3)$
0673 7520 cu : $\sqrt{2} - \sqrt{5} + \sqrt{7}$
0673 8733 $\Gamma : 3\zeta(3)/4$
0674 0273 $\Gamma : 10\sqrt[3]{5}$
0674 0317 rt : (6,6,8,9)
0674 2172 $J_2 : \ln(\sqrt{2}/3)$
0674 2269 rt : (5,3,5,8)
0674 4224 rt : (4,0,2,7)
0674 6183 rt : (3,6,2,5)
0674 6677 id : $e\pi/8$
0674 7164 rt : (5,6,-8,2)
0674 8041 $e^x : 3\sqrt{2} - \sqrt{3}/3$
0674 8146 rt : (2,6,-4,-5)
0674 9707 $\Gamma : 5\sqrt[3]{3}/8$
0675 0111 rt : (1,9,-7,-4)
0675 1172 $\Gamma : 3/7$
0675 1283 $J_1 : (e)^{-2}$
0675 6370 sq : $3\ln 2/8$
0675 6762 sq : $4e - 2\pi$
0675 8313 cr : $\sqrt{2}/3 + \sqrt{5}/3$
0675 9277 sr : $5\sqrt[3]{5}/2$
0676 1767 rt : (5,7,-4,6)
0676 3778 sr : $(\sqrt{2}\pi/3)^{1/3}$
0676 3807 sr : $3\sqrt{3}/2 + 3\sqrt{5}/4$
0676 3997 $\sqrt{2} - \ln(\sqrt{2})$
0676 5394 $J_0 : 23/2$
0676 5898 $J_1 : 11/3$
0676 6764 $1/2 \div \exp(2)$
0676 7026 $\theta_3 : (2e)^{-2}$
0676 8687 $2^x : 3/4 + e$
0677 0031 sr : $2\sqrt[3]{5}/3$
0677 0654 at : $\sqrt[3]{6}$
0677 1493 rt : (5,4,7,9)
0677 3133 $\Psi : 7\pi$
0677 3490 rt : (4,1,4,8)
0677 4801 $J : (3\pi)^{-1/2}$
0677 5491 rt : (3,7,6,9)
0677 6377 sr : $\sqrt{2}/2 + \sqrt{3}/4$
0677 7496 rt : (2,7,2,-2)
0677 8529 $J_2 : (\ln 3)^{-3}$
0677 9504 rt : (1,8,-5,-5)
0678 0539 $\exp(\sqrt{2}) \times s\pi(\sqrt{3})$
0678 1198 $Li_2 : 1/15$
0678 3055 sr : $4\sqrt{2}/3 - \sqrt{5}/3$
0678 3490 $\theta_3 : 9(1/\pi)/4$
0678 4116 $\ell_2 : (\ln 3)^{1/2}$
0678 7179 $Ei : 9\pi^2/5$
0678 7226 $2^x : (\ln 3)^{1/2}$
0678 7905 $Ei : 9(1/\pi)$
0678 8663 sinh : 13/14
0679 2359 rt : (4,3,-8,6)
0679 2443 $J_2 : (\sqrt[3]{5}/3)^{1/2}$
0679 3143 $2^x : 2\sqrt{2} - 3\sqrt{5}$
0679 3148 cr : $\sqrt{2} + \sqrt{6} - \sqrt{7}$
0679 3405 $\ell_{10} : \sqrt{2} - \sqrt{5}/4$
0679 5022 $J_2 : \sqrt{13}\pi$
0679 5064 sq : $(e/3)^{-1/3}$
0679 5311 at : 20/11
0679 6220 $\zeta : \ln(2e)$
0679 6311 $e^x : 3/2 - 4\pi/3$
0679 6523 $\lambda : \exp(-19\pi/11)$
0679 7465 at : $9\sqrt{2}/7$
0679 7490 $2^x : \exp(\sqrt[3]{3}/3)$
0679 9854 $Ei : 4\sqrt[3]{5}/3$
0680 2346 $J_1 : 3/22$
0680 2388 $\Psi : e/8$
0680 3258 rt : (4,2,6,9)

0680 3999 ℓ_{10} : $\sqrt[3]{5}/2$
0680 4138 cu : $1/\sqrt{6}$
0680 4664 ζ : $\exp(\sqrt[3]{3})$
0680 5002 cu : $3/2 + 3\sqrt{5}/4$
0680 5071 J_1 : $3(1/\pi)/7$
0680 5293 sq : $6/23$
0680 5302 rt : (3,1,3,-8)
0680 5525 e^x : $2/3 - 3\sqrt{5}/2$
0680 5653 $\exp(3/4) + s\pi(2/5)$
0680 5957 10^x : $1/2 - \sqrt{2}/3$
0680 6225 id : $\sqrt{3}/3 + 2\sqrt{5}/3$
0680 6729 ζ : $3\pi^2/7$
0680 7350 rt : (2,8,8,7)
0680 9143 J_2 : $3\sqrt[3]{2}/5$
0680 9402 rt : (1,7,-3,-6)
0680 9556 2^x : $7\ln 2/3$
0681 0816 $\ln(4) \times \exp(2/5)$
0681 3385 rt : (9,3,-8,-1)
0681 4303 Ψ : 22
0681 4695 sr : $3/4 - \sqrt{5}/3$
0681 5387 sinh : $19/9$
0681 5474 Ei : $(\sqrt{2}\pi/2)^3$
0681 7898 ζ : $\ln(1/3)$
0681 8282 e^x : $3\sqrt{2} - 4\sqrt{3}$
0682 0527 tπ : $9/19$
0682 0790 cu : $1/3 - 3\sqrt{5}$
0682 1393 e^x : $\exp(-e)$
0682 2888 J_1 : $8e/3$
0682 3508 rt : (3,8,-5,-4)
0682 4169 2^x : $2/21$
0682 7223 J_1 : $7\pi/6$
0682 8641 sq : $(\ln 2/2)^{-3}$
0682 9020 cr : $(2e/3)^{1/3}$
0683 0997 Ei : $5\pi/9$
0683 3737 rt : (1,7,-8,5)
0683 4159 e^x : $3\zeta(3)/2$
0683 4772 rt : (1,5,-5,9)
0683 5632 rt : (3,0,5,9)
0683 7725 rt : (2,3,2,9)
0683 8773 rt : (3,9,1,5)
0683 9265 e^x : $4\sqrt{3}/3$
0683 9822 rt : (1,6,-1,-7)
0684 0674 cr : $\sqrt{2}/4 + \sqrt{3}/2$
0684 1923 rt : (9,4,-6,0)
0684 2764 cu : $3\sqrt{2}/2 - 4\sqrt{3}$
0684 2953 $4/5 \times \exp(3)$
0684 3042 2^x : $3(1/\pi)/10$
0684 3893 rt : (8,1,-9,-1)
0684 5856 Γ : $\sqrt{3} - \sqrt{5} + \sqrt{7}$
0684 6356 cu : $9/22$
0684 8598 at : $1/\ln(\sqrt{3})$
0684 9470 J_1 : $7\sqrt{2}$
0684 9502 id : $8\zeta(3)/9$
0684 9742 sr : $\sqrt{2}/3 + 4\sqrt{5}$
0685 1512 10^x : $1/4 - \sqrt{2}$
0685 1864 cr : $e\pi/7$
0685 4626 cu : $9(1/\pi)/7$
0685 4671 cu : $1/4 - 3e/2$
0685 4996 sq : $3\sqrt{3}/4 - 2\sqrt{5}$
0685 5564 ℓ_{10} : $e\pi/10$
0685 5965 at : $1/4 + \pi/2$
0685 6844 rt : (9,4,-8,-7)
0685 8347 id : $9\pi/4$
0685 9098 cu : $2\sqrt{3}/3 - \sqrt{5}/3$
0686 1699 e^x : $2\zeta(3)$
0686 2870 Γ : $9/10$
0686 4941 J_1 : $10\zeta(3)/3$
0686 7420 Γ : $5\sqrt[3]{2}/7$
0686 7739 ℓ_2 : $\sqrt{2}/3 + \sqrt{3}/3$
0686 8638 rt : (2,2,4,-9)
0686 9191 at : $2/3 + 2\sqrt{3}/3$
0686 9604 sq : $((e+\pi)/3)^{-2}$
0687 0782 rt : (1,7,9,-4)
0687 0920 rt : (9,5,-4,1)
0687 1891 rt : (9,3,-6,-8)
0687 2076 e^x : $3\pi^2/5$
0687 2879 ln : $2\zeta(3)/7$
0687 2930 rt : (8,6,-9,-7)
0687 3462 2^x : $4/3 - 3\sqrt{3}$
0687 3496 id : $5\sqrt[3]{5}/8$
0687 4351 2^x : $\sqrt{2}/3 + \sqrt{3}/3$
0687 5860 J_2 : $(\sqrt{3})^{-1/2}$
0687 6930 ℓ_{10} : $4 - 2\sqrt{2}$
0687 8506 J_2 : $4\sqrt[3]{5}/9$
0687 8690 J_2 : $19/25$
0687 8694 $\ln(3/4) + \exp(\sqrt{5})$
0688 2631 tπ : $e\pi/6$
0688 3214 cr : $(\ln 3/2)^{-1/3}$
0688 3403 Ψ : $5e/4$
0688 4338 cu : $3e - \pi/3$
0688 5655 cu : $1/3 + 2\sqrt{5}/3$
0688 6470 rt : (2,9,4,2)
0688 7067 rt : (9,2,-4,-9)
0688 8117 rt : (8,5,-7,-8)
0688 8943 θ_3 : $12/25$
0689 1830 ℓ_{10} : $2(e+\pi)$
0689 2775 sr : $3\sqrt{3}/4 + 4\sqrt{5}/3$
0689 3910 e^x : $1/15$
0689 3961 cr : $3/4 + \sqrt{2}/3$
0689 4162 cr : $5\sqrt[3]{5}/7$
0689 5963 e^x : $3\sqrt{3} + 3\sqrt{5}/2$
0689 7703 e^x : $7\sqrt[3]{5}$
0689 9287 ln : $14/15$
0689 9801 e^x : $3e/4 - 3\pi/2$
0689 9848 e^x : $7\sqrt[3]{2}/4$
0690 0389 rt : (9,6,-2,2)
0690 1953 $\ln(4) \div \exp(3)$
0690 2299 rt : (1,5,6,9)
0690 2441 rt : (8,3,-5,1)
0690 3434 rt : (8,4,-5,-9)
0690 4496 sr : $8/7$
0690 7150 Γ : $15/7$
0690 8350 sq : $\sqrt{2} - 3\sqrt{5}/4$
0690 8628 2^x : $1/3 - 4\pi/3$
0691 1522 J_2 : $16/21$
0691 2011 ℓ_{10} : $\exp(1/(2\pi))$
0691 3273 $\ln(3/5) \div \exp(2)$
0691 3808 10^x : $6/19$
0691 4833 J_1 : $\ln 2/5$
0691 6840 Γ : $\exp(-1/(3\pi))$
0691 7291 ℓ_2 : $(\sqrt{3}/2)^{1/3}$
0691 7810 cr : $11/9$
0691 8856 J_2 : $\pi^2/2$
0691 9528 e^x : $3/2 + 4e/3$
0691 9958 rt : (7,6,-6,-9)
0692 0392 tanh : $\ln 2/10$
0692 0402 at : $\ln 2/10$
0692 1033 rt : (6,9,-9,-4)
0692 3403 tπ : $\sqrt{2}/3 - \sqrt{3}$
0692 5207 sq : $5/19$
0692 5376 rt : (7,9,-7,9)
0692 6019 2^x : $(\sqrt{3}/2)^{-1/3}$
0692 6814 sq : $4e/3 - 2\pi$
0692 6899 rt : (1,7,-9,-3)
0692 7722 at : $\sqrt{2} - \sqrt{5} + \sqrt{7}$
0692 7918 Ei : $10e\pi/9$
0692 8189 e^x : $1/2 - \sqrt{3}/4$
0693 0345 rt : (9,7,0,3)
0693 1053 ℓ_{10} : $\sqrt{3} - \sqrt{5}/4$
0693 1152 at : $1/3 + 2\sqrt{5}/3$
0693 1471 id : $\ln 2/10$
0693 2440 rt : (8,4,-3,2)
0693 3646 ln : $1/2 + \sqrt{3}/4$
0693 3662 10^x : $9\pi/2$
0693 4081 ζ : $7\zeta(3)/2$
0693 4538 rt : (7,1,-6,1)
0693 4703 Ψ : $7\zeta(3)$
0693 5559 10^x : $(e+\pi)^{-2}$
0693 6212 e^x : $17/3$
0693 6641 rt : (6,8,-7,-9)
0693 6790 Ψ : $-\sqrt{2} + \sqrt{3} - \sqrt{5}$
0693 7023 sinh : $\ln 2/10$
0693 7034 as : $\ln 2/10$
0693 8893 rt : (1,9,1,-8)
0693 8964 cu : $\sqrt{2}/2 - \sqrt{5}/2$
0694 0704 $3 \times \exp(\sqrt{5})$
0694 2395 J_1 : $5(e+\pi)/8$
0694 2557 Ψ : $\sqrt{3} + \sqrt{6} - \sqrt{7}$
0694 2900 e^x : $8/11$
0694 4160 10^x : $3\sqrt[3]{3}/4$
0694 4456 10^x : $3\pi/8$
0694 8073 id : $\sqrt{2}/3 + 3\sqrt{3}/2$
0694 8309 ζ : $1/\sqrt[3]{19}$
0694 8345 sq : $\exp(-4/3)$
0695 1799 rt : (1,4,-7,-8)
0695 2762 e^x : $\sqrt{3}/3 + 3\sqrt{5}$
0695 5011 $\ln(5) \div \exp(\pi)$
0695 6055 $e - \exp(1/2)$
0695 6450 2^x : $1/2 + \sqrt{5}/2$
0695 6456 at : $\sqrt{2}/2 + \sqrt{5}/2$
0695 6783 Ψ : $17/5$
0695 9500 J_0 : $4\pi^2$
0695 9924 J_2 : $13/17$
0695 9997 rt : (2,9,-4,9)
0695 9999 ln : $3/2 + \sqrt{2}$
0696 0194 10^x : $1/3 - 2\sqrt{5}/3$
0696 0246 J_2 : $(\sqrt[3]{5})^{-1/2}$
0696 0802 rt : (9,8,2,4)
0696 2942 rt : (8,5,-1,3)
0696 4309 sπ : $7e\pi/10$
0696 4552 e^x : $\exp(-1/\pi)$
0696 5085 rt : (7,2,-4,2)
0696 5642 ζ : $22/13$
0696 7091 e^x : P_{73}
0696 7233 rt : (6,1,-7,-1)
0696 9384 rt : (5,6,-9,0)
0696 9878 $1/2 + s\pi(\pi)$
0697 0331 at : $\sqrt{10/3}$
0697 1690 rt : (1,8,-1,-9)
0697 2501 e^x : $(\sqrt[3]{4}/3)^{1/2}$
0697 5097 ℓ_2 : $2 - \pi/3$
0697 5147 ζ : $21/5$
0697 5777 sq : $4 + e/3$
0697 6436 J_1 : $9(e+\pi)/2$
0697 6654 ζ : $10\sqrt[3]{2}/3$
0697 7280 tπ : $\sqrt{3}/2 + 2\sqrt{5}/3$
0697 7288 10^x : $16/7$
0697 7812 sinh : $5\zeta(3)/2$
0698 1188 ζ : $3\sqrt[3]{3}/7$
0698 1260 tπ : $7e\pi/10$
0698 1477 cu : $7/17$
0698 2428 J_1 : $\sqrt[3]{2}/9$
0698 2991 sπ : $(\ln 2/2)^{-3}$
0698 3303 tπ : $\ln(2\pi/3)$
0698 5424 rt : (5,7,-5,4)
0698 6662 e^x : $1 + 2\sqrt{5}/3$
0698 9252 J_2 : $(\sqrt{2}\pi/2)^{-1/3}$
0699 0227 $\ln(\sqrt{3}) \times \exp(2/3)$
0699 0776 rt : (1,8,7,3)
0699 1319 sr : $\sqrt[3]{3/2}$
0699 1776 rt : (9,9,4,5)
0699 2050 sr : $\ln(\pi)$
0699 2973 at : $2\sqrt{3}/3 - 4\sqrt{5}/3$
0699 3962 rt : (8,6,1,4)
0699 4032 $\ln(3/4) \div \exp(\sqrt{2})$
0699 4606 cu : $1/3 - \sqrt{5}/3$
0699 6152 rt : (7,3,-2,3)
0699 7084 cu : $2\sqrt[3]{3}/7$
0699 8012 θ_3 : $(e)^{-1/3}$
0699 8012 id : $(3\pi)^{1/2}$
0699 8346 rt : (6,0,-5,2)
0700 0079 tπ : $(\ln 2/2)^{-3}$
0700 0545 rt : (5,3,-8,-1)
0700 1608 rt : (5,8,-1,8)
0700 2829 tπ : $e - \pi$
0700 3154 e^x : $4e/3 - 2\pi$
0700 3308 ln : $\sqrt{17/2}$
0700 8792 ζ : $3/8$
0701 0594 rt : (1,9,8,4)
0701 2864 J_1 : $7(e+\pi)$
0701 6331 cu : $4/3 - 3\pi/4$
0701 6860 e^x : $3 - 4\sqrt{2}$
0701 8066 sr : $4 + 4\pi$
0701 8649 cr : $2/3 + \sqrt{5}/4$
0701 9091 rt : (4,3,-9,4)
0702 1023 tπ : $2\pi^2$
0702 1216 id : $3\sqrt{3}/2 + 2\sqrt{5}$
0702 1389 ℓ_{10} : $\sqrt{14}\pi$
0702 2013 θ_3 : $(\ln 2)^2$
0702 2136 e^x : $7\sqrt[3]{3}/9$

0702 2409 as : $(\sqrt{2}\pi/3)^{-1/3}$
0702 5518 rt : (8,7,3,5)
0702 7756 rt : (7,4,0,4)
0702 9892 ℓ_2 : $5\sqrt[3]{2}/3$
0702 9999 rt : (6,1,-3,3)
0703 1343 at : $3/4 - e/4$
0703 1805 J_1 : 13
0703 2229 at : $1-2\sqrt{2}$
0703 2245 rt : (5,2,-6,-2)
0703 2899 J_0 : 25/11
0703 4243 sinh : $1/\ln((e+\pi)/2)$
0703 4453 $s\pi(\sqrt{3}) + s\pi(\sqrt{5})$
0703 4496 rt : (4,7,-7,-9)
0703 5585 rt : (4,4,-5,8)
0703 6028 J : $\exp(-14\pi/15)$
0703 6193 sq : $5(1/\pi)/6$
0703 6330 10^x : $1/\ln(\sqrt[3]{2}/3)$
0703 8411 J_2 : 10/13
0703 8932 ℓ_2 : 5/21
0703 9632 ζ : $4\pi/3$
0704 0954 $\sqrt{2} \div \exp(3)$
0704 2643 ln : $1+4\sqrt{3}$
0704 2954 id : $3/4 - e/4$
0704 3197 erf : 1/16
0704 3539 J_2 : $3(e+\pi)$
0704 3542 2^x : $5\sqrt[3]{2}/6$
0704 4188 2^x : $1/4 - 3e/2$
0704 4584 ζ : $3\sqrt{3}/5$
0704 7853 J_1 : $3e\pi/7$
0704 8317 J_2 : $4\sqrt{3}/9$
0704 8789 as : $3/4 - e/4$
0704 9649 sr : $(e+\pi)^{-3}$
0705 1235 tπ : $7\sqrt{5}/9$
0705 2984 2^x : 21/20
0705 3404 J_1 : $\sqrt{2}/10$
0705 3411 rt : (3,8,-6,-7)
0705 4636 J_2 : $10\ln 2/9$
0705 4708 Li_2 : $(\pi/2)^{-1/2}$
0705 5297 $\ln(2/5) - \exp(e)$
0705 5436 Li_2 : $\ln 2/10$
0705 5871 J_1 : $4(1/\pi)/9$
0705 7426 Γ : $2\pi/7$
0705 7627 rt : (8,8,5,6)
0705 7962 ζ : $5(e+\pi)/7$
0705 8107 ℓ_{10} : 17/20
0705 9915 rt : (7,5,2,5)
0706 1779 tanh : $2(1/\pi)/9$
0706 1790 at : $2(1/\pi)/9$
0706 2207 rt : (6,2,-1,4)
0706 4504 rt : (5,1,-4,-3)
0706 6805 rt : (4,8,-3,-5)
0706 7206 Γ : $\exp(-3\pi)$
0706 7840 at : $3/2 - \pi/2$
0706 9326 λ : $\exp(-12\pi/7)$
0707 0226 rt : (3,9,0,2)
0707 1327 ln : $10(e+\pi)$
0707 1357 $\sqrt{5} \div \ln(\sqrt{3})$
0707 2247 sr : $3/2 - \sqrt{2}/4$
0707 2334 Ei : $6\pi/7$
0707 2519 J_1 : $\exp(-(e+\pi)/3)$
0707 3549 cu : P_{56}
0707 3553 id : $2(1/\pi)/9$
0707 3855 tπ : 6/23
0707 3865 θ_3 : $(1/\pi)/9$
0707 4238 2^x : $\exp(\sqrt{2}\pi/3)$
0707 5877 Ei : $10e/9$
0707 6848 $2/5 \div \exp(\sqrt{3})$
0707 8167 J_0 : $9\sqrt{2}/5$
0707 8624 e^x : $1/3 - 4\sqrt{5}/3$
0707 9453 sinh : $2(1/\pi)/9$
0707 9465 as : $2(1/\pi)/9$
0707 9632 id : $1/2+\pi/2$
0708 0042 rt : (2,9,-5,4)
0708 0220 rt : (7,9,-8,7)
0708 2011 rt : (9,4,-9,-6)
0708 2326 Ψ : $3\sqrt[3]{2}/5$
0708 4924 sinh : 11
0708 5560 as : $3/2 - \pi/2$
0708 6376 sinh : $(\sqrt{3}/2)^{1/2}$
0708 6382 e^x : $\sqrt{2}/2 - 3\sqrt{5}/2$
0708 7000 cr : $2\sqrt{3} - \sqrt{5}$
0708 7453 θ_3 : $(\ln 2/3)^{1/2}$
0708 7604 ℓ_{10} : $2\sqrt{2}+4\sqrt{5}$
0709 0305 rt : (8,9,7,7)
0709 1908 sq : $9e\pi/10$
0709 2645 rt : (7,6,4,6)
0709 4370 rt : (2,6,-7,7)
0709 4990 rt : (6,3,1,5)
0709 6857 θ_3 : $(\ln 2)^{1/3}$
0709 7339 rt : (5,0,-2,4)
0709 8515 rt : (9,3,-7,-7)
0709 9311 J_2 : 17/22
0709 9478 ℓ_{10} : $(\sqrt[3]{3}/2)^{1/2}$
0709 9692 rt : (4,9,1,-1)
0710 0251 cu : $(\sqrt{5}/3)^3$
0710 0311 cr : $9\pi^2/10$
0710 0897 sr : $e+\pi/2$
0710 1455 $\exp(3/5) \times s\pi(1/5)$
0710 2051 rt : (3,6,-8,-2)
0710 2543 J_2 : $8\sqrt{2}$
0710 3654 Γ : $\exp(\sqrt[3]{3})$
0710 4643 2^x : $\ln 2/7$
0710 6781 id : $5\sqrt{2}$
0710 9176 Li_2 : $\ln(\sqrt{2}\pi/2)$
0710 9280 θ_3 : $\sqrt[3]{3}/3$
0710 9655 J_1 : $9e\pi/4$
0711 0271 $\exp(1/3) + s\pi(\sqrt{5})$
0711 1111 sq : 4/15
0711 3372 ℓ_2 : $(e/3)^{1/2}$
0711 5168 rt : (9,2,-5,-8)
0711 6359 rt : (8,5,-8,-7)
0711 8114 ℓ_{10} : $3\pi/8$
0712 0749 sr : $e - \pi/2$
0712 1593 J_0 : $8(1/\pi)$
0712 2129 ℓ_2 : $1/4+3e$
0712 2732 10^x : $1/\sqrt{10}$
0712 2868 rt : (2,9,3,-2)
0712 4077 J_0 : $1/\ln(\sqrt{2}\pi/3)$
0712 4571 cr : $\sqrt{8}\pi$
0712 4651 J_1 : 1/7
0712 4770 e^x : $1/2 - \pi$
0712 5195 $2/3 \div \exp(\sqrt{5})$
0712 5425 ζ : $\exp(-\sqrt[3]{3}/3)$
0712 5968 rt : (7,7,6,7)
0712 6680 Γ : $(\sqrt[3]{3}/2)^{1/3}$
0712 8119 ℓ_{10} : $\sqrt{3}/4+\sqrt{5}/3$
0712 8366 rt : (6,4,3,6)
0712 8374 cu : $3\sqrt{2}/2 - 3\sqrt{3}$
0713 0721 θ_3 : $2\zeta(3)/5$
0713 0734 tanh : 1/14
0713 0746 at : 1/14
0713 0769 rt : (5,1,0,5)
0713 0774 2^x : $(e/3)^{-1/2}$
0713 1961 J_2 : $(5/3)^{-1/2}$
0713 1972 rt : (9,1,-3,-9)
0713 2927 2^x : $1+2\sqrt{3}$
0713 3177 rt : (8,4,-6,-8)
0713 3293 rt : (1,9,7,-4)
0713 3375 ℓ_{10} : $3\sqrt{2}/5$
0713 4383 rt : (7,7,-9,-7)
0713 5510 sq : $2\zeta(3)/9$
0713 5590 rt : (3,5,-6,-3)
0713 8007 rt : (2,8,-9,-2)
0714 0900 sπ : $6(e+\pi)/7$
0714 2763 cu : $\sqrt{3}/3+4\sqrt{5}/3$
0714 2857 id : 1/14
0714 3930 ln : $2\sqrt{3}+2\sqrt{5}$
0714 4960 at : 11/6
0714 8106 2^x : $e/3 - 3\pi/2$
0714 8314 sr : $2/3+4e/3$
0714 8932 sinh : 1/14
0714 8944 as : 1/14
0715 0150 rt : (8,3,-4,-9)
0715 0694 10^x : $10\sqrt{2}/3$
0715 1369 rt : (7,6,-7,-8)
0715 1787 sinh : $\sqrt{11}\pi$
0715 2199 2^x : $2/3 - 2\sqrt{5}$
0715 2296 tπ : $3e+2\pi$
0715 2844 10^x : $2/3 - 2e/3$
0715 2989 id : $e/2+3\pi/2$
0715 3603 ln : $\sqrt[3]{2}/10$
0715 3907 rt : (1,8,6,-4)
0715 5133 sπ : $(\pi/3)^{-1/2}$
0715 5566 J_1 : $(2e)^2$
0715 6441 sq : P_{66}
0715 6470 J_2 : $3e$
0715 6808 sq : P_{65}
0715 9177 tπ : $6(e+\pi)/7$
0715 9670 J_2 : $\exp(e)$
0715 9901 rt : (7,8,8,8)
0715 9948 ln : $\sqrt{2}/4+\sqrt{3}/3$
0716 2356 rt : (6,5,5,7)
0716 2522 sr : $(\pi/3)^3$
0716 4815 rt : (5,2,2,6)
0716 5176 10^x : $3/2+2e$
0716 5889 10^x : $\ln(1/\pi)$
0716 6457 cr : 16/13
0716 7234 cu : $\sqrt{2}+\sqrt{5}/3$
0716 7279 rt : (4,1,-1,-5)
0716 7929 J_2 : $2e/7$
0716 8514 rt : (7,5,-5,-9)
0716 8857 ζ : 25/6
0716 9749 rt : (6,8,-8,-8)
0717 0553 10^x : 18/5
0717 2223 rt : (2,7,-7,-3)
0717 2388 ℓ_{10} : $1/2+e/4$
0717 2906 J : $(\ln 2/3)^2$
0717 3467 J_1 : $\ln(\sqrt{3}/2)$
0717 3519 tπ : $(\pi/3)^{-1/2}$
0717 3537 $\ln(2/3) \div \exp(\sqrt{3})$
0717 3792 erf : $(1/\pi)/5$
0717 4210 sq : $3/4 - 3e/2$
0717 5871 sπ : $2/3+3\pi/4$
0717 7346 2^x : 1/10
0717 8920 sπ : $4\sqrt[3]{5}/7$
0717 9676 sq : $2-\sqrt{3}$
0718 0024 2^x : $10\sqrt[3]{5}$
0718 4394 Ψ : 20/13
0718 4598 sinh : $((e+\pi)/3)^3$
0718 4616 cu : $\sqrt{2}/4+\sqrt{3}$
0718 5961 2^x : P_{18}
0718 7068 rt : (6,7,-6,-9)
0718 7295 Ei : $(\ln 2/2)^{1/3}$
0718 7586 2^x : P_{48}
0718 7664 J_2 : 7/9
0718 9056 J_1 : $8e\pi/5$
0718 9903 sr : $4+2e$
0719 1477 e^x : $2/3+\pi$
0719 2051 ln : $(\sqrt{3}/2)^{1/2}$
0719 2514 J_1 : $\sqrt[3]{3}/10$
0719 3332 cu : $3\ln 2/5$
0719 4389 10^x : $3\ln 2$
0719 4418 tπ : $1/3 - 3\pi/4$
0719 5580 J : $\exp(-(e+\pi)/2)$
0719 6329 J_2 : $(\sqrt{2}/3)^{1/3}$
0719 6541 e^x : $3e+2\pi/3$
0719 6980 rt : (6,6,7,8)
0719 7491 tπ : $4\sqrt[3]{5}/7$
0719 9498 rt : (5,3,4,7)
0720 2021 rt : (4,0,1,6)
0720 2737 Li_2 : $2(1/\pi)/9$
0720 4187 rt : (2,9,-6,-1)
0720 4549 rt : (3,6,1,3)
0720 4935 sq : $1/4+\pi/4$
0720 5295 sinh : $\sqrt{14}$
0720 5815 rt : (5,9,-7,-9)
0720 7083 rt : (2,6,-5,-4)
0720 9622 rt : (1,9,-8,-3)
0721 1778 erf : $(5/2)^{-3}$
0721 1880 $1/4+\exp(3/5)$
0721 1920 Ei : 9/23
0721 3177 $1/3+\ln(2/3)$
0721 4949 ζ : $6\ln 2$
0721 7841 10^x : $2-\pi$
0721 9104 tπ : $1/\sqrt[3]{22}$
0721 9763 2^x : $2/3+e/2$
0722 3002 J_2 : $9\ln 2/8$
0722 3211 e^x : $\sqrt{2}/4 - 4\sqrt{5}/3$
0722 3479 rt : (5,7,-6,2)
0722 6188 sr : $3\sqrt{2}+3\sqrt{3}$

0723 0299 ln : $\ln((e+\pi)/2)$
0723 1100 Ψ : $9\sqrt[3]{5}/10$
0723 1416 2^x : P_{61}
0723 2262 rt : (6,7,9,9)
0723 3796 cu : 5/12
0723 4840 rt : (5,4,6,8)
0723 6062 e^x : $e/2+3\pi/4$
0723 7425 rt : (4,1,3,7)
0723 8052 sr : 23/20
0723 9006 J_1 : $10\pi/3$
0724 0014 rt : (3,7,5,7)
0724 0184 J : $10(1/\pi)/9$
0724 0931 e^x : $(\sqrt[3]{2})^{1/2}$
0724 1311 rt : (5,8,-2,6)
0724 1736 2^x : P_{45}
0724 1743 rt : (7,9,-9,5)
0724 2609 rt : (2,7,1,-5)
0724 4359 $\exp(1/4)-\exp(\sqrt{5})$
0724 5210 rt : (1,8,-6,-4)
0724 7245 10^x : $(\sqrt[3]{5}/3)^{-1/3}$
0724 8499 e^x : $1-4e/3$
0724 8949 at : $\ln(2\pi)$
0724 9401 $t\pi$: $9(e+\pi)$
0724 9526 $\ln(3)\div\exp(e)$
0725 1469 cr : $\pi^2/8$
0725 1676 2^x : $\sqrt{17\pi}$
0725 1678 at : $\sqrt{2}/3-4\sqrt{3}/3$
0725 2690 rt : (9,3,-9,-2)
0725 4701 J_2 : $(2\pi/3)^{-1/3}$
0725 5066 ℓ_{10} : 11/13
0725 6541 $s\pi$: $2\sqrt{3}+\sqrt{5}/4$
0725 9595 J_0 : $(e+\pi)^2$
0726 0846 rt : (2,6,-8,3)
0726 1463 Ψ : 4/23
0726 2354 $4/5\div s\pi(\sqrt{3})$
0726 7537 sinh : $\sqrt{24}$
0726 8540 J_0 : $\sqrt{13/2}$
0726 8895 10^x : $(\sqrt[3]{3}/3)^{1/3}$
0726 9808 2^x : $\sqrt{2}-3\sqrt{3}$
0727 0866 rt : (5,5,8,9)
0727 2600 J_2 : 18/23
0727 2711 cr : $3/4+3e$
0727 3513 rt : (4,2,5,8)
0727 4191 J_0 : $(3)^3$
0727 4627 Li_2 : 1/14
0727 5204 Ψ : $6\pi^2$
0727 5541 2^x : $(\pi)^{-2}$
0727 5723 $t\pi$: $2\sqrt{3}+\sqrt{5}/4$
0727 6167 rt : (3,1,2,-7)
0727 8826 rt : (6,6,-9,9)
0727 9636 Γ : $\sqrt[3]{5}/4$
0728 0331 10^x : $(\pi/3)^3$
0728 1490 rt : (1,7,-4,-5)
0728 2202 sq : $2/3-3e$
0728 3887 Ψ : 25/18
0728 6011 ℓ_2 : $7\zeta(3)$
0728 6217 J_2 : $(\sqrt[3]{3}/3)^{1/3}$
0728 6655 rt : (9,4,-7,-1)
0728 7318 Ψ : $8\sqrt{3}/9$
0728 7880 cu : $3+4e$
0728 8868 rt : (1,9,0,6)
0728 9917 sr : $2\sqrt{2}/3+3\sqrt{5}/2$
0729 1178 Γ : 17/19
0729 1238 $\ln(4)\times s\pi(e)$
0729 1911 10^x : $\sqrt{2}/4-2\sqrt{5}/3$
0729 5860 ln : $\sqrt[3]{5}/5$
0729 6896 e^x : $3/4-e/4$
0729 7628 cr : 21/17
0729 8298 J_0 : $9\sqrt[3]{2}/5$
0729 8940 ℓ_{10} : $2-2\sqrt{3}/3$
0730 2196 sr : $2\sqrt{2}-3\sqrt{5}/4$
0730 3267 ln : $1/4+e/4$
0730 4685 Ei : $(2\pi)^{-2}$
0730 6262 2^x : $3\sqrt{2}/4+\sqrt{5}/4$
0730 8226 $t\pi$: $3\sqrt{2}-4\sqrt{5}/3$
0730 8949 as : $3-3\sqrt{2}/2$
0731 0313 rt : (4,3,7,9)
0731 0632 θ_3 : $(\sqrt{3})^{-1/3}$
0731 1448 2^x : $7\zeta(3)/8$
0731 3032 rt : (3,0,4,8)
0731 5341 sr : $1/2+4\sqrt{5}$
0731 5758 rt : (2,3,1,7)
0731 6609 Γ : $2\sqrt{5}/5$
0731 6676 sinh : $(3\pi)^{1/3}$
0731 7123 rt : (3,9,-1,-1)
0731 7823 10^x : $3/4-4\sqrt{2}/3$
0731 8490 rt : (1,6,-2,-6)
0732 0206 at : $4\sqrt{3}/3-\sqrt{5}$
0732 1228 rt : (9,5,-5,0)
0732 1411 cu : $1-e$
0732 1743 $s\pi$: $(\pi/3)^{1/2}$
0732 1832 J_2 : $\pi/4$
0732 2657 sq : $e/4+3\pi/2$
0732 3784 rt : (8,2,-8,-1)
0732 4253 cr : $6\sqrt[3]{3}/7$
0732 4653 rt : (1,5,-8,-9)
0732 6255 $4\div\exp(4)$
0732 6711 J_2 : $5\sqrt{2}/9$
0732 7236 Ψ : $(\sqrt[3]{5}/3)^{1/2}$
0732 7420 J_2 : 11/14
0732 8473 cr : $(\sqrt[3]{4}/3)^{-1/3}$
0732 9733 e^x : $2(1/\pi)/9$
0733 0374 $s\pi$: $(e+\pi)/6$
0733 2640 rt : (2,9,-7,-6)
0733 3309 id : $4\sqrt{3}/3-\sqrt{5}$
0733 4014 sq : $8\sqrt[3]{2}/7$
0733 9371 rt : (9,3,-8,-6)
0733 9898 as : $4\sqrt{3}/3-\sqrt{5}$
0734 1218 sr : $3+3\sqrt{3}/4$
0734 1448 $t\pi$: $(\pi/3)^{1/2}$
0734 3642 e^x : $5e\pi$
0734 4113 at : $3\sqrt{3}-3\sqrt{5}/2$
0734 8099 ℓ_{10} : $6\pi^2/5$
0734 8351 sinh : $7\pi/4$
0735 0148 $t\pi$: $(e+\pi)/6$
0735 0638 rt : (3,1,6,9)
0735 2178 cr : $5\sqrt{3}/7$
0735 3410 2^x : $9\sqrt[3]{2}/7$
0735 3433 rt : (2,2,3,-8)
0735 5518 2^x : $\sqrt{2}/2-2\sqrt{5}$
0735 6235 rt : (1,7,8,-9)
0735 6427 rt : (9,6,-3,1)
0735 7588 $1/5\div e$
0735 7688 rt : (9,2,-6,-7)
0735 8144 J_2 : $5\sqrt[3]{2}/8$
0735 9042 rt : (8,5,-9,-6)
0735 9632 cr : $7\sqrt{2}/8$
0736 1509 ℓ_{10} : $4\sqrt{2}-2\sqrt{5}$
0736 1664 rt : (7,0,-9,1)
0736 3435 ℓ_2 : $2\sqrt{3}+\sqrt{5}/3$
0736 8680 e^x : $4\sqrt{2}/3+3\sqrt{3}$
0736 8681 ζ : $(\sqrt[3]{2})^{1/3}$
0737 5395 rt : (2,9,2,-6)
0737 6182 rt : (9,1,-4,-8)
0737 6589 Ei : 5/2
0737 7051 rt : (1,9,-1,-3)
0737 7553 rt : (8,4,-7,-7)
0737 8653 J_1 : $7(e+\pi)/3$
0738 1567 λ : $\exp(-17\pi/10)$
0738 7103 J_1 : $(\sqrt[3]{4}/3)^3$
0738 8205 e^x : $1+2\pi/3$
0738 9056 sq : $e/10$
0738 9641 ζ : $\ln(\ln 2/2)$
0739 1819 Γ : $1/\ln(\sqrt{2}/3)$
0739 1879 rt : (2,1,5,-9)
0739 2274 rt : (9,7,-1,2)
0739 3889 tanh : $(\sqrt[3]{2}/3)^3$
0739 3903 at : $(\sqrt[3]{2}/3)^3$
0739 4008 J_2 : 15/19
0739 4484 at : $\sqrt{2}/4+2\sqrt{5}/3$
0739 4753 rt : (2,4,-4,8)
0739 4856 rt : (9,0,-2,9)
0739 4951 rt : (8,4,-4,1)
0739 4958 ℓ_{10} : $(5/3)^{-1/3}$
0739 5905 sq : $1/3+4\sqrt{5}$
0739 6244 rt : (8,3,-5,-8)
0739 6543 cu : $\sqrt{2}/2+3\sqrt{3}/4$
0739 6953 e^x : $3/4-3\sqrt{5}/2$
0739 7335 Ei : $\exp(-e)$
0739 7431 at : $1/4-2\pi/3$
0739 7634 rt : (7,8,-6,-7)
0740 0058 ℓ_2 : 19/20
0740 0114 10^x : $9\zeta(3)/10$
0740 0691 10^x : $8(e+\pi)/3$
0740 1502 J : $\exp(-13\pi/19)$
0740 2087 ℓ_2 : $5\sqrt[3]{5}/9$
0740 4143 e^x : 1/14
0740 4600 sq : P_{30}
0740 7407 id : $(\sqrt[3]{2}/3)^3$
0740 7783 θ_3 : $(3)^{-3}$
0740 8213 $\ln(2)\div\exp(\sqrt{5})$
0741 0797 ln : 13/14
0741 3368 Ψ : $1/\sqrt{23}$
0741 4183 sinh : $(\sqrt[3]{2}/3)^3$
0741 4198 as : $(\sqrt[3]{2}/3)^3$
0741 4280 Γ : 24/7
0741 4593 ln : P_{53}
0741 5066 sr : $4\sqrt[3]{3}/5$
0741 5119 rt : (8,2,-3,-9)
0741 6526 rt : (7,5,-6,-8)
0741 7164 cr : $1/3+e/3$
0741 7231 sr : 15/13
0741 7303 at : $(2\pi)^{1/3}$
0741 7935 rt : (6,8,-9,-7)
0741 7969 at : $3-2\sqrt{3}/3$
0742 0965 ln : P_{17}
0742 3533 $t\pi$: $\sqrt{15/2}$
0742 5538 J : $\exp(-\sqrt{2}\pi/3)$
0742 7357 $\exp(2/5)\div\exp(3)$
0742 8792 rt : (9,8,1,3)
0743 0646 e^x : $1/4+2\sqrt{3}/3$
0743 1008 2^x : 20/19
0743 1117 ζ : $\sqrt{17}$
0743 1533 rt : (8,5,-2,2)
0743 2349 Ei : $\sqrt[3]{16/3}$
0743 2967 J_2 : 19/24
0743 4050 $\ln(\sqrt{3})\div\exp(2)$
0743 4075 rt : (1,3,4,-9)
0743 4280 rt : (7,2,-5,1)
0743 5060 rt : (2,6,-9,-1)
0743 5065 sq : $4/3-3\sqrt{2}/4$
0743 5158 erf : $\exp(-e)$
0743 5586 K : $(\sqrt[3]{5})^{-1/3}$
0743 5606 rt : (7,4,-4,-9)
0743 7033 rt : (6,7,-7,-8)
0743 7357 at : 24/13
0743 8016 sq : 3/11
0743 8118 $1/3\times\ln(4/5)$
0743 8150 $\sqrt{5}\times\exp(3/5)$
0743 9997 rt : (1,9,2,-7)
0744 1428 J_0 : $5e/6$
0744 1660 cu : $\exp(-\sqrt{3}/2)$
0744 1888 J_2 : $8\ln 2/7$
0744 2238 2^x : $3\sqrt{3}-4\sqrt{5}$
0744 4005 sq : $6(1/\pi)/7$
0744 6788 cu : $5\sqrt{5}/7$
0744 7631 id : $e+3\pi/4$
0745 0176 J_1 : $9\sqrt[3]{3}$
0745 0457 ln : $1/2+\sqrt{3}/3$
0745 0855 θ_3 : $\ln(\sqrt{3}/3)$
0745 0970 $3/4-s\pi(\sqrt{5})$
0745 1618 $t\pi$: $5\sqrt{2}/3$
0745 3142 e^x : P_{23}
0745 3781 cr : $2+4\sqrt{3}$
0745 5595 J : $5(1/\pi)/6$
0745 5825 10^x : $7\sqrt{2}$
0745 6322 rt : (6,6,-5,-9)
0745 6512 ℓ_{10} : $1+4e$
0745 6993 id : $(\sqrt{3}/2)^{-1/2}$
0745 7357 Ei : $10(1/\pi)/9$
0745 7768 rt : (5,9,-8,-8)
0745 9143 $t\pi$: $2\sqrt{2}+\sqrt{3}/4$
0746 1673 at : $\sqrt{2}+\sqrt{3}/4$
0746 2529 at : $(e/2)^2$
0746 3057 2^x : $1/3-3e/2$
0746 3361 ℓ_{10} : 16/19
0746 4644 cu : 8/19
0746 6003 rt : (9,9,3,4)
0746 8810 rt : (8,6,0,3)
0746 9174 J_2 : $\sqrt[3]{1/2}$

0747 0555 $2^x : 1/2 - 3\sqrt{2}$
0747 1624 rt : (7,3,-3,2)
0747 4444 rt : (6,0,-6,1)
0747 4775 $\ell_{10} : 3 - 2e/3$
0747 4887 $Li_2 : ((e+\pi)/3)^{-1/3}$
0747 5084 $e + \exp(\sqrt{5})$
0747 5751 rt : (1,5,1,7)
0747 7270 rt : (5,8,-6,-9)
0747 8926 $J_1 : 3/20$
0747 9460 $Li_2 : 4/5$
0747 9998 $t\pi : 10\sqrt[3]{2}/9$
0748 0315 rt : (1,8,0,-8)
0748 0930 at : $3/4 - 3\sqrt{3}/2$
0748 1232 $J_0 : \sqrt{2} - \sqrt{3} - \sqrt{5}$
0748 2338 sr : $5\ln 2/3$
0748 3207 id : $3\sqrt{2}/2 - 3\sqrt{3}$
0748 4870 $\sqrt{3} \div \exp(\pi)$
0748 5283 $10^x : 3/4 - \sqrt{3}/4$
0748 6545 sinh : 14/15
0748 8092 at : $8\ln 2/3$
0748 8890 $2^x : P_{51}$
0748 9385 $Ei : e\pi/6$
0748 9503 $J_0 : 7\sqrt{3}$
0748 9858 $2 \div s\pi(\sqrt{2})$
0749 1673 $J_1 : \zeta(3)/8$
0749 1928 $\zeta : \exp(\sqrt{2})$
0749 3918 $\Psi : 11/23$
0749 4366 $e^x : 7\sqrt{5}/8$
0749 6972 rt : (5,8,-3,4)
0749 7693 $\ell_{10} : 7\zeta(3)/10$
0749 8100 id : $\ln((e+\pi)/2)$
0749 8164 $\Psi : 2e\pi/5$
0749 8935 cr : $5\sqrt{5}/9$
0749 8971 cu : $1 - 4e/3$
0750 1731 $10^x : 8\sqrt{2}/9$
0750 4225 $\Psi : 6\pi^2/7$
0750 4771 $s\pi : 3\sqrt{3}/4 + 3\sqrt{5}/4$
0750 4940 sq : $\sqrt{2}/3 - \sqrt{5}/3$
0750 5625 $10^x : 3e/2 - \pi/3$
0750 5715 $\Psi : \sqrt[3]{11/3}$
0750 5764 $t\pi : \sqrt{3} - 3\sqrt{5}$
0750 6173 $J_2 : 5(1/\pi)/2$
0750 6809 rt : (8,7,2,4)
0750 7404 $Ei : \sqrt{2}/4$
0750 7909 id : $(\ln 2/3)^{-3}$
0750 8468 cu : $(4/3)^{-3}$
0750 8657 $\Psi : \sqrt[3]{8/3}$
0750 9692 rt : (7,4,-1,3)
0751 0757 cr : $1/\ln(\sqrt{5})$
0751 1061 sq : $1/3 + \pi$
0751 1398 $erf : 1/15$
0751 1458 $s\pi : 2/3 + 4\sqrt{3}/3$
0751 2582 rt : (6,1,-4,2)
0751 2633 $10^x : 3 - 2\sqrt{2}/3$
0751 2756 sq : $\sqrt{2}/2 - \sqrt{3}/4$
0751 3716 $J_1 : 8\sqrt[3]{5}$
0751 4870 cu : $(\ln 2)^{-3}$
0751 5478 rt : (5,2,-7,-1)
0751 6876 rt : (5,9,1,8)
0752 1453 ln : $4\sqrt{2} + 4\sqrt{3}/3$
0752 1509 rt : (1,7,-2,-9)
0752 1809 sr : $1/4 + e/3$
0752 1839 $J_0 : 8\pi/3$
0752 1916 sq : $\zeta(5)$
0752 2510 $\ell_2 : 3/2 + 4\sqrt{3}$
0752 4648 $t\pi : 1/3 + 4\sqrt{3}$
0752 5749 $\ell_{10} : (\sqrt{2})^{1/2}$
0752 5995 $t\pi : 3\sqrt{3}/4 + 3\sqrt{5}/4$
0752 6671 sq : $4\sqrt{3}/3 + 3\sqrt{5}/2$
0752 7683 $Ei : \sqrt{8\pi}$
0752 8172 $e^x : 4\sqrt{2}/3 - 2\sqrt{5}$
0752 8226 at : $\sqrt[3]{19/3}$
0753 1292 $2^x : (\sqrt{2}/3)^3$
0753 1853 $\Psi : (\ln 3)^{-3}$
0753 2139 $\ell_2 : \exp(\sqrt{5}/3)$
0753 2739 $t\pi : 1/3 - 4\sqrt{3}/3$
0753 3005 $\ln(\sqrt{3}) \div \ln(3/5)$
0753 3680 $\exp(\sqrt{2}) \div \exp(4)$
0753 4132 $\ell_2 : (\sqrt[3]{5}/2)^{1/3}$
0753 6473 $2^x : 1/2 + 3\pi/2$
0753 6864 ln : P_{34}
0753 8512 rt : (4,4,-7,4)
0754 0062 $e^x : 5e$
0754 2911 $\zeta : 5\sqrt{2}/7$
0754 3885 $J_2 : (\pi/2)^{-1/2}$
0754 4893 $\Gamma : 4(1/\pi)/7$
0754 5554 rt : (8,8,4,5)
0754 5781 $\Gamma : \sqrt[3]{10}$
0754 6995 cr : $3e + \pi/4$
0754 8020 $J_1 : 9\sqrt{5}/5$
0754 8338 ln : $1/3 + 3\sqrt{3}/2$
0754 8509 rt : (7,5,1,4)
0754 8748 $J_2 : \ln(\sqrt{2\pi}/2)$
0754 9029 $2/5 + s\pi(\sqrt{5})$
0754 9295 $Li_2 : (\sqrt[3]{2}/3)^3$
0754 9591 cr : $2/3 + \sqrt{3}/3$
0754 9910 cu : $1 - \sqrt{3}/3$
0755 0477 $10^x : 1/2 + \sqrt{2}$
0755 0585 $J : 3(1/\pi)/7$
0755 1472 rt : (6,2,-2,3)
0755 2130 $2^x : 9\sqrt{2}/4$
0755 3140 $\Psi : 2\ln 2$
0755 3847 $\ln(\pi) \div \exp(e)$
0755 4441 rt : (5,1,-5,-2)
0755 4815 $J_2 : 7e\pi/5$
0755 7335 $2^x : 4/3 + 4e/3$
0755 7417 rt : (4,8,-4,-7)
0755 8578 $J_0 : 8\sqrt{5}/7$
0755 8854 rt : (4,5,-3,8)
0756 0958 $J_0 : 23/9$
0756 2580 $\zeta : 7(e+\pi)/10$
0756 2748 $10^x : 1 - 3\sqrt{2}/2$
0756 3446 $t\pi : 3e\pi/10$
0756 6841 rt : (2,9,-1,-9)
0756 9058 $2^x : 2/19$
0757 0591 $\ell_{10} : 3\sqrt{3} + 3\sqrt{5}$
0757 1774 $2^x : (\sqrt[3]{5}/2)^{-1/3}$
0757 2071 $\ell_{10} : 21/25$
0757 3383 $\ell_2 : e\pi/9$
0757 4264 rt : (3,6,-7,9)
0757 4671 ln : $1/3 + \sqrt{5}/3$
0757 4792 $\ell_{10} : 2\sqrt[3]{2}/3$
0757 4806 $\ell_2 : (\sqrt{2\pi})^{1/2}$
0757 7491 $2^x : 3/4 - 2\sqrt{5}$
0757 7577 $J_0 : 8\sqrt{2}/5$
0757 8163 cr : $4\sqrt{5}$
0758 0978 rt : (3,9,-2,-4)
0758 1368 $J_2 : ((e+\pi)/3)^{-1/3}$
0758 1776 $J_2 : 4/5$
0758 3532 $2^x : P_{57}$
0758 4103 $\Gamma : 9\ln 2/7$
0758 5073 rt : (8,9,6,6)
0758 5393 cu : $e - \pi$
0758 5904 at : $3\sqrt{3}/2 - \sqrt{5}/3$
0758 6220 as : 22/25
0758 6418 $J_1 : \pi^2$
0758 8104 rt : (7,6,3,5)
0758 8166 cr : $1/2 + \sqrt{5}/3$
0758 9328 $e^x : 1/4 - 2\sqrt{2}$
0759 0969 sr : $2 + 4\sqrt{3}/3$
0759 1143 rt : (6,3,0,4)
0759 3013 $Ei : (\ln 2/2)^{-1/3}$
0759 4188 rt : (5,0,-3,3)
0759 5713 rt : (9,3,-9,-5)
0759 5814 $10^x : 3\sqrt{2}/2 - \sqrt{5}/2$
0759 6954 $e^x : 3/2 - 3e/2$
0759 7240 rt : (4,9,0,-3)
0759 7533 $J_0 : 23/4$
0759 8531 $2^x : P_{38}$
0760 0300 rt : (3,6,-9,-1)
0760 1212 $\Psi : (\sqrt[3]{2}/3)^{-1/2}$
0760 1776 rt : (3,1,-7,8)
0760 4197 $J_2 : 5\sqrt[3]{3}/9$
0760 4559 $\ell_{10} : 4\sqrt{3}/3 - \sqrt{5}/2$
0760 4637 $10^x : (1/\pi)/10$
0760 4694 at : $3/2 + \sqrt{2}/4$
0760 5517 sr : 22/19
0760 6379 $J_2 : 2\zeta(3)/3$
0760 7343 $e^x : 4 - \sqrt{2}/3$
0760 7878 $10^x : 3/4 + 4\pi/3$
0760 8243 $\Gamma : (\sqrt[3]{2})^{-1/2}$
0760 8892 $e^x : 4\sqrt{3}/3 - \sqrt{5}$
0760 9000 $2^x : 2/3 + 3\sqrt{5}/4$
0761 0804 cu : $4\sqrt{2}/3 - 4\sqrt{3}/3$
0761 5752 sinh : $6\sqrt[3]{3}$
0761 5909 rt : (9,2,-7,-6)
0761 7058 at : $3/2 - 3\sqrt{5}/2$
0761 7295 $2^x : (e+\pi)$
0761 8367 $\sqrt{3} \div \ln(5)$
0762 0584 $\zeta : 13/21$
0762 1088 $\Psi : 4\sqrt{3}/5$
0762 1991 at : $9\sqrt[3]{3}/7$
0762 2025 $2^x : 10\zeta(3)/3$
0762 3325 $e^x : 1 + \pi/2$
0762 3983 sr : $(\sqrt{5}/3)^{-1/2}$
0762 4538 $10^x : \sqrt{2} + \sqrt{3}/3$
0762 5461 sq : $2/3 - 2\sqrt{2}/3$
0762 8397 $\exp(3/5) + s\pi(\sqrt{3})$
0762 8505 rt : (7,7,5,6)
0762 8809 $\ell_{10} : 4e + \pi/3$
0762 8858 $2^x : 1 - 3\pi/2$
0763 1623 rt : (6,4,2,5)
0763 1718 $2^x : (1/\pi)/3$
0763 1723 $2^x : 1/4 + 3e/2$
0763 2662 $\Psi : (2e)^3$
0763 4747 rt : (5,1,-1,4)
0763 4972 at : $\sqrt{2} - 2\sqrt{5}/3$
0763 5422 $J_2 : 10\sqrt[3]{3}$
0763 5932 $\Psi : 2\sqrt[3]{5}$
0763 6313 rt : (9,1,-5,-7)
0763 6753 cu : $3\sqrt{2}/10$
0763 7880 rt : (8,4,-8,-6)
0764 1019 rt : (3,5,-7,-2)
0764 3015 rt : (6,7,-7,9)
0764 3179 cu : $3 + 3\pi$
0764 3202 sq : $2 - \sqrt{5}/4$
0764 3926 $\lambda : (e+\pi)^{-3}$
0764 4808 cu : $4(1/\pi)/3$
0764 6662 $Ei : 16/15$
0764 8373 $\ell_{10} : 3\sqrt{5}/8$
0764 8692 $J_0 : 4\sqrt{5}$
0764 9842 id : $\sqrt{2} - 2\sqrt{5}/3$
0765 1020 $2^x : 3 - 3\sqrt{5}$
0765 1168 $K : (e+\pi)/7$
0765 2139 $t\pi : 5/14$
0765 2244 $\Psi : \sqrt{2} - \sqrt{5} - \sqrt{6}$
0765 3615 rt : (1,8,-9,7)
0765 4615 cr : $9\ln 2/5$
0765 6068 ln : $\ln(\pi/3)$
0765 6927 rt : (9,0,-3,8)
0765 7092 $\ell_{10} : 1/4 + 2\sqrt{2}/3$
0765 7323 as : $\sqrt{2} - 2\sqrt{5}/3$
0765 7737 cr : $2e - 4\pi/3$
0765 8049 cu : $8\pi^2/7$
0765 8515 rt : (8,3,-6,-7)
0766 0105 rt : (7,6,-9,-6)
0766 2812 $Ei : 7(e+\pi)/10$
0766 3327 at : $2\sqrt{2}/3 - \sqrt{3}/2$
0766 4170 at : $1/2 - 3\pi/4$
0766 4918 rt : (1,8,0,5)
0766 5548 $\ell_2 : 3 + 2e$
0766 6579 $J_2 : \ln(\sqrt{5})$
0766 7193 rt : (2,4,-5,5)
0766 7287 $\ln(\pi) - \exp(1/5)$
0766 9571 $J_1 : 2/13$
0766 9680 $2^x : \sqrt[3]{25/3}$
0766 9744 rt : (7,8,7,7)
0767 1421 cu : $\sqrt{2}/2 - \sqrt{3}$
0767 1836 $10^x : \exp(\sqrt{5}/2)$
0767 2944 rt : (6,5,4,6)
0767 6151 rt : (5,2,1,5)
0767 7171 tanh : 1/13
0767 7189 at : 1/13
0767 7758 rt : (9,1,-1,9)
0767 8363 id : $2\sqrt{2}/3 - \sqrt{3}/2$
0767 8816 $J_1 : 2\ln 2/9$
0767 9366 rt : (8,2,-4,-8)
0768 0977 rt : (7,5,-7,-7)
0768 2590 rt : (3,5,-4,-3)
0768 5495 at : 13/7

0768 5821 rt : (2,7,-8,-2)
0768 5928 as : $2\sqrt{2}/3 - \sqrt{3}/2$
0768 6443 tanh : $\ln 2/9$
0768 6461 at : $\ln 2/9$
0768 7248 sq : $2\ln 2/5$
0768 8657 e^x : $(\sqrt[3]{2}/3)^3$
0768 8707 Ei : $3(e+\pi)/5$
0768 9525 2^x : $\exp(-\sqrt{5})$
0768 9717 10^x : $2 + \sqrt{2}/3$
0768 9838 ln : $6\sqrt[3]{2}/7$
0769 0685 2^x : $2(e+\pi)/5$
0769 1279 10^x : $4e - 3\pi$
0769 2195 cu : $\exp(-\sqrt[3]{5}/2)$
0769 2307 id : $1/13$
0769 2623 J_0 : $10(e+\pi)/7$
0769 5962 $\ln(3/5) + s\pi(1/5)$
0769 6429 sq : $7\sqrt[3]{3}/5$
0769 8546 e^x : $4\sqrt{2}/3 - 2\sqrt{3}/3$
0769 9896 sinh : $1/13$
0769 9914 as : $1/13$
0770 0438 rt : (8,1,-2,-9)
0770 1635 id : $\ln 2/9$
0770 2070 rt : (7,4,-5,-8)
0770 3704 rt : (6,7,-8,-7)
0770 3919 cu : $4\sqrt{3}/3 - 4\sqrt{5}$
0770 5992 at : $3\sqrt{3}/4 + \sqrt{5}/4$
0770 8886 10^x : $(\pi)^{-3}$
0770 9251 sinh : $\ln 2/9$
0770 9269 as : $\ln 2/9$
0771 1852 rt : (7,9,9,8)
0771 4220 $s\pi$: $2\sqrt{2} + 3\sqrt{3}$
0771 5138 rt : (6,6,6,7)
0771 5201 2^x : $3/2 - 3\sqrt{3}$
0771 5291 $\exp(\sqrt{2}) - s\pi(\sqrt{2})$
0771 5968 ℓ_{10} : $3 + 4\sqrt{5}$
0771 6049 sq : $5/18$
0771 6593 e^x : $(\sqrt[3]{5}/2)^2$
0771 7744 ζ : $3e/2$
0771 8432 rt : (5,3,3,6)
0771 9655 at : $1/2 - \sqrt{3}/3$
0771 9938 sr : $4/3 + 4\sqrt{5}/3$
0772 0972 ℓ_{10} : $(e+\pi)/7$
0772 1553 sinh : $7\zeta(3)/9$
0772 1708 cr : $1/\ln(\sqrt{5}/2)$
0772 1734 cr : $5/4$
0772 1862 sq : $3/4 - 4\sqrt{2}$
0772 2754 cr : $((e+\pi)/3)^{1/3}$
0772 3388 rt : (7,3,-3,-9)
0772 5045 rt : (6,6,-6,-8)
0772 6703 rt : (5,9,-9,-7)
0772 6859 at : $4 - 3e/2$
0772 7743 Ψ : $18/13$
0772 7974 J_2 : $4\sqrt{2}/7$
0772 8363 rt : (2,6,-6,-3)
0772 8654 ℓ_2 : $\sqrt{3} + 3\sqrt{5}$
0773 0368 at : $1/2 + e/2$
0773 1179 Li_2 : $5\sqrt[3]{3}/9$
0773 1691 rt : (1,9,-9,-2)
0773 3566 $\ln(4/5) \times \ln(\sqrt{2})$
0773 4393 2^x : $19/2$
0773 4548 ℓ_{10} : $\sqrt{2} - \sqrt{3}/3$
0773 5026 id : $1/2 + \sqrt{3}/3$
0773 5503 Ei : $\ln(\sqrt{2}\pi/3)$
0773 7277 $t\pi$: $2\sqrt{2} + 3\sqrt{3}$
0773 7909 J_2 : $7\ln 2/6$
0773 8480 Ei : $\pi/8$
0773 8542 cu : $16/11$
0774 0700 J_2 : $(\sqrt[3]{4}/3)^{1/3}$
0774 2274 id : $3e/2$
0774 2760 as : $1/2 - \sqrt{3}/3$
0774 6615 rt : (6,5,-4,-9)
0774 7147 2^x : $2/3 - \sqrt{5}/4$
0774 8084 ζ : $1/\sqrt{7}$
0774 8296 rt : (5,8,-7,-8)
0775 0030 as : $4 - 3e/2$
0775 1052 ζ : $3\sqrt[3]{2}/10$
0775 3322 J_2 : $17/21$
0775 4143 Ei : $6/17$
0775 4673 2^x : $1/2 - 4\pi/3$
0775 5683 Li_2 : $2\zeta(3)/3$
0775 7239 ℓ_{10} : $7e\pi/5$
0775 7403 sq : $7(1/\pi)/8$
0775 7422 rt : (1,5,0,2)
0775 8240 rt : (6,7,8,8)
0776 0783 Γ : $2/11$
0776 1113 $\sqrt{2} - \exp(2/5)$
0776 1544 sq : $3/4 - \sqrt{2}/3$
0776 1624 rt : (5,4,5,7)
0776 2246 $t\pi$: $\sqrt[3]{5/2}$
0776 2334 rt : (3,6,-8,6)
0776 5018 rt : (4,1,2,6)
0776 6333 ℓ_{10} : $(\sqrt[3]{5})^{1/3}$
0776 8353 $t\pi$: $2/5$
0776 8419 rt : (3,7,4,5)
0777 0124 rt : (5,8,-4,2)
0777 0521 Ψ : $9\pi^2/4$
0777 0916 J_0 : $\exp(5/3)$
0777 1830 rt : (2,7,0,-8)
0777 4755 e^x : $\sqrt{2}/4 + 4\sqrt{5}/3$
0777 5249 rt : (1,8,-7,-3)
0777 5883 Γ : $8/9$
0777 7726 2^x : $((e+\pi)/3)^2$
0777 7733 e^x : $\sqrt{2} - \sqrt{3} - \sqrt{5}$
0777 7841 10^x : $1/4 - e/2$
0778 1417 sq : $10(e+\pi)/7$
0778 1959 rt : (1,8,-1,-3)
0778 5063 rt : (9,4,-8,-2)
0778 7883 $s\pi$: $\sqrt{5}\pi$
0779 0135 at : $\sqrt{2}/2 + 2\sqrt{3}/3$
0779 2191 rt : (5,9,0,6)
0779 2391 J_0 : $1/\ln(\pi/3)$
0779 3364 J_1 : $\sqrt{19}\pi$
0779 3921 rt : (4,9,-6,-9)
0779 5670 id : $9\sqrt[3]{5}/5$
0779 6511 2^x : $\sqrt{2} - \sqrt{6} - \sqrt{7}$
0779 7073 $t\pi$: $6e/5$
0779 7448 $\exp(1/5) \div s\pi(1/5)$
0779 8549 cu : $4/3 - e/3$
0779 9595 sq : P_{72}
0779 9996 sq : P_{69}
0780 0251 ℓ_2 : $9/19$
0780 3968 Ei : $10\ln 2$
0780 4614 cr : $1/\ln(\sqrt{2}\pi/2)$
0780 4849 ln : $3\sqrt{2}/4 + 4\sqrt{3}$
0780 5766 rt : (5,5,7,8)
0780 7135 2^x : $\exp(-\sqrt{2}\pi/2)$
0780 7237 cu : $2/3 - 3\sqrt{2}/2$
0780 7251 J_2 : $13/16$
0780 8804 ℓ_{10} : $7\sqrt[3]{5}$
0780 8820 erf : $\ln 2/10$
0780 9154 Ei : $\pi^2/3$
0780 9254 rt : (4,2,4,7)
0780 9384 J_1 : $\ln(\sqrt[3]{5}/2)$
0781 1608 $t\pi$: $\sqrt{5}\pi$
0781 1843 id : $3\sqrt{3} - \sqrt{5}/2$
0781 2177 J_1 : $3\pi^2$
0781 2500 sq : $\sqrt{5}/8$
0781 2752 rt : (3,8,8,9)
0781 4624 Ψ : $17/11$
0781 6258 rt : (4,4,-8,2)
0781 6825 cr : $3e/4 - \pi/4$
0781 6851 sr : $(\pi/2)^{1/3}$
0781 7574 $t\pi$: $7\sqrt[3]{3}/8$
0781 8500 ln : $4 - 3\sqrt{2}/4$
0781 8864 Ψ : $9\zeta(3)/7$
0781 9774 rt : (1,7,-5,-4)
0782 0581 e^x : $\sqrt{3}/4 - 4\sqrt{5}/3$
0782 1593 $s\pi$: $9\sqrt{5}/5$
0782 2527 e^x : $\sqrt{2} + 3\sqrt{3}/4$
0782 4243 rt : (2,7,-3,7)
0782 4650 id : $\sqrt{3}/2 - 4\sqrt{5}$
0782 4808 cu : $1/3 + 2\sqrt{2}/3$
0782 5168 Ψ : $6\sqrt{2}$
0782 5386 sr : $2\sqrt{2} + 2\sqrt{5}/3$
0782 5547 at : $5\sqrt{5}/6$
0782 6295 $\sqrt{5} \times e$
0782 6577 rt : (9,5,-6,-1)
0782 9863 rt : (8,2,-9,-2)
0783 1288 rt : (6,7,-8,7)
0783 2513 2^x : $3e + 4\pi$
0783 2517 J_1 : $\sqrt{2}/9$
0783 3393 ln : $(\sqrt[3]{5}/2)^{1/2}$
0783 3754 Γ : $(3\pi/2)^{-3}$
0783 4692 ln : $7\sqrt[3]{2}/3$
0783 6750 2^x : $(2\pi/3)^{-3}$
0783 7781 e^x : $4/3 + \sqrt{2}/3$
0783 8845 rt : (4,5,-4,6)
0783 9017 sq : $2\sqrt[3]{2}/9$
0783 9291 at : $1/\ln(\sqrt[3]{5})$
0784 0000 sq : $7/25$
0784 1903 e^x : $13/8$
0784 2712 id : $1/4 + 2\sqrt{2}$
0784 5524 Li_2 : $1/13$
0784 5628 $t\pi$: $9\sqrt{5}/5$
0784 6096 id : $6\sqrt{3}/5$
0784 7416 2^x : $1/4 + 2\pi/3$
0785 0117 J_1 : $\sqrt[3]{2}/8$
0785 0896 rt : (5,6,9,9)
0785 1845 2^x : $19/18$
0785 2202 ln : $3\sqrt[3]{3}/4$
0785 2477 $t\pi$: $\sqrt{2}/2 - \sqrt{3}$
0785 2751 at : $2/3 - \sqrt{5}/3$
0785 2942 Ψ : $5/24$
0785 4484 rt : (4,3,6,8)
0785 4972 cr : $2 - \sqrt{5}/3$
0785 5230 Li_2 : $\ln 2/9$
0785 7119 $t\pi$: $1/3 + 4\sqrt{3}/3$
0785 7187 J_2 : $7\sqrt[3]{5}$
0785 8081 rt : (3,0,3,7)
0786 0495 ζ : $e/3$
0786 0761 10^x : $9\pi^2/4$
0786 1484 J_2 : $3e/10$
0786 1688 rt : (2,3,0,5)
0786 2439 $\sqrt{3} + \ln(\sqrt{2})$
0786 3496 rt : (3,9,-3,-7)
0786 4623 Ei : $e\pi/8$
0786 5305 rt : (1,6,-3,-5)
0786 7365 ln : $9\zeta(3)/10$
0786 7757 J_0 : $3e\pi/10$
0786 8932 id : $1/3 + \sqrt{5}/3$
0787 0159 J_1 : $3/19$
0787 1035 Γ : $\sqrt{14/3}$
0787 1522 2^x : $3/2 + 2\pi/3$
0787 1578 2^x : $\exp(1/\sqrt{2})$
0787 1720 cu : $3/7$
0787 2303 rt : (8,3,-7,-1)
0787 3621 cr : $3\sqrt{2}/2 - \sqrt{3}/2$
0787 3805 Γ : $\exp(-2e)$
0787 4089 $\exp(3/5) \div \exp(\pi)$
0787 7076 as : $2/3 - \sqrt{5}/3$
0787 8645 J_2 : $8\pi^2/7$
0787 9651 $\pi \div \ln(4/5)$
0787 9909 J_2 : $\sqrt{2/3}$
0788 1736 J_0 : $\sqrt[3]{23/2}$
0788 1954 ℓ_2 : $(3\pi)^{1/3}$
0788 3084 ln : $4\ln 2/3$
0788 3849 $\ln(5) \div \exp(2/5)$
0788 3876 θ_3 : $1/\ln(1/(2\pi))$
0788 4411 at : $1 + \sqrt{3}/2$
0788 5001 J_1 : $21/2$
0788 6626 rt : (3,1,-8,6)
0788 6643 10^x : $\sqrt{3}/4 + \sqrt{5}/3$
0788 8141 $2/3 - s\pi(1/5)$
0788 8842 sr : $(2e)^{-3}$
0788 9546 at : $\sqrt[3]{13/2}$
0789 0757 10^x : $3e/4 - \pi$
0789 1255 rt : (9,2,-8,-5)
0789 2131 sq : $1 - 3e/4$
0789 4910 ℓ_{10} : $1/3 + \sqrt{3}/2$
0789 5062 $\ln(\sqrt{5}) \div s\pi(\sqrt{3})$
0790 0211 sq : $3\sqrt{2} - 3\sqrt{5}$
0790 0748 rt : (4,4,8,9)
0790 2248 $\exp(1/2) - s\pi(\pi)$
0790 3174 2^x : P_{76}
0790 4149 rt : (1,6,1,9)
0790 4450 rt : (3,1,5,8)
0790 6568 $t\pi$: $((e+\pi)/3)^{-2}$
0790 6609 cu : $2 - \pi/2$
0790 8163 rt : (2,4,4,9)
0791 0628 J_2 : $9/11$

0791 1192 sinh : $\sqrt{19}$
0791 1886 rt : (1,5,-1,-6)
0791 2052 cr : $2\pi/5$
0791 2107 Ψ : $(2e)^{-2}$
0791 2159 rt : (9,7,-2,1)
0791 3200 cu : $3\sqrt{2}-\sqrt{5}$
0791 3822 rt : (9,1,-6,-6)
0791 4704 2^x : $9\sqrt[3]{3}/8$
0791 5619 rt : (8,4,-5,0)
0791 6816 $1/3 - s\pi(\sqrt{3})$
0791 8124 ℓ_{10} : 12
0791 8422 Ψ : $3(1/\pi)/2$
0791 9088 rt : (7,1,-8,-1)
0792 0759 Ei : $7\sqrt{3}$
0792 2797 2^x : $2\sqrt{3}-3\sqrt{5}/2$
0792 4398 J_2 : $(e+\pi)^2$
0792 4693 cr : $8\sqrt{2}/9$
0792 5132 J_2 : $(\ln 3/2)^{1/3}$
0792 6897 cu : $1/4 - e/4$
0792 7928 Ψ : 24/7
0792 8247 2^x : $2-4\sqrt{2}$
0792 9188 at : $3/4+\sqrt{5}/2$
0792 9271 at : $\sqrt{2}/4-\sqrt{3}/4$
0793 0002 sq : $\sqrt{3}/3+3\sqrt{5}$
0793 0251 ln : $5\sqrt{3}/8$
0793 2577 J_1 : $(1/\pi)/2$
0793 4056 $\exp(\sqrt{3}) \div \exp(1)$
0793 4143 sr : $3e/7$
0793 5067 10^x : $10\sqrt[3]{2}/7$
0793 6458 ln : $2+2\sqrt{2}/3$
0793 6512 sq : $3-e$
0793 6639 rt : (9,0,-4,7)
0793 6839 id : $8\sqrt[3]{2}$
0793 8461 rt : (8,3,-7,-6)
0794 0791 sq : P_6
0794 0991 tanh : $(1/\pi)/4$
0794 1013 at : $(1/\pi)/4$
0794 2526 $s\pi$: $3\sqrt{2}+\sqrt{3}$
0794 3802 cu : $1/3-e/2$
0794 4154 id : $3\ln 2$
0794 5784 10^x : 17/20
0794 5931 id : $\sqrt{2}/4-\sqrt{3}/4$
0794 6771 ℓ_{10} : $\sqrt{2}+\sqrt{5}-\sqrt{6}$
0794 6844 id : $2e\pi$
0794 7912 J_2 : $(2e/3)^{-1/3}$
0794 8758 10^x : $\sqrt{3}/3-3\sqrt{5}/4$
0794 8884 ln : $\sqrt{2}/2+\sqrt{5}$
0794 9350 $s\pi$: $(2\pi)^{-2}$
0795 0398 id : $4\sqrt{2}-\sqrt{3}/3$
0795 1292 sinh : $\sqrt{2}-\sqrt{3}+\sqrt{7}$
0795 1904 rt : (3,2,7,9)
0795 2020 ℓ_{10} : $(\sqrt[3]{3})^{1/2}$
0795 2109 θ_3 : $((e+\pi)/2)^{-3}$
0795 2177 ζ : $7\sqrt{3}/3$
0795 3113 $\ln(5)\div s\pi(e)$
0795 3574 cu : $3\sqrt{2}/4-2\sqrt{5}/3$
0795 4068 Γ : $-\sqrt{2}+\sqrt{6}+\sqrt{7}$
0795 4316 as : $\sqrt{2}/4-\sqrt{3}/4$
0795 5728 rt : (2,1,4,-8)
0795 5960 10^x : $3e+4\pi$
0795 6292 rt : (9,8,0,2)
0795 7747 id : $(1/\pi)/4$
0795 8248 θ_3 : $(1/\pi)/8$
0795 8726 ℓ_{10} : $(\ln 2)^{1/2}$
0795 9562 rt : (3,6,-9,3)
0795 9709 rt : (9,1,-2,8)
0795 9845 rt : (8,5,-3,1)
0796 0994 $\exp(2/5)+s\pi(1/5)$
0796 1557 rt : (8,2,-5,-7)
0796 2222 J_2 : $(e/3)^2$
0796 3407 rt : (7,9,-4,-6)
0796 3419 ln : $3\sqrt{2}-3\sqrt{3}/4$
0796 4091 sq : $2e/3-2\pi/3$
0796 4187 ζ : $(\sqrt[3]{5}/2)^{1/3}$
0796 4312 10^x : $2\sqrt{2}-\sqrt{5}/3$
0796 4713 $\ln(4)\div\exp(1/4)$
0796 6148 sinh : $(1/\pi)/4$
0796 6170 as : $(1/\pi)/4$
0796 6978 rt : (6,1,-9,1)
0796 7180 ζ : $9\pi/7$
0796 7698 $t\pi$: $3\sqrt{2}+\sqrt{3}$
0796 8052 Γ : $\sqrt{5}/10$
0796 8054 sr : $3e/2+4\pi$
0796 8281 2^x : $7\sqrt[3]{3}/2$
0796 8357 erf : $2(1/\pi)/9$
0796 8704 10^x : $\ln(1/3)$
0797 0632 sq : $10e/7$
0797 3236 sr : $(e/2)^{1/2}$
0797 4427 J_1 : 4/25
0797 4587 $t\pi$: $(2\pi)^{-2}$
0797 5432 J_2 : $(\sqrt{2}\pi/3)^{-1/2}$
0797 8546 10^x : $3/2-3\sqrt{3}/2$
0797 9728 cu : $\sqrt{2}/2+\sqrt{5}/2$
0798 0271 rt : (1,7,0,4)
0798 0484 sq : $4+3\sqrt{3}/4$
0798 0842 e^x : $2\sqrt{2}/3-\sqrt{3}/2$
0798 1170 $\ln(4)\times\exp(\pi)$
0798 2576 $1/3-\exp(5)$
0798 2633 Ei : $5\sqrt[3]{5}/6$
0798 2976 tanh : 2/25
0798 2998 at : 2/25
0798 3038 rt : (9,2,0,9)
0798 4913 rt : (8,1,-3,-8)
0798 5917 ℓ_{10} : $5\sqrt[3]{3}/6$
0798 6789 rt : (7,4,-6,-7)
0798 6804 J_1 : $\sqrt[3]{3}/9$
0798 7987 10^x : $7(e+\pi)/8$
0798 8400 ℓ_{10} : $\exp(1/(2e))$
0798 8669 rt : (6,7,-9,-6)
0799 1364 at : $\sqrt{7/2}$
0799 2502 ℓ_{10} : $\zeta(3)$
0799 3232 id : $6\sqrt[3]{2}/7$
0799 5899 e^x : 1/13
0799 7226 as : $7\sqrt[3]{2}/10$
0799 8757 e^x : $1/2+e/3$
0799 9329 ℓ_{10} : $6\ln 2/5$
0799 9520 K : $3\sqrt{5}/8$
0800 0000 id : 2/25
0800 0512 θ_3 : 1/25
0800 1366 rt : (9,9,2,3)
0800 1740 as : $2-\sqrt{5}/2$
0800 2288 at : $1/4-3\sqrt{2}/2$
0800 4270 ln : 12/13
0800 4431 rt : (2,0,6,9)
0800 5016 rt : (8,6,-1,2)
0800 5624 sq : $8(1/\pi)/9$
0800 5973 id : $(\sqrt[3]{2})^{1/3}$
0800 8382 id : $\sqrt[3]{9}$
0800 8428 J_2 : 14/17
0800 8534 rt : (8,0,-1,9)
0800 8536 sinh : 2/25
0800 8558 as : 2/25
0800 8675 rt : (7,3,-4,1)
0800 9081 Ψ : 17/2
0800 9815 10^x : $2\sqrt[3]{5}/7$
0800 9835 at : $3e-2\pi$
0801 0438 rt : (7,3,-4,-8)
0801 2344 sr : 7/6
0801 2402 $\exp(1/3)\times s\pi(e)$
0801 2496 ℓ_2 : $7e/9$
0801 2919 $\ln(5)\div\exp(3)$
0801 4089 J_0 : $3\sqrt[3]{5}/2$
0801 5844 $3/4\div\exp(\sqrt{5})$
0801 6319 rt : (1,9,3,-6)
0801 6334 ℓ_{10} : P_{25}
0801 9673 J_2 : $4\sqrt[3]{3}/7$
0801 9684 cr : $e+2\pi$
0802 0364 10^x : $4\sqrt{2}/3-4\sqrt{5}/3$
0802 1014 J_1 : $\sqrt{17}\pi$
0802 1684 e^x : 9/8
0802 1804 10^x : 12/17
0802 3599 ℓ_2 : $2-2\sqrt{2}/3$
0802 4216 e^x : $(e+\pi)/8$
0802 4583 sr : $1/4+3e/2$
0802 6251 rt : (2,7,-4,3)
0802 8488 rt : (6,7,-9,5)
0802 8875 sr : $4/3+3e$
0803 0812 e^x : $3\sqrt{2}/4+4\sqrt{3}/3$
0803 2278 J_1 : $(\sqrt[3]{4}/3)^{-3}$
0803 4358 rt : (7,2,-2,-9)
0803 4638 ln : $3\sqrt{3}/4+3\sqrt{5}$
0803 5245 $t\pi$: $e/3+3\pi/4$
0803 5525 $t\pi$: $3\sqrt{2}/4+3\sqrt{5}/4$
0803 5567 e^x : $2\sqrt{2}/3-2\sqrt{3}$
0803 5608 sr : $10e\pi/9$
0803 6292 rt : (6,5,-5,-8)
0803 6916 id : $2e-3\pi/4$
0803 8229 rt : (5,8,-8,-7)
0804 0466 2^x : $\ln(\sqrt{5}/2)$
0804 0503 id : $(\ln 2/3)^{-1/2}$
0804 2984 cr : $3\sqrt{2}-4\sqrt{5}/3$
0804 4434 2^x : $4\sqrt{3}+3\sqrt{5}$
0804 4758 rt : (6,8,-5,9)
0804 6164 erf : 1/14
0804 6460 10^x : $1-2\pi/3$
0804 7231 sq : $\exp(-\sqrt[3]{2})$
0804 7342 Γ : $3\sqrt[3]{3}/2$
0804 8202 ℓ_2 : $(\sqrt{5}/2)^{1/2}$
0804 8723 sr : $3/2+2\sqrt{2}$
0805 0560 ln : $3/2-\sqrt{3}/3$
0805 1169 rt : (8,7,1,3)
0805 2440 $\ln(3/5)-s\pi(\pi)$
0805 4383 J_1 : $(e)^3$
0805 4931 rt : (7,4,-2,2)
0805 5373 J_2 : 19/23
0805 5778 $\sqrt{2}\div\ln(\sqrt{2})$
0805 6753 10^x : 7/22
0805 8398 rt : (1,2,5,-9)
0805 8702 rt : (6,1,-5,1)
0805 9796 $\exp(1/5)\div\exp(e)$
0805 9983 ℓ_2 : $3\pi^2/7$
0806 0440 rt : (1,5,-1,-3)
0806 0519 rt : (6,4,-3,-9)
0806 2425 $t\pi$: $3\sqrt{2}/2+\sqrt{5}$
0806 2484 rt : (5,8,-5,0)
0806 2582 ζ : $9\sqrt{5}/5$
0806 3069 at : $3\sqrt{3}/2-2\sqrt{5}$
0806 4444 cr : $7\sqrt[3]{3}/8$
0806 5165 cu : $3+\sqrt{5}/4$
0806 5889 Γ : $9\zeta(3)/5$
0806 6583 rt : (1,8,1,-7)
0807 0455 at : $4/3-\sqrt{2}$
0807 1474 sq : $6\zeta(3)/5$
0807 2630 ℓ_2 : $\exp(\sqrt[3]{3})$
0807 7297 cu : $3\pi/4$
0807 9027 rt : (5,4,-9,9)
0807 9550 $\ln(2)-s\pi(e)$
0808 1104 $\ln(\pi)-\exp(4/5)$
0808 3720 J_1 : $9\pi/7$
0808 3900 at : 15/8
0808 4944 J_2 : $\exp(2\pi/3)$
0808 5128 Γ : $1/\ln(\sqrt[3]{4})$
0808 5536 $\sqrt{3}\times\exp(\pi)$
0808 7024 rt : (5,9,-1,4)
0808 7594 2^x : $2-2\sqrt{2}/3$
0808 8022 id : $1/3-\sqrt{2}$
0808 9022 rt : (4,9,-7,-8)
0808 9572 sr : $5\sqrt{3}/2$
0809 0254 $\exp(3/4)-s\pi(\sqrt{2})$
0809 2754 cr : $4\sqrt{2}+3\sqrt{5}/2$
0809 5540 $\ln(2/3)-s\pi(\sqrt{5})$
0809 5956 Γ : $(\sqrt[3]{3})^{-1/3}$
0809 6867 as : $4/3-\sqrt{2}$
0809 8345 rt : (8,8,3,4)
0809 8386 cr : 24/19
0809 8810 $s\pi$: $2/3+e/2$
0809 9191 sinh : 15/16
0809 9643 e^x : $3\sqrt[3]{5}/7$
0810 0000 cu : $3\sqrt[3]{3}/10$
0810 0420 J_2 : $(\ln 3)^{-2}$
0810 1928 sq : $3\ln 2/2$
0810 2022 10^x : $(2e)^{-2}$
0810 2213 rt : (7,5,0,3)
0810 3678 Γ : $(\ln 2)^{1/3}$
0810 3891 $1/3\div\exp(\sqrt{2})$
0810 6092 rt : (6,2,-3,2)
0810 9302 $1/5\times\ln(2/3)$
0810 9982 rt : (5,1,-6,-1)
0811 1384 J_2 : $(\sqrt[3]{5}/3)^{1/3}$
0811 2476 rt : (1,6,0,3)

0811 3227 e^x : $\sqrt[3]{15/2}$
0811 3602 2^x : $(\sqrt{5}/2)^{1/2}$
0811 3866 J_0 : $1/\ln(\sqrt{5}/2)$
0811 3883 rt : (4,8,-5,-9)
0811 4073 J_2 : $\sqrt{2}+\sqrt{3}+\sqrt{5}$
0811 5219 cu : $\sqrt{14/3}$
0811 8112 rt : (1,7,-1,-8)
0811 8116 10^x : $(1/\pi)$
0811 8988 cu : $\sqrt{3}/4$
0811 9413 sinh : $7e/9$
0812 0116 $3/5 \div \exp(2)$
0812 1924 Li_2 : $(1/\pi)/4$
0812 1924 ln : $6\pi^2$
0812 2271 sq : $\sqrt[3]{5}/6$
0812 2321 cu : $7\pi/5$
0812 2422 cu : P_{10}
0812 4306 cu : $1-3\sqrt{3}/2$
0812 5502 tπ : $1/3-e/2$
0812 6665 e^x : $\sqrt{2}/3-4\sqrt{5}/3$
0813 0484 cu : $5\ln 2/8$
0813 3605 2^x : $\sqrt{11/2}$
0813 3750 2^x : $4(e+\pi)/9$
0813 6743 id : $2\sqrt{3}/3-\sqrt{5}$
0813 6898 id : $(\ln 2)^{-2}$
0813 7638 J_1 : $\exp(-2e/3)$
0813 7708 ℓ_{10} : $(\sqrt[3]{5}/3)^{1/3}$
0813 7865 Γ : $\exp(-1/(3e))$
0813 8881 cu : $\sqrt{3}/3-2\sqrt{5}$
0813 9000 rt : (1,7,-1,-3)
0813 9042 rt : (4,5,-5,4)
0814 0011 at : $\exp(\sqrt[3]{2}/2)$
0814 1785 $\ln(3/5)\times\exp(3/4)$
0814 2943 Γ : $5\sqrt{3}/4$
0814 3923 cr : $4\sqrt{3}/3+3\sqrt{5}$
0814 4748 Ei : $8\zeta(3)/9$
0814 5211 sq : $1/2-\pi/4$
0814 5255 2^x : $4/3+2\sqrt{5}/3$
0814 5486 J_2 : $\sqrt{21}\pi$
0814 6585 rt : (8,9,5,5)
0814 8374 id : $(\sqrt[3]{5}/2)^{-1/2}$
0815 0556 Γ : $7/24$
0815 0565 rt : (7,6,2,4)
0815 1688 10^x : $4\pi/3$
0815 3590 J : $\exp(-13\pi/14)$
0815 4542 sπ : $3\sqrt[3]{5}/5$
0815 4557 rt : (6,3,-1,3)
0815 7341 e^x : $3/4+e$
0815 8559 rt : (5,0,-4,2)
0816 0207 J_2 : $6\ln 2/5$
0816 2388 λ : $\exp(-5\pi/3)$
0816 2574 rt : (4,9,-1,-5)
0816 3265 sq : $2/7$
0816 3861 2^x : $4\sqrt{2}+4\sqrt{3}/3$
0816 4074 J_2 : $\exp(-1/(2e))$
0816 4506 rt : (4,6,-1,8)
0816 4653 J_2 : $\sqrt{24}$
0816 5262 at : $4-3\sqrt{2}/2$
0816 5575 10^x : $4e/3-3\pi/2$
0816 5958 Li_2 : $2/25$
0816 6599 id : $\sqrt{13/3}$
0816 8309 J_1 : $7\sqrt{3}/3$
0816 8717 id : $3\sqrt[3]{3}/4$
0816 8890 ℓ_{10} : $(\ln 3)^2$
0816 9793 sq : $1/2+2\sqrt{2}/3$
0817 0416 ℓ_2 : $3\sqrt[3]{2}/4$
0817 0863 $\exp(2/3)-s\pi(1/3)$
0817 0967 rt : (1,6,-3,-9)
0817 2258 $\exp(5)\div\exp(\sqrt{2})$
0817 4568 ℓ_{10} : $1/2+\sqrt{2}/2$
0817 4584 J_2 : $(\ln 2)^{1/2}$
0817 4788 rt : (1,7,-8,6)
0817 5349 rt : (2,9,0,-8)
0817 6748 10^x : $1+4e/3$
0817 6777 2^x : $3e/4+\pi/4$
0817 6961 J_2 : $(\sqrt[3]{3})^{-1/2}$
0817 7050 id : $4\sqrt{2}/3+3\sqrt{3}$
0817 9633 2^x : P_{37}
0818 0837 Ei : $12/17$
0818 1790 tπ : $3\sqrt[3]{5}/5$
0818 1871 Ei : $\sqrt{10}$
0818 2925 cr : $\sqrt{2}/2+\sqrt{5}/4$
0818 3834 sπ : $\pi^2/5$
0818 3999 rt : (3,7,-4,9)
0818 5121 id : $9\zeta(3)/10$
0818 5227 ζ : $10\zeta(3)/3$
0818 5303 rt : (9,2,-9,-4)
0818 7368 sq : $4/3-\pi/3$
0818 8288 2^x : $2+2\pi/3$
0818 8984 J_2 : $5/6$
0818 9726 cu : $1/2+4\pi/3$
0819 0113 tanh : $\exp(-5/2)$
0819 0137 at : $\exp(-5/2)$
0819 2408 rt : (3,1,-9,4)
0819 8247 Γ : $5\sqrt{2}/8$
0819 8385 cr : $19/15$
0819 9171 J_0 : $4\sqrt[3]{3}$
0820 0032 rt : (7,7,4,5)
0820 0908 e^x : $11/15$
0820 2128 $3/4\div\ln(2)$
0820 4142 rt : (6,4,1,4)
0820 7015 sq : $9(1/\pi)/10$
0820 7258 J_1 : $10e\pi/3$
0820 8263 rt : (5,1,-2,3)
0820 8499 id : $\exp(-5/2)$
0820 8992 $\ln(4/5)\div\exp(1)$
0821 0118 Ei : $5\sqrt[3]{5}/8$
0821 0328 rt : (9,1,-7,-5)
0821 1196 ℓ_{10} : $4e/9$
0821 1378 tπ : $\pi^2/5$
0821 2068 2^x : $\sqrt{3}/2-2\sqrt{5}$
0821 2396 rt : (4,2,-5,-2)
0821 3051 10^x : $3\sqrt{2}/2+2\sqrt{3}/3$
0821 6542 rt : (3,5,-8,-1)
0821 6606 cu : $3\sqrt{3}+3\sqrt{5}/2$
0821 6934 10^x : $4/3-3\sqrt{3}/4$
0821 7721 sinh : $\exp(-5/2)$
0821 7745 as : $\exp(-5/2)$
0821 8537 rt : (3,2,-5,8)
0821 8952 cu : $10/23$
0822 0199 sq : $\sqrt[3]{15}$
0822 0327 $\ln(2)+\exp(2)$
0822 0700 id : $\sqrt{19/2}$
0822 5213 rt : (1,8,8,-9)
0822 5245 Ψ : $7e\pi$
0822 6785 ln : $3\sqrt{3}/2-3\sqrt{5}/4$
0822 9742 rt : (2,8,-2,-9)
0823 1058 ℓ_{10} : $1/4+\sqrt{3}/3$
0823 2323 ζ : 4
0823 2500 sq : $\sqrt{3}/3+3\sqrt{5}/4$
0823 3922 Γ : $13/6$
0823 4890 cr : $3-\sqrt{3}$
0823 5648 rt : (9,0,-5,6)
0823 5742 ln : $\sqrt{2}/4+3\sqrt{3}/2$
0823 7719 at : $3-\sqrt{5}/2$
0823 7747 rt : (8,3,-8,-5)
0823 8560 rt : (2,7,-5,-1)
0823 9220 sr : $4-2\sqrt{2}$
0823 9708 Ψ : $\ln(3\pi/2)$
0824 0054 at : $\sqrt[3]{20/3}$
0824 1494 cr : $4e/3-3\pi/4$
0824 1668 e^x : $1/\ln(5/3)$
0824 3031 J_2 : $(\sqrt[3]{5})^{-1/3}$
0824 6216 ℓ_2 : $17/18$
0824 8871 e^x : $1/3-2\sqrt{2}$
0824 9795 at : $\sqrt{2}/4-\sqrt{5}$
0825 0313 10^x : $1/\ln(2e/3)$
0825 0663 rt : (7,8,6,6)
0825 0927 ln : $2\sqrt{2}+3\sqrt{3}$
0825 1117 2^x : $2-4\sqrt{2}/3$
0825 1563 sq : $1/3+\sqrt{2}/2$
0825 2725 rt : (6,8,-6,7)
0825 3175 id : $5\sqrt{3}/8$
0825 3489 $\exp(\sqrt{3})-s\pi(\pi)$
0825 4146 cu : $2e/3-\pi/4$
0825 4896 rt : (6,5,3,5)
0825 5918 2^x : $10(1/\pi)$
0825 6912 sq : $3e/4+2\pi/3$
0825 7789 sr : $(e+\pi)/5$
0825 8021 e^x : $\sqrt[3]{17}$
0825 8087 ζ : $\ln(\ln 3/3)$
0825 8590 ζ : $2\sqrt[3]{5}/9$
0825 8649 ζ : $-\sqrt{2}-\sqrt{5}+\sqrt{7}$
0825 8880 J_1 : $(\ln 3/2)^3$
0825 9142 rt : (5,2,0,4)
0825 9247 J_2 : $(e+\pi)/7$
0825 9736 10^x : $7e/9$
0825 9750 Ψ : $5\zeta(3)/8$
0826 1269 rt : (9,1,-3,7)
0826 3400 rt : (8,2,-6,-6)
0826 4250 ℓ_2 : $\sqrt[3]{19/2}$
0826 5276 sq : $1/\ln(2/3)$
0826 5534 rt : (7,5,-9,-5)
0826 7671 rt : (3,5,-5,-5)
0826 8086 2^x : $6e$
0827 1955 rt : (2,7,-9,-1)
0827 2574 ℓ_2 : $2+\sqrt{5}$
0827 3203 cu : $4/3+2\sqrt{2}$
0827 4017 rt : (2,4,-7,-1)
0827 4985 J_0 : $9/4$
0827 5470 ln : $4-\pi/3$
0827 6097 sq : $\ln(4/3)$
0828 1572 sq : $3/2+4\sqrt{5}$
0828 1945 Ei : $1/\sqrt[3]{23}$
0828 2944 e^x : $(1/\pi)/4$
0828 5262 J_2 : $3\sqrt{5}/8$
0828 5268 cr : $5(e+\pi)$
0828 5529 rt : (1,7,-9,-2)
0828 6749 cu : P_{29}
0828 7198 rt : (9,2,-1,8)
0828 7559 rt : (4,9,-8,7)
0828 9361 rt : (8,1,-4,-7)
0829 1489 10^x : $2\sqrt{3}/3-\sqrt{5}$
0829 1527 rt : (7,4,-7,-6)
0829 7262 ℓ_2 : $(1/(3e))^3$
0829 7423 ℓ_{10} : $19/23$
0829 7985 2^x : $4e-2\pi$
0829 8421 J : $\exp(-3\pi)$
0829 8962 sq : $1/\ln(4/3)$
0829 9324 ln : $-\sqrt{3}+\sqrt{5}+\sqrt{6}$
0830 0796 10^x : $1/3-\sqrt{2}$
0830 1188 rt : (2,5,-1,8)
0830 1951 ζ : $7\sqrt[3]{5}/3$
0830 2508 rt : (7,9,8,7)
0830 2632 J : $3/17$
0830 3461 at : $3\pi/5$
0830 4431 J_1 : $1/6$
0830 5556 cu : $2e+\pi/2$
0830 5876 e^x : $4\sqrt{3}+\sqrt{5}/4$
0830 6685 sr : $\sqrt{3}-\sqrt{5}/4$
0830 6871 rt : (6,6,5,6)
0830 6959 e^x : $\sqrt{3}/2-3\sqrt{5}/2$
0830 7113 id : $2\sqrt{2}-\sqrt{5}/3$
0831 0213 2^x : P_{78}
0831 0676 2^x : $\sqrt{2}-3\sqrt{3}/4$
0831 1247 rt : (5,3,2,5)
0831 1706 J_2 : $2\sqrt[3]{2}/3$
0831 2685 J_2 : $21/25$
0831 3440 rt : (9,3,1,9)
0831 4096 tanh : $1/12$
0831 4123 at : $1/12$
0831 5637 rt : (8,0,-2,8)
0831 6079 sq : $3/2+4\sqrt{5}/3$
0831 7037 J_0 : $\sqrt[3]{17}$
0831 7837 rt : (7,3,-5,-7)
0831 8008 at : $4\sqrt{2}/3$
0831 8919 e^x : $e/2+\pi$
0832 0040 rt : (6,6,-8,-6)
0832 0076 ℓ_2 : $4/3+\pi/4$
0832 0335 sq : $\sqrt[3]{3}/5$
0832 3202 2^x : $18/17$
0832 4067 e^x : $25/9$
0832 4073 J_0 : $18/7$
0832 4457 rt : (2,6,-7,-2)
0832 4514 cu : $4+4e$
0832 6216 θ_3 : $(\ln 2/2)^3$
0832 8706 e^x : $2/25$
0832 8794 Γ : $15/17$
0832 9375 J_2 : $(\sqrt{2})^{-1/2}$
0833 0711 sπ : $8\pi/5$
0833 1770 rt : (1,6,-1,-3)

0833 2953 $\sqrt{3} - \exp(1/2)$
0833 3333 id : $1/12$
0833 3936 θ_3 : $1/24$
0833 5555 10^x : $5e/3$
0833 6886 2^x : $\ln 2/6$
0833 8160 ln : $23/25$
0833 9499 J_2 : $7\zeta(3)/10$
0834 1101 J_0 : $10\sqrt{3}/3$
0834 1960 sq : $3 - \sqrt{5}/3$
0834 2235 rt : (8,1,0,9)
0834 2981 sinh : $1/12$
0834 3008 as : $1/12$
0834 3101 erf : $(\sqrt[3]{2}/3)^3$
0834 4320 tπ : $8\sqrt{2}/5$
0834 4469 rt : (7,2,-3,-8)
0834 5712 rt : (9,4,-9,-3)
0834 6707 rt : (6,5,-6,-7)
0834 8949 rt : (5,8,-9,-6)
0835 0160 sπ : $2\sqrt{3} - 2\sqrt{5}/3$
0835 0586 sq : $4/3 - 3\sqrt{5}/2$
0835 1062 ζ : $25/23$
0835 1140 2^x : $1/2 + \sqrt{5}/4$
0835 1903 J_2 : $16/19$
0835 6173 J : $5\ln 2/8$
0835 6814 K : $2\sqrt[3]{2}/3$
0835 6820 ℓ_{10} : $6\sqrt{2}/7$
0835 9159 J : $(2e)^{-2}$
0835 9770 tπ : $8\pi/5$
0836 0121 rt : (6,7,7,7)
0836 2406 e^x : $1/2 - 4\sqrt{5}/3$
0836 3723 Γ : $7\sqrt[3]{2}/10$
0836 4635 rt : (5,4,4,6)
0836 5365 Ei : $7/4$
0836 5866 ℓ_{10} : $7\sqrt{3}$
0836 7144 10^x : $3 - 3e/2$
0836 9163 rt : (4,1,1,5)
0836 9558 Γ : $1/\sqrt[3]{13}$
0837 0111 K : $21/25$
0837 0585 Γ : $\exp(-\sqrt[3]{5}/2)$
0837 0676 cr : $14/11$
0837 0926 ln : $2e/5$
0837 1433 rt : (7,1,-1,-9)
0837 1950 J_1 : $(2e/3)^{-3}$
0837 2519 cr : $9\sqrt{2}/10$
0837 2762 $\ln(\sqrt{2}) + s\pi(\pi)$
0837 3706 rt : (6,4,-4,-8)
0837 3837 2^x : $4 + 2\pi$
0837 4023 cu : $7/16$
0837 5896 J_2 : $9e/5$
0837 5982 rt : (5,8,-6,-2)
0837 6668 J_2 : $(5/3)^{-1/3}$
0837 6998 2^x : $1/2 - 3e/2$
0837 8263 rt : (2,5,-5,-3)
0837 9424 tπ : $2\sqrt{3} - 2\sqrt{5}/3$
0838 1863 10^x : $5\sqrt{2}/6$
0838 2834 rt : (1,8,-8,-2)
0838 3393 Li_2 : $\exp(-5/2)$
0838 5214 cr : $4(1/\pi)$
0838 5414 2^x : $2/3 - 3\sqrt{2}$
0838 6350 ℓ_{10} : $7\ln 2/4$
0838 7235 rt : (1,5,-2,-8)
0838 9214 cr : $3\sqrt{2}/2 + 4\sqrt{3}$
0838 9709 at : $17/9$
0839 2757 e^x : 9
0839 5609 $\ln(\pi) \div \ln(\sqrt{3})$
0839 5921 rt : (9,5,-7,-2)
0839 6224 Γ : $23/2$
0839 6989 sq : P_{82}
0839 8564 J_0 : $\sqrt{15}\pi$
0839 9763 ℓ_{10} : $4\sqrt[3]{3}/7$
0839 9958 sπ : $\sqrt{2} + \sqrt{5}/4$
0840 0832 cu : $(\sqrt{5}/3)^{-3}$
0840 1043 rt : (6,3,-2,-9)
0840 2481 10^x : $\sqrt{3} + \sqrt{5}/2$
0840 2541 $1/5 - \exp(1/4)$
0840 2944 rt : (3,7,-5,6)
0840 3356 rt : (5,9,-2,2)
0840 3360 cu : $4/3 - 2e$
0840 3409 e^x : $\sqrt{2}/3 + 2\sqrt{3}/3$
0840 4467 e^x : $3\sqrt{2}/2 + 4\sqrt{5}$
0840 5673 rt : (4,9,-8,-7)
0840 6307 sπ : $2e/3 - \pi/4$
0840 8754 ζ : $1/\ln(\ln 3/3)$
0840 9080 2^x : P_{77}
0840 9304 2^x : $((e+\pi)/2)^{-2}$
0840 9712 tπ : $\sqrt{2} - 3\sqrt{5}/4$
0841 0137 as : $5\sqrt{2}/8$
0841 0627 ζ : $23/21$
0841 1432 at : $3\sqrt[3]{2}/2$
0841 4707 rt : (6,8,9,8)
0841 6921 e^x : $2/3 - \pi$
0841 9365 rt : (5,5,6,7)
0842 3865 10^x : $11/18$
0842 4039 rt : (4,2,3,6)
0842 5282 θ_3 : $7\ln 2/10$
0842 5783 $\ln(\sqrt{2}) \div \exp(\sqrt{2})$
0842 7528 J_2 : $11/13$
0842 8728 rt : (3,8,7,7)
0842 9751 tπ : $\sqrt{2} + \sqrt{5}/4$
0843 1078 rt : (5,5,-3,-9)
0843 1307 2^x : $1/3 + 3e$
0843 2088 ℓ_{10} : $14/17$
0843 2813 ℓ_{10} : $e + 3\pi$
0843 3016 Li_2 : $\ln(\sqrt{5})$
0843 3432 rt : (4,8,-6,-8)
0843 5217 at : $\sqrt{2}/2 - 3\sqrt{3}/2$
0843 6167 tπ : $2e/3 - \pi/4$
0843 6508 2^x : $2\sqrt{3} - \sqrt{5}/2$
0843 8152 rt : (1,7,-6,-3)
0844 0438 e^x : $2 - 2\sqrt{5}$
0844 1522 J_2 : $\exp(\sqrt[3]{4})$
0844 2165 2^x : $13/8$
0844 3393 ℓ_{10} : $2 - \pi/4$
0844 7267 rt : (9,6,-5,-1)
0844 7506 sr : $7e/2$
0845 0008 2^x : $e - 2\pi$
0845 0324 e^x : $18/7$
0845 1661 rt : (8,3,-8,-2)
0845 4632 rt : (4,9,-9,4)
0846 1549 rt : (4,7,-4,-9)
0846 1942 rt : (2,7,-6,-5)
0846 2026 ln : $4/3 + 3\sqrt{5}$
0846 5228 sr : $20/17$
0846 5378 at : $4/3 + \sqrt{5}/4$
0846 7519 cr : $1/3 + 2\sqrt{2}/3$
0846 9807 Γ : $(3\pi/2)^{1/2}$
0847 1002 rt : (6,8,-7,5)
0847 2005 J_2 : $3\sqrt{2}/5$
0847 3031 e^x : $1/4 - e$
0847 4944 ℓ_{10} : $4 + 3e$
0847 5501 rt : (5,6,8,8)
0847 6804 cr : $10e/3$
0847 7598 J_2 : $8(1/\pi)/3$
0847 7673 Ψ : $10\sqrt{5}$
0847 8161 cr : $(\sqrt[3]{3}/3)^{-1/3}$
0847 8905 10^x : $3\sqrt{5}/7$
0848 0114 10^x : $1 + 3\sqrt{3}/4$
0848 0114 cu : $3/2 - 3\sqrt{2}/4$
0848 0328 rt : (4,3,5,7)
0848 1369 rt : (1,9,0,7)
0848 2641 ζ : $21/19$
0848 3258 J_0 : $9(e+\pi)/10$
0848 4432 J_2 : $(\sqrt[3]{3}/2)^{1/2}$
0848 4518 10^x : $(1/\pi)/9$
0848 5172 rt : (3,0,2,6)
0848 9810 e^x : $1/\ln(2/3)$
0849 0031 rt : (6,9,-3,9)
0849 2467 rt : (3,9,-5,-9)
0849 3049 J_2 : 12
0849 3778 $3/4 \times \exp(\sqrt{2})$
0849 4907 rt : (1,6,-4,-4)
0849 5788 10^x : $1/2 - \pi/2$
0849 6097 e^x : $3\sqrt{2} - 3\sqrt{5}$
0849 6250 ℓ_2 : $3\sqrt{2}$
0849 6391 2^x : $2/17$
0849 7909 ℓ_{10} : $3\sqrt{2}/2 - 3\sqrt{3}/4$
0849 9623 J_2 : $17/20$
0849 9800 rt : (9,7,-3,0)
0850 0618 ζ : $8/21$
0850 3111 $s\pi(1/3) - s\pi(2/5)$
0850 3317 2^x : $4\sqrt{3} - 3\sqrt{5}/4$
0850 3891 Ψ : $e\pi$
0850 4324 rt : (8,4,-6,-1)
0850 6944 sq : $7/24$
0850 8524 e^x : $1 - 2\sqrt{3}$
0850 8862 rt : (7,1,-9,-2)
0851 3136 cr : $4\sqrt{5}/7$
0851 3692 Li_2 : $1/12$
0851 3834 cr : $23/18$
0851 6903 sr : $(\sqrt[3]{3}/2)^{-1/2}$
0851 7491 $\sqrt{5} \times \exp(4)$
0851 7517 10^x : $\sqrt{3} + 3\sqrt{5}/4$
0851 7558 Ψ : $4e/7$
0851 7575 10^x : $23/24$
0851 8400 cu : $11/25$
0851 8667 J_1 : $\sqrt[3]{5}/10$
0851 9550 ℓ_{10} : $\sqrt{15}\pi$
0852 0455 ln : $4\sqrt{3} + \sqrt{5}/2$
0852 0510 ℓ_{10} : $\sqrt{2}/3 + \sqrt{5}/3$
0852 1779 ln : $3\sqrt{3} - \sqrt{5}$
0852 5483 $\ln(4) \times \exp(4/5)$
0852 6235 e^x : $e/4 - \pi$
0852 7180 ℓ_{10} : $(\sqrt{2}\pi/3)^{1/2}$
0852 7626 rt : (9,1,-8,-4)
0852 7864 J_1 : $4e/3$
0853 0252 rt : (5,5,-7,9)
0853 0778 Γ : $22/25$
0853 2115 Ei : $\sqrt{2}/2$
0853 8099 rt : (4,4,7,8)
0854 0188 sr : $3\pi/8$
0854 1219 tπ : $e/2 + 2\pi$
0854 3106 rt : (3,1,4,7)
0854 5504 ℓ_2 : $1/3 + 3e$
0854 6946 ℓ_2 : $3\pi/10$
0854 8130 rt : (2,4,3,7)
0854 8577 sr : $8e/5$
0854 8696 ln : $3\pi^2/10$
0855 0234 $4/5 \div \exp(\sqrt{5})$
0855 0648 rt : (3,2,-6,6)
0855 2692 sr : $\sqrt{3}/4 + \sqrt{5}/3$
0855 3171 rt : (1,5,-2,-5)
0855 3470 J_2 : $\exp(-1/(2\pi))$
0855 3569 rt : (9,8,-1,1)
0855 3692 id : $(e)^3$
0855 4510 J_1 : $\zeta(3)/7$
0855 4807 e^x : $\exp(-5/2)$
0855 5802 rt : (9,0,-6,5)
0855 7691 at : $3/2 - \sqrt{2}$
0855 8230 rt : (8,5,-4,0)
0855 9260 sr : $5\sqrt{2}/6$
0856 0419 id : $\sqrt{2}/4 + \sqrt{3}$
0856 0836 cr : $(2\pi/3)^{1/3}$
0856 2275 as : $\exp(-1/(3e))$
0856 2838 e^x : $\sqrt{15}$
0856 2895 rt : (1,6,-2,-9)
0856 2905 rt : (7,2,-7,-1)
0856 3017 ℓ_{10} : $\sqrt{2} + \sqrt{6} - \sqrt{7}$
0856 3569 $3/5 \div \ln(3/4)$
0856 5354 ℓ_{10} : $(e/3)^2$
0856 7978 cu : $\sqrt{2}/3 + 3\sqrt{5}$
0857 0784 Ei : $\exp(-\pi/3)$
0857 1258 sr : $\sqrt{3}/3 + 4\sqrt{5}$
0857 1463 ζ : $19/17$
0857 3620 ℓ_{10} : $\exp(5/2)$
0857 4360 J_2 : $e\pi/10$
0857 5620 cu : $1 - \sqrt{5}/4$
0857 5975 2^x : $\sqrt{2}/4 + 3\sqrt{5}/4$
0857 8643 id : $3/2 - \sqrt{2}$
0857 8691 Ei : $8\sqrt[3]{3}$
0858 0619 cu : $\sqrt{10/3}$
0858 2307 at : $2e\pi/9$
0858 3384 $\ln(\pi) \times \exp(3/5)$
0858 3439 rt : (1,9,-1,-2)
0858 4336 rt : (9,1,-4,6)
0858 5718 e^x : $4\sqrt{3} - \sqrt{5}$
0858 6087 Ψ : $3/4$
0858 6803 rt : (8,2,-7,-5)
0858 8579 2^x : $3\sqrt{2}/4$
0858 8633 sq : $3e\pi/2$
0858 9200 as : $3/2 - \sqrt{2}$

0859 0892 id : $3e/4+\pi/3$
0859 1345 sinh : $(2\pi)^{1/3}$
0859 3484 J_2 : $\sqrt[3]{5}/2$
0859 5870 ℓ_{10} : $3/2-e/4$
0859 7369 K : $(\sqrt{2})^{-1/2}$
0859 7427 rt : (4,5,9,9)
0859 7545 sπ : $\sqrt[3]{25/3}$
0860 0535 Ψ : $10/21$
0860 0669 $\ln(\sqrt{5})\div\exp(\sqrt{5})$
0860 0776 ζ : $2\pi^2/5$
0860 2325 10^x : $2/3-\sqrt{3}$
0860 2606 rt : (3,2,6,8)
0860 4588 J_0 : $5\pi/7$
0860 5169 10^x : $4e-4\pi/3$
0860 6024 $1/5\times s\pi(\pi)$
0860 6774 ℓ_{10} : $(2e/3)^{1/3}$
0860 7308 10^x : $\sqrt{11}\pi$
0860 7803 rt : (2,1,3,-7)
0860 8031 sr : $1/2+e/4$
0860 8629 rt : (9,9,1,2)
0861 2755 cu : $4\sqrt{5}/7$
0861 3019 rt : (2,4,-8,-4)
0861 3235 rt : (9,2,-2,7)
0861 3434 rt : (8,6,-2,1)
0861 5277 id : $\sqrt{3}+3\sqrt{5}/2$
0861 5283 e^x : $3\sqrt{2}/4+\sqrt{5}/3$
0861 5742 rt : (8,1,-5,-6)
0861 8254 rt : (7,4,-8,-5)
0862 0986 ℓ_{10} : $\sqrt{2}/4+\sqrt{3}/2$
0862 3090 rt : (6,0,-8,1)
0862 3373 $\sqrt{3}\div\exp(3)$
0862 3644 2^x : $7\pi/5$
0862 4828 cu : $23/18$
0862 5264 cr : $4-e$
0862 5755 2^x : $3(1/\pi)/8$
0862 7818 J_1 : $\sqrt{3}/10$
0862 8957 tπ : $5/19$
0862 9498 tπ : $\sqrt[3]{25/3}$
0862 9842 Ψ : $5\sqrt[3]{5}$
0863 1260 at : $10\sqrt[3]{5}/9$
0863 1839 at : $19/10$
0863 1858 J_1 : $\ln 2/4$
0863 3066 at : $2\sqrt{3}/3+\sqrt{5}/3$
0863 3214 cr : $8\sqrt[3]{3}/9$
0863 3335 rt : (3,7,-6,3)
0863 3835 cu : $3\sqrt{2}/4-4\sqrt{5}/3$
0863 4160 J_2 : $6/7$
0863 4631 ℓ_{10} : $e\pi/7$
0863 5958 $\ln(2/3)+\exp(2/5)$
0863 6070 J_2 : $(\sqrt[3]{4})^{-1/3}$
0863 6139 10^x : $4-\pi/4$
0863 7329 sq : $\ln(\sqrt{5}/3)$
0864 0088 ln : $-\sqrt{2}+\sqrt{3}+\sqrt{7}$
0864 0440 J_0 : $\ln(3\pi)$
0864 1319 sinh : $16/17$
0864 1857 e^x : $4\sqrt{2}-3\sqrt{3}/4$
0864 1975 sq : $13/9$
0864 2484 2^x : $10(1/\pi)/3$
0864 2509 rt : (9,3,0,8)
0864 2723 tanh : $\ln 2/8$
0864 2755 at : $\ln 2/8$
0864 2783 $2\div\exp(\pi)$
0864 3481 θ_3 : $\exp(-\pi)$
0864 3563 rt : (2,5,-2,5)
0864 5058 rt : (8,0,-3,7)
0864 5896 J_2 : $(e/2)^{-1/2}$
0864 6837 cr : $3\sqrt[3]{5}/4$
0864 7611 rt : (7,3,-6,-6)
0864 7948 sq : $4\sqrt{2}+\sqrt{3}/4$
0864 8766 as : $(\ln 2)^{1/3}$
0864 8865 10^x : $2+3\sqrt{5}$
0865 0168 rt : (6,6,-9,-5)
0865 0519 sq : $5/17$
0865 1704 sr : $4+\sqrt{2}/4$
0865 3057 2^x : $4+3e$
0865 3482 tπ : $(5/2)^{-1/3}$
0865 7196 e^x : $2+3\sqrt{3}/4$
0865 7466 ζ : $1/\sqrt[3]{18}$
0865 8975 ℓ_{10} : $4/3+4e$
0865 9557 10^x : $\sqrt{5}/7$
0865 9835 J_1 : $9\sqrt{5}$
0866 0973 2^x : $2\sqrt{2}/3-2\sqrt{5}$
0866 1936 J_2 : $5\zeta(3)/7$
0866 2750 erf : $1/13$
0866 2817 J_1 : $4/23$
0866 3752 rt : (3,3,8,9)
0866 4339 id : $\ln 2/8$
0866 4850 sr : $1+3\sqrt{5}/2$
0866 5868 10^x : $3/4-2e/3$
0866 6249 cu : $10\sqrt{5}/3$
0866 8337 as : $(\sqrt[3]{3})^{-1/3}$
0866 9133 rt : (2,0,5,8)
0866 9537 J_2 : $(\sqrt[3]{5}/3)^{-3}$
0866 9993 rt : (8,7,0,2)
0867 2167 rt : (9,4,2,9)
0867 2851 ℓ_{10} : $(\ln 3/2)^{1/3}$
0867 3213 erf : $\ln 2/9$
0867 3486 2^x : $3/25$
0867 3801 tanh : $2/23$
0867 3833 at : $2/23$
0867 4533 rt : (1,3,2,7)
0867 4758 rt : (8,1,-1,8)
0867 4965 rt : (7,4,-3,1)
0867 5184 sinh : $\ln 2/8$
0867 5217 as : $\ln 2/8$
0867 6159 tπ : $4\sqrt{2}/3+2\sqrt{5}$
0867 7354 rt : (7,2,-4,-7)
0867 8512 2^x : $\sqrt{2}/3+2\sqrt{3}/3$
0867 8970 e^x : $4+3\sqrt{5}/4$
0867 9954 rt : (6,7,-5,-6)
0868 1435 J_0 : $9e\pi/5$
0868 2017 2^x : $(\ln 2/2)^2$
0868 2041 θ_3 : $(\ln 3)^{-2}$
0868 2592 cu : $1/2-2\sqrt{2}/3$
0868 3752 ζ : $17/15$
0868 4316 2^x : $8e/3$
0868 4960 rt : (5,2,-9,1)
0868 5395 rt : (1,9,4,-5)
0868 5952 ℓ_{10} : $3/4+\sqrt{2}/3$
0868 6196 ℓ_{10} : $5\sqrt[3]{5}/7$
0868 8099 ℓ_{10} : $4\sqrt{3}/3-2\sqrt{5}/3$
0868 8981 2^x : $\zeta(3)/10$
0868 9129 Γ : $4e/5$
0868 9917 Ei : $\sqrt[3]{3}/9$
0869 0404 e^x : $1/12$
0869 1092 at : $1/3-\sqrt{5}$
0869 2448 at : $7e/10$
0869 3019 J_2 : $(\pi/2)^{-1/3}$
0869 5652 id : $2/23$
0869 6366 θ_3 : $1/23$
0869 7250 rt : (2,7,-7,-9)
0869 9495 10^x : $3+3e$
0870 0316 rt : (6,8,-8,3)
0870 1137 ln : $11/12$
0870 2582 cr : $\sqrt{2}/2+\sqrt{3}/3$
0870 4326 2^x : $2/3-4\pi/3$
0870 4853 rt : (8,2,1,9)
0870 6522 sq : $\exp(1/e)$
0870 6614 sinh : $2/23$
0870 6648 as : $2/23$
0870 7097 e^x : $2\sqrt{2}+\sqrt{3}/4$
0870 7493 rt : (7,1,-2,-8)
0871 0137 rt : (6,4,-5,-7)
0871 1424 J_0 : $25/3$
0871 1461 sr : $13/11$
0871 2606 J_2 : $5(e+\pi)/6$
0871 2786 rt : (5,8,-7,-4)
0871 4787 J_0 : $\exp(5/2)$
0871 4984 sr : $2+3\pi/4$
0871 5017 ℓ_{10} : $11/9$
0871 5028 cr : $(\sqrt{2}/3)^{-1/3}$
0871 5968 $\ln(5)\times s\pi(\sqrt{5})$
0871 6455 10^x : $\ln(\ln 2/2)$
0871 8335 Ψ : $14/9$
0871 8360 rt : (2,8,1,7)
0871 8690 J_0 : $8e$
0872 0863 rt : (6,9,-4,7)
0872 1268 at : $1/3+\pi/2$
0872 3080 id : $7\pi^2$
0872 3194 θ_3 : $\exp(-\sqrt[3]{3}/2)$
0872 3545 rt : (1,9,8,5)
0872 3922 2^x : $\sqrt{3}/2-\sqrt{5}/3$
0872 4542 Ei : $\exp(\sqrt[3]{4})$
0872 5715 sq : $1/3-4e$
0872 5909 2^x : $e/3-\pi/4$
0872 7353 rt : (1,8,-9,8)
0872 7970 rt : (8,8,2,3)
0872 9124 cr : $1/2+\pi/4$
0873 1273 id : $2e/5$
0873 2205 rt : (2,1,7,9)
0873 2440 $\ln(3/5)+\exp(4)$
0873 3103 rt : (7,5,-1,2)
0873 3238 Ψ : $11/8$
0873 3606 at : $3/4+2\sqrt{3}/3$
0873 5866 K : $7\zeta(3)/10$
0873 6420 cu : $2e-\pi$
0873 7802 rt : (1,2,4,-8)
0873 8037 cr : $9/7$
0873 8253 rt : (6,2,-4,1)
0873 8933 at : $\sqrt{2}/3-\sqrt{5}/4$
0873 9424 as : P_{54}
0874 0726 rt : (6,3,-3,-8)
0874 2157 ln : $\ln(5/2)$
0874 2572 sπ : $1/2+2\sqrt{5}$
0874 2816 $\exp(1/5)+s\pi(1/3)$
0874 3420 rt : (5,9,-3,0)
0874 3582 sr : $3\sqrt{2}/2+\sqrt{5}$
0874 6088 e^x : $3-2e$
0874 6119 rt : (4,9,-9,-6)
0874 6284 ℓ_2 : 17
0874 9060 rt : (1,8,2,-6)
0875 0699 ℓ_{10} : $9e/2$
0875 1559 J_2 : $(\sqrt{5}/3)^{1/2}$
0875 2820 sq : $4\sqrt{2}/3+\sqrt{3}$
0875 3140 at : $4-2\pi/3$
0875 3261 ζ : $5\pi/4$
0875 3829 sr : $4\sqrt{2}-3\sqrt{3}/4$
0875 4019 Ψ : $\exp(1/\pi)$
0875 5898 10^x : $\sqrt{3}/4-2\sqrt{5}/3$
0875 7188 J_2 : $19/22$
0875 7939 sq : $\sqrt[3]{23}$
0875 9103 ζ : $6(1/\pi)/5$
0876 1247 id : $\sqrt{2}/3-\sqrt{5}/4$
0876 1648 Ψ : $(\sqrt[3]{4}/3)^{-1/2}$
0876 3698 at : $1+e/3$
0876 3793 Γ : $(\sqrt{2}\pi/3)^{-1/3}$
0876 4370 rt : (5,5,-8,7)
0876 5565 cu : $2/3+4\sqrt{3}$
0876 6387 sr : $3/4+\sqrt{3}/4$
0876 8207 $1/5+\ln(3/4)$
0876 8778 cu : $3\sqrt{2}/2-3\sqrt{5}/4$
0876 9014 Ei : $8e/5$
0877 0307 sq : $4\sqrt{2}+2\sqrt{5}/3$
0877 0504 $s\pi(1/5)-s\pi(\sqrt{5})$
0877 1732 rt : (6,2,-1,-9)
0877 2233 sq : $\sqrt{2}-\sqrt{5}/2$
0877 2494 as : $\sqrt{2}/3-\sqrt{5}/4$
0877 2915 id : $4e-\pi/4$
0877 4473 rt : (5,5,-4,-8)
0877 5942 ℓ_2 : $3/2-\sqrt{5}/4$
0877 6176 tπ : $1/2+2\sqrt{5}$
0877 6607 sπ : $7e$
0877 7219 rt : (4,8,-7,-7)
0877 7647 ln : $\beta(2)$
0877 7879 e^x : $3+4\pi$
0877 8525 $1/2+s\pi(1/5)$
0877 9149 sq : $(3/2)^{-3}$
0877 9762 sr : $\sqrt{19}$
0878 0675 tπ : $4\sqrt{2}-3\sqrt{3}/4$
0878 2321 as : $1-4\sqrt{2}/3$
0878 2529 2^x : $e/4-4\pi/3$
0878 3198 e^x : $-\sqrt{5}+\sqrt{6}-\sqrt{7}$
0878 3572 $\ln(4/5)\times\exp(\sqrt{5})$
0878 4645 J_1 : $(\sqrt[3]{2}/3)^2$
0878 5557 sr : $3+e/2$
0878 7431 rt : (8,9,4,4)
0878 9062 cu : $21/8$
0878 9226 J_1 : $3/17$
0878 9765 J_2 : $3\sqrt[3]{3}/5$
0879 2732 rt : (7,6,1,3)

0879 3212 rt : (2,9,-5,5)
0879 4014 id : $3\sqrt{2}-2\sqrt{3}/3$
0879 8052 rt : (6,3,-2,2)
0879 8327 ln : $1/4+e$
0880 0000 cu : 21/5
0880 0149 Ψ : $3\ln 2/10$
0880 0527 cr : $8e\pi$
0880 1320 id : $4e/3-3\pi/2$
0880 2189 J_1 : $6e\pi/7$
0880 2625 J_2 : $\sqrt{3}/2$
0880 2920 rt : (1,1,6,-9)
0880 3353 2^x : $7e\pi/6$
0880 3391 rt : (5,0,-5,1)
0880 4352 J_1 : $\sqrt{2}/8$
0880 4562 ℓ_{10} : $\sqrt{2/3}$
0880 5954 rt : (5,4,-2,-9)
0880 6818 sq : $\pi^2/4$
0880 6853 Γ : $4(1/\pi)/3$
0880 7423 J_1 : $5(1/\pi)/9$
0880 8748 rt : (4,9,-2,-7)
0880 9433 10^x : $2\sqrt{3}/3-\sqrt{5}/2$
0881 0404 J_2 : $5\ln 2/4$
0881 0606 $t\pi$: $7e$
0881 1493 J_1 : $\exp(-\sqrt{3})$
0881 4362 sr : $2\sqrt{2}+3\sqrt{5}$
0881 4600 rt : (1,7,0,-7)
0881 4836 J_2 : 13/15
0881 5936 10^x : $3\pi/4$
0882 1101 $s\pi$: $\exp(1/\sqrt{2})$
0882 4647 sq : $3\ln 2/7$
0882 4893 10^x : $6\pi^2/5$
0882 6292 cu : $3/2-\sqrt{2}/3$
0882 8021 J_0 : $\sqrt{20/3}$
0882 8303 at : 21/11
0883 0214 J_2 : $4e\pi/7$
0883 0223 ζ : 15/13
0883 1921 J_2 : $\ln(\sqrt[3]{2}/3)$
0883 2638 sq : $3/4-\pi/3$
0883 3562 sinh : $3\pi/10$
0883 7354 2^x : $2/3+4\sqrt{2}$
0883 7839 ℓ_{10} : $2/3+\sqrt{5}/4$
0883 8578 e^x : $\sqrt{2}+2\sqrt{5}$
0883 9184 $s\pi$: $\sqrt[3]{23/3}$
0884 0719 rt : (4,6,-3,4)
0884 2667 $\exp(2/3)+\exp(\pi)$
0884 3568 rt : (3,9,-6,-8)
0884 4765 sr : $4\sqrt{2}-2\sqrt{5}$
0884 4841 at : $6(1/\pi)$
0884 6060 $1/2\div\exp(\sqrt{3})$
0884 9282 cu : $4+e/3$
0884 9821 cu : $7(1/\pi)/5$
0885 1234 $t\pi$: $(\ln 2/2)^{-1/3}$
0885 1982 $\sqrt{3}+\exp(\sqrt{5})$
0885 3926 rt : (7,7,3,4)
0885 4756 2^x : 17/16
0885 5622 $t\pi$: $\exp(1/\sqrt{2})$
0885 6403 ζ : $7\sqrt{5}/4$
0885 7546 cu : $1/\ln(3\pi)$
0885 8426 ℓ_{10} : $3e/10$
0885 9425 rt : (6,4,0,3)
0885 9616 Li_2 : $\ln 2/8$
0886 1568 Ψ : $8\zeta(3)/7$
0886 2571 at : $4/3+\sqrt{3}/3$
0886 4939 $\exp(5)+s\pi(\sqrt{5})$
0886 4944 rt : (5,1,-3,2)
0886 7711 rt : (9,1,-9,-3)
0886 8696 sinh : $(e)^2$
0887 0116 J_2 : 20/23
0887 0172 Ei : $(1/(3e))^{1/2}$
0887 0281 $\ln(\pi)\times s\pi(2/5)$
0887 0484 rt : (4,2,-6,-1)
0887 3123 Ei : 4/25
0887 3143 rt : (4,7,1,8)
0887 3252 Ψ : $(\sqrt[3]{2}/3)^{1/3}$
0887 3919 $t\pi$: $\sqrt[3]{23/3}$
0887 4437 $\ln(\sqrt{5})+\exp(1/4)$
0887 4831 rt : (1,9,7,-3)
0887 5960 $\ln(2)+\exp(1/3)$
0887 6043 rt : (3,8,-4,-9)
0887 8128 sinh : 24/13
0887 8819 id : $3\sqrt{3}/2+2\sqrt{5}/3$
0887 9304 $\pi\times\ln(\sqrt{2})$
0888 1384 Γ : $3\sqrt{2}/10$
0888 2119 rt : (1,6,-2,-8)
0888 2526 sinh : $2\sqrt{2}/3$
0888 3824 Ei : 17/24
0888 3941 2^x : $3\sqrt{2}+\sqrt{5}/2$
0888 4037 2^x : $3/4-3\sqrt{2}$
0888 4688 sq : 24/23
0888 4854 10^x : $4\sqrt{2}-3\sqrt{5}$
0888 6688 sr : $\sqrt[3]{5/3}$
0888 7917 $t\pi$: $4\sqrt{3}+2\sqrt{5}$
0888 8220 rt : (2,9,1,-7)
0889 0164 Γ : $\exp(-3/2)$
0889 2372 Li_2 : 2/23
0889 3543 J_2 : $(\pi/3)^{-3}$
0889 5432 2^x : $1/4+3\pi/4$
0889 7523 J_1 : $9\pi^2/4$
0889 7860 ζ : $7\zeta(3)/5$
0889 8268 rt : (3,8,-1,9)
0889 9153 rt : (9,0,-7,4)
0890 2296 10^x : $(3)^{-3}$
0890 5478 ζ : $2(e+\pi)/3$
0890 6218 K : 16/19
0890 7285 ln : $4+3e/2$
0890 8002 at : $2/3-\sqrt{3}/3$
0890 8984 rt : (3,2,-7,4)
0891 0625 2^x : $\exp(-2\pi/3)$
0891 0854 at : $\sqrt[3]{7}$
0891 6760 rt : (7,8,5,5)
0891 6895 $\ln(4/5)-s\pi(1/3)$
0891 7437 $3/5+\ln(3/5)$
0892 0496 tanh : $(\sqrt{5})^{-3}$
0892 0534 at : $(\sqrt{5})^{-3}$
0892 1026 ℓ_{10} : $2\sqrt{3}-\sqrt{5}$
0892 1399 J_2 : $10\zeta(3)$
0892 1461 J_0 : $\sqrt{2}+\sqrt{3}+\sqrt{7}$
0892 1612 J_0 : $(\ln 2/2)^{-2}$
0892 2449 rt : (6,5,2,4)
0892 5240 e^x : $(5/2)^{-1/3}$
0892 5603 e^x : $2\sqrt{3}/3+2\sqrt{5}/3$
0892 5742 $2/5\times\ln(4/5)$
0892 8124 $\exp(\sqrt{2})\div s\pi(\sqrt{5})$
0892 8159 rt : (5,2,-1,3)
0892 9613 10^x : 8/25
0893 0236 $t\pi$: $\exp(-4/3)$
0893 1022 rt : (9,1,-5,5)
0893 1639 id : $2/3-\sqrt{3}/3$
0893 2271 10^x : $1/4-3\sqrt{3}/4$
0893 2669 $\pi+\exp(2/3)$
0893 2721 e^x : 14/19
0893 3890 rt : (8,2,-8,-4)
0893 5647 $\exp(\sqrt{5})\div s\pi(e)$
0893 6084 cr : $4\sqrt{2}+2\sqrt{3}$
0893 7388 e^x : $\exp(\sqrt[3]{4})$
0893 8074 e^x : $3/2+\sqrt{2}/2$
0893 8363 at : $1/2+\sqrt{2}$
0893 9643 rt : (3,5,-6,-7)
0894 0045 cr : $2-\sqrt{2}/2$
0894 0176 λ : $\exp(-18\pi/11)$
0894 1228 J_1 : $5(e+\pi)/4$
0894 1457 rt : (6,8,-9,1)
0894 2378 sq : $1-3\sqrt{3}/4$
0894 2403 rt : (3,3,-3,8)
0894 3400 Ei : 7/20
0894 3557 as : $2/3-\sqrt{3}/3$
0894 4037 Ei : 15/14
0894 4271 id : $(\sqrt{5})^{-3}$
0894 5417 rt : (2,9,-6,0)
0894 6688 J_2 : $1/\ln(\pi)$
0894 6917 J_2 : $\sqrt[3]{2}/3$
0895 0919 ζ : $7\sqrt[3]{3}/6$
0895 1732 rt : (1,8,9,-7)
0895 2095 at : $\sqrt{11/3}$
0895 6202 sinh : $(\sqrt{5})^{-3}$
0895 6240 as : $(\sqrt{5})^{-3}$
0895 7361 e^x : $3/2-\sqrt{2}$
0895 7650 Ψ : $\exp(-\pi/2)$
0895 8075 rt : (2,8,-1,-8)
0895 8898 J_2 : $9\zeta(3)/2$
0896 0437 erf : $(1/\pi)/4$
0896 3042 J_1 : $\sqrt[3]{2}/7$
0896 3328 rt : (9,2,-3,6)
0896 3678 rt : (6,9,-5,5)
0896 5071 sr : $(\sqrt{2}/3)^{-3}$
0896 5235 Γ : 7/8
0896 6247 rt : (8,1,-6,-5)
0896 7101 cu : $\sqrt{2}/2-2\sqrt{3}/3$
0896 7893 J_1 : $9(e+\pi)/5$
0896 8336 rt : (2,8,0,3)
0896 9170 rt : (7,4,-9,-4)
0896 9924 e^x : $2\sqrt{2}/3-3\sqrt{5}/2$
0897 0385 $\sqrt{2}+s\pi(\sqrt{5})$
0897 1456 $s\pi$: $1/2+\sqrt{2}/3$
0897 2473 sr : 19/16
0897 3069 10^x : $3e/2-\pi/2$
0897 4130 J_2 : 7/8
0897 4425 cr : 22/17
0897 4470 cu : $1/3+2e$
0897 5482 ln : $2-e/3$
0897 7121 ζ : $9\sqrt{3}/4$
0897 7488 J_1 : $\sqrt[3]{3}/8$
0897 9385 rt : (2,4,-9,-7)
0898 1318 rt : (7,9,7,6)
0898 1537 J_0 : $3(e+\pi)$
0898 1701 cu : $\sqrt{3}/2+2\sqrt{5}/3$
0898 6579 $1/5\div\exp(4/5)$
0898 6695 id : $4\sqrt{2}+\sqrt{3}/4$
0898 6793 sr : $3-2e/3$
0898 7207 rt : (6,6,4,5)
0898 7551 sq : P_{79}
0899 0113 at : $6\sqrt{5}/7$
0899 0904 at : 23/12
0899 1267 sq : $3+3e/2$
0899 3119 rt : (5,3,1,4)
0899 3415 ℓ_2 : P_{46}
0899 5356 10^x : $1+3e/2$
0899 6083 rt : (9,3,-1,7)
0899 7473 $\exp(2/5)+\exp(4)$
0899 9053 rt : (8,0,-4,6)
0900 0000 sq : 3/10
0900 2029 rt : (7,3,-7,-5)
0900 5010 rt : (3,6,-2,-3)
0900 6102 2^x : $\sqrt{2}-\sqrt{5}+\sqrt{6}$
0900 6576 J_1 : $\exp(-\sqrt[3]{5})$
0900 6799 $\sqrt{3}-\exp(3/5)$
0900 7149 θ_3 : $(3\pi/2)^{-2}$
0900 7779 $t\pi$: $1/2+\sqrt{2}/3$
0900 7812 erf : 2/25
0900 9387 sinh : $6e/7$
0900 9618 ζ : $\exp(e/2)$
0901 0821 rt : (5,5,-9,5)
0901 0823 J_0 : $9\sqrt[3]{5}$
0901 0991 rt : (2,6,-8,-1)
0901 1028 Ψ : $9\sqrt{3}/10$
0901 1206 2^x : $1-2\sqrt{5}$
0901 1742 Ei : $4\ln 2/7$
0901 3554 2^x : $10e/3$
0901 3860 rt : (2,5,-3,2)
0901 4112 sinh : $1/\ln(\ln 2/2)$
0901 7663 ℓ_{10} : 13/16
0901 8447 J_2 : $(\sqrt{2}\pi/3)^{-1/3}$
0902 1922 10^x : $4\sqrt{3}/3-3\sqrt{5}/2$
0902 2352 $2/3\div\exp(2)$
0902 4349 at : $\sqrt{2}-\sqrt{3}+\sqrt{5}$
0902 4493 10^x : $3+2\sqrt{3}/3$
0902 5434 J_0 : $\sqrt{24}\pi$
0902 6403 ζ : $4\sqrt[3]{3}/5$
0902 8091 ℓ_2 : $2-3\sqrt{2}/4$
0902 8753 2^x : $3/4+3\pi/2$
0902 9278 ζ : 13/11
0902 9298 rt : (9,4,1,8)
0902 9892 2^x : $3/2+2\sqrt{2}$
0903 0166 rt : (2,7,-3,-9)
0903 0879 sq : $\zeta(3)/4$
0903 1108 Ei : 10/7
0903 1326 rt : (1,6,-9,-9)
0903 2322 rt : (8,1,-2,7)
0903 3361 J_0 : 9
0903 3397 ℓ_2 : $4e-3\pi/4$

0903 3723 rt : (5,6,-5,9)
0903 4914 $e^x : \sqrt{3}/3 - 4\sqrt{5}/3$
0903 5351 rt : (7,2,-5,-6)
0903 7208 $10^x : 4\sqrt{3}/3 + \sqrt{5}$
0903 7588 $\ln(3/5) \div \exp(\sqrt{3})$
0903 8386 rt : (6,5,-8,-5)
0903 9546 $2^x : \sqrt{2} - \sqrt{5} - \sqrt{7}$
0903 9717 $J_1 : 16$
0903 9724 rt : (9,5,-8,-3)
0903 9730 cu : $\pi/7$
0904 0532 $J_0 : \sqrt{5}$
0904 0628 $e^x : 3\sqrt{2}/4 - 2\sqrt{3}$
0904 3160 sq : $3(e+\pi)/10$
0904 7692 $\ln(2/3) \times \ln(4/5)$
0905 0773 $2^x : 1/8$
0905 3395 $J_1 : 2/11$
0905 3789 rt : (6,7,6,6)
0905 4615 cu : P_{41}
0905 5608 $e^x : 3/2 + e/3$
0905 5882 $J_1 : 10\pi^2/3$
0905 7008 $J_1 : 4(1/\pi)/7$
0905 8349 sq : $3/4 + 2\pi/3$
0905 8844 $\Psi : ((e+\pi)/2)^2$
0905 8893 $10^x : 10\sqrt{5}$
0905 9914 rt : (5,4,3,5)
0906 0287 ln : $2/3 + 4\sqrt{3}/3$
0906 0806 cr : $1/4 + \pi/3$
0906 1005 ln : $3\sqrt{3}/4 + 3\sqrt{5}/4$
0906 2986 rt : (9,5,3,9)
0906 4003 ln : 7π
0906 4415 $t\pi : 7e\pi/9$
0906 5947 tanh : 1/11
0906 5988 at : 1/11
0906 6064 rt : (8,2,0,8)
0906 6448 $\Psi : 1/\sqrt{15}$
0906 6990 $t\pi : \ln(\ln 2/3)$
0906 7770 $10^x : \sqrt{21}$
0906 9148 rt : (7,1,-3,-7)
0906 9190 $J_0 : 10e\pi/7$
0906 9576 tanh : $2(1/\pi)/7$
0906 9617 at : $2(1/\pi)/7$
0907 0241 $J_2 : 22/25$
0907 1546 $10^x : \sqrt{2}/2 + 4\sqrt{5}$
0907 1795 $\exp(3/5) \div \exp(3)$
0907 2238 rt : (9,6,-9,-9)
0907 5334 rt : (5,8,-8,-6)
0907 6340 $\Psi : 9\ln 2/4$
0907 8422 at : $3\sqrt{2}/4 - 4\sqrt{5}/3$
0907 8437 rt : (2,5,-6,-2)
0908 0353 $\zeta : 7\zeta(3)/6$
0908 1537 sq : $3\sqrt{2} + 3\sqrt{3}$
0908 2346 rt : (4,2,-9,9)
0908 3132 $\zeta : 10e/7$
0908 3644 $\zeta : 1/\ln(2e/3)$
0908 3980 cr : $9\sqrt[3]{3}/10$
0908 4660 rt : (1,8,-9,-1)
0908 4925 $e^x : 2/23$
0908 6091 $2^x : \sqrt{2}/4 + 4\sqrt{5}/3$
0908 9892 $\Gamma : \sqrt[3]{2/3}$
0909 0909 id : 1/11
0909 0943 $\Gamma : 1/\ln(\pi)$
0909 1762 $\theta_3 : 1/22$
0909 2755 $J_1 : 7\pi/3$
0909 2991 sq : $2\sqrt{2}/3 - 3\sqrt{3}$
0909 4568 id : $2(1/\pi)/7$
0909 5423 $\theta_3 : (1/\pi)/7$
0909 5796 $10^x : 4 + 4e$
0909 6892 tanh : $(\sqrt{2}\pi/2)^{-3}$
0909 6934 at : $(\sqrt{2}\pi/2)^{-3}$
0909 7177 ln : 21/23
0909 8418 sinh : $(2\pi)^{1/2}$
0909 8449 $\Gamma : 25/3$
0910 0293 rt : (8,3,2,9)
0910 0953 $2^x : P_{53}$
0910 2428 rt : (9,6,-6,-2)
0910 2799 rt : (2,9,-7,-5)
0910 3433 rt : (7,0,-1,8)
0910 3436 sinh : 1/11
0910 3477 as : 1/11
0910 4245 ln : 22
0910 4536 sq : $(\ln 3/2)^2$
0910 5622 $\Gamma : 24/11$
0910 6580 rt : (6,3,-4,-7)
0910 7110 sinh : $2(1/\pi)/7$
0910 7152 as : $2(1/\pi)/7$
0910 7730 $J_2 : 7\sqrt[3]{2}/10$
0910 8219 rt : (8,3,-9,-3)
0910 8945 sr : 25/21
0910 9734 rt : (5,9,-4,-2)
0911 0492 $10^x : (2\pi)^{1/2}$
0911 1539 rt : (3,9,-2,-9)
0911 2363 cr : $3\sqrt{3}/4$
0911 2500 cu : 9/20
0911 5607 $J_2 : 15/17$
0911 8050 $J_1 : \sqrt{5} + \sqrt{6} + \sqrt{7}$
0912 0975 $\ell_{10} : \pi^2/8$
0912 2090 $s\pi : 7\pi^2/3$
0912 2111 id : $(\sqrt{2}\pi/2)^{-3}$
0912 2290 rt : (6,8,8,7)
0912 2291 sq : $e/9$
0912 2706 $\Psi : 2\sqrt{3}$
0912 4443 sinh : 17/18
0912 5790 $2^x : \sqrt[3]{2}/10$
0912 5969 $Li_2 : 4\sqrt{2}/7$
0912 6053 at : $4\sqrt[3]{3}/3$
0912 6454 id : $\sqrt{2} + 3\sqrt{5}/4$
0912 7703 at : 25/13
0912 8643 rt : (5,5,5,6)
0912 9268 ln : $2e/3 + 2\pi$
0913 2024 rt : (3,7,-8,-3)
0913 2260 at : $8\zeta(3)/5$
0913 2649 sq : $\sqrt{2}/4 + 4\sqrt{3}/3$
0913 3551 sq : $3e/2 + 2\pi/3$
0913 3809 $\sqrt{5} - \ln(\pi)$
0913 4768 sinh : $(\sqrt{2}\pi/2)^{-3}$
0913 4810 as : $(\sqrt{2}\pi/2)^{-3}$
0913 5021 rt : (4,2,2,5)
0913 5666 id : $1/\ln(5/2)$
0913 6233 $s\pi : (e+\pi)^{-2}$
0913 6329 $2^x : 7\sqrt{2}$
0913 6359 $10^x : 1 - e/4$
0913 8220 rt : (7,1,1,9)
0913 9288 cr : 13/10
0913 9994 sq : $4\sqrt{3}/3 - 3\sqrt{5}/2$
0914 0920 $\zeta : (\pi/2)^3$
0914 1426 rt : (6,2,-2,-8)
0914 1450 at : $9\sqrt[3]{5}/8$
0914 1487 at : $\sqrt{3}/4 + 2\sqrt{5}/3$
0914 4309 $\ell_{10} : 3e + 4\pi/3$
0914 4462 id : $1/3 - 3\pi$
0914 4638 rt : (5,5,-5,-7)
0914 5163 $J_2 : 5\sqrt{2}/8$
0914 7209 $10^x : 1 - 3e/4$
0914 7856 rt : (4,8,-8,-6)
0914 7884 $\ln(4) \div \exp(e)$
0914 8386 $e^x : 3/4 - \pi$
0914 8627 sq : $4/3 + 3\sqrt{2}$
0914 9006 ln : $\sqrt{2}/4 + \sqrt{5}/4$
0914 9763 sr : $4\sqrt{3}/3 - \sqrt{5}/2$
0915 0084 sr : $3/4 + 4e/3$
0915 2646 $Li_2 : (\sqrt{5})^{-3}$
0915 4313 rt : (1,7,-7,-2)
0915 5494 cu : $1/2 + 4e$
0915 6161 rt : (3,8,-2,6)
0915 6633 $\exp(4/5) + s\pi(1/3)$
0915 7819 $5 \div \exp(4)$
0915 7994 at : $10\sqrt{3}/9$
0915 8897 $J_2 : \exp(-1/(3e))$
0915 9439 cu : $3\zeta(3)/8$
0916 0282 $t\pi : 7\pi^2/3$
0916 0513 rt : (1,6,6,7)
0916 0543 $t\pi : 1/2 + \sqrt{5}$
0916 2684 $\zeta : \sqrt{15}$
0916 3856 rt : (1,8,0,6)
0916 4584 $J : \exp(-15\pi/8)$
0916 6691 $J_2 : (\ln 2)^{1/3}$
0916 6760 rt : (9,7,-4,-1)
0916 6866 $2^x : 3 + \sqrt{2}/3$
0916 8453 $J_2 : (\sqrt[3]{3})^{-1/3}$
0917 0244 $10^x : 2\sqrt[3]{3}/9$
0917 2123 $J_2 : 7e\pi/2$
0917 2746 rt : (8,4,-7,-2)
0917 3300 $J_1 : 22/3$
0917 4604 $t\pi : (e+\pi)^{-2}$
0917 4914 $\exp(\pi) + s\pi(2/5)$
0917 6788 rt : (6,1,0,-9)
0917 7037 $\ell_{10} : 17/21$
0917 7323 $\Psi : 5\ln 2$
0918 0060 rt : (5,4,-3,-8)
0918 1998 $e^x : \sqrt{2} - \sqrt{5} + \sqrt{6}$
0918 2469 $2/5 - \exp(2/5)$
0918 3340 rt : (4,7,-6,-7)
0918 3488 $J_2 : 1/\ln(\pi/3)$
0918 3506 sq : $(2e)^{1/3}$
0918 6497 cu : $\sqrt{2}/4 + 4\sqrt{5}/3$
0918 8749 $\zeta : 1/\ln(5)$
0919 0810 sq : $9e\pi$
0919 2140 $\ell_2 : P_{62}$
0919 3312 $\exp(5) \times \exp(e)$
0919 4537 cr : $1 + 3e$
0919 5817 cu : $2/3 - \sqrt{5}/2$
0919 6725 sq : $1/3 - 3\pi/4$
0919 6986 $1/4 \div e$
0919 7226 $\ell_2 : \sqrt{2} - \sqrt{3} + \sqrt{6}$
0919 7922 sinh : $3\sqrt[3]{2}/4$
0919 8739 $\exp(4) \times s\pi(1/5)$
0919 9410 rt : (5,6,7,7)
0920 0660 cr : $2(e+\pi)/9$
0920 1048 ln : $2\sqrt{2} - \sqrt{3}$
0920 3830 $\Gamma : 1/\ln(\sqrt[3]{4}/2)$
0920 4235 $\ell_{10} : 1/4 + \sqrt{5}/4$
0920 4399 at : $3\sqrt{2}/4 + \sqrt{3}/2$
0920 6031 rt : (4,3,4,6)
0920 6471 $2^x : \sqrt{2}/2 + 4\sqrt{3}/3$
0920 8271 $s\pi(1/3) - s\pi(e)$
0920 8987 cu : $\ln(\pi/2)$
0920 9362 $\ell_{10} : 6\sqrt[3]{3}/7$
0920 9372 $e^x : 1/3 - e$
0921 2152 at : $1/4 + 3\sqrt{5}/4$
0921 2681 rt : (3,0,1,5)
0921 4486 $\ell_{10} : (\sqrt[3]{4}/3)^{1/3}$
0921 5797 sr : $1/4 + 2\sqrt{2}/3$
0921 6016 rt : (5,3,-1,-9)
0921 6845 $\theta_3 : 4\sqrt[3]{2}/7$
0921 9358 rt : (6,9,-6,3)
0921 9616 $\ell_{10} : 4\sqrt{2} + 3\sqrt{5}$
0921 9624 $J_1 : (\sqrt[3]{5}/3)^3$
0922 1912 $2^x : 3/4 - 4\pi/3$
0922 2707 rt : (3,9,-7,-7)
0922 2774 $\ell_{10} : 7\ln 2/6$
0922 3519 ln : $\pi^2/9$
0922 4424 $\theta_3 : \ln(\pi/3)$
0922 6064 rt : (1,6,-5,-3)
0922 6578 $2^x : 2(1/\pi)/5$
0922 6842 $e^x : 23/11$
0922 8707 $3/5 \div \ln(\sqrt{3})$
0923 2407 rt : (2,8,-1,-1)
0923 2798 rt : (9,8,-2,0)
0923 3871 $t\pi : 8\sqrt[3]{5}/5$
0923 6155 $2^x : 2 - 2e$
0923 6642 sinh : 13/3
0923 6870 $2^x : 4\sqrt{2}/3 + 3\sqrt{3}/4$
0923 8320 $Li_2 : 7\ln 2/6$
0923 8514 cu : $1/3 - \pi/4$
0923 8752 sinh : $(e/2)^2$
0923 8899 at : $\sqrt{3} - \sqrt{6} + \sqrt{7}$
0923 8989 rt : (8,5,-5,-1)
0923 9776 $\Psi : -\sqrt{2} + \sqrt{5} + \sqrt{7}$
0924 0102 ln : $4\sqrt{5}/3$
0924 0841 $\ln(\sqrt{2}) - s\pi(\sqrt{3})$
0924 1518 sq : $1/4 - e$
0924 1539 $erf : \exp(-5/2)$
0924 2083 $J_2 : 8/9$
0924 2547 $10^x : \sqrt{2}/3 - \sqrt{3}/4$
0924 2780 $10^x : 9\pi/8$
0924 3259 $\ell_{10} : 5\sqrt{3}/7$
0924 4868 cu : $\exp(-\sqrt[3]{4}/2)$
0924 5202 rt : (7,2,-8,-2)
0924 6334 $\theta_3 : 18/25$
0924 6392 $\ln(\pi) + \exp(2/3)$

0924 7948 $\ell_{10} : 3/2 + 4e$
0924 8555 $K : (5/3)^{-1/3}$
0924 9630 $\lambda : \exp(-13\pi/8)$
0925 2305 $\ell_{10} : 4\sqrt{2}/7$
0925 3508 sq : $1/3 + 2\sqrt{5}$
0925 5734 $3/5 \times \exp(e)$
0925 5926 rt : (4,7,0,6)
0925 6700 rt : (1,8,7,4)
0925 7127 $t\pi : 9(1/\pi)/8$
0925 9341 rt : (3,8,-5,-8)
0926 2759 sq : $7/23$
0926 4916 $\pi + s\pi(2/5)$
0926 6307 cr : $4\sqrt{3} + \sqrt{5}$
0926 7839 $Ei : 16/7$
0926 7985 rt : (9,0,-8,3)
0926 8151 $J_2 : 8\pi^2/9$
0926 8667 sinh : $\sqrt{5}\pi$
0926 9890 $Li_2 : (\sqrt[3]{4}/3)^{1/3}$
0927 2292 sq : $(2e/3)^{-2}$
0927 2331 rt : (5,7,9,8)
0927 3533 cr : $(\sqrt{2}\pi/2)^{1/3}$
0927 6145 rt : (7,9,-8,8)
0927 8426 cu : $3e/2 + 2\pi/3$
0927 9210 rt : (4,4,6,7)
0927 9563 cu : $\sqrt{17}$
0928 0431 $10^x : (e/3)^{-3}$
0928 1110 $J_2 : (\sqrt[3]{2})^{-1/2}$
0928 3811 cu : $3/2 - \pi/3$
0928 3824 $s\pi : 7\ln 2/5$
0928 4657 $J_0 : \sqrt{7}\pi$
0928 6120 rt : (3,1,3,6)
0928 6556 $e^x : 2e/3 - 4\pi/3$
0928 6756 $J_2 : 9\ln 2/7$
0928 8496 $J_1 : 3\zeta(3)$
0928 8511 $J_1 : 3\sqrt{5}$
0928 9321 $4/5 - s\pi(1/4)$
0929 2405 $J : \sqrt{2}/6$
0929 2551 $2^x : 1/3 + 4e$
0929 3060 rt : (2,4,2,5)
0929 3600 $10^x : 5\sqrt{3}/8$
0929 5381 rt : (5,6,-6,7)
0929 6542 rt : (3,7,-3,-9)
0929 7520 sq : $23/22$
0929 8859 cu : $e/6$
0929 9016 $Ei : 1/\sqrt[3]{16}$
0930 0031 rt : (1,5,-3,-4)
0930 0626 rt : (9,9,0,1)
0930 2744 rt : (1,8,-1,-2)
0930 3684 rt : (9,1,-6,4)
0930 4457 at : $2/3 - 3\sqrt{3}/2$
0930 5498 $\zeta : 1/\ln(\ln 2/2)$
0930 6322 $Li_2 : 1/11$
0930 7033 rt : (8,6,-3,0)
0930 8098 $J_0 : 21/4$
0931 0158 $Li_2 : 2(1/\pi)/7$
0931 0940 $\ell_2 : 15/16$
0931 1358 $J_1 : 5\sqrt[3]{3}/2$
0931 2906 $\ell_{10} : \exp(3\pi)$
0931 3465 rt : (7,3,-6,-1)
0931 4396 $J_1 : \sqrt{13}$
0931 4718 $2/5 + \ln(2)$
0931 5417 $\zeta : 11/9$
0931 6533 sinh : $(\sqrt{5}/2)^{-1/2}$
0931 9922 rt : (6,0,-9,2)
0932 2104 $\ell_{10} : 1/3 + e/3$
0932 2478 $Ei : \pi/9$
0932 4093 $t\pi : 7\ln 2/5$
0932 5261 $J_0 : 4e\pi/3$
0932 6730 cu : $e/2 + 3\pi$
0932 6830 $e^x : e/2 + \pi/3$
0932 7128 $3/5 \times \exp(3/5)$
0932 7937 ln : P_{11}
0932 9446 cu : $5\sqrt[3]{3}/7$
0933 0427 cu : $\sqrt{2} - 3\sqrt{3}$
0933 0788 $10^x : 3e + 3\pi/4$
0933 2117 $\sqrt{2} \div \exp(e)$
0933 3815 sinh : $6\sqrt{3}/7$
0933 3861 $J_1 : 3/16$
0933 4327 rt : (3,3,-4,6)
0933 5158 $\Psi : 25/16$
0933 5601 at : $e - \pi/4$
0933 6519 $\Gamma : (\pi/3)^{-3}$
0933 7886 rt : (2,9,-4,-9)
0933 8326 at : $3/2 + \sqrt{3}/4$
0933 9037 $Li_2 : (\sqrt{2}\pi/2)^{-3}$
0933 9625 $e^x : 2 + 4\sqrt{2}$
0933 9907 rt : (9,2,-4,5)
0934 0114 $\ln(4) + s\pi(1/4)$
0934 2655 $e^x : 2/3 - \sqrt{3}/3$
0934 3116 at : $3\sqrt{2} - 4\sqrt{3}/3$
0934 3319 rt : (8,1,-7,-4)
0934 8369 $e^x : e/4 + 3\pi$
0934 9779 $J_2 : 2\sqrt{5}/5$
0935 3242 sr : $(\sqrt[3]{5})^{1/3}$
0935 3950 rt : (9,5,-8,-9)
0935 4129 cr : $17/13$
0935 4300 $\ell_2 : 3 - \sqrt{3}/2$
0935 4684 rt : (4,5,8,8)
0935 5815 $J_2 : 17/19$
0935 5974 ln : $1/3 + \sqrt{3}/3$
0935 6469 $e^x : (\sqrt{5})^{-3}$
0935 8750 $\zeta : 6\sqrt[3]{2}/7$
0936 1048 $\sqrt{2} \div s\pi(\sqrt{5})$
0936 1871 rt : (3,2,5,7)
0936 3147 at : $1 - e/3$
0936 3655 rt : (1,7,7,9)
0936 3831 $\zeta : 9\sqrt[3]{5}/4$
0936 5294 sq : $3/2 + 3\sqrt{5}/4$
0936 5376 $5 \div \exp(1/5)$
0936 6772 cr : $3\pi^2$
0936 6894 rt : (2,6,-7,8)
0936 7108 cu : $3/4 + \sqrt{2}/2$
0936 8716 $\Psi : (\pi/2)^{-3}$
0936 9092 rt : (2,5,6,9)
0936 9360 $e^x : 2\sqrt{2} - 3\sqrt{3}$
0937 0169 cr : $4\sqrt{2}/3 - \sqrt{3}/3$
0937 1485 $\ell_{10} : 2\sqrt{3} + 4\sqrt{5}$
0937 3875 sinh : $23/4$
0937 5064 $10^x : 2e/3 + 3\pi/4$
0937 5378 $\zeta : 8\sqrt[3]{3}/3$
0937 6345 rt : (1,5,2,-7)
0937 6461 at : $3\sqrt{2}/4 - 2\sqrt{3}/3$
0937 6671 rt : (9,3,-2,6)
0937 6969 rt : (8,7,-1,1)
0937 7152 tanh : $\ln(\ln 3)$
0937 7200 at : $\ln(\ln 3)$
0937 9192 cu : $6\zeta(3)/7$
0938 0147 rt : (8,0,-5,5)
0938 0727 $\ln(2) \div \exp(2)$
0938 1273 rt : (8,7,-9,-9)
0938 1438 $erf : 1/12$
0938 2491 $e^x : 19/7$
0938 3632 rt : (7,4,-4,0)
0938 4123 $2^x : 10e$
0939 0322 rt : (6,1,-7,-1)
0939 0605 id : $1 - e/3$
0939 1042 cr : $3/4 + \sqrt{5}/4$
0939 1435 cu : $5/11$
0939 4444 cu : $\sqrt{2} + 2\sqrt{3}$
0939 5040 $J_2 : (\sqrt[3]{3}/2)^{1/3}$
0939 5602 $2^x : 3 + 2e/3$
0939 5746 ln : $3/4 + \sqrt{5}$
0939 8173 $2^x : 3\sqrt{2}/4 - 2\sqrt{5}$
0940 1730 $J_1 : \exp(-5/3)$
0940 2329 rt : (3,7,-9,-6)
0940 2779 cu : $10(1/\pi)/7$
0940 2863 $2^x : 2/3 - 3e/2$
0940 3010 $e^x : \sqrt{7}$
0940 4036 id : $3\sqrt{2}/4 - 2\sqrt{3}/3$
0940 4462 as : $1 - e/3$
0940 4782 id : $\ln(\ln 3)$
0940 6725 sr : $7\sqrt[3]{5}/10$
0940 6910 cu : $\sqrt{2}/3 + \sqrt{5}/4$
0940 7925 $\exp(\pi) \div \exp(\sqrt{3})$
0940 9488 ln : $\sqrt{2} + \sqrt{3} - \sqrt{5}$
0941 0902 $2^x : \sqrt{2}/2 - \sqrt{3}/3$
0941 1375 $e^x : 17/23$
0941 1622 at : $2 - 2\pi/3$
0941 1652 $J_2 : 2\pi/7$
0941 2721 $Li_2 : 17/21$
0941 3989 rt : (9,4,0,7)
0941 4488 sq : $4e + \pi/3$
0941 5028 as : $3\sqrt{2} - 3\sqrt{5}/2$
0941 5426 rt : (2,5,-4,-1)
0941 5865 sq : $\ln(e/2)$
0941 7532 rt : (8,1,-3,6)
0941 7953 as : $3\sqrt{2}/4 - 2\sqrt{3}/3$
0941 8653 sinh : $\ln(\ln 3)$
0941 8702 as : $\ln(\ln 3)$
0941 9117 $\ell_2 : e\pi$
0942 1084 rt : (7,2,-6,-5)
0942 1249 $\ell_{10} : 5\sqrt{5}/9$
0942 1365 $\Psi : \ln 2/4$
0942 3070 $10^x : 1/3 - e/2$
0942 4643 rt : (6,5,-9,-4)
0942 5416 sq : $\sqrt{3}/2 - \sqrt{5}/4$
0942 7309 $\theta_3 : 2\sqrt[3]{5}/7$
0942 7952 sq : $3/2 + 4e$
0942 8594 rt : (3,8,-3,3)
0942 8863 $\ell_{10} : 3 + 3\pi$
0942 9189 $J_0 : 7\pi^2/2$
0943 0009 cu : $3/2 + 2\sqrt{2}$
0943 1821 ln : $\sqrt{2}/4 + \sqrt{5}/3$
0943 2915 $2^x : 3e\pi$
0943 3471 $\zeta : 5/13$
0943 4556 $\ell_{10} : 1/3 + \sqrt{2}/3$
0943 5576 $\ell_{10} : \ln(\sqrt{5})$
0943 6795 cr : $1/4 + 3\sqrt{2}/4$
0943 8528 ln : $3\sqrt{2}/2 + \sqrt{3}/2$
0943 9510 id : $2\pi/3$
0944 0072 rt : (3,3,7,8)
0944 1345 $2^x : (\ln 2/3)^{-1/3}$
0944 1825 $10^x : \sqrt{2}/2 - \sqrt{3}$
0944 2570 $e^x : 9\sqrt{2}/4$
0944 5552 $J_2 : \exp(-1/(3\pi))$
0944 5670 $\Gamma : 20/23$
0944 7331 $1/5 \div \exp(3/4)$
0944 7592 rt : (2,0,4,7)
0944 8894 rt : (8,8,1,2)
0945 1877 rt : (9,5,2,8)
0945 2024 $\sqrt{2} \times s\pi(e)$
0945 2738 $2^x : 3e/2 + 2\pi/3$
0945 3489 $1/2 + \ln(2/3)$
0945 3585 as : $2 - 2\pi/3$
0945 5148 rt : (2,6,2,8)
0945 5490 rt : (8,2,-1,7)
0945 5802 rt : (7,5,-2,1)
0945 6117 $10^x : \sqrt{2}/2$
0945 6203 $2^x : 1/\ln(\sqrt{5}/3)$
0945 7002 cu : $5\sqrt{2}/3$
0945 7063 $J_1 : \sqrt[3]{5}/9$
0945 7526 $J_2 : 5\sqrt[3]{2}/7$
0945 7706 $\theta_3 : (\sqrt{3}/2)^{1/3}$
0945 8630 $J_2 : 9/10$
0945 8824 $2^x : 16/15$
0945 9110 rt : (7,1,-4,-6)
0946 2353 $2^x : 3/23$
0946 2739 rt : (6,8,-3,-5)
0946 2783 ln : $\sqrt{2} + 3\sqrt{5}$
0946 3491 $s\pi : 7\sqrt[3]{5}$
0946 4286 $\theta_3 : (\sqrt[3]{5}/3)^{1/3}$
0946 5367 $\ell_2 : 1/4 + 4\sqrt{2}/3$
0946 6376 rt : (5,8,-9,-8)
0946 6584 $s\pi : 5\sqrt[3]{3}/7$
0946 7455 sq : $4/13$
0946 9690 rt : (1,8,6,-3)
0946 9705 rt : (5,1,-8,1)
0947 0366 rt : (1,9,5,-4)
0947 0469 $t\pi : \sqrt{2}/2 - 3\sqrt{5}/4$
0947 1563 at : $3 - 3\sqrt{2}/4$
0947 3290 $\Psi : 4\sqrt[3]{5}/5$
0947 3698 cu : $4 - \sqrt{3}/3$
0947 4986 $\Psi : \exp(-\sqrt{5}/3)$
0947 5019 cu : $\sqrt{15}$
0947 5910 $J_0 : 7(1/\pi)$
0947 7145 $\zeta : 23/6$
0947 8022 sr : $\exp(-3\pi/2)$
0947 8181 $\lambda : (2e)^{-3}$
0947 8471 sq : $(\sqrt{2}\pi/3)^{-3}$
0948 0226 $Ei : 19/5$

0948 0580 rt : (7,9,-9,6)
0948 0682 J_1 : 4/21
0948 1260 sπ : $3e/2 - \pi/3$
0948 2629 ℓ_{10} : $2/3 + \sqrt{3}/3$
0948 3082 at : $\sqrt{2} - 3\sqrt{5}/2$
0948 5576 sq : $\sqrt[3]{23/2}$
0948 6168 J_2 : $5\sqrt[3]{3}/8$
0948 7217 10^x : $4/3 - 3\pi/4$
0948 7978 cr : 21/16
0948 8414 id : $e/3 + 4\pi/3$
0948 8847 J_2 : $3\zeta(3)/4$
0948 8861 rt : (6,9,-7,1)
0949 0351 rt : (9,6,4,9)
0949 0429 sq : $4\ln 2/9$
0949 1407 as : 8/9
0949 4035 rt : (8,3,1,8)
0949 4269 J_0 : $9\sqrt[3]{3}/5$
0949 4305 Ei : $(1/\pi)/2$
0949 4414 cr : $1/4 + 4\sqrt{5}$
0949 4414 cu : $2 + \sqrt{2}/3$
0949 5119 tanh : 2/21
0949 5170 at : 2/21
0949 5733 ℓ_{10} : $4e + \pi/2$
0949 7728 rt : (7,0,-2,7)
0949 7876 e^x : $1 - 3\sqrt{5}/2$
0949 8377 Γ : $7(1/\pi)/10$
0950 1428 rt : (6,3,-5,-6)
0950 1865 ln : $2/3 + \sqrt{3}/4$
0950 2003 J_1 : $3e/2$
0950 5137 rt : (5,9,-5,-4)
0950 5380 2^x : $3/2 - \sqrt{3}/4$
0950 5823 J_1 : $3(1/\pi)/5$
0950 6154 tπ : $7\sqrt[3]{5}$
0950 6480 ζ : $2\sqrt{3}/9$
0950 7597 rt : (1,7,-8,7)
0950 8140 sr : $2e - \pi/3$
0950 8378 Γ : $8e\pi/7$
0950 8695 cr : $4 + 3\sqrt{3}$
0950 9289 tπ : $5\sqrt[3]{3}/7$
0951 0799 10^x : $2/3 + \sqrt{3}/2$
0951 1739 rt : (2,8,-2,-5)
0951 2971 J_2 : $(e/2)^{-1/3}$
0951 3143 10^x : $3\sqrt{2}/2 + 4\sqrt{3}/3$
0951 3326 e^x : $(\sqrt[3]{3})^{1/3}$
0951 5238 sr : $1/3 + \sqrt{3}/2$
0951 5442 sπ : $6\zeta(3)/7$
0951 6943 e^x : 1/11
0951 9618 at : $5e/7$
0952 0375 tanh : $3(1/\pi)/10$
0952 0427 at : $3(1/\pi)/10$
0952 0873 rt : (3,4,9,9)
0952 0951 e^x : $2(1/\pi)/7$
0952 1111 ζ : $4\sqrt[3]{2}/3$
0952 1653 Ψ : $1/\ln(4/3)$
0952 2915 rt : (8,9,3,3)
0952 3809 id : 2/21
0952 4166 tπ : $3e/2 - \pi/3$
0952 4837 θ_3 : 1/21
0952 5916 ln : $7\sqrt{2}/9$
0952 8714 rt : (2,1,6,8)

0952 8730 as : P_{43}
0952 9140 e^x : $3\sqrt{2}/2 - 2\sqrt{5}$
0952 9351 ℓ_{10} : $1/2 + \sqrt{5}/3$
0952 9707 J : $1/\sqrt{24}$
0953 0082 rt : (7,6,0,2)
0953 1017 ln : 10/11
0953 3187 rt : (8,4,3,9)
0953 6206 e^x : $e/3 + 2\pi/3$
0953 6593 rt : (1,2,3,-7)
0953 6953 rt : (7,1,0,8)
0953 7281 rt : (6,3,-3,1)
0953 7319 id : $2e/3 + 2\pi$
0953 8213 sinh : 2/21
0953 8265 as : 2/21
0954 0728 rt : (6,2,-3,-7)
0954 2658 sπ : $\sqrt{2}/3 + \sqrt{5}/4$
0954 4327 at : $1 + 2\sqrt{2}/3$
0954 4511 sr : 6/5
0954 4673 at : $\sqrt[3]{22/3}$
0954 4717 J_0 : $9\sqrt{3}/7$
0954 5442 cu : $1 + 3\sqrt{2}$
0954 5650 10^x : $3/2 + 3e/4$
0954 6554 J_1 : 18/5
0954 6803 Γ : $(e)^2$
0954 8303 rt : (4,8,-9,-5)
0954 9150 sq : $1/4 - \sqrt{5}/4$
0954 9296 id : $3(1/\pi)/10$
0954 9329 e^x : $(\ln 2)^{-1/3}$
0955 0736 Ψ : $(\sqrt{2}\pi)^{-1/2}$
0955 1091 cu : $1/4 - \sqrt{2}/2$
0955 1121 e^x : $(\sqrt{2}\pi/2)^{-3}$
0955 1148 ln : $3/2 + 2\sqrt{5}/3$
0955 1773 rt : (4,9,-3,-9)
0955 2023 J_2 : 19/21
0955 2468 rt : (1,8,3,-5)
0955 3067 ζ : $5\ln 2/9$
0955 3420 cr : $3e + \pi/3$
0955 3497 2^x : $6\sqrt[3]{5}$
0955 4120 ln : P_7
0955 5280 e^x : $1/4 - 3\sqrt{3}/2$
0955 7700 sinh : 18/19
0955 8019 ℓ_2 : $8\zeta(3)/9$
0955 8815 tπ : $6\zeta(3)/7$
0956 2115 cu : $1/4 + 4\sqrt{2}$
0956 2720 cu : $2\sqrt{3}/3 - 4\sqrt{5}/3$
0956 3816 sinh : $3(1/\pi)/10$
0956 3869 as : $3(1/\pi)/10$
0956 3976 J : $\exp(-12\pi/13)$
0956 5166 10^x : $\sqrt{2}/3 - 2\sqrt{5}/3$
0956 7333 10^x : $2\sqrt{3} + 3\sqrt{5}$
0956 8147 ℓ_2 : $3e/2 - \pi$
0957 1132 Γ : $\sqrt[3]{21/2}$
0957 1523 rt : (5,6,-7,5)
0957 2393 e^x : $\pi^2/7$
0957 2902 sq : $2 - 4\sqrt{3}/3$
0957 3444 10^x : 9/5
0957 4699 id : $7\sqrt[3]{3}$
0957 5021 2^x : $e\pi/8$
0957 6804 rt : (7,2,2,9)
0957 7074 $4 \times s\pi(e)$

0957 8024 J_1 : $\sqrt{3}/9$
0957 8070 2^x : P_{49}
0957 8207 Ψ : $7\sqrt{5}/10$
0957 8211 J_2 : $e/3$
0957 8640 $\ln(\pi) + s\pi(2/5)$
0957 9370 cr : 25/19
0958 0588 id : $4\sqrt{2}/3 - 4\sqrt{5}/3$
0958 0645 sr : $\sqrt{2} + \sqrt{5} - \sqrt{6}$
0958 0655 rt : (6,1,-1,-8)
0958 1595 e^x : $\sqrt[3]{20}$
0958 1968 cu : $2\sqrt{3} - \sqrt{5}/3$
0958 2399 rt : (2,6,-8,4)
0958 2606 2^x : $\sqrt[3]{13/3}$
0958 4516 rt : (5,4,-4,-7)
0958 5679 2^x : $3\sqrt{3}/2 + 3\sqrt{5}$
0958 6406 tπ : $\sqrt{2}/3 + \sqrt{5}/4$
0958 6750 10^x : $((e+\pi)/2)^{-3}$
0958 7269 sr : $(\sqrt[3]{3})^{1/2}$
0958 8385 rt : (4,7,-7,-6)
0958 9402 ln : $(4/3)^{-1/3}$
0958 9763 J_2 : $(\sqrt{5}/3)^{1/3}$
0958 9868 J_0 : $3\sqrt{3}/2$
0959 0412 ℓ_2 : $5\sqrt[3]{5}$
0959 1887 2^x : P_{64}
0959 4493 10^x : $(1/\pi)/8$
0959 5730 sr : $(\ln 2)^{-1/2}$
0959 6679 2^x : $4\sqrt{2} - \sqrt{5}/3$
0959 6734 ℓ_2 : $3\sqrt{3} + 3\sqrt{5}/2$
0959 7976 e^x : $3\sqrt{3}/4 - \sqrt{5}/4$
0959 8002 10^x : $2 + \pi/4$
0959 8388 e^x : $9\sqrt[3]{5}/2$
0959 8832 rt : (5,7,-3,9)
0959 9841 sπ : $\sqrt{2}/4 + 3\sqrt{5}/4$
0960 0000 cu : 8/5
0960 0053 at : $\sqrt{3} - \sqrt{5} + \sqrt{6}$
0960 1894 $\ln(4/5) \times s\pi(\pi)$
0960 5703 sinh : $8\ln 2/3$
0960 6588 rt : (7,7,2,3)
0960 7685 J : $(1/(3e))^{1/2}$
0960 9509 ln : $7\sqrt[3]{5}/4$
0960 9796 ℓ_{10} : $9\ln 2/5$
0961 1759 2^x : $4/3 - 3\pi/2$
0961 2625 rt : (2,2,8,9)
0961 3574 ℓ_{10} : $2e - 4\pi/3$
0961 4065 rt : (6,4,-1,2)
0961 4125 cu : $\sqrt{2}/4 - 2\sqrt{3}$
0961 6288 $\exp(2) + s\pi(1/4)$
0961 6623 ℓ_{10} : $2\zeta(3)/3$
0961 7434 $\exp(4/5) \div \exp(\pi)$
0961 9090 ζ : $\zeta(11)$
0962 0851 rt : (1,1,5,-8)
0962 1227 $\exp(1/3) - \exp(2/5)$
0962 1230 rt : (6,0,1,9)
0962 1575 rt : (5,1,-4,1)
0962 3208 ℓ_{10} : $5\sqrt[3]{3}/9$
0962 3832 ζ : $((e+\pi)/3)^2$
0962 3996 J_1 : $8(e+\pi)/7$
0962 4200 tπ : $4\sqrt{2}/3 - 4\sqrt{3}/3$
0962 5170 rt : (5,3,-2,-8)
0962 6525 rt : (9,5,-9,-8)

0962 6678 J_0 : 25
0962 6770 J_2 : $(4/3)^{-1/3}$
0962 8182 cu : 11/24
0962 9120 rt : (4,6,-5,0)
0963 0043 sr : $5\sqrt[3]{3}/6$
0963 0098 J_1 : $7\pi^2/2$
0963 0923 e^x : $9\pi^2/4$
0963 1350 ζ : $\exp(-1/(3e))$
0963 1581 e^x : $\sqrt{2} + 4\sqrt{5}/3$
0963 3079 rt : (3,9,-8,-6)
0963 3177 sr : $\exp(1/(2e))$
0963 5590 rt : (1,7,-9,-1)
0963 5670 Li_2 : $\ln(\ln 3)$
0963 7230 J_2 : 10/11
0963 7430 rt : (1,7,1,-6)
0963 7434 sr : $3/2 - 2\sqrt{5}/3$
0963 7631 id : $2\sqrt{2} - \sqrt{3}$
0963 8324 tπ : $\sqrt{22}\pi$
0963 8355 sr : $\zeta(3)$
0964 0307 sr : $3(e+\pi)/4$
0964 2989 2^x : $1/4 - 4e/3$
0964 4383 tπ : $\sqrt{2}/4 + 3\sqrt{5}/4$
0964 7272 at : $\exp(2/3)$
0964 7819 10^x : 1/25
0964 8527 Ei : $(e+\pi)$
0964 9047 e^x : $1/2 + 2e/3$
0965 0974 sq : $3/4 - 3\sqrt{2}/4$
0965 3600 rt : (9,4,-7,-9)
0965 5183 rt : (1,7,6,3)
0965 7080 rt : (4,3,-7,9)
0965 7359 $1/4 - \ln(\sqrt{2})$
0965 7757 sr : $\sqrt{2} + 4\sqrt{5}/3$
0965 9034 J_2 : $\sqrt{2} + \sqrt{3} - \sqrt{5}$
0965 9880 J_2 : $1/\ln(3)$
0966 2271 id : $\pi^2/9$
0966 2925 $\sqrt{5} \div \exp(\pi)$
0966 4437 at : $9\sqrt{3}/8$
0966 4836 rt : (9,0,-9,2)
0966 6494 rt : (5,2,0,-9)
0966 7727 rt : (1,9,0,8)
0967 0526 rt : (4,7,-1,4)
0967 2296 10^x : $4/3 + 4e$
0967 3797 sinh : $\sqrt[3]{19/2}$
0967 4568 rt : (3,8,-6,-7)
0967 5858 10^x : $5e\pi/4$
0967 8364 J_2 : $3(e+\pi)/2$
0968 0495 J_0 : 13/5
0968 0707 sπ : $(\ln 3)^{-1/3}$
0968 2497 2^x : 2/15
0968 3633 rt : (8,6,-8,-9)
0968 4335 10^x : $9e/5$
0968 5442 rt : (7,8,4,4)
0968 7345 Ei : $5(1/\pi)/4$
0969 0349 tπ : $7\ln 2/9$
0969 1001 ℓ_{10} : 4/5
0969 1466 Ψ : $7\pi^2/8$
0969 1916 $\ln(5) \times \exp(\sqrt{3})$
0969 2235 ℓ_{10} : $((e+\pi)/3)^{1/3}$
0969 2409 rt : (1,5,-8,-8)
0969 2489 J_2 : $(\pi/3)^{-2}$

0969 3216 rt : (6,5,1,3)
0969 4283 sr : $\exp(\sqrt{2}\pi/3)$
0969 6788 $1/3 + s\pi(\pi)$
0969 7196 $\exp(2/3) \div \exp(3)$
0969 7591 2^x : $2\sqrt{2} + 4\sqrt{5}/3$
0969 9874 2^x : $\zeta(3)/9$
0970 1025 rt : (5,2,-2,2)
0970 4307 Ψ : $\sqrt{3}/10$
0970 4459 ζ : $7e/5$
0970 4944 rt : (9,1,-7,3)
0970 5446 Γ : 13/15
0970 5954 Γ : $\exp(\sqrt{2}\pi/3)$
0970 6488 ℓ_2 : $7\zeta(3)/9$
0970 7201 J_0 : $\sqrt[3]{11}$
0970 8028 cr : $2 - e/4$
0970 8108 rt : (1,0,7,9)
0970 8730 e^x : $\exp(1/(3e))$
0970 8855 sq : $3/4 + 4e$
0970 8871 rt : (4,1,-5,-1)
0971 2625 rt : (4,8,3,8)
0971 4042 rt : (7,8,-9,-9)
0971 5277 J_2 : 21/23
0971 5812 e^x : $4e - \pi$
0971 6754 rt : (3,8,-4,0)
0971 8043 2^x : $\sqrt{3} - \sqrt{6} - \sqrt{7}$
0971 8390 $\ln(\sqrt{3}) \div \exp(\sqrt{3})$
0971 9567 sr : $7\pi/5$
0972 0385 $t\pi$: $1/3 - 3\sqrt{3}/2$
0972 0976 2^x : $3e/5$
0972 1282 Ei : $((e+\pi)/2)^{1/3}$
0972 2416 $t\pi$: $e/2 - 2\pi/3$
0972 2755 10^x : $3e/4 + 3\pi/2$
0972 4444 2^x : $8\zeta(3)/9$
0972 4669 $2/5 \div \exp(\sqrt{2})$
0972 5445 rt : (1,6,-1,-7)
0972 6390 $t\pi$: $(\ln 3)^{-1/3}$
0972 6408 Γ : $5\ln 2/4$
0973 0866 sq : $e/2 - \pi/3$
0973 1261 2^x : $1 - \sqrt{3}/2$
0973 4184 rt : (2,9,2,-6)
0973 4942 sinh : 23/6
0973 5966 J_0 : $9e/2$
0973 6329 e^x : $\sqrt{3} + 2\sqrt{5}/3$
0974 1213 2^x : $(\ln 3/3)^2$
0974 4890 e^x : $1/2 - 2\sqrt{2}$
0974 5172 2^x : $3e/2 + \pi/4$
0974 5685 rt : (9,2,-5,4)
0974 5726 rt : (3,9,2,9)
0974 6127 sq : $3/2 - 2e/3$
0974 7145 2^x : $((e+\pi)/3)^{-3}$
0974 7439 10^x : $10\sqrt{3}/7$
0974 9692 rt : (8,1,-8,-3)
0974 9891 e^x : $-\sqrt{2} + \sqrt{3} - \sqrt{7}$
0975 0566 sq : 22/21
0975 0957 id : $2\sqrt{2}/3 + 2\sqrt{3}/3$
0975 1057 ζ : 19/5
0975 1349 $s\pi$: $7\sqrt{3}/4$
0975 1766 ln : $7\sqrt[3]{2}/8$
0975 2250 erf : $\ln 2/8$
0975 2786 at : $3 - \pi/3$
0975 4848 10^x : $3\sqrt{3}/2 + 4\sqrt{5}/3$
0975 5457 ℓ_2 : $2/3 - \sqrt{3}/4$
0975 5833 sq : $\sqrt{3}/4 - \sqrt{5}/3$
0975 9327 2^x : $5\sqrt[3]{5}/8$
0975 9656 rt : (3,6,-2,-9)
0976 0712 Li_2 : 2/21
0976 1217 $\ln(\sqrt{5}) - s\pi(1/4)$
0976 1673 cr : $\sqrt{3} + \sqrt{5} - \sqrt{7}$
0976 1731 ζ : 9/7
0976 1769 sr : 22/5
0976 2327 $2 \div \exp(3/5)$
0976 2944 at : $(e+\pi)/3$
0976 3254 Γ : $\sqrt{3}/2$
0976 3728 10^x : $2/3 - 3\sqrt{5}/4$
0976 3885 rt : (2,9,-5,-8)
0976 5625 sq : 5/16
0976 6782 rt : (7,9,6,5)
0976 6849 J_1 : 17
0977 0083 rt : (9,5,-9,-4)
0977 1046 cr : $\sqrt{3}/3 + \sqrt{5}/3$
0977 1371 $s\pi$: $9\sqrt{5}/4$
0977 1584 sq : $4e - 3\pi$
0977 2290 Ψ : $\sqrt{5}/3$
0977 3131 J_2 : $\beta(2)$
0977 3233 rt : (6,9,-8,-1)
0977 4871 rt : (6,6,3,4)
0977 5484 sr : $(3\pi/2)^{-3}$
0977 6654 ℓ_2 : $3\sqrt{3}/4 + 4\sqrt{5}/3$
0977 7354 Γ : $2\pi^2/9$
0977 7761 cr : $3 - 3\sqrt{5}/4$
0977 8219 cu : $4\sqrt{2} - 3\sqrt{3}$
0977 8978 sinh : $e\pi/9$
0977 9576 J_2 : $\ln(5/2)$
0977 9876 $t\pi$: $1/2 + \pi$
0978 2999 rt : (5,3,0,3)
0978 3012 at : $8\sqrt[3]{5}/7$
0978 3467 rt : (1,9,-1,-1)
0978 3725 id : $4\sqrt{2} - \sqrt{5}/4$
0978 5246 2^x : $e/2 - 3\pi/2$
0978 5694 sq : $e/4 + 3\pi$
0978 7030 J_2 : 11/12
0978 7078 rt : (9,3,-3,5)
0978 7317 erf : 2/23
0978 7498 Li_2 : $3(1/\pi)/10$
0978 7712 $s\pi(1/3) + s\pi(\sqrt{2})$
0978 9249 ζ : $4e\pi/9$
0979 1167 rt : (8,0,-6,4)
0979 1205 ℓ_{10} : $\ln(\sqrt{2}\pi/2)$
0979 1256 J_0 : $5\pi/3$
0979 3125 $\ln(2/5) \div \exp(\sqrt{5})$
0979 5265 rt : (7,3,-9,-3)
0979 6355 rt : (1,8,-9,9)
0979 6444 ln : $(\sqrt{5}/3)^{1/3}$
0979 7777 J_2 : $7\ln 2$
0979 8045 $t\pi$: $7\sqrt{3}/4$
0979 9374 rt : (3,6,-3,-5)
0980 0120 Ei : $(\sqrt{3}/2)^{-1/2}$
0980 2788 sq : $2 + 4\sqrt{2}/3$
0980 3300 rt : (3,4,-1,8)
0980 3422 J : $(\sqrt{2}/3)^3$
0980 3992 sq : $2/3 - \sqrt{2}/4$
0980 4107 ζ : $\exp(4/3)$
0980 4336 J_0 : 20/9
0980 5769 J_2 : $2e$
0980 5961 ℓ_{10} : $3e/4 - \pi/4$
0980 5993 ℓ_{10} : $(\pi/2)^{1/2}$
0980 7621 id : $1/2 + 3\sqrt{3}/2$
0981 0240 2^x : $9\pi/10$
0981 5003 $1/2 + \exp(4)$
0981 5973 at : $\sqrt{2}/4 - 4\sqrt{3}/3$
0981 6192 e^x : $3\sqrt{5}$
0981 6703 10^x : $2\sqrt{3} - 2\sqrt{5}$
0981 6724 rt : (1,5,-3,-8)
0981 7321 sinh : $(\sqrt[3]{5}/2)^{1/3}$
0981 8356 $t\pi$: $9\sqrt{5}/4$
0981 8547 $s\pi(2/5) \div s\pi(1/3)$
0982 1796 sr : $(\sqrt[3]{5}/3)^{-1/3}$
0982 3155 θ_3 : $(\ln 3/3)^3$
0982 4296 Γ : $3\sqrt[3]{3}/5$
0982 5877 rt : (2,8,0,-7)
0982 6595 Ei : 8/23
0982 6982 Ψ : $\sqrt{2} - \sqrt{3} - \sqrt{7}$
0982 7545 e^x : $4e/3 - \pi/4$
0982 9143 rt : (9,4,-1,6)
0982 9387 tanh : $\ln(e/3)$
0982 9449 at : $\ln(e/3)$
0983 0720 at : $7\sqrt{5}/8$
0983 1588 cu : 6/13
0983 3316 rt : (8,1,-4,5)
0983 4804 2^x : $(e)^{-2}$
0983 5122 e^x : $(\ln 3/2)^{1/2}$
0983 7500 rt : (7,2,-7,-4)
0983 7675 rt : (2,9,5,7)
0984 0394 Ei : $(e/2)^{-3}$
0984 2266 ζ : $\pi/4$
0984 4258 $\exp(2/5) \div \exp(e)$
0984 5239 $s\pi(\sqrt{5}) - s\pi(e)$
0984 5654 e^x : $1 - e/3$
0984 6922 sq : $3\sqrt{2} + \sqrt{3}/2$
0984 7190 rt : (9,6,-7,-3)
0984 8823 at : $\sqrt[3]{15/2}$
0985 2051 ℓ_{10} : $2 - \sqrt{5}/3$
0985 2124 rt : (2,5,-5,-4)
0985 2577 at : $1/\ln(5/3)$
0985 3219 J_2 : 23/25
0985 3532 ln : $3/4 + \sqrt{2}/4$
0985 3635 J_1 : $2\ln 2/7$
0985 7535 $\ln(2/3) \div \exp(\sqrt{2})$
0985 8460 10^x : $4\sqrt{2}/3 + 3\sqrt{5}/2$
0985 9179 rt : (6,7,5,5)
0986 1087 $\ln(3/5) - s\pi(1/5)$
0986 1228 id : $\ln(3)$
0986 3284 rt : (5,6,-8,3)
0986 5667 Ei : $3e/4$
0986 6596 sq : $3\sqrt{2}/4 + 4\sqrt{5}$
0986 7397 cu : $2\ln 2/3$
0986 7648 rt : (5,4,2,4)
0986 8411 sr : $1/2 + \sqrt{2}/2$
0986 9604 sq : $\pi/10$
0986 9865 tanh : $\ln 2/7$
0986 9927 at : $\ln 2/7$
0987 1817 cr : $\sqrt[3]{7/3}$
0987 1898 rt : (9,5,1,7)
0987 2602 cu : $4\sqrt{2}/3 + \sqrt{5}/2$
0987 4577 ℓ_{10} : $3\sqrt{2}/2 - \sqrt{3}/2$
0987 6159 rt : (8,2,-2,6)
0987 6543 sq : $2\sqrt{2}/9$
0987 7219 sinh : $\ln(e/3)$
0987 7281 as : $\ln(e/3)$
0987 7529 ln : $e/3 + 2\pi/3$
0987 7940 ℓ_2 : $\sqrt{3} - \sqrt{5} + \sqrt{7}$
0987 8684 cr : $4\sqrt{2}/3 - \sqrt{5}/4$
0987 9015 e^x : $2/3 - 4\sqrt{5}/3$
0988 0431 rt : (7,1,-5,-5)
0988 4713 rt : (6,4,-8,-4)
0988 5812 sinh : $(2e)^{1/2}$
0988 6853 J_0 : $4(e+\pi)/9$
0989 0938 id : $\sqrt{2}/4 + \sqrt{5}/3$
0989 3044 rt : (5,7,-4,7)
0989 3310 rt : (2,5,-7,-1)
0989 5449 10^x : $\ln(\ln 3/3)$
0989 6229 10^x : $-\sqrt{2} - \sqrt{5} + \sqrt{7}$
0989 7244 id : $2\sqrt{3}/3 + 4\sqrt{5}$
0989 7421 rt : (2,6,1,5)
0989 8053 J_1 : $5(1/\pi)/8$
0989 9364 cr : $3/4 + \sqrt{3}/3$
0990 0691 J_0 : $7e\pi/2$
0990 2102 id : $\ln 2/7$
0990 3036 ζ : 20/23
0990 5504 e^x : $(\ln 3/2)^{-3}$
0990 7208 Ψ : $6\sqrt[3]{3}$
0991 1499 rt : (1,7,6,-9)
0991 2159 Ei : $\ln((e+\pi)/2)$
0991 2501 $t\pi$: $\sqrt{2} + 4\sqrt{5}$
0991 3122 2^x : 3/22
0991 4465 J_2 : 12/13
0991 4751 sr : $4e/9$
0991 5216 ln : $3 - 2\pi/3$
0991 5367 rt : (9,6,3,8)
0991 7303 2^x : $3(1/\pi)/7$
0991 8292 sinh : $\ln 2/7$
0991 8356 as : $\ln 2/7$
0991 9639 2^x : $5e/2$
0991 9718 rt : (8,3,0,7)
0992 0877 rt : (1,7,0,5)
0992 0986 ℓ_{10} : 4π
0992 1102 rt : (2,7,-2,-8)
0992 1256 sq : $\sqrt[3]{2}/4$
0992 1512 10^x : $9\pi/10$
0992 1763 ζ : $3\sqrt[3]{2}$
0992 3783 ζ : $1/\ln(1/3)$
0992 4081 rt : (7,0,-3,6)
0992 4595 ζ : $-\sqrt{2} - \sqrt{3} + \sqrt{5}$
0992 5720 e^x : $2 + 4e/3$
0992 6431 sr : $2\sqrt{2}/3 + 2\sqrt{3}$
0992 6550 rt : (9,7,-5,-2)
0992 8455 rt : (6,3,-6,-5)
0992 8483 at : $4 - 3e/4$
0993 1965 sq : $\sqrt{2}/2 + 4\sqrt{3}/3$
0993 2002 as : $(\sqrt[3]{2})^{-1/2}$

0993 2840 rt : (5,9,-6,-6)
0993 3344 cu : $(\pi)^3$
0993 3915 Ψ : 15/11
0993 4562 rt : (8,4,-8,-3)
0993 5118 at : $1/3-\sqrt{3}/4$
0993 5677 cu : $3e/2-3\pi/4$
0993 6247 ℓ_{10} : $8\sqrt{2}/9$
0993 6780 J_2 : $4\ln 2/3$
0993 7664 e^x : $1+4\sqrt{2}$
0994 1512 $\ln(\sqrt{3})-\exp(1/2)$
0994 1672 sq : $3\sqrt{2}/4-\sqrt{5}/3$
0994 2117 J : 5/12
0994 3200 rt : (9,4,-8,-8)
0994 3485 id : $10\pi^2/7$
0994 4731 J_1 : $8(e+\pi)$
0994 5559 at : $\sqrt{2}/3+2\sqrt{5}/3$
0994 5944 J_2 : $(\sqrt[3]{5}/2)^{1/2}$
0994 6302 rt : (6,8,7,6)
0994 6445 sinh : $5\sqrt[3]{5}/9$
0994 6917 J_1 : 10π
0994 8075 sπ : $1/4+e$
0994 8431 sinh : 19/20
0994 8586 J_1 : $8\pi/7$
0995 0083 J_1 : 1/5
0995 1252 Ψ : 9/19
0995 1439 at : $2\sqrt{2}-\sqrt{3}/2$
0995 3567 ℓ_2 : 7/15
0995 3876 ln : $(\ln 2)^3$
0995 4610 J_2 : $\sqrt{3}\pi$
0995 4851 sr : $9e\pi/8$
0995 5136 rt : (5,5,4,5)
0995 6670 rt : (4,3,-8,7)
0995 7413 $2\div\exp(3)$
0995 8642 θ_3 : $(e)^{-3}$
0995 9570 rt : (9,7,5,9)
0996 1058 K : 11/13
0996 1986 2^x : $3/4-3e/2$
0996 4015 rt : (8,4,2,8)
0996 4358 2^x : P_{24}
0996 4772 rt : (1,7,5,-3)
0996 6534 λ : $\exp(-8\pi/5)$
0996 6799 tanh : 1/10
0996 6865 at : 1/10
0996 6967 J_1 : $\zeta(3)/6$
0996 7936 id : $1/3-\sqrt{3}/4$
0996 8472 rt : (7,1,-1,7)
0997 0071 id : $\sqrt{3}/3-3\sqrt{5}/4$
0997 0279 J_2 : $(\sqrt[3]{2})^{-1/3}$
0997 0300 $\ln(3/4)\times\ln(\sqrt{2})$
0997 1929 cr : $7\sqrt[3]{5}/9$
0997 2299 sq : 6/19
0997 2941 rt : (9,3,-6,-9)
0997 3249 2^x : $7\sqrt{5}/6$
0997 3974 at : $5\pi/8$
0997 5926 rt : (8,6,-9,-8)
0997 5946 id : $10\sqrt[3]{5}$
0997 7128 id : $\sqrt{2}+\sqrt{5}+\sqrt{6}$
0997 7421 rt : (5,5,-7,-5)
0997 7610 id : $2\sqrt{2}-4\sqrt{3}$
0997 8346 at : P_{18}
0997 9700 Γ : 19/22
0998 0512 at : P_{48}
0998 1015 id : $5e\pi/7$
0998 1913 rt : (2,8,3,-8)
0998 2890 ln : $4\sqrt{2}/3+\sqrt{5}/2$
0998 3341 sπ : $(1/\pi)/10$
0998 3700 θ_3 : $\sqrt[3]{3}/2$
0998 4517 as : $1/3-\sqrt{3}/4$
0998 4842 $\sqrt{5}\times\ln(4)$
0998 5520 ℓ_{10} : $4\sqrt{2}+4\sqrt{3}$
0998 6457 at : $3/2-2\sqrt{3}$
0998 6750 J_2 : $7\pi^2/2$
0998 6841 id : $5\sqrt[3]{2}/3$
0998 7983 sπ : $(\ln 3)^{1/3}$
0998 8660 sq : $\sqrt{2}/3+\sqrt{3}/3$
0998 8924 sinh : $(\pi/2)^3$
0998 9090 sπ : $6(e+\pi)/5$
0998 9939 Ψ : $5\sqrt{3}$
0999 0309 e^x : $4\sqrt{2}/5$
0999 0933 rt : (1,7,-8,-1)
0999 2070 e^x : 2/21
0999 3420 sr : $7\sqrt[3]{2}/2$
0999 4388 id : $7\sqrt{2}/9$
0999 4617 sr : $4e/3+\pi/4$
0999 4715 ζ : $8\sqrt{2}/3$
0999 5244 rt : (1,4,-9,-8)
0999 6003 as : $9\ln 2/7$
0999 6299 cu : $3+2\sqrt{3}$
0999 7366 $\exp(3/5)\div s\pi(1/5)$
0999 7669 tπ : $1/4+e$
0999 9041 sπ : $\sqrt{3}+\sqrt{5}$
1000 0000 id : 1/10
1000 1250 θ_3 : 1/20
1000 3137 ℓ_2 : $1/2+\sqrt{3}/4$
1000 4080 as : $e/4-\pi/2$
1000 6052 rt : (8,5,-7,-9)
1000 6105 ζ : $6\pi/5$
1000 6587 Ψ : $10\pi/9$
1000 6668 Γ : $(\sqrt{5}/3)^{1/2}$
1000 7159 tπ : $2\sqrt{2}+3\sqrt{5}/4$
1000 7516 e^x : $6e/5$
1000 8283 rt : (9,8,-3,-1)
1000 8345 ln : 19/21
1000 9032 rt : (9,9,9,-1)
1000 9073 rt : (8,5,4,9)
1001 1596 id : P_{18}
1001 3628 rt : (7,2,1,8)
1001 3784 id : P_{48}
1001 5137 sr : $2/3+4\sqrt{5}$
1001 6606 rt : (8,5,-6,-2)
1001 6675 sinh : 1/10
1001 6742 as : 1/10
1001 8001 sr : $1/3+3e/2$
1001 8112 rt : (8,9,9,-1)
1001 8195 rt : (6,1,-2,-7)
1001 9438 at : $2/3+3\sqrt{3}/4$
1002 0108 e^x : $3(1/\pi)/10$
1002 0120 rt : (2,6,-4,-9)
1002 1947 rt : (3,8,-5,-3)
1002 2181 2^x : $1/3+3\sqrt{3}/4$
1002 2774 rt : (5,4,-5,-6)
1002 3920 sr : 23/19
1002 4126 Li_2 : 13/16
1002 4187 J_2 : 13/14
1002 4968 rt : (7,2,-9,-3)
1002 5099 J_0 : 10π
1002 6486 $\ln(4/5)\div\exp(4/5)$
1002 7240 rt : (7,9,9,-1)
1002 7365 rt : (4,7,-8,-5)
1002 7810 Γ : $3(e+\pi)/8$
1002 8397 as : P_{18}
1003 0273 rt : (3,9,-1,-8)
1003 0595 as : P_{48}
1003 1029 2^x : $\sqrt{2}/4+2\sqrt{3}$
1003 1189 2^x : $2\sqrt{3}/3-2\sqrt{5}$
1003 2490 ln : $5\zeta(3)/2$
1003 3467 tπ : $(1/\pi)/10$
1003 4333 ℓ_{10} : $\sqrt[3]{2}$
1003 6416 rt : (6,9,9,7)
1003 8179 tπ : $(\ln 3)^{1/3}$
1003 8904 at : P_{61}
1003 8998 ln : $9\pi^2/4$
1003 9091 tπ : $\zeta(3)/3$
1003 9302 tπ : $6(e+\pi)/5$
1003 9607 rt : (7,7,-8,-9)
1004 0191 J_2 : $7(e+\pi)/2$
1004 2879 2^x : P_4
1004 3046 J : $\sqrt[3]{3}/5$
1004 3198 rt : (1,6,4,-3)
1004 4345 sq : $5\sqrt[3]{5}$
1004 5642 rt : (5,9,9,-1)
1004 6449 ln : $\sqrt{2}/3+\sqrt{3}/4$
1004 8094 sq : $3/4-\sqrt{3}/4$
1004 8119 ζ : 17/20
1004 9405 tπ : $\sqrt{3}+\sqrt{5}$
1005 0074 J_1 : $\sqrt{2}/7$
1005 2648 at : P_{45}
1005 3627 rt : (3,9,1,6)
1005 4347 J_2 : $8e$
1005 4916 rt : (4,9,9,-1)
1005 5610 θ_3 : $3\zeta(3)/5$
1005 6968 cu : $3\sqrt{2}/2+4\sqrt{3}$
1005 7524 $\pi\times\exp(3)$
1005 7777 J_2 : $1/\ln((e+\pi)/2)$
1005 8570 e^x : $3\sqrt{2}/2+3\sqrt{3}/2$
1005 9572 rt : (7,3,3,9)
1005 9638 10^x : $(\ln 2/2)^3$
1006 1592 tπ : $5e/6$
1006 2332 sq : $10e/3$
1006 2718 sinh : $3e$
1006 4241 id : $\sqrt[3]{4/3}$
1006 4908 J_2 : $(\sqrt{3}/2)^{1/2}$
1006 5681 erf : $(\sqrt{5})^{-3}$
1006 6760 ℓ_2 : $2e+\pi$
1006 7337 Ψ : $\pi/2$
1006 8074 10^x : $e/2-3\pi/4$
1006 8922 rt : (5,3,-3,-7)
1006 9417 10^x : 1/24
1007 2365 at : $3/4-e$
1007 2764 id : P_{61}
1007 3616 rt : (6,9,-9,-3)
1007 7744 e^x : $3\sqrt{3}/7$
1007 7836 tπ : $5(1/\pi)/6$
1007 8323 rt : (3,9,-9,-5)
1007 8529 ℓ_{10} : $3/2-\sqrt{2}/2$
1007 8971 ℓ_{10} : $3\sqrt{2}-4\sqrt{5}/3$
1007 9288 2^x : $4\sqrt{2}/3-3\sqrt{3}$
1007 9553 sq : $3e/4-3\pi/4$
1008 0199 J_1 : $(\sqrt{2}\pi/2)^{-2}$
1008 1725 Ei : $(2\pi)^{-1/2}$
1008 3043 rt : (1,9,9,-1)
1008 3341 Γ : $(4/3)^{-3}$
1008 5880 2^x : $\ln 2/5$
1008 6649 id : P_{45}
1008 9876 as : P_{61}
1009 0964 J_1 : $5(e+\pi)/3$
1009 1205 J_0 : $7\sqrt{5}/6$
1009 1852 e^x : $3\sqrt{2}/4-3\sqrt{5}/2$
1009 2193 J_2 : $5\sqrt{5}$
1009 2521 rt : (9,9,-1,0)
1009 2824 Γ : $9\sqrt[3]{5}/7$
1009 5980 10^x : $\pi/4$
1009 6514 Ψ : $(3\pi)^{-1/3}$
1009 7588 tanh : $(\pi)^{-2}$
1009 7658 at : $(\pi)^{-2}$
1009 9184 sr : $6\sqrt{2}/7$
1010 0298 sr : $3+\sqrt{2}$
1010 1095 rt : (9,8,9,-1)
1010 1174 rt : (8,6,-4,-1)
1010 2051 sq : $\sqrt{2}-\sqrt{3}$
1010 3831 as : P_{45}
1010 4099 sr : $8\zeta(3)$
1010 4847 ℓ_{10} : $7\sqrt[3]{3}/8$
1010 5172 10^x : $1/\ln(\ln 3/3)$
1010 9870 rt : (7,3,-7,-2)
1011 0583 rt : (8,8,9,-1)
1011 0651 sr : $7\sqrt{3}/10$
1011 1106 rt : (6,1,2,9)
1011 1634 rt : (2,9,-9,-1)
1011 3353 sr : $3\sqrt{2}/4+3\sqrt{5}/2$
1011 4791 sπ : $(\pi)^{-3}$
1011 5369 $\ln(3/5)\times\exp(\sqrt{2})$
1011 5879 Ψ : $10\sqrt{2}/9$
1011 5893 rt : (5,2,-1,-8)
1011 6035 rt : (1,7,-1,-2)
1011 6095 e^x : $4/3-4e/3$
1011 8259 ℓ_{10} : $8\ln 2/7$
1011 9495 cr : $(\sqrt[3]{2}/3)^{-1/3}$
1011 9644 J_2 : 14/15
1012 0124 rt : (7,8,9,-1)
1012 0148 ζ : $3\sqrt{5}/4$
1012 0694 rt : (4,7,-2,2)
1012 1270 rt : (3,9,-9,-1)
1012 2926 Ψ : 11/7
1012 3966 sq : 7/22
1012 5437 cr : $1/3+4\sqrt{5}$
1012 5508 rt : (3,8,-7,-6)
1012 6261 $\ln(5)+\exp(2/5)$
1012 7092 cr : $10\zeta(3)/9$
1012 7204 Γ : $7\pi/10$

1012 9700 10^x : 15/11
1012 9717 rt : (6,8,9,-1)
1012 9869 sq : $2/3+3\sqrt{2}$
1013 0206 θ_3 : $1/\ln(4)$
1013 0960 rt : (4,9,-9,-1)
1013 2118 id : $(\pi)^{-2}$
1013 2816 ℓ_2 : $1/3+2e/3$
1013 3435 θ_3 : $(\sqrt{2}\pi)^{-2}$
1013 3855 sq : $1/3-2\pi/3$
1013 3917 rt : (3,8,-3,-9)
1013 6351 at : $3/2+\sqrt{2}/3$
1013 6622 sr : $7\ln 2/4$
1013 6627 $1/4\times\ln(2/3)$
1013 7402 $\ln(3/4)+\exp(2)$
1013 7719 rt : (9,1,-8,2)
1013 9062 $\sqrt{2}\div\exp(1/4)$
1013 9363 rt : (5,8,9,-1)
1013 9582 ln : $3\sqrt{3}+4\sqrt{5}/3$
1014 0411 10^x : $(4/3)^{-1/3}$
1014 0664 J_1 : $7e$
1014 0705 rt : (5,9,-9,-1)
1014 1312 $s\pi$: $(e/3)^{1/3}$
1014 1994 $t\pi$: $3(1/\pi)/2$
1014 4316 Ψ : $(e/3)^3$
1014 5000 at : $\sqrt[3]{23/3}$
1014 5764 ℓ_{10} : 19/24
1014 5887 J_2 : $10\pi^2/3$
1014 8158 Ψ : $9e/7$
1014 8432 Ei : $9(1/\pi)/2$
1014 9062 rt : (4,8,9,-1)
1014 9463 sinh : $(\pi)^{-2}$
1014 9534 as : $(\pi)^{-2}$
1015 0146 $3/4\div\exp(2)$
1015 0505 rt : (6,9,-9,-1)
1015 1327 2^x : 15/14
1015 1460 $t\pi$: $\sqrt{2}/3$
1015 1789 J_2 : $7\zeta(3)/9$
1015 3142 J_1 : $1/\sqrt{24}$
1015 4419 rt : (2,9,4,3)
1015 5812 2^x : $\sqrt{8/3}$
1015 7215 10^x : 17/12
1015 8661 Li_2 : $\ln 2/7$
1015 8817 rt : (3,8,9,-1)
1015 9615 Ψ : $(5/2)^{-3}$
1016 0361 rt : (7,9,-9,-1)
1016 2962 cu : 7/15
1016 3714 rt : (5,1,1,-9)
1016 3942 J_1 : $7\pi^2/5$
1016 5066 e^x : $7(1/\pi)/3$
1016 6933 $t\pi$: $(\pi)^{-3}$
1016 7555 cr : $3+2\pi$
1016 8626 rt : (4,8,2,6)
1016 9490 J_1 : $\exp(-\sqrt[3]{4})$
1017 0273 rt : (8,9,-9,-1)
1017 1908 rt : (5,6,-9,1)
1017 3551 rt : (3,7,-5,-7)
1017 3693 at : $\sqrt{2}+\sqrt{5}/4$
1017 3952 ℓ_2 : $(e+\pi)^2$
1017 4014 10^x : $9\sqrt{3}/2$
1017 6222 Γ : $4\pi^2/3$
1017 6848 cr : $\sqrt{2}/3+\sqrt{3}/2$
1017 6943 at : $2\sqrt{3}-2\sqrt{5}/3$
1017 7647 ℓ_2 : P_{77}
1017 8490 rt : (1,8,9,-1)
1017 9086 ζ : 15/4
1018 0242 rt : (9,9,-9,-1)
1018 0249 Γ : 11/5
1018 3333 sinh : $\sqrt[3]{19/3}$
1018 3684 rt : (9,2,-6,3)
1018 4679 cr : $2\sqrt{2}-2\sqrt{5}/3$
1018 5439 $s\pi$: $4e+2\pi/3$
1018 7795 at : $\pi^2/5$
1018 8411 rt : (8,9,-1,0)
1018 9841 2^x : $\sqrt[3]{2}/9$
1019 0760 e^x : $(2e/3)^{-1/2}$
1019 1043 cu : $3\sqrt{2}+\sqrt{5}/3$
1019 1213 id : $7(e+\pi)/10$
1019 3867 $t\pi$: $(e/3)^{1/3}$
1019 4633 sr : 17/14
1019 5891 e^x : $4-2\pi$
1019 6461 rt : (9,7,9,-1)
1019 7463 rt : (7,4,-5,-1)
1019 8262 ln : $3\sqrt{3}/2-2\sqrt{5}/3$
1019 8388 rt : (1,8,-9,-1)
1019 8418 10^x : $e+\pi/2$
1019 9628 $\exp(2/3)+\exp(e)$
1020 1161 $\exp(\sqrt{3})\div\exp(3/5)$
1020 3445 J_2 : 15/16
1020 4081 sq : $\sqrt{5}/7$
1020 4401 rt : (5,7,-5,5)
1020 4916 cu : $4/3-\sqrt{3}/2$
1020 6389 rt : (8,7,9,-1)
1020 6560 rt : (6,1,-8,-2)
1020 6775 ln : $4/3+3\sqrt{5}/4$
1020 8423 rt : (2,8,-9,-1)
1020 8975 sr : $2-\pi/4$
1020 9338 id : $\sqrt{3}/2+\sqrt{5}$
1021 0469 rt : (3,6,-8,7)
1021 0564 J_0 : $7(e+\pi)$
1021 0861 $\exp(e)\div\exp(5)$
1021 6167 10^x : $1/2-2\sqrt{5}/3$
1021 6374 rt : (7,7,9,-1)
1021 6391 2^x : $1/3-4e/3$
1021 6512 $1/5\times\ln(3/5)$
1021 6843 $e\times\ln(2/3)$
1021 7017 e^x : $(\ln 2/3)^{-1/3}$
1021 7440 rt : (4,3,0,-9)
1021 8516 rt : (3,8,-9,-1)
1022 0514 λ : $\exp(-5)$
1022 2481 rt : (3,6,-3,-8)
1022 5868 J : $\exp(-\sqrt[3]{5})$
1022 6416 rt : (6,7,9,-1)
1022 6894 $\ln(4)-\exp(1/4)$
1022 7536 rt : (2,9,-6,-7)
1022 7871 cu : $1/4+3\sqrt{3}/2$
1022 8427 ln : $(e/2)^{1/3}$
1022 8507 ζ : $(3\pi/2)^{1/3}$
1022 8668 rt : (4,8,-9,-1)
1022 9475 sinh : $(e/3)^{1/2}$
1022 9615 sq : $3\sqrt{2}/4-3\sqrt{3}$
1022 9803 erf : 1/11
1022 9837 J_0 : $10\sqrt[3]{3}$
1023 0384 $\exp(\sqrt{5})-s\pi(\sqrt{3})$
1023 0439 rt : (9,3,-4,4)
1023 1604 at : $1/3-4\sqrt{3}/3$
1023 3898 erf : $2(1/\pi)/7$
1023 5263 J_0 : $1/\ln(\pi/2)$
1023 5271 rt : (8,0,-7,3)
1023 5686 cu : $2/3-2\sqrt{2}$
1023 6184 sq : $5\sqrt[3]{2}/6$
1023 6515 rt : (5,7,9,-1)
1023 7298 rt : (5,8,-1,9)
1023 8688 $t\pi$: $4e+2\pi/3$
1023 8880 rt : (5,8,-9,-1)
1023 9226 $5\div\ln(\sqrt{3})$
1024 0000 sq : 8/25
1024 2092 2^x : $4\sqrt{2}-4\sqrt{5}$
1024 3091 id : $7\sqrt[3]{2}/8$
1024 4238 rt : (4,9,-8,8)
1024 4742 cu : $3e/4-\pi/2$
1024 4976 ζ : 14/17
1024 6673 rt : (4,7,9,-1)
1024 7214 J_1 : $\sqrt[3]{3}/7$
1024 7288 rt : (6,7,-8,8)
1024 7544 sr : $2e+4\pi/3$
1024 7616 ℓ_{10} : $\sqrt{2}/2+\sqrt{5}/4$
1024 8077 $\exp(3)\times s\pi(2/5)$
1024 8818 rt : (9,4,-9,-7)
1024 9151 rt : (6,8,-9,-1)
1025 0000 sq : 21/20
1025 2582 ζ : $\sqrt{14}$
1025 2606 as : $1/3+\sqrt{5}/4$
1025 2626 J_1 : $8e\pi/7$
1025 2862 cu : $3/2+4\pi/3$
1025 6146 2^x : $3/2-e/2$
1025 6890 rt : (3,7,9,-1)
1025 9484 rt : (7,8,-9,-1)
1026 1779 Li_2 : 1/10
1026 4720 erf : $(\sqrt{2}\pi/2)^{-3}$
1026 6234 ℓ_{10} : 15/19
1026 7167 rt : (2,7,9,-1)
1026 7509 sq : $1-e/4$
1026 8342 $1/5\div\exp(2/3)$
1026 9878 rt : (8,8,-9,-1)
1027 0898 10^x : $1+\sqrt{3}/4$
1027 1283 2^x : P_{70}
1027 1560 2^x : P_{67}
1027 1920 $\exp(1/3)+s\pi(1/4)$
1027 2018 sq : $2\sqrt[3]{3}/9$
1027 2164 E : 19/20
1027 2182 2^x : $3-2\pi$
1027 2328 rt : (3,5,-1,-9)
1027 4052 rt : (4,3,-9,5)
1027 4256 E : $5\sqrt[3]{5}/9$
1027 4302 2^x : $3\sqrt{2}/4+3\sqrt{5}$
1027 4431 id : $3\sqrt{2}/2+4\sqrt{5}/3$
1027 5665 2^x : $e/3-4\pi/3$
1027 6617 ln : $4+4\pi/3$
1027 7504 rt : (2,8,-4,-8)
1027 7599 J_2 : 16/17
1027 7978 cr : $\sqrt{2}/4+4\sqrt{5}$
1027 8011 rt : (9,4,-2,5)
1027 8331 Ei : $\sqrt{5}+\sqrt{6}-\sqrt{7}$
1027 8471 rt : (8,8,0,1)
1027 9159 sq : $e/3-3\pi/4$
1028 0334 rt : (9,8,-9,-1)
1028 1193 rt : (9,3,-7,-8)
1028 2950 rt : (8,1,-5,4)
1028 4149 $2/3\times\exp(e)$
1028 6712 e^x : $\sqrt{2}/2-4\sqrt{5}/3$
1028 7903 rt : (7,9,-1,0)
1028 8128 id : $3e/4-\pi$
1028 9073 Γ : $(\pi/2)^{-1/3}$
1029 2356 id : $(\sqrt{5}/3)^{-1/3}$
1029 2444 $2\div s\pi(2/5)$
1029 2752 $\ln(\sqrt{3})-\exp(\sqrt{3})$
1029 3405 Ei : $1/\sqrt[3]{24}$
1029 3985 $\exp(\sqrt{5})\times s\pi(1/3)$
1029 4086 rt : (1,7,4,-9)
1029 5330 sinh : $10e\pi/9$
1029 5356 rt : (9,6,9,-1)
1029 6125 2^x : $\sqrt[3]{13}$
1029 7385 rt : (6,2,-6,-1)
1029 8364 rt : (1,7,-9,-1)
1029 9125 2^x : $\sqrt{2}/10$
1029 9330 2^x : $2\sqrt{2}$
1030 0162 2^x : $\exp(\sqrt[3]{5}/2)$
1030 0673 sq : $e\pi/3$
1030 1212 at : $8\sqrt{3}/7$
1030 2384 10^x : $3+3\sqrt{5}/4$
1030 2606 sinh : 20/21
1030 2925 2^x : $4(1/\pi)/9$
1030 3895 J_2 : $3\pi/10$
1030 5759 rt : (8,6,9,-1)
1030 6868 sr : $\sqrt{2}/3+\sqrt{5}/3$
1030 6916 rt : (5,1,-9,2)
1030 7837 rt : (4,4,-5,9)
1030 8887 rt : (2,7,-9,-1)
1030 9394 sq : $2e+\pi/2$
1030 9595 at : $7\sqrt{2}/5$
1031 0185 ℓ_{10} : $3-\sqrt{3}$
1031 0592 J_2 : $2\sqrt{2}/3$
1031 1939 cr : $4e-\pi/2$
1031 3960 rt : (9,2,-5,-9)
1031 4021 J : 1/12
1031 5339 sr : $(\sqrt{2}\pi/3)^{1/2}$
1031 5664 Ψ : 7/2
1031 5927 rt : (2,7,-3,8)
1031 6222 rt : (7,6,9,-1)
1031 7364 rt : (8,5,-8,-8)
1031 8133 ℓ_{10} : $4e/3-3\pi/4$
1031 8560 at : $1/4-\sqrt{2}/4$
1031 8793 rt : (1,7,-2,-9)
1031 8994 e^x : 24/17
1031 9474 rt : (3,7,-9,-1)
1032 3032 id : $1/3-2e$
1032 6425 rt : (9,5,0,6)
1032 6748 rt : (6,6,9,-1)
1032 8353 J : $3\sqrt[3]{2}/10$
1032 8428 rt : (2,7,-2,-9)

1032 8578 $2^x : \exp(-(e+\pi)/3)$
1032 8590 $J_2 : 1/\ln(\ln 2/2)$
1032 8977 $1/4 \times \exp(5)$
1032 9362 cr : $7\pi^2$
1032 9622 $\ell_2 : \sqrt{2}/4+\sqrt{3}/3$
1032 9864 cu : $4\sqrt{2}/3+2\sqrt{3}/3$
1032 9989 $\ell_2 : 4\sqrt{2}/3+3\sqrt{5}$
1033 0125 rt : (4,7,-9,-1)
1033 0495 id : $9\pi^2/8$
1033 1476 rt : (8,2,-3,5)
1033 2952 $e^x : 2\sqrt{2}/3+\sqrt{3}/2$
1033 6226 $10^x : 4-2e/3$
1033 6318 $10^x : 23/22$
1033 6542 rt : (7,1,-6,-4)
1033 7082 $Ei : \ln 2/2$
1033 7337 rt : (5,6,9,-1)
1033 7934 $J_1 : \exp(-\pi/2)$
1033 9793 cr : $7e\pi/2$
1034 0571 at : $1-4\sqrt{5}/3$
1034 0803 $\ln(\pi) \times s\pi(\sqrt{2})$
1034 0841 rt : (5,7,-9,-1)
1034 1110 $J_1 : 3\ln 2/10$
1034 1621 rt : (6,4,-9,-3)
1034 1697 cr : $3\sqrt{3}/2+3\sqrt{5}$
1034 3681 $J_2 : 17/18$
1034 4229 $\Psi : 8e\pi/3$
1034 4421 $\ell_{10} : 9\pi^2/7$
1034 5620 rt : (3,8,-6,-6)
1034 6932 $t\pi : e/4-3\pi/2$
1034 7990 rt : (4,6,9,8)
1034 8094 $2^x : e+2\pi/3$
1034 9070 ln : $8\ln 2/5$
1035 0584 rt : (8,4,-6,-9)
1035 1452 sq : $4\sqrt{2}/3+3\sqrt{3}/2$
1035 1623 rt : (6,7,-9,-1)
1035 2427 $\exp(\sqrt{3}) \div \exp(4)$
1035 2615 $J_1 : 4(e+\pi)$
1035 3148 $10^x : 2-3\sqrt{5}/4$
1035 3293 at : $1/4+\sqrt{3}$
1035 3731 $J_2 : 3\sqrt[3]{2}/4$
1035 4047 rt : (7,7,-9,-8)
1035 5339 id : $1/4-\sqrt{2}/4$
1035 6840 $2^x : \sqrt{2}-\sqrt{5}-\sqrt{6}$
1035 7630 sinh : $(\ln 2/3)^{-2}$
1035 8086 $J_0 : 9\sqrt[3]{2}$
1035 8708 rt : (3,6,9,-1)
1035 9958 id : $7\zeta(3)/4$
1036 0254 $J_1 : 5/24$
1036 0862 sr : $\sqrt{2}+\sqrt{6}-\sqrt{7}$
1036 2472 rt : (7,7,-9,-1)
1036 3832 $3 \div e$
1036 4789 ln : $3\zeta(3)/4$
1036 6482 ln : $3+3\sqrt{3}$
1036 9161 $J_1 : 1/\sqrt{23}$
1036 9491 rt : (2,6,9,-1)
1036 9954 $J_2 : (\sqrt{5}/2)^{-1/2}$
1037 1525 rt : (8,9,2,2)
1037 3105 at : $3e/2-2\pi/3$
1037 3388 rt : (8,7,-9,-1)
1037 3936 as : $1/4-\sqrt{2}/4$
1037 5710 rt : (9,6,2,7)
1037 5937 $\ell_2 : (\sqrt{3}/2)^{1/2}$
1037 7665 $\ell_{10} : 5\sqrt[3]{2}/8$
1037 7947 $t\pi : 3\sqrt[3]{5}/8$
1037 9433 $e \times s\pi(e)$
1037 9953 ln : $5\sqrt[3]{3}/8$
1038 0340 rt : (3,9,0,3)
1038 0719 $10^x : 3/4-\sqrt{2}/2$
1038 0877 rt : (8,3,-1,6)
1038 1301 at : P_{51}
1038 1364 rt : (7,6,-1,1)
1038 3307 $e^x : 3-\sqrt{2}/4$
1038 4372 rt : (9,7,-9,-1)
1038 5165 $\zeta : 5\sqrt{5}/3$
1038 5673 $e^x : 1/3-3\sqrt{3}/2$
1038 5733 $t\pi : 10\sqrt[3]{2}$
1038 5789 rt : (1,9,-3,-9)
1038 6059 rt : (7,0,-4,5)
1038 7732 rt : (7,6,-7,-9)
1038 9635 sq : $2/3+e/2$
1039 0196 $\zeta : e/7$
1039 0485 $3 \div s\pi(1/5)$
1039 1256 rt : (6,9,-1,0)
1039 2416 $J_2 : 7\sqrt[3]{2}$
1039 4014 at : $\sqrt{3}/2+\sqrt{5}/2$
1039 4202 $J_0 : \exp(\sqrt[3]{4}/2)$
1039 6468 rt : (5,9,-7,-8)
1039 6794 $J_1 : 7(e+\pi)/10$
1039 8027 rt : (9,5,9,-1)
1040 0155 $\exp(3/5) \div s\pi(1/3)$
1040 1029 $Li_2 : (\pi)^{-2}$
1040 1202 rt : (5,0,-7,1)
1040 1389 $\ell_2 : 3+3\sqrt{3}/4$
1040 2240 rt : (1,9,6,-3)
1040 2876 $J : 1/18$
1040 2939 $J_2 : 18/19$
1040 3780 $Ei : \sqrt{3}/5$
1040 6596 $s\pi : 3/4+2\pi$
1040 8942 rt : (8,5,9,-1)
1040 8951 $2^x : 1/7$
1040 9945 $E : (\sqrt[3]{5}/2)^{1/3}$
1040 9982 ln : $\sqrt{2}/2+4\sqrt{3}/3$
1041 0196 id : $1/4-3\sqrt{5}/2$
1041 2660 $\Psi : e/2$
1041 3292 rt : (2,6,-9,-1)
1041 3848 $s\pi(\sqrt{3}) \div s\pi(\sqrt{5})$
1041 4408 $e^x : (e/3)^3$
1041 6472 sr : $(2e/3)^{1/3}$
1041 6977 $10^x : 3/2+\sqrt{2}/2$
1041 8163 cr : $2/3+e/4$
1041 8416 $\Psi : (2e/3)^{-1/2}$
1041 8756 id : P_{51}
1041 9564 $e^x : 3\sqrt{3}/2+\sqrt{5}/2$
1041 9924 rt : (7,5,9,-1)
1042 1331 cu : 8/17
1042 1359 rt : (1,9,8,6)
1042 1954 $10^x : 3/4-\sqrt{3}$
1042 2247 sinh : $(\sqrt{3}/2)^{1/3}$
1042 2530 cr : $3\pi/7$
1042 4415 rt : (3,6,-9,-1)
1042 4851 $e^x : 2/3+\sqrt{5}/3$
1042 5418 rt : (6,8,-8,-9)
1042 5728 $t\pi : 2\sqrt{2}/3-3\sqrt{5}/4$
1042 5826 $Ei : 3/19$
1042 5893 rt : (9,7,4,8)
1042 6477 $\ell_{10} : \sqrt{2}/4+\sqrt{3}/4$
1042 7736 $\Psi : 5\sqrt[3]{2}/4$
1042 8032 $J_0 : 7\sqrt{5}/3$
1042 9606 sq : $2-3\sqrt{5}/4$
1043 0974 rt : (6,5,9,-1)
1043 1116 $\ell_2 : \ln((e+\pi)/2)$
1043 1181 rt : (8,4,1,7)
1043 3203 $J_2 : e\pi/9$
1043 4541 sr : $\sqrt{2}/4+\sqrt{3}/2$
1043 4670 at : $1/4-\sqrt{5}$
1043 4841 id : $e/4+3\pi$
1043 5607 rt : (4,6,-9,-1)
1043 6484 rt : (7,1,-2,6)
1043 7008 $10^x : 2-4\sqrt{5}/3$
1043 7503 tanh : $(\sqrt{2}/3)^3$
1043 7586 at : $(\sqrt{2}/3)^3$
1043 7698 as : P_{51}
1043 8447 $J_2 : (\sqrt[3]{5}/2)^{1/3}$
1044 0351 $\zeta : 5\sqrt{2}/9$
1044 0506 $\Gamma : 5\zeta(3)/7$
1044 1521 $Ei : 14/13$
1044 1577 $J_1 : \sqrt[3]{2}/6$
1044 1803 rt : (6,2,-5,-5)
1044 2093 rt : (5,5,9,-1)
1044 3133 cr : $7\sqrt{3}/9$
1044 6235 rt : (4,9,-9,5)
1044 6871 rt : (5,6,-9,-1)
1044 7138 rt : (5,5,-8,-4)
1044 9847 $s\pi : (\ln 3/2)^{-3}$
1044 9964 $\Psi : 7(1/\pi)/3$
1045 0116 $E : e\pi/9$
1045 1891 sr : $e\pi/7$
1045 3282 rt : (4,5,9,-1)
1045 6106 $J_2 : 5\sqrt[3]{5}/9$
1045 6359 $10^x : (\ln 3)^2$
1045 6378 $J_2 : 19/20$
1045 7618 $e+\ln(4)$
1045 8207 rt : (6,6,-9,-1)
1045 9516 $\exp(\pi)-s\pi(\sqrt{2})$
1046 0042 ln : $7(e+\pi)/5$
1046 2258 $10^x : \exp(-\pi)$
1046 2969 rt : (3,6,-9,4)
1046 3408 $t\pi : 1/4-2\pi$
1046 3703 cu : $\sqrt{3}/4-3\sqrt{5}$
1046 4542 rt : (3,5,9,-1)
1046 5958 at : $8\sqrt{5}/9$
1046 7487 $s\pi : 3\sqrt{2}/2-2\sqrt{3}/3$
1046 8105 $J_1 : 4/19$
1046 9616 rt : (7,6,-9,-1)
1047 1206 $J_1 : 9\pi$
1047 2948 $\Gamma : 7\sqrt[3]{2}/4$
1047 3284 $\lambda : \exp(-19\pi/12)$
1047 3535 $\ell_{10} : 11/14$
1047 5656 id : $(\sqrt{2}/3)^3$
1047 5750 $\ell_{10} : 9\sqrt{2}$
1047 5874 rt : (2,5,9,-1)
1047 6022 $e^x : \exp(e)$
1047 6521 $Ei : 2/5$
1047 6556 $2^x : \sqrt{2}+\sqrt{6}+\sqrt{7}$
1047 7006 rt : (9,8,6,9)
1047 8032 rt : (7,7,1,2)
1047 8978 $s\pi : (e/3)^{-1/3}$
1047 9603 $J_1 : 8\sqrt{5}/5$
1048 0089 $10^x : \sqrt[3]{20}$
1048 0668 $\ln(\sqrt{3}) \times \exp(\sqrt{3})$
1048 1100 rt : (8,6,-9,-1)
1048 2419 rt : (8,5,3,8)
1048 3880 $10^x : 5\sqrt{2}/9$
1048 4165 $10^x : 4+\pi/2$
1048 5135 rt : (2,9,-5,6)
1048 5832 $10^x : \sqrt{2}+\sqrt{5}-\sqrt{7}$
1048 7020 $2^x : P_{75}$
1048 7278 rt : (1,5,9,-1)
1048 7608 tanh : 2/19
1048 7693 at : 2/19
1048 7848 rt : (7,2,0,7)
1048 8364 rt : (6,4,-2,1)
1049 1011 $\ell_{10} : \pi/4$
1049 1096 rt : (2,9,3,-1)
1049 2658 rt : (9,6,-9,-1)
1049 2788 sr : $3\sqrt{2}/2+4\sqrt{3}/3$
1049 3294 rt : (6,1,-3,-6)
1049 4433 cu : $4\sqrt{3}+3\sqrt{5}/2$
1049 4812 $J_2 : (e/3)^{1/2}$
1049 4826 sinh : $(\sqrt{2}/3)^3$
1049 4911 as : $(\sqrt{2}/3)^3$
1049 5953 rt : (6,7,-9,6)
1049 7443 $Ei : 10e/7$
1049 7744 $J : 1/\sqrt[3]{14}$
1049 8756 rt : (5,9,-1,0)
1050 0432 ln : $4/3-\sqrt{3}/4$
1050 0503 sr : $(\ln 3/2)^{-1/3}$
1050 0516 $\Gamma : 8/19$
1050 1988 $10^x : 2\sqrt{2}/3+4\sqrt{5}/3$
1050 3068 $\exp(1/4)-\exp(2)$
1050 4234 rt : (4,7,-9,-4)
1050 4748 rt : (9,4,9,-1)
1050 4813 $J_2 : 20/21$
1050 5424 $2^x : 2\sqrt{2}/3+4\sqrt{3}$
1050 5586 $J_0 : 10e/3$
1050 6891 at : P_{57}
1050 7375 $t\pi : (\ln 3/2)^{-3}$
1050 9206 rt : (4,2,-8,1)
1051 0308 rt : (1,8,4,-4)
1051 3413 ln : $3/2+3\sqrt{5}$
1051 3679 $2^x : \sqrt[3]{3}/10$
1051 5286 $e^x : 3+2\pi/3$
1051 6218 rt : (8,4,9,-1)
1051 6437 cu : $4\sqrt{3}/3+4\sqrt{5}/3$
1051 7091 $e^x : 1/10$
1051 7171 sr : $3/4+\sqrt{2}/3$
1051 7482 sr : $5\sqrt[3]{5}/7$
1051 9055 $\Gamma : (e/2)^{-1/2}$
1052 1174 $J_2 : (\sqrt{3}/2)^{1/3}$
1052 1781 $\ell_2 : 4\sqrt{3}+3\sqrt{5}/4$

1052 1935 rt : (2,5,-9,-1)
1052 2882 cu : $1/2+2\pi$
1052 3152 $J_0 : 5\pi/6$
1052 4493 cu : $4+2\sqrt{5}$
1052 5308 tπ : $3\sqrt{2}/2-2\sqrt{3}/3$
1052 5910 sq : $2\sqrt{2}/3+\sqrt{5}$
1052 6315 id : 2/19
1052 6782 at : P_{38}
1052 7330 $\zeta : 11/14$
1052 7762 rt : (7,4,9,-1)
1052 7850 $\theta_3 : 1/19$
1052 8418 at : $1/2+2\sqrt{5}/3$
1052 8983 $J_1 : 20/3$
1052 9514 $10^x : 1/23$
1052 9908 $e^x : P_{18}$
1053 0552 $Ei : (2\pi/3)^3$
1053 1875 $\exp(e)+s\pi(2/5)$
1053 2326 $e^x : P_{48}$
1053 3615 sq : $3\sqrt{2}-4\sqrt{5}$
1053 3639 rt : (3,5,-9,-1)
1053 4308 rt : (5,7,-6,3)
1053 4624 rt : (8,6,5,9)
1053 6051 ln : 9/10
1053 6192 $\ln(2)-s\pi(1/5)$
1053 6383 $\zeta : 7/18$
1053 6387 $\ell_2 : 1/4+e/4$
1053 6991 tπ : $(e/3)^{-1/3}$
1053 9113 $J_2 : (\ln 3)^{-1/2}$
1053 9381 rt : (6,4,9,-1)
1053 9922 $\exp(3/4)\div\exp(3)$
1054 0184 rt : (7,3,2,8)
1054 0229 $10^x : 11/14$
1054 0555 $\ln(4/5)\div\exp(3/4)$
1054 1234 $e^x : 17/4$
1054 2161 $\zeta : 1/\sqrt[3]{17}$
1054 2317 ln : $5\sqrt[3]{2}/7$
1054 3526 at : $e/4-\pi/4$
1054 5421 rt : (4,5,-9,-1)
1054 5577 at : $\sqrt{2}+\sqrt{3}/3$
1054 5726 id : P_{57}
1054 5761 rt : (6,0,-1,7)
1054 5765 sinh : 2/19
1054 5852 as : 2/19
1054 6875 cu : $3\sqrt[3]{2}/8$
1054 7208 sr : $4+\sqrt{3}/4$
1054 8261 rt : (1,6,3,-8)
1054 8917 $J_2 : 21/22$
1054 9518 $\pi-s\pi(\sqrt{2})$
1055 0716 $J_1 : 2(1/\pi)/3$
1055 1076 rt : (5,4,9,-1)
1055 1356 rt : (5,3,-4,-6)
1055 2495 $J_2 : 7(e+\pi)$
1055 3033 $\ln(4)-\exp(2/5)$
1055 3076 $e^x : \sqrt[3]{13/3}$
1055 3430 sinh : $(\ln 3)^{-1/2}$
1055 4159 sr : 11/9
1055 4262 $2^x : 2\sqrt{3}-3\sqrt{5}$
1055 4727 $\zeta : 7/5$
1055 5200 at : $\sqrt{2}/3-\sqrt{3}/3$
1055 5407 sq : $(\sqrt[3]{5}/3)^2$
1055 6752 $J_2 : 3(1/\pi)$
1055 6967 rt : (4,6,-7,-4)
1055 7272 $\Psi : 3\sqrt{3}/7$
1055 7280 rt : (5,5,-9,-1)
1056 0849 cu : $4-e$
1056 2524 $J_0 : 9\pi$
1056 2848 rt : (4,4,9,-1)
1056 4237 sq : $2e+4\pi$
1056 4960 $2^x : 1-3\sqrt{2}$
1056 5098 cu : $2e/3+4\pi/3$
1056 5371 as : P_{57}
1056 5838 id : P_{38}
1056 6526 $10^x : 4/3-4\sqrt{3}/3$
1056 7311 $\Gamma : (\sqrt[3]{4})^{-1/3}$
1056 9220 rt : (6,5,-9,-1)
1057 0143 $J_1 : 19$
1057 0326 rt : (5,8,-2,7)
1057 0691 tanh : $(1/\pi)/3$
1057 0779 at : $(1/\pi)/3$
1057 2197 $\Gamma : \exp(-(e+\pi))$
1057 3690 $\Gamma : \exp(-\sqrt[3]{5})$
1057 4697 rt : (3,4,9,-1)
1057 6317 rt : (2,7,-4,4)
1057 6707 $\Gamma : 6/7$
1057 8112 rt : (7,8,3,3)
1057 9151 $s\pi(\pi)-s\pi(\sqrt{5})$
1058 0955 $erf : \ln(\ln 3)$
1058 1240 rt : (7,5,-9,-1)
1058 1356 sπ : $4\sqrt{2}+4\sqrt{3}/3$
1058 2770 id : $e/4-\pi/4$
1058 3232 $\ln(3/4)\div\exp(1)$
1058 4056 cr : $9\zeta(3)/8$
1058 5597 as : P_{38}
1058 6625 rt : (2,4,9,-1)
1058 7059 sπ : P_{50}
1058 7253 $\pi\div\exp(2/5)$
1058 8144 $Ei : 5/7$
1058 8919 rt : (6,5,0,2)
1058 9244 $J_2 : 22/23$
1058 9912 $\ell_{10} : 1/3+2\sqrt{2}/3$
1059 0169 sπ : $10\pi^2/9$
1059 1013 $e^x : 3\sqrt{2}+3\sqrt{3}/4$
1059 3341 rt : (8,5,-9,-1)
1059 3525 rt : (7,4,4,9)
1059 4448 cr : $1/\ln(2\pi/3)$
1059 4574 id : $\sqrt{2}/3-\sqrt{3}/3$
1059 4832 $K : 3\sqrt{2}/5$
1059 6736 at : $e-3\pi/2$
1059 7538 $e^x : P_{61}$
1059 7809 $e^x : \sqrt{2}/4-3\sqrt{3}/2$
1059 8633 rt : (1,4,9,-1)
1059 9240 rt : (6,1,1,8)
1059 9257 $e^x : 17/15$
1059 9582 $J_1 : 1/\sqrt{22}$
1059 9788 rt : (5,2,-3,1)
1060 0462 rt : (1,9,7,-2)
1060 1148 cr : 23/17
1060 2624 as : $e/4-\pi/4$
1060 2694 $\ell_{10} : (\sqrt[3]{3}/3)^{1/3}$
1060 4972 rt : (5,2,-2,-7)
1060 5247 rt : (9,1,-9,1)
1060 5525 rt : (9,5,-9,-1)
1060 6950 rt : (9,3,-8,-7)
1060 8962 $\zeta : 10\sqrt[3]{2}/9$
1060 9204 sπ : $(2e)^{-2}$
1060 9415 $2^x : (\sqrt{3}/2)^{-1/2}$
1060 9854 $\Psi : 3\sqrt[3]{2}/8$
1061 0329 id : $(1/\pi)/3$
1061 0722 rt : (4,9,-1,0)
1061 1913 $\theta_3 : (1/\pi)/6$
1061 2308 sq : $3\sqrt{2}/2-4\sqrt{3}$
1061 2895 $e^x : P_{45}$
1061 2981 $Ei : (2\pi)^{1/2}$
1061 4107 at : $7\sqrt[3]{5}/6$
1061 4421 $\exp(1/4)+\exp(3/5)$
1061 4495 as : $\sqrt{2}/3-\sqrt{3}/3$
1061 5272 $3/5\div\exp(\sqrt{3})$
1061 5823 rt : (9,3,9,-1)
1061 6490 rt : (3,8,-8,-5)
1061 6680 tπ : $\sqrt{19}$
1061 7775 $e^x : 2-3\sqrt{2}$
1061 7828 cr : $1+\sqrt{2}/4$
1061 8671 $\zeta : 3\pi^2/8$
1061 9083 $\Psi : \sqrt[3]{5/2}$
1061 9449 id : $1/4-3\pi/4$
1062 0364 $\ln(5)\div\exp(e)$
1062 0378 ln : $2/3+3\pi/4$
1062 1278 $J_0 : \sqrt{13}\pi$
1062 1721 rt : (3,4,-9,1)
1062 2893 rt : (1,7,2,-5)
1062 5126 sinh : 21/22
1062 5882 $J_2 : 3\sqrt{5}/7$
1062 6136 $\Psi : 19/14$
1062 6260 $J_2 : 23/24$
1062 6430 $J_1 : \sqrt[3]{5}/8$
1062 7897 rt : (8,3,9,-1)
1062 8279 sq : $7\zeta(3)/8$
1062 8371 cu : 9/19
1062 9480 $J_2 : 25$
1063 0249 sinh : $(1/\pi)/3$
1063 0339 as : $(1/\pi)/3$
1063 0628 $\ln(2)+\exp(5)$
1063 0745 cu : $4\sqrt{2}/3+2\sqrt{3}$
1063 2533 $\exp(2/5)-\exp(4)$
1063 4413 $e^x : 3\pi^2/2$
1063 5147 rt : (2,4,-9,-1)
1063 8486 $\zeta : 1/\ln(1/\pi)$
1064 0053 rt : (7,3,9,-1)
1064 0831 $Li_2 : 3e/10$
1064 1095 tπ : $4\sqrt{2}+4\sqrt{3}/3$
1064 2704 rt : (9,2,-6,-8)
1064 3847 sπ : $3e-4\pi/3$
1064 4695 $\ell_{10} : 4\sqrt{5}/7$
1064 5533 $\ell_{10} : 18/23$
1064 6537 rt : (8,5,-9,-7)
1064 6710 sq : $1/3+\sqrt{5}/2$
1064 6895 tπ : P_{50}
1064 6901 $1/4-\exp(\sqrt{5})$
1064 7282 tanh : $\exp(-\sqrt{5})$
1064 7374 at : $\exp(-\sqrt{5})$
1064 7485 rt : (3,4,-9,-1)
1064 7773 rt : (4,4,-6,7)
1065 0059 tπ : $10\pi^2/9$
1065 1017 $J_0 : \sqrt[3]{18}$
1065 1951 rt : (1,8,0,7)
1065 2291 rt : (6,3,9,-1)
1065 2905 $J_1 : 3/14$
1065 3766 rt : (6,2,3,9)
1065 7267 rt : (9,2,-7,2)
1065 7956 $\Gamma : 2/9$
1065 8339 $J_2 : 10\pi$
1065 9445 cu : $\sqrt{2}/2+3\sqrt{3}/2$
1065 9641 rt : (5,1,0,-8)
1065 9909 rt : (4,4,-9,-1)
1066 0356 $J_2 : 24/25$
1066 0780 id : $8e/7$
1066 1779 $Ei : (5/2)^{-3}$
1066 3201 $e^x : (\pi)^{-2}$
1066 3354 $Ei : \sqrt[3]{17/2}$
1066 4342 rt : (2,9,-6,1)
1066 4612 rt : (5,3,9,-1)
1066 5536 rt : (4,8,1,4)
1066 6016 cu : $1/4-4\pi/3$
1066 6681 sr : $\sqrt{2}/2+4\sqrt{5}$
1066 7467 $5\div\ln(5)$
1066 8191 sr : $\sqrt{3/2}$
1066 9419 tπ : $(2e)^{-2}$
1066 9433 $2^x : \ln((e+\pi)/2)$
1067 1449 rt : (3,7,-6,-6)
1067 1454 cu : $8e/9$
1067 2419 rt : (5,4,-9,-1)
1067 2976 $J_1 : (e)^2$
1067 3199 $10^x : 2\sqrt{2}+3\sqrt{5}/2$
1067 5254 $K : 8(1/\pi)/3$
1067 7019 rt : (4,3,9,-1)
1067 8314 sr : $(\sqrt{2}\pi)^{-3}$
1067 8357 $\theta_3 : (\ln 2/3)^2$
1067 8918 rt : (9,1,-4,-9)
1068 0868 $E : 18/19$
1068 1045 tπ : $\sqrt{2}/2+\sqrt{5}/4$
1068 1704 $\theta_3 : \exp(-(e+\pi)/2)$
1068 1831 rt : (7,9,5,4)
1068 2129 $\Psi : (\sqrt[3]{4}/3)^3$
1068 2429 sinh : $3(1/\pi)$
1068 2814 rt : (8,4,-7,-8)
1068 3994 $2^x : 1/2-\sqrt{2}/4$
1068 4093 sinh : $4\sqrt[3]{3}$
1068 4365 sr : $1/4+4\pi/3$
1068 5011 sq : $3\sqrt{3}+3\sqrt{5}/2$
1068 5017 rt : (6,4,-9,-1)
1068 5281 $4/5-\ln(2)$
1068 7792 id : $\exp(-\sqrt{5})$
1068 9178 sq : $\ln(\sqrt[3]{3}/2)$
1068 9372 rt : (3,8,-7,-9)
1068 9510 rt : (3,3,9,-1)
1069 1071 $J_2 : 2\sqrt[3]{3}/3$
1069 1090 sinh : $(\sqrt{2}\pi/2)^{1/2}$
1069 1520 $\ell_2 : 13/14$
1069 3147 rt : (6,6,2,3)
1069 3878 $e^x : e/3-\pi$

1069 4059 $J_2 : 4\zeta(3)/5$
1069 6149 rt : (9,6,-8,-4)
1069 6419 $J_0 : 10\pi^2/3$
1069 7704 rt : (7,4,-9,-1)
1069 9588 cu : $8\sqrt[3]{3}/9$
1070 1379 $5 \times \exp(1/5)$
1070 1953 $\ell_{10} : (2\pi/3)^{1/3}$
1070 2089 rt : (2,3,9,9)
1070 2650 cu : $2\sqrt{2} + 2\sqrt{5}$
1070 4532 rt : (5,3,-1,2)
1070 4657 t$\pi : 3e - 4\pi/3$
1070 6190 $\ell_2 : P_{17}$
1070 6458 $J_2 : 5\sqrt{3}/9$
1070 7126 $\theta_3 : 13/18$
1070 8151 sinh : $\exp(-\sqrt{5})$
1070 8245 as : $\exp(-\sqrt{5})$
1070 9946 sq : $(e + \pi)^2$
1071 0189 $J_1 : 4\pi^2$
1071 0250 rt : (9,3,-5,3)
1071 0482 rt : (8,4,-9,-1)
1071 0598 sr : $2/3 + \sqrt{5}/4$
1071 0678 $2/5 + s\pi(1/4)$
1071 1524 $J_0 : \sqrt{21}\pi$
1071 3821 ln : $5\pi^2/6$
1071 4065 $erf : 2/21$
1071 4756 rt : (1,3,9,-1)
1071 4871 at : 2
1071 5293 $\zeta : 2(e + \pi)/7$
1071 5402 rt : (5,0,2,9)
1071 5524 cr : 19/14
1071 5819 sq : $1/4 - \sqrt{3}/3$
1071 5987 rt : (8,0,-8,2)
1071 6569 $10^x : \sqrt{2}/2 - 3\sqrt{5}/4$
1071 7055 $10^x : 13/5$
1071 7317 cr : $\sqrt[3]{5/2}$
1071 9144 id : $\exp(\sqrt{5}/3)$
1071 9565 rt : (8,3,-5,-9)
1071 9630 sinh : $9\pi^2/8$
1072 0682 $\Gamma : \exp(-\sqrt{3}/2)$
1072 1447 rt : (4,9,5,8)
1072 2472 sr : $3/2 + 3e$
1072 3351 rt : (9,4,-9,-1)
1072 3534 rt : (7,6,-8,-8)
1072 3670 at : $2/3 - \sqrt{5}/4$
1072 4315 rt : (1,6,0,4)
1072 5078 $2^x : 3e + 2\pi/3$
1072 6231 $Ei : \sqrt[3]{2}/8$
1072 6700 $\Gamma : \ln(\sqrt[3]{3}/2)$
1072 7512 rt : (3,9,-1,0)
1072 7601 $1/3 + s\pi(e)$
1072 7619 $J_0 : 8(e + \pi)/9$
1073 0404 ln : $10\sqrt{5}$
1073 0936 $\ell_2 : \sqrt[3]{10}$
1073 1591 rt : (9,2,9,-1)
1073 1933 $J_2 : (\sqrt{5}/2)^{-1/3}$
1073 2102 $Ei : \zeta(3)/3$
1073 3298 $J_1 : (\sqrt[3]{4}/3)^{-2}$
1073 3597 rt : (2,9,-7,-6)
1073 3792 $e + \exp(2)$
1073 6074 $\sqrt{2} + \ln(2)$
1073 6422 id : $3\sqrt{3}/2 - 2\sqrt{5}/3$
1073 7136 $J_1 : (5/3)^{-3}$
1073 7542 $e^x : 3/4 - 4\sqrt{5}/3$
1074 0359 rt : (1,6,0,-6)
1074 2563 $erf : 3(1/\pi)/10$
1074 2674 $J_1 : 25/7$
1074 3396 t$\pi : 3e/2 + 4\pi/3$
1074 4325 rt : (8,2,9,-1)
1074 5187 $e^x : e/4 + 4\pi/3$
1074 8737 $\ell_2 : 1 + 2\sqrt{3}/3$
1075 1010 cu : $4 + 4\sqrt{3}$
1075 2176 sinh : $2\sqrt{5}/3$
1075 2594 sq : $2\sqrt{3}/3 + 4\sqrt{5}/3$
1075 3298 rt : (2,9,3,-5)
1075 3330 s$\pi : 2/3 + 3\sqrt{3}/4$
1075 3561 $\ell_{10} : 3e\pi/2$
1075 4794 sr : $4 + 4\sqrt{2}$
1075 7147 rt : (7,2,9,-1)
1075 7446 t$\pi : \ln(\ln 3/2)$
1075 7530 $1/4 \times s\pi(\pi)$
1075 8475 $\zeta : (\sqrt{2}/3)^{-1/3}$
1075 9376 $\zeta : \ln(\sqrt[3]{2}/3)$
1076 0836 rt : (7,5,-6,-9)
1076 0876 $\Gamma : (5/2)^{-3}$
1076 2011 $J_1 : \sqrt{3}/8$
1076 3584 $Li_2 : (\sqrt{2}/3)^3$
1076 4230 rt : (9,4,-3,4)
1076 4326 sinh : 7/3
1076 4445 sq : $3/2 + 3e/2$
1076 4793 id : $3e - \pi/3$
1076 4881 rt : (6,8,-9,-8)
1076 4967 id : $2/3 - \sqrt{5}/4$
1076 6095 $J_0 : 7\sqrt[3]{2}/4$
1076 6330 rt : (3,3,-9,-1)
1076 9831 id : $(e/2)^{1/3}$
1077 0061 rt : (6,2,9,-1)
1077 0105 rt : (8,1,-6,3)
1077 1368 cu : $\sqrt{2}/4 + 4\sqrt{3}$
1077 3745 $K : (\sqrt[3]{3}/2)^{1/2}$
1077 5998 rt : (7,2,-9,-2)
1077 6035 $2^x : 1/4 - 2\sqrt{3}$
1077 6505 $\ell_{10} : 1/4 + 4\pi$
1077 7671 $\Gamma : \sqrt[3]{5}/2$
1077 7992 $J_1 : 5e\pi$
1077 8498 rt : (4,2,1,-9)
1077 9254 $\ell_{10} : 4 - e$
1077 9457 rt : (4,3,-9,-1)
1077 9879 cu : $3(e + \pi)/2$
1078 1473 id : $(\sqrt{2}\pi)^{1/2}$
1078 1838 $\theta_3 : \exp(-1/\sqrt{2})$
1078 3067 rt : (5,2,9,-1)
1078 3341 at : $5\zeta(3)/3$
1078 4030 sq : $2 + \sqrt{2}/3$
1078 4159 as : 17/19
1078 4308 $e^x : 9\sqrt[3]{2}/10$
1078 4721 rt : (6,8,-6,8)
1078 5648 rt : (1,6,-9,-8)
1078 5867 as : $2/3 - \sqrt{5}/4$
1078 6200 cu : $2\sqrt{2} + 4\sqrt{5}/3$
1078 8789 $\ell_{10} : 8\sqrt[3]{3}/9$
1078 8835 $2^x : 3/2 - 3\pi/2$
1078 9701 t$\pi : 4\sqrt[3]{3}/9$
1079 0964 rt : (2,8,-5,-7)
1079 2663 $10^x : (\ln 2/3)^{1/3}$
1079 2679 rt : (5,3,-9,-1)
1079 3035 $\ln(3/4) + \exp(1/3)$
1079 4692 rt : (9,7,-6,-3)
1079 6167 rt : (4,2,9,-1)
1079 7969 cu : 10/21
1079 8496 $e^x : 2 - \sqrt{3}/2$
1080 1301 rt : (6,7,4,4)
1080 2201 $\ell_{10} : 9 \ln 2/8$
1080 2749 rt : (6,7,-7,-9)
1080 2944 tanh : $\exp(-\sqrt{2}\pi/2)$
1080 3043 at : $\exp(-\sqrt{2}\pi/2)$
1080 3308 ln : $2\pi/7$
1080 3324 sq : 20/19
1080 5126 $\ell_{10} : 3\sqrt[3]{5}/4$
1080 5434 $\zeta : -\sqrt{2} + \sqrt{6} + \sqrt{7}$
1080 5480 $J_1 : 5/23$
1080 5536 rt : (8,4,-9,-4)
1080 5999 rt : (6,3,-9,-1)
1080 7310 $J_1 : \exp(\sqrt{2})$
1080 7477 $e^x : 7\sqrt{2}/3$
1080 9361 rt : (3,2,9,-1)
1081 0130 $2^x : \sqrt{2}/2 - \sqrt{5}/4$
1081 2206 rt : (1,8,-1,-1)
1081 3242 rt : (5,4,1,3)
1081 4613 $2^x : (\sqrt[3]{4}/3)^3$
1081 6047 t$\pi : 1/3 - 3\sqrt{3}/4$
1081 6679 sr : $2\sqrt{3} - \sqrt{5}$
1081 7106 $Li_2 : 2/19$
1081 7645 $J_2 : (e/3)^{1/3}$
1081 9241 rt : (9,5,-1,5)
1081 9404 cu : $1/\ln(1/(3e))$
1081 9418 rt : (7,3,-9,-1)
1082 1693 $10^x : 1/3 - 3\sqrt{3}/4$
1082 1901 $10^x : 4 + 2e/3$
1082 2652 rt : (2,2,9,-1)
1082 3289 $Ei : 10\sqrt{3}/3$
1082 5260 rt : (8,2,-4,4)
1082 5557 $10^x : 4\sqrt{3}/3 + 2\sqrt{5}$
1082 6822 $4/5 \div \exp(2)$
1082 9991 ln : $3 - 4\sqrt{2}/3$
1083 1298 rt : (7,1,-7,-3)
1083 2296 rt : (1,7,-8,8)
1083 2937 rt : (8,3,-9,-1)
1083 3442 $J_1 : 1/\ln(\ln 3)$
1083 5055 cu : $2 + 2\sqrt{5}$
1083 6040 rt : (1,2,9,-1)
1083 6477 cu : $3\sqrt{2}/2 - 3\sqrt{3}/2$
1083 7355 rt : (3,7,0,-3)
1083 7468 at : $\sqrt{2}/2 + 3\sqrt{3}/4$
1084 0381 $\Psi : 3(e + \pi)/5$
1084 2107 tanh : $(2\pi/3)^{-3}$
1084 2207 at : $(2\pi/3)^{-3}$
1084 3118 rt : (3,5,1,8)
1084 5266 id : $\exp(-\sqrt{2}\pi/2)$
1084 5324 rt : (5,9,-8,-9)
1084 5981 at : $3\sqrt{2} - \sqrt{5}$
1084 6077 $J_1 : 1/\sqrt{21}$
1084 6559 rt : (9,3,-9,-1)
1084 7981 $J_2 : (\ln 3)^{-1/3}$
1084 8024 rt : (1,9,0,9)
1084 8135 $e^x : 4 - 3e/4$
1084 8780 $J_0 : 21/8$
1084 9227 sq : $2/3 + \pi/4$
1084 9528 rt : (2,9,2,-5)
1084 9986 $\lambda : \exp(-11\pi/7)$
1085 0777 $Li_2 : \sqrt{2/3}$
1085 2434 rt : (9,1,9,-1)
1085 4091 $\exp(3/4) - \exp(4/5)$
1085 6155 $\Psi : (\ln 3/2)^{1/2}$
1085 6371 t$\pi : 4e - 3\pi$
1086 0440 $J_1 : ((e + \pi)/2)^3$
1086 3116 rt : (1,5,-2,-7)
1086 3241 $\ln(\pi) - s\pi(\sqrt{2})$
1086 3693 ln : $3e/2 - \pi/3$
1086 4157 cr : $8(e + \pi)/5$
1086 5888 rt : (8,1,9,-1)
1086 6539 sinh : $\exp(-\sqrt{2}\pi/2)$
1086 6609 id : $3\sqrt{2} + \sqrt{3}/2$
1086 6640 as : $\exp(-\sqrt{2}\pi/2)$
1086 7288 $Ei : \sqrt[3]{12}$
1087 1712 $J_2 : 6e\pi$
1087 1958 $\ell_{10} : 3\sqrt{2} - 2\sqrt{3}$
1087 2784 $\Gamma : e\pi/10$
1087 3397 $\ell_2 : P_{34}$
1087 4134 $J_2 : 7 \ln 2/5$
1087 5293 cr : $\sqrt{3}/4 + 4\sqrt{5}$
1087 5320 rt : (9,6,1,6)
1087 6760 $J_1 : 8\sqrt{3}$
1087 6806 rt : (2,8,1,-6)
1087 8366 $2^x : 3e/4$
1087 9441 rt : (7,1,9,-1)
1087 9587 $e^x : 1/2 - e$
1088 0248 $\Gamma : \exp(\sqrt[3]{4}/2)$
1088 1256 $2^x : 18/11$
1088 1488 rt : (8,3,-2,5)
1088 2326 $e^x : 3e/5$
1088 4304 rt : (5,7,-7,1)
1088 4391 $e - \ln(5)$
1088 4892 id : $(2\pi/3)^{-3}$
1088 5175 as : $\sqrt{2} - 4\sqrt{3}/3$
1088 6875 $\ell_{10} : (\sqrt{2}/3)^{1/3}$
1088 7677 rt : (7,0,-5,4)
1088 8116 $J_1 : 7e\pi/2$
1088 9959 rt : (2,4,-4,9)
1089 0192 sinh : $\ln(\sqrt{2}\pi)$
1089 0600 rt : (3,2,-9,-1)
1089 0686 $\ln(\sqrt{5}) \div \exp(2)$
1089 1823 cr : 15/11
1089 2193 ln : $7\sqrt{3}/4$
1089 3095 rt : (6,1,9,-1)
1089 3885 rt : (6,3,-8,-3)
1089 6560 rt : (9,8,-4,-2)
1089 6977 $10^x : 17/24$
1089 7565 $\ln(3/5) - \exp(4)$
1089 8102 ln : $(\sqrt[3]{3}/2)^{1/3}$
1090 2850 $J_1 : 9(e + \pi)$

1090 3535 $10^x : 4/3 + 2\sqrt{2}$
1090 3548 id : $8\ln 2/5$
1090 3767 $\ell_{10} : 1/2 + \pi/4$
1090 4499 rt : (4,2,-9,-1)
1090 5899 $Li_2 : (1/\pi)/3$
1090 6399 sinh : $(2\pi/3)^{-3}$
1090 6502 as : $(2\pi/3)^{-3}$
1090 6852 rt : (5,1,9,-1)
1090 7881 rt : (8,5,-7,-3)
1090 8036 $J : \sqrt{2}/3$
1090 9335 rt : (2,6,-1,-1)
1091 0942 $2^x : 2 - 3\sqrt{3}$
1091 2268 $J_0 : 8\sqrt{2}$
1091 3658 rt : (6,8,6,5)
1091 4091 id : $1/4 - e/2$
1091 4446 $\ell_{10} : 7/9$
1091 5940 rt : (1,9,-4,-8)
1091 8504 rt : (5,2,-9,-1)
1091 8729 $Ei : (\ln 3/3)^{1/3}$
1092 0054 sinh : 22/23
1092 0712 rt : (4,1,9,-1)
1092 1698 at : $1/3 + 3\sqrt{5}/4$
1092 1802 $e^x : e/3 + 3\pi/4$
1092 3063 $\zeta : 9\ln 2/10$
1092 3825 rt : (5,8,-3,5)
1092 5536 2^x : 25/4
1092 5561 $10^x : (3\pi/2)^{-2}$
1092 6201 rt : (5,5,3,4)
1092 7074 cu : $2/3 - 2\sqrt{5}$
1092 7941 $\ell_2 : 1/3 + \sqrt{5}/3$
1092 9997 at : P_{76}
1093 0326 $E : (\sqrt{5}/2)^{-1/2}$
1093 2076 ln : $4 + 3\sqrt{2}$
1093 2505 rt : (9,7,3,7)
1093 2618 rt : (6,2,-9,-1)
1093 4677 rt : (3,1,9,-1)
1093 4792 $2^x : 4\sqrt{3} - \sqrt{5}/2$
1093 7500 cu : 23/4
1093 8541 $Ei : ((e+\pi)/3)^{-1/2}$
1093 8713 sr : $3\sqrt{3} + 2\sqrt{5}$
1093 8829 rt : (8,4,0,6)
1093 9426 cu : 11/23
1093 9523 $10^x : 7\sqrt[3]{5}/3$
1094 0039 sr : 16/13
1094 1369 $\zeta : 11/3$
1094 3791 $1/2 + \ln(5)$
1094 3840 $e^x : 9\sqrt{5}/4$
1094 5174 rt : (7,1,-3,5)
1094 6843 rt : (7,2,-9,-1)
1094 7259 $Ei : (\sqrt{5}/2)^{-3}$
1094 8521 $\ell_{10} : 3e + 3\pi/2$
1094 8750 rt : (2,1,9,-1)
1095 1541 rt : (9,3,-9,-6)
1095 3182 $J_1 : 3e\pi/2$
1095 3215 2^x : 14/13
1095 5444 $\zeta : 7\pi/6$
1095 5917 at : $2\sqrt{3} - 3\sqrt{5}/2$
1095 5956 $J_0 : 6e\pi$
1095 6545 cr : $1/2 + \sqrt{3}/2$
1095 6947 2^x : 3/20
1095 7402 $2^x : 1 + 3\sqrt{3}/2$
1095 7638 $\zeta : 5\sqrt[3]{5}/7$
1095 7929 rt : (5,5,-9,-3)
1096 1179 rt : (8,2,-9,-1)
1096 2858 at : $9\sqrt{5}/10$
1096 2931 rt : (1,1,9,-1)
1096 3407 $e^x : 9e/4$
1096 3901 rt : (5,9,1,9)
1096 4338 rt : (2,4,-6,-1)
1096 4810 sq : $\sqrt{2}/2 + \sqrt{5}/3$
1096 6276 $\zeta : 8/11$
1096 6764 $2^x : 1 - 4\pi/3$
1096 8406 $\ell_{10} : 2 + 4e$
1096 8538 $\Psi : \ln(\sqrt{3}/2)$
1096 9107 sr : $3\sqrt{3} - \sqrt{5}/3$
1096 9991 at : $(1/(3e))^{-1/3}$
1097 0434 rt : (2,7,5,8)
1097 1621 at : $4/3 + e/4$
1097 3002 $\Psi : \sqrt{5/2}$
1097 3731 id : P_{76}
1097 4516 cu : $2/3 + 3\sqrt{5}$
1097 5629 rt : (9,2,-9,-1)
1097 5872 $\sqrt{2} \times \exp(2/5)$
1097 6232 $1/5 \div \exp(3/5)$
1097 6458 $10^x : (e/2)^3$
1097 6723 $2^x : \zeta(3)/8$
1097 7137 $\Gamma : \exp(-1/(2\pi))$
1097 7222 rt : (1,9,-1,0)
1097 7356 $\ell_{10} : 2e/7$
1097 7362 ln : $1/4 + \sqrt{3}/2$
1097 7444 $2^x : 3\sqrt{3}/2 - \sqrt{5}/4$
1097 8779 rt : (9,0,9,1)
1097 9817 $10^x : 4 + \sqrt{5}$
1098 0087 $\ln(3/4) - \exp(3/5)$
1098 0215 cu : $4/3 - 2e/3$
1098 0859 $e^x : P_{51}$
1098 1929 $\zeta : 5(e+\pi)/8$
1098 4992 cu : $4 + 4\pi/3$
1098 5656 $t\pi : 1/3 + 3\sqrt{2}$
1098 6122 $1/5 \times \ln(\sqrt{3})$
1098 7805 $Li_2 : \exp(-\sqrt{5})$
1098 8140 $Ei : 3(e+\pi)/10$
1099 0558 $Ei : \sqrt{2}/9$
1099 0623 rt : (9,2,-7,-7)
1099 0836 rt : (9,8,5,8)
1099 1470 rt : (9,9,8,-1)
1099 1626 rt : (1,7,7,-7)
1099 3024 rt : (8,0,9,1)
1099 3546 K : 17/20
1099 5876 as : P_{76}
1099 6641 $10^x : 7(1/\pi)$
1099 7322 rt : (8,5,2,7)
1099 9793 cu : $1/4 + \pi/4$
1099 9964 id : $2\sqrt{3} - 3\sqrt{5}/2$
1100 1957 rt : (9,9,-2,-1)
1100 3052 $J_2 : (e+\pi)/6$
1100 3831 rt : (7,2,-1,6)
1100 4400 rt : (8,9,8,-1)
1100 5489 $J_0 : (e+\pi)$
1100 6143 rt : (2,7,-1,-7)
1100 6237 $\zeta : 3e\pi/7$
1100 6239 sq : $4\sqrt{2}/3 - \sqrt{3}/4$
1100 6368 id : $\sqrt{3}/4 + 3\sqrt{5}/4$
1100 6379 sinh : $7\pi^2/8$
1100 7380 rt : (7,0,9,1)
1100 9056 rt : (4,4,-7,5)
1100 9207 rt : (1,6,-1,-2)
1100 9461 sq : $(\sqrt[3]{5}/2)^{-1/3}$
1101 0362 rt : (6,1,-4,-5)
1101 0901 cr : $1/4 + \sqrt{5}/2$
1101 2276 $J_2 : 7\sqrt{3}$
1101 2772 id : $3\sqrt{3}/2 - 3\sqrt{5}$
1101 3061 $J_2 : 4\sqrt[3]{5}/7$
1101 3791 rt : (8,6,-5,-2)
1101 4634 $J_2 : (\pi/3)^{-1/2}$
1101 5689 $2^x : 1/2 + \sqrt{3}/3$
1101 6648 $e^x : 3e/2 - 2\pi$
1101 6916 rt : (5,4,-7,-4)
1101 7418 rt : (7,9,8,-1)
1101 8978 $1/3 + \ln(4/5)$
1101 9304 $s\pi : P_{22}$
1102 0593 $J_1 : 3\pi^2/4$
1102 0776 rt : (3,1,-9,-1)
1102 1848 rt : (6,0,9,1)
1102 2269 as : $2\sqrt{3} - 3\sqrt{5}/2$
1102 5696 rt : (7,3,-8,-3)
1102 9702 $e^x : 3e/2 + \pi/3$
1103 0236 rt : (9,1,-5,-8)
1103 0341 at : $8\sqrt[3]{2}/5$
1103 0462 $3/4 \div s\pi(\sqrt{5})$
1103 0525 rt : (6,9,8,6)
1103 0568 $t\pi : e/2$
1103 3631 10^x : 1/22
1103 4029 sinh : $8\pi/7$
1103 4191 at : $4\sqrt{3} - 4\sqrt{5}$
1103 4640 rt : (8,4,-8,-7)
1103 4718 $\ell_2 : 7\pi^2$
1103 5345 sq : $\sqrt[3]{25/3}$
1103 5527 rt : (4,1,-9,-1)
1103 5925 $J_1 : 2e\pi$
1103 6226 $J_0 : 11/5$
1103 6366 rt : (3,7,-5,7)
1103 6429 rt : (5,0,9,1)
1103 8309 $10^x : (1/\pi)/7$
1103 9020 $2^x : 5e/3$
1104 0065 $J_2 : 9e\pi/5$
1104 0983 rt : (2,9,-8,-9)
1104 1697 $10^x : 1/2 + \sqrt{2}/2$
1104 2664 $J_1 : 2/9$
1104 3051 sr : $1/4 + 3\pi$
1104 3723 rt : (5,9,8,-1)
1104 4025 $e^x : (\sqrt{2}/3)^3$
1104 6343 ln : $e/4 + 3\pi/4$
1104 7683 $t\pi : 3\sqrt{2} + 2\sqrt{5}/3$
1105 0356 rt : (9,9,7,9)
1105 0369 rt : (4,5,-3,9)
1105 0397 rt : (5,1,-9,-1)
1105 0545 $\Gamma : \sqrt[3]{2}/3$
1105 1126 rt : (4,0,9,1)
1105 2256 $E : 3\sqrt[3]{2}/4$
1105 4822 id : $\sqrt{2}/4 - 2\sqrt{3}$
1105 5867 $\Gamma : 1/\ln(\pi/2)$
1105 7012 rt : (8,6,4,8)
1105 7974 $\ell_{10} : 1/3 + 4\pi$
1105 8925 rt : (6,8,-7,6)
1105 9200 cu : 12/25
1105 9400 $5 \times \exp(3/5)$
1106 1251 $t\pi$: 4/15
1106 2561 rt : (1,5,1,8)
1106 3249 $e^x : 4 + e/3$
1106 3691 rt : (7,3,1,7)
1106 5388 rt : (6,1,-9,-1)
1106 5611 tanh : 1/9
1106 5722 at : 1/9
1106 5941 rt : (3,0,9,1)
1107 0394 rt : (9,0,-3,9)
1107 1304 $J_2 : 7\sqrt[3]{2}/9$
1107 1849 $J_1 : 7(1/\pi)/10$
1107 2073 sr : $\pi^2/8$
1107 4378 $\ell_2 : \sqrt{2} + \sqrt{5}/3$
1107 4876 rt : (8,3,-6,-8)
1107 6448 $\Psi : 23/17$
1107 7120 rt : (5,3,-5,-5)
1107 9368 rt : (7,6,-9,-7)
1108 0315 $\exp(4/5) \div \exp(3)$
1108 0503 rt : (7,1,-9,-1)
1108 0876 rt : (2,0,9,1)
1108 1065 ln : $4\sqrt{2} + 3\sqrt{3}/2$
1108 1380 cu : P_{21}
1108 1507 $2^x : 2 + \sqrt{2}/4$
1108 2434 cu : $3/4 - 4\sqrt{5}/3$
1108 2562 $2/5 + \ln(3/5)$
1108 3870 rt : (4,6,-8,-4)
1108 5442 $J_0 : 7\pi/10$
1108 6701 $2^x : 3\sqrt{3}/4 - 2\sqrt{5}$
1108 6820 $t\pi : P_{22}$
1108 6950 $e^x : 2\sqrt{3}/3 - 3\sqrt{5}/2$
1108 7220 $J_1 : \exp(-3/2)$
1108 8192 $J_1 : 8\pi$
1108 8262 $s\pi : (1/\pi)/9$
1109 0541 sq : $(\ln 2)^3$
1109 1043 at : $3e/2 - 4\pi/3$
1109 1241 $erf : \ln(e/3)$
1109 2437 $\ell_{10} : \sqrt{5/3}$
1109 4091 $\ell_2 : 6\sqrt[3]{2}/7$
1109 4946 at : $7\sqrt[3]{3}/5$
1109 5743 rt : (8,1,-9,-1)
1109 5931 rt : (1,0,9,1)
1109 6973 rt : (3,9,-2,-3)
1109 7442 rt : (1,9,8,-1)
1109 7753 $s\pi : 1/4 + \pi/4$
1110 0294 $e^x : 2/19$
1110 1260 $e^x : 4/3 + 2\sqrt{5}$
1110 2857 $\Psi : 1/\ln(2\pi/3)$
1110 3905 rt : (1,3,-7,-8)
1110 5965 sr : $(\ln 2/3)^3$
1110 6200 $\zeta : \exp(-1/\pi)$
1110 8157 $\ell_{10} : 4/3 - \sqrt{5}/4$
1110 8163 cu : $4/3 + 3\pi$
1110 9459 rt : (1,9,-2,-9)

1111 0608 J_1 : $\sqrt{5}/10$
1111 1111 id : $1/9$
1111 1224 at : $\ln(\sqrt{5}/2)$
1111 3016 θ_3 : $1/18$
1111 3573 sinh : $3\sqrt{2}/2$
1111 5672 rt : (8,2,-4,-9)
1111 6166 cu : $2\zeta(3)/5$
1111 7384 J_0 : $9\sqrt[3]{5}/7$
1111 7699 at : $10\sqrt{2}/7$
1111 7944 rt : (8,7,6,9)
1111 8521 2^x : P_{71}
1111 8822 2^x : P_{68}
1111 9621 ℓ_2 : $\sqrt{14/3}$
1111 9848 Ei : $(2e)^{1/3}$
1112 0243 rt : (7,5,-7,-8)
1112 1861 e^x : P_{57}
1112 2563 ln : $17/19$
1112 3030 e^x : $3-3\sqrt{3}$
1112 3495 rt : (8,7,-3,-1)
1112 4725 rt : (8,8,8,-1)
1112 4801 rt : (7,4,3,8)
1112 4877 rt : (1,9,-8,-1)
1112 5980 at : $7\sqrt{3}/6$
1112 6227 rt : (8,1,9,1)
1112 6415 rt : (1,3,-6,-9)
1112 6819 at : $4/3-3\sqrt{5}/2$
1112 7544 E : $17/18$
1112 7768 ζ : $\exp(-\sqrt{2}/3)$
1112 9823 10^x : $3\sqrt{2}-3\sqrt{3}$
1113 0606 sq : $7\sqrt[3]{2}/5$
1113 0830 10^x : $10/11$
1113 1683 rt : (6,1,0,7)
1113 2726 $\sqrt{5}\div\exp(3)$
1113 3937 10^x : $19/17$
1113 3987 sinh : $1/9$
1113 4101 as : $1/9$
1113 4599 Ei : $9(1/\pi)/4$
1113 5957 rt : (7,4,-6,-2)
1113 6475 rt : (2,7,-6,-4)
1113 6746 id : $3e/2-4\pi/3$
1113 6914 erf : $\ln 2/7$
1113 8437 rt : (7,8,8,-1)
1113 8589 rt : (5,2,-3,-6)
1113 8744 rt : (2,9,-8,-1)
1113 9596 rt : (3,1,-7,9)
1114 1467 rt : (7,1,9,1)
1114 1847 rt : (2,6,-3,-8)
1114 2886 Ψ : $\exp(\sqrt[3]{2})$
1114 3353 ln : $1+3e/4$
1114 3786 sr : $21/17$
1114 3827 Ψ : $9\zeta(3)/8$
1114 4213 e^x : P_{38}
1114 5521 rt : (4,7,-4,-2)
1114 5618 sq : $4-\sqrt{5}$
1114 8199 10^x : $8\sqrt[3]{5}/9$
1114 8479 Γ : $9e/8$
1114 8498 rt : (6,1,-9,-3)
1115 2246 rt : (6,8,8,-1)
1115 2477 rt : (3,8,-9,-4)
1115 2711 rt : (3,9,-8,-1)
1115 3810 sq : $1/4+3\sqrt{3}/2$
1115 4419 Li_2 : $\exp(-\sqrt{2}\pi/2)$
1115 6265 ℓ_{10} : $2-\sqrt{2}/2$
1115 6833 rt : (6,1,9,1)
1115 7062 tπ : $(1/\pi)/9$
1115 7177 id : $\ln(\sqrt{5}/2)$
1115 7409 rt : (3,9,0,-7)
1115 7595 ζ : $(\sqrt[3]{4}/3)^{1/2}$
1115 8468 Γ : $9e\pi/4$
1115 8700 $\exp(1/3)-\exp(1/4)$
1115 9896 as : $3e/2-4\pi/3$
1116 0625 Γ : $((e+\pi)/2)^{-2}$
1116 1059 ln : $4/3+4\sqrt{3}$
1116 1068 J_0 : $3\sqrt{3}$
1116 1697 rt : (7,4,-5,-9)
1116 2155 J_1 : $\sqrt{17}$
1116 2684 cr : $\sqrt{2}/3+4\sqrt{5}$
1116 4167 Ψ : $19/12$
1116 5643 10^x : $\sqrt{3}-\sqrt{5}/2$
1116 5930 cr : $2/3+\sqrt{2}/2$
1116 6088 cr : $8\zeta(3)/7$
1116 6155 rt : (5,8,8,-1)
1116 6362 rt : (6,7,-8,-8)
1116 6572 Γ : $7\pi/5$
1116 6731 tπ : $1/4+\pi/4$
1116 6781 rt : (4,9,-8,-1)
1116 6832 ζ : $9/23$
1116 7948 at : $1/3-3\pi/4$
1116 9422 rt : (1,4,1,6)
1116 9844 $e\times\ln(\pi)$
1117 0166 rt : (9,2,-8,1)
1117 1199 J_2 : $(\pi/3)^{-1/3}$
1117 2326 rt : (5,1,9,1)
1117 3103 rt : (4,0,-9,1)
1117 3407 10^x : $1/3+3\pi/2$
1117 7880 J_0 : $3(e+\pi)/8$
1117 9567 sq : P_7
1118 0165 rt : (4,8,8,-1)
1118 0339 $1/4\div\sqrt{5}$
1118 0456 as : $\ln(\sqrt{5}/2)$
1118 0955 rt : (5,9,-8,-1)
1118 1315 $\exp(\sqrt{2})\div\exp(2/3)$
1118 5157 sr : $6\sqrt[3]{3}/7$
1118 7208 rt : (7,5,5,9)
1118 7949 rt : (4,1,9,1)
1118 8020 sinh : $3\sqrt{5}/7$
1118 8931 rt : (5,0,-9,1)
1119 0350 2^x : $\sqrt{5}+\sqrt{6}-\sqrt{7}$
1119 0593 Ψ : $\sqrt{2}/3$
1119 0784 sinh : $23/24$
1119 1715 sr : $(\sqrt[3]{4}/3)^{-1/3}$
1119 1734 10^x : $17/11$
1119 3672 id : $\exp(1/(3\pi))$
1119 4277 rt : (6,2,2,8)
1119 5235 rt : (6,9,-8,-1)
1119 6231 ln : $4\sqrt{2}/3+2\sqrt{3}/3$
1119 6368 Li_2 : $(2\pi/3)^{-3}$
1119 6783 J_0 : 8π
1119 7086 cu : $25/13$
1119 7375 ℓ_{10} : $17/22$
1119 9004 cr : $11/8$
1119 9862 $\exp(1/4)\times s\pi(1/3)$
1120 1319 Li_2 : $9/11$
1120 1372 rt : (5,1,-1,-7)
1120 3279 10^x : $\ln(\pi/3)$
1120 3703 rt : (3,1,9,1)
1120 4247 sq : $10\sqrt[3]{3}/3$
1120 4895 rt : (6,0,-9,1)
1120 5222 e^x : $2-4\pi/3$
1120 7762 ℓ_{10} : $4+4\sqrt{5}$
1120 8493 rt : (6,6,-6,-9)
1120 9313 Ψ : $\exp(-e/2)$
1120 9622 rt : (7,9,-8,-1)
1121 1637 2^x : $1/3+\sqrt{5}/3$
1121 2263 sπ : $e/4+3\pi/4$
1121 2657 2^x : $22/5$
1121 3255 rt : (5,9,-9,-8)
1121 5641 rt : (3,7,-7,-5)
1121 6037 J_0 : $9\pi^2/2$
1121 7351 J_2 : $\pi^2/10$
1121 7647 ln : $3e/2+4\pi/3$
1121 9545 ln : $1/3+\pi/4$
1121 9592 rt : (2,1,9,1)
1122 0854 10^x : $\exp(-1/\sqrt{2})$
1122 0999 rt : (7,0,-9,1)
1122 1152 cr : $2\sqrt{2}/3+\sqrt{3}/4$
1122 2815 rt : (1,8,8,-1)
1122 4118 rt : (8,9,-8,-1)
1122 4810 J_2 : $5\pi^2/9$
1122 4994 cr : $3\sqrt{2}/2-\sqrt{5}/3$
1122 5745 at : $2/3+e/2$
1122 6243 Ei : $(1/\pi)/5$
1122 8555 sr : $5\sqrt{3}/7$
1123 0057 Ei : $(e)^{-1/3}$
1123 0399 rt : (9,3,-6,2)
1123 0702 id : $(3\pi)^{1/3}$
1123 1644 2^x : $4\sqrt{2}/3-\sqrt{3}$
1123 1751 rt : (2,9,1,-9)
1123 1792 $2/5+\ln(3/4)$
1123 3224 $1/4\div\exp(4/5)$
1123 4787 $3\div s\pi(\sqrt{2})$
1123 5617 rt : (1,1,9,1)
1123 5797 rt : (9,7,8,-1)
1123 6323 2^x : $3\sqrt{2}/4+\sqrt{3}/3$
1123 6415 as : $(\sqrt[3]{3}/2)^{1/3}$
1123 7243 rt : (8,8,-1,0)
1123 7433 ℓ_{10} : $\sqrt{17}\pi$
1123 8724 rt : (9,9,-8,-1)
1123 8886 J_0 : $\exp((e+\pi)/2)$
1124 0139 sr : $7\sqrt{2}/8$
1124 0242 E : $1/\ln(\ln 2/2)$
1124 1313 Ei : $10\sqrt[3]{2}$
1124 1501 J_2 : $\exp(\pi/2)$
1124 5313 cu : $2\sqrt{3}-4\sqrt{5}/3$
1124 6291 erf : $1/10$
1124 8154 2^x : $2/3+4\pi$
1124 8373 Γ : $17/20$
1124 8497 cu : $1+4\sqrt{5}/3$
1124 9978 rt : (9,2,9,1)
1125 0158 rt : (8,7,8,-1)
1125 0305 rt : (7,5,-4,-1)
1125 1445 J_2 : $5e\pi/3$
1125 1780 rt : (1,9,1,0)
1125 2686 cu : $4e/3-\pi$
1125 3147 2^x : $2/13$
1125 3631 rt : (9,0,-9,1)
1125 3918 sπ : $1/2+2\sqrt{3}$
1125 6085 rt : (5,8,-7,-9)
1125 6510 Ψ : $\sqrt{3}-\sqrt{5}-\sqrt{6}$
1125 6938 at : $\sqrt[3]{25/3}$
1125 7188 $\ln(2/3)-s\pi(1/4)$
1125 8229 rt : (6,3,4,9)
1125 9772 10^x : $(\sqrt[3]{5}/3)^{-1/2}$
1126 0546 J_1 : $7e\pi/9$
1126 1916 Ei : $6\sqrt[3]{2}/7$
1126 3267 cu : $5\zeta(3)$
1126 3453 rt : (6,2,-7,-2)
1126 4625 rt : (7,7,8,-1)
1126 5481 $\exp(2/5)\times s\pi(\sqrt{3})$
1126 5521 rt : (5,0,1,8)
1126 6059 rt : (8,2,9,1)
1126 6427 rt : (2,8,-8,-1)
1126 7438 2^x : $\sqrt{3}-\sqrt{5}-\sqrt{7}$
1126 7534 2^x : $2\ln 2/9$
1126 8085 rt : (1,1,-9,1)
1126 9837 cu : $1+2\sqrt{2}$
1127 0912 at : $\exp(1/\sqrt{2})$
1127 1091 J_1 : $5\pi^2$
1127 2840 rt : (4,9,4,6)
1127 5419 J_2 : $4\sqrt{3}/7$
1127 6443 rt : (2,4,-5,6)
1127 7046 $1/2\times\exp(4/5)$
1127 8485 10^x : $3\sqrt{2}-3\sqrt{3}/2$
1127 9200 rt : (6,7,8,-1)
1127 9725 J_2 : $7\sqrt{2}/10$
1127 9840 e^x : $\exp(-\sqrt{5})$
1128 0187 rt : (3,6,-5,-6)
1128 1187 rt : (3,8,-8,-1)
1128 2279 rt : (7,2,9,1)
1128 3412 tπ : $e/4+3\pi/4$
1128 4207 sr : $1+2\sqrt{3}$
1128 4533 rt : (2,5,-5,-9)
1128 5173 J_2 : $10\ln 2/7$
1128 5700 ℓ_{10} : $4e+2\pi/3$
1128 7563 rt : (2,9,-8,-5)
1128 9561 cu : $3\sqrt{2}/4-\sqrt{3}/3$
1129 0369 sq : $2\sqrt{3}/3-2\sqrt{5}/3$
1129 0423 J_1 : $5/22$
1129 1854 rt : (9,4,-4,3)
1129 2330 cu : $3e/4-3\pi/2$
1129 3886 rt : (5,8,9,7)
1129 4617 at : P_{37}
1129 4909 J_1 : $5(1/\pi)/7$
1129 5797 10^x : $(\sqrt[3]{5}/3)^2$
1129 6059 rt : (4,8,-8,-1)
1129 6655 Ei : $(\sqrt[3]{2})^{1/3}$
1129 7597 J_1 : $\exp(-\sqrt{2}\pi/3)$
1129 8639 rt : (6,2,9,1)
1129 8877 rt : (8,1,-7,2)
1129 9172 J_2 : 8

1129 9545 rt : (5,8,-4,3)
1130 0612 ℓ_{10} : $1/4+\pi/3$
1130 1128 rt : (3,8,-2,-8)
1130 1198 ℓ_2 : $(\sqrt[3]{5}/2)^{1/2}$
1130 1782 Γ : $\exp(\sqrt{5}/2)$
1130 1877 $s\pi(\sqrt{2}) \div s\pi(1/3)$
1130 3210 J_2 : 9π
1130 8307 10^x : 21/11
1130 8683 rt : (4,7,8,-1)
1130 9058 rt : (1,6,-2,-8)
1130 9739 cu : P_{32}
1131 1047 rt : (5,8,-8,-1)
1131 3632 Ei : $5\sqrt{5}$
1131 5144 rt : (5,2,9,1)
1131 7258 Γ : $(5/2)^3$
1131 7871 rt : (4,1,-9,1)
1131 7987 J : 1/8
1131 8713 cr : $4+2e$
1131 9549 at : $\sqrt{2}/4+3\sqrt{5}/4$
1131 9582 J_2 : $4\zeta(3)$
1132 0008 Γ : $(2e)^{-3}$
1132 1590 Γ : $\exp(\pi/2)$
1132 2662 J_0 : $(\ln 2/3)^{-2}$
1132 3593 rt : (3,7,8,-1)
1132 4742 J_1 : $(2\pi/3)^{-2}$
1132 5037 id : $\exp(\sqrt{2})$
1132 5482 Γ : $(\sqrt[3]{3}/2)^{1/2}$
1132 5868 $t\pi$: $1/2+2\sqrt{3}$
1132 6152 rt : (6,8,-8,-1)
1132 6602 ln : $8\sqrt[3]{2}/9$
1132 7568 sq : $2/3+4\pi$
1132 7937 θ_3 : $2\sqrt{3}/7$
1132 8292 ℓ_{10} : $9\sqrt[3]{3}$
1132 8333 ℓ_2 : $3\sqrt[3]{3}$
1132 9568 sr : $1/3+e/3$
1132 9997 rt : (3,7,-6,4)
1133 1091 rt : (5,1,3,9)
1133 1794 rt : (4,2,9,1)
1133 3969 cu : $\sqrt{3}\pi$
1133 4766 rt : (5,1,-9,1)
1133 5353 cr : $3\sqrt{2}+3\sqrt{3}$
1133 7447 10^x : $\sqrt{2}/4-3\sqrt{3}/4$
1133 8619 rt : (4,2,0,-8)
1133 9127 2^x : $4+2\sqrt{5}$
1134 0981 2^x : $4\sqrt{2}+\sqrt{3}/4$
1134 1377 rt : (7,8,-8,-1)
1134 1704 ℓ_{10} : $\ln 2/9$
1134 2876 J_2 : $9\sqrt[3]{5}$
1134 2891 id : P_{37}
1134 3653 rt : (5,9,0,7)
1134 3670 2^x : $10e/9$
1134 3953 ζ : $\pi^2/10$
1134 4883 J_2 : $10\sqrt[3]{3}/3$
1134 5820 Ψ : $\ln(2\pi/3)$
1134 5980 10^x : $3+\sqrt{2}/4$
1134 6175 rt : (6,8,-8,4)
1134 7902 cr : $1/3+\pi/3$
1134 8212 J_1 : $10\sqrt[3]{5}$
1134 8592 rt : (3,2,9,1)
1135 0209 ℓ_2 : $9\zeta(3)/5$
1135 0806 ℓ_2 : $1/2+3e$
1135 1815 rt : (6,1,-9,1)
1135 1847 cu : P_{42}
1135 3313 ζ : $4e/3$
1135 3408 2^x : $4/3-2\sqrt{5}$
1135 3761 rt : (2,8,-6,-6)
1135 4577 rt : (9,5,-2,4)
1135 4860 sq : $(\sqrt{2}\pi)^3$
1135 5311 rt : (8,9,1,1)
1135 6257 sr : $3/4+4\sqrt{5}$
1135 6286 J_2 : $\sqrt{24}\pi$
1135 6722 rt : (8,8,-8,-1)
1135 9205 rt : (9,2,-8,-6)
1135 9758 Ψ : 23
1136 0222 Ψ : $8\pi^2/9$
1136 0259 Γ : $8(1/\pi)/3$
1136 0424 J_2 : $4\sqrt{5}/9$
1136 1786 rt : (8,2,-5,3)
1136 2189 ℓ_{10} : $3\sqrt{3}/4$
1136 4546 cu : $\sqrt{2}/2-4\sqrt{3}/3$
1136 5106 2^x : $4+3e/2$
1136 5542 rt : (2,2,9,1)
1136 5753 e^x : $2/3-\sqrt{5}/4$
1136 5932 rt : (9,6,8,-1)
1136 7047 Li_2 : $(\ln 3/2)^{1/3}$
1136 7356 as : P_{37}
1136 9021 rt : (7,8,-1,0)
1136 9521 $\exp(1/2) \times s\pi(\sqrt{5})$
1137 1006 2^x : $4/3+\sqrt{2}/2$
1137 1541 Γ : $7(e+\pi)/2$
1137 2050 λ : $\exp(-14\pi/9)$
1137 2178 cu : $3\sqrt{2}/2+3\sqrt{3}/2$
1137 2189 rt : (9,8,-8,-1)
1137 2887 ℓ_2 : $\ln 2/3$
1137 4110 E : $2\sqrt{2}/3$
1137 6236 cr : $1/2+4\sqrt{5}$
1137 8933 J_0 : $7(e+\pi)/2$
1137 9111 10^x : $10\sqrt[3]{2}/9$
1137 9727 rt : (6,9,-4,8)
1138 0754 Ψ : 17/23
1138 0800 J : $\exp(-11\pi/12)$
1138 1108 rt : (8,6,8,-1)
1138 2481 Ei : $7(e+\pi)/9$
1138 2486 $t\pi$: $2\zeta(3)/9$
1138 2645 rt : (1,2,9,1)
1138 2825 rt : (6,3,-5,-1)
1138 4164 10^x : $1/\ln(\ln 2/2)$
1138 4286 ℓ_2 : $3+4\sqrt{2}$
1138 4402 rt : (1,7,-8,-1)
1138 6388 rt : (8,1,-9,1)
1138 7892 10^x : $5\pi/2$
1138 8323 $t\pi$: $5\ln 2/2$
1138 8696 at : $2-4\sqrt{2}/3$
1138 8758 Γ : $3\sqrt{2}/5$
1138 9425 $e+\exp(1/3)$
1138 9647 ln : $1/3+\sqrt{5}/4$
1139 0787 ζ : $\sqrt[3]{14/3}$
1139 3570 rt : (4,4,-8,3)
1139 3692 2^x : $6\sqrt[3]{2}/7$
1139 3868 erf : $(\pi)^{-2}$
1139 4335 ℓ_{10} : 13
1139 4958 J_2 : $1/\ln(\ln 3/3)$
1139 5486 J_1 : $1/\sqrt{19}$
1139 6008 rt : (9,3,9,1)
1139 6103 sq : $3/2+3\sqrt{2}/2$
1139 6402 rt : (7,6,8,-1)
1139 6724 rt : (5,0,-8,2)
1139 8480 2^x : $5\sqrt{3}/2$
1139 8744 $\ln(3/5) \times \ln(4/5)$
1139 8817 θ_3 : $5\ln 2/7$
1139 8915 J_2 : $9\pi^2/2$
1139 9906 rt : (2,9,1,0)
1140 1592 rt : (7,9,-8,9)
1140 2616 rt : (9,1,-6,-7)
1140 3190 $t\pi$: $3\sqrt[3]{5}/7$
1140 3711 sq : $\sqrt{2}/3-\sqrt{5}$
1140 3917 rt : (9,1,-9,1)
1140 5887 rt : (4,1,2,-9)
1140 7608 rt : (8,4,-9,-6)
1140 8460 id : $7(1/\pi)/2$
1140 9685 J_0 : $((e+\pi)/2)^3$
1141 1816 rt : (6,6,8,-1)
1141 2079 J_0 : $2\pi^2/9$
1141 2362 2^x : $(\sqrt[3]{2})^{1/3}$
1141 3162 rt : (8,3,9,1)
1141 3612 cu : $3\sqrt{2}+3\sqrt{3}/2$
1141 3663 rt : (3,5,0,6)
1141 4040 J_1 : $(e+\pi)^2$
1141 4062 $s\pi$: $-\sqrt{2}+\sqrt{3}+\sqrt{7}$
1141 5034 J_2 : $10\sqrt{3}$
1141 5534 rt : (3,7,-8,-1)
1141 6811 2^x : $1/3-2\sqrt{3}$
1141 7326 rt : (1,2,-9,1)
1141 8329 sq : $3e/2+\pi$
1141 8617 rt : (9,6,0,5)
1141 9753 sq : 19/18
1142 1469 rt : (2,7,-4,-7)
1142 1919 id : $7e/9$
1142 2233 ln : $2+\pi/3$
1142 2745 cu : $7\ln 2/10$
1142 2759 ln : $2+2\pi$
1142 3774 E : $3\pi/10$
1142 4794 at : $3\sqrt{2}/2-\sqrt{5}$
1142 6020 rt : (8,3,-3,4)
1142 7352 rt : (5,6,8,-1)
1142 8527 2^x : $3\sqrt{3}/4+4\sqrt{5}$
1142 8969 as : $2\pi/7$
1142 9463 $\sqrt{3} \div \exp(e)$
1143 0472 rt : (7,3,9,1)
1143 0617 sinh : $1/\ln((e+\pi)/3)$
1143 1288 rt : (4,7,-8,-1)
1143 3450 rt : (7,0,-6,3)
1143 4674 cu : $8\zeta(3)/5$
1143 4909 rt : (2,2,-9,1)
1143 6020 Li_2 : 1/9
1143 6874 e^x : $\sqrt{2}/3+2\sqrt{5}/3$
1143 8161 rt : (2,7,-7,-8)
1143 8191 id : $2-4\sqrt{2}/3$
1143 9127 rt : (4,5,-4,7)
1144 0179 sinh : 24/25
1144 0909 rt : (6,3,-9,-2)
1144 0934 ℓ_2 : $5\sqrt{3}$
1144 2203 Ψ : $9\pi/8$
1144 3012 rt : (4,6,8,7)
1144 3133 $s\pi$: $5\pi/8$
1144 4046 $s\pi$: $(\sqrt{5}/2)^{-1/3}$
1144 6658 rt : (9,0,-4,8)
1144 7170 rt : (5,7,-8,-1)
1144 7941 rt : (6,3,9,1)
1144 8211 ln : $e/3-\pi/4$
1144 8270 e^x : $3/4+3\sqrt{2}/4$
1144 8877 cu : $4\sqrt{3}/3-3\sqrt{5}$
1145 1743 rt : (8,3,-7,-7)
1145 1946 rt : (5,5,-8,8)
1145 2657 rt : (3,7,-4,-9)
1145 5214 e^x : $\exp(-\sqrt{2}\pi/2)$
1145 6571 $\ln(4/5) \div \exp(2/3)$
1145 6717 sr : $5\sqrt{5}/9$
1145 6958 sr : $4+\sqrt{2}/3$
1145 7607 cr : 18/13
1145 8797 rt : (3,6,8,-1)
1146 1253 e^x : $(\sqrt[3]{2}/3)^{1/3}$
1146 1822 J_1 : 3/13
1146 3183 rt : (6,7,-8,-1)
1146 3281 as : $2-4\sqrt{2}/3$
1146 5572 rt : (5,3,9,1)
1146 5680 Ei : 19/6
1146 6411 at : P_{78}
1146 7018 at : $\sqrt{2}-3\sqrt{3}/4$
1146 7511 2^x : $1/2-4e/3$
1146 7579 ℓ_{10} : $2(e+\pi)/9$
1146 7914 rt : (1,5,0,3)
1147 0061 ln : $\sqrt{3}/2-\sqrt{5}/3$
1147 0574 rt : (4,2,-9,1)
1147 1331 cu : $7e/5$
1147 1705 10^x : $1/\ln(\sqrt[3]{5})$
1147 3131 ℓ_{10} : $2\sqrt{2}/3-\sqrt{3}/2$
1147 3505 Ψ : $(\ln 3/2)^{-1/2}$
1147 3767 rt : (2,8,1,8)
1147 3904 sinh : $\sqrt[3]{10/3}$
1147 4252 sr : $2\sqrt{5}$
1147 4710 rt : (2,6,8,9)
1147 4763 id : $3\sqrt{2}/2-\sqrt{5}$
1147 5103 sr : $1/\ln(\sqrt{5})$
1147 5408 at : $2\sqrt{2}/3-4\sqrt{5}/3$
1147 5535 J_1 : $\ln 2/3$
1147 7278 at : $3e/4$
1147 9032 J_1 : $(\sqrt[3]{3}/3)^2$
1147 9310 $t\pi$: $1/\sqrt{14}$
1147 9329 rt : (7,7,-8,-1)
1148 1494 2^x : $\sqrt[3]{17/2}$
1148 2716 rt : (3,3,1,-9)
1148 2775 Ψ : $5\sqrt{2}/2$
1148 3368 rt : (4,3,9,1)
1148 4023 at : $3\sqrt{3}/2-\sqrt{5}/4$
1148 4840 Γ : $\sqrt{24}\pi$
1148 4859 Li_2 : $\ln(\sqrt{5}/2)$
1148 5111 cr : $4\sqrt{3}/5$
1148 5985 e^x : $1/\ln(\sqrt[3]{2}/2)$
1148 7403 ℓ_{10} : $4\sqrt{2}/3-\sqrt{5}/2$

1148 8664 rt : (5,2,-9,1)
1148 8884 $\Psi : 7\pi^2/3$
1148 8966 ln : $7\sqrt[3]{3}/9$
1148 9148 $t\pi : \sqrt{2}-\sqrt{3}-\sqrt{7}$
1149 0348 $J_2 : 1$
1149 0754 rt : (3,9,-3,-6)
1149 1348 rt : (9,1,-2,9)
1149 1629 rt : (8,4,-1,5)
1149 2347 rt : (1,8,7,5)
1149 2415 rt : (7,7,0,1)
1149 2459 cu : $\zeta(5)$
1149 3662 cu : $\exp(-\sqrt[3]{3}/2)$
1149 3678 $s\pi : 2\sqrt{3}/3-\sqrt{5}/2$
1149 5609 rt : (8,7,-8,-1)
1149 6528 rt : (8,2,-5,-8)
1149 8506 at : $\sqrt{5}+\sqrt{6}-\sqrt{7}$
1149 8825 rt : (1,9,-5,-7)
1149 9264 rt : (7,1,-4,4)
1149 9389 $e^x : (2\pi/3)^{-3}$
1150 0095 as : $3\sqrt{2}/2-\sqrt{5}$
1150 0740 $J_2 : \zeta(11)$
1150 1015 $e^x : 4\sqrt{3}/3-2\sqrt{5}$
1150 1201 $\exp(3/5)-s\pi(1/4)$
1150 1333 rt : (3,3,9,1)
1150 1467 at : $\ln 2/6$
1150 1722 rt : (7,5,-8,-7)
1150 1968 rt : (9,5,8,-1)
1150 2640 cr : $2\ln 2$
1150 6929 rt : (6,8,-1,0)
1150 7043 $2^x : \sqrt{2}/9$
1150 7282 $2/5\times\ln(3/4)$
1151 0782 at : $4/3+\sqrt{2}/2$
1151 1978 $\Gamma : \sqrt{3}/6$
1151 2026 rt : (9,7,-8,-1)
1151 2227 $e^x : 2/3-2\sqrt{2}$
1151 4119 id : $(\sqrt[3]{3}/2)^{-1/3}$
1151 6930 id : P_{78}
1151 7545 id : $\sqrt{2}-3\sqrt{3}/4$
1151 7590 $\Psi : \sqrt[3]{4}$
1151 8038 rt : (8,5,8,-1)
1151 8284 at : $\sqrt[3]{17/2}$
1151 8798 $t\pi : 5\pi/8$
1151 9469 rt : (2,3,9,1)
1151 9729 $t\pi : (\sqrt{5}/2)^{-1/3}$
1151 9849 ln : $9\ln 2/7$
1152 0313 $4\div\exp(1/4)$
1152 1547 rt : (5,1,-6,-1)
1152 2633 cu : $10\sqrt[3]{3}/9$
1152 3238 rt : (1,9,7,-2)
1152 5331 $10^x : e\pi$
1152 5373 rt : (7,2,-9,1)
1152 6405 $J_2 : 5\zeta(3)/6$
1152 7658 $10^x : 13/12$
1153 0383 $Ei : 6\sqrt[3]{2}$
1153 1142 $\zeta : \ln(\sqrt{2}\pi/3)$
1153 2600 $J_2 : \zeta(9)$
1153 3220 $\zeta : \pi/8$
1153 4239 rt : (7,5,8,-1)
1153 4301 $\zeta : (\sqrt[3]{3}/2)^{1/3}$
1153 4507 $2^x : \sqrt[3]{2}/8$
1153 5507 sr : $2/3+\sqrt{3}/3$
1153 6157 $\zeta : 3\zeta(3)$
1153 6271 rt : (4,2,-9,2)
1153 6728 $\Gamma : 20/9$
1153 7779 rt : (1,3,9,1)
1153 7852 rt : (3,1,-8,7)
1153 9684 rt : (2,6,-8,-1)
1153 9856 id : $\sqrt{3}/2-4\sqrt{5}/3$
1153 9986 sinh : $9\sqrt[3]{3}/7$
1154 0208 sr : $4/3+\pi$
1154 1632 $J : \exp(-15\pi/2)$
1154 1702 $\zeta : 5\sqrt[3]{3}/2$
1154 1894 $e^x : 25/22$
1154 1982 rt : (8,1,-3,-9)
1154 2438 $\zeta : \sqrt{13}$
1154 2543 as : P_{78}
1154 3024 $e^x : \sqrt{3}/3+\sqrt{5}/4$
1154 3163 as : $\sqrt{2}-3\sqrt{3}/4$
1154 3999 rt : (8,2,-9,1)
1154 4062 cu : $e-4\pi$
1154 5195 $e^x : 7/2$
1154 5400 sq : $e/8$
1154 7005 $1/5\div\sqrt{3}$
1154 7274 rt : (7,4,-6,-8)
1154 7721 $\ell_2 : 3/13$
1154 9926 rt : (9,4,9,1)
1155 0574 rt : (6,5,8,-1)
1155 0850 rt : (9,8,4,7)
1155 0892 at : $3/4-\sqrt{3}/2$
1155 2453 id : $\ln 2/6$
1155 2581 rt : (6,7,-9,-7)
1155 3163 $\sqrt{3}\times\exp(1/5)$
1155 4495 $\ell_{10} : (\sqrt{2}\pi/2)^{1/3}$
1155 6268 rt : (3,9,1,0)
1155 7508 $\ell_{10} : 1/3+\sqrt{3}/4$
1155 7735 $\ln(2)\times\ln(5)$
1155 8234 $\ell_{10} : 4e/3+3\pi$
1155 8669 rt : (8,5,1,6)
1156 0327 $10^x : 3\sqrt{3}/4-\sqrt{5}$
1156 1682 rt : (2,7,4,5)
1156 2812 rt : (9,2,-9,1)
1156 3204 $\Psi : 3(e+\pi)/2$
1156 5791 $2^x : 3/19$
1156 6518 rt : (7,2,-2,5)
1156 7044 rt : (5,5,8,-1)
1156 7101 ln : $1/3+e$
1156 8277 rt : (8,4,9,1)
1156 9998 rt : (1,8,-3,-8)
1157 0247 sq : $16/11$
1157 0357 $t\pi : 2\sqrt{3}/3-\sqrt{5}/2$
1157 1775 $5\times\ln(4/5)$
1157 2158 cr : $25/18$
1157 2994 rt : (4,6,-8,-1)
1157 3254 $10^x : 9\zeta(3)/7$
1157 4398 rt : (6,1,-5,-4)
1157 4939 rt : (1,3,-9,1)
1157 5190 $s\pi : \zeta(5)$
1157 7950 $2^x : \sqrt{2}/4-2\sqrt{3}$
1157 8166 sinh : $\ln 2/6$
1157 8304 as : $\ln 2/6$
1157 9908 sr : $3+3\sqrt{5}$
1157 9957 cu : $4+\sqrt{3}/4$
1158 1323 $10^x : 1/2+3\pi/4$
1158 2309 rt : (5,4,-8,-3)
1158 2343 $J_0 : 5\pi^2/6$
1158 3653 rt : (4,5,8,-1)
1158 3984 sq : $\sqrt[3]{11/2}$
1158 4478 id : $2\sqrt{2}-4\sqrt{5}$
1158 5731 $J_2 : \sqrt{2}+\sqrt{5}-\sqrt{7}$
1158 6453 $J_2 : \ln(\ln 3/3)$
1158 6806 rt : (7,4,9,1)
1158 7240 at : $3\sqrt{3}/4+\sqrt{5}/3$
1158 8399 $10^x : 1/21$
1158 9863 rt : (5,6,-8,-1)
1159 0476 $s\pi : 7e\pi/6$
1159 1536 sq : $3/2+2e$
1159 2505 sq : $3\sqrt{3}/4+3\sqrt{5}$
1159 3516 rt : (7,3,-4,-9)
1159 3795 rt : (2,3,-9,1)
1159 5519 sr : $1/2+\sqrt{5}/3$
1159 6298 at : P_{77}
1159 6452 tanh : $((e+\pi)/2)^{-2}$
1159 6593 at : $((e+\pi)/2)^{-2}$
1159 7061 rt : (2,6,-7,9)
1159 7354 $10^x : 8\sqrt{5}/3$
1159 8487 $e^x : P_{76}$
1159 8907 $\zeta : 18/5$
1159 8926 rt : (6,6,-7,-8)
1159 9260 $e^x : 1/3+3\sqrt{3}/4$
1160 0402 rt : (3,5,8,-1)
1160 1731 sq : $2/3-3\sqrt{2}/2$
1160 2403 cu : $4\sqrt{3}/3-2\sqrt{5}$
1160 2540 id : $1/4+\sqrt{3}/2$
1160 5038 $J_2 : 24/5$
1160 5516 rt : (6,4,9,1)
1160 6025 $t\pi : 1/2+4\sqrt{5}/3$
1160 6296 $J_0 : \sqrt[3]{21/2}$
1160 6827 cu : $\sqrt{3}/2+2\sqrt{5}$
1160 6878 rt : (6,6,-8,-1)
1160 9291 $s\pi : (3)^{-3}$
1161 1871 sinh : $(e/2)^3$
1161 1925 cu : $2\sqrt{3}+\sqrt{5}/4$
1161 2840 rt : (3,3,-9,1)
1161 3411 id : $9e/4$
1161 4503 $\ell_2 : 3/2-\sqrt{3}/3$
1161 5497 $\ell_{10} : 1/2+4\pi$
1161 6662 $\Gamma : 11/13$
1161 7293 rt : (2,5,8,-1)
1161 7582 $Ei : 2\zeta(3)/7$
1161 7998 $E : 16/17$
1161 9157 rt : (9,9,6,8)
1162 0828 rt : (7,8,2,2)
1162 1139 $2^x : (\sqrt[3]{5}/2)^{-1/2}$
1162 3847 cu : $2/3-2\sqrt{3}/3$
1162 4041 rt : (7,6,-8,-1)
1162 4410 rt : (5,4,9,1)
1162 7197 rt : (8,6,3,7)
1162 7614 $Li_2 : (2e/3)^{-1/3}$
1162 7767 $e^x : 2\sqrt{3}-3\sqrt{5}/2$
1162 8731 as : $3/4-\sqrt{3}/2$
1162 9073 $1/5-\ln(2/5)$
1162 9798 $2^x : 1/4-3\sqrt{5}/2$
1163 2078 rt : (4,9,-3,-9)
1163 4329 rt : (1,5,8,-1)
1163 4405 $\ell_{10} : 2e/3-\pi/3$
1163 5270 rt : (7,3,0,6)
1163 6114 rt : (6,5,-1,1)
1163 6877 $\Psi : 8/17$
1163 6998 sq : $7e\pi/9$
1163 8611 rt : (3,7,-7,1)
1163 9815 $e^x : 2\sqrt{2}-\sqrt{3}/2$
1164 1305 at : $1+\pi/3$
1164 1356 rt : (8,6,-8,-1)
1164 2579 $Ei : 23/16$
1164 2903 rt : (1,7,-1,-9)
1164 3063 $J_0 : 8\pi^2/7$
1164 3376 rt : (6,0,-3,5)
1164 3492 rt : (4,4,9,1)
1164 4412 rt : (9,4,8,-1)
1164 4518 id : $5e\pi/6$
1164 4778 rt : (7,9,-9,7)
1164 5181 $s\pi : 8\sqrt{5}/3$
1164 5983 rt : (6,5,-5,-9)
1164 6078 $\Gamma : \sqrt[3]{11}$
1164 7325 rt : (6,8,-9,2)
1164 7497 $e^x : 3e/4-4\pi/3$
1164 7585 $s\pi : 1/\ln(\sqrt{5}/2)$
1164 8560 id : P_{77}
1164 8858 id : $((e+\pi)/2)^{-2}$
1164 8994 $4/5\times\exp(1/3)$
1164 9500 $\ell_{10} : (\sqrt[3]{5})^{1/2}$
1165 0556 $\ell_{10} : 13/17$
1165 0981 $2^x : 3\sqrt[3]{3}/4$
1165 1513 rt : (5,8,-1,0)
1165 3524 $t\pi : \zeta(5)$
1165 5706 $e^x : 3-e/3$
1165 7637 $J_2 : 4\sqrt[3]{2}/5$
1165 8443 $t\pi : P_{66}$
1165 8750 $t\pi : (e+\pi)/8$
1165 8824 rt : (9,6,-8,-1)
1165 9684 rt : (4,6,-9,-3)
1166 1466 rt : (8,4,8,-1)
1166 1520 $\exp(4/5)\times s\pi(2/5)$
1166 1807 cu : $2\sqrt[3]{5}/7$
1166 2765 rt : (3,4,9,1)
1166 3288 $2^x : (1/\pi)/2$
1166 3294 $t\pi : P_{65}$
1166 4504 $\ln(2/3)\times\ln(3/4)$
1166 4856 sinh : $2\sqrt[3]{3}/3$
1166 5394 $J_2 : 7\pi/4$
1166 5557 rt : (9,6,-9,-5)
1166 6353 $J_2 : \zeta(7)$
1166 7029 rt : (4,1,-7,1)
1166 8187 $2^x : \sqrt{3}/3+2\sqrt{5}$
1166 8848 rt : (1,8,5,-3)
1166 9122 $t\pi : 7e\pi/6$
1166 9665 $\ell_{10} : 4\sqrt{2}/3-\sqrt{3}/3$
1167 0007 ln : $4\sqrt{3}/3+\sqrt{5}/3$
1167 1149 rt : (6,3,-9,1)
1167 3942 $s\pi : e/4+2\pi$

1167 5047 $2^x : 9\zeta(3)/10$
1167 5065 as : P_{77}
1167 5221 sinh : $((e+\pi)/2)^{-2}$
1167 5365 as : $((e+\pi)/2)^{-2}$
1167 6736 at : $3/4+3\sqrt{3}/4$
1167 8667 rt : (7,4,8,-1)
1168 2235 rt : (2,4,9,1)
1168 3459 rt : (6,9,-5,6)
1168 3476 $J_1 : 4/17$
1168 4941 $Ei : (\sqrt[3]{5}/2)^{-1/2}$
1168 6335 rt : (2,5,-8,-1)
1168 6713 sinh : $4\zeta(3)/5$
1168 8323 $t\pi : (3)^{-3}$
1168 9584 rt : (2,4,-6,3)
1169 0991 rt : (7,3,-9,1)
1169 2263 $J_2 : 7\sqrt[3]{3}/10$
1169 3080 at : $\sqrt{2}-2\sqrt{3}$
1169 3740 sr : $3/2+4\sqrt{5}/3$
1169 4528 $\ell_{10} : 3-\sqrt{5}$
1169 6017 rt : (6,4,8,-1)
1169 6070 sq : $\sqrt[3]{5}/5$
1169 7275 rt : (8,7,5,8)
1169 7407 $\zeta : 8\pi/7$
1169 8922 sr : $9\ln 2/5$
1169 9414 rt : (5,8,-5,1)
1170 0001 $e^x : 3/4$
1170 0175 $\Psi : 7\sqrt{3}/9$
1170 1904 rt : (1,4,9,1)
1170 2714 $2^x : 3/2+3e$
1170 3296 cr : $4/3+3e$
1170 3461 $J_1 : \sqrt{2}/6$
1170 3780 sr : $2e-4\pi/3$
1170 3977 rt : (3,5,-8,-1)
1170 4488 $J_2 : 5\sqrt{2}/7$
1170 5583 rt : (7,4,2,7)
1170 6381 cr : $10e\pi/9$
1170 6897 $10^x : (3\pi)^{-1/2}$
1170 8679 cr : $1/3+3\sqrt{2}/4$
1170 9881 cu : $4\zeta(3)/3$
1171 0726 tanh : $2/17$
1171 0874 at : $2/17$
1171 1041 rt : (8,3,-9,1)
1171 2491 $t\pi : \ln(\sqrt[3]{3}/3)$
1171 2567 rt : (9,5,9,1)
1171 3519 rt : (5,4,8,-1)
1171 3924 rt : (6,1,-1,6)
1171 6166 $J_1 : 5e\pi/4$
1171 6406 $\lambda : \exp(-17\pi/11)$
1171 7617 rt : (1,3,-8,-6)
1171 7778 $J_1 : 22$
1172 1768 $\zeta : (\ln 3/2)^{-1/3}$
1172 1778 rt : (4,9,1,0)
1172 2300 rt : (5,2,-4,-5)
1172 2491 cu : $4e/3-\pi/3$
1172 4954 $t\pi : 8\sqrt{5}/3$
1172 6258 at : $2/3-e$
1172 6501 rt : (9,2,-9,0)
1172 7408 $t\pi : 1/\ln(\sqrt{5}/2)$
1172 8713 $2^x : 4/25$
1172 8748 sinh : $7\sqrt[3]{2}/2$
1173 0711 rt : (4,7,-5,-4)
1173 1175 rt : (4,4,8,-1)
1173 1306 rt : (9,3,-9,1)
1173 1548 $J_2 : 11/2$
1173 2264 rt : (8,5,9,1)
1173 3091 at : $6\sqrt[3]{5}/5$
1173 4079 $10^x : 4/3+3\pi/2$
1173 6052 $2/3-\ln(\sqrt{3})$
1173 7395 $J_1 : \exp(-\sqrt[3]{3})$
1173 7894 $10^x : 4\sqrt[3]{2}$
1173 8892 $10^x : \sqrt{2}/4+\sqrt{3}/4$
1173 9739 rt : (5,5,-8,-1)
1174 0414 $Ei : 3\sqrt[3]{3}/4$
1174 1740 $\ln(3) \div \exp(\sqrt{5})$
1174 1828 $\zeta : 5/8$
1174 1859 rt : (1,4,-9,1)
1174 3938 $\ln(4/5) \times \exp(5)$
1174 5098 $e^x : -\sqrt{3}+\sqrt{5}-\sqrt{7}$
1174 5577 $2^x : P_{20}$
1174 6760 $e^x : 1-\pi$
1174 7356 sr : $4\sqrt{2}/3+3\sqrt{3}/2$
1174 8072 $2^x : \sqrt[3]{3}/9$
1174 8072 rt : (5,9,-1,5)
1174 8988 rt : (3,5,9,8)
1174 9010 $\ell_{10} : 1/4+3\sqrt{2}/4$
1174 9037 at : $1-\sqrt{5}/2$
1174 9603 $J_2 : \sqrt{23}$
1174 9650 ln : $4-2\sqrt{2}/3$
1175 0063 rt : (9,2,-9,-5)
1175 1906 $e^x : 1/9$
1175 2164 rt : (7,5,9,1)
1175 3166 rt : (1,8,6,-2)
1175 3842 rt : (4,9,-7,-9)
1175 4311 $t\pi : e/4+2\pi$
1175 4642 rt : (7,9,4,3)
1175 5705 $1/5 \times s\pi(1/5)$
1175 6325 $\Psi : 1/\ln(\sqrt{3}/3)$
1175 7865 rt : (6,5,-8,-1)
1175 8855 $Ei : 25/2$
1175 9076 ln : P_{43}
1175 9143 $J_2 : 9\pi^2/8$
1176 1321 $\ln(5)-\exp(2/5)$
1176 1413 rt : (5,5,-9,6)
1176 2153 rt : (2,4,-9,1)
1176 3938 at : $4e/3-\pi/2$
1176 4705 id : $2/17$
1176 5272 sq : $2-2\sqrt{2}/3$
1176 6961 rt : (2,4,8,-1)
1176 7100 $\theta_3 : 1/17$
1176 8002 $\Gamma : \sqrt[3]{3}/5$
1176 8496 ln : $\sqrt{7}\pi$
1176 8969 rt : (8,8,7,9)
1176 9145 id : $9\sqrt{3}/5$
1176 9279 cr : $4\pi/9$
1177 0767 rt : (6,6,1,2)
1177 0783 $\ln(3) \times \exp(2)$
1177 2270 rt : (6,5,9,1)
1177 4208 $t\pi : (\ln 2/3)^{-2}$
1177 4921 $2^x : 5\sqrt{3}/8$
1177 6158 rt : (7,5,-8,-1)
1177 7414 $erf : (\sqrt{2}/3)^3$
1177 7420 sinh : $5\sqrt{3}/9$
1177 7522 rt : (7,5,4,8)
1177 7736 $10^x : 3e/4+\pi$
1177 8303 ln : $8/9$
1178 1637 $\Psi : 3\pi/7$
1178 2613 $K : \exp(-1/(2\pi))$
1178 2665 rt : (3,4,-9,1)
1178 4400 $s\pi : 2\sqrt{2}-\sqrt{3}/2$
1178 5097 rt : (1,4,8,-1)
1178 5145 $Ei : 9\zeta(3)/10$
1178 6111 rt : (6,2,1,7)
1178 7019 rt : (5,3,-2,1)
1178 8375 $s\pi : 1/4+3\pi/2$
1178 8576 rt : (9,7,-7,-4)
1178 9838 $J_0 : 5\pi^2/4$
1179 1179 id : $\sqrt[3]{19/2}$
1179 1484 $Li_2 : (e/3)^2$
1179 1863 sinh : $2/17$
1179 2015 as : $2/17$
1179 2373 at : $3\sqrt{3}/4-3\sqrt{5}/2$
1179 2587 rt : (5,5,9,1)
1179 2970 $s\pi : (\sqrt{2}\pi/2)^3$
1179 3842 rt : (9,3,8,-1)
1179 4227 rt : (2,8,0,4)
1179 4621 rt : (8,5,-8,-1)
1179 4736 rt : (5,1,-2,-6)
1179 4747 $2/3 \div \exp(\sqrt{3})$
1179 5192 rt : (9,3,-7,1)
1179 5435 sq : $2\zeta(3)/7$
1179 7725 rt : (9,1,-7,-6)
1180 1550 ln : $4\sqrt{2}-3\sqrt{3}/2$
1180 3398 id : $\sqrt{5}/2$
1180 4987 sr : $((e+\pi)/3)^{1/3}$
1180 5002 $J_0 : \sqrt{7}$
1180 5594 $\Psi : (2e/3)^{1/2}$
1180 8597 cr : $(\ln 3/3)^{-1/3}$
1180 9163 $1/4 \div \exp(3/4)$
1180 9477 $\ell_2 : 4\sqrt{2}-3\sqrt{3}$
1180 9931 $\ell_{10} : 16/21$
1181 1984 rt : (8,3,8,-1)
1181 2097 rt : (3,7,-8,-4)
1181 3120 rt : (4,5,9,1)
1181 3258 rt : (9,5,-8,-1)
1181 4182 sq : $4/3-3\sqrt{5}/4$
1181 6258 cu : $1-2\sqrt{5}/3$
1181 6477 $\exp(e)-s\pi(\sqrt{2})$
1181 7189 $\ell_2 : P_{78}$
1181 7643 $J_2 : (\pi/3)^{1/3}$
1181 7705 ln : $3\sqrt{2}-3\sqrt{5}/2$
1181 7824 at : $2/3-\pi/4$
1181 9908 rt : (3,6,-5,-9)
1182 0601 $J_1 : 5/21$
1182 0885 $10^x : 8\sqrt{3}/3$
1182 1869 rt : (1,7,3,-4)
1182 1917 ln : $3/4+4\sqrt{3}/3$
1182 2931 $\ell_2 : (3\pi/2)^{1/2}$
1182 3885 as : $\exp(-1/(3\pi))$
1182 4359 rt : (5,4,-9,1)
1182 4543 at : $\exp(\sqrt[3]{3}/2)$
1182 8182 $2/5+e$
1182 8451 $\Psi : 9\sqrt{2}/8$
1182 8850 $J_1 : 7\pi$
1182 8917 $\zeta : 8\sqrt{5}/5$
1182 9026 $10^x : 3/4-3\sqrt{5}/4$
1182 9789 $\Gamma : 9\sqrt{3}/7$
1183 0290 rt : (7,3,8,-1)
1183 0979 as : $1-\sqrt{5}/2$
1183 1591 $s\pi : 5\sqrt{3}/9$
1183 3063 at : $3-2\sqrt{2}/3$
1183 3873 rt : (3,5,9,1)
1183 3951 $erf : 2/19$
1183 4516 $\sqrt{3}+\ln(4)$
1183 6087 sinh : $(\sqrt{5})^{1/2}$
1184 0456 $J_2 : \sqrt{15}\pi$
1184 0511 rt : (2,4,-8,-1)
1184 0917 rt : (1,7,0,6)
1184 5551 rt : (6,4,-9,1)
1184 6119 rt : (9,0,-5,7)
1184 6865 $2^x : 1-3e/2$
1184 8755 sr : $10\pi/7$
1184 8764 rt : (6,3,8,-1)
1185 1157 rt : (7,6,6,9)
1185 1401 cu : $\sqrt{2}/2-3\sqrt{5}$
1185 1475 $2^x : 3-e/2$
1185 1784 $J_1 : 3(1/\pi)/4$
1185 1904 rt : (8,3,-8,-6)
1185 3762 rt : (4,5,-5,5)
1185 4851 rt : (2,5,9,1)
1185 5091 ln : $\zeta(3)/10$
1185 8256 id : $\sqrt{2}/4-2\sqrt{5}$
1185 8819 rt : (2,6,-8,5)
1185 9329 rt : (3,4,-8,-1)
1185 9612 sr : $3e+\pi/2$
1186 0006 rt : (6,3,3,8)
1186 1521 $e^x : 3/4+3\pi$
1186 3063 ln : $2+3\sqrt{2}/4$
1186 4107 $e^x : -\sqrt{2}+\sqrt{3}-\sqrt{6}$
1186 5388 rt : (9,4,-5,2)
1186 5876 sinh : $10e/3$
1186 6557 cr : $10\sqrt[3]{2}/9$
1186 6980 rt : (7,4,-9,1)
1186 7088 $t\pi : 2\sqrt{2}-\sqrt{3}/2$
1186 7407 rt : (5,3,8,-1)
1186 7939 at : $3/2+\sqrt{5}/4$
1186 8157 $2^x : 3\sqrt{2}/2-3\sqrt{3}$
1186 8742 $\ell_2 : 3\sqrt{3}/2-3\sqrt{5}/4$
1186 8894 cr : $7/5$
1187 1147 $t\pi : 1/4+3\pi/2$
1187 2336 $10^x : 17/10$
1187 2777 sq : $\sqrt{3}/4-2\sqrt{5}/3$
1187 3149 id : $1/3+\pi/4$
1187 3374 $s\pi : \sqrt{2}/3+2\sqrt{5}/3$
1187 3826 rt : (8,1,-8,1)
1187 4602 $e^x : 4/3-2\sqrt{3}$
1187 5268 at : $1/4-4\sqrt{3}/3$
1187 5658 $s\pi : (\sqrt{5}/2)^{1/3}$
1187 5840 $t\pi : (\sqrt{2}\pi/2)^3$
1187 6059 rt : (1,5,9,1)
1187 6574 $10^x : \sqrt{5/2}$

1187 6798 $\Psi : 5(1/\pi)$
1187 7532 $2^x : 1/3 + \pi$
1187 7549 $J_2 : 9\pi^2/10$
1187 7820 rt : (4,9,3,4)
1187 8325 rt : (4,4,-8,-1)
1188 0073 $\Gamma : (5/3)^{-1/3}$
1188 0249 tanh : $3(1/\pi)/8$
1188 0408 at : $3(1/\pi)/8$
1188 2300 $1/3 \times \exp(\sqrt{5})$
1188 4909 rt : (9,6,9,1)
1188 6225 rt : (4,3,8,-1)
1188 6431 $10^x : 7/5$
1188 6785 rt : (3,6,-6,-5)
1188 6852 $3/4 \times \exp(2/5)$
1188 8653 rt : (8,4,-9,1)
1188 9019 ln : $3e/4 + 2\pi$
1189 2323 $\zeta : (\sqrt[3]{4}/3)^{-2}$
1189 2618 $2^x : 13/12$
1189 3491 at : $10\sqrt[3]{3}/7$
1189 4673 $\zeta : 25/7$
1189 5064 cr : $7e/2$
1189 5266 rt : (9,1,-3,8)
1189 5790 rt : (2,9,-9,-4)
1189 7503 rt : (5,9,1,0)
1189 7873 $s\pi(1/3) \div s\pi(e)$
1189 8695 $\pi \times \exp(2/3)$
1189 9279 at : $1 + 3\sqrt{2}/4$
1189 9782 $J_2 : 3e/8$
1190 0492 $\ln(\sqrt{3}) + s\pi(\pi)$
1190 0580 $t\pi : 9\sqrt[3]{2}/5$
1190 1166 rt : (8,2,-6,-7)
1190 1224 as : $2/3 - \pi/4$
1190 2667 rt : (1,5,-1,-2)
1190 2765 at : $2\sqrt{2}/3 + \sqrt{5}/2$
1190 4434 $Li_2 : \ln 2/6$
1190 4958 rt : (4,6,-1,9)
1190 5219 rt : (3,3,8,-1)
1190 5277 sq : $4\zeta(3)$
1190 6126 rt : (8,6,9,1)
1190 6796 $\sqrt{5} - \exp(3/4)$
1190 7083 rt : (7,5,-9,-6)
1190 7886 $\Gamma : 7(1/\pi)$
1191 0573 rt : (9,4,-9,1)
1191 1113 rt : (3,9,-4,-9)
1191 1322 rt : (6,7,3,3)
1191 3000 sr : $3 + 2\sqrt{5}/3$
1191 3566 cu : $\sqrt{2}/2 + \sqrt{3}/3$
1191 3807 $J_1 : 6/25$
1191 4062 cu : $9\sqrt[3]{5}/8$
1191 5284 $t\pi : 5\sqrt{3}/9$
1191 6001 cu : $\sqrt{3}/4 + 2\sqrt{5}/3$
1191 6002 $\ln(2/5) \times \exp(1/5)$
1191 6231 id : $9 \ln 2/2$
1191 6268 rt : (9,8,-5,-3)
1191 6866 rt : (6,4,-8,-1)
1191 7432 $e^x : \sqrt{8/3}$
1191 8332 $\exp(3) \div s\pi(2/5)$
1191 8640 $\ell_{10} : 19/25$
1191 9102 $\theta_3 : (1/(3e))^{1/3}$
1191 9187 rt : (1,5,-9,1)
1191 9251 $\ell_{10} : 4\sqrt[3]{5}/9$
1191 9632 $\Psi : 7\sqrt[3]{2}$
1192 0111 cr : $4 - 3\sqrt{3}/2$
1192 3171 sq : $3/2 - 2\sqrt{3}/3$
1192 3348 $e^x : \sqrt{2}/3 - 3\sqrt{3}/2$
1192 3848 $\ell_{10} : 4\pi^2/3$
1192 4394 rt : (2,3,8,8)
1192 4475 sinh : $(2e/3)^3$
1192 7577 rt : (7,6,9,1)
1192 7697 $erf : (1/\pi)/3$
1192 8031 $\ell_{10} : (\sqrt{3})^{1/2}$
1192 8382 at : $1/4 + 2e/3$
1192 8505 rt : (5,4,0,2)
1193 1831 rt : (8,5,-8,-4)
1193 2020 ln : $(\ln 2/2)^2$
1193 2149 $J_1 : \sqrt[3]{3}/6$
1193 2152 $s\pi(1/4) - s\pi(1/5)$
1193 2454 sr : $1/\ln(\sqrt{2}\pi/2)$
1193 2775 cr : $7\zeta(3)/6$
1193 3932 $J_1 : \zeta(3)/5$
1193 5686 rt : (6,4,5,9)
1193 6418 rt : (7,4,-8,-1)
1193 6620 id : $3(1/\pi)/8$
1193 7149 rt : (9,5,-3,3)
1193 7957 $2^x : 2\sqrt{3}/3 + 3\sqrt{5}/4$
1194 1116 rt : (2,5,-9,1)
1194 2729 tanh : $3/25$
1194 2882 $Li_2 : (\sqrt{2}\pi/3)^{-1/2}$
1194 2892 at : $3/25$
1194 3752 rt : (1,3,8,-1)
1194 4849 rt : (5,1,2,8)
1194 5187 rt : (9,2,-1,9)
1194 5829 rt : (8,2,-6,2)
1194 6218 $\theta_3 : 2\sqrt{5}/9$
1194 6296 $J : (\pi)^{-3}$
1194 8545 $t\pi : 10(e + \pi)$
1194 9265 rt : (6,6,9,1)
1194 9520 $J_2 : \sqrt{8\pi}$
1195 0754 $e^x : 3/2 - 4e/3$
1195 0914 rt : (9,2,8,-1)
1195 1207 rt : (8,1,-4,-8)
1195 1471 sr : $3e/4 - \pi/4$
1195 1513 sr : $(\pi/2)^{1/2}$
1195 1737 cr : $\sqrt{3}/3 + 4\sqrt{5}$
1195 3893 tanh : $(\ln 2/2)^2$
1195 4053 rt : (4,2,-1,-7)
1195 4057 at : $(\ln 2/2)^2$
1195 4545 rt : (7,1,-9,-1)
1195 6163 rt : (8,4,-8,-1)
1195 7243 rt : (7,4,-7,-7)
1195 7963 $t\pi : \sqrt{2}/3 + 2\sqrt{5}/3$
1195 8502 sr : $1/4 + 3\sqrt{2}$
1196 0296 $t\pi : (\sqrt{5}/2)^{1/3}$
1196 0478 $\lambda : \exp(-20\pi/13)$
1196 0936 $J_0 : 7\pi$
1196 1298 $\zeta : (\ln 3/3)^{-1/3}$
1196 3005 tanh : $\zeta(3)/10$
1196 3169 at : $\zeta(3)/10$
1196 3298 rt : (3,8,-1,0)
1196 3787 sinh : $(\sqrt{5}/2)^{-1/3}$
1196 4015 as : $5\sqrt[3]{2}/7$
1196 4987 sinh : $3(1/\pi)/8$
1196 5150 as : $3(1/\pi)/8$
1196 5201 $t\pi : \sqrt{3}$
1197 0263 rt : (8,2,8,-1)
1197 0729 $Ei : 5\sqrt{3}/8$
1197 1197 rt : (5,6,9,1)
1197 2135 $10^x : (\ln 3/3)^3$
1197 2330 $e^x : 5\zeta(3)/8$
1197 2584 rt : (2,8,-7,-5)
1197 6105 rt : (9,4,-8,-1)
1197 6951 as : $9/10$
1197 7332 $\Psi : 14/19$
1198 0913 rt : (2,5,-9,1)
1198 3010 $\Psi : \exp(\pi)$
1198 3034 rt : (1,6,1,-5)
1198 4299 $2^x : \exp(-2e/3)$
1198 4390 $J_2 : (\pi/3)^{1/2}$
1198 5736 rt : (4,5,-9,1)
1198 6525 $Ei : 2\sqrt{2}/7$
1198 6689 $\Psi : (5/2)^{-1/3}$
1198 7325 cu : $\exp(-1/\sqrt{2})$
1198 7620 $10^x : 3\sqrt{3}/2 + \sqrt{5}/3$
1198 8014 sq : $1/3 - e/4$
1198 8633 $10^x : 4/3 + 3\sqrt{5}$
1198 9004 $t\pi : 3\sqrt{2}/4 + 3\sqrt{3}/4$
1198 9156 at : $1/3 + \sqrt{3}$
1198 9737 $10^x : 3\sqrt{3}/4 + \sqrt{5}/4$
1198 9797 rt : (7,2,8,-1)
1199 2982 id : $8\sqrt[3]{2}/9$
1199 3379 rt : (4,6,9,1)
1199 3955 cr : $1/4 + 2\sqrt{3}/3$
1199 7831 sq : $4/3 + \sqrt{3}/4$
1200 0000 id : $3/25$
1200 0821 $Ei : \exp(-e/3)$
1200 2049 rt : (8,0,-2,9)
1200 2140 $2^x : 1/\ln(\sqrt[3]{3}/2)$
1200 2264 rt : (6,9,-6,4)
1200 2965 rt : (2,9,4,-4)
1200 3407 rt : (3,8,-2,7)
1200 4958 $J_1 : (\ln 3/3)^{-3}$
1200 6905 $Li_2 : ((e + \pi)/2)^{-2}$
1200 7083 $\ell_{10} : 9(e + \pi)/4$
1200 8081 $2^x : \sqrt{2} - 2\sqrt{5}$
1200 8209 rt : (7,3,-5,-8)
1200 8439 rt : (5,5,-9,1)
1200 8878 at : $\sqrt{3}/2 - \sqrt{5}/3$
1200 9429 $\Gamma : 16/19$
1200 9519 rt : (6,2,8,-1)
1201 0539 rt : (9,6,-1,4)
1201 0839 $10^x : 24/25$
1201 0892 sr : $2 - \sqrt{5}/3$
1201 1226 $e^x : P_{37}$
1201 1325 id : $(\ln 2/2)^2$
1201 1390 $J_2 : 3\pi^2$
1201 1477 at : $e/3 - \pi/4$
1201 2188 $e^x : \sqrt{3}/3 + 4\sqrt{5}/3$
1201 3898 $\Psi : (e/3)^{-3}$
1201 4119 $erf : \exp(-\sqrt{5})$
1201 4312 $\ell_{10} : 3\sqrt{2} + 4\sqrt{5}$
1201 4388 rt : (6,6,-8,-7)
1201 5623 $\lambda : (5)^{-3}$
1201 5815 rt : (3,6,9,1)
1201 6372 rt : (3,2,-5,9)
1201 7289 ln : $4/3 + \sqrt{3}$
1201 7986 $2^x : 3\pi/4$
1201 8746 sq : $1/\sqrt[3]{24}$
1201 9052 cu : $1/2 + \sqrt{3}$
1201 9057 $s\pi(1/3) + s\pi(\sqrt{3})$
1201 9473 rt : (8,3,-4,3)
1202 0278 $s\pi : 4\zeta(3)/5$
1202 0569 id : $\zeta(3)/10$
1202 1985 cu : $\sqrt{23\pi}$
1202 2683 rt : (5,2,4,9)
1202 3094 rt : (3,3,-8,-1)
1202 3804 $\exp(3/5) \div \exp(e)$
1202 5649 rt : (1,8,5,-9)
1202 6353 ln : $3/25$
1202 6588 $\exp(5) + s\pi(1/4)$
1202 8445 rt : (7,0,-7,2)
1202 8820 sinh : $3/25$
1202 8988 as : $3/25$
1202 9423 $\ell_2 : 23/25$
1202 9435 rt : (5,2,8,-1)
1203 0974 cr : $1/2 + e/3$
1203 1285 sq : $1/2 + 4\sqrt{5}/3$
1203 1411 rt : (6,5,-9,1)
1203 1924 $s\pi : \sqrt{2}/3 - \sqrt{3}/4$
1203 2179 rt : (4,1,1,-8)
1203 5082 sinh : $3(e + \pi)/7$
1203 5502 $1/4 \div \ln(4/5)$
1203 6735 sq : $\sqrt[3]{19}$
1203 7832 $J_1 : 1/\sqrt{17}$
1203 8512 rt : (2,6,9,1)
1203 9945 sr : $3\sqrt{2}/2 - \sqrt{3}/2$
1203 9961 at : $\sqrt{3}/3 + 2\sqrt{5}/3$
1204 0103 $\sqrt{5} - \exp(\sqrt{5})$
1204 0227 sinh : $(\ln 2/2)^2$
1204 0396 as : $(\ln 2/2)^2$
1204 1155 $J_2 : 3\sqrt[3]{5}/5$
1204 1718 rt : (3,5,-1,4)
1204 2352 sinh : $7\sqrt[3]{5}/8$
1204 3071 $2^x : 4 + \sqrt{2}/2$
1204 3425 rt : (4,3,-8,-1)
1204 5983 $\zeta : ((e + \pi)/3)^{1/2}$
1204 8958 rt : (9,9,-3,-2)
1204 9383 $J_0 : 24/11$
1204 9538 sinh : $\zeta(3)/10$
1204 9546 rt : (4,2,8,-1)
1204 9707 as : $\zeta(3)/10$
1205 0565 as : $4/3 - \sqrt{3}/4$
1205 1302 rt : (2,7,-5,-6)
1205 1627 $\ln(\sqrt{2}) + s\pi(e)$
1205 2739 id : $\exp(2\pi/3)$
1205 4660 rt : (7,5,-9,1)
1205 7621 $\theta_3 : -\sqrt{2} - \sqrt{3} + \sqrt{7}$
1205 8264 rt : (6,8,5,4)
1205 8521 $e^x : \sqrt{3}/2 - 4\sqrt{5}/3$
1206 0004 rt : (7,2,-3,-9)
1206 1475 rt : (1,6,9,1)

1206 3045 cr : $2\sqrt{2}+3\sqrt{5}$
1206 3964 rt : (5,3,-8,-1)
1206 4987 10^x : $1/2+e/4$
1206 5147 J_0 : 22
1206 5322 rt : (8,6,-6,-3)
1206 5332 cu : $2+\sqrt{3}/3$
1206 5752 $s\pi$: $2\sqrt[3]{3}/3$
1206 6247 J_1 : $\exp(-\sqrt{2})$
1206 6311 rt : (6,5,-6,-8)
1206 6941 id : $\sqrt{3}/2-\sqrt{5}/3$
1206 7522 ln : $4e/3+3\pi/2$
1206 8100 rt : (9,7,9,1)
1206 8425 $\sqrt{5}\times\exp(1/3)$
1206 8783 cu : $1-3e/4$
1206 9577 id : $e/3-\pi/4$
1206 9858 rt : (3,2,8,-1)
1207 1327 rt : (1,7,-1,-1)
1207 1523 ℓ_{10} : $2-e/4$
1207 2638 rt : (5,8,-9,-7)
1207 3032 at : $2-3\sqrt{2}/2$
1207 4322 e^x : $2/3+\sqrt{2}/3$
1207 4490 Γ : $7\zeta(3)/10$
1207 4515 $t\pi$: $\sqrt{2}/2-\sqrt{5}/3$
1207 6467 rt : (5,5,2,3)
1207 6694 ℓ_2 : $2e/5$
1207 7697 sinh : $9\pi/8$
1207 8193 rt : (8,5,-9,1)
1207 9153 J_2 : 8π
1208 1810 rt : (7,3,-9,-4)
1208 3258 ζ : $5\sqrt{5}/8$
1208 4713 rt : (6,9,1,0)
1208 5627 rt : (9,7,1,5)
1208 8531 cu : $2e+2\pi/3$
1208 8683 sq : $2+3e$
1209 0373 rt : (2,2,8,-1)
1209 0916 J_1 : $\sqrt{3}+\sqrt{5}+\sqrt{7}$
1209 1048 rt : (8,7,9,1)
1209 1718 at : $1/2+\pi/2$
1209 2990 K : $e\pi/10$
1209 3876 10^x : $\sqrt{3}/4+3\sqrt{5}/2$
1209 4228 e^x : $10\sqrt[3]{3}$
1209 4828 rt : (8,4,-2,4)
1209 5586 id : $4\sqrt{2}+2\sqrt{3}$
1209 5978 J_2 : $(\ln 3/3)^{-3}$
1209 6419 as : $\sqrt{3}/2-\sqrt{5}/3$
1209 8298 sq : $8/23$
1209 9075 as : $e/3-\pi/4$
1209 9824 sr : $2\pi/5$
1210 2015 rt : (9,5,-9,1)
1210 4068 rt : (7,1,-5,3)
1210 4379 rt : (1,9,8,7)
1210 5677 rt : (7,3,-8,-1)
1210 5968 cu : $3/4+3e$
1210 8069 $t\pi$: $4\zeta(3)/5$
1210 8230 rt : (1,6,-9,1)
1211 0690 $\sqrt{3}+\exp(2)$
1211 0726 sq : $18/17$
1211 0925 cu : $3\sqrt{3}/4+\sqrt{5}/2$
1211 1097 rt : (1,2,8,-1)
1211 1764 $t\pi$: $\exp(-\sqrt{3}/3)$
1211 2287 rt : (4,0,3,9)
1211 3349 rt : (6,2,-8,-2)
1211 4266 rt : (7,7,9,1)
1211 6380 rt : (9,1,8,-1)
1211 8024 e^x : $2-4\sqrt{2}/3$
1211 9094 rt : (6,4,-4,-9)
1211 9305 cu : $2\sqrt{3}/7$
1211 9522 sr : $8\sqrt{2}/9$
1211 9972 $t\pi$: $\sqrt{2}/3-\sqrt{3}/4$
1212 0170 sq : $\sqrt{2}/3+3\sqrt{3}$
1212 2137 rt : (3,6,3,8)
1212 2798 J_0 : $7(e+\pi)/5$
1212 3232 J_1 : $\sqrt[3]{5}/7$
1212 3528 10^x : $\sqrt{3}/4+2\sqrt{5}$
1212 5554 rt : (5,8,-6,-1)
1212 6839 J_2 : $\exp(5/2)$
1212 6860 rt : (8,3,-8,-1)
1212 7080 rt : (5,6,-6,8)
1212 7727 Γ : $(\sqrt{2})^{-1/2}$
1212 8399 e^x : $11/3$
1212 9405 rt : (1,8,0,8)
1212 9653 $t\pi$: $4e/3-3\pi/4$
1213 0114 Li_2 : $2/17$
1213 0613 $1/5\div\exp(1/2)$
1213 0852 J_2 : $5\sqrt[3]{3}/7$
1213 1579 $s\pi$: $3e/4$
1213 2034 id : $3\sqrt{2}/2$
1213 3118 cr : $\pi^2/7$
1213 4198 J_2 : $6\zeta(3)/7$
1213 5215 ℓ_{10} : $\sqrt{3}+\sqrt{5}-\sqrt{7}$
1213 5295 ln : $\sqrt{2}+4\sqrt{3}$
1213 6466 cu : $5\ln 2/7$
1213 7074 rt : (8,1,8,-1)
1213 7760 rt : (6,7,9,1)
1213 8669 e^x : $10(1/\pi)$
1214 1978 rt : (1,9,-6,-6)
1214 6340 ℓ_{10} : $\sqrt{3}/3+\sqrt{5}/3$
1214 6489 $t\pi$: $8\sqrt[3]{2}/3$
1214 6847 10^x : $(e)^{-3}$
1214 8268 rt : (9,3,-8,-1)
1214 8793 ℓ_{10} : $4/3-\sqrt{3}/3$
1214 9012 rt : (1,7,-8,9)
1214 9324 sr : $e/2+\pi$
1215 0541 ℓ_{10} : $3\sqrt[3]{2}/5$
1215 0840 ln : $\sqrt{2}/3+3\sqrt{3}/2$
1215 1699 sq : $1/2+\sqrt{5}/4$
1215 3187 rt : (1,5,-1,-6)
1215 3758 ℓ_{10} : $4e+3\pi/4$
1215 4310 ℓ_{10} : $3-3\sqrt{5}/4$
1215 4551 $t\pi$: $2\sqrt[3]{3}/3$
1215 4924 cu : $1/4-\sqrt{5}/3$
1215 5186 rt : (9,9,7,-1)
1215 5836 $1/2\div\exp(\sqrt{2})$
1215 6132 rt : (3,6,-9,1)
1215 6488 rt : (2,9,-5,7)
1215 7019 10^x : $3+3\pi/2$
1215 7978 rt : (7,1,8,-1)
1215 8483 J_2 : $4(e+\pi)$
1215 9317 E : $15/16$
1215 9944 rt : (4,9,-8,9)
1216 0215 Ei : $9\sqrt{5}/4$
1216 1538 rt : (5,7,9,1)
1216 1994 as : $2-3\sqrt{2}/2$
1216 2486 rt : (9,8,3,6)
1216 3439 $\exp(3)+s\pi(\sqrt{2})$
1216 3518 sinh : $4\sqrt{2}$
1216 4790 rt : (1,9,-1,1)
1216 5953 ℓ_{10} : $2/3+4\pi$
1216 6564 10^x : $\sqrt{11}$
1216 6590 J_2 : $(\ln 3)^{1/3}$
1216 7108 ln : $(3\pi)^{1/2}$
1216 7761 J_2 : $7\pi^2/4$
1216 8606 id : $4\sqrt{2}/3+\sqrt{5}$
1216 9454 ln : P_{54}
1216 9741 sq : $9\sqrt[3]{3}/2$
1217 0021 e^x : $1/4-3\pi/4$
1217 0340 $\ln(2)\times\exp(2)$
1217 1966 rt : (8,5,0,5)
1217 4002 rt : (8,9,7,-1)
1217 4562 rt : (2,8,2,-5)
1217 4606 2^x : $(\ln 3/2)^3$
1217 4966 id : $7\sqrt[3]{3}/9$
1217 9097 rt : (6,1,8,-1)
1217 9157 2^x : $4+4\sqrt{2}/3$
1217 9360 rt : (5,9,-2,3)
1218 0532 rt : (4,6,-9,1)
1218 1378 cr : $24/17$
1218 1444 e^x : $7e\pi/3$
1218 1488 rt : (7,2,-3,4)
1218 2791 sr : $4\sqrt{2}-2\sqrt{3}/3$
1218 3903 10^x : $1/2-\sqrt{2}$
1218 4576 Ψ : $1/\ln(1/\sqrt{2})$
1218 4696 sq : $\pi/9$
1218 5606 rt : (4,7,9,1)
1218 5979 rt : (4,9,-8,-8)
1218 8104 rt : (2,5,-1,9)
1218 8210 cr : $2/3+\sqrt{5}/3$
1218 9332 Ei : $13/12$
1218 9762 erf : $\exp(-\sqrt{2}\pi/2)$
1219 1053 rt : (6,1,-6,-3)
1219 1100 $t\pi$: $2\sqrt{2}/3-4\sqrt{5}/3$
1219 2985 rt : (7,9,7,-1)
1219 3691 ln : $3/2+\pi/2$
1219 6164 rt : (3,2,-8,-1)
1219 7501 sq : $3\sqrt{3}+\sqrt{5}/4$
1219 8452 cu : $19/13$
1219 8549 2^x : P_1
1220 0293 J_2 : $(e/3)^{-1/3}$
1220 0434 rt : (5,1,8,-1)
1220 0661 rt : (5,4,-9,-2)
1220 1845 10^x : $1/20$
1220 4237 rt : (8,7,-4,-2)
1220 4649 rt : (3,2,2,-9)
1220 5241 rt : (5,6,-9,1)
1220 6338 e^x : P_{78}
1220 6562 ℓ_{10} : $(\sqrt[3]{5}/3)^{1/2}$
1220 6803 ln : $(\sqrt[3]{3})^{1/3}$
1220 7029 e^x : $\sqrt{2}-3\sqrt{3}/4$
1220 7399 sq : $\sqrt{2}/4+4\sqrt{5}/3$
1220 7604 Ei : $4\sqrt[3]{2}/7$
1220 7824 10^x : $\sqrt{3}/3-2\sqrt{5}/3$
1220 8281 ℓ_{10} : $e/4+4\pi$
1220 9762 cu : $3/2-3e/2$
1220 9971 rt : (3,7,9,1)
1221 0699 2^x : $\sqrt{3}/2+2\sqrt{5}/3$
1221 1534 $\pi\times s\pi(\sqrt{5})$
1221 2138 rt : (6,9,7,5)
1221 3467 Ei : $\sqrt[3]{5}/5$
1221 4876 rt : (2,7,3,2)
1221 5336 at : $3/2+\sqrt{3}/3$
1221 5757 Γ : $21/25$
1221 6327 ζ : $4\sqrt[3]{3}/7$
1221 6699 at : $2-3e/2$
1221 7097 ln : $(\ln 2)^{1/3}$
1221 7370 rt : (9,1,-8,-5)
1221 7996 rt : (4,2,-8,-1)
1221 9341 ℓ_2 : $4+\sqrt{2}/4$
1222 0477 Ei : $18/25$
1222 0934 Γ : $2\sqrt[3]{2}/3$
1222 1606 rt : (7,4,-7,-3)
1222 1850 $t\pi$: $3e/4$
1222 1994 rt : (4,1,8,-1)
1222 5153 rt : (1,8,-4,-7)
1222 5365 e^x : $4+3\sqrt{3}$
1222 5787 ℓ_{10} : $3/2-\sqrt{5}/3$
1222 7149 sq : $5\sqrt[3]{5}/3$
1222 9148 sq : $10e/9$
1223 0267 rt : (6,6,-9,1)
1223 1464 rt : (5,9,7,-1)
1223 2988 $\exp(2)\times s\pi(\sqrt{2})$
1223 3951 erf : $(2\pi/3)^{-3}$
1223 4635 $\ln(\pi)\div\exp(\sqrt{5})$
1223 4642 rt : (2,7,9,1)
1223 5587 at : $3/4-2\sqrt{2}$
1223 6223 at : $6\sqrt{3}/5$
1223 6533 10^x : $1/4+3e/2$
1223 6892 cu : $3\sqrt{3}/5$
1223 7519 ℓ_2 : $2+3\sqrt{5}$
1223 9811 e^x : $6(e+\pi)/5$
1224 0043 $s\pi$: $3\sqrt{3}/2-\sqrt{5}/4$
1224 0064 rt : (5,2,-8,-1)
1224 0515 at : $4/3+\sqrt{5}/3$
1224 0848 rt : (4,8,-6,-9)
1224 1192 rt : (9,9,5,7)
1224 1749 id : $\sqrt{2}+3\sqrt{5}$
1224 2018 2^x : $1+3\sqrt{5}$
1224 3002 Ei : $7e\pi/6$
1224 3782 rt : (3,1,8,-1)
1224 4897 sq : $10\sqrt{3}/7$
1224 6204 id : $(\sqrt[3]{2})^{1/2}$
1224 6377 $\ln(4/5)\div\exp(3/5)$
1224 7395 J_1 : $5\sqrt{2}/2$
1224 9623 10^x : $8\sqrt{2}/5$
1225 0000 sq : $7/20$
1225 0966 rt : (8,6,2,6)
1225 2598 tanh : $\exp(-2\pi/3)$
1225 2651 10^x : $2/3+e/2$
1225 2783 at : $\exp(-2\pi/3)$
1225 3335 ℓ_{10} : $(\ln 3)^3$
1225 4648 at : $3\ln 2$

1225 5618 rt : (7,6,-9,1)
1225 7412 $\ln(4/5) \times \ln(\sqrt{3})$
1225 7797 $10^x : 1 + 4\sqrt{3}$
1225 8583 $s\pi : 3\pi^2/10$
1225 9114 $10^x : \ln(3\pi)$
1225 9247 $e^x : 1/3 + 3\sqrt{2}$
1225 9625 rt : (1,7,9,1)
1226 0784 rt : (7,3,-1,5)
1226 1115 sr : $2\sqrt{2} + 3\sqrt{5}/4$
1226 1289 $\sqrt{3} - \ln(5)$
1226 2373 rt : (6,2,-8,-1)
1226 2648 $1/3 \div e$
1226 3243 $\ln(2) \div \exp(\sqrt{3})$
1226 3508 rt : (9,8,9,1)
1226 5802 rt : (2,1,8,-1)
1226 5892 $\ell_{10} : \sqrt[3]{7/3}$
1226 6708 at : $\sqrt[3]{9}$
1226 9223 $e^x : 1/2 - 3\sqrt{3}/2$
1226 9234 cu : $2\sqrt{5}/9$
1227 0647 rt : (9,0,-6,6)
1227 2738 at : $(\ln 2/3)^{-1/2}$
1227 3265 $J_2 : ((e + \pi)/2)^3$
1227 4034 $\ell_{10} : 4\sqrt{2}/3 - \sqrt{5}/4$
1227 4235 cu : $4/3 + 2e/3$
1227 5719 $J_2 : \zeta(5)$
1227 7248 rt : (8,3,-9,-5)
1227 7350 $J_1 : \sqrt{3}/7$
1227 8786 $\zeta : 5\sqrt{2}/2$
1227 9046 $\ln(3) - \exp(1/5)$
1228 0556 rt : (5,3,-7,-3)
1228 0993 $s\pi : 9\pi/7$
1228 1304 rt : (8,6,-9,1)
1228 1512 sq : $4\sqrt{2} - \sqrt{5}/3$
1228 1655 at : $2/3 + \sqrt{2}$
1228 3672 $2^x : 23/14$
1228 4118 $J_1 : 7(1/\pi)/9$
1228 4264 $\Gamma : 3e$
1228 4929 rt : (7,9,1,0)
1228 6264 $10^x : \exp(-\sqrt{5}/2)$
1228 7952 sq : $\ln(\ln 2/2)$
1228 8061 rt : (1,2,9,9)
1228 8440 rt : (8,8,9,1)
1229 0512 rt : (2,9,7,-1)
1229 0823 at : $(\ln 2)^{-2}$
1229 1106 rt : (9,0,8,1)
1229 2336 $\zeta : 9\pi/8$
1229 2557 $2^x : P_{60}$
1229 3442 $s\pi : 3\sqrt{3}/5$
1229 4574 $2^x : 1/3 + 2\pi$
1229 5912 $10^x : 1/\ln(1/3)$
1229 6392 at : $\sqrt{13/3}$
1229 6437 $J_2 : (\sqrt{5}/2)^{1/3}$
1229 6669 rt : (4,5,-6,3)
1229 7067 $\Psi : 3\sqrt{5}/5$
1229 7125 $10^x : -\sqrt{2} - \sqrt{3} + \sqrt{5}$
1229 8334 sq : $\sqrt{3} + \sqrt{5}/2$
1229 8554 $\ell_{10} : 3/4 + \sqrt{3}/3$
1229 8978 cu : $\sqrt{2}/2 + 4\sqrt{3}$
1229 9015 $e^x : 6e\pi$
1229 9490 $J_1 : 9\pi/8$
1229 9933 rt : (2,4,1,-9)
1230 0587 as : $5\sqrt[3]{3}/8$
1230 2598 $10^x : \sqrt{5}/2$
1230 3905 sr : $3\sqrt{2} - 4\sqrt{5}/3$
1230 5241 rt : (3,7,-9,-5)
1230 6836 $\ell_2 : 2 + 3\pi/4$
1230 7334 rt : (9,6,-9,1)
1230 7577 $\ln(\sqrt{5}) \times \exp(1/3)$
1230 7737 rt : (8,2,-8,-1)
1231 0503 $2^x : P_{44}$
1231 0562 id : $\sqrt{17}$
1231 1068 cr : 17/12
1231 1971 $2^x : 5\sqrt{2}/3$
1231 2731 id : $1/4 + 4e$
1231 3011 rt : (9,8,7,-1)
1231 3099 $Li_2 : 3(1/\pi)/8$
1231 3308 rt : (8,0,8,1)
1231 3688 rt : (7,8,9,1)
1231 4471 id : $\exp(-2\pi/3)$
1231 6017 sq : $3/4 + \sqrt{2}/2$
1232 1342 $2^x : 1/3 - 3\sqrt{5}/2$
1232 1573 $Li_2 : 14/17$
1232 2724 at : $2\sqrt{2} - \sqrt{5}/3$
1232 4803 rt : (9,1,-4,7)
1232 4839 $10^x : \sqrt{3} + 4\sqrt{5}/3$
1232 5276 $J_2 : 3\sqrt{3}/5$
1232 6703 $J_1 : ((e + \pi)/3)^3$
1232 6992 $J_1 : \sqrt{5}/9$
1232 7635 at : 25/12
1232 7986 $t\pi : 3\sqrt{2}/10$
1233 0739 cr : $9\sqrt[3]{2}/8$
1233 0802 rt : (9,7,-1,0)
1233 1296 $t\pi : 4e - \pi$
1233 1543 rt : (8,2,-7,-6)
1233 1911 rt : (8,7,4,7)
1233 2169 as : $3\zeta(3)/4$
1233 2776 $t\pi : 3\sqrt{3}/2 - \sqrt{5}/4$
1233 2828 $J : 1/4$
1233 3006 rt : (8,8,7,-1)
1233 3312 rt : (1,7,8,-5)
1233 5299 $e^x : 17/12$
1233 5412 sq : $\sqrt{2}/4 + 3\sqrt{5}/4$
1233 5750 rt : (7,0,8,1)
1233 5836 $J_2 : 3\ln 2/2$
1233 6363 $\Gamma : \sqrt[3]{3}/8$
1233 6533 rt : (2,7,-9,1)
1233 7165 rt : (6,9,-7,2)
1233 7365 sr : $7\sqrt[3]{3}/8$
1233 7593 cu : $2\sqrt{2}/3 + \sqrt{5}$
1233 9260 rt : (6,8,9,1)
1233 9410 sr : $2\sqrt{3}/3 + 3\sqrt{5}/2$
1234 0531 $e^x : 3/2 + 4\sqrt{3}/3$
1234 0916 $\Gamma : 9\ln 2$
1234 0979 $4/5 \times \exp(e)$
1234 2041 rt : (7,4,1,6)
1234 3003 ln : $4\sqrt{2}/5$
1234 4839 $\ln(2) - s\pi(\pi)$
1234 5618 sinh : $\exp(-2\pi/3)$
1234 5809 as : $\exp(-2\pi/3)$
1234 6368 $\ell_2 : 3\sqrt{2} + 2\sqrt{5}$
1234 8614 $Ei : 8\sqrt[3]{2}/7$
1234 8726 rt : (3,8,-3,4)
1234 8980 rt : (8,8,-2,-1)
1235 1235 rt : (1,9,-7,-1)
1235 1741 $t\pi : 3\pi^2/10$
1235 1996 sq : $5\zeta(3)$
1235 2218 rt : (6,1,-2,5)
1235 2296 $\zeta : 5/3$
1235 2432 $2^x : (2e/3)^{-3}$
1235 3188 rt : (7,8,7,-1)
1235 3200 $t\pi : \sqrt{3}/4 - 2\sqrt{5}$
1235 3433 $J_0 : 9\sqrt{3}$
1235 3473 rt : (4,6,-2,7)
1235 4133 $e^x : P_{77}$
1235 4177 $J_1 : 10\sqrt{5}/3$
1235 4468 $e^x : ((e + \pi)/2)^{-2}$
1235 5454 rt : (1,4,-4,9)
1235 5720 $J_0 : 10e\pi/3$
1235 6052 sr : $2\sqrt{2} + 4\sqrt{3}$
1235 6315 rt : (2,7,0,-6)
1235 8436 rt : (6,0,8,1)
1236 0533 $\ell_2 : 4\sqrt{2} - 3\sqrt{3}/4$
1236 0964 $10^x : \sqrt{2}/4 + 3\sqrt{3}$
1236 1126 $\Gamma : 3\sqrt{5}/8$
1236 1699 rt : (2,9,-6,2)
1236 2444 rt : (5,2,-5,-4)
1236 2852 rt : (3,7,-9,1)
1236 3369 $\Psi : 5\sqrt{5}/7$
1236 4639 sq : $1/\sqrt[3]{23}$
1236 5164 rt : (5,8,9,1)
1236 6321 $e^x : 4\sqrt{3} + 4\sqrt{5}$
1236 7307 rt : (7,5,-5,-2)
1236 9090 $2^x : 1/4 + 2\sqrt{3}$
1236 9961 rt : (1,5,-2,-7)
1237 0120 at : $\sqrt{2}/4 + \sqrt{3}$
1237 1865 rt : (2,9,-7,-1)
1237 2642 $10^x : (\sqrt{2}\pi)^{-2}$
1237 2718 rt : (4,7,-6,-6)
1237 3560 rt : (6,8,7,-1)
1237 4667 $t\pi : 9\pi/7$
1237 4693 $e - \ln(2/3)$
1237 7777 $J_2 : 25/24$
1237 8994 $\ln(3/4) \times s\pi(\pi)$
1237 9190 $10^x : \sqrt[3]{14}$
1237 9275 $K : \sqrt[3]{5}/2$
1237 9578 rt : (3,1,-8,-1)
1237 9710 cu : $1/2 + \pi/4$
1237 9757 sr : $1/3 + 3\pi$
1237 9863 rt : (9,2,-2,8)
1238 0604 $Li_2 : 3/25$
1238 1372 rt : (5,0,8,1)
1238 3102 id : $\exp(2e/3)$
1238 4053 id : $3e\pi/5$
1238 4553 $\ell_{10} : 7\sqrt[3]{5}/9$
1238 5786 rt : (6,2,-8,-3)
1238 6747 rt : (8,1,-5,-7)
1238 7403 $t\pi : 3\sqrt{3}/5$
1238 9527 rt : (4,7,-9,1)
1239 0297 sr : 24/19
1239 1411 rt : (4,8,9,1)
1239 2668 $\zeta : \exp(\sqrt[3]{2})$
1239 2669 $Li_2 : (\ln 2/2)^2$
1239 2696 rt : (3,9,-7,-1)
1239 2698 $e^x : 2\sqrt{2} + 2\sqrt{5}/3$
1239 3654 rt : (7,4,-8,-6)
1239 4127 rt : (5,8,7,-1)
1239 4644 rt : (4,9,-9,6)
1239 5820 cu : $3\ln 2/2$
1239 6375 $\ell_2 : \sqrt{19}$
1239 7607 $\exp(e) \times s\pi(1/3)$
1239 8317 rt : (1,6,0,-9)
1239 9167 sq : $\sqrt{3}/4 + \sqrt{5}$
1240 0646 $\ln(2/5) \div \exp(2)$
1240 2517 $Li_2 : \zeta(3)/10$
1240 2597 $J_1 : 1/4$
1240 3106 rt : (4,1,-8,-1)
1240 4384 $\ell_2 : 3 + e/2$
1240 4563 rt : (4,0,8,1)
1240 5125 sr : $8e\pi/7$
1240 6233 $\Gamma : \sqrt{5}$
1240 9377 rt : (9,3,-8,0)
1240 9980 $t\pi : 1/4 - 4\sqrt{5}/3$
1241 3731 rt : (4,9,-7,-1)
1241 4891 rt : (8,8,6,8)
1241 6569 rt : (5,7,-9,1)
1241 6709 sinh : $5\pi^2/4$
1241 6867 $J_2 : 24/23$
1241 8007 rt : (3,8,9,1)
1241 8553 $e^x : 5\pi/8$
1241 9025 $\ln(3/5) \div \exp(\sqrt{2})$
1241 9495 $\ell_{10} : 5\zeta(3)/8$
1242 5348 rt : (7,5,3,7)
1242 5435 $2^x : 25/23$
1242 6596 ln : $9\sqrt[3]{5}/5$
1242 6698 $1/5 \div \ln(5)$
1242 6906 rt : (5,1,-8,-1)
1242 8015 rt : (3,0,8,1)
1242 9700 $\zeta : 4\ln 2/7$
1243 4470 $J_1 : 6\ln 2$
1243 4887 $s\pi : 4\sqrt[3]{2}$
1243 4975 rt : (5,9,-7,-1)
1243 5300 tanh : 1/8
1243 5499 at : 1/8
1243 5565 id : $7\sqrt{3}$
1243 5856 rt : (9,3,0,9)
1243 5864 $\sqrt{3} \times \exp(\sqrt{2})$
1243 7455 $J_1 : (\ln 3/3)^{-2}$
1243 7577 id : $1/2 + 4e/3$
1243 8587 $\ell_{10} : 3/4 + 4\pi$
1243 9761 $Ei : (\pi/2)^{-2}$
1244 1879 ln : $1/4 + 2\sqrt{2}$
1244 2890 rt : (8,0,-3,8)
1244 3987 rt : (6,7,-9,1)
1244 4119 $\Gamma : 9\sqrt[3]{3}$
1244 4964 rt : (2,8,9,1)
1244 6287 $s\pi : 3\ln 2/2$
1244 6415 rt : (5,1,-3,-5)
1244 6720 $J_2 : 10e\pi/7$
1244 8546 $J_2 : (2e)^2$
1244 9947 rt : (7,3,-6,-7)

1245 0813 $Li_2 : 4\sqrt[3]{3}/7$
1245 0970 $4/5 - s\pi(\sqrt{5})$
1245 0984 rt : (6,1,-8,-1)
1245 1437 $J_0 : 4e/5$
1245 1714 rt : (9,5,-9,-9)
1245 1735 rt : (2,0,8,1)
1245 3730 $10^x : 1/4 - 2\sqrt{3}/3$
1245 6431 rt : (6,9,-7,-1)
1245 6747 sq : 6/17
1245 7026 rt : (6,6,-9,-6)
1245 7907 $s\pi : ((e+\pi)/2)^{-3}$
1245 9557 $J_2 : 23/22$
1246 0669 id : $3\sqrt{2} - \sqrt{5}/2$
1246 1179 id : $9\sqrt{5}$
1246 2029 id : $3e/2 + \pi/3$
1246 2638 $10^x : 8/13$
1246 4569 sinh : 13/7
1246 7473 $s\pi : (1/\pi)/8$
1246 7690 rt : (3,7,-9,-3)
1246 8955 cr : $2 - \sqrt{3}/3$
1246 9077 at : 23/11
1247 0111 $10^x : 2/3 - \pi/2$
1247 1733 $t\pi : 2\sqrt[3]{2}/7$
1247 1794 rt : (7,7,-9,1)
1247 2278 rt : (5,6,-7,6)
1247 2292 rt : (1,8,9,1)
1247 2772 rt : (9,9,9,1)
1247 2982 $s\pi : \sqrt{5} + \sqrt{6} - \sqrt{7}$
1247 5346 rt : (7,1,-8,-1)
1247 5727 rt : (1,0,8,1)
1247 6097 rt : (9,1,8,1)
1247 7581 rt : (3,2,-6,7)
1247 7890 $2^x : 2e/5$
1247 8105 rt : (7,9,-7,-1)
1247 8406 rt : (1,8,7,-1)
1247 8612 $\ell_{10} : \sqrt{18}\pi$
1247 8700 rt : (9,7,7,-1)
1247 9166 $2^x : e/2 + 3\pi$
1248 0312 ln : $10(e+\pi)/7$
1248 4703 $e^x : 2/17$
1248 5800 cr : $e\pi/6$
1248 6141 $erf : 1/9$
1248 8541 rt : (2,8,-2,-4)
1248 9022 $2^x : 4 - 3\pi/4$
1248 9811 rt : (9,4,-6,1)
1249 3873 $\ell_{10} : 3/4$
1249 4737 $\zeta : 3(e+\pi)/5$
1249 5159 cr : $(\ln 2/2)^{-1/3}$
1249 5247 $2^x : \sqrt{2}/4 - 3\sqrt{5}/2$
1249 5673 cu : $1/2 - 4\sqrt{5}$
1249 7042 rt : (1,8,-2,-8)
1249 7248 $J_2 : \pi/3$
1249 9474 $J_1 : \sqrt[3]{2}/5$
1249 9865 $\Gamma : (e+\pi)/7$
1249 9993 $2^x : 3\sqrt{3}/4 + \sqrt{5}/3$
1250 0000 id : 1/8
1250 0843 at : P_{53}
1250 3051 $\theta_3 : 1/16$
1250 4313 sq : $1/3 - 3\sqrt{5}/2$
1250 4944 ln : $2e - 3\pi/4$
1250 6368 $J_2 : 22/21$
1250 6498 $Ei : \ln(3/2)$
1250 7002 $s\pi : 3\sqrt{3} - \sqrt{5}$
1250 7212 rt : (7,2,-4,-8)
1250 8788 sq : $10(1/\pi)/9$
1251 0231 $10^x : 4/3 - \sqrt{5}$
1251 0802 rt : (7,6,5,8)
1251 0817 $t\pi : 9/25$
1251 2056 $Ei : 1/\sqrt[3]{15}$
1251 2686 $e^x : 3/4 - 2\sqrt{2}$
1251 3319 sr : $5(e+\pi)/3$
1251 4448 rt : (6,5,-7,-7)
1251 5569 $10^x : 2\sqrt{3}/7$
1251 6314 ln : 15/17
1251 7796 $J_2 : (\ln 3)^{1/2}$
1251 9373 rt : (7,6,-3,-1)
1252 0062 $2^x : (\sqrt{2}\pi)^{1/3}$
1252 1509 rt : (7,7,7,-1)
1252 1659 rt : (6,3,2,7)
1252 1810 rt : (1,8,-7,-1)
1252 2120 rt : (9,9,-7,-1)
1252 2165 sr : $\sqrt{2}/2 + \sqrt{5}/4$
1252 3653 rt : (1,9,6,-9)
1252 4176 rt : (7,1,8,1)
1252 4558 rt : (1,6,4,-9)
1252 4802 at : $3 - e/3$
1252 4951 rt : (9,1,-8,-1)
1252 5260 $e^x : 2 - 3e/2$
1252 5362 $e^x : 4e - \pi/4$
1252 5864 sq : $2/3 - \pi$
1252 5992 cu : $1/3 + 3\sqrt{5}/4$
1252 6114 $10^x : 2\sqrt{2}/3 + \sqrt{3}/3$
1252 7479 cu : $5(e+\pi)/8$
1252 7600 rt : (7,9,9,1)
1252 8099 rt : (1,8,-9,1)
1252 8616 rt : (9,7,-9,1)
1253 0328 cr : $5\sqrt[3]{5}/6$
1253 1041 $E : 7\zeta(3)/9$
1253 1971 cr : $\sqrt{3}/2 + \sqrt{5}/4$
1253 2155 $t\pi : 4\sqrt[3]{2}$
1253 2570 rt : (5,0,-1,6)
1253 2577 sinh : 1/8
1253 2783 as : 1/8
1253 2964 tanh : $\sqrt[3]{2}/10$
1253 3171 at : $\sqrt[3]{2}/10$
1253 3323 $s\pi : 1/25$
1253 3883 at : $2\pi/3$
1253 4316 $\Gamma : 5\zeta(3)$
1253 4371 at : $\sqrt{3}/4 - \sqrt{5}/4$
1253 5024 $J_1 : 25$
1253 6443 cu : 9/7
1253 6467 sq : $8\sqrt[3]{3}$
1253 7481 $erf : \ln(\sqrt{5}/2)$
1253 7544 $10^x : 23/20$
1253 8226 cu : $3\sqrt{2}/2 + 2\sqrt{5}/3$
1253 8518 cu : $\sqrt{2} + \sqrt{3} - \sqrt{7}$
1253 8756 $J_2 : (\sqrt{3}/2)^{-1/3}$
1253 8820 sq : $3 - 3\sqrt{5}/2$
1253 8914 rt : (6,3,-6,-2)
1253 9082 $J_0 : 3\pi^2$
1254 0008 $J_1 : 9\sqrt{2}$
1254 1351 $Ei : \sqrt[3]{3}/2$
1254 3233 rt : (6,7,7,-1)
1254 3536 rt : (4,9,2,2)
1254 3825 $t\pi : 3\ln 2/2$
1254 3844 rt : (2,8,-7,-1)
1254 6286 sr : 19/15
1254 6599 $t\pi : 9\sqrt[3]{5}/10$
1254 7703 $\ln(3/4) + \exp(5)$
1254 8634 rt : (6,1,8,1)
1254 9409 rt : (2,6,-2,-7)
1255 2381 $2^x : 3/2 + 3\sqrt{3}/2$
1255 3088 $\ell_2 : 11/12$
1255 4558 rt : (3,6,-7,-4)
1255 5585 rt : (6,9,9,1)
1255 5720 $t\pi : ((e+\pi)/2)^{-3}$
1255 5913 ln : $8\pi/3$
1255 6502 $J_2 : 5\sqrt[3]{2}/6$
1255 6601 rt : (2,8,-9,1)
1255 6825 cu : $10(e+\pi)$
1255 7928 $J_2 : 21/20$
1255 8627 rt : (5,0,-9,3)
1255 9264 $\ell_{10} : (\sqrt[3]{2}/3)^{1/3}$
1256 0175 ln : $2 - \sqrt{5}/2$
1256 1524 $\exp(2/3) - \exp(3/5)$
1256 1756 $10^x : 2\sqrt{2} + 3\sqrt{3}/4$
1256 2588 ln : $7\sqrt[3]{2}/10$
1256 3435 $2^x : \sqrt{2}/3 - 2\sqrt{3}$
1256 4589 ln : $\sqrt{19/2}$
1256 5175 rt : (5,8,8,6)
1256 5481 rt : (7,1,-2,-9)
1256 5513 $t\pi : (1/\pi)/8$
1256 6103 rt : (3,8,-7,-1)
1256 6370 id : P_{53}
1256 8251 $\ell_{10} : 10\zeta(3)/9$
1256 8854 ln : $9\sqrt[3]{2}/10$
1256 9673 $J_2 : (e/3)^{-1/2}$
1256 9897 $\ln(3/4) \times \exp(2)$
1257 0180 $e^x : 9e\pi/10$
1257 1153 $t\pi : \sqrt{5} + \sqrt{6} - \sqrt{7}$
1257 2184 rt : (9,5,-4,2)
1257 2540 $Ei : 3\zeta(3)/5$
1257 2880 ln : $2 - \sqrt{3}/2$
1257 2880 rt : (6,4,-5,-8)
1257 3253 $2^x : 18/5$
1257 3379 rt : (5,1,8,1)
1257 3627 rt : (2,9,-7,-3)
1257 4560 rt : (3,9,1,-6)
1257 9092 sq : $10(1/\pi)/3$
1258 0304 rt : (5,8,-7,-3)
1258 0714 rt : (6,7,-8,9)
1258 2690 rt : (8,2,-7,1)
1258 3670 $2^x : \sqrt[3]{5}/10$
1258 3670 $e^x : (\ln 3)^{-3}$
1258 3964 rt : (5,9,9,1)
1258 4840 cr : $9e\pi/8$
1258 5518 rt : (3,8,-9,1)
1258 6302 $J_2 : \exp(\sqrt[3]{5})$
1258 6758 $\Gamma : (\sqrt[3]{5})^{-1/3}$
1258 7340 rt : (4,7,7,-1)
1258 7821 $J_1 : 4e\pi$
1258 8594 rt : (4,8,-7,-1)
1259 0931 sinh : $(e/3)^{1/3}$
1259 1633 at : $2\sqrt{2}/3 + 2\sqrt{3}/3$
1259 3236 $10^x : 10\zeta(3)/3$
1259 3301 $10^x : 8e\pi/7$
1259 4298 $t\pi : P_{42}$
1259 6707 rt : (3,6,-8,8)
1259 6949 $J_2 : 7\zeta(3)/8$
1259 8419 rt : (4,1,8,1)
1259 8507 rt : (7,7,7,9)
1259 9210 id : $\sqrt[3]{2}/10$
1259 9681 as : P_{53}
1260 0019 rt : (4,0,-8,1)
1260 0429 id : $\sqrt{3}/4 - \sqrt{5}/4$
1260 2125 at : $1/2 - 3\sqrt{3}/2$
1260 3250 sr : $3 - \sqrt{3}$
1260 3619 $Ei : (e/2)^3$
1260 4745 $\Psi : 8/5$
1260 4875 $Ei : 1/\ln(4)$
1260 5986 $t\pi : 3\sqrt{3} - \sqrt{5}$
1260 6771 rt : (2,4,-8,-3)
1260 7885 $J_1 : 8\pi^2/5$
1260 8442 as : $4/3 - \sqrt{5}$
1260 9731 rt : (6,4,4,8)
1261 1321 rt : (5,8,-7,-1)
1261 2047 at : $\ln(1/(3e))$
1261 2266 $\ell_2 : \ln(5/2)$
1261 2405 cr : $2/3 + 4\sqrt{5}$
1261 2749 rt : (4,9,9,1)
1261 3555 sr : $4e/3 - 3\pi/4$
1261 4862 rt : (4,8,-9,1)
1261 4996 $J_2 : 20/19$
1261 6609 cu : $2\sqrt{3}/3 + 3\sqrt{5}$
1261 6980 $Ei : \exp(-3\pi/2)$
1261 7332 as : $(e/2)^{-1/3}$
1261 8414 $\ln(\sqrt{3}) - s\pi(\sqrt{5})$
1262 1014 rt : (5,1,1,7)
1262 3643 cu : $3\sqrt{2}/4 - \sqrt{5}/4$
1262 3759 rt : (3,1,8,1)
1262 4788 cr : 10/7
1262 5792 rt : (5,0,-8,1)
1262 7103 $\ell_{10} : \sqrt{2}/3 + \sqrt{3}/2$
1262 7727 sq : $\ln(e+\pi)$
1262 8573 $2^x : 3 - 2\sqrt{2}$
1262 9268 cu : $1/3 + \sqrt{2}/2$
1262 9469 $J_1 : 4(1/\pi)/5$
1263 0076 $e^x : (\sqrt{2}\pi/3)^{1/3}$
1263 0167 $10^x : \exp(-1/(2\pi))$
1263 2355 rt : (6,3,-3,-9)
1263 2570 sinh : $\sqrt[3]{2}/10$
1263 2625 $s\pi : 4\sqrt{2}/3 + 2\sqrt{3}/3$
1263 2784 as : $\sqrt[3]{2}/10$
1263 2937 $t\pi : 1/25$
1263 4012 as : $\sqrt{3}/4 - \sqrt{5}/4$
1263 4046 $\zeta : 1/\sqrt[3]{16}$
1263 4289 rt : (6,8,-7,-1)
1263 5279 at : $5\sqrt[3]{2}/3$
1263 6242 $J_2 : (\sqrt[3]{5}/2)^{-1/3}$
1263 6364 $\ell_{10} : 2\sqrt{2} - 2\sqrt{5}/3$

1263 7711 at : 21/10
1263 9948 rt : (5,9,-3,1)
1264 0247 2^x : $\zeta(3)/7$
1264 1400 cr : $3-\pi/2$
1264 1953 rt : (3,9,9,1)
1264 2403 e^x : $9\sqrt[3]{2}/8$
1264 3755 rt : (3,5,-5,-5)
1264 4014 Γ : $\ln(4/3)$
1264 4647 rt : (5,8,-9,1)
1264 7568 rt : (4,9,-9,-7)
1264 8179 Ψ : 25/7
1264 9409 rt : (2,1,8,1)
1265 1035 cr : $3/4+e/4$
1265 1889 rt : (6,0,-8,1)
1265 1919 10^x : $1/3+3\sqrt{2}$
1265 2851 sr : $1/3+4\pi/3$
1265 2993 rt : (9,6,7,-1)
1265 3077 cr : $8\zeta(3)$
1265 3148 $\sqrt{2}+\ln(3/4)$
1265 3458 2^x : $3\sqrt{2}-3\sqrt{3}/2$
1265 5215 rt : (2,8,-8,-4)
1265 5402 Ψ : $(\sqrt[3]{4}/3)^{-2}$
1265 6583 rt : (9,6,-2,3)
1265 7504 rt : (7,8,-7,-1)
1266 2702 $\sqrt{5}\times s\pi(2/5)$
1266 3468 ℓ_2 : $\beta(2)$
1266 3511 rt : (9,1,-9,-4)
1266 4035 tanh : $2(1/\pi)/5$
1266 4253 at : $2(1/\pi)/5$
1266 5061 ζ : $(\ln 2)^{1/3}$
1266 5252 10^x : $1/3+4e/3$
1266 5340 J_0 : $(3\pi/2)^{1/2}$
1266 7169 id : $\sqrt{2}/3-3\sqrt{3}/2$
1266 7423 rt : (8,3,-5,2)
1266 9388 rt : (2,5,-2,6)
1267 0246 10^x : $e/3+2\pi/3$
1267 0487 $s\pi$: $1/3+\sqrt{2}/2$
1267 1587 rt : (2,9,9,1)
1267 1822 e^x : $2\sqrt[3]{5}/3$
1267 2531 rt : (9,2,8,1)
1267 3225 sinh : 17/8
1267 3386 ζ : 7/2
1267 4885 rt : (6,8,-9,1)
1267 5375 rt : (1,1,8,1)
1267 5486 Ψ : $9\pi^2/10$
1267 5742 rt : (8,6,7,-1)
1267 8247 e^x : $3(1/\pi)/8$
1267 8317 rt : (7,7,-1,0)
1267 8506 J_2 : 19/18
1268 0970 rt : (8,8,-7,-1)
1268 3193 J_1 : $\exp(\sqrt[3]{2})$
1268 3618 10^x : $5\ln 2/7$
1268 4624 ln : $3e/4+\pi/3$
1268 5424 e^x : $3/2-\sqrt{5}/3$
1268 6825 cr : $((e+\pi)/2)^{1/3}$
1268 7451 sq : $2-3\pi/4$
1268 9263 rt : (6,9,-8,0)
1268 9914 sq : $2\sqrt{2}/3-3\sqrt{3}/4$
1269 2863 $\ln(3)-\exp(4/5)$
1269 4589 Ei : $7e$
1269 8198 2^x : $2e-\pi/4$
1269 8361 rt : (8,2,8,1)
1269 8728 rt : (7,6,7,-1)
1269 9021 rt : (6,4,-4,-1)
1269 9935 cu : $3\sqrt{2}/4-3\sqrt{5}$
1270 0185 rt : (6,5,6,9)
1270 0676 rt : (5,5,-4,-9)
1270 1665 rt : (1,9,9,1)
1270 4154 at : $7\zeta(3)/4$
1270 4223 Γ : $\sqrt{15}$
1270 4693 rt : (9,8,-7,-1)
1270 5084 rt : (8,0,-8,1)
1270 5592 rt : (7,8,-9,1)
1270 7891 J_1 : 25/6
1270 8473 rt : (4,8,-7,-8)
1270 8768 $t\pi$: $(\sqrt[3]{5}/2)^2$
1271 1860 rt : (5,2,3,8)
1271 2672 2^x : $3/4-\sqrt{3}/3$
1271 2678 Ψ : $\sqrt{8}\pi$
1271 2962 rt : (3,8,-4,1)
1271 4198 e^x : $\sqrt{2}/2+\sqrt{3}/4$
1271 4874 sr : $3\sqrt{2}/4+2\sqrt{3}$
1271 5897 Li_2 : $\exp(-2\pi/3)$
1271 7996 J_2 : $(\sqrt{5}/2)^{1/2}$
1271 8650 cr : $2e+4\pi/3$
1271 8974 2^x : 17/6
1271 9916 rt : (5,1,-7,-2)
1272 1659 J_1 : $9(e+\pi)/8$
1272 1956 rt : (6,6,7,-1)
1272 2276 rt : (9,0,-7,5)
1272 2936 J : $\exp(-15\pi/19)$
1272 3598 rt : (4,1,0,-7)
1272 4509 rt : (7,2,8,1)
1272 4888 e^x : 16/3
1272 5165 cr : $9(1/\pi)/2$
1272 5265 rt : (2,7,-7,-1)
1272 8287 rt : (1,2,-7,-9)
1272 9533 10^x : $\sqrt{2}-4\sqrt{3}/3$
1273 0847 ζ : $9e/7$
1273 2066 rt : (1,7,6,4)
1273 2200 rt : (9,9,1,0)
1273 2395 id : $2(1/\pi)/5$
1273 2776 sq : $2-3\pi$
1273 3564 2^x : $\pi^2/6$
1273 4646 $t\pi$: $4\sqrt{2}/3+2\sqrt{3}/3$
1273 5398 rt : (3,5,-2,2)
1273 5680 θ_3 : $(1/\pi)/5$
1273 6542 sq : $1/\sqrt[3]{22}$
1273 6781 rt : (8,8,-9,1)
1273 8414 J_1 : $\exp(-e/2)$
1273 9017 id : $7(e+\pi)/8$
1274 0889 J_1 : 6π
1274 1380 cr : $1+\sqrt{3}/4$
1274 3100 rt : (9,7,0,4)
1274 3307 2^x : $7e\pi/4$
1274 3770 2^x : $3/2-2\sqrt{5}$
1274 5431 rt : (5,6,7,-1)
1274 6523 id : $2\sqrt{2}+3\sqrt{3}/4$
1274 7262 rt : (2,7,-6,-5)
1274 9123 rt : (3,7,-7,-1)
1274 9611 J_2 : 18/17
1274 9685 e^x : 3/25
1274 9729 $\ln(\sqrt{2})\div\exp(1)$
1275 0425 ln : $3\sqrt{2}-2\sqrt{3}/3$
1275 0982 rt : (6,2,8,1)
1275 1466 2^x : $1/\ln(4/3)$
1275 1937 at : $1/4-3\pi/4$
1275 3032 e^x : $1/4-4\sqrt{3}/3$
1275 4291 rt : (8,4,-3,3)
1275 4823 rt : (2,7,-3,9)
1275 5102 sq : 5/14
1275 5251 rt : (2,5,-4,-8)
1275 5705 Γ : 5/12
1275 6068 2^x : $\sqrt{3}/10$
1275 6098 J_0 : 8/3
1275 6500 e^x : $(\sqrt[3]{5}/3)^{1/2}$
1275 8796 e^x : $4\sqrt{2}/3-\sqrt{5}/3$
1276 0242 E : 14/15
1276 2455 2^x : $\ln 2/4$
1276 3206 rt : (1,9,-9,1)
1276 3625 ℓ_{10} : $6\sqrt{5}$
1276 5539 rt : (7,1,-6,2)
1276 6323 $\sqrt{3}+\exp(1/3)$
1276 6824 sinh : $2(1/\pi)/5$
1276 6900 10^x : $2\sqrt{2}+3\sqrt{3}/2$
1276 7050 as : $2(1/\pi)/5$
1276 7834 J_2 : $\ln(\ln 2/2)$
1276 8037 J_2 : $\exp((e+\pi)/2)$
1276 8469 rt : (9,8,-9,1)
1276 9157 rt : (4,6,7,6)
1277 0270 at : $\exp(\sqrt{5}/3)$
1277 0464 $5\times\exp(4/5)$
1277 0505 rt : (4,7,-5,-9)
1277 0640 $1/4\times\ln(3/5)$
1277 0761 $t\pi$: $4\sqrt{2}/3-2\sqrt{3}/3$
1277 2879 e^x : $\zeta(3)/10$
1277 3243 rt : (4,7,-7,-1)
1277 3435 $t\pi$: $1/3+\sqrt{2}/2$
1277 5926 10^x : $\pi^2/8$
1277 6844 rt : (6,2,-9,-1)
1277 7788 rt : (5,2,8,1)
1277 7889 2^x : $2/3+3\pi/4$
1277 8110 cu : $10\sqrt{3}/9$
1277 9575 ζ : $10\pi/9$
1278 1724 at : $(\sqrt{2}\pi)^{1/2}$
1278 2067 rt : (9,1,-5,6)
1278 2564 rt : (3,8,-1,-7)
1278 3337 ln : 22/25
1278 3656 ln : $\sqrt{3}/3+\sqrt{5}/4$
1278 9630 J_2 : $3\sqrt{2}/4$
1278 9790 rt : (8,2,-8,-5)
1279 0711 ℓ_{10} : $4+3\pi$
1279 1213 $s\pi$: $\sqrt[3]{17/2}$
1279 2590 rt : (2,9,-8,-8)
1279 3142 rt : (3,6,7,-1)
1279 3508 Γ : $7(e+\pi)/8$
1279 3561 J_1 : $(\pi/2)^{-3}$
1279 4700 rt : (2,9,-9,1)
1279 7114 Ψ : $\exp(\sqrt{2}/3)$
1279 7633 rt : (5,7,-7,-1)
1279 7757 J_2 : $10(1/\pi)/3$
1280 0000 cu : $2\sqrt[3]{2}/5$
1280 2660 J_1 : $1/\sqrt{15}$
1280 3355 θ_3 : $(5/2)^{-3}$
1280 3714 cu : $\sqrt{3}-\sqrt{5}$
1280 4937 rt : (4,2,8,1)
1280 5232 rt : (5,3,5,9)
1280 6077 at : $3/4+e/2$
1280 6910 ℓ_2 : $1-2\sqrt{2}/3$
1280 6918 cu : $3/4+3\sqrt{5}/2$
1280 8465 e^x : $3\sqrt{3}/4-3\sqrt{5}/2$
1281 0237 rt : (4,1,-8,1)
1281 1413 2^x : 4/23
1281 2931 sinh : $(\ln 3)^{-1/3}$
1281 4937 2^x : $1/2-2\sqrt{3}$
1281 5214 sr : 14/11
1281 7389 rt : (4,0,2,8)
1281 8092 sr : $9\sqrt{2}/10$
1281 9092 Ψ : $10\sqrt[3]{3}/9$
1281 9495 2^x : $\sqrt{2}-\sqrt{3}-\sqrt{7}$
1282 2297 rt : (6,7,-7,-1)
1282 3007 at : $\sqrt{3}/4+3\sqrt{5}/4$
1282 3462 sq : $9(1/\pi)/8$
1282 4310 sq : $3/4-2e/3$
1282 5186 e^x : $\sqrt{3}/2-\sqrt{5}/3$
1282 6695 rt : (3,9,-9,1)
1282 8161 e^x : $e/3-\pi/4$
1282 9614 rt : (3,6,2,6)
1282 9759 J_2 : 17/16
1282 9921 J_1 : $7e\pi/8$
1283 1737 2^x : $e/3+3\pi/2$
1283 1832 rt : (9,8,2,5)
1283 2435 rt : (3,2,8,1)
1283 3703 rt : (4,6,-3,5)
1283 4414 J_0 : $\sqrt[3]{19}$
1283 5427 $1/4\div\exp(2/3)$
1283 5751 rt : (5,6,-8,4)
1283 6738 rt : (9,5,7,-1)
1283 7916 sr : $4(1/\pi)$
1283 8278 rt : (5,1,-8,1)
1283 9909 Ψ : $4\zeta(3)/3$
1284 1524 J_0 : $3\pi^2/5$
1284 1906 rt : (3,9,-6,-9)
1284 2210 at : 19/9
1284 2918 rt : (9,2,-3,7)
1284 3394 rt : (8,5,-1,4)
1284 3792 J_2 : $(\ln 2/3)^{-2}$
1284 4714 rt : (7,8,1,1)
1284 6183 10^x : $e/4-\pi/2$
1284 7243 rt : (7,7,-7,-1)
1284 8031 ℓ_{10} : $(e/3)^3$
1284 8073 J_0 : $9e\pi/2$
1284 8655 rt : (9,4,-8,-9)
1284 9000 sr : $5e/3$
1285 0702 Ψ : $8\sqrt{5}/5$
1285 0817 rt : (8,1,-6,-6)
1285 2714 e^x : $2/3-e$
1285 4267 rt : (1,9,-7,-5)
1285 4289 Ei : 13/18
1285 5016 rt : (7,2,-4,3)

1285 8024 ln : $\sqrt{2}+3\sqrt{5}/4$
1285 8743 rt : (7,4,-9,-5)
1285 8935 cr : 23/16
1285 9210 rt : (4,9,-9,1)
1286 0294 rt : (2,2,8,1)
1286 1104 rt : (8,5,7,-1)
1286 1691 Li_2 : 19/23
1286 4116 at : $(3\pi)^{1/3}$
1286 4850 e^x : $3+\sqrt{5}/4$
1286 5235 rt : (2,8,-3,-8)
1286 6223 J_2 : $10\sqrt{2}$
1286 6697 rt : (6,7,-1,0)
1287 0636 Γ : $\ln(3\pi)$
1287 1846 2^x : $2/3-4e/3$
1287 2476 rt : (8,7,-7,-1)
1287 4217 10^x : $1/4+3e$
1287 4931 e^x : $\sqrt{2}-2\sqrt{3}$
1287 5524 rt : (6,7,-9,7)
1287 8702 Γ : 5/6
1287 8847 J_2 : $9e\pi/2$
1288 0838 ln : $1/4+3e$
1288 1795 rt : (9,3,8,1)
1288 3789 10^x : 1/19
1288 3977 rt : (5,7,-4,8)
1288 4781 cu : P_{35}
1288 5740 rt : (7,5,7,-1)
1288 6408 ζ : $4\pi/9$
1288 7056 J_1 : $3\ln 2/8$
1288 8523 rt : (1,2,8,1)
1288 8893 rt : (5,2,-5,-1)
1288 8991 e^x : $\sqrt{3}/4+3\sqrt{5}/2$
1289 0625 sq : 17/16
1289 1770 rt : (1,9,8,-1)
1289 2261 rt : (5,9,-9,1)
1289 3206 2^x : $10(e+\pi)/7$
1289 3682 θ_3 : 1/2
1289 5506 rt : (7,1,-8,1)
1289 5582 Γ : 16/3
1289 6625 10^x : $4\sqrt{5}/3$
1289 7111 J_0 : 13/6
1289 7157 tπ : $\sqrt[3]{17/2}$
1289 7161 $\ln(2)-\exp(3/5)$
1289 7760 Ψ : 11/15
1289 8003 rt : (9,7,-7,-1)
1289 8109 rt : (4,7,1,9)
1289 8219 sq : $1-e/2$
1289 9020 id : $4\sqrt{2}+2\sqrt{5}$
1289 9100 at : $7e/9$
1290 0238 rt : (3,6,-9,5)
1290 1697 rt : (8,6,-9,-9)
1290 2074 at : $4-4\sqrt{2}/3$
1290 3149 tπ : $4\pi^2/7$
1290 3555 at : $\sqrt{2}/2-\sqrt{3}/3$
1290 4863 rt : (9,3,-1,8)
1290 6030 sq : $1/3-3\sqrt{3}/2$
1290 6351 J_0 : 6π
1290 7715 ℓ_2 : $3/4+4e/3$
1290 9393 2^x : $\sqrt{3}-\sqrt{5}-\sqrt{6}$
1290 9828 rt : (8,3,8,1)
1291 0161 ℓ_{10} : $(2e/3)^{1/2}$
1291 0192 cr : $\sqrt{2}/2+4\sqrt{5}$
1291 0649 rt : (6,5,7,-1)
1291 1305 rt : (4,1,-8,2)
1291 1509 ζ : $5(1/\pi)/4$
1291 2156 ℓ_{10} : $2/3+e/4$
1291 2555 Γ : $5\pi/7$
1291 2944 rt : (8,0,-4,7)
1291 3368 $\exp(1/4)-\exp(5)$
1291 3502 tπ : $1/\ln(\sqrt[3]{3})$
1291 3863 rt : (4,1,4,9)
1291 3986 Li_2 : 1/8
1291 4431 J_0 : $(5/2)^3$
1291 5442 rt : (2,9,5,8)
1291 7130 rt : (2,8,1,0)
1291 7308 ℓ_{10} : $3\pi/7$
1292 0651 at : $\sqrt{3}/2-4\sqrt{5}/3$
1292 0690 10^x : 11/6
1292 0792 J_2 : 16/15
1292 1053 rt : (7,3,-7,-6)
1292 1964 cr : $8\sqrt[3]{2}/7$
1292 2866 $\ln(\sqrt{5})-s\pi(\sqrt{5})$
1292 2885 rt : (9,9,4,6)
1292 4715 rt : (8,1,-8,1)
1292 5866 rt : (6,9,-9,1)
1292 6388 $\ln(3/4)\div\exp(4/5)$
1292 6507 tπ : $6\sqrt[3]{3}/5$
1292 6537 rt : (3,2,1,-8)
1292 6939 rt : (9,7,-8,-5)
1292 7544 Γ : $\sqrt{2}+\sqrt{3}+\sqrt{6}$
1292 7945 sinh : 3/2
1293 2361 J_2 : $10e\pi/3$
1293 2746 J_0 : $(2e)^2$
1293 2836 J_1 : 6/23
1293 3488 ln : $3-3\sqrt{2}/2$
1293 3827 as : $2/3-\pi/2$
1293 4534 J_0 : $8(e+\pi)/3$
1293 4837 rt : (8,6,1,5)
1293 5681 cr : $3/2+3e$
1293 5840 rt : (5,5,7,-1)
1293 8231 rt : (7,3,8,1)
1293 8296 J_2 : $e\pi/8$
1293 8493 Ei : $3(e+\pi)/7$
1293 9283 rt : (2,7,2,-1)
1294 1615 ℓ_{10} : $3\sqrt{3}/7$
1294 1701 $\ln(2/5)\div s\pi(\pi)$
1294 2784 rt : (3,6,-7,-1)
1294 3438 ln : $4-e/3$
1294 4143 Γ : $(\sqrt[3]{3})^{-1/2}$
1294 4704 ℓ_{10} : $e/3+4\pi$
1294 6128 rt : (1,2,-8,1)
1294 6370 Ei : $7\sqrt[3]{2}/5$
1294 6853 rt : (7,3,-2,4)
1294 7392 2^x : $4/3+2e$
1294 7625 ζ : $1/\ln(4/3)$
1294 8884 ℓ_2 : $2-e/3$
1294 9066 Γ : 6π
1295 0013 cr : $2-\sqrt{5}/4$
1295 0448 cr : $4+4\sqrt{2}$
1295 2044 J_2 : $9\pi/2$
1295 2103 rt : (1,8,-5,-6)
1295 3277 Ψ : $10\zeta(3)/9$
1295 4335 rt : (9,1,-8,1)
1295 7100 Γ : $(\ln 2)^{1/2}$
1295 7648 e^x : $3\sqrt{2}+\sqrt{5}$
1295 8375 sq : $2\sqrt[3]{2}/7$
1295 8933 rt : (6,0,-5,3)
1295 9244 ln : $1+2\pi/3$
1296 0000 sq : 9/25
1296 0046 rt : (7,9,-9,1)
1296 0699 cu : $9\zeta(3)/5$
1296 0803 J_2 : $8\zeta(3)/9$
1296 1318 rt : (4,7,9,7)
1296 3931 e^x : $3\sqrt[3]{2}/5$
1296 4175 $3\div\exp(\pi)$
1296 5478 2^x : P_{14}
1296 5684 sinh : $9\sqrt{2}$
1296 6057 J_2 : $5\sqrt[3]{5}/8$
1296 6471 sr : $1/3+2\sqrt{2}/3$
1296 6511 at : $\sqrt[3]{19/2}$
1296 7014 rt : (6,3,8,1)
1296 7939 rt : (9,4,1,9)
1296 8245 e^x : $\sqrt{3}/2+2\sqrt{5}$
1296 8714 sq : $1/3-2\sqrt{5}$
1296 8738 at : $1+\sqrt{5}/2$
1296 8740 rt : (4,6,-7,-1)
1297 0007 tanh : 3/23
1297 0253 at : 3/23
1297 0413 e^x : $4/3-\sqrt{3}/3$
1297 1078 rt : (5,3,-8,-2)
1297 2375 rt : (3,2,-7,5)
1297 5526 rt : (2,4,-6,-9)
1297 5651 id : $\sqrt{2}/2-\sqrt{3}/3$
1297 6208 rt : (8,1,-2,8)
1297 7788 erf : $\ln 2/6$
1298 0695 id : $2e-4\pi$
1298 1449 at : $4/3+\pi/4$
1298 3096 id : $(\sqrt[3]{3})^{1/3}$
1298 3490 Ψ : $(\sqrt[3]{2}/3)^{-1/3}$
1298 4508 rt : (7,2,-5,-7)
1298 5717 sπ : $7\sqrt{3}/3$
1298 6141 10^x : $5\sqrt[3]{2}/8$
1298 6985 J_0 : $5\sqrt{3}/4$
1298 7090 rt : (3,5,7,-1)
1298 8807 cr : $6\zeta(3)/5$
1298 9043 rt : (4,3,-8,8)
1299 0693 cu : $1+e/2$
1299 0730 θ_3 : $6\ln 2/5$
1299 2754 J_1 : $((e+\pi)/3)^{-2}$
1299 2784 id : $3\sqrt[3]{5}$
1299 2839 rt : (6,5,-8,-6)
1299 3028 10^x : $(1/\pi)/6$
1299 4432 ln : $7\zeta(3)$
1299 4727 id : $(\ln 2)^{-1/3}$
1299 4820 rt : (8,9,-9,1)
1299 5003 rt : (5,6,-7,-1)
1299 5096 K : 6/7
1299 6186 rt : (5,3,8,1)
1299 7344 10^x : $9\sqrt[3]{3}$
1299 7703 cr : $1/2+2\sqrt{2}/3$
1299 9600 e^x : $\exp(\sqrt[3]{4}/2)$
1300 0523 sq : $\sqrt[3]{3}/4$
1300 1179 as : $\sqrt{2}/3+\sqrt{3}/4$
1300 1553 $\ln(\pi)\div\ln(4/5)$
1300 4317 2^x : $(\sqrt[3]{2}/3)^2$
1300 4333 sinh : $7\ln 2/5$
1300 4467 sq : $3\zeta(3)/10$
1300 4889 tπ : $2\sqrt{2}-\sqrt{5}/4$
1300 5335 rt : (3,7,-3,-8)
1300 5388 e^x : $-\sqrt{5}-\sqrt{6}+\sqrt{7}$
1300 8217 2^x : 12/11
1300 9277 ℓ_{10} : $(\ln 3/2)^{1/2}$
1301 1578 2^x : 3/17
1301 2313 sπ : $1/3+3\sqrt{5}$
1301 2341 as : $\sqrt{2}/2-\sqrt{3}/3$
1301 2494 cr : $5\sqrt{3}/6$
1301 3134 10^x : $1-4\sqrt{2}/3$
1301 3163 rt : (2,5,7,-1)
1301 4102 10^x : $4-e$
1301 4477 Γ : $\exp(-1/(2e))$
1301 7751 sq : 23/13
1301 8814 $\sqrt{2}-\exp(1/4)$
1301 9213 rt : (7,9,3,2)
1301 9932 sr : $8\pi^2/3$
1301 9997 Li_2 : $\sqrt[3]{2}/10$
1302 0893 e^x : $2\sqrt{2}/3-4\sqrt{5}/3$
1302 1254 ζ : $(e/2)^{-3}$
1302 1582 rt : (6,6,-7,-1)
1302 4259 K : $(\sqrt[3]{4})^{-1/3}$
1302 5084 J_2 : 15/14
1302 5759 rt : (4,3,8,1)
1302 5926 Ψ : $3\sqrt[3]{5}/7$
1302 6081 ln : $2-3\sqrt{5}/4$
1302 6314 rt : (3,1,3,-9)
1302 8067 cu : 25/24
1302 8281 Ei : $9\sqrt{5}/7$
1302 8566 at : $3\sqrt{2}/2$
1302 8735 rt : (8,7,3,6)
1302 8858 $\exp(1/2)\div s\pi(e)$
1303 0210 rt : (9,9,-9,1)
1303 0910 rt : (9,4,7,-1)
1303 3376 ℓ_{10} : $(\sqrt[3]{2}/3)^3$
1303 4462 cr : $3\sqrt{3}+2\sqrt{5}$
1303 4644 Ei : $\sqrt[3]{3}$
1303 5559 2^x : $\sqrt{2}/8$
1303 5569 rt : (4,2,-8,1)
1303 5626 Γ : $6\ln 2/5$
1303 6636 Ei : $1/\ln(\ln 3)$
1303 6719 J_2 : 7π
1303 7742 sr : $4\sqrt{5}/7$
1303 8833 sr : 23/18
1303 9208 Γ : $1/\ln(\ln 2/3)$
1303 9543 rt : (3,3,-3,9)
1304 0381 cr : 13/9
1304 0427 2^x : $5(1/\pi)/9$
1304 0455 ℓ_{10} : $3e/2+3\pi$
1304 0600 sπ : $(\ln 2/2)^3$
1304 0649 rt : (8,2,0,9)
1304 1168 rt : (7,4,0,5)
1304 1170 $\exp(\sqrt{5})+s\pi(e)$
1304 1831 J_0 : $4(e+\pi)$

1304 2607 rt : (6,6,0,1)
1304 3172 10^x : $4e/3 - \pi/2$
1304 3478 id : 3/23
1304 4237 J_0 : $1/\ln(\sqrt[3]{4})$
1304 4281 10^x : $2\sqrt{2} + 3\sqrt{5}$
1304 4321 J_1 : 5/19
1304 4673 $1/5 \times \exp(\sqrt{3})$
1304 5071 sr : $4\sqrt{2} - \sqrt{5}/2$
1304 5273 cu : $3e/2 + 2\pi$
1304 5360 J_2 : $9e/2$
1304 5822 J : $(1/\pi)/10$
1304 6123 ζ : $-\sqrt{2} + \sqrt{5} + \sqrt{7}$
1304 6209 id : $\exp(1/(3e))$
1304 6882 2^x : $\exp(-\sqrt{3})$
1304 7227 sinh : $8\pi/9$
1304 8098 ln : $4\sqrt{3} + 2\sqrt{5}/3$
1304 8484 rt : (7,6,-7,-1)
1304 9146 rt : (7,1,-3,-8)
1304 9516 id : $7\sqrt{5}/5$
1304 9657 $e - s\pi(1/5)$
1305 1352 θ_3 : $\sqrt{2} + \sqrt{3} - \sqrt{7}$
1305 2619 $s\pi$: 1/24
1305 2851 rt : (1,7,-3,-7)
1305 3669 rt : (6,1,-3,4)
1305 4099 $\exp(1/5) \div \exp(\sqrt{5})$
1305 5744 rt : (3,3,8,1)
1305 7091 rt : (8,4,7,-1)
1305 7676 rt : (6,4,-6,-7)
1305 8377 $s\pi$: $3\sqrt{5}/7$
1305 9746 rt : (6,9,-9,-2)
1306 1610 2^x : $2 - \sqrt{2}/4$
1306 2472 Ei : 23/8
1306 3313 J_0 : $9\zeta(3)/5$
1306 4751 2^x : $1 + e/2$
1306 5072 J_2 : $8\ln 2$
1306 5715 J_1 : $\exp(-4/3)$
1306 6238 rt : (5,8,0,-3)
1306 7631 as : $1/4 - 2\sqrt{3}/3$
1306 8027 ζ : $5\ln 2$
1306 8938 10^x : $\exp(e)$
1306 9833 rt : (2,7,-4,5)
1306 9848 J_1 : $3(e+\pi)/5$
1307 1431 ζ : $(\pi/3)^{1/3}$
1307 4311 2^x : $1/\ln(5/2)$
1307 5589 ℓ_{10} : $3\sqrt{3}/4 - \sqrt{5}/4$
1307 5718 rt : (8,6,-7,-1)
1307 6828 id : $1/3 - 2\sqrt{3}$
1307 7247 ℓ_2 : $\sqrt[3]{21/2}$
1307 7414 J_0 : $10\sqrt{3}$
1307 7542 Ei : $10\sqrt[3]{5}/3$
1307 8134 ln : $1/2 + 3\sqrt{3}/2$
1307 8144 rt : (9,3,-9,-1)
1307 8877 rt : (4,7,-7,-8)
1307 9444 10^x : $(\ln 2/3)^2$
1308 0495 sinh : 3/23
1308 0749 as : 3/23
1308 1715 J_0 : $3\sqrt[3]{3}/2$
1308 2037 as : 19/21
1308 2116 cr : $1/4 + 3\pi$
1308 2227 sq : $3\sqrt{3}/2 + 4\sqrt{5}/3$
1308 3350 ζ : $5\sqrt[3]{5}/8$
1308 3578 rt : (7,4,7,-1)
1308 3798 10^x : $\exp(-(e+\pi)/2)$
1308 4054 at : $3/2 - 4e/3$
1308 5115 erf : $((e+\pi)/2)^{-2}$
1308 5366 id : $\sqrt{3}/3 - 3\sqrt{5}$
1308 6153 rt : (2,3,8,1)
1308 7208 ζ : $2\sqrt{3}$
1308 7624 rt : (9,8,-6,-4)
1308 8250 sinh : 25/3
1308 9706 ln : $(\sqrt{2}\pi/3)^{1/3}$
1309 0066 J_1 : $2\pi^2/3$
1309 0112 rt : (4,0,-6,1)
1309 2808 $\exp(4/5) - \exp(\sqrt{5})$
1309 2922 $s\pi$: $1/4 + 3\sqrt{5}$
1309 3256 rt : (1,8,6,-2)
1309 4032 J_2 : $(\sqrt{3}/2)^{-1/2}$
1309 4206 J_2 : $7e/4$
1309 5374 at : 17/8
1309 6610 $t\pi$: $7\sqrt{3}/3$
1309 7356 rt : (6,2,-8,1)
1309 7575 rt : (3,8,-5,-2)
1309 8796 Ei : $6\zeta(3)/5$
1310 1419 ln : $2\sqrt[3]{5}/3$
1310 3063 J_2 : $\ln((e+\pi)/2)$
1310 3290 rt : (9,6,-7,-1)
1310 4808 e^x : $\exp(-2\pi/3)$
1310 4996 sq : $3\sqrt{3}/2 - \sqrt{5}$
1310 5545 rt : (9,4,8,1)
1310 7754 ℓ_{10} : $9\zeta(3)/8$
1310 9202 rt : (8,5,-9,-5)
1310 9988 Ψ : $-\sqrt{2} + \sqrt{3} - \sqrt{6}$
1311 0379 rt : (6,4,7,-1)
1311 0590 sq : $4 + 4e/3$
1311 0724 sq : $e - 3\pi/4$
1311 2283 sr : $(2\pi/3)^{1/3}$
1311 3306 ln : $\sqrt{2}/2 + \sqrt{3}/4$
1311 4233 rt : (3,3,-9,2)
1311 4513 2^x : $7e/2$
1311 5009 rt : (7,0,-1,9)
1311 6048 $\ln(4/5) \times s\pi(1/5)$
1311 6997 rt : (1,3,8,1)
1311 7154 rt : (2,4,-9,-6)
1311 9998 ℓ_{10} : $\ln(2\pi/3)$
1312 0107 at : P_{49}
1312 0604 rt : (2,5,-7,-1)
1312 1519 Ψ : $(e+\pi)/8$
1312 3746 rt : (6,3,-4,-8)
1312 3894 $t\pi$: $1/3 + 3\sqrt{5}$
1312 4453 ℓ_2 : 21/23
1312 5095 Ei : $e/8$
1312 5210 rt : (8,8,5,7)
1312 5663 at : $\sqrt{2}/3 - 3\sqrt{3}/2$
1312 5814 ln : $4\sqrt{2}/3 - \sqrt{5}/3$
1312 7891 ℓ_{10} : 17/23
1312 8895 sq : $5e/6$
1312 8936 rt : (7,2,-8,1)
1312 9230 sq : $3\sqrt{2} + 2\sqrt{3}/3$
1312 9682 sq : $\sqrt{22}\pi$
1313 0400 J_2 : 22
1313 1251 rt : (1,7,5,-2)
1313 2516 rt : (5,9,-4,-1)
1313 4448 10^x : $4 - 4\sqrt{2}/3$
1313 6113 rt : (8,4,8,1)
1313 6114 sq : $8(e+\pi)/9$
1313 7084 id : $4\sqrt{2}/5$
1313 7502 rt : (5,4,7,-1)
1313 7734 sq : $1/\sqrt[3]{21}$
1313 7810 2^x : $4 - 4\sqrt{3}$
1313 7984 at : P_{64}
1313 8083 rt : (7,5,2,6)
1314 2154 cr : $4e - 3\pi$
1314 2846 rt : (2,3,2,-9)
1314 4093 $e + \exp(5)$
1314 4102 Γ : $(2\pi/3)^3$
1314 5756 J_2 : 14/13
1314 6602 J_1 : $5(1/\pi)/6$
1314 6904 E : $(\sqrt{3}/2)^{1/2}$
1314 7539 ℓ_{10} : $1 + \sqrt{2}/4$
1314 8290 rt : (3,8,1,0)
1314 9749 rt : (3,9,1,7)
1315 1029 rt : (6,2,-1,5)
1315 1465 Γ : $-\sqrt{3} - \sqrt{5} + \sqrt{7}$
1315 2917 $t\pi$: $(\ln 2/2)^3$
1315 5805 10^x : $3\sqrt{2} + 4\sqrt{3}/3$
1315 6673 rt : (1,6,-1,-8)
1315 8360 ln : $3/2 + 4\sqrt{3}$
1315 9156 e^x : 22
1316 0993 rt : (8,2,-8,1)
1316 1327 Γ : $3\ln 2/5$
1316 2404 Li_2 : $2(1/\pi)/5$
1316 4048 rt : (5,1,-4,-4)
1316 4954 rt : (4,4,7,-1)
1316 5249 id : $\sqrt{2} - \sqrt{3} + \sqrt{6}$
1316 5312 Ei : $1/\ln(2)$
1316 7117 rt : (7,4,8,1)
1317 0586 rt : (9,4,-7,0)
1317 1157 $t\pi$: $3\sqrt{5}/7$
1317 1947 10^x : $8\ln 2/9$
1317 2899 $t\pi$: $4(1/\pi)/3$
1317 4566 K : $(e/2)^{-1/2}$
1317 5219 $\exp(\pi) \div \exp(2)$
1317 6324 rt : (4,5,-7,-1)
1317 7142 rt : (4,8,-3,-4)
1317 7603 Ei : 25/23
1318 0045 rt : (1,3,-8,1)
1318 1078 2^x : $4/3 - 2\sqrt{3}/3$
1318 2639 $s\pi$: $3\sqrt{2}/4 + 4\sqrt{5}/3$
1318 7316 rt : (2,5,-3,3)
1318 8277 Ei : 1/16
1319 1091 rt : (6,2,-2,-9)
1319 2743 rt : (3,5,8,7)
1319 3540 rt : (9,2,-8,1)
1319 4753 ζ : $(2\pi)^{-1/2}$
1319 5911 id : P_{49}
1319 6949 E : $1/\ln((e+\pi)/2)$
1319 7607 $2 \div \exp(e)$
1319 8169 10^x : $\sqrt{2} + \sqrt{3} - \sqrt{5}$
1319 8569 rt : (6,4,8,1)
1319 9225 ℓ_2 : $\sqrt{2}/4 + \sqrt{5}/4$
1319 9723 at : $4/3 - 2\sqrt{3}$
1320 0078 rt : (5,5,-5,-8)
1320 1056 sr : $4\sqrt{3}/3 + \sqrt{5}$
1320 1162 e^x : $\sqrt{2}/3 + 4\sqrt{3}/3$
1320 1399 θ_3 : $\exp(-e)$
1320 2753 2^x : $\sqrt{3}/4 - 3\sqrt{5}/2$
1320 3204 rt : (9,0,-8,4)
1320 3378 Ψ : 1/3
1320 4713 rt : (5,5,-7,-1)
1320 5080 $2/5 + \sqrt{3}$
1320 6609 $t\pi$: $1/4 + 3\sqrt{5}$
1320 7492 2^x : $3e/4 + 4\pi$
1320 7716 ln : $\sqrt{3}/2 + \sqrt{5}$
1320 9101 rt : (4,8,-8,-7)
1320 9621 cu : $2 - 2\sqrt{5}/3$
1321 0228 cu : $8(1/\pi)/5$
1321 1836 sq : $10(1/\pi)$
1321 1997 J_2 : $6\sqrt[3]{2}/7$
1321 2275 rt : (2,3,-8,1)
1321 2919 $\ln(4) - s\pi(\sqrt{3})$
1321 2992 sr : $4 - e$
1321 3941 sr : $\sqrt{3}/2 + 4\sqrt{5}$
1321 4056 erf : 2/17
1321 4067 sinh : $4\pi/5$
1321 4100 id : P_{64}
1321 4804 J_2 : $(\sqrt[3]{2})^{1/3}$
1321 5165 J_1 : 4/15
1321 5677 at : $\sqrt{2} - \sqrt{3} + \sqrt{6}$
1321 6098 Γ : $\sqrt[3]{2}/7$
1321 8820 rt : (5,6,-9,2)
1322 0315 J_0 : $6(e+\pi)$
1322 0646 rt : (9,4,-9,-8)
1322 0685 cr : $1/3 + \sqrt{5}/2$
1322 0879 rt : (2,4,7,-1)
1322 3140 sq : 4/11
1322 4393 rt : (8,9,7,8)
1322 4760 rt : (6,8,-6,9)
1322 5135 cr : $3/4 + 4\sqrt{5}$
1322 5421 sr : $8\sqrt[3]{3}/9$
1322 7503 rt : (6,7,2,2)
1322 7645 e^x : $1/3 - 3\pi/4$
1323 0482 rt : (5,4,8,1)
1323 3467 rt : (6,5,-7,-1)
1323 3787 sq : $8(1/\pi)/7$
1323 4511 as : P_{49}
1323 6640 rt : (9,3,7,-1)
1323 6841 2^x : $4/3 + 4\sqrt{2}$
1323 7412 J_1 : $2\zeta(3)/9$
1323 7733 rt : (7,6,4,7)
1323 8820 cr : $2/3 + \pi/4$
1324 3153 J : $10(1/\pi)/7$
1324 3927 cu : $4\sqrt{3} + 4\sqrt{5}/3$
1324 4103 J_1 : $1/\sqrt{14}$
1324 4996 rt : (3,6,-5,-9)
1324 6189 J_2 : $(\sqrt[3]{5}/2)^{-1/2}$
1324 6367 ℓ_{10} : $1 + 4\pi$
1324 6720 sr : $3\sqrt[3]{5}/4$
1324 7023 $s\pi$: $1/3 + 4e/3$
1324 9163 cr : $\sqrt{2}/2 + \sqrt{5}/3$
1324 9214 cu : $2\sqrt{2}/3 - \sqrt{3}/4$

1324 9368 rt : (1,4,7,-1)
1324 9539 $\sqrt{2}+e$
1325 0674 J_2 : $3\sqrt[3]{3}/4$
1325 1151 rt : (6,3,1,6)
1325 1903 10^x : 19/4
1325 2726 rt : (5,4,-1,1)
1325 2861 as : P_{64}
1325 2869 cr : $4\sqrt{2}/3-\sqrt{3}/4$
1325 4291 J_2 : $9\zeta(3)/10$
1325 4878 tanh : 2/15
1325 5153 at : 2/15
1325 5354 e^x : $4/3-3\sqrt{5}/2$
1325 5384 rt : (9,9,-4,-3)
1325 7204 J_0 : $\sqrt{14/3}$
1325 7507 ln : $3+2e$
1325 7525 at : $3-\sqrt{3}/2$
1325 8462 Ψ : $8\pi/7$
1326 1964 J_1 : $9\pi^2/7$
1326 2556 ℓ_{10} : 14/19
1326 2595 rt : (7,5,-7,-1)
1326 2869 rt : (4,4,8,1)
1326 3004 2^x : $3\sqrt{2}/4+3\sqrt{3}/4$
1326 3749 rt : (1,7,9,-3)
1326 4646 rt : (5,0,-2,5)
1326 4666 ℓ_{10} : $(5/2)^{-1/3}$
1326 4866 rt : (8,3,7,-1)
1326 5439 rt : (9,5,-5,1)
1326 7338 J_1 : $(\pi)^3$
1326 8971 rt : (5,4,-3,-9)
1326 9090 $3/4 \div \exp(\sqrt{3})$
1326 9302 J_2 : $5\sqrt{3}/8$
1326 9346 rt : (9,1,-6,5)
1327 0091 10^x : $8\sqrt{2}$
1327 0395 sq : $3e-\pi$
1327 1206 rt : (5,7,-5,6)
1327 1424 rt : (9,3,-7,-9)
1327 2515 J_2 : $4\sqrt{5}$
1327 3135 θ_3 : $\exp(-1/(2e))$
1327 3135 sr : $2e\pi$
1327 4122 id : 8π
1327 4306 ℓ_2 : $2\sqrt{2}-\sqrt{3}$
1327 4778 Ei : $2e/5$
1327 4786 at : $e\pi/4$
1327 5105 J_0 : $(\ln 3/3)^{-3}$
1327 6267 $s\pi$: $1/\ln(5/3)$
1327 7331 tanh : $\zeta(3)/9$
1327 7607 at : $\zeta(3)/9$
1327 8221 rt : (8,6,-7,-4)
1327 8378 10^x : $1/\ln(3)$
1328 0379 as : $3-2\pi/3$
1328 6989 J_2 : 13/12
1328 7099 at : $1/4+4\sqrt{2}/3$
1328 7502 2^x : $\sqrt[3]{2}/7$
1329 1877 rt : (3,6,-8,-3)
1329 2107 rt : (8,5,-7,-1)
1329 3322 rt : (1,6,0,5)
1329 3445 rt : (7,3,7,-1)
1329 3937 Ψ : $4(e+\pi)$
1329 3963 Γ : $(\sqrt{2}\pi)^{-3}$
1329 5744 rt : (3,4,8,1)
1329 7178 ζ : $(e)^{1/3}$
1329 8699 $t\pi$: $3\sqrt{2}/4+4\sqrt{5}/3$
1330 0309 Γ : 9/4
1330 1249 ln : $\sqrt{3}+3\sqrt{5}$
1330 2345 2^x : $1+\pi/3$
1330 3266 cr : 16/11
1330 3996 rt : (3,7,-2,-7)
1330 4228 Γ : $(\sqrt[3]{5}/3)^{1/3}$
1330 6725 ℓ_2 : $\pi^2/9$
1330 7447 rt : (1,7,4,-3)
1330 7646 e^x : $5\sqrt[3]{5}$
1331 0471 2^x : $\sqrt[3]{3}/8$
1331 0647 id : $3e/4+2\pi/3$
1331 1016 2^x : $4/3-3\sqrt{2}$
1331 1969 rt : (5,3,-8,1)
1331 4845 e^x : 1/8
1331 4971 ζ : $\ln(\sqrt{2}/3)$
1331 7800 e^x : $4\sqrt{3}-4\sqrt{5}$
1331 8154 at : $1-\sqrt{3}/2$
1331 8566 rt : (2,9,4,4)
1332 0138 Ei : $2\ln 2/9$
1332 0376 at : $5\sqrt[3]{5}/4$
1332 2014 rt : (9,5,-7,-1)
1332 2385 rt : (6,3,7,-1)
1332 4067 J_0 : $7\sqrt{5}$
1332 6448 ℓ_{10} : $5e$
1332 7231 cr : $3+3\sqrt{5}$
1332 9124 rt : (2,4,8,1)
1332 9627 cu : $\ln(5/3)$
1332 9863 rt : (8,5,-8,-9)
1333 0725 tanh : $(\ln 3/3)^2$
1333 1007 at : $(\ln 3/3)^2$
1333 2319 $t\pi$: $e\pi/2$
1333 2546 $\ln(4/5)+\exp(\sqrt{5})$
1333 2739 $s\pi$: $\sqrt[3]{15/2}$
1333 3333 id : 2/15
1333 3889 sr : $\sqrt{2}/2+\sqrt{3}/3$
1333 5796 J_2 : 19/4
1333 6732 rt : (9,2,-4,6)
1333 7055 rt : (2,4,-7,-1)
1333 7283 θ_3 : 1/15
1333 7515 J_1 : 14
1333 8383 tanh : $((e+\pi)/3)^{-3}$
1333 8666 at : $((e+\pi)/3)^{-3}$
1333 9834 ln : $3/4+3\pi/4$
1334 0261 rt : (7,7,6,8)
1334 0338 e^x : $4\sqrt{3}+3\sqrt{5}/2$
1334 3209 cu : P_{13}
1334 5771 rt : (9,5,8,1)
1334 5822 rt : (8,1,-7,-5)
1334 6252 rt : (1,9,-1,2)
1334 6254 rt : (6,3,-8,1)
1334 8730 rt : (4,6,-4,3)
1335 1696 rt : (5,3,7,-1)
1335 2103 J_2 : $6(e+\pi)$
1335 2335 sq : $9\pi^2/4$
1335 3139 ln : 7/8
1335 3354 sr : $(\sqrt{2}/3)^{-1/3}$
1335 3803 ζ : $(\sqrt[3]{3})^{-1/3}$
1335 4178 rt : (6,4,3,7)
1335 4551 $\ln(\sqrt{3}) \div \exp(\sqrt{2})$
1335 4967 Ψ : $1/\ln(\sqrt[3]{3}/2)$
1335 6011 sq : $2\sqrt{2}/3-\sqrt{3}/3$
1335 6187 id : $\zeta(3)/9$
1335 6737 2^x : $\exp(-\sqrt[3]{5})$
1335 8257 rt : (3,9,-7,-8)
1336 2820 rt : (9,6,-3,2)
1336 3024 rt : (1,4,8,1)
1336 4535 e^x : $3\sqrt{2}-\sqrt{3}/4$
1336 4728 10^x : $3/4+4\sqrt{2}/3$
1336 4807 $t\pi$: $1/3+4e/3$
1336 4962 Γ : $(\ln 3)^{-2}$
1336 4994 Ei : $5\sqrt{3}/6$
1336 7027 J_2 : 25/23
1336 7056 rt : (3,4,-7,-1)
1336 7264 2^x : $3\sqrt{3}/4-\sqrt{5}/2$
1336 8180 rt : (5,1,0,6)
1336 8650 J_1 : $5(e+\pi)/7$
1336 8920 e^x : $2\sqrt{2}+2\sqrt{5}$
1336 9733 cr : $3/4+\sqrt{2}/2$
1337 1477 Ei : $\sqrt[3]{11/2}$
1337 2874 sinh : 2/15
1337 3158 as : 2/15
1337 3321 cu : $3\sqrt{2}+4\sqrt{5}$
1337 4271 rt : (1,2,-1,8)
1337 4905 J_2 : $2e/5$
1337 5401 sr : $1/2+\pi/4$
1337 6043 rt : (8,3,-6,1)
1337 6852 J_0 : $3e$
1337 9240 10^x : $1/\ln(1/\pi)$
1337 9240 $t\pi$: $3e-3\pi$
1337 9286 rt : (8,5,8,1)
1338 0191 cu : $4/3-4\sqrt{3}$
1338 1095 rt : (7,3,-8,1)
1338 1388 rt : (4,3,7,-1)
1338 1473 Li_2 : $(\ln 3)^{-2}$
1338 2266 rt : (4,2,-3,-5)
1338 2988 Ei : $\sqrt{23}\pi$
1338 5687 rt : (4,3,-9,6)
1338 8613 $3/5 \times \ln(4/5)$
1338 9341 sr : 9/7
1338 9412 λ : $\exp(-3\pi/2)$
1339 0078 e^x : P_{53}
1339 1844 J_0 : $(2\pi/3)^3$
1339 2894 id : $9\sqrt[3]{2}/10$
1339 4292 at : $1-\pi$
1339 4840 $t\pi$: $1/\ln(5/3)$
1339 5301 2^x : $2e/3-3\pi/2$
1339 5828 as : $e/3$
1339 5932 sinh : $\zeta(3)/9$
1339 6219 as : $\zeta(3)/9$
1339 6438 rt : (3,5,-6,-4)
1339 6825 at : $\sqrt{3}-\sqrt{5}+\sqrt{7}$
1339 7459 id : $1-\sqrt{3}/2$
1339 9838 rt : (2,7,-5,1)
1340 1260 ℓ_2 : $2e-\pi/3$
1340 1988 $\exp(3/4) \div s\pi(\sqrt{5})$
1340 5336 erf : $3(1/\pi)/8$
1340 5405 rt : (9,3,-2,7)
1340 6400 $1/5 \div \exp(2/5)$
1340 7124 Ψ : $4\sqrt{5}$
1341 0544 id : $(\ln 3/3)^2$
1341 0697 rt : (2,8,-9,-3)
1341 1469 rt : (3,3,7,-1)
1341 1807 Ψ : $\ln(5)$
1341 3319 rt : (7,5,8,1)
1341 4718 rt : (8,0,-5,6)
1341 6511 rt : (8,3,-8,1)
1341 6916 at : 15/7
1341 8340 id : $((e+\pi)/3)^{-3}$
1341 9262 10^x : $2/3+2\pi$
1342 0107 $2/3 \div s\pi(1/5)$
1342 0680 rt : (4,7,0,7)
1342 1149 K : $5\zeta(3)/7$
1342 1263 J_0 : $3\sqrt[3]{5}$
1342 2244 rt : (6,8,4,3)
1342 4067 rt : (7,3,-8,-5)
1342 7321 e^x : $\sqrt[3]{2}/10$
1342 8277 rt : (5,4,-7,-1)
1343 0456 rt : (3,8,-5,-9)
1343 1252 2^x : 2/11
1343 1700 E : 13/14
1343 2448 rt : (1,4,-8,1)
1343 3172 sq : $\ln(\ln 2)$
1343 4557 $s\pi$: $1/4+\sqrt{2}/2$
1343 7006 2^x : $4(1/\pi)/7$
1343 7865 as : $1-\sqrt{3}/2$
1343 8488 cu : $1/\ln(\sqrt[3]{4})$
1344 1951 rt : (2,3,7,7)
1344 5821 rt : (7,8,8,9)
1344 6709 sq : $\sqrt{2}/3+3\sqrt{3}/4$
1344 6975 at : $e/2+\pi/4$
1344 7885 rt : (6,5,8,1)
1344 8040 $\exp(3)-s\pi(2/5)$
1344 9239 rt : (5,5,1,2)
1345 0776 sinh : $(\ln 3/3)^2$
1345 1069 as : $(\ln 3/3)^2$
1345 1168 2^x : $2-e/3$
1345 1504 tanh : $(e)^{-2}$
1345 1799 at : $(e)^{-2}$
1345 2520 rt : (9,3,-8,1)
1345 2845 $t\pi$: $\sqrt[3]{15/2}$
1345 4515 J_2 : 12/11
1345 4881 cr : $3e+\pi/2$
1345 4963 2^x : $9\sqrt[3]{2}/4$
1345 5002 rt : (8,7,-5,-3)
1345 5248 rt : (9,2,7,-1)
1345 7272 J : $(\ln 3/3)^2$
1345 7369 ℓ_2 : P_{11}
1345 8643 sinh : $((e+\pi)/3)^{-3}$
1345 8912 $t\pi$: $4\sqrt{2}/3-4\sqrt{3}$
1345 8927 cu : $\sqrt{2}/4-\sqrt{3}/2$
1345 8937 as : $((e+\pi)/3)^{-3}$
1345 9519 rt : (6,4,-7,-1)
1345 9809 10^x : $4+\sqrt{2}/4$
1346 0270 rt : (6,5,5,8)
1346 0840 2^x : $8\sqrt[3]{3}/7$
1346 1492 Ei : 2/13
1346 2851 rt : (9,7,-1,3)
1346 3391 sr : $4e-\pi/3$

1346 3879 $2^x : e/4+\pi$
1346 4434 $J_2 : 1/\ln(5/2)$
1346 4510 at : $1/3+2e/3$
1346 6260 $J_1 : e/10$
1346 8009 rt : (2,4,-8,1)
1346 9857 $\ell_{10} : 11/15$
1346 9996 $\ell_{10} : 4\sqrt{3}+3\sqrt{5}$
1347 0975 $\ln(3)+s\pi(\sqrt{2})$
1347 2845 rt : (1,3,7,-1)
1347 4809 rt : (5,2,2,7)
1347 5410 rt : (9,4,0,8)
1347 5468 $2^x : 1/4-\pi$
1347 5700 $\ln(5)\times\exp(2/3)$
1347 5835 $erf : 3/25$
1347 6542 rt : (8,4,-4,2)
1347 8256 $J_1 : 4\pi/3$
1347 9459 rt : (7,4,-8,-4)
1347 9778 $\zeta : 2/5$
1348 3002 rt : (5,5,8,1)
1348 3756 $J_2 : 9(e+\pi)/2$
1348 4552 cr : 19/13
1348 4955 rt : (8,1,-3,7)
1348 5796 rt : (8,2,7,-1)
1348 7239 sr : $7(e+\pi)/9$
1348 8431 $erf : (\ln 2/2)^2$
1348 9077 $t\pi : \sqrt[3]{3}/4$
1348 9092 $\exp(2)-s\pi(\sqrt{3})$
1348 9439 rt : (4,1,-1,-6)
1349 0311 rt : (7,1,-7,1)
1349 1199 rt : (7,4,-7,-1)
1349 1905 $2^x : e/2+3\pi/4$
1349 3355 id : $e\pi/4$
1349 3674 rt : (1,6,6,8)
1349 3906 $\exp(\sqrt{5})\div\exp(3/5)$
1349 4538 rt : (7,2,-6,-6)
1349 4617 $\Psi : \sqrt[3]{3}/7$
1349 5441 $Li_2 : 3/23$
1349 7818 $\ell_2 : 1/3+\sqrt{3}/3$
1349 8540 sq : $e/3+\pi/2$
1349 8678 $\ell_{10} : 3\sqrt[3]{5}/7$
1349 8712 $erf : \zeta(3)/10$
1350 0000 cu : $3\sqrt[3]{5}/10$
1350 0124 $\Psi : \sqrt{2}-\sqrt{6}+\sqrt{7}$
1350 0182 $\ln(4)\div\exp(1/5)$
1350 1035 $10^x : \sqrt{3}-3\sqrt{5}/4$
1350 2393 $J_0 : 6\sqrt{5}/5$
1350 4161 rt : (6,5,-9,-5)
1350 4336 sq : $2/3-\sqrt{3}$
1350 5350 rt : (1,9,-2,-7)
1350 6625 $\exp(1/5)-\exp(\sqrt{5})$
1350 7341 $J_0 : 7(e+\pi)/8$
1350 8207 $Li_2 : (\sqrt[3]{5}/3)^{1/3}$
1350 9805 $1/5\times s\pi(\sqrt{5})$
1350 9971 $J_1 : 3/11$
1351 1783 sr : $9(e+\pi)/2$
1351 2513 $\zeta : 24/7$
1351 2514 $\Psi : (\sqrt[3]{5}/2)^2$
1351 4968 sq : $e/4-\pi/3$
1351 5273 $\Psi : (\ln 2)^3$
1351 5307 $J_1 : 6(1/\pi)/7$
1351 5503 ln : $\sqrt[3]{2/3}$
1351 6338 sq : $3/4+4e/3$
1351 6752 rt : (7,2,7,-1)
1351 6819 at : $\sqrt{2}/3+3\sqrt{5}/4$
1351 6870 ln : $\ln(\pi)$
1351 8547 sr : $4+\sqrt{5}/4$
1351 8554 sr : $10\sqrt[3]{5}$
1351 8687 rt : (4,5,8,1)
1351 8976 rt : (2,7,-7,-4)
1351 9096 $s\pi : 8\pi^2$
1351 9981 rt : (1,9,1,-9)
1352 0167 $\ell_{10} : (e+\pi)/8$
1352 0353 sinh : $6\pi^2/5$
1352 3329 rt : (8,4,-7,-1)
1352 4842 $J_1 : 15/2$
1352 7075 cu : $3\sqrt{3}/2+3\sqrt{5}/4$
1353 0909 ln : $6\pi^2/7$
1353 1072 sq : $4+e$
1353 2103 rt : (2,4,-8,1)
1353 2688 $J_2 : 5e\pi/9$
1353 3484 sinh : $5\zeta(3)/4$
1353 3528 id : $(e)^{-2}$
1353 4034 $Ei : 1/\sqrt{6}$
1353 4340 $\ln(4/5)\div\exp(1/2)$
1353 4388 $s\pi : \exp(-\pi)$
1353 4828 as : $(\sqrt{5}/3)^{1/3}$
1353 5911 rt : (1,6,2,-4)
1354 0042 sr : $\sqrt{3}/3-\sqrt{5}/4$
1354 0063 sr : $8\sqrt[3]{5}/3$
1354 0924 rt : (4,8,-4,-9)
1354 2919 rt : (9,9,6,-1)
1354 4901 sq : $3/4-\sqrt{5}/2$
1354 5707 at : $3e/4-4\pi/3$
1354 5877 $\ell_{10} : 1/2+\sqrt{3}/2$
1354 6003 $\Gamma : (3\pi)^{1/2}$
1354 6795 rt : (9,5,2,9)
1354 7445 $e^x : 2+\sqrt{5}$
1354 8127 rt : (6,2,7,-1)
1354 8744 rt : (6,8,-7,7)
1354 9225 id : $3\sqrt{2}/4-3\sqrt{3}$
1354 9421 ln : 23
1355 0541 $J_2 : 23/21$
1355 0833 $10^x : 1/4-\sqrt{5}/2$
1355 2464 tanh : 3/22
1355 2755 sr : $2\sqrt{2}+\sqrt{3}$
1355 2771 at : 3/22
1355 2797 $\exp(\sqrt{3})\div s\pi(\pi)$
1355 4522 $e^x : \sqrt[3]{14}$
1355 4960 rt : (3,5,8,1)
1355 4992 rt : (1,7,4,-8)
1355 5596 $2^x : (e)^{1/2}$
1355 5921 rt : (9,4,-7,-1)
1355 6433 $\ell_{10} : 8e\pi/5$
1355 6581 rt : (8,2,-1,8)
1355 7462 $t\pi : 1/4+\sqrt{2}/2$
1355 7852 tanh : $3(1/\pi)/7$
1355 8159 at : $3(1/\pi)/7$
1355 8810 $2^x : \sqrt{2}/3-3\sqrt{5}/2$
1356 0853 $\Gamma : -\sqrt{5}-\sqrt{6}+\sqrt{7}$
1356 1084 rt : (3,9,0,4)
1356 1808 id : $1/4+4\sqrt{2}/3$
1356 3527 cr : $(\pi)^{1/3}$
1356 5530 $10^x : \ln(\sqrt[3]{2}/3)$
1356 5666 rt : (9,8,1,4)
1356 6283 $\Psi : 18/5$
1356 6408 rt : (7,1,-4,-7)
1356 8107 rt : (2,9,5,-3)
1356 8233 $\ell_2 : \ln(3)$
1356 8820 $t\pi : 3e/2+2\pi$
1356 9596 rt : (6,6,7,9)
1357 0804 $10^x : 9\pi^2/8$
1357 0887 rt : (8,9,6,-1)
1357 1258 $\Psi : 7\sqrt[3]{5}/9$
1357 1476 $e^x : 8/7$
1357 2088 sq : $1/\sqrt[3]{20}$
1357 2635 $2^x : P_{63}$
1357 3264 cr : $(e+\pi)/4$
1357 3407 sq : 7/19
1357 4878 sinh : $(e)^{-2}$
1357 5022 $\ell_2 : \sqrt{2}+\sqrt{3}-\sqrt{5}$
1357 5185 as : $(e)^{-2}$
1357 5638 $Ei : 2\pi^2$
1357 6052 $\pi\div\exp(\pi)$
1357 6276 rt : (6,4,-7,-6)
1357 8319 rt : (5,4,-8,1)
1357 8490 $e^x : 2(1/\pi)/5$
1357 9431 rt : (3,3,-4,7)
1357 9851 rt : (8,5,-2,3)
1357 9934 rt : (5,2,7,-1)
1358 1299 $J_2 : \pi^2/9$
1358 3150 $J_1 : 8\zeta(3)$
1358 3226 $\ell_2 : 3(e+\pi)$
1358 3670 $J_0 : \sqrt[3]{10}$
1358 4582 $t\pi : 3\zeta(3)/10$
1358 4707 rt : (5,3,4,8)
1358 6185 rt : (5,8,-9,-7)
1358 6509 $\ell_{10} : 7(e+\pi)/3$
1358 8976 $Ei : 9\zeta(3)/4$
1358 9138 $\ln(3/4)\div\exp(3/4)$
1359 1839 rt : (2,5,8,1)
1359 1997 cr : $6\sqrt[3]{5}/7$
1359 2561 $2^x : P_{47}$
1359 2716 $\Gamma : 5e/3$
1359 4120 rt : (7,2,-5,2)
1359 4366 $J_2 : 10\pi^2/7$
1359 5180 $J_0 : 5e\pi/3$
1359 6297 rt : (3,7,-5,8)
1359 6416 rt : (1,8,0,9)
1359 9183 rt : (7,9,6,-1)
1359 9916 rt : (4,0,1,7)
1360 0763 rt : (3,3,-7,-1)
1360 4905 rt : (9,6,8,1)
1360 4918 $s\pi : 7\sqrt{5}/8$
1360 5228 $t\pi : 1/4+4\pi/3$
1360 7242 $\ell_2 : \sqrt{2}+4\sqrt{5}/3$
1360 7998 $\ell_{10} : 8\sqrt[3]{5}$
1360 8474 rt : (6,1,-8,-1)
1360 9467 sq : 19/13
1360 9688 $\ell_{10} : 1/4+\sqrt{5}/2$
1360 9973 sq : $4e/3$
1361 0166 $4\times\exp(1/4)$
1361 0185 $\Gamma : 22/5$
1361 2182 rt : (4,2,7,-1)
1361 2450 id : $2\sqrt{3}/3+4\sqrt{5}/3$
1361 3161 $\ell_{10} : 4\sqrt{2}/3-2\sqrt{3}/3$
1361 3514 $e^x : e-3\pi/2$
1361 5223 rt : (3,6,1,4)
1361 6113 $\Gamma : 19/23$
1361 6368 rt : (6,4,-8,1)
1361 6664 $s\pi : 1/23$
1361 7128 cr : 22/15
1361 8056 $t\pi : P_{23}$
1361 8133 $\zeta : 2\sqrt[3]{5}$
1361 8773 at : P_{24}
1361 8804 cu : 24/23
1362 1936 sr : $\sqrt{5/3}$
1362 3045 at : $\sqrt[3]{10}$
1362 4850 $\exp(2/5)\times\exp(\sqrt{2})$
1362 5434 cu : $1/\ln(\ln 3)$
1362 5535 $J_2 : \ln(3)$
1362 6520 $\Psi : 1/\ln(\sqrt{5}/2)$
1362 7670 $J_0 : 3e\pi/5$
1362 7757 at : $1+2\sqrt{3}/3$
1362 7814 rt : (6,9,6,4)
1362 9347 rt : (1,5,8,1)
1362 9391 sr : $3-4\sqrt{5}/3$
1362 9436 $1/4-\ln(4)$
1362 9647 rt : (8,3,1,9)
1363 0217 $t\pi : 3\sqrt{2}+3\sqrt{3}$
1363 0631 rt : (2,6,-5,-5)
1363 3890 rt : (4,3,-7,-1)
1363 5743 rt : (1,7,-9,2)
1363 5903 $10^x : 3-e/4$
1363 6363 id : 3/22
1363 6726 id : $\sqrt{3}/3+\sqrt{5}/4$
1363 7245 $2^x : 3/4-4e/3$
1363 7375 sq : $9\pi^2$
1363 9728 rt : (7,0,-2,8)
1364 0192 rt : (8,8,-3,-2)
1364 0529 rt : (1,6,-1,-1)
1364 1852 id : $3(1/\pi)/7$
1364 1880 rt : (8,6,8,1)
1364 4357 $t\pi : 8\pi^2$
1364 4886 rt : (3,2,7,-1)
1364 5748 $e^x : 18/11$
1364 6140 rt : (1,9,-8,-4)
1364 6366 $10^x : 1/18$
1364 8338 $2^x : 23/21$
1364 8776 $e^x : 2\sqrt{2}+3\sqrt{3}$
1364 9852 rt : (6,3,-5,-7)
1365 0479 $10^x : 3-\sqrt{3}/2$
1365 2667 $\ell_{10} : \sqrt{19}\pi$
1365 4611 $\ell_2 : 9\sqrt[3]{5}/7$
1365 5096 rt : (7,4,-8,1)
1365 5166 $J_2 : 7\sqrt{2}/9$
1365 6415 $J_2 : 11/10$
1365 6789 rt : (5,9,6,-1)
1365 7506 $\ell_2 : \exp(\sqrt{2}\pi/3)$
1365 8472 $J_1 : (3)^3$
1366 0019 rt : (5,9,-5,-3)

1366 0078 $t\pi : \exp(-\pi)$
1366 5090 $\zeta : \zeta(3)/3$
1366 5923 $10^x : 1/4 + 3\sqrt{2}/2$
1366 6193 rt : (7,5,-6,-3)
1366 6725 $\ln : 3/2 - \sqrt{2}/4$
1366 6744 rt : (9,3,-8,-8)
1366 7504 rt : (5,8,1,0)
1366 8198 $\zeta : 2e\pi/5$
1367 0718 $J_2 : \sqrt[3]{4/3}$
1367 1408 rt : (9,9,3,5)
1367 1776 $\Gamma : 1/\sqrt[3]{14}$
1367 2399 cu : $3\zeta(3)/7$
1367 5170 at : $2/3 + 2\sqrt{5}/3$
1367 5763 $\ln : 7\pi^2/3$
1367 8058 rt : (2,4,9,8)
1367 8265 $\ln : 4 + 2\sqrt{5}$
1367 8631 $Ei : 13/9$
1367 8664 sinh : 3/22
1367 8751 cu : $3/4 + \sqrt{2}$
1367 8982 as : 3/22
1367 9476 rt : (7,6,8,1)
1367 9913 rt : (5,7,-6,4)
1368 1349 cr : $2\sqrt{2} + 4\sqrt{3}$
1368 1788 $J_2 : 1/\ln(\sqrt{5}/2)$
1368 3464 $\ell_{10} : P_{23}$
1368 4204 sinh : $3(1/\pi)/7$
1368 4523 as : $3(1/\pi)/7$
1368 5274 $t\pi : 3\sqrt{2} + \sqrt{5}/2$
1368 6116 rt : (8,6,0,4)
1368 6333 $\ln(3/5) \times \exp(4/5)$
1368 8013 $2^x : \sqrt{2} + 3\sqrt{3}/2$
1368 8288 rt : (9,1,7,-1)
1369 2160 cr : $1/3 + 3\pi$
1369 2295 $\ell_2 : 7\pi/5$
1369 2468 rt : (6,2,-9,-4)
1369 3350 $J_0 : (2e/3)^3$
1369 3943 $10^x : 6\sqrt[3]{3}$
1369 4191 $t\pi : 10\sqrt{5}$
1369 4527 rt : (8,4,-8,1)
1369 6290 $2^x : (\sqrt[3]{5}/3)^3$
1369 7667 cu : $1/2 - 4\sqrt{2}$
1369 7725 $J_0 : 5\sqrt{5}$
1369 8062 rt : (5,4,6,9)
1369 8754 rt : (1,8,7,6)
1370 0761 cr : $7\sqrt[3]{2}/6$
1370 0914 rt : (7,3,-3,3)
1370 1618 rt : (6,3,-7,-1)
1370 3598 id : P_{24}
1370 3729 cr : $8e\pi/7$
1370 5462 sr : $2 - \sqrt{2}/2$
1370 6333 rt : (1,5,-8,1)
1370 8294 $\ell_2 : 2/3 + \sqrt{3}/4$
1370 9139 $\ln(\sqrt{3}) + s\pi(1/5)$
1370 9280 $\ln : 9\sqrt{3}/5$
1371 0561 $J_2 : 7\sqrt[3]{2}/8$
1371 1421 $e^x : \sqrt{2}/4 + 3\sqrt{3}/2$
1371 1711 rt : (1,2,7,-1)
1371 2868 $J : \ln(\sqrt{5}/2)$
1371 3524 sq : $8\sqrt[3]{5}$
1371 3885 at : $\sqrt{2} + \sqrt{5}/3$
1371 3888 rt : (4,1,3,8)
1371 4553 rt : (7,1,0,9)
1371 5804 rt : (9,0,-9,3)
1371 5859 id : $\sqrt{2}/4 - 2\sqrt{5}/3$
1371 6694 id : $9\pi/2$
1371 7717 rt : (6,6,8,1)
1371 8300 cr : 25/17
1371 9842 at : P_4
1372 0267 $\ln : 2 + \sqrt{5}/2$
1372 0290 $1/4 \div \exp(3/5)$
1372 0370 at : $1/3 - \sqrt{2}/3$
1372 1489 rt : (8,1,7,-1)
1372 1543 $J_2 : (\sqrt{5}/3)^{-1/3}$
1372 2584 rt : (9,2,-6,-9)
1372 3397 $e^x : 1/4 - \sqrt{5}$
1372 4943 rt : (6,2,-3,-8)
1372 5841 at : $\sqrt{14/3}$
1372 8507 $J_1 : 7(e+\pi)/2$
1372 9820 rt : (3,7,5,8)
1373 0159 $J_1 : 2\ln 2/5$
1373 0318 rt : (8,5,-9,-8)
1373 0785 rt : (5,3,-9,-1)
1373 2603 $t\pi : 7\sqrt{5}/8$
1373 2653 $1/4 \times \ln(\sqrt{3})$
1373 3051 $e^x : 18$
1373 4590 $J_1 : 1/\sqrt{13}$
1373 4688 rt : (9,4,-8,1)
1373 5379 rt : (5,5,-6,-7)
1373 6248 rt : (7,3,-7,-1)
1373 7056 $10^x : 19/8$
1373 7752 $J_1 : 7/2$
1373 7827 rt : (5,8,-2,8)
1373 9337 cr : $1 + \sqrt{2}/3$
1374 0502 rt : (7,9,-9,8)
1374 0799 $s\pi(1/4) - s\pi(\pi)$
1374 0951 rt : (1,7,0,7)
1374 2992 $\ell_2 : 7\sqrt{2}/9$
1374 3700 $\Psi : \sqrt{13}$
1374 4016 $\zeta : 3\sqrt[3]{5}/4$
1374 4049 cr : $\sqrt{2}/4 + \sqrt{5}/2$
1374 4206 $Ei : \exp(1/e)$
1374 4510 $t\pi : \exp(2\pi)$
1374 4683 $t\pi : 1/23$
1374 5568 $\sqrt{5} - \ln(3)$
1374 5860 rt : (4,8,-9,-6)
1374 6020 $\Psi : 5\sqrt[3]{3}/2$
1374 6855 as : P_{24}
1374 6993 id : $5\sqrt[3]{5}/4$
1375 0352 $\ell_2 : 5/11$
1375 1591 $\sqrt{3} - \ln(2/3)$
1375 2536 at : $2/3 - 2\sqrt{2}$
1375 3066 cr : $5(e+\pi)/3$
1375 4058 rt : (9,8,6,-1)
1375 4445 sq : $\sqrt{3} + \sqrt{5}/3$
1375 5167 rt : (7,1,7,-1)
1375 5359 $J_1 : 5/18$
1375 6447 $\ln : 9\ln 2/2$
1375 6624 rt : (5,6,8,1)
1375 7302 $\ln : e - \pi/2$
1375 9291 sr : 22/17
1376 1947 $10^x : 1/2 + \sqrt{2}/4$
1376 2010 rt : (1,8,-6,-5)
1376 3478 $\Psi : 3\zeta(3)$
1376 3726 $J_0 : 7\pi^2/4$
1376 4480 $\ln(2/3) \times \exp(4)$
1376 8580 $\Gamma : \sqrt[3]{23/2}$
1376 8957 sq : $2/3 + 3\pi/4$
1376 9349 $\ln(2/5) - \exp(1/5)$
1376 9705 at : $4\sqrt{3}/3 - 2\sqrt{5}$
1377 1409 rt : (8,3,-7,-1)
1377 1549 $e^x : 2\sqrt{2} + \sqrt{3}/3$
1377 2940 rt : (1,9,8,8)
1377 3730 $J_2 : 21/19$
1377 4814 tanh : $\ln 2/5$
1377 5147 at : $\ln 2/5$
1377 6295 rt : (1,9,6,-1)
1377 7777 sq : 16/15
1377 9576 $J : \exp(-10\pi/11)$
1378 0288 $\exp(2/3) - \exp(3)$
1378 0520 rt : (1,5,0,-5)
1378 0966 at : $3\sqrt[3]{3}/2$
1378 1711 $2^x : \sqrt{5} - \sqrt{6} - \sqrt{7}$
1378 2601 $J_1 : 7\pi^2/4$
1378 2820 id : $2\sqrt{2} + 4\sqrt{3}/3$
1378 4110 rt : (8,8,6,-1)
1378 6110 rt : (3,5,-8,1)
1378 6741 at : $9\zeta(3)/5$
1378 7171 rt : (8,4,-7,-9)
1378 7277 $e^x : 1 - 4\sqrt{5}/3$
1378 7330 $\ln : 2\sqrt[3]{3}/9$
1378 9112 rt : (9,1,-7,4)
1378 9335 rt : (6,1,7,-1)
1378 9564 $\ln(3/5) + \exp(1/2)$
1379 0938 sq : $7(1/\pi)/6$
1379 1233 $J_1 : 9e\pi/8$
1379 1455 $J_1 : 7(1/\pi)/8$
1379 1511 $\ell_{10} : 2/3 + \sqrt{2}/2$
1379 1697 $\ell_{10} : 8\zeta(3)/7$
1379 2489 $e^x : (\sqrt{3})^{-1/2}$
1379 2727 at : $1/\ln(\sqrt[3]{4})$
1379 4207 sr : $3 + \pi/2$
1379 5494 rt : (8,7,2,5)
1379 5735 at : $3/4 + \sqrt{2}$
1379 6222 rt : (4,6,8,1)
1379 6271 $\zeta : \ln(\sqrt[3]{3}/3)$
1379 8025 rt : (1,8,-1,1)
1379 9656 $\Psi : 7/15$
1380 0549 $10^x : 3\sqrt{2}/2 - 4\sqrt{5}/3$
1380 1338 cu : $4\sqrt{2}/3 + 3\sqrt{3}/2$
1380 1608 rt : (6,1,-1,-9)
1380 4446 $\ln(5) \times s\pi(1/4)$
1380 4488 $e^x : 4/3 + 3e/4$
1380 5112 $E : (\sqrt[3]{2})^{-1/3}$
1380 5228 $1/3 + \ln(\sqrt{5})$
1380 6276 $Li_2 : 2/15$
1380 6580 id : P_4
1380 7118 id : $1/3 - \sqrt{2}/3$
1380 7912 rt : (2,9,-5,8)
1380 9286 $\ln : 1 - e/4$
1381 0684 at : $5\sqrt{3}/4$
1381 0854 rt : (7,4,-1,4)
1381 0979 $10^x : 1 + \sqrt{5}$
1381 2325 rt : (5,4,-4,-8)
1381 2523 $s\pi : 4\sqrt{2} + 3\sqrt{3}/4$
1381 4536 rt : (7,8,6,-1)
1381 5328 $2^x : 7\sqrt{2}/6$
1381 5705 rt : (2,8,3,-4)
1381 6963 $2^x : 2\sqrt{2} - \sqrt{3}$
1381 7012 $\Gamma : 4\sqrt[3]{3}/7$
1381 7256 $2^x : 4/3 - 4\pi/3$
1381 8255 $\theta_3 : 23/25$
1382 1729 $\ell_{10} : (\sqrt[3]{4}/3)^{1/2}$
1382 3089 rt : (4,7,-7,-7)
1382 3290 $\ell_{10} : P_{73}$
1382 4008 rt : (5,1,7,-1)
1382 4022 $\ell_{10} : \exp(1/\pi)$
1382 5336 $e^x : 4\sqrt[3]{5}/9$
1382 5471 $erf : \exp(-2\pi/3)$
1382 5609 $\ln : 1 + 3\sqrt{2}/2$
1382 6312 rt : (6,1,-4,3)
1382 6529 rt : (2,6,2,9)
1382 7110 rt : (4,5,-8,1)
1382 7622 $e^x : 19/25$
1382 8114 $J_2 : (e/2)^{1/3}$
1382 8843 cu : $2\sqrt{2}/3 + \sqrt{3}$
1382 9962 $2^x : 3/4 + 3\sqrt{3}/4$
1383 0239 $2^x : 1/2 - 3\sqrt{5}/2$
1383 0269 $\ell_{10} : 11/8$
1383 0806 $Li_2 : \zeta(3)/9$
1383 1175 $10^x : 1/2 - e/2$
1383 1563 rt : (4,2,5,9)
1383 3305 $\ln : 6\sqrt{2}$
1383 4523 rt : (8,9,-1,-1)
1383 4583 $\ell_2 : (4/3)^{-1/3}$
1383 5279 $\ln : (\pi/3)^3$
1383 6535 rt : (3,6,8,1)
1383 7531 $10^x : 3 - e/3$
1383 8339 rt : (1,9,-6,-1)
1383 8855 at : 13/6
1383 9389 $J_1 : \sqrt{5}/8$
1384 1868 rt : (5,2,-7,-2)
1384 3876 cu : $2 + 2\sqrt{3}$
1384 4305 $Ei : 9/22$
1384 5345 rt : (6,8,6,-1)
1384 6189 sq : $3/2 - \sqrt{3}/4$
1384 8163 rt : (3,1,2,-8)
1385 0824 as : P_4
1385 1368 as : $1/3 - \sqrt{2}/3$
1385 1432 rt : (3,2,-7,-1)
1385 2986 rt : (7,6,-8,-9)
1385 3484 $2^x : \pi^2/9$
1385 5112 $e^x : \sqrt{2}/2 - \sqrt{3}/3$
1385 5652 $J_1 : 10\sqrt[3]{2}/3$
1385 6216 $\ell_{10} : 2\sqrt{2}/3 + \sqrt{3}/4$
1385 8005 $J_2 : 8\ln 2/5$
1385 9201 rt : (4,1,7,-1)
1386 0718 $\ell_{10} : 3\sqrt{2}/2 - \sqrt{5}/3$
1386 1258 $2^x : 5\pi^2/6$
1386 1322 $e^x : 1/3 - 4\sqrt{3}/3$
1386 1484 $Ei : 9e/5$

1386 2228 rt : (7,6,-4,-2)
1386 2395 J_1 : $2\sqrt[3]{2}/9$
1386 2943 id : $\ln 2/5$
1386 3247 J_1 : $7/25$
1386 3884 rt : (9,2,-5,5)
1386 4688 $\ln(4/5) \div \ln(5)$
1386 4694 J_1 : $21/5$
1386 4880 rt : (4,4,-6,8)
1386 6377 ζ : $17/5$
1386 7446 ln : $1/3 + 3e$
1386 7560 θ_3 : $\ln 2/10$
1386 7966 10^x : $1/2 + 3\sqrt{2}/2$
1386 8889 rt : (5,5,-8,1)
1386 9968 rt : (2,9,-6,-1)
1387 0754 J_2 : $8\pi^2/3$
1387 0924 Ei : $4\pi/5$
1387 1814 cu : $1/3 + 3\sqrt{5}/2$
1387 4376 rt : (8,1,-8,-4)
1387 6548 rt : (5,8,6,-1)
1387 7590 rt : (2,6,8,1)
1387 8863 2^x : $3/16$
1388 0149 $e \div s\pi(1/3)$
1388 0262 id : $1/3 - 2\sqrt{5}$
1388 0640 Γ : $14/17$
1388 1715 rt : (1,7,-4,-6)
1388 3705 2^x : P_9
1388 5133 cu : $9e/5$
1388 5955 rt : (9,7,8,1)
1388 7286 rt : (6,8,-8,5)
1388 7718 rt : (4,2,-7,-1)
1388 8888 sq : $\sqrt{5}/6$
1388 9161 Li_2 : $(\ln 3/3)^2$
1389 0240 rt : (6,3,-7,-3)
1389 0918 rt : (5,3,-2,-9)
1389 3417 ζ : $5e/4$
1389 4580 sr : $1/4 + \pi/3$
1389 4799 $s\pi$: $7e\pi/5$
1389 4929 rt : (3,1,7,-1)
1389 5833 2^x : $5\pi/6$
1389 6146 id : $2\sqrt{2}/3 + 3\sqrt{3}$
1389 7533 Li_2 : $((e+\pi)/3)^{-3}$
1390 1647 $s\pi$: $3\sqrt{3}/4 + \sqrt{5}/3$
1390 1978 rt : (4,6,-5,1)
1390 2015 rt : (3,9,-6,-1)
1390 2509 J_1 : 5π
1390 3370 K : $(\pi/2)^{-1/3}$
1390 3502 2^x : $2 - 2e/3$
1390 4442 J_2 : $10/9$
1390 4751 J_2 : 11
1390 4892 Ei : $9(1/\pi)/7$
1390 5609 $1/4 - \exp(2)$
1390 7389 sinh : $\ln 2/5$
1390 7735 as : $\ln 2/5$
1390 7908 cu : $5\zeta(3)/2$
1390 8153 rt : (8,8,4,6)
1390 8384 tanh : $\sqrt[3]{2}/9$
1390 8733 at : $\sqrt[3]{2}/9$
1391 1392 at : $(3\pi/2)^{1/2}$
1391 1478 rt : (6,5,-8,1)
1391 2901 2^x : $3/2 + \sqrt{5}/2$
1391 3089 rt : (3,9,-8,-7)
1391 3652 rt : (9,4,-8,-1)
1391 5264 sr : $1/3 + 3\sqrt{2}$
1391 7150 Ψ : $21/13$
1391 7164 10^x : $(1/(3e))^{1/3}$
1391 9414 rt : (1,6,8,1)
1392 0073 rt : (6,8,-9,-9)
1392 1522 cu : $3\sqrt{2}/4 + 3\sqrt{3}/4$
1392 2926 J_2 : $\exp(1/(3\pi))$
1392 3368 ζ : $e\pi/7$
1392 4110 rt : (7,5,1,5)
1392 4580 rt : (5,2,-7,-1)
1392 5330 10^x : $3/2 - 3\pi/4$
1392 5913 J_0 : $6\pi/7$
1392 7038 rt : (8,7,8,1)
1392 8266 2^x : $1/2 + 2\pi$
1392 9553 $\ln(\pi) - \exp(1/4)$
1393 0841 ln : $3\sqrt{2} - \sqrt{5}/2$
1393 0883 sr : $9\sqrt[3]{3}/10$
1393 1209 rt : (2,1,7,-1)
1393 1726 e^x : $7\sqrt{2}/2$
1393 2363 e^x : $3/23$
1393 4466 rt : (3,8,-7,-8)
1393 4492 rt : (4,9,-6,-1)
1393 5669 rt : (6,9,-4,9)
1393 7609 rt : (9,0,7,1)
1394 0172 rt : (9,3,-3,6)
1394 3428 ln : $8/25$
1394 6200 $t\pi$: $4\sqrt{2} + 3\sqrt{3}/4$
1394 6236 2^x : P_{36}
1394 7429 sr : $4 + \sqrt{3}/3$
1394 7809 rt : (3,7,-6,5)
1394 8092 sinh : $(e+\pi)/6$
1394 8532 sq : $e\pi/8$
1394 9123 sr : $1/2 + 3e/2$
1395 0210 J_1 : $9e/7$
1395 0939 rt : (8,0,-6,5)
1395 4909 rt : (7,5,-8,1)
1395 6340 rt : (5,1,-5,-3)
1395 6592 $\ln(\sqrt{3}) \times \exp(\sqrt{5})$
1395 7151 J : $\exp(-4/3)$
1396 0507 $t\pi$: $\sqrt{2}/4 - 4\sqrt{3}/3$
1396 1468 cr : $3\sqrt{3}/2 - \sqrt{5}/2$
1396 1753 rt : (7,3,-9,-4)
1396 2038 rt : (6,8,1,0)
1396 5512 $t\pi$: $\exp(4)$
1396 7411 rt : (5,9,-6,-1)
1396 8058 rt : (1,2,8,8)
1396 8882 rt : (7,7,8,1)
1396 9667 e^x : $3/4 - e$
1397 0484 rt : (3,0,4,9)
1397 0659 sq : $1/3 - \sqrt{2}/2$
1397 0953 Ψ : $\sqrt[3]{7/3}$
1397 1049 J_2 : $7(1/\pi)/2$
1397 2363 E : $(\sqrt[3]{5}/2)^{1/2}$
1397 2615 rt : (4,8,-4,-6)
1397 3867 rt : (8,0,7,1)
1397 5352 sr : $3\sqrt{3}/4$
1397 5454 $\ln(3) \div s\pi(\sqrt{2})$
1397 6194 ln : $20/23$
1397 6764 ℓ_2 : $\sqrt{2}/3 + \sqrt{3}$
1397 7659 rt : (9,7,6,-1)
1397 8010 ζ : $\sqrt{23/2}$
1397 8197 at : $4e/5$
1397 9255 cu : $7(e+\pi)/5$
1398 0455 $\ln(3) \times \exp(2/3)$
1398 1453 10^x : $2\sqrt[3]{5}/9$
1398 1940 cu : $2\sqrt{2} - 4\sqrt{3}/3$
1398 2474 as : $(4/3)^{-1/3}$
1398 2594 rt : (4,7,-1,5)
1398 5044 id : $(\sqrt{2}\pi/3)^{1/3}$
1398 7498 2^x : $\exp(-5/3)$
1398 7925 rt : (2,3,1,8)
1398 9984 Ψ : $8\sqrt{2}/7$
1399 3841 at : $3/2 - e/2$
1399 4017 rt : (3,8,-6,-8)
1399 4066 J_2 : $7\pi/2$
1399 4285 e^x : $2/3 + 3\sqrt{3}/4$
1399 4741 J_2 : $(\sqrt[3]{3}/2)^{-1/3}$
1399 6693 rt : (3,9,-1,1)
1399 7031 J_2 : $4\pi^2/5$
1399 8396 id : $2\sqrt[3]{5}/3$
1399 9122 id : $\sqrt[3]{2}/9$
1399 9216 rt : (8,5,-8,1)
1400 0111 rt : (7,2,-7,-1)
1400 0783 rt : (6,9,-6,-1)
1400 0879 sq : P_{74}
1400 1184 J_1 : $\sqrt{2}/5$
1400 4612 ℓ_{10} : $1/3 + \pi/3$
1400 5494 rt : (1,9,7,0)
1400 6006 J_1 : $8(1/\pi)/9$
1400 6616 ln : $2/17$
1400 8030 cu : $4\sqrt{3} - \sqrt{5}/2$
1401 0057 rt : (8,7,6,-1)
1401 0685 rt : (7,0,7,1)
1401 1513 rt : (6,7,8,1)
1401 1948 id : $\sqrt{2}/2 + \sqrt{3}/4$
1401 2887 J_0 : $9(e+\pi)/2$
1401 3260 at : P_{70}
1401 3274 Ψ : $(\ln 3)^3$
1401 3615 at : P_{67}
1401 3966 ℓ_2 : P_{37}
1401 5666 ℓ_{10} : $2e - 3\pi/2$
1401 7542 sr : $13/10$
1401 7817 at : $e/3 - \pi/3$
1401 8034 rt : (9,4,-1,7)
1401 8649 cu : $4\sqrt{3}/3 - 3\sqrt{5}/2$
1402 0495 at : $1/2 + 3\sqrt{5}/4$
1402 1258 Li_2 : $(e)^{-2}$
1402 1386 sinh : $4\sqrt[3]{5}/7$
1402 3192 rt : (9,5,-6,0)
1402 3212 J_0 : $8\sqrt{5}/3$
1402 4276 rt : (8,9,6,7)
1402 5155 ℓ_{10} : $3\sqrt{3} - 2\sqrt{5}$
1402 6209 id : $4\sqrt{2}/3 - \sqrt{5}/3$
1402 6220 cu : $3/4 - 3e/4$
1402 6571 Γ : $9(1/\pi)/10$
1402 6852 sq : $4e/3 + 4\pi$
1402 8185 e^x : $3/2 - 2\sqrt{3}$
1402 9087 rt : (8,1,-4,6)
1402 9611 cu : $3\sqrt{3}/10$
1403 0903 $t\pi$: $7e\pi/5$
1403 1620 erf : $1/8$
1403 2899 sinh : $(\pi/3)^{-1/2}$
1403 4621 rt : (7,9,-6,-1)
1403 5135 E : $4\ln 2/3$
1403 7954 $t\pi$: $3\sqrt{3}/4 + \sqrt{5}/3$
1403 8182 rt : (2,9,-6,3)
1403 8820 rt : (8,6,-1,0)
1403 9528 sinh : $9\sqrt[3]{2}$
1404 0189 rt : (7,2,-7,-5)
1404 0869 rt : (7,6,3,6)
1404 1419 rt : (5,5,-8,9)
1404 1623 J_1 : $\exp(-\sqrt[3]{2})$
1404 2433 id : $3\sqrt{3} + 4\sqrt{5}$
1404 2778 ℓ_{10} : $7\pi^2/5$
1404 2850 θ_3 : $(\ln 2)^{1/2}$
1404 2885 rt : (7,7,6,-1)
1404 3329 2^x : $\sqrt[3]{9/2}$
1404 3536 rt : (1,7,9,-3)
1404 4101 Ψ : $1/\ln(\sqrt[3]{5}/2)$
1404 4219 sq : $1/\sqrt[3]{19}$
1404 4437 rt : (9,5,-8,1)
1404 4892 sinh : $\sqrt[3]{2}/9$
1404 5255 as : $\sqrt[3]{2}/9$
1404 8082 rt : (6,0,7,1)
1404 8602 tanh : $\sqrt{2}/10$
1404 8970 at : $\sqrt{2}/10$
1404 9107 at : $3\sqrt{2}/4 + \sqrt{5}/2$
1404 9477 cu : $3\ln 2/4$
1404 9707 10^x : $(e)^2$
1404 9814 rt : (2,6,-8,1)
1405 0977 J_2 : $19/17$
1405 2349 $\ln(2/3) \times \ln(\sqrt{2})$
1405 3475 tanh : $4(1/\pi)/9$
1405 3843 at : $4(1/\pi)/9$
1405 4596 $\exp(1/2) + \exp(2/5)$
1405 4964 rt : (5,7,8,1)
1405 7574 rt : (6,3,0,5)
1405 9234 sr : $2\sqrt{3} + \sqrt{5}/2$
1405 9666 J_2 : $\sqrt{5}/2$
1406 0800 cu : $13/25$
1406 2500 sq : $3/8$
1406 4147 Ψ : 9
1406 4350 at : $3/2 + e/4$
1406 5006 rt : (4,8,3,9)
1406 5763 at : $3 - \pi$
1406 6650 2^x : $e/2 - 4\pi/3$
1406 8293 Γ : $(\sqrt{2}\pi/3)^{-1/2}$
1406 8413 rt : (7,7,-2,-1)
1406 8825 ℓ_2 : $7\sqrt[3]{2}$
1406 8939 rt : (8,9,-6,-1)
1406 9263 id : $\exp(\pi)$
1406 9514 sr : $\sqrt{21}$
1406 9603 cu : $5(e+\pi)$
1407 0469 ℓ_2 : $4e/3 + \pi/4$
1407 2397 id : $1/\ln(e/3)$
1407 2422 cr : $\sqrt{3}/2 + 4\sqrt{5}$
1407 2605 rt : (1,8,-6,-1)
1407 3659 Li_2 : $6\ln 2/5$

1407 4393 rt : (5,0,-3,4)
1407 5126 10^x : $1-2\sqrt{2}/3$
1407 6153 rt : (6,7,6,-1)
1407 6160 2^x : $\sqrt[3]{5}/9$
1407 6756 rt : (3,7,-4,-9)
1407 8185 rt : (9,2,-7,-1)
1407 8271 Ψ : $\exp(\sqrt[3]{3}/3)$
1407 8712 cr : $6\sqrt{3}/7$
1408 1606 J_2 : $5(e+\pi)$
1408 2205 rt : (2,7,1,-5)
1408 2846 rt : (9,3,-9,-7)
1408 3073 Ψ : $7\ln 2/3$
1408 5908 id : $3/2-e/2$
1408 6076 rt : (5,0,7,1)
1408 6359 tanh : $\exp(-(e+\pi)/3)$
1408 6731 at : $\exp(-(e+\pi)/3)$
1408 8304 θ_3 : $8/11$
1409 1328 rt : (4,9,0,-2)
1409 2334 rt : (2,6,-8,6)
1409 5034 rt : (3,6,-8,1)
1409 5787 ln : $2\sqrt{2}-3\sqrt{5}/4$
1409 7240 Ei : $5e/2$
1409 7431 ζ : $\pi/5$
1409 7530 rt : (9,5,1,8)
1409 8352 rt : (6,4,-5,-2)
1409 9267 rt : (4,7,8,1)
1409 9963 $s\pi$: $(3\pi/2)^{-2}$
1410 2265 J_2 : $8\sqrt[3]{2}/9$
1410 2597 ℓ_2 : $2/3+3e$
1410 3404 at : 24/11
1410 3749 rt : (9,9,-6,-1)
1410 5612 J_1 : $\sqrt[3]{5}/6$
1410 5713 id : P_{70}
1410 6076 id : P_{67}
1410 6166 e^x : P_{49}
1410 6561 Γ : $(\sqrt{5}/3)^3$
1410 6862 rt : (2,8,-6,-1)
1410 8380 rt : (3,6,-9,-2)
1410 8881 rt : (8,2,-2,7)
1410 9011 rt : (2,7,-7,-7)
1410 9356 sq : $4-4e/3$
1410 9666 as : 10/11
1410 9874 rt : (5,8,7,5)
1411 0360 id : $e/3-\pi/3$
1411 0961 10^x : $1/3+\sqrt{3}/3$
1411 1465 J_1 : $7\zeta(3)/2$
1411 1705 rt : (5,7,-7,2)
1411 2000 $s\pi$: $3(1/\pi)$
1411 3729 sr : $2(e+\pi)/9$
1411 4030 2^x : 4/21
1411 5254 rt : (2,2,3,-9)
1411 5259 id : $2\sqrt{3}+3\sqrt{5}/4$
1411 8176 θ_3 : $2\sqrt[3]{2}/5$
1411 8205 J_0 : 5π
1411 8529 Li_2 : $\exp(-1/(2e))$
1411 9119 ln : $\sqrt{5}/7$
1411 9535 J_0 : $\exp(\sqrt{2}\pi/2)$
1412 0284 rt : (7,1,-5,-6)
1412 1198 J_1 : $\sqrt{20}\pi$
1412 1521 rt : (3,1,-7,-1)
1412 1663 id : $\sqrt{3}/4+3\sqrt{5}$
1412 4687 rt : (4,0,7,1)
1412 4718 cu : 23/11
1412 6922 e^x : P_{64}
1412 8494 2^x : $2e-2\pi/3$
1412 8644 rt : (5,1,-8,-3)
1412 9482 J_1 : $10\pi/9$
1413 1741 rt : (6,4,-8,-5)
1413 1783 Li_2 : 3/22
1413 2909 as : $3/2-e/2$
1413 2915 ℓ_{10} : 13/18
1413 3281 ℓ_2 : $(\sqrt{5}/3)^{1/3}$
1413 3611 rt : (1,7,8,-8)
1413 5033 Γ : $8\sqrt{2}/5$
1413 5894 rt : (9,6,-4,1)
1413 6248 $t\pi$: $4\sqrt{3}/3-3\sqrt{5}/2$
1413 7684 Li_2 : $3(1/\pi)/7$
1414 0436 J_1 : 2/7
1414 0805 $\ln(3/5)+\exp(\sqrt{3})$
1414 1057 Γ : $\ln(\ln 3)$
1414 1193 rt : (4,6,-8,1)
1414 1605 rt : (3,8,-6,-1)
1414 1818 erf : $\sqrt[3]{2}/10$
1414 2135 id : $\sqrt{2}/10$
1414 3194 J_2 : $7\sqrt[3]{3}/9$
1414 3453 rt : (9,2,-7,-8)
1414 3825 Γ : $(e/3)^2$
1414 4059 rt : (4,7,6,-1)
1414 4452 e^x : $\sqrt{2}/4-4\sqrt{3}/3$
1414 4456 rt : (3,7,8,1)
1414 4725 $\exp(1/3)-s\pi(\sqrt{3})$
1414 6664 e^x : $(\ln 2)^{-3}$
1414 7106 id : $4(1/\pi)/9$
1414 8095 $s\pi$: $2\sqrt{3}+2\sqrt{5}/3$
1414 8606 2^x : ln(3)
1415 2109 rt : (8,3,-7,0)
1415 2113 θ_3 : $2(1/\pi)/9$
1415 2914 as : P_{70}
1415 3280 as : P_{67}
1415 3696 $4/5\div\exp(\sqrt{3})$
1415 4126 2^x : $3/4-\sqrt{5}/4$
1415 4350 e^x : $\sqrt[3]{3/2}$
1415 4357 2^x : $3(1/\pi)/5$
1415 7172 2^x : 10π
1415 7608 as : $e/3-\pi/3$
1415 8110 cu : $2\sqrt{3}-3\sqrt{5}$
1415 9225 J_2 : $(\sqrt[3]{2})^{1/2}$
1415 9265 id : π
1416 1330 rt : (7,7,5,7)
1416 1383 rt : (3,3,-5,5)
1416 1506 rt : (4,1,-7,-1)
1416 1892 rt : (2,8,1,9)
1416 3304 e^x : 13/2
1416 3934 rt : (3,0,7,1)
1416 5061 ℓ_{10} : $8\sqrt{3}$
1416 5166 sq : $4+2\pi/3$
1416 6215 cr : $1/3+2\sqrt{3}/3$
1416 8161 sq : $8\zeta(3)/9$
1417 0951 Ψ : $(\sqrt[3]{5}/3)^{-1/2}$
1417 1519 e^x : 11
1417 3414 id : $\sqrt{3}-\sqrt{5}+\sqrt{7}$
1417 3535 rt : (2,9,-5,-9)
1417 4878 rt : (5,8,-3,6)
1417 6847 rt : (4,8,-6,-1)
1417 7500 J_1 : $9(1/\pi)/10$
1417 8722 rt : (9,6,3,9)
1417 9549 2^x : P_{52}
1417 9714 λ : $(3\pi/2)^{-3}$
1418 0654 id : $\exp(-(e+\pi)/3)$
1418 2114 ℓ_{10} : $1/4+\sqrt{2}/3$
1418 3166 $\ln(2/5)-\exp(4/5)$
1418 5024 Γ : $\ln(\sqrt[3]{5}/2)$
1418 5025 ζ : $(3/2)^3$
1418 5325 θ_3 : $(\sqrt[3]{4}/3)^{1/2}$
1418 5545 ℓ_{10} : $2\ln 2$
1418 6068 cr : $4e-\pi/3$
1418 7535 E : 12/13
1418 7681 10^x : $e/2+3\pi$
1418 8317 10^x : $4e-\pi/3$
1418 8329 rt : (5,6,-8,1)
1418 9319 tanh : 1/7
1418 9323 sinh : $\sqrt{2}/10$
1418 9705 at : 1/7
1419 0385 rt : (8,3,0,8)
1419 0567 rt : (2,7,8,1)
1419 2372 ℓ_{10} : $3\zeta(3)/5$
1419 2705 rt : (9,8,8,1)
1419 2710 2^x : $\sqrt{2}/4+\sqrt{5}/3$
1419 4343 sinh : $4(1/\pi)/9$
1419 4726 as : $4(1/\pi)/9$
1419 6237 rt : (5,1,-1,5)
1419 7502 2^x : $3\sqrt{3}/2+\sqrt{5}/2$
1419 8957 ℓ_{10} : $\sqrt[3]{3}/2$
1419 9385 rt : (2,9,2,-4)
1420 1794 sr : $4e+2\pi$
1420 1946 J_0 : $9\pi^2/4$
1420 2103 rt : (7,0,-3,7)
1420 2183 rt : (5,1,-7,-1)
1420 2350 cu : 12/23
1420 3504 sq : $4\sqrt{2}/3+3\sqrt{3}/4$
1420 3839 rt : (2,0,7,1)
1420 4979 sq : $10\pi/7$
1420 5041 rt : (9,1,-5,-9)
1420 5432 rt : (9,1,7,1)
1420 5627 ln : $4e-3\pi/4$
1420 7009 as : $3-\pi$
1420 7224 at : $4-2e/3$
1420 8197 sinh : 14
1420 9336 sinh : $5\sqrt{2}/2$
1421 2208 cu : $1/4-2\sqrt{2}$
1421 2602 rt : (5,8,-6,-1)
1421 3562 id : $10\sqrt{2}$
1421 3876 rt : (8,4,-8,-8)
1421 5108 rt : (9,6,6,-1)
1421 5642 ℓ_2 : $3+\sqrt{2}$
1421 6382 J_2 : 9/8
1421 7214 10^x : $1/\ln(\sqrt{3})$
1421 7397 θ_3 : $(\sqrt[3]{3})^{-1/2}$
1421 7437 ℓ_{10} : $3+4e$
1421 7499 $t\pi$: $3\sqrt{3}/2$
1421 9442 sq : $5\sqrt[3]{5}/8$
1422 4081 $\sqrt{5}\times\ln(3/5)$
1422 4121 at : $2-4\pi/3$
1422 5706 rt : (5,9,-6,-5)
1422 5909 Γ : $(2e/3)^{-1/3}$
1422 6533 $t\pi$: $3\sqrt{2}/4+2\sqrt{5}/3$
1422 6745 ℓ_2 : $e/3$
1422 7975 sr : $(\sqrt{2}\pi/2)^{1/3}$
1422 8197 cu : $1/3+3\sqrt{5}$
1422 8229 sinh : $\exp(-(e+\pi)/3)$
1422 8616 as : $\exp(-(e+\pi)/3)$
1422 8837 rt : (9,7,-9,-6)
1423 0152 Ei : $5\sqrt{2}/4$
1423 0203 ln : $4-\sqrt{3}/2$
1423 0484 sq : $2+3\sqrt{3}/2$
1423 1483 $s\pi$: 1/22
1423 1642 rt : (3,5,-7,-3)
1423 3570 ℓ_2 : $3\sqrt{2}/2+3\sqrt{5}$
1423 4665 cr : $2\sqrt{5}/3$
1423 4804 $\ln(2/5)+s\pi(e)$
1423 5221 J_0 : 15/7
1423 5300 e^x : 16/21
1423 5684 10^x : $2e-2\pi$
1423 5810 J_1 : ln(4/3)
1423 6488 rt : (6,6,-8,1)
1423 7172 $s\pi$: $(1/\pi)/7$
1423 7184 $\ln(\sqrt{5})\div\exp(\sqrt{3})$
1423 7639 rt : (1,7,8,1)
1423 8738 rt : (8,8,8,1)
1424 0518 Li_2 : $(\ln 2)^{1/2}$
1424 0854 at : $\sqrt[3]{21/2}$
1424 1044 J_0 : $\exp(2\pi/3)$
1424 1264 rt : (6,8,-9,3)
1424 1499 sr : $4e-2\pi$
1424 2248 $t\pi$: $(3\pi/2)^{-2}$
1424 3577 rt : (6,1,-7,-1)
1424 4425 rt : (1,0,7,1)
1424 5239 rt : (8,1,7,1)
1424 5736 rt : (1,9,6,-8)
1424 7144 J_1 : $(\ln 2/3)^{-2}$
1424 7978 rt : (3,4,-1,9)
1424 8886 rt : (6,8,-6,-1)
1424 9535 rt : (1,7,6,-1)
1424 9772 cr : $\ln(\sqrt{2}\pi)$
1425 0163 rt : (8,6,6,-1)
1425 1926 rt : (9,7,-2,2)
1425 2331 cu : $2\sqrt{3}/3-3\sqrt{5}/4$
1425 4241 10^x : $8\sqrt[3]{3}/9$
1425 4654 $t\pi$: $3(1/\pi)$
1425 5694 ζ : $4\sqrt{2}/9$
1425 6043 Ei : 12/11
1425 8202 J_2 : 6π
1426 1914 $\ln(3/4)-s\pi(\pi)$
1426 3081 e^x : 2/15
1426 5589 cu : 23/22
1426 5854 2^x : $2+2\sqrt{3}$
1426 6061 Γ : $8\pi^2/7$
1426 6355 rt : (1,5,5,6)
1426 6750 ℓ_{10} : 18/25
1426 8119 Li_2 : $(\sqrt[3]{3})^{-1/2}$

1426 8764 rt : (8,4,-5,1)
1426 9471 ℓ_{10} : $4\sqrt[3]{2}/7$
1427 0268 2^x : $\sqrt{3}/9$
1427 3014 J_1 : $\sqrt[3]{3}/5$
1427 3667 rt : (8,4,2,9)
1427 5851 rt : (2,9,-7,-2)
1427 6328 Ei : $6\sqrt[3]{5}$
1427 6660 rt : (8,3,-6,-9)
1427 6761 2^x : $1/3-\pi$
1428 0004 rt : (1,7,-2,-7)
1428 0650 $\ln(2)\times\exp(1/2)$
1428 3926 J_1 : $\sqrt{3}/6$
1428 5714 id : 1/7
1428 6113 at : $\ln(\sqrt{3}/2)$
1428 6604 2^x : $4\pi^2/5$
1428 6609 sq : $3\sqrt[3]{2}/10$
1428 9198 e^x : $\zeta(3)/9$
1428 9373 sq : $3\sqrt{3}/4-3\sqrt{5}/4$
1428 9627 at : P_{75}
1428 9709 erf : $2(1/\pi)/5$
1429 0920 θ_3 : 1/14
1429 1857 $t\pi$: $2\sqrt{3}+2\sqrt{5}/3$
1429 1966 e^x : $-\sqrt{3}+\sqrt{5}-\sqrt{6}$
1429 2444 at : $4\sqrt{2}-2\sqrt{3}$
1429 2765 Ei : 8/11
1429 3173 rt : (6,9,-5,7)
1429 7819 rt : (6,2,-4,-7)
1429 8508 J_0 : $\sqrt{3}-\sqrt{5}+\sqrt{7}$
1429 9110 Ψ : $9\sqrt[3]{2}/7$
1430 0285 Γ : $5e/6$
1430 0555 ℓ_{10} : $4/3+4\pi$
1430 0927 at : $2\pi^2/9$
1430 1508 $\ln(\sqrt{5})-\exp(2/3)$
1430 2228 rt : (4,9,-9,7)
1430 3840 rt : (6,5,4,7)
1430 4633 ℓ_2 : $3-2\pi/3$
1430 5978 id : $e+3\pi$
1430 6054 ln : $4\sqrt[3]{3}/5$
1430 6559 e^x : $\sqrt{11}\pi$
1430 7058 2^x : $2/3+\sqrt{3}/4$
1430 8809 cr : $1/\ln((e+\pi)/3)$
1430 9983 rt : (5,5,-7,-6)
1431 0084 ln : 13/15
1431 0247 cr : $3\sqrt{2}/4+\sqrt{3}/4$
1431 1658 ζ : $7\sqrt[3]{3}/3$
1431 3359 J_2 : $9e\pi/7$
1431 3530 cr : $\sqrt[3]{10/3}$
1431 3959 rt : (4,4,-7,6)
1431 4744 $s\pi$: $1/3+3\pi/2$
1431 6480 rt : (3,7,-7,2)
1431 7786 J_1 : $\exp((e+\pi)/2)$
1431 7799 rt : (6,5,-3,-1)
1431 7940 sq : $9\sqrt[3]{2}/5$
1432 0795 $s\pi$: $8\sqrt[3]{5}/7$
1432 1775 rt : (6,6,6,-1)
1432 2098 rt : (5,2,1,6)
1432 2427 rt : (1,7,-6,-1)
1432 2535 Ei : $\exp(-1/\pi)$
1432 3103 rt : (8,8,-6,-1)
1432 3320 tanh : $\sqrt[3]{3}/10$
1432 3725 at : $\sqrt[3]{3}/10$
1432 5370 J_2 : $(\sqrt[3]{3})^{1/3}$
1432 5669 rt : (1,8,5,-8)
1432 6657 ζ : $5\sqrt[3]{2}/8$
1432 6879 rt : (6,1,7,1)
1432 7732 rt : (1,6,5,-7)
1432 7997 J_2 : $(\ln 2)^{-1/3}$
1432 8620 rt : (8,1,-7,-1)
1433 0626 $1/5\div\exp(1/3)$
1433 2969 J_0 : $5(e+\pi)$
1433 3463 Ei : $(\sqrt[3]{4}/3)^{1/2}$
1433 3673 rt : (6,8,8,1)
1433 4354 sinh : 1/7
1433 4756 as : 1/7
1433 4836 rt : (1,7,-8,1)
1433 6057 rt : (8,6,-8,1)
1433 6377 e^x : $1-\sqrt{3}/2$
1433 6936 ln : $5\ln 2/4$
1433 8565 ζ : $-\sqrt{3}+\sqrt{6}+\sqrt{7}$
1433 9627 J_2 : $\exp(1/(3e))$
1434 0252 cu : $3/4-4\sqrt{2}$
1434 0491 rt : (4,1,-2,-5)
1434 3010 Ψ : $4e/3$
1434 3373 $1/3\times s\pi(\pi)$
1434 4030 rt : (9,1,-8,3)
1434 6141 ζ : $8\sqrt[3]{3}/9$
1434 6354 2^x : $7\sqrt{2}/9$
1434 7447 J_0 : $8\pi^2/3$
1434 9522 rt : (2,5,-5,-3)
1434 9738 rt : (7,5,-7,-9)
1434 9860 Ei : $\ln(e+\pi)$
1435 0285 rt : (5,2,-6,-2)
1435 0904 at : $3-3\sqrt{3}$
1435 1338 e^x : $(\ln 3/3)^2$
1435 2983 sr : $(\sqrt[3]{5})^{1/2}$
1435 3167 cr : $3/4+\sqrt{5}/3$
1435 4374 sr : 17/13
1435 4692 2^x : 11/10
1435 4757 cu : $\pi/6$
1435 4796 J_2 : $9\sqrt{3}$
1435 7140 Γ : $(\ln 3/2)^{1/3}$
1435 7214 e^x : $1/3+3\sqrt{5}$
1435 7582 2^x : $2+4\sqrt{3}$
1435 8361 rt : (5,9,9,6)
1435 9020 rt : (3,5,-4,-2)
1435 9353 sq : $4-\sqrt{3}$
1435 9427 J_2 : $5\sqrt{5}/2$
1435 9690 rt : (2,7,-6,-1)
1436 0162 J_2 : $4\sqrt{2}/5$
1436 0254 e^x : $((e+\pi)/3)^{-3}$
1436 1070 rt : (9,8,-6,-1)
1436 2366 J_1 : $9(e+\pi)/7$
1436 8759 rt : (5,1,7,1)
1437 0504 rt : (2,6,-1,-6)
1437 1190 rt : (7,2,1,9)
1437 1466 rt : (9,8,0,3)
1437 1998 e^x : $\sqrt{2}-3\sqrt{5}/2$
1437 2098 cu : 11/21
1437 2326 rt : (9,1,-7,-1)
1437 3227 at : $3(e+\pi)/8$
1437 5399 2^x : $(\ln 3/3)^{-1/2}$
1437 5415 cr : $7\sqrt[3]{5}/8$
1437 5533 Li_2 : $\ln 2/5$
1437 5799 as : $\sqrt{2}+\sqrt{3}-\sqrt{5}$
1437 5810 cu : $3/4-3\pi/4$
1437 7687 rt : (2,7,-8,-3)
1437 7829 $t\pi$: 1/22
1437 8184 Ei : $1/\ln(5/2)$
1437 9502 at : $\sqrt{2}/2+2\sqrt{5}/3$
1437 9535 sr : $4\sqrt{2}/3-\sqrt{3}/3$
1438 1000 $s\pi$: $(\ln 3)^{-1/2}$
1438 1115 ℓ_{10} : $e/2+4\pi$
1438 2657 rt : (5,8,8,1)
1438 3144 ζ : $8\sqrt[3]{2}/3$
1438 3182 rt : (4,1,-9,3)
1438 3642 rt : (6,1,-2,-8)
1438 3695 $t\pi$: $(1/\pi)/7$
1438 4103 id : $\ln(\sqrt{3}/2)$
1438 4281 2^x : $2/3-2\sqrt{3}$
1438 4734 2^x : $4e/3+3\pi/4$
1438 5052 rt : (2,7,-8,1)
1438 6142 as : $1/\ln(3)$
1438 7570 rt : (9,6,-8,1)
1438 7690 id : P_{75}
1438 7691 sq : $4\sqrt{2}+3\sqrt{3}/2$
1438 8967 rt : (8,5,-3,2)
1439 1186 Ei : $\exp(\sqrt[3]{5}/3)$
1439 1878 at : $9\sqrt[3]{5}/7$
1439 5487 rt : (4,6,6,5)
1439 6157 rt : (5,4,-5,-7)
1439 6629 J_1 : $10\sqrt{3}$
1439 7521 rt : (3,7,-6,-1)
1439 7881 2^x : $\sqrt{2}/4+3\sqrt{3}/4$
1439 7966 $t\pi$: $\sqrt{2}-\sqrt{5}-\sqrt{6}$
1439 8439 $\ln(2/3)\times\exp(3)$
1440 1464 $t\pi$: $4\sqrt{2}-4\sqrt{3}$
1440 1721 at : $7\pi/10$
1440 1758 cu : $(e+\pi)/4$
1440 4539 J_2 : 17/15
1440 5837 $\exp(2/5)+\exp(\sqrt{3})$
1440 5900 at : $4/3+\sqrt{3}/2$
1440 6588 rt : (7,2,-6,1)
1440 6631 at : $2\sqrt{3}/3-3\sqrt{5}/2$
1440 6932 rt : (5,5,-9,7)
1440 7812 Li_2 : 5/6
1440 8734 rt : (4,7,-8,-6)
1441 0787 rt : (2,6,-9,2)
1441 1378 rt : (4,1,7,1)
1441 1589 Ψ : $\sqrt{3}+\sqrt{5}-\sqrt{7}$
1441 1943 10^x : $1/3+\pi/4$
1441 2280 sr : $3/4+\sqrt{5}/4$
1441 4060 rt : (3,9,2,-5)
1441 4258 rt : (7,9,9,9)
1441 5416 Ei : $7\sqrt{3}/4$
1441 5628 J_2 : $\sqrt{20}\pi$
1441 6883 at : 11/5
1441 8015 J_2 : $9\sqrt[3]{2}/10$
1441 9352 10^x : $4\sqrt{2}-\sqrt{3}$
1441 9978 e^x : $4+3\sqrt{3}/4$
1442 2495 id : $\sqrt[3]{3}/10$
1442 4332 rt : (6,7,-8,-9)
1442 4737 10^x : $\sqrt{17/2}$
1442 6075 ℓ_{10} : $1/3+3\sqrt{2}/4$
1442 6555 $\sqrt{3}-s\pi(1/5)$
1442 6950 $1/5\div\ln(4)$
1442 7156 rt : (9,2,-6,4)
1442 7196 e^x : $((e+\pi)/3)^3$
1442 7264 e^x : $3e/2+\pi$
1442 8807 J_1 : 7/24
1443 1252 sr : $2+3\sqrt{3}/2$
1443 1270 ln : $5\ln 2/3$
1443 1299 rt : (9,8,-7,-5)
1443 2713 rt : (4,8,8,1)
1443 3169 rt : (6,6,6,8)
1443 3756 $1/4\div\sqrt{3}$
1443 4172 as : $\ln(\sqrt{3}/2)$
1443 5936 rt : (4,7,-6,-1)
1443 6413 rt : (3,7,-8,1)
1443 7290 sr : $2\sqrt{2}/3+4\sqrt{5}$
1443 7636 $s\pi$: $\ln(\pi/3)$
1443 7797 as : P_{75}
1443 8990 ℓ_2 : 19/21
1443 9306 rt : (8,1,-9,-3)
1443 9593 sq : $2\sqrt[3]{5}/9$
1444 0716 rt : (2,6,1,6)
1444 1063 Γ : 9/11
1444 3439 2^x : $3\sqrt{2}/4-\sqrt{3}/2$
1444 4790 rt : (1,9,2,-8)
1444 5874 J_0 : $7e\pi/10$
1444 7924 sinh : $7\sqrt[3]{2}/9$
1444 8983 J_0 : $9\zeta(3)/4$
1444 9338 sq : $3/4+\pi$
1445 0163 2^x : $\sqrt[3]{4/3}$
1445 2224 rt : (5,3,3,7)
1445 2457 $\ln(2/3)\times\exp(e)$
1445 2602 10^x : $e\pi/10$
1445 3907 id : $e/2+\pi/4$
1445 4763 rt : (3,1,7,1)
1445 7403 Γ : $\exp(2\pi/3)$
1445 8067 ζ : $3\sqrt{5}/2$
1445 8427 rt : (4,0,-7,1)
1445 8567 rt : (3,9,-2,-2)
1445 9131 Γ : $(\pi/2)^3$
1446 2152 ln : $3\sqrt[3]{3}/5$
1446 3700 $t\pi$: $1/3+3\pi/2$
1446 4568 $t\pi$: $\sqrt{19/3}$
1446 5869 Ψ : $(\sqrt[3]{4}/3)^{1/2}$
1446 8026 rt : (9,5,6,-1)
1446 9301 $s\pi$: $2e/3+\pi$
1446 9942 $t\pi$: $8\sqrt[3]{5}/7$
1447 1424 id : $\sqrt[3]{3/2}$
1447 1989 sq : $2/3+2e/3$
1447 2370 $t\pi$: $2\sqrt{5}/7$
1447 2547 sinh : $\sqrt[3]{3}/10$
1447 2969 as : $\sqrt[3]{3}/10$
1447 2988 id : $\ln(\pi)$
1447 3141 J_2 : 25/22
1447 4821 $s\pi$: $4\sqrt{3}+\sqrt{5}/2$
1447 4827 Γ : $9\sqrt[3]{2}/5$
1447 4954 rt : (5,7,-6,-1)

1447 5973 at : $\sqrt{2}/3+\sqrt{3}$
1447 6105 sr : 23/5
1447 6113 Ψ : $\exp(-1/\pi)$
1447 6482 ℓ_{10} : $(e)^{1/3}$
1448 0375 sq : $2/3-\pi/3$
1448 0645 cu : $\sqrt{2}+\sqrt{3}$
1448 2667 cu : $1-3e/2$
1448 3887 rt : (3,8,8,1)
1448 4067 sr : $1/4+3\sqrt{2}/4$
1448 4306 rt : (5,3,-3,-8)
1448 4734 J_2 : 9
1448 4934 $\exp(2/3)\times s\pi(1/5)$
1448 7130 sq : $9\sqrt[3]{2}/2$
1448 8476 Ei : $(2\pi/3)^{1/2}$
1448 8974 rt : (4,7,-8,1)
1449 0771 rt : (3,6,0,2)
1449 2059 e^x : $(e)^{-2}$
1449 2319 e^x : $3\sqrt{2}/4+\sqrt{3}/3$
1449 3218 2^x : $2/3-\sqrt{2}/3$
1449 3584 as : $1/3+\sqrt{3}/3$
1449 3962 ℓ_2 : $\sqrt{2}/3+\sqrt{3}/4$
1449 4374 e^x : $2/3-3\sqrt{3}/2$
1449 4452 rt : (9,2,7,1)
1449 4712 rt : (9,9,2,4)
1449 6735 ℓ_{10} : $4\pi/9$
1449 7254 rt : (4,6,-6,-1)
1449 8943 rt : (2,1,7,1)
1449 9981 at : $7\sqrt[3]{2}/4$
1450 0501 rt : (5,9,-9,-9)
1450 2587 cu : $\ln(\ln 2/3)$
1450 2939 sq : $(\pi)^{1/3}$
1450 3637 rt : (5,0,-7,1)
1450 4019 Ψ : 8/11
1450 4703 ln : $1/4+e/3$
1450 4756 10^x : 1/17
1450 6109 rt : (8,5,6,-1)
1450 6781 ℓ_2 : $\exp(\sqrt[3]{4}/2)$
1451 0269 rt : (3,9,-9,-6)
1451 0362 sinh : $((e+\pi)/2)^2$
1451 2076 rt : (9,3,-4,5)
1451 2471 sq : 8/21
1451 2918 rt : (8,6,-1,3)
1451 3230 ln : 7/22
1451 3978 e^x : $\exp(\sqrt{2})$
1451 4593 rt : (6,7,-6,-1)
1451 5235 rt : (7,9,2,1)
1451 5303 cu : $e/3-3\pi/2$
1451 5343 at : $3e/2-2\pi$
1451 7628 J_2 : $\sqrt{2}+\sqrt{3}+\sqrt{6}$
1451 7856 cr : $2\sqrt{2}/3+\sqrt{5}/4$
1452 0213 sinh : $5\sqrt{5}/6$
1452 0453 $\exp(4/5)\times s\pi(\sqrt{2})$
1452 0991 $\sqrt{3}+\exp(5)$
1452 1209 Ψ : $9\sqrt[3]{3}/8$
1452 1260 rt : (2,9,-8,-7)
1452 1693 rt : (1,9,-1,3)
1452 2183 Li_2 : $\sqrt[3]{2}/9$
1452 2774 J_1 : $8e$
1452 4563 rt : (8,0,-7,4)
1452 5110 $\ln(\sqrt{2})-\exp(2/5)$
1452 6185 cu : $2\sqrt{2}-3\sqrt{5}/2$
1452 9918 rt : (1,9,-9,-3)
1453 0008 rt : (9,9,8,1)
1453 1255 rt : (7,3,-4,2)
1453 2056 $t\pi$: $(\ln 3)^{-1/2}$
1453 2320 ζ : $4(e+\pi)/7$
1453 6232 rt : (2,8,8,1)
1453 6501 ℓ_{10} : $5\sqrt{5}/8$
1453 6583 J_1 : $\ln(\sqrt{5}/3)$
1453 6791 cr : $5\zeta(3)/4$
1453 8249 at : $3/2+\sqrt{2}/2$
1453 8352 ℓ_{10} : $((e+\pi)/3)^{1/2}$
1453 8425 rt : (8,2,7,1)
1453 8953 J_0 : $5\sqrt[3]{5}/4$
1453 9336 e^x : $-\sqrt{3}+\sqrt{6}-\sqrt{7}$
1454 1295 at : $1/2-\sqrt{2}/4$
1454 2560 ℓ_{10} : $(\ln 3/3)^{1/3}$
1454 2793 rt : (5,7,-8,1)
1454 3945 rt : (1,1,7,1)
1454 4769 rt : (7,5,6,-1)
1454 7437 J_1 : 5/17
1454 7595 rt : (2,8,0,5)
1454 8873 10^x : 23/15
1454 9722 rt : (6,6,-1,0)
1455 2121 id : $1/3+2e/3$
1455 2196 J_2 : $(\sqrt{2}\pi/3)^{1/3}$
1455 4354 $s\pi$: $-\sqrt{3}+\sqrt{5}+\sqrt{6}$
1455 4875 rt : (7,7,-6,-1)
1455 5226 J_2 : $2\sqrt[3]{5}/3$
1455 5443 as : P_{11}
1455 6890 ζ : $(\sqrt{5}/3)^{1/3}$
1455 7652 rt : (1,5,1,9)
1455 9674 sq : $1/\sqrt[3]{18}$
1456 1242 $1/4-\exp(1/3)$
1456 3438 Ψ : $\exp(-4)$
1456 4392 sr : 21/16
1456 5723 J_1 : $9\zeta(3)$
1456 6976 rt : (6,7,8,9)
1456 8322 rt : (5,7,-8,0)
1457 1511 $t\pi$: $1/\ln(\ln 2)$
1457 4514 rt : (5,2,-1,-9)
1457 4851 J_2 : $10\sqrt{2}/3$
1457 4952 $s\pi(e)\div s\pi(\sqrt{5})$
1457 9384 cu : 10/19
1458 0439 J_2 : $10\pi^2/9$
1458 2104 rt : (8,9,8,1)
1458 3201 rt : (7,2,7,1)
1458 3482 $2/5-s\pi(\sqrt{3})$
1458 3786 ζ : $2\sqrt{2}/7$
1458 4025 rt : (6,5,6,-1)
1458 4666 rt : (5,3,-4,-1)
1458 5802 e^x : $2\sqrt{2}/3+2\sqrt{3}/3$
1458 6119 2^x : $4\sqrt{2}+3\sqrt{3}/4$
1458 6885 rt : (5,4,5,8)
1458 6974 rt : (9,2,-8,-7)
1458 7004 $3/5\div\exp(\sqrt{2})$
1458 7851 rt : (4,7,-2,3)
1458 9803 sq : $3/2-\sqrt{5}/2$
1458 9804 10^x : P_{11}
1459 0250 sq : $6(1/\pi)/5$
1459 0503 $t\pi$: $\ln(\pi/3)$
1459 1721 ln : $5\sqrt[3]{5}$
1459 1987 2^x : $\sqrt{3}/3-3\sqrt{5}/2$
1459 3587 ℓ_{10} : $3/2-\pi/4$
1459 4344 ζ : $\exp(-e/3)$
1459 5819 rt : (8,7,-6,-1)
1459 6103 ln : $3\sqrt{3}+3\sqrt{5}/2$
1459 6718 rt : (7,0,-7,1)
1459 7174 Γ : $e\pi$
1459 7798 J_0 : $\sqrt{22/3}$
1459 7932 rt : (6,7,-8,1)
1459 7960 ln : $4/3+2e/3$
1459 8858 rt : (9,4,-2,6)
1460 1259 rt : (3,8,-7,-7)
1460 2471 E : 23/25
1460 3968 J_2 : $10\sqrt[3]{5}$
1460 6952 rt : (4,1,2,7)
1460 9213 sq : $3/4-4\sqrt{5}$
1460 9858 e^x : 3/22
1461 0082 ℓ_{10} : $10\sqrt[3]{2}/9$
1461 1699 rt : (8,1,-5,5)
1461 2803 ℓ_{10} : 14
1461 3305 sq : $(e+\pi)/4$
1461 4059 at : $\exp(\sqrt[3]{4}/2)$
1461 6148 e^x : $3(1/\pi)/7$
1461 9751 Γ : $\sqrt{2/3}$
1462 0079 rt : (4,0,-7,2)
1462 0467 J_2 : 8/7
1462 1305 $s\pi$: $(e+\pi)/3$
1462 1583 ln : $\sqrt{2}+\sqrt{3}$
1462 3187 $t\pi$: $2e/3+\pi$
1462 3894 rt : (5,5,6,-1)
1462 4605 rt : (7,2,-8,-4)
1462 6436 id : $\sqrt{2}+\sqrt{3}$
1462 7179 rt : (3,7,4,6)
1462 7680 Γ : $(e+\pi)^{-1/2}$
1462 8812 rt : (6,2,7,1)
1462 8885 $t\pi$: $4\sqrt{3}+\sqrt{5}/2$
1462 9475 cr : $2\sqrt{2}/3+4\sqrt{5}$
1463 0532 rt : (2,6,-6,-1)
1463 1792 J_2 : $3\pi/2$
1463 3756 $\ln(\sqrt{5})-s\pi(2/5)$
1463 4628 sr : $7\sqrt{2}$
1463 4946 erf : 3/23
1463 5393 rt : (7,9,8,1)
1463 5666 10^x : $9\sqrt[3]{2}/8$
1463 6548 rt : (1,2,-6,-8)
1463 7039 10^x : $7\sqrt[3]{3}/4$
1463 7353 rt : (5,8,-4,4)
1463 7448 rt : (9,7,-6,-1)
1463 7787 10^x : $10\zeta(3)/7$
1464 0833 rt : (8,7,1,4)
1464 3443 10^x : $\sqrt{17/3}$
1464 3722 rt : (9,9,-5,-4)
1464 3787 sq : $6\sqrt[3]{2}$
1464 4240 rt : (1,4,-9,-9)
1464 4472 10^x : $e/3+\pi/4$
1464 4660 rt : (8,8,1,0)
1464 4945 J_0 : $7\sqrt[3]{5}/2$
1464 6178 sinh : $9\sqrt{5}/8$
1464 7569 rt : (2,7,0,-7)
1464 9502 e^x : $3\sqrt{2}/4-4\sqrt{5}/3$
1465 0879 2^x : $3\sqrt{2}+2\sqrt{3}/3$
1465 2831 J_1 : $(3/2)^{-3}$
1465 3961 rt : (9,1,-6,-8)
1465 4461 rt : (7,7,-8,1)
1465 4621 $s\pi$: $(\sqrt{3}/2)^{1/3}$
1465 6646 J_2 : $(\sqrt{2}\pi/2)^3$
1465 7359 $1/5-\ln(\sqrt{2})$
1465 8031 J_1 : $10e$
1465 8442 ℓ_{10} : $4e+\pi$
1465 9187 sq : $\ln(\ln 2/3)$
1465 9932 rt : (7,4,-2,3)
1466 0347 ln : 19/22
1466 0477 sq : $1/2-\pi/2$
1466 2681 J_2 : $\sqrt[3]{3/2}$
1466 2700 at : $1/\ln(\pi/2)$
1466 2850 $\exp(3/5)-s\pi(\sqrt{5})$
1466 2943 J_0 : 19
1466 3037 J_2 : $\ln(\pi)$
1466 3880 rt : (8,4,-9,-7)
1466 4395 rt : (4,7,8,6)
1466 5546 at : $3-\pi/4$
1466 7221 2^x : $2\sqrt{2}/3-\sqrt{5}/3$
1466 7372 rt : (6,9,-6,5)
1466 7690 at : $3\sqrt{3}-4\sqrt{5}/3$
1466 8123 rt : (1,8,-7,-4)
1466 9781 10^x : $1/3+4\sqrt{5}$
1467 0052 e^x : $3-\sqrt{5}$
1467 1929 rt : (3,6,-6,-1)
1467 2143 2^x : $6\sqrt[3]{5}/5$
1467 2440 ℓ_{10} : $4-3\sqrt{3}/2$
1467 5116 $s\pi$: $\sqrt{17}\pi$
1467 5288 rt : (5,2,7,1)
1467 5786 rt : (8,6,-8,-5)
1467 6313 Li_2 : $\sqrt{2}/10$
1467 8395 at : $3\sqrt{2}/4+2\sqrt{3}/3$
1467 9172 rt : (6,1,-5,2)
1468 0617 rt : (4,8,2,7)
1468 1004 cr : $\sqrt{2}/4+2\sqrt{3}/3$
1468 1419 rt : (2,9,1,-8)
1468 1672 Li_2 : $4(1/\pi)/9$
1468 2074 J_0 : $e\pi/4$
1468 3279 cu : $1/3-\pi$
1468 4217 rt : (2,5,-3,-7)
1468 5907 $\exp(4/5)\div\exp(e)$
1468 6657 e^x : $-\sqrt{2}+\sqrt{3}-\sqrt{5}$
1468 6755 10^x : $(\pi)^{1/3}$
1468 6942 e^x : P_{24}
1468 6975 2^x : $4\sqrt{3}/3+4\sqrt{5}/3$
1468 7181 ℓ_{10} : $7\zeta(3)/6$
1468 7580 rt : (9,5,0,7)
1468 7624 2^x : $\sqrt{2}-\sqrt{3}-\sqrt{6}$
1468 8405 J_0 : 25/2
1468 9912 J_1 : $3\ln 2/7$
1468 9934 rt : (6,9,8,1)
1469 0430 2^x : $2\sqrt{2}/3+3\sqrt{5}/4$
1469 3590 rt : (9,0,-7,1)
1469 3786 ℓ_2 : $\ln(\pi/2)$
1469 3908 rt : (1,9,-2,-6)

1469 4437 rt : (3,7,-5,-8)
1469 4631 $1/4 \times s\pi(1/5)$
1469 4666 ln : $(\sqrt{5}/3)^{1/2}$
1469 5537 $\ln(3/4) \times \ln(3/5)$
1469 5550 sq : $3\sqrt{3}/2 - 4\sqrt{5}/3$
1469 7519 as : $1/2 - \sqrt{2}/4$
1469 8556 rt : (5,2,-8,-1)
1469 9053 e^x : $3/2 - \sqrt{2}/4$
1470 0791 rt : (8,2,-3,6)
1470 0869 rt : (1,8,-8,1)
1470 2124 at : $\sqrt{2}/2 - \sqrt{5}/4$
1470 3480 rt : (3,7,-8,-1)
1470 4733 ℓ_2 : $3 - \pi/4$
1470 5550 rt : (3,5,6,-1)
1470 7207 cr : $3 - 2\sqrt{5}/3$
1470 7373 tanh : $(\sqrt[3]{4}/3)^3$
1470 7634 rt : (5,9,0,8)
1470 7835 at : $(\sqrt[3]{4}/3)^3$
1470 7866 sr : 25/19
1470 8382 ζ : $-\sqrt{2} - \sqrt{3} + \sqrt{6}$
1470 9451 $t\pi$: $3\sqrt{2} - 3\sqrt{3}$
1471 0999 $t\pi$: $\sqrt{3} - \sqrt{5} - \sqrt{6}$
1471 2452 rt : (8,7,-8,1)
1471 2982 ℓ_2 : $3\sqrt{3} - 4\sqrt{5}/3$
1471 4018 rt : (4,6,-6,-1)
1471 4046 2^x : $2\ln 2/7$
1471 4070 rt : (7,1,-6,-5)
1471 4715 J_1 : $7\sqrt{5}$
1471 5177 $2/5 \div e$
1471 5363 Ei : 23/13
1471 6181 2^x : $7\sqrt[3]{2}/8$
1471 7848 Li_2 : $\exp(-(e+\pi)/3)$
1471 9265 cr : $7\sqrt{2}$
1472 0269 sr : $(\sqrt{3})^{1/2}$
1472 0955 10^x : $3 - 3e/4$
1472 2101 rt : (9,0,-4,9)
1472 2663 rt : (4,2,7,1)
1472 3185 sinh : $1/\ln(\sqrt[3]{5})$
1472 5381 rt : (9,4,-9,-2)
1472 6379 rt : (5,5,7,9)
1472 7417 rt : (6,4,-9,-4)
1472 7541 Γ : $3e/10$
1472 7786 at : $1/2 - e$
1472 8476 Γ : $1/\ln(\ln 3)$
1472 9123 Ψ : 13/8
1472 9207 cr : $5e/9$
1472 9579 cu : $\sqrt{2}/4 - 4\sqrt{5}/3$
1472 9633 sq : $2\sqrt{2}/3 - \sqrt{5}/4$
1473 2240 rt : (8,3,-7,-8)
1473 2846 rt : (3,8,0,-6)
1473 2895 ln : $-\sqrt{3} + \sqrt{5} + \sqrt{7}$
1473 3979 ln : $5\sqrt[3]{2}/2$
1473 4734 $t\pi$: $7\pi^2/4$
1473 5623 ζ : 10/3
1473 5666 Γ : $\sqrt{12}\pi$
1473 8329 rt : (9,4,6,-1)
1473 9016 $\sqrt{2} \times \exp(4/5)$
1473 9092 at : $1/3 + 4\sqrt{2}/3$
1474 1504 J_1 : $1/\ln(4/3)$
1474 3023 rt : (1,6,4,-2)
1474 5617 $\ln(\sqrt{3}) + \exp(4)$
1474 5786 rt : (5,9,8,1)
1474 6126 J_2 : $(\pi/3)^3$
1474 7378 rt : (4,2,4,8)
1474 8550 id : $e - \pi/2$
1474 9639 $t\pi$: $e/10$
1475 4162 ℓ_2 : $3\sqrt{2}/2 + 4\sqrt{3}/3$
1475 5107 rt : (2,4,-5,7)
1475 5373 $\sqrt{5} \div \exp(e)$
1475 5577 $\ln(3) - s\pi(2/5)$
1475 6501 ℓ_2 : $(e/2)^{1/3}$
1475 6623 id : $4\sqrt{2} + 2\sqrt{5}/3$
1475 6820 rt : (5,6,-6,-1)
1475 8045 θ_3 : $1/\ln(1/(3e))$
1475 8374 ℓ_{10} : $1/4 + 2\sqrt{3}/3$
1475 8497 rt : (2,8,-8,1)
1475 9354 Ei : $\sqrt{2} + \sqrt{3} + \sqrt{7}$
1476 0166 sinh : $\sqrt{12}\pi$
1476 0878 cu : $9\pi/8$
1476 2238 rt : (3,8,-2,8)
1476 2784 sr : $4e/3 + 2\pi$
1476 6487 ℓ_{10} : $\sqrt{20}\pi$
1476 8552 rt : (3,1,1,-7)
1476 9654 cu : $1 - \sqrt{2}/3$
1477 0090 $\ln(3/4) \div \exp(2/3)$
1477 0384 sq : $3\sqrt{2}/4 + 3\sqrt{3}$
1477 0972 rt : (3,2,7,1)
1477 1985 rt : (9,7,-8,1)
1477 2943 rt : (8,8,3,5)
1477 3275 $s\pi$: $\pi/3$
1477 3404 $1/5 + \exp(2/3)$
1477 3473 cr : $6\sqrt[3]{2}/5$
1477 6108 e^x : $e/3 + 2\pi$
1477 7567 cr : $4e/3 + 2\pi$
1477 7710 Ψ : $10e/3$
1477 8320 rt : (9,6,2,8)
1477 9887 rt : (8,4,6,-1)
1478 0146 $t\pi$: $(e+\pi)/3$
1478 2476 rt : (4,1,-7,1)
1478 3026 J_2 : 23/20
1478 5760 2^x : $5(1/\pi)/8$
1478 6722 as : $(\pi/3)^{-2}$
1478 9514 2^x : $(\sqrt{5}/3)^{-1/3}$
1478 9903 rt : (4,4,-8,4)
1479 1810 sq : $4 + 2\sqrt{5}/3$
1479 1919 rt : (8,3,-1,7)
1479 2496 cr : $4\sqrt{3} + 4\sqrt{5}/3$
1479 2858 rt : (7,5,0,4)
1479 2899 sq : 5/13
1479 3190 2^x : $\sqrt{2}/2 - 2\sqrt{3}$
1479 4240 at : 20/9
1479 5436 rt : (6,7,1,1)
1479 5603 sr : $4\sqrt{3} + 4\sqrt{5}/3$
1479 5918 sq : 15/14
1480 0362 rt : (6,6,-6,-1)
1480 0709 Γ : 25/11
1480 1433 ℓ_{10} : $1/2 + e/3$
1480 1517 10^x : $\pi/3$
1480 1742 10^x : $1/4 + 3\pi/2$
1480 1798 rt : (8,2,-5,-9)
1480 3015 rt : (4,9,8,1)
1480 3557 $\sqrt{5} \div \exp(2/3)$
1480 4112 ζ : $\ln(1/2)$
1480 4237 rt : (2,9,-6,-8)
1480 5108 e^x : P_4
1480 5589 rt : (7,0,-4,6)
1480 7993 rt : (9,3,7,1)
1480 8978 id : $\sqrt{2}/2 - \sqrt{5}/4$
1480 9723 e^x : $2 - \sqrt{3}/3$
1481 1434 rt : (1,7,-5,-5)
1481 2206 rt : (7,5,-8,-8)
1481 2925 rt : (6,2,-3,3)
1481 4561 $t\pi$: $(\sqrt{3}/2)^{1/3}$
1481 4814 id : $(\sqrt[3]{4}/3)^3$
1481 5971 ζ : $4\sqrt[3]{5}/7$
1481 6426 $\ln(4) \div \exp(\sqrt{5})$
1481 7620 rt : (3,8,-8,1)
1481 8205 ℓ_{10} : $3/2 + 4\pi$
1481 8474 cu : P_3
1481 9332 rt : (6,3,-7,-5)
1482 0253 rt : (2,2,7,1)
1482 0836 θ_3 : $(\sqrt[3]{2}/3)^3$
1482 2123 rt : (7,4,6,-1)
1482 3822 at : $\sqrt[3]{11}$
1482 4229 10^x : $4e/9$
1482 5793 sq : $6\sqrt[3]{5}/7$
1482 6317 sq : $1/3 - 4\sqrt{5}$
1482 6368 Ei : 7/17
1482 7599 K : $(\sqrt{5}/3)^{1/2}$
1482 8749 ℓ_2 : $4 + \sqrt{3}/4$
1482 8796 sq : $5\ln 2/9$
1483 1179 Li_2 : 1/7
1483 1881 J_1 : 3/10
1483 3147 rt : (5,9,2,-2)
1483 5736 $t\pi$: $\sqrt{17}\pi$
1483 6241 e^x : 13/17
1483 8061 id : $(\pi/3)^3$
1483 8184 cu : 9/17
1483 9989 2^x : $e\pi$
1484 0238 e^x : $(\sqrt[3]{5})^{-1/2}$
1484 3552 J_2 : $1/\ln(\sqrt[3]{2}/3)$
1484 4668 rt : (7,6,-6,-1)
1484 4749 10^x : $2 - 2\sqrt{2}$
1484 5550 id : $\sqrt{2}/3 + 3\sqrt{5}/4$
1484 6529 Ei : $4e$
1484 7410 J_2 : $(5/2)^3$
1484 8324 rt : (5,6,-6,9)
1485 2204 rt : (9,5,-7,-1)
1485 3527 rt : (4,8,-5,-8)
1485 5195 sq : $3e/4 - 3\pi/2$
1485 6729 J_1 : $\zeta(3)/4$
1485 6909 rt : (8,3,7,1)
1485 7828 2^x : $1/2 + 2\sqrt{3}/3$
1485 8259 ln : $4/3 - \sqrt{2}/3$
1485 8304 $e - s\pi(\pi)$
1486 1092 J_2 : $5\pi^2/4$
1486 1692 rt : (3,9,8,1)
1486 3066 rt : (4,5,-4,8)
1486 3648 as : $\sqrt{2}/2 - \sqrt{5}/4$
1486 4979 J_2 : $8(e+\pi)/3$
1486 5060 rt : (6,4,6,-1)
1486 5487 cr : $7\sqrt{3}/8$
1486 6066 J_2 : $2e\pi$
1486 7327 J_0 : $\sqrt{2} - \sqrt{3} + \sqrt{6}$
1486 8100 $\ln(3) \div \exp(2)$
1486 8804 cu : $6\sqrt[3]{5}/7$
1486 9066 sinh : $(\sqrt[3]{4}/3)^3$
1486 9549 as : $(\sqrt[3]{4}/3)^3$
1486 9710 J_2 : $4\sqrt[3]{3}/5$
1486 9835 2^x : 1/5
1487 0130 sq : $3/2 - 4\sqrt{2}/3$
1487 0158 J_1 : $10\pi^2/7$
1487 0546 rt : (1,2,7,1)
1487 0772 J_2 : 15/13
1487 1164 rt : (9,7,4,9)
1487 1398 rt : (4,1,-5,-1)
1487 1898 10^x : $\sqrt{2}/3 - 3\sqrt{3}/4$
1487 2127 $1/2 + \exp(1/2)$
1487 2660 J_0 : 19/7
1487 3252 at : $9\sqrt{3}/7$
1487 3520 cu : $5\sqrt{3}/4$
1487 5209 cu : $e/4 + \pi/3$
1487 6236 $2/3 \times \ln(4/5)$
1487 7135 rt : (1,8,7,-1)
1487 8140 Ei : $6\sqrt[3]{5}/5$
1487 8320 rt : (4,8,-8,1)
1487 8427 J_0 : $\sqrt[3]{20}$
1487 9873 cu : $3/2 + \sqrt{5}$
1487 9935 J_1 : $3\pi^2/7$
1488 0341 ln : P_5
1488 3240 rt : (7,4,-6,-9)
1488 3303 2^x : $3/4 + \sqrt{2}/4$
1488 3331 Ψ : $\zeta(3)/7$
1488 4512 J_2 : $9\sqrt{5}$
1488 4906 rt : (6,1,-7,1)
1488 5169 rt : (8,4,1,8)
1488 7761 rt : (3,4,-2,7)
1488 8000 J_1 : $10\sqrt[3]{2}$
1488 8303 2^x : P_{16}
1488 8403 2^x : P_{40}
1488 8503 tanh : 3/20
1488 8994 at : 3/20
1488 9356 Ψ : $4(1/\pi)/5$
1488 9357 ln : $2 + 2\sqrt{3}/3$
1488 9766 rt : (8,6,-6,-1)
1489 0284 J_2 : $2\sqrt{3}/3$
1489 2433 J_1 : $\exp(\sqrt[3]{3})$
1489 2896 Ei : $4(e+\pi)/3$
1489 3037 rt : (4,3,6,9)
1489 3929 rt : (6,7,-9,-8)
1489 4164 at : $7(1/\pi)$
1489 5263 id : $2e + 3\pi/2$
1489 7134 2^x : $\zeta(3)/6$
1489 7360 e^x : $2e/3 - \pi/3$
1489 9250 rt : (7,1,-2,7)
1490 1243 rt : (8,7,-6,-4)
1490 1905 cr : $1/2 + 3\pi$
1490 2555 rt : (2,8,-4,-9)
1490 2729 J_2 : $5\ln 2/3$
1490 4153 ln : $3/4 - \sqrt{2}/2$

1490 4226 $s\pi : 1/21$
1490 4901 $J_1 : 1/\sqrt{11}$
1490 6806 rt : (7,3,7,1)
1490 7119 $1/3 \div \sqrt{5}$
1490 8720 rt : (5,4,6,-1)
1490 9506 rt : (8,9,5,6)
1490 9944 sr : $2 - e/4$
1491 0203 rt : (3,7,-3,-9)
1491 1064 sq : $\sqrt{2}/2 + 4\sqrt{5}$
1491 1511 $\Gamma : 2/7$
1491 3103 $\ln(\sqrt{2}) \times s\pi(\pi)$
1491 3242 cu : $10\sqrt{3}/7$
1491 3407 rt : (6,2,-5,-6)
1491 3643 tanh : $\zeta(3)/8$
1491 3986 sr : $8\sqrt{3}/3$
1491 4139 at : $\zeta(3)/8$
1491 5232 rt : (3,0,3,8)
1491 5533 cu : $3\sqrt{2}/8$
1491 5813 $J_1 : (\ln 3/2)^2$
1491 6016 $e^x : 1/3 - \sqrt{5}$
1491 6227 $\ln(2/3) \div \exp(1)$
1491 6696 $1/5 \times s\pi(\sqrt{3})$
1491 7611 $K : 19/22$
1491 9504 $\zeta : (\pi/2)^{-2}$
1492 0170 $\ell_{10} : \pi^2/7$
1492 1272 $s\pi(\sqrt{5}) \div s\pi(1/5)$
1492 1894 rt : (2,9,8,1)
1492 1900 ln : $2e + \pi$
1492 4975 rt : (3,6,-8,9)
1492 5210 $s\pi : e/3 + \pi$
1492 6114 $2^x : 4/3 - 3e/2$
1492 6625 cr : $3 + 4\sqrt{3}$
1492 7641 rt : (5,5,-8,-5)
1493 0017 $J_1 : e/9$
1493 0206 $\exp(5) \times s\pi(2/5)$
1493 0293 rt : (7,6,2,5)
1493 0553 $\ell_2 : 8\ln 2/5$
1493 0614 $2/5 - \ln(\sqrt{3})$
1493 1265 cu : $5(1/\pi)/3$
1493 1582 $t\pi : 9\sqrt{2}$
1493 4784 $\exp(\sqrt{2}) + s\pi(\sqrt{2})$
1493 4921 cr : $\sqrt[3]{7/2}$
1493 5685 rt : (9,6,-6,-1)
1493 5928 rt : (7,4,-9,-5)
1493 6120 $3 \div \exp(3)$
1493 6958 rt : (9,1,-9,2)
1493 7176 $t\pi : \pi/3$
1493 7396 $\Gamma : 8(e+\pi)/7$
1493 7621 rt : (2,3,0,6)
1493 7801 rt : (7,1,-7,1)
1493 8528 $10^x : 3\pi^2/2$
1493 8739 $Ei : 2\sqrt[3]{3}/7$
1494 0686 rt : (5,8,-8,1)
1494 1110 $2^x : 3/2 - 3\sqrt{2}$
1494 3363 cr : $\sqrt{3} + \sqrt{5} - \sqrt{6}$
1494 3677 $e^x : 4\pi^2/3$
1494 6448 $2^x : 3/2 - 3\sqrt{3}/4$
1494 7667 $e^x : 1/3 + e$
1494 8664 at : $3/4 - 4\sqrt{5}/3$
1494 8889 rt : (3,9,-3,-5)
1495 0568 $\Psi : \sqrt{2} - \sqrt{5} + \sqrt{6}$
1495 1246 rt : (6,3,-1,4)
1495 2630 $J_0 : 7e$
1495 2641 as : $\sqrt{2}/4 + \sqrt{5}/4$
1495 2894 rt : (1,7,-9,3)
1495 3128 rt : (4,4,6,-1)
1495 3230 $\ell_2 : 3\zeta(3)/4$
1495 3377 rt : (2,8,-1,1)
1495 6373 $erf : 2/15$
1495 7725 rt : (6,3,7,1)
1495 7759 $\ln(4/5) \div \exp(2/5)$
1495 7851 $t\pi : P_{30}$
1495 9145 at : $1/2 + \sqrt{3}$
1495 9393 $3 \div \exp(1/3)$
1496 0206 rt : (1,6,-3,-6)
1496 2706 $10^x : 3/4 + 2\sqrt{2}$
1496 3296 $J_2 : 22/19$
1496 3531 $\zeta : \sqrt{11}$
1496 4658 $Ei : 7e/6$
1496 5173 ln : $3\sqrt{3}/2 + \sqrt{5}/4$
1496 6498 rt : (6,6,-7,-9)
1496 7215 $s\pi : 3\sqrt{3}/2 + 3\sqrt{5}/2$
1496 7449 rt : (3,5,-6,-1)
1496 7600 $2^x : \sqrt{3} - 2\sqrt{5}$
1496 9108 $\zeta : \ln(3/2)$
1497 2367 rt : (5,0,-4,3)
1497 2388 $J_2 : (\sqrt{5}/3)^{-1/2}$
1497 2689 cr : $\sqrt{10\pi}$
1497 2761 $J_1 : 6\sqrt[3]{2}$
1497 3241 $\zeta : 1/\sqrt[3]{15}$
1497 4160 $\lambda : \exp(-19\pi/13)$
1497 4341 rt : (1,2,-7,1)
1497 5107 $\ell_2 : 5\sqrt[3]{3}/8$
1497 6232 $\ell_{10} : 17/24$
1497 6784 cu : 22/21
1497 6848 id : $\sqrt{3} - \sqrt{5} - \sqrt{7}$
1497 7402 cr : $8\sqrt[3]{5}/9$
1497 7412 $2^x : 8e\pi/5$
1497 7632 $1/3 \div \exp(4/5)$
1497 7799 $2^x : P_8$
1497 8831 $Li_2 : \sqrt[3]{3}/10$
1497 9716 $4 \div s\pi(\sqrt{2})$
1498 0262 id : $5\sqrt[3]{2}/2$
1498 0633 rt : (8,5,3,9)
1498 1689 $J : ((e+\pi)/3)^{-2}$
1498 1706 $erf : \zeta(3)/9$
1498 1958 cr : $2\sqrt{2}/3 + \sqrt{3}/3$
1498 2991 rt : (9,6,-5,0)
1498 3266 $s\pi : 7\sqrt[3]{3}/2$
1498 3702 rt : (1,9,8,1)
1498 4166 $\ell_{10} : 2/3 + \sqrt{5}/3$
1498 7808 $\ell_2 : 1/3 + 4\sqrt{2}/3$
1499 1883 rt : (8,1,-7,1)
1499 2177 $J : 7/22$
1499 3658 rt : (4,9,-1,-4)
1499 4237 sr : $\sqrt{3} + \sqrt{5} - \sqrt{7}$
1499 5144 rt : (7,2,0,8)
1499 5283 $10^x : 2/3 - 2\sqrt{5}/3$
1499 6200 ln : $((e+\pi)/2)^2$
1499 6278 $J_2 : \exp(\pi)$
1499 7757 $\zeta : (\ln 3/2)^{-2}$
1499 8311 rt : (3,5,7,6)
1499 8761 ln : P_{77}
1500 0000 id : 3/20
1500 2631 $Ei : 9\sqrt{5}/8$
1500 2949 $\pi \div \ln(3/5)$
1500 2968 rt : (8,3,-8,-1)
1500 3287 $J_2 : 9\pi^2/4$
1500 3403 rt : (2,5,-6,-6)
1500 4812 rt : (6,8,-8,1)
1500 5082 $\exp(1/4) + s\pi(1/3)$
1500 6485 cu : $3/2 - 3\pi/2$
1500 7883 id : $3e/4 - 4\pi/3$
1500 8233 sq : P_{27}
1500 8967 sr : $\sqrt{3}/3 + \sqrt{5}/3$
1500 9708 rt : (5,3,7,1)
1500 9737 rt : (6,1,-3,-7)
1501 3830 rt : (4,5,-6,-1)
1501 6652 $\ell_2 : 1/4 + 4\pi/3$
1501 6709 at : $e/3 - \pi$
1501 7153 $2^x : 3\sqrt{3}/2 + \sqrt{5}/3$
1501 8208 sinh : $8e/7$
1501 9520 sr : $3 - 3\sqrt{5}/4$
1502 0026 $\Psi : -\sqrt{2} - \sqrt{5} + \sqrt{7}$
1502 2233 sq : $2\sqrt{2} - \sqrt{5}/4$
1502 4155 $e^x : 4\sqrt{2}/3 + 4\sqrt{5}$
1502 4411 rt : (5,4,-6,-6)
1502 5164 rt : (3,1,-8,8)
1502 5613 $\Gamma : \ln(\pi/3)$
1502 5711 id : $\zeta(3)/8$
1502 6161 $e^x : e - \pi/2$
1502 6199 at : $\sqrt{5}$
1502 6370 $e^x : \sqrt[3]{2}/9$
1502 7936 rt : (2,4,-5,-8)
1502 8297 rt : (9,3,6,-1)
1502 9544 rt : (9,2,-7,3)
1503 1621 $2^x : \sqrt{2}/7$
1503 4146 $10^x : \pi^2/6$
1503 6156 sr : $1/2 + 3\pi$
1503 6238 $\ell_{10} : 9\pi/2$
1503 7130 $Li_2 : (\sqrt[3]{5})^{-1/3}$
1503 9167 rt : (4,7,-9,-5)
1504 0310 $J_0 : 3e\pi$
1504 1876 $J_1 : 7/23$
1504 1952 $erf : (\ln 3/3)^2$
1504 2514 cr : $1 + 4\sqrt{5}$
1504 3618 sr : $1 + 4e/3$
1504 4295 rt : (2,4,6,-1)
1504 5574 $E : 11/12$
1504 6244 $\ell_{10} : 3\sqrt{3} + 4\sqrt{5}$
1504 6777 cu : $9\sqrt{5}/10$
1504 7008 $J_2 : 5\pi/2$
1504 7124 $J_0 : e$
1504 7207 rt : (9,8,1,0)
1504 7544 $\Gamma : 13/16$
1504 9433 $J_1 : (2e/3)^{-2}$
1504 9437 at : $2/3 + \pi/2$
1505 0378 cu : $7\sqrt{2}$
1505 0593 $erf : ((e+\pi)/3)^{-3}$
1505 1499 $\ell_{10} : \sqrt{2}$
1505 1651 rt : (5,8,-8,-9)
1505 2756 ln : $(\pi/2)^{1/3}$
1505 3511 cu : $4\sqrt{2} - 4\sqrt{5}/3$
1505 4230 $10^x : 4(e+\pi)/9$
1505 4667 rt : (9,2,-9,-6)
1505 5578 $J_0 : 10\sqrt{2}$
1505 6250 $10^x : 3\sqrt[3]{3}/7$
1505 6313 sinh : 3/20
1505 6465 rt : (6,8,3,2)
1505 6826 $s\pi : (e/3)^{1/2}$
1505 6827 as : 3/20
1505 6853 sinh : $1/\ln(4/3)$
1505 7728 $2^x : 1/4 - 4\sqrt{5}/3$
1505 9297 rt : (6,9,-7,3)
1506 1067 rt : (5,5,-6,-1)
1506 2801 rt : (4,3,7,1)
1506 4525 $J_0 : 6$
1506 6010 $\ln(3) \div \ln(3/5)$
1506 6070 $2^x : \sqrt[3]{18}$
1506 7533 $J_2 : (\pi/2)^{1/3}$
1506 7596 rt : (3,1,5,9)
1506 8237 $s\pi : (\ln 3)^{1/2}$
1506 8484 as : 21/23
1507 0806 rt : (7,8,-8,1)
1507 1121 $10^x : \sqrt{2} - \sqrt{5}$
1507 2160 $J_1 : 10(e+\pi)/9$
1507 2521 rt : (7,7,4,6)
1507 2574 $t\pi : 1/21$
1507 2828 $4 \times \ln(3/4)$
1507 2946 $2^x : 3/2 + 3\pi/2$
1507 3130 $\ln(4/5) \times s\pi(\sqrt{5})$
1507 3876 rt : (8,3,6,-1)
1507 4998 $e^x : e\pi/6$
1507 7024 rt : (1,5,0,4)
1507 9706 sq : $e/7$
1508 0418 $2^x : (\sqrt{2}\pi/2)^{-2}$
1508 2062 $J_1 : 1/\ln(\pi/3)$
1508 2117 at : $4/3 + e/3$
1508 2314 sinh : $\zeta(3)/8$
1508 2731 rt : (3,7,-2,-7)
1508 2833 as : $\zeta(3)/8$
1508 2894 $J_1 : 8(e+\pi)/3$
1508 7301 $\zeta : 1/\ln(\ln 2/3)$
1508 8038 $10^x : 1/3 - 2\sqrt{3}/3$
1508 8734 sr : $(\sqrt[3]{5}/3)^{-1/2}$
1508 9147 $2^x : 1/\ln(\ln 2)$
1509 0514 sr : $3 + 4\sqrt{3}$
1509 1108 rt : (2,2,2,-8)
1509 2542 $e^x : \sqrt{2}/2 - 3\sqrt{3}/2$
1509 3372 rt : (7,3,2,9)
1509 4243 at : P_{71}
1509 4278 $t\pi : e/3 + \pi$
1509 4425 rt : (6,4,1,5)
1509 4624 at : P_{68}
1509 5141 $E : \ln(5/2)$
1509 5769 $J_1 : \sqrt{2} - \sqrt{5} - \sqrt{7}$
1509 7312 rt : (5,5,0,1)
1509 8916 $\ell_2 : 9\pi^2/5$
1510 1395 $J_1 : (5/2)^3$
1510 3051 $\theta_3 : 5/6$

1510 3465 at : $3/4+2\sqrt{5}/3$
1510 4357 ln : $4\sqrt{2}/3+3\sqrt{5}$
1510 5023 Ψ : $(4)^3$
1510 5651 $1/5+s\pi(2/5)$
1510 6285 e^x : $7\pi/4$
1510 6902 rt : (2,6,0,3)
1510 7885 10^x : $3/4-\pi/2$
1510 8219 10^x : $2+2e/3$
1510 8422 rt : (6,0,-1,8)
1510 8970 cu : $\exp(-\sqrt[3]{2}/2)$
1510 9193 rt : (6,5,-6,-1)
1511 0072 rt : (3,7,-9,-4)
1511 1111 sq : 22/15
1511 1206 Ψ : $(\sqrt{3})^{1/2}$
1511 2507 rt : (1,9,-8,1)
1511 2897 ln : $1/3+2\sqrt{2}$
1511 4821 $\sqrt{2}\div\exp(\sqrt{5})$
1511 4835 rt : (1,5,-1,-7)
1511 5371 J_1 : $\sqrt{2}+\sqrt{6}+\sqrt{7}$
1511 5715 cu : $1/3-\sqrt{3}/2$
1511 6511 rt : (5,1,-2,4)
1511 7052 rt : (3,3,7,1)
1511 7967 rt : (9,7,-3,1)
1511 9133 cu : $\sqrt{3}+\sqrt{5}/2$
1512 0048 Ψ : $1/\ln(1/\pi)$
1512 0258 rt : (7,3,6,-1)
1512 0882 Ψ : $(\ln 2/3)^{-1/3}$
1512 2311 at : $2\sqrt{2}/3+3\sqrt{3}/4$
1512 3456 sq : 7/18
1512 3558 rt : (5,3,-4,-7)
1512 4253 rt : (9,3,-5,4)
1512 5185 sq : $1/\sqrt[3]{17}$
1512 5324 Γ : $3(e+\pi)/2$
1512 5561 J_2 : $3e/7$
1512 6240 e^x : $3/2-e/2$
1512 6767 ℓ_{10} : 12/17
1512 7217 rt : (5,8,-5,2)
1512 7623 rt : (9,1,-7,-7)
1512 9254 ln : $\sqrt{10}$
1512 9993 $\ln(2/3)+s\pi(\sqrt{3})$
1513 2326 J_0 : $9\pi/2$
1513 3403 sq : $3\sqrt{2}/4-4\sqrt{5}$
1513 4675 2^x : $4e/3-\pi/2$
1513 5476 at : $2-3\sqrt{2}$
1513 7592 sπ : $\sqrt{2}/4+3\sqrt{3}/2$
1513 7614 id : $2\sqrt{2}-3\sqrt{5}/4$
1513 7731 tπ : $3\sqrt{3}/2+3\sqrt{5}/2$
1513 7931 J_1 : $5\pi^2/2$
1513 7944 E : $\beta(2)$
1513 8117 2^x : 21/19
1513 8781 rt : (8,8,-4,-3)
1514 2945 Ψ : 25/19
1514 3388 J_1 : $(\sqrt{2}/3)^{-3}$
1514 3475 2^x : $3e/2+\pi/2$
1514 3642 sinh : $7e/3$
1514 4930 J_2 : $(e/2)^{1/2}$
1514 5474 sinh : $5\sqrt{5}/4$
1514 6578 as : $\sqrt{3}/3-2\sqrt{5}/3$
1514 7107 at : $\ln(3\pi)$
1514 7349 sq : $4\sqrt{2}/3+3\sqrt{3}$
1514 8921 ℓ_2 : $4/3-\sqrt{3}/4$
1514 9043 e^x : P_{70}
1514 9461 e^x : P_{67}
1514 9585 ℓ_{10} : $9\sqrt[3]{2}/8$
1514 9612 ℓ_2 : $\sqrt{2}\pi$
1515 0212 rt : (9,4,7,1)
1515 0608 $5\times s\pi(\pi)$
1515 0725 sr : $(\ln 3)^3$
1515 3077 sq : $3\sqrt{2}/2-\sqrt{3}$
1515 4338 tπ : $7\sqrt[3]{3}/2$
1515 6751 tπ : $1/\ln(\sqrt{3}/2)$
1515 7922 at : $5\pi/7$
1515 8241 rt : (7,5,-6,-1)
1515 9756 rt : (7,1,-9,-1)
1516 1161 J_2 : $(\sqrt[3]{4})^{1/3}$
1516 1240 rt : (3,5,-8,-2)
1516 1600 Ei : $(3\pi)^{-3}$
1516 2075 2^x : $3/4+e/3$
1516 2767 J_1 : $\ln(e/2)$
1516 3266 $1/4\div\exp(1/2)$
1516 4321 J_2 : 7/6
1516 5655 10^x : $2\sqrt[3]{3}/3$
1516 5741 sr : $(5/3)^3$
1516 6670 at : $\sqrt{2}/4-3\sqrt{3}/2$
1516 7373 sr : $\sqrt[3]{7/3}$
1516 7474 rt : (6,3,6,-1)
1516 8854 10^x : $2+4\pi/3$
1517 0370 cu : 8/15
1517 2513 rt : (2,3,7,1)
1517 3065 Γ : $9\sqrt[3]{5}/5$
1517 3563 rt : (3,8,-3,5)
1517 4110 J_1 : $5\ln 2$
1517 4112 cu : $1+3\sqrt{3}/4$
1517 4505 10^x : $4+\pi/3$
1517 4552 Γ : $2\sqrt[3]{3}/7$
1517 4930 rt : (6,7,-9,8)
1517 4937 tπ : $3\pi^2/4$
1517 5128 e^x : $3\sqrt{2}/2+\sqrt{3}$
1517 5957 rt : (7,5,-7,-4)
1517 6048 Ψ : $\sqrt[3]{13/3}$
1517 7342 2^x : $\sqrt{2}+2\sqrt{5}$
1517 7594 cr : $7e\pi/6$
1517 8170 sr : $4\sqrt{2}/3-\sqrt{5}/4$
1517 8229 erf : $(e)^{-2}$
1517 9709 rt : (2,9,-8,1)
1517 9736 sinh : $(\pi/3)^{-1/3}$
1518 0465 at : $3/2+\sqrt{5}/3$
1518 0894 rt : (3,8,1,-5)
1518 1955 at : $2\sqrt{2}-4\sqrt{5}/3$
1518 5375 cu : $\sqrt{3}-4\sqrt{5}$
1518 7347 rt : (1,7,5,-2)
1518 7983 J : $\exp(-10\pi)$
1518 8894 e^x : $1/3+\sqrt{3}/4$
1519 0991 e^x : $\sqrt{2}/10$
1519 1818 sr : $\sqrt{10}\pi$
1519 4264 cr : $2-\sqrt{2}/3$
1519 5336 2^x : $4+\sqrt{3}$
1519 6146 rt : (5,2,-7,1)
1519 6716 e^x : $4(1/\pi)/9$
1519 6839 ζ : $7\sqrt{2}/3$
1519 7485 cu : $(\ln 2/3)^{-3}$
1519 8676 2^x : $1/\sqrt{24}$
1519 8780 tπ : $4\sqrt{2}+\sqrt{5}/3$
1519 9245 Ψ : $1/\sqrt[3]{10}$
1520 0309 ℓ_2 : 9/10
1520 0334 e^x : $(\sqrt{2}\pi/2)^{-1/3}$
1520 1899 rt : (9,0,-5,8)
1520 1988 J_2 : 14
1520 3265 J_1 : 4/13
1520 4227 rt : (5,9,-1,6)
1520 4244 J_2 : $7\sqrt{5}$
1520 5072 rt : (8,4,7,1)
1520 5474 rt : (9,9,5,-1)
1520 6743 tπ : $2\sqrt{3}/3+3\sqrt{5}/2$
1520 8247 rt : (8,5,-6,-1)
1520 8287 cu : $3\sqrt{2}/4+\sqrt{3}/2$
1520 8866 rt : (9,8,-8,1)
1520 9349 ℓ_2 : $5\sqrt[3]{2}/7$
1520 9568 rt : (6,1,1,9)
1520 9931 id : P_{71}
1521 0322 id : P_{68}
1521 0688 sr : $3/4+\sqrt{3}/3$
1521 0724 $\exp(\pi)\div\exp(3)$
1521 1897 J_1 : $(\sqrt{2}\pi/3)^{-3}$
1521 3318 rt : (8,3,-8,-7)
1521 3522 J_0 : $9\pi^2/8$
1521 4069 id : $((e+\pi)/2)^3$
1521 4194 cr : $2\sqrt{2}-3\sqrt{3}/4$
1521 5551 rt : (5,3,6,-1)
1521 9852 rt : (7,8,6,7)
1522 0688 e^x : $\sqrt{2}/4-\sqrt{5}$
1522 1174 rt : (9,4,-3,5)
1522 1262 J_1 : $4\ln 2/9$
1522 2509 e^x : $(\ln 2/2)^{-1/3}$
1522 3367 $1/2+\exp(\sqrt{3})$
1522 3668 rt : (2,4,-9,2)
1522 4329 J_0 : $8\pi^2/5$
1522 4513 ℓ_{10} : $3/4-e/4$
1522 5191 rt : (5,2,-2,-8)
1522 5201 2^x : $\exp(-\sqrt[3]{4})$
1522 6397 Li_2 : $(e+\pi)/7$
1522 9049 cr : $4e/3-2\pi/3$
1522 9240 rt : (1,3,7,1)
1523 0032 Ψ : $3e/5$
1523 0459 tπ : $(e/3)^{1/2}$
1523 5370 e^x : $\exp(-(e+\pi)/3)$
1523 5870 $\sqrt{5}-\ln(2/5)$
1523 6141 rt : (8,1,-6,4)
1523 6735 $\sqrt{3}\div\ln(\sqrt{5})$
1523 6838 rt : (2,4,-6,-1)
1523 7297 ln : $4\sqrt{3}+3\sqrt{5}/4$
1523 7688 at : $4\sqrt{2}/3-\sqrt{3}$
1523 9615 2^x : $3/4-2\sqrt{3}$
1524 0906 rt : (4,7,-3,1)
1524 2270 cu : $10\sqrt{3}$
1524 2577 J_1 : $2\sqrt{3}$
1524 2779 rt : (6,5,3,6)
1524 3319 J_0 : 17/8
1524 3806 ln : $5\zeta(3)/7$
1524 7845 rt : (8,9,5,-1)
1524 8514 cu : $4\zeta(3)/9$
1524 8929 rt : (3,9,-8,1)
1524 8956 e^x : $6\sqrt[3]{3}$
1524 9190 sπ : $(2e/3)^3$
1525 0792 rt : (2,1,4,-9)
1525 0922 sq : $4/3-2\sqrt{2}/3$
1525 1192 rt : (7,2,-9,-3)
1525 1867 2^x : $7\pi^2/10$
1525 3687 tπ : 3π
1525 4409 rt : (5,6,-7,7)
1525 4889 rt : (6,2,-7,1)
1525 5216 e^x : $1/2+\pi$
1525 6715 rt : (3,8,-8,-6)
1525 6949 sπ : $\sqrt{2}/3+\sqrt{3}/3$
1525 7199 at : 9/4
1525 7379 rt : (9,8,-1,2)
1525 7717 2^x : $2-3\pi/2$
1525 9177 cu : $4e/3-3\pi$
1525 9249 rt : (9,5,-6,-1)
1525 9490 10^x : $\exp(-\sqrt[3]{3}/3)$
1526 1150 rt : (7,4,7,1)
1526 3065 e^x : $1/4+4e/3$
1526 3184 cu : $1/4+e$
1526 3248 at : $e/4+\pi/2$
1526 4375 tanh : 2/13
1526 4523 rt : (4,3,6,-1)
1526 4932 at : 2/13
1526 5374 id : $1/\ln(\sqrt[3]{2}/3)$
1526 5903 rt : (5,2,0,5)
1526 7652 ln : $4-\pi$
1526 7950 ln : 6/19
1526 9195 as : P_{71}
1526 9590 as : P_{68}
1526 9724 J_2 : $(e)^3$
1527 0213 ln : $3e/7$
1527 4150 cu : $4\sqrt{2}+2\sqrt{5}/3$
1527 4801 rt : (1,8,-1,2)
1527 4955 at : $2\sqrt{2}-\sqrt{3}/3$
1527 5760 rt : (3,6,-9,6)
1527 6586 10^x : 25/12
1527 7541 rt : (9,1,-3,9)
1527 8769 ζ : $\sqrt[3]{2}/2$
1527 8970 sq : $1+\pi$
1527 9087 rt : (8,5,-4,1)
1528 0058 λ : $\exp(-16\pi/11)$
1528 0493 e^x : $\sqrt{24}$
1528 2596 tanh : $2\ln 2/9$
1528 3156 at : $2\ln 2/9$
1528 6333 J_2 : $(e+\pi)/5$
1528 7291 rt : (3,7,1,0)
1528 9229 rt : (8,2,-6,-8)
1528 9594 cu : $2\sqrt{3}/3+2\sqrt{5}$
1528 9786 Γ : $4\sqrt[3]{5}/3$
1529 0134 cu : 7π
1529 0393 10^x : $(\ln 2)^3$
1529 0395 10^x : $3+2\sqrt{2}/3$
1529 0878 rt : (7,9,5,-1)
1529 1032 sinh : $\sqrt{19/3}$
1529 2144 erf : 3/22
1529 3510 Ei : $2e\pi/5$

1529 4876 rt : (4,4,-9,2)
1529 7078 cr : $2/3+\sqrt{3}/2$
1529 8223 erf : $3(1/\pi)/7$
1529 8812 $\sqrt{5}-\exp(2)$
1529 8868 J_2 : $\exp(1/(2\pi))$
1529 9684 id : $2\sqrt{2}-4\sqrt{5}/3$
1530 0968 rt : (7,9,-5,-7)
1530 1426 tπ : $(\sqrt[3]{4}/3)^{1/2}$
1530 2041 2^x : $4-3\sqrt{5}$
1530 7469 ℓ_2 : $\exp(1/(3\pi))$
1530 8272 e^x : $(\pi/3)^3$
1530 9748 $\exp(3/4)+s\pi(\sqrt{2})$
1530 9798 ℓ_{10} : $2-\sqrt{3}/3$
1530 9941 J_1 : $3\sqrt{2}$
1531 1909 sq : 9/23
1531 2742 10^x : $7\sqrt[3]{3}/6$
1531 2759 rt : (3,5,-5,-4)
1531 2944 2^x : $1/2+3\sqrt{2}/2$
1531 3156 cr : 23/15
1531 4067 tπ : $\sqrt{2}/4+3\sqrt{3}/2$
1531 4387 ℓ_2 : $\sqrt[3]{11}$
1531 4423 rt : (3,6,9,7)
1531 5079 rt : (7,2,-7,1)
1531 5630 Ei : $(e+\pi)^{-1/2}$
1531 6117 $\sqrt{3}\div\ln(\sqrt{3})$
1531 8501 rt : (6,4,7,1)
1532 0296 rt : (4,9,-8,1)
1532 0401 rt : (9,5,-1,6)
1532 0560 tπ : P_{73}
1532 1265 $\ln(5)\div\exp(1/3)$
1532 4814 sr : $7\sqrt[3]{5}/9$
1532 6821 cu : 19/11
1532 8210 sinh : $5e/9$
1532 9310 ℓ_{10} : $e\pi/6$
1532 9426 rt : (5,1,0,-9)
1532 9537 tπ : $\exp(-1/\pi)$
1532 9679 at : $\sqrt{3}/3+3\sqrt{5}/4$
1533 3670 at : $3-\sqrt{5}/3$
1533 4593 rt : (6,9,5,3)
1533 5829 rt : (8,2,-4,5)
1533 6498 rt : (2,7,-9,-2)
1533 7316 ζ : $\pi^2/3$
1533 8021 Γ : 7/17
1533 8743 rt : (4,4,-6,-1)
1534 0089 sr : $3\sqrt{3}-\sqrt{5}/4$
1534 0151 ℓ_{10} : $(\ln 2/2)^{1/3}$
1534 0678 rt : (9,2,6,-1)
1534 1459 sinh : $7\sqrt{2}$
1534 1800 10^x : $8\sqrt{3}$
1534 2640 ln : $(e/2)^{1/2}$
1534 4907 sπ : $3/4+3\sqrt{3}/4$
1534 5396 sr : $1+4\sqrt{5}$
1534 5659 rt : (4,8,1,5)
1534 6730 rt : (1,3,-7,1)
1534 7252 cu : $1/4-\pi/4$
1534 8036 J_0 : $5\zeta(3)$
1534 8385 at : $1-2\sqrt{3}/3$
1534 9922 e^x : $3\sqrt{3}/2-2\sqrt{5}$
1535 0927 sq : $\sqrt{2}/4-\sqrt{5}/3$
1535 1135 cu : $\sqrt{2}/3+\sqrt{3}/3$
1535 1345 rt : (7,1,-7,-4)
1535 1411 2^x : $\sqrt[3]{3}/7$
1535 3872 Ei : $(\sqrt[3]{5}/2)^2$
1535 6294 as : $1/2-\sqrt{2}$
1535 6499 e^x : 1/7
1535 6727 id : $4\sqrt{2}/3-\sqrt{3}$
1536 0011 as : $2\sqrt{2}-4\sqrt{5}/3$
1536 1992 rt : (3,7,-6,-7)
1536 3394 sq : $4/3-4\pi/3$
1536 4894 cr : $3/4+\pi/4$
1536 5285 rt : (2,3,6,6)
1536 6519 tπ : $1/3-3\sqrt{2}/4$
1536 6562 rt : (8,1,-4,-9)
1536 7214 sπ : $(\ln 3/3)^3$
1536 8798 sπ : $(\sqrt{3}/2)^{-1/3}$
1536 9838 Γ : 17/21
1537 0747 $\ln(3/4)-s\pi(1/3)$
1537 2627 rt : (7,9,8,8)
1537 4688 cr : $\sqrt{3}+\sqrt{6}-\sqrt{7}$
1537 5239 rt : (4,5,-5,6)
1537 5255 cu : $2/3+3\sqrt{2}/4$
1537 5296 at : $\sqrt[3]{23/2}$
1537 6790 rt : (8,2,-7,1)
1537 7183 rt : (5,4,7,1)
1537 8584 rt : (7,4,-7,-8)
1537 9012 rt : (5,9,5,-1)
1537 9965 id : $4\sqrt[3]{3}/5$
1538 0721 rt : (2,8,-2,-3)
1538 0801 sπ : $4e/3+3\pi$
1538 0875 ℓ_{10} : $5\sqrt[3]{5}/6$
1538 1313 sπ : $3\sqrt{2}+3\sqrt{5}$
1538 2778 ℓ_{10} : $\sqrt{3}/2+\sqrt{5}/4$
1538 4130 J_1 : 13/2
1538 4615 id : 2/13
1538 5114 J : $1/\sqrt{21}$
1538 5588 J_2 : $3e\pi$
1538 5955 J_2 : $\sqrt{22}$
1538 7648 2^x : $10\sqrt{5}/7$
1538 9571 rt : (8,9,-2,-2)
1538 9880 J_2 : 20/17
1539 0309 cu : $4-2\sqrt{3}$
1539 0951 rt : (8,2,6,-1)
1539 1235 rt : (5,4,-6,-1)
1539 1617 θ_3 : 1/13
1539 2029 e^x : $1/4-3\sqrt{2}/2$
1539 3952 rt : (5,9,-8,1)
1539 4615 e^x : $e/2+4\pi$
1539 4815 J_0 : $10\sqrt{5}$
1539 6339 ζ : 23/7
1539 6656 rt : (6,6,5,7)
1539 9342 Ψ : $\sqrt{8/3}$
1540 1492 rt : (9,9,1,3)
1540 2968 Li_2 : $(\sqrt[3]{4}/3)^3$
1540 3270 id : $2\ln 2/9$
1540 3806 J_1 : 24
1540 5595 10^x : $2e\pi/3$
1540 6152 tπ : 3/11
1540 6351 ℓ_2 : $3\sqrt{3}-\sqrt{5}/3$
1540 7626 rt : (2,3,-7,1)
1541 0002 2^x : P_{39}
1541 0307 θ_3 : $\ln 2/9$
1541 0336 10^x : $1-2e/3$
1541 1217 2^x : P_{33}
1541 1271 e^x : $\sqrt[3]{6}$
1541 1466 tπ : $2\sqrt{3}/3+\sqrt{5}/2$
1541 1507 J_1 : $9\pi/2$
1541 5067 ln : 6/7
1541 5603 $4/5\div\ln(2)$
1541 5735 J_2 : $(\sqrt[3]{3}/2)^{-1/2}$
1541 7149 rt : (1,3,6,9)
1541 7736 as : $4\sqrt{2}/3-\sqrt{3}$
1542 0422 ζ : $(2e/3)^2$
1542 0636 sq : $\ln(\sqrt{2}\pi/3)$
1542 0706 $\exp(4/5)\times\exp(\sqrt{2})$
1542 0901 rt : (5,3,2,6)
1542 1256 sq : $\pi/8$
1542 1745 ln : $3\sqrt{2}/4-\sqrt{5}/3$
1542 2037 rt : (9,6,1,7)
1542 2833 tπ : P_{21}
1542 4157 rt : (8,6,-2,2)
1542 6224 id : $\exp(e)$
1542 7362 rt : (1,9,8,9)
1542 7393 J_2 : $3\pi/8$
1542 9530 rt : (7,6,-5,-3)
1542 9645 tπ : $(2e/3)^3$
1543 0559 ℓ_{10} : $2-3\sqrt{3}/4$
1543 1285 Γ : $3\pi^2/2$
1543 5039 J_1 : 5/16
1543 6946 J_2 : $5\sqrt{2}/6$
1543 7260 rt : (4,4,7,1)
1543 7600 Ei : 23/21
1543 7682 tπ : $\sqrt{2}/3+\sqrt{3}/3$
1543 7948 rt : (8,3,-2,6)
1543 8079 Ei : 23/10
1543 8667 $3\div s\pi(2/5)$
1543 8717 sinh : $\sqrt{2}-\sqrt{5}-\sqrt{6}$
1544 0019 rt : (2,9,-5,9)
1544 0102 rt : (9,2,-7,1)
1544 0413 cu : $\ln(\sqrt[3]{5})$
1544 1567 cr : 20/13
1544 2187 rt : (7,2,6,-1)
1544 2254 ln : $4(e+\pi)$
1544 2477 K : $3\sqrt[3]{3}/5$
1544 3469 id : $\sqrt[3]{10}$
1544 3557 sr : $3/2+\pi$
1544 4644 cu : $1/4-3\sqrt{3}/4$
1544 4808 10^x : $1/2+2\sqrt{3}/3$
1544 4809 rt : (6,4,-6,-1)
1544 4812 10^x : $1/4+\sqrt{2}$
1544 5365 rt : (4,0,-1,5)
1544 5376 sinh : 2/13
1544 5862 Γ : $(\sqrt[3]{4}/3)^{1/3}$
1544 5959 as : 2/13
1544 7010 rt : (7,3,-5,1)
1545 1506 J_0 : $3\sqrt{2}/2$
1545 1634 10^x : $\sqrt{3}/2-3\sqrt{5}/4$
1545 1661 2^x : $3\sqrt{3}/2-2\sqrt{5}/3$
1545 2300 rt : (1,6,3,-7)
1545 2680 cu : $3\sqrt{2}/2+3\sqrt{5}/2$
1545 3081 J_1 : 4π
1545 3167 rt : (4,9,5,9)
1545 3731 tπ : $9\sqrt{2}/10$
1545 3954 rt : (7,0,-5,5)
1545 4492 cr : $9\sqrt[3]{5}/10$
1545 5476 e^x : $4\sqrt{2}/3-\sqrt{5}/2$
1545 7681 rt : (7,3,-5,-9)
1545 8681 $\exp(2/5)\times s\pi(e)$
1545 9057 ℓ_2 : $3/4+3e$
1546 1770 Ei : $\sqrt{22/3}$
1546 2704 Γ : $7\ln 2/6$
1546 3791 10^x : $10\sqrt{5}/9$
1546 3932 Ψ : $3e\pi/7$
1546 4252 sinh : $2\ln 2/9$
1546 4640 10^x : $1/4-3\sqrt{2}/4$
1546 4840 as : $2\ln 2/9$
1546 5038 2^x : $3\sqrt{2}+2\sqrt{5}$
1546 6382 at : $8\sqrt{2}/5$
1546 7132 $\ln(2)\times\ln(4/5)$
1546 7253 sinh : $\sqrt[3]{13/2}$
1546 8787 2^x : 21/2
1547 0053 id : $2\sqrt{3}/3$
1547 2157 J_2 : $4(e+\pi)/5$
1547 2929 2^x : $((e+\pi)/3)^3$
1547 2962 10^x : $4\zeta(3)/5$
1547 4195 e^x : P_{75}
1547 8198 10^x : 1/16
1547 8853 $1/5\times s\pi(e)$
1547 8882 cr : $3\sqrt{2}/4+4\sqrt{5}$
1548 1076 J_1 : $10\sqrt{2}$
1548 4005 sπ : $\sqrt{3}/3+2\sqrt{5}$
1548 4440 at : $3/4-e/3$
1548 4548 id : $3e$
1548 6249 rt : (5,9,-8,-9)
1548 6600 tπ : $6(1/\pi)/7$
1548 6921 J : $1/\sqrt{15}$
1548 7046 rt : (2,9,-7,-7)
1548 7368 J_0 : $1/\ln(\ln 2)$
1548 7521 sq : $2\sqrt{2}/3+\sqrt{3}$
1549 0195 ℓ_{10} : 10/7
1549 1204 rt : (9,8,5,-1)
1549 2234 $\ln(\pi)\div\exp(2)$
1549 2917 ℓ_{10} : $5\sqrt[3]{2}/9$
1549 4423 rt : (6,2,6,-1)
1549 4969 rt : (2,6,-7,-3)
1549 5598 sπ : $3\sqrt{2}/2+4\sqrt{3}$
1549 6296 cu : 22/15
1549 6650 10^x : 19/9
1549 6909 rt : (1,9,3,-7)
1549 8801 rt : (3,4,7,1)
1549 8937 2^x : $\exp(-\pi/2)$
1549 9057 at : $1/3-3\sqrt{3}/2$
1549 9511 rt : (7,4,-6,-1)
1550 1560 2^x : $(e/2)^{1/3}$
1550 2798 rt : (1,8,-2,-6)
1550 4107 2^x : $3\ln 2/10$
1550 5132 ℓ_2 : $9\sqrt{3}/7$
1550 6761 ℓ_2 : P_{41}
1550 7082 at : $5e/6$
1550 7138 sinh : $\sqrt{2}-\sqrt{3}+\sqrt{6}$
1550 9346 2^x : $3/2-4\pi/3$

1550 9412 ℓ_{10} : $3 - \pi/2$
1550 9540 10^x : $3/2 - 4\sqrt{3}/3$
1550 9965 rt : (2,7,-4,6)
1551 0557 Ψ : 21/16
1551 0752 J_1 : $(\ln 2/2)^{-3}$
1551 1892 Ei : $\sqrt{19/3}$
1551 2815 2^x : $4e/3 - \pi/4$
1551 3298 J_2 : 13/11
1551 4393 e^x : $\sqrt[3]{3}/10$
1551 4932 $\ln(3) \div s\pi(2/5)$
1551 4969 J_1 : $\pi/10$
1551 5114 2^x : $3\pi^2/4$
1551 5256 Ei : $10\zeta(3)$
1551 6724 rt : (2,9,5,-1)
1551 7944 sinh : $\pi^2/10$
1551 8049 ζ : $(\ln 3/3)^{-1/2}$
1551 8929 2^x : $3/2 + \sqrt{3}/2$
1551 9951 E : 21/23
1552 0287 J_1 : $2\sqrt{2}/9$
1552 0556 ℓ_{10} : $3/4 + e/4$
1552 2750 Γ : $4\sqrt{2}/7$
1552 3088 sq : $2/3 - 3\sqrt{2}/4$
1552 3918 2^x : $2/3 - 3\sqrt{5}/2$
1552 4041 rt : (1,6,3,-3)
1552 4212 Ei : $5\pi^2/8$
1552 4530 id : $5\ln 2/3$
1552 6187 rt : (9,7,3,8)
1552 6402 rt : (9,5,7,1)
1552 8461 $s\pi$: $\sqrt{20}\pi$
1552 8638 J_0 : $\exp(\pi)$
1552 8816 e^x : $1/2 + 3\sqrt{3}/2$
1552 8822 $t\pi$: $1/4 - 3\sqrt{3}/4$
1552 9267 $t\pi$: $4\sqrt{2} + 4\sqrt{5}/3$
1553 0488 Li_2 : $3\sqrt{5}/8$
1553 0545 cr : $\sqrt[3]{11/3}$
1553 2433 as : $1 - 2\sqrt{3}/3$
1553 3373 $1/4 \div \ln(5)$
1553 4091 rt : (4,9,-1,-7)
1553 4288 J : 1/7
1553 5158 $\sqrt{5} \times s\pi(\sqrt{2})$
1553 5269 2^x : 5/24
1553 5430 J : 1/5
1553 6756 cr : 10π
1553 7327 rt : (8,8,5,-1)
1553 9854 tanh : $\ln(\sqrt[3]{5}/2)$
1554 0454 ζ : $(2e/3)^{1/3}$
1554 0463 at : $\ln(\sqrt[3]{5}/2)$
1554 2608 rt : (8,4,0,7)
1554 3024 erf : $\ln 2/5$
1554 3426 J : $\sqrt{5}/6$
1554 3492 Ψ : $5(e + \pi)/8$
1554 4083 2^x : $3\sqrt{2} - 4\sqrt{3}$
1554 4177 J : $(1/\pi)/2$
1554 5213 $1/3 + \exp(3/5)$
1554 6510 $1/4 + \ln(2/3)$
1554 7700 rt : (5,2,6,-1)
1554 8773 rt : (7,9,-8,1)
1554 9772 2^x : $1/\sqrt{23}$
1554 9822 at : $9\sqrt[3]{2}/5$
1555 0993 rt : (6,5,-6,-9)
1555 1308 at : $4 - \sqrt{3}$
1555 1942 $t\pi$: $(\ln 3/3)^3$
1555 3583 $t\pi$: $(\sqrt{3}/2)^{-1/3}$
1555 3991 2^x : $4\sqrt{2}/3 - 3\sqrt{5}/4$
1555 4054 rt : (3,1,-9,6)
1555 4506 J_1 : $\sqrt[3]{2}/4$
1555 5319 J_1 : 17/4
1555 5390 rt : (8,4,-6,-1)
1555 5761 $s\pi$: $5\pi^2/7$
1555 6441 rt : (6,7,7,8)
1555 7016 sr : $(\sqrt[3]{2}/3)^{-1/3}$
1555 8415 sq : $\ln((e + \pi)/2)$
1555 9128 rt : (7,1,-3,6)
1556 1143 Γ : $9\sqrt{5}/4$
1556 1878 rt : (2,4,7,1)
1556 1943 ℓ_{10} : $((e + \pi)/2)^{1/3}$
1556 2862 J_1 : $4e$
1556 3733 rt : (5,8,-9,-8)
1556 4197 rt : (1,9,5,-1)
1556 6025 $t\pi$: $4e/3 + 3\pi$
1556 6556 $t\pi$: $3\sqrt{2} + 3\sqrt{5}$
1556 8974 sr : $10\zeta(3)/9$
1557 2555 $\ln(\sqrt{2}) \div \exp(4/5)$
1557 2564 sq : $\sqrt{2}/3 - \sqrt{3}/2$
1557 2734 $\pi \div e$
1557 4290 rt : (8,7,0,3)
1557 5075 at : $2\sqrt{2} - \sqrt{5}/4$
1557 5748 rt : (6,2,-6,-5)
1557 6015 $1/5 \div \exp(1/4)$
1557 7372 $s\pi$: $(e)^{-3}$
1557 9156 rt : (2,9,6,-2)
1557 9184 cr : $3\sqrt{2}/2 - \sqrt{3}/3$
1558 1035 rt : (3,5,-2,-9)
1558 1898 rt : (5,4,4,7)
1558 2479 e^x : $\sqrt{2} + 3\sqrt{5}$
1558 4104 $2/3 + \ln(3/5)$
1558 4224 rt : (7,8,5,-1)
1558 4668 cu : $4/3 + e/4$
1558 4923 2^x : $2\sqrt{3} + 4\sqrt{5}/3$
1558 5419 tanh : $\sqrt{2}/9$
1558 5879 ℓ_2 : $\pi/7$
1558 6037 at : $\sqrt{2}/9$
1558 6868 J_0 : $1/\ln(\sqrt[3]{3})$
1558 8524 rt : (8,5,7,1)
1559 2076 $\exp(\pi) \div \exp(5)$
1559 2469 rt : (5,5,-9,-4)
1559 2837 ln : $7\pi^2/8$
1559 3468 J_1 : 6/19
1559 3644 sr : $7\pi^2/4$
1559 5082 sq : $2\sqrt{2}/3 - 4\sqrt{5}/3$
1559 5340 ln : $3/2 + 3\sqrt{5}/4$
1559 8185 rt : (7,4,-3,2)
1559 8620 rt : (2,8,-5,-8)
1559 9075 J_1 : $9\sqrt{3}$
1559 9825 rt : (5,3,-7,1)
1560 0233 $\sqrt{3} \times \exp(3/5)$
1560 1461 J_2 : $\sqrt[3]{5/3}$
1560 2058 rt : (4,2,6,-1)
1560 2636 ζ : $-\sqrt{2} + \sqrt{5} + \sqrt{6}$
1560 3503 Li_2 : 3/20
1560 6264 ℓ_{10} : $2\pi/9$
1560 6335 rt : (3,8,-4,2)
1560 7594 rt : (4,1,1,6)
1560 9394 id : $1/4 + e/3$
1560 9830 at : $4/3 - 2\sqrt{5}/3$
1561 2198 cu : 7/13
1561 2498 rt : (9,5,-1,0)
1561 4567 J_1 : $1/\sqrt{10}$
1561 5294 cu : $1/4 + 3\sqrt{5}/4$
1561 6214 cr : 17/11
1561 7391 cr : $9\zeta(3)/7$
1562 0083 tanh : $\sqrt[3]{2}/8$
1562 0328 ln : $4\sqrt{2} + 4\sqrt{5}/3$
1562 0708 at : $\sqrt[3]{2}/8$
1562 1517 rt : (1,7,0,8)
1562 2287 rt : (6,1,-6,1)
1562 2326 e^x : $1/\ln(\pi/2)$
1562 3031 $s\pi$: $5\sqrt[3]{2}/6$
1562 4375 ℓ_2 : $3 - 4\sqrt{2}/3$
1562 5003 ℓ_{10} : $1 + \sqrt{3}/4$
1562 5855 Ei : $4e/3$
1562 6572 rt : (1,4,7,1)
1562 6617 e^x : $1/2 - 3\pi/4$
1562 7709 rt : (9,1,-8,-6)
1562 8945 at : 25/11
1562 9062 at : $2\sqrt{3}/3 + \sqrt{5}/2$
1563 0050 e^x : $4\sqrt{3}/3 + 3\sqrt{5}/2$
1563 0306 rt : (8,9,-8,1)
1563 0335 Ψ : $7\pi/6$
1563 1262 id : $4e + 2\pi$
1563 1362 Li_2 : $\zeta(3)/8$
1563 1918 rt : (6,8,5,-1)
1563 2517 rt : (1,5,-1,-1)
1563 2967 rt : (9,8,5,9)
1563 3535 rt : (3,7,3,4)
1563 3939 cu : $3/2 - 3e/4$
1563 4405 sq : $3e/4$
1563 5448 rt : (3,3,-6,-1)
1564 1068 $\pi \div \exp(3)$
1564 2393 ln : $\sqrt{2} - \sqrt{5}/4$
1564 2570 sr : $7e\pi/6$
1564 3446 $s\pi$: 1/20
1564 4241 J : 5/16
1564 4490 J_0 : $\sqrt[3]{19/2}$
1564 4491 sq : $4\sqrt{2}/3 - 3\sqrt{5}/2$
1564 4723 J_2 : 19/16
1564 6599 rt : (5,8,-6,0)
1564 6901 $1/5 - \exp(\sqrt{5})$
1564 7305 sr : $\sqrt{2}/3 + \sqrt{3}/2$
1564 7592 $s\pi$: $5\sqrt[3]{5}/9$
1564 9924 rt : (8,5,2,8)
1565 1564 K : $\sqrt{3}/2$
1565 2184 rt : (7,5,7,1)
1565 2733 Γ : $7\pi/4$
1565 2799 ln : $2\sqrt{2}/3 + \sqrt{5}$
1565 4718 at : $\sqrt{2}/2 - 4\sqrt{5}/3$
1565 7542 rt : (3,2,6,-1)
1565 9350 e^x : $3/2 - 3\sqrt{5}/2$
1565 9554 tanh : 3/19
1565 9635 sr : $2\sqrt{2} - 2\sqrt{5}/3$
1565 9663 cr : $1/2 + \pi/3$
1565 9726 rt : (2,5,-5,-4)
1566 0187 at : 3/19
1566 1657 rt : (1,9,-5,-1)
1566 3761 sq : $4e + 2\pi/3$
1566 3876 J_1 : $9\sqrt[3]{3}/2$
1566 5609 sr : $2e - \pi/4$
1566 6787 id : $\ln(\sqrt[3]{5}/2)$
1566 6986 rt : (7,2,-1,7)
1566 7350 rt : (6,3,-7,1)
1566 7763 2^x : $\sqrt[3]{2}/6$
1566 9061 cu : $7\ln 2/9$
1566 9353 E : $(\pi/3)^{-2}$
1567 2664 2^x : $3e/4 - 3\pi/2$
1567 3029 $t\pi$: $\sqrt{3}/3 + 2\sqrt{5}$
1567 3487 as : $3/4 - e/3$
1567 4246 rt : (9,2,-8,2)
1567 4989 rt : (1,8,1,-8)
1567 6163 Ei : $(\sqrt{5}/3)^3$
1567 6558 Ψ : 11/3
1567 8819 rt : (9,1,6,-1)
1568 0437 rt : (5,8,5,-1)
1568 0670 Ψ : 18/11
1568 0709 rt : (5,6,-8,5)
1568 1779 Γ : 16/7
1568 4154 rt : (6,1,-4,-6)
1568 4267 at : $4/3 + 2\sqrt{2}/3$
1568 4268 J_2 : $(\sqrt{2})^{1/2}$
1568 5051 $t\pi$: $3\sqrt{2}/2 + 4\sqrt{3}$
1568 5424 id : $1/2 + 4\sqrt{2}$
1568 5606 rt : (1,9,7,1)
1568 6170 rt : (1,8,-8,-3)
1568 8261 sq : $4\ln 2/7$
1568 9376 $s\pi$: $2/3 + 2\pi$
1568 9657 $\exp(1/2) - \exp(2/5)$
1569 0767 ℓ_{10} : $\sqrt{2} + \sqrt{3} - \sqrt{6}$
1569 1190 rt : (1,9,-1,4)
1569 2966 rt : (4,7,1,0)
1569 3293 10^x : $2 + e/4$
1569 3732 erf : $\sqrt[3]{2}/9$
1569 4453 rt : (2,9,-6,4)
1569 5086 rt : (3,5,1,9)
1569 8033 rt : (7,7,-3,-2)
1570 1392 2^x : $8\ln 2/5$
1570 1430 rt : (5,4,-7,-5)
1570 3693 J_0 : $10\pi^2/7$
1570 5508 cr : $1/4 + 3\sqrt{3}/4$
1570 7584 Ψ : 24
1570 8609 J_1 : 7/22
1570 8783 rt : (9,0,-6,7)
1570 9320 id : $3\sqrt{3}/2 + \sqrt{5}/4$
1570 9672 $s\pi$: $(2\pi)^3$
1571 0127 J_1 : $8e\pi/9$
1571 1023 2^x : 4/19
1571 1382 sr : $3\sqrt{3}/4 + 3\sqrt{5}/2$
1571 1704 rt : (2,9,-5,-1)
1571 3297 rt : (2,7,-3,-9)
1571 3484 id : $\sqrt{2}/9$
1571 3684 J_2 : 25/21
1571 4199 rt : (2,4,8,7)

1571 4771 $J_1 : (1/\pi)$
1571 4868 rt : (9,9,-8,1)
1571 5224 sq : $3+2e/3$
1571 6991 sq : $3/4-\sqrt{2}/4$
1571 7462 rt : (6,5,7,1)
1571 9137 tπ : $\sqrt{20}\pi$
1572 1671 rt : (8,3,-9,-6)
1572 1799 sπ : $7\sqrt{2}/2$
1572 2559 rt : (6,8,9,9)
1572 2638 $\ell_2 : (\sqrt[3]{3}/2)^{1/3}$
1572 9810 rt : (8,8,2,4)
1573 0956 sinh : $\ln(\sqrt[3]{5}/2)$
1573 1139 rt : (5,9,-2,4)
1573 1595 as : $\ln(\sqrt[3]{5}/2)$
1573 1928 $\lambda : \exp(-13\pi/9)$
1573 4163 $J_1 : \sqrt{12}\pi$
1573 4305 cr : $\ln(3\pi/2)$
1573 4628 rt : (8,1,6,-1)
1573 5159 $J_2 : 7\sqrt{5}/2$
1573 5978 $J_2 : 7\pi^2/3$
1573 6768 rt : (7,3,-7,1)
1573 7865 id : $1/3-2\sqrt{5}/3$
1573 8454 $\zeta : 1/\sqrt{6}$
1573 9461 $\zeta : 6e/5$
1574 0740 cu : $5\sqrt[3]{2}/6$
1574 1762 rt : (6,4,-6,-3)
1574 3185 cu : $1/4+4e/3$
1574 3440 cu : $3\sqrt[3]{2}/7$
1574 3706 sinh : $6\sqrt[3]{2}/5$
1574 4607 rt : (4,9,-8,-9)
1574 4705 rt : (1,7,6,5)
1574 5088 $J_2 : 4\pi^2/7$
1574 5134 $J_0 : 9\sqrt{5}$
1574 5551 $1/3 \div \exp(3/4)$
1574 6074 at : $4\sqrt[3]{5}/3$
1574 6491 sq : $1/2-4\sqrt{5}/3$
1574 7370 $\Gamma : \sqrt[3]{5}/6$
1574 7457 tπ : $5\pi^2/7$
1574 8403 $s\pi(1/5)+s\pi(\pi)$
1574 9013 id : $\sqrt[3]{2}/8$
1574 9329 rt : (5,5,6,8)
1574 9779 rt : (5,7,-4,9)
1575 0380 ln : $2\sqrt{2}/9$
1575 0447 at : $\sqrt{2}+\sqrt{3}/2$
1575 0739 $J_0 : 4\pi$
1575 1767 rt : (5,3,-6,-1)
1575 2152 $\ell_2 : P_{31}$
1575 2849 $J_2 : 10e/3$
1575 3994 $e^x : 3/4-3\sqrt{3}/2$
1575 4826 rt : (7,5,-1,3)
1575 5496 cr : $\sqrt{3}/4+\sqrt{5}/2$
1575 9302 cu : $1/2+4e/3$
1575 9635 rt : (9,5,-8,-2)
1576 0021 rt : (8,6,4,9)
1576 0069 at : $\sqrt{2}/2-\sqrt{3}/2$
1576 0785 $\ell_{10} : 16/23$
1576 1155 rt : (1,4,-7,1)
1576 2500 cu : 21/20
1576 2672 rt : (3,9,-5,-1)
1576 7785 tπ : $5\sqrt{5}/3$
1576 9055 $J_1 : \sqrt{5}/7$
1576 9878 tπ : $(e)^{-3}$
1577 1311 $e^x : 1/2-\sqrt{2}/4$
1577 1534 $\ell_{10} : 2e/3+4\pi$
1577 2080 rt : (1,2,6,-1)
1577 3124 rt : (7,9,-9,9)
1577 4268 $Ei : (e+\pi)/8$
1577 5590 $J_2 : 19$
1577 6209 $e^x : 3-\pi/4$
1577 6249 sq : $3\sqrt{3}/2-\sqrt{5}/4$
1577 6362 rt : (4,2,3,7)
1577 6574 tπ : $e/4+\pi/3$
1577 7450 $e^x : 5\sqrt[3]{5}/6$
1577 7655 rt : (7,3,1,8)
1577 8228 sinh : $\sqrt{2}/9$
1577 8578 $2^x : 7e/3$
1577 8586 $K : 5\ln 2/4$
1577 8631 $\ln(4/5)\times s\pi(1/4)$
1577 8877 as : $\sqrt{2}/9$
1578 0063 $\ln(\sqrt{3})-s\pi(1/4)$
1578 0067 rt : (9,3,-6,3)
1578 0162 $\theta_3 : 8(1/\pi)/5$
1578 2128 tπ : $4(1/\pi)$
1578 2460 tanh : $(1/\pi)/2$
1578 3119 at : $(1/\pi)/2$
1578 4090 $\zeta : 1/\ln(e/2)$
1578 4442 rt : (5,5,7,1)
1578 4696 sq : $1/2+3\pi/4$
1578 4808 $Ei : (\pi)^{1/2}$
1578 5493 $10^x : e/2+4\pi/3$
1578 5520 ln : $\pi/10$
1578 6013 $\sqrt{2}\div\exp(1/5)$
1578 6158 rt : (5,1,-9,-4)
1578 7580 $10^x : (1/\pi)/5$
1578 8326 $\ln(3/4)\div\exp(3/5)$
1578 8329 tπ : $7\sqrt{3}/5$
1578 9473 id : 3/19
1579 0125 $\ln(5)\div s\pi(\sqrt{3})$
1579 0777 as : $\beta(2)$
1579 1425 rt : (9,1,-4,8)
1579 1605 rt : (7,1,6,-1)
1579 2718 sq : $4\sqrt{2}-4\sqrt{5}/3$
1579 5401 rt : (6,0,-2,7)
1579 6071 $J_1 : 8/25$
1579 6356 ln : $6\sqrt[3]{3}$
1579 6870 rt : (9,7,5,-1)
1579 7021 rt : (8,0,-9,2)
1579 7921 at : $4-2\pi$
1580 1736 $10^x : (2\pi/3)^3$
1580 2187 $\Psi : (\ln 2/2)^{-3}$
1580 2394 tπ : $\sqrt{2}-2\sqrt{3}$
1580 2553 $\ln(3/4)\times\ln(\sqrt{3})$
1580 3413 $e^x : \sqrt{3}/2+\sqrt{5}/4$
1580 3566 as : $4/3-2\sqrt{5}/3$
1580 3663 rt : (3,8,7,8)
1580 4633 rt : (8,2,-7,-7)
1580 4957 $s\pi(1/5)+s\pi(\sqrt{3})$
1580 5536 rt : (5,1,-7,-1)
1580 6287 cu : $\sqrt{3}/3-\sqrt{5}/2$
1580 7123 $10^x : \sqrt{2}/4-2\sqrt{3}/3$
1580 8189 rt : (8,3,-7,1)
1580 9210 $Ei : 1/3$
1581 0067 sr : $6e\pi$
1581 0553 $e^x : 10/13$
1581 1576 sπ : $(e/3)^{-1/2}$
1581 1622 cr : $4e/7$
1581 1908 rt : (6,3,-6,-1)
1581 1933 ln : $1/2+3e$
1581 2093 $e^x : 1/4-2\pi/3$
1581 3010 $J_0 : 8\sqrt[3]{5}/5$
1581 3233 $10^x : 3/2+3\pi/4$
1581 3262 rt : (5,3,-5,-6)
1581 4198 sinh : $\sqrt[3]{2}/8$
1581 4594 rt : (4,9,-5,-1)
1581 4854 as : $\sqrt[3]{2}/8$
1581 5610 $Ei : \pi^2/9$
1581 7257 tπ : $5\sqrt[3]{2}/6$
1581 9290 $e^x : 23/20$
1582 0112 $J_1 : 2\sqrt[3]{3}/9$
1582 0312 cu : $3\sqrt[3]{3}/8$
1582 0576 $\exp(e)\div\exp(2/5)$
1582 1288 rt : (1,6,0,6)
1582 5297 cu : $\sqrt{3}/2+3\sqrt{5}/2$
1582 5952 $\ell_{10} : \sqrt{21}\pi$
1582 7238 ln : $\ln 2/6$
1582 9218 id : $(\sqrt{5}/3)^{-1/2}$
1583 0131 $J_1 : 7e/2$
1583 1239 rt : (4,6,-8,-5)
1583 1434 sq : $5(1/\pi)/4$
1583 3509 at : $3/2+\pi/4$
1583 3527 $\ell_{10} : 8\sqrt[3]{2}/7$
1583 4153 $\zeta : 5(e+\pi)/9$
1583 4433 ln : $4\sqrt{2}/3+3\sqrt{3}/4$
1583 4718 ln : $4-2\sqrt{2}$
1583 5018 $2^x : 4e/3-2\pi$
1583 5140 ln : $3+4\sqrt{2}$
1583 6384 $J_2 : 17$
1583 6661 $\exp(e)\div\exp(3/4)$
1583 6986 $\ell_2 : 1+2\sqrt{3}$
1583 7549 $J_2 : (\sqrt[3]{5})^{1/3}$
1583 8444 tπ : 1/20
1583 8588 at : 16/7
1584 0166 $Li_2 : 2\sqrt[3]{2}/3$
1584 2747 tπ : $5\sqrt[3]{5}/9$
1584 5163 cu : $2\sqrt{2}+2\sqrt{3}$
1584 5868 $2^x : 2(1/\pi)/3$
1584 7305 rt : (8,7,5,-1)
1584 8211 $10^x : 15/19$
1584 8388 sπ : $(\sqrt{2}\pi)^{-2}$
1584 9136 $1/3-\exp(2/5)$
1584 9800 rt : (6,1,6,-1)
1585 0244 rt : (1,9,-4,-9)
1585 0927 tπ : $7\zeta(3)/6$
1585 1128 $K : 13/15$
1585 1648 $Li_2 : 21/25$
1585 1941 $erf : \sqrt{2}/10$
1585 3214 rt : (4,5,7,1)
1585 3702 $J_0 : 7e/9$
1585 5162 sinh : 3/19
1585 5828 as : 3/19
1585 6494 $2^x : 3-4\sqrt{2}$
1585 7439 $erf : 4(1/\pi)/9$
1585 8766 $\Psi : 13/18$
1585 9094 rt : (1,7,-6,-4)
1586 0162 cu : $7\sqrt{2}/4$
1586 1315 rt : (1,7,-1,-9)
1586 1340 cu : $2-3e$
1586 4013 sq : $(e/2)^{-3}$
1586 4738 $J_2 : 7\sqrt[3]{5}/10$
1586 4850 tanh : 4/25
1586 5526 at : 4/25
1586 5534 cu : $\sqrt{2}/4+3\sqrt{5}$
1586 5885 $\ell_{10} : 2-\sqrt{5}/4$
1586 6929 $\zeta : \sqrt[3]{9/2}$
1586 7508 rt : (5,9,-5,-1)
1586 7554 cr : 14/9
1586 8649 $1/3\div\ln(3/4)$
1586 9027 ln : $(e+\pi)/5$
1587 1882 as : $\ln(5/2)$
1587 3451 rt : (7,3,-6,-1)
1587 3679 $2^x : 13/3$
1587 4405 ln : $5\sqrt{3}$
1587 5689 rt : (9,2,-2,9)
1587 6322 rt : (1,9,-2,-5)
1587 6902 rt : (1,8,7,7)
1587 7217 $Ei : 3\sqrt[3]{5}/7$
1587 7735 $10^x : (5/2)^{-3}$
1588 1734 rt : (9,3,-7,1)
1588 2113 $E : 1/\ln(3)$
1588 3360 rt : (1,8,5,-1)
1588 3992 $10^x : 22/21$
1588 4073 $10^x : 1/2-3\sqrt{3}/4$
1588 4655 $10^x : 3e-4\pi/3$
1588 6117 tπ : $1/3-2\pi$
1588 6685 at : $1/4+3e/4$
1588 6755 at : P_{20}
1588 7612 $E : \sqrt{2}+\sqrt{3}-\sqrt{5}$
1588 8308 id : $6\ln 2$
1588 8327 $J_2 : 5\pi$
1588 8336 $10^x : 8/3$
1588 8517 rt : (9,4,-4,4)
1588 8850 tπ : $4\sqrt{2}-\sqrt{5}/2$
1588 9215 tanh : $\sqrt[3]{3}/9$
1588 9231 rt : (8,1,-5,-8)
1588 9896 at : $\sqrt[3]{3}/9$
1588 9989 cr : $9\sqrt{5}/2$
1589 1073 rt : (8,9,4,5)
1589 1270 rt : (7,4,3,9)
1589 1732 rt : (1,5,1,-4)
1589 1862 id : $\sqrt{2}/2-\sqrt{3}/2$
1589 2169 $e^x : 3\sqrt{3}-4\sqrt{5}/3$
1589 2239 $2^x : 3\sqrt{2}+4\sqrt{5}$
1589 2650 cu : 13/24
1589 3055 rt : (2,7,-5,2)
1589 3689 id : $4e\pi$
1589 4543 $erf : \exp(-(e+\pi)/3)$
1589 6964 $\Gamma : \ln(\sqrt{5})$
1589 7562 sq : $4/3-\sqrt{3}$
1589 7631 $J_2 : 23$
1589 8085 $e^x : 4\sqrt{2}/3+3\sqrt{3}/4$

1589 8177 at : $\sqrt[3]{12}$
1589 8649 rt : (7,7,5,-1)
1589 9393 $\exp(5) - s\pi(\sqrt{3})$
1590 0004 $t\pi$: $2e/3 + \pi/4$
1590 0818 rt : (6,9,-9,-1)
1590 2485 e^x : $4\sqrt{2}/3 + \sqrt{5}/2$
1590 2839 rt : (7,4,-8,-7)
1590 2916 rt : (4,3,-8,9)
1590 3327 rt : (3,5,-6,-9)
1590 4041 ℓ_{10} : $\sqrt[3]{3}$
1590 5145 rt : (2,9,5,9)
1590 6018 rt : (8,1,-7,3)
1590 7188 $t\pi$: $(2\pi)^3$
1590 9266 rt : (5,1,6,-1)
1590 9627 rt : (6,1,0,8)
1591 0627 ℓ_{10} : $6\zeta(3)/5$
1591 1744 rt : (9,6,-6,-1)
1591 3592 e^x : $\sqrt{2}/3 - 4\sqrt{3}/3$
1591 5198 ζ : $13/4$
1591 5494 id : $(1/\pi)/2$
1591 7204 $s\pi$: $(\sqrt[3]{5}/2)^{1/3}$
1591 7297 rt : (7,6,1,4)
1591 7453 ℓ_{10} : $\ln 2$
1591 8357 $\ln(\sqrt{5}) + s\pi(\sqrt{2})$
1591 9124 rt : (1,7,-1,1)
1591 9690 rt : (4,5,-6,4)
1591 9779 $t\pi$: $7\sqrt{2}/2$
1592 0885 ℓ_{10} : $1/2 + 2\sqrt{2}/3$
1592 1450 rt : (6,9,-5,-1)
1592 2599 sq : P_{59}
1592 3514 θ_3 : $(1/\pi)/4$
1592 3677 rt : (5,6,8,9)
1592 3878 rt : (3,5,7,1)
1592 4021 at : $4/3 - 4e/3$
1592 4427 at : $\sqrt{3} + \sqrt{5}/4$
1592 4689 id : $6(e + \pi)$
1592 5722 2^x : $1/\sqrt{22}$
1592 7016 J_0 : $\sqrt{15/2}$
1592 8090 cu : $3\sqrt{2}/2 + 2\sqrt{3}/3$
1592 8106 rt : (5,2,-3,-7)
1593 0875 e^x : $3\sqrt{2} + 3\sqrt{3}/2$
1593 1263 J_0 : $10\sqrt[3]{5}$
1593 3511 e^x : $4\sqrt{3}/9$
1593 4901 J_2 : $6/5$
1593 6012 10^x : $3 + 3\pi/4$
1593 6465 rt : (8,5,-1,0)
1593 6970 Γ : $\sqrt[3]{12}$
1593 7938 ℓ_{10} : $2\sqrt{3}/5$
1593 8490 Ei : $(\ln 2)^3$
1594 1832 sr : $(e/3)^{-3}$
1594 3222 ζ : $(\sqrt{2}\pi/3)^3$
1594 3452 rt : (9,6,7,1)
1594 3584 sinh : $4\sqrt{3}/7$
1594 3766 rt : (6,3,-2,3)
1594 4312 $\exp(2/5) \div \exp(\sqrt{5})$
1594 6708 rt : (4,7,-4,-1)
1594 7253 id : $4\pi^2/3$
1594 9188 cr : $9\sqrt{3}/10$
1594 9591 rt : (6,8,-7,8)
1595 0796 ℓ_{10} : $3e + 2\pi$
1595 0935 rt : (6,7,5,-1)
1595 1533 10^x : $1/4 - \pi/3$
1595 2164 rt : (4,3,5,8)
1595 2274 rt : (1,9,6,-7)
1595 2622 rt : (2,7,5,9)
1595 2680 10^x : P_7
1595 3107 2^x : $1/3 - 4\sqrt{5}/3$
1595 3330 J_2 : $\sqrt{2} + \sqrt{5} - \sqrt{6}$
1595 3434 cr : $1 + \sqrt{5}/4$
1595 3618 at : $3 - \sqrt{2}/2$
1595 3622 rt : (2,8,4,-3)
1595 4689 $\ln(3/5) - \exp(1/2)$
1595 6178 cu : $2/3 - \pi$
1595 6216 10^x : $13/21$
1595 6476 cu : $\sqrt{2}/3 + 4\sqrt{3}$
1595 6709 J_2 : $(\sqrt[3]{3})^{1/2}$
1595 6915 rt : (2,9,-7,-1)
1595 6955 id : $\sqrt{2} + \sqrt{5}/3$
1595 7228 e^x : $4 - \pi/3$
1595 9339 ln : $\sqrt{3} - \sqrt{5}/4$
1595 9525 as : $\sqrt{2}/2 - \sqrt{3}/2$
1596 0014 e^x : $3 - e/2$
1596 1026 J_2 : $(\ln 2)^{-1/2}$
1596 1701 e^x : $\sqrt{2}/2 - \sqrt{5}/4$
1596 2099 J_0 : $(3\pi)^{1/3}$
1596 2384 at : $3\sqrt{2}/4 - 3\sqrt{5}/2$
1596 3365 10^x : $3e/4 + 2\pi$
1596 5225 2^x : $\sqrt{2}/2 - 3\sqrt{5}/2$
1596 5435 rt : (3,8,-9,-5)
1596 5826 Ei : $7\sqrt{2}/3$
1596 5846 as : $11/12$
1596 6569 rt : (9,8,-8,-6)
1596 7419 cr : $9\ln 2/4$
1596 8468 e^x : $(\sqrt[3]{4}/3)^3$
1596 9625 2^x : $\sqrt[3]{5}/8$
1597 0058 rt : (4,1,6,-1)
1597 0084 ℓ_{10} : $13/9$
1597 0485 rt : (5,0,-5,2)
1597 2848 ζ : $9/22$
1597 5154 sinh : $7\sqrt{2}/10$
1597 5470 J_0 : $(\ln 3/2)^{-3}$
1597 5525 rt : (8,0,-3,9)
1597 6331 sq : $14/13$
1597 6461 rt : (7,9,-5,-1)
1597 6652 sq : $4e/3 + 2\pi$
1597 6801 ℓ_{10} : $\exp(1/e)$
1597 7540 rt : (4,8,-3,-8)
1597 8011 Ei : $1/\sqrt[3]{14}$
1597 8541 J_2 : $5\sqrt[3]{3}/6$
1598 0142 J_2 : $\exp(1/(2e))$
1598 0947 rt : (3,0,2,7)
1598 1503 J_2 : $10\sqrt{5}$
1598 2770 sinh : $(1/\pi)/2$
1598 2786 J_2 : $\zeta(3)$
1598 3070 rt : (7,8,-1,-1)
1598 3462 as : $(1/\pi)/2$
1598 6799 at : $2e - \pi$
1598 8583 ζ : $9\sqrt[3]{3}/4$
1598 8934 $\ln(4/5) \div \exp(1/3)$
1598 9171 cu : $6e\pi/5$
1598 9484 rt : (7,3,-6,-8)
1599 0591 rt : (1,8,-5,-1)
1599 4121 Ei : $6e\pi/7$
1599 4157 cr : $1/2 + 3\sqrt{2}/4$
1599 4963 J_1 : $9e$
1599 5123 e^x : $4e/3 - \pi/3$
1599 6542 rt : (2,5,7,1)
1599 7276 $s\pi$: $e\pi/9$
1599 7458 rt : (4,9,-2,-6)
1599 9718 rt : (9,5,-2,5)
1600 0000 id : $4/25$
1600 1020 rt : (9,3,-6,-1)
1600 3513 rt : (6,6,-9,-7)
1600 4197 rt : (5,8,6,4)
1600 4839 $\sqrt{2} - s\pi(\sqrt{3})$
1600 5040 $t\pi$: $3\sqrt{2}/4 + \sqrt{3}/3$
1600 5214 J : $(4/3)^{-3}$
1600 6828 $t\pi$: $\exp((e + \pi)/2)$
1600 8192 θ_3 : $2/25$
1600 8397 Γ : $5(e + \pi)/6$
1601 0032 rt : (4,6,-2,8)
1601 0712 erf : $1/7$
1601 1947 2^x : $10/9$
1601 2938 2^x : $3/14$
1601 3010 $t\pi$: $(e/3)^{-1/2}$
1601 4611 rt : (8,6,7,1)
1601 5090 sinh : $10\ln 2/7$
1601 5351 Ei : $11/15$
1601 6974 rt : (3,2,-6,-1)
1601 7794 rt : (8,2,-5,4)
1601 8500 J_2 : $7e$
1601 8732 ζ : $9(1/\pi)/7$
1602 0678 Li_2 : $2/13$
1602 1773 id : P_{20}
1602 3704 $\ln(4) + s\pi(e)$
1602 4689 id : $\sqrt{14/3}$
1602 4793 sr : $(2e/3)^{1/2}$
1602 4828 ζ : $(\sqrt[3]{5}/2)^2$
1602 4995 id : $\sqrt[3]{3}/9$
1602 5123 2^x : $1/2 - \pi$
1602 5827 2^x : $1/4 + 3\sqrt{5}/2$
1602 6040 rt : (9,4,-9,-9)
1602 6980 rt : (6,2,2,9)
1602 7458 sr : $2/3 + e/4$
1602 9194 E : $10/11$
1602 9913 J_0 : $19/9$
1603 0420 rt : (6,5,-4,-2)
1603 1148 J_1 : $(\sqrt[3]{5}/3)^2$
1603 1414 ℓ_2 : $3e + \pi/4$
1603 2238 rt : (3,1,6,-1)
1603 2584 rt : (8,9,-5,-1)
1603 4135 J_2 : 20
1603 4341 sr : $3\pi/7$
1603 4852 cr : $3\sqrt{2}/2 - \sqrt{5}/4$
1603 5985 rt : (7,1,-8,-3)
1603 8285 $t\pi$: $3\sqrt{3}/2 - \sqrt{5}$
1603 8362 2^x : $1 - \pi/4$
1603 9425 rt : (1,6,-4,-5)
1603 9606 2^x : $4\sqrt{3} + \sqrt{5}/3$
1603 9720 cr : $25/16$
1604 0897 $\exp(2/5) - \exp(\sqrt{3})$
1604 0936 Li_2 : $2\ln 2/9$
1604 5778 rt : (2,8,-5,-1)
1604 6112 rt : (5,1,-1,-8)
1604 6467 ℓ_2 : $17/19$
1604 6651 2^x : $\exp(\sqrt[3]{3}/2)$
1604 6853 rt : (9,0,6,1)
1604 7249 Γ : $\sqrt{19/2}$
1604 7409 Li_2 : $(\sqrt{2})^{-1/2}$
1604 9116 cu : $4\sqrt{2}/3 - 3\sqrt{5}$
1604 9382 sq : $16/9$
1605 0113 Ψ : $17/13$
1605 1251 $t\pi$: $(\sqrt{2\pi})^{-2}$
1605 1600 at : $1 + 3\sqrt{3}/4$
1605 1989 sq : $7(e + \pi)/4$
1605 2930 ℓ_{10} : $(2\pi/3)^{1/2}$
1605 3694 Ψ : $(\sqrt[3]{5})^{1/2}$
1605 4008 rt : (5,4,-7,1)
1605 4897 sq : $\zeta(3)/3$
1605 5902 Γ : $6\pi^2/7$
1605 6341 J_2 : $4(e + \pi)/3$
1605 7621 sq : $3/4 + \sqrt{3}$
1605 8471 rt : (4,7,5,-1)
1605 8670 ζ : $\sqrt{21/2}$
1606 2092 rt : (3,8,-5,-1)
1606 2920 sq : $7\sqrt[3]{2}/6$
1606 4708 e^x : $20/11$
1606 5376 rt : (4,8,0,3)
1606 6607 e^x : $1 - 2\sqrt{2}$
1606 6818 sr : $7\sqrt{3}/9$
1606 6898 at : $23/10$
1606 8354 sinh : $4/25$
1606 8360 sq : $1/2 + \sqrt{3}/3$
1606 8429 cu : $e/5$
1606 9065 as : $4/25$
1606 9133 rt : (9,7,-4,0)
1607 0980 cu : $7e/3$
1607 1324 rt : (1,5,7,1)
1607 1657 cr : $4e - \pi/4$
1607 1693 J_1 : $(3\pi)^{-1/2}$
1607 1732 J_0 : $(e)^3$
1607 2807 ℓ_2 : $4 + \sqrt{2}/3$
1607 3123 sinh : $9e\pi/5$
1607 3833 id : $4e - 3\pi/2$
1607 4955 Ei : $7e/5$
1607 6614 J_2 : $(\sqrt[3]{5}/3)^{-1/3}$
1607 7410 ℓ_{10} : $3 - 4\sqrt{3}/3$
1607 7902 rt : (7,2,-4,-9)
1607 7957 cr : $4\sqrt{3}/3 - \sqrt{5}/3$
1607 8540 rt : (5,2,-7,-3)
1608 1018 sq : $\sqrt{5} + \sqrt{6} - \sqrt{7}$
1608 1858 rt : (4,2,-6,-1)
1608 3976 sq : $3 - 3e/2$
1608 4773 rt : (3,7,-7,-6)
1608 6000 rt : (7,7,3,5)
1608 6037 J_2 : $\exp(\sqrt{3})$
1608 7332 ℓ_{10} : $4e - 3\pi$
1608 7760 rt : (7,6,7,1)
1608 7854 rt : (3,7,-5,9)
1608 8942 rt : (2,9,0,-9)

1608 9865 rt : (9,9,-5,-1)
1609 0985 $s\pi$: $9\sqrt{3}/8$
1609 1123 as : P_{20}
1609 2301 rt : (6,5,-7,-8)
1609 3670 sinh : $\sqrt[3]{3}/9$
1609 4003 e^x : $2\sqrt{3}/3 - 4\sqrt{5}/3$
1609 4379 $1/5 \times \ln(\sqrt{5})$
1609 4387 as : $\sqrt[3]{3}/9$
1609 4477 2^x : $4/3 - \sqrt{5}/2$
1609 4915 Ψ : $1/\ln(4)$
1609 4980 rt : (8,4,-7,-1)
1609 5869 rt : (2,1,6,-1)
1609 6404 ℓ_2 : $\sqrt{5}$
1609 6823 J_2 : $(\ln 3)^2$
1609 6921 E : $(4/3)^{-1/3}$
1609 8182 cr : $2/3 + 3\pi$
1609 9504 cu : $2e/3 - 3\pi/4$
1610 2067 rt : (3,8,-5,-1)
1610 2837 $\sqrt{3} \div \exp(2/5)$
1610 6333 Ei : $8\sqrt[3]{5}/3$
1610 7660 ζ : $7\sqrt{2}/6$
1610 7697 cr : $7\sqrt{5}/10$
1610 8333 cu : $1/\ln(2\pi)$
1610 9267 rt : (8,0,6,1)
1611 0611 at : $4\sqrt{2} - 3\sqrt{5}/2$
1611 3796 rt : (9,6,0,6)
1611 4164 2^x : $4e\pi$
1611 4905 $s\pi(1/3) \div s\pi(\sqrt{3})$
1611 6523 cu : $2 - \sqrt{2}/2$
1611 8797 $\ln(\sqrt{5}) + \exp(\sqrt{5})$
1611 9024 ln : $2/3 - \sqrt{2}/4$
1611 9475 erf : $\ln(\sqrt{3}/2)$
1611 9594 J_0 : $7\pi^2/3$
1612 1632 sinh : 8/3
1612 1858 e^x : $9\sqrt{2}/7$
1612 2424 cu : $2e/3 - 3\pi$
1612 2584 cu : $4\sqrt{2}/3 + 3\sqrt{3}$
1612 2755 $t\pi$: $(\sqrt[3]{5}/2)^{1/3}$
1612 3578 10^x : $\sqrt[3]{5}/2$
1612 4275 J_2 : $4e/9$
1612 5076 rt : (9,6,5,-1)
1612 5410 Ψ : $-\sqrt{2} + \sqrt{6} + \sqrt{7}$
1612 5557 Ψ : $3\zeta(3)/5$
1612 6443 J_1 : 19/2
1612 8554 rt : (5,6,-9,3)
1612 8685 J_1 : $\exp(-\sqrt{5}/2)$
1612 8888 cr : $7\sqrt[3]{3}$
1612 9168 e^x : 9π
1612 9233 e^x : $2\sqrt{3} - \sqrt{5}/3$
1612 9702 J_1 : $\ln(\sqrt[3]{3}/2)$
1612 9860 ln : $\sqrt{2} - 3\sqrt{3}/4$
1613 1991 K : $\ln(\sqrt[3]{2}/3)$
1613 2478 rt : (8,3,-3,5)
1613 2874 rt : (6,4,-7,1)
1613 5205 ln : P_{78}
1613 5555 rt : (4,4,7,9)
1613 5601 2^x : $\exp(1/(3\pi))$
1613 6818 10^x : $1/\ln((e + \pi)/3)$
1613 7906 rt : (3,9,1,8)
1613 8227 sr : $4e/3 + \pi/3$
1613 9390 ln : $10\sqrt{5}/7$
1613 9558 e^x : $-\sqrt{2} + \sqrt{5} - \sqrt{7}$
1613 9616 2^x : $3\sqrt{3}/4 + 3\sqrt{5}/2$
1614 1866 rt : (5,1,-3,3)
1614 4413 $s\pi$: $1/3 + e$
1614 8351 rt : (5,7,1,0)
1615 0698 cr : $2 - \sqrt{3}/4$
1615 0873 2^x : $(5/3)^{-3}$
1615 1284 rt : (7,0,-6,4)
1615 1902 cr : $2\sqrt{3}/3 + 4\sqrt{5}$
1615 4273 sq : $3 - 3\sqrt{3}/2$
1615 5120 Ψ : $\sqrt[3]{3}/2$
1615 6004 rt : (9,1,-9,-5)
1615 7267 sr : $(\ln 3/2)^{-1/2}$
1615 7716 J_2 : $9\pi/5$
1615 8370 J_0 : $2e\pi$
1615 9504 rt : (4,8,-5,-1)
1616 1020 rt : (1,2,7,7)
1616 1906 erf : $\sqrt[3]{3}/10$
1616 2305 $\exp(1/5) \times s\pi(2/5)$
1616 3013 rt : (6,6,7,1)
1616 5955 rt : (3,1,4,8)
1616 6272 Li_2 : $7\zeta(3)/10$
1616 6793 cu : $4\zeta(3)$
1616 7443 rt : (5,0,1,9)
1617 0213 rt : (6,3,-9,-3)
1617 1151 rt : (4,9,-9,8)
1617 3120 rt : (7,0,6,1)
1617 5115 e^x : $4\sqrt{2}/3 + \sqrt{3}/4$
1617 5188 2^x : $3 - 2\sqrt{2}/3$
1617 6045 id : $1/3 + 2\sqrt{2}$
1617 7198 2^x : $1 + e$
1617 7734 ℓ_{10} : $1/3 + \sqrt{5}/2$
1617 8271 2^x : $\sqrt{2}/4 - 4\sqrt{5}/3$
1617 9145 sr : $10\sqrt{3}$
1617 9200 $t\pi$: $e - 3\pi/4$
1617 9214 J_0 : $7\sqrt[3]{3}/2$
1618 0347 J_2 : 23/19
1618 0503 rt : (8,6,5,-1)
1618 2944 rt : (6,4,-5,-9)
1618 3424 e^x : 3/20
1618 5924 sq : $e + 2\pi/3$
1618 6031 tanh : $\exp(-2e/3)$
1618 6382 ℓ_2 : $4/3 + \pi$
1618 6778 at : $\exp(-2e/3)$
1618 7379 at : $8\sqrt[3]{3}/5$
1618 7407 rt : (4,9,4,7)
1618 7940 J_2 : $\sqrt{12}\pi$
1618 9116 e^x : $\sqrt{2}/4 + 3\sqrt{3}$
1619 0250 cr : $e/4 + 3\pi$
1619 1646 2^x : $\sqrt{3}/8$
1619 4824 e^x : $1/\ln(\sqrt{3}/3)$
1619 6690 rt : (2,4,3,8)
1619 7672 $1/2 \div s\pi(\pi)$
1619 7806 rt : (5,8,-7,-2)
1619 8055 cu : $\sqrt{3}/2 + 4\sqrt{5}$
1619 8193 J_2 : 14/3
1619 8600 ℓ_{10} : $2/3 + \pi/4$
1619 8845 rt : (3,5,-9,-1)
1620 1652 $s\pi$: $7\zeta(3)/8$
1620 3243 rt : (5,7,-5,7)
1620 3703 cu : 11/6
1620 5986 $t\pi$: $e\pi/9$
1620 7514 rt : (3,6,-5,-7)
1620 7782 $2/3 \div \exp(\sqrt{2})$
1620 9481 cu : $4\sqrt{2} - 3\sqrt{5}$
1620 9638 Ei : $\zeta(3)/8$
1621 0501 ℓ_{10} : $\sqrt{2}/2 + \sqrt{5}/3$
1621 1126 $\ln(2/5) \div \exp(\sqrt{3})$
1621 1621 J_1 : $e\pi/2$
1621 1732 J_2 : $4\sqrt{2}$
1621 2185 rt : (3,2,-6,8)
1621 3300 e^x : $\zeta(3)/8$
1621 4299 rt : (7,4,-7,1)
1621 4764 ℓ_{10} : $4\sqrt{2}/3 - \sqrt{3}/4$
1621 5447 2^x : $4 + \sqrt{3}/2$
1621 5847 at : $4\sqrt{3}/3$
1621 6129 Ei : $\sqrt{11}\pi$
1621 6535 rt : (6,2,-6,-1)
1621 6913 J_n : $(\sqrt{2}\pi)^{1/2}$
1621 7509 2^x : $1 - 4e/3$
1621 8135 rt : (5,8,-5,-1)
1621 8604 $2/5 \times \ln(2/3)$
1621 9066 J_2 : $6\sqrt{2}/7$
1621 9386 $\exp(\sqrt{2}) - s\pi(2/5)$
1622 0497 $s\pi$: $\exp((e + \pi)/3)$
1622 0547 Ψ : 23/14
1622 1474 $3/4 - s\pi(1/5)$
1622 1558 e^x : $2 + \sqrt{3}/3$
1622 4969 J_2 : $7\sqrt{3}/10$
1622 7433 Γ : $(\ln 3/3)^{-3}$
1622 7766 id : $\sqrt{10}$
1622 8399 cu : 6/11
1623 0889 rt : (9,7,2,7)
1623 2133 rt : (9,8,-2,1)
1623 2592 sr : $3\sqrt{2} + \sqrt{3}/4$
1623 3043 at : $10 \ln 2/3$
1623 5024 rt : (9,9,-6,-5)
1623 5202 $\sqrt{3} - s\pi(\pi)$
1623 7010 rt : (7,6,5,-1)
1623 8344 J_2 : $7 \ln 2/4$
1623 8369 Ψ : $(2\pi/3)^3$
1623 8477 rt : (6,0,6,1)
1623 9488 sq : $e - 4\pi/3$
1624 0496 rt : (5,6,7,1)
1624 1278 J_0 : 23
1624 4614 rt : (9,0,-7,6)
1624 4735 id : $(\pi/2)^{1/3}$
1624 6440 10^x : $2\sqrt{2}/3 - \sqrt{3}$
1624 8978 J_0 : $10\sqrt[3]{2}$
1625 0211 rt : (8,4,-1,6)
1625 1892 ln : 17/20
1625 2285 J_0 : $\exp(\sqrt{5}/3)$
1625 4202 e^x : $2\sqrt{2} - 3\sqrt{5}/4$
1625 4769 $s\pi$: $6\sqrt[3]{5}/5$
1625 8352 cr : $10\sqrt{2}/9$
1625 9190 rt : (8,5,-5,0)
1625 9355 $t\pi$: $9(e + \pi)/10$
1625 9804 at : $1/2 + 2e/3$
1626 0329 cr : 11/7
1626 1343 θ_3 : $\ln(5/3)$
1626 1376 rt : (7,8,5,6)
1626 2143 rt : (1,7,-9,4)
1626 2940 2^x : 5/23
1626 2975 sq : 25/17
1626 4906 sr : $3 + 3\sqrt{5}/4$
1626 8241 J_2 : 17/14
1626 9263 Γ : $2\zeta(3)/3$
1626 9664 rt : (7,1,-4,5)
1627 2729 ℓ_{10} : 11/16
1627 3390 $t\pi$: $4\sqrt{2} - 3\sqrt{5}$
1627 3487 id : $4\sqrt{3}/3 - 2\sqrt{5}$
1627 7357 10^x : $3\sqrt[3]{3}$
1627 8010 rt : (6,8,-5,-1)
1628 0129 rt : (8,6,-9,-6)
1628 2855 Γ : $5\sqrt[3]{3}/9$
1628 4822 2^x : $\sqrt{3}/4 + 3\sqrt{5}$
1628 5000 rt : (2,6,-8,7)
1628 6499 rt : (7,9,1,0)
1628 8111 $\ln(\sqrt{3}) \times \exp(3/4)$
1628 9037 sr : $9\zeta(3)/8$
1628 9250 rt : (6,2,-7,-4)
1629 0611 rt : (6,5,2,5)
1629 0784 rt : (5,9,-3,2)
1629 3169 cr : $2/3 + e/3$
1629 3447 10^x : $4/3 - 3\sqrt{2}/2$
1629 4119 rt : (2,7,-6,-2)
1629 4639 rt : (6,6,5,-1)
1629 6111 cu : $4\sqrt{2} + \sqrt{5}/4$
1629 6185 rt : (1,4,-1,-5)
1629 7190 J : $\sqrt{3}/8$
1629 7871 J : $\exp(-13\pi/8)$
1629 8457 rt : (8,4,-7,1)
1629 9878 $\exp(2) + s\pi(e)$
1629 9972 cu : $2\sqrt{2}/3 - \sqrt{5}$
1630 0254 at : $2/3 - 4\sqrt{5}/3$
1630 2327 cr : $\sqrt{2}/2 + \sqrt{3}/2$
1630 3433 $t\pi$: $9\sqrt{3}/8$
1630 5411 rt : (5,0,6,1)
1630 5430 sr : $1/\ln(2\pi/3)$
1630 5739 sr : $3\sqrt{2}/4 + 4\sqrt{5}$
1630 5929 $\ln(3/5) - \exp(\sqrt{3})$
1630 6135 rt : (4,9,-9,-8)
1630 6972 $t\pi$: $\sqrt{2}/3 - \sqrt{5}/3$
1630 9721 rt : (2,5,-7,1)
1631 1036 Ψ : $(\sqrt{2}\pi)^{1/3}$
1631 1302 ℓ_{10} : $4\zeta(3)/7$
1631 1487 ℓ_{10} : $1/3 + \sqrt{2}/4$
1631 2012 10^x : $4 + 3\sqrt{5}/2$
1631 2021 Li_2 : 16/19
1631 2765 $s\pi$: $2\pi^2/5$
1631 2980 rt : (4,2,0,-9)
1631 2981 ℓ_2 : $4/3 + e/3$
1631 3388 10^x : $4\sqrt{2}/3 + 3\sqrt{5}/4$
1631 5080 ln : 5/16
1631 5220 $t\pi$: $\pi/7$
1631 5910 $1/4 - \exp(5)$
1631 5999 sr : 23/17
1631 8673 sinh : $2(e + \pi)/5$
1632 0145 rt : (5,2,-1,4)

1632 0347 rt : (4,6,7,1)
1632 1230 $\ln(\sqrt{5}) \times \exp(3)$
1632 3148 sr : $4 + e/4$
1632 6530 sq : $20/7$
1632 9572 2^x : $1/\sqrt{21}$
1632 9647 id : $\exp(-2e/3)$
1633 0312 ℓ_{10} : $(\sqrt{2}/3)^{1/2}$
1633 0544 J_2 : $(\sqrt{2}\pi/3)^{1/2}$
1633 3612 Ei : $3\ln 2/5$
1633 3841 rt : (6,8,-8,6)
1633 5026 rt : (9,1,-5,7)
1633 5135 ℓ_{10} : $2 + 4\pi$
1633 5594 E : $(\sqrt{5}/3)^{1/3}$
1633 7435 id : $3\sqrt[3]{3}/2$
1633 7965 rt : (6,6,-2,-1)
1633 9184 rt : (7,8,-5,-1)
1633 9586 sq : $4/3 + \sqrt{2}/2$
1634 0833 ℓ_2 : $8\sqrt[3]{2}/9$
1634 2196 e^x : $4 + 2\sqrt{2}/3$
1634 2313 sr : $1 + \sqrt{2}/4$
1634 5908 cr : $5\sqrt[3]{2}/4$
1634 6151 sπ : $\exp(2/3)$
1634 6462 rt : (1,7,-5,-1)
1634 7154 ln : $(\sqrt[3]{3}/2)^{1/2}$
1634 7656 J : $1/17$
1634 9137 ℓ_{10} : $3/4 + \sqrt{2}/2$
1634 9985 rt : (8,2,-8,-6)
1635 1143 rt : (9,8,4,8)
1635 1517 $\exp(2) - \exp(4/5)$
1635 2102 cu : $1 - 3\pi/2$
1635 3434 rt : (5,9,8,5)
1635 4075 J_2 : $\sqrt{2} + \sqrt{6} - \sqrt{7}$
1635 4855 rt : (2,8,-6,-7)
1635 5610 J_2 : $(e/3)^{-2}$
1635 7066 sq : $1/3 + \sqrt{5}/3$
1635 8337 rt : (8,2,-6,-1)
1635 8801 cu : $7\zeta(3)/8$
1635 8859 Ei : $\ln(3)$
1635 9013 tπ : $1/3 + e$
1635 9174 Ψ : $2\ln 2/3$
1635 9327 rt : (3,2,6,9)
1636 1090 at : $4\sqrt{2}/3 + \sqrt{3}/4$
1636 1530 rt : (1,4,1,7)
1636 4659 rt : (9,2,-9,1)
1636 5850 cr : $4\sqrt{2} + 2\sqrt{5}$
1636 6110 J_1 : $10e\pi/9$
1636 6365 rt : (2,7,2,-4)
1636 6992 J_0 : $7\pi/8$
1636 8183 J_2 : $4e$
1636 9419 rt : (4,3,-9,7)
1636 9633 $\ln(4/5) \times \exp(\pi)$
1637 0242 id : $9\zeta(3)/5$
1637 0977 $\ln(\sqrt{2}) \div \exp(3/4)$
1637 1141 rt : (8,5,1,7)
1637 1692 ζ : $\ln(\sqrt[3]{4}/3)$
1637 4000 rt : (4,0,6,1)
1637 4615 $1/5 \div \exp(1/5)$
1637 5218 e^x : $1/2 - 4\sqrt{3}/3$
1637 5680 Ψ : $(\sqrt{2}\pi/2)^{1/3}$
1637 6576 2^x : $3 + 2\pi/3$
1637 6797 10^x : $(\pi/3)^{-2}$
1637 7749 J_1 : $5\sqrt[3]{5}/2$
1637 8117 2^x : $4 + 3\sqrt{5}/4$
1637 8118 Li_2 : $\sqrt{2}/9$
1637 8252 sq : $3/4 - 2\sqrt{3}/3$
1638 0136 rt : (3,5,-6,-6)
1638 1457 10^x : $3\sqrt[3]{5}/4$
1638 2844 J_2 : $(2e/3)^{1/3}$
1638 3082 ζ : $(e)^{1/2}$
1638 4012 10^x : $6\sqrt{2}$
1638 4471 cu : $1/2 - \pi/3$
1638 5543 rt : (9,4,-7,1)
1638 7116 rt : (2,9,4,5)
1638 9283 at : $3 - e/4$
1639 0063 ln : $3\pi/8$
1639 0253 $1/5 - s\pi(\sqrt{2})$
1639 0324 rt : (5,3,-5,-2)
1639 1279 rt : (7,2,-2,6)
1639 1871 rt : (2,1,3,-8)
1639 2638 Ψ : $\pi^2/6$
1639 3925 e^x : $4/3 - \pi$
1639 4389 rt : (3,5,-7,1)
1639 4997 at : $\sqrt[3]{25/2}$
1639 5027 at : $3/4 + \pi/2$
1639 5598 ℓ_2 : $\ln(\sqrt{5}/2)$
1639 5722 ℓ_2 : $3/2 + 4\sqrt{5}/3$
1640 1110 rt : (9,9,0,2)
1640 1182 J_2 : $e\pi/7$
1640 1564 rt : (4,8,-7,-9)
1640 1713 rt : (8,8,-5,-1)
1640 2317 sinh : $\exp(-2e/3)$
1640 2714 rt : (3,6,7,1)
1640 3105 as : $\exp(-2e/3)$
1640 4256 id : $1/\ln(\sqrt[3]{4})$
1640 6182 cr : $1 + \sqrt{3}/3$
1640 7706 rt : (2,7,-5,-1)
1640 8641 Ei : $3/20$
1640 9395 10^x : $\exp(-e)$
1640 9800 E : $e/3$
1641 0561 rt : (9,7,7,1)
1641 1558 rt : (6,1,-5,-5)
1641 2432 tπ : $\sqrt{2}/2 - \sqrt{3}/4$
1641 3101 ln : $\sqrt{3}/4 + \sqrt{5}/3$
1641 3442 rt : (4,6,5,4)
1641 4142 J_0 : $11/4$
1641 4456 tπ : $5\sqrt[3]{5}/6$
1641 6774 Li_2 : $\sqrt[3]{2}/8$
1641 7209 10^x : $1/3 - \sqrt{5}/2$
1641 8574 tπ : $7\zeta(3)/8$
1642 1356 id : $1/4 - \sqrt{2}$
1642 1457 J_1 : $(\ln 2)^3$
1642 2971 Γ : $4/5$
1642 4483 tanh : $(\ln 3/2)^3$
1642 4793 rt : (2,4,-8,-2)
1642 5203 ln : $3\sqrt{2}/5$
1642 5287 at : $(\ln 3/2)^3$
1642 5525 Γ : $((e + \pi)/3)^{-1/3}$
1642 5571 sq : $(\pi/2)^{-2}$
1642 5816 J_0 : $\sqrt{20}\pi$
1642 6079 θ_3 : $\exp(-5/2)$
1642 6368 J_2 : $(\ln 3/2)^{-1/3}$
1642 6556 ln : $2 + 3\sqrt{5}$
1642 7120 id : $4\sqrt{3} + \sqrt{5}$
1642 7308 rt : (9,2,-3,8)
1642 7585 e^x : P_{71}
1642 8040 e^x : P_{68}
1642 8710 at : $4 - 3\sqrt{5}/4$
1642 9470 rt : (8,6,-3,1)
1643 1668 Ei : $7\sqrt{5}$
1643 1788 ℓ_2 : $1/3 + \sqrt{5}/4$
1643 1980 rt : (5,4,-8,-4)
1643 2155 rt : (9,2,-6,-1)
1643 2696 2^x : $(5/2)^3$
1643 5170 J_2 : $5\sqrt[3]{5}/7$
1643 6254 J_1 : $1/3$
1643 8187 tπ : $\exp((e + \pi)/3)$
1643 8398 rt : (3,1,-6,-1)
1643 8635 ℓ_{10} : $4\sqrt{2} + 4\sqrt{5}$
1644 0195 sq : $\ln(2/3)$
1644 1413 sq : $1/\sqrt[3]{15}$
1644 2665 rt : (8,1,-6,-7)
1644 3609 rt : (4,0,-8,3)
1644 3917 rt : (7,9,7,7)
1644 4326 rt : (3,0,6,1)
1644 7019 2^x : $3/4 - 3\sqrt{5}/2$
1644 7916 cu : $2\sqrt{2}/3 - 2\sqrt{5}/3$
1644 9961 rt : (9,1,6,1)
1645 0399 ℓ_{10} : $3e/4 + 4\pi$
1645 0819 cu : $4/3 - \pi/4$
1645 1231 J_0 : $4\sqrt[3]{2}$
1645 3564 ζ : $12/19$
1645 4192 J_2 : $11/9$
1645 5252 at : P_1
1645 6204 J_0 : $7\zeta(3)/4$
1645 6282 θ_3 : $8/9$
1645 6937 e^x : $4/3 + 4\sqrt{5}$
1645 7621 2^x : $7(1/\pi)/2$
1645 8104 rt : (7,4,-9,-6)
1645 8672 10^x : $\sqrt{2}/2 - 2\sqrt{5}/3$
1645 9459 sπ : $1/19$
1645 9600 Ψ : $18/25$
1646 0806 Li_2 : $3/19$
1646 2213 sq : $1/\ln(\sqrt[3]{5}/3)$
1646 3948 rt : (3,5,0,7)
1646 5404 10^x : $e/2 + 4\pi$
1646 5657 rt : (9,8,-5,-1)
1646 6457 λ : $\exp(-10\pi/7)$
1646 6467 $\ln(3/5) + s\pi(\sqrt{5})$
1646 7620 10^x : $\sqrt{2} - \sqrt{5}/4$
1646 8791 ℓ_2 : $2\sqrt{2}/3 + 3\sqrt{3}/4$
1647 0288 rt : (3,7,-5,-1)
1647 0616 cu : $\sqrt{2}/3 + 4\sqrt{5}/3$
1647 1432 e^x : $3\sqrt{2}/4 + 2\sqrt{3}/3$
1647 1828 Ψ : $4\sqrt[3]{2}/7$
1647 3301 2^x : $4\sqrt{3} - 3\sqrt{5}$
1647 3616 cu : $\sqrt{2} - \sqrt{3}/2$
1647 3860 tπ : $6\sqrt[3]{5}/5$
1647 4603 ln : $4 + 3\pi/2$
1647 4714 rt : (9,9,6,9)
1647 7314 J_2 : $17/3$
1647 8950 rt : (9,5,5,-1)
1647 9462 e^x : $\sqrt{3}/4 - \sqrt{5}$
1648 1024 ℓ_{10} : $13/19$
1648 1958 rt : (4,7,-5,-9)
1648 3094 rt : (9,3,-7,2)
1648 3878 ζ : $8\sqrt[3]{3}/7$
1648 5774 rt : (2,7,-4,-8)
1648 6090 λ : $(3\pi)^{-2}$
1648 6342 ln : $8\zeta(3)/3$
1648 7018 rt : (6,0,-9,1)
1648 7765 rt : (2,6,7,1)
1648 9160 $\ln(4) - \exp(1/5)$
1649 1229 sq : $1/2 - e/3$
1649 2529 rt : (3,7,-6,6)
1649 3225 rt : (8,7,7,1)
1649 3363 ln : $e/2 - \pi/3$
1649 5000 ℓ_{10} : $2\sqrt[3]{5}/5$
1649 5428 rt : (8,6,3,8)
1649 6474 sr : $19/14$
1649 7709 sq : $\sqrt[3]{17/2}$
1649 7792 id : $3e/7$
1649 7973 2^x : $2\sqrt{3} + 4\sqrt{5}$
1649 9117 rt : (4,5,-7,2)
1649 9305 sr : $\sqrt[3]{5/2}$
1650 1328 J : $((e + \pi)/3)^{-3}$
1650 2004 ln : $3\sqrt{2} + 2\sqrt{5}$
1650 2234 2^x : $3 - 4\sqrt{2}/3$
1650 3126 sq : $1 + \sqrt{2}/3$
1650 4856 sr : $4/3 + 3\sqrt{5}/2$
1650 5841 rt : (5,3,1,5)
1650 6196 at : $\sqrt{2} - \sqrt{3} + \sqrt{7}$
1650 6350 id : $5\sqrt{3}/4$
1650 7282 cu : $4e - \pi$
1650 7897 cr : $2e + 3\pi/2$
1650 8502 θ_3 : $\ln(\sqrt[3]{2}/2)$
1650 8958 cu : P_{12}
1650 9110 $\exp(3/5) \times \exp(\pi)$
1650 9544 e^x : $5\sqrt[3]{2}/3$
1651 2385 rt : (4,1,-6,-1)
1651 3409 J_2 : $\sqrt{3/2}$
1651 4041 tanh : $1/6$
1651 4187 at : $1/2 - 2\sqrt{2}$
1651 4867 at : $1/6$
1651 5035 ln : $1/2 + e/4$
1651 5578 sr : $4(e + \pi)/5$
1651 6291 rt : (7,3,0,7)
1651 6480 rt : (2,0,6,1)
1651 8027 J_2 : $9\sqrt{3}/2$
1652 0366 rt : (8,1,6,1)
1652 1530 rt : (9,3,-1,9)
1652 4854 10^x : $1/4 + 4\sqrt{2}$
1652 5805 tπ : $1/\ln(\sqrt{5}/3)$
1652 6800 rt : (2,5,-2,7)
1652 8925 sq : $25/11$
1652 9888 $\exp(1/5) \div \exp(2)$
1653 2223 sq : $2\sqrt{3}/3 + \sqrt{5}/2$
1653 4045 2^x : $2 + 3\pi$
1653 4241 tπ : $2\pi^2/5$
1653 4266 rt : (4,7,-5,-1)
1653 6144 sr : $1/2 + 4\pi/3$

1653 7304 rt : (8,0,-4,8)
1653 7980 at : $6e/7$
1653 9863 Ei : $\exp(\sqrt[3]{2})$
1654 0207 rt : (8,5,5,-1)
1654 1208 10^x : $3\sqrt{2}/4+\sqrt{3}/4$
1654 2503 rt : (3,8,-6,-4)
1654 8380 J_1 : $25/2$
1654 9793 cr : $2+3e$
1654 9985 rt : (9,3,-8,-9)
1655 0360 cu : $3/4-3\sqrt{3}/4$
1655 1633 $\exp(2/3)-\exp(\sqrt{2})$
1655 3164 rt : (7,3,-7,-7)
1655 3177 cr : $19/12$
1655 6941 sq : $\sqrt{2}/4+\sqrt{5}/2$
1655 7121 ℓ_{10} : $1/4+\sqrt{3}/4$
1655 7553 J_2 : $9\sqrt[3]{2}/2$
1655 8469 rt : (5,3,-6,-5)
1655 9066 Γ : $4(e+\pi)/5$
1656 4223 at : $(2e)^{1/2}$
1656 4812 J_0 : $9e\pi/4$
1656 4968 ℓ_2 : $\ln(3\pi)$
1656 6456 e^x : $17/22$
1656 6889 sinh : $4\sqrt{5}/9$
1656 7237 rt : (8,7,-7,-5)
1656 9009 tπ : $\exp(2/3)$
1656 9110 rt : (3,6,-2,-2)
1657 0995 e^x : $4/3+4\pi/3$
1657 1662 ℓ_{10} : $(\pi)^{1/3}$
1657 2639 rt : (5,5,-7,1)
1657 2816 at : P_{60}
1657 3677 sr : $\sqrt{22}$
1657 4612 id : $(\ln 3/2)^3$
1657 5075 ℓ_2 : $7\sqrt[3]{3}/9$
1657 5681 rt : (1,6,7,1)
1657 7482 E : $19/21$
1657 7647 tπ : $\sqrt{2}/2-4\sqrt{5}/3$
1657 8535 rt : (7,7,7,1)
1657 9790 rt : (4,6,-9,-4)
1658 2199 id : $(e/2)^{1/2}$
1658 2832 ℓ_{10} : $(e+\pi)/4$
1658 8434 rt : (5,1,-6,-1)
1658 9613 sπ : $(1/\pi)/6$
1659 0019 tπ : 9π
1659 0454 at : $7/3$
1659 0558 rt : (1,3,9,8)
1659 1254 $\ln(4/5)+\exp(2)$
1659 1440 10^x : $1/15$
1659 1615 ζ : $8\zeta(3)/3$
1659 2570 rt : (7,1,6,1)
1659 3688 rt : (1,8,5,-7)
1659 4350 $\exp(3/4)-s\pi(2/5)$
1659 5230 erf : $(\sqrt[3]{4}/3)^3$
1659 5243 at : P_{44}
1659 6316 rt : (2,0,5,9)
1659 8015 Li_2 : $(1/\pi)/2$
1659 8622 e^x : $4\sqrt{2}/3-\sqrt{3}$
1659 8882 rt : (4,6,-3,6)
1659 9703 rt : (5,7,-5,-1)
1660 1267 rt : (2,6,-8,-2)
1660 1566 10^x : $\sqrt{2}+\sqrt{3}-\sqrt{7}$
1660 2761 rt : (7,5,5,-1)
1660 3411 Li_2 : $(5/3)^{-1/3}$
1660 4108 J_2 : $7e\pi/3$
1660 4321 ℓ_{10} : $6\sqrt[3]{5}/7$
1660 4661 rt : (9,4,-5,3)
1660 5402 id : P_1
1660 6230 rt : (8,7,-1,2)
1660 6930 ℓ_2 : $5\pi/7$
1660 8036 tπ : $3e+\pi/2$
1661 0511 sinh : $7\sqrt{3}/8$
1661 3339 cr : $3-\sqrt{2}$
1661 3371 sr : $4\sqrt{3}-\sqrt{5}$
1661 4226 id : $4\sqrt{2}-2\sqrt{5}/3$
1661 6207 2^x : $(\sqrt[3]{3}/2)^{-1/3}$
1661 6991 e^x : $21/10$
1661 7132 Ei : $5/12$
1661 9629 ℓ_2 : $9\ln 2/7$
1662 0739 rt : (2,6,-2,-3)
1662 3245 rt : (8,7,5,9)
1662 3611 Ψ : $\exp(\sqrt{2}\pi/2)$
1662 4885 rt : (3,9,0,5)
1662 5186 rt : (8,1,-8,2)
1662 5383 sq : $6\sqrt[3]{2}/7$
1662 9073 $1/4-\ln(2/5)$
1663 0750 Γ : $\ln(\sqrt{2}\pi/2)$
1663 1144 e^x : $2/13$
1663 1274 rt : (1,3,2,8)
1663 3142 ℓ_{10} : $15/22$
1663 3976 rt : (8,1,-2,9)
1663 5077 cu : $20/19$
1663 6296 rt : (7,4,-4,1)
1663 7500 cu : $11/20$
1663 8166 Ei : $7(e+\pi)/8$
1663 9758 ζ : $3e\pi/8$
1664 0454 rt : (8,5,-9,-9)
1664 0761 2^x : $3e+\pi$
1664 1996 sπ : $4\sqrt{3}-4\sqrt{5}/3$
1664 2823 $\ln(4/5)\times s\pi(\sqrt{3})$
1664 3050 θ_3 : $(\sqrt[3]{5}/2)^2$
1664 3611 rt : (1,9,-5,-8)
1664 4875 rt : (7,4,2,8)
1664 5125 at : $\sqrt{3}/3-\sqrt{5}/3$
1664 6164 rt : (2,6,-9,3)
1664 6712 Ei : $8\pi/3$
1664 6758 tanh : $(2e/3)^{-3}$
1664 7618 at : $(2e/3)^{-3}$
1664 7949 Ei : $16/11$
1664 8707 2^x : $4\sqrt{2}/3-2\sqrt{5}$
1664 9413 sq : $3+2\sqrt{5}/3$
1665 0276 rt : (7,2,-5,-8)
1665 0605 sinh : $(\ln 3/2)^3$
1665 1454 as : $(\ln 3/2)^3$
1665 2903 id : $(\sqrt[3]{4})^{1/3}$
1665 4509 rt : (1,6,-2,-6)
1665 5039 J_2 : $16/13$
1665 8039 sq : $2/3+3\sqrt{5}/2$
1666 0196 sq : $1+3\sqrt{5}/4$
1666 0405 cu : $1/4+e/2$
1666 0698 J_0 : $21/10$
1666 1483 Γ : $(\pi/2)^{-1/2}$
1666 2859 10^x : $7\pi/10$
1666 3591 sinh : $14/3$
1666 6666 id : $1/6$
1666 7720 Ei : $((e+\pi)/3)^3$
1666 8175 J_0 : $5\sqrt[3]{2}/3$
1666 9713 Ψ : $8\sqrt[3]{3}/7$
1667 0460 cu : $4\sqrt{2}/3-3\sqrt{5}/2$
1667 0484 Ψ : $2(e+\pi)/9$
1667 0630 e^x : $10\zeta(3)$
1667 1190 Γ : $23/10$
1667 3832 cr : $2\sqrt{3}+3\sqrt{5}$
1667 4803 J : $2/19$
1667 5732 Ψ : $6/13$
1667 6311 θ_3 : $1/12$
1667 7337 ζ : $16/5$
1667 7549 10^x : $1/2-\sqrt{3}/4$
1667 9063 J_1 : $5e\pi/3$
1667 9357 λ : $(\sqrt{2}\pi)^{-3}$
1668 0248 rt : (5,7,-6,5)
1668 2677 as : P_1
1668 3863 10^x : $4\sqrt{2}-\sqrt{5}/4$
1668 3902 cr : $3/4-\sqrt{5}/3$
1668 4511 sq : $2/3+4e$
1668 5798 Γ : $9(1/\pi)/7$
1668 6219 cu : $1/2-4e$
1668 7048 tπ : $1/19$
1668 8620 rt : (5,2,-4,-6)
1669 0079 Li_2 : $4/25$
1669 1226 Ei : $\exp(\sqrt[3]{3}/2)$
1669 1385 cr : $3/4+3\pi$
1669 2466 sπ : $(\ln 2/3)^2$
1669 4419 ℓ_2 : $3+2\sqrt{5}/3$
1669 7506 10^x : 6π
1669 7645 sπ : $\exp(-(e+\pi)/2)$
1669 8738 J_2 : $8e\pi/3$
1669 9529 rt : (5,4,3,6)
1670 0000 cu : $23/10$
1670 0999 2^x : $7(1/\pi)/10$
1670 2280 2^x : $3/2+\sqrt{5}/4$
1670 2466 J_0 : 20
1670 2503 rt : (1,9,4,-6)
1670 2791 cr : $\sqrt{2}/3+\sqrt{5}/2$
1670 5408 ln : $11/13$
1670 5711 Ψ : $(e)^{1/2}$
1671 0740 rt : (4,7,-5,-3)
1671 4343 rt : (2,7,-7,-6)
1671 6829 $3\times\exp(2)$
1671 7145 J_0 : $8e\pi/3$
1671 7318 Li_2 : $\sqrt[3]{3}/9$
1671 7545 e^x : $1/4-3e/4$
1671 8427 sq : $2-3\sqrt{5}$
1672 1199 ζ : $7/17$
1672 2501 Ei : $7\sqrt{2}/9$
1672 3885 rt : (5,4,-3,-1)
1672 4061 J_2 : $\pi^2/8$
1672 6231 id : P_{60}
1672 6340 2^x : $\exp(-3/2)$
1672 9012 ℓ_{10} : $7\sqrt[3]{2}/6$
1672 9512 rt : (9,5,-3,4)
1672 9589 J_1 : $24/7$
1672 9659 e^x : $1/3-3\sqrt{2}/2$
1673 1121 cu : $22/17$
1673 1499 $\exp(2/5)+s\pi(\sqrt{5})$
1673 1980 rt : (5,5,5,-1)
1673 2137 cu : $3-\sqrt{2}/3$
1673 2763 rt : (4,1,0,5)
1673 3565 rt : (1,6,-5,-1)
1673 3602 id : $\sqrt{15}\pi$
1673 3902 J_0 : $9\sqrt{5}/4$
1673 4658 rt : (6,8,-9,4)
1673 4813 10^x : $3e/2-2\pi/3$
1673 5229 rt : (7,7,-5,-1)
1673 5537 sq : $9/22$
1673 5766 $3/4\times\ln(4/5)$
1673 7827 Ei : $11/10$
1673 9554 rt : (1,7,4,-7)
1673 9558 cu : $7e/2$
1673 9570 J_0 : $\ln(1/(3e))$
1674 0543 sπ : $2\sqrt{2}+\sqrt{5}/2$
1674 0757 cr : $9\sqrt{2}/8$
1674 1088 rt : (1,8,-1,3)
1674 2337 2^x : $1/4-2\sqrt{2}$
1674 2755 rt : (5,1,6,1)
1674 3736 J : $\exp(-20\pi/3)$
1674 3934 sinh : $1/6$
1674 4807 as : $1/6$
1674 4919 rt : (1,6,6,-5)
1674 5253 J_1 : $e/8$
1674 5919 J_0 : $7\sqrt{3}/2$
1674 7210 rt : (7,1,-6,-1)
1674 9007 10^x : $5\sqrt{3}/9$
1674 9009 2^x : $1/4+\sqrt{3}/2$
1674 9011 sq : $9(1/\pi)/7$
1674 9108 ℓ_{10} : $17/25$
1674 9286 id : P_{44}
1674 9503 Ψ : $3\pi^2/8$
1674 9522 rt : (7,1,-3,-9)
1675 0758 rt : (8,2,-6,3)
1675 0986 at : $2/3+3\sqrt{5}/4$
1675 2899 rt : (9,5,-9,-3)
1675 3105 sπ : $4e/3-\pi/2$
1675 3997 2^x : $3/2-3e/2$
1675 4432 cr : $5(1/\pi)$
1675 4561 at : $2(e+\pi)/5$
1675 6295 sq : $2\sqrt{3}/3-\sqrt{5}/3$
1675 6368 ℓ_2 : $1/4+3\sqrt{2}$
1675 7814 rt : (5,7,7,1)
1675 8001 $1/4\div\exp(2/5)$
1675 9984 rt : (2,7,4,6)
1676 0946 rt : (1,6,-7,1)
1676 1417 at : $1/4+2\pi/3$
1676 1613 J_2 : $21/17$
1676 2621 rt : (5,8,-2,9)
1676 2654 sπ : $(\sqrt[3]{5}/2)^{-1/3}$
1676 4303 rt : (7,5,-7,1)
1676 4911 2^x : $\sqrt{5}/10$
1676 5334 $\exp(1/5)-\exp(2)$
1676 5677 ζ : $10\sqrt{5}/7$
1676 6377 rt : (6,4,-6,-8)
1676 7079 K : $20/23$

1676 9292 sinh : $5e\pi/4$
1677 0350 ln : $4-\pi/4$
1677 2156 rt : (7,1,-9,-2)
1677 3208 ℓ_{10} : $1+\sqrt{2}/3$
1677 3925 at : $\sqrt{11/2}$
1677 4526 tπ : $\sqrt{3}/2+\sqrt{5}/4$
1677 4841 sr : $15/11$
1677 4870 Γ : $9/22$
1677 6550 ℓ_{10} : $e/4$
1677 7221 rt : (7,5,4,9)
1677 8605 ℓ_{10} : $\sqrt{2}/4+\sqrt{5}/2$
1678 2066 e^x : $\sqrt{14}$
1678 2190 rt : (4,1,-6,-2)
1678 3297 J_2 : $6\sqrt[3]{3}/7$
1678 3329 rt : (5,8,-8,-4)
1678 3792 rt : (1,8,-3,-9)
1678 4081 tπ : $\sqrt{2}/2+\sqrt{3}$
1678 4881 $\exp(4)+s\pi(\pi)$
1678 6736 J_2 : $(\sqrt[3]{4}/3)^{-1/3}$
1678 7147 at : $2\sqrt{3}-\sqrt{5}/2$
1678 9920 rt : (8,8,1,3)
1679 3014 J_0 : $\sqrt[3]{21}$
1679 5818 J : $3(1/\pi)/4$
1679 8767 rt : (4,7,7,5)
1679 9597 erf : $3/20$
1679 9838 rt : (6,2,1,8)
1680 0380 rt : (2,5,-6,-3)
1680 0572 id : $\sqrt{3}/3-\sqrt{5}/3$
1680 2034 rt : (2,6,-5,-1)
1680 3135 id : $(2e/3)^{-3}$
1680 3701 rt : (6,9,-5,8)
1680 4266 Ψ : $7\sqrt{2}/6$
1680 5220 as : P_{60}
1680 5469 rt : (8,7,-5,-1)
1680 5742 cu : $4+4\sqrt{3}/3$
1680 6060 J_2 : $5\sqrt{3}/7$
1680 6432 ln : $2-2\sqrt{3}/3$
1680 7459 ζ : $2\sqrt[3]{3}/7$
1680 8048 as : $23/25$
1680 8155 2^x : P_{31}
1681 1364 rt : (9,0,-8,5)
1681 2139 J_2 : $7\sqrt{2}/8$
1681 4893 e^x : $7\zeta(3)/2$
1681 8007 at : $1/4-3\sqrt{3}/2$
1682 0197 sinh : $1/\ln(\ln 3/3)$
1682 0947 rt : (4,1,6,1)
1682 1137 J_2 : $8\pi^2/5$
1682 1228 ln : $2-4\sqrt{2}/3$
1682 1536 rt : (7,5,-2,2)
1682 2519 e^x : $e/2-\pi$
1682 2629 rt : (5,1,-2,-7)
1682 2721 tπ : $(1/\pi)/6$
1682 2907 2^x : $\exp(\sqrt[3]{5})$
1682 3245 rt : (3,2,-7,6)
1682 5440 rt : (2,6,0,-5)
1682 6208 E : $(e/2)^{-1/3}$
1682 7962 erf : $\zeta(3)/8$
1682 8605 as : P_{44}
1682 9094 cu : $8(e+\pi)$
1683 0206 sq : $1/3-\sqrt{2}$
1683 0207 rt : (8,1,-6,-1)
1683 4137 $4\div s\pi(e)$
1683 4780 rt : (1,8,-9,-2)
1683 6121 ℓ_{10} : $\sqrt{22\pi}$
1683 6144 sπ : $3/4+3\sqrt{3}$
1683 7933 10^x : $19/2$
1683 8237 id : $2e/3+3\pi/4$
1683 8297 sq : $1-4\pi/3$
1684 1621 rt : (3,2,-9,3)
1684 4739 cu : $3\sqrt{3}-\sqrt{5}/3$
1684 4997 sq : $(\ln 2/3)^{-3}$
1684 5316 tπ : $1/\sqrt[3]{21}$
1684 5599 rt : (4,8,-1,1)
1684 5703 cu : $4/3-4\sqrt{2}/3$
1684 8078 rt : (1,2,4,-9)
1685 0275 id : $5\pi^2/8$
1685 0974 J_1 : $\sqrt[3]{5}/5$
1685 1567 $\ln(2)\div\exp(\sqrt{2})$
1685 2195 rt : (4,7,7,1)
1685 3485 rt : (6,2,-5,1)
1685 4823 rt : (1,9,-1,5)
1685 7808 rt : (9,6,-1,5)
1685 8722 tπ : $(\ln 3/3)^{-3}$
1685 8772 rt : (2,6,-7,1)
1686 0023 at : $3\sqrt{2}/2-2\sqrt{5}$
1686 1021 e^x : $1+3e$
1686 2284 rt : (9,4,5,-1)
1686 2615 2^x : $4/3+3e$
1686 4042 $3/5\times\exp(2/3)$
1686 5845 rt : (8,5,-7,1)
1686 7092 rt : (6,8,8,8)
1686 7975 at : $\sqrt[3]{13}$
1686 8136 J_0 : $\exp(\sqrt{5})$
1686 8150 at : $\exp(\sqrt[3]{5}/2)$
1686 8331 rt : (6,3,-4,-9)
1686 8748 rt : (3,7,-8,-5)
1686 9239 $\ln(\sqrt{2})+\exp(3/5)$
1687 1494 rt : (8,8,-5,-4)
1687 2146 rt : (3,6,-5,-1)
1687 7089 sr : $1/2+\sqrt{3}/2$
1687 7351 tπ : $4\sqrt{3}-4\sqrt{5}/3$
1687 7469 rt : (9,7,-5,-1)
1687 9817 rt : (8,3,-4,4)
1688 0628 as : $\sqrt{3}/3-\sqrt{5}/3$
1688 1725 e^x : $1/\ln(\sqrt[3]{5}/3)$
1688 2318 sinh : $(2e/3)^{-3}$
1688 3228 as : $(2e/3)^{-3}$
1688 3627 10^x : $\sqrt{15\pi}$
1688 4330 2^x : $21/8$
1688 4762 ln : $1/2+e$
1688 5414 10^x : $4+4\pi/3$
1688 5775 rt : (5,9,-4,0)
1688 6116 sq : $\sqrt{2}/2-\sqrt{5}/2$
1688 6167 J_0 : $8\pi/5$
1688 7257 $1/4\times s\pi(\sqrt{5})$
1689 0637 Ψ : $\sqrt[3]{9/2}$
1689 1156 cu : $\ln(5)$
1689 2233 cr : $5\sqrt{5}/7$
1689 3542 e^x : $3e\pi$
1689 4761 rt : (9,2,6,1)
1689 6685 rt : (2,9,3,1)
1689 6789 2^x : $1/2+4\sqrt{5}/3$
1689 7568 $e-\ln(\sqrt{3})$
1689 7951 e^x : $1+3\sqrt{2}$
1689 9888 sq : $1/2+3\pi/2$
1690 1053 Γ : $5(1/\pi)/2$
1690 1361 rt : (3,1,6,1)
1690 1853 rt : (5,5,5,7)
1690 1932 at : $2+\sqrt{2}/4$
1690 1987 rt : (7,0,-7,3)
1690 5565 ln : $4/3+4\sqrt{2}/3$
1690 8365 rt : (3,9,3,-4)
1690 9527 $\ln(3/4)\times s\pi(1/5)$
1691 0197 sr : $1/2-\sqrt{2}/3$
1691 0319 at : $1-3\sqrt{5}/2$
1691 0407 rt : (9,1,-6,6)
1691 0868 e^x : $4/3-\sqrt{5}/4$
1691 1607 sinh : $e\pi/4$
1691 3280 Ei : $\sqrt[3]{4/3}$
1691 3679 $\exp(1/5)+\exp(2/3)$
1691 5814 rt : (9,1,-6,-1)
1691 6144 rt : (3,7,-7,3)
1691 9088 Ψ : $13/10$
1692 0290 J_1 : $2\zeta(3)/7$
1692 0527 cu : $2/3+2\sqrt{2}/3$
1692 3972 rt : (1,8,2,-7)
1692 4619 $\ln(2/5)+\exp(3)$
1692 4813 rt : (7,5,-8,-5)
1692 6003 J_2 : $5\sqrt{5}/9$
1692 6597 $\exp(2/3)-\exp(3/4)$
1692 7384 rt : (8,2,-9,-5)
1692 9999 tπ : $(\ln 2/3)^2$
1693 0092 rt : (9,6,-7,-2)
1693 0419 rt : (8,4,5,-1)
1693 5019 tanh : $\sqrt[3]{5}/10$
1693 5403 tπ : $\exp(-(e+\pi)/2)$
1693 5553 sq : $2\sqrt{3}/3-\sqrt{5}$
1693 5688 J_2 : $1/\ln(\sqrt{5})$
1693 5957 at : $\sqrt[3]{5}/10$
1693 7026 rt : (6,3,3,9)
1693 7207 2^x : $3e/2-\pi/3$
1694 0403 rt : (9,8,7,1)
1694 0837 2^x : $1/4+\sqrt{2}$
1694 2282 at : $3\pi/4$
1694 3984 rt : (4,6,-5,-1)
1694 4853 ζ : $10(1/\pi)$
1694 6048 θ_3 : $3\sqrt[3]{5}/10$
1694 6727 rt : (4,4,-6,9)
1694 7036 id : $(3\pi)^3$
1695 0049 rt : (3,7,7,1)
1695 0502 ln : $4\sqrt{2}-2\sqrt{5}$
1695 0567 at : $\sqrt{3}/2+2\sqrt{5}/3$
1695 4918 at : $5\sqrt{2}/3$
1695 5017 sq : $7/17$
1695 5511 $\exp(1/3)+s\pi(e)$
1695 7151 cu : $e/2+\pi$
1695 7253 rt : (5,5,-9,8)
1695 8677 sr : $4+\sqrt{2}/2$
1696 0428 rt : (3,6,-7,1)
1696 0709 cr : $8/5$
1696 2736 ζ : $9\sqrt{2}/4$
1696 2985 sr : $1/4+\sqrt{5}/2$
1696 3349 sπ : $(\sqrt{5}/2)^{-1/2}$
1696 4761 ℓ_2 : P_{43}
1696 5034 10^x : $13/8$
1696 7997 $e\times s\pi(\pi)$
1696 9878 $3/5+s\pi(\pi)$
1697 1630 rt : (9,5,-7,1)
1697 1683 e^x : $(5/3)^{-1/2}$
1697 2621 rt : (3,8,6,6)
1697 3893 cu : $3\sqrt{3}/4-\sqrt{5}/3$
1697 4987 rt : (8,2,6,1)
1697 6267 sq : $1/3-\sqrt{5}/3$
1697 9064 E : $3\zeta(3)/4$
1697 9382 J_0 : $2\pi/3$
1698 0165 tπ : $2\sqrt{2}+\sqrt{5}/2$
1698 0276 sq : $2\sqrt[3]{3}/7$
1698 0727 sinh : $\sqrt{11/2}$
1698 1041 rt : (8,9,3,4)
1698 3972 sinh : $\sqrt{7/2}$
1698 4125 rt : (2,1,6,1)
1698 4582 rt : (4,9,3,5)
1698 5425 J_0 : 17
1698 7207 at : $1+e/2$
1698 9169 rt : (1,8,-2,-5)
1698 9725 rt : (9,7,1,6)
1699 0580 rt : (5,5,-5,-9)
1699 1845 at : $3-2\sqrt{2}$
1699 1962 e^x : $23/14$
1699 2500 ℓ_2 : 9
1699 2782 2^x : $19/17$
1699 3274 tπ : $4e/3-\pi/2$
1699 3844 rt : (4,0,-6,1)
1699 4466 2^x : $3e\pi/2$
1699 5694 at : $3\sqrt{2}/4+3\sqrt{3}/4$
1699 6003 E : $5\sqrt[3]{3}/8$
1699 7229 e^x : $22/7$
1699 8350 $\sqrt{3}\times s\pi(\sqrt{5})$
1700 0129 rt : (7,4,5,-1)
1700 0744 Ei : $(5/2)^{-1/3}$
1700 3240 tπ : $(\sqrt[3]{5}/2)^{-1/3}$
1700 4715 sq : $3\sqrt[3]{3}/4$
1700 5413 tanh : $\zeta(3)/7$
1700 6371 at : $\zeta(3)/7$
1700 8648 rt : (4,8,-8,-8)
1700 9064 sq : $2e-2\pi/3$
1701 0896 Ei : $14/19$
1701 1607 rt : (9,2,-4,7)
1701 1614 J_0 : $6(e+\pi)/7$
1701 2541 rt : (8,4,-2,5)
1701 4337 rt : (7,6,0,3)
1701 5338 e^x : $\sqrt{2}/9$
1701 6016 ℓ_2 : $e+2\pi$
1701 6191 Ψ : $(\ln 3/3)^{-1/2}$
1701 7632 rt : (5,6,-5,-1)
1701 8001 10^x : $\pi^2/2$
1701 9102 e^x : $3\sqrt{2}+3\sqrt{5}$
1701 9210 rt : (6,8,2,1)
1701 9515 sq : $3/4-4\sqrt{3}$
1702 0527 10^x : $2\sqrt{2}-3\sqrt{5}/4$

1702 1479 cu : $\sqrt{3}-\sqrt{6}+\sqrt{7}$
1702 1578 e^x : $4\sqrt[3]{3}/5$
1702 1583 cr : $10\sqrt[3]{3}/9$
1702 3900 ln : $\sqrt{3}+2\sqrt{5}/3$
1702 4454 sinh : $9e/2$
1702 5102 rt : (9,3,-9,-8)
1702 7410 ℓ_{10} : $3\sqrt{3}/2-\sqrt{5}/2$
1702 7498 cr : $4\zeta(3)/3$
1702 7520 ln : $(5/3)^{-1/3}$
1702 7589 J : $1/\sqrt[3]{22}$
1702 8247 Ψ : $3\sqrt{3}/4$
1702 9061 rt : (8,1,-7,-6)
1702 9344 2^x : $\sqrt{2}-\sqrt{3}-\sqrt{5}$
1703 2911 rt : (2,9,-9,-5)
1703 4635 J_1 : $3(e+\pi)$
1703 5530 rt : (7,1,-5,4)
1703 6319 e^x : 15/13
1703 7447 λ : $\exp(-17\pi/12)$
1703 8132 rt : (8,8,7,1)
1703 8286 10^x : 24/13
1703 9100 ℓ_{10} : $3\pi^2/2$
1704 0177 tπ : $5\sqrt[3]{5}/2$
1704 0204 sq : $9\zeta(3)/10$
1704 3826 J : $\exp(-10\pi/9)$
1704 5117 rt : (1,7,-7,-3)
1704 5482 sq : $3\pi^2/2$
1704 5590 $\ln(2/5)-s\pi(\sqrt{3})$
1704 6638 sπ : $\sqrt{3}-\sqrt{5}+\sqrt{6}$
1704 8000 rt : (6,3,-3,2)
1704 9344 ℓ_2 : $3\sqrt{2}-3\sqrt{5}/2$
1704 9376 rt : (3,8,-7,-7)
1704 9641 Li_2 : $\exp(-2e/3)$
1705 0862 cu : $4\ln 2/5$
1705 0987 2^x : $\sqrt{5}/2$
1705 1536 J_1 : $5e\pi/2$
1705 1645 rt : (2,7,7,1)
1705 2674 rt : (1,9,-2,-4)
1705 3821 J_2 : $9\ln 2/5$
1705 6920 e^x : $\sqrt[3]{2}/8$
1705 7508 rt : (7,2,6,1)
1705 8695 rt : (6,2,-8,-3)
1706 0032 rt : (2,4,-9,-5)
1706 1531 ℓ_2 : $2\sqrt{2}-\sqrt{3}/3$
1706 1991 2^x : 5/22
1706 1996 J_1 : $\sqrt{3}/5$
1706 2926 cu : $6\zeta(3)$
1706 5211 sr : $(e+\pi)^{-2}$
1706 6234 rt : (4,6,-7,1)
1706 7697 tπ : $1/\ln(2\pi)$
1706 8016 e^x : $4\sqrt{3}+4\sqrt{5}/3$
1706 9383 rt : (1,1,6,1)
1706 9414 2^x : $5(1/\pi)/7$
1706 9802 J_1 : $\ln 2/2$
1707 0449 as : $3\sqrt{3}/2-3\sqrt{5}/4$
1707 1486 rt : (6,4,5,-1)
1707 3859 2^x : $(4/3)^3$
1707 3863 2^x : $\exp(-\sqrt{2}\pi/3)$
1707 4863 cu : $\sqrt{3}/2-4\sqrt{5}$
1707 4916 J_1 : $1/\sqrt[3]{24}$
1707 8244 rt : (1,3,-8,-7)
1707 9955 tπ : $1/4-3\sqrt{3}$
1708 0376 id : $(3\pi/2)^{1/2}$
1708 0537 θ_3 : $\exp(-2/3)$
1708 1596 J_1 : $(\sqrt{2}\pi/2)^3$
1708 2039 rt : (5,5,-1,0)
1708 4391 id : $3\sqrt{2}+4\sqrt{3}$
1708 7759 $\ln(4/5)-\exp(2/3)$
1708 9365 2^x : $10\sqrt[3]{3}/7$
1708 9801 J_1 : $2\sqrt[3]{5}$
1709 1806 at : $3/2+\sqrt{3}/2$
1709 3184 rt : (6,6,-5,-1)
1709 4233 ζ : $(5/2)^{-1/2}$
1709 4317 2^x : $\sqrt{3}/4-4\sqrt{5}/3$
1709 6432 at : $3/4-\sqrt{3}/3$
1709 7726 ζ : $(e+\pi)^{-1/2}$
1709 8656 Ei : $(\ln 3/3)^{-2}$
1709 9620 tπ : $5\sqrt[3]{2}/4$
1709 9759 id : $\sqrt[3]{5}/10$
1710 1631 e^x : $\sqrt{19}$
1710 3089 rt : (5,1,2,9)
1710 3527 Ei : $2\sqrt[3]{2}$
1710 3759 sq : P_{56}
1710 4292 e^x : 3/19
1710 4841 10^x : $2+\pi/2$
1710 5384 sr : $\sqrt{3}+4\sqrt{5}/3$
1710 7347 J_0 : 14
1710 7384 $2\times\exp(3)$
1710 8669 rt : (9,2,-7,-9)
1710 9113 J_2 : 5/4
1710 9954 J_2 : $((e+\pi)/3)^{1/3}$
1711 1880 2^x : $7(e+\pi)/10$
1711 2027 J_1 : 21
1711 3536 rt : (5,6,7,8)
1711 3903 $1/3\div\exp(2/3)$
1711 3932 rt : (9,7,-5,-1)
1711 5049 rt : (9,3,-2,8)
1711 5238 cu : $2/3-4\sqrt{5}$
1711 6460 rt : (4,9,-3,-8)
1711 7553 at : $2\sqrt{2}-3\sqrt{3}$
1711 7722 sπ : $4\sqrt{3}/3+\sqrt{5}/3$
1711 8520 sr : $10\sqrt{2}/3$
1711 8798 2^x : $(2\pi/3)^{-2}$
1711 9030 J_0 : 11
1712 0839 id : $\sqrt{2}/2+2\sqrt{3}$
1712 4440 e^x : $\sqrt{2}/3-\sqrt{5}$
1712 5449 rt : (9,8,3,7)
1712 7925 rt : (4,2,-1,-8)
1712 9620 J_1 : 8/23
1712 9629 cu : 13/6
1713 0827 10^x : $1/4+4e/3$
1713 2879 id : $7e/6$
1713 3003 rt : (8,0,-5,7)
1713 4219 ζ : $7e/6$
1713 6970 $\exp(\sqrt{5})\div\exp(4)$
1713 8233 Γ : $\sqrt[3]{1/2}$
1713 9045 rt : (3,9,-1,2)
1713 9507 rt : (7,8,7,1)
1714 2457 rt : (6,2,6,1)
1714 3381 Ψ : $9\sqrt[3]{3}/10$
1714 3933 J_1 : $10\pi^2/9$
1714 4569 rt : (5,4,5,-1)
1714 4881 10^x : $1/3+3\sqrt{3}$
1714 6154 rt : (8,4,-8,-2)
1714 6776 cu : 5/9
1714 7240 ℓ_{10} : $\exp(5)$
1714 7413 J_0 : $\sqrt{2}-\sqrt{3}-\sqrt{6}$
1714 9121 rt : (8,5,0,6)
1714 9356 tanh : $\sqrt{3}/10$
1715 0355 at : $\sqrt{3}/10$
1715 0840 rt : (4,3,4,7)
1715 1064 rt : (7,3,-8,-6)
1715 1475 J_0 : $7e\pi/3$
1715 2969 rt : (3,5,-4,-7)
1715 5643 $\exp(3)\div s\pi(1/5)$
1715 5952 2^x : $1/3+\pi/4$
1715 7287 id : $3-2\sqrt{2}$
1715 7555 at : $(4/3)^3$
1715 8288 at : $\ln 2/4$
1716 1383 ℓ_{10} : $6\sqrt{3}/7$
1716 1595 K : $(\pi/3)^{-3}$
1716 5080 J_0 : 8
1716 7673 ln : $8\pi^2/9$
1716 8020 ℓ_2 : $2\sqrt{2}+3\sqrt{5}/4$
1716 9705 E : 9/10
1716 9758 ζ : $(\pi/3)^{-3}$
1717 0741 rt : (7,6,-5,-1)
1717 1739 sq : $4\pi^2/3$
1717 1903 at : $1/4+3\sqrt{2}/2$
1717 2241 id : $\zeta(3)/7$
1717 2979 rt : (7,2,-3,5)
1717 3130 rt : (6,0,-6,1)
1717 3606 rt : (2,5,-3,4)
1717 4599 Ψ : $(3\pi/2)^{-1/2}$
1717 6553 rt : (5,9,-4,-9)
1717 6652 E : $5\sqrt[3]{2}/7$
1717 6804 cr : $3\sqrt{3}/4+4\sqrt{5}$
1717 6941 J : $\exp(-15\pi/4)$
1717 7159 2^x : $1+3\sqrt{2}/4$
1717 7542 J_2 : $1/\ln(\sqrt{2}\pi/2)$
1717 7752 J_0 : 23/11
1717 8224 rt : (2,8,-7,-6)
1717 9740 tπ : $3\sqrt{3}/2+3\sqrt{5}/4$
1718 0186 10^x : $\pi^2/5$
1718 0306 cu : $e/3-\pi$
1718 1784 id : $3e/2+2\pi/3$
1718 2245 $\exp(2)\div s\pi(\pi)$
1718 2386 rt : (5,7,-7,3)
1718 2837 at : $1/3+3e/4$
1718 3021 cr : $1/4+e/2$
1718 3214 sinh : $\sqrt[3]{5}/10$
1718 3894 sq : $2/3-4e$
1718 4208 as : $\sqrt[3]{5}/10$
1718 4999 J_2 : $2e\pi/3$
1718 5025 ln : 16/19
1718 6228 e^x : $1/3-2\pi/3$
1718 7500 sq : $5\sqrt{3}/8$
1718 7659 J_2 : $(\pi/2)^{1/2}$
1718 8613 rt : (3,0,1,6)
1718 8809 J_1 : $\pi/9$
1719 0230 cr : $\ln(5)$
1719 1148 cr : $2/3+2\sqrt{2}/3$
1719 2855 sπ : $\sqrt{3}-3\sqrt{5}/4$
1719 4727 rt : (8,9,-3,-3)
1719 7025 rt : (6,1,-6,-4)
1719 7117 J_0 : $7\pi/2$
1719 7489 id : $(e+\pi)/5$
1719 8496 J : $\sqrt[3]{2}/7$
1719 9340 rt : (9,9,4,-1)
1719 9641 10^x : $\sqrt{2}/4-\sqrt{5}/2$
1720 1338 $\ln(5)\div\exp(\sqrt{5})$
1720 3508 Li_2 : 11/13
1720 5353 $\exp(4/5)\div s\pi(\pi)$
1720 5801 ℓ_{10} : $2e+3\pi$
1720 6246 Γ : $8\sqrt[3]{3}/5$
1720 7691 cr : $6e\pi/5$
1720 8081 sr : $2/3+\sqrt{2}/2$
1720 8332 sr : $8\zeta(3)/7$
1720 8561 rt : (8,4,-8,-9)
1720 9934 ζ : 19/6
1721 1234 sπ : $3\sqrt[3]{2}/4$
1721 1305 ln : $3-2e/3$
1721 1379 sr : $9\sqrt{5}/2$
1721 1607 rt : (3,8,0,-7)
1721 2048 $2/5\times s\pi(\pi)$
1721 2438 J_0 : $\sqrt{23/3}$
1721 2811 tπ : $(\sqrt{5}/2)^{-1/2}$
1721 5269 rt : (7,7,2,4)
1721 5301 sq : $1/\sqrt[3]{14}$
1721 5400 cr : $\sqrt{2}-\sqrt{6}+\sqrt{7}$
1721 6063 sr : $2+e$
1721 6226 10^x : $(e+\pi)/4$
1721 6319 Ei : $(\sqrt{2}/3)^{-1/2}$
1721 8062 tanh : 4/23
1721 8709 cr : $3e+2\pi/3$
1721 9081 at : 4/23
1721 9461 rt : (4,8,9,6)
1722 0825 rt : (9,4,0,9)
1722 1262 rt : (5,4,-9,-3)
1722 3686 erf : 2/13
1722 6868 rt : (6,9,-6,6)
1722 7388 at : 19/8
1722 7492 tπ : $\sqrt{3}/4-3\sqrt{5}$
1722 9982 rt : (5,2,6,1)
1723 0039 2^x : $2\sqrt{2}/3+\sqrt{5}/2$
1723 0554 id : $2\sqrt{3}+3\sqrt{5}$
1723 1887 J_0 : $8(e+\pi)/5$
1723 1906 Γ : $1/\sqrt{6}$
1723 3112 rt : (2,8,0,6)
1723 3200 ln : $4e-2\pi/3$
1723 3395 J_1 : 7/20
1723 4488 rt : (2,5,-5,-1)
1723 6006 2^x : $1/\sqrt{19}$
1723 7107 rt : (9,3,-8,1)
1723 7382 cr : $\sqrt{3}/2+\sqrt{5}/3$
1723 9302 rt : (8,1,-3,8)
1724 0229 ℓ_{10} : $4+4e$
1724 1537 10^x : $(\ln 3)^{1/2}$
1724 1721 sr : $3\sqrt{2}/2+3\sqrt{3}/2$
1724 2015 J_1 : $(1/(3e))^{1/2}$
1724 2599 as : $3-2\sqrt{2}$

1724 4243 erf : $2\ln 2/9$
1724 4813 rt : (6,8,7,1)
1724 5927 $\exp(1/5) + s\pi(2/5)$
1724 6180 rt : (3,9,-7,-9)
1724 6887 $\ln(4/5) + \exp(1/3)$
1724 8009 rt : (1,5,2,-9)
1724 8042 2^x : $\sqrt{13}$
1724 8865 sr : $(\sqrt[3]{4}/3)^{-1/2}$
1725 0411 rt : (8,6,-5,-1)
1725 0798 rt : (6,4,-1,3)
1725 1503 at : $2e/3 - 4\pi/3$
1725 1867 sq : $\sqrt{2}/2 - 4\sqrt{5}/3$
1725 1931 E : $\exp(-1/(3\pi))$
1725 1960 id : $\exp(1/(2\pi))$
1725 2563 rt : (7,6,-6,-4)
1725 6763 sinh : $\zeta(3)/7$
1725 7667 e^x : $3/4 + 2e$
1725 7779 as : $\zeta(3)/7$
1725 7838 cr : $7(e+\pi)/4$
1725 7893 rt : (7,2,-6,-7)
1725 9308 J_1 : $2e\pi/5$
1726 0393 sr : 11/8
1726 0924 $3/5 \times \ln(3/4)$
1726 1281 ℓ_{10} : $1/3 + 2\sqrt{3}/3$
1726 3205 ζ : $\pi^2/6$
1726 4076 ln : $7\zeta(3)/10$
1726 4139 at : $\exp(\sqrt{3}/2)$
1726 4203 sq : $6(e+\pi)$
1726 4449 rt : (8,9,4,-1)
1726 4538 10^x : $7\pi^2/6$
1726 4973 id : $3/4 - \sqrt{3}/3$
1726 5182 rt : (9,9,5,8)
1726 5617 rt : (1,6,-5,-4)
1726 6499 J_2 : $2\pi/5$
1726 6555 $t\pi$: $5\sqrt{5}/7$
1726 7316 rt : (7,7,1,0)
1726 7431 sinh : $\sqrt[3]{7/2}$
1726 8841 10^x : $2\sqrt{2}/3 + 2\sqrt{3}/3$
1727 0031 $\ln(4/5) \times s\pi(e)$
1727 0373 10^x : $9\sqrt[3]{3}/4$
1727 1783 rt : (5,8,-3,7)
1727 2902 $t\pi$: $\sqrt{2}/3 - 3\sqrt{3}$
1727 3388 e^x : 10/7
1727 4431 ℓ_2 : $4\sqrt{3}/3 + 3\sqrt{5}$
1727 4542 sinh : $4\pi^2/9$
1727 5884 rt : (4,1,1,-9)
1727 6598 rt : (6,5,-9,-6)
1727 6984 J_2 : $8\sqrt{2}/9$
1727 7285 J_1 : $\exp(-\pi/3)$
1727 9211 J_2 : $8\sqrt{3}$
1727 9743 rt : (9,3,5,-1)
1727 9977 at : $7e/8$
1728 1585 ζ : $\sqrt{10}$
1728 3950 sq : $10\sqrt{5}/9$
1728 4806 cu : $7(1/\pi)/4$
1728 5567 $4 \div \exp(\pi)$
1728 6742 rt : (5,1,-4,2)
1728 7822 $\ln(2) - s\pi(1/3)$
1728 9764 rt : (8,6,2,7)
1728 9966 ℓ_2 : $3 - \sqrt{5}/3$
1729 1798 rt : (6,6,-7,1)
1729 2289 ln : $1/4 + 4\sqrt{5}/3$
1729 3674 cr : $6\sqrt[3]{5}$
1729 4453 Γ : $\exp(-\sqrt[3]{2})$
1729 5428 sr : $2\sqrt{2}/3 + \sqrt{3}/4$
1729 5919 $t\pi$: $\sqrt{2}/4 - 3\sqrt{3}/4$
1729 6255 rt : (3,5,6,5)
1729 6308 sq : $3\ln 2/5$
1729 8648 cu : $3/2 - 2\sqrt{2}/3$
1729 9848 $t\pi$: $\sqrt{3} - \sqrt{5} + \sqrt{6}$
1730 1506 sr : $3\sqrt{2}/2 - \sqrt{5}/3$
1730 1972 rt : (3,5,-1,5)
1730 3381 id : $\sqrt{3} - \sqrt{5}/4$
1730 4512 10^x : $\ln 2/10$
1730 4761 sr : $1/4 + 2\sqrt{5}$
1730 4846 J_2 : $9e\pi/4$
1730 4872 rt : (9,8,-3,0)
1730 5508 Γ : $\ln(\sqrt[3]{2}/3)$
1730 7306 e^x : $2\sqrt{3}/3$
1730 7426 e^x : $-\sqrt{3} + \sqrt{5} + \sqrt{6}$
1730 8101 sinh : $2\sqrt[3]{2}$
1730 9341 2^x : $5\sqrt[3]{3}/2$
1730 9691 at : $\sqrt{17/3}$
1730 9784 id : $3\sqrt{3}/4 - 2\sqrt{5}$
1731 0512 at : $4/3 + \pi/3$
1731 0811 rt : (7,6,-9,-9)
1731 1351 J_1 : $1/\sqrt[3]{23}$
1731 1438 rt : (6,3,-9,-5)
1731 1686 ln : $3/2 + \sqrt{3}$
1731 2025 2^x : $2\sqrt{2} - 3\sqrt{3}/2$
1731 3401 J_0 : $8\sqrt{3}/5$
1731 3449 e^x : $9(e+\pi)/10$
1731 3746 rt : (3,5,-5,-1)
1731 4515 Γ : $8\ln 2/7$
1731 4544 rt : (7,3,-1,6)
1731 6266 Γ : $5\sqrt[3]{5}$
1731 7281 Li_2 : $(\ln 3/2)^3$
1732 0241 rt : (4,2,6,1)
1732 0508 id : $\sqrt{3}/10$
1732 1263 10^x : $1/4 + 3\sqrt{5}$
1732 3109 rt : (4,2,-7,-1)
1732 7496 $\exp(2/3) + \exp(4/5)$
1732 8288 rt : (2,7,-5,-7)
1732 8679 id : $\ln 2/4$
1733 0917 rt : (7,9,4,-1)
1733 1138 rt : (2,9,-6,5)
1733 1743 ℓ_{10} : $(\sqrt{2}\pi/2)^{1/2}$
1733 2312 rt : (9,6,-5,-1)
1733 4003 Γ : $4\sqrt{3}/3$
1733 4388 cr : 21/13
1733 4659 $t\pi$: $\sqrt{2}/4 + \sqrt{5}/2$
1733 5216 10^x : $\sqrt{17}$
1733 5394 rt : (5,7,9,9)
1733 5574 10^x : $4/3 - 2\pi/3$
1733 6146 J_2 : 25/2
1733 6400 2^x : $8\sqrt[3]{2}/9$
1733 6834 $s\pi$: $7e\pi/4$
1733 7208 cu : $1/4 + 2\sqrt{2}$
1733 8037 e^x : $2/3 + 3\sqrt{3}/2$
1733 8737 rt : (8,5,-6,-1)
1733 9374 ℓ_{10} : $2\sqrt{5}/3$
1733 9526 rt : (6,0,-4,5)
1733 9752 $\exp(1/2) - \exp(3/5)$
1733 9950 θ_3 : $\ln 2/8$
1733 9987 rt : (4,7,0,8)
1734 1735 rt : (2,5,-2,-6)
1734 4497 J_2 : $\sqrt[3]{2}$
1734 6046 2^x : 3/13
1734 6392 at : $2\sqrt{2}/3 - \sqrt{5}/2$
1734 7461 sinh : $\sqrt{3} + \sqrt{5} - \sqrt{6}$
1734 8057 rt : (8,2,-1,9)
1734 9901 cr : $3/4 + \sqrt{3}/2$
1735 1087 e^x : 4/25
1735 1233 rt : (1,9,7,2)
1735 1916 as : $3/4 - \sqrt{3}/3$
1735 4369 rt : (5,8,7,1)
1735 5193 cr : $8\sqrt{2}/7$
1735 6081 rt : (8,3,5,-1)
1735 6603 ℓ_{10} : $\ln(\sqrt{2}\pi)$
1735 7023 2^x : $5\sqrt{5}/2$
1735 9091 J_1 : $9e\pi/7$
1735 9329 ln : $3(e+\pi)/2$
1735 9909 rt : (3,7,-8,0)
1736 1036 J_2 : $9\zeta(3)$
1736 1111 sq : 5/12
1736 3493 J_2 : $10\sqrt[3]{5}/3$
1736 3753 sinh : $8e/3$
1736 4083 θ_3 : $(\ln 3)^{-1/2}$
1736 4713 rt : (5,3,-7,-4)
1736 4817 $s\pi$: 1/18
1736 4842 at : $\sqrt{2}/2 + 3\sqrt{5}/4$
1736 4895 10^x : $4/3 + 2\sqrt{3}$
1736 4919 sq : $(\sqrt{2}\pi/2)^3$
1736 7202 rt : (7,1,-4,-8)
1736 8171 J_0 : $4\ln 2$
1736 8808 2^x : $\ln 2/3$
1737 2410 Γ : 19/24
1737 2967 rt : (7,2,-9,-2)
1737 3594 rt : (9,4,-6,2)
1737 3700 J_1 : 6/17
1737 4160 $t\pi$: $4\sqrt{3}/3 + \sqrt{5}/3$
1737 4614 2^x : $(\sqrt[3]{3}/3)^2$
1737 5046 rt : (4,4,6,8)
1737 5533 Ei : $9\pi/7$
1737 6641 e^x : P_{20}
1737 6669 at : $1/3 - e$
1737 6777 ζ : $(\sqrt{5}/3)^3$
1737 7224 sinh : $\sqrt[3]{19}$
1737 7394 $\exp(1/4) \div \exp(2)$
1737 8437 $\ln(4/5) \div \exp(1/4)$
1737 9135 $\exp(2/3) - s\pi(e)$
1737 9327 Ei : $3e$
1738 0422 e^x : $\sqrt[3]{3}/9$
1738 1148 e^x : $2/3 + 4\sqrt{5}$
1738 1795 cr : $7\ln 2/3$
1738 2121 rt : (1,7,-7,1)
1738 5655 cu : $1/2 + 4\sqrt{3}/3$
1738 6468 rt : (6,4,-7,-7)
1738 6678 at : $1/2 + 4\sqrt{2}/3$
1738 9880 rt : (9,3,6,1)
1739 0466 10^x : $4/3 + \sqrt{2}$
1739 1304 id : 4/23
1739 3164 J_2 : $7\sqrt[3]{3}/8$
1739 4268 $2/5 + s\pi(e)$
1739 5171 rt : (4,5,-5,-1)
1739 6818 sinh : $2\pi^2/3$
1739 7734 rt : (8,1,-9,1)
1739 8499 cr : $1/2 + \sqrt{5}/2$
1739 8799 rt : (6,9,4,2)
1740 0753 ℓ_{10} : $1/2 - \sqrt{3}/4$
1740 0849 rt : (6,7,-9,9)
1740 1800 Ei : $7\sqrt[3]{2}/8$
1740 2739 θ_3 : 2/23
1740 2890 J_1 : $\sqrt{2}/4$
1740 3325 Ψ : $(e)^{-1/3}$
1740 3681 rt : (1,6,-2,-9)
1740 5856 rt : (5,8,-9,-6)
1740 7240 sinh : $\sqrt{3}/10$
1740 8301 as : $\sqrt{3}/10$
1740 8815 J_1 : $10(1/\pi)/9$
1741 0129 at : P_{14}
1741 1113 rt : (9,0,-9,4)
1741 1424 Γ : $10\ln 2/3$
1741 2448 rt : (7,6,-7,1)
1741 3412 rt : (3,2,6,1)
1741 4449 2^x : $1/4 + 3\sqrt{2}/2$
1741 5223 rt : (3,1,3,7)
1741 5534 sinh : $\ln 2/4$
1741 6583 J_1 : 23/3
1741 6597 as : $\ln 2/4$
1741 7956 Li_2 : 1/6
1741 8080 e^x : $2e/7$
1741 8195 $s\pi$: $4\sqrt{5}$
1741 8682 2^x : $2\sqrt{2}/3 - 2\sqrt{3}$
1741 8908 cr : $4/3 + 4\sqrt{5}$
1742 0006 $\ln(4/5) - s\pi(2/5)$
1742 0400 10^x : $9(e+\pi)/7$
1742 0966 $\exp(1/3) - \exp(1/5)$
1742 1266 Γ : $7\sqrt{3}$
1742 1454 J_2 : 24/19
1742 1875 sinh : $\sqrt{8\pi}$
1742 3910 ℓ_{10} : $\ln((e+\pi)/3)$
1742 4963 rt : (7,8,4,5)
1742 5549 ℓ_{10} : $3\sqrt{2}/4 + \sqrt{3}/4$
1742 6408 cu : $7\pi/4$
1742 6442 2^x : $7\sqrt{3}/4$
1742 6497 10^x : $3\sqrt{2}/4 - \sqrt{5}/4$
1742 9291 ℓ_{10} : $\sqrt[3]{10/3}$
1743 4358 rt : (7,3,5,-1)
1743 4374 J : 7/18
1743 4691 rt : (8,7,4,8)
1743 5338 ln : 21/25
1743 5900 ln : $1 + \sqrt{5}$
1743 6054 rt : (2,9,2,-3)
1743 8227 cr : $3\sqrt{2}/4 + \sqrt{5}/4$
1743 8673 rt : (3,8,1,-5)
1743 9458 rt : (3,8,-2,9)
1744 1604 ln : $2\sqrt[3]{2}/3$
1744 2760 Ψ : $\exp(-\sqrt[3]{4})$
1744 3576 cr : $9\sqrt[3]{2}/7$

1744 6161 sq : $\sqrt{14}\pi$
1744 6747 Ψ : $\sqrt[3]{5}/10$
1744 7212 $\ln(2/3) \times s\pi(\pi)$
1744 8115 rt : (2,9,1,-8)
1744 8240 ℓ_{10} : $7e\pi/4$
1744 8522 10^x : $e\pi/2$
1744 8799 $\ln(3/4) \div \exp(1/2)$
1745 1185 10^x : $(\sqrt{2}\pi/2)^{1/3}$
1745 2045 cr : $4\sqrt{3} + 3\sqrt{5}/2$
1745 2065 ℓ_2 : $\sqrt[3]{23/2}$
1745 2737 $t\pi$: $\sqrt{3} - 3\sqrt{5}/4$
1745 3966 cu : $1/3 + 3\pi$
1745 5941 rt : (3,6,3,9)
1745 6911 $3/5 \div \ln(3/5)$
1745 7138 tanh : $(\sqrt[3]{2}/3)^2$
1745 8108 J_2 : $(5/3)^3$
1745 8230 at : $(\sqrt[3]{2}/3)^2$
1745 8249 cr : $4 + 2\pi$
1746 0453 rt : (7,4,1,7)
1746 2531 rt : (6,5,1,4)
1746 2546 id : $4e/5$
1746 4284 E : $2\pi/7$
1746 4970 10^x : $\sqrt[3]{10/3}$
1746 5879 rt : (9,0,-6,1)
1746 6124 tanh : 3/17
1746 6234 cu : $2/3 - 2\pi$
1746 6635 ℓ_{10} : $\sqrt{2} - \sqrt{5}/3$
1746 7219 at : 3/17
1746 8156 rt : (5,9,4,-1)
1746 8540 rt : (4,8,7,1)
1746 8676 sq : $3/2 + e/2$
1746 9281 cu : $\sqrt{5}/4$
1747 1370 $1/5 \div \ln(\pi)$
1747 1963 $t\pi$: $3\sqrt[3]{2}/4$
1747 2382 rt : (4,4,-7,7)
1747 3468 $t\pi$: $3\sqrt{3}/4 - 3\sqrt{5}/2$
1747 4250 ℓ_{10} : $(\sqrt{5})^{1/2}$
1747 4459 ℓ_{10} : $3/4 + \sqrt{5}/3$
1747 5405 J_0 : $9e\pi/7$
1747 5779 at : $3/4 - \pi$
1747 6506 rt : (3,2,-8,4)
1747 7796 id : $1/4 - 3\pi$
1747 8885 rt : (5,5,-5,-1)
1747 9099 rt : (7,0,-2,9)
1747 9105 sinh : 4/23
1748 0188 as : 4/23
1748 0210 2^x : 5/3
1748 1210 e^x : $1/\ln(\sqrt{3})$
1748 2409 rt : (8,3,6,1)
1748 2493 e^x : $4 - 3\pi/4$
1748 3127 rt : (1,7,0,9)
1748 3392 rt : (2,6,-3,-6)
1748 6431 rt : (6,1,-2,6)
1748 7068 sq : $\sqrt{5} - \sqrt{6} - \sqrt{7}$
1748 7214 ζ : $\ln(\ln 3/2)$
1748 7221 ζ : $5\sqrt[3]{2}/2$
1748 7788 ζ : $-\sqrt{3} + \sqrt{5} + \sqrt{7}$
1748 9598 as : $3/2 - \sqrt{3}/3$
1748 9907 J_0 : 16
1749 3916 ζ : $(\sqrt{2}\pi)^{1/3}$
1749 4595 Ψ : $9(1/\pi)/4$
1749 5800 tanh : $\sqrt{2}/8$
1749 5995 sr : $1/3 + \pi/3$
1749 6904 at : $\sqrt{2}/8$
1749 7212 rt : (1,5,-3,-5)
1749 8954 rt : (6,3,-5,-8)
1749 9805 ℓ_{10} : $7\sqrt[3]{5}/8$
1750 0557 rt : (5,2,-2,3)
1750 1822 tanh : $5(1/\pi)/9$
1750 2147 rt : (2,7,-7,1)
1750 2929 at : $5(1/\pi)/9$
1750 3408 rt : (9,9,-1,1)
1750 4965 J_2 : 19/15
1750 7155 cr : $9\sqrt[3]{3}/8$
1750 9687 rt : (2,2,6,1)
1750 9808 tanh : $\exp(-\sqrt{3})$
1751 0146 ln : $4\sqrt{3}/3 - \sqrt{5}/2$
1751 0917 at : $\exp(-\sqrt{3})$
1751 1073 at : $7\sqrt[3]{5}/5$
1751 1095 rt : (1,9,-6,-7)
1751 1477 at : $4/3 + 3\sqrt{2}/4$
1751 2630 rt : (5,2,-5,-5)
1751 4006 rt : (9,5,-4,3)
1751 4673 rt : (6,3,5,-1)
1751 8938 rt : (5,9,-5,-2)
1751 9745 rt : (9,1,-7,5)
1752 0119 sinh : 1
1752 0451 J_0 : $9\pi^2/7$
1752 2494 id : $2\sqrt{2}/3 - \sqrt{5}/2$
1752 3500 at : $3/2 - 3\sqrt{5}/4$
1753 2576 at : $2\sqrt{2} - \sqrt{3}/4$
1753 2774 sr : $3 + \sqrt{3}$
1753 4500 erf : $\ln(\sqrt[3]{5}/2)$
1753 4747 ℓ_{10} : $5\zeta(3)/9$
1753 4857 2^x : $e/3 + 4\pi/3$
1753 6379 Ei : $(\sqrt{5}/3)^{-1/3}$
1753 7290 e^x : $3 - \pi/2$
1753 7402 ln : $3\sqrt{2}/2 + \sqrt{5}/2$
1753 7764 10^x : $3\sqrt{3}/2 - 3\sqrt{5}/2$
1753 9052 rt : (8,6,-4,0)
1754 0780 id : $3\sqrt{2}/4 - \sqrt{5}$
1754 2231 Ψ : $5\sqrt{5}/3$
1754 4467 sq : $\sqrt{2}/2 - 2\sqrt{5}$
1754 9029 $1/2 + s\pi(\sqrt{5})$
1755 1139 rt : (9,9,7,1)
1755 4627 Ei : 16/9
1755 5324 2^x : $\sqrt{2}/3 - 4\sqrt{5}/3$
1755 6654 e^x : $(\sqrt{2}\pi)^{1/3}$
1755 6811 ℓ_2 : P_{54}
1755 7050 $2 \times s\pi(1/5)$
1755 8299 cu : $4\sqrt[3]{2}/9$
1756 0633 sq : $3 + 2e$
1756 1559 J_1 : $1/\sqrt[3]{22}$
1756 1600 cu : 14/25
1756 3612 rt : (1,7,-9,5)
1756 5020 rt : (6,5,-5,-1)
1756 6734 cr : 13/8
1756 7302 Li_2 : $(2e/3)^{-3}$
1756 7929 E : $(\sqrt[3]{3}/2)^{1/3}$
1756 8762 ln : $\sqrt{21/2}$
1757 0555 ℓ_{10} : P_{26}
1757 1843 $t\pi$: $3\sqrt{3} - \sqrt{5}/4$
1757 3939 J_1 : 5/14
1757 5099 rt : (7,3,-7,-1)
1757 7833 ζ : $(2\pi/3)^{1/3}$
1757 7934 rt : (7,3,6,1)
1757 8029 rt : (3,5,-7,-8)
1758 0919 2^x : $2/3 - \sqrt{3}/4$
1758 1539 at : $2/3 + \sqrt{3}$
1758 4144 rt : (8,8,6,9)
1758 5000 J_0 : 25/9
1758 5910 erf : $\sqrt{2}/9$
1758 7737 rt : (3,8,7,1)
1758 8196 id : P_{14}
1758 8378 Ψ : 22/17
1758 8576 rt : (9,8,4,-1)
1759 0200 10^x : $3\sqrt{2}/4 + 2\sqrt{3}$
1759 2592 cu : $7\sqrt[3]{2}/6$
1759 3378 cr : $\sqrt{2}/3 + 2\sqrt{3}/3$
1759 3470 rt : (3,6,-9,7)
1759 6105 rt : (3,3,-4,8)
1759 6391 sinh : $\zeta(11)$
1759 7133 rt : (5,3,5,-1)
1760 0520 at : 12/5
1760 1862 Γ : $3\pi^2/8$
1760 2023 rt : (7,7,-4,-3)
1760 2996 ζ : 22/7
1760 3401 $t\pi$: $7e\pi/4$
1760 5234 sq : $1/3 - 3e$
1760 6017 10^x : $3/4 - e/4$
1760 8507 sq : $2e/3 + \pi/3$
1760 9125 ℓ_{10} : 15
1760 9266 J_0 : 25/12
1760 9282 rt : (1,2,6,1)
1760 9693 J_1 : $7\pi/2$
1760 9739 cu : 19/18
1761 0601 rt : (4,5,8,9)
1761 0694 ℓ_2 : $(\sqrt[3]{3})^{1/3}$
1761 0725 2^x : $7\sqrt[3]{3}/9$
1761 0951 rt : (7,5,3,8)
1761 1029 ln : $3\sqrt{5}/8$
1761 1450 ζ : $1/\sqrt[3]{14}$
1761 1555 rt : (3,9,4,-1)
1761 1856 sr : $4e - \pi/4$
1761 2835 sq : $1/3 - 3\pi/2$
1761 3424 as : $2\sqrt{2}/3 - \sqrt{5}/2$
1761 3598 rt : (9,2,-8,-8)
1761 3604 $s\pi(\pi) + s\pi(\sqrt{3})$
1761 4170 rt : (6,2,-3,-9)
1761 7497 rt : (2,9,-7,0)
1761 9154 2^x : $1/4 + 2e/3$
1761 9456 J_1 : $9(1/\pi)/8$
1761 9585 2^x : $1 + 4e$
1762 1077 ζ : 25/18
1762 1190 e^x : $1/2 - \sqrt{5}$
1762 3782 cu : $1/2 - 3\sqrt{2}/4$
1762 4173 ζ : π
1762 4505 rt : (1,7,6,-1)
1762 5020 erf : $\sqrt[3]{2}/8$
1762 5217 $s\pi$: $3\sqrt{2} - 3\sqrt{3}/4$
1762 5545 ℓ_2 : $(\ln 2)^{1/3}$
1762 6707 Γ : 15/19
1762 6709 J_1 : $7\pi^2/9$
1762 7905 rt : (3,7,-7,1)
1762 8330 e^x : $4\sqrt{2} + 4\sqrt{3}$
1763 0254 cr : $\sqrt{2} - \sqrt{5} + \sqrt{6}$
1763 0869 rt : (9,2,-5,6)
1763 1106 ln : $1/4 + 2\sqrt{2}/3$
1763 2109 ln : $4\sqrt{2}/3 + 4\sqrt{3}$
1763 2698 $t\pi$: 1/18
1763 4780 rt : (5,5,-6,-8)
1763 5760 $\ln(2/3) \times \exp(5)$
1763 6696 θ_3 : $3\zeta(3)/7$
1763 7789 id : $(\sqrt[3]{2}/3)^2$
1763 7992 rt : (6,2,0,7)
1764 3129 rt : (5,1,-6,1)
1764 3631 rt : (1,9,6,-6)
1764 3856 rt : (2,7,3,3)
1764 4127 rt : (7,9,6,6)
1764 4423 $s\pi$: $\sqrt{2}/3 + 2\sqrt{5}$
1764 6039 J_2 : $2\pi^2$
1764 7058 id : 3/17
1764 9422 J_2 : 14/11
1765 0749 rt : (8,1,-8,-5)
1765 0971 J_2 : $9\sqrt{2}/10$
1765 1367 at : $3\sqrt{2}/4 - 2\sqrt{3}$
1765 1484 sq : $5e\pi$
1765 3443 rt : (3,2,5,8)
1765 3720 rt : (7,5,-5,-1)
1765 4058 $\ln(2/5) \times \exp(1/4)$
1765 4129 2^x : $1/4 + 3\pi/2$
1765 5062 sq : $1/4 - 3\pi$
1765 5528 rt : (4,8,-9,-7)
1765 5542 10^x : $\sqrt{2}/4 + \sqrt{5}/4$
1765 5909 at : $5\sqrt[3]{3}/3$
1765 7895 ln : P_{37}
1765 8549 rt : (9,6,-2,4)
1765 9436 10^x : $2\sqrt{2} - 2\sqrt{3}/3$
1765 9492 Ei : 17/23
1765 9744 $\exp(4/5) + s\pi(2/5)$
1766 0498 rt : (8,8,4,-1)
1766 0695 at : $\sqrt{3}/3 - 4\sqrt{5}/3$
1766 1287 at : $2\zeta(3)$
1766 1644 J_2 : $4(1/\pi)$
1766 1962 ℓ_{10} : $2\sqrt{2}/3 + \sqrt{5}/4$
1766 2993 e^x : 7/9
1766 4586 rt : (1,8,9,-8)
1766 4762 Ψ : $9e$
1766 5270 rt : (5,1,-3,-6)
1766 6204 rt : (6,4,-7,-4)
1766 7209 cu : $2 + 3e$
1766 8924 rt : (6,9,-7,4)
1766 9328 rt : (8,9,7,1)
1766 9554 erf : 3/19
1766 9681 sr : 18/13
1766 9894 θ_3 : $(e + \pi)/8$
1767 0342 sr : $2/3 + 3\pi$
1767 0543 sinh : $8\sqrt[3]{5}/9$
1767 0616 10^x : $4/3 + \pi/3$
1767 2257 rt : (9,6,-7,1)

1767 2519 J_1 : 3π
1767 2798 rt : (1,8,-4,-8)
1767 3218 e^x : $4\sqrt{2}+\sqrt{5}/3$
1767 3878 Ψ : $(\sqrt{5}/2)^{-3}$
1767 4220 tπ : $3e/4-2\pi/3$
1767 4302 2^x : $1/3+3e/4$
1767 4898 10^x : $3+4e/3$
1767 6655 rt : (6,3,6,1)
1767 6681 J : $(\sqrt{2}\pi/2)^{-2}$
1767 6832 at : $4/3-2\sqrt{3}/3$
1767 7203 10^x : $5e/6$
1767 7522 e^x : $1/4+\pi/2$
1767 7669 id : $\sqrt{2}/8$
1767 8529 J_1 : 11
1767 9009 $\exp(\pi)+s\pi(\sqrt{2})$
1767 9152 cr : $(\ln 2/3)^{-1/3}$
1768 0162 as : P_{14}
1768 1372 e^x : $1/3+2\sqrt{5}$
1768 1439 rt : (1,6,6,9)
1768 1856 rt : (4,3,5,-1)
1768 2226 Ψ : $((e+\pi)/3)^{-1/2}$
1768 2440 rt : (3,9,-2,-1)
1768 3278 J_1 : $\sqrt{3}+\sqrt{5}+\sqrt{6}$
1768 3503 ℓ_{10} : $5\zeta(3)/4$
1768 3882 id : $5(1/\pi)/9$
1768 3928 rt : (6,6,3,5)
1768 4558 rt : (8,3,-5,3)
1768 5644 $2/5+\ln(4/5)$
1768 5738 rt : (2,9,4,-1)
1768 8591 tπ : $4\sqrt{5}$
1768 8907 10^x : $2(1/\pi)/9$
1769 0464 e^x : $3/4+e/4$
1769 0473 at : $3/2+e/3$
1769 1261 ℓ_2 : $\exp(1/(3e))$
1769 1453 sq : $3+3\sqrt{3}$
1769 2120 id : $\exp(-\sqrt{3})$
1769 2789 rt : (4,8,-2,-1)
1769 4074 at : $e/2+\pi/3$
1769 4999 cr : $\sqrt[3]{13/3}$
1769 5920 ln : $7\sqrt[3]{2}$
1769 6893 rt : (2,1,2,-7)
1769 7553 Ei : $\ln(2\pi/3)$
1769 7684 cu : $\sqrt{2}-\sqrt{6}+\sqrt{7}$
1769 9512 10^x : $\ln(\sqrt{2}/3)$
1770 0040 ℓ_2 : $1/3+4\pi/3$
1770 1199 Ψ : $(\ln 3/3)^{1/3}$
1770 1561 tπ : $2e-3\pi/2$
1770 1751 rt : (2,8,-1,2)
1770 1762 ℓ_{10} : P_{81}
1770 3867 $\ln(3/5)\times\ln(\sqrt{2})$
1770 4312 J_0 : $\sqrt{13/3}$
1770 5098 id : $1/2+3\sqrt{5}/4$
1770 5571 rt : (9,1,-6,-9)
1770 6058 sinh : $10\sqrt{5}/7$
1770 7117 J_0 : $10\pi^2/9$
1770 8793 rt : (9,8,-9,-7)
1770 8896 J_1 : $2\sqrt[3]{2}/7$
1770 9112 $\sqrt{5}-\exp(5)$
1770 9970 J_1 : 9/25
1771 0510 cr : $3e/5$
1771 0781 rt : (7,0,-8,2)
1771 1365 rt : (5,7,-8,1)
1771 1383 $s\pi(2/5)-s\pi(e)$
1771 2434 rt : (2,8,7,1)
1771 3238 sr : $4\sqrt{3}/5$
1771 3420 sq : $e/2-\pi$
1771 3431 ln : $9\sqrt[3]{3}/4$
1771 4669 2^x : 4/17
1771 7430 10^x : $\sqrt{13}\pi$
1771 8205 2^x : $(\sqrt[3]{2})^{1/2}$
1771 9330 ln : $2/3+3e$
1772 0553 rt : (3,7,-9,-4)
1772 1247 J_0 : $(\ln 2)^{-2}$
1772 2054 rt : (8,4,-9,-8)
1772 3289 J_0 : $\sqrt{2}+\sqrt{5}+\sqrt{6}$
1772 4238 rt : (5,3,0,4)
1772 4284 J_0 : $5e\pi/7$
1772 5664 2^x : $\sqrt{2}/3+3\sqrt{5}/2$
1772 8531 sq : 8/19
1772 9381 sinh : $(\sqrt[3]{2}/3)^2$
1773 0169 Li_2 : $3\sqrt{2}/5$
1773 0543 as : $(\sqrt[3]{2}/3)^2$
1773 1321 $\exp(1/5)\times s\pi(\sqrt{2})$
1773 2788 sπ : $\exp(\sqrt[3]{3}/2)$
1773 4045 rt : (7,8,4,-1)
1773 6733 J_1 : $\sqrt[3]{3}/4$
1773 7137 rt : (9,2,5,-1)
1773 7888 2^x : $1/3-2\sqrt{2}$
1773 8052 sr : $7\sqrt[3]{3}$
1773 8569 e^x : $\exp(-2e/3)$
1773 8795 sinh : 3/17
1773 9335 J_1 : $3\zeta(3)/10$
1773 9911 J_2 : $(\sqrt[3]{3}/3)^{-1/3}$
1773 9960 as : 3/17
1774 0968 rt : (1,5,2,-6)
1774 1002 sr : $2\ln 2$
1774 4237 cr : $1/3+3\sqrt{3}/4$
1774 4313 ln : $4\ln 2/9$
1774 4587 rt : (9,3,-3,7)
1774 5142 rt : (8,5,-5,-1)
1774 7666 rt : (8,7,-2,1)
1774 7975 2^x : $\sqrt{2}/6$
1774 9229 $1/3+\ln(3/5)$
1774 9553 ζ : $1/\ln(1/(2e))$
1774 9700 cu : $1/4-\pi$
1774 9751 e^x : $1/4+e/3$
1775 0015 at : $\sqrt[3]{14}$
1775 0392 ζ : 23/14
1775 1189 rt : (6,1,-6,1)
1775 2944 sq : $3+2\pi$
1775 3506 rt : (5,4,-4,-9)
1775 7273 sinh : $5\sqrt[3]{2}$
1775 7502 sπ : $\sqrt{2}/2+\sqrt{5}$
1775 7639 id : $3\sqrt{3}+4\sqrt{5}/3$
1775 8609 $\pi\times\ln(2)$
1775 9184 id : $(\sqrt[3]{3}/2)^{-1/2}$
1775 9677 J_0 : 5
1775 9989 rt : (4,7,-7,1)
1776 1677 rt : (1,9,4,-1)
1776 3036 cu : $\ln(\sqrt[3]{5}/3)$
1776 5069 rt : (8,0,-6,6)
1776 6307 rt : (7,6,5,9)
1776 6908 at : $2\sqrt{2}/3-3\sqrt{5}/2$
1776 7684 at : $1/2-e/4$
1776 8965 rt : (3,6,8,6)
1776 9012 $\pi+s\pi(\sqrt{2})$
1776 9416 J_2 : $4\sqrt{5}/7$
1776 9884 sinh : $\sqrt{2}/8$
1777 0005 J_2 : 23/18
1777 0531 e^x : $(\sqrt{2}/3)^{1/3}$
1777 0791 2^x : $3\zeta(3)$
1777 1060 as : $\sqrt{2}/8$
1777 3050 J_0 : $(\sqrt{2}\pi/2)^3$
1777 4922 rt : (4,7,-7,-8)
1777 4966 $\exp(\sqrt{2})\div\exp(\pi)$
1777 6047 sr : $1/2+3\sqrt{2}$
1777 6195 sinh : $5(1/\pi)/9$
1777 6212 J_0 : $(\ln 2/3)^{-1/2}$
1777 7372 as : $5(1/\pi)/9$
1777 8196 ln : $(e+\pi)/7$
1777 8792 rt : (5,3,6,1)
1777 9050 10^x : $\sqrt{3}/3+4\sqrt{5}$
1778 4184 ℓ_2 : $3\sqrt{2}/2+4\sqrt{3}$
1778 4562 sinh : $\exp(-\sqrt{3})$
1778 4789 sinh : $5\zeta(3)/6$
1778 5685 rt : (7,4,-5,0)
1778 5742 as : $\exp(-\sqrt{3})$
1778 7895 J_2 : $7\pi^2/5$
1778 8026 ℓ_{10} : $4e+4\pi/3$
1778 8119 $2-\exp(3/5)$
1778 8804 sr : $2\sqrt{3}/3+4\sqrt{5}$
1779 2777 rt : (7,9,7,1)
1779 3681 $\ln(\sqrt{2})\div\exp(2/3)$
1779 3873 cr : $e/3+3\pi$
1779 4482 rt : (6,3,2,8)
1779 4530 J_0 : $\sqrt[3]{9}$
1779 6525 Li_2 : $8(1/\pi)/3$
1779 7851 sq : $(4/3)^{-3}$
1779 8928 as : $3/2-3\sqrt{5}/4$
1779 9191 e^x : $(e)^2$
1780 1379 ℓ_{10} : $\sqrt{23}\pi$
1780 2985 rt : (3,4,-5,-1)
1780 4558 2^x : $\exp(-\sqrt[3]{3})$
1780 5383 ln : 24
1780 6437 rt : (3,6,-3,-4)
1780 6994 tanh : $\sqrt[3]{2}/7$
1780 7190 ℓ_2 : $4\sqrt{2}/5$
1780 7356 ln : $(\sqrt{2}\pi/3)^3$
1780 7447 rt : (9,7,0,5)
1780 8201 at : $\sqrt[3]{2}/7$
1780 8224 erf : $(1/\pi)/2$
1780 8521 rt : (5,8,-4,5)
1780 9293 rt : (6,8,4,-1)
1780 9306 sr : $2/3+3e/2$
1780 9455 ln : $\sqrt{2}-\sqrt{3}/3$
1780 9704 J_2 : $(2\pi/3)^{1/3}$
1780 9724 id : $3\pi/8$
1781 0028 rt : (2,2,-1,9)
1781 0113 ln : $3\sqrt{2}/2+3\sqrt{5}$
1781 1369 ζ : $7\sqrt{5}/5$
1781 1625 E : 17/19
1781 2780 rt : (1,7,-1,2)
1781 3005 10^x : $\sqrt{3}+2\sqrt{5}/3$
1781 3241 cu : $3\sqrt{3}/4+\sqrt{5}$
1781 4067 sr : $5e\pi/9$
1781 4355 J_1 : $1/\ln(\sqrt{5}/3)$
1781 6220 rt : (8,3,-7,-9)
1781 8298 Ei : $\sqrt[3]{2}/3$
1781 9408 rt : (1,2,-6,1)
1782 0189 tπ : $3\sqrt{3}-2\sqrt{5}$
1782 0323 id : $1/4+4\sqrt{3}$
1782 0556 at : $(\sqrt{5}/3)^{-3}$
1782 2913 rt : (5,0,-1,7)
1782 3386 rt : (8,2,5,-1)
1782 4156 rt : (6,1,-8,-1)
1782 4375 2^x : $\sqrt{3}/2-3\sqrt{5}/2$
1782 4710 J_2 : $8\sqrt{3}/3$
1782 5128 rt : (3,7,-9,-3)
1782 7011 J_1 : $1/\sqrt[3]{21}$
1782 8730 J_1 : $9e\pi/10$
1782 8761 e^x : $3\sqrt{2}-2\sqrt{3}$
1783 0267 sinh : $\zeta(9)$
1783 1160 J_0 : $3\ln 2$
1783 4477 rt : (8,4,-3,4)
1783 5313 tanh : $\sqrt[3]{3}/8$
1783 6529 at : $\sqrt[3]{3}/8$
1783 6586 10^x : $(\sqrt{2}\pi)^{1/2}$
1783 6869 id : $\sqrt{3}/4+\sqrt{5}/3$
1783 8380 sr : $4+\sqrt{5}/3$
1783 9458 rt : (9,5,-5,-1)
1783 9533 at : $8e/9$
1783 9611 cu : $4/3+3e$
1784 0146 cr : 18/11
1784 0302 rt : (1,7,-2,-9)
1784 1241 sr : $(1/\pi)/10$
1784 3174 rt : (1,8,7,1)
1784 7239 sq : $\sqrt[3]{17/3}$
1784 7446 ℓ_{10} : $\sqrt{2}/4+2\sqrt{3}/3$
1784 8215 tπ : $3\sqrt{2}/2+2\sqrt{3}/3$
1784 8353 rt : (2,6,-9,-1)
1784 8992 E : $2\sqrt{5}/5$
1785 0077 2^x : $4\sqrt{2}-4\sqrt{3}/3$
1785 0594 Ei : $(\sqrt[3]{4}/3)^3$
1785 1130 id : $5\sqrt{2}/6$
1785 1484 $4/5\times\ln(4/5)$
1785 1544 at : $3\sqrt{3}/4+\sqrt{5}/2$
1785 1604 rt : (4,9,2,3)
1785 3890 sq : $3\sqrt{3}/2+\sqrt{5}/3$
1785 8597 rt : (2,3,5,5)
1785 8781 $\exp(3/5)+\exp(\sqrt{5})$
1786 1012 rt : (9,4,-1,8)
1786 1739 rt : (7,1,-6,3)
1786 2964 Γ : $5\sqrt[3]{2}/8$
1786 3279 id : $4/3-2\sqrt{3}/3$
1786 3555 rt : (7,1,-6,1)
1786 5451 sπ : $\sqrt[3]{22/3}$
1786 5499 ln : 4/13
1786 6580 λ : $\exp(-7\pi/5)$
1786 7781 rt : (2,5,-4,1)
1786 9416 id : $3\sqrt{2}/4+\sqrt{5}/2$

1786 9810 $J_0 : 2\pi^2$
1787 0557 $s\pi : 2\sqrt{2}/3$
1787 0942 $J_2 : 8\sqrt[3]{3}/9$
1787 3377 sr : $e/4 + 3\pi$
1787 4322 $t\pi : 1/\ln(\ln 2/2)$
1787 4361 $\ell_{10} : P_{55}$
1787 4692 $\ln(3/4) \div \ln(5)$
1787 5394 $e^x : 3\sqrt{2} - 3\sqrt{3}/2$
1787 6863 $10^x : 1/14$
1787 7212 $\ell_{10} : 3 - 2\sqrt{5}/3$
1787 7646 $Li_2 : (\sqrt[3]{3}/2)^{1/2}$
1787 9663 cr : $3\sqrt{2}/4 + \sqrt{3}/3$
1788 0559 rt : (3,6,-7,-5)
1788 2125 rt : (8,1,-4,7)
1788 2482 $J_2 : 3\sqrt[3]{5}/4$
1788 2643 ln : $(\sqrt[3]{5})^{1/3}$
1788 2943 $J_1 : 4/11$
1788 3491 $\Gamma : 3\sqrt[3]{3}$
1788 3495 sq : $e + \pi/3$
1788 3524 $10^x : 2/3 - \sqrt{2}$
1788 4590 rt : (4,3,6,1)
1788 6317 rt : (5,8,4,-1)
1788 7091 $J_0 : 6\sqrt{3}/5$
1788 7701 id : $2\sqrt{2}/3 + \sqrt{5}$
1788 8543 $2/5 \div \sqrt{5}$
1788 9144 $\zeta : 3\ln 2/5$
1788 9235 rt : (6,2,-9,-2)
1788 9901 $J_1 : 8(1/\pi)/7$
1789 2327 $Li_2 : \sqrt[3]{5}/10$
1789 2328 tanh : $\exp(-\sqrt[3]{5})$
1789 2567 as : $4\ln 2/3$
1789 3564 at : $\exp(-\sqrt[3]{5})$
1789 6318 rt : (4,4,-5,-1)
1789 7369 rt : (4,6,-5,2)
1789 8030 ln : $(\ln 2/2)^3$
1789 9095 rt : (5,7,-7,1)
1789 9603 rt : (2,7,-4,7)
1790 1181 $erf : 4/25$
1790 1343 $\theta_3 : (\sqrt{5})^{-3}$
1790 2197 $\ell_{10} : 5e/9$
1790 3380 rt : (7,2,-7,-6)
1790 4406 rt : (3,3,7,9)
1790 4757 at : $(e + \pi)^{1/2}$
1790 4957 $\zeta : 25/8$
1790 5527 $t\pi : 3\sqrt{2} - 3\sqrt{3}/4$
1790 6537 at : $3\sqrt{3}/4 - \sqrt{5}/2$
1790 6757 $2^x : 1/2 - 4\sqrt{5}/3$
1790 7694 rt : (5,9,0,9)
1790 8932 $\ln(2/5) \times \exp(\sqrt{3})$
1791 0824 rt : (3,8,-3,6)
1791 2060 rt : (7,2,5,-1)
1791 2790 $\lambda : (\ln 2/3)^3$
1791 2841 rt : (2,9,-8,-5)
1791 3282 $1/4 \div \exp(1/3)$
1791 3669 $\ln(5) + s\pi(\pi)$
1791 4455 $t\pi : (\sqrt{2}\pi/3)^{-2}$
1791 5287 sq : $2 + 4\pi$
1791 5809 rt : (6,7,5,6)
1791 9169 rt : (1,9,-4,-1)
1791 9205 sq : $e - \pi$
1791 9645 rt : (3,9,-8,-8)
1792 1981 rt : (6,9,7,1)
1792 2585 $J_1 : 17/5$
1792 2647 $\theta_3 : 3\sqrt[3]{5}/7$
1792 4273 id : $\sqrt{2}/2 + 2\sqrt{5}$
1792 5241 $\ln(2) \div s\pi(1/5)$
1792 5665 $t\pi : \sqrt{2}/3 + 2\sqrt{5}$
1792 6287 $s\pi : (\sqrt{5}/2)^{1/2}$
1792 8671 $erf : \sqrt[3]{3}/9$
1792 9665 rt : (7,5,-8,-9)
1792 9777 rt : (5,6,-7,8)
1793 0499 rt : (2,4,-4,-7)
1793 2901 at : $3 - \sqrt{3}/3$
1793 3149 $10^x : 3e/2 + 3\pi/4$
1793 3480 ln : $2\sqrt{2}/3 + 4\sqrt{3}/3$
1793 6600 $J_1 : \sqrt{24}\pi$
1793 7407 $e^x : 1 - e$
1793 9433 $t\pi : 1/3 + 2\sqrt{2}/3$
1794 0316 $J_2 : (\sqrt{2}/3)^{-1/3}$
1794 0779 $\ln(2/3) - s\pi(e)$
1794 1841 $2^x : \sqrt[3]{23}$
1794 3444 $2^x : 5/21$
1794 4346 $J_0 : 10(e + \pi)/3$
1794 4894 $J_1 : 9\sqrt[3]{5}$
1794 4947 sr : 19/4
1794 5435 $\Psi : \sqrt{5/3}$
1794 6856 rt : (9,4,6,1)
1794 7467 $10^x : 5/7$
1794 7958 cr : $3 - e/2$
1795 0902 rt : (2,0,4,8)
1795 1983 $\exp(2) \times s\pi(\pi)$
1795 2457 $\ell_{10} : 6\sqrt[3]{2}/5$
1795 4722 sq : $3\sqrt{3}/4 + \sqrt{5}/3$
1795 6198 rt : (6,4,4,9)
1795 6784 $\ln(4/5) \times \ln(\sqrt{5})$
1795 7045 id : $1/2 + e/4$
1795 8631 rt : (5,4,2,5)
1795 8712 sr : $(\pi)^{-3}$
1795 9202 sq : $4\sqrt{2}/3 - 4\sqrt{3}/3$
1795 9672 as : $4/3 - 2\sqrt{3}/3$
1795 9856 $J_2 : 9/7$
1796 0845 id : $\sqrt{2}/3 + 3\sqrt{5}$
1796 0937 rt : (9,8,2,6)
1796 1063 $\pi \times \exp(1/2)$
1796 4389 $\zeta : (e/3)^{-2}$
1796 5204 rt : (8,8,0,2)
1796 5214 at : $7\sqrt{3}/5$
1796 6063 $\ell_2 : 5e/3$
1797 1834 $Li_2 : \zeta(3)/7$
1797 2936 $s\pi : 3\pi/10$
1797 3158 $2/5 \div \exp(4/5)$
1797 5778 rt : (7,8,-2,-2)
1798 0192 cu : $4e/3 - 4\pi/3$
1798 0260 rt : (9,5,1,9)
1798 0436 ln : $7\sqrt[3]{5}/10$
1798 0589 rt : (8,7,1,0)
1798 1836 at : $7\ln 2/2$
1798 1979 cu : $e - 4\pi/3$
1798 4081 tanh : 2/11
1798 5349 at : 2/11
1798 5863 rt : (5,1,1,8)
1798 6185 $e^x : 3/2 + 2\sqrt{3}$
1798 7592 $2^x : 2/3 - \pi$
1798 7749 cr : $\sqrt{2} + 4\sqrt{5}$
1798 9056 rt : (8,5,-1,5)
1799 1163 tanh : $4(1/\pi)/7$
1799 2433 at : $4(1/\pi)/7$
1799 2609 rt : (5,4,-5,-1)
1799 4320 rt : (3,3,6,1)
1799 5546 $2^x : 3(1/\pi)/4$
1799 5813 cr : 23/14
1799 6874 at : $3/4 + 3\sqrt{5}/4$
1799 8107 rt : (1,3,1,5)
1799 8661 $10^x : \sqrt{7}\pi$
1799 8683 $\Psi : \sqrt{14}$
1799 8872 id : $\sqrt[3]{2}/7$
1800 0000 sq : $3\sqrt{2}/10$
1800 0388 sinh : $5\sqrt[3]{5}/4$
1800 0908 rt : (2,9,-4,-1)
1800 2037 rt : (8,2,-2,8)
1800 2637 cr : $3e/2 + 2\pi$
1800 3299 rt : (6,2,5,-1)
1800 3396 rt : (4,9,-9,9)
1800 4613 $e^x : 3 - e/4$
1800 4977 $J_1 : \ln(\sqrt[3]{3})$
1800 5035 $\zeta : 9\ln 2/2$
1800 5373 rt : (7,5,-3,1)
1800 6703 at : $1/3 + 2\pi/3$
1800 7635 rt : (2,9,1,-7)
1800 8290 ln : $4 - \sqrt{5}/3$
1800 9626 rt : (9,7,4,-1)
1800 9783 sq : $2 + e/4$
1801 1139 $10^x : 2\sqrt{2} + \sqrt{5}/2$
1801 2137 $J_1 : 5e/4$
1801 2654 sq : $4(1/\pi)/3$
1801 2668 rt : (1,9,-1,6)
1801 3018 as : $(\sqrt[3]{5}/2)^{1/2}$
1801 4003 rt : (1,3,-9,-5)
1801 5809 rt : (4,2,-2,-7)
1801 6298 $J_0 : 9\sqrt{2}$
1801 7423 rt : (7,2,-4,4)
1801 8144 $\Psi : 5/7$
1801 8346 $t\pi : \exp(\sqrt[3]{3}/2)$
1801 8514 cr : $4 - 3\pi/4$
1801 8928 at : 17/7
1801 9649 $J_1 : \ln(\ln 2)$
1802 1943 cr : $(\sqrt{2}\pi)^{1/3}$
1802 2178 cu : $(3\pi)^3$
1802 2412 $2^x : 2 - 2\sqrt{5}$
1802 3015 rt : (4,7,-1,6)
1802 3213 $J_1 : 9\sqrt[3]{5}/2$
1802 3963 rt : (7,1,-5,-7)
1802 7341 $e^x : (\ln 3/2)^3$
1802 7447 rt : (4,4,-8,5)
1802 7720 rt : (1,8,7,8)
1802 8091 at : $4 - \pi/2$
1802 8119 id : $\sqrt[3]{3}/8$
1803 0121 $1/4 + s\pi(\pi)$
1803 0241 $e^x : 2/3 + 2\sqrt{3}/3$
1803 0358 $\zeta : 9\sqrt{3}/5$
1803 0402 id : $3e/4 + \pi$
1803 1847 $J_1 : \sqrt{6}\pi$
1803 3166 $J_1 : 3e\pi/4$
1803 3688 $1/4 \div \ln(4)$
1803 3787 cu : 25/17
1803 3988 id : $5\sqrt{5}$
1803 4151 ln : $5(e + \pi)/9$
1803 6675 cr : $3\sqrt{2} - 3\sqrt{3}/2$
1803 9821 $J_1 : 3\sqrt[3]{3}$
1804 3422 $e^x : 3 - 3\pi/2$
1804 3737 $10^x : 8\sqrt[3]{2}/9$
1804 4275 $t\pi : \sqrt{2}/2 + \sqrt{5}$
1804 4631 cu : $\sqrt{3}/4 + 3\sqrt{5}$
1804 5516 cr : $\pi^2/6$
1804 6042 rt : (9,9,-7,-6)
1805 2419 $Ei : \exp(-\sqrt{3}/2)$
1805 3398 sq : $1/4 + \sqrt{5}$
1805 3479 $\ell_{10} : \exp(e)$
1805 4979 as : $1/2 - e/4$
1805 5154 rt : (8,4,6,1)
1805 6868 $\ell_{10} : 7\sqrt{3}/8$
1805 7039 cu : 13/23
1805 7224 $\ell_2 : 15/17$
1805 7514 rt : (5,9,7,1)
1805 8112 $Li_2 : 17/20$
1806 1086 cu : $3 - e/3$
1806 3687 $e^x : P_1$
1806 5617 $Ei : 10\sqrt[3]{3}/7$
1806 6831 $e^x : \pi^2/6$
1806 7030 $\Gamma : 11/14$
1806 7502 sr : $1/3 + 3\sqrt{2}/4$
1807 0190 $\zeta : (e + \pi)/8$
1807 0623 $2^x : 1/4 - e$
1807 1102 $K : 1/\ln(\pi)$
1807 1755 $\Gamma : 5\sqrt{2}/9$
1807 3337 at : $\sqrt{5} - \sqrt{6} + \sqrt{7}$
1807 5060 $K : \sqrt[3]{2/3}$
1807 5492 $e^x : 3/2 + \sqrt{2}/3$
1807 6566 rt : (2,8,-8,-5)
1808 1687 cr : $2 - \sqrt{2}/4$
1808 1843 $J_1 : 1/\ln(\sqrt[3]{2})$
1808 2411 sq : $4/3 + 2\sqrt{2}/3$
1808 2735 $\ln(\pi) + s\pi(\sqrt{2})$
1808 3213 at : $2 + \sqrt{3}/4$
1808 4552 $J_1 : \exp(-1)$
1808 4777 rt : (3,9,-4,-1)
1808 6404 $J_2 : \sqrt{5/3}$
1808 7014 id : $\exp(-\sqrt[3]{5})$
1808 7031 cu : $9\pi/4$
1808 7189 sq : $1/\sqrt[3]{13}$
1808 7222 rt : (5,1,-9,1)
1808 8013 ln : $\sqrt{3}/2 - \sqrt{5}/4$
1808 9548 rt : (8,7,4,-1)
1809 0558 at : $2\sqrt{2}/3 + 2\sqrt{5}/3$
1809 0662 rt : (2,5,-7,-2)
1809 0841 cu : $1/\ln(e + \pi)$
1809 2053 rt : (6,4,-5,-1)
1809 4730 rt : (1,9,5,-5)
1809 5420 $2^x : 1/\ln(2/3)$
1809 6211 sinh : $\sqrt[3]{2}/7$

1809 7253 rt : (5,2,5,-1)
1809 7376 $e^x : 1+\sqrt{2}$
1809 7418 $J_2 : 10(e+\pi)/3$
1809 7498 as : $\sqrt[3]{2}/7$
1809 9266 $2^x : 6/25$
1810 0411 id : $3\sqrt{3}/4-\sqrt{5}/2$
1810 0955 at : $1/4-\sqrt{3}/4$
1810 1546 $2^x : 9/8$
1810 1886 $\zeta : \sqrt{2}+\sqrt{6}-\sqrt{7}$
1810 1933 cu : $2\sqrt{2}/5$
1810 2706 rt : (9,1,-6,1)
1810 4311 $\Gamma : \pi/4$
1810 4707 $2^x : 3\sqrt{2}-3\sqrt{5}$
1810 5323 sr : $7e/4$
1810 8287 rt : (2,3,6,1)
1810 9419 $J_1 : 1/\sqrt[3]{20}$
1811 0269 $J_1 : 7/19$
1811 1507 $\zeta : 5/12$
1811 2655 sq : $4/3-4\pi$
1811 4823 rt : (2,4,-5,8)
1811 7783 rt : (5,2,-8,-4)
1811 9281 rt : (9,9,4,7)
1812 0503 $\ell_2 : 1-\sqrt{5}/4$
1812 1145 $J_0 : 3\pi$
1812 1207 $e^x : 3/2+3e$
1812 2115 $t\pi : \sqrt{2}+4\sqrt{3}/3$
1812 2350 $2^x : 3+4\sqrt{5}/3$
1812 3058 $2^x : 1-2\sqrt{3}$
1812 3983 $\ell_2 : 7\sqrt[3]{2}/10$
1812 4933 rt : (8,3,0,9)
1812 5420 rt : (1,8,-7,1)
1812 5625 cr : $8\sqrt[3]{3}/7$
1812 5934 sinh : $\sqrt[3]{3}/8$
1812 7232 as : $\sqrt[3]{3}/8$
1812 7862 $J_1 : 9e\pi/5$
1812 8928 rt : (2,8,4,-1)
1812 9961 $2^x : \sqrt[3]{3}/6$
1813 0945 rt : (6,9,-8,2)
1813 1402 rt : (4,5,-4,9)
1813 2946 $2^x : \zeta(3)/5$
1813 3023 $\ell_2 : 9\sqrt[3]{2}/5$
1813 4466 at : $3-2e$
1813 4577 $Li_2 : \sqrt{3}/10$
1813 5601 $\ell_{10} : \sqrt[3]{7/2}$
1813 6041 $e^x : 1/6$
1813 6357 sinh : $\exp(\sqrt[3]{2}/3)$
1813 8706 ln : $\ln(e/2)$
1813 8832 $\ell_2 : 4-\sqrt{3}$
1813 9089 $J_0 : 9e/4$
1814 0081 $2^x : 2\sqrt{2}/3+4\sqrt{5}/3$
1814 0400 $10^x : 3\sqrt{3}/2+3\sqrt{5}/2$
1814 1293 $s\pi : 9e/8$
1814 2766 $10^x : 7\zeta(3)/5$
1814 3551 $Li_2 : \ln 2/4$
1814 3606 rt : (9,2,-9,-7)
1814 4194 rt : (1,6,4,-2)
1814 5170 $\ell_{10} : \sqrt{3}+\sqrt{5}-\sqrt{6}$
1814 5965 $J_1 : 5\sqrt{3}/2$
1814 7250 $e^x : 2\sqrt{2}+3\sqrt{5}$
1814 7448 sq : 25/23
1814 7566 rt : (7,0,-3,8)
1814 8560 rt : (8,6,1,6)
1814 9199 $2^x : e/4-\pi$
1814 9880 $\Gamma : \sqrt[3]{25/2}$
1815 2728 $10^x : 1/\ln(5)$
1815 4453 rt : (5,2,3,9)
1815 4464 rt : (1,7,-2,-5)
1815 7235 cu : $2-2\sqrt{2}/3$
1815 7573 $t\pi : \sqrt[3]{22/3}$
1815 9099 rt : (6,8,7,7)
1815 9252 at : P_{63}
1815 9793 $t\pi : \sqrt{3}-\sqrt{6}-\sqrt{7}$
1816 0061 sq : $5e/2$
1816 1346 $J_2 : 22/17$
1816 1893 $\Gamma : 8(1/\pi)/9$
1816 2933 $t\pi : 2\sqrt{2}/3$
1816 3407 $\ell_{10} : P_{58}$
1816 3506 $\ell_{10} : P_{19}$
1816 3590 sr : $4\pi/9$
1816 4565 cr : $7\sqrt{2}/6$
1816 5331 rt : (9,1,-8,4)
1816 6411 rt : (4,2,-6,1)
1816 7200 $10^x : 3/4-2\sqrt{5}/3$
1816 7479 rt : (7,4,6,1)
1816 7818 $\Psi : \exp(\sqrt{5})$
1816 9653 $J_2 : (2\pi/3)^3$
1817 0360 rt : (6,3,-6,-7)
1817 0368 $e^x : 1/3-3e/4$
1817 0890 rt : (4,9,-4,-1)
1817 1438 rt : (7,7,4,-1)
1817 1826 at : $\sqrt{2}/2+\sqrt{3}$
1817 5577 $Ei : 21/19$
1817 5835 rt : (3,2,-9,2)
1817 6287 $e \div \ln(4/5)$
1817 6558 $\Psi : 5/3$
1817 8082 $\zeta : 8\zeta(3)/9$
1817 8105 rt : (7,3,-2,5)
1817 8702 ln : $1/3+\sqrt{3}/2$
1818 1818 id : 2/11
1818 2617 sq : $8\pi^2$
1818 2852 at : $2e\pi/7$
1818 3735 at : P_{47}
1818 3747 $\ell_{10} : 8\sqrt[3]{5}/9$
1818 5746 rt : (4,1,0,-8)
1818 5792 sinh : $\exp(-\sqrt[3]{5})$
1818 7111 as : $\exp(-\sqrt[3]{5})$
1818 8910 $\ell_{10} : 2\sqrt{2}/3+\sqrt{3}/3$
1818 9136 id : $4(1/\pi)/7$
1818 9574 cr : $\sqrt[3]{9/2}$
1819 1088 $e^x : 3+\sqrt{2}/3$
1819 2362 rt : (8,9,2,3)
1819 3317 rt : (5,9,-6,-4)
1819 4085 rt : (4,2,5,-1)
1819 4333 rt : (2,8,-2,-2)
1819 4861 rt : (7,4,-5,-1)
1819 5478 $\theta_3 : 1/11$
1819 6495 $J_0 : 8\pi/9$
1819 7079 rt : (1,8,-1,4)
1819 8051 id : $9\sqrt{2}/4$
1819 8076 at : $3-\sqrt{5}/4$
1819 8278 $t\pi : \exp(-\sqrt[3]{5}/2)$
1819 9486 rt : (9,4,-7,1)
1820 0041 rt : (4,9,7,1)
1820 0733 as : $3\sqrt{3}/4-\sqrt{5}/2$
1820 1809 rt : (7,7,-7,1)
1820 2818 $\theta_3 : 2(1/\pi)/7$
1820 4697 rt : (5,5,4,6)
1820 4704 $2^x : 3/4+2\pi/3$
1820 4784 $1/5 \div \ln(3)$
1820 6429 $e^x : P_{60}$
1820 6709 $t\pi : 1/2+3e/4$
1820 7923 rt : (6,0,-5,4)
1820 7999 $e^x : \sqrt[3]{23}$
1820 8913 $2^x : \sqrt{3}-2\sqrt{5}/3$
1820 9002 $Ei : 8/19$
1820 9737 $10^x : (\ln 3)^3$
1821 0173 $J_2 : 4\pi$
1821 0269 $t\pi : 1/\sqrt[3]{13}$
1821 1370 $t\pi : \sqrt{3}/4+\sqrt{5}/2$
1821 2336 $Li_2 : 4/23$
1821 3234 $2^x : 2e/3-\pi/2$
1821 3966 rt : (1,8,4,-1)
1821 4119 $2^x : 7\zeta(3)$
1821 5214 $J_1 : 9\pi^2/5$
1821 6876 ln : $2\sqrt{2}+\sqrt{3}/4$
1821 7350 rt : (3,5,-2,3)
1821 7371 $Ei : 8e\pi/9$
1821 7701 sr : $5\sqrt{5}/8$
1821 8721 $\ln(2/3) \div \exp(4/5)$
1821 9325 rt : (2,1,6,9)
1822 0220 sr : $((e+\pi)/3)^{1/2}$
1822 0378 sinh : $\sqrt{2}+\sqrt{5}-\sqrt{7}$
1822 1453 $t\pi : (\sqrt{5}/2)^{1/2}$
1822 1981 $t\pi : (\sqrt[3]{3}/3)^{-1/3}$
1822 2282 $t\pi : 3\sqrt{2}/4-\sqrt{5}/2$
1822 2579 $\zeta : 8e/7$
1822 3045 rt : (1,9,-2,-3)
1822 3068 $10^x : 4+4\sqrt{3}/3$
1822 4301 at : $3/2+2\sqrt{2}/3$
1822 4499 at : $10\sqrt[3]{5}/7$
1822 4727 cr : $6\sqrt{3}$
1822 4897 sq : $2e/5$
1822 5027 sq : $8(e+\pi)/3$
1822 5459 $2^x : \sqrt{2}/2+4\sqrt{5}$
1822 5679 sinh : $\ln(\ln 3/3)$
1822 5946 sr : $(\ln 3/3)^{-1/3}$
1822 6471 sq : $3e/2-\pi/3$
1822 6832 rt : (1,3,6,1)
1822 7201 cu : $1-\sqrt{3}/4$
1822 8406 cr : $\sqrt{2}/4+3\sqrt{3}/4$
1823 1735 $\ell_2 : 2\sqrt{2}-\sqrt{5}/4$
1823 2013 $Ei : (\ln 3/2)^{1/2}$
1823 2014 $\ln(2/3)+s\pi(1/5)$
1823 2155 ln : 5/6
1823 3345 cu : P_{80}
1823 3656 at : $7\pi/9$
1823 3684 $e^x : P_{44}$
1823 3755 $3/4 \div \exp(\sqrt{2})$
1823 4881 rt : (7,6,-1,2)
1823 4916 $s\pi : 5e/7$
1823 7145 $E : 9\ln 2/7$
1823 8017 rt : (5,3,-8,-3)
1823 9225 $\ell_{10} : \exp(\sqrt[3]{2}/3)$
1824 1829 rt : (9,1,5,-1)
1824 2892 ln : $e/3+3\pi/4$
1824 3113 rt : (9,1,-7,-8)
1824 5556 rt : (2,9,7,-1)
1824 6068 $\Gamma : \ln 2/6$
1824 6448 $J_1 : 13/3$
1824 7760 at : 22/9
1824 9271 rt : (2,7,-6,-6)
1824 9396 id : $\exp(5/2)$
1824 9821 $J_1 : 7(1/\pi)/6$
1825 0493 at : $4\sqrt{2}/3+\sqrt{5}/4$
1825 0949 rt : (4,2,1,5)
1825 2909 id : $2\sqrt{2}+3\sqrt{5}/2$
1825 3349 sq : $4/3-e/3$
1825 3738 $\Psi : 15/4$
1825 5132 $J : \exp(-5)$
1825 5185 $J_2 : 9\sqrt[3]{3}/10$
1825 5393 rt : (6,7,4,-1)
1825 7295 rt : (3,9,-3,-4)
1825 8071 $\theta_3 : (\sqrt{2}\pi/2)^{-3}$
1825 9367 rt : (5,9,-4,-1)
1826 0166 $e^x : ((e+\pi)/2)^{1/3}$
1826 0745 sr : $4\sqrt{2}+2\sqrt{5}$
1826 0982 rt : (9,7,-6,-2)
1826 2590 $\theta_3 : 11/15$
1826 3554 $erf : \exp(-2e/3)$
1826 9010 rt : (5,7,-9,-1)
1826 9291 rt : (6,8,-7,9)
1826 9448 $\ln(4/5) \div \exp(1/5)$
1827 0043 rt : (1,2,3,-8)
1827 0451 $t\pi : 3\pi/10$
1827 1193 $2^x : 4\sqrt{2}/3+4\sqrt{5}/3$
1827 1215 sq : $1/\ln(\pi/3)$
1827 1333 $2 \div \ln(2/5)$
1827 1742 $E : (\sqrt[3]{2})^{-1/2}$
1827 2433 $s\pi : 3\sqrt{3}+\sqrt{5}/3$
1827 3038 rt : (9,9,9,-2)
1827 4326 rt : (7,1,-1,9)
1827 4466 $\sqrt{3}-\ln(\sqrt{3})$
1827 5110 sq : $\sqrt[3]{5}/4$
1827 6545 at : $9e/10$
1827 7954 rt : (6,3,-4,1)
1827 8543 rt : (2,8,-7,1)
1827 8680 cr : $1/2+2\sqrt{3}/3$
1827 9544 $J_2 : 3\sqrt{3}/4$
1827 9688 $\Gamma : \sqrt{2}/5$
1828 0506 $\Psi : 1/\sqrt{24}$
1828 1828 id : $10e$
1828 2159 sinh : 2/11
1828 2219 cu : $1/4+\pi/3$
1828 3513 as : 2/11
1828 4142 rt : (6,4,6,1)
1828 4723 id : $4\sqrt{3}-\sqrt{5}/3$
1828 7499 rt : (9,2,-6,5)
1828 9598 sinh : $4(1/\pi)/7$
1828 9738 $2^x : 3/4+3\sqrt{3}/2$
1829 0848 $J_1 : \sqrt{23/2}$

1829 0955 as : $4(1/\pi)/7$
1829 2036 rt : (5,2,-6,1)
1829 3721 $1/3 \div \exp(3/5)$
1829 3975 rt : (3,2,5,-1)
1829 7370 e^x : $(2e/3)^{-3}$
1829 7874 rt : (6,2,-4,-8)
1829 8117 ln : $\sqrt{2}+\sqrt{5}-\sqrt{6}$
1830 1270 id : $1/4-\sqrt{3}/4$
1830 1830 J_0 : $5\sqrt{5}/4$
1830 2669 J_2 : 13/10
1830 7015 2^x : $1/\sqrt{17}$
1830 9194 rt : (3,3,-5,6)
1830 9693 tanh : $(\sqrt[3]{5}/3)^3$
1830 9886 id : $10(1/\pi)$
1831 0190 rt : (8,1,-9,-4)
1831 0204 ln : $(\sqrt[3]{3})^{1/2}$
1831 1081 at : $(\sqrt[3]{5}/3)^3$
1831 1871 cr : $3/4+e/3$
1831 2261 J_1 : $\sqrt{5}/6$
1831 3276 rt : (8,7,3,7)
1831 4841 2^x : P_{15}
1831 7888 sr : $10\sqrt[3]{2}/9$
1831 8100 ln : $2/3+3\sqrt{3}/2$
1831 9626 rt : (8,9,9,-2)
1831 9964 at : $\sqrt{6}$
1832 1595 sr : 7/5
1832 2038 J : 4/11
1832 2469 e^x : 22/19
1832 2509 $\ln(4/5)\times\exp(4)$
1832 4618 10^x : $3/2-\pi/4$
1832 5646 ln : $(\ln 2)^{1/2}$
1832 5814 $1/5\times\ln(2/5)$
1832 7756 cu : $\sqrt{3}/4-2\sqrt{5}/3$
1832 7945 tπ : $\sqrt{3}/4-2\sqrt{5}/3$
1833 0839 $\ln(3/4)\times\exp(\sqrt{2})$
1833 2417 rt : (2,7,-5,3)
1833 2621 J_0 : $\exp(2e/3)$
1833 3733 2^x : $4+\sqrt{2}/3$
1833 4415 as : $(\sqrt[3]{2})^{-1/3}$
1833 7951 J : $\ln 2/3$
1834 0232 rt : (8,1,5,-1)
1834 1516 rt : (5,8,5,3)
1834 1657 cu : $\sqrt{2}/2+4\sqrt{5}/3$
1834 2004 $\exp(\sqrt{2})\times s\pi(e)$
1834 3130 Γ : $(\sqrt[3]{3}/3)^{1/3}$
1834 4082 rt : (7,4,0,6)
1834 4635 rt : (9,0,-5,9)
1834 5492 rt : (2,8,3,-7)
1834 6273 $\exp(\pi)\times\exp(\sqrt{2})$
1834 6768 at : $4/3+\sqrt{5}/2$
1834 7732 10^x : $6e$
1835 0341 rt : (6,9,-4,-1)
1835 4678 2^x : $\exp(-\sqrt{2})$
1835 5446 J_2 : $2(e+\pi)/9$
1835 7628 rt : (9,5,-5,2)
1836 0903 sπ : $4\sqrt{2}-3\sqrt{3}/2$
1836 1290 J : $e/8$
1836 1527 id : P_{63}
1836 1789 rt : (4,0,2,9)
1836 2509 10^x : $3/2-\sqrt{5}$
1836 3503 rt : (8,3,-8,-8)
1836 4729 sr : $\sqrt{2}+3\sqrt{5}/2$
1836 6484 rt : (3,1,-8,9)
1836 6889 rt : (7,9,9,-2)
1836 6927 cu : $e/3-3\pi$
1836 7346 sq : 3/7
1836 7581 rt : (8,7,-7,1)
1837 3184 2^x : $3\sqrt{2}+\sqrt{5}$
1837 3207 sinh : 15/8
1837 4010 rt : (1,8,3,-6)
1837 4746 rt : (5,8,-5,3)
1837 4951 sπ : 1/17
1837 5052 tπ : $1/2+3\sqrt{2}/4$
1837 5181 rt : (6,1,-3,5)
1837 6124 e^x : $\sqrt[3]{25/2}$
1837 6874 rt : (7,9,0,-1)
1837 7274 sq : $4e/3-3\pi/2$
1837 8069 e^x : $3/4+\pi/2$
1837 8365 sπ : $\exp(\sqrt{5}/2)$
1837 9096 e^x : $\sqrt[3]{23/3}$
1838 0201 $\sqrt{5}+\exp(2/3)$
1838 2034 sq : $1/4+3\sqrt{2}$
1838 5351 sq : $8\pi/3$
1838 5611 Ψ : $8(e+\pi)/5$
1838 6264 id : $\sqrt{2}-3\sqrt{3}/2$
1838 6836 id : P_{47}
1838 7237 rt : (8,2,-8,1)
1838 8253 ln : $5\sqrt[3]{3}/6$
1839 0959 rt : (1,8,-4,-1)
1839 2204 e^x : $4e-4\pi$
1839 2582 $\ln(2/5)\div s\pi(e)$
1839 3601 at : $1/3+3\sqrt{2}/2$
1839 3647 $\sqrt{2}-\exp(4)$
1839 3822 rt : (1,7,-8,-2)
1839 3972 $1/2\div e$
1839 4869 rt : (3,6,2,7)
1839 4928 10^x : $4\sqrt{3}/3-\sqrt{5}$
1839 6949 10^x : $e/2-2\pi/3$
1839 7120 rt : (2,4,7,6)
1839 8839 ℓ_{10} : $\sqrt{7/3}$
1840 0000 cu : 22/5
1840 0400 tπ : $\sqrt{2}-2\sqrt{5}$
1840 2862 sr : $4-3\sqrt{3}/2$
1840 3237 cu : $\sqrt{3}/2+\sqrt{5}/3$
1840 3417 ln : $\zeta(3)$
1840 5004 as : $1/4-\sqrt{3}/4$
1840 5490 rt : (5,4,6,1)
1840 5640 rt : (3,8,-4,3)
1840 6579 rt : (5,2,-6,-4)
1840 9925 ln : $9\pi^2/10$
1841 0780 J_1 : $1/\sqrt[3]{19}$
1841 1537 rt : (5,6,-8,6)
1841 1543 rt : (9,4,-5,-1)
1841 2674 rt : (9,3,-4,6)
1841 3712 at : $4\sqrt{3}-2\sqrt{5}$
1841 4804 cr : $\sqrt{11}\pi$
1841 4849 rt : (6,9,9,8)
1841 4921 $\ln(3)+\exp(3)$
1841 8230 J_2 : $(\sqrt{2}\pi/2)^{1/3}$
1841 9136 ln : $6\ln 2/5$
1842 0327 2^x : $5\sqrt[3]{3}$
1842 0824 sq : $3\sqrt{2}/2+\sqrt{5}/4$
1842 1012 rt : (1,8,6,1)
1842 1579 sq : $2-\pi/2$
1842 2335 J_1 : 3/8
1842 2958 sr : $7\zeta(3)/6$
1842 3403 rt : (6,2,-6,1)
1842 5556 ℓ_{10} : $e+4\pi$
1842 5873 cu : $7(e+\pi)/2$
1842 7283 tπ : $4e/3-4\pi$
1842 8293 rt : (2,6,-4,-7)
1842 8737 rt : (6,1,-2,-9)
1842 9257 ℓ_{10} : $2-\sqrt{2}/3$
1842 9920 rt : (4,7,4,-1)
1843 2493 $4/5\div s\pi(\sqrt{5})$
1843 4692 sπ : $1/2+\sqrt{5}/4$
1843 5002 Γ : 18/23
1843 6090 rt : (8,0,-7,5)
1843 8282 rt : (4,7,-7,-7)
1843 8717 rt : (2,6,-8,8)
1843 9878 10^x : $2\sqrt{2}/3-3\sqrt{5}/4$
1844 0262 e^x : $e+\pi/3$
1844 1002 rt : (3,8,-7,1)
1844 1717 e^x : $4/3+2\sqrt{2}$
1844 1727 rt : (7,1,5,-1)
1844 2457 ℓ_2 : 11/25
1844 2917 ℓ_2 : $4\sqrt{3}-3\sqrt{5}$
1844 3263 10^x : $4\sqrt[3]{2}/5$
1844 3955 rt : (7,9,-4,-1)
1844 5065 ln : $\sqrt{8}\pi$
1844 7391 tπ : $9e/8$
1844 9011 e^x : $(\sqrt{5}/3)^{-1/2}$
1845 0334 2^x : $\sqrt[3]{5}/7$
1845 1451 $\exp(1/4)\div s\pi(1/5)$
1845 1788 cr : $1+3\pi$
1845 1796 ℓ_{10} : $2\sqrt{2}-3\sqrt{3}/4$
1845 3077 sq : $1/4-e/4$
1845 3264 rt : (5,4,-5,-8)
1845 3841 2^x : $e-\pi/3$
1845 4159 rt : (6,6,-3,-2)
1845 7160 2^x : $\sqrt[3]{14/3}$
1845 7974 ℓ_{10} : $4/3-e/4$
1845 8291 ln : P_{25}
1845 8790 rt : (4,3,-5,-1)
1845 9170 J : $\ln(4/3)$
1845 9310 J_2 : 23/5
1846 0667 e^x : $2/3-3\pi/4$
1846 2068 rt : (1,9,-7,-6)
1846 3524 rt : (5,9,9,-2)
1846 4815 rt : (1,8,-2,-4)
1846 5618 id : $4\sqrt{2}/3+3\sqrt{3}/4$
1846 5852 e^x : $4+\sqrt{3}/4$
1846 6299 as : P_{63}
1846 7304 rt : (9,6,4,-1)
1846 7614 rt : (8,2,-6,-9)
1846 8594 ℓ_{10} : $4e/3-2\pi/3$
1846 9679 sq : $10\pi/9$
1847 1089 J_1 : $\sqrt{21}\pi$
1847 1758 id : $9(e+\pi)/4$
1847 1829 id : $4\sqrt{2}-2\sqrt{5}$
1847 2311 2^x : $3-2e$
1847 2785 rt : (8,7,-8,-6)
1847 5015 rt : (7,7,1,3)
1847 7978 rt : (4,7,-8,-7)
1847 9253 rt : (1,3,-6,1)
1848 2790 rt : (5,9,-1,7)
1848 3167 rt : (2,8,-4,-1)
1848 3312 Li_2 : $(\sqrt[3]{2}/3)^2$
1848 3519 rt : (8,8,5,8)
1848 3594 10^x : $(3\pi/2)^{1/2}$
1848 4885 rt : (9,8,-4,-1)
1848 7050 J_2 : $(\sqrt[3]{5})^{1/2}$
1848 7099 rt : (7,5,-9,-8)
1848 7817 J_2 : 17/13
1849 2047 as : P_{47}
1849 2160 e^x : $3+2\sqrt{5}/3$
1849 2316 sinh : $(\sqrt[3]{2}/3)^{-3}$
1849 3510 Li_2 : 3/17
1849 3816 e^x : $e/3+4\pi/3$
1849 4456 sq : $3\sqrt{2}/4-2\sqrt{5}/3$
1849 4933 e^x : $(2\pi/3)^{-1/3}$
1849 7187 $\ln(2)+\exp(2/5)$
1849 8217 at : $e/4-\pi$
1850 1517 Ψ : 11/24
1850 3603 J_0 : 14/5
1850 3737 rt : (1,2,5,-1)
1850 3986 tπ : $2(1/\pi)$
1850 4912 cr : $1/4+\sqrt{2}$
1850 6405 Ei : $(4/3)^{-3}$
1850 6770 Ψ : $(1/\pi)/5$
1850 7016 sq : $2+2\sqrt{5}/3$
1850 8110 $\ln(2/5)\div\ln(3/4)$
1850 9336 rt : (2,9,7,1)
1851 0084 E : 8/9
1851 0224 sr : $(e+\pi)^3$
1851 1028 Ei : $\exp(-\sqrt{5}/2)$
1851 1611 at : $4\sqrt{2}/3+\sqrt{3}/3$
1851 1799 $\sqrt{3}\div\exp(\sqrt{5})$
1851 2937 rt : (4,9,9,-2)
1851 5642 sq : $2/3+\sqrt{5}/2$
1851 5673 rt : (7,5,2,7)
1851 7903 10^x : $\ln(\sqrt[3]{3}/3)$
1851 8518 id : $(\sqrt[3]{5}/3)^3$
1851 8970 sr : $4/3-3\sqrt{3}/4$
1852 0063 sr : $1/4+2\sqrt{3}/3$
1852 0095 ln : $-\sqrt{2}+\sqrt{5}+\sqrt{6}$
1852 0726 rt : (9,6,-3,3)
1852 6605 2^x : $-\sqrt{5}+\sqrt{6}-\sqrt{7}$
1852 7191 Li_2 : $\sqrt{2}/8$
1852 7406 rt : (8,5,-7,-2)
1852 7535 Ei : 5
1852 7642 at : $1-2\sqrt{3}$
1852 8275 e^x : $(e+\pi)^2$
1853 0529 $\exp(1/5)-s\pi(\sqrt{2})$
1853 1915 rt : (4,4,6,1)
1853 2586 erf : $(\ln 3/2)^3$
1853 2706 rt : (9,8,9,-2)
1853 3147 10^x : $1-\sqrt{3}$
1853 3196 rt : (5,3,-6,-3)
1853 3319 tanh : 3/16

1853 4028 $Li_2 : 5(1/\pi)/9$
1853 4632 cu : $8(e+\pi)/5$
1853 4794 at : $3/16$
1853 4940 $\Psi : 5\pi^2/2$
1853 5883 cu : $1+3\sqrt{5}/4$
1853 6698 $\Psi : \sqrt[3]{14/3}$
1854 0365 rt : (8,9,-4,-1)
1854 0720 at : P_9
1854 0991 rt : (9,4,-2,7)
1854 1312 $\ln(3) - \exp(1/4)$
1854 3094 $Li_2 : \exp(-\sqrt{3})$
1854 4809 rt : (9,7,-7,1)
1854 5405 $2^x : 1/3+\sqrt{3}$
1854 5491 $\ell_{10} : 2/3+\sqrt{3}/2$
1854 5859 $t\pi : 5e/7$
1854 6513 rt : (6,1,5,-1)
1854 6730 $10^x : 21/23$
1854 7375 $2^x : 7\pi/4$
1854 8145 rt : (6,2,-1,6)
1854 8300 at : $3\sqrt{2}-3\sqrt{5}$
1854 8334 sq : $2\sqrt{3}+\sqrt{5}/4$
1854 8846 $K : 7/8$
1855 1551 rt : (8,3,-6,2)
1855 1798 $\Psi : 9/7$
1855 3273 $s\pi : 3/4+4\sqrt{3}/3$
1855 4965 cu : $\sqrt{2}/3-4\sqrt{3}$
1855 6230 $\Gamma : (2\pi/3)^{-1/3}$
1855 6716 rt : (8,6,4,-1)
1855 7464 at : $\sqrt[3]{15}$
1855 7851 $t\pi : \sqrt{2}/4-4\sqrt{3}$
1855 8421 rt : (1,1,5,-9)
1855 8754 at : $1/\ln(2/3)$
1855 9117 $Ei : 3\sqrt{3}/7$
1855 9261 $\zeta : 19/23$
1856 1074 rt : (7,2,-6,1)
1856 2463 id : $5(e+\pi)/7$
1856 2758 $J_1 : 1/\sqrt{7}$
1856 3110 id : $\sqrt[3]{5/3}$
1856 3318 $J_1 : 3\sqrt[3]{2}/10$
1856 3657 $\ell_{10} : 15/23$
1856 3678 cu : $1+\sqrt{2}/3$
1856 4944 at : $2-2e/3$
1856 5170 rt : (8,1,-5,6)
1856 7071 rt : (5,1,-5,1)
1856 9921 sq : $\sqrt{3}+\sqrt{5}+\sqrt{6}$
1857 1168 rt : (5,3,-5,-1)
1857 4208 sr : $2e+3\pi/2$
1857 4246 at : $\pi^2/4$
1857 4715 $\zeta : 3\sqrt[3]{5}/7$
1857 8037 rt : (3,8,-4,-1)
1857 8832 sr : $1/2+e/3$
1857 9019 rt : (3,2,1,-9)
1858 0941 rt : (5,1,-4,-5)
1858 1598 rt : (9,5,6,1)
1858 2452 rt : (8,8,9,-2)
1858 2663 rt : (7,0,-9,1)
1858 5102 $2^x : 4/3+4e$
1858 5333 $t\pi : 3\sqrt{3}+\sqrt{5}/3$
1858 6667 at : $1/4-e$
1858 7087 $J_0 : \sqrt[3]{22}$
1858 7869 cr : $3/2+4\sqrt{5}$
1858 8488 rt : (5,3,-3,-9)
1858 8694 $2^x : 4(e+\pi)/7$
1858 9528 rt : (7,2,-8,-5)
1859 0890 at : $3\sqrt{2}/2-4\sqrt{3}/3$
1859 1691 rt : (1,5,0,5)
1859 2984 $e^x : 3\sqrt{3}/4-4\sqrt{5}/3$
1859 3528 $\exp(4)+s\pi(1/5)$
1859 3929 rt : (7,4,-7,-9)
1859 7026 cu : $1-\pi/2$
1859 7101 $10^x : (\sqrt[3]{2}/3)^3$
1860 3726 $J_2 : 21/16$
1860 5546 $\theta_3 : 1/\ln(1/(3\pi))$
1860 8535 cu : $3\sqrt{2}+3\sqrt{3}/4$
1861 2301 $J_2 : 10\sqrt[3]{2}$
1861 4066 rt : (6,9,-9,0)
1861 5743 $s\pi(1/5)-s\pi(e)$
1861 6792 $10^x : 10\sqrt{3}$
1861 7211 at : P_{36}
1861 9872 $J_2 : 16$
1862 2101 $\ell_{10} : 3/4+\pi/4$
1862 2308 $e^x : 1/2+3\sqrt{5}/2$
1862 4544 sinh : $(\sqrt[3]{5}/3)^3$
1862 5557 $J_1 : 1/\ln(\sqrt[3]{5}/2)$
1862 5944 sinh : $\exp(3/2)$
1862 6030 as : $(\sqrt[3]{5}/3)^3$
1862 6289 $e^x : \sqrt{5}\pi$
1862 9436 $1/5-\ln(4)$
1863 0648 at : $2+\sqrt{2}/3$
1863 1591 $\ln(3/5)-s\pi(\sqrt{5})$
1863 2385 $t\pi : 1/\ln(\sqrt[3]{3}/2)$
1863 2959 rt : (7,8,9,-2)
1863 3087 $J_1 : 8(e+\pi)/5$
1863 3162 $\ell_{10} : \sqrt{3}+\sqrt{6}-\sqrt{7}$
1863 3628 $erf : 1/6$
1863 5419 $\ell_{10} : (e+\pi)/9$
1863 5435 $\Gamma : 1/\sqrt{21}$
1863 8068 $t\pi : 4\sqrt{2}/3-4\sqrt{5}$
1863 9742 rt : (9,9,-4,-1)
1863 9845 rt : (3,9,-9,-7)
1864 0936 at : $2-2\sqrt{5}$
1864 5810 $\Psi : (\sqrt{2}/3)^{-1/3}$
1864 5870 $1/4\times s\pi(\sqrt{3})$
1864 6935 as : $3/4-3\sqrt{5}/4$
1864 8211 $J : 1/\sqrt[3]{20}$
1864 8540 rt : (7,6,4,-1)
1864 8789 $e^x : \sqrt[3]{5}/10$
1864 9496 rt : (1,8,-5,-7)
1865 0475 $J_1 : 8\sqrt{5}$
1865 4638 $\zeta : \sqrt{19/2}$
1865 4805 rt : (5,1,5,-1)
1865 4945 cu : $6e\pi/7$
1865 5054 $2^x : 4-2\sqrt{3}/3$
1865 6365 id : $1/4-2e$
1865 6473 $J_2 : 23/4$
1865 7039 $\zeta : \ln(\sqrt[3]{5}/3)$
1865 8072 sinh : $10\pi^2/7$
1865 8854 $J_1 : 2\sqrt[3]{5}/9$
1865 8892 cu : $4/7$
1865 9080 $\ell_2 : 3-3\sqrt{2}/2$
1865 9404 at : $4-4\pi/3$
1865 9556 ln : $3/4+3e$
1865 9629 rt : (8,9,7,9)
1866 1471 $10^x : 2\sqrt{2}+3\sqrt{5}/4$
1866 2937 $s\pi : 3e+\pi/4$
1866 2954 ln : $3\sqrt{2}/2+2\sqrt{3}/3$
1866 3560 $\lambda : \exp(-18\pi/13)$
1866 3863 rt : (3,4,6,1)
1866 5829 rt : (1,9,9,-2)
1866 6121 tanh : $\exp(-5/3)$
1866 6682 sr : $2+3e$
1866 6795 $\Gamma : \sqrt{2}-\sqrt{3}+\sqrt{7}$
1866 7650 at : $\exp(-5/3)$
1866 7770 cr : $e-\pi/3$
1866 7997 $10^x : e/8$
1866 8126 cr : $\sqrt[3]{14/3}$
1867 0704 rt : (1,7,1,-7)
1867 1685 $\ell_{10} : 9e\pi/5$
1867 2162 at : $10\sqrt{3}/7$
1867 2593 rt : (9,5,0,8)
1867 5723 rt : (4,8,-4,-1)
1867 6083 at : $\exp(e/3)$
1867 6720 $Ei : 7(1/\pi)/3$
1867 8127 rt : (1,9,7,1)
1867 8449 $t\pi : 4\sqrt{2}-3\sqrt{3}/2$
1867 9398 at : $7\sqrt{2}/4$
1868 0131 at : $2/3-\pi$
1868 2462 cu : $\sqrt{2}/4+\sqrt{5}/2$
1868 3114 $J_2 : 25/19$
1868 3727 $3\times\exp(1/3)$
1868 4252 rt : (6,8,9,-2)
1868 7769 rt : (6,3,-5,-1)
1868 8961 $10^x : \sqrt{14}\pi$
1868 9066 rt : (9,7,-1,4)
1868 9984 $J_2 : (\sqrt{3})^{1/2}$
1869 1259 id : $3\sqrt{2}+4\sqrt{5}$
1869 3230 rt : (7,6,4,8)
1869 3239 $t\pi : 1/17$
1869 4507 rt : (6,8,-8,7)
1869 5699 $t\pi : \ln(\sqrt[3]{4}/3)$
1869 6498 rt : (1,7,6,6)
1869 6834 $t\pi : \exp(\sqrt{5}/2)$
1869 7576 rt : (8,2,-3,7)
1869 9531 $Li_2 : \exp(-1/(2\pi))$
1869 9858 $J_0 : 7\zeta(3)/3$
1870 0000 cu : $9\sqrt[3]{3}/10$
1870 0423 $e^x : e/4-3\pi/4$
1870 1078 sq : $1/3-2e/3$
1870 4167 $J_1 : 8/21$
1870 4494 id : $(2\pi/3)^3$
1870 4608 $\ln(4/5)+s\pi(\sqrt{2})$
1870 4945 $10^x : e+\pi/4$
1870 5475 cu : $18/17$
1870 5702 rt : (8,2,-6,1)
1870 7680 at : $e/3+\pi/2$
1870 8664 $\ell_{10} : 13/20$
1870 8742 $t\pi : 2e/3-3\pi/4$
1870 8965 $J_1 : 10\sqrt[3]{3}$
1870 9120 $Ei : 19/13$
1870 9535 $2^x : \sqrt{3}/7$
1871 0080 rt : (1,7,4,-1)
1871 0178 $\exp(5)+s\pi(e)$
1871 0693 rt : (8,5,6,1)
1871 1017 sq : $3\sqrt{2}+4\sqrt{5}/3$
1871 1311 $Ei : (2e/3)^{-1/2}$
1871 2449 $\sqrt{5}+s\pi(2/5)$
1871 2551 rt : (2,8,-3,-6)
1871 3497 $\ln(3/5)\div s\pi(\pi)$
1871 4918 at : $\sqrt{3}+\sqrt{5}/3$
1871 7049 $e^x : 18/23$
1871 7065 $e^x : 3-2\sqrt{2}$
1871 8427 rt : (9,9,-2,0)
1872 0754 sq : $3\sqrt[3]{3}/10$
1872 0934 $2^x : 7(1/\pi)/9$
1872 1188 rt : (8,4,-4,3)
1872 2749 rt : (7,1,-6,-6)
1872 3251 $\ell_{10} : 9\sqrt[3]{5}$
1872 3752 rt : (6,6,-8,-9)
1872 4645 $5\div s\pi(\sqrt{2})$
1872 5549 $\ell_{10} : \sqrt{24}\pi$
1872 6679 rt : (7,8,3,4)
1872 7175 rt : (6,3,1,7)
1872 7617 rt : (4,5,-5,7)
1872 8372 rt : (1,5,2,-3)
1873 0283 cr : $2\sqrt{2}-2\sqrt{3}/3$
1873 1171 $\zeta : 9\sqrt[3]{5}/5$
1873 1767 rt : (2,4,-6,5)
1873 1834 $Ei : \exp(\sqrt{3}/3)$
1873 3448 $J_1 : 1/\sqrt[3]{18}$
1873 3465 $2^x : 23/3$
1873 4819 $e^x : \zeta(3)/7$
1873 5192 at : $2/3+2e/3$
1873 5821 $2^x : 19/8$
1873 5839 $10^x : 1/3-3\sqrt{2}/4$
1873 6356 rt : (5,8,9,-2)
1873 7766 ln : $(\sqrt[3]{5}/3)^{1/3}$
1874 0460 $2^x : 4\sqrt{2}-\sqrt{3}$
1874 0541 as : P_{34}
1874 0691 $Ei : 8\sqrt[3]{3}/5$
1874 0810 $\ell_{10} : 3\sqrt{3}/8$
1874 1041 sr : $\pi^2/7$
1874 1545 $J_1 : 8e/5$
1874 2368 cu : $9\sqrt{3}/4$
1874 2655 cr : $2(e+\pi)/7$
1874 2905 rt : (6,6,4,-1)
1874 5291 $e\times\ln(\sqrt{5})$
1874 6375 $e^x : (\sqrt{2}\pi/2)^3$
1874 6632 $10^x : 3e/4+\pi/2$
1874 7838 $\Gamma : 1/\sqrt[3]{15}$
1874 8113 rt : (6,4,-9,-5)
1874 8552 sinh : $4\sqrt[3]{2}/5$
1874 9361 $\Gamma : -\sqrt{3}+\sqrt{5}-\sqrt{7}$
1875 0000 id : $3/16$
1875 0708 $\Psi : \ln(\pi/3)$
1875 2387 $J_1 : 6(1/\pi)/5$
1875 3618 rt : (7,1,-7,2)
1875 3938 rt : (4,7,-2,4)
1875 5287 sq : P_{10}
1875 5355 rt : (3,6,-3,-7)
1875 5529 sr : $4+\pi/4$

1875 6128 $\Gamma : \ln(3/2)$
1875 6133 $\mathrm{id} : P_9$
1875 6149 $\mathrm{t}\pi : 1/2 + \sqrt{5}/4$
1875 9920 $\ln(\pi) \div s\pi(\sqrt{2})$
1876 0295 $\mathrm{rt} : (2,6,1,7)$
1876 1454 $\ln(4) \div \exp(2)$
1876 1471 $\mathrm{rt} : (5,0,-2,6)$
1876 1817 $\exp(4/5) - \exp(5)$
1876 2635 $J_2 : \exp(\sqrt{2}\pi/2)$
1876 3380 $\mathrm{rt} : (8,6,-5,-1)$
1876 4660 $\mathrm{sq} : 1 + \pi/4$
1876 6844 $\mathrm{rt} : (4,1,5,-1)$
1876 7695 $\mathrm{sq} : 5\ln 2/8$
1877 0555 $\mathrm{cu} : 1/2 + \sqrt{5}/4$
1877 1122 $\mathrm{at} : 1/2 - 4\sqrt{5}/3$
1877 1884 $\mathrm{rt} : (1,9,-9,-2)$
1877 2263 $\Gamma : 9\ln 2/8$
1877 2527 $10^x : \sqrt{3}/3 + 2\sqrt{5}$
1877 2806 $\mathrm{rt} : (1,6,-1,1)$
1877 4362 $\tanh : \sqrt[3]{5}/9$
1877 5295 $\mathrm{rt} : (6,5,0,3)$
1877 5936 $\mathrm{at} : \sqrt[3]{5}/9$
1877 6397 $\mathrm{rt} : (5,8,-4,-1)$
1877 8525 $2/5 - s\pi(1/5)$
1877 9878 $\mathrm{at} : 3/4 + \sqrt{3}$
1878 0644 $\mathrm{rt} : (3,8,-7,-8)$
1878 0967 $\mathrm{cr} : 10\pi/3$
1878 1211 $\mathrm{id} : 2 - 2e/3$
1878 1572 $\mathrm{t}\pi : 2\ln 2/5$
1878 3364 $erf : (2e/3)^{-3}$
1878 4883 $\mathrm{rt} : (2,7,-6,-1)$
1878 6138 $10^x : \sqrt{2} + \sqrt{5} + \sqrt{7}$
1878 7390 $\ell_2 : P_{76}$
1878 9295 $\mathrm{rt} : (4,8,9,-2)$
1878 9681 $\mathrm{cu} : 1/3 - e/3$
1879 1205 $2^x : 4 + 2e/3$
1879 1916 $\Psi : 2(e + \pi)/7$
1879 2224 $\ln(3/5) \div \exp(1)$
1879 3178 $2^x : \sqrt{5}/9$
1879 4976 $s\pi : 7\sqrt[3]{2}/3$
1879 6125 $\mathrm{rt} : (4,9,1,1)$
1879 6752 $\mathrm{cr} : (3\pi/2)^{1/3}$
1879 8729 $2^x : 2\sqrt{2}/3 - 3\sqrt{5}/2$
1879 9288 $\mathrm{rt} : (5,8,-7,1)$
1879 9488 $J_1 : 18$
1880 1840 $\mathrm{rt} : (2,4,6,1)$
1880 3324 $\mathrm{rt} : (9,0,5,1)$
1880 5151 $e^x : \sqrt{2}/3 + 2\sqrt{3}$
1880 6008 $10^x : \sqrt{2}/3 + \sqrt{3}/3$
1880 6275 $\mathrm{rt} : (9,7,9,-2)$
1880 6319 $\mathrm{sr} : (1/\pi)/9$
1880 7069 $\mathrm{rt} : (2,8,8,9)$
1880 7635 $\mathrm{rt} : (9,6,2,9)$
1880 7898 $\mathrm{rt} : (9,1,-8,-7)$
1880 8073 $\mathrm{id} : 3\sqrt{2}/2 - 4\sqrt{3}/3$
1880 8839 $\mathrm{cr} : 3\sqrt{5}/4$
1880 8925 $\mathrm{rt} : (7,4,-1,0)$
1880 9047 $\ell_{10} : \sqrt[3]{11/3}$
1880 9094 $\mathrm{cu} : 9(1/\pi)/5$
1880 9565 $\ln : (\ln 3)^2$
1881 2591 $\sinh : \zeta(7)$
1881 4331 $\mathrm{at} : 10\sqrt{5}/9$
1881 5624 $\sqrt{2} + s\pi(e)$
1881 6048 $\Gamma : 6e/7$
1881 7705 $\mathrm{sr} : 24/17$
1881 9005 $10^x : 3\sqrt{3} - \sqrt{5}$
1882 0557 $\tanh : 4/21$
1882 2150 $\mathrm{at} : 4/21$
1882 2640 $\ln : 1/2 + \sqrt{2}/2$
1882 4620 $\mathrm{rt} : (5,2,-3,2)$
1882 5212 $\theta_3 : \ln(\ln 3)$
1882 6230 $\mathrm{rt} : (2,9,-9,-2)$
1882 8559 $\mathrm{sr} : 2/3 + \sqrt{5}/3$
1882 9968 $\ell_2 : 7(e + \pi)/9$
1883 0115 $\mathrm{id} : e/3 - 2\pi/3$
1883 0853 $\mathrm{rt} : (1,8,5,-6)$
1883 1144 $\mathrm{rt} : (3,6,-8,-4)$
1883 2534 $\zeta : 15/19$
1883 3098 $2^x : (\sqrt[3]{3})^{1/3}$
1883 3466 $\mathrm{rt} : (8,3,-1,8)$
1883 3643 $\mathrm{t}\pi : 3\sqrt{2} + 3\sqrt{5}/2$
1883 5327 $\mathrm{id} : P_{36}$
1883 5424 $Ei : 2\pi^2/3$
1883 5901 $\mathrm{at} : 1/4 + \sqrt{5}$
1883 8896 $\ell_{10} : (\sqrt[3]{2}/3)^{1/2}$
1883 9952 $\mathrm{rt} : (5,9,7,4)$
1884 0268 $\mathrm{rt} : (2,6,-9,4)$
1884 0295 $e^x : 1/\ln(\ln 3/2)$
1884 1037 $Ei : (e/2)^{1/3}$
1884 2067 $J_2 : \sqrt{3} + \sqrt{5} - \sqrt{7}$
1884 3097 $\mathrm{rt} : (3,8,9,-2)$
1884 3609 $10^x : 4\sqrt{2} - \sqrt{3}/4$
1884 4292 $\mathrm{rt} : (1,7,-3,-8)$
1884 4975 $e^x : 3/4 - \sqrt{3}/3$
1884 5401 $\mathrm{rt} : (7,5,6,1)$
1884 6952 $\mathrm{rt} : (2,8,5,-2)$
1884 8142 $s\pi : 10\sqrt[3]{3}/7$
1884 9136 $J_0 : 8\sqrt{3}$
1884 9555 $\mathrm{rt} : (9,1,-9,3)$
1885 0691 $\mathrm{t}\pi : 1/\sqrt{13}$
1885 0742 $2^x : (\ln 2)^{-1/3}$
1885 1022 $e^x : 2 - \sqrt{2}/4$
1885 4457 $J_2 : \sqrt{6}\pi$
1885 5692 $\Gamma : (\pi/2)^{-2}$
1885 7271 $\mathrm{rt} : (5,8,-9,-9)$
1885 7422 $\mathrm{rt} : (1,7,-9,6)$
1885 8048 $\mathrm{rt} : (9,7,1,0)$
1885 9498 $\mathrm{rt} : (7,0,-4,7)$
1885 9538 $\mathrm{rt} : (8,7,9,-2)$
1885 9879 $\Psi : 6\pi/5$
1886 0056 $\sinh : 3/16$
1886 0142 $\mathrm{rt} : (8,8,-6,-5)$
1886 1638 $\mathrm{as} : 3/16$
1886 2952 $\mathrm{rt} : (9,8,1,5)$
1886 3261 $\mathrm{at} : 4/3 + 2\sqrt{3}/3$
1886 3795 $J_2 : 9\sqrt[3]{5}/2$
1886 3854 $\mathrm{at} : \sqrt{3}/2 - 3\sqrt{5}/2$
1886 3887 $\ell_{10} : 3\sqrt{2}/2 - \sqrt{3}/3$
1886 4315 $10^x : 12/5$
1886 6028 $\mathrm{rt} : (3,9,-4,-7)$
1886 7883 $\mathrm{as} : P_9$
1886 8306 $\mathrm{cr} : 1 + e/4$
1886 9720 $\tanh : 3(1/\pi)/5$
1887 1053 $\mathrm{at} : 3/4 - \sqrt{5}/4$
1887 1183 $e^x : 3e/2 + 4\pi/3$
1887 1335 $\mathrm{at} : 3(1/\pi)/5$
1887 1702 $e^x : 9(1/\pi)/2$
1887 2158 $s\pi : P_{46}$
1887 2278 $\ell_2 : 4 + \sqrt{5}/4$
1887 4631 $10^x : 8\sqrt{5}/9$
1887 4674 $\mathrm{rt} : (4,1,-6,-1)$
1887 5841 $\zeta : (3\pi)^{1/2}$
1887 5956 $\mathrm{cr} : 4\sqrt[3]{2}/3$
1887 7098 $s\pi : 3e - 2\pi/3$
1887 7135 $\mathrm{rt} : (7,7,6,9)$
1887 7355 $J_1 : 5/13$
1887 7551 $\mathrm{sq} : 25/14$
1887 9020 $\mathrm{id} : 4\pi/3$
1887 9620 $\mathrm{sr} : 4\sqrt{2} - \sqrt{3}/2$
1888 0245 $\mathrm{rt} : (6,8,-4,-1)$
1888 0982 $Li_2 : \sqrt[3]{2}/7$
1888 1085 $\mathrm{t}\pi : 1/4 - 4\sqrt{3}/3$
1888 1495 $\mathrm{rt} : (3,9,-9,-2)$
1888 2895 $\mathrm{rt} : (3,1,5,-1)$
1888 4454 $\ell_2 : (\sqrt{2}\pi/3)^{1/3}$
1888 5735 $\mathrm{rt} : (6,3,-7,-6)$
1888 6038 $e^x : (\sqrt[3]{3}/3)^{1/3}$
1888 6506 $\mathrm{cu} : e/2 - \pi/4$
1888 7560 $\mathrm{id} : \exp(-5/3)$
1889 0241 $\mathrm{t}\pi : 4\sqrt{3} + 3\sqrt{5}$
1889 0813 $J_1 : 2\sqrt{3}/9$
1889 3033 $2^x : \sqrt{3}/3 - 4\sqrt{5}/3$
1889 3414 $\mathrm{as} : 2 - 2e/3$
1889 4376 $\mathrm{rt} : (6,7,-1,-1)$
1889 4662 $J_2 : (\sqrt[3]{5}/3)^{-1/2}$
1889 4662 $2/5 \div \exp(3/4)$
1889 6461 $\mathrm{rt} : (8,5,-2,4)$
1889 7789 $\mathrm{rt} : (4,8,2,8)$
1889 9392 $J_1 : 5\ln 2/9$
1890 0085 $\Psi : 8\sqrt{2}/3$
1890 0472 $\mathrm{at} : 1 + 2\sqrt{5}/3$
1890 1315 $2^x : 3\sqrt{2}/4 - 2\sqrt{3}$
1890 1353 $\ell_2 : 2\sqrt[3]{5}/3$
1890 2047 $\mathrm{at} : P_{52}$
1890 3591 $\mathrm{sq} : 10/23$
1890 3787 $\mathrm{rt} : (4,9,0,-6)$
1890 5609 $1/5 - \exp(2)$
1890 5623 $\ell_{10} : 11/17$
1890 6949 $\ell_{10} : 9\zeta(3)/7$
1890 6978 $\ln : 4e/9$
1890 8794 $e^x : 5\zeta(3)/2$
1891 0769 $\mathrm{rt} : (1,7,-4,-1)$
1891 0994 $e^x : \sqrt{3}/10$
1891 2178 $\mathrm{rt} : (5,9,-7,-6)$
1891 2664 $\mathrm{rt} : (6,4,3,8)$
1891 3230 $Li_2 : \sqrt[3]{3}/8$
1891 3664 $\mathrm{rt} : (7,7,9,-2)$
1891 4671 $\mathrm{cu} : \sqrt{22}$
1891 5580 $Ei : 2\sqrt[3]{5}$
1891 6223 $\mathrm{rt} : (5,6,-9,4)$
1891 6897 $\mathrm{rt} : (8,0,5,1)$
1891 7937 $\mathrm{rt} : (9,0,-6,8)$
1891 8502 $\ell_2 : 2\sqrt{2} + \sqrt{3}$
1892 0711 $\mathrm{id} : (\sqrt{2})^{1/2}$
1892 0763 $\mathrm{as} : 3\sqrt{2}/2 - 4\sqrt{3}/3$
1892 1372 $J_0 : 10\sqrt[3]{3}/7$
1892 1516 $\ln : 2 + 4\sqrt{3}$
1892 4685 $\Psi : 3\sqrt[3]{5}/4$
1892 4910 $e^x : (\sqrt{5}/3)^{-3}$
1892 5289 $\mathrm{cr} : 1/\ln(2e/3)$
1892 5478 $\mathrm{rt} : (3,8,-5,0)$
1892 6651 $Ei : 9e/4$
1892 6730 $2^x : 1/3 + 3\sqrt{3}$
1892 7924 $\mathrm{id} : e/3 + 2\pi$
1892 8851 $2^x : \exp(1/(3e))$
1892 9204 $J_2 : (\ln 3)^3$
1893 0302 $\mathrm{rt} : (7,2,-5,3)$
1893 5009 $\mathrm{rt} : (8,3,-5,-1)$
1893 5341 $\Psi : 3\pi$
1893 6399 $\mathrm{rt} : (7,5,-9,-6)$
1893 6546 $\ell_2 : 4\sqrt{2}/3 - \sqrt{5}/3$
1893 7228 $Ei : 2\sqrt[3]{3}$
1893 7700 $\mathrm{at} : 3\sqrt{3}/4 - 2\sqrt{5}/3$
1893 7709 $\mathrm{rt} : (4,9,-9,-2)$
1893 8486 $J_2 : \sqrt[3]{7/3}$
1893 9041 $\mathrm{rt} : (8,3,-9,-7)$
1893 9791 $\Gamma : (2e)^{1/2}$
1893 9835 $\mathrm{rt} : (4,6,4,3)$
1894 0310 $\mathrm{cr} : 7\sqrt[3]{3}/6$
1894 0520 $\mathrm{sr} : 2\sqrt{3} + 3\sqrt{5}$
1894 1816 $s\pi : 3\sqrt{2}/4$
1894 3165 $Ei : 19/7$
1894 4346 $\exp(4/5) - s\pi(\sqrt{2})$
1894 5037 $J_1 : 5\pi^2/4$
1894 6322 $\mathrm{cr} : 7\zeta(3)/5$
1894 6428 $\mathrm{rt} : (1,4,6,1)$
1894 6733 $\ell_{10} : 1 - \sqrt{2}/4$
1894 8513 $\mathrm{as} : P_{36}$
1894 8575 $\mathrm{rt} : (5,1,0,7)$
1894 8833 $\mathrm{rt} : (2,9,-6,6)$
1894 9103 $Li_2 : e\pi/10$
1895 2713 $\ln : 1/4 + \sqrt{3}/3$
1895 3402 $\mathrm{rt} : (1,8,9,-2)$
1895 4576 $\ell_{10} : 1/2 + \pi/3$
1895 5369 $\Gamma : (\sqrt{2}/3)^{1/3}$
1895 6062 $E : (\sqrt[3]{3})^{-1/3}$
1895 8406 $\ln : 7/23$
1895 8683 $2^x : 16/5$
1896 1208 $\mathrm{at} : 1/3 - 2\sqrt{2}$
1896 3611 $\exp(\pi) - s\pi(2/5)$
1896 3733 $J_1 : (3/2)^3$
1896 4484 $\mathrm{rt} : (6,1,-8,-2)$
1896 5175 $Ei : 7\sqrt[3]{2}/2$
1896 5374 $\mathrm{t}\pi : \ln(\ln 2/2)$
1896 5818 $10^x : 19/24$
1896 6678 $E : (\ln 2)^{1/3}$
1896 7532 $\mathrm{rt} : (9,5,4,-1)$
1896 8683 $\mathrm{rt} : (6,7,9,-2)$
1896 8710 $\Psi : (1/(3e))^{-2}$
1896 9263 $\mathrm{rt} : (3,7,-6,7)$

1896 9456 $t\pi : 4/11$
1897 0138 rt : (3,2,-5,-1)
1897 2503 rt : (5,8,-6,1)
1897 3008 rt : (8,4,1,9)
1897 3979 sq : $3 + 4\sqrt{3}/3$
1897 5956 id : $\sqrt[3]{21/2}$
1897 6213 $J_0 : 3e\pi/2$
1897 6980 $\Psi : (3\pi/2)^{1/3}$
1897 8184 $Li_2 : \exp(-\sqrt[3]{5})$
1897 9277 sr : $3/4 + 3\pi$
1898 0440 $\Psi : 8\sqrt[3]{3}/9$
1898 1927 rt : (5,4,-4,-2)
1898 3603 $\ell_{10} : 4 - 3\sqrt{5}/2$
1898 4015 rt : (9,2,-7,4)
1898 4875 rt : (4,2,-3,-6)
1898 5639 rt : (3,1,-9,7)
1898 6231 rt : (6,5,6,1)
1898 6937 rt : (1,6,-1,-8)
1898 7063 cu : $\ln(\ln 2/2)$
1898 7470 rt : (7,8,-4,-1)
1898 7638 $J : \exp(-11\pi/14)$
1898 8160 cu : $5\sqrt{3}/2$
1899 0891 rt : (7,9,5,5)
1899 3005 as : $4 - 4\pi/3$
1899 3370 $e^x : 3 + e/2$
1899 3870 sr : $\sqrt{23}$
1899 4906 rt : (5,9,-9,-2)
1899 5208 $e^x : 4/23$
1899 5984 $s\pi(\sqrt{2}) + s\pi(e)$
1899 6007 $J : 2\zeta(3)/9$
1899 6701 $t\pi : 3e + \pi/4$
1899 8220 $s\pi : 2\sqrt{2}/3 + \sqrt{5}/2$
1899 8610 rt : (6,8,-7,1)
1899 9732 id : $\sqrt[3]{5}/9$
1899 9948 rt : (7,1,-2,8)
1900 0059 sinh : $\exp(-5/3)$
1900 1515 $J_1 : \exp(\sqrt{5})$
1900 1555 $\ln(4/5) + \exp(5)$
1900 1676 cu : $e/2 + 2\pi/3$
1900 1701 as : $\exp(-5/3)$
1900 2698 rt : (1,9,7,3)
1900 2982 sinh : $7\sqrt[3]{3}/10$
1900 3251 rt : (2,5,9,7)
1900 5228 $Ei : 7e\pi/9$
1900 6210 $\ell_{10} : 1/4 + 3\sqrt{3}/4$
1900 7404 sq : P_{29}
1900 7686 cu : $1/3 + 3\sqrt{3}/2$
1900 8264 sq : 12/11
1900 9966 rt : (8,9,-2,0)
1901 0884 tanh : $\sqrt{3}/9$
1901 2166 $Ei : (e/3)^3$
1901 2560 at : $\sqrt{3}/9$
1901 3597 $E : \exp(-1/(3e))$
1901 3772 cu : $8e\pi$
1901 4895 rt : (5,7,-5,8)
1901 5083 $Ei : (3\pi)^{-1/2}$
1901 5118 $\Gamma : 7/9$
1901 5796 rt : (2,7,-4,-1)
1901 6518 $Ei : \sqrt[3]{20}$
1901 8307 $\Psi : 3\sqrt{5}/4$
1901 9180 $\Psi : 1/\ln(\ln 2/2)$
1902 0012 $J_2 : 1/\ln(\sqrt{2}/3)$
1902 0361 $\ln(\sqrt{2}) \div \exp(3/5)$
1902 1130 $1/5 \times s\pi(2/5)$
1902 1573 rt : (3,5,-6,-5)
1902 3807 sr : 17/12
1902 4623 rt : (5,7,9,-2)
1902 6371 $J_2 : 7\sqrt[3]{5}/9$
1902 7106 rt : (6,2,-5,-7)
1902 7313 $\Gamma : (1/\pi)/5$
1902 7361 rt : (3,9,1,9)
1902 7951 $s\pi : 10e/3$
1902 8994 at : 5/2
1902 9080 as : P_{17}
1903 0344 rt : (9,1,-4,9)
1903 0476 sr : $4/3 + 2\sqrt{3}$
1903 1254 sq : $4/3 + 2\sqrt{3}/3$
1903 4499 rt : (7,0,5,1)
1903 5970 $\exp(e) + s\pi(\sqrt{2})$
1903 7500 $\ln(\sqrt{2}) \times \ln(\sqrt{3})$
1903 7627 $2^x : 2\sqrt{2} - 2\sqrt{3}/3$
1903 8295 $e^x : \sqrt{3}/3 - \sqrt{5}$
1903 8575 $J_1 : \sqrt{19}$
1903 8633 $\ell_{10} : \ln(3\pi/2)$
1903 8988 cu : $3 + 2\sqrt{5}$
1904 0654 cr : $4/3 + \sqrt{2}/4$
1904 2719 rt : (9,9,3,6)
1904 4256 $t\pi : 1/3 + 4\sqrt{5}$
1904 4996 $J_2 : \sqrt{21}$
1904 6607 $\zeta : \sqrt[3]{8/3}$
1904 7003 rt : (1,6,-1,-9)
1904 7171 at : $3/4 - 2\sqrt{2}/3$
1904 7619 id : 4/21
1904 9595 rt : (2,9,2,-7)
1905 0737 rt : (1,9,-7,1)
1905 2036 rt : (8,2,-7,-8)
1905 2498 sq : $3\sqrt{3}/2 - \sqrt{5}/2$
1905 2601 $J_1 : e/7$
1905 2675 cr : $3e + 3\pi/4$
1905 3120 rt : (6,9,-9,-2)
1905 3510 $s\pi(1/3) - s\pi(\sqrt{5})$
1905 3613 $\pi - s\pi(2/5)$
1905 4484 rt : (5,5,-8,-6)
1905 4512 as : 13/14
1905 5078 sr : $9\sqrt[3]{2}/8$
1905 6284 ln : $3e + \pi/4$
1905 6796 $s\pi : 10(1/\pi)/3$
1905 7190 $2^x : 3/4 - \pi$
1905 8234 rt : (7,4,-6,-1)
1905 8677 rt : (2,8,-9,-4)
1906 0136 $10^x : 1/4 + \pi/2$
1906 0139 rt : (5,3,-6,1)
1906 1002 $\theta_3 : 3\sqrt{3}/10$
1906 2487 $\ell_{10} : \sqrt{3}/4 + \sqrt{5}/2$
1906 3103 $\zeta : \sqrt[3]{2}/3$
1906 3934 $\Gamma : 7/3$
1906 4073 $\theta_3 : 2/21$
1906 6028 $e^x : 23/4$
1906 6442 rt : (9,3,-5,-1)
1906 6798 $2^x : 4\sqrt{2}/5$
1906 7514 rt : (1,8,-9,-2)
1906 8330 rt : (8,5,4,-1)
1907 0716 cr : $2\sqrt{2}/3 + \sqrt{5}/3$
1907 1680 rt : (4,1,-7,-3)
1907 2201 $10^x : 21/17$
1907 2521 $J_2 : 9e\pi/10$
1907 3805 $e^x : 4 - 4\sqrt{2}$
1907 4000 $e^x : 1/4 + 3\pi$
1907 4069 cr : $3\sqrt{2}/2 - \sqrt{3}/4$
1907 4563 rt : (3,3,-6,4)
1907 4638 id : $4e/3 + 4\pi$
1907 7712 rt : (8,6,0,5)
1907 8570 sq : $1/\sqrt[3]{12}$
1907 9169 $J_1 : 7/18$
1907 9507 $J_0 : \sqrt{12}\pi$
1907 9558 $\zeta : \ln(\sqrt{3}/3)$
1908 0218 $J_1 : 1/\sqrt[3]{17}$
1908 0283 $\zeta : \exp(\sqrt{5}/2)$
1908 1515 rt : (4,7,9,-2)
1908 2788 $Li_2 : 2/11$
1908 4381 $2^x : \sqrt[3]{2}/5$
1908 4748 ln : $\pi^2/3$
1908 6643 rt : (9,4,-8,0)
1908 6834 $10^x : \sqrt{2}/2 + 3\sqrt{5}/4$
1908 9023 sr : 24/5
1909 0380 rt : (5,9,-2,5)
1909 0741 $\theta_3 : (\sqrt{5}/2)^{-1/3}$
1909 0866 $Li_2 : 4(1/\pi)/7$
1909 1860 $10^x : (2e/3)^{1/2}$
1909 2822 sinh : $5\sqrt{2}/7$
1909 4462 $\zeta : 9e/8$
1909 5138 rt : (9,6,9,-2)
1909 5344 rt : (5,3,-1,3)
1909 5840 $10^x : ((e + \pi)/3)^3$
1909 6150 $E : 5\sqrt{2}/8$
1909 8300 id : $3/4 - \sqrt{5}/4$
1909 8593 id : $3(1/\pi)/5$
1910 1331 ln : $4\sqrt{5}$
1910 1781 sq : $1 + \pi/3$
1910 3546 rt : (2,9,-6,-9)
1910 4100 $e^x : 2\sqrt{2}/3 - 3\sqrt{3}/2$
1910 5046 rt : (6,5,5,9)
1910 5523 ln : 19/23
1910 5937 sq : $1/\ln(5/2)$
1910 7001 $\sqrt{2} + \ln(4/5)$
1910 7177 rt : (6,1,-9,-2)
1910 7619 $J_2 : 4/3$
1910 8587 $erf : \sqrt[3]{5}/10$
1911 1008 $s\pi : \sqrt{2}/2 + 3\sqrt{5}/2$
1911 1331 $s\pi : 3/4 + 4\pi/3$
1911 2357 $t\pi : \sqrt{2} - 3\sqrt{5}/2$
1911 2388 rt : (7,9,-9,-2)
1911 3066 rt : (7,3,-3,4)
1911 4251 sinh : $\sqrt[3]{5}/9$
1911 5224 $\theta_3 : 3(1/\pi)/10$
1911 5663 $e^x : 3/4 + 2\pi/3$
1911 5942 as : $\sqrt[3]{5}/9$
1912 0185 at : $3e/2 - \pi/2$
1912 0210 at : $(2\pi)^{1/2}$
1912 1962 rt : (9,3,-5,5)
1912 4210 rt : (3,7,-4,-1)
1912 4531 $\ell_{10} : 3 - 3\pi/4$
1912 5304 $Ei : 7\sqrt[3]{3}/4$
1912 5643 $\ell_{10} : 4e/7$
1912 6081 rt : (2,8,-9,-2)
1912 6750 $\Gamma : 2\sqrt{5}$
1912 6850 sr : $3\sqrt{2} + \sqrt{5}/4$
1912 7911 cr : $4 - 4\sqrt{3}/3$
1912 8030 $J_0 : \exp(\sqrt[3]{3}/2)$
1912 8244 rt : (1,2,6,6)
1913 0141 $10^x : 2 - e$
1913 0427 id : P_{52}
1913 0578 $10^x : 2\sqrt[3]{2}/5$
1913 0734 $e^x : 1 + \sqrt{3}/4$
1913 1090 $\ln(\sqrt{5}) \div s\pi(\sqrt{5})$
1913 3454 $J_2 : 4\sqrt[3]{3}$
1913 3767 rt : (5,5,6,1)
1913 3931 $J_1 : 9\sqrt{2}/2$
1913 4794 $\Gamma : 9\sqrt{5}/5$
1913 6006 $t\pi : 7\sqrt[3]{2}/3$
1913 6708 id : $4\sqrt{3}/3 - \sqrt{5}/2$
1913 7117 rt : (6,8,-9,5)
1913 9299 $\sqrt{2} \div \exp(2)$
1913 9392 rt : (3,7,9,-2)
1913 9669 $J : \exp(-17\pi/19)$
1913 9956 $\theta_3 : 3\ln 2/4$
1913 9969 $\theta_3 : (\sqrt[3]{5})^{-1/3}$
1914 0625 sq : 7/16
1914 2697 rt : (5,2,2,8)
1914 4282 rt : (7,2,0,9)
1914 5158 $2^x : 1/3 - e$
1914 8511 sr : $2\sqrt{3}/3 - \sqrt{5}/2$
1914 8788 rt : (8,0,-8,4)
1914 8887 cr : $e/3 + \pi/4$
1915 0274 ln : $3e/2 - \pi/4$
1915 1677 $\Gamma : 2e/7$
1915 2337 rt : (8,6,9,-2)
1915 2467 $\exp(e) \div \exp(1/2)$
1915 2815 $\ln(2/3) \div \exp(3/4)$
1915 3363 $2^x : 2(e + \pi)/7$
1915 6347 $J_2 : (\sqrt[3]{2}/3)^{-1/3}$
1915 6423 rt : (6,0,5,1)
1915 6619 $t\pi : 4\sqrt{5}/7$
1915 6696 at : $4 - 2\sqrt{5}/3$
1915 8451 $\Psi : 3\sqrt[3]{2}$
1916 2349 cu : $3e/2 + \pi/2$
1916 3006 sinh : 4/21
1916 3052 $J_2 : 10\zeta(3)/9$
1916 4719 as : 4/21
1916 4799 rt : (1,9,-1,7)
1916 6718 at : $\sqrt{2}/3 - 4\sqrt{5}/3$
1916 7387 id : $3\sqrt{3}/4 - 2\sqrt{5}/3$
1916 7520 rt : (8,1,-5,-9)
1916 8034 cr : 22/13
1917 1943 cr : $1/3 + e/2$
1917 2126 rt : (7,5,4,-1)
1917 2415 rt : (6,1,-3,-8)
1917 2749 rt : (8,9,-9,-2)
1917 4428 $10^x : 1/3 + \pi/2$
1917 4530 at : $3\sqrt{2} - \sqrt{3}$

1917 5359 $t\pi : 5/18$
1917 5634 $\exp(1/5) - \exp(5)$
1917 5930 $\mathrm{at} : (e/2)^3$
1917 6657 $\ln : 4 - \sqrt{2}/2$
1917 7463 $\Gamma : 10e/7$
1917 7466 $\lambda : \exp(-11\pi/8)$
1917 7510 $J_1 : 9\pi^2/8$
1917 7955 $Li_2 : \sqrt[3]{5}/2$
1917 8804 $2/3 \times \ln(3/4)$
1917 9801 $\mathrm{cr} : 3/4 + 2\sqrt{2}/3$
1918 0774 $\mathrm{rt} : (4,1,-1,-7)$
1918 1391 $\mathrm{sq} : 1 - 3e$
1918 3588 $\mathrm{sq} : 2\sqrt{2} + 4\sqrt{3}$
1918 4886 $\mathrm{rt} : (5,3,-9,-2)$
1918 4944 $\theta_3 : 13/25$
1918 5275 $\mathrm{at} : 3(e+\pi)/7$
1918 5703 $\mathrm{rt} : (3,8,-9,-2)$
1918 7736 $\mathrm{cr} : \ln(2e)$
1918 8007 $erf : \zeta(3)/7$
1918 8552 $\ell_{10} : 14/9$
1918 9974 $\mathrm{rt} : (7,4,-8,-8)$
1919 1698 $\theta_3 : (\sqrt[3]{3}/2)^2$
1919 1835 $\sinh : 14/5$
1919 2127 $t\pi : 10\sqrt[3]{3}/7$
1919 2626 $10^x : 4\sqrt{2} + 3\sqrt{5}/4$
1919 3122 $J_1 : 9/23$
1919 4407 $\zeta : 5\sqrt{2}/6$
1919 8288 $\mathrm{rt} : (2,7,9,-2)$
1920 0000 $\mathrm{cu} : 2\sqrt[3]{3}/5$
1920 0782 $\mathrm{rt} : (5,4,-6,-7)$
1920 3078 $\mathrm{rt} : (3,6,-4,-6)$
1920 6515 $Ei : 8\ln 2/5$
1920 7561 $J_0 : 10e\pi/9$
1921 0520 $\mathrm{rt} : (7,6,9,-2)$
1921 1251 $\mathrm{at} : 4\pi/5$
1921 2680 $\mathrm{cu} : 3\sqrt{3}/4$
1921 2983 $\mathrm{rt} : (9,8,-4,-1)$
1921 3806 $\mathrm{sr} : 1/3 + 2\sqrt{5}$
1921 4505 $\mathrm{rt} : (7,8,-7,1)$
1921 4910 $\sinh : 3(1/\pi)/5$
1921 6348 $\mathrm{as} : 3/4 - \sqrt{5}/4$
1921 6646 $\mathrm{as} : 3(1/\pi)/5$
1921 7484 $t\pi : P_{46}$
1921 8736 $\ell_{10} : 3 + 4\pi$
1921 9285 $\mathrm{rt} : (3,5,-3,1)$
1922 0672 $\sinh : \exp(\sqrt[3]{2}/2)$
1922 2700 $t\pi : 3e - 2\pi/3$
1922 3140 $\mathrm{at} : 3\sqrt{2}/4 - \sqrt{3}/2$
1922 4274 $\mathrm{rt} : (6,9,-5,9)$
1922 5325 $\mathrm{rt} : (6,3,-6,1)$
1922 8130 $\mathrm{rt} : (3,4,-2,8)$
1922 9390 $\mathrm{rt} : (4,7,-9,-6)$
1922 9731 $e^x : P_{14}$
1923 0811 $J_0 : 4e$
1923 1551 $\mathrm{cu} : \gamma$
1923 1860 $\mathrm{rt} : (5,2,-5,-1)$
1923 1884 $J_2 : \sqrt{19}\pi$
1923 4243 $\mathrm{rt} : (9,9,-9,-2)$
1923 6176 $t\pi : 8(1/\pi)/7$
1923 6232 $\mathrm{rt} : (4,7,-4,-1)$
1923 6965 $Ei : \sqrt[3]{17/3}$
1923 8136 $\mathrm{rt} : (4,3,-9,8)$
1924 0205 $\mathrm{cr} : \sqrt{3}/3 + \sqrt{5}/2$
1924 0973 $t\pi : 1/4 + 2\sqrt{5}$
1924 1695 $e^x : 4/3 - 4\sqrt{5}/3$
1924 2291 $\ln : 6\sqrt{2}/7$
1924 2694 $\mathrm{at} : 9\sqrt{5}/8$
1924 4553 $\ell_2 : 3 + \pi/2$
1924 5008 $\mathrm{id} : \sqrt{3}/9$
1924 5429 $J_2 : 10\sqrt{3}/3$
1924 6309 $\mathrm{cu} : \sqrt{2} - \sqrt{3} - \sqrt{6}$
1924 6418 $\mathrm{rt} : (4,8,-9,-2)$
1924 7640 $\Psi : 4\sqrt[3]{2}/3$
1924 9079 $\mathrm{as} : P_{52}$
1924 9358 $2^x : 9\sqrt{5}/2$
1925 0393 $\zeta : \exp(-\sqrt{3}/2)$
1925 3375 $\zeta : 1/\ln(1/(2\pi))$
1925 3945 $\zeta : 11/15$
1925 6236 $2^x : 2e/3 - 4\pi/3$
1925 6813 $\mathrm{at} : \sqrt{19/3}$
1925 8240 $\mathrm{rt} : (2,9,-7,1)$
1925 8511 $J_1 : \ln(\sqrt{2}\pi/3)$
1925 8883 $J_1 : \pi/8$
1925 9070 $J_0 : 7\pi^2/5$
1926 0086 $\mathrm{sr} : 10\sqrt[3]{3}/3$
1926 3120 $\ln : 7\sqrt{3}/10$
1926 3559 $\mathrm{rt} : (9,4,-3,6)$
1926 4507 $\ell_2 : 7/8$
1926 4957 $\mathrm{rt} : (9,5,-6,1)$
1926 5316 $\mathrm{rt} : (8,7,2,6)$
1926 7299 $2^x : 2e + \pi$
1926 7354 $\mathrm{rt} : (9,6,-9,-4)$
1926 8027 $\mathrm{rt} : (8,8,-1,1)$
1926 8492 $\ln(2/3) + \exp(4)$
1926 9722 $\mathrm{rt} : (6,6,9,-2)$
1927 1008 $s\pi : P_{62}$
1927 2119 $\ell_{10} : 4e + 3\pi/2$
1927 2259 $\mathrm{cu} : 5\ln 2/6$
1927 3268 $E : 15/17$
1927 4104 $\mathrm{rt} : (8,9,-4,-4)$
1927 4881 $\mathrm{sr} : 2 - \sqrt{3}/3$
1927 5263 $\mathrm{id} : 4\sqrt{2} - 2\sqrt{3}$
1927 6711 $\mathrm{sr} : 4\zeta(3)$
1927 6929 $\Psi : (2\pi/3)^{1/3}$
1927 8029 $\exp(3) \div s\pi(1/3)$
1927 8050 $1/5 \times s\pi(\sqrt{2})$
1927 9096 $\mathrm{rt} : (6,5,4,-1)$
1927 9391 $\mathrm{cr} : 6\sqrt{2}/5$
1927 9604 $10^x : 5\pi$
1928 0143 $s\pi : \sqrt{2}/4 + 3\sqrt{5}$
1928 0313 $\ell_{10} : 9\sqrt{3}$
1928 0615 $\mathrm{sr} : 3(e+\pi)$
1928 0904 $\mathrm{id} : 1/4 + 2\sqrt{2}/3$
1928 1080 $\sqrt{2} - \exp(1/5)$
1928 2993 $\mathrm{rt} : (5,0,5,1)$
1928 3580 $\mathrm{at} : 2/3 - \sqrt{2}/3$
1928 3906 $\ln(3/4) \div \exp(2/5)$
1928 4748 $\Gamma : 10\pi/3$
1928 5084 $\ell_{10} : 1 + \sqrt{5}/4$
1928 6147 $J_2 : 7\pi^2/9$
1928 6736 $\mathrm{as} : 3\sqrt{3}/4 - 2\sqrt{5}/3$
1928 8512 $\mathrm{sq} : 4\sqrt{2} + 4\sqrt{5}$
1928 8682 $\mathrm{rt} : (4,5,6,1)$
1928 8875 $e^x : (\sqrt[3]{2}/3)^2$
1929 1051 $t\pi : 3\sqrt{2}/4$
1929 1309 $\mathrm{rt} : (8,1,-6,5)$
1929 1868 $10^x : 8\pi/9$
1929 2257 $2^x : e\pi/3$
1929 3050 $\ln : 3\sqrt{2}/4 + \sqrt{5}$
1929 4646 $\zeta : 18/11$
1929 5859 $\exp(2/3) - \exp(\pi)$
1929 7873 $\ell_{10} : 3\sqrt[3]{5}/8$
1929 9933 $e^x : 3/17$
1930 0046 $\mathrm{rt} : (8,2,-9,0)$
1930 0797 $\ell_{10} : 9\ln 2/4$
1930 0818 $\mathrm{at} : 2\sqrt[3]{2}$
1930 1677 $\mathrm{sr} : e\pi/6$
1930 1948 $\mathrm{sq} : 3/2 - 3\sqrt{2}/4$
1930 2293 $\mathrm{rt} : (7,4,-1,5)$
1930 4131 $2^x : 1 - \sqrt{5}/3$
1930 4454 $2^x : 4(1/\pi)/5$
1930 7561 $\mathrm{cr} : 9(e+\pi)/5$
1930 8268 $\mathrm{rt} : (5,8,-9,-2)$
1930 8697 $\mathrm{rt} : (7,3,-6,-9)$
1930 8716 $\ln : \ln(\sqrt{5}/2)$
1930 9253 $J_2 : 3\sqrt{5}/5$
1931 0286 $\ln : 7\ln 2/4$
1931 1807 $J_0 : 2\pi^2/7$
1931 3516 $t\pi : 3\sqrt{2}/2 - \sqrt{3}/3$
1931 4209 $\ell_{10} : 4\sqrt[3]{3}/9$
1931 4718 $1/2 + \ln(2)$
1931 6568 $\mathrm{sr} : (\ln 2/2)^{-1/3}$
1931 6898 $\mathrm{rt} : (9,6,6,1)$
1931 9285 $\mathrm{rt} : (7,9,-2,0)$
1932 0158 $\mathrm{rt} : (1,9,6,-5)$
1932 0344 $E : 7\sqrt[3]{2}/10$
1932 0544 $\mathrm{at} : 2\sqrt{2}/3 - 2\sqrt{3}$
1932 1864 $\mathrm{rt} : (6,0,-1,9)$
1932 3084 $2^x : \sqrt{3}/3 + 2\sqrt{5}/3$
1932 3183 $\mathrm{cr} : 5e/8$
1932 3745 $\ell_{10} : 2 - e/2$
1932 4200 $\mathrm{rt} : (6,7,4,5)$
1932 4269 $\mathrm{cu} : 3\sqrt{2}/4$
1932 4425 $\mathrm{sq} : 9e\pi/4$
1932 4542 $\mathrm{id} : 2\pi^2/9$
1932 4593 $\mathrm{as} : 1/4 + e/4$
1932 4798 $\ln(4/5) \times s\pi(1/3)$
1932 5242 $\mathrm{rt} : (1,5,-3,-9)$
1932 5724 $10^x : 1/3 - \pi/3$
1932 6193 $e^x : 2\sqrt{2} - 2\sqrt{5}$
1932 8005 $e^x : \pi/4$
1932 8248 $10^x : (\ln 3/3)^{1/3}$
1932 9523 $s\pi : 4e + 4\pi/3$
1932 9978 $\mathrm{rt} : (5,6,9,-2)$
1933 0281 $\zeta : 2\ln 2$
1933 0834 $\ell_{10} : 1/2 + 3\sqrt{2}/4$
1933 1960 $\mathrm{rt} : (6,6,-9,-8)$
1933 2408 $K : (\sqrt{2}\pi/3)^{-1/3}$
1933 4102 $10^x : \sqrt[3]{19/2}$
1933 6457 $e^x : \sqrt{2}/8$
1933 9341 $10^x : 2\sqrt{2}/3 - \sqrt{3}/2$
1933 9669 $\mathrm{rt} : (6,1,-4,4)$
1934 1169 $\ln : 4\sqrt[3]{3}/7$
1934 1501 $\sinh : 7e/4$
1934 3872 $e^x : 5(1/\pi)/9$
1934 4329 $\mathrm{rt} : (5,3,4,9)$
1934 4825 $t\pi : P_{41}$
1934 5034 $e^x : 2/3 - 4\sqrt{3}/3$
1934 5975 $4/5 \times \exp(2/5)$
1934 7580 $Ei : (\sqrt[3]{5}/3)^{-3}$
1934 8319 $\mathrm{cr} : 17/10$
1934 8567 $1/4 \times s\pi(e)$
1935 0405 $erf : \sqrt{3}/10$
1935 0642 $t\pi : 2\sqrt{2}/3 + \sqrt{5}/2$
1935 1280 $\mathrm{rt} : (5,3,-4,-8)$
1935 1586 $J : 3\ln 2/8$
1935 2108 $\mathrm{rt} : (5,7,-4,-1)$
1935 2154 $10^x : 2e$
1935 3704 $e^x : \exp(-\sqrt{3})$
1935 3913 $\mathrm{cu} : 2\sqrt{2} + 2\sqrt{3}/3$
1935 4996 $\mathrm{sq} : 2\sqrt{3} - 3\sqrt{5}/4$
1935 5567 $e^x : 1 + e$
1935 6452 $\mathrm{at} : 7\sqrt[3]{3}/4$
1935 7530 $\mathrm{cu} : 10\pi^2$
1935 7572 $\mathrm{rt} : (7,6,-7,-5)$
1935 8302 $\Psi : \sqrt[3]{2}/5$
1935 8780 $\mathrm{rt} : (4,5,-6,5)$
1935 9353 $erf : \ln 2/4$
1936 0000 $\mathrm{sq} : 11/25$
1936 3094 $\ln : 2 + 3\sqrt{3}/4$
1936 4025 $\sinh : \sqrt{3}/9$
1936 4995 $2^x : 17/15$
1936 5830 $\mathrm{as} : \sqrt{3}/9$
1936 6995 $\mathrm{sr} : 3 + 2e/3$
1936 7134 $e^x : 3/2 - \pi$
1936 7266 $\mathrm{rt} : (2,9,4,6)$
1936 7336 $\ln(\pi) - s\pi(2/5)$
1936 8499 $Ei : 3\sqrt{2}/10$
1936 9010 $J_1 : 7\sqrt[3]{3}/3$
1937 0086 $10^x : \sqrt[3]{2}$
1937 0824 $\mathrm{cr} : 3 - 3\sqrt{3}/4$
1937 1294 $\mathrm{rt} : (6,8,1,0)$
1937 2524 $\mathrm{sr} : 5\sqrt[3]{5}/6$
1937 2682 $J_2 : (e/3)^{-3}$
1937 4335 $Ei : (\sqrt[3]{5}/3)^2$
1937 4563 $10^x : (\sqrt{5}/2)^{-1/3}$
1937 5087 $2^x : 2\sqrt{2} - 3\sqrt{3}$
1937 5139 $\mathrm{sr} : \sqrt{3}/2 + \sqrt{5}/4$
1937 5988 $\zeta : 8/19$
1937 6536 $\ell_{10} : 3\sqrt{2}/2 - \sqrt{5}/4$
1937 7454 $\mathrm{rt} : (5,2,-7,-3)$
1937 7505 $\ln(\sqrt{5}) + \exp(2)$
1937 7664 $10^x : 1/13$
1937 8142 $\mathrm{sr} : e + 2\pi/3$
1937 9341 $J_2 : 8\sqrt[3]{5}$
1938 0491 $\mathrm{rt} : (5,4,1,4)$
1938 0952 $\mathrm{rt} : (4,6,-7,-2)$
1938 1463 $\mathrm{rt} : (1,7,-9,-2)$
1938 2002 $\ell_{10} : 16/25$
1938 2063 $t\pi : 10e/3$

1938 4327 rt : (4,0,1,8)
1938 5704 ℓ_{10} : $8(e+\pi)/3$
1938 5861 $\pi - \exp(2/3)$
1938 7145 ln : $7\sqrt{2}/3$
1938 7514 rt : (1,9,-2,-2)
1938 7762 rt : (2,4,-7,2)
1938 7922 e^x : $\sqrt{2}+\sqrt{5}/4$
1938 8395 10^x : $4\sqrt{2}/3-3\sqrt{3}/2$
1938 8556 e^x : $5\sqrt{2}/9$
1938 9438 rt : (5,5,4,-1)
1939 0981 J_0 : $7\sqrt{2}/2$
1939 1327 rt : (4,6,9,9)
1939 1482 10^x : $4-3\pi/2$
1939 1816 J_0 : $6e$
1939 2874 J_0 : $19/2$
1939 3656 rt : (2,8,0,-8)
1939 7350 e^x : $11/14$
1940 0437 rt : (7,3,-6,1)
1940 0712 cu : $1/2-4\pi/3$
1940 0899 rt : (9,5,9,-2)
1940 1414 J_0 : $6\sqrt[3]{5}/5$
1940 1504 rt : (9,1,-9,-6)
1940 2320 Γ : $(5/3)^{-1/2}$
1940 2410 as : $3/4-2\sqrt{2}/3$
1940 3306 10^x : $\ln 2/9$
1940 5161 cu : $11/19$
1940 7880 10^x : $2/3+e/4$
1940 7886 ℓ_2 : $1/3+3\sqrt{2}$
1940 8982 rt : (9,5,-1,7)
1940 9753 $s\pi$: $7\sqrt{3}/2$
1941 0348 10^x : $4/3+4\pi/3$
1941 0910 10^x : $\zeta(7)$
1941 1136 10^x : $((e+\pi)/3)^{-1/2}$
1941 2550 $t\pi$: $10(1/\pi)/3$
1941 2853 $s\pi$: $1/4+2e/3$
1941 4572 rt : (4,0,5,1)
1941 5601 ln : $14/17$
1941 5655 rt : (4,7,-8,-9)
1941 6054 Ψ : $1/\ln(2e/3)$
1941 7382 cu : $5\sqrt{2}/2$
1941 7743 rt : (2,5,-6,-5)
1941 8368 J_1 : $4\ln 2/7$
1941 9473 J_2 : $(2e/3)^{1/2}$
1941 9562 at : $3-\sqrt{2}/3$
1942 0821 rt : (5,5,-9,9)
1942 1409 Ei : $\sqrt{5}/3$
1942 2213 Ei : $4(1/\pi)/3$
1942 4801 rt : (3,6,1,5)
1942 4861 J_2 : $3\pi/7$
1942 4926 ℓ_{10} : $4\sqrt{3}/3-\sqrt{5}/3$
1942 7169 rt : (2,4,-6,1)
1942 7190 id : $1/4+4\sqrt{5}$
1942 7920 erf : $4/23$
1942 9327 J_0 : $5\pi^2/8$
1943 2644 $\ln(3/4)\times s\pi(\sqrt{5})$
1943 4168 rt : (9,1,5,1)
1943 5542 rt : (7,8,-9,-2)
1943 6781 $\ln(3)\div\exp(\sqrt{3})$
1943 7709 rt : (8,2,-4,6)
1943 7752 cu : $4\sqrt{3}-4\sqrt{5}$
1943 8282 id : $10\sqrt{5}/7$
1944 1472 cu : $4/3-4e$
1944 1631 ln : $2-\pi/4$
1944 3195 J_2 : $7\sqrt{3}/9$
1944 3322 rt : (3,7,-7,4)
1944 4091 2^x : $4\sqrt{2}/3+\sqrt{5}/3$
1944 4817 rt : (2,7,-9,-2)
1944 6601 sq : $1-\sqrt{5}/4$
1944 7610 10^x : $(\sqrt{5}/2)^{-3}$
1944 8943 J_2 : $7(e+\pi)/3$
1944 9221 $s\pi$: $9\sqrt{5}/2$
1944 9338 $4/5\div\exp(\sqrt{2})$
1945 0127 cu : $10(1/\pi)/3$
1945 0196 rt : (8,8,-7,1)
1945 1269 ℓ_2 : $4+\sqrt{3}/3$
1945 1758 rt : (3,5,6,1)
1945 3489 $3/5+\ln(2/3)$
1945 3553 ℓ_2 : $1+3e$
1945 3807 rt : (3,6,9,-2)
1945 4101 rt : (6,5,-7,-9)
1945 4443 J_1 : $1/\sqrt[3]{16}$
1945 4513 J_1 : $\sqrt{3}-\sqrt{6}-\sqrt{7}$
1945 4771 Li_2 : $(\sqrt[3]{5}/3)^3$
1945 5578 2^x : $9\sqrt[3]{2}/10$
1945 8304 ℓ_{10} : $7\sqrt{5}$
1945 8695 2^x : $2\sqrt{5}$
1945 9679 rt : (8,8,4,7)
1946 1697 sq : $3/2-4\sqrt{5}/3$
1946 2523 2^x : $2-\sqrt{3}/2$
1946 2526 rt : (8,5,9,-2)
1946 3476 id : $3\sqrt{2}/4-\sqrt{3}/2$
1946 5762 rt : (2,6,-5,-6)
1946 6676 rt : (7,1,-7,-5)
1946 7308 Ψ : $7\sqrt[3]{3}/6$
1946 8868 rt : (5,5,-2,-1)
1946 8894 at : $\sqrt{2}+\sqrt{5}/2$
1946 9690 Ψ : $23/18$
1946 9865 $t\pi$: $\sqrt{2}/2+3\sqrt{5}/2$
1947 0019 $1/4\div\exp(1/4)$
1947 0206 $t\pi$: $1/4-4\pi/3$
1947 2026 rt : (3,8,-6,-3)
1947 2111 rt : (6,7,-4,-1)
1947 2552 Ψ : $4\sqrt{5}/7$
1947 4427 $\exp(3/5)\div\exp(\sqrt{5})$
1947 4436 rt : (8,6,6,1)
1947 8685 rt : (1,2,-1,9)
1948 0600 $t\pi$: $3\sqrt{2}/4-2\sqrt{3}$
1948 3016 2^x : $2/3+3\sqrt{3}$
1948 3160 rt : (3,9,-7,1)
1948 3191 e^x : $1/4-4\sqrt{2}/3$
1948 4491 sinh : $\sqrt{7/3}$
1948 5850 rt : (8,3,-7,1)
1948 6056 rt : (4,6,-2,9)
1948 7818 Ψ : $7\zeta(3)/5$
1948 8191 rt : (1,6,-4,-1)
1948 9294 2^x : $\exp(-e/2)$
1949 0103 Ei : $1/\ln(\sqrt{2})$
1949 2209 $\ln(\sqrt{5})\times\exp(e)$
1949 3118 $\ln(4/5)\div\ln(\pi)$
1949 4299 J_2 : $(\ln 3/2)^{-1/2}$
1949 4537 at : $2\sqrt{2}/3-\sqrt{5}/3$
1949 5886 rt : (3,7,0,-2)
1949 8402 rt : (7,5,1,6)
1949 8750 ℓ_2 : $\sqrt[3]{12}$
1949 9577 Γ : $2\sqrt{3}$
1950 0721 ℓ_2 : $\ln(\pi)$
1950 1061 rt : (8,8,-9,-2)
1950 1915 id : $4\sqrt{2}/3+4\sqrt{3}/3$
1950 2367 e^x : $2\sqrt{3}-2\sqrt{5}/3$
1950 3260 J_1 : $5(1/\pi)/4$
1950 3363 rt : (4,7,6,4)
1950 4984 J_2 : $23/3$
1950 6200 rt : (5,2,-2,-9)
1950 6547 ℓ_{10} : $2-\sqrt{3}/4$
1950 7221 rt : (1,9,-8,-5)
1950 9032 $s\pi$: $1/16$
1950 9085 as : $1/\ln((e+\pi)/2)$
1950 9392 rt : (3,7,-9,-2)
1951 2278 Γ : $\exp(-e/3)$
1951 4398 cr : $1+\sqrt{2}/2$
1951 7147 rt : (7,2,-5,-1)
1951 7209 id : $4e/3+\pi/2$
1951 7462 rt : (2,6,9,-2)
1951 7699 rt : (9,4,4,-1)
1951 8650 rt : (9,7,-7,-3)
1952 0882 rt : (9,0,-7,7)
1952 2435 $\sqrt{2}-\ln(5)$
1952 2517 J_1 : $(e/2)^{-3}$
1952 2860 sr : $10/7$
1952 4654 $4\div\exp(3/5)$
1952 5282 rt : (7,5,9,-2)
1952 5813 Ei : $4\sqrt{3}/3$
1952 6214 id : $2/3-\sqrt{2}/3$
1952 8104 $1-\ln(\sqrt{5})$
1952 8755 J : $\exp(-4)$
1952 9570 10^x : $3\sqrt{2}/4+2\sqrt{3}/3$
1953 0637 J_0 : $4\pi^2/5$
1953 3712 Γ : $2\sqrt{2}/7$
1953 3999 cr : $e\pi/5$
1953 4848 Γ : $1/\ln(\sqrt[3]{2})$
1953 4963 rt : (2,7,3,-3)
1953 7003 rt : (4,7,-3,2)
1953 7559 rt : (6,2,-2,5)
1953 8532 rt : (8,9,1,2)
1954 3697 E : $22/25$
1954 4778 sr : $3\sqrt{2}+\sqrt{3}/3$
1954 4895 J_2 : $8e\pi/5$
1954 9184 ln : $\sqrt{2}/2+3\sqrt{3}/2$
1954 9292 tanh : $2\ln 2/7$
1954 9306 sr : $3-\pi/2$
1955 1221 at : $2\ln 2/7$
1955 1568 rt : (3,0,5,1)
1955 1613 rt : (5,7,-6,6)
1955 2901 J_1 : $(2\pi)^{-1/2}$
1955 5052 rt : (2,5,-8,-1)
1955 5906 at : $1/2+3e/4$
1955 8163 e^x : $4/3-2\sqrt{3}/3$
1955 8422 rt : (9,6,1,8)
1955 8912 2^x : $1-3\sqrt{5}/2$
1956 1242 $1/5-\exp(1/3)$
1956 1766 e^x : $7\zeta(3)/4$
1956 1828 $t\pi$: $9\sqrt{2}/2$
1956 4064 $\ln(\sqrt{5})\div\exp(\sqrt{2})$
1956 4645 sr : $3/4+e/4$
1956 7076 rt : (8,1,5,1)
1956 7158 ln : $3\sqrt{2}/2-3\sqrt{3}/4$
1956 7902 rt : (9,8,-9,-2)
1956 8085 rt : (3,8,-8,-7)
1956 8566 cu : $e/3+\pi/2$
1956 8862 J_2 : $9\zeta(3)/8$
1956 9119 rt : (4,2,-5,-2)
1957 0845 10^x : $1/2+\pi$
1957 1609 ζ : $2\pi/7$
1957 5235 rt : (4,7,-9,-2)
1957 7161 rt : (5,1,-5,-4)
1957 7182 rt : (2,9,-8,-4)
1957 8148 J_2 : $1/\ln(2\pi/3)$
1957 8346 Ψ : $\exp(4/3)$
1958 0035 rt : (9,9,3,-1)
1958 1317 id : $(\sqrt[3]{5})^{1/3}$
1958 2334 rt : (3,9,0,6)
1958 2997 2^x : $(\pi/2)^{-3}$
1958 3349 ζ : $(\ln 3)^{-1/2}$
1958 4136 J_2 : $23/17$
1958 4149 2^x : $9\pi/2$
1958 4314 e^x : $\sqrt{2}/4+\sqrt{3}/4$
1958 6786 rt : (8,3,-6,1)
1958 8182 rt : (8,3,-2,7)
1958 8508 as : $3\sqrt{2}/4-\sqrt{3}/2$
1958 9208 rt : (6,5,9,-2)
1959 1041 cr : $2\sqrt{2}-\sqrt{5}/2$
1959 2594 2^x : P_{28}
1959 2841 $1/3\times s\pi(1/5)$
1959 3082 rt : (7,6,-2,1)
1959 3830 sr : $\sqrt{2}/3-\sqrt{3}/4$
1959 4771 cu : $1/2+4\sqrt{5}/3$
1959 4884 cu : $\sqrt{2}/4-4\sqrt{3}$
1959 5646 e^x : $\sqrt{2}-\sqrt{5}+\sqrt{7}$
1959 5703 J_1 : $8\sqrt[3]{2}/3$
1959 6094 rt : (4,1,3,9)
1959 6542 rt : (7,7,-4,-1)
1959 8466 2^x : $1/\sqrt{15}$
1960 1138 ζ : $7\sqrt{3}/4$
1960 2015 ℓ_2 : $4\sqrt{3}+\sqrt{5}$
1960 2657 J_1 : $2/5$
1960 3516 2^x : $3\sqrt{2}/2-2\sqrt{5}$
1960 3974 rt : (5,8,-7,-1)
1960 6182 rt : (3,5,-5,-8)
1960 6189 as : $(\sqrt{3}/2)^{1/2}$
1960 7984 sq : $1/2-2\sqrt{2}/3$
1960 9197 rt : (2,6,-4,-1)
1961 0536 Ei : $(\pi)^{1/3}$
1961 0679 J_2 : $9\pi^2/7$
1961 1987 ℓ_{10} : 5π
1961 3796 e^x : $4/3+4\sqrt{3}/3$
1961 4448 at : $\sqrt{3}/2+3\sqrt{5}/4$
1961 4854 2^x : $\exp(\sqrt{3}/2)$
1961 5024 J_0 : $9\pi/10$
1961 5242 id : $3\sqrt{3}$
1961 5274 ζ : $(4/3)^{-3}$

1961 5871 $\ell_2 : \sqrt{21}$
1961 6110 $\Psi : (\sqrt[3]{3}/3)^{-1/3}$
1961 6356 $J_0 : 7e/2$
1961 7217 rt : (4,8,-4,-5)
1961 8197 rt : (7,0,-5,6)
1961 9201 $\ln : \sqrt{2}/3 + \sqrt{5}/3$
1962 1109 rt : (6,8,6,6)
1962 1631 $\mathrm{sr} : ((e+\pi)/2)^{1/3}$
1962 3911 rt : (2,5,6,1)
1962 7248 $\ell_{10} : 10\sqrt{2}/9$
1962 8016 $\exp(3/4) \div s\pi(\sqrt{2})$
1962 8115 $\ln(3/5) + s\pi(1/4)$
1962 9464 $\ell_{10} : 11/7$
1962 9766 $Ei : 1/17$
1963 0006 $2 \times \exp(4)$
1963 1657 $\Psi : 4e\pi/9$
1963 1812 $\Gamma : 17/22$
1963 2339 rt : (8,4,4,-1)
1963 4559 $\ln : (\sqrt{2}\pi/3)^{1/2}$
1963 4901 $J_1 : \zeta(3)/3$
1963 5993 $\tanh : 5(1/\pi)/8$
1963 7966 $\mathrm{at} : 5(1/\pi)/8$
1963 8699 rt : (3,7,5,9)
1963 9129 $t\pi : P_{62}$
1963 9343 $e^x : -\sqrt{2} + \sqrt{5} - \sqrt{6}$
1963 9795 rt : (9,7,-2,3)
1964 0171 rt : (7,6,6,1)
1964 0775 $2^x : 1/4 - 3\sqrt{3}/2$
1964 2083 $J_0 : 5\pi^2/3$
1964 2398 rt : (5,7,-9,-2)
1964 2948 rt : (9,1,-5,8)
1964 2963 rt : (1,8,-1,5)
1964 7099 $\Psi : 10e\pi/9$
1964 7459 $\mathrm{sq} : 1/4 - \sqrt{3}$
1964 8007 $\mathrm{at} : 9\sqrt{2}/5$
1964 8472 rt : (6,9,-2,0)
1964 8798 $t\pi : \sqrt{2}/4 + 3\sqrt{5}$
1965 1790 $\sqrt{5} \div s\pi(\pi)$
1965 2473 $\mathrm{as} : 2/3 - \sqrt{2}/3$
1965 2868 $\ln(\sqrt{3}) + s\pi(\sqrt{3})$
1965 4350 rt : (5,5,9,-2)
1965 4809 $J_0 : 2\sqrt{2}$
1965 9964 $\mathrm{at} : 8(1/\pi)$
1966 0214 $2^x : (3\pi/2)^{1/3}$
1966 0648 $\mathrm{at} : 1/\ln(\sqrt{2}\pi/3)$
1966 1250 rt : (8,9,6,8)
1966 2744 $\Psi : 17/24$
1966 3046 $\mathrm{sq} : 2 - e/3$
1966 5889 $Li_2 : 6/7$
1966 6262 $\ell_{10} : 2/3 + e/3$
1966 6848 $10^x : \sqrt{11/3}$
1966 7281 rt : (8,2,-8,-7)
1966 8472 $\mathrm{cr} : 2\sqrt{3}/3 + \sqrt{5}/4$
1966 9561 $\mathrm{at} : 3/2 + \pi/3$
1967 0045 rt : (8,2,-5,-1)
1967 1886 $\mathrm{cr} : 2/3 + \pi/3$
1967 2191 rt : (3,8,-1,-9)
1967 4517 $\mathrm{sr} : \sqrt{2}/4 + 2\sqrt{5}$
1967 6521 $\ell_{10} : \sqrt{2}/2 + \sqrt{3}/2$
1967 6722 $\ln : 1 + 4\sqrt{3}/3$
1967 7888 $\mathrm{at} : 2/3 - \sqrt{3}/2$
1967 8072 rt : (8,4,-5,2)
1967 9011 rt : (5,9,-8,-8)
1968 0681 rt : (8,9,3,-1)
1968 1036 $10^x : 8\ln 2/7$
1968 1494 rt : (5,5,3,5)
1968 1696 cr : 12/7
1968 1839 rt : (2,7,1,-3)
1968 2684 $\mathrm{sr} : 9(1/\pi)/2$
1968 2697 rt : (3,1,-5,-1)
1968 5673 $10^x : 6/7$
1968 5763 $\mathrm{at} : \sqrt{3}/4 - 4\sqrt{5}/3$
1968 6514 $J_2 : 19/14$
1968 7001 rt : (1,7,-1,3)
1968 7615 $e^x : 1/3 + 2\sqrt{5}/3$
1968 7908 $\mathrm{as} : \sqrt{2}/4 + \sqrt{3}/3$
1968 8121 $J_2 : \sqrt[3]{5/2}$
1968 8850 $Li_2 : (\sqrt[3]{4})^{-1/3}$
1968 9754 $\mathrm{sq} : 1/3 - 3\sqrt{2}/2$
1969 1011 rt : (6,9,-6,7)
1969 2966 $J_2 : 5\pi^2/3$
1969 3845 $\mathrm{sq} : 3\sqrt{2}/2 + 2\sqrt{3}$
1969 4443 rt : (2,0,5,1)
1969 4842 $2^x : 5(e+\pi)/7$
1969 5964 $\ln(2/5) + \exp(\sqrt{2})$
1969 7646 $erf : (\sqrt[3]{2}/3)^2$
1969 7789 rt : (4,8,1,6)
1969 8316 $\mathrm{id} : 7\sqrt[3]{5}/10$
1969 8671 $J_1 : 7e/3$
1970 0000 cu : 13/10
1970 0415 $\mathrm{at} : \sqrt{13/2}$
1970 1073 $t\pi : 4e + 4\pi/3$
1970 1102 $J_2 : \sqrt{2} + \sqrt{3} + \sqrt{7}$
1970 1855 rt : (7,6,3,7)
1970 3463 $e^x : 2 - 4e/3$
1970 4729 rt : (5,0,-8,1)
1970 5375 rt : (7,1,5,1)
1970 6166 $\zeta : 23/18$
1970 7065 $10^x : 4/3 - 3e/4$
1970 7456 $2^x : \sqrt{2} - 2\sqrt{3}/3$
1970 7785 $erf : 3/17$
1970 8508 $\mathrm{sr} : 1 + \sqrt{3}/4$
1971 0016 rt : (9,8,-7,1)
1971 0933 rt : (6,7,-9,-2)
1971 0940 $Li_2 : 3/16$
1971 2079 rt : (9,7,3,9)
1971 3967 $\mathrm{sr} : 3/4 + 3e/2$
1971 5925 rt : (1,9,6,-4)
1971 6764 rt : (7,1,-8,1)
1971 6916 $\ell_2 : 3 - \sqrt{2}/2$
1971 7075 $\ln : \sqrt{2} + \sqrt{6} - \sqrt{7}$
1972 0218 $2^x : 3/4 + 3\sqrt{5}/2$
1972 0386 $e^x : \sqrt[3]{2}/7$
1972 0752 rt : (4,5,9,-2)
1972 1813 $Ei : (e+\pi)/4$
1972 2316 $\ln(5) + s\pi(1/5)$
1972 2457 ln : 9
1972 3032 rt : (9,2,-8,3)
1972 5231 $\mathrm{at} : 3\sqrt{2}/4 + 2\sqrt{5}/3$
1972 5334 $\ell_{10} : 5\sqrt[3]{2}/4$
1972 5422 rt : (9,4,9,-2)
1972 5528 rt : (1,8,-6,-6)
1972 5741 rt : (8,7,-4,-1)
1972 8894 rt : (4,9,-7,1)
1973 0094 rt : (3,7,-6,-8)
1973 1259 $10^x : \sqrt{3} + \sqrt{5}/3$
1973 2762 rt : (5,9,-3,3)
1973 3192 $\mathrm{sq} : 4\sqrt{2} + 3\sqrt{5}/2$
1973 4510 $10^x : (\sqrt{3}/2)^{-1/3}$
1973 4608 rt : (3,6,-4,-1)
1973 5230 $J_2 : e/2$
1973 6822 $\mathrm{sr} : 2 + 2\sqrt{2}$
1973 7237 $Ei : \exp(-\sqrt[3]{5}/2)$
1973 7264 $\mathrm{cu} : 1/4 + 4e$
1973 7380 $\mathrm{at} : 2/3 + 4\sqrt{2}/3$
1973 7526 $\mathrm{sq} : 3\sqrt{2}/2 - 3\sqrt{5}/4$
1973 7532 tanh : 1/5
1973 7939 $\zeta : 4\sqrt{5}/7$
1973 7980 $Ei : 1/\sqrt[3]{13}$
1973 8757 $\ln : e + 2\pi$
1973 9555 at : 1/5
1974 1265 $erf : \sqrt{2}/8$
1974 2076 $2^x : 3\ln 2/8$
1974 2162 $t\pi : 7(1/\pi)/8$
1974 2933 rt : (8,4,0,8)
1974 5293 $\mathrm{id} : 3(e+\pi)/8$
1974 5304 $\mathrm{id} : 2\sqrt{2}/3 - \sqrt{5}/3$
1974 6341 $10^x : 7\sqrt{3}$
1974 6659 $\theta_3 : 12/23$
1974 6907 rt : (9,9,8,-2)
1974 8060 $erf : 5(1/\pi)/9$
1974 9319 $e^x : \sqrt{3} - 3\sqrt{5}/2$
1975 0038 $\mathrm{cr} : 10\zeta(3)/7$
1975 0745 $\sqrt{5} \times \exp(\sqrt{2})$
1975 0763 rt : (7,4,4,-1)
1975 2588 $J_2 : 5e\pi/4$
1975 3086 sq : 4/9
1975 3815 rt : (2,7,-8,-9)
1975 5406 $e^x : \sqrt[3]{3}/8$
1975 7069 $erf : \exp(-\sqrt{3})$
1975 8579 rt : (4,9,-9,-9)
1975 8933 $\mathrm{at} : \sqrt{2} - \sqrt{3} - \sqrt{5}$
1976 0896 $\ln : 3/2 + 2e/3$
1976 1855 $\mathrm{at} : P_{16}$
1976 1976 $\mathrm{at} : P_{40}$
1976 2952 $t\pi : 4(e+\pi)$
1976 3757 $e^x : 1/2 - 3\sqrt{2}/2$
1976 3790 rt : (2,7,-2,-9)
1976 4595 $\zeta : 4\sqrt{3}/5$
1976 4620 $\mathrm{at} : 3\sqrt{2}/2 + \sqrt{3}/4$
1976 7812 rt : (9,2,-3,9)
1976 8127 $e^x : 1/3 + 4\sqrt{2}/3$
1976 8328 $\ell_2 : P_{29}$
1976 8984 $\Psi : 19/5$
1977 0064 $\mathrm{sq} : 1 - 2\pi/3$
1977 0475 $\tanh : \zeta(3)/6$
1977 2517 $\mathrm{at} : \zeta(3)/6$
1977 3404 $1/4 + \exp(2/3)$
1977 3632 $2^x : 3\sqrt{5}/4$
1977 3787 $\Psi : 19/2$
1977 3923 $Ei : 10/9$
1977 4060 rt : (7,1,-3,7)
1977 4972 $e^x : (\pi/2)^{1/3}$
1977 7161 rt : (3,2,-6,9)
1977 8636 $e^x : 5\sqrt[3]{2}/8$
1977 9782 $J_2 : 6e$
1978 0205 $\mathrm{at} : 8\sqrt{5}/7$
1978 0860 at : 23/9
1978 0898 rt : (7,7,-9,-2)
1978 1221 rt : (9,8,-5,-2)
1978 1876 $\mathrm{id} : \sqrt{2}/2 + 2\sqrt{5}/3$
1978 3552 $10^x : \sqrt[3]{5}/5$
1978 4120 rt : (7,9,3,-1)
1978 4497 rt : (5,0,-3,5)
1978 4746 rt : (2,6,-9,-2)
1978 5657 $e^x : 13/3$
1978 5978 rt : (9,3,-6,1)
1978 6039 $t\pi : 7\sqrt{3}/2$
1978 8464 rt : (3,5,9,-2)
1978 9323 $t\pi : 1/4 + 2e/3$
1979 1587 $\Psi : \exp(-\sqrt{5}/2)$
1979 2059 rt : (8,4,9,-2)
1979 2724 $\ln : 3/2 - e/4$
1979 2814 $\ell_{10} : 1 + \sqrt{3}/3$
1979 2868 rt : (8,1,-6,-8)
1979 3539 $J_1 : 2\sqrt{2}/7$
1979 5350 $J_1 : \exp(-e/3)$
1979 6410 $3 \div \exp(e)$
1979 6593 rt : (4,8,-2,-7)
1979 7456 $s\pi : 1/2 + 2e$
1979 9011 $t\pi : 1/4 + \sqrt{2}/3$
1980 0793 $\lambda : \exp(-15\pi/11)$
1980 2299 $J_2 : 8\sqrt[3]{5}/3$
1980 2428 $\Gamma : \ln(\sqrt[3]{4}/2)$
1980 2741 $\ln(\sqrt{3}) + \exp(1/2)$
1980 3878 $t\pi : (3\pi/2)^{-1/3}$
1980 4205 $\mathrm{id} : 2\ln 2/7$
1980 4483 $\mathrm{cu} : \sqrt{11}\pi$
1980 5464 rt : (9,3,-9,-9)
1980 5867 $e^x : 8\sqrt[3]{3}/7$
1980 6226 rt : (1,5,6,1)
1980 6983 $Li_2 : (e/2)^{-1/2}$
1980 8128 $\zeta : 10e/9$
1980 8514 rt : (8,9,8,-2)
1980 9142 $t\pi : 7\pi^2/8$
1980 9645 rt : (7,7,-5,-4)
1981 2968 $e^x : 4\sqrt{2} + 3\sqrt{5}$
1981 5003 $2/5 - \exp(4)$
1981 5005 rt : (6,6,6,1)
1981 7829 $\ln : (2e/3)^{1/3}$
1981 8104 rt : (7,4,-9,-7)
1981 9870 $\ln : (\ln 3/2)^2$
1982 0068 $2^x : 6/23$
1982 1132 $J : 2/9$
1982 3433 $\theta_3 : \ln 2/7$
1982 5375 $\mathrm{sq} : 4\sqrt[3]{5}/3$
1982 5664 $\sinh : \exp(4/3)$
1982 5956 $e^x : \exp(-\sqrt[3]{5})$
1982 6244 $2^x : 25/22$
1982 6556 rt : (4,9,0,-1)

1982 6769 at : $2+\sqrt{5}/4$
1982 6797 2^x : $\sqrt{3}/3+\sqrt{5}/4$
1982 7853 tπ : $9\sqrt{5}/2$
1983 0713 rt : (9,2,-5,-1)
1983 1688 J_1 : $3\sqrt{5}/2$
1983 1857 at : $1/4+4\sqrt{3}/3$
1983 2028 at : $3/2-3\sqrt{3}/4$
1983 4718 rt : (8,5,-8,-3)
1983 6173 cu : $4\sqrt{2}+2\sqrt{5}$
1983 6932 sinh : $\sqrt{3}-\sqrt{5}+\sqrt{7}$
1983 7069 rt : (9,8,0,4)
1983 7150 rt : (5,5,-9,-5)
1983 7473 rt : (4,1,-5,-1)
1983 8440 ln : $1/3+4\sqrt{5}/3$
1984 0615 cu : $3/4-2e/3$
1984 1972 ℓ_{10} : $8\pi^2/5$
1984 2011 e^x : $3\sqrt{3}/4-\sqrt{5}/2$
1984 2268 J_1 : $\exp(e)$
1984 2627 tπ : $1/\ln(4)$
1984 3041 cr : $3e/2-3\pi/4$
1984 3721 rt : (1,3,8,7)
1984 4119 ln : $4\sqrt{2}+3\sqrt{5}/2$
1984 4910 J_2 : $15/11$
1984 7592 ℓ_2 : $2e-\pi$
1984 8525 at : $3/2+3\sqrt{2}/4$
1984 9513 rt : (6,1,5,1)
1984 9537 cu : $7/12$
1985 0374 E : $(\sqrt{2}\pi/3)^{-1/3}$
1985 0554 ln : $\sqrt{2}/4+\sqrt{3}/2$
1985 0745 J_0 : $17/6$
1985 1009 J_1 : $(\pi/2)^{-2}$
1985 1229 Γ : $2(e+\pi)/5$
1985 2320 rt : (1,7,4,-6)
1985 2354 rt : (8,7,-9,-2)
1985 3271 10^x : $(\sqrt[3]{4})^{-1/3}$
1985 3896 sq : $3/4-3\sqrt{2}$
1985 4050 id : $9\sqrt[3]{5}/7$
1985 4978 rt : (2,8,0,7)
1985 4992 rt : (3,6,-9,-2)
1985 5715 sq : $3/4+3\sqrt{3}/4$
1985 6910 rt : (3,6,-9,8)
1985 7539 rt : (2,5,9,-2)
1985 8951 sq : $7(1/\pi)/5$
1985 9479 J_1 : $\ln(3/2)$
1985 9999 rt : (7,4,9,-2)
1986 0084 rt : (9,7,-4,-1)
1986 0184 J_1 : $1/\sqrt[3]{15}$
1986 1621 rt : (3,1,2,-9)
1986 3338 Li_2 : $\exp(-5/3)$
1986 3933 θ_3 : $16/17$
1986 4753 rt : (4,6,-4,-1)
1986 5193 at : $3e\pi/10$
1986 6873 sr : $3\sqrt{3}/2+\sqrt{5}$
1986 6933 sπ : $(1/\pi)/5$
1986 8731 at : $3/4+2e/3$
1986 9124 rt : (3,6,-9,-3)
1986 9841 at : P_8
1987 0046 J_0 : $\pi^2/2$
1987 0507 sq : $1/\ln(3\pi)$
1987 1206 rt : (7,9,8,-2)
1987 1418 10^x : $\exp(3/2)$
1987 3224 rt : (6,4,4,-1)
1987 3592 ζ : $4\sqrt{2}/5$
1987 3893 sinh : $7\sqrt[3]{3}/4$
1987 4251 Ψ : 25
1987 5089 J_1 : $4\pi^2/9$
1987 5211 rt : (9,3,-6,4)
1987 5913 as : $2\sqrt{2}/3-\sqrt{5}/3$
1987 6276 Ψ : $(\sqrt{2}\pi/3)^{-2}$
1987 7116 rt : (8,5,-3,3)
1987 7513 J_2 : $7(e+\pi)/9$
1987 8730 $\exp(3)+\exp(\sqrt{2})$
1988 0759 rt : (7,7,0,2)
1988 1973 ln : $e\pi/7$
1988 4704 e^x : $\pi^2/5$
1988 6962 J_2 : $1/\ln(\sqrt[3]{3}/3)$
1989 0504 rt : (6,9,3,1)
1989 1236 tπ : $1/16$
1989 2687 10^x : $4/3+3\sqrt{5}/4$
1989 3147 $\exp(2/5)+s\pi(1/4)$
1989 4367 id : $5(1/\pi)/8$
1989 4763 ln : $\sqrt{11}$
1989 5788 sr : $23/16$
1989 5872 cr : $1/\ln(\ln 3)$
1989 7000 ℓ_{10} : $\sqrt{5/2}$
1989 7184 rt : (3,6,-4,-9)
1989 7784 $\ln(3)\div\ln(2/5)$
1989 7861 sπ : $2\sqrt{3}+2\sqrt{5}$
1990 2187 rt : (8,5,2,9)
1990 3649 rt : (6,4,-8,-5)
1990 4241 ℓ_{10} : $4\sqrt{3}/3-3\sqrt{5}/4$
1990 5393 at : $3\sqrt[3]{5}/2$
1990 6003 rt : (8,0,-9,3)
1990 6070 rt : (2,9,-9,-9)
1990 6331 10^x : $1/3+e$
1990 7136 rt : (2,8,7,6)
1990 7928 sinh : $4\sqrt[3]{5}$
1990 9703 sr : $9e\pi/2$
1991 0050 J_0 : $9\sqrt[3]{2}/4$
1991 0106 2^x : $1/2-2\sqrt{2}$
1991 1082 10^x : $13/11$
1991 1485 id : $7\pi/10$
1991 3157 rt : (7,7,5,8)
1991 3643 e^x : $4/3+2\sqrt{3}$
1991 3648 at : $3\sqrt{2}-3\sqrt{5}/4$
1991 4827 $4\div\exp(3)$
1991 7188 2^x : $-\sqrt{2}+\sqrt{3}-\sqrt{7}$
1991 7614 rt : (7,2,-6,2)
1991 7877 ln : $4\sqrt{3}/3+3\sqrt{5}$
1992 1407 rt : (8,0,-4,9)
1992 2246 2^x : $((e+\pi)/3)^{-2}$
1992 4748 sinh : $(\pi/3)^{1/3}$
1992 5364 rt : (9,7,-9,-2)
1992 6722 rt : (4,6,-9,-2)
1992 8031 rt : (1,5,9,-2)
1992 8912 Ψ : $7e/2$
1992 9295 rt : (6,4,9,-2)
1992 9440 tπ : $3\zeta(3)/5$
1992 9935 rt : (1,8,-2,-3)
1993 1120 rt : (2,9,-7,-8)
1993 2072 at : $3-\sqrt{3}/4$
1993 2594 tanh : $\sqrt{2}/7$
1993 2890 tπ : $4\sqrt{3}-2\sqrt{5}$
1993 3914 sinh : $2\ln 2/7$
1993 3951 tπ : $3e/2-3\pi/4$
1993 3955 rt : (1,7,-9,-1)
1993 4492 rt : (7,2,-1,8)
1993 4720 at : $\sqrt{2}/7$
1993 5021 rt : (6,9,8,7)
1993 5873 id : $1/3+\sqrt{3}/2$
1993 6000 as : $2\ln 2/7$
1993 6011 Ei : $6\sqrt[3]{5}/7$
1993 7117 Ψ : $7e/5$
1993 8020 2^x : $1+4\sqrt{5}$
1993 8119 2^x : $4e-2\pi/3$
1993 8334 $1/3\times\exp(4)$
1993 9003 rt : (3,7,-8,1)
1993 9420 sr : $((e+\pi)/2)^{-3}$
1993 9610 e^x : $2/11$
1993 9892 rt : (6,4,-3,1)
1994 0142 id : $2\sqrt{3}/3-3\sqrt{5}/2$
1994 0601 $\ln(2)\times\ln(3/4)$
1994 3262 tπ : $3e+4\pi$
1994 3716 J_1 : $19/3$
1994 5346 Γ : $\sqrt{11/2}$
1994 6289 cu : $17/16$
1994 7114 sr : $(1/\pi)/8$
1994 7411 rt : (7,3,-7,-8)
1994 7455 10^x : P_{35}
1994 8387 e^x : $4(1/\pi)/7$
1994 8836 sq : $1/4-3e/4$
1994 8916 Γ : $10\ln 2/9$
1994 8974 sq : $\sqrt{2}+\sqrt{3}/2$
1995 1000 J_2 : $4\sqrt[3]{5}/5$
1995 1853 rt : (2,9,3,2)
1995 3257 rt : (1,7,-4,-7)
1995 4648 sq : $23/21$
1995 6313 rt : (6,4,2,7)
1995 7235 ℓ_{10} : $12/19$
1995 7443 at : $\sqrt{2}+2\sqrt{3}/3$
1995 8572 rt : (6,1,-9,-1)
1996 0088 ℓ_2 : $(\pi/3)^3$
1996 0966 Γ : $\exp(-(e+\pi)/2)$
1996 2031 id : $10e\pi/7$
1996 3853 at : $1/3+\sqrt{5}$
1996 4523 2^x : $4+\pi$
1996 7095 rt : (6,1,-4,-7)
1996 9978 ln : $(\ln 3/2)^{1/3}$
1997 0498 rt : (1,5,-2,-5)
1997 0998 sπ : $(5/2)^{-3}$
1997 1202 10^x : $2/3+2e/3$
1997 1474 cr : $e/4+\pi/3$
1997 3683 id : $10\sqrt[3]{2}/3$
1997 3705 sinh : $8\pi/5$
1997 5494 tπ : $2\sqrt{2}+\sqrt{3}$
1997 5908 Ψ : $\ln(\sqrt{5}/3)$
1997 6020 Ei : $23/2$
1997 9266 e^x : $-\sqrt{2}+\sqrt{6}-\sqrt{7}$
1998 2196 at : $1+\pi/2$
1998 2554 10^x : $5\sqrt[3]{3}$
1998 3162 cr : $19/11$
1998 4415 cr : $2/3+3\sqrt{2}/4$
1998 6770 J_2 : $9\sqrt{2}$
1998 7702 Li_2 : $\sqrt[3]{5}/9$
1998 8531 ζ : $7/11$
1998 8572 at : $\sqrt[3]{17}$
1999 0098 J_1 : $1/\sqrt{6}$
1999 0503 at : $18/7$
1999 1341 tanh : $(\sqrt{2}\pi/2)^{-2}$
1999 3500 at : $(\sqrt{2}\pi/2)^{-2}$
1999 3847 2^x : $3\sqrt{2}+2\sqrt{5}/3$
1999 4060 Γ : $4\sqrt{3}/9$
1999 4585 2^x : $7e/8$
1999 6240 sr : $8\sqrt[3]{2}/7$
1999 7051 $\sqrt{5}-s\pi(\sqrt{2})$
1999 8096 Ψ : $4(1/\pi)$
1999 9653 Ei : $\exp(1/(3\pi))$
2000 0000 id : $1/5$
2000 0033 Li_2 : $5\zeta(3)/7$
2000 0144 ln : $3/4+\sqrt{2}/3$
2000 0706 ln : $5\sqrt[3]{5}/7$
2000 0964 Ei : $10\ln 2/3$
2000 4355 Ψ : $\sqrt{2}/2$
2000 5088 ln : $4\sqrt{3}/3-2\sqrt{5}/3$
2000 7965 cu : $4\sqrt{2}/3+4\sqrt{5}$
2000 9415 ln : $4-e/4$
2001 0271 2^x : $5/19$
2001 3244 tπ : $\sqrt[3]{3}/2$
2002 0000 θ_3 : $1/10$
2002 3193 id : P_{16}
2002 3298 sπ : $3e/2-\pi$
2002 3318 id : P_{40}
2002 4027 ℓ_{10} : $3\sqrt{2}/2-2\sqrt{5}/3$
2002 4469 ℓ_{10} : $3-\sqrt{2}$
2002 5859 sinh : $5(1/\pi)/8$
2002 5896 J_1 : $9\sqrt{3}/2$
2002 7504 rt : (4,5,-7,3)
2002 7993 as : $5(1/\pi)/8$
2002 9621 J_1 : $9/22$
2003 1614 $\ln(2/5)-\exp(1/4)$
2003 1634 rt : (9,4,-4,5)
2003 2576 e^x : $(e)^{1/2}$
2003 3138 $\ln(\sqrt{5})+\exp(1/3)$
2003 3915 cu : $1/4+3\pi/2$
2003 4017 sq : $\sqrt{2}/2-2\sqrt{3}/3$
2003 4281 id : $\zeta(3)/6$
2003 7343 J_1 : $9(1/\pi)/7$
2003 7485 $\ln(5)\times s\pi(\sqrt{3})$
2003 9425 rt : (9,4,-9,-1)
2003 9579 2^x : $1/4+3\sqrt{3}/2$
2004 0278 J_0 : $\sqrt[3]{17/2}$
2004 0818 Li_2 : $4/21$
2004 0951 sr : $2-\sqrt{5}/4$
2004 1393 2^x : $\sqrt{12}\pi$
2004 1472 rt : (9,9,2,5)
2004 4230 rt : (4,2,-4,-5)
2004 4362 2^x : $4/3+3\sqrt{3}/4$
2004 4820 tπ : $3\sqrt{3}/4-\sqrt{5}$
2004 6815 2^x : $\exp(-4/3)$
2004 7100 rt : (3,8,-7,-6)

2004 9961 $\ln(\sqrt{5}) \times \exp(2/5)$
2005 0323 $\Psi : 9\sqrt{2}/10$
2005 1714 sr : $3\sqrt{3}/4 + 4\sqrt{5}$
2005 1882 cu : $3 - 3\pi$
2005 3776 $J_0 : 9\zeta(3)$
2005 5971 rt : (9,9,-3,-1)
2005 6613 $\sqrt{3} \times \ln(2)$
2005 7906 $\Psi : 14/11$
2006 0687 rt : (1,7,-2,-4)
2006 1115 rt : (4,2,-9,1)
2006 1492 $2^x : 10(e+\pi)/9$
2006 2455 cr : $6\sqrt[3]{3}/5$
2006 3548 rt : (8,1,-7,4)
2006 3955 $J_1 : 4(e+\pi)/7$
2006 4465 $\ell_{10} : 4\sqrt{3} + 4\sqrt{5}$
2006 4973 $\Gamma : 10/13$
2006 4973 $e^x : 3/4 - 3\pi/4$
2006 5100 rt : (9,8,8,-2)
2006 5888 at : $4e/3 - \pi/3$
2006 6183 rt : (4,9,8,-2)
2006 7069 ln : 9/11
2006 8140 at : $2 + \sqrt{3}/3$
2006 8666 $\ell_{10} : \sqrt[3]{4}$
2006 9088 at : $3/2 - 3e/2$
2006 9385 $3/4 - \ln(\sqrt{3})$
2006 9653 rt : (3,4,-3,6)
2007 0017 $\Gamma : 5\ln 2$
2007 0347 as : $2/3 - \sqrt{3}/2$
2007 0884 rt : (9,3,9,-2)
2007 0928 $\ln(2/5) + \exp(3/4)$
2007 1124 sq : $4\sqrt{3} + 4\sqrt{5}/3$
2007 2172 rt : (4,4,9,-2)
2007 3507 rt : (1,8,4,-5)
2007 3533 $\ell_2 : 1/4 + 4\sqrt{5}$
2007 4893 rt : (6,6,-9,-2)
2007 6862 $\theta_3 : 1/\sqrt[3]{7}$
2007 9054 sq : $4\sqrt{2}/3 - 4\sqrt{5}/3$
2007 9179 id : $\sqrt{2} + \sqrt{5} - \sqrt{6}$
2007 9835 rt : (7,2,-5,-9)
2008 0166 $2^x : e/4 + 3\pi/4$
2008 0749 $10^x : \sqrt[3]{7/3}$
2008 0986 $10^x : \sqrt{3} - \sqrt{5} + \sqrt{6}$
2008 2225 at : $1/4 - 2\sqrt{2}$
2008 3424 rt : (8,6,-1,4)
2008 5404 $2^x : 4e - \pi/4$
2008 6020 rt : (9,9,-8,-7)
2008 6166 rt : (2,4,-8,-1)
2008 6581 $2^x : 2/3 + \sqrt{2}/3$
2008 7377 cu : $9\zeta(3)$
2008 9017 rt : (3,5,-7,-4)
2009 2351 $erf : \sqrt[3]{2}/7$
2009 2743 $J_2 : 8\zeta(3)/7$
2009 3695 id : $(\sqrt[3]{3})^{1/2}$
2009 6189 id : $3/2 - 3\sqrt{3}/4$
2009 7376 $Li_2 : 3(1/\pi)/5$
2009 8871 $J_0 : \sqrt{5} + \sqrt{6} - \sqrt{7}$
2009 9655 $2^x : 2/3 - 4\sqrt{5}/3$
2009 9741 rt : (7,3,1,9)
2010 1012 cu : $2 - \sqrt{2}$
2010 1356 $10^x : -\sqrt{2} - \sqrt{3} + \sqrt{6}$

2010 2801 sr : $6\zeta(3)/5$
2010 3037 $\ell_2 : 4 + 3\sqrt{3}$
2010 4124 $\zeta : 5\zeta(3)/2$
2010 5005 rt : (2,8,-5,-9)
2010 5651 $1/4 + s\pi(2/5)$
2010 6850 rt : (6,5,-8,-8)
2010 9017 $\zeta : \sqrt{8/3}$
2010 9531 $10^x : (1/\pi)/4$
2010 9668 rt : (9,8,3,-1)
2011 1063 $E : 7/8$
2011 1273 $J_0 : (\sqrt{2}/3)^{-3}$
2011 2240 id : $(\ln 2)^{-1/2}$
2011 2578 cr : $5\ln 2/2$
2011 2789 rt : (8,6,-6,-2)
2011 3254 $J : 4/19$
2011 4843 rt : (5,7,-7,4)
2011 5988 $J_2 : (\sqrt[3]{4}/3)^{-1/2}$
2011 6986 sr : $1/2 + 2\sqrt{2}/3$
2011 7764 $J_2 : \exp(1/\pi)$
2011 7973 $1/4 \times \ln(\sqrt{5})$
2011 8502 $2^x : 1/2 + \pi/2$
2011 9994 sr : $6e\pi/5$
2012 1712 cu : $(e+\pi)/10$
2012 2602 $J_2 : 11/8$
2012 3000 $s\pi : \sqrt{2}/3 + 2\sqrt{3}$
2012 3381 $s\pi : 2\sqrt{2} + \sqrt{5}$
2012 4300 $erf : \sqrt[3]{3}/8$
2012 4373 $\ell_{10} : \sqrt{2}/3 + \sqrt{5}/2$
2012 5865 rt : (7,3,-4,3)
2012 6341 $J_1 : 3(e+\pi)/4$
2012 7171 rt : (9,3,4,-1)
2012 7818 cu : $2/3 + \pi/2$
2012 8738 at : $\sqrt{20/3}$
2013 1397 rt : (8,8,8,-2)
2013 2869 rt : (7,8,7,9)
2013 3555 tanh : $1/\sqrt{24}$
2013 3600 sinh : 1/5
2013 3617 rt : (6,0,-2,8)
2013 5056 $10^x : \sqrt{15}$
2013 5532 id : P_8
2013 5792 at : $1/\sqrt{24}$
2013 5916 rt : (1,8,9,1)
2014 0570 sr : $5\sqrt{3}/6$
2014 0761 rt : (6,6,-4,-1)
2014 2049 sq : $\pi/7$
2014 2394 rt : (2,3,1,9)
2014 3225 rt : (8,3,9,-2)
2014 3383 $e^x : \sqrt{2}/2 - 4\sqrt{3}/3$
2014 3690 rt : (1,7,-9,7)
2014 4351 sr : $3e + 2\pi/3$
2014 5872 rt : (3,4,9,-2)
2014 7233 rt : (1,6,3,-6)
2014 8620 rt : (2,5,-9,-2)
2014 9749 $10^x : 4\sqrt[3]{5}$
2015 0689 $\zeta : (\ln 2)^{-3}$
2015 1367 rt : (9,2,5,1)
2015 1476 rt : (7,6,-9,-2)
2015 2089 rt : (1,8,7,9)
2015 3981 $\ln(\sqrt{3}) + \exp(\sqrt{3})$
2015 4171 $e^x : 3e$

2015 4964 $J_1 : 7/17$
2015 5187 rt : (9,0,-8,6)
2015 5346 $e^x : P_{63}$
2015 5540 $J_1 : 5\sqrt{5}$
2015 7411 rt : (4,1,5,1)
2015 9463 as : P_{16}
2015 9591 as : P_{40}
2016 1714 $J_0 : 3e/4$
2016 3386 $\ell_2 : 5/23$
2016 3967 rt : (1,6,7,-3)
2016 4153 sq : P_{41}
2016 4870 $\lambda : \exp(-19\pi/14)$
2016 5421 tanh : $\exp(-\sqrt[3]{4})$
2016 6129 $2^x : 9\sqrt[3]{5}/2$
2016 6701 rt : (3,9,-1,3)
2016 6752 $\ell_{10} : 4\sqrt{2}/9$
2016 7165 rt : (4,6,-3,7)
2016 7676 at : $\exp(-\sqrt[3]{4})$
2016 7818 rt : (5,3,-5,-7)
2016 8571 sinh : $\zeta(3)/6$
2016 8807 rt : (6,0,-7,2)
2016 9330 $J_1 : 2\sqrt[3]{3}/7$
2017 0474 sq : $5\pi^2/4$
2017 0480 rt : (7,3,-9,-3)
2017 0781 as : $\zeta(3)/6$
2017 0845 sinh : $\sqrt{17/2}$
2017 1103 rt : (6,1,-5,-1)
2017 3024 $10^x : \sqrt{3}/2 + 3\sqrt{5}/4$
2017 6983 sq : $4/3 + 4\pi$
2017 7452 rt : (6,9,-7,5)
2017 8208 at : $4 - \sqrt{2}$
2017 8276 rt : (6,5,4,8)
2018 0419 $\zeta : 2(1/\pi)$
2018 2013 $\ell_{10} : \pi/5$
2018 2813 cu : $4\sqrt{2}/3 - 3\sqrt{3}/4$
2018 3509 cr : $5e\pi/4$
2018 3604 rt : (7,8,2,3)
2018 5042 sr : 13/9
2018 5115 $2^x : 5(1/\pi)/6$
2018 5760 $e^x : P_{47}$
2018 7045 rt : (9,7,6,1)
2018 7464 id : $5\sqrt[3]{3}/6$
2018 7722 at : $4\sqrt{2}/3 - 2\sqrt{5}$
2018 8623 $erf : \exp(-\sqrt[3]{5})$
2018 9651 $\exp(2/5) \div \exp(2)$
2018 9921 $\exp(2/3) + s\pi(\sqrt{3})$
2019 1177 rt : (1,6,-2,-8)
2019 2517 rt : (9,5,-2,6)
2019 2541 $10^x : 4\sqrt{2}/3 + 4\sqrt{3}/3$
2019 3609 $J_1 : \exp(\sqrt{2}\pi/3)$
2019 3705 $10^x : 3 - \sqrt{5}/2$
2019 4336 id : $\exp(1/(2e))$
2019 4887 cr : $\sqrt{3} + 4\sqrt{5}$
2019 5420 $\ell_2 : 3e + \pi/3$
2019 5822 $\Gamma : \sqrt{5} + \sqrt{6} + \sqrt{7}$
2019 6418 rt : (4,6,6,1)
2019 7216 $t\pi : 1/2 + 2e$
2019 8934 rt : (7,8,8,-2)
2020 0607 $Ei : 25/14$
2020 2349 rt : (4,6,-8,-4)

2020 3050 id : $\sqrt{2}/7$
2020 4102 sq : $2\sqrt{2} - \sqrt{3}$
2020 4303 id : $3e + \pi/3$
2020 5690 id : $\zeta(3)$
2020 6803 rt : (1,5,-6,1)
2020 7843 $\ln(\sqrt{3}) \div \exp(1)$
2021 3682 $10^x : (2e/3)^2$
2021 6603 sinh : $\sqrt[3]{13}$
2021 7083 rt : (7,3,9,-2)
2021 7549 $J_1 : (e+\pi)^{-1/2}$
2021 7688 $1/3 \div \exp(1/2)$
2021 8000 sq : $1/\sqrt[3]{11}$
2021 8377 rt : (6,4,-6,1)
2021 9373 $\zeta : (\sqrt{2}\pi/3)^{1/2}$
2022 1167 rt : (2,4,9,-2)
2022 1593 $J_2 : 5e$
2022 2576 rt : (8,8,3,-1)
2022 2649 sinh : $\exp(\sqrt[3]{5}/2)$
2022 3538 $Ei : 22/15$
2022 3704 $e^x : 15/19$
2022 4266 rt : (5,2,1,7)
2022 4299 $10^x : 3\pi/7$
2022 5080 $J_1 : 7\pi/5$
2022 5411 rt : (3,5,-9,-2)
2022 5619 rt : (8,2,-5,5)
2022 6018 ln : $\zeta(3)/4$
2022 6443 $10^x : 2/25$
2022 8037 at : $\sqrt{2}/4 + \sqrt{5}$
2022 8421 $\ell_{10} : 3\sqrt{2}/4 - \sqrt{3}/4$
2022 8529 $e^x : 1 - 3\sqrt{3}/2$
2022 9070 rt : (3,9,3,-1)
2022 9355 $J_1 : 5(e+\pi)/2$
2022 9558 $\Psi : ((e+\pi)/3)^2$
2022 9825 rt : (8,6,-9,-2)
2023 0009 $t\pi : 2\sqrt{2} - 2\sqrt{3}$
2023 0859 sr : $7(e+\pi)/4$
2023 2166 rt : (5,8,-3,8)
2023 2217 $10^x : 9(1/\pi)/4$
2023 2955 rt : (2,5,-2,8)
2023 3974 as : $3/2 - 3\sqrt{3}/4$
2024 0033 id : $3e\pi/8$
2024 0649 rt : (9,5,-7,0)
2024 0698 $J_1 : 9e/2$
2024 1389 rt : (2,7,-4,8)
2024 2814 $\Psi : 22/13$
2024 3388 rt : (6,4,-6,-9)
2024 5337 $10^x : 3\sqrt{3}/2 - \sqrt{5}/2$
2024 6894 rt : (6,5,-1,2)
2024 9078 rt : (3,9,-1,-9)
2024 9985 ln : $1/2 + 2\sqrt{2}$
2025 0000 sq : 9/20
2025 2699 $\ln(\pi) \div \exp(\sqrt{3})$
2025 3693 $\exp(3) + \exp(3/4)$
2025 3760 $\exp(1/3) - \exp(4)$
2025 7870 cr : $7\sqrt{5}/9$
2025 8137 sq : $\pi^2/9$
2025 8889 rt : (8,3,4,-1)
2025 9026 rt : (7,1,-8,-4)
2025 9926 $Li_2 : \sqrt{3}/9$
2026 1524 $e^x : \sqrt{2}/2 + 4\sqrt{5}$

2026 1936 $\exp(3) \times s\pi(1/4)$
2026 3726 $J_1 : (\sqrt{5}/3)^3$
2026 4236 id : $(\sqrt{2}\pi/2)^{-2}$
2026 4346 at : $\sqrt{2}/4 - \sqrt{5}/4$
2026 7696 $K : 22/25$
2026 7757 rt : (6,8,8,-2)
2026 8660 $J_0 : \sqrt[3]{23}$
2026 9624 as : $1/2 + \sqrt{3}/4$
2026 9956 $10^x : \ln(1/2)$
2027 0375 $E : \sqrt[3]{2/3}$
2027 0853 rt : (4,1,-2,-6)
2027 1003 $t\pi : (1/\pi)/5$
2027 1211 ln : $3\sqrt{2}/2 + 4\sqrt{3}$
2027 1481 rt : (5,8,-8,-3)
2027 1712 $E : 1/\ln(\pi)$
2027 2429 rt : (1,9,8,-2)
2027 3245 sr : $7\ln 2$
2027 3255 ln : $\sqrt{2/3}$
2027 3366 $J : 2\sqrt[3]{5}/7$
2027 4138 as : P_8
2027 4333 cu : $7(e+\pi)/3$
2027 7271 rt : (2,5,-4,-1)
2027 7552 $J_1 : 22/5$
2028 0157 $J_1 : \sqrt{23}\pi$
2028 2431 $\Psi : (\sqrt{2}/3)^{-3}$
2028 4691 rt : (2,5,-7,-8)
2028 5315 $\theta_3 : (\pi)^{-2}$
2028 5744 $10^x : 1/4 - 2\sqrt{2}/3$
2028 7505 rt : (7,6,-4,-1)
2028 7859 rt : (9,1,-6,7)
2028 9885 at : $1/2 + 2\pi/3$
2029 2137 $erf : 2/11$
2029 2521 rt : (6,3,9,-2)
2029 3540 cr : $1/4 + 2\sqrt{5}/3$
2029 6622 rt : (7,8,-3,-3)
2029 7478 rt : (8,7,1,5)
2029 8125 rt : (1,4,9,-2)
2029 9295 $s\pi : 7\zeta(3)/9$
2029 9730 sr : $(2\pi/3)^{1/2}$
2030 0126 $erf : 4(1/\pi)/7$
2030 0191 $s\pi(1/4) \div s\pi(1/5)$
2030 1388 $10^x : 2/3 - e/2$
2030 2425 $J_1 : 1/\sqrt[3]{14}$
2030 2503 $2^x : 4/15$
2030 2825 rt : (6,9,-7,1)
2030 3130 $e^x : 1/2 - 2\pi/3$
2030 3860 $t\pi : 2\sqrt{3} + 2\sqrt{5}$
2030 3955 rt : (4,5,-9,-2)
2030 5305 rt : (6,1,0,9)
2030 9552 rt : (8,2,5,1)
2030 9776 $\Psi : \ln(2e)$
2031 0024 rt : (9,6,-9,-2)
2031 0094 sr : $6\sqrt[3]{5}$
2031 1269 at : $9\sqrt[3]{3}/5$
2031 1287 rt : (5,2,-4,1)
2031 1291 rt : (1,9,-1,8)
2031 1343 cu : $1/4 - 2\sqrt{3}$
2031 1401 $\ln(4/5) \div \ln(3)$
2031 4410 $\zeta : 3\sqrt{2}/10$
2031 5196 rt : (8,2,-9,-6)
2031 6886 tanh : $\sqrt[3]{3}/7$
2031 7878 cu : $1/4 - 4\sqrt{5}$
2031 8053 rt : (3,5,-4,-1)
2031 9227 at : $\sqrt[3]{3}/7$
2031 9479 sq : $3\zeta(3)/8$
2031 9862 $\Gamma : 5/23$
2032 0266 sr : $3/2 + 3\sqrt{5}/2$
2032 0572 $\exp(e) - s\pi(2/5)$
2032 1057 sinh : 5
2032 1864 sinh : 15/7
2032 2173 $t\pi : P_{72}$
2032 2410 rt : (3,1,5,1)
2032 6852 $\ell_{10} : 4/3 - \sqrt{2}/2$
2032 7397 cu : $4/3 - \sqrt{5}/3$
2032 7690 $t\pi : P_{69}$
2033 1116 at : $2e/3 + \pi/4$
2033 2562 $2^x : 1 + e/4$
2033 3719 $t\pi : 1/4 - 4\sqrt{2}/3$
2033 5460 $Ei : 5\sqrt{3}/3$
2033 5696 $\ell_{10} : 5\sqrt{5}/7$
2033 5922 $\ell_2 : 2\sqrt{2} - 3\sqrt{5}/4$
2033 6421 rt : (2,6,1,-4)
2033 7442 at : $3\sqrt{3}/2$
2033 7916 rt : (5,8,8,-2)
2033 8914 rt : (7,8,3,-1)
2033 9575 $s\pi : \pi^2/2$
2033 9709 $J_0 : \sqrt{17}\pi$
2034 0261 $\exp(\sqrt{2}) \div \exp(1/4)$
2034 0625 $2^x : 2\zeta(3)/9$
2034 0767 sinh : $\sqrt{2}/7$
2034 0886 rt : (5,2,-3,-8)
2034 2015 rt : (7,4,-2,4)
2034 2194 $\ln(3/4) \times s\pi(1/4)$
2034 2206 cu : $4e\pi/7$
2034 3074 as : $\sqrt{2}/7$
2034 3909 rt : (9,8,-2,0)
2034 4127 $e^x : (\sqrt[3]{5}/3)^3$
2034 5532 id : $\sqrt{2}/3 + \sqrt{3}$
2034 5912 cu : $7\pi^2/3$
2034 6313 $\Psi : 12/17$
2034 7386 sr : $4e - 3\pi$
2034 8060 $J_1 : 3\ln 2/5$
2034 9059 rt : (2,9,3,-1)
2034 9877 ln : $2/3 + \sqrt{5}/4$
2035 0009 $\theta_3 : \pi/6$
2035 1012 sr : $2/3 + 4\pi/3$
2035 1018 cu : $5\ln 2/2$
2035 1752 rt : (7,1,-5,-1)
2035 2093 $2^x : 1/\sqrt{14}$
2035 4162 cu : 10/17
2035 5803 $\zeta : 7\sqrt[3]{5}/4$
2035 8043 $J_2 : 18/13$
2035 8092 rt : (9,6,0,7)
2035 8178 $2^x : (\sqrt{2}\pi/3)^{1/3}$
2035 8244 $\zeta : 4(1/\pi)/3$
2035 8830 as : 14/15
2036 0218 $\ln(2/5) \times \exp(\pi)$
2036 2249 at : 13/5
2036 4023 $\ln(\pi) \div s\pi(2/5)$
2036 4619 sq : $1 + 2\sqrt{5}/3$
2036 5881 rt : (2,7,-8,-4)
2036 6824 $e^x : \sqrt{2}/2 + \sqrt{5}/2$
2036 9390 $\theta_3 : (e+\pi)/7$
2036 9609 rt : (5,3,9,-2)
2037 0110 $J_0 : \sqrt{19}\pi$
2037 1084 $\ln(\sqrt{2}) \times s\pi(1/5)$
2037 2637 $J : \exp(-13\pi/3)$
2037 3245 sq : $2/3 - \sqrt{5}/2$
2037 3285 $2^x : P_{66}$
2037 3859 $2^x : P_{65}$
2037 4222 cu : $9\sqrt{5}/5$
2037 4886 $Li_2 : (\pi/2)^{-1/3}$
2037 5608 $J_0 : e\pi/3$
2037 6822 rt : (4,9,-2,0)
2037 7207 $e^x : 8e/9$
2037 8573 $2^x : 2\sqrt[3]{5}/3$
2038 1140 $J_2 : 1/\ln(\ln 3)$
2038 1583 $t\pi : (5/2)^{-3}$
2038 1904 rt : (2,8,-1,3)
2038 2427 id : $7(e+\pi)/5$
2038 3172 $J_2 : 4\sqrt{3}/5$
2038 3371 $J_1 : 4(e+\pi)/3$
2038 4334 rt : (5,5,-9,-2)
2038 4479 $J_1 : 5/12$
2038 5389 rt : (8,7,6,1)
2038 7100 rt : (6,1,-5,3)
2038 9495 at : P_{39}
2039 0952 at : P_{33}
2039 2455 rt : (8,3,-3,6)
2039 2693 sq : $\ln(\pi/2)$
2039 5479 rt : (7,3,4,-1)
2039 6888 $s\pi : 1/3 + \sqrt{3}$
2039 7280 ln : 3/10
2039 8828 $2^x : 3\sqrt{2}/4 - 3\sqrt{5}/2$
2039 9197 $J_2 : 2\ln 2$
2039 9275 $2^x : \sqrt{2}/2 + \sqrt{3}/4$
2039 9806 cr : $5\pi/9$
2040 0421 cr : $1 + \sqrt{5}/3$
2040 0670 rt : (6,5,-6,-4)
2040 2345 rt : (9,7,8,-2)
2040 2960 sr : $(\ln 2/2)^3$
2040 3209 sinh : $(\sqrt{2}\pi/2)^{-2}$
2040 4382 rt : (8,7,-4,-1)
2040 4571 $2^x : 4\sqrt[3]{2}/3$
2040 5375 $s\pi : \sqrt{10}\pi$
2040 5552 as : $(\sqrt{2}\pi/2)^{-2}$
2040 5781 rt : (3,6,6,1)
2040 6190 $\Gamma : \sqrt[3]{13}$
2040 7053 $\Gamma : \exp(\sqrt[3]{5}/2)$
2040 7667 $\exp(4) \div s\pi(\sqrt{3})$
2040 8163 sq : $6\sqrt{3}/7$
2040 9463 rt : (6,6,6,9)
2040 9497 $2^x : 2 - \sqrt{3}$
2040 9694 $J_2 : \sqrt[3]{8/3}$
2041 1998 $\ell_{10} : 16$
2041 2384 rt : (5,9,-4,1)
2041 2414 id : $1/\sqrt{24}$
2041 3121 rt : (3,8,-9,-6)
2041 3253 $\exp(\sqrt{5}) \div \exp(4/5)$
2041 3848 cr : $4 + 3\sqrt{5}$
2041 4262 $\ln(3/4) + \exp(2/5)$
2041 4515 $\Psi : 8\pi$
2041 4710 $Ei : (\sqrt[3]{2}/3)^{1/3}$
2041 5037 at : $3/4 - 3\sqrt{5}/2$
2041 6847 rt : (1,9,-8,-2)
2041 8563 $\theta_3 : 11/21$
2041 8676 id : $\sqrt{3} + 2\sqrt{5}$
2041 8721 at : $4(e+\pi)/9$
2041 9544 $\Gamma : (\sqrt{2}\pi/2)^{-1/3}$
2042 1062 $2^x : 4\sqrt{2}/3 - \sqrt{5}/3$
2042 1615 $10^x : 4e/3 - \pi/4$
2042 1957 at : $1/2 - \sqrt{2}/2$
2042 3715 rt : (9,2,-4,8)
2042 4390 $s\pi(\pi) - s\pi(e)$
2042 4513 rt : (3,5,-4,-1)
2042 4830 cr : $10\pi^2/3$
2042 6788 $J_0 : 5\pi/2$
2042 7143 rt : (7,0,-6,5)
2042 7848 $Ei : 9\sqrt{3}/2$
2042 8564 rt : (2,5,-6,1)
2043 2744 rt : (5,2,-8,-2)
2043 2782 $2^x : 4/3 - 4e/3$
2043 3024 $2/5 \times \ln(3/5)$
2043 6261 sq : $1/3 - \pi/4$
2043 7012 $\Gamma : 10e/3$
2043 7186 $t\pi : 3e/2 - \pi$
2043 9853 rt : (9,2,9,-2)
2044 0155 rt : (1,7,8,-7)
2044 0760 rt : (8,6,-4,-1)
2044 1909 at : $1/4 + 3\pi/4$
2044 1910 cr : $\sqrt[3]{16/3}$
2044 5629 id : $\exp(-\sqrt[3]{4})$
2044 6283 $t\pi : (2\pi/3)^{1/3}$
2044 6436 $\ln(2/5) \times \ln(4/5)$
2044 6589 ln : $\sqrt{2}/4 + 4\sqrt{5}/3$
2044 8418 rt : (4,3,9,-2)
2044 8760 $\zeta : 18/13$
2044 9008 rt : (9,6,-5,1)
2045 0778 cr : $1/3 + \sqrt{2}$
2045 1904 rt : (8,1,-7,-7)
2045 4208 rt : (7,4,-6,1)
2045 7337 rt : (1,7,2,-6)
2045 7620 rt : (3,7,-9,-2)
2045 7959 rt : (5,3,3,8)
2045 8889 rt : (6,8,3,-1)
2046 0391 $e^x : 4\sqrt{3}/3 + \sqrt{5}$
2046 0890 $s\pi : 3\sqrt{2}/2 + 4\sqrt{5}$
2046 2164 rt : (6,3,-9,-4)
2046 2816 $J_2 : 25/18$
2046 3480 $10^x : 2e - \pi/2$
2046 5830 rt : (7,4,-7,-2)
2046 6485 ln : P_{79}
2046 6635 rt : (6,5,-9,-2)
2046 8268 $1/4 \div \exp(1/5)$
2046 8731 rt : (9,2,-8,-9)
2047 0542 rt : (2,6,-1,1)
2047 1041 $\ln(\sqrt{2}) \div \ln(3/4)$
2047 1895 $2/5 + \ln(\sqrt{5})$
2047 2707 sr : $1/3 + \sqrt{5}/2$
2047 2838 $\ell_{10} : \exp(\sqrt{2}/3)$

2047 2992 rt : (1,9,3,-1)
2047 3929 rt : (8,7,8,-2)
2047 4625 at : $7\sqrt{5}/6$
2047 5189 rt : (7,2,5,1)
2047 6721 $\Gamma : 3\sqrt{2}$
2047 7853 $J_0 : 8\sqrt[3]{5}$
2047 8667 $\exp(\sqrt{2}) \div \exp(3)$
2047 9790 $\ell_{10} : 10\sqrt[3]{3}/9$
2048 1380 rt : (1,4,-8,-9)
2048 1974 cu : $2/3 - 3\sqrt{3}/2$
2048 2452 rt : (3,8,8,-2)
2048 6183 id : $7\sqrt[3]{2}/4$
2048 6259 $\ell_2 : \exp(\sqrt{2}\pi/2)$
2048 6376 $\ell_{10} : 4\zeta(3)/3$
2048 7811 $\Gamma : 1/\ln(\pi/3)$
2048 8382 rt : (3,2,-7,7)
2048 8905 $2^x : 3e\pi/8$
2048 9425 $\ell_{10} : e - 2\pi/3$
2049 1302 rt : (2,9,-8,-2)
2049 2628 rt : (8,3,-8,0)
2049 3202 $\ell_{10} : 9\ln 2/10$
2049 3228 $\Psi : ((e+\pi)/2)^3$
2049 3601 tanh : $\exp(-\pi/2)$
2049 3843 $\zeta : 3\pi/8$
2049 4664 $\Gamma : 3/23$
2049 5761 rt : (2,1,5,1)
2049 6046 at : $\exp(-\pi/2)$
2049 6640 $J_0 : \exp(\pi/3)$
2049 7500 $10^x : 1/2 + 2e/3$
2049 9788 tanh : $3\ln 2/10$
2049 9911 $t\pi : \sqrt{5}/8$
2050 0117 rt : (4,4,-7,8)
2050 1652 sr : $2/3 + \pi/4$
2050 2236 at : $3\ln 2/10$
2050 3005 sq : $3/2 - \pi/3$
2050 3035 $10^x : 4e/3 + \pi/2$
2050 3078 rt : (7,9,4,4)
2050 5661 $10^x : \sqrt{3} + \sqrt{5}$
2050 7113 cr : 7/4
2050 7243 rt : (9,2,-9,2)
2050 7528 rt : (5,2,-9,-5)
2050 8421 rt : (5,6,-7,9)
2050 8664 $J_0 : 5\sqrt[3]{5}/3$
2051 0403 $J_1 : 15$
2051 5928 cr : $4\sqrt{3}/3 - \sqrt{5}/4$
2051 8047 sr : $3e/2 + \pi/4$
2051 8163 sr : $\sqrt{2}/2 + \sqrt{5}/3$
2051 8292 rt : (3,9,4,-3)
2051 8398 $10^x : 2\zeta(3)/7$
2051 8735 rt : (8,2,9,-2)
2051 9436 cu : $\sqrt{3} + \sqrt{5}/3$
2051 9506 rt : (5,1,-1,-9)
2051 9808 rt : (8,8,3,6)
2052 2168 $s\pi(2/5) + s\pi(\sqrt{3})$
2052 4079 sr : $4\sqrt{2}/3 - \sqrt{3}/4$
2052 5155 sq : $e/6$
2052 8243 $J_0 : 7(e+\pi)/3$
2052 8608 rt : (9,7,2,8)
2052 9026 rt : (3,3,9,-2)
2052 9058 sinh : $3e/8$
2052 9583 $Ei : \sqrt[3]{5}/4$
2053 0259 $t\pi : 8\pi^2/7$
2053 1631 $J_1 : \sqrt{22}\pi$
2053 6684 $2/5 \div \exp(2/3)$
2053 7067 tanh : 5/24
2053 7306 rt : (6,3,4,-1)
2053 9111 $J_1 : \sqrt[3]{2}/3$
2053 9538 at : 5/24
2053 9756 rt : (2,4,-9,-2)
2054 1422 ln : $2\sqrt{3} - \sqrt{5}$
2054 3093 rt : (8,1,-5,-1)
2054 3232 $t\pi : \sqrt{2}/3 + 2\sqrt{3}$
2054 3637 $t\pi : 2\sqrt{2} + \sqrt{5}$
2054 4274 sq : $4\sqrt{2} - 4\sqrt{3}/3$
2054 4365 $2^x : 4 - 2\pi$
2054 4951 $\Gamma : \pi^2$
2054 6024 $\Gamma : (\ln 2/3)^2$
2054 6159 rt : (1,9,-2,-1)
2054 6360 id : $\sqrt{2}/4 - \sqrt{5}/4$
2054 6510 $1/5 + \ln(2/3)$
2054 6939 rt : (7,7,8,-2)
2054 8104 rt : (2,9,-6,7)
2054 8507 id : $8\zeta(3)/3$
2054 9569 $J_0 : 9\sqrt[3]{3}$
2055 0953 rt : (7,5,-9,-2)
2055 4411 tanh : $1/\sqrt{23}$
2055 4463 sinh : $1/\sqrt{24}$
2055 5243 rt : (2,6,-8,9)
2055 6893 at : $1/\sqrt{23}$
2055 6943 rt : (4,8,0,4)
2055 7072 $Ei : 7\sqrt{3}/3$
2055 7304 $Ei : 8e\pi/7$
2056 1942 at : $4\sqrt{2}/3 - 3\sqrt{5}/4$
2056 2893 rt : (9,3,-2,9)
2056 4309 $J_1 : 7\sqrt[3]{2}/2$
2056 4310 rt : (8,4,-1,7)
2056 4356 $10^x : 9\ln 2/10$
2056 4646 id : $\sqrt{2}/3 - 3\sqrt{5}/4$
2056 4891 sinh : $2e\pi/5$
2056 6304 $10^x : 7\sqrt{5}/6$
2056 6612 rt : (7,5,0,5)
2056 7255 rt : (3,8,-3,7)
2056 7338 rt : (3,9,-8,-2)
2056 7778 $t\pi : 5\sqrt[3]{3}/3$
2056 9091 rt : (2,9,2,-2)
2056 9313 $J_1 : \exp(-\sqrt{3}/2)$
2057 1008 rt : (6,6,1,3)
2057 2683 $e^x : 4 - 3\sqrt{5}/4$
2057 5050 ln : $5\pi^2/2$
2057 6256 id : $3e/2 - 2\pi$
2057 7136 sq : $3/2 - 3\sqrt{3}/2$
2057 7907 $\zeta : 4\sqrt{5}/3$
2057 8149 rt : (4,5,-4,-1)
2058 1166 $E : (\pi/3)^{-3}$
2058 1659 $\Psi : 10(1/\pi)/7$
2058 2582 $s\pi : \exp(-e)$
2058 2727 rt : (5,8,3,-1)
2058 4889 $4 \div s\pi(2/5)$
2058 5211 $e^x : 3e/7$
2058 6209 rt : (1,8,6,2)
2058 6985 $Ei : 7(1/\pi)/2$
2058 7043 sr : $4/3 + 4\sqrt{5}$
2058 7173 rt : (4,8,-8,-9)
2058 7971 $\sqrt{3} \times \exp(\sqrt{5})$
2058 8373 sinh : $\exp(-\sqrt[3]{4})$
2058 9524 $J_1 : 8/19$
2059 0693 sr : $4 + \sqrt{3}/2$
2059 0784 $2^x : 3e/4 + \pi/2$
2059 0823 as : $\exp(-\sqrt[3]{4})$
2059 1979 rt : (8,0,-5,8)
2059 2739 at : $5\pi/6$
2059 3249 at : $3/2 + \sqrt{5}/2$
2059 4770 rt : (3,7,-7,-7)
2059 6488 rt : (7,7,6,1)
2059 7190 $J_0 : 8e\pi/5$
2059 7343 $\ell_{10} : (2e)^3$
2059 8232 $e^x : 10e\pi/9$
2059 8660 $2^x : 1 + 2\sqrt{2}$
2059 9396 rt : (7,2,9,-2)
2060 0361 rt : (7,1,-4,6)
2060 1132 rt : (9,6,-4,-1)
2060 2288 $J_1 : 10e\pi/7$
2060 2307 cu : $1/\ln(2e)$
2060 2360 $\zeta : 3e/5$
2060 3565 id : $\sqrt[3]{3}/7$
2060 4537 sr : 16/11
2060 7467 rt : (2,6,-6,-5)
2060 8269 id : $(\sqrt[3]{5}/3)^{-1/3}$
2061 0848 rt : (8,4,-9,-9)
2061 0876 $Ei : 2(e+\pi)$
2061 1518 rt : (2,3,9,-2)
2061 3321 $\ln(3/4) \div \exp(1/3)$
2061 3736 sr : $4\sqrt{2}/3 + 4\sqrt{5}/3$
2061 4015 rt : (6,2,-3,4)
2061 5971 $\zeta : \exp(-\sqrt[3]{5}/2)$
2061 6485 at : $2\sqrt{2}/3 + 3\sqrt{5}/4$
2061 6580 $\zeta : 1/\sqrt[3]{13}$
2061 6907 $10^x : e - 2\pi/3$
2062 1019 $\Psi : 6\sqrt{2}/5$
2062 1296 rt : (7,3,-8,-7)
2062 1434 rt : (6,7,8,-2)
2062 2971 $\Gamma : ((e+\pi)/3)^3$
2062 3024 $e^x : 3/16$
2062 4170 rt : (3,4,-9,-2)
2062 5439 $J_1 : 7\sqrt{5}/2$
2062 6933 $e^x : 2 + \pi/4$
2062 7690 at : $\sqrt[3]{18}$
2062 7741 $10^x : 9\pi^2/7$
2062 7815 $J_2 : (e)^{1/3}$
2062 7937 $J_1 : (4/3)^{-3}$
2062 8768 $e^x : 4\sqrt{2}/3 - 2\sqrt{3}$
2062 9388 rt : (8,7,-9,-7)
2062 9947 rt : (2,6,6,1)
2063 0424 $e^x : P_9$
2063 0862 $\Gamma : (\sqrt[3]{5})^{-1/2}$
2063 2542 $10^x : (e)^{-1/3}$
2063 2676 $e^x : 4\sqrt{2}/3 + 3\sqrt{5}$
2063 2982 cu : 13/22
2063 2996 rt : (1,8,8,-2)
2063 3217 $\Gamma : 13/17$
2063 3909 $J_1 : 7e\pi/4$
2063 5046 at : $1/2 + 3\sqrt{2}/2$
2063 6651 cu : $2\sqrt{3} + \sqrt{5}/2$
2063 6768 rt : (6,2,-7,-5)
2063 7389 rt : (8,5,-9,-2)
2063 9273 $\ell_2 : 4\sqrt[3]{3}/5$
2064 0192 rt : (5,3,-2,2)
2064 0314 rt : (1,9,7,4)
2064 3619 sr : $e/4 + 4\pi/3$
2064 3800 $J_2 : 4\pi/9$
2064 5022 rt : (4,9,-8,-2)
2064 5087 $\ell_2 : 13/15$
2064 6064 id : $1/3 + 4e$
2064 6634 rt : (7,9,-7,1)
2064 9003 rt : (6,2,5,1)
2064 9870 $10^x : 3/2 + e/4$
2065 2649 rt : (3,8,-8,-9)
2065 2789 $10^x : \sqrt{3}/4 - \sqrt{5}/2$
2065 2820 $\exp(2/3) - \exp(e)$
2065 3201 $\Psi : \exp(\sqrt[3]{4}/3)$
2065 4153 $e^x : 7\sqrt{2}/6$
2065 7955 $1/2 \times \exp(5)$
2065 8038 $t\pi : 2\sqrt[3]{2}/5$
2065 8705 rt : (1,9,-9,-4)
2065 9407 $\ell_{10} : 1/4 + e/2$
2065 9483 $erf : (\sqrt[3]{5}/3)^3$
2066 0337 sr : $4\sqrt{3} + 3\sqrt{5}/2$
2066 0678 $e^x : 2 - 2e/3$
2066 0990 id : $9\pi^2/4$
2066 1157 sq : 5/11
2066 1986 ln : $2e - 2\pi/3$
2066 2032 rt : (5,1,-6,-3)
2066 4959 rt : (9,7,-3,2)
2066 5139 $J_1 : 3\pi^2/2$
2066 7422 $\ell_{10} : \ln(5)$
2066 7693 rt : (3,5,-6,1)
2066 8442 $\ell_{10} : 2/3 + 2\sqrt{2}/3$
2067 1039 $2^x : \sqrt{2}/2 - 4\sqrt{5}/3$
2067 1064 $e^x : 4 - 2\sqrt{3}/3$
2067 3647 $2^x : 7\sqrt{5}/3$
2067 3824 at : $1 - 4e/3$
2067 4060 sr : $4 + 2\pi$
2067 5214 $J_2 : 5\sqrt{5}/8$
2067 5269 rt : (9,3,-7,3)
2067 6678 $J_2 : ((e+\pi)/3)^{1/2}$
2067 6826 id : P_{39}
2067 7792 sq : $10(1/\pi)/7$
2067 8346 id : P_{33}
2067 8349 rt : (1,1,5,1)
2068 0003 $J_2 : (\ln 3/3)^{-1/3}$
2068 1737 at : 21/8
2068 1917 rt : (6,2,9,-2)
2068 3827 $\ell_2 : 5\ln 2/4$
2068 4013 sq : $2/3 - 4\sqrt{3}$
2068 4519 sr : $(\sqrt{2}/3)^{-1/2}$
2068 4625 rt : (6,9,-8,3)
2068 4776 rt : (5,3,4,-1)
2068 6365 $J_2 : 3e\pi/2$
2068 7992 $\Psi : 5/11$
2068 8305 cu : $3e + \pi/3$

2068 9073 rt : (9,7,3,-1)
2068 9602 cr : $3(e+\pi)/10$
2069 1801 J_1 : 10/3
2069 2610 J_0 : 13
2069 3739 as : $\sqrt{2}/4-\sqrt{5}/4$
2069 4896 id : $(\ln 3)^2$
2069 5390 tanh : $\sqrt[3]{2}/6$
2069 5402 ℓ_{10} : $\sqrt{2}-\sqrt{6}+\sqrt{7}$
2069 5981 rt : (1,3,9,-2)
2069 7411 sinh : $\sqrt{2}+\sqrt{5}+\sqrt{7}$
2069 7472 rt : (5,8,9,8)
2069 7959 at : $\sqrt[3]{2}/6$
2069 8735 cr : $(2e)^{1/3}$
2069 9788 ln : $3\sqrt{3}/2+\sqrt{5}/3$
2070 0184 sπ : $1/2+4\pi$
2070 1168 rt : (1,5,5,7)
2070 1587 sq : $9(1/\pi)$
2070 1815 rt : (5,4,5,9)
2070 4334 rt : (9,8,4,9)
2070 6217 rt : (5,7,-8,2)
2070 7181 e^x : 19/24
2070 7843 $3/4\times\ln(5)$
2070 8566 2^x : $\sqrt{17/3}$
2071 0678 id : $1/2+\sqrt{2}/2$
2071 1164 $1/3\div\ln(5)$
2071 2196 $\ln(2/3)\times\ln(3/5)$
2071 3435 tπ : $3\sqrt{3}/10$
2071 4737 θ_3 : $(5/2)^{-1/3}$
2071 5261 e^x : $(\ln 2/3)^{-2}$
2071 5537 10^x : $(e/2)^{-1/2}$
2071 5766 $\ln(4/5)-s\pi(\pi)$
2071 6398 10^x : $3/2-\sqrt{2}/2$
2071 6888 E : 20/23
2071 8081 at : $\sqrt{2}/4-4\sqrt{5}/3$
2071 9834 ℓ_{10} : $\sqrt{3}/2+\sqrt{5}/3$
2071 9916 id : $7\zeta(3)/2$
2072 0869 rt : (2,7,-5,4)
2072 0970 sq : $\sqrt{2}+\sqrt{5}+\sqrt{6}$
2072 2810 tπ : $3\sqrt{2}-3\sqrt{5}$
2072 2970 id : $e/2-4\pi$
2072 4270 Γ : $-\sqrt{2}+\sqrt{5}+\sqrt{7}$
2072 4427 rt : (5,9,-8,-2)
2072 5992 e^x : P_{36}
2072 6053 rt : (9,5,-9,-2)
2072 7810 ln : $2\sqrt{3}-3\sqrt{5}/2$
2072 8325 2^x : $4/3+\pi/3$
2073 0752 $\pi\div\exp(e)$
2073 0908 tπ : $7\zeta(3)/9$
2073 1536 ℓ_2 : P_{10}
2073 2393 J_0 : $(1/(3e))^{-1/2}$
2073 3281 Ei : 3/4
2073 3434 J_2 : $10\sqrt[3]{2}/9$
2073 3679 2^x : $e/10$
2073 3767 rt : (1,9,-3,-1)
2073 4997 10^x : $\sqrt{2}/3-2\sqrt{3}/3$
2073 5571 rt : (8,1,-3,9)
2073 5589 J_2 : 7/5
2073 6600 rt : (4,5,-8,1)
2073 6711 J : $2\zeta(3)/5$
2073 8261 sinh : $4\sqrt[3]{2}$
2073 8756 rt : (5,5,-4,-1)
2073 9472 J_1 : $3\sqrt{2}/10$
2073 9832 e^x : $\sqrt{10/3}$
2074 0249 cu : $\sqrt{2}/2-3\sqrt{3}/4$
2074 0253 θ_3 : 14/19
2074 1196 sinh : $\sqrt[3]{20/3}$
2074 1461 rt : (8,5,1,8)
2074 3050 rt : (4,9,4,8)
2074 5620 $e+\ln(3/5)$
2074 6431 J_1 : $4(1/\pi)/3$
2074 6503 rt : (9,1,-5,-1)
2074 7021 tanh : 4/19
2074 7137 Ψ : $(\ln 2/2)^{-1/2}$
2074 8985 $1/5\div s\pi(\sqrt{2})$
2074 9622 at : 4/19
2074 9647 sinh : $\sqrt[3]{3}/7$
2075 1001 rt : (8,9,5,7)
2075 1874 ℓ_2 : $\sqrt{3}/2$
2075 2195 as : $\sqrt[3]{3}/7$
2075 3773 rt : (4,1,2,8)
2075 5034 sπ : $\sqrt{23}\pi$
2075 5743 cr : $4/3+3\pi$
2075 6955 rt : (7,8,-8,-9)
2075 7297 at : $4\sqrt{2}/3+\sqrt{5}/3$
2075 7597 2^x : P_{30}
2075 8794 $\ln(5)+\exp(4)$
2076 0176 ζ : $\sqrt[3]{13/3}$
2076 0183 sq : $\sqrt{2}/4+\sqrt{5}/3$
2076 0680 rt : (1,6,0,8)
2076 0768 rt : (9,6,8,-2)
2076 3936 ln : 13/16
2076 5845 rt : (7,2,-6,-8)
2076 6385 rt : (5,2,9,-2)
2076 6630 Ψ : 23/6
2076 8033 Ψ : 19/15
2076 8397 sq : $5e\pi/7$
2076 9570 Ψ : $5e/8$
2077 0116 ζ : $-\sqrt{2}-\sqrt{3}+\sqrt{7}$
2077 0313 J_0 : $\exp(1/\sqrt{2})$
2077 1338 10^x : $1/\ln(\ln 2/3)$
2077 3818 tπ : $\pi^2/2$
2077 4696 Γ : $3\pi/4$
2077 4928 at : $4/3+3\sqrt{3}/4$
2077 5118 rt : (4,7,8,-2)
2077 6279 rt : (7,5,-5,-1)
2077 8928 Ψ : $(3\pi/2)^{-3}$
2077 8963 rt : (7,2,-2,7)
2077 9928 $\exp(1/4)-\exp(2/5)$
2078 1236 J_2 : $5e/3$
2078 2512 rt : (3,9,-2,0)
2078 2540 e^x : $1/\ln(\sqrt[3]{4}/3)$
2078 3650 J_1 : $\exp(5/2)$
2078 5252 cu : $2\sqrt{2}-\sqrt{5}$
2078 6038 cu : $4+3e/4$
2078 7262 J_1 : $\exp(-\sqrt[3]{5}/2)$
2078 7358 J_1 : $1/\sqrt[3]{13}$
2078 7957 id : $\exp(-\pi/2)$
2078 8614 sq : $(\sqrt{2}\pi/3)^{-2}$
2078 9068 e^x : $\exp(-5/3)$
2078 9305 Γ : $10(e+\pi)/7$
2078 9968 J_0 : 20/7
2079 0059 rt : (1,8,-8,-2)
2079 1169 sπ : 1/15
2079 1230 10^x : $4\sqrt{2}-\sqrt{3}/3$
2079 2263 2^x : $2\sqrt{2}+4\sqrt{3}/3$
2079 4415 id : $3\ln 2/10$
2079 4568 J_2 : $7\zeta(3)/6$
2079 6369 rt : (6,5,-9,-7)
2079 6665 cu : $\ln(2\pi)$
2079 9386 rt : (5,4,-9,-2)
2080 0262 rt : (7,6,2,6)
2080 4413 2^x : $4/3-3\sqrt{2}/4$
2080 5024 10^x : $\exp(-5/2)$
2080 5631 rt : (6,9,-8,-2)
2080 6479 rt : (3,7,4,7)
2080 7060 as : $7\zeta(3)/9$
2080 8028 rt : (9,2,4,-1)
2080 8676 2^x : $1/3-3\sqrt{3}/2$
2080 8944 2^x : 3/11
2080 9848 cu : $\sqrt{14}\pi$
2081 1370 J_0 : $\sqrt[3]{25/3}$
2081 2525 id : $4e/9$
2081 4403 cu : $2(e+\pi)/9$
2081 5829 at : $3/4+4\sqrt{2}/3$
2081 6695 rt : (8,7,3,-1)
2081 6847 rt : (6,1,-5,-6)
2081 6977 e^x : $2/3-\sqrt{5}$
2081 7272 $\ln(2/3)\div\exp(2/3)$
2081 7902 2^x : 8/7
2081 8136 2^x : $6(1/\pi)/7$
2081 8831 J_1 : $5\sqrt[3]{2}$
2081 9922 ℓ_2 : $5\ln 2/3$
2081 9968 sπ : $3\sqrt{2}-4\sqrt{3}/3$
2081 9980 ln : $4\sqrt{2}-4\sqrt{3}/3$
2082 0393 id : $1/2+3\sqrt{5}$
2082 1865 Ei : $5(e+\pi)/4$
2082 2103 rt : (6,7,6,1)
2082 2406 rt : (2,9,-8,-7)
2082 2631 rt : (9,0,-9,5)
2082 3273 rt : (7,9,-1,-2)
2082 5088 cr : $7\sqrt[3]{2}/5$
2082 6060 cr : $4-\sqrt{5}$
2082 7068 as : P_{39}
2082 7428 2^x : $\exp(\pi/3)$
2082 7594 ℓ_{10} : 13/21
2082 8621 as : P_{33}
2083 0663 rt : (2,4,-9,-4)
2083 1831 rt : (5,2,5,1)
2083 2152 10^x : $\sqrt{2}/4+4\sqrt{3}/3$
2083 3119 rt : (5,8,-4,6)
2083 3333 id : 5/24
2083 4892 tπ : $1/3+\sqrt{3}$
2083 5382 rt : (9,1,9,-2)
2083 7752 Γ : $5\sqrt{2}/3$
2083 8347 rt : (8,6,8,-2)
2083 8591 ln : $3/4+3\sqrt{3}/2$
2083 9803 sq : $3/4-\sqrt{5}$
2084 1316 $\ln(4)+\exp(3/5)$
2084 1403 sinh : $10\sqrt[3]{3}/3$
2084 3014 rt : (6,2,-8,-2)
2084 3938 tπ : $\sqrt{10}\pi$
2084 4818 ℓ_{10} : $3/4+\sqrt{3}/2$
2084 7794 10^x : $\exp(-\sqrt{2}/3)$
2084 8233 rt : (9,4,-5,4)
2085 0179 rt : (6,3,-1,5)
2085 0648 $\exp(2/5)\div s\pi(\sqrt{5})$
2085 0694 ℓ_{10} : $8\sqrt{2}/7$
2085 1239 ln : $4(e+\pi)/7$
2085 1441 id : $1/\sqrt{23}$
2085 2893 rt : (4,2,9,-2)
2085 4415 cr : $\sqrt[3]{11/2}$
2085 4440 tπ : $4\sqrt[3]{5}/3$
2085 4441 rt : (3,7,8,-2)
2085 4787 Ψ : 17/10
2085 5121 rt : (5,4,-8,-5)
2085 5920 e^x : $(e/2)^{1/2}$
2085 6710 id : $4\sqrt{2}/3-3\sqrt{5}/4$
2085 7715 at : $e/2-\pi/2$
2085 9954 rt : (2,7,0,-6)
2086 1669 as : $1/2-\sqrt{2}/2$
2086 2635 e^x : $\sqrt{2}-4\sqrt{5}/3$
2086 3581 $\exp(1/3)\times s\pi(1/3)$
2086 4476 ℓ_2 : $3\sqrt[3]{3}/5$
2086 5670 tπ : $2\sqrt[3]{2}/9$
2086 9082 sr : $\sqrt{2}+2\sqrt{3}$
2086 9156 J_0 : $\sqrt{5}-\sqrt{6}-\sqrt{7}$
2086 9590 2^x : $1/\ln(2e/3)$
2087 1215 rt : (2,9,-3,-1)
2087 5955 Ei : $(\sqrt[3]{3}/2)^{-1/3}$
2087 6544 sinh : $\sqrt[3]{22}$
2087 7731 id : $8(e+\pi)/9$
2087 7984 cu : $1/4+4\pi$
2087 8700 ℓ_{10} : $\exp(\sqrt[3]{3}/3)$
2087 9235 tπ : 7/25
2087 9754 rt : (2,9,-7,2)
2088 0224 ℓ_{10} : $7\ln 2/3$
2088 0310 at : $2\sqrt{2}/3-2\sqrt{3}/3$
2088 1153 cr : $4/3+\sqrt{3}/4$
2088 1668 at : $4-e/2$
2088 2054 Li_2 : $2\ln 2/7$
2088 5004 rt : (8,1,-8,3)
2088 8715 rt : (7,9,-8,-2)
2088 9000 rt : (9,8,-1,3)
2088 9067 sinh : 23/15
2088 9686 sπ : $1/2+\sqrt{3}/4$
2089 0104 rt : (4,6,-4,5)
2089 0122 J_1 : $\sqrt[3]{5}/4$
2089 0410 rt : (6,4,-9,-2)
2089 0865 at : $1/2-\pi$
2089 3636 ℓ_{10} : $3\sqrt[3]{3}/7$
2089 4104 sr : 19/13
2089 4607 rt : (9,7,-8,-4)
2089 4660 sq : $1/4-\sqrt{2}/2$
2089 8764 ℓ_{10} : $1/2+\sqrt{5}/2$
2090 0771 rt : (5,0,-4,4)
2090 1568 tπ : $\sqrt{2}/2+4\sqrt{3}$
2090 3120 tπ : $3\sqrt{2}/2+4\sqrt{5}$
2090 3782 sr : $4e\pi/7$
2090 5171 at : $1/3+4\sqrt{3}/3$
2090 6362 Γ : $\zeta(3)/8$

2090 6994 rt : (6,5,-4,-1)
2090 7760 tanh : $2(1/\pi)/3$
2090 9393 $\exp(2) - \exp(4)$
2091 0459 J_1 : $\sqrt{2} + \sqrt{5} + \sqrt{7}$
2091 0464 at : $2(1/\pi)/3$
2091 1767 erf : 3/16
2091 2477 sq : $2/3 - e$
2091 3558 cr : $5\sqrt{2}/4$
2091 3991 rt : (6,7,3,4)
2091 4123 tπ : $4\sqrt[3]{2}/7$
2091 4124 rt : (7,1,-4,-9)
2091 4223 rt : (1,8,-7,-5)
2091 4884 rt : (5,0,-5,1)
2091 5780 Ei : 3/7
2091 7578 rt : (7,6,8,-2)
2091 7756 cu : $\sqrt{2}/3 - 4\sqrt{3}/3$
2091 7808 tπ : $e/2 - 3\pi$
2091 8395 $\ln(2/5) + s\pi(1/4)$
2092 1703 ζ : $(\ln 2/3)^{-1/3}$
2092 1794 cr : $\ln(e + \pi)$
2092 1823 rt : (8,1,9,-2)
2092 3452 2^x : $\sqrt{2}/2 - \sqrt{3}/4$
2092 4206 rt : (8,6,3,9)
2092 4636 e^x : $\sqrt[3]{5}/9$
2092 5720 cu : $2/3 - \sqrt{3}$
2092 5863 ℓ_2 : $1 + 4e/3$
2092 6686 ℓ_{10} : $4e/3 + 4\pi$
2092 7256 rt : (4,7,-4,-8)
2092 9328 sπ : $e - \pi/4$
2092 9471 sq : $2/3 + \sqrt{3}/4$
2092 9821 J_2 : $\exp(\sqrt{5})$
2093 1303 10^x : $7\sqrt{3}/4$
2093 2125 cr : $e/2 + 3\pi$
2093 4166 sq : $\sqrt{3}/3 - 3\sqrt{5}/4$
2093 4758 rt : (2,8,-2,-1)
2093 5223 J_2 : $(\sqrt[3]{2}/3)^{-3}$
2093 5513 rt : (2,7,8,-2)
2093 6982 rt : (8,5,-4,2)
2093 7289 E : $\ln(\sqrt[3]{2}/3)$
2093 8002 sinh : $\exp(-\pi/2)$
2093 8684 at : $2\sqrt{3}/3 + 2\sqrt{5}/3$
2093 9086 rt : (6,6,-4,-3)
2093 9328 J_1 : $\sqrt{15}\pi$
2093 9682 2^x : $e/4 + 4\pi/3$
2094 0339 J_1 : 3/7
2094 0667 as : $\exp(-\pi/2)$
2094 1541 rt : (3,2,9,-2)
2094 2845 ℓ_{10} : $3\sqrt{2}/4 + \sqrt{5}/4$
2094 2920 at : $\sqrt{7}$
2094 3295 ζ : $-\sqrt{2} + \sqrt{3} + \sqrt{7}$
2094 3791 $2/5 - \ln(5)$
2094 4600 sinh : $3\ln 2/10$
2094 5671 rt : (6,4,-7,-8)
2094 5760 id : $2\sqrt{3} + \sqrt{5}/3$
2094 6924 cr : 23/13
2094 7269 as : $3\ln 2/10$
2094 8516 cu : $3/2 - e/3$
2094 8587 rt : (7,7,3,-1)
2094 8780 ℓ_{10} : $9\sqrt[3]{2}/7$
2095 1609 at : $3 - \sqrt{2}/4$
2095 2056 rt : (4,9,-1,-3)
2095 3068 cu : $2\sqrt{3} + \sqrt{5}$
2095 4225 rt : (3,8,-8,-2)
2095 4808 J_1 : $(2e)^{-1/2}$
2095 5089 2^x : $9\sqrt{3}$
2095 6673 Ψ : $9e\pi/8$
2095 6888 Ei : $\sqrt[3]{3}/10$
2095 8294 K : $7\sqrt[3]{2}/10$
2095 8462 at : $\sqrt{2}/2 - 3\sqrt{5}/2$
2096 1191 rt : (8,2,4,-1)
2096 1337 $\ln(3) \times \exp(\sqrt{3})$
2096 1431 sq : $3/4 + 3\sqrt{3}/2$
2096 2200 rt : (2,9,3,-6)
2096 3254 rt : (7,3,0,8)
2096 6581 ln : P_{76}
2096 6947 rt : (9,1,-7,6)
2096 7229 10^x : $1/3 + 4e$
2096 7682 e^x : $1/4 - 2e/3$
2097 1814 cu : $6\ln 2/7$
2097 2139 at : $1/3 - 4\sqrt{5}/3$
2097 3770 rt : (8,9,-8,-2)
2097 4532 cr : $\sqrt{2}/3 + 3\sqrt{3}/4$
2097 5397 θ_3 : $(\sqrt{2}/3)^3$
2097 6729 rt : (6,8,-8,8)
2097 7488 rt : (5,8,-9,-5)
2097 8056 e^x : $3/2 - \sqrt{2}/2$
2097 8722 rt : (9,3,5,1)
2098 0038 rt : (2,8,3,-1)
2098 0121 J_2 : $\pi^2/7$
2098 0287 sr : $5(e + \pi)/6$
2098 0335 2^x : $5\sqrt[3]{5}/3$
2098 1711 e^x : $8\sqrt{2}$
2098 2192 rt : (1,9,6,-4)
2098 2556 e^x : 4/21
2098 2559 Li_2 : $5(1/\pi)/8$
2098 3873 rt : (7,4,-9,-2)
2098 4364 sinh : 5/24
2098 5573 rt : (7,2,-7,1)
2098 7059 as : 5/24
2098 7654 sq : $7\sqrt{2}/9$
2098 8558 rt : (5,4,0,3)
2098 9174 Γ : $(\sqrt{2}\pi)^{-2}$
2098 9391 Γ : 16/21
2099 1609 sq : $4 + 4e$
2099 1981 rt : (9,4,-6,1)
2099 3182 rt : (2,5,-3,5)
2099 5040 rt : (2,6,-9,5)
2099 7541 10^x : $2\sqrt{3}/3 + 4\sqrt{5}/3$
2099 7763 J_0 : $9e\pi/8$
2099 8532 rt : (6,6,8,-2)
2099 8684 id : $\sqrt[3]{2}/6$
2099 9458 ζ : $3\pi^2/10$
2100 0000 sq : 11/10
2100 0308 cu : $3/2 - 2\pi/3$
2100 1823 ln : $\pi^2/8$
2100 2049 ln : $3 + \sqrt{2}/4$
2100 2714 rt : (6,0,-3,7)
2100 2807 tanh : $1/\sqrt{22}$
2100 2867 sinh : $1/\sqrt{23}$
2100 4311 10^x : $3\sqrt{2} - \sqrt{3}/2$
2100 5573 at : $1/\sqrt{22}$
2100 5940 J_0 : $\sqrt{24}$
2100 6003 J_0 : $7\sqrt{5}/2$
2100 6944 sq : 11/24
2100 9493 10^x : $3/2 + 2\sqrt{2}/3$
2101 0370 rt : (7,1,9,-2)
2101 0652 $e + \exp(2/5)$
2101 0960 as : $4\sqrt{2}/3 - 3\sqrt{5}/4$
2101 1392 2^x : $7\sqrt[3]{3}/6$
2101 1691 rt : (4,2,4,9)
2101 3831 rt : (3,9,-3,-1)
2101 5128 cu : $\ln(2e/3)$
2101 5725 e^x : 23/16
2101 7917 tπ : $\sqrt{3}/2 + 3\sqrt{5}$
2101 8406 ln : $3\sqrt{5}/2$
2101 8412 rt : (2,8,-6,-8)
2101 9294 ℓ_{10} : $9\sqrt[3]{3}/8$
2102 0324 sr : $(\pi)^{1/3}$
2102 0399 cr : $3/2 - 2\sqrt{5}/3$
2102 0750 $\ln(\sqrt{2}) \div \exp(1/2)$
2102 1107 $s\pi(2/5) \div s\pi(\pi)$
2102 4651 rt : (4,2,5,1)
2102 4950 J_2 : 24/17
2102 6196 e^x : $3/4 - 4\sqrt{3}/3$
2102 6397 rt : (9,5,-3,5)
2102 6977 Ei : $(2e)^{-1/2}$
2102 8658 10^x : $2\sqrt{2}/3 + 3\sqrt{5}/2$
2102 9244 $1/5 \div s\pi(2/5)$
2103 0959 Γ : $7(e + \pi)/6$
2103 1058 10^x : $10(e + \pi)/7$
2103 2437 rt : (2,2,9,-2)
2103 2706 ℓ_{10} : $8\ln 2/9$
2103 2926 tπ : $\exp(-e)$
2103 3011 rt : (8,9,0,1)
2103 4343 sr : $3 + 4\sqrt{2}/3$
2103 5314 10^x : $1 - 3\sqrt{5}/4$
2103 5520 E : 13/15
2103 5888 sr : $(e + \pi)/4$
2103 6972 sπ : $e\pi/8$
2103 7116 Ψ : $(3\pi)^{-1/2}$
2103 8245 rt : (1,8,5,-5)
2103 9176 rt : (4,8,-8,-2)
2103 9468 rt : (5,6,-8,7)
2104 2594 rt : (5,3,-6,-6)
2104 2749 2^x : $\sqrt{3} + \sqrt{5}/2$
2104 3633 rt : (7,7,4,7)
2104 3888 e^x : $3/4 - \sqrt{5}/4$
2104 4242 e^x : $3(1/\pi)/5$
2104 4880 2^x : $5\pi/4$
2104 5843 rt : (8,9,-7,1)
2104 7892 rt : (4,7,0,9)
2104 9617 J_0 : $9\ln 2$
2105 0160 rt : (9,2,-9,-8)
2105 1348 2^x : $\sqrt{2} + 4\sqrt{3}/3$
2105 2631 id : 4/19
2105 4455 at : $3/2 + 2\sqrt{3}/3$
2105 5019 tanh : $\sqrt[3]{5}/8$
2105 5587 rt : (3,3,-9,-2)
2105 7157 $\exp(1/2) \div s\pi(\sqrt{3})$
2105 7460 ln : $4\sqrt{2} + 2\sqrt{3}$
2105 7820 at : $\sqrt[3]{5}/8$
2105 8134 Ψ : $8\zeta(3)$
2105 8298 rt : (5,3,-7,-4)
2105 8554 sq : $6\sqrt[3]{5}/5$
2106 0693 as : $3e/2 - \pi$
2106 0889 rt : (9,9,-8,-2)
2106 0968 E : $5\ln 2/4$
2106 1586 erf : $\exp(-5/3)$
2106 3211 J_0 : $5e$
2106 4172 tπ : $\sqrt{2} + \sqrt{3}/2$
2106 4398 rt : (5,7,6,1)
2106 4616 rt : (8,2,-6,4)
2106 5836 sr : $6\sqrt[3]{5}/7$
2106 7552 rt : (3,1,1,-8)
2106 8173 2^x : $7\zeta(3)/5$
2107 3102 cu : $4\sqrt{2} - 3\sqrt{5}/2$
2107 3521 tπ : $2(e + \pi)$
2107 5588 rt : (1,5,3,-7)
2107 7186 θ_3 : 2/19
2107 8921 rt : (1,8,-1,6)
2107 9910 rt : (8,4,-9,-2)
2108 0773 ln : $1 + 3\pi/4$
2108 1199 at : $3 - 4\sqrt{2}$
2108 1285 rt : (5,6,8,-2)
2108 2782 e^x : P_{52}
2108 2861 e^x : $(\sqrt[3]{4})^{1/3}$
2108 2904 rt : (4,6,-9,-5)
2108 3563 Li_2 : $(\sqrt{5}/3)^{1/2}$
2108 3623 rt : (7,5,-4,-1)
2108 5036 rt : (6,7,3,-1)
2108 5247 J_2 : $\sqrt{2}$
2108 5336 ℓ_{10} : 13/8
2108 7459 J_0 : $9(1/\pi)$
2108 9678 ℓ_2 : $(5/3)^3$
2109 5245 rt : (1,9,2,-9)
2109 6076 Ei : $5\zeta(3)/8$
2109 6286 rt : (6,4,1,6)
2109 8937 rt : (6,3,-5,-9)
2110 0377 Li_2 : 1/5
2110 1128 rt : (6,1,9,-2)
2110 1295 2^x : $4\sqrt{2}/3 + 4\sqrt{5}$
2110 2312 $\ln(4) \div \ln(\pi)$
2110 2332 2^x : $\sqrt[3]{3/2}$
2110 2373 2^x : $\sqrt{2}/4 - 3\sqrt{3}/2$
2110 3222 rt : (7,8,-2,0)
2110 4729 2^x : $\ln(\pi)$
2110 4888 K : 15/17
2110 5446 at : $4e/3 - 2\pi$
2110 5602 E : $\sqrt{3}/2$
2110 6014 sr : 22/15
2110 6497 tanh : 3/14
2110 9333 at : 3/14
2111 3935 rt : (1,3,-7,-8)
2111 4860 ℓ_{10} : $\sqrt{2}/3 + 2\sqrt{3}/3$
2111 4886 rt : (9,2,-5,7)
2111 7141 rt : (8,8,-7,-6)
2111 7489 $\exp(2) + \exp(3/5)$
2111 8489 rt : (3,8,-4,4)
2111 9649 2^x : $\ln(1/(3\pi))$
2112 0768 rt : (7,2,4,-1)

2112 1111 e^x : $4\sqrt{2} - 2\sqrt{3}/3$
2112 1680 rt : (9,9,1,4)
2112 2085 rt : (1,8,3,-1)
2112 2606 cu : $3/2 + 4\pi$
2112 4270 rt : (4,4,-8,6)
2112 4428 J_0 : $8\zeta(3)$
2112 4785 id : $5\sqrt[3]{3}$
2112 5696 rt : (1,2,9,-2)
2112 6160 rt : (5,8,-8,-2)
2112 6343 J_1 : $5\pi/2$
2112 7054 e^x : $7/6$
2112 9921 2^x : $2 - 3\sqrt{2}$
2113 0909 ln : $17/21$
2113 1427 J_1 : $3\sqrt[3]{3}/10$
2113 1850 rt : (5,9,-5,-1)
2113 2486 rt : (6,6,1,0)
2113 5919 sq : $9\sqrt[3]{5}/2$
2113 7168 cu : $2\sqrt{3}/3 - \sqrt{5}/4$
2113 8630 Li_2 : $\zeta(3)/6$
2113 9555 at : $1 - \pi/4$
2114 1372 cr : $16/9$
2114 2224 rt : (3,7,-1,-4)
2114 2853 rt : (9,5,8,-2)
2114 3126 Ψ : $8\sqrt[3]{3}/3$
2114 3427 10^x : $9\sqrt{3}/10$
2114 5671 J_2 : $17/12$
2114 6649 rt : (8,1,-8,-6)
2114 7147 J_1 : $\sqrt{3}/4$
2114 9364 Ei : $9\pi^2/7$
2114 9682 rt : (5,1,-2,5)
2115 0410 ℓ_2 : $19/22$
2115 1507 rt : (4,3,-9,-2)
2115 1904 Li_2 : $19/22$
2115 1919 sinh : $(\pi/3)^{1/2}$
2115 2765 10^x : $1/12$
2115 3345 sinh : $\sqrt[3]{2}/6$
2115 3393 ζ : $-\sqrt{3} + \sqrt{5} + \sqrt{6}$
2115 3569 rt : (7,4,2,9)
2115 5710 ℓ_{10} : $\sqrt{2} - \sqrt{5} + \sqrt{6}$
2115 6150 as : $\sqrt[3]{2}/6$
2115 6525 id : $\exp(\sqrt[3]{4}/2)$
2115 6652 J_1 : $5\ln 2/8$
2115 6742 at : $\sqrt{2}/4 + 4\sqrt{3}/3$
2115 8465 tπ : $1/2 + 4\pi$
2115 9291 cu : $3/4 - 3\pi/2$
2116 1185 Γ : $21/4$
2116 2011 rt : (4,9,-3,-1)
2116 2124 J : $(2e)^{-1/2}$
2116 4014 J_2 : $9\sqrt[3]{2}/8$
2116 4581 Γ : $8e/7$
2116 5291 J_2 : $8e\pi/9$
2116 5541 id : $e/2 - \pi/2$
2116 5917 rt : (4,6,8,8)
2116 6723 cu : $4\sqrt{2}/3 + \sqrt{5}/3$
2116 8112 cr : $9\zeta(3)$
2116 8165 Ei : e
2116 8526 ln : $2 + e/2$
2117 1015 rt : (8,3,5,1)
2117 2107 rt : (8,6,-2,3)
2117 3913 $\ln(\pi) - \exp(\sqrt{5})$
2117 4713 rt : (6,3,-6,-1)
2117 6359 e^x : $1/3 - 4\sqrt{2}/3$
2117 6665 rt : (9,1,-7,-9)
2117 7838 rt : (3,5,0,8)
2117 8668 rt : (9,4,-9,-2)
2117 9284 E : $3\sqrt[3]{3}/5$
2118 1309 Ψ : $24/19$
2118 1871 rt : (1,7,-5,-6)
2118 1973 ℓ_{10} : $\sqrt{3} - \sqrt{5}/2$
2118 1984 Ei : $7\sqrt[3]{2}/6$
2118 3698 erf : $\sqrt[3]{5}/9$
2118 5533 ζ : $(e/3)^{1/2}$
2118 5567 $\ln(4) - \exp(4)$
2118 6051 Ψ : $9\sqrt[3]{5}/4$
2118 6266 $\ln(3) + \exp(\sqrt{2})$
2118 7831 ln : $8\sqrt[3]{2}/3$
2118 9009 2^x : $2\ln 2/5$
2118 9149 id : $2\sqrt{2}/3 - 2\sqrt{3}/3$
2119 0031 rt : (1,7,-8,-2)
2119 3535 ln : $1/4 + \sqrt{5}/4$
2119 4210 rt : (5,1,9,-2)
2119 4713 rt : (6,1,-1,8)
2119 5205 J_0 : $7\sqrt{3}/6$
2119 6668 J_0 : $9e/5$
2119 6673 2^x : $1/\sqrt{13}$
2119 9170 sr : $9e/5$
2119 9922 ℓ_2 : $(\sqrt{5}/3)^{1/2}$
2119 9981 e^x : $\sqrt[3]{9/2}$
2120 2178 rt : (9,8,-6,-3)
2120 2565 at : $8/3$
2120 3842 rt : (4,2,-5,-4)
2120 5341 ln : $6\sqrt[3]{3}/7$
2120 6226 at : $4/3 - \sqrt{5}/2$
2120 6607 J_2 : $8(e + \pi)/5$
2120 8490 sinh : $4/19$
2120 9859 ℓ_{10} : $(\ln 2/3)^{1/3}$
2121 0044 rt : (9,6,-1,6)
2121 1252 rt : (5,5,-6,1)
2121 1331 as : $4/19$
2121 3609 rt : (6,9,-9,1)
2121 5276 rt : (6,8,-8,-2)
2121 5701 10^x : $2 + 3\pi/2$
2121 7049 tπ : $\sqrt{23}\pi$
2121 7138 ln : $(\sqrt[3]{4}/3)^{1/3}$
2121 8281 $\pi \div s\pi(\sqrt{3})$
2121 8305 id : $6\sqrt{2}/7$
2121 9417 J_0 : $10\sqrt{2}/7$
2121 9823 sπ : $\sqrt{3}/3 + 2\sqrt{5}/3$
2122 0659 id : $2(1/\pi)/3$
2122 1445 rt : (2,9,1,-6)
2122 1474 $4/5 - s\pi(1/5)$
2122 1600 e^x : $\sqrt{3}/9$
2122 2110 id : $\sqrt{3} - 4\sqrt{5}$
2122 2212 $\ln(4/5) \times s\pi(2/5)$
2122 3033 sq : $6\pi/5$
2122 3201 rt : (7,3,-5,2)
2122 3943 at : $\sqrt[3]{19}$
2122 4617 sq : $3\sqrt{2} - 4\sqrt{3}$
2122 6356 rt : (5,8,4,2)
2122 7273 rt : (8,5,8,-2)
2122 7403 ℓ_{10} : $\sqrt[3]{13/3}$
2122 8610 rt : (3,2,5,1)
2122 9476 J_1 : $10/23$
2123 0469 J_1 : $8\pi^2/7$
2123 1792 $1/2 + \ln(3/4)$
2123 2303 at : $\sqrt{3}/4 + \sqrt{5}$
2123 2606 2^x : $5/18$
2123 3537 Γ : $19/25$
2123 4141 id : $6\zeta(3)$
2123 4783 2^x : $e/3 - \pi$
2123 4912 Γ : $4\sqrt[3]{5}/9$
2123 5811 erf : $4/21$
2123 6224 ln : $7\ln 2/6$
2123 6310 rt : (5,2,-4,-7)
2123 6628 θ_3 : $10/19$
2123 7141 tπ : $3\sqrt{3}/4 + 4\sqrt{5}/3$
2123 8253 J_1 : 2π
2123 8898 id : $1/2 + 3\pi/2$
2123 9757 sr : $7\sqrt[3]{2}/6$
2124 1461 cu : $7e\pi/8$
2124 3306 sinh : $4(e + \pi)/7$
2124 3556 id : $7\sqrt{3}/10$
2124 4573 ℓ_{10} : $6e$
2124 5026 sq : $4(e + \pi)/7$
2124 5268 rt : (3,2,-8,5)
2124 5620 cr : $4\sqrt{2}/3 + 4\sqrt{5}$
2124 6007 θ_3 : $(1/\pi)/3$
2124 7085 rt : (9,8,6,1)
2124 7530 rt : (5,0,-9,2)
2124 9804 rt : (8,3,-4,5)
2125 0046 Ei : $4\sqrt{5}/5$
2125 0101 rt : (5,3,-9,-2)
2125 2513 rt : (3,6,8,-2)
2125 3288 Ei : $2\sqrt[3]{3}/9$
2125 4673 Γ : $(\sqrt{3})^{-1/2}$
2125 5656 tπ : $1/15$
2125 5810 J_0 : $\exp(\sqrt[3]{4})$
2125 5954 cr : $\sqrt[3]{17/3}$
2125 6743 $\ln(4) \div \ln(4/5)$
2125 8787 rt : (3,5,-8,-3)
2125 9899 $\ln(\sqrt{2}) + s\pi(1/3)$
2126 1141 rt : (9,0,9,2)
2126 6614 rt : (9,3,-3,8)
2126 7812 sr : $25/17$
2126 8088 $\ln(5) - \exp(3/5)$
2126 9142 ζ : $\sqrt[3]{5}/4$
2126 9526 rt : (8,5,-4,-1)
2127 0229 tanh : $(5/3)^{-3}$
2127 3178 at : $(5/3)^{-3}$
2127 4040 J_0 : $4(e + \pi)/3$
2127 4979 rt : (6,0,-8,1)
2127 7409 rt : (1,6,-1,2)
2127 7872 rt : (6,8,5,5)
2127 8371 rt : (4,7,-5,-2)
2127 8934 rt : (2,7,-8,-2)
2128 0268 10^x : $4\sqrt{3}/3 - 4\sqrt{5}/3$
2128 1898 ℓ_{10} : $1/3 + 3\sqrt{3}/4$
2128 3022 tπ : $9\sqrt[3]{3}/5$
2128 3390 ln : $5\sqrt{3}/7$
2128 5913 J_0 : $7\sqrt[3]{3}/5$
2128 6432 tπ : $3\sqrt{2} - 4\sqrt{3}/3$
2128 7304 rt : (6,2,4,-1)
2128 8803 at : $3e/4 - 3\pi/2$
2128 9000 ln : $-\sqrt{3} + \sqrt{6} + \sqrt{7}$
2128 9331 rt : (9,5,-8,-1)
2128 9735 rt : (4,1,9,-2)
2128 9980 rt : (7,0,-7,4)
2129 0432 rt : (5,5,-6,-9)
2129 1275 erf : $3(1/\pi)/5$
2129 2937 rt : (2,5,-1,-5)
2129 3514 sq : $4 - 2\pi$
2129 4843 2^x : $3/4 - 4\sqrt{5}/3$
2129 5089 2^x : $7(1/\pi)/8$
2129 5100 10^x : $4\sqrt{2}/3 - \sqrt{5}/4$
2129 7471 rt : (7,8,6,8)
2129 8065 e^x : $3\sqrt{3}/4 + \sqrt{5}/2$
2129 8415 cr : $2/3 + \sqrt{5}/2$
2129 8436 ℓ_{10} : $\sqrt{8/3}$
2129 9437 rt : (8,0,-6,7)
2130 0756 id : $7\ln 2/4$
2130 1339 2^x : $3/4 - \sqrt{2}/3$
2130 1464 sr : $1 + \sqrt{2}/3$
2130 1775 sq : $6/13$
2130 3306 at : $2\sqrt{2}/3 + \sqrt{3}$
2130 4219 ln : $4\sqrt{2}/7$
2130 6131 $2 \div \exp(1/2)$
2130 6629 rt : (7,8,-8,-2)
2130 7310 J_2 : $(e + \pi)$
2130 8899 J_2 : $e\pi/6$
2130 9001 sr : $\sqrt{2}/4 + \sqrt{5}/2$
2130 9005 J_0 : $25/4$
2130 9154 cu : $3e/2 + \pi$
2131 0299 at : $4\sqrt{2} - 4\sqrt{5}/3$
2131 1271 10^x : $3\sqrt{2}/4 - \sqrt{3}$
2131 3621 rt : (7,5,8,-2)
2131 4215 cr : $1 + \pi/4$
2131 5910 $1/5 - \exp(5)$
2131 6199 rt : (5,9,-3,-1)
2131 7655 Ψ : $(\ln 2/2)^{1/3}$
2131 7661 J_2 : $(\ln 2/2)^{-1/3}$
2131 8568 tanh : $\sqrt{3}/8$
2131 9818 rt : (1,5,-2,-8)
2132 0071 id : $1/\sqrt{22}$
2132 1375 cr : $25/14$
2132 1550 at : $\sqrt{3}/8$
2132 1775 Ψ : $7\sqrt[3]{3}/8$
2132 2802 J_1 : $1/\sqrt[3]{12}$
2132 3386 ℓ_2 : $3\sqrt{3} - \sqrt{5}/4$
2132 4361 sr : $(1/\pi)/7$
2132 4815 rt : (4,8,-9,-8)
2132 5931 cu : $\sqrt{20/3}$
2132 6075 rt : (4,7,6,1)
2132 6733 rt : (9,6,3,-1)
2132 6843 as : $e/2 - \pi/2$
2132 7063 Li_2 : $\sqrt{2}/7$
2132 7310 ℓ_2 : $4\sqrt{2}/3 + \sqrt{3}/4$
2132 7459 rt : (5,7,-9,0)
2133 0155 at : $1 + 3\sqrt{5}/4$
2133 1072 rt : (1,4,1,8)
2133 2545 rt : (7,3,-9,-6)

2133 3664 cu : $3/4 - 4\sqrt{5}$
2133 3894 rt : (8,3,-8,-9)
2133 4934 $5 \div \ln(\sqrt{5})$
2133 6383 sr : $\sqrt{24}$
2133 7441 ζ : $(\sqrt[3]{3}/3)^{-1/3}$
2133 9539 cu : $4 + 4e/3$
2134 0435 $\pi \times \exp(2)$
2134 1166 rt : (3,0,3,9)
2134 4425 $\sqrt{2} \times \exp(4)$
2134 6892 tπ : $3\sqrt{2}/2 + 3\sqrt{3}/2$
2134 7520 ℓ_2 : $\exp(5)$
2134 9841 2^x : $2\sqrt{2}/3 + 2\sqrt{3}$
2135 0195 ln : $7\sqrt[3]{3}/3$
2135 0601 J_2 : $5\sqrt[3]{5}/6$
2135 0999 as : $2\sqrt{2}/3 - 2\sqrt{3}/3$
2135 1523 rt : (6,3,-9,-2)
2135 1631 cr : $2\sqrt{3} - 3\sqrt{5}/4$
2135 2666 sπ : $8\zeta(3)/9$
2135 3107 rt : (6,5,3,7)
2135 3467 sq : $2\ln 2/3$
2135 5676 J_1 : $7\sqrt{3}$
2135 5780 J_1 : $7/16$
2135 6394 rt : (8,0,9,2)
2135 7319 2^x : $3/2 + 2\sqrt{3}$
2135 8519 rt : (5,5,2,4)
2135 8702 2^x : P_{72}
2135 9305 2^x : P_{69}
2136 0959 tπ : $1/2 + \sqrt{3}/4$
2136 0981 at : $2 + e/4$
2136 2846 sπ : $10\ln 2$
2136 2962 cu : $16/15$
2136 5230 $2/3 \div \ln(\sqrt{3})$
2136 5402 E : $19/22$
2136 5404 rt : (7,0,-5,1)
2136 5629 rt : (2,4,-5,9)
2136 6031 cu : 8π
2136 7988 2^x : $3/2 - \sqrt{2}/4$
2136 8853 sπ : $\sqrt{3}/3 + 3\sqrt{5}/2$
2137 0035 rt : (3,7,-8,-2)
2137 0353 at : $3\sqrt{2}/2 + \sqrt{5}/4$
2137 1444 sr : $10e$
2137 2896 rt : (4,7,3,-1)
2137 4036 rt : (7,3,5,1)
2137 4699 id : $\sqrt[3]{5}/8$
2137 5246 ℓ_2 : $1/3 + 4\sqrt{5}$
2137 5585 $2 \div \exp(\sqrt{5})$
2137 8129 2^x : $\sqrt{5}/8$
2137 8967 10^x : $8\ln 2$
2137 9811 sπ : $9\pi/4$
2138 0284 sinh : $2(1/\pi)/3$
2138 1078 ln : $3\ln 2/7$
2138 1097 10^x : $\sqrt{5}$
2138 1772 J_1 : $\sqrt{11}$
2138 1954 sπ : $1/3 + 3\sqrt{3}/2$
2138 2548 $\ln(5) - \exp(1/3)$
2138 2555 rt : (3,7,-6,8)
2138 3242 as : $2(1/\pi)/3$
2138 3290 Ei : $25/17$
2138 4323 cu : $4e\pi/3$
2138 4726 rt : (1,8,-2,-2)
2138 6409 Γ : $8e\pi/5$
2138 7834 rt : (3,1,9,-2)
2138 7981 ℓ_{10} : $11/18$
2138 9791 λ : $(1/(3e))^2$
2139 0253 $1/4 - s\pi(\sqrt{2})$
2139 2446 cr : $4\sqrt{5}/5$
2139 2896 cu : $2 + 3\sqrt{3}/2$
2139 3215 rt : (6,2,1,9)
2139 5425 Li_2 : $(\sqrt{2}\pi/2)^{-2}$
2139 7488 E : $(\sqrt{5}/3)^{1/2}$
2139 9482 rt : (9,7,1,7)
2140 0336 rt : (8,8,-8,-2)
2140 1459 rt : (2,3,0,7)
2140 1681 θ_3 : $\exp(-\sqrt{5})$
2140 1987 rt : (6,5,8,-2)
2140 2657 as : $3\sqrt{3}/4 - \sqrt{5}$
2140 2926 cr : $4\sqrt{2} + 3\sqrt{3}$
2140 3026 tanh : $5/23$
2140 3350 tπ : $e - \pi/4$
2140 4147 10^x : $8e\pi/3$
2140 6068 at : $5/23$
2140 6114 J_1 : $\exp(\sqrt{2}\pi/2)$
2140 6293 at : $6\sqrt{5}/5$
2140 9565 rt : (5,2,0,6)
2141 0161 id : $1/4 - 2\sqrt{3}$
2141 3031 cu : $3 + 2\pi/3$
2141 3505 rt : (9,9,7,-2)
2141 3635 rt : (3,3,-4,9)
2141 6678 rt : (8,7,0,4)
2141 6737 sr : $e/3 + 3\pi$
2141 7499 rt : (1,6,-6,1)
2141 8011 2^x : $2\sqrt[3]{2}/9$
2141 9488 2^x : $7/25$
2142 0964 rt : (2,8,1,-7)
2142 1356 $1/5 - \sqrt{2}$
2142 2044 ζ : $\sqrt{2} - \sqrt{5} + \sqrt{6}$
2142 2308 rt : (9,4,-1,9)
2142 2530 rt : (1,7,-9,8)
2142 2777 rt : (1,8,-3,-1)
2142 4480 sq : $1/3 + 2\sqrt{3}/3$
2142 6302 sπ : $5\sqrt[3]{5}/8$
2142 7127 Γ : $\zeta(3)/3$
2142 8330 ln : $1 + 3e$
2142 8571 id : $3/14$
2143 1666 ℓ_{10} : $3\sqrt{2}/4 + \sqrt{3}/3$
2143 1974 rt : (3,9,-3,-3)
2143 4089 at : $3\sqrt{2} - 4\sqrt{3}$
2143 5817 cu : $8\pi^2/9$
2143 5937 ℓ_2 : $4/3 - \sqrt{2}/3$
2143 6652 rt : (5,1,-2,-8)
2143 8909 cu : $5e\pi$
2143 9189 ℓ_2 : $3 - e/4$
2143 9208 J_2 : $10/7$
2144 0200 id : $9e\pi/4$
2144 0766 e^x : $9\pi^2/8$
2144 0885 rt : (8,4,-2,6)
2144 1282 2^x : $3/4 + 4\sqrt{2}/3$
2144 2043 J_2 : $6\sqrt{5}$
2144 2918 ζ : $7\sqrt[3]{2}/3$
2144 3378 id : $1/\ln(\pi/2)$
2144 5068 rt : (2,2,5,1)
2144 8243 rt : (7,7,-1,1)
2145 0525 erf : $\sqrt{3}/9$
2145 2215 rt : (1,9,-1,9)
2145 4170 rt : (7,0,9,2)
2145 5939 rt : (7,3,-9,-2)
2145 6212 sq : $9e/7$
2145 6249 rt : (8,1,-4,8)
2145 6330 $\ln(3/4) \times s\pi(\sqrt{3})$
2145 6837 rt : (6,8,-9,6)
2145 6880 at : $2/3 - 3\sqrt{5}/2$
2145 9952 e^x : $2 + 3\sqrt{2}$
2146 0183 id : $1 - \pi/4$
2146 1417 rt : (5,2,4,-1)
2146 1872 ℓ_2 : $\sqrt[3]{25/2}$
2146 1991 ℓ_2 : $3 + 2\pi$
2146 2124 10^x : $3/2 - 2\sqrt{3}/3$
2146 2740 rt : (1,6,-3,-7)
2146 3443 rt : (4,7,-8,-2)
2146 4875 rt : (5,8,-5,4)
2146 4939 ln : $1/3 + e/3$
2146 5737 rt : (9,5,-4,-1)
2146 5753 e^x : $1/2 - 3e/4$
2146 6869 rt : (4,1,-3,-5)
2146 7795 ℓ_2 : P_5
2146 8309 Ei : $8/25$
2147 0485 rt : (7,4,-3,3)
2147 1877 J_1 : $11/25$
2147 1904 cu : $e/4 - 4\pi/3$
2147 2226 rt : (8,6,3,-1)
2147 2438 cu : $3/2 - \sqrt{3}/4$
2147 2845 id : $3\sqrt{3} - 4\sqrt{5}/3$
2147 3351 at : $3/2 - 4\pi/3$
2147 4442 J_0 : $8\sqrt[3]{2}/5$
2147 4586 $2/3 \times \exp(3/5)$
2147 5008 sr : $\ln(\pi/3)$
2147 5701 sr : $\sqrt{3}/4 + 2\sqrt{5}$
2147 5855 Γ : $\ln(1/2)$
2147 6895 rt : (6,9,-3,-1)
2147 8219 rt : (1,9,-5,-9)
2147 9454 J_0 : $9\sqrt{5}/7$
2148 0209 θ_3 : $(\sqrt[3]{2})^{-1/2}$
2148 0591 rt : (4,8,-1,2)
2148 1585 J_0 : $23/8$
2148 1884 tanh : $1/\sqrt{21}$
2148 1954 sinh : $1/\sqrt{22}$
2148 2312 at : $1/3 + 3\pi/4$
2148 2735 rt : (7,1,-5,5)
2148 3257 Ψ : $e\pi/5$
2148 4143 J_1 : $(\ln 3/2)^{-2}$
2148 4983 at : $1/\sqrt{21}$
2148 5219 sinh : $\sqrt{3} + \sqrt{6} - \sqrt{7}$
2148 6719 e^x : $3\sqrt{2}/4 - \sqrt{3}/2$
2148 6866 sq : $e + 3\pi/2$
2148 7636 $\pi - \exp(\sqrt{5})$
2148 8650 rt : (2,1,9,-2)
2148 8883 sr : $9\pi^2/5$
2148 9075 rt : (4,7,-7,-9)
2148 9713 2^x : $1/2 - e$
2149 0492 rt : (7,2,-7,-7)
2149 2466 rt : (5,5,8,-2)
2149 3091 ln : $3\sqrt{2}/4 + 4\sqrt{3}/3$
2149 3577 e^x : $3\sqrt{2}/4 - 3\sqrt{3}/2$
2149 5558 rt : (8,9,7,-2)
2149 5845 rt : (7,5,-9,-9)
2149 6520 rt : (9,8,-8,-2)
2149 7041 sr : $4 + e/3$
2149 7505 J_2 : $((e + \pi)/2)^{1/3}$
2149 8916 J_0 : $(\sqrt[3]{2}/3)^{-3}$
2149 9543 J_0 : $5(e + \pi)/6$
2150 2876 cu : $\ln(\ln 3/2)$
2150 3294 e^x : $2\sqrt{2} + 3\sqrt{5}/2$
2150 7128 ℓ_{10} : $3 - e/2$
2150 7985 rt : (8,8,6,1)
2150 8863 $\ln(4/5) \times s\pi(\sqrt{2})$
2150 9248 rt : (1,5,-9,-9)
2151 5060 $1/2 \times s\pi(\pi)$
2151 5274 rt : (2,8,-3,-5)
2151 7112 10^x : $e + 2\pi$
2151 8063 at : $4/3 + e/2$
2151 8516 tπ : $e\pi/8$
2152 1935 at : $6\pi/7$
2152 3477 J_1 : $\sqrt{2}\pi$
2152 5043 rt : (9,9,-4,-2)
2152 7453 2^x : $e - \pi/2$
2152 8571 e^x : $-\sqrt{3} - \sqrt{6} + \sqrt{7}$
2152 9358 cu : $21/13$
2152 9934 id : $4/3 - \sqrt{5}/2$
2153 1235 ln : $4\sqrt{3} + \sqrt{5}$
2153 3575 J_2 : $9(1/\pi)/2$
2153 3827 $\ln(\sqrt{2}) \div \ln(5)$
2153 5393 sq : $7\sqrt[3]{2}/8$
2153 5720 cr : $4\pi/7$
2153 6071 id : $3\sqrt{2}/4 + 2\sqrt{3}/3$
2153 7180 sπ : $\sqrt{2}/4 + \sqrt{3}/3$
2153 7512 as : $15/16$
2153 7831 sinh : $\sqrt[3]{5}/8$
2153 8098 tπ : $3e/2 - 3\pi/2$
2153 9030 sq : $2 - 3\sqrt{3}$
2153 9663 cr : $4e$
2154 0899 as : $\sqrt[3]{5}/8$
2154 0948 J_0 : $4\pi^2/3$
2154 1211 10^x : $4 + 2e$
2154 2299 rt : (1,7,-1,4)
2154 4346 sq : $1/\sqrt[3]{10}$
2154 8043 Li_2 : $3\sqrt[3]{3}/5$
2154 9970 λ : $\exp(-4\pi/3)$
2155 1061 J_1 : $8\sqrt{2}$
2155 1531 rt : (9,4,8,-2)
2155 1929 ln : $4/3 + 3e/4$
2155 4603 rt : (6,0,9,2)
2155 8367 $5 \div \exp(\sqrt{2})$
2155 8416 ℓ_2 : $2/3 - \sqrt{5}/4$
2155 9278 rt : (5,7,-8,-2)
2155 9980 ℓ_{10} : $14/23$
2156 1088 Li_2 : $1/\sqrt{24}$
2156 1230 $\ln(3) + \exp(3/4)$
2156 1418 Ei : $19/17$
2156 1707 Ψ : $e/6$
2156 1724 J_2 : $9(e + \pi)/5$

2156 2606 rt : (7,9,8,9)

2156 2961 e^x : $2/3 - \sqrt{2}/3$

2156 3167 e^x : $3\sqrt{3}+4\sqrt{5}/3$

2156 3273 $\exp(\sqrt{3}) \times s\pi(\sqrt{3})$

2156 3534 rt : (8,3,-9,-2)

2156 3698 ζ : $7\ln 2/5$

2156 3966 Ψ : $\sqrt[3]{2}$

2156 4178 2^x : $3-e$

2156 5259 $e \div \sqrt{5}$

2156 8061 sq : $2/3+2\sqrt{2}$

2156 8323 $\sqrt{3}-\exp(2/3)$

2156 9568 sr : $2/3+3\sqrt{2}$

2157 0033 sinh : $3\sqrt[3]{5}/5$

2157 0195 10^x : $4+3\pi/2$

2157 0575 2^x : P_6

2157 3599 rt : (2,9,-9,-8)

2157 6155 $3/4 \times \ln(3/4)$

2157 6182 rt : (9,1,4,-1)

2157 7078 rt : (8,3,-9,-1)

2157 8177 J_2 : $\sqrt{17}\pi$

2157 9368 rt : (7,9,7,-2)

2158 0374 rt : (5,2,-9,-1)

2158 0877 cr : $3/4+\pi/3$

2158 2305 rt : (2,8,-3,-1)

2158 3430 tπ : $4\sqrt{2}+3\sqrt{5}$

2158 4115 rt : (2,7,-9,-3)

2158 5043 ℓ_{10} : $4-3\pi/4$

2158 5158 rt : (4,5,8,-2)

2158 6881 sq : $e/4+3\pi/4$

2158 7124 id : $4\sqrt{2}+\sqrt{5}/4$

2158 8829 ℓ_{10} : $(\sqrt{2}\pi)^{1/3}$

2158 9035 rt : (6,3,5,1)

2159 0089 ζ : 3/7

2159 0560 rt : (1,7,6,7)

2159 2334 rt : (1,1,9,-2)

2159 2942 sinh : 3/14

2159 3024 at : $3\sqrt{2}/2+\sqrt{3}/3$

2159 4397 rt : (5,6,-9,5)

2159 5042 rt : (9,8,3,8)

2159 5315 rt : (5,4,-4,-1)

2159 5749 ℓ_2 : $4-3\sqrt{5}/4$

2159 6049 as : 3/14

2159 6908 rt : (8,6,-7,-3)

2159 7281 $1/3-\ln(\sqrt{3})$

2159 8242 Li_2 : $\exp(-\sqrt[3]{4})$

2160 0000 id : $(5/3)^{-3}$

2160 3087 e^x : $4\sqrt{3}/3+\sqrt{5}/3$

2160 3353 rt : (5,1,-7,-1)

2160 4134 sq : $3/2+3e$

2160 4170 sπ : $\ln 2/10$

2160 5001 J_0 : $4e\pi/7$

2160 5090 ℓ_{10} : $3\sqrt{2}-3\sqrt{3}/2$

2160 5311 cr : $\sqrt{12}\pi$

2160 5600 rt : (5,9,-1,8)

2160 6959 $5 \div \exp(\pi)$

2160 9204 $\exp(1/5)\div\exp(\sqrt{3})$

2161 0570 rt : (3,7,6,1)

2161 1073 rt : (2,4,-7,-1)

2161 4849 ℓ_{10} : $\pi^2/6$

2161 5722 e^x : $5(1/\pi)/2$

2161 6022 rt : (8,0,-5,1)

2161 7278 rt : (8,2,-2,9)

2161 9003 sr : $4\sqrt{2}-\sqrt{5}/3$

2162 0488 rt : (1,9,7,-3)

2162 0666 at : $4-3\sqrt{3}/4$

2162 1505 rt : (6,6,5,8)

2162 2367 cr : $1/2+3\sqrt{3}/4$

2162 3122 rt : (7,6,3,-1)

2162 4181 2^x : $3\pi/2$

2162 4674 J_0 : $9(e+\pi)/4$

2162 5714 10^x : $4\sqrt{3}/3+2\sqrt{5}/3$

2162 8415 as : $1-\pi/4$

2162 8839 sπ : $(\sqrt{3}/2)^{1/2}$

2163 2231 rt : (4,2,-9,-2)

2163 3621 e^x : 23/6

2163 4712 sq : $3e/4-\pi$

2163 6272 cu : $e\pi/8$

2163 8198 rt : (8,5,0,7)

2163 9532 ln : $(3/2)^3$

2164 1106 rt : (8,9,-5,-5)

2164 1499 cr : $10\sqrt[3]{2}/7$

2164 2011 10^x : $e/3-\pi/2$

2164 3121 Ψ : $\sqrt[3]{5}$

2164 3812 rt : (8,4,8,-2)

2164 4039 sq : $(\sqrt{5}/3)^{-1/3}$

2164 4666 rt : (7,9,-3,-1)

2165 0178 J_0 : $(1/(3e))^{-1/3}$

2165 0520 rt : (2,7,4,7)

2165 0635 id : $\sqrt{3}/8$

2165 2720 rt : (7,1,-5,-8)

2165 3154 ζ : $1/\ln(1/(3e))$

2165 4765 ℓ_{10} : $2-\sqrt{2}/4$

2165 4882 ζ : $(e+\pi)/2$

2165 4963 at : $4\sqrt{3}-3\sqrt{5}$

2165 5091 sπ : $\sqrt{2}/3+3\sqrt{3}/2$

2165 6674 J_0 : $9\sqrt{3}/2$

2165 6985 sr : $3\sqrt{3}/2-\sqrt{5}/2$

2165 7631 J : $\exp(-8\pi/9)$

2165 7669 rt : (6,7,-8,-2)

2165 7840 rt : (5,0,9,2)

2165 8071 rt : (4,6,-5,3)

2165 8919 rt : (5,4,-5,-3)

2165 8971 2^x : $\sqrt{2}/5$

2165 9670 J_2 : 23/16

2165 9692 K : $5\sqrt{2}/8$

2166 1772 e^x : $2\sqrt{2}/3+4\sqrt{3}$

2166 4810 at : $9\zeta(3)/4$

2166 4942 2^x : $(\pi/3)^3$

2166 5011 rt : (6,9,7,6)

2166 6296 cu : $3\sqrt{2}/2+\sqrt{3}$

2166 6384 10^x : $3/4-\sqrt{2}$

2166 7277 Ei : $\sqrt{5}/2$

2166 7354 2^x : $8(1/\pi)/9$

2166 9285 J : 7/16

2166 9576 10^x : $4\sqrt{2}+3\sqrt{3}/4$

2167 0332 J_1 : $\sqrt{13}\pi$

2167 0924 J_0 : $9\sqrt{5}/10$

2167 1383 rt : (8,8,2,5)

2167 2409 cu : $4e/3+\pi/2$

2167 3827 at : $2/3+3e/4$

2167 4508 rt : (9,3,-9,-2)

2167 5597 tπ : $2\sqrt{2}/7$

2167 5657 rt : (1,2,5,1)

2167 6051 id : $\sqrt{2}/3+\sqrt{5}/3$

2167 6072 e^x : $\sqrt{2}/2-\sqrt{5}$

2167 7006 2^x : $3e/2-2\pi$

2167 7348 10^x : 25/23

2167 8022 J_1 : 4/9

2167 8934 ln : $\sqrt{2}-\sqrt{5}/2$

2168 0175 rt : (3,5,8,-2)

2168 0475 e^x : $4\sqrt[3]{3}$

2168 1317 rt : (5,3,2,7)

2168 1837 rt : (7,2,-3,6)

2168 2135 rt : (9,1,-8,5)

2168 2816 ζ : $(2e)^{-1/2}$

2168 3231 rt : (2,7,3,-1)

2168 3484 Ψ : $8e\pi$

2168 3612 e^x : $4+4\pi$

2168 4447 sq : $\sqrt{3}+4\sqrt{5}/3$

2168 4971 cu : $\sqrt{2}+\sqrt{3}+\sqrt{6}$

2168 5170 Γ : 7/25

2168 7381 ln : $3\sqrt{2}-\sqrt{3}/2$

2168 8490 rt : (6,4,-8,-7)

2168 9331 ℓ_2 : $\sqrt{2}/4+4\sqrt{5}$

2168 9995 rt : (4,9,3,6)

2169 2708 rt : (2,2,2,-9)

2169 3229 ln : $5\sqrt{5}/9$

2169 3370 $\ln(\sqrt{5})-s\pi(1/5)$

2169 4596 at : $2+\sqrt{2}/2$

2169 4740 id : $(\sqrt{2}\pi/3)^{1/2}$

2169 5384 rt : (3,9,-8,-9)

2169 6329 rt : (3,8,-5,1)

2169 6503 10^x : 5/8

2169 7494 J_2 : $9\sqrt[3]{3}$

2169 8987 at : $\sqrt{2}/3+\sqrt{5}$

2169 9056 rt : (1,9,-2,0)

2169 9344 10^x : $\sqrt{2}/2+3\sqrt{3}/2$

2169 9835 as : $4/3-\sqrt{5}/2$

2170 3236 ℓ_{10} : $8\sqrt[3]{3}/7$

2170 4734 Li_2 : $\sqrt{3}/2$

2170 5472 at : $\sqrt{22/3}$

2170 6412 Γ : $2\sqrt[3]{2}/9$

2170 7351 cr : $5\sqrt[3]{3}/4$

2170 7765 at : $4-3\sqrt{5}$

2170 8503 cr : $(e+\pi)^{1/3}$

2171 0190 Ei : $\sqrt{5}/7$

2171 0484 10^x : $1+\pi/2$

2171 0569 2^x : $e/3+4\pi$

2171 1263 cu : $\zeta(3)/2$

2171 1308 sinh : $3\pi/5$

2171 2556 id : $(e+\pi)^3$

2171 3503 cr : $3\zeta(3)/2$

2171 4332 tπ : $\sqrt{3}/3+2\sqrt{5}/3$

2171 4724 ℓ_{10} : $(e)^{1/2}$

2171 5289 tπ : $4\sqrt{3}/3-4\sqrt{5}$

2171 5635 J : $1/\sqrt[3]{12}$

2171 6440 rt : (9,4,-6,3)

2171 6537 ℓ_2 : $(\pi/2)^{1/3}$

2171 6648 ℓ_{10} : $7\ln 2/8$

217i 8202 θ_3 : $\exp(-\sqrt{2}\pi/2)$

2171 8203 rt : (5,7,-5,9)

2171 8432 rt : (2,6,-8,-2)

2171 8853 sq : $4e/3-\pi/2$

2171 9214 J_2 : $8\sqrt[3]{2}/7$

2172 0817 tπ : $\sqrt{3}/3-4\sqrt{5}/3$

2172 1590 rt : (9,1,9,2)

2172 3870 ln : $1/3+\sqrt{2}/3$

2172 5829 tπ : $9\sqrt[3]{3}/2$

2172 5970 rt : (6,1,-6,-5)

2172 6218 ln : $\ln(\sqrt{5})$

2172 8139 rt : (7,5,-1,4)

2172 9296 2^x : $\exp(-\sqrt[3]{2})$

2172 9793 E : $(\pi/2)^{-1/3}$

2173 1197 rt : (2,6,-6,1)

2173 2051 cu : $3\sqrt{3}/4+2\sqrt{5}$

2173 3136 J_1 : $7(1/\pi)/5$

2173 5850 rt : (1,3,-3,9)

2173 6508 rt : (3,6,-6,-7)

2173 7976 sπ : $1/\ln((e+\pi)/2)$

2173 8364 rt : (7,4,8,-2)

2173 9130 id : 5/23

2173 9141 J_1 : $1/\ln(3\pi)$

2174 2068 rt : (4,0,-1,6)

2174 2370 J_0 : $6\sqrt{5}$

2174 2719 rt : (5,2,-9,-2)

2174 4383 rt : (2,7,-7,-4)

2174 6178 ℓ_{10} : $3\sqrt{2}/7$

2174 8634 rt : (3,8,-3,-1)

2174 9194 id : $7\sqrt{5}/3$

2175 0674 cr : $4/3+\sqrt{2}/3$

2175 2565 rt : (5,9,7,-2)

2175 5271 cu : $9\sqrt{2}/4$

2175 6732 2^x : $1+3e/4$

2175 6837 rt : (8,1,4,-1)

2175 7609 Γ : $3\sqrt[3]{2}/5$

2175 7911 at : $2-3\pi/2$

2175 8070 rt : (1,5,-1,-8)

2175 8754 rt : (7,7,-8,-2)

2175 8877 rt : (8,1,-9,2)

2175 9227 ℓ_2 : $2e-\pi/4$

2176 0117 as : P_{62}

2176 0357 rt : (7,6,-8,-6)

2176 2640 id : $6\pi^2$

2176 3019 e^x : $4-\sqrt{5}/4$

2176 4042 rt : (4,0,9,2)

2176 6056 10^x : $4e-3\pi/2$

2176 8231 at : $\sqrt{2}+3\sqrt{3}/4$

2176 8353 sinh : $(5/3)^{-3}$

2176 9876 rt : (9,7,-4,1)

2177 0608 rt : (5,4,-9,-4)

2177 1589 as : $(5/3)^{-3}$

2177 2622 tπ : $4\sqrt{3}-3\sqrt{5}/2$

2177 2795 2^x : $2\sqrt{3}/3-3\sqrt{5}/2$

2177 3122 $\exp(1/3)+\exp(3/5)$

2177 3750 ℓ_{10} : $\sqrt[3]{9/2}$

2177 4088 rt : (4,1,-5,1)

2177 4203 J_1 : $9\sqrt[3]{2}$

2177 4426 J_2 : 13

2177 5013 Li_2 : $\sqrt[3]{3}/7$

2177 7046 J_2 : $\sqrt[3]{3}$

2177 7632 rt : (2,5,8,-2)
2177 7777 sq : 7/15
2177 8393 at : $3/4-2\sqrt{3}$
2177 8945 θ_3 : 21/22
2177 9226 Ψ : $((e+\pi)/2)^{1/2}$
2177 9413 cr : $3\sqrt{2}/4+\sqrt{5}/3$
2177 9433 ln : $(2\pi/3)^3$
2177 9822 rt : (6,6,3,-1)
2178 0009 $\exp(e)\div s\pi(\pi)$
2178 0593 at : 19/7
2178 1373 $\ln(5)\div\exp(2)$
2178 1661 rt : (4,4,-9,4)
2178 2170 at : $\sqrt[3]{20}$
2178 2320 10^x : $6\pi^2/7$
2178 2453 J_2 : $6\zeta(3)/5$
2178 3008 sq : $3/4+\sqrt{2}/4$
2178 3147 ζ : $\sqrt[3]{25}$
2178 3321 sq : $3\sqrt{2}/2+\sqrt{5}/3$
2178 3592 rt : (2,9,-9,-6)
2178 4159 10^x : $4-\pi$
2178 4898 J_2 : 9/2
2178 5722 rt : (6,2,-4,3)
2178 7805 $\ln(\sqrt{5})+\exp(5)$
2178 8059 J_2 : $1/\ln(2)$
2178 8958 at : $1/4-\sqrt{2}/3$
2178 9353 cu : $3\sqrt{2}+\sqrt{3}/4$
2179 0609 rt : (7,8,6,1)
2179 3240 $\ln(3/5)-s\pi(1/4)$
2179 4450 rt : (9,1,-8,-8)
2179 5199 id : $\sqrt{2}+\sqrt{6}-\sqrt{7}$
2179 7085 rt : (9,9,5,9)
2179 7217 rt : (4,1,-8,-4)
2179 7861 θ_3 : $(2\pi/3)^{-3}$
2179 9616 Li_2 : $5\ln 2/4$
2180 1337 rt : (6,4,-4,-1)
2180 1571 10^x : $(\pi)^{1/2}$
2180 1754 id : $(e/3)^{-2}$
2180 3789 rt : (3,6,-1,1)
2180 4807 rt : (9,8,7,-2)
2180 4886 J_2 : $5\sqrt{3}/6$
2180 5215 10^x : $5(e+\pi)/3$
2180 5654 rt : (2,5,-4,2)
2180 6301 J_1 : $\sqrt{5}/5$
2180 6770 $s\pi(\sqrt{2})-s\pi(\sqrt{3})$
2180 7510 Ψ : $-\sqrt{3}-\sqrt{6}-\sqrt{7}$
2180 8230 $s\pi$: $(3\pi)^{1/2}$
2180 8998 rt : (1,7,7,-8)
2181 3127 $t\pi$: $\exp(-e/3)$
2181 3153 rt : (7,8,1,2)
2181 3840 ℓ_{10} : $(\ln 3/3)^{1/2}$
2181 5954 Γ : 2/5
2181 6551 ℓ_{10} : $\sqrt{2}/4+3\sqrt{3}/4$
2181 7499 rt : (5,3,5,1)
2181 8670 rt : (1,6,-2,-4)
2181 9007 rt : (3,6,-8,-2)
2181 9434 J_2 : $\sqrt{18}\pi$
2181 9440 rt : (7,0,-3,9)
2182 0155 rt : (8,9,-3,-1)
2182 0177 sinh : $\sqrt{3}/8$
2182 0461 ℓ_2 : $3\sqrt{3}/2+3\sqrt{5}$
2182 1789 id : $1/\sqrt{21}$
2182 1884 2^x : $3-3\sqrt{3}$
2182 3451 as : $\sqrt{3}/8$
2182 4188 J_2 : $6\sqrt[3]{2}$
2182 4583 rt : (8,4,-7,0)
2182 7133 $1/5\div\ln(2/5)$
2182 7222 rt : (8,1,9,2)
2182 8182 id : $1/2+e$
2182 8290 at : e
2182 9586 e^x : $2\sqrt{2}/3-\sqrt{5}/3$
2183 1311 J_2 : 13/9
2183 2054 rt : (4,7,-1,7)
2183 3565 $s\pi$: $(e+\pi)/2$
2183 3818 at : $2\sqrt{3}-\sqrt{5}/3$
2183 4059 $t\pi$: $4e/3+2\pi/3$
2183 4560 ln : $2/3+\sqrt{3}/3$
2183 5300 rt : (6,4,8,-2)
2183 5664 $t\pi$: $2\sqrt{3}/3-3\sqrt{5}/4$
2183 5672 J_0 : $\sqrt{18}\pi$
2183 6836 J_2 : $\exp(1/e)$
2183 7670 sq : $4/3-\sqrt{3}/2$
2183 9031 10^x : $3/2-\sqrt{2}$
2183 9213 $1/4\div\ln(\pi)$
2183 9959 cu : P_2
2184 0687 2^x : $\sqrt[3]{5}/6$
2184 2005 $\exp(1/2)+s\pi(\pi)$
2184 2116 rt : (8,6,2,8)
2184 2725 cr : $2\sqrt{2}/3+\sqrt{3}/2$
2184 3354 rt : (9,2,-6,6)
2184 4158 rt : (5,1,-7,-2)
2184 4777 sr : $6\sqrt{3}/7$
2184 4972 J_0 : $2\sqrt[3]{3}$
2184 5799 J_1 : $4\pi^2/5$
2184 6010 sr : $\sqrt{2}+4\sqrt{5}$
2184 7945 rt : (1,7,3,-1)
2184 8491 10^x : $5\sqrt[3]{2}/6$
2185 0000 10^x : $3\sqrt{2}-3\sqrt{3}/4$
2185 3698 Li_2 : 13/15
2185 3769 $t\pi$: $2\sqrt{3}-\sqrt{5}/3$
2185 6500 rt : (6,3,-6,-8)
2185 6717 rt : (6,2,-9,-2)
2185 6743 $t\pi$: $8\zeta(3)/9$
2185 8067 ln : $1/4+4\sqrt{5}$
2186 1838 cu : $4-\pi/4$
2186 2686 rt : (8,7,-8,-2)
2186 3508 tanh : 2/9
2186 3885 $t\pi$: $2\zeta(3)$
2186 4717 Γ : $(4/3)^3$
2186 6510 rt : (2,6,-7,-4)
2186 6894 at : 2/9
2186 7662 $t\pi$: $10\ln 2$
2187 1316 10^x : $\sqrt[3]{1/2}$
2187 1370 J_2 : 3π
2187 1940 ℓ_{10} : $1/2+2\sqrt{3}/3$
2187 3381 rt : (3,0,9,2)
2187 3843 cu : $4+\sqrt{3}/2$
2187 4104 $t\pi$: $\sqrt{3}/3+3\sqrt{5}/2$
2187 5822 ln : $2e/3+\pi/2$
2187 6086 rt : (7,5,-6,1)
2187 7658 rt : (1,5,8,-2)
2187 8518 ln : $4+3\sqrt{3}$
2187 8893 J_0 : $1/\ln(\sqrt{2})$
2187 8984 sr : $3e/2+2\pi$
2187 9196 rt : (8,1,-9,-5)
2187 9368 $s\pi$: $3\sqrt{3}/2+2\sqrt{5}$
2187 9686 J_1 : $\pi/7$
2187 9746 J : $(e+\pi)^{-2}$
2188 3016 ζ : $2\sqrt{5}/7$
2188 3705 cr : $3/4+3\sqrt{2}/4$
2188 4876 Γ : $(\sqrt[3]{5}/3)^{1/2}$
2188 5858 $t\pi$: $9\pi/4$
2188 6492 10^x : $10\sqrt[3]{5}/7$
2188 7347 rt : (9,0,-5,1)
2188 7582 ln : 25
2188 7669 rt : (7,3,-1,7)
2188 8158 $t\pi$: $1/3+3\sqrt{3}/2$
2188 9902 cr : $3/2+3\pi$
2189 0512 Ei : π^2
2189 1803 ℓ_{10} : $2e/9$
2189 3251 $\sqrt{2}+\ln(\sqrt{5})$
2189 3859 rt : (8,8,7,-2)
2189 3916 rt : (5,9,-6,-3)
2189 4448 sq : $3e/4-\pi/2$
2189 5141 id : $1/3+4\sqrt{2}/3$
2189 5908 rt : (6,5,-2,1)
2189 9520 J_2 : $(2\pi/3)^{1/2}$
2190 1107 Ψ : $8\sqrt{2}/9$
2190 1193 e^x : $-\sqrt{3}-\sqrt{5}+\sqrt{6}$
2190 1365 2^x : 2/7
2190 1539 id : $3e/2+\pi$
2190 2447 rt : (6,7,7,9)
2190 3082 rt : (3,7,-7,5)
2190 3469 rt : (4,8,-6,-9)
2190 3780 ℓ_2 : $\sqrt{2}-\sqrt{3}+\sqrt{7}$
2190 5260 E : $5\zeta(3)/7$
2190 7057 rt : (4,2,-1,-9)
2190 8244 Ei : $(\ln 3)^{-3}$
2190 8496 ℓ_{10} : $3/4+e/3$
2190 9264 cu : $3-\sqrt{3}/3$
2191 0173 Ψ : $\exp(-(e+\pi)/2)$
2191 0763 sinh : 5/23
2191 3090 cr : $4+4\sqrt{3}$
2191 3894 2^x : 23/20
2191 3899 rt : (9,5,-4,4)
2191 4105 as : 5/23
2191 4804 2^x : $2+3\sqrt{3}/2$
2191 5148 10^x : $3\sqrt{2}/2+\sqrt{5}/2$
2191 7975 id : $(2e/3)^{1/3}$
2191 8800 J_1 : $1/\sqrt[3]{11}$
2192 0069 K : $\exp(-1/(3e))$
2192 0128 tanh : $7(1/\pi)/10$
2192 2359 rt : (4,8,-3,-1)
2192 2737 ℓ_{10} : $4+4\pi$
2192 3558 at : $7(1/\pi)/10$
2192 8024 Ψ : $(\sqrt[3]{5}/3)^2$
2192 8430 ℓ_{10} : $1/4+\sqrt{2}/4$
2193 0488 rt : (4,5,-5,8)
2193 0667 J_0 : $5\sqrt{3}/3$
2193 3528 2^x : $2-4\pi/3$
2193 3756 rt : (9,0,-6,9)
2193 3869 ln : $2/3+e$
2193 4568 sinh : $4\sqrt{2}/3$
2193 4740 rt : (5,4,8,-2)
2193 5253 J_1 : 9/20
2193 5737 $t\pi$: $5\sqrt[3]{5}/8$
2193 5910 rt : (7,1,9,2)
2193 6985 2^x : $4/3-\pi/3$
2193 6987 rt : (8,9,4,6)
2193 8179 $t\pi$: $7\sqrt[3]{3}/3$
2194 0603 Ψ : $\sqrt{15}$
2194 1417 rt : (8,7,-5,-2)
2194 2142 ln : $1/2+\sqrt{5}/3$
2194 2552 ln : $3e+\pi/3$
2194 2773 rt : (5,9,6,3)
2194 5999 $s\pi$: $1/4+e/4$
2194 6161 rt : (7,1,4,-1)
2194 6975 J_1 : $(2\pi/3)^3$
2194 6999 rt : (1,7,-2,-3)
2194 7723 J_0 : $5e\pi/4$
2194 8706 at : $1/\ln(\ln 2)$
2194 9944 tanh : $\exp(-3/2)$
2195 0900 rt : (9,4,5,1)
2195 1151 $\exp(3)-s\pi(1/3)$
2195 1371 rt : (1,9,-3,-9)
2195 2465 $2/5\div\exp(3/5)$
2195 3399 at : $\exp(-3/2)$
2195 3589 Ψ : $2\pi/5$
2195 3648 ln : $3/2+4\sqrt{2}/3$
2195 5791 rt : (3,5,-4,-8)
2195 7213 cu : $4\sqrt{2}/3+3\sqrt{5}/4$
2195 7879 id : $\sqrt{2}/4+\sqrt{3}/2$
2195 8103 rt : (8,2,-7,3)
2196 1174 2^x : $4/3+\sqrt{2}/4$
2196 1794 sinh : $7(e+\pi)/10$
2196 4784 rt : (8,3,-9,-8)
2196 5944 rt : (5,4,4,8)
2196 5988 2^x : $9(1/\pi)/10$
2196 9630 rt : (9,7,-8,-2)
2196 9906 ζ : $\sqrt{17/2}$
2197 0928 J_1 : $3\zeta(3)/8$
2197 2245 $1/5\times\ln(3)$
2197 4459 rt : (7,2,-9,-2)
2197 5934 at : $1/\ln(\sqrt[3]{3})$
2198 0390 rt : (5,8,-2,0)
2198 0888 $\ln(3/5)\times s\pi(\pi)$
2198 1609 Li_2 : $\exp(-\pi/2)$
2198 1855 Ψ : 12/7
2198 4284 at : $1/4-4\sqrt{5}/3$
2198 4949 rt : (7,8,7,-2)
2198 4993 sr : $1/3+2\sqrt{3}/3$
2198 5300 cu : $2/3-3\pi/2$
2198 6043 cu : $1/4+\sqrt{2}/4$
2198 6049 rt : (2,0,9,2)
2198 8112 cu : $8\zeta(3)/9$
2198 8848 Li_2 : $3\ln 2/10$
2199 0287 rt : (9,3,8,-2)
2199 1333 Γ : $(\ln 3)^{-3}$
2199 1691 at : $1+\sqrt{3}$
2199 2163 ℓ_2 : $5\zeta(3)/7$
2199 5264 e^x : $(\ln 3/3)^{-1/2}$

2199 5307 tanh : $\sqrt{5}/10$
2199 5390 sinh : $1/\sqrt{21}$
2199 5543 E : $(e/2)^{-1/2}$
2199 6203 id : $e\pi/7$
2199 6957 rt : (7,6,1,5)
2199 8797 at : $\sqrt{5}/10$
2199 9929 id : $4\sqrt{3} - 3\sqrt{5}$
2200 0091 cu : $\sqrt{3} - 3\sqrt{5}$
2200 2355 Ψ : 7/10
2200 2737 rt : (4,3,-9,9)
2200 4103 rt : (9,9,-3,-1)
2200 5146 rt : (2,8,-7,-7)
2200 7568 ln : $2\sqrt{2} + \sqrt{5}/4$
2200 8443 J_1 : $\ln(\pi/2)$
2200 8895 rt : (9,3,-4,7)
2201 1324 e^x : $5(1/\pi)/8$
2201 2737 id : $\sqrt{3}/2 + 3\sqrt{5}/2$
2201 4786 Ψ : $5\sqrt[3]{2}/9$
2201 6899 at : $\sqrt{2}/4 - \sqrt{3}/3$
2201 8113 rt : (1,5,0,6)
2201 8454 10^x : 21/20
2201 9732 rt : (7,4,-4,-1)
2202 1802 cr : $2 + 4\sqrt{5}$
2202 3028 Ψ : $(\pi/2)^3$
2202 4463 rt : (9,8,-2,2)
2202 4939 ℓ_{10} : P_2
2202 6567 ℓ_2 : $4 - \pi$
2202 7287 Γ : 3/20
2202 7583 rt : (2,9,7,-2)
2202 7690 J_0 : 2π
2202 8493 cr : $\sqrt[3]{6}$
2202 8644 rt : (5,6,-8,-2)
2202 9234 10^x : $\sqrt{3}/5$
2202 9308 rt : (6,2,-4,-9)
2203 0261 ℓ_2 : $3e/7$
2203 0490 rt : (4,1,1,7)
2203 0519 e^x : $7\pi^2/8$
2203 1192 cu : $3/4 + \sqrt{3}/2$
2203 1378 2^x : $3/2 + \sqrt{3}/3$
2203 1500 e^x : $8\sqrt[3]{2}/7$
2203 2484 Li_2 : 5/24
2203 3300 rt : (9,5,3,-1)
2203 5518 J_1 : $\exp(-\sqrt[3]{4}/2)$
2203 6816 rt : (4,4,8,-2)
2203 7836 at : $8\sqrt[3]{5}/5$
2203 9062 2^x : $\sqrt{2}/2 + 3\sqrt{5}/4$
2203 9091 at : $1/2 + \sqrt{5}$
2203 9360 rt : (3,1,-9,-2)
2203 9774 $s\pi$: $2(1/\pi)/9$
2204 1741 cu : $2e/9$
2204 2641 10^x : $4\sqrt{3} + 4\sqrt{5}/3$
2204 2795 rt : (5,1,-5,1)
2204 4247 Γ : $\sqrt{22}$
2204 5724 e^x : $3\sqrt{3} - 3\sqrt{5}$
2204 5992 rt : (8,0,-7,6)
2204 7834 rt : (6,1,9,2)
2204 9119 e^x : $\sqrt{2}/4 + 3\sqrt{3}/4$
2204 9668 at : P_{31}
2204 9837 Ψ : $(\sqrt{2}\pi/2)^{-2}$
2205 0762 E : $(\sqrt[3]{4})^{-1/3}$
2205 1561 rt : (3,2,-9,3)
2205 1654 e^x : $(\pi/2)^3$
2205 2244 cr : 20/11
2205 2790 Li_2 : $1/\sqrt{23}$
2205 3042 rt : (8,7,4,9)
2205 4320 cr : $9\sqrt{2}/7$
2205 4640 10^x : $\sqrt[3]{20/3}$
2205 4758 $t\pi$: $\sqrt{2}/4 + \sqrt{3}/3$
2205 6274 sq : $3/2 + 4\sqrt{2}$
2205 6889 Ei : 1/7
2205 7841 rt : (6,3,-2,4)
2205 7908 erf : $2\ln 2/7$
2205 8404 $s\pi$: $1/2 + \pi/2$
2205 8444 at : $3\sqrt{2}/4 + 3\sqrt{5}/4$
2206 1212 rt : (4,3,5,1)
2206 1251 id : $\exp(\sqrt{2}\pi/2)$
2206 1492 E : 6/7
2206 2685 e^x : $\exp(\sqrt{2}\pi/3)$
2206 4513 Γ : $5\sqrt{3}/2$
2206 6239 cr : $3\sqrt{2} + 3\sqrt{5}$
2206 6670 rt : (4,1,-8,1)
2206 6893 Γ : $\sqrt{13}\pi$
2206 7675 2^x : $\sqrt{5}\pi$
2206 7748 2^x : $\ln(4/3)$
2206 8114 rt : (5,5,-7,-8)
2206 8530 rt : (6,9,-6,8)
2206 8565 K : $(\ln 2)^{1/3}$
2206 9054 at : $\sqrt{15/2}$
2206 9281 rt : (3,6,-9,9)
2207 0043 as : $2 - 3\sqrt{2}/4$
2207 0312 cu : $5\sqrt[3]{5}/8$
2207 2087 J_1 : 25/4
2207 2685 e^x : $3/2 + 2\sqrt{3}/3$
2207 2766 $3/5 \div e$
2207 3267 J_1 : $7\sqrt{2}/3$
2207 3602 sq : $1/3 + 3\pi$
2207 5271 10^x : $1/4 - e/3$
2207 6118 J_1 : $e/6$
2207 6290 Ψ : $\exp(-\sqrt[3]{4}/2)$
2207 6348 ζ : 13/8
2207 8171 rt : (6,8,7,-2)
2207 9684 10^x : $\ln 2/8$
2208 0329 rt : (3,6,-6,1)
2208 1219 J_2 : 16/11
2208 2064 rt : (8,5,-5,1)
2208 3790 e^x : $(\pi/2)^{-1/2}$
2208 3991 sr : $(\sqrt{2}\pi/2)^{1/2}$
2208 5938 J_2 : $9(e + \pi)/4$
2208 5976 rt : (1,8,5,-4)
2208 6368 at : $\sqrt{3} - 2\sqrt{5}$
2208 6876 J_2 : $4\pi^2/3$
2208 8178 ζ : $(\sqrt[3]{3}/2)^{-1/2}$
2208 9713 cu : $(\pi/2)^3$
2209 0542 10^x : $6\sqrt[3]{5}/7$
2209 1675 rt : (8,3,8,-2)
2209 4716 sr : $2\sqrt{5}/3$
2209 6119 rt : (3,7,3,5)
2209 6199 rt : (8,2,-9,-2)
2209 8948 rt : (6,8,6,1)
2210 0647 rt : (7,4,1,8)
2210 2215 K : $(\sqrt[3]{3})^{-1/3}$
2210 2253 rt : (1,0,9,2)
2210 3612 id : $(\ln 3/2)^{-1/3}$
2210 4157 rt : (5,8,-3,-1)
2210 5472 cu : $1/2 + 2e$
2210 5659 ln : $2\sqrt{3}/3 + \sqrt{5}$
2210 8266 rt : (8,2,-7,-9)
2210 9644 id : $\sqrt{2}/2 - 4\sqrt{3}$
2211 0496 ℓ_{10} : $\zeta(3)/2$
2211 0719 cr : $1/4 + \pi/2$
2211 2478 rt : (4,6,3,2)
2211 2801 10^x : $\ln 2/2$
2211 4707 10^x : $8\sqrt{2}/3$
2211 6380 at : $3/2 - 3\sqrt{2}$
2211 6568 cu : $4\sqrt{2}/3 - 4\sqrt{3}$
2211 6762 10^x : $5e\pi/8$
2211 7351 ln : $\sqrt{23/2}$
2211 7454 rt : (3,9,-4,-6)
2211 7768 rt : (9,6,-2,5)
2211 8073 $\exp(1/4) \times s\pi(2/5)$
2211 8936 sr : $\ln(\sqrt{2}\pi)$
2211 9668 rt : (5,0,-5,3)
2212 0905 ℓ_{10} : $1/4 + \sqrt{2}$
2212 2518 $\sqrt{3} + \ln(3/5)$
2212 3479 cr : $2/3 + 2\sqrt{3}/3$
2212 3700 rt : (1,9,7,-2)
2212 4489 10^x : $5\zeta(3)/7$
2212 5231 ℓ_{10} : $3e/2 + 4\pi$
2212 5322 rt : (2,8,-4,-9)
2212 5672 2^x : $2\sqrt{2} - 3\sqrt{5}/4$
2212 6260 λ : $(4)^{-3}$
2212 6712 $t\pi$: $\ln 2/10$
2212 7374 ln : $9\ln 2/5$
2212 8997 J_2 : $(\sqrt{2}/3)^{-1/2}$
2212 9351 rt : (5,8,-6,2)
2212 9491 rt : (2,9,-6,8)
2212 9962 ln : $1/4 + \pi$
2213 2733 2^x : $\sqrt[3]{3}/5$
2213 2926 cu : $e - \pi/4$
2213 3372 at : $4/3 - 3e/2$
2213 4752 ℓ_2 : $(2e)^{1/2}$
2213 5416 sinh : 20/13
2213 5936 as : P_{46}
2213 6072 ln : $2e - 4\pi/3$
2213 7287 Ei : $(\sqrt[3]{5}/3)^{1/2}$
2213 7907 Γ : $\sqrt{3}/8$
2213 8032 rt : (6,6,-8,-2)
2214 0275 e^x : 1/5
2214 0452 id : $1/4 - \sqrt{2}/3$
2214 0491 rt : (7,9,-6,-8)
2214 1139 id : $5\sqrt[3]{5}/7$
2214 1584 $s\pi$: $5\sqrt{2}$
2214 1668 rt : (3,5,9,9)
2214 2819 $t\pi$: $1/\ln(\sqrt[3]{3}/3)$
2214 3093 ln : $2\zeta(3)/3$
2214 4146 $\pi \div \sqrt{2}$
2214 5005 rt : (6,1,4,-1)
2214 5328 sq : 8/17
2214 5339 J_1 : 5/11
2214 6161 J_1 : $4e\pi/3$
2214 6212 10^x : $1/2 - 2\sqrt{3}/3$
2214 8214 at : $2 + \sqrt{5}/3$
2214 8831 rt : (6,1,-2,7)
2215 1800 2^x : $\sqrt{3}/6$
2215 3216 $t\pi$: $(\sqrt{3}/2)^{1/2}$
2215 3788 J_1 : $10(1/\pi)/7$
2215 5715 erf : $5(1/\pi)/8$
2215 8256 ln : $5\sqrt[3]{3}/9$
2216 0242 rt : (4,1,-9,-2)
2216 0664 sq : 21/19
2216 0767 sr : $(\ln 3/3)^3$
2216 2213 cu : $2e + \pi$
2216 2868 rt : (2,7,-1,-9)
2216 2914 J_2 : 21/2
2216 3014 2^x : $3 + \sqrt{2}/4$
2216 3190 rt : (5,1,9,2)
2216 3858 rt : (8,3,-5,4)
2216 6045 rt : (6,8,0,-1)
2216 7555 10^x : $1/\sqrt[3]{24}$
2216 7734 10^x : 2/23
2216 7741 rt : (1,3,-8,-8)
2216 8357 $\ln(4) \div s\pi(\pi)$
2216 8607 e^x : P_{16}
2216 8760 e^x : P_{40}
2217 0168 cr : $10\pi^2/9$
2217 1048 10^x : $\pi/2$
2217 2090 cu : $2e\pi$
2217 3625 rt : (5,8,7,-2)
2217 3860 at : $4/3 + \sqrt{2}$
2217 3971 Ψ : $\exp(-2e)$
2217 7958 J_2 : $10\pi/7$
2217 8969 rt : (9,4,-2,8)
2217 9992 cr : $\sqrt{2} - \sqrt{5} + \sqrt{7}$
2218 1375 as : $4\sqrt{3} - 3\sqrt{5}$
2218 1426 $t\pi$: $\sqrt{2}/3 + 3\sqrt{3}/2$
2218 2154 e^x : $\zeta(3)/6$
2218 2466 rt : (4,9,-2,-5)
2218 3306 cr : $1/3 + 2\sqrt{5}/3$
2218 3429 rt : (3,5,-1,6)
2218 4874 ℓ_{10} : 3/5
2218 6011 Ei : $8\sqrt[3]{2}/9$
2218 7616 rt : (1,2,-5,1)
2218 9605 at : $7\pi/8$
2219 1095 rt : (8,4,5,1)
2219 1404 J_0 : $5\zeta(3)/3$
2219 3245 $\exp(\sqrt{3}) + s\pi(\pi)$
2219 4579 rt : (2,5,-8,-2)
2219 5762 rt : (7,3,8,-2)
2219 6182 id : $2e + \pi/4$
2219 6506 at : $1/3 - \sqrt{5}/4$
2219 6965 Ei : $(1/\pi)$
2219 7607 rt : (5,2,-5,-6)
2219 9212 rt : (1,7,-3,-1)
2220 0037 rt : (7,9,3,3)
2220 0262 ζ : $\ln(\sqrt[3]{2}/2)$
2220 0806 rt : (8,5,3,-1)
2220 2532 at : 11/4
2220 3752 J_1 : $10\zeta(3)$
2220 7761 cr : $\sqrt{2}/2 + \sqrt{5}/2$
2220 9971 J_1 : $(\sqrt{2}\pi/3)^{-2}$

2221 0481 rt : (7,0,-8,3)
2221 1880 $2/5 + \exp(3/5)$
2221 2177 $\Psi : 10\zeta(3)/7$
2221 3595 id : $1/4 - 2\sqrt{5}$
2221 3604 sr : $1/\ln((e+\pi)/3)$
2221 4499 $e^x : 4/3 + 2e$
2221 5910 sr : $3\sqrt{2}/4 + \sqrt{3}/4$
2221 6467 $t\pi : e$
2221 7393 rt : (8,1,-5,7)
2221 7641 $\theta_3 : 9\ln 2/7$
2221 7993 $Li_2 : \sqrt[3]{2}/6$
2222 1175 sr : $\sqrt[3]{10/3}$
2222 1716 at : $4\sqrt{2}/3 + \sqrt{3}/2$
2222 2222 id : 2/9
2222 5043 at : $3\sqrt{2} - 2\sqrt{5}/3$
2222 5615 $\Gamma : 19/8$
2222 6252 rt : (1,7,5,1)
2223 0775 rt : (1,8,-8,-4)
2223 0848 sq : $2/3 - 3\sqrt{5}/2$
2223 1094 sinh : $5\sqrt[3]{3}/7$
2223 3890 sr : $3/4 + 4\pi/3$
2223 5239 rt : (2,7,-5,-8)
2223 7981 rt : (3,3,-5,7)
2223 9242 $\ell_2 : 3/7$
2224 2651 $\Psi : 10e/7$
2224 5012 $2^x : P_{82}$
2224 6577 $2^x : 2\sqrt{2}/3 + \sqrt{5}/3$
2224 6593 $t\pi : \sqrt{2}/4 + 4\sqrt{3}$
2224 6990 rt : (4,7,-6,-4)
2224 7446 rt : (5,4,-5,-9)
2224 9442 $\exp(\sqrt{5}) + s\pi(1/3)$
2224 9451 rt : (2,4,8,-2)
2225 0330 $\ln(\sqrt{3}) \times \exp(4/5)$
2225 0711 rt : (7,6,-8,-2)
2225 1726 cu : $4\sqrt{3} + 3\sqrt{5}/4$
2225 2093 $s\pi : 1/14$
2225 2286 $Ei : 7/22$
2225 2396 $\ln(2/3) \div \exp(3/5)$
2225 2705 $\theta_3 : 1/9$
2225 4392 $J_2 : 19/13$
2225 4402 $\theta_3 : 9/17$
2225 5767 sinh : $6\zeta(3)/7$
2225 6189 rt : (7,2,-8,-6)
2225 7818 $e^x : 3/2 - 3\sqrt{3}/4$
2225 9399 $10^x : P_{22}$
2226 0834 cr : $9e\pi/7$
2226 1344 sinh : $9\sqrt[3]{5}/10$
2226 4380 rt : (1,9,7,5)
2226 4583 rt : (5,5,6,9)
2226 4586 cu : $3\sqrt{2}/7$
2226 4943 $\ln(3/4) \times s\pi(e)$
2226 5371 rt : (2,9,4,7)
2226 7431 rt : (1,7,6,1)
2227 0258 $erf : 1/5$
2227 0529 $t\pi : 1/\ln((e+\pi)/2)$
2227 1417 rt : (4,8,7,-2)
2227 2447 $\ln(2/3) \times \ln(\sqrt{3})$
2227 2796 sq : $4\pi/7$
2227 5244 rt : (8,5,-6,1)
2227 5304 rt : (3,1,-9,8)
2227 5961 $t\pi : P_6$
2227 6279 id : $\sqrt{3} + 2\sqrt{5}/3$
2227 6561 $\ln(2/5) \div \exp(\sqrt{2})$
2227 7561 cu : $3/2 + \sqrt{3}/4$
2227 7823 rt : (7,7,3,6)
2227 8432 $2^x : 3\sqrt{2}/2 - \sqrt{3}/4$
2227 8558 $Li_2 : 4/19$
2228 1011 $s\pi : P_{17}$
2228 1314 $\Gamma : \sqrt{5}/8$
2228 1692 id : $7(1/\pi)/10$
2228 2070 rt : (5,9,-2,6)
2228 2194 rt : (4,1,9,2)
2228 3125 $s\pi : e/2 + 3\pi/2$
2228 4040 at : $\sqrt{2}/2 - 2\sqrt{3}$
2228 4454 sr : $(\sqrt{5})^{1/2}$
2228 4749 sr : $3/4 + \sqrt{5}/3$
2228 5274 rt : (5,1,-9,-2)
2228 8143 rt : (4,7,-8,-8)
2228 8394 rt : (7,4,-8,-9)
2228 9325 $E : \sqrt[3]{5}/2$
2228 9503 rt : (9,9,0,3)
2229 1042 rt : (2,7,-8,-8)
2229 1236 sq : $2 - 4\sqrt{5}$
2229 1380 $10^x : 7\sqrt[3]{3}/10$
2229 4806 rt : (6,8,-3,-1)
2229 8447 $J_0 : \sqrt{2} + \sqrt{5} + \sqrt{7}$
2229 9835 $\ell_{10} : e - \pi/3$
2230 0226 $\ell_{10} : \sqrt[3]{14/3}$
2230 2450 rt : (3,8,-6,-2)
2230 2693 rt : (6,3,8,-2)
2230 4476 sq : $3/2 + \pi/4$
2230 4646 rt : (8,8,-3,-1)
2230 5927 $e^x : P_8$
2230 6375 rt : (3,5,-8,-2)
2230 6458 at : $\sqrt[3]{21}$
2230 7422 $erf : \zeta(3)/6$
2230 7526 rt : (5,7,-6,7)
2231 0049 cu : $7\ln 2/8$
2231 3016 id : $\exp(-3/2)$
2231 4355 ln : 4/5
2231 6752 cu : $4\sqrt{3}/3 + 4\sqrt{5}$
2231 7196 ln : $((e+\pi)/3)^{1/3}$
2231 7509 rt : (5,5,-3,-2)
2231 9265 rt : (8,7,7,-2)
2232 0149 $J_1 : 11/24$
2232 0437 sr : $7\sqrt[3]{5}/8$
2232 1226 rt : (7,5,3,9)
2232 2346 rt : (3,3,5,1)
2232 2827 sq : $3\sqrt[3]{2}/8$
2232 3268 rt : (1,9,-7,-2)
2232 4291 sr : $4 + 2\sqrt{2}/3$
2232 5450 as : $1/4 - \sqrt{2}/3$
2232 8422 rt : (9,7,0,6)
2232 8747 sq : $3\sqrt{3}/4 - 3\sqrt{5}/2$
2232 9269 $t\pi : 2\sqrt{2}/3 - \sqrt{3}/3$
2233 0044 $J_2 : (\pi)^{1/3}$
2233 0255 $\Psi : ((e+\pi)/2)^{-1/3}$
2233 1263 at : $2/3 + 2\pi/3$
2233 1429 $s\pi : \sqrt{3} - \sqrt{6} + \sqrt{7}$
2233 3250 $2^x : 4\sqrt{3}/3 - 2\sqrt{5}$
2233 3321 rt : (4,2,3,8)
2233 4628 rt : (6,1,-5,1)
2233 5877 $t\pi : \sqrt{2} + \sqrt{3} + \sqrt{6}$
2233 8177 rt : (7,7,-6,-5)
2233 9380 $J_2 : (e+\pi)/4$
2234 0218 rt : (8,2,9,2)
2234 0740 sr : $\sqrt{2}/3 + 2\sqrt{5}$
2234 1101 $10^x : 9(e+\pi)/8$
2234 1203 $e^x : \sqrt{20/3}$
2234 2320 rt : (6,4,0,5)
2234 3260 $J_2 : \ln(\ln 2/3)$
2234 3882 tanh : 5/22
2234 4001 $1/3 \div \exp(2/5)$
2234 4710 $J_1 : 12$
2234 5173 $J_1 : 2\sqrt{5}$
2234 5347 $\theta_3 : \ln(\sqrt{5}/2)$
2234 5476 $J_2 : 9(e+\pi)/7$
2234 6092 $t\pi : (3\pi)^{1/2}$
2234 6207 rt : (1,0,-9,2)
2234 7660 at : 5/22
2234 8339 $2^x : 2/3 - 2\sqrt{2}$
2234 9269 $\Psi : (\pi/2)^{1/2}$
2235 0245 rt : (8,6,-3,2)
2235 2573 tanh : $5(1/\pi)/7$
2235 3802 rt : (9,5,0,9)
2235 4050 rt : (9,9,-9,-8)
2235 4351 rt : (5,1,4,-1)
2235 4531 $e^x : 2\sqrt{2} + \sqrt{3}/2$
2235 6358 at : $5(1/\pi)/7$
2235 6416 $2^x : 6\sqrt{3}/5$
2235 7348 $J_2 : 6\sqrt[3]{5}/7$
2235 7782 tanh : $\exp(-\sqrt{2}\pi/3)$
2235 9786 $\theta_3 : 3\sqrt{5}/8$
2236 0330 rt : (1,4,8,-2)
2236 0679 id : $\sqrt{5}/10$
2236 0908 $J_1 : 9\ln 2$
2236 1572 at : $\exp(-\sqrt{2}\pi/3)$
2236 4851 $s\pi : 4\sqrt{3}$
2236 6596 sq : $2\sqrt{2} + 3\sqrt{5}/2$
2236 6883 rt : (8,6,-8,-2)
2236 8178 cr : $7\pi/2$
2236 8451 $\ell_{10} : 2\sqrt{2} - 2\sqrt{3}/3$
2236 9861 $\Psi : 3e\pi$
2237 0979 sr : $6\sqrt{3}$
2237 0985 $\Gamma : 5\zeta(3)/8$
2237 1663 rt : (6,2,0,8)
2237 2690 $e^x : \sqrt{2} + 4\sqrt{3}$
2237 2984 $J_0 : 5\sqrt[3]{2}$
2237 3350 $t\pi : (e+\pi)/2$
2237 5267 rt : (7,5,3,-1)
2237 6529 rt : (8,4,-3,5)
2237 7543 ln : 5/17
2237 9687 id : $\sqrt{2}/4 - \sqrt{3}/3$
2238 0956 rt : (5,3,-3,1)
2238 1122 $\theta_3 : 17/23$
2238 1462 $J_2 : 22/15$
2238 1565 $Ei : 3\sqrt[3]{3}/10$
2238 2026 $\ell_{10} : 2(e+\pi)/7$
2238 3546 $J_2 : \exp(3/2)$
2238 4154 id : $4\sqrt{2} - \sqrt{3}/4$
2238 7090 rt : (2,7,-3,-1)
2238 7550 $\sqrt{3} + \exp(2/5)$
2238 8534 $e^x : \sqrt{2}/7$
2238 8713 rt : (7,5,-6,-2)
2238 9077 $J_0 : 2$
2238 9996 rt : (9,9,2,-1)
2239 0341 cr : 11/6
2239 0851 $\ell_2 : 4e/3 + \pi/3$
2239 1320 $e^x : 2/3 + \sqrt{5}$
2239 3629 rt : (8,2,-3,8)
2239 5285 rt : (9,7,-9,-5)
2239 6058 $E : e\pi/10$
2239 8009 id : $\sqrt[3]{11}$
2239 8605 sq : $\ln(\sqrt{2}\pi)$
2239 9725 $\sqrt{3} \times \exp(1/4)$
2240 0183 $\Psi : 1/\ln(\sqrt{2}\pi/2)$
2240 3274 rt : (3,8,7,9)
2240 3771 at : $\sqrt{2} - \sqrt{3} - \sqrt{6}$
2240 4702 $\ln(3/4) \div \exp(1/4)$
2240 5083 rt : (3,1,9,2)
2240 5354 $2^x : 7/24$
2240 5573 sinh : 2/9
2240 6465 id : $3\sqrt{2} + 4\sqrt{5}/3$
2240 7863 rt : (5,1,-3,4)
2240 9309 as : 2/9
2241 0371 tanh : $(2\pi/3)^{-2}$
2241 1862 $Ei : 3\sqrt[3]{2}/5$
2241 1898 rt : (7,3,-6,1)
2241 2273 sq : $(3\pi)^{-1/3}$
2241 2386 cu : $4/3 + 3\pi/4$
2241 2625 rt : (5,3,8,-2)
2241 3071 sinh : $8\sqrt{3}/9$
2241 4100 id : P_{31}
2241 4205 at : $(2\pi/3)^{-2}$
2241 4752 rt : (6,1,-9,-2)
2241 5473 rt : (9,5,-9,-2)
2241 6987 rt : (2,8,0,8)
2241 7014 $\Gamma : \exp(\sqrt{3}/2)$
2241 7142 ln : 8π
2241 8407 $10^x : 9\sqrt{2}/5$
2241 8625 $\Gamma : (2\pi)^{-1/2}$
2241 8692 rt : (7,7,7,-2)
2241 8864 $J_0 : 1/\ln(\ln 3)$
2242 0332 $\Psi : \ln(\pi/2)$
2242 1556 rt : (4,5,-8,-2)
2242 1637 at : $\sqrt{23/3}$
2242 2618 $e^x : 2\pi^2$
2242 2643 $t\pi : 3\sqrt{3}/2 + 2\sqrt{5}$
2242 2745 rt : (5,1,-3,-7)
2242 2757 $e^x : 4/3 - 2\sqrt{2}$
2242 3274 $2^x : 4/3 + \sqrt{5}/3$
2242 5213 rt : (7,1,-6,4)
2242 5314 $2^x : (\sqrt[3]{5}/3)^{-3}$
2242 6971 rt : (2,9,-7,-2)
2242 7353 $J_1 : (3\pi/2)^{-1/2}$
2242 7456 rt : (9,3,-9,1)
2242 8378 $e^x : (3/2)^3$
2243 1722 $e^x : 3/2 + 3\sqrt{3}/2$
2243 3855 rt : (7,1,-6,-7)
2243 5393 rt : (9,1,-9,4)

2243 6364 $10^x : 5e\pi/7$
2243 7195 $t\pi : 8\sqrt[3]{3}/9$
2243 7673 sq : 9/19
2243 8235 rt : (5,8,6,1)
2244 0190 $\ln(2/5) + \exp(\pi)$
2244 1038 rt : (9,1,-9,-7)
2244 1371 $\ell_{10} : (3\pi/2)^{1/3}$
2244 2544 rt : (3,9,0,7)
2244 6172 sinh : $10e/9$
2244 6414 rt : (3,7,-8,2)
2244 7583 rt : (7,4,5,1)
2244 8979 sq : 16/7
2244 9385 at : $8\sqrt{3}/5$
2245 3101 rt : (9,0,4,1)
2245 4626 $\ell_{10} : 3\sqrt{5}/4$
2245 9182 $2^x : 8\zeta(3)/3$
2246 0746 $t\pi : 7\ln 2/2$
2246 1808 $J_2 : 7\sqrt[3]{2}/6$
2246 1969 rt : (7,2,9,2)
2246 3311 rt : (9,2,8,-2)
2246 3442 $e^x : (\sqrt{2}\pi/2)^{-2}$
2246 3654 $\ln(\sqrt{3}) - s\pi(e)$
2246 4441 at : $4\ln 2$
2246 6064 $J_0 : 7\ln 2$
2246 6329 sq : $4 + 3\pi$
2246 6448 $1/2 \div \exp(4/5)$
2246 6521 sinh : $7(1/\pi)/10$
2246 7033 id : $5\pi^2/6$
2246 7324 $Li_2 : 2(1/\pi)/3$
2246 7878 $J_1 : 6/13$
2247 0308 as : $7(1/\pi)/10$
2247 4487 id : $\sqrt{3/2}$
2247 8079 $\theta_3 : \ln(2\pi/3)$
2247 8676 $J_2 : 25/17$
2247 9643 $\ln(\sqrt{3}) + s\pi(\sqrt{5})$
2248 0279 sr : $7\sqrt{2}/2$
2248 0283 at : $e/4 + 2\pi/3$
2248 0728 $10^x : 10\pi/7$
2248 2536 $J_1 : \pi^2/3$
2248 2619 rt : (2,9,-7,3)
2248 4682 $e^x : 2 - \sqrt{5}/4$
2248 5167 $\exp(2) \times s\pi(1/4)$
2248 6774 rt : (9,6,-8,-2)
2248 9084 $\Psi : \exp(-3\pi)$
2248 9758 sq : $4e$
2249 0303 $erf : \sqrt{2}/7$
2249 1368 cr : $\ln(2\pi)$
2249 3656 $J_1 : 2\ln 2/3$
2249 4210 $J_2 : 10\pi/3$
2249 4301 ln : $((e + \pi)/2)^3$
2249 4378 $t\pi : 1/4 + e/4$
2249 4616 sinh : $(\ln 3)^{1/3}$
2249 4621 $e^x : (\ln 2/2)^{-3}$
2249 5202 rt : (7,8,-3,-1)
2249 8439 $\ell_{10} : 2\sqrt{3}/3 - \sqrt{5}/4$
2249 8627 sinh : $\exp(-3/2)$
2249 9124 $2^x : 4\sqrt[3]{3}/5$
2249 9286 $\Psi : (\ln 2/3)^2$
2249 9806 $\Gamma : 7e/8$
2250 0352 rt : (9,4,-4,-1)
2250 1889 $Ei : \sqrt{3}/4$
2250 2441 as : $\exp(-3/2)$
2250 5130 rt : (1,8,-1,7)
2250 6295 $2^x : 15/13$
2250 7907 sq : $(\sqrt{2}\pi)^{-1/2}$
2250 9465 $2^x : 1 - \sqrt{2}/2$
2251 0815 $e^x : \exp(\sqrt{5}/3)$
2251 2298 at : $\sqrt{3}/3 - 3\sqrt{5}/2$
2251 2363 $E : \exp(-1/(2\pi))$
2251 6343 $1/3 \times s\pi(\sqrt{5})$
2251 6798 $\ell_2 : 3\sqrt{2} + \sqrt{3}/4$
2251 7237 rt : (3,9,-9,-8)
2251 8120 $t\pi : 3\sqrt{2} - \sqrt{5}/3$
2251 9822 $\ell_{10} : 1 + e/4$
2252 0619 rt : (6,7,7,-2)
2252 1225 sq : $\exp(-\sqrt{5}/3)$
2252 3644 $\Psi : \exp(e/2)$
2252 4074 at : 25/9
2252 5363 $e^x : \sqrt{2}/2 + 3\sqrt{5}$
2252 5730 rt : (4,3,8,-2)
2252 6906 rt : (3,7,-9,-5)
2252 8206 $\ell_{10} : 4\sqrt[3]{2}/3$
2252 9721 rt : (1,9,-6,-8)
2253 0030 $2^x : 3e/4 - 4\pi/3$
2253 0108 $Li_2 : 20/23$
2253 0192 $\pi + \ln(2/5)$
2253 1947 cu : $3\sqrt{2} - 4\sqrt{3}/3$
2253 2118 rt : (2,1,9,2)
2253 2966 $\Psi : 2\pi/9$
2253 3471 rt : (3,9,-7,-2)
2253 4873 $J_1 : 7\sqrt[3]{5}$
2253 5292 $Ei : \exp(e/2)$
2253 7746 rt : (2,7,-4,9)
2254 0333 rt : (5,5,-8,-2)
2254 1670 $\Gamma : 3/4$
2254 2226 rt : (3,5,-9,-2)
2254 4355 cu : $2 + 3\pi$
2254 4695 rt : (8,9,2,-1)
2254 5084 ln : $\ln(\sqrt{2}\pi/2)$
2254 6271 rt : (9,8,2,7)
2254 7004 rt : (1,7,-6,-5)
2254 7322 ln : $2\sqrt{2} + \sqrt{3}/3$
2254 7386 tanh : $1/\sqrt{19}$
2254 7485 sinh : $\sqrt{5}/10$
2254 9008 rt : (7,1,-9,-2)
2254 9012 sr : $2\sqrt{2}/3 + \sqrt{5}/4$
2254 9034 $e^x : ((e + \pi)/3)^{-1/3}$
2254 9302 sq : $1/3 - 2\sqrt{2}$
2255 0048 cu : $3e - \pi/4$
2255 1340 at : $1/\sqrt{19}$
2255 1759 $2^x : 3\sqrt{5}/2$
2255 2307 as : $3/2 - \sqrt{5}/4$
2255 2509 rt : (1,9,3,-8)
2255 2806 cu : 14/23
2255 4092 $e^x : 4/5$
2255 6575 $erf : (\sqrt{2}\pi/2)^{-2}$
2255 7265 rt : (6,5,3,-1)
2255 7846 $\theta_3 : 3\sqrt{2}/8$
2255 8779 at : $\sqrt{2}/3 + 4\sqrt{3}/3$
2255 8896 at : $3\sqrt{2} - 2\sqrt{5}$
2255 9915 $\ell_2 : 3 + 3\sqrt{5}/4$
2256 1633 $10^x : \sqrt{2}/3 - \sqrt{5}/2$
2256 6347 sq : $7\pi/4$
2256 7203 $\ell_2 : \sqrt{2} - \sqrt{5}/4$
2256 7746 $2^x : 1/2 + 4\sqrt{2}/3$
2256 8366 id : $1/3 - \sqrt{5}/4$
2257 0187 rt : (3,3,-9,1)
2257 0842 as : $\sqrt{2}/4 - \sqrt{3}/3$
2257 0899 sr : $2e/3 + \pi$
2257 1145 $e^x : 2/3 + 2\sqrt{3}$
2257 1728 rt : (7,8,5,7)
2257 1736 $10^x : 1 + \sqrt{5}/4$
2257 4622 $Ei : 5\ln 2/8$
2257 4950 rt : (8,3,-1,9)
2257 5341 rt : (8,2,8,-2)
2257 7470 cu : $2/3 - 4e$
2257 7977 $10^x : 2 + 4\sqrt{3}$
2257 9061 ln : $3e/4 - \pi/4$
2257 9097 $Li_2 : 1/\sqrt{22}$
2257 9135 ln : $(\pi/2)^{1/2}$
2257 9408 sr : $5\zeta(3)/4$
2258 0015 sq : $3\sqrt{3}/2 + 4\sqrt{5}$
2258 0021 rt : (1,8,7,-2)
2258 2265 $\ell_{10} : \ln(2e/3)$
2258 4014 cr : $3\sqrt{3} - 3\sqrt{5}/2$
2258 4176 rt : (3,7,-3,-1)
2258 7720 rt : (6,2,9,2)
2258 8529 $J_1 : 1/\sqrt[3]{10}$
2258 8573 $t\pi : 2\sqrt{3}/3 - \sqrt{5}/4$
2259 1243 rt : (9,0,-7,8)
2259 2119 rt : (1,7,3,-5)
2259 3821 rt : (9,9,6,1)
2259 4105 sr : $2\sqrt{3} + 2\sqrt{5}/3$
2259 4513 rt : (6,9,-7,6)
2259 4554 $\exp(3/4) \div s\pi(2/5)$
2259 5393 $t\pi : 2(1/\pi)/9$
2259 6418 $\ell_2 : \exp(\sqrt{5})$
2259 6539 rt : (8,5,-1,6)
2259 7067 $10^x : 1/\ln(\pi/3)$
2259 8720 $\ell_{10} : 7\sqrt[3]{3}/6$
2259 8962 $e^x : 3/2 + 4\pi$
2260 0313 $t\pi : 2e/3 - 2\pi/3$
2260 2051 sq : $\sqrt{2}/4 - 4\sqrt{3}$
2260 2335 cu : $3/4 - e/2$
2260 2396 $\ell_2 : \sqrt[3]{5}/2$
2260 2769 rt : (6,3,2,9)
2260 3585 rt : (2,3,5,1)
2260 5306 $\ell_{10} : 7\zeta(3)/5$
2260 5731 $1 - s\pi(e)$
2260 6152 as : P_{31}
2260 7371 rt : (3,9,0,-8)
2260 8454 $J_1 : \exp(3/2)$
2260 9514 as : 16/17
2261 0258 $s\pi : P_{34}$
2261 0629 rt : (5,3,-4,-1)
2261 1196 rt : (9,2,-7,5)
2261 1292 at : $2 + \pi/4$
2261 2132 $\ell_{10} : 6\ln 2/7$
2261 2784 $\zeta : 5\sqrt{3}/3$
2261 3484 $2^x : 5/17$
2261 5468 $t\pi : 1/2 + \pi/2$
2261 5621 $\Psi : 8e\pi/7$
2261 6691 rt : (7,0,-4,8)
2261 7523 $J_1 : 7e\pi/5$
2261 8255 $2^x : 3e + \pi/4$
2261 9524 $\theta_3 : 5(1/\pi)/3$
2262 1861 $\Gamma : \exp(-\sqrt{3})$
2262 2955 $\exp(3) + \exp(\pi)$
2262 5173 rt : (5,8,8,7)
2262 5552 sq : $3\sqrt{3}/2 - 2\sqrt{5}/3$
2262 6057 $e^x : 3/4 - \sqrt{5}$
2262 8224 $\ell_{10} : 3/2 - e/3$
2262 8242 $10^x : 1/3 + 4\sqrt{5}/3$
2262 9930 rt : (8,7,-1,3)
2263 0058 rt : (1,9,6,-3)
2263 3127 cr : $\sqrt{2}/4 + 2\sqrt{5}/3$
2263 3529 $\theta_3 : 7\ln 2/5$
2263 6214 rt : (1,6,5,5)
2263 6474 rt : (4,5,-6,6)
2263 7376 rt : (3,1,-9,4)
2263 7610 $\ell_2 : 4 + e/4$
2263 8105 $2^x : 2\sqrt{3}/3$
2263 9364 rt : (9,4,-7,2)
2263 9623 cu : $1/3 - 2\sqrt{2}/3$
2264 0164 at : $2\sqrt{2} - 3\sqrt{3}/2$
2264 0190 rt : (6,5,2,6)
2264 0546 $Li_2 : \sqrt[3]{5}/8$
2264 0826 at : $2/3 + 3\sqrt{2}/2$
2264 2193 rt : (3,3,8,-2)
2264 2922 rt : (4,9,-7,-2)
2264 3581 $2^x : 3\ln 2$
2264 3830 $\zeta : 1/\ln(\sqrt{2})$
2264 4885 ln : $\sqrt{3} + 3\sqrt{5}/4$
2264 5040 $e^x : 1/\sqrt{24}$
2264 5713 $e - \exp(2/5)$
2264 6150 $10^x : 8e/7$
2264 7432 rt : (7,2,-4,5)
2264 8439 $\Psi : 9\sqrt{3}/4$
2264 9081 at : $3/4 + 3e/4$
2265 0075 $2 \times \exp(\sqrt{2})$
2265 1911 rt : (4,3,5,9)
2265 2932 $J_1 : 23/7$
2265 4091 rt : (7,6,1,0)
2265 5394 cr : $(2\pi)^{1/3}$
2265 5653 $\Gamma : \sqrt{17/3}$
2265 6043 cr : $3 - 2\sqrt{3}/3$
2265 6336 rt : (5,0,-1,8)
2265 7200 $J_2 : 10e\pi/9$
2265 8867 rt : (6,3,-7,-7)
2265 9136 sr : $4/3 + 4e/3$
2266 0723 $2^x : -\sqrt{3} + \sqrt{5} - \sqrt{7}$
2266 0911 at : $3\sqrt{3}/4 + 2\sqrt{5}/3$
2266 2936 rt : (6,5,-8,-2)
2266 2946 $2^x : 1 - \pi$
2266 3588 rt : (1,1,9,2)
2266 4178 $\zeta : 2\sqrt[3]{3}$
2266 6648 rt : (4,7,-2,5)
2266 8159 rt : (1,6,3,-1)
2266 8971 rt : (9,6,7,-2)
2266 9882 rt : (8,0,4,1)

2267 2729 2^x : $4\sqrt{2}/3 + 2\sqrt{3}/3$
2267 4970 cr : $24/13$
2267 5121 rt : (2,5,-5,-1)
2267 5736 sq : $10/21$
2267 5815 tanh : $3/13$
2267 7390 tπ : $4\sqrt{3} + 3\sqrt{5}/2$
2267 9139 J_0 : $7\sqrt[3]{5}/6$
2267 9783 10^x : $2e/5$
2267 9884 at : $3/13$
2268 1074 J_2 : $3\pi^2/5$
2268 2121 rt : (9,6,-7,-1)
2268 4052 Ei : $7\sqrt[3]{3}/9$
2268 5187 ln : $2 - \sqrt{5}/3$
2268 5783 e^x : $\exp(-\sqrt[3]{4})$
2268 8342 ℓ_2 : $4/3 - e/3$
2268 8411 rt : (8,9,-1,0)
2268 8541 Ψ : $5(e+\pi)/3$
2269 0614 rt : (7,2,8,-2)
2269 2109 Γ : $(\sqrt[3]{2}/3)^{1/3}$
2269 2246 id : $9\sqrt{3}/7$
2269 2499 at : $8\pi/9$
2269 4053 rt : (1,7,-9,9)
2269 4462 J_2 : $2\sqrt{5}$
2269 4568 ζ : $9\sqrt[3]{3}/8$
2269 4592 at : $3\sqrt{2}/4 + \sqrt{3}$
2269 5097 rt : (7,4,-4,2)
2269 7293 J : $\exp(-16\pi/3)$
2269 8719 cr : $\sqrt{2} + \sqrt{3}/4$
2269 8964 rt : (6,1,-7,-4)
2269 9555 sq : $(e/2)^{1/3}$
2270 0592 ℓ_2 : $\exp((e+\pi)/2)$
2270 1165 Li_2 : $3/14$
2270 1485 rt : (5,9,-7,-5)
2270 2358 tanh : $\ln 2/3$
2270 3885 J_1 : $7/15$
2270 5013 rt : (7,9,2,-1)
2270 5138 tπ : $5\sqrt{2}$
2270 5735 sq : $1/\ln(1/(3e))$
2270 6055 10^x : $3\sqrt{2}/2$
2270 6388 rt : (8,8,-3,-1)
2270 6441 e^x : $4 - 2\sqrt{2}$
2270 6451 at : $\ln 2/3$
2270 8374 rt : (6,7,2,3)
2270 8592 ℓ_{10} : $4/3 + \sqrt{2}/4$
2270 9126 tanh : $(\sqrt[3]{3}/3)^2$
2270 9819 rt : (5,2,-1,5)
2271 0623 rt : (4,9,2,4)
2271 1135 rt : (3,8,-7,-9)
2271 2957 $\pi + \exp(3)$
2271 3225 at : $(\sqrt[3]{3}/3)^2$
2271 3444 rt : (2,4,-8,-2)
2271 6912 10^x : $6\sqrt[3]{3}/7$
2271 7000 erf : $1/\sqrt{24}$
2271 7132 e^x : $1/4 - \sqrt{3}$
2271 7523 sπ : $1/4 + 3\sqrt{5}/4$
2271 7738 rt : (5,2,9,2)
2271 8012 $\sqrt{5} \div \exp(3/5)$
2272 1552 at : $5\sqrt{5}/4$
2272 2189 2^x : $5\ln 2/3$
2272 2191 J_1 : $(2e/3)^2$
2272 2704 rt : (6,4,5,1)
2272 3573 rt : (2,7,3,4)
2272 4279 rt : (1,8,6,3)
2272 4535 cr : $8\ln 2/3$
2272 6668 rt : (3,0,2,8)
2272 7272 id : $5/22$
2272 7585 sq : $5\pi^2$
2272 8615 2^x : $3\sqrt{2}/2 + 2\sqrt{5}/3$
2272 9615 sq : $3\sqrt{2}/2 - 3\sqrt{3}/2$
2273 1376 e^x : $3/2 - 4\sqrt{5}/3$
2273 2489 rt : (4,7,7,-2)
2273 5290 e^x : $3/2 + 2e$
2273 5930 sπ : $5\pi/4$
2273 6420 id : $5(1/\pi)/7$
2273 7055 ln : $3\sqrt{2}/2 - \sqrt{3}/2$
2273 8611 ζ : $16/25$
2273 9357 rt : (9,5,-6,1)
2274 1100 Ψ : $((e+\pi)/3)^{1/3}$
2274 1490 ℓ_{10} : $2\sqrt{2} - \sqrt{5}$
2274 1903 id : $\exp(-\sqrt{2}\pi/3)$
2274 3210 sinh : $(e/3)^{-1/3}$
2274 5083 rt : (9,1,-5,9)
2274 5158 ℓ_{10} : $3\sqrt{2}/2 - \sqrt{3}/4$
2274 5204 rt : (4,0,-9,2)
2274 5353 Ψ : $1/4$
2274 5563 cu : $7\ln 2$
2274 5774 ℓ_2 : $(\ln 2/3)^2$
2274 7455 rt : (5,5,3,-1)
2274 8198 at : $2/3 - 2\sqrt{3}$
2275 0157 id : $3e/4 + 4\pi/3$
2275 2947 erf : $\exp(-\sqrt[3]{4})$
2275 3603 2^x : $7(e+\pi)/2$
2275 4295 10^x : $8/23$
2275 4701 rt : (9,8,-7,-4)
2275 5489 rt : (5,9,-7,-2)
2275 5980 $1/4 \div \ln(3)$
2275 8390 $\ln(2/3) - \exp(3/5)$
2275 9996 sq : $4\sqrt{3} - \sqrt{5}/3$
2276 0345 ln : $1/3 + 4\sqrt{5}$
2276 1299 rt : (6,1,-7,1)
2276 2217 rt : (2,3,8,9)
2276 3782 cr : $\sqrt[3]{19/3}$
2276 4202 sr : $1/4 + 3\pi/2$
2276 4474 as : $1/3 - \sqrt{5}/4$
2276 5650 rt : (3,6,-7,-6)
2276 6354 at : $3/2 + 3\sqrt{3}/4$
2276 9902 rt : (9,3,9,2)
2277 1761 rt : (9,9,4,8)
2277 3692 ℓ_2 : $e\pi/10$
2277 5203 rt : (8,6,7,-2)
2277 6601 10^x : $5/2$
2277 7238 at : $14/5$
2277 8017 cu : $1/2 - \pi/2$
2277 8909 rt : (7,6,-4,-1)
2278 0654 rt : (8,2,-8,-8)
2278 0715 J_1 : $10\pi/7$
2278 1435 e^x : $20/9$
2278 1773 rt : (1,8,-4,-9)
2278 2223 sq : $4\sqrt{3}/3 + \sqrt{5}/4$
2278 6882 rt : (3,7,0,-5)
2278 7141 rt : (1,5,-1,1)
2278 7915 J_2 : $19/2$
2278 8253 Γ : $(e/2)^{-3}$
2278 8851 2^x : $\sqrt{2} - \sqrt{5}/2$
2278 9622 rt : (7,5,-8,-2)
2278 9858 e^x : $1/3 - 2e/3$
2279 0679 2^x : $4 - 4\sqrt{3}/3$
2279 1404 rt : (4,7,-3,-1)
2279 1931 rt : (9,3,-5,6)
2279 2206 sq : $3 + 3\sqrt{2}/2$
2279 2256 sr : $\sqrt{11}\pi$
2279 4717 ln : $2 + \sqrt{2}$
2279 7192 sinh : $3\pi/4$
2279 7266 id : $(2\pi/3)^{-2}$
2279 8785 2^x : $(3/2)^{-3}$
2279 9812 rt : (1,9,2,0)
2280 0293 at : $\sqrt[3]{22}$
2280 1252 ln : $\sqrt{3}/4 + 4\sqrt{5}/3$
2280 1310 sq : $6(e+\pi)/7$
2280 1528 at : $3/2 - \sqrt{3}$
2280 2639 sr : $3/2 + 2\sqrt{3}$
2280 2849 rt : (2,3,-1,5)
2280 3363 id : $2\sqrt{3} - \sqrt{5}$
2280 3876 cu : $\sqrt{23/3}$
2280 4058 ℓ_{10} : $4 - 4\sqrt{3}/3$
2280 4973 rt : (7,1,-2,9)
2280 7749 rt : (5,4,-1,2)
2280 9080 cu : $2\sqrt{2}/3 + 2\sqrt{3}/3$
2280 9320 rt : (6,2,8,-2)
2281 0831 E : $17/20$
2281 0990 sr : $\sqrt{2}/4 + 2\sqrt{3}/3$
2281 1523 Ψ : $7\pi^2$
2281 5217 rt : (2,4,-7,3)
2281 5549 rt : (4,8,6,1)
2281 6591 tπ : $3\sqrt[3]{5}/4$
2281 6920 id : $7(1/\pi)$
2281 7799 Li_2 : $(\pi/3)^{-3}$
2281 9633 2^x : $-\sqrt{2} + \sqrt{3} - \sqrt{6}$
2282 0053 sq : $8\pi^2/7$
2282 0241 cr : $3\sqrt{3}/2 - \sqrt{5}/3$
2282 0473 ln : $3 + 2\pi$
2282 2357 rt : (9,4,3,-1)
2282 2359 cu : $11/18$
2282 3882 ζ : $3\sqrt[3]{3}/10$
2282 4347 tπ : $1/14$
2282 4349 rt : (8,6,1,7)
2282 6661 J_2 : $6\sqrt{3}/7$
2282 6997 ℓ_{10} : $e/3 + \pi/4$
2282 8568 rt : (5,8,-7,0)
2282 8587 $\sqrt{5} \times \ln(\sqrt{3})$
2282 9303 sπ : $3\sqrt{2}/4 + \sqrt{3}/2$
2282 9376 rt : (1,8,-2,-1)
2283 1448 at : $7\zeta(3)/3$
2283 1782 2^x : $\sqrt[3]{9}$
2283 2406 e^x : $5\sqrt[3]{3}/9$
2283 3378 10^x : $2/3 - \sqrt{3}/3$
2283 3623 2^x : $4/3 - 2\sqrt{3}$
2283 3863 J_0 : $8e\pi/7$
2283 3917 rt : (8,0,-8,5)
2283 5015 sπ : $4\sqrt{3}/3 - \sqrt{5}$
2283 6202 e^x : $(e+\pi)/5$
2283 6750 ζ : $1/\ln(1/(3\pi))$
2283 7862 J_1 : $23/2$
2283 8649 cr : $3/2 + \sqrt{2}/4$
2283 8699 rt : (3,4,-8,-2)
2283 9181 cr : $3\sqrt{2}/2 + 4\sqrt{5}$
2284 1490 rt : (3,9,-5,-9)
2284 2714 rt : (3,7,7,-2)
2284 3915 ln : $2\pi/5$
2284 3961 sinh : $6\pi^2/7$
2284 4669 ℓ_2 : $4 - 2\sqrt{2}$
2284 6255 rt : (2,9,-8,-2)
2284 6279 rt : (1,9,-2,1)
2284 7681 2^x : $8(e+\pi)$
2284 7932 ℓ_{10} : $13/22$
2284 8788 rt : (6,2,-5,-8)
2285 2206 ℓ_{10} : $1/3 + e/2$
2285 2324 rt : (4,2,9,2)
2285 2737 rt : (3,5,-6,-5)
2285 3083 sr : $3 - 2\sqrt{5}/3$
2285 3241 2^x : $1/4 + e/3$
2285 4376 $\ln(2/5) \div s\pi(\sqrt{3})$
2285 5557 tπ : P_{17}
2285 5601 cr : $9\sqrt[3]{3}/7$
2285 7839 tπ : $e/2 + 3\pi/2$
2285 8335 rt : (9,5,-5,3)
2285 9483 e^x : $2\zeta(3)/3$
2286 0797 ℓ_{10} : $3/4 + 2\sqrt{2}/3$
2286 1944 cu : $2e - 3\pi/4$
2286 4069 2^x : $3\ln 2/7$
2286 4964 rt : (4,8,2,9)
2286 9111 10^x : $(\sqrt{5})^{-3}$
2286 9471 ℓ_{10} : $\ln(2e)$
2287 0426 at : $1/3 - \pi$
2287 1355 rt : (6,9,2,0)
2287 2543 e^x : $e\pi/3$
2287 2667 10^x : $1/2 + \sqrt{3}/2$
2287 3345 sq : $11/23$
2287 4247 sr : $1 + 3\pi$
2287 7614 rt : (7,3,-2,6)
2287 9038 Ei : $(\sqrt[3]{2})^{1/2}$
2287 9055 Γ : $5(1/\pi)/9$
2287 9055 ln : $8\sqrt{2}/9$
2287 9701 e^x : $\sqrt[3]{3}/7$
2287 9782 rt : (7,9,7,8)
2287 9877 ℓ_2 : $2/3 + 3\sqrt{5}/4$
2288 0507 rt : (1,6,-4,-6)
2288 2561 ζ : $23/8$
2288 2626 rt : (2,6,1,8)
2288 3269 at : $1/2 + 4\sqrt{3}/3$
2288 3843 ζ : $9\sqrt{5}/7$
2288 4060 J_1 : $8/17$
2288 4246 rt : (7,6,7,-2)
2288 4618 $s\pi(2/5) \div s\pi(e)$
2288 6024 rt : (1,3,8,-2)
2288 7309 ln : $2/3 - \sqrt{5}/4$
2288 8372 rt : (5,0,-9,2)
2288 8427 sr : $5e/9$
2288 9159 10^x : $4/3 - \sqrt{2}/2$
2289 0786 sq : $4 + 3\sqrt{5}/4$

2289 3078 rt : (5,5,-8,-7)
2289 4166 $\ell_2 : (e+\pi)/5$
2289 4195 $Li_2 : (5/3)^{-3}$
2289 4597 $1/5 \times \ln(\pi)$
2289 4690 $E : (\sqrt[3]{3}/2)^{1/2}$
2289 7409 rt : (1,7,4,-5)
2289 7624 $Ei : 10$
2289 8192 $\Psi : 3\zeta(3)/8$
2289 8552 $2^x : \sqrt{2}/3 - 3\sqrt{3}/2$
2289 8766 rt : (7,0,4,1)
2289 8967 rt : (5,1,1,9)
2290 2637 $J_0 : 5(e+\pi)/3$
2290 2826 rt : (8,3,9,2)
2290 3012 cu : $(5/3)^3$
2290 4907 $J_1 : 7\pi^2/6$
2290 7241 $e^x : 3 + 4\sqrt{3}/3$
2290 7268 $1/4 \times \ln(2/5)$
2290 7927 $Ei : \exp(-(e+\pi)/3)$
2290 8300 rt : (1,3,5,1)
2290 9538 rt : (8,2,-8,2)
2290 9958 $\Gamma : 6\sqrt{5}$
2290 9986 $t\pi : \sqrt{3} - \sqrt{6} + \sqrt{7}$
2291 0471 $10^x : 2e - \pi$
2291 0745 $e - \ln(3/5)$
2291 2764 rt : (1,8,1,-9)
2291 4029 $e^x : \sqrt{2}/2 + 3\sqrt{3}$
2291 4395 rt : (2,8,-7,-2)
2291 4580 at : $1 + 2e/3$
2291 7022 $\ell_2 : \sqrt{3}/4 + 4\sqrt{5}$
2291 7892 cr : 13/7
2291 8623 rt : (2,9,3,3)
2291 9803 $\Psi : \sqrt{2} + \sqrt{3} - \sqrt{6}$
2292 0671 rt : (8,5,-8,-2)
2292 1531 $J_1 : \sqrt{2}/3$
2292 1572 $\ell_2 : 1/2 + 4\pi/3$
2292 2011 rt : (8,8,1,4)
2292 3433 sinh : 5/22
2292 3807 $erf : \sqrt[3]{3}/7$
2292 4257 rt : (5,7,-7,5)
2292 4831 $\Psi : 2(e+\pi)/3$
2292 5936 $2^x : (\ln 2/3)^{-1/2}$
2292 6275 $\zeta : \sqrt{3}/4$
2292 6814 $\ell_{10} : \sqrt{3}/3 + \sqrt{5}/2$
2292 7620 as : 5/22
2292 8055 rt : (5,9,-3,-8)
2292 8386 at : $\sqrt{3}/3 + \sqrt{5}$
2292 9586 rt : (9,8,-3,-1)
2293 0168 sq : $4/3 - 2e/3$
2293 0542 sq : $7e/3$
2293 1575 rt : (6,0,-5,5)
2293 1666 rt : (5,2,8,-2)
2293 2180 id : $4e + 3\pi/4$
2293 2369 $E : 8(1/\pi)/3$
2293 2818 sinh : $5(1/\pi)/7$
2293 5022 $2^x : 3/2 - 4e/3$
2293 7013 as : $5(1/\pi)/7$
2293 8014 cr : $3\sqrt{3}/4 + \sqrt{5}/4$
2293 8444 sinh : $\exp(-\sqrt{2}\pi/3)$
2293 8639 rt : (3,8,-7,-5)
2293 9434 rt : (8,1,-6,-9)
2294 1149 rt : (1,1,-9,2)
2294 1573 id : $1/\sqrt{19}$
2294 2644 as : $\exp(-\sqrt{2}\pi/3)$
2294 3489 $e^x : 3 - 2\sqrt{5}$
2294 4116 $J_2 : 7e/2$
2294 6078 $t\pi : 4\sqrt{3}$
2294 6582 rt : (4,7,5,3)
2294 8505 sr : $8\sqrt{5}$
2294 9526 id : $3\sqrt{2} - 2\sqrt{5}$
2295 1250 $Li_2 : \sqrt{3}/8$
2295 2613 rt : (6,6,4,7)
2295 2678 $10^x : \sqrt{2}/4 + 2\sqrt{3}/3$
2295 2874 $\ln(3/5) \div \exp(4/5)$
2295 3598 at : $2/3 - \sqrt{3}/4$
2295 6007 rt : (4,2,-2,-8)
2295 6430 cu : $5e/4$
2295 7311 $10^x : 1 + 3e/4$
2295 7905 $Ei : 5e\pi/4$
2295 9276 sq : $3/2 - 4\pi/3$
2295 9556 sr : $6\sqrt[3]{2}/5$
2295 9699 rt : (9,7,-5,0)
2296 0523 rt : (8,9,6,1)
2296 1204 $\ell_2 : \exp(1/(2\pi))$
2296 1957 cr : $1/2 + e/2$
2296 2648 ln : $2\sqrt[3]{5}$
2296 2936 $\ln(\pi) \times \exp(2/3)$
2296 3182 $E : 3\sqrt{2}/5$
2296 4512 rt : (7,8,-4,-4)
2296 7173 sr : $(3\pi/2)^3$
2296 7374 $\exp(1/2) \times s\pi(\sqrt{3})$
2296 8147 rt : (4,4,-8,-2)
2296 9624 $\ell_{10} : 6\sqrt{2}/5$
2296 9757 $e^x : P_{39}$
2297 0295 $Ei : 4\pi^2/5$
2297 0442 $J_1 : 3\sqrt[3]{2}/8$
2297 1536 $J_2 : (\sqrt{2}\pi/2)^{1/2}$
2297 1580 $\ell_2 : \sqrt{22}$
2297 1626 $e^x : P_{33}$
2297 4535 ln : $3\sqrt{2}/2 + 3\sqrt{3}/4$
2297 5694 rt : (9,1,8,-2)
2297 6478 rt : (7,4,-9,-8)
2297 7847 rt : (9,4,-3,7)
2297 8040 $J_2 : 2\sqrt{5}/3$
2297 8053 ln : $\sqrt{2}/4 + 4\sqrt{5}$
2297 9983 $\ell_{10} : \exp(\sqrt[3]{4}/3)$
2298 0862 $e^x : e - 4\pi/3$
2298 2868 sr : $1/2 + 2\sqrt{5}$
2298 3045 id : $3\pi^2/7$
2298 5151 $10^x : 3 - 3\sqrt{3}/4$
2298 5608 $t\pi : e/3 - 4\pi/3$
2298 6248 rt : (5,3,-8,-4)
2298 7042 ln : $4\sqrt{2} - \sqrt{5}$
2298 8253 $\zeta : 5\ln 2/8$
2299 0583 $2^x : e/3 + \pi/4$
2299 0716 rt : (7,9,-7,-2)
2299 1018 rt : (7,5,-2,3)
2299 1807 rt : (3,2,9,2)
2299 1903 sq : $3/4 + \pi/3$
2299 2729 $J_2 : \ln(\sqrt{2}\pi)$
2299 4483 rt : (5,9,-3,4)
2299 5248 sinh : $(2\pi/3)^{-2}$
2299 5626 cu : 15/14
2299 5971 sq : $8\ln 2/5$
2299 5980 $\Psi : 19/11$
2299 6247 rt : (6,6,7,-2)
2299 6707 as : $3\pi/10$
2299 8102 rt : (2,8,-1,4)
2299 9500 as : $(2\pi/3)^{-2}$
2300 0795 at : $2\pi^2/7$
2300 0924 $Ei : 9\sqrt{2}/4$
2300 2041 cu : $4/3 + 3\sqrt{2}/2$
2300 2782 $J_1 : 8\sqrt[3]{3}$
2300 4096 sinh : $\sqrt[3]{11/3}$
2300 4453 $\zeta : \exp(-1/(2\pi))$
2300 7217 rt : (8,1,-5,1)
2300 7914 $J_0 : \sqrt{17/2}$
2300 9858 rt : (5,7,-3,-1)
2301 0226 $\ell_{10} : (\ln 2/2)^{1/2}$
2301 2102 $e^x : \exp(1/(2\pi))$
2301 3819 $J_1 : (3\pi)^{-1/3}$
2301 3858 rt : (3,8,-2,0)
2301 4565 $4/5 \times \ln(3/4)$
2301 6255 cu : $5e\pi/3$
2301 7410 rt : (8,4,3,-1)
2301 7449 $\ell_{10} : 5e/8$
2301 9350 rt : (5,4,5,1)
2301 9356 sq : $1/4 - e/2$
2302 0125 id : $\exp(\sqrt[3]{3})$
2302 0149 $t\pi : 5\pi^2/3$
2302 0714 cr : $\sqrt{2}/2 + 2\sqrt{3}/3$
2302 1410 rt : (8,1,-6,6)
2302 3150 $\Gamma : 5(1/\pi)/4$
2302 3633 $e^x : (\sqrt{2}\pi)^{1/2}$
2302 4460 $\ell_2 : 4\sqrt{3} - \sqrt{5}$
2302 4721 $s\pi : 7e\pi/3$
2302 5039 $\exp(3/4) + \exp(\sqrt{2})$
2302 5243 rt : (8,4,-8,-1)
2302 5301 $\Psi : 9\ln 2/5$
2302 5984 $10^x : 1 + \sqrt{5}/2$
2302 6118 $J_1 : 9/19$
2302 6501 ln : $4e - \pi/2$
2302 6680 $\ln(\pi) + \exp(3)$
2302 6750 rt : (5,3,1,6)
2302 7341 $J_2 : 15/2$
2302 7434 $e^x : 4\sqrt{2}/3 - 3\sqrt{5}/2$
2303 0121 $1/5 + s\pi(\pi)$
2303 2344 rt : (3,8,-7,-2)
2303 2943 sr : $(1/\pi)/6$
2303 5091 id : $2\sqrt{2} - 3\sqrt{3}/2$
2303 7264 $10^x : 3/2 + 3\sqrt{5}$
2303 7309 rt : (6,0,-9,2)
2303 7983 $10^x : 4 - 3\sqrt{3}/2$
2303 8907 sinh : 17/9
2304 0000 sq : 12/25
2304 0439 rt : (7,3,9,2)
2304 1502 ln : $4 - \sqrt{3}/3$
2304 3903 $2^x : 3\sqrt{3} + 3\sqrt{5}/4$
2304 4170 rt : (5,9,2,-1)
2304 4557 rt : (6,1,-3,-9)
2304 4892 $\ell_{10} : 17$
2304 7187 at : $4/3 + 2\sqrt{5}/3$
2304 7903 at : $3e/4 + \pi/4$
2304 8451 $t\pi : 10\zeta(3)/7$
2305 0177 $J_2 : 1/\ln((e+\pi)/3)$
2305 1007 $Li_2 : 5/23$
2305 4774 $J_2 : \sqrt[3]{10/3}$
2305 4871 $10^x : -\sqrt{2} + \sqrt{3} + \sqrt{6}$
2305 5563 cr : $5\sqrt{5}/6$
2305 6397 rt : (9,5,-8,-2)
2305 7198 cu : $2/3 + \pi$
2305 7880 rt : (4,2,8,-2)
2306 0085 $J_1 : (\sqrt{2}\pi)^{-1/2}$
2306 0456 $J_0 : 19/3$
2306 0461 rt : (8,7,3,8)
2306 1105 rt : (2,7,-5,5)
2306 1173 rt : (6,2,-5,2)
2306 1587 $s\pi : (\sqrt[3]{2}/3)^3$
2306 3590 $10^x : 2 - 2\sqrt{5}/3$
2306 3941 $\ell_{10} : 4/3 - \sqrt{5}/3$
2306 4650 cu : $7\ln 2/3$
2306 5261 $2^x : 2/3 + \sqrt{2}$
2306 5433 rt : (7,2,-9,-5)
2306 6518 $J_1 : \exp(-\sqrt{5}/3)$
2306 7922 rt : (3,1,4,9)
2306 8945 ln : $3\sqrt{3}/2 + 3\sqrt{5}$
2306 9095 cr : $1/\ln(\sqrt[3]{5})$
2306 9394 $10^x : 8(1/\pi)/5$
2306 9458 $\ell_{10} : 3 - 3\sqrt{3}/4$
2307 0592 rt : (5,8,-3,9)
2307 0795 sq : P_{21}
2307 1058 $J_2 : \sqrt{11}\pi$
2307 2542 rt : (2,8,-8,-6)
2307 3189 $\Gamma : \sqrt{2}/8$
2307 6923 id : 3/13
2307 7242 $s\pi : (\sqrt[3]{2})^{-1/3}$
2307 8181 $2^x : \sqrt{3}/2 - 4\sqrt{5}/3$
2308 0510 cu : $4/3 + e/3$
2308 2356 $\exp(3/5) \times s\pi(\sqrt{5})$
2308 3509 sq : $(\ln 2)^2$
2308 4793 rt : (9,6,-3,4)
2308 4896 at : $9\pi/10$
2308 5036 rt : (1,6,-3,-1)
2308 7999 rt : (2,1,-9,2)
2308 9030 sq : $1/\ln((e+\pi)/3)$
2309 0006 rt : (3,9,-1,4)
2309 0627 rt : (5,4,-6,-8)
2309 2047 cr : $9\pi^2/8$
2309 3207 $J_2 : (\sqrt{5})^{1/2}$
2309 3291 $10^x : \ln(\sqrt[3]{4}/3)$
2309 4010 $2/5 \div \sqrt{3}$
2309 4374 rt : (1,7,-6,1)
2309 5941 at : $2\sqrt{2}$
2309 6731 $2^x : P_{79}$
2309 8261 $Ei : 1/\sqrt{10}$
2310 0023 sq : $3\sqrt{2} - \sqrt{5}/3$
2310 0286 rt : (8,1,8,-2)
2310 1273 $t\pi : \sqrt{2}/5$
2310 2077 rt : (5,4,-8,-2)
2310 2239 rt : (4,0,-2,5)
2310 2444 rt : (9,8,2,-1)

2310 3776 $J_0 : 8\sqrt{5}/9$
2310 4594 tanh : $4/17$
2310 4906 id : $\ln 2/3$
2310 5865 sq : $3\sqrt{2}/4 + \sqrt{3}/4$
2310 6043 $J_1 : 9/2$
2310 6491 $e^x : \exp(-\pi/2)$
2310 7451 sr : $7\sqrt{3}/8$
2310 8524 sr : $\exp(-(e+\pi)/2)$
2310 9066 at : $4/17$
2310 9516 at : $e/2 - 4\pi/3$
2311 0610 rt : (1,9,-3,-8)
2311 1368 rt : (5,6,7,-2)
2311 2042 id : $(\sqrt[3]{3}/3)^2$
2311 2278 rt : (2,6,-9,6)
2311 2619 $J_1 : 6\pi^2/5$
2311 2655 $e^x : 6\zeta(3)/5$
2311 3548 cr : $1 + \sqrt{3}/2$
2311 3789 rt : (8,9,-7,-2)
2311 4441 $2^x : 3/10$
2311 5072 $J_2 : 7\sqrt[3]{5}/8$
2311 5175 $\Psi : 7\sqrt{5}/4$
2311 5470 $\ln(4) \times \ln(5)$
2311 6281 rt : (7,4,0,7)
2311 6922 rt : (6,5,-7,-5)
2311 8123 rt : (3,3,-6,5)
2311 8610 cr : $\sqrt[3]{13/2}$
2311 9052 sq : $2\zeta(3)/5$
2312 0259 $\zeta : 9(1/\pi)$
2312 0999 rt : (9,5,5,1)
2312 3148 $erf : \exp(-\pi/2)$
2312 6294 $2^x : 1/2 + 4\sqrt{3}$
2312 6711 cu : $4\pi^2/5$
2312 8568 $e^x : 2 - 2\sqrt{3}$
2312 9921 rt : (4,2,-5,1)
2313 0127 $erf : 3\ln 2/10$
2313 1399 $Ei : 10/23$
2313 1583 $2^x : 22/19$
2313 2152 $\lambda : \exp(-17\pi/13)$
2313 2514 rt : (5,1,-8,-1)
2313 2840 at : $2\sqrt{3}/3 + 3\sqrt{5}/4$
2313 3160 rt : (9,9,-5,-3)
2313 6555 rt : (2,2,9,2)
2313 7218 rt : (4,7,-9,-7)
2313 7983 $10^x : 2\sqrt{2} - 2\sqrt{3}$
2313 8307 rt : (8,3,-6,3)
2314 0506 rt : (7,3,-7,-9)
2314 0529 $\theta_3 : \ln 2/6$
2314 0997 $J_1 : 10/21$
2314 1144 rt : (6,0,4,1)
2314 1726 rt : (6,9,-8,4)
2314 2396 id : $1/4 + 4\sqrt{5}/3$
2314 3226 tanh : $\sqrt{2}/6$
2314 3345 sinh : $1/\sqrt{19}$
2314 4233 $Li_2 : 1/\sqrt{21}$
2314 4316 sq : $\sqrt[3]{10/3}$
2314 4383 cu : $2e + 3\pi/4$
2314 5988 rt : (1,6,1,-6)
2314 7736 at : $\sqrt{2}/6$
2314 8796 rt : (3,8,-3,8)
2314 8878 rt : (9,5,7,-2)
2314 9456 cu : $\sqrt{3} - \sqrt{5}/2$
2314 9659 rt : (2,4,3,9)
2315 0011 $\ln(4/5) \div s\pi(\sqrt{2})$
2315 0321 $\ln(3/4) \times \ln(\sqrt{5})$
2315 0371 at : $17/6$
2315 1305 $e^x : 1/2 + 2\sqrt{3}/3$
2315 1629 rt : (7,3,-4,-1)
2315 2490 cu : $3\sqrt{2} + 2\sqrt{3}/3$
2315 3860 rt : (4,8,-7,-2)
2315 5422 $J_1 : 1/\ln(1/(3e))$
2315 5500 rt : (6,8,4,4)
2315 5907 as : $3\sqrt{2} - 2\sqrt{5}$
2315 7706 cr : $3/4 + \sqrt{5}/2$
2315 8331 $2^x : \zeta(3)/4$
2316 2133 sinh : $7\zeta(3)/3$
2316 2364 $e^x : 5/24$
2316 2567 rt : (1,8,-3,-9)
2316 4564 $\Gamma : \sqrt{5}/3$
2316 6857 at : $9\sqrt[3]{2}/4$
2316 7276 $\ell_{10} : 4\sqrt{2}/3 - 3\sqrt{3}/4$
2316 9216 rt : (9,5,-1,8)
2317 1068 $2^x : 2/3 + 2\sqrt{5}$
2317 2179 $erf : 5/24$
2317 2857 rt : (6,1,-3,6)
2317 3228 $2^x : 22/13$
2317 3787 cu : $4 + e$
2317 5135 rt : (1,6,0,9)
2317 5748 $10^x : 3e/2 - 3\pi/2$
2317 5945 $\exp(1/4) + \exp(2/3)$
2317 5987 sr : $3/2 + 4\sqrt{5}$
2317 6514 sq : $e\pi/2$
2317 8116 $e^x : 4/3 + 2e/3$
2317 8240 $e^x : \sqrt{3} - \sqrt{5}/4$
2317 9130 rt : (3,6,-2,-1)
2317 9859 $t\pi : 8(1/\pi)/9$
2318 0829 $Ei : 4(1/\pi)/9$
2318 1218 ln : $4\sqrt{3}/3 + \sqrt{5}/2$
2318 2380 at : $2 - \sqrt{5}$
2318 3081 rt : (6,3,9,2)
2318 4531 $s\pi : e + 3\pi/4$
2318 4668 $e^x : 1/\sqrt{23}$
2318 5049 $\Gamma : (5/3)^{-3}$
2318 6687 cu : $2/3 + 2e/3$
2318 6955 rt : (6,8,-8,9)
2318 7341 $2^x : 9\pi/4$
2318 8211 rt : (3,2,8,-2)
2318 8245 $s\pi : e/2 + 4\pi$
2318 9777 ln : $4/3 + 2\pi/3$
2318 9894 $10^x : 4\sqrt{3}/3 + 4\sqrt{5}/3$
2319 1038 sr : $2 + 4\sqrt{5}/3$
2319 1158 $e^x : 4\sqrt{2}/3 - 3\sqrt{5}/4$
2319 1743 $erf : 1/\sqrt{23}$
2319 2505 rt : (7,7,-2,0)
2319 3062 $2^x : (\sqrt{5}/3)^{-1/2}$
2319 9367 $J_1 : 3(1/\pi)/2$
2319 9738 $2^x : 2/3 + 3\sqrt{5}/2$
2320 0000 cu : $7\sqrt[3]{3}/5$
2320 3463 sq : $3 - 2\sqrt{2}/3$
2320 4537 $\sqrt{2} \times \exp(\sqrt{5})$
2320 5080 id : $1/2 + \sqrt{3}$
2320 6672 ln : $3/2 - \sqrt{2}/2$
2320 7497 $E : 11/13$
2320 7690 ln : $3\sqrt{2} - 4\sqrt{5}/3$
2320 8610 $e^x : 1/\ln(2)$
2320 8767 $J_2 : 3/2$
2320 8815 tanh : $\exp(-\sqrt[3]{3})$
2321 0540 $2^x : 1/3 + e/2$
2321 1168 $\ell_{10} : (e+\pi)/10$
2321 1312 rt : (4,1,0,-9)
2321 1351 $t\pi : P_{34}$
2321 2781 at : $4e/3 - \pi/4$
2321 3138 $s\pi : (\sqrt{3}/2)^{-1/2}$
2321 3389 at : $\exp(-\sqrt[3]{3})$
2321 4368 ln : $7/24$
2321 4451 rt : (8,2,-4,7)
2321 6886 rt : (1,8,5,-4)
2321 9092 sr : $\sqrt[3]{7/2}$
2322 0830 rt : (2,9,-9,-7)
2322 1308 $Ei : \sqrt{2}/10$
2322 1606 rt : (7,4,3,-1)
2322 3961 rt : (8,6,-8,-4)
2322 4170 $2^x : 2 + 2e$
2322 4728 cr : $1/4 + 4e$
2322 5885 $2^x : 1/4 - 3\pi/4$
2322 6068 $\ell_{10} : 2 - \sqrt{2}$
2322 6822 id : $9e/2$
2322 7481 rt : (8,9,3,5)
2322 8822 rt : (7,1,8,-2)
2322 9785 rt : (4,6,7,7)
2323 0210 rt : (5,2,-6,-5)
2323 1281 sq : $3\sqrt{2} + \sqrt{3}/3$
2323 1522 $\ln(\sqrt{2}) \div \exp(2/5)$
2323 2667 sr : $\sqrt{3} + \sqrt{5} - \sqrt{6}$
2323 3078 rt : (1,5,-2,-7)
2323 5814 $J_1 : 11/23$
2323 7345 cr : $3e - 2\pi$
2323 8156 sinh : $5\sqrt{2}/3$
2323 9087 cu : $1/2 - \sqrt{5}$
2324 0668 $\Psi : 16/23$
2324 0812 rt : (9,9,-7,-2)
2324 1322 $J_1 : \sqrt{2} - \sqrt{5} - \sqrt{6}$
2324 3481 $2^x : 1/\sqrt{11}$
2324 3827 as : $2\sqrt{2} - 3\sqrt{3}/2$
2324 5275 $J_0 : 7e/3$
2324 5790 $10^x : 6\pi/5$
2324 7435 $\ell_{10} : 2e\pi$
2324 9008 rt : (9,8,-3,1)
2325 2311 $\zeta : -\sqrt{5} + \sqrt{6} + \sqrt{7}$
2325 3273 $t\pi : 8(1/\pi)/5$
2325 6859 $e^x : 1/2 + 2\sqrt{2}/3$
2325 7474 rt : (2,6,-8,-3)
2325 8688 $\ln(3) - s\pi(1/3)$
2326 0179 rt : (7,1,-7,-6)
2326 1926 $\Psi : 6\sqrt[3]{3}/5$
2326 2158 $J_0 : 9(e+\pi)/5$
2326 2780 $2^x : (\ln 3/2)^2$
2326 3327 sinh : $(\sqrt[3]{2}/3)^{-1/2}$
2326 3933 rt : (5,5,1,3)
2326 5757 rt : (8,5,7,-2)
2326 6667 at : $\sqrt[3]{23}$
2326 7270 ln : $7\sqrt[3]{3}/8$
2327 2228 rt : (3,5,-2,4)
2327 2477 at : $3/4 + 2\pi/3$
2327 2661 $J_2 : 5\zeta(3)/4$
2327 5984 sinh : $7\sqrt{5}/3$
2327 8162 id : $3\sqrt{2}/2 - 3\sqrt{5}/2$
2327 8561 rt : (8,8,2,-1)
2327 9165 rt : (5,8,-7,-2)
2327 9882 rt : (9,0,-8,7)
2328 0900 rt : (6,7,6,8)
2328 2293 sinh : $3/13$
2328 2423 at : $4 - 2\sqrt{3}/3$
2328 2712 rt : (2,9,4,-5)
2328 4012 $J_2 : (\sqrt{2}/3)^{-3}$
2328 4673 $10^x : 1/11$
2328 5553 $2^x : 3/4 + 2\sqrt{2}/3$
2328 6817 as : $3/13$
2328 6984 rt : (1,2,9,2)
2328 7412 sr : $8\sqrt[3]{5}/9$
2328 7470 sinh : 10
2328 7908 $2^x : e/9$
2328 8362 rt : (4,7,-7,-6)
2328 8503 $Ei : 6/19$
2328 9037 id : $5e\pi/3$
2329 2463 rt : (6,4,-5,-1)
2329 4470 rt : (5,3,-4,-9)
2329 4740 sr : $2\sqrt{2}/3 + \sqrt{3}/3$
2329 5061 $10^x : 2(1/\pi)/7$
2329 5547 $10^x : (\sqrt[3]{4}/3)^{-1/3}$
2329 5748 $2^x : \sqrt{13/3}$
2329 6182 rt : (9,9,6,-2)
2329 6474 at : $e\pi/3$
2329 7770 sq : $2\sqrt{3} - 4\sqrt{5}/3$
2329 8152 ln : $8\ln 2/7$
2329 9000 $\ell_{10} : \sqrt[3]{5}$
2329 9802 sinh : $\zeta(5)$
2330 0283 $\zeta : 20/7$
2330 0698 rt : (7,6,0,4)
2330 2954 $J_1 : \sqrt{14}\pi$
2330 3728 id : $1/3 - 4\pi$
2330 4313 $J_2 : (2e/3)^3$
2330 4506 cu : $8/13$
2330 5428 rt : (8,8,5,9)
2330 7953 sq : $4e/3 - \pi$
2330 8130 $10^x : 1/2 + 2\sqrt{2}$
2330 9594 $\ell_{10} : 2\sqrt{2} - \sqrt{5}/2$
2331 0321 rt : (2,6,-3,-1)
2331 0603 cr : $15/8$
2331 1026 sinh : $\ln 2/3$
2331 2924 at : $1/4 + 3\sqrt{3}/2$
2331 3978 $t\pi : \sqrt{3} - \sqrt{5}$
2331 4219 cu : $\sqrt{3} + \sqrt{6} + \sqrt{7}$
2331 4476 at : $4/3 - \pi/2$
2331 4631 $10^x : 2/3 - 3\sqrt{3}/4$
2331 5173 $1/2 \div \ln(2/3)$
2331 5261 $\theta_3 : \exp(-\sqrt[3]{2}/2)$
2331 5403 $J_1 : 12/25$
2331 5578 as : $\ln 2/3$
2331 6102 rt : (1,7,-7,-2)
2331 7229 sinh : $(5/3)^3$

2331 8249 rt : (8,5,-6,0)
2331 8354 sinh : $(\sqrt[3]{3}/3)^2$
2331 9200 rt : (9,7,-1,5)
2331 9369 rt : (1,3,-9,-6)
2332 2595 J_0 : $\sqrt[3]{25}$
2332 2912 as : $(\sqrt[3]{3}/3)^2$
2332 2933 rt : (2,2,8,-2)
2332 6066 2^x : $\sqrt{2}/2 + 2\sqrt{5}$
2332 7445 $t\pi$: $1/4 + 3\sqrt{5}/4$
2332 8255 J_1 : $2(e+\pi)$
2332 9542 Ei : $\exp(e)$
2333 0231 at : $\exp(\pi/3)$
2333 1133 rt : (5,3,9,2)
2333 2001 rt : (3,4,-2,9)
2333 3586 at : $5\sqrt[3]{5}/3$
2333 4059 rt : (2,7,-6,-7)
2333 4544 θ_3 : $((e+\pi)/2)^{-2}$
2333 4955 at : $\sqrt{3} + \sqrt{5}/2$
2333 5186 ζ : $(1/(3e))^{-1/2}$
2333 6126 J_1 : $(\ln 2)^2$
2333 8234 sr : $3\sqrt{2} + \sqrt{5}/3$
2333 8733 $s\pi$: $\ln((e+\pi)/2)$
2334 1158 rt : (4,4,5,1)
2334 2915 ℓ_{10} : $((e+\pi)/2)^{1/2}$
2334 2999 rt : (4,6,-7,-1)
2334 3821 ln : $3 + \sqrt{3}/4$
2334 6308 J_1 : $(\ln 2/3)^{1/2}$
2334 6878 J_2 : $6\sqrt{3}$
2334 6998 10^x : $2(e+\pi)/7$
2334 7376 $t\pi$: $5\pi/4$
2334 7990 ℓ_2 : $\sqrt[3]{13}$
2334 8120 e^x : $2/3 - 3\sqrt{2}/2$
2334 8272 2^x : $\ln(1/(3e))$
2334 8690 ℓ_2 : $\exp(\sqrt[3]{5}/2)$
2334 8747 rt : (6,5,-8,-9)
2334 9703 J_1 : $\sqrt[3]{3}/3$
2335 0771 erf : $\sqrt[3]{2}/6$
2335 1687 rt : (3,6,7,-2)
2335 3038 J_1 : $2\zeta(3)/5$
2335 3864 Ψ : 9/20
2335 4518 rt : (8,0,-9,2)
2335 5991 sq : $1/3 + 3\pi/4$
2335 6949 2^x : $1/2 - 3\sqrt{3}/2$
2335 8846 sq : $3\sqrt{2}/4 - \sqrt{3}/3$
2335 9170 ζ : $9\sqrt[3]{2}/7$
2335 9569 sinh : $3\pi^2/8$
2336 0030 rt : (5,4,3,7)
2336 0716 Γ : $(e/3)^3$
2336 1334 2^x : $\ln(2e)$
2336 1485 ln : 19/24
2336 1557 rt : (6,1,8,-2)
2336 3894 as : $1/\ln(\ln 2/2)$
2336 4002 rt : (7,5,2,8)
2336 5396 id : $2/3 - \sqrt{3}/4$
2336 6182 e^x : $\sqrt[3]{2}/6$
2336 6325 rt : (9,6,1,9)
2336 6614 Ψ : $\sqrt{3}$
2336 7088 10^x : $\sqrt[3]{11/2}$
2336 8696 $s\pi$: $(\ln 2/3)^{-3}$
2336 9581 tanh : 5/21
2336 9843 rt : (7,9,6,1)
2337 0055 id : $\pi^2/8$
2337 2362 $t\pi$: 2π
2337 3281 10^x : $(\sqrt{2}\pi/2)^{-3}$
2337 3365 rt : (9,4,9,2)
2337 3826 Ei : $4\pi/7$
2337 4318 at : 5/21
2337 4756 sinh : $3\sqrt[3]{2}/2$
2337 5192 rt : (8,4,-4,4)
2337 6679 sq : $1 + 3e/4$
2337 7008 Ei : $4e\pi/3$
2337 7381 cu : $1/4 - \sqrt{3}/2$
2337 8933 at : $1/2 - 3\sqrt{5}/2$
2337 9169 rt : (1,7,-1,5)
2338 1615 rt : (4,5,-7,4)
2338 4709 rt : (7,4,-8,-2)
2338 5981 rt : (7,5,7,-2)
2338 6670 sq : P_{32}
2338 7555 e^x : $3 + 4e/3$
2338 8611 rt : (7,2,-9,-1)
2338 9385 cu : $8\ln 2/9$
2339 1091 10^x : $\pi/9$
2339 1216 cr : $4 - 3\sqrt{2}/2$
2339 1407 $t\pi$: $1/2 + \sqrt{3}/2$
2339 3741 at : $4/3 - 4\pi/3$
2339 3923 ℓ_{10} : $2\sqrt{3}/3 + \sqrt{5}/4$
2339 4758 rt : (5,1,-8,-2)
2339 6059 at : $(1/(3e))^{-1/2}$
2339 7641 ℓ_{10} : $2/3 + \pi/3$
2339 8661 rt : (5,0,4,1)
2340 0003 $2 \times \exp(3/4)$
2340 0096 rt : (4,1,-9,2)
2340 1069 at : $3\sqrt{2}/4 - 3\sqrt{3}/4$
2340 1798 at : $1/2 + 3\pi/4$
2340 2362 rt : (9,1,-5,1)
2340 4576 $s\pi$: $4\sqrt{2} - \sqrt{3}$
2340 6009 rt : (8,9,6,-2)
2340 7287 sq : $1/4 - 4\sqrt{2}$
2340 8320 ℓ_{10} : 12/7
2340 8477 10^x : $4/3 + \sqrt{5}$
2340 8499 rt : (6,8,-7,-2)
2340 8959 cu : $\sqrt{21}$
2340 9012 erf : 4/19
2341 0361 rt : (4,4,-7,9)
2341 0709 $\ln(\sqrt{2}) \times s\pi(\sqrt{5})$
2341 1292 rt : (3,9,2,-1)
2341 2150 at : 20/7
2341 2242 $t\pi$: $8\sqrt[3]{2}/7$
2341 2359 $s\pi$: $1/2 + 3\pi$
2341 3251 10^x : 7/4
2341 3338 rt : (8,3,-2,8)
2341 8551 as : $3/2 - \sqrt{3}$
2341 8620 10^x : $5e\pi/3$
2341 9006 cu : $3\sqrt{2}/4 - 3\sqrt{5}/4$
2341 9399 10^x : $(e/3)^{-1/2}$
2342 0451 id : $4\sqrt{2} + \sqrt{3}/3$
2342 0521 rt : (9,2,-8,4)
2342 3403 rt : (6,2,-1,7)
2342 4358 rt : (3,3,-8,-2)
2342 4536 ζ : $(\sqrt[3]{5}/2)^{-1/3}$
2342 6787 J : $(\sqrt[3]{5}/3)^2$
2342 7085 cu : $1/3 - 2e/3$
2342 9810 tanh : $3(1/\pi)/4$
2342 9868 Ψ : $5\ln 2/2$
2343 1928 rt : (7,1,-7,3)
2343 2125 rt : (8,5,5,1)
2343 2753 e^x : 4/19
2343 3354 $\exp(3/5) - s\pi(1/5)$
2343 3942 at : $3/2 + e/2$
2343 4292 at : $\sqrt{5} - \sqrt{6} - \sqrt{7}$
2343 4608 at : $3(1/\pi)/4$
2343 5808 rt : (6,4,3,-1)
2343 6343 cu : $1/4 + 2\sqrt{3}$
2343 6607 at : $2e/3 + \pi/3$
2343 6705 cr : $3 + 3e$
2344 0683 rt : (4,1,0,6)
2344 0758 $\sqrt{3} \div \exp(2)$
2344 1944 10^x : $2\sqrt{2}/3 - \sqrt{3}/4$
2344 2144 Γ : $\sqrt{3} + \sqrt{6} + \sqrt{7}$
2344 2177 sr : $3/4 + 3\sqrt{2}$
2344 2395 ℓ_{10} : $4e + 2\pi$
2344 2878 cu : $1/4 + 4\sqrt{5}$
2344 3556 rt : (2,9,2,0)
2344 4682 sq : P_{42}
2344 6011 rt : (9,1,-6,8)
2344 6525 ℓ_2 : 17/20
2344 8521 $t\pi$: $3\sqrt{2}/4 + \sqrt{3}/2$
2344 9730 $s\pi$: $(\sqrt[3]{5}/2)^{1/2}$
2345 0770 $\sqrt{2} - \exp(1/2)$
2345 1009 rt : (5,0,-6,2)
2345 1928 $\exp(3/5) \div s\pi(\pi)$
2345 2041 sr : $\sqrt{3} - 3\sqrt{5}/4$
2345 2749 sinh : $(\sqrt{5}/2)^{1/3}$
2345 4670 e^x : $e/3 - 3\pi/4$
2345 4672 rt : (8,3,-4,-1)
2345 4710 $t\pi$: $4\sqrt{3}/3 - \sqrt{5}$
2345 6790 sq : 10/9
2346 0879 rt : (7,0,-5,7)
2346 1216 J_2 : $5e/9$
2346 1746 rt : (7,8,2,-1)
2346 2350 rt : (1,2,8,-2)
2346 2701 $\ln(4/5) \div s\pi(2/5)$
2346 3123 cr : $3 - \sqrt{5}/2$
2346 3634 J_1 : 8
2346 4251 rt : (6,6,-6,1)
2346 4771 ζ : 10/23
2346 5442 cr : $\sqrt[3]{20/3}$
2346 8604 rt : (9,1,4,1)
2346 8680 10^x : $4e\pi/7$
2347 1008 ζ : $5\sqrt[3]{5}/3$
2347 2813 Li_2 : $1/\ln(\pi)$
2347 5639 Li_2 : $\sqrt[3]{2/3}$
2347 6990 J : $1/\sqrt{22}$
2347 7282 rt : (2,6,7,8)
2347 8316 ζ : $\exp(\pi/3)$
2348 2698 ℓ_{10} : $10\zeta(3)/7$
2348 3652 rt : (5,1,-4,-6)
2348 4058 ℓ_2 : $2 + \sqrt{2}/4$
2348 5017 rt : (4,3,9,2)
2348 5340 E : $(5/3)^{-1/3}$
2348 5509 Γ : $7(1/\pi)/8$
2348 5639 rt : (8,2,-9,-7)
2348 5772 rt : (7,7,-3,-1)
2348 6028 2^x : 7/23
2348 9427 rt : (3,8,2,-2)
2349 5302 10^x : $4\sqrt{2}/3 + 4\sqrt{5}$
2349 5395 at : $9(1/\pi)$
2349 6758 e^x : $5\sqrt{3}/6$
2349 6973 $s\pi$: $10\sqrt{3}/9$
2349 8769 rt : (5,1,8,-2)
2349 9430 2^x : $(2e/3)^{-2}$
2350 0337 at : $2/3 - e/3$
2350 0920 $\ln(3/5) - s\pi(\sqrt{3})$
2350 2017 $\ln(\sqrt{5}) - s\pi(\pi)$
2350 4604 ζ : $4\sqrt[3]{3}/9$
2350 4694 J_2 : $6\sqrt[3]{2}/5$
2350 5100 Ei : $(\sqrt{3})^{-1/2}$
2350 5478 Γ : $(2e/3)^{-1/2}$
2350 6474 ℓ_2 : $\sqrt{2}/3 + 4\sqrt{5}$
2350 6616 $\exp(5) \times s\pi(1/5)$
2350 8193 $\exp(1/4) + s\pi(2/5)$
2350 8819 at : $2 + \sqrt{3}/2$
2350 8961 rt : (6,3,-8,-6)
2350 9664 e^x : $10(e+\pi)$
2350 9741 rt : (6,5,7,-2)
2351 0503 rt : (5,2,-5,1)
2351 1410 $2/5 \times s\pi(1/5)$
2351 5882 at : $3\sqrt{2}/2 + \sqrt{5}/3$
2351 6886 rt : (2,1,3,-9)
2351 7971 Ψ : $5\pi/4$
2351 8695 rt : (7,9,6,-2)
2352 2153 Γ : $7(1/\pi)/3$
2352 2918 rt : (3,7,2,3)
2352 3987 rt : (9,0,-9,2)
2352 4592 rt : (8,4,9,2)
2352 7790 $\exp(3/5) + \exp(5)$
2352 8194 rt : (4,9,-3,-7)
2352 8463 cr : $3\pi/5$
2352 9411 id : 4/17
2353 0544 sq : $4/3 - 2\sqrt{2}$
2353 3418 K : 8/9
2353 3699 rt : (9,0,8,2)
2353 4170 rt : (8,4,-8,-2)
2353 4767 at : $4\sqrt{3}/3 + \sqrt{5}/4$
2353 6992 ln : $7\pi^2$
2353 9135 J_0 : $(e+\pi)/2$
2353 9772 $t\pi$: $2e\pi/7$
2354 0915 J_0 : $\sqrt{6}\pi$
2354 1245 $\sqrt{2} \div \ln(\pi)$
2354 1535 rt : (3,6,2,8)
2354 1611 $\ln(3/4) - \exp(2/3)$
2354 2129 rt : (7,8,-7,-2)
2354 2197 sq : $7\ln 2/10$
2354 2934 cr : $4\sqrt{2}/3$
2354 3125 J_2 : $8\sqrt{5}/3$
2354 3473 cr : $3\sqrt{2} + 4\sqrt{3}$
2354 4775 rt : (4,8,-3,-2)
2354 5600 Γ : $9\sqrt{3}/5$
2354 8257 J_0 : $9\sqrt[3]{5}/2$
2354 8324 Ei : $4\sqrt[3]{5}/9$

2354 8520 rt : (3,6,-3,-1)
2354 9574 tanh : 6/25
2354 9601 J_0 : $7\sqrt{2}/5$
2354 9871 id : $e/3 - \pi$
2355 0931 rt : (9,9,-1,2)
2355 1331 Ei : 19/25
2355 1932 ζ : $e\pi/3$
2355 3198 J_1 : $7\ln 2/10$
2355 3415 $\ln(3/4) \div \exp(1/5)$
2355 4498 at : 6/25
2355 5257 θ_3 : 8/15
2355 5578 rt : (4,7,-3,3)
2355 7158 cu : $1/2 - 3\sqrt[3]{3}/2$
2355 7597 rt : (5,9,-8,-7)
2356 2059 rt : (9,8,1,6)
2356 3175 rt : (4,6,-3,8)
2356 4446 Ψ : $1/\sqrt[3]{11}$
2356 4648 rt : (5,8,-8,-2)
2356 5426 J_2 : $7e\pi/8$
2356 6190 at : $3/4 + 3\sqrt{2}/2$
2356 6432 rt : (5,1,-9,2)
2356 7726 θ_3 : 2/17
2356 9932 rt : (5,7,-8,3)
2357 0226 id : $\sqrt{2}/6$
2357 1295 rt : (4,3,-8,-2)
2357 3433 J_0 : $8\sqrt{3}/7$
2357 3824 Ei : 9/8
2357 4038 rt : (5,6,-8,8)
2357 4095 tπ : 6/13
2357 5113 rt : (3,7,-7,-2)
2357 5718 ln : $4 - \sqrt{5}/4$
2357 8383 10^x : $7\sqrt[3]{3}/9$
2357 8767 sπ : $2\sqrt{2}/3 + 4\sqrt{5}/3$
2357 8891 Γ : $3\sqrt{3}/7$
2358 3398 as : $2/3 - \sqrt{3}/4$
2358 3846 ℓ_{10} : $3e/2 - 3\pi/4$
2358 3958 ℓ_2 : $(\sqrt[3]{3}/2)^{1/2}$
2358 4353 10^x : $e/2 + \pi$
2358 4985 tanh : $\sqrt[3]{3}/6$
2358 8427 tanh : $\zeta(3)/5$
2358 9945 at : $\sqrt[3]{3}/6$
2358 9993 sπ : $4\ln 2/3$
2359 0016 as : 17/18
2359 0326 erf : $2(1/\pi)/3$
2359 3091 sr : $\sqrt{7/3}$
2359 3392 at : $\zeta(3)/5$
2359 5189 J_2 : $7\sqrt{3}/8$
2359 6008 ln : $\sqrt{2}/2 + \sqrt{5}/4$
2359 6532 Li_2 : 2/9
2359 7142 tπ : $(e)^{-1/3}$
2359 8004 2^x : $\sqrt{3} + 4\sqrt{5}/3$
2359 8775 id : $5\pi/3$
2359 8940 J_1 : $\exp(-\sqrt[3]{3}/2)$
2359 9210 at : $3/4 - 4e/3$
2359 9975 e^x : $\exp(5/3)$
2360 4199 Ψ : $(\sqrt{2}\pi)^{-2}$
2360 4318 sr : $10\pi/3$
2360 5350 at : $9\sqrt{5}/7$
2360 5948 at : 23/8
2360 6797 id : $\sqrt{5}$
2360 8954 sq : $3/4 + \sqrt{5}/3$
2361 1023 e^x : $1/3 + \sqrt{2}/3$
2361 4325 cr : 17/9
2361 5160 cu : $3\sqrt[3]{3}/7$
2361 6409 rt : (9,2,-4,9)
2361 7060 ζ : $\sqrt[3]{23}$
2361 7953 rt : (9,3,-6,5)
2361 8327 $1/2 \div \exp(3/4)$
2361 8393 rt : (8,4,0,9)
2361 8531 ln : $3e - 3\pi/2$
2362 0045 E : 16/19
2362 0087 rt : (9,4,-8,1)
2362 0717 rt : (8,5,-2,5)
2362 0842 Γ : $1/\sqrt[3]{16}$
2362 1391 id : $6\sqrt[3]{3}/7$
2362 1408 rt : (7,6,4,9)
2362 1444 cu : $3\sqrt{2} + 3\sqrt{5}$
2362 1698 Ψ : $\sqrt{2}/7$
2362 2673 Ψ : $-\sqrt{2} + \sqrt{5} - \sqrt{7}$
2362 3122 cu : 17/13
2362 4205 rt : (8,6,-4,1)
2362 5301 rt : (7,7,2,5)
2362 5500 J_1 : $6e/5$
2362 5585 λ : $\exp(-13\pi/10)$
2362 6544 rt : (6,8,8,9)
2362 6596 Ψ : $1/\ln(\sqrt{5})$
2362 9008 e^x : $1/\ln(1/2)$
2363 0027 rt : (8,7,-6,-3)
2363 2020 rt : (7,8,0,1)
2363 3236 $\sqrt{2} + \exp(3/5)$
2363 4388 rt : (6,9,6,5)
2363 5976 id : $(\sqrt[3]{4}/3)^{-1/3}$
2363 6381 sr : $2 - \sqrt{2}/3$
2363 6544 $2/3 + s\pi(\pi)$
2363 6710 $\ln(\sqrt{3}) \times s\pi(\pi)$
2363 7247 rt : (5,5,7,-2)
2363 8877 ln : 15/19
2363 9537 id : $\exp(-\sqrt[3]{3})$
2364 0218 Ei : $\sqrt[3]{2}/4$
2364 0261 J_2 : $\sqrt{2}\pi$
2364 0328 e^x : $2(1/\pi)/3$
2364 0768 rt : (4,1,8,-2)
2364 1695 rt : (8,8,-8,-7)
2364 4485 sπ : $\sqrt[3]{25}$
2364 5210 rt : (3,3,9,2)
2364 5863 ℓ_2 : 3π
2364 6396 rt : (7,9,-2,-3)
2364 9762 cu : $3/4 + 3\sqrt{3}$
2365 0652 $\exp(1/2) + s\pi(1/5)$
2365 2483 10^x : $5\sqrt{2}/7$
2365 2564 rt : (6,8,2,-1)
2365 3810 sq : $1/2 + 3\sqrt{3}/4$
2365 4095 $3/4 \times \exp(1/2)$
2365 5706 $\exp(e) \times s\pi(\sqrt{5})$
2365 7366 rt : (8,1,-7,-8)
2365 8531 tπ : $2\sqrt{2} + \sqrt{5}/3$
2365 8920 J_0 : $9\sqrt{2}/2$
2365 9474 10^x : $5\sqrt[3]{5}/4$
2366 0425 tπ : $7e\pi/3$
2366 1010 rt : (5,4,3,-1)
2366 3582 J_2 : $\sqrt[3]{7/2}$
2366 3800 Li_2 : $7(1/\pi)/10$
2366 4819 J_0 : 21/2
2366 5531 rt : (8,0,-9,4)
2366 5697 sinh : $3\sqrt{3}/5$
2366 6524 rt : (9,4,7,-2)
2366 7865 rt : (7,1,-3,8)
2366 8468 sr : $2\sqrt{2} - 3\sqrt{3}/4$
2367 1903 J_2 : $\sqrt{3} + \sqrt{5} - \sqrt{6}$
2367 3102 rt : (1,9,-7,-7)
2367 3290 rt : (8,0,8,2)
2367 6086 Ψ : $5\sqrt{5}/9$
2367 7240 2^x : $3/4 - 2\sqrt{2}$
2367 9100 ℓ_2 : $\sqrt{3} + 4\sqrt{5}/3$
2367 9941 id : $\sqrt{2}/3 - 3\sqrt{5}$
2368 0183 rt : (7,2,-5,4)
2368 0350 rt : (8,8,-7,-2)
2368 0968 at : $4 - \sqrt{5}/2$
2368 1050 at : $\sqrt{3} - 2\sqrt{5}/3$
2368 1790 rt : (7,4,9,2)
2368 1898 cu : $7\pi/6$
2368 2879 e^x : $3 + \pi/3$
2368 3865 rt : (6,3,1,8)
2368 6033 at : $2e/3 - \pi/2$
2368 7371 E : $7\zeta(3)/10$
2368 8826 at : $\sqrt{2}/3 - 3\sqrt{5}/2$
2368 9647 rt : (9,4,-8,-2)
2368 9788 cr : $4/3 + \sqrt{5}/4$
2369 0020 ζ : $3\sqrt[3]{5}/8$
2369 2017 2^x : $4\pi/3$
2369 2386 rt : (1,8,-9,-3)
2369 2387 sr : $4e/3 - 2\pi/3$
2369 2792 rt : (3,4,5,1)
2369 3508 rt : (2,9,-6,9)
2369 3730 2^x : $2 - 3e/2$
2369 6559 ℓ_2 : $3\sqrt{2}/5$
2369 6584 id : $2e/3 + 3\pi$
2369 7539 erf : $1/\sqrt{22}$
2369 8721 rt : (7,3,-7,0)
2369 9241 Li_2 : $\exp(-3/2)$
2370 0435 tπ : $(\sqrt[3]{2}/3)^3$
2370 0627 2^x : $\ln(e/2)$
2370 0873 rt : (6,8,-9,7)
2370 2951 2^x : $4 - e/2$
2370 3697 Γ : $1/\ln(4/3)$
2370 4575 rt : (6,4,-1,4)
2370 5470 J_2 : $8\sqrt[3]{5}/9$
2370 6577 J_1 : $2\sqrt[3]{5}/7$
2370 8167 at : $2\sqrt[3]{3}$
2370 8224 2^x : $2\sqrt{2} - \sqrt{5}/3$
2371 0413 sinh : 22/3
2371 1044 rt : (4,7,-7,-2)
2371 1184 rt : (6,9,-9,2)
2371 1227 rt : (5,5,5,8)
2371 1524 rt : (9,3,3,-1)
2371 1939 J_0 : $9e\pi/10$
2371 3895 Γ : $4\pi^2/5$
2371 3968 2^x : $\sqrt{3}/2 - \sqrt{5}/4$
2371 6512 id : $\sqrt{2}/2 - 4\sqrt{5}$
2371 7429 tπ : $(\sqrt[3]{2})^{-1/3}$
2371 7724 at : $1/\ln(\sqrt{2})$
2371 7914 id : $5\sqrt{3}/7$
2371 7918 rt : (6,2,-6,-7)
2372 0169 at : $1 + 4\sqrt{2}/3$
2372 0773 10^x : $7e\pi/6$
2372 1470 sinh : $8\pi/3$
2372 3140 cu : 13/21
2372 3399 ℓ_{10} : $e/4 + \pi/3$
2372 3932 rt : (5,3,-8,-2)
2372 7557 rt : (5,8,-4,7)
2372 8116 tπ : $\ln(\sqrt[3]{3})$
2372 8621 rt : (2,8,6,1)
2372 8810 J_0 : $\exp(\pi/2)$
2372 9719 Γ : $7\sqrt[3]{5}/5$
2372 9758 rt : (1,5,-9,-8)
2372 9768 rt : (7,4,-9,-4)
2373 0856 10^x : $\sqrt{3}/2 - 2\sqrt{5}/3$
2373 2315 at : $5\sqrt{3}/3$
2373 3779 ℓ_{10} : $7\pi^2/4$
2373 3929 sπ : $\sqrt{3}/4 + 2\sqrt{5}/3$
2373 4304 rt : (8,1,4,1)
2373 4462 sπ : $9\sqrt[3]{5}/8$
2373 6091 ℓ_{10} : 11/19
2373 6479 rt : (3,7,-6,9)
2373 7030 2^x : $7\sqrt[3]{5}$
2373 7088 Γ : $(\ln 3/2)^{1/2}$
2373 7452 ℓ_{10} : $2/3 + 3\sqrt{2}/4$
2374 0078 ln : $3 - \sqrt{3}$
2374 0481 rt : (6,7,-2,0)
2374 1361 as : $3\sqrt[3]{2}/4$
2374 2252 E : $(\sqrt{2})^{-1/2}$
2374 3218 rt : (1,6,-1,3)
2374 3363 e^x : $10e/3$
2374 3686 id : $7\sqrt{2}/8$
2374 5088 rt : (5,9,-4,2)
2374 6299 id : $1/3 - \pi/2$
2374 6542 rt : (8,7,-3,-1)
2374 7124 sinh : 4/17
2374 7252 rt : (5,0,-2,7)
2374 7440 cu : $\sqrt{2} + \sqrt{5}/2$
2374 7846 10^x : $3 - 4e/3$
2374 9778 J_1 : $1/\ln(e/2)$
2375 0296 2^x : $(\sqrt[3]{2}/3)^{-3}$
2375 0448 $3 \times s\pi(\sqrt{3})$
2375 1535 rt : (3,8,-4,5)
2375 2117 as : 4/17
2375 3182 Li_2 : $\sqrt{5}/10$
2375 3248 rt : (5,9,6,-2)
2375 3798 J_2 : $\exp(\sqrt[3]{2}/3)$
2375 4105 e^x : $e + \pi/4$
2375 5065 2^x : $4/3 + \pi$
2375 6434 erf : $\sqrt[3]{5}/8$
2375 6650 rt : (1,5,-1,-7)
2375 8381 ln : $4e/3 - 3\pi/4$
2376 1190 sinh : $4e\pi/9$
2376 2231 tπ : $\exp(-\sqrt[3]{2})$
2376 3303 e^x : $1/\sqrt{22}$
2376 4601 $2/3 \times \exp(\sqrt{5})$
2376 5952 $\exp(2/5) - s\pi(\sqrt{3})$
2376 7535 $\exp(1/4) \times s\pi(\sqrt{2})$

2376 8567 rt : (5,5,-9,-6)
2376 8728 rt : (4,7,9,8)
2376 9841 rt : (7,5,5,1)
2377 0545 rt : (3,9,-2,1)
2377 2565 J_1 : $10e/3$
2377 2628 2^x : $4/13$
2377 4332 e^x : $4-2e$
2377 5370 rt : (4,8,1,7)
2377 6412 $1/4 \times s\pi(2/5)$
2377 6739 10^x : $23/19$
2377 7306 rt : (2,2,-9,2)
2377 8088 rt : (9,8,6,-2)
2378 0800 cr : $1/3+4e$
2378 4109 at : $1/4-\pi$
2378 4597 rt : (9,3,-4,-1)
2378 5237 2^x : $25/12$
2378 5328 rt : (5,1,-4,3)
2378 6119 $\exp(4/5) \div \exp(\sqrt{5})$
2378 7893 rt : (3,1,8,-2)
2378 7982 2^x : $(\sqrt{2}\pi/3)^{-3}$
2378 8931 tanh : $1/\sqrt{17}$
2378 9075 sinh : $\sqrt{2}/6$
2379 0007 sq : $3\sqrt{3}+\sqrt{5}$
2379 4112 at : $1/\sqrt{17}$
2379 4455 Ψ : π^2
2379 5746 $\sqrt{3} \div s\pi(e)$
2379 5847 rt : (8,4,7,-2)
2379 6338 J_0 : $4\zeta(3)$
2379 7150 $\ln(2/5)+\exp(e)$
2379 7635 rt : (4,2,2,7)
2379 7686 cu : $4+\pi$
2379 9980 rt : (9,2,-9,-9)
2380 1180 rt : (4,6,-3,-1)
2380 1941 sr : $2/3+\sqrt{3}/2$
2380 2011 $\sqrt{2}-\exp(\sqrt{3})$
2380 2332 rt : (2,5,-2,9)
2380 3125 at : P_{15}
2380 4034 at : $4-3\sqrt{2}$
2380 4642 2^x : $4\ln 2/9$
2380 6642 cr : $2e\pi/9$
2380 7194 e^x : $\sqrt{3}+3\sqrt{5}/4$
2380 9523 id : $5/21$
2381 1196 rt : (1,9,2,-1)
2381 1597 rt : (4,9,1,2)
2381 2253 rt : (2,3,9,2)
2381 2348 Li_2 : $7/8$
2381 3181 ln : $8(e+\pi)/5$
2381 3920 rt : (9,9,3,7)
2381 3936 rt : (1,7,-2,-2)
2381 4501 erf : $3/14$
2381 4649 10^x : $\sqrt{2}/3+3\sqrt{5}/2$
2381 4895 J_0 : π^2
2381 6131 ζ : $19/24$
2381 7705 sq : $2/3-2\sqrt{3}/3$
2381 7747 rt : (7,0,8,2)
2381 7898 rt : (6,6,-5,-4)
2381 8201 J_0 : $10\sqrt[3]{3}/3$
2381 9230 cu : $3/2-4\pi$
2381 9974 sq : $1/4+3e/4$
2382 1309 rt : (9,4,-4,6)
2382 2166 ℓ_{10} : $6\sqrt[3]{3}/5$
2382 3488 rt : (9,8,-7,-2)
2382 6159 ℓ_2 : $2+e$
2382 7837 sr : $23/15$
2382 8857 rt : (3,7,-6,-9)
2382 9023 ln : $\sqrt{3}/4+4\sqrt{5}$
2383 0931 e^x : $\sqrt[3]{5}/8$
2383 1746 as : $2-\sqrt{5}$
2383 2462 id : $9\ln 2$
2383 2588 E : $21/25$
2383 2641 $\ln(2/3) \times s\pi(1/5)$
2383 3224 rt : (6,9,6,1)
2383 3635 rt : (8,0,-5,9)
2383 3942 $t\pi$: $e+3\pi/4$
2383 5578 ℓ_{10} : $5\ln 2/6$
2383 5984 ζ : $9\sqrt[3]{2}/4$
2383 6816 2^x : $(\pi/2)^{1/3}$
2383 7793 id : $3\sqrt{2}/4-3\sqrt{3}/4$
2383 7885 E : $2\sqrt[3]{2}/3$
2383 7977 $t\pi$: $e/2+4\pi$
2383 8697 rt : (2,6,0,5)
2383 9083 2^x : $(1/(3e))^{-1/2}$
2384 0533 id : $6e\pi$
2384 3709 Ei : $1/\sqrt[3]{12}$
2384 3746 tanh : $\exp(-\sqrt{2})$
2384 4403 sinh : $17/11$
2384 5447 rt : (6,4,9,2)
2384 6991 J_2 : $7e\pi/10$
2384 7470 θ_3 : $(\ln 3/2)^{1/2}$
2384 8077 Ei : $10(1/\pi)$
2384 8987 at : $\exp(-\sqrt{2})$
2385 0561 J : $3/10$
2385 1648 rt : (5,8,2,-1)
2385 5652 cr : $10\sqrt[3]{5}/9$
2385 5972 sinh : $9\zeta(3)/7$
2385 6062 ℓ_{10} : $\sqrt{3}$
2385 6232 cr : $19/10$
2385 7461 cr : $2\sqrt{3}/3+\sqrt{5}/3$
2385 7836 Γ : $\sqrt{6}\pi$
2385 8751 J_2 : $(\ln 3/3)^{-2}$
2386 0240 ℓ_2 : $3\sqrt{2}+3\sqrt{3}$
2386 0327 sinh : $\exp(-\sqrt[3]{3})$
2386 0412 θ_3 : $4\zeta(3)/9$
2386 2761 $t\pi$: $9(1/\pi)/4$
2386 2993 rt : (9,5,-6,2)
2386 3135 2^x : $\sqrt{3}/3+\sqrt{5}/2$
2386 4275 Γ : $9\ln 2/2$
2386 4844 ln : P_{82}
2386 5024 $t\pi$: $(\sqrt{3}/2)^{-1/2}$
2386 5439 as : $\exp(-\sqrt[3]{3})$
2386 6189 ℓ_{10} : γ
2386 6247 $s\pi$: $8\zeta(3)/5$
2386 7617 rt : (4,2,-7,-2)
2386 8783 J_2 : $9e\pi/8$
2386 9532 sq : $2\sqrt[3]{5}/7$
2387 0108 sq : $7\sqrt[3]{5}/8$
2387 0762 rt : (8,1,-7,5)
2387 2113 10^x : $7/20$
2387 2270 ζ : $17/6$
2387 3241 id : $3(1/\pi)/4$
2387 3858 $\ln(3/4)-s\pi(2/5)$
2387 5188 rt : (1,9,7,6)
2387 5450 rt : (8,6,0,6)
2387 5822 at : $2e/3-3\pi/2$
2387 6417 rt : (7,3,-8,-8)
2387 6546 ℓ_{10} : $5\ln 2/2$
2387 6758 J_0 : $10\pi/3$
2387 6930 rt : (3,6,-8,-5)
2387 8607 rt : (1,9,8,-2)
2387 8613 J_0 : $7\pi^2/9$
2387 9356 e^x : $10\sqrt{3}$
2388 0774 e^x : $3/4+e/3$
2388 1495 rt : (7,2,-1,9)
2388 2631 10^x : $22/13$
2388 2732 rt : (6,3,-8,-2)
2388 4377 $1/3 \div \exp(1/3)$
2388 5837 rt : (3,8,6,7)
2388 8588 J_1 : $5(e+\pi)/9$
2388 9030 $s\pi$: $2\sqrt{2}/3-\sqrt{3}/2$
2388 9628 rt : (9,7,2,-1)
2388 9959 ℓ_{10} : $2\sqrt[3]{3}/5$
2389 1658 rt : (2,7,2,1)
2389 3205 J_2 : $\sqrt{7/3}$
2389 4957 Ei : $7\sqrt{5}/3$
2389 5199 J_0 : $\pi^2/5$
2389 5456 ln : $5\sqrt[3]{2}/8$
2389 7659 e^x : $3/14$
2389 7690 J_0 : $7\sqrt[3]{2}/3$
2389 8110 Ψ : $\sqrt[3]{1/3}$
2389 8362 rt : (8,8,6,-2)
2389 8630 $\sqrt{5} \div \exp(\sqrt{5})$
2389 8958 $t\pi$: $e/2-\pi/4$
2390 1286 sinh : $10(e+\pi)/9$
2390 2578 e^x : $8\sqrt{3}/7$
2390 2722 at : $2/3+\sqrt{5}$
2390 2943 sr : $3e-\pi$
2390 4434 rt : (3,5,7,-2)
2390 7978 as : $\sqrt{2}/4-3\sqrt{3}/4$
2390 8987 10^x : $4\pi^2/3$
2391 0129 10^x : $5\sqrt{3}/4$
2391 1184 sr : $3/4+\pi/4$
2391 1777 J_1 : $\exp(-1/\sqrt{2})$
2391 2327 rt : (2,8,2,-6)
2391 2880 10^x : $8\pi^2/7$
2391 3215 J_1 : $5e/3$
2391 3845 θ_3 : $3(1/\pi)/8$
2391 4631 sr : $1-2\sqrt{2}/3$
2391 5510 10^x : $3/2-3\sqrt{2}/2$
2391 5947 Ψ : $7\sqrt{5}/9$
2391 6985 cr : $7e/10$
2391 7437 $3/4+\ln(3/5)$
2391 7516 at : $4/3+\pi/2$
2391 7525 $3/4 \times \exp(\sqrt{3})$
2391 8497 cu : $4\sqrt{2}/3-\sqrt{3}/3$
2391 9374 ln : $\sqrt{2}/3+4\sqrt{5}/3$
2392 0038 J_2 : $10\sqrt{5}/3$
2392 1761 rt : (1,8,-1,8)
2392 2376 rt : (8,2,-9,1)
2392 3008 rt : (7,1,-9,2)
2392 4703 rt : (9,6,-8,-2)
2392 5306 Ψ : $2(1/\pi)/5$
2392 5950 at : $\sqrt{2}+2\sqrt{5}/3$
2392 6963 sr : $\sqrt{3}+\sqrt{6}-\sqrt{7}$
2392 7635 at : $2\sqrt{3}-\sqrt{5}/4$
2392 9192 rt : (7,4,7,-2)
2392 9486 2^x : $4e+2\pi$
2393 0492 $\ln(3)+\exp(\pi)$
2393 1566 $s\pi$: $1/13$
2393 2084 $t\pi$: $3\sqrt{3}/2+\sqrt{5}/2$
2393 3652 rt : (6,1,-4,-8)
2393 3966 at : $1/3-\sqrt{3}/3$
2393 5433 id : $3\sqrt{2}/2+\sqrt{5}/2$
2393 6832 e^x : $1-\pi/4$
2393 7862 rt : (7,3,-3,5)
2393 8279 rt : (6,2,-5,1)
2393 8324 at : $2+e/3$
2393 9439 10^x : $3\sqrt{3}/2-\sqrt{5}/3$
2393 9863 ln : $e/2+2\pi/3$
2394 0184 J_2 : $((e+\pi)/3)^3$
2394 0526 rt : (2,1,8,-2)
2394 1691 rt : (8,3,3,-1)
2394 2727 id : $1/3+e/3$
2394 3957 ℓ_2 : $1/2+4\sqrt{5}$
2394 4140 rt : (8,7,-2,2)
2394 5902 cr : $1/3+\pi/2$
2394 6899 $\exp(3/4)-\exp(\sqrt{5})$
2394 7730 $3 \times \exp(5)$
2394 9409 J_2 : $8\zeta(3)$
2394 9576 Ei : $2\sqrt{2}/9$
2394 9621 e^x : $13/9$
2395 2088 $t\pi$: $3\sqrt{2}/2-3\sqrt{3}$
2395 2963 2^x : $5\sqrt{5}/3$
2395 3644 tanh : $\sqrt[3]{5}/7$
2395 4965 rt : (6,4,3,9)
2395 5098 sq : $3/2+\sqrt{5}/4$
2395 5204 $s\pi$: $4\sqrt[3]{3}/3$
2395 5764 rt : (3,2,-9,2)
2395 7803 cu : $2+e/4$
2395 8287 cr : $3/4+2\sqrt{3}/3$
2395 9008 at : $\sqrt[3]{5}/7$
2395 9398 sq : $10\sqrt[3]{2}/7$
2395 9727 Ψ : $\ln 2$
2396 0017 $s\pi$: $\ln 2/9$
2396 1827 rt : (1,8,-5,-8)
2396 5250 10^x : $(1/(3e))^{1/2}$
2396 5654 rt : (4,1,-9,2)
2396 6138 rt : (7,8,4,6)
2396 7417 rt : (6,0,8,2)
2397 0276 $t\pi$: $1/4+2\sqrt{3}/3$
2397 2004 id : $4\sqrt{2}/3+3\sqrt{5}/2$
2397 2221 ln : $4/3+3\sqrt{2}/2$
2397 2312 at : $4/3-3\sqrt{2}$
2397 5330 as : $4/3-\pi/2$
2397 5909 rt : (2,8,4,-3)
2397 7824 sq : $2\sqrt{2}/3-3\sqrt{5}$
2397 7903 cr : $4-2\pi/3$
2397 9916 $\exp(3)+\exp(e)$
2398 0657 E : $3\sqrt{5}/8$
2398 3668 cr : $2e/3+3\pi$
2398 5314 e^x : $2/3-2\pi/3$

2398 5386 cu : $3/2 - 3\sqrt{2}/2$
2398 6451 rt : (5,4,-7,-7)
2398 6758 rt : (1,3,9,2)
2398 6951 as : $(\sqrt{5}/2)^{-1/2}$
2398 7076 cu : $1/\ln(5)$
2398 7901 rt : (1,9,-2,2)
2398 7907 sq : $4 - 4e$
2398 8507 cr : $1 + e/3$
2398 9264 J_2 : $7\sqrt[3]{5}/2$
2398 9478 rt : (3,2,-7,8)
2399 0865 cu : $1 + 2e/3$
2399 1560 2^x : $1/4 - 4\sqrt{3}/3$
2399 2324 ζ : $2\sqrt{2}$
2399 3816 J_1 : $2\sqrt{3}/7$
2399 5802 rt : (5,3,-8,-5)
2399 6788 Γ : $\ln(2\pi/3)$
2399 6814 e^x : $2e + 4\pi$
2399 7254 rt : (6,7,-7,-2)
2399 7689 Ei : $\pi/10$
2399 9188 erf : $(5/3)^{-3}$
2400 0000 id : $6/25$
2400 1180 rt : (9,0,-9,6)
2400 1564 $t\pi$: $\ln((e+\pi)/2)$
2400 1568 e^x : $1/4 - 3\sqrt{5}/4$
2400 1892 $\exp(\sqrt{3}) + s\pi(1/5)$
2400 4098 rt : (1,3,2,9)
2400 4435 J_1 : $5\ln 2/7$
2400 6462 $s\pi(\sqrt{2}) \div s\pi(\pi)$
2400 6572 10^x : $3 + 2\sqrt{3}$
2400 7852 ln : $\sqrt{2}/4 + \sqrt{3}/4$
2400 9835 ζ : $7\ln 2/3$
2401 3939 rt : (2,9,7,-7)
2401 5341 Γ : $17/23$
2401 6112 rt : (5,4,9,2)
2401 6280 J_0 : $\sqrt[3]{23/3}$
2401 6733 ζ : $9\pi/10$
2401 7047 J_0 : $1/\ln(\sqrt[3]{5}/2)$
2401 7588 rt : (7,1,4,1)
2402 0494 $\pi + \ln(3)$
2402 1477 rt : (2,6,-7,-2)
2402 1760 rt : (5,1,0,8)
2402 2004 rt : (7,8,6,-2)
2402 2141 2^x : $4 + 4\sqrt{2}$
2402 2610 $\sqrt{3} - \exp(2/5)$
2402 2650 $\ln(2) \times \ln(\sqrt{2})$
2402 3206 rt : (7,4,-5,1)
2402 3309 e^x : $4/3 - \sqrt{5}/2$
2402 4074 at : $3/2 + \sqrt{2}$
2402 4347 ζ : $\exp(\sqrt[3]{3}/3)$
2402 5307 rt : (9,7,-3,-1)
2402 9205 10^x : $3\sqrt{3}/4 + 2\sqrt{5}/3$
2403 0654 J_0 : $7\sqrt{2}$
2403 0909 rt : (9,5,-2,7)
2403 1696 Γ : $3/17$
2403 4053 ℓ_{10} : $7\sqrt{5}/9$
2403 4157 $t\pi$: $(\ln 2/3)^{-3}$
2403 4734 sr : $20/13$
2403 5120 sinh : $5/21$
2403 5960 J_1 : $5\pi^2/8$
2403 5972 2^x : $1 + 3\pi/4$
2403 7034 id : $\sqrt{21/2}$
2403 7367 at : $\sqrt{17/2}$
2403 7492 id : $\sqrt[3]{3}/6$
2403 7721 J_2 : $23/15$
2403 9798 $t\pi$: $(2e/3)^2$
2404 0421 as : $5/21$
2404 1138 id : $\zeta(3)/5$
2404 1473 θ_3 : $3/25$
2404 2532 J_0 : $24/5$
2404 3943 rt : (9,5,9,2)
2404 4349 e^x : $\exp(1/e)$
2404 4642 rt : (2,5,7,-2)
2404 4917 $1/3 \div \ln(4)$
2404 8217 rt : (7,3,-8,-2)
2404 8440 J_0 : $23/3$
2405 0695 rt : (6,5,1,5)
2405 3456 cr : $21/11$
2405 3511 sinh : $25/24$
2405 3713 erf : $\sqrt{3}/8$
2405 3857 Ψ : $2\sqrt{3}/5$
2405 5566 sr : $9\sqrt[3]{5}/10$
2405 8750 rt : (3,3,-7,3)
2405 8951 rt : (7,2,-6,-9)
2405 9715 rt : (8,8,-4,-2)
2406 0307 at : $1/2 - \sqrt{5}/3$
2406 1851 $s\pi$: $1/2 + \sqrt{3}/3$
2406 2725 $\exp(2/3) - s\pi(1/4)$
2406 3795 2^x : $3\sqrt{3}/4 - 3\sqrt{5}/2$
2406 4280 θ_3 : $(\ln 2/2)^2$
2406 4560 Γ : $4\ln 2/7$
2406 5433 rt : (5,3,-9,-3)
2406 6099 Ψ : $\pi/7$
2406 6811 rt : (6,4,7,-2)
2406 7388 rt : (2,9,-7,4)
2406 7930 rt : (1,7,-7,-4)
2406 8853 2^x : $10\pi/9$
2406 9529 as : $3\sqrt{2}/4 - 3\sqrt{3}/4$
2407 0040 cu : $4/3 + 3\sqrt{3}/4$
2407 0098 cr : $6(1/\pi)$
2407 0151 rt : (5,6,-3,-1)
2407 1005 $1/2 \div \ln(4/5)$
2407 1092 10^x : $2/3 + 4\sqrt{5}/3$
2407 1198 id : $1/4 - 2\sqrt{5}/3$
2407 2691 ℓ_{10} : $1/4 + 2\sqrt{5}/3$
2407 2827 10^x : $\beta(2)$
2407 3195 $t\pi$: $4\sqrt{2} - \sqrt{3}$
2407 4299 rt : (3,2,-8,-2)
2407 9797 ζ : $1/\sqrt[3]{12}$
2407 9825 sq : $1 - 2\sqrt{5}/3$
2408 0043 $\ln(2/5) + s\pi(\sqrt{5})$
2408 0384 rt : (2,4,5,1)
2408 0647 sr : $8\sqrt{3}/9$
2408 0842 cu : $3\sqrt{2} - \sqrt{3}/3$
2408 1143 10^x : $7\sqrt{3}/9$
2408 1664 $t\pi$: $1/2 + 3\pi$
2408 2135 rt : (5,6,7,9)
2408 2179 J_1 : $(1/(3e))^{1/3}$
2408 2357 rt : (8,2,-5,6)
2408 2896 θ_3 : $\zeta(3)/10$
2408 3950 $s\pi$: $3e/2$
2408 6221 J_1 : $2\sqrt{5}/9$
2408 7080 Ei : $16/21$
2408 7944 cr : $4/3 + \sqrt{3}/3$
2408 9609 rt : (4,2,-3,-7)
2409 1914 rt : (8,7,2,-1)
2409 2505 sq : $3\sqrt{3} + 3\sqrt{5}/4$
2409 4061 $\ln(\pi) \div \ln(3/5)$
2409 4620 cr : $4\sqrt{3}/3 + 4\sqrt{5}$
2409 5094 Ei : $7/16$
2409 5310 rt : (4,4,-8,7)
2409 6353 at : $\sqrt{3}/4 - 3\sqrt{5}/2$
2409 7168 10^x : $1/2 - \sqrt{5}/2$
2409 7328 10^x : $3/4 + \sqrt{3}/4$
2409 7563 $1/4 \times s\pi(\sqrt{2})$
2409 8836 J_2 : $\sqrt{3} + \sqrt{6} - \sqrt{7}$
2409 9092 rt : (1,1,8,-2)
2410 0377 rt : (7,9,2,2)
2410 0657 sinh : $3(1/\pi)/4$
2410 0809 ℓ_2 : $11/13$
2410 2780 2^x : $1/2 + 3\pi/4$
2410 3699 rt : (6,7,-7,-8)
2410 6031 as : $3(1/\pi)/4$
2410 6824 sr : $10e\pi$
2410 6863 $\sqrt{3} \div \exp(1/3)$
2410 9205 e^x : $(2e/3)^3$
2411 0237 e^x : $(5/3)^{-3}$
2411 0523 2^x : $2 + 2\sqrt{5}/3$
2411 1315 rt : (1,5,-3,-1)
2411 1968 J_1 : $13/4$
2411 2663 Ψ : $7\sqrt{2}$
2411 3279 sq : $1/4 - 4\sqrt{3}/3$
2411 4635 rt : (9,6,-4,3)
2411 4897 rt : (8,1,-9,2)
2411 4915 sr : $6(e+\pi)/7$
2411 5418 rt : (2,9,-7,-9)
2411 5698 sinh : $9\pi/4$
2411 5838 e^x : $3/4 + e/2$
2411 6205 ln : $11/14$
2411 8449 sq : $7(1/\pi)/2$
2412 0401 Ψ : $2\pi^2/5$
2412 0699 E : $(e+\pi)/7$
2412 1166 $\ln(\sqrt{2}) - s\pi(1/5)$
2412 1307 ln : $5\sqrt{2}/9$
2412 1388 2^x : $2/3 - e$
2412 2339 $t\pi$: $(\sqrt[3]{5}/2)^{1/2}$
2412 2682 rt : (5,0,8,2)
2412 4854 rt : (7,6,-6,1)
2412 7046 at : $\sqrt[3]{25}$
2412 8278 ℓ_{10} : $e/2 - \pi/4$
2412 9693 $\ln(3/5) \div \exp(3/4)$
2413 0193 sq : $4\pi^2/9$
2413 0645 rt : (2,9,6,-2)
2413 3882 id : $\sqrt{3} - 2\sqrt{5}/3$
2413 4417 rt : (7,1,-8,-5)
2413 6581 cr : $\sqrt[3]{7}$
2413 6915 rt : (5,2,-2,4)
2413 7075 $\exp(\sqrt{5}) \times s\pi(e)$
2413 7875 2^x : $e/2 - \pi/3$
2413 8375 10^x : $1 - e/3$
2413 9059 rt : (6,5,5,1)
2413 9155 id : $2e/3 - \pi/2$
2414 0018 rt : (8,7,2,7)
2414 1546 10^x : $e + 3\pi$
2414 2961 rt : (4,2,-9,2)
2414 5147 rt : (9,1,8,2)
2414 8230 rt : (7,7,-7,-2)
2414 8278 sq : $\sqrt[3]{12}$
2414 8978 erf : $5/23$
2414 9203 rt : (6,8,6,-2)
2415 0282 2^x : $\sqrt{2} - 2\sqrt{3}$
2415 5938 Γ : $3\sqrt[3]{5}$
2415 6447 ln : $\pi/4$
2415 6566 rt : (6,0,-2,9)
2415 8546 rt : (3,5,-6,-6)
2416 4314 cr : $1/2 + \sqrt{2}$
2416 5339 J_2 : $20/13$
2416 6806 J_0 : $\sqrt{23}$
2416 7367 rt : (7,5,-7,-3)
2416 8101 $\exp(4) - \exp(\sqrt{5})$
2416 8533 rt : (4,5,-8,2)
2416 8592 Li_2 : $5/22$
2416 9426 rt : (3,9,2,0)
2417 0817 at : $4 - 4\sqrt{3}$
2417 3097 e^x : $\sqrt{3}/8$
2417 3666 rt : (5,6,-9,6)
2417 3773 $t\pi$: $10\sqrt{3}/9$
2417 4986 rt : (4,3,4,8)
2417 6096 ζ : $(5/2)^{-1/3}$
2417 6108 rt : (2,7,-7,-3)
2417 6648 Ei : $5(e+\pi)/3$
2417 6847 rt : (8,3,-7,2)
2417 7592 as : $2/3 - e/3$
2417 8165 sr : $\sqrt[3]{11/3}$
2417 8200 J_2 : $9\sqrt[3]{5}/10$
2417 8905 10^x : $\ln(\ln 3)$
2417 8970 Li_2 : $5(1/\pi)/7$
2418 0669 rt : (9,1,-7,7)
2418 2230 10^x : $2 + 4\sqrt{3}/3$
2418 2469 $1/4 - \exp(2/5)$
2418 4166 rt : (7,3,3,-1)
2418 4705 sq : $3 - 4\sqrt{2}/3$
2418 4714 id : $2\sqrt{2}/3 + 3\sqrt{3}/4$
2418 5191 Li_2 : $\exp(-\sqrt{2}\pi/3)$
2418 5781 2^x : $5/16$
2418 7596 cu : $3 + \pi/4$
2418 7736 ℓ_{10} : $5\pi/9$
2418 8402 ℓ_{10} : $1 + \sqrt{5}/3$
2418 8513 10^x : $3\sqrt{2}/4 - 3\sqrt{5}/4$
2418 8827 J_1 : $(\sqrt{2}\pi/3)^3$
2418 8919 at : $(e+\pi)/2$
2418 9656 rt : (2,2,-2,8)
2419 3338 sq : $\sqrt{3}/2 + 4\sqrt{5}$
2419 3689 J_2 : $8\sqrt{3}/9$
2419 4412 rt : (4,4,9,2)
2419 7622 rt : (6,2,-4,-1)
2419 9648 sr : $8\pi/5$
2420 1141 e^x : $7\sqrt{2}/5$
2420 1630 Ψ : $9/13$
2420 3123 ζ : $2\pi^2/7$
2420 4273 at : $1/3 + 3\sqrt{3}/2$

2420 5613 rt : (7,4,-1,6)
2420 5743 rt : (1,9,4,-7)
2420 6946 cu : $3\sqrt{3}/2-\sqrt{5}/2$
2420 7340 sr : $3e+3\pi/4$
2420 7786 E : $(\sqrt[3]{5})^{-1/3}$
2420 8874 10^x : $1/4-\sqrt{3}/2$
2420 8981 rt : (5,4,7,-2)
2420 9415 10^x : $\ln(5/3)$
2421 1665 rt : (5,3,-5,-8)
2421 6533 cr : $6\sqrt{5}/7$
2421 6934 Γ : 12/5
2421 7331 cr : 23/12
2421 7864 10^x : $2\sqrt{3}-3\sqrt{5}/4$
2421 8033 rt : (8,5,9,2)
2421 8358 $t\pi$: 11/25
2421 8516 rt : (6,6,-1,1)
2422 0845 sr : $(\sqrt[3]{2}/3)^{-1/2}$
2422 0978 rt : (8,3,-8,-2)
2422 2347 rt : (3,7,-6,1)
2422 3101 cu : $\ln((e+\pi)/2)$
2422 4582 J_0 : $\sqrt{11}\pi$
2422 4864 rt : (4,1,-5,-3)
2422 5998 id : $5\sqrt{5}/9$
2422 6845 J_1 : 1/2
2422 7457 rt : (9,3,7,-2)
2422 7522 ζ : 14/19
2422 8606 rt : (2,8,-9,-5)
2422 9460 sinh : 25/6
2422 9784 2^x : $3e/7$
2423 1064 sinh : 6/25
2423 2514 Ψ : $\exp(-1/e)$
2423 3101 10^x : $3+2\pi/3$
2423 3290 ℓ_{10} : $\sqrt[3]{16/3}$
2423 6585 as : 6/25
2423 7600 ln : $\sqrt{2}/3+4\sqrt{5}$
2423 7927 erf : $1/\sqrt{21}$
2423 8582 2^x : $2/3-\sqrt{2}/4$
2423 8695 2^x : $6\sqrt{2}/5$
2423 9228 rt : (9,7,-6,-1)
2424 2883 ℓ_{10} : $1/3+\sqrt{2}$
2424 3097 rt : (4,2,-8,-2)
2424 4823 rt : (2,8,-3,-4)
2424 5332 ln : $2\sqrt{3}$
2424 6157 rt : (5,7,-9,1)
2424 6556 ℓ_2 : $3+\sqrt{3}$
2424 6727 $s\pi$: $9\sqrt[3]{5}/5$
2424 7101 rt : (9,6,0,8)
2424 8013 Li_2 : $(2\pi/3)^{-2}$
2425 0068 rt : (1,9,-2,-1)
2425 0132 J_1 : $\sqrt{2}+\sqrt{3}-\sqrt{7}$
2425 0554 $t\pi$: $6\sqrt[3]{5}/7$
2425 0679 tanh : $\sqrt{3}/7$
2425 1111 cr : $\sqrt{2}-\sqrt{3}+\sqrt{5}$
2425 2077 ℓ_2 : P_{38}
2425 3446 θ_3 : $\ln(2/3)$
2425 3562 id : $1/\sqrt{17}$
2425 3940 cu : $\sqrt{3}/3+4\sqrt{5}$
2425 3942 J_2 : $\sqrt[3]{11/3}$
2425 5772 J_0 : $\sqrt{10}\pi$
2425 6387 at : $\sqrt{3}/7$
2425 6592 sr : $3\sqrt{2}/2-\sqrt{3}/3$
2426 0144 Ψ : $7\sqrt{2}/8$
2426 1226 $2/5\div\exp(1/2)$
2426 2868 $t\pi$: $2\sqrt{2}/3+4\sqrt{5}/3$
2426 3105 id : P_{15}
2426 3717 tanh : $7(1/\pi)/9$
2426 4068 id : $3\sqrt{2}$
2426 5270 cu : $\ln(1/(3e))$
2426 6237 cr : $8\pi^2/7$
2426 6986 id : $1/\ln(\sqrt{5})$
2426 7262 ℓ_{10} : $1-2\sqrt{2}/3$
2426 7474 rt : (4,6,-8,-3)
2426 9440 at : $7(1/\pi)/9$
2426 9643 sinh : $\sqrt[3]{3}/6$
2426 9885 rt : (3,0,1,7)
2427 0505 $\ln(\sqrt{2})\times\exp(\sqrt{5})$
2427 2201 rt : (6,1,-4,5)
2427 3395 sinh : $\zeta(3)/5$
2427 3445 rt : (9,2,-9,3)
2427 5027 10^x : $3\sqrt{2}/4-\sqrt{3}/4$
2427 5100 $t\pi$: $4\ln 2/3$
2427 5208 as : $\sqrt[3]{3}/6$
2427 7497 cu : $9\ln 2/10$
2427 7580 rt : (3,8,2,-1)
2427 8190 10^x : $10e\pi/7$
2427 8805 sq : $1/2+2e$
2427 8963 as : $\zeta(3)/5$
2427 9199 rt : (8,8,0,3)
2428 0164 rt : (5,8,6,-2)
2428 0328 J_2 : $(\sqrt[3]{2}/3)^{-1/2}$
2428 3033 e^x : 5/23
2428 3832 cu : $e-2\pi/3$
2428 3973 rt : (4,0,8,2)
2428 6739 at : $4-3\sqrt{2}/4$
2428 7320 J_2 : 6
2428 7643 $\sqrt{3}-\ln(3/5)$
2429 0842 e^x : 20/17
2429 1442 Ψ : $5\sqrt{3}/7$
2429 1493 ζ : $8\sqrt{2}/7$
2429 1676 at : $7\sqrt[3]{2}/3$
2429 2499 ln : $5\ln 2$
2429 2939 rt : (6,4,-7,-9)
2429 3955 rt : (8,9,-6,-6)
2429 5605 rt : (9,7,6,-2)
2429 6017 $s\pi$: $3\sqrt{3}-\sqrt{5}/2$
2429 6821 e^x : $e/4-2\pi/3$
2429 6989 rt : (3,4,-3,7)
2429 7780 e^x : $7(e+\pi)/6$
2429 8308 ζ : 7/16
2429 8648 sr : $2e+4\pi$
2429 9936 rt : (3,7,-7,6)
2430 0675 rt : (8,3,-3,7)
2430 1882 rt : (1,8,6,1)
2430 2317 sr : $9\sqrt{5}/4$
2430 2885 rt : (8,1,8,2)
2430 2909 ζ : $3e/8$
2430 3231 rt : (7,7,2,-1)
2430 3398 cu : $3/4+\sqrt{5}/4$
2430 3804 ℓ_{10} : 4/7
2430 4989 rt : (8,7,-7,-2)
2430 8010 rt : (5,2,2,9)
2430 8204 Ei : $(5)^{-3}$
2431 1673 id : $\exp(-\sqrt{2})$
2431 2031 $t\pi$: $\ln(\ln 2)$
2431 2680 K : $(\sqrt[3]{2})^{-1/2}$
2431 2821 J_1 : $9\sqrt[3]{3}/4$
2431 3335 ℓ_{10} : $4\sqrt{3}/3-\sqrt{5}/4$
2431 4612 rt : (3,5,-7,-7)
2431 6312 sr : 17/11
2431 7188 rt : (9,1,-9,2)
2431 7797 rt : (8,4,-9,-2)
2431 8209 sr : $9\zeta(3)/7$
2431 9636 2^x : $-\sqrt{5}-\sqrt{6}+\sqrt{7}$
2432 0874 rt : (6,1,4,1)
2432 1863 $\exp(\sqrt{2})-\exp(\sqrt{5})$
2432 2616 Γ : $(\sqrt[3]{2}/3)^2$
2432 2690 at : $2+2\sqrt{2}/3$
2432 2823 rt : (4,6,-7,-2)
2432 3381 ln : $\sqrt[3]{3}/5$
2432 4094 ζ : $(e+\pi)/6$
2432 4692 rt : (7,9,6,7)
2432 5616 $\ln(2/3)+\exp(1/2)$
2432 5622 $t\pi$: $e/2-2\pi$
2432 6475 at : $\sqrt{2}/2+\sqrt{5}$
2432 7906 $3/5\times\ln(2/3)$
2432 8691 2^x : $\pi/10$
2432 9698 2^x : $\exp(\sqrt[3]{4}/3)$
2433 0903 at : $3\sqrt{2}-3\sqrt{3}/4$
2433 1001 id : $3\sqrt{3}/4+4\sqrt{5}$
2433 2256 $t\pi$: $3e\pi/4$
2433 2257 $s\pi$: $3\pi^2/5$
2433 2365 rt : (2,9,1,-5)
2433 2677 Γ : 14/19
2433 3831 $\exp(1/3)\div s\pi(\pi)$
2433 4217 id : $\ln(3\pi)$
2433 4491 $t\pi$: $\sqrt[3]{25}$
2433 7644 rt : (1,6,-2,-3)
2433 7662 Γ : $(5/2)^{-1/3}$
2433 8207 2^x : $2\sqrt{2}/9$
2433 8915 10^x : P_{13}
2433 9383 J_2 : 17/11
2433 9730 2^x : $2\sqrt{2}/3-4\sqrt{5}/3$
2433 9811 rt : (5,8,-1,-4)
2434 0557 J_2 : $9\zeta(3)/7$
2434 0680 $\ln(4)\times\exp(2)$
2434 0983 cu : $3-4e/3$
2434 2374 sinh : 24/23
2434 6296 tanh : $\sqrt{5}/9$
2434 6436 ln : $-\sqrt{2}+\sqrt{5}+\sqrt{7}$
2434 6762 10^x : $\exp(-\pi/3)$
2435 0804 $\ln(5)\times\exp(\pi)$
2435 2118 at : $\sqrt{5}/9$
2435 2920 $\ln(4/5)\div\ln(2/5)$
2435 3732 sq : $4\sqrt{2}/3+2\sqrt{3}/3$
2435 3988 sq : $(\sqrt[3]{3}/2)^{-1/3}$
2435 4907 rt : (7,0,-6,6)
2435 5658 cr : 25/13
2435 6007 rt : (4,4,7,-2)
2435 7672 rt : (6,6,-3,-1)
2436 0273 cr : $8\zeta(3)/5$
2436 0635 $5\times\exp(1/2)$
2436 1011 2^x : $(e/2)^{1/2}$
2436 1064 rt : (1,3,-9,2)
2436 2587 Li_2 : $(\sqrt{2}\pi/3)^{-1/3}$
2436 3265 rt : (7,5,-3,2)
2436 4961 rt : (9,2,-5,8)
2436 5569 10^x : $2/3+3\pi/2$
2436 5932 ln : $3/4+e$
2436 6922 rt : (6,9,-6,9)
2436 7639 rt : (5,9,6,1)
2436 9045 e^x : $4\sqrt{2}/7$
2436 9581 cr : $9\sqrt[3]{5}/8$
2436 9619 cr : $\sqrt{3}/4+2\sqrt{5}/3$
2437 1464 rt : (8,3,7,-2)
2437 4521 as : $\sqrt{3}-2\sqrt{5}/3$
2437 5577 rt : (9,7,-2,4)
2437 7352 cu : $\sqrt{3}/2-2\sqrt{5}/3$
2437 8122 cr : $3e+\pi$
2437 9156 J_0 : $6\sqrt{3}$
2437 9680 rt : (4,1,-1,-8)
2437 9956 as : $2e/3-\pi/2$
2438 1061 rt : (3,4,9,2)
2438 2106 rt : (3,8,-5,2)
2438 4175 ln : $1/3+2\sqrt{2}/3$
2438 5003 $t\pi$: $\sqrt{2}/2+\sqrt{3}/3$
2438 5806 e^x : $1/\sqrt{21}$
2438 5960 $t\pi$: $(\sqrt{5}/2)^{-3}$
2438 6344 cr : $10\sqrt{3}/9$
2438 6396 sr : $1/2+\pi/3$
2438 7674 rt : (2,5,-3,-1)
2438 8786 J_0 : $\sqrt{3}-\sqrt{5}-\sqrt{6}$
2438 9738 Ψ : $5\pi/9$
2439 0077 $s\pi$: $1/4+2\sqrt{2}$
2439 1008 Ψ : $(\sqrt[3]{4}/3)^{-1/3}$
2439 1273 Ei : $\sqrt[3]{2}/9$
2439 1908 e^x : $4+2\sqrt{3}/3$
2439 4973 $s\pi(1/4)-s\pi(2/5)$
2439 8045 Γ : 5/18
2439 9463 2^x : $\sqrt[3]{2}/4$
2439 9475 id : $5\pi/7$
2439 9904 rt : (7,5,9,2)
2440 0389 $s\pi$: $6\sqrt{3}/5$
2440 0544 Ei : $(\sqrt[3]{4}/3)^{-3}$
2440 1693 id : $1/3-\sqrt{3}/3$
2440 3828 id : $\sqrt{3}/4-3\sqrt{5}/4$
2440 4990 e^x : $\sqrt[3]{11}$
2440 6195 θ_3 : $2\sqrt[3]{2}/3$
2440 6842 J_1 : $2\sqrt[3]{2}/5$
2440 8475 rt : (4,7,-8,-8)
2440 8741 Ψ : $6\sqrt[3]{3}/7$
2440 9814 rt : (6,3,-9,-5)
2441 0231 id : $2\sqrt{3}-3\sqrt{5}$
2441 0367 $t\pi$: $((e+\pi)/3)^{-1/2}$
2441 0703 rt : (8,1,-8,-7)
2441 1867 Li_2 : $1/\sqrt{19}$
2441 3125 ℓ_{10} : $\sqrt[3]{5}/3$
2441 3606 ln : $(\sqrt[3]{3}/3)^{1/3}$
2441 3754 at : $\sqrt{2}/4+3\sqrt{3}/2$
2441 4062 cu : 5/8
2441 5111 rt : (8,8,4,8)

2441 5735 $J_0 : 7e\pi/6$
2441 6094 rt : (5,8,-5,5)
2441 7945 $J_0 : 3e\pi/4$
2441 9324 rt : (5,2,-8,-2)
2442 5516 $\exp(\sqrt{3}) \div \exp(\pi)$
2442 5826 at : $4 - \pi/3$
2442 6652 $K : 9\ln 2/7$
2442 7259 sq : $e/2 - 3\pi/2$
2442 7396 $2^x : 3\sqrt{2}/4 - \sqrt{5}/3$
2442 7400 rt : (7,2,-5,1)
2442 7418 $10^x : 1/4 + \sqrt{5}$
2442 7964 rt : (8,7,6,-2)
2442 8024 cu : $\sqrt{7}\pi$
2442 8055 $1/5 \times \exp(1/5)$
2442 8227 id : $\sqrt[3]{5}/7$
2442 9752 rt : (1,7,6,8)
2443 2030 $t\pi : \sqrt{3}/4 + 2\sqrt{5}/3$
2443 2611 $t\pi : 9\sqrt[3]{5}/8$
2443 2621 $\ln(\pi) - \exp(2)$
2443 3073 at : $\sqrt{3} - \sqrt{5} - \sqrt{6}$
2443 3394 cr : $3\sqrt{2}/4 + \sqrt{3}/2$
2443 3657 $e^x : 4 + 3\sqrt{5}$
2443 4138 $\exp(3/4) \times s\pi(1/5)$
2443 5992 rt : (1,7,8,-6)
2444 0286 rt : (6,3,3,-1)
2444 0837 rt : (8,4,-5,3)
2444 1260 cr : $1/4 + 3\sqrt{5}/4$
2444 3734 rt : (9,8,-8,-5)
2444 4662 rt : (5,2,-3,-9)
2444 5475 $Ei : \exp(\sqrt{2})$
2444 6872 $e^x : \sqrt{3}/2 + \sqrt{5}$
2444 7393 id : $3e/4 - 2\pi$
2444 7673 rt : (2,4,-9,-3)
2444 8051 sq : $3/4 + 3\sqrt{2}/2$
2444 8925 rt : (6,2,-6,1)
2445 1770 rt : (3,0,8,2)
2445 2207 rt : (2,9,-8,-1)
2445 2282 id : $\sqrt{2}/4 - 3\sqrt{3}/2$
2445 2553 rt : (3,5,-3,2)
2445 2819 $2^x : \sqrt{2}/4 + \sqrt{3}$
2445 4405 $\ell_2 : 4\sqrt{2} - 2\sqrt{5}$
2445 4467 $2^x : 4\sqrt{2} - \sqrt{5}/2$
2445 5927 ln : $3 + \sqrt{2}/3$
2445 7374 $J_2 : \ln(3\pi/2)$
2445 9189 ln : $4 + 2e$
2445 9422 $J_2 : 6\sqrt[3]{5}$
2446 0359 sr : $1/4 + 3\sqrt{3}/4$
2446 1432 rt : (7,6,-9,-7)
2446 2017 $10^x : 18/13$
2446 2027 $t\pi : 2\sqrt{2}/3 - 4\sqrt{3}/3$
2446 4428 rt : (4,0,-7,1)
2446 5193 cr : $4e\pi$
2446 5412 rt : (5,9,-9,-9)
2446 5859 $t\pi : (\ln 3/3)^{1/3}$
2446 6194 rt : (1,6,-5,-5)
2446 6725 rt : (7,1,8,2)
2446 7257 cu : $4 + 3\sqrt{3}/2$
2446 7989 rt : (9,7,-7,-2)
2446 8406 cr : $\sqrt{3} - \sqrt{6} + \sqrt{7}$
2446 9258 $2^x : 6/19$
2446 9954 $s\pi : 1/3 + \sqrt{5}/3$
2447 0260 rt : (9,7,2,9)
2447 0741 sr : $3 + 3e/4$
2447 0995 $2^x : (\sqrt[3]{4})^{1/3}$
2447 3033 rt : (3,6,-2,-6)
2447 5816 sq : $4 + 3e/2$
2447 6236 at : $2/3 - 4e/3$
2448 1696 rt : (1,5,-3,-9)
2448 2084 rt : (5,6,-7,-2)
2448 2452 $\theta_3 : 21/25$
2448 2812 ln : $3\sqrt{2} + 3\sqrt{3}$
2448 4036 $2^x : 2 + 3\sqrt{5}$
2448 4154 rt : (7,5,1,7)
2448 4272 $\zeta : \ln(2/3)$
2448 4972 $\Psi : \sqrt{10}\pi$
2448 6304 rt : (3,9,-3,-2)
2448 6563 rt : (2,9,-2,-1)
2448 7299 $J_2 : 5\zeta(3)$
2448 9205 rt : (9,3,-7,4)
2448 9795 sq : $2\sqrt{3}/7$
2449 1336 as : 18/19
2449 1866 tanh : 1/4
2449 2042 sinh : $1/\sqrt{17}$
2449 2409 $2^x : 7/6$
2449 4215 $J_2 : 7(e+\pi)/4$
2449 4521 $t\pi : 1/4 + 3e$
2449 5481 rt : (1,8,6,-3)
2449 5691 $t\pi : 1/4 + 3\sqrt{2}$
2449 6925 $E : 5/6$
2449 7866 at : 1/4
2449 8358 $J_0 : 5\pi/8$
2450 0586 at : $3\sqrt{3} - \sqrt{5}$
2450 0897 cu : $1 + 4\sqrt{3}/3$
2450 0956 $\ell_{10} : 3(e+\pi)$
2450 2995 rt : (4,7,-4,1)
2450 3183 $10^x : e/4 + 2\pi/3$
2450 3402 $J_1 : \sqrt{21/2}$
2450 6153 id : $9\sqrt[3]{3}/4$
2450 6173 rt : (2,8,2,-1)
2450 6318 $t\pi : \sqrt{2} - 2\sqrt{5}/3$
2450 6453 $\ln(\sqrt{2}) \times s\pi(1/4)$
2450 6826 sr : $\ln(3\pi/2)$
2450 6902 sq : $10\sqrt[3]{3}/7$
2450 7057 rt : (7,1,-8,2)
2450 7078 $2^x : 1/\sqrt{10}$
2450 7409 rt : (5,3,0,5)
2450 7703 as : P_{15}
2450 8226 rt : (3,5,8,8)
2450 8696 as : $4 - 3\sqrt{2}$
2450 8747 at : $3\pi^2/10$
2450 9113 $Ei : \sqrt[3]{25/2}$
2451 0316 ln : $4\sqrt{5}/7$
2451 0815 $\ell_{10} : (2e)^{1/3}$
2451 1249 $\ell_2 : (4/3)^3$
2451 2233 rt : (1,4,5,1)
2451 2245 ln : 18/23
2451 2407 $\zeta : 21/13$
2451 2908 sq : $5\ln 2/7$
2451 4214 $2^x : 1/3 + 4\sqrt{3}/3$
2451 4426 cu : $3/2 + 3\sqrt{3}$
2451 4824 $\ln(3) \times \ln(4/5)$
2451 5444 rt : (8,9,-2,-1)
2451 8907 $s\pi(\pi) + s\pi(\sqrt{5})$
2451 9052 cu : $3/2 + \sqrt{3}/2$
2451 9708 $10^x : 2/21$
2452 0366 rt : (7,3,7,-2)
2452 0637 $\Psi : 21/17$
2452 1829 $\Gamma : 5\sqrt[3]{3}/3$
2452 4384 rt : (6,7,2,-1)
2452 5296 $2/3 \div e$
2452 6101 rt : (8,4,-1,8)
2452 6487 $\ln(4) \div \exp(\sqrt{3})$
2452 6668 rt : (2,7,-7,-6)
2452 6959 $\ell_2 : P_{57}$
2452 7062 $e^x : (\sqrt[3]{4}/3)^{1/3}$
2453 0084 $\Psi : \sqrt[3]{16/3}$
2453 1178 $J_1 : 9$
2453 2494 $t\pi : 4\sqrt[3]{2}/9$
2453 4758 $J_2 : 4e/7$
2453 5599 id : $1/2 + \sqrt{5}/3$
2453 6240 rt : (5,9,-5,0)
2453 6443 at : $\sqrt{2} - \sqrt{3} - \sqrt{7}$
2453 7504 $2^x : 4\sqrt{2} - \sqrt{5}/4$
2453 7755 sq : $1/4 - \sqrt{5}/3$
2454 0841 ln : $1/2 + 4\sqrt{5}$
2454 1024 sr : $\sqrt{3}/4 + \sqrt{5}/2$
2454 1478 $t\pi : e + 4\pi$
2454 1686 at : $1/2 - 2\sqrt{3}$
2454 1893 $J_2 : 6e\pi/5$
2454 3788 rt : (1,9,-6,-2)
2454 4433 $s\pi(\sqrt{2}) \div s\pi(e)$
2454 4828 sinh : $(5/2)^3$
2454 6216 rt : (5,5,5,1)
2454 7383 rt : (6,2,-9,2)
2454 8511 $t\pi : e/5$
2454 9148 $\Gamma : -\sqrt{2} - \sqrt{3} + \sqrt{6}$
2455 1218 $t\pi : e/2 + 3\pi/4$
2455 1270 sq : $1/4 + \sqrt{3}/2$
2455 1560 $\Gamma : 2\zeta(3)$
2455 1875 sinh : $\exp(-\sqrt{2})$
2455 4287 rt : (9,3,-3,9)
2455 6864 sinh : $\sqrt{21}\pi$
2455 6894 $s\pi : 3\sqrt{3}/2 - 3\sqrt{5}/4$
2455 7318 ln : $1/3 + \pi$
2455 7770 as : $\exp(-\sqrt{2})$
2455 8081 cu : $4/3 - \sqrt{2}/2$
2455 9391 $2^x : 4(e+\pi)$
2456 1018 $E : (\sqrt[3]{3})^{-1/2}$
2456 1269 $\ln(5) \times s\pi(e)$
2456 2119 rt : (3,8,-9,-7)
2456 2609 rt : (2,3,-9,2)
2456 4328 rt : (7,7,6,-2)
2456 5399 $\lambda : \exp(-9\pi/7)$
2456 5519 $\ell_2 : (5/3)^{-1/3}$
2456 5683 $Li_2 : 3/13$
2456 6684 cr : $e - \pi/4$
2456 7806 rt : (9,8,-4,0)
2456 9057 $\ell_2 : 1/2 + 3\sqrt{2}$
2456 9456 cr : $3/2 + \sqrt{3}/4$
2456 9936 $e^x : 1/4 + \sqrt{5}/4$
2457 2644 $2^x : 3/4 - \sqrt{3}/4$
2457 3678 $E : (\ln 2)^{1/2}$
2457 4330 cr : $3\sqrt{2} - 4\sqrt{3}/3$
2457 4925 rt : (4,4,6,9)
2457 5924 rt : (9,6,5,1)
2457 6443 $t\pi : 8\zeta(3)/5$
2457 6740 $10^x : 1/3 - 2\sqrt{2}/3$
2457 6880 rt : (2,4,9,2)
2457 7827 rt : (7,2,-4,-1)
2457 8931 $2^x : 20/7$
2458 1282 $\zeta : 7\zeta(3)/3$
2458 1433 cr : $\sqrt{13}\pi$
2458 2653 rt : (7,1,-4,7)
2458 2764 $\Psi : (2\pi/3)^{-1/2}$
2458 3482 $1/2 + s\pi(\sqrt{3})$
2458 4349 at : $1/4 + e$
2458 6697 rt : (6,7,-3,-3)
2458 8399 sinh : 16/5
2458 9148 $2^x : 4/3 + 2\pi$
2458 9932 ln : $\ln(4/3)$
2458 9982 $e^x : 9\sqrt{5}$
2459 0279 rt : (6,5,9,2)
2459 0447 rt : (5,9,-1,9)
2459 0805 $J_2 : 14/9$
2459 2701 $\ln(2/3) \div \exp(1/2)$
2459 2805 $10^x : 3(1/\pi)/10$
2459 3576 $J_0 : 10$
2459 3577 $10^x : \exp(4/3)$
2459 4107 id : $e/4 + 4\pi$
2459 5635 $2^x : (\ln 2/2)^{-1/2}$
2459 5694 $10^x : 3/4 - e/2$
2459 7499 $Li_2 : \ln 2/3$
2460 1323 $t\pi : 2\sqrt{2}/3 - \sqrt{3}/2$
2460 1665 rt : (8,0,-6,8)
2460 3668 rt : (6,2,-8,-2)
2460 5615 $Li_2 : (\sqrt[3]{3}/3)^2$
2460 6389 rt : (2,5,-7,-7)
2460 6968 $2^x : 1/3 - 3\pi/4$
2460 7585 $e^x : 4\sqrt{3} - 3\sqrt{5}$
2460 8454 $\theta_3 : \ln(\sqrt[3]{5})$
2461 1697 $J_1 : 7(e+\pi)/9$
2461 3119 $\ell_2 : 4/3 + 3e$
2461 4322 cu : $10\sqrt{2}/7$
2461 5154 $\ln(5) \times \exp(1/3)$
2461 5294 rt : (9,1,-8,-9)
2461 5474 rt : (2,7,0,-7)
2461 5722 rt : (1,8,-3,-8)
2461 9426 $\ell_2 : 5e\pi/9$
2462 1318 sq : $e/2 - 2\pi$
2462 3588 at : $3/2 - 2\sqrt{5}$
2462 5544 rt : (5,1,-5,-5)
2462 6185 $J : \sqrt{2}/9$
2462 6382 $10^x : 21/13$
2462 6617 rt : (2,0,8,2)
2462 6853 sr : $1/3 + 3\pi/2$
2462 9387 $J : 1/16$
2462 9615 $E : \exp(-1/(2e))$
2463 0670 rt : (8,9,2,4)
2463 1613 sr : $4e/7$
2463 1881 $2^x : 3\sqrt{2} + \sqrt{3}/3$

2463 2008 rt : (7,6,-5,-2)
2463 2034 sq : $3 + \sqrt{2}/4$
2463 5559 2^x : $10\sqrt{2}/3$
2463 7150 rt : (6,1,8,2)
2463 7787 ℓ_{10} : P_{80}
2463 9933 rt : (6,2,-7,-6)
2464 2159 ln : $(2\pi/3)^{1/3}$
2464 2667 ℓ_{10} : $1 - \sqrt{3}/4$
2464 2685 2^x : $4/3 - 3\sqrt{5}/2$
2464 2944 Ψ : $1/\ln(1/2)$
2464 4437 2^x : $\exp(5/3)$
2464 6438 rt : (9,8,0,5)
2464 6673 J_0 : $\sqrt{3} + \sqrt{5} + \sqrt{6}$
2464 7020 tπ : $9\pi/8$
2464 7120 rt : (5,1,4,1)
2464 7136 ℓ_{10} : $7\sqrt[3]{2}/5$
2464 7490 rt : (1,8,2,-8)
2464 7674 rt : (6,6,-7,-2)
2464 7786 tπ : $1/13$
2464 8026 J_1 : $8(1/\pi)/5$
2464 8184 ℓ_{10} : $4 - \sqrt{5}$
2464 9552 rt : (7,3,-9,-7)
2465 0184 E : $6\ln 2/5$
2465 0476 J_0 : $3\pi^2/10$
2465 0585 as : $1/3 - \sqrt{3}/3$
2465 1088 rt : (8,5,-7,-1)
2465 1269 ζ : $\sqrt[3]{22}$
2465 1632 ℓ_2 : $4 + \sqrt{5}/3$
2465 4658 e^x : $(\sqrt[3]{3}/2)^{-1/2}$
2465 5136 rt : (1,6,3,-5)
2465 7082 rt : (2,5,-7,-2)
2465 7963 sinh : $23/22$
2465 9211 2^x : $5e/8$
2465 9688 sr : $4 + \pi/3$
2465 9696 $\exp(3/5) \div \exp(2)$
2466 0013 Ψ : $1/\ln(1/(3e))$
2466 0513 cu : $4e/3 + 4\pi$
2466 0822 10^x : $1/3 + 3\pi/4$
2466 1479 cr : $9\sqrt[3]{2}$
2466 1605 rt : (9,4,-9,0)
2466 2775 10^x : $\sqrt{2}/2 + 4\sqrt{5}/3$
2466 3086 rt : (2,5,-3,6)
2466 3524 at : $2/3 + 4\sqrt{3}/3$
2466 3741 at : $3\sqrt{3}/4 + 3\sqrt{5}/4$
2466 4482 $5 \div \exp(4/5)$
2466 5540 rt : (9,2,4,1)
2466 6013 rt : (2,4,7,-2)
2466 6483 rt : (7,6,-3,-1)
2466 8377 erf : $2/9$
2466 8642 J_1 : $8\sqrt[3]{5}/3$
2467 1908 sinh : $\sqrt[3]{5}/7$
2467 2702 J_2 : $9\sqrt{3}/10$
2467 3350 e^x : $7\sqrt[3]{2}/2$
2467 3612 tπ : $4\sqrt[3]{3}/3$
2467 4501 rt : (6,3,7,-2)
2467 4731 sr : $7\sqrt[3]{3}/2$
2467 4936 θ_3 : $\exp(-2\pi/3)$
2467 5830 2^x : $7/22$
2467 6385 e^x : $4\sqrt{2}$
2467 6750 sπ : $8\sqrt[3]{2}$
2467 6856 $\ln(2/3) + \exp(\sqrt{3})$
2467 7379 rt : (3,1,3,8)
2467 7947 as : $\sqrt[3]{5}/7$
2467 8294 tanh : $\sqrt[3]{2}/5$
2467 8756 ℓ_{10} : $\sqrt[3]{11/2}$
2467 8872 tπ : $\ln 2/9$
2468 1916 $\ln(\sqrt{3}) \div \exp(4/5)$
2468 2520 sq : $(1/(3e))^{1/3}$
2468 2780 rt : (4,0,1,9)
2468 2982 rt : (3,5,-3,-1)
2468 3780 e^x : $17/21$
2468 4528 at : $\sqrt[3]{2}/5$
2468 4641 2^x : $1/2 + 2e$
2468 6820 at : $\sqrt{3}/2 - \sqrt{5}/2$
2468 6898 2^x : $(1/\pi)$
2468 7130 $\exp(1/2) + \exp(4)$
2469 0006 10^x : $\ln(5/2)$
2469 0648 rt : (2,9,-6,-2)
2469 1008 J_2 : $9\ln 2/4$
2469 1343 e^x : $1/3 - \sqrt{3}$
2469 1358 sq : $2\sqrt{5}/9$
2469 2936 θ_3 : $3\sqrt{3}/7$
2469 7960 rt : (4,8,-4,-4)
2469 9017 sπ : $3\ln 2$
2470 0308 J_2 : $7\sqrt[3]{2}/2$
2470 1490 rt : (8,9,6,9)
2470 3142 ζ : $14/5$
2470 4941 rt : (6,7,6,-2)
2470 5221 cr : $3 - 3\sqrt{2}/4$
2470 7207 sπ : $2e\pi$
2470 7399 sq : $2\sqrt{2}/3 + \sqrt{5}/2$
2470 7578 ℓ_{10} : $4/3 + \sqrt{3}/4$
2471 0618 sr : $\sqrt{3}/3 + 2\sqrt{5}$
2471 1613 rt : (5,3,3,-1)
2471 1735 rt : (9,4,-5,5)
2471 4774 Ψ : $\pi^2/8$
2471 4986 rt : (8,5,-3,4)
2471 5943 10^x : $1/\sqrt[3]{23}$
2471 6529 Γ : $5\pi^2/3$
2471 7188 J_1 : $\ln(5/3)$
2471 7776 at : $4\sqrt{5}/3$
2471 8025 sπ : $4\sqrt{2} - \sqrt{3}/3$
2472 0247 rt : (7,6,-1,3)
2472 1912 sr : $14/9$
2472 2095 J : $\exp(-15\pi/17)$
2472 2112 Ei : $5/16$
2472 2497 $\exp(\sqrt{2}) - s\pi(1/3)$
2472 3095 2^x : $4\sqrt{3} - 4\sqrt{5}$
2472 4107 rt : (9,2,3,-1)
2472 5628 cu : $3\sqrt{2}/4 - \sqrt{3}/4$
2473 0213 rt : (6,7,1,2)
2473 2240 erf : $7(1/\pi)/10$
2473 6049 rt : (3,9,-2,-1)
2473 8702 tπ : $(\sqrt{2}/3)^{-1/3}$
2474 0395 sπ : $(1/\pi)/4$
2474 2501 ℓ_{10} : $2\sqrt{2}/5$
2474 3582 id : $\sqrt{3}/7$
2474 4139 ln : $1/2 + 4\sqrt{5}/3$
2474 4253 10^x : $(1/(3e))^{-1/2}$
2474 6046 J_0 : $\sqrt{2} - \sqrt{3} - \sqrt{7}$
2474 6344 $\pi - \exp(2)$
2474 6566 rt : (1,8,2,-1)
2474 6983 rt : (4,8,0,5)
2474 7245 Ψ : $7/4$
2475 1374 ℓ_{10} : $\ln(e + \pi)$
2475 1522 rt : (6,3,-8,-3)
2475 2946 10^x : $4\sqrt[3]{2}/7$
2475 4288 cr : $5e/7$
2475 4413 ln : P_{38}
2475 4437 ℓ_{10} : $\exp(\sqrt[3]{5}/3)$
2475 6290 rt : (5,8,3,1)
2475 7435 id : $7(1/\pi)/9$
2475 7845 Ei : $\sqrt{17}\pi$
2475 9057 rt : (8,5,1,9)
2475 9963 rt : (3,6,1,6)
2476 3663 J_2 : $25/16$
2476 3894 10^x : $1/3 + \sqrt{3}$
2476 4672 at : $3/4 + \sqrt{5}$
2476 5306 rt : (9,6,2,-1)
2476 5871 erf : $\exp(-3/2)$
2476 6492 id : $9\ln 2/5$
2476 6942 rt : (7,2,-9,2)
2476 7940 rt : (8,1,-8,4)
2476 8106 id : $6e\pi/5$
2477 0122 sπ : $4e + \pi/3$
2477 0158 10^x : $1/4 + 2\pi$
2477 0403 rt : (2,7,-8,-7)
2477 3368 sr : $9(e + \pi)/5$
2477 4279 rt : (7,6,3,8)
2477 5156 rt : (3,3,-9,2)
2477 7345 id : $2e - 4\pi/3$
2477 7552 at : $3\sqrt{2}/2 + \sqrt{3}/2$
2477 8448 ℓ_{10} : $13/23$
2477 9537 cr : $1 + 2\sqrt{2}/3$
2477 9891 cr : $\sqrt[3]{22/3}$
2478 2295 rt : (2,4,2,7)
2478 2822 rt : (1,4,9,2)
2478 4460 2^x : $\sqrt{5}/7$
2478 4584 rt : (7,2,-6,3)
2478 5630 Ψ : $7e\pi/6$
2478 8689 as : $1/2 - \sqrt{5}/3$
2478 9033 rt : (5,4,-6,-4)
2478 9573 rt : (4,8,-9,-9)
2478 9990 rt : (5,5,9,2)
2479 0193 tπ : $1/2 + \sqrt{3}/3$
2479 2751 ℓ_2 : 19
2479 2794 $\exp(1/4) - s\pi(\sqrt{2})$
2479 2871 tπ : $8\sqrt[3]{5}/3$
2479 5215 $\sqrt{3} \times \exp(e)$
2479 6384 rt : (2,6,-9,-2)
2479 6708 rt : (9,6,9,2)
2479 6921 rt : (7,2,-8,-2)
2479 8005 rt : (8,1,-4,9)
2479 9601 rt : (6,7,5,7)
2480 3210 rt : (3,1,-8,-2)
2480 5021 cu : $\pi/5$
2480 6070 10^x : $3\sqrt{2} - 4\sqrt{5}/3$
2480 6510 cu : $3\sqrt{2}/2 + 2\sqrt{3}$
2480 7460 10^x : $18/25$
2480 8185 ℓ_{10} : $\sqrt{2}/3 + 3\sqrt{3}/4$
2480 8889 rt : (3,7,-7,-8)
2480 9127 rt : (1,0,8,2)
2481 1088 id : $(\sqrt{2}\pi/3)^3$
2481 1437 at : $3/2 + 2\sqrt{5}/3$
2481 1614 rt : (6,3,-4,2)
2481 4361 tπ : $3e/2$
2481 4580 rt : (4,4,-9,5)
2481 4703 rt : (5,1,8,2)
2481 5090 J_1 : $3\sqrt[3]{5}/10$
2481 5799 J_0 : $9\sqrt{5}/2$
2481 6180 e^x : $3\sqrt{3} - \sqrt{5}/4$
2481 7036 erf : $\sqrt{5}/10$
2481 7182 tπ : $\sqrt[3]{5}/6$
2481 8091 rt : (7,2,-2,8)
2481 8781 e^x : $3\pi/8$
2481 9967 rt : (9,2,8,2)
2482 0115 rt : (7,6,-7,-2)
2482 0149 ln : $4 - e$
2482 4158 Γ : $11/15$
2482 5080 rt : (3,5,-7,-2)
2482 6111 rt : (8,3,-9,-9)
2482 6811 e^x : $8\pi/7$
2482 7923 e^x : $2/3 + 3\sqrt{2}/2$
2482 8542 Γ : $\sqrt{2} - \sqrt{6} - \sqrt{7}$
2482 8730 ζ : $5\sqrt{5}/4$
2482 8985 at : $7\sqrt[3]{5}/4$
2482 9782 rt : (1,4,7,-2)
2483 0665 ℓ_2 : $3 - 2e/3$
2483 1184 cu : $4\sqrt{2}/9$
2483 1387 at : $\sqrt{2}/3 - 2\sqrt{3}$
2483 2053 J_2 : $7\sqrt{5}/10$
2483 3054 2^x : $8/25$
2483 3082 $\ln(\sqrt{2}) \div \exp(1/3)$
2483 3856 e^x : $5\pi^2/4$
2483 4240 rt : (5,3,7,-2)
2483 4249 J_1 : $\exp(-2/3)$
2483 5713 rt : (3,2,-8,6)
2483 6095 rt : (1,8,6,4)
2483 6534 cr : $\sqrt{3} - \sqrt{5} + \sqrt{6}$
2483 8474 rt : (9,2,7,-2)
2483 8528 rt : (2,6,-5,-7)
2483 8797 J_0 : $1/\ln(5/3)$
2484 2106 ln : $8\sqrt[3]{3}/9$
2484 2598 rt : (3,9,-6,-2)
2484 5199 id : $\sqrt{5}/9$
2484 6423 rt : (1,8,6,-2)
2484 7657 rt : (7,2,-7,-8)
2484 8267 rt : (6,3,0,7)
2484 8396 sπ : $6\sqrt[3]{2}/7$
2484 8401 rt : (2,9,-9,-6)
2484 8476 10^x : $5(1/\pi)/2$
2484 9300 J_0 : $\sqrt[3]{15/2}$
2485 0068 rt : (5,8,7,6)
2485 1757 10^x : $1/2 + 3\sqrt{2}$
2485 3397 $2/5 \div \ln(5)$
2485 3545 rt : (9,6,6,-2)
2485 3743 sr : $9\sqrt{3}/10$
2485 5263 rt : (6,1,-9,-2)
2485 7493 ℓ_{10} : $(\pi)^{1/2}$
2485 7753 J_2 : $3\pi^2/4$

2485 9672 $J_0 : 8\sqrt[3]{2}$
2485 9798 $10^x : (\pi/2)^{-1/3}$
2486 0602 sr : $1+\sqrt{5}/4$
2486 1160 $1/3 \times s\pi(\sqrt{3})$
2486 3146 rt : (5,4,-2,1)
2486 3911 $Ei : 3(e+\pi)/2$
2486 6206 rt : (2,6,-1,2)
2486 8988 $s\pi$: 2/25
2487 0227 $2^x : 1-e/4$
2487 2127 $2/5-\exp(1/2)$
2487 2988 ln : $9\ln 2/8$
2487 4090 id : $3\sqrt{2}/4-4\sqrt{3}/3$
2487 4367 Ei : 13/17
2487 5154 id : $8e/3$
2487 6317 $2^x : 2\sqrt[3]{3}/9$
2487 6659 $e^x : \sqrt{3}/4+3\sqrt{5}/4$
2487 7632 $J_0 : 6\sqrt[3]{5}$
2487 7912 rt : (6,1,-5,-7)
2487 8212 sinh : 15/4
2487 9595 $Ei : (\sqrt[3]{5})^{-1/2}$
2487 9723 ln : $3\sqrt[3]{5}/4$
2488 3191 sr : $9\ln 2/4$
2488 4729 cr : $1/2+4e$
2488 4886 e^x : 2/9
2488 5608 rt : (9,9,-6,-4)
2488 7149 rt : (3,7,-8,3)
2488 7166 $s\pi : (\sqrt[3]{2})^{1/3}$
2488 7545 sq : $4/3-4e/3$
2488 8203 $J_0 : 7(e+\pi)/4$
2488 9792 rt : (1,5,-3,-6)
2489 0260 e^x : 23
2489 0284 $t\pi : \sqrt{6}$
2489 2775 $\Gamma : 3\sqrt[3]{5}/7$
2489 3534 $5 \div \exp(3)$
2489 4416 $\zeta : 8\pi/9$
2489 4494 $s\pi : \sqrt[3]{9}$
2489 4914 $J_0 : 7\sqrt[3]{3}$
2489 6887 $Ei : (\sqrt[3]{3})^{1/3}$
2489 7587 cu : 14/13
2489 8376 cu : $\sqrt{3}-\sqrt{5}-\sqrt{7}$
2489 8677 cu : $3\sqrt{2}/4+\sqrt{5}/4$
2489 8687 rt : (5,4,2,6)
2489 9167 sq : $\sqrt{3}+\sqrt{5}/4$
2489 9935 $J_0 : 7\sqrt{5}/8$
2490 0958 2^x : 17/10
2490 1922 $J_0 : 6e\pi/5$
2490 2476 cr : $9\sqrt{3}/8$
2490 4577 at : 3
2490 4839 rt : (5,0,-7,1)
2490 4954 $\exp(2/3) \div s\pi(1/3)$
2490 5230 $e^x : \sqrt{2}+\sqrt{3}$
2490 6798 $\ell_2 : 7\zeta(3)/10$
2490 6965 $e^x : \sqrt{3}/4+\sqrt{5}/3$
2490 9100 $E : (\sqrt[3]{5}/3)^{1/3}$
2490 9466 at : $e/3+2\pi/3$
2491 0062 at : $\sqrt{2}/4-3\sqrt{5}/2$
2491 1551 rt : (9,9,-2,1)
2491 3234 $J_1 : 3\zeta(3)/7$
2491 3494 sq : 19/17
2491 3998 $\ln(3/4) \times s\pi(1/3)$
2491 4053 rt : (2,8,-4,-8)
2491 8522 rt : (1,9,-8,-6)
2492 2359 sq : $3+3\sqrt{5}$
2492 2574 rt : (2,8,0,9)
2492 3493 rt : (5,0,-3,6)
2492 4058 id : $3e+2\pi/3$
2492 5363 $\Gamma : 1/\sqrt{13}$
2492 5672 $J_2 : (\ln 3/2)^{-3}$
2492 6385 sr : $1/2+3\sqrt{2}/4$
2492 7894 rt : (9,9,2,6)
2492 8276 tanh : $4(1/\pi)/5$
2492 8329 rt : (6,9,-7,7)
2492 8751 $Ei : (\ln 2)^{-1/3}$
2492 9870 $t\pi : 2e+2\pi/3$
2493 0859 $\ln(\pi) \div \ln(2/5)$
2493 1069 rt : (9,5,-7,1)
2493 1112 $10^x : \pi/5$
2493 2361 at : $(\ln 2)^{-3}$
2493 3100 $e^x : 2+2e/3$
2493 4469 at : $1-\sqrt{5}/3$
2493 4835 at : $4(1/\pi)/5$
2493 4877 $J_2 : 8e\pi/7$
2493 5359 $e^x : 4/3+\pi/2$
2493 6705 sinh : $\pi/3$
2493 7316 id : $\sqrt{3}-4\sqrt{5}/3$
2493 8541 rt : (5,4,-8,-6)
2493 9079 ζ : 9/14
2494 1055 $\ell_2 : \exp(\sqrt{3}/2)$
2494 1057 at : $4\sqrt{2}/3+\sqrt{5}/2$
2494 1415 rt : (9,5,-3,6)
2494 1757 sq : $4/3+e/2$
2494 2038 sq : $2/3+2e$
2494 3043 $10^x : 1/\ln(\sqrt{5}/2)$
2494 3981 $\Gamma : (e+\pi)/8$
2494 4946 ln : P_{57}
2494 8850 $Ei : 10\sqrt[3]{2}/7$
2495 0100 $10^x : 4\sqrt[3]{3}/5$
2495 0665 rt : (9,1,-8,6)
2495 2198 $J_1 : 1/\ln(\sqrt{5}/2)$
2495 3302 $e^x : 5\sqrt{2}/6$
2495 5920 at : $5\zeta(3)/2$
2495 7088 $J_0 : 1/\ln(e/3)$
2495 7225 $\ell_{10} : 9\pi^2/5$
2495 7601 $10^x : 3/4+\sqrt{3}/3$
2495 9111 sr : $4+3\sqrt{2}/4$
2495 9177 $e^x : 7(1/\pi)/10$
2495 9338 rt : (5,7,-6,8)
2496 0122 rt : (1,6,-3,-9)
2496 0298 as : $e\pi/9$
2496 2406 $1/3-\ln(2/5)$
2496 2798 $s\pi : 3\sqrt{2}+3\sqrt{5}/4$
2496 5057 $10^x : \sqrt[3]{9}$
2496 7057 $E : (\ln 3)^{-2}$
2496 9090 $J_2 : 5(e+\pi)/3$
2497 0162 $J_2 : \pi/2$
2497 0716 $\zeta : 8\ln 2/7$
2497 1600 cr : $4e\pi/3$
2497 6649 $J_2 : 1/\ln(\sqrt[3]{4}/3)$
2497 8724 Ei : 11/25
2498 1305 $s\pi : 2e-3\pi/4$
2498 3903 $J_2 : 10\sqrt{2}/9$
2498 5899 J_2 : 11/7
2498 6755 Ei : 9/5
2498 7747 ℓ_{10} : 16/9
2498 9953 $Ei : 3\sqrt[3]{5}$
2499 0765 $J : \sqrt[3]{5}/6$
2499 2133 sr : $3\sqrt{2}/2-\sqrt{5}/4$
2499 2217 $s\pi : (\ln 2/3)^{-1/2}$
2499 2514 $t\pi : 9\sqrt[3]{5}/5$
2499 3118 cr : $3-\pi/3$
2499 4681 $Ei : 6\sqrt{3}/5$
2499 5388 $J_1 : \exp(2\pi/3)$
2499 6842 sinh : $\sqrt{3}/7$
2499 8287 $\theta_3 : 7(1/\pi)/3$
2499 8326 $e^x : \exp(-3/2)$
2500 0000 id : 1/4
2500 0056 $t\pi : 3\sqrt{2}+2\sqrt{5}$
2500 3289 as : $\sqrt{3}/7$
2500 3551 id : $((e+\pi)/3)^{1/3}$
2500 4167 sinh : 22/21
2500 4261 sinh : $9\sqrt{5}/2$
2500 4668 ln : $4/3+3e$
2500 4996 $\ell_2 : 7e$
2500 8097 at : $4/3+3\sqrt{5}/4$
2500 8609 J_2 : 22/5
2500 8758 Li_2 : 22/25
2500 9040 ln : $10\pi/9$
2500 9119 sinh : $\ln(3\pi/2)$
2501 0572 ln : $2+2\sqrt{5}/3$
2501 1121 sinh : $7(1/\pi)/9$
2501 1323 $\Psi : \sqrt{5}/5$
2501 3097 sq : $5\sqrt[3]{3}/4$
2501 4477 cu : $3/4+2\sqrt{5}/3$
2501 6912 $e^x : 1/2-4\sqrt{2}/3$
2501 7586 as : $7(1/\pi)/9$
2502 0436 $\sqrt{2} \div \exp(\sqrt{3})$
2502 4170 cr : $8\sqrt[3]{5}/7$
2503 1567 sq : $(e+\pi)^{1/3}$
2503 3224 $J_0 : 8\sqrt[3]{5}/7$
2503 3610 ln : $3\sqrt{2}-2\sqrt{3}$
2503 3669 $e^x : 4/3-e$
2503 5329 cu : $3\sqrt{3}-3\sqrt{5}/2$
2503 6678 id : $e/4+\pi/2$
2503 8062 $\Gamma : 2\ln 2/5$
2503 8370 $\exp(4)+\exp(\sqrt{3})$
2504 0476 $e^x : 4/3+4\pi$
2504 2148 rt : (3,8,-6,-1)
2504 2182 as : $(\sqrt[3]{5}/2)^{1/3}$
2504 4331 sr : $2\sqrt{2}+\sqrt{5}$
2504 4738 $2^x : e+4\pi$
2504 5523 $\Gamma : \sqrt[3]{14}$
2504 6279 cu : $1/2+\sqrt{3}/3$
2504 6319 $J : \exp(-11\pi/2)$
2504 6503 $t\pi : 3\sqrt{3}-\sqrt{5}/2$
2504 8829 θ_3 : 1/8
2505 1332 sq : $\sqrt{2}+\sqrt{3}-\sqrt{7}$
2505 1522 rt : (7,1,-5,-9)
2505 7919 $e^x : \sqrt{5}/10$
2505 8018 $J_1 : \exp(2e/3)$
2505 9345 rt : (7,1,-9,-4)
2506 1693 rt : (7,3,0,9)
2506 1788 sr : $4\sqrt{3}/3-\sqrt{5}/3$
2506 3124 $e^x : 4\sqrt{2}/3+\sqrt{3}$
2506 3943 cr : $4\sqrt{3}+2\sqrt{5}$
2506 4610 $J_2 : 7\pi/5$
2506 4904 rt : (3,6,-5,-9)
2506 5998 cu : $3/4+4e$
2506 7956 ln : $(\sqrt{2}/3)^{1/3}$
2506 8842 at : $\sqrt{2}/2+4\sqrt{3}/3$
2506 9385 $4/5-\ln(\sqrt{3})$
2506 9789 $Ei : \exp(1/(3e))$
2507 1516 cu : $3-3e/2$
2507 2330 $J_2 : 5\sqrt[3]{2}/4$
2507 2494 ζ : 11/25
2507 2795 Ψ : 16/13
2507 2886 cu : $9\sqrt[3]{2}/7$
2507 3226 cr : $7\sqrt{5}/8$
2507 3255 rt : (7,3,-4,4)
2507 4003 $\ell_{10} : \exp(\sqrt{3}/3)$
2507 6867 rt : (7,7,5,9)
2507 7207 cu : $3\sqrt{2}/2-2\sqrt{5}/3$
2507 7829 rt : (3,6,-9,-4)
2508 0848 Li_2 : 4/17
2508 1920 rt : (3,2,0,-9)
2508 6212 $t\pi : 3\pi^2/5$
2508 6510 $e^x : \exp(\sqrt{2}\pi/2)$
2508 6733 $e^x : 4e/3+\pi/4$
2508 7073 rt : (2,9,4,8)
2508 8240 $\theta_3 : (2e/3)^{-1/2}$
2508 8253 $\ln(\pi)-\exp(1/3)$
2508 8487 $2^x : 2-3\sqrt{5}/4$
2508 8712 sinh : $(\ln 3)^{1/2}$
2508 8744 $t\pi : \sqrt{3}/3-4\sqrt{5}$
2508 9152 $J_2 : (e)^2$
2508 9224 $J_0 : (e+\pi)/3$
2508 9267 $t\pi : \sqrt{3}/2-4\sqrt{5}$
2509 0143 rt : (7,3,-8,-1)
2509 1853 cr : $\sqrt[3]{15/2}$
2509 5677 rt : (7,7,1,4)
2509 5717 cr : $1/\ln(5/3)$
2509 7461 $\exp(3/4)-s\pi(1/3)$
2509 7989 rt : (2,9,0,-9)
2509 8138 $J_2 : \exp(\sqrt{2}\pi/3)$
2510 0832 sq : $\sqrt{3}/4-\sqrt{5}$
2510 1598 sinh : $\sqrt{5}/9$
2510 2324 $2^x : e-3\pi/2$
2510 2670 10^x : 15/13
2510 3639 rt : (6,4,-8,-8)
2510 4684 Ei : 19
2510 4950 $\pi \div \exp(1/3)$
2510 6478 at : $10e/9$
2510 6852 ln : $1/2+\pi/4$
2510 7685 id : $2\sqrt{2}-\sqrt{3}/3$
2510 8181 as : $\sqrt{5}/9$
2510 9855 sr : $7\sqrt{5}/10$
2511 0922 $\ell_{10} : \sqrt[3]{17/3}$
2511 0977 at : $1/3-3\sqrt{5}/2$
2511 1029 rt : (3,2,5,9)
2511 1679 sq : $3\zeta(3)/2$
2511 3720 $J_1 : 3\sqrt{3}/10$

2511 3792 sq : $6\pi/7$
2511 5224 id : $4\sqrt{3}-3\sqrt{5}/4$
2511 5819 cu : $3/2-4\sqrt{5}/3$
2511 6650 10^x : $\sqrt{2}/2+\sqrt{3}/3$
2511 7101 Ψ : $8\pi^2/3$
2511 7653 2^x : $3-3\sqrt{3}/4$
2511 8577 J_2 : $1/\ln(e/3)$
2512 0515 e^x : $(2\pi/3)^{1/2}$
2512 0738 ℓ_2 : $\sqrt{3}/3+4\sqrt{5}$
2512 1913 rt : (1,7,0,-9)
2512 3992 rt : (1,9,-2,3)
2512 4239 rt : (6,0,-3,8)
2512 4743 e^x : P_{31}
2512 4759 J_1 : $3\ln 2/4$
2512 5017 ℓ_2 : $\sqrt{17/3}$
2512 6652 rt : (7,7,-3,-1)
2512 7387 Li_2 : $\sqrt{2}/6$
2512 8334 ℓ_2 : $4/3+\pi/3$
2512 8850 10^x : $4\pi^2/5$
2512 8881 ln : $4/3-\pi/3$
2512 9179 ln : $2/19$
2512 9759 cu : $7\sqrt{3}/6$
2513 0771 rt : (2,9,-8,-8)
2513 1000 $\ln(3/4)\div\ln(\pi)$
2513 1045 J_1 : $13/25$
2513 1442 ln : $7/9$
2513 1631 at : $2/3+3\pi/4$
2513 1989 J_1 : $(\sqrt[3]{3}/2)^2$
2513 2029 Ψ : $(2e)^{-3}$
2513 5073 rt : (1,7,5,2)
2513 5364 Γ : $25/8$
2513 6021 ln : $2/3+2\sqrt{2}$
2513 6737 $s\pi$: $2/3+\sqrt{2}$
2513 7626 tanh : $\exp(-e/2)$
2513 7977 rt : (5,8,-6,3)
2514 0877 $t\pi$: $1/2+\pi/4$
2514 1004 rt : (3,7,-3,-8)
2514 4468 at : $\exp(-e/2)$
2514 6506 2^x : $23/5$
2514 7237 10^x : $4\sqrt{2}/9$
2514 7696 rt : (6,0,-7,3)
2514 9147 cu : $1/4+3\sqrt{2}/4$
2514 9590 $t\pi$: $1/4+2\sqrt{2}$
2515 0015 sinh : $9e\pi/7$
2515 0071 rt : (9,2,-6,7)
2515 0500 J_1 : $4\sqrt{5}$
2515 1265 rt : (7,6,6,-2)
2515 3443 cu : $10(1/\pi)$
2515 3682 Γ : $(\sqrt[3]{5}/2)^2$
2515 3876 ℓ_2 : $21/25$
2515 5058 rt : (6,4,2,8)
2515 6016 sq : $1/3+\pi/4$
2515 6401 e^x : $1+4\pi/3$
2515 6538 ℓ_{10} : $2/3+\sqrt{5}/2$
2515 7644 $\ln(\sqrt{2})-\exp(4)$
2515 8460 $\ln(3/4)+s\pi(\sqrt{2})$
2515 9048 rt : (1,8,-6,-2)
2516 0895 $t\pi$: $6\sqrt{3}/5$
2516 2916 ℓ_2 : $\sqrt[3]{2}/3$
2516 3252 rt : (5,9,-6,-2)
2516 4587 sq : $3\sqrt{2}/4-\sqrt{5}/4$
2516 4851 $\exp(5)\times s\pi(\sqrt{5})$
2516 5381 rt : (2,7,1,-2)
2516 7097 sq : $8e\pi/7$
2516 7566 rt : (7,2,7,-2)
2516 9397 ζ : $\exp(1/(3e))$
2516 9649 J_2 : $3(e+\pi)/4$
2516 9681 cr : $9e\pi$
2517 1010 10^x : $\ln(\ln 3/2)$
2517 2241 rt : (3,3,7,-2)
2517 3508 ℓ_{10} : $1+\pi/4$
2517 3915 cr : $4-3e/4$
2517 5258 Ψ : 10
2517 7189 rt : (1,7,4,-4)
2517 7266 sq : $3/4-3\pi$
2517 8659 rt : (9,6,-1,7)
2517 8896 10^x : $3\sqrt[3]{5}/2$
2517 9363 sr : $2-\sqrt{3}/4$
2518 1197 ℓ_{10} : $14/25$
2518 2056 cu : $4/3-3\sqrt{5}/2$
2518 2433 rt : (5,5,-7,-2)
2518 3918 ℓ_{10} : $4\sqrt[3]{2}/9$
2518 7158 rt : (7,7,-7,-6)
2518 7662 rt : (5,3,-6,-7)
2518 7831 rt : (7,2,8,2)
2518 8002 rt : (9,6,-7,-2)
2519 0349 10^x : $(\sqrt[3]{4}/3)^{-3}$
2519 1525 cr : $\sqrt{2}/3+2\sqrt{5}/3$
2519 2963 $\ln(2/3)\div\ln(5)$
2519 3176 cu : $12/19$
2519 3369 rt : (6,4,-2,3)
2519 3647 rt : (2,6,-9,7)
2519 3739 rt : (3,1,8,2)
2519 5685 e^x : $\sqrt{2}/4-\sqrt{3}$
2519 7590 cr : $2\sqrt{2}-\sqrt{3}/2$
2519 8083 rt : (1,7,-1,6)
2519 8420 id : $\sqrt[3]{2}/5$
2519 8686 ln : $3\sqrt{2}-\sqrt{5}/3$
2519 8961 J : $\sqrt[3]{3}/9$
2520 0038 rt : (1,3,-5,-9)
2520 0858 id : $\sqrt{3}/2-\sqrt{5}/2$
2520 1207 rt : (8,1,-9,-6)
2520 1266 2^x : $4\sqrt{3}+2\sqrt{5}/3$
2520 3080 10^x : P_{50}
2520 4112 at : $3e/2-\pi/3$
2520 4473 E : $19/23$
2520 6203 rt : (6,8,7,8)
2520 6447 Li_2 : $\exp(-\sqrt[3]{3})$
2520 6707 $\ln(\sqrt{2})\times\exp(e)$
2520 6767 rt : (5,1,-8,-2)
2520 9320 J_1 : $12/23$
2520 9586 J : $(\ln 2/3)^3$
2520 9714 rt : (2,5,-7,-3)
2521 0194 erf : $5/22$
2521 0711 rt : (9,6,-5,2)
2521 2592 at : $7\sqrt{3}/4$
2521 3422 rt : (3,9,0,8)
2521 3685 ℓ_{10} : $2\sqrt{3}-3\sqrt{5}/4$
2521 3721 rt : (7,6,9,2)
2521 3973 rt : (9,2,-8,-2)
2521 7047 rt : (4,6,-5,4)
2521 9996 erf : $5(1/\pi)/7$
2522 0844 cr : $5\pi/8$
2522 1011 id : $2\sqrt{2}/3+4\sqrt{3}/3$
2522 1033 cu : $e/3+4\pi/3$
2522 1429 rt : (3,5,9,2)
2522 3367 $2/5-\exp(\sqrt{3})$
2522 4651 cr : $2+3\pi$
2522 4915 rt : (2,1,2,-8)
2522 5871 erf : $\exp(-\sqrt{2}\pi/3)$
2522 7559 J_2 : $\sqrt{5/2}$
2522 9725 rt : (1,4,-9,2)
2523 0002 J_1 : $9e/4$
2523 0379 10^x : $2-3\sqrt{3}/2$
2523 1325 sr : $(1/\pi)/5$
2523 1973 J_1 : $\sqrt{21}$
2523 1976 $\ln(4)+s\pi(1/3)$
2523 4619 rt : (5,1,-1,7)
2523 7191 $t\pi$: $1/3+\sqrt{5}/3$
2523 7583 $t\pi$: $7\pi^2/2$
2523 8682 rt : (5,3,-9,2)
2523 9156 J : $2/11$
2523 9566 rt : (3,9,-4,-5)
2523 9787 rt : (8,3,-4,6)
2524 2001 $\exp(3/5)-s\pi(\pi)$
2524 2593 2^x : $1/4-\sqrt{5}$
2524 3540 tanh : $(\pi/2)^{-3}$
2524 3791 sinh : $(\sqrt{3}/2)^{-1/3}$
2524 6440 rt : (7,6,2,-1)
2524 7832 2^x : $\sqrt{2}+3\sqrt{5}/2$
2524 8382 rt : (9,2,-9,2)
2524 8819 θ_3 : $\sqrt[3]{2}/10$
2525 0441 rt : (1,8,-6,-7)
2525 0529 at : $(\pi/2)^{-3}$
2525 1215 $t\pi$: $\sqrt{8/3}$
2525 1986 rt : (4,6,-9,-5)
2525 5149 J_1 : $1/\sqrt[3]{7}$
2525 5377 Γ : $7\ln 2$
2525 6725 rt : (9,6,-9,-3)
2525 6967 2^x : $(\sqrt[3]{5}/3)^2$
2525 7146 2^x : $4-2\sqrt{2}$
2525 7498 ℓ_{10} : $8\sqrt{5}$
2525 8428 at : $e/4+3\pi/4$
2526 1012 tanh : $1/\sqrt{15}$
2526 1231 sinh : $1/4$
2526 1384 at : P_{28}
2526 1800 ℓ_2 : $4\sqrt{3}/3-\sqrt{5}/2$
2526 1887 sq : $1/4+2e/3$
2526 5171 10^x : $\pi^2/3$
2526 5932 Ψ : $\exp(-(e+\pi))$
2526 5983 sr : $e+3\pi/4$
2526 6743 $t\pi$: $\sqrt{3}/3+3\sqrt{5}$
2526 7794 cr : $2/3+3\sqrt{3}/4$
2526 7922 rt : (1,8,-2,-1)
2526 8025 at : $1/\sqrt{15}$
2527 6296 ln : $2/7$
2527 6536 θ_3 : $7/13$
2527 7456 ζ : $25/9$
2527 9876 rt : (6,8,3,3)
2528 0160 rt : (5,9,-2,-1)
2528 0471 J_2 : $7\sqrt[3]{3}$
2528 2166 J_2 : $19/12$
2528 3106 rt : (8,3,-8,1)
2528 5185 erf : $(2\pi/3)^{-2}$
2528 5323 $s\pi$: $(\ln 2)^{-2}$
2528 5570 cu : $4\sqrt{3}/3-3\sqrt{5}/4$
2528 7244 at : $1+3e/4$
2528 7453 $\ln(3)+\exp(e)$
2528 8131 cu : $2/3-3\sqrt{3}/4$
2528 8744 id : $1/\ln(\sqrt{2}\pi/2)$
2528 9492 $t\pi$: $\sqrt[3]{20}$
2529 2470 $s\pi(1/5)\times s\pi(\pi)$
2529 2943 rt : (7,2,3,-1)
2529 2956 J_1 : $\pi/6$
2529 3159 rt : (5,1,-5,2)
2529 6654 rt : (8,7,1,6)
2529 7660 e^x : $1/2+e/4$
2529 8221 sr : $(5/2)^{-3}$
2529 8364 rt : (2,4,-6,7)
2529 8639 sq : $4/3+4\sqrt{3}$
2530 1736 rt : (7,0,-7,5)
2530 2430 J_1 : $11/21$
2530 2942 at : $4\sqrt{2}/3+2\sqrt{3}/3$
2530 4505 J_0 : $8e\pi/9$
2530 5312 Γ : $\exp(e/2)$
2530 5569 rt : (9,1,-9,-8)
2530 5994 sinh : $4\zeta(3)$
2530 7658 rt : (9,9,9,-3)
2530 9375 J_0 : $7e/4$
2531 0884 $\exp(1/2)-\exp(1/3)$
2531 1193 $t\pi$: $\sqrt{3}/4-3\sqrt{5}/2$
2531 1934 10^x : $1/2+\pi/3$
2531 2440 Ψ : $9(e+\pi)/2$
2531 2905 rt : (5,5,4,7)
2531 2912 e^x : $4+4e$
2531 4802 2^x : $e/2+3\pi/2$
2531 5597 rt : (6,3,-6,-9)
2531 8771 Ei : $4\sqrt{2}/5$
2531 9310 as : $5\sqrt[3]{5}/9$
2531 9924 2^x : $(e+\pi)/5$
2532 0205 $s\pi$: $(\sqrt[3]{5}/2)^{-1/2}$
2532 3589 as : $19/20$
2532 3967 rt : (2,8,-6,-2)
2532 3979 2^x : $1-4\sqrt{5}/3$
2532 5945 Ψ : $7\sqrt[3]{5}/3$
2532 6314 rt : (5,9,-2,7)
2532 7400 J_2 : $8\sqrt[3]{2}$
2532 8981 rt : (1,8,-1,9)
2533 0279 2^x : $(3\pi)^{-1/2}$
2533 0809 rt : (3,4,-4,5)
2533 1320 id : $3e/4-\pi/4$
2533 1372 sr : $1+3e/2$
2533 1413 id : $(\pi/2)^{1/2}$
2533 2600 $t\pi$: $3\sqrt{3}/2-3\sqrt{5}/4$
2533 2804 rt : (6,9,-6,-2)
2533 4338 id : $2\sqrt{2}/3-3\sqrt{3}$
2534 1631 rt : (6,2,7,-2)
2534 1803 rt : (1,5,2,-5)
2534 2149 rt : (5,5,-3,-1)
2534 2640 $3/5-\ln(\sqrt{2})$

2534 3150 $\ln(5) \div \exp(1/4)$
2534 4897 $J : 1/\sqrt{11}$
2534 7970 $\ell_2 : 2\sqrt{2} + 3\sqrt{5}$
2534 9963 $Ei : (\sqrt{2}\pi/2)^{-1/3}$
2535 1485 rt : (2,3,7,8)
2535 2225 $J_2 : \pi^2$
2535 2269 sinh : $\sqrt[3]{10}$
2535 2832 $e^x : \sqrt{2}/2 + 3\sqrt{3}/2$
2535 3436 sr : $10\sqrt{2}/9$
2535 3478 $e^x : 13/16$
2535 3875 $\Psi : 3(e + \pi)/10$
2535 5150 rt : (9,3,-4,8)
2535 5591 rt : (2,7,-5,6)
2535 6523 ln : $\sqrt{3}/3 + 4\sqrt{5}$
2535 6546 $\theta_3 : 3(1/\pi)$
2535 6634 sr : 11/7
2535 8791 $e^x : 2/3 - 3e/4$
2536 1327 rt : (7,2,4,1)
2536 1852 rt : (1,5,0,7)
2536 1940 rt : (2,4,-7,-2)
2536 2013 $Ei : \sqrt{12}\pi$
2536 2557 rt : (9,6,-3,-1)
2536 3125 $J_1 : 3e$
2536 3534 $J_0 : 9\sqrt{3}/8$
2536 5834 rt : (8,7,-3,1)
2536 7298 id : $4\sqrt{3}/3 + 4\sqrt{5}$
2536 7740 rt : (1,8,5,-3)
2536 8940 $J_0 : 4\sqrt{5}/3$
2536 9033 $\pi \times \exp(5)$
2536 9423 $J_2 : 9\sqrt{5}/2$
2536 9959 at : $2 + \pi/3$
2537 0285 $t\pi : 2\sqrt{2} + 3\sqrt{3}/2$
2537 0395 rt : (5,9,-6,-2)
2537 1829 $Ei : 3\ln 2$
2537 3054 rt : (6,5,-7,-2)
2537 3879 rt : (7,4,-2,5)
2537 4181 sq : $3e\pi/5$
2537 5117 sinh : $5\sqrt[3]{2}/6$
2537 5591 $s\pi : \sqrt{13/3}$
2537 7167 $e^x : 3/4 - 3\sqrt{2}/2$
2537 7549 sr : $4\sqrt{2} - \sqrt{3}/3$
2537 9452 rt : (6,1,-1,9)
2538 1384 ln : $e + \pi/4$
2538 1872 $Ei : 9\pi/8$
2538 2027 $s\pi : 3\sqrt[3]{3}/4$
2538 3378 $J_2 : \sqrt[3]{4}$
2538 4157 rt : (6,2,8,2)
2538 4190 $\Psi : (2e)^{1/3}$
2538 5668 sinh : 21/20
2538 8752 cr : $3/2 + \sqrt{2}/3$
2539 0401 rt : (5,1,-9,-3)
2539 1186 at : $\sqrt{2} - 2\sqrt{3}/3$
2539 1819 $E : 4\sqrt[3]{3}/7$
2539 3390 $10^x : 6/17$
2539 3406 $e^x : (e + \pi)^{1/2}$
2539 4301 $\exp(\pi) + \exp(\sqrt{2})$
2539 5138 rt : (4,2,-8,-1)
2539 6033 $t\pi : 2/7$
2539 6432 sq : $3e/4 + 2\pi$
2539 6723 rt : (2,1,8,2)
2539 7424 $10^x : 2/3 + 4\pi$
2539 7714 cr : $\sqrt[3]{23/3}$
2539 8043 rt : (3,5,-7,-5)
2539 8416 sq : $2\sqrt[3]{2}/5$
2539 8658 $t\pi : 5\ln 2$
2539 9064 $Ei : 9e\pi/8$
2539 9593 $\ell_{10} : 3/2 - 2\sqrt{2}/3$
2540 0007 rt : (2,8,-6,-9)
2540 0489 $Li_2 : 5/21$
2540 3330 sq : $\sqrt{3} - \sqrt{5}$
2540 4185 $10^x : 11/12$
2540 5013 $2^x : \exp(1/(2\pi))$
2540 6971 $s\pi : \sqrt{2} - \sqrt{3} + \sqrt{5}$
2540 7345 $\ell_2 : 3\sqrt{5}/8$
2540 7347 $s\pi : 4\sqrt{2}/3 + 3\sqrt{3}$
2540 7916 rt : (9,3,-8,3)
2540 8195 rt : (8,0,-7,7)
2540 9614 $e^x : \sqrt{3}/2 - \sqrt{5}$
2540 9752 sr : $2/3 + e/3$
2541 0168 rt : (2,6,-2,-8)
2541 1182 $\ell_{10} : 4\pi/7$
2541 1236 $J_0 : \exp(2/3)$
2541 1312 $\lambda : \exp(-4)$
2541 1821 rt : (1,9,-3,-6)
2541 2853 at : $1/3 + e$
2541 3997 $\zeta : 4\ln 2$
2541 4456 $J_2 : 7\sqrt{2}$
2541 4941 rt : (8,9,9,-3)
2541 5034 $J_1 : 10/19$
2541 6517 $1 + s\pi(\sqrt{3})$
2541 6545 rt : (9,9,5,-2)
2541 8173 $2^x : 1/3 - 4\sqrt{3}/3$
2542 0020 $t\pi : 3\sqrt{2}/4 - 4\sqrt{5}/3$
2542 2156 rt : (4,2,1,6)
2542 3023 tanh : $3\ln 2/8$
2542 3095 $\Psi : \sqrt{5}/9$
2542 3916 rt : (9,7,1,8)
2542 4280 sq : $8\sqrt[3]{2}/9$
2542 4566 sr : $\sqrt{2}/2 + \sqrt{3}/2$
2542 4590 rt : (6,1,-8,-2)
2542 6283 rt : (5,5,0,2)
2542 7455 cr : $\sqrt{2} + \sqrt{5}/4$
2543 0267 at : $3\ln 2/8$
2543 0579 $\Gamma : 15/2$
2543 0825 cr : $2\sqrt{3} - 2\sqrt{5}/3$
2543 1869 $s\pi : 9\zeta(3)/10$
2543 3429 $2^x : \exp(-\sqrt{5}/2)$
2543 4743 rt : (6,8,-1,-2)
2543 4944 $J_2 : 4\pi^2/9$
2543 6309 $\ell_2 : 1/2 + 4\sqrt{2}/3$
2543 7446 rt : (3,1,-9,9)
2543 7949 $2^x : 23/6$
2543 8995 rt : (6,6,9,2)
2543 9721 $erf : 1/\sqrt{19}$
2544 0103 rt : (5,1,-4,-1)
2544 0125 id : $\sqrt{3}/3 + 3\sqrt{5}/4$
2544 1099 rt : (4,7,-1,8)
2544 2080 cr : $\pi^2/5$
2544 2823 $\Gamma : (\sqrt{5}/3)^{-3}$
2544 3293 at : $4\sqrt{3}/3 + \sqrt{5}/3$
2544 5143 $10^x : 3/2 - 2\pi/3$
2544 5191 $1/3 - s\pi(1/5)$
2544 5880 rt : (5,2,-4,-8)
2544 6557 $\lambda : \exp(-14\pi/11)$
2544 8547 $\zeta : 8\sqrt{3}/5$
2545 0688 $E : 14/17$
2545 2387 $t\pi : 1/\ln(\ln 3)$
2545 2409 rt : (6,1,-5,4)
2545 5577 $e^x : 1/4 + 3\sqrt{3}/2$
2545 5599 rt : (2,5,9,2)
2545 5734 $J_2 : 7\sqrt{3}/2$
2545 7698 rt : (9,5,6,-2)
2545 8686 $\ln(2) - \exp(2/3)$
2545 9529 $J_2 : \sqrt{10}\pi$
2545 9581 $\ell_{10} : 3/4 + \pi/3$
2546 1856 rt : (7,4,-6,0)
2546 3031 $J_2 : 10$
2546 3679 cu : $3/2 - 2\sqrt{5}$
2546 4239 $t\pi : 8\sqrt[3]{2}$
2546 4400 id : $1 - \sqrt{5}/3$
2546 4790 id : $4(1/\pi)/5$
2546 5935 sinh : $\sqrt[3]{2}/5$
2546 6833 at : $4 - 2\sqrt{2}/3$
2546 7194 $2^x : 2e + 3\pi$
2546 9458 $\zeta : 23/20$
2546 9529 rt : (5,9,9,7)
2547 0095 $e^x : 11/6$
2547 2180 rt : (7,6,5,1)
2547 2599 sinh : $(e/3)^{-1/2}$
2547 2674 $J_2 : 9\sqrt{2}/8$
2547 3008 as : $\sqrt[3]{2}/5$
2547 3016 rt : (1,9,7,7)
2547 3277 $Li_2 : 3(1/\pi)/4$
2547 3563 rt : (2,4,-9,2)
2547 3901 at : $\sqrt{2} - 2\sqrt{5}$
2547 3908 $J_2 : 7e\pi/6$
2547 4962 rt : (9,2,-4,-1)
2547 5299 at : $9e/8$
2547 5527 as : $\sqrt{3}/2 - \sqrt{5}/2$
2547 5685 rt : (6,5,4,9)
2547 6741 rt : (8,7,-7,-4)
2547 8034 id : $7(e + \pi)/4$
2547 8601 $Li_2 : 7\sqrt[3]{2}/10$
2547 8745 cr : $3e/4 + 3\pi$
2548 0797 at : $1/\ln(\sqrt[3]{3}/2)$
2548 0947 cu : $3/2 - \sqrt{3}/2$
2548 2011 rt : (1,7,6,-2)
2548 2166 at : $4\sqrt{2} - 3\sqrt{3}/2$
2548 2712 at : $\exp(\sqrt{5}/2)$
2548 3399 $2^x : 11/2$
2548 4303 $\pi + \exp(\sqrt{2})$
2548 5367 $2^x : \sqrt{3} - \sqrt{5}/4$
2548 6184 rt : (9,0,-7,9)
2548 6584 $J_2 : 5(1/\pi)$
2548 7766 rt : (8,4,-2,7)
2548 8181 at : $3/4 + 4\sqrt{3}/3$
2548 8709 $t\pi : 3\ln 2$
2548 8750 $\ell_2 : (\sqrt{2}/3)^3$
2549 0159 $2^x : 7\sqrt[3]{2}/2$
2549 1280 rt : (7,8,3,5)
2549 3046 id : $4\sqrt{2} + 3\sqrt{3}/2$
2549 3067 rt : (6,3,-9,2)
2549 4654 $t\pi : \sqrt{2}/4 - \sqrt{3}/4$
2549 5072 sr : $5\sqrt[3]{2}/4$
2549 5198 rt : (3,8,-6,-2)
2549 7365 $\theta_3 : 7\ln 2/9$
2549 7442 $e^x : 2\sqrt{2}/3 - 4\sqrt{3}/3$
2549 7710 $t\pi : 2e\pi$
2549 9401 rt : (3,7,-9,0)
2549 9459 $\ln(2) \div \exp(1)$
2550 0330 at : $2 + 3\sqrt{2}/4$
2550 1157 cu : $3/2 + 4e$
2550 1218 rt : (9,7,-3,3)
2550 2893 $s\pi : \exp(-5/2)$
2550 4036 $\ell_{10} : 1/2 + 3\sqrt{3}/4$
2550 5805 rt : (6,6,2,-1)
2550 8014 rt : (9,1,7,-2)
2550 8150 $\exp(2) + s\pi(1/3)$
2550 8583 rt : (2,8,7,7)
2550 9156 rt : (7,9,-6,-2)
2550 9600 $t\pi : 4\sqrt{2} - \sqrt{3}/3$
2551 0445 sq : P_{35}
2551 0868 tanh : 6/23
2551 1262 rt : (6,9,-8,5)
2551 2082 $J_0 : 19/4$
2551 2302 $\zeta : \sqrt{23/3}$
2551 2549 rt : (3,5,5,1)
2551 3125 cu : $1/3 + \sqrt{5}/3$
2551 3288 rt : (4,7,-5,-1)
2551 4028 ln : $2\sqrt{2} + 3\sqrt{5}$
2551 4925 $Ei : \ln 2/5$
2551 7214 $e^x : 5/22$
2551 7354 $\theta_3 : 2(1/\pi)/5$
2551 8239 at : 6/23
2552 2709 rt : (5,2,7,-2)
2552 4529 $\ell_{10} : 10\sqrt[3]{2}/7$
2552 4770 rt : (7,9,9,-3)
2552 5004 sr : $\sqrt{3} + 3\sqrt{5}/2$
2552 7250 $\ell_{10} : 18$
2552 7878 rt : (2,7,2,-1)
2552 8016 ln : $\sqrt[3]{5}/6$
2552 8450 $t\pi : 1/3 - 4\sqrt{3}$
2552 8694 cu : $1/4 - \sqrt{3}$
2552 8696 $e^x : 5(1/\pi)/7$
2552 8905 $e^x : 1/\ln(\sqrt[3]{3}/3)$
2552 9493 id : $3\sqrt{2}/2 - \sqrt{3}/2$
2553 2287 rt : (5,2,-8,-3)
2553 4078 rt : (3,8,5,5)
2553 4192 $t\pi : (1/\pi)/4$
2553 4329 $\ell_{10} : 2e + 4\pi$
2553 5579 $e^x : \exp(-\sqrt{2}\pi/3)$
2553 6789 sq : $3e\pi/8$
2553 8315 rt : (1,3,7,-2)
2553 9952 $s\pi : \sqrt{19/2}$
2554 1281 ln : $\sqrt{5/3}$
2554 2309 $J_0 : \sqrt{3} - \sqrt{5} + \sqrt{6}$
2554 3505 $Ei : 6\sqrt{3}/7$
2554 3909 $\ln(4/5) \times \ln(\pi)$
2554 4921 $t\pi : 1/4 + 2\sqrt{3}$
2554 5831 at : $4/3 + \sqrt{3}$

2554 7257 10^x : $4/3+\sqrt{3}/2$
2554 8144 sq : $2\sqrt{2}/3+\sqrt{5}/4$
2554 8582 id : $5(e+\pi)/9$
2555 0406 Γ : $8e/9$
2555 2040 rt : (2,8,-2,-1)
2555 2505 rt : (7,1,-5,6)
2555 3408 ζ : $-\sqrt{2}+\sqrt{3}+\sqrt{6}$
2555 3778 sq : $9(e+\pi)/5$
2555 3968 J_1 : 9/17
2555 4189 rt : (2,8,-1,5)
2555 4922 rt : (3,4,-7,-2)
2555 5720 2^x : $3/4-e$
2555 9850 rt : (6,1,-9,-1)
2555 9870 cr : $8\sqrt{3}/7$
2555 9994 rt : (5,2,1,8)
2556 0874 rt : (1,4,-5,1)
2556 3833 rt : (8,9,5,-2)
2556 4207 2^x : $4e/3+\pi/4$
2556 5269 10^x : $\sqrt{2}+4\sqrt{3}$
2556 6193 rt : (9,4,-2,9)
2556 6876 tπ : $4e+\pi/3$
2556 8588 cr : $7\sqrt{2}/5$
2557 1005 rt : (9,3,8,2)
2557 2635 rt : (7,5,-7,-2)
2557 7357 rt : (8,4,-6,2)
2557 7478 ln : $4/3-\sqrt{5}/4$
2557 7551 Li_2 : 15/17
2557 8614 rt : (6,9,-2,-1)
2557 9581 rt : (2,5,-4,3)
2558 3130 2^x : $3/2+e/2$
2558 3884 $\exp(4)\times s\pi(e)$
2558 4571 erf : 3/13
2558 5187 at : $\sqrt{2}/3+3\sqrt{3}/2$
2558 9771 rt : (5,2,8,2)
2558 9905 10^x : $\sqrt{2}/2-3\sqrt{3}/4$
2558 9978 at : $(3\pi)^{1/2}$
2559 0057 cu : $4\sqrt{2}-2\sqrt{3}/3$
2559 0582 tπ : $10\sqrt{2}/3$
2559 0638 J_1 : $5e\pi/7$
2559 1081 rt : (5,5,-7,-9)
2559 1491 J_1 : $\sqrt{2}+\sqrt{5}+\sqrt{6}$
2559 2606 sr : $1+\sqrt{3}/3$
2559 4022 10^x : 22/9
2559 5043 ℓ_{10} : $5\sqrt[3]{3}/4$
2559 5144 J_1 : $3\sqrt{2}/8$
2559 6013 rt : (6,5,0,4)
2559 6047 Ψ : 11/16
2559 6277 ℓ_{10} : $(e+\pi)^{1/3}$
2559 7805 at : $3/2+\pi/2$
2559 9250 rt : (1,7,-4,-8)
2559 9306 2^x : $-\sqrt{5}+\sqrt{6}+\sqrt{7}$
2560 0382 2^x : $7\sqrt[3]{5}/5$
2560 0703 cu : $3e/2-3\pi/2$
2560 1628 ℓ_{10} : $3\zeta(3)/2$
2560 3499 J_1 : $5(1/\pi)/3$
2560 5086 rt : (6,2,3,-1)
2560 5098 e^x : $(2\pi/3)^{-2}$
2560 5808 rt : (8,8,3,7)
2560 7314 rt : (3,5,0,9)
2560 8342 e^x : $4e-3\pi$
2560 8455 ℓ_{10} : $4\ln 2/5$
2560 9077 10^x : $\ln 2/7$
2560 9874 rt : (1,1,8,2)
2561 0160 cu : $3\sqrt{2}/2+\sqrt{5}/4$
2561 0284 2^x : $4/3+3\sqrt{2}/4$
2561 1609 ln : $(\sqrt{2}/3)^3$
2561 1766 Ψ : 4
2561 2685 rt : (9,7,-7,-2)
2561 4062 cr : $1/4+\sqrt{3}$
2561 4507 erf : $\ln 2/3$
2561 4905 10^x : $9\sqrt[3]{3}/2$
2561 8158 rt : (6,2,-8,-5)
2561 8166 Li_2 : 6/25
2561 8968 Ei : $\sqrt[3]{9}$
2561 9925 rt : (5,3,-5,1)
2562 0742 rt : (8,5,6,-2)
2562 1262 rt : (8,1,-5,8)
2562 2141 erf : $(\sqrt[3]{3}/3)^2$
2562 3016 E : $(\sqrt{2}\pi/3)^{-1/2}$
2562 3264 rt : (5,7,-7,6)
2562 5146 Ei : $5\sqrt{5}/3$
2562 5804 tanh : $((e+\pi)/3)^{-2}$
2562 6919 e^x : $2+4\pi/3$
2562 6925 J_2 : $5\sqrt{5}/7$
2562 7705 $2/5\times\exp(\pi)$
2562 9875 2^x : $3/2-2\sqrt{3}$
2563 1468 rt : (3,9,1,-7)
2563 3343 at : $((e+\pi)/3)^{-2}$
2563 4588 rt : (2,9,-7,5)
2563 4694 cr : $3e/2-2\pi/3$
2563 4933 e^x : $4+\sqrt{3}/3$
2563 6454 at : $3\sqrt{2}/2-3\sqrt{3}$
2563 7266 rt : (6,9,9,9)
2563 8580 sπ : $5\sqrt{3}/8$
2564 1292 $\ln(\sqrt{3})+s\pi(1/4)$
2564 1412 ℓ_{10} : $4/3+\sqrt{2}/3$
2564 3078 rt : (9,9,1,-1)
2564 5877 rt : (4,1,-2,-7)
2564 8515 ℓ_2 : $(e+\pi)/7$
2564 8802 rt : (2,8,3,-6)
2565 1493 rt : (9,4,-6,4)
2565 2287 rt : (9,7,9,2)
2565 2476 e^x : $\sqrt{2}-\sqrt{3}+\sqrt{7}$
2565 2976 tπ : $6\sqrt[3]{2}/7$
2565 3600 ζ : $4(1/\pi)$
2565 4736 rt : (7,8,-1,0)
2565 5283 J_1 : 23/5
2565 6479 cr : $\sqrt{3}/2+\sqrt{5}/2$
2565 7951 ζ : 20/17
2566 0264 sq : $e/2+\pi$
2566 0708 rt : (3,8,-3,9)
2566 1044 Li_2 : $\sqrt[3]{3}/6$
2566 1076 $3\times\exp(3)$
2566 1215 at : $1-3e/2$
2566 2012 rt : (1,7,-2,-1)
2566 2124 $\exp(1/3)-\exp(\sqrt{3})$
2566 3706 id : $2\pi/5$
2566 5213 Li_2 : $\zeta(3)/5$
2566 6045 $\ln(3/5)+s\pi(\sqrt{3})$
2566 6314 at : $9\sqrt[3]{5}/5$
2566 6694 Γ : $(\sqrt[3]{4}/3)^{1/2}$
2566 6840 rt : (5,5,-4,-3)
2566 7518 10^x : $1/\ln(1/(2e))$
2567 0805 at : $1/4+2\sqrt{2}$
2567 0855 $1/2\div\exp(2/3)$
2567 2163 ℓ_{10} : $3\sqrt{2}/4+\sqrt{5}/3$
2567 2220 Γ : $\exp(-1/\pi)$
2567 2550 10^x : $6(1/\pi)$
2567 3238 rt : (4,8,-6,-2)
2567 3950 ℓ_{10} : $3\sqrt{3}/4-\sqrt{5}/3$
2567 4521 sinh : $7\zeta(3)/8$
2567 4593 2^x : $2\sqrt{3}/3+2\sqrt{5}/3$
2567 5636 tπ : 2/25
2567 6695 J_0 : $5e\pi/9$
2567 6991 rt : (7,2,-8,-7)
2567 7140 rt : (5,6,9,2)
2567 7676 rt : (9,8,3,9)
2567 8630 cr : $7\pi^2/2$
2567 9575 $s\pi(1/4)+s\pi(\sqrt{2})$
2568 0508 $1/5\times\exp(1/4)$
2568 1151 Γ : $7\sqrt{2}/2$
2568 1259 id : $3\sqrt{2}/4+3\sqrt{3}$
2568 1930 Ei : 1/18
2568 5709 Γ : $7\pi^2/10$
2568 6465 cu : $2\sqrt{2}-2\sqrt{3}$
2568 7274 Γ : 8/11
2568 7353 rt : (7,5,0,6)
2568 8136 id : $\exp(-e/2)$
2568 8251 ln : $2-\sqrt{2}/2$
2568 8982 rt : (1,8,-2,1)
2568 9331 at : $2e-3\pi/4$
2569 0704 rt : (8,1,7,-2)
2569 1842 E : $(e/3)^2$
2569 2873 rt : (8,9,-6,-2)
2569 3612 ℓ_2 : $\sqrt{2}-\sqrt{3}/3$
2569 4031 rt : (9,2,-5,1)
2569 5641 tπ : $(\sqrt[3]{2})^{1/3}$
2569 5756 J_0 : $\sqrt[3]{22/3}$
2569 6775 J_2 : 8/5
2569 7426 J_1 : $\exp(-\sqrt[3]{2}/2)$
2569 8195 rt : (2,0,4,9)
2570 2397 at : $\sqrt{2}-3\sqrt{5}/4$
2570 3126 rt : (5,2,-3,3)
2570 3707 tπ : $\sqrt[3]{9}$
2570 4079 rt : (1,5,9,2)
2570 6120 2^x : $2e/3+\pi/3$
2570 6844 at : $\sqrt{19/2}$
2570 7667 sr : $3+2\pi/3$
2570 7872 id : $8\sqrt{2}/9$
2571 0721 2^x : $4/3+3\sqrt{3}/2$
2571 1345 10^x : $\sqrt{2}/4$
2571 1377 rt : (4,2,7,-2)
2571 3753 rt : (6,5,-3,-1)
2571 3804 rt : (5,1,-2,-9)
2571 5433 rt : (8,1,-9,3)
2571 5767 rt : (7,9,5,-2)
2571 7362 10^x : $2+\sqrt{5}/4$
2571 7541 e^x : $4\sqrt{2}/3+3\sqrt{5}/4$
2571 7871 id : $\sqrt[3]{23/2}$
2571 8500 sr : $e/3+4\pi/3$
2571 9823 rt : (8,2,-8,-9)
2572 0532 K : $2\sqrt{5}/5$
2572 1588 J_1 : $\sqrt{8}\pi$
2572 4683 tanh : 5/19
2572 9689 J_1 : 8/15
2573 0256 ln : $3(e+\pi)/5$
2573 1112 $5\div s\pi(2/5)$
2573 1518 cr : $8\sqrt{5}/9$
2573 2371 at : 5/19
2573 3395 rt : (3,8,1,-4)
2573 3427 sq : $4\sqrt{2}/3-3\sqrt{5}$
2573 5931 sq : $3-\sqrt{2}/2$
2573 7232 tπ : $1/3-\pi/3$
2573 9169 rt : (9,8,9,-3)
2573 9881 ℓ_{10} : $2\sqrt{2}/3+\sqrt{3}/2$
2573 9967 $\exp(\sqrt{2})\div\exp(3/5)$
2574 0897 sinh : $4(1/\pi)/5$
2574 1109 e^x : 19/9
2574 2062 at : $3e/4+\pi/3$
2574 2597 Ei : $(\ln 2/3)^{-1/2}$
2574 3342 sr : $\sqrt{5/2}$
2574 3514 $\exp(5)\div\exp(\sqrt{3})$
2574 4444 rt : (8,5,0,8)
2574 6920 rt : (9,5,2,-1)
2574 7955 as : $1-\sqrt{5}/3$
2574 8359 as : $4(1/\pi)/5$
2574 8414 J_0 : $7\sqrt[3]{5}/4$
2574 9738 J_1 : $9\pi^2/10$
2575 0022 $\exp(1/5)+s\pi(\sqrt{2})$
2575 0054 rt : (8,8,-1,2)
2575 1259 10^x : $(\sqrt[3]{2})^{1/2}$
2575 2552 J_2 : $\exp(\sqrt{2}/3)$
2575 2559 rt : (5,9,9,-3)
2575 4067 rt : (6,2,4,1)
2575 5869 2^x : $\sqrt{3}/3+3\sqrt{5}/2$
2575 6902 rt : (4,4,-7,-2)
2575 6989 sq : $3+e/3$
2575 7609 rt : (9,1,-9,5)
2575 8931 J_2 : $10\sqrt[3]{3}/9$
2576 0777 Ψ : $(\sqrt{2}\pi)^{-3}$
2576 1169 e^x : $1/4+\sqrt{3}$
2576 1353 at : $3\sqrt{2}-2\sqrt{3}/3$
2576 1431 $\ln(3)\times\ln(\pi)$
2576 2417 $\exp(\pi)\div s\pi(\sqrt{5})$
2576 3936 e^x : $1-3\pi/4$
2576 4974 J_2 : $4\zeta(3)/3$
2576 5079 J_0 : $5e/7$
2576 5438 rt : (7,3,-9,2)
2576 5696 tanh : $\exp(-4/3)$
2576 6293 E : $(2e/3)^{-1/3}$
2576 6397 $\ln(4/5)\div s\pi(1/3)$
2576 6470 rt : (1,7,-8,-3)
2576 9264 $\exp(3/4)+\exp(\pi)$
2577 0097 cu : 7/11
2577 0339 rt : (6,2,-3,5)
2577 0610 J_1 : $4\zeta(3)/9$
2577 1131 rt : (4,5,-6,7)
2577 1340 10^x : $1/3+e/2$
2577 1999 sq : $5\zeta(3)/4$
2577 2465 rt : (2,9,1,-9)

2577 3446 at : $\exp(-4/3)$
2577 4183 Ψ : $4\zeta(3)/7$
2577 5713 sinh : $4e/7$
2577 5935 10^x : $10(1/\pi)/9$
2577 6929 2^x : $\sqrt{2}/4-4\sqrt{3}/3$
2577 7280 rt : (3,9,6,1)
2577 8066 ζ : $\sqrt[3]{21}$
2577 8912 tπ : $3\sqrt{2}+3\sqrt{5}/4$
2577 8927 rt : (8,3,8,2)
2577 9457 rt : (5,9,5,2)
2577 9771 10^x : $3+4\sqrt{3}$
2578 2062 rt : (8,5,-7,-2)
2578 2910 ln : 17/22
2578 3694 ℓ_{10} : $3/4+3\sqrt{2}/4$
2578 3906 sr : $4\sqrt{2}-\sqrt{5}/4$
2578 4541 rt : (4,8,-1,3)
2578 5484 J_0 : $6\sqrt[3]{2}$
2578 5672 $\ln(3)-\exp(\sqrt{5})$
2578 5989 θ_3 : $3\sqrt[3]{2}/7$
2578 6486 e^x : $1/\sqrt{19}$
2578 6559 ζ : $\sqrt{2}-\sqrt{6}+\sqrt{7}$
2578 6807 θ_3 : $(\sqrt{2})^{-1/2}$
2578 8389 cu : $\ln(\sqrt[3]{4}/3)$
2578 9630 rt : (7,5,6,-2)
2578 9701 rt : (1,3,-7,-7)
2579 0933 $\ln(2)-s\pi(2/5)$
2579 2877 at : $\sqrt{2}+3\sqrt{5}/4$
2579 4535 e^x : $1+2e$
2579 6766 cr : $1/2+2\sqrt{5}/3$
2579 6866 rt : (6,5,-4,-1)
2579 8026 $\ln(\pi)+\exp(\sqrt{2})$
2579 8089 $1/3\times s\pi(e)$
2579 9201 ℓ_2 : $(\sqrt[3]{5})^{1/3}$
2579 9295 tπ : $2e-3\pi/4$
2580 1227 id : $(\pi/2)^{-3}$
2580 2329 sπ : $2\sqrt{2}-\sqrt{5}/3$
2580 3157 Ψ : $7\sqrt[3]{2}/5$
2580 3344 rt : (9,8,-1,4)
2580 3394 rt : (2,9,3,4)
2580 4428 10^x : $7\zeta(3)/6$
2580 5587 rt : (4,2,8,2)
2580 6177 Ψ : $10\zeta(3)/3$
2580 6661 sq : $\sqrt{3}/2-4\sqrt{5}$
2580 8154 sinh : 20/19
2581 1298 rt : (3,6,6,-2)
2581 1315 tπ : $(\ln 2/3)^{-1/2}$
2581 2177 rt : (7,2,-3,7)
2581 2327 Ψ : $\sqrt{3/2}$
2581 2727 rt : (4,1,-6,-2)
2581 2805 id : P_{28}
2581 2950 cr : $7\pi^2/6$
2581 3707 sq : $10\pi^2/9$
2581 4706 cr : $\sqrt{2}+\sqrt{3}/3$
2581 5021 e^x : $\sqrt{2}/4+4\sqrt{5}$
2581 5112 rt : (1,7,2,-1)
2581 5918 $\ln(5)+\exp(1/2)$
2581 7882 at : $4-e/3$
2581 7904 e^x : $2-3\sqrt{5}/2$
2581 8073 $3/5\times s\pi(\pi)$
2581 9888 id : $1/\sqrt{15}$
2582 0322 ℓ_{10} : $2e/3$
2582 0568 ln : $3\sqrt{3}-3\sqrt{5}/4$
2582 1538 2^x : $\sqrt{7}$
2582 2507 at : $1+2\pi/3$
2582 2610 rt : (2,7,-8,-5)
2582 4224 cu : $7\sqrt{3}/5$
2582 5888 rt : (1,3,1,6)
2582 8371 Ψ : $9\sqrt{5}/2$
2583 0274 Ei : $10e/3$
2583 0573 sr : 19/12
2583 1287 10^x : $3\sqrt[3]{5}/10$
2583 2231 sq : $7\sqrt[3]{3}/9$
2583 4261 rt : (1,8,2,0)
2583 4371 J_1 : $7(e+\pi)/5$
2584 0732 rt : (8,2,-3,9)
2584 1706 $1/5\div s\pi(e)$
2584 5890 2^x : $2\sqrt{3}+\sqrt{5}/4$
2584 6592 K : 17/19
2584 8665 $\ln(\sqrt{2})\times s\pi(\sqrt{3})$
2585 0271 tπ : $4\zeta(3)/9$
2585 2573 rt : (7,5,-4,1)
2585 4384 rt : (3,8,-2,-1)
2585 4814 rt : (5,1,-6,-4)
2585 4976 rt : (8,8,9,-3)
2585 6130 J_0 : $9\sqrt[3]{3}/2$
2585 6239 tπ : $2\sqrt{3}/3+\sqrt{5}/4$
2585 6380 rt : (3,7,-8,-7)
2585 6544 Ei : 17/15
2585 7278 at : $1/2+3\sqrt{3}/2$
2585 8646 rt : (5,8,-6,-2)
2585 9267 $\pi+\exp(3/4)$
2586 2013 Ψ : $(\sqrt{2}/3)^{1/2}$
2586 3735 Ψ : $-\sqrt{3}+\sqrt{5}-\sqrt{6}$
2586 3896 ℓ_2 : $4+\pi/4$
2586 5202 rt : (4,0,-8,2)
2586 6592 λ : $\exp(-19\pi/15)$
2587 0442 J_1 : $\ln(\sqrt[3]{5})$
2587 0788 rt : (4,9,9,-3)
2587 2639 rt : (6,9,5,4)
2587 3308 θ_3 : $(e/3)^3$
2587 5351 rt : (6,8,-9,8)
2587 6324 rt : (8,9,1,-1)
2587 8567 rt : (1,7,4,-4)
2587 8930 rt : (4,7,-8,-9)
2588 0902 rt : (6,1,-6,-6)
2588 0966 rt : (7,1,7,-2)
2588 1038 Ψ : $1/\ln(3\pi)$
2588 1904 sr : $1/2-\sqrt{3}/4$
2588 3537 rt : (8,5,-4,3)
2588 4526 rt : (1,9,6,-1)
2588 4580 E : $(\ln 3/2)^{1/3}$
2588 4597 rt : (9,9,-6,-2)
2588 6433 cr : $7\sqrt[3]{5}/6$
2588 8417 e^x : $2\sqrt{3}+4\sqrt{5}$
2588 9135 id : $1/\ln(e/2)$
2588 9686 $\exp(5)-\exp(e)$
2589 1614 rt : (9,1,3,-1)
2589 2541 10^x : 1/10
2589 2547 sr : $3\sqrt{2}/2+4\sqrt{5}/3$
2589 3123 sπ : $6\sqrt{5}/7$
2589 3917 rt : (8,7,9,2)
2589 4727 rt : (2,4,-9,1)
2589 5054 rt : (5,8,-7,1)
2589 5138 at : $\sqrt{3}/2+\sqrt{5}$
2589 7710 e^x : $4\sqrt{2}+\sqrt{5}$
2589 7884 rt : (7,9,-2,-1)
2589 8658 rt : (7,1,-6,-8)
2589 8709 rt : (8,1,-8,-2)
2590 0283 Ψ : $\sqrt[3]{11/2}$
2590 0822 rt : (5,3,3,9)
2590 1328 rt : (1,8,-6,1)
2590 3251 rt : (9,5,-4,5)
2590 4174 e^x : $2\sqrt{2}-3\sqrt{3}/2$
2590 7494 sq : $3\sqrt{3}/4-3\sqrt{5}$
2590 8287 rt : (3,2,7,-2)
2590 8343 Li_2 : $1/\sqrt{17}$
2590 9024 rt : (7,9,5,6)
2590 9934 rt : (4,2,-7,-4)
2591 3454 $\exp(\pi)\times s\pi(\sqrt{3})$
2591 3737 J_1 : $8\zeta(3)/3$
2591 4035 at : $1/4-3\sqrt{5}/2$
2591 7867 rt : (3,9,-1,5)
2592 0712 tanh : $5(1/\pi)/6$
2592 0838 rt : (1,5,-9,-7)
2592 2461 Γ : $(e+\pi)^{1/2}$
2592 2704 sq : $3/2+4e/3$
2592 4307 $\pi\div s\pi(\sqrt{2})$
2592 4932 rt : (3,5,-8,-9)
2592 5925 cu : $5\sqrt[3]{2}/3$
2592 6007 ζ : 17/14
2592 6161 10^x : P_{18}
2592 8012 sr : $3-\sqrt{2}$
2592 8067 tπ : $\sqrt{3}/4-4\sqrt{5}$
2592 8417 rt : (8,9,5,8)
2592 8700 at : $5(1/\pi)/6$
2592 9719 rt : (4,6,9,2)
2593 1430 J_2 : $\ln(5)$
2593 1665 e^x : $9\sqrt{2}$
2593 2423 Ei : $5\sqrt[3]{3}/4$
2593 2504 10^x : P_{48}
2593 3698 at : $3/4+3\pi/4$
2593 4448 as : $(e/3)^{1/2}$
2593 5140 rt : (6,1,-4,-1)
2593 5572 sq : $1/4+4\pi$
2593 5582 J : $1/\sqrt[3]{17}$
2593 6232 $1/4\div s\pi(\sqrt{2})$
2593 7371 cu : $1/2-2e/3$
2593 7428 sq : $2-2\sqrt{5}/3$
2593 7480 rt : (1,6,2,-5)
2593 7579 at : $8e/7$
2593 8223 sq : $8(1/\pi)/5$
2593 8375 ℓ_{10} : $\sqrt[3]{6}$
2593 9014 rt : (5,2,3,-1)
2594 0285 ℓ_2 : $7\sqrt[3]{5}/5$
2594 0834 rt : (4,8,-5,-6)
2594 1923 ℓ_2 : $4/3+3\sqrt{2}/4$
2594 3218 rt : (5,2,-7,-2)
2594 3370 $\ln(\sqrt{3})\times\exp(\sqrt{2})$
2594 3675 rt : (2,7,-6,2)
2594 7384 $\ln(\sqrt{3})\div\exp(3/4)$
2594 7521 cu : $6\sqrt[3]{2}/7$
2594 9578 Li_2 : $5\sqrt{2}/8$
2594 9662 Ei : $(e+\pi)^{1/3}$
2595 0184 10^x : $3\sqrt{3}/4+\sqrt{5}/2$
2595 0648 rt : (5,4,-9,-5)
2595 1302 id : $\sqrt{2}-2\sqrt{3}/3$
2595 6853 e^x : 3/13
2595 7212 J_2 : $\sqrt{2}-\sqrt{6}+\sqrt{7}$
2595 7308 rt : (6,8,-5,-7)
2595 9026 Ψ : $7(1/\pi)/5$
2595 9031 J_1 : 7/13
2595 9752 E : 9/11
2596 0026 $1/3\div\exp(1/4)$
2596 2125 rt : (7,4,-9,-9)
2596 2971 2^x : $-\sqrt{3}+\sqrt{5}-\sqrt{6}$
2596 3731 ℓ_{10} : 11/20
2596 3869 ℓ_{10} : $\sqrt{3}-3\sqrt{5}/4$
2596 4782 rt : (6,5,6,-2)
2596 5093 rt : (7,2,-7,2)
2596 5150 2^x : $(\ln 2)^3$
2596 5501 sinh : $(\sqrt[3]{5}/2)^{-1/3}$
2596 5946 ℓ_{10} : $9\sqrt{2}/7$
2596 6192 cr : $8\sqrt[3]{3}$
2596 6576 rt : (1,6,-2,-9)
2596 8757 rt : (5,9,2,0)
2597 0031 rt : (8,2,-7,4)
2597 0608 tπ : $1/3-\sqrt{2}$
2597 1588 sinh : $\exp(-e/2)$
2597 1945 rt : (2,6,-2,-1)
2597 3442 rt : (9,2,-7,6)
2597 3710 rt : (7,8,9,-3)
2597 4532 at : $\sqrt{2}/4-2\sqrt{3}$
2597 4910 Li_2 : $\exp(-\sqrt{2})$
2597 5448 rt : (3,3,-5,8)
2597 7591 rt : (7,8,-5,-5)
2597 7926 cr : $2/3+4e$
2597 9389 as : $\exp(-e/2)$
2598 2934 rt : (8,8,-5,-3)
2598 4117 10^x : $4+\sqrt{5}/2$
2598 5100 $\ln(2/5)\div\ln(2/3)$
2598 5402 rt : (9,8,-5,-1)
2598 5568 id : $6\sqrt[3]{5}$
2598 7526 sq : $3\sqrt{2}/4-3\sqrt{5}/2$
2598 8195 J_1 : $7\ln 2/9$
2598 9230 sq : $2\sqrt{2}/3-\sqrt{3}/4$
2599 1089 ln : $\sqrt{2}/3-\sqrt{3}/4$
2599 2104 id : $\sqrt[3]{2}$
2599 2472 2^x : $3e/4+\pi$
2599 3019 id : $3\ln 2/8$
2599 4319 10^x : $\exp(\sqrt[3]{2}/3)$
2599 4596 cr : $3\sqrt{3}/2+4\sqrt{5}$
2599 5965 sπ : $\ln(5/2)$
2599 7048 rt : (7,3,8,2)
2600 1096 e^x : $(\sqrt[3]{3}/3)^2$
2600 2366 rt : (9,5,-7,-2)
2600 4650 rt : (6,6,5,1)
2600 5116 sq : $9\zeta(3)/2$
2600 5195 J_0 : 3
2600 5312 e^x : $3/4+3e/4$
2600 6486 Ψ : $8\sqrt[3]{2}$

2600 7297 $Ei : \sqrt{19}$
2600 9247 rt : (9,8,5,-2)
2600 9460 $J_0 : 13/2$
2600 9479 $e^x : 9\sqrt{3}/4$
2601 0374 rt : (8,6,2,9)
2601 1055 $\Psi : -\sqrt{2} - \sqrt{5} - \sqrt{7}$
2601 1525 rt : (4,4,-9,2)
2601 4676 rt : (7,6,2,7)
2601 5512 $2^x : \sqrt{6}\pi$
2601 5663 $J_1 : 5\pi^2/6$
2601 5821 $t\pi : 9(1/\pi)/10$
2601 6435 $2^x : 23/11$
2601 8554 $\ell_{10} : \ln(\sqrt{3})$
2601 8802 $10^x : 3\sqrt{2}/2 - 2\sqrt{3}/3$
2601 9426 $\zeta : 11/4$
2601 9782 $Ei : 9\sqrt[3]{2}/10$
2601 9967 $10^x : 2\sqrt{2} + \sqrt{3}/2$
2602 0620 $\ln : 1/4 + \pi/3$
2602 0692 $\ln(3/5) \times \exp(3)$
2602 1135 rt : (2,7,-6,-2)
2602 1629 rt : (6,6,2,5)
2602 2361 rt : (8,5,2,-1)
2602 3156 $2^x : 20/17$
2602 3421 rt : (5,6,-8,9)
2602 3584 $sr : 3\sqrt{2} + \sqrt{3}/2$
2602 4429 $Ei : 3\zeta(3)/2$
2602 4501 $\Gamma : (3\pi)^{-2}$
2602 5388 sinh : 16
2602 6136 $\ell_{10} : 1/4 + \pi/2$
2602 6224 $J_1 : 3\sqrt[3]{2}/7$
2602 7061 $e^x : 3e/10$
2602 7527 $\ell_2 : 2\sqrt{2} - \sqrt{3}/4$
2602 9341 $J_2 : 22/3$
2602 9662 $e^x : 13/11$
2603 2661 rt : (3,2,8,2)
2603 2750 rt : (3,9,-5,-8)
2603 4771 rt : (5,9,5,-2)
2603 6776 $10^x : 6\sqrt{5}/5$
2603 7459 $\sinh : 2e\pi/9$
2603 7833 $J_1 : 3e\pi/8$
2603 9751 $\ell_{10} : 2/3 + 2\sqrt{3}/3$
2603 9864 rt : (2,9,-8,0)
2604 0564 $cu : 2 + 2e$
2604 1306 $at : 9\sqrt{3}/5$
2604 1532 $e^x : 2e/3 + 2\pi$
2604 3426 $cu : 7\sqrt{5}/9$
2604 4438 $J : \exp(-19\pi/8)$
2604 4501 $at : 2 + \sqrt{5}/2$
2604 7348 $\Gamma : \pi/8$
2604 9550 $\zeta : 7\pi/8$
2605 2015 $\Gamma : \ln(\sqrt{2}\pi/3)$
2605 2044 tanh : 4/15
2605 2051 rt : (6,8,-6,-2)
2605 2826 $\Gamma : 13/3$
2605 3492 $t\pi : \sqrt{3} + 4\sqrt{5}/3$
2605 5020 $at : 9\ln 2/2$
2605 7087 $2^x : 3\sqrt{3} + 3\sqrt{5}$
2605 7412 $\Gamma : 10\pi^2/7$
2605 8438 $\zeta : \ln(5)$
2605 8635 rt : (8,3,-9,2)
2605 8711 rt : (1,8,-3,-7)
2606 0239 at : 4/15
2606 1309 $2^x : \exp((e + \pi)/2)$
2606 2236 $J_2 : \sqrt{5} + \sqrt{6} + \sqrt{7}$
2606 3660 $2^x : \sqrt{2} - 3\sqrt{5}/2$
2606 4050 $cr : 5\zeta(3)/3$
2606 4628 $id : \sqrt{2}/3 - \sqrt{3}$
2606 5387 $J_1 : 3\sqrt[3]{3}/8$
2606 5588 rt : (9,5,-8,0)
2606 6443 rt : (4,9,3,7)
2606 8162 $erf : 4/17$
2606 8985 rt : (4,6,2,1)
2606 9735 $e^x : 3/4 - 2\pi/3$
2607 1138 rt : (1,1,-8,2)
2607 2935 $sr : \sqrt{2}/3 + \sqrt{5}/2$
2607 5121 $at : 1 + 3\sqrt{2}/2$
2607 6010 $\ell_{10} : P_{12}$
2607 6594 $cu : 2\sqrt{5}/7$
2607 7257 $J_2 : 7\pi/3$
2607 9012 $sq : 3(e + \pi)/2$
2607 9224 $J_2 : 21/13$
2607 9438 rt : (6,1,7,-2)
2608 0766 rt : (5,8,-5,-9)
2608 1091 $sr : 1/\ln(\ln 3)$
2608 1326 rt : (7,3,-1,8)
2608 1674 $2^x : P_7$
2608 2562 $1/4 + \ln(3/5)$
2608 4017 $\theta_3 : 3\sqrt[3]{3}/8$
2608 4358 $\ln : 9\sqrt[3]{3}/10$
2608 5214 rt : (3,6,0,4)
2608 5551 rt : (8,5,-8,-2)
2608 6494 $\Gamma : 9\zeta(3)/2$
2608 6956 id : 6/23
2608 7099 $10^x : (\ln 2/3)^{-2}$
2608 7402 $e^x : 1/3 - 3\sqrt{5}/4$
2608 7588 $J_1 : (e/2)^{-2}$
2608 7793 $10^x : 2\sqrt{3} + 3\sqrt{5}/4$
2608 8447 $\sinh : (\pi/2)^{-3}$
2608 9346 $2/5 \times \exp(\sqrt{3})$
2608 9991 $\ln : 4 - \sqrt{2}/3$
2609 2678 rt : (1,9,5,-6)
2609 2706 $t\pi : (\pi/2)^{-2}$
2609 4281 $sq : \ln(5/3)$
2609 4585 $s\pi : \beta(2)$
2609 4646 $\tanh : 2\zeta(3)/9$
2609 5168 as : 20/21
2609 5517 rt : (6,8,9,-3)
2609 6027 $\Psi : 5\sqrt{2}/4$
2609 6425 $as : (\pi/2)^{-3}$
2609 8632 cu : 21/16
2609 8669 rt : (5,9,-3,5)
2609 9333 $J_0 : (\ln 2)^{-3}$
2610 0000 cu : 21/10
2610 0028 $\ell_{10} : \sqrt{2} - \sqrt{5} + \sqrt{7}$
2610 0580 $J_2 : 8\sqrt{2}/7$
2610 1180 rt : (1,4,1,9)
2610 1637 rt : (2,5,5,1)
2610 1960 $J_1 : 8\sqrt{3}/3$
2610 2111 $J_1 : 13/24$
2610 2530 rt : (6,3,-1,6)
2610 2908 $at : 2\zeta(3)/9$
2610 3538 $at : 1/2 - 4e/3$
2610 3562 $\ell_{10} : 1/3 + 2\sqrt{5}/3$
2610 3647 $10^x : P_{61}$
2610 5684 $at : 3\sqrt{2} - \sqrt{5}/2$
2610 7034 $\ell_{10} : \sqrt{2} - \sqrt{3}/2$
2610 7458 $\tanh : 1/\sqrt{14}$
2610 7734 $\sinh : 1/\sqrt{15}$
2610 8114 rt : (9,3,4,1)
2610 8409 $as : P_{28}$
2610 8499 $Li_2 : \sqrt[3]{5}/7$
2610 9338 at : 25/8
2611 1731 $erf : \sqrt{2}/6$
2611 1841 $e^x : 3\sqrt{2}/2 - 2\sqrt{3}$
2611 2003 $sq : P_{13}$
2611 2868 rt : (5,0,-8,2)
2611 3683 $J_1 : (2\pi)^{-1/3}$
2611 3723 $Ei : (\ln 2)^{-2}$
2611 4182 rt : (2,4,9,9)
2611 4256 $J_2 : \sqrt{2} + \sqrt{5} + \sqrt{6}$
2611 4918 $J_2 : 5e\pi/7$
2611 5105 rt : (9,4,6,-2)
2611 5240 $\ln : 10\ln 2/9$
2611 5741 $at : 1/\sqrt{14}$
2611 6669 rt : (6,9,-9,3)
2611 6837 $5 \times \exp(\sqrt{3})$
2611 7918 rt : (9,9,1,5)
2611 8594 $E : \sqrt{2/3}$
2611 9705 $sq : 1/2 + \pi$
2612 0155 $2/3 + \ln(2/3)$
2612 0387 rt : (7,9,1,-1)
2612 0518 $J : 5/11$
2612 1007 $cr : \sqrt{2}/2 + 3\sqrt{3}/4$
2612 1671 $id : 3\sqrt{2} - 4\sqrt{5}/3$
2612 2844 $Li_2 : \exp(-1/(3e))$
2612 3221 $2^x : 3 - \sqrt{2}/4$
2612 3269 $\Psi : \ln(e + \pi)$
2612 3507 $\Psi : 11/9$
2612 5217 rt : (2,4,-7,4)
2612 5949 $\ln(4/5) \times \exp(\sqrt{3})$
2612 6486 $J_2 : \exp(\sqrt[3]{3}/3)$
2612 7082 $\ell_{10} : 4/3 - \pi/4$
2612 7217 rt : (7,0,-4,9)
2612 7896 $J_2 : 7\ln 2/3$
2612 8215 $2^x : 2\sqrt{2} - \sqrt{3}/4$
2612 9503 $2^x : 2 + e/2$
2612 9636 $\ell_{10} : \sqrt{2}/2 + \sqrt{5}/2$
2612 9972 $cr : 3\sqrt{2} - \sqrt{5}$
2613 0795 $t\pi : 2\sqrt{5}/3$
2613 2671 $\Psi : \exp(\sqrt[3]{5}/3)$
2613 4058 $t\pi : 2\sqrt{3}/3 - \sqrt{5}$
2613 4324 $J_1 : 16/5$
2613 4462 $sr : 9\sqrt{2}/8$
2613 4576 $t\pi : (\ln 2)^{-2}$
2613 6516 $sq : 7\pi^2/2$
2613 6948 rt : (5,3,-1,4)
2613 9064 $sq : 2(e + \pi)/3$
2613 9447 $at : P_{66}$
2614 0088 $at : P_{65}$
2614 1649 $\ell_2 : 4\sqrt{2}/3 - 3\sqrt{5}/4$
2614 1987 rt : (3,6,-6,-8)
2614 2090 rt : (1,7,-3,-8)
2614 3937 $\ell_{10} : \sqrt{10/3}$
2614 3968 $10^x : P_{45}$
2614 3982 $id : 2\sqrt{2} + \sqrt{3}/4$
2614 4710 $Ei : 10/13$
2614 4848 $\theta_3 : 3/23$
2614 6669 rt : (5,5,6,-2)
2614 9801 $10^x : \exp(-2/3)$
2615 0030 rt : (7,7,9,2)
2615 1436 rt : (2,9,5,-4)
2615 3358 rt : (6,0,-4,7)
2615 3641 $J_0 : \sqrt{2} + \sqrt{6} + \sqrt{7}$
2615 3656 $id : 1/3 + 4\sqrt{3}$
2615 4309 rt : (8,9,1,3)
2615 6251 $\ln : P_{51}$
2615 6626 $sr : 5(1/\pi)$
2615 8093 $J_0 : 9(e + \pi)/7$
2615 8318 rt : (9,1,-8,-2)
2616 0300 $at : 7\sqrt{5}/5$
2616 1536 $e^x : 4(e + \pi)$
2616 1749 $\Gamma : \sqrt{14}\pi$
2616 2407 $\ln : 3\sqrt{3}/4$
2616 2549 $2^x : 3\sqrt{2}/2 + 4\sqrt{5}$
2616 2829 $at : 1/3 - 2\sqrt{3}$
2616 3684 rt : (9,6,-2,6)
2616 3782 $\exp(1/3) + s\pi(1/3)$
2616 3839 $\exp(4/5) + s\pi(\sqrt{2})$
2616 6706 $J_2 : 5(e + \pi)/4$
2616 7158 $\ln(5) + \exp(\sqrt{3})$
2616 8427 $10^x : \sqrt[3]{3}/2$
2616 9242 $sq : \sqrt{21}\pi$
2616 9438 $sq : 3\sqrt{2}/4 + \sqrt{5}/3$
2617 1769 rt : (1,6,-1,4)
2617 2037 rt : (8,8,5,-2)
2617 2990 $\ln(\pi) + \exp(3/4)$
2617 3097 $t\pi : (\sqrt[3]{5}/2)^{-1/2}$
2617 3315 $sq : 3e/2 + \pi/3$
2617 6539 $J_0 : 10(e + \pi)/9$
2617 7163 $\Psi : 7\sqrt[3]{3}$
2617 7398 rt : (4,8,-2,-1)
2617 8097 $\ell_2 : \sqrt{23}$
2617 9054 $J_0 : 5\zeta(3)/2$
2617 9938 $at : 2 - \sqrt{3}$
2618 0045 $Ei : (\ln 3/2)^{-3}$
2618 0224 rt : (3,9,-9,-9)
2618 3247 $sq : 7e\pi/6$
2618 4034 rt : (5,2,4,1)
2618 5702 $erf : \exp(-\sqrt[3]{3})$
2618 5950 $\ln(4) \div \ln(3)$
2618 8260 rt : (6,3,-7,-8)
2618 8580 $\zeta : \ln(\ln 2)$
2618 8994 rt : (1,8,9,-9)
2619 0833 $J_1 : e/5$
2619 1373 $J_2 : 9\sqrt[3]{2}/7$
2619 1506 rt : (6,4,-7,-2)
2619 2037 rt : (5,0,-4,5)
2619 2485 $at : 4 - \sqrt{3}/2$
2619 3180 $cu : 1/2 - 3\sqrt{2}/2$
2619 3819 $id : 6e/5$

2619 5666 rt : (9,3,-5,7)
2619 6515 rt : (6,5,-8,-6)
2619 6837 id : $7\sqrt[3]{3}/8$
2619 7448 rt : (9,7,9,-3)
2619 8606 rt : (3,6,9,2)
2619 8888 2^x : $(\sqrt[3]{3}/2)^{-1/2}$
2620 0537 e^x : 25/3
2620 2515 rt : (8,6,-2,4)
2620 3607 sq : $3/4 - 4e/3$
2620 6375 Ψ : 23/13
2620 9785 cr : $1/3 + 3\sqrt{5}/4$
2620 9931 id : $((e+\pi)/3)^{-2}$
2621 0873 J_1 : $1/\ln(2\pi)$
2621 3624 E : $3e/10$
2621 4400 cu : 16/25
2621 4759 rt : (2,8,-2,1)
2621 5328 sr : $9(e+\pi)$
2621 5978 rt : (3,7,-6,-2)
2621 7289 2^x : $2/3 - 3\sqrt{3}/2$
2621 8341 sq : $2 + e$
2621 8777 rt : (2,9,-9,-7)
2621 9279 rt : (9,0,-8,8)
2621 9784 sinh : $10\sqrt[3]{2}$
2622 0275 rt : (7,9,1,1)
2622 0559 rt : (5,8,9,-3)
2622 1281 Li_2 : $(\ln 2)^{1/3}$
2622 1897 rt : (2,4,-5,1)
2622 3716 Ψ : $5\sqrt[3]{5}/7$
2622 5723 rt : (1,5,-1,2)
2622 6324 ℓ_2 : $2/3 + \sqrt{3}$
2622 6404 rt : (6,3,8,2)
2622 6501 rt : (5,3,-7,-6)
2622 6662 $\ln(3/5) \div \exp(2/3)$
2622 8114 Ei : $\sqrt{13/3}$
2622 8571 rt : (8,1,3,-1)
2622 8843 id : $e/3 + 3\pi/4$
2623 0722 sr : $4 + \sqrt{5}/2$
2623 3507 rt : (8,3,-5,5)
2623 3704 as : $2 - \pi/3$
2623 4284 tπ : $\sqrt{13/3}$
2623 5117 tπ : $\sqrt{2} + 3\sqrt{3}/4$
2623 6426 ln : 10/13
2623 7036 at : $4/3 - 2\sqrt{5}$
2624 1064 rt : (8,9,-2,-1)
2624 1396 tπ : $3\sqrt[3]{3}/4$
2624 2690 $\ln(4) - \exp(1/2)$
2624 3051 sπ : $\sqrt{17/2}$
2624 3549 Li_2 : $(\sqrt[3]{3})^{-1/3}$
2624 4658 rt : (1,9,9,-3)
2624 6715 rt : (9,0,7,2)
2624 7993 J_2 : $6e\pi/7$
2625 0100 rt : (5,9,-7,-4)
2625 0616 ζ : $9\sqrt{2}/10$
2625 1220 sinh : $7\sqrt[3]{2}$
2625 1292 ln : $9\pi/8$
2625 1791 as : $\sqrt{2} - 2\sqrt{3}/3$
2625 1931 rt : (1,5,-9,2)
2625 3231 cr : $9\sqrt{5}/10$
2625 3324 θ_3 : $(e/2)^{-2}$
2625 4172 rt : (7,8,-6,-2)
2625 4624 rt : (1,9,-2,4)
2625 5112 rt : (8,0,-8,6)
2625 5924 e^x : $\sqrt{2/3}$
2625 6758 J_2 : $9\sqrt[3]{3}/8$
2626 0764 cr : $(1/(3e))^{-1/3}$
2626 2486 cr : $4/3 + e/4$
2626 2725 at : π
2626 2756 sq : $\sqrt{2}/4 - \sqrt{3}/2$
2626 4033 rt : (7,6,-2,2)
2626 7944 J_1 : $7\sqrt[3]{2}$
2626 8963 tπ : $\sqrt{2} - \sqrt{3} + \sqrt{5}$
2626 9379 tπ : $4\sqrt{2}/3 + 3\sqrt{3}$
2627 0126 Ψ : $(\ln 3/2)^{-1/3}$
2627 0516 Γ : $7\sqrt{3}/5$
2627 0948 J_1 : 6/11
2627 1604 rt : (9,8,-9,-6)
2627 2227 rt : (2,2,8,2)
2627 3633 2^x : $-\sqrt{3} + \sqrt{6} - \sqrt{7}$
2627 4100 sq : $4 - 3e$
2627 4169 id : $8\sqrt{2}/5$
2627 4330 ln : $3\sqrt{3}/4 + \sqrt{5}$
2627 4354 at : 22/7
2627 4502 ℓ_2 : P_{51}
2627 5918 cu : $\sqrt{2}/3 - 4\sqrt{5}$
2627 6106 10^x : $(\pi)^{-2}$
2627 6385 rt : (2,7,4,8)
2627 7212 rt : (1,9,-9,-5)
2627 8144 2^x : $3\pi/8$
2627 8604 sinh : 19/18
2627 9460 cu : $1/3 - \sqrt{2}$
2628 1009 tπ : $(5/2)^{-1/2}$
2628 2420 10^x : $4/3 + 4\sqrt{5}/3$
2628 3259 rt : (4,9,-1,-2)
2628 3742 id : $\sqrt{2} - 3\sqrt{5}/4$
2628 4596 $\ln(2) + s\pi(\pi)$
2628 6432 ln : $\sqrt{2}/5$
2628 6555 $1/4 \div s\pi(2/5)$
2628 6707 sinh : $3\ln 2/8$
2628 6847 rt : (5,1,7,-2)
2628 8611 rt : (7,3,-5,3)
2629 0188 ln : $2/3 + 4\sqrt{5}$
2629 4991 as : $3\ln 2/8$
2629 6485 tπ : $9\zeta(3)/10$
2629 7874 rt : (8,4,6,-2)
2629 8113 10^x : $2\sqrt{3}/3 + 3\sqrt{5}/2$
2629 8296 J_2 : $\sqrt{19}$
2629 8827 at : $4/3 + 2e/3$
2629 9650 $\ln(2) \times \exp(3/5)$
2630 0935 10^x : $3\zeta(3)/5$
2630 3343 sr : $3 + 3\sqrt{2}/2$
2630 3440 ℓ_2 : 5/6
2630 4356 rt : (7,0,-8,4)
2630 4497 Ei : $4\sqrt{3}/9$
2630 5647 at : $\sqrt{2} + \sqrt{3}$
2630 8631 rt : (1,9,5,-9)
2631 0638 rt : (3,8,-4,6)
2631 0810 rt : (5,4,-9,2)
2631 1706 $\ln(3/5) + s\pi(e)$
2631 3393 rt : (7,5,2,-1)
2631 4608 cu : $4\sqrt{3}/3 + \sqrt{5}/2$
2631 5789 id : 5/19
2631 6159 sq : $3\sqrt[3]{5}/10$
2631 6574 rt : (5,7,-8,4)
2631 6646 Ψ : $9\sqrt{5}/5$
2631 7993 e^x : $4\sqrt{2}/3 + 2\sqrt{5}/3$
2631 8083 J_2 : 13/8
2631 8219 Ψ : $-\sqrt{2} - \sqrt{3} - \sqrt{7}$
2632 0106 cu : $2 - e/2$
2632 0723 2^x : $\sqrt{3}/4 + \sqrt{5}/3$
2632 0730 e^x : $2/3 - \sqrt{3}/4$
2632 1986 rt : (2,5,-4,-9)
2632 2902 rt : (8,7,9,-3)
2632 4143 ℓ_{10} : 6/11
2632 4553 cr : $8\sqrt[3]{2}/5$
2632 7554 $s\pi(1/5) + s\pi(\sqrt{5})$
2632 7688 sr : $\ln 2/10$
2632 9908 rt : (1,2,7,-2)
2633 1600 ζ : $\sqrt{15/2}$
2633 2288 sinh : 14/9
2633 5815 rt : (4,7,8,7)
2633 7439 ζ : 14/11
2633 7448 cu : $4\sqrt[3]{3}/9$
2633 7765 at : $\sqrt{3} - \sqrt{5} - \sqrt{7}$
2633 8078 at : $5\sqrt[3]{2}/2$
2633 9810 e^x : $1/2 + 3e/2$
2634 0388 rt : (7,8,5,-2)
2634 1120 Γ : $7\sqrt{5}/5$
2634 1812 $\exp(3) - \exp(3/5)$
2634 3096 2^x : $5\sqrt{2}/6$
2634 6444 rt : (1,2,3,-9)
2634 7571 ln : $8\zeta(3)$
2634 8828 e^x : $3e/2 - \pi/2$
2634 9006 rt : (4,8,9,-3)
2635 1108 e^x : $(2\pi)^{1/2}$
2635 3255 ℓ_2 : $3\sqrt{2} + \sqrt{5}/4$
2635 4284 rt : (1,7,8,-6)
2635 4386 J_1 : $(5/3)^3$
2635 5709 rt : (4,1,2,9)
2635 6960 rt : (7,7,4,8)
2635 9018 sr : $3e\pi/5$
2635 9713 id : $\exp(-4/3)$
2635 9729 as : $(\sqrt{3}/2)^{1/3}$
2635 9880 J_1 : $10\sqrt{5}/7$
2636 0041 $\ln(2/5) \times \ln(3/4)$
2636 0581 rt : (7,4,1,9)
2636 2484 rt : (4,7,-2,6)
2636 2524 sq : $e/4 - 3\pi/2$
2636 4220 θ_3 : 13/24
2636 6326 e^x : $3e - 3\pi/2$
2636 6748 Γ : $7\ln 2/2$
2636 7014 erf : 5/21
2636 7187 cu : $3\sqrt[3]{5}/8$
2636 8171 cr : $\sqrt{3}/3 - \sqrt{5}/4$
2637 0841 sr : $3/2 + 4e/3$
2637 5025 tπ : $\exp(-5/2)$
2637 6241 sr : $3e/2 + \pi/3$
2637 6261 rt : (9,9,-3,0)
2637 6264 ln : $3/2 + 3e/4$
2638 0038 sr : $5\sqrt{5}/7$
2638 0151 Ei : 17/2
2638 2862 at : $2 + 2\sqrt{3}/3$
2638 3521 J_2 : $\sqrt{2} - \sqrt{5} + \sqrt{6}$
2638 3847 sinh : 6/23
2638 7137 tπ : $(\ln 2)^2$
2638 8380 e^x : $\sqrt{2} + 2\sqrt{5}/3$
2638 9399 J_2 : $9e/4$
2639 2284 as : 6/23
2639 2940 cr : $7\sqrt[3]{3}/5$
2639 3542 $\exp(2/5) \div \exp(\sqrt{3})$
2639 5214 $4 \div \exp(e)$
2639 8602 ℓ_2 : $\sqrt{2} + \sqrt{5} - \sqrt{6}$
2640 0223 J_1 : $7\sqrt{3}/2$
2640 3072 Ψ : $e\pi/7$
2640 4145 sq : $1 + 3\pi/4$
2640 4694 at : $3\sqrt{3}/2 + \sqrt{5}/4$
2640 4969 ζ : $8\sqrt[3]{5}/5$
2640 5078 ln : $2(e+\pi)/9$
2640 6364 Ei : $10\ln 2/9$
2641 0580 $\exp(1/5) \times \exp(\pi)$
2641 1464 2^x : $3\sqrt{2}/4 - 4\sqrt{5}/3$
2641 1617 ℓ_2 : $9e\pi$
2641 5267 rt : (9,4,8,2)
2641 6024 tπ : $\sqrt{19/2}$
2641 6041 ℓ_2 : $(\sqrt[3]{3})^{1/2}$
2641 7050 cr : $10\sqrt{2}/7$
2641 8217 rt : (9,1,-6,9)
2641 8258 rt : (8,8,-9,-8)
2641 9344 e^x : $3/4 + \sqrt{3}/4$
2641 9490 rt : (4,7,-6,-2)
2641 9797 2^x : $5e\pi/3$
2642 0550 Γ : 13/18
2642 2469 rt : (6,7,9,2)
2642 4620 rt : (9,4,-3,8)
2642 5827 cr : $7\sqrt{3}/6$
2642 6336 rt : (7,4,-7,-2)
2642 6652 rt : (6,2,-5,-9)
2643 1478 sπ : $\sqrt{11/3}$
2643 1645 ℓ_{10} : $\ln(2\pi)$
2643 3328 rt : (9,7,0,7)
2643 3935 J_2 : $(\ln 2/3)^{-1/3}$
2643 4939 erf : $3(1/\pi)/4$
2643 6180 rt : (6,1,-2,8)
2643 7412 sr : $7(e+\pi)/8$
2643 7676 ln : $2\sqrt{2}/3 + 3\sqrt{3}/2$
2643 8318 ℓ_2 : $(\ln 2)^{1/2}$
2643 8351 10^x : $1/\ln(4)$
2643 9333 rt : (3,5,-4,-9)
2643 9971 tπ : $e/4 - \pi/3$
2644 0064 ln : $2e + 4\pi/3$
2644 2336 J_1 : $\ln(\sqrt{3})$
2644 2795 tπ : $1/3 + 3\sqrt{3}/4$
2644 6044 rt : (1,7,-2,-1)
2644 7193 at : $1/3 + 2\sqrt{2}$
2644 8932 Ψ : $(\pi)^{1/2}$
2645 0150 rt : (6,4,1,7)
2645 0281 J_2 : $\sqrt[3]{13/3}$
2645 0300 cu : $2\sqrt{3}/3 - \sqrt{5}$
2645 0723 ln : $4\sqrt{2}/3 - \sqrt{5}/2$
2645 1741 rt : (7,7,9,-3)
2645 1895 at : $\sqrt{10}$

2645 2657 θ_3 : $(2\pi)^{-1/3}$
2645 5007 rt : (3,7,4,8)
2645 5164 e^x : $1/\ln(\sqrt{2}/3)$
2645 5347 rt : (8,0,7,2)
2645 6917 rt : (2,9,-9,-5)
2645 7683 rt : (3,8,-8,-7)
2645 7878 2^x : $-\sqrt{2}+\sqrt{3}-\sqrt{5}$
2646 0549 rt : (8,3,-9,0)
2646 3473 e^x : $2e/3-\pi$
2646 3857 sq : $6\sqrt[3]{5}$
2646 5829 rt : (8,8,-6,-2)
2646 6284 J_2 : $3e/5$
2646 6474 ζ : $4/9$
2646 7552 ℓ_{10} : $e/5$
2646 7732 ℓ_2 : $1/3+2\sqrt{5}$
2646 8199 as : $3\sqrt{2}-3\sqrt{3}$
2646 8201 rt : (5,3,8,2)
2647 0437 Li_2 : $\sqrt{3}/7$
2647 0463 e^x : $3e/2-2\pi/3$
2647 1386 rt : (9,7,5,1)
2647 1770 rt : (8,2,-9,-8)
2647 2612 rt : (6,7,4,6)
2647 3180 J_1 : $11/20$
2647 4287 id : $1/3-3\sqrt{3}/2$
2647 6602 J_1 : $3(e+\pi)/2$
2648 1046 rt : (3,8,9,-3)
2648 1669 $1/3-\exp(4)$
2648 2703 e^x : $e/3+\pi$
2648 3003 rt : (8,6,-6,-1)
2648 3892 $t\pi$: $4\sqrt{3}/3-3\sqrt{5}/4$
2648 4044 cr : $3-4\sqrt{5}/3$
2648 6074 rt : (2,6,9,2)
2648 6352 Li_2 : $7(1/\pi)/9$
2648 6435 rt : (8,1,-6,7)
2648 7455 J_1 : $1/\sqrt[3]{6}$
2648 7831 rt : (7,4,6,-2)
2648 9357 e^x : $3/2-2\sqrt{2}$
2648 9866 θ_3 : $12/13$
2649 1106 sr : $8/5$
2649 1745 at : $19/6$
2649 2316 E : $13/16$
2649 3440 sr : $3\sqrt[3]{5}$
2650 0001 Ψ : $(2e/3)^{1/3}$
2650 1386 rt : (7,1,-4,-1)
2650 4018 rt : (4,1,7,-2)
2650 5383 2^x : $1+\sqrt{2}/2$
2650 5452 rt : (5,8,-4,8)
2650 6418 rt : (8,4,-3,6)
2650 7224 $t\pi$: $\sqrt{3}+3\sqrt{5}$
2650 7450 10^x : $1/4+\sqrt{2}/3$
2650 9326 2^x : $1/2+e/4$
2651 1050 sinh : $((e+\pi)/3)^{-2}$
2651 1685 rt : (1,9,-9,-3)
2651 2992 rt : (5,2,-5,-7)
2651 4679 rt : (6,8,5,-2)
2651 6393 J_2 : $\exp(2e/3)$
2651 6522 J_2 : $\sqrt{8/3}$
2651 9199 10^x : $4/3+3\sqrt{2}$
2651 9692 as : $((e+\pi)/3)^{-2}$
2652 0270 rt : (2,8,-7,-8)
2652 0580 cr : $3/4+4e$
2652 1740 rt : (8,9,-3,-2)
2652 3485 id : $5e/6$
2652 4059 rt : (5,8,-2,-1)
2652 5197 $t\pi$: $5\sqrt{3}/8$
2652 5738 rt : (1,2,8,2)
2652 5823 id : $5(1/\pi)/6$
2652 8085 e^x : $4/17$
2652 8642 ℓ_2 : $5\sqrt[3]{3}/3$
2653 0152 ℓ_{10} : $3\sqrt{3}-3\sqrt{5}/2$
2653 0343 sinh : $10(e+\pi)/3$
2653 0612 sq : $9\sqrt{5}/7$
2653 1715 cr : $2/3+e/2$
2653 2514 tanh : $e/10$
2653 2808 rt : (3,5,6,-2)
2653 3965 at : $7e/6$
2653 5249 rt : (8,7,0,5)
2653 6892 ℓ_2 : $\exp(1/(2e))$
2653 7234 rt : (5,3,-5,-1)
2653 7580 J_0 : $\sqrt{3}-\sqrt{6}+\sqrt{7}$
2653 8564 sr : $3/4-e/4$
2653 9047 J_0 : $10\sqrt{2}/3$
2653 9728 sq : $3\zeta(3)/7$
2654 0746 10^x : $\sqrt[3]{2}/2$
2654 1501 2^x : $7\sqrt[3]{3}$
2654 1501 at : $e/10$
2654 2714 Ei : $4/9$
2654 3887 rt : (8,3,4,1)
2654 5180 rt : (5,1,-2,6)
2654 5277 e^x : $3/2+3\sqrt{5}/2$
2654 6270 sr : $8e\pi$
2654 8862 Γ : $1/\ln(4)$
2654 9050 rt : (7,2,-9,-6)
2654 9954 10^x : $5\sqrt{3}/7$
2654 9956 at : $3\sqrt{3}/4-2\sqrt{5}$
2654 9968 $s\pi$: $(e)^3$
2655 0519 ℓ_2 : $\zeta(3)$
2655 0852 ζ : $1/\ln(\sqrt[3]{3})$
2655 1202 cu : $2/3-4\sqrt{2}$
2655 3942 rt : (1,9,-3,-5)
2655 6443 rt : (2,9,5,-2)
2655 6634 Ei : $3\sqrt{2}$
2655 6817 2^x : $e/8$
2655 9120 Ψ : $13/19$
2656 1959 cu : $1/4-2\sqrt{5}$
2656 2500 sq : $9/8$
2656 4875 cr : $\sqrt[3]{25/3}$
2656 5523 Γ : $3\zeta(3)/5$
2656 7055 cu : $9/14$
2656 8110 at : P_{30}
2656 9430 rt : (8,5,-3,-1)
2657 0005 erf : $6/25$
2657 0344 $s\pi$: $\sqrt{2}/4+\sqrt{3}$
2657 1284 sinh : $(\sqrt{5}/2)^{1/2}$
2657 3196 ℓ_2 : $3\ln 2/5$
2657 6340 Γ : $1/\ln(\ln 2/2)$
2657 6738 $\exp(2/5)+s\pi(e)$
2657 9737 e^x : $\sqrt{2}/6$
2657 9927 rt : (9,8,1,-1)
2658 0623 rt : (7,1,-6,5)
2658 0900 rt : (1,9,9,-1)
2658 1164 sq : $1/3+\sqrt{3}$
2658 1599 Γ : $\sqrt[3]{3}/2$
2658 2322 rt : (4,5,-7,5)
2658 2341 ℓ_{10} : $\sqrt{2}/4+2\sqrt{5}/3$
2658 2350 Γ : $17/7$
2658 2716 θ_3 : $7\zeta(3)/10$
2658 4148 rt : (6,7,9,-3)
2658 5193 J_1 : $8\pi^2/9$
2658 5630 at : $3/2+3\sqrt{5}/4$
2658 5724 J_0 : $3\pi/2$
2658 6896 Ei : $5\sqrt{2}/2$
2658 7218 Li_2 : $\sqrt{5}/9$
2658 8572 rt : (3,8,-1,-8)
2658 9870 sr : $10\sqrt[3]{3}/9$
2658 9961 rt : (7,1,3,-1)
2659 0895 2^x : $8\sqrt[3]{2}/3$
2659 1057 rt : (4,7,-6,-3)
2659 2053 rt : (7,3,-9,-2)
2659 3931 $\ln(4)-\exp(\sqrt{3})$
2659 6152 sr : $2(1/\pi)/9$
2659 6184 as : $\sqrt{2}-3\sqrt{5}/4$
2659 7173 rt : (5,4,1,5)
2659 7760 J_2 : $8e/5$
2659 7978 e^x : $\sqrt{2}/3+2\sqrt{5}$
2659 9468 sr : $4\zeta(3)/3$
2660 0162 J_2 : $18/11$
2660 2082 at : $2\sqrt{2}/3+\sqrt{5}$
2660 2379 cu : $\sqrt{3}+3\sqrt{5}$
2660 2540 $2/5+s\pi(1/3)$
2660 4691 2^x : $2+\pi/3$
2660 5209 ln : $(\sqrt{2}\pi/2)^{1/3}$
2660 5995 ℓ_{10} : $(2\pi)^{1/3}$
2660 6685 ℓ_{10} : $3-2\sqrt{3}/3$
2660 7855 sinh : $5\pi^2/7$
2660 8699 10^x : $3e$
2660 9940 erf : $\sqrt[3]{3}/6$
2661 2010 rt : (9,9,-2,-1)
2661 2146 ln : $1/3+\sqrt{3}/4$
2661 2377 id : $\sqrt{2}/2+\sqrt{5}/4$
2661 3119 rt : (4,3,-7,-2)
2661 3822 erf : $\zeta(3)/5$
2661 5005 rt : (5,6,5,1)
2661 6073 tanh : $3/11$
2661 6881 rt : (2,8,9,-3)
2661 8007 ℓ_2 : $\exp(\pi/2)$
2661 8726 sq : $8\pi/5$
2662 0087 cu : $9\zeta(3)/10$
2662 0168 at : $4/3-3\sqrt{2}/4$
2662 0260 e^x : $\sqrt{2}/4-3\sqrt{5}/4$
2662 0580 sinh : $5/19$
2662 0672 $t\pi$: $10e\pi/3$
2662 1294 id : $3e/2+4\pi/3$
2662 1790 rt : (6,5,2,-1)
2662 3838 Ψ : $2\sqrt[3]{5}/5$
2662 5204 at : $3/11$
2662 5534 $s\pi$: $1/2+\sqrt{2}$
2662 6038 $\exp(\sqrt{5})\div\exp(2)$
2662 6273 tanh : $6(1/\pi)/7$
2662 6297 rt : (7,4,-3,4)
2662 6788 ℓ_{10} : $13/24$
2662 9147 $t\pi$: $2\sqrt{2}-3\sqrt{3}$
2662 9401 as : $5/19$
2662 9685 ℓ_2 : P_{25}
2663 0003 at : $9\sqrt{2}/4$
2663 1296 Ψ : $2\sqrt[3]{3}/9$
2663 1506 cr : $\sqrt{2}/4+3\sqrt{5}/4$
2663 2474 rt : (5,7,-6,-2)
2663 3965 J_0 : $15/2$
2663 4860 rt : (6,4,-9,2)
2663 5421 at : $6(1/\pi)/7$
2663 7540 e^x : $9/11$
2663 7926 rt : (3,7,-7,7)
2663 9947 rt : (9,8,9,2)
2664 0052 at : $10(1/\pi)$
2664 0492 rt : (1,8,3,-7)
2664 2123 rt : (4,2,-5,-5)
2664 2899 rt : (9,4,-7,3)
2664 3376 Ei : $24/7$
2664 4165 Ψ : $(e/3)^{-2}$
2664 5082 rt : (5,9,1,-1)
2664 8002 J_1 : $\sqrt{7}\pi$
2664 9975 rt : (9,7,5,-2)
2665 0361 e^x : $-\sqrt{3}-\sqrt{5}+\sqrt{7}$
2665 1156 rt : (2,9,-9,-3)
2665 2009 ℓ_{10} : $\sqrt{2}+\sqrt{3}/4$
2665 2188 as : $(\ln 3)^{-1/2}$
2665 2302 Ψ : $\sqrt{2}+\sqrt{6}-\sqrt{7}$
2665 2526 10^x : $4/3+4e/3$
2665 2897 ℓ_{10} : $(e/2)^2$
2665 3481 sq : $2e-\pi/3$
2665 4036 at : $4\sqrt{2}/3+3\sqrt{3}/4$
2665 5976 rt : (6,2,-9,-4)
2665 6576 rt : (3,9,-2,2)
2665 7559 rt : (7,0,-8,2)
2665 7864 ln : P_6
2665 8841 rt : (8,4,8,2)
2666 0111 rt : (1,8,-7,-6)
2666 0632 10^x : $\sqrt{2}+3\sqrt{3}$
2666 0671 rt : (9,5,-1,9)
2666 2622 $s\pi$: $3e/4+\pi/3$
2666 3041 rt : (5,6,-9,7)
2666 3083 rt : (1,4,-8,-7)
2666 6036 sinh : $\exp(-4/3)$
2666 6666 id : $4/15$
2666 7501 e^x : $\exp(-\sqrt[3]{3})$
2666 7778 sr : $2\sqrt{2}+4\sqrt{3}/3$
2666 9297 ℓ_2 : $3+2e/3$
2667 1306 Ei : $4\ln 2/9$
2667 1409 rt : (7,6,-6,-3)
2667 3643 rt : (7,0,7,2)
2667 3774 J_1 : $4\ln 2/5$
2667 4646 rt : (8,4,-7,-2)
2667 4933 as : $\exp(-4/3)$
2667 5329 K : $(\sqrt[3]{3}/2)^{1/3}$
2667 5902 ζ : $1/\ln(3\pi/2)$
2667 6418 Ψ : $4/9$
2667 7937 10^x : $7\zeta(3)/8$
2667 7981 rt : (8,1,-7,-9)
2667 8933 e^x : $2\sqrt{3}-\sqrt{5}/4$

2667 9419 ℓ_{10} : $8\ln 2/3$
2668 0672 J_1 : $(e+\pi)^{-1/3}$
2668 2364 10^x : $4e/3+2\pi/3$
2668 3958 ℓ_2 : $e+2\pi/3$
2668 4807 cu : $3-3\pi/4$
2668 4809 ln : $3-e$
2668 5549 rt : (9,6,9,-3)
2668 6209 J_0 : $10e/9$
2668 7172 Ei : 25/22
2668 7965 rt : (9,8,-6,-2)
2668 9190 sq : $2e-\pi$
2668 9246 rt : (5,8,-8,-1)
2668 9272 sr : $2/3+2\sqrt{5}$
2669 1095 at : $1-4\pi/3$
2669 2265 e^x : $1/4-\pi/2$
2669 2831 ℓ_{10} : $3\sqrt[3]{3}/8$
2669 3359 10^x : $e/3+4\pi/3$
2669 5330 rt : (5,8,5,-2)
2669 5559 2^x : $e\pi/5$
2669 7461 rt : (1,6,5,6)
2669 8214 rt : (7,7,0,3)
2669 9595 rt : (9,7,-4,2)
2670 3509 rt : (4,0,-8,2)
2670 6651 $t\pi$: $2\sqrt{2}-\sqrt{5}/3$
2670 9103 $\ln(3)\div\exp(\sqrt{2})$
2671 0077 ln : $\sqrt{2}/2+4\sqrt{5}$
2671 1055 cu : $8\pi^2$
2671 2375 id : $2\zeta(3)/9$
2671 2768 rt : (9,8,2,8)
2671 3460 rt : (5,7,9,2)
2671 4342 $\exp(1/5)\div s\pi(\sqrt{2})$
2671 5450 rt : (7,8,6,9)
2671 7982 Ψ : $9\pi/7$
2671 8830 cu : $3\sqrt{3}/4-3\sqrt{5}$
2671 9061 rt : (6,1,-6,3)
2671 9797 J_1 : 5/9
2672 0320 rt : (5,7,9,-3)
2672 1078 ℓ_{10} : $\sqrt[3]{19/3}$
2672 1100 sr : $5e\pi/4$
2672 3855 rt : (4,3,8,2)
2672 5007 rt : (8,2,-4,8)
2672 5167 cu : $2\sqrt{2}/3+3\sqrt{3}/4$
2672 5292 10^x : $4+e/3$
2672 6124 id : $1/\sqrt{14}$
2672 8192 J_1 : $(\ln 2/2)^{-2}$
2672 8869 $\exp(\sqrt{2})\div s\pi(\sqrt{2})$
2672 9211 e^x : $(3\pi)^{1/3}$
2672 9879 θ_3 : 2/15
2673 0088 rt : (6,2,0,9)
2673 1901 rt : (3,1,7,-2)
2673 6450 $s\pi$: $\sqrt{3}+3\sqrt{5}/2$
2673 8313 rt : (2,5,6,-2)
2673 9114 at : $\sqrt{2}/3-\sqrt{5}/3$
2674 0171 rt : (5,4,-6,-9)
2674 1055 rt : (3,5,-8,-4)
2674 1090 at : $10\sqrt{5}/7$
2674 1099 sr : $2\sqrt{3}+3\sqrt{5}/4$
2674 3230 sinh : $\sqrt{3}+\sqrt{5}+\sqrt{6}$
2674 3543 rt : (1,9,5,-2)
2674 3595 2^x : $1/3-\sqrt{5}$
2674 4802 10^x : $1/3-e/3$
2674 5991 ln : $3/2+3e$
2674 6426 sr : $\sqrt{3}+4\sqrt{5}$
2674 7302 Γ : 18/25
2674 9313 rt : (3,5,-1,7)
2675 0232 e^x : $4/3+4e/3$
2675 0302 cr : $4e+\pi/4$
2675 0802 sr : $2+\pi$
2675 0829 2^x : $\sqrt[3]{5}/5$
2675 1262 $\exp(\sqrt{2})+\exp(e)$
2675 1525 id : P_{66}
2675 1961 ζ : $(\ln 3)^{-2}$
2675 2212 id : P_{65}
2675 2378 at : $\sqrt{2}/2-\sqrt{3}/4$
2675 2537 e^x : $4\sqrt{3}/3-2\sqrt{5}/3$
2675 3159 $t\pi$: $\ln(2/3)$
2675 3962 Γ : $4\sqrt[3]{2}/7$
2675 6733 rt : (1,8,9,-3)
2675 6877 at : $2-3\sqrt{3}$
2675 7086 Ei : $(\sqrt{2}\pi/3)^{-3}$
2675 8790 J_0 : $10\sqrt{3}/9$
2675 8797 $e+\ln(\sqrt{3})$
2676 1235 J_2 : 23/14
2676 1307 cu : $4\sqrt{2}+4\sqrt{3}$
2676 1522 sinh : $2\sqrt[3]{5}$
2676 3345 ℓ_{10} : $3\sqrt[3]{2}/7$
2676 4533 rt : (1,5,5,8)
2676 5263 Li_2 : 1/4
2676 6794 ln : $4+4\sqrt{2}$
2676 7941 E : 17/21
2676 8028 Γ : 3/14
2676 8271 $t\pi$: $(\sqrt{2}/3)^{-1/2}$
2676 9763 cu : $\sqrt{3}-3\sqrt{5}/2$
2677 2233 J_1 : 25/3
2677 3123 e^x : $4-2\sqrt{2}/3$
2677 6022 θ_3 : $\zeta(3)/9$
2677 7080 Ψ : $(\sqrt{2}\pi/3)^{1/2}$
2678 0984 ℓ_{10} : $3\sqrt{3}/2-\sqrt{5}/3$
2678 1442 Ψ : $7\sqrt{3}/3$
2678 1926 $t\pi$: $4\sqrt{2}-4\sqrt{5}$
2678 4749 rt : (9,9,-7,-5)
2678 5691 J_1 : $7(1/\pi)/4$
2678 5788 id : $9\sqrt[3]{2}/5$
2678 8305 J_2 : $(\sqrt{2}\pi)^{1/3}$
2678 8476 rt : (8,4,-7,1)
2678 9207 ln : $2e/3-\pi/3$
2678 9373 rt : (8,5,-1,7)
2679 0204 rt : (7,1,-7,-7)
2679 1145 at : 16/5
2679 4919 id : $2-\sqrt{3}$
2679 7206 10^x : $9\sqrt{3}/5$
2679 9803 cr : $3e/4$
2680 0510 ℓ_{10} : $3/2+\sqrt{2}/4$
2680 3947 J_0 : $9\sqrt[3]{5}/8$
2680 5680 sinh : 18/17
2680 6894 rt : (4,6,-3,9)
2680 6988 e^x : $3\sqrt{3}+2\sqrt{5}$
2680 7014 cr : $3\sqrt{3}/2-\sqrt{5}/4$
2680 7157 sinh : $10\sqrt[3]{5}/9$
2680 7368 $t\pi$: $6\sqrt{5}/7$
2680 8284 $t\pi$: $1/\sqrt[3]{15}$
2680 8373 rt : (4,7,-9,-8)
2681 0268 cu : $2e+3\pi$
2681 0872 rt : (5,1,-3,-8)
2681 1908 J_1 : $10(1/\pi)$
2681 2486 at : $3e\pi/8$
2681 2734 J_2 : $\pi^2/6$
2681 2801 $2/5\div\exp(2/5)$
2681 3105 as : 21/22
2681 3608 rt : (5,2,-9,-2)
2681 4509 rt : (6,5,3,8)
2681 5588 $1/5\div s\pi(\sqrt{3})$
2681 6291 sinh : 19/10
2681 6883 rt : (1,6,-2,-2)
2681 7736 e^x : $(\ln 3/2)^{1/3}$
2681 8128 id : $4e/3-3\pi/4$
2681 8488 ℓ_{10} : $9\sqrt[3]{3}/7$
2681 8803 rt : (6,4,-3,2)
2682 1985 rt : (8,6,9,-3)
2682 2502 cr : $\sqrt{5}+\sqrt{6}-\sqrt{7}$
2682 2809 $\ln(\sqrt{2})\times s\pi(e)$
2682 3965 ln : $(\sqrt[3]{5})^{1/2}$
2682 5284 10^x : $4e/3+\pi/3$
2682 5506 rt : (6,3,-9,-4)
2682 6398 ln : 13/17
2682 9018 J_0 : $8\zeta(3)/5$
2682 9436 rt : (2,6,1,9)
2682 9858 cu : $(\sqrt{2}\pi/3)^3$
2683 0929 rt : (8,7,5,-2)
2683 1900 ℓ_{10} : $7\ln 2/9$
2683 2233 E : $(\sqrt[3]{4}/3)^{1/3}$
2683 2815 $3/5\div\sqrt{5}$
2683 4431 rt : (9,3,6,-2)
2683 4772 rt : (2,7,-5,-9)
2683 5634 cr : $4/3+\sqrt{2}/2$
2683 6184 Ei : 4/13
2683 6771 rt : (9,2,-8,5)
2683 7988 sinh : $5(1/\pi)/6$
2683 9869 at : $8\zeta(3)/3$
2683 9941 erf : $1/\sqrt{17}$
2684 0959 $t\pi$: $1/\sqrt[3]{11}$
2684 1448 J_0 : 25/13
2684 2403 rt : (7,9,-3,-4)
2684 2921 $\ln(\pi)-\exp(5)$
2684 3411 ζ : $7(1/\pi)/5$
2684 3615 cu : $1/\ln(3\pi/2)$
2684 3661 cr : $\sqrt[3]{17/2}$
2684 4489 rt : (3,1,1,-9)
2684 5946 J_0 : $4\sqrt[3]{3}/3$
2684 6439 E : $7\ln 2/6$
2684 7174 as : $5(1/\pi)/6$
2684 8354 Ψ : 16/9
2684 8688 rt : (8,8,1,-1)
2685 1918 sr : $1/4+e/2$
2685 5848 rt : (6,7,-6,-2)
2685 6285 rt : (3,7,0,-1)
2685 6629 J_1 : $9\sqrt{2}/4$
2685 6820 $\ln(3)\div s\pi(1/3)$
2685 6891 rt : (9,4,2,-1)
2685 9190 cu : $5\sqrt{3}/8$
2686 0474 rt : (4,7,9,-3)
2686 2504 2^x : 13/11
2686 3375 rt : (5,3,-7,-2)
2686 3624 sr : ln(5)
2686 5113 sr : $2/3+2\sqrt{2}/3$
2686 6041 10^x : $1-\pi/2$
2686 7191 rt : (7,2,-4,6)
2686 8114 $\ln(4/5)+\exp(2/5)$
2687 0398 ln : $4\sqrt{2}/3-\sqrt{3}/3$
2687 0589 Ei : 25/12
2687 3030 rt : (4,1,-8,2)
2687 3143 J_1 : $\sqrt{5}/4$
2687 5081 Γ : 9/23
2687 8254 2^x : $2\zeta(3)/7$
2688 1297 cu : $7\sqrt{2}/2$
2688 2808 rt : (8,8,2,6)
2688 3002 e^x : 5/21
2688 3721 $t\pi$: $4\sqrt{2}/3-3\sqrt{3}/2$
2688 4531 ℓ_{10} : 7/13
2688 4594 ζ : $1/\ln(3\pi)$
2688 4941 $s\pi$: $\ln 2/8$
2688 5129 ln : $3\sqrt{3}+2\sqrt{5}$
2688 5777 θ_3 : $(\ln 3/3)^2$
2688 5871 rt : (3,5,-9,2)
2688 9000 $s\pi(\sqrt{3})\div s\pi(1/5)$
2689 1668 rt : (5,4,6,-2)
2689 3029 rt : (4,8,-2,1)
2689 4755 e^x : $1/3+\sqrt{5}/2$
2689 5809 J_2 : $8\sqrt[3]{3}/7$
2689 6979 E : $4\sqrt{2}/7$
2689 8053 rt : (6,8,-2,-1)
2690 0748 Γ : $\sqrt{5}-\sqrt{6}+\sqrt{7}$
2690 0841 rt : (8,9,-3,0)
2690 0960 ζ : e
2690 0980 at : $3/2-3\pi/2$
2690 1522 θ_3 : $((e+\pi)/3)^{-3}$
2690 1757 erf : $\exp(-\sqrt{2})$
2690 2336 2^x : $3-e/3$
2690 2521 rt : (6,0,7,2)
2690 3043 ℓ_2 : $3\sqrt{2}+\sqrt{3}/3$
2690 4496 sr : $\sqrt{2}-\sqrt{6}+\sqrt{7}$
2690 5686 10^x : $10\sqrt{2}/9$
2690 5858 ℓ_{10} : $3\sqrt{3}/4+\sqrt{5}/4$
2690 6029 rt : (2,8,-3,-3)
2690 6616 J_2 : $(e)^{1/2}$
2690 7128 10^x : $3e\pi/2$
2690 9108 sq : $6\pi^2/5$
2690 9551 rt : (5,9,-4,3)
2691 1199 cu : $\sqrt{3}/3+\sqrt{5}$
2691 1830 ℓ_2 : $\sqrt[3]{14}$
2691 2447 2^x : $7\pi^2/3$
2691 5096 J_1 : $4\sqrt[3]{2}/9$
2691 5624 rt : (5,2,0,7)
2691 6103 at : $1/4-2\sqrt{3}$
2691 6127 rt : (7,4,8,2)
2691 6648 J_1 : 14/25
2691 7508 cr : $3\sqrt{3}/4+\sqrt{5}/3$
2691 8887 rt : (9,5,-5,4)
2692 0517 at : $4-\pi/4$
2692 1542 $t\pi$: ln(5/2)

2692 1677 rt : (5,4,-3,-1)
2692 2486 rt : (1,8,6,5)
2692 3036 e^x : $1/2 - 2e/3$
2692 6616 tπ : $\exp(-1)$
2692 7646 ln : $3 - \sqrt{5}$
2692 8159 rt : (4,9,1,-1)
2693 1230 ℓ_{10} : $1/2 + e/2$
2693 3692 rt : (8,8,9,2)
2693 6222 J_2 : $7\sqrt{2}/6$
2693 6411 Γ : $8e$
2693 8098 rt : (9,5,-2,0)
2693 8803 sq : $2\sqrt{2} - 4\sqrt{3}/3$
2694 0143 sq : $2e/3 + 3\pi$
2694 0198 sr : $\sqrt{3}/2 + \sqrt{5}/3$
2694 0434 tπ : $3\pi/2$
2694 0824 sinh : $\ln(\ln 2/2)$
2694 1013 id : $2\sqrt{2} - \sqrt{5}/4$
2694 1617 ln : $\sqrt{3}/3 + 4\sqrt{5}/3$
2694 2166 10^x : $2\sqrt{2} - \sqrt{3}/3$
2694 2271 as : $3(1/\pi)$
2694 2370 rt : (5,1,-6,1)
2694 2795 at : $2/3 - 2\sqrt{2}/3$
2694 3244 rt : (4,9,-9,-3)
2694 5849 cu : $4 - 3\sqrt{5}/2$
2694 6260 rt : (6,1,-7,-5)
2694 8078 rt : (7,3,-8,-9)
2694 8367 $\sqrt{2} - \ln(\pi)$
2694 8438 ln : $3 + \sqrt{5}/4$
2694 9042 θ_3 : $\sqrt{5}/3$
2694 9645 rt : (5,5,2,-1)
2695 1398 sq : $4/3 + 4\sqrt{3}/3$
2695 1730 rt : (6,8,6,7)
2695 1871 cu : $1 - 3e$
2695 2228 ln : $1/4 + 3\pi$
2695 2953 at : $1/2 + e$
2695 3084 rt : (4,4,-8,8)
2695 5229 Ψ : $7(1/\pi)/9$
2695 6842 sπ : $7\sqrt{5}/4$
2695 7952 10^x : $17/7$
2695 8847 at : $4/3 + 4\sqrt{2}/3$
2695 9443 rt : (8,0,-8,2)
2695 9737 $3/5 \div \exp(4/5)$
2695 9782 Ei : $7(1/\pi)/5$
2696 1264 J_1 : $10(e+\pi)/7$
2696 2187 J_2 : $\sqrt[3]{9/2}$
2696 2365 rt : (7,6,9,-3)
2696 2582 2^x : $\sqrt{2}/2 - 3\sqrt{3}/2$
2696 3875 e^x : $3(1/\pi)/4$
2696 4026 J : $3\sqrt{2}/10$
2697 1239 rt : (8,3,-2,9)
2697 1592 rt : (2,1,7,-2)
2697 5495 cr : $1 + \pi/3$
2697 6558 e^x : $4\sqrt{2} - 2\sqrt{5}$
2697 6605 rt : (3,7,-9,-6)
2697 7863 J_1 : $\exp(-\sqrt{3}/3)$
2697 9054 Ψ : $(\pi)^{-1/3}$
2697 9267 rt : (7,5,-1,5)
2697 9426 rt : (6,1,3,-1)
2697 9677 sπ : $2/23$
2698 0176 sq : $4/3 - \pi$
2698 1384 sinh : $9\sqrt{2}/2$
2698 3841 sinh : $4/15$
2698 4231 rt : (9,0,-9,7)
2698 6711 id : $e\pi/2$
2698 8343 rt : (7,4,-9,2)
2698 9033 J_1 : $8\pi/3$
2698 9337 rt : (3,8,-5,3)
2698 9438 2^x : $22/3$
2699 1178 $\ln(\sqrt{2}) \div \exp(1/4)$
2699 1193 rt : (2,5,-8,-2)
2699 2357 at : $\sqrt{3} + 2\sqrt{5}/3$
2699 3247 id : $3e - 3\pi$
2699 3279 as : $4/15$
2699 3473 ℓ_{10} : $\sqrt{2}/2 + 2\sqrt{3}/3$
2699 3736 Li_2 : $\sqrt[3]{2}/5$
2699 5039 rt : (3,3,8,2)
2699 5067 rt : (8,2,-8,3)
2699 8170 e^x : $1 - 4\sqrt{3}/3$
2699 8742 tπ : $\ln(4/3)$
2699 9437 J_0 : $7e\pi/8$
2699 9492 rt : (1,7,-1,7)
2699 9733 J_1 : $(\ln 3/2)^{-3}$
2699 9966 J_2 : $(\ln 3/3)^{-1/2}$
2700 0000 sq : $3\sqrt{3}/10$
2700 0017 rt : (6,7,0,1)
2700 0113 rt : (2,4,-8,1)
2700 2518 $\ln(2/3) + s\pi(\sqrt{5})$
2700 2644 rt : (9,9,4,9)
2700 4844 rt : (3,7,9,-3)
2700 5201 Ei : $1/\ln(3\pi)$
2701 0952 ζ : $\sqrt[3]{20}$
2701 0980 J_1 : $\ln(\sqrt[3]{5}/3)$
2701 1717 2^x : $1/3 + 3e/2$
2701 3537 cr : $3/4 + 3\sqrt{3}/4$
2701 3645 sπ : $\sqrt[3]{7}$
2701 4565 cu : $1 - \sqrt{2}/4$
2701 4717 ζ : $19/7$
2701 7463 rt : (7,4,-7,-1)
2701 8607 rt : (7,7,5,-2)
2701 8880 rt : (4,1,-3,-6)
2701 9208 rt : (3,6,-6,-2)
2701 9611 $2/5 \times s\pi(\sqrt{5})$
2702 0000 $3/5 \times \exp(3/4)$
2702 0004 rt : (1,3,-8,-9)
2702 3494 rt : (7,3,4,1)
2702 4867 $\ln(2/5) \times \ln(4)$
2702 5482 sq : $3\ln 2/4$
2702 5691 erf : $\sqrt[3]{5}/7$
2702 5730 rt : (4,7,9,2)
2702 7202 J_1 : $9/16$
2702 8902 ln : $8\pi^2/3$
2702 9743 rt : (3,5,-4,-7)
2703 0375 ℓ_{10} : $5\sqrt{5}/6$
2703 1007 $2/3 \times \ln(2/3)$
2703 1120 tπ : $\beta(2)$
2703 1188 sinh : $2\zeta(3)/9$
2703 1325 K : $2\pi/7$
2703 2882 ℓ_2 : $(\sqrt[3]{5}/3)^{1/3}$
2703 4624 $\exp(1/5) - s\pi(2/5)$
2703 5923 cu : $8e/5$
2703 6505 rt : (9,8,-2,3)
2703 6621 tanh : $2\ln 2/5$
2703 7496 cu : $\sqrt{2}/3 - \sqrt{5}/2$
2703 9191 rt : (7,0,-5,8)
2703 9655 sr : $4 + 2\sqrt{3}/3$
2704 0000 sq : $13/25$
2704 0709 as : $2\zeta(3)/9$
2704 0753 rt : (5,7,-9,2)
2704 0923 rt : (8,3,6,-2)
2704 1176 J_0 : $7\sqrt{3}/4$
2704 1425 e^x : $6\pi^2/5$
2704 1465 2^x : $3/2 - 2\sqrt{3}/3$
2704 2179 sq : $(\sqrt[3]{3}/2)^2$
2704 2193 $\exp(1/5) - \exp(2/5)$
2704 4132 θ_3 : $e/5$
2704 4702 ℓ_{10} : $\ln(\sqrt[3]{5})$
2704 5076 tanh : $1/\sqrt{13}$
2704 5430 sinh : $1/\sqrt{14}$
2704 5518 rt : (9,9,8,-3)
2704 6505 at : $2\ln 2/5$
2704 6658 rt : (2,8,3,-5)
2704 7072 2^x : $2\pi/3$
2704 9073 ζ : $11/8$
2704 9470 rt : (1,8,-9,-3)
2705 0419 2^x : $3/4 + \sqrt{3}/4$
2705 0969 λ : $\exp(-5\pi/4)$
2705 1000 ln : $4\sqrt{2}/3 + 3\sqrt{5}/4$
2705 2401 ζ : $(\sqrt[3]{3}/2)^{-1/3}$
2705 3095 ln : $1/4 + 3\sqrt{2}/4$
2705 3402 rt : (1,6,2,-1)
2705 4976 at : $1/\sqrt{13}$
2705 8137 rt : (1,9,7,8)
2706 0556 10^x : $1 + 3\sqrt{2}/2$
2706 2641 sπ : $7\pi^2$
2706 4207 rt : (1,7,-5,-7)
2706 7056 $2 \div \exp(2)$
2706 8223 rt : (9,1,7,2)
2706 8238 at : $1/4 + 4\sqrt{5}/3$
2706 9063 rt : (4,6,-7,0)
2707 0393 sq : $3/4 + 2\sqrt{5}$
2707 0963 rt : (9,5,-3,-1)
2707 3336 e^x : $9\pi/8$
2707 3504 ℓ_2 : $\sqrt{2}/2 + 4\sqrt{5}$
2707 3562 10^x : $(\sqrt{2}/3)^{-1/3}$
2707 3716 at : $3/2 + \sqrt{3}$
2707 4111 cr : $6\sqrt[3]{5}/5$
2707 5535 $\ln(2) + s\pi(\sqrt{2})$
2707 5839 sinh : $\exp((e+\pi)/2)$
2707 7632 rt : (9,3,-6,6)
2707 9293 tπ : $1/\ln(4/3)$
2708 0084 sr : $4\sqrt{3}/3 - \sqrt{5}$
2708 1337 as : P_{66}
2708 1848 rt : (1,2,-8,2)
2708 2050 as : P_{65}
2708 3038 rt : (8,6,1,8)
2708 3485 rt : (9,7,-8,-3)
2708 4711 tanh : $5/18$
2708 6759 rt : (5,5,3,6)
2708 7416 sπ : $2e/5$
2708 8997 J_2 : $13/3$
2708 9853 10^x : $3 + \sqrt{3}/4$
2709 0675 rt : (7,7,-6,-2)
2709 0963 rt : (6,6,-6,-5)
2709 1390 cu : $11/17$
2709 1755 ℓ_{10} : $4 - 2\sqrt{3}$
2709 4685 at : $5/18$
2709 5516 rt : (8,7,5,1)
2709 6427 rt : (5,9,-9,-3)
2709 7111 ℓ_{10} : $\sqrt[3]{13/2}$
2709 7781 sr : $21/13$
2710 0756 e^x : $(2e/3)^{-1/3}$
2710 1847 J_1 : $(\pi)^{-1/2}$
2710 2497 sinh : $3\sqrt{2}/4$
2710 4290 rt : (1,8,-2,2)
2710 6666 ζ : $\ln(1/\sqrt{2})$
2710 6916 rt : (6,6,9,-3)
2710 7302 cr : $4e/3 - \pi/2$
2710 8095 rt : (6,3,-8,-7)
2710 8376 Ψ : $17/14$
2710 8772 at : $1 + \sqrt{5}$
2710 8873 e^x : $3 + 3e/4$
2711 2295 10^x : P_{51}
2711 2671 Ψ : $\exp(\sqrt{3}/3)$
2712 0286 ζ : $7\sqrt[3]{2}/9$
2712 0969 rt : (5,0,-1,9)
2712 1058 2^x : $\sqrt{2}/4 - \sqrt{5}$
2712 2495 tπ : $4\sqrt[3]{5}/5$
2712 2762 sπ : $\sqrt{2}/4 + \sqrt{5}/4$
2712 2804 J_0 : $\sqrt{2} - \sqrt{3} + \sqrt{5}$
2712 2987 sr : $3/4 + \sqrt{3}/2$
2712 4196 J_1 : $5\sqrt{3}$
2712 4368 rt : (3,5,-5,-2)
2712 4915 e^x : $6/25$
2712 5156 Ei : $17/22$
2712 5317 ℓ_2 : $3/2 + 3e$
2712 6225 $\exp(3/4) + \exp(e)$
2712 6375 as : $2 - \sqrt{3}$
2712 8644 rt : (6,9,2,0)
2712 8942 e^x : $9\sqrt{3}/7$
2712 8973 J_1 : $7\zeta(3)$
2713 0381 rt : (8,5,-5,2)
2713 1302 rt : (7,8,1,-1)
2713 1588 sr : $8\sqrt{2}/7$
2713 2678 J : $4\ln 2/7$
2713 4152 θ_3 : $(e)^{-2}$
2713 4415 id : $\sqrt{2} - \sqrt{5} - \sqrt{6}$
2713 4898 id : $4\sqrt{2} - 4\sqrt{3}$
2713 6466 ℓ_2 : $(\ln 3)^2$
2713 7392 at : $3\sqrt{2}/2 + \sqrt{5}/2$
2713 8477 ℓ_{10} : $3/4 + \sqrt{5}/2$
2713 9309 2^x : $\sqrt{3}/5$
2714 0715 rt : (1,9,-5,-2)
2714 1203 cu : $13/12$
2714 2531 $\exp(\sqrt{2}) \div \exp(e)$
2714 3036 rt : (5,0,7,2)
2714 4291 rt : (1,5,-2,-3)
2714 4305 rt : (8,0,-9,5)
2714 5075 J_0 : $2\pi^2/3$
2714 5952 J_1 : $6\sqrt[3]{3}$
2714 6230 at : $\sqrt{21/2}$

2714 6528 cr : $2(e+\pi)$
2714 7227 J_1 : $13/23$
2714 7532 e^x : $3/2-e/4$
2714 7633 sinh : $9\sqrt{3}/10$
2714 8437 cu : $9\sqrt[3]{3}/8$
2714 9051 $\ln(2/3)-s\pi(1/3)$
2715 2570 Γ : $7\sqrt[3]{2}/2$
2715 2742 e^x : $3\sqrt{2}-\sqrt{3}/2$
2715 3518 ln : $3/4+4\sqrt{5}$
2715 3581 tanh : $7(1/\pi)/8$
2715 3688 rt : (2,7,9,-3)
2715 3712 2^x : $\ln 2/2$
2715 5330 ℓ_2 : $1+\sqrt{2}$
2715 5368 2^x : $\sqrt[3]{5}$
2715 6644 $\pi-\exp(5)$
2715 7371 rt : (7,8,2,4)
2716 1024 J_1 : $\exp(-\sqrt[3]{5}/3)$
2716 2784 J_1 : $1/\ln(e+\pi)$
2716 2794 sinh : $10(1/\pi)/3$
2716 3148 2^x : $1/\sqrt[3]{24}$
2716 3329 rt : (2,6,-3,-4)
2716 3684 at : $7(1/\pi)/8$
2716 4090 rt : (8,1,-4,-1)
2716 4755 rt : (7,3,-2,7)
2716 5396 e^x : $3e-\pi/3$
2716 5959 at : $1-3\sqrt{2}$
2716 7885 J_1 : $2\sqrt{2}/5$
2717 0489 rt : (3,7,-2,-1)
2717 0581 at : $3/4-\sqrt{2}/3$
2717 1164 10^x : $6\ln 2$
2717 2134 ln : $4-\sqrt{3}/4$
2717 2586 e^x : $\sqrt[3]{3}/6$
2717 4235 Ψ : $8/25$
2717 4817 sr : $7\ln 2/3$
2717 6290 $\sqrt{3}\div\ln(2/3)$
2717 7222 e^x : $\zeta(3)/5$
2717 7488 10^x : $3\sqrt{2}/2-2\sqrt{5}/3$
2717 7807 Li_2 : $8/9$
2717 7978 rt : (5,9,1,-3)
2717 8647 at : $2\sqrt{3}-3\sqrt{5}$
2717 9138 $\ln(2/3)\div\exp(2/5)$
2718 0603 rt : (8,9,8,-3)
2718 0770 id : $7\pi^2/4$
2718 1771 cr : $3-2\sqrt{2}/3$
2718 2818 id : $e/10$
2718 2874 $\exp(\sqrt{5})\div\exp(2/5)$
2718 4450 rt : (8,4,2,-1)
2718 4719 rt : (2,9,-7,6)
2718 5982 rt : (6,9,-7,8)
2718 6969 at : $9\sqrt[3]{3}/4$
2718 7515 J_2 : $5\sqrt{3}/2$
2718 8205 sq : $2\sqrt{2}/3+\sqrt{3}/2$
2718 8420 ζ : $\exp(1/\pi)$
2718 8737 rt : (6,4,8,2)
2719 1031 J_1 : $14/3$
2719 2621 e^x : $2/3+\pi/4$
2719 3371 ln : $16/21$
2719 4533 ζ : $\sqrt{22/3}$
2719 4612 J_0 : $(\ln 3/3)^{-2}$
2719 5086 10^x : $4\sqrt{2}/3+\sqrt{3}/4$
2719 5262 rt : (2,7,-8,-6)
2719 6256 tπ : $\sqrt{17/2}$
2719 6938 J_1 : $7\pi^2/8$
2719 7084 rt : (6,6,5,9)
2719 7958 rt : (9,1,-7,8)
2719 7994 Ψ : $\sqrt{3}/7$
2719 8294 θ_3 : $1/\ln(2\pi)$
2719 9535 ℓ_2 : $(\sqrt{5}/3)^3$
2719 9587 J_0 : $\sqrt{22}$
2720 1964 sr : $1/2+\sqrt{5}/2$
2720 2907 rt : (2,8,-9,-3)
2720 3402 ℓ_{10} : $\sqrt{7/2}$
2720 4811 rt : (9,6,-3,5)
2720 5608 $\exp(2)-\exp(3/4)$
2720 7050 rt : (9,5,9,-3)
2720 7572 rt : (4,9,2,5)
2720 7666 Ei : $\ln(e/2)$
2720 8188 E : $\ln(\sqrt{5})$
2720 9915 rt : (2,8,-6,1)
2721 0304 10^x : $2\sqrt{2}/3+3\sqrt{3}$
2721 1396 id : P_{30}
2721 3324 sπ : $4e-\pi/4$
2721 3392 at : $(\sqrt{2}\pi/3)^3$
2721 3510 rt : (6,7,5,-2)
2721 3626 J_0 : $23/12$
2721 3993 rt : (8,9,-7,-7)
2721 5773 J_0 : $6\sqrt{5}/7$
2721 6544 $\sqrt{5}+s\pi(\sqrt{2})$
2721 6552 sr : $(\sqrt[3]{2}/3)^3$
2721 6715 rt : (1,7,6,9)
2721 9390 cr : $3/2+\sqrt{5}/4$
2722 0387 e^x : $\sqrt{3}/2+\sqrt{5}/2$
2722 0558 ln : $9(e+\pi)/2$
2722 1172 sq : $12/23$
2722 1260 cu : $6\sqrt{3}/7$
2722 2453 J_2 : $5\pi^2/8$
2722 2700 ℓ_{10} : $3e-2\pi$
2722 3961 sinh : $7\sqrt{3}$
2722 4187 $\exp(1/5)\times\exp(5)$
2722 4365 rt : (1,2,8,9)
2722 4374 $\exp(2/3)-s\pi(\sqrt{5})$
2722 5523 tπ : $1/4+\sqrt{5}/2$
2722 5576 rt : (4,2,0,5)
2722 5749 ℓ_{10} : $4\zeta(3)/9$
2722 5757 Ψ : $\sqrt[3]{17/3}$
2722 6026 rt : (7,2,-8,1)
2722 7007 rt : (3,9,1,-1)
2722 9301 rt : (6,5,-9,-9)
2722 9739 at : $13/4$
2723 2184 rt : (2,7,-9,-4)
2723 3866 at : P_{72}
2723 3921 sr : $4+3\sqrt{5}$
2723 4531 at : P_{69}
2723 5011 J_0 : $10\sqrt{5}/3$
2723 5027 rt : (1,7,1,-8)
2723 9603 ζ : $(\sqrt[3]{4}/3)^{-1/2}$
2723 9788 ln : $4/3+\sqrt{5}$
2724 0696 rt : (2,6,-9,8)
2724 3596 rt : (5,8,-5,6)
2724 4810 rt : (4,5,-9,2)
2724 5012 tanh : $\sqrt{5}/8$
2724 5557 ℓ_{10} : $\exp((e+\pi)/2)$
2724 6380 rt : (8,9,4,7)
2724 6646 $\exp(5)-\exp(\pi)$
2724 6972 cr : $10\sqrt[3]{3}/7$
2724 7840 rt : (6,0,-5,6)
2724 8262 J_0 : $((e+\pi)/3)^3$
2724 8487 rt : (7,8,9,2)
2724 8842 at : $2\sqrt{2}/3+4\sqrt{3}/3$
2724 9987 2^x : $2\sqrt{2}-\sqrt{5}/2$
2725 0179 J_2 : $1/\ln(\sqrt[3]{2})$
2725 0606 J_1 : $6\pi^2/7$
2725 3223 cr : $1+3\sqrt{2}/4$
2725 4795 rt : (6,9,-9,-3)
2725 4814 $\ln(4/5)\times\exp(1/5)$
2725 5287 at : $\sqrt{5}/8$
2725 5468 rt : (1,9,-6,-9)
2725 5890 rt : (5,6,9,-3)
2725 6388 rt : (7,3,6,-2)
2725 6987 cr : $2\sqrt{2}/3+\sqrt{5}/2$
2725 7722 rt : (9,0,3,1)
2725 9158 ℓ_{10} : $(\ln 2/3)^2$
2726 0406 rt : (7,7,-4,-2)
2726 1132 10^x : $\sqrt{7/2}$
2726 2429 Γ : $(e)^{-1/3}$
2726 3748 10^x : $4e/3-4\pi/3$
2726 4151 2^x : $8/23$
2726 4195 Ψ : $15/22$
2726 5668 rt : (3,7,-8,4)
2726 6537 sr : $3\sqrt{2}/4+\sqrt{5}/4$
2726 6695 tπ : $1/2+2\pi/3$
2726 6942 J : $9/22$
2726 7316 id : $4/3-3\sqrt{2}/4$
2726 7826 Ψ : $7\ln 2/4$
2726 9231 J_0 : $4(e+\pi)/5$
2726 9313 rt : (9,5,-9,-1)
2726 9851 at : $4-\sqrt{5}/3$
2727 0241 10^x : $1/3+\sqrt{5}/2$
2727 0384 sq : $3\sqrt{2}/2+3\sqrt{3}/2$
2727 2727 id : $3/11$
2727 3452 id : $2\sqrt{3}/3+\sqrt{5}/2$
2727 5160 e^x : $\sqrt[3]{5/3}$
2727 5232 sr : $9\sqrt[3]{2}/7$
2727 5974 $\ln(2/5)-\exp(\sqrt{5})$
2727 7003 ℓ_2 : $4e/9$
2727 7104 sπ : $3\sqrt{2}-2\sqrt{3}/3$
2727 7111 at : $5(e+\pi)/9$
2727 7703 tπ : $1/\sqrt[3]{6}$
2727 8563 e^x : $(e/3)^2$
2727 8863 ln : $2+\pi/2$
2727 8943 10^x : $(\sqrt{2}/3)^3$
2727 9220 id : $9\sqrt{2}/10$
2727 9746 Ei : $(\sqrt{2}\pi/2)^{1/2}$
2728 0385 rt : (4,8,-6,-8)
2728 0467 2^x : $3\sqrt{3}/2-2\sqrt{5}$
2728 0626 10^x : $9\ln 2/4$
2728 1265 $\ln(\sqrt{3})-\exp(3/5)$
2728 1834 J_1 : $7e/6$
2728 3704 id : $6(1/\pi)/7$
2728 3760 rt : (2,3,8,2)
2728 3916 $1/4\div\ln(2/5)$
2728 4632 rt : (8,3,-6,4)
2728 4663 cr : $1/4+2e/3$
2728 4996 J : $1/\sqrt[3]{16}$
2728 4998 sq : $2\sqrt{3}/3-3\sqrt{5}/4$
2728 5268 rt : (9,0,-8,2)
2728 7332 rt : (6,5,-1,3)
2728 7354 e^x : $2/3+4e$
2728 8271 tπ : $7\sqrt[3]{3}/4$
2728 8884 tanh : $2\sqrt[3]{2}/9$
2728 9885 10^x : $1/3+4\sqrt{3}/3$
2729 0508 tanh : $7/25$
2729 1025 $1/3\div\exp(1/5)$
2729 1159 J_2 : $3\sqrt[3]{3}$
2729 2197 ζ : $9\zeta(3)/4$
2729 2707 cu : $\sqrt{2}/4+2\sqrt{5}/3$
2729 4830 J_1 : $6\sqrt{2}$
2729 5226 e^x : $\sqrt{3}-2\sqrt{5}/3$
2729 5738 rt : (3,6,-7,-7)
2729 6472 rt : (7,1,-3,9)
2729 6567 ln : $7/25$
2729 7129 ln : $3+3\sqrt{5}$
2729 8725 J_1 : $((e+\pi)/2)^2$
2729 9243 at : $2\sqrt[3]{2}/9$
2729 9454 rt : (4,7,4,2)
2730 0127 ℓ_{10} : $8/15$
2730 0870 at : $7/25$
2730 0903 Li_2 : $4(1/\pi)/5$
2730 1329 J_1 : $(2e)^{-1/3}$
2730 1939 e^x : $2e/3-\pi/2$
2730 2833 ln : $2\sqrt[3]{2}/9$
2730 4016 rt : (7,8,-2,-1)
2730 4927 rt : (5,3,2,8)
2730 6445 at : $1/\ln(e/2)$
2730 7286 rt : (1,7,9,-3)
2730 7503 J : $\exp(-3\pi/8)$
2730 7981 J_0 : $9(e+\pi)/8$
2730 9191 rt : (8,1,7,2)
2731 2196 J_1 : $17/2$
2731 2371 Γ : $9(1/\pi)/4$
2731 3222 rt : (3,2,-7,-2)
2731 8880 J_0 : $\sqrt{11/3}$
2731 9003 2^x : $4\sqrt{2}-2\sqrt{5}$
2731 9431 rt : (7,9,8,-3)
2732 3954 id : $4(1/\pi)$
2732 5732 J_2 : $8e/3$
2732 6050 ℓ_2 : $3\sqrt{3}+2\sqrt{5}$
2732 6084 rt : (9,4,-4,7)
2732 7516 J_1 : $5\sqrt[3]{5}$
2732 7588 sq : $1/\sqrt[3]{7}$
2732 8239 sπ : $(\pi/3)^{-2}$
2732 8361 at : $2\sqrt{2}+\sqrt{3}/4$
2732 8909 e^x : $\sqrt{2}+3\sqrt{3}/2$
2732 9836 J_1 : $e\pi$
2733 0117 rt : (4,6,5,1)
2733 0217 sinh : $9\ln 2/4$
2733 2119 rt : (3,5,7,7)
2733 2316 2^x : $1/4-3\sqrt{2}/2$
2733 2337 rt : (5,3,-8,-5)
2733 2643 at : $6e/5$

2733 4166 2^x : $2/3+\sqrt{3}$
2733 4386 10^x : $\exp(\sqrt{2})$
2733 5651 at : $e/3+3\pi/4$
2733 8194 rt : (8,7,-6,-2)
2733 9071 rt : (4,7,-3,4)
2733 9259 Ψ : $7\sqrt{3}/10$
2734 0913 rt : (8,8,-2,1)
2734 1721 10^x : P_5
2734 1885 θ_3 : 3/22
2734 2985 ℓ_2 : $1/4+\sqrt{3}/3$
2734 3745 2^x : $4/3+4\sqrt{5}$
2734 4375 rt : (3,2,-7,9)
2734 5736 rt : (9,6,5,-2)
2734 6629 ζ : $\sqrt{5}/5$
2734 8590 cu : $4/3+2\pi/3$
2734 8661 rt : (2,9,1,-4)
2734 8772 rt : (7,6,1,6)
2735 0390 cr : $1/3+\sqrt{3}$
2735 0990 J_2 : 5/3
2735 1052 rt : (5,3,-9,-6)
2735 2074 rt : (2,9,-5,-2)
2735 2974 θ_3 : $3(1/\pi)/7$
2735 4292 10^x : $7\sqrt[3]{2}/4$
2735 6070 rt : (8,5,9,-3)
2735 6470 rt : (7,3,-5,1)
2735 6719 at : $2/3+3\sqrt{3}/2$
2735 6988 sq : $3/4+2e$
2735 7752 Ei : $2\sqrt{5}/3$
2735 7754 J_1 : $\sqrt[3]{5}/3$
2735 8837 ℓ_{10} : $\exp(\sqrt[3]{2}/2)$
2735 9632 Ei : $3(1/\pi)/7$
2736 0190 $\ln(3/4)\times s\pi(2/5)$
2736 0659 erf : $\sqrt{3}/7$
2736 1471 rt : (3,8,-9,-3)
2736 2289 rt : (4,1,-6,-1)
2736 2473 ln : $2\sqrt{2}+\sqrt{5}/3$
2736 2654 e^x : $\sqrt{2}/2+\sqrt{5}/3$
2736 2671 rt : (3,7,9,2)
2736 4032 $\exp(4)+s\pi(\sqrt{5})$
2736 5736 rt : (7,0,-9,3)
2736 8396 rt : (6,2,-6,-8)
2737 0809 Ψ : $6\sqrt{2}/7$
2737 0815 e^x : $3\sqrt{2}+2\sqrt{5}$
2737 1374 ln : $4\sqrt{3}-3\sqrt{5}/2$
2737 2233 rt : (2,3,0,8)
2737 2394 e^x : $4-\sqrt{2}$
2737 2513 cr : $\sqrt{14}\pi$
2737 3561 2^x : $\pi/9$
2737 4868 rt : (1,4,0,5)
2737 5361 erf : $7(1/\pi)/9$
2737 5881 rt : (9,5,8,2)
2737 5945 cu : $2\sqrt{2}-3\sqrt{3}$
2737 6941 J : $\exp(-5\pi/7)$
2737 7450 rt : (8,4,-9,2)
2737 8599 sr : $9\sqrt[3]{3}/8$
2737 8813 e^x : $3/2+4\sqrt{3}$
2737 9671 e^x : $4\sqrt{3}+3\sqrt{5}/4$
2737 9863 rt : (1,9,-2,5)
2738 0620 $\pi\times\ln(2/3)$
2738 0729 ℓ_2 : P_{33}
2738 0730 ℓ_2 : P_{56}
2738 2827 e^x : $3\sqrt{3}/4+4\sqrt{5}/3$
2738 5273 ℓ_{10} : $4-3\sqrt{2}/2$
2738 8005 rt : (9,9,0,4)
2738 8146 rt : (8,7,3,9)
2738 8703 id : $9(e+\pi)/10$
2738 8774 $\ln(2/3)\times s\pi(\sqrt{5})$
2738 9055 rt : (2,2,-8,2)
2739 1333 ℓ_2 : P_{39}
2739 3225 sq : $1/2-4\sqrt{3}/3$
2739 4235 e^x : $3\sqrt{2}/4+2\sqrt{3}$
2739 4268 $1/2+s\pi(e)$
2739 5147 id : $\sqrt{2}/3-\sqrt{5}/3$
2739 5529 rt : (8,1,-7,6)
2739 6414 rt : (4,0,7,2)
2740 0259 sinh : 17/16
2740 1495 rt : (5,1,3,-1)
2740 1585 cu : $3\sqrt{3}/8$
2740 4854 2^x : $9e/7$
2740 5409 cr : $\sqrt{3}/3+2\sqrt{5}/3$
2740 5754 Ei : 3/22
2740 6137 tπ : $\sqrt{11/3}$
2740 6473 10^x : $3/4+3e/2$
2740 7000 $3\div\ln(2/5)$
2740 7348 10^x : $\ln(\sqrt[3]{5}/3)$
2740 9407 id : $\sqrt{2}/2-\sqrt{3}/4$
2740 9558 rt : (4,6,9,-3)
2741 0536 Γ : $(\sqrt{5}/2)^{-3}$
2741 0863 rt : (7,3,-7,-2)
2741 2965 J_2 : $1/\ln(\ln 3/2)$
2741 3236 at : $\sqrt{2}-\sqrt{5}-\sqrt{6}$
2741 3984 rt : (1,8,7,-2)
2741 4801 rt : (3,8,-8,-9)
2741 5108 Γ : $((e+\pi)/3)^{-1/2}$
2741 5177 rt : (4,0,-1,7)
2741 5567 sq : $\pi/6$
2741 6197 rt : (5,8,6,5)
2741 7675 rt : (9,2,-5,9)
2741 7692 e^x : $3e+\pi$
2741 8709 rt : (7,9,-9,-3)
2741 9984 sq : $3\sqrt{3}/2-4\sqrt{5}$
2742 0702 rt : (5,2,-4,2)
2742 1006 J_1 : 4/7
2742 2855 ℓ_2 : $1/4+3\pi$
2742 3603 e^x : $4\sqrt{2}/3-\sqrt{3}/4$
2742 4724 cu : $3\sqrt{2}+4\sqrt{3}/3$
2742 5089 cu : $1/4-2\pi/3$
2742 5502 Γ : $(\ln 3/3)^{1/3}$
2742 7498 10^x : 2/19
2742 9188 rt : (6,8,1,-1)
2743 0566 J_0 : $\sqrt[3]{7}$
2743 1608 rt : (3,9,-3,-1)
2743 1718 id : $\sqrt{2}/2-4\sqrt{5}/3$
2743 3388 id : 9π
2743 5990 rt : (4,5,-8,3)
2743 7515 e^x : $2\sqrt{2}/3-\sqrt{5}$
2743 7641 sq : 11/21
2744 0581 $1/2\div\exp(3/5)$
2744 0942 Ψ : 25/14
2744 1378 rt : (5,5,-9,-7)
2744 2711 e^x : $(\sqrt{2}\pi/3)^{-1/2}$
2744 3684 ln : 19/25
2744 5091 ln : $4\sqrt[3]{5}/9$
2744 6801 sπ : $4\sqrt{2}-\sqrt{5}/3$
2744 7665 e^x : $1/\sqrt{17}$
2744 8421 10^x : $1/\sqrt[3]{22}$
2744 8441 $\exp(4/5)-s\pi(2/5)$
2744 9346 cu : $1/4+2\sqrt{5}/3$
2745 3150 at : $3\sqrt{2}/2+2\sqrt{3}/3$
2745 3289 rt : (6,3,-2,5)
2745 5350 cu : $3/2-3\sqrt{5}$
2745 6062 2^x : 7/20
2745 6950 cu : $4/3+2e$
2745 9828 e^x : P_{15}
2746 0125 at : $3-e$
2746 0295 J : $2\sqrt{5}/9$
2746 0620 J_2 : $\sqrt[3]{14/3}$
2746 1177 ℓ_{10} : $3-\sqrt{5}/2$
2746 1530 cr : $1/2+\pi/2$
2746 2209 rt : (6,9,8,8)
2746 2237 ln : $3+\sqrt{3}/3$
2746 2500 cu : 13/20
2746 2875 2^x : $\sqrt[3]{5/3}$
2746 3624 ℓ_{10} : $\sqrt[3]{20/3}$
2746 3690 10^x : $3(e+\pi)/10$
2746 4783 J_1 : $5\zeta(3)$
2746 5307 ln : $(\sqrt{3})^{1/2}$
2746 5946 rt : (7,9,-3,0)
2746 6722 $\exp(\pi)-s\pi(1/3)$
2746 6776 10^x : $3\zeta(3)/7$
2746 7158 at : P_6
2746 7516 J_1 : 19/6
2746 8485 erf : $\sqrt{5}/9$
2747 1419 $\exp(\sqrt{5})\div\exp(\sqrt{2})$
2747 2022 2^x : $(1/(3e))^{1/2}$
2747 2258 ln : $\sqrt{5}/8$
2747 4080 rt : (7,4,0,8)
2747 5487 sr : 13/8
2747 6588 ln : $3e+\pi/2$
2747 8593 rt : (5,4,8,2)
2747 9140 rt : (8,1,-8,-8)
2748 1617 rt : (6,3,6,-2)
2748 2991 sq : $\sqrt{2}/4+2\sqrt{3}/3$
2748 3340 as : 22/23
2748 3981 cu : $3\sqrt{2}/4+4\sqrt{3}/3$
2748 4465 10^x : P_{57}
2748 4804 rt : (8,7,-8,-5)
2748 7326 rt : (8,6,-3,3)
2748 8098 cr : $2\sqrt{2}+4\sqrt{5}$
2748 8291 J_1 : $9(1/\pi)/5$
2748 9298 sinh : $2(e+\pi)$
2749 1721 rt : (1,9,6,0)
2749 2301 $4\div\exp(1/5)$
2749 2335 ln : $3/4+2\sqrt{2}$
2749 3986 id : $5\sqrt[3]{5}/2$
2749 5024 sinh : $9\sqrt{3}/5$
2749 9706 10^x : 13/18
2750 0074 Ei : $9e\pi/5$
2750 0451 10^x : $1/2-3\sqrt{2}/4$
2750 1310 rt : (9,7,-1,6)
2750 2322 tπ : $((e+\pi)/2)^{1/2}$
2750 2870 rt : (5,6,-6,-2)
2750 5427 at : $2e/3-2\pi/3$
2750 5448 rt : (9,8,-6,-2)
2750 6672 rt : (6,1,-8,2)
2750 7120 rt : (2,7,3,5)
2750 9722 rt : (7,5,9,-3)
2750 9942 at : $e/3-4\pi/3$
2751 0922 cu : $9\zeta(3)/2$
2751 1086 sq : $5\sqrt[3]{5}/2$
2751 1531 E : $2\zeta(3)/3$
2751 2719 id : $3\sqrt{3}/2+3\sqrt{5}/4$
2751 2778 Ei : $\sqrt{5}/5$
2751 3456 sq : $8\zeta(3)/3$
2751 4094 at : $3-2\pi$
2751 5120 2^x : $4\sqrt{2}/3+\sqrt{3}$
2751 5450 rt : (2,8,2,-9)
2751 5701 $s\pi(2/5)\div s\pi(\sqrt{3})$
2751 6689 rt : (5,7,-6,9)
2751 8816 sinh : $e/10$
2751 8824 sr : $\sqrt{2}/3+2\sqrt{3}/3$
2751 9123 id : $\sqrt{3}/4-3\sqrt{5}$
2752 1220 at : $(2e/3)^2$
2752 1747 e^x : $\exp(-\sqrt{2})$
2752 2489 E : $5\sqrt[3]{3}/9$
2752 2535 rt : (5,8,-9,-3)
2752 5512 rt : (4,8,-9,-3)
2752 7943 rt : (3,6,-1,2)
2752 9223 as : $e/10$
2752 9903 2^x : $2/3+2\sqrt{2}$
2753 0112 ℓ_{10} : 6π
2753 2991 cu : $1/2-3\sqrt{5}$
2753 3185 rt : (7,4,2,-1)
2753 3326 sπ : $3\sqrt{3}/2+2\sqrt{5}/3$
2753 3951 Ei : $\ln(\sqrt{2}\pi)$
2753 5548 at : 23/7
2753 5653 rt : (4,2,-2,-9)
2753 7349 2^x : $\exp(-\pi/3)$
2753 8291 tπ : $(e)^3$
2753 8505 J_2 : $2(e+\pi)/7$
2754 1135 sinh : $3e\pi/8$
2754 3400 rt : (2,9,1,-1)
2754 3488 ℓ_2 : $(e+\pi)^{1/2}$
2754 3517 10^x : P_{38}
2754 3630 ζ : 17/23
2754 5373 ℓ_{10} : $3\sqrt{2}/8$
2754 7910 2^x : $((e+\pi)/2)^{1/2}$
2754 8166 rt : (8,6,5,-2)
2754 9029 $2/5-s\pi(\sqrt{5})$
2754 9407 Γ : $2e\pi/7$
2754 9982 at : $4\sqrt{2}-4\sqrt{5}$
2755 2344 ln : P_{69}
2755 3402 tanh : $\sqrt{2}/5$
2755 3403 rt : (3,2,-9,2)
2755 4172 rt : (9,8,8,-3)
2755 4913 ln : P_{72}
2755 6399 rt : (6,3,4,1)
2755 6622 $s\pi(2/5)-s\pi(\sqrt{5})$
2755 6724 $\pi-s\pi(1/3)$
2755 7659 rt : (4,1,-7,-1)

2755 8858 $Li_2 : \exp(-e/2)$
2755 8921 as : P_{30}
2756 0016 rt : (5,0,-5,4)
2756 1030 $t\pi : \sqrt{2}/4 + \sqrt{3}$
2756 1512 rt : (6,1,-3,7)
2756 2573 $e^x : 3/4 - 3e/4$
2756 2588 tanh : $8(1/\pi)/9$
2756 2732 $\theta_3 : 16/19$
2756 2763 rt : (7,1,7,2)
2756 3444 $\ell_2 : 19/23$
2756 4279 at : $\sqrt{2}/5$
2756 7333 $t\pi : \sqrt{2}/2 + \sqrt{3}/2$
2756 8220 rt : (3,6,9,-3)
2756 8757 $e^x : 3\sqrt{2}/2 - 3\sqrt{3}/4$
2756 9732 $e^x : 3 - \sqrt{3}/3$
2757 0722 at : $\pi^2/3$
2757 2951 rt : (3,9,-5,-2)
2757 3484 at : $8(1/\pi)/9$
2757 4826 $10^x : 7\sqrt{2}/8$
2757 4863 $J_0 : \sqrt{3} + \sqrt{5} + \sqrt{7}$
2757 7092 rt : (4,7,-2,-1)
2757 7348 sq : $e + \pi/4$
2757 8034 $\Psi : 23/19$
2757 8812 sr : $\sqrt{2} - \sqrt{5} + \sqrt{6}$
2757 9496 sr : $\sqrt{2}/2 + 2\sqrt{5}$
2758 0572 $\exp(2) - \exp(\sqrt{2})$
2758 0651 at : $1/3 - 4e/3$
2758 2554 rt : (9,5,-2,8)
2758 4592 $10^x : 5/14$
2758 5052 rt : (8,4,-4,5)
2758 7583 rt : (6,8,9,2)
2758 8571 rt : (8,9,-9,-3)
2758 8603 rt : (7,3,-6,2)
2758 8950 at : $3e/2 - \pi/4$
2758 9423 cu : $4/3 - 3\sqrt{5}$
2759 0958 $3/4 \div e$
2759 2444 rt : (1,3,8,2)
2759 3720 $10^x : 11/7$
2759 4053 rt : (6,2,-8,-1)
2759 5082 $J_2 : (3\pi/2)^{1/3}$
2759 5858 cr : $3/2 + \sqrt{3}/3$
2759 6286 at : $4 - \sqrt{2}/2$
2759 6471 rt : (4,2,-7,-2)
2759 7621 $3/4 \div s\pi(1/5)$
2759 8684 rt : (2,5,-6,-3)
2759 9267 $\Gamma : 5/7$
2759 9861 rt : (9,7,-6,-2)
2760 0491 $2^x : 1/\sqrt[3]{23}$
2760 0524 $10^x : 1/2 + 3e$
2760 1576 $\exp(1/2) \times s\pi(e)$
2760 1800 cu : $(e + \pi)/9$
2760 1982 cr : $6(e + \pi)$
2760 2088 id : $3\sqrt{2}/2 + 2\sqrt{3}/3$
2760 2812 sr : $3e/4 + \pi$
2760 6598 rt : (2,7,-5,7)
2760 7728 $J_2 : 3\sqrt{5}/4$
2760 8892 $\sqrt{2} \times \ln(5)$
2760 8998 $J_0 : 6(1/\pi)$
2760 9163 rt : (5,9,8,-3)
2760 9765 rt : (2,9,-8,1)
2761 0457 $1/3 - \ln(5)$
2761 0606 $J : (1/\pi)/5$
2761 2078 sinh : $3/11$
2761 4237 id : $1/3 + 2\sqrt{2}/3$
2761 5716 $s\pi : P_{11}$
2761 7038 as : $4/3 - 3\sqrt{2}/4$
2761 8594 cr : $6\sqrt{3}/5$
2761 9012 rt : (4,8,1,8)
2762 0109 sinh : $9(e + \pi)/5$
2762 0388 $2^x : 1/2 - 3\pi/4$
2762 0641 $\ell_{10} : 9/17$
2762 2339 rt : (9,7,1,-1)
2762 2641 $t\pi : 1/2 + \sqrt{2}$
2762 2663 as : $3/11$
2762 3268 cr : $4/3 + \sqrt{5}/3$
2762 3466 sinh : $6(1/\pi)/7$
2762 5742 rt : (8,5,-9,-3)
2762 6484 rt : (9,2,6,-2)
2762 7294 rt : (4,7,5,-2)
2762 8131 sq : $4/3 + 3e/2$
2762 8633 at : $3\sqrt{2}/4 + \sqrt{5}$
2762 8825 rt : (3,5,-5,-8)
2762 9856 rt : (2,8,-4,-7)
2763 0010 rt : (1,7,-9,-3)
2763 0426 tanh : $\exp(-\sqrt[3]{2})$
2763 0968 $t\pi : \sqrt[3]{3}/5$
2763 2639 $erf : 1/4$
2763 3403 sq : $2\sqrt{2} - 3\sqrt{5}/2$
2763 4073 as : $6(1/\pi)/7$
2763 4994 $E : 4/5$
2763 6921 $Ei : ((e + \pi)/3)^2$
2763 7037 $E : ((e + \pi)/3)^{-1/3}$
2763 7224 $\zeta : 6\pi/7$
2763 7708 rt : (6,1,-4,-9)
2763 8660 cr : $3\ln 2$
2763 9320 rt : (5,5,1,0)
2763 9594 $Ei : \sqrt{2} - \sqrt{3} + \sqrt{7}$
2763 9883 $\ell_{10} : P_3$
2764 1458 at : $\exp(-\sqrt[3]{2})$
2764 2078 $J_1 : \sqrt{10}$
2764 3305 $Ei : (\sqrt{2}\pi/3)^{1/3}$
2764 3339 rt : (2,6,-7,-5)
2764 3459 $\ell_{10} : 3\sqrt[3]{2}/2$
2764 4092 rt : (7,9,4,5)
2764 8084 at : $2 + 3\sqrt{3}/4$
2764 8934 $Ei : (5/3)^{-1/2}$
2764 9429 cu : $4/3 - 2\pi$
2765 0494 rt : (5,2,-6,-6)
2765 1749 $J_1 : 4(e + \pi)/5$
2765 1800 id : $(\sqrt[3]{3}/3)^{-1/3}$
2765 3635 $J_0 : 21/11$
2765 4760 at : $7\sqrt{2}/3$
2765 4958 rt : (8,2,-5,7)
2765 6191 $\ln(\sqrt{3}) \div s\pi(\pi)$
2765 6362 cu : $4 + 4\sqrt{5}/3$
2765 7357 $\Gamma : 2e$
2765 8371 sr : $(\ln 2/3)^{-1/3}$
2765 8583 $\ell_2 : 3 - \sqrt{3}/3$
2766 0478 $2^x : 3/2 - 3\sqrt{5}/2$
2766 0520 $10^x : 4/3 - \sqrt{2}/3$
2766 1517 $J_1 : 2\sqrt[3]{3}/5$
2766 1734 $\theta_3 : 6/11$
2766 2241 rt : (2,5,-6,-2)
2766 4061 $t\pi : 3e/4 + \pi/3$
2766 4084 rt : (3,0,7,2)
2766 6249 rt : (8,5,8,2)
2766 6993 rt : (1,7,-9,-2)
2766 8029 cu : 10π
2766 8090 cr : $2/3 + \sqrt{2}$
2766 8294 rt : (6,5,9,-3)
2766 8385 $J_1 : 6$
2766 9047 rt : (4,9,-2,-4)
2767 0192 rt : (7,1,-7,4)
2767 0466 $e^x : \sqrt[3]{5}/7$
2767 1252 sq : $3\sqrt{3}/2 + 3\sqrt{5}/4$
2767 4227 rt : (6,8,2,2)
2767 4244 $10^x : (1/\pi)/3$
2767 4469 $\Psi : 4\sqrt{5}/5$
2767 5382 $J_1 : \gamma$
2767 6106 $Li_2 : (\sqrt[3]{2})^{-1/2}$
2767 7979 $J_2 : 4\sqrt[3]{2}/3$
2767 8084 sq : $(\ln 2)^{-1/3}$
2767 8689 rt : (8,0,3,1)
2767 9927 $Ei : 2\sqrt[3]{5}/3$
2768 0129 rt : (4,6,-4,7)
2768 0556 $s\pi(1/4) + s\pi(\pi)$
2768 1290 $J_1 : \sqrt{3}/3$
2768 4157 sr : $\sqrt[3]{13/3}$
2768 4594 as : $1/4 + \sqrt{2}/2$
2768 5644 $1/2 + \ln(4/5)$
2768 6917 sq : $7\sqrt[3]{5}$
2768 7655 $\ell_{10} : 1 - \sqrt{2}/3$
2768 8148 sq : $\sqrt{3}/3 + 2\sqrt{5}/3$
2768 8173 rt : (9,4,-8,2)
2768 9614 $Li_2 : (\pi/2)^{-3}$
2769 0503 rt : (1,9,-3,-4)
2769 1914 rt : (6,0,-9,1)
2769 2813 $s\pi : 1/3 + \sqrt{3}/3$
2769 3245 $J_1 : 5\ln 2/6$
2769 5153 $e^x : 3\sqrt{3}/4 + 2\sqrt{5}/3$
2769 5421 rt : (5,8,-9,-3)
2769 8396 rt : (3,8,-6,0)
2769 9704 at : $\sqrt{2}/2 + 3\sqrt{3}/2$
2770 0153 $\ell_{10} : 4/3 + \sqrt{5}/4$
2770 0831 sq : $10/19$
2770 0956 rt : (8,8,8,-3)
2770 5423 $J_1 : \sqrt{22}$
2770 7460 rt : (8,0,-6,9)
2770 8734 $10^x : \sqrt{2}/3 + 2\sqrt{3}/3$
2770 9400 sr : $3e/5$
2770 9860 sr : $2\sqrt{2}/3 - \sqrt{3}/2$
2771 1200 $Li_2 : 1/\sqrt{15}$
2771 2382 rt : (8,5,-2,0)
2771 2880 cr : $2\sqrt{2} - \sqrt{5}/3$
2771 3254 $\ell_2 : 3/4 + 4\sqrt{5}$
2771 4014 rt : (5,4,-7,-8)
2771 4858 rt : (5,4,4,9)
2771 6168 $2^x : 6/17$
2771 7509 rt : (5,3,6,-2)
2771 8238 cr : $25/12$
2771 9672 rt : (3,7,-3,-9)
2772 0405 $\Psi : (3)^3$
2772 3789 rt : (2,8,-8,-7)
2772 5887 id : $2\ln 2/5$
2772 6863 rt : (4,4,-9,6)
2772 8411 rt : (7,1,-8,-6)
2772 8582 rt : (2,7,9,2)
2772 9376 $\ln(4/5) \div \ln(\sqrt{5})$
2772 9652 $J_2 : 1/\ln(2e/3)$
2772 9747 $\ln(3/4) \times s\pi(\sqrt{2})$
2773 0944 $s\pi : (\sqrt{5})^{-3}$
2773 2206 rt : (2,6,9,-3)
2773 5009 id : $1/\sqrt{13}$
2773 5037 at : $1 + 4\sqrt{3}/3$
2773 6214 rt : (7,7,3,7)
2773 8966 cu : $15/23$
2774 0130 $\zeta : \ln(2\pi/3)$
2774 1081 rt : (4,7,-7,-5)
2774 1722 rt : (9,2,-9,4)
2774 3992 rt : (5,8,1,-1)
2774 4516 at : $4\sqrt{2}/3 - 3\sqrt{3}$
2774 5391 $J_2 : 7\sqrt[3]{3}/6$
2774 6550 $t\pi : \sqrt{3} + 3\sqrt{5}/2$
2774 6713 $\Psi : 6e\pi/5$
2774 7906 rt : (8,8,-2,-1)
2774 8162 cu : $1/3 - 3\pi/4$
2774 8383 $Li_2 : 9\ln 2/7$
2774 8780 $e^x : 6e/7$
2774 9929 as : $\sqrt{2}/3 - \sqrt{5}/3$
2775 0384 $10^x : 4e - 3\pi/4$
2775 1366 $J_1 : 11/19$
2775 1691 $J_2 : 7\zeta(3)/5$
2775 1820 sr : $\ln 2/9$
2775 2265 tanh : $\sqrt[3]{5}/6$
2775 5999 $\exp(e) \div \exp(4)$
2775 7726 $2^x : 19/16$
2775 8335 at : $3/2 + 2e/3$
2775 8824 rt : (7,6,5,-2)
2776 0524 id : $1/3 + 4\sqrt{5}$
2776 0537 rt : (4,9,8,-3)
2776 0757 $\ell_2 : 6\sqrt{2}/7$
2776 1072 rt : (5,9,-5,1)
2776 3544 at : $\sqrt[3]{5}/6$
2776 4292 sr : $1/3 + 3\sqrt{3}/4$
2776 4626 cr : $\sqrt{2}/4 + \sqrt{3}$
2776 4757 as : $\sqrt{2}/2 - \sqrt{3}/4$
2776 4827 rt : (9,9,-9,-3)
2776 6016 $10^x : 7/2$
2776 6178 rt : (9,4,9,-3)
2776 6583 $10^x : 2/3 + 2\sqrt{3}/3$
2776 7425 $K : \exp(-1/(3\pi))$
2777 0376 $2^x : \sqrt{2}/4$
2777 2053 rt : (7,3,-9,-8)
2777 3606 sinh : $7e/10$
2777 4649 at : $(\ln 3/2)^{-2}$
2777 5312 id : $4\sqrt{5}/7$
2777 5436 rt : (1,7,9,-4)
2777 6250 $2^x : 3/4 - 3\sqrt{3}/2$
2777 7184 sq : $3 - 3\pi$
2777 7777 id : $5/18$

2777 9785 at : $1/3+4\sqrt{5}/3$
2778 0454 Ei : $(\ln 3/2)^{-2}$
2778 0688 $\exp(2/3)-\exp(4/5)$
2778 1382 2^x : $10(1/\pi)/9$
2778 1929 cu : $2\sqrt{3}/3-3\sqrt{5}$
2778 4610 rt : (6,6,1,4)
2778 5554 Ψ : $3e/2$
2778 6163 Ψ : $\sqrt{5}/7$
2778 8009 rt : (4,4,8,2)
2778 8620 sr : $\sqrt{8/3}$
2778 9161 sr : $1+4\pi/3$
2779 0807 ℓ_2 : $7\sqrt{3}/5$
2779 1698 rt : (9,9,4,-2)
2779 4454 sq : $\exp(1/(3e))$
2779 4709 ln : $2\sqrt{2}+4\sqrt{3}$
2779 5031 sq : $3-2\sqrt{5}/3$
2779 5355 at : $\sqrt{11}$
2779 5709 ln : $2-e/4$
2779 6008 rt : (7,5,2,9)
2779 8847 Γ : $10\sqrt[3]{5}/7$
2779 9444 ln : $3/4-\sqrt{2}/3$
2779 9758 θ_3 : $\ln 2/5$
2779 9854 rt : (2,7,-9,-3)
2780 0275 E : $\ln(\sqrt{2}\pi/2)$
2780 0737 at : $1/2-\pi/4$
2780 0951 rt : (6,9,-8,6)
2780 2109 at : $2\sqrt{3}/3-2\sqrt{5}$
2780 2703 Ψ : $17/25$
2780 2886 cu : $25/19$
2780 3147 rt : (9,9,9,2)
2780 3164 2^x : $12/5$
2780 3957 rt : (1,6,4,1)
2780 4228 rt : (4,9,-5,-2)
2780 5312 rt : (8,9,0,2)
2780 7005 2^x : $3-2e/3$
2780 9083 rt : (9,8,1,7)
2780 9886 ln : $1/3+3\pi$
2781 0677 rt : (9,4,-9,2)
2781 6307 rt : (1,4,6,-2)
2781 7149 tπ : $\sqrt{3}/6$
2781 8549 tanh : $2/7$
2781 9005 10^x : $9\sqrt{5}/2$
2781 9453 Ψ : $7(e+\pi)/4$
2781 9458 cu : $3\sqrt{2}+\sqrt{3}$
2782 1728 rt : (7,7,5,1)
2782 3184 ℓ_{10} : $2e\pi/9$
2782 4015 $s\pi(1/3)-s\pi(1/5)$
2782 4233 10^x : $2\pi/3$
2782 4566 E : $(\pi/2)^{-1/2}$
2782 6127 ln : $8\pi/7$
2782 6345 sinh : $9\sqrt{3}/2$
2782 6931 sπ : $1/\ln(3)$
2782 7028 at : $4-e/4$
2782 7063 2^x : $(\sqrt{2}\pi/2)^3$
2782 9965 at : $2/7$
2783 0298 rt : (6,1,7,2)
2783 0299 $\ln(\pi)\div\exp(\sqrt{2})$
2783 2106 rt : (5,5,9,-3)
2783 2890 rt : (4,1,1,8)
2783 3121 J_2 : $6\zeta(3)$
2783 4879 rt : (1,6,-2,-1)
2783 5405 J_0 : $3\pi^2/4$
2783 6508 ζ : $4\zeta(3)/3$
2783 7408 rt : (2,9,4,9)
2783 9856 sπ : $\sqrt{2}+\sqrt{3}-\sqrt{5}$
2784 2863 erf : $\sqrt[3]{2}/5$
2784 3195 cu : $7\ln 2/2$
2784 7136 tπ : $1/4+3e/4$
2784 7213 2^x : $1/4-2\pi/3$
2784 7509 rt : (9,6,0,9)
2784 7875 J_2 : $5\sqrt[3]{3}$
2784 9025 sq : $3/4+3\sqrt{2}/4$
2785 1492 J_0 : $14/3$
2785 2090 rt : (7,8,8,-3)
2785 2115 id : $7(1/\pi)/8$
2785 2752 e^x : $14/17$
2785 3901 Γ : $7\pi/9$
2785 4604 ζ : $\pi/7$
2785 4805 cu : $1+3e$
2785 6533 2^x : $3/2+2\pi$
2785 8854 ℓ_2 : $7\ln 2$
2785 8969 sπ : $4\sqrt{2}+\sqrt{3}/4$
2785 9547 id : $3/4-\sqrt{2}/3$
2786 0490 rt : (8,7,-1,4)
2786 1884 rt : (8,2,6,-2)
2786 3243 rt : (7,1,-8,2)
2786 8930 at : $4/3-\pi/3$
2787 0448 $\ln(\pi)-s\pi(1/3)$
2787 1423 Ψ : $6\sqrt[3]{5}$
2787 1628 rt : (6,8,-9,-3)
2787 2093 rt : (5,5,-1,1)
2787 2861 cr : $23/11$
2787 4749 ℓ_{10} : $10\sqrt[3]{5}/9$
2787 5360 ℓ_{10} : 19
2787 6652 ℓ_{10} : $2\sqrt{3}/3+\sqrt{5}/3$
2787 8673 $\ln(2/5)\times\exp(1/3)$
2787 9063 Ψ : $4e/9$
2787 9425 rt : (1,9,1,-1)
2787 9898 2^x : $9\sqrt{5}/5$
2788 4219 rt : (1,6,-1,-9)
2788 5070 at : $3/4-3e/2$
2788 7376 e^x : $19/16$
2788 8160 rt : (6,2,-1,8)
2788 9078 tanh : $9(1/\pi)/10$
2788 9433 $\ln(\sqrt{2})\times\ln(\sqrt{5})$
2788 9997 rt : (2,6,-9,2)
2789 1437 10^x : $(\pi/2)^{-1/2}$
2789 1953 rt : (7,8,-2,-1)
2789 3388 at : $1/2+2\sqrt{2}$
2789 5142 2^x : $3e+3\pi/4$
2789 5380 ln : $5(e+\pi)/3$
2789 7057 rt : (8,5,-2,6)
2789 7551 rt : (9,4,4,1)
2789 8649 rt : (5,2,-7,-2)
2789 9108 sinh : $\sqrt{14/3}$
2790 0642 at : $9(1/\pi)/10$
2790 1806 ζ : $10\sqrt[3]{3}/9$
2790 1878 rt : (3,9,0,9)
2790 2173 10^x : $\exp(-\sqrt{5})$
2790 2421 tπ : $2\sqrt{2}+\sqrt{3}/3$
2790 3408 ℓ_2 : $\sqrt[3]{3}/7$
2790 3494 J_0 : $7e\pi/9$
2790 5437 Ei : $9(e+\pi)/10$
2790 5860 rt : (6,4,2,-1)
2790 6204 Γ : $\sqrt[3]{5}/8$
2790 9024 10^x : $2\sqrt{3}/3$
2791 0552 J_0 : $9e/8$
2791 0890 tπ : $\sqrt{3}/3-2\sqrt{5}/3$
2791 1518 ζ : $8\zeta(3)/7$
2791 1581 Li_2 : $3\ln 2/8$
2791 2248 $1/5\times\exp(1/3)$
2791 2621 tπ : $\ln 2/8$
2791 3184 $\ln(3)\times\exp(\sqrt{5})$
2791 3572 J : $\sqrt{2}/5$
2791 6598 rt : (3,9,8,7)
2791 8450 $\exp(5)+s\pi(1/3)$
2791 8622 ζ : $6\sqrt{5}/5$
2791 9173 rt : (3,7,-9,1)
2791 9973 Ei : $9\sqrt{3}/4$
2792 0429 sr : $18/11$
2792 0441 ℓ_2 : $3+3\sqrt{5}$
2792 2061 cu : $2+3\sqrt{2}$
2792 3075 rt : (8,3,-3,8)
2792 4077 rt : (3,8,2,0)
2792 6604 rt : (2,4,-9,-2)
2792 7756 id : P_{72}
2792 8104 cu : $1/2-4\sqrt{5}/3$
2792 8473 id : P_{69}
2792 8599 J_0 : $1/\ln(\sqrt[3]{3}/2)$
2792 9053 rt : (5,3,-2,3)
2792 9201 sr : $4e/3+\pi/2$
2792 9724 rt : (8,4,9,-3)
2793 0261 Ψ : $e/4$
2793 3927 cr : $3-e/3$
2793 3953 at : $10/3$
2793 4880 J_0 : $\exp(\sqrt{5}/2)$
2793 4988 rt : (8,7,1,-1)
2793 7571 ln : $3/2+2\pi/3$
2793 8415 rt : (1,9,5,-8)
2793 8843 Γ : $22/9$
2793 9252 ℓ_{10} : $7e$
2793 9727 rt : (7,6,-3,1)
2794 1003 10^x : $13/4$
2794 1318 sq : $1-\sqrt{2}/3$
2794 1549 sπ : $6(1/\pi)$
2794 2185 tπ : $1/\sqrt[3]{20}$
2794 2210 cu : $4/3-e/4$
2794 2367 ln : $\sqrt{3}+\sqrt{5}-\sqrt{7}$
2794 3207 rt : (1,8,-5,-2)
2794 3386 cr : $6\pi^2/5$
2794 3535 J_1 : $7/12$
2794 3725 sq : $3/2-4\sqrt{2}$
2794 3886 id : $(2\pi/3)^{1/3}$
2794 5231 rt : (1,5,-1,-9)
2794 7521 at : $\sqrt{2}/4+4\sqrt{5}/3$
2794 7734 rt : (2,0,7,2)
2794 8975 rt : (9,9,-4,-1)
2795 0705 sr : $3\sqrt{3}$
2795 0849 id : $\sqrt{5}/8$
2795 3772 rt : (3,7,3,6)
2795 4473 10^x : $1/4+2\sqrt{3}$
2795 4788 id : $8\pi^2/7$
2795 5069 rt : (5,8,9,2)
2795 8105 Ψ : $\ln(\sqrt[3]{4}/3)$
2795 9222 10^x : $4\sqrt{3}-\sqrt{5}$
2795 9719 10^x : $3\sqrt{3}/2+\sqrt{5}/2$
2796 0705 ℓ_2 : $1/3+2\pi/3$
2796 1281 rt : (5,1,-3,5)
2796 2481 J_1 : $7\sqrt[3]{5}/2$
2796 5098 rt : (4,3,6,-2)
2796 5530 e^x : $3/4+3\sqrt{3}$
2796 5977 rt : (9,1,-4,-1)
2796 7095 10^x : $7\sqrt[3]{3}/8$
2796 7935 rt : (7,4,-4,3)
2796 7983 ln : $\sqrt{3}/3+\sqrt{5}/3$
2796 9651 ℓ_{10} : $1/3+\pi/2$
2796 9973 2^x : $2\sqrt{2}/3+2\sqrt{3}/3$
2797 0792 ζ : $\exp(\sqrt{2}/3)$
2797 0994 2^x : $\sqrt{2}/3-4\sqrt{3}/3$
2797 3311 2^x : $\ln(1/(2\pi))$
2797 3630 ln : $4/3-\sqrt{3}/3$
2797 3870 rt : (9,7,-5,1)
2797 5694 rt : (7,5,8,2)
2797 5895 rt : (3,7,-9,-3)
2797 6999 rt : (3,5,-2,5)
2797 7036 tπ : $7/19$
2797 7656 ln : $3\sqrt[3]{2}/5$
2797 7809 as : $2/3-2\sqrt{2}/3$
2797 8013 J_0 : $(e)^2$
2797 8380 rt : (6,6,5,-2)
2798 0518 Ei : $\exp(-(e+\pi))$
2798 1767 10^x : $3/4+4\pi$
2798 2007 J_1 : $((e+\pi)/2)^{-1/2}$
2798 2016 rt : (5,1,-4,-7)
2798 2670 ℓ_{10} : $3/4+2\sqrt{3}/3$
2798 4320 J_2 : $22/13$
2798 4781 sr : $3\sqrt{2}/4+\sqrt{3}/3$
2798 5625 rt : (6,8,-9,9)
2798 6334 ln : $3-3\sqrt{5}/4$
2798 6558 rt : (7,2,-5,5)
2798 7704 rt : (1,7,5,3)
2798 8864 rt : (8,9,4,-2)
2798 9731 e^x : $3-2e/3$
2799 0395 rt : (9,2,7,2)
2799 0689 rt : (9,5,-6,3)
2799 1956 tπ : $9\ln 2/4$
2799 2514 e^x : $4\sqrt[3]{3}/7$
2799 2890 rt : (3,3,-7,4)
2799 3112 tπ : $7\sqrt{5}/4$
2799 5598 sr : $4/3+3\pi$
2799 6477 rt : (5,3,-5,-9)
2799 6792 id : $4\sqrt[3]{5}/3$
2799 8041 10^x : $4\sqrt{2}-2\sqrt{5}/3$
2799 8245 id : $2\sqrt[3]{2}/9$
2799 9774 rt : (7,0,-6,7)
2800 0000 id : $7/25$
2800 0828 rt : (1,7,-3,-7)
2800 1506 rt : (4,5,9,-3)
2800 1780 rt : (7,4,-3,-1)
2800 2814 rt : (9,3,-7,5)

2800 2856 sq : P_3
2800 3286 ℓ_{10} : $4-2\pi/3$
2800 4937 2^x : $2\sqrt{3}/3+\sqrt{5}/4$
2800 4974 Ψ : $\zeta(3)/6$
2800 5019 J_2 : $\ln(2e)$
2800 6727 at : $2e-2\pi/3$
2800 7274 cr : $2\sqrt{2}/3+2\sqrt{3}/3$
2800 7831 rt : (6,8,8,-3)
2800 7863 J_1 : $1/\sqrt[3]{5}$
2800 9640 rt : (7,2,-7,-9)
2801 0791 ℓ_2 : $7/17$
2801 1559 sπ : $4\sqrt{3}+4\sqrt{5}/3$
2801 1798 at : $\ln(4/3)$
2801 1971 rt : (9,1,-8,7)
2801 2569 E : $5(1/\pi)/2$
2801 4430 ℓ_{10} : $1+e/3$
2801 5323 rt : (5,8,-6,4)
2801 7107 at : $3\sqrt{3}/2+\sqrt{5}/3$
2801 8685 tπ : $2/23$
2801 9130 rt : (9,1,-9,-9)
2801 9299 J_0 : $7e/10$
2802 0233 rt : (5,7,-2,-1)
2802 0397 Li_2 : $6/23$
2802 3896 id : $\sqrt{2}+\sqrt{3}/2$
2802 6301 cu : $3/2+4\sqrt{3}/3$
2802 6844 Ψ : $(\ln 3)^2$
2802 7388 2^x : $(\sqrt{2})^{1/2}$
2802 7681 sq : $9/17$
2802 8098 cu : $7\sqrt{3}$
2803 0570 K : $5\sqrt[3]{2}/7$
2803 1251 cu : $3e+3\pi/4$
2803 4704 $\ln(3/5) \div \exp(3/5)$
2803 5948 10^x : $4/3-4\sqrt{2}/3$
2803 6089 rt : (9,3,-7,-2)
2803 8289 2^x : $2/3+\pi/3$
2803 8782 e^x : $2e-2\pi/3$
2803 9931 ln : $1+3\sqrt{3}/2$
2804 1312 sq : $4e/3+3\pi$
2804 2314 rt : (7,6,-6,-2)
2804 2521 ζ : $\ln(\sqrt[3]{3}/2)$
2804 5304 e^x : $4\sqrt{2}-4\sqrt{3}$
2804 6207 id : $3\sqrt{3}/4+4\sqrt{5}/3$
2804 6919 rt : (5,9,-5,-2)
2804 6956 rt : (2,9,-9,-4)
2804 8345 ℓ_2 : $2-\pi/4$
2804 8522 e^x : $e+3\pi$
2805 0087 at : $4\sqrt{2}-4\sqrt{3}/3$
2805 0548 cu : $2/3+4\pi$
2805 1047 sinh : $21/4$
2805 1114 2^x : $4\sqrt{2}+\sqrt{3}/3$
2805 2649 cu : $6\pi^2/5$
2805 3650 rt : (2,8,-1,6)
2805 4618 rt : (7,8,-9,-3)
2805 4913 K : $9/10$
2805 5191 at : $3/4+3\sqrt{3}/2$
2805 5241 cr : $5\sqrt[3]{2}/3$
2805 6089 sinh : $25/16$
2805 6737 tπ : $\sqrt[3]{7}$
2805 7287 sq : $5e/9$
2805 7916 cr : $21/10$
2805 8659 at : $4(e+\pi)/7$
2805 9630 J_1 : $(e+\pi)/10$
2805 9965 $\ln(3/5) \times \ln(\sqrt{3})$
2806 0083 Γ : $9\sqrt{3}/4$
2806 2612 cu : $1/2-2\sqrt{3}/3$
2806 4150 e^x : $\sqrt{2}+\sqrt{3}+\sqrt{6}$
2806 5593 2^x : $1/\sqrt[3]{22}$
2806 6650 Ei : $\pi/7$
2807 0749 tanh : $\sqrt[3]{3}/5$
2807 3717 e^x : $\sqrt{3}/7$
2807 5062 θ_3 : $\sqrt[3]{2}/9$
2807 6198 sinh : $16/15$
2807 6739 Ei : $7\sqrt{3}/2$
2807 7640 rt : (8,8,-6,-4)
2808 0245 e^x : $9\sqrt[3]{2}$
2808 1796 $\ln(\sqrt{5}) - \exp(3)$
2808 2481 sinh : $2\ln 2/5$
2808 2660 ℓ_{10} : $11/21$
2808 2699 at : $\sqrt[3]{3}/5$
2808 3496 rt : (1,8,4,-9)
2808 5058 e^x : $3e-3\pi$
2808 5996 10^x : $9(1/\pi)/8$
2808 6233 10^x : $10\sqrt{2}$
2808 7004 sinh : $10e/7$
2808 8668 2^x : $5/14$
2809 1460 e^x : $7(1/\pi)/9$
2809 1495 tanh : $\sqrt{3}/6$
2809 1956 sinh : $1/\sqrt{13}$
2809 2209 cu : $2\sqrt{2}/3-4\sqrt{5}$
2809 3243 $\ln(2) + s\pi(1/5)$
2809 3384 ln : $5/18$
2809 3995 as : $2\ln 2/5$
2809 6022 sr : $3-e/2$
2809 6651 cu : $\sqrt{19/2}$
2809 7018 rt : (9,7,8,-3)
2809 7490 at : $e/2-3\pi/2$
2809 8758 rt : (7,4,9,-3)
2809 9063 J_1 : $7e\pi/10$
2809 9983 at : $3+\sqrt{2}/4$
2810 0137 ℓ_{10} : $\pi/6$
2810 1372 J_2 : $6\sqrt{2}/5$
2810 3490 at : $\sqrt{3}/6$
2810 4462 at : $3\sqrt{5}/2$
2810 4699 $\ln(2) \times \ln(2/3)$
2810 5055 rt : (9,5,5,-2)
2810 6649 ln : $(\sqrt[3]{5}/3)^{1/2}$
2810 6737 $1/3 + \exp(2/3)$
2810 8005 sπ : $2/3+3\sqrt{2}$
2810 8672 rt : (7,2,6,-2)
2811 0695 sinh : $\sqrt[3]{25}$
2811 0933 as : $23/24$
2811 1348 J_2 : $\exp(\sqrt[3]{4}/3)$
2811 1642 tπ : $7\pi^2$
2811 2743 Γ : $9e/10$
2811 3391 rt : (5,1,7,2)
2811 8877 ℓ_{10} : $4/3+\sqrt{3}/3$
2811 9805 rt : (3,4,8,2)
2812 1534 at : $1+3\pi/4$
2812 2261 rt : (1,5,-3,-8)
2812 4475 rt : (9,3,2,-1)
2812 4748 erf : $4(1/\pi)/5$
2812 5000 sq : $3\sqrt{2}/8$
2812 5794 tπ : $3\sqrt{3}+3\sqrt{5}/2$
2812 6699 ζ : $1/\sqrt[3]{11}$
2812 8425 Ei : $1/\ln(\sqrt[3]{3})$
2812 8899 rt : (9,9,3,8)
2812 9030 rt : (1,7,9,2)
2812 9659 10^x : $2/3-\sqrt{5}/4$
2813 1042 cr : $7\zeta(3)/4$
2813 3138 cr : $1+4e$
2813 3239 10^x : $12/19$
2813 4116 sq : $4\sqrt[3]{3}$
2813 4142 2^x : $12/7$
2813 4817 cu : $4/3+3\sqrt{5}/4$
2813 5695 $\sqrt{3}+\ln(\sqrt{3})$
2813 5805 Ψ : $(\sqrt[3]{5}/3)^{-1/3}$
2813 6128 rt : (7,0,3,1)
2813 6383 sinh : $5/18$
2813 6938 J_1 : $5\sqrt[3]{2}/2$
2813 8289 J_1 : $\sqrt{3}-\sqrt{5}-\sqrt{7}$
2813 8526 $2 \times \exp(\pi)$
2813 9415 tπ : $2e/5$
2814 0481 J_2 : $(\ln 2/2)^{-1/2}$
2814 0814 $\exp(\sqrt{3}) \div \exp(3)$
2814 3171 rt : (7,8,5,8)
2814 4480 Ψ : $4\pi/7$
2814 4773 sq : $5(1/\pi)/3$
2814 5540 at : $2+e/2$
2814 7442 J_2 : $5e/8$
2814 8007 as : $5/18$
2815 0818 at : $8\sqrt[3]{2}/3$
2815 0915 ln : $3/2-\sqrt{5}/3$
2815 1152 tπ : $e+\pi/2$
2815 1803 e^x : $1/4+\sqrt{2}$
2815 4829 2^x : $9\pi^2/5$
2815 5522 rt : (5,3,4,1)
2815 6708 rt : (5,9,-9,-8)
2815 7143 2^x : $1-2\sqrt{2}$
2815 7838 J_1 : $10/17$
2815 8598 rt : (4,7,-9,-3)
2815 8804 ζ : $e\pi/10$
2815 9982 Ψ : $(2e/3)^{-3}$
2816 2950 Li_2 : $((e+\pi)/3)^{-2}$
2816 5790 J_1 : $3\pi/2$
2816 8463 rt : (5,8,8,-3)
2816 9934 ℓ_{10} : $\sqrt[3]{7}$
2817 1025 rt : (8,9,9,2)
2817 1817 id : $3-e$
2817 2070 J_0 : $20/3$
2817 3255 sπ : $1/11$
2817 3554 2^x : $9(1/\pi)/8$
2817 3894 J_2 : $17/10$
2817 3988 sr : $23/14$
2817 5662 id : $\sqrt{2}/4+4\sqrt{3}$
2817 6884 rt : (3,5,9,-3)
2817 8347 J_1 : $(\ln 2/2)^{1/2}$
2817 8468 at : $\sqrt{3}-\sqrt{6}-\sqrt{7}$
2817 9047 tπ : $\sqrt{2}/4+\sqrt{5}/4$
2817 9345 ℓ_2 : $3e+\pi/2$
2817 9409 id : P_6
2817 9971 e^x : $\exp(\pi/3)$
2818 1855 J_0 : $19/10$
2818 2657 rt : (2,9,0,-8)
2818 2959 cu : $3/4-\sqrt{5}$
2818 3409 J_0 : $10\sqrt[3]{5}/9$
2818 4285 sπ : $2(1/\pi)/7$
2818 5161 rt : (2,8,-5,-2)
2818 9292 10^x : $2\sqrt{2}+2\sqrt{5}$
2818 9536 e^x : $4\sqrt{2}+2\sqrt{3}/3$
2819 0413 2^x : $2\sqrt{3}/3-4\sqrt{5}/3$
2819 1381 $4 \div \ln(\sqrt{3})$
2819 3519 rt : (7,9,4,-2)
2819 4283 rt : (3,6,5,1)
2819 5181 at : $7\sqrt[3]{3}/3$
2819 6249 J_1 : $\exp(-\sqrt[3]{4}/3)$
2819 6364 E : $\sqrt[3]{1/2}$
2819 6704 rt : (3,5,-9,-3)
2819 7696 rt : (2,5,-3,7)
2819 9038 ℓ_{10} : $1/2+\sqrt{2}$
2819 9312 10^x : $3/2+\sqrt{3}$
2819 9880 J_1 : $10\sqrt{2}/3$
2819 9961 id : $8\sqrt[3]{3}/9$
2820 0403 rt : (8,4,-1,9)
2820 1056 rt : (4,5,-6,-2)
2820 2317 $\ln(\sqrt{3}) \div \exp(2/3)$
2820 2404 ζ : $(\ln 2)^{-1/3}$
2820 3927 e^x : $\sqrt{5}/9$
2820 4644 rt : (1,8,-8,-5)
2820 5041 at : P_{82}
2820 6250 sinh : $e\pi/8$
2820 6935 $s\pi(1/3) \div s\pi(\sqrt{5})$
2820 7263 ln : $1/4+3\sqrt{5}/2$
2820 7592 rt : (5,9,8,6)
2820 9479 sr : $(1/\pi)/4$
2821 0126 θ_3 : $3\pi/10$
2821 0978 sr : $4-3\pi/4$
2821 3571 ℓ_{10} : $\sqrt{11/3}$
2821 3614 sinh : $7(1/\pi)/8$
2821 4348 ln : $(\ln 3)^3$
2821 6567 sr : $(\sqrt{2}\pi)^{1/3}$
2821 8499 J_2 : $9\ln 2$
2821 9232 rt : (5,0,-9,1)
2822 0721 id : $2e/3-2\pi/3$
2822 1448 rt : (8,6,0,7)
2822 1608 2^x : $3/4+4\sqrt{5}/3$
2822 2500 rt : (6,2,-7,-2)
2822 2514 rt : (5,7,-7,7)
2822 3314 Ei : $1/\ln((e+\pi)/3)$
2822 4466 Ei : $2e/7$
2822 5129 Ψ : $\exp(2\pi)$
2822 5398 as : $7(1/\pi)/8$
2822 5575 rt : (3,6,9,8)
2822 5666 cu : $4\sqrt{2}/3-3\sqrt{3}$
2822 7459 sr : $8(e+\pi)/9$
2822 8061 2^x : $25/21$
2822 8528 $\pi + \exp(\pi)$
2822 8760 rt : (6,3,1,9)
2822 8893 10^x : $\ln(\sqrt{3}/3)$
2822 9442 ℓ_2 : $3\sqrt{2}/2-3\sqrt{3}/4$
2823 0519 id : $4\sqrt{3}+3\sqrt{5}/2$

2823 0609 rt : (3,6,-5,-7)
2823 1187 e^x : $4/3 - 3\sqrt{3}/2$
2823 2512 2^x : $3/2 + 4\sqrt{5}$
2823 3137 as : $3/4 - \sqrt{2}/3$
2823 3870 ℓ_2 : $\sqrt{5} - \sqrt{6} + \sqrt{7}$
2823 4175 at : $3\sqrt{2}/4 + 4\sqrt{3}/3$
2823 4581 rt : (2,7,-6,3)
2823 4771 sinh : $6\sqrt{5}/5$
2823 5007 cu : $4/3 - 4\pi/3$
2823 7569 rt : (9,8,-2,-1)
2823 9567 rt : (5,9,-2,8)
2824 0434 ζ : 11/17
2824 0573 sr : $3\sqrt{2} - 3\sqrt{3}/2$
2824 1065 tπ : $\sqrt{2}/3 - \sqrt{5}/4$
2824 1439 ζ : 9/20
2824 2171 cu : $1/4 - e/3$
2824 3262 ln : $\sqrt[3]{7/3}$
2824 3451 cr : $3/4 + e/2$
2824 3984 rt : (1,9,8,-3)
2824 4934 rt : (8,8,-9,-3)
2824 5135 rt : (3,9,-4,-4)
2824 5698 2^x : $-\sqrt{2} + \sqrt{5} - \sqrt{7}$
2824 6321 10^x : $3/4 - 3\sqrt{3}/4$
2824 7118 10^x : $(e/3)^{1/3}$
2824 7467 ln : $\sqrt{13}$
2824 8196 id : $3\sqrt[3]{5}/4$
2824 8306 rt : (9,2,-6,8)
2824 9374 rt : (1,0,7,2)
2824 9482 ln : $5\sqrt[3]{3}/2$
2825 0217 at : $4/3 + 3e/4$
2825 0630 Ei : $(2e/3)^{-2}$
2825 1192 rt : (8,8,1,5)
2825 2226 rt : (7,5,-9,-5)
2825 3050 rt : (8,1,-8,2)
2825 3641 e^x : 16/11
2825 3821 ℓ_{10} : $6\sqrt{5}/7$
2825 4658 ℓ_{10} : 12/23
2825 4983 sr : $\pi^2/6$
2825 7166 rt : (8,7,8,-3)
2825 8874 tπ : $6\ln 2/7$
2825 9939 rt : (1,9,6,-5)
2826 1741 J_1 : $1/\ln(2e)$
2826 1963 rt : (6,5,2,7)
2826 2009 ln : $4\sqrt{2}/3 - \sqrt{5}/4$
2826 2151 cr : $\sqrt{3}/4 + 3\sqrt{5}/4$
2826 4646 ln : $3\zeta(3)$
2826 5670 rt : (7,7,1,-1)
2826 7296 sπ : $(\sqrt{2}\pi/2)^{-3}$
2826 8359 rt : (3,2,-8,7)
2826 8398 sq : $1/3 - 3\sqrt{2}$
2826 9049 at : $1/4 - 4e/3$
2826 9626 id : $e/3 - 4\pi/3$
2826 9855 rt : (5,2,-8,-3)
2827 2460 rt : (9,4,-5,6)
2827 2479 rt : (8,2,7,2)
2827 2886 Ei : $(e)^{-2}$
2827 3651 rt : (6,4,9,-3)
2827 4087 at : $(3/2)^3$
2827 4169 rt : (4,2,3,9)
2827 4385 ℓ_2 : $2 + \sqrt{3}/4$
2827 4518 J_1 : 13/22
2827 5508 rt : (2,6,-2,-1)
2827 7374 10^x : $3(e+\pi)/2$
2827 8492 Ei : $\sqrt[3]{10/3}$
2828 0365 ln : $2\ln 2/5$
2828 0384 sπ : $\sqrt{2} + 3\sqrt{5}/4$
2828 0582 rt : (7,1,-4,8)
2828 0648 tπ : $4e - \pi/4$
2828 2307 sq : $3\sqrt{2}/2 + \sqrt{3}/3$
2828 2386 rt : (1,3,-8,2)
2828 3371 cr : 19/9
2828 4112 sπ : $7\pi^2/10$
2828 4271 id : $\sqrt{2}/5$
2828 4427 10^x : $\ln(\sqrt{2}\pi/2)$
2828 4818 at : $4\sqrt{2}/3 + 2\sqrt{5}/3$
2828 5549 e^x : $7(1/\pi)$
2828 5605 J_0 : $(3\pi)^{1/2}$
2828 5751 Li_2 : 5/19
2828 7118 at : $3\sqrt{2} - \sqrt{3}/2$
2829 0084 ℓ_{10} : $\sqrt{2} - \sqrt{3} + \sqrt{5}$
2829 1017 10^x : $3\sqrt{2}/2 + \sqrt{3}/2$
2829 1468 rt : (3,8,0,-6)
2829 2605 10^x : $4\sqrt{2}/3 + 2\sqrt{3}/3$
2829 4212 id : $8(1/\pi)/9$
2829 9058 rt : (3,4,-3,8)
2830 2197 rt : (6,9,-5,-2)
2830 4165 as : P_{72}
2830 4524 ℓ_2 : $\sqrt{2}/3 + \sqrt{5}/3$
2830 4568 rt : (9,6,-4,4)
2830 4912 as : P_{69}
2830 5819 rt : (5,4,2,-1)
2830 6569 sr : $1/2 + 3\pi/2$
2830 6783 at : $4/3 - 3\pi/2$
2830 6864 rt : (6,5,8,2)
2830 6896 2^x : $4 - \pi/4$
2830 7589 cr : $(3\pi)^{1/3}$
2830 8142 sπ : $1/\ln(5/2)$
2831 0300 Γ : $1/\sqrt[3]{17}$
2831 1578 e^x : $7e/9$
2831 1858 rt : (1,9,6,1)
2831 2705 2^x : $1/\ln(\sqrt{3}/3)$
2831 3837 rt : (6,7,3,5)
2831 3935 sr : $2 - \sqrt{2}/4$
2831 4440 J_0 : $2e\pi/9$
2831 5672 rt : (7,3,-3,6)
2831 6217 sinh : $\sqrt{5}/8$
2831 6487 rt : (8,1,-9,-7)
2831 7598 sq : $3e/2 - \pi/2$
2831 8467 ln : $3/4 + \sqrt{3}/3$
2831 8530 id : 2π
2831 8599 rt : (1,7,5,-2)
2832 0248 10^x : $2\sqrt{2}/3 - 2\sqrt{5}/3$
2832 0359 Ei : 7/23
2832 3721 Γ : 7/18
2832 5237 rt : (6,4,-8,-9)
2832 6681 ℓ_2 : $(\sqrt{2}\pi/3)^{1/2}$
2832 7929 rt : (5,0,-2,8)
2832 8216 as : $\sqrt{5}/8$
2832 9800 10^x : $\sqrt[3]{5}$
2833 1340 rt : (3,6,-9,2)
2833 1493 E : $8\ln 2/7$
2833 2977 cr : $3\sqrt{3} + 3\sqrt{5}$
2833 3143 rt : (8,5,5,-2)
2833 3160 sπ : $(4/3)^{-1/3}$
2833 4293 rt : (4,8,8,-3)
2833 4644 sπ : $2/3 + 3\pi$
2833 4659 rt : (4,5,-9,1)
2833 5965 rt : (7,7,-8,-7)
2833 6730 Li_2 : $\exp(-4/3)$
2833 7298 rt : (8,6,-6,-2)
2833 8083 sq : $9\pi^2/8$
2833 8385 at : $2e/3 + \pi/2$
2833 8492 $\exp(4) \times s\pi(1/3)$
2833 8879 10^x : $\sqrt{21/2}$
2834 0583 2^x : $2\sqrt[3]{2}/7$
2834 1528 rt : (3,6,2,9)
2834 1993 $3/5 \div \exp(3/4)$
2834 2175 cu : $8\pi/7$
2834 2589 2^x : 9/25
2834 3258 ln : $\sqrt{3}/2 + 4\sqrt{5}$
2834 3605 ℓ_2 : $e/4 + 4\pi/3$
2834 4032 cu : $8\sqrt{3}/5$
2834 6295 cr : $7e/9$
2834 7076 e^x : $\sqrt{2}/3 - \sqrt{3}$
2834 8474 cu : $3 - 2\sqrt{3}/3$
2834 8486 rt : (5,7,-9,-3)
2834 9319 rt : (9,8,-3,2)
2834 9587 cr : $4 - 4\sqrt{2}/3$
2834 9727 10^x : $4\sqrt{3} + 2\sqrt{5}/3$
2835 0531 rt : (8,1,-8,5)
2835 1283 cu : $e + 3\pi/2$
2835 2250 tπ : $3\sqrt{2} - 2\sqrt{3}/3$
2835 2581 J : $(1/\pi)/3$
2835 3694 rt : (1,7,5,-3)
2835 4161 at : $2/3 + e$
2835 6187 rt : (4,8,9,2)
2835 6605 2^x : $3 + 3\sqrt{3}$
2835 6705 rt : (1,3,-9,-7)
2835 7139 ln : $3e/4 + \pi/2$
2835 8667 rt : (2,5,9,-3)
2835 9535 at : $3/2 + 4\sqrt{2}/3$
2836 0752 tπ : $7(e+\pi)/2$
2836 0808 erf : $\exp(-e/2)$
2836 1551 Ei : $1/\sqrt[3]{11}$
2836 2025 rt : (6,2,-7,-7)
2836 2687 ℓ_{10} : $9e\pi/4$
2836 4275 θ_3 : $\sqrt{2}/10$
2836 5408 rt : (7,5,-2,4)
2836 5479 sinh : $2\sqrt[3]{2}/9$
2836 6259 ζ : $\sqrt[3]{19}$
2836 6284 10^x : $1/2 - \pi/3$
2836 6785 10^x : $\exp(-\sqrt{2}\pi/2)$
2836 6809 tanh : 7/24
2836 7303 sinh : 7/25
2836 7632 sq : $9e\pi/5$
2836 7642 id : $\exp(-\sqrt[3]{2})$
2836 7971 rt : (9,3,9,-3)
2836 9041 2^x : $4\sqrt{3}/3 - \sqrt{5}/2$
2836 9513 J_2 : $e\pi/5$
2837 2073 J_2 : 25/4
2837 3505 sinh : $8\zeta(3)/9$
2837 4180 at : $2\sqrt{2} + \sqrt{5}/4$
2837 4329 θ_3 : $4(1/\pi)/9$
2837 4539 rt : (4,7,-4,2)
2837 5045 $\ln(\sqrt{2}) \div \exp(1/5)$
2837 5603 E : 19/24
2837 6084 sq : $1/3 - \sqrt{3}/2$
2837 6466 Γ : $\sqrt{6}$
2837 6556 rt : (5,4,-7,-5)
2837 7049 tπ : $4e/3 - 3\pi/2$
2837 7583 as : $2\sqrt[3]{2}/9$
2837 9410 at : 7/24
2837 9490 cu : $4e/5$
2838 2768 sq : $2 + e/2$
2838 2896 J_1 : $8\sqrt{5}/3$
2838 3543 rt : (5,2,-1,6)
2838 5560 sr : $8\sqrt[3]{3}/7$
2838 8699 10^x : $4\sqrt{3}/3 - \sqrt{5}/4$
2838 8776 sr : $e/2 + 3\pi$
2839 2630 2^x : $\sqrt[3]{3}/4$
2839 4958 Γ : $\sqrt{2} - \sqrt{5} - \sqrt{7}$
2839 5871 rt : (8,3,-7,3)
2839 7496 2^x : $3\zeta(3)/10$
2839 7915 ℓ_{10} : $4\sqrt[3]{3}/3$
2839 9665 ℓ_{10} : 13/25
2840 0805 at : $2\sqrt{3}/3 + \sqrt{5}$
2840 0919 rt : (3,1,-7,-2)
2840 2493 id : $(2e/3)^2$
2840 2541 e^x : 1/4
2840 3977 at : $\sqrt{23/2}$
2840 4142 rt : (3,1,0,-8)
2840 4500 ℓ_{10} : $8\zeta(3)/5$
2840 6229 rt : (6,9,4,3)
2840 7398 at : $1/4 + \pi$
2840 7835 Ψ : $10e$
2840 9687 tπ : $(\pi/3)^{-2}$
2841 1327 ℓ_{10} : $3\ln 2/4$
2841 1597 J_1 : 22/7
2841 1631 rt : (6,0,-6,5)
2841 2553 sinh : $5\sqrt[3]{5}/8$
2841 3929 rt : (4,1,7,2)
2841 4252 ℓ_{10} : $9\sqrt[3]{5}/8$
2841 4291 ℓ_{10} : $\sqrt{3}/4 + 2\sqrt{5}/3$
2841 4706 J_1 : $6\ln 2/7$
2841 5983 rt : (9,8,-3,0)
2841 6477 e^x : 14
2841 8299 sr : $7\sqrt{5}/3$
2841 9054 ζ : 8/3
2841 9422 J_2 : $\sqrt[3]{5}$
2842 0973 cr : $\sqrt[3]{19/2}$
2842 1139 cu : 25/23
2842 2408 rt : (7,7,8,-3)
2842 3015 J : $\exp(-11\pi/3)$
2842 3442 cr : $1 + \sqrt{5}/2$
2842 5886 $\exp(1/5) \div s\pi(2/5)$
2842 7058 ln : $3\sqrt{2}/2 + 2\sqrt{5}/3$
2842 7753 rt : (2,4,-6,8)
2842 9052 rt : (4,9,1,3)
2842 9311 2^x : $9(1/\pi)$
2843 1811 ℓ_{10} : $10\sqrt{3}/9$

2843 2408 rt : (3,8,1,-1)
2843 2501 J_1 : $\ln(2e/3)$
2843 5027 cr : $4e+\pi/3$
2843 6241 e^x : 19/23
2843 7538 cr : $4/3+\pi/4$
2843 8470 rt : (6,9,-9,4)
2843 8984 $1/3+s\pi(2/5)$
2843 9351 rt : (3,8,-5,-2)
2843 9484 rt : (3,8,-7,-3)
2844 1343 rt : (7,7,-1,2)
2844 2189 θ_3 : $\exp(-(e+\pi)/3)$
2844 2370 tπ : $\sqrt[3]{12}$
2844 3186 rt : (9,8,-9,-3)
2844 4444 sq : 8/15
2844 5705 id : $\sqrt{2}/2+\sqrt{3}/3$
2844 5727 ℓ_2 : $\sqrt{2}+\sqrt{6}-\sqrt{7}$
2844 6369 e^x : $4-4\sqrt{2}/3$
2844 7320 rt : (4,6,5,5)
2844 7596 e^x : $(\sqrt{2})^{1/2}$
2844 8414 rt : (2,6,-4,-7)
2844 9049 sr : $7\sqrt{2}/6$
2844 9274 rt : (2,6,-9,-3)
2845 3491 ℓ_2 : $(e/3)^2$
2845 3926 cu : $9(e+\pi)$
2845 4815 rt : (5,4,9,-3)
2845 6489 rt : (6,0,-4,1)
2845 7378 at : $5e/4$
2845 9227 rt : (8,5,-6,1)
2846 1109 rt : (3,8,-9,-8)
2846 1534 J_1 : π
2846 1962 J_2 : $((e+\pi)/2)^{1/2}$
2846 3477 Ψ : $7(e+\pi)/10$
2846 5244 id : $e+4\pi$
2846 7394 10^x : 23/11
2846 8028 rt : (5,2,-8,2)
2846 8115 Ei : 8/7
2847 0859 rt : (5,4,0,4)
2847 1512 λ : $\exp(-16\pi/13)$
2847 4369 $\ln(5)\div\exp(\sqrt{3})$
2847 4488 at : 17/5
2847 6830 $\ln(3/5)-s\pi(e)$
2847 7476 rt : (2,4,8,2)
2847 7730 J_0 : $8(e+\pi)/7$
2848 0007 10^x : $7\sqrt{2}/2$
2848 0234 erf : $(\pi/2)^{-3}$
2848 1084 ℓ_{10} : $2\sqrt{2}-4\sqrt{3}/3$
2848 2149 rt : (6,2,-5,3)
2848 3964 10^x : $(2\pi/3)^{-3}$
2848 5351 rt : (9,6,8,2)
2848 5540 Ei : 9/20
2848 9320 ℓ_{10} : $1/4+3\sqrt{5}/4$
2848 9829 id : $(\sqrt{2}/3)^{-1/3}$
2849 0079 tπ : $6\sqrt[3]{2}$
2849 0787 ζ : $3\zeta(3)/8$
2849 0830 Ψ : $10\sqrt[3]{2}/7$
2849 1657 rt : (9,3,-4,9)
2849 2014 rt : (8,4,4,1)
2849 2412 at : $1-\sqrt{2}/2$
2849 2820 $\ln(5)+s\pi(\sqrt{5})$
2849 5028 rt : (5,5,-6,-2)
2849 5135 at : $1/\ln(\sqrt{5}/3)$
2849 8670 10^x : $3/2+4e$
2849 9143 Ψ : 9/5
2849 9599 id : $\sqrt[3]{5}/6$
2849 9777 rt : (1,9,-2,6)
2849 9934 erf : $1/\sqrt{15}$
2850 0326 rt : (2,3,6,7)
2850 1146 $\sqrt{5}-s\pi(2/5)$
2850 2478 ln : $4e-\pi/3$
2850 4939 rt : (9,1,6,-2)
2850 5069 Γ : 17/24
2850 5662 rt : (3,8,8,-3)
2850 6476 e^x : $3/2+3\pi/4$
2850 7061 rt : (6,7,-2,-1)
2850 7955 J : $\exp(-7\pi/8)$
2850 8041 rt : (4,1,-4,-5)
2850 8514 Γ : 17/4
2850 9388 rt : (5,3,-9,-4)
2851 0160 rt : (1,8,-2,3)
2851 1633 J : $(5/2)^{-3}$
2851 4248 rt : (1,5,2,-1)
2851 4486 2^x : $10\sqrt{5}/3$
2851 6406 cu : P_{19}
2851 6487 ln : $7\sqrt[3]{5}/9$
2851 6565 J_1 : $(3\pi/2)^{-1/3}$
2851 6602 cu : P_{58}
2851 7273 rt : (9,8,4,-2)
2851 7738 ℓ_{10} : $\sqrt{3}-\sqrt{6}+\sqrt{7}$
2851 9932 sr : $3/4+2\sqrt{5}$
2851 9995 rt : (8,3,2,-1)
2852 0415 at : $2\sqrt{2}+\sqrt{3}/3$
2852 1367 2^x : $3\sqrt{3}/2-\sqrt{5}$
2852 3571 e^x : $2+e/3$
2852 5375 J_2 : 12/7
2852 8414 2^x : $e-3\pi/4$
2852 9256 sinh : $3\sqrt{3}$
2852 9651 Li_2 : $5(1/\pi)/6$
2853 0934 2^x : $1/2-4\sqrt{3}/3$
2853 1151 J_0 : 22/3
2853 1708 2^x : $3+2\sqrt{3}$
2853 2205 rt : (3,6,-8,-6)
2853 2774 sinh : $7(e+\pi)/8$
2853 3133 ζ : $\sqrt[3]{1/2}$
2853 3677 J_0 : $9\sqrt[3]{5}/5$
2853 7528 rt : (2,8,5,1)
2853 8018 10^x : $4\sqrt{2}+\sqrt{5}/4$
2853 9582 Ei : 7/9
2853 9706 Ei : $6e/7$
2853 9816 id : $1/2+\pi/4$
2854 1291 sq : $2+\pi/3$
2854 2040 sq : $4\zeta(3)/9$
2854 2251 $\ln(\pi)+\exp(\pi)$
2854 2958 tπ : $4\sqrt{2}-\sqrt{5}/3$
2854 3910 cu : $e/2+3\pi/4$
2854 5536 sinh : 18/5
2854 6143 rt : (6,7,-9,-3)
2854 6776 at : $\sqrt{3}+3\sqrt{5}/4$
2854 7335 rt : (1,5,9,-3)
2854 7436 cu : $2e/5$
2854 8420 rt : (8,3,9,-3)
2854 9147 sr : $(\ln 3/3)^{-1/2}$
2854 9664 J_0 : $\sqrt{5}+\sqrt{6}+\sqrt{7}$
2855 1160 rt : (9,5,-3,7)
2855 1263 10^x : $\sqrt{13}$
2855 3160 sr : $\sqrt{2}/4+3\sqrt{3}/4$
2855 3523 2^x : $4/3-\pi$
2855 3692 $1/5+\exp(3)$
2855 3938 rt : (8,7,-5,-1)
2855 4865 ℓ_2 : $3/2-e/4$
2855 5420 id : $\sqrt{3}/3+3\sqrt{5}$
2855 7247 sr : $4\sqrt{2}-\sqrt{3}/4$
2855 7611 rt : (1,9,-7,-8)
2855 7621 sq : $1+3\sqrt{3}/4$
2855 8097 J_0 : $7\pi/3$
2855 8288 Li_2 : $2\sqrt{5}/5$
2855 8433 as : $3-e$
2855 9111 rt : (8,7,2,8)
2855 9243 Γ : $\sqrt{5}\pi$
2855 9877 at : $2/3-3e/2$
2856 1300 $\ln(4)\times\exp(1/2)$
2856 1629 2^x : $1/\sqrt[3]{21}$
2856 2504 rt : (8,0,-7,8)
2856 4079 cr : 17/8
2856 4503 rt : (1,6,-1,5)
2856 5573 at : $3\sqrt{2}/4-2\sqrt{5}$
2856 6346 as : P_6
2856 7786 E : 15/19
2856 8929 10^x : $1/\ln(1/(2\pi))$
2856 9545 cu : $\sqrt{20}\pi$
2857 1428 id : 2/7
2857 5470 10^x : $2e/3-3\pi/4$
2857 6566 ζ : 8/5
2857 9485 sq : $9\sqrt[3]{2}/10$
2858 2986 ln : $4\sqrt{2}/3+\sqrt{3}$
2858 3684 J_0 : $3\sqrt{5}$
2858 4497 at : $\ln(\sqrt{5}/3)$
2858 5751 sq : $6\sqrt[3]{2}/5$
2858 7220 at : $2+\sqrt{2}$
2858 7894 cr : $3+4\sqrt{5}$
2858 8982 at : $\sqrt{3}/4+4\sqrt{5}/3$
2858 9838 sq : $2-\sqrt{3}/2$
2859 0270 rt : (6,4,-4,1)
2859 0924 10^x : $(\sqrt{5})^{1/2}$
2859 0967 tπ : $2\sqrt{2}-\sqrt{5}/2$
2859 1084 ℓ_2 : $(2e/3)^{1/3}$
2859 2027 tanh : 5/17
2859 3069 rt : (6,7,8,-3)
2859 4029 rt : (1,9,-8,-3)
2859 6945 ln : $5\zeta(3)/8$
2859 7407 2^x : $1/4+2\sqrt{2}/3$
2859 7572 J_2 : $10\zeta(3)/7$
2859 8720 J_1 : $(2e/3)^3$
2860 0321 rt : (4,6,-5,5)
2860 0488 at : $2e\pi/5$
2860 1525 rt : (5,1,-8,-2)
2860 3662 e^x : $3/2+3\sqrt{3}$
2860 5144 at : 5/17
2860 7246 cu : $2-4e/3$
2860 8049 J_0 : $5(e+\pi)/4$
2860 9406 as : $2e/3-2\pi/3$
2861 1844 rt : (2,9,3,5)
2861 3578 id : $4/3-\pi/3$
2861 4207 2^x : $4\sqrt{3}/3+3\sqrt{5}$
2861 6455 rt : (6,7,1,-1)
2861 8247 $1/4\times\ln(\pi)$
2862 0571 ℓ_{10} : $e-\pi/4$
2862 0826 rt : (1,9,-1,-1)
2862 3470 ℓ_{10} : $3/2+\sqrt{3}/4$
2862 6167 tπ : $9(1/\pi)/5$
2862 7631 rt : (5,9,4,-2)
2862 7919 rt : (1,5,0,8)
2862 8568 ℓ_{10} : $3\sqrt{2}-4\sqrt{3}/3$
2862 9976 sq : $3\sqrt{3}-3\sqrt{5}$
2863 0567 rt : (9,7,-2,5)
2863 0810 rt : (1,9,7,9)
2863 0986 J_1 : $\ln(\ln 3/2)$
2863 2253 $\pi+\ln(\pi)$
2863 2487 at : $2\sqrt[3]{5}$
2863 2702 rt : (8,2,-6,6)
2863 4070 sπ : $4e/3+2\pi$
2863 5163 sr : $1/2+2\sqrt{3}/3$
2863 5598 rt : (3,5,-6,-4)
2863 5689 at : $3\sqrt{2}/2+3\sqrt{3}/4$
2863 6163 Li_2 : 17/19
2863 6611 rt : (6,0,3,1)
2863 7202 e^x : $9e/8$
2863 8297 ℓ_2 : $\sqrt{2}+2\sqrt{3}$
2863 9058 at : $4\sqrt{2}-\sqrt{5}$
2863 9098 cu : $4\sqrt{3}/3+3\sqrt{5}$
2864 0307 tπ : $3\sqrt{3}/2+2\sqrt{5}/3$
2864 1078 rt : (5,2,6,-2)
2864 2707 rt : (4,4,9,-3)
2864 2874 10^x : $3/4+\sqrt{5}/3$
2864 3614 Ψ : $\zeta(3)$
2864 4564 10^x : 22/15
2864 6206 rt : (3,6,-9,-3)
2864 7065 10^x : $2e-\pi/3$
2864 7889 id : $9(1/\pi)/10$
2865 0479 sr : $\exp(-5/2)$
2865 1726 sq : $4\sqrt{2}/3+4\sqrt{5}$
2865 2087 e^x : 25/2
2865 2154 rt : (9,6,-6,-2)
2865 2199 $\ln(4/5)\times\exp(1/4)$
2865 3037 J_0 : $6e\pi/7$
2865 3721 at : $4-\sqrt{3}/3$
2865 3993 rt : (9,7,-9,-4)
2865 4732 θ_3 : 1/7
2865 5405 rt : (5,9,-6,-1)
2865 6649 $\sqrt{2}\div s\pi(\pi)$
2865 6706 2^x : $\sqrt{3}/4-\sqrt{5}$
2865 6715 rt : (6,1,-5,-8)
2865 7572 e^x : $\sqrt[3]{2}/5$
2865 7970 Ψ : $\exp(1/(2e))$
2865 8014 rt : (3,9,-1,6)
2865 8462 rt : (7,4,-1,7)
2865 9137 rt : (8,4,-3,-1)
2866 0193 ℓ_2 : $1/3+3\pi$
2866 0314 cr : $7e\pi/5$
2866 0904 rt : (8,9,3,6)
2866 0946 sq : $3\sqrt[3]{2}$

2866 1252 $2/5 \div \exp(1/3)$
2866 2906 sinh : $\sqrt{2}/5$
2866 3013 rt : (5,5,8,2)
2866 4110 Γ : $e/7$
2866 4859 J_0 : $\sqrt{19/2}$
2866 5119 sq : $1/4 - \pi/4$
2866 6489 2^x : $4/11$
2866 6661 Ψ : $5\sqrt[3]{3}/6$
2866 6829 K : $5\sqrt[3]{3}/8$
2866 8444 e^x : $\sqrt{3} - 4\sqrt{5}/3$
2867 0098 J_1 : $3/5$
2867 0285 2^x : $5\sqrt[3]{2}/3$
2867 0712 $\ln(2/3) \times s\pi(1/4)$
2867 2114 rt : (2,3,-8,2)
2867 3247 sinh : $8(1/\pi)/9$
2867 4506 rt : (1,7,-6,-6)
2867 5655 as : $\sqrt{2}/5$
2867 7802 Ei : $(\sqrt{2}/3)^{1/3}$
2867 8242 rt : (9,6,8,-3)
2867 9375 10^x : $7(e+\pi)/3$
2867 9543 2^x : $8(1/\pi)/7$
2867 9597 $t\pi$: $1/4 - 3\sqrt{5}/4$
2868 2539 $\exp(\sqrt{5}) \div \exp(1/4)$
2868 2610 erf : $3\ln 2/8$
2868 2943 rt : (4,8,0,6)
2868 3244 rt : (9,1,-8,2)
2868 3624 ℓ_2 : $e\pi/7$
2868 3869 ln : $3/2 + 3\sqrt{2}/2$
2868 3931 10^x : $2e - 3\pi/4$
2868 5816 10^x : $3 + 4\pi$
2868 6019 as : $8(1/\pi)/9$
2868 6576 e^x : $3\sqrt{2}/4 - 4\sqrt{3}/3$
2868 6747 $2/3 \times s\pi(\pi)$
2868 9313 sr : $3/4 + e/3$
2868 9377 rt : (5,1,0,9)
2868 9596 rt : (6,7,5,1)
2869 1309 at : $4\sqrt{3}/3 + \sqrt{5}/2$
2869 3388 Li_2 : $4/15$
2869 3611 at : $4/3 + 2\pi/3$
2869 4088 Γ : $\sqrt{2}/2$
2869 5436 rt : (7,4,-8,-2)
2869 6843 sq : $3/2 - 3e$
2869 8100 cr : $\sqrt{2} - \sqrt{3} + \sqrt{6}$
2869 8544 rt : (6,6,4,8)
2870 0221 at : $24/7$
2870 5842 10^x : $3\sqrt{2}/4 + \sqrt{5}$
2870 6170 Ψ : $5\sqrt[3]{3}/4$
2870 7062 rt : (4,8,-5,-2)
2870 8880 e^x : $5\sqrt[3]{5}/3$
2870 9385 2^x : $21/10$
2870 9938 Ψ : $(e+\pi)^{1/3}$
2871 0266 e^x : $2 + 3\sqrt{3}$
2871 0689 rt : (8,9,-4,-3)
2871 1712 rt : (2,5,-7,-6)
2871 4739 J_1 : $\zeta(3)/2$
2871 5165 2^x : $3 + 4\sqrt{3}$
2871 5523 rt : (7,1,-9,-5)
2871 6393 sinh : $7e\pi/10$
2871 7196 ζ : $7\ln 2/4$
2871 8262 rt : (2,9,-7,7)
2871 8707 sq : $2 - 4\sqrt{3}$
2872 1635 2^x : $7(e+\pi)/6$
2872 5010 e^x : $1/4 + \sqrt{3}/3$
2872 5773 rt : (1,5,-5,1)
2872 6277 Ψ : $3\zeta(3)/2$
2872 6572 rt : (8,4,-5,4)
2872 6673 10^x : $3/2 + \sqrt{3}/2$
2872 6840 K : $3\zeta(3)/4$
2872 7265 $\sqrt{2} \div \ln(3)$
2872 7766 sq : $\sqrt{2}/4 + 2\sqrt{5}$
2873 1728 $t\pi$: P_{82}
2873 3073 $t\pi$: P_{11}
2873 4170 rt : (3,1,7,2)
2873 4999 at : $3 + \sqrt{3}/4$
2873 5437 rt : (7,3,9,-3)
2873 7153 rt : (8,8,4,-2)
2873 9931 Ei : $(\sqrt{5})^{1/2}$
2874 1147 sinh : $7\sqrt{5}/10$
2874 1766 id : $4\sqrt{2} - 4\sqrt{5}$
2874 1809 rt : (9,9,-1,3)
2874 3420 rt : (5,9,-3,0)
2874 4091 E : $5\sqrt[3]{2}/8$
2874 4815 cr : $3 - \sqrt{3}/2$
2874 6225 ℓ_2 : $3 - \sqrt{5}/4$
2874 6559 Li_2 : $2\zeta(3)/9$
2874 7058 10^x : P_{76}
2874 8457 rt : (4,1,-7,-2)
2874 9644 sinh : $\exp(-\sqrt[3]{2})$
2875 0831 cr : $7\sqrt[3]{5}$
2875 2231 rt : (7,7,-9,-3)
2875 3850 ζ : $\ln(\pi/2)$
2875 3952 sr : $(1/(3e))^{-3}$
2875 4236 Ei : $3\zeta(3)/8$
2875 5032 $4/5 \times \ln(5)$
2875 7511 rt : (3,6,-2,-1)
2875 9721 rt : (6,1,-4,6)
2876 1281 10^x : $18/11$
2876 1831 Ψ : $(\ln 2)^{-1/2}$
2876 2526 rt : (2,8,-2,2)
2876 2555 Li_2 : $1/\sqrt{14}$
2876 2586 as : $\exp(-\sqrt[3]{2})$
2876 2746 at : $2 - 2e$
2876 4097 cr : $e\pi/4$
2876 4878 e^x : $3\sqrt{2} + 2\sqrt{3}$
2876 5391 ℓ_{10} : $3 - 3\sqrt{2}/4$
2876 8207 id : $\ln(4/3)$
2876 9506 rt : (5,8,9,9)
2877 0011 J_0 : $3\sqrt[3]{2}/2$
2877 0176 10^x : $\sqrt{3} + \sqrt{6} + \sqrt{7}$
2877 3928 rt : (3,7,-7,-9)
2877 5805 Ψ : $\exp(\sqrt{2})$
2877 5993 10^x : $3 - \sqrt{2}/2$
2877 6100 rt : (8,1,6,-2)
2877 6710 rt : (6,6,-3,-1)
2877 7857 cr : $1/4 + 4\sqrt{2}/3$
2877 7861 2^x : $3e + 2\pi$
2877 7935 rt : (4,5,-6,8)
2877 8554 rt : (2,9,-8,-3)
2877 8714 ζ : $(\sqrt[3]{5}/3)^{1/3}$
2878 0117 at : $3/4 - 4\pi/3$
2878 0235 e^x : $4 + 2\sqrt{3}$
2878 1004 sq : $\ln(\sqrt[3]{5})$
2878 1241 $s\pi$: $e + 4\pi/3$
2878 1657 erf : $6/23$
2878 2582 rt : (7,6,0,5)
2878 3535 ℓ_2 : $5(e+\pi)/3$
2878 3832 rt : (1,7,-1,8)
2878 5306 Ψ : $(\sqrt[3]{3})^{1/2}$
2878 7864 sinh : $9e\pi/4$
2878 8036 J_2 : 2π
2878 9394 rt : (9,4,-9,1)
2879 0217 rt : (9,6,1,-1)
2879 0975 rt : (1,3,6,-2)
2879 1957 tanh : $(3/2)^{-3}$
2879 4814 at : $\sqrt{2} - \sqrt{5}/2$
2879 4927 rt : (8,3,-5,1)
2879 4946 10^x : $\sqrt{3}/3 - \sqrt{5}/2$
2879 6038 cu : $4e/3 - 3\pi/2$
2879 6520 rt : (4,2,-3,-8)
2879 7204 at : $4 - \sqrt{5}/4$
2879 7827 rt : (3,8,9,2)
2879 8061 $e + s\pi(\pi)$
2879 8834 2^x : $2 + \sqrt{3}$
2879 9213 rt : (1,7,4,-3)
2879 9661 rt : (3,7,-1,-3)
2880 0000 cu : $8\sqrt[3]{3}/5$
2880 0983 rt : (3,8,-4,7)
2880 3156 2^x : $10\zeta(3)/7$
2880 3685 Ψ : $\sqrt{2} + \sqrt{5} - \sqrt{6}$
2880 3779 Γ : π
2880 5175 ℓ_{10} : $3\zeta(3)/7$
2880 5544 at : $(3/2)^{-3}$
2880 7385 J_1 : $5e\pi/9$
2880 8220 rt : (6,5,-6,-2)
2880 8675 at : $3e - 3\pi/2$
2880 9002 $t\pi$: $\pi^2/3$
2881 0588 ℓ_2 : $(\ln 3/2)^{1/3}$
2881 1008 rt : (2,8,1,-1)
2881 5069 cr : $5\sqrt[3]{5}/4$
2881 6080 rt : (8,1,-5,9)
2881 6644 ℓ_{10} : $5e/7$
2881 7718 rt : (2,8,0,-9)
2881 8525 sq : $8\sqrt[3]{2}/3$
2881 9742 sq : $1/2 + \pi/2$
2881 9943 $t\pi$: $1/3 + \sqrt{3}/3$
2882 0829 rt : (6,5,5,-2)
2882 1575 rt : (1,9,-3,-3)
2882 1795 e^x : $\sqrt{3}/4 - 3\sqrt{5}/4$
2882 2314 rt : (5,8,-7,2)
2882 2808 sr : $5\pi/3$
2882 3683 $\exp(e) - s\pi(1/3)$
2882 4331 rt : (7,1,-8,3)
2882 4561 sr : $3 + \sqrt{5}$
2882 4851 10^x : $2\sqrt{3} - 3\sqrt{5}/2$
2882 6736 rt : (4,6,-9,2)
2882 7725 J_0 : $17/9$
2882 8812 rt : (5,3,1,7)
2882 9123 2^x : $2\sqrt{2}/3 - \sqrt{3}/3$
2882 9505 rt : (9,9,-8,-6)
2882 9917 2^x : $3\sqrt{3}/4 + 3\sqrt{5}$
2883 2051 $\ln(4/5) \div s\pi(e)$
2883 3393 $\sqrt{5} - \exp(2/3)$
2883 5434 J : $\exp(-19\pi/4)$
2883 7332 rt : (1,7,-5,-2)
2883 7623 rt : (2,7,2,2)
2883 7836 rt : (3,4,9,-3)
2883 7891 10^x : $20/19$
2883 9189 rt : (8,6,8,2)
2883 9458 rt : (9,6,-1,8)
2884 1223 $s\pi(\sqrt{2}) + s\pi(\sqrt{5})$
2884 3011 ℓ_{10} : $1 + 2\sqrt{2}/3$
2884 3203 J_0 : $(5/3)^3$
2884 3255 $\exp(3/5) \times s\pi(1/4)$
2884 3380 ℓ_{10} : $\sqrt[3]{22/3}$
2884 4191 J_2 : $19/11$
2884 4991 id : $\sqrt[3]{3}/5$
2884 6309 J_1 : $2e/9$
2884 6758 $s\pi$: $1/4 + 4\sqrt{2}$
2884 7520 $4 \times \exp(3/5)$
2885 1365 rt : (4,6,-9,-3)
2885 1422 sinh : $15/14$
2885 2051 sq : $4\sqrt{2} - \sqrt{3}/4$
2885 3762 rt : (8,6,8,-3)
2885 3900 $1/5 \div \ln(2)$
2885 4109 ℓ_2 : $3 + 4\sqrt{2}/3$
2885 4920 ℓ_2 : $5\sqrt[3]{5}/7$
2885 6220 rt : (8,9,-5,-2)
2885 8434 rt : (8,6,-4,2)
2885 9531 J_2 : $5\sqrt[3]{5}/2$
2886 1242 ℓ_2 : $2\sqrt{3}/3 - \sqrt{5}/3$
2886 1574 rt : (9,1,-9,6)
2886 2105 e^x : $3 - 3\sqrt{2}$
2886 2263 tanh : $3\ln 2/7$
2886 2929 $t\pi$: $(\sqrt{5})^{-3}$
2886 2962 rt : (5,2,-7,-5)
2886 4685 e^x : $25/21$
2886 5426 rt : (1,4,8,2)
2886 6549 rt : (1,8,8,-3)
2886 7513 id : $\sqrt{3}/6$
2887 1137 id : $1/4 + 3e/4$
2887 5162 rt : (4,7,-1,9)
2887 6019 at : $3\ln 2/7$
2887 8743 rt : (6,8,5,6)
2888 1534 $t\pi$: $\ln(3\pi/2)$
2888 1702 rt : (3,0,2,9)
2888 3595 Γ : $12/17$
2888 4386 sinh : $2e\pi$
2888 4663 rt : (9,1,3,1)
2888 4700 10^x : $4 - \sqrt{2}$
2888 6971 sinh : $\sqrt[3]{5}/6$
2888 8154 rt : (2,7,-7,-1)
2888 8376 at : $3/4 - \pi/3$
2888 9164 at : $\sqrt{2}/3 + 4\sqrt{5}/3$
2888 9337 rt : (6,2,7,2)
2889 0806 rt : (7,2,-8,-8)
2889 0987 ζ : $(\sqrt[3]{3})^{1/3}$
2889 2605 ℓ_2 : $7\pi/9$
2889 3325 J_1 : $(\ln 3/3)^{1/2}$
2889 4121 10^x : $4\sqrt{3}/3 - 3\sqrt{5}/4$
2889 4639 at : $e/2 + 2\pi/3$

2889 4669 $E: 11/14$
2889 5170 $10^x: 4\sqrt{2}-\sqrt{3}/2$
2889 5696 $2^x: \ln(\sqrt[3]{3})$
2889 7517 rt : (6,2,-8,2)
2889 8137 $E: 5\sqrt{2}/9$
2889 8826 $s\pi: (\sqrt{5}/3)^{1/3}$
2889 8896 $J_1: 7\sqrt{5}/5$
2890 0224 as : $\sqrt[3]{5}/6$
2890 0673 cr : $\sqrt{3}-\sqrt{5}+\sqrt{7}$
2890 1786 rt : (7,6,-7,-4)
2890 2511 $\ell_{10}: \sqrt{3}-\sqrt{5}+\sqrt{6}$
2890 3282 at : $4/3+3\sqrt{2}/2$
2890 3989 $\Psi: 1/5$
2890 4347 sr : $4\sqrt{2}/3+3\sqrt{5}/2$
2890 5920 rt : (3,5,-6,-7)
2890 7090 rt : (6,3,-3,4)
2890 7815 id : $e+\pi/2$
2891 1023 rt : (8,8,4,9)
2891 1249 $erf: ((e+\pi)/3)^{-2}$
2891 3223 sinh : $9\sqrt{3}$
2891 4566 $2^x: 2\sqrt{3}+3\sqrt{5}/4$
2891 4930 rt : (1,8,-3,-5)
2891 5066 sr : $9\zeta(3)$
2891 6854 rt : (1,6,5,-8)
2891 7437 $4/5+\ln(3/5)$
2891 7825 $e^x: 1/4-2\sqrt{5}/3$
2891 8679 $J_1: 19/4$
2891 8774 ln : $(\sqrt[3]{2}/3)^{1/3}$
2892 0686 rt : (3,7,-7,8)
2892 1751 $10^x: \exp(e/3)$
2892 2009 $E: \pi/4$
2892 2422 $\ln(3/4)\div\ln(4/5)$
2892 3198 cr : 15/7
2892 4719 rt : (8,3,-4,7)
2892 4784 $e^x: \sqrt{3}+\sqrt{5}/2$
2892 6016 $10^x: 3/2-3e/4$
2892 7199 rt : (9,6,-8,-1)
2892 8193 $J_2: 6\sqrt[3]{3}/5$
2892 8345 $s\pi: 2(e+\pi)/3$
2892 9506 rt : (6,3,9,-3)
2893 1531 $\theta_3: \sqrt[3]{3}/10$
2893 2188 cu : $4-\sqrt{2}$
2893 2671 $\Gamma: 10\pi/9$
2893 4176 $J_1: 3\sqrt{2}/7$
2893 8424 rt : (9,4,5,-2)
2893 9340 $t\pi: \sqrt[3]{5}$
2893 9469 ln : $10\zeta(3)/9$
2894 0871 $J_2: \sqrt{2}+\sqrt{5}+\sqrt{7}$
2894 2184 as : $1/2-\pi/4$
2894 2848 id : $\sqrt[3]{12}$
2894 3045 $2^x: 1/4-3e/4$
2894 4253 $\zeta: \exp(-\sqrt[3]{4}/2)$
2894 5047 rt : (7,3,2,-1)
2894 5678 $J_0: (\sqrt[3]{4}/3)^{-3}$
2894 5977 ln : π^2
2894 6494 $\Psi: \ln(\sqrt[3]{4}/2)$
2894 7524 rt : (1,8,4,-6)
2894 9740 rt : (8,2,-7,-2)
2895 0661 $\ell_2: 9/11$
2895 2019 $J_1: 7\ln 2/8$
2895 2113 rt : (4,7,8,-3)
2895 2965 $\ell_{10}: \exp(2/3)$
2895 3182 $J_1: (e)^{-1/2}$
2895 3996 rt : (1,8,-5,-9)
2895 5804 rt : (9,9,7,-3)
2895 6920 cr : $e/2+\pi/4$
2895 7249 rt : (3,4,-6,-2)
2895 7580 $2^x: 1/3-3\sqrt{2}/2$
2895 7903 rt : (5,7,-8,5)
2896 1305 $J_2: \sqrt{3}$
2896 1745 sinh : 2/7
2896 1912 $\ell_2: 4\sqrt{2}/3+\sqrt{5}/4$
2896 2390 rt : (6,0,-3,9)
2896 2843 sq : $3/4-4\sqrt{2}/3$
2896 6129 rt : (7,8,4,-2)
2896 6253 $10^x: 1+4\sqrt{5}/3$
2896 7504 rt : (8,7,-9,-3)
2896 8135 sr : $1+3\sqrt{2}$
2896 8262 rt : (4,7,-8,-7)
2896 9521 rt : (9,8,0,6)
2897 0042 rt : (3,9,-8,-3)
2897 1204 $t\pi: 1/\ln(3)$
2897 1314 $\ell_{10}: 9\sqrt{3}/8$
2897 2062 rt : (2,5,-9,-1)
2897 2925 rt : (9,3,-8,4)
2897 3707 rt : (7,8,1,3)
2897 5170 as : 2/7
2897 6142 at : $2\sqrt{3}$
2897 6604 cr : $1/3+2e/3$
2897 7560 id : P_{82}
2897 7609 rt : (7,3,-7,1)
2897 7716 $Ei: \sqrt[3]{3}/2$
2898 0818 cu : $\ln(3\pi)$
2898 1323 $J_2: 5\ln 2/2$
2898 2009 $Ei: \ln(\pi)$
2898 2551 $J_2: 5\sqrt[3]{2}$
2898 4240 rt : (1,8,6,6)
2898 5358 $\Gamma: 6\zeta(3)$
2898 5791 $t\pi: \sqrt{2}+\sqrt{3}-\sqrt{5}$
2898 6813 id : $\pi^2/3$
2898 8708 at : $5\ln 2$
2898 8874 $\ell_{10}: 3\sqrt[3]{5}/10$
2898 9794 rt : (5,8,2,0)
2899 1719 sinh : 4
2899 4082 sq : 7/13
2899 4508 rt : (1,6,3,-4)
2899 4774 $\theta_3: \ln(\sqrt{3})$
2899 6251 $e^x: (\ln 3)^{-2}$
2899 8261 $10^x: (5/2)^{-1/2}$
2899 9455 rt : (9,0,-8,9)
2899 9556 sinh : $8(e+\pi)/3$
2900 0000 sq : 23/10
2900 0230 $e^x: 1-\sqrt{5}/3$
2900 0734 $e^x: 4(1/\pi)/5$
2900 2541 $Ei: 7\sqrt[3]{5}/8$
2900 3072 at : $\sqrt{2}-\sqrt{5}-\sqrt{7}$
2900 4401 sr : $1/4+\sqrt{2}$
2900 5879 cr : $8\pi^2$
2900 7364 $t\pi: 4\sqrt{2}+\sqrt{3}/4$
2900 8262 at : $3/4+e$
2901 0692 rt : (6,9,9,2)
2901 0902 $J_2: e\pi/2$
2901 1367 rt : (7,0,-7,6)
2901 1931 $e^x: 1/3-\pi/2$
2901 2997 sq : $7(e+\pi)/8$
2901 3562 $\ell_2: \exp(\sqrt[3]{4})$
2901 3672 $2^x: -\sqrt{3}+\sqrt{6}+\sqrt{7}$
2901 4197 cu : $3\sqrt{2}-\sqrt{3}/4$
2901 5952 rt : (7,5,1,8)
2901 6557 $J_2: 5e\pi/6$
2901 7672 rt : (2,8,-9,-6)
2901 7899 $J_0: 4\sqrt{2}/3$
2901 8491 rt : (9,2,9,-3)
2901 9156 as : $4/3-\pi/3$
2902 0994 sq : $3/2-3e/4$
2902 1396 $\Psi: (1/\pi)$
2902 2736 $erf: 5/19$
2902 5837 rt : (2,3,-1,6)
2902 9350 $J_1: (\sqrt{2}\pi)^{-1/3}$
2903 2209 at : $3+\sqrt{2}/3$
2903 2844 rt : (2,9,-1,-1)
2903 3839 $\zeta: (\sqrt[3]{2}/3)^{1/2}$
2903 4557 rt : (5,0,-6,3)
2903 5297 rt : (7,6,8,-3)
2903 5374 cr : $\sqrt{2}/3+3\sqrt{5}/4$
2903 6603 $Ei: \ln(\pi/2)$
2903 7158 rt : (9,3,7,2)
2903 7813 at : $1-2\sqrt{5}$
2903 9422 $e^x: ((e+\pi)/2)^2$
2903 9579 sinh : $5\sqrt[3]{3}$
2904 0771 rt : (2,4,9,-3)
2904 1357 sinh : $9(1/\pi)/10$
2904 1853 cu : $10e\pi/9$
2904 3297 $\Psi: \ln(1/(3\pi))$
2904 4381 $\ln(3)-\exp(2)$
2904 5464 $2^x: \exp(-1)$
2904 6497 $\Psi: \sqrt{17}$
2904 6825 $J_1: 14/23$
2904 7108 rt : (7,7,-2,-1)
2904 8195 rt : (4,5,8,2)
2904 9126 rt : (2,6,-1,3)
2905 0146 rt : (5,5,2,5)
2905 2432 $e^x: 1-\sqrt{5}$
2905 2844 $\ln(2/3)\div\exp(1/3)$
2905 3362 rt : (5,3,-6,-8)
2905 3738 $J_1: 7e/4$
2905 3861 $2^x: 2+\sqrt{3}/2$
2905 4966 as : $9(1/\pi)/10$
2905 6421 $J_0: 3\pi/5$
2905 6941 rt : (8,8,-3,0)
2905 7408 at : $1-3\sqrt{3}/4$
2905 8819 cu : $3/2+4\sqrt{5}$
2905 9167 at : $1/3+\pi$
2906 0030 $4/5\times\exp(\sqrt{2})$
2906 1996 rt : (7,1,6,-2)
2906 3715 rt : (5,9,-3,6)
2906 4441 sq : $7\ln 2/9$
2906 5465 rt : (5,6,-9,-3)
2906 5600 $\zeta: \sqrt{7}$
2906 5831 $\ell_{10}: 3-\pi/3$
2906 5968 rt : (1,7,-9,2)
2906 6385 $\Gamma: 1/\sqrt{22}$
2906 6429 $s\pi(\pi)\times s\pi(\sqrt{5})$
2906 7834 at : $1/\ln(4/3)$
2906 8897 $2^x: e/2-\pi$
2906 8978 $erf: \exp(-4/3)$
2906 9194 $\ell_2: 9e/5$
2907 0270 rt : (5,6,-9,8)
2907 0616 rt : (8,5,-3,5)
2907 3583 cr : $10\zeta(3)$
2907 3913 $2^x: (\sqrt[3]{5})^{1/3}$
2907 4866 $10^x: 2\sqrt[3]{2}/7$
2907 4944 $t\pi: 1/2+e/3$
2907 4980 ln : $\sqrt{2}/3+\sqrt{3}/2$
2907 5379 $s\pi: e/3$
2907 6705 $\ell_{10}: (e+\pi)/3$
2907 6850 rt : (2,1,7,2)
2908 0909 $e^x: (\sqrt{2}/3)^{-1/2}$
2908 2395 rt : (5,5,5,-2)
2908 2504 id : $4\sqrt{3}/3+4\sqrt{5}/3$
2908 5936 rt : (1,9,6,1)
2908 6316 cu : $3/4+\sqrt{3}$
2908 6750 $\Gamma: 22/7$
2908 6765 $10^x: 9/25$
2908 7982 sr : $4\sqrt{2}/3+4\sqrt{5}$
2909 0365 $3\times s\pi(\pi)$
2909 1702 cu : P_{55}
2909 2317 $2^x: 1/\sqrt[3]{20}$
2909 3919 $2^x: 7/19$
2909 4146 $Ei: 2e/3$
2909 5932 $E: (\sqrt[3]{3}/3)^{1/3}$
2909 6303 ln : $2\sqrt{2}-2\sqrt{5}/3$
2909 7081 cu : $4\sqrt{2}/3+4\sqrt{5}/3$
2909 8194 $\ell_{10}: 8\sqrt[3]{5}/7$
2909 9444 id : $\sqrt{5/3}$
2910 0619 $t\pi: 6(1/\pi)$
2910 1562 cu : 13/8
2910 2708 $Ei: 9\ln 2/8$
2910 4243 id : $1/3-4e/3$
2910 6780 id : $\sqrt{3}+\sqrt{5}/4$
2910 8779 at : $1/2+4\sqrt{5}/3$
2911 3157 $\exp(3/5)-\exp(\sqrt{2})$
2911 4822 $2^x: e\pi/2$
2911 4858 $J_1: 25/8$
2911 5819 rt : (9,2,-7,7)
2911 7711 rt : (2,7,-5,-2)
2911 7991 $s\pi: \ln(\ln 3)$
2911 8609 $2^x: \exp(\sqrt{3})$
2911 8950 $1/3\div\ln(\pi)$
2912 0873 rt : (9,5,-7,2)
2912 1604 $\lambda: \exp(-11\pi/9)$
2912 2895 ln : $2\sqrt{2}/3+4\sqrt{5}$
2912 2989 rt : (6,2,-2,7)
2912 4241 rt : (5,1,-7,-2)
2912 5413 rt : (9,8,5,1)
2912 6155 rt : (8,9,7,-3)
2912 6637 at : P_{79}
2912 7191 $J_0: 8\sqrt{3}/3$
2912 8784 sr : 21/4
2912 9721 $\ln(2)+\exp(4)$

2913 0436 rt : (9,8,-7,-3)
2913 1169 rt : (5,3,9,-3)
2913 1261 tanh : 3/10
2913 2231 sq : 25/22
2913 2567 e^x : $(\sqrt[3]{5}/3)^{1/3}$
2913 3055 sq : $\sqrt{3}/3+\sqrt{5}/4$
2913 3511 2^x : 3π
2913 3739 J_1 : $3\pi^2/5$
2913 5067 rt : (6,5,-2,2)
2913 5462 J_2 : $7\sqrt{5}/9$
2913 7944 rt : (9,7,1,9)
2913 8054 rt : (8,5,-9,2)
2913 9678 2^x : $1/3+e$
2913 9772 2^x : $1/\ln(\sqrt[3]{5}/3)$
2914 1327 rt : (3,7,8,-3)
2914 3333 rt : (7,5,-6,-2)
2914 3647 Li_2 : $(\sqrt[3]{3}/2)^{1/3}$
2914 5679 at : 3/10
2914 5933 rt : (4,6,-9,-4)
2914 6731 Ψ : $(\pi)^{-2}$
2914 7535 10^x : $1/4-\pi/4$
2914 8065 e^x : $3\sqrt{2}/2-3\sqrt{5}/2$
2914 9305 ℓ_{10} : $7\sqrt{5}/8$
2915 1181 J_1 : 11/18
2915 3927 sr : $4\sqrt{3}-3\sqrt{5}/4$
2915 4966 10^x : 1/9
2915 5455 rt : (7,1,-6,-9)
2915 6345 sq : $3\sqrt[3]{2}/7$
2915 7800 e^x : $4\sqrt{3}/3-\sqrt{5}/2$
2915 7818 rt : (8,6,1,-1)
2915 7989 ℓ_{10} : P_{13}
2915 8480 rt : (9,9,-5,-2)
2915 9411 rt : (6,7,6,9)
2915 9820 $t\pi$: $4\sqrt{2}/3+\sqrt{5}/3$
2916 0278 cr : $1+2\sqrt{3}/3$
2916 0324 rt : (2,5,-9,-3)
2916 1141 sinh : $9\ln 2/2$
2916 2096 Ψ : 7/22
2916 2283 E : 18/23
2916 2421 rt : (2,9,-8,2)
2916 4137 as : $3-3e/4$
2916 4571 rt : (7,4,4,1)
2916 5125 rt : (4,9,-3,-6)
2916 6666 id : 7/24
2916 8708 ℓ_{10} : $\sqrt[3]{15/2}$
2916 9032 rt : (4,9,-8,-3)
2916 9375 e^x : $1/2-\sqrt{3}$
2916 9598 $t\pi$: $3e/5$
2917 0337 J : $(\sqrt[3]{4}/3)^3$
2917 1127 sq : $1/3+2e$
2917 1649 10^x : $4/3+3e/4$
2917 2559 rt : (6,8,-2,-3)
2917 2732 ℓ_{10} : $\ln(5/3)$
2917 3636 rt : (7,2,-6,4)
2917 4116 rt : (1,6,-4,-7)
2917 6605 2^x : $5\sqrt[3]{3}/3$
2917 8244 rt : (5,8,-4,9)
2917 8312 tanh : $\zeta(3)/4$
2917 8353 $\ln(2/3)\times\exp(\sqrt{3})$
2917 8990 at : $10\pi/9$
2917 9396 at : $2+2\sqrt{5}/3$
2917 9732 $t\pi$: $4\sqrt{3}+4\sqrt{5}/3$
2918 0571 as : $\ln(4/3)$
2918 1397 rt : (4,1,-1,-9)
2918 2381 $\sqrt{3}\times s\pi(\sqrt{3})$
2918 2469 $1/5-\exp(2/5)$
2918 7777 cu : $1/2+\pi$
2918 8653 rt : (5,0,3,1)
2919 2822 rt : (9,7,-9,-3)
2919 2849 at : $\zeta(3)/4$
2919 4016 at : $3/4-3\sqrt{2}$
2919 5808 $\exp(2/5)\times s\pi(1/3)$
2919 7015 rt : (2,9,-8,-9)
2919 7519 rt : (8,4,5,-2)
2920 2457 id : $3e/2-\pi/4$
2920 4952 rt : (6,8,4,-2)
2920 6231 10^x : $\sqrt{2}/3+2\sqrt{5}$
2920 6311 rt : (2,5,-4,4)
2920 6905 10^x : 17/6
2920 8232 J_0 : $8e/3$
2920 8534 $1\div s\pi(e)$
2920 9273 rt : (7,5,-6,-1)
2921 1272 rt : (4,0,-9,3)
2921 1379 at : $9e/7$
2921 2590 at : $2/3+2\sqrt{2}$
2921 3768 cr : $2/3+2\sqrt{5}/3$
2921 4356 rt : (3,8,-8,-6)
2921 6712 rt : (1,8,1,-1)
2921 8193 $t\pi$: $9(e+\pi)/7$
2921 8219 Ψ : $6\sqrt{3}$
2921 8789 rt : (8,2,9,-3)
2922 0748 rt : (7,6,8,2)
2922 1213 rt : (7,7,2,6)
2922 1665 cu : 24/13
2922 2346 $s\pi$: $2\pi/3$
2922 3260 rt : (6,6,8,-3)
2922 4098 J_0 : $\sqrt[3]{20/3}$
2922 7199 rt : (8,4,-2,8)
2922 8721 rt : (5,2,7,2)
2922 9159 at : $3\sqrt{2}-\sqrt{5}/3$
2922 9559 cu : $1/2+2\sqrt{3}$
2923 0622 ζ : $e/6$
2923 1890 cu : $3+\pi/3$
2923 2217 rt : (5,1,-5,-6)
2923 3888 sq : $\sqrt{3}/3-\sqrt{5}/2$
2923 4433 rt : (6,4,-9,-8)
2923 5035 rt : (3,2,6,-2)
2923 6410 θ_3 : 11/20
2923 8069 $s\pi(\sqrt{2})\div s\pi(\sqrt{3})$
2924 0158 rt : (2,6,-8,-4)
2924 0272 Ei : $\exp(-\sqrt[3]{4}/2)$
2924 0846 as : $2\sqrt[3]{3}/3$
2924 2112 ln : $1/2+\pi$
2924 3492 2^x : $3\sqrt{2}-\sqrt{5}/3$
2924 3755 erf : $5(1/\pi)/6$
2924 4934 rt : (3,2,-9,5)
2924 6658 sinh : $\sqrt[3]{3}/5$
2924 6990 rt : (2,2,-2,9)
2924 7514 sinh : $3\sqrt{5}/2$
2924 8125 ℓ_2 : $\sqrt{6}$
2924 8373 ln : $7\sqrt{2}$
2924 9368 E : $(2\pi/3)^{-1/3}$
2924 9666 at : 7/2
2925 0964 Ei : $((e+\pi)/3)^{-3}$
2925 1178 sq : $3\sqrt[3]{3}/8$
2925 2085 10^x : $9\sqrt[3]{2}/5$
2925 2154 rt : (1,4,9,-3)
2925 2873 id : $2\sqrt{2}+2\sqrt{3}$
2925 4151 ℓ_{10} : $4-3e/4$
2925 4629 rt : (1,8,-8,-3)
2925 4869 rt : (2,6,-9,9)
2925 7494 cr : $\sqrt{2}+\sqrt{5}/3$
2925 7994 rt : (1,6,-2,-1)
2925 9556 J_1 : $(\ln 2/3)^{1/3}$
2925 9761 2^x : $7\sqrt[3]{5}/10$
2926 0328 ℓ_{10} : $2\sqrt{2}/3-\sqrt{3}/4$
2926 0754 as : $\sqrt[3]{3}/5$
2926 1064 rt : (1,9,3,-9)
2926 5569 rt : (9,4,-6,5)
2926 9004 2^x : $4+4\pi$
2926 9509 tanh : $1/\sqrt{11}$
2927 0125 sinh : $\sqrt{3}/6$
2927 0424 sr : $e-\pi/3$
2927 1006 sr : $\sqrt[3]{14/3}$
2927 1119 K : $(e/2)^{-1/3}$
2927 2479 ℓ_{10} : $\sqrt{2}/3+2\sqrt{5}/3$
2927 3461 ln : $4/3+4\sqrt{3}/3$
2927 5497 sinh : $3\sqrt[3]{3}/2$
2927 6704 rt : (8,7,-2,3)
2927 7413 at : $e+\pi/4$
2927 8653 rt : (6,1,-9,-3)
2927 8791 ℓ_{10} : $2\sqrt{2}-\sqrt{3}/2$
2928 1266 J_0 : $5e/2$
2928 4277 at : $1/\sqrt{11}$
2928 4702 Ei : $(2e)^{1/2}$
2928 6221 J_2 : $5\pi/9$
2928 7010 Ψ : $7\sqrt[3]{5}/10$
2928 8803 $t\pi$: $1/3-3\sqrt{2}$
2928 9173 e^x : $\exp(-e/2)$
2928 9321 id : $1-\sqrt{2}/2$
2929 0161 tanh : $(\ln 3/2)^2$
2929 0630 sq : $e/2+3\pi$
2929 2036 10^x : $\ln(\sqrt{5}/2)$
2929 2345 J : 5/23
2929 2888 10^x : $1/2+\pi/4$
2929 2934 sq : $2\pi^2/3$
2929 3955 as : $4\zeta(3)/5$
2929 4719 Li_2 : $e/10$
2929 7670 rt : (3,5,-3,3)
2929 9162 rt : (8,2,-9,-9)
2929 9392 rt : (5,4,3,8)
2930 1302 rt : (5,2,-5,1)
2930 2008 rt : (7,9,7,-3)
2930 2801 rt : (9,5,8,-3)
2930 2989 ℓ_{10} : $5\pi/8$
2930 3534 rt : (1,7,-2,1)
2930 3654 ℓ_{10} : $2-2\sqrt{5}/3$
2930 3751 cu : $3/4-\sqrt{2}$
2930 4982 at : $(\ln 3/2)^2$
2930 5022 sq : $(e/2)^{-2}$
2930 6459 rt : (9,7,4,-2)
2931 0328 2^x : $2\zeta(3)$
2931 0527 rt : (7,8,-6,-6)
2931 1282 e^x : $2+3\sqrt{3}/2$
2931 2841 rt : (9,9,0,-1)
2931 2966 sq : $\sqrt{2}/4-2\sqrt{5}/3$
2931 4718 $2/5-\ln(2)$
2931 7044 tanh : $e/9$
2931 7431 Ei : $(\ln 3/3)^2$
2931 7959 rt : (7,1,-5,7)
2931 9080 at : $e/4-4\pi/3$
2932 2345 rt : (9,9,2,7)
2932 5893 id : $2\sqrt{2}/3-\sqrt{5}$
2932 6781 rt : (9,7,-6,0)
2932 7760 rt : (2,4,-7,5)
2932 9720 10^x : $1/3-\sqrt{3}/2$
2932 9821 ln : $4e/3+2\pi$
2933 0975 J_2 : $\sqrt[3]{16/3}$
2933 1934 at : $e/9$
2933 3359 cr : $3\sqrt[3]{3}/2$
2933 4344 cu : $8\zeta(3)$
2933 5501 J_1 : 8/13
2933 7633 rt : (2,7,8,-3)
2933 9896 cr : $9\zeta(3)/5$
2934 0277 sq : 13/24
2934 1031 rt : (4,3,9,-3)
2934 1658 ζ : $5\sqrt{5}/7$
2934 3725 rt : (4,8,-4,-3)
2934 3833 J_1 : $9\ln 2/2$
2934 4179 id : $3\sqrt{2}/4-3\sqrt{5}/2$
2934 6674 cr : $1/\ln(\sqrt[3]{4})$
2934 8427 θ_3 : $1/\sqrt[3]{6}$
2935 0080 cr : $3/4+\sqrt{2}$
2935 0672 ln : $4\sqrt{3}+4\sqrt{5}/3$
2935 1573 rt : (2,9,4,-2)
2935 1829 ℓ_{10} : $2/3+3\sqrt{3}/4$
2935 1944 e^x : $3/4+\sqrt{2}/2$
2935 3382 rt : (8,1,-9,4)
2935 4326 rt : (6,4,-1,5)
2935 5013 Ei : $e/9$
2935 7305 2^x : $7(1/\pi)/6$
2935 7333 rt : (4,7,7,6)
2935 8471 J_2 : 19/3
2935 8667 ℓ_2 : $2/3+\sqrt{5}/4$
2935 9276 Li_2 : $2\pi/7$
2935 9391 $s\pi$: $\sqrt{3}/4+2\sqrt{5}$
2936 2649 $t\pi$: 1/11
2936 3001 rt : (9,2,-7,-2)
2936 4268 rt : (6,1,6,-2)
2936 4425 sinh : $(\sqrt{3}/2)^{-1/2}$
2936 5850 rt : (6,2,-9,-2)
2936 7011 cr : $5\sqrt{3}/4$
2936 7645 J_1 : $8\ln 2/9$
2936 8386 sq : $(2\pi)^{-1/3}$
2936 8496 cu : $e/3-\pi/2$
2936 9243 $s\pi$: $e/3+4\pi/3$
2936 9349 at : $3(e+\pi)/5$
2937 2585 sr : $2\sqrt{2}-2\sqrt{3}/3$
2937 2781 rt : (8,3,7,2)
2937 3622 rt : (6,9,-7,9)

2937 5136 $\text{t}\pi : 2(1/\pi)/7$
2937 5199 rt : (7,2,-8,2)
2937 5316 $\ln : 4 - \sqrt{2}/4$
2937 5376 rt : (5,2,-4,-9)
2937 6129 rt : (5,9,-8,-3)
2937 6207 $\zeta : (\pi/3)^{-1/2}$
2937 7885 rt : (9,3,-5,8)
2937 8622 $\text{s}\pi : 2\sqrt{3} - \sqrt{5}/4$
2938 2588 rt : (3,5,-9,-3)
2938 2677 $J : (\ln 2/3)^{1/2}$
2938 2879 $2^x : 3\sqrt{2}/2 + \sqrt{5}/3$
2938 3614 $10^x : \sqrt[3]{3}/4$
2938 4024 $10^x : 2\sqrt{3} - \sqrt{5}/3$
2938 6994 $\Psi : 11/25$
2938 7530 $\text{t}\pi : 4\sqrt{2}/3 + 3\sqrt{5}$
2938 8609 rt : (4,7,1,-1)
2938 9262 $1/2 \times s\pi(1/5)$
2938 9284 rt : (7,6,3,9)
2938 9333 id : $\ln(\sqrt{5}/3)$
2938 9660 cu : $3 + 2\sqrt{2}/3$
2939 1338 $e^x : 2/3 + 4\pi$
2939 1825 $erf : 4/15$
2939 2263 rt : (5,6,-9,2)
2939 2807 sr : $2(e+\pi)/7$
2939 3104 at : $3\sqrt{3} - 3\sqrt{5}/4$
2939 3266 cu : $3 - 3\sqrt{2}/4$
2939 3489 $\Psi : 2e/3$
2939 5751 $\ln : P_{45}$
2939 8934 cr : 13/6
2939 9037 id : $\sqrt{2} - 3\sqrt{5}$
2939 9236 as : P_{82}
2939 9671 $Li_2 : 3/11$
2940 0312 $J_2 : 7/4$
2940 1451 $J_1 : 9\sqrt{3}/5$
2940 3258 $E : 9\ln 2/8$
2940 3546 rt : (7,4,-5,2)
2940 4114 rt : (6,3,2,-1)
2940 4427 $J_0 : 8e/7$
2941 0274 rt : (6,2,-8,-6)
2941 1764 id : 5/17
2941 2490 $Li_2 : 6(1/\pi)/7$
2941 2496 $10^x : 3\zeta(3)/10$
2941 4205 rt : (3,7,-5,-2)
2941 5666 at : $2/3 - 4\pi/3$
2941 6572 rt : (2,7,-7,-7)
2941 7585 $2^x : 4 + e$
2941 8111 rt : (5,6,8,-3)
2942 0123 sinh : $9\zeta(3)/5$
2942 0391 $\ln : 2/3 + 4\sqrt{5}/3$
2942 3513 $\Gamma : (\ln 2/2)^{1/3}$
2942 4433 rt : (2,9,2,1)
2942 5368 rt : (8,1,3,1)
2942 6387 $\text{s}\pi : \sqrt{2} + 2\sqrt{5}/3$
2942 6616 cu : $4\sqrt{2} + \sqrt{3}/3$
2942 7037 rt : (7,2,9,-3)
2942 7055 $\ell_2 : 3e/10$
2942 8036 rt : (1,6,3,-4)
2942 8387 $\ln : \sqrt{2}/2 - \sqrt{3}/4$
2942 8963 rt : (8,6,-1,6)
2942 9532 $2^x : \sqrt{2}/3 - \sqrt{5}$
2942 9685 $\ell_2 : \sqrt{3}/2 + 4\sqrt{5}$
2943 0355 $4/5 \div e$
2943 0763 $\ln(\sqrt{2}) + \exp(2/3)$
2943 1650 sinh : $\ln((e+\pi)/2)$
2943 2007 rt : (9,4,-3,-1)
2943 5250 sr : $\ln 2/8$
2943 5470 $e^x : (\pi/2)^{-3}$
2943 5917 $\Psi : (\sqrt[3]{5})^{1/3}$
2943 7834 $J_0 : 6\zeta(3)$
2943 8174 at : $\exp(\sqrt[3]{2})$
2943 8612 rt : (3,4,-4,6)
2943 9018 sr : $4\sqrt{2} + 3\sqrt{3}$
2943 9856 $erf : 2\zeta(3)/9$
2944 0632 cu : P_{81}
2944 3144 rt : (4,1,0,7)
2944 4145 $J_0 : 5\sqrt[3]{3}$
2944 5326 rt : (1,1,7,2)
2944 7883 rt : (8,4,-9,-1)
2944 8749 cu : $2 + \pi/3$
2944 9005 id : $\exp(5/3)$
2944 9256 rt : (3,9,-2,3)
2945 0425 $\text{s}\pi : 3/4 + 3e$
2945 0457 $e^x : P_{28}$
2945 2709 $J_1 : 3\sqrt[3]{3}/7$
2945 4301 $erf : 1/\sqrt{14}$
2945 4471 rt : (5,8,4,-2)
2945 5212 cu : $9e/2$
2945 5235 sinh : $7e\pi/8$
2945 6634 rt : (8,0,-8,7)
2945 6763 $J_2 : 7e/3$
2945 7485 $\ell_2 : 4 + e/3$
2945 9627 $e^x : 1/\sqrt{15}$
2946 0187 rt : (2,8,-8,-3)
2946 1856 $J_1 : \exp(-\sqrt[3]{3}/3)$
2946 3863 at : $4 - \sqrt{2}/3$
2946 5319 rt : (9,6,-5,3)
2946 7567 rt : (3,5,8,2)
2946 7888 cu : $4 + \sqrt{2}/2$
2946 9156 $\text{t}\pi : (\sqrt{2}\pi/2)^{-3}$
2946 9300 at : $2\sqrt{2}/3 - 2\sqrt{5}$
2946 9441 rt : (7,4,5,-2)
2947 0452 $\exp(4/5) \div s\pi(\sqrt{5})$
2947 2708 rt : (4,9,-3,0)
2947 5027 $\text{t}\pi : 4\sqrt{3} - \sqrt{5}/4$
2947 5398 $2^x : \sqrt{5}/6$
2947 5517 $\text{s}\pi : 2/21$
2947 6275 $J_0 : \sqrt{3} + \sqrt{6} + \sqrt{7}$
2947 7574 $\ell_{10} : 3/2 + \sqrt{2}/3$
2947 8139 rt : (3,9,-1,-1)
2948 0429 $\ln : \sqrt{2} + \sqrt{5}$
2948 1241 sr : $(3\pi/2)^{1/3}$
2948 3716 rt : (6,9,7,7)
2948 3986 $\text{t}\pi : \sqrt{2} + 3\sqrt{5}/4$
2948 5217 $e^x : (2e)^{1/2}$
2948 6501 rt : (7,0,-4,1)
2948 6755 $Ei : (\ln 3/2)^2$
2948 6886 $\ell_{10} : \sqrt[3]{23/3}$
2948 7126 cu : $1/3 + 2\sqrt{3}/3$
2948 7406 rt : (9,0,6,2)
2948 7870 $J_0 : \exp(\sqrt[3]{2}/2)$
2948 8121 $\exp(3/4) - \exp(3/5)$
2948 8211 $\text{t}\pi : 7\pi^2/10$
2948 8391 sr : 2/23
2948 8904 $10^x : 7\pi^2/10$
2948 9203 $s\pi(1/4) + s\pi(1/5)$
2949 0778 rt : (5,1,-4,4)
2949 1459 $\Psi : (1/(3e))^{-3}$
2949 2362 $\Psi : \sqrt{11}\pi$
2949 3171 $J_1 : 13/21$
2949 3939 $\text{s}\pi : 3/4 + 2\sqrt{3}/3$
2949 5487 $\exp(\pi) + \exp(e)$
2949 6134 rt : (8,5,8,-3)
2949 6651 $J_2 : 5\sqrt{2}$
2949 7100 id : $2e - \pi$
2949 7260 rt : (5,9,9,2)
2950 1003 sr : $3\sqrt{5}/4$
2950 1571 $J_0 : 4e\pi/5$
2950 2635 rt : (6,3,0,8)
2950 3095 $2^x : 1/3 - 2\pi/3$
2950 3221 rt : (2,8,-3,-2)
2950 3445 $\ln : 1/2 + 3\pi$
2950 3701 rt : (8,5,-6,-2)
2950 6148 at : $9\pi/8$
2951 1113 rt : (6,3,-7,-9)
2951 1672 rt : (1,5,-2,-1)
2951 2183 at : $3\sqrt{3}/4 + \sqrt{5}$
2951 4097 $\ln(\sqrt{3}) - s\pi(\sqrt{3})$
2951 4519 $\ln(5) \times \ln(\sqrt{5})$
2951 5352 at : $5\sqrt{2}/2$
2951 5445 $\text{t}\pi : 1/\ln(5/2)$
2951 6071 $\text{s}\pi : 2e/3 + 2\pi$
2951 7783 $\ell_{10} : \sqrt{2} + \sqrt{5}/4$
2951 9257 rt : (5,5,-5,-4)
2951 9515 sq : $4 - 3\sqrt{3}/4$
2951 9820 $e^x : 4/3 + 3\sqrt{5}/4$
2952 0602 sq : $2/3 + \sqrt{2}/3$
2952 1284 $\ell_{10} : 2\sqrt{3} - 2\sqrt{5}/3$
2952 1416 $J_2 : 9\pi/4$
2952 2210 $e^x : \sqrt{12}\pi$
2952 3770 rt : (3,8,-5,4)
2952 3865 rt : (7,6,-9,-3)
2952 4548 sinh : $5\pi^2/9$
2952 7541 $J_0 : 7(e+\pi)/6$
2952 8641 tanh : 7/23
2953 1152 $2^x : 3e - \pi$
2953 2405 $E : (\sqrt{2}/3)^{1/3}$
2953 2974 $\ell_{10} : 2\pi^2$
2953 3364 rt : (6,7,-1,0)
2953 3496 $\ln : P_{61}$
2953 5661 $\theta_3 : (5/3)^{-1/3}$
2953 7369 $2^x : 3\sqrt{2}/4 + 2\sqrt{5}$
2953 7700 rt : (7,3,-4,5)
2953 7951 $\ln : 3 + 4\sqrt{3}$
2953 8869 at : $3/2 + 3e/4$
2954 0183 rt : (7,9,3,4)
2954 0814 rt : (8,5,0,9)
2954 1572 rt : (1,7,8,-3)
2954 2931 tanh : $(2e/3)^{-2}$
2954 3807 $\text{t}\pi : (4/3)^{-1/3}$
2954 3816 $e^x : 1/3 + \pi$
2954 4020 rt : (3,3,-5,9)
2954 4083 at : 7/23
2954 5485 $Ei : e/6$
2954 5488 $\text{t}\pi : 1/3 - 3\pi$
2954 5691 $J_0 : 4\sqrt[3]{5}$
2954 5887 rt : (2,2,1,-9)
2954 9319 rt : (7,6,1,-1)
2955 1949 $\ell_2 : 2/3 + 3\sqrt{2}$
2955 2020 $\text{s}\pi : 3(1/\pi)/10$
2955 3255 rt : (8,7,4,-2)
2955 3990 $e^x : 2/3 - 4\sqrt{2}/3$
2955 4656 rt : (1,9,5,-7)
2955 4936 at : $2\sqrt{2}/3 + 3\sqrt{3}/2$
2955 5038 rt : (9,8,7,-3)
2955 6224 sq : $e/5$
2955 7180 cr : $4e/5$
2955 8411 at : $(2e/3)^{-2}$
2955 9473 $\theta_3 : (\pi/3)^{-1/3}$
2955 9774 rt : (3,3,9,-3)
2956 0435 $2^x : 10\pi/3$
2956 8425 rt : (9,5,-4,6)
2956 8671 rt : (2,7,-8,-5)
2956 9784 rt : (8,3,-8,2)
2957 0040 $10^x : 3/4 + 2\sqrt{2}/3$
2957 0117 sinh : $1/\ln(\sqrt[3]{4})$
2957 0286 rt : (4,6,-6,3)
2957 2660 rt : (4,9,-2,-9)
2957 3116 $10^x : 2 + 3e/4$
2957 3778 $\Psi : \sqrt[3]{3}/10$
2957 3785 $e^x : 3/2 - e$
2957 4295 $E : 7/9$
2957 8974 $\text{s}\pi : \sqrt{2}/3 + \sqrt{3}/4$
2957 9807 rt : (1,5,-1,3)
2958 1382 $4 \div s\pi(\pi)$
2958 1961 sinh : 7/24
2958 3078 $\sqrt{2} \times \ln(2/5)$
2958 3541 $e^x : -\sqrt{2} - \sqrt{6} + \sqrt{7}$
2958 3686 $\ln : (3)^3$
2958 5489 $\pi + \exp(e)$
2958 5964 rt : (3,7,-8,5)
2958 8014 $Ei : 1/\sqrt{11}$
2958 8446 rt : (9,2,2,-1)
2958 9270 rt : (8,9,-1,1)
2958 9694 rt : (1,8,5,-1)
2959 0872 rt : (3,7,2,4)
2959 1477 $J_1 : 1/\ln(5)$
2959 2017 rt : (6,9,-8,-3)
2959 2641 rt : (4,2,7,2)
2959 4318 sq : $2e/3 - 3\pi/4$
2959 4556 $J_2 : 3(e+\pi)/10$
2959 4770 rt : (5,9,-7,-3)
2959 6880 as : 7/24
2959 7240 $\text{s}\pi : 3\sqrt{3} + 3\sqrt{5}$
2959 8242 sr : $1 + e/4$
2959 8324 $\theta_3 : (\sqrt[3]{2}/3)^{1/3}$
2959 8644 rt : (7,7,-5,-3)
2959 9786 rt : (2,4,-7,-2)
2959 9826 $J_2 : 17/4$
2960 1797 sq : $e/4 - 4\pi$
2960 2243 $\ln : \sqrt{10}\pi$

2960 3285 id : $\sqrt{2}+\sqrt{5}+\sqrt{7}$
2960 3955 $\ln(\sqrt{5})\div\exp(1)$
2960 4284 J_2 : $(2e)^{1/3}$
2960 5137 sq : $1/\ln(2\pi)$
2960 5331 cr : $1/2+3\sqrt{5}/4$
2960 6604 Ei : $(2\pi/3)^{-1/3}$
2960 7755 rt : (7,8,-5,-2)
2960 8123 $1/5\div s\pi(\sqrt{5})$
2961 0041 2^x : P_{74}
2961 0303 rt : (5,4,-5,1)
2961 0753 sr : $4\sqrt[3]{2}/3$
2961 1522 $t\pi$: $4\sqrt{3}/3+4\sqrt{5}/3$
2961 1541 cu : $3\sqrt{3}$
2961 3781 J_0 : $23/5$
2961 4432 rt : (1,9,-2,7)
2961 5192 rt : (4,5,-9,-3)
2961 5863 rt : (1,9,4,-2)
2961 5908 $\exp(3/4)-\exp(5)$
2961 6318 ℓ_2 : $4\sqrt{2}-\sqrt{5}/3$
2961 6746 rt : (6,5,-9,-7)
2961 7225 e^x : $4-2\sqrt{5}/3$
2961 7957 id : $\sqrt{2}-\sqrt{5}/2$
2961 8417 rt : (5,0,-3,7)
2961 9066 rt : (2,9,-9,-3)
2962 0361 rt : (4,6,8,9)
2962 7280 rt : (2,7,-9,2)
2962 7311 sq : $3/4+3e$
2962 8251 $s\pi$: $7\sqrt[3]{3}$
2962 9629 id : $(3/2)^{-3}$
2962 9866 e^x : $\sqrt{2}-2\sqrt{3}/3$
2963 0559 $t\pi$: $1/4+4\sqrt{3}/3$
2963 0737 rt : (4,2,-7,-3)
2963 0850 sq : $6\ln 2$
2963 2313 rt : (9,9,-5,-2)
2963 2662 $s\pi(1/3)-s\pi(\pi)$
2963 2773 e^x : $1/4+2\sqrt{2}/3$
2963 4689 rt : (6,6,8,2)
2963 5007 ℓ_2 : $2\sqrt{3}-\sqrt{5}$
2963 5394 J_0 : $15/8$
2963 6002 rt : (7,2,-3,8)
2963 7576 2^x : $1/3+\sqrt{3}/2$
2963 7930 cr : $3\sqrt{2}/4+\sqrt{5}/2$
2964 2555 10^x : $\sqrt{2}/2+2\sqrt{3}$
2964 3813 id : $3e+\pi$
2964 3878 rt : (6,2,9,-3)
2964 5476 rt : (5,4,-6,-2)
2964 6708 rt : (4,5,-7,6)
2964 7039 rt : (3,5,5,-2)
2964 8201 rt : (9,4,-3,9)
2964 8639 rt : (6,0,-7,4)
2964 9450 sr : $9(e+\pi)/10$
2965 2634 sq : $4-2\sqrt{5}/3$
2965 3516 rt : (8,9,0,-1)
2965 4141 rt : (1,7,3,-9)
2965 4365 $\ln(\sqrt{5})+\exp(2/5)$
2965 5257 ℓ_{10} : $8\sqrt{3}/7$
2965 5308 cr : $3/2+e/4$
2965 6995 2^x : $6\zeta(3)$
2965 7287 sq : $4-\sqrt{2}/4$
2965 9392 ℓ_2 : $4e-\pi/3$
2966 0308 rt : (8,2,-7,5)
2966 2040 2^x : $1/\sqrt[3]{19}$
2966 2184 cu : $3e/2+3\pi/4$
2966 4098 ℓ_{10} : P_{35}
2966 4303 ℓ_{10} : $7\sqrt{2}/5$
2966 4876 e^x : P_{25}
2966 5260 $s\pi$: $1/3+\pi/2$
2966 6295 rt : (5,8,-8,0)
2966 9575 E : $2e/7$
2967 1670 rt : (5,9,7,-3)
2967 2135 J_2 : $9\sqrt{2}/2$
2967 2814 id : $3\sqrt{2}/4+\sqrt{5}$
2967 2942 cu : $3/4+3\sqrt{5}/4$
2967 4259 rt : (3,8,-8,-3)
2968 3288 rt : (7,8,4,7)
2968 3528 10^x : $3\sqrt{3}-2\sqrt{5}$
2968 3955 2^x : $3/8$
2968 3976 J_2 : $\exp((e+\pi)/3)$
2968 4849 rt : (5,1,6,-2)
2968 6461 at : $\sqrt{3}/3+4\sqrt{5}/3$
2968 6959 cu : $1/4+2\sqrt{5}$
2968 7500 sq : $7\sqrt{3}/8$
2968 8238 at : $3+\sqrt{5}/4$
2969 0596 e^x : $3\sqrt{2}+3\sqrt{5}/4$
2969 1100 id : $2\sqrt{2}/3+3\sqrt{5}/2$
2969 1444 sr : $1/\ln(2e/3)$
2969 3461 rt : (3,9,2,-6)
2969 3833 ζ : $\ln(\sqrt{5}/3)$
2969 4344 $\exp(1/5)\div\exp(\sqrt{2})$
2969 6634 rt : (7,5,8,-3)
2969 8685 J_1 : $9\ln 2/10$
2969 9669 ln : $1+4\sqrt{5}$
2969 9863 cr : $24/11$
2970 0013 $\ln(3)-\exp(1/3)$
2970 1263 10^x : $5\ln 2/3$
2970 3258 rt : (1,7,2,-7)
2970 4286 J_2 : $5\pi^2/7$
2970 6307 id : $3\ln 2/7$
2970 7054 rt : (1,5,-2,-8)
2970 8196 rt : (4,4,-8,9)
2970 8677 cu : P_{26}
2970 9070 rt : (3,6,1,7)
2971 1235 J_1 : $\exp(-\sqrt{2}/3)$
2971 1478 ℓ_{10} : $1/4+\sqrt{3}$
2971 1783 ln : $3\sqrt{2}/4+3\sqrt{3}/2$
2971 4935 at : $4\sqrt{2}/3+3\sqrt{5}/4$
2971 5650 rt : (8,8,0,4)
2971 5656 2^x : $3/2+2\sqrt{3}/3$
2971 6016 sr : $7\sqrt[3]{3}/6$
2971 6995 2^x : $3e/2-3\pi/4$
2971 9755 id : $1/4+\pi/3$
2972 1250 ζ : $5/11$
2972 2530 10^x : $3/4+2\sqrt{3}/3$
2972 3586 $1/3-s\pi(\sqrt{2})$
2972 4222 rt : (5,7,-9,3)
2972 5115 rt : (9,1,9,-3)
2972 5136 as : $1-\sqrt{2}/2$
2972 5851 sr : $7\zeta(3)/5$
2972 5978 θ_3 : $(\sqrt[3]{4}/3)^3$
2972 6019 ζ : $21/8$
2972 6744 ln : $(2e/3)^{1/2}$
2972 8754 rt : (4,7,-5,-2)
2973 0226 rt : (8,1,-6,8)
2973 0975 cr : $7\sqrt{3}$
2973 1244 at : $e-2\pi$
2973 1338 ln : $2/3+e/4$
2973 1683 rt : (7,3,7,2)
2973 2705 J_0 : $9\sqrt{3}/5$
2973 2876 ℓ_{10} : $3e/2-2\pi/3$
2973 2886 rt : (6,1,-6,-7)
2973 5653 λ : $\exp(-17\pi/14)$
2973 6735 rt : (4,9,0,1)
2973 8938 J_2 : $7\sqrt[3]{2}/5$
2973 9670 2^x : $6/5$
2974 0079 2^x : $4-4e/3$
2974 1393 rt : (8,8,7,-3)
2974 2796 rt : (2,5,-6,-5)
2974 3202 ln : $3\pi/7$
2974 4254 sr : $4e$
2974 4712 rt : (9,8,-4,1)
2974 6436 at : $4-\sqrt{3}/4$
2974 8755 J_1 : $5/8$
2974 8900 rt : (6,3,-3,-1)
2974 9543 sinh : $14/13$
2975 2066 sq : $6/11$
2975 3269 Ψ : $\sqrt[3]{6}$
2975 3750 rt : (7,8,-3,0)
2975 4243 $t\pi$: $\sqrt{5/3}$
2975 5466 ℓ_{10} : $\sqrt{3}/2+\sqrt{5}/2$
2975 5475 rt : (6,4,5,-2)
2975 6641 rt : (1,4,-8,2)
2975 7128 tanh : $\ln(e/2)$
2975 9667 ℓ_{10} : $2\sqrt[3]{2}/5$
2976 3010 $\ln(\sqrt{2})+s\pi(2/5)$
2976 4016 at : $4/3+\sqrt{5}$
2976 8070 rt : (5,7,5,1)
2977 0174 rt : (8,6,-9,-3)
2977 0208 J_2 : $\sqrt[3]{11/2}$
2977 0947 rt : (5,4,-4,-1)
2977 3186 at : $\ln(e/2)$
2977 4165 at : $2+\pi/2$
2977 5407 J_1 : $\sqrt{23}$
2977 5908 J_0 : $9\ln 2/2$
2977 8762 at : $25/7$
2977 9533 rt : (2,9,6,-3)
2978 0367 2^x : $7\zeta(3)/4$
2978 0389 at : $(\sqrt[3]{4}/3)^{-2}$
2978 1342 ζ : $10(1/\pi)/7$
2978 1741 Ei : $\zeta(3)/9$
2978 2254 cu : $5\zeta(3)/9$
2978 2530 id : $\sqrt{2}/4+4\sqrt{5}$
2978 7062 rt : (8,6,-8,-3)
2978 7406 at : $\sqrt{3}/2-\sqrt{5}/4$
2978 7769 e^x : $\exp(-1/(2e))$
2978 8170 rt : (2,3,9,-3)
2979 0527 id : $4e+3\pi$
2979 1256 rt : (9,7,8,2)
2979 4710 $t\pi$: $1/3-4e/3$
2979 5869 at : $2\sqrt{2}+\sqrt{5}/3$
2979 5897 sq : $\sqrt{2}/2+4\sqrt{3}$
2979 6790 Ei : $23/11$
2979 7929 rt : (5,5,5,9)
2979 8179 at : $4\sqrt{3}-3\sqrt{5}/2$
2979 9171 ln : $3\sqrt{3}/7$
2980 0327 rt : (9,0,-9,8)
2980 0688 Li_2 : $\exp(-1/(3\pi))$
2980 2461 id : $9\sqrt[3]{3}/10$
2980 2679 rt : (8,7,1,7)
2980 3581 rt : (8,0,6,2)
2980 5310 rt : (9,5,-9,2)
2980 5533 Γ : $7/10$
2980 5834 e^x : $6/23$
2980 6699 10^x : $4\sqrt{2}$
2980 7472 10^x : $2\sqrt{2}-3\sqrt{5}/2$
2981 1225 rt : (7,7,4,-2)
2981 1415 e^x : $8\sqrt{5}/9$
2981 1768 at : $2/3-3\sqrt{2}$
2981 2455 ln : $5(e+\pi)/8$
2981 2479 Γ : $5\sqrt[3]{2}/9$
2981 2816 rt : (4,8,-1,4)
2981 3499 rt : (7,2,-9,-7)
2981 4239 $2/3\div\sqrt{5}$
2981 6105 $t\pi$: $\sqrt{3}+\sqrt{5}/4$
2981 6214 rt : (2,7,-5,8)
2981 6375 J_2 : $3\sqrt{2}$
2981 6428 rt : (6,4,-8,-4)
2981 7024 $s\pi$: $2\sqrt{2}-\sqrt{3}$
2981 7471 rt : (7,9,-8,-3)
2981 8526 cr : $4-2e/3$
2982 1747 at : $3+\sqrt{3}/3$
2982 2128 sq : $4\sqrt{2}+\sqrt{5}$
2982 2272 at : $1/2-3e/2$
2982 3629 rt : (9,7,-3,4)
2982 4345 at : $8\sqrt{5}/5$
2982 5973 rt : (6,5,1,6)
2982 6207 $\ln(2)\times s\pi(\pi)$
2982 7197 cu : $12/11$
2982 9415 $t\pi$: $1/3-e/3$
2982 9549 at : $3/4+2\sqrt{2}$
2982 9751 as : $\ln(\sqrt{5}/3)$
2983 0509 Ψ : $20/11$
2983 0580 rt : (3,6,8,-3)
2983 1106 sq : $9e\pi/8$
2983 1526 rt : (5,4,-9,-6)
2983 3247 ℓ_{10} : $8\sqrt{5}/9$
2983 3291 J_2 : $5\sqrt{2}/4$
2983 3393 $2/5\times s\pi(\sqrt{3})$
2983 3624 tanh : $4/13$
2983 4667 J_1 : $8e/7$
2983 5886 rt : (2,6,-5,-8)
2983 6493 $1/5\times\exp(2/5)$
2983 7257 Ψ : $9\sqrt{2}/7$
2983 7647 sinh : $5/17$
2983 9495 J_0 : $7\pi^2/10$
2984 0794 rt : (1,7,-3,-6)
2984 2077 J_2 : $\ln(e+\pi)$
2984 2480 Ei : $\sqrt{11}$
2984 2803 E : $(5/3)^{-1/2}$
2984 3788 rt : (1,8,9,2)
2984 5110 J_2 : $\exp(\sqrt[3]{5}/3)$

2984 5556 $\ln(3/4) \div s\pi(\sqrt{2})$
2984 6100 $10^x : P_{37}$
2984 8936 $2/3 \times \exp(2/3)$
2984 9048 cr : $e + 3\pi$
2984 9525 $\Gamma : \sqrt[3]{15}$
2984 9893 at : $4/13$
2984 9926 tanh : $(\sqrt{2}\pi/3)^{-3}$
2984 9985 $J_1 : 24/5$
2985 0951 rt : (7,5,-3,3)
2985 1513 $10^x : 4 - 3\sqrt{3}/4$
2985 3100 $J_2 : 1/\ln(\sqrt[3]{5}/2)$
2985 3220 as : $5/17$
2985 6797 sq : $1/4 + 2\sqrt{5}$
2985 7033 cr : $\sqrt[3]{21/2}$
2985 7116 $e^x : 2e + \pi$
2985 7672 $\Gamma : 1/\ln(3/2)$
2985 7699 rt : (3,6,-9,-5)
2985 8915 rt : (9,3,5,-2)
2985 9145 rt : (5,5,-9,-3)
2985 9782 $J_0 : 5e\pi/6$
2985 9974 $Ei : \exp(-(e + \pi)/2)$
2986 1228 $1/5 + \ln(3)$
2986 3952 $\zeta : \sqrt[3]{18}$
2986 5303 $e^x : 3\sqrt{3}/4 + \sqrt{5}$
2986 5794 $2^x : \sqrt{2} + \sqrt{5} - \sqrt{6}$
2986 6240 at : $(\sqrt{2}\pi/3)^{-3}$
2986 6301 rt : (4,9,7,-3)
2986 7610 tanh : $4\ln 2/9$
2986 8890 $J_2 : 23/13$
2986 8959 $J_2 : 6(e + \pi)/5$
2987 0034 rt : (5,2,9,-3)
2987 0099 rt : (4,7,-2,7)
2987 0110 $\Psi : \ln(1/3)$
2987 1998 $2^x : 7\pi$
2987 3336 rt : (8,5,-7,0)
2987 6521 $10^x : 2e - 3\pi/2$
2987 8021 $J_0 : \sqrt{7/2}$
2987 9964 $t\pi : 3\sqrt{2}/2 - 2\sqrt{5}/3$
2988 0203 sr : $4/3 + \sqrt{2}/4$
2988 2124 rt : (9,6,-2,7)
2988 2303 rt : (5,6,-2,-1)
2988 3973 at : $4\ln 2/9$
2988 4571 rt : (5,5,-8,-9)
2988 5438 $t\pi : 4e/3 + 2\pi$
2988 5849 sr : $2/3 - \sqrt{3}/3$
2988 6759 rt : (4,2,-8,-5)
2988 8056 ln : $7\pi/6$
2988 8923 $2^x : (\sqrt[3]{3})^{1/2}$
2988 9466 cu : $4\sqrt{2}/3 + 3\sqrt{3}/4$
2989 0602 $Ei : 18/23$
2989 0757 ln : $3\sqrt{2} - \sqrt{3}/3$
2989 0833 sr : $\sqrt{12}\pi$
2989 0901 $s\pi : \pi^2/9$
2989 0905 $J_1 : \pi/5$
2989 3009 $Ei : (\ln 2/3)^2$
2989 3508 rt : (9,5,-6,-2)
2989 6586 $J_0 : 4\sqrt{3}$
2989 6786 rt : (5,3,-3,2)
2989 7571 rt : (4,8,-8,-3)
2989 7669 $Ei : 9\sqrt{2}/5$
2989 9157 rt : (1,8,-9,-4)
2989 9212 $\ln(\pi) + \exp(e)$
2990 0002 rt : (6,4,2,9)
2990 0356 $J_1 : 4\sqrt{2}/9$
2990 0843 $\ell_{10} : 1/2 + 2\sqrt{5}/3$
2990 1277 $\exp(1/2) - \exp(2/3)$
2990 1337 rt : (2,5,-8,-9)
2990 2775 rt : (5,3,2,-1)
2990 3810 id : $3\sqrt{3}/4$
2990 4124 $\exp(3/5) \times \exp(\sqrt{3})$
2990 4838 rt : (6,5,8,-3)
2990 5039 $J_0 : 10\ln 2$
2990 6257 rt : (1,2,6,-2)
2990 6755 rt : (1,8,-2,4)
2990 6975 sr : $(\sqrt{5})^{-3}$
2990 7501 $1/2 \times \exp(4)$
2990 8765 $J : 5/24$
2991 3939 rt : (8,2,-8,2)
2991 5948 at : $8\pi/7$
2991 6171 cr : $4\sqrt{2} - 2\sqrt{3}$
2991 7305 rt : (3,7,-8,-8)
2991 7644 rt : (7,4,-2,6)
2991 8476 $e^x : (\ln 2)^{1/2}$
2991 8628 $\ln(4/5) \div s\pi(\sqrt{3})$
2991 9423 $\ell_{10} : \sqrt{2} + \sqrt{3}/3$
2992 2711 cu : $\sqrt{2} - \sqrt{5}/3$
2992 3098 rt : (6,3,-7,-1)
2992 3386 cr : $4 + 3e$
2992 5902 sq : $(\sqrt{2}\pi/3)^{1/3}$
2992 6301 $s\pi : \sqrt{2}/2 + 3\sqrt{3}$
2992 6321 rt : (5,9,-4,4)
2992 7827 rt : (2,5,8,2)
2992 8298 ln : $3/11$
2992 9405 sr : $2\sqrt{2}/3 + \sqrt{5}/3$
2992 9593 $Li_2 : 2\ln 2/5$
2993 0614 $1/4 - \ln(\sqrt{3})$
2993 1502 $J_2 : \sqrt{5}\pi$
2993 3298 sq : $2\sqrt{2} + 3\sqrt{5}/4$
2993 3511 $erf : e/10$
2993 3725 rt : (8,4,-6,3)
2993 4213 rt : (7,8,7,-3)
2993 4892 sr : $3\sqrt{2}/2 - \sqrt{3}/4$
2993 6867 sinh : $21/11$
2993 7241 id : $5(e + \pi)$
2993 7871 rt : (6,4,4,1)
2994 0277 $Li_2 : 1/\sqrt{13}$
2994 2113 $\ell_{10} : 7e\pi/3$
2994 2516 sq : $1/2 - \pi/3$
2994 2789 cu : $1/\ln(\pi/3)$
2994 4739 at : $3/2 + 2\pi/3$
2994 4745 rt : (2,4,-9,-3)
2994 6540 $J_0 : 25/8$
2994 7226 rt : (1,9,-3,-2)
2994 7232 $J_2 : (\pi)^{1/2}$
2994 7641 rt : (8,8,-5,-2)
2994 8038 $e^x : (\sqrt[3]{3})^{-1/2}$
2994 8141 ln : $4/3 - 3\sqrt{2}/4$
2994 8961 rt : (8,1,9,-3)
2994 9832 $e^x : \sqrt{2}/3 - 3\sqrt{5}/4$
2995 0706 $2^x : 1/\sqrt{7}$
2995 0966 $J_0 : 1/\ln(\sqrt{3}/2)$
2995 1772 $2^x : 3\sqrt[3]{2}/10$
2995 3144 rt : (4,2,2,8)
2995 3383 $\zeta : 5\pi/6$
2995 4967 ln : $(\ln 3/2)^{1/2}$
2995 5264 $2/3 \div \exp(4/5)$
2995 5573 $10^x : 3/4 + e/3$
2995 5599 $\Gamma : \pi^2/4$
2995 5990 $\exp(4/5) \times \exp(5)$
2995 6028 $\ell_2 : 13/16$
2995 6343 sq : $2\sqrt[3]{5}/3$
2995 7045 $Li_2 : 5\sqrt[3]{2}/7$
2995 8326 $Ei : \pi^2/2$
2995 9670 $Ei : 4\pi$
2996 0041 rt : (5,8,-5,7)
2996 0144 $e^x : 3\sqrt{3} - \sqrt{5}$
2996 0508 $\ell_{10} : 3\sqrt{2}/4 - \sqrt{5}/4$
2996 0524 id : $5\sqrt[3]{2}$
2996 1150 $J_1 : \sqrt[3]{2}/2$
2996 2391 rt : (4,9,-1,-1)
2996 5517 $J_0 : 5\sqrt{2}$
2996 5560 $e^x : ((e + \pi)/3)^{-2}$
2996 5796 rt : (1,9,-8,-7)
2996 5929 $t\pi : 9(e + \pi)/2$
2996 6269 rt : (7,3,-1,9)
2996 6351 cu : $5\pi/7$
2996 7043 $10^x : 3 + e$
2996 7826 rt : (6,6,1,-1)
2996 9544 $J_0 : 9\pi/4$
2997 0859 at : $1/4 - \sqrt{5}/4$
2997 1160 at : $1 + 3\sqrt{3}/2$
2997 1472 $Li_2 : 9/10$
2997 2444 rt : (5,2,-2,5)
2997 4850 rt : (1,7,-8,-3)
2997 5449 as : $(\sqrt{5}/2)^{-1/3}$
2997 6612 rt : (9,4,8,-3)
2997 7230 $\zeta : 7\sqrt{3}/10$
2997 7452 $Ei : 2/15$
2997 7572 cu : $\sqrt{2}/3 + 2\sqrt{3}/3$
2997 8381 rt : (8,3,-5,6)
2997 9245 id : P_{79}
2998 2489 rt : (7,1,-7,-2)
2998 2702 rt : (1,6,-3,-7)
2998 2730 $J_1 : 10\sqrt[3]{3}/3$
2998 2919 $10^x : 6\sqrt{2}/7$
2998 3164 id : $7\sqrt{2}/3$
2998 4148 $Ei : (\pi/3)^3$
2998 4871 rt : (3,2,7,2)
2998 4947 at : $18/5$
2998 5942 $t\pi : e/2 + \pi/3$
2998 7059 cu : $1/\ln(5/2)$
2998 7243 sq : $\sqrt{2}/2 + \sqrt{3}/4$
2998 8897 $\Gamma : ((e + \pi)/2)^{-1/3}$
2998 9592 rt : (9,3,-9,3)
2998 9940 $t\pi : 3e - \pi/4$
2999 0373 $Li_2 : 5/18$
2999 0872 $\exp(5) - \exp(\sqrt{2})$
2999 1130 $2^x : 2 + \pi$
2999 1866 $J_0 : \exp((e + \pi)/3)$
2999 3679 $\ell_{10} : 7\sqrt[3]{5}/6$
2999 4124 $J_0 : 5\pi^2/7$
2999 5539 $J_1 : 4\zeta(3)$
2999 7470 $\Psi : 1/\ln(\sqrt{3})$
2999 9571 $E : 17/22$
3000 0000 id : $3/10$
3000 2117 cr : $\sqrt{15}\pi$
3000 5916 at : $2 - 4\sqrt{3}/3$
3000 7609 $J_0 : 6(e + \pi)/5$
3000 7927 $J_0 : 7$
3000 8932 cr : $3(e + \pi)/8$
3001 0297 $J_0 : \sqrt{5}\pi$
3001 0317 $\ln(3/4) + s\pi(1/5)$
3001 1152 cu : $\ln((e + \pi)/3)$
3001 2406 cu : $\sqrt{2} - 3\sqrt{5}/2$
3001 2753 rt : (8,2,-4,9)
3001 3713 rt : (7,9,0,-1)
3001 4153 $\ln(\sqrt{2}) \times s\pi(1/3)$
3001 4300 at : $1/4 + 3\sqrt{5}/2$
3001 6147 cr : $\sqrt{2}/2 + 2\sqrt{5}/3$
3001 7265 $10^x : (\ln 3/2)^{-3}$
3001 7457 rt : (6,9,-8,7)
3001 7935 sr : $4\sqrt{3}/3 + 4\sqrt{5}/3$
3001 8671 $2^x : 1/2 - \sqrt{5}$
3001 8721 rt : (4,1,-7,-2)
3001 9698 $\Psi : 10\pi/3$
3001 9763 sq : $4\sqrt{2}/3 - \sqrt{5}/3$
3002 0637 sinh : $5\sqrt{3}/4$
3002 1338 $Ei : 5\pi/2$
3002 1580 rt : (9,2,-8,6)
3002 2984 rt : (5,1,-1,8)
3002 3033 sr : $4 - 4\sqrt{3}/3$
3002 3110 $\Psi : 6\ln 2$
3002 3295 sq : $4/3 - \pi/4$
3002 3303 rt : (7,1,3,1)
3002 4656 at : $\sqrt{13}$
3002 4851 rt : (3,8,-9,-9)
3002 5175 at : $5\sqrt[3]{3}/2$
3002 5576 $\ln(3/5) \times s\pi(1/5)$
3002 6037 rt : (4,1,6,-2)
3002 7097 rt : (1,3,9,-3)
3002 7714 $erf : 3/11$
3002 8650 at : $\sqrt{3}/2 - 2\sqrt{5}$
3002 8846 $10^x : (\sqrt{5}/3)^{1/2}$
3002 9080 at : $3\zeta(3)$
3002 9500 rt : (9,6,-9,-3)
3003 0327 $J_1 : 12/19$
3003 0377 cr : $9\sqrt[3]{5}/7$
3003 0958 $Ei : 7/3$
3003 4959 $J_1 : \exp(\pi/2)$
3003 5044 rt : (6,1,-5,5)
3003 5615 $Ei : \zeta(3)/4$
3003 6372 $e^x : 2/3 + 2e$
3003 6515 $10^x : 2\sqrt{3}/3 - 3\sqrt{5}/4$
3003 7319 $2^x : 3/2 + e/3$
3003 8389 $2^x : 5\sqrt[3]{3}/6$
3003 9212 $erf : 6(1/\pi)/7$
3003 9755 cu : $1/2 + 3\pi/4$
3003 9809 $\ell_2 : 4\sqrt{2}/3 + \sqrt{3}/3$
3003 9947 $J : \sqrt{3}/10$
3004 1699 cr : $7\pi/10$

3004 2756 rt : (4,9,9,2)
3004 3566 10^x : $1/2+e/2$
3004 4245 sq : $2/3+2\pi$
3004 6021 rt : (7,1,-9,2)
3004 6506 cr : $4/3+\sqrt{3}/2$
3004 7161 rt : (9,1,-7,9)
3004 9085 J_2 : $3e\pi/4$
3004 9196 2^x : $\sqrt{2}/3+2\sqrt{3}$
3004 9348 2^x : $\exp(1/(2e))$
3004 9407 rt : (2,6,8,-3)
3004 9632 e^x : $2\sqrt{2}+\sqrt{5}$
3005 1025 sq : $\sqrt{2}-\sqrt{3}/2$
3005 1422 id : $\zeta(3)/4$
3005 2875 tπ : $e+4\pi/3$
3005 2887 at : $3e/4+\pi/2$
3005 2994 rt : (6,6,-9,2)
3005 3376 rt : (8,9,-8,-3)
3005 6307 id : $2\sqrt{2}+2\sqrt{5}$
3005 7111 rt : (5,4,5,-2)
3005 7132 J_0 : $\sqrt{21}$
3005 7376 sr : $e/3+\pi/4$
3005 8458 ℓ_{10} : $\sqrt{2}+\sqrt{3}-\sqrt{7}$
3005 9144 cr : $11/5$
3006 0600 ζ : $(\sqrt{5}/3)^{-1/3}$
3006 3681 rt : (5,7,-5,-2)
3006 4736 sπ : $7e/10$
3006 4991 e^x : $3\sqrt{2}/2+2\sqrt{5}$
3006 5074 sinh : $(3/2)^{-3}$
3006 5328 Ei : $5/11$
3006 7326 sq : $10e\pi/3$
3006 7452 2^x : $\zeta(3)$
3006 7771 J_1 : $(5/2)^{-1/2}$
3006 8089 rt : (6,0,-4,8)
3006 9026 as : $\sqrt{2}-\sqrt{5}/2$
3007 0872 at : $3\sqrt{2}/2+2\sqrt{5}/3$
3007 2047 sπ : $(e/2)^{-1/3}$
3007 4531 rt : (8,2,2,-1)
3007 6333 rt : (7,0,-8,5)
3007 6512 J_2 : $16/9$
3007 7300 e^x : $7\sqrt{3}/5$
3007 7482 Li_2 : $7(1/\pi)/8$
3008 1247 as : $(3/2)^{-3}$
3008 1382 rt : (6,7,4,-2)
3008 3503 sπ : $2/3+\sqrt{5}$
3008 3956 rt : (4,0,-6,1)
3008 6976 rt : (5,6,8,2)
3008 8727 sr : $22/13$
3008 9628 rt : (2,5,-2,-1)
3009 1510 $\pi \div s\pi(\pi)$
3009 3990 sq : P_{12}
3009 4642 10^x : $4\sqrt{2}-2\sqrt{5}$
3009 5128 sr : $1/3+e/2$
3009 5494 $\exp(1/2)+\exp(\sqrt{3})$
3009 5582 e^x : $-\sqrt{2}-\sqrt{5}+\sqrt{6}$
3009 7589 e^x : $5/6$
3009 9620 rt : (7,1,-7,-8)
3010 1257 θ_3 : $3/20$
3010 2389 Γ : $2\pi/9$
3010 2999 ℓ_{10} : 2
3010 3212 e^x : $5/19$
3010 3257 rt : (5,3,-8,2)
3010 4676 J_2 : $1/\ln(\sqrt[3]{5}/3)$
3010 5937 J_0 : $7\sqrt{5}/5$
3010 6326 rt : (4,2,9,-3)
3010 6427 Ei : $(\sqrt[3]{3}/3)^{1/3}$
3010 7656 ln : $3\sqrt{3}/4-\sqrt{5}/4$
3010 7995 sr : $3/4+2\sqrt{2}/3$
3010 8624 rt : (9,5,1,-1)
3011 0941 at : $4\sqrt{2}/3+\sqrt{3}$
3011 2419 sq : $2+4\pi/3$
3011 2931 rt : (3,8,6,8)
3011 4930 10^x : $\sqrt{2}/4+3\sqrt{5}/4$
3011 5612 rt : (6,5,-9,-3)
3011 6152 cu : $1/4-3\pi$
3011 7253 rt : (6,3,7,2)
3011 7765 Ψ : $25/21$
3011 9421 $\exp(4/5)\div\exp(2)$
3012 0763 ln : P_{48}
3012 0784 at : $3/4-3\sqrt{2}/4$
3012 0989 sr : $\ln(2e)$
3012 1344 rt : (5,5,8,-3)
3012 2902 rt : (4,1,-5,-4)
3012 6377 rt : (1,8,-1,-1)
3012 7163 2^x : $e/2+\pi/3$
3012 7199 cr : $\sqrt{2}/3+\sqrt{3}$
3012 7488 tπ : $1/4+4\sqrt{2}$
3012 7634 cu : $2+4e$
3012 8544 Ei : $3/2$
3012 8733 Ei : $10(1/\pi)/7$
3012 9890 sinh : $\pi/2$
3013 0952 rt : (5,8,-8,-3)
3013 1344 10^x : $2-4\sqrt{2}/3$
3013 1738 rt : (5,1,-9,-1)
3013 3703 2^x : $2\sqrt[3]{5}/9$
3013 3948 rt : (6,8,7,-3)
3013 5246 rt : (4,0,-7,2)
3013 6836 at : $3/2+3\sqrt{2}/2$
3013 9261 rt : (7,0,6,2)
3014 1722 J_2 : 7
3014 2611 ln : P_{18}
3014 4045 J_0 : $\sqrt[3]{13/2}$
3014 4284 sq : $3/4-3\sqrt{3}/4$
3014 4596 sinh : $9\sqrt[3]{3}/2$
3014 5152 sinh : $3\ln 2/7$
3014 6560 $\ln(\sqrt{3})\div\exp(3/5)$
3014 8545 10^x : $3\sqrt{3}/2-\sqrt{5}$
3014 8794 rt : (4,2,-4,-7)
3015 1013 cu : $1/\ln(\sqrt{2}\pi)$
3015 1134 id : $1/\sqrt{11}$
3015 1161 e^x : $4\sqrt{2}-3\sqrt{3}/2$
3015 2224 sq : $3\sqrt{3}+\sqrt{5}/3$
3015 3376 θ_3 : $\zeta(3)/8$
3015 3432 ln : P_{30}
3015 4881 cr : $7\sqrt[3]{2}/4$
3015 4981 rt : (5,2,-8,-4)
3015 5972 rt : (8,3,5,-2)
3015 6665 sπ : $2\sqrt{2}/3+2\sqrt{3}/3$
3015 7201 sr : $2(1/\pi)/7$
3015 7357 rt : (3,3,-6,-2)
3015 7535 J_2 : $\sqrt{3}+\sqrt{5}+\sqrt{6}$
3015 7883 10^x : $1/3+e/4$
3015 8468 at : $4e/3$
3016 0372 e^x : $\exp(-4/3)$
3016 1540 as : $3\ln 2/7$
3016 2249 J_2 : $\exp(\sqrt{3}/3)$
3016 4167 $\ln(2/3)+s\pi(1/4)$
3016 7256 rt : (4,9,3,8)
3016 7519 K : $19/21$
3016 9089 rt : (9,6,4,-2)
3016 9587 rt : (8,8,5,1)
3017 1069 rt : (7,9,6,8)
3017 1829 rt : (5,3,-7,-7)
3017 3540 rt : (2,6,-5,-2)
3017 3724 id : $(\ln 3/2)^2$
3017 4959 $\sqrt{3}+s\pi(\pi)$
3017 5624 as : $3/4-\pi/3$
3017 7089 rt : (8,9,2,5)
3017 7377 ℓ_{10} : $5\zeta(3)/3$
3018 0400 J_2 : $\exp(\sqrt[3]{3})$
3018 1721 ln : $9\zeta(3)/8$
3018 2093 ζ : $(\sqrt{2}\pi/3)^{-2}$
3018 2466 rt : (9,9,-2,2)
3018 2555 rt : (7,1,9,-3)
3018 3688 J_1 : $(e+\pi)$
3018 5892 rt : (1,9,-4,-2)
3018 6818 tπ : $(\sqrt{5}/3)^{1/3}$
3018 6917 cu : $3\sqrt{5}/10$
3018 9761 Ψ : $(e+\pi)^3$
3018 9816 cu : $\sqrt{15}\pi$
3019 0469 10^x : $e-3\pi/4$
3019 0757 rt : (8,4,8,-3)
3019 1212 J_2 : $3\pi^2/7$
3019 2331 10^x : $4\sqrt{3}/5$
3019 2724 sq : $\sqrt[3]{6}$
3019 2941 rt : (8,8,3,8)
3019 3245 Li_2 : $\sqrt{5}/8$
3019 5289 sinh : $1/\ln(\sqrt[3]{4}/3)$
3019 6396 sr : $4+3\sqrt{3}/4$
3019 6789 rt : (9,8,-1,5)
3019 7659 rt : (9,7,7,-3)
3019 8252 rt : (3,4,-9,-3)
3019 8975 J_2 : $\sqrt[3]{17/3}$
3019 9039 cr : $3/2+\sqrt{2}/2$
3020 1880 sinh : $6(1/\pi)$
3020 2833 sr : $(\sqrt{2}\pi/2)^{-3}$
3020 3131 id : $e/9$
3020 3704 rt : (9,5,4,1)
3020 5363 cr : $10e\pi/7$
3020 5643 rt : (2,7,-8,-3)
3020 6922 sr : $\sqrt{3}/3+\sqrt{5}/2$
3020 7360 rt : (9,7,0,8)
3020 9913 ln : $\ln(2\pi/3)$
3021 0598 ζ : $3\sqrt{3}/8$
3021 1106 ln : $2+3\sqrt{5}/4$
3021 3350 E : $10\ln 2/9$
3021 3506 tπ : $3\sqrt{2}/4+\sqrt{3}/4$
3021 4768 $\ln(\sqrt{2})-\exp(1/2)$
3021 7470 Γ : $(\sqrt[3]{4}/3)^3$
3021 8681 rt : (1,8,4,-8)
3021 9432 id : $2(e+\pi)/9$
3022 0120 2^x : $8/21$
3022 0467 tπ : $2(e+\pi)/3$
3022 1403 cu : $2-e/4$
3022 3824 10^x : $2e/3+2\pi$
3022 8087 ln : $17/23$
3022 9686 ℓ_2 : $\sqrt[3]{15}$
3023 3098 id : $4e-\pi/2$
3023 4328 Ψ : $25/6$
3023 4499 J_1 : $7/11$
3023 4879 rt : (8,7,8,2)
3023 5032 ℓ_2 : $\ln(3/2)$
3023 5531 e^x : $4-3\sqrt{3}$
3023 5671 rt : (2,9,-7,8)
3023 6229 ℓ_{10} : $\sqrt{2}/2+3\sqrt{3}/4$
3023 7767 at : $e/2-\pi/3$
3023 8153 cu : $1/2-2e$
3023 9684 cu : $9\pi^2/7$
3024 0914 J_1 : $\ln(\sqrt[3]{4}/3)$
3024 0999 $\ln(2/3)\times s\pi(\sqrt{3})$
3024 1478 rt : (9,4,7,2)
3024 3088 sinh : $6\sqrt[3]{2}/7$
3024 3520 E : $4\sqrt{3}/9$
3024 5414 J_1 : $2(1/\pi)$
3024 5487 Ψ : $\sqrt{2}-\sqrt{5}+\sqrt{7}$
3024 5490 ℓ_{10} : $3\sqrt{2}-\sqrt{5}$
3024 7115 10^x : $4\sqrt{2}+\sqrt{3}/2$
3024 7539 Γ : $1/\ln(\sqrt[3]{4}/3)$
3024 7902 sq : $2+2e$
3024 8380 cr : $4/3+4e$
3024 8683 $\ln(3/4)\div s\pi(2/5)$
3024 8842 Li_2 : $2\sqrt[3]{2}/9$
3024 9287 10^x : $\sqrt{23/2}$
3025 0000 sq : $11/20$
3025 0901 Li_2 : $7/25$
3025 4799 sπ : $4\sqrt{2}-\sqrt{5}/4$
3025 6916 sq : $4/3-4\sqrt{3}$
3025 6973 ζ : $7\sqrt{5}/6$
3025 7659 $\exp(e)\times s\pi(\sqrt{3})$
3025 8509 ln : 10
3025 9841 10^x : $2+\sqrt{5}/2$
3026 0049 at : $3/2-2e/3$
3026 1889 $\sqrt{5}\div\exp(2)$
3026 3965 cu : $3\sqrt{2}/4-\sqrt{3}$
3026 3979 2^x : $4\sqrt{3}/3+\sqrt{5}/4$
3026 4011 sinh : $(\sqrt[3]{2})^{1/3}$
3026 5456 e^x : $3\sqrt{2}+\sqrt{3}/4$
3026 6807 sq : $4\sqrt{2}-3\sqrt{5}/2$
3026 6853 Ei : $3/10$
3026 7485 cu : $4\sqrt{3}-\sqrt{5}$
3026 8236 e^x : $4/3+3\pi/2$
3026 8450 sinh : $10\sqrt{2}/9$
3026 8934 J_2 : $25/14$
3027 0972 tanh : $5/16$
3027 1112 sr : $6\sqrt{2}/5$
3027 1617 e^x : $\exp(\sqrt{5}/2)$
3027 3273 sq : $7(e+\pi)/5$
3027 3327 ln : $1+\sqrt{2}/4$
3027 4209 at : $\sqrt{3}/4-\sqrt{5}/3$
3027 4983 J_0 : $1/\ln(\sqrt[3]{5})$
3027 5228 id : $4\sqrt{2}-3\sqrt{5}/2$

3027 5578 $\ln(\sqrt{2}) \div \ln(\pi)$
3027 6007 $2^x : 1/\sqrt[3]{18}$
3027 7563 rt : (4,7,-9,-9)
3027 9602 ln : $3 + e/4$
3027 9726 at : $1/2 + \pi$
3028 0532 $\Psi : (\sqrt{2})^{1/2}$
3028 2570 $\theta_3 : \ln(\ln 2)$
3028 2643 $10^x : (\ln 3/2)^{-2}$
3028 5343 sq : $1/\sqrt[3]{6}$
3028 6650 sr : $\exp(\sqrt[3]{4}/3)$
3028 7730 at : $4/3 + 4\sqrt{3}/3$
3028 8344 $\ell_{10} : (e)^3$
3028 8486 at : 5/16
3028 8577 sinh : 11/7
3028 8837 cu : $1/2 + 3e$
3029 0782 $E : 10/13$
3029 9225 $\ell_2 : \pi^2$
3030 0103 $\Psi : 21/2$
3030 0748 rt : (9,9,-8,-3)
3030 3812 rt : (9,3,-6,7)
3030 6154 sq : $3\sqrt{2} - \sqrt{3}$
3030 7142 rt : (9,4,-7,4)
3030 7817 ln : $3\sqrt{2}/4 + 4\sqrt{5}$
3030 8325 rt : (8,4,-3,7)
3031 0751 $J_0 : 5\sqrt{5}/6$
3031 1440 rt : (9,5,-8,1)
3031 1644 $2^x : 3/2 - \sqrt{5}/2$
3031 2092 rt : (9,8,-5,-2)
3031 2173 $2^x : 6(1/\pi)/5$
3031 3001 rt : (8,5,-4,4)
3031 3720 at : $4 - \sqrt{2}/4$
3031 4712 rt : (7,5,0,7)
3031 6828 $\exp(\sqrt{3}) \div s\pi(e)$
3031 7198 rt : (9,6,-9,-2)
3031 8834 cu : $\sqrt{2} + \sqrt{3}/4$
3031 9192 ln : $-\sqrt{2} + \sqrt{6} + \sqrt{7}$
3031 9254 $\Gamma : \sqrt{2} + \sqrt{3} - \sqrt{6}$
3031 9352 rt : (8,6,-5,1)
3032 0699 cu : 24/7
3032 1753 rt : (7,6,-1,4)
3032 2537 $t\pi : 7/24$
3032 3378 sq : $2 - \pi$
3032 4447 rt : (6,6,3,7)
3032 5215 at : $2/3 + 4\sqrt{5}/3$
3032 6342 $s\pi : 1/2 + 3\sqrt{3}/2$
3032 6532 $1/2 \div \exp(1/2)$
3032 6599 $\pi \div s\pi(2/5)$
3032 7908 $\ell_{10} : 1/3 + 3\sqrt{5}/4$
3032 8475 rt : (8,7,-6,-2)
3032 8513 rt : (1,8,-3,-4)
3033 2004 $Li_2 : 5\sqrt[3]{3}/8$
3033 2022 sr : $(\ln 2/2)^{-1/2}$
3033 2085 rt : (7,7,-2,1)
3033 6247 rt : (6,7,2,4)
3033 6937 $J : 5(1/\pi)/7$
3034 0076 $\Gamma : 2\pi^2/3$
3034 0518 at : $\sqrt{2} + \sqrt{5}$
3034 1097 rt : (5,8,7,-3)
3034 1307 $1/3 \div \ln(3)$
3034 1524 $J_1 : 2\sqrt{5}/7$
3034 2689 rt : (8,8,-7,-5)
3034 2861 sr : $5e/8$
3034 2956 cr : $1/\ln(\pi/2)$
3034 4348 at : $2/3 - \sqrt{2}/4$
3034 4964 $J_2 : 4\sqrt{5}/5$
3034 5185 rt : (9,7,-2,-1)
3034 5929 $Ei : 8(1/\pi)$
3034 6253 cr : $3 - \pi/4$
3034 6623 $e^x : 4e + 2\pi$
3034 6824 rt : (4,5,8,-3)
3034 8716 rt : (7,8,-3,-2)
3034 8737 cr : $3\sqrt{3} - 4\sqrt{5}/3$
3034 9546 cu : $4\sqrt{3}/3 - 4\sqrt{5}/3$
3035 1156 $\Gamma : 7e$
3035 3688 rt : (3,2,9,-3)
3035 4682 $\Gamma : 5\sqrt{5}/3$
3035 5975 rt : (6,8,1,1)
3035 7489 sq : $(e/2)^3$
3036 1140 cr : $3\sqrt{2}/4 + 2\sqrt{3}/3$
3036 4886 rt : (5,8,5,4)
3036 6494 $\ell_{10} : (e + \pi)^3$
3036 7144 $Li_2 : 3\zeta(3)/4$
3036 7488 $10^x : P_{78}$
3036 7894 rt : (8,9,-8,-8)
3036 8447 as : $1 - 3\sqrt{3}/4$
3036 9336 $10^x : \sqrt{2} - 3\sqrt{3}/4$
3037 1585 $Ei : 1/\ln(\sqrt{2}\pi/3)$
3037 2751 $\ell_{10} : 9\sqrt{5}$
3037 5346 rt : (6,8,-8,-3)
3037 6088 rt : (4,8,9,7)
3037 6761 $e^x : 5(1/\pi)/6$
3037 9097 $\Psi : \sqrt{10/3}$
3037 9478 $\theta_3 : 2\sqrt{2}/3$
3037 9792 rt : (8,1,-8,-9)
3037 9919 rt : (7,9,-4,-5)
3038 0524 $\ell_{10} : (1/(3e))^{1/3}$
3038 2301 $\ell_{10} : 4/3 + e/4$
3038 4048 sr : 17/10
3038 5952 rt : (7,5,-9,-5)
3038 6326 at : $2 - 4\sqrt{2}$
3038 8143 $10^x : 1/\sqrt[3]{21}$
3038 8214 $t\pi : e/3$
3038 9324 $J_1 : 16/25$
3038 9356 $10^x : 2\sqrt{2} - 4\sqrt{3}/3$
3038 9696 ln : $3\sqrt{2} - \sqrt{5}/4$
3039 0155 rt : (2,9,-9,-8)
3039 0598 rt : (3,1,6,-2)
3039 0630 $10^x : 4\sqrt{2} + 2\sqrt{5}$
3039 1044 $2^x : 1 - e$
3039 1386 rt : (8,0,-9,6)
3039 5445 sinh : $(4/3)^3$
3039 5449 rt : (7,0,-5,9)
3039 5644 rt : (6,9,0,-1)
3039 6092 cu : $9\sqrt{3}/5$
3039 9415 at : $3\sqrt{2}/4 + 3\sqrt{3}/2$
3039 9777 cu : $4\sqrt{2}/3 - 2\sqrt{5}$
3040 0000 cu : 17/5
3040 1033 sq : $3/4 + 3e/2$
3040 1402 $\zeta : 4(e + \pi)/9$
3040 2466 rt : (8,7,7,-3)
3040 6826 $\sqrt{2} - e$
3040 7404 at : $3e\pi/7$
3040 9883 $3/4 \times \ln(2/3)$
3041 0145 rt : (2,2,7,2)
3041 0534 cr : $9e/2$
3041 1137 rt : (7,1,-6,6)
3041 2503 cu : $\sqrt{23}$
3041 2751 $\Gamma : 9\pi/7$
3041 3307 cu : $1/4 - \pi/2$
3041 3534 rt : (7,4,8,-3)
3041 5768 $\ell_2 : 3/4 + 4\pi/3$
3041 6674 rt : (1,8,-6,-8)
3041 6771 rt : (6,1,-2,9)
3041 7091 rt : (5,9,4,1)
3042 0305 $\exp(2/3) + \exp(\sqrt{5})$
3042 0929 sr : $3 - 3\sqrt{3}/4$
3042 1376 sr : $3 + 4\sqrt{3}/3$
3042 1619 tanh : $\pi/10$
3042 1793 rt : (6,7,-5,-2)
3042 2302 sinh : $\exp(2e/3)$
3042 4217 $J_0 : \pi$
3042 5006 at : $5(e + \pi)/8$
3042 5877 $\Psi : \ln(\ln 2)$
3042 6764 rt : (6,1,9,-3)
3042 6890 $s\pi(1/4) \times s\pi(\pi)$
3042 8825 $Ei : 23/20$
3043 1039 rt : (9,9,3,-2)
3043 1532 cr : $1/3 + 4\sqrt{2}/3$
3043 1636 rt : (7,2,-7,3)
3043 1638 tanh : $2\sqrt{2}/9$
3043 1841 $J_1 : 4\sqrt[3]{3}/9$
3043 2055 $e^x : 3e + 3\pi/2$
3043 2291 $t\pi : 3\sqrt{5}$
3043 4311 $t\pi : 3\sqrt{2}/4 - 2\sqrt{3}/3$
3043 4782 id : 7/23
3043 6872 $t\pi : \ln(\ln 3)$
3043 7256 $10^x : 2/3 + 4e/3$
3043 7449 $10^x : 2 - \sqrt{2}/4$
3043 7923 rt : (1,5,8,2)
3043 8455 ln : $(e + \pi)^3$
3043 9184 rt : (6,2,-3,6)
3043 9579 at : $\pi/10$
3044 0412 $s\pi : 3\zeta(3)/4$
3044 1713 $t\pi : \sqrt[3]{13/3}$
3044 2093 $J_1 : 3\sqrt[3]{5}/8$
3044 4210 at : $7\pi/6$
3044 4896 at : $3\sqrt{2} - \sqrt{3}/3$
3044 6181 rt : (7,4,-6,-2)
3044 6331 $\ell_{10} : 8\sqrt[3]{2}/5$
3044 6734 $2 \times \exp(\sqrt{3})$
3044 7009 rt : (3,7,-8,-3)
3044 7337 rt : (1,7,-7,-5)
3044 7509 as : P_{79}
3044 7513 rt : (5,2,1,9)
3044 8096 rt : (8,6,4,-2)
3044 9366 $Ei : \sqrt{24}$
3044 9629 at : $2\sqrt{2}/9$
3045 0438 id : $(2e/3)^{-2}$
3045 2029 sinh : 3/10
3045 4427 at : 11/3
3045 8953 $e^x : 3 - 4\pi/3$
3045 9553 rt : (7,3,-8,0)
3045 9724 $\theta_3 : 3/4$
3046 0175 $J_0 : 22/7$
3046 3247 $e^x : 7e/3$
3046 5036 rt : (4,4,-9,-3)
3046 9265 as : 3/10
3046 9439 rt : (7,3,5,-2)
3047 0184 rt : (6,3,-4,3)
3047 0709 cu : $e - 2\pi$
3047 1895 $1/2 + \ln(\sqrt{5})$
3047 3852 $e^x : e/3 - 2\pi/3$
3047 4166 $10^x : \ln 2/6$
3047 6686 $2^x : 2\sqrt{2}/3 - \sqrt{5}/4$
3048 0303 id : $(\sqrt{2}\pi/2)^{1/3}$
3048 0589 rt : (8,9,2,0)
3048 1315 $s\pi : 5\sqrt[3]{3}/8$
3048 1813 rt : (2,7,-6,4)
3048 2141 rt : (5,3,0,6)
3048 3141 $2^x : 7\sqrt[3]{3}/3$
3048 4313 $\Gamma : 9e/7$
3048 5458 $\ell_2 : 17/21$
3048 5721 sq : $\sqrt{13}\pi$
3048 6782 $s\pi : \ln(3)$
3048 7575 $J : \exp(-14\pi/5)$
3048 7662 rt : (6,3,-8,-8)
3048 8376 as : P_{22}
3049 2896 rt : (5,9,-1,-1)
3049 3117 ln : $1/3 + 3\sqrt{5}/2$
3049 4628 rt : (2,9,-4,-2)
3049 5319 rt : (1,9,7,-3)
3049 5588 cr : 20/9
3049 5691 rt : (4,3,4,9)
3049 6102 tanh : $\sqrt[3]{2}/4$
3049 6928 rt : (9,0,9,3)
3049 7996 sinh : $(\sqrt[3]{5}/2)^{-1/2}$
3049 8329 $J_2 : 4\pi/7$
3049 8826 $2^x : (\ln 2/3)^{-2}$
3049 9416 $\Gamma : 16/23$
3049 9648 rt : (2,8,-1,7)
3049 9795 rt : (7,4,-9,-3)
3049 9865 $\Psi : 19/16$
3050 0744 $10^x : e/4 + \pi/3$
3050 1065 $J_2 : 1/\ln(\sqrt{3}/2)$
3050 1427 $10^x : 3/2 - \sqrt{3}/2$
3050 1807 $erf : 2\ln 2/5$
3050 1844 sq : $4/3 - 4\sqrt{2}/3$
3050 5262 $10^x : 2\sqrt{2} + \sqrt{5}/4$
3050 5787 sinh : $\zeta(3)/4$
3050 6038 rt : (9,8,0,-1)
3050 7679 $2/3 \div \ln(3/5)$
3051 0763 $J_1 : 9/14$
3051 1202 $\ln(2/5) + \exp(1/5)$
3051 1339 $erf : 1/\sqrt{13}$
3051 4286 at : $\sqrt[3]{2}/4$
3051 5434 $2^x : 3 - 3\pi/2$
3051 5864 rt : (6,8,-3,0)
3051 6845 $\ell_{10} : 7\sqrt[3]{3}/5$
3051 8299 id : $\sqrt{2}/2 + 3\sqrt{3}/2$
3052 1373 $\Psi : (\sqrt{2}\pi/2)^{-1/2}$

3052 1814 sq : $\sqrt[3]{7/2}$
3052 2050 ζ : $(\ln 3/2)^{1/2}$
3052 2820 ln : $\sqrt{2}/2 + 4\sqrt{5}/3$
3052 3175 as : $\zeta(3)/4$
3052 4719 E : $(\sqrt{2}\pi/2)^{-1/3}$
3052 5153 rt : (3,6,-5,-2)
3052 5681 $\exp(4) + s\pi(1/4)$
3052 6044 rt : (5,2,-5,-8)
3052 6130 at : $2 + 3\sqrt{5}/4$
3052 6518 sr : $3/2 + 3\pi$
3052 7154 10^x : $19/22$
3052 9988 cr : $\sqrt[3]{11}$
3053 0257 ℓ_{10} : $5\ln 2/7$
3053 0309 rt : (4,4,-9,7)
3053 1444 sinh : $3\sqrt[3]{3}/4$
3053 2910 rt : (9,2,-8,2)
3053 3117 ℓ_2 : $2 + \sqrt{2}/3$
3053 3687 rt : (5,3,7,2)
3053 4455 rt : (5,4,-1,3)
3053 4683 $\ln(2/5) - \exp(2)$
3053 5068 $1/4 \times \exp(1/5)$
3053 5646 ζ : $6\sqrt{2}/7$
3053 7224 $\exp(\pi) \times s\pi(\sqrt{2})$
3053 8164 ln : $14/19$
3054 0728 rt : (1,6,0,-8)
3054 1695 ℓ_{10} : $10\sqrt{2}/7$
3054 3024 ln : $(5/2)^{-1/3}$
3054 3470 at : $3 + e/4$
3054 3752 at : $3\sqrt{2}/4 - \sqrt{5}/3$
3054 4889 rt : (9,8,2,9)
3054 6022 rt : (8,5,1,-1)
3054 7780 ζ : $13/5$
3054 9032 rt : (5,8,-9,-2)
3054 9848 ln : $4/3 + 3\pi/4$
3055 0118 Ei : $(\sqrt{2}\pi/3)^{-2}$
3055 0741 ℓ_{10} : $2\sqrt{3}/7$
3055 1004 Ei : $(1/\pi)/6$
3055 1169 2^x : $5/13$
3055 1520 rt : (1,7,-1,9)
3055 2329 rt : (6,6,-2,-1)
3055 3487 at : $\sqrt{2} - \sqrt{6} - \sqrt{7}$
3055 4464 ℓ_2 : $\sqrt{2}/3 + 2\sqrt{5}$
3055 6019 erf : $5/18$
3055 6118 $t\pi$: $2\pi/3$
3055 6214 rt : (4,8,7,-3)
3055 7125 Ψ : $3\sqrt{5}/10$
3055 8174 rt : (4,5,-8,4)
3055 8419 sinh : $9\zeta(3)/10$
3055 8513 2^x : $(\ln 2/2)^{-3}$
3056 0517 e^x : $4/15$
3056 5772 rt : (4,3,-6,-2)
3056 6008 $\ln(\sqrt{2}) - \exp(\sqrt{3})$
3056 6474 rt : (5,1,-6,-5)
3056 9479 tanh : $6/19$
3057 1319 at : $3\sqrt{2} - \sqrt{5}/4$
3057 2809 sq : $4 - \sqrt{5}/2$
3057 5660 $s\pi$: $\sqrt{2}/4 + \sqrt{5}/3$
3057 5808 ℓ_2 : $1/4 + \sqrt{5}/4$
3057 5843 sq : 6π
3057 6943 2^x : $2\sqrt{3}/9$
3057 7468 10^x : $7(e + \pi)/5$
3057 8330 sr : $4 + 4\sqrt{3}$
3057 8512 sq : $20/11$
3057 9646 cr : $1/2 - \sqrt{2}/3$
3057 9834 Γ : $10\sqrt{3}/7$
3058 1029 rt : (1,8,4,-2)
3058 1425 rt : (5,9,-8,-5)
3058 1598 rt : (4,1,-2,-8)
3058 2036 rt : (3,5,8,-3)
3058 4467 2^x : $\exp(2\pi/3)$
3058 4728 Li_2 : $\sqrt{2}/5$
3058 5427 rt : (4,6,8,2)
3058 5890 rt : (9,9,1,6)
3058 7411 e^x : $5e\pi/2$
3058 7527 cr : $9\sqrt{3}/7$
3058 7887 at : $6/19$
3058 9039 ln : $2\sqrt{3}/3 - \sqrt{5}/2$
3058 9752 rt : (1,5,-2,-2)
3058 9830 rt : (6,5,-6,-3)
3059 2841 ℓ_2 : $6\sqrt[3]{3}/7$
3059 2935 rt : (4,6,-7,1)
3059 3380 2^x : $5\ln 2/9$
3059 4523 $s\pi$: $2\sqrt{3}/3 + 4\sqrt{5}$
3059 4617 $t\pi$: $\sqrt{22/3}$
3059 5034 Γ : $6(1/\pi)/7$
3059 5248 rt : (9,4,-4,8)
3059 6413 Li_2 : $8(1/\pi)/9$
3059 6999 rt : (9,1,6,2)
3059 7456 at : $1/3 + 3\sqrt{5}/2$
3059 9010 ζ : $13/20$
3060 1065 rt : (8,9,5,9)
3060 3076 rt : (7,2,2,-1)
3060 4282 Γ : $8\ln 2$
3060 4958 at : $\sqrt{2}/2 + 4\sqrt{5}/3$
3060 5018 Γ : $\exp(e/3)$
3060 5119 J_1 : $1/\ln(3\pi/2)$
3060 5360 rt : (3,3,-6,7)
3060 6708 $e + s\pi(1/5)$
3060 6735 at : $1/2 - 4\pi/3$
3060 9052 $s\pi$: $\ln 2/7$
3060 9207 tanh : $1/\sqrt{10}$
3060 9411 e^x : $\sqrt{2} - 3\sqrt{3}/2$
3060 9861 ℓ_2 : $(\sqrt[3]{4}/3)^{1/3}$
3061 0050 sinh : $1/\sqrt{11}$
3061 1631 J_2 : $10\sqrt[3]{2}/7$
3061 1784 at : $4/3 + 3\pi/4$
3061 1892 cr : $7(1/\pi)$
3061 2235 ζ : $3\sqrt{3}/2$
3061 2244 sq : $8/7$
3061 3190 rt : (2,2,9,-3)
3061 4353 J_2 : $9/5$
3061 4938 rt : (7,7,7,-3)
3062 0208 e^x : $2\zeta(3)/9$
3062 2031 J_0 : $8\sqrt[3]{5}/3$
3062 2577 rt : (5,5,-2,0)
3062 3416 sq : $\sqrt{3} + \sqrt{5} - \sqrt{6}$
3062 4523 e^x : $(\sqrt[3]{5})^{1/3}$
3062 5889 2^x : $3/2 + 3\sqrt{2}/2$
3062 6325 Γ : $7\sqrt{2}/4$
3062 6394 $t\pi$: $3e/2 - \pi/4$
3062 7736 at : $1/\sqrt{10}$
3062 8014 id : $3\sqrt{3}/2 + 3\sqrt{5}$
3063 1841 rt : (7,8,-8,-3)
3063 3434 rt : (1,6,5,7)
3063 3655 erf : $7(1/\pi)/8$
3063 3675 sinh : $(\ln 3/2)^2$
3063 5901 10^x : $3 - 2\sqrt{5}/3$
3063 6190 rt : (9,5,-5,5)
3063 6727 $t\pi$: 5π
3063 6988 J_2 : $10\ln 2$
3063 7287 θ_3 : $2\sqrt{5}/5$
3063 7395 ℓ_2 : $7\ln 2/6$
3063 7506 rt : (4,7,-6,-2)
3063 7860 Ψ : $1/\ln(\sqrt{2}\pi)$
3063 8168 e^x : $1/\sqrt{14}$
3064 1546 rt : (2,8,-1,-1)
3064 2017 Γ : $-\sqrt{3} + \sqrt{5} + \sqrt{7}$
3064 3651 $\sqrt{3} \div \exp(\sqrt{3})$
3064 4095 sr : $1 - e/3$
3064 4393 rt : (4,0,-3,5)
3064 5003 sq : $3\sqrt{3}/4 + 2\sqrt{5}$
3064 5443 at : $2\sqrt{2} + \sqrt{3}/2$
3064 5645 rt : (6,4,8,-3)
3064 6137 J : $2/17$
3064 6735 Γ : $5\ln 2/9$
3064 8862 rt : (8,4,7,2)
3064 9537 $3/5 \times \ln(3/5)$
3064 9737 Γ : $5\sqrt[3]{2}/2$
3065 0693 $t\pi$: $e\pi/5$
3065 0932 rt : (8,5,-1,8)
3065 1429 as : $(\ln 3/2)^2$
3065 3801 rt : (3,4,-5,4)
3065 5593 J_0 : $\sqrt{3} - \sqrt{5} - \sqrt{7}$
3065 6296 sr : $1 + \sqrt{2}/2$
3065 6388 sq : $3\sqrt{3}/4 - \sqrt{5}/3$
3065 6554 J_0 : $5\sqrt[3]{2}/2$
3065 7043 at : $3/2 - 3\sqrt{3}$
3065 7445 J_2 : $4\sqrt{3}$
3065 7897 rt : (7,3,-3,-1)
3065 9819 ℓ_{10} : $2/3 + e/2$
3065 9950 2^x : $\exp(\sqrt{2})$
3066 2288 E : $(\sqrt[3]{5})^{-1/2}$
3066 3068 rt : (4,7,4,-2)
3066 3814 E : $13/17$
3066 3872 rt : (3,9,9,2)
3066 4088 2^x : $1/3 - 3e/4$
3066 4433 sinh : $e/9$
3066 7218 sr : $\ln(\ln 3)$
3066 7443 rt : (1,9,6,2)
3066 9186 rt : (3,0,1,8)
3066 9880 Ei : $\pi/4$
3067 0368 sinh : $5\sqrt{3}/8$
3067 1356 e^x : P_{66}
3067 1659 rt : (2,7,4,9)
3067 1765 rt : (8,5,-9,-3)
3067 2254 e^x : P_{65}
3067 4202 J_0 : $13/7$
3067 7486 rt : (1,5,1,-5)
3067 7612 J_0 : $7(e + \pi)/9$
3067 7822 2^x : $1/2 + e$
3068 0340 ζ : $9\sqrt[3]{3}/5$
3068 0497 e^x : $7\pi/5$
3068 2197 J_2 : $5\sqrt[3]{3}/4$
3068 2276 as : $e/9$
3068 2576 rt : (5,1,9,-3)
3068 2751 Li_2 : $\exp(-\sqrt[3]{2})$
3068 3119 J_2 : $9\sqrt[3]{3}/2$
3068 3237 ln : $2\sqrt{2} + \sqrt{3}/2$
3068 3433 J_2 : $(e + \pi)^{1/3}$
3068 3808 rt : (6,9,-9,5)
3068 4106 Li_2 : $(e/2)^{-1/3}$
3068 5281 id : $\ln(e/2)$
3068 6607 rt : (8,1,-7,7)
3068 7693 cr : $1/2 + \sqrt{3}$
3068 7938 e^x : $7\sqrt{2}$
3068 8440 sr : $e\pi/5$
3068 8791 J_2 : $3\zeta(3)/2$
3068 9011 J_1 : $11/17$
3068 9200 rt : (9,6,-6,2)
3069 0757 at : $3\pi^2/8$
3069 1238 rt : (6,1,3,1)
3069 3357 rt : (8,9,3,-2)
3069 3958 ℓ_{10} : $\sqrt[3]{25/3}$
3069 3968 $\exp(1/5) + \exp(3)$
3069 6712 rt : (4,8,-5,-5)
3069 6770 at : $3/4 - \sqrt{3}/4$
3069 7728 10^x : $9/7$
3069 8322 rt : (2,9,-8,3)
3069 9953 rt : (4,7,-8,-3)
3070 0327 sq : $3(e + \pi)/7$
3070 0840 id : $\sqrt{3}/2 - \sqrt{5}/4$
3070 2421 cu : $1/4 - 4\pi$
3070 5442 ℓ_2 : $5\sqrt{3}/7$
3070 5607 rt : (5,1,-8,-1)
3070 6805 rt : (9,3,8,-3)
3070 7928 rt : (6,3,-8,2)
3070 8759 rt : (8,6,-2,5)
3070 9257 ℓ_{10} : $\exp(1/\sqrt{2})$
3071 0344 2^x : $(\sqrt[3]{5}/3)^{-1/3}$
3071 0678 $2/5 - s\pi(1/4)$
3071 2902 $t\pi$: $\sqrt{3}/4 + 2\sqrt{5}$
3071 4456 rt : (3,5,6,6)
3071 6517 10^x : $1/2 + 2e$
3071 6636 10^x : $3/4 + \sqrt{3}/2$
3071 9006 rt : (3,5,-4,1)
3072 0244 at : $3\sqrt{3} - 2\sqrt{5}/3$
3072 1156 rt : (7,7,8,2)
3072 1723 ℓ_2 : $\exp(e/3)$
3072 2022 rt : (9,7,-3,0)
3072 3891 rt : (1,9,-2,8)
3072 4180 $t\pi$: $e/3 + 4\pi/3$
3072 7309 rt : (7,3,-9,-9)
3072 7553 10^x : $\sqrt{2}/4 - \sqrt{3}/2$
3072 8071 e^x : $2 - \sqrt{3}$
3072 8975 sinh : $13/6$
3072 9807 rt : (6,6,-7,-6)
3073 0293 rt : (7,6,2,8)
3073 1079 rt : (5,7,-7,8)
3073 1160 J_1 : $(\sqrt[3]{2}/3)^{1/2}$
3073 1552 at : $3 + \sqrt{2}/2$

3073 2796 $s\pi : 10\pi^2/7$
3073 3125 cu : $\sqrt{2}+\sqrt{3}+\sqrt{7}$
3073 3457 $J_1 : 7\ln 2$
3073 4918 $t\pi : 2\sqrt{3}-\sqrt{5}/4$
3073 5272 at : $\sqrt{2}/4+3\sqrt{5}/2$
3073 5378 $\Gamma : 3/11$
3073 5492 $\ell_2 : 7\sqrt{2}$
3073 6721 $erf : \sqrt{5}/8$
3073 6904 rt : (3,5,-2,-1)
3073 8992 at : $3-3\sqrt{5}$
3073 9761 rt : (8,2,-8,4)
3074 0491 $\Psi : \sqrt[3]{5/3}$
3074 1009 rt : (7,6,4,-2)
3074 1818 at : $3e/4-3\pi/4$
3074 3039 $J_2 : 13/2$
3074 3839 rt : (2,8,7,8)
3074 5122 $\ell_{10} : 4e+3\pi$
3074 6566 rt : (5,4,-9,-3)
3074 6929 $\Psi : 5(e+\pi)/7$
3074 6962 $Ei : 5\sqrt{2}/9$
3074 8992 sq : $4\ln 2/5$
3074 9038 rt : (8,0,9,3)
3075 2553 $\exp(\sqrt{5})+s\pi(2/5)$
3075 3581 rt : (2,5,-3,-8)
3075 4145 $t\pi : 3\sqrt{2}/2+2\sqrt{5}$
3075 8155 $Ei : 11/14$
3075 8771 $\Gamma : 2\sqrt{3}/9$
3075 8993 rt : (7,2,-4,7)
3075 9931 id : $8\sqrt[3]{3}/5$
3076 0503 rt : (9,7,-7,-1)
3076 2531 $\ell_{10} : \sqrt{2}/4+3\sqrt{5}/4$
3076 2876 $Ei : \sqrt[3]{6}$
3076 3218 $10^x : P_{77}$
3076 4116 $10^x : ((e+\pi)/2)^{-2}$
3076 4267 cu : $\sqrt{2}/4-3\sqrt{3}/2$
3076 4718 $10^x : 7\ln 2$
3076 5240 rt : (1,9,7,-4)
3076 6048 id : $(\sqrt[3]{5})^{1/2}$
3076 6242 sq : $(e+\pi)^{-1/3}$
3076 7334 at : $1-3\pi/2$
3076 8816 $10^x : (\sqrt[3]{5})^{1/2}$
3076 9230 id : 4/13
3076 9924 $e^x : (\sqrt[3]{5})^{-1/3}$
3077 0870 $10^x : 9e/4$
3077 2994 rt : (2,8,-7,-9)
3077 3329 $J_2 : 7\pi^2/10$
3077 3985 at : $\sqrt{2}-\sqrt{3}$
3077 5388 rt : (5,8,-6,5)
3077 6312 $\Psi : 9(e+\pi)/5$
3077 7369 rt : (1,7,1,-1)
3077 8848 sinh : $4(e+\pi)/5$
3077 8915 at : $1/4+2\sqrt{3}$
3077 9142 rt : (4,9,-4,-8)
3077 9531 $K : e/3$
3077 9918 rt : (3,8,7,-3)
3078 1902 rt : (2,1,6,-2)
3078 1999 sr : $2\sqrt{2}-\sqrt{5}/2$
3078 2190 $J_1 : \sqrt{19/2}$
3078 5559 $2^x : 4e/3+3\pi/2$
3078 6176 $erf : 2\sqrt[3]{2}/9$
3078 6198 tanh : 7/22
3078 6904 rt : (1,7,5,4)
3078 7126 id : $(\sqrt{2}\pi/3)^{-3}$
3078 7252 at : $e/2+3\pi/4$
3078 7678 rt : (8,7,-3,2)
3078 8006 $erf : 7/25$
3078 9622 $t\pi : \sqrt{2}+2\sqrt{5}/3$
3079 2484 at : $3\sqrt{3}/2+\sqrt{5}/2$
3079 3194 $J_1 : 3\sqrt{3}/8$
3079 3236 cr : $2/3+\pi/2$
3079 7770 $J_2 : \sqrt{2}+\sqrt{6}+\sqrt{7}$
3079 7791 tanh : $(1/\pi)$
3079 7852 rt : (1,6,5,1)
3079 8039 rt : (6,2,-6,-9)
3080 1160 rt : (6,3,5,-2)
3080 1915 rt : (5,9,0,-1)
3080 2305 sinh : 13/12
3080 3814 $2^x : P_{27}$
3080 4820 $Ei : \sqrt{10}\pi$
3080 5027 $3/5 \div \exp(2/3)$
3080 5278 at : 7/22
3080 5884 $s\pi : 2/3+\sqrt{3}/4$
3080 6042 $2^x : 2e$
3080 6312 $J_2 : 10(e+\pi)/9$
3080 6540 rt : (7,7,-5,-2)
3080 6541 id : $4\ln 2/9$
3080 7140 at : $1+e$
3080 8718 rt : (8,3,-9,1)
3081 1466 rt : (3,6,-3,-2)
3081 3545 $J_1 : 13/20$
3081 4013 $\exp(4/5)\times s\pi(1/5)$
3081 4420 as : $1/3-3\sqrt{3}/4$
3081 6823 cu : $2-3\pi$
3081 6907 at : $(1/\pi)$
3081 7161 $t\pi : 1/4-3e$
3081 7941 rt : (2,9,3,-7)
3081 8137 rt : (7,7,1,5)
3081 8499 $10^x : \exp(\sqrt[3]{3})$
3081 8624 $e^x : 1/2-3\sqrt{5}/4$
3082 1069 $\theta_3 : 4\ln 2/5$
3082 1279 sr : $2+4\sqrt{5}$
3082 1611 rt : (3,9,-4,-2)
3082 1819 $10^x : 3/4+\pi/4$
3082 6781 id : $4\sqrt{2}/3-\sqrt{3}/3$
3082 7536 cu : $3/4+4\sqrt{2}/3$
3082 7840 rt : (2,5,8,-3)
3082 9267 rt : (5,9,-5,2)
3082 9823 $s\pi : 10\sqrt[3]{5}$
3083 1206 $4/5 \div \ln(\sqrt{2})$
3083 1500 cr : $4/3+e/3$
3083 2180 sr : $((e+\pi)/2)^{1/2}$
3083 2269 $\Psi : 4\pi/3$
3083 3111 at : $3/4-2\sqrt{5}$
3083 3354 $s\pi : \sqrt{2}+\sqrt{5}+\sqrt{6}$
3083 4307 rt : (7,3,-5,4)
3083 5652 rt : (6,7,7,-3)
3083 6173 $2^x : 4+4\sqrt{3}/3$
3083 7889 $10^x : 3\sqrt{3}/10$
3083 8009 $Li_2 : \sqrt[3]{5}/6$
3083 8391 $J_0 : 9\sqrt[3]{3}/7$
3084 0893 rt : (5,2,-9,-4)
3084 1980 sq : $3e/4+4\pi$
3084 3062 at : $\sqrt{2}+4\sqrt{3}/3$
3084 4972 $s\pi : 5e\pi/7$
3084 5914 $t\pi : 2/21$
3084 6141 rt : (5,4,4,1)
3084 6755 $J_2 : 7\zeta(3)/2$
3084 8315 rt : (4,6,-4,8)
3084 8916 $2^x : (\ln 3)^2$
3085 1720 $2^x : \exp(\sqrt{5}/3)$
3085 1757 rt : (7,6,-9,2)
3085 2448 $2\times\exp(e)$
3085 2522 rt : (6,7,5,8)
3085 2995 $\ell_2 : 2e/3+\pi$
3085 3142 $\ln(\sqrt{5})-\exp(\sqrt{2})$
3085 6513 cr : $3/4+2\sqrt{5}/3$
3085 6529 sq : $2/3+4\sqrt{5}/3$
3085 9903 rt : (2,7,-8,-6)
3085 9949 $J_1 : (e+\pi)/9$
3086 0669 sr : 2/21
3086 1456 rt : (9,8,-8,-4)
3086 2381 $s\pi : 5\sqrt[3]{2}/3$
3086 2527 rt : (6,3,-1,7)
3086 3051 ln : $3\pi^2/8$
3086 3807 rt : (9,2,3,1)
3086 4197 sq : 5/9
3086 4339 at : $5\sqrt{5}/3$
3086 6259 $\lambda : \exp(-6\pi/5)$
3086 6492 rt : (3,4,-8,2)
3086 7029 $t\pi : 1/4-2\sqrt{3}/3$
3086 8047 $\ln(4/5)-\exp(3)$
3086 9150 rt : (6,1,-7,-6)
3086 9306 $e^x : 3\sqrt{2}/4-\sqrt{5}$
3086 9363 $\Gamma : \sqrt[3]{1/3}$
3087 0412 cu : $7\zeta(3)/4$
3087 0452 cu : $1/3+2\pi/3$
3087 4171 $2^x : 1/2+\sqrt{2}/2$
3087 4521 rt : (1,2,7,2)
3087 5383 sq : $3e-\pi/4$
3087 5971 $\theta_3 : (e+\pi)^{-1/3}$
3087 8607 cr : $2\sqrt{2}/3+3\sqrt{3}/4$
3087 9570 rt : (5,0,6,2)
3087 9635 ln : $9\sqrt{5}/2$
3088 1281 $\theta_3 : 2/13$
3088 3038 rt : (9,2,5,-2)
3088 3040 $\ell_{10} : (\ln 3/3)^3$
3088 3078 $\ell_2 : \sqrt{3}+\sqrt{5}/3$
3088 4808 $s\pi : 2\sqrt{3}/3+\sqrt{5}/3$
3088 4931 $s\pi : 7\sqrt{2}/9$
3088 6070 rt : (1,2,9,-3)
3088 7377 $2^x : e/7$
3088 7884 rt : (5,4,8,-3)
3088 8585 $\exp(4/5)\div s\pi(\sqrt{2})$
3088 9320 rt : (9,6,7,-3)
3089 2401 $t\pi : 2e/3+2\pi$
3089 2604 $E : 16/21$
3089 3251 rt : (1,7,-4,-9)
3089 5204 rt : (5,1,-3,-9)
3089 5494 at : $3/4+4\sqrt{5}/3$
3089 6179 rt : (5,5,-9,-8)
3089 6424 rt : (9,5,-2,9)
3089 7074 $Ei : 5\zeta(3)/4$
3089 7086 rt : (9,1,-8,8)
3089 8876 $t\pi : \sqrt{2}/4+3\sqrt{5}/2$
3089 9693 at : $2+\sqrt{3}$
3089 9892 tanh : $\sqrt{5}/7$
3090 0131 cu : $2-e/3$
3090 1067 $\zeta : 5(1/\pi)$
3090 1699 id : $1/4-\sqrt{5}/4$
3090 1936 sr : $3(1/\pi)/10$
3090 4091 $\Gamma : \ln 2$
3090 5464 $J_1 : 15/23$
3090 6813 sinh : 7/23
3090 7694 cr : $\ln(3\pi)$
3090 8047 $J_2 : 2e/3$
3090 9034 sr : $2\sqrt{3}/3+\sqrt{5}/4$
3090 9684 $s\pi : 10\sqrt[3]{5}/9$
3091 1259 $e^x : \sqrt{2}-\sqrt{3}/3$
3091 4636 sr : $2/3+\pi/3$
3091 7620 $10^x : 17/13$
3091 8548 $s\pi : 5\sqrt[3]{2}/7$
3091 9136 $\theta_3 : 2\ln 2/9$
3091 9332 at : $\sqrt{5}/7$
3092 0269 rt : (4,7,-3,5)
3092 0386 cr : $5\pi/7$
3092 0604 sr : $3\sqrt{2}+3\sqrt{5}$
3092 2579 $Li_2 : 2/7$
3092 2770 rt : (1,6,-1,6)
3092 3096 $2^x : 4e-4\pi$
3092 3200 sinh : $(2e/3)^{-2}$
3092 4321 $10^x : ((e+\pi)/3)^{-1/3}$
3092 5019 $\ln(2/5)+\exp(4/5)$
3092 5156 $2^x : \ln(1/(2e))$
3092 5370 as : 7/23
3092 6576 at : $3/2+\sqrt{5}$
3092 7058 $\Psi : 11/6$
3093 0734 sr : 12/7
3093 0915 $10^x : 7\sqrt{3}/10$
3093 2782 $\Psi : 7/16$
3093 3624 $t\pi : 3(1/\pi)/10$
3093 4264 $\ln(4)\times\ln(4/5)$
3093 4713 $\Gamma : 5/13$
3093 5574 $\ell_{10} : 3e/4$
3093 6345 $s\pi : P_{18}$
3093 7385 rt : (7,4,-6,1)
3093 7574 $2^x : 23/2$
3093 8084 $e^x : 7\sqrt{3}/2$
3093 8457 $2^x : 7/18$
3094 0107 id : $4\sqrt{3}/3$
3094 0475 $2^x : 1/\sqrt[3]{17}$
3094 1806 as : $(2e/3)^{-2}$
3094 2383 sq : $3e/2+3\pi$
3094 2880 $s\pi : P_{48}$
3094 2983 $\ell_{10} : 3\sqrt{3}/2-\sqrt{5}/4$
3094 5149 $\Gamma : 5\pi^2/6$
3094 5496 rt : (8,3,8,-3)
3094 6228 $e^x : 3\sqrt{3}-3\sqrt{5}/2$
3094 6336 $J_1 : 9\sqrt[3]{5}/5$
3094 6853 cr : $3/2+\sqrt{5}/3$
3094 7350 $2^x : 4\sqrt{3}/3+\sqrt{5}/3$

3094 7988 rt : (7,8,0,2)
3094 8112 ℓ_2 : P_{45}
3094 9692 id : $(1/(3e))^{-3}$
3095 0692 tanh : 8/25
3095 1127 rt : (4,1,9,-3)
3095 2677 rt : (3,7,-2,-5)
3095 3458 Ψ : $\ln((e+\pi)/3)$
3095 7160 Γ : $2\sqrt{3}/5$
3095 7344 10^x : 4/5
3095 7707 $2/5 \times s\pi(e)$
3095 8243 cu : $3+3\sqrt{5}/4$
3095 8424 rt : (1,9,-7,-3)
3095 8895 ℓ_{10} : $\sqrt{5}+\sqrt{6}-\sqrt{7}$
3096 2252 cu : $1/3-4\pi/3$
3096 2650 rt : (6,0,-8,3)
3096 3891 at : $\sqrt{14}$
3096 4542 $t\pi$: $\sqrt{2}/3+\sqrt{3}/4$
3096 5634 rt : (5,7,-8,-3)
3096 6902 rt : (9,2,-9,5)
3096 7362 ℓ_2 : $1/3+e/3$
3096 7554 rt : (7,9,3,-2)
3096 9097 id : $6e$
3097 0294 at : 8/25
3097 0445 at : $1/2-3\sqrt{2}$
3097 1006 rt : (8,1,6,2)
3097 1681 e^x : $(e+\pi)/7$
3097 2385 ℓ_{10} : $4/3+\sqrt{2}/2$
3097 3009 2^x : $3\sqrt{2}/2-\sqrt{3}$
3097 3864 ζ : 11/24
3097 4953 rt : (9,5,-9,-3)
3097 7473 rt : (9,6,-3,6)
3098 0095 at : $1/3-3e/2$
3098 0212 ln : $3\sqrt{3}-2\sqrt{5}/3$
3098 0630 ℓ_{10} : $\sqrt[3]{17/2}$
3098 3140 $\ln(3/5) \div \exp(1/2)$
3098 3649 rt : (2,6,-9,-3)
3098 5498 2^x : $e/4+\pi/3$
3098 5499 $t\pi$: $3\sqrt{3}+3\sqrt{5}$
3098 6270 rt : (4,3,7,2)
3098 6932 $3/4 \times \exp(5)$
3098 8526 at : $3+\sqrt{5}/3$
3099 0141 $s\pi$: $1/3+4\pi$
3099 0923 rt : (1,8,8,-1)
3099 1830 rt : (8,2,-5,8)
3099 4456 sq : $\sqrt{2}/3-3\sqrt{5}/2$
3099 5886 tanh : $2\sqrt[3]{3}/9$
3099 7907 rt : (5,0,-4,6)
3100 0667 Ei : $7\sqrt{5}/2$
3100 1055 $\ln(\sqrt{5}) \times \exp(\sqrt{2})$
3100 1823 rt : (6,8,4,5)
3100 2837 2^x : $2/3-3\pi/4$
3100 4478 J_0 : $\sqrt{10}$
3100 6902 at : $3\sqrt{3}-4\sqrt{5}$
3100 7072 rt : (8,6,1,9)
3100 9253 at : $1-e/4$
3101 0205 rt : (5,8,2,0)
3101 1576 e^x : $7\sqrt[3]{5}/10$
3101 2646 Li_2 : $9(1/\pi)/10$
3101 2904 rt : (4,8,-2,2)
3101 2970 10^x : 4/11
3101 3277 rt : (7,0,9,3)
3101 5124 rt : (9,9,-9,-7)
3101 5492 ln : 11/15
3101 5634 at : $2\sqrt[3]{3}/9$
3101 5735 rt : (7,5,1,-1)
3101 9393 at : 15/4
3102 0220 as : P_{50}
3102 1087 $t\pi$: $7\sqrt[3]{3}$
3102 2104 rt : (1,8,6,7)
3102 2849 rt : (5,4,2,7)
3102 4688 10^x : $3\ln 2/4$
3102 5172 ln : $3+\sqrt{2}/2$
3102 6637 J_2 : $\sqrt[3]{6}$
3102 6706 sr : $5e\pi/8$
3102 6757 rt : (2,6,-8,-3)
3102 8883 $\sqrt{5} \div s\pi(\sqrt{5})$
3102 9612 sq : $7(1/\pi)/4$
3102 9796 $t\pi$: $4\sqrt{3}/3+\sqrt{5}/2$
3103 3449 rt : (1,6,-5,-6)
3103 3500 sq : $8\sqrt[3]{5}/9$
3103 5494 rt : (5,9,-2,9)
3103 5883 cu : $1-3\sqrt{5}/4$
3103 7069 cr : 9/4
3103 7215 2^x : $4e/9$
3103 7900 2^x : $(\sqrt{2}\pi)^{1/2}$
3103 7915 Γ : $8(e+\pi)/5$
3103 9969 ln : $\sqrt{2}/4+3\sqrt{5}/2$
3104 0200 rt : (9,1,-7,-2)
3104 0517 Γ : 9/13
3104 0651 sq : $\ln(\pi)$
3104 1371 $t\pi$: $4\sqrt{2}/3-4\sqrt{5}/3$
3104 2899 sr : $10\zeta(3)/7$
3104 4189 cr : $e/4+\pi/2$
3104 4576 rt : (6,4,-9,-3)
3104 4613 sr : $\sqrt{3}/2+2\sqrt{5}$
3104 6176 sq : $3/2-2\sqrt{2}/3$
3104 6796 $\ln(3/4)+\exp(4)$
3104 7261 E : 19/25
3104 7329 rt : (3,5,-1,8)
3104 7612 $\ln(\sqrt{2})-s\pi(\sqrt{2})$
3104 8127 E : $4\sqrt[3]{5}/9$
3104 9060 id : $10\ln 2/3$
3104 9220 rt : (9,9,6,-3)
3105 1971 2^x : $2/3+4e$
3105 2130 J_2 : 20/11
3105 2386 K : $(\sqrt{5}/3)^{1/3}$
3105 2564 $s\pi$: $7\sqrt{2}$
3105 3240 rt : (4,7,-9,2)
3105 3433 id : $4\sqrt{2}/3-3\sqrt{3}$
3105 3470 J_2 : 21/5
3105 4358 J_2 : $9\sqrt{2}/7$
3105 4872 $t\pi$: $\sqrt{2}/3+\sqrt{5}$
3105 5929 $t\pi$: $2\sqrt{2}+2\sqrt{3}$
3105 6461 ℓ_{10} : $3\sqrt{3}/4+\sqrt{5}/3$
3105 7279 ζ : $9\sqrt{2}/8$
3105 7946 Γ : $\exp(-1/e)$
3105 7971 cr : $2\sqrt{2}-\sqrt{3}/3$
3105 8946 10^x : $3\sqrt[3]{2}$
3106 0567 E : $(\sqrt{3})^{-1/2}$
3106 1009 J_2 : $10\sqrt[3]{2}/3$
3106 3572 $t\pi$: $1/3+\pi/2$
3106 3962 rt : (3,8,-7,-2)
3106 5250 rt : (5,8,8,8)
3106 5378 cu : $4\sqrt{2}-2\sqrt{5}/3$
3106 6017 id : $1/4+3\sqrt{2}/4$
3106 6287 cr : $5\pi^2/4$
3106 6746 $1/2 \div \ln(5)$
3106 7525 rt : (1,9,-3,-1)
3106 7704 rt : (8,5,4,1)
3107 4655 rt : (2,4,5,-2)
3107 5771 cu : $1-2\pi/3$
3107 9251 rt : (6,9,-1,-1)
3107 9838 Ei : $8\sqrt[3]{5}/5$
3108 0117 J_0 : $\sqrt[3]{19/3}$
3108 0573 rt : (8,9,-5,-4)
3108 1855 ln : $3\sqrt[3]{5}/7$
3108 2562 $1/5+\ln(3/5)$
3108 2774 rt : (9,7,-4,3)
3108 4348 erf : $\sqrt{2}/5$
3108 5216 rt : (1,5,8,-3)
3108 6183 rt : (8,3,-6,5)
3108 6873 rt : (7,5,-7,-2)
3108 6970 2^x : $4/3-2\sqrt{2}/3$
3108 8433 sq : $2\sqrt{2}/3+\sqrt{3}/3$
3108 8852 rt : (7,1,-8,-7)
3108 9643 rt : (7,4,7,2)
3109 0353 10^x : $1/3+2\sqrt{2}$
3109 0837 10^x : $8(1/\pi)/7$
3109 0998 rt : (6,7,-8,-9)
3109 3579 $s\pi$: $\sqrt[3]{4/3}$
3109 4702 erf : $8(1/\pi)/9$
3109 4763 sr : $(\sqrt{2}\pi/2)^3$
3109 7903 rt : (1,7,-2,2)
3109 9165 J_2 : $4\sqrt[3]{5}$
3110 1294 $s\pi$: $\exp(-1/(3\pi))$
3110 1311 2^x : 19/11
3110 3482 ℓ_2 : $1/2+3\pi$
3110 3862 cu : $1/\ln(\ln 2)$
3110 4863 sq : $\sqrt{3}-\sqrt{6}-\sqrt{7}$
3110 7274 J_2 : $1/\ln(\sqrt{3})$
3110 7673 cr : $3e+4\pi/3$
3110 7785 $\exp(3)+\exp(4/5)$
3110 7804 ℓ_{10} : $2\sqrt[3]{5}/7$
3110 7995 Γ : $4\sqrt[3]{3}$
3110 9871 2^x : $4/3+4\sqrt{2}/3$
3111 0781 rt : (3,3,-9,-3)
3111 1797 J_2 : $7(e+\pi)/6$
3111 1836 J_1 : $\exp(-\sqrt[3]{2}/3)$
3111 3393 10^x : 2/17
3111 3730 2^x : $2/3+3\sqrt{2}/4$
3111 5565 rt : (8,6,7,-3)
3111 5975 ℓ_{10} : $1+\pi/3$
3111 7204 rt : (9,5,4,-2)
3111 8800 at : $\sqrt{2}/2-2\sqrt{5}$
3111 8954 sq : $2+3e/4$
3111 9027 $s\pi$: P_{61}
3112 0815 rt : (7,3,-2,8)
3112 1766 at : $e+\pi/3$
3112 2275 rt : (8,7,0,6)
3112 2371 Ei : 20/11
3112 2455 cr : $\sqrt{3}/3+3\sqrt{5}/4$
3112 3978 J_2 : $2\pi^2/3$
3112 5417 J_0 : 19/6
3112 6818 Ψ : $10\sqrt[3]{2}/3$
3112 7161 cr : $3-\sqrt{5}/3$
3112 8931 J_2 : $4e\pi/5$
3112 9317 $\ln(3/5)+\exp(3/5)$
3113 0056 10^x : $4e+\pi/2$
3113 1121 10^x : 13/25
3113 1335 ln : $(e+\pi)/8$
3113 1974 ln : $4e-\pi/4$
3113 3120 rt : (3,7,-9,-7)
3113 3889 Ψ : 21/5
3113 6581 rt : (4,9,-1,-1)
3114 0592 rt : (3,6,8,2)
3114 1152 rt : (4,4,8,-3)
3114 1311 rt : (5,1,-5,3)
3114 2191 $\ln(4/5) \times \exp(1/3)$
3114 4626 J_2 : $\sqrt{3}+\sqrt{6}+\sqrt{7}$
3114 4991 rt : (3,9,-4,-3)
3114 5110 $\ln(2) \div \exp(4/5)$
3114 6837 ℓ_2 : P_{61}
3114 7099 10^x : $(\sqrt[3]{3}/2)^2$
3114 7179 Ei : $2\pi/3$
3114 8164 rt : (6,5,-3,1)
3115 0931 at : $6\pi/5$
3115 1496 rt : (3,7,-7,9)
3115 2031 $2/5 \div \exp(1/4)$
3115 2942 sr : $10\pi^2/9$
3115 3264 ℓ_2 : $3+4\sqrt{3}$
3115 3309 rt : (5,3,5,-2)
3115 3803 Ei : $9\sqrt{2}/7$
3115 4467 Li_2 : $\ln(4/3)$
3115 5003 ℓ_{10} : $3/4+3\sqrt{3}/4$
3115 6695 2^x : $3\sqrt{3}/4-4\sqrt{5}/3$
3115 7648 rt : (1,9,5,-6)
3115 7867 2^x : 9/23
3115 8238 ζ : $\sqrt{20/3}$
3115 8980 rt : (7,9,-1,-1)
3115 9423 $\exp(3)-s\pi(e)$
3115 9638 at : $8\sqrt{2}/3$
3116 0477 $s\pi$: P_{45}
3116 1846 sinh : $5\sqrt[3]{2}/4$
3116 5014 $t\pi$: $\sqrt[3]{2}/2$
3116 5943 rt : (7,7,4,9)
3116 6245 rt : (6,0,-7,2)
3116 7498 J_2 : $9(e+\pi)/8$
3116 8799 ln : $2/3+3\pi$
3116 8930 rt : (9,2,-6,9)
3116 9090 rt : (4,9,-4,-2)
3116 9101 sinh : $\ln(e/2)$
3116 9182 as : $3\sqrt{2}/2-2\sqrt{3}/3$
3116 9510 J_1 : $4e\pi/7$
3117 1169 erf : $\exp(-\sqrt[3]{2})$
3117 1189 $t\pi$: $\sqrt{3}/2-\sqrt{5}$
3117 3769 rt : (2,9,0,-7)
3117 6281 cr : $\sqrt[3]{23/2}$
3118 1676 rt : (6,2,2,-1)
3118 3331 J_0 : $8\ln 2/3$
3118 4078 rt : (3,6,0,5)

3118 6394 rt : (9,8,-8,-3)
3118 6849 10^x : 19/12
3118 7288 id : $\sqrt{7}\pi$
3118 7908 θ_3 : 5/9
3118 8455 as : $\ln(e/2)$
3118 9287 J_2 : $\sqrt{2}-\sqrt{5}+\sqrt{7}$
3118 9303 cu : $3/4-3\sqrt{3}/2$
3118 9410 sq : $3+4\pi$
3118 9687 Ψ : $(\sqrt{5})^{-1/2}$
3119 0315 $\ln(2/5)-\exp(1/3)$
3119 0535 ln : $1/2+\sqrt{3}/2$
3119 2105 Γ : $6(e+\pi)/5$
3119 2623 cu : $\sqrt{2}-\sqrt{5}+\sqrt{6}$
3119 3468 Γ : $2(1/\pi)/3$
3119 4336 id : $e/2-\pi/3$
3119 4523 rt : (3,8,-1,-8)
3119 4701 rt : (7,3,8,-3)
3119 5588 sr : $3e/2-3\pi/4$
3119 5635 ℓ_{10} : $7(e+\pi)/2$
3119 5705 rt : (4,7,5,1)
3119 6967 rt : (7,0,-9,4)
3119 8418 rt : (4,1,-1,6)
3119 9317 Li_2 : 19/21
3120 4116 rt : (1,2,7,8)
3120 4803 as : $\sqrt{3}/2-\sqrt{5}/4$
3120 5374 rt : (2,9,-7,-3)
3120 6452 $s\pi$: $\sqrt{24}$
3120 7075 rt : (3,8,-1,-1)
3120 8414 $t\pi$: $3\sqrt{2}/4+4\sqrt{3}/3$
3120 9023 rt : (8,4,-7,2)
3121 1424 ln : $7\sqrt[3]{3}$
3121 3682 ln : $1/4+2\sqrt{3}$
3121 3945 10^x : $\sqrt[3]{25/2}$
3121 5537 at : $3\sqrt[3]{2}$
3121 6806 e^x : $1/4-\sqrt{2}$
3121 7124 ℓ_{10} : $6\sqrt[3]{5}/5$
3121 7925 J_1 : $\sqrt{2}+\sqrt{3}+\sqrt{7}$
3121 8449 ℓ_2 : P_8
3121 8788 id : $1/2+2e/3$
3121 9062 e^x : $3\sqrt{2}-\sqrt{3}$
3121 9728 rt : (5,5,1,4)
3122 0766 J_1 : $5(e+\pi)/6$
3122 1886 sq : $2/3-2e/3$
3122 2222 rt : (8,7,-5,-2)
3122 2534 rt : (5,4,-7,-9)
3122 4996 rt : (9,8,-5,0)
3122 5850 e^x : 7/3
3122 7193 rt : (3,8,-4,8)
3122 7267 rt : (8,2,5,-2)
3122 8106 $\ln(\sqrt{5})-\exp(3/4)$
3122 9762 at : $\sqrt{2}-3\sqrt{3}$
3123 0251 rt : (1,8,-9,2)
3123 1093 J_2 : $\sqrt{3}+\sqrt{5}+\sqrt{7}$
3123 1792 $3/5+\ln(3/4)$
3123 3071 Ψ : 13/11
3123 3516 J_2 : $\sqrt{10/3}$
3123 3729 rt : (3,1,9,-3)
3123 4329 id : $\sqrt{3}/4-\sqrt{5}/3$
3123 5509 sq : $1/3+e$
3123 5594 cu : $\sqrt{3}+\sqrt{5}+\sqrt{7}$
3123 5628 rt : (4,9,0,-1)
3123 6149 e^x : $e/10$
3123 6546 cu : $\sqrt{3}+\sqrt{5}-\sqrt{7}$
3123 6856 rt : (6,0,-5,7)
3123 7573 at : $2-3\sqrt{5}/4$
3123 7954 $t\pi$: $2\sqrt{2}-\sqrt{3}$
3123 8934 J_2 : $5e/2$
3124 1292 cr : $4\sqrt{2}+3\sqrt{5}$
3124 1587 rt : (1,8,-4,-2)
3124 2306 ln : $\ln 2/7$
3124 2893 Ei : $5\sqrt[3]{2}/8$
3124 3368 ln : $2\sqrt{3}/3+4\sqrt{5}$
3124 5033 Li_2 : $\sqrt[3]{3}/5$
3124 5401 rt : (6,7,-8,-3)
3124 6017 ℓ_2 : $\sqrt{10}\pi$
3124 6409 sq : $\sqrt{2}+\sqrt{3}+\sqrt{6}$
3124 6895 ln : $e/2+3\pi/4$
3124 8991 J_0 : $(e/2)^2$
3125 0000 id : 5/16
3125 1150 ℓ_{10} : $4e/3-\pi/2$
3125 2348 at : $3+\pi/4$
3125 3041 J_0 : $7e/6$
3125 3609 rt : (2,8,-2,3)
3125 3648 Ψ : $\ln(2\pi)$
3125 3651 J_1 : $(3\pi)^{1/2}$
3125 4702 rt : (6,9,3,2)
3125 5382 rt : (7,4,-3,5)
3125 5585 rt : (8,1,-9,-8)
3125 5760 Γ : $(2\pi/3)^{-1/2}$
3125 5957 rt : (1,8,7,-3)
3125 7044 sinh : 4/13
3125 8917 e^x : 19/13
3125 9071 rt : (6,7,8,2)
3126 0398 rt : (4,8,1,9)
3126 2786 rt : (5,1,-4,1)
3126 3270 sinh : $\sqrt[3]{7}$
3126 3541 at : $\sqrt{3}/4+3\sqrt{5}/2$
3126 3836 rt : (3,1,3,9)
3126 3979 rt : (8,9,6,-3)
3126 7747 ln : $3\sqrt{3}/2+\sqrt{5}/2$
3126 9329 cu : $2\sqrt{5}/3$
3127 0831 rt : (9,3,-7,6)
3127 1606 Li_2 : $\sqrt{3}/6$
3127 2539 e^x : $3\sqrt{3}+4\sqrt{5}$
3127 3659 e^x : P_{30}
3127 5794 sinh : $(\sqrt{2}\pi/3)^{-3}$
3127 6672 as : 4/13
3128 0279 rt : (8,8,-1,3)
3128 0577 rt : (2,4,-9,-1)
3128 1101 rt : (3,8,0,-7)
3128 1377 2^x : $e/4-3\pi/4$
3128 1672 $e+\ln(2/3)$
3128 3957 cr : $8\sqrt{2}/5$
3128 4007 2^x : $\ln(\sqrt{2}\pi/3)$
3128 4726 2^x : $\pi/8$
3128 6986 θ_3 : $4\ln 2/3$
3129 0840 rt : (6,0,9,3)
3129 1582 cr : $3/2+4e$
3129 3206 2^x : $4e-3\pi/4$
3129 4196 J_2 : $7e\pi/9$
3129 4235 rt : (1,8,-2,5)
3129 4610 sr : $4\sqrt{2}/3+2\sqrt{3}$
3129 5399 e^x : $3\sqrt{5}/8$
3129 5427 $s\pi$: $10\pi^2/3$
3129 5480 as : $(\sqrt{2}\pi/3)^{-3}$
3129 6137 sinh : $4\ln 2/9$
3129 6179 $s\pi$: $(\pi)^{-2}$
3129 6586 ln : $e/4+3\pi$
3129 6714 ℓ_2 : $5\sqrt{5}/9$
3129 8964 rt : (1,9,4,-8)
3130 0992 rt : (9,1,2,-1)
3130 1645 rt : (3,6,-8,-3)
3130 4457 rt : (4,7,7,-3)
3130 5999 10^x : $3/4+\pi/2$
3130 6186 at : $\exp(4/3)$
3130 6726 rt : (6,4,1,8)
3130 6964 rt : (5,6,-5,-2)
3130 7227 e^x : $2e+3\pi/2$
3130 7661 rt : (8,3,-3,9)
3130 8502 erf : $\sqrt[3]{5}/6$
3130 9903 rt : (4,2,-8,-2)
3131 1327 id : $2/3-\sqrt{2}/4$
3131 1912 rt : (9,8,3,-2)
3131 3588 J_0 : 24/13
3131 3665 e^x : 4π
3131 5887 as : $4\ln 2/9$
3131 7207 rt : (3,9,-1,7)
3131 7677 at : $4e\pi/9$
3131 8051 ℓ_{10} : $\exp(\sqrt[3]{3}/2)$
3132 1826 Ψ : $1/\sqrt{10}$
3132 2605 rt : (7,8,0,-1)
3132 2935 $t\pi$: $\pi^2/9$
3132 5037 $1/5+\exp(\sqrt{2})$
3132 5613 J_2 : $(\sqrt[3]{4}/3)^{-3}$
3132 6168 ln : $1+e$
3132 7119 Ψ : $7\zeta(3)/2$
3132 7460 ℓ_{10} : $3-2\sqrt{2}/3$
3133 0637 at : $1/3+2\sqrt{3}$
3133 0765 J_2 : 20/3
3133 1618 10^x : $\sqrt{3}-\sqrt{5}$
3133 1799 rt : (9,7,-1,7)
3133 2157 cr : $5e/6$
3133 2991 e^x : $5\sqrt{2}/2$
3133 3575 ln : $4\sqrt[3]{5}/5$
3133 5033 rt : (2,6,-6,-7)
3133 6708 at : $3\sqrt{2}/2+3\sqrt{5}/4$
3133 7466 ln : $1/4+\sqrt{5}/2$
3133 7472 J_1 : $\exp(\sqrt[3]{4})$
3133 7852 J_0 : $5e/3$
3134 0919 ℓ_2 : $1/3+\sqrt{2}/3$
3134 2589 rt : (7,8,3,6)
3134 3589 cu : $3/2+3e/4$
3134 4307 ℓ_2 : $\ln(\sqrt{5})$
3134 4943 e^x : $(e/2)^3$
3134 5273 rt : (9,1,9,3)
3134 5462 ln : $4\sqrt{2}/3-2\sqrt{3}/3$
3134 6172 sinh : $8\zeta(3)/3$
3134 7088 e^x : $4/3-3\sqrt{2}/4$
3134 7261 at : 19/5
3134 7993 at : $4\sqrt{3}/3+2\sqrt{5}/3$
3134 9035 rt : (2,9,3,6)
3135 0983 rt : (7,6,7,-3)
3135 1852 J_2 : $8(e+\pi)/7$
3135 2983 J_2 : $3\sqrt{5}$
3135 4195 e^x : 3/11
3135 4936 sq : $(\ln 2/2)^{-2}$
3135 5287 rt : (5,8,-3,0)
3135 5674 sr : $9e\pi/7$
3135 6070 sq : $4\sqrt[3]{2}/9$
3135 7772 cu : $4\ln 2$
3136 0000 sq : 14/25
3136 1133 rt : (7,4,-9,-3)
3136 3677 $t\pi$: $\sqrt{2}/2+3\sqrt{3}$
3136 5007 J_0 : $(2\pi)^{1/3}$
3136 5424 J_1 : $9e/5$
3136 5993 ℓ_{10} : $3/2+\sqrt{5}/4$
3136 6932 rt : (9,8,8,2)
3136 6939 Ei : $6\sqrt[3]{3}$
3136 7011 10^x : $4+\pi$
3136 7570 $t\pi$: $\sqrt{2}/2$
3136 8307 $\exp(2/3)\div s\pi(1/5)$
3136 8615 e^x : $6(1/\pi)/7$
3136 9451 cu : $\sqrt{3}+\sqrt{5}+\sqrt{6}$
3137 0434 cu : $2/3+e/2$
3137 0849 id : $8\sqrt{2}$
3137 1535 rt : (7,1,6,2)
3137 3439 E : $3\sqrt[3]{2}/5$
3137 3568 J_2 : $4\pi/3$
3137 4343 rt : (5,9,7,5)
3137 4457 Ψ : $1/\sqrt[3]{12}$
3137 5151 $t\pi$: 9/20
3137 5245 Ei : 11/24
3137 5373 rt : (8,5,-8,-1)
3137 5849 e^x : $\sqrt[3]{19/2}$
3137 8900 cu : 23/21
3137 9553 rt : (3,7,1,2)
3137 9695 J : $\exp(-13\pi/6)$
3138 0675 $\ln(2/3)\times s\pi(e)$
3138 2636 at : $2/3-2\sqrt{5}$
3138 2829 cr : $9\sqrt[3]{2}/5$
3138 3215 erf : 2/7
3138 3445 at : $7e/5$
3138 3651 cu : $e/4$
3138 4593 cr : $4-\sqrt{3}$
3138 4636 10^x : $(\sqrt[3]{5}/2)^{-1/3}$
3138 5933 rt : (2,9,9,2)
3138 6574 ℓ_2 : $1+4\sqrt{5}$
3138 7969 at : $e/3-3\pi/2$
3138 9103 Ei : 1/19
3138 9324 sr : $2+3\sqrt{5}/2$
3139 1594 rt : (6,1,-6,4)
3139 2254 tanh : $(\sqrt[3]{5}/3)^2$
3139 2499 $t\pi$: $\sqrt[3]{5}/4$
3139 4237 ℓ_{10} : $10\sqrt[3]{3}/7$
3139 4349 $\exp(2/5)+\exp(3/5)$
3139 5035 cu : $2+\sqrt{5}/2$
3139 6374 $\ln(3/4)\div\ln(2/5)$
3139 8789 rt : (9,4,-8,3)
3139 9267 e^x : $3e/4+4\pi$
3139 9718 sinh : 25/23

3140 0637 ℓ_{10} : $1+3\sqrt{2}/4$
3140 0645 at : $2/3+\pi$
3140 1681 10^x : $(\ln 3)^{-1/3}$
3140 2067 rt : (5,6,-9,9)
3140 4244 rt : (9,4,-6,-2)
3140 4366 cu : $4+2e$
3140 4491 ℓ_{10} : $2\sqrt{2}/3+\sqrt{5}/2$
3140 4972 rt : (4,9,2,6)
3140 6478 rt : (3,4,8,-3)
3140 6545 sr : $e/4+\pi/3$
3140 7649 ℓ_{10} : $7\ln 2/10$
3140 8008 at : $3/2+4\sqrt{3}/3$
3140 9153 rt : (7,0,-6,8)
3140 9471 at : $3\sqrt{2}-\sqrt{3}/4$
3141 2412 rt : (7,3,-8,2)
3141 2461 $\exp(1/4)-\exp(4)$
3141 2798 cr : $2\sqrt{2}-\sqrt{5}/4$
3141 3318 at : $(\sqrt[3]{5}/3)^2$
3141 3397 rt : (6,8,7,9)
3141 4179 id : $(\ln 3/2)^{-2}$
3141 4319 cu : $\sqrt{3}/3+\sqrt{5}/3$
3141 4910 10^x : $10\pi^2/7$
3141 5261 J_2 : $11/6$
3141 5926 id : $\pi/10$
3141 6195 rt : (2,5,-6,-2)
3141 8641 id : $3\sqrt{3}+\sqrt{5}/2$
3142 1117 rt : (4,3,-9,-3)
3142 2047 2^x : $23/19$
3142 5748 sr : $19/11$
3142 5962 at : $2+2e/3$
3142 6968 id : $2\sqrt{2}/9$
3142 7335 rt : (9,9,-6,-3)
3142 7806 sr : $2/3+3\sqrt{2}/4$
3143 0901 e^x : $1/3-2\sqrt{5}/3$
3143 2778 $\sqrt{5}\times s\pi(1/5)$
3143 2785 10^x : $3\sqrt{2}-\sqrt{5}/2$
3143 2823 ℓ_{10} : $1/4+2e/3$
3143 3982 sq : $3/2-\sqrt{2}/4$
3143 4310 2^x : $3/4+e/2$
3143 4536 sr : $3+3\pi/4$
3143 5190 rt : (8,5,4,-2)
3144 0517 rt : (7,5,-4,2)
3144 3200 cu : $17/25$
3144 3370 2^x : $1/\ln(\ln 3/2)$
3144 6055 rt : (8,4,-4,6)
3144 6204 rt : (9,9,9,-4)
3144 6289 at : $((e+\pi)/3)^2$
3144 6728 rt : (5,1,3,1)
3144 7779 θ_3 : $5\zeta(3)/8$
3144 8149 e^x : $4\sqrt{2}+3\sqrt{3}/4$
3145 1033 rt : (5,1,-2,7)
3145 1482 E : $(\sqrt[3]{5}/3)^{1/2}$
3145 1665 sq : $\sqrt{3}/4-2\sqrt{5}$
3145 4599 J_1 : $\sqrt{24}$
3145 4681 $\ln(3)-\exp(5)$
3145 5361 rt : (6,3,8,-3)
3145 6313 as : $2-4\sqrt{3}/3$
3145 8545 sinh : $2e/5$
3145 9621 rt : (2,4,2,8)
3146 1112 at : $\sqrt{2}/4+2\sqrt{3}$
3146 2323 sq : $e/4-4\pi/3$
3146 2332 θ_3 : $17/19$
3146 2711 erf : $9(1/\pi)/10$
3146 3567 J_2 : $5(e+\pi)/7$
3146 3843 rt : (3,9,-7,-3)
3146 5094 rt : (2,4,-6,9)
3146 7052 $\exp(\sqrt{2})\div s\pi(e)$
3146 7606 sr : $\ln 2/7$
3146 8330 tanh : $(3\pi)^{-1/2}$
3146 9487 ln : $\sqrt{2}+4\sqrt{3}/3$
3147 2177 rt : (4,5,-2,-1)
3147 2453 $s\pi$: $7(e+\pi)/10$
3147 2568 Ei : $4\sqrt[3]{3}/5$
3147 3672 10^x : $e/3+3\pi/4$
3147 3806 Ψ : $5\zeta(3)/9$
3147 5730 id : $1/3+4\sqrt{5}/3$
3147 6794 cr : $25/11$
3147 6934 cr : $2\sqrt{3}/3+\sqrt{5}/2$
3147 7446 e^x : $1+\sqrt{5}/2$
3147 7975 rt : (1,5,-2,-9)
3147 8084 J_1 : $10\sqrt{3}/3$
3147 8741 e^x : $3/4+4\sqrt{3}/3$
3148 1798 rt : (3,3,7,2)
3148 3617 at : $e/4+\pi$
3148 5343 Ei : $15/13$
3148 5717 2^x : $4\sqrt{2}/3-2\sqrt{5}/3$
3148 6793 rt : (7,9,6,-3)
3148 9126 ζ : $(e+\pi)/9$
3148 9654 at : $(3\pi)^{-1/2}$
3149 0149 rt : (1,9,-9,-6)
3149 1103 $2/3\div\exp(3/4)$
3149 3327 rt : (9,8,-2,4)
3149 7611 rt : (7,4,0,9)
3149 7710 rt : (6,3,-6,-2)
3149 7781 ℓ_{10} : P_{42}
3149 8026 id : $\sqrt[3]{2}/4$
3149 8183 sr : $4+e/2$
3149 8694 e^x : $3\pi^2/10$
3150 0083 ℓ_{10} : $1/3+\sqrt{3}$
3150 0434 rt : (7,8,5,1)
3150 0612 ℓ_2 : $3-3\sqrt{3}/2$
3150 1279 sq : $9\zeta(3)/4$
3150 2069 rt : (9,2,8,-3)
3150 2659 Γ : $10\sqrt{5}/9$
3150 3866 rt : (5,7,-8,6)
3150 6033 sq : $9\pi/7$
3150 7340 ln : P_{23}
3150 9966 rt : (8,9,-2,0)
3151 0993 cr : $2\sqrt{3}+4\sqrt{5}$
3151 1427 at : $\sqrt{2}/3+3\sqrt{5}/2$
3151 2457 rt : (6,3,-9,-7)
3151 3414 cu : $8\ln 2/3$
3151 4126 rt : (6,5,0,5)
3151 4617 rt : (4,5,-9,2)
3151 5189 sq : $\exp(-\sqrt{3}/3)$
3151 5459 J_1 : $2/3$
3151 6341 10^x : $e+4\pi/3$
3151 6419 E : $(\ln 3)^{-3}$
3151 9753 sq : $3(e+\pi)/4$
3152 2560 rt : (6,5,1,-1)
3152 3143 $t\pi$: $7e/10$
3152 3590 ζ : $18/7$
3152 3678 at : $1/4-3e/2$
3152 3847 J_2 : $\ln(2\pi)$
3152 6493 $s\pi$: $\sqrt{3}/2+\sqrt{5}$
3152 6736 2^x : $\sqrt[3]{14}$
3152 8474 rt : (4,3,5,-2)
3152 8715 ζ : $\sqrt[3]{17}$
3152 9926 sq : $1/4+\pi/2$
3153 0095 at : $1+2\sqrt{2}$
3153 0417 id : $3\sqrt{2}/4-\sqrt{5}/3$
3153 1054 rt : (5,2,-9,-3)
3153 1308 sr : $3\sqrt{2}+\sqrt{5}/2$
3153 1570 $t\pi$: $(e/2)^{-1/3}$
3153 1910 rt : (2,1,9,-3)
3153 2270 rt : (3,6,-7,-8)
3153 3854 e^x : $\sqrt{2}/2-\sqrt{3}/4$
3153 4129 sq : $7\pi^2/4$
3153 7065 cu : $2\sqrt{2}+3\sqrt{3}/4$
3153 8414 sq : $3/2+\sqrt{3}/3$
3153 8793 $1/3-\exp(1/2)$
3153 9268 rt : (8,0,-7,9)
3153 9768 rt : (5,9,-4,-2)
3153 9929 J_1 : $4\sqrt[3]{3}$
3154 0162 ln : $4\sqrt{2}+2\sqrt{5}$
3154 0828 rt : (7,7,-8,-3)
3154 1376 J_0 : $9\sqrt{2}/4$
3154 1756 cu : $3-3\sqrt{5}/4$
3154 2616 cr : $4/3+2\sqrt{2}/3$
3154 4777 $t\pi$: $1/3-\sqrt{5}$
3154 4880 rt : (4,4,-8,2)
3154 6277 Li_2 : $e/3$
3154 6487 $\ln(\sqrt{2})\div\ln(3)$
3154 8912 θ_3 : $\sqrt{2}/9$
3155 1288 $t\pi$: $2\sqrt{2}/3+\sqrt{5}/4$
3155 1580 ℓ_{10} : P_{32}
3155 3361 $s\pi(\pi)-s\pi(\sqrt{3})$
3155 4088 rt : (3,7,7,-3)
3155 4457 ln : $5\sqrt{5}/3$
3155 4964 rt : (8,8,2,7)
3155 6025 rt : (5,9,3,-2)
3155 6052 sr : $6\sqrt[3]{3}/5$
3155 6360 ℓ_{10} : $\sqrt{3}/3+2\sqrt{5}/3$
3155 7263 $\sqrt{2}\times\ln(4/5)$
3155 8526 10^x : $2e/3+\pi$
3156 0127 $\sqrt{2}-\ln(3)$
3156 0134 rt : (3,2,-8,8)
3156 1393 at : $23/6$
3156 2579 rt : (7,1,-7,5)
3156 3347 J_1 : $5\zeta(3)/9$
3156 4120 rt : (9,5,-9,0)
3156 6609 rt : (2,5,-7,-4)
3156 7544 $\exp(2/3)\times s\pi(\sqrt{5})$
3156 8008 e^x : $1/4+4\sqrt{5}/3$
3156 8525 sq : $2-3e/2$
3156 9644 rt : (6,4,7,2)
3156 9804 cu : $\sqrt{3}/4+3\sqrt{5}/2$
3156 9838 rt : (4,1,-9,1)
3157 1386 J_0 : $10(1/\pi)$
3157 4690 J : $e/9$
3157 5225 tanh : $\exp(-\sqrt{5}/2)$
3157 7133 tanh : $\ln(\sqrt[3]{3}/2)$
3157 7430 ℓ_{10} : $3\sqrt{2}/4-\sqrt{3}/3$
3157 7654 $\ln(2/3)\div\exp(1/4)$
3157 8947 id : $6/19$
3157 9757 as : $(e/3)^{1/3}$
3157 9761 e^x : $1/\ln(\sqrt[3]{2}/3)$
3158 2477 rt : (8,2,3,1)
3158 2509 $s\pi$: $2e\pi/9$
3158 2981 2^x : $4\sqrt{2}/3+4\sqrt{3}/3$
3158 3114 rt : (5,0,9,3)
3158 3334 cu : $4\sqrt{2}/3-4\sqrt{5}/3$
3158 3528 rt : (3,2,-6,-2)
3158 5018 rt : (9,4,-5,7)
3158 5440 10^x : $-\sqrt{2}-\sqrt{3}+\sqrt{7}$
3158 6729 Ψ : $\sqrt{2}-\sqrt{3}-\sqrt{6}$
3158 7735 erf : $\ln(4/3)$
3158 8749 as : $3/4-3\sqrt{2}/4$
3159 1334 rt : (4,6,-8,-3)
3159 2890 Ei : $3\ln 2/7$
3159 2901 rt : (7,2,5,-2)
3159 3123 2^x : $4\ln 2/7$
3159 4042 rt : (2,6,-3,-3)
3159 5752 sr : $7\pi/2$
3159 6322 rt : (6,6,7,-3)
3159 6920 at : $\exp(-\sqrt{5}/2)$
3159 8023 rt : (5,5,4,8)
3159 8834 at : $\ln(\sqrt[3]{3}/2)$
3159 9040 rt : (4,2,-5,-6)
3159 9349 sq : $\ln(\sqrt[3]{5}/3)$
3160 2336 J_1 : $(\sqrt{5})^{-1/2}$
3160 3086 rt : (2,8,-4,-2)
3160 5106 $\ln(3)\times\ln(3/4)$
3160 5166 rt : (6,2,-7,1)
3160 5765 rt : (7,9,2,3)
3160 6849 rt : (8,8,3,-2)
3160 7401 id : $(\sqrt{3})^{1/2}$
3160 9266 rt : (2,5,-3,8)
3160 9405 rt : (9,4,1,-1)
3161 2327 sinh : 7
3161 2852 rt : (8,6,-9,-4)
3161 2971 cr : $3+3\pi$
3161 3738 ℓ_{10} : $1/2+\pi/2$
3161 4690 e^x : $3\sqrt{3}/2+\sqrt{5}/3$
3161 5964 sq : $3\sqrt[3]{5}$
3161 6270 cr : $4\sqrt[3]{5}/3$
3161 7637 rt : (5,9,-9,-7)
3161 8519 $s\pi$: $2\pi/7$
3161 9147 Ψ : $1/\ln(\ln 3)$
3162 0046 rt : (6,1,-3,8)
3162 1077 θ_3 : $\sqrt[3]{2}/8$
3162 1485 cr : $\sqrt{2}+\sqrt{3}/2$
3162 2776 id : $1/\sqrt{10}$
3162 4506 e^x : $2\sqrt[3]{2}/3$
3162 4792 ℓ_{10} : $4e/3-\pi$
3162 4946 Li_2 : $7/24$
3162 5805 rt : (5,8,-7,3)
3162 6193 2^x : $3/4-\sqrt{2}/4$
3162 6909 rt : (8,5,-5,3)
3162 7108 $s\pi$: $7\sqrt[3]{2}/8$

3162 8439 rt : (8,9,9,-4)
3162 9073 $2/5 - \ln(2/5)$
3162 9157 $t\pi : 2\sqrt{2}/3 + 2\sqrt{3}/3$
3163 1201 $10^x : 3 + 3\sqrt{5}/2$
3163 1765 rt : (8,1,9,3)
3163 3432 $10^x : 3(1/\pi)/8$
3163 4281 $\ell_{10} : 2\sqrt{3} - 4\sqrt{5}/3$
3163 4805 $J_1 : \ln((e+\pi)/3)$
3163 5320 cu : $e/3 + \pi$
3163 5616 at : $1/4 - \sqrt{3}/3$
3163 6697 $e^x : 21/25$
3163 6773 rt : (9,5,7,-3)
3163 7061 id : $1/4 - 4\pi$
3163 8442 sr : $5\ln 2/2$
3163 9352 rt : (1,8,5,-5)
3164 0625 sq : 9/16
3164 1107 sr : P_{18}
3164 1523 $\ell_{10} : 3e + 4\pi$
3164 1845 at : $8\sqrt[3]{3}/3$
3164 3186 $e + \exp(4)$
3164 4563 sr : P_{48}
3165 0314 rt : (9,5,7,2)
3165 1005 at : $9\sqrt[3]{5}/4$
3165 1027 at : $\sqrt{3}/2 + 4\sqrt{5}/3$
3165 4469 $\ln(5) + s\pi(1/4)$
3165 4681 sq : $4\pi/5$
3165 5820 $\ell_2 : 1 + 2\sqrt{5}/3$
3165 7693 cu : $5\pi/9$
3165 8359 $\ell_2 : 7e\pi/3$
3166 1477 cu : $\sqrt{17}\pi$
3166 2089 $\Psi : 5\sqrt{2}/6$
3166 2479 id : $\sqrt{11}$
3166 3026 $2^x : 1/\sqrt[3]{16}$
3166 3813 cu : $\ln(\sqrt{2}\pi)$
3166 7477 $erf : \sqrt[3]{3}/5$
3166 9609 rt : (4,5,-6,9)
3167 0994 $s\pi(e) \div s\pi(1/5)$
3167 1263 rt : (4,6,-8,-1)
3167 1925 $\ln(\sqrt{3}) - s\pi(1/3)$
3167 2052 $2^x : \sqrt{3}/3 - \sqrt{5}$
3167 2297 sq : $e - \pi/2$
3167 4307 rt : (9,7,-5,-2)
3167 5124 rt : (3,7,4,9)
3167 8254 $J_1 : 1/\ln(\sqrt{2}\pi)$
3167 8992 ln : $3/4 + 4\sqrt{5}/3$
3168 0150 $t\pi : 3\sqrt{2} + 2\sqrt{3}$
3168 0574 cu : $4\sqrt{3}/3$
3168 1214 $t\pi : 2\sqrt{2}/3 - \sqrt{5}$
3168 1410 $J_1 : \exp(\sqrt{5}/2)$
3168 1460 rt : (3,8,2,-1)
3168 2130 cu : $1 + \sqrt{5}/3$
3168 3545 $\exp(e) \div \exp(3/5)$
3168 5042 rt : (2,4,8,-3)
3168 5740 $E : \ln(\sqrt{2}/3)$
3168 6811 rt : (8,1,-8,6)
3168 7956 $2^x : 6\sqrt{2}/7$
3168 8446 at : $3\sqrt{2}/2 + \sqrt{3}$
3168 9001 $J_1 : 1/\ln(\sqrt[3]{3}/2)$
3168 9382 $J_1 : 3\sqrt{5}/10$
3169 0804 rt : (5,2,-3,4)
3169 0860 $erf : \sqrt{3}/6$
3169 2028 cr : $(1/\pi)/10$
3169 2772 $\Psi : 5(1/\pi)/8$
3169 3056 at : $1/2 + 3\sqrt{5}/2$
3169 4312 $J_1 : (\sqrt{2}\pi/2)^{-1/2}$
3169 4786 $\zeta : \sqrt{2} + \sqrt{5} - \sqrt{7}$
3169 5789 ln : $2 + \sqrt{3}$
3169 6093 cu : 15/22
3169 6364 rt : (7,5,-1,6)
3169 8729 id : $3/4 - \sqrt{3}/4$
3169 8762 $e^x : e + 4\pi/3$
3169 9725 $Li_2 : (\sqrt{5}/3)^{1/3}$
3170 0001 $1/5 + \exp(3/4)$
3170 0051 rt : (9,8,1,8)
3170 0208 $J_2 : (2\pi)^{1/3}$
3170 0513 rt : (3,9,0,-1)
3170 1600 at : $1/3 - 4\pi/3$
3170 1883 $1/3 \times s\pi(2/5)$
3170 3268 $\theta_3 : 3/19$
3170 4596 $J : 5(1/\pi)/8$
3170 6248 at : $3/2 + 3\pi/4$
3170 7708 rt : (5,3,-7,-3)
3171 0351 $2^x : \sqrt{3}/4 + 3\sqrt{5}/4$
3171 0753 rt : (7,6,-5,-1)
3171 0796 $J_1 : 9e/8$
3171 2434 rt : (9,9,-3,1)
3171 2988 $2^x : 4 - 4\sqrt{2}$
3171 5292 $\theta_3 : 7(1/\pi)/4$
3171 5926 $\Psi : 3\pi/8$
3171 6044 $e \div s\pi(\pi)$
3171 8264 rt : (6,9,6,6)
3172 0010 sq : $4\sqrt{3}/3 + 3\sqrt{5}$
3172 0112 $Ei : 2\sqrt{3}/3$
3172 0495 $s\pi : 3\sqrt{2}/2 + 4\sqrt{5}/3$
3172 0678 cr : $3/2 + \pi/4$
3172 1262 $J_2 : 24/13$
3172 3777 as : $e/2 - \pi/3$
3172 6751 cr : 16/7
3172 8512 $2^x : 7\sqrt{3}/10$
3172 8549 rt : (5,3,8,-3)
3172 9540 rt : (3,3,-7,5)
3173 1876 cr : $4e + \pi/2$
3173 2639 rt : (1,8,-3,-3)
3173 4462 rt : (7,9,-1,-1)
3173 4531 $2^x : 3/4 + 3\sqrt{2}/2$
3173 4942 rt : (4,9,-7,-3)
3173 5261 rt : (3,0,6,2)
3173 6155 sr : 9π
3173 7051 ln : $2e + 3\pi/2$
3173 7299 $2/3 \div \ln(3/4)$
3173 7386 rt : (2,9,-7,9)
3173 7619 sr : P_{61}
3173 7849 sq : $2/3 + 2\sqrt{3}/3$
3173 9380 $\ln(3/5) \div \ln(5)$
3174 1055 rt : (7,1,-4,9)
3174 2435 $t\pi : 4\sqrt{2} - \sqrt{5}/4$
3174 3541 id : $2\sqrt{3}/3 - 2\sqrt{5}$
3174 5115 rt : (5,2,-6,-7)
3174 5485 $E : 5\zeta(3)/8$
3174 5688 rt : (6,6,-5,-2)
3174 6342 rt : (7,7,-9,-8)
3174 7276 id : $3\sqrt{2}/2 + 3\sqrt{3}$
3174 7359 $2^x : 16/3$
3174 7703 $J_2 : (e/2)^2$
3174 7894 $2^x : 2\sqrt{2}/3 - 3\sqrt{3}/2$
3174 7923 at : $4/3 - 3\sqrt{3}$
3174 8311 id : $3e/4 - 3\pi/4$
3174 9515 as : $3/2 - 2e/3$
3175 0016 $\zeta : 3\sqrt[3]{5}/2$
3175 0935 rt : (3,5,-8,-5)
3175 0973 $\ell_{10} : 3/2 + \sqrt{3}/3$
3175 1183 rt : (5,3,-9,-3)
3175 1822 $\zeta : (3\pi/2)^{-1/2}$
3175 3097 cu : $3\sqrt{3}/2 + \sqrt{5}/2$
3175 3295 $t\pi : (\ln 2/3)^{-1/3}$
3175 3594 rt : (9,8,6,-3)
3175 6129 ln : $2/3 + \sqrt{2}/2$
3175 6169 rt : (9,5,-6,4)
3175 6216 $\ln(2/5) \times \ln(\sqrt{2})$
3175 6556 ln : $8\zeta(3)/7$
3175 7707 $2^x : 5(1/\pi)/4$
3175 9133 $2^x : 4 + 4e/3$
3175 9485 sr : P_{45}
3176 1115 sinh : 5/16
3176 2611 $s\pi : 9\sqrt{3}/4$
3176 3540 rt : (7,2,-8,2)
3176 3624 $Ei : 5\pi^2/3$
3176 3771 $10^x : 23/25$
3176 5873 as : $\sqrt{3}/4 - \sqrt{5}/3$
3176 6428 at : $2e - \pi/2$
3176 7219 rt : (2,6,8,2)
3176 8046 at : $3 + \sqrt{3}/2$
3176 9631 $\ln(5) \div \exp(1/5)$
3177 0011 rt : (8,2,8,-3)
3177 0731 rt : (7,5,4,-2)
3177 1951 rt : (6,8,0,-1)
3177 3644 rt : (6,5,3,9)
3177 3897 $s\pi : (\sqrt{5}/3)^{-1/3}$
3177 4187 $\ell_{10} : 6\sqrt{3}/5$
3177 4468 rt : (5,9,-6,0)
3177 4571 $J_2 : 8\ln 2/3$
3177 5943 rt : (1,8,-7,-3)
3177 7848 id : $\exp(3\pi/2)$
3177 8958 $\ell_{10} : 4/3 + \sqrt{5}/3$
3178 1165 rt : (9,1,-9,7)
3178 1589 rt : (7,0,-7,2)
3178 1675 $\Psi : (\sqrt[3]{3}/2)^{-1/2}$
3178 2370 as : 5/16
3178 2848 $\Psi : (2\pi)^{1/3}$
3178 3724 id : $\sqrt{2} - \sqrt{3}$
3178 4300 cr : $1/4 + 3e/4$
3178 6426 $J : \ln 2/8$
3178 7378 id : $8e\pi$
3178 8717 rt : (4,2,1,7)
3178 8931 cu : $2\sqrt{2} - \sqrt{3}$
3179 3078 $e^x : e - \pi/3$
3179 3128 rt : (2,6,0,7)
3179 4452 $J_1 : 23/4$
3179 4671 $\ell_{10} : 3\ln 2$
3179 5082 $2^x : (e/2)^{-3}$
3179 5097 ln : $2 + 3e$
3179 5184 $J_0 : \ln(2\pi)$
3179 5464 rt : (8,3,-3,-1)
3179 6240 cu : $1/\ln(\ln 2/3)$
3179 8062 cr : $\sqrt[3]{12}$
3179 8853 $e^x : 1/3 + \sqrt{3}/2$
3179 9370 rt : (8,9,1,4)
3180 0691 rt : (2,8,1,-8)
3180 1073 $e^x : \sqrt[3]{14/3}$
3180 1498 $\ell_{10} : 2\zeta(3)/5$
3180 2497 rt : (6,1,6,2)
3180 3371 ln : $3/2 + \sqrt{5}$
3180 5809 $e^x : 2/3 - 2e/3$
3180 6647 $J : \sqrt[3]{2}/5$
3180 7369 $Ei : 15/19$
3180 8083 $\ell_{10} : \sqrt[3]{9}$
3180 9612 $\Psi : 6/19$
3181 1607 at : $\sqrt{15}$
3181 1723 rt : (3,5,-5,-2)
3181 4688 $10^x : 2/3 + \pi/4$
3181 4712 $J : 7/19$
3181 4789 $\ell_{10} : (\ln 2/3)^{1/2}$
3181 5067 rt : (2,7,7,-3)
3181 6510 rt : (7,9,9,-4)
3181 6786 $J_2 : \sqrt[3]{19/3}$
3181 8181 id : 7/22
3182 0141 rt : (5,2,2,-1)
3182 0306 at : $1/4 + 4e/3$
3182 0406 $2^x : 7\ln 2/4$
3182 0850 rt : (1,5,0,9)
3182 3441 $\Psi : 2/3$
3182 4412 cu : $1/2 + 3\sqrt{5}/4$
3182 4676 rt : (6,6,-1,2)
3182 4708 $\ell_{10} : 2/3 + \sqrt{2}$
3182 5089 $t\pi : 1/2 + 3\sqrt{3}/2$
3182 5673 $10^x : 3/25$
3182 5708 ln : $(\sqrt[3]{4}/3)^{1/2}$
3182 6057 $t\pi : (4/3)^3$
3182 8182 $2/5 - e$
3182 8218 rt : (1,9,-2,9)
3182 8856 rt : (4,6,-5,6)
3182 9103 at : $(\pi/2)^3$
3182 9302 ln : P_{73}
3182 9497 cu : $(4/3)^3$
3182 9512 cr : $\sqrt{3} + \sqrt{5}/4$
3182 9742 $\Gamma : 11/16$
3182 9910 cu : $\sqrt{2}/4 - 3\sqrt{5}/4$
3183 0988 id : $(1/\pi)$
3183 3588 rt : (4,8,-1,-1)
3183 4907 $\ell_{10} : (\ln 2)^2$
3183 7338 rt : (2,7,0,-4)
3183 7771 rt : (9,7,0,-1)
3183 8392 $t\pi : 3\sqrt{2}/4 - 3\sqrt{5}/2$
3183 9039 $Ei : 5e\pi/9$
3184 1104 $\ell_{10} : \sqrt{13/3}$
3184 1223 rt : (6,2,-4,5)
3184 3476 $10^x : 3/2 + 3\sqrt{5}/4$
3184 4428 $\ln(4) \times s\pi(2/5)$
3184 4433 $e^x : (\sqrt{2})^{-1/2}$
3184 5373 ln : 8/11

3184 5863 $\Gamma : -\sqrt{2}+\sqrt{6}-\sqrt{7}$
3184 5923 $\Psi : 24/13$
3184 6050 $\ell_2 : 3\sqrt{2}+\sqrt{5}/3$
3184 6081 $\Gamma : -\sqrt{3}-\sqrt{6}+\sqrt{7}$
3184 6871 $\ell_{10} : P_{21}$
3184 6938 as : $2/3-\sqrt{2}/4$
3184 7210 $E : 3/4$
3184 7469 rt : (1,1,9,-3)
3184 9319 rt : (6,2,-7,-8)
3185 1627 rt : (3,7,-8,6)
3185 2314 $\exp(e) \div s\pi(\sqrt{3})$
3185 2428 rt : (5,6,7,-3)
3185 3781 rt : (8,7,-8,-3)
3185 3996 at : $2\sqrt{2}-3\sqrt{5}$
3185 4086 $2^x : (2\pi)^{-1/2}$
3185 5299 ln : P_{65}
3185 6365 sq : $4e/3-4\pi/3$
3185 7359 $\zeta : 3e\pi/10$
3185 7383 $\exp(3/5)-\exp(\pi)$
3185 7868 ln : P_{66}
3186 0055 $10^x : (\ln 2/2)^2$
3186 0844 rt : (5,7,8,2)
3186 2224 $2^x : P_{59}$
3186 2821 rt : (1,6,4,2)
3186 2976 cu : $1/4+\sqrt{3}/4$
3186 3078 id : $4\sqrt{2}/3+\sqrt{3}/4$
3186 4515 cr : $3-\sqrt{2}/2$
3186 4737 rt : (9,2,6,2)
3186 4881 sinh : $7e\pi/9$
3186 5258 $s\pi : 2/3+2e$
3186 6196 $Ei : \sqrt{13/2}$
3186 9816 $Ei : 5\ln 2/3$
3187 0410 $\ell_{10} : 2\sqrt{2}-\sqrt{5}/3$
3187 1382 $J_0 : 10\sqrt{5}/7$
3187 1882 $s\pi : (\sqrt[3]{3}/2)^{1/3}$
3187 2930 rt : (8,6,-6,0)
3187 5675 at : $10e/7$
3187 5876 $\ell_{10} : 12/25$
3187 7366 sr : $7\sqrt{5}/9$
3187 7804 sq : $(\pi/3)^3$
3188 2328 $\lambda : \exp(-19\pi/16)$
3188 4634 rt : (3,8,-8,-5)
3188 7480 $s\pi : 9\pi^2/8$
3188 8123 $10^x : \zeta(3)/10$
3188 8131 rt : (8,5,7,-3)
3188 8694 $2^x : 6\sqrt[3]{3}/5$
3188 9311 $\Gamma : e/10$
3189 0034 $1/3-\exp(\sqrt{3})$
3189 0333 at : $2+4\sqrt{2}/3$
3189 1712 rt : (4,0,9,3)
3189 1802 $t\pi : e-3\pi$
3189 2676 cu : $2\sqrt{2}+3\sqrt{5}/2$
3189 3417 rt : (2,7,-8,-4)
3189 4275 cu : $2+2\pi$
3189 4506 id : $8\pi^2/3$
3189 5788 $e^x : 1/2+\sqrt{3}$
3189 7503 rt : (5,6,-8,-3)
3189 7701 $4/5 \times \exp(1/2)$
3189 7952 rt : (7,9,5,7)
3190 0955 $Ei : 1/\ln(\sqrt{3})$
3190 2643 cu : $\sqrt{2}/3-2\sqrt{3}/3$
3190 4334 cr : $2e-\pi$
3190 4760 rt : (5,9,-3,7)
3190 5118 ln : $2\sqrt{2}/3+\sqrt{3}/4$
3190 6214 $2^x : 4/3-4\sqrt{5}/3$
3190 7032 sq : $2/3+3e/4$
3191 0076 rt : (9,5,-3,8)
3191 0724 rt : (8,1,2,-1)
3191 3910 id : $2\sqrt{2}+2\sqrt{5}/3$
3191 4965 $Li_2 : 5/17$
3191 5483 ln : $3\sqrt{2}/2-\sqrt{5}/3$
3191 5562 $J_2 : 9\sqrt[3]{3}/7$
3191 6397 rt : (7,8,3,-2)
3191 8269 rt : (6,3,-8,-2)
3192 1982 rt : (1,7,-1,-1)
3192 2146 rt : (9,9,0,5)
3192 2785 $t\pi : 1/4-4\sqrt{2}$
3192 3053 $\ell_2 : 9\ln 2/5$
3192 3189 $\ell_{10} : \sqrt{2}/4+\sqrt{3}$
3192 5108 $\Psi : (e/2)^2$
3192 6678 rt : (2,4,-8,-1)
3192 7393 at : $3/4+\pi$
3192 7394 rt : (7,2,-8,-9)
3192 7685 $\Psi : 20/17$
3192 7842 rt : (5,2,0,8)
3192 7912 sr : $2/3+3\pi/2$
3192 9306 sinh : $\sqrt{11/3}$
3192 9773 rt : (3,6,8,7)
3193 0908 $\Gamma : 4\zeta(3)/7$
3193 1006 $e^x : 2-\pi$
3193 2152 $\Psi : 3\pi^2/7$
3193 3603 rt : (7,1,9,3)
3193 5032 rt : (1,7,8,-5)
3193 5164 at : $\exp(e/2)$
3193 5253 sinh : $\pi/10$
3193 5601 $\ell_2 : 2e-4\pi/3$
3193 6044 sr : $1/4+2\sqrt{5}/3$
3193 6232 $E : (\sqrt[3]{2}/3)^{1/3}$
3193 6927 rt : (6,9,-4,-2)
3193 8532 $K : (4/3)^{-1/3}$
3193 8925 rt : (2,5,-8,-3)
3194 0610 $Ei : (3/2)^{-3}$
3194 2035 $\Psi : \exp(\sqrt[3]{3})$
3194 3047 rt : (7,2,-5,6)
3194 3828 id : $\sqrt{5}/7$
3194 4296 sq : $3/2-3\pi/2$
3194 4845 rt : (8,8,8,2)
3194 5730 $\ell_2 : \zeta(3)/3$
3194 6845 sinh : $2\sqrt{2}/9$
3194 7069 sq : 13/23
3194 7156 at : $\sqrt{3}/3-2\sqrt{5}$
3194 7385 $\pi-\exp(3/5)$
3195 0136 rt : (3,5,-5,-9)
3195 0791 $2^x : 2/5$
3195 2866 ln : $\sqrt{14}$
3195 3503 $e^x : (2e)^2$
3195 5080 rt : (1,2,-7,2)
3195 5780 $J_1 : \pi^2/2$
3195 6995 $t\pi : 3\zeta(3)/4$
3195 7095 as : $\pi/10$
3195 7321 $\Gamma : (\sqrt[3]{2}/3)^{-3}$
3195 9068 rt : (5,9,6,-3)
3195 9326 $\theta_3 : (1/\pi)/2$
3196 1445 $s\pi : 3/4+\sqrt{2}/4$
3196 1549 at : $9\sqrt{3}/4$
3196 1811 rt : (6,7,-5,-5)
3196 2732 cu : $6\sqrt{5}/5$
3196 2740 $10^x : 1/4-\sqrt{5}/3$
3196 2828 $e^x : 1/\sqrt{13}$
3196 5027 rt : (5,2,-3,-1)
3196 6888 ln : $2\sqrt{3}+3\sqrt{5}$
3196 7607 $\ell_2 : 5\sqrt[3]{3}/9$
3196 8725 as : $2\sqrt{2}/9$
3196 9878 $3/4+s\pi(\pi)$
3197 0455 $e^x : 7\zeta(3)/10$
3197 1319 rt : (7,6,-2,3)
3197 4718 $10^x : 4e-2\pi$
3197 5194 $s\pi : 7\zeta(3)/4$
3197 7125 $10^x : \sqrt{23/3}$
3197 7219 id : $6e\pi/7$
3197 7724 $\sqrt{5}+\ln(2/5)$
3197 8048 rt : (9,6,-7,1)
3197 8207 rt : (1,4,8,-3)
3198 0208 cu : $5e/7$
3198 0308 $\ell_2 : 3/4+3\sqrt{2}$
3198 0713 $\sqrt{5} \div s\pi(\sqrt{2})$
3198 0817 $\Gamma : (\sqrt{2}/3)^{1/2}$
3198 2208 cr : $1+3\sqrt{3}/4$
3198 2415 sq : $\exp(-\sqrt[3]{5}/3)$
3198 2582 $J_2 : 13/7$
3198 2664 rt : (6,2,5,-2)
3198 3605 $\sqrt{3}+s\pi(1/5)$
3198 4395 $10^x : 2\sqrt{2}/3-\sqrt{3}/3$
3198 5882 rt : (3,8,5,6)
3198 5931 sq : $3/4-2\sqrt{2}$
3198 6371 rt : (2,7,-5,9)
3198 6926 sq : $1/\ln(e+\pi)$
3198 9070 rt : (3,8,-4,-2)
3198 9712 rt : (5,1,-7,-4)
3199 0100 rt : (8,8,6,-3)
3199 0496 rt : (3,8,-5,5)
3199 1190 ln : $3/4+3\pi$
3199 4084 $\ell_2 : P_{48}$
3199 8541 sr : $\sqrt{2}+\sqrt{3}+\sqrt{5}$
3199 8858 $J_2 : 25/6$
3200 0000 id : 8/25
3200 0612 cr : 23/10
3200 1159 $erf : 7/24$
3200 1173 rt : (2,6,1,-1)
3200 1504 $\ln(3)+\exp(1/5)$
3200 1604 $e^x : 2\sqrt{3}+\sqrt{5}/3$
3200 3268 $2^x : 2\sqrt{2}-2\sqrt{5}$
3200 4300 rt : (1,8,-7,-7)
3200 4322 ln : $2\zeta(3)/9$
3200 4332 $t\pi : 5\sqrt[3]{3}/8$
3200 5210 $2^x : 3/4+3e$
3200 5527 rt : (1,8,2,-9)
3200 5535 $\Psi : 8\ln 2/3$
3200 8413 rt : (8,9,4,8)
3200 9307 $2/3-\ln(\sqrt{2})$
3201 0660 $t\pi : \ln(3)$
3201 0792 rt : (6,9,9,-4)
3201 1692 $e^x : 6/5$
3201 2288 $\exp(1/4)+s\pi(\sqrt{2})$
3201 3515 $2^x : \zeta(3)/3$
3201 5497 rt : (4,3,8,-3)
3201 6553 rt : (8,2,-6,7)
3201 6902 at : $3+e/3$
3201 7368 cu : $4+3\pi/2$
3201 8816 $J_0 : 16/5$
3201 9278 $e^x : 5/18$
3201 9898 rt : (6,2,-1,9)
3201 9910 at : $2(e+\pi)/3$
3201 9955 rt : (5,9,-7,-3)
3202 0400 $\exp(\sqrt{5}) \times s\pi(\sqrt{5})$
3202 1449 sinh : $\sqrt[3]{2}/4$
3202 1465 rt : (5,7,-9,2)
3202 3895 $2^x : 19/9$
3202 3961 cu : $3/2+\sqrt{5}/3$
3202 4700 $\ell_2 : P_{40}$
3202 4891 $2^x : 2/3-4\sqrt{3}/3$
3202 5011 sq : $4/3+2\sqrt{2}$
3202 5603 $\ell_2 : P_{16}$
3202 5877 $2^x : 17/14$
3202 8910 rt : (4,7,-4,3)
3202 9023 $10^x : \sqrt{3}/2-\sqrt{5}/3$
3202 9223 rt : (2,3,7,2)
3203 0288 $\Psi : 5e\pi/4$
3203 0908 cu : 13/19
3203 1969 rt : (9,1,5,-2)
3203 3515 $\ell_{10} : 11/23$
3203 3529 sq : $2\sqrt[3]{3}$
3203 4859 rt : (9,8,9,-4)
3203 5588 rt : (2,7,3,6)
3203 6653 at : $1/3-3\sqrt{2}$
3203 7039 $10^x : e/3-\pi/4$
3203 7154 $\exp(\sqrt{5})-s\pi(\sqrt{2})$
3203 7724 rt : (7,4,-2,0)
3204 0299 rt : (2,8,-3,-1)
3204 2954 id : $1-e/4$
3204 3585 as : $\sqrt[3]{2}/4$
3204 7866 $\zeta : 6/13$
3204 9990 id : $2\sqrt[3]{3}/9$
3205 0246 $2^x : 3/2-\pi$
3205 0807 id : $10\sqrt{3}$
3205 0932 rt : (7,2,8,-3)
3205 1667 ln : $3+\sqrt{5}/3$
3205 1741 rt : (5,3,-4,1)
3205 3243 cr : $4\sqrt{2}-3\sqrt{5}/2$
3205 3410 sinh : 12/11
3205 3685 rt : (3,9,-5,-6)
3205 4250 $J_0 : 9/2$
3205 4427 $J_1 : e/4$
3205 5052 rt : (2,8,-7,-3)
3205 7611 $e^x : 4/3+\pi/4$
3205 9565 $J_0 : 11/6$
3206 0040 at : $7\sqrt{5}/4$
3206 0415 rt : (9,4,-9,-3)
3206 0841 rt : (3,5,-8,-8)
3206 1938 $J : \exp(-18\pi/5)$

3206 3377 $3 \div \exp(\sqrt{5})$
3206 3738 rt : (8,6,-3,4)
3206 7205 $\zeta : \sqrt[3]{4}$
3206 8291 cu : $2/3 + 3\sqrt{2}$
3206 8377 rt : (6,1,-8,-5)
3207 2114 rt : (9,2,-7,8)
3207 2300 $J_1 : 17/25$
3207 2434 rt : (5,5,1,-1)
3207 2904 $e^x : \sqrt{2}/4 - 2\sqrt{5}/3$
3207 4416 $e^x : 1/2 + 2\sqrt{5}/3$
3207 6724 $2^x : 2 - \pi/4$
3207 7716 as : $3\sqrt{2}/4 - \sqrt{5}/3$
3207 9441 id : $\sqrt[3]{25/2}$
3207 9632 id : $1/4 - \pi/2$
3208 1432 $J_0 : 3e\pi/8$
3208 3202 rt : (9,9,3,9)
3208 3371 $\zeta : 23/9$
3208 5130 $\zeta : 8\sqrt{5}/7$
3208 6211 rt : (4,1,-3,-7)
3208 7248 $\zeta : 4\sqrt[3]{5}/5$
3208 8449 rt : (2,8,-8,-8)
3208 8890 rt : (7,5,4,1)
3208 9151 $2^x : 1 + 3\sqrt{3}$
3209 2070 $2^x : \exp(5/2)$
3209 3363 $s\pi(\pi) \times s\pi(\sqrt{3})$
3209 4931 sq : $4/3 + \sqrt{5}/3$
3209 5719 $\ell_{10} : 3 - e/3$
3209 6373 rt : (5,4,7,2)
3210 0635 $1/4 \times \exp(1/4)$
3210 3681 rt : (6,3,-9,-3)
3210 4449 rt : (3,4,-3,9)
3210 5861 $\ell_{10} : 2\pi/3$
3210 6428 sinh : 6/19
3211 0909 sr : $5\pi/9$
3211 1921 sr : $1 + \sqrt{5}/3$
3211 3589 $t\pi : \sqrt{2}/4 + \sqrt{5}/3$
3211 5314 $2^x : 3 + \sqrt{2}$
3211 6942 rt : (2,1,2,-9)
3211 7454 $e^x : 7(1/\pi)/8$
3211 8901 $t\pi : (\sqrt{2}\pi/2)^{1/2}$
3212 0258 rt : (4,6,7,8)
3212 1570 $10^x : 1 + 3\pi/2$
3212 1850 cu : $8(e + \pi)/9$
3212 2351 $e^x : 3/4 - 4\sqrt{2}/3$
3212 3593 tanh : $(\ln 2)^3$
3212 4698 $s\pi : 3/4 + 3\sqrt{5}/2$
3212 4867 $e^x : 16/19$
3212 4970 rt : (7,1,-9,-6)
3212 5784 rt : (6,5,4,-2)
3212 6860 $2^x : 3 - 3\sqrt{3}/2$
3212 7275 $e^x : 3/4 - \sqrt{2}/3$
3212 7330 rt : (9,6,-4,5)
3212 7562 sinh : $1/\ln(5/2)$
3212 7991 at : $2\sqrt{2}/3 + 4\sqrt{5}/3$
3212 8858 as : 6/19
3213 0084 $1/3 \times s\pi(\sqrt{2})$
3213 0729 $J_2 : 5\sqrt{5}/6$
3213 0752 rt : (1,4,0,6)
3213 1085 $\theta_3 : 4/25$
3213 1469 at : $4\sqrt{2} - \sqrt{3}$
3213 1837 $\Psi : \sqrt[3]{19/3}$
3213 4979 $e \div \ln(3/5)$
3213 5447 $t\pi : 2\sqrt{3}/3 + 4\sqrt{5}$
3213 5867 $t\pi : 2\sqrt{2}/3 + 2\sqrt{3}$
3213 6833 rt : (8,4,1,-1)
3214 0236 rt : (7,7,-6,-4)
3214 2010 rt : (4,8,-6,-7)
3214 2318 $\zeta : -\sqrt{2} + \sqrt{3} + \sqrt{5}$
3214 4317 rt : (3,2,-9,-3)
3214 4796 at : $5\pi/4$
3214 4909 cu : $\sqrt{3}/4 - \sqrt{5}/2$
3214 5293 $J_2 : 1/\ln(\sqrt[3]{5})$
3214 5671 rt : (2,8,-9,2)
3214 5831 cr : $8\sqrt[3]{3}/5$
3214 6267 $e^x : 9e\pi/5$
3214 6945 rt : (3,8,-9,-9)
3214 7271 at : $(\ln 2)^3$
3214 7459 sq : $1 - \sqrt{3}/4$
3214 7905 $J_1 : 15/22$
3214 8273 rt : (6,3,-5,2)
3215 0162 $s\pi : P_{51}$
3215 0581 rt : (7,5,7,-3)
3215 1273 tanh : 1/3
3215 2177 at : $3 - 4\sqrt{3}$
3215 2210 rt : (5,1,-4,-8)
3215 2235 $J_1 : 7\sqrt{2}/2$
3215 2284 $t\pi : \ln 2/7$
3215 2464 sinh : $1/\sqrt{10}$
3215 4683 sq : P_{80}
3215 6096 rt : (3,9,-2,4)
3215 7397 rt : (8,5,7,2)
3215 8935 rt : (7,6,1,7)
3215 9082 cu : $3 + 3\sqrt{2}/2$
3216 0490 $\sqrt{5} + \exp(3)$
3216 1560 $J_0 : 8\zeta(3)/3$
3216 2931 sq : $\sqrt{2} + 4\sqrt{5}/3$
3216 3717 rt : (1,4,-1,-6)
3216 5721 rt : (9,4,4,-2)
3216 6586 rt : (6,9,-8,8)
3217 0395 $\ell_{10} : 2\sqrt{2}/3 + 2\sqrt{3}/3$
3217 0575 as : $(\ln 3)^{-1/3}$
3217 1676 at : $4/3 + 3\sqrt{3}/2$
3217 1936 at : $\sqrt{3}/3 + 3\sqrt{5}/2$
3217 3158 $Li_2 : (3/2)^{-3}$
3217 4691 $\Gamma : 4/23$
3217 5055 at : 1/3
3217 5241 $\Gamma : 8\sqrt[3]{5}$
3217 5583 ln : 4/15
3217 6550 $Ei : (3\pi/2)^{-1/2}$
3217 7712 $J_1 : 1/\ln(\ln 2/3)$
3217 9676 sq : $1/2 - 4\sqrt{3}$
3218 0215 sr : $\sqrt[3]{16/3}$
3218 1897 $\theta_3 : \sqrt[3]{3}/9$
3218 2021 $\ln(\pi) \div s\pi(1/3)$
3218 2538 rt : (1,9,-3,0)
3218 2805 $3 \times s\pi(e)$
3218 3249 $2^x : 1/4 - 4\sqrt{2}/3$
3218 5824 $e^x : 3(e + \pi)/7$
3218 6488 rt : (9,7,-8,-3)
3218 8037 $J_1 : (\pi)^{-1/3}$
3218 8510 $s\pi : 4\pi^2/5$
3218 8758 $1/5 \times \ln(5)$
3218 9667 id : $3e/4 + 2\pi$
3218 9960 rt : (2,8,3,-5)
3219 2403 $K : 10/11$
3219 2809 $\ell_2 : 5$
3219 2872 $Li_2 : (4/3)^{-1/3}$
3219 4814 sr : $1/3 + \sqrt{2}$
3219 6546 at : $\sqrt{2}/3 + 2\sqrt{3}$
3219 6908 $\ell_2 : ((e + \pi)/3)^{1/3}$
3219 8009 $s\pi : e/4 + 3\pi$
3219 8583 $J_2 : \sqrt[3]{13/2}$
3219 8922 rt : (6,1,-5,-9)
3219 9708 $2^x : \sqrt{3}$
3220 0998 cr : $10\ln 2/3$
3220 1114 rt : (2,9,0,-1)
3220 2959 at : $3/2 - 2e$
3220 5930 sr : $e/4 + 3\pi/2$
3220 7042 rt : (3,9,3,-2)
3220 8713 rt : (7,3,-6,3)
3220 9960 rt : (8,6,0,8)
3221 1696 rt : (5,9,9,-4)
3221 1880 $1/2 + \exp(3/5)$
3221 1996 $E : \sqrt{5}/3$
3221 6446 rt : (2,9,2,2)
3221 6448 at : $1/4 - 4\pi/3$
3221 6779 $J_2 : 6\ln 2$
3221 7428 $e^x : P_{72}$
3221 7934 rt : (2,9,-8,4)
3221 8376 $e^x : P_{69}$
3221 8535 rt : (3,0,9,3)
3221 9208 $\ell_{10} : 5\sqrt[3]{2}/3$
3222 1075 $e^x : 8\pi/9$
3222 1929 $\ell_{10} : 21$
3222 2139 rt : (6,6,-8,-3)
3222 2251 $\Gamma : 7\sqrt{3}/3$
3222 3925 rt : (7,6,-5,-2)
3222 6041 rt : (8,7,-7,-3)
3222 7538 $t\pi : \ln(\sqrt{5}/3)$
3222 9213 rt : (8,6,-2,-1)
3222 9810 $\sqrt{3} \div s\pi(\sqrt{3})$
3222 9959 $\exp(\sqrt{3}) \times s\pi(1/5)$
3223 1771 $erf : \ln(\sqrt{5}/3)$
3223 2669 rt : (8,8,9,-4)
3223 2988 $10^x : 2/3 + e$
3223 3362 cr : $1/2 + 2e/3$
3223 5558 sq : $3\sqrt{3}/4 + 4\sqrt{5}/3$
3223 6178 rt : (7,8,6,-3)
3223 6503 rt : (1,9,6,3)
3223 6745 $\zeta : 2\ln 2/3$
3223 6747 id : $\sqrt{3} + \sqrt{5} - \sqrt{7}$
3223 7145 $\exp(3/5) \div \exp(\sqrt{3})$
3223 8128 $J_1 : 2\sqrt[3]{5}/5$
3223 9642 $\Gamma : (1/\pi)/6$
3224 0761 at : $3 + 2\sqrt{2}/3$
3224 2003 rt : (6,8,3,-2)
3224 5783 $\zeta : 3\sqrt{3}/7$
3224 6674 rt : (9,6,-1,9)
3224 6812 ln : $1/3 + \pi/3$
3224 7265 $J_1 : 13/19$
3224 7965 $e^x : \sqrt{5}/8$
3224 8979 rt : (3,5,-5,-1)
3225 0000 sq : 23/20
3225 0120 rt : (8,3,-7,4)
3225 0405 rt : (1,9,9,2)
3225 2485 rt : (6,1,9,3)
3225 3569 rt : (5,8,0,-1)
3225 4987 $erf : 5/17$
3225 5127 as : $3/4 - \sqrt{3}/4$
3225 5780 rt : (3,5,-8,-3)
3225 9946 rt : (8,3,-8,2)
3226 2814 at : $2\sqrt{2} + \sqrt{5}/2$
3226 3946 $\ell_2 : 3\sqrt{2}/4 + 4\sqrt{5}$
3226 4119 $Li_2 : 3\ln 2/7$
3226 4734 at : $4\sqrt{3} - 4\sqrt{5}/3$
3226 6940 $e - \exp(1/3)$
3226 7301 at : P_7
3226 8742 rt : (5,1,6,2)
3227 0626 id : $\sqrt{3}/3 + \sqrt{5}/3$
3227 1142 at : $2\pi^2/5$
3227 2264 ln : $2e - 3\pi/2$
3227 2591 $3/4 \times s\pi(\pi)$
3227 3019 $\Psi : 3\sqrt{2}$
3227 3459 rt : (4,5,-5,-2)
3227 3842 sinh : $(2e/3)^2$
3227 4357 rt : (5,3,-1,5)
3227 4682 $e^x : \sqrt{2} + \sqrt{5} - \sqrt{6}$
3227 4964 rt : (9,7,3,-2)
3227 6534 rt : (9,7,-8,-2)
3227 6835 $t\pi : \sqrt{2}/3 - 4\sqrt{3}$
3227 8098 sr : P_{51}
3228 0263 rt : (9,3,-8,5)
3228 4545 rt : (9,2,9,3)
3228 7405 $\ln(\sqrt{3}) \times s\pi(1/5)$
3228 7565 sr : 7/4
3229 4114 ln : $3\sqrt{3} - 2\sqrt{5}$
3229 4901 id : $2 - 3\sqrt{5}/4$
3229 5731 $10^x : 3e/4$
3229 5794 $t\pi : 10\pi^2/7$
3229 6307 $\ell_{10} : 7\zeta(3)/4$
3230 0012 $\zeta : \sqrt{13/2}$
3230 0383 $\Gamma : -\sqrt{2} - \sqrt{5} + \sqrt{7}$
3230 1577 rt : (8,7,0,-1)
3230 2081 sr : $4\sqrt{3}/3 - \sqrt{5}/4$
3230 2133 $1/4 \div s\pi(e)$
3230 5618 $2^x : 2/3 + 4\sqrt{3}$
3230 6359 $10^x : 1 - 2\sqrt{5}/3$
3230 6735 $J_2 : \sqrt{7/2}$
3230 7410 as : $3e/4 - 3\pi/4$
3230 7461 rt : (5,7,-9,4)
3231 0659 $e^x : 2\sqrt[3]{2}/9$
3231 1358 $t\pi : \sqrt{2} - 3\sqrt{5}$
3231 2981 $e^x : 7/25$
3231 3330 rt : (6,3,-2,6)
3231 4761 rt : (5,4,-8,-8)
3231 4860 rt : (8,2,6,2)
3231 7251 $2^x : 2 + 4\pi$
3231 7633 rt : (3,3,8,-3)
3231 9009 rt : (1,7,6,-2)
3231 9376 $s\pi : (\sqrt{2}/3)^3$

3232 0377 rt : (6,9,-7,-3)
3232 1786 sr : $3\sqrt{2}+2\sqrt{3}/3$
3232 2740 2^x : $2\sqrt{2}/7$
3232 2918 e^x : $(\sqrt[3]{3})^{1/2}$
3232 3091 rt : (5,5,-2,-1)
3232 3371 sq : $10\pi^2/3$
3232 4273 sq : $\sqrt{2}/3-3\sqrt{3}$
3232 4822 E : $(e/3)^3$
3232 4882 $\ln(\sqrt{3})+s\pi(e)$
3232 6277 2^x : $\exp(-e/3)$
3232 7297 rt : (1,8,4,-7)
3232 9816 J_0 : $10\pi/7$
3233 0496 at : $1/3+4e/3$
3233 0522 rt : (5,0,-8,1)
3233 2701 Li_2 : 10/11
3233 4962 rt : (6,4,-9,-9)
3233 5097 Γ : 7/2
3233 8137 rt : (6,7,-2,-1)
3233 9452 rt : (7,3,-3,7)
3234 0427 rt : (1,5,2,-4)
3234 3251 sq : $(2e)^{-1/3}$
3234 4757 as : $\sqrt{2}-\sqrt{3}$
3234 5272 e^x : $\sqrt{2}/4+2\sqrt{5}/3$
3234 5915 J_1 : $(\sqrt{2}/3)^{1/2}$
3234 6112 rt : (6,2,8,-3)
3234 6981 rt : (5,4,-8,2)
3234 8638 rt : (3,8,-7,-3)
3234 9759 id : $\sqrt{2}/4-3\sqrt{5}/4$
3235 0377 2^x : $4-2\sqrt{2}/3$
3235 2091 id : $1/3-4\sqrt{2}$
3235 2455 λ : $\exp(-13\pi/11)$
3235 4396 2^x : $2+e$
3235 6070 cr : $4\sqrt{2}/3+\sqrt{3}/4$
3235 7232 rt : (6,0,-9,2)
3235 7783 sinh : 7/22
3235 8181 rt : (8,3,-4,8)
3235 8373 J_1 : $4\zeta(3)/7$
3235 8550 at : $3/4-3\pi/2$
3236 1007 rt : (2,3,5,6)
3236 1151 2^x : $5(e+\pi)/4$
3236 1516 cu : 17/7
3236 1820 2^x : $-\sqrt{2}+\sqrt{5}-\sqrt{6}$
3236 2464 J : $2\sqrt{3}/9$
3236 4618 rt : (7,9,-4,-2)
3236 6118 sr : $(\sqrt{2}/3)^3$
3236 8191 ζ : 15/23
3236 8800 at : $1/2+2\sqrt{3}$
3236 9724 rt : (2,8,6,5)
3237 0399 ℓ_{10} : $\exp(\sqrt{5}/3)$
3237 1243 sinh : $(1/\pi)$
3237 2266 rt : (9,3,-5,9)
3237 3117 rt : (9,1,8,-3)
3237 5447 rt : (7,7,-3,0)
3237 7610 Γ : 13/19
3237 9000 sr : $(\sqrt[3]{5}/3)^{-3}$
3238 0152 2^x : $4\sqrt{3}+3\sqrt{5}/2$
3238 0484 at : $3e-4\pi/3$
3238 0642 tπ : $1/3-\sqrt{3}/4$
3238 1101 as : 7/22
3238 2166 2^x : $(3\pi)^{1/3}$
3238 2862 10^x : $\ln(\sqrt[3]{3})$
3238 3243 ℓ_{10} : $(\sqrt{2}\pi)^{1/2}$
3238 3661 J_1 : 11/16
3238 4554 e^x : $(\ln 2)^{-1/2}$
3238 7920 2^x : $5\ln 2/2$
3238 8051 tπ : $\sqrt{3}/3-3\sqrt{5}/4$
3238 8339 rt : (3,5,-2,6)
3238 9244 Γ : $1/\ln(\sqrt{5}/3)$
3238 9649 rt : (5,4,-8,-6)
3239 0288 cr : $3-e/4$
3239 2812 at : $\sqrt{3}+\sqrt{5}$
3239 4610 as : $(1/\pi)$
3239 5032 10^x : $2e+2\pi/3$
3239 5345 $1\div s\pi(\pi)$
3239 5639 Γ : $2e\pi$
3239 6468 rt : (8,7,-4,1)
3239 6582 rt : (7,2,3,1)
3239 7227 cr : $\sqrt[3]{25/2}$
3239 7263 cr : $3/4+\pi/2$
3239 9833 rt : (5,2,5,-2)
3240 0907 rt : (3,6,7,-3)
3240 1212 rt : (4,8,0,7)
3240 1442 rt : (1,7,-8,-4)
3240 2885 cu : $3/2-3\sqrt{3}/2$
3240 3039 rt : (4,8,-4,-2)
3240 5237 J_2 : 15/8
3240 7158 e^x : $3+\sqrt{5}/3$
3240 7474 E : $(2e/3)^{-1/2}$
3240 7483 ζ : $1/\ln(\sqrt{2}\pi/3)$
3240 7712 sq : $3\ln 2$
3240 8234 cu : $1/3+\sqrt{2}/4$
3240 8446 tπ : $10\sqrt[3]{5}$
3240 8649 cu : $4\zeta(3)/7$
3240 9334 ζ : $8(1/\pi)$
3240 9959 rt : (9,7,-5,2)
3241 0529 10^x : $(\ln 3/3)^{-2}$
3241 0559 ℓ_{10} : $3/4+e/2$
3241 2161 Ψ : $\sqrt[3]{2}/10$
3241 2548 tπ : $\sqrt{2}+\sqrt{5}+\sqrt{6}$
3241 4223 tπ : $2\sqrt{2}-4\sqrt{3}$
3241 4508 Γ : $2\sqrt[3]{5}/5$
3241 5321 $\ln(3)+\exp(4/5)$
3241 5704 J_1 : $10\sqrt[3]{5}/3$
3241 5891 at : $2\sqrt{3}/3-2\sqrt{5}/3$
3241 6391 rt : (2,7,-9,-5)
3241 6635 Ψ : 9π
3241 6778 at : $1/2-2\sqrt{5}$
3241 6961 E : $7(1/\pi)/3$
3241 8910 rt : (6,4,-9,-5)
3241 8924 Ei : 19/24
3241 9176 θ_3 : $\sqrt{5}/4$
3241 9678 rt : (4,9,9,-4)
3242 1156 $\exp(1/2)+s\pi(\sqrt{5})$
3242 1400 tπ : 5/17
3242 1906 rt : (2,5,-8,2)
3242 2138 tπ : $2e/3-3\pi/2$
3242 4235 2^x : $\sqrt{2}/3+\sqrt{5}/3$
3242 5111 rt : (6,5,7,-3)
3242 6045 tπ : $5e\pi/7$
3242 7040 Ψ : $9\sqrt[3]{3}/7$
3242 9556 ℓ_{10} : $\sqrt{3}/4+3\sqrt{5}/4$
3243 0590 rt : (5,3,2,9)
3243 1527 rt : (2,4,-7,6)
3243 2208 Ei : $10\sqrt{5}/7$
3243 2936 $\ln(\sqrt{3})\times\exp(e)$
3243 3195 e^x : $(5/3)^{-1/3}$
3243 3571 cu : $3e/4+2\pi$
3243 4738 rt : (1,6,-3,-6)
3243 5019 2^x : $2-4e/3$
3243 5026 2^x : $(\pi/2)^{-2}$
3243 5814 rt : (8,1,5,-2)
3243 7208 $4/5\times\ln(2/3)$
3243 7247 rt : (7,8,9,-4)
3243 8186 cr : $4-3\sqrt{5}/4$
3243 8672 rt : (1,7,-4,-2)
3244 1699 ζ : $9\sqrt{2}/5$
3244 3747 Ψ : $\exp(1/(2\pi))$
3244 4284 sr : 2/19
3244 6273 tπ : $5\sqrt[3]{2}/3$
3244 8147 rt : (9,4,7,-3)
3244 9191 E : $3\sqrt{3}/7$
3245 0630 2^x : $\sqrt{13}\pi$
3245 0970 $1-s\pi(\sqrt{5})$
3245 1109 ℓ_{10} : 9/19
3245 1585 2^x : $\ln(3/2)$
3245 2964 2^x : $1/\sqrt[3]{15}$
3245 4167 id : $(\sqrt[3]{5}/3)^{-1/2}$
3245 4346 2^x : $(\sqrt{2}\pi/3)^{1/2}$
3245 5532 sq : $\sqrt{2}+\sqrt{5}$
3245 9793 10^x : 12/23
3246 2394 at : $4\sqrt{2}-3\sqrt{5}/4$
3246 5065 J_2 : $\exp(\sqrt[3]{2}/2)$
3246 6960 $\ln(3)-s\pi(e)$
3246 8321 Ψ : 17/4
3246 9946 sπ : 2/19
3247 1795 rt : (9,0,-9,9)
3247 2015 at : $1+4\sqrt{5}/3$
3247 2336 tπ : $2\sqrt{3}/3+\sqrt{5}/3$
3247 2479 tπ : $7\sqrt{2}/9$
3247 2548 e^x : $\sqrt{2}/4+\sqrt{5}$
3247 2802 J_0 : $\exp(3/2)$
3247 2856 10^x : $4-2\sqrt{3}/3$
3247 4184 sr : P_{57}
3247 5704 ℓ_{10} : $(3\pi)^{1/3}$
3247 6823 rt : (8,9,-1,-1)
3247 9061 Ei : 6/13
3248 0304 erf : $(3/2)^{-3}$
3248 1928 rt : (7,3,-9,-3)
3248 2121 at : $2\sqrt{2}+2\sqrt{3}/3$
3248 4310 id : $5(e+\pi)/4$
3248 5180 Ei : 3/23
3248 6912 rt : (1,6,8,2)
3248 7324 2^x : $\sqrt{3}-3\sqrt{5}/2$
3248 9085 id : $(\sqrt[3]{5}/3)^2$
3248 9470 cr : 4π
3248 9614 cr : $4/3-3\sqrt{3}/4$
3248 9869 sinh : $\sqrt{5}/7$
3249 0081 sq : $7\sqrt[3]{3}/3$
3249 1602 rt : (8,0,-7,2)
3249 1969 tπ : 1/10
3249 2605 rt : (6,8,6,-3)
3249 2739 $\exp(\sqrt{2})\div s\pi(2/5)$
3249 4007 J_1 : $2e\pi/3$
3249 4937 e^x : $3/4+3\sqrt{2}$
3249 5117 cu : 11/16
3249 7731 10^x : $1/4+e/3$
3250 0139 as : $\sqrt{2}/2-3\sqrt{5}/4$
3250 1252 tπ : $10\sqrt[3]{5}/9$
3250 1264 J_0 : $\sqrt{10/3}$
3250 1458 sq : $8\sqrt[3]{3}/5$
3250 1817 at : $4\sqrt{3}/3+3\sqrt{5}/4$
3250 2652 rt : (5,5,4,-2)
3250 3785 2^x : $1/2-3\sqrt{2}/2$
3250 4017 2^x : $3/2+\sqrt{5}$
3250 5136 sr : P_{38}
3250 6194 10^x : $2/3-2\sqrt{3}/3$
3250 7258 rt : (1,5,5,9)
3251 1557 tπ : $5\sqrt[3]{2}/7$
3251 1763 rt : (4,2,-9,-3)
3251 2245 at : $2e-3\pi$
3251 2473 rt : (9,7,6,-3)
3251 2843 rt : (5,8,-8,1)
3251 3207 rt : (9,7,-2,6)
3251 3665 as : $\sqrt{5}/7$
3251 5001 ℓ_{10} : $7e/9$
3251 5087 Ψ : $(e+\pi)/5$
3251 8343 ℓ_{10} : $4-4\sqrt{2}/3$
3251 9034 10^x : 23/4
3252 1026 rt : (8,7,-1,5)
3252 1719 e^x : $3\sqrt{2}/4+\sqrt{3}$
3252 2472 at : $7\sqrt[3]{5}/3$
3252 5681 ℓ_2 : $\ln(\sqrt{2}\pi/2)$
3252 7617 sπ : P_{57}
3252 8015 J_1 : $(2\pi/3)^{-1/2}$
3253 1316 cu : $1-2e$
3253 1459 rt : (8,4,4,-2)
3253 2252 tπ : P_{18}
3253 2476 cr : $\sqrt{2}-\sqrt{3}+\sqrt{7}$
3253 2486 rt : (7,7,0,4)
3253 3396 sinh : $8\pi^2/5$
3253 5536 rt : (5,8,-1,-1)
3253 5595 rt : (2,7,-6,-9)
3253 8401 at : $1/4-3\sqrt{2}$
3253 8640 E : $(\ln 3/2)^{1/2}$
3253 9677 rt : (1,7,7,-9)
3253 9851 tπ : P_{48}
3254 0513 e^x : $3-e$
3254 1249 $\ln(5)-\exp(1/4)$
3254 1962 rt : (9,4,-9,2)
3254 2240 ln : 13/18
3254 2609 ℓ_2 : P_{59}
3254 3388 e^x : $3/4+3\sqrt{5}/4$
3254 3732 rt : (4,7,8,2)
3254 4281 rt : (1,9,-4,-9)
3254 5327 sinh : $6\sqrt{5}/7$
3254 7568 rt : (5,1,-7,2)
3254 7592 id : $(\ln 2/2)^{-2}$
3254 8936 sinh : 8/25
3255 0056 rt : (1,7,-5,-8)
3255 0575 e^x : P_6

3255 0898 rt : (6,7,1,3)
3255 4670 rt : (8,4,-8,1)
3255 8166 sinh : 23/12
3255 8738 Ei : $8\ln 2/7$
3255 9536 erf : $3\ln 2/7$
3256 0258 rt : (2,9,3,-2)
3256 1940 $s\pi$: $2\sqrt{5}/5$
3256 2028 sinh : $(3\pi/2)^{1/2}$
3256 2540 ℓ_{10} : $3\sqrt[3]{2}/8$
3256 5840 rt : (2,0,9,3)
3256 6701 sq : $2\sqrt{2}-3\sqrt{5}/4$
3256 7623 rt : (7,6,-8,-3)
3257 1213 cr : $6e/7$
3257 1916 J_2 : $\sqrt[3]{20/3}$
3257 2633 sq : $3/4+4\pi$
3257 2948 as : 8/25
3257 2998 rt : (7,4,-7,0)
3257 3500 id : $(3\pi)^{-1/2}$
3257 4699 ℓ_2 : $3e-\pi$
3257 4806 ℓ_2 : $(2\pi)^{1/2}$
3257 5808 Ei : $\sqrt{15/2}$
3257 6107 rt : (5,8,-5,8)
3257 6223 $\ln(3)\times\exp(3/4)$
3257 7747 e^x : $(\pi)^{1/3}$
3257 7964 rt : (5,0,-2,9)
3257 8074 sr : $9\zeta(3)/2$
3257 8115 J_1 : $\exp(-1/e)$
3257 8500 rt : (3,7,-9,3)
3258 0844 sq : $1-\pi/2$
3258 1302 Γ : $6(1/\pi)/5$
3258 1766 at : 4
3258 2538 J_1 : 9/13
3258 3525 rt : (1,7,3,-8)
3258 3659 rt : (7,1,2,-1)
3258 4927 e^x : $1-3\sqrt{2}/2$
3258 5340 rt : (5,8,3,-2)
3258 6068 at : $\sqrt{2}/3-2\sqrt{5}$
3258 7005 rt : (2,7,-1,-1)
3258 7359 $s\pi$: P_{38}
3258 7517 ln : $e+\pi/3$
3258 8172 sr : $3(e+\pi)/10$
3258 8956 rt : (4,5,-7,7)
3259 0405 rt : (5,1,9,3)
3259 0786 ℓ_{10} : $\sqrt[3]{19/2}$
3259 1081 Ei : $\sqrt{24}\pi$
3259 2026 rt : (4,5,-8,-3)
3259 3292 ℓ_{10} : $1+\sqrt{5}/2$
3259 3743 rt : (7,8,8,2)
3259 4827 $t\pi$: $1/3+4\pi$
3259 6896 id : $(\ln 3)^3$
3259 6915 rt : (2,4,-5,-9)
3259 8054 Ei : 22/19
3260 0517 rt : (4,0,-1,8)
3260 1312 10^x : $3\sqrt{2}/2+\sqrt{5}$
3260 1512 sinh : $2\sqrt[3]{3}/9$
3260 1729 rt : (6,4,-6,-1)
3260 3222 sr : $(2e)^{1/3}$
3260 3711 J_1 : $2\sqrt{3}/5$
3260 4654 cr : $4\sqrt{2}+4\sqrt{3}$
3260 5416 rt : (3,2,-9,6)
3260 7114 rt : (5,3,-9,-5)
3260 7592 ℓ_{10} : $4/3+\pi/4$
3260 8606 J_0 : $\sqrt{2}-\sqrt{5}+\sqrt{7}$
3260 8898 rt : (8,7,3,-2)
3261 0319 sr : $4/3+3e/2$
3261 2951 Li_2 : 3/10
3261 3507 rt : (8,2,9,3)
3261 4985 rt : (1,7,3,-6)
3261 6148 rt : (8,7,2,9)
3261 6259 ln : $4\sqrt{3}/5$
3261 6267 2^x : $\sqrt{2}+\sqrt{6}-\sqrt{7}$
3261 7207 J_1 : $\ln 2$
3261 7207 Γ : $(\pi)^{-1/3}$
3261 8290 as : $1-e/4$
3261 8911 rt : (9,4,-6,6)
3262 2032 at : $10\zeta(3)/3$
3262 2880 2^x : $9e\pi/7$
3262 4858 Li_2 : $\sqrt{2}+\sqrt{3}-\sqrt{5}$
3262 5718 as : $2\sqrt[3]{3}/9$
3262 6000 $t\pi$: $(\ln 3/3)^{-2}$
3262 6046 J_1 : $\sqrt[3]{1/3}$
3262 6837 2^x : $(e/3)^{-2}$
3262 7078 Ψ : 13/7
3262 7261 rt : (2,5,-7,-5)
3262 8545 $\ln(2/3)\times\ln(\sqrt{5})$
3263 0582 Ψ : 10/23
3263 2447 rt : (1,6,1,-1)
3263 4546 rt : (4,9,-2,-3)
3263 4679 e^x : $5\sqrt[3]{3}/6$
3263 5240 id : $\sqrt[3]{7/3}$
3263 6201 Li_2 : $1/\ln(3)$
3263 6619 rt : (2,3,8,-3)
3263 7960 rt : (7,9,-7,-3)
3263 9758 J_2 : $3\pi/5$
3264 0589 Γ : $-\sqrt{2}+\sqrt{3}-\sqrt{7}$
3264 0694 rt : (1,3,7,2)
3264 0743 Γ : $8\sqrt[3]{3}$
3264 2162 rt : (2,7,-9,-8)
3264 2461 $\ln(5)\times\exp(3)$
3264 2523 rt : (8,4,-5,5)
3264 3303 rt : (8,3,-6,-2)
3264 4990 rt : (7,7,3,8)
3264 9059 rt : (6,8,9,-4)
3264 9849 sr : $3\sqrt{2}/2+4\sqrt{5}$
3265 2105 2^x : $4\sqrt{2}+2\sqrt{3}/3$
3265 3061 sq : 4/7
3265 3218 rt : (5,4,-5,-2)
3265 3850 at : $\sqrt{2}+3\sqrt{3}/2$
3265 5335 J_2 : $4\sqrt{2}/3$
3265 5524 ln : $1/4+\sqrt{2}/3$
3265 7036 rt : (5,2,8,-3)
3265 7540 e^x : $\exp(1/(2e))$
3265 8264 rt : (4,8,-7,-3)
3265 8569 $\sqrt{3}+\ln(2/3)$
3265 9988 sq : $\sqrt{2}-\sqrt{5}+\sqrt{7}$
3266 0108 id : $4\sqrt{2}/3-\sqrt{5}/4$
3266 0625 ℓ_{10} : $\sqrt{2}/3$
3266 2361 rt : (9,7,9,-4)
3266 2481 ln : $3\sqrt{3}/4+4\sqrt{5}$
3266 3425 ln : $2\ln 2$
3266 7481 $t\pi$: $7\sqrt{2}$
3266 8521 $t\pi$: $e/3-3\pi/4$
3266 9721 2^x : $3/4+2\sqrt{5}$
3267 1378 Ei : $2\ln 2/3$
3267 2753 rt : (1,8,-2,6)
3267 2829 rt : (4,7,3,1)
3267 4098 Li_2 : $\zeta(3)/4$
3267 4159 J : $\exp(-11\pi/18)$
3267 4871 id : $3\sqrt[3]{3}$
3267 5160 rt : (7,4,-4,4)
3267 5726 $\ln(\sqrt{5})\times\exp(1/2)$
3267 5930 rt : (3,6,-4,-4)
3267 6330 rt : (8,1,8,-3)
3267 7458 rt : (1,7,-7,-3)
3267 9144 ln : $3\zeta(3)/5$
3267 9172 sq : $1+4\sqrt{2}/3$
3267 9758 rt : (4,4,7,2)
3268 1960 rt : (5,2,-6,-2)
3268 2847 e^x : $\sqrt{2}/3+3\sqrt{3}$
3268 3959 rt : (5,3,-6,-9)
3268 4170 E : $\ln(2\pi/3)$
3268 4626 sr : $4+\sqrt{2}$
3268 7489 rt : (2,7,-6,5)
3268 7528 J_0 : $2\sqrt{5}$
3268 8235 rt : (6,7,4,7)
3268 9209 sinh : $10\sqrt[3]{2}/3$
3268 9644 e^x : $\sqrt{2}/5$
3269 0528 cu : $3/2+2\sqrt{3}$
3269 0669 as : $7\ln 2/5$
3269 1816 cr : $10\sqrt[3]{2}$
3269 2189 id : $\exp(-\sqrt{5}/2)$
3269 4308 id : $\ln(\sqrt[3]{3}/2)$
3269 4505 E : 17/23
3269 5311 e^x : $\zeta(3)$
3269 5628 rt : (2,6,7,9)
3269 8018 rt : (9,8,-9,-5)
3269 8297 e^x : $\sqrt{2}+\sqrt{3}/3$
3269 9970 rt : (2,5,-4,5)
3270 0343 sr : $2\sqrt{2}/3+2\sqrt{5}$
3270 2835 e^x : $8(1/\pi)/9$
3270 3340 at : $2/3+3\sqrt{5}/2$
3270 4498 cu : $\sqrt{2}/4+\sqrt{5}/3$
3270 5144 ln : $6\pi/5$
3270 7322 Ei : $(\sqrt{5}/3)^{-1/2}$
3270 7351 2^x : $1/\sqrt{6}$
3270 8929 rt : (7,4,1,-1)
3271 1562 rt : (2,9,-9,-1)
3271 2457 rt : (8,4,-2,9)
3271 2851 rt : (5,5,7,-3)
3271 3472 Ψ : $\sqrt[3]{2}/4$
3271 4132 sq : $1/3+2\sqrt{5}/3$
3271 4801 rt : (7,5,7,2)
3271 5244 $t\pi$: $\sqrt[3]{4/3}$
3271 6675 rt : (2,7,2,-5)
3271 7022 at : $2\sqrt{3}+\sqrt{5}/4$
3271 7339 id : $\sqrt{13}\pi$
3271 9469 $s\pi$: $(1/\pi)/3$
3272 0361 ln : $3e+2\pi/3$
3272 0557 J_1 : 16/23
3272 1586 rt : (4,9,1,4)
3272 3219 rt : (6,4,-3,3)
3272 4230 $t\pi$: $\exp(-1/(3\pi))$
3272 6624 K : $\sqrt{2}+\sqrt{3}-\sqrt{5}$
3272 7513 at : $9\sqrt{5}/5$
3272 7807 ℓ_2 : $2-\sqrt{5}/3$
3272 8564 tanh : $e/8$
3272 9195 rt : (8,4,7,-3)
3272 9788 10^x : $(2e)^{1/3}$
3273 0629 $\exp(1/3)\times s\pi(2/5)$
3273 2186 J_2 : 17/9
3273 2355 rt : (8,1,-9,5)
3273 3707 ζ : $(\pi/3)^3$
3273 5026 id : $1/4-\sqrt{3}/3$
3273 5893 ℓ_{10} : 8/17
3273 7363 sinh : $\sqrt{5/2}$
3273 7664 J_1 : $7\sqrt{3}/4$
3274 0284 ln : $8\sqrt{2}/3$
3274 0744 e^x : $(e+\pi)/4$
3274 1317 rt : (3,8,-9,-8)
3274 1954 $\ln(2)\div\exp(3/4)$
3274 2274 id : $1/4+3e/2$
3274 3039 rt : (1,9,0,-1)
3274 4048 $t\pi$: $\exp(5/3)$
3274 4888 $t\pi$: P_{61}
3274 5540 rt : (2,9,6,-3)
3274 6541 $s\pi$: $3/4+3\pi/4$
3274 7469 K : $1/\ln(3)$
3274 7671 rt : (1,9,5,-5)
3274 8000 rt : (1,3,9,9)
3274 9047 2^x : $-\sqrt{2}+\sqrt{6}-\sqrt{7}$
3274 9230 $2/5\div\exp(1/5)$
3274 9588 rt : (8,6,-5,-2)
3275 2988 rt : (7,4,-1,8)
3275 3153 2^x : $4+2\pi/3$
3275 4597 at : $e/8$
3275 5493 J_2 : $3\sqrt[3]{2}/2$
3275 5678 J_1 : $9\sqrt[3]{2}/2$
3275 7913 J_1 : 5
3275 8459 θ_3 : $4\sqrt[3]{2}/9$
3276 0255 rt : (5,8,6,-3)
3276 2217 $\exp(3)-\exp(5)$
3276 3861 rt : (5,9,-7,-2)
3276 6817 J_1 : $\sqrt{2}+\sqrt{3}-\sqrt{6}$
3277 1035 θ_3 : 14/25
3277 1725 sinh : 23/21
3277 1974 $e+\ln(5)$
3277 2119 rt : (8,8,-8,-6)
3277 2396 $\exp(\sqrt{3})+s\pi(\sqrt{5})$
3277 3336 at : $e/4-3\pi/2$
3277 4272 rt : (8,7,6,-3)
3277 4397 ln : $7(e+\pi)/4$
3277 4621 rt : (7,1,-8,4)
3277 5446 rt : (3,6,-1,3)
3277 5653 10^x : $5\sqrt[3]{3}/9$
3277 6392 rt : (4,1,6,2)
3277 7764 rt : (5,5,-5,-2)
3278 0140 Γ : 15/22
3278 3683 10^x : $\exp(-2\pi/3)$
3278 4882 2^x : 9/22
3278 5261 rt : (3,8,-6,2)

3278 5942 rt : (8,1,-6,9)
3278 7649 J_1 : 17/3
3279 1406 id : $\sqrt{2}-\sqrt{3}+\sqrt{7}$
3279 1731 rt : (9,8,-6,-1)
3279 2728 Li_2 : $1/\sqrt{11}$
3279 3191 tπ : P_{45}
3279 4615 Ψ : $\sqrt[3]{5}/7$
3279 6978 10^x : $8\sqrt{2}/7$
3279 8527 rt : (7,7,0,-1)
3279 9655 λ : $\exp(-20\pi/17)$
3280 0038 2^x : $9(1/\pi)/7$
3280 0314 e^x : $\exp(-\sqrt[3]{2})$
3280 0999 rt : (5,4,-2,2)
3280 1523 θ_3 : $\exp(-2e/3)$
3280 2233 rt : (7,2,6,2)
3280 2637 ℓ_2 : $3\sqrt{2}-\sqrt{3}$
3280 3426 rt : (2,4,-5,-2)
3280 4384 tπ : $3\sqrt{3}-2\sqrt{5}/3$
3280 5471 sq : $1/3-e/3$
3280 7424 at : $2+3e/4$
3280 7488 J_0 : $1/\ln(\sqrt{3})$
3280 8229 2^x : $2\sqrt{3}/3-\sqrt{5}/3$
3280 8319 rt : (4,6,-9,-3)
3280 8512 id : $1/\ln(\sqrt[3]{2})$
3280 9803 at : $\sqrt{3}/4-2\sqrt{5}$
3280 9935 rt : (6,4,0,7)
3281 0009 sr : $2/3-\sqrt{5}/4$
3281 0192 at : $9\pi/7$
3281 1500 sr : $7\sqrt[3]{2}/5$
3281 2144 rt : (5,9,-4,5)
3281 2500 sq : $9\sqrt{5}/8$
3281 3102 sr : $4-\sqrt{5}$
3281 4312 2^x : $(2e/3)^{1/3}$
3281 9615 Li_2 : $(\ln 3/2)^2$
3282 2705 J_1 : $2\pi/9$
3282 3245 at : $7\sqrt{3}/3$
3282 3877 ln : $6\sqrt[3]{5}$
3282 4834 cr : $(1/\pi)/9$
3282 6700 10^x : $2\pi^2/5$
3282 6885 rt : (1,3,5,8)
3282 6892 at : $3\sqrt{2}/4+4\sqrt{5}/3$
3282 7920 rt : (8,9,-4,-2)
3282 8063 sq : $9(1/\pi)/5$
3283 1670 rt : (7,1,-5,8)
3283 1709 cr : $2/3+3\sqrt{5}/4$
3283 2871 Γ : $1/\sqrt[3]{18}$
3283 3248 rt : (6,1,-7,3)
3283 6095 cr : $2(e+\pi)/5$
3283 6382 rt : (5,1,-9,-2)
3283 6477 cu : $3\sqrt{2}+\sqrt{3}/2$
3284 0016 cu : $4/3+4\sqrt{5}/3$
3284 1461 tπ : $1/3+2\pi/3$
3284 2712 id : $1/2+2\sqrt{2}$
3284 4507 cr : $1/4+2\pi/3$
3284 6362 2^x : $3/4-3\pi/4$
3284 6739 10^x : 12/11
3284 6791 tπ : $\sqrt{24}$
3284 7734 ℓ_2 : $3(e+\pi)/7$
3284 7857 at : $2/3-3\pi/2$
3284 8407 rt : (4,2,5,-2)
3284 9317 rt : (5,8,-4,-2)
3284 9867 10^x : $1/2+4e$
3285 0406 ln : 18/25
3285 0547 $\ln(2/5)+s\pi(1/5)$
3285 2043 J_1 : $((e+\pi)/2)^{-1/3}$
3285 4622 Li_2 : $e/9$
3285 4623 rt : (1,5,-1,4)
3285 5988 2^x : $9e/8$
3285 6348 at : $3+\pi/3$
3285 6672 ln : $4\sqrt[3]{2}/7$
3285 8576 tπ : $1/3-3e/4$
3285 8941 rt : (6,9,-9,6)
3285 9162 at : $e/3+\pi$
3285 9857 sr : $\sqrt[3]{11/2}$
3286 1064 sq : $1/\ln(\sqrt[3]{2}/3)$
3286 1216 rt : (9,8,-3,3)
3286 1505 rt : (2,8,-4,-5)
3286 2675 erf : 3/10
3286 2853 J : $\ln 2/2$
3286 4881 id : $\sqrt{18}\pi$
3286 7505 e^x : $2-2\sqrt{3}/3$
3286 8215 rt : (7,1,5,-2)
3286 8618 rt : (5,8,9,-4)
3286 8752 rt : (3,6,-8,-7)
3286 9020 rt : (2,7,-4,-2)
3286 9200 sπ : $8e/7$
3286 9488 ζ : $7(1/\pi)/3$
3287 0016 E : 14/19
3287 0370 cu : 23/6
3287 1640 ℓ_{10} : $\sqrt{2}-\sqrt{3}+\sqrt{6}$
3287 2755 E : $(5/2)^{-1/3}$
3287 3157 sinh : $4\sqrt{3}$
3287 5261 rt : (1,7,-2,3)
3287 6090 cr : $2\sqrt{3}-\sqrt{5}/2$
3287 6128 10^x : 7/11
3287 6760 rt : (8,8,-5,-2)
3287 7918 rt : (8,7,9,-4)
3287 8716 2^x : $\sqrt{2}/4+\sqrt{3}/2$
3288 1740 at : $4/3+e$
3288 2517 ζ : $\beta(2)$
3288 4382 as : $2-3\sqrt{5}/4$
3288 7218 rt : (1,6,4,-2)
3289 0059 rt : (8,3,-9,-3)
3289 1116 J_1 : $4\sqrt{2}$
3289 1423 tπ : $10\pi/3$
3289 1551 rt : (7,2,-9,-8)
3289 1670 $\ln(\sqrt{2})-s\pi(\sqrt{5})$
3289 2978 rt : (6,1,-4,7)
3289 5071 rt : (2,8,-1,8)
3289 5804 rt : (5,4,1,6)
3289 7770 J_1 : $5\sqrt[3]{2}/9$
3289 9574 J_1 : 7/10
3289 9934 sinh : 19/8
3290 3951 sr : $4/3+\sqrt{3}/4$
3290 4073 rt : (4,7,6,5)
3290 6608 rt : (5,2,-9,-3)
3291 1721 J_1 : $9\pi/5$
3291 4830 rt : (9,8,0,7)
3291 5697 erf : $\zeta(3)/4$
3291 6543 cu : $4e+3\pi/4$
3291 7982 rt : (7,8,-7,-7)
3291 8074 sq : $e/2-\pi/4$
3291 8280 Ei : $7\pi^2/9$
3291 8924 ℓ_{10} : $3-\sqrt{3}/2$
3291 9725 rt : (7,4,4,-2)
3292 0009 rt : (5,1,-6,2)
3292 0676 rt : (3,3,-8,3)
3292 1997 rt : (2,7,-1,-9)
3292 2782 tπ : $5\sqrt[3]{5}$
3292 3925 sq : $3+\pi/4$
3292 5745 tanh : $\sqrt[3]{5}/5$
3292 6191 2^x : $3\sqrt{2}/4+3\sqrt{5}/2$
3292 8698 $3/5\div\exp(3/5)$
3292 9905 rt : (9,5,-7,3)
3293 0002 sq : $2\sqrt{3}/3+3\sqrt{5}/2$
3293 1826 $\ln(3/4)\times\ln(\pi)$
3293 2056 Ei : 5/17
3293 2152 sr : $\exp(-\sqrt{2}\pi/2)$
3293 3568 at : $3+3\sqrt{2}/4$
3293 4038 Γ : 5/2
3293 4806 rt : (1,9,3,-2)
3293 5279 2^x : $\sqrt{2}/2-4\sqrt{3}/3$
3293 5576 rt : (2,3,0,9)
3293 5702 J_0 : $9\sqrt{2}/7$
3293 5713 ζ : $1/\sqrt[3]{10}$
3293 6339 rt : (1,0,9,3)
3293 6518 cu : $3-4\sqrt{3}/3$
3293 6704 at : $\sqrt{2}/2+3\sqrt{5}/2$
3293 6839 rt : (8,6,-8,-3)
3293 8436 ℓ_{10} : $e\pi/4$
3293 8837 J_1 : $\exp(\sqrt{3})$
3293 9198 J_2 : $2e\pi/9$
3293 9783 2^x : $2+4e/3$
3294 0476 id : $2e/3-\pi$
3294 0586 2^x : $e\pi/7$
3294 1099 J_0 : 20/11
3294 2406 ζ : $\sqrt[3]{5}/2$
3294 4983 $\ln(2/5)-\exp(5)$
3294 8354 rt : (8,8,-2,2)
3294 8563 sr : $2\sqrt{2}+3\sqrt{3}/2$
3294 9334 sπ : $\exp(-\sqrt{5})$
3294 9737 rt : (4,1,9,3)
3295 0188 rt : (5,5,-8,-3)
3295 0595 tπ : $10\pi^2/3$
3295 1473 tπ : $(\pi)^{-2}$
3295 2358 ℓ_{10} : $1/4+4\sqrt{2}/3$
3295 2586 at : $\sqrt[3]{5}/5$
3295 2717 10^x : $2\zeta(3)/3$
3295 3266 rt : (6,2,-8,-7)
3295 5635 2^x : $7e/9$
3295 5670 rt : (2,6,-7,-6)
3295 6803 ℓ_2 : $2\pi/5$
3295 7397 sr : $5\sqrt{2}/4$
3295 8368 $3/5\times\ln(\sqrt{3})$
3296 1025 rt : (7,7,3,-2)
3296 1107 $\ln(\sqrt{2})\times s\pi(2/5)$
3296 1732 e^x : $3/4+2\sqrt{5}$
3296 2278 rt : (7,2,9,3)
3296 2687 rt : (2,4,-8,-3)
3296 5276 tπ : $5e/9$
3296 6134 ln : $3\sqrt[3]{2}$
3296 6894 cu : $10\zeta(3)/3$
3296 7257 rt : (8,5,-9,-2)
3296 9679 rt : (9,5,-4,7)
3297 0981 sr : $\ln(e+\pi)$
3297 1880 id : $1/\ln(\sqrt{2}/3)$
3297 4412 rt : (1,3,8,-3)
3297 4425 $1/5\times\exp(1/2)$
3297 4782 rt : (8,9,-7,-3)
3297 5454 cr : $\sqrt[3]{13}$
3297 5669 e^x : $\sqrt[3]{5}/6$
3297 9467 rt : (5,1,-3,6)
3298 0710 10^x : 24/19
3298 0954 e^x : $(2\pi)^{1/3}$
3298 3299 ℓ_{10} : $3e/4-\pi/2$
3298 3644 cu : $2/3+\sqrt{3}/4$
3298 4220 e^x : $1/4-e/2$
3298 4400 rt : (4,1,-8,-3)
3298 5431 rt : (4,2,8,-3)
3298 5625 cu : $5(e+\pi)/7$
3298 5769 rt : (5,8,-7,-3)
3298 5788 Ei : $\sqrt[3]{1/2}$
3299 0001 ℓ_{10} : $5\sqrt[3]{5}/4$
3299 1388 cu : $\sqrt{3}/3-3\sqrt{5}/4$
3299 1947 J_2 : $10\sqrt[3]{5}/9$
3299 2261 sr : $(2\pi/3)^{-3}$
3299 2482 2^x : $4\sqrt{2}+\sqrt{5}/4$
3299 2572 J_2 : 19/10
3299 3671 Ψ : $e\pi/2$
3299 4017 $5\div\exp(e)$
3299 5585 id : $6e/7$
3299 5806 rt : (7,1,8,-3)
3299 6115 rt : (2,7,-7,-3)
3299 6727 ln : $4/3+4\sqrt{5}$
3299 8129 id : $7\sqrt[3]{5}/9$
3299 9121 J_1 : $(\ln 2/2)^{1/3}$
3299 9509 e^x : $3-2\sqrt{3}/3$
3300 0469 rt : (8,8,1,6)
3300 2000 sinh : $\pi^2/9$
3300 2836 J_0 : $\sqrt[3]{6}$
3300 3276 10^x : $4\sqrt{2}+3\sqrt{3}/2$
3300 4167 J_1 : $6(e+\pi)/7$
3300 4469 2^x : $4-4\sqrt{2}/3$
3300 5873 rt : (3,9,-6,-9)
3300 6252 sq : $2/3+\sqrt{2}$
3300 7499 ℓ_2 : $8\sqrt{2}/9$
3300 8606 sπ : $4\sqrt{2}+\sqrt{5}$
3301 1807 rt : (9,2,-8,7)
3301 2434 sr : 23/13
3301 2701 id : $5\sqrt{3}/2$
3301 3035 sπ : $9e/5$
3301 4699 rt : (8,5,-6,2)
3301 5092 rt : (4,7,9,9)
3301 7266 cr : $2+\sqrt{2}/4$
3301 8464 erf : $1/\sqrt{11}$
3302 0084 $\sqrt{3}+\exp(4)$
3302 0978 10^x : $2/3+2e$
3302 3088 rt : (7,8,-4,-3)
3302 3814 rt : (7,4,7,-3)
3302 4960 Ei : $9e/8$

3302 7649 sq : $3e/2 + 4\pi/3$
3302 8828 2^x : $1 + \sqrt{2}$
3302 9043 at : $3e/2$
3302 9410 $\exp(2/5) - \exp(3/5)$
3303 0743 $s\pi$: $\exp(e/2)$
3303 1205 2^x : $7/17$
3303 1715 2^x : $1 - 3\sqrt{3}/2$
3303 2298 rt : (1,9,6,-3)
3303 2519 rt : (3,9,-3,1)
3303 2989 at : $3\sqrt{3} - \sqrt{5}/2$
3303 4309 $\ln(4/5) \div s\pi(\sqrt{5})$
3303 6260 rt : (5,2,-7,-6)
3303 6790 rt : (1,8,6,8)
3303 8285 id : $7\pi/3$
3303 9684 ℓ_{10} : $4/3 - \sqrt{3}/2$
3304 0121 rt : (4,8,6,-3)
3304 0732 sq : $\sqrt{2}/3 + \sqrt{5}$
3304 1004 J_0 : $\sqrt{21/2}$
3304 1737 erf : $(\ln 3/2)^2$
3304 2313 $s\pi$: $\exp(\sqrt{5}/3)$
3304 2447 ln : $4\sqrt{3} + 3\sqrt{5}/2$
3304 3584 J_1 : $8\pi/5$
3304 4277 rt : (2,6,-1,4)
3304 4919 sq : $1/4 - 2\sqrt{3}$
3304 7554 rt : (7,7,6,-3)
3305 0429 $\sqrt{2} - \ln(2/5)$
3305 1006 ln : $4 + 2\pi$
3305 1224 rt : (8,5,-3,6)
3305 2287 10^x : $3\sqrt{2} + 3\sqrt{3}/2$
3305 3344 ζ : $(2e/3)^{-1/2}$
3305 4616 e^x : $6\sqrt[3]{5}/7$
3305 4931 tanh : $2\zeta(3)/7$
3305 7957 J_2 : $7e/10$
3305 7980 sr : $\sqrt{2}/3 + 3\sqrt{3}/4$
3305 9480 2^x : $2\sqrt[3]{3}/7$
3305 9744 Ei : 18
3306 1338 Ei : $\sqrt{2} - \sqrt{5} + \sqrt{7}$
3306 1520 rt : (4,1,-5,1)
3306 3669 sinh : $(\sqrt[3]{5}/3)^2$
3306 4314 J_1 : $4\pi^2/7$
3306 6246 e^x : $-\sqrt{3} + \sqrt{5} + \sqrt{7}$
3306 6549 e^x : $11/13$
3306 7003 cr : $3\pi/4$
3306 8091 2^x : $3 + 3\pi$
3306 8541 Ψ : $5\sqrt{5}/6$
3307 1219 e^x : $2/7$
3307 2029 erf : $e/9$
3307 4608 rt : (1,3,1,7)
3307 4832 Li_2 : $(\pi/3)^{-2}$
3307 6555 ℓ_{10} : $\sqrt{3} - \sqrt{5} + \sqrt{7}$
3307 7223 cr : $\sqrt{3}/2 + 2\sqrt{5}/3$
3307 9312 2^x : 4π
3307 9630 cu : $7\sqrt{2}/9$
3308 0398 10^x : $7\ln 2/4$
3308 1301 10^x : $3/2 + 2\sqrt{2}$
3308 2313 at : $2\zeta(3)/7$
3308 2591 cr : $5\sqrt{2}/3$
3308 2605 e^x : $4\sqrt{2} + 3\sqrt{3}$
3308 5602 rt : (7,8,-1,1)
3308 7190 id : $e/3 + 3\pi$
3308 8386 Γ : $17/25$
3308 8540 rt : (8,2,-7,6)
3308 8904 ℓ_2 : $9\sqrt{5}$
3308 9632 as : $(\sqrt[3]{5}/3)^2$
3309 0711 J_1 : $9\sqrt{5}/4$
3309 1405 rt : (1,9,9,-4)
3309 3357 at : $3\sqrt{3}/2 + 2\sqrt{5}/3$
3309 3540 $s\pi$: $3\sqrt{3}/2 - 2\sqrt{5}/3$
3309 4450 10^x : $10\sqrt[3]{2}/3$
3309 6497 rt : (4,8,9,-4)
3309 9321 ℓ_{10} : $7/15$
3310 0000 cu : $11/10$
3310 1076 rt : (7,5,-8,-3)
3310 1381 rt : (7,7,9,-4)
3310 1722 sinh : $\sqrt{2} - \sqrt{3} + \sqrt{5}$
3310 3491 Ψ : $10e\pi/3$
3310 6714 at : $4/3 - 3\sqrt{5}/4$
3310 7644 as : $1/2 + \sqrt{2}/3$
3311 0665 rt : (4,1,-2,5)
3311 1886 Ψ : $1/\ln(\sqrt[3]{5})$
3311 3883 sq : $\sqrt{2}/2 + \sqrt{5}/2$
3311 4077 2^x : $(\ln 3/2)^{-1/3}$
3311 5018 cu : $2\sqrt{2} + 3\sqrt{5}$
3311 5107 ln : $3 + \pi/4$
3311 6104 2^x : $1/2 - 2\pi/3$
3311 6175 J : $2\sqrt[3]{2}/9$
3312 0412 ζ : $\ln(\sqrt[3]{4}/2)$
3312 2447 cr : $1 + e/2$
3312 3240 rt : (4,2,-9,-1)
3312 4961 at : $2 + 2\pi/3$
3312 5364 sq : $4\sqrt[3]{3}/5$
3312 5443 rt : (2,9,1,-2)
3312 6623 sr : P_{76}
3312 7099 rt : (7,8,2,5)
3312 7320 e^x : $4/3 - \pi/3$
3312 7372 Ψ : $5\sqrt[3]{5}/2$
3312 7471 rt : (1,8,-3,-2)
3313 0628 Li_2 : $7/23$
3313 0903 id : $\sqrt{5} + \sqrt{6} + \sqrt{7}$
3313 1024 cu : $(\ln 3)^3$
3313 1090 rt : (1,6,-3,-9)
3313 1502 10^x : $2(1/\pi)$
3313 2930 cr : $3\sqrt{2}/4 + 3\sqrt{3}/4$
3313 3396 ℓ_{10} : $e/2 + \pi/4$
3313 3536 sr : $(\pi)^{1/2}$
3313 6094 sq : $15/13$
3313 6561 cu : $\sqrt{13}\pi$
3313 7236 E : $11/15$
3313 7939 rt : (7,5,-5,1)
3314 1035 J_1 : $12/17$
3314 4463 J_1 : $10e/9$
3314 4738 rt : (4,1,1,9)
3314 5666 at : $3/2 + 3\sqrt{3}/2$
3314 5907 e^x : $5\sqrt[3]{2}/2$
3314 7937 10^x : $4 - \sqrt{3}$
3314 8250 ℓ_2 : $\sqrt{19/3}$
3314 9297 Li_2 : $(2e/3)^{-2}$
3315 1731 $5 \div \ln(2/3)$
3315 2590 sinh : $(3\pi)^{-1/2}$
3315 3281 ℓ_{10} : $1/3 + 2e/3$
3315 3841 rt : (4,2,-6,-5)
3315 4075 sq : $4\sqrt{3}/3 + \sqrt{5}/3$
3315 4447 2^x : $(e + \pi)^{-1/2}$
3315 5215 at : $2\sqrt{2} - 4\sqrt{3}$
3315 5387 e^x : $3/2 + \sqrt{3}$
3315 5508 J_0 : $9\sqrt[3]{3}/4$
3315 6414 $\exp(\pi) \div s\pi(2/5)$
3315 6669 rt : (7,8,5,9)
3315 6758 rt : (5,7,-7,9)
3315 7411 $t\pi$: $7(e + \pi)/10$
3315 8471 $t\pi$: $\sqrt{5}/4$
3315 9580 rt : (1,3,-9,-8)
3316 0442 ln : $\sqrt{3}/4 + 3\sqrt{5}/2$
3316 1504 Γ : $e/4$
3316 2195 sq : $(e + \pi)^3$
3316 2229 $t\pi$: $2e - \pi$
3316 4398 id : $(2e)^{1/2}$
3316 4685 rt : (7,5,-2,5)
3316 6194 sr : $2\sqrt{3} - 3\sqrt{5}/2$
3316 6219 cu : $\exp(-1/e)$
3316 7204 at : $7(e + \pi)/10$
3316 7335 rt : (1,9,-7,-9)
3316 8283 Ei : $5e/9$
3316 8899 e^x : $1/3 + 3e/2$
3317 2008 rt : (7,2,-9,1)
3317 3006 e^x : $9(1/\pi)/10$
3317 3610 2^x : $3/4 + \sqrt{2}/3$
3317 4106 E : $3\sqrt[3]{5}/7$
3317 4579 2^x : $1/4 + 3\pi$
3317 4596 at : $4/3 - 2e$
3317 4720 2^x : $5\sqrt[3]{5}/7$
3317 6130 J_1 : $4\sqrt[3]{2}$
3317 6652 rt : (4,2,-3,-9)
3317 7634 $s\pi$: $3e - \pi/3$
3317 8150 $s\pi$: $1/3 + \sqrt{5}/4$
3317 8903 as : $(3\pi)^{-1/2}$
3317 9482 at : $3/4 + 3\sqrt{5}/2$
3318 1031 rt : (1,6,3,-3)
3318 1611 cu : $9/13$
3318 2990 rt : (4,7,-8,-6)
3318 4288 sr : $4\sqrt{3} - 2\sqrt{5}/3$
3318 4347 10^x : $1/2 + 3\sqrt{5}/4$
3318 4890 sinh : $6\sqrt[3]{5}$
3318 4982 rt : (7,5,1,9)
3318 7800 sq : $8e\pi$
3319 0679 rt : (7,2,-6,5)
3319 1189 J_1 : $\sqrt{2}/2$
3319 1336 e^x : $3e/4 - \pi$
3319 2565 $s\pi$: $(e/2)^{1/3}$
3319 5850 sq : $1/\ln(\sqrt{3}/2)$
3319 6675 $\ln(2/3) \div \exp(1/5)$
3319 7492 2^x : P_{56}
3319 8746 $e - \ln(4)$
3319 9328 rt : (4,7,-5,1)
3320 0098 e^x : $2\sqrt{2} - 2\sqrt{3}/3$
3320 0562 10^x : $4/3 - 2e/3$
3320 1553 E : $(e + \pi)/8$
3320 2192 10^x : $3/4 + 3\sqrt{5}/4$
3320 4243 J_2 : $\sqrt{17}$
3320 4814 J_2 : $21/11$
3320 5080 $2/5 - \sqrt{3}$
3320 5472 rt : (7,2,-3,9)
3320 5565 cu : $2/3 - e/2$
3320 6611 e^x : $1/\ln(4/3)$
3320 7312 E : $\ln(\sqrt[3]{3}/3)$
3320 9460 cu : $1/\ln((e + \pi)/3)$
3320 9683 sq : $(\ln 2)^{-2}$
3321 1125 $\exp(1/4) \div s\pi(\sqrt{2})$
3321 2460 rt : (4,7,-2,8)
3321 2636 ℓ_{10} : $\sqrt{2}/3 + 3\sqrt{5}/4$
3321 2750 Ψ : $1/6$
3321 3205 at : $3\sqrt{3}/2 - 3\sqrt{5}$
3321 4344 $\exp(3/5) - \exp(e)$
3321 6220 sr : $9\pi^2/8$
3321 7265 ln : $1/3 + 3\sqrt{2}/4$
3321 8664 $\ln(3/4) \div s\pi(1/3)$
3322 0622 $t\pi$: $\sqrt{3}/2 + \sqrt{5}$
3322 0989 rt : (7,1,-7,-9)
3322 1814 cr : $2\sqrt{3}/3 - \sqrt{5}/2$
3322 2720 J_2 : $6(1/\pi)$
3322 2877 Γ : $8/21$
3322 7062 $s\pi$: $(\sqrt{2}\pi)^{1/2}$
3322 7139 ζ : $19/12$
3322 9457 J_0 : $(\sqrt{2}\pi/3)^3$
3323 0644 at : $\exp(\sqrt{2})$
3323 0898 Ψ : $(\sqrt[3]{4})^{1/3}$
3323 2571 cu : $2e - 2\pi/3$
3323 4943 rt : (4,4,-9,8)
3323 4945 2^x : $4e/3$
3323 7960 ζ : $7\sqrt[3]{3}/4$
3324 0639 10^x : $1/\sqrt[3]{7}$
3324 1392 J_1 : $17/24$
3324 5486 2^x : $(\sqrt{5}/3)^3$
3324 6286 cr : $9\pi^2/7$
3324 7179 cu : $3\sqrt{2}/4 + \sqrt{3}/4$
3324 7841 rt : (1,6,-1,7)
3324 8113 at : $3/2 - 2\sqrt{3}/3$
3325 0508 sq : $1/\ln(\sqrt{5}/2)$
3325 1815 cr : $3/2 + \sqrt{3}/2$
3325 3751 cu : $1/4 - 2\sqrt{2}/3$
3325 5375 cu : $2\sqrt{3}/5$
3325 5810 J_1 : $7\sqrt[3]{3}/2$
3325 7311 at : $3 + \sqrt{5}/2$
3325 8512 10^x : $2\sqrt{2} - \sqrt{5}/2$
3326 0366 at : $\sqrt{2}/4 - 2\sqrt{5}$
3326 6533 2^x : $4\sqrt{2} - 3\sqrt{3}/2$
3326 7893 J : $\exp(-13\pi/15)$
3326 8045 sr : $\sqrt{3}\pi$
3326 8062 θ_3 : $\exp(-\sqrt{3}/3)$
3327 0413 Ψ : $\sqrt[3]{13/2}$
3327 2369 Ei : $5\sqrt[3]{2}/3$
3327 5080 J_0 : $13/4$
3327 5597 at : $2 + 3\sqrt{2}/2$
3327 7631 at : $4\sqrt{2}/3 + \sqrt{5}$
3327 7655 sinh : $\exp(-\sqrt{5}/2)$
3327 9888 sinh : $\ln(\sqrt[3]{3}/2)$
3328 1039 10^x : $\exp(-1)$
3328 1341 sq : $2\sqrt[3]{3}/5$
3328 3244 cu : $1/2 + e$
3328 5392 J : $\exp(-7\pi/9)$

3328 5520 at : $\sqrt{17}$
3328 6100 θ_3 : $(\sqrt[3]{5}/2)^{1/2}$
3328 6181 tπ : $2e\pi/9$
3328 9106 $\ln(4/5)\times\exp(2/5)$
3328 9781 J_0 : $2e/3$
3329 2329 rt : (1,9,-3,1)
3329 2574 at : $1/2+4e/3$
3329 3817 sinh : 19/12
3329 4250 J_2 : $\sqrt[3]{7}$
3329 4442 $\ln(3/5)-\exp(3/5)$
3329 6890 $\exp(1/4)-s\pi(2/5)$
3329 7072 e^x : $\sqrt{3}/3-3\sqrt{5}/4$
3329 9194 2^x : $\exp(\sqrt{5}/2)$
3330 0182 θ_3 : $(\ln 3/2)^3$
3330 2250 2^x : $(\sqrt{5}/3)^{-3}$
3330 2465 id : $(\ln 2)^3$
3330 2847 $s\pi(\pi)\times s\pi(e)$
3330 4465 as : $\exp(-\sqrt{5}/2)$
3330 5481 ℓ_2 : $3+3e/4$
3330 5807 2^x : 11/9
3330 6707 as : $\ln(\sqrt[3]{3}/2)$
3330 9716 at : $2\sqrt{2}+3\sqrt{3}/4$
3331 0463 erf : 7/23
3331 0708 ℓ_{10} : $3/4-\sqrt{5}/3$
3331 0893 e^x : $1/3+2\pi/3$
3331 3342 E : $(\sqrt[3]{5}/2)^2$
3331 4001 10^x : $\sqrt[3]{5/3}$
3331 4627 cu : $1+2\sqrt{2}/3$
3331 7101 $\ln(\sqrt{3})\div\exp(1/2)$
3331 7792 sq : γ
3331 8805 tanh : $\sqrt{3}/5$
3332 1852 2^x : $1/\sqrt[3]{14}$
3332 3535 Ei : 21/10
3332 4178 Ψ : $(e/2)^{1/2}$
3332 6565 erf : $(2e/3)^{-2}$
3332 7490 $\exp(\sqrt{2})\times\exp(e)$
3332 8015 at : $2/3+2\sqrt{3}$
3332 8347 tπ : $2\pi/7$
3332 8704 2^x : $\sqrt{2}/4+4\sqrt{3}/3$
3332 9229 J_0 : $\sqrt{2}\pi$
3333 1866 at : $1/3-e/4$
3333 3333 id : 1/3
3333 8407 tπ : $7\sqrt[3]{2}/8$
3333 8692 ℓ_{10} : $1+2\sqrt{3}/3$
3333 8985 J_2 : $\sqrt{11/3}$
3334 0953 at : $3e/4+2\pi/3$
3334 2848 tanh : $1/\sqrt[3]{24}$
3334 4435 λ : $(2\pi)^{-2}$
3334 7317 at : $\sqrt{3}/5$
3334 9797 as : $1/4-\sqrt{3}/3$
3335 1142 cr : $1/4+3\sqrt{2}/2$
3335 1207 e^x : $3/2-3\sqrt{3}/2$
3335 2143 10^x : 1/8
3335 4140 at : $3\sqrt{2}/4-3\sqrt{3}$
3335 4674 10^x : $4\sqrt[3]{5}/5$
3335 7632 at : $2\sqrt{3}/3+4\sqrt{5}/3$
3336 1389 10^x : $3\sqrt{2}/2-3\sqrt{3}/2$
3336 1908 at : $\ln 2/2$
3336 2008 $s\pi(1/5)-s\pi(\sqrt{3})$
3336 4719 cr : $1/3+3e/4$
3336 4792 sq : $5\ln 2/6$
3336 6496 sq : $4-\pi/4$
3336 9929 sr : $1/4+3\sqrt{3}$
3337 1464 at : $1/\sqrt[3]{24}$
3337 2413 at : $1/3-2\sqrt{5}$
3337 8446 Ei : $1/\sqrt[3]{10}$
3337 9966 ln : $4\pi/9$
3338 0258 J_2 : $6\sqrt{5}/7$
3338 0638 10^x : $1/\ln(1/(3e))$
3338 0909 10^x : $4e$
3338 1117 J_2 : 23/12
3338 2569 Li_2 : 21/23
3338 3611 2^x : $4\sqrt{2}+3\sqrt{5}/4$
3338 7792 at : $1+\pi$
3339 0523 id : $4\sqrt{2}+3\sqrt{5}/4$
3339 1462 ζ : $2\sqrt[3]{2}$
3339 2637 ℓ_{10} : $2/3+2\sqrt{5}/3$
3340 2239 cu : $8\sqrt{5}$
3340 2361 sr : $\ln(\sqrt{5}/2)$
3340 2603 sπ : $3/4+\pi$
3340 2840 $\ln(4)+\exp(2/3)$
3340 6316 $\ln(\sqrt{2})\times s\pi(\sqrt{2})$
3340 6741 J_0 : $5(e+\pi)/9$
3340 9667 10^x : $\ln(2e)$
3341 1987 2^x : $3\ln 2/5$
3341 4922 10^x : $3\sqrt[3]{3}/5$
3341 5609 rt : (2,7,-7,1)
3341 6022 sπ : $\exp(-\sqrt{2}\pi/2)$
3341 7450 J_2 : $\sqrt{2}-\sqrt{3}+\sqrt{5}$
3342 0082 cr : 19/8
3342 4636 cu : $10\ln 2/3$
3342 7011 e^x : $4\sqrt{2}/3-4\sqrt{5}/3$
3342 7216 10^x : $4\sqrt{2}+3\sqrt{5}/2$
3342 9577 Li_2 : $\ln(e/2)$
3342 9886 rt : (5,8,-6,6)
3343 1411 cu : $1/4+3\sqrt{2}/2$
3343 2582 ln : $1/3+2\sqrt{3}$
3343 3010 2^x : $7\sqrt[3]{5}/2$
3343 3027 Ψ : $9\zeta(3)$
3343 5640 Ψ : $3e/7$
3343 5751 e^x : $\sqrt[3]{3}/5$
3343 5860 id : P_7
3343 6719 ℓ_{10} : $\sqrt{2}+\sqrt{5}/3$
3343 6854 sq : $3-4\sqrt{5}$
3343 8071 rt : (5,8,-9,-1)
3344 0751 ζ : $5(1/\pi)/2$
3344 3229 rt : (8,0,-9,7)
3344 4619 tanh : 8/23
3344 7842 tπ : $3\sqrt{2}/2+4\sqrt{5}/3$
3344 9007 rt : (2,4,-8,3)
3345 0339 ℓ_{10} : $\sqrt{14/3}$
3345 2954 ℓ_2 : $4e-\pi/4$
3345 7237 ln : $3\sqrt{2}/2+3\sqrt{5}/4$
3345 8025 rt : (8,3,-5,7)
3345 9118 cu : $4+\sqrt{3}$
3345 9170 sq : $5\ln 2/3$
3345 9786 at : $3+2\sqrt{3}/3$
3346 0579 at : $4-3e$
3346 3907 rt : (5,4,-9,-7)
3346 4605 cu : $4\sqrt{2}/3-\sqrt{5}/4$
3346 5313 $\exp(1/5)+\exp(\sqrt{2})$
3346 5807 e^x : $\sqrt{3}/6$
3346 6352 rt : (1,7,-3,-4)
3346 8026 $\sqrt{5}+\ln(3)$
3346 9806 J_1 : $\sqrt{2}+\sqrt{3}+\sqrt{6}$
3347 1532 ln : $5\sqrt{5}/8$
3347 1574 at : $3/2-4\sqrt{2}$
3347 2012 rt : (8,3,-8,3)
3347 2067 J_2 : $\exp(\sqrt{2})$
3347 3683 at : 8/23
3347 4203 e^x : $1-2\pi/3$
3347 5794 ln : $((e+\pi)/3)^{1/2}$
3347 6182 e^x : 22/15
3347 7423 rt : (8,6,-1,7)
3347 9176 e^x : $2(e+\pi)/7$
3347 9210 sπ : $3\sqrt{2}+\sqrt{3}/2$
3348 0150 ℓ_2 : $3/2-\sqrt{2}/2$
3348 0493 cu : $4-\pi/2$
3348 1514 cr : $9\sqrt{2}$
3348 1619 ℓ_2 : $3\sqrt{2}-4\sqrt{5}/3$
3348 2667 at : $6\ln 2$
3348 3365 2^x : $4\sqrt{2}/3-2\sqrt{3}$
3348 3985 2^x : 5/12
3348 4497 J_1 : 5/7
3348 5110 e^x : $1/2+2\sqrt{5}$
3348 5340 rt : (5,1,-5,-7)
3348 5482 ln : $(\ln 3/3)^{1/3}$
3348 5526 cr : $7e/8$
3348 5973 10^x : $9\sqrt[3]{2}$
3348 7674 θ_3 : 1/6
3348 7858 J_0 : $1/\ln(e/2)$
3349 4183 rt : (2,8,-9,-7)
3349 4448 Γ : $\sqrt{10}$
3349 5693 tπ : $3e/4-\pi$
3349 6426 rt : (8,6,-4,3)
3349 7230 tπ : $9\sqrt{3}/4$
3349 7736 id : $\sqrt{2}/4+4\sqrt{5}/3$
3349 8384 at : $4/3+2\sqrt{2}$
3350 0106 ln : 5/19
3350 0666 10^x : $\sqrt{7}$
3350 1516 rt : (2,7,2,3)
3350 2217 cu : $1+2\pi$
3350 3081 ln : $4\sqrt{3}/3+2\sqrt{5}/3$
3350 3964 rt : (8,9,3,7)
3350 6081 ℓ_2 : $2/3+3\pi$
3350 6648 rt : (5,1,-8,-3)
3350 9139 cu : $3/4+4\pi$
3350 9732 sq : $\exp(\sqrt{2}\pi/3)$
3351 0469 tπ : $(\sqrt{5}/3)^{-1/3}$
3351 0683 Ψ : $2\sqrt{2}/9$
3351 2149 tπ : $9\zeta(3)/4$
3351 3167 ℓ_{10} : $3\sqrt[3]{3}/2$
3351 3233 ζ : $\sqrt{19/3}$
3351 3527 sr : $1/4+4e$
3351 3668 ln : $e/3+3\pi$
3351 5189 rt : (5,2,-1,7)
3351 6002 $1/2\div\exp(2/5)$
3351 6992 J_1 : $5\sqrt{5}/2$
3351 8005 sq : 11/19
3351 9485 $1/4\div s\pi(\sqrt{3})$
3351 9753 ℓ_{10} : $9\zeta(3)/5$
3352 1190 rt : (8,6,-7,-1)
3352 1432 10^x : 9/8
3352 2035 sπ : $9\ln 2/7$
3352 2278 at : $4\sqrt{2}-2\sqrt{5}/3$
3352 2547 sr : $\sqrt[3]{17/3}$
3352 3571 cr : $4/3+\pi/3$
3352 5134 at : 25/6
3352 6579 ℓ_{10} : $2\ln 2/3$
3352 8005 rt : (2,7,-1,-7)
3352 8089 J_2 : $4\sqrt[3]{3}/3$
3352 9876 Li_2 : 4/13
3352 9886 J_2 : 25/13
3353 0011 ℓ_{10} : $3/4+\sqrt{2}$
3353 1262 rt : (8,9,0,3)
3353 1618 J_1 : $(\ln 3/3)^{1/3}$
3353 2455 $3/5\times\exp(4/5)$
3353 3294 Γ : $9\pi/5$
3353 3323 sr : $10e\pi/3$
3353 3330 sπ : $(2\pi/3)^{-3}$
3353 4240 id : $(\sqrt[3]{2}/3)^{-1/3}$
3353 4441 J_1 : $((e+\pi)/3)^{-1/2}$
3353 4475 at : $2e/3+3\pi/4$
3353 4848 J_2 : $8\zeta(3)/5$
3353 5220 rt : (1,7,5,5)
3353 5683 J_1 : $(\sqrt{5}/2)^{-3}$
3354 0565 e^x : 24/13
3354 0897 e^x : $5e\pi/4$
3354 1019 $3/4\div\sqrt{5}$
3354 3481 Ψ : $9e\pi$
3354 4857 J_2 : $9\sqrt[3]{5}/8$
3354 6536 rt : (5,2,-4,3)
3354 7064 ℓ_{10} : $5\sqrt{3}/4$
3354 8255 ℓ_2 : $4+\pi/3$
3354 9844 at : $\sqrt{2}/2+2\sqrt{3}$
3355 0877 rt : (4,5,-8,5)
3355 1265 Li_2 : $(\sqrt{2}\pi/3)^{-3}$
3355 2369 ζ : $9\sqrt{5}/8$
3355 3692 $1/4+\exp(3)$
3355 4683 tanh : $\pi/9$
3355 5053 rt : (8,9,8,-4)
3355 5383 sinh : $7\sqrt{2}/9$
3355 6093 10^x : P_{53}
3355 6301 θ_3 : $\ln(1/\sqrt{2})$
3355 6489 Γ : $(2\pi)^{1/2}$
3355 9353 Ei : $\exp(\sqrt{5}/2)$
3355 9620 rt : (5,5,3,7)
3355 9629 K : $(\pi/3)^{-2}$
3356 0031 J_0 : $6e/5$
3356 1878 id : $10\zeta(3)/9$
3356 2380 J_1 : $9(1/\pi)/4$
3356 2518 10^x : $1/\sqrt[3]{20}$
3356 2881 J_2 : $10\sqrt{3}/9$
3356 3585 Ei : $5(1/\pi)/2$
3356 4380 rt : (2,5,-8,-3)
3356 4747 sinh : 11/10
3356 6773 sinh : 21/5
3356 7575 ℓ_2 : $7\sqrt[3]{3}$
3356 7918 erf : $\ln(e/2)$
3356 8924 rt : (8,9,-3,-1)

3356 9241 rt : (8,6,9,-4)
3357 1858 $s\pi$: $\exp(\sqrt[3]{4})$
3357 2146 10^x : $7/19$
3357 2971 2^x : $\sqrt{3}+2\sqrt{5}/3$
3357 4423 rt : (5,7,9,-4)
3357 4472 Li_2 : $4\ln 2/9$
3357 5035 2^x : $9\sqrt{5}/7$
3357 5965 sinh : $7\sqrt{3}/4$
3357 5982 J_1 : $(e)^{-1/3}$
3357 6690 $\pi \div \exp(\sqrt{5})$
3357 7666 cu : $2\sqrt{2}/3+3\sqrt{5}/2$
3357 9210 ℓ_{10} : $6/13$
3357 9841 rt : (2,8,9,-4)
3358 2143 $\sqrt{5}\times\exp(2/5)$
3358 2769 E : $(\sqrt[3]{4}/3)^{1/2}$
3358 4237 at : $\pi/9$
3358 5252 rt : (7,0,-8,6)
3358 5511 rt : (1,9,-9,-4)
3358 5639 E : $\exp(-1/\pi)$
3358 8304 e^x : $3/4+3\pi/4$
3358 8540 $s\pi$: $8\ln 2/5$
3359 0897 rt : (5,2,-7,-1)
3359 1347 cr : $\sqrt{2}/2+3\sqrt{5}/4$
3359 1452 2^x : $4\sqrt{3}-2\sqrt{5}/3$
3359 1762 Ψ : $\sqrt{7/2}$
3359 2688 sr : $2/3+\sqrt{5}/2$
3359 3401 rt : (3,8,-4,9)
3359 3458 E : $8/11$
3359 7367 $\ln(\sqrt{5})-\exp(\pi)$
3359 7426 $t\pi$: $10(e+\pi)/7$
3360 0000 cu : $23/5$
3360 1144 id : $2\sqrt{3}/3-2\sqrt{5}/3$
3360 2976 ln : $3/2-\pi/4$
3360 3234 2^x : $23/8$
3360 3427 rt : (8,9,5,-3)
3360 4068 rt : (1,9,5,-7)
3360 7126 rt : (4,8,-1,5)
3360 8008 $s\pi$: $(\sqrt[3]{2})^{-1/2}$
3361 0247 rt : (5,5,0,3)
3361 0679 10^x : $4\sqrt{2}/3-\sqrt{3}/3$
3361 2129 ℓ_2 : $\ln 2/7$
3361 2517 sinh : $9\sqrt{2}/5$
3361 2762 e^x : P_{82}
3361 2778 10^x : $23/3$
3361 3011 rt : (3,8,-7,-1)
3361 3332 ℓ_{10} : $\ln(\pi/3)$
3361 3660 ℓ_2 : $\sqrt{3}/3+2\sqrt{5}$
3361 7216 Ψ : $5\ln 2/8$
3361 7695 $t\pi$: $1/3-2e$
3361 8617 cr : $1/2+4\sqrt{2}/3$
3361 8792 sr : $1+\pi/4$
3361 9734 $s\pi$: $3/4+e/2$
3362 0574 e^x : $3\sqrt{2}/5$
3362 4193 rt : (8,9,-6,-5)
3362 4790 rt : (8,6,6,-3)
3362 4891 rt : (7,3,-4,6)
3362 5264 $\ln(4)\times s\pi(\sqrt{2})$
3362 5474 $t\pi$: $(\sqrt[3]{3}/2)^{1/3}$
3362 6463 2^x : $3/4+4\sqrt{3}/3$
3362 6805 J : $\zeta(3)/5$
3362 7942 at : $5(e+\pi)/7$
3363 0125 rt : (9,1,-8,9)
3363 0620 sr : $25/14$
3363 1419 λ : $\exp(-7\pi/6)$
3363 1550 Ei : $3e\pi/2$
3363 1897 Γ : $-\sqrt{3}-\sqrt{5}+\sqrt{6}$
3363 2760 rt : (5,8,7,7)
3363 4856 Ψ : $\pi/10$
3363 7554 tanh : $7/20$
3363 9393 $\ln(4/5)-\exp(\sqrt{2})$
3363 9644 ζ : $4\pi/5$
3364 0234 $3\div\exp(1/4)$
3364 0690 10^x : $1/4+\sqrt{5}/2$
3364 0957 ln : $10\sqrt[3]{2}/9$
3364 1185 rt : (4,8,-4,-2)
3364 3792 $t\pi$: $9\pi^2/8$
3364 3838 $\sqrt{3}-\exp(1/3)$
3364 5023 at : $4\pi/3$
3364 7223 ln : $5/7$
3364 9846 rt : (8,3,7,-3)
3365 0105 rt : (1,9,-6,-3)
3365 1106 J_2 : $\sqrt{3}-\sqrt{6}+\sqrt{7}$
3365 2740 sinh : $(e+\pi)/2$
3365 3573 tanh : $(1/(3e))^{1/2}$
3365 4110 erf : $4/13$
3365 4305 cu : $10\sqrt{5}/9$
3365 5320 sq : $1/4+e/3$
3365 6173 2^x : $1/\ln(\sqrt[3]{4}/3)$
3365 7025 θ_3 : $11/13$
3365 7121 10^x : $\sqrt[3]{2}/10$
3365 8006 ℓ_{10} : $4\sqrt{2}-3\sqrt{3}$
3365 9285 rt : (5,4,7,-3)
3366 0413 sq : $2-\sqrt{2}/3$
3366 2056 ℓ_{10} : $(3\pi/2)^{1/2}$
3366 3784 rt : (7,3,-7,2)
3366 4774 rt : (1,9,-4,-8)
3366 4831 cu : $16/23$
3366 5062 cu : $3/2+2e/3$
3366 5605 rt : (9,4,-4,9)
3366 7481 at : $7/20$
3366 9314 rt : (2,5,7,-3)
3366 9923 10^x : $8/11$
3367 0328 θ_3 : $9/16$
3367 1962 sinh : $\sqrt[3]{4/3}$
3367 2477 erf : $(\sqrt{2}\pi/3)^{-3}$
3367 6475 id : $4e/3+3\pi/2$
3367 8563 at : $4\sqrt{2}/3+4\sqrt{3}/3$
3367 8652 rt : (8,9,2,-2)
3367 9013 $\exp(2)+\exp(2/3)$
3367 9209 sr : P_{37}
3367 9319 rt : (7,6,0,6)
3367 9543 rt : (6,3,-8,-9)
3367 9652 rt : (8,0,8,3)
3367 9992 rt : (1,9,8,-3)
3368 0613 sr : $2\sqrt{3}-3\sqrt{5}/4$
3368 3573 at : $(1/(3e))^{1/2}$
3368 4655 at : $1-3\sqrt{3}$
3368 4889 rt : (4,8,-7,-9)
3368 6868 cu : $1/3+\sqrt{2}$
3368 7369 $\pi-\ln(\sqrt{5})$
3368 7489 θ_3 : $(\ln 3)^{-3}$
3368 8351 Ei : $\sqrt{10/3}$
3368 8514 e^x : $4e/3-3\pi/2$
3368 9150 Γ : $1/\ln(\sqrt[3]{2}/3)$
3368 9908 rt : (5,5,-3,-1)
3369 0256 e^x : $8(1/\pi)/3$
3369 0438 ℓ_2 : $e/4+3\pi$
3369 1011 rt : (5,1,8,-3)
3369 1198 rt : (2,8,-2,4)
3369 1385 rt : (4,7,-7,-3)
3369 1849 sq : $2\sqrt{3}/3-4\sqrt{5}/3$
3369 2403 erf : $4\ln 2/9$
3369 2694 Ei : $6\sqrt[3]{2}/5$
3369 4819 2^x : $2/3-\sqrt{5}$
3369 8227 rt : (9,4,-7,5)
3369 9672 $3/4\div\exp(4/5)$
3370 0550 id : $5\pi^2/4$
3370 0622 sq : $1/4+2e$
3370 2544 rt : (4,1,-4,-6)
3370 3135 $\ln(4)\div\exp(\sqrt{2})$
3370 3155 rt : (2,2,8,-3)
3370 3498 ℓ_2 : $19/24$
3370 3570 rt : (7,8,-7,-3)
3370 3902 at : $10\sqrt[3]{2}/3$
3370 5314 at : $21/5$
3370 7555 at : $4\sqrt{2}/3-\sqrt{5}$
3370 8452 $\ln(2/5)\div\exp(1)$
3371 0711 rt : (9,7,0,9)
3371 0854 rt : (1,8,5,1)
3371 1868 ζ : $3(e+\pi)/7$
3371 3040 rt : (8,9,-9,-9)
3371 3819 J_1 : $5\zeta(3)/2$
3371 4115 2^x : $\sqrt{3/2}$
3371 4367 rt : (8,6,3,-2)
3371 5215 J_1 : $4\sqrt[3]{2}/7$
3371 5714 rt : (8,3,9,3)
3371 6172 rt : (1,6,1,-7)
3371 7047 J_1 : $18/25$
3371 7221 ln : $2/3+\pi$
3371 7542 sr : $3/4+3\pi/2$
3371 8173 10^x : $4-2\sqrt{5}$
3371 9102 tanh : $\exp(-\pi/3)$
3371 9107 rt : (2,8,-5,-9)
3372 1646 J : $6/19$
3372 1701 rt : (2,9,-8,5)
3372 8151 rt : (5,8,4,3)
3372 8816 rt : (6,0,-4,9)
3372 8925 e^x : $1/3+4e/3$
3372 9151 rt : (1,8,-8,-6)
3372 9614 10^x : $3\sqrt{3}/2-3\sqrt{5}/4$
3372 9650 rt : (5,2,9,3)
3373 0161 rt : (4,4,-8,-3)
3373 0698 $t\pi$: $1/4-\sqrt{2}/4$
3373 1518 cu : $4/3+4\sqrt{3}/3$
3373 3388 id : $5e\pi/8$
3373 4910 cr : $\sqrt{2}/3-\sqrt{3}/4$
3373 5087 $\ln(4)+s\pi(2/5)$
3373 5431 rt : (7,6,-3,2)
3373 7349 Γ : $-\sqrt{2}+\sqrt{5}-\sqrt{6}$
3373 7877 ζ : $(e/2)^3$
3373 8446 ℓ_{10} : $8e$
3374 1691 $t\pi$: $1/4+3\sqrt{2}/2$
3374 2989 rt : (2,8,3,-2)
3374 2992 id : $\sqrt{2}/3+\sqrt{3}/2$
3374 3874 at : $7\zeta(3)/2$
3374 4666 rt : (2,1,9,3)
3374 5238 rt : (7,5,-8,-3)
3374 6027 2^x : $\sqrt{2}-4\sqrt{5}/3$
3374 6414 Ψ : $\sqrt{3}/4$
3374 6861 $t\pi$: $7\zeta(3)/4$
3374 7197 ln : $3/2+4\sqrt{3}/3$
3374 8060 sr : $4\sqrt{5}/5$
3374 8591 2^x : $4\sqrt{3}+\sqrt{5}/2$
3374 8769 rt : (3,7,-8,-9)
3374 9401 at : $\exp(-\pi/3)$
3374 9844 10^x : $9\sqrt{2}/2$
3375 0602 $t\pi$: $8\sqrt[3]{3}$
3375 1728 rt : (1,2,-8,-5)
3375 3154 ln : $3\sqrt{2}-\sqrt{3}/4$
3375 3329 rt : (9,7,-3,5)
3375 4178 sr : $2+2\sqrt{3}$
3375 5944 at : $2\sqrt{3}+\sqrt{5}/3$
3375 8383 rt : (8,3,4,-2)
3375 8697 rt : (7,9,4,6)
3375 9012 rt : (1,9,-3,-2)
3375 9321 rt : (5,9,-5,3)
3376 0895 rt : (1,0,-9,3)
3376 2725 J_1 : $\sqrt[3]{3}/2$
3376 5730 θ_3 : $(2e/3)^{-3}$
3376 7165 J_1 : $3\zeta(3)/5$
3376 7179 rt : (4,1,-7,-2)
3376 8510 $\sqrt{2}\div\ln(4/5)$
3377 0494 Ψ : $(\pi/2)^{1/3}$
3377 1513 id : $2\sqrt{2}-2\sqrt{5}/3$
3377 1594 at : $1/2-3\pi/2$
3377 1767 J_1 : $1/\ln(4)$
3377 2233 sq : $\sqrt{2}/2-\sqrt{5}$
3377 4297 cr : $7\sqrt[3]{5}/5$
3377 4514 $1/2\times s\pi(\sqrt{5})$
3377 4803 cr : $4/3+3\sqrt{2}/4$
3377 5405 e^x : $(\sqrt[3]{3}/2)^{1/2}$
3377 5659 rt : (5,4,4,-2)
3377 5895 rt : (6,0,-7,5)
3377 7911 J_2 : $7(e+\pi)/10$
3377 8495 rt : (4,1,-9,-3)
3377 8525 $1/4-s\pi(1/5)$
3378 0387 rt : (8,1,-7,8)
3378 0445 J_0 : $\sqrt{2}-\sqrt{5}-\sqrt{6}$
3378 0603 ln : $\sqrt{2}+4\sqrt{5}$
3378 0727 at : $3/4+2\sqrt{3}$
3378 1947 2^x : $8e/9$
3378 2376 tanh : $1/\sqrt[3]{23}$
3378 4542 ln : $4-3\sqrt{3}/2$
3378 4847 rt : (7,9,8,-4)
3378 6859 ℓ_{10} : $1/2+3\sqrt{5}/4$
3379 0314 2^x : $\sqrt[3]{2}/3$
3379 3371 rt : (1,9,6,4)
3379 3579 2^x : $3/2+4\sqrt{5}/3$
3379 3759 ζ : $7/15$
3379 4433 rt : (6,3,0,9)

3379 6134 $s\pi : P_{76}$
3379 6296 cu : $25/6$
3379 7652 rt : (7,2,-9,-3)
3379 8500 $10^x : \exp(-1/\pi)$
3379 9384 rt : (5,9,-8,-4)
3380 1092 ln : $3e/2 + 2\pi$
3380 1265 cr : $2\sqrt{2} - \sqrt{3}/4$
3380 2087 $J_1 : (\ln 2)^{-3}$
3380 2990 at : $3/2 + e$
3380 6734 rt : (9,7,-6,1)
3380 7262 $J_1 : 13/18$
3380 9095 $10^x : 1 + 4e$
3381 1347 rt : (8,9,-1,-1)
3381 2806 at : $\sqrt{3}/2 + 3\sqrt{5}/2$
3381 2894 id : $(e + \pi)^2$
3381 2965 at : $1/\sqrt[3]{23}$
3381 3117 rt : (7,6,-6,-2)
3381 3575 $10^x : P_{73}$
3381 4002 rt : (8,0,5,2)
3381 4457 rt : (7,6,9,-4)
3381 4914 rt : (1,6,-4,-2)
3381 6135 id : $\sqrt{3}/2 + 2\sqrt{5}$
3381 8485 ln : $7\zeta(3)/6$
3381 9214 $J_0 : 3\zeta(3)/2$
3381 9314 $e^x : 4 - \sqrt{2}/4$
3381 9626 $\ell_{10} : 3\sqrt{2}/4 + \sqrt{5}/2$
3382 0327 ln : $2 + 2e/3$
3382 0395 sr : $2 - 4\sqrt{2}/3$
3382 1843 rt : (3,5,-3,4)
3382 3480 at : $1/4 - 2\sqrt{5}$
3382 5356 $t\pi : 7(1/\pi)/6$
3382 5381 rt : (4,7,9,-4)
3382 8051 cu : $\sqrt{2} + \sqrt{3} - \sqrt{6}$
3382 8957 $\Gamma : 2\sqrt[3]{5}/9$
3382 9612 $\ln(5) - \exp(2/3)$
3383 2135 $J_0 : (e + \pi)^{1/3}$
3383 3263 rt : (5,5,-6,-5)
3383 3625 $\ell_2 : 3 - \sqrt{2}/3$
3383 5114 $J_0 : 5\sqrt[3]{3}/4$
3383 6305 rt : (5,1,5,-2)
3383 6474 cu : $1 + 4\sqrt{3}$
3383 6682 sq : $10 \ln 2/3$
3383 6827 rt : (1,8,9,-4)
3383 7091 $\ell_{10} : 3/2 + e/4$
3383 7351 rt : (4,7,-4,-2)
3383 8451 sq : $7\pi^2/3$
3384 0277 $\exp(1/5) + \exp(3/4)$
3384 1656 $2^x : 7\sqrt{5}/9$
3384 1690 rt : (8,4,-3,8)
3384 4420 sq : $3\sqrt{2}/4 + 4\sqrt{5}/3$
3384 5710 $10^x : (\sqrt[3]{4}/3)^{1/2}$
3384 5726 sq : $3 + 3\sqrt{3}/2$
3384 6630 rt : (7,9,1,2)
3384 7597 cu : $3e/5$
3384 8833 rt : (2,9,-9,-4)
3385 0027 $Ei : (\pi/2)^{1/3}$
3385 0267 $2^x : \exp(-\sqrt{3}/2)$
3385 3095 $10^x : 2 \ln 2$
3385 4768 rt : (2,5,-5,2)
3385 8947 $e^x : \sqrt{2}/4 + 4\sqrt{3}/3$
3385 9659 rt : (6,3,-3,5)
3385 9809 $\zeta : \sqrt{5/2}$
3386 0366 cu : $3/4 + \sqrt{3}/3$
3386 0839 rt : (2,4,7,7)
3386 1780 $J_1 : 8 \ln 2$
3386 2049 rt : (7,8,-4,-2)
3386 2312 $10^x : (\ln 2/3)^{-1/2}$
3386 2736 cr : $2/3 + \sqrt{3}$
3386 3171 $J : \exp(-11\pi/19)$
3386 3650 sinh : $(e + \pi)$
3386 3713 $2^x : 1/4 - 2e/3$
3386 4280 at : $3\pi^2/7$
3386 5672 $e^x : 7/24$
3386 6243 at : $\exp(\sqrt[3]{3})$
3386 6247 $2^x : 2/3 + \sqrt{5}/4$
3386 6443 rt : (5,3,-7,-8)
3387 2072 $10^x : 2/3 + 4\sqrt{2}$
3387 3688 $s\pi : 2\sqrt{3} - 3\sqrt{5}/2$
3387 5580 rt : (9,7,-9,-3)
3387 9130 $\exp(4/5) + \exp(\sqrt{2})$
3388 0798 $e^x : 1/3 + 2\pi$
3388 1855 $\ell_{10} : 11/24$
3388 2534 rt : (8,6,0,-1)
3388 4007 $\Psi : 15/8$
3388 4734 $J : 3(1/\pi)/2$
3388 5225 rt : (7,9,5,-3)
3388 6133 rt : (6,6,4,9)
3388 6435 $10^x : \pi/6$
3388 6573 rt : (8,3,6,2)
3388 6590 cr : $12/5$
3388 6592 $J_1 : 3e\pi/5$
3388 6864 rt : (4,6,-4,9)
3388 7968 rt : (1,6,4,-3)
3389 0410 $2^x : 8/19$
3389 2615 $s\pi : \sqrt{3}/4 + 3\sqrt{5}/4$
3389 3257 $\zeta : (2\pi)^{1/2}$
3389 5729 rt : (3,5,-6,-3)
3389 7252 at : $2 + \sqrt{5}$
3389 8134 tanh : $6/17$
3389 9674 rt : (8,4,-6,4)
3390 1394 rt : (3,9,-1,8)
3390 3077 sq : $2\sqrt{2} - 3\sqrt{3}/4$
3390 3179 ln : $((e + \pi)/3)^2$
3390 5895 $J_1 : 3$
3390 6275 sq : $4e + 2\pi$
3390 8794 $s\pi : 3\sqrt[3]{2}/2$
3391 1276 $J_1 : 7(e + \pi)/8$
3391 1725 $1/4 \times \exp(\sqrt{5})$
3391 1773 rt : (5,8,1,-1)
3391 2306 sq : $7e\pi/4$
3391 2615 $10^x : \sqrt{3} + \sqrt{6} - \sqrt{7}$
3391 3146 $e^x : 2\sqrt{3}/3 - \sqrt{5}$
3391 6116 rt : (2,5,-8,-8)
3391 6465 cu : $7e$
3391 6502 rt : (5,2,6,2)
3391 8141 rt : (4,4,-5,-2)
3391 8534 sq : $2\sqrt{2} - \sqrt{5}/3$
3392 1458 sinh : $(\ln 2)^3$
3392 2448 rt : (8,7,1,8)
3392 2759 $t\pi : 1/4 - 3\sqrt{5}/2$
3392 5970 sr : $1 + 2\sqrt{5}$
3392 6120 rt : (3,1,0,-9)
3392 6158 $2^x : 3\sqrt{2}/4 + 4\sqrt{3}/3$
3392 7622 rt : (7,6,-9,-6)
3392 8539 $10^x : 4\sqrt{2} + 2\sqrt{3}/3$
3392 8660 $J_1 : 3\sqrt[3]{5}$
3392 8944 id : $9\sqrt[3]{2}$
3392 9190 $2^x : 3/4 - 4\sqrt{3}/3$
3392 9261 at : $6/17$
3392 9673 $e^x : 1/3 - \sqrt{2}$
3393 0167 rt : (7,6,6,-3)
3393 0769 $\exp(\sqrt{2}) - s\pi(e)$
3393 1896 at : $3\sqrt{2}$
3393 2559 $\ell_2 : 4 + 3\sqrt{2}/4$
3393 6602 sr : P_{78}
3393 7509 sr : $\sqrt{2} - 3\sqrt{3}/4$
3394 0686 sinh : $19/5$
3394 0981 $\Gamma : (e/2)^3$
3394 1151 rt : (4,6,-7,2)
3394 1541 at : $3e/4 - 2\pi$
3394 4386 rt : (2,8,0,-1)
3394 6227 rt : (6,3,-6,1)
3394 7155 rt : (4,7,6,-3)
3394 8408 rt : (3,9,-4,-2)
3394 9225 rt : (1,5,-2,-1)
3394 9975 rt : (2,1,6,2)
3395 0952 as : $(\ln 2)^3$
3395 1920 rt : (7,5,-5,-2)
3395 2309 tanh : $\sqrt{2}/4$
3395 2534 sq : $1/3 - 2\sqrt{5}/3$
3395 2747 $t\pi : P_{51}$
3395 4055 sinh : $1/3$
3395 6029 rt : (5,0,-3,8)
3395 6273 cr : $5\sqrt[3]{3}/3$
3395 8281 $J_2 : 5e/7$
3396 0188 $J : 2/23$
3396 0305 $\Psi : 3\sqrt[3]{3}/10$
3396 0515 $\Psi : 4e$
3396 2146 $e^x : 2\sqrt{3} + 3\sqrt{5}$
3396 2962 rt : (5,7,-8,7)
3396 3044 cr : $2\zeta(3)$
3396 3303 tanh : $10(1/\pi)/9$
3396 3636 ln : $\sqrt{2}/4 + 2\sqrt{3}$
3396 4685 $e^x : 17/20$
3396 5172 rt : (1,8,6,-3)
3396 6752 $2^x : (4/3)^{-3}$
3396 8563 $E : 13/18$
3397 0209 $\exp(3) + s\pi(\sqrt{3})$
3397 0565 at : $17/4$
3397 0745 sinh : $7\sqrt[3]{2}/8$
3397 1326 $J_1 : \exp(\sqrt[3]{5})$
3397 2473 rt : (8,4,-9,0)
3397 5113 $3 \times \exp(\sqrt{2})$
3397 5438 rt : (7,1,-6,7)
3397 8522 id : $e/8$
3397 8802 rt : (8,3,1,-1)
3397 9892 cr : $3e\pi/2$
3397 9902 rt : (7,9,-2,-2)
3398 1010 rt : (1,9,0,-1)
3398 2414 ln : $1/4 + 2\sqrt{3}/3$
3398 2456 rt : (6,6,1,5)
3398 3184 rt : (7,3,7,-3)
3398 3690 at : $\sqrt{2}/4$
3398 4315 rt : (2,9,-6,-3)
3398 4375 cu : $7\sqrt[3]{2}/8$
3398 4917 sr : $4\pi/7$
3398 5388 rt : (8,6,7,2)
3398 5777 $J_2 : \sqrt[3]{22/3}$
3398 7494 sq : $1/3 - 4\sqrt{2}$
3398 7688 rt : (1,6,6,-6)
3398 8091 at : $2\sqrt{2}/3 - 3\sqrt{3}$
3398 8704 sr : $3 + 3e$
3398 8749 cu : $10\sqrt{5}$
3398 8899 sr : $\ln 2/6$
3398 9049 rt : (9,2,-9,6)
3399 3751 rt : (2,7,-7,-8)
3399 4735 at : $10(1/\pi)/9$
3399 6203 sr : $3\sqrt{2}/2 + 3\sqrt{5}/2$
3399 7922 $t\pi : 4\pi^2/5$
3399 8641 $J_0 : 9/5$
3399 9813 cr : $3/2 + e/3$
3400 1002 $\ell_{10} : 4 - 2e/3$
3400 1092 $\ln(\pi) - \ln(\sqrt{5})$
3400 1647 rt : (8,7,-2,4)
3400 3457 rt : (4,4,7,-3)
3400 4352 cr : $e/2 + \pi/3$
3400 5199 $J_0 : 10\sqrt[3]{2}/7$
3400 5785 $\Gamma : 3(e + \pi)/7$
3400 7221 $\exp(4/5) \div s\pi(2/5)$
3400 8839 rt : (9,8,8,-4)
3400 9116 $t\pi : e/4 + 3\pi$
3401 1261 $\exp(2) + s\pi(2/5)$
3401 1834 $J_1 : 8/11$
3401 1924 $\Gamma : \exp(3/2)$
3401 4356 $\ln(4/5) - \exp(3/4)$
3401 6065 $J_1 : \exp(-1/\pi)$
3401 7618 $J_1 : (\sqrt[3]{4}/3)^{1/2}$
3401 9505 rt : (2,9,3,7)
3401 9795 rt : (5,4,1,-1)
3402 1101 cr : $1/4 + 4\pi$
3402 3302 rt : (6,9,8,9)
3402 5083 rt : (1,5,7,-3)
3402 6032 $1/5 \div s\pi(1/5)$
3402 6092 cu : $2\pi/9$
3402 6865 $t\pi : 4\sqrt{2}/9$
3402 7110 $J_0 : 7\sqrt[3]{2}/2$
3402 7746 rt : (5,5,7,2)
3402 7777 sq : $7/12$
3402 9966 $e^x : 1 - \sqrt{2}/2$
3403 0537 rt : (4,1,-6,-2)
3403 1747 rt : (1,6,-2,1)
3403 2444 at : $3 - 3\sqrt{5}/2$
3403 3030 rt : (4,9,3,9)
3403 3065 $E : 1/\ln(4)$
3403 6972 $\ln(2/3) - s\pi(\sqrt{3})$
3403 7371 $e^x : (\sqrt[3]{5}/3)^{-1/3}$
3403 8421 rt : (3,8,4,4)
3403 8712 $2^x : 1 - \sqrt{3}/3$
3403 9643 $\ell_{10} : \sqrt[3]{21/2}$
3404 1416 $E : 3\zeta(3)/5$

3404 1844 $\ell_2 : 2\sqrt{2}+\sqrt{5}$
3404 2317 rt : (8,1,-9,-9)
3404 2370 rt : (7,9,2,-2)
3404 2461 rt : (1,8,-2,7)
3404 5265 rt : (5,0,-6,4)
3404 6596 $J_2 : \sqrt{3}-\sqrt{5}+\sqrt{6}$
3404 6707 rt : (7,0,8,3)
3404 7187 $e^x : 3-3e/2$
3404 8208 rt : (2,6,-7,-3)
3404 8498 $10^x : 11/21$
3404 9396 $e^x : 9e/2$
3404 9429 rt : (7,1,-9,3)
3404 9469 $E : \sqrt[3]{3}/2$
3405 0805 $\sqrt{3}\times s\pi(e)$
3405 2509 $10^x : 2+e/2$
3405 3121 sinh : $(\sqrt{5}/3)^{-1/3}$
3405 3217 rt : (9,5,9,-4)
3405 3610 $\Psi : \sqrt{12}\pi$
3405 3649 $Ei : 7/24$
3405 5041 $2/3\times\ln(3/5)$
3405 5485 ln : $e/4+\pi$
3405 7631 $10^x : 8\sqrt{5}/7$
3405 9596 sr : $3/4+\pi/3$
3406 1310 $\Psi : \exp(\sqrt[3]{2}/2)$
3406 4009 $2\div\exp(2/5)$
3406 5208 $2^x : \sqrt[3]{19/2}$
3406 5902 rt : (3,7,-8,7)
3406 6532 rt : (6,3,-9,-3)
3406 7124 $e^x : 9e/5$
3406 7636 $10^x : 2(1/\pi)/5$
3406 9263 $1/5+\exp(\pi)$
3406 9719 rt : (6,6,9,-4)
3407 1334 rt : (4,1,8,-3)
3407 1873 $J_0 : (2e/3)^2$
3407 2022 sq : 22/19
3407 3000 rt : (5,7,-7,-3)
3407 4327 at : $e\pi/2$
3407 4924 cr : $\sqrt[3]{14}$
3407 6066 $Ei : -\sqrt{2}+\sqrt{3}+\sqrt{5}$
3407 6265 rt : (2,4,7,2)
3407 6611 rt : (9,5,-5,6)
3407 7943 $1\div s\pi(\sqrt{3})$
3407 8471 rt : (2,9,0,-6)
3407 8667 rt : (7,4,-2,7)
3407 9702 rt : (7,2,-6,-2)
3408 1560 ln : $1/2+e/3$
3408 1776 rt : (5,3,1,8)
3408 4084 sq : $4e/3-2\pi/3$
3408 6010 $2^x : 3\sqrt{3}/4+\sqrt{5}/2$
3408 6885 cu : $7\pi^2/9$
3408 7027 rt : (3,7,9,-4)
3409 1476 $t\pi : \sqrt{2}+\sqrt{5}+\sqrt{7}$
3409 2457 as : P_7
3409 5683 $\Psi : 2\ln 2/7$
3409 6001 rt : (3,3,-6,8)
3409 6965 $2^x : 1/3-4\sqrt{2}/3$
3409 7798 rt : (1,2,8,-3)
3409 8501 $J_2 : \exp(2/3)$
3409 8964 $\ell_{10} : 4\sqrt{2}-2\sqrt{3}$
3409 9467 $e^x : 2\sqrt{3}+3\sqrt{5}/2$
3409 9654 rt : (8,8,-7,-3)
3410 0674 at : $5\sqrt[3]{5}/2$
3410 1646 at : $3\sqrt{3}/2+3\sqrt{5}/4$
3410 1975 $2^x : 1+\sqrt{5}/2$
3410 3691 $\ell_2 : 15/19$
3410 5201 rt : (8,9,-4,0)
3410 5386 $Li_2 : 5/16$
3410 6561 ln : $6\sqrt{3}$
3410 8723 $\ell_{10} : 2\pi^2/9$
3411 0200 $J_0 : 23/7$
3411 2072 rt : (6,2,-9,-6)
3411 6390 rt : (8,0,2,1)
3411 7381 $J_2 : 9\sqrt{3}/8$
3411 7946 $10^x : 1/\ln(5/2)$
3411 8364 rt : (7,6,3,-2)
3411 8662 $\ell_2 : (\ln 2/3)^3$
3411 9026 rt : (6,6,-2,1)
3412 0358 rt : (1,6,-1,-1)
3412 1912 rt : (4,9,0,2)
3412 4265 rt : (7,3,9,3)
3412 4313 rt : (3,9,-9,-4)
3412 6223 $J : \exp(-e)$
3412 6321 rt : (2,9,4,-6)
3412 6632 $e^x : 3/4+4e$
3412 7274 $10^x : 7\ln 2/5$
3412 7874 $Ei : \exp(\sqrt{2}\pi/2)$
3412 8226 sr : $1/2+3\sqrt{3}/4$
3412 8252 rt : (8,9,8,2)
3412 9275 rt : (8,2,-5,9)
3412 9288 at : $3\sqrt{3}/4+4\sqrt{5}/3$
3412 9986 sr : P_{77}
3413 0309 $\ln(\sqrt{3})\div\ln(5)$
3413 0423 sq : $((e+\pi)/2)^{-1/2}$
3413 2168 $E : 18/25$
3413 2468 rt : (1,3,-7,2)
3413 2519 $10^x : 11/4$
3413 5481 $E : 4\sqrt[3]{2}/7$
3413 5834 $\Gamma : 8\zeta(3)$
3413 6599 $K : 21/23$
3413 7179 rt : (2,8,5,2)
3413 9203 cr : $(1/\pi)/8$
3414 3374 at : $2-2\pi$
3414 3821 $J_1 : 11/2$
3414 6863 $erf : 5/16$
3414 8238 rt : (4,7,3,-2)
3414 8607 $s\pi : 7e\pi/2$
3414 8647 cu : $3e/4-\pi$
3414 8725 $\zeta : 5/2$
3414 8872 $\sqrt{3}+\ln(5)$
3414 9026 $\zeta : 1/\ln(\pi)$
3415 0376 cr : $1+\sqrt{2}$
3415 0963 $t\pi : 9e\pi/5$
3415 0998 $Ei : (\pi/2)^{-1/2}$
3415 1551 $s\pi : P_{43}$
3415 2231 $t\pi : (\sqrt{2}/3)^3$
3415 4854 rt : (4,2,9,3)
3415 5966 $J_1 : 7\pi/4$
3415 7163 rt : (5,4,-8,-3)
3415 7958 rt : (9,8,5,-3)
3415 8727 rt : (9,5,-8,2)
3415 9265 $1/5+\pi$
3415 9875 sr : $10\sqrt[3]{2}/7$
3416 0329 $10^x : 3/2+e$
3416 0491 sr : $5\pi^2/9$
3416 0752 sr : 6π
3416 2553 cu : $3\sqrt{2}/4+\sqrt{5}/2$
3416 2573 $J_1 : (\sqrt[3]{5}/2)^2$
3416 3244 $e^x : 4e+\pi/4$
3416 4078 id : $3\sqrt{5}/5$
3416 4166 $e^x : 2+3\sqrt{2}/4$
3416 6366 $t\pi : e/4-\pi$
3416 8399 $Li_2 : \beta(2)$
3416 9087 ln : $\sqrt{2}/3+3\sqrt{5}/2$
3417 1850 $t\pi : \sqrt{2}-4\sqrt{3}/3$
3417 2043 rt : (2,7,-8,-3)
3417 3795 at : $e+\pi/2$
3417 8083 rt : (5,1,2,-1)
3417 9201 cu : $2/3+2e$
3417 9739 rt : (6,9,5,5)
3418 0072 rt : (7,4,-5,3)
3418 0572 rt : (1,8,3,-2)
3418 1826 sq : $4\pi^2/5$
3418 3028 rt : (3,3,-9,1)
3418 3091 rt : (4,7,-1,-1)
3418 3917 at : $2/3+4e/3$
3418 4807 $\ln(3)\times\exp(1/5)$
3418 6410 $\Gamma : 4\pi/5$
3418 6666 sinh : $8(1/\pi)$
3418 6764 $J_1 : 7\sqrt[3]{5}/4$
3418 6845 $\ln(2)+\exp(1/2)$
3418 8026 rt : (1,1,9,3)
3418 8094 cr : $8e/9$
3418 8781 $2^x : 3\sqrt{2}/10$
3419 0636 rt : (8,5,-8,-3)
3419 1085 rt : (9,8,-1,6)
3419 1958 $\ell_{10} : 3(e+\pi)/8$
3419 3078 rt : (5,8,8,2)
3419 4176 $e^x : 5/17$
3419 5446 ln : 9π
3419 8456 rt : (4,2,-7,2)
3419 9129 cu : $4/3-2\sqrt{2}$
3419 9187 $\ell_{10} : \sqrt{2}/2+2\sqrt{5}/3$
3419 9518 id : $\sqrt[3]{5}/5$
3420 0058 $2^x : 1/4+2\sqrt{5}/3$
3420 2014 $s\pi : 1/9$
3420 2347 $Ei : \sqrt{19}\pi$
3420 2652 $2^x : 4(1/\pi)/3$
3420 2752 rt : (3,8,1,-3)
3420 3301 cr : $3\sqrt{3}/4+\sqrt{5}/2$
3420 3943 $J_0 : \pi^2/3$
3420 4084 $2^x : \exp(3/2)$
3420 4849 rt : (7,9,-5,-6)
3420 5994 $\Gamma : \ln 2/4$
3420 7838 $J : (3/2)^{-3}$
3420 9305 rt : (8,2,-8,5)
3421 2881 rt : (7,3,4,-2)
3421 3447 $\ell_{10} : 9\sqrt[3]{5}/7$
3421 4108 at : $2\sqrt{2}/3+3\sqrt{5}/2$
3421 4285 rt : (5,3,-2,4)
3421 4471 $e^x : 3\sqrt{2}/2$
3421 4769 $4\times\exp(3)$
3421 5696 $1/3-s\pi(\sqrt{5})$
3421 5698 rt : (2,9,-3,-2)
3421 6855 id : $2e-2\pi/3$
3421 8260 at : $2-3\pi/4$
3421 9184 $J_1 : \ln(\sqrt[3]{3}/3)$
3421 9533 sinh : $1/\ln(\sqrt{2}\pi/3)$
3421 9916 sq : $3e/2+2\pi$
3422 0350 $e^x : \sqrt{2}+\sqrt{3}/4$
3422 0530 $t\pi : \sqrt{2}-\sqrt{5}/2$
3422 1328 at : $2\sqrt{2}/3-3\sqrt{3}/4$
3422 1353 rt : (7,7,2,7)
3422 2253 $J_1 : (e+\pi)/8$
3422 3627 $\zeta : \sqrt[3]{2/3}$
3422 3750 sq : $\sqrt{3}/4+3\sqrt{5}/2$
3422 4143 rt : (2,0,-9,3)
3422 4791 $\ell_{10} : 7\pi$
3422 5032 at : $3+3\sqrt{3}/4$
3422 5154 rt : (9,5,6,-3)
3422 5679 $J_0 : 22/5$
3422 5841 $J_2 : (e+\pi)/3$
3422 5947 sinh : $\exp(\sqrt{3}/2)$
3422 6631 $Ei : \ln(\sqrt{2}\pi/2)$
3422 7807 $2/3\div\exp(2/3)$
3422 8124 sr : $3\sqrt{2}+4\sqrt{3}$
3422 9607 $\ell_{10} : 4/3+\sqrt{3}/2$
3423 2522 $J_1 : 5\pi^2/9$
3423 2756 $\zeta : 23/19$
3423 2921 rt : (2,9,-9,0)
3423 5279 $\Gamma : 2(e+\pi)/3$
3423 6444 rt : (4,9,-3,-5)
3423 6627 $Ei : 7/15$
3423 6864 $J_1 : 3\sqrt[3]{5}/7$
3423 8807 rt : (6,1,-5,6)
3424 0502 rt : (8,5,-1,9)
3424 1665 $\ln(3/5)\div\exp(2/5)$
3424 1679 id : $\sqrt{2}+4\sqrt{3}$
3424 2268 $\ell_{10} : 22$
3424 4299 $e^x : (e/2)^2$
3424 5404 ln : $1+2\sqrt{2}$
3424 5979 cu : $4/3+3e/4$
3424 6200 rt : (8,7,-8,-4)
3424 6548 tanh : $1/\sqrt[3]{22}$
3424 7497 $2^x : 2\sqrt{3}-\sqrt{5}$
3424 7961 $J_2 : 8\sqrt[3]{5}/7$
3424 9693 $\ell_2 : 3-\sqrt{3}$
3425 0794 sinh : $7\pi^2/4$
3425 0803 rt : (8,8,8,-4)
3425 1600 rt : (7,1,-8,-8)
3425 2192 cu : $3/2+3\pi/4$
3425 3429 rt : (6,9,-8,9)
3425 4094 $\exp(3/4)+\exp(4/5)$
3425 6361 $Li_2 : \ln(5/2)$
3425 6467 $J_1 : 11/15$
3425 7263 rt : (7,8,-1,-1)
3426 0407 $2^x : e+\pi/4$
3426 1529 $J_0 : 7\pi/5$
3426 1914 sinh : $4e/5$
3426 3636 rt : (5,1,-9,-3)
3426 5759 id : $7e/3$

3426 6669 e^x : 17/7
3426 7539 e^x : 14/3
3426 7890 as : $2\sqrt{3}/3 - 2\sqrt{5}/3$
3426 8833 sr : $5\sqrt[3]{3}/4$
3426 9490 tanh : 5/14
3427 0189 rt : (2,7,8,2)
3427 0740 sr : $(e+\pi)^{1/3}$
3427 1075 rt : (5,9,8,-4)
3427 3448 Ψ : $(1/\pi)/6$
3427 3547 e^x : $1/2 - \pi/2$
3427 6098 ℓ_2 : $4e/3 - 3\pi/4$
3427 7225 rt : (7,1,-7,2)
3427 7778 Γ : 11/2
3427 7993 J_0 : $4\pi/7$
3427 8042 rt : (3,7,6,-3)
3427 8104 at : $2 + 4\sqrt{3}/3$
3427 8127 id : $3\sqrt{2}/2 - 2\sqrt{3}$
3427 9013 sr : $3\zeta(3)/2$
3427 9339 at : $1/\sqrt[3]{22}$
3428 0296 e^x : $3/2 + 2\sqrt{5}$
3428 1745 θ_3 : $(\pi)^{-1/2}$
3428 2961 J_0 : $\exp(\sqrt{2}\pi/3)$
3428 3695 rt : (4,0,-2,7)
3428 4082 2^x : $\exp(-\sqrt[3]{5}/2)$
3428 4268 e^x : $\sqrt{2}/3 + 4\sqrt{3}$
3428 4274 2^x : $1/\sqrt[3]{13}$
3428 6139 rt : (5,2,-5,-9)
3429 2308 rt : (3,6,-9,-6)
3429 3552 cu : $5\sqrt[3]{2}/9$
3429 9717 sr : 2/17
3430 0000 cu : 7/10
3430 0577 J_2 : $7\sqrt{5}/8$
3430 2096 rt : (9,8,-4,2)
3430 2394 at : 5/14
3430 3863 $s\pi$: $8\sqrt{5}$
3430 4447 Li_2 : $\pi/10$
3430 4498 rt : (9,2,7,-3)
3430 5010 J_1 : $3\sqrt{3}$
3430 5367 $s\pi$: $3\sqrt{2} - 3\sqrt{5}/2$
3430 5440 at : $4/3 + 4\sqrt{5}/3$
3430 5733 $\exp(3/4) - s\pi(e)$
3430 6451 cr : $3 - \sqrt{3}/3$
3430 6612 J : $2\ln 2/7$
3430 7033 rt : (8,6,-3,0)
3431 0425 ℓ_{10} : $\sqrt{2}/3 + \sqrt{3}$
3431 0922 $\pi \times s\pi(\sqrt{3})$
3431 1904 2^x : $4/3 + \pi/4$
3431 2626 rt : (8,5,9,-4)
3431 2891 $t\pi$: $\sqrt{3}/3 - 2\sqrt{5}$
3431 3021 rt : (7,4,-8,-1)
3431 4403 e^x : $e/2 + 3\pi/2$
3431 4575 sq : $2 - \sqrt{2}$
3431 6583 erf : $\pi/10$
3431 7702 Li_2 : $2\sqrt{2}/9$
3431 7820 $\exp(2) \times s\pi(1/5)$
3431 8366 rt : (5,8,-7,4)
3431 9508 cu : $1/4 - 3\pi/4$
3432 0554 J_2 : $\sqrt[3]{15/2}$
3432 2015 rt : (7,9,-1,-1)
3432 2555 Ψ : $(\sqrt{5}/3)^{-1/2}$
3432 2683 sr : $4 + 2\sqrt{5}/3$
3432 3311 $t\pi$: $(3/2)^{-3}$
3432 4697 J_2 : $1/\ln(5/3)$
3432 4999 rt : (1,9,5,-4)
3432 5109 rt : (4,7,-3,6)
3432 5894 ℓ_2 : $4e + 3\pi$
3432 6190 θ_3 : $(\sqrt[3]{5}/3)^{1/2}$
3432 6863 e^x : $(\ln 3)^2$
3432 7216 rt : (6,6,-5,-3)
3432 7755 at : $2\sqrt{2} + 2\sqrt{5}/3$
3432 7871 erf : $2\sqrt{2}/9$
3432 8229 sinh : $\sqrt[3]{4}$
3432 8590 J_0 : $3(e+\pi)/4$
3432 9681 rt : (8,3,3,1)
3433 0043 $t\pi$: 2/19
3433 0276 cr : $3e + 3\pi/2$
3433 2218 $t\pi$: $(\sqrt[3]{2}/3)^{-1/2}$
3433 3405 rt : (9,3,-7,7)
3433 3784 rt : (7,0,5,2)
3433 4646 $\exp(1/3) + \exp(2/3)$
3433 5167 rt : (6,3,7,-3)
3433 5862 rt : (5,6,9,-4)
3433 6558 rt : (3,6,1,8)
3433 6757 rt : (2,9,-9,-9)
3433 7956 rt : (3,9,-6,-3)
3433 7972 $s\pi$: $\ln(\sqrt{5}/2)$
3433 8128 sq : $(e+\pi)/10$
3433 8138 ℓ_{10} : $7\sqrt[3]{2}/4$
3434 0532 sr : $4/3 + \sqrt{2}/3$
3434 3220 id : $3\sqrt{3}/2 + \sqrt{5}/3$
3434 4482 id : $2\zeta(3)/7$
3434 5767 rt : (8,5,-4,5)
3434 7489 cr : $7\sqrt{3}/5$
3434 7599 e^x : $8(e+\pi)/5$
3434 7757 10^x : $3 - 2\sqrt{3}$
3434 9388 10^x : $3\pi^2/4$
3435 0027 at : $4/3 - 4\sqrt{2}$
3435 3815 tanh : $9(1/\pi)/8$
3435 4318 2^x : $2 + 3\sqrt{2}/4$
3435 4534 sq : $3e\pi$
3435 4962 ln : $\pi^2/7$
3435 5745 rt : (1,7,-6,-7)
3435 8198 Li_2 : 11/12
3435 9866 J_1 : $8(e+\pi)/9$
3436 0287 rt : (2,7,9,-4)
3436 3527 rt : (6,1,-8,2)
3436 3569 id : $3e + 4\pi/3$
3436 3733 rt : (1,6,-8,2)
3436 4146 $s\pi(\pi) + s\pi(e)$
3436 6043 cr : $2 + 4e$
3436 6406 at : $3\sqrt[3]{3}$
3436 7752 ln : $\sqrt{11}\pi$
3436 7944 rt : (3,5,8,9)
3436 8613 cr : $7\ln 2/2$
3436 8763 rt : (7,7,-1,3)
3436 9824 at : $1/4 + 3e/2$
3436 9978 rt : (4,2,-9,-6)
3437 0152 sr : $5\sqrt{5}$
3437 0541 θ_3 : $\sqrt[3]{5}/10$
3437 1764 id : $1/3 - 3\sqrt{5}/4$
3437 3181 at : $1/\ln(\sqrt[3]{2})$
3437 3474 ln : 6/23
3437 4914 at : $3/2 + 2\sqrt{2}$
3437 4989 cu : $3/4 + \sqrt{5}/3$
3437 5507 Ψ : 22/19
3437 7600 Ψ : $\sqrt[3]{20/3}$
3437 8178 rt : (2,6,-2,1)
3437 8668 sinh : $\sqrt{3} + \sqrt{6} + \sqrt{7}$
3437 9630 e^x : $1/2 + \sqrt{2}/2$
3438 1307 rt : (7,2,-4,8)
3438 1898 $\sqrt{2} \div \exp(\sqrt{2})$
3438 2334 ℓ_{10} : $3/2 + \sqrt{2}/2$
3438 3524 at : $5\sqrt{3}/2$
3438 3589 rt : (4,1,5,-2)
3438 5676 ℓ_{10} : $e/6$
3438 5801 E : $(e)^{-1/3}$
3438 6002 rt : (1,8,-9,-4)
3438 7108 10^x : $\sqrt{2} + \sqrt{3}/4$
3438 7133 at : $9(1/\pi)/8$
3438 7734 cr : $3/4 + 3\sqrt{5}/4$
3438 7777 rt : (8,8,3,9)
3438 8100 sr : $3\sqrt{2}/4 + \sqrt{5}/3$
3438 8437 rt : (5,7,-4,-2)
3438 9122 2^x : $3\sqrt{2}/2 + 3\sqrt{3}/2$
3438 9507 e^x : $1/\ln(\sqrt[3]{3})$
3439 0126 rt : (9,8,2,-2)
3439 2274 sinh : 11/2
3439 4050 rt : (3,8,-5,6)
3439 4052 cu : $3/4 + \sqrt{2}/4$
3439 4580 J_1 : $7\sqrt{5}/3$
3439 4979 id : $2(e+\pi)/5$
3439 6230 J_1 : $(5/2)^{-1/3}$
3439 6963 rt : (1,9,-3,2)
3439 7669 J_1 : 14/19
3439 8225 $t\pi$: P_{57}
3439 8486 $\exp(4) - s\pi(\sqrt{3})$
3439 9693 rt : (9,1,8,3)
3439 9747 at : 13/3
3440 0237 cr : $1/3 + 2\pi/3$
3440 0495 erf : $\sqrt[3]{2}/4$
3440 0854 J : $2\sqrt[3]{5}/9$
3440 3026 Li_2 : $\sqrt[3]{2}/4$
3440 3065 rt : (5,8,-2,-5)
3440 6003 Γ : $9\sqrt{5}/8$
3440 7263 $t\pi$: 10/21
3440 7606 sq : $4\sqrt{2}/3 - 3\sqrt{3}/4$
3440 9123 ℓ_{10} : $3/2 - \pi/3$
3440 9638 ℓ_2 : $2 + 3e$
3441 0116 E : $9(1/\pi)/4$
3441 0195 rt : (1,8,4,-6)
3441 0666 10^x : $\sqrt{2}/2 + \sqrt{5}/3$
3441 2207 J_1 : $\sqrt{3}\pi$
3441 3115 rt : (4,9,-9,-4)
3441 4690 rt : (6,4,-1,6)
3441 5791 cr : 17/7
3441 7122 rt : (1,5,4,5)
3441 8546 ln : $1 + 3\pi$
3441 9275 2^x : $1/2 - 3e/4$
3442 4097 $4/5 \times s\pi(\pi)$
3442 5084 id : $(e/3)^{-3}$
3442 5910 J_0 : $7\sqrt{2}/3$
3442 6395 $t\pi$: $\sqrt{3}/2 + 3\sqrt{5}/4$
3442 6442 rt : (6,9,2,1)
3442 7454 cr : $4 - \pi/2$
3442 9377 J_1 : $2e$
3443 0009 cu : $3/2 - 2\sqrt{2}$
3443 0042 J_2 : $3e/2$
3443 3354 rt : (5,2,3,1)
3443 4869 rt : (4,7,-6,-1)
3443 7054 rt : (6,0,8,3)
3443 8824 $t\pi$: $2\sqrt{5}/5$
3443 8923 rt : (1,2,5,-2)
3443 9510 id : $1/4 + 2\pi/3$
3444 0805 rt : (3,6,-7,-3)
3444 1592 cu : $2 - 3\sqrt{3}/4$
3444 2462 sq : $7\pi/5$
3444 4609 rt : (8,8,-4,-2)
3444 4768 sinh : 21/19
3444 5600 rt : (9,8,-7,-2)
3444 5632 $s\pi$: $\exp(1/(3\pi))$
3444 8347 $t\pi$: $3e + \pi$
3445 0192 2^x : $3\sqrt{2}/4 - 3\sqrt{3}/2$
3445 0212 rt : (5,1,-1,9)
3445 0217 10^x : $2\sqrt{2}/3 + 2\sqrt{5}/3$
3445 3489 $3/4 + \ln(2/3)$
3445 3736 rt : (1,6,5,8)
3445 4766 rt : (4,0,-5,3)
3445 7298 rt : (8,8,5,-3)
3445 7771 E : $(\sqrt{5}/2)^{-3}$
3445 8790 cu : $4 + 2e/3$
3445 8835 J_2 : $5\pi/8$
3445 9545 e^x : $2/3 - \sqrt{3}$
3445 9737 J_1 : $5\pi/3$
3445 9987 E : $((e+\pi)/3)^{-1/2}$
3446 5021 E : $(\ln 3/3)^{1/3}$
3446 5476 rt : (3,1,-9,3)
3446 5809 2^x : $4/3 - e/3$
3446 8762 ℓ_2 : $4\sqrt{2} - \sqrt{3}/3$
3446 8901 $t\pi$: P_{38}
3446 8977 rt : (1,8,3,-8)
3446 9362 rt : (9,6,-3,7)
3446 9975 ℓ_{10} : $\exp(\sqrt[3]{4}/2)$
3447 1160 e^x : $\sqrt{2} - \sqrt{5}/2$
3447 1236 rt : (1,7,4,-1)
3447 2149 Γ : $\sqrt{3}/10$
3447 2988 $1/5 + \ln(\pi)$
3447 3604 sr : $7\pi/4$
3447 3857 ℓ_2 : $5\sqrt[3]{2}/8$
3447 5326 rt : (3,8,-8,-4)
3447 5960 rt : (7,6,0,-1)
3447 6933 $t\pi$: $\pi/5$
3447 7297 rt : (3,1,8,-3)
3447 8293 rt : (8,5,-7,1)
3447 9485 at : $1 - e/2$
3447 9666 $\pi \div s\pi(1/5)$
3447 9950 at : $8e/5$
3448 0134 rt : (4,5,-8,2)
3448 1488 rt : (6,7,-7,-3)
3448 2798 Ψ : $3\sqrt[3]{3}$
3448 3159 erf : 6/19

3448 4048 ln : $17/24$
3448 5090 cr : $\sqrt{5}-\sqrt{6}+\sqrt{7}$
3448 6857 e^x : $(3/2)^{-3}$
3448 8124 rt : (5,9,5,-3)
3448 9061 2^x : $-\sqrt{3}-\sqrt{6}+\sqrt{7}$
3448 9541 2^x : $\sqrt[3]{5}/4$
3448 9588 J_1 : $17/23$
3449 2295 rt : (3,6,-2,1)
3449 2915 sr : $2\sqrt{2}/3+\sqrt{3}/2$
3449 3499 2^x : $1/4+3\sqrt{5}$
3449 3750 J : $\sqrt{5}/8$
3449 4239 rt : (7,3,6,2)
3449 4980 J_1 : $\ln(2\pi/3)$
3449 6898 J_0 : $4\pi^2/9$
3449 7679 cr : $2+\sqrt{3}/4$
3449 9702 $\sqrt{2}\times s\pi(2/5)$
3450 0243 Li_2 : $6/19$
3450 0822 rt : (2,9,7,-2)
3450 1397 J_1 : $21/4$
3450 1533 at : $4+\sqrt{2}/4$
3450 2213 rt : (7,8,8,-4)
3450 2318 ln : $2/3+\sqrt{5}/3$
3450 3128 rt : (9,5,3,-2)
3450 4215 rt : (7,2,-7,4)
3450 4282 at : $1+3\sqrt{5}/2$
3450 4950 Γ : $\sqrt{19/3}$
3450 5775 $\ln(3/5)\times s\pi(\sqrt{5})$
3450 6576 10^x : $\ln(\sqrt[3]{2}/2)$
3450 6831 Γ : $19/6$
3450 7045 cr : $2\sqrt{2}/3+2\sqrt{5}/3$
3451 1667 e^x : $8e/7$
3451 2018 J_1 : $9\zeta(3)/2$
3451 3171 rt : (1,8,-3,-1)
3451 4761 at : $2+3\pi/4$
3451 6154 rt : (9,4,9,3)
3451 7521 Ψ : $1/\ln(\sqrt[3]{2})$
3451 8424 θ_3 : $\zeta(3)/7$
3451 9416 tanh : $2\sqrt[3]{2}/7$
3452 0736 at : $3\sqrt{2}/2+\sqrt{5}$
3452 0787 id : $\sqrt{11/2}$
3452 1403 tanh : $9/25$
3452 2876 at : $4\sqrt{2}-3\sqrt{3}/4$
3452 3435 $\exp(1/3)\times s\pi(\sqrt{2})$
3452 5277 rt : (6,1,-7,-7)
3452 5495 rt : (9,8,-7,-3)
3452 6269 ℓ_{10} : $\ln(\pi/2)$
3452 7215 cr : $1/3+4\pi$
3452 7915 erf : $1/\sqrt{10}$
3452 8292 at : $\sqrt{19}$
3452 8393 e^x : $4\pi/5$
3452 8425 $t\pi$: $e/4-\pi/4$
3452 8899 rt : (4,9,8,-4)
3452 9501 at : $3+e/2$
3452 9565 ℓ_{10} : $3-\pi/4$
3452 9946 id : $3/2-2\sqrt{3}/3$
3453 2048 ℓ_{10} : $3\sqrt{3}-4\sqrt{5}/3$
3453 2372 rt : (8,8,0,5)
3453 2756 sinh : $23/7$
3453 3489 e^x : $2\sqrt{2}+4\sqrt{3}$
3453 4479 J_1 : $(\sqrt[3]{5}/3)^{-3}$
3453 4576 cu : $e/4+2\pi/3$
3453 8085 $\ln(2)+\exp(\sqrt{3})$
3454 1038 e^x : $2\sqrt{2}+4\sqrt{3}/3$
3454 1043 rt : (4,7,0,-1)
3454 2448 rt : (9,1,-9,8)
3454 4444 ℓ_{10} : $3\sqrt{2}/4+2\sqrt{3}/3$
3454 6281 Ei : $3e/7$
3454 8106 rt : (6,6,3,-2)
3454 8878 E : $5/7$
3454 9150 $s\pi(1/5)\times s\pi(4/5)$
3454 9414 sr : $3(1/\pi)/8$
3455 1390 sinh : $3\pi^2/7$
3455 1699 rt : (8,5,6,-3)
3455 2921 Li_2 : $1/\sqrt{10}$
3455 3561 at : $2\sqrt[3]{2}/7$
3455 3578 rt : (7,5,0,8)
3455 4820 $s\pi$: $(3\pi)^{1/3}$
3455 5558 at : $9/25$
3455 6834 10^x : $\sqrt{3}/2$
3455 6909 rt : (1,7,-9,-3)
3455 7135 J_1 : $9(e+\pi)/10$
3455 7435 rt : (4,5,-9,3)
3456 0773 sr : $3/4+3\sqrt{2}/4$
3456 2739 rt : (6,3,9,3)
3456 4470 10^x : $4e\pi/3$
3456 7637 sq : $2-2\pi$
3456 7963 rt : (3,3,-8,-3)
3456 9513 J_1 : $\sqrt{2}+\sqrt{3}+\sqrt{5}$
3456 9863 e^x : $3/4-2e/3$
3456 9935 $t\pi$: $\sqrt{2}/3-\sqrt{3}/3$
3457 0203 10^x : $2/3+\sqrt{5}$
3457 0551 Ψ : $5\sqrt{3}/2$
3457 0610 rt : (5,4,-5,-2)
3457 0675 J_1 : $(\ln 3/2)^{1/2}$
3457 0930 tanh : $\sqrt[3]{3}/4$
3457 1650 rt : (4,7,-9,-8)
3457 1735 sq : $4/3-\sqrt{5}/3$
3457 3835 rt : (7,7,-4,-1)
3457 5644 $\sqrt{3}-\ln(4)$
3457 5745 tanh : $3\zeta(3)/10$
3457 8182 Ψ : $3\pi/5$
3457 8691 rt : (7,5,-2,-1)
3458 1643 $1/3\div s\pi(\sqrt{2})$
3458 3087 rt : (7,5,9,-4)
3458 3482 $2/5+s\pi(\sqrt{3})$
3458 8241 rt : (5,9,9,8)
3458 9773 J_1 : $\exp(5/3)$
3459 0019 2^x : $3/7$
3459 5825 J_1 : $4\sqrt{5}/3$
3459 6329 rt : (1,6,2,-8)
3459 6892 rt : (9,6,-6,3)
3460 2076 sq : $10/17$
3460 4370 ℓ_{10} : $3\zeta(3)/8$
3460 5336 at : $\sqrt[3]{3}/4$
3460 5368 ln : $3/2+4\sqrt{5}$
3460 5416 at : $3/4+4e/3$
3460 6762 id : $2\sqrt{3}-\sqrt{5}/2$
3460 8807 rt : (8,3,-6,6)
3461 0174 at : $3\zeta(3)/10$
3461 0814 cr : $\sqrt{2}/2+\sqrt{3}$
3461 2040 J_1 : $5e\pi/8$
3461 2560 J_1 : $16/3$
3461 3421 rt : (3,2,9,3)
3461 3584 2^x : $3/2+4\pi$
3461 3824 rt : (4,6,9,-4)
3461 3996 rt : (5,1,-4,5)
3461 4419 rt : (1,8,0,-1)
3461 4779 ℓ_{10} : $1/3+4\sqrt{2}/3$
3461 6982 J_1 : $3\sqrt{3}/7$
3461 7528 id : $(2e/3)^{1/2}$
3461 8756 rt : (7,0,-6,9)
3461 8990 2^x : $(2e)^{-1/2}$
3461 9348 rt : (6,4,-8,-3)
3461 9569 id : $3\sqrt{3}/2-4\sqrt{5}$
3462 1281 sq : $1/2+2e/3$
3462 3712 id : $1/3-e/4$
3462 4210 Ψ : $4\sqrt{2}/3$
3462 4913 cr : $2e\pi/7$
3462 5354 $t\pi$: $(1/\pi)/3$
3462 7154 $\ln(\sqrt{5})\times s\pi(\pi)$
3462 7190 rt : (4,2,-4,-8)
3462 8266 ζ : $19/15$
3462 8634 at : $1/3-3\pi/2$
3463 0065 rt : (7,4,-8,2)
3463 3627 J_1 : $7(1/\pi)/3$
3463 4363 cu : $4\sqrt{2}/3+\sqrt{3}$
3463 4430 rt : (5,9,-3,8)
3463 5768 e^x : $\exp(-1/(2\pi))$
3463 6009 ℓ_2 : $\sqrt{2}/4+\sqrt{3}/4$
3463 6026 rt : (1,7,-2,4)
3463 6133 sinh : $e/8$
3463 7679 rt : (3,4,-5,5)
3463 8522 J_1 : $(2e/3)^{-1/2}$
3463 9685 id : $3\pi/7$
3464 0483 rt : (1,1,6,2)
3464 1016 id : $\sqrt{3}/5$
3464 1655 sq : $4-2\sqrt{2}/3$
3464 3756 rt : (9,2,4,-2)
3464 4386 cr : $3-\sqrt{5}/4$
3464 6218 rt : (1,7,9,-4)
3464 6663 J_0 : $4\sqrt{5}/5$
3464 7920 rt : (9,9,1,7)
3464 8226 ℓ_{10} : $9\pi^2/4$
3464 8288 J_2 : $\sqrt[3]{23/3}$
3464 9014 2^x : $2-\pi/2$
3465 0023 rt : (8,5,-5,-2)
3465 2679 2^x : $\sqrt{2}/2-\sqrt{5}$
3465 3743 cu : $1/3+3e/4$
3465 3764 Ψ : $13/3$
3465 5136 θ_3 : $13/23$
3465 5187 Γ : $(\sqrt{2}\pi/2)^{-1/2}$
3465 7359 id : $\ln 2/2$
3465 7442 $t\pi$: $1/4-3\pi/4$
3465 7479 ℓ_2 : $2\sqrt{3}+3\sqrt{5}$
3465 9194 $\ln(2/5)+s\pi(\pi)$
3466 0193 ζ : $(e/3)^3$
3466 2759 10^x : $4\sqrt{2}+4\sqrt{3}/3$
3466 3869 rt : (8,2,7,-3)
3466 5426 at : $4\pi^2/9$
3466 6894 $\exp(1/3)+s\pi(2/5)$
3466 8063 id : $1/\sqrt[3]{24}$
3466 8806 rt : (1,9,3,0)
3466 8853 as : $e/8$
3467 0692 sr : $\zeta(3)/10$
3467 2971 e^x : $-\sqrt{2}+\sqrt{5}+\sqrt{6}$
3467 3281 e^x : $3\sqrt{2}+\sqrt{3}$
3467 5051 rt : (6,7,3,6)
3467 5575 rt : (9,5,-8,-3)
3467 5909 Γ : $3\sqrt{5}/10$
3467 7952 cr : $3/2+2\sqrt{2}/3$
3467 8205 cr : $10\sqrt[3]{5}/7$
3467 8748 ℓ_{10} : $9/20$
3467 9622 at : $2e-\pi/3$
3468 0415 rt : (2,8,-9,-4)
3468 1092 J_1 : $(e/3)^3$
3468 1164 rt : (6,6,-8,-7)
3468 2463 rt : (5,4,3,9)
3468 7141 cu : $9\ln 2/2$
3468 7512 rt : (7,3,1,-1)
3468 9628 rt : (9,4,-5,8)
3468 9932 cr : $7\pi/9$
3469 0031 10^x : $13/3$
3469 1803 $t\pi$: $\sqrt[3]{10/3}$
3469 2092 2^x : $16/13$
3469 2540 ℓ_2 : $3/4+3\pi$
3469 2655 rt : (5,3,-8,-4)
3469 5779 J_2 : $\pi^2/5$
3469 7878 rt : (2,9,0,-1)
3469 9740 rt : (4,0,-8,1)
3470 0987 rt : (6,3,4,-2)
3470 1891 e^x : $(3\pi/2)^{1/3}$
3470 3334 ln : $8\sqrt[3]{3}/3$
3470 4463 $t\pi$: $3\sqrt{2}/4+\sqrt{5}$
3470 4546 rt : (3,6,-5,-6)
3470 5661 sq : $\exp(-\sqrt[3]{4}/3)$
3470 6924 at : $3(e+\pi)/4$
3470 7286 rt : (8,6,4,1)
3470 7929 rt : (5,3,7,-3)
3470 8000 cr : $22/9$
3470 8021 rt : (2,6,-8,-5)
3470 8143 rt : (3,9,-3,-2)
3470 9481 $\ln(\pi)-\exp(2/5)$
3471 0525 at : $\sqrt{2}+4\sqrt{5}/3$
3471 0650 rt : (1,9,-8,-8)
3471 1114 sq : $9(e+\pi)/2$
3471 1502 cr : $4\sqrt{2}/3+\sqrt{5}/4$
3471 3089 ℓ_{10} : $\sqrt[3]{11}$
3471 3393 rt : (4,9,-6,-3)
3471 5062 id : $7\sqrt{3}/9$
3471 6581 rt : (5,9,-9,-4)
3471 8060 at : $\exp(\sqrt{2}\pi/3)$
3471 8277 rt : (7,6,7,2)
3472 0362 e^x : $4e/9$
3472 2109 rt : (8;8,-3,1)
3472 2713 Γ : $1/\ln(\sqrt{2}\pi)$
3472 3026 rt : (7,5,-3,4)
3472 3274 at : $7\pi/5$
3472 3744 rt : (6,2,-3,7)
3472 4608 Ei : $7\zeta(3)/4$
3472 5192 rt : (3,2,-8,9)

3472 5382 $e^x : \sqrt{3}/4 - 2\sqrt{5}/3$
3472 6091 at : $4\sqrt{3}/3 - 3\sqrt{5}$
3472 7300 $erf : 7/22$
3472 9147 $\Psi : 5\ln 2/3$
3472 9635 rt : (3,9,0,-9)
3473 0032 $\zeta : 7\sqrt[3]{2}/8$
3473 1647 rt : (9,7,8,-4)
3473 1972 at : 22/5
3473 3226 at : $3\sqrt{3}/2 - \sqrt{5}$
3473 3331 $Ei : ((e+\pi)/3)^{-1/3}$
3473 4217 ln : $3\ln 2/8$
3473 5186 sπ : $2\sqrt{2}/3 + 4\sqrt{5}$
3473 5521 id : $3e/2 - 3\pi$
3473 7221 $Ei : 2(e+\pi)/5$
3473 7502 sr : $\sqrt{3}/2 - \sqrt{5}/3$
3473 7568 $J_0 : (\ln 3/2)^{-2}$
3473 7913 tanh : $1/\sqrt[3]{21}$
3473 8977 $J_1 : \sqrt{5}/3$
3473 9654 $Ei : 4/5$
3474 0220 at : $e - 3\pi/4$
3474 0359 $erf : (1/\pi)$
3474 0952 ln : $9\sqrt[3]{5}/4$
3474 1042 ln : $\sqrt{3}/2 + 4\sqrt{5}/3$
3474 1298 sr : $e/3 - \pi/4$
3474 2296 $\lambda : \exp(-15\pi/13)$
3474 4898 cr : $9e/10$
3474 5317 id : $4\sqrt{2} - 4\sqrt{3}/3$
3474 6289 tπ : $9\sqrt{3}/10$
3474 7979 rt : (5,9,-6,1)
3474 9502 rt : (6,3,-9,-8)
3475 1301 rt : (7,0,-9,5)
3475 3458 rt : (8,3,-9,2)
3475 3688 $\ln(\pi) \times \exp(e)$
3475 5230 rt : (2,4,7,-3)
3475 5378 $\zeta : 10\sqrt{5}/9$
3475 6092 rt : (9,6,-9,-1)
3475 8046 $10^x : 3\sqrt{3}/2$
3476 0521 sr : $1/3 + 4e$
3476 3810 rt : (6,8,8,-4)
3476 4013 sinh : $4\sqrt[3]{3}/3$
3476 5757 $10^x : 7e/3$
3476 5864 at : $2\sqrt{2}/3 + 2\sqrt{3}$
3476 8913 $\theta_3 : \exp(-\sqrt[3]{5}/3)$
3477 0509 $\ell_{10} : 9\sqrt{3}/7$
3477 0691 rt : (7,8,5,-3)
3477 0999 $\ell_{10} : P_{41}$
3477 3174 at : $1/\sqrt[3]{21}$
3477 3404 $2/5 + \exp(2/3)$
3477 8569 $Ei : (e/2)^{1/2}$
3477 9631 at : $7\sqrt[3]{2}/2$
3477 9876 at : $4e/3 + \pi/4$
3477 9956 rt : (1,9,-4,-7)
3478 1038 rt : (8,8,2,-2)
3478 2608 id : 8/23
3478 3436 $\theta_3 : 1/\ln(e+\pi)$
3478 3673 rt : (3,9,-2,5)
3478 4049 $\ell_2 : (\ln 3/3)^3$
3478 4679 at : $1/3 + 3e/2$
3478 5042 $Ei : 7\sqrt{3}/8$
3478 5293 $\Gamma : 6e\pi/5$
3478 7634 rt : (2,3,-1,7)
3478 7971 $Li_2 : 7/22$
3478 8438 at : $3e - 4\pi$
3479 0786 $J_0 : \sqrt{11}$
3479 1089 sinh : 25/13
3479 2330 $\ell_2 : 11/14$
3479 3388 rt : (7,8,4,8)
3479 4816 $\ell_{10} : \pi/7$
3479 6822 rt : (6,1,-9,-3)
3479 6999 cr : $4 + 4\sqrt{5}$
3479 7531 $e^x : 1/2 + \sqrt{2}/4$
3479 7599 rt : (3,9,8,8)
3479 8344 rt : (9,8,-1,-1)
3479 8613 $e^x : 8e/3$
3479 8689 $3 \div \exp(4/5)$
3479 9057 sinh : $7e/8$
3479 9514 rt : (8,1,8,3)
3479 9690 $\ell_2 : 5\sqrt{2}/9$
3479 9723 rt : (5,7,-9,5)
3480 0061 $\Psi : \exp(-\sqrt[3]{2}/3)$
3480 0615 sr : $\sqrt[3]{6}$
3480 1569 at : $3 + \sqrt{2}$
3480 2003 $\Psi : 2\sqrt{3}/3$
3480 2200 id : $5\pi^2$
3480 2489 sq : $1/2 + 4e$
3480 2950 tπ : $8e/7$
3480 3380 rt : (1,9,-9,2)
3480 3387 $Li_2 : (1/\pi)$
3480 3984 rt : (7,3,-2,9)
3480 4247 at : $3\sqrt{2}/4 + 3\sqrt{5}/2$
3480 6169 rt : (1,8,6,-4)
3480 7621 id : $1/4 - 3\sqrt{3}/2$
3480 9246 rt : (3,5,-7,-7)
3481 0819 $J : 3\ln 2/7$
3481 1693 rt : (8,6,-2,6)
3481 4718 $\Gamma : 2\sqrt[3]{2}$
3481 5003 $1/4 - \exp(4)$
3481 6546 $\Psi : (\sqrt{2}\pi/2)^3$
3481 6569 rt : (8,1,-8,7)
3481 8966 rt : (4,5,7,2)
3481 9638 rt : (4,2,-7,-4)
3482 0679 $10^x : \sqrt{2}/2 - \sqrt{3}/3$
3482 1044 $\theta_3 : \sqrt{3}/10$
3482 1182 rt : (9,9,-2,3)
3482 1806 $J_2 : 8\sqrt{3}/7$
3482 3842 rt : (9,1,5,2)
3482 4137 rt : (9,4,-8,4)
3482 5518 rt : (3,7,-9,4)
3482 5534 $\theta_3 : 2\sqrt{2}/5$
3482 8465 rt : (9,4,9,-4)
3482 9186 $J_0 : 25/14$
3483 0669 ln : 12/17
3483 1130 $J_2 : 7\sqrt{2}/5$
3483 2269 sπ : $5\sqrt{3}/3$
3483 2777 sπ : $\exp(\sqrt{2})$
3483 3147 rt : (5,9,2,-2)
3483 5049 cr : $4/3 + \sqrt{5}/2$
3483 5638 $\Gamma : 4/19$
3483 7727 $\theta_3 : \ln 2/4$
3483 9972 sr : 20/11
3484 0839 $2^x : 4e + 3\pi/4$
3484 1301 tanh : 4/11
3484 3412 sr : $9\sqrt{2}/7$
3484 9393 rt : (1,6,-3,-5)
3484 9970 id : $4(e+\pi)/7$
3485 0387 $\ell_2 : \pi/4$
3485 0475 cr : $\sqrt{17}\pi$
3485 1159 $\Psi : 17/9$
3485 1417 rt : (6,4,-7,-2)
3485 3195 sinh : $(e/2)^{1/3}$
3485 3287 $\ell_2 : \ln(\sqrt{2}\pi/3)$
3485 3337 $\Psi : 10\pi^2/9$
3485 3562 rt : (2,7,-6,6)
3485 3761 rt : (5,0,8,3)
3485 4160 tanh : $8(1/\pi)/7$
3485 5375 $erf : \sqrt{5}/7$
3485 7584 $e^x : \sqrt{8}\pi$
3486 1218 rt : (4,6,-7,-3)
3486 1228 $1/4 + \ln(3)$
3486 3853 $10^x : (\ln 3/2)^{-1/2}$
3486 5553 rt : (6,5,9,-4)
3486 5913 sinh : $8\zeta(3)/5$
3486 6274 tπ : $2\sqrt{2}/3 + 3\sqrt{5}/2$
3486 6742 $\zeta : \zeta(7)$
3486 8410 rt : (9,2,-7,9)
3486 9740 rt : (6,9,-1,-1)
3487 0095 sinh : $\sqrt[3]{5}/5$
3487 0269 ln : $10\pi/3$
3487 0407 $\ell_{10} : 1/2 + \sqrt{3}$
3487 1516 rt : (9,2,-9,-3)
3487 7100 at : 4/11
3487 7440 rt : (9,7,-1,8)
3487 7494 rt : (7,7,-7,-5)
3487 9367 $J_1 : (\sqrt[3]{2}/3)^{1/3}$
3488 1783 rt : (4,8,-2,3)
3488 1868 at : $3\sqrt{2}/2 + 4\sqrt{3}/3$
3488 2475 $Ei : 8\sqrt{5}/7$
3488 2738 sq : $1/\ln(2e)$
3488 2961 rt : (5,9,-9,-6)
3488 3209 ln : $9\sqrt[3]{2}/8$
3488 3455 $e^x : 7\sqrt[3]{2}/6$
3488 5347 sπ : P_{37}
3488 9421 rt : (1,5,1,-1)
3489 0026 at : $8(1/\pi)/7$
3489 0310 $1/4 \times \exp(1/3)$
3489 1085 $\ell_2 : 3 + 2\pi/3$
3489 2252 $\sqrt{3} \div \exp(1/4)$
3489 2968 at : $4 + \sqrt{3}/4$
3489 4838 ln : $3\sqrt{2}/2 + \sqrt{3}$
3489 5276 cr : $1/3 + 3\sqrt{2}/2$
3489 5311 rt : (7,5,6,-3)
3489 6175 $e^x : e\pi/10$
3489 7705 $\Psi : \exp(-\sqrt{2})$
3489 8119 tπ : $\exp(-\sqrt{5})$
3489 8504 rt : (5,1,-6,-6)
3489 9286 rt : (6,0,5,2)
3490 0808 rt : (4,1,0,8)
3490 1742 $2^x : -\sqrt{3} - \sqrt{5} + \sqrt{6}$
3490 3929 as : $\sqrt[3]{5}/5$
3490 4678 rt : (3,6,9,-4)
3490 4934 $\ell_2 : e/3 + 4\pi/3$
3490 5448 $10^x : 1/4 - \sqrt{2}/2$
3490 5766 rt : (3,9,-5,-5)
3490 5821 $\Gamma : \ln((e+\pi)/3)$
3490 6585 id : $\pi/9$
3490 7316 $Ei : 23/9$
3490 8342 rt : (2,5,-3,9)
3491 0147 rt : (3,6,-4,-2)
3491 0150 at : $1 - 2e$
3491 2534 rt : (2,1,8,-3)
3491 2599 $erf : 8/25$
3491 3802 ln : $1/2 + 3\sqrt{5}/2$
3491 4498 sq : $2/3 + \sqrt{3}/2$
3491 6364 $\Psi : 15/13$
3491 7045 tπ : $2\sqrt{2} - \sqrt{5}$
3491 7355 sq : 13/22
3491 8407 tπ : $9(e+\pi)/8$
3491 9343 rt : (6,2,-6,3)
3491 9779 rt : (5,5,4,1)
3491 9823 rt : (4,6,-5,7)
3491 9940 $\Psi : 3\sqrt[3]{2}/2$
3492 0910 at : $1/4 + 4\pi/3$
3492 0935 rt : (7,7,-7,-3)
3492 1166 cr : $4\sqrt{3} - 2\sqrt{5}$
3492 2591 $\Psi : 4\sqrt[3]{3}/5$
3492 4360 $J_1 : 3/4$
3492 5092 id : $8e/5$
3492 5107 id : $(\ln 3/2)^{-1/2}$
3492 8142 rt : (5,4,0,5)
3493 0614 $1/5 - \ln(\sqrt{3})$
3493 1835 $\sqrt{5} + \exp(\sqrt{2})$
3493 2458 $e^x : 2\pi^2/3$
3493 3555 id : $5e\pi/2$
3493 6886 sr : $1/4 + \pi/2$
3493 7080 rt : (1,4,7,2)
3493 7496 cr : $4e + 2\pi/3$
3493 7540 rt : (2,9,2,3)
3493 7979 $10^x : 7\sqrt[3]{5}/8$
3493 9276 $Li_2 : \sqrt{5}/7$
3494 0661 at : $\sqrt{2}\pi$
3494 1689 rt : (8,5,3,-2)
3494 2741 $2/5 \div \ln(\pi)$
3494 4274 rt : (4,6,6,7)
3494 6028 rt : (7,9,-4,0)
3494 6576 $e^x : 4\sqrt{2} - 3\sqrt{5}$
3494 6656 sq : $4\sqrt{3}/3 - 3\sqrt{5}$
3494 7969 cu : $2e/3 - \pi$
3494 8500 $\ell_{10} : \sqrt{5}$
3495 0573 sπ : $\sqrt{2} + 2\sqrt{5}$
3495 4328 $\ln(\sqrt{5}) - \exp(e)$
3495 4522 $Ei : 6\pi^2/7$
3495 5185 rt : (8,2,-6,-2)
3495 5926 rt : (6,0,-5,8)
3495 6415 at : $3/4 - 3\sqrt{3}$
3495 7868 $e^x : P_{79}$
3495 8038 sr : $2/3 + 2\sqrt{3}/3$
3495 8168 rt : (2,7,-9,-7)
3496 0420 $2^x : 8/3$
3496 0937 cu : $7\sqrt[3]{5}/8$
3496 3509 $erf : 2\sqrt[3]{3}/9$

3496 5598 $\theta_3 : 4/23$
3496 5825 rt : (6,7,0,2)
3496 7172 rt : (8,4,9,3)
3496 7654 cu : $1/3 + 3\sqrt[3]{3}/4$
3496 8081 ln : $3/2 + 3\pi/4$
3496 8564 $t\pi : 4\sqrt{2} + \sqrt{5}$
3496 9331 rt : (6,9,-9,7)
3497 0332 sq : $3e - \pi/2$
3497 1969 id : $4\sqrt{2}/3 + 2\sqrt{3}$
3497 3172 $Ei : (\sqrt[3]{4})^{1/3}$
3497 3377 $2^x : 3\sqrt[3]{3}/10$
3497 3830 $t\pi : 9e/5$
3497 4487 sq : $\sqrt{2}/4 + \sqrt{3}$
3497 5579 $J_1 : 5\zeta(3)/8$
3497 5586 $\ell_{10} : 2/3 + \pi/2$
3497 5616 $e^x : 3\sqrt{5}/4$
3497 6218 $2^x : 8(e + \pi)/5$
3497 6473 rt : (1,5,-1,-8)
3497 6935 $E : 17/24$
3497 8754 at : $3\sqrt{3} - \sqrt{5}/3$
3497 9636 rt : (7,3,-5,5)
3498 1407 rt : (8,8,-6,-3)
3498 1486 $\Psi : 9e\pi/7$
3498 3460 rt : (3,1,5,-2)
3498 4029 $e^x : 4 - \pi/2$
3498 4524 $e^x : 4 + 2e/3$
3498 5880 $e^x : 3/10$
3498 8531 $\ell_2 : 4\sqrt{2} - \sqrt{5}/4$
3498 9812 rt : (3,8,-9,-4)
3499 1990 sr : $4/3 + 4\pi/3$
3499 4887 $t\pi : \exp(e/2)$
3499 5982 rt : (1,8,-4,-9)
3499 6265 rt : (6,7,-4,-2)
3499 6671 rt : (1,7,6,-3)
3499 6843 rt : (8,7,8,-4)
3499 6991 $J_0 : \sqrt[3]{17/3}$
3499 7266 rt : (7,0,2,1)
3499 8752 $10^x : (e/2)^2$
3500 0000 id : 7/20
3500 1031 $t\pi : 3\ln 2/7$
3500 1231 rt : (6,5,1,7)
3500 4956 cr : $3/4 - \sqrt{2}/2$
3500 4987 $2^x : \sqrt{3}/4$
3500 5285 $J_2 : 8\sqrt{5}/9$
3500 5509 $\zeta : (e/3)^{1/3}$
3500 5565 $J_1 : \ln(\sqrt{2}/3)$
3500 6635 $\ln(2) \times \exp(2/3)$
3500 6915 rt : (5,9,6,4)
3500 6961 $Li_2 : 8/25$
3500 8647 $t\pi : \exp(\sqrt{5}/3)$
3500 8872 rt : (8,6,-5,2)
3501 0422 $\exp(1/4) \div s\pi(2/5)$
3501 0700 $2^x : P_{10}$
3501 1054 $Ei : 7/6$
3501 3696 $\ell_{10} : 4/3 + e/3$
3501 4311 cr : $9\sqrt[3]{3}$
3501 6794 rt : (2,6,-1,-1)
3501 6882 sinh : $9\sqrt[3]{5}/8$
3501 8063 id : $(1/(3e))^{1/2}$
3501 9682 cu : 21/19
3501 9689 $\ln(4) - s\pi(\sqrt{2})$
3502 1985 $\ell_2 : \sqrt{13/2}$
3502 3656 sinh : $2\zeta(3)/7$
3502 3858 rt : (9,5,0,-1)
3502 4105 $2^x : 5\ln 2/8$
3502 4182 rt : (8,4,-4,7)
3502 6174 rt : (1,6,-4,-8)
3502 7451 $e^x : 1/4 - 3\sqrt[3]{3}/4$
3502 8976 rt : (1,8,6,9)
3503 1403 $10^x : 3/23$
3503 3793 $J_0 : \sqrt{19}$
3503 4324 at : $1/4 - 3\pi/2$
3503 5762 rt : (5,3,9,3)
3503 6270 rt : (6,9,-9,-4)
3503 6425 rt : (5,8,8,-4)
3503 7963 at : $2\sqrt{2}/3 - \sqrt{3}/3$
3503 8269 sq : $\sqrt{2}/2 - 3\sqrt{3}/4$
3503 8603 $\ell_{10} : 3/4 + 2\sqrt{5}/3$
3503 8947 $e^x : 4\sqrt{2} + 3\sqrt{5}/4$
3503 9783 $Ei : 5(e + \pi)/7$
3504 0440 $\zeta : (\ln 3)^{1/2}$
3504 1711 rt : (7,8,1,4)
3504 2510 at : $1 + 2\sqrt{3}$
3504 3033 $\Gamma : (\sqrt{5})^{-1/2}$
3504 3233 rt : (3,7,-9,-8)
3504 4412 rt : (7,2,7,-3)
3504 4989 id : $4\sqrt{2}/3 - \sqrt{5}$
3504 5808 rt : (9,9,-5,-1)
3504 6352 rt : (4,3,-8,-3)
3504 7415 cr : $4\sqrt{2}/3 + \sqrt{3}/3$
3504 8740 $1/3 \div s\pi(2/5)$
3505 0129 rt : (9,7,5,-3)
3505 1382 rt : (7,6,2,9)
3505 1726 sr : $\sqrt{2} - \sqrt{5} + \sqrt{7}$
3505 2907 rt : (2,8,-6,-3)
3505 3659 rt : (9,7,-4,4)
3505 5311 $e^x : \zeta(3)/4$
3505 5599 rt : (7,9,8,2)
3505 7221 sr : $1/3 + 2\sqrt{5}/3$
3505 8238 as : $2\zeta(3)/7$
3505 8439 rt : (9,5,-3,9)
3506 0599 $\ell_{10} : 2\sqrt{2}/3 + 3\sqrt{3}/4$
3506 1229 $2^x : 3\sqrt{3} - 3\sqrt{5}$
3506 1608 $2^x : 3/4 + 4e$
3506 2932 $2^x : 1/2 + 4\sqrt{2}$
3506 3354 rt : (9,4,6,2)
3506 4388 $E : \sqrt{2}/2$
3506 5859 $\Psi : 8e/5$
3506 6702 tanh : $\ln(\sqrt[3]{3})$
3506 7219 $Li_2 : 2\sqrt[3]{3}/9$
3506 8288 cu : $e/3 + 4\pi$
3506 9592 $t\pi : 3\sqrt{3}/2 - 2\sqrt{5}/3$
3507 5412 at : $2e/3 - 2\pi$
3507 7349 sq : $3/2 + 3\pi$
3507 7351 at : $4 + \sqrt{2}/3$
3507 7807 sinh : $8\ln 2/5$
3507 8105 rt : (4,8,-5,-4)
3507 8322 $s\pi : 7(1/\pi)/2$
3508 0834 at : $2\sqrt{5}$
3508 1561 id : $3\sqrt{2}/2 - 2\sqrt{5}$
3508 5029 $J_0 : \exp(\sqrt{3}/3)$
3508 7275 $Ei : 5\sqrt[3]{3}/9$
3508 7289 as : $4/3 - 3\sqrt{5}/4$
3508 7941 at : $1/2 - \sqrt{3}/2$
3508 8410 $\zeta : 7(1/\pi)/2$
3508 8935 sq : $2\sqrt{2} - \sqrt{5}$
3508 9552 $\ell_{10} : \ln(3\pi)$
3509 0163 $J_1 : (\ln 3)^{-3}$
3509 0861 rt : (8,9,2,6)
3509 1955 $2^x : 3\sqrt{2}/2$
3509 1980 id : $\exp(-\pi/3)$
3509 3385 rt : (9,8,-4,-2)
3509 3784 tanh : $\ln(\ln 2)$
3509 4112 at : $4/3 + \pi$
3509 6020 $\theta_3 : 3\sqrt[3]{2}/5$
3509 6661 cu : $4/3 - 3e/4$
3509 7770 sr : $\sqrt{2}/2 + \sqrt{5}/2$
3509 8058 $2 \times s\pi(\sqrt{5})$
3509 9459 rt : (6,8,5,-3)
3510 0004 $3 \times \exp(3/4)$
3510 0894 rt : (5,8,-5,9)
3510 2183 $\ell_{10} : 5\pi/7$
3510 3698 at : $\ln(\sqrt[3]{3})$
3510 3994 rt : (4,3,7,-3)
3510 5651 $2/5 + s\pi(2/5)$
3510 5903 cu : $9\sqrt{5}/4$
3510 6675 cr : $\sqrt[3]{15}$
3510 8343 cr : $1/\ln(3/2)$
3510 9344 rt : (1,5,-3,-6)
3511 0976 rt : (2,2,9,3)
3511 1111 sq : 23/15
3511 1511 rt : (5,3,-8,-7)
3511 3454 rt : (5,9,-6,-3)
3511 4100 $4 \times s\pi(1/5)$
3511 4308 $\ln(2/3) \times s\pi(1/3)$
3511 4487 rt : (3,5,-1,9)
3511 4510 cu : $1/\ln(\sqrt{2}/3)$
3511 5220 rt : (8,4,9,-4)
3511 7027 $\Gamma : 3\sqrt[3]{2}/10$
3511 7917 $s\pi : 7e/9$
3511 8597 $e^x : 4\sqrt{2} + 3\sqrt{5}/2$
3512 0015 sr : $\sqrt{10/3}$
3512 0340 $t\pi : 1/4 + \pi/3$
3512 0719 rt : (6,6,0,-1)
3512 0987 cu : $e/2 + 2\pi$
3512 1066 $Ei : 2\zeta(3)/3$
3512 1322 $\Gamma : 6\ln 2$
3512 1656 rt : (5,1,-9,-2)
3512 2307 $s\pi : \sqrt{8\pi}$
3512 2458 $J_1 : (\sqrt[3]{5}/3)^{1/2}$
3512 3124 rt : (7,4,-8,-3)
3512 4628 $\Gamma : 1/\sqrt{7}$
3512 4976 at : $3/2 + 4\sqrt{5}/3$
3512 6233 at : $\exp(3/2)$
3512 6856 rt : (3,7,2,5)
3512 7335 $\ell_2 : 3\sqrt{2}/2 + 4\sqrt{5}/3$
3512 7818 cu : $4\sqrt{2} + 3\sqrt{3}$
3512 7872 $2 - \exp(1/2)$
3512 8384 sq : $(\pi/2)^{1/3}$
3512 8520 $\ell_{10} : 3/2 + \sqrt{5}/3$
3513 0926 at : $\ln(\ln 2)$
3513 3132 $\Psi : 7\pi/2$
3513 3468 id : $\sqrt[3]{13}$
3513 4610 id : $\exp(\sqrt[3]{5}/2)$
3513 5739 at : $4\sqrt{2}/3 + 3\sqrt[3]{3}/2$
3513 6839 rt : (6,2,-7,-9)
3513 7525 ln : 2/21
3513 7567 $\ln(3/4) \times \exp(1/5)$
3514 0660 $\zeta : 7\sqrt{2}/4$
3514 2283 $J_0 : 10/3$
3514 3627 rt : (8,2,4,-2)
3514 8586 sr : $1/3 + 3\sqrt{3}$
3515 0164 $\zeta : \exp(e/3)$
3515 1435 $E : 12/17$
3515 1960 rt : (6,0,-8,4)
3515 2211 $\zeta : 8/17$
3515 6074 at : $10\pi/7$
3515 6186 rt : (1,8,-3,-2)
3515 7803 as : $4/3 - 4\sqrt{3}/3$
3515 8961 $10^x : 7(1/\pi)/6$
3515 9310 rt : (4,1,-2,-9)
3516 0023 $5 \div \exp(2/5)$
3516 1068 at : $\sqrt{2}/2 - 3\sqrt{3}$
3516 1104 rt : (5,5,9,-4)
3516 1161 $J_1 : 3\sqrt[3]{2}/5$
3516 1407 $\zeta : 10\sqrt{3}/7$
3516 3388 id : $1/\sqrt[3]{23}$
3516 5776 $s\pi : 4\sqrt{2}/3$
3516 7742 rt : (6,3,6,2)
3516 8288 rt : (1,4,7,-3)
3516 8943 at : $3 + 2\sqrt{5}/3$
3516 9108 rt : (5,7,-1,-1)
3516 9432 $2^x : \pi^2/8$
3516 9715 $t\pi : 3e - \pi/3$
3517 0328 $t\pi : 1/3 + \sqrt{5}/4$
3517 0688 $J_2 : 7\sqrt[3]{5}/6$
3517 0713 $2^x : 10/23$
3517 1992 cu : 12/17
3517 3980 $t\pi : \sqrt{2}/4 + 3\sqrt{3}$
3517 4265 rt : (7,1,-7,6)
3517 4983 $J : (\sqrt[3]{3}/3)^2$
3517 5258 $\Psi : 11$
3517 7412 sq : $9e/8$
3517 7558 rt : (6,2,-9,-1)
3517 8051 at : $1/4 + 3\sqrt{2}$
3517 8929 $\ell_2 : 1/3 + 2\sqrt{2}/3$
3517 9426 $e^x : 25/17$
3517 9702 $e^x : 4\sqrt{3}/3 - 3\sqrt{5}/2$
3518 3114 $\pi \times s\pi(\pi)$
3518 3663 $e^x : \sqrt{2} - \sqrt{5}/4$
3518 5502 rt : (3,3,-5,-2)
3518 7501 $t\pi : (e/2)^{1/3}$
3518 9894 cu : $4\sqrt{3}/3 + 3\sqrt{5}/4$
3519 0045 $e^x : 1/\sqrt{11}$
3519 1368 rt : (8,2,-6,8)
3519 1653 $2^x : 4\sqrt{3}/3 + \sqrt{5}$
3519 3920 rt : (1,1,-9,3)
3519 4806 rt : (7,8,2,-2)
3519 7849 rt : (8,9,-7,1)
3519 8816 rt : (9,4,6,-3)

3519 9671 $e^x : 7\sqrt[3]{5}/6$
3520 1428 cr : $2+\sqrt{2}/3$
3520 1923 rt : (7,3,-8,1)
3520 3518 $2^x : \sqrt{3}+3\sqrt{5}$
3520 5935 $2^x : 9(e+\pi)/7$
3520 8236 $\Gamma : 5\zeta(3)/9$
3520 9345 $\Gamma : 7\sqrt[3]{3}/4$
3520 9647 rt : (2,6,9,-4)
3521 0051 rt : (2,9,-8,6)
3521 1410 $10^x : 2\sqrt{2}/3+4\sqrt{3}/3$
3521 2738 at : 9/2
3521 3549 tanh : $\exp(-1)$
3521 4321 10^x : 17/8
3521 5412 $10^x : 2+2e$
3521 5838 sr : $2e/3+3\pi$
3521 5874 rt : (5,2,-2,6)
3521 6019 ln : $2e-\pi/2$
3521 6047 $J_0 : 8e/5$
3521 6189 at : $e/2+\pi$
3521 8251 ℓ_{10} : 4/9
3521 8964 sr : $3\sqrt{2}/4+2\sqrt{5}$
3521 9211 rt : (8,4,-7,3)
3522 0112 rt : (4,8,8,2)
3522 0428 tπ : $1/3+3e/4$
3522 0587 $e^x : (\ln 3/2)^2$
3522 1388 $\ell_2 : (\sqrt[3]{3}/3)^{1/3}$
3522 2695 ln : $3+\sqrt{3}/2$
3522 2868 at : $4\sqrt{2}-2\sqrt{3}/3$
3522 2964 $J_0 : 1/\ln(\sqrt[3]{5}/3)$
3522 4382 sπ : P_{54}
3522 5330 $\ell_{10} : e/4+\pi/2$
3522 6309 rt : (7,1,8,3)
3522 6463 rt : (3,8,-6,3)
3522 7275 sq : $\pi^2/2$
3522 7806 rt : (5,3,4,-2)
3522 8608 tπ : $(\sqrt{2}\pi)^{1/2}$
3522 9113 at : $e/4-\pi/3$
3522 9716 $e^x : 4+3e/4$
3523 0256 rt : (3,7,0,-1)
3523 1068 $Ei : 2(1/\pi)/5$
3523 1401 id : $9\zeta(3)/8$
3523 2114 rt : (9,5,-6,5)
3523 2586 rt : (6,3,-1,8)
3523 2821 sπ : $1/\ln(\sqrt{2})$
3523 5363 $\ell_{10} : 3\sqrt{2}/2-3\sqrt{5}/4$
3523 7927 rt : (2,5,-7,-3)
3523 8487 at : $2\sqrt{2}+3\sqrt{5}/4$
3523 9032 $\ell_{10} : 2\sqrt{2}-\sqrt{3}/3$
3523 9171 $2^x : 2+4e$
3524 2297 rt : (4,9,-3,-2)
3524 2560 rt : (2,8,-1,9)
3524 2613 ln : $3e+3\pi/4$
3524 2938 rt : (5,8,-8,2)
3524 4176 sinh : $9\sqrt{2}/8$
3524 8209 $10^x : 5\ln 2/4$
3524 9007 $e^x : 3\sqrt{3}+\sqrt{5}/2$
3525 0379 rt : (5,2,-7,2)
3525 1342 at : $\exp(-1)$
3525 1578 sq : $2(e+\pi)$
3525 1873 rt : (7,5,-9,-4)
3525 2114 ln : $2-\sqrt{3}/3$
3525 2308 $J_1 : \sqrt{2}-\sqrt{3}-\sqrt{7}$
3525 2866 rt : (1,7,-3,-3)
3525 3772 cu : $7\sqrt[3]{5}/9$
3525 4084 at : $2\sqrt{3}/3+3\sqrt{5}/2$
3525 5270 cr : $10\sqrt{3}/7$
3525 5785 as : $3/2-2\sqrt{3}/3$
3525 7794 rt : (6,5,6,-3)
3525 9417 tanh : $1/\sqrt[3]{20}$
3525 9891 sinh : $6\pi/7$
3526 0358 $e^x : e/9$
3526 0985 tanh : 7/19
3526 1108 rt : (1,7,-9,-4)
3526 4661 cr : $7\sqrt{2}/4$
3526 4954 at : $3/4-\sqrt{5}/2$
3526 5170 rt : (8,6,-8,-2)
3526 7115 $3/5\times s\pi(1/5)$
3526 7516 Li_2 : 23/25
3526 9531 id : $1/\ln(2\pi/3)$
3527 0498 $\ell_2 : -\sqrt{2}+\sqrt{3}+\sqrt{5}$
3527 1360 $2^x : 5\pi/9$
3527 2440 sq : $3/2-e/3$
3527 3179 rt : (7,7,8,-4)
3527 4175 rt : (9,7,-7,0)
3527 7575 $2^x : 1+\sqrt{5}/3$
3527 7745 $\ln(3)+s\pi(\sqrt{3})$
3528 0261 $Ei : 3(e+\pi)$
3528 0678 rt : (5,4,-3,1)
3528 2461 $2^x : P_{29}$
3528 2562 rt : (1,8,-6,-9)
3528 2787 rt : (7,2,-3,-1)
3528 3402 J_2 : 2
3528 5981 rt : (4,0,-9,3)
3528 6987 rt : (3,2,6,2)
3528 7732 $e^x : 2\sqrt{2}+3\sqrt{3}/2$
3528 8494 $Ei : \sqrt{3}+\sqrt{6}+\sqrt{7}$
3528 8975 sinh : $10\sqrt{3}/9$
3528 9443 rt : (6,5,-2,3)
3529 0262 sinh : $9\pi/2$
3529 0407 J_0 : 16/9
3529 2699 rt : (3,3,-7,6)
3529 4117 id : 6/17
3529 4664 $\ell_2 : 3\sqrt{2}+\sqrt{3}/2$
3529 5077 rt : (7,6,-1,5)
3529 5863 $\exp(2)-s\pi(\sqrt{2})$
3529 6828 $Ei : \sqrt{11/2}$
3529 7041 ln : $e\pi/6$
3529 7462 at : $1/\sqrt[3]{20}$
3529 8588 sq : $6\ln 2/7$
3529 9004 rt : (8,7,0,7)
3529 9038 at : 7/19
3529 9670 $10^x : 1+e$
3530 0548 rt : (4,0,8,3)
3530 1389 cr : $e/3+\pi/2$
3530 1897 rt : (9,8,1,9)
3530 3121 $\ell_{10} : \sqrt{3}/3+3\sqrt{5}/4$
3530 3866 ζ : 15/11
3530 6799 $\sqrt{5}+\exp(3/4)$
3530 7797 $\ell_{10} : 3-\sqrt{5}/3$
3530 8705 rt : (6,4,-5,-2)
3531 0793 cr : $\sqrt{3}+\sqrt{5}/3$
3531 3045 $10^x : 1/3-\pi/4$
3531 3674 rt : (5,6,-7,-3)
3531 4819 $\Psi : \sqrt{19}$
3531 4854 $10^x : 4-3\sqrt{5}/4$
3531 5133 $J_1 : (\sqrt{3})^{-1/2}$
3531 5782 rt : (4,8,-9,-4)
3531 6363 at : $1/3+4\pi/3$
3531 9914 $\ln(2/3)-\exp(2/3)$
3532 0996 rt : (4,8,8,-4)
3532 1217 $J_1 : 4\sqrt[3]{5}/9$
3532 1577 sπ : $(\sqrt[3]{3})^{-1/3}$
3532 1640 J_1 : 19/25
3532 2003 ln : $(\ln 2/2)^{1/3}$
3532 4806 id : $e/2-3\pi/2$
3532 6397 rt : (8,0,-8,9)
3532 7033 rt : (7,1,-9,-7)
3532 8656 at : $3\sqrt{2}/4+2\sqrt{3}$
3533 0553 sq : $3/2-2\pi/3$
3533 2856 rt : (9,2,1,-1)
3533 7146 cr : $2/3+2e/3$
3533 8005 sinh : $\sqrt{3}/5$
3533 8291 rt : (1,9,6,5)
3534 1543 $10^x : 4\sqrt{2}/3-\sqrt{3}/4$
3534 7173 sq : $\ln(2e/3)$
3534 7731 rt : (9,9,-8,-5)
3534 8352 sπ : $(\ln 2)^{1/3}$
3534 8605 $2^x : 4/3+3e/4$
3534 8611 $\sqrt{3}-\exp(3)$
3534 9500 rt : (4,9,2,7)
3535 0633 at : $2/3-3\sqrt{3}$
3535 1435 $J_1 : 3\pi^2/10$
3535 4338 rt : (3,5,-4,2)
3535 5205 at : $5e/3$
3535 5339 id : $\sqrt{2}/4$
3535 5715 as : $1/3-e/4$
3535 6594 $\ell_{10} : \sqrt[3]{23/2}$
3535 8945 $2^x : 1/\sqrt[3]{12}$
3536 0090 $J_2 : 5\zeta(3)/3$
3536 0534 sπ : $3\pi/5$
3536 0912 $\ell_2 : 4\sqrt{5}/7$
3536 1155 rt : (2,5,-6,-6)
3536 3695 ℓ_2 : 9/23
3536 4222 cu : $4/3+4\sqrt{2}/3$
3536 6442 $J_2 : 7\sqrt{3}/3$
3536 6693 sinh : $1/\sqrt[3]{24}$
3536 7765 id : $10(1/\pi)/9$
3536 9547 rt : (9,9,7,-4)
3536 9806 $10^x : 2/3-\sqrt{5}/2$
3537 1476 rt : (1,9,8,-4)
3537 3504 rt : (9,3,-8,6)
3537 3843 rt : (3,8,-9,-7)
3537 4011 rt : (7,9,-9,-4)
3537 4160 as : $\sqrt{3}/5$
3537 5110 rt : (8,9,-1,2)
3537 6799 rt : (9,1,7,-3)
3537 9443 $10^x : \sqrt{2}/2+\sqrt{5}$
3538 0231 rt : (8,2,-9,4)
3538 1528 rt : (1,1,8,-3)
3538 3744 rt : (8,7,5,-3)
3538 4241 $2\div\exp(\sqrt{3})$
3538 5578 rt : (2,8,-4,-4)
3538 7147 sinh : $5(1/\pi)$
3538 8110 $10^x : 2\sqrt{5}/7$
3538 8936 rt : (7,8,-2,0)
3538 8974 sπ : $(\sqrt[3]{3}/2)^{-1/3}$
3539 1044 $e^x : 1-3e/4$
3539 1582 as : $\ln 2/2$
3539 3931 at : $4\sqrt{2}-\sqrt{5}/2$
3539 4404 tπ : $\sqrt{2}/4-4\sqrt{5}/3$
3539 5289 cr : $3/4+\sqrt{3}$
3539 6031 $E : (\ln 2/2)^{1/3}$
3539 6456 rt : (8,7,-7,-3)
3539 7015 J_1 : 16/21
3539 7232 sπ : P_{78}
3539 8385 rt : (7,9,6,9)
3539 9041 sπ : $\sqrt{2}-3\sqrt{3}/4$
3539 9175 rt : (9,7,7,2)
3540 0640 sr : 11/6
3540 2510 ln : $\sqrt{15}$
3540 2994 as : $1/\sqrt[3]{24}$
3540 3503 rt : (1,8,-2,8)
3540 5337 rt : (6,1,-3,9)
3540 5768 rt : (8,1,5,2)
3540 7705 sr : $3\sqrt{2}+3\sqrt{3}/4$
3540 7734 $\ln(2)\times\ln(3/5)$
3540 8177 $Ei : (\sqrt{2}\pi)^{-2}$
3540 9669 rt : (7,5,3,-2)
3540 9976 $erf : (\sqrt[3]{5}/3)^2$
3541 0111 $e\times s\pi(1/3)$
3541 0196 id : $3\sqrt{5}/2$
3541 0912 cu : $3/2+e/4$
3541 1756 tπ : $\sqrt{3}/3+4\sqrt{5}/3$
3541 1793 Γ : 2/3
3541 1797 cu : $3\sqrt{2}/2+3\sqrt{5}$
3541 2247 rt : (6,7,-3,-2)
3541 5229 rt : (7,4,9,-4)
3541 5774 ln : $5\sqrt[3]{5}/6$
3541 6334 rt : (6,8,5,7)
3541 8936 rt : (5,0,-4,7)
3542 0155 $\ln(2/3)\div\ln(\pi)$
3542 0156 ln : $\sqrt{3}/2+\sqrt{5}/4$
3542 0991 rt : (4,5,-7,8)
3542 4439 $J_2 : 9\pi/7$
3542 4669 at : $4\sqrt{3}/3+\sqrt{5}$
3542 4868 rt : (1,8,9,0)
3542 5198 as : $(e+\pi)/6$
3542 5554 2^x : 7/16
3542 6224 $1/5+\exp(e)$
3542 6939 sinh : 10/9
3542 7491 $3/4\div\exp(3/4)$
3542 9338 2^x : 21/17
3542 9447 cr : $4e/3+3\pi$
3542 9947 ln : P_{28}
3543 1691 $Ei : \sqrt{3}/6$
3543 2667 Ψ : 23/20
3543 3299 $\exp(3/5)\div s\pi(e)$
3543 7343 $2^x : (e+\pi)^{1/2}$
3543 7569 $\zeta : \sqrt{2}/3$
3543 7762 rt : (4,1,-5,-5)

3543 8027 $\mathrm{t}\pi : 1/4 - \pi$
3543 8455 $\ln : 1/4 + 4e/3$
3544 0172 $\mathrm{cr} : 10\sqrt{5}/9$
3544 3039 $\zeta : \pi^2/4$
3544 3714 rt : (9,5,-9,1)
3544 4109 rt : (2,4,-7,7)
3544 5121 rt : (5,8,5,-3)
3544 8678 rt : (6,2,7,-3)
3544 9077 $\mathrm{sr} : P_{53}$
3545 1343 $J_0 : 4(e+\pi)/7$
3545 1934 rt : (9,5,-5,-2)
3545 3751 rt : (7,4,9,3)
3545 4053 $\mathrm{t}\pi : \exp(-\sqrt{2}\pi/2)$
3545 4483 rt : (1,7,3,-7)
3545 6544 rt : (8,5,-2,8)
3545 8072 $\mathrm{t}\pi : (\ln 2/2)^{1/3}$
3545 9869 rt : (5,9,-1,-1)
3546 1350 $\Psi : 2e\pi/9$
3546 2221 rt : (3,8,-6,-3)
3546 3498 $\ell_{10} : 8\sqrt{2}/5$
3546 4843 $\mathrm{cu} : \sqrt{8/3}$
3546 4945 $\mathrm{sr} : 4\sqrt{3}/3 + 4\sqrt{5}$
3546 5054 $\mathrm{id} : \sqrt{2}/4 - 3\sqrt{5}$
3546 5546 rt : (3,7,-1,-2)
3546 6766 $\mathrm{s}\pi : \exp(-1/(3e))$
3546 7446 rt : (8,1,-7,2)
3546 8295 $\mathrm{cr} : 1/4 + \sqrt{5}$
3546 9167 $\theta_3 : (\sqrt[3]{2}/3)^2$
3547 0205 $10^x : \exp(\sqrt[3]{5})$
3547 0263 $\mathrm{cu} : 2e + 3\pi/2$
3547 0310 rt : (2,8,2,-7)
3547 0977 rt : (4,5,9,-4)
3547 2047 rt : (6,3,1,-1)
3547 4811 $\ln : (\pi/2)^3$
3547 5449 $10^x : 2\sqrt{3} + \sqrt{5}/2$
3547 5564 $\ln(\sqrt{3}) \div \ln(2/3)$
3547 5777 $2^x : 4/3 - 2\sqrt{2}$
3547 7349 rt : (7,4,-3,6)
3548 0897 $\mathrm{at} : 7(e+\pi)/9$
3548 2004 $\mathrm{sr} : 8\ln 2$
3548 3888 $\mathrm{sq} : 2\sqrt{3}/3 - \sqrt{5}/4$
3548 7036 $\mathrm{at} : 4 + \sqrt{5}/4$
3548 7746 $\zeta : 1/\ln(3/2)$
3548 8114 $\theta_3 : 3/17$
3548 8214 $\sinh : 8/23$
3549 1253 $\mathrm{at} : 8\sqrt[3]{5}/3$
3549 1472 $\zeta : \sqrt[3]{15}$
3549 3740 $\mathrm{at} : 2\sqrt{2} + \sqrt{3}$
3549 4632 $\mathrm{s}\pi : 2\sqrt[3]{3}$
3549 5333 rt : (4,7,-4,4)
3549 5366 $\mathrm{sr} : \sqrt[3]{2}/10$
3549 5662 $erf : (3\pi)^{-1/2}$
3549 6501 rt : (1,9,-3,3)
3549 6590 $10^x : 9\pi/5$
3549 7237 rt : (7,3,3,1)
3549 8797 rt : (4,6,3,3)
3549 9840 rt : (3,9,0,-1)
3550 0292 $e^x : 4\sqrt{2}/3 + 4\sqrt{3}/3$
3550 1584 $\mathrm{s}\pi : \ln 2/6$
3550 3993 $\mathrm{cr} : 4/3 + 2\sqrt{3}/3$
3550 5696 rt : (6,3,-4,4)
3550 6182 $10^x : P_{49}$
3550 7661 rt : (9,6,8,-4)
3550 7688 $J_1 : 13/17$
3550 8423 $J_1 : (\sqrt[3]{5})^{-1/2}$
3551 0435 $J_0 : 13/3$
3551 0557 $\Psi : 5/16$
3551 0622 $10^x : 1/3 + e/3$
3551 0679 rt : (2,9,5,-3)
3551 1323 $\ell_{10} : 5e/6$
3551 3663 $10^x : 1/3 + 2\sqrt{5}$
3551 4381 $\mathrm{cu} : e/3 - 4\pi$
3551 4540 $e^x : 1/4 + 4e$
3551 4729 $\mathrm{cu} : 3/4 + 3\pi$
3551 7218 $\pi \times \ln(4)$
3551 8205 $\tanh : 7(1/\pi)/6$
3551 8945 rt : (5,0,5,2)
3551 9978 $\ln(3/5) + s\pi(1/3)$
3552 0972 rt : (3,6,-7,-9)
3552 4947 $\mathrm{sq} : 3\sqrt{2}/4 + 3\sqrt{5}$
3552 5010 $e^x : 23/19$
3552 5140 $\mathrm{as} : 8/23$
3552 5156 rt : (7,7,-9,2)
3552 5628 $\mathrm{sq} : 1/\ln(\sqrt[3]{3}/2)$
3552 6382 rt : (3,3,7,-3)
3552 6571 $\sqrt{5} \times \exp(2/3)$
3552 7274 $\ln(5) - s\pi(\sqrt{3})$
3552 9548 $\mathrm{t}\pi : 3\sqrt{2} + \sqrt{3}/2$
3553 0139 rt : (2,6,0,8)
3553 0175 $\ln : 2 - 3\sqrt{3}/4$
3553 0404 rt : (9,8,-2,5)
3553 2942 $\mathrm{cr} : 1/2 + 4\pi$
3553 4802 $e^x : 1 + \sqrt{2}/3$
3553 5848 $Ei : \sqrt[3]{3}/5$
3553 7735 $\exp(1/5) - s\pi(1/3)$
3553 9321 $\mathrm{sq} : 1/4 - \sqrt{2}$
3553 9641 $\mathrm{cu} : 17/24$
3554 0466 $\mathrm{sr} : e/2 + 4\pi/3$
3554 0910 rt : (1,6,-1,8)
3554 0975 $\mathrm{at} : 3 + \pi/2$
3554 1085 $\mathrm{cu} : 4\sqrt{3} - \sqrt{5}/3$
3554 1572 rt : (6,9,-6,-3)
3554 6591 rt : (5,2,-5,2)
3554 9141 rt : (4,3,9,3)
3554 9852 rt : (9,7,2,-2)
3555 0685 $\theta_3 : \sqrt{2}/8$
3555 1120 $\ell_2 : (2\pi/3)^{1/3}$
3555 1947 $3/4 \times \exp(\pi)$
3555 2595 $\mathrm{cr} : 1 + 2\sqrt{5}/3$
3555 4241 rt : (6,6,7,2)
3555 5455 rt : (2,7,-8,-7)
3555 7697 $\mathrm{at} : 7(1/\pi)/6$
3555 7814 $\ell_{10} : 1 - \sqrt{5}/4$
3555 7970 rt : (4,2,2,9)
3555 8972 $\ell_2 : 2 + \sqrt{5}/4$
3556 1520 $J_2 : 9\sqrt{5}/10$
3556 1583 $\ell_{10} : 9\sqrt[3]{2}/5$
3556 1598 rt : (6,7,8,-4)
3556 2947 $10^x : P_{64}$
3556 3096 $J_0 : 3\sqrt{5}/2$
3556 3332 $\ell_{10} : 4 - \sqrt{3}$
3556 3386 $\theta_3 : 5(1/\pi)/9$
3556 4600 $\mathrm{at} : 1/3 + 3\sqrt{2}$
3556 5975 $\sinh : \exp(1/(3\pi))$
3556 6113 $E : 7/10$
3556 6318 rt : (8,7,-3,3)
3556 7286 rt : (8,4,6,-3)
3556 7779 $\mathrm{t}\pi : e/4 - \pi/2$
3556 8212 rt : (5,3,-8,-3)
3556 8324 $\mathrm{sr} : \ln(2\pi)$
3556 8778 $J_0 : 5\sqrt{3}/2$
3556 9183 $E : 5\sqrt[3]{2}/9$
3556 9534 $J_2 : (1/(3e))^{-1/3}$
3556 9990 rt : (5,2,-6,-8)
3557 0871 $\mathrm{at} : 4 + \sqrt{3}/3$
3557 1201 $\mathrm{at} : 1/2 + 3e/2$
3557 3702 $Ei : 8/17$
3557 4053 $e^x : 7/23$
3557 4491 $J_1 : (\sqrt{2}\pi/2)^{-1/3}$
3557 6294 $\mathrm{cu} : 6\pi$
3557 7266 $\mathrm{sq} : (3\pi/2)^{-1/3}$
3557 8135 $\mathrm{sr} : \sqrt{2}/4 + 3\sqrt{3}$
3557 9486 $2^x : 6\sqrt[3]{3}/7$
3557 9991 rt : (8,3,-4,9)
3558 0227 $\theta_3 : \exp(-\sqrt{3})$
3558 0405 rt : (8,9,5,1)
3558 0624 $\ell_2 : 1/4 + 4\sqrt{3}/3$
3558 0746 $\mathrm{t}\pi : 9\ln 2/7$
3558 0877 rt : (5,9,-4,6)
3558 2805 $\exp(1/2) + s\pi(1/4)$
3558 6708 rt : (3,9,3,-5)
3558 7362 $\ln(2) \div \exp(2/3)$
3558 8866 $\Gamma : 7e/6$
3558 9999 $\mathrm{cu} : 4 + 2\pi/3$
3559 0260 $10^x : 2\sqrt{3}/3 + 3\sqrt{5}$
3559 1298 $\ell_{10} : 2\sqrt{2} - \sqrt{5}/4$
3559 1484 $\ln : 9(e+\pi)/5$
3559 2648 $\mathrm{at} : 2\sqrt{3} + \sqrt{5}/2$
3559 3073 rt : (2,7,-9,-4)
3559 4097 rt : (9,2,8,3)
3559 4253 $\mathrm{t}\pi : (2\pi/3)^{-3}$
3559 4649 $\mathrm{at} : \sqrt{21}$
3559 5280 $e^x : (2e/3)^{-2}$
3559 6771 rt : (1,7,3,-2)
3559 7400 $Li_2 : (\sqrt[3]{5}/3)^2$
3559 8369 $2^x : 3/2 - 3\sqrt{2}/4$
3559 9710 $J_0 : (\pi)^{1/2}$
3560 3303 $2^x : (\sqrt[3]{4}/3)^{-1/3}$
3560 4315 rt : (1,8,-9,-5)
3560 5765 $J_0 : 1/\ln(\sqrt[3]{2})$
3561 0722 rt : (4,9,-1,0)
3561 1393 $Ei : \sqrt[3]{7/2}$
3561 2308 $\mathrm{sq} : 4\sqrt{2} - 3\sqrt{3}/2$
3561 4213 $\mathrm{at} : 3\sqrt{2}/2 - 3\sqrt{5}$
3561 4450 $e^x : \sqrt{2}/4 + \sqrt{5}/2$
3561 4861 rt : (7,2,-5,7)
3561 5451 $10^x : \sqrt{2}/4 + 2\sqrt{3}$
3561 5810 rt : (2,7,-7,2)
3561 6060 $erf : \exp(-\sqrt{5}/2)$
3561 6577 $\Psi : 10\sqrt[3]{5}/9$
3561 7082 rt : (7,6,-4,1)
3561 8208 $erf : \ln(\sqrt[3]{3}/2)$
3561 8416 $\Psi : 19/10$
3561 8586 rt : (3,8,8,-4)
3561 9449 $\mathrm{id} : 3\pi/4$
3561 9793 $\sinh : \pi/9$
3562 0747 $J_1 : \sqrt{3} - \sqrt{5} - \sqrt{6}$
3562 2491 rt : (6,6,3,8)
3562 2906 $\mathrm{id} : 2\sqrt{2}/3 - 3\sqrt{3}/4$
3562 4378 $\sqrt{5} \div \exp(1/2)$
3562 8081 $\mathrm{at} : 4e - 2\pi$
3562 9019 $10^x : \sqrt{2}/3 + 3\sqrt{3}$
3562 9901 $J_0 : 3\sqrt[3]{3}$
3563 3190 $\tanh : \sqrt{5}/6$
3563 4099 rt : (6,8,2,-2)
3563 7373 $J_2 : 8\sqrt[3]{2}/5$
3563 7650 $K : \beta(2)$
3563 9084 rt : (8,9,7,-4)
3564 0340 $\mathrm{t}\pi : \exp(\sqrt[3]{4})$
3564 1254 rt : (5,5,6,-3)
3564 1700 $\mathrm{cu} : 9e$
3564 1844 $e^x : 6/7$
3564 2254 $10^x : 13/15$
3564 6772 rt : (2,5,9,8)
3564 6892 $E : ((e+\pi)/2)^{-1/3}$
3564 6901 $\mathrm{id} : \exp(\sqrt{5})$
3564 7479 $e^x : 1/3 + 3\pi/2$
3564 8245 $\ln(2/5) \div s\pi(\sqrt{5})$
3564 8644 $\mathrm{s}\pi : 1/4 + \sqrt{3}/2$
3564 9227 $\mathrm{sq} : 7e\pi/2$
3565 0771 $\sinh : 8\sqrt{5}/3$
3565 0782 $\Psi : (\pi/3)^3$
3565 1332 $\mathrm{as} : 4\sqrt[3]{5}/7$
3565 1580 $\ell_2 : 3 + 3\sqrt{2}/2$
3565 2162 $\ln(2/3) - s\pi(2/5)$
3565 4727 rt : (1,2,9,3)
3565 4732 $\ell_{10} : 11/25$
3565 4870 $\ell_{10} : 2\sqrt{3}/3 + \sqrt{5}/2$
3565 6077 rt : (9,6,-4,6)
3565 7208 $\mathrm{cu} : e + 2\pi$
3565 7406 $\mathrm{as} : \pi/9$
3565 8560 $\mathrm{t}\pi : \sqrt{2}/2 - 3\sqrt{3}/2$
3566 0188 rt : (5,8,-9,-4)
3566 0302 $\mathrm{t}\pi : 8\ln 2/5$
3566 0432 $2^x : 11/25$
3566 0574 rt : (6,1,-6,5)
3566 1007 $\ell_2 : 3\sqrt{3}/4 + 4\sqrt{5}$
3566 4879 $\mathrm{at} : 2 + 3\sqrt{3}/2$
3566 5674 $e^x : (\sqrt[3]{4})^{-1/3}$
3566 6797 rt : (9,3,9,-4)
3566 7385 rt : (3,5,-7,-5)
3566 7494 $\ln : 7/10$
3566 8955 rt : (2,8,7,9)
3567 2863 $\mathrm{sq} : 3/4 + 3\sqrt{3}$
3567 3338 $\mathrm{at} : \sqrt{5}/6$
3567 3564 $\mathrm{at} : 23/5$
3567 3738 $\mathrm{id} : \sqrt{3}/2 + 2\sqrt{5}/3$
3567 3760 $\ln : 5\sqrt[3]{2}/9$
3567 3931 rt : (7,7,-4,-2)
3567 4136 $\mathrm{sq} : 4\sqrt{2} - 2\sqrt{5}/3$

3567 5218 J_0 : $8\sqrt[3]{2}/3$
3567 5388 $s\pi$: $5\sqrt{2}/8$
3567 6896 rt : (8,4,-8,-3)
3567 8471 10^x : $\sqrt{2}/2 - 2\sqrt{3}/3$
3567 8917 rt : (9,8,-4,0)
3567 9105 $t\pi$: $\sqrt{2}/4 + 4\sqrt{5}$
3568 0550 $s\pi$: $9e/4$
3568 1296 rt : (8,5,0,-1)
3568 2482 sr : $2(1/\pi)/5$
3568 2807 rt : (7,2,4,-2)
3568 3601 $t\pi$: $(\sqrt[3]{2})^{-1/2}$
3568 3852 rt : (6,1,8,3)
3568 5029 ζ : $5\sqrt[3]{2}/4$
3568 6029 J_1 : $10/13$
3568 6784 rt : (5,4,-8,-9)
3568 7048 Ψ : $1/\ln(\sqrt{5}/3)$
3568 7059 as : $(\pi/3)^{-1/2}$
3568 8292 id : $1/\sqrt[3]{22}$
3568 9586 rt : (2,6,-8,2)
3569 1544 $\ln(2) \div \ln(3/5)$
3569 1624 sq : $2\sqrt{2}/3 - 2\sqrt{3}$
3569 3080 cr : $(1/\pi)/7$
3569 3854 rt : (5,5,2,6)
3569 4286 rt : (6,5,-5,-1)
3569 6614 E : $2\pi/9$
3569 7299 2^x : $\sqrt[3]{16/3}$
3569 7638 $t\pi$: $1/4 - e/2$
3569 8250 e^x : $5\sqrt[3]{2}$
3569 8417 2^x : $3/4 - \sqrt{5}$
3569 9510 Li_2 : $(3\pi)^{-1/2}$
3570 2260 id : $5\sqrt{2}/3$
3570 2683 rt : (3,5,-7,-3)
3570 3737 rt : (3,9,-3,2)
3570 5115 rt : (2,8,-3,-2)
3570 5398 sq : $2 - 3\pi/2$
3570 5754 rt : (8,5,-5,4)
3570 7267 cu : $7\sqrt{5}/2$
3570 8440 J_1 : $4\sqrt{3}/9$
3570 9873 e^x : $1/4 + 3\sqrt{2}$
3571 0059 J_2 : $7\sqrt[3]{3}/5$
3571 1741 ln : $3 - \pi/2$
3571 1976 Ei : $\sqrt{3} + \sqrt{5} - \sqrt{6}$
3571 3349 cu : $3\sqrt{3}/2 + 2\sqrt{5}/3$
3571 4285 id : $5/14$
3571 5305 λ : $\exp(-8\pi/7)$
3571 7356 sq : $3e/7$
3571 8972 sinh : $7/20$
3571 9942 ℓ_{10} : $3/2 - 3\sqrt{2}/4$
3572 0880 id : $\sqrt[3]{5/2}$
3572 2159 sr : $3\sqrt{3} - 3\sqrt{5}/2$
3572 2337 rt : (7,0,-7,8)
3572 2557 ℓ_2 : $3e\pi/5$
3572 2725 J_1 : $10\ln 2/9$
3572 3118 rt : (5,0,-7,3)
3572 4482 2^x : $\sqrt[3]{19}$
3572 5700 rt : (3,9,2,-2)
3572 8206 ln : $2 + 4\sqrt{2}/3$
3572 9733 rt : (6,4,9,-4)
3573 1281 sq : $3\sqrt{2}/4 + 4\sqrt{3}/3$
3573 1965 rt : (8,9,-9,-4)
3573 3376 e^x : $2e/3 + 3\pi$
3573 3792 sq : $1/2 + e$
3573 4401 rt : (7,7,5,-3)
3573 5562 rt : (1,8,5,2)
3573 5673 J_2 : $10\sqrt{2}/7$
3573 7152 2^x : $5\sqrt{3}/7$
3573 7401 ln : $3/4 + e/4$
3573 7627 ℓ_2 : $3/2 + 4e/3$
3573 8155 sinh : $(1/(3e))^{1/2}$
3573 8177 rt : (3,5,7,2)
3573 8300 rt : (2,9,-9,1)
3573 8832 id : $3\sqrt{2}/2 + \sqrt{5}$
3573 8873 rt : (4,7,-7,-3)
3574 1308 ℓ_{10} : $8e\pi/3$
3574 1619 J_0 : $\sqrt{3} - \sqrt{6} - \sqrt{7}$
3574 2379 Γ : $5\pi/2$
3574 4511 ℓ_2 : $3e + 2\pi/3$
3574 4752 sq : $3/4 + \pi/4$
3574 4997 J_2 : $7\sqrt{3}/6$
3574 6253 rt : (1,5,-2,-9)
3574 9384 $\ln(3/4) \div \ln(\sqrt{5})$
3575 0505 $t\pi$: $4e + 3\pi$
3575 1597 rt : (2,5,-9,-2)
3575 2704 $t\pi$: $2\sqrt{3} - 4\sqrt{5}/3$
3575 2898 2^x : $1 - \sqrt{5}/4$
3575 5288 rt : (9,4,-6,7)
3575 7110 as : $7/20$
3575 8081 at : $8\sqrt{3}/3$
3575 8629 rt : (7,9,3,5)
3576 1261 rt : (8,4,6,2)
3576 5069 rt : (8,9,-4,-2)
3576 7092 sinh : $2\pi^2/7$
3576 7449 rt : (6,4,1,9)
3576 9490 at : $1/3 - \sqrt{2}/2$
3576 9515 sq : $2 - 3\sqrt{3}/2$
3577 1827 $s\pi$: $5e\pi/6$
3577 2450 rt : (2,1,-9,3)
3577 2699 rt : (7,4,-6,2)
3577 5276 2^x : $5\pi/2$
3577 5403 id : $4\sqrt{2}/3 + 2\sqrt{5}$
3577 6394 as : $(1/(3e))^{1/2}$
3577 6995 10^x : $2 + 2\pi$
3577 7021 rt : (6,1,-8,-6)
3577 7087 $4/5 \div \sqrt{5}$
3577 9267 2^x : $7\sqrt{2}/8$
3577 9802 rt : (5,9,-7,-1)
3578 0324 rt : (6,1,-6,-2)
3578 0372 2^x : $3e - \pi/4$
3578 1583 J_0 : $7\sqrt[3]{3}/3$
3578 1614 id : $4\sqrt{2} - 3\sqrt{3}/4$
3578 1993 rt : (3,0,8,3)
3578 3009 at : $1 + 4e/3$
3578 3546 rt : (9,1,2,1)
3578 3683 $s\pi$: P_{77}
3578 4558 $s\pi$: $((e + \pi)/2)^{-2}$
3578 5010 $\exp(2/5) + s\pi(1/3)$
3578 6923 J_0 : $23/13$
3578 7120 2^x : $1/3 + \sqrt{2}$
3578 7963 J_2 : $9\sqrt{5}/5$
3578 8229 e^x : $(e/2)^{-1/2}$
3578 8641 rt : (8,1,7,-3)
3579 1045 E : $\sqrt{2} + \sqrt{3} - \sqrt{6}$
3579 1150 cu : $3\sqrt{3}/2 - 2\sqrt{5}/3$
3579 2636 rt : (1,5,2,-4)
3579 2873 ℓ_{10} : $4\sqrt[3]{5}/3$
3579 3517 e^x : $4\pi^2/9$
3579 4739 rt : (1,6,4,3)
3579 6591 rt : (3,5,9,-4)
3579 7907 sinh : $25/8$
3579 7957 2^x : $1/4 - \sqrt{3}$
3579 8036 ℓ_{10} : $\sqrt{2} + \sqrt{3}/2$
3579 9702 rt : (8,6,8,-4)
3580 0091 2^x : $10\sqrt{2}$
3580 1094 rt : (2,6,-3,-2)
3580 3473 rt : (6,6,-7,-3)
3580 3732 sr : $\sqrt{2}/4 + 2\sqrt{5}/3$
3580 3886 rt : (3,8,6,9)
3580 4285 rt : (9,4,3,-2)
3580 4313 $t\pi$: $1/2 + 4\sqrt{3}$
3580 4937 at : P_{74}
3580 5141 as : $4\sqrt{2}/3 - \sqrt{5}$
3580 6454 at : $(5/3)^3$
3580 6913 rt : (1,7,-6,-3)
3580 7907 ℓ_2 : $4 - e$
3580 8211 K : $\ln(5/2)$
3580 8671 rt : (7,2,-9,-9)
3580 9432 rt : (8,8,2,8)
3580 9862 id : $9(1/\pi)/8$
3581 0137 sq : $2/3 - 4\sqrt{5}/3$
3581 0672 cu : $2\sqrt{2}/3 + 3\sqrt{3}$
3581 0682 Ψ : $7e/10$
3581 0738 2^x : $\sqrt{2}/4 + 2\sqrt{5}$
3581 1358 rt : (9,9,4,-3)
3581 1538 ζ : $3\sqrt[3]{2}/8$
3581 2001 rt : (8,3,-7,5)
3581 3515 2^x : $3/2 - 4\sqrt{5}/3$
3581 3871 rt : (4,1,-8,-1)
3581 4486 tanh : $1/\sqrt[3]{19}$
3581 6660 sinh : $\exp(-\pi/3)$
3581 7077 rt : (9,8,-5,1)
3582 2468 ℓ_2 : $7(e + \pi)$
3582 3466 J_1 : $17/22$
3582 5694 sinh : $\exp(\sqrt[3]{3})$
3582 5756 rt : (5,9,3,0)
3582 6565 $1/2 \div \exp(1/3)$
3582 9151 10^x : $1/\ln(1/(3\pi))$
3583 0769 rt : (3,2,-9,7)
3583 2700 ln : $((e + \pi)/2)^{1/3}$
3583 5618 rt : (9,2,-8,8)
3583 5739 tanh : $3/8$
3583 9583 ℓ_2 : $8\sqrt[3]{3}/9$
3583 9860 at : $3\sqrt{3} - \sqrt{5}/4$
3584 0687 cr : $3e/2 - \pi/2$
3584 0721 sr : $(2\pi)^{1/3}$
3584 1799 sr : $3 - 2\sqrt{3}/3$
3584 3180 Li_2 : $\exp(-\sqrt{5}/2)$
3584 3718 J_0 : $\exp(\sqrt[3]{5}/3)$
3584 4014 Γ : $\sqrt{3} - \sqrt{6} - \sqrt{7}$
3584 5187 e^x : $3 + 2\sqrt{5}$
3584 8547 id : $\sqrt{2} + 4\sqrt{5}$
3584 8945 rt : (5,2,-9,-4)
3584 9367 rt : (4,4,-9,9)
3585 0381 sr : $8\pi^2/7$
3585 0483 $\ln(4) \div s\pi(1/5)$
3585 0682 e^x : $1/3 - e/2$
3585 0959 J_0 : $\ln(e + \pi)$
3585 1274 Ei : $\sqrt{2}/3$
3585 2378 10^x : $3\sqrt{2}/4 + 3\sqrt{5}/2$
3585 3290 $t\pi$: $2\ln 2/3$
3585 3637 $t\pi$: $3\sqrt{2}/4 - \sqrt{3}/4$
3585 4386 rt : (5,3,0,7)
3585 5319 as : $\exp(-\pi/3)$
3585 5689 at : $1/\sqrt[3]{19}$
3585 7141 $t\pi$: $9\sqrt[3]{3}/10$
3585 8374 $s\pi$: $10e/7$
3585 9649 at : $3/2 + \pi$
3586 0985 cr : $\ln(\pi/3)$
3586 3166 rt : (5,7,8,-4)
3586 4858 sq : $\sqrt{3} + \sqrt{6} - \sqrt{7}$
3586 6140 rt : (9,5,9,3)
3586 7470 $s\pi$: $5(e + \pi)/6$
3586 7569 Ei : $7\pi/3$
3586 7789 rt : (6,9,-4,0)
3586 8921 E : $16/23$
3587 1938 J_0 : $5\sqrt{2}/4$
3587 2872 10^x : $\sqrt{5}/6$
3587 3244 sr : $24/13$
3587 7067 at : $3/8$
3587 7112 sinh : $\sqrt{17/3}$
3587 7355 2^x : $1/3 - 2e/3$
3587 8673 $t\pi$: $\sqrt{2} - \sqrt{5} + \sqrt{6}$
3587 9712 rt : (5,2,7,-3)
3588 1654 e^x : $\sqrt{2}/2 - \sqrt{3}$
3588 1849 ln : $3/4 + \pi$
3588 2072 rt : (6,8,2,3)
3588 4137 ℓ_2 : $9\ln 2/8$
3588 5728 rt : (7,2,-8,3)
3588 8750 cr : $4 - 2\sqrt{5}/3$
3588 9894 id : $\sqrt{19}$
3589 1386 Ei : $\ln(4/3)$
3589 2003 sq : $\ln(\ln 3/2)$
3589 2519 sinh : $1/\sqrt[3]{23}$
3589 2536 e^x : $\sqrt{3}/2 + 3\sqrt{5}$
3589 2593 J_2 : $\sqrt[3]{25/3}$
3589 3853 ℓ_2 : $3\sqrt[3]{5}$
3589 5677 rt : (8,1,-9,6)
3589 6187 ℓ_{10} : $3/2 + \pi/4$
3589 6813 J_1 : $(5/3)^{-1/2}$
3589 7014 cu : $2/3 + 4\sqrt{3}/3$
3589 7943 rt : (2,9,1,-1)
3589 8978 rt : (7,7,1,6)
3589 9966 $\exp(3/5) \times s\pi(\sqrt{3})$
3590 0246 rt : (4,8,-6,-3)
3590 2028 at : $2e - \pi/4$
3590 2194 ℓ_{10} : $7/16$
3590 4831 rt : (5,0,-9,3)
3590 5556 rt : (7,8,-5,-4)
3590 6661 rt : (3,9,-6,-8)
3590 7122 rt : (8,6,0,9)
3590 8358 J_2 : $\exp(1/\sqrt{2})$

3590 9021 $t\pi$: P_{76}
3590 9122 cu : $8(e+\pi)/7$
3590 9849 2^x : $5\sqrt[3]{5}/2$
3591 0749 at : $3\sqrt{3}/4+3\sqrt{5}/2$
3591 1057 rt : (6,5,3,-2)
3591 2246 cr : $3\sqrt{2}-\sqrt{3}$
3591 2702 sr : $\sqrt{2}+\sqrt{3}/4$
3591 3569 $\ln(4/5)\times\ln(5)$
3591 4091 id : $e/2$
3591 4597 rt : (9,7,-7,-3)
3591 4903 rt : (8,7,-6,-1)
3591 9140 rt : (9,6,-7,2)
3591 9744 rt : (7,9,7,-4)
3592 2520 rt : (5,3,6,2)
3592 6408 cr : $3(e+\pi)/7$
3592 7133 at : $1-4\sqrt{2}$
3592 8114 sinh : $7(1/\pi)/2$
3592 9046 ℓ_2 : $3\sqrt{2}-3\sqrt{5}/4$
3593 0408 rt : (4,8,0,8)
3593 1588 as : $1/\sqrt[3]{23}$
3593 1782 at : $4-4e/3$
3593 2750 θ_3 : $(2e)^{-1/3}$
3593 4045 $\exp(\sqrt{3})+s\pi(1/4)$
3593 4751 ln : $2\pi/9$
3593 5239 e^x : $\sqrt{3}/2-\sqrt{5}/4$
3593 5639 10^x : 2/15
3593 7500 cu : 7/4
3593 9999 e^x : $4-\pi$
3594 3791 $1/4-\ln(5)$
3594 3813 rt : (3,7,-9,-4)
3594 6317 $4/5\div\exp(4/5)$
3594 7781 rt : (4,9,-4,-7)
3594 8702 cu : $5\pi^2/6$
3595 1495 $\exp(1/3)-s\pi(\sqrt{2})$
3595 1552 $\exp(4/5)-s\pi(1/3)$
3595 3365 Ψ : $1/\sqrt{17}$
3595 3731 cu : $3\sqrt{3}/4+3\sqrt{5}/4$
3595 4619 rt : (7,8,-1,-1)
3595 5252 $\ln(\sqrt{2})\div s\pi(\sqrt{2})$
3595 5598 sr : $8\ln 2/3$
3595 6469 e^x : $4/3-3\pi/4$
3595 6952 rt : (7,4,6,-3)
3595 7432 rt : (3,7,-4,-9)
3595 9102 ℓ_{10} : $1/4+3e/4$
3595 9593 cu : $3\sqrt{3}/2-\sqrt{5}/3$
3596 0168 Ei : $8(e+\pi)/3$
3596 0667 cr : $4\pi/5$
3596 0956 $\ln(3/4)\times\exp(e)$
3596 1179 rt : (4,7,2,0)
3596 6634 rt : (5,1,-2,8)
3596 8474 J_0 : $(3/2)^3$
3596 9827 id : $3\sqrt{2}/4+3\sqrt{3}/4$
3597 0299 at : 14/3
3597 2708 ℓ_{10} : $\sqrt[3]{12}$
3597 7345 J_1 : $2e/7$
3597 7901 ln : $1+\sqrt{3}/4$
3597 8745 rt : (2,3,7,9)
3597 8946 id : $8\sqrt[3]{2}/3$
3598 0812 cu : $1-4\pi$
3598 1384 $1/3-\ln(2)$
3598 1827 rt : (6,4,9,3)
3598 1828 10^x : 10/19
3598 3342 rt : (4,2,-1,5)
3598 5460 rt : (6,1,-9,1)
3598 5646 rt : (8,3,9,-4)
3598 8307 e^x : $5\zeta(3)/7$
3598 8738 rt : (5,8,-6,7)
3599 1197 rt : (2,7,0,-1)
3599 1819 at : $4e/3+\pi/3$
3599 3494 sq : $3/4+4\sqrt{3}/3$
3599 4285 $s\pi$: $\sqrt{12}\pi$
3599 4878 $\exp(2/3)-s\pi(1/5)$
3599 5586 rt : (8,7,2,-2)
3599 7657 rt : (9,4,-9,3)
3599 7744 id : $2\sqrt[3]{2}/7$
3599 9207 rt : (7,5,-1,7)
3600 0000 id : 9/25
3600 1477 rt : (6,0,2,1)
3600 1771 rt : (4,6,-6,5)
3600 1816 rt : (1,9,-8,-4)
3600 1967 rt : (7,9,-6,-3)
3600 2090 $t\pi$: $2\sqrt{3}-3\sqrt{5}/2$
3600 2171 cr : $9\sqrt{5}/8$
3600 3794 ℓ_{10} : $\sqrt{3}+\sqrt{5}/4$
3600 6328 K : 11/12
3600 7193 10^x : $\zeta(3)/9$
3600 7615 ℓ_2 : $3-\sqrt{3}/4$
3600 9282 rt : (3,2,-8,-3)
3600 9680 at : $3\sqrt{2}+\sqrt{3}/4$
3601 2048 10^x : $2+\sqrt{5}/3$
3601 3420 rt : (5,2,-8,-2)
3601 3733 10^x : $\pi^2/4$
3601 5792 at : $3+3\sqrt{5}/4$
3601 7712 Γ : $\sqrt[3]{2}/6$
3601 7870 rt : (9,2,-6,-2)
3601 9091 rt : (8,5,-8,0)
3601 9466 rt : (9,6,5,-3)
3602 0820 sr : $\sqrt[3]{19/3}$
3602 1411 J_1 : 7/9
3602 1731 sr : $\sqrt{2}/2-\sqrt{3}/3$
3602 2304 rt : (3,4,-6,3)
3602 2506 J_0 : $\sqrt[3]{11/2}$
3602 3331 e^x : $1+3e/2$
3602 3424 Ψ : $4\pi^2/9$
3602 3636 ln : $9\sqrt{3}/4$
3602 3767 2^x : $\sqrt{3}/4+\sqrt{5}$
3602 4817 $t\pi$: $\sqrt{3}/4+3\sqrt{5}/4$
3602 5234 rt : (6,8,-9,-4)
3602 5344 sr : $4+\pi/2$
3602 6800 at : $4+e/4$
3602 7225 E : $\sqrt[3]{1/3}$
3602 8237 e^x : 4/13
3602 8648 rt : (9,9,0,6)
3603 1444 sinh : 6/17
3603 8382 ℓ_{10} : $3-\sqrt{2}/2$
3603 8898 $s\pi$: $9\pi^2/10$
3603 9464 2^x : $\sqrt{2}+\sqrt{3}+\sqrt{6}$
3604 0382 Ψ : $5(e+\pi)$
3604 0570 e^x : $3/2+4\sqrt{5}/3$
3604 0733 J_1 : $(\sqrt{2}/3)^{1/3}$
3604 1976 E : $\ln 2$
3604 2944 rt : (7,1,5,2)
3604 4247 $t\pi$: $3\sqrt[3]{2}/2$
3604 4824 2^x : $3-2\sqrt{5}$
3604 7547 $t\pi$: $3\sqrt{3}/2-3\sqrt{5}$
3604 8173 rt : (4,7,8,8)
3604 9328 cu : $5\pi^2/7$
3604 9556 rt : (5,1,-4,-9)
3604 9991 $\exp(3)\times s\pi(\sqrt{2})$
3605 1667 Ei : $\ln(\sqrt{5})$
3605 2232 rt : (5,5,-8,2)
3605 2582 e^x : $(\sqrt{2}\pi/3)^{-3}$
3605 3319 $\ln(\sqrt{2})-s\pi(1/4)$
3605 3859 ℓ_{10} : P_{29}
3605 4894 rt : (1,5,-1,5)
3605 5227 rt : (8,1,-9,-3)
3605 6239 id : $\sqrt[3]{3}/4$
3605 6989 rt : (3,6,-1,-1)
3605 7297 at : $3/4-2e$
3605 9427 rt : (2,8,4,-1)
3606 0160 rt : (5,4,9,-4)
3606 0558 rt : (7,6,-7,-3)
3606 0804 id : $3e/2+2\pi$
3606 1092 at : $4/3+3\sqrt{5}/2$
3606 1522 rt : (8,2,8,3)
3606 1707 id : $3\zeta(3)/10$
3606 1891 cu : $e/3+3\pi/2$
3606 2486 2^x : $3\sqrt{2}/2-3\sqrt{5}/4$
3606 3113 at : $4(e+\pi)/5$
3606 4481 E : $2\sqrt{3}/5$
3606 6988 at : $1/2+4\pi/3$
3606 7347 cu : $1/3+4e$
3606 7376 $1/2\div\ln(4)$
3606 7467 id : $3\sqrt{2}+\sqrt{5}/2$
3606 7977 id : $10\sqrt{5}$
3607 1272 as : 6/17
3607 2600 rt : (2,5,-4,6)
3607 3722 10^x : $1/2-2\sqrt{2}/3$
3607 4058 at : $\sqrt{22}$
3607 4395 rt : (7,3,-3,8)
3607 7720 ℓ_{10} : $2e-\pi$
3607 7835 rt : (2,5,-4,-2)
3607 8200 rt : (2,8,-2,5)
3607 9000 2^x : 4/9
3608 1348 e^x : $6\sqrt{2}/7$
3608 1532 at : $4\sqrt{3}-\sqrt{5}$
3608 4475 e^x : $\sqrt{2}/3-2\sqrt{5}/3$
3608 5511 2^x : $e-4\pi/3$
3608 6962 rt : (1,7,4,-5)
3608 7767 e^x : $\sqrt[3]{19/3}$
3608 9314 $\exp(2/3)+\exp(5)$
3609 1033 cr : $4\pi^2/3$
3609 2084 rt : (3,8,-7,0)
3609 3842 tanh : $1/\sqrt{7}$
3609 4614 rt : (4,0,-3,6)
3609 4872 tanh : $3\sqrt[3]{2}/10$
3609 6525 sinh : $\sqrt{2}/4$
3609 7103 J_0 : $7\sqrt[3]{2}/5$
3609 7123 rt : (9,7,-2,7)
3609 8971 at : $1/2-3\sqrt{3}$
3609 8972 rt : (4,9,-1,-1)
3609 9742 E : 9/13
3610 0109 J_1 : $9\ln 2/8$
3610 1742 sr : $3e+\pi$
3610 2081 rt : (5,8,2,-2)
3610 3852 rt : (6,7,5,-3)
3610 3919 $\ln(3/4)+\exp(1/2)$
3610 4129 $4/5\div s\pi(1/5)$
3610 4784 2^x : $1/3+e/3$
3610 4995 rt : (7,6,8,-4)
3610 6490 sinh : $(\sqrt[3]{3}/2)^{-1/3}$
3610 6508 $3/4\div\ln(4/5)$
3610 6959 ℓ_{10} : $\exp(2e)$
3610 7102 E : $\exp(-1/e)$
3610 9737 sinh : $10(1/\pi)/9$
3610 9936 10^x : $1/2+3\sqrt{3}/2$
3611 0308 rt : (3,3,9,3)
3611 1111 sq : 7/6
3611 2721 rt : (9,9,-9,-4)
3611 4665 sr : $3\sqrt{3}/2-\sqrt{5}/3$
3611 5755 sr : 3/23
3611 5865 ℓ_2 : $3\sqrt{2}-2\sqrt{3}$
3611 7333 Li_2 : 12/13
3611 7758 rt : (1,3,-8,-9)
3611 7838 Ei : $8\sqrt[3]{5}/9$
3611 7966 J_1 : $7\sqrt[3]{2}/3$
3611 8274 Ψ : $1/\ln(1/(2\pi))$
3612 0826 $\ln(3/5)\times s\pi(1/4)$
3612 2710 at : $3\sqrt{2}-4\sqrt{5}$
3612 3519 sq : $\zeta(3)/2$
3612 4166 $s\pi$: 2/17
3612 5318 Ei : $\exp(\sqrt{5}/3)$
3612 5566 rt : (6,4,-2,5)
3612 7488 rt : (9,1,4,-2)
3612 9326 ln : $\sqrt{2}+\sqrt{3}-\sqrt{6}$
3613 0807 sq : $1/4-\pi$
3613 2895 rt : (7,1,-5,9)
3613 4019 cu : $1/4-3\sqrt{5}$
3613 5004 sr : $4/3+3\sqrt{2}$
3613 5338 sinh : $\sqrt{13/2}$
3613 5708 rt : (8,2,1,-1)
3613 6185 2^x : $4\sqrt{2}/3-3\sqrt{5}/2$
3613 6504 10^x : $1-\sqrt{3}/2$
3613 6712 as : $\sqrt{2}/4$
3613 7171 $s\pi$: $9\sqrt{3}/5$
3613 7748 at : $3\sqrt[3]{2}/10$
3613 9591 rt : (2,5,9,-4)
3614 0947 at : $3\sqrt{3}/4-3\sqrt{5}/4$
3614 1494 J_2 : $3e/4$
3614 2363 rt : (6,3,-8,-3)
3614 3055 rt : (9,8,7,-4)
3614 3224 ℓ_2 : $1/3+\sqrt{5}$
3614 4099 Ψ : $\ln(\pi)$
3614 4617 ζ : $(3\pi)^{-1/3}$
3614 4851 cu : $1/2+2e/3$
3614 5267 sr : $3/2+\sqrt{2}/4$
3614 6217 Ψ : $\sqrt[3]{3/2}$
3614 6382 at : $4+\sqrt{2}/2$
3614 6876 rt : (2,6,-6,-8)
3614 9462 rt : (8,9,4,-3)

3614 9996 as : $10(1/\pi)/9$
3615 0973 rt : (9,5,-4,8)
3615 1119 at : $2-3\sqrt{5}$
3615 2667 cr : $7\sqrt[3]{3}/4$
3615 2998 rt : (8,8,-1,4)
3615 3576 Ψ : $9\pi^2/8$
3615 3602 cu : $4-3\pi/2$
3615 4616 ℓ_{10} : $1+3\sqrt{3}/4$
3615 4848 2^x : $3-\sqrt{3}/3$
3615 8353 $\ln(\sqrt{5}) \div \exp(4/5)$
3615 8406 10^x : $8e$
3615 8983 rt : (7,4,-5,-2)
3616 1235 2^x : $3\sqrt{3}/2+3\sqrt{5}/4$
3616 4131 cu : $4\sqrt{2}/3-3\sqrt{3}/2$
3616 4822 sq : $4/3+4\sqrt{2}/3$
3616 5279 tπ : $3\sqrt{2}-4\sqrt{5}$
3616 5416 ℓ_2 : $(\sqrt{2}/3)^{1/3}$
3616 5510 J_2 : $\sqrt{5}+\sqrt{6}-\sqrt{7}$
3616 5677 sr : $3/2+3e/2$
3616 6223 e^x : $7\sqrt{3}/10$
3616 7957 rt : (6,9,8,2)
3616 8871 J : $1/15$
3616 9168 at : $3\pi/2$
3617 0475 J_1 : $(2\pi/3)^{-1/3}$
3617 1952 rt : (4,2,-5,-7)
3617 2642 sq : $3(e+\pi)/5$
3617 2783 ℓ_{10} : 23
3617 2793 $s\pi(1/5)+s\pi(e)$
3617 3451 sr : $9\sqrt[3]{3}/7$
3617 3461 sq : $\sqrt{2}+3\sqrt{3}/4$
3617 3846 at : $\sqrt{3}+4\sqrt{5}/3$
3617 4615 rt : (6,7,-6,-6)
3617 6302 at : $10\sqrt{2}/3$
3617 6780 rt : (5,1,8,3)
3617 7526 10^x : $(\ln 3/3)^2$
3617 9097 rt : (4,7,8,-4)
3618 2071 ζ : $\sqrt{6}$
3618 3065 rt : (1,8,0,-1)
3618 3276 $5 \div \exp(3/4)$
3618 6311 rt : (9,8,-8,-3)
3618 7891 J_2 : $\sqrt[3]{17/2}$
3619 0302 E : $(2\pi/3)^{-1/2}$
3619 1233 2^x : $7(1/\pi)/5$
3619 4438 rt : (9,3,-6,9)
3619 4530 at : $2+e$
3619 4760 $\sqrt{2}+\exp(2/3)$
3619 5364 as : $3-3\sqrt{5}/2$
3619 6889 tπ : $\sqrt{2}/4-2\sqrt{3}$
3619 9321 at : $3\sqrt{2}/2+3\sqrt{3}/2$
3619 9775 e^x : $7(e+\pi)/9$
3620 0823 id : $3\sqrt{3}/2-\sqrt{5}$
3620 1706 sπ : $\sqrt[3]{19/2}$
3620 1976 10^x : $((e+\pi)/3)^{-3}$
3620 3093 2^x : $17/8$
3620 3471 2^x : $1/\ln(3\pi)$
3620 3777 rt : (4,0,5,2)
3620 4993 rt : (5,8,-9,0)
3620 5262 2^x : $3\sqrt{2}/2+4\sqrt{5}/3$
3620 5876 rt : (4,5,-7,-3)
3620 6435 sπ : $\sqrt[3]{20/3}$
3620 6687 e^x : $\sqrt{3}+\sqrt{5}+\sqrt{7}$
3620 7683 θ_3 : $\sqrt[3]{2}/7$
3620 8205 $\sqrt{5}-\exp(4)$
3620 8733 id : $e-3\pi/4$
3620 9183 ℓ_2 : $2\sqrt{3}+3\sqrt{5}/4$
3620 9539 $\exp(1/5)+\exp(\pi)$
3620 9656 sr : $3\sqrt{3}/2+4\sqrt{5}/3$
3621 0063 rt : (2,8,-8,-9)
3621 0110 J_1 : $18/23$
3621 0646 sr : $7e$
3621 1086 at : $1/4+2\sqrt{5}$
3621 1453 rt : (8,5,-2,-1)
3621 2448 rt : (6,9,7,8)
3621 3335 Ei : $3\sqrt[3]{2}/8$
3621 3550 cu : $1/2+3e/4$
3621 4711 rt : (8,6,-3,5)
3621 8573 rt : (2,9,2,-2)
3621 9643 Ei : $7\pi^2/4$
3622 0410 2^x : $\ln(\ln 2/3)$
3622 1531 ℓ_2 : $2+\pi$
3622 2291 at : $\sqrt{2}/3-3\sqrt{3}$
3622 4505 rt : (1,9,6,-6)
3622 4722 ℓ_{10} : $4\sqrt{2}-3\sqrt{5}/2$
3622 6898 e^x : $2+3\sqrt{5}$
3622 7585 rt : (7,1,7,-3)
3622 7653 ℓ_{10} : $7\pi^2/3$
3622 8836 rt : (4,8,-3,1)
3623 1727 rt : (3,7,-8,8)
3623 3081 rt : (5,5,-1,2)
3623 3399 sq : $3/2+4\pi/3$
3623 3639 erf : $(\ln 2)^3$
3623 5004 rt : (1,7,5,6)
3623 6387 cr : $3-\sqrt{2}/3$
3623 7489 sπ : $\sqrt{5}/2$
3623 8545 Ψ : $3(e+\pi)/4$
3623 9302 ζ : $9/19$
3623 9652 rt : (3,6,-4,-9)
3624 0221 J_1 : $(\sqrt[3]{3}/3)^{1/3}$
3624 0326 10^x : $3/2+4\pi/3$
3624 1909 cr : $9(e+\pi)/4$
3624 1958 Ψ : $21/11$
3624 2208 J_2 : $10\zeta(3)/3$
3624 3718 sπ : $7\sqrt[3]{2}/10$
3624 4700 2^x : $4+4\sqrt{5}/3$
3624 5866 rt : (5,6,-4,-2)
3624 5910 rt : (6,0,-6,7)
3624 6012 id : $1/\sqrt[3]{21}$
3624 6117 2^x : $2-2\sqrt{3}$
3624 6760 id : $(\ln 3/3)^{-3}$
3624 8761 ℓ_2 : $\sqrt[3]{17}$
3625 3556 at : $3+\sqrt{3}$
3625 3788 ln : $3+e/3$
3625 3907 10^x : $\sqrt[3]{8/3}$
3625 4139 rt : (2,9,3,0)
3625 5018 cr : $3\sqrt{2}+4\sqrt{5}$
3625 5537 Ψ : $8\pi^2$
3625 7007 ℓ_2 : $7/9$
3625 7856 rt : (1,8,8,-4)
3626 1504 rt : (8,4,-5,6)
3626 4138 rt : (4,1,-9,-4)
3626 4811 erf : $1/3$
3626 5084 rt : (7,9,0,1)
3626 6130 sq : P_2
3626 6307 ln : $2(e+\pi)/3$
3626 7546 θ_3 : $\sqrt[3]{3}/8$
3626 7640 rt : (9,3,6,-3)
3626 8454 e^x : $4\sqrt{2}+\sqrt{3}/4$
3626 8754 $\exp(2/5) \div \exp(\sqrt{2})$
3626 9586 J_0 : $\sqrt{23/2}$
3627 0282 tanh : $2\sqrt[3]{5}/9$
3627 2436 rt : (4,6,-9,-2)
3627 2482 cu : $5e\pi/4$
3627 3845 Ei : $11/6$
3627 4917 rt : (8,7,7,2)
3627 7028 sr : $13/7$
3627 7357 10^x : $1/2+\sqrt{2}/3$
3628 2165 rt : (6,7,5,9)
3628 5842 rt : (6,2,-8,-8)
3628 7391 rt : (2,8,-5,-8)
3629 0453 cu : $4\sqrt{3}/3-\sqrt{5}/4$
3629 0549 ln : $16/23$
3629 2014 rt : (9,5,-3,0)
3629 4068 $\exp(\pi) \times s\pi(1/4)$
3629 4491 Ψ : $6(1/\pi)$
3629 6327 Ψ : $\exp(\sqrt{2}\pi/3)$
3629 8210 rt : (8,2,-7,7)
3629 8730 at : $1/2+3\sqrt{2}$
3629 9252 rt : (5,8,5,1)
3629 9541 $\pi+\exp(1/5)$
3630 0085 rt : (6,5,-8,-5)
3630 1945 cr : $\sqrt{2}+\sqrt{5}/2$
3630 3818 rt : (2,0,8,3)
3630 3900 10^x : $1/2+3\sqrt{2}/4$
3630 4518 rt : (3,8,-3,-2)
3630 4895 at : $2/3+3e/2$
3630 5777 at : $5e\pi/9$
3630 9375 ζ : $9e/10$
3630 9653 rt : (5,3,-3,3)
3631 0282 at : $4+\sqrt{5}/3$
3631 0494 sr : $3\sqrt{3}/4+\sqrt{5}/4$
3631 1773 $4 \div s\pi(\sqrt{3})$
3631 3838 10^x : $(\pi/3)^{1/3}$
3631 3969 rt : (8,4,3,-2)
3631 4233 at : $2\sqrt[3]{5}/9$
3631 5563 rt : (4,7,-9,-4)
3631 6040 ℓ_{10} : $8\sqrt[3]{3}/5$
3631 6403 $\sqrt{2} \times s\pi(\sqrt{2})$
3631 8783 J_1 : $\pi/4$
3631 9024 id : $\sqrt{3}-\sqrt{6}-\sqrt{7}$
3632 0562 rt : (7,3,9,-4)
3632 3389 Ψ : $7\pi/5$
3632 5037 $1/4+\exp(\sqrt{2})$
3632 5172 e^x : $1+e/4$
3632 5358 rt : (1,7,-7,-6)
3632 6176 sr : P_{49}
3632 7126 $s\pi(1/5)-s\pi(2/5)$
3632 7775 rt : (8,0,-9,8)
3632 9157 rt : (8,9,-7,-6)
3632 9452 ℓ_{10} : $5\ln 2/8$
3632 9526 J_1 : $5\sqrt{2}/9$
3633 0010 at : $19/4$
3633 0154 rt : (7,7,-2,2)
3633 1086 J_1 : $11/14$
3633 2680 tπ : $7e\pi/2$
3633 2823 rt : (1,2,-5,-8)
3633 4819 cu : $\sqrt{3}-\sqrt{5}+\sqrt{6}$
3633 4973 sr : $3\sqrt{2}/2+2\sqrt{3}$
3633 6225 tπ : P_{43}
3633 7388 rt : (7,6,-7,-3)
3634 0048 rt : (1,8,9,-9)
3634 0444 2^x : $\sqrt{5}/5$
3634 1186 rt : (4,2,7,-3)
3634 1530 rt : (2,7,3,7)
3634 1780 rt : (2,9,-8,-4)
3634 2524 rt : (5,7,-8,8)
3634 3020 Ψ : $15/23$
3634 3388 rt : (6,7,-1,-1)
3634 3775 rt : (3,6,-3,-1)
3634 3813 ℓ_{10} : P_{10}
3634 3869 rt : (9,5,8,-4)
3634 4914 rt : (8,9,4,9)
3634 9087 rt : (5,3,1,-1)
3634 9936 ℓ_{10} : $\sqrt{3}/4$
3635 0317 sr : $1/2+e/2$
3635 1203 sr : P_{64}
3635 3427 tanh : $8/21$
3635 8566 2^x : $7/4$
3635 9033 10^x : $4-e/4$
3635 9647 at : $7e/4$
3635 9910 rt : (5,1,-5,4)
3636 1082 at : $2/3-\pi/3$
3636 3636 id : $4/11$
3636 4807 sq : $3\sqrt{2}/4-3\sqrt{3}/2$
3636 7120 rt : (1,6,-2,2)
3636 8548 Ψ : $22/5$
3636 8648 Ei : $(\sqrt{2}\pi)^{1/2}$
3637 0291 rt : (6,4,6,-3)
3637 0420 ℓ_{10} : $10\ln 2/3$
3637 0749 rt : (1,9,-9,-7)
3637 1275 rt : (5,8,-6,-3)
3637 4161 rt : (7,5,-4,3)
3637 4664 rt : (2,5,-7,-4)
3637 6104 e^x : $(\pi/2)^{-1/3}$
3637 7056 J : $1/\sqrt{19}$
3637 8156 ζ : $\exp(-1/(3\pi))$
3637 8272 id : $8(1/\pi)/7$
3637 8555 $1/3 \div \ln(2/5)$
3637 8671 cu : $1/3-\pi/3$
3637 8674 J : $\exp(-17\pi/4)$
3638 0598 rt : (1,7,-2,5)
3638 1961 Ψ : $(2e)^{-1/2}$
3638 3618 $\ln(4/5)-\exp(\pi)$
3638 3783 rt : (3,5,-8,-6)
3638 3832 ℓ_{10} : $3\sqrt[3]{3}/10$
3638 6344 rt : (8,7,-9,-5)
3638 6782 2^x : $6\sqrt{2}$
3638 6837 rt : (9,9,-3,2)
3638 8111 θ_3 : $\exp(-\sqrt[3]{5})$
3638 8171 rt : (1,2,6,7)
3638 8667 rt : (1,8,-4,-8)

3639 0253 $2/5 - s\pi(\sqrt{2})$
3639 0762 $e^x : 4 + 4\pi/3$
3639 0970 $J_0 : 5e/4$
3639 1839 $3/5 \div \exp(1/2)$
3639 2369 $J : 6(1/\pi)/7$
3639 3589 rt : (8,6,5,-3)
3639 3951 $\zeta : 22/9$
3639 5182 ln : $\ln(\ln 3)$
3639 5435 rt : (5,1,-7,-5)
3639 6103 id : $9\sqrt{2}/2$
3639 7023 $t\pi : 1/9$
3639 7895 at : 8/21
3639 8024 $\Psi : 1/7$
3639 8611 $J_1 : 5\sqrt[3]{2}/8$
3639 8757 rt : (7,5,0,-1)
3640 1917 $e^x : 5\sqrt{5}/4$
3640 2312 $\ell_{10} : 1/2 + 2e/3$
3640 4327 $\theta_3 : \sqrt[3]{5}/3$
3640 5915 rt : (7,3,-6,4)
3640 6886 rt : (2,7,-8,-2)
3640 6897 cu : $8 \ln 2/5$
3640 7137 tanh : $1/\sqrt[3]{18}$
3640 7424 $\Gamma : \sqrt{2} + \sqrt{3} + \sqrt{5}$
3640 7455 at : $\sqrt{2} + 3\sqrt{5}/2$
3640 7630 rt : (8,5,9,3)
3640 7902 $2^x : 3 + \sqrt{5}/2$
3640 7903 $e^x : 2/3 - 3\sqrt{5}/4$
3640 8160 rt : (4,4,9,-4)
3640 8301 rt : (4,7,-2,9)
3640 8887 $10^x : 4 \ln 2$
3640 9071 rt : (4,9,0,-1)
3640 9569 $2/5 \div \ln(3)$
3641 2814 $J_2 : 4$
3641 3551 rt : (7,8,-9,-4)
3641 4116 at : $2/3 - 2e$
3641 5389 rt : (9,7,-5,3)
3641 5847 rt : (3,9,-1,9)
3641 6692 rt : (3,1,-9,3)
3641 7821 cr : $1/2 + 3e/4$
3641 8029 $J_0 : (2e)^{1/3}$
3641 9209 as : $2 - 3\pi/4$
3642 1561 $\exp(5) + s\pi(2/5)$
3642 1882 rt : (5,3,-9,-6)
3642 2909 as : $2\sqrt{2}/3 - 3\sqrt{3}/4$
3642 4597 $Ei : \sqrt[3]{2}/10$
3642 4754 rt : (6,6,8,-4)
3642 7271 rt : (4,5,-8,6)
3642 7669 sq : $1/4 + \sqrt{2}/4$
3642 8377 $E : 11/16$
3642 9054 $Li_2 : 4 \ln 2/3$
3642 9559 $J_0 : 17/5$
3642 9911 rt : (7,1,-8,5)
3643 0707 $\zeta : \exp(-\sqrt[3]{2}/3)$
3643 1153 $10^x : 3\sqrt{2} + \sqrt{3}/4$
3643 1309 $J_2 : 6\sqrt[3]{5}/5$
3643 2180 rt : (9,7,-1,-1)
3643 3474 ln : $7\sqrt{5}/4$
3643 5402 sr : $5\sqrt{5}/2$
3643 5446 $\zeta : 7\pi/9$
3643 6408 $e^x : 1 + \sqrt{3}$
3643 7260 rt : (9,5,-7,4)
3643 7635 $\ell_{10} : \exp(\pi)$
3643 8747 rt : (8,8,7,-4)
3643 9834 rt : (8,7,-4,-2)
3644 0903 $\ln(\sqrt{2}) \div s\pi(2/5)$
3644 1197 $J_0 : 3(e + \pi)/10$
3644 1636 $s\pi : 1/3 + \pi/4$
3644 1869 tanh : $6(1/\pi)/5$
3644 2111 $t\pi : \exp(3/2)$
3644 2376 $1/5 \times \exp(3/5)$
3644 3148 cu : 5/7
3644 4805 rt : (1,7,8,-7)
3644 5801 cu : $1/4 - e/2$
3644 8060 sr : $\sqrt{2}/2 + 2\sqrt{3}/3$
3644 8125 $2^x : 4\sqrt{3}/3 - \sqrt{5}/4$
3644 9860 $Ei : 9(1/\pi)/10$
3645 0705 rt : (5,5,3,-2)
3645 0707 sinh : $1/\sqrt[3]{22}$
3645 1942 at : $1/\sqrt[3]{18}$
3645 4551 rt : (9,3,-9,5)
3645 5160 $\ln(3/4) + \exp(\sqrt{3})$
3645 6976 rt : (7,6,4,1)
3645 8045 ln : $8\sqrt[3]{2}/7$
3646 2406 $\zeta : 10\sqrt[3]{5}/7$
3646 3666 $10^x : 19/18$
3646 4170 $\ln(4/5) + s\pi(1/5)$
3646 4874 $10^x : 9\pi/7$
3646 5560 rt : (1,3,7,-3)
3646 5988 $\ell_2 : e/7$
3646 6298 $t\pi : 1/3 + 4\pi/3$
3646 8186 rt : (1,8,4,-5)
3646 8453 rt : (6,9,-3,-2)
3646 9393 sinh : $1/\ln(\ln 3)$
3646 9585 $\exp(1/2) - \exp(1/4)$
3646 9832 $E : 4\zeta(3)/7$
3646 9913 at : $3/2 - 2\pi$
3647 0061 rt : (7,7,2,-2)
3647 2713 $Ei : (e + \pi)/5$
3647 3620 $\pi - \ln(4/5)$
3647 4443 $J_1 : (e + \pi)/2$
3647 5929 $J_0 : 1/\ln(\sqrt{5}/3)$
3647 7181 $J_1 : 15/19$
3647 7410 sr : $(\sqrt{2\pi})^3$
3647 8123 $J : 1/2$
3647 8374 sinh : 5/14
3647 9176 at : $4 + \pi/4$
3648 1503 rt : (3,5,6,-3)
3648 4265 $2^x : 2/3 - 3\sqrt{2}/2$
3648 5049 rt : (9,9,1,-2)
3648 6382 at : $3/2 - \sqrt{5}/2$
3648 6893 at : $6(1/\pi)/5$
3648 8084 rt : (1,4,-9,-9)
3648 8254 $t\pi : 3e/2 - 4\pi/3$
3648 9165 sq : $2e/9$
3649 0228 $E : (\sqrt{2}/3)^{1/2}$
3649 0349 $2^x : \pi/7$
3649 0979 $2^x : 9e/4$
3649 1556 cu : $3/2 - \pi/4$
3649 1922 $10^x : 4 - \sqrt{3}/3$
3649 2895 as : $1/\sqrt[3]{22}$
3649 3561 $e^x : 2\sqrt{3} - 2\sqrt{5}$
3649 4130 rt : (5,8,6,6)
3649 5961 cr : $\sqrt{3}/2 + 3\sqrt{5}/4$
3649 7577 rt : (3,8,8,2)
3649 9118 $e^x : 4\sqrt[3]{2}/3$
3649 9914 rt : (8,9,-6,-3)
3650 1357 $\zeta : (\sqrt{2\pi})^{-1/2}$
3650 1874 at : $4\sqrt{2} - \sqrt{3}/2$
3650 1899 rt : (1,5,9,-4)
3650 4234 $\Psi : \sqrt[3]{7}$
3650 4376 rt : (7,9,4,-3)
3650 4793 $10^x : 4 - \sqrt{3}/2$
3650 5818 id : $4\sqrt{2} + 3\sqrt{5}$
3650 6043 sr : $5\sqrt{5}/6$
3650 6461 sinh : $5\sqrt{5}$
3650 6957 rt : (9,4,3,1)
3650 8013 cr : $4e + 3\pi/4$
3650 8699 $2^x : 9\sqrt{2}/2$
3651 0786 rt : (3,7,8,-4)
3651 3643 $2^x : P_{41}$
3651 4837 sr : 2/15
3651 5715 $Ei : 7\pi/2$
3651 5832 $10^x : 16/25$
3651 8235 rt : (5,9,7,-4)
3651 9839 $t\pi : 8\sqrt{5}$
3652 0722 as : 5/14
3652 1238 $\ln(3) \div \ln(\sqrt{5})$
3652 1655 $t\pi : 3\sqrt{2} - 3\sqrt{5}/2$
3652 1744 $J_0 : 5\sqrt[3]{5}/2$
3652 2355 $2^x : 4e - \pi/2$
3652 2739 at : $\sqrt{23}$
3652 2763 cu : $1/4 + \sqrt{5}$
3652 3001 rt : (2,4,-8,4)
3652 3159 $\ell_{10} : 4\sqrt{2}/3 + \sqrt{3}/4$
3652 4899 id : $7\sqrt[3]{3}/3$
3652 6283 $t\pi : \sqrt{5}/6$
3652 6295 rt : (5,6,7,2)
3652 6364 $2/5 \times \exp(5)$
3652 7007 $\sqrt{2} + s\pi(2/5)$
3652 8561 sr : $1/\ln(\sqrt[3]{5})$
3652 8660 rt : (3,0,1,9)
3652 9418 at : $4/3 + 2\sqrt{3}$
3653 0142 sinh : 19/17
3653 0152 cr : $2/3 + 4\pi$
3653 1973 cu : $4/3 + 3\sqrt{2}$
3653 2552 ln : $2 - \sqrt{5}/4$
3653 3568 rt : (5,4,2,8)
3653 4191 $Ei : (3\pi)^{-1/3}$
3653 4205 sq : $3e + 4\pi/3$
3653 5124 $J_2 : \exp(\sqrt[3]{3}/2)$
3653 5883 id : $1/\ln(\sqrt[3]{3}/3)$
3653 7866 rt : (7,8,3,7)
3653 7935 $2^x : 4\sqrt{2}/3 + 3\sqrt{5}$
3654 0093 at : 24/5
3654 0818 cr : $9\sqrt{2}/5$
3654 2667 $4 \div \ln(2/5)$
3654 2694 rt : (7,4,6,2)
3654 4181 rt : (5,2,0,9)
3654 5877 id : $2\sqrt{2}/3 - \sqrt{3}/3$
3654 6038 ln : $1 - e/3$
3654 6118 sr : $\zeta(3)/9$
3654 6986 at : $3\sqrt{2} + \sqrt{5}/4$
3654 7288 rt : (7,6,1,8)
3654 9372 rt : (3,5,-2,7)
3655 1084 $\zeta : \exp(-\sqrt{5}/3)$
3655 3294 rt : (9,2,5,2)
3655 3463 sr : $\sqrt{2} + \sqrt{3} + \sqrt{6}$
3655 4314 rt : (7,4,-1,9)
3655 5909 rt : (3,3,-8,4)
3655 6457 rt : (9,8,0,8)
3655 6812 cr : $8(1/\pi)$
3655 6838 $\ell_{10} : 3 - e/4$
3655 7388 $Li_2 : (\sqrt[3]{5}/2)^{1/2}$
3655 7727 cr : $1/\ln(\sqrt{2}\pi/3)$
3655 8663 sr : $\sqrt{13}\pi$
3655 8970 rt : (5,4,9,3)
3656 1003 $t\pi : \ln(\sqrt{5}/2)$
3656 1450 rt : (9,6,-2,9)
3656 1828 $10^x : 3/4 + 3\sqrt{5}$
3656 2226 $J_1 : 19/24$
3656 2819 at : $1/3 + 2\sqrt{5}$
3656 3667 $\ell_{10} : \sqrt[3]{25/2}$
3656 3702 $10^x : (e)^{-2}$
3656 3702 $\ell_{10} : 3/4 + \pi/2$
3656 4873 rt : (7,2,8,3)
3656 5127 $\zeta : (4/3)^{-1/3}$
3656 6415 rt : (5,9,-5,4)
3656 6482 $t\pi : \sqrt{2}/2 - 3\sqrt{3}/4$
3656 7641 $s\pi : 9 \ln 2/2$
3656 8685 at : $3\sqrt{2}/2 - 4\sqrt{3}$
3656 8828 $2^x : 5\sqrt{5}/9$
3656 9369 rt : (9,0,7,3)
3656 9654 cr : $3/2 + \pi/3$
3656 9668 $3/4 \times \exp(e)$
3657 0351 $2^x : 1/\sqrt[3]{11}$
3657 1239 at : $10\sqrt[3]{3}/3$
3657 4041 $\lambda : \exp(-17\pi/15)$
3657 4262 at : $4\zeta(3)$
3657 7315 $e^x : 4 + 2\pi/3$
3657 7532 rt : (9,2,9,-4)
3657 9230 $\ell_2 : 4e/3 - \pi/3$
3658 0128 sinh : $9(1/\pi)/8$
3658 1658 $J_1 : 8 \ln 2/7$
3658 1875 $\Psi : 3/7$
3658 2244 $\theta_3 : 2/11$
3658 3493 $\Psi : -\sqrt{3} - \sqrt{6} + \sqrt{7}$
3658 3585 at : $\exp(\pi/2)$
3658 3662 $10^x : 2 + 2\sqrt{5}/3$
3658 3779 $Li_2 : (\ln 2)^3$
3658 5198 $\zeta : 2e\pi/7$
3658 7442 $2/3 \div \exp(3/5)$
3658 8861 $\ell_2 : 2 + \sqrt{3}/3$
3659 0669 at : $3 + 2e/3$
3659 1006 rt : (1,9,-3,4)
3659 2693 at : $e + 2\pi/3$
3659 5634 sinh : $\sqrt{5}/2$
3659 7232 $\theta_3 : 4(1/\pi)/7$
3659 8246 $J : 1/\sqrt[3]{23}$
3659 8556 $s\pi(1/5) \div s\pi(\pi)$
3659 9591 $2^x : e/3 - 3\pi/4$

3660 2255 $\ln(3/5) \div \exp(1/3)$
3660 2540 id : $1/2 + \sqrt{3}/2$
3660 3967 ℓ_{10} : $4 - 3\sqrt{5}/4$
3660 4025 2^x : $9/20$
3660 4403 J_0 : $e\pi/2$
3660 6909 at : $3\sqrt{3}/2 - 4\sqrt{5}/3$
3660 7009 cr : $e/4 + 4\pi$
3660 7732 e^x : $e/2 - \pi/3$
3661 0965 sr : $\sqrt[3]{13/2}$
3661 2157 sinh : $9\zeta(3)$
3661 3264 $\ln(\pi) + \exp(1/5)$
3661 3461 J_2 : $10\sqrt[3]{3}/7$
3661 6231 Ψ : $7\sqrt[3]{2}/2$
3662 0409 id : $\ln(\sqrt[3]{3})$
3662 1322 Li_2 : $1/3$
3662 1663 e^x : $-\sqrt{2} - \sqrt{5} + \sqrt{7}$
3662 2766 Ei : $\exp(1/(2\pi))$
3662 2920 at : $3\sqrt{2} + \sqrt{3}/3$
3662 3066 as : $9(1/\pi)/8$
3662 3356 $\exp(4/5) + \exp(\pi)$
3662 4989 sinh : $\sqrt{3} - \sqrt{6} + \sqrt{7}$
3662 5125 Ei : $9/19$
3662 7252 $s\pi$: $3(1/\pi)/8$
3662 7338 rt : (2,9,3,8)
3663 1053 sr : $((e+\pi)/3)^{-3}$
3663 3623 at : $4\sqrt{2}/3 - 3\sqrt{5}$
3663 4757 sq : $3/4 - 2\sqrt{3}$
3663 5320 Ψ : $\sqrt{11/3}$
3663 5573 ln : $6\zeta(3)/5$
3663 5737 cu : $(\sqrt{5}/2)^{-3}$
3663 6041 cu : $\sqrt{2}/4 + \sqrt{5}$
3663 6048 2^x : $1/\ln(\sqrt{5})$
3664 0984 J_1 : $\sqrt[3]{1/2}$
3664 1194 $t\pi$: $3\pi^2/8$
3664 4012 ζ : $(e/3)^{-1/2}$
3664 4221 cr : $3\sqrt{2}/4 + 2\sqrt{5}/3$
3664 5639 at : $2\sqrt{2}/3 - \sqrt{5}/4$
3664 6409 at : $\sqrt{2}/4 + 2\sqrt{5}$
3665 1100 E : $13/19$
3665 1292 id : $\ln(\ln 2)$
3665 1629 $2/5 \times \ln(2/5)$
3665 3543 at : $3/4 + 3e/2$
3665 5350 10^x : $5\sqrt[3]{3}/3$
3665 7675 at : $2 + 2\sqrt{2}$
3665 8818 e^x : $3e/2 + 3\pi$
3665 8910 $3/4 \times \exp(3/5)$
3665 9193 ln : $1/2 + 2\sqrt{2}/3$
3665 9203 id : $2\sqrt{2}/3 - 4\sqrt{3}/3$
3666 0511 cr : $2/3 + 4\sqrt{2}/3$
3666 1244 J_2 : $7\sqrt[3]{5}/3$
3666 3399 rt : (3,9,-4,-1)
3666 5944 E : $2\sqrt[3]{5}/5$
3666 6566 Ψ : $(\ln 3/2)^3$
3666 7164 rt : (8,5,8,-4)
3667 0894 tanh : $5/13$
3667 3192 rt : (6,3,9,-4)
3667 4994 $\exp(\pi) - s\pi(e)$
3667 6039 sr : $3/4 + \sqrt{5}/2$
3667 7080 2^x : $3\zeta(3)/8$
3668 1162 at : $3\sqrt{3}/2 + \sqrt{5}$
3668 3390 rt : (2,9,-8,7)
3668 3794 e^x : $5/16$
3668 5267 Ψ : $(e+\pi)/9$
3668 6390 sq : $20/13$
3668 6758 rt : (8,3,6,-3)
3668 7124 J_1 : $\sqrt[3]{25}$
3668 8681 10^x : P_{74}
3668 9427 cr : $-\sqrt{2} + \sqrt{3} + \sqrt{5}$
3669 1037 $t\pi$: $\exp(1/(3\pi))$
3669 2164 id : $\sqrt{3}/3 - 4\sqrt{5}$
3669 4514 $\ln(2/5) \times \exp(2/5)$
3669 5541 tanh : $2\sqrt{3}/9$
3669 6694 ℓ_{10} : $\sqrt{2} - \sqrt{3} + \sqrt{7}$
3669 7059 cr : $3\sqrt{2}/2 + \sqrt{3}/4$
3669 7300 rt : (6,1,7,-3)
3669 8458 ln : $2\sqrt{3}/5$
3669 9231 10^x : $19/11$
3670 0001 $1/4 + \exp(3/4)$
3670 0309 $e + \exp(1/2)$
3670 0683 rt : (3,9,-8,-4)
3670 2456 Ei : $\exp(\sqrt[3]{2}/3)$
3670 4588 10^x : P_{23}
3670 4745 rt : (1,4,0,7)
3670 4858 rt : (8,1,-7,9)
3670 9145 J_0 : $2e\pi/5$
3671 0870 rt : (4,1,8,3)
3671 0986 rt : (5,7,-9,-4)
3671 1253 tanh : $5\ln 2/9$
3671 1305 rt : (9,8,-1,-1)
3671 3081 $t\pi$: $(\sqrt[3]{4}/3)^{-2}$
3671 3358 $\pi + \exp(4/5)$
3671 5457 Ei : $7\pi/4$
3671 5799 at : $\sqrt{2}/4 - 3\sqrt{3}$
3671 5831 sinh : $5\pi/4$
3671 6062 rt : (8,3,-5,8)
3671 7092 ln : $2\sqrt{2}/3 + 4\sqrt{5}/3$
3671 7383 at : $5/13$
3671 7985 cr : $8\sqrt{5}/7$
3671 8864 cr : $23/9$
3671 8866 ζ : $11/7$
3671 9903 rt : (1,9,-5,-3)
3672 0818 rt : (6,1,-6,-9)
3672 1187 J_1 : $5(1/\pi)/2$
3672 3392 rt : (8,1,4,-2)
3672 5509 $e - \exp(3)$
3672 5991 rt : (4,7,-5,2)
3672 7262 e^x : P_5
3672 7304 sq : $8\pi^2/5$
3672 8994 rt : (2,3,9,3)
3673 0021 rt : (8,5,-3,7)
3673 0265 Ei : $16/5$
3673 1627 ln : $4\sqrt{2} - \sqrt{3}$
3673 3656 ℓ_{10} : $2 - \pi/2$
3673 4693 sq : $19/7$
3673 4723 as : $1 - e/2$
3673 4769 ℓ_{10} : $6e/7$
3673 4801 e^x : $-\sqrt{2} + \sqrt{3} + \sqrt{7}$
3673 4950 rt : (3,7,-6,-3)
3673 5539 10^x : $4 + 4\sqrt{2}/3$
3673 6051 rt : (6,1,-4,8)
3673 6513 cu : $9(1/\pi)/4$
3673 8493 rt : (3,8,-5,7)
3673 8614 sr : $9\sqrt[3]{2}$
3674 1554 $\exp(\sqrt{2}) + s\pi(\sqrt{3})$
3674 2190 at : $2\sqrt{3}/9$
3674 2682 rt : (4,9,1,5)
3674 2884 ζ : $10\sqrt{2}/9$
3674 2934 rt : (7,8,-8,-8)
3674 6374 rt : (6,1,5,2)
3674 7149 E : $(\pi)^{-1/3}$
3674 7568 rt : (7,8,7,-4)
3674 7891 rt : (8,7,-1,6)
3674 8454 rt : (1,9,2,-2)
3675 0801 10^x : $\sqrt{2} + 4\sqrt{5}$
3675 3236 cu : $3\sqrt{2}$
3675 3969 2^x : $\ln(\pi/2)$
3675 4300 at : $7\ln 2$
3675 4446 rt : (5,5,-7,-3)
3675 5785 rt : (6,3,-2,7)
3675 6022 rt : (1,8,-2,9)
3675 6200 Ψ : $6\sqrt{5}/7$
3675 8005 at : $5\ln 2/9$
3675 8716 Ψ : $23/12$
3676 0161 $\exp(\sqrt{3}) \div s\pi(\sqrt{5})$
3676 0366 rt : (5,6,8,-4)
3676 1136 ℓ_{10} : $3/4 - \sqrt{2}/2$
3676 2415 cu : $2 + 3\sqrt{5}$
3676 2709 id : $e/4 - \pi/3$
3676 2738 at : $3/2 + 3\sqrt{5}/2$
3676 3268 sq : $1/2 - 3e/4$
3676 3851 E : $1/\ln(\ln 2/3)$
3676 4068 sq : $2 + 3\sqrt{2}/4$
3676 4458 $\exp(\sqrt{2}) \div \exp(1/5)$
3676 4591 rt : (4,1,-3,-8)
3676 4980 rt : (5,4,-9,-7)
3676 4996 $\pi - s\pi(e)$
3676 6223 ℓ_{10} : $(2e)^{1/2}$
3676 7151 $t\pi$: $3\sqrt{3}/4$
3676 7644 e^x : $2/3 - \sqrt{2}/4$
3676 8252 at : $2/3 + 4\pi/3$
3677 1074 rt : (9,8,4,-3)
3677 1582 rt : (8,9,1,5)
3677 2318 e^x : $4/3 - \sqrt{2}/3$
3677 2469 rt : (1,7,-3,-2)
3677 2478 ln : $9/13$
3677 2529 id : $2\sqrt{2} - 3\sqrt{3}$
3677 3468 $\ln(2/5) + \exp(1/4)$
3677 4474 rt : (2,7,0,-3)
3677 5654 rt : (3,4,9,-4)
3677 8239 sr : $\sqrt{7/2}$
3677 8843 J_0 : $2\sqrt[3]{5}$
3678 0207 rt : (4,3,6,2)
3678 0250 sinh : $2\sqrt[3]{2}/7$
3678 0563 cr : $2 + \sqrt{5}/4$
3678 0722 rt : (7,3,-8,-3)
3678 2360 rt : (6,5,0,6)
3678 2654 sinh : $9/25$
3678 4263 $5 \div \ln(\pi)$
3678 4683 sq : $7\ln 2/8$
3678 6903 sq : $2/3 - 3\pi/2$
3678 7343 ln : $5\pi/4$
3678 7406 cr : $1/4 + 4\sqrt{3}/3$
3678 7944 id : $\exp(-1)$
3678 8459 Ei : $1/20$
3678 8610 rt : (7,6,5,-3)
3678 8745 Ψ : $2\sqrt[3]{5}/3$
3678 8757 e^x : $17/14$
3678 8875 ln : $1 - \sqrt{5}/3$
3679 0151 $\ln(4/5) \times \exp(1/2)$
3679 1955 $s\pi$: $8\sqrt[3]{2}/9$
3679 4614 rt : (4,1,-1,7)
3679 5263 rt : (3,9,-1,-1)
3679 7678 ℓ_{10} : $3/7$
3679 8075 id : $4\sqrt[3]{5}/5$
3679 8166 at : $1/3 - 3\sqrt{3}$
3679 8174 at : $3e/2 + \pi/4$
3680 2649 J_1 : $(\pi/2)^{-1/2}$
3680 2845 ln : $\sqrt{3} + 4\sqrt{5}$
3680 3398 id : $1/4 + \sqrt{5}/2$
3680 4702 at : $3/2 - 4\sqrt{2}/3$
3680 5637 Ei : $2/7$
3680 6006 rt : (3,5,-4,-2)
3680 6299 sq : $\sqrt[3]{20}$
3680 6937 Ψ : $(\sqrt{2}\pi/3)^{1/3}$
3680 8631 sr : $3e - 2\pi$
3680 8670 rt : (2,3,-2,5)
3680 8695 ℓ_2 : P_{27}
3680 9501 2^x : $\exp(-\sqrt[3]{4}/2)$
3680 9833 cr : $3/2 + 3\sqrt{2}/4$
3681 0238 rt : (5,4,6,-3)
3681 1167 at : $4 + \sqrt{3}/2$
3681 2003 E : $15/22$
3681 2455 $s\pi$: $3/25$
3681 2812 rt : (9,7,-8,-1)
3681 3132 J_1 : $\ln(\sqrt{2}\pi/2)$
3681 5286 at : $4\sqrt{2}/3 + 4\sqrt{5}/3$
3681 6293 rt : (8,1,2,1)
3681 7956 cu : $3/2 + \sqrt{2}/4$
3681 8102 rt : (9,1,-9,-3)
3682 0074 rt : (6,7,2,5)
3682 0626 at : $e/4 + 4\pi/3$
3682 0825 $\exp(1/4) - \exp(\sqrt{3})$
3682 1091 $\ln(\sqrt{3}) \div \exp(2/5)$
3682 3085 $t\pi$: $(3\pi)^{1/3}$
3682 3766 10^x : $2\sqrt{2}/3 + 4\sqrt{3}$
3682 4371 as : $2\sqrt[3]{2}/7$
3682 4960 rt : (7,3,-9,0)
3682 6789 as : $9/25$
3682 8324 rt : (8,8,-9,-4)
3682 9092 rt : (9,6,2,-2)
3683 2271 cr : $3e\pi/10$
3683 3827 sinh : $5\sqrt{5}/7$
3683 5045 rt : (2,5,7,2)
3683 7035 cr : $3/4 + 2e/3$
3683 7593 rt : (3,2,7,-3)
3683 8283 rt : (4,9,7,-4)
3683 8468 rt : (9,2,-9,7)
3683 9485 rt : (1,7,0,-1)
3684 0314 id : $1/\sqrt[3]{20}$
3684 2105 id : $7/19$

3684 2583 sinh : $\sqrt[3]{3}/4$
3684 5436 sq : $9\sqrt[3]{5}/10$
3684 5534 $s\pi : (\ln 2/2)^2$
3684 7757 $\ln(2/3) + s\pi(e)$
3684 8279 $\ell_2 : \sqrt{5/3}$
3684 8410 sinh : $3\zeta(3)/10$
3684 9420 rt : (9,9,-6,-2)
3685 4066 $10^x : 3/2 - \sqrt{3}/3$
3685 6193 rt : (5,3,-5,-2)
3685 6461 rt : (5,9,-8,-3)
3685 6812 $s\pi : 4e\pi/7$
3685 7102 $2^x : 2/3 + \sqrt{3}/3$
3685 9841 rt : (2,7,8,-4)
3686 0847 at : $\sqrt{2} + 2\sqrt{3}$
3686 1985 rt : (7,4,3,-2)
3686 3559 rt : (9,4,-7,6)
3686 3747 $\ln(2) + s\pi(\sqrt{5})$
3686 5082 $\Psi : \sqrt{2} - \sqrt{3} + \sqrt{5}$
3686 6344 rt : (7,5,-7,-1)
3686 7027 at : $4e\pi/7$
3686 8554 $\zeta : \sqrt{5}/3$
3686 9638 $2^x : 3/2 - \pi/3$
3687 1502 rt : (1,9,6,6)
3687 2181 rt : (2,6,-1,5)
3687 2529 $s\pi : \zeta(3)/10$
3687 2879 $e^x : 1 + 4\sqrt{3}/3$
3687 3318 rt : (1,0,8,3)
3687 4210 rt : (7,0,-8,7)
3687 6166 $Ei : (\sqrt{2}\pi)^{-1/2}$
3687 6937 $\Psi : 5\sqrt{5}$
3687 7772 rt : (6,9,4,4)
3687 9665 sq : $3e + 4\pi$
3688 0642 at : $5(e+\pi)/6$
3688 0666 rt : (6,8,-6,-3)
3688 3328 $J_1 : ((e+\pi)/3)^{-1/3}$
3688 4204 $J_1 : 4/5$
3688 5775 rt : (5,9,-4,0)
3688 6437 cr : $3\sqrt[3]{5}/2$
3688 7077 as : $\sqrt[3]{3}/4$
3688 7450 $10^x : 3/22$
3688 8020 $\theta_3 : \ln(\sqrt[3]{3}/2)$
3688 9011 $\ell_2 : e/3 + 3\pi$
3689 0131 ln : $8\pi^2$
3689 0254 at : $3 + 4\sqrt{2}/3$
3689 2330 cu : $6\sqrt{3}$
3689 2838 $2^x : e/6$
3689 2939 as : $3\zeta(3)/10$
3689 3817 cu : $3\sqrt{2} - 4\sqrt{3}$
3689 4612 rt : (9,6,-5,5)
3689 5003 sq : $3\sqrt{3}/2 + \sqrt{5}$
3689 5240 $e^x : 2 - \pi/4$
3689 6496 $e^x : e/2 - 3\pi/4$
3689 7568 cr : $3\sqrt{2} - 3\sqrt{5}/4$
3689 9146 $Li_2 : (\sqrt[3]{2})^{-1/3}$
3689 9802 ln : $4/3 + 3\sqrt{3}/2$
3690 0500 $\ell_2 : 4/3 - \sqrt{5}/4$
3690 0696 rt : (6,3,3,1)
3690 0888 ln : $\sqrt{3}/3 + 3\sqrt{5}/2$
3690 3060 rt : (2,9,0,-5)
3690 3094 cu : $3\sqrt{2}/4 + 2\sqrt{5}$
3690 3253 $J_0 : 7/4$
3690 4751 $10^x : 3(1/\pi)/7$
3690 6130 rt : (5,2,4,-2)
3690 6966 $\ln(\sqrt{5}) \div s\pi(1/5)$
3690 6998 $\ell_{10} : \sqrt[3]{5}/4$
3690 7024 $\ln(2/3) \div \ln(3)$
3690 7041 rt : (9,9,9,-5)
3690 7519 $\theta_3 : (\sqrt[3]{3}/2)^{1/3}$
3690 7621 rt : (4,7,5,-3)
3690 8023 rt : (7,2,-6,6)
3690 8318 rt : (3,6,-6,-8)
3690 8469 $\zeta : \pi/2$
3690 9544 $\zeta : \sqrt{5} - \sqrt{6} + \sqrt{7}$
3691 0238 rt : (1,5,-1,-1)
3691 0777 $e^x : \pi/10$
3691 0796 rt : (5,8,-7,5)
3691 0890 $e^x : 4\sqrt{3} - \sqrt{5}/3$
3691 1956 at : $\exp(\sqrt[3]{4})$
3691 4862 $erf : e/8$
3691 6821 sinh : $8\sqrt[3]{2}/9$
3691 7581 rt : (2,9,2,-9)
3691 8623 id : $4\sqrt{3} - \sqrt{5}/4$
3691 9522 at : $9e/5$
3691 9977 $J_0 : 17/4$
3692 2002 rt : (1,2,-9,3)
3692 2422 cr : $3 - \sqrt{3}/4$
3692 2734 rt : (5,2,-7,-7)
3692 3698 $Ei : \exp(-\sqrt{5}/3)$
3692 4281 rt : (8,6,-7,-3)
3692 4597 $J_0 : 24/7$
3692 5894 $e^x : 2\sqrt{2}/9$
3692 6597 rt : (7,7,-5,-2)
3692 7447 sr : 3/22
3692 8206 $\Psi : (2e)^2$
3692 9778 $\exp(\pi) \div s\pi(1/5)$
3693 0639 sr : 15/8
3693 1803 cu : $4 + 3\sqrt{2}/2$
3693 2328 $J_1 : 5\sqrt[3]{3}/9$
3693 2871 $\ell_{10} : 4/3 - e/3$
3693 3013 rt : (8,9,1,-2)
3693 3687 $E : 17/25$
3693 4037 rt : (9,8,-3,4)
3693 4411 rt : (8,2,9,-4)
3693 4878 sr : $3(1/\pi)/7$
3693 5412 $\theta_3 : 4/7$
3693 5864 as : $4/3 - \sqrt{2}/4$
3693 6764 $\ln(2/3) + s\pi(\sqrt{2})$
3693 7004 $J_1 : 2\zeta(3)/3$
3693 8947 rt : (3,6,-8,-8)
3693 9109 $\ln(3/4) \times \exp(1/4)$
3693 9806 rt : (7,8,2,0)
3694 3516 id : $9(e+\pi)/2$
3694 3840 at : $\sqrt{24}$
3694 4623 $2^x : 4 - 2e$
3694 4732 id : $3e - \pi/4$
3694 4796 rt : (2,5,6,-3)
3694 5346 rt : (9,3,8,3)
3694 7706 rt : (9,4,0,-1)
3694 7887 rt : (2,8,-3,1)
3695 1433 rt : (7,4,-4,5)
3695 3054 $e^x : 2/3 + \pi/2$
3695 5478 $e^x : 1/\ln(\ln 3/3)$
3695 6233 $\exp(1/4) + \exp(3)$
3695 6447 rt : (5,3,-6,-1)
3695 6671 cr : $\sqrt{2} + 2\sqrt{3}/3$
3695 8755 $e^x : 8e\pi/7$
3696 0135 at : P_{27}
3696 2359 $E : e/4$
3696 3238 ln : $(2\pi/3)^{1/2}$
3696 3827 $10^x : 2/3 + 2\sqrt{5}$
3696 4430 rt : (4,8,-3,-2)
3696 5329 cr : $1/3 + \sqrt{5}$
3696 6073 sq : $1/2 - 2e$
3696 6335 rt : (1,8,-8,-4)
3696 6456 $s\pi : \exp(2\pi/3)$
3696 8487 at : $\sqrt{3}/4 + 2\sqrt{5}$
3696 8560 rt : (3,0,5,2)
3696 9878 $4/5 + s\pi(\pi)$
3697 0774 rt : (5,0,-5,6)
3697 2258 at : $4 + e/3$
3697 3682 rt : (9,7,7,-4)
3697 5291 at : $3/4 - 4\sqrt{2}$
3697 6994 rt : (6,7,2,-2)
3697 9409 sinh : $3e\pi/4$
3698 1032 $2^x : 7\sqrt{3}/5$
3698 1843 rt : (2,7,-6,7)
3698 4770 $e^x : (\sqrt[3]{2}/3)^{-3}$
3698 5068 at : $2/3 + 3\sqrt{2}$
3699 0112 cr : $1 + \pi/2$
3699 0529 $\ell_{10} : 2/3 + 3\sqrt{5}/4$
3699 1616 tanh : $e/7$
3699 2407 rt : (3,8,0,-5)
3699 2476 $e^x : 3e + \pi/3$
3699 2698 $Ei : 2e\pi$
3699 2775 $J_1 : \sqrt{17/2}$
3699 3793 at : $4\sqrt{2} - \sqrt{5}/3$
3699 4651 rt : (1,9,-4,-5)
3699 4830 $\ell_{10} : 4(e+\pi)$
3699 6891 $Ei : 4\sqrt{2}/7$
3699 8731 cr : $\sqrt[3]{17}$
3699 9131 rt : (7,5,9,3)
3700 1342 cr : 18/7
3700 1848 sq : $(\sqrt{2}\pi)^{-1/3}$
3700 3080 $\ell_{10} : 1/4 + 2\pi/3$
3700 3689 $J_2 : 6\sqrt{3}/5$
3700 3949 ln : $\sqrt{2}/3 + 2\sqrt{3}$
3700 4257 id : $\sqrt{3}/2 - \sqrt{5}$
3700 4273 $10^x : 1/\sqrt[3]{19}$
3700 4578 rt : (3,6,-9,-4)
3700 6124 id : $3\sqrt{2}/4 + 4\sqrt{3}/3$
3700 6373 rt : (7,5,8,-4)
3700 7907 $s\pi : \sqrt{3}/2 - \sqrt{5}/3$
3700 9195 rt : (7,6,-2,4)
3701 0154 $1/4 \div s\pi(\sqrt{5})$
3701 1914 $Ei : 10\pi^2/9$
3701 3283 $\Psi : \ln(\sqrt{5}/2)$
3701 4032 sq : $2/3 + 4\sqrt{5}$
3701 5314 as : $7\sqrt[3]{2}/9$
3701 5602 $s\pi : e/3 - \pi/4$
3701 6771 rt : (1,8,-7,-8)
3701 8134 $\ell_{10} : \sqrt{11/2}$
3701 8372 sr : P_{24}
3701 9605 ln : $3 - 4\sqrt{3}/3$
3701 9781 rt : (6,9,-9,8)
3702 1810 rt : (3,7,-9,5)
3702 2356 rt : (7,9,-3,-3)
3702 2452 rt : (1,7,-3,-2)
3702 2564 $t\pi : \sqrt{3}\pi$
3702 3226 $e^x : \sqrt[3]{2}/4$
3702 4704 $J_2 : 3\ln 2$
3702 5619 cr : $3/4 + 4\pi$
3702 5670 sq : $7\sqrt[3]{3}/4$
3702 7081 $\ln(\pi) + \exp(4/5)$
3703 0403 rt : (7,2,1,-1)
3703 3392 sr : $e/3 + 3\pi/2$
3703 3551 $J_0 : 3\sqrt{2}$
3703 4052 $\ell_{10} : 2\sqrt{3} - \sqrt{5}/2$
3703 4603 rt : (4,7,5,4)
3703 4946 $\Psi : 13/20$
3703 5098 $2^x : 5/11$
3703 5406 rt : (9,9,2,9)
3703 5746 rt : (5,2,-3,5)
3703 6290 $2^x : 4 + 3\sqrt[3]{3}/4$
3703 6588 rt : (1,8,5,-3)
3703 7037 id : $(4/3)^3$
3703 8462 $J_2 : \sqrt[3]{9}$
3704 0195 tanh : 7/18
3704 0223 at : $e/7$
3704 1453 rt : (3,4,-4,8)
3704 1585 $t\pi : 2\sqrt{2}/3 + 4\sqrt{5}$
3704 2113 tanh : $1/\sqrt[3]{17}$
3704 2134 as : $3\sqrt{3}/2 - \sqrt{5}$
3704 2138 rt : (9,9,-6,-3)
3704 2453 ln : $4e - 3\pi$
3704 3610 at : $e/2 - 2\pi$
3704 3618 $\Gamma : 3/8$
3704 4892 sinh : $1/\sqrt[3]{21}$
3704 5340 $J_2 : (\ln 2/3)^{-1/2}$
3704 5453 rt : (5,3,9,-4)
3704 6415 rt : (8,0,7,3)
3704 6802 $e^x : 3\sqrt{2}/4 + 4\sqrt{5}$
3705 0620 as : $e - 3\pi/4$
3705 1039 sq : 14/23
3705 1550 $e + \exp(\sqrt{3})$
3705 2477 $2^x : 10(1/\pi)/7$
3705 5645 rt : (2,7,0,-8)
3705 8227 cu : $2 - e$
3706 0069 at : $2 - 4\sqrt{3}$
3706 0312 $\ell_2 : 4 - \sqrt{2}$
3706 3831 rt : (1,2,6,2)
3706 4935 sr : $4 - 3\sqrt{2}/2$
3706 5689 $J_1 : \ln(\sqrt{5})$
3706 5974 $J_2 : (\ln 2)^{-2}$
3706 6049 rt : (3,7,1,3)
3706 7617 $e^x : 3\sqrt{2}/4 - \sqrt{5}/3$
3706 7948 $J_0 : \sqrt[3]{16/3}$
3706 8549 sq : $4/3 + 3e/4$
3706 8903 rt : (8,3,-8,4)
3706 9061 $t\pi : 5(e+\pi)$
3707 0687 rt : (6,8,7,-4)

3707 1471 ζ : 17/7
3707 2329 J_2 : $\sqrt{13/3}$
3707 3913 rt : (5,5,-4,-2)
3707 7046 2^x : $1/2+\sqrt{5}/3$
3707 7555 rt : (9,5,5,-3)
3707 8720 $\ln(\pi) - s\pi(e)$
3708 0731 rt : (4,9,-8,-4)
3708 3689 $t\pi$: $2\sqrt{2}-3\sqrt{5}/2$
3708 5193 2^x : $4\sqrt{2}-\sqrt{3}/4$
3708 6132 at : $\pi^2/2$
3708 6584 rt : (8,7,-1,-1)
3708 9128 at : 7/18
3708 9810 rt : (7,8,0,3)
3709 0616 as : $1/\sqrt[3]{21}$
3709 0687 rt : (6,1,-9,-5)
3709 0811 2^x : $4/3+\sqrt{3}$
3709 1059 at : $1/\sqrt[3]{17}$
3709 1494 $s\pi$: $4\sqrt{2}+2\sqrt{3}$
3709 2218 rt : (1,9,9,-2)
3709 3078 at : $1/2-2e$
3709 6277 cu : $\sqrt{2}/2-2\sqrt{5}$
3709 7206 rt : (4,5,4,1)
3709 8446 cr : $\sqrt{18}\pi$
3709 9537 10^x : P_{24}
3710 0972 rt : (1,6,-5,-7)
3710 1017 ln : $4+3\sqrt{5}$
3710 1850 at : $3/4+4\pi/3$
3710 3376 cr : $4e/3-\pi/3$
3710 3395 10^x : $5e/9$
3710 4798 rt : (8,6,-9,-3)
3710 5096 $\ln(\sqrt{3}) \times s\pi(\sqrt{5})$
3710 5265 sq : $3/4-e/2$
3710 5579 $s\pi$: $8(e+\pi)$
3710 6427 cr : $2+\sqrt{3}/3$
3710 6689 e^x : $(\sqrt{5}/3)^{1/2}$
3710 7984 J_2 : 25/12
3710 8345 $1/3 \times \exp(\sqrt{2})$
3710 8828 rt : (9,1,-7,2)
3710 9015 Ψ : $3\pi^2$
3710 9930 rt : (6,2,8,3)
3711 2828 10^x : $4\sqrt{3}-3\sqrt{5}/4$
3711 3433 rt : (4,6,8,-4)
3711 4251 cu : $\sqrt{2}-3\sqrt{5}$
3711 4632 rt : (4,7,-8,-5)
3711 5422 10^x : $3\sqrt{2}/2+\sqrt{3}/4$
3711 6518 $\exp(3/4) + s\pi(\sqrt{3})$
3711 7665 at : $4+2\sqrt{2}/3$
3711 9596 rt : (2,8,0,-1)
3712 0541 at : $\sqrt{2}/3+2\sqrt{5}$
3712 1790 10^x : $3/4+\sqrt{5}/4$
3712 2188 at : $3\sqrt{2}/2-\sqrt{3}$
3712 3416 at : $4-4\sqrt{5}$
3712 3619 e^x : $6(e+\pi)$
3712 3932 rt : (7,7,3,9)
3712 4926 id : $10(e+\pi)/7$
3712 8493 rt : (5,4,-1,4)
3712 8953 ζ : 10/21
3713 0204 rt : (7,9,-7,1)
3713 0803 at : $1/4-3\sqrt{3}$
3713 1222 e^x : $1/2-2\sqrt{5}/3$
3713 1445 ℓ_{10} : $\sqrt[3]{13}$
3713 1655 ℓ_{10} : $\exp(\sqrt[3]{5}/2)$
3713 2034 id : $1/4+3\sqrt{2}/2$
3713 2583 rt : (7,3,6,-3)
3713 2673 rt : (8,5,-6,3)
3713 3302 rt : (6,7,-9,-4)
3713 4152 e^x : 6/19
3713 4991 rt : (4,6,-7,3)
3713 6153 id : $7(1/\pi)/6$
3713 6872 erf : $\sqrt[3]{5}/5$
3713 7370 10^x : 3/8
3713 7734 e^x : $3e/4+2\pi/3$
3714 0063 rt : (3,8,2,-2)
3714 0483 Ei : $\sqrt[3]{5}/6$
3714 0632 Ei : $10\sqrt[3]{3}$
3714 1936 $t\pi$: $e/4-4\pi$
3714 2494 $\ln(\sqrt{3}) + \exp(3/5)$
3714 3633 rt : (9,5,6,2)
3714 3917 rt : (4,1,-9,3)
3714 4911 at : $7\sqrt{2}/2$
3714 5270 10^x : $4/3+2\pi$
3714 5321 at : $4/3-2\pi$
3714 6063 sq : $1/3-2\sqrt{2}/3$
3714 7960 rt : (8,8,4,-3)
3714 9090 10^x : $3\sqrt{2}/4-2\sqrt{5}/3$
3714 9494 Ei : $7\ln 2/6$
3715 0264 rt : (8,4,-3,9)
3715 2160 id : $9e\pi/5$
3715 2392 $\ln(2/3) \times \ln(2/5)$
3715 3981 2^x : $4-\sqrt{5}/2$
3715 6312 Γ : $\exp(-\sqrt[3]{2}/3)$
3715 7206 sr : P_4
3715 7664 sr : $2+4e/3$
3715 9387 $t\pi$: $5\sqrt{3}/3$
3716 0004 $t\pi$: $\exp(\sqrt{2})$
3716 0715 at : $2e/3+\pi$
3716 1039 rt : (6,1,-7,4)
3716 3380 rt : (8,4,-5,-2)
3716 4497 rt : (3,4,-7,-3)
3716 4759 at : $2\sqrt{3}+2\sqrt{5}/3$
3716 4904 rt : (2,4,9,-4)
3716 5775 rt : (5,0,2,1)
3716 5973 rt : (9,1,-9,9)
3716 8130 2^x : $(\sqrt{2}\pi/3)^{-2}$
3717 0354 sinh : 4/11
3717 0823 $e - \ln(\sqrt{2})$
3717 0978 $\ln(3/4) \div s\pi(e)$
3717 1570 2^x : $2/3-2\pi/3$
3717 2192 rt : (8,9,9,-5)
3717 2405 ℓ_{10} : $2+\sqrt{2}/4$
3717 3942 rt : (3,9,7,6)
3717 4187 J_0 : $5\pi/9$
3717 4211 cu : 10/9
3717 6085 at : $4/3+4e/3$
3717 6964 e^x : 19/22
3717 7004 rt : (2,9,-5,-3)
3718 2159 ζ : $7\ln 2/2$
3718 3739 rt : (7,9,-3,-2)
3718 4766 sr : $3-\sqrt{5}/2$
3718 5969 sinh : $8(1/\pi)/7$
3718 7227 rt : (6,5,0,-1)
3718 8530 Ψ : $3\sqrt{3}/8$
3718 8631 sr : $\sqrt[3]{20/3}$
3718 8709 Ψ : $4\sqrt[3]{3}/3$
3718 9029 2^x : $1/4-3\sqrt{5}/4$
3718 9342 ln : $3+2\sqrt{2}/3$
3719 0082 sq : 23/11
3719 0288 10^x : $1/4-e/4$
3719 2342 Ei : $(\sqrt[3]{4}/3)^{1/3}$
3719 2657 rt : (4,9,-2,-2)
3719 3960 Ψ : 25/13
3719 4270 e^x : $1/\sqrt{10}$
3719 4365 at : $1/4+3\pi/2$
3719 5062 rt : (4,4,9,3)
3719 6187 J_1 : $4\sqrt{2}/7$
3719 6877 ℓ_2 : 17/22
3719 7807 $s\pi$: $3\sqrt{2}/2$
3719 9792 rt : (4,8,-1,6)
3720 1046 at : $3/2+2\sqrt{3}$
3720 2258 rt : (5,1,7,-3)
3720 2901 Γ : $1/\sqrt[3]{19}$
3720 4470 id : $1/3+3e/4$
3720 6882 rt : (6,6,5,-3)
3720 8467 Ψ : $8\zeta(3)/5$
3720 8944 rt : (5,7,-9,6)
3720 9271 $\exp(4) + s\pi(e)$
3720 9538 sr : $2\sqrt{3}/3+2\sqrt{5}$
3721 2479 rt : (8,8,-7,-4)
3721 5325 rt : (7,9,5,8)
3721 6570 rt : (8,7,-4,2)
3721 6853 as : 4/11
3721 7235 J_1 : $7\ln 2/6$
3721 7760 ln : $3/4-\sqrt{5}/3$
3721 8966 rt : (8,6,-1,8)
3722 0995 rt : (9,3,-7,8)
3722 1113 ℓ_{10} : $3\pi/4$
3722 1300 10^x : $\sqrt{3}/2+2\sqrt{5}/3$
3722 1405 J_0 : $\exp(\sqrt[3]{3})$
3722 3144 J_1 : $(\sqrt[3]{4}/3)^{1/3}$
3722 3850 $t\pi$: P_{37}
3722 5597 sinh : $7\sqrt[3]{3}/9$
3722 5711 e^x : $3+4\sqrt{2}/3$
3722 6924 J_0 : $3\pi^2/7$
3722 8132 rt : (2,9,-9,2)
3723 1106 at : $4/3-2\sqrt{2}/3$
3723 1119 ℓ_{10} : $\sqrt{3}/2+2\sqrt{5}/3$
3723 1838 ζ : $7\sqrt{3}/5$
3723 2329 at : $1/2+2\sqrt{5}$
3723 2565 as : $8(1/\pi)/7$
3723 2974 sr : $\ln 2/5$
3723 5944 e^x : $((e+\pi)/2)^3$
3723 6374 ℓ_{10} : $5\sqrt{2}/3$
3723 7065 Ei : $(e)^{-3}$
3723 7722 Ψ : $9\sqrt[3]{5}/8$
3724 0675 sr : $1/2+4e$
3724 4331 rt : (7,1,-8,-9)
3724 7211 e^x : 15
3724 7935 at : $\sqrt{3}-3\sqrt{5}$
3724 8415 tanh : 9/23
3724 8605 rt : (9,4,8,-4)
3724 9166 rt : (6,2,-2,9)
3724 9847 J_1 : 17/21
3725 0609 ln : $1/3+\sqrt{5}/2$
3725 3064 e^x : $8\sqrt{3}/3$
3725 4744 rt : (4,7,-6,-3)
3725 5011 rt : (6,3,-5,3)
3725 5856 cr : $4-\sqrt{2}$
3725 7790 rt : (1,8,-3,1)
3725 8115 rt : (5,5,4,9)
3725 8139 cu : $3/2+e/2$
3725 8300 sq : $2-4\sqrt{2}$
3726 0910 $\exp(4) - \exp(4/5)$
3726 1699 rt : (9,3,3,-2)
3726 2432 rt : (5,2,-8,-3)
3726 3388 rt : (5,9,-3,9)
3726 3430 $t\pi$: $1/\sqrt{6}$
3726 4767 sq : $2/3+3\sqrt{3}$
3726 5178 $s\pi(1/3) \times s\pi(\pi)$
3726 5352 rt : (6,4,-8,-3)
3726 7799 id : $\sqrt{5}/6$
3726 8374 at : $2+4\sqrt{5}/3$
3726 9525 J_2 : 23/11
3727 1582 rt : (5,9,4,-3)
3727 2200 rt : (8,2,5,2)
3727 2297 θ_3 : $(\sqrt[3]{5}/3)^3$
3727 3457 rt : (9,8,-9,-4)
3727 4117 ℓ_2 : $\sqrt{2}+4\sqrt{5}$
3727 4906 ln : $4/3-3\sqrt{3}/4$
3727 5388 ℓ_{10} : $1+e/2$
3727 5471 $\exp(5) \div s\pi(1/3)$
3727 7525 Ψ : $\sqrt[3]{5}/4$
3727 9487 λ : $(e+\pi)^{-2}$
3727 9852 $t\pi$: $5\sqrt[3]{2}$
3728 0096 $3/5 \div \ln(5)$
3728 0318 rt : (4,4,6,-3)
3728 1925 ln : $2\sqrt{2}+\sqrt{5}/2$
3728 2319 erf : $2\zeta(3)/7$
3728 3062 Ψ : 25/22
3728 5647 ℓ_{10} : $3\sqrt{2}/4+3\sqrt{3}/4$
3728 7893 rt : (9,5,-5,7)
3728 9748 e^x : 17/8
3728 9986 ln : $4\sqrt{3}-4\sqrt{5}/3$
3729 0091 cu : $\sqrt{2}/4+3\sqrt{5}/4$
3729 0397 Ψ : $10\sqrt{3}/9$
3729 0488 2^x : $((e+\pi)/2)^3$
3729 1741 $1/2 \times s\pi(\sqrt{3})$
3729 2192 2^x : $8\sqrt{5}/3$
3729 2279 rt : (7,7,7,2)
3729 3459 rt : (3,1,8,3)
3729 3684 sr : $3\pi/5$
3729 3803 at : $3\sqrt{2}+\sqrt{5}/3$
3729 3909 rt : (1,8,4,-7)
3729 4915 rt : (4,6,-1,-1)
3729 5617 $1/4 \times \exp(2/5)$
3729 7472 cu : $3e+3\pi/2$
3729 7979 $\exp(\sqrt{3}) \div \exp(e)$
3729 8513 e^x : $3/4-\sqrt{3}/4$
3729 8656 ln : $2/3+\pi/4$
3729 8670 rt : (4,1,-6,-4)
3729 8772 at : 9/23

3729 9589 rt : (8,7,7,-4)
3729 9940 sq : $\sqrt{2}+\sqrt{6}+\sqrt{7}$
3730 0889 rt : (9,6,-8,1)
3730 1803 rt : (2,5,-5,3)
3730 2265 at : $2/3-4\sqrt{2}$
3730 2869 10^x : $2e/3+\pi/2$
3730 3124 $t\pi$: $\sqrt{2}+2\sqrt{5}$
3730 3677 ℓ_2 : $3e+4\pi$
3730 4070 $s\pi$: $\sqrt{2}+2\sqrt{3}$
3730 4431 $s\pi$: $4\sqrt{2}/3+\sqrt{5}$
3730 6978 rt : (5,8,-1,-1)
3730 7311 rt : (8,8,1,7)
3731 1004 rt : (7,2,9,-4)
3731 1598 2^x : $4\sqrt{2}/3+3\sqrt{3}/2$
3731 1731 at : $3/4+3\sqrt{2}$
3731 2731 id : $1/2+4e$
3731 6219 e^x : $10\pi^2/7$
3731 6709 rt : (7,1,-6,8)
3731 6903 ln : $2\pi^2/5$
3731 7784 cu : $4\sqrt[3]{2}/7$
3731 7809 sr : $4\sqrt{2}/3$
3731 9595 rt : (2,6,-4,-5)
3732 0840 e^x : $1/2+4e$
3732 0852 Ei : 1/8
3732 1099 rt : (9,7,-4,-2)
3732 1899 $\ln(3/4)-\exp(3)$
3732 2970 $s\pi$: $7\sqrt[3]{3}/9$
3732 3676 cr : $\sqrt{2}/4+\sqrt{5}$
3732 4800 cu : 18/25
3732 6058 ln : $\sqrt{2}/2+\sqrt{5}/3$
3732 8412 rt : (4,0,-9,2)
3733 1850 rt : (8,9,-2,1)
3733 1993 e^x : $1/2+3\pi$
3733 3437 Γ : $9\sqrt{2}/5$
3733 5875 ln : $4\sqrt{2}/3-\sqrt{3}/4$
3733 6976 rt : (7,2,-9,2)
3733 7015 cu : $\sqrt{5}-\sqrt{6}-\sqrt{7}$
3733 7129 λ : $\exp(-9\pi/8)$
3733 7776 rt : (3,9,-2,6)
3734 0076 at : 5
3734 0360 θ_3 : $3\sqrt{2}/5$
3734 1390 rt : (8,6,2,-2)
3734 1974 at : $\sqrt{2}/4-\sqrt{5}/3$
3734 3602 J_2 : $2\pi/3$
3734 5679 sq : 11/18
3734 6592 sinh : $(\sqrt[3]{2})^{1/2}$
3734 7143 rt : (6,4,0,8)
3734 8100 rt : (5,3,-7,-9)
3734 8709 rt : (2,5,-7,-5)
3735 1447 rt : (2,8,-8,-4)
3735 2515 sq : $(e+\pi)/5$
3735 7431 $\sqrt{3}\times\exp(2/3)$
3735 7457 rt : (6,5,-7,-3)
3736 3084 rt : (6,5,8,-4)
3736 3597 J_1 : 13/16
3736 4201 10^x : $2/3+3\sqrt{2}/4$
3736 7795 tanh : $\ln(\sqrt{2}\pi/3)$
3736 8474 tanh : $\pi/8$
3737 0566 10^x : $(3/2)^3$
3737 0959 rt : (9,7,-3,6)
3737 3039 rt : (7,1,4,-2)
3737 3987 e^x : $9\sqrt{5}/8$
3737 4510 rt : (2,2,7,-3)
3737 7344 id : $1/3-\sqrt{2}/2$
3737 7931 id : $8\zeta(3)/7$
3738 0166 10^x : $2-e/2$
3738 1348 rt : (5,1,-3,-1)
3738 3539 cu : $3/2+3\sqrt{5}/2$
3738 6018 Ei : 17/21
3738 6833 rt : (6,5,-3,2)
3739 0919 at : $3e-\pi$
3739 1841 rt : (4,0,-1,9)
3739 4268 $2/5-s\pi(e)$
3739 5364 2^x : 11/24
3740 1940 ℓ_{10} : $1-\sqrt{3}/3$
3740 3222 rt : (7,0,-9,3)
3740 6669 rt : (9,8,-6,0)
3740 7006 2^x : $7\ln 2/2$
3740 7432 rt : (7,3,-4,7)
3740 7906 Li_2 : $e/8$
3740 8006 cr : $1/2+2\pi/3$
3740 8627 rt : (7,9,1,-2)
3740 9440 rt : (5,8,7,-4)
3741 1405 rt : (2,5,-2,-7)
3741 2562 as : $2\sqrt{2}/3-\sqrt{3}/3$
3741 2951 ζ : $(e+\pi)^{1/2}$
3741 5259 $\exp(\sqrt{3})\div\exp(\sqrt{2})$
3741 5401 sr : $\sqrt[3]{2}/9$
3741 7274 2^x : $4+3\sqrt{2}/4$
3741 7749 id : P_{74}
3741 8138 sq : $6e\pi$
3741 8318 rt : (1,3,9,3)
3741 8984 at : $\ln(\sqrt{2}\pi/3)$
3741 9392 sinh : $8\sqrt[3]{2}/3$
3741 9668 at : $\pi/8$
3742 2605 Γ : $8(1/\pi)$
3742 5017 10^x : P_4
3742 7193 at : $6(e+\pi)/7$
3742 7710 Γ : $1/\ln(\sqrt{2}\pi/3)$
3742 8809 10^x : $3+\sqrt{3}$
3742 9353 rt : (9,6,9,3)
3743 2998 J_2 : $\ln(1/(3e))$
3743 3856 rt : (4,6,-9,-4)
3743 5232 rt : (7,8,-6,-3)
3743 5767 $\exp(1/3)\div s\pi(1/5)$
3743 6102 sr : $4e\pi/3$
3743 6854 sr : 17/9
3743 7202 cr : $9\sqrt[3]{3}/5$
3743 7577 id : $1/4-4e/3$
3743 9592 rt : (4,3,9,-4)
3744 1666 at : $8\pi/5$
3744 1773 $\ln(\sqrt{5})+s\pi(\pi)$
3744 2376 rt : (3,3,-6,9)
3744 2448 cu : $2e/3-2\pi$
3744 2610 rt : (1,9,5,-2)
3744 4414 sinh : $\ln(\sqrt[3]{3})$
3744 7239 rt : (2,7,-3,-2)
3744 7613 rt : (5,1,-5,-8)
3744 7761 rt : (7,9,9,-5)
3744 8103 rt : (9,2,6,-3)
3744 9640 $t\pi$: P_{79}
3745 0490 $\exp(\sqrt{3})\times s\pi(e)$
3745 0957 rt : (8,2,-8,6)
3745 4186 rt : (6,4,3,-2)
3745 4219 e^x : $3/4-\sqrt{3}$
3745 4980 rt : (7,4,-7,1)
3745 6311 Ψ : $\sqrt{2}\pi$
3745 6775 2^x : $9\ln 2/5$
3745 6852 cu : $1+4\sqrt{5}$
3745 7400 cr : $9e\pi/2$
3745 7899 rt : (5,9,0,-1)
3745 8486 2^x : $2e+3\pi/2$
3745 8564 $t\pi$: $7(1/\pi)/2$
3745 9182 at : $9\sqrt{5}/4$
3745 9579 J_2 : $5\sqrt[3]{2}/3$
3746 0259 $t\pi$: $1/2+4e$
3746 0545 $e\div\ln(\pi)$
3746 1179 sq : $3+3\sqrt{5}/2$
3746 2362 J_2 : 21/10
3746 2617 e^x : 7/22
3746 4317 cr : $2e/3+\pi/4$
3746 4594 cu : $e/3-4\pi/3$
3746 6105 rt : (9,9,-9,-6)
3746 8695 cu : $3\sqrt{3}/2+\sqrt{5}/3$
3746 9344 ln : 11/16
3747 1643 2^x : $(3/2)^3$
3747 1686 10^x : $2e\pi/7$
3747 2298 10^x : $5\ln 2$
3747 2963 id : $(\sqrt[3]{4}/3)^{-1/2}$
3747 3443 as : $1/2-\sqrt{3}/2$
3747 3629 Ei : 10/21
3747 4639 2^x : $2e-4\pi/3$
3747 5617 id : $1/\sqrt[3]{19}$
3747 6831 rt : (9,9,-1,5)
3747 7393 sinh : $\ln(\ln 2)$
3747 7420 J_1 : $3e/10$
3747 7635 rt : (4,5,-9,4)
3747 7704 e^x : $2-4\sqrt{5}/3$
3747 8444 sinh : 24/11
3747 8462 10^x : $4-4e/3$
3747 9996 rt : (6,6,2,7)
3748 0222 id : $\exp(1/\pi)$
3748 1620 ℓ_{10} : $(4/3)^3$
3748 2082 sr : $4\pi^2/7$
3748 4092 $t\pi$: $2\sqrt{3}+\sqrt{5}$
3748 4548 rt : (5,9,-8,-4)
3748 5216 $t\pi$: $7\sqrt{2}/3$
3748 5810 rt : (3,6,8,-4)
3748 7059 id : $1/3-3\sqrt{5}$
3748 7204 rt : (6,2,-9,-7)
3748 7866 at : $3+3e/4$
3749 0583 $e\times\ln(5)$
3749 1552 at : $4\sqrt[3]{2}$
3749 2113 rt : (3,5,-6,-9)
3749 2595 J_0 : $2\sqrt{3}$
3749 2645 as : $\ln(\sqrt[3]{3})$
3749 5673 cu : $1/2+4\sqrt{5}$
3749 9022 ℓ_{10} : $1/4+3\sqrt{2}/2$
3750 0000 id : 3/8
3750 0648 $\ln(4/5)+\exp(4)$
3750 2194 ζ : $\ln(5/2)$
3750 2535 at : $4\sqrt{2}/3-4\sqrt{3}$
3750 3214 θ_3 : $9(1/\pi)/5$
3750 5533 2^x : $e/4-2\pi/3$
3750 6790 $t\pi$: $7e/9$
3750 6886 cr : 13/5
3750 9592 rt : (5,0,-8,2)
3751 0737 sq : $3+3\pi$
3751 2138 $t\pi$: $\sqrt{8}\pi$
3751 2286 ℓ_{10} : $1/3+3e/4$
3751 2294 2^x : $10e\pi/9$
3751 4398 at : $1/3+3\pi/2$
3751 5960 J_1 : $\sqrt{2/3}$
3751 7062 cu : $3\zeta(3)/5$
3751 7450 J_0 : $5\ln 2$
3751 7531 $s\pi$: $\sqrt{2}+3\sqrt{5}$
3751 9972 at : $4+\pi/3$
3752 0862 rt : (5,8,3,2)
3752 1912 10^x : $4\sqrt[3]{3}/9$
3752 2525 at : $7\sqrt[3]{3}/2$
3752 4455 rt : (7,5,-2,6)
3752 5835 as : $\ln(\ln 2)$
3752 6761 rt : (2,9,7,-2)
3752 8613 at : $\sqrt{3}/3+2\sqrt{5}$
3753 0506 $s\pi$: $(\sqrt[3]{2})^{1/2}$
3753 0705 rt : (5,1,5,2)
3753 0939 $s\pi$: $\exp(\sqrt[3]{2}/2)$
3753 1198 sr : $3/2-e/2$
3753 1682 J_0 : $7\sqrt{5}/9$
3753 1766 at : $2/3-3\sqrt{2}/4$
3753 2491 E : $(\sqrt{2}\pi/2)^{-1/2}$
3753 4758 cu : $1/\ln(4)$
3753 5764 $t\pi$: $2\sqrt{3}/3+2\sqrt{5}$
3753 7317 rt : (3,2,-5,-2)
3753 8408 J_2 : $7\zeta(3)/4$
3753 9820 ℓ_2 : $1+4\pi/3$
3754 0197 E : $3\sqrt{5}/10$
3754 1163 sq : $2e/3+3\pi/4$
3754 2940 rt : (4,2,1,8)
3754 3657 cu : $1/4+\sqrt{2}/3$
3754 5210 rt : (7,8,4,-3)
3754 5750 J_0 : $\sqrt{2}-\sqrt{5}-\sqrt{7}$
3754 8081 Ψ : $\sqrt{3}-\sqrt{6}+\sqrt{7}$
3755 5002 J_0 : $7\zeta(3)/2$
3755 5035 rt : (8,4,-6,5)
3755 6795 sinh : 8/5
3755 7574 sr : P_{70}
3755 7583 E : $1/\ln(\sqrt{2}\pi)$
3755 8056 sr : P_{67}
3755 8161 ln : $4\zeta(3)/7$
3755 8173 rt : (5,9,8,2)
3755 8588 ln : $1/3+\sqrt{2}/4$
3755 8842 rt : (7,0,7,3)
3755 8977 rt : (9,9,6,-4)
3755 9181 rt : (2,9,-1,-1)
3755 9366 $\exp(\sqrt{3})\times s\pi(2/5)$
3756 2422 id : $4-4e/3$
3756 2724 sr : $4/3+\sqrt{5}/4$
3756 5111 $t\pi$: $4\sqrt{2}/3$
3756 6000 ln : $4/3+3\pi$

3756 6361 ℓ_{10} : 8/19
3756 6362 rt : (1,6,2,-6)
3756 6499 $\exp(1/5) + \exp(e)$
3756 6534 ln : $1/3 + 4e/3$
3756 8157 cu : $1/4 - 4\sqrt{2}/3$
3757 0694 at : $4 + 3\sqrt{2}/4$
3757 1853 rt : (3,5,-8,-7)
3757 2045 rt : (9,1,9,-4)
3757 4514 rt : (4,7,-3,7)
3757 5055 cr : $(1/\pi)/6$
3757 6095 rt : (9,6,-7,-3)
3757 6603 rt : (4,2,-7,-3)
3757 7694 rt : (5,1,-3,7)
3757 8590 rt : (1,4,9,-4)
3757 9388 erf : $\sqrt{3}/5$
3757 9917 id : $8(e+\pi)/5$
3758 0073 J_1 : 9/11
3758 0558 cu : $4\sqrt{3} - \sqrt{5}/4$
3758 2137 2^x : $10\sqrt[3]{2}/3$
3758 2174 id : $2\sqrt{2}/3 + \sqrt{3}/4$
3758 3687 e^x : $3\sqrt[3]{3}/5$
3758 4211 cr : $4(e+\pi)/9$
3758 4658 ζ : $3(1/\pi)/2$
3758 4941 rt : (6,7,-1,1)
3758 5095 at : $2\sqrt{2} + \sqrt{5}$
3758 6062 at : $\sqrt{2}/3 - \sqrt{3}/2$
3758 6451 rt : (7,7,-9,-4)
3758 7735 rt : (3,6,-8,2)
3758 9914 2^x : $4\sqrt{2} + 3\sqrt{5}$
3759 2250 rt : (2,9,2,4)
3759 2529 rt : (5,9,-6,2)
3759 5743 erf : $\ln 2/2$
3759 6435 id : $3\sqrt{2}/2 - \sqrt{5}/3$
3759 7434 $s\pi$: $(\sqrt{2}\pi/3)^{-1/3}$
3760 0832 sq : $\sqrt{3} - \sqrt{5}/4$
3760 1062 rt : (6,0,-7,6)
3760 1175 $t\pi$: $1/4 - \sqrt{2}/2$
3760 1934 ln : $(\sqrt{2}/3)^{1/2}$
3760 3485 10^x : $\ln 2/5$
3760 4871 sq : $4\sqrt{2}/3 + \sqrt{3}/4$
3760 5600 rt : (3,8,-6,4)
3760 6030 sr : $\sqrt{2}/10$
3760 6454 erf : $1/\sqrt[3]{24}$
3760 8588 rt : (8,4,8,-4)
3760 9075 ζ : $8e/9$
3760 9329 cu : $9\sqrt[3]{3}/7$
3761 0282 J_1 : $(\ln 3/2)^{1/3}$
3761 0367 10^x : $4\sqrt{3}/3 + \sqrt{5}/3$
3761 0607 sinh : $7\pi^2/9$
3761 0738 at : $4\sqrt{2}/3 - 2\sqrt{5}/3$
3761 1005 ℓ_{10} : $\exp(\sqrt{3}/2)$
3761 1728 $s\pi(1/5) + s\pi(\sqrt{2})$
3761 2638 sr : $4(1/\pi)/9$
3761 3968 rt : (9,9,-2,-1)
3761 4115 J_2 : $\exp(\sqrt{5}/3)$
3761 5873 rt : (2,4,2,9)
3761 6003 cr : $1/4 + 3\pi/4$
3761 6750 Ψ : $9\sqrt[3]{2}/10$
3761 6870 J : $\exp(-14\pi/19)$
3761 6901 rt : (9,4,-5,9)
3761 7015 2^x : $(3\pi/2)^{-1/2}$
3761 7679 cr : $6\sqrt{5}$
3761 7851 rt : (7,8,-4,0)
3761 9470 10^x : $3/2 + \sqrt{2}/4$
3762 1113 2^x : $4\sqrt{2} - 3\sqrt{3}$
3762 1217 e^x : $6\pi/5$
3762 1302 rt : (2,6,-6,-3)
3762 2478 at : $e + 3\pi/4$
3762 2829 e^x : $1/\ln(2e/3)$
3762 3273 e^x : $\sqrt{2}/3 + \sqrt{5}/3$
3762 3361 sinh : $\exp(-1)$
3762 5336 E : $\ln((e+\pi)/3)$
3762 7234 J_2 : $(\sqrt{2}\pi)^{1/2}$
3762 8852 rt : (4,5,-4,-2)
3763 0253 ℓ_{10} : $7e/8$
3763 0536 Ψ : $\ln(\sqrt[3]{2}/2)$
3763 1774 ℓ_2 : $9\sqrt[3]{3}/5$
3763 3006 id : $4\sqrt{2}/3 + 2\sqrt{5}/3$
3763 3487 at : $1 + 3e/2$
3763 4903 rt : (9,8,9,-5)
3763 5442 e^x : $\sqrt{5}/7$
3763 6576 $t\pi$: P_{54}
3763 7629 rt : (3,6,-1,4)
3763 7886 rt : (8,5,-9,-1)
3763 8192 $t\pi$: 3/10
3764 1120 rt : (7,7,7,-4)
3764 1255 at : $4\sqrt{2} - \sqrt{3}/3$
3764 1809 rt : (5,8,-5,-9)
3764 3883 sr : $4\sqrt{3} + 2\sqrt{5}$
3764 5281 ln : $3/4 + \sqrt{2}/2$
3764 5614 as : $e/4 - \pi/3$
3764 6870 $t\pi$: $1/\ln(\sqrt{2})$
3765 0551 rt : (6,5,9,3)
3765 1184 rt : (5,9,8,7)
3765 1883 rt : (4,8,-4,-1)
3765 3279 sq : $(\ln 2/3)^{1/3}$
3765 4904 e^x : $\sqrt{2}/3 + 4\sqrt{5}$
3765 5705 J_0 : 21/5
3765 6878 Ei : 19/9
3765 6881 rt : (2,4,-9,1)
3765 7210 sr : $\exp(-(e+\pi)/3)$
3765 7584 Ψ : $(\sqrt[3]{2}/3)^{1/2}$
3765 7644 J_1 : $(2e/3)^{-1/3}$
3765 8508 rt : (6,4,-2,-1)
3765 8851 as : $\sqrt{2} - \sqrt{3}/4$
3765 9339 tanh : $4\ln 2/7$
3765 9352 J_0 : $10\sqrt[3]{2}/3$
3765 9821 sr : $3e/2 + \pi/2$
3766 0231 id : $2e/3 - 4\pi/3$
3766 0897 cr : $7\sqrt{5}/6$
3766 1260 Li_2 : 13/14
3766 1528 id : $3\sqrt{2} - \sqrt{3}/2$
3766 4525 rt : (3,9,-5,-4)
3766 6031 at : $\sqrt{3} + 3\sqrt{5}/2$
3766 6279 ζ : $(\sqrt{5}/3)^{-3}$
3766 6383 ℓ_{10} : $\sqrt{17/3}$
3766 7083 cr : $4 + 3\pi$
3766 7382 ℓ_{10} : $4/3 + \pi/3$
3766 7509 rt : (3,9,-5,-3)
3767 0262 rt : (6,8,4,6)
3767 1467 cu : 13/18
3767 1868 J_0 : $1/\ln(4/3)$
3767 2750 as : $\exp(-1)$
3767 4030 Γ : $5\pi^2/8$
3767 4727 $\exp(1/5) - \exp(4)$
3767 5438 rt : (9,1,1,-1)
3767 5832 E : $(\sqrt{5})^{-1/2}$
3767 6061 $\sqrt{5} + \exp(\pi)$
3767 6254 rt : (2,2,-9,3)
3767 6327 ℓ_2 : $5\ln 2/9$
3767 6528 2^x : $\sqrt{14}$
3767 7640 10^x : 12/13
3767 7774 J_2 : $2\pi^2/5$
3767 7792 ℓ_{10} : $\sqrt[3]{2}/3$
3767 8083 $t\pi$: $5\sqrt[3]{2}/9$
3767 8156 Li_2 : $\sqrt[3]{5}/5$
3767 8639 J : $\exp(-11\pi/20)$
3767 8762 sinh : 17
3767 8982 e^x : $4/3 - 4\sqrt{3}/3$
3767 9321 sinh : $1/\sqrt[3]{20}$
3767 9426 J : $(2e)^{-3}$
3768 0537 rt : (1,9,-3,5)
3768 1085 rt : (7,7,0,5)
3768 1234 sinh : 7/19
3768 3646 sinh : $\sqrt{20}\pi$
3768 5014 rt : (1,7,-4,-9)
3768 5102 $\ln(3/5) - s\pi(1/3)$
3768 5644 $3/5 + \ln(4/5)$
3768 6377 e^x : $(\sqrt{2}\pi/3)^{1/2}$
3768 6691 rt : (4,6,7,2)
3768 7347 J_1 : $(e/3)^2$
3768 8058 rt : (8,6,-4,4)
3768 9207 10^x : $4\sqrt{2}/3 - 4\sqrt{3}/3$
3768 9371 as : $3/4 - \sqrt{5}/2$
3769 2924 rt : (9,5,-8,3)
3769 6274 $t\pi$: $3\sqrt{2}/2 - \sqrt{5}$
3769 6514 rt : (5,2,-6,1)
3769 6529 J_2 : 19/9
3769 6660 at : $3 + 2\pi/3$
3769 8277 rt : (5,8,-3,-2)
3769 8474 at : $e/3 + 4\pi/3$
3769 8536 Ψ : 17/15
3770 0945 2^x : 6/13
3770 1665 sq : $\sqrt{3} - \sqrt{5}/2$
3770 3998 rt : (5,2,8,3)
3770 5646 rt : (1,7,-5,-9)
3770 6163 $\exp(5) - s\pi(\sqrt{2})$
3770 8311 rt : (2,8,-9,-8)
3770 9136 id : $e/3 - 2\pi$
3770 9422 at : $4\sqrt{2} - \sqrt{5}/4$
3770 9536 rt : (6,2,9,-4)
3771 1123 $t\pi$: $10\sqrt[3]{5}/3$
3771 1523 Ei : 20/17
3771 2605 at : $4\ln 2/7$
3771 2776 e^x : 8/25
3771 4317 e^x : $3\sqrt{3}/2 - \sqrt{5}/3$
3771 4733 J_1 : $(\sqrt{2}\pi/3)^{-1/2}$
3771 6769 cu : $\sqrt{2}/4 + 2\sqrt{5}$
3771 7154 ℓ_2 : $2e/3 + \pi/4$
3771 7835 $s\pi$: $\sqrt{17}$
3772 0285 rt : (2,8,2,-2)
3772 1022 erf : 8/23
3772 1631 J_2 : $(3\pi)^{1/3}$
3772 3796 J : $(2\pi)^{-1/2}$
3772 4145 $s\pi$: $1/4 + 4e$
3772 5070 tanh : $1/\sqrt[3]{16}$
3772 5480 Γ : $\sqrt{13/2}$
3772 5594 $\exp(2/3) \times s\pi(1/4)$
3772 7588 at : $3\sqrt{2}/2 + 4\sqrt{5}/3$
3772 7925 ln : $1/2 + 2\sqrt{3}$
3772 8461 id : $\sqrt{3}/4 + 4\sqrt{5}$
3772 8558 rt : (9,6,-3,8)
3772 9077 as : $1/\sqrt[3]{20}$
3772 9207 $s\pi$: $\exp(-2\pi/3)$
3772 9385 at : $1/3 - 2e$
3772 9537 rt : (2,5,-9,-4)
3772 9907 rt : (8,7,1,9)
3773 0205 10^x : $e - \pi$
3773 0519 $t\pi$: $\sqrt{13/2}$
3773 1003 as : 7/19
3773 2705 ζ : $\ln(\sqrt[3]{5}/2)$
3773 3499 ℓ_{10} : $\sqrt{2}/2 + 3\sqrt{5}/4$
3773 4244 rt : (9,8,1,-2)
3773 4568 rt : (6,9,9,-5)
3773 5040 rt : (3,1,-8,-3)
3773 6063 $\sqrt{5} \div \ln(3/5)$
3773 6321 E : $5\zeta(3)/9$
3773 7306 rt : (1,8,5,3)
3773 9132 rt : (5,5,8,-4)
3774 0579 $\ln(4/5) - \exp(e)$
3774 3883 rt : (8,4,0,-1)
3774 3936 at : $3/4 - \sqrt{2}/4$
3774 3956 rt : (7,1,-9,4)
3774 4267 id : $\exp(\sqrt{3}/2)$
3774 4375 ℓ_2 : $3\sqrt{3}$
3774 5147 $5 \times s\pi(\sqrt{5})$
3774 7249 θ_3 : 3/16
3774 7985 rt : (4,1,7,-3)
3774 8407 rt : (6,2,-5,5)
3774 9349 rt : (2,5,-8,-7)
3774 9465 at : $3\sqrt{2} + \sqrt{3}/2$
3775 0094 10^x : $1 - \sqrt{2}/3$
3775 0941 $t\pi$: $2\sqrt{3} + 4\sqrt{5}$
3775 4373 2^x : $2\ln 2/3$
3775 4919 rt : (5,3,-1,6)
3775 5164 rt : (1,3,-9,-9)
3775 5206 $t\pi$: $(\sqrt[3]{3})^{-1/3}$
3775 6609 ln : $5(e+\pi)$
3775 6718 rt : (4,9,-1,-8)
3775 7201 Ei : $\exp(-\sqrt[3]{2})$
3775 7711 sr : $2e\pi/9$
3775 8040 id : $8\pi/3$
3776 0091 ℓ_{10} : $1/2 + 4\sqrt{2}/3$
3776 0490 rt : (3,8,-8,-4)
3776 1728 J_2 : $7e/9$
3776 4943 10^x : $3\sqrt[3]{5}/8$
3776 5357 rt : (4,8,7,-4)
3776 6063 $\pi + \sqrt{5}$
3777 0083 rt : (8,1,-8,8)
3777 1942 e^x : $1 - e/4$

3777 2208 rt : (1,8,-4,-7)
3777 2528 rt : (3,3,6,2)
3777 3562 rt : (4,9,-5,-9)
3777 4800 as : $2-4\sqrt{5}/3$
3777 6344 rt : (2,7,-7,3)
3777 7196 ln : $3e-4\pi/3$
3777 7380 rt : (7,2,-4,9)
3777 8219 sinh : 9/8
3777 8814 at : $1/\sqrt[3]{16}$
3777 9211 sq : $\ln(2\pi)$
3778 1013 2^x : $4+\sqrt{5}$
3778 1636 e^x : $2\sqrt[3]{3}/9$
3778 2003 rt : (4,4,-7,-3)
3778 2992 J_1 : 14/17
3778 3082 sq : $\sqrt[3]{11/3}$
3778 3974 at : $4+\sqrt{5}/2$
3778 4804 rt : (3,5,7,8)
3778 7912 tπ : $(\ln 2)^{1/3}$
3778 8753 rt : (3,7,-2,-4)
3778 9324 $4/5 \div \exp(3/4)$
3779 2692 $e \div \ln(\sqrt{5})$
3779 3002 2^x : $3/4+3\sqrt{5}/4$
3779 6051 at : $3+3\sqrt{2}/2$
3779 6447 id : $1/\sqrt{7}$
3779 7631 id : $3\sqrt[3]{2}/10$
3779 9227 10^x : $4\sqrt{2}/3+\sqrt{3}$
3779 9637 sr : $9\pi/5$
3780 0866 Ψ : $8\pi^2/7$
3780 1287 id : $3\sqrt{3}/4-3\sqrt{5}/4$
3780 2498 sq : $\sqrt{3}/4-3\sqrt{5}$
3780 2795 tπ : $3\pi/5$
3780 3105 rt : (1,6,-1,9)
3780 3525 $\exp(2/3)-s\pi(\pi)$
3780 5303 at : $3e\pi/5$
3780 5603 ln : $e/2+3\pi$
3780 6209 J_1 : $4\sqrt[3]{3}/7$
3780 7117 K : 23/25
3780 7267 at : $3/2+4e/3$
3780 8164 at : $3e/2+\pi/3$
3780 8935 θ_3 : $8(1/\pi)/3$
3780 8962 J_0 : $4\pi/3$
3781 0393 E : 2/3
3781 0717 10^x : $4\sqrt{2}/3+\sqrt{5}/4$
3781 1609 $\sqrt{2}-s\pi(\sqrt{2})$
3781 3082 rt : (9,7,4,-3)
3781 3923 2^x : $3+2\sqrt{2}/3$
3781 3985 tanh : $5(1/\pi)/4$
3781 4964 Γ : $9\sqrt{3}$
3781 6067 rt : (6,3,-9,-9)
3781 8319 at : $7(e+\pi)/8$
3782 0360 rt : (5,7,-6,-3)
3782 3133 sq : $\sqrt{2}/3-4\sqrt{3}/3$
3782 3373 cr : $5\pi/6$
3782 3539 $\ln(\sqrt{2}) \div \ln(2/5)$
3782 4077 cr : $3/2+\sqrt{5}/2$
3782 4281 rt : (9,0,4,2)
3782 5356 rt : (8,7,-7,-2)
3782 5613 Ei : $\ln(2\pi)$
3782 7172 cu : $1/4-4\sqrt{5}/3$
3782 7613 at : $3\sqrt[3]{5}$
3782 9212 ln : $\sqrt{3}+\sqrt{5}$
3783 3794 rt : (2,0,5,2)
3783 7550 tπ : $(\sqrt[3]{3}/2)^{-1/3}$
3783 8080 rt : (6,8,-4,-5)
3783 8885 ln : $\sqrt[3]{2}/5$
3783 9014 J_2 : $\sqrt[3]{19/2}$
3783 9356 rt : (9,7,-6,2)
3783 9518 sr : $10\sqrt[3]{5}/9$
3783 9938 rt : (3,8,5,7)
3784 0487 sr : 19/10
3784 1423 2^x : 5/4
3784 2538 sr : $2\sqrt{3}/3+\sqrt{5}/3$
3784 3014 $\exp(3)-s\pi(1/4)$
3784 4920 erf : $\pi/9$
3784 7278 2^x : $((e+\pi)/3)^{1/3}$
3784 7644 tπ : P_{78}
3784 9047 tanh : $(e/2)^{-3}$
3784 9659 id : $7e/8$
3784 9741 id : $\sqrt{2}/4-\sqrt{3}$
3784 9856 tπ : $\sqrt{2}-3\sqrt{3}/4$
3785 0582 10^x : $\ln(\sqrt{5})$
3785 1162 ℓ_2 : 5/13
3785 1455 J_0 : $5(e+\pi)/7$
3785 3000 Γ : $5e\pi/6$
3785 3215 rt : (5,0,-6,2)
3785 3305 rt : (2,8,-4,-3)
3785 3461 2^x : $4/3+3\sqrt{3}$
3785 5658 Li_2 : $2\zeta(3)/7$
3785 5849 id : $2e/3+4\pi$
3785 6113 cr : $2\sqrt{2}/3+3\sqrt{5}/4$
3785 6492 at : $2\sqrt{2}+4\sqrt{3}/3$
3785 6850 rt : (7,8,-3,-1)
3785 7859 cu : $2e/3+\pi/3$
3785 8271 rt : (3,3,9,-4)
3786 0048 at : $2/3+2\sqrt{5}$
3786 0506 rt : (8,3,3,-2)
3786 3946 rt : (8,8,-2,3)
3786 7675 rt : (9,8,-1,7)
3786 8312 rt : (5,8,-8,3)
3786 8380 at : $5(1/\pi)/4$
3786 8619 at : $2\sqrt{3}+3\sqrt{5}/4$
3786 8838 rt : (9,6,7,-4)
3786 9822 sq : 8/13
3786 9994 10^x : $7\sqrt[3]{5}/9$
3787 0223 at : $2+\pi$
3787 0475 ζ : 11/23
3787 1570 cr : $\sqrt[3]{18}$
3787 5717 Γ : $1/\sqrt{14}$
3787 8338 10^x : $1/3+\sqrt{2}/3$
3787 9485 ζ : $\sqrt[3]{14}$
3787 9670 rt : (2,6,8,-4)
3787 9757 J_1 : 19/23
3788 1628 rt : (9,7,4,1)
3788 1722 cr : $1/2+3\sqrt{2}/2$
3788 2613 J_0 : $10\pi/9$
3788 3022 10^x : $7e/5$
3788 5931 rt : (8,9,6,-4)
3788 7493 rt : (4,6,-5,8)
3788 9389 rt : (3,3,-9,2)
3789 0321 rt : (7,6,2,-2)
3789 2499 rt : (2,6,-7,-7)
3789 3706 10^x : 17/14
3789 4327 rt : (5,6,-9,-4)
3789 5236 rt : (8,9,3,8)
3789 6594 J_0 : $5\ln 2/2$
3789 6772 e^x : 13/15
3789 8459 $2/3+\ln(3/4)$
3789 8534 rt : (1,9,7,-4)
3790 0062 rt : (8,3,-6,7)
3790 1334 rt : (7,9,2,4)
3790 3363 rt : (3,4,9,3)
3790 3702 at : $(e/2)^{-3}$
3790 4168 Ei : $3(1/\pi)/2$
3790 4357 tanh : $(2\pi)^{-1/2}$
3790 5564 id : $1/3-3\pi/2$
3790 5605 rt : (7,3,-7,3)
3790 6734 rt : (9,1,7,3)
3791 0186 J_2 : $3\sqrt{2}/2$
3791 0419 e^x : $\sqrt{2}/2-3\sqrt{5}/4$
3791 0786 10^x : $7(e+\pi)/9$
3791 1803 ℓ_{10} : $7\sqrt[3]{5}/5$
3791 2296 ℓ_{10} : $4/3+3\sqrt{2}/4$
3791 5286 rt : (6,9,1,0)
3791 7367 2^x : 21/5
3791 7882 at : $4+2\sqrt{3}/3$
3791 8408 at : $3-3e$
3792 1847 rt : (2,7,2,4)
3792 3507 rt : (8,8,9,-5)
3792 5690 at : $1/2-4\sqrt{2}$
3792 6615 2^x : $1/3-\sqrt{3}$
3792 8797 sπ : $\exp(2e/3)$
3792 8969 rt : (3,8,-9,-6)
3792 9165 rt : (8,2,6,-3)
3792 9815 rt : (7,4,-2,8)
3792 9870 cu : $4/3+3\sqrt{3}$
3793 0010 tπ : $\sqrt{3}/2-4\sqrt{5}/3$
3793 1108 sr : P_{75}
3793 1560 sπ : $3e\pi/5$
3793 2353 2^x : 6π
3793 2677 tπ : $\exp(-1/(3e))$
3793 3199 10^x : 25/2
3793 4069 rt : (2,1,8,3)
3793 6495 rt : (5,3,-9,-5)
3793 8066 ℓ_{10} : $2\sqrt{2}-\sqrt{3}/4$
3793 8205 erf : 7/20
3793 8985 rt : (1,5,-3,-8)
3794 0026 at : $4/3-\sqrt{3}$
3794 1918 sr : $7e/10$
3794 2637 J_0 : $9e/7$
3794 3730 rt : (1,9,-2,-2)
3794 3942 J_0 : $\sqrt{3}$
3794 6208 cr : 21/8
3794 6450 rt : (6,4,-3,4)
3794 8159 cr : $e/3+4\pi$
3794 8383 rt : (7,5,5,-3)
3794 8729 sq : $1/4-\sqrt{3}/2$
3794 8962 ln : 13/19
3795 1283 2^x : $1/\sqrt[3]{10}$
3795 2932 cu : $3\sqrt{3}-2\sqrt{5}$
3795 5347 rt : (2,9,-1,-9)
3795 6237 erf : $(1/(3e))^{1/2}$
3795 7266 rt : (6,1,-7,-8)
3795 8469 rt : (7,6,-4,-2)
3795 8980 rt : (1,2,7,-3)
3795 9421 at : $(2\pi)^{-1/2}$
3795 9807 rt : (8,4,3,1)
3796 1473 Γ : $\sqrt{14}$
3796 1719 sq : $8\ln 2/9$
3796 3448 $\ln(4/5) \div s\pi(1/5)$
3796 4587 ζ : $6\ln 2/5$
3796 5030 rt : (6,8,4,-3)
3796 5636 e^x : $7\sqrt[3]{3}/6$
3796 6775 tπ : $2\sqrt[3]{3}$
3796 7103 at : P_{59}
3796 8447 Ψ : $4\sqrt{2}/5$
3796 9750 rt : (1,2,-2,9)
3797 1379 rt : (6,5,2,9)
3797 2223 J_1 : $(\ln 3)^{-2}$
3797 3269 10^x : $7e\pi/8$
3797 4468 rt : (8,1,9,-4)
3797 5283 tπ : $\ln 2/6$
3797 6961 sr : $\sqrt[3]{3}/10$
3797 7055 Ψ : 11/17
3797 7817 cu : $2e-3\pi/2$
3797 9589 rt : (5,9,0,-4)
3798 0316 $\exp(4/5)+\exp(e)$
3798 0801 sq : $3+\pi/3$
3798 1142 ln : $2\sqrt[3]{5}/5$
3798 1155 sr : $4\sqrt{3}/3+3\sqrt{5}/2$
3798 5203 e^x : $\sqrt{5}+\sqrt{6}+\sqrt{7}$
3798 6836 J_2 : 17/8
3798 7889 rt : (7,4,8,-4)
3798 9953 Ei : 8
3799 0204 sr : $1/3+\pi/2$
3799 0210 rt : (8,9,-3,-2)
3799 3763 sq : $3\sqrt{2}/4-3\sqrt{5}/4$
3799 4671 J_1 : $(\sqrt[3]{5}/3)^{1/3}$
3799 4896 tanh : 2/5
3799 5629 sinh : $7(1/\pi)/6$
3799 6919 rt : (1,8,-5,-3)
3799 7909 ℓ_{10} : $2/3+\sqrt{3}$
3799 8591 rt : (5,0,-3,9)
3799 9465 id : $2\sqrt[3]{5}/9$
3799 9785 rt : (6,7,7,-4)
3800 5591 sr : $2+3\pi$
3800 6242 rt : (6,2,-8,-3)
3800 6333 tπ : $\sqrt[3]{7/2}$
3800 6490 at : $\sqrt{2}/2+2\sqrt{5}$
3800 6927 rt : (7,1,2,1)
3800 7060 2^x : $8\sqrt[3]{3}/3$
3800 9600 J_1 : $5\sqrt{3}/3$
3801 0303 at : $3e/4+\pi$
3801 0401 rt : (5,5,1,5)
3801 0888 sr : $3/4+2\sqrt{3}/3$
3801 2231 $\ln(\sqrt{5}) \div \exp(3/4)$
3801 2545 rt : (9,8,-9,-4)
3801 2744 sinh : $\sqrt{3}\pi$
3801 2773 J_0 : 7/2
3801 5291 rt : (9,4,-8,5)
3801 6669 10^x : $4-e/2$

3801 9438 $\lambda : \exp(-19\pi/17)$
3801 9932 rt : (4,9,-4,0)
3802 0620 $Ei : (\sqrt[3]{3}/2)^{-1/2}$
3802 1124 $\ell_{10} : 24$
3802 2236 $J_0 : 6\sqrt[3]{3}/5$
3802 5785 $e^x : \sqrt{2} + \sqrt{6} - \sqrt{7}$
3802 7089 rt : (7,5,-7,-3)
3802 7866 $Ei : \sqrt[3]{13}$
3802 8053 rt : (9,8,7,2)
3802 9484 cr : $\sqrt{3} - 3\sqrt{5}/4$
3802 9708 $\theta_3 : \exp(-5/3)$
3802 9748 $5 \div \ln(3/4)$
3802 9998 $erf : \exp(-\pi/3)$
3803 0778 rt : (9,9,-4,1)
3803 1593 rt : (1,7,-8,-4)
3803 1955 $\exp(e) - s\pi(e)$
3803 2765 at : $1/4 - 2e$
3803 2962 $Ei : \exp(\sqrt[3]{5}/2)$
3803 3535 rt : (5,9,9,-5)
3803 5638 $10^x : \sqrt[3]{2}/9$
3803 5700 cu : $3/2 + 3\sqrt{2}$
3803 8418 rt : (6,2,1,-1)
3804 0505 s$\pi : (\pi/2)^3$
3804 0695 $\Psi : -\sqrt{2} + \sqrt{6} - \sqrt{7}$
3804 0723 $\ln(2) \div \exp(3/5)$
3804 0742 at : $1 + 4\pi/3$
3804 0891 rt : (4,0,-4,5)
3804 2260 $2/5 \times s\pi(2/5)$
3804 3648 sr : $4 - 2\pi/3$
3804 3836 rt : (8,8,-6,-3)
3804 5595 $2^x : 1/3 + 2\pi/3$
3804 7498 as : $7(1/\pi)/6$
3804 7614 id : $\sqrt{17/3}$
3804 7945 $e^x : (e/3)^{-2}$
3805 0637 at : 2/5
3805 0715 rt : (8,4,-9,1)
3805 0777 cr : $4\sqrt{2}/3 + \sqrt{5}/3$
3805 1500 t$\pi : 5(1/\pi)$
3805 3088 id : $1/3 + \pi/3$
3805 3354 rt : (8,5,6,2)
3805 3546 tanh : $\zeta(3)/3$
3805 5996 rt : (1,9,4,-9)
3805 7357 $J_1 : 1/\ln(\sqrt{2})$
3805 9242 $\Gamma : 2\zeta(3)/9$
3806 0945 rt : (5,5,0,-1)
3806 1360 $\ell_{10} : (\ln 2/2)^3$
3806 1361 sr : $1 + e/3$
3806 1778 cu : $10\pi/3$
3806 2399 rt : (8,5,-4,6)
3806 2795 rt : (3,5,-9,-5)
3806 3570 at : $4e/3 + \pi/2$
3806 3992 $J : \pi/8$
3806 6313 sr : $\sqrt{2}/3 + 3\sqrt{3}$
3806 7072 at : $3\sqrt{3}$
3806 7106 $\Psi : 7e\pi/2$
3806 8630 $\Gamma : 15/23$
3807 1478 $e^x : 1/3 - 3\sqrt{3}/4$
3807 2060 rt : (7,2,5,2)
3807 5000 $\ln(2) \times \ln(\sqrt{3})$
3807 5165 rt : (3,5,-3,5)
3807 5213 cr : $4/3 + 3\sqrt{3}/4$
3807 5335 rt : (8,7,-9,-4)
3807 5758 rt : (7,0,-9,6)
3807 7207 $\sqrt{3} + \exp(1/2)$
3807 7859 rt : (4,8,-1,-1)
3807 9786 $\sqrt{5} + \ln(\pi)$
3808 1246 s$\pi : 7\sqrt{3}$
3808 6451 rt : (6,1,4,-2)
3808 6976 $\Gamma : 9\sqrt{2}/4$
3808 7090 s$\pi : 1/2 + 4e/3$
3808 8593 $J_1 : 2\sqrt[3]{3}$
3808 8916 $\ell_{10} : 5\sqrt[3]{3}/3$
3809 0448 t$\pi : 2\sqrt{2} - 4\sqrt{5}$
3809 0957 $10^x : 2 + 4e$
3809 3659 $\Psi : \exp(1/(3e))$
3809 3873 rt : (8,6,9,3)
3809 4290 rt : (1,9,-4,-4)
3809 4364 $J_1 : 6\ln 2/5$
3809 4476 $\ell_2 : 2(e+\pi)/9$
3809 4559 rt : (8,9,-5,-3)
3809 5238 id : 8/21
3809 5502 $\ell_{10} : 2\zeta(3)$
3809 7863 rt : (5,4,3,-2)
3809 8646 $J_0 : 25/6$
3809 9154 $\ln(3/5) \times s\pi(\sqrt{3})$
3810 1220 $erf : 1/\sqrt[3]{23}$
3810 1276 $Ei : 8(1/\pi)/9$
3810 2243 $J_1 : \exp(-1/(2e))$
3810 2328 $\ell_{10} : 3\ln 2/5$
3810 2923 $e^x : 7\zeta(3)/5$
3810 3864 rt : (7,3,8,3)
3810 4308 as : $3/4 - \sqrt{3}$
3810 6825 t$\pi : \zeta(3)/4$
3810 7135 rt : (4,7,2,-2)
3810 9361 rt : (1,7,-2,6)
3810 9730 at : $\zeta(3)/3$
3810 9798 $10^x : 2 + \sqrt{5}$
3811 0017 at : $3/2 - 3\sqrt{5}$
3811 0157 sq : $(\sqrt[3]{2}/3)^{-1/2}$
3811 0419 $10^x : 4\sqrt{2}/3 + \sqrt{3}/3$
3811 2055 at : $8(e+\pi)/9$
3811 2086 rt : (6,0,7,3)
3811 7042 $\Psi : 8\sqrt{2}$
3811 7553 $\Gamma : 9\pi^2/8$
3811 8970 rt : (5,3,6,-3)
3811 9493 $e^x : 2 - 3\sqrt{5}/4$
3811 9957 $\sqrt{3} - \exp(\sqrt{2})$
3812 0143 rt : (7,5,-5,2)
3812 0568 sinh : $9\sqrt[3]{5}$
3812 0970 $Ei : 3e\pi/10$
3812 1193 $\ell_2 : \sqrt{11}\pi$
3812 3095 cr : $3e/2 + 3\pi$
3812 3238 ln : $4\sqrt{2} - 3\sqrt{5}/4$
3812 3638 $J_1 : (\ln 2)^{1/2}$
3812 4182 ln : $1/4 + \sqrt{3}/4$
3812 4794 $Ei : (3\pi)^{1/3}$
3812 4818 rt : (4,6,5,6)
3812 4885 at : $1/2 + 3\pi/2$
3812 4925 $J_2 : \sqrt{2} - \sqrt{3} + \sqrt{6}$
3812 5875 ln : $9\zeta(3)$
3812 7424 rt : (7,6,0,7)
3812 8474 $J_1 : (\sqrt[3]{3})^{-1/2}$
3812 8587 $\Gamma : 23/5$
3812 8626 t$\pi : 3/7$
3812 9848 rt : (3,9,4,-1)
3813 1257 $\ell_{10} : 3/2 + e/3$
3813 1955 cr : $3/4 + 4\sqrt{2}/3$
3813 2637 rt : (5,2,9,-4)
3813 5465 rt : (6,1,-5,7)
3813 5517 sinh : $\exp(\sqrt{2}/3)$
3813 5671 $\ell_{10} : e/2 + \pi/3$
3813 5773 $s\pi(2/5) - s\pi(\pi)$
3813 6490 sinh : $\sqrt{5}/6$
3813 6638 $10^x : 23/9$
3813 6661 rt : (4,5,8,-4)
3813 7593 sinh : $9\pi/7$
3813 9740 $Li_2 : 1/\ln((e+\pi)/2)$
3814 0209 rt : (3,8,7,-4)
3814 2109 rt : (2,9,-8,8)
3814 2984 at : $7\sqrt{5}/3$
3814 7905 $Ei : \sqrt{2}/5$
3814 8932 $\zeta : 2\zeta(3)$
3814 9379 $e^x : 3 + e$
3815 0164 rt : (2,8,3,-4)
3815 1456 t$\pi : 2\sqrt{2} + 2\sqrt{5}$
3815 2923 $J_1 : 5/6$
3815 2952 rt : (4,5,-7,9)
3815 4160 s$\pi : 3\sqrt{2} - \sqrt{5}/2$
3815 5437 t$\pi : 1/4 + \sqrt{3}/2$
3815 5640 s$\pi : 9\sqrt{5}$
3815 5648 $10^x : 3e/2 + 4\pi$
3815 7141 id : $1/\sqrt[3]{18}$
3815 7589 $\ln(4/5) \times \exp(e)$
3815 7662 ln : $(\pi)^{1/3}$
3815 8108 s$\pi : 3e/2 + \pi/3$
3815 8350 sq : $3\sqrt{2}/4 - \sqrt{5}$
3815 9425 at : $3/4 + 2\sqrt{5}$
3815 9961 $Ei : 3\pi/8$
3816 0327 $\ell_2 : 2\sqrt{2}/3 - \sqrt{5}/4$
3816 3953 ln : $1 + 4\sqrt{5}/3$
3816 4655 $\Psi : (\ln 2)^{-1/3}$
3816 5307 $\zeta : 5\sqrt[3]{3}/3$
3816 5456 at : $4\sqrt{2} - \sqrt{3}/4$
3816 6791 cu : 18/11
3816 6829 rt : (1,6,5,9)
3816 6842 $J_2 : 5\pi/4$
3816 7016 $\Gamma : -\sqrt{2} + \sqrt{3} + \sqrt{5}$
3816 7596 $10^x : 4\sqrt{2}/3 - 2\sqrt{3}/3$
3816 9855 sr : 21/11
3817 2804 $Ei : 11/23$
3817 4228 $10^x : 25/11$
3818 0702 $\Psi : (\sqrt[3]{3})^{1/3}$
3818 1939 $10^x : 5(e+\pi)/4$
3818 3382 ln : $(e+\pi)/4$
3818 4304 $10^x : \exp(\pi/3)$
3818 5884 rt : (9,6,-6,4)
3818 6651 rt : (9,7,-1,9)
3818 7548 rt : (8,2,-6,9)
3818 8235 t$\pi : 5\sqrt{2}/8$
3818 9324 as : $\sqrt{5}/6$
3819 1287 $2^x : 7/15$
3819 1643 $\Psi : 2\sqrt{5}$
3819 2800 $J_2 : e\pi/4$
3819 4475 $\ell_2 : 1 + 3\pi$
3819 4567 t$\pi : 9e/4$
3819 6495 $J_0 : 6\ln 2$
3819 6601 id : $3/2 - \sqrt{5}/2$
3819 7186 id : $6(1/\pi)/5$
3819 7659 sr : $6(1/\pi)$
3820 1781 sinh : $10\sqrt[3]{3}/9$
3820 2839 $10^x : P_3$
3820 4267 $\ell_{10} : \sqrt[3]{14}$
3820 5621 sq : $3\sqrt[3]{3}/7$
3820 6735 ln : $2\sqrt{2} + 2\sqrt{3}/3$
3820 7088 $\theta_3 : (\sqrt{3})^{-1/2}$
3820 8297 at : $5\pi/3$
3820 8579 at : $3 + \sqrt{5}$
3821 1701 $Ei : 4\sqrt[3]{2}$
3821 1774 $Ei : 13/16$
3821 6371 at : $3 - 3\sqrt{3}/2$
3821 6530 rt : (4,9,0,3)
3821 8469 rt : (4,8,-7,-8)
3821 9139 $2^x : \sqrt{2} - \sqrt{3} + \sqrt{6}$
3821 9314 rt : (5,2,-1,8)
3821 9318 $Li_2 : \sqrt{3}/5$
3822 0087 $2^x : 3(e+\pi)/10$
3822 0740 $J_0 : 19/11$
3822 1422 at : $4\sqrt{2}/3 + 3\sqrt{5}/2$
3822 1866 rt : (5,1,-6,3)
3822 3001 rt : (6,7,4,8)
3822 3454 cr : $4 - e/2$
3822 4241 rt : (7,8,9,-5)
3822 6225 $J_0 : 3(e+\pi)/5$
3822 6513 rt : (6,6,-1,3)
3822 7479 sr : $4/3 + \sqrt{3}/3$
3822 8116 rt : (7,9,6,-4)
3822 9900 rt : (8,6,7,-4)
3823 0266 $Ei : 2\sqrt{5}$
3823 0990 ln : $4\sqrt{2}/3 + 4\sqrt{5}$
3823 1103 rt : (6,5,-6,-2)
3823 1515 $erf : 6/17$
3823 1681 at : $1 + 3\sqrt{2}$
3823 1899 rt : (9,3,8,-4)
3823 1908 sq : $\exp(-\sqrt[3]{3}/3)$
3823 2862 ln : $6\sqrt[3]{5}/7$
3823 3234 id : $\sqrt{2} + \sqrt{3} + \sqrt{5}$
3823 5689 rt : (8,7,4,-3)
3823 8437 $2^x : 8e/5$
3823 8472 rt : (9,4,5,-3)
3823 9096 $\Gamma : 3(e+\pi)/5$
3823 9383 $Li_2 : \ln 2/2$
3824 1393 $\Psi : \sqrt{13}\pi$
3824 1684 $Li_2 : (\sqrt{3}/2)^{1/2}$
3824 3912 rt : (8,8,1,-2)
3824 4139 rt : (5,9,-4,7)
3824 5163 $J_2 : 5\sqrt[3]{5}/4$
3824 5864 $e^x : 3/2 + \sqrt{2}/4$
3824 6146 cu : $3/4 + e/2$
3824 8050 rt : (9,5,2,-2)
3825 0152 $2^x : 9\pi$

3825 0972 rt : (7,9,-6,-7)
3825 2527 $Li_2 : 1/\sqrt[3]{24}$
3825 2726 $2^x : 4/3 - \sqrt{3}/2$
3825 6165 cr : $1/3 + 4\sqrt{3}/3$
3825 6509 rt : (8,9,-2,-1)
3825 7482 at : 21/4
3825 8033 cu : $10\zeta(3)/9$
3825 9707 $s\pi(1/4) + s\pi(\sqrt{5})$
3826 0157 $\theta_3 : \sqrt[3]{5}/9$
3826 1515 at : $4\sqrt{3} - 3\sqrt{5}/4$
3826 2181 $\ln(5) \div s\pi(\sqrt{5})$
3826 2481 rt : (7,2,-7,5)
3826 2491 $J_1 : (\sqrt[3]{5})^{-1/3}$
3826 3301 rt : (9,6,-1,-1)
3826 3770 $\ln(3) + \exp(1/4)$
3826 4568 sinh : $4\zeta(3)/3$
3826 5628 sr : $4 + 3\sqrt{5}/4$
3826 8343 sr : $1/2 - \sqrt{2}/4$
3826 9993 rt : (1,7,3,-6)
3827 0550 sq : $4 - 4\pi$
3827 0784 rt : (8,7,-2,5)
3827 1728 $\Gamma : (e + \pi)/9$
3827 2549 $2^x : 1/2 - 4\sqrt{2}/3$
3827 4126 $Ei : 5\sqrt{2}/6$
3827 4344 $\ln(2/3) \times \exp(\pi)$
3827 5278 $10^x : (\sqrt[3]{5}/2)^2$
3827 7568 $\ell_{10} : 1 + \sqrt{2}$
3827 8221 rt : (8,6,-7,0)
3827 8453 cu : $7(1/\pi)/2$
3828 4438 $s\pi : 9\sqrt{5}/7$
3828 5362 rt : (9,3,-8,7)
3828 6085 $t\pi : 3\sqrt{2}/4 - 4\sqrt{5}$
3829 0162 ln : $4\sqrt{3}/3 + 3\sqrt{5}/4$
3829 0316 $2^x : 4/3 - e$
3829 0875 $\ell_{10} : (\sqrt{5}/3)^3$
3829 1355 rt : (9,7,-4,0)
3829 2492 $erf : \sqrt{2}/4$
3829 2943 id : $1/\ln(\sqrt[3]{5}/2)$
3829 5262 $J_1 : (e + \pi)/7$
3829 7576 rt : (3,9,-3,3)
3829 8421 id : $2e/3 + \pi/2$
3829 9225 ln : 15/22
3830 2851 cr : $2\sqrt{3}/3 + 2\sqrt{5}/3$
3830 4092 $\Gamma : 5e\pi/8$
3830 4674 rt : (2,3,9,-4)
3830 4865 $erf : 10(1/\pi)/9$
3830 5773 $2^x : 9\pi/5$
3830 6591 $t\pi : 5e\pi/6$
3830 7043 rt : (2,7,-7,-9)
3830 7402 as : $1/3 - \sqrt{2}/2$
3830 8755 sr : $\sqrt[3]{7}$
3831 0917 $2^x : 3e/4 - \pi/2$
3831 1753 $10^x : 3/2 - e/2$
3831 1921 $3/4 \times \ln(3/5)$
3831 3669 $2^x : (2e)^{1/3}$
3831 4195 $\ell_{10} : 8e/9$
3831 7920 $2^x : 1/\ln(\sqrt{2}\pi/2)$
3831 8467 rt : (1,7,5,-3)
3832 0341 cu : $\sqrt{14}$
3832 0870 cr : $3 - \sqrt{2}/4$
3832 1152 $t\pi : P_{77}$
3832 1774 rt : (4,6,-8,1)
3832 1995 sq : 13/21
3832 2226 $t\pi : ((e + \pi)/2)^{-2}$
3832 3496 rt : (3,7,3,8)
3832 4780 rt : (9,3,-8,-3)
3832 8513 $\Gamma : 8\sqrt{5}/7$
3832 8958 rt : (2,4,6,-3)
3832 8960 $\ell_{10} : 3\sqrt{3}/4 + \sqrt{5}/2$
3832 9745 $\exp(5) \times \exp(\pi)$
3833 1626 $\theta_3 : 4\sqrt[3]{5}/9$
3833 2928 $\Gamma : 10\zeta(3)$
3833 2989 $J_2 : \sqrt{3} - \sqrt{5} + \sqrt{7}$
3833 3491 $\Gamma : 23/9$
3833 4285 $\zeta : 12/5$
3833 4775 id : $3\sqrt{3}/2 - 4\sqrt{5}/3$
3833 5646 $\ell_2 : 7\sqrt{5}/3$
3833 8059 $\ln(5) + s\pi(e)$
3833 8806 rt : (3,6,-4,-3)
3834 0297 $\theta_3 : 19/25$
3834 0747 at : $9(e + \pi)/10$
3834 1431 rt : (3,1,7,-3)
3834 1835 tanh : $2\sqrt{2}/7$
3834 2354 sr : $9e\pi/4$
3834 4839 rt : (8,0,-9,3)
3834 4927 $J_0 : \exp(\sqrt[3]{2})$
3834 5124 tanh : $\exp(-e/3)$
3834 5420 $\ell_{10} : P_{56}$
3834 5703 rt : (4,9,9,-5)
3834 7183 rt : (5,2,-8,-6)
3834 7737 $J_1 : 3\sqrt{5}/8$
3834 9682 $2^x : 3\sqrt{3}/2 + 2\sqrt{5}$
3835 0230 rt : (1,8,2,-2)
3835 0494 $Ei : \sqrt{17/2}$
3835 0967 as : P_{74}
3835 1369 $\Gamma : 10(1/\pi)$
3835 1752 rt : (7,1,-9,-8)
3835 1779 rt : (3,7,-8,9)
3835 3155 $\Psi : 9\sqrt[3]{2}$
3835 5106 sr : $1/2 + \sqrt{2}$
3835 6074 $J_2 : 15/7$
3835 6507 rt : (4,2,8,3)
3835 8568 $\theta_3 : 4/21$
3835 8991 sinh : $1/\sqrt[3]{19}$
3835 9718 rt : (9,8,-7,1)
3836 0077 $2^x : 17/7$
3836 7176 sq : $4\sqrt{2} + 4\sqrt{3}$
3836 7287 rt : (8,4,-4,8)
3836 9151 sq : $5(e + \pi)/3$
3837 2192 rt : (2,4,-7,8)
3837 2962 at : $1 - 2\pi$
3837 4842 $10^x : P_{70}$
3837 5100 rt : (5,5,9,3)
3837 5316 $10^x : 4 - e/3$
3837 5996 $10^x : P_{67}$
3837 6678 sq : $e/3 + \pi$
3837 6790 rt : (7,9,-8,-4)
3837 7322 rt : (5,8,8,9)
3837 7716 ln : $7\sqrt[3]{5}/3$
3837 8257 sr : $\sqrt{11/3}$
3837 9204 id : $2\sqrt{2}/3 - \sqrt{5}/4$
3838 0597 rt : (8,3,-9,3)
3838 3153 $\theta_3 : (\sqrt[3]{3}/2)^{1/2}$
3838 3204 $\ell_2 : (\sqrt{2}\pi/2)^{1/3}$
3838 3317 $e^x : \exp(3/2)$
3838 4359 sq : $3\sqrt{2}/2 - \sqrt{3}/3$
3838 5106 sinh : 3/8
3838 5507 rt : (6,2,-8,1)
3838 5515 $10^x : 9/17$
3838 6274 rt : (9,9,1,8)
3838 7960 $e^x : (\sqrt[3]{5}/3)^2$
3838 8263 $2^x : 3e/4 - \pi/4$
3838 8416 $2^x : (\pi/2)^{1/2}$
3838 8614 rt : (6,4,8,-4)
3838 9190 cu : $3 - 4\sqrt{2}/3$
3839 0817 $Ei : \sqrt{7/3}$
3839 0894 rt : (6,6,-9,-4)
3839 3212 $\ell_2 : 1/3 + \sqrt{3}/4$
3839 3233 rt : (1,9,6,7)
3839 3234 $Li_2 : 8/23$
3839 3576 rt : (7,8,2,6)
3839 4415 $\ell_{10} : (e + \pi)^{1/2}$
3839 4718 rt : (9,4,-5,-2)
3839 6153 rt : (4,1,-7,-1)
3839 9349 at : $4\sqrt{3}/3 + 4\sqrt{5}/3$
3840 0234 at : $2\sqrt{2}/7$
3840 0582 $Ei : 7\pi/8$
3840 0953 $J_1 : 2\sqrt[3]{2}/3$
3840 1264 $e^x : 3e/2 + 2\pi$
3840 1416 rt : (7,1,9,-4)
3840 2846 rt : (9,8,-4,3)
3840 2922 $J_1 : 21/25$
3840 3218 $10^x : 3(e + \pi)/5$
3840 3549 at : $\exp(-e/3)$
3840 4294 rt : (1,9,-1,-1)
3840 7494 sq : $3\sqrt{3} - 3\sqrt{5}/4$
3840 8304 sq : 20/17
3841 0006 rt : (5,8,4,-3)
3841 0261 at : $\sqrt{2} - 3\sqrt{5}$
3841 1943 cu : $4 + 2\pi$
3841 1982 at : $\exp(5/3)$
3841 1983 rt : (9,7,9,-5)
3841 2927 $t\pi : 10e/7$
3841 3377 as : $1/\sqrt[3]{19}$
3841 5776 id : $\sqrt{2}/2 + 3\sqrt{5}/4$
3841 6054 rt : (4,1,5,2)
3841 7273 rt : (2,8,-2,6)
3841 8228 rt : (7,7,-3,1)
3841 9634 $10^x : 23/16$
3842 0042 cu : $4 - \sqrt{3}/4$
3842 0347 $\Gamma : \ln(\sqrt{5}/3)$
3842 0565 $J_1 : 23/8$
3842 1218 $\Gamma : (\pi)^{-2}$
3842 1218 rt : (3,4,-5,6)
3842 2498 $J_1 : 9\sqrt{5}/7$
3842 2580 rt : (9,7,-9,-2)
3842 4110 $t\pi : 5(e + \pi)/6$
3842 5108 $E : \exp(-\sqrt[3]{2}/3)$
3842 5931 rt : (6,5,5,-3)
3842 7635 at : $4 + 3\sqrt{3}/4$
3842 8448 $\sqrt{3} + \exp(\sqrt{3})$
3842 8564 $2^x : 4\sqrt{2}/3 + 2\sqrt{5}/3$
3842 9062 $\ell_{10} : 3 - \sqrt{3}/3$
3843 0618 $\Psi : \exp(3/2)$
3843 1841 sinh : 17/2
3843 6441 $J_1 : (\sqrt{2})^{-1/2}$
3843 6715 rt : (2,6,0,-1)
3843 9326 sq : $3 - e/4$
3843 9677 as : 3/8
3843 9987 $J : 1/\sqrt{10}$
3844 0663 rt : (6,7,-6,-3)
3844 1053 $e^x : (2e/3)^{1/3}$
3844 1403 rt : (3,6,-9,-7)
3844 1802 rt : (9,8,6,-4)
3844 2395 sr : $6\sqrt{5}/7$
3844 3421 $\Psi : 5e/7$
3844 3731 sr : 23/12
3844 3845 sq : $4\sqrt{2} + 3\sqrt{3}/4$
3844 4128 cu : $3e - 4\pi/3$
3844 4215 ln : $4\sqrt{2} + 3\sqrt{3}$
3844 4257 sq : $3 - e/3$
3844 4432 rt : (7,2,6,-3)
3844 4574 $\pi \times \exp(1/3)$
3844 6168 tanh : $(\pi/2)^{-2}$
3844 7403 rt : (8,1,-6,-2)
3844 7654 rt : (6,0,-5,9)
3844 9437 $2^x : 2\sqrt[3]{3}$
3844 9448 $s\pi : 1/4 + 4e/3$
3844 9888 $10^x : 22/19$
3844 9899 rt : (7,6,-8,-4)
3845 0812 $\Gamma : 3\sqrt{3}$
3845 2686 rt : (5,4,6,2)
3845 5198 at : $3/4 - 2\sqrt{3}/3$
3845 6749 $J_1 : 7\zeta(3)/10$
3845 7050 rt : (5,9,1,-2)
3845 7256 cu : $4\pi^2/7$
3845 9205 $J_0 : 9\pi/8$
3846 0432 rt : (5,8,-6,8)
3846 0897 $s\pi : P_{53}$
3846 1538 id : 5/13
3846 1917 $2^x : \sqrt{2}/4 - \sqrt{3}$
3846 2818 tanh : $1/\sqrt[3]{15}$
3846 3204 at : $3 + 4\sqrt{3}/3$
3846 3348 $\theta_3 : 3(1/\pi)/5$
3846 4001 $\ell_2 : 3/2 + 4\sqrt{5}$
3846 4522 cr : $3/2 + 2\sqrt{3}/3$
3846 5168 sr : $1/4 + 2e$
3846 6191 rt : (8,1,7,3)
3846 7317 cu : 8/11
3846 7330 rt : (5,4,-7,-3)
3846 7855 $e^x : 9e$
3846 8866 $\ell_{10} : 7\sqrt{3}/5$
3846 9090 rt : (1,5,-3,-5)
3847 0094 rt : (9,5,-6,6)
3847 0220 cu : $e/4 + 4\pi/3$
3847 2464 rt : (2,7,-8,-4)
3847 4450 $J_0 : 5\sqrt{2}/2$
3847 4818 rt : (1,7,-8,-5)
3847 5904 cu : $1/3 - 3\sqrt{2}/4$
3847 6168 rt : (4,1,-4,-7)

3847 7044 rt : (7,5,0,9)
3847 7168 10^x : $2/3+4\sqrt{3}/3$
3847 9510 Γ : 13/20
3848 1156 rt : (9,9,3,-3)
3848 1606 J_1 : 16/19
3848 2435 sr : $\sqrt{2}/2-\sqrt{5}/4$
3848 3921 cu : $\exp(-1/\pi)$
3848 5127 sq : $1+4e/3$
3848 5868 cu : P_{73}
3848 6839 J_2 : $7\sqrt{5}/4$
3848 7231 10^x : $2\sqrt{3}/3+\sqrt{5}/2$
3848 9350 ℓ_{10} : $7\ln 2/2$
3849 0017 id : $2\sqrt{3}/9$
3849 0937 10^x : $\sqrt{2}/10$
3849 2158 rt : (9,4,8,3)
3849 2837 rt : (1,7,2,-8)
3849 4849 id : $1/3-e$
3849 4920 $\sqrt{3}-\exp(3/4)$
3849 8017 ζ : 12/25
3849 9730 cr : $1+4\pi$
3850 0206 sr : $\sqrt{2}-\sqrt{3}+\sqrt{5}$
3850 0593 rt : (4,1,-8,-3)
3850 2038 rt : (1,8,4,-4)
3850 2762 ℓ_{10} : $2\sqrt[3]{3}/7$
3850 2927 rt : (7,4,-5,4)
3850 4096 rt : (6,7,7,2)
3850 4829 e^x : $(3\pi)^{-1/2}$
3850 5386 at : $(\pi/2)^{-2}$
3850 6017 rt : (3,8,-7,1)
3850 6283 $3/4\div\exp(2/3)$
3850 6788 10^x : $4(1/\pi)/9$
3850 7023 as : $4-4e/3$
3850 7889 ℓ_{10} : $3/4+3\sqrt{5}/4$
3850 8095 rt : (4,2,-9,-2)
3850 8176 id : $5\ln 2/9$
3851 0192 rt : (7,3,3,-2)
3851 1113 ℓ_2 : $4\sqrt{2}-\sqrt{3}/4$
3851 1453 at : $1/3-4\sqrt{2}$
3851 1848 sr : $3/2+4\pi/3$
3851 5096 rt : (9,2,-8,9)
3851 9974 ln : $7\sqrt[3]{2}/6$
3852 0010 ℓ_{10} : $1/3+2\pi/3$
3852 0878 at : $\ln(2/3)$
3852 2168 at : $1/\sqrt[3]{15}$
3852 2419 rt : (5,4,1,7)
3852 3429 Ψ : $\sqrt[3]{22/3}$
3852 4641 rt : (6,8,-3,-2)
3852 7940 cr : $1-2\sqrt{2}/3$
3852 8757 Ψ : $(3\pi)^{-2}$
3852 8884 10^x : $1-\sqrt{2}$
3853 0484 rt : (5,5,-7,-6)
3853 1153 J_1 : $(5/3)^{-1/3}$
3853 2448 rt : (2,8,-5,-3)
3853 5088 ℓ_{10} : 7/17
3853 6054 rt : (4,8,-2,4)
3853 6403 ζ : $(\pi/2)^{-1/2}$
3853 7294 e^x : $2/3+3\sqrt{5}$
3853 7396 e^x : $\sqrt{5}-\sqrt{6}+\sqrt{7}$
3853 8131 rt : (6,8,9,-5)
3853 8527 e^x : $3\sqrt{2}-3\sqrt{3}$
3853 9105 rt : (7,2,-7,2)
3854 0840 $t\pi$: $\sqrt[3]{15}$
3854 3273 rt : (3,2,-9,3)
3854 4837 at : 16/3
3854 4901 sq : $1/2-3\sqrt{5}/4$
3854 5656 Li_2 : $\pi/9$
3854 5849 rt : (8,0,4,2)
3854 6392 ℓ_{10} : $4-\pi/2$
3854 7361 sinh : $6e$
3854 7721 $\sqrt{3}-\ln(\sqrt{2})$
3854 7786 rt : (8,9,0,4)
3854 8143 Γ : $7\sqrt{5}/4$
3855 2521 rt : (7,1,-7,7)
3855 6101 $2/5\times s\pi(\sqrt{2})$
3855 6110 $s\pi$: $\sqrt[3]{2}/10$
3855 6958 $t\pi$: $(\sqrt{2}\pi)^3$
3855 8193 rt : (3,5,8,-4)
3855 8414 at : $5e\pi/8$
3856 1220 at : $\sqrt{3}/2+2\sqrt{5}$
3856 1808 id : $1/2+4\sqrt{2}/3$
3856 2023 $\ln(2/3)\times s\pi(2/5)$
3856 3156 rt : (1,8,7,-3)
3856 4064 id : $4\sqrt{3}/5$
3856 5327 rt : (4,7,-9,-7)
3856 6248 ln : 17/25
3856 6565 rt : (5,3,-4,2)
3856 7433 2^x : 8/17
3856 8849 J : $\exp(-19\pi/6)$
3856 9589 Γ : $\sqrt{5}/6$
3857 0858 Γ : $3\sqrt{3}/8$
3857 1099 rt : (6,7,-1,-1)
3857 1820 $\ln(3/4)\div s\pi(\sqrt{3})$
3857 1980 Ei : $3e\pi/8$
3857 4797 rt : (5,5,-4,-2)
3857 4871 at : $1/2-e/3$
3857 6133 e^x : $\sqrt{2}/4+\sqrt{3}/2$
3857 6671 2^x : $2\sqrt{2}/3+\sqrt{3}$
3857 7750 sr : $3e/4+3\pi$
3858 0144 $t\pi$: $\sqrt{12}\pi$
3858 3441 rt : (4,2,9,-4)
3858 6954 rt : (6,9,6,7)
3858 7252 e^x : 25
3858 7329 e^x : 20/23
3858 7916 Ψ : $10\pi/7$
3858 9243 rt : (8,8,-5,-1)
3859 0693 rt : (9,3,5,2)
3859 1053 ζ : $7\sqrt{5}/10$
3859 2338 at : $3e/2-3\pi$
3859 2594 $t\pi$: $8(e+\pi)/3$
3859 2618 rt : (8,6,-2,7)
3859 2999 J_2 : $\sqrt[3]{10}$
3859 3104 10^x : $5\pi/7$
3859 6058 2^x : $4-\pi/2$
3859 6213 sinh : $\sqrt{14}\pi$
3859 7068 ζ : $\ln(\sqrt{3}/2)$
3860 0051 $t\pi$: $5e$
3860 0093 rt : (2,9,4,-3)
3860 0324 at : $4\sqrt{2}/3+2\sqrt{3}$
3860 2241 ℓ_{10} : $\sqrt{5}-\sqrt{6}+\sqrt{7}$
3860 2270 sinh : $(\sqrt[3]{3})^{1/3}$
3860 3355 sr : $2e\pi/3$
3860 3896 sq : $3/2-3\sqrt{2}/2$
3860 5710 sq : $1/\ln(5)$
3860 5898 ℓ_2 : P_{52}
3860 6166 rt : (9,7,-9,-4)
3860 6474 e^x : $(1/(3e))^{-1/2}$
3860 6658 id : $3e/4-3\pi$
3860 7456 ζ : $\exp(1/(2\pi))$
3860 7877 cr : $\sqrt{2}/4+4\sqrt{3}/3$
3860 8263 2^x : $2-\sqrt{5}/3$
3860 8343 ζ : $7\sqrt[3]{5}/5$
3860 8672 sq : $\sqrt[3]{25/2}$
3860 9559 sq : $3/4+\pi/2$
3860 9732 rt : (7,6,7,-4)
3861 3824 10^x : $\exp(-(e+\pi)/3)$
3861 4437 ℓ_{10} : $2+\sqrt{3}/4$
3861 4768 Ψ : $1/\ln(3\pi/2)$
3861 5108 at : $2+3\sqrt{5}/2$
3861 5328 rt : (5,0,-6,5)
3861 5786 $t\pi$: $\sqrt{2}/3-3\sqrt{5}/2$
3861 5937 10^x : $4+3\pi$
3861 7007 rt : (1,8,-3,2)
3861 7580 cu : 24/11
3861 8684 J_0 : $\sqrt{17}$
3861 9221 $t\pi$: $5e/8$
3862 0001 sq : $\sqrt{3}+2\sqrt{5}/3$
3862 1740 ln : $1+\sqrt{2}/3$
3862 2150 sinh : $(\ln 2)^{-1/3}$
3862 2159 at : $3+3\pi/4$
3862 2972 rt : (7,4,0,-1)
3862 3510 ℓ_{10} : $2\sqrt{2}/3+2\sqrt{5}/3$
3862 3561 Γ : $8\sqrt{2}$
3862 3653 erf : $1/\sqrt[3]{22}$
3862 4445 rt : (5,2,-9,-3)
3862 5198 2^x : $2\sqrt{2}+2\sqrt{3}$
3862 5583 rt : (2,9,-2,-2)
3862 7627 rt : (8,5,-7,2)
3862 9436 id : $2\ln 2$
3863 1229 id : $4e\pi/3$
3863 1766 rt : (3,6,5,-3)
3863 2078 at : $4+e/2$
3863 2566 J_1 : 11/13
3863 2962 λ : $\exp(-10\pi/9)$
3863 3877 2^x : $3\sqrt{2}-\sqrt{3}/2$
3863 4166 ln : $\sqrt{2}/4+\sqrt{5}/2$
3863 4330 2^x : $2/3-3e/4$
3863 4442 rt : (9,5,-2,-1)
3863 5095 $t\pi$: $9\pi^2/10$
3863 5192 sq : $3\sqrt{2}/4+3\sqrt{3}/2$
3863 5350 rt : (8,3,8,-4)
3863 6108 J_2 : $2(e+\pi)/3$
3863 7237 at : $3\sqrt{2}+\sqrt{5}/2$
3863 7832 $t\pi$: $10\sqrt{2}/9$
3863 9948 e^x : $4/3+3\sqrt{5}$
3864 1184 sq : $1/4-4e/3$
3864 2928 rt : (5,3,-8,-8)
3864 3277 rt : (4,2,-5,-2)
3864 3929 10^x : $8\sqrt{3}/7$
3864 5387 rt : (1,1,8,3)
3864 5858 2^x : $\sqrt{2}/3$
3864 6367 cr : $5e$
3864 8657 ℓ_2 : $2e/3-\pi/3$
3864 8820 rt : (5,3,3,1)
3864 8925 rt : (5,7,-8,9)
3864 9084 id : $4\pi^2/9$
3864 9473 erf : 5/14
3864 9596 $\sqrt{3}\times\ln(4/5)$
3865 0634 rt : (8,2,-9,5)
3865 1937 rt : (6,8,1,2)
3865 3733 2^x : $3/4-3\sqrt{2}/2$
3865 5336 e^x : $\sqrt{19/3}$
3865 9803 cu : $3e/4+\pi/3$
3866 0594 Li_2 : 7/20
3866 0687 rt : (5,9,-7,0)
3866 2020 Ei : 17/4
3866 2186 ζ : $(\ln 2)^2$
3866 3590 rt : (5,8,-8,-4)
3866 4375 rt : (9,0,9,4)
3866 6135 sr : $1/2+3\sqrt{3}$
3866 8869 rt : (4,3,6,-3)
3866 9316 e^x : $\exp(-\sqrt{5}/2)$
3866 9843 rt : (1,6,-2,3)
3867 0811 Γ : 4/15
3867 2254 id : $\sqrt[3]{8/3}$
3867 5049 sr : 25/13
3867 5816 Ei : $(\ln 3/3)^3$
3867 7015 rt : (1,6,-3,-2)
3867 9668 10^x : $3/2+3\sqrt{2}/2$
3867 9974 $s\pi$: $\sqrt[3]{2/3}$
3868 0211 θ_3 : $(\sqrt[3]{2})^{-1/3}$
3868 0962 rt : (6,5,-1,5)
3868 1813 $t\pi$: $\sqrt{2}/4-\sqrt{5}$
3868 2769 sr : $8\zeta(3)/5$
3868 2828 Li_2 : $(1/(3e))^{1/2}$
3868 3223 rt : (7,7,4,-3)
3868 3433 $s\pi$: $1/\ln(\pi)$
3868 4854 at : $4/3-3\sqrt{5}$
3868 7983 2^x : $\sqrt{3}/2-\sqrt{5}$
3868 9384 $t\pi$: $10\sqrt{3}/7$
3868 9879 e^x : $3/2+3\sqrt{2}/2$
3869 1073 rt : (3,6,1,9)
3869 1545 rt : (8,1,1,-1)
3869 2281 at : $e/3-2\pi$
3869 4168 $t\pi$: $((e+\pi)/2)^{-1/3}$
3869 5906 rt : (6,2,-3,8)
3869 5955 cu : $3e/4+4\pi$
3869 6314 rt : (6,9,0,-1)
3869 7134 $1/2\times s\pi(e)$
3869 7667 cu : $1/2+4\sqrt{2}$
3869 8339 sr : $9\sqrt[3]{5}/8$
3869 8402 sr : $\sqrt{3}/4+2\sqrt{5}/3$
3869 8431 tanh : $1/\sqrt{6}$
3869 8801 ℓ_2 : $(\sqrt[3]{5})^{1/2}$
3869 8845 at : $2/3+3\pi/2$
3869 9910 rt : (9,7,-4,5)
3869 9947 10^x : $3e+\pi/3$
3869 9976 rt : (4,5,-9,-4)
3870 0329 Ψ : $\sqrt{3}-\sqrt{5}+\sqrt{6}$
3870 1946 rt : (3,3,-7,7)
3870 2275 rt : (2,4,9,3)

3870 2312 $\ell_2 : 13/17$
3870 2322 cr : $\sqrt[3]{19}$
3870 2627 $\Psi : \ln(1/(2e))$
3870 2812 sinh : $1/\sqrt{7}$
3870 2831 rt : (2,9,-9,3)
3870 3682 $\zeta : \exp(-1/(2e))$
3870 4082 sinh : $3\sqrt[3]{2}/10$
3870 5913 $e^x : e\pi/7$
3870 7654 rt : (9,4,-6,8)
3870 7690 rt : (1,9,-7,-4)
3870 9029 rt : (3,9,-6,-7)
3870 9243 $10^x : 4e/3 + 3\pi/4$
3870 9785 at : $\sqrt{2} + \sqrt{3} + \sqrt{5}$
3870 9908 $\Psi : 1/\sqrt[3]{13}$
3871 0164 sinh : $\exp(1/(3e))$
3871 1196 $J_2 : \sqrt{14/3}$
3871 1256 $\Psi : \exp(-\sqrt[3]{5}/2)$
3871 2944 rt : (5,0,7,3)
3871 4086 cr : $\sqrt{3}/4 + \sqrt{5}$
3871 4996 $\zeta : e/2$
3871 5945 $2^x : 3\sqrt{2}/2 - \sqrt{3}/2$
3871 8321 ln : $\sqrt{12}\pi$
3871 8329 rt : (9,8,-6,-3)
3871 9149 rt : (8,4,5,-3)
3872 0878 $J_1 : 3\sqrt{2}/5$
3872 2778 $Ei : 3e\pi/4$
3872 3744 $10^x : 1/3 - \sqrt{5}/3$
3872 3986 $\ell_{10} : \sqrt{2}/2 + \sqrt{3}$
3872 6381 sr : $10\sqrt{3}/9$
3872 6943 $J_0 : \exp(\sqrt{2})$
3872 7375 rt : (8,7,9,-5)
3872 9833 sr : 3/20
3873 1959 $J_1 : 8(1/\pi)/3$
3873 2919 $E : 15/23$
3873 4075 cu : $2\sqrt{2}/3 + \sqrt{5}/4$
3873 5896 $e^x : 9\sqrt[3]{3}/7$
3873 7631 $\ell_{10} : 2e\pi/7$
3873 9691 sinh : $\sqrt{13}$
3874 0206 $t\pi : 2/17$
3874 0461 id : P_{27}
3874 1852 at : $e/4 + 3\pi/2$
3874 2588 rt : (9,9,-7,-3)
3874 2968 $\zeta : (\ln 2/3)^{1/2}$
3874 4372 $erf : 9(1/\pi)/8$
3874 4411 id : $2\sqrt{2} + \sqrt{5}/4$
3874 5048 sr : $10\sqrt[3]{5}/3$
3874 5492 $J_1 : (\sqrt[3]{3}/2)^{1/2}$
3874 8328 $2^x : 3\sqrt[3]{2}/8$
3875 0279 sr : $2\sqrt{3} + \sqrt{5}$
3875 6247 $t\pi : 9\sqrt{3}/5$
3875 6475 $\ell_{10} : 3 - \sqrt{5}/4$
3875 8892 $Ei : 12/25$
3875 9668 at : $1/\sqrt{6}$
3875 9731 at : $3\sqrt{2} + 2\sqrt{3}/3$
3876 0838 rt : (9,1,6,-3)
3876 0947 as : $3\sqrt[3]{2}/10$
3876 1595 $10^x : 1/\sqrt{7}$
3876 1905 rt : (1,6,2,-7)
3876 3012 sr : $\zeta(3)/8$
3876 3099 rt : (7,7,2,8)
3876 4438 $\theta_3 : \sqrt{3}/9$
3876 4896 as : $3\sqrt{3}/4 - 3\sqrt{5}/4$
3876 5153 rt : (1,9,-3,6)
3876 5790 $\ell_2 : 4\sqrt{2}/3 - \sqrt{3}/3$
3876 6178 rt : (6,3,8,3)
3876 8106 $10^x : 3\sqrt[3]{2}/10$
3876 8550 at : $(\sqrt[3]{5}/3)^{-3}$
3876 9919 $\zeta : \sqrt[3]{3}/3$
3877 0050 tanh : 9/22
3877 0790 rt : (5,9,-5,-3)
3877 3841 $Li_2 : \exp(-\pi/3)$
3877 4587 $J_2 : 3\sqrt[3]{3}/2$
3877 5237 $J_1 : 9(1/\pi)$
3877 5525 rt : (1,7,-3,-1)
3877 5543 $J_1 : 17/20$
3877 5749 rt : (4,7,7,-4)
3877 8525 $1/5 - s\pi(1/5)$
3877 8918 $J : 5/17$
3878 0304 rt : (8,5,-7,-3)
3878 0625 $2^x : 2\sqrt{2}/3 - 4\sqrt{3}/3$
3878 0822 $\ln(3/4) + s\pi(\sqrt{5})$
3878 1229 $J_2 : 9\zeta(3)/5$
3878 2654 rt : (1,3,9,-4)
3878 4040 tanh : $9(1/\pi)/7$
3878 5731 $\Psi : 4e\pi/3$
3878 6442 rt : (9,2,2,1)
3878 8113 $J_2 : 1/\ln(\sqrt[3]{4})$
3878 8140 rt : (7,8,1,-2)
3878 8588 $2^x : 1/2 + 2\sqrt{5}$
3878 8951 $\ell_{10} : 3/2 + 2\sqrt{2}/3$
3878 9183 $s\pi : 3\sqrt{3} + 3\sqrt{5}/4$
3878 9196 $\ell_{10} : 10\sqrt[3]{5}/7$
3879 1099 $\ell_{10} : 2\sqrt{3}/3 - \sqrt{5}/3$
3879 1249 at : $1/4 - 4\sqrt{2}$
3879 1311 sq : $3\pi/8$
3879 3120 rt : (7,0,-7,9)
3879 5304 $e^x : 4/3 + e/3$
3879 6214 rt : (2,8,-5,-7)
3879 6397 $\zeta : 2\zeta(3)/5$
3879 7121 rt : (4,7,-4,5)
3879 8891 at : $3\sqrt{3}/4 - 3\sqrt{5}$
3879 9190 at : $9\zeta(3)/2$
3880 0225 $E : (e + \pi)/9$
3880 0540 $\ell_{10} : 7\pi/9$
3880 0697 ln : $10\zeta(3)/3$
3880 2271 rt : (8,8,6,-4)
3880 2507 $J_0 : 10\zeta(3)/7$
3880 3515 $\ln(5) - \exp(1/5)$
3880 4145 at : $4/3 + 3e/2$
3880 4904 rt : (7,3,-5,6)
3880 5099 sr : $3\sqrt{2}/4 + \sqrt{3}/2$
3880 7206 ln : $3\pi^2$
3880 8765 $J_2 : 5\sqrt{3}/4$
3881 1104 rt : (4,2,-4,-9)
3881 1216 $s\pi : 4e$
3881 3234 rt : (5,4,8,-4)
3881 3788 $2^x : 1/\ln(\sqrt[3]{3}/3)$
3881 4130 cr : $2\sqrt{2}/3 + \sqrt{3}$
3881 4378 $t\pi : 1/3 - \sqrt{2}/2$
3881 5558 at : $4 + \sqrt{2}$
3881 7971 at : $2\sqrt{2}/3 + 2\sqrt{5}$
3881 8017 $\ell_{10} : 9/22$
3881 8261 sr : $1/4 + 3\sqrt{5}/4$
3882 1403 $\ell_{10} : 4\sqrt{2}/3 + \sqrt{5}/4$
3882 1542 $10^x : \sqrt{2}/2 - \sqrt{5}/2$
3882 3289 rt : (5,6,-1,-1)
3882 3850 rt : (7,6,-3,3)
3882 3997 cr : $4\sqrt{2} - 4\sqrt{5}/3$
3882 6741 $t\pi : 8\zeta(3)/7$
3882 6896 rt : (4,6,-6,-3)
3882 9144 rt : (1,3,8,8)
3882 9209 $2^x : 4\sqrt{2} - 4\sqrt{5}/3$
3883 1794 rt : (7,6,9,3)
3883 1871 at : 9/22
3883 2597 id : $e/7$
3883 4727 rt : (4,1,-2,6)
3883 5887 $t\pi : \sqrt[3]{19/2}$
3883 7325 rt : (9,5,7,-4)
3883 8268 $10^x : \sqrt{2} + 2\sqrt{5}/3$
3883 8826 rt : (6,6,4,1)
3883 9303 $2^x : (3\pi)^{-1/3}$
3883 9820 $Ei : 15/2$
3884 1162 $J_2 : 13/6$
3884 1726 $t\pi : \sqrt[3]{20/3}$
3884 1809 $\ln(\sqrt{3}) \times s\pi(1/4)$
3884 2019 rt : (8,5,2,-2)
3884 2975 sq : 17/11
3884 4457 sq : $9(e + \pi)$
3884 5976 at : $9(1/\pi)/7$
3884 6172 $\ell_2 : 5\pi/3$
3884 6187 $\Psi : \ln(e/3)$
3884 7154 $J_0 : 7(e + \pi)/10$
3884 8382 $\ell_2 : 3 + \sqrt{5}$
3884 8395 $10^x : 1 + 2\sqrt{2}$
3884 8624 $e^x : 1/2 + 2\pi/3$
3884 8974 rt : (1,5,2,-3)
3884 9121 $e^x : \sqrt{2}/4 - 3\sqrt{3}/4$
3884 9266 $\Psi : 1/8$
3885 0171 $\Gamma : (\sqrt[3]{2}/3)^{1/2}$
3885 0343 at : $4 - 3\pi$
3885 0634 $J_2 : 9\sqrt{3}/4$
3885 1223 $\Psi : \exp(2/3)$
3885 1748 cu : $(\sqrt{2}\pi/2)^3$
3885 2023 cr : $1 + 3\sqrt{5}/4$
3885 2895 $s\pi : \sqrt{15}$
3885 3184 sinh : $10\pi/9$
3885 3554 $t\pi : 5e\pi$
3885 3597 at : $2\sqrt{3}/3 - \sqrt{5}/3$
3885 3699 $\ell_{10} : 9e$
3885 3915 rt : (8,5,-2,9)
3885 4014 $t\pi : 3\sqrt{2}/2 + \sqrt{3}/3$
3885 4476 $Ei : 9\sqrt{5}/2$
3885 5277 sq : $\sqrt{3}/4 + \sqrt{5}/3$
3885 5960 rt : (6,1,9,-4)
3885 6011 at : $2\sqrt{2} + 3\sqrt{3}/2$
3885 6801 $e \times \ln(3/5)$
3885 7560 sq : $9\zeta(3)/7$
3885 9332 rt : (7,9,-1,0)
3885 9399 cu : P_{23}
3886 1228 sq : $3 + 3e/4$
3886 1593 at : $3/2 - 4\sqrt{3}$
3886 1811 $Li_2 : 1/\sqrt[3]{23}$
3886 2524 rt : (5,8,-9,1)
3886 3687 sr : $\sqrt{3} - \sqrt{6} + \sqrt{7}$
3886 5114 $2^x : 9/19$
3886 5618 sinh : $4\sqrt{2}/5$
3886 6337 rt : (5,8,9,-5)
3886 8110 $e^x : 3e/2 + \pi/4$
3886 8635 $E : 13/20$
3887 1635 sq : $2/3 + 3\sqrt{5}$
3887 2106 rt : (4,9,-3,-4)
3887 3375 $2^x : (\ln 3/3)^{-3}$
3887 3478 sinh : $5\sqrt[3]{3}/2$
3887 3573 $Ei : 7e/9$
3887 3808 rt : (8,9,-8,-4)
3887 5972 rt : (8,1,-9,7)
3887 6720 rt : (5,1,4,-2)
3887 7525 rt : (6,6,-9,-8)
3887 8865 $t\pi : (\ln 2/2)^{-1/2}$
3888 0067 $e^x : (\pi/3)^{-3}$
3888 0073 $t\pi : \sqrt{5}/2$
3888 0801 rt : (9,9,8,-5)
3888 1740 $J_1 : \exp(-1/(2\pi))$
3888 3190 rt : (8,8,0,6)
3888 7056 $J_0 : 25/7$
3888 7092 $\Psi : 9/2$
3888 7768 $t\pi : 7\sqrt[3]{2}/10$
3888 8031 rt : (1,8,-8,-7)
3888 8437 rt : (1,7,5,7)
3888 8888 id : 7/18
3888 8994 at : $2e$
3888 9459 $J_0 : (\sqrt[3]{4}/3)^{-2}$
3889 0505 id : $4\sqrt{2} + \sqrt{3}$
3889 1111 id : $1/\sqrt[3]{17}$
3889 2029 at : $4\sqrt{3} - 2\sqrt{5}/3$
3889 2324 rt : (1,7,0,-1)
3889 4878 $\Gamma : 21/2$
3889 5341 rt : (3,3,-7,-3)
3889 5569 cr : $2 + e/4$
3889 8568 $E : 3\sqrt{3}/8$
3889 9017 rt : (7,5,-8,-2)
3889 9206 rt : (5,3,1,9)
3890 0393 rt : (2,4,-7,2)
3890 1467 rt : (8,9,3,-3)
3890 4802 at : $\sqrt{3}\pi$
3890 5609 id : $(e)^2$
3890 5919 rt : (8,4,-7,4)
3890 6090 $\Psi : 9\sqrt{3}/8$
3890 6499 $e^x : 7\sqrt[3]{5}/2$
3890 6747 rt : (9,2,3,-2)
3890 7562 $\ell_{10} : \sqrt{6}$
3890 8818 cr : $3\sqrt{2}/2 + \sqrt{5}/4$
3890 9445 cr : $4\sqrt{3} + 3\sqrt{5}$
3891 1376 $Ei : (\ln 2)^2$
3891 4973 rt : (9,9,-3,-2)
3891 6694 sq : $9\ln 2/10$
3891 7390 $2^x : 1 + 2\sqrt{5}$
3891 8563 $e^x : 1/\ln(\ln 2/2)$
3891 9357 sq : $(\pi)^3$
3891 9359 rt : (1,4,6,-3)

3891 9582 $\Psi : (\ln 2/3)^{-3}$
3892 0275 $t\pi : P_{32}$
3892 0321 at : $1/4 + 3\sqrt{3}$
3892 0585 sinh : $2\sqrt[3]{5}/9$
3892 1555 rt : (5,1,-6,-7)
3892 2804 $J_1 : e\pi/10$
3892 2855 rt : (4,0,-7,1)
3892 3464 sq : $e - 2\pi/3$
3892 4587 $J_2 : (3\pi/2)^{1/2}$
3892 5046 $2^x : 3 - \sqrt{3}/2$
3892 6953 id : $3\sqrt{2}/2 - \sqrt{3}$
3892 7930 rt : (3,8,-1,-1)
3892 9754 rt : (7,6,-9,-4)
3893 0734 $erf : 2\sqrt[3]{2}/7$
3893 2763 $e^x : 3/2 + 3e/2$
3893 2970 $erf : 9/25$
3893 4349 rt : (3,2,-9,8)
3893 4715 $\sqrt{5} \div \ln(5)$
3893 6209 ln : $\sqrt{2} + 3\sqrt{3}/2$
3893 6436 $2^x : (\sqrt{2}\pi)^{-1/2}$
3893 6507 rt : (5,5,5,-3)
3893 6610 id : $2e - \pi/3$
3893 8123 $2^x : 2\pi/5$
3894 0039 $1/2 \div \exp(1/4)$
3894 0517 rt : (2,6,-5,-8)
3894 0839 $\ell_{10} : 4/3 + \sqrt{5}/2$
3894 1834 $s\pi : 2(1/\pi)/5$
3894 2892 rt : (9,6,-4,7)
3894 3007 cu : $e/2 - 2\pi$
3894 4222 rt : (3,7,-8,-4)
3894 6635 $J_2 : \exp(e/2)$
3894 8873 rt : (2,3,6,2)
3894 8973 $\ell_2 : 2\sqrt{2}/3 + 3\sqrt{5}/4$
3894 9359 rt : (9,6,4,-3)
3894 9549 $10^x : 1/7$
3894 9630 rt : (7,9,-2,-1)
3894 9951 $2^x : \exp(-\sqrt{5}/3)$
3895 3183 rt : (8,7,-5,1)
3895 3208 sq : $\exp(-\sqrt{2}/3)$
3895 3696 $J_0 : 8\sqrt{5}/5$
3895 4680 $1/5 \times \exp(2/3)$
3895 5310 rt : (1,8,7,-4)
3895 9662 cr : $6\sqrt{5}/5$
3896 0275 $10^x : e\pi/3$
3896 0331 $J_1 : \sqrt[3]{5}/2$
3896 0994 $s\pi : 7(e+\pi)/8$
3896 3792 rt : (6,1,-9,3)
3896 4075 rt : (5,9,6,-4)
3896 5901 $10^x : \sqrt{24}$
3896 6294 $10^x : \sqrt{19/2}$
3896 9326 $J_1 : \sqrt{5} - \sqrt{6} - \sqrt{7}$
3897 1941 sinh : $\sqrt{2} - \sqrt{3} - \sqrt{5}$
3897 2249 rt : (6,2,5,2)
3897 2533 $J_0 : 12/7$
3897 3125 at : $3/4 + 3\pi/2$
3897 8276 $\Gamma : 2(1/\pi)/5$
3897 8352 id : $9\sqrt[3]{5}$
3897 8677 at : $2 + 2\sqrt{3}$
3897 8815 $\theta_3 : 2\sqrt[3]{3}/5$
3897 9051 as : $2\sqrt[3]{5}/9$
3897 9096 $\Gamma : 3e\pi/10$
3897 9456 rt : (7,8,-6,-5)
3897 9941 rt : (5,2,-4,4)
3898 0784 rt : (2,6,-2,-9)
3898 2709 $s\pi : 2\sqrt{2} + 3\sqrt{3}/4$
3898 3394 rt : (1,9,5,-1)
3898 4418 $\ln(5) \times \exp(e)$
3898 4510 sq : $3 - 4e/3$
3898 6281 $Ei : (\ln 2/3)^{1/2}$
3898 8704 $erf : \sqrt[3]{3}/4$
3898 9074 at : $\sqrt{2}/2 - \sqrt{5}/2$
3898 9557 $E : (\sqrt[3]{2}/3)^{1/2}$
3899 1509 rt : (2,1,7,-3)
3899 3354 rt : (9,9,-2,4)
3899 4121 $erf : 3\zeta(3)/10$
3899 7004 tanh : 7/17
3899 7500 $\ell_2 : \sqrt[3]{18}$
3899 8892 rt : (6,2,6,-3)
3899 9022 $\ell_{10} : 1/3 + 3\sqrt{2}/2$
3899 9757 rt : (9,5,-9,2)
3899 9911 sr : P_{71}
3899 9960 $s\pi : 4\sqrt{3} + 4\sqrt{5}$
3900 0412 sr : P_{68}
3900 1422 $J_2 : 4e/5$
3900 2123 rt : (2,4,-9,-4)
3900 2381 cu : $1/4 + \sqrt{3}/2$
3900 4678 at : $1 + 2\sqrt{5}$
3900 6045 sr : $1 + 3\pi/2$
3900 6727 rt : (2,5,8,-4)
3901 0305 rt : (6,6,7,-4)
3901 1066 cu : $e + \pi/3$
3901 1253 $Ei : \sqrt[3]{3}/3$
3901 1282 $2^x : 8\sqrt{2}/9$
3901 3738 $Ei : \exp(-2\pi/3)$
3901 4538 rt : (6,9,-9,9)
3901 5292 at : $3\sqrt{2}/2 + 3\sqrt{5}/2$
3901 6644 rt : (6,8,6,9)
3902 0677 rt : (6,4,-1,7)
3902 1377 cu : $\sqrt{5} - \sqrt{6} + \sqrt{7}$
3902 1525 $t\pi : 1/\sqrt{11}$
3902 2394 $2^x : 1 + 4\sqrt{2}/3$
3902 2977 $Li_2 : 6/17$
3902 2998 tanh : $2\sqrt[3]{3}/7$
3902 3332 sq : $\sqrt{3}/2 - 2\sqrt{5}/3$
3902 3374 sinh : 8/21
3902 3469 rt : (6,0,-8,5)
3902 4025 $\ell_{10} : 4\sqrt{3} - 2\sqrt{5}$
3902 6639 $e^x : 3/4 + 4e/3$
3902 7475 rt : (3,8,-5,8)
3902 8186 sr : $e - \pi/4$
3902 8591 $Li_2 : 14/15$
3902 9367 $\ell_2 : 1 + 3\sqrt{2}$
3902 9722 $Ei : 11/4$
3903 0056 $10^x : e + 4\pi$
3903 0605 at : $2\sqrt{3} - 4\sqrt{5}$
3903 2827 sr : $3/2 + \sqrt{3}/4$
3903 3021 $\sqrt{5} + \exp(e)$
3903 5778 $Ei : 2\zeta(3)/5$
3903 5794 $2/3 \times \exp(3)$
3903 7030 as : $2/3 - \pi/3$
3903 8820 rt : (8,6,-1,-1)
3903 9280 $Ei : 3e/10$
3903 9321 $J_1 : 20/7$
3903 9940 $J_1 : 6/7$
3904 0085 at : $5\pi^2/9$
3904 0693 $\Gamma : 11/17$
3904 0987 sr : $3\sqrt{2} - 4\sqrt{3}/3$
3904 2379 $2/5 \div \ln(3/4)$
3904 3673 $J_1 : (\sqrt[3]{4})^{-1/3}$
3904 4289 $10^x : 2/3 + e/3$
3904 5829 cu : $3 - 2\pi$
3904 8572 cu : $4\sqrt{2}/3 - 2\sqrt{3}/3$
3904 9511 rt : (7,2,-5,8)
3905 1164 $E : 11/17$
3905 2033 rt : (3,5,-7,-8)
3905 2429 id : $4/3 - 2\sqrt{2}/3$
3905 6208 $2 - \ln(5)$
3905 6995 rt : (7,7,9,-5)
3905 9796 id : $\sqrt{24}\pi$
3906 0704 at : 7/17
3906 1127 $2^x : 1 - 3\pi/4$
3906 2500 sq : 5/8
3906 2528 rt : (7,3,8,-4)
3906 2859 $J_1 : (e/2)^{-1/2}$
3906 3059 $t\pi : \sqrt[3]{17}$
3906 4512 at : $4 + 2\sqrt{5}/3$
3906 5727 rt : (3,2,9,-4)
3906 6986 rt : (8,0,-9,9)
3906 7403 cr : $1/3 + 3\pi/4$
3906 7998 cr : $8e\pi/5$
3906 8896 $\sqrt{3} \div \ln(3/5)$
3906 9263 $1/4 + \exp(\pi)$
3906 9911 $e^x : (\ln 3/2)^{-1/3}$
3907 0462 $t\pi : \sqrt{2}/3 + 2\sqrt{3}/3$
3907 0722 rt : (9,8,-7,-1)
3907 2903 rt : (7,1,7,3)
3907 4007 rt : (2,7,-6,8)
3907 6851 id : $2\sqrt{3}/3 + \sqrt{5}$
3907 7065 rt : (4,9,2,8)
3907 8598 $t\pi : \sqrt{2}/4 - 2\sqrt{5}$
3907 9785 cu : $4/3 + 4\pi/3$
3907 9934 rt : (3,2,8,3)
3908 1340 rt : (5,1,-9,-1)
3908 2613 as : 8/21
3908 2758 at : $1/3 - \sqrt{5}/3$
3908 2884 $\ln(2/3) \times s\pi(\sqrt{2})$
3908 3240 $J_0 : 8\pi/7$
3908 5520 $J : 1/\ln(3\pi)$
3908 6361 $10^x : 2 - \pi/4$
3908 6916 at : $2\sqrt[3]{3}/7$
3908 7198 at : $7\pi/4$
3908 9831 sinh : $1/\sqrt[3]{18}$
3909 0076 $J_1 : (1/(3e))^{-1/2}$
3909 0630 rt : (1,3,-9,3)
3909 0831 $J_0 : 3e/2$
3909 4142 $J_1 : 5\zeta(3)/7$
3909 4282 at : 11/2
3909 4719 rt : (5,5,-2,1)
3909 7792 $\theta_3 : \gamma$
3909 8504 $Li_2 : \sqrt{2}/4$
3909 9428 rt : (7,5,6,2)
3910 1643 $10^x : 4/3 + \sqrt{3}/4$
3910 1792 $10^x : 3\sqrt{2}/8$
3910 2561 rt : (1,9,4,-3)
3910 3023 $10^x : 1 + 3\sqrt{5}/4$
3910 3341 ln : $3/2 + 3\pi$
3910 3863 sinh : $\exp(\pi/2)$
3910 6561 $2^x : 10/21$
3910 7967 rt : (9,7,1,-2)
3911 0220 tanh : $(e+\pi)^{-1/2}$
3911 0379 rt : (6,3,-6,2)
3911 2968 rt : (3,8,-5,-3)
3911 3838 $Li_2 : 10(1/\pi)/9$
3911 6499 id : $\sqrt{23/2}$
3911 7824 $2^x : 2 - 3\sqrt{5}/2$
3911 8168 cr : $4/3 + e/2$
3912 0379 rt : (5,9,5,3)
3912 1769 $J_0 : ((e+\pi)/2)^{1/2}$
3912 2004 rt : (8,0,9,4)
3912 3192 cr : $7(e+\pi)/3$
3912 3670 cr : $6\pi/7$
3912 5971 rt : (3,6,7,2)
3912 7450 rt : (7,7,-6,-3)
3912 7821 rt : (2,6,0,9)
3912 8128 $\ln(5) \div \exp(\sqrt{2})$
3912 8425 rt : (6,7,1,4)
3912 9291 sr : $3 + e$
3913 0434 id : 9/23
3913 1473 rt : (7,5,-3,5)
3913 2542 $t\pi : 1/3 + \pi/4$
3913 2830 sinh : $6(1/\pi)/5$
3913 3076 rt : (8,3,-7,6)
3913 4020 $3/5 \times \exp(\sqrt{3})$
3913 4689 ln : $4 + 4\sqrt{3}$
3913 8238 rt : (2,4,-8,-2)
3913 8646 sq : $1/2 + e/4$
3913 8939 at : $\sqrt{2} - 4\sqrt{3}$
3913 9517 sr : $4e/3 + 2\pi/3$
3914 3069 sq : $3e/2 + 3\pi/4$
3914 5426 $J_2 : 24/11$
3914 5505 rt : (9,5,-4,9)
3914 5883 $\ell_{10} : 4\sqrt{2}/3 + \sqrt{3}/3$
3914 6763 rt : (1,8,-4,-6)
3914 7308 ln : $2/3 + 3\sqrt{5}/2$
3914 7491 rt : (8,7,0,8)
3914 8533 $\theta_3 : \sqrt{3}/3$
3914 9574 as : $1/\sqrt[3]{18}$
3915 0471 rt : (2,9,-7,-4)
3915 0490 $Ei : \sqrt{17}$
3915 0550 $\ln(3) - s\pi(1/4)$
3915 0970 rt : (7,9,4,7)
3915 1980 rt : (9,3,0,-1)
3915 2303 $1/5 \div \ln(3/5)$
3915 4413 rt : (4,8,-5,-3)
3915 4641 $J_1 : (\pi/2)^{-1/3}$
3915 7776 sq : $2\sqrt{3} - \sqrt{5}/3$
3915 8640 rt : (6,7,4,-3)
3915 9265 id : $1/4 + \pi$
3916 0836 $J_2 : 10e/7$
3916 1042 rt : (3,7,0,1)

3916 2171 rt : (8,4,8,3)
3916 2634 cr : $8\sqrt[3]{5}$
3916 4803 at : $4/3+4\pi/3$
3916 4868 sq : $3/4+4\pi/3$
3916 6153 rt : (6,8,-8,-4)
3916 6919 rt : (5,2,-6,-9)
3917 0271 rt : (3,7,-9,6)
3917 3044 cu : $3\sqrt{3}/4+3\sqrt{5}$
3917 3054 10^x : $3/2+4\pi$
3917 3261 E : $1/\ln(3\pi/2)$
3917 4874 at : $(e+\pi)^{-1/2}$
3917 6227 rt : (2,9,3,9)
3917 6606 erf : $1/\sqrt[3]{21}$
3917 6877 rt : (4,0,-2,8)
3917 6898 J_0 : $18/5$
3917 8796 rt : (8,9,-8,-7)
3917 8977 e^x : $3\sqrt{3}/4-\sqrt{5}$
3918 0260 id : $\sqrt{2}/4-\sqrt{5}/3$
3918 0859 rt : (7,8,6,-4)
3918 1758 $\exp(3/5)+s\pi(\pi)$
3918 4260 10^x : $3/2+3\sqrt{3}$
3918 4902 rt : (1,5,-1,6)
3918 5683 $2/3\times s\pi(1/5)$
3918 6036 rt : (6,9,-7,1)
3918 6068 at : $\exp(\sqrt[3]{5})$
3918 6169 Ei : $13/11$
3918 7660 sr : $4\sqrt{2}/3-\sqrt{3}$
3918 8149 at : $1/3+3\sqrt{3}$
3918 8873 rt : (1,9,-4,-3)
3918 9543 rt : (5,2,1,-1)
3919 0971 rt : (4,5,9,3)
3919 1609 e^x : $8/3$
3919 1898 cu : $4+2\sqrt{2}$
3919 2267 as : $3/2-\sqrt{5}/2$
3919 2900 as : $6(1/\pi)/5$
3919 3710 tanh : $(\sqrt{5}/3)^3$
3919 4359 rt : (1,7,-8,2)
3919 4390 10^x : $9\sqrt{2}$
3919 4506 2^x : $1/4+4\sqrt{5}/3$
3919 4845 e^x : $3/4+\sqrt{2}/3$
3919 5378 rt : (8,9,8,-5)
3919 5943 id : $e/4+3\pi/2$
3919 7175 e^x : $5\sqrt[3]{5}/7$
3919 7427 rt : (3,7,7,-4)
3919 8472 rt : (4,5,-8,7)
3919 8627 at : $3\sqrt{2}/4+2\sqrt{5}$
3919 9106 rt : (1,9,-8,-9)
3919 9859 at : $3/4-2\pi$
3920 0458 ζ : $\ln(\sqrt{2}\pi/2)$
3920 0892 e^x : 13
3920 1695 rt : (2,5,-6,-3)
3920 1715 sinh : 17/15
3920 2133 Ψ : $(\sqrt[3]{2})^{1/2}$
3920 2387 sinh : $8\sqrt[3]{3}/3$
3920 3041 ℓ_{10} : $\sqrt[3]{15}$
3920 3857 $t\pi$: $\ln(5/3)$
3920 4651 ℓ_{10} : $\ln(3/2)$
3920 5737 ln : $2\sqrt{3}+\sqrt{5}/4$
3920 7061 ln : $3\sqrt{3}/2-\sqrt{5}/2$
3920 8802 sq : $\sqrt{20}\pi$
3921 0181 rt : (4,8,9,-5)
3921 1189 rt : (9,4,-9,4)
3921 4701 at : P_{56}
3921 5005 e^x : $3\sqrt{2}/2+3\sqrt{3}$
3921 5969 sq : $4/3-\sqrt{2}/2$
3921 6051 rt : (7,3,-5,-2)
3921 6896 2^x : $e\pi/4$
3921 8395 rt : (1,9,7,-5)
3921 9584 rt : (6,3,3,-2)
3922 0011 2^x : $3/4+3e/2$
3922 1103 Ψ : $(e+\pi)/3$
3922 1217 10^x : $1/4+2\sqrt{3}/3$
3922 1828 J_0 : $\sqrt[3]{5}$
3922 3227 sr : 2/13
3922 3975 ℓ_{10} : $\pi^2/4$
3922 4807 cr : $3\sqrt{2}/2+\sqrt{3}/3$
3922 6683 at : $3\sqrt{2}+3\sqrt{3}/4$
3922 8095 rt : (7,4,-8,0)
3922 8605 cu : $\sqrt{2}/3+\sqrt{3}/2$
3922 9249 J_0 : $\sqrt{13}$
3922 9491 2^x : $3(1/\pi)/2$
3922 9856 $t\pi$: $(\ln 3/2)^2$
3922 9926 J_0 : $5\sqrt[3]{3}/2$
3923 0484 id : $6\sqrt{3}$
3923 0720 ζ : $\sqrt{17/3}$
3923 1742 ℓ_2 : 21
3923 2462 rt : (7,4,5,-3)
3923 2882 rt : (8,6,-5,3)
3923 4434 rt : (1,7,7,-9)
3923 5011 J_0 : $3\zeta(3)$
3923 5586 $s\pi$: $3e-2\pi$
3923 6560 rt : (9,8,-2,6)
3923 7478 rt : (6,4,-7,-3)
3923 7526 10^x : $2\sqrt{2}/3+\sqrt{3}/2$
3923 7709 at : $8\ln 2$
3923 8488 cu : $5\zeta(3)/4$
3923 8974 $\ln(2)-\exp(3)$
3923 9618 rt : (8,5,7,-4)
3924 4172 rt : (9,6,9,-5)
3924 4644 cr : $\sqrt{19}\pi$
3924 5799 rt : (5,5,-9,-4)
3924 6379 at : $e/2+4\pi/3$
3924 7000 sr : $2\ln 2/9$
3924 7360 10^x : $5(1/\pi)/3$
3925 0420 $\ln(\sqrt{5})+s\pi(1/5)$
3925 0433 rt : (6,1,-8,-7)
3925 0447 J : $\exp(-17\pi/8)$
3925 0562 ln : $\sqrt{5}/9$
3925 1050 10^x : $1/3+3e$
3925 1272 θ_3 : $5\ln 2/6$
3925 1961 at : $\sqrt{2}/4+3\sqrt{3}$
3925 3483 e^x : $((e+\pi)/3)^2$
3925 6001 10^x : $1/2-e/3$
3925 6648 $\exp(\sqrt{5})+s\pi(\sqrt{2})$
3925 7944 θ_3 : $2\pi/7$
3925 9076 at : $(\sqrt{5}/3)^3$
3926 0013 $4\times\exp(4)$
3926 0181 sr : $3-3\sqrt{2}/4$
3926 3183 rt : (5,9,-5,5)
3926 3402 ℓ_2 : $4\sqrt{3}-3\sqrt{5}/4$
3926 3649 tanh : $1/\sqrt[3]{14}$
3926 3896 at : $2\sqrt{3}/3-3\sqrt{5}$
3926 4199 cr : $4-3\sqrt{3}/4$
3926 4678 rt : (4,4,8,-4)
3926 8136 J_1 : $(\sqrt{5}/3)^{1/2}$
3926 9118 id : $\ln(\sqrt{2}\pi/3)$
3926 9908 id : $\pi/8$
3927 2047 rt : (9,7,9,3)
3927 2924 rt : (7,8,-1,2)
3927 3753 Γ : $9\sqrt{3}/2$
3927 4810 rt : (6,6,-4,-1)
3927 5178 sq : $10(e+\pi)/9$
3927 6030 Γ : $1/\sqrt{23}$
3927 6197 10^x : P_{75}
3927 7465 rt : (8,8,7,2)
3927 7814 rt : (5,4,-7,-4)
3927 9020 J_1 : 19/22
3928 1620 ln : $2+4\sqrt{5}$
3928 3723 Ψ : $4(1/\pi)/3$
3928 4645 rt : (6,3,-1,9)
3928 5309 Ψ : $8\sqrt[3]{5}/7$
3928 6226 J_1 : $5\sqrt[3]{5}/3$
3928 7551 rt : (5,1,-4,6)
3928 7969 cu : $\ln(\sqrt[3]{3}/3)$
3928 8695 $t\pi$: $9\ln 2/2$
3928 9843 Γ : $3\sqrt[3]{5}/2$
3929 1448 rt : (4,1,-7,-3)
3929 2941 erf : 4/11
3929 3499 rt : (4,8,0,9)
3929 4383 ℓ_{10} : $2+\sqrt{2}/3$
3929 6718 J_1 : $\exp(\pi/3)$
3929 6944 rt : (3,9,-4,0)
3930 0219 cu : $(e+\pi)/8$
3930 1384 Ψ : $7\sqrt[3]{3}/9$
3930 3493 J_2 : $\sqrt[3]{21/2}$
3930 3779 e^x : $2+\sqrt{2}$
3930 3826 sinh : $9\sqrt[3]{2}/10$
3930 5936 rt : (9,2,8,-4)
3930 6335 2^x : 11/23
3930 7410 erf : $8(1/\pi)/7$
3930 9766 E : 9/14
3930 9871 rt : (1,5,0,-7)
3931 2882 10^x : $\ln(2/3)$
3931 2981 rt : (7,9,5,1)
3931 3116 e^x : $3\sqrt{3}/2-\sqrt{5}/2$
3931 4496 e^x : $3e-\pi$
3931 4988 sq : $3\sqrt{3}-3\sqrt{5}/2$
3931 5460 e^x : $2+\sqrt{3}/4$
3931 8042 at : $4+\pi/2$
3931 8707 rt : (8,1,6,-3)
3931 9752 Ei : $\sqrt{2/3}$
3932 1240 $\ln(3)-\exp(2/5)$
3932 2156 rt : (8,2,-7,8)
3932 3186 ζ : $7e/8$
3932 4511 rt : (2,8,-3,2)
3932 6001 rt : (9,7,-7,1)
3932 6370 J_2 : $(\pi/2)^3$
3932 7180 cr : $9\zeta(3)/4$
3932 9616 at : $1/\sqrt[3]{14}$
3933 0789 sr : $7\pi^2/6$
3933 3765 $s\pi$: $3/4+3\sqrt{2}/2$
3933 4191 at : $4/3+3\sqrt{2}$
3933 4562 rt : (2,5,-4,7)
3933 5007 Ψ : 9/14
3933 8704 at : $3/2+3e/2$
3933 9164 J_0 : $e\pi/5$
3934 0057 cr : $2/3+3e/4$
3934 0090 rt : (2,7,3,-4)
3934 1658 ln : $3\sqrt{2}+3\sqrt{5}$
3934 1788 rt : (5,1,9,-4)
3934 1825 as : $3\sqrt{3}/2-4\sqrt{5}/3$
3934 1899 J_1 : $3\sqrt[3]{3}/5$
3934 2379 sr : $5e/7$
3934 2925 rt : (7,0,4,2)
3934 4092 rt : (5,5,3,8)
3934 5172 at : $3\sqrt{3}/2+4\sqrt{5}/3$
3934 5550 rt : (7,9,3,-3)
3934 6090 tanh : $3\ln 2/5$
3934 6258 ℓ_{10} : $10\sqrt{3}/7$
3934 6266 Γ : $\zeta(3)/7$
3934 8680 rt : (9,8,-2,-1)
3935 0913 $t\pi$: $2\pi/9$
3935 1159 ℓ_{10} : $\exp(e/3)$
3935 5304 ℓ_{10} : $2\sqrt{2}/7$
3935 6144 rt : (2,8,5,3)
3935 8600 cu : $3\sqrt[3]{5}/7$
3935 9505 $\ln(\sqrt{3})\div\exp(1/3)$
3936 1363 rt : (4,9,6,-4)
3936 2657 $t\pi$: $3(1/\pi)/8$
3936 3583 at : $3\sqrt{2}/2+2\sqrt{3}$
3936 3679 rt : (1,4,-9,-8)
3936 6671 J_1 : $\sqrt{3}/2$
3936 7887 e^x : $3/2+2\pi/3$
3936 9731 cr : $2+\sqrt{2}/2$
3937 0073 rt : (4,0,7,3)
3937 1607 rt : (6,8,1,-2)
3937 2535 ζ : $\exp(\sqrt{3}/2)$
3937 2566 J_2 : $2\pi^2/9$
3937 3159 rt : (7,4,-3,7)
3937 4104 rt : (1,9,9,-5)
3937 4538 $\exp(4/5)\div\exp(\sqrt{3})$
3937 6006 cr : $\sqrt{2}/3+\sqrt{5}$
3937 7212 $e+s\pi(\sqrt{5})$
3937 8003 rt : (5,1,-8,-3)
3937 8318 at : $5\sqrt{5}/2$
3937 8748 rt : (3,9,-2,-2)
3937 8813 J : $\exp(-6\pi/7)$
3937 9509 sq : $3-\sqrt{2}/3$
3937 9953 rt : (4,7,-7,-2)
3938 1411 $\ln(5)\times s\pi(1/3)$
3938 1473 Ψ : $3\sqrt{2}/10$
3938 1644 J_1 : $5\ln 2/4$
3938 1688 cu : $2\sqrt{2}-2\sqrt{5}/3$
3938 1907 2^x : $3/4-2\pi/3$
3938 2026 sq : $1/2+\pi/3$
3938 2153 10^x : $3/4-2\sqrt{3}/3$
3938 2655 rt : (2,8,-6,1)
3938 3351 ℓ_2 : $3e+3\pi/4$
3938 4685 sr : $1+2\sqrt{2}/3$
3938 5277 sr : $\sqrt[3]{22/3}$

3938 7691 sq : $2\sqrt{2}+3\sqrt{3}$
3938 7862 10^x : $\sqrt[3]{3}/10$
3938 8068 J_2 : $\sqrt{15}$
3938 8230 rt : (9,3,-9,6)
3938 9933 as : $2\sqrt{2}/3-\sqrt{5}/4$
3939 0170 J_1 : $13/15$
3939 0676 ℓ_{10} : $e/3+\pi/2$
3939 2876 rt : (9,7,6,-4)
3939 2879 at : $4/3-4\sqrt{3}$
3939 4134 sq : $3\sqrt{2}/4-\sqrt{3}/4$
3939 5616 at : $\sqrt{2}+\sqrt{3}+\sqrt{6}$
3939 6072 2^x : $1/4+2\sqrt{5}$
3939 6632 id : $7\sqrt[3]{5}/5$
3939 7056 10^x : $9/14$
3939 7183 J_0 : $4e/3$
3939 8428 rt : (6,9,-5,-3)
3939 9350 id : $1/3+3\sqrt{2}/4$
3939 9732 ℓ_{10} : $\sqrt{3}+\sqrt{5}/3$
3940 0404 2^x : $1/3-3\sqrt{5}/4$
3940 0841 rt : (8,9,-3,0)
3940 2072 J_1 : $e\pi/3$
3940 2137 rt : (6,7,9,-5)
3940 2800 at : $3+3\sqrt{3}/2$
3940 5383 Γ : $\exp(\sqrt[3]{4})$
3940 6441 rt : (6,1,2,1)
3940 6525 cu : $1/\ln(\sqrt{5}/3)$
3940 7397 cu : $2e/3-3\pi/2$
3940 8164 rt : (5,0,-9,1)
3940 8567 E : $3\sqrt[3]{5}/8$
3940 8707 J_0 : $7\sqrt{3}/3$
3941 1266 cu : $7\pi/2$
3941 1856 tanh : $5/12$
3941 2277 rt : (5,8,-7,6)
3941 2319 rt : (9,9,-8,-4)
3941 2771 at : $3\ln 2/5$
3941 6839 sinh : $5/13$
3941 7017 sr : $4+\sqrt{3}$
3941 7041 rt : (4,9,8,2)
3941 9910 at : $e/4-2\pi$
3942 0374 $s\pi(\pi)+s\pi(\sqrt{2})$
3942 0849 rt : (9,2,7,3)
3942 1432 $\pi\times\exp(\sqrt{5})$
3942 1690 Γ : $1/\ln(3\pi/2)$
3942 2389 rt : (5,8,0,-2)
3942 2729 sπ : $2\sqrt{2}/3+4\sqrt{3}$
3942 3004 J_1 : $\ln(\sqrt[3]{2}/3)$
3942 3278 E : $4\sqrt[3]{3}/9$
3942 3444 sπ : $4\sqrt{2}+2\sqrt{5}$
3942 5104 ℓ_{10} : $2/3+2e/3$
3942 5344 2^x : $1/4+4\sqrt{2}/3$
3942 5985 2^x : $3\sqrt{2}/2-2\sqrt{3}$
3942 6330 rt : (8,6,4,-3)
3942 7082 J_0 : $9\pi/7$
3942 8101 $\ln(2/5)\times s\pi(\pi)$
3942 9396 e^x : $\sqrt{2}/3+3\sqrt{5}$
3942 9774 rt : (7,7,-6,-3)
3943 1142 rt : (3,1,5,2)
3943 3918 rt : (5,6,7,-4)
3943 4609 rt : (7,8,4,9)
3943 4923 rt : (9,7,-2,8)
3943 5760 rt : (2,7,2,-2)
3943 6226 10^x : $2e-\pi/4$
3943 6770 ζ : $25/16$
3943 7037 cu : $11/15$
3943 7207 $3/5\div s\pi(\pi)$
3943 7777 rt : (6,2,-8,-9)
3943 7988 Ψ : $7\sqrt{5}/8$
3943 9940 J : $\exp(-\sqrt{5})$
3944 1149 rt : (3,8,-8,-2)
3944 3825 Γ : $7(1/\pi)/6$
3944 4203 sr : $3\sqrt{2}+2\sqrt{5}/3$
3944 4872 rt : (1,8,3,0)
3944 4915 $4\times\ln(3)$
3944 6888 ln : $(\sqrt{2}\pi/2)^3$
3944 7453 sinh : $2\sqrt{3}/9$
3944 8422 tπ : $2\sqrt{5}$
3945 0107 2^x : $9\sqrt[3]{5}/4$
3945 1482 rt : (4,7,-8,-4)
3945 3489 $4/5+\ln(2/3)$
3945 3593 2^x : $\sqrt{3}/2+4\sqrt{5}/3$
3945 3960 ℓ_{10} : $(2\pi)^3$
3945 4396 rt : (2,7,-9,-5)
3945 4891 rt : (8,5,-5,5)
3945 5610 $\exp(1/2)-s\pi(\sqrt{3})$
3945 5686 J_2 : $3(e+\pi)/8$
3945 7089 2^x : $10\sqrt{5}$
3945 8522 $\exp(3)\times s\pi(1/3)$
3945 9648 at : $2/3-2\pi$
3946 2088 id : $\sqrt{2}/3-\sqrt{3}/2$
3946 2205 rt : (4,8,0,-1)
3946 2882 rt : (1,7,-4,-8)
3946 3284 rt : (6,6,1,6)
3946 5565 rt : (3,6,-2,2)
3946 5681 at : $e/3+3\pi/2$
3946 6974 sinh : $5\ln 2/9$
3946 8350 e^x : $4+e$
3947 0367 rt : (7,8,-3,-2)
3947 1142 2^x : $\sqrt{2}/3+3\sqrt{3}/2$
3947 2318 e^x : $11/9$
3947 2988 $1/4+\ln(\pi)$
3947 4288 rt : (6,5,-9,-6)
3947 4366 2^x : $12/25$
3947 5103 cr : $\sqrt{2}+3\sqrt{3}/4$
3947 5724 sπ : $\sqrt{7/2}$
3947 6623 sq : $3e\pi/7$
3947 7126 J_2 : $9\sqrt[3]{5}/7$
3947 8162 cu : $\sqrt{3}/4+3\sqrt{5}/4$
3947 8417 sq : $\pi/5$
3947 9111 at : $5/12$
3948 0159 rt : (7,3,-8,2)
3948 0198 sr : $\sqrt{3}-\sqrt{5}+\sqrt{6}$
3948 1066 ℓ_{10} : $3/4+\sqrt{3}$
3948 2028 ln : $10\pi^2/9$
3948 2635 2^x : $\sqrt[3]{2}$
3948 3368 rt : (7,5,2,-2)
3948 3757 at : $2+4e/3$
3948 4180 E : $16/25$
3948 4802 rt : (4,7,7,7)
3948 5780 $\exp(\pi)+s\pi(\sqrt{3})$
3948 5867 rt : (1,5,8,-4)
3948 6344 Ei : $7/25$
3948 6480 sπ : $(\pi/3)^{-3}$
3948 7086 cu : $3/4-4e$
3948 7235 ζ : $19/8$
3948 8441 J_2 : $7\pi/10$
3949 0586 id : $3(e+\pi)/4$
3949 0609 id : $4\sqrt{2}/3-2\sqrt{5}/3$
3949 1011 cu : $3\sqrt{2}/4+\sqrt{3}/3$
3949 1294 at : $2\sqrt{3}/3+2\sqrt{5}$
3949 2819 cr : $19/7$
3949 3084 rt : (9,0,-9,3)
3949 4448 Li_2 : $7\zeta(3)/9$
3949 4635 Ei : $2\sqrt[3]{2}/9$
3949 4756 J_1 : $\sqrt[3]{23}$
3949 5079 cr : $\sqrt[3]{20}$
3949 5938 Ψ : $\sqrt[3]{15/2}$
3949 6237 J_1 : $20/23$
3949 6855 $e-\exp(\sqrt{2})$
3949 9066 rt : (5,3,8,3)
3950 0564 rt : (8,3,5,2)
3950 1681 tπ : $e/9$
3950 2639 rt : (2,5,-8,-4)
3950 3590 rt : (5,4,-2,3)
3950 5372 2^x : P_{21}
3950 5868 J_2 : $11/5$
3950 6172 sq : $4\sqrt{2}/9$
3950 6906 cu : $4-3\sqrt{2}/4$
3950 7030 rt : (2,4,-8,5)
3950 7954 Ψ : $1/\ln(5/3)$
3950 9842 Li_2 : $1/\sqrt[3]{22}$
3950 9966 as : $2\sqrt{3}/9$
3951 0863 Γ : $(e)^3$
3951 1299 e^x : $10\sqrt{5}/7$
3951 5323 cu : $2\sqrt{2}/3+3\sqrt{3}/2$
3951 5546 ln : $6\sqrt{3}/7$
3951 6280 rt : (6,3,8,-4)
3951 8169 e^x : $(\ln 2)^3$
3951 8899 rt : (8,6,-9,-4)
3951 9473 rt : (9,3,5,-3)
3952 0320 e^x : $1/2+3\pi/4$
3952 3739 rt : (7,9,8,-5)
3952 3758 10^x : $1/4+2e/3$
3952 3803 rt : (5,6,-6,-3)
3952 4249 ℓ_{10} : $10\sqrt{5}/9$
3952 5701 rt : (5,0,-4,8)
3952 7905 rt : (8,9,2,7)
3952 9642 as : $5\ln 2/9$
3953 0563 rt : (9,6,-7,3)
3953 0817 at : $4\pi^2/7$
3953 1003 rt : (2,6,-3,-2)
3953 4975 $\ln(3/5)\times s\pi(e)$
3953 7562 ln : $(\ln 2/3)^3$
3953 8605 J_0 : $9\sqrt{5}/5$
3953 8724 rt : (9,2,-9,8)
3953 9699 2^x : $(\ln 2/3)^{1/2}$
3954 1033 J_1 : $(\pi/3)^{-3}$
3954 1442 id : $2\sqrt{2}-\sqrt{3}/4$
3954 1991 rt : (5,7,-9,7)
3954 1995 Li_2 : $5/14$
3954 2143 e^x : $7e/4$
3954 2603 rt : (1,6,0,-1)
3954 3901 K : $12/13$
3954 4012 cu : $1/2-3\pi/4$
3954 4382 e^x : $1/\ln(\pi)$
3954 4669 e^x : 6π
3954 6553 erf : $\ln(\sqrt[3]{3})$
3954 6878 2^x : $\sqrt[3]{3}/3$
3954 7242 e^x : $\sqrt[3]{2/3}$
3954 7264 2^x : $1+3\sqrt{5}/4$
3955 1299 ℓ_{10} : $1/4+\sqrt{5}$
3955 2444 E : $2\sqrt{5}/7$
3955 3826 rt : (3,4,-9,-4)
3955 3930 2^x : $2\zeta(3)/5$
3955 4469 at : $3\sqrt{2}/4-3\sqrt{5}$
3955 5337 Ψ : $8\sqrt[3]{2}/9$
3955 6522 at : $3e/2+\pi/2$
3955 9703 ℓ_2 : $4\sqrt{2}/3+\sqrt{5}/3$
3956 0784 $\sqrt{5}\div\exp(\sqrt{3})$
3956 1242 id : $(e)^{1/3}$
3956 3753 id : $\sqrt{2}+4\sqrt{5}/3$
3956 4392 rt : (8,8,-8,-5)
3956 7300 tπ : $8\sqrt[3]{2}/9$
3956 8240 rt : (7,3,-3,9)
3956 8285 sinh : $5e/3$
3956 8715 at : $\exp(\sqrt{3})$
3956 9179 cr : $2\sqrt{3}-\sqrt{5}/3$
3956 9972 rt : (7,7,-1,4)
3957 0642 sinh : 12
3957 1181 rt : (3,8,9,-5)
3957 1897 e^x : $3/4-3\sqrt{5}/4$
3957 2327 rt : (1,6,-3,-3)
3957 5782 $\pi+s\pi(\sqrt{3})$
3957 6703 at : $9\pi/5$
3957 7024 erf : $\ln(\ln 2)$
3957 8289 Γ : $\sqrt{2}+\sqrt{5}+\sqrt{6}$
3957 9394 rt : (6,8,6,-4)
3958 2728 at : $4\sqrt{2}$
3958 3807 cu : $2\sqrt{2}/3-3\sqrt{5}/4$
3958 4118 rt : (2,2,9,-4)
3958 5628 ℓ_{10} : $3-3\sqrt{3}/2$
3958 6582 rt : (2,9,-8,9)
3958 7762 as : $3/2-4\sqrt{2}/3$
3958 9532 rt : (9,8,3,-3)
3959 0148 rt : (1,9,4,-9)
3959 0251 as : $(\pi/3)^{-1/3}$
3959 0355 rt : (8,6,9,-5)
3959 0728 sr : $9\sqrt{3}/8$
3959 1814 e^x : $2+4\sqrt{3}/3$
3959 2567 ln : $2+3e/4$
3959 2800 tπ : $3/25$
3959 2867 ℓ_2 : $19/25$
3959 3186 rt : (1,2,-6,2)
3959 4896 ℓ_2 : $\sqrt[3]{5}/9$
3959 6283 J : $\sqrt[3]{2}/10$
3959 8714 rt : (5,2,6,-3)
3960 0717 cu : $2+2\sqrt{3}/3$
3960 1371 J_2 : $7\sqrt[3]{2}/4$
3960 2356 rt : (6,4,0,-1)
3960 2666 2^x : $3/2+\sqrt{3}$
3960 2853 at : $4\sqrt{3}/3+3\sqrt{5}/2$

3960 3138 $2^x : 5\sqrt{3}/3$
3960 3640 $\zeta : 24/19$
3960 4431 $\ln : 9\pi/7$
3960 5832 rt : (6,1,-6,6)
3960 8410 id : $4\ln 2/7$
3960 8423 $10^x : \sqrt{3}/2 + \sqrt{5}/2$
3960 8964 $\mathrm{t}\pi : 9e\pi/2$
3960 9143 $2^x : 3\sqrt{2} + 3\sqrt{5}$
3960 9200 cu : 19/17
3960 9213 sq : $4 - 3\sqrt{5}/4$
3960 9612 rt : (1,6,4,4)
3961 0623 rt : (7,0,9,4)
3961 1982 $10^x : 3\sqrt{2} + 2\sqrt{5}$
3961 2412 at : 17/3
3961 2473 rt : (2,7,1,1)
3961 2478 $2^x : 7\sqrt[3]{2}/5$
3961 5101 at : $\sqrt{2}/3 + 3\sqrt{3}$
3961 6539 rt : (3,5,-9,-9)
3961 8484 rt : (1,4,9,3)
3961 8861 rt : (8,2,3,-2)
3961 9142 sq : $e + \pi/2$
3961 9643 cu : $e/4 + \pi/2$
3962 0021 rt : (7,6,-6,1)
3962 1402 at : $9\sqrt[3]{2}/2$
3962 2495 $2^x : 4 - \sqrt{5}$
3962 3222 rt : (3,9,-7,-4)
3962 4062 $\ell_2 : (\sqrt{3})^{1/2}$
3962 6340 id : $4\pi/9$
3962 8013 rt : (6,5,-4,1)
3963 2351 $\ell_{10} : 1 + 2\sqrt{5}/3$
3963 3963 $\mathrm{t}\pi : (\ln 2/2)^2$
3963 5561 rt : (8,4,-5,7)
3963 6307 $\ell_2 : 4/3 + 3\sqrt{3}/4$
3963 8074 sr : $3/2 + 3\sqrt{2}$
3964 0237 sr : $\sqrt{2}/9$
3964 1826 rt : (9,5,-7,-3)
3964 2107 $\Psi : 5e/3$
3964 2325 $J_1 : 1/\ln(\pi)$
3964 2760 $J_1 : \sqrt[3]{2/3}$
3964 3719 at : $4 + 3\sqrt{5}/4$
3964 4574 $\mathrm{s}\pi : \sqrt{2}/2 - \sqrt{3}/3$
3964 4660 id : $3/4 - \sqrt{2}/4$
3964 5145 rt : (2,7,7,-4)
3964 6163 $J : (e/2)^{-3}$
3964 8002 $\mathrm{t}\pi : 4e\pi/7$
3964 8028 $2^x : \sqrt{2}/4 + 4\sqrt{5}$
3964 8437 cu : 19/8
3965 2666 $\mathrm{t}\pi : 2(e+\pi)/9$
3965 3113 rt : (2,9,0,-4)
3965 4038 rt : (5,1,-9,-3)
3965 7993 sq : $4/3 + \sqrt{3}$
3965 8613 id : $\sqrt{21}\pi$
3966 0277 $Li_2 : 9(1/\pi)/8$
3966 0400 $\ln : \sqrt{3}/7$
3966 0825 rt : (6,6,9,3)
3966 2175 $Ei : 3\sqrt[3]{5}/2$
3966 4690 rt : (7,5,7,-4)
3966 5455 rt : (5,8,5,5)
3966 5964 rt : (8,8,-3,2)
3966 6046 $\mathrm{s}\pi : (\sqrt[3]{3})^{1/3}$
3966 6078 $\theta_3 : 17/20$
3966 6598 rt : $(3,9,1,-2)$
3966 7569 $\mathrm{t}\pi : \zeta(3)/10$
3966 7831 $\mathrm{t}\pi : P_{74}$
3966 8805 $J_0 : 10\zeta(3)/3$
3966 9421 sq : 13/11
3966 9429 cu : $8e\pi/9$
3967 0835 rt : (1,7,-6,-8)
3967 2300 at : $1/4 + 2e$
3967 3314 $\ln(\pi) \times \ln(\sqrt{2})$
3967 3716 $J_0 : 3e\pi/7$
3967 3932 $\mathrm{t}\pi : 2\sqrt{2} - 3\sqrt{5}$
3967 5318 sinh : $\exp(2\pi/3)$
3967 6043 $\ln : 3\sqrt{2}/4 + 4\sqrt{5}/3$
3967 6265 sr : $8\sqrt[3]{3}$
3967 8976 at : $3/2 + 4\pi/3$
3967 9389 rt : (4,3,-7,-3)
3967 9677 rt : (7,2,-8,4)
3968 0010 $\Gamma : 5/24$
3968 5026 id : $1/\sqrt[3]{16}$
3968 6689 rt : (4,2,0,7)
3968 7914 rt : (6,5,-4,-2)
3968 9039 $E : 2(1/\pi)$
3969 0826 tanh : $\sqrt[3]{2}/3$
3969 0975 $\lambda : \exp(-11\pi/10)$
3969 1452 $J_0 : 5(e+\pi)/8$
3969 1980 rt : (8,7,1,-2)
3969 2053 at : $2e\pi/3$
3969 3982 $\mathrm{s}\pi : 3\sqrt[3]{5}$
3969 4448 $J_1 : 7/8$
3969 5413 $E : \ln(\sqrt[3]{4}/3)$
3969 5656 $\zeta : 4e/9$
3969 7349 rt : (3,5,-4,3)
3969 7808 $2^x : 3\sqrt{2} - 4\sqrt{5}/3$
3969 7937 $10^x : 4 - \sqrt{5}/3$
3969 8313 rt : (9,5,-7,5)
3969 9587 $\mathrm{s}\pi : (\ln 2)^{-1/3}$
3969 9951 rt : (9,8,8,-5)
3970 1016 at : $1/2 + 3\sqrt{3}$
3970 1182 rt : (6,9,-2,-1)
3970 2309 $10^x : \exp(\sqrt[3]{4})$
3970 2722 sr : $2/3 + 4e$
3970 4323 $\mathrm{s}\pi(1/5) \times \mathrm{s}\pi(\sqrt{5})$
3970 4403 $e^x : P_7$
3970 4496 $E : 7/11$
3970 5871 $\zeta : (4/3)^3$
3970 9414 rt : (9,6,6,2)
3970 9921 rt : (1,1,7,-3)
3971 0173 rt : (7,8,-8,-4)
3971 0522 $J_0 : 7\pi/6$
3971 0925 rt : (8,7,-1,-1)
3971 1769 $erf : \exp(-1)$
3971 1995 rt : (6,5,1,8)
3971 2273 at : $10\sqrt[3]{5}/3$
3971 2277 $10^x : 3\sqrt{3} + 4\sqrt{5}$
3971 3019 at : $2\sqrt{3} + \sqrt{5}$
3971 3229 rt : (5,3,-7,-2)
3971 4980 $J_0 : 4$
3971 6757 rt : (1,8,5,4)
3971 6924 id : $\exp(\sqrt{2}\pi/3)$
3971 8749 $Ei : \sqrt{5}/8$
3971 9622 $\ell_{10} : \zeta(3)/3$
3972 0549 $J_0 : 11/3$
3972 1741 sinh : 25/22
3972 7579 $\exp(1/4) + \exp(\sqrt{2})$
3972 7862 sq : $2\sqrt{2} + 4\sqrt{3}/3$
3972 9024 rt : (8,8,2,9)
3972 9506 rt : (2,6,-8,-4)
3972 9877 cu : $4/3 - 4\pi$
3973 0810 cu : $10\pi/7$
3973 2432 $J_2 : \exp(\sqrt[3]{4}/2)$
3973 3471 $2^x : 2\sqrt{3} - 4\sqrt{5}/3$
3973 4122 id : $3\sqrt{2} + 2\sqrt{3}/3$
3973 4222 id : $10e\pi$
3973 5205 rt : (6,1,7,3)
3973 5970 sr : 3/19
3973 6975 at : $1 - 3\sqrt{5}$
3973 7685 rt : (7,6,-6,-1)
3974 0314 sr : $3\sqrt{3}/2 + 4\sqrt{5}$
3974 2708 sr : $3 - \pi/3$
3974 3687 $2^x : 4e/3 - \pi$
3974 5264 tanh : $\exp(-\sqrt{3}/2)$
3974 5569 $\ln : 1/3 + 2\sqrt{3}/3$
3974 5940 rt : (7,4,3,1)
3974 6449 rt : (3,4,8,-4)
3974 6618 rt : (4,8,-5,-3)
3974 7747 cu : $e/2 - 2\pi/3$
3974 9285 $\ln : 7\pi/2$
3974 9428 at : $1 + 3\pi/2$
3975 2144 $\theta_3 : 11/19$
3975 2746 rt : (9,9,-5,0)
3975 4248 id : $5\sqrt{5}/8$
3975 9976 $e^x : 3\sqrt{3}/2 + 2\sqrt{5}$
3976 0205 id : $((e+\pi)/3)^{1/2}$
3976 0562 at : $\sqrt[3]{2}/3$
3976 1414 rt : (8,2,8,-4)
3976 1777 $2^x : (3\pi)^{1/2}$
3976 3323 $J : 7(1/\pi)/8$
3976 3373 $erf : 1/\sqrt[3]{20}$
3976 4281 rt : (5,7,9,-5)
3976 4294 $2^x : \sqrt{5} - \sqrt{6} + \sqrt{7}$
3976 5137 $erf : 7/19$
3976 6154 rt : (2,6,-8,-6)
3976 6690 sq : $3\sqrt{2}/2 - 2\sqrt{5}/3$
3976 6901 $10^x : (\sqrt{5}/3)^{-1/2}$
3976 6932 at : $3 + e$
3976 8383 at : $4e/3 + 2\pi/3$
3977 2329 rt : (5,9,-8,-2)
3977 3744 id : $(\ln 3/3)^{-1/3}$
3977 4556 $e^x : 8\pi^2/9$
3977 4878 rt : (4,9,-1,1)
3977 5080 rt : (3,5,-1,-1)
3977 8344 $\exp(\pi) \div \exp(4/5)$
3977 8362 $J_1 : (\sqrt{2}\pi/3)^{-1/3}$
3977 9404 $\mathrm{t}\pi : 4e - \pi/2$
3977 9467 $J_0 : 7\sqrt[3]{5}/3$
3978 1008 rt : (6,0,-6,8)
3978 1469 as : P_{27}
3978 1684 tanh : 8/19
3978 2724 rt : (1,8,9,-8)
3978 2917 rt : (6,4,5,-3)
3978 3139 $\ln(3/5) \div \exp(1/4)$
3978 4577 $2^x : 1/\ln(\sqrt{2}/3)$
3978 4586 $\mathrm{t}\pi : \exp(2\pi/3)$
3978 5228 id : $5e/4$
3978 5550 rt : (8,3,-5,9)
3978 6641 rt : (9,1,4,2)
3978 6653 $Ei : 9/11$
3978 8295 $J_2 : 1/\ln(\pi/2)$
3978 8735 id : $5(1/\pi)/4$
3978 9052 rt : (5,3,-2,5)
3978 9527 $\ln : 11$
3979 1576 sr : 23/4
3979 2282 rt : (8,7,6,-4)
3979 3238 $2^x : 2e/3 - \pi$
3979 4000 $\ell_{10} : 25$
3979 4096 rt : (7,6,-1,6)
3979 4722 $2^x : 3\sqrt{2}/4 - \sqrt{3}/3$
3979 4786 sr : $8\sqrt[3]{5}/7$
3979 6486 cr : $1 + \sqrt{3}$
3979 7554 rt : (9,9,0,7)
3979 7628 rt : (1,8,-2,-2)
3979 8485 $J_0 : 17/10$
3980 2467 $\ln : 3 + \pi/3$
3980 2759 $J_1 : 9\sqrt[3]{2}/4$
3980 3234 cu : $7(e+\pi)/4$
3980 6850 $\exp(\sqrt{3}) - \mathrm{s}\pi(\sqrt{3})$
3980 7696 at : $4 + \sqrt{3}$
3981 1540 at : $3\sqrt{2} + 2\sqrt{5}/3$
3981 2007 rt : (2,6,-3,-1)
3981 3429 $J_0 : \sqrt{2} - \sqrt{6} - \sqrt{7}$
3981 4133 $\ell_2 : 3/4 + 4\sqrt{2}/3$
3981 4550 $\ln : e/3 + \pi$
3981 5003 $1/5 - \exp(4)$
3981 5492 at : $\exp(-\sqrt{3}/2)$
3981 5956 sinh : $e/7$
3981 6095 rt : (6,9,3,3)
3981 7623 $\ln(\pi) \times \exp(1/5)$
3982 0213 $2^x : 3/2 - 2\sqrt{2}$
3982 0519 $\Psi : \sqrt{5}/2$
3982 0576 cu : $2\sqrt{2}/3 + 4\sqrt{3}/3$
3982 2609 $2^x : P_{32}$
3982 2686 rt : (1,7,-2,7)
3982 2725 $2^x : 7\sqrt[3]{3}/8$
3982 2971 id : $7\pi/5$
3982 3537 rt : (3,9,-2,7)
3982 3912 $\Psi : 23/2$
3982 4006 $10^x : 2\sqrt{3}$
3982 8817 $\mathrm{s}\pi : \pi^2$
3982 9654 id : $(e/2)^{-3}$
3983 1491 rt : (9,4,2,-2)
3983 5018 rt : (8,7,-8,-3)
3983 6272 $\mathrm{t}\pi : \sqrt{3}/2 - \sqrt{5}/3$
3983 8455 rt : (9,4,-7,7)
3983 8919 at : $3/2 + 3\sqrt{2}$
3983 9391 $10^x : 4\ln 2/3$
3984 0108 $\mathrm{s}\pi : 3/23$
3984 0491 rt : (6,4,-9,-4)
3984 1662 rt : (7,1,-8,6)
3984 1683 $2^x : 14/3$

3984 1951 rt : (6,5,-9,-4)
3984 2789 $\ln(\sqrt{3}) - \exp(2/3)$
3984 2796 cu : $9\sqrt{3}/8$
3984 2815 rt : (1,3,1,8)
3984 4521 rt : (7,1,1,-1)
3984 4911 rt : (1,9,-3,7)
3984 5871 $t\pi : e/3 - \pi/4$
3984 5943 rt : (5,2,-9,-5)
3984 7978 $s\pi : \exp(1/(3e))$
3985 0698 $e^x : 7e\pi/2$
3985 0884 tanh : $(4/3)^{-3}$
3985 2244 at : 8/19
3985 3085 $\Gamma : 9/14$
3985 3257 $\Psi : \zeta(3)/5$
3985 3287 $J_1 : 17/6$
3985 3444 rt : (1,6,0,-9)
3985 7508 $s\pi : 7\sqrt{5}/5$
3986 0526 $J_0 : 5e/8$
3986 0551 at : 23/4
3986 0778 rt : (9,9,0,-2)
3986 1174 $\Psi : 3\sqrt[3]{5}/8$
3986 1364 rt : (8,2,-8,-3)
3986 3157 cr : $8\sqrt[3]{5}/5$
3986 3388 rt : (4,1,9,-4)
3986 4385 $10^x : 6(e+\pi)/5$
3986 4404 $J : 1/\sqrt{7}$
3986 4971 cr : $1/2+\sqrt{5}$
3986 7098 rt : (6,9,8,-5)
3986 8880 rt : (8,7,-3,4)
3987 1747 id : $1/3 - \sqrt{3}$
3987 2076 rt : (9,1,9,4)
3987 2127 $1/4 - \exp(1/2)$
3987 3954 rt : (2,8,-1,-1)
3987 4720 $\Psi : 19/17$
3987 5108 $10^x : 7e\pi/5$
3987 5714 at : $3\sqrt{3}+\sqrt{5}/4$
3987 6034 $J_1 : 22/25$
3987 6040 $2^x : 2+4\sqrt{5}$
3987 6080 rt : (1,8,-7,-4)
3987 6551 sinh : 7/18
3987 6668 $2/3 \times \exp(4)$
3987 6848 $J_0 : (\ln 2/2)^{-1/2}$
3987 7070 sr : $7\sqrt{5}/8$
3987 7834 $t\pi : 5(1/\pi)/3$
3987 8130 rt : (3,9,1,-8)
3987 8420 $\Psi : (\sqrt{2}\pi/2)^{-3}$
3987 8945 sinh : $1/\sqrt[3]{17}$
3987 9784 rt : (3,4,-6,4)
3987 9873 cu : $3/2 - \sqrt{5}$
3988 0285 id : $4\sqrt{3}/3 - 3\sqrt{5}$
3988 0326 rt : (6,4,-4,3)
3988 0339 $10^x : 2\sqrt[3]{5}/9$
3988 0705 sinh : $3\zeta(3)$
3988 0717 $2^x : P_{42}$
3988 1203 $\ln(3/4)\times\ln(4)$
3988 1430 as : $e/7$
3988 3274 rt : (4,6,7,9)
3988 4022 cu : $4+\sqrt{2}/3$
3988 6666 $\ell_2 : 9(e+\pi)/5$
3988 7529 $e^x : 7/8$
3988 8016 rt : (7,6,-1,-1)
3988 9196 sq : 12/19
3988 9505 $\ln(\pi) + s\pi(\sqrt{3})$
3988 9683 rt : (9,4,7,-4)
3989 0063 $2^x : 2\sqrt{2}+3\sqrt{3}$
3989 1038 $3 \div \ln(2/3)$
3989 1297 rt : (2,2,8,3)
3989 2965 cr : $3\sqrt{2}/4+3\sqrt{5}/4$
3989 3030 $Li_2 : 2\sqrt[3]{2}/7$
3989 3929 $\theta_3 : 16/21$
3989 4228 id : $(2\pi)^{-1/2}$
3989 4750 $e^x : 2\sqrt{2}/3+2\sqrt{5}/3$
3989 5826 $Li_2 : 9/25$
3989 6795 $\Psi : 5\pi/8$
3989 6942 rt : (8,7,-6,-3)
3989 7948 sq : $\sqrt{2}/2+2\sqrt{3}$
3989 8114 rt : (4,1,-5,2)
3989 9033 rt : (2,7,-7,4)
3989 9301 $\ell_{10} : P_{59}$
3989 9341 sr : $3\sqrt{3}+\sqrt{5}/4$
3990 1908 rt : (7,4,8,3)
3990 3035 rt : (9,9,5,-4)
3990 3132 id : P_{59}
3990 3707 rt : (1,9,6,8)
3990 5630 at : $2\sqrt{2}/3 - 3\sqrt{5}$
3990 6527 rt : (3,5,-6,-3)
3990 7814 ln : $(\sqrt{2}\pi/2)^{1/2}$
3990 8320 sr : $\sqrt[3]{15/2}$
3990 8961 $\ell_{10} : 3e/2 - \pi/2$
3990 8993 $\ell_{10} : (2\pi)^{1/2}$
3990 9918 $e^x : 4\pi^2/7$
3991 1029 $\exp(2)\times s\pi(1/3)$
3991 1560 ln : $4/3+e$
3991 4487 rt : (2,3,6,8)
3991 4560 $J : \sqrt{5}/7$
3991 4802 sr : $1/\ln(5/3)$
3991 5033 $2^x : \sqrt[3]{11/2}$
3991 6147 at : $4\sqrt[3]{3}$
3991 6155 $\theta_3 : 2\ln 2/7$
3991 8647 $e^x : 4e/3+\pi/2$
3991 8768 at : $1/3+2e$
3992 0747 $\Psi : \sqrt[3]{3}/6$
3992 1202 rt : (7,1,6,-3)
3992 2076 at : $(4/3)^{-3}$
3992 2492 at : $3\sqrt{3}/4+2\sqrt{5}$
3992 2734 rt : (4,1,0,9)
3992 2877 $e^x : 4\sqrt[3]{5}$
3992 5384 ln : $2\sqrt{5}/3$
3992 6123 $\ln(\sqrt{2})+s\pi(\sqrt{3})$
3992 6526 $10^x : 19/16$
3992 6644 rt : (3,8,-6,5)
3992 8456 $10^x : 4/3-\sqrt{3}$
3992 8925 $J_0 : 3\pi^2/8$
3992 9277 at : $10\sqrt{3}/3$
3992 9454 cr : $(1/\pi)/5$
3992 9532 rt : (2,8,4,-3)
3993 0746 $Ei : 2(e+\pi)/3$
3993 2495 sq : $1/4 - 2\pi$
3993 4159 rt : (7,6,4,-3)
3993 5608 rt : (1,7,4,1)
3993 6185 $s\pi : 2/3+2\sqrt{3}$
3993 7258 rt : (9,8,-5,2)
3993 9044 $\Gamma : \sqrt[3]{17}$
3993 9213 $E : (5/2)^{-1/2}$
3993 9302 $J_2 : 20/9$
3993 9875 $\Psi : 4\sqrt[3]{3}/9$
3994 0582 $t\pi : 4\sqrt{2}+2\sqrt{3}$
3994 2107 $J_2 : 9\sqrt[3]{5}/4$
3994 2254 rt : (4,8,-3,2)
3994 2523 as : 7/18
3994 2764 $\zeta : 2\sqrt[3]{3}/3$
3994 4936 as : $1/\sqrt[3]{17}$
3994 5153 $J_0 : \exp(\sqrt[3]{4}/3)$
3994 6462 $J_1 : 7\sqrt[3]{2}/10$
3994 7548 rt : (9,8,0,9)
3994 8965 $e^x : 8e$
3994 9510 $\lambda : (1/\pi)/10$
3995 1085 rt : (2,8,9,-5)
3995 1905 sq : $3/4+\sqrt{3}/4$
3995 3361 rt : (7,6,9,-5)
3995 4215 $\Gamma : 18/7$
3995 4643 sq : $2\sqrt{2}/3+2\sqrt{3}/3$
3995 5051 $\ell_{10} : 4-2\sqrt{5}/3$
3995 5902 $\Psi : 7\pi^2/6$
3995 6507 $2^x : \sqrt{2}/4 - 3\sqrt{5}/4$
3995 7339 rt : (9,3,-7,9)
3995 7433 at : $1/2 - 2\pi$
3995 8170 $t\pi : 8(e+\pi)$
3995 9542 $J : \exp(-13\pi/4)$
3996 0472 sq : $2\sqrt{2}/3+4\sqrt{5}/3$
3996 0775 id : $\sqrt{2}/3+4\sqrt{3}$
3996 1231 $J_1 : 15/17$
3996 2066 cr : $7\pi^2/5$
3996 2176 rt : (7,5,-6,1)
3996 5055 ln : $\ln(\sqrt{2}\pi)$
3996 5559 $Li_2 : \sqrt[3]{3}/4$
3996 5745 rt : (2,7,-8,-8)
3996 7687 rt : (1,8,-3,3)
3996 8538 $J_0 : 6\sqrt{2}/5$
3996 8724 cu : $\sqrt{3}/2 - 3\sqrt{5}$
3996 9410 rt : (7,5,-1,8)
3997 1786 rt : (2,5,-7,-3)
3997 2341 $Li_2 : 3\zeta(3)/10$
3997 2487 $t\pi : \exp(\sqrt[3]{4}/3)$
3997 2529 $10^x : 10(1/\pi)$
3997 3042 $J_2 : 8\sqrt[3]{3}/3$
3997 3248 $J_2 : \sqrt[3]{11}$
3997 5869 rt : (7,0,-8,8)
3997 7577 $\ell_{10} : 3\sqrt{2}-\sqrt{3}$
3997 8283 $2^x : 7\ln 2/10$
3997 9345 $\ell_{10} : (e/2)^3$
3998 1953 $e^x : \sqrt{3}/4+4\sqrt{5}/3$
3998 2735 $t\pi : 3\sqrt{3}-3\sqrt{5}$
3998 3031 at : $\sqrt{2}+\sqrt{3}+\sqrt{7}$
3998 3844 as : $3\sqrt{2}/2-\sqrt{3}$
3998 6664 sq : $4\sqrt{3}/3-3\sqrt{5}/4$
3998 7819 $2^x : -\sqrt{3}-\sqrt{5}+\sqrt{7}$
3998 7826 at : $1-\sqrt{3}/3$
3998 9363 sq : $2/3-3\sqrt{3}/4$
3998 9755 $2^x : 5\sqrt[3]{5}/4$
3999 1152 $\ell_{10} : 3(e+\pi)/7$
3999 1227 id : $10\sqrt[3]{2}/9$
3999 1569 $E : 12/19$
4000 0000 id : 2/5
4000 2157 $e^x : 3/4+2\sqrt{5}/3$
4000 2216 at : $3e - 3\pi/4$
4000 5831 cu : 14/19
4000 6575 sinh : 24/7
4000 6785 $Ei : (\ln 3/2)^{1/3}$
4000 7272 at : $4e/3 - 3\pi$
4000 7312 $J_0 : 2\pi^2/5$
4000 8231 sinh : $8\sqrt{5}/7$
4001 5977 cu : $1/3+\pi/4$
4001 6561 $J_1 : 5\sqrt{2}/8$
4001 7260 $Ei : 4\pi/3$
4001 8871 $\ln(\sqrt{2}) \div s\pi(1/3)$
4001 9413 $J_1 : 2\sqrt{2}$
4001 9948 $2^x : 2+\sqrt{3}/4$
4002 0543 $2^x : 24/19$
4002 1886 at : $4/3+2\sqrt{5}$
4002 3057 cr : $2+\sqrt{5}/3$
4002 3985 $\ell_{10} : 8\pi$
4002 5396 $\ln(2)+s\pi(1/4)$
4002 7169 rt : (3,3,-8,5)
4002 7207 sr : P_{20}
4002 9359 rt : (8,6,-3,6)
4002 9955 $J_2 : 9\sqrt{3}/7$
4003 1187 $10^x : \sqrt{17}\pi$
4003 1231 sr : $\sqrt[3]{3}/9$
4003 1391 $2^x : 1/4 - \pi/2$
4003 3918 id : $4\sqrt{3}+2\sqrt{5}$
4003 4503 at : $2\sqrt{2}+4\sqrt{5}/3$
4003 5418 at : $4\sqrt{3}-\sqrt{5}/2$
4003 5955 rt : (8,6,-8,-1)
4004 0164 sinh : 23/9
4004 1224 at : $4+2e/3$
4004 2224 $J_1 : \exp(-1/(3e))$
4004 3841 rt : (8,8,8,-5)
4004 3903 at : $e - \pi$
4004 5855 rt : (3,8,4,5)
4004 6014 sr : $4-3e/4$
4004 7437 cu : $3\sqrt{3}+4\sqrt{5}$
4004 9938 sq : $e/4 - 2\pi$
4005 1658 tanh : $3\sqrt{2}/10$
4005 2370 $J_0 : 5\sqrt{5}/3$
4005 2994 $J_1 : 9\pi/10$
4005 3940 $J_2 : 7(1/\pi)$
4005 4507 $erf : 7(1/\pi)/6$
4005 6774 $J_1 : (\ln 2)^{1/3}$
4005 7495 $\ell_{10} : ((e+\pi)/2)^3$
4006 0063 $J_1 : (\sqrt[3]{3})^{-1/3}$
4006 0295 cr : $4/3+\sqrt{2}$
4006 0374 rt : (8,8,3,-3)
4006 3752 $\ell_{10} : 9\sqrt{5}/8$
4006 4177 tanh : $4(1/\pi)/3$
4006 4263 rt : (3,8,-1,-7)
4006 5724 $10^x : 4+2\sqrt{3}$
4006 8563 id : $\zeta(3)/3$
4006 9882 rt : (6,3,-4,5)
4007 1117 rt : (8,3,5,-3)

4007 3414 tπ : $3\sqrt{2}/2$
4007 3851 sq : $1/3 - 2\pi$
4007 5541 2^x : $\exp(-\sqrt[3]{3}/2)$
4007 5569 sr : $\sqrt{2}/3 + 2\sqrt{5}/3$
4007 6580 rt : (3,5,5,-3)
4007 8328 10^x : $\sqrt{18}\pi$
4007 8523 10^x : $\sqrt{2}+\sqrt{5}/3$
4007 8592 rt : (4,7,-5,3)
4008 1617 ℓ_{10} : $\sqrt{19/3}$
4008 2954 rt : (2,7,-5,-3)
4008 3173 e^x : $1/2 - \sqrt{2}$
4008 3174 cr : $7\pi/8$
4008 3938 at : $4\sqrt{2}/3 - 4\sqrt{3}/3$
4008 4667 rt : (9,7,-5,4)
4008 5749 sr : $2\sqrt{2} - \sqrt{3}/2$
4008 6518 rt : (8,2,7,3)
4008 7787 at : $3+2\sqrt{2}$
4008 7952 E : $\sqrt[3]{2}/2$
4008 8460 rt : (1,5,4,6)
4008 8560 tπ : $1/4 - 4e/3$
4008 9386 rt : (6,3,-9,-2)
4009 0492 rt : (7,9,-5,-3)
4009 3245 $\ln(3/4) - \exp(\sqrt{2})$
4009 4727 J_0 : $5\pi/4$
4009 4748 rt : (3,0,7,3)
4009 6886 rt : (8,8,-2,-1)
4009 9250 cu : $2\sqrt{2} - 3\sqrt{5}$
4009 9529 sq : $(\ln 3/2)^{-3}$
4009 9922 $\ln(5)\times\exp(2/5)$
4010 0645 rt : (6,8,3,5)
4010 0732 ℓ_2 : $2 - e/4$
4010 1744 sinh : $8\sqrt[3]{3}$
4010 1966 cr : $11/4$
4010 2126 θ_3 : $5(1/\pi)/8$
4010 2543 $\exp(1/4)+\exp(3/4)$
4010 2733 10^x : $1/2 - \sqrt{2}/4$
4010 3272 sinh : $\pi^2/3$
4010 4717 rt : (2,8,3,-6)
4010 7187 tπ : $(2e)^{-1/2}$
4010 9664 sπ : $4\sqrt{2}/5$
4011 0453 rt : (8,7,9,3)
4011 1455 J_0 : $\sqrt{14}$
4011 2127 e^x : $4\sqrt{2} - \sqrt{3}/2$
4011 2601 10^x : $(e+\pi)/8$
4011 3226 $\sqrt{3}\times\ln(4)$
4011 5157 rt : (6,5,7,-4)
4011 7047 rt : (7,4,-7,-3)
4011 7332 e^x : $\sqrt{3}/3 - 2\sqrt{5}/3$
4012 0037 ln : $\ln((e+\pi)/3)$
4012 0103 as : $4/3 - 2\sqrt{2}/3$
4012 0839 sq : $4+3\pi/4$
4012 2230 rt : (1,7,7,-4)
4012 3810 ln : $3\sqrt{2}/4+\sqrt{3}/4$
4012 4257 rt : (3,5,9,3)
4012 4714 at : $3\sqrt{2}/10$
4012 4780 sr : $5\pi/8$
4012 6455 rt : (5,4,-9,-9)
4012 7020 at : $\sqrt{3}/2 - 3\sqrt{5}$
4012 7402 e^x : $2/3+4\sqrt{5}/3$
4012 7874 rt : (8,3,0,-1)
4012 9437 rt : (4,0,-5,4)
4012 9870 cr : $4\sqrt{2}/3+\sqrt{3}/2$
4013 0235 rt : (4,9,-7,-4)
4013 1477 sq : $\sqrt{2}/4+2\sqrt{5}/3$
4013 2426 ln : $\sqrt[3]{10/3}$
4013 3367 rt : (7,4,-6,3)
4013 4556 ln : $3+3\sqrt{2}/4$
4013 4570 rt : (6,0,9,4)
4013 4712 cr : $3\sqrt{2} - 2\sqrt{5}/3$
4013 6710 sinh : $9/23$
4013 7332 ℓ_{10} : $2\sqrt[3]{2}$
4013 7350 at : $4(1/\pi)/3$
4013 7617 tanh : $\exp(-\sqrt[3]{5}/2)$
4013 7790 tanh : $1/\sqrt[3]{13}$
4013 8393 rt : (9,5,9,-5)
4014 0605 J_0 : $15/4$
4014 2037 rt : (2,3,-9,3)
4014 3467 J_0 : $7\sqrt{5}/4$
4014 4206 Ei : $23/15$
4014 4344 rt : (1,2,9,-4)
4014 5127 rt : (4,7,9,-5)
4014 6718 rt : (6,8,-2,-2)
4014 8065 ln : $\sqrt{2}/2+3\sqrt{5}/2$
4014 9267 $\ln(3/4)\times\exp(1/3)$
4015 2556 rt : (1,9,-9,-5)
4015 3077 sq : $3\sqrt{2} - \sqrt{3}/2$
4015 5661 rt : (4,4,-9,-4)
4015 9544 rt : (8,5,-3,8)
4016 1102 10^x : $4\pi/7$
4016 2387 rt : (5,1,-2,9)
4016 2795 rt : (2,9,-9,4)
4016 3811 J_0 : $2(e+\pi)/3$
4016 5823 rt : (7,1,-9,3)
4016 6848 rt : (2,3,-1,8)
4016 7442 cu : $2+2e/3$
4016 8505 Ψ : $8\sqrt[3]{3}$
4016 8921 rt : (9,6,-9,-4)
4016 8968 Γ : $3\sqrt[3]{5}/8$
4017 0400 sπ : $4\sqrt{3}/3+\sqrt{5}/4$
4017 0729 cu : $4/3+2\sqrt{3}/3$
4017 0753 cu : 4π
4017 1080 tπ : $4\sqrt{2} - 3\sqrt{5}/2$
4017 1522 10^x : $3\sqrt{2}/4+4\sqrt{3}$
4017 2120 rt : (2,7,0,-1)
4017 2290 E : $4\sqrt{2}/9$
4017 5037 $\ln(\sqrt{3}) - s\pi(2/5)$
4017 7243 at : $(e+\pi)$
4017 7382 sq : $10\sqrt[3]{5}$
4017 9329 sq : $1/4 - 2\pi/3$
4017 9477 sq : $(2\pi/3)^3$
4017 9735 rt : (6,5,2,-2)
4018 1529 ℓ_{10} : $3/4 - \sqrt{2}/4$
4018 3854 erf : $\sqrt{5}/6$
4018 4652 rt : (2,9,2,5)
4018 5370 E : $\pi/5$
4018 5572 at : $2/3+3\sqrt{3}$
4018 6503 Ei : $7(1/\pi)/8$
4018 6916 sπ : $e/4+4\pi/3$
4018 7391 sr : $4\sqrt[3]{3}$
4018 8044 rt : (8,2,2,1)
4018 8252 rt : (7,9,1,3)
4019 0298 J_0 : $9\sqrt{3}/4$
4019 0701 sπ : $\sqrt{2}-\sqrt{3}+\sqrt{6}$
4019 1250 2^x : $4/3+\sqrt{3}/4$
4019 1275 rt : (8,5,-8,1)
4019 1824 rt : (4,6,-7,4)
4019 2378 id : $3 - 3\sqrt{3}/2$
4019 4279 J_0 : $\ln(2e)$
4019 4320 rt : (9,5,4,-3)
4019 4625 K : $4\ln 2/3$
4019 6401 $\exp(\sqrt{3})\div\exp(1/4)$
4019 7026 J_1 : $8/9$
4019 7300 10^x : $4\sqrt{2}+\sqrt{5}$
4019 7623 rt : (1,7,2,-2)
4019 8646 J_0 : $6\pi/5$
4019 8916 at : $3\sqrt{2}/4 - 4\sqrt{3}$
4020 0981 Ei : $3\pi/4$
4020 1076 J_0 : $\exp(e/2)$
4020 1076 Li_2 : $1/\sqrt[3]{21}$
4020 1933 J_0 : $8\sqrt{2}/3$
4020 3085 $5\div s\pi(\sqrt{5})$
4020 3144 ℓ_2 : $1/3+4\sqrt{3}/3$
4020 3593 sr : $2/3+3\sqrt{3}/4$
4020 3658 rt : (5,1,-7,2)
4020 4853 as : $9/23$
4020 5224 rt : (9,6,-5,6)
4020 5305 J_2 : $\sqrt{5}$
4020 6098 sr : $1/3+2e$
4020 6367 tπ : $\sqrt{2}+2\sqrt{3}$
4020 6819 tπ : $4\sqrt{2}/3+\sqrt{5}$
4020 7686 ζ : $\sqrt[3]{5/2}$
4020 7846 ℓ_{10} : $7\sqrt[3]{3}/4$
4020 7890 rt : (4,7,4,-3)
4020 8267 rt : (7,9,0,-1)
4020 8676 e^x : $\sqrt{2}+4\sqrt{5}$
4021 0254 2^x : $2\sqrt{2}/3+2\sqrt{5}/3$
4021 1481 at : $\exp(-\sqrt[3]{5}/2)$
4021 1656 at : $1/\sqrt[3]{13}$
4021 3434 rt : (7,7,6,-4)
4021 6269 Γ : $4\sqrt[3]{3}/9$
4021 7640 rt : (6,2,-4,7)
4021 8414 ln : $\sqrt{2} - \sqrt{5}/3$
4021 9202 $3/5\div\exp(2/5)$
4021 9502 rt : (4,9,-2,-2)
4021 9902 Ψ : $-\sqrt{2}-\sqrt{6}+\sqrt{7}$
4022 0330 id : $3\pi^2/4$
4022 1024 id : $4\sqrt{2}+\sqrt{5}/3$
4022 1258 ℓ_{10} : $4\ln 2/7$
4022 1370 J_0 : $3\sqrt[3]{2}$
4022 2666 J_0 : $10e/7$
4022 3367 $1/4 - \exp(\sqrt{3})$
4022 3517 rt : (1,9,-4,-3)
4022 5578 rt : (2,9,6,-4)
4022 6387 Ψ : $(\sqrt[3]{3}/2)^{-1/3}$
4022 6833 rt : (5,9,8,-5)
4023 0034 tπ : $7\sqrt[3]{3}/9$
4023 1513 J : $\ln 2/7$
4023 1993 rt : (9,0,6,3)
4023 2680 sr : $3\sqrt{3}/4+2\sqrt{5}$
4023 4230 cu : $2+\sqrt{3}/4$
4023 5947 ln : $(\sqrt{5})^{1/2}$
4023 6193 sq : $1/3+4\pi$
4023 6430 ln : $3/4+\sqrt{5}/3$
4023 6749 rt : (3,7,2,6)
4023 6968 J_0 : $(\pi/2)^3$
4023 7005 2^x : $3/2+\pi/2$
4023 8354 Ei : $\sqrt[3]{5/3}$
4023 9972 id : $7\zeta(3)/6$
4024 1696 J_2 : $23/6$
4024 1755 J_0 : $\sqrt{15}$
4024 2745 J_0 : $22/13$
4024 2794 e^x : $1/\ln(1/3)$
4024 3684 rt : (8,8,-7,1)
4024 4517 e^x : $-\sqrt{2}-\sqrt{3}+\sqrt{5}$
4024 5308 sinh : $9(e+\pi)/7$
4024 6425 rt : (7,2,8,-4)
4024 6708 J_0 : $\exp(4/3)$
4024 7744 tπ : $9\sqrt{2}/8$
4024 9060 Li_2 : $15/16$
4024 9368 J_0 : $4e\pi/9$
4024 9783 at : $4+4\sqrt{2}/3$
4025 0011 cu : $\sqrt{3}/2 - 3\sqrt{5}/2$
4025 1647 rt : (4,2,6,-3)
4025 1835 at : $\sqrt{2}+2\sqrt{5}$
4025 3353 cr : $\sqrt[3]{21}$
4025 5641 J_0 : $19/5$
4025 8926 ζ : $19/14$
4025 9002 as : $\sqrt{2}/4 - \sqrt{5}/3$
4025 9505 id : $1/\ln(\sqrt{5}/3)$
4026 0323 $2\div s\pi(1/5)$
4026 2178 J_0 : $7e/5$
4026 2794 rt : (2,4,8,-4)
4026 3460 e^x : $4/3+\sqrt{2}/4$
4026 6347 rt : (7,9,-4,-4)
4026 6654 Ψ : $16/25$
4026 7882 rt : (2,8,-4,-2)
4026 8347 rt : (7,3,-6,5)
4026 9279 J_1 : $(\sqrt[3]{2})^{-1/2}$
4026 9805 sinh : $\sqrt{2}-\sqrt{6}+\sqrt{7}$
4026 9974 rt : (6,8,-7,-9)
4027 0428 cu : $2/3 - 4\sqrt{5}/3$
4027 0543 J_0 : $((e+\pi)/3)^2$
4027 0957 J_0 : $9\sqrt[3]{5}/4$
4027 0969 2^x : $1/2 - 2e/3$
4027 1830 J_0 : $8\sqrt[3]{3}/3$
4027 4986 rt : (1,7,5,-4)
4027 5886 J_0 : $23/6$
4027 8462 rt : (1,9,-4,-2)
4027 8886 sπ : P_{49}
4027 9713 J_1 : $9\ln 2/7$
4028 0870 sπ : $3/4+\sqrt{5}/2$
4028 1141 sr : $10\sqrt{3}/3$
4028 1523 rt : (3,6,-8,-4)
4028 3220 10^x : $2\sqrt{3}/3+4\sqrt{5}$
4028 3965 cr : $3+4e$
4028 4027 10^x : $1/4+3\sqrt{3}/4$
4028 6187 sinh : $\ln(\sqrt{2}\pi/3)$
4028 6898 $\ln(\sqrt{5})+\exp(4)$
4028 7038 sinh : $\pi/8$
4028 7384 rt : (4,5,-9,5)

4028 7881 rt : (9,5,8,3)
4028 7935 ℓ_{10} : $3-\sqrt{2}/3$
4028 9566 cr : $2/3+2\pi/3$
4029 4791 ln : $7\sqrt[3]{5}/8$
4029 8219 rt : (8,9,5,-4)
4029 8914 rt : (6,7,3,7)
4029 9136 at : $\sqrt{2}/2+3\sqrt{3}$
4029 9438 rt : (4,1,-5,-6)
4030 3171 rt : (8,8,-8,-4)
4030 4818 2^x : $2\sqrt[3]{5}/7$
4030 5699 rt : (9,5,-5,8)
4030 6873 10^x : $\sqrt{2}/3-\sqrt{3}/2$
4030 7258 J_1 : $2\pi^2/7$
4030 8178 cu : $(2\pi/3)^3$
4030 9159 at : $1/4+4\sqrt{2}$
4031 0451 sq : $\ln(3\pi/2)$
4031 1598 Ψ : $7(e+\pi)/9$
4031 2314 ℓ_2 : $\sqrt{3}+\sqrt{5}-\sqrt{7}$
4031 4015 rt : (3,7,-3,-6)
4031 5035 rt : (8,4,-8,3)
4031 6268 rt : (5,9,3,-3)
4031 8044 cu : $1+3\pi/2$
4031 8212 sq : $3e/2-3\pi/2$
4031 8903 rt : (4,3,8,3)
4031 9702 rt : (7,7,1,-2)
4032 0102 θ_3 : $1/5$
4032 1671 Ei : $\sqrt[3]{19/2}$
4032 1695 sinh : $(\sqrt{2}\pi/3)^{1/3}$
4032 1828 rt : (5,9,-6,3)
4032 2487 tanh : $\sqrt[3]{5}/4$
4032 5576 rt : (9,5,-1,-1)
4032 7263 rt : (2,2,-3,9)
4032 7381 rt : (6,5,6,2)
4032 9423 10^x : $4\sqrt{3}-\sqrt{5}/4$
4032 9769 e^x : $\sqrt{3/2}$
4033 1182 $s\pi$: P_{64}
4033 2150 10^x : $3\sqrt{2}+4\sqrt{5}$
4033 4842 rt : (6,6,9,-5)
4033 6117 rt : (6,1,-4,9)
4033 7615 rt : (3,9,-5,-3)
4034 0753 rt : (8,4,7,-4)
4034 2212 $\exp(3/5)-\exp(4/5)$
4034 2640 $3/4-\ln(\sqrt{2})$
4034 3786 J_2 : $\ln(3\pi)$
4034 4144 id : $3\sqrt{2}/4-2\sqrt{3}$
4034 4489 ζ : $9\ln 2/4$
4034 4703 sinh : $2\sqrt[3]{5}/3$
4034 4851 at : $3\sqrt{2}+3\sqrt{5}/4$
4034 4937 e^x : $\sqrt{3}/2+4\sqrt{5}$
4034 5481 ζ : $5\sqrt{2}/3$
4034 7216 Li_2 : $4/11$
4034 7838 rt : (2,9,1,-2)
4034 8834 2^x : $1-4\sqrt{3}/3$
4034 9271 ℓ_2 : $\sqrt{3}/3+\sqrt{5}/3$
4034 9543 $t\pi$: $6\sqrt{2}/5$
4035 0595 at : $3\pi^2/5$
4035 0615 ℓ_{10} : $\sqrt{2}+\sqrt{5}/2$
4035 1927 rt : (1,8,9,-5)
4035 2101 Ei : $(2e/3)^{-1/3}$
4035 3817 e^x : $8\pi^2/3$
4035 3868 λ : $(\pi)^{-3}$
4035 5422 Ei : $21/4$
4035 5602 as : $\ln(\sqrt{2}\pi/3)$
4035 5658 rt : (8,7,4,1)
4035 5743 sq : $2\sqrt{2}-4\sqrt{5}$
4035 5954 rt : (5,0,-7,4)
4035 6166 J_2 : $5\pi/7$
4035 6460 as : $\pi/8$
4035 7371 Ei : $(2\pi)^{1/3}$
4035 7417 ℓ_2 : $4/3-\sqrt{3}/3$
4035 7562 rt : (1,1,-8,3)
4035 7613 10^x : $3-3\pi/4$
4035 8529 rt : (9,6,-4,-2)
4036 1588 rt : (3,6,7,-4)
4036 2944 rt : (7,8,1,5)
4036 3226 ℓ_2 : $3\sqrt[3]{2}/5$
4036 5142 10^x : $2/3-3\sqrt{2}/4$
4036 5409 Li_2 : $8(1/\pi)/7$
4036 5675 rt : (4,7,-1,-1)
4036 6898 Ψ : $8\sqrt[3]{5}/3$
4036 7018 cr : $10e\pi$
4036 7746 ℓ_2 : $\sqrt{7}$
4036 7822 2^x : $2+3\sqrt{2}/2$
4036 8434 at : $1-4\sqrt{3}$
4037 4928 id : $5\sqrt[3]{3}/3$
4037 5005 Ψ : $7(1/\pi)/2$
4037 5248 ln : $5\zeta(3)/9$
4037 5746 ℓ_2 : $3-3\sqrt{5}/4$
4037 6022 rt : (5,4,5,-3)
4037 6590 at : $4/3-e/3$
4037 7031 cu : $5e\pi/6$
4037 7567 rt : (7,2,-6,7)
4037 9715 cu : $17/23$
4038 0961 rt : (3,6,0,7)
4038 1126 E : $5/8$
4038 4049 ln : $3\sqrt{2}/2+4\sqrt{5}$
4038 5185 rt : (2,8,-7,-4)
4038 5622 ζ : $3\pi/4$
4038 7785 erf : $1/\sqrt[3]{19}$
4039 0867 θ_3 : $\zeta(3)/6$
4039 1533 at : $1/2+2e$
4039 2836 rt : (1,3,6,2)
4039 5598 cr : $-\sqrt{2}+\sqrt{3}+\sqrt{6}$
4039 5834 J_1 : $2\sqrt{5}/5$
4039 7538 rt : (8,1,9,4)
4039 8117 at : $\sqrt[3]{5}/4$
4039 8885 rt : (5,8,-8,4)
4039 9914 rt : (7,2,3,-2)
4040 1338 ζ : $7\ln 2/10$
4040 1737 cu : $\ln(2\pi/3)$
4040 2612 $t\pi$: $8(e+\pi)/7$
4040 3896 rt : (7,8,8,-5)
4040 4699 rt : (1,3,-8,-8)
4040 5165 at : $3\sqrt{3}+\sqrt{5}/3$
4040 5417 cu : $4\sqrt{2}+\sqrt{3}$
4040 5654 ℓ_2 : $3-\sqrt{2}/4$
4040 6101 id : $2\sqrt{2}/7$
4040 6713 sr : $3/2+\sqrt{2}/3$
4040 6923 J_1 : $17/19$
4040 6933 $t\pi$: $\ln(\sqrt{3})$
4040 7148 rt : (6,7,-2,0)
4040 7370 id : $\sqrt{3}/3-4\sqrt{5}/3$
4040 8205 sq : $4\sqrt{2}-\sqrt{3}$
4040 9918 10^x : $8/21$
4040 9958 id : $\exp(-e/3)$
4041 1380 id : $2\zeta(3)$
4041 1690 erf : $3/8$
4041 2439 rt : (3,8,-9,-5)
4041 2673 tanh : $3/7$
4041 2775 at : $3-4\sqrt{5}$
4041 2817 Γ : $16/25$
4041 5687 $\ln(3)\div\exp(1)$
4041 6024 rt : (8,3,-8,5)
4041 7208 rt : (1,9,1,-1)
4041 7353 sq : $\ln(1/(3e))$
4041 7949 at : $3/4+3\sqrt{3}$
4041 9132 $\ln(2/5)\times\exp(3)$
4042 0177 rt : (8,9,-1,3)
4042 1303 rt : (9,6,6,-4)
4042 1612 ζ : $e\pi/8$
4042 1765 sr : $\sqrt[3]{23/3}$
4042 2786 10^x : $3\sqrt{2}-\sqrt{5}/4$
4042 3217 rt : (2,8,2,-7)
4042 6224 $1/4+\exp(e)$
4042 6300 rt : (3,1,9,-4)
4042 8118 at : $1/3-2\pi$
4042 9736 rt : (4,2,-5,-8)
4042 9931 J : $\sqrt{2}/10$
4043 1358 rt : (1,5,-2,1)
4043 1637 rt : (8,9,0,-2)
4043 2013 at : $(2e/3)^3$
4043 2487 E : $\exp(-\sqrt{2}/3)$
4043 2968 $\exp(\sqrt{5})\div\exp(\pi)$
4043 4506 at : $3\sqrt{3}/2+3\sqrt{5}/2$
4043 5377 $2/3\div\exp(1/2)$
4043 7304 cr : $4/3+4\pi$
4043 8448 Ei : $\exp(-4)$
4043 8650 tanh : $(2e)^{-1/2}$
4043 9432 rt : (7,4,-2,-1)
4043 9537 ζ : $(e+\pi)/5$
4044 1847 e^x : $(\sqrt{2}\pi/3)^{-1/3}$
4044 2730 e^x : $2+3e$
4044 4011 rt : (5,8,-5,-3)
4044 5153 rt : (6,6,3,9)
4044 5252 Ψ : $\sqrt[3]{23/3}$
4044 7055 rt : (9,0,1,1)
4044 9173 ℓ_2 : $\exp(5/3)$
4044 9640 E : $9\ln 2/10$
4045 0022 2^x : $4\sqrt{2}+4\sqrt{3}$
4045 1031 at : $3/4-3\sqrt{5}$
4045 1239 rt : (2,7,3,8)
4045 4093 10^x : $3\sqrt{3}$
4045 7023 10^x : $10\ln 2/3$
4045 7700 ln : P_{26}
4045 9650 rt : (2,9,3,-8)
4046 0783 J : $\sqrt{5}/10$
4046 0984 e^x : $5\sqrt{2}$
4046 1332 ℓ_{10} : $1/2+3e/4$
4046 1719 sr : $7(e+\pi)$
4046 2438 cr : $8\sqrt{3}/5$
4046 3745 at : $8\sqrt{5}/3$
4046 3859 rt : (5,1,7,3)
4046 4588 e^x : $e/8$
4046 4718 K : $(\sqrt[3]{5}/2)^{1/2}$
4046 6304 e^x : $1/4-2\sqrt{3}/3$
4046 6392 cu : $8\sqrt[3]{2}/9$
4046 7792 rt : (7,1,-6,9)
4046 8208 $\ln(3/4)-\exp(3/4)$
4046 9757 J_2 : $9/4$
4047 0053 id : $1/4+2\sqrt{3}/3$
4047 1725 sr : $\sqrt{2}+\sqrt{5}/4$
4047 1895 $2/5-\ln(\sqrt{5})$
4047 4001 $t\pi$: $\sqrt{2}+3\sqrt{5}$
4047 4225 rt : (5,1,-7,-6)
4047 6876 rt : (6,6,4,-3)
4047 7387 sr : $2\sqrt{3}-2\sqrt{5}/3$
4047 7403 cu : $3/4-2\pi$
4047 8851 J_1 : $(\sqrt[3]{3}/2)^{1/3}$
4048 0066 id : $3e\pi/4$
4048 4397 rt : (7,8,-4,-2)
4048 4523 cr : $4\ln 2$
4048 4548 id : $1/4+3e$
4048 5675 rt : (9,1,8,-4)
4048 5762 rt : (3,6,-5,-5)
4048 9114 at : $3/2+2\sqrt{5}$
4048 9178 at : $3/7$
4048 9412 e^x : $2/3-\pi/2$
4049 0291 $t\pi$: $(\sqrt[3]{2})^{1/2}$
4049 0836 $t\pi$: $\exp(\sqrt[3]{2}/2)$
4049 2101 rt : (4,1,-6,2)
4049 3788 rt : (3,5,-2,8)
4049 4026 rt : (7,7,1,7)
4049 4720 $t\pi$: $7\sqrt[3]{5}/8$
4049 5867 sq : $7/11$
4049 6081 at : $3\sqrt{2}+\sqrt{3}$
4049 6294 $\pi\div\sqrt{5}$
4049 7252 as : $2/3-3\sqrt{2}/4$
4049 8639 $t\pi$: $1/\ln(\sqrt[3]{4}/3)$
4049 8811 rt : (7,5,-9,-4)
4050 0026 rt : (8,2,-8,7)
4050 2192 sq : $(2\pi)^{1/3}$
4050 2198 rt : (3,6,-3,-2)
4050 3660 $s\pi$: 8π
4050 4586 at : $7e\pi/10$
4050 5764 Ei : $7\ln 2/10$
4050 6827 sinh : $\sqrt{15}\pi$
4050 7774 cr : $e/4+2\pi/3$
4050 9240 J_1 : $2\pi/7$
4051 0642 $s\pi$: $3e+3\pi/2$
4051 2086 at : $4e/3+3\pi/4$
4051 2483 rt : (1,9,5,0)
4051 3010 sq : $3-2\sqrt{3}/3$
4051 4407 at : $3+4\sqrt{5}/3$
4051 4483 2^x : $\sqrt{2}/2+\sqrt{5}/4$
4051 4957 rt : (6,0,-9,4)
4051 5007 sq : $e/2+2\pi$
4051 5028 sq : $\ln(\sqrt[3]{4}/3)$
4051 5408 at : $(2e)^{-1/2}$
4051 6702 rt : (5,5,0,4)
4051 7778 rt : (4,3,8,-4)

4052 0212 rt : (8,5,9,-5)
4052 0950 rt : (5,9,-2,-1)
4052 3872 $\sqrt{2}\times\exp(3)$
4052 3896 at : $7\sqrt[3]{5}/2$
4052 4781 cu : 13/7
4052 5097 10^x : $3+e/2$
4052 5196 at : $2\sqrt{2}/3-4\sqrt{3}$
4052 5868 cu : $3\sqrt{3}/4-\sqrt{5}/4$
4052 6471 2^x : $5\sqrt{2}/4$
4052 6785 rt : (8,4,2,-2)
4052 8473 id : $(\pi/2)^{-2}$
4052 9417 $3/5\times s\pi(\sqrt{5})$
4053 2727 Γ : $5e\pi/7$
4053 2827 rt : (4,9,1,6)
4053 4308 rt : (7,3,5,2)
4053 4530 sq : $4e\pi/9$
4053 5421 sq : $2/3-4\pi/3$
4053 5940 ℓ_{10} : $\sqrt{3}/2+3\sqrt{5}/4$
4053 8202 at : $1/3+4\sqrt{2}$
4053 8311 rt : (8,8,-1,5)
4053 9464 Ei : 5/18
4054 0943 e^x : 13/7
4054 1250 $\exp(\sqrt{5})-s\pi(2/5)$
4054 2580 at : $2-\pi/2$
4054 5935 e^x : $4/3-\sqrt{5}$
4054 6510 id : $\ln(2/3)$
4054 6650 rt : (3,7,9,-5)
4054 8013 id : $1/\sqrt[3]{15}$
4054 9323 ℓ_2 : $(\sqrt[3]{5}/3)^{1/2}$
4054 9942 rt : (2,6,-1,6)
4055 0689 Γ : $3\ln 2/10$
4055 0834 rt : (8,9,-6,-4)
4055 4542 rt : (1,8,5,-6)
4055 8552 rt : (3,5,4,1)
4056 0219 rt : (7,8,3,-3)
4056 3571 ln : $3\sqrt{3}-\sqrt{5}/2$
4056 3644 10^x : $3\sqrt[3]{5}/7$
4056 4637 rt : (2,7,-2,-9)
4056 4764 at : 6
4056 5521 as : $\sqrt{2}/3-\sqrt{3}/2$
4056 5900 sπ : $4\sqrt{2}/3+4\sqrt{5}/3$
4056 6361 e^x : $3/4+4\sqrt{5}$
4056 7408 at : $2e/3+4\pi/3$
4056 7729 at : $\sqrt{2}/2-3\sqrt{5}$
4056 8469 10^x : 20/23
4056 8723 tπ : 9/22
4056 8767 rt : (5,2,-7,-8)
4056 8776 Ei : $(e/3)^2$
4056 9034 Γ : $5(1/\pi)/6$
4056 9289 10^x : $\sqrt{2}/4-\sqrt{5}/3$
4057 0516 λ : $\exp(-12\pi/11)$
4057 1014 rt : (8,1,-8,9)
4057 1126 J_1 : $\exp(-1/(3\pi))$
4057 1888 Ei : $5\sqrt{2}/3$
4057 2074 10^x : $1/3+4\sqrt{2}$
4057 2110 sq : $\sqrt[3]{5}/3$
4057 2848 rt : (5,3,-7,-3)
4057 3050 ℓ_2 : $4+3\sqrt{3}/4$
4057 3548 at : $1/4-e/4$
4057 4373 tπ : $(\sqrt{2}\pi/3)^{-1/3}$
4057 4583 sq : $\sqrt{3}/4+\sqrt{5}/2$
4057 4724 ζ : $9\sqrt{3}/10$
4057 5693 rt : (6,1,6,-3)
4057 5993 rt : (2,9,-9,-5)
4057 6523 rt : (8,8,-3,-2)
4057 7040 rt : (9,7,8,-5)
4057 7739 id : $2\sqrt{2}+\sqrt{3}/3$
4057 8616 rt : (9,9,-3,3)
4057 8750 ℓ_{10} : $9\sqrt{2}/5$
4058 2612 Ψ : $\pi^2/5$
4058 4490 rt : (3,4,-4,9)
4058 6018 cr : $e/2+4\pi$
4059 2514 at : $5\zeta(3)$
4059 2944 J_1 : $5\sqrt[3]{2}/7$
4059 4011 ℓ_{10} : $\pi/8$
4059 4109 rt : (5,5,7,-4)
4059 4884 ℓ_{10} : $\ln(\sqrt{2}\pi/3)$
4059 4954 J_1 : 9/10
4059 5430 $\ln(5)\div\ln(\pi)$
4059 5579 E : $1/\ln(5)$
4059 6084 rt : (7,6,1,9)
4059 6563 as : $4\sqrt{2}/3-2\sqrt{5}/3$
4059 7093 rt : (6,7,-8,-4)
4059 9186 10^x : $2e/3+3\pi/2$
4060 0382 $\sqrt{2}\div s\pi(1/5)$
4060 0584 $3\div\exp(2)$
4060 1957 rt : (6,1,-9,-6)
4060 3553 e^x : $8\pi/5$
4060 4525 rt : (4,9,8,-5)
4060 4745 J_2 : $\sqrt[3]{23/2}$
4060 5007 2^x : 19/15
4060 6043 rt : (5,6,9,3)
4060 6262 ℓ_{10} : $3/2+\pi/3$
4060 7514 rt : (8,3,-5,1)
4060 8545 sπ : $3e/4+2\pi/3$
4060 9394 id : $1/2+e/3$
4061 1224 rt : (3,5,-7,-4)
4061 1745 2^x : $\ln(e+\pi)$
4061 1798 tπ : $\sqrt{3}+3\sqrt{5}/4$
4061 2644 tπ : $\sqrt{2}+\sqrt{3}-\sqrt{6}$
4061 2914 cr : $1/2-\sqrt{3}/4$
4061 3186 ℓ_2 : $3/2-\sqrt{5}/3$
4061 4178 at : $3\sqrt{2}/4-2\sqrt{5}/3$
4061 4359 $\exp(\sqrt{2})-s\pi(1/4)$
4061 4527 Ψ : $4\ln 2/9$
4061 6034 Ψ : 6/25
4061 6449 rt : (4,8,-1,7)
4061 6634 J_2 : $((e+\pi)/3)^2$
4061 9105 rt : (2,1,5,2)
4062 1732 ζ : $\exp(\sqrt[3]{5}/2)$
4062 1926 Γ : $9\pi^2/7$
4062 2290 ζ : $\sqrt[3]{13}$
4062 2802 $\ln(4/5)\div\ln(\sqrt{3})$
4062 3167 cr : $\sqrt{2}/3+4\sqrt{3}/3$
4062 5886 rt : (9,6,1,-2)
4062 8757 rt : (1,3,6,-3)
4063 0393 rt : (2,5,-5,4)
4063 1382 rt : (8,7,-1,7)
4063 2696 rt : (7,7,-4,0)
4063 3824 10^x : $\sqrt{2}/2-\sqrt{5}/4$
4063 3846 id : $e/2+\pi/3$
4063 4545 Γ : $2\sqrt{5}/7$
4063 4829 Ψ : $2\sqrt{5}/7$
4063 6644 rt : (6,7,-7,-7)
4063 7409 Ei : 11
4063 9477 cu : $3/4-2\sqrt{5}/3$
4064 1185 2^x : $\exp(\sqrt[3]{5}/3)$
4064 1620 e^x : $(\sqrt[3]{5}/3)^{-3}$
4064 2065 rt : (2,8,-2,-2)
4064 5043 J_1 : $5\sqrt[3]{3}/8$
4064 5667 ℓ_{10} : $\sqrt{13/2}$
4064 9417 e^x : $2/3+\sqrt{5}/4$
4064 9910 J_1 : $3\zeta(3)/4$
4065 2209 sinh : $4\ln 2/7$
4065 2724 10^x : $(\sqrt[3]{4}/3)^3$
4065 3076 rt : (4,9,-4,-6)
4065 3974 at : $1/4-2\pi$
4065 4344 at : $(\ln 3/2)^{-3}$
4065 5321 rt : (5,4,0,6)
4065 8716 rt : (6,7,6,-4)
4065 9406 $\ln(4/5)\times\exp(3/5)$
4066 0566 Ei : 24/13
4066 2616 rt : (9,1,-6,-2)
4066 3823 rt : (9,8,-3,5)
4066 4278 tπ : $2e+4\pi/3$
4066 4448 rt : (7,3,5,-3)
4066 5124 2^x : $2+e/4$
4066 6069 sr : $\sqrt{2}+\sqrt{3}+\sqrt{7}$
4066 6673 Li_2 : $\ln(\sqrt[3]{3})$
4066 8172 ln : $2\sqrt{2}/3+\sqrt{5}/4$
4066 8737 at : $4+3e/4$
4066 9203 Ei : $6\pi^2/5$
4067 0883 rt : (2,6,-6,-9)
4067 0889 sπ : $3\sqrt{2}/2+\sqrt{5}/3$
4067 3663 rt : (6,2,-9,-8)
4067 3664 sπ : 2/15
4067 6197 rt : (8,1,4,2)
4067 6276 at : $2/3-3\sqrt{5}$
4067 6403 10^x : $1/4+3e/4$
4067 6814 rt : (5,9,-7,-4)
4067 7381 ℓ_{10} : $3\sqrt{2}/4+2\sqrt{5}/3$
4067 7738 Ψ : $\exp(1/(3\pi))$
4067 9473 θ_3 : 22/23
4068 0325 rt : (8,8,-6,-2)
4068 1700 cu : $4/3+3e/2$
4068 4388 $\sqrt{2}\times\ln(3/4)$
4068 5424 id : $1/4-4\sqrt{2}$
4068 5651 rt : (4,7,-3,8)
4068 7429 at : $4/3+3\pi/2$
4069 1065 id : $2\sqrt{2}/3+2\sqrt{3}$
4069 2912 ℓ_{10} : $2/3+4\sqrt{2}/3$
4069 2966 rt : (4,9,3,0)
4069 3683 J_1 : $(e/2)^{-1/3}$
4069 4230 sr : $8\sqrt{3}/7$
4069 5488 Γ : $\exp(-\pi/2)$
4069 5505 10^x : $4+4\sqrt{5}$
4069 7231 rt : (2,4,-9,2)
4069 7899 10^x : $2/3+3\sqrt{3}/4$
4069 8118 rt : (1,9,6,-4)
4069 8757 cu : $2\sqrt{2}-\sqrt{3}/3$
4069 9166 rt : (5,0,9,4)
4070 0246 rt : (6,5,-2,4)
4070 0539 cr : $2+\pi/4$
4070 1990 erf : $1/\sqrt{7}$
4070 3148 erf : $3\sqrt[3]{2}/10$
4070 4241 tπ : $8e\pi/9$
4070 4700 ℓ_2 : $(\ln 3)^3$
4070 5369 rt : (5,4,0,-1)
4070 6631 rt : (8,6,-1,9)
4070 6934 rt : (3,5,-8,-7)
4070 8748 J_2 : $8\sqrt{2}/5$
4070 8883 sr : $7\sqrt{2}/5$
4070 9650 sq : $9e/4$
4071 0058 $5\div\ln(4/5)$
4071 0853 rt : (2,8,-2,7)
4071 1929 sr : $(\ln 3/2)^3$
4071 3425 $1/3\times\exp(1/5)$
4071 3889 rt : (9,5,3,1)
4071 4075 rt : (7,9,5,-4)
4071 7772 ln : $5\zeta(3)/4$
4071 8008 $\ln(5)\times\exp(3/4)$
4071 9042 rt : (2,7,9,9)
4072 0477 ℓ_{10} : $-\sqrt{2}+\sqrt{3}+\sqrt{5}$
4072 0755 ζ : $(\ln 2)^{1/2}$
4072 2580 rt : (9,9,-8,-4)
4072 4280 ln : $9\pi^2/8$
4072 4500 rt : (1,9,5,-8)
4072 4821 as : $4\ln 2/7$
4072 5818 tπ : $\sqrt{17}$
4072 6053 $\exp(e)\div\exp(1/5)$
4072 6116 rt : (7,8,-9,-9)
4072 6558 at : $3e-2\pi/3$
4072 6934 rt : (6,4,8,3)
4072 7752 ℓ_{10} : $3\sqrt{2}/2+\sqrt{3}/4$
4072 8165 10^x : $(\sqrt[3]{5}/2)^{1/2}$
4072 8494 E : 13/21
4072 9238 sq : $2/3+2\sqrt{5}$
4072 9348 Γ : 17/2
4073 1135 at : $7\sqrt{3}/2$
4073 3311 rt : (9,7,-3,7)
4073 3762 tπ : $1/4+4e$
4073 4926 sinh : $1/\sqrt[3]{16}$
4073 6526 rt : (8,5,4,-3)
4073 6708 rt : (5,6,9,-5)
4073 7213 rt : (3,1,-5,-2)
4073 9245 sπ : $\zeta(3)/9$
4073 9408 θ_3 : $\sqrt{2}/7$
4073 9725 rt : (4,8,-6,-5)
4074 0134 tπ : $\exp(-2\pi/3)$
4074 0250 rt : (7,6,-4,2)
4074 0435 sπ : $5\ln 2/4$
4074 0740 cu : $7\sqrt[3]{2}/3$
4074 2169 $\ln(2)\times s\pi(1/5)$
4074 2840 Ei : $1/\sqrt{13}$
4074 3538 2^x : $\exp(-1/\sqrt{2})$
4074 3875 rt : (4,6,-5,9)
4074 4116 cr : $2/3+3\sqrt{2}/2$
4074 6414 ℓ_2 : $\sqrt[3]{7/3}$
4074 7694 ℓ_{10} : $8\sqrt{5}/7$
4074 8532 ℓ_{10} : 9/23

4074 8806 as : $\sqrt{3}-\sqrt{5}/3$
4074 9726 sr : P_1
4075 1263 10^x : $2-2\sqrt{2}/3$
4075 2837 10^x : $1/\sqrt[3]{18}$
4075 3645 rt : (2,9,-1,-8)
4075 4110 $\sqrt{3}+s\pi(\sqrt{5})$
4075 4310 Ei : 19/16
4075 5070 cu : $1+e$
4075 5187 J_2 : $5e/6$
4075 5431 tanh : $3\sqrt[3]{3}/10$
4075 5872 at : $e/2+3\pi/2$
4075 6304 cr : $3/4+3e/4$
4075 6982 rt : (5,3,0,8)
4075 9814 ln : P_{81}
4076 0001 10^x : $3\sqrt{3}/2+2\sqrt{5}/3$
4076 0316 rt : (5,5,-5,-3)
4076 4304 as : $3/4-\sqrt{2}/4$
4076 4360 J_1 : 19/21
4076 4896 rt : (6,2,8,-4)
4076 7084 J : $2\sqrt[3]{2}/7$
4076 8622 Ei : $(\sqrt{2}\pi/3)^{-1/2}$
4077 0653 E : $\exp(-\sqrt[3]{3}/3)$
4077 1420 at : $2+3e/2$
4077 1859 e^x : $3-3\sqrt{2}/2$
4077 2401 rt : (9,7,-6,-3)
4077 2420 ζ : $\exp(-\sqrt[3]{3}/2)$
4077 3461 ℓ_2 : $4\sqrt{2}/3-\sqrt{5}/4$
4077 3778 cr : $3\sqrt{3}/4+2\sqrt{5}/3$
4077 4243 2^x : $\sqrt{2}+\sqrt{3}+\sqrt{7}$
4077 5352 e^x : $\sqrt[3]{5}/5$
4077 6091 rt : (8,9,-5,0)
4077 9299 rt : (9,1,3,-2)
4078 1646 rt : (6,8,8,-5)
4078 2375 rt : (3,6,-8,-9)
4078 2951 E : $3\sqrt[3]{3}/7$
4078 3598 tanh : $\sqrt{3}/4$
4078 3761 rt : (3,7,-5,-3)
4078 5326 sr : $1/4+\sqrt{3}$
4078 6251 Ei : $2\ln 2/5$
4078 6753 J_0 : $7\zeta(3)/5$
4078 9227 $\exp(4)\div s\pi(2/5)$
4078 9589 rt : (1,8,8,2)
4079 0523 $\sqrt{2}-\exp(3/5)$
4079 1079 rt : (9,6,-3,9)
4079 1612 $s\pi$: $\sqrt[3]{13/2}$
4079 4327 Ψ : 1/9
4079 4993 $s\pi(2/5)\div s\pi(\sqrt{5})$
4079 6184 rt : (9,7,3,-3)
4080 0628 tanh : $5\ln 2/8$
4080 1082 e^x : $7e$
4080 1468 J_0 : $7\sqrt[3]{3}/6$
4080 2828 2^x : $2\sqrt{2}/3-\sqrt{5}$
4080 3727 sinh : $1/\ln(e/3)$
4080 3882 sr : $3e-3\pi/4$
4080 3930 J_2 : $9\sqrt[3]{2}/5$
4080 4159 at : $4\sqrt{2}+\sqrt{3}/4$
4080 7317 ℓ_{10} : $2+\sqrt{5}/4$
4080 7329 rt : (7,8,7,2)
4080 8276 as : $1/\sqrt[3]{16}$
4080 8383 rt : (9,8,-8,-2)
4080 8764 rt : (4,7,-8,-4)
4080 9054 2^x : $4\sqrt{2}/3+\sqrt{5}$
4080 9506 e^x : $1/4+2\sqrt{5}$
4081 0565 rt : (5,2,-9,3)
4081 1542 $\ln(2/5)-\exp(2/5)$
4081 1615 J_1 : $7\zeta(3)/3$
4081 1625 J_1 : $e/3$
4081 1780 rt : (5,9,-4,8)
4081 3834 ℓ_{10} : $1/4+4\sqrt{3}/3$
4081 4717 rt : (1,2,8,3)
4081 6039 at : $4+2\pi/3$
4081 6326 sq : $2\sqrt{5}/7$
4081 6704 J_2 : $7e/5$
4081 6997 rt : (7,2,7,3)
4081 7719 rt : (5,4,-9,-4)
4081 8927 rt : (1,4,8,-4)
4081 8995 2^x : $3-\sqrt{3}$
4081 9646 rt : (3,9,-3,4)
4081 9784 rt : (7,4,7,-4)
4082 0014 sr : $3e/2-2\pi/3$
4082 0468 cr : $8\pi/9$
4082 0662 rt : (1,7,-9,-4)
4082 1882 rt : (7,5,-4,4)
4082 3563 cr : $3\sqrt{2}/4+\sqrt{3}$
4082 4829 id : $1/\sqrt{6}$
4082 4864 ln : $3\sqrt{3}/2+2\sqrt{5}/3$
4082 8402 sq : 24/13
4083 0123 at : $\sqrt{2}+\sqrt{5}+\sqrt{6}$
4083 0224 at : $5e\pi/7$
4083 2443 J_1 : $(\sqrt{5}/3)^{1/3}$
4083 5194 ℓ_{10} : $3/2+3\sqrt{2}/4$
4083 5333 at : $3\sqrt[3]{3}/10$
4083 7352 id : $2\sqrt{3}+4\sqrt{5}$
4083 8227 J_0 : $1/\ln(2e/3)$
4083 9171 at : $2/3+2e$
4084 0410 sinh : 8/7
4084 1070 Ψ : $(\sqrt{2}\pi/3)^{-3}$
4084 1080 Ei : $\exp(-\sqrt[3]{3}/2)$
4084 2556 $t\pi$: $(\sqrt[3]{4}/3)^{-1/2}$
4084 2741 $\exp(e)+s\pi(\sqrt{3})$
4084 4532 $s\pi$: $9\sqrt[3]{2}/10$
4084 6932 sinh : $5(1/\pi)/4$
4084 8701 ℓ_2 : P_{36}
4084 9842 rt : (8,6,-6,2)
4085 2292 rt : (5,8,-6,9)
4085 3098 e^x : $\sqrt{2}-4\sqrt{3}/3$
4085 4679 rt : (2,9,-4,-3)
4085 4912 ℓ_2 : $3+4\sqrt{3}/3$
4085 6560 ℓ_{10} : $3e\pi$
4085 6643 sr : $\sqrt{3}/2+\sqrt{5}/2$
4085 7623 $s\pi$: $\sqrt{3}/2$
4085 7739 2^x : $4e/3-3\pi/4$
4085 9122 rt : (3,8,-7,2)
4086 1097 ℓ_{10} : $3/4+2e/3$
4086 1528 rt : (6,3,-2,8)
4086 1838 rt : (8,9,-5,-3)
4086 3455 cr : $5\sqrt{5}/4$
4086 3785 at : $\sqrt{3}/4$
4086 4182 $\sqrt{3}-\exp(\pi)$
4086 5486 $t\pi$: $3\sqrt{3}/4-3\sqrt{5}$
4086 5837 θ_3 : $(\sqrt{2}\pi/2)^{-2}$
4086 6049 $4/5\times\ln(3/5)$
4086 6284 rt : (8,6,6,-4)
4086 8926 at : P_{10}
4087 2084 at : $2\sqrt{2}-4\sqrt{5}$
4087 2158 2^x : 23/13
4087 2718 rt : (2,6,7,-4)
4087 2838 at : $9e/4$
4087 3497 $\exp(2/3)-\exp(\sqrt{5})$
4087 5331 rt : (9,7,-8,0)
4087 5428 Li_2 : $\exp(-1)$
4087 7914 cu : 19/9
4087 7935 rt : (7,7,-9,-7)
4088 0987 at : $5\ln 2/8$
4088 1521 rt : (4,6,-8,-4)
4088 5983 rt : (7,4,-4,6)
4088 6330 at : $4+3\sqrt{2}/2$
4088 6583 rt : (8,0,6,3)
4088 6676 Ei : $\sqrt{3}+\sqrt{6}-\sqrt{7}$
4088 8872 10^x : $\exp(-\sqrt[3]{2}/2)$
4089 1136 sinh : $(e/2)^{-3}$
4089 2853 at : $\exp(2e/3)$
4089 4101 J_2 : 25/11
4089 5138 $s\pi$: $(\ln 3/3)^2$
4089 6059 Ψ : $\ln(\sqrt{3}/3)$
4089 6968 E : $8\ln 2/9$
4089 7715 sr : P_{60}
4089 8662 rt : (5,4,-5,-1)
4089 8997 J_1 : $(4/3)^{-1/3}$
4089 9207 Γ : $5\pi^2/7$
4089 9268 $t\pi$: $1/\sqrt[3]{19}$
4090 0423 erf : $2\sqrt[3]{5}/9$
4090 2050 rt : (2,0,7,3)
4090 2648 ln : $1/4+4e$
4090 2657 cu : $3\sqrt{3}/7$
4090 3299 J_1 : $\sqrt[3]{22}$
4090 6167 rt : (9,8,5,-4)
4090 7630 e^x : $\sqrt{2}+2\sqrt{3}$
4090 7646 rt : (1,8,-9,-6)
4090 8126 ℓ_{10} : $3\sqrt[3]{5}/2$
4090 8409 rt : (8,5,-6,4)
4090 9090 id : 9/22
4090 9103 sq : π^2
4091 0179 id : $\sqrt{3}+3\sqrt{5}/4$
4091 0256 rt : (7,8,-2,-1)
4091 1073 at : $\sqrt{3}/3-3\sqrt{5}$
4091 3002 as : $\pi^2/10$
4091 4554 rt : (8,3,-5,-2)
4091 6582 id : $3\sqrt{3}/4-3\sqrt{5}$
4091 7489 $s\pi$: $((e+\pi)/3)^{-3}$
4091 7770 J_1 : 10/11
4091 7778 rt : (3,8,6,-4)
4091 8720 ℓ_{10} : $3\sqrt{2}-3\sqrt{5}/4$
4091 9538 2^x : $2\sqrt{3}/7$
4091 9869 rt : (1,9,-3,8)
4092 0250 Ψ : $\sqrt{21}$
4092 1292 as : $5(1/\pi)/4$
4092 2097 rt : (7,5,9,-5)
4092 3354 ζ : $\sqrt{11/2}$
4092 4077 $s\pi(2/5)\times s\pi(\pi)$
4092 5556 id : $9(1/\pi)/7$
4092 5606 id : $9\zeta(3)/2$
4092 5891 sr : P_{44}
4092 7078 sr : $3/4+4e$
4092 7902 rt : (4,0,-8,3)
4092 8849 rt : (5,9,-9,-4)
4092 9047 rt : (9,6,-8,2)
4092 9832 cr : $3/2+3\sqrt{3}/4$
4093 0430 J_2 : 19/5
4093 1044 tanh : 10/23
4093 1375 10^x : $4+3\sqrt{3}$
4093 1610 $s\pi$: $2e-\pi/2$
4093 1647 2^x : $3/4-3e/4$
4093 2058 at : $2\sqrt{2}/3+3\sqrt{3}$
4093 3270 rt : (3,8,-7,-4)
4093 4454 id : $2\sqrt{3}/3-\sqrt{5}/3$
4093 6537 $1/2\div\exp(1/5)$
4093 7667 rt : (7,3,-4,8)
4093 8583 rt : (1,8,-1,-1)
4093 8856 at : $3+\pi$
4093 9850 rt : (6,5,-7,-3)
4093 9886 E : 8/13
4094 0737 Li_2 : $1/\sqrt[3]{20}$
4094 0742 rt : (5,5,2,-2)
4094 1084 rt : (1,6,-2,4)
4094 1912 rt : (9,2,5,-3)
4094 2083 $\ln(2)\div\ln(3/4)$
4094 2345 2^x : $5\ln 2/7$
4094 2371 ℓ_{10} : $3-\sqrt{3}/4$
4094 2970 Li_2 : 7/19
4094 3084 ln : $\sqrt[3]{5}/7$
4094 3292 Ei : $8\zeta(3)/3$
4094 3791 $1/5-\ln(5)$
4094 4736 sq : $5\sqrt[3]{2}/3$
4094 5414 sr : $4/3+2\sqrt{5}$
4094 5856 rt : (8,6,6,2)
4094 5974 cr : 14/5
4094 6882 Ψ : $8\sqrt{3}/7$
4094 8677 cu : $2/3-3\sqrt{5}/2$
4095 4184 $\pi-\sqrt{3}$
4095 4526 e^x : $2\sqrt{2}/3+\sqrt{5}/3$
4095 4589 rt : (8,7,8,-5)
4095 4992 rt : (9,8,-8,-4)
4095 5876 Ψ : $(4/3)^{-3}$
4095 6057 rt : (8,4,-6,6)
4095 6686 2^x : $4+4e$
4095 6855 J_1 : $\sqrt{2}+\sqrt{3}-\sqrt{5}$
4095 8369 J_1 : $1/\ln(3)$
4095 8794 sq : $1/3-4\sqrt{2}/3$
4095 8867 J_0 : $4\sqrt[3]{2}/3$
4096 0000 sq : 16/25
4096 0908 sinh : $(2\pi)^{-1/2}$
4096 1899 ln : $2+2\pi/3$
4096 2869 rt : (7,1,9,4)
4096 5897 as : $(e/2)^{-3}$
4096 7621 rt : (3,2,6,-3)
4096 9165 $\ln(\sqrt{3})\times s\pi(\sqrt{3})$
4097 0044 rt : (9,5,-6,1)
4097 0924 J_1 : 14/5
4097 1154 rt : (2,7,9,-5)

4097 1393 cu : $7(1/\pi)/3$
4097 1683 10^x : $3/2-\sqrt{5}/2$
4097 2367 id : $7\sqrt[3]{2}/2$
4097 2518 ℓ_2 : $\exp(\sqrt{2}\pi)$
4097 2601 rt : (1,9,-9,-8)
4097 2741 cu : $\sqrt{3}/3-3\sqrt{5}/2$
4097 3012 at : $2-3e$
4097 3110 rt : (9,5,-8,4)
4097 3819 Ψ : $7\sqrt{2}/5$
4097 4328 10^x : $3\sqrt{3}/2+3\sqrt{5}$
4097 4930 10^x : $6(1/\pi)/5$
4097 4958 ℓ_{10} : $3\sqrt{2}/2-\sqrt{3}$
4097 5057 rt : (7,6,-9,-5)
4097 5326 e^x : $4-\sqrt{3}/4$
4097 6056 rt : (1,5,-3,-2)
4097 6323 2^x : 5π
4097 7393 id : $4e/3+\pi/4$
4097 8177 at : $1/2+4\sqrt{2}$
4097 9027 e^x : 20
4097 9574 e^x : $2\zeta(3)/7$
4098 0184 cr : $\sqrt[3]{22}$
4098 2835 sr : $8\sqrt{5}/9$
4098 3140 rt : (9,9,7,-5)
4098 3194 ℓ_{10} : $1/3+\sqrt{5}$
4098 5551 ζ : $2(e+\pi)/5$
4098 8154 at : $4e-3\pi/2$
4098 8966 rt : (5,3,-5,1)
4098 9323 $\exp(3/4)-s\pi(1/4)$
4099 0257 cu : $1+3\sqrt{2}/2$
4099 0405 $\exp(3/5)+s\pi(1/5)$
4099 1627 sr : $(2e/3)^{-3}$
4099 1755 tπ : $\exp(2e/3)$
4099 3926 erf : 8/21
4099 4348 id : $\pi^2/7$
4099 4806 10^x : $3/2+3\sqrt{5}/2$
4099 5171 rt : (5,8,0,-1)
4099 5243 tπ : $3e\pi/5$
4099 5585 rt : (8,3,-6,8)
4099 7685 rt : (6,7,1,-2)
4099 7727 tπ : $\exp(1/\pi)$
4100 0000 sq : 21/10
4100 1994 rt : (3,9,8,9)
4100 3892 10^x : 25/4
4100 4138 rt : (8,1,8,-4)
4100 4662 $\exp(3)-s\pi(\sqrt{5})$
4100 5767 2^x : $3\sqrt{2}/2+\sqrt{5}/4$
4100 6312 sq : $4(e+\pi)$
4100 6767 ℓ_{10} : $1+\pi/2$
4100 7899 rt : (3,2,-7,-3)
4100 8061 Γ : $\sqrt{2}+\sqrt{5}+\sqrt{7}$
4100 8061 at : $5\pi^2/8$
4100 9915 rt : (9,4,-8,6)
4101 1792 as : $4/3-\sqrt{3}$
4101 2734 at : 10/23
4101 4226 id : $\sqrt[3]{14}$
4101 4964 ℓ_{10} : $\sqrt[3]{17}$
4101 5265 e^x : $e/4-\pi/2$
4101 5625 sq : 19/16
4101 6546 at : $3e/2+2\pi/3$
4101 6685 J_1 : $(\pi/3)^{-2}$
4101 7446 ℓ_{10} : 7/18
4101 7556 rt : (6,4,-7,-1)
4101 8604 rt : (4,8,9,8)
4101 8923 10^x : $1/4+3\sqrt{3}$
4102 1551 $\exp(3/5)\times s\pi(e)$
4102 2616 J_2 : $4e\pi/9$
4102 4548 rt : (2,6,7,2)
4102 5981 rt : (3,9,-9,-5)
4102 6459 cr : $7\zeta(3)/3$
4102 7420 J_2 : $4\sqrt[3]{5}/3$
4102 8910 rt : (4,1,-3,5)
4103 1675 e^x : $3\sqrt{2}/2-\sqrt{3}/4$
4103 2793 rt : (1,7,3,-5)
4103 2864 at : $3/4-4\sqrt{3}$
4103 3622 10^x : $3-2e/3$
4103 3716 rt : (9,4,5,2)
4103 5442 $\exp(1/5)\div s\pi(1/3)$
4103 6306 as : $(2\pi)^{-1/2}$
4103 6327 sr : $2\sqrt{2}+4\sqrt{5}/3$
4103 7470 rt : (2,1,9,-4)
4103 8638 rt : (9,3,7,-4)
4104 0928 E : $(\ln 2/3)^{1/3}$
4104 1121 rt : (9,3,-8,8)
4104 1641 Ei : $(e/2)^2$
4104 1747 Γ : $10\sqrt{5}/7$
4104 2619 rt : (7,9,0,-2)
4104 2926 sr : $4\sqrt{3}-\sqrt{5}/2$
4104 3900 at : $2\sqrt{2}+3\sqrt{5}/2$
4104 4712 at : $4\sqrt{3}-\sqrt{5}/3$
4104 6017 as : P_{59}
4104 6157 ℓ_2 : $\ln(\ln 3)$
4105 0113 Ψ : 4/13
4105 1270 sπ : $3\sqrt[3]{3}/5$
4105 1764 ln : $3/2+3\sqrt{3}/2$
4105 2328 Γ : $\sqrt{20/3}$
4105 2601 rt : (5,2,-5,3)
4105 4180 at : $3/4+2e$
4105 4327 erf : $1/\sqrt[3]{18}$
4105 4860 10^x : $\sqrt{2}-\sqrt{3}+\sqrt{6}$
4105 5397 10^x : $3\sqrt{3}/2-\sqrt{5}/4$
4105 6070 sinh : $\sqrt[3]{21/2}$
4105 7345 J_1 : 21/23
4105 8233 J_2 : $\exp(4/3)$
4105 9351 rt : (5,9,7,6)
4105 9847 at : $2+4\pi/3$
4106 2049 rt : (3,1,-1,-9)
4106 2625 rt : (7,7,9,3)
4106 4610 $\ln(3)\times\exp(1/4)$
4106 5079 e^x : $9\sqrt[3]{3}/5$
4106 5784 rt : (2,4,-6,-3)
4106 9818 e^x : $2\sqrt{2}/3+3\sqrt{3}/4$
4107 0036 sq : $2-e/2$
4107 3369 $4/5\div\exp(2/3)$
4107 3610 rt : (6,3,-7,1)
4107 4612 rt : (3,3,8,-4)
4107 5232 sinh : 2/5
4107 5607 id : $1/3+3e/2$
4107 7923 cr : $\ln 2/10$
4107 8558 at : $1+3\sqrt{3}$
4108 0270 rt : (9,2,-8,-3)
4108 0355 ℓ_{10} : $e/7$
4108 2008 rt : (4,0,-3,7)
4108 2787 Γ : $2(1/\pi)$
4108 4795 sr : $4+2e/3$
4108 5143 J : $(2\pi)^{-2}$
4108 7981 Ψ : $8\ln 2/5$
4108 8075 sq : $4\sqrt[3]{3}/9$
4108 9761 sq : $3-2e/3$
4108 9970 e^x : 22/25
4109 1963 rt : (1,9,1,-2)
4109 2366 rt : (7,4,-9,-1)
4109 2593 sr : $1/2+2\sqrt{5}/3$
4109 2650 rt : (6,8,3,-3)
4109 2720 id : $\sqrt{2}/2-\sqrt{5}/2$
4109 3384 erf : $6(1/\pi)/5$
4109 4297 rt : (5,1,2,1)
4109 5264 ln : $\sqrt{2}/4+2\sqrt{3}/3$
4109 8034 tanh : $1/\sqrt[3]{12}$
4109 8666 cr : $4e+\pi$
4109 8928 at : $\sqrt{3}+2\sqrt{5}$
4109 9879 rt : (5,1,-5,5)
4110 0727 Ei : 17/3
4110 2832 rt : (9,4,9,-5)
4110 3541 cr : $1/2+4\sqrt{3}/3$
4110 5243 rt : (4,5,7,-4)
4110 5442 tπ : $1/\ln(2/3)$
4110 7106 $\ln(3/5)\times\ln(\sqrt{5})$
4110 9094 at : $1/2-3\sqrt{5}$
4110 9494 2^x : $(1/(3e))^{1/3}$
4110 9554 rt : (1,6,-5,-3)
4110 9619 10^x : $4/3+\sqrt{3}/3$
4111 0300 2^x : $3+\sqrt{5}/3$
4111 1769 10^x : $4\sqrt{3}+\sqrt{5}/4$
4111 2957 at : P_{29}
4111 3886 rt : (1,4,0,8)
4111 4445 ℓ_{10} : $4e/3-\pi/3$
4111 5012 rt : (8,5,8,3)
4111 5971 rt : (6,2,-7,3)
4111 6638 sq : $2e/3-3\pi/2$
4111 7344 ℓ_{10} : $2+\sqrt{3}/3$
4111 8193 2^x : $2\sqrt{5}/9$
4111 8999 sq : $3\sqrt[3]{5}/8$
4111 9466 rt : (6,7,-3,-2)
4111 9671 at : $3/2+3\pi/2$
4112 2434 rt : (4,1,-3,-9)
4112 2777 sr : $\sqrt{2}+\sqrt{3}/3$
4112 2786 J_0 : $3\sqrt{5}/4$
4112 5303 rt : (3,3,-9,-4)
4112 5538 e^x : $3\sqrt{3}/4+\sqrt{5}/4$
4112 8461 at : $4\sqrt{2}+\sqrt{5}/4$
4112 9292 id : $2\sqrt{2}/3-3\sqrt{5}/2$
4113 0537 rt : (7,3,-9,1)
4113 0929 rt : (5,8,7,8)
4113 1616 rt : (2,7,-6,9)
4113 2352 cu : $3/4+2\sqrt{5}$
4113 2410 sπ : $e\pi/4$
4113 2573 J_2 : 16/7
4113 2875 tπ : $(\pi/2)^3$
4113 3183 rt : (3,4,6,2)
4113 3334 $2/3\times\exp(3/4)$
4113 3500 J_1 : $5\sqrt{5}/4$
4113 3900 Γ : 7/11
4113 6396 rt : (5,0,-5,7)
4113 7135 cu : $2\sqrt{2}/3+2\sqrt{5}/3$
4113 8182 rt : (1,9,4,-8)
4114 0595 ℓ_2 : $7\sqrt[3]{5}/9$
4114 1633 at : $\sqrt{2}/2-4\sqrt{3}$
4114 2900 $e+\ln(2)$
4114 3782 rt : (8,4,-3,0)
4114 3812 at : $2e+\pi/4$
4114 4035 tπ : $1/3-3\sqrt{5}$
4114 5323 ln : $7(e+\pi)/10$
4114 7412 rt : (1,6,3,-1)
4114 7578 id : $3\sqrt{2}/4-2\sqrt{5}$
4114 9118 rt : (6,1,-7,5)
4114 9364 sinh : $\zeta(3)/3$
4115 0182 cr : $1+2e/3$
4115 1144 10^x : $3/2-4\sqrt{2}/3$
4115 1684 as : 2/5
4115 2263 cu : $7\sqrt[3]{3}/9$
4115 2281 J_0 : $(3\pi/2)^{1/3}$
4115 2512 id : $3e-4\pi$
4115 2772 rt : (6,9,5,6)
4115 4116 rt : (9,3,7,3)
4115 4733 10^x : $2e/3+\pi/3$
4115 6299 rt : (2,5,-8,-4)
4115 7005 tanh : 7/16
4115 7239 ln : P_{55}
4115 7750 at : $3e/4+4\pi/3$
4115 8609 2^x : $\sqrt{2}/3+3\sqrt{3}/4$
4115 8766 sq : $3e/4+3\pi$
4116 0226 J_1 : $\beta(2)$
4116 0757 rt : (7,2,-9,3)
4116 0857 $\pi\div\exp(4/5)$
4116 0872 $3/4\div\exp(3/5)$
4116 1180 rt : (4,6,9,-5)
4116 1432 sinh : $\sqrt[3]{3/2}$
4116 3696 2^x : $\sqrt{2}+2\sqrt{3}$
4116 3802 ln : $3-2\sqrt{5}/3$
4116 4138 sinh : $\ln(\pi)$
4116 5455 rt : (5,1,-5,-9)
4116 7931 cu : $(e/3)^3$
4116 9526 10^x : 11/15
4117 0543 rt : (5,9,-2,-2)
4117 0764 cr : $\sqrt{3}/3+\sqrt{5}$
4117 0806 e^x : 20/7
4117 1596 $\ln(2/5)\div\exp(4/5)$
4117 1657 J_1 : $\ln(5/2)$
4117 2174 θ_3 : $1/\sqrt{24}$
4117 2295 rt : (6,1,1,-1)
4117 4581 at : $4\sqrt{2}+\sqrt{3}/3$
4117 5765 rt : (6,0,-7,7)
4117 6470 id : 7/17
4117 8695 ζ : $(\sqrt[3]{3})^{-1/2}$
4117 8846 rt : (5,8,8,-5)
4117 9254 at : $4+\sqrt{5}$
4118 0885 cu : $3\sqrt{2}+3\sqrt{5}/2$
4118 1088 at : $\sqrt{2}/3-3\sqrt{5}$
4118 1452 rt : (1,7,-7,-4)
4118 1454 at : $1/\sqrt[3]{12}$

4118 3521 ℓ_{10} : P_{27}
4118 4247 E : 11/18
4118 4395 tπ : $7\sqrt{3}$
4118 4871 J_1 : 11/12
4118 4910 at : $9\ln 2$
4118 5280 rt : (7,1,-9,5)
4118 7397 ln : $3+3e$
4118 7502 rt : (7,6,-6,-3)
4119 0335 K : $(\sqrt[3]{2})^{-1/3}$
4119 1788 tπ : $1/2+4e/3$
4119 4276 rt : (6,8,5,8)
4119 5358 rt : (4,2,5,2)
4119 5437 ℓ_{10} : $\sqrt{20/3}$
4119 5715 at : $2+3\sqrt{2}$
4119 6204 tπ : $1/2+3\sqrt{3}$
4119 7320 2^x : $2+2\sqrt{2}$
4119 7656 rt : (6,1,-7,-9)
4119 7960 $3/4\times\ln(\sqrt{3})$
4119 8695 ln : $3/4+3\sqrt{5}/2$
4120 0230 J_2 : $\sqrt[3]{12}$
4120 1811 10^x : $5\sqrt{5}/2$
4120 2265 id : $1/3-\sqrt{5}/3$
4120 2693 sq : $2e\pi/3$
4120 5580 rt : (7,0,-9,7)
4120 5964 sq : $e/3-2\pi/3$
4120 7130 id : $2\sqrt[3]{3}/7$
4120 7866 rt : (7,9,3,6)
4121 1848 sπ : $9(1/\pi)$
4121 4106 at : 25/4
4121 4427 rt : (2,5,9,3)
4121 7402 rt : (9,2,9,4)
4121 7887 J_1 : $8\pi/9$
4121 8031 $1/4\times\exp(1/2)$
4121 9388 rt : (6,8,-5,-3)
4122 0306 rt : (7,3,0,-1)
4122 1333 ln : $5e/9$
4122 1474 $1-s\pi(1/5)$
4122 1593 tπ : $9(1/\pi)/7$
4122 1696 rt : (2,4,-7,9)
4122 2662 rt : (7,1,-9,-9)
4122 3565 tπ : $9\zeta(3)/2$
4122 3925 rt : (1,7,-4,-7)
4122 4487 sq : $\sqrt{2}+\sqrt{3}/4$
4122 4964 10^x : $(\sqrt{5}/2)^{1/2}$
4122 5735 Ei : $(\sqrt{2})^{1/2}$
4122 6401 rt : (1,8,-9,-5)
4122 6505 as : $\zeta(3)/3$
4122 7766 sq : $\sqrt{2}+\sqrt{5}/2$
4123 0473 rt : (8,5,-9,-4)
4123 1093 at : $3\sqrt{2}/4+3\sqrt{3}$
4123 4940 rt : (7,8,3,8)
4123 5541 rt : (5,0,4,2)
4123 6397 rt : (2,9,1,1)
4123 7576 sq : $4e+\pi$
4123 8437 sq : $(e/2)^2$
4124 0871 θ_3 : $\exp(-\sqrt[3]{4})$
4124 1044 at : 7/16
4124 1281 e^x : $3/2-2\sqrt{3}/3$
4124 2851 at : $2/3-4\sqrt{3}$
4124 3475 sr : $7\sqrt[3]{5}/6$
4124 4204 rt : (8,9,1,6)
4124 5916 e^x : $1-4\sqrt{2}/3$
4124 7413 sπ : $(e)^{-2}$
4124 8588 rt : (3,3,8,3)
4124 9080 rt : (2,8,-5,-6)
4125 0149 $1/3+s\pi(\sqrt{3})$
4125 2278 rt : (7,7,-8,-4)
4125 2817 rt : (8,5,-1,-1)
4125 3754 10^x : 3/20
4125 5985 $\exp(e)\times s\pi(2/5)$
4125 6671 ℓ_2 : $5\zeta(3)/8$
4125 9265 ℓ_{10} : $4-\sqrt{2}$
4126 3255 rt : (8,8,1,8)
4126 3621 rt : (6,2,3,-2)
4126 3797 ℓ_2 : $1-e/3$
4126 4295 rt : (3,8,-5,9)
4126 7075 Ei : 14/17
4126 9630 rt : (6,9,-7,-4)
4126 9972 rt : (9,9,-1,6)
4127 1159 10^x : $(\ln 2/2)^{-3}$
4127 3101 rt : (4,1,7,3)
4127 3537 rt : (3,7,-9,7)
4127 5301 sq : $4e/7$
4127 6285 rt : (9,7,-2,-1)
4127 6671 tπ : $3\sqrt{2}-\sqrt{5}/2$
4127 6740 at : $\sqrt{3}/4-3\sqrt{5}$
4127 7576 Ψ : $(e/2)^{1/3}$
4127 8546 tπ : $9\sqrt{5}$
4127 8876 cr : $2\pi^2/7$
4128 0558 sinh : $2e\pi/3$
4128 1670 tπ : $3e/2+\pi/3$
4128 4102 rt : (9,8,-1,8)
4128 4195 J_0 : $2(e+\pi)/7$
4128 4342 rt : (7,4,2,-2)
4128 8946 sr : $5(e+\pi)$
4129 1538 rt : (5,1,6,-3)
4129 2167 2^x : $\sqrt{3}-\sqrt{5}+\sqrt{7}$
4129 6513 at : 2π
4129 6979 cr : $3/4-e/4$
4129 8213 tπ : $10\pi^2$
4129 8685 cr : $\sqrt{20}\pi$
4130 0258 rt : (8,6,1,-2)
4130 1841 J_1 : 23/25
4130 2776 ℓ_2 : $\sqrt{2}/4+4\sqrt{3}/3$
4130 4520 sq : $2/3+4e/3$
4130 5707 10^x : π^2
4130 5841 $1/4\times\exp(\sqrt{3})$
4130 5906 rt : (6,3,5,-3)
4130 9238 10^x : $\sqrt{13/2}$
4130 9970 rt : (1,8,-3,4)
4131 0061 id : $(e+\pi)^{-1/2}$
4131 0132 Li_2 : $7(1/\pi)/6$
4131 1043 rt : (4,0,9,4)
4131 2865 rt : (9,8,0,-2)
4131 2876 λ : $\exp(-13\pi/12)$
4131 5648 $\ln(\sqrt{5})\div\exp(2/3)$
4131 5910 id : $\exp(5)$
4131 7414 rt : (7,5,4,-3)
4131 9562 at : $2\sqrt{2}+2\sqrt{3}$
4132 0756 ζ : $7\sqrt[3]{3}/8$
4132 1295 E : 14/23
4132 1593 rt : (5,2,8,-4)
4132 1980 id : $9\pi^2/2$
4132 2215 sq : $3-4\pi/3$
4132 3429 10^x : $(\sqrt{2}/3)^{-3}$
4132 3626 ℓ_{10} : $\sqrt{2}/4+\sqrt{5}$
4132 4547 tπ : $\sqrt{2}/4-4\sqrt{5}$
4132 5190 10^x : 15/4
4132 6530 sq : 9/14
4132 6588 e^x : $6\sqrt{3}/7$
4132 6839 rt : (8,7,3,-3)
4132 8189 at : $\sqrt{2}+\sqrt{5}+\sqrt{7}$
4132 8331 sπ : $1/4+4\sqrt{2}/3$
4133 0304 rt : (6,4,7,-4)
4133 0716 ln : $3\sqrt{2}+4\sqrt{3}$
4133 1610 sinh : $5e/7$
4133 3161 sq : $5(e+\pi)/8$
4133 4699 rt : (9,9,2,-3)
4133 5447 J_2 : $3\sqrt[3]{2}$
4133 6975 at : $5\sqrt[3]{2}$
4133 7061 ln : $6\sqrt[3]{2}/5$
4133 7404 10^x : $\zeta(3)/8$
4133 7618 rt : (7,6,6,-4)
4133 8892 J : $\sqrt[3]{2}/6$
4134 3846 rt : (8,8,5,-4)
4134 4061 Ψ : 23/5
4134 4153 E : $(\sqrt{2}\pi)^{-1/3}$
4134 5960 Li_2 : 16/17
4134 6186 rt : (6,5,9,-5)
4134 8288 cr : $4/3+2\sqrt{5}/3$
4134 9359 cr : $3e/4+\pi/4$
4135 0673 Ei : $(3\pi)^{1/2}$
4135 0918 erf : 5/13
4135 1243 rt : (7,7,8,-5)
4135 1412 $\exp(5)\div\exp(\pi)$
4135 1855 sr : $\sqrt[3]{5}/10$
4135 5663 rt : (8,9,7,-5)
4135 6692 id : P_{56}
4135 6699 cr : $2(1/\pi)/9$
4136 0795 2^x : $2\sqrt{3}/3+2\sqrt{5}$
4136 1015 at : $4+4\sqrt{3}/3$
4136 1683 as : $3-3\sqrt{3}/2$
4136 3114 rt : (1,9,-4,-1)
4136 4444 tanh : 11/25
4136 6829 10^x : $3\sqrt{3}/2-4\sqrt{5}/3$
4136 9733 θ_3 : 17/18
4137 2733 at : $3\sqrt{3}+\sqrt{5}/2$
4137 7613 Ψ : $2(1/\pi)$
4137 8631 erf : $2\sqrt{3}/9$
4138 1989 e^x : $\sqrt{2}+4\sqrt{3}/3$
4138 3753 10^x : $2-e/3$
4139 1459 J_2 : 23/10
4139 4489 cr : $3/2+4\pi$
4139 4917 $\sqrt{5}-\exp(3/5)$
4139 5364 at : $3/2-3\sqrt{2}/4$
4139 5540 at : $2/3+4\sqrt{2}$
4139 6299 erf : $5\ln 2/9$
4139 6898 Γ : $7e\pi/6$
4139 8245 e^x : $\sqrt{3}/5$
4140 3140 rt : (1,9,6,9)
4140 3611 ℓ_{10} : $1/2+2\pi/3$
4140 4792 cr : $9\pi/10$
4140 8666 id : $(\sqrt{5}/3)^3$
4140 9511 J_1 : 12/13
4141 5686 ln : $5\sqrt{5}$
4141 9444 at : 19/3
4142 0707 Γ : 7/19
4142 1356 id : $\sqrt{2}$
4142 2328 $2/3\div\ln(5)$
4142 4851 θ_3 : 7/12
4142 7222 2^x : $4\sqrt{2}-4\sqrt{3}$
4142 9188 Ei : $8\ln 2/3$
4143 1189 tπ : $3\zeta(3)/8$
4143 1292 ℓ_{10} : $9\sqrt[3]{3}/5$
4143 2847 Γ : $1/\sqrt[3]{20}$
4143 6184 sπ : $1/\ln(\sqrt[3]{5})$
4143 6495 e^x : $1/\sqrt[3]{24}$
4143 6723 Ei : $4\sqrt[3]{3}/7$
4143 9403 sr : $\zeta(3)/7$
4143 9832 id : $7\zeta(3)$
4144 1769 tπ : $9\sqrt{5}/7$
4144 2093 at : $7e/3$
4144 3052 e^x : $5\zeta(3)/3$
4144 3479 E : $(e)^{-1/2}$
4144 3667 id : $\sqrt{3}/4+4\sqrt{5}/3$
4144 4704 ℓ_{10} : $5\ln 2/9$
4144 4992 E : $7\ln 2/8$
4144 5829 sr : $4e+\pi/4$
4144 8553 sq : $3-3\pi/4$
4144 8608 J_1 : $4\ln 2/3$
4145 0669 at : $3\sqrt{3}/2-4\sqrt{5}$
4145 0687 at : 11/25
4145 0876 e^x : $2\sqrt{3}-\sqrt{5}$
4145 4887 10^x : 8/15
4145 6561 ℓ_2 : P_9
4145 6994 ℓ_{10} : $2e/3+\pi/4$
4145 9638 $\ln(\sqrt{3})+s\pi(\sqrt{2})$
4146 2186 Ψ : 7/11
4146 4641 J_1 : $(\sqrt[3]{5}/2)^{1/2}$
4146 5188 ℓ_{10} : $2\sqrt{3}/9$
4146 5674 J_0 : $\sqrt[3]{14/3}$
4146 7918 2^x : $3e-3\pi$
4146 8187 E : $3\sqrt{2}/7$
4146 9137 cu : $3\sqrt{3}/4+\sqrt{5}/4$
4146 9802 at : $3+3\sqrt{5}/2$
4147 1127 at : $\sqrt{2}/4-3\sqrt{5}$
4147 1658 2^x : $\sqrt{2}+\sqrt{3}-\sqrt{7}$
4147 3154 sπ : $2\sqrt{3}/3+4\sqrt{5}/3$
4147 3482 2^x : $\sqrt{3}+4\sqrt{5}$
4147 4768 Li_2 : $\sqrt{5}/6$
4147 4858 at : $4+3\pi/4$
4147 6213 id : $3\sqrt{2}/4+3\sqrt{5}/2$
4147 6631 Ψ : $8\sqrt{5}/9$
4147 6741 cr : $2\sqrt{3}/3+3\sqrt{5}/4$
4147 6843 $s\pi(\pi)\times s\pi(\sqrt{2})$
4147 8624 at : $4\sqrt{2}/3+2\sqrt{5}$
4147 9836 cu : $4/3+2\sqrt{3}$
4148 2464 id : $e/4-2\pi/3$
4148 3582 tπ : $5e/3$
4149 1326 id : $1/\sqrt[3]{14}$

4149 3432 rt : (9,8,9,3)
4149 3594 at : $9\sqrt{2}/2$
4149 4499 id : $2\sqrt{2}/3+2\sqrt{5}$
4149 4724 10^x : $3e-2\pi$
4149 5341 id : $(\sqrt{5}/3)^{-3}$
4149 7334 ℓ_{10} : 5/13
4149 7546 rt : (1,7,5,8)
4149 7971 $2/5 \div s\pi(\sqrt{2})$
4150 1449 Ψ : 8/19
4150 1536 rt : (9,4,-6,9)
4150 3079 cr : 17/6
4150 3102 J_2 : $8\sqrt{2}/3$
4150 3749 ℓ_2 : 3/4
4150 6029 rt : (4,9,-9,-5)
4150 6175 at : $4\sqrt{3}-\sqrt{5}/4$
4150 7162 J_1 : $(\sqrt[3]{2})^{-1/3}$
4150 8225 rt : (9,5,-6,7)
4151 1341 rt : (8,4,-4,9)
4151 4597 sinh : $2\sqrt{2}/7$
4151 5560 ζ : 7/3
4151 6131 rt : (9,6,-6,5)
4151 8495 ζ : $3\sqrt{5}/7$
4151 8773 sinh : $\exp(-e/3)$
4151 9222 rt : (9,6,8,-5)
4151 9839 at : $1/3-3\sqrt{5}$
4151 9853 rt : (8,5,-4,7)
4152 0927 rt : (1,7,-2,8)
4152 1190 E : $(\ln 3/3)^{1/2}$
4152 1636 Ei : $10\sqrt{5}/3$
4152 3879 rt : (7,4,-2,9)
4152 5619 rt : (9,7,-6,3)
4152 6207 rt : (8,4,9,-5)
4152 6727 rt : (4,8,-7,-4)
4152 7687 J_2 : $8\sqrt[3]{3}/5$
4152 7863 cr : $9\sqrt[3]{2}/4$
4152 8993 J_2 : $6\pi/5$
4153 0141 rt : (8,6,-4,5)
4153 1071 id : $\sqrt{2}/2+3\sqrt{5}$
4153 3013 at : $1-\sqrt{5}/4$
4153 3154 $\ln(\sqrt{3})+s\pi(1/3)$
4153 5075 rt : (7,5,-2,7)
4153 6471 $t\pi$: $\sqrt{2}+\sqrt{3}-\sqrt{7}$
4153 7216 rt : (9,8,-6,1)
4153 7586 ln : $3+\sqrt{5}/2$
4153 7920 rt : (9,5,6,-4)
4153 8568 sq : $3/4-3\sqrt{3}/2$
4153 8636 rt : (5,6,-8,-4)
4153 9169 at : $1/\ln(\sqrt[3]{5}/2)$
4154 0478 rt : (6,4,0,9)
4154 1501 $s\pi$: 3/22
4154 2205 rt : (9,9,7,2)
4154 2508 sr : $5\zeta(3)/3$
4154 2537 $s\pi$: $\sqrt{3}/3+\sqrt{5}/4$
4154 2828 rt : (8,7,-4,3)
4154 2857 10^x : $\sqrt{11/2}$
4154 4295 rt : (3,9,-4,-3)
4154 7969 rt : (8,3,7,-4)
4154 9012 rt : (7,6,-2,5)
4155 0395 sinh : 21/13
4155 1140 sr : $3/4-\sqrt{3}/3$
4155 1712 rt : (9,9,-6,-1)
4155 3592 rt : (6,4,-9,-4)
4155 5862 rt : (6,5,0,7)
4155 6778 sq : $5\pi/3$
4155 7018 ln : $2/3-\sqrt{3}/3$
4155 7185 $s\pi$: $3(1/\pi)/7$
4155 8541 sq : $4/3+e$
4155 8860 rt : (8,8,-4,1)
4155 9059 sinh : $3e\pi/7$
4155 9282 e^x : $7\sqrt[3]{2}/10$
4155 9833 J_2 : $4\sqrt{3}/3$
4155 9918 rt : (7,1,8,-4)
4156 0820 rt : (4,7,-5,-3)
4156 2693 $s\pi(1/4)\times s\pi(1/5)$
4156 3491 rt : (5,4,2,9)
4156 4422 e^x : $2-\sqrt{5}/2$
4156 5439 rt : (9,4,4,-3)
4156 6839 rt : (7,7,-2,3)
4156 7643 id : $\sqrt{2}/3+4\sqrt{5}$
4156 7676 θ_3 : $\sqrt[3]{3}/7$
4156 8834 cu : $4\sqrt{2}+\sqrt{5}/3$
4157 0579 ℓ_{10} : $4(e+\pi)/9$
4157 4366 rt : (6,1,9,4)
4157 5350 $t\pi$: $2\sqrt{3}/3-\sqrt{5}/3$
4157 5805 rt : (6,6,0,5)
4157 6131 ℓ_{10} : $\sqrt{2}/3-\sqrt{3}/4$
4157 6277 Ei : 25/21
4157 6644 J_0 : $1/\ln(\ln 3/2)$
4157 7475 ln : $7\sqrt{3}/8$
4157 9183 e^x : $\sqrt{2}+\sqrt{5}+\sqrt{7}$
4157 9244 J_2 : $10\ln 2/3$
4157 9759 rt : (8,9,-4,-1)
4158 1115 rt : (8,2,5,-3)
4158 2017 E : $2e/9$
4158 2503 rt : (5,5,-6,-3)
4158 2521 2^x : $3\sqrt{2}/4-\sqrt{5}/4$
4158 3481 cr : $10\pi^2/7$
4158 5224 at : $4\sqrt{2}+\sqrt{5}/3$
4158 5952 rt : (5,5,2,7)
4158 8830 id : $3\ln 2/5$
4158 8833 10^x : 5/3
4158 9368 id : $2e\pi/5$
4159 0402 ℓ_{10} : $2\sqrt{2}/3-\sqrt{5}/4$
4159 0445 rt : (7,8,-2,1)
4159 1391 at : $3e\pi/4$
4159 2653 id : 10π
4159 3320 rt : (3,8,-2,-2)
4159 5208 as : $2\sqrt{2}/7$
4159 6276 at : $3/4+4\sqrt{2}$
4159 6979 cr : $4e/3-\pi/4$
4159 7530 rt : (4,8,8,-5)
4159 7583 ℓ_2 : $\sqrt[3]{19}$
4159 8596 e^x : 8/23
4159 9424 as : $\exp(-e/3)$
4160 0632 rt : (7,0,6,3)
4160 0682 ℓ_{10} : $1/4+3\pi/4$
4160 0759 ζ : $(2e)^{1/2}$
4160 1058 J_1 : 13/14
4160 2681 rt : (6,7,0,3)
4160 3502 Ei : $\exp(-5)$
4160 8403 rt : (2,9,-9,5)
4160 9927 rt : (9,3,2,-2)
4161 0814 Ei : $(5/2)^3$
4161 0868 rt : (3,6,9,-5)
4161 1897 $s\pi$: $5\sqrt{5}/6$
4161 1926 ℓ_2 : $5e\pi$
4161 1971 Ψ : $\sqrt{5}-\sqrt{6}-\sqrt{7}$
4161 2221 rt : (6,3,-7,-3)
4161 2366 ζ : 14/9
4161 2838 sq : $1/\ln(3\pi/2)$
4161 6569 $\sqrt{5}-\exp(\sqrt{3})$
4161 6833 rt : (5,9,5,-4)
4161 7353 2^x : $4/3-3\sqrt{3}/2$
4161 7358 ln : $2+3\sqrt{2}/2$
4161 7888 2^x : 14/11
4161 7914 sr : $\sqrt{3}/10$
4162 1326 ln : P_{15}
4162 1811 at : $\sqrt{3}+\sqrt{5}+\sqrt{6}$
4162 3175 rt : (7,9,-2,-1)
4162 3693 Ψ : 21/19
4162 5051 id : $8e/9$
4162 5604 rt : (6,2,7,3)
4162 6168 $s\pi$: $(\sqrt{5}/3)^{1/2}$
4162 6231 ln : $4\sqrt{2}/3+\sqrt{5}$
4162 7540 rt : (4,6,-3,-2)
4162 7730 sr : $\ln 2/4$
4162 8763 2^x : $9\sqrt{2}/10$
4162 9073 $1/2+\ln(2/5)$
4162 9240 Ei : $2\sqrt[3]{5}/7$
4163 3390 rt : (4,7,6,-4)
4163 4291 ℓ_2 : $\sqrt{3}+4\sqrt{5}$
4163 4542 2^x : $(\pi)^{1/2}$
4163 5805 2^x : 15/7
4163 6011 10^x : $2/3-\pi/3$
4163 7719 rt : (8,1,3,-2)
4163 8444 sr : $\sqrt{2}/2+3\sqrt{3}/4$
4163 8787 at : $3-3\pi$
4163 9111 rt : (2,7,-9,-4)
4163 9377 cu : $\sqrt{2}+3\sqrt{5}/2$
4164 0786 id : $6\sqrt{5}$
4164 0858 rt : (6,8,0,1)
4164 3179 ℓ_{10} : $7\sqrt{5}/6$
4164 5456 rt : (2,9,1,-1)
4164 6885 at : $1/2-4\sqrt{3}$
4164 7126 sinh : $(\pi/2)^{-2}$
4164 9045 ln : $1/3+4e$
4165 1297 $t\pi$: $1/4+4e/3$
4165 1935 10^x : $1/\ln(3\pi/2)$
4165 2061 ζ : $2\sqrt[3]{5}/7$
4165 3026 rt : (3,5,7,-4)
4165 3324 rt : (2,8,-3,3)
4165 3546 sr : $3\sqrt{2}-\sqrt{5}$
4165 5476 sq : $8\ln 2/3$
4165 5512 rt : (7,1,-8,-3)
4165 7768 cu : $3e-\pi/2$
4165 7909 e^x : 15/17
4165 8707 rt : (5,4,8,3)
4165 9355 J_1 : $1/\ln((e+\pi)/2)$
4165 9667 at : $3e/2+3\pi/4$
4166 0196 sq : $1/4+3\sqrt{5}$
4166 0667 ln : $\sqrt{17}$
4166 0886 Ei : $3\sqrt{2}/2$
4166 1950 rt : (5,8,3,-3)
4166 5132 as : $3/4-2\sqrt{3}/3$
4166 5369 $\ln(2/3)+\exp(3/5)$
4166 5851 $t\pi$: P_{53}
4166 6614 at : $1+2e$
4166 6666 id : 5/12
4166 7725 $t\pi$: 7/23
4166 8294 sinh : $1/\sqrt[3]{15}$
4166 8380 $\ln(\sqrt{5})-\exp(1/5)$
4167 0891 10^x : $\sqrt{2}+3\sqrt{5}/2$
4167 1711 J_1 : $(\sqrt{3}/2)^{1/2}$
4167 4554 rt : (1,8,5,5)
4167 5160 rt : (7,1,4,2)
4167 6694 rt : (2,3,8,-4)
4167 8220 cr : $\sqrt[3]{23}$
4167 8904 rt : (5,4,-4,-2)
4168 5785 at : $1/2-2\sqrt{2}/3$
4168 6134 ζ : $6e/7$
4168 6989 cr : $3/4+2\pi/3$
4168 7547 rt : (4,6,4,5)
4168 7707 at : $2\sqrt{3}+4\sqrt{5}/3$
4169 0748 e^x : $\sqrt[3]{19}$
4169 1468 ln : $1/2+4e/3$
4169 3973 rt : (9,2,0,-1)
4169 5290 Ei : 20/13
4169 8174 rt : (2,8,-9,-9)
4169 8512 rt : (8,1,-9,8)
4169 9329 rt : (6,9,0,-1)
4170 1324 J_1 : 25/9
4170 2004 cr : $4-2\sqrt{3}/3$
4170 2882 sr : 4/23
4170 3697 2^x : $4(1/\pi)$
4170 4720 rt : (4,6,9,3)
4170 5786 rt : (1,1,9,-4)
4170 6230 rt : (2,8,-9,-5)
4170 7209 id : $3\sqrt{3}/4+\sqrt{5}/2$
4170 7304 rt : (3,7,0,-1)
4170 9965 $3\div\exp(3/4)$
4171 1523 erf : $e/7$
4171 1712 2^x : $2\sqrt{2}+\sqrt{5}/2$
4171 4159 at : $\sqrt{2}/3-4\sqrt{3}$
4171 7450 at : $1/4-3\sqrt{5}$
4171 7637 Γ : $\exp(\sqrt[3]{2})$
4171 8427 sq : $3/2-4\sqrt{5}$
4171 9276 rt : (3,5,6,7)
4172 0886 J_0 : 5/3
4172 0972 ℓ_2 : $(\sqrt[3]{2}/3)^{1/3}$
4172 1916 rt : (8,2,-9,6)
4172 3227 cr : $e\pi/3$
4172 3356 sq : 25/21
4172 4629 rt : (8,1,-9,3)
4172 6790 Ψ : $\exp(-(e+\pi)/3)$
4172 7162 $\sqrt{3}\times\exp(1/3)$
4172 8364 rt : (6,3,5,2)
4172 9026 as : $(\pi/2)^{-2}$
4173 1247 at : $3+2\sqrt{3}$
4173 1992 $t\pi$: $1/\ln(2e)$
4173 2165 tanh : 4/9

4173 2413 rt : (2,7,-9,-7)
4173 2893 rt : (7,1,-7,8)
4173 3346 rt : (9,9,-5,-3)
4173 3546 $s\pi$: P_{24}
4173 4054 rt : (5,8,2,1)
4173 4990 Li_2 : $1/\sqrt[3]{19}$
4173 5695 2^x : $\sqrt{2}/3 - \sqrt{3}$
4173 6489 10^x : $1/3 + 2\sqrt{3}$
4173 8399 Γ : $1/\ln(\ln 3/2)$
4173 8965 Li_2 : $3\pi/10$
4174 0401 J : $\exp(-17\pi/2)$
4174 1118 id : $9\sqrt[3]{2}/8$
4174 1849 Ei : $\zeta(3)/10$
4174 2430 rt : (1,9,-6,-4)
4174 5917 e^x : $1/\ln(1/\pi)$
4174 6334 rt : (7,9,7,-5)
4174 7294 rt : (1,5,-3,-4)
4174 8085 cr : $1/4 + 3\sqrt{3}/2$
4174 8758 as : $\ln(2/3)$
4174 9347 rt : (1,8,-7,-9)
4174 9628 rt : (8,3,-9,4)
4175 0003 at : $2 + 2\sqrt{5}$
4175 0401 as : $1/\sqrt[3]{15}$
4175 0543 $s\pi$: $2\sqrt{3}/3 + 3\sqrt{5}$
4175 0828 ℓ_2 : $10\zeta(3)/9$
4175 1248 E : $\zeta(3)/2$
4175 6117 rt : (8,0,1,1)
4175 7133 sinh : $\sqrt[3]{22/3}$
4175 8765 ln : $\sqrt[3]{7/2}$
4175 9657 rt : (2,4,8,9)
4176 0414 Γ : $\sqrt[3]{5}/10$
4176 0852 id : $\sqrt{3} + \sqrt{5} + \sqrt{6}$
4176 1698 sq : $(2e/3)^3$
4176 1743 J_2 : $\sqrt[3]{25/2}$
4176 2771 rt : (7,2,-7,6)
4176 2839 θ_3 : $((e+\pi)/2)^{-1/2}$
4176 3562 $t\pi$: $3\sqrt{3} - \sqrt{5}/3$
4176 4855 rt : (6,2,-5,-2)
4176 5313 at : $3\sqrt{2} + \sqrt{5}$
4176 5546 Li_2 : $3/8$
4176 6138 erf : $7/18$
4176 6313 J_1 : $14/15$
4176 6347 ln : $2\sqrt{2} + 3\sqrt{3}/4$
4176 7852 $t\pi$: $3\sqrt{3}/2 - 2\sqrt{5}$
4176 8294 erf : $1/\sqrt[3]{17}$
4176 8952 rt : (6,7,8,-5)
4177 0791 $\exp(\sqrt{2}) \times s\pi(1/5)$
4177 0928 $s\pi$: $9\pi/2$
4177 1229 cu : $4 - e/2$
4177 2400 cu : $1/3 + e$
4177 4254 e^x : $\pi/9$
4177 4397 $s\pi$: $3e/2 + \pi/4$
4177 4592 rt : (2,7,-7,-4)
4177 4891 cu : $2/3 - \sqrt{2}$
4177 4916 $s\pi$: $2/3 + 3\sqrt{3}$
4177 5173 sinh : $8\sqrt{2}/7$
4177 6304 rt : (2,6,-9,-5)
4177 7017 rt : (6,1,-5,8)
4177 7061 at : $2e + \pi/3$
4177 8747 rt : (4,7,4,3)
4177 9329 cr : $5\sqrt[3]{5}/3$
4178 0798 ln : $\sqrt{3} + \sqrt{5} - \sqrt{6}$
4178 1400 cr : $\sqrt{3} + \sqrt{5}/2$
4178 2954 rt : (8,4,-9,2)
4178 6984 $t\pi$: $\sqrt[3]{2}/10$
4178 7942 cu : $3 - e/2$
4178 8021 sr : $1/3 + 3\sqrt{5}/4$
4178 8473 2^x : $2/3 + 3e$
4178 8522 Γ : $\exp(-1)$
4178 9199 Ψ : $\exp(-\sqrt{3}/2)$
4178 9321 sq : $1 - \sqrt{2}/4$
4178 9685 rt : (9,0,8,4)
4178 9714 ζ : $\sqrt{2} - \sqrt{3} + \sqrt{7}$
4179 1481 $t\pi$: $\sqrt{3}/4 - \sqrt{5}/4$
4179 1829 at : $9\sqrt[3]{3}/2$
4179 4980 rt : (5,5,9,-5)
4179 6639 sinh : $(\pi/3)^3$
4179 6862 ℓ_{10} : $5\pi/6$
4179 7528 ℓ_{10} : $3/2 + \sqrt{5}/2$
4179 7721 Ψ : $(\pi)^3$
4179 8545 rt : (1,7,-7,-7)
4179 8793 cr : $9\pi/2$
4179 8972 rt : (7,3,-7,4)
4179 9389 Ψ : $8\sqrt{3}/3$
4180 0001 $\exp(1/4) - s\pi(1/3)$
4180 0002 rt : (6,8,-2,-1)
4180 0606 Ψ : $2(e + \pi)$
4180 1706 sq : $4/3 - 3\pi/2$
4180 4849 rt : (5,7,-9,8)
4180 6287 rt : (7,8,5,-4)
4180 7811 at : $3\sqrt{2}/2 - 3\sqrt{5}/4$
4180 8326 $\sqrt{3} \div \exp(1/5)$
4180 8339 E : $3/5$
4181 0061 rt : (5,5,6,2)
4181 0709 e^x : $3\pi^2$
4181 0899 2^x : $2\sqrt[3]{2}/5$
4181 2829 rt : (1,0,7,3)
4181 2966 sq : $\sqrt{2}/3 - \sqrt{5}/2$
4181 3988 rt : (3,5,-8,-4)
4181 4699 at : $13/2$
4181 5726 $t\pi$: $(2e/3)^{-2}$
4181 6486 rt : (6,2,-5,6)
4181 6817 rt : (4,5,-1,-1)
4181 7362 cr : $3\sqrt{3} + 4\sqrt{5}$
4181 7560 as : $1/2 - e/3$
4182 0532 ℓ_2 : P_{74}
4182 1675 J_1 : $7\zeta(3)/9$
4182 2060 e^x : $1/2 + e/2$
4182 2306 10^x : $3 + 4\sqrt{3}/3$
4182 2432 at : $4/9$
4182 2790 ln : P_{58}
4182 2864 $\ln(3/5) \div \exp(1/5)$
4182 3020 ln : P_{19}
4182 3787 rt : (8,5,-9,0)
4182 6235 rt : (3,8,-8,-1)
4182 6302 ζ : $\ln(e/3)$
4182 7117 cr : $10\sqrt{2}$
4182 7516 rt : (8,2,9,4)
4182 7808 ℓ_{10} : $2\sqrt{2}/3 + 3\sqrt{5}/4$
4182 8600 Ei : $(\ln 2/2)^2$
4183 0341 tanh : $7(1/\pi)/5$
4183 0695 rt : (5,8,-7,7)
4183 1896 rt : (1,8,-4,-3)
4183 4591 rt : (2,5,-9,-3)
4183 5717 rt : (5,1,-3,8)
4183 6529 at : $\sqrt{2} + \sqrt{6} + \sqrt{7}$
4183 8459 rt : (9,9,9,-6)
4183 9412 Li_2 : $2\sqrt{2}/3$
4184 0026 at : $10(e + \pi)/9$
4184 1036 tanh : $1/\ln(3\pi)$
4184 2416 ℓ_{10} : $\sqrt[3]{18}$
4184 3735 rt : (7,4,-7,2)
4184 6341 ln : $2/3 + 2\sqrt{3}$
4185 1391 λ : $(2e)^{-2}$
4185 1760 Ei : $9\sqrt[3]{5}/10$
4185 2009 ℓ_{10} : $1/2 + 3\sqrt{2}/2$
4185 2500 rt : (8,9,2,-3)
4185 5126 rt : (7,2,2,1)
4185 5167 rt : (3,9,-4,1)
4185 6178 J : $2\zeta(3)/7$
4185 6863 cr : $4\sqrt{3}/3 - \sqrt{5}$
4185 7395 $s\pi$: $5\sqrt[3]{5}/4$
4185 8223 E : $\ln(\ln 3/2)$
4185 9283 rt : (9,9,1,9)
4186 0272 rt : (5,9,-5,6)
4186 1241 sr : $9\sqrt{5}/10$
4186 3406 rt : (4,3,-9,-4)
4186 5085 rt : (1,9,8,-5)
4186 5781 rt : (6,3,-5,4)
4186 7419 at : $4\sqrt{2} + \sqrt{3}/2$
4186 7460 rt : (5,9,2,-1)
4186 8512 sq : $11/17$
4186 9516 rt : (1,8,-4,-4)
4186 9626 ln : $8\sqrt[3]{5}/9$
4187 0143 J_1 : $4\ln 2$
4187 1315 at : $2e/3 + 3\pi/2$
4187 3937 sr : $(1/(3e))^{-1/3}$
4187 3964 rt : (4,5,-8,8)
4187 4973 rt : (8,6,-9,-2)
4187 6551 rt : (5,4,7,-4)
4187 6840 sr : $4/3 + e/4$
4187 8114 10^x : $3\sqrt{3}/4 - 3\sqrt{5}/4$
4188 0959 $\exp(2/5) \times s\pi(2/5)$
4188 1513 ln : $2\sqrt{2}/3 + \sqrt{3}/3$
4188 2574 at : $4/3 + 3\sqrt{3}$
4188 2639 cr : $1/2 + 3\pi/4$
4188 3267 rt : (1,6,-3,-2)
4188 6116 rt : (2,8,3,0)
4188 6709 J_2 : $\sqrt{2} - \sqrt{3} + \sqrt{7}$
4188 7025 $1/3 + \exp(3)$
4188 7692 cu : $4e/3 + 3\pi/2$
4189 0275 rt : (5,2,-3,6)
4189 1048 at : $1/4 + 2\pi$
4189 1521 id : $4\sqrt{3} + 2\sqrt{5}/3$
4189 3248 rt : (7,7,3,-3)
4189 5157 rt : (9,1,6,3)
4189 8273 $t\pi$: $9\sqrt{5}/8$
4189 8341 cr : $20/7$
4190 0491 rt : (7,9,-5,0)
4190 2258 $\ln(4) \div \ln(2/3)$
4190 2556 rt : (9,8,7,-5)
4190 2615 e^x : $\exp(4/3)$
4190 2929 ln : $3e/4 + 2\pi/3$
4190 5227 cu : $1 - 4e$
4190 6754 e^x : $7/20$
4191 0335 J_1 : $15/16$
4191 0404 rt : (4,6,-6,7)
4191 1131 rt : (9,8,-3,-2)
4191 1477 Ei : $1/21$
4191 2573 J_1 : $8\sqrt{3}/5$
4191 2837 J_2 : $15/4$
4191 2930 ℓ_{10} : $8/21$
4191 5724 rt : (2,4,-9,-1)
4191 7239 rt : (7,9,-7,-4)
4191 7654 rt : (4,1,-1,8)
4191 8389 sq : $\sqrt{2} - \sqrt{3} + \sqrt{7}$
4191 9910 Γ : $(5/2)^{-1/2}$
4192 0073 2^x : P_{35}
4192 0886 ln : $2e/3 + 3\pi$
4192 1119 $t\pi$: $\sqrt{3}/3 + \sqrt{5}/2$
4192 1709 at : $7(1/\pi)/5$
4192 2375 J_2 : $6e/7$
4192 2383 rt : (4,2,8,-4)
4192 7487 rt : (3,2,-9,9)
4192 7547 rt : (8,9,3,9)
4192 9075 rt : (6,4,-5,2)
4193 1410 cr : $3/2 + e/2$
4193 1943 cr : $-\sqrt{5} + \sqrt{6} + \sqrt{7}$
4193 2391 e^x : $(1/(3e))^{1/2}$
4193 2524 at : $1/\ln(3\pi)$
4193 3993 $t\pi$: $(\sqrt{2}\pi)^{-1/2}$
4193 4093 at : $3\sqrt{2} + 4\sqrt{3}/3$
4193 4966 Ei : $3/25$
4193 5456 cr : $2e/3 + \pi/3$
4193 5553 sq : $4\sqrt{3}/3 - \sqrt{5}/2$
4193 5678 rt : (8,6,8,-5)
4193 8206 10^x : P_{71}
4193 8283 sr : P_{14}
4193 9483 10^x : P_{68}
4193 9849 rt : (7,0,-6,2)
4194 0997 rt : (8,7,-9,-4)
4194 1034 10^x : $1/3 + 4\pi/3$
4194 2115 rt : (1,8,3,-9)
4194 2470 $\ln(3/4) + s\pi(1/4)$
4194 2575 rt : (6,5,4,-3)
4194 4801 $t\pi$: $\sqrt[3]{2/3}$
4194 6330 ℓ_2 : $2\sqrt{2}/3 + \sqrt{3}$
4194 7615 λ : $\exp(-14\pi/13)$
4194 8407 id : $\sqrt{11}\pi$
4194 9211 $t\pi$: $1/\ln(\pi)$
4194 9548 θ_3 : $\exp(-\pi/2)$
4195 0083 $\ln(3) \div s\pi(e)$
4195 1804 J_2 : $(2e)^{1/2}$
4195 2065 rt : (4,7,7,2)
4195 3215 rt : (4,7,-4,6)
4195 3644 Ψ : $7\sqrt[3]{5}/6$
4195 7078 rt : (3,4,-6,-3)
4195 7252 Ψ : $(\sqrt{5}/3)^{-1/3}$
4195 9587 $s\pi$: $2\sqrt{2} + 4\sqrt{3}/3$
4195 9773 rt : (8,3,7,3)

4196 0077 rt : (1,8,6,-4)
4196 0574 sr : $10(e+\pi)/3$
4196 0588 tanh : $\sqrt{5}/5$
4196 1322 rt : (5,3,-3,4)
4196 2929 θ_3 : $3\ln 2/10$
4196 4337 rt : (8,8,0,-2)
4196 8301 $\exp(\sqrt{5}) \div \exp(3/4)$
4196 8339 sinh : $1/\sqrt{6}$
4197 3498 rt : (3,3,-7,8)
4197 3750 rt : (7,4,9,-5)
4197 4499 Ei : 19/23
4197 4773 rt : (7,6,-7,-2)
4197 5308 sq : 14/9
4197 5924 ln : $2\sqrt{3}/3+4\sqrt{5}/3$
4197 7093 ℓ_2 : $2\sqrt{2}-2\sqrt{5}/3$
4197 7602 e^x : $1/4-\sqrt{5}/2$
4197 8273 rt : (8,7,-8,-4)
4197 8614 rt : (3,0,9,4)
4198 1202 J_2 : 7/3
4198 1466 sr : $8\sqrt[3]{2}/5$
4198 1665 rt : (5,3,-8,-9)
4198 5385 at : $\sqrt{2}/4-4\sqrt{3}$
4198 5567 rt : (2,7,-7,5)
4198 5798 e^x : $\sqrt{2}/2+4\sqrt{3}/3$
4198 7006 10^x : $2\sqrt{2}/3-\sqrt{5}/4$
4198 9268 10^x : $1/3+3\sqrt{5}/4$
4198 9821 rt : (9,7,5,-4)
4199 0081 rt : (1,9,-3,9)
4199 0219 e^x : $2/3+3\sqrt{5}/4$
4199 0313 rt : (1,7,-2,-2)
4199 0567 J_1 : $\sqrt{23/3}$
4199 0579 θ_3 : $1/\sqrt[3]{5}$
4199 3895 rt : (1,4,-9,3)
4199 4454 e^x : $3+4\sqrt{2}$
4199 5189 id : $2\sqrt[3]{5}$
4199 5409 rt : (2,5,-6,1)
4199 6877 at : $2\pi^2/3$
4199 7368 id : $\sqrt[3]{2}/3$
4199 7601 cu : $1/2-2\pi$
4199 7990 rt : (4,2,-1,6)
4200 0218 erf : 9/23
4200 1900 tπ : $\sqrt{6}\pi$
4200 3432 E : $(3\pi/2)^{-1/3}$
4200 3541 rt : (5,3,5,-3)
4200 4222 rt : (4,8,-2,5)
4200 4658 sinh : $10\sqrt[3]{5}/3$
4200 6608 at : $3e-\pi/2$
4200 8402 sr : 3/17
4201 0319 $\ln(2)-\exp(\sqrt{2})$
4201 1656 ℓ_{10} : $4\sqrt{2}/3+\sqrt{5}/3$
4201 2708 $5\times\exp(1/4)$
4201 3288 rt : (6,7,-4,0)
4202 0410 rt : (5,9,-9,-5)
4202 1038 rt : (7,9,5,9)
4202 2251 ℓ_{10} : $2\sqrt[3]{5}/9$
4202 4808 cr : $9(1/\pi)$
4202 5322 at : $9(e+\pi)/8$
4202 6362 rt : (9,6,-4,8)
4202 6847 ℓ_{10} : $8\pi^2/3$
4202 6883 $\exp(3/5)+\exp(4)$
4202 7331 sπ : P_4
4202 7619 Ψ : $7\sqrt[3]{2}/8$
4202 7790 at : $3\sqrt{2}/2+2\sqrt{5}$
4202 8058 e^x : $5\sqrt{2}/8$
4202 8695 rt : (3,4,-5,7)
4202 8866 sπ : $2/3+\sqrt{2}/3$
4202 9350 rt : (8,8,-9,-6)
4203 0108 rt : (1,9,5,1)
4203 0608 rt : (7,6,1,-2)
4203 0968 at : $1/3-4\sqrt{3}$
4203 3742 Ψ : $\ln(e/2)$
4203 4717 ℓ_{10} : $4/3+3\sqrt{3}/4$
4203 5844 id : $3\sqrt{2}/2+3\sqrt{3}/4$
4203 6343 rt : (2,9,3,-3)
4203 6678 rt : (8,5,6,-4)
4203 6967 J_1 : 16/17
4203 7341 e^x : $\exp(-\pi/3)$
4203 7986 tπ : $\sqrt{2}/3-3\sqrt{3}/2$
4203 8060 rt : (6,4,3,1)
4203 8171 at : $4+3\sqrt{3}/2$
4203 8518 rt : (5,2,-8,-7)
4203 8803 Ei : $\sqrt[3]{19/3}$
4204 0067 rt : (3,8,8,-5)
4204 0225 Ei : $8\sqrt{3}/9$
4204 0646 J_1 : $\sqrt{2}-\sqrt{3}-\sqrt{6}$
4204 1501 $\ln(2)\div\exp(1/2)$
4204 3574 θ_3 : 5/24
4204 4751 e^x : $8e/5$
4204 4820 sr : $\sqrt{2}/8$
4204 5083 rt : (7,5,8,3)
4204 5237 cr : $2+\sqrt{3}/2$
4204 8093 sinh : $\exp(\sqrt[3]{3}/3)$
4204 9303 rt : (1,1,5,2)
4205 1218 sq : $1/3+2\sqrt{3}$
4205 1732 2^x : $4\sqrt{3}-4\sqrt{5}/3$
4205 2208 sr : $5(1/\pi)/9$
4205 3433 at : $\sqrt{5}/5$
4205 4133 rt : (4,2,-7,-3)
4205 5344 rt : (2,6,-2,3)
4205 5990 cr : $3\sqrt{2}/2+\sqrt{5}/3$
4205 6820 rt : (9,5,-9,-4)
4205 7624 rt : (5,4,-3,2)
4205 8488 $2/5\div s\pi(2/5)$
4205 9736 sinh : 9/22
4206 0074 rt : (9,9,-3,-1)
4206 0662 Ψ : $(5/3)^3$
4206 1258 Γ : $10\sqrt{3}$
4206 2002 id : $\exp(-\sqrt{3}/2)$
4206 2953 sinh : $7\ln 2/3$
4206 3092 2^x : $\sqrt{3}-4\sqrt{5}/3$
4206 3497 sπ : $\sqrt{2}/2+2\sqrt{3}/3$
4206 4461 ℓ_2 : $4+3\sqrt{5}$
4206 4948 $\ln(2/3)\div s\pi(\sqrt{2})$
4206 5717 at : $\sqrt{2}+3\sqrt{3}$
4206 6023 rt : (4,9,0,4)
4206 6245 rt : (5,3,-2,-1)
4206 9456 ln : $4\sqrt{3}/3+4\sqrt{5}$
4207 0733 J_2 : $\sqrt{14}$
4207 1776 id : $(e+\pi)^{1/2}$
4207 3552 at : $\sqrt{3}+\sqrt{5}+\sqrt{7}$
4207 6115 rt : (7,7,-7,-4)
4207 7600 sinh : $9(1/\pi)/7$
4207 7807 sinh : 23/20
4207 8353 rt : (1,7,3,-7)
4207 8627 id : $4\sqrt{2}-\sqrt{5}$
4207 9287 10^x : $\sqrt{14}$
4207 9469 at : $1/3+2\pi$
4208 0633 sinh : $7\sqrt{5}/5$
4208 0877 rt : (4,1,6,-3)
4208 1103 θ_3 : $1/\sqrt{23}$
4208 1530 2^x : $3\sqrt{2}/4-4\sqrt{3}/3$
4208 1689 J_1 : $3\pi/10$
4208 2197 $\ln(\sqrt{5})-\exp(4/5)$
4208 2409 rt : (9,7,-4,6)
4208 3110 sπ : P_5
4208 4168 tπ : $3\sqrt{3}+3\sqrt{5}/4$
4208 4391 Γ : $\sqrt{23}\pi$
4208 4730 rt : (9,2,4,2)
4208 4755 cr : $4\sqrt{3}/3+\sqrt{5}/4$
4208 5109 at : $\sqrt{2}/2-2\sqrt{3}/3$
4208 6153 sq : $2\sqrt{2}/3+2\sqrt{3}$
4208 8247 ℓ_{10} : $3/4+4\sqrt{2}/3$
4208 8865 rt : (2,6,9,-5)
4209 1124 tanh : $\pi/7$
4209 2056 rt : (7,3,7,-4)
4209 2528 rt : (7,8,-5,-3)
4209 3064 J_1 : $2\sqrt{2}/3$
4209 5308 tπ : $\sqrt{3}/2-2\sqrt{5}/3$
4209 6130 rt : (3,5,-3,6)
4209 6776 sr : $7\sqrt[3]{3}/5$
4209 6996 at : $3+4e/3$
4209 7803 Γ : 12/19
4209 8620 tπ : $(\sqrt{2}\pi/2)^{1/3}$
4210 0519 rt : (2,5,-3,-2)
4210 3640 sq : $4\sqrt{2}/3+2\sqrt{5}$
4210 5263 id : 8/19
4210 6661 rt : (1,7,4,-3)
4210 8041 ln : $1+\pi$
4210 8163 rt : (5,1,-8,-5)
4210 8556 rt : (9,2,-9,9)
4210 9053 $\sqrt{3}\div\exp(\sqrt{2})$
4210 9228 rt : (8,6,-2,8)
4210 9431 E : $\ln(2e/3)$
4210 9442 ζ : $5\sqrt[3]{5}/9$
4211 0082 ℓ_{10} : $9(e+\pi)/2$
4211 1727 cu : $3\sqrt{2}/2-2\sqrt{3}$
4211 1993 Ψ : $\sqrt{14}\pi$
4211 2259 $\ln(\pi)\div\exp(1)$
4211 2309 tπ : $4e$
4211 5481 rt : (9,6,3,-3)
4211 7226 10^x : $\sqrt{12}\pi$
4212 0343 at : $4\sqrt{3}/3-4\sqrt{5}$
4212 0834 ℓ_2 : $3+3\pi/4$
4212 3604 J_1 : $1/\ln(\ln 2/2)$
4212 4103 e^x : $4\sqrt[3]{2}$
4212 5687 sq : $5\pi/4$
4212 8031 cu : $1/3+4\sqrt{5}/3$
4212 9232 rt : (2,7,2,5)
4213 0749 rt : (6,6,-5,-2)
4213 1795 E : $6\ln 2/7$
4213 2379 e^x : $3e/2+3\pi/4$
4213 2512 $s\pi(\sqrt{3})-s\pi(\sqrt{5})$
4213 2589 sr : $2/3+3\sqrt{3}$
4213 2659 cr : $3/4+3\sqrt{2}/2$
4213 3493 ln : $2e/3-\pi/2$
4213 4416 erf : $\ln(\sqrt{2}\pi/3)$
4213 5180 erf : $\pi/8$
4213 6196 at : $7e\pi/9$
4213 7436 sr : $10\sqrt{2}/7$
4213 7504 Li_2 : $1/\sqrt{7}$
4213 8803 e^x : $1/\sqrt[3]{23}$
4213 8991 Li_2 : $3\sqrt[3]{2}/10$
4214 0275 $1/5+\exp(1/5)$
4214 2436 rt : (4,2,-6,-7)
4214 2983 sinh : $9\pi/10$
4214 5756 as : 9/22
4214 7252 rt : (9,3,9,-5)
4214 9176 J_1 : 17/18
4214 9840 rt : (9,8,-4,4)
4215 0977 2^x : $e/2+\pi/4$
4215 2240 sr : $7\sqrt{3}/6$
4215 3516 rt : (8,9,0,-1)
4215 3658 ζ : $\sqrt[3]{25/2}$
4215 5343 ln : $\sqrt{3}-2\sqrt{5}/3$
4215 6926 rt : (6,9,7,9)
4215 7287 sq : $1/2-2\sqrt{2}$
4215 8224 Γ : $\ln(\ln 2/2)$
4215 8552 rt : (6,1,8,-4)
4215 9820 rt : (2,8,1,-9)
4216 0658 tanh : $1/\sqrt[3]{11}$
4216 1190 rt : (9,3,-9,7)
4216 3147 rt : (6,7,9,3)
4216 3446 J : $4(1/\pi)/7$
4216 3802 as : $9(1/\pi)/7$
4216 4580 at : $3/2-3e$
4216 4730 J_2 : $2(e+\pi)/5$
4216 5569 tπ : $\sqrt{15}$
4216 6189 J_1 : $3\sqrt[3]{2}/4$
4216 9014 at : $1+4\sqrt{2}$
4217 0073 rt : (3,7,6,-4)
4217 1196 sq : $\sqrt{2}/3+3\sqrt{3}/2$
4217 3554 as : $2\sqrt{3}/3-\sqrt{5}/3$
4217 3694 rt : (5,8,-7,-4)
4217 4398 10^x : $4\zeta(3)/9$
4217 4522 ℓ_{10} : $4-e/2$
4217 5222 tπ : $\pi/2$
4218 0366 rt : (3,6,-1,5)
4218 2336 rt : (3,1,7,3)
4218 2569 rt : (8,7,-2,6)
4218 3037 rt : (1,4,-9,-7)
4218 5468 at : $\pi/7$
4218 5881 rt : (8,2,-7,9)
4218 6339 J_2 : $\sqrt{11/2}$
4218 6713 rt : (8,4,4,-3)
4218 7500 id : $(4/3)^{-3}$
4218 7939 sπ : $\ln 2/5$
4218 9900 tanh : 9/20
4219 0516 2^x : $1/3+2\sqrt{2}/3$
4219 0571 rt : (8,9,9,-6)
4219 0637 at : 20/3

4219 2434 cr : $9\sqrt{5}/7$
4219 3166 rt : (3,8,-6,6)
4219 3348 cr : 23/8
4219 3622 J_1 : $(\sqrt{5}/2)^{-1/2}$
4219 5444 rt : (5,5,-3,0)
4219 6184 $s\pi$: $2e+3\pi$
4220 0089 e^x : $\exp(-1/(3e))$
4220 0175 ℓ_2 : $2+e/4$
4220 3225 cu : $1/4-4e/3$
4220 5351 ℓ_{10} : $1/3+4\sqrt{3}/3$
4220 5958 at : P_{41}
4220 7789 $3\times\exp(\pi)$
4220 9980 rt : (5,7,8,-5)
4221 0421 rt : (8,6,-6,-3)
4221 0829 θ_3 : 13/17
4221 1880 $2/5-\exp(3/5)$
4221 5981 at : $1/4-4\sqrt{3}$
4221 6323 10^x : $4e+3\pi/2$
4221 8421 rt : (7,6,0,8)
4221 8924 2^x : $\sqrt{3}/4-3\sqrt{5}/4$
4221 9060 rt : (8,4,5,2)
4221 9663 rt : (3,8,-9,-5)
4222 0089 Ψ : $\sqrt[3]{2}/3$
4222 1143 $\sqrt{3}-\exp(e)$
4222 2394 rt : (7,8,-7,-6)
4222 2510 rt : (9,4,-9,5)
4222 2579 rt : (2,8,6,7)
4222 2626 2^x : $1+\sqrt{5}$
4222 6351 θ_3 : $(\sqrt[3]{5})^{-1/2}$
4222 8183 rt : (5,2,3,-2)
4222 9420 at : $4e-4\pi/3$
4223 2138 e^x : $2(e+\pi)/5$
4223 2487 rt : (9,9,-4,2)
4223 3711 rt : (1,7,-3,1)
4223 4282 rt : (2,8,1,-2)
4223 4951 at : $3\sqrt{3}+2\sqrt{5}/3$
4223 7262 $s\pi$: $2/3+2\sqrt{5}$
4223 9182 $\ln(4)+s\pi(\sqrt{2})$
4223 9316 rt : (4,1,-6,-5)
4223 9953 rt : (5,1,9,4)
4224 1080 $e-\exp(\pi)$
4224 1458 ℓ_2 : $3\sqrt{2}+\sqrt{5}/2$
4224 2932 rt : (2,5,7,-4)
4224 3906 sinh : $4e\pi/5$
4224 4725 10^x : $\sqrt{2}/2+\sqrt{3}/2$
4224 5529 rt : (3,9,-2,8)
4224 5778 e^x : $4\sqrt{3}/3+2\sqrt{5}$
4224 8540 rt : (1,5,-1,7)
4224 9290 J_1 : 18/19
4224 9340 ℓ_{10} : $2\sqrt{3}/3+2\sqrt{5}/3$
4224 9570 id : $10\sqrt[3]{3}$
4225 0000 sq : 13/20
4225 1567 rt : (8,3,-7,7)
4225 2133 $\ln(\pi)\div\ln(\sqrt{5})$
4225 3273 e^x : $3/2+3e/4$
4225 3284 tanh : $3\zeta(3)/8$
4225 3541 ℓ_{10} : $3\sqrt[3]{2}/10$
4225 3583 2^x : $(\sqrt[3]{3}/3)^{-1/3}$
4225 4902 ℓ_{10} : $\sqrt{7}$
4225 5241 at : $3/2+3\sqrt{3}$
4225 5809 at : $1/\sqrt[3]{11}$
4225 7089 at : $8(e+\pi)/7$
4225 9843 2^x : $3-3\sqrt{2}$
4226 4929 $\ln(3)\times\exp(\pi)$
4226 4973 id : $1-\sqrt{3}/3$
4226 6313 ℓ_{10} : $3-\sqrt{2}/4$
4226 6395 $\exp(1/2)+s\pi(e)$
4226 9860 rt : (1,7,8,-3)
4227 1427 rt : (4,5,9,-5)
4227 2753 e^x : $4-4\sqrt{3}/3$
4227 3304 rt : (8,8,-2,4)
4227 4472 rt : (7,2,5,-3)
4227 7773 at : $e-3\pi$
4227 8433 Ψ : 2
4227 9186 rt : (7,5,-1,-1)
4227 9321 $t\pi$: $2(1/\pi)/5$
4227 9340 sq : $1/4+2\sqrt{2}/3$
4228 1486 at : $3\sqrt{5}$
4228 2489 $s\pi$: $2\sqrt{2}/3+3\sqrt{3}$
4228 2845 rt : (7,2,-5,9)
4228 3454 Ψ : $\sqrt[3]{4/3}$
4228 5392 at : 9/20
4228 8551 rt : (3,7,1,4)
4228 8915 10^x : $1/3-\sqrt{2}/2$
4228 9995 2^x : $4\sqrt{3}+\sqrt{5}/4$
4229 0578 at : $2+3\pi/2$
4229 1146 J_2 : $\sqrt[3]{13}$
4229 1340 J_2 : $\exp(\sqrt[3]{5}/2)$
4229 3168 rt : (7,8,-7,1)
4229 3239 cu : $6\pi^2/7$
4229 4845 rt : (9,5,-9,3)
4229 6305 rt : (6,8,5,-4)
4229 7723 e^x : $(\ln 2)^{1/3}$
4229 8841 Γ : 15/4
4229 9116 ln : $8\pi^2/7$
4229 9699 rt : (9,2,7,-4)
4230 0236 J_1 : $e\pi/9$
4230 0728 Ei : $\sqrt{14}$
4230 3362 at : $4+e$
4230 3845 $t\pi$: $7(e+\pi)/8$
4230 4086 rt : (4,9,-4,-3)
4230 5718 rt : (6,7,-5,-4)
4230 7081 E : 13/22
4230 7157 cu : $2\pi^2/7$
4230 7983 rt : (5,1,-8,1)
4230 8099 cr : $4-\sqrt{5}/2$
4230 8932 rt : (4,7,-9,-6)
4230 9051 J_1 : $(\sqrt[3]{5}/2)^{1/3}$
4231 1247 rt : (1,6,3,-5)
4231 1471 rt : (8,8,7,-5)
4231 1506 J_1 : $\sqrt[3]{21}$
4231 1558 sq : $\sqrt[3]{19/3}$
4231 1649 rt : (9,5,1,-2)
4231 1823 e^x : $3\sqrt{2}/2-4\sqrt{5}/3$
4231 2199 $\ln(3)-s\pi(\sqrt{5})$
4231 4355 $1/5-\ln(4/5)$
4231 5260 2^x : $6\sqrt{5}/5$
4231 6376 2^x : $1/4-2\sqrt{5}/3$
4231 8514 rt : (7,7,0,6)
4231 9799 e^x : $(\sqrt[3]{3})^{-1/3}$
4231 9912 tanh : $\ln(\pi/2)$
4232 1621 rt : (2,3,8,3)
4232 2979 E : $1/\ln(2e)$
4232 4022 rt : (2,9,-6,-4)
4232 4742 e^x : 6/17
4232 4966 2^x : $\sqrt{2}/2+\sqrt{3}$
4232 7876 sinh : $\sqrt{3}-\sqrt{6}-\sqrt{7}$
4232 8903 id : $e\pi/6$
4232 9527 rt : (8,4,-7,5)
4232 9741 ζ : $4e/7$
4233 0593 $\ln(\sqrt{2})\times\exp(1/5)$
4233 0867 sr : $2/3+e/2$
4233 1010 10^x : $4+2\pi/3$
4233 1082 id : $e-\pi$
4233 1648 $t\pi$: $2\sqrt{2}+3\sqrt{3}/4$
4233 1834 rt : (1,3,8,-4)
4233 3601 rt : (2,9,0,-3)
4233 4758 2^x : $2-2\sqrt{5}/3$
4233 5528 2^x : $8(1/\pi)/5$
4233 5668 10^x : $(\ln 2)^{-3}$
4233 6107 rt : (9,8,-5,0)
4233 7817 Li_2 : 17/18
4233 8709 J_1 : $5\sqrt[3]{5}/9$
4233 9165 J_1 : 19/20
4233 9318 $\ln(\pi)\times\exp(3/4)$
4233 9688 ζ : $(\sqrt[3]{2}/3)^{1/3}$
4234 0000 $1/5\times\exp(3/4)$
4234 1188 sr : $2(e+\pi)$
4234 3309 cr : $5e\pi/3$
4234 4215 rt : (2,3,-2,6)
4234 6526 rt : (1,6,0,-1)
4234 7087 as : $\sqrt{2}/2-\sqrt{5}/2$
4234 7985 J_2 : $5\sqrt{5}/3$
4234 9427 rt : (3,6,-9,-8)
4234 9518 at : $3\zeta(3)/8$
4234 9781 cr : $2\sqrt[3]{3}$
4234 9954 sinh : 7/17
4235 2577 e^x : $1/2-e/2$
4235 3740 $t\pi$: $4\sqrt{3}+4\sqrt{5}$
4236 1691 cu : $2e/3-4\pi/3$
4236 2213 rt : (1,6,2,-2)
4236 4435 id : $(\ln 2/2)^{-1/3}$
4236 4893 ln : $\sqrt{7/3}$
4236 6105 rt : (4,0,-6,3)
4236 7127 rt : (7,3,-5,7)
4236 7982 tanh : $\exp(-\sqrt[3]{4}/2)$
4236 8185 cr : $1+4\sqrt{2}/3$
4236 8877 rt : (6,6,2,8)
4237 1797 at : $(\sqrt[3]{4}/3)^{-3}$
4237 2563 rt : (5,9,9,9)
4237 3773 J_2 : $3\pi/4$
4237 4157 Ei : $8\sqrt{3}/3$
4237 4159 at : $3e/4+3\pi/2$
4237 4903 rt : (7,6,8,-5)
4237 5494 Ψ : 1/10
4237 8299 id : $4\sqrt{2}/3-4\sqrt{3}/3$
4237 8523 $t\pi$: $1/4-3\pi/2$
4238 0525 rt : (9,4,-7,-3)
4238 0728 cu : $3/2-\pi$
4238 1430 rt : (9,6,-9,1)
4238 1829 10^x : $e/2+2\pi/3$
4238 2259 rt : (7,4,-9,-4)
4238 2812 cu : 9/8
4238 3252 sinh : $2\sqrt[3]{3}/7$
4238 3537 Ψ : $7\sqrt{2}/9$
4238 3656 $\exp(\pi)\div\exp(4)$
4238 3979 10^x : $5e/7$
4238 4918 2^x : $2\sqrt{2}/3-\sqrt{3}/4$
4238 5283 $\ln(2)-\exp(3/4)$
4238 5379 rt : (8,0,8,4)
4238 6226 e^x : 16/13
4238 6391 θ_3 : $\sqrt[3]{2}/6$
4238 6820 sr : $\sqrt[3]{25/3}$
4238 7071 Γ : $6\sqrt{2}$
4238 7808 J_2 : $5\sqrt{2}/3$
4238 8397 rt : (8,9,-2,2)
4238 8505 $t\pi$: $9\sqrt[3]{5}/2$
4239 2705 Li_2 : $2\sqrt[3]{5}/9$
4239 2751 sq : $(e+\pi)/9$
4239 6547 Γ : 16/5
4239 9041 at : $2/3-\sqrt{5}/2$
4239 9845 ℓ_2 : $\sqrt{5}/3$
4239 9930 e^x : P_{54}
4239 9944 rt : (9,9,4,-4)
4240 1553 ℓ_{10} : $3/2+2\sqrt{3}/3$
4240 3287 rt : (7,9,2,-3)
4240 3568 J_1 : $(e/3)^{1/2}$
4240 4257 E : $\exp(-\sqrt[3]{4}/3)$
4240 4811 cu : $5\zeta(3)/8$
4240 6012 at : $4e/3+\pi$
4240 7684 rt : (2,4,-8,6)
4240 8031 e^x : $\exp(5/2)$
4240 8033 rt : (6,2,-3,9)
4240 8121 rt : (9,6,8,3)
4240 8624 rt : (5,9,0,-2)
4241 0853 rt : (4,8,-7,-7)
4241 1782 2^x : $1/3-\pi/2$
4241 1901 e^x : $\sqrt{2}/4$
4241 2453 rt : (1,7,-9,-5)
4241 4406 at : $4/3+2e$
4241 6329 J : $\sqrt[3]{2}/4$
4241 6932 at : $\ln(\pi/2)$
4241 8376 ζ : $(\ln 3)^2$
4241 8489 rt : (3,9,-2,-1)
4241 8784 10^x : $4\sqrt{2}/3-\sqrt{3}$
4241 9817 rt : (3,7,-7,-4)
4242 0293 J_1 : 20/21
4242 2188 rt : (6,2,-9,3)
4242 3055 rt : (7,6,6,2)
4242 3532 rt : (8,5,-7,3)
4242 4035 ln : $3+2\sqrt{3}/3$
4242 5056 at : $4\sqrt{2}+\sqrt{5}/2$
4242 5077 sr : $\sqrt[3]{2}/7$
4242 6406 id : $3\sqrt{2}/10$
4242 6414 E : $(\ln 2/2)^{1/2}$
4242 7792 Γ : $\sqrt[3]{2}/2$
4242 9599 e^x : $10(1/\pi)/9$
4243 2402 rt : (8,3,2,-2)
4243 2538 rt : (3,8,3,3)
4243 2685 rt : (5,7,-5,-3)

4243 4932 ln : $2-\sqrt{2}/3$
4243 8970 as : $7/17$
4243 9219 at : $4\sqrt{3}/3+2\sqrt{5}$
4244 0039 tanh : $e/6$
4244 1318 id : $4(1/\pi)/3$
4244 2726 at : $1/2+2\pi$
4244 2890 rt : (1,9,-4,0)
4244 6201 10^x : $5/13$
4244 7132 rt : (7,8,0,4)
4244 7626 J_1 : $(\sqrt{3}/2)^{1/3}$
4244 7994 θ_3 : $(e+\pi)/10$
4244 8206 ζ : $5\sqrt[3]{2}/7$
4244 8266 rt : (6,4,9,-5)
4244 8746 ln : $3e+\pi$
4245 1288 cu : $1/2-3\sqrt{2}$
4245 1323 rt : (7,9,-7,-8)
4245 1766 E : $10/17$
4245 2082 2^x : $1/3+2e/3$
4245 2811 2^x : $1-\sqrt{5}$
4245 6725 e^x : $8\sqrt[3]{5}/5$
4245 6971 at : $1/3-\pi/4$
4245 7411 rt : (6,3,0,-1)
4245 8492 rt : (5,2,-8,-1)
4245 8557 e^x : $4e+\pi/3$
4245 9533 sr : $\sqrt[3]{3}/8$
4245 9626 tπ : $\sqrt{2}/2+3\sqrt{3}/2$
4246 0919 rt : (7,7,-3,-2)
4246 1070 2^x : $4\sqrt{5}/7$
4246 1360 10^x : $3/4+\sqrt{3}$
4246 1533 rt : (3,5,-9,-6)
4246 2117 erf : $4\ln 2/7$
4246 3686 rt : (4,4,7,-4)
4246 5213 2^x : $23/18$
4246 5347 2^x : $\sqrt{20}\pi$
4246 5573 at : $\exp(-\sqrt[3]{4}/2)$
4246 7279 as : $1/3-\sqrt{5}/3$
4246 9058 rt : (2,7,1,-7)
4246 9308 at : $5e/2$
4246 9566 rt : (7,4,-5,5)
4247 1804 $\exp(1/4)+\exp(\pi)$
4247 2618 as : $2\sqrt[3]{3}/7$
4247 3761 Ψ : $\exp(3\pi)$
4247 7095 rt : (8,7,5,-4)
4247 7549 e^x : $3/2-3\pi/4$
4247 7553 J_1 : $(\ln 3)^{-1/2}$
4247 7796 id : 3π
4248 2558 J : $10/21$
4248 4001 cu : $1-4\pi/3$
4248 4471 10^x : $4-3\sqrt{5}/2$
4248 6538 2^x : $\ln(5/3)$
4248 6832 ln : $2\sqrt{2}-3\sqrt{3}/4$
4248 6894 rt : (9,7,-9,-1)
4248 7113 id : $7\sqrt{3}/5$
4248 9125 rt : (2,4,5,-3)
4248 9913 Li_2 : $3\sqrt[3]{2}/4$
4248 9998 rt : (7,2,9,4)
4249 3892 J_1 : $21/22$
4249 4845 rt : (2,5,-4,8)
4249 5069 sinh : $(e+\pi)^{-1/2}$
4249 5145 rt : (5,6,4,1)
4249 6419 λ : $\exp(-15\pi/14)$
4249 6975 10^x : $\sqrt{2}/2+3\sqrt{3}/4$
4249 7995 id : $5\sqrt[3]{5}/6$
4249 8301 θ_3 : $4/19$
4249 9276 sr : $\sqrt{2}/4+3\sqrt{5}/4$
4249 9290 Ψ : $5\zeta(3)/3$
4249 9633 e^x : $10\sqrt[3]{3}/3$
4250 0216 rt : (1,8,4,-2)
4250 0256 rt : (6,7,3,-3)
4250 1056 ln : $4/3-e/4$
4250 2826 at : $4\sqrt{2}+2\sqrt{3}/3$
4250 3323 cr : $2\sqrt{2}/3-\sqrt{3}/2$
4250 3668 2^x : P_{13}
4250 4239 id : $\sqrt{3}/2+\sqrt{5}/4$
4250 5709 sπ : $(\pi/2)^{-1/3}$
4250 5929 cu : $3\sqrt{3}/2+2\sqrt{5}$
4250 6940 J_1 : $3(1/\pi)$
4250 9246 rt : (2,8,8,-5)
4251 0267 10^x : $2/13$
4251 3119 sq : $7\zeta(3)/4$
4251 3147 $\ln(\sqrt{3})\times s\pi(e)$
4251 3228 Li_2 : $8/21$
4251 3606 rt : (6,7,2,6)
4251 5356 $\exp(2)+s\pi(\sqrt{2})$
4251 6833 $\pi\div\exp(2)$
4251 6840 at : $2\sqrt{3}+3\sqrt{5}/2$
4251 7807 rt : (2,6,-7,-8)
4251 8199 at : $3/2-\pi/3$
4251 9798 $\sqrt{3}+\ln(2)$
4252 0120 rt : (6,3,-3,7)
4252 2740 rt : (9,3,2,1)
4252 2797 2^x : $\sqrt{17}$
4252 3804 at : $4/3-3e$
4252 3979 rt : (6,1,-8,-8)
4252 4640 ℓ_{10} : $4-4e/3$
4252 4654 ln : $6\ln 2$
4252 5510 ln : $4e/3-2\pi/3$
4252 5692 rt : (1,5,9,3)
4252 6831 at : $3\sqrt{2}/2-4\sqrt{5}$
4252 8830 id : $\exp(-\sqrt[3]{5}/2)$
4252 9037 id : $1/\sqrt[3]{13}$
4253 0233 rt : (5,2,7,3)
4253 1855 Ei : $3(1/\pi)/8$
4253 2466 cu : $\ln(\sqrt{2}/3)$
4253 2470 rt : (9,1,5,-3)
4253 2540 $1/4\div s\pi(1/5)$
4253 3081 sq : $15/23$
4253 4929 rt : (9,5,8,-5)
4253 5411 sπ : $(\sqrt{2}\pi/3)^{1/3}$
4253 5953 at : $\sqrt{3}+\sqrt{6}+\sqrt{7}$
4253 5994 erf : $1/\sqrt[3]{16}$
4253 6248 J_0 : $(\ln 3/3)^{-1/2}$
4253 6374 ℓ_{10} : $\sqrt{2}/4+4\sqrt{3}/3$
4253 8337 at : $4+2\sqrt{2}$
4253 8489 at : $e/6$
4253 8860 $e+s\pi(1/4)$
4253 9052 rt : (8,6,-4,0)
4253 9372 sinh : $1/\ln(\sqrt[3]{2}/3)$
4253 9404 Γ : $9\sqrt[3]{3}/5$
4254 0373 ln : $\zeta(3)/5$
4254 4578 sr : $3\sqrt{3}/4-\sqrt{5}/2$
4254 5389 at : $4e\pi/5$
4254 6220 cr : $\ln 2/9$
4254 9029 $1/4-s\pi(\sqrt{5})$
4254 9626 2^x : $3\sqrt{2}/2-3\sqrt{5}/2$
4254 9959 rt : (3,9,-7,-9)
4255 0677 cu : $2\sqrt{2}+\sqrt{3}/2$
4255 4092 $1/5+\exp(4/5)$
4255 4395 $\ln(\sqrt{2})\times\exp(\sqrt{2})$
4255 5310 at : $7(e+\pi)/6$
4255 5537 ln : $\sqrt[3]{3}/6$
4255 7771 $\ln(4/5)+\exp(1/2)$
4255 8745 rt : (7,9,9,-6)
4255 9349 sinh : $5(e+\pi)/2$
4256 0962 J_1 : $22/23$
4256 1624 cu : $4/3-3e$
4256 1807 $\pi+\exp(1/4)$
4256 2394 at : $4\sqrt[3]{5}$
4256 2584 sq : $4+4\sqrt{3}$
4256 2819 tanh : $5/11$
4256 3009 rt : (5,9,-8,-1)
4256 4096 at : $3\sqrt{2}+3\sqrt{3}/2$
4256 5686 rt : (9,7,-2,9)
4256 6177 10^x : $1/4+4\sqrt{5}$
4256 7714 rt : (7,5,6,-4)
4257 0469 ζ : $24/25$
4257 1186 2^x : $1/2-\sqrt{3}$
4257 1496 10^x : $2\ln 2/9$
4257 2705 E : $(e+\pi)/10$
4257 3224 rt : (9,7,-1,-1)
4257 3370 sπ : $2\sqrt[3]{5}/3$
4257 4395 rt : (6,9,-9,-5)
4257 4574 Ψ : $\ln(3)$
4257 4608 rt : (3,2,8,-4)
4257 5435 sπ : $\sqrt[3]{2}/9$
4257 5909 rt : (5,2,-1,9)
4257 7799 tanh : $10(1/\pi)/7$
4257 8225 Ψ : $1/\ln(2/3)$
4258 2671 ζ : $23/24$
4258 5615 as : $(e+\pi)^{-1/2}$
4258 6841 sq : $2/3+\sqrt{5}$
4258 7606 $4\times\exp(\sqrt{5})$
4258 7760 rt : (2,8,-4,-3)
4258 8631 $\ln(3/4)\div s\pi(\sqrt{5})$
4258 9488 rt : (5,9,7,-5)
4258 9873 rt : (8,8,9,3)
4259 1175 Li_2 : $1/\sqrt[3]{18}$
4259 1680 rt : (7,1,3,-2)
4259 2815 sinh : $9\sqrt[3]{5}/4$
4259 3026 rt : (6,8,-5,-6)
4259 3817 ln : $4/3+2\sqrt{2}$
4259 4383 cu : $1/2+e/2$
4259 5925 rt : (1,7,4,2)
4259 5937 ζ : $22/23$
4259 6683 rt : (7,5,-5,3)
4259 6873 ℓ_{10} : $3/8$
4259 6910 sq : $7e\pi$
4259 7230 J_1 : $11/4$
4259 8875 rt : (1,6,9,9)
4260 1513 id : $7\ln 2/2$
4260 2204 rt : (1,9,6,1)
4260 2228 sinh : $(\sqrt{5}/3)^3$
4260 2928 sr : $4+4\sqrt{2}/3$
4260 3011 rt : (9,8,2,-3)
4260 4276 rt : (3,5,-8,-6)
4260 4966 Γ : $25/6$
4260 5241 10^x : $2\sqrt{3}/9$
4260 6758 rt : (6,5,-6,-3)
4261 0412 ζ : $21/22$
4261 1890 sπ : $\sqrt{2}/2+\sqrt{3}/4$
4261 2218 J_2 : $(4/3)^3$
4261 2263 $4\div\exp(1/2)$
4261 3169 2^x : $2e\pi/7$
4261 3605 sπ : $(e+\pi)$
4261 5969 rt : (4,9,2,9)
4261 7349 rt : (2,7,-1,-1)
4261 7635 $\exp(\sqrt{3})+s\pi(e)$
4261 8002 sr : $\sqrt{2}+2\sqrt{5}$
4261 8087 rt : (9,8,-9,-3)
4261 8349 rt : (7,9,0,2)
4261 8584 cu : $1/3-3\sqrt{5}/4$
4261 8900 rt : (5,5,4,-3)
4261 9490 rt : (8,3,9,-5)
4262 0806 rt : (3,2,5,2)
4262 0948 e^x : $1/2+\sqrt{5}$
4262 1709 J_1 : $3\sqrt{5}/7$
4262 2334 J_1 : $23/24$
4262 3839 J_0 : $\sqrt[3]{9/2}$
4262 5120 ℓ_{10} : $\sqrt[3]{19}$
4262 6270 ζ : $20/21$
4262 8342 tπ : $8\pi/7$
4263 0897 rt : (8,9,-7,-4)
4263 1593 at : $4-4e$
4263 1651 10^x : $\sqrt{2}-\sqrt{5}+\sqrt{6}$
4263 1751 at : $3\sqrt{3}+3\sqrt{5}/4$
4263 2219 rt : (2,8,-4,-1)
4263 2514 J_1 : $7\pi/8$
4263 3124 $\ln(2/3)\div s\pi(2/5)$
4263 3481 rt : (6,0,-8,6)
4263 5232 sinh : $2\pi^2/9$
4263 5924 erf : $5(1/\pi)/4$
4263 6146 E : $1/\sqrt[3]{5}$
4263 6170 ℓ_{10} : $\sqrt{3}/4+\sqrt{5}$
4263 6824 as : P_{56}
4263 6826 10^x : $17/9$
4264 0143 sr : $2/11$
4264 0742 $\exp(3/5)\times\exp(5)$
4264 0919 cu : $2\sqrt{2}/3-4\sqrt{3}$
4264 1617 Li_2 : $6(1/\pi)/5$
4264 3719 ζ : $19/20$
4264 3992 rt : (1,8,-3,5)
4264 5324 rt : (1,2,6,-3)
4264 8723 sr : $4(1/\pi)/7$
4264 9103 10^x : $2\sqrt{3}+2\sqrt{5}/3$
4264 9125 cr : $2/3+\sqrt{5}$
4265 0045 rt : (9,4,-7,8)
4265 0333 id : $2\sqrt{2}+3\sqrt{3}/2$
4265 0637 $\exp(1/3)-\exp(3/5)$
4265 1403 Ei : $(\ln 3)^{-2}$
4265 2416 sπ : $4\sqrt{2}/3-\sqrt{5}/3$

4265 3335 2^x : $9(e+\pi)$
4265 5770 rt : (1,9,8,-4)
4265 6023 tπ : $3e-2\pi$
4265 8999 rt : (9,4,-1,-1)
4266 0714 rt : (6,4,-3,5)
4266 2387 10^x : $(\pi/3)^{-3}$
4266 2749 at : 5/11
4266 3011 ζ : 18/19
4266 3434 e^x : $2\sqrt[3]{2}$
4266 6087 rt : (7,8,0,-2)
4266 7753 E : $((e+\pi)/2)^{-1/2}$
4266 9274 $\exp(4/5)-\exp(\sqrt{3})$
4266 9805 rt : (5,3,-8,-3)
4267 0033 e^x : $\sqrt{3}/4+\sqrt{5}$
4267 1973 cr : $4/3+\pi/2$
4267 2709 J : 1/3
4267 3652 rt : (9,8,-2,7)
4267 3665 rt : (8,1,6,3)
4267 4309 rt : (1,9,4,-7)
4267 5329 erf : $(e/2)^{-3}$
4267 5750 rt : (6,3,7,-4)
4267 6978 rt : (4,7,8,-5)
4267 7378 tanh : $(\sqrt{2}\pi/3)^{-2}$
4267 7788 rt : (2,5,-7,-6)
4267 7910 at : $10(1/\pi)/7$
4267 8706 J_1 : 24/25
4268 0237 ℓ_2 : $(e/3)^3$
4268 1725 rt : (4,8,-2,-2)
4268 4023 J_0 : $7\sqrt{2}/6$
4268 4335 rt : (8,7,-7,-1)
4268 4455 ζ : 17/18
4268 5005 cr : $\sqrt{2}+2\sqrt{5}/3$
4268 7091 rt : (5,1,-6,-8)
4268 7609 cr : $2\sqrt{3}-\sqrt{5}/4$
4268 8086 ζ : $10\ln 2/3$
4268 8809 rt : (9,8,9,-6)
4268 9253 J_2 : 19/8
4269 1904 rt : (5,3,-9,-4)
4269 2092 sinh : $1/\sqrt[3]{14}$
4269 2234 ℓ_{10} : P_{74}
4269 3916 as : $(\sqrt{5}/3)^3$
4269 5040 $5\div\ln(\sqrt{2})$
4269 6357 e^x : $1/4+2\pi/3$
4269 6725 $s\pi(\sqrt{2})\div s\pi(\sqrt{5})$
4269 8046 rt : (3,8,-1,-9)
4269 8512 sπ : $3\sqrt{3}+4\sqrt{5}$
4269 9049 ln : $4\sqrt{2}-2\sqrt{5}/3$
4270 0736 2^x : $3\sqrt[3]{5}/10$
4270 0898 sinh : $\sqrt{3}-\sqrt{5}+\sqrt{6}$
4270 1621 at : $e+4\pi/3$
4270 1642 rt : (1,9,3,-3)
4270 2571 ln : $2/3+\sqrt{3}/2$
4270 4132 cr : $2+e/3$
4270 5010 at : $7\pi^2/10$
4270 5098 id : $1/4-3\sqrt{5}/4$
4270 6700 10^x : $5\ln 2/9$
4270 7858 as : $1-\sqrt{2}$
4270 8431 ζ : 16/17
4271 0004 rt : (6,8,2,4)
4271 0131 θ_3 : $\ln(\sqrt{5}/3)$
4271 1635 ln : 6/25
4271 2474 cu : $10\sqrt{2}$
4271 2789 rt : (2,0,9,4)
4271 2842 $2/3\times\exp(\pi)$
4271 4685 E : 7/12
4271 7517 rt : (8,6,3,-3)
4271 8432 rt : (2,9,2,6)
4271 9227 Γ : $4\sqrt{2}/9$
4272 0456 ln : $\sqrt{13}\pi$
4272 2895 rt : (1,5,-7,2)
4272 3939 id : $4/3-e/3$
4272 6196 rt : (8,7,0,9)
4272 8027 rt : (4,4,8,3)
4272 9011 rt : (9,3,9,4)
4272 9356 J_1 : $2\sqrt[3]{3}/3$
4272 9730 J_2 : $\exp(\sqrt{3}/2)$
4273 0103 ℓ_{10} : $2\sqrt{2}/3+\sqrt{3}$
4273 0462 rt : (5,1,1,-1)
4273 0492 10^x : $\exp(\sqrt[3]{3}/3)$
4273 1851 $\exp(1/2)-\exp(1/5)$
4273 2300 rt : (5,3,-1,7)
4273 2931 sinh : $9\sqrt[3]{2}/7$
4273 4276 J_1 : $4\zeta(3)/5$
4273 5291 ℓ_2 : $4/3+3\pi$
4273 5416 ζ : 15/16
4273 5961 cu : $2e/3+3\pi/4$
4273 6142 Li_2 : $(\sqrt{5}/2)^{-1/2}$
4273 6909 $\ln(\sqrt{2})-s\pi(e)$
4273 7291 rt : (6,6,-6,1)
4273 7488 erf : $(2\pi)^{-1/2}$
4273 8992 sinh : $4\sqrt[3]{3}/5$
4273 9363 ℓ_{10} : $4\sqrt{2}-4\sqrt{5}/3$
4274 0589 sq : $4/3-e/4$
4274 1807 rt : (7,8,7,-5)
4274 2715 2^x : $\exp(-2/3)$
4274 3506 id : $4\sqrt{3}/3+\sqrt{5}/2$
4274 4401 ln : 15/23
4274 4542 2^x : $(2\pi/3)^{1/3}$
4274 4875 as : $4\sqrt{3}/7$
4274 5137 ζ : $4\sqrt{3}/3$
4274 5792 rt : (2,7,6,-4)
4274 7096 sinh : 15/13
4274 7158 J_2 : $7e/8$
4274 9398 id : $\sqrt[3]{5}/4$
4275 1170 $4\div\exp(\sqrt{5})$
4275 1543 at : $10\ln 2$
4275 2545 sπ : $2e/3+\pi/3$
4275 2625 J_0 : $(e)^{1/2}$
4275 2804 ln : $2e/3+3\pi/4$
4275 3579 rt : (9,5,-7,6)
4275 4094 rt : (3,6,-5,-3)
4275 4683 J_1 : $5\sqrt{3}/9$
4275 5714 rt : (9,4,6,-4)
4275 6168 Γ : $3\sqrt{3}/2$
4275 7318 e^x : $e/3+\pi/4$
4275 8508 rt : (7,6,-5,1)
4275 9485 Ψ : $(5/2)^{-1/2}$
4276 1321 rt : (1,6,-5,-8)
4276 1917 at : $3/2+2e$
4276 2473 sq : $3\sqrt{2}/2+3\sqrt{5}/4$
4276 4653 Γ : $\pi/5$
4276 5664 ℓ_{10} : $1+3\sqrt{5}/4$
4276 6015 ζ : 14/15
4276 6496 sq : $4\sqrt{2}/3-4\sqrt{3}$
4276 6832 tπ : $4e/3$
4276 6934 rt : (6,1,-8,4)
4276 7603 rt : (4,3,5,-3)
4276 8461 $5\times\exp(3)$
4277 0864 J : $\pi/9$
4277 1283 J : $\ln(e/2)$
4277 1712 rt : (4,8,-9,-5)
4277 2618 rt : (6,8,7,2)
4277 2843 id : $1/3+2\pi/3$
4277 3729 Ei : $\sqrt[3]{11/3}$
4277 4715 sπ : $\exp(\pi)$
4277 6208 e^x : $2e+\pi/3$
4277 7594 at : $2-4\sqrt{5}$
4277 7664 J_0 : $8\sqrt[3]{3}/7$
4277 8499 Ei : $7(e+\pi)/3$
4277 8704 at : $(\sqrt{2}\pi/3)^{-2}$
4277 9042 rt : (3,5,9,-5)
4277 9835 J_2 : $\sqrt{17/3}$
4278 0005 $\ln(\sqrt{3})\div\exp(1/4)$
4278 0432 rt : (5,8,-2,-1)
4278 2031 rt : (8,5,-4,-2)
4278 2269 tπ : $1/4-3\sqrt{2}/2$
4278 3450 sr : $3e/4$
4278 4747 as : $1/\sqrt[3]{14}$
4278 5568 rt : (9,9,-9,-5)
4278 7887 rt : (1,6,2,-6)
4278 8625 rt : (4,2,-6,-1)
4278 8922 at : $2/3+2\pi$
4279 2096 rt : (3,6,-6,-7)
4279 2668 rt : (9,7,-8,-4)
4279 3518 at : $1/\ln(\sqrt{3}/2)$
4279 3543 10^x : $10\sqrt[3]{5}/9$
4279 5630 sr : $3\sqrt{3}/2-\sqrt{5}/4$
4279 6547 J_1 : $(\sqrt{5}/2)^{-1/3}$
4279 8130 sinh : $3\ln 2/5$
4280 0673 rt : (2,4,-8,-4)
4280 0847 $1/3\times\exp(1/4)$
4280 1005 ζ : 13/14
4280 1163 at : $4\sqrt{2}+3\sqrt{3}/4$
4280 1706 $\exp(1/5)\times\exp(\sqrt{5})$
4280 3376 rt : (8,4,-5,8)
4280 3886 rt : (9,9,-2,5)
4280 5843 at : $1/4+3\sqrt{5}$
4280 6154 sq : $3\sqrt{2}/4-4\sqrt{3}$
4280 6518 ℓ_{10} : $2+e/4$
4280 6913 rt : (5,1,8,-4)
4280 7081 tπ : $2\sqrt{2}+\sqrt{3}/2$
4281 1017 e^x : $8e\pi/5$
4281 1913 rt : (6,1,4,2)
4281 2846 sπ : $-\sqrt{5}+\sqrt{6}+\sqrt{7}$
4281 2899 rt : (4,2,1,9)
4281 3374 J_2 : $3\pi^2/8$
4281 4820 cu : $3\sqrt{3}/4+4\sqrt{5}/3$
4281 5048 at : $e/4+2\pi$
4281 5870 Ei : $(\sqrt[3]{5}/3)^{1/3}$
4281 7222 rt : (5,8,5,-4)
4281 7849 $3\div\ln(3/4)$
4281 8945 ℓ_{10} : $3\sqrt{2}/2+\sqrt{5}/4$
4282 0323 id : $1/2+4\sqrt{3}$
4282 0577 ln : $\sqrt{2}/2+2\sqrt{3}$
4282 1416 rt : (1,9,-1,-2)
4282 1592 sq : $4-3\pi$
4282 1791 sr : $\sqrt{5}+\sqrt{6}-\sqrt{7}$
4282 1977 sπ : $1/2+e/2$
4282 3147 ℓ_2 : $\sqrt{2}+\sqrt{3}+\sqrt{5}$
4282 5301 $\exp(\sqrt{3})\div\exp(1/2)$
4282 6274 rt : (6,6,1,-2)
4282 7363 ln : $9\sqrt[3]{2}$
4282 7762 rt : (4,7,-7,-1)
4282 9331 rt : (3,8,-9,-4)
4283 3446 rt : (5,0,-6,6)
4283 4936 Γ : $\exp(-4/3)$
4283 6913 cr : $3/2+\sqrt{2}$
4283 7470 $\ln(3/4)-\exp(\pi)$
4283 7913 rt : (1,9,-8,-5)
4283 8019 e^x : $1/3+2\sqrt{3}/3$
4283 9235 erf : 2/5
4283 9416 rt : (6,6,8,-5)
4283 9713 ζ : $8\sqrt[3]{3}/5$
4284 1406 ζ : 12/13
4284 2150 rt : (6,5,-3,3)
4284 2505 sq : $3\sqrt{3}/2+3\sqrt{5}/2$
4284 2698 K : 13/14
4284 3129 rt : (5,8,-8,5)
4284 3974 sr : $4/3+\sqrt{2}/2$
4284 7062 θ_3 : $2(1/\pi)/3$
4285 0352 sr : P_{63}
4285 1498 10^x : $3/4-\sqrt{5}/2$
4285 2203 sr : $\sqrt{14}\pi$
4285 2678 at : $4+4\sqrt{5}/3$
4285 7142 id : 3/7
4285 7535 sr : $\sqrt[3]{17/2}$
4285 9809 $\pi\div\ln(2/5)$
4286 0625 Ψ : $\pi^2/9$
4286 3279 sq : $1/2-2\sqrt{3}/3$
4286 3948 rt : (1,6,-4,-8)
4286 6624 ℓ_{10} : $\sqrt{5}/6$
4286 7206 10^x : $\exp(\sqrt[3]{2}/2)$
4286 8302 rt : (8,8,0,7)
4286 8336 10^x : $4\sqrt{2}/7$
4286 9373 rt : (2,7,-8,1)
4286 9427 sq : $6e/7$
4287 0275 at : $4/3+4\sqrt{2}$
4287 0388 Ei : $\sqrt[3]{17}$
4287 2385 rt : (8,8,-7,-3)
4287 2485 tanh : 11/24
4287 4818 at : $1/4-\sqrt{2}/2$
4287 5222 rt : (8,9,4,-4)
4287 5424 2^x : $10(e+\pi)$
4287 5530 $\ln(\pi)+\exp(1/4)$
4287 5973 rt : (9,6,-7,4)
4287 8195 sπ : P_{70}
4287 8385 e^x : $\sqrt{2}-\sqrt{3}+\sqrt{6}$
4287 8973 ln : $3/4+\pi/4$
4287 9223 sπ : P_{67}
4287 9349 e^x : 6

4287 9355 rt : (8,2,7,-4)
4287 9874 sr : P_{47}
4288 0400 id : $9e\pi/2$
4288 1770 rt : (1,5,7,-4)
4288 2808 sinh : 5/12
4288 3636 e^x : $3/2+3\sqrt{3}/4$
4288 4774 rt : (6,8,-7,-4)
4288 6133 e^x : $2e-2\pi$
4288 6627 ℓ_2 : $(2e/3)^{1/2}$
4288 6857 e^x : $1/\sqrt[3]{22}$
4288 8194 id : $(2e)^{-1/2}$
4288 8576 ζ : 11/12
4288 9080 rt : (8,9,-3,-1)
4288 9599 rt : (9,7,7,-5)
4288 9927 at : 7
4289 1093 sinh : $8\sqrt{5}$
4289 1665 10^x : $e/4-\pi/3$
4289 1937 as : $3\ln 2/5$
4289 2104 rt : (2,4,-7,-4)
4289 3254 ℓ_2 : $2/3+e/4$
4289 4333 cu : $(\ln 3)^{-3}$
4289 6058 sinh : $2\sqrt{3}/3$
4289 6814 tπ : $2\sqrt{2}/3+4\sqrt{3}$
4289 7049 Ψ : $6\pi^2/5$
4289 7593 2^x : 16/9
4289 7736 tπ : $4\sqrt{2}+2\sqrt{5}$
4290 2916 Ei : $6(1/\pi)/7$
4290 4441 ln : $\sqrt{3}+\sqrt{6}-\sqrt{7}$
4290 4632 at : $2e+\pi/2$
4290 4683 rt : (1,9,-5,-9)
4290 4716 rt : (8,2,0,-1)
4290 5143 erf : $\zeta(3)/3$
4290 5286 sπ : $2\sqrt{3}+3\sqrt{5}/4$
4290 7073 rt : (4,1,-4,-8)
4290 7517 cu : $(e/3)^{-3}$
4290 9012 $\ln(4/5)+\exp(\sqrt{3})$
4290 9639 ln : $(e+\pi)/9$
4291 0300 rt : (9,7,0,-2)
4291 0370 ℓ_2 : $3\pi/7$
4291 5661 rt : (7,0,-8,9)
4291 5980 2^x : $3\zeta(3)/7$
4291 7122 $\ln(3/4)\times\exp(2/5)$
4291 7372 rt : (1,5,0,-1)
4291 8745 J : 2/13
4292 0367 id : $2-\pi/2$
4292 3457 sπ : $\sqrt{3}/4+3\sqrt{5}$
4292 4003 e^x : 5/14
4292 7472 $\pi-\ln(3/4)$
4292 8015 10^x : $1/4+3\sqrt{2}$
4293 0325 rt : (9,0,3,2)
4293 1710 rt : (3,9,-5,-2)
4293 3071 rt : (6,2,-8,2)
4293 3636 Γ : $\ln(\sqrt[3]{3})$
4293 5170 sq : $1+2e$
4293 5628 rt : (8,5,-5,6)
4293 6781 J_1 : $(e/3)^{1/3}$
4293 7248 rt : (5,4,-1,5)
4293 9182 rt : (7,7,2,9)
4293 9239 rt : (6,8,0,-1)
4293 9384 at : $\sqrt{5}\pi$
4294 2682 10^x : $3/4+\sqrt{5}$
4294 2805 2^x : $5e\pi/8$
4294 4373 ζ : 10/11
4294 4453 rt : (6,9,9,-6)
4294 5174 Ei : 18/7
4294 7358 rt : (5,9,-6,4)
4294 7384 E : 11/19
4294 7563 10^x : $9\sqrt[3]{5}/2$
4294 9285 Ψ : 14/3
4295 0366 rt : (8,5,3,1)
4295 3085 rt : (5,4,9,-5)
4295 3343 at : $6(e+\pi)/5$
4295 3398 rt : (2,5,-9,-9)
4295 3791 rt : (3,9,-6,-4)
4295 5779 Ei : 3/11
4295 5990 at : $3/4+2\pi$
4295 7045 id : $1/4-e/4$
4295 8877 J_1 : $\sqrt{15/2}$
4295 9484 2^x : $2/3-4\sqrt{2}/3$
4295 9857 rt : (3,1,6,-3)
4296 1183 rt : (2,8,-2,8)
4296 1995 rt : (2,7,-9,-5)
4296 2455 Γ : 13/5
4296 4353 J : $4(1/\pi)/9$
4296 5117 tπ : $\sqrt{7/2}$
4296 5377 rt : (5,6,-3,-2)
4296 6236 sr : $\sqrt{2}/2+3\sqrt{3}$
4296 7604 ℓ_{10} : $1/3+3\pi/4$
4296 8558 rt : (1,7,-4,-6)
4296 9799 rt : (4,1,9,4)
4297 0045 J_0 : $\pi^2/6$
4297 1260 rt : (7,7,-5,-1)
4297 2054 sπ : $5\zeta(3)/7$
4297 2521 at : $1/3+3\sqrt{5}$
4297 4327 sinh : $7\sqrt[3]{2}/3$
4297 4593 rt : (9,7,6,2)
4297 4986 Li_2 : 5/13
4297 5177 e^x : $4+2\sqrt{5}/3$
4297 5547 λ : $\exp(-16\pi/15)$
4297 6038 cu : $3/2-\sqrt{5}/3$
4297 6169 rt : (4,8,-5,-2)
4297 6227 at : 11/24
4297 7543 as : 5/12
4297 8231 rt : (4,4,-6,-3)
4297 8658 Γ : $3e\pi/8$
4297 8985 tπ : $(\pi/3)^{-3}$
4297 9008 sq : $3/2+\pi/2$
4297 9427 2^x : $3/2-e$
4297 9685 tπ : $1/4-4\sqrt{5}$
4298 1537 sπ : $\sqrt{2}/10$
4298 1969 Ψ : $3(1/\pi)/4$
4298 2310 sr : $3\sqrt{3}/4+\sqrt{5}/3$
4298 5727 cu : $3/4-2\pi/3$
4298 5877 rt : (3,2,-9,-4)
4298 6146 sinh : $2\sqrt{2}$
4298 6185 J_1 : $(\ln 3)^{-1/3}$
4298 7223 rt : (5,4,-8,-5)
4298 8674 at : $5\pi^2/7$
4298 9056 rt : (3,7,3,9)
4298 9254 2^x : $-\sqrt{2}-\sqrt{6}+\sqrt{7}$
4299 1092 sinh : $5\ln 2/3$
4299 1117 ℓ_2 : $3\sqrt{3}/7$
4299 1260 rt : (6,9,2,2)
4299 1878 $3/5\div\exp(1/3)$
4299 2900 at : $\exp((e+\pi)/3)$
4299 4302 rt : (7,7,5,-4)
4299 5636 sπ : $4(1/\pi)/9$
4299 6678 rt : (8,5,8,-5)
4299 6914 sq : $(\sqrt[3]{5})^{1/3}$
4299 6996 rt : (4,5,-9,6)
4300 0000 sq : $9\sqrt{3}/10$
4300 0479 J_2 : $7\sqrt[3]{5}/5$
4300 0529 rt : (7,4,-3,8)
4300 1296 $\exp(3/4)\times s\pi(\sqrt{5})$
4300 1844 10^x : $\ln(\ln 2)$
4300 3413 rt : (9,7,-7,1)
4300 4042 cu : $8\sqrt{3}$
4300 5181 id : $3\sqrt{2}/4-2\sqrt{5}/3$
4300 6279 at : $4\sqrt{2}/3-4\sqrt{5}$
4300 8376 rt : (1,8,8,-5)
4301 0940 Li_2 : $2\sqrt{3}/9$
4301 1401 ζ : 9/10
4301 1595 2^x : $3-\sqrt{5}/4$
4301 2383 at : $\sqrt{2}/4+3\sqrt{5}$
4301 2700 cr : $(1/\pi)/4$
4301 5155 ℓ_{10} : $4/3+e/2$
4301 5970 rt : (3,6,9,3)
4301 7197 E : $5\ln 2/6$
4301 7715 rt : (5,1,-6,4)
4302 0308 ℓ_{10} : $6\pi/7$
4302 2826 rt : (9,7,-7,2)
4302 3446 e^x : $4\sqrt{2}-4\sqrt{3}/3$
4302 5789 at : $9\pi/4$
4302 6589 J_0 : $(\sqrt{2}\pi)^{1/3}$
4302 8681 J_1 : $7\ln 2/5$
4302 8984 at : $3\sqrt{3}/2+2\sqrt{5}$
4302 9035 rt : (7,0,8,4)
4303 0121 sπ : π
4303 0662 at : $5\sqrt{2}$
4303 1418 rt : (6,2,5,-3)
4303 1525 E : $\sqrt{3}/3$
4303 3148 sr : $(\sqrt[3]{5}/3)^3$
4303 3868 Li_2 : $5\ln 2/9$
4303 4866 rt : (5,4,2,-2)
4303 8602 E : γ
4304 0020 rt : (2,9,-9,6)
4304 0207 sr : $1/4+4\sqrt{2}$
4304 0292 rt : (4,6,-1,-1)
4304 2602 J_1 : $8\sqrt[3]{5}/5$
4304 3112 at : $3+3e/2$
4304 3744 10^x : $1-\sqrt{2}/4$
4304 4730 rt : (8,9,0,5)
4304 5563 e^x : $7\pi^2/6$
4304 5926 sq : $1/4-e/3$
4304 6400 rt : (4,9,7,-5)
4304 8138 rt : (7,1,-8,7)
4305 0087 rt : (3,8,-4,0)
4305 0142 10^x : $1/2-\sqrt{3}/2$
4305 1234 Ei : $(\sqrt[3]{5})^{1/3}$
4305 1617 at : $4\sqrt{2}/3+3\sqrt{3}$
4305 2259 Ψ : 12/19
4305 3398 sq : $1+\sqrt{5}/4$
4305 3548 θ_3 : $1/\sqrt{22}$
4305 5199 E : $2\sqrt[3]{3}/5$
4305 7239 rt : (1,8,8,-2)
4305 7780 sq : $3+3e$
4305 9975 $e+\ln(3/4)$
4306 0136 Ψ : 23/21
4306 0670 e^x : $9(1/\pi)/8$
4306 2105 tanh : $(3\pi/2)^{-1/2}$
4306 3920 Γ : $(5)^{-3}$
4306 5015 rt : (4,3,-9,3)
4306 7080 id : $e+3\pi/2$
4306 7655 $\ln(2)\div\ln(5)$
4306 8184 rt : (2,7,-2,-2)
4306 9511 $1/3\div s\pi(e)$
4307 0241 sπ : $\sqrt{3}-\sqrt{5}+\sqrt{7}$
4307 2142 id : $3\sqrt{2}/2+4\sqrt{3}/3$
4307 2520 rt : (8,8,9,-6)
4307 5521 as : $10\ln 2/7$
4307 7080 rt : (7,6,-8,-4)
4307 8061 sq : $3-4\sqrt{3}$
4307 8291 ln : 13/20
4307 8975 Ψ : $9\sqrt{5}/10$
4307 9761 e^x : $3e+\pi/2$
4308 0311 sr : $1+\pi/3$
4308 0964 ℓ_2 : $e/2+3\pi$
4308 3091 $\exp(1/5)-\exp(\sqrt{3})$
4308 4051 rt : (9,3,4,-3)
4308 4956 rt : (6,6,-3,1)
4308 7021 rt : (3,9,1,-1)
4308 8248 $\ln(\sqrt{5})\times\exp(5)$
4308 8693 cr : 2/25
4308 9231 J_0 : 23/14
4309 0284 tπ : $7\sqrt[3]{2}/2$
4309 0769 sπ : $\exp(-(e+\pi)/3)$
4309 1502 10^x : $(\sqrt[3]{2})^{-1/3}$
4309 2147 rt : (8,8,-5,-3)
4309 3037 e^x : $5\pi/7$
4309 3092 tπ : $(\sqrt{5})^{1/2}$
4309 3346 id : $((e+\pi)/2)^{1/3}$
4309 3431 ζ : 8/9
4309 5139 Ei : $(\sqrt[3]{2}/3)^{-1/2}$
4309 5543 rt : (8,6,-5,4)
4309 6837 rt : (8,5,1,-2)
4309 8004 J_2 : 12/5
4309 8077 rt : (3,4,7,-4)
4310 2026 Ψ : $(1/(3e))^{-1/3}$
4310 2026 at : $3e-\pi/3$
4310 2115 cu : $\sqrt{2}/4+2\sqrt{3}/3$
4310 3230 rt : (5,1,-6,2)
4310 6071 rt : (6,5,8,3)
4310 9329 Ei : 17/8
4311 1878 ln : $9\sqrt[3]{5}/10$
4311 2714 rt : (6,0,-6,9)
4311 3669 sr : $2\sqrt{2}+4\sqrt{5}$
4311 4987 ℓ_{10} : $3\sqrt{2}/2+\sqrt{3}/3$
4311 5522 rt : (7,7,-2,-1)
4311 7310 cr : $1/3+3\sqrt{3}/2$
4311 9081 at : $5e\pi/6$

4312 1150 rt : (7,3,9,-5)
4312 4892 rt : (8,9,-7,-5)
4312 5331 ln : $1/2+4e$
4312 6042 rt : (4,6,-7,5)
4312 7069 rt : (4,7,-7,-4)
4312 7642 e^x : $2e/3+\pi/4$
4312 8357 2^x : $4-e$
4312 8417 Ei : $\exp(-1/\sqrt{2})$
4312 8720 10^x : $3\sqrt{3}/4+3\sqrt{5}/4$
4312 9987 sπ : $3\sqrt{3}/4+\sqrt{5}/4$
4313 1505 Ψ : 10π
4313 1647 rt : (2,6,-5,-7)
4313 3690 rt : (1,8,1,-2)
4313 3726 tanh : 6/13
4313 4665 rt : (7,8,2,7)
4313 4902 $\sqrt{5}-\ln(\sqrt{5})$
4313 5268 rt : (6,5,6,-4)
4313 6122 sq : $4+2\sqrt{2}/3$
4313 6318 $\sqrt{2}\times\exp(e)$
4313 6376 ℓ_{10} : $(3)^3$
4313 7529 rt : (1,8,9,-7)
4313 7834 rt : (5,3,5,2)
4314 1265 $\pi\times s\pi(e)$
4314 4615 sr : $3/4+3\sqrt{3}/4$
4314 4906 at : $2e-4\pi$
4314 5003 sq : $2\sqrt{3}+3\sqrt{5}/4$
4314 5390 rt : (6,9,-5,0)
4314 6207 tπ : $8(e+\pi)/5$
4314 9771 rt : (5,9,-4,-3)
4315 1845 ℓ_{10} : $4-3\sqrt{3}/4$
4315 2310 ln : $3\sqrt{3}/8$
4315 3755 rt : (5,8,4,4)
4315 4568 2^x : $2e\pi$
4315 5267 rt : (3,8,-7,3)
4315 5671 $\ln(5)+\exp(3/5)$
4315 7384 e^x : $3\sqrt{2}-3\sqrt{5}/2$
4315 7477 rt : (9,5,-5,9)
4315 8240 J_2 : $5\sqrt[3]{3}/3$
4315 9298 rt : (4,9,-3,-3)
4316 2711 rt : (1,2,-8,3)
4316 3724 cu : $2\sqrt{2}/3+4\sqrt{5}/3$
4316 3891 rt : (9,4,7,3)
4316 4082 J_2 : $2\zeta(3)$
4316 4611 rt : (1,8,-6,-4)
4316 4919 rt : (9,6,5,-4)
4316 5493 J_2 : $\sqrt{2}-\sqrt{6}-\sqrt{7}$
4316 5594 ln : $5(e+\pi)/7$
4316 6884 at : $\sqrt{3}/4+3\sqrt{5}$
4316 7059 θ_3 : $\sqrt[3]{5}/8$
4316 7268 cr : $2e/3+4\pi$
4316 7607 at : $4+\pi$
4316 8240 at : $(3\pi/2)^{-1/2}$
4317 0989 $\exp(1/3)+s\pi(\sqrt{2})$
4317 1784 at : $4\sqrt{2}-3\sqrt{3}$
4317 3082 rt : (3,7,8,-5)
4317 3171 sq : $3/4-4\sqrt{3}/3$
4317 3324 sq : $\exp(-\sqrt[3]{2}/3)$
4317 4604 rt : (7,9,-9,-5)
4317 5385 rt : (7,5,-3,6)
4317 5792 2^x : $8\sqrt[3]{3}/9$
4317 9003 rt : (4,9,0,-2)
4317 9085 at : $4\sqrt{2}+2\sqrt{5}/3$
4317 9269 tanh : $2\ln 2/3$
4318 1948 rt : (1,6,-2,5)
4318 3075 tπ : $\sqrt{2}/2-\sqrt{3}/3$
4318 4038 sπ : $10\sqrt{2}$
4318 5326 rt : (8,8,2,-3)
4318 7824 2^x : $2\pi^2/5$
4319 3046 at : $1-3e$
4319 3416 J : $7\ln 2/10$
4319 3816 rt : (2,9,5,-4)
4319 4562 $1/3+\ln(3)$
4319 5231 tπ : $2\sqrt{2}/3+\sqrt{3}/4$
4319 5822 rt : (6,8,7,-5)
4319 6132 ζ : 7/8
4319 6893 at : $3/2+4\sqrt{2}$
4320 0000 cu : $3\sqrt[3]{2}/5$
4320 0164 e^x : 22/13
4320 1880 tπ : $2e-4\pi$
4320 2160 rt : (9,8,-7,0)
4320 4080 rt : (8,2,4,2)
4320 4429 rt : (1,7,-2,9)
4320 5217 cu : $4/3-\sqrt{3}/3$
4320 6111 rt : (7,2,-8,5)
4320 7665 J_1 : $1/\ln(\sqrt[3]{3})$
4321 0754 ℓ_{10} : $9\zeta(3)/4$
4321 0830 tπ : $(\sqrt[3]{3})^{1/3}$
4321 1051 $\ln(\sqrt{2})+\exp(3)$
4321 2510 sπ : $(e/2)^{-1/2}$
4321 2537 cu : $3\sqrt{3}/2-3\sqrt{5}/2$
4321 4179 $\exp(2/5)\times\exp(\sqrt{3})$
4321 4283 rt : (6,2,9,4)
4321 4825 rt : (9,9,6,-5)
4321 5883 ℓ_2 : $(\ln 3/2)^{1/2}$
4321 6302 $\exp(\sqrt{3})\times s\pi(\pi)$
4321 7174 rt : (5,5,-1,3)
4321 8006 rt : (1,8,-4,-3)
4321 9030 rt : (2,1,7,3)
4322 0159 rt : (3,8,5,8)
4322 1791 10^x : $2/3+3e/4$
4322 2040 id : $3\sqrt{3}+\sqrt{5}$
4322 2796 ℓ_{10} : $2/3+3e/4$
4322 4889 ℓ_2 : $3\sqrt{2}+2\sqrt{3}/3$
4322 9083 erf : $2\sqrt{2}/7$
4322 9171 2^x : $2e/3+\pi/2$
4322 9338 sq : $9\ln 2/4$
4323 2044 rt : (2,6,1,-5)
4323 2779 erf : $\exp(-e/3)$
4323 2954 id : $\sqrt{5}-\sqrt{6}+\sqrt{7}$
4323 5489 rt : (6,7,4,9)
4323 6883 J_1 : $(e+\pi)/6$
4323 7656 at : $1/4+4\sqrt{3}$
4323 8363 Ψ : $2(1/\pi)/7$
4323 9448 id : $9(1/\pi)/2$
4323 9563 Li_2 : 18/19
4324 0331 at : $\sqrt{2}/3+3\sqrt{5}$
4324 0777 at : 6/13
4324 1195 ln : $4\pi/3$
4324 1298 rt : (9,1,8,4)
4324 1779 ζ : 23/10
4324 2545 e^x : 8/9
4324 2869 sinh : $\sqrt[3]{2}/3$
4324 3035 sinh : $8(e+\pi)/9$
4324 6251 cr : $4-3\sqrt{2}/4$
4324 6951 tπ : $3\sqrt[3]{5}$
4324 7029 sr : $6\sqrt[3]{5}/5$
4324 8363 2^x : $1+3\pi/2$
4325 0538 ℓ_{10} : $2+\sqrt{2}/2$
4325 2456 rt : (3,5,-6,-1)
4325 2960 J_1 : $4\sqrt[3]{5}/7$
4325 3619 rt : (8,1,-8,-3)
4325 3982 cr : $7\sqrt[3]{2}/3$
4325 4200 tπ : $(\ln 2)^{-1/3}$
4325 5484 J_1 : $(\pi/3)^{-1/2}$
4325 5574 rt : (5,0,6,3)
4325 6405 ℓ_{10} : $\sqrt{2}/3+\sqrt{5}$
4325 6611 rt : (5,2,-6,2)
4325 6668 sπ : $9e\pi$
4325 6916 10^x : $10(e+\pi)/9$
4325 7107 2^x : $3\sqrt[3]{5}/4$
4325 7220 sq : $3\sqrt{3}/2-\sqrt{5}/3$
4325 7783 at : $3+4\pi/3$
4325 8711 at : $e/3+2\pi$
4326 0304 J_2 : $\sqrt[3]{14}$
4326 1317 E : $9(1/\pi)/5$
4326 2888 rt : (7,8,-5,-3)
4326 3486 rt : (6,4,-1,8)
4326 4462 rt : (1,1,-9,-4)
4326 5071 ℓ_{10} : $\sqrt{22/3}$
4326 7487 id : $3\sqrt[3]{3}/10$
4326 7849 λ : $(1/\pi)/9$
4326 8852 cr : $\sqrt{21}\pi$
4327 1744 at : $2+3\sqrt{3}$
4327 1773 10^x : $7\ln 2/3$
4327 4079 rt : (7,0,1,1)
4327 4920 rt : (3,9,-3,5)
4327 6869 sq : $7\sqrt[3]{5}/10$
4327 7964 rt : (4,7,-5,4)
4327 9032 θ_3 : 3/14
4327 9867 J_1 : $1/\ln(\ln 2)$
4327 9969 sinh : $7\sqrt[3]{5}/5$
4328 0851 $3/5\div\ln(4)$
4328 1175 rt : (8,1,5,-3)
4328 2347 10^x : 19/10
4328 2509 $\ln(2/5)\div\exp(3/4)$
4328 3392 rt : (9,2,9,-5)
4328 4124 e^x : P_{43}
4328 4304 rt : (6,1,-6,7)
4328 6239 cu : $3(e+\pi)/10$
4328 6905 at : $2\ln 2/3$
4328 7625 rt : (2,2,8,-4)
4329 0130 10^x : $\exp(\pi/2)$
4329 0644 e^x : $1/3+e/2$
4329 1747 rt : (9,9,0,8)
4329 2045 tπ : $4e/3+\pi/4$
4329 2236 rt : (5,9,-4,9)
4329 2671 rt : (8,7,-5,2)
4329 4538 at : $1/2+3\sqrt{5}$
4329 5940 ℓ_2 : $(\sqrt[3]{5}/3)^3$
4329 7122 Ψ : $8\sqrt[3]{2}/5$
4329 8752 2^x : $2\sqrt{2}-4\sqrt{3}/3$
4329 8913 rt : (5,2,-7,-3)
4330 0284 at : $5\sqrt[3]{3}$
4330 0363 cu : $3\sqrt{3}/2-3\sqrt{5}$
4330 1270 id : $\sqrt{3}/4$
4330 2120 at : $\sqrt{3}-4\sqrt{5}$
4330 2347 at : $6\zeta(3)$
4330 2585 cr : $2+2\sqrt{2}/3$
4330 3155 sr : $4e/3-\pi/2$
4330 4164 sr : $3\sqrt{2}+3\sqrt{5}/4$
4330 4421 cu : $2/3-4\sqrt{3}/3$
4330 4990 rt : (5,3,7,-4)
4330 6262 $2\div\exp(1/3)$
4330 6274 rt : (9,6,-5,7)
4330 6456 cu : $1/2-\pi$
4330 7375 id : P_{10}
4330 8352 sr : P_9
4330 8521 cr : $\sqrt{2}/2+\sqrt{5}$
4330 9432 ln : $\sqrt[3]{11/3}$
4331 0841 tπ : $\exp(-\sqrt{2}/3)$
4331 3297 sinh : $\exp(-\sqrt{3}/2)$
4331 4757 10^x : $(2e/3)^3$
4331 4924 at : $3e/2+\pi$
4331 5465 cr : $3\sqrt{2}-3\sqrt{3}/4$
4331 6142 rt : (1,6,4,5)
4331 8487 rt : (8,4,6,-4)
4332 1093 θ_3 : 10/17
4332 1698 id : $5\ln 2/8$
4332 2062 rt : (2,5,9,-5)
4332 4130 sq : P_{19}
4332 4330 sq : P_{58}
4332 4423 at : $3\sqrt{2}+4\sqrt{5}/3$
4332 5403 rt : (2,9,6,-4)
4332 8444 ζ : 6/7
4332 9708 e^x : $2\sqrt[3]{2}/7$
4333 0083 ζ : $\ln(3\pi/2)$
4333 0324 $\exp(3)-\exp(\sqrt{3})$
4333 1472 rt : (8,4,-9,-4)
4333 2177 rt : (5,6,8,-5)
4333 2941 e^x : 9/25
4333 4013 rt : (3,7,-9,8)
4333 4138 2^x : $3\sqrt{3}/4+4\sqrt{5}/3$
4333 6573 J_2 : $(\sqrt{5}/3)^{-3}$
4333 7295 sr : $2-2e/3$
4333 8101 rt : (7,3,2,-2)
4334 0386 cu : $e/2+4\pi$
4334 0498 rt : (8,7,7,-5)
4334 0763 E : 4/7
4334 1632 as : $\sqrt[3]{2}/3$
4334 2766 rt : (1,5,8,9)
4334 3365 $3/5\times\exp(2)$
4334 6255 J_1 : $7\sqrt[3]{2}/9$
4334 6307 erf : $(\pi/2)^{-2}$
4334 6720 sr : $3\pi^2/5$
4334 6774 tanh : $1/\sqrt[3]{10}$
4334 6954 rt : (5,5,-8,-4)
4334 7050 sq : $(5/3)^3$
4334 7328 ℓ_2 : $4-3\sqrt{3}/4$
4334 9007 ℓ_{10} : $\sqrt{2}+3\sqrt{3}/4$
4334 9379 rt : (5,9,9,-6)

4335 0467 $t\pi : \sqrt{19}\pi$
4335 1970 $e^x : 1 + \sqrt{5}$
4335 2102 id : $2\sqrt{2}/3 + 2\sqrt{5}/3$
4335 2519 rt : (3,3,-8,6)
4335 2897 $2^x : \sqrt{2}/3 + 3\sqrt{5}/4$
4335 4472 $\zeta : \exp(-1/\sqrt{2})$
4335 5019 rt : (4,7,6,6)
4335 5109 rt : (1,8,3,-9)
4335 6584 rt : (2,7,-1,-5)
4335 7056 $J_2 : 8e/9$
4335 7287 $2^x : 3\sqrt{3}/10$
4335 7501 $2^x : \sqrt{2}/3 - 3\sqrt{5}/4$
4335 8082 rt : (2,6,-7,-4)
4335 8585 $\exp(\pi) - s\pi(1/4)$
4335 9261 at : $3 + 3\sqrt{2}$
4335 9748 $s\pi : (\sqrt[3]{4})^{-1/3}$
4336 0445 sinh : 8/19
4336 1597 $\Gamma : 10\sqrt[3]{3}$
4336 1723 id : $3e/2 + 3\pi/4$
4336 2821 sq : $7\pi/6$
4336 3575 $erf : \ln(2/3)$
4336 4126 rt : (7,6,3,-3)
4336 4310 ln : $4\sqrt{3} + 2\sqrt{5}$
4336 5014 $erf : 1/\sqrt[3]{15}$
4336 5556 $\ell_{10} : 7/19$
4336 6462 rt : (1,7,6,-4)
4336 7666 $\ell_{10} : \sqrt[3]{20}$
4336 8627 rt : (5,8,-9,-5)
4336 9922 rt : (5,0,-4,9)
4337 0684 at : $8e/3$
4337 2296 $e^x : 8\sqrt{2}/3$
4337 3001 rt : (4,8,5,-4)
4337 4721 rt : (2,9,-8,-5)
4337 4749 $Ei : 7\sqrt[3]{5}/10$
4337 6545 rt : (5,3,-9,-8)
4337 8161 ln : $(\sqrt[3]{2}/3)^{1/2}$
4337 8243 rt : (7,9,4,-4)
4337 8954 rt : (8,5,-3,9)
4338 1648 $2^x : 3\ln 2/4$
4338 3504 cu : $3/2 - 2\pi$
4338 3701 rt : (8,2,-8,8)
4338 7105 rt : (7,9,2,5)
4338 8373 $s\pi : 1/7$
4338 9329 $Ei : e/10$
4338 9790 ln : $4\sqrt{2}/3 + 4\sqrt{3}/3$
4339 0798 rt : (4,2,-9,-3)
4339 1204 $Ei : \sqrt{5}\pi$
4339 1342 $e^x : \pi^2/8$
4339 3185 rt : (7,6,-3,4)
4339 4520 at : $1/3 + 4\sqrt{3}$
4339 5524 $2^x : 13/25$
4339 6908 sq : $4/3 + \pi/2$
4339 7418 $\lambda : \exp(-17\pi/16)$
4339 7607 $2^x : (\sqrt[3]{3}/2)^2$
4339 7616 rt : (7,3,-8,3)
4339 9685 sr : P_{36}
4340 3790 cu : 23/14
4341 1039 $J_2 : 11/3$
4341 2863 as : $\exp(-\sqrt{3}/2)$
4341 3573 $e^x : \sqrt[3]{3}/4$
4341 4702 cr : $10\sqrt[3]{3}$
4341 5136 $E : \sqrt[3]{5}/3$
4341 8965 $3 \times \ln(\pi)$
4342 1415 $e^x : 3\zeta(3)/10$
4342 1497 $t\pi : \pi^2$
4342 3512 $\Gamma : \exp(2e/3)$
4342 4810 sinh : $9\sqrt[3]{3}/8$
4342 5641 $e - \exp(1/4)$
4342 5854 rt : (9,9,-1,-2)
4342 7339 $J_2 : (e + \pi)^{1/2}$
4342 9109 sr : $3 - 2\sqrt{2}/3$
4342 9448 $\ell_{10} : e$
4343 0926 $e^x : 3/2 + \sqrt{2}$
4343 2050 at : $\sqrt{2}/4 + 4\sqrt{3}$
4343 3579 rt : (2,5,-7,-2)
4343 4694 at : $1 + 2\pi$
4343 5096 $\Gamma : 4(e + \pi)/9$
4343 5384 sq : $3\sqrt{2} - \sqrt{3}/3$
4343 5706 ln : $3\sqrt{2}/2 - \sqrt{3}/3$
4343 5962 $J_2 : 7\pi/6$
4343 6129 $t\pi : 3/23$
4343 6166 $\ell_2 : 3\sqrt{3}/4 - \sqrt{5}/4$
4343 6857 $\ell_{10} : 2\sqrt{3} - \sqrt{5}/3$
4343 8486 $\Psi : \exp(2e)$
4343 9076 at : $\sqrt{3}/3 + 3\sqrt{5}$
4343 9082 $\Gamma : 5/19$
4343 9670 as : $1/2 - 2\sqrt{5}/3$
4344 4032 $Li_2 : e/7$
4344 4689 $\Psi : 4(1/\pi)/9$
4344 5617 cr : $\sqrt{2}/4 + 3\sqrt{3}/2$
4344 5723 $s\pi : e + 3\pi$
4344 6328 $t\pi : \exp(1/(3e))$
4344 6333 $\exp(e) \div s\pi(\sqrt{5})$
4344 6666 rt : (8,3,9,4)
4344 8587 $\pi - s\pi(1/4)$
4345 0094 sinh : $(4/3)^{-3}$
4345 1097 cu : $2 + 3\pi/2$
4345 1875 at : $3 - 2\sqrt{3}$
4345 1884 $\Gamma : 5/8$
4345 3892 sinh : 22/19
4345 5200 $\Psi : 4(e + \pi)/5$
4345 6587 at : $1/\sqrt[3]{10}$
4345 6923 rt : (8,6,8,3)
4345 8681 $t\pi : 7\sqrt{5}/5$
4345 9387 rt : (4,8,-3,3)
4345 9820 sr : $\exp(-5/3)$
4346 0080 $e^x : \sqrt{2}/2 + 3\sqrt{3}/4$
4346 0552 as : 8/19
4346 1636 $J_0 : 18/11$
4346 2429 $\ln(4/5) \times \exp(2/3)$
4346 3173 $10^x : 4 + 3\sqrt{2}$
4346 4615 cr : $4 - \pi/3$
4346 5393 rt : (5,7,9,3)
4346 6773 at : $2\sqrt{2} + 2\sqrt{5}$
4346 8489 $Ei : 9\sqrt[3]{3}/7$
4346 9003 $10^x : 10/3$
4347 2565 $e^x : 3/4 + 2\sqrt{2}/3$
4347 4961 rt : (7,8,9,-6)
4347 6023 cr : $-\sqrt{3} + \sqrt{5} + \sqrt{6}$
4347 7572 $10^x : 4 - 2\sqrt{3}$
4347 8260 id : 10/23
4347 8481 rt : (2,9,-1,-7)
4348 1227 $E : (2e)^{-1/3}$
4348 2572 rt : (9,7,-5,5)
4348 2603 $J_2 : 5(e + \pi)/8$
4348 5815 rt : (7,5,8,-5)
4348 7120 rt : (8,9,7,2)
4349 2193 rt : (4,4,9,-5)
4349 2320 $J_2 : 7\sqrt{3}/5$
4349 2752 sr : $3/2 + \sqrt{5}/4$
4349 4166 rt : (6,2,-6,5)
4349 7845 at : $3\sqrt{2}/2 + 3\sqrt{3}$
4350 2059 at : $6e\pi/7$
4350 2068 cr : $3e + 2\pi$
4350 2186 ln : $10\sqrt[3]{2}/3$
4350 2830 rt : (7,2,7,-4)
4350 3645 $2^x : -\sqrt{2} - \sqrt{5} + \sqrt{6}$
4350 4308 $\theta_3 : (\ln 2/2)^{1/2}$
4350 5277 $J_1 : (\pi/3)^{-1/3}$
4350 5323 $\zeta : 5/6$
4350 5344 $t\pi : 3\sqrt{2}/2 - \sqrt{5}/3$
4350 5515 rt : (6,5,-1,6)
4350 6091 $\Psi : 7\sqrt[3]{3}/5$
4350 7213 cu : $3 + 3\sqrt{3}/2$
4350 8111 $Ei : 6e\pi/5$
4350 8452 ln : 5/21
4350 8982 $t\pi : 3\sqrt{3}/2 + 3\sqrt{5}$
4350 9103 sinh : $\exp(2/3)$
4351 0157 $J_2 : 7\ln 2/2$
4351 1344 at : $5(e + \pi)/4$
4351 1721 rt : (3,6,-4,-2)
4351 3465 $2^x : 2 + 3e/4$
4351 3674 rt : (4,1,8,-4)
4351 3882 rt : (5,5,-8,-7)
4351 4985 $\Psi : \sqrt{22}$
4351 5302 $Li_2 : 7/18$
4351 7594 $t\pi : 5\sqrt{5}/2$
4351 7845 rt : (1,9,-4,1)
4351 8117 $Li_2 : 1/\sqrt[3]{17}$
4352 1384 rt : (6,8,4,7)
4352 1472 at : $7\pi/3$
4352 3164 at : $\sqrt{5} + \sqrt{6} + \sqrt{7}$
4352 3404 sinh : $(\sqrt{5}/3)^{-1/2}$
4352 3784 $\ln(4) - s\pi(2/5)$
4352 5122 $J_2 : 3e\pi/7$
4352 5281 sq : $\sqrt{12}\pi$
4352 6861 at : 22/3
4352 7905 at : $4\sqrt{2} + 3\sqrt{5}/4$
4352 8084 rt : (1,0,9,4)
4353 0455 rt : (3,9,7,7)
4353 1807 ln : 11/17
4353 3326 rt : (7,1,6,3)
4353 3334 cu : $3e/4 + 3\pi$
4353 3492 rt : (9,8,4,-4)
4353 3708 rt : (2,9,-2,-1)
4353 4860 ln : $9\zeta(3)/7$
4353 5701 $e^x : \sqrt{2}/2 + 2\sqrt{3}/3$
4353 6491 rt : (1,9,5,2)
4353 9420 sr : $10\sqrt[3]{3}/7$
4354 1434 rt : (8,8,-5,0)
4354 3019 $\ell_2 : 9\zeta(3)$
4354 4182 $e^x : \sqrt{11/2}$
4354 5148 rt : (6,7,5,-4)
4354 6920 $Ei : 6\ln 2/5$
4354 9875 rt : (3,4,-6,5)
4354 9918 $J_2 : 17/7$
4354 9997 sr : $1 + 3\sqrt{2}/4$
4355 0211 tanh : 7/15
4355 1237 as : $(4/3)^{-3}$
4355 4047 rt : (7,9,-2,-2)
4355 4136 rt : (8,3,-8,6)
4355 4998 rt : (3,7,-9,-5)
4355 5738 rt : (4,2,7,3)
4355 6366 sr : $2\sqrt{2}/3 + \sqrt{5}/2$
4355 9644 rt : (3,6,6,-4)
4356 0725 $t\pi : 1/3 - 2\sqrt{3}$
4356 3157 $\zeta : 23/17$
4356 3322 rt : (1,5,-2,2)
4356 4628 rt : (9,5,3,-3)
4356 4670 at : $4 + 3\sqrt{5}/2$
4356 5093 rt : (1,8,1,-1)
4356 6017 sq : $3/2 + \sqrt{2}/4$
4356 8488 $2^x : 12/23$
4356 9972 rt : (8,6,-3,7)
4357 2418 $s\pi(1/3) + s\pi(\pi)$
4357 2995 $\ell_{10} : 2\sqrt{3}/3 - \sqrt{5}/2$
4357 8166 rt : (5,8,-3,-6)
4357 8325 $J_1 : \pi^2/10$
4357 8450 ln : $2 + 3\pi$
4357 9715 $\Psi : 10\sqrt{2}/7$
4358 0353 $2^x : 3\pi^2$
4358 2029 $\Gamma : \ln(\ln 2/3)$
4358 2443 $\ln(4) - \exp(3/5)$
4358 2452 cr : $3\sqrt{3} - \sqrt{5}$
4358 3020 $\ell_2 : 2/3 + 3e/4$
4358 3691 $\ell_2 : \ln(2\pi/3)$
4358 4642 rt : (6,4,4,-3)
4358 8682 sr : $\sqrt[3]{5}/9$
4359 0026 rt : (1,3,8,3)
4359 0269 rt : (8,9,2,8)
4359 0361 $2^x : \sqrt{2}/2 + \sqrt{3}/3$
4359 2472 at : $3e - \pi/4$
4359 2642 $t\pi : \sqrt{3}/3 - 3\sqrt{5}$
4359 3519 $10^x : \sqrt{2}/9$
4359 4540 $\Psi : \sqrt[3]{2}/2$
4359 5177 cu : $2e - 3\pi$
4359 5335 cr : $3\pi^2/10$
4359 6258 $J_1 : e$
4359 6692 $s\pi : 8\sqrt{3}$
4359 6991 rt : (9,7,9,-6)
4359 7482 id : P_{29}
4359 7501 sq : $2 + \pi$
4359 8041 $e - \exp(e)$
4360 1025 rt : (5,1,-4,7)
4360 2270 at : $2/3 + 3\sqrt{5}$
4360 2675 rt : (3,4,-8,-4)
4360 3199 sr : $1/4 + 2e/3$
4360 3833 rt : (7,4,5,2)
4360 4739 $Ei : \exp(-1/(2e))$
4360 5523 sq : $1/4 - 3\pi/4$

4360 6513 $\Psi : 7\sqrt{3}/6$
4360 6797 $1/5 + \sqrt{5}$
4360 8130 $J_2 : \sqrt{5} - \sqrt{6} + \sqrt{7}$
4360 8139 $\ln(\sqrt{2}) \times \exp(5)$
4360 9911 $\ell_2 : 17/23$
4361 1308 rt : (1,8,5,6)
4361 1433 rt : (3,6,8,8)
4362 1074 rt : (9,2,2,-2)
4362 1076 $e^x : 3\sqrt{3}/2 - \sqrt{5}$
4362 1344 $\Psi : 1/\ln(5/2)$
4362 1465 rt : (5,9,2,-3)
4362 2374 rt : (4,7,0,-1)
4362 2454 at : $3e/4 - 3\pi$
4362 6466 ln : $1 - \sqrt{2}/4$
4362 7561 at : $4\sqrt{2} + \sqrt{3}$
4362 7680 $\ell_{10} : \ln(\sqrt[3]{3})$
4362 7833 at : $(e)^2$
4362 9713 $erf : 1/\sqrt{6}$
4363 0411 $\theta_3 : (\sqrt{2}\pi/2)^{-1/3}$
4363 1956 rt : (5,4,1,8)
4363 2438 $e^x : e - 3\pi/4$
4363 4159 rt : (9,4,8,-5)
4363 4396 rt : (7,4,-8,1)
4363 4520 $\Gamma : \exp(-\sqrt{2}/3)$
4363 5561 $\theta_3 : (5/3)^{-3}$
4363 6587 $E : 2\sqrt{2}/5$
4363 6657 rt : (6,3,-9,-4)
4363 6703 as : $1 - \sqrt{3}/3$
4363 8019 rt : (7,2,-6,8)
4363 9087 cr : $-\sqrt{2} + \sqrt{3} + \sqrt{7}$
4364 0311 rt : (4,9,-6,-4)
4364 2499 $E : 1/\ln(e + \pi)$
4364 3578 sr : 4/21
4364 3768 $2^x : 4 - 3\sqrt{3}$
4364 4526 ln : $1/2 + \pi/3$
4364 4538 $E : \exp(-\sqrt[3]{5}/3)$
4364 6784 at : $\sqrt{2}/3 + 4\sqrt{3}$
4364 8877 $\ell_{10} : 1 + \sqrt{3}$
4364 9167 rt : (2,8,3,-3)
4365 0644 sr : $1/2 + 2e$
4365 1438 at : $3\pi^2/4$
4365 3269 $J : \ln 2/4$
4365 4266 $2/5 \div \ln(2/5)$
4365 4726 rt : (8,9,6,-5)
4365 4794 $J_0 : \sqrt{8/3}$
4365 5610 sinh : $7e\pi/6$
4365 5975 rt : (6,3,9,-5)
4365 6172 at : $3/4 - 3e$
4365 6365 id : $2e$
4365 6614 $s\pi : 1/2 + 3\pi/4$
4365 6750 cu : $3 + 2e/3$
4365 8756 rt : (8,7,0,-2)
4366 0519 $E : 13/23$
4366 0792 rt : (9,3,-8,9)
4366 2715 at : 7/15
4366 3195 rt : (6,1,3,-2)
4366 4487 $\theta_3 : \exp(-\sqrt[3]{4}/3)$
4366 4874 $2^x : (\sqrt{2}/3)^{-1/3}$
4366 5417 rt : (7,5,-1,9)
4366 8733 $10^x : 11/17$
4366 9859 $J_1 : 4\sqrt{3}/7$
4366 9912 $2^x : 1/\sqrt[3]{7}$
4367 1639 rt : (7,7,-3,2)
4367 2800 rt : (4,7,-3,9)
4367 3553 $10^x : 3 + 4\sqrt{2}/3$
4367 3567 $e^x : 2 - 2\sqrt{2}$
4367 4175 rt : (7,8,-7,-4)
4367 4891 at : $\sqrt{2}/2 + 3\sqrt{5}$
4367 5179 $\ell_2 : 4 + \sqrt{2}$
4367 5295 rt : (4,5,6,2)
4367 6151 rt : (5,8,7,-5)
4367 6630 $J_1 : 7\sqrt{2}/10$
4367 6790 $s\pi : P_{75}$
4367 8426 $1/2 \div \ln(\pi)$
4367 9023 id : $1/\sqrt[3]{12}$
4367 9714 ln : $7\zeta(3)/2$
4367 9737 rt : (4,9,-1,2)
4368 1565 $\exp(1/2) - \exp(3)$
4368 2512 $10^x : 7\ln 2/6$
4368 2920 $\ln(4/5) \div \ln(3/5)$
4368 3983 rt : (8,5,-6,1)
4368 5193 $J_1 : 10\ln 2/7$
4368 5577 rt : (7,9,-5,-5)
4368 5992 $e^x : 1/\sqrt[3]{21}$
4368 6069 rt : (1,7,-4,-3)
4368 6242 $\Psi : 1/11$
4369 0841 $\ln(3/5) + \exp(2/3)$
4369 1779 at : $2 - 3\pi$
4369 1940 rt : (9,1,7,-4)
4369 2436 rt : (4,1,-9,-1)
4369 4625 rt : (9,8,-5,3)
4369 4668 $\ell_2 : \sqrt{2}/3 + \sqrt{5}$
4369 5732 $\Gamma : 9\ln 2/10$
4369 6754 rt : (6,2,-9,-9)
4369 7191 $\zeta : ((e + \pi)/3)^{-1/3}$
4369 7879 at : $1/2 + 4\sqrt{3}$
4369 9041 $\Psi : \exp(\sqrt{2}\pi)$
4369 9107 rt : (5,9,4,2)
4369 9326 $2^x : 3/2 + 2\sqrt{2}/3$
4369 9882 $\Psi : \sqrt{2}/10$
4370 0832 $10^x : 4 - 4\sqrt{5}/3$
4370 1602 sr : $3/4 - \sqrt{5}/4$
4370 1937 sr : $3(1/\pi)/5$
4370 2031 rt : (2,7,8,-5)
4370 2270 at : $e + 3\pi/2$
4370 4504 $2^x : 10\sqrt[3]{5}/7$
4370 4833 $Li_2 : e\pi/9$
4370 5026 at : $3\sqrt{3} + \sqrt{5}$
4370 6358 rt : (3,1,-7,-3)
4370 8238 $e^x : \sqrt{2}/3 - 3\sqrt{3}/4$
4370 9660 as : $e - \pi$
4371 0169 $erf : 9/22$
4371 0508 rt : (8,6,5,-4)
4371 0703 sinh : $3\sqrt{2}/10$
4371 0998 $\ell_{10} : 8\sqrt[3]{5}/5$
4371 1039 $10^x : \sqrt[3]{2}/8$
4371 1363 ln : $4 - 3\sqrt{5}/2$
4371 2688 $\ell_{10} : 1/2 + \sqrt{5}$
4371 2745 at : $2 + 2e$
4371 2980 $E : (\pi)^{-1/2}$
4371 4443 sr : $1/3 + \sqrt{3}$
4371 4871 cr : $1/4 + e$
4371 5361 at : $4/3 - \sqrt{3}/2$
4371 6160 $\ell_{10} : 2\sqrt{2}/3 - \sqrt{3}/3$
4371 6182 $J_1 : \sqrt[3]{20}$
4372 0006 rt : (4,0,-2,9)
4372 0269 $J_1 : 19/7$
4372 3455 $\ell_2 : \sqrt{22/3}$
4372 4882 $J_2 : 2e\pi/7$
4372 5884 $erf : 9(1/\pi)/7$
4372 6421 at : $3/2 - 4\sqrt{5}$
4372 6576 $Ei : 2\sqrt{3}/7$
4372 6977 sinh : $4(1/\pi)/3$
4372 8665 rt : (6,0,8,4)
4372 9554 $\Gamma : 8\zeta(3)/3$
4372 9842 $e^x : 1 + 4\sqrt{3}$
4373 1913 $e^x : (\sqrt[3]{2})^{-1/2}$
4373 3380 ln : $2\sqrt{3} + \sqrt{5}/3$
4373 3932 rt : (1,9,6,-7)
4373 7720 rt : (4,6,-5,-3)
4373 8016 $10^x : 1 - e/2$
4373 8344 sr : P_{52}
4373 8673 $2^x : \exp(\sqrt{3}/3)$
4373 8762 $\ell_{10} : 3\sqrt{2}/4 + 3\sqrt{5}/4$
4374 0932 $\theta_3 : \sqrt{3}/8$
4374 0965 at : $((e + \pi)/3)^3$
4374 1938 rt : (1,8,-8,-8)
4374 2864 at : $10\sqrt{5}/3$
4374 3798 $10^x : e\pi/4$
4374 4470 rt : (5,5,6,-4)
4374 4945 $10^x : 4 - \sqrt{2}/4$
4374 5405 rt : (1,7,3,-4)
4374 7979 rt : (5,9,-9,-3)
4374 8625 at : $(\ln 3/3)^{-2}$
4374 9124 id : $4\sqrt{3} - 2\sqrt{5}/3$
4374 9314 $2^x : 1/2 + \pi/4$
4375 0000 id : 7/16
4375 1071 at : $3/4 + 3\sqrt{5}$
4375 2095 sr : $3\sqrt{3} + \sqrt{5}/3$
4375 3063 $\ell_{10} : \sqrt{15/2}$
4375 3668 $2^x : \pi/6$
4375 3841 $\zeta : 4/5$
4375 4612 rt : (9,1,1,1)
4375 6489 rt : (8,4,-8,4)
4375 8965 $\zeta : 1/\ln(2\pi/3)$
4376 0947 rt : (7,6,-1,-1)
4376 1478 at : $4 + 2\sqrt{3}$
4376 1783 as : $4\sqrt{2}/3 - 4\sqrt{3}/3$
4376 1788 $Ei : (\ln 2)^{1/2}$
4376 2310 $\ln(\pi) - s\pi(1/4)$
4376 3416 ln : $1/4 + 3\sqrt{3}/4$
4376 5179 at : $3e/4 - \pi/2$
4376 5674 cu : $2e - 4\pi$
4376 9131 $J_2 : 10\sqrt[3]{5}/7$
4376 9410 sq : $3/2 - 3\sqrt{5}/2$
4377 0745 $J_0 : 3e/5$
4377 1678 $\lambda : \exp(-18\pi/17)$
4377 3429 $Ei : \sqrt[3]{25}$
4377 4669 $2^x : 11/21$
4377 5128 $s\pi : \sqrt[3]{3}/10$
4377 5164 $4 \times \ln(5)$
4377 5463 rt : (4,9,9,-6)
4377 5630 at : $3 + 2\sqrt{5}$
4377 5861 at : $7e\pi/8$
4377 6008 $\exp(\sqrt{2}) - s\pi(\sqrt{5})$
4377 7268 rt : (3,1,9,4)
4377 7517 rt : (8,4,-1,-1)
4377 7744 $\Gamma : 4\sqrt{5}$
4377 8217 $e^x : 3\sqrt{2} - \sqrt{5}$
4377 8850 $J_2 : 7\pi/9$
4378 4261 rt : (2,7,-9,-3)
4378 5792 $Li_2 : (\sqrt[3]{5}/2)^{1/3}$
4378 6134 $t\pi : 4\sqrt{2}/5$
4378 6154 $e^x : 3\sqrt{3}/2$
4378 7718 rt : (2,4,7,-4)
4379 1137 rt : (7,9,-3,-1)
4379 1816 rt : (3,5,-4,4)
4379 3810 $J_2 : 22/9$
4379 4048 rt : (9,2,6,3)
4379 5366 rt : (1,9,5,-4)
4379 5648 sq : 9π
4379 7293 $Ei : (\sqrt[3]{3})^{-1/2}$
4379 7360 $\exp(2/5) \times s\pi(\sqrt{2})$
4379 8944 $E : 9/16$
4379 9958 $\exp(2) - s\pi(2/5)$
4380 1625 rt : (8,7,-3,5)
4380 2046 $\Gamma : (4)^{-3}$
4380 2119 at : $4\sqrt{3} + \sqrt{5}/4$
4380 2195 rt : (8,3,4,-3)
4380 2730 $2^x : 9/7$
4380 3079 id : $3e + 2\pi$
4380 3104 $J_1 : 4\sqrt{5}/9$
4380 3651 rt : (6,0,-8,3)
4380 3705 sinh : $9\sqrt{3}/8$
4380 3799 at : $2/3 - 3e$
4380 3952 $Ei : 5\ln 2/7$
4380 5076 $s\pi : (1/(3e))^{-1/2}$
4380 6939 rt : (1,5,-3,-9)
4380 7588 sr : $\sqrt{3}/3 + 2\sqrt{5}/3$
4380 7669 $J_0 : \sqrt[3]{13/3}$
4380 8086 rt : (7,8,2,-3)
4380 8310 cu : $3 - 2\sqrt{5}/3$
4380 9760 $\zeta : \sqrt[3]{12}$
4381 0904 rt : (2,9,1,2)
4381 1716 $Ei : 10/3$
4381 2543 rt : (6,5,-8,-4)
4381 3612 rt : (8,2,9,-5)
4381 4899 $Ei : 17/11$
4381 4903 as : $3\sqrt{2}/10$
4381 5332 rt : (7,5,-6,-3)
4381 6886 rt : (7,7,7,-5)
4381 7572 $E : \ln(\sqrt[3]{5}/3)$
4381 9536 rt : (6,6,-1,4)
4382 1010 $\ln(4) \div s\pi(\sqrt{2})$
4382 1466 $Li_2 : 9/23$
4382 2507 sinh : $\exp(-\sqrt[3]{5}/2)$
4382 2732 sinh : $1/\sqrt[3]{13}$
4382 3560 $\zeta : (\sqrt[3]{4})^{-1/3}$
4382 4312 $J_2 : 9e/10$
4382 4479 at : 15/2

4382 5533 rt : (5,1,-9,-4)

4382 8765 $\exp(1/4)+\exp(e)$

4382 9082 10^x : $\exp(\sqrt{3}/3)$

4382 9219 Ei : $9\zeta(3)/7$

4382 9524 rt : (8,9,-9,-5)

4382 9609 2^x : $3+3\sqrt{3}/2$

4382 9913 tπ : $e-2\pi/3$

4383 1370 as : $4(1/\pi)/3$

4383 3499 e^x : $3+\sqrt{5}/2$

4383 6705 rt : (9,4,-8,7)

4383 8072 ln : $\ln(3\pi/2)$

4383 9071 $s\pi(1/5)\times s\pi(\sqrt{3})$

4384 0418 cr : $2/3+4\sqrt{3}/3$

4384 0762 cr : $3\sqrt{3}/4+3\sqrt{5}/4$

4384 1036 $5\times\ln(3/4)$

4384 2068 cu : $3/2+2\pi/3$

4384 3642 ln : $3/4+2\sqrt{3}$

4384 4718 rt : (2,9,-1,-2)

4384 4988 10^x : 3/19

4384 5377 J_0 : $(\ln 2/3)^{-1/3}$

4384 6138 sq : $1/3+\sqrt{3}/2$

4384 7337 sr : $3/4+3\sqrt{3}$

4385 0591 sq : $9\sqrt[3]{3}/7$

4385 0631 rt : (2,5,-5,5)

4385 3884 ℓ_2 : $\ln(\pi/3)$

4385 4197 tπ : $\sqrt[3]{1/3}$

4385 5100 e^x : 4/11

4385 5454 rt : (3,7,-2,-3)

4385 5541 E : $\exp(-\sqrt{3}/3)$

4385 6701 rt : (4,6,8,-5)

4385 6985 J_1 : $1/\ln(\ln 3/3)$

4385 7663 rt : (7,3,-6,6)

4385 9243 sq : $\sqrt{2}+2\sqrt{5}/3$

4385 9866 ℓ_{10} : $2+\sqrt{5}/3$

4386 1016 rt : (4,7,3,-3)

4386 2414 rt : (9,9,8,-6)

4386 3833 sπ : $e/2+\pi/4$

4386 4074 rt : (5,2,5,-3)

4386 4595 rt : (8,9,-5,-2)

4386 4977 sπ : $2/3+4\pi/3$

4386 5183 tπ : $4\sqrt{3}/3+\sqrt{5}/4$

4386 5356 e^x : $2/3-2\sqrt{5}/3$

4386 7056 2^x : $3-4\pi/3$

4386 7447 tanh : 8/17

4386 8543 rt : (6,3,7,3)

4386 9133 sr : $\sqrt{3}/9$

4387 0187 rt : (4,8,-2,-1)

4387 0244 J_2 : $\sqrt{6}$

4387 1266 e^x : $\sqrt{3}+4\sqrt{5}/3$

4387 5661 2^x : $5\pi^2$

4387 5698 rt : (7,8,9,3)

4387 6157 e^x : $8(1/\pi)/7$

4387 6777 at : $\sqrt{2}-4\sqrt{5}$

4387 7822 rt : (4,8,-1,8)

4387 8337 at : $2e+2\pi/3$

4387 8532 rt : (5,3,0,-1)

4387 9020 id : $1/4+4\pi/3$

4387 9310 id : $3\sqrt{2}+3\sqrt{3}$

4387 9821 Ψ : $7e\pi/5$

4388 1928 2^x : $e/3-2\pi/3$

4388 2109 rt : (9,7,-3,8)

4388 3819 at : $9(e+\pi)/7$

4388 5821 cu : $3/2-4\pi/3$

4388 6691 tπ : $e/4+4\pi/3$

4389 0572 rt : (5,8,-9,3)

4389 1621 tπ : $\sqrt{2}-\sqrt{3}+\sqrt{6}$

4389 2998 ln : $\sqrt{3}/4+\sqrt{5}/2$

4389 3448 rt : (5,2,-4,5)

4389 4511 ℓ_{10} : $4/3+\sqrt{2}$

4389 5014 rt : (6,2,2,1)

4389 5747 cu : $4\sqrt[3]{5}/9$

4389 7600 cu : 19/25

4389 7973 rt : (6,8,9,-6)

4389 8200 $\ln(2)-s\pi(\sqrt{3})$

4390 0823 e^x : $4+\pi$

4390 2617 sr : $1/2+\pi/2$

4390 2890 rt : (4,9,-7,1)

4390 4876 sq : P_{55}

4390 4930 $4/5\div\exp(3/5)$

4390 5204 K : $1/\ln((e+\pi)/2)$

4390 5373 rt : (6,1,-9,-7)

4390 5659 rt : (1,5,9,-5)

4390 7187 Γ : $7\sqrt{5}/6$

4390 9305 e^x : $(\sqrt{2}\pi/2)^{1/2}$

4390 9513 rt : (5,7,-7,-4)

4390 9683 e^x : $4e\pi$

4391 1322 10^x : $(\sqrt[3]{4}/3)^{1/3}$

4391 2254 rt : (6,9,4,5)

4391 3280 sπ : $\sqrt[3]{3/2}$

4391 4084 J_1 : $\sqrt{22/3}$

4391 5758 id : $\sqrt{2}/2+\sqrt{3}$

4391 5792 ℓ_{10} : $7\pi/8$

4391 6207 rt : (2,7,1,2)

4391 6714 ln : $3e/4+3\pi$

4391 7681 2^x : $4\sqrt{3}+3\sqrt{5}/4$

4391 7695 sπ : $\ln(\pi)$

4391 9639 rt : (7,6,-1,7)

4392 0813 10^x : $7\sqrt[3]{3}$

4392 1268 rt : (7,4,6,-4)

4392 2876 rt : (8,3,-6,9)

4392 3406 rt : (4,2,-9,-4)

4392 5159 θ_3 : 5/23

4392 5513 E : 14/25

4392 6659 cr : $4\sqrt{5}/3$

4392 7284 E : $4\sqrt[3]{2}/9$

4392 7650 at : $6\sqrt[3]{2}$

4392 8041 as : $\exp(-\sqrt[3]{5}/2)$

4392 8269 as : $1/\sqrt[3]{13}$

4392 8272 2^x : $3(e+\pi)/5$

4393 1167 rt : (1,6,4,-3)

4393 3061 rt : (8,1,8,4)

4393 3269 ℓ_{10} : 4/11

4393 3344 tanh : $\sqrt{2}/3$

4393 3982 id : $3/2-3\sqrt{2}/4$

4393 4270 rt : (7,5,-8,-1)

4393 5808 e^x : $2/3+4\pi/3$

4393 6192 e^x : $9\pi^2/5$

4393 6766 rt : (2,8,-3,4)

4393 6985 rt : (6,8,-3,-3)

4393 7321 10^x : $\ln(\sqrt[3]{5})$

4393 8452 sπ : $\sqrt{2}-\sqrt{5}/4$

4393 8995 e^x : 21/17

4393 9577 rt : (1,7,-3,2)

4394 0478 2^x : $3e/2+3\pi/4$

4394 2789 ln : $3/2+e$

4394 2866 rt : (7,7,4,1)

4394 4491 $2/5\times\ln(3)$

4394 5063 2^x : $7\pi/9$

4394 8361 tπ : $9\ln 2/10$

4394 9792 id : $4(e+\pi)$

4395 0645 rt : (2,1,6,-3)

4395 1398 Γ : $10\pi/7$

4395 1665 sq : $2\sqrt{3}-\sqrt{5}/4$

4395 1690 rt : (3,9,-8,-5)

4395 2253 sr : $(2e/3)^3$

4395 2819 rt : (9,6,7,-5)

4395 2881 rt : (4,2,-7,-6)

4395 2888 at : $\sqrt{3}/2+3\sqrt{5}$

4395 3058 rt : (5,8,-2,-2)

4395 3314 10^x : $5(e+\pi)/8$

4395 4273 rt : (9,9,-5,1)

4395 4668 rt : (7,5,1,-2)

4395 5873 10^x : $\sqrt{2}+\sqrt{3}$

4395 5873 $\exp(2/3)+\exp(2/5)$

4395 9216 ℓ_{10} : $4\sqrt{2}/3+\sqrt{3}/2$

4396 1566 e^x : $\sqrt{2}-\sqrt{5}$

4396 1646 J_0 : $\sqrt{2}-\sqrt{5}+\sqrt{6}$

4396 3717 ℓ_{10} : $3\sqrt{2}-2\sqrt{5}/3$

4396 3821 tπ : $1/4+3\sqrt{3}/4$

4396 5101 erf : 7/17

4396 8018 tπ : $4e-4\pi$

4396 8414 rt : (6,2,-4,8)

4396 9884 rt : (1,8,-3,6)

4397 0862 sr : $3\sqrt{3}/2+3\sqrt{5}/2$

4397 4296 tπ : $\pi^2/7$

4397 5075 E : $\sqrt{5}/4$

4397 6824 Ei : 5/6

4397 7490 rt : (5,4,-5,1)

4398 0567 10^x : $3\sqrt{2}/2+3\sqrt{3}/4$

4398 0743 rt : (8,0,3,2)

4398 1345 at : $8e\pi/9$

4398 1486 rt : (3,4,8,3)

4398 2477 2^x : $10\pi/7$

4398 2990 e^x : $1/3-2\sqrt{3}/3$

4398 4258 at : 8/17

4398 5876 cu : $2/3+e/4$

4398 6133 rt : (9,7,2,-3)

4398 6531 ln : $\sqrt{3}/2+3\sqrt{5}/2$

4398 7159 rt : (4,3,7,-4)

4398 8156 at : $2/3+4\sqrt{3}$

4398 8861 J : $\ln(\sqrt[3]{3})$

4398 9153 rt : (5,5,1,6)

4398 9196 rt : (1,9,-5,-8)

4399 0977 id : $8\sqrt[3]{2}/7$

4399 1347 at : $3\sqrt{2}+3\sqrt{5}/2$

4399 2406 id : $2e\pi/7$

4399 2654 sinh : $5(e+\pi)/3$

4399 3310 rt : (1,7,-1,-1)

4399 4298 erf : $2\sqrt[3]{3}/7$

4399 7309 sπ : $\sqrt[3]{5}/2$

4399 7503 Li_2 : $\ln(\sqrt{2}\pi/3)$

4399 8505 Li_2 : $\pi/8$

4399 8517 rt : (8,6,-8,-4)

4400 0000 id : 11/25

4400 0467 rt : (8,5,-8,2)

4400 1349 cr : $3/4+\sqrt{5}$

4400 2287 ℓ_2 : $4\sqrt{2}+3\sqrt{3}$

4400 2367 rt : (7,9,4,8)

4400 2773 rt : (5,7,-9,9)

4400 3655 Ψ : $7\sqrt[3]{5}$

4400 5058 J_1 : 1

4400 5640 rt : (6,5,8,-5)

4400 6552 tπ : P_{49}

4400 8106 e^x : $3/4-\pi/2$

4400 8308 10^x : P_{27}

4400 8853 rt : (9,1,-9,3)

4400 9140 tπ : $1/4-\sqrt{5}/2$

4400 9642 rt : (5,2,-7,-9)

4401 1749 ζ : 16/7

4401 2617 rt : (5,2,9,4)

4401 5482 at : $4/3-4\sqrt{5}$

4401 6800 rt : (8,7,9,-6)

4401 7141 2^x : $\sqrt{2}-3\sqrt{3}/2$

4401 7828 $\ln(2/5)+\exp(\sqrt{5})$

4401 8231 J_1 : $9\zeta(3)/4$

4401 8284 at : $2e/3-3\pi$

4401 8350 rt : (6,8,-9,-5)

4401 9318 tanh : $3\sqrt[3]{2}/8$

4402 1123 J_1 : $\zeta(11)$

4402 1189 rt : (3,9,0,-2)

4402 1886 cr : $3\sqrt{2}/2+\sqrt{3}/2$

4402 3057 rt : (9,5,5,2)

4402 3089 $\ln(3/5)+s\pi(2/5)$

4402 4653 2^x : 10/19

4402 4945 at : $4/3+2\pi$

4402 5474 id : $\sqrt{3}+3\sqrt{5}$

4402 5578 e^x : $2\sqrt{5}/3$

4402 8741 rt : (2,5,-5,-3)

4403 0592 Ψ : $\sqrt[3]{25/3}$

4403 2255 10^x : $2\sqrt{2}/3-3\sqrt{3}/4$

4403 2965 rt : (5,4,-6,-2)

4403 4769 $\sqrt{5}\div\ln(2/5)$

4403 5561 Ψ : $3\pi/2$

4403 5761 10^x : $2-3\pi/4$

4403 5860 ln : $3-3\pi/4$

4403 7517 rt : (2,7,-7,6)

4403 8246 at : $4+4e/3$

4403 8421 ln : $4e/7$

4403 9464 rt : (7,8,-3,0)

4404 2093 rt : (9,5,-8,5)

4404 2601 as : $3e/4-\pi/3$

4404 3099 sq : $1/4-2e/3$

4404 4621 ζ : $2\sqrt{3}/7$

4404 4956 rt : (2,9,7,-1)

4404 8293 rt : (4,1,-2,7)

4404 8981 rt : (9,8,-3,-1)

4405 0536 rt : (6,5,1,9)

4405 0910 rt : (3,9,-6,-5)

4405 1066 at : $\sqrt{2}/3$

4405 1694 10^x : $\sqrt{2}/2+2\sqrt{5}$

4405 2180 $\mathrm{t}\pi : 2e\pi/3$
4405 6711 $\mathrm{at} : \sqrt{2}/2 + 4\sqrt{3}$
4405 7259 $\ell_2 : 7/19$
4405 7586 $2^x : 1/2 + 4e/3$
4405 9165 $Li_2 : 5\sqrt[3]{5}/9$
4405 9439 $2^x : 9\pi/7$
4406 0743 $J_1 : 5\zeta(3)/6$
4406 0844 $\mathrm{t}\pi : \ln 2$
4406 0935 $1/3 - s\pi(e)$
4406 1137 $2^x : 1/3 + 4\sqrt{3}$
4406 1970 rt : (8,8,4,-4)
4406 3178 $\zeta : 9\zeta(3)/8$
4406 3379 $Li_2 : 19/20$
4406 3431 $\sinh : \sqrt[3]{5}/4$
4406 4213 rt : (1,7,5,9)
4406 4269 $\ell_2 : (5/2)^{-1/3}$
4406 8533 $\mathrm{at} : e/2 + 2\pi$
4407 0295 $J_1 : \zeta(9)$
4407 0423 rt : (3,4,9,-5)
4407 2600 $\Psi : 4\sqrt{2}/9$
4407 2731 $\mathrm{cu} : 3\pi/7$
4407 3425 $\mathrm{at} : 3\sqrt{3}/4 - 4\sqrt{5}$
4407 3632 rt : (1,2,8,-4)
4407 3976 $\ell_{10} : \sqrt[3]{21}$
4407 4158 rt : (5,9,-7,2)
4407 4289 $E : 7(1/\pi)/4$
4407 4690 $\Psi : 10\sqrt{2}/3$
4407 4778 $\mathrm{t}\pi : P_{64}$
4407 5368 $\sinh : 13/8$
4407 5873 $\Psi : \exp(1/\sqrt{2})$
4407 5962 $\mathrm{cr} : 3/2 + 2\sqrt{5}/3$
4407 7714 $\exp(1/2) \div s\pi(\sqrt{5})$
4407 9175 $\mathrm{sq} : 3\sqrt{2}/2 - \sqrt{5}/4$
4408 0771 rt : (1,6,5,-9)
4408 1840 $\mathrm{cu} : 4/3 - 2\pi/3$
4408 3139 $\mathrm{at} : 2\sqrt{2}/3 + 3\sqrt{5}$
4408 3893 $3/4 \times s\pi(1/5)$
4408 5928 $\mathrm{sq} : 4\sqrt{3}/3 + 2\sqrt{5}/3$
4408 5973 $e^x : 1/3 + \sqrt{5}/4$
4408 8118 rt : (8,8,-3,3)
4408 9573 $\mathrm{at} : 1/2 - 3e$
4409 0370 rt : (9,3,6,-4)
4409 1019 $10^x : 3\sqrt{2} + 4\sqrt{3}$
4409 1985 rt : (2,6,-1,7)
4409 2264 $erf : (e+\pi)^{-1/2}$
4409 2943 $\mathrm{at} : 2 + 4\sqrt{2}$
4409 2999 $\sinh : 9\zeta(3)/4$
4409 3295 rt : (6,4,-6,1)
4409 3753 $\mathrm{cu} : 2\sqrt{2} - 2\sqrt{5}$
4409 5207 rt : (3,6,-2,3)
4409 5526 $\tanh : (3\pi)^{-1/3}$
4409 7317 $\theta_3 : 1/\sqrt{21}$
4409 8300 $\mathrm{id} : 1 - \sqrt{5}/4$
4409 9806 $2^x : \sqrt[3]{17/3}$
4410 1027 rt : (4,0,-9,1)
4410 1566 rt : (7,1,5,-3)
4410 2303 rt : (1,7,-8,-6)
4410 3022 $J_2 : 4e/3$
4410 3173 rt : (6,9,-4,-3)
4410 3481 $\mathrm{cr} : 3/2 - \sqrt{2}$
4410 3993 $\mathrm{cr} : 7\sqrt[3]{5}/4$
4410 5927 $\lambda : \exp(-19\pi/18)$
4410 7612 $\ell_{10} : 2/3 + 2\pi/3$
4410 9378 $\mathrm{at} : 23/3$
4411 0477 rt : (3,7,-1,-7)
4411 0894 $\mathrm{at} : 4 - 2\sqrt{5}$
4411 2483 $J_0 : 13/8$
4411 3252 rt : (1,8,-8,-5)
4411 6545 rt : (9,8,-3,6)
4411 7127 $\tanh : 9/19$
4411 7430 $\mathrm{sr} : 3\sqrt{2}/4 - \sqrt{3}/2$
4411 7502 $e^x : 2\sqrt{2}/3 - \sqrt{3}/3$
4411 7965 $\mathrm{sq} : 3/4 - \sqrt{2}$
4411 8352 rt : (7,9,6,-5)
4411 8666 $\ell_{10} : e - 3\pi/4$
4411 9436 $10^x : 4/3 + \pi/4$
4411 9782 $J_2 : \sqrt[3]{15}$
4412 0449 $Ei : \sqrt[3]{21}$
4412 0899 $\mathrm{at} : 4\sqrt{3} + \sqrt{5}/3$
4412 1130 $J_2 : 1/\ln(2/3)$
4412 1851 $3 \div s\pi(\sqrt{5})$
4412 3524 rt : (4,9,1,7)
4412 4766 rt : (7,4,-6,4)
4412 5572 $\mathrm{at} : 7\pi^2/9$
4412 7339 rt : (5,1,4,2)
4412 8155 $\ell_{10} : 3\sqrt{3}/2 - \sqrt{5}$
4412 8538 $\zeta : 3/4$
4412 8649 $\mathrm{at} : 3/4 + 4\sqrt{3}$
4413 0158 $\mathrm{sr} : 3/2 + \sqrt{3}/3$
4413 2205 $\exp(2) - \exp(2/3)$
4413 2368 rt : (3,6,-7,-4)
4413 3341 rt : (2,7,-8,-9)
4413 3850 $e^x : 1/3 + 4e$
4413 4069 rt : (5,8,6,7)
4413 4162 $K : (\sqrt{3}/2)^{1/2}$
4413 4358 $\zeta : 5\ln 2/7$
4413 5903 rt : (5,0,-9,2)
4413 7302 $J_2 : \pi^2/4$
4413 8238 $\mathrm{at} : 3\sqrt[3]{2}/8$
4413 9110 rt : (8,9,-1,-2)
4413 9809 $\theta_3 : \exp(-1/(3\pi))$
4413 9809 $\mathrm{id} : \sqrt{3}\pi$
4414 0625 $\mathrm{sq} : 25/16$
4414 0906 $s\pi : 1/3 + 2e/3$
4414 1242 $\mathrm{at} : 9e\pi/10$
4414 3726 $\mathrm{as} : 4/3 - e/3$
4414 4092 rt : (8,4,-6,7)
4414 4232 $\mathrm{sr} : 6\pi^2/5$
4414 4856 $\Psi : 10e\pi$
4414 5372 $\Psi : \ln 2/9$
4414 7042 $\Psi : \pi/5$
4414 8681 $E : 5/9$
4414 9047 rt : (8,4,8,-5)
4415 0972 $\exp(3/4) - s\pi(\sqrt{5})$
4415 1844 rt : (3,8,-8,0)
4415 2023 $J_1 : \sqrt{2} + \sqrt{5} - \sqrt{7}$
4415 3131 $J_1 : \ln(\ln 3/3)$
4415 3330 $\mathrm{sq} : \sqrt{3}/4 - 4\sqrt{5}$
4415 4776 $10^x : \sqrt{2} + \sqrt{3} + \sqrt{7}$
4415 5458 rt : (6,0,-9,5)
4415 6425 $\mathrm{at} : 9\sqrt[3]{5}/2$
4415 7101 $\mathrm{at} : \sqrt{6\pi}$
4415 8882 rt : (8,4,7,3)
4416 1938 rt : (3,5,4,-3)
4416 2165 rt : (1,6,-3,-1)
4416 3912 $\ln(\sqrt{5}) \div \exp(3/5)$
4416 5073 $Ei : 2/17$
4416 7971 rt : (7,0,-9,8)
4416 8684 $\mathrm{sr} : 6\sqrt{3}/5$
4417 0478 rt : (4,8,-8,-9)
4417 1887 $\mathrm{as} : \sqrt[3]{5}/4$
4417 2120 $\mathrm{cu} : 4 - 3\pi/4$
4417 2879 rt : (5,8,-7,8)
4417 4243 rt : (5,9,3,0)
4417 6077 $\mathrm{at} : 3\sqrt{2} + 2\sqrt{3}$
4417 6248 $\sinh : 3e\pi/10$
4417 6604 $\mathrm{sr} : 4/3 + \sqrt{5}/3$
4417 6768 $\tanh : (\sqrt{2\pi})^{-1/2}$
4417 6931 rt : (6,2,7,-4)
4417 8496 $\mathrm{at} : 1 + 3\sqrt{5}$
4417 8531 rt : (6,7,-2,-1)
4418 0232 $\Psi : 5/21$
4418 1197 $\sinh : 3/7$
4418 2925 $\mathrm{sq} : e/3 - \pi/2$
4418 3275 $\ln : 9/14$
4418 5413 $s\pi : 9\sqrt[3]{3}/7$
4418 5420 $\mathrm{at} : 3 + 3\pi/2$
4418 5793 $\mathrm{t}\pi : \ln(\sqrt[3]{2}/3)$
4418 5893 rt : (4,8,7,-5)
4418 6033 $erf : (\sqrt{5}/3)^3$
4418 8062 $\tanh : \exp(-\sqrt{5}/3)$
4418 8476 $\mathrm{sr} : 2/3 - \sqrt{2}/3$
4418 9433 $\mathrm{sr} : 8\sqrt{5}/3$
4419 0093 $\mathrm{sq} : \sqrt{2} + \sqrt{5} - \sqrt{6}$
4419 2398 rt : (2,1,-9,-4)
4419 2704 $E : (e+\pi)^{-1/3}$
4419 3367 $\mathrm{t}\pi : 3\sqrt{2}/4 - 4\sqrt{3}$
4419 3741 rt : (9,4,1,-2)
4419 4431 $\ln(3/4) - \exp(e)$
4419 4472 $10^x : 1/4 + \sqrt{5}/4$
4419 8768 rt : (1,9,4,-6)
4419 8875 rt : (4,7,-9,-5)
4420 0454 $E : 4\ln 2/5$
4420 0593 $\exp(3) + \exp(\sqrt{5})$
4420 2688 $\mathrm{sr} : 3\ln 2$
4420 3706 $\ln(\sqrt{3}) \times \ln(\sqrt{5})$
4420 5259 $\exp(\sqrt{5}) \div \exp(1)$
4420 6047 $\ell_{10} : -\sqrt{2} + \sqrt{3} + \sqrt{6}$
4420 9096 $\Psi : 2e/5$
4420 9228 $Ei : 6/5$
4421 0087 $\mathrm{sq} : 1/4 + 4\sqrt{5}/3$
4421 0815 rt : (9,9,-8,1)
4421 1273 $\mathrm{t}\pi : \sqrt{3}/2 - \sqrt{5}/4$
4421 2530 rt : (2,6,0,-1)
4421 5146 $\sinh : (2e)^{-1/2}$
4421 5516 $\mathrm{at} : (3\pi)^{-1/3}$
4421 6191 $\ln : 3\pi^2/7$
4421 6998 $\mathrm{at} : 4e - \pi$
4422 0005 rt : (1,9,-3,-3)
4422 1003 $\Gamma : 1/\ln(5)$
4422 1438 $\mathrm{cr} : 2 + 4\pi$
4422 4957 $\mathrm{id} : \sqrt[3]{3}$
4422 5458 $2^x : 1/2 - 3\sqrt{5}/4$
4422 7214 $\mathrm{cu} : 3\sqrt{2} + 3\sqrt{5}/4$
4422 8477 rt : (5,3,9,-5)
4422 8485 $\mathrm{cu} : 16/21$
4423 0329 $\ell_{10} : \sqrt{23/3}$
4423 1110 rt : (4,0,6,3)
4423 2793 $\mathrm{cr} : e/3 + 2\pi/3$
4423 4703 $\ln : 2/23$
4423 4782 rt : (7,3,9,4)
4423 4828 rt : (8,7,-1,8)
4423 6092 $\mathrm{sr} : (\ln 2/3)^{-1/2}$
4423 7422 $\mathrm{at} : 9/19$
4423 7566 rt : (7,7,-1,5)
4423 8796 $\ln(3/5) \times s\pi(1/3)$
4423 9213 $J_2 : 10\sqrt{3}/7$
4424 2407 rt : (6,7,-1,2)
4424 3202 rt : (9,8,6,-5)
4424 3282 $J_2 : \exp(e/3)$
4424 5650 $\mathrm{id} : 3e - 3\pi/2$
4424 6723 $J_2 : 7\sqrt{2}/4$
4424 6828 $\mathrm{id} : 6\zeta(3)/5$
4424 7098 $s\pi : 1/2 + 3\sqrt{5}/2$
4424 8447 $10^x : 3 - 3\sqrt{5}/2$
4424 8481 rt : (3,5,-2,9)
4424 9333 rt : (9,9,9,3)
4424 9852 $\mathrm{cr} : \ln 2/8$
4425 0704 $\ln(2/3) \div \ln(2/5)$
4425 1490 $\sinh : (\pi/2)^{1/3}$
4425 2415 $2^x : 1 - \sqrt{2}/3$
4425 2511 $\theta_3 : 1/\ln(2e)$
4425 2564 $\mathrm{sr} : 2/3 + \sqrt{2}$
4425 2594 rt : (8,5,3,-3)
4425 3280 rt : (5,8,0,-2)
4425 3757 $J_0 : 9\sqrt[3]{3}/8$
4425 4759 rt : (7,3,-4,9)
4425 5245 $\mathrm{sq} : P_{81}$
4425 5464 rt : (2,8,5,4)
4425 5496 $e^x : 6\sqrt[3]{3}/7$
4425 9374 rt : (6,5,-8,-4)
4426 1018 $\Psi : 25/23$
4426 2096 $J_1 : 4\sqrt[3]{2}/5$
4426 2259 rt : (6,3,-4,6)
4426 2489 rt : (9,8,-5,-3)
4426 2994 $10^x : (1/\pi)/2$
4426 4446 $\mathrm{cu} : \sqrt{3}/3 - 3\sqrt{5}$
4426 4582 $erf : 1/\sqrt[3]{14}$
4426 6167 $\Psi : 12$
4426 8061 $\ell_{10} : 8\sqrt{3}/5$
4426 8344 rt : (1,7,8,-5)
4426 9504 $\mathrm{id} : 1/\ln(2)$
4426 9742 $e^x : 4/3 + \sqrt{3}$
4427 1909 $\mathrm{cu} : 2\sqrt{5}$
4427 2205 rt : (8,9,8,-6)
4427 4801 rt : (5,3,-4,3)
4427 5397 $J_1 : \zeta(7)$
4427 5858 $2^x : 3\sqrt{2}/4 - \sqrt{5}$
4427 8129 $\mathrm{at} : 3\sqrt{2}/4 + 3\sqrt{5}$
4427 8353 $\mathrm{sq} : (2e)^3$

4427 9797 sr : $\sqrt{13/3}$
4428 0551 $2 \times \exp(1/5)$
4428 0904 id : $1/2+2\sqrt{2}/3$
4428 0996 $s\pi$: $5\pi/2$
4428 1542 J : $\exp(-\sqrt[3]{2})$
4428 1902 rt : (7,2,0,-1)
4428 2278 id : $10\sqrt[3]{5}/7$
4428 2757 rt : (5,1,-7,-7)
4428 3309 $s\pi$: $e\pi/10$
4428 3457 cr : $4\sqrt{2}/3+\sqrt{5}/2$
4428 4874 rt : (9,6,-8,3)
4428 8121 $\exp(2/5)+s\pi(2/5)$
4428 8293 θ_3 : $\exp(-1/(2\pi))$
4428 8293 id : $\sqrt{2}\pi$
4428 8545 ℓ_{10} : $4\ln 2$
4429 0011 rt : (3,1,8,-4)
4429 0519 2^x : $4e\pi/7$
4429 1104 as : 3/7
4429 2642 rt : (7,6,5,-4)
4429 4117 ℓ_2 : $2\sqrt{3}-\sqrt{5}/3$
4429 5371 ℓ_{10} : $3\zeta(3)/10$
4429 7908 at : $(\sqrt{2}\pi)^{-1/2}$
4429 8757 $\exp(3/5) \div \exp(\sqrt{2})$
4430 0046 rt : (8,6,-8,0)
4430 0164 $t\pi$: 8π
4430 1428 at : $3/2+2\pi$
4430 1957 ℓ_{10} : $\sqrt[3]{3}/4$
4430 3093 cr : 2/23
4430 3730 $t\pi$: $4\sqrt{2}/3+2\sqrt{5}/3$
4430 5709 e^x : $(\sqrt[3]{4}/3)^{-1/3}$
4430 5941 $\exp(3/5) \div s\pi(\sqrt{3})$
4430 7314 cr : $5\zeta(3)/2$
4430 9298 $t\pi$: $3e+3\pi/2$
4430 9362 at : $\exp(-\sqrt{5}/3)$
4431 0106 ℓ_{10} : $e/4+2\pi/3$
4431 0597 2^x : P_3
4431 1790 at : $2\sqrt{3}/3-4\sqrt{5}$
4431 3122 ζ : $\sqrt[3]{2}$
4431 4718 $1/4-\ln(2)$
4431 4883 J_1 : $7\sqrt[3]{3}/10$
4431 6000 2^x : 22/9
4431 6369 10^x : $2\sqrt{3}/3+\sqrt{5}/3$
4431 6654 rt : (3,7,-2,-2)
4431 6955 at : $2e+3\pi/4$
4431 8750 tanh : 10/21
4431 9337 at : $9\sqrt{3}/2$
4432 1670 rt : (6,7,7,-5)
4432 3946 10^x : $4\sqrt{3}-2\sqrt{5}/3$
4432 4804 rt : (9,0,5,3)
4432 5474 as : $(2e)^{-1/2}$
4432 5480 ℓ_2 : P_{47}
4432 5502 rt : (7,6,-8,-3)
4432 6029 2^x : $4+\sqrt{2}/4$
4432 6784 ζ : $4\sqrt[3]{5}/3$
4432 8484 sr : $2\sqrt{2}-\sqrt{5}/3$
4433 2090 rt : (3,8,0,-4)
4433 3485 J_1 : $5\sqrt{2}/7$
4433 4057 2^x : 9/17
4433 5937 cu : $(3/2)^3$
4433 7567 id : $5\sqrt{3}/6$
4433 9197 rt : (5,5,8,3)
4434 0416 rt : (7,2,-5,-2)
4434 3284 rt : (1,8,6,-5)
4434 3471 rt : (3,9,-4,2)
4434 3683 rt : (5,8,9,-6)
4434 4050 tanh : $1/\ln(1/(3e))$
4434 4061 e^x : $4e\pi/3$
4434 4210 rt : (6,3,2,-2)
4434 6095 id : $7\pi/9$
4434 6900 10^x : 7/3
4434 6924 10^x : $8\sqrt[3]{5}/3$
4434 9347 10^x : $2\sqrt{2}/3+3\sqrt{3}/2$
4434 9931 at : $4(e+\pi)/3$
4435 0277 $s\pi$: $\sqrt{2}+\sqrt{3}$
4435 2311 $2/3+\ln(4/5)$
4435 7166 erf : $3\ln 2/5$
4435 7484 rt : (8,1,7,-4)
4435 8202 rt : (9,9,1,-3)
4436 0469 e^x : $1/3+4\sqrt{3}$
4436 1045 ln : $7\pi^2/6$
4436 1092 as : $2-\pi/2$
4436 3369 at : $1/3-3e$
4436 3547 ln : $2+\sqrt{5}$
4436 3570 rt : (1,8,-6,1)
4436 5957 rt : (4,8,-4,-3)
4436 6750 rt : (5,9,-5,7)
4436 7610 θ_3 : 13/22
4436 9749 ℓ_{10} : 9/25
4437 0094 Ei : $(1/(3e))^{1/3}$
4437 0965 at : $7\sqrt{5}/2$
4437 1387 2^x : $3\sqrt{2}/2+\sqrt{5}/2$
4437 1400 rt : (2,6,-9,-5)
4437 2471 ℓ_{10} : $2\sqrt[3]{2}/7$
4437 5600 rt : (9,5,-6,8)
4437 7030 rt : (6,6,-8,-6)
4437 7105 $t\pi$: $2\sqrt{3}/5$
4437 7687 e^x : $3/2+\sqrt{5}/3$
4437 9195 rt : (5,4,4,-3)
4437 9539 sr : $3/2+2\sqrt{5}$
4437 9957 J_1 : $6\pi/7$
4438 0080 rt : (7,2,9,-5)
4438 1634 $t\pi$: $4\sqrt{2}/3+4\sqrt{5}/3$
4438 2191 rt : (2,2,5,2)
4438 3341 rt : (8,1,-9,9)
4438 6217 J_2 : $10\sqrt{5}/9$
4438 6308 rt : (2,7,3,9)
4438 6457 J_2 : $3\zeta(3)$
4438 7928 2^x : $4\sqrt{2}/3+\sqrt{5}/4$
4438 8038 $t\pi$: $1/4-2\sqrt{2}/3$
4438 8583 e^x : $1-2e/3$
4438 8779 $\exp(4)-\exp(e)$
4439 1174 cr : $4/3+3\sqrt{5}/4$
4439 2364 id : $4e+\pi/2$
4439 2542 rt : (4,6,-8,3)
4439 4562 ln : $9\sqrt{3}/10$
4439 4725 rt : (5,9,-6,-4)
4439 4834 J_2 : $5\sqrt[3]{3}/2$
4439 5946 J_2 : $\sqrt{13}$
4439 7737 ℓ_2 : $\sqrt{3}\pi$
4439 9254 rt : (9,9,-3,4)
4439 9512 Ei : $2\sqrt{5}/9$
4440 0794 2^x : $3+3\sqrt{2}$
4440 1584 $s\pi$: $1/2+\sqrt{2}/4$
4440 1612 rt : (4,7,8,9)
4440 1705 as : $1/4-e/4$
4440 2210 cu : $5e/9$
4440 2903 $t\pi$: $6\pi/7$
4440 4299 J_0 : $9\sqrt[3]{2}/7$
4440 5549 ln : $1+\sqrt{5}/4$
4440 5648 rt : (6,6,1,7)
4440 6237 λ : $\exp(-20\pi/19)$
4440 7393 rt : (8,5,-6,5)
4440 8392 rt : (3,8,-6,7)
4440 8509 E : $1/\sqrt[3]{6}$
4441 1800 rt : (2,7,1,-2)
4441 4359 rt : (7,1,-9,6)
4441 4487 e^x : $2e\pi/5$
4441 5274 rt : (5,2,-2,8)
4441 5378 at : $5\pi/2$
4441 5522 cr : $4\sqrt{2}+4\sqrt{5}$
4441 6210 sr : $\sqrt{2}/4+\sqrt{3}$
4441 6609 rt : (8,3,-7,-3)
4441 7062 ℓ_{10} : $\sqrt{2}/3+4\sqrt{3}/3$
4441 7206 rt : (3,6,8,-5)
4441 7632 2^x : $9\sqrt[3]{5}/5$
4442 1098 tanh : $3(1/\pi)/2$
4442 3185 rt : (4,2,-5,-9)
4442 3981 rt : (1,1,7,3)
4442 4348 E : 11/20
4442 5808 sq : $1/\ln(\ln 2)$
4442 5960 2^x : $3\sqrt{2}/8$
4442 6035 $3 \div \ln(4/5)$
4442 6936 id : $3\sqrt{2}/2-3\sqrt{5}/4$
4442 7190 id : $1/2+4\sqrt{5}$
4442 8314 Ei : $\sqrt{2}+\sqrt{5}-\sqrt{6}$
4442 8957 Li_2 : $4\ln 2/7$
4442 9597 at : $2\sqrt{3}/3+3\sqrt{5}$
4443 1020 erf : 5/12
4443 1627 rt : (1,8,3,-3)
4443 1820 sr : $3\sqrt{2}+\sqrt{3}$
4443 4994 ln : $3\sqrt[3]{5}/8$
4443 5689 sr : $2\sqrt{2}/3-\sqrt{5}/3$
4443 6475 $t\pi$: $4e-2\pi$
4443 6873 $\sqrt{3}+\ln(3/4)$
4443 7482 rt : (1,7,7,-8)
4443 7497 $t\pi$: $3e/4+2\pi/3$
4443 7593 cr : $3e/4+4\pi$
4444 0206 e^x : $\sqrt{3}/2-3\sqrt{5}/4$
4444 1176 Ei : 13/7
4444 1729 ln : $9\ln 2/4$
4444 1751 e^x : $2\sqrt{3}-\sqrt{5}/2$
4444 1920 at : 10/21
4444 2489 at : $2\sqrt{2}/3+4\sqrt{3}$
4444 3125 Ψ : 5/12
4444 3428 ζ : $(\sqrt[3]{5}/3)^{-1/3}$
4444 4444 id : 4/9
4444 4620 2^x : $5(1/\pi)/3$
4444 4757 Ψ : $10\zeta(3)$
4444 5619 as : $\sqrt{3}/4+\sqrt{5}/4$
4444 5848 at : $3-4e$
4445 0105 $\exp(3/4) \div s\pi(1/3)$
4445 0265 sq : $5\sqrt[3]{3}/6$
4445 0953 rt : (8,9,-3,1)
4445 2477 rt : (8,6,7,-5)
4445 2875 $s\pi$: $3\sqrt{2}/2+\sqrt{3}$
4445 4630 $\ln(4/5)-\exp(1/5)$
4445 5016 as : $3\sqrt{2}/4-2\sqrt{5}/3$
4445 5590 $\exp(1/3)-s\pi(2/5)$
4445 6182 rt : (7,5,-6,2)
4445 6203 J : $\ln(\sqrt{2}\pi/3)$
4445 6448 e^x : $1/4-3\sqrt{2}/4$
4445 7996 rt : (2,9,-9,7)
4445 8218 sq : $2e/3+\pi/2$
4445 8522 E : $\ln(\sqrt{3})$
4445 8606 rt : (7,7,9,-6)
4446 0475 rt : (6,1,-9,3)
4446 1427 rt : (8,6,-2,-1)
4446 1912 rt : (4,5,-8,9)
4446 2472 at : $3\sqrt{2}/4-4\sqrt{5}$
4446 3156 sinh : $\sqrt{22}$
4446 3507 id : $4\sqrt{2}/3+\sqrt{5}/4$
4446 3533 sq : $4+4\pi$
4446 4677 e^x : 14/5
4446 5898 rt : (1,8,4,-1)
4446 6304 10^x : $4\sqrt{2}/3+\sqrt{5}/2$
4446 6532 cu : $3\sqrt{2}-2\sqrt{3}/3$
4446 6786 id : $\exp(1/e)$
4446 7585 at : $1/\ln(1/(3e))$
4446 8482 Ei : $(\sqrt[3]{3})^{1/2}$
4447 0888 rt : (7,7,0,-2)
4447 2611 ln : $4\sqrt[3]{3}/9$
4447 6919 rt : (6,8,2,-3)
4447 7198 at : $4\sqrt{2}+\sqrt{5}$
4447 8093 rt : (8,3,2,1)
4447 9223 J : $\exp(-17\pi/20)$
4448 0539 J_2 : 18/5
4448 1100 rt : (5,9,4,-4)
4448 1559 at : $4\pi^2/5$
4448 2744 rt : (9,8,-1,9)
4448 2891 cu : $7\sqrt{3}/9$
4448 4169 rt : (4,2,-2,5)
4448 4971 tanh : 11/23
4448 6518 rt : (3,7,0,2)
4448 7489 rt : (7,2,4,2)
4448 7999 at : $3\sqrt{2}/2-3\sqrt{3}/2$
4448 8373 rt : (2,6,-8,-7)
4448 8728 ℓ_{10} : $2+\pi/4$
4448 9011 cr : $\sqrt{2}/2+4\sqrt{3}/3$
4449 1909 rt : (6,1,6,3)
4449 4079 sq : $\zeta(3)$
4449 4479 rt : (5,0,8,4)
4449 4568 ln : $2-e/2$
4449 5684 sr : $7e\pi/10$
4449 6007 at : $1/4-3e$
4450 0661 $\ln(3) \times \exp(4/5)$
4450 0929 $\ln(\sqrt{2}) \times \exp(1/4)$
4450 1915 sr : $2\ln 2/7$
4450 4186 rt : (4,4,-8,-4)
4450 4838 J_1 : $(\pi/3)^{1/3}$
4450 5542 sq : $1/2+3e/4$

4450 9581 cu : $5(e+\pi)/6$
4451 0658 e^x : $3/2+2e/3$
4451 0818 $1/5\times\exp(4/5)$
4451 0891 ln : $1/2+3\sqrt{2}/4$
4451 2458 e^x : $3/2-4\sqrt{3}/3$
4451 8587 ln : $3\sqrt{2}$
4451 9228 tπ : $3\sqrt{2}/2+\sqrt{5}/3$
4451 9801 Ei : $(\ln 2)^{-1/2}$
4452 0931 cu : $7\pi^2/4$
4452 2868 tπ : $2/15$
4452 3229 sq : $1/2-2\pi$
4452 3457 sq : P_{26}
4452 3735 ℓ_2 : $1/4+3\sqrt{3}$
4452 4205 ℓ_2 : P_{63}
4452 5491 rt : (9,1,9,-5)
4452 6524 10^x : $e/7$
4452 6537 Li_2 : $1/\sqrt[3]{16}$
4452 7325 at : $3/2-3\pi$
4452 9074 ℓ_{10} : $2/3+3\sqrt{2}/2$
4453 2691 at : $1+4\sqrt{3}$
4453 6240 rt : (5,9,-5,0)
4453 8200 sq : $2+e/3$
4453 8645 rt : (6,0,-7,8)
4453 9230 rt : (1,3,-8,-7)
4454 0356 ℓ_{10} : $3/4+3e/4$
4454 0412 Γ : $1/\ln(\sqrt{3}/3)$
4454 0531 rt : (6,5,-6,-1)
4454 1439 rt : (7,4,-4,7)
4454 2464 e^x : $1/\sqrt[3]{20}$
4454 2835 rt : (1,4,7,-4)
4454 3977 10^x : $4/25$
4454 4726 rt : (8,8,-1,6)
4454 4895 $\ln(2/3)\times\ln(3)$
4454 5052 e^x : $7/19$
4454 5261 at : $2\sqrt{3}+2\sqrt{5}$
4454 5749 at : $3(1/\pi)/2$
4454 5798 sq : $1/2-3\pi/4$
4454 6147 rt : (7,9,-3,-2)
4454 6966 2^x : $2/3+\sqrt{5}/2$
4454 8483 sinh : $3(e+\pi)/8$
4454 9743 cr : $10e/9$
4455 0293 rt : (9,9,-9,-5)
4455 0365 J_0 : $7\ln 2/3$
4455 2044 sr : $4e/3+3\pi/4$
4455 2558 id : $2\sqrt{3}+4\sqrt{5}/3$
4455 2845 rt : (2,3,5,7)
4455 3608 J_0 : $\exp(\sqrt[3]{3}/3)$
4455 5007 e^x : $5\sqrt{5}/6$
4455 5415 sπ : $4\sqrt{2}+3\sqrt{3}$
4455 5516 tπ : $1/4-e/4$
4455 6133 rt : (6,8,-7,1)
4455 6530 ℓ_{10} : $3\sqrt{3}/4+2\sqrt{5}/3$
4455 7805 at : $1-4\sqrt{5}$
4455 8092 rt : (1,8,-4,-2)
4456 0083 rt : (5,5,8,-5)
4456 1596 rt : (8,8,-7,-4)
4456 3384 id : $7(1/\pi)/5$
4456 4563 ln : $8\sqrt[3]{3}$
4456 6288 sq : $7\sqrt[3]{2}/2$
4456 6803 sinh : $7\sqrt{3}/3$
4456 9498 sr : $3+4\sqrt{5}/3$
4456 9849 rt : (9,6,9,-6)
4457 4045 sr : $1+4e$
4457 5235 rt : (4,9,-8,-5)
4457 5999 rt : (9,7,-8,1)
4457 6347 id : $1/\ln(3\pi)$
4457 9006 rt : (3,6,0,8)
4458 0140 ln : $2/3+4e$
4458 0243 rt : (7,7,-5,-3)
4458 2472 sq : $4-4\sqrt{5}$
4458 3746 rt : (7,3,4,-3)
4458 7942 e^x : $5\sqrt{3}/7$
4458 8035 rt : (1,9,-4,2)
4458 9613 rt : (3,8,-6,-4)
4459 0380 cr : $2/3+3\pi/4$
4459 2001 at : $4\sqrt{2}+4\sqrt{3}/3$
4459 3433 e^x : $2\sqrt{5}/5$
4459 3852 rt : (1,9,7,-5)
4459 5446 ln : $\ln 2/8$
4459 5480 sπ : $\exp(-1/(2\pi))$
4459 6268 rt : (5,5,3,9)
4459 6938 sq : $5\zeta(3)/9$
4459 9735 ℓ_{10} : $8\pi/9$
4459 9761 sr : $23/11$
4460 1485 e^x : $2+2\sqrt{3}/3$
4460 2271 ln : $3\sqrt{3}/2+4\sqrt{5}$
4460 2599 ℓ_{10} : $3\sqrt{2}/4+\sqrt{3}$
4460 3102 sr : $5(1/\pi)/8$
4460 5090 rt : (8,2,2,-2)
4460 5905 J_2 : $5/2$
4460 7847 rt : (3,9,-2,9)
4460 8272 rt : (6,9,6,8)
4460 8600 Γ : $8(1/\pi)/7$
4460 8928 tπ : $\zeta(3)/9$
4460 9293 rt : (4,8,-6,-4)
4461 0489 tπ : $5\ln 2/4$
4461 0554 at : $11/23$
4461 0615 sq : $4e/3+\pi/4$
4461 1281 rt : (2,7,-4,-3)
4461 3193 J_0 : $8\sqrt{2}/7$
4461 5242 id : $1/4+3\sqrt{3}$
4461 6123 ln : $3\sqrt{2}/2-\sqrt{5}/4$
4461 6462 10^x : P_{20}
4461 6834 rt : (6,6,-3,-2)
4461 9194 rt : (5,0,-7,5)
4462 0742 2^x : $1/4-\sqrt{2}$
4462 2110 10^x : $4\sqrt{2}/3-\sqrt{5}$
4462 2922 rt : (8,2,-9,7)
4462 3702 sq : $4\sqrt{3}/3-\sqrt{5}/3$
4462 3869 rt : (7,8,4,-4)
4462 4118 $\ln(3/5)\div\ln(\pi)$
4462 4361 tanh : $12/25$
4462 5023 J_2 : $8\pi/7$
4462 5725 2^x : $4e/3+2\pi$
4462 6976 at : $3\sqrt{2}/4+4\sqrt{3}$
4462 7191 10^x : $\sqrt[3]{3}/9$
4462 8284 J_1 : $3e/8$
4462 8710 ln : $16/25$
4463 0180 sinh : $3\sqrt[3]{3}/10$
4463 1151 rt : (9,6,-6,6)
4463 5963 sπ : $3\sqrt{3}/2-\sqrt{5}/3$
4463 9501 ℓ_{10} : $5\sqrt{5}/4$
4464 0079 rt : (7,7,-9,2)
4464 0875 sr : $7\sqrt[3]{5}/2$
4464 1569 $s\pi(\pi)\div s\pi(\sqrt{2})$
4464 2040 rt : (9,1,0,-1)
4464 2621 rt : (2,6,9,3)
4464 4133 at : 8
4464 4954 rt : (8,7,2,-3)
4464 5364 id : $9e/10$
4464 5877 rt : (7,8,-1,3)
4464 6206 J : $1/\sqrt[3]{13}$
4464 6383 at : $2\sqrt{2}/3-4\sqrt{5}$
4464 7197 E : $6/11$
4464 7658 rt : (1,6,-2,-2)
4465 0534 2^x : $2/3+e$
4465 3242 e^x : $3/2+e/2$
4465 4662 2^x : $\exp(-\sqrt[3]{2}/2)$
4465 5264 at : $3\sqrt{3}/4+3\sqrt{5}$
4465 8371 $1/3+\exp(\sqrt{2})$
4465 8716 Li_2 : $5(1/\pi)/4$
4465 8860 at : $4/3-2e/3$
4466 0634 tanh : $(\ln 2)^2$
4466 0961 10^x : $4e-2\pi/3$
4466 1973 rt : (4,7,-6,2)
4466 2304 J_0 : $21/13$
4466 2478 Li_2 : $(e/3)^{1/2}$
4466 4459 rt : (4,6,1,1)
4466 5241 Ei : $8\sqrt[3]{2}$
4466 5621 rt : (7,6,8,3)
4466 7177 sinh : $\sqrt{3}/4$
4466 7481 $\pi\div\exp(1/4)$
4466 7899 cu : $\sqrt{3}/2+3\sqrt{5}/4$
4466 7968 J_1 : $6\sqrt{5}/5$
4466 9183 e^x : $17/19$
4467 5027 rt : (9,6,0,-2)
4467 6031 rt : (8,7,-8,-2)
4467 6760 e^x : $7\sqrt{2}/8$
4467 7705 tπ : $\sqrt[3]{13/2}$
4467 8453 tanh : $(\ln 2/3)^{1/2}$
4467 8552 rt : (1,7,-6,-9)
4467 8584 cu : $\sqrt{2}/4-\sqrt{5}/2$
4468 0436 rt : (2,1,9,4)
4468 1833 at : $2\sqrt{2}+3\sqrt{3}$
4468 3691 rt : (4,1,-9,-3)
4468 4394 tanh : $\sqrt[3]{3}/3$
4468 6179 cr : $5(e+\pi)/2$
4468 6316 10^x : $(\sqrt[3]{2}/3)^{1/2}$
4468 7670 rt : (7,1,8,4)
4468 8003 $2/3\div\exp(2/5)$
4468 8806 rt : (6,3,-2,9)
4468 8990 cu : $3-\pi/3$
4468 9552 sinh : $5\ln 2/8$
4469 0229 tanh : $2\zeta(3)/5$
4469 0568 sq : $6(e+\pi)/5$
4469 1898 ln : $4/17$
4469 2647 $1/3\div s\pi(\sqrt{3})$
4469 3017 2^x : $1/4+2\sqrt{2}$
4469 3718 rt : (2,4,9,-5)
4469 3787 sπ : $e-\pi/2$
4469 3845 sq : $2\sqrt{2}+3\sqrt{3}/2$
4469 4368 $\exp(1/5)+\exp(4/5)$
4469 6128 sinh : $3e/7$
4469 6249 cu : $9\pi/2$
4469 6666 2^x : $\sqrt{5/3}$
4469 7176 rt : (1,7,-4,-5)
4469 7220 rt : (7,4,8,-5)
4469 8388 J_2 : $(2\pi)^{1/2}$
4470 0296 cr : $2/3-\sqrt{3}/3$
4470 0491 rt : (4,6,-6,8)
4470 0881 ℓ_{10} : $3/2+3\sqrt{3}/4$
4470 2552 rt : (7,9,8,-6)
4470 3353 sr : $3-e/3$
4470 5847 rt : (2,8,-6,-9)
4470 5954 rt : (7,7,1,8)
4470 6466 Γ : $13/21$
4470 6635 cu : $\sqrt{2}+4\sqrt{5}$
4470 7697 cr : $3e/2-\pi/3$
4470 7711 at : $4/3+3\sqrt{5}$
4470 8314 rt : (7,2,-9,4)
4470 9355 e^x : $-\sqrt{5}+\sqrt{6}+\sqrt{7}$
4471 0610 Γ : $4/11$
4471 0896 Li_2 : $(e/2)^{-3}$
4471 2846 E : $1/\ln(2\pi)$
4471 2985 tπ : $1/3+e/2$
4471 3586 2^x : $1+\pi/4$
4471 4864 at : $4\sqrt{3}+\sqrt{5}/2$
4471 5546 $\exp(e)-s\pi(1/4)$
4471 5803 ℓ_{10} : $5/14$
4471 6474 sπ : $4\sqrt{2}+2\sqrt{5}/3$
4471 8094 cu : $13/17$
4472 0250 id : $(2\pi/3)^{1/2}$
4472 1359 id : $\sqrt{5}/5$
4472 1445 cr : $7\sqrt{3}/4$
4472 5733 rt : (8,6,-6,3)
4472 6923 2^x : $8/15$
4472 7119 rt : (3,5,-9,-8)
4472 7546 ln : $4\sqrt{3}/3-\sqrt{5}/3$
4472 7912 rt : (8,8,6,-5)
4472 7978 Ei : $5\sqrt[3]{3}/6$
4472 8181 ζ : $25/11$
4472 8733 rt : (3,8,7,-5)
4472 9790 rt : (2,8,-8,-5)
4473 0085 rt : (8,3,6,-4)
4473 1201 rt : (7,8,-9,-5)
4473 1497 ln : 10π
4473 1605 rt : (3,3,7,-4)
4473 3192 sq : $3\sqrt{2}/2+4\sqrt{5}$
4473 3767 rt : (2,9,4,-7)
4473 4415 rt : (8,2,6,3)
4473 4690 E : $e/5$
4473 7044 sq : $\sqrt{2}-\sqrt{5}/3$
4473 7085 rt : (7,3,-9,-4)
4473 8141 rt : (3,2,7,3)
4474 0285 cu : $4/3+4\pi$
4474 2881 rt : (7,1,-7,9)
4474 4272 erf : $\sqrt[3]{2}/3$
4474 4304 at : $e/2-3\pi$
4474 4332 10^x : $3\sqrt{2}-2\sqrt{3}/3$
4474 4845 rt : (2,2,-8,3)

4474 5251 $\Psi : 3e/4$
4474 5764 as : $3\sqrt[3]{3}/10$
4474 5897 $\ell_2 : 11/15$
4474 6007 $\exp(1/5) - s\pi(e)$
4474 7000 $Ei : \exp(1/(2e))$
4474 7264 tπ : $9\sqrt[3]{2}/10$
4474 7277 rt : (8,7,6,2)
4474 7282 sr : P_{16}
4474 7422 $\ell_{10} : \sqrt[3]{22}$
4474 7422 sr : P_{40}
4474 8597 sr : $1/3 + 4\sqrt{2}$
4475 0089 $3/5 \times s\pi(\sqrt{3})$
4475 1997 at : 12/25
4475 4545 $\ln(2) - \exp(\pi)$
4475 4617 $J_2 : (e/2)^3$
4475 4625 $J_1 : (\pi/3)^{1/2}$
4475 5325 rt : (4,0,-7,2)
4475 8073 $\zeta : 2/3$
4475 8637 cu : $1/3 - 3\pi$
4475 9375 id : $\sqrt{2}/2 - 2\sqrt{3}/3$
4475 9671 sr : $\zeta(3)/6$
4476 0102 rt : (8,5,-4,8)
4476 0494 rt : (7,2,-9,3)
4476 0681 rt : (9,7,4,-4)
4476 2120 at : $4 + 3e/2$
4476 3364 at : $\sqrt{3}/2 - 4\sqrt{5}$
4476 4027 $J_2 : 3(e+\pi)/7$
4476 4479 tπ : $\sqrt{3}/2$
4476 6441 rt : (4,7,5,-4)
4476 7728 rt : (3,7,7,2)
4476 8015 cu : $2e/3 - \pi/3$
4477 0005 rt : (3,5,-8,-8)
4477 0766 $\lambda : (3)^{-3}$
4477 1327 sinh : $\sqrt{2} - \sqrt{5} + \sqrt{6}$
4477 3404 $1/2 + \exp(2/3)$
4477 4181 rt : (1,7,-6,-4)
4477 6386 rt : (9,2,4,-3)
4477 7344 tπ : $\sqrt[3]{14}$
4477 8057 at : P_{21}
4477 8422 $Ei : \zeta(3)$
4477 8648 $J : \ln 2/10$
4477 8919 rt : (6,9,-3,-1)
4477 9992 rt : (9,9,-1,7)
4478 0265 $\ln(\sqrt{2}) \div s\pi(e)$
4478 2217 $Ei : (\sqrt[3]{5})^{-1/3}$
4478 3239 as : $\sqrt{3}/4$
4478 3251 at : $4/3 - 3\pi$
4478 4514 tπ : $8\sqrt[3]{3}/5$
4478 5137 rt : (6,7,-7,-4)
4478 6731 $\Psi : 5e\pi/9$
4478 7644 cu : $3\sqrt{2} - 3\sqrt{3}/2$
4478 7712 $\exp(1/3) \div s\pi(\sqrt{2})$
4478 8373 rt : (4,2,5,-3)
4478 8809 at : $(\ln 2)^2$
4478 9128 $2^x : 25/14$
4478 9158 at : $2e/3 + 2\pi$
4479 0012 as : P_{10}
4479 0155 $J_2 : 4\pi/5$
4479 0181 $\ell_{10} : 7\zeta(3)/3$
4479 0532 $\Psi : 1/12$
4479 1673 rt : (5,4,-4,1)
4479 3274 $Li_2 : (2\pi)^{-1/2}$
4479 4180 $\zeta : (1/(3e))^{1/3}$
4479 4515 rt : (9,3,-9,8)
4479 4598 $10^x : 2/3 + \sqrt{3}$
4479 5076 $2^x : 4e + 3\pi$
4479 5311 $e^x : 3/4 + 3\sqrt{3}/2$
4479 5827 cr : $e/4 + 3\pi/4$
4479 6228 rt : (1,9,-2,-1)
4479 6471 rt : (4,1,-5,-7)
4479 8987 rt : (3,3,-9,4)
4480 0750 $\theta_3 : 3\sqrt[3]{2}/4$
4480 2539 $\sqrt{3} - \exp(1/4)$
4480 2613 $\Gamma : 3e\pi/2$
4480 4401 ln : $2\sqrt{5}/7$
4480 5394 $erf : \exp(-\sqrt{3}/2)$
4480 5486 rt : (9,8,8,-6)
4480 5904 as : $5\ln 2/8$
4480 6894 at : $(\ln 2/3)^{1/2}$
4480 7056 rt : (1,5,1,-6)
4481 1882 $\Psi : (2e/3)^{-2}$
4481 1973 $J_2 : 8\sqrt{5}/5$
4481 2923 at : $\sqrt[3]{3}/3$
4481 2974 rt : (9,3,4,2)
4481 3831 tπ : $(\ln 3/3)^2$
4481 4177 $\Psi : \sqrt{5} + \sqrt{6} - \sqrt{7}$
4481 4567 rt : (4,8,9,-6)
4481 5468 cu : $4\sqrt{2}/5$
4481 6603 $10^x : 3\sqrt{3} + \sqrt{5}$
4481 6924 cu : $\sqrt{2}/2 + 4\sqrt{3}/3$
4481 8526 $E : (2\pi)^{-1/3}$
4481 8660 $2^x : 4\zeta(3)/9$
4481 8845 at : $2\zeta(3)/5$
4481 9304 $Li_2 : 20/21$
4481 9308 rt : (5,9,6,5)
4482 0235 $\exp(\sqrt{3}) \times s\pi(\sqrt{2})$
4482 1721 $J_2 : 9\sqrt{5}/8$
4482 2696 $\zeta : (\pi/3)^{1/2}$
4482 5153 sq : $\ln((e+\pi)/3)$
4482 6849 at : $\exp(2\pi/3)$
4482 7814 sr : $2\sqrt{2}/3 + 2\sqrt{3}/3$
4482 8466 rt : (4,9,1,-1)
4482 8622 $\zeta : 2\sqrt{5}/9$
4482 8770 rt : (5,3,-2,6)
4482 8773 sr : $3/2 - 3\sqrt{3}/4$
4482 8858 rt : (9,3,8,-5)
4482 9672 at : $\sqrt{2} + 3\sqrt{5}$
4482 9823 sπ : $7\ln 2$
4483 1065 $E : 13/24$
4483 2639 $2^x : 1/3 - 2\sqrt{5}/3$
4483 4935 id : $4e - 3\pi$
4483 5874 $J_2 : \sqrt{19/3}$
4483 7906 $e^x : 3 - \sqrt{3}/2$
4483 8927 $J_1 : 3\sqrt[3]{5}/5$
4484 0262 rt : (6,8,-1,0)
4484 1638 $\ell_2 : 3\sqrt[3]{5}/7$
4484 2657 cr : $1 + 3e/4$
4484 3248 tπ : $((e+\pi)/3)^{-3}$
4484 3674 $10^x : 7/18$
4484 3954 $10^x : 1/4 + 3\sqrt{2}/4$
4484 4036 tπ : $(\sqrt[3]{5})^{1/2}$
4484 4158 rt : (4,3,9,-5)
4484 4547 $\pi - \ln(2)$
4484 4641 sinh : $(e/2)^{1/2}$
4484 6286 $erf : 8/19$
4484 6789 $E : (e/2)^{-2}$
4484 7539 sq : $3/4 + e/2$
4485 1897 rt : (4,3,5,2)
4485 5391 rt : (7,7,-8,-5)
4485 5871 $10^x : 5(e+\pi)/2$
4485 6207 $10^x : 1/\sqrt[3]{17}$
4485 8285 rt : (5,7,7,-5)
4485 9367 rt : (6,1,-7,6)
4486 1092 sinh : 10/23
4486 1374 $\ell_{10} : 1/2 + 4\sqrt{3}/3$
4486 1837 tπ : $2e - \pi/2$
4486 1881 $\Gamma : \exp(-\sqrt[3]{3}/3)$
4486 2247 $\zeta : 9\zeta(3)/7$
4486 3532 sπ : $\sqrt{2}/2 - \sqrt{5}/4$
4486 7515 $10^x : 2/3 + 3\sqrt{5}$
4486 8189 cr : $4\sqrt{2}/3 + 2\sqrt{3}/3$
4486 8447 $\ln(4/5) - \exp(4/5)$
4486 9060 rt : (6,7,1,5)
4486 9346 $\Gamma : \sqrt[3]{3}/7$
4487 0220 rt : (9,2,8,4)
4487 0798 $E : 3\sqrt[3]{3}/8$
4487 2127 $1/5 - \exp(1/2)$
4487 2633 sr : P_8
4487 2681 rt : (5,6,-5,-3)
4487 3928 $e^x : \sqrt{3}/3 + \sqrt{5}/2$
4487 5011 tπ : 4/13
4487 5321 cr : $7(e+\pi)$
4487 7548 rt : (7,8,-5,0)
4487 7788 $\zeta : 17/11$
4487 7901 at : $3e$
4487 8410 $\Psi : \sqrt[3]{17/2}$
4487 8582 tπ : $7e\pi/8$
4487 9895 id : $\pi/7$
4487 9897 $J_2 : 2\sqrt[3]{2}$
4487 9918 sr : $(\sqrt[3]{4}/3)^3$
4488 1381 $e^x : \sqrt{2}/4 - 2\sqrt{3}/3$
4488 2729 rt : (5,1,3,-2)
4488 6352 cu : $3/4 + 3\sqrt{2}$
4489 0936 $e \div \ln(3/4)$
4489 1218 rt : (7,5,-4,5)
4489 7959 sq : 13/7
4489 8704 rt : (6,5,1,-2)
4489 9730 $J_2 : (\sqrt[3]{4}/3)^{-2}$
4490 0989 tπ : $2e/3 - 4\pi/3$
4490 1315 rt : (4,2,9,4)
4490 2954 $J_2 : 25/7$
4490 3158 rt : (5,7,-9,-5)
4490 3738 $\Gamma : ((e+\pi)/3)^{-2}$
4490 3756 rt : (8,3,-9,5)
4490 4433 $\ell_{10} : 1 + 2e/3$
4490 4514 id : P_{41}
4490 7347 $\Gamma : 3\sqrt[3]{3}/7$
4490 8023 $\Psi : 5\sqrt{3}/8$
4490 8474 rt : (7,9,-1,-2)
4490 9227 sr : $5\sqrt[3]{2}/3$
4490 9878 rt : (9,9,0,-1)
4491 0183 rt : (5,2,7,-4)
4491 1483 at : $3\sqrt{3} + 4\sqrt{5}/3$
4491 3023 $\ell_2 : (e+\pi)/8$
4491 3068 $E : 3\sqrt[3]{2}/7$
4491 3767 sr : 21/10
4491 5485 ln : $2 - \sqrt{3}/4$
4491 6269 rt : (6,6,5,-4)
4491 8545 $\Gamma : 5\pi/6$
4492 0405 $\Psi : 19/4$
4492 3429 $\ell_{10} : \sqrt{3}/3 + \sqrt{5}$
4492 3978 $erf : (4/3)^{-3}$
4492 4737 rt : (6,7,9,-6)
4492 5656 $\exp(5) + s\pi(\sqrt{2})$
4492 7983 at : $4 + 4\pi/3$
4492 8016 $\ell_2 : \ln(\sqrt[3]{3})$
4492 8297 $Li_2 : 2/5$
4492 9311 rt : (8,9,-1,4)
4492 9492 tπ : $3\sqrt{2} - \sqrt{3}/2$
4493 1006 rt : (9,8,-8,-1)
4493 2439 $\theta_3 : 2/9$
4493 2724 rt : (8,8,1,9)
4493 2896 $1 \div \exp(4/5)$
4493 3899 rt : (9,7,-6,4)
4493 4066 id : $5\pi^2/3$
4493 5365 $J_2 : 7\sqrt[3]{3}/4$
4493 5990 rt : (9,6,-4,9)
4493 6032 at : $3/4 - 4\sqrt{5}$
4493 8791 at : $3 + 3\sqrt{3}$
4494 2706 $J_0 : \sqrt{2} - \sqrt{6} + \sqrt{7}$
4494 3297 $s\pi(\sqrt{5}) + s\pi(e)$
4494 5170 sπ : $(\pi/3)^3$
4494 6677 $10^x : 17/21$
4494 7804 sr : $\sqrt{2}/7$
4494 8974 id : $\sqrt{6}$
4495 0034 at : $7(e+\pi)/5$
4495 1300 $e^x : 3\sqrt{2} + \sqrt{3}/2$
4495 1415 $e^x : 1/\ln(\sqrt[3]{5})$
4495 3205 $\ell_2 : 3/2 + 3\pi$
4495 4013 $E : 7\ln 2/9$
4495 4290 $J : \sqrt[3]{5}/9$
4495 5532 sq : $\sqrt{2}/4 + 4\sqrt{5}$
4495 6442 at : $3/2 + 3\sqrt{5}$
4495 9992 sq : $3\sqrt{3}/4 - 4\sqrt{5}$
4496 0129 sq : $1/3 + 4\pi/3$
4496 2301 $\Gamma : -\sqrt{2} + \sqrt{3} - \sqrt{6}$
4496 3366 rt : (9,7,8,3)
4496 4310 sq : $1/\ln(\sqrt{2}\pi)$
4496 4431 id : $1/\sqrt[3]{11}$
4496 6185 sπ : $\sqrt{2}/3 + 3\sqrt{5}/4$
4496 8938 sr : $2e/3 + 4\pi/3$
4496 9126 sinh : $(\sqrt[3]{4})^{1/3}$
4496 9394 at : $2\sqrt{3} - 4\sqrt{5}/3$
4497 0334 $\sqrt{2} \times \exp(2)$
4497 0463 cr : $2(1/\pi)/7$
4497 0709 $e^x : 7(1/\pi)/6$
4497 0874 $\Psi : 3\ln 2/5$
4497 1385 $J_1 : 5\sqrt[3]{3}/7$
4497 3383 $\ln(\sqrt{3}) \div \exp(1/5)$
4497 4729 $\Gamma : 5e\pi/9$

4497 6137 e^x : $1/2 - 3\sqrt{3}/4$
4497 6310 J_1 : $6\zeta(3)/7$
4497 7365 cr : $2 + \pi/3$
4497 7947 at : $4e/3 - \pi$
4497 9244 rt : (4,7,-4,7)
4497 9294 tπ : $\exp(-1/e)$
4497 9686 as : 10/23
4497 9709 2^x : $1/\ln(\sqrt[3]{2}/3)$
4498 0478 at : $5\pi^2/6$
4498 2732 rt : (7,6,7,-5)
4498 4389 rt : (4,8,-4,1)
4498 4471 2^x : $4 - 2\sqrt{3}$
4498 5347 E : 7/13
4498 7884 rt : (6,2,9,-5)
4498 9253 rt : (8,9,1,-3)
4499 0882 J : $2\sqrt{3}/7$
4499 1641 rt : (4,9,-4,-5)
4499 3365 sinh : 7/6
4499 4486 rt : (9,8,4,1)
4499 4984 rt : (4,6,-7,-4)
4499 5971 rt : (3,7,2,7)
4499 6809 rt : (7,5,3,-3)
4499 8431 ℓ_2 : $1 + \sqrt{3}$
4499 8493 Ψ : 7/23
4499 8652 at : $\sqrt{2}/2 - 4\sqrt{5}$
4500 0000 id : 9/20
4500 0947 ζ : $9\sqrt[3]{2}/5$
4500 1917 J_0 : $\ln(5)$
4500 6448 cu : $1/3 + \sqrt{3}/4$
4500 6600 at : $4 + 3\sqrt{2}$
4500 7856 rt : (6,1,5,-3)
4500 7871 Ψ : $9\zeta(3)/10$
4500 8051 rt : (5,2,-9,-4)
4500 8972 cu : $10\sqrt{3}/3$
4500 9658 rt : (3,8,2,1)
4501 0054 id : $e/3 - 3\pi/4$
4501 0249 sinh : $7e\pi/5$
4501 0882 e^x : $4 + 4\sqrt{5}/3$
4501 1035 rt : (8,8,-3,-1)
4501 2022 rt : (3,9,-1,-2)
4501 5268 sq : $(1/(3e))^{-2}$
4501 5815 $\sqrt{2} \div \pi$
4501 5880 Li_2 : $\zeta(3)/3$
4501 8814 rt : (2,6,8,-5)
4501 8930 2^x : $2 - \sqrt{2}/2$
4501 9522 tπ : $3\sqrt[3]{3}/5$
4502 0660 at : $3\sqrt{2}/4 - \sqrt{3}/3$
4502 2168 10^x : $\ln(1/\sqrt{2})$
4502 3170 ℓ_{10} : $2\pi^2/7$
4502 3354 Ei : $(e + \pi)/7$
4502 3906 J_1 : $(\ln 3)^{1/3}$
4502 4400 at : $4\sqrt{2} + 3\sqrt{3}/2$
4502 5550 rt : (5,3,7,3)
4502 5901 rt : (5,0,-5,8)
4502 6902 sr : $3\sqrt{3} + 3\sqrt{5}$
4503 0991 rt : (7,4,-1,-1)
4503 1102 $\sqrt{2} + s\pi(\sqrt{2})$
4503 1144 rt : (8,6,9,-6)
4503 1849 rt : (1,9,5,3)
4503 1951 Ψ : $3\sqrt[3]{3}/4$
4503 2686 $2/3 \times s\pi(\sqrt{5})$
4503 3263 $\sqrt{3} + e$
4503 3334 $1/3 + \exp(3/4)$
4503 3947 at : $4/3 + 4\sqrt{3}$
4503 4352 sinh : $7(e + \pi)/5$
4503 4585 rt : (3,9,2,-3)
4503 5255 J : $1/\sqrt[3]{21}$
4503 6695 2^x : $\sqrt{21/2}$
4503 7911 sr : $7\zeta(3)/4$
4503 9448 rt : (4,5,-3,-2)
4503 9665 cu : $7e/9$
4504 0085 tanh : $7\ln 2/10$
4504 0696 at : $3e/2 + 4\pi/3$
4504 2862 2^x : $\ln(\sqrt[3]{5})$
4504 3020 tπ : $4\sqrt{3} - \sqrt{5}$
4504 3309 rt : (5,1,-7,3)
4504 3985 at : P_{32}
4504 4306 sinh : $9\sqrt[3]{5}/7$
4504 4489 rt : (9,9,3,-4)
4504 6627 rt : (7,2,-7,7)
4504 7390 cr : $1/3 + e$
4504 7569 10^x : 18/17
4504 9373 tπ : $(\sqrt{2}\pi/3)^{-3}$
4505 4017 rt : (4,2,0,8)
4505 6627 θ_3 : $7(1/\pi)/10$
4505 7062 10^x : $1/3 - e/4$
4505 7105 at : $2/3 - 4\sqrt{5}$
4505 8367 10^x : $3\sqrt{2}/2 - \sqrt{3}$
4505 8490 rt : (4,7,9,3)
4505 8995 e^x : $1/4 - \pi/3$
4505 9153 rt : (8,5,5,-4)
4506 1822 Ψ : $(\sqrt[3]{5}/2)^{-1/2}$
4506 1827 rt : (9,4,-9,6)
4506 2712 2^x : $3/4 + 4\sqrt{5}$
4506 4640 rt : (9,6,-8,-4)
4506 4695 rt : (7,3,-9,2)
4506 4696 e^x : $\exp(\sqrt{5})$
4506 5127 at : $2 + 2\pi$
4506 5812 Γ : $(2e/3)^3$
4506 6580 rt : (3,6,-9,-5)
4506 8815 rt : (1,9,-5,-7)
4506 9385 $1 - \ln(\sqrt{3})$
4506 9882 rt : (6,0,1,1)
4507 0889 rt : (5,2,-9,-2)
4507 3305 J_1 : $(e/3)^{-1/3}$
4507 4581 2^x : $9e/10$
4507 5221 rt : (6,4,-2,7)
4507 6538 sq : $3\sqrt{2}/4 - \sqrt{3}$
4507 6719 Li_2 : $(\sqrt{3}/2)^{1/3}$
4507 7133 id : $3\zeta(3)/8$
4507 9147 rt : (7,1,7,-4)
4507 9643 id : $3\sqrt{3} - \sqrt{5}/3$
4507 9902 e^x : $2e/3 + \pi/3$
4508 0172 E : $\ln(\sqrt[3]{5})$
4508 1221 sinh : $1/\sqrt[3]{12}$
4508 1235 θ_3 : $7\sqrt[3]{2}/9$
4508 3833 rt : (8,7,-7,1)
4508 4072 Ψ : $7e/4$
4508 4513 rt : (9,8,-1,-2)
4508 4892 rt : (1,2,7,9)
4508 6866 e^x : 22/5
4508 7166 ℓ_{10} : $4/3 + 2\sqrt{5}/3$
4508 7589 cu : $\pi^2/6$
4508 8153 ℓ_{10} : $3e/4 + \pi/4$
4508 9390 rt : (8,6,-4,6)
4509 2555 at : P_{42}
4509 4187 rt : (3,8,-2,-1)
4509 7153 cr : $4\sqrt{3}/3 + \sqrt{5}/3$
4509 8572 rt : (2,9,-3,-3)
4509 8605 sq : $4 - \sqrt{2}/3$
4509 9019 Γ : $5\pi/3$
4510 2238 2^x : $1/3 + 3\pi/4$
4510 2295 rt : (9,5,7,-5)
4510 3519 2^x : $1 + 3\sqrt{5}/2$
4510 5629 sπ : $2e + 3\pi/2$
4510 5651 $1/2 + s\pi(2/5)$
4510 6198 at : $\sqrt{7}\pi$
4510 7137 rt : (6,5,-4,2)
4510 8034 rt : (6,3,9,4)
4510 8185 $2 \times \exp(4/5)$
4510 8649 2^x : $2\sqrt{3} - 3\sqrt{5}/4$
4510 9410 rt : (5,1,-5,-2)
4511 2117 as : P_{29}
4511 3556 J_1 : $\sqrt[3]{19}$
4511 4509 Γ : $9e/5$
4511 4531 rt : (3,5,6,-4)
4511 6068 rt : (8,4,1,-2)
4511 7803 rt : (8,7,-6,1)
4511 9925 tanh : $\exp(-\sqrt[3]{3}/2)$
4512 0483 at : $3e/4 + 2\pi$
4512 0925 tπ : $\sqrt[3]{3}$
4512 0959 rt : (3,3,-7,9)
4512 2056 θ_3 : $\exp(-3/2)$
4512 4098 rt : (8,9,-2,-2)
4512 5213 rt : (8,1,9,-5)
4512 5575 at : $(\ln 2/2)^{-2}$
4512 6604 tπ : $e\pi/4$
4512 7494 rt : (5,9,6,-5)
4513 4764 rt : (7,8,1,6)
4513 5493 2^x : $3/2 + 4\sqrt{2}/3$
4513 5678 cr : $4 - 2\sqrt{2}/3$
4513 6732 id : $1/3 + \sqrt{5}/2$
4513 6740 at : 25/3
4513 9229 $\pi \times \ln(3)$
4513 9238 ℓ_{10} : 9π
4514 1609 at : $4e/3 + 3\pi/2$
4514 1622 rt : (3,8,3,0)
4514 7813 10^x : $7\sqrt{2}/4$
4514 8832 ζ : $5e/6$
4514 8953 rt : (1,9,-5,-4)
4514 9376 erf : $3\sqrt{2}/10$
4514 9541 cr : $9e/8$
4514 9592 cu : $1 + 2\sqrt{5}/3$
4514 9620 at : $\sqrt{2} + 4\sqrt{3}$
4515 0683 rt : (2,1,8,-4)
4515 1640 rt : (3,4,-7,3)
4515 3887 rt : (4,5,8,-5)
4515 4499 ℓ_{10} : $2\sqrt{2}$
4515 5526 rt : (6,9,8,-6)
4515 7832 e^x : $3\sqrt{2}/4 + 3\sqrt{5}/4$
4515 8270 id : $\ln(\pi/2)$
4515 8816 sr : $5\zeta(3)$
4515 9088 sinh : 7/16
4515 9824 $\exp(4/5) - s\pi(e)$
4516 0596 rt : (8,8,-8,-4)
4516 0789 cr : $4\sqrt{2} - 3\sqrt{3}/2$
4516 1476 sq : $4\sqrt{3}/3 - 4\sqrt{5}/3$
4516 1607 e^x : $(\sqrt[3]{3}/2)^{1/3}$
4516 1683 e^x : $\sqrt{5}/6$
4516 2328 cr : $\sqrt{22}\pi$
4516 3429 erf : $4(1/\pi)/3$
4516 4336 rt : (9,4,3,-3)
4516 5114 J_1 : 8/3
4516 6820 rt : (7,0,3,2)
4516 9881 tπ : $(\ln 2/3)^{1/2}$
4517 0329 rt : (2,8,-2,9)
4517 0644 cr : $3/4 + 4\sqrt{3}/3$
4517 4011 at : $7\ln 2/10$
4517 9124 rt : (5,4,3,1)
4518 0100 sr : $1/\sqrt{24}$
4518 1606 rt : (1,6,-4,-7)
4518 1921 rt : (9,7,-5,0)
4518 3150 sr : $(\sqrt{2}\pi)^{1/2}$
4518 3393 J_1 : $\zeta(5)$
4518 4161 rt : (3,8,-9,-3)
4518 4231 at : $\sqrt{3}/3 - 4\sqrt{5}$
4518 6395 E : $4\zeta(3)/9$
4518 7944 rt : (7,9,-1,1)
4519 0323 at : $10(e + \pi)/7$
4519 0556 cr : $2 + 3\sqrt{2}/4$
4519 1944 rt : (2,5,-7,-4)
4519 2836 $\sqrt{5} \div \ln(\sqrt{2})$
4519 3410 ln : $10\sqrt{2}/9$
4519 4166 2^x : $\sqrt[3]{10}$
4519 5625 rt : (3,5,-7,-3)
4519 5944 rt : (7,9,-4,-3)
4519 6711 rt : (8,3,-7,8)
4519 6933 rt : (2,9,2,7)
4519 7084 10^x : 23/21
4519 8512 ln : 7/11
4519 8706 rt : (5,8,2,-3)
4519 9224 at : $8\pi/3$
4520 2527 rt : (6,1,-5,9)
4520 2635 2^x : $2/3 - 2e/3$
4520 2742 as : $1/\sqrt[3]{12}$
4520 4793 ℓ_2 : $4\sqrt[3]{5}/5$
4520 5175 rt : (1,9,9,-6)
4520 5514 ℓ_{10} : $2\sqrt{3}/3 + 3\sqrt{5}/4$
4520 6483 id : $1/3 - \pi/4$
4520 6704 10^x : $3\sqrt{2}/4 + \sqrt{3}/3$
4520 8345 rt : (5,8,-8,6)
4521 0088 10^x : $5\sqrt[3]{5}$
4521 0407 ℓ_2 : $1 + 2\sqrt{5}$
4521 1179 rt : (1,7,4,3)
4521 2486 rt : (8,0,5,3)
4521 3520 J_1 : $(\sqrt{5}/2)^{1/3}$
4521 4408 rt : (3,9,-7,-8)
4521 4671 Γ : $9\pi/8$
4521 6843 id : $\exp(-\sqrt[3]{4}/2)$
4521 9767 rt : (4,1,-7,-3)

4522 1495 $\Gamma : \sqrt[3]{18}$
4522 1639 $\theta_3 : \sqrt{5}/10$
4522 1889 rt : (1,8,4,-8)
4522 1943 $\ell_2 : 2\sqrt{2}/3 - \sqrt{3}/3$
4522 2435 $J_2 : 9\sqrt{2}/5$
4522 2456 rt : (2,3,-1,9)
4522 3367 $1/5 - \exp(\sqrt{3})$
4522 3439 rt : (6,8,4,-4)
4522 3532 $2^x : 8\pi^2/9$
4522 4929 cu : $4\sqrt{2}/3 - \sqrt{5}/2$
4522 6968 $2^x : 22/17$
4522 7435 $2^x : \ln(1/\pi)$
4522 8816 sr : $3/4 + e/2$
4522 9741 $E : 8/15$
4522 9767 $\ell_{10} : 6/17$
4523 0799 rt : (2,4,-8,7)
4523 1565 tπ : $9(1/\pi)$
4523 1676 $\Gamma : \exp(e)$
4523 1900 rt : (6,9,-6,-4)
4523 3397 rt : (7,3,-5,1)
4523 4075 $J_2 : 8(1/\pi)$
4523 4674 $\ln(4/5) + s\pi(\sqrt{5})$
4523 4740 $J_2 : 1/\ln(\sqrt{2}\pi/3)$
4523 6875 sq : $3\sqrt{3}/4 + \sqrt{5}/4$
4523 7195 rt : (8,4,-9,3)
4523 7404 at : $1/4 + 3e$
4523 8863 tπ : $4\ln 2/9$
4523 9065 rt : (9,0,7,4)
4524 0789 rt : (7,8,6,-5)
4524 1746 rt : (5,9,-8,0)
4524 2285 $2^x : 7/13$
4524 4547 $s\pi(\pi) \div s\pi(2/5)$
4524 5099 rt : (2,8,1,-1)
4524 5418 rt : (9,7,-4,7)
4524 5734 $10^x : 3\pi^2/5$
4524 5868 $erf : \exp(-\sqrt[3]{5}/2)$
4524 6063 $erf : 1/\sqrt[3]{13}$
4524 6277 id : $\sqrt{2}/2 + \sqrt{5}/3$
4524 8613 id : $((e+\pi)/3)^3$
4524 9362 rt : (1,5,-1,8)
4525 0723 at : $7\zeta(3)$
4525 1383 $2^x : 4\sqrt{2}$
4525 1934 rt : (6,2,-7,4)
4525 2584 $\ell_{10} : 9\sqrt[3]{2}/4$
4525 2956 rt : (9,5,-4,-2)
4525 3478 rt : (5,9,-8,-5)
4525 3633 rt : (8,8,8,-6)
4525 4723 $Ei : \ln(3\pi/2)$
4525 5088 at : $\exp(-\sqrt[3]{3}/2)$
4525 5377 $J_1 : 3\sqrt{3}/5$
4525 7010 at : $4\sqrt{3} + 2\sqrt{5}/3$
4525 7882 cu : $4 - 4\sqrt{2}/3$
4525 8234 sr : $4e + \pi/3$
4525 8687 rt : (3,1,-9,-4)
4526 0202 rt : (7,5,-2,8)
4526 0538 id : $4\sqrt{2}/3 - \sqrt{3}/4$
4526 0582 sr : $\sqrt{3}/4 + 3\sqrt{5}/4$
4526 2075 rt : (6,4,5,2)
4526 2333 $2^x : \sqrt{3}/2 + 2\sqrt{5}$
4526 3837 $E : \exp(-\sqrt[3]{2}/2)$
4526 3941 rt : (3,6,-7,-9)
4526 5161 at : $1 - 3\pi$
4526 5216 cr : $4/3 + \sqrt{3}$
4526 8332 rt : (6,7,-8,-8)
4526 8788 $\exp(2/3) \times s\pi(\sqrt{3})$
4526 9021 $Ei : \ln(\pi/3)$
4526 9918 at : $3/2 + 4\sqrt{3}$
4527 0680 $J_1 : 3\ln 2/2$
4527 0876 $\Psi : 5/8$
4527 1143 $\Psi : (\sqrt[3]{2})^{1/3}$
4527 1380 rt : (5,4,6,-4)
4527 2497 $e^x : 3 + 3\sqrt{5}/4$
4527 3369 $J_2 : \sqrt{13/2}$
4527 3700 rt : (4,0,-5,5)
4527 5370 rt : (2,5,-6,2)
4527 6211 $2^x : 1 + 2\sqrt{3}/3$
4527 7003 ln : $5\sqrt[3]{5}/2$
4527 7558 $Ei : ((e+\pi)/2)^{-2}$
4527 7560 rt : (1,8,-1,-2)
4527 8597 tπ : $(e)^{-2}$
4528 0244 id : $3/2 - \pi/3$
4528 1233 rt : (5,3,0,9)
4528 1385 ln : $3\sqrt{3}/2 + 3\sqrt{5}/4$
4528 1513 at : $3 + 2e$
4528 1659 as : $7/16$
4528 2849 id : $\sqrt{2}/3 + 4\sqrt{5}/3$
4528 3062 rt : (6,4,8,-5)
4528 3242 ln : $2/3 + e/3$
4528 6625 at : $\sqrt{3} + 3\sqrt{5}$
4528 7775 rt : (1,8,-3,7)
4528 8526 rt : (7,4,7,3)
4528 9891 $\Psi : 6\sqrt[3]{2}/7$
4529 0017 $\zeta : 8\sqrt{2}/5$
4529 2183 at : $1/2 - 4\sqrt{5}$
4529 3442 rt : (8,5,-6,-3)
4529 4161 $s\pi(1/4) - s\pi(\sqrt{3})$
4529 6631 sr : $19/9$
4529 7024 $\ell_2 : 3\sqrt{2} + 3\sqrt{5}$
4529 7808 rt : (9,8,-6,2)
4529 9684 ln : $3/4 + 4e$
4530 0151 $4 \times \exp(\sqrt{2})$
4530 1539 rt : (8,4,9,4)
4530 2132 $e^x : 8\pi^2/7$
4530 2863 rt : (1,8,5,-4)
4530 4697 id : $e/6$
4530 6366 sinh : $7\pi/10$
4530 6865 ln : $\sqrt{2}/2 + \sqrt{3}/2$
4530 7457 sinh : $7\sqrt[3]{3}/3$
4530 7639 tanh : $2\sqrt[3]{5}/7$
4530 7690 $\ln(3/5) - s\pi(\sqrt{2})$
4530 8034 $2^x : 7\ln 2/9$
4530 8183 rt : (7,5,-8,-4)
4530 8757 sinh : $(\ln 2/3)^{-1/3}$
4530 9099 rt : (4,8,-2,6)
4531 0136 rt : (9,3,-1,-1)
4531 3428 at : $6\pi^2/7$
4531 3532 rt : (3,8,9,-6)
4531 6196 $\ell_{10} : 4e/3 - \pi/4$
4531 7140 $e^x : 1/\ln((e+\pi)/3)$
4531 9212 $10^x : 4/3 - 3\sqrt{5}/4$
4531 9684 rt : (9,4,-7,9)
4532 2363 sq : $5(e+\pi)$
4532 5170 rt : (6,3,-9,-1)
4532 5892 $2^x : 2 - \pi$
4532 6836 rt : (7,6,-4,3)
4532 8161 rt : (8,9,1,7)
4532 9898 cr : $\sqrt{2}/3 + 3\sqrt{3}/2$
4532 9992 $J_2 : \sqrt{2} - \sqrt{3} - \sqrt{5}$
4533 0595 at : $4 + 2\sqrt{5}$
4533 1339 $J_1 : 25/24$
4533 1600 at : $\sqrt{2}/3 - 4\sqrt{5}$
4533 1715 $\Gamma : 8\ln 2/9$
4533 4231 sπ : $-\sqrt{3} + \sqrt{5} + \sqrt{7}$
4533 6319 rt : (1,9,-9,-9)
4533 6664 rt : (2,6,-4,-3)
4533 7779 sr : $(3\pi)^{1/3}$
4533 8033 tπ : $3\sqrt{2}/4 - 3\sqrt{3}$
4533 9429 $10^x : ((e+\pi)/2)^2$
4533 9765 rt : (4,0,8,4)
4534 1047 $\Psi : 7\sqrt{3}$
4534 2014 $\ln(\pi) - \exp(4)$
4534 2237 $e^x : 3\sqrt{2}/4 + \sqrt{3}/4$
4534 2640 $4/5 - \ln(\sqrt{2})$
4534 2810 $2^x : \sqrt{3}\pi$
4534 3792 sπ : $5\sqrt[3]{2}/2$
4534 4022 $\ln(\sqrt{5}) + \exp(1/2)$
4534 4529 $\ell_2 : \sqrt{15/2}$
4534 8630 at : $6\sqrt{2}$
4534 9741 rt : (9,7,6,-5)
4535 0243 rt : (1,8,3,-8)
4535 0508 $J_2 : 8\sqrt{5}/7$
4535 0658 cr : $3/2 + \pi/2$
4535 1139 $J_2 : 23/9$
4535 1449 rt : (8,7,4,-4)
4535 2474 sπ : $\sqrt[3]{19/3}$
4535 2597 at : $1/3 + 3e$
4535 3601 id : $e/2 + 2\pi/3$
4535 3650 at : $3e/2 - 4\pi$
4535 3863 rt : (7,7,2,-3)
4535 3903 rt : (3,7,-9,9)
4535 5992 id : $10\sqrt{5}/3$
4535 7533 rt : (6,7,0,-2)
4535 8339 sq : $\sqrt{2}/3 - 3\sqrt{5}/4$
4536 0202 $Li_2 : (\ln 3)^{-1/2}$
4536 2705 $E : 5(1/\pi)/3$
4536 3204 cu : $\sqrt{2}/2 + 2\sqrt{3}/3$
4536 3491 $e^x : 1/3 + e/3$
4536 3785 rt : (5,8,-1,-3)
4536 5565 cu : $3 + 3\sqrt{3}/4$
4536 7446 rt : (6,7,3,8)
4536 8758 at : $17/2$
4536 9501 rt : (1,4,9,-5)
4537 0199 $e^x : 2\pi/7$
4537 1455 rt : (5,4,-2,4)
4537 1471 $E : 3\sqrt{2}/8$
4537 2586 rt : (9,9,-8,-3)
4537 2760 rt : (1,4,0,9)
4537 3929 rt : (9,5,-9,4)
4537 4208 $2^x : 3\sqrt{2}/2 + \sqrt{3}$
4537 5312 sπ : $\sqrt{3} + \sqrt{5}/2$
4537 6822 rt : (4,9,-2,0)
4537 9077 $10^x : 7(e+\pi)/6$
4537 9516 $e^x : P_{74}$
4538 1065 $s\pi(1/3) + s\pi(1/5)$
4538 3883 $J_0 : 4\zeta(3)/3$
4538 4109 at : $\sqrt{3}/4 - 4\sqrt{5}$
4538 5699 tπ : $1/4 + 4\sqrt{2}/3$
4538 7698 $J_1 : 24/23$
4538 8926 rt : (1,7,6,-3)
4539 0104 sq : $e + 3\pi$
4539 0927 $\ell_{10} : \sqrt[3]{23}$
4539 1150 sr : $\sqrt[3]{3}/7$
4539 1404 ln : $2/3 - \sqrt{3}/4$
4539 1829 at : $4e - 3\pi/4$
4539 3477 rt : (1,6,-2,6)
4539 3841 $2^x : 3\sqrt[3]{2}/7$
4539 4210 at : $e/3 - 3\pi$
4539 7739 $J_0 : 10\sqrt[3]{3}/9$
4539 7929 rt : (3,8,-7,4)
4539 8523 rt : (3,8,-8,-5)
4539 8885 $\Gamma : \exp(-\sqrt{2}\pi)$
4539 8992 $\ell_{10} : 3/4 + 2\pi/3$
4539 9049 sπ : $3/20$
4539 9560 $e^x : \sqrt[3]{10/3}$
4540 1185 $\ell_2 : \exp(e+\pi)$
4540 2624 $J_2 : 5\sqrt{2}/2$
4540 2896 at : $2/3 - 2\sqrt{3}/3$
4540 3548 sr : $7e/9$
4540 4150 rt : (8,6,-2,9)
4540 6096 ln : $3\sqrt{3}/4 + 4\sqrt{5}/3$
4540 6627 rt : (8,3,8,-5)
4540 7266 rt : (5,3,-9,3)
4540 9144 sr : $4 - 4\sqrt{2}/3$
4540 9313 $10^x : 2 + 3\sqrt{3}$
4540 9733 $Ei : 3\sqrt{5}/8$
4541 0027 rt : (7,3,-7,5)
4541 0271 sπ : $5\sqrt[3]{5}/3$
4541 2364 $J_0 : \exp(\sqrt{2}/3)$
4541 2776 rt : (4,8,-6,-4)
4541 2798 $\ell_{10} : 4 - 2\sqrt{3}/3$
4541 4604 $E : 9/17$
4541 7911 rt : (5,7,9,-6)
4541 8904 $e^x : 2\sqrt{2}/3 - \sqrt{3}$
4541 9261 ln : $5\sqrt[3]{2}/4$
4541 9293 $J_2 : 9\pi/8$
4542 0033 rt : (7,3,6,-4)
4542 0693 rt : (3,7,-6,1)
4542 1990 $Ei : 1/2$
4542 2753 at : $e\pi$
4542 5144 rt : (9,1,3,2)
4542 6241 rt : (2,7,-2,-8)
4542 6955 rt : (6,8,9,3)
4542 9585 rt : (7,7,-6,-2)
4543 0829 rt : (4,7,7,-5)
4543 2119 $J_2 : 3e\pi/10$
4543 2309 $\ell_{10} : e\pi/3$
4543 3539 sinh : $11/25$
4543 4264 $2^x : 15/4$
4543 6160 rt : (5,8,-4,-3)
4543 6461 at : $5\sqrt[3]{5}$

4543 6735 10^x : $1+4\sqrt{2}/3$
4543 6762 $t\pi$: $4\sqrt{2}/3-\sqrt{3}/3$
4543 6967 at : $3\sqrt{3}+3\sqrt{5}/2$
4544 0158 rt : (6,3,4,-3)
4544 0348 10^x : $8\sqrt[3]{5}/5$
4544 5747 at : $2\sqrt[3]{5}/7$
4544 7720 Li_2 : $2\sqrt{2}/7$
4544 8838 ℓ_2 : $(\sqrt{2}\pi)^3$
4544 8863 rt : (3,7,5,-4)
4544 9055 J_1 : 23/22
4545 2661 Li_2 : $\exp(-e/3)$
4545 2677 rt : (3,9,-5,-1)
4545 3207 cr : $1-e/3$
4545 3378 erf : $\sqrt[3]{5}/4$
4545 3828 e^x : $4\sqrt{3}+3\sqrt{5}$
4545 4545 id : 5/11
4545 5160 ℓ_{10} : $1/4+3\sqrt{3}/2$
4545 5483 ℓ_2 : P_{23}
4545 8674 at : $4-4\pi$
4545 9231 sq : $3\sqrt{2}/2-3\sqrt{3}$
4546 0851 rt : (5,1,-5,6)
4546 3545 sq : $(\sqrt[3]{5}/3)^{-1/3}$
4546 3669 e^x : $1/\sqrt[3]{19}$
4546 5367 id : $1/3+3\sqrt{2}/2$
4546 5414 2^x : $\sqrt{2}/4-2\sqrt{5}/3$
4546 6619 sq : $4\sqrt{2}/3+3\sqrt{5}/2$
4546 8760 2^x : $8\sqrt{5}$
4546 9419 rt : (8,1,1,1)
4547 0520 J_2 : $3\sqrt[3]{5}/2$
4547 0955 cr : $1/4+2\sqrt{2}$
4547 1005 $s\pi$: $\zeta(3)/8$
4547 1778 sr : P_{39}
4547 2101 10^x : $\sqrt{2}+\sqrt{6}+\sqrt{7}$
4547 2840 id : $10(1/\pi)/7$
4547 3449 sr : P_{33}
4547 3740 rt : (5,3,2,-2)
4547 4497 at : $2e+\pi$
4547 5481 J : $\exp(-13\pi/5)$
4547 5798 rt : (2,7,3,-3)
4547 5930 e^x : $4/3-3\sqrt{2}/2$
4547 6069 cr : $\ln(\ln 3)$
4547 6916 sq : $1/3+3e/2$
4547 8843 rt : (5,9,-6,5)
4547 9211 ℓ_{10} : $\exp(\pi/3)$
4547 9935 rt : (3,4,-5,8)
4548 1601 rt : (6,8,-2,-2)
4548 2246 rt : (1,5,4,7)
4548 2294 2^x : $3\sqrt[3]{3}/8$
4548 3039 at : $((e+\pi)/2)^2$
4548 3237 sinh : $\sqrt[3]{13/3}$
4548 3875 ℓ_{10} : $5\sqrt[3]{5}/3$
4548 5778 ℓ_{10} : $\sqrt{3}+\sqrt{5}/2$
4548 6065 cu : $1/3+3\pi/4$
4548 6364 as : $3/2-3\sqrt{2}/4$
4548 8666 rt : (8,7,-4,4)
4548 9799 $3/4\div\exp(1/2)$
4549 0409 rt : (1,0,-9,4)
4549 1209 $s\pi(1/5)\times s\pi(e)$
4549 1315 at : $\sqrt{2}/4-4\sqrt{5}$
4549 2798 Γ : 8/13
4549 4774 rt : (2,4,8,3)
4549 5463 at : $4\sqrt{2}/3+3\sqrt{5}$
4549 5486 sq : $e/4+4\pi$
4549 5900 $s\pi$: $\exp(\pi/3)$
4549 6597 $t\pi$: $1/4-4e$
4549 7294 rt : (6,8,1,3)
4549 9141 e^x : 3/8
4549 9535 ℓ_2 : $5\pi^2/9$
4550 0633 rt : (9,6,2,-3)
4550 1539 cr : $2e-3\pi/4$
4550 2701 sinh : $(e+\pi)/3$
4550 3061 J_1 : $\pi/3$
4550 5039 rt : (7,4,-9,0)
4550 7678 J : $\exp(-16\pi/5)$
4550 7720 rt : (5,8,7,2)
4550 8130 $\ln(3)+\exp(\sqrt{5})$
4550 9827 rt : (3,3,9,-5)
4551 0716 at : $4\sqrt{3}+3\sqrt{5}/4$
4551 0729 10^x : 7/13
4551 1961 $1/2\div\ln(3)$
4551 3988 2^x : $3/4-4\sqrt{2}/3$
4551 4501 10^x : $(5/2)^{-1/3}$
4551 5528 rt : (5,2,-8,-8)
4551 5693 Li_2 : 21/22
4551 6105 J_1 : 22/21
4551 6613 cu : 10/13
4551 7071 rt : (6,1,8,4)
4551 8285 at : $1/3-4\sqrt{5}$
4551 8475 rt : (7,6,9,-6)
4552 0376 rt : (1,7,1,-2)
4552 0962 rt : (8,8,-9,-5)
4552 2375 rt : (9,1,-8,-3)
4552 3311 $s\pi$: 6π
4552 7601 rt : (1,8,5,7)
4552 8524 rt : (5,5,-4,-1)
4552 8654 $t\pi$: $1/\ln(\sqrt[3]{5})$
4552 9317 rt : (9,5,7,3)
4553 0470 sr : $\sqrt[3]{19/2}$
4553 2438 J_1 : $(\ln 3)^{1/2}$
4553 2476 2^x : $(e/2)^{-2}$
4553 3603 Γ : $1/\sqrt[3]{21}$
4553 3770 rt : (5,4,-8,-4)
4553 3963 rt : (8,4,-7,6)
4553 4669 sr : $1+\sqrt{5}/2$
4553 5194 10^x : $2+\sqrt{2}$
4553 6516 rt : (1,7,-8,-5)
4553 7751 rt : (3,8,4,6)
4553 8453 rt : (7,8,3,9)
4554 0216 J_0 : 8/5
4554 0396 2^x : $4\sqrt{5}/5$
4554 0773 J_2 : $\exp(\sqrt[3]{2})$
4554 1004 rt : (8,6,0,-2)
4554 1330 10^x : $\sqrt{2}/2+\sqrt{5}/4$
4554 4570 cr : $3\pi^2/2$
4554 4919 sq : $2-\sqrt{3}/4$
4554 7376 rt : (6,6,7,-5)
4554 9369 Γ : $\sqrt{5}-\sqrt{6}-\sqrt{7}$
4554 9555 J_2 : $\sqrt[3]{17}$
4555 0407 rt : (2,3,7,-4)
4555 1383 J_2 : 18/7
4555 1411 at : $7\pi^2/8$
4555 2970 $1/4\times\exp(3/5)$
4555 4552 at : $4\sqrt{2}+4\sqrt{5}/3$
4555 4602 erf : 3/7
4555 7767 10^x : $1/2+\sqrt{5}/4$
4555 8630 sr : $4/3+\pi/4$
4555 8940 rt : (2,6,-1,-8)
4555 9316 E : 10/19
4555 9478 10^x : 14/19
4555 9867 as : 11/25
4556 0980 rt : (9,2,6,-4)
4556 1309 rt : (2,5,-4,9)
4556 2355 J_1 : $(\sqrt{3}/2)^{-1/3}$
4556 2572 rt : (6,5,-2,5)
4556 5318 2^x : 13/24
4556 8343 10^x : $1+\sqrt{3}/2$
4556 8800 Ψ : $\exp(-\sqrt{2}/3)$
4556 9459 rt : (2,4,-8,-3)
4557 0370 cu : 17/15
4557 0786 ℓ_{10} : $(1/(3e))^{1/2}$
4557 1005 2^x : $7e\pi/2$
4557 2723 $\exp(2/5)-s\pi(\sqrt{2})$
4557 3136 10^x : π
4557 3699 rt : (5,1,6,3)
4557 4055 cu : $7e/8$
4557 4526 rt : (9,9,5,-5)
4557 4639 ln : $1+\sqrt{3}/3$
4557 4643 at : $6\sqrt[3]{3}$
4557 5165 rt : (4,2,-9,-5)
4557 5599 rt : (9,8,-7,-4)
4557 6191 10^x : $3e/8$
4557 6419 at : $1/2+3e$
4557 6782 Ψ : $6\sqrt[3]{5}/5$
4557 7690 e^x : $7(e+\pi)/10$
4557 7708 $t\pi$: $2\sqrt{3}/3+4\sqrt{5}/3$
4557 7824 10^x : $4/3+3e/2$
4557 8777 ℓ_{10} : $1/2+3\pi/4$
4557 9065 at : $3+4\sqrt{2}$
4558 2808 rt : (2,7,-6,-4)
4558 3540 at : $5\sqrt{3}$
4558 3757 erf : $(2e)^{-1/2}$
4558 4412 sq : $3+3\sqrt{2}$
4558 5340 rt : (3,5,-5,2)
4558 7416 rt : (5,9,8,8)
4558 7648 J_1 : $5\sqrt[3]{2}/6$
4558 8711 cr : $3e/4+\pi/3$
4558 9435 ln : $3/2-\sqrt{2}$
4558 9678 J_1 : 21/20
4558 9994 e^x : $4-4e/3$
4559 0934 $\exp(2/3)-\exp(2/5)$
4559 1125 Ei : $\sqrt{2}+\sqrt{3}-\sqrt{7}$
4559 1493 2^x : $(2\pi)^{-1/3}$
4559 3195 ℓ_{10} : 7/20
4559 3348 rt : (5,2,-7,1)
4559 3812 sr : $\exp(-\pi/2)$
4559 4532 id : $(\sqrt{2}\pi/3)^{-2}$
4559 8100 sq : $4/3+3\sqrt{3}/2$
4559 8666 10^x : $3\sqrt{2}/4+\sqrt{3}$
4560 0000 cu : $6\sqrt[3]{2}/5$
4560 0406 rt : (6,2,-5,7)
4560 0894 sr : $3\ln 2/10$
4560 2619 at : $3/4-3\pi$
4560 3771 ln : $4e+\pi/4$
4560 4557 Li_2 : $(\pi/2)^{-2}$
4560 4859 sr : $3+4\sqrt{5}$
4560 6398 J_1 : $(e/3)^{-1/2}$
4560 6727 id : $4\sqrt{3}-2\sqrt{5}$
4560 7182 ln : $e+\pi/2$
4560 7252 ζ : $\sqrt[3]{23/2}$
4560 7854 Ei : $1/\sqrt{14}$
4560 8320 rt : (7,8,0,-1)
4561 1458 rt : (8,8,-6,-1)
4561 1704 10^x : $2+\sqrt{2}/2$
4561 2502 rt : (8,9,3,-4)
4561 2845 rt : (9,8,-4,5)
4561 4584 rt : (8,2,4,-3)
4561 4846 sq : $\sqrt{2}-4\sqrt{5}/3$
4561 5262 2^x : $8\sqrt{2}$
4561 5300 rt : (4,5,-9,7)
4561 5386 rt : (9,5,9,-6)
4561 6424 rt : (1,3,5,9)
4561 7799 cu : $4\sqrt{3}/9$
4561 8697 J : $\sqrt[3]{3}/6$
4561 8962 at : $1-2\sqrt{5}/3$
4561 9225 $3\div\exp(1/5)$
4561 9797 rt : (6,4,-6,-3)
4562 0647 cr : $3\sqrt{2}-2\sqrt{3}/3$
4562 0794 rt : (8,5,5,2)
4562 3556 ℓ_{10} : $3/2+e/2$
4562 4044 ℓ_{10} : $-\sqrt{5}+\sqrt{6}+\sqrt{7}$
4562 5101 rt : (6,1,-8,-9)
4562 5450 $s\pi$: $(\sqrt[3]{3}/2)^{1/2}$
4562 5625 e^x : $1/3-\sqrt{5}/2$
4562 7270 ℓ_{10} : $2e/3+\pi/3$
4562 7686 Li_2 : $\ln(3/2)$
4562 8128 at : $1/4-4\sqrt{5}$
4562 8250 sr : $(\ln 3/2)^{-3}$
4562 9613 Li_2 : $1/\sqrt[3]{15}$
4563 0434 rt : (1,7,-3,3)
4563 1098 rt : (1,4,6,2)
4563 1667 sq : $\sqrt{3}/3+3\sqrt{5}/2$
4563 3528 rt : (5,9,8,-6)
4563 3670 θ_3 : $4\sqrt{3}/7$
4563 3731 rt : (9,5,-7,7)
4563 4438 Ψ : $1/\sqrt[3]{14}$
4563 8276 $4/5\div\ln(\sqrt{3})$
4563 8441 θ_3 : $6\ln 2/7$
4563 9024 rt : (8,5,-9,1)
4564 0253 Li_2 : $3(1/\pi)$
4564 0493 sq : $1/2+e/2$
4564 3263 rt : (5,2,9,-5)
4564 3546 sr : 5/24
4564 4413 2^x : $4\sqrt{2}+4\sqrt{5}/3$
4564 5167 J_1 : $7\zeta(3)/8$
4564 5299 10^x : $\exp(-2e/3)$
4564 6289 at : $2+3\sqrt{5}$
4564 6559 rt : (4,6,3,4)
4564 7531 id : $(\sqrt{2}/3)^{-1/2}$
4564 8460 e^x : $e\pi/4$
4565 1734 at : $4+3\pi/2$

4565 2969 ln : $2/3+4e/3$
4565 3517 rt : (1,9,-4,3)
4565 4173 sinh : $3e/5$
4565 4837 at : $3\sqrt{2}+2\sqrt{5}$
4565 5737 $\pi-\exp(4)$
4565 7175 $\sqrt{5}\times\ln(3)$
4565 7602 rt : (3,7,-4,-3)
4565 8994 rt : (8,5,7,-5)
4566 0883 J_2 : $3(e+\pi)/5$
4566 2796 ζ : $(\sqrt[3]{2}/3)^{-1/2}$
4566 3378 sr : $1/\sqrt{23}$
4566 4861 tanh : $\exp(-1/\sqrt{2})$
4566 5094 rt : (7,9,1,-3)
4566 7114 cr : $2/21$
4566 7834 rt : (3,9,-3,6)
4566 8469 tπ : $3/22$
4566 8582 Ψ : $9\ln 2/10$
4566 9147 sr : $4\sqrt{2}/3-3\sqrt{5}/4$
4566 9363 as : $1-\sqrt{5}/4$
4566 9846 tπ : $\sqrt{3}/3+\sqrt{5}/4$
4567 0775 J_1 : $20/19$
4567 1977 tπ : $e/3+\pi/4$
4567 2579 sq : $(\ln 3)^2$
4567 2885 cr : $\sqrt{2}+3\sqrt{5}/4$
4567 3316 sq : $3e+2\pi$
4567 5070 Ei : $2\zeta(3)/9$
4567 5633 e^x : $\sqrt{2}/2-2\sqrt{5}/3$
4567 5671 E : $11/21$
4567 6124 ℓ_2 : $9e\pi/7$
4567 8333 $5\div\ln(2/5)$
4567 9012 sq : $19/9$
4567 9574 rt : (4,9,6,-5)
4567 9870 id : $\sqrt{2}/3-4\sqrt{3}$
4568 0418 rt : (6,7,-9,-5)
4568 1422 J_2 : $\sqrt{20/3}$
4568 1904 id : $(\ln 3/3)^{-2}$
4568 2393 cu : $10\ln 2/9$
4568 3421 rt : (8,2,8,4)
4568 5423 E : $\pi/6$
4568 5523 cu : $1/4+4\pi/3$
4568 5770 Ei : $19/4$
4568 8135 rt : (7,9,1,4)
4568 9310 tπ : $3(1/\pi)/7$
4569 2970 Γ : $5\sqrt{2}/2$
4569 3320 Γ : $21/8$
4569 4249 at : $e/4-3\pi$
4569 5263 $\ln(\sqrt{5})+\exp(\sqrt{3})$
4569 5710 rt : (4,0,-3,8)
4569 7647 cu : $1/3+4\sqrt{3}/3$
4569 9323 ℓ_2 : $2+\sqrt{5}/3$
4570 0237 J_0 : $5\sqrt{5}/7$
4570 0301 cu : $3\sqrt{3}-3\sqrt{5}$
4570 0625 rt : (7,6,-2,6)
4570 0875 J_1 : $(\sqrt[3]{5}/2)^{-1/3}$
4570 2362 rt : (7,2,2,-2)
4570 4648 rt : (1,1,9,4)
4570 5277 rt : (4,9,0,5)
4570 7814 cr : $3(1/\pi)/10$
4570 8463 rt : (7,3,-5,8)
4570 9107 e^x : $6\pi^2/7$
4570 9263 ℓ_{10} : $\pi/9$
4570 9884 Ψ : $\sqrt{15}\pi$
4571 0517 sinh : $11/5$
4571 0678 id : $1/4-\sqrt{2}/2$
4571 0880 at : $2/3-3\pi$
4571 1607 rt : (4,6,-9,1)
4571 1789 rt : (1,9,4,-5)
4571 1992 Γ : $(1/\pi)/7$
4571 2678 2^x : $3\sqrt{3}+\sqrt{5}/3$
4571 3462 rt : (4,2,7,-4)
4571 4367 cr : $4-e/3$
4571 8963 rt : (7,5,5,-4)
4572 2044 cr : $1+2\pi/3$
4572 4280 E : $1/\sqrt[3]{7}$
4572 5885 rt : (7,8,8,-6)
4572 7262 sπ : $8(1/\pi)/3$
4572 8003 ℓ_{10} : $2+\sqrt{3}/2$
4572 9985 at : $8\pi^2/9$
4573 3616 Ψ : $1/13$
4573 5163 Ei : $\sqrt{2}-\sqrt{3}+\sqrt{6}$
4573 6762 tπ : $\sqrt{2}/4+\sqrt{5}$
4573 7356 at : $4e-2\pi/3$
4573 7866 ℓ_{10} : $3\sqrt{2}/2+\sqrt{5}/3$
4573 7896 sr : $4+3e/4$
4574 2710 rt : (9,9,-6,0)
4574 5740 $\exp(4)-\exp(\pi)$
4574 7073 $\sqrt{5}+\exp(1/5)$
4575 1044 2^x : $1/4+\pi/3$
4575 1531 at : $3(e+\pi)/2$
4575 3375 rt : (8,9,9,3)
4575 9515 e^x : $\sqrt{17/2}$
4575 9985 cu : $\sqrt{3}/3+3\sqrt{5}/4$
4576 0608 J_1 : $19/18$
4576 2049 tπ : $5\sqrt{5}/6$
4576 3615 $\ln(4)\div s\pi(2/5)$
4576 4244 ℓ_{10} : $4\sqrt{3}/3+\sqrt{5}/4$
4576 6213 2^x : $e/5$
4576 7409 10^x : $4/3-2\sqrt{2}/3$
4576 9112 sr : $7e\pi/5$
4576 9504 $4/5\times\exp(3/5)$
4577 0195 rt : (7,1,9,-5)
4577 1258 E : $12/23$
4577 3797 sr : $17/8$
4577 9521 rt : (7,7,-7,-4)
4577 9805 cr : $1/2+3\sqrt{3}/2$
4578 0059 J_1 : $\sqrt{7}$
4578 1033 tπ : $(\sqrt{5}/3)^{1/2}$
4578 1388 ζ : $5\zeta(3)/9$
4578 2127 at : $4\sqrt{2}/3+4\sqrt{3}$
4578 3831 sπ : $2\sqrt{2}-3\sqrt{5}/4$
4578 5127 rt : (6,8,6,-5)
4578 5728 e^x : $6\sqrt{2}/5$
4578 6577 2^x : $5\zeta(3)$
4578 7631 sq : $2/3+e$
4578 8621 rt : (7,2,6,3)
4578 9195 $2/5\times\ln(\pi)$
4578 9273 at : $7\sqrt[3]{2}$
4578 9639 ln : $2\sqrt{2}/3+3\sqrt{5}/2$
4579 1893 at : $2/3+3e$
4579 2840 rt : (3,5,8,-5)
4579 4573 rt : (9,8,1,-3)
4579 4599 rt : (8,4,-5,9)
4579 6002 e^x : $\exp(-1/(3\pi))$
4579 8976 rt : (4,7,-2,-2)
4579 9427 cu : $4\sqrt{2}-\sqrt{3}$
4579 9888 tπ : $7(e+\pi)/9$
4580 0000 cu : $9\sqrt[3]{2}/10$
4580 0829 θ_3 : $\ln(2e/3)$
4580 1883 Ei : $2\sqrt[3]{2}/3$
4580 2049 at : $3\sqrt{2}/2+3\sqrt{5}$
4580 5022 rt : (8,7,-2,7)
4580 5750 rt : (8,9,-8,-6)
4580 6015 cu : $2/3+\sqrt{5}$
4580 6591 rt : (6,5,3,-3)
4580 7443 tanh : $2\sqrt{3}/7$
4580 8164 ℓ_{10} : $3/4+3\sqrt{2}/2$
4580 8720 at : $\exp(-1/\sqrt{2})$
4580 8966 rt : (4,5,8,3)
4581 0563 sπ : $3\sqrt{2}/5$
4581 1336 rt : (5,1,-3,9)
4581 1654 2^x : $1/\ln(2\pi)$
4581 4410 ℓ_2 : $4/3+\sqrt{2}$
4581 4536 ln : $\sqrt{5/2}$
4581 4965 rt : (9,7,8,-6)
4581 5024 rt : (1,9,-7,-5)
4581 5027 ℓ_2 : $8\zeta(3)/7$
4581 5275 $\ln(3/4)-s\pi(\sqrt{3})$
4581 6247 J_1 : $(\sqrt{5}/2)^{1/2}$
4581 6397 Ei : $21/25$
4581 7483 rt : (5,3,-9,-4)
4581 7610 cu : $2-\sqrt{3}/2$
4581 9077 rt : (2,7,0,-1)
4582 0034 cu : $2\sqrt{2}+3\sqrt{5}/4$
4582 0393 id : $1/4-3\sqrt{5}$
4582 1930 Γ : $(3)^{-3}$
4582 4321 sr : $\sqrt[3]{2}/6$
4582 5870 rt : (3,2,5,-3)
4582 5890 tanh : $5\ln 2/7$
4583 0704 rt : (9,5,-2,-1)
4583 0830 rt : (5,4,0,7)
4583 0841 2^x : $2e-3\pi/4$
4583 0917 rt : (4,6,-9,-5)
4583 1209 ln : $4\sqrt{3}/3-3\sqrt{5}/4$
4583 3333 id : $11/24$
4583 4590 rt : (2,9,2,-3)
4583 9130 ln : $3+3\sqrt{3}/4$
4583 9638 Ψ : $\exp(5/2)$
4583 9803 sq : $1-3\sqrt{5}/4$
4584 0292 rt : (4,7,-9,-5)
4584 2787 cr : $\sqrt{3}/2+\sqrt{5}$
4584 3926 rt : (9,4,5,-4)
4584 4022 rt : (2,8,9,-6)
4584 4446 rt : (7,4,-4,-2)
4584 5788 e^x : $1/4+e$
4584 7334 $\ln(\pi)\times\exp(2)$
4584 8442 sπ : $8\ln 2/3$
4585 0218 E : $(\sqrt[3]{3}/2)^2$
4585 0578 J_2 : $9\sqrt[3]{3}/5$
4585 0580 sinh : $8\sqrt[3]{5}/7$
4585 1179 E : $13/25$
4585 1974 rt : (7,4,-7,3)
4585 5064 e^x : $2e/3+\pi/2$
4585 5966 $\ln(3/5)-\exp(2/3)$
4585 7002 at : $2-4e$
4585 7581 E : $3\ln 2/4$
4585 8940 cr : $2e+3\pi$
4586 0668 J_1 : $18/17$
4586 1873 rt : (1,7,3,0)
4586 2666 rt : (2,9,-9,8)
4586 2918 J_2 : $7/2$
4586 2947 ℓ_{10} : $9\sqrt{5}/7$
4586 3784 ℓ_{10} : $8/23$
4586 4299 rt : (4,2,-8,-5)
4586 5784 rt : (5,8,4,-4)
4586 6717 rt : (6,1,7,-4)
4586 8613 rt : (5,9,-3,-1)
4586 8815 E : $3\sqrt{3}/10$
4586 8926 at : $9\pi^2/10$
4587 1422 Γ : $9e/2$
4587 1957 rt : (6,2,0,-1)
4587 2831 at : $\sqrt{8}\pi$
4587 2919 as : $1/2-2\sqrt{2}/3$
4587 4535 Ψ : $\exp(\sqrt[3]{3}/2)$
4587 4587 J_2 : $3\sqrt{3}/2$
4587 4979 Γ : $(\ln 2/3)^{1/3}$
4587 9768 rt : (1,8,-9,-7)
4588 0505 sr : $4/3+3\pi/2$
4588 1893 rt : (7,5,3,1)
4588 2971 rt : (7,7,-4,1)
4588 3146 sr : $4/19$
4588 5105 ℓ_2 : 7π
4588 6225 J_1 : $\ln(\ln 2/2)$
4588 6340 rt : (8,7,6,-5)
4588 6732 rt : (9,8,-2,8)
4588 7764 sinh : $9e/7$
4588 9911 rt : (1,8,-4,-1)
4589 0990 rt : (9,0,9,5)
4589 1681 rt : (2,7,-9,-2)
4589 1968 2^x : $9\sqrt[3]{3}/10$
4589 2198 θ_3 : $5\sqrt[3]{2}/7$
4589 3025 Ei : $(\sqrt[3]{5}/3)^{-1/3}$
4589 3165 rt : (2,5,6,-4)
4589 6643 at : $3/4+3e$
4589 7285 J_2 : $13/5$
4589 8722 Ei : $4/15$
4589 9397 rt : (8,8,-1,-2)
4590 2806 rt : (3,2,9,4)
4590 3963 rt : (4,1,-5,3)
4590 7029 cr : $3/4+3\pi/4$
4591 1568 rt : (9,1,2,-2)
4591 1885 rt : (5,4,8,-5)
4591 2325 rt : (4,3,-6,-3)
4591 2348 $5\times\exp(2/5)$
4591 2907 sinh : $5e\pi/9$
4591 3500 cr : $8e/7$
4591 4791 ℓ_2 : $(\sqrt[3]{4}/3)^{1/2}$
4591 5053 rt : (3,5,-3,7)
4591 6745 J_1 : $3\sqrt{2}/4$
4591 8969 rt : (2,9,-5,-4)
4591 9444 sq : $9(e+\pi)/8$

4591 9976 $\ell_2 : P_{73}$
4592 1152 $\sqrt{5} - \ln(4/5)$
4592 1412 at : $1/2 - 3\pi$
4592 2155 sinh : $4/9$
4592 2409 $\ell_2 : \exp(1/\pi)$
4592 3915 cr : $4 + 4e$
4592 3996 tπ : P_{24}
4592 5658 at : $2 + 4\sqrt{3}$
4592 5741 $J_2 : 9e/7$
4592 6129 rt : (2,6,-1,-1)
4592 8113 $J_1 : 10(1/\pi)/3$
4592 9912 $J : 1/\sqrt{6}$
4593 0371 sinh : $(e + \pi)/5$
4593 1109 $e^x : 1/\sqrt{7}$
4593 1366 sq : $1/3 + 3\pi/2$
4593 2374 rt : (1,8,7,-5)
4593 2837 $e^x : 3\sqrt[3]{2}/10$
4593 3106 $K : 14/15$
4593 5380 rt : (8,5,-7,4)
4593 6294 rt : (1,5,-1,-9)
4593 6720 sπ : $1/4 + 3\sqrt{3}/2$
4593 7631 rt : (6,5,0,8)
4593 9255 $erf : 3\sqrt[3]{3}/10$
4594 0555 at : $3e + \pi/4$
4594 1330 rt : (4,7,9,-6)
4594 2246 rt : (8,9,-8,1)
4594 2719 at : $4e/3 - 4\pi$
4594 3161 $\ell_2 : 11$
4594 5531 at : $4\sqrt{5}$
4594 6441 $e^x : 5\sqrt[3]{2}/7$
4594 6621 rt : (5,5,1,-2)
4594 6647 tπ : $2\sqrt{3}/3 + 3\sqrt{5}$
4594 8010 $2^x : 6/11$
4594 8742 $J_2 : 4(e + \pi)/9$
4594 9722 rt : (4,6,-7,6)
4595 3232 ln : $12/19$
4595 3441 rt : (3,4,-9,3)
4595 3652 at : $2\sqrt{3}/7$
4595 4321 rt : (5,1,-8,-6)
4595 6663 sq : $4e/9$
4595 6704 rt : (8,4,3,-3)
4596 0311 $e^x : 9/10$
4596 0522 rt : (8,7,-5,-3)
4596 0868 tanh : $(1/(3e))^{1/3}$
4596 3711 rt : (6,4,-9,-3)
4596 6091 id : $6\pi^2/7$
4596 7294 $2^x : 1 - 3\sqrt{2}/2$
4596 7883 tanh : $2\sqrt{5}/9$
4596 8408 at : $1/\ln(\sqrt{5}/2)$
4596 8757 rt : (5,6,-7,-4)
4596 8885 $\ell_{10} : 4 - \sqrt{5}/2$
4597 0645 tπ : $\sqrt{2}/4 - 2\sqrt{5}/3$
4597 0862 $erf : \sqrt{3}/4$
4597 2405 at : $5\ln 2/7$
4597 2804 $J_1 : 17/16$
4597 3008 rt : (2,5,-9,-5)
4597 3821 tπ : $9\pi/2$
4597 4444 sr : $7\sqrt[3]{5}$
4597 5745 $\Psi : 1/\ln(\sqrt[3]{4}/2)$
4597 7412 rt : (2,7,-9,-8)
4597 8210 $J_2 : 10\pi/9$
4597 8336 $10^x : e + 3\pi/2$
4597 8446 tπ : $3e/2 + \pi/4$
4597 9137 tπ : $1/3 - 3\sqrt{3}$
4598 0156 tπ : $e/3 - 2\pi$
4598 0522 rt : (1,9,0,-2)
4598 1978 rt : (8,8,-4,2)
4598 2396 $\ln(\sqrt{2}) + \exp(\sqrt{2})$
4598 2850 rt : (7,7,4,-4)
4598 4989 $Li_2 : 1/\sqrt{6}$
4598 5693 sπ : P_{71}
4598 6040 $\Psi : 10e\pi/7$
4598 6326 rt : (9,3,6,3)
4598 6783 sπ : P_{68}
4598 6910 $2^x : 4 + 3\pi/2$
4598 8215 $\Psi : \sqrt{23}$
4598 8473 $\theta_3 : 5/22$
4598 9970 $erf : 5\ln 2/8$
4599 1547 $10^x : (\sqrt{2})^{1/2}$
4599 2550 at : $1/4 - \sqrt{5}/3$
4599 6920 rt : (6,9,-8,-5)
4599 7238 sπ : $((e + \pi)/2)^3$
4599 7980 rt : (9,9,7,-6)
4599 9367 rt : (2,8,-5,-4)
4599 9388 $J_2 : 7\sqrt{5}/6$
4599 9911 rt : (6,8,3,6)
4600 0316 $e^x : 4(e + \pi)/7$
4600 1138 rt : (9,6,-7,5)
4600 1797 sr : $\sqrt{2} - \sqrt{3} + \sqrt{6}$
4600 3283 cu : $3 + \sqrt{3}/4$
4600 4756 $\Gamma : (\pi)^{-3}$
4600 6739 $e^x : \exp(\sqrt[3]{4}/3)$
4600 7041 $\ell_{10} : 2\sqrt[3]{3}$
4600 7629 $\theta_3 : 5(1/\pi)/7$
4601 0861 rt : (8,0,7,4)
4601 1425 $e^x : \sqrt{2} + \sqrt{6} + \sqrt{7}$
4601 1922 rt : (5,2,-5,4)
4601 3910 at : 9
4601 5699 at : $e + 2\pi$
4601 5820 $\zeta : \sqrt[3]{11/3}$
4601 7058 rt : (7,6,0,9)
4601 8910 rt : (5,1,5,-3)
4601 9112 $\theta_3 : \exp(-\sqrt{2}\pi/3)$
4602 0453 $\ell_{10} : \ln 2/2$
4602 0927 $\Psi : \ln((e + \pi)/2)$
4602 1186 $\zeta : 9/4$
4602 1452 $J_0 : 5(1/\pi)$
4602 3651 rt : (6,0,-8,7)
4602 3885 $\ell_{10} : 1 + 4\sqrt{2}/3$
4602 4650 $10^x : 10\sqrt{5}/3$
4602 5304 rt : (7,3,8,-5)
4602 6755 sinh : $\exp(1/(2\pi))$
4602 7255 at : $4\sqrt{2} + 3\sqrt{5}/2$
4602 9356 $\ell_2 : 2\sqrt{2}/3 + \sqrt{3}/4$
4603 0341 rt : (9,2,-5,1)
4603 0574 $10^x : 7\ln 2/9$
4603 4820 rt : (6,6,9,-6)
4603 5338 at : $4\sqrt{3}/3 + 3\sqrt{5}$
4603 5450 $\zeta : 1/2$
4603 5855 as : $3\sqrt{2}/2 - 3\sqrt{5}/4$
4603 7045 $\theta_3 : 10/13$
4603 8055 rt : (7,0,-8,3)
4603 9186 $e^x : 4/3 - \sqrt{3}/4$
4604 0937 $\ell_{10} : \sqrt{3}/5$
4604 1124 $\Psi : \exp(-2e/3)$
4604 2671 $5 \div s\pi(e)$
4604 4249 rt : (3,7,7,-5)
4604 4309 $\ell_2 : 3\sqrt{2} - 2\sqrt{5}/3$
4604 8289 $\exp(5) \times \exp(\sqrt{2})$
4605 0487 rt : (3,6,-9,-9)
4605 3069 sinh : $7(1/\pi)/5$
4605 3270 $J_0 : 9\sqrt{2}/8$
4605 4502 $\theta_3 : 9/10$
4605 5399 as : $4/9$
4605 5609 rt : (9,9,-4,3)
4605 6029 $\ln(2/3) + s\pi(1/3)$
4605 6345 rt : (2,7,-7,7)
4605 8230 rt : (8,5,-4,0)
4606 0734 $10^x : P_{34}$
4606 1607 rt : (7,5,-9,-2)
4606 2706 cu : $1/3 + 3\pi/2$
4606 3570 $Ei : (\sqrt{2})^{-1/2}$
4606 4768 $2^x : 3\sqrt{3}/4$
4606 5886 id : $(3\pi/2)^{-1/2}$
4606 6339 rt : (7,8,-3,-1)
4606 7341 sinh : $1/\ln(3\pi)$
4606 7690 sq : $1/4 - 4\sqrt{5}/3$
4607 0182 id : $4\sqrt{2} - 3\sqrt{3}$
4607 0204 rt : (6,7,-6,-5)
4607 1497 $E : 3\zeta(3)/7$
4607 3978 at : $3\sqrt{2}/2 + 4\sqrt{3}$
4607 8684 rt : (7,1,-8,8)
4607 9893 ln : $2 + 4\sqrt{3}/3$
4608 0053 $3/4 \times \exp(2/3)$
4608 1112 rt : (3,9,-3,-3)
4608 1299 sr : $3 - \sqrt{3}/2$
4608 1837 $\Psi : (\sqrt{3}/2)^{-1/2}$
4608 4774 $\Psi : 24/5$
4608 6223 rt : (5,3,9,4)
4608 6822 cr : $9\sqrt{3}/5$
4608 7733 at : $10e/3$
4608 8764 rt : (6,6,8,3)
4608 9178 tπ : $5\sqrt[3]{5}/4$
4608 9215 $e^x : (\sqrt{5})^{1/2}$
4608 9258 rt : (6,3,-5,5)
4609 0442 rt : (7,9,3,7)
4609 2028 rt : (8,9,5,-5)
4609 2172 cr : $2 + \sqrt{5}/2$
4609 2432 $e^x : 3/4 + \sqrt{5}/3$
4609 3315 $Li_2 : 9/22$
4609 4265 rt : (9,6,4,-4)
4609 5502 $\Gamma : 1/25$
4609 6326 rt : (7,9,7,2)
4609 9186 $\Psi : 10\sqrt[3]{3}/7$
4609 9338 $J_1 : 16/15$
4610 3557 $2^x : 2/3 + 2\sqrt{5}/3$
4610 5297 $10^x : e + \pi/3$
4610 5403 $J_2 : 5\pi/6$
4610 6566 $2^x : 1/2 + 3\sqrt{5}/2$
4610 7027 ln : $3\sqrt{2}/2 - 2\sqrt{5}/3$
4610 7219 rt : (4,9,-5,0)
4610 8045 ln : $3 - \sqrt{2}$
4610 9641 at : $(1/(3e))^{1/3}$
4610 9792 cr : $9\ln 2/2$
4611 0836 tπ : $7\sqrt{3}/8$
4611 1506 $\ln(2) - \exp(e)$
4611 1550 sπ : $\sqrt{3}/2 + 4\sqrt{5}/3$
4611 2518 rt : (5,5,-2,2)
4611 2524 sπ : $9\sqrt[3]{5}/4$
4611 3617 $\theta_3 : e\pi/10$
4611 4118 sr : $e\pi/4$
4611 4138 cu : $\sqrt{2}/2 - 3\sqrt{3}$
4611 4492 $Li_2 : 9(1/\pi)/7$
4611 5761 rt : (1,1,8,-4)
4611 6773 at : $2\sqrt{5}/9$
4612 0516 rt : (6,7,2,-3)
4612 2392 sq : $4\sqrt{2} + 4\sqrt{3}/3$
4612 3428 rt : (8,2,-8,9)
4612 3568 $J_1 : e\pi/8$
4612 4320 at : $1/3 - 3\pi$
4612 4673 rt : (8,5,9,-6)
4612 7536 ln : $2(e + \pi)$
4612 9008 rt : (4,8,-8,-5)
4613 0154 cu : $3 - \sqrt{5}/3$
4613 0541 $10^x : 2\sqrt{3}/3 - 2\sqrt{5}/3$
4613 1077 rt : (3,6,-3,1)
4613 1751 rt : (8,6,-9,-1)
4613 2615 rt : (2,2,7,3)
4613 2966 $Ei : (\ln 3)^2$
4613 3415 rt : (9,2,8,-5)
4613 3976 rt : (7,4,1,-2)
4613 4742 rt : (2,9,-1,-6)
4613 5073 $\theta_3 : (2\pi/3)^{-2}$
4613 6303 $erf : 10/23$
4613 6513 $J_2 : \sqrt[3]{18}$
4613 7183 rt : (5,9,-3,-8)
4613 7540 sr : $1/4 + 4\sqrt{2}/3$
4613 9367 rt : (4,9,8,-6)
4614 0120 cu : $17/22$
4614 3481 cr : $1 + 3\sqrt{2}/2$
4614 3535 $3 \div \ln(\sqrt{3})$
4614 3596 rt : (1,9,-5,-6)
4614 4459 rt : (3,5,-7,-4)
4614 6801 $\Gamma : 8e/5$
4614 6916 $10^x : 4\sqrt{2}/3 - \sqrt{3}/2$
4614 7381 rt : (7,4,-5,6)
4614 8476 rt : (2,7,2,6)
4615 0643 $E : \exp(-2/3)$
4615 0953 rt : (5,6,7,-5)
4615 3602 $J_2 : 1/\ln(4/3)$
4615 3846 id : $6/13$
4615 4676 $J_1 : 8\zeta(3)/9$
4615 5396 sq : $3e - 4\pi$
4615 9485 at : $4\sqrt{2} + 2\sqrt{3}$
4615 9560 $Li_2 : 22/23$
4616 0890 $\sqrt{5} + \exp(4/5)$
4616 1386 rt : (8,1,0,-1)
4616 1929 $J_1 : 5\sqrt[3]{5}/8$
4616 3212 sπ : $(e/2)^2$
4616 3367 sinh : $3\sqrt[3]{5}/2$

4616 3631 rt : (3,8,0,-2)
4616 4876 cu : $1/3 - 2\pi/3$
4616 7160 $\ln(\sqrt{3}) \div \ln(4/5)$
4616 7262 tπ : $(2\pi/3)^{-1/2}$
4616 7781 rt : (5,7,0,-1)
4616 8289 rt : (6,3,6,-4)
4616 9782 E : $3\sqrt[3]{5}/10$
4617 2359 rt : (7,9,-6,-4)
4617 2478 tπ : $1/4 - \sqrt{5}/4$
4617 3663 sr : $1/\sqrt{22}$
4617 3734 sπ : $\sqrt{2} + \sqrt{3}/4$
4617 4636 rt : (2,8,4,-5)
4617 7083 rt : (2,8,-3,5)
4617 9820 sr : $3e - 2\pi/3$
4618 0310 rt : (7,8,-6,-4)
4618 0409 ℓ_{10} : $3/2 - 2\sqrt{3}/3$
4618 1419 rt : (3,5,6,2)
4618 1600 sq : $e/4$
4618 4017 sinh : $\exp(\sqrt{3})$
4618 4094 sq : $(3\pi)^{1/3}$
4618 4380 J_2 : 21/8
4618 4488 rt : (9,7,-9,0)
4618 6411 ζ : $5\zeta(3)/8$
4618 6497 10^x : $(e + \pi)^{1/2}$
4618 8021 $4/5 \div \sqrt{3}$
4618 8214 $\ln(3/5) - s\pi(2/5)$
4618 8216 as : $7(1/\pi)/5$
4618 9198 10^x : $3\sqrt{3}/8$
4618 9564 rt : (4,2,-6,-8)
4619 1516 sinh : $\sqrt{8/3}$
4619 2660 rt : (8,5,-5,7)
4619 4744 rt : (6,6,-2,3)
4619 4753 cr : $3\sqrt{2} - \sqrt{5}/2$
4619 7069 rt : (7,0,5,3)
4619 7961 Ei : $5\pi^2/7$
4619 9140 Ψ : $(\sqrt{5}/3)^3$
4619 9174 2^x : $4/3 - \pi/4$
4619 9460 Γ : 19/2
4619 9590 at : $1 + 3e$
4620 0313 Ei : $4e/7$
4620 0886 sr : $5\sqrt[3]{5}/4$
4620 2219 id : $e/4 - \pi$
4620 2697 as : $1/\ln(3\pi)$
4620 4107 ln : $4/3 + 4\sqrt{5}/3$
4620 4478 rt : (1,3,1,9)
4620 5426 rt : (4,7,1,-1)
4620 9240 10^x : 9/23
4620 9812 id : $2\ln 2/3$
4621 0258 $\exp(5) - s\pi(2/5)$
4621 0693 at : $4\sqrt{3} + \sqrt{5}$
4621 0817 Ei : $\ln 2/6$
4621 1715 tanh : 1/2
4621 2529 $\sqrt{5} - s\pi(e)$
4621 3388 Ei : $7\zeta(3)/10$
4621 4902 sr : $7\sqrt{3}/2$
4621 4962 rt : (6,6,-5,-3)
4621 7252 rt : (7,9,3,-4)
4622 1683 10^x : $5\sqrt{3}/3$
4622 2878 2^x : $\sqrt{6}$
4622 3042 at : $1/4 - 3\pi$
4622 4814 2^x : $\sqrt{2} - \sqrt{3}/2$
4622 4853 rt : (6,8,8,-6)
4622 5711 tπ : $2\sqrt{2} + 4\sqrt{3}/3$
4622 7049 sinh : $\sqrt{5}/5$
4622 7528 e^x : $1/4 + 4\sqrt{2}/3$
4622 7677 e^x : $2\sqrt[3]{5}/9$
4622 8882 2^x : 13/10
4622 8927 rt : (9,6,-5,8)
4623 1792 $3/4 + \ln(3/4)$
4623 2779 sr : $\sqrt[3]{5}/8$
4623 3939 rt : (5,1,-6,-9)
4623 4054 rt : (2,3,9,-5)
4623 6916 ζ : $\sqrt{2} + \sqrt{3} - \sqrt{7}$
4623 7425 at : $(2\pi/3)^3$
4623 8898 id : $1/4 - 3\pi/2$
4624 0000 sq : 17/25
4624 0980 sq : $3/2 + 4\sqrt{2}/3$
4624 3225 J_1 : 15/14
4624 5880 at : $1/4 + 4\sqrt{5}$
4624 7529 rt : (9,7,-3,-1)
4624 8079 at : $4 + 3\sqrt{3}$
4624 8769 Γ : 1/24
4624 9753 rt : (8,9,-6,-3)
4625 0862 rt : (8,6,2,-3)
4625 1296 rt : (7,7,-2,4)
4625 1861 rt : (1,8,-3,-3)
4625 2055 tanh : $\sqrt{2} + \sqrt{3} - \sqrt{7}$
4625 2489 J_2 : $\sqrt{2} - \sqrt{5} - \sqrt{7}$
4625 2588 2^x : $3/2 + 4\sqrt{3}$
4625 2775 rt : (7,5,7,-5)
4625 3938 rt : (2,7,-8,-5)
4625 4177 rt : (8,3,4,2)
4625 4958 at : $3e + \pi/3$
4625 5922 e^x : $1 + \sqrt{3}/2$
4625 6163 rt : (7,4,9,4)
4625 7424 J_0 : $\sqrt[3]{4}$
4625 7753 cu : $1/2 + 2\pi/3$
4625 8237 Ψ : $10\sqrt[3]{3}/3$
4625 9842 rt : (1,5,3,-8)
4626 3924 cr : $\ln 2/7$
4626 4398 cu : $4\sqrt{2} + 3\sqrt{5}/4$
4626 4525 2^x : P_{12}
4626 4533 Ψ : $9e/2$
4626 7216 E : $\ln(5/3)$
4626 8758 rt : (3,9,6,5)
4626 9122 Ei : 6π
4627 1087 2^x : $1/4 + 4e$
4627 1145 rt : (4,2,-7,-2)
4627 4096 J_2 : $5\ln 2$
4627 5086 Ψ : $4\zeta(3)$
4627 6589 at : $\exp(\sqrt{2}\pi/2)$
4627 7966 rt : (5,8,-9,4)
4627 8672 $3 \times \exp(e)$
4628 0733 ℓ_{10} : $2/3 + \sqrt{5}$
4628 2210 e^x : $3e - \pi/2$
4628 2288 rt : (3,0,8,4)
4628 6532 cr : $7\sqrt{5}/5$
4628 9912 cu : $3(e + \pi)/5$
4629 1004 sr : 3/14
4629 2577 rt : (8,8,-2,5)
4629 2902 J_2 : $2\sqrt{3}$
4629 4146 rt : (8,2,6,-4)
4629 4410 cu : $3\sqrt{2} + 2\sqrt{5}/3$
4629 6835 id : $4\sqrt{2}/3 + \sqrt{3}/3$
4629 6841 rt : (6,2,-9,-4)
4630 0502 2^x : $2\sqrt{2} + \sqrt{5}$
4630 0643 $\ln(3/4) \times \ln(5)$
4630 1600 ℓ_{10} : $4/3 + \pi/2$
4630 1869 $1/3 \times \exp(2)$
4630 2047 sq : $2/3 - \sqrt{5}$
4630 5610 ln : $2\sqrt{2} + 2\sqrt{5}/3$
4630 6369 e^x : $5\sqrt[3]{3}/8$
4630 6453 rt : (9,3,1,-2)
4630 7103 rt : (8,7,8,-6)
4630 7524 tπ : $3\sqrt{2}/8$
4630 7721 rt : (1,7,4,-6)
4631 3500 ℓ_{10} : $\sqrt{2} + 2\sqrt{5}/3$
4631 3515 rt : (4,6,5,7)
4631 5704 $\sqrt{2} - s\pi(2/5)$
4631 5878 ℓ_{10} : $2\sqrt{3} - \sqrt{5}/4$
4631 6336 tπ : P_4
4631 6513 cu : $2 - \sqrt{2}/4$
4631 6857 cr : $7e\pi/4$
4631 7113 rt : (9,8,3,-4)
4631 7564 rt : (1,9,-9,-6)
4631 8392 tπ : $1/3 - \sqrt{2}/3$
4631 9884 rt : (4,1,-3,6)
4632 2174 2^x : $5\sqrt{2}$
4632 3053 rt : (4,8,-7,-6)
4632 3656 erf : $1/\sqrt[3]{12}$
4632 4037 rt : (9,9,-2,6)
4632 5137 sr : $1 - \pi/4$
4632 7063 Ψ : $\exp(\pi/2)$
4632 7274 rt : (2,9,1,3)
4633 0964 ℓ_{10} : $2 + e/3$
4633 2495 rt : (5,8,1,0)
4633 5684 E : $8(1/\pi)/5$
4633 7723 J_1 : $(\sqrt{3}/2)^{-1/2}$
4633 7979 e^x : $\sqrt{2}/2 + \sqrt{3}$
4633 8081 ln : $\sqrt{2}/3 + \sqrt{5}/2$
4633 8172 2^x : $\ln(\sqrt{3})$
4633 8416 rt : (8,7,8,3)
4634 0041 e^x : $3\zeta(3)/4$
4634 0709 cr : $4 - \sqrt{3}/2$
4634 1158 rt : (4,7,-5,5)
4634 2442 at : $1/3 + 4\sqrt{5}$
4634 3193 e^x : $5\sqrt{5}/9$
4634 4420 sq : $2\sqrt{2}/3 + 3\sqrt{5}/2$
4634 4974 rt : (9,9,9,-7)
4634 6648 sr : $\sqrt{3} - \sqrt{5} + \sqrt{7}$
4634 8846 at : $3 + 2\pi$
4634 8933 id : $3e/4 + 3\pi$
4634 9989 rt : (9,4,7,-5)
4635 0064 J_1 : $\ln((e + \pi)/2)$
4635 3994 rt : (4,2,9,-5)
4635 4266 sπ : $e\pi/3$
4635 5615 rt : (5,2,-3,7)
4635 5685 cu : $8\sqrt[3]{5}/7$
4635 6999 2^x : $1/4 - e/2$
4635 7360 $\ln(\sqrt{2}) + \exp(3/4)$
4636 1742 rt : (9,1,5,3)
4636 4757 tπ : $\sqrt{2}/2 + 2\sqrt{3}/3$
4636 4760 at : 1/2
4636 4832 rt : (5,8,6,-5)
4636 4919 10^x : $2 - \sqrt{3}/3$
4636 5614 at : $\sqrt{2}/4 + 4\sqrt{5}$
4636 6040 10^x : $4\zeta(3)$
4636 6146 sq : $3 + 4e$
4636 7790 e^x : 8/21
4636 8519 tπ : $\pi/6$
4636 9343 rt : (4,9,-1,-2)
4637 0763 at : $4e - \pi/2$
4637 1464 ζ : $5\pi/7$
4637 2166 rt : (6,8,-5,0)
4637 3803 e^x : 13/5
4637 4975 rt : (4,3,7,3)
4637 5273 at : $3\sqrt{3}/2 + 3\sqrt{5}$
4637 7974 J_1 : 21/8
4638 0160 ζ : $(e/2)^{-1/2}$
4638 1993 rt : (6,3,-3,8)
4638 5010 sr : 15/7
4638 6004 sr : $7e\pi/3$
4638 6269 rt : (5,3,4,-3)
4638 9813 erf : 7/16
4639 0253 $1/2 + s\pi(\sqrt{2})$
4639 1027 tπ : P_5
4639 2075 rt : (6,9,1,1)
4639 4722 sπ : $4\sqrt{2}/3 - \sqrt{3}$
4639 6825 Ei : 16/19
4639 8292 rt : (6,8,5,9)
4639 9925 rt : (7,5,-7,1)
4640 0000 cu : $(\sqrt[3]{5}/3)^{-3}$
4640 0111 rt : (9,7,-3,-2)
4640 0360 sr : $4/3 - \sqrt{5}/2$
4640 1763 sinh : $\pi/7$
4640 2266 rt : (7,4,-3,9)
4640 4691 e^x : $\sqrt[3]{13/2}$
4640 4746 sr : $e/2 + 3\pi/2$
4640 5797 at : $\sqrt{2} + \sqrt{3} - \sqrt{7}$
4640 6673 rt : (9,8,-9,-5)
4640 6927 sq : $4/3 - 3\sqrt{2}$
4640 7268 as : $\sqrt{2}/2 - 2\sqrt{3}/3$
4640 8289 J_1 : 14/13
4640 8569 2^x : 11/20
4640 8843 tπ : $\sqrt{3}/4 + 4\sqrt{5}$
4640 9750 ζ : $\ln(3\pi)$
4641 0161 id : $2\sqrt{3}$
4641 0199 Γ : $-\sqrt{2} - \sqrt{6} + \sqrt{7}$
4641 0580 ℓ_2 : $\sqrt[3]{21}$
4641 1474 J_2 : $\sqrt{7}$
4641 1665 $e - s\pi(\sqrt{3})$
4641 3798 rt : (9,5,9,4)
4641 4301 ℓ_{10} : $2\zeta(3)/7$
4641 4802 $\ln(2/3) \times \ln(\pi)$
4641 4863 Γ : 1/23
4641 5888 id : $1/\sqrt[3]{10}$
4641 7543 $\exp(4) + s\pi(1/3)$
4642 0438 rt : (8,6,-7,2)
4642 1080 2^x : $4/3 + e/2$
4642 1559 rt : (9,6,3,1)

4642 1725 rt : (3,8,-8,1)
4642 4254 Γ : 11/18
4642 5358 cu : $4/3-\sqrt{5}/4$
4642 5855 ln : $\sqrt{14}\pi$
4643 0592 rt : (6,5,5,-4)
4643 2258 at : $\exp(\sqrt{5})$
4643 3823 cr : P_{18}
4643 5211 rt : (2,8,4,1)
4643 5662 ln : $4\sqrt{2}/9$
4643 6215 rt : (9,7,-7,3)
4643 6561 $3/5\times s\pi(e)$
4643 6970 rt : (5,1,8,4)
4643 7205 cr : P_{48}
4643 7515 θ_3 : $1/\sqrt{19}$
4643 7523 Li_2 : 7/17
4643 7693 cu : $2\sqrt{2}+\sqrt{5}/2$
4644 1170 2^x : $1/\sqrt[3]{6}$
4644 2378 sq : $1/4-3\sqrt{3}$
4644 2448 sr : $e/2+\pi/4$
4644 4065 rt : (4,5,-5,-3)
4644 4787 rt : (5,8,-7,9)
4644 5206 2^x : $4\sqrt{2}/3+3\sqrt{3}$
4644 5905 10^x : $\sqrt{3}+\sqrt{5}/4$
4644 7923 rt : (2,9,4,-4)
4644 8341 2^x : $3+2e$
4644 9927 2^x : $3\sqrt{3}-3\sqrt{5}/4$
4645 1039 at : $3/2-4e$
4645 1143 rt : (5,9,-9,-2)
4645 1655 cu : $1-2\pi$
4645 2048 rt : (6,4,-1,-1)
4645 2137 ℓ_{10} : $3/2+\sqrt{2}$
4645 2536 cu : $3/4-4\sqrt{2}/3$
4645 2910 $\exp(1/2)-\exp(\sqrt{2})$
4645 2968 rt : (3,6,-1,6)
4645 3645 id : $9e$
4645 4045 at : $8(e+\pi)/5$
4645 5715 at : $\sqrt{3}/4+4\sqrt{5}$
4645 5857 sq : $3/2-4\pi$
4645 6235 e^x : $1/2+3\sqrt{5}$
4645 6655 rt : (7,7,6,-5)
4645 8425 e^x : $1/\sqrt[3]{18}$
4645 8835 Ei : $4e/9$
4645 9171 2^x : $2\sqrt{2}+\sqrt{5}/4$
4645 9188 id : $(\pi)^{1/3}$
4645 9382 $s\pi$: $4\sqrt[3]{3}/5$
4645 9570 rt : (1,3,7,-4)
4646 0758 10^x : $4-2\pi/3$
4646 3044 $\ln(2)\div\exp(2/5)$
4646 3228 10^x : $3\sqrt{3}+3\sqrt{5}/4$
4646 7488 rt : (6,1,9,-5)
4646 7874 $\ln(\sqrt{2})\div s\pi(\sqrt{3})$
4646 9136 10^x : $(\ln 3/2)^3$
4647 0247 rt : (3,7,-3,-5)
4647 0802 ln : $\pi/5$
4647 0946 ℓ_{10} : $\sqrt{17/2}$
4647 2317 $s\pi$: 2/13
4647 2527 rt : (5,1,-8,3)
4647 3815 rt : (8,9,7,-6)
4647 5800 sr : $(5/3)^{-3}$
4647 5978 sr : $1/3+2e/3$
4647 6630 ℓ_{10} : $4/3-3\sqrt{3}/4$
4647 6772 rt : (8,9,-4,-3)
4647 7038 Li_2 : $2\sqrt[3]{3}/7$
4647 8836 cr : 22/7
4647 9069 $1/5\div s\pi(\pi)$
4648 1624 rt : (3,7,-6,-4)
4648 1638 ln : $3\sqrt[3]{3}$
4648 2023 e^x : $7\sqrt[3]{5}/8$
4648 2373 $t\pi$: $\sqrt{2}/3+\sqrt{5}/2$
4648 4122 rt : (2,5,8,-5)
4648 4994 rt : (4,1,-8,-3)
4648 5182 e^x : $1/\ln(\sqrt{5})$
4648 6141 sinh : $7e/5$
4648 6798 rt : (2,8,-2,-1)
4648 7603 sq : 15/22
4648 8512 rt : (7,6,0,-2)
4648 8633 J_0 : 19/12
4649 1926 rt : (5,2,2,1)
4649 4568 sinh : $5(e+\pi)/8$
4649 4973 sinh : $1/\sqrt[3]{11}$
4649 5460 rt : (5,0,-8,4)
4649 6128 at : $3\sqrt{2}/4-\sqrt{5}/4$
4649 6862 id : $(e+\pi)/4$
4649 7215 ln : $1/4+3e/2$
4649 7352 $\ln(5)\div\ln(3)$
4649 8111 10^x : $3e/4+2\pi/3$
4649 8254 J_1 : $6\sqrt[3]{2}/7$
4649 8711 at : $\sqrt{2}/3+4\sqrt{5}$
4649 8803 rt : (3,7,9,-6)
4649 9154 J_1 : $\sqrt[3]{18}$
4650 0456 rt : (2,9,-7,-5)
4650 2056 J_1 : $(\sqrt[3]{2})^{1/3}$
4650 3274 rt : (6,1,-8,5)
4650 6184 rt : (8,8,1,-3)
4650 6514 rt : (1,8,-5,-9)
4650 7205 cu : $4\sqrt{2}/3+\sqrt{3}/4$
4650 8853 at : 3π
4650 9331 rt : (7,2,-8,6)
4651 0534 10^x : $\sqrt{5}\pi$
4651 2520 id : $\ln(\ln 2/3)$
4651 2999 rt : (4,1,-9,-4)
4651 4163 rt : (8,3,-8,7)
4651 5426 $s\pi$: $8\sqrt[3]{3}/3$
4651 5953 2^x : $3\pi^2/2$
4651 6228 e^x : $3/2-\sqrt{5}/2$
4651 6390 rt : (1,9,5,4)
4651 7085 e^x : $6(1/\pi)/5$
4651 8107 rt : (9,4,-8,8)
4651 9141 rt : (2,4,-9,4)
4652 0212 cr : $4/3+2e/3$
4652 0233 $\exp(\pi)-s\pi(\sqrt{5})$
4652 0414 $1/3\times\exp(1/3)$
4652 0422 ln : $3/2+2\sqrt{2}$
4652 1957 at : $4+2e$
4652 2316 ℓ_2 : $1/3+\pi/3$
4652 3238 tanh : $2\sqrt[3]{2}/5$
4652 3422 Ψ : $1/\ln(5)$
4652 3561 cu : $3/4+4\pi/3$
4652 4203 $s\pi$: $2\ln 2/9$
4652 4293 sr : $2+3e/2$
4652 4432 at : $3\sqrt{2}+3\sqrt{3}$
4652 5126 rt : (6,0,3,2)
4652 6207 θ_3 : $4\sqrt{3}/9$
4652 8197 cr : P_{61}
4652 8200 e^x : $9\zeta(3)/2$
4653 0242 sr : $\sqrt{3}/8$
4653 0510 at : $1/2+4\sqrt{5}$
4653 1062 rt : (9,3,-7,-3)
4653 1217 ζ : $(\ln 3/2)^{-1/2}$
4653 1570 $t\pi$: $\ln 2/5$
4653 1750 cr : $\sqrt{2}+\sqrt{3}$
4653 4201 sinh : 9/20
4653 5921 rt : (7,2,4,-3)
4653 7396 sq : 23/19
4653 9030 sq : $3/2-4\sqrt{3}$
4654 0919 10^x : $3\sqrt{2}+3\sqrt{5}/4$
4654 1270 rt : (8,4,5,-4)
4654 2087 as : $\pi/7$
4654 2506 $\exp(2/3)-\exp(5)$
4654 2633 $t\pi$: $2e+3\pi$
4654 3336 rt : (8,5,-8,-4)
4654 4463 rt : (9,6,6,-5)
4654 4514 J_1 : $(\sqrt[3]{5}/2)^{-1/2}$
4654 5733 rt : (2,5,-9,-8)
4654 6986 rt : (3,3,-8,7)
4654 7751 rt : (7,7,-9,-5)
4654 8411 Ψ : 1/14
4654 9456 cu : $3-2e$
4654 9566 cr : P_{45}
4655 0574 J_1 : $3\sqrt[3]{3}/4$
4655 1424 Ei : 11/2
4655 1623 $t\pi$: $4\sqrt{3}/3$
4655 1625 e^x : $\sqrt{15/2}$
4655 5459 J_1 : $9\zeta(3)/10$
4655 6324 id : $3\sqrt{2}-3\sqrt{5}$
4655 6585 Γ : 25/2
4655 7123 rt : (9,9,0,9)
4655 7568 e^x : $\sqrt{2}/4-\sqrt{5}/2$
4655 8572 2^x : $3e/4-\pi$
4655 9036 ℓ_2 : $e-3\pi/4$
4655 9687 ln : $5\sqrt{3}/2$
4656 0222 sinh : $10\pi/7$
4656 4233 2^x : $6\pi/7$
4656 6882 e^x : $4+3\sqrt{2}/2$
4656 9366 id : $6\sqrt[3]{5}/7$
4656 9556 sq : $3/4+2\pi$
4656 9638 as : P_{41}
4657 1230 θ_3 : $(3\pi/2)^{-1/3}$
4657 2114 E : $2\sqrt[3]{2}/5$
4657 3014 10^x : P_1
4657 3590 id : $5\ln 2$
4657 4757 Ei : $(4/3)^3$
4657 4962 rt : (8,0,9,5)
4657 5248 rt : (3,8,-6,8)
4657 5717 J_1 : $5\sqrt{3}/8$
4657 6106 sr : $\sqrt{2}/3+3\sqrt{5}/4$
4657 7057 J_1 : $5\pi/6$
4657 7662 ln : $3\sqrt{2}/4-\sqrt{3}/4$
4657 8074 id : $10e\pi/3$
4657 8320 ln : $2\sqrt{2}+4\sqrt{5}$
4657 8826 rt : (5,3,-9,-9)
4657 8968 at : $4/3+3e$
4657 9423 at : $10e\pi/9$
4657 9480 e^x : 10
4658 0823 rt : (7,2,8,4)
4658 1391 e^x : $4\pi^2$
4658 1500 rt : (4,8,-5,-1)
4658 3116 $\exp(\sqrt{5})\div\exp(3)$
4658 3697 rt : (5,6,9,-6)
4658 4927 rt : (3,5,-9,-7)
4658 5473 sq : $1/\ln(\ln 2/3)$
4658 6129 cr : $-\sqrt{3}+\sqrt{5}+\sqrt{7}$
4658 6659 cr : $5\sqrt[3]{2}/2$
4658 7995 rt : (2,8,-1,-2)
4658 8022 $s\pi$: $\exp(e)$
4658 8718 rt : (9,9,-4,-1)
4658 9474 rt : (2,6,-9,-6)
4658 9631 Ei : $5(1/\pi)/6$
4659 0212 rt : (4,4,8,-5)
4659 0558 ℓ_2 : $3\sqrt{3}-2\sqrt{5}$
4659 1938 at : 19/2
4659 2421 rt : (6,4,7,3)
4659 2451 Γ : $10\sqrt[3]{2}$
4659 5465 Γ : 1/22
4659 5800 rt : (1,7,-9,-5)
4659 7775 $t\pi$: $1/3-2\sqrt{5}$
4659 7791 rt : (1,8,-3,8)
4659 8000 ℓ_{10} : $\sqrt[3]{5}/5$
4659 8964 Ei : $5\sqrt{5}/6$
4659 9557 J_1 : 13/12
4660 0120 $3/4\div\ln(5)$
4660 0960 rt : (3,2,7,-4)
4660 2540 $2/5-s\pi(1/3)$
4660 3502 sq : $4+3e/4$
4660 3680 2^x : $2(e+\pi)/9$
4660 4046 rt : (3,9,-6,-4)
4660 5201 rt : (8,9,-4,0)
4660 7244 at : $7e/2$
4660 8220 10^x : $4/3+3\pi/4$
4660 9767 Li_2 : $(e+\pi)^{-1/2}$
4661 0235 $\pi-s\pi(\sqrt{5})$
4661 3296 J_0 : $\sqrt{5/2}$
4661 4757 rt : (7,8,-4,-1)
4661 5581 at : $\sqrt{3}/3+4\sqrt{5}$
4661 7814 rt : (3,6,0,-1)
4661 8617 e^x : $1/3+3\sqrt{5}/4$
4661 9291 sinh : $3\zeta(3)/8$
4661 9407 sq : $(\pi)^{-1/3}$
4662 0290 ζ : $(\sqrt{5})^{-1/2}$
4662 0643 rt : (1,9,9,-3)
4662 0743 Γ : 6/23
4662 1207 id : $\sqrt[3]{15}$
4662 2051 cu : $\sqrt{3}/2-4\sqrt{5}/3$
4662 2052 rt : (2,0,6,3)
4662 2511 erf : 11/25
4662 2921 sinh : 12/5
4662 5240 sr : 5/23
4662 6258 rt : (5,6,6,2)
4662 7794 e^x : $4\sqrt{2}-2\sqrt{5}/3$
4662 8037 rt : (1,5,-2,3)

4662 9053 rt : (6,7,-4,-2)
4662 9353 $\ln(\sqrt{5}) \times \exp(3/5)$
4662 9496 rt : (5,9,-7,3)
4662 9934 $J_2 : 8/3$
4663 0346 id : $1/\ln(2/3)$
4663 1929 at : $2\sqrt{2} + 3\sqrt{5}$
4663 2649 sq : $\sqrt{2}/2 + 2\sqrt{3}/3$
4663 2935 rt : (9,5,0,-2)
4663 3706 ln : 3/13
4663 5368 at : $4/3 - 4e$
4663 5640 $3 \times \exp(3/5)$
4663 5953 $s\pi : \sqrt[3]{10}$
4663 6708 as : $1/\sqrt[3]{11}$
4663 8415 $2^x : 7\sqrt{3}/3$
4663 8492 sinh : $\sqrt{22/3}$
4664 0282 rt : (7,9,5,-5)
4664 2046 at : $(\sqrt{2}/3)^{-3}$
4664 3276 $e^x : (e/2)^{-1/3}$
4664 6449 $t\pi : (1/(3e))^{-3}$
4664 7582 $J_2 : \sqrt[3]{19}$
4665 0635 sq : $1/4 + \sqrt{3}/4$
4665 0707 rt : (2,4,-5,-3)
4665 2756 $e^x : (\ln 2/2)^{-1/2}$
4665 2774 rt : (5,9,-1,-2)
4665 5740 rt : (9,7,-5,6)
4665 7058 cu : $3e/4 - 2\pi$
4665 7186 rt : (4,1,-1,9)
4665 7517 rt : (7,5,-6,1)
4665 8526 $t\pi : 2\sqrt{2}/3 + 3\sqrt{3}$
4666 0516 rt : (6,7,4,-4)
4666 1328 $2^x : \sqrt{3}/3 - 3\sqrt{5}/4$
4666 2600 cr : $2 + 2\sqrt{3}/3$
4666 4174 at : $3 - 4\pi$
4666 5042 rt : (7,5,9,-6)
4666 6666 id : 7/15
4666 7687 rt : (8,6,-5,5)
4667 0883 rt : (8,3,-1,-1)
4667 0940 $10^x : 3\sqrt{2}/4 + \sqrt{3}/2$
4667 4886 rt : (9,1,4,-3)
4667 6348 rt : (3,9,8,-6)
4667 6533 as : 9/20
4667 8445 rt : (6,8,-4,-3)
4667 9738 rt : (5,6,-9,-5)
4667 9988 sinh : $7\sqrt{5}/8$
4668 1730 at : $2\sqrt[3]{2}/5$
4668 2316 rt : (1,2,5,2)
4668 2926 rt : (7,5,-5,4)
4668 3524 $J_2 : 24/7$
4668 3592 $10^x : 13/20$
4668 5617 at : $\sqrt{3} - \sqrt{5}$
4668 5831 $\ell_{10} : (e + \pi)/2$
4668 5998 rt : (1,6,2,-5)
4668 9344 sq : $\sqrt{2}/3 - 2\sqrt{3}/3$
4669 0816 rt : (6,3,8,-5)
4669 3116 $\exp(2/5) \div s\pi(\pi)$
4669 3758 rt : (5,5,3,-3)
4669 5243 $\Gamma : 22/3$
4669 6544 $2^x : \ln(1/3)$
4669 7381 $\ell_2 : \exp(\sqrt[3]{5})$
4669 7819 $10^x : 4/3 + 4\sqrt{5}$
4669 9669 cr : $3\sqrt{3}/2 + \sqrt{5}/4$
4669 9814 rt : (4,7,-2,-1)
4670 3038 rt : (6,4,-5,3)
4670 4592 rt : (2,7,7,-5)
4670 6922 sr : $10\zeta(3)$
4670 7042 $J_1 : 25/23$
4670 7649 $\ell_{10} : 1/3 + 3\sqrt{3}/2$
4670 8122 at : $9e\pi/8$
4670 8827 sinh : $\ln(\pi/2)$
4670 8969 sr : $7e\pi/2$
4670 8986 $\ln(2) + s\pi(e)$
4670 9037 rt : (2,0,-9,4)
4670 9817 $s\pi : 2\sqrt{3}/3$
4670 9837 $10^x : 3\sqrt[3]{2}/7$
4671 2127 at : $2/3 + 4\sqrt{5}$
4671 2217 rt : (1,9,2,-3)
4671 3797 sr : $1/\sqrt{21}$
4671 3899 $2^x : 3/2 - 3\sqrt{3}/2$
4671 4343 rt : (1,9,-4,4)
4671 4531 $\ell_2 : 1/3 + 3\sqrt{3}$
4671 4781 rt : (9,3,8,4)
4671 5543 $10^x : 3\sqrt{3} - 3\sqrt{5}/4$
4671 7013 $e^x : 4/3 - 2\pi/3$
4671 7485 $\sqrt{2} \div s\pi(\sqrt{2})$
4671 7585 $J_1 : 2e/5$
4671 7959 $s\pi : (2\pi)^{1/3}$
4671 8032 at : $8\zeta(3)$
4671 9129 rt : (7,2,-6,9)
4671 9467 rt : (3,9,-7,1)
4671 9929 rt : (9,8,5,-5)
4672 1424 $\theta_3 : 3/13$
4672 2155 rt : (9,8,-9,-2)
4672 3750 sq : $\sqrt{3}/2 + 3\sqrt{5}/4$
4672 3807 $E : \sqrt{2} + \sqrt{3} - \sqrt{7}$
4672 7162 sinh : 20/17
4672 7543 at : $2e + 4\pi/3$
4672 7974 $Ei : 2\sqrt[3]{2}/5$
4672 8046 $3/5 \div \exp(1/4)$
4673 0792 id : $4/3 - \sqrt{3}/2$
4673 0804 rt : (5,3,-5,2)
4673 1608 rt : (8,1,3,2)
4673 2773 rt : (5,1,7,-4)
4673 2904 $t\pi : \sqrt{22}$
4673 4952 rt : (6,4,-8,-4)
4673 7016 $Li_2 : (\sqrt{5}/3)^3$
4673 8243 rt : (6,1,-6,8)
4673 9259 $\ln(2) \times \exp(3/4)$
4674 0110 id : $\pi^2/4$
4674 1172 cu : 25/22
4674 2577 cu : $\sqrt{3}/3 + \sqrt{5}/4$
4674 2931 $Ei : (1/\pi)/7$
4674 3023 rt : (9,4,9,-6)
4674 3374 $\exp(1/3) \div s\pi(2/5)$
4674 6220 $E : 1/2$
4674 6296 rt : (7,9,-8,-9)
4674 8389 rt : (8,7,-9,-3)
4674 9782 cu : $3\sqrt{3}/2 + \sqrt{5}/4$
4675 0075 $s\pi : 3e$
4675 0385 rt : (2,8,6,8)
4675 0709 $Li_2 : 3\sqrt{5}/7$
4675 3597 rt : (5,8,8,-6)
4675 5259 at : $\sqrt{2}/2 + 4\sqrt{5}$
4675 6839 $Li_2 : 23/24$
4675 8495 rt : (8,6,4,-4)
4675 8940 at : $3/2 + 3e$
4676 0435 $t\pi : \exp(\sqrt{3}/2)$
4676 0572 id : $\sqrt{2} - \sqrt{5} - \sqrt{7}$
4676 1071 at : $4 + 4\sqrt{2}$
4676 2472 rt : (5,0,-6,7)
4676 2675 $Ei : (5/3)^{-1/3}$
4676 2734 $10^x : 3\sqrt{2}/4 + 3\sqrt{3}$
4676 2926 as : $3\zeta(3)/8$
4676 4421 $e^x : \exp(\pi)$
4676 6656 rt : (3,9,-4,3)
4676 7634 rt : (3,9,-5,-4)
4677 0227 at : P_{35}
4677 1616 rt : (4,2,-5,1)
4677 1924 cr : $1/3 + 2\sqrt{2}$
4677 2706 rt : (2,5,-7,-1)
4677 3020 $J_2 : 2\sqrt[3]{5}$
4677 3188 at : $3\sqrt{3} + 2\sqrt{5}$
4677 3483 sinh : $\exp(-\sqrt[3]{4}/2)$
4677 6557 sr : $4\sqrt{2} + \sqrt{3}/4$
4677 7798 cu : $9\sqrt{5}$
4677 9418 $\Gamma : \sqrt{3} + \sqrt{5} + \sqrt{7}$
4677 9926 $10^x : 1/6$
4678 0053 at : $1/4 + 3\pi$
4678 0150 $\theta_3 : \ln 2/3$
4678 1532 $2^x : \sqrt{2} + \sqrt{5}/3$
4678 2717 $2^x : 3\sqrt{3}/4 - \sqrt{5}/3$
4678 4016 $e^x : 2\sqrt{2}/3 - \sqrt{5}/4$
4678 4283 sq : $2\sqrt[3]{5}/5$
4678 4531 rt : (7,8,-1,-2)
4678 5045 $Ei : 1/22$
4678 5704 rt : (8,9,9,-7)
4678 5921 rt : (1,2,-9,-5)
4678 6290 rt : (8,2,8,-5)
4678 7469 $2^x : 4\sqrt{2}/3 - 4\sqrt{5}/3$
4678 8042 rt : (8,5,7,3)
4678 8779 sinh : $6\sqrt[3]{2}$
4678 8982 sr : $1 + 2\sqrt{3}/3$
4679 0414 rt : (5,9,-5,8)
4679 1111 rt : (1,6,9,3)
4679 1504 id : $3e/4 - \pi/2$
4679 2565 $\Gamma : 1/21$
4679 4187 rt : (4,1,-6,-6)
4679 5022 $3/5 \times \exp(\sqrt{2})$
4679 5129 $\theta_3 : (\sqrt[3]{3}/3)^2$
4679 5931 $J_2 : 6\sqrt{5}/5$
4679 8644 $2^x : 3\sqrt{2} + 4\sqrt{3}$
4679 8970 rt : (4,6,7,-5)
4680 0006 $4 \times \exp(3/4)$
4680 0618 at : $3/4 + 4\sqrt{5}$
4680 0876 $\ell_2 : 3\sqrt{2}/2 + 4\sqrt{5}$
4680 3328 rt : (3,5,-9,-5)
4680 4306 ln : $4/3 - \sqrt{2}/2$
4680 5237 rt : (4,8,-3,4)
4680 6478 $\ln(2/5) \times \ln(3/5)$
4680 7230 $e^x : 5e/8$
4680 9290 $t\pi : 3\sqrt{2}/4 - 3\sqrt{3}/2$
4681 1664 $Ei : 1/\ln(\sqrt[3]{5})$
4681 3142 rt : (4,1,6,3)
4681 4404 sq : 13/19
4681 4441 $J_2 : 2e\pi/5$
4681 5142 ln : $2\sqrt{2} - 3\sqrt{3}/2$
4681 5266 at : $3 + 3\sqrt{5}$
4681 9077 $\ln(2/3) \div s\pi(1/3)$
4681 9507 rt : (7,9,-8,-5)
4681 9595 $t\pi : 9\sqrt[3]{3}/8$
4681 9827 $\Psi : (\sqrt{2}\pi)^3$
4682 0987 rt : (7,4,3,-3)
4682 1139 $3/4 \div \ln(3/5)$
4682 1419 $\ln(2) \times s\pi(\sqrt{5})$
4682 1990 sq : $1/\ln(\sqrt[3]{4}/3)$
4682 3782 $J_1 : 12/11$
4682 4671 ln : $5\sqrt{5}/7$
4682 4979 $\ell_{10} : 4 - 3\sqrt{2}/4$
4682 7265 $\zeta : 8\sqrt{3}/9$
4682 7716 rt : (7,7,8,-6)
4682 7874 $\exp(1/2) - \exp(3/4)$
4682 8182 id : $1/4 - e$
4682 9703 as : $2/3 - \sqrt{5}/2$
4683 2011 $\ell_{10} : 7\sqrt[3]{2}/3$
4683 3244 $2^x : 1 - 2\pi/3$
4683 3541 at : $3e + \pi/2$
4683 4687 $\Gamma : 3\zeta(3)/10$
4683 6968 $J_1 : 1/\ln(5/2)$
4683 7494 rt : (6,5,-9,-5)
4683 7663 $J_1 : 7\sqrt{5}/6$
4683 8680 $t\pi : 6e$
4683 8996 $\theta_3 : 10\ln 2/9$
4683 9509 $\zeta : \sqrt{5}$
4684 3764 $Li_2 : 1/\sqrt[3]{14}$
4684 4238 cu : $e/3 - 2\pi$
4684 6733 cu : $2e/7$
4684 7801 cr : 19/6
4684 8388 id : $4\sqrt{2}/3 - 3\sqrt{5}/2$
4684 9036 rt : (9,7,-3,9)
4684 9312 $\ell_2 : -\sqrt{2} + \sqrt{3} + \sqrt{6}$
4685 1293 $\Gamma : (\sqrt{2}\pi/2)^{-3}$
4685 3842 as : $\ln(\pi/2)$
4685 3908 $10^x : 9(1/\pi)$
4685 4743 rt : (6,6,2,9)
4685 5815 rt : (7,0,7,4)
4686 1073 $s\pi : 5\ln 2/3$
4686 1138 rt : (4,7,-4,-3)
4686 3499 rt : (6,9,3,4)
4686 5858 at : $2\sqrt{2} + 4\sqrt{3}$
4686 7397 at : $1/3 + 3\pi$
4686 7590 $10^x : 5\sqrt{5}/9$
4686 7761 $2^x : 4\ln 2/5$
4686 8286 sr : $4 + 2\pi/3$
4686 9044 at : $8e\pi/7$
4687 0490 sinh : $e/6$
4687 1617 rt : (3,5,7,9)
4687 3504 $\Gamma : \sqrt[3]{3}/4$
4687 5370 $\Psi : (e + \pi)^{-1/2}$
4687 6064 at : $5(e + \pi)/3$
4687 6208 $\ell_{10} : 2 + 2\sqrt{2}/3$
4687 9402 rt : (2,6,-7,-9)

4687 9550 $\ell_{10} : e/8$
4687 9862 $Ei : \sqrt{20}\pi$
4688 0177 sr : $2/3 + 2\sqrt{5}/3$
4688 0866 $E : 2\sqrt{5}/9$
4688 1447 rt : (7,2,-7,-3)
4688 1605 $\ell_{10} : \sqrt{2}/2 + \sqrt{5}$
4688 2053 rt : (8,6,-3,8)
4688 3593 $2^x : (e+\pi)^{-1/3}$
4688 4719 $E : (1/(3e))^{1/3}$
4688 5818 $Ei : 14/9$
4688 7112 rt : (1,9,4,0)
4688 7917 $\ell_{10} : 3\sqrt{2} - 3\sqrt{3}/4$
4688 7931 $J_2 : 6\pi/7$
4688 8555 $\zeta : 5\sqrt[3]{3}/9$
4688 8654 rt : (6,5,7,-5)
4689 0294 rt : (6,6,-2,-1)
4689 3456 rt : (7,3,2,1)
4689 6964 rt : (8,9,-2,3)
4689 8868 rt : (9,6,8,-6)
4690 0174 rt : (2,9,6,-4)
4690 0318 $10^x : 16/3$
4690 4082 sr : $4\sqrt{3} - 3\sqrt{5}$
4690 4919 $e^x : 5/13$
4690 7105 rt : (9,7,-1,-2)
4690 7885 as : $1/3 - \pi/4$
4690 8035 rt : (5,5,2,8)
4690 9023 $2^x : \sqrt{19/2}$
4690 9344 rt : (4,3,-8,-4)
4691 3580 sq : $10\sqrt{2}/9$
4691 9166 rt : (1,7,2,-9)
4691 9500 as : $\exp(-\sqrt[4]{4}/2)$
4691 9832 cr : $7e/6$
4692 0823 rt : (4,9,-5,-7)
4692 1347 at : $\sqrt{3}/2 + 4\sqrt{5}$
4692 2022 rt : (1,6,4,6)
4692 2548 rt : (5,4,-9,-6)
4692 3493 rt : (7,5,-3,7)
4692 3959 $\ln(\pi) - s\pi(\sqrt{5})$
4692 5416 sq : $\sqrt{3}/4 - \sqrt{5}/2$
4692 6349 sinh : $(\sqrt[3]{3}/2)^{-1/2}$
4692 9972 $\ell_2 : \sqrt{23/3}$
4693 0977 rt : (8,7,-7,-4)
4693 2042 rt : (9,5,-8,6)
4693 3761 rt : (5,8,-8,-5)
4693 4217 $2^x : 4/3 + \sqrt{5}/2$
4693 5191 rt : (2,9,-9,-6)
4693 6062 rt : (1,7,-7,-8)
4693 7398 at : $4e - \pi/3$
4693 8775 sq : $11/7$
4693 9635 tanh : $8(1/\pi)/5$
4693 9642 rt : (9,3,3,-3)
4694 6763 $e^x : 2\sqrt{3}/9$
4694 6969 $J_2 : 1/\ln(\sqrt{5}/3)$
4694 8528 $\ell_2 : 13/18$
4695 0050 $\Psi : 5\sqrt[3]{5}/8$
4695 0310 $2^x : 3\sqrt{2}/4 + 3\sqrt{3}$
4695 0771 rt : (8,8,3,-4)
4695 1023 $J_1 : 23/21$
4695 2235 $e^x : 2/3 + \sqrt{3}/3$
4695 2874 $e^x : 3\sqrt{3}/2 - 3\sqrt{5}/2$
4695 4446 rt : (5,8,3,3)
4695 4739 sr : $\sqrt{2} + \sqrt{5}/3$
4695 7222 $Ei : \sqrt{21}\pi$
4695 7373 rt : (7,8,-2,2)
4695 7547 $\Gamma : 14/23$
4695 8624 $E : 5\ln 2/7$
4695 8624 rt : (9,8,-7,1)
4695 8993 $t\pi : (\pi/2)^{-1/3}$
4695 9587 $J_1 : 4(e+\pi)/9$
4696 0063 at : $e - 4\pi$
4696 0078 cr : $4e + 4\pi/3$
4696 0842 $s\pi : 3/4 + 2\pi/3$
4696 1743 rt : (2,6,1,-2)
4696 2775 $Ei : 9\sqrt[3]{5}/5$
4696 3300 cu : $7\zeta(3)/2$
4696 3620 cu : $2/3 + 4\pi/3$
4696 7347 $J_0 : 5\sqrt[3]{2}/4$
4696 8690 $E : 2\sqrt{3}/7$
4696 9770 $Li_2 : 3\ln 2/5$
4697 0825 rt : (9,1,7,4)
4697 1279 rt : (2,5,-5,6)
4697 1939 $\exp(\sqrt{2}) + \exp(\sqrt{5})$
4697 2256 $J_2 : 17/5$
4697 3252 rt : (8,4,-8,5)
4697 3449 $2^x : 5/9$
4697 3943 rt : (8,4,7,-5)
4697 5809 rt : (7,9,7,-6)
4697 7050 rt : (6,4,-3,6)
4697 7150 sr : $\sqrt{2} + \sqrt{5} + \sqrt{6}$
4697 7784 sr : $\sqrt{14/3}$
4697 7937 sr : $5e\pi/7$
4698 0770 $10^x : 4e + 3\pi$
4698 1376 $10^x : P_{60}$
4698 1974 at : π^2
4698 3228 rt : (1,5,0,-1)
4698 3417 $10^x : 4 - \sqrt{5}/4$
4698 4436 rt : (7,0,-9,9)
4698 4587 rt : (4,8,6,-5)
4698 5039 rt : (5,3,6,-4)
4698 5389 cr : $\sqrt{23}\pi$
4698 5552 at : $1 - 4e$
4698 5898 $\Psi : 8\zeta(3)/9$
4698 5942 rt : (6,2,6,3)
4698 6226 $\ln(\pi) \times \exp(1/4)$
4698 8641 rt : (7,7,6,2)
4699 0596 as : $3/2 - \pi/3$
4699 0789 id : $7\sqrt[3]{2}/6$
4699 1345 $2^x : \sqrt{14/3}$
4699 1584 $J_1 : \pi^2/9$
4699 2282 sq : $7e/9$
4699 3042 $J_2 : 5e/4$
4699 4024 rt : (2,7,-8,3)
4699 5596 sinh : $\sqrt[3]{15/2}$
4699 6720 $10^x : \ln(\sqrt{2}\pi/3)$
4699 6820 $s\pi : \sqrt{2}/4 + 2\sqrt{5}/3$
4699 8995 rt : (1,7,5,-4)
4699 9051 $t\pi : (\sqrt{2}\pi/3)^{1/3}$
4699 9286 $J_2 : 9\zeta(3)/4$
4699 9702 at : $2\sqrt{2}/3 + 4\sqrt{5}$
4700 0000 sq : $7\sqrt{3}/10$
4700 0362 ln : $5/8$
4700 0942 rt : (4,8,-1,9)
4700 1210 $10^x : \pi/8$
4700 1736 cu : $2 + 3e/2$
4700 6185 $\ell_{10} : \sqrt{2}/4 + 3\sqrt{3}/2$
4700 6751 rt : (2,4,7,8)
4700 8144 cr : $3/2 + 3\sqrt{5}/4$
4700 8531 $\Gamma : 1/20$
4701 0410 cu : $5\sqrt{5}/6$
4701 2257 at : $7\sqrt{2}$
4701 6192 sinh : $3\pi/8$
4701 6670 $t\pi : 3\sqrt{2}/2 - 4\sqrt{5}/3$
4701 6947 rt : (8,9,-6,0)
4701 8023 as : $e/6$
4702 0398 at : $4e/3 + 2\pi$
4702 0897 $t\pi : 4e/3 - \pi$
4702 2481 at : $4\sqrt{3} + 4\sqrt{5}/3$
4702 2820 $4/5 \times s\pi(1/5)$
4702 3439 $\ell_{10} : 4 - \pi/3$
4702 3467 $e^x : 3\sqrt{2}/2 + 2\sqrt{3}/3$
4702 4475 rt : (5,9,1,-1)
4702 4641 rt : (7,3,-8,4)
4702 5385 $K : 7\zeta(3)/9$
4702 7009 rt : (2,6,-4,-3)
4702 7851 rt : (1,3,9,-5)
4702 8080 $\ln(\pi) \times \exp(\sqrt{3})$
4702 9531 rt : (3,3,5,2)
4703 0330 rt : (6,1,-9,-8)
4703 0499 $J_2 : \sqrt{22/3}$
4703 2309 $\ell_2 : 3\sqrt{2} + 3\sqrt{3}/4$
4703 3660 $Ei : e\pi/4$
4703 3799 $\ell_{10} : -\sqrt{3} + \sqrt{5} + \sqrt{6}$
4703 4932 $erf : 4/9$
4703 6034 sinh : $5/11$
4703 6303 cr : $2\sqrt{2}/3 + \sqrt{5}$
4703 6579 $10^x : 2/3 + \sqrt{2}$
4703 7067 $\zeta : 9\sqrt[3]{5}/10$
4703 7731 at : $1/2 + 3\pi$
4704 0874 $2^x : 4e/3 - 3\pi/2$
4704 0955 rt : (9,8,7,-6)
4704 1046 rt : (3,7,-8,-5)
4704 1172 at : $3 + 4\sqrt{3}$
4704 3872 rt : (6,7,-2,1)
4704 3934 $\ln(2/5) \div \exp(2/3)$
4704 6231 $E : \exp(-1/\sqrt{2})$
4704 7168 sr : $2/3 + 2e$
4704 7275 $\Gamma : (\sqrt{2}\pi)^{-1/3}$
4704 7581 at : $\sqrt{10}\pi$
4704 7621 $2^x : 4\pi^2/3$
4704 8949 rt : (5,3,-3,5)
4704 9175 $e^x : \sqrt{2}/3 + \sqrt{3}/4$
4704 9184 rt : (2,2,9,4)
4704 9351 cu : $\sqrt{2}/4 - 2\sqrt{5}/3$
4704 9750 $J_1 : \ln(3)$
4704 9996 $2^x : (\sqrt{2}\pi/2)^{1/3}$
4705 0275 $t\pi : 2\sqrt[3]{5}/3$
4705 0754 cu : $7/9$
4705 0837 id : $e - 4\pi/3$
4705 0956 $\Gamma : 7\sqrt[3]{5}$
4705 2281 rt : (4,1,-4,-9)
4705 3062 $t\pi : \sqrt[3]{2}/9$
4705 3073 sq : $1 - 3e/2$
4705 3385 rt : (9,4,-6,1)
4705 4564 cu : $e + 2\pi/3$
4705 4636 rt : (3,4,-6,6)
4705 4947 cr : P_{51}
4705 5315 $\ell_2 : 4\sqrt{3}/5$
4705 6253 sinh : $10(1/\pi)/7$
4705 6274 sq : $4 + 3\sqrt{2}/2$
4705 6940 $J_2 : \sqrt{23/2}$
4705 7284 at : $1 + 4\sqrt{5}$
4705 8823 id : $8/17$
4705 9423 $10^x : P_{44}$
4705 9762 $10^x : 1/4 - \sqrt{3}/3$
4706 1089 sq : $4 - 4\sqrt{2}/3$
4706 1097 sinh : $1/\ln(5/3)$
4706 1361 rt : (8,5,-2,-1)
4706 2574 $2^x : 4\pi/7$
4706 3788 rt : (7,6,2,-3)
4706 4895 $\ell_{10} : (2e)^2$
4706 4963 rt : (6,7,6,-5)
4706 7510 $t\pi : 8\pi/3$
4706 7798 $e^x : 5\pi^2/8$
4706 8037 cu : $3 - 4\pi$
4706 8341 rt : (2,8,2,-8)
4707 0427 $Li_2 : 5/12$
4707 6022 at : $7e\pi/6$
4708 1352 rt : (4,9,-3,-1)
4708 1826 $J_1 : 13/5$
4708 4007 $\exp(1/2) + \exp(3/5)$
4708 4137 sr : $3\sqrt[3]{3}/2$
4708 7027 rt : (6,6,-7,-4)
4708 7277 rt : (5,2,-9,-7)
4708 7595 $J_2 : 19/7$
4708 8538 sinh : $18/11$
4708 8603 $J_1 : 7\sqrt{2}/9$
4708 8785 $J_2 : \sqrt[3]{20}$
4708 9551 rt : (2,8,-6,-8)
4708 9825 sinh : $5\sqrt{2}/6$
4709 0239 $J_1 : 11/10$
4709 0509 rt : (6,2,-8,3)
4709 1219 $e^x : 4e - \pi/3$
4709 1294 rt : (7,2,6,-4)
4709 4059 $e^x : 7\sqrt{5}/3$
4709 4911 rt : (2,7,9,-6)
4709 4998 rt : (9,5,-6,9)
4709 5070 rt : (7,6,-7,-1)
4709 5289 sr : $9\zeta(3)/5$
4709 5766 rt : (5,8,-7,1)
4709 6430 $s\pi : 1/4 + e/3$
4709 9238 ln : $4 + \sqrt{2}/4$
4709 9742 id : $2e/3 - 2\pi$
4710 1367 cr : $10(1/\pi)$
4710 2284 $t\pi : \sqrt{2}/2 + \sqrt{3}/4$
4710 3170 rt : (8,1,2,-2)
4710 3894 rt : (3,6,5,-4)
4710 3905 $10^x : \ln(\sqrt[3]{3}/2)$
4710 4601 $t\pi : (e+\pi)$
4710 5038 at : $2 - 2\sqrt{5}/3$
4710 5261 rt : (5,5,-3,-2)

4710 5303 $1/3 \times \exp(5)$
4710 5623 $\theta_3 : \pi^2/10$
4710 5657 at : $8(1/\pi)/5$
4710 6851 sr : $1/\ln(\sqrt[3]{4})$
4710 7259 $s\pi : \sqrt[3]{23}$
4710 8960 $J_1 : \sqrt[3]{4/3}$
4710 9669 rt : (3,7,9,3)
4711 1640 $2^x : 7\zeta(3)/2$
4711 1838 ln : $1 + 3\sqrt{5}/2$
4711 1963 $\ell_2 : 1/4 + \sqrt{2}/3$
4711 2663 sr : $3/4 + \sqrt{2}$
4711 2767 at : 10
4711 5517 $e^x : e/3 + 3\pi/2$
4711 7648 at : $3\sqrt{2}/4 + 4\sqrt{5}$
4711 7990 $2^x : 2\sqrt{2}/3 + 3\sqrt{3}$
4711 9020 $10^x : 9e/2$
4712 2049 rt : (4,7,3,2)
4712 3144 rt : (2,6,-5,-6)
4712 3362 $\ell_2 : 2\ln 2$
4712 3458 $J_2 : e$
4712 4205 $Ei : 23/19$
4712 4867 $t\pi : 3e - 3\pi/2$
4712 4988 $2^x : 7(1/\pi)/4$
4712 5352 cr : $4\sqrt{2}/3 + 3\sqrt{3}/4$
4712 9937 rt : (3,2,9,-5)
4713 0410 $\ell_{10} : 3\sqrt{3} - \sqrt{5}$
4713 4112 rt : (3,7,1,5)
4713 4343 $e^x : 19/21$
4713 5232 $J_1 : 3\sqrt{3}/2$
4713 7256 $\lambda : ((e+\pi)/2)^{-3}$
4713 8562 $1/3 - \ln(\sqrt{5})$
4713 8735 sq : $7\ln 2/4$
4713 9652 $\Psi : e\pi/8$
4714 0149 $2^x : 3/2 - 2\sqrt{2}/3$
4714 0275 $1/4 + \exp(1/5)$
4714 0452 id : $\sqrt{2}/3$
4714 1186 rt : (8,9,0,6)
4714 1547 sr : $5\sqrt{3}/4$
4714 1553 rt : (8,6,6,-5)
4714 1892 rt : (1,9,7,-6)
4714 2100 $\ell_{10} : 3\pi^2$
4714 2651 rt : (8,3,6,3)
4714 4604 rt : (1,3,-8,3)
4714 5027 $erf : 7(1/\pi)/5$
4714 5393 at : $2\sqrt{2}/3 - \sqrt{3}/4$
4714 6040 $\ell_2 : 3\zeta(3)/5$
4714 7615 sq : $4/3 - 3\pi$
4715 0564 rt : (8,4,-6,8)
4715 1776 $4 \div e$
4715 2768 $\Psi : 5\pi^2/4$
4715 6461 ln : $10\sqrt[3]{3}/9$
4715 7019 $erf : 1/\ln(3\pi)$
4715 7022 $s\pi : 4e + 2\pi$
4715 7037 $t\pi : 4\sqrt{2}/3 - \sqrt{5}/3$
4715 8737 id : $\sqrt{2}/4 + \sqrt{5}/2$
4715 9821 rt : (4,9,-3,-2)
4715 9885 ln : $2 + 3\pi/4$
4716 0530 rt : (4,1,5,-3)
4716 1005 $J_1 : 7\sqrt[3]{2}/8$
4716 1173 sinh : 17/6
4716 2315 rt : (9,8,-5,4)
4716 3370 $\lambda : (1/\pi)/8$
4716 4282 $J_0 : 11/7$
4716 5760 $\Gamma : 5\sqrt{5}$
4716 6168 rt : (9,5,2,-3)
4716 7916 $\ell_2 : \sqrt[3]{3}/2$
4716 8826 $J_0 : 10\sqrt{2}/9$
4716 9307 rt : (4,6,9,-6)
4716 9428 cu : $\sqrt{2}/3 - 3\sqrt{3}$
4717 1624 ln : $4\zeta(3)/3$
4717 4078 at : $9\sqrt{5}/2$
4717 4151 $\Gamma : \sqrt[3]{2}/10$
4717 5321 $J_1 : (\sqrt{5}/3)^{-1/3}$
4717 7978 rt : (5,9,1,-3)
4717 8132 ln : $6\pi^2/5$
4717 8645 ln : $e - 2\pi/3$
4718 1337 sq : $1/3 + \sqrt{2}/4$
4718 1740 sq : $4\zeta(3)/7$
4718 1790 $\ell_{10} : -\sqrt{2} + \sqrt{3} + \sqrt{7}$
4718 3128 $\ln(4) + \exp(3)$
4718 5349 $J_0 : 1/\ln(\sqrt[4]{4}/3)$
4718 6183 as : 5/11
4718 7286 ln : $3\sqrt{2}/2 + \sqrt{5}$
4718 7343 ln : $9\ln 2/10$
4718 9050 cu : $3\sqrt{2} - 2\sqrt{3}$
4719 0737 at : $8\sqrt[3]{2}$
4719 0779 sinh : $(\sqrt{2}\pi/3)^{-2}$
4719 0967 rt : (9,1,6,-4)
4719 1384 $J_1 : 9\sqrt[3]{3}/5$
4719 4987 $\ell_2 : e/2 + 4\pi/3$
4719 5014 rt : (4,9,-3,-3)
4719 5436 $\zeta : 2\zeta(3)/3$
4719 6014 sr : 13/6
4719 7011 rt : (4,3,9,4)
4719 7104 ln : $4\sqrt{2} - 3\sqrt{3}/4$
4719 7551 id : $10\pi/3$
4719 8879 at : $4e - \pi/4$
4720 0121 $J_0 : \pi/2$
4720 1048 cu : $4 - 2e/3$
4720 1923 $s\pi : 6\pi^2/5$
4720 2371 rt : (5,5,5,-4)
4720 2494 at : $2/3 + 3\pi$
4720 3487 sr : $7(1/\pi)/10$
4720 5113 rt : (6,5,-7,-2)
4720 6242 $J_2 : (3/2)^3$
4720 6676 at : $7\sqrt[3]{3}$
4720 6724 as : $10(1/\pi)/7$
4720 8005 rt : (6,4,-1,9)
4720 9557 $10^x : 4 - 3e/4$
4720 9808 at : $2\sqrt{3}/3 + 4\sqrt{5}$
4720 9915 rt : (9,5,-6,-3)
4721 0621 $t\pi : 3\sqrt{3}/2 + 4\sqrt{5}$
4721 1663 $\zeta : 20/13$
4721 2629 $J_2 : 1/\ln(\ln 2)$
4721 3595 id : $2\sqrt{5}$
4721 5026 at : $e/4 + 3\pi$
4721 6319 cr : 2/19
4721 6955 $e^x : 2e\pi/7$
4721 7675 rt : (9,4,2,1)
4721 8101 $\Psi : 6\sqrt{3}/5$
4721 8938 rt : (7,3,-6,7)
4721 9356 $t\pi : 3\sqrt{3} + 4\sqrt{5}$
4722 0582 rt : (2,8,-4,1)
4722 1948 ln : $\sqrt{19}$
4722 2222 sq : $5\sqrt{5}/6$
4722 2956 rt : (6,9,5,7)
4722 5897 cu : $\sqrt{3} + 2\sqrt{5}/3$
4722 5931 rt : (5,1,9,-5)
4722 6744 id : $7e\pi/8$
4722 7055 at : $\ln(5/3)$
4722 7374 tanh : $3\sqrt[3]{5}/10$
4722 7499 ln : $3 + e/2$
4722 7572 $s\pi : (5/3)^{-1/3}$
4722 9804 rt : (4,5,-7,-4)
4723 0320 cu : 23/7
4723 2516 $J_2 : 1/\ln(\sqrt[3]{3})$
4723 2724 rt : (6,8,-3,-1)
4723 3207 at : $3/4 - 4e$
4723 5201 rt : (7,8,0,5)
4723 6655 sr : $\exp(-3/2)$
4723 6786 rt : (1,5,8,-5)
4723 8339 cu : $2\sqrt{2}/3 - 4\sqrt{5}/3$
4723 8420 $E : 2\sqrt[3]{5}/7$
4723 8869 at : $4\sqrt{2} + 2\sqrt{5}$
4723 9361 $t\pi : 6\zeta(3)/5$
4723 9490 $\ln(4/5) \times \exp(3/4)$
4724 0346 rt : (6,5,9,-6)
4724 0808 at : P_{13}
4724 1129 $\ell_2 : \sqrt{2}/4 + 3\sqrt{3}$
4724 1880 $10^x : (2e/3)^{-3}$
4724 3195 $J_1 : 21/19$
4724 3320 rt : (8,7,-5,3)
4724 5324 cr : P_{57}
4724 5652 rt : (8,9,-6,-4)
4724 6209 $\Gamma : 1/19$
4724 6455 id : $e/3 + 4\pi$
4724 7039 id : $3\sqrt[3]{2}/8$
4724 7476 rt : (3,9,-7,-5)
4724 8411 rt : (4,6,-8,4)
4724 8686 $e^x : 2\sqrt{2}/3 + 3\sqrt{5}/2$
4724 8980 rt : (7,9,9,-7)
4725 0038 rt : (7,9,-6,-6)
4725 0183 at : $1/\ln(e/3)$
4725 0513 $\ell_{10} : 1/4 + e$
4725 1313 rt : (5,0,1,1)
4725 3781 $10^x : \sqrt{2}/3 + \sqrt{5}/3$
4725 4352 rt : (2,9,-9,9)
4725 5395 rt : (3,5,-8,-5)
4725 8068 $2^x : 2\sqrt{3}/3 - \sqrt{5}$
4725 8101 at : $2e + 3\pi/2$
4725 9440 $\Psi : 1/15$
4726 0337 tanh : $\exp(-2/3)$
4726 3764 at : $2 + 3e$
4726 5305 $e^x : 3\sqrt{2}/2 + 4\sqrt{5}/3$
4726 5625 sq : 11/16
4726 7538 rt : (6,4,1,-2)
4727 2863 rt : (7,8,1,-3)
4727 3420 $\ln(3) \times s\pi(\pi)$
4727 3481 $\Gamma : 9/25$
4727 4030 $2^x : 1/3 - \sqrt{2}$
4727 4985 cr : $10\sqrt{5}/7$
4727 5340 cr : P_{38}
4727 6536 $\ln(2/5) + \exp(2)$
4727 7343 $J_2 : 8\sqrt[3]{5}/5$
4727 8357 $\Psi : 3\ln 2$
4727 9068 $t\pi : 1/3 + 3e/2$
4728 0505 at : $2\sqrt{3} + 3\sqrt{5}$
4728 2258 $\Psi : 7\ln 2$
4728 2871 at : $3/4 + 3\pi$
4728 3390 rt : (2,8,0,-2)
4728 4100 $e^x : 3 + 4\sqrt{3}$
4728 6738 id : $\sqrt{2}/3 - 4\sqrt{5}$
4728 6977 rt : (9,8,6,2)
4728 7080 sr : $\sqrt{5}/10$
4728 9551 $\Gamma : 2\sqrt[3]{2}/7$
4729 1074 $erf : \sqrt{5}/5$
4729 1726 $\ell_2 : 9\pi^2$
4729 2068 rt : (8,8,5,-5)
4729 2465 $t\pi : 2e/3 + \pi/3$
4729 2773 $J_2 : 7\sqrt[3]{3}/3$
4729 4935 rt : (7,4,5,-4)
4729 9676 $\ln(3) \div s\pi(\sqrt{3})$
4729 9744 $J_2 : \sqrt{15/2}$
4729 9760 rt : (4,0,-8,1)
4730 0193 $\ln(\sqrt{5}) \times s\pi(1/5)$
4730 0632 rt : (6,0,5,3)
4730 3333 rt : (9,8,9,-7)
4730 4858 cu : $9\zeta(3)/8$
4730 5128 rt : (9,3,-9,9)
4730 6626 rt : (1,7,-3,4)
4730 7004 $s\pi : 1/2 + 4\sqrt{2}$
4730 7312 cu : $5\pi/2$
4730 8190 sr : $9e/4$
4730 8226 rt : (8,4,9,-6)
4730 9392 rt : (5,3,-1,8)
4731 0702 $J_2 : \sqrt{3} - \sqrt{6} - \sqrt{7}$
4731 1244 $\Psi : 13/21$
4731 2452 $\zeta : 7(1/\pi)$
4731 2586 $Li_2 : 24/25$
4731 3088 at : $2/3 - 4e$
4731 3638 $J_1 : (e/2)^{1/3}$
4731 4038 rt : (3,8,1,-1)
4731 4355 $1/4 - \ln(4/5)$
4731 5242 $e^x : P_{27}$
4731 5736 rt : (4,8,8,-6)
4731 6301 rt : (7,0,9,5)
4731 7428 rt : (5,4,5,2)
4731 7806 $\Psi : \sqrt[3]{9}$
4732 2480 $t\pi : \exp(\pi)$
4732 2860 $\zeta : \ln((e+\pi)/3)$
4732 2922 $\exp(\pi) \div \exp(\sqrt{5})$
4732 2951 rt : (1,8,3,-7)
4732 4082 $Ei : 2\pi$
4732 6178 rt : (3,4,8,-5)
4732 6504 $2^x : \sqrt{5}/4$
4733 0514 rt : (2,8,-3,-3)
4733 1063 rt : (2,6,-3,1)
4733 2160 rt : (6,4,9,4)
4733 4532 $t\pi : 1/\ln(e/3)$
4733 5914 rt : (1,8,-7,-5)

4733 6477 sr : $(3\pi/2)^{1/2}$
4733 7041 $s\pi(1/3) \div s\pi(1/5)$
4733 7529 $\Psi : (\ln 2/3)^{-1/2}$
4733 8234 $E : \exp(-\sqrt[3]{3}/2)$
4733 9761 rt : (9,8,-3,7)
4734 0058 $J_2 : 8\sqrt[3]{2}/3$
4734 0202 rt : (1,4,-1,5)
4734 0742 $10^x : 4\sqrt{2}/3 + \sqrt{3}/2$
4734 1602 id : $(3\pi)^{-1/3}$
4734 2764 $e^x : 3 - 2\pi/3$
4734 2805 $\lambda : 1/25$
4734 3408 as : $(\sqrt{2}\pi/3)^{-2}$
4734 3531 sr : P_{31}
4734 6580 rt : (2,0,8,4)
4734 6903 $\exp(3/4) + \exp(\sqrt{5})$
4734 6922 sq : $3\sqrt{2}/4 + 2\sqrt{3}$
4734 7120 sinh : $4(e + \pi)/3$
4734 8001 at : $3\sqrt{3}/4 + 4\sqrt{5}$
4734 8869 rt : (8,9,2,9)
4734 9336 rt : (7,6,-5,2)
4734 9618 $\Gamma : 9e/4$
4735 2125 at : $6e\pi/5$
4735 2232 $J_1 : 8\ln 2/5$
4735 2673 $e^x : 2/3 - \sqrt{2}$
4735 3596 at : $3e + 2\pi/3$
4735 4562 $10^x : 7\pi/2$
4735 7536 at : $\sqrt{2}/4 - \sqrt{3}/2$
4735 7593 sq : $2/3 + e/3$
4735 8588 cu : $3e/4$
4735 8817 at : $7(e + \pi)/4$
4736 0209 $2^x : 3\sqrt{2}/4 + 4\sqrt{5}/3$
4736 0701 rt : (2,4,-7,-4)
4736 1259 cr : 16/5
4736 2030 rt : (9,7,1,-3)
4736 3596 at : $6\sqrt[3]{5}$
4736 3718 rt : (4,9,-1,3)
4736 4281 $\ell_{10} : 2/3 + 4\sqrt{3}/3$
4736 4592 $\ell_{10} : 3\sqrt{3}/4 + 3\sqrt{5}/4$
4736 5261 rt : (6,7,0,4)
4736 6596 sq : $3\sqrt{2}/2 + 2\sqrt{5}$
4736 8421 id : 9/19
4737 0037 rt : (8,7,-3,-1)
4737 1593 rt : (5,4,-7,-3)
4737 3129 $s\pi : 3\sqrt{3}/2 + \sqrt{5}/4$
4737 3740 $J : 7(1/\pi)/9$
4737 4127 $t\pi : \sqrt{5} - \sqrt{6} - \sqrt{7}$
4737 4533 rt : (6,8,-6,-4)
4737 6253 rt : (8,2,-9,8)
4737 7558 rt : (5,8,9,3)
4738 0133 rt : (6,7,8,-6)
4738 0270 at : $4/3 + 4\sqrt{5}$
4738 0445 $E : 7\ln 2/10$
4738 1220 $10^x : 1/2 + e/3$
4738 1747 sq : $9\sqrt[3]{5}/5$
4738 2054 rt : (9,3,5,-4)
4738 4035 $2^x : 1 + 4\pi/3$
4738 4648 $s\pi : \sqrt{2}/9$
4738 4676 at : $4\sqrt{3} + 3\sqrt{5}/2$
4738 4937 $J_2 : 7\pi/8$
4738 5501 at : $4 + 2\pi$
4738 6499 $t\pi : 1/2 + e/2$
4738 7062 rt : (2,5,-9,-4)
4738 7460 $2^x : 3 - 3e/2$
4738 7707 $\zeta : 9\sqrt{3}/7$
4738 8430 $J_2 : 3\sqrt{5}/2$
4738 8602 rt : (9,9,-9,-4)
4739 1042 ln : $(\sqrt{2}\pi)^3$
4739 1330 rt : (5,8,5,6)
4739 3118 $\ell_2 : 9/25$
4739 3946 $J_2 : 11/4$
4739 4296 rt : (4,6,-6,9)
4739 4739 $e^x : 17/10$
4739 5245 rt : (7,7,-7,1)
4739 6147 tanh : $3\zeta(3)/7$
4739 6695 $\Psi : (\ln 2)^{-2}$
4739 8028 $e^x : 2 + 3\sqrt{5}/2$
4739 8096 cr : $3e\pi/8$
4739 8772 at : $3\sqrt[3]{5}/10$
4740 0006 rt : (8,3,1,-2)
4740 1376 rt : (7,8,-2,-2)
4740 1474 rt : (9,1,9,5)
4740 2158 $\ell_2 : \sqrt[3]{2}/7$
4740 2596 $1/3 + \exp(\pi)$
4740 3666 cu : $2/3 + \sqrt{2}/3$
4740 4251 rt : (1,4,8,3)
4740 6806 $\Psi : \exp(-\sqrt[3]{3})$
4740 6991 cu : $7(e + \pi)$
4740 7963 $Ei : \exp(-4/3)$
4741 0480 rt : (5,3,8,-5)
4741 1018 $10^x : 3\sqrt[3]{3}/8$
4741 2015 $J_1 : 10/9$
4741 3021 sr : $4 + 3\sqrt{2}/2$
4741 3738 rt : (4,7,0,-2)
4741 4919 $\Psi : \sqrt{13/3}$
4741 6180 rt : (9,6,-8,4)
4741 6986 cu : $9\ln 2/8$
4741 7135 $e^x : 1/2 + \sqrt{5}/3$
4741 7144 rt : (5,1,-6,5)
4741 9365 rt : (1,7,7,-5)
4742 0312 sq : $4 + \pi/3$
4742 0598 $\Gamma : 7\sqrt{5}/3$
4742 3336 $2^x : 4\sqrt[3]{2}/9$
4742 3990 rt : (1,8,5,8)
4742 4703 $J : \exp(-11\pi/13)$
4742 6560 $10^x : 1/\ln(\pi)$
4742 6898 $Ei : \sqrt{2} + \sqrt{3} + \sqrt{5}$
4742 6921 $2^x : 14/25$
4742 7221 $t\pi : \sqrt[3]{5}/3$
4742 7763 ln : $1 + 4e$
4742 8628 rt : (8,1,5,3)
4742 8674 $e \div \ln(3)$
4742 9925 rt : (3,9,2,-7)
4742 9970 at : $e/3 + 3\pi$
4743 0755 $\ln(3/4) \times \exp(1/2)$
4743 2360 at : $\exp(-2/3)$
4743 5191 $J_2 : 4(e + \pi)/7$
4743 5247 $\exp(3/5) + \exp(\sqrt{3})$
4743 5753 $J_1 : \exp(1/(3\pi))$
4743 5829 id : $10\sqrt{3}/7$
4743 6308 cu : $4e$
4743 6946 rt : (8,7,-3,6)
4743 7431 $erf : \pi/7$
4743 8464 as : $1/4 + \sqrt{5}/3$
4743 9786 $\Gamma : (e)^{-1/2}$
4744 1768 rt : (8,6,8,-6)
4744 2373 $\ell_{10} : 4\sqrt{5}/3$
4744 2499 id : $(\sqrt{2}\pi)^{-1/2}$
4744 5408 cr : $8\zeta(3)/3$
4744 5798 $\Gamma : 7\ln 2/8$
4744 7106 $10^x : \sqrt[3]{2/3}$
4744 8240 rt : (4,2,-1,7)
4744 8979 sq : 17/14
4745 1036 $e^x : e/7$
4745 1390 sq : $1/\ln(\sqrt[3]{5})$
4745 2085 $s\pi : 2/3 + 2\sqrt{5}/3$
4745 4648 rt : (7,4,-6,-3)
4745 4969 sinh : 11/24
4745 5536 at : $\sqrt{2} + 4\sqrt{5}$
4745 6014 rt : (8,5,9,4)
4745 6532 id : $\exp(-\sqrt{5}/3)$
4745 7495 at : $3e/2 + 2\pi$
4745 7714 $\Psi : \sqrt{3} - \sqrt{5} - \sqrt{7}$
4745 9302 $\exp(2) + \exp(3)$
4746 2335 rt : (7,1,-9,7)
4746 2713 $t\pi : P_{70}$
4746 3755 id : $\exp(e/3)$
4746 4108 $t\pi : P_{67}$
4746 5415 rt : (7,8,2,8)
4746 5482 $J_2 : \sqrt[3]{21}$
4746 5937 $e^x : 8\pi/3$
4746 6113 sr : $4e/5$
4746 6709 rt : (1,6,-4,-6)
4746 8609 rt : (4,1,8,4)
4746 9037 at : $1/2 - 4e$
4747 1304 $\ln(2/5) \times \ln(5)$
4747 2611 rt : (7,6,4,-4)
4747 2815 $\ln(3/5) + s\pi(\sqrt{2})$
4747 3951 as : $1/4 - \sqrt{2}/2$
4747 4487 sq : $\sqrt{2}/2 + \sqrt{3}/2$
4747 7766 sq : $2\sqrt{2} + \sqrt{5}/4$
4748 0602 $t\pi : e/3 - \pi/3$
4748 0963 rt : (3,8,-1,-6)
4748 2910 $s\pi : \sqrt[3]{2}/8$
4748 6166 $e^x : 22/3$
4748 6663 at : $6\sqrt{3}$
4748 7373 id : $7\sqrt{2}/4$
4748 9639 rt : (6,2,-4,9)
4748 9993 $10^x : \exp(\sqrt{3}/2)$
4749 0072 rt : (7,2,8,-5)
4749 1109 sq : $\sqrt{3}/2 - 4\sqrt{5}/3$
4749 2598 id : $1/3 + \pi$
4749 3085 rt : (4,7,-6,-4)
4749 4043 rt : (6,5,-5,1)
4749 7397 $J_1 : 7(1/\pi)/2$
4749 8196 rt : (3,6,7,-5)
4749 8774 $Li_2 : \sqrt[3]{2}/3$
4749 9464 $t\pi : 2\sqrt{3} + 3\sqrt{5}/4$
4750 0882 rt : (5,8,-8,7)
4750 1053 cu : $4 + e/4$
4750 3235 rt : (2,9,0,-1)
4750 6851 rt : (6,9,7,-6)
4750 7516 rt : (6,8,-6,-7)
4750 8809 rt : (7,1,1,1)
4750 9050 $\Gamma : 1/18$
4750 9969 $\ell_{10} : 3/4 + \sqrt{5}$
4751 0607 rt : (2,6,-1,8)
4751 1533 at : $\sqrt{11}\pi$
4751 2390 $Ei : 11/13$
4751 4106 cu : $1/\ln(2\pi/3)$
4751 4465 $J_0 : 7\sqrt{5}/10$
4751 5389 $erf : 1/\sqrt[3]{11}$
4751 6363 at : $1 + 3\pi$
4751 7159 $\Psi : 25/12$
4751 9510 rt : (8,5,-8,3)
4752 1311 $\Psi : -\sqrt{2} - \sqrt{3} + \sqrt{5}$
4752 2775 rt : (1,9,5,-9)
4752 4123 $t\pi : \sqrt{3}/4 + 3\sqrt{5}$
4752 5274 rt : (2,3,-2,7)
4752 5762 sq : $2 - \pi/4$
4752 7662 $J_1 : (\sqrt[3]{3}/2)^{-1/3}$
4752 8548 $\ell_{10} : 3\sqrt{2}/2 + \sqrt{3}/2$
4752 9314 rt : (8,7,-9,-5)
4753 0990 $J_2 : \sqrt{2} - \sqrt{3} - \sqrt{6}$
4753 2958 rt : (9,9,0,-3)
4753 4061 $e^x : 7/18$
4753 4104 at : $3/2 + 4\sqrt{5}$
4753 7341 $e^x : 1/\sqrt[3]{17}$
4753 8094 $\Gamma : 3\sqrt{2}/7$
4753 8852 rt : (8,8,-9,-5)
4753 9796 $2^x : (\sqrt[3]{5})^{1/2}$
4753 9837 rt : (4,9,1,8)
4754 2230 id : $3\sqrt{2}/2 + 3\sqrt{5}/2$
4754 2646 $10^x : 4\sqrt{3} + \sqrt{5}/3$
4754 2860 $J_2 : \sqrt{23/3}$
4754 3211 as : $1/\ln(\ln 3/3)$
4754 4460 $2^x : 3/4 + \pi/3$
4754 5257 $2^x : 17/13$
4754 6331 $\ln(5) + s\pi(1/3)$
4754 7409 $3 \times \exp(2/5)$
4754 8171 $erf : 9/20$
4754 8330 sr : $1/2 + 3\sqrt{5}/4$
4754 9029 $1/5 - s\pi(\sqrt{5})$
4754 9055 $\ell_2 : 1/4 + 4e$
4754 9236 cu : $4 - \sqrt{5}/3$
4754 9345 rt : (9,5,4,-4)
4754 9467 $Ei : (e + \pi)/2$
4755 0289 $\ell_2 : \sqrt{2}/3 + 4\sqrt{3}/3$
4755 1239 $\ell_{10} : 7e\pi/2$
4755 2825 $1/2 \times s\pi(2/5)$
4755 4092 $1/4 + \exp(4/5)$
4755 4521 $\ln(\sqrt{2}) - \exp(3/5)$
4755 5286 $e^x : 3/4 + \sqrt{5}/2$
4755 6204 rt : (1,7,1,-1)
4755 6792 $\exp(1/5) + s\pi(\sqrt{3})$
4755 6985 rt : (5,9,3,-4)
4755 7364 $J_2 : 10/3$
4755 7375 rt : (6,2,4,-3)
4755 8542 rt : (7,5,5,2)
4755 9204 at : $10\pi/3$
4756 1196 $J_2 : 8\sqrt{3}/5$

4756 1678 rt : (7,6,-3,5)
4756 2335 cu : $1/3+4\pi/3$
4756 3006 Ψ : $\exp(-\sqrt[3]{3}/3)$
4756 3039 rt : (8,8,7,-6)
4756 3679 rt : (9,6,-6,7)
4756 3715 E : $2\zeta(3)/5$
4756 5299 rt : (9,1,8,-5)
4756 5975 ζ : $\sqrt[3]{11}$
4756 6748 E : $\sqrt[3]{3}/3$
4756 8409 2^x : $\exp(-\sqrt{3}/3)$
4756 8651 rt : (6,0,-9,6)
4756 9836 E : $(\ln 2/3)^{1/2}$
4757 0044 ln : $1/4+e/2$
4757 0484 cr : $2/3-\sqrt{5}/4$
4757 0764 at : $3\zeta(3)/7$
4757 1095 J_2 : $4\ln 2$
4757 1307 $\ln(\sqrt{3})\times s\pi(1/3)$
4757 2743 rt : (5,5,7,-5)
4757 6382 ln : $3/4+4e/3$
4757 6655 rt : (1,6,-2,7)
4757 7459 ℓ_{10} : $3/2+2\sqrt{5}/3$
4757 7537 J_1 : $\sqrt{20/3}$
4757 8750 ℓ_{10} : P_7
4757 9092 E : $(\ln 2)^2$
4758 0015 sq : $2\sqrt{3}+3\sqrt{5}$
4758 0098 rt : (1,9,6,-5)
4758 0465 10^x : $1+3\sqrt{3}/2$
4758 0841 rt : (6,2,8,4)
4758 2622 Li_2 : $\exp(-\sqrt{3}/2)$
4758 3474 Ei : $6\sqrt{2}/7$
4758 4314 Ψ : $2\sqrt[3]{3}/7$
4758 4462 at : $21/2$
4758 5059 cr : $4-\pi/4$
4758 8499 ln : $\ln(5)$
4759 0106 $t\pi$: $5\zeta(3)/7$
4759 0230 e^x : $3\sqrt{2}/2-\sqrt{3}$
4759 0255 rt : (3,8,-7,5)
4759 0846 Ei : $\sqrt[3]{13/2}$
4759 0847 ln : $2/3+2\sqrt{2}/3$
4759 1694 rt : (2,2,7,-4)
4759 1897 sinh : $5e\pi/3$
4759 4375 at : $3e+3\pi/4$
4759 4739 $s\pi$: $3/19$
4759 5962 rt : (9,9,-7,-1)
4759 7540 cu : $6e$
4759 7913 E : $12/25$
4759 9284 J_1 : $19/17$
4760 2360 rt : (3,9,-9,-6)
4760 2805 ℓ_{10} : $7\sqrt[3]{5}/4$
4760 2988 $t\pi$: $\sqrt{2}/10$
4760 4002 sr : $3\sqrt{2}/4+\sqrt{5}/2$
4760 4484 rt : (7,9,9,3)
4760 4593 rt : (8,7,-1,9)
4760 4827 ζ : $2\sqrt[3]{2}/5$
4760 5615 2^x : $1/2-\pi/2$
4760 5949 id : $1/\ln(4/3)$
4760 8780 cu : $4+3\pi$
4760 9108 e^x : $(\sqrt{5}/3)^{1/3}$
4760 9259 at : $3e/4-4\pi$
4760 9881 $s\pi$: $3\sqrt{3}-3\sqrt{5}/2$
4760 9901 $\ln(5)-\exp(3)$
4760 9954 J_2 : $25/9$
4761 0296 e^x : $(5/3)^3$
4761 0324 J_1 : $\sqrt{5}/2$
4761 1906 as : $11/24$
4761 3733 cu : $(\ln 2/2)^{-3}$
4761 4071 rt : (7,9,-4,-3)
4761 5554 rt : (4,7,-8,-2)
4761 6077 $t\pi$: $1/3+3\pi/4$
4761 8208 rt : (5,1,-4,8)
4761 9047 id : $10/21$
4761 9228 erf : $3\zeta(3)/8$
4762 0098 at : $1/3-4e$
4762 0125 rt : (8,9,-4,-1)
4762 1109 sq : $8\zeta(3)$
4762 2143 $t\pi$: $4(1/\pi)/9$
4762 2199 rt : (6,7,2,7)
4762 3149 rt : (5,7,-2,-2)
4762 3195 rt : (2,9,2,8)
4762 4989 Ei : $5/19$
4762 5184 rt : (3,5,8,3)
4762 5935 Γ : $7\sqrt{3}/2$
4762 7212 at : $9(e+\pi)/5$
4762 8874 rt : (7,8,3,-4)
4763 3683 sr : $3/2+e/4$
4763 4281 rt : (6,6,-9,-5)
4763 4481 $t\pi$: $10\ln 2/3$
4763 6617 Ψ : $3\sqrt[3]{3}/7$
4763 8219 rt : (2,5,-7,-7)
4763 8766 Li_2 : $8/19$
4763 8949 10^x : $4/3+\sqrt{3}$
4763 9124 rt : (9,6,-7,1)
4764 0613 cu : $4/3+\sqrt{5}$
4764 1355 cr : $1/2+e$
4764 2319 rt : (8,5,0,-2)
4764 3312 rt : (7,4,7,-5)
4764 3749 at : $2-4\pi$
4764 4025 2^x : $4\sqrt{2}/3-\sqrt{3}/3$
4764 5156 10^x : $(e+\pi)/6$
4764 5203 rt : (1,1,-7,3)
4764 9094 cu : $23/17$
4765 0068 rt : (3,8,6,-5)
4765 0535 id : $1/\ln(1/(3e))$
4765 1061 cu : $1/\ln(\sqrt[3]{5})$
4765 1074 rt : (7,4,-8,2)
4765 1594 cr : $4/3+4\sqrt{2}/3$
4765 2488 $t\pi$: $9\pi^2/7$
4765 2603 rt : (4,9,-1,-2)
4765 2924 ln : $\sqrt{2}-\sqrt{6}+\sqrt{7}$
4765 3485 Ei : $7\sqrt{3}/10$
4765 4831 cu : $4/3-4\sqrt{5}/3$
4765 9263 10^x : $2/3+\sqrt{2}/4$
4766 3692 rt : (8,1,4,-3)
4766 4085 rt : (1,7,2,-9)
4766 4338 J_1 : $8\sqrt[3]{2}/9$
4766 4584 rt : (5,3,-6,-3)
4766 9014 $t\pi$: π
4766 9966 E : $11/23$
4766 9990 J_0 : $25/16$
4767 0524 rt : (1,8,-9,-6)
4767 1356 sq : $1/4+2\sqrt{2}$
4767 1748 rt : (3,5,-4,5)
4767 2285 θ_3 : $4/17$
4767 2924 ζ : $20/9$
4767 3129 sr : $5/22$
4767 3609 rt : (4,7,5,5)
4767 5586 id : $3\sqrt{2}/2-3\sqrt{3}/2$
4767 6878 rt : (8,5,-6,6)
4767 7433 e^x : $3/4-2\sqrt{5}/3$
4767 8179 $t\pi$: $4\sqrt{2}/3-3\sqrt{3}$
4767 8803 sinh : $13/11$
4768 2614 2^x : $9/16$
4768 2722 sr : $5(1/\pi)/7$
4768 3362 rt : (9,7,-5,-3)
4768 4301 rt : (8,3,8,4)
4768 4480 J_2 : $\sqrt{11}$
4768 8472 sr : $\exp(-\sqrt{2}\pi/3)$
4768 9026 id : $e/3+\pi/2$
4768 9361 10^x : $-\sqrt{3}+\sqrt{5}+\sqrt{6}$
4769 0441 ln : $3\sqrt{3}+3\sqrt{5}$
4769 2687 sq : $3-4\sqrt{3}/3$
4769 3866 at : $1/4-4e$
4769 3875 rt : (4,1,7,-4)
4769 3919 erf : $\ln(\pi/2)$
4769 6934 rt : (9,7,3,-4)
4769 7041 Γ : $(2\pi)^{-3}$
4770 2345 10^x : $2\sqrt[3]{3}$
4770 2432 at : $1/\ln(\ln 3)$
4770 2701 J_2 : $(\ln 3/2)^{-2}$
4770 2840 E : $3(1/\pi)/2$
4770 3296 rt : (5,9,-1,-5)
4770 4515 $s\pi$: $(\sqrt{5}/3)^{-1/2}$
4770 4898 10^x : $7\sqrt{2}/5$
4770 9182 ln : $\sqrt{3}/2+\sqrt{5}/3$
4770 9789 sr : $24/11$
4770 9846 cr : $\sqrt{3}+2\sqrt{5}/3$
4771 0189 rt : (9,3,7,-5)
4771 2125 ℓ_{10} : 3
4771 2508 sinh : $(3\pi/2)^{-1/2}$
4771 2567 rt : (4,9,-5,-4)
4771 6069 J_1 : $7\sqrt[3]{3}/9$
4771 6406 rt : (5,8,7,9)
4771 6595 $t\pi$: $3\sqrt{2}/4-2\sqrt{5}/3$
4771 6633 J_2 : $8\pi/9$
4771 9204 ℓ_{10} : $e/3+2\pi/3$
4771 9490 rt : (6,0,-7,9)
4772 0903 rt : (9,3,-9,-4)
4772 3576 $t\pi$: $\sqrt{3}-\sqrt{5}+\sqrt{7}$
4772 4360 rt : (9,2,1,1)
4772 7600 rt : (6,5,-3,4)
4773 1526 rt : (4,5,-9,-5)
4773 1654 J : $(3\pi/2)^{-1/2}$
4773 4567 J_2 : $5\sqrt{5}/4$
4773 5234 rt : (1,7,9,-6)
4773 6287 J_1 : $(\sqrt[3]{2})^{1/2}$
4773 7121 rt : (6,9,9,-7)
4773 8920 at : $5e\pi/4$
4774 0300 tanh : $3\sqrt{3}/10$
4774 0359 at : $\sqrt{3}+4\sqrt{5}$
4774 0680 id : $\sqrt{3}+\sqrt{5}/3$
4774 1868 rt : (7,6,-1,8)
4774 2375 E : $1/\ln(1/(3e))$
4774 2918 rt : (5,2,0,-1)
4774 5548 Li_2 : $(4/3)^{-3}$
4774 6482 id : $3(1/\pi)/2$
4774 7804 erf : $\exp(-\sqrt[3]{4}/2)$
4774 9573 Ei : $9e/2$
4774 9888 Γ : $(\ln 3/3)^{1/2}$
4775 0143 e^x : $4/3+2\pi$
4775 1507 $t\pi$: $\exp(-(e+\pi)/3)$
4775 1973 rt : (6,8,-1,-2)
4775 2361 ℓ_{10} : $(\ln 2)^3$
4775 2621 rt : (9,4,-9,7)
4775 3368 cu : $3+2e$
4775 4960 rt : (7,7,-9,-6)
4775 5329 E : $10/21$
4775 8185 θ_3 : $\sqrt{2}/6$
4775 9225 tanh : $3\ln 2/4$
4776 0316 rt : (4,7,-6,3)
4776 0887 rt : (1,6,-1,-1)
4776 2135 e^x : $7\sqrt[3]{3}/4$
4776 2541 rt : (9,6,5,2)
4776 3060 Ei : $(3\pi/2)^{-2}$
4776 4376 $s\pi(1/4)\times s\pi(\sqrt{5})$
4776 4962 ℓ_{10} : $4\sqrt{2}/3+\sqrt{5}/2$
4776 5701 rt : (8,8,-7,-2)
4776 8004 at : $4+3\sqrt{5}$
4776 8549 J_2 : $14/5$
4776 8754 $2/3\div\exp(1/3)$
4776 9277 sr : $2\sqrt{2}/3+3\sqrt{3}$
4777 0001 tanh : $13/25$
4777 0473 rt : (6,4,3,-3)
4777 0568 rt : (1,9,-4,5)
4777 0661 sq : $3/4-4e$
4777 1618 tanh : $(\sqrt[3]{3}/2)^2$
4777 1811 Ψ : $1/\ln(\sqrt{2}/3)$
4777 2047 rt : (9,9,-5,2)
4777 2658 2^x : $3/4+\sqrt{5}/4$
4777 3919 2^x : $4/3+3\sqrt{5}$
4777 5536 e^x : $4/3-2\sqrt{2}/3$
4777 8490 $s\pi$: $7\zeta(3)/10$
4777 9097 rt : (2,8,2,-3)
4777 9363 rt : (7,6,6,-5)
4777 9474 sinh : $10\zeta(3)/3$
4777 9892 rt : (5,2,4,2)
4778 0430 Ψ : $e/9$
4778 1121 $1/5\times\exp(2)$
4778 2466 J_2 : $\sqrt[3]{22}$
4778 3317 rt : (1,7,4,4)
4778 3754 rt : (8,8,9,-7)
4778 4540 2^x : $2/3-\sqrt{3}$
4778 5036 rt : (3,8,1,-1)
4778 6503 ℓ_{10} : $5\zeta(3)/2$
4778 7981 rt : (4,0,-4,7)
4778 8357 ℓ_2 : $2+\pi/4$
4778 8371 rt : (6,0,7,4)
4778 8563 $\sqrt{3}-s\pi(\sqrt{3})$
4779 2392 cu : $3\sqrt{3}/2-\sqrt{5}/4$
4779 6066 e^x : $5\sqrt[3]{5}/4$
4779 6091 Ψ : $7/17$

4779 6399 $2^x : 2\sqrt{2} + 3\sqrt{3}/4$
4779 6724 rt : (3,6,9,-6)
4779 9088 rt : (3,9,-7,-7)
4780 0895 $\theta_3 : \sqrt[3]{5}/2$
4780 1130 $J_2 : 7\zeta(3)/3$
4780 1275 $\Gamma : 1/17$
4780 3997 $t\pi : 1/4 + 3\sqrt{2}/4$
4780 4330 $J_2 : 7\sqrt{2}/3$
4780 4893 $t\pi : 3\sqrt{3}/4 + \sqrt{5}/4$
4780 4954 rt : (2,8,-5,-4)
4780 5062 $e^x : 7\pi^2/5$
4780 6321 $1/3 + \ln(\pi)$
4780 7587 at : $2e/3 - 4\pi$
4780 8173 $J_1 : 9/8$
4780 8342 $t\pi : \exp(-\sqrt[3]{4}/3)$
4780 9254 $10^x : (e/2)^{-2}$
4780 9985 sinh : 6/13
4781 0954 at : $4/3 + 3\pi$
4781 2089 $Ei : 7\ln 2/4$
4781 3064 rt : (3,6,-2,-2)
4781 3411 cu : $8\sqrt[3]{3}/7$
4781 3482 $10^x : (e+\pi)/9$
4781 4534 rt : (5,6,8,3)
4781 5099 rt : (8,5,-4,9)
4781 5741 $\Psi : 10\pi^2/3$
4781 8009 $Li_2 : 2\sqrt[3]{3}/3$
4781 8764 rt : (7,4,-6,5)
4781 9639 rt : (5,9,-8,1)
4782 1706 $e^x : \sqrt[3]{22}$
4782 1781 rt : (2,4,-9,-5)
4782 2020 $E : \exp(-\sqrt{5}/3)$
4782 2368 sr : $3 + \pi$
4782 4257 rt : (6,3,-8,1)
4782 4491 ln : $4e + \pi/3$
4782 4688 $\Gamma : \ln 2/9$
4782 6086 id : 11/23
4782 7765 $E : (\sqrt{2}\pi)^{-1/2}$
4782 8045 rt : (9,9,2,-4)
4782 8574 $erf : e/6$
4782 9649 rt : (6,8,-8,-5)
4783 0357 rt : (1,2,7,3)
4783 3009 at : $e/2 + 3\pi$
4783 5115 $J_0 : 9\ln 2/4$
4783 9170 rt : (9,5,6,-5)
4783 9966 $\Gamma : 5\pi/4$
4784 1760 id : $4\pi^2$
4784 2051 cr : $1/4 + 4\sqrt{5}/3$
4784 2594 $2^x : 2/3 + \sqrt{5}$
4784 3717 rt : (8,7,-1,-2)
4784 4060 rt : (1,8,-5,-8)
4784 4412 rt : (5,9,5,-5)
4784 4752 rt : (2,7,-1,-4)
4784 5742 rt : (3,2,-6,-3)
4784 9797 $10^x : 19/15$
4785 1610 cr : $3/2 + \sqrt{3}$
4785 2955 ln : $4\pi^2/9$
4785 5206 rt : (5,5,9,-6)
4785 5671 $2^x : (\pi)^{-1/2}$
4785 7942 rt : (7,9,-8,1)
4785 8062 $E : 9/19$
4786 0079 rt : (8,3,3,-3)
4786 0296 rt : (1,9,8,-6)
4786 0711 rt : (7,6,-5,-3)
4786 2194 $\ell_{10} : 4/3 + 3\sqrt{5}/4$
4786 2408 at : $9\zeta(3)$
4786 3258 rt : (5,1,-7,-8)
4786 3684 $J_1 : 18/7$
4786 4295 sq : $1 - 4e$
4786 5428 sq : $e/4 - 3\pi$
4786 7427 $Li_2 : 4\zeta(3)/5$
4786 7584 $s\pi : 6\ln 2$
4786 7643 $J_1 : \sqrt[3]{17}$
4786 7735 rt : (4,7,2,-3)
4786 9016 $E : (3\pi)^{-1/3}$
4787 0866 id : $3\sqrt{2} + \sqrt{5}$
4787 1686 $J_2 : \pi^2/3$
4787 2026 sinh : $2\ln 2/3$
4787 2037 at : $4\sqrt{2}/3 + 4\sqrt{5}$
4787 3737 as : $(3\pi/2)^{-1/2}$
4787 5233 at : $2\sqrt{2} - 4\sqrt{3}/3$
4787 5506 rt : (8,3,-9,6)
4787 5949 rt : (8,1,7,4)
4787 6027 cr : P_{76}
4787 6705 $J_0 : 9\sqrt{3}/10$
4787 7194 $\exp(e) - s\pi(\sqrt{5})$
4787 8525 $t\pi : 10\sqrt{2}$
4787 8577 as : $4\sqrt{2} - 3\sqrt{3}$
4787 9349 rt : (4,2,-9,-4)
4788 2425 $s\pi : 4e\pi$
4788 2578 sinh : 8
4788 3382 rt : (4,3,6,-4)
4788 5349 $\Psi : 1/16$
4788 5455 id : $1/3 - 2e/3$
4788 8280 rt : (4,7,-4,8)
4788 8487 $Ei : 9\sqrt{3}/10$
4789 0523 $2^x : 3/4 - 2e/3$
4789 0713 rt : (7,9,-2,0)
4789 0854 $e^x : 9/23$
4789 1539 at : $4\sqrt{2} + 3\sqrt{3}$
4789 2158 $t\pi : \sqrt{3} - 3\sqrt{5}/2$
4789 7362 sr : $1/\sqrt{19}$
4789 8941 $J_2 : 23/7$
4789 9363 $e^x : 3/2 - \sqrt{5}$
4789 9661 $J_2 : 2\pi^2/7$
4790 0054 rt : (1,8,-3,9)
4790 0894 rt : (7,8,5,-5)
4790 4112 $\theta_3 : \exp(-\sqrt[3]{3})$
4790 4116 tanh : 12/23
4790 5701 rt : (3,9,-5,0)
4790 7582 $E : 3\sqrt[3]{2}/8$
4790 8446 at : $4e$
4790 8870 $\ln(\pi) \div s\pi(e)$
4790 9887 $J_2 : (2e/3)^2$
4791 0878 rt : (7,4,9,-6)
4791 2545 sr : $4 - 2e/3$
4791 2732 rt : (4,7,-8,-5)
4791 2840 cr : $1 + \sqrt{5}$
4791 4147 cr : $2\sqrt{3} - 3\sqrt{5}/2$
4791 4170 sq : $\exp(-1/e)$
4791 5595 rt : (3,8,8,-6)
4791 6445 cu : $4\sqrt{2} - \sqrt{5}/3$
4791 6549 at : $\sqrt{12}\pi$
4791 6680 $\Psi : 4e\pi/7$
4791 7338 $t\pi : (e/2)^{-1/2}$
4791 8482 ln : $2e - \pi/3$
4792 0576 rt : (6,5,-1,7)
4792 1515 rt : (7,1,0,-1)
4792 1528 $e^x : 3 - 3\sqrt{3}/4$
4792 1638 at : $3\sqrt{3}/10$
4792 2383 $\ell_2 : 4/3 + 3\sqrt{2}$
4792 2549 $2^x : 4\sqrt{2} + 3\sqrt{3}/2$
4792 3574 rt : (9,9,-3,5)
4792 4311 cu : $4\sqrt{3} - 4\sqrt{5}/3$
4792 4605 rt : (5,9,-6,6)
4792 7250 $t\pi : 1/\sqrt{7}$
4792 8393 $s\pi : (\sqrt{2})^{-1/2}$
4792 8527 $t\pi : 3e/4 - 4\pi$
4792 8994 sq : 9/13
4792 9557 $J : 6/13$
4793 0646 $\ln(\sqrt{5}) - \exp(1/4)$
4793 1176 $10^x : 9\sqrt{2}/4$
4793 2169 $\ln(2/5) + \exp(1/3)$
4793 2751 $J : 5/13$
4793 2933 cu : 18/23
4793 5646 rt : (6,8,-4,-4)
4793 6636 sinh : $7\sqrt[3]{2}/4$
4793 8358 $e^x : e/2 - 2\pi/3$
4794 0938 at : $3\ln 2/4$
4794 2553 $s\pi : (1/\pi)/2$
4794 3699 $\Gamma : 3\ln 2/8$
4794 4381 $J_1 : (\sqrt[3]{3})^{1/3}$
4794 6721 $J_2 : 9\pi/10$
4794 7650 $J_1 : (\ln 2)^{-1/3}$
4794 9402 rt : (1,9,2,-1)
4794 9982 rt : (6,6,2,-3)
4795 0355 $e^x : 4 + \sqrt{2}/3$
4795 0446 $\ell_{10} : \sqrt{2}/2 + 4\sqrt{3}/3$
4795 0561 rt : (7,3,4,2)
4795 0939 $E : \sqrt{2}/3$
4795 1276 $e^x : 10(e+\pi)/9$
4795 1565 at : $3/2 + 3\pi$
4795 1929 at : 13/25
4795 2058 sq : $2/3 - e/2$
4795 2808 $J_2 : 2\sqrt{2}$
4795 3360 rt : (8,8,-5,1)
4795 3578 at : $(\sqrt[3]{3}/2)^2$
4795 3841 $t\pi : 3\sqrt[3]{2}/10$
4795 4410 at : $4 + 4\sqrt{3}$
4795 4672 rt : (9,7,5,-5)
4795 7308 ln : 13/21
4795 9130 rt : (5,4,-3,3)
4795 9862 $\ell_2 : 3 + 3e$
4796 1044 $2^x : 13/23$
4796 1378 $2^x : 3\sqrt[3]{3}/2$
4796 2115 $J_1 : \exp(1/(3e))$
4796 2792 $10^x : (\sqrt{2}\pi/3)^{1/2}$
4796 2894 cr : $3\sqrt{2}/2 + \sqrt{5}/2$
4796 3068 rt : (2,7,-9,-1)
4796 3699 rt : (7,4,-4,8)
4796 3923 rt : (9,3,9,-6)
4796 6191 $erf : 5/11$
4796 7734 at : $2 + 4\sqrt{5}$
4796 7902 $s\pi : 6(e+\pi)$
4796 8302 rt : (5,7,8,-6)
4797 0477 rt : (2,6,5,6)
4797 0609 $\exp(4/5) + s\pi(\sqrt{3})$
4797 2863 as : 6/13
4797 3173 at : $3\sqrt{2} + 3\sqrt{5}$
4797 4507 $2^x : \ln(\ln 2/2)$
4797 6948 rt : (1,5,-2,-2)
4797 7568 $t\pi : 9e\pi$
4797 7883 $s\pi : 3\sqrt{2} + 3\sqrt{3}/2$
4797 8031 rt : (7,7,8,3)
4797 8207 $3/5 \div \ln(2/3)$
4797 8362 sr : $\sqrt[3]{21/2}$
4798 0424 $\Psi : 23/11$
4798 1574 rt : (2,4,-8,8)
4798 2478 $J_2 : 17/6$
4798 2535 $t\pi : \sqrt{2}/2 - \sqrt{5}/2$
4798 2580 tanh : $1/\sqrt[3]{7}$
4798 2698 at : $(\sqrt{2}\pi/2)^3$
4798 2980 $erf : 10(1/\pi)/7$
4798 3872 rt : (2,2,9,-5)
4798 4064 $E : 8/17$
4798 5434 cu : $1 + \sqrt{2}/4$
4798 5655 $\ln(\sqrt{3}) \div \ln(\pi)$
4798 5876 at : $10\pi^2/9$
4798 6908 $e^x : 2\sqrt{2}/3 - 3\sqrt{5}/4$
4798 7547 sinh : $6\sqrt{2}$
4798 7625 $J_1 : 4\sqrt{2}/5$
4798 8132 $2^x : 1/2 + 3\sqrt{3}/4$
4798 9058 rt : (9,7,-8,2)
4798 9480 $J_2 : \sqrt{2} - \sqrt{5} - \sqrt{6}$
4798 9650 rt : (5,9,-1,-2)
4799 0175 rt : (2,6,-8,-5)
4799 0318 rt : (1,9,5,5)
4799 0699 rt : (3,3,7,3)
4799 1358 $J_2 : 9\sqrt[3]{2}/4$
4799 1451 rt : (8,3,-7,9)
4799 2532 $\theta_3 : 3/5$
4799 3106 $2^x : \exp(-\sqrt[3]{5}/3)$
4799 3507 $\Psi : 5(e+\pi)/6$
4799 4480 $\Gamma : 2e/9$
4799 4886 sr : $2\sqrt{2} - 3\sqrt{3}/2$
4799 6783 sq : $e/4 + 2\pi$
4799 6939 at : $9e\pi/7$
4799 6968 ln : $3/4 + \sqrt{3}/2$
4799 7197 $2^x : 1/\ln(e+\pi)$
4799 8436 sq : $1/4 - 2\sqrt{2}/3$
4800 0000 id : 12/25
4800 1400 rt : (7,7,-7,-3)
4800 1825 rt : (6,3,-6,4)
4800 2361 rt : (3,9,-3,7)
4800 3062 rt : (1,4,-1,-7)
4800 4222 id : $3\sqrt{3}/2 - \sqrt{5}/2$
4800 5197 $\ell_{10} : 10e/9$
4800 9053 $2^x : 2\sqrt{2}/5$
4801 0014 at : $7\pi/2$
4801 0498 ln : $8\sqrt{2}/7$
4801 1440 $10^x : 3 + 3\sqrt{3}/4$

4801 3588 $\ell_2 : 3\sqrt{3}/2 + 4\sqrt{5}/3$

4801 3643 at : 11

4801 3670 rt : (8,9,-2,-2)

4801 4400 $J : (5)^{-3}$

4801 5133 rt : (5,5,-5,-3)

4801 5174 $10^x : 1 + \pi/3$

4801 7029 id : $2\sqrt{3} - 4\sqrt{5}$

4801 7703 rt : (2,7,1,3)

4801 8579 rt : (7,6,8,-6)

4802 0925 $\ln(\sqrt{5}) + s\pi(\sqrt{5})$

4802 2616 rt : (3,8,3,4)

4802 4923 rt : (9,8,-8,1)

4802 5182 ln : $3 + 4\sqrt{5}$

4802 6463 rt : (8,5,2,-3)

4802 6558 rt : (7,2,-9,5)

4802 6923 rt : (9,9,-4,-3)

4803 0090 $s\pi : 4\pi^2/3$

4803 1414 sq : $5\sqrt[3]{2}/4$

4803 2068 id : P_{21}

4803 2176 $\Gamma : \sqrt{7}$

4803 2490 rt : (4,9,1,-3)

4803 5441 cu : $3\sqrt{2}/4 + 4\sqrt{5}$

4803 5868 $10^x : 16/11$

4803 5961 as : $2\ln 2/3$

4803 5970 $t\pi : 3\sqrt{3}/4 - 3\sqrt{5}/4$

4803 7196 $J_1 : 3\sqrt[3]{5}/2$

4803 8446 sr : 3/13

4803 8829 rt : (3,3,-9,5)

4803 8929 rt : (8,1,6,-4)

4803 8999 $Ei : 5\sqrt[3]{5}/4$

4803 9375 rt : (5,1,-9,-4)

4803 9755 $2^x : \sqrt{3}/4 - 2\sqrt{5}/3$

4804 0333 $2^x : 1/2 + \pi$

4804 0617 $1 \div s\pi(\sqrt{5})$

4804 1820 $\ell_{10} : 2/3 + 3\pi/4$

4804 2615 $J_1 : 17/15$

4804 3414 rt : (2,7,-7,8)

4804 3683 $e^x : 3\sqrt{2}/2 + 4\sqrt{3}$

4804 4049 $J_2 : \sqrt[3]{23}$

4804 4611 ln : $3(e + \pi)/4$

4804 4806 rt : (4,7,7,8)

4804 5301 id : $(\ln 2)^2$

4804 5591 $J_2 : 6e/5$

4804 7277 tanh : $\pi/6$

4804 7719 rt : (4,2,-7,-7)

4804 9738 cr : $9\sqrt[3]{3}/4$

4805 0614 sq : $\sqrt{2}/3 + \sqrt{5}/3$

4805 5023 $2^x : 1/4 + 3\sqrt{2}/4$

4805 5435 rt : (9,9,-1,8)

4805 6151 $Li_2 : 3\sqrt{2}/10$

4805 6816 rt : (4,1,9,-5)

4805 6821 $s\pi : \sqrt{2} + \sqrt{5}/3$

4805 8663 rt : (9,9,4,-5)

4805 9277 $J_1 : 9\sqrt[3]{2}/10$

4805 9425 $J_2 : e\pi/3$

4806 1257 ln : $\sqrt{2} + 4\sqrt{5}/3$

4806 2484 rt : (5,8,-5,0)

4806 2690 $J_0 : 14/9$

4806 3231 $J_2 : 1/\ln(e/2)$

4806 3257 $2^x : 9\zeta(3)/5$

4806 3485 tanh : 11/21

4806 4089 $\Psi : 1/\ln(1/3)$

4806 6579 rt : (9,5,8,-6)

4806 7092 at : $3\sqrt{2}/2 + 4\sqrt{5}$

4806 7562 id : $(\ln 2/3)^{1/2}$

4806 7722 at : $3/2 - 4\pi$

4806 8088 rt : (4,9,-7,-5)

4807 0058 $10^x : 13/24$

4807 0339 rt : (5,9,7,-6)

4807 1944 $2^x : 2 + 3\pi/4$

4807 2988 $Li_2 : 5\sqrt{3}/9$

4807 4841 $e^x : (4/3)^{-1/3}$

4807 4985 id : $\sqrt[3]{3}/3$

4807 5557 $Li_2 : 4(1/\pi)/3$

4807 6634 $J_2 : \exp(\pi/3)$

4807 8333 $J_2 : 5\sqrt[3]{5}/3$

4807 8493 ln : $7\ln 2/3$

4807 8662 rt : (1,5,-3,-2)

4807 9459 sr : $4\sqrt{2} - 2\sqrt{3}$

4808 1738 rt : (1,6,-5,-9)

4808 2276 id : $2\zeta(3)/5$

4808 2639 $J_2 : 5(e + \pi)/9$

4808 2741 $J : 9(1/\pi)/8$

4808 3278 $J : 1/\sqrt{23}$

4808 6271 rt : (5,4,-1,-1)

4808 8728 at : 12/23

4808 9834 $\ell_2 : (e)^{1/3}$

4809 1816 rt : (6,6,-9,-7)

4809 2170 $e^x : 1 - \sqrt{3}$

4809 3332 cu : $3/2 - 4e$

4809 4378 sq : $3 + 3\pi/2$

4809 4583 $erf : (\sqrt{2}\pi/3)^{-2}$

4809 6097 sr : $2\pi^2/9$

4809 7267 $e^x : \pi/8$

4809 7298 rt : (1,8,-2,-1)

4809 7538 at : $9\pi^2/8$

4809 8159 rt : (9,4,4,2)

4810 0600 sinh : $1/\sqrt[3]{10}$

4810 1617 cu : $7\sqrt{3}/8$

4810 1958 $10^x : \sqrt{2} - \sqrt{3}$

4810 2683 sq : $7\sqrt[3]{3}/2$

4810 3142 rt : (6,8,1,-3)

4810 4483 $\Psi : 9\pi^2$

4810 4946 $s\pi(1/4) + s\pi(e)$

4810 6567 $\Psi : 10(1/\pi)/3$

4810 7878 rt : (1,7,-4,-3)

4810 8634 rt : (7,9,0,3)

4810 8913 rt : (5,0,3,2)

4810 9376 ln : $3\sqrt[3]{3}/7$

4810 9569 $J_2 : (1/(3e))^{-1/2}$

4811 1356 rt : (8,8,-3,4)

4811 2769 $Ei : 9\ln 2/4$

4811 2982 rt : (9,7,-6,5)

4811 3214 $J_2 : 13/4$

4811 3459 at : $1/4 + 4e$

4811 4200 rt : (6,4,5,-4)

4811 4654 $2^x : 9\sqrt[3]{3}/4$

4811 4934 $\exp(1/3) + \exp(3)$

4811 5001 $\exp(3/4) - \exp(4)$

4811 5950 rt : (7,8,7,-6)

4811 6267 cu : $\sqrt{2}/2 - 2\sqrt{5}/3$

4811 7493 $J_2 : 20/7$

4811 8330 $t\pi : (\sqrt[3]{4})^{-1/3}$

4811 8450 $J_1 : 3e\pi/10$

4811 9430 rt : (4,8,-8,-8)

4811 9927 $\zeta : \sqrt{3} + \sqrt{6} - \sqrt{7}$

4812 0212 ln : $7\pi/5$

4812 0242 rt : (9,8,8,3)

4812 1182 ln : $1/2 + \sqrt{5}/2$

4812 3547 $J_2 : (\sqrt{2}\pi/3)^3$

4812 4803 cr : 13/4

4812 5003 rt : (6,0,9,5)

4812 7259 $J_1 : 25/22$

4812 8108 $\Gamma : 1/16$

4812 8315 $J_2 : \sqrt{5} - \sqrt{6} - \sqrt{7}$

4812 9719 rt : (3,8,-1,-2)

4813 0091 sr : $1/2 + 4\sqrt{2}$

4813 0115 rt : (2,4,8,-5)

4813 0509 rt : (5,4,7,3)

4813 5172 rt : (2,8,-7,-5)

4813 6402 $\Psi : (\ln 3/2)^2$

4813 6767 at : $2\sqrt{3}/3 - 3\sqrt{5}/4$

4813 8818 at : $3 + 3e$

4814 0013 $J_2 : 9\sqrt[3]{3}/4$

4814 0560 rt : (1,0,6,3)

4814 1196 $10^x : 3e/4 - 3\pi/4$

4814 1333 cr : $\ln(\sqrt{5}/2)$

4814 2011 $1 \div \ln(4/5)$

4814 2236 $E : 7/15$

4814 2396 id : $1/2 + 4\sqrt{5}/3$

4814 2675 $2^x : 1 - \sqrt{3}/4$

4814 3512 $10^x : 7\sqrt{5}/8$

4814 5697 rt : (7,2,-7,8)

4814 7490 $\ell_{10} : 3e/2 - \pi/3$

4814 8148 cu : $2\sqrt[3]{5}/3$

4814 8446 $\Gamma : 9e\pi/5$

4814 9121 $Ei : ((e + \pi)/3)^{-2}$

4814 9216 $2^x : P_{80}$

4814 9922 rt : (4,5,-9,8)

4815 0401 rt : (3,4,-5,-3)

4815 0826 $\zeta : 1/\ln(\pi/2)$

4815 1555 at : $3\sqrt{2} + 4\sqrt{3}$

4815 2156 rt : (6,3,-4,7)

4815 7112 $\ell_2 : 4\pi/9$

4815 7461 $t\pi : 1/7$

4815 7681 $J_2 : 9(1/\pi)$

4815 8372 cr : $2\sqrt{2}/3 + 4\sqrt{3}/3$

4815 8482 $10^x : P_{17}$

4815 9098 at : $5\sqrt{5}$

4815 9362 sinh : 19

4815 9578 rt : (9,7,7,-6)

4815 9650 sq : $2/3 - 2\sqrt{5}$

4815 9867 $\ell_{10} : 7\sqrt{3}/4$

4816 0040 rt : (7,8,-4,-3)

4816 0454 ln : 5/22

4816 2828 rt : (5,4,-1,6)

4816 2850 $\Psi : 3\sqrt{2}/4$

4816 4833 $J_2 : \sqrt{21/2}$

4816 6230 $Ei : 3\sqrt{2}/5$

4816 6550 $Ei : 17/14$

4816 6627 $\ell_2 : 3\sqrt{2} + 4\sqrt{3}$

4816 8777 at : $1/\sqrt[3]{7}$

4816 8907 id : $\exp(3/2)$

4816 9144 rt : (8,7,1,-3)

4817 0197 rt : (3,0,-9,4)

4817 0365 $\Psi : \exp(\sqrt[3]{4})$

4817 0853 $e^x : 9\sqrt{5}/10$

4817 1471 cr : $e + 4\pi$

4817 2863 sq : $3 + 3\sqrt{3}/4$

4817 5367 $s\pi : 4/25$

4817 8272 cu : $\sqrt{2}/4 - 4\sqrt{3}/3$

4817 8926 rt : (8,6,-8,1)

4817 9105 rt : (8,3,5,-4)

4817 9439 rt : (7,4,-8,-4)

4817 9781 at : $1/3 + 4e$

4818 0388 at : $e/2 - 4\pi$

4818 1513 $2^x : 1/3 + 3\sqrt{2}/2$

4818 3828 rt : (4,7,4,-4)

4818 6622 cu : $\sqrt{3} + \sqrt{5}$

4818 8469 sinh : $10\sqrt[3]{5}$

4818 8851 rt : (8,1,9,5)

4818 9492 $Li_2 : \exp(-\sqrt[3]{5}/2)$

4818 9761 $Li_2 : 1/\sqrt[3]{13}$

4818 9856 $s\pi : 2\sqrt[3]{2}/3$

4818 9897 $J_0 : 4e/7$

4819 0621 rt : (1,5,-1,9)

4819 2897 $\Psi : 2\pi/3$

4819 3003 $2^x : 10\sqrt[3]{2}/7$

4819 3031 cu : $2/3 + 3e$

4819 3474 rt : (4,3,8,-5)

4819 5126 $1/2 \times s\pi(\sqrt{2})$

4819 5322 cr : $4 - \sqrt{5}/3$

4819 5803 $\ln(4/5) \times \exp(3)$

4819 6136 sq : $3e + 3\pi/4$

4819 8050 rt : (7,7,-5,0)

4820 0454 sr : $7\sqrt{3}$

4820 0726 at : $4/3 - 4\pi$

4820 0988 cu : $\sqrt{2}/2 + \sqrt{3}/4$

4820 1852 $s\pi : 4\sqrt[3]{5}$

4820 3814 at : $2e/3 + 3\pi$

4820 5080 id : $1/4 - \sqrt{3}$

4820 5411 $J : 6/25$

4820 5560 $2^x : 3/2 + 3\pi/4$

4820 6508 $e^x : 10/11$

4820 8098 cr : $5(e + \pi)/9$

4820 8346 sr : $4e - 3\pi/2$

4820 8664 $J_2 : 9\sqrt{5}/7$

4820 8935 $J_2 : 23/8$

4821 2551 $2^x : 1/2 + 4e$

4821 3134 $\Psi : 9e/5$

4821 3599 rt : (3,8,5,9)

4821 6722 rt : (6,1,-9,4)

4821 6922 at : $4\sqrt{3}/3 + 4\sqrt{5}$

4821 7227 rt : (7,3,-1,-1)

4822 0225 $2^x : 9/5$

4822 1047 $2^x : 3 + 2\sqrt{5}/3$

4822 1385 $\sqrt{5} - e$

4822 2031 $2^x : 3/4 + \sqrt{2}$

4822 2275 rt : (9,7,-4,8)

4822 2684 ln : $3\sqrt{2}/4 + \sqrt{5}/4$

4822 4251 $J_1 : (\sqrt{2}\pi/3)^{1/3}$
4822 6184 rt : (3,7,-2,-1)
4822 6814 $\ell_{10} : e/4 + 3\pi/4$
4822 7957 $J_1 : 2\sqrt[3]{5}/3$
4823 1257 cu : $4\sqrt{2} - \sqrt{5}/4$
4823 4790 at : $\pi/6$
4823 5282 rt : (6,8,-2,-1)
4823 5298 $s\pi : P_{20}$
4823 5599 rt : (7,1,3,2)
4823 5913 $t\pi : e + 3\pi$
4823 6348 ln : $9\sqrt[3]{2}/7$
4823 7147 at : $8\pi^2/7$
4823 8083 sr : $3(e + \pi)/8$
4823 8945 ln : $7\sqrt[3]{5}$
4823 9087 cu : $1/2 + \sqrt{5}$
4824 2550 $E : 1/\sqrt[3]{10}$
4824 3324 $s\pi : \sqrt{14/3}$
4824 4164 $s\pi : \sqrt[3]{3}/9$
4824 4194 rt : (9,9,6,-6)
4824 6317 rt : (8,8,-1,7)
4824 7022 rt : (5,2,-6,3)
4824 7735 $t\pi : 2\sqrt{2}/3 - \sqrt{3}/4$
4824 8337 $Ei : 8(1/\pi)/3$
4824 9082 $\zeta : 7\sqrt{3}/9$
4825 0300 at : $3e + \pi$
4825 0422 sr : $\sqrt{2}/2 + 2\sqrt{5}/3$
4825 0987 $10^x : \sqrt[3]{5}/10$
4825 1329 at : 11/21
4825 1654 $2^x : 4\sqrt{2} - 3\sqrt{5}$
4825 2625 rt : (1,7,-1,-2)
4825 2776 rt : (4,8,-6,-3)
4825 2826 rt : (5,9,9,-7)
4825 2927 rt : (6,2,8,-5)
4825 3229 rt : (7,5,7,3)
4825 4128 $J_2 : 2\sqrt[3]{3}$
4825 4995 sq : $7e/5$
4825 5982 tanh : 10/19
4825 6607 cu : $4\sqrt{2}/3 - \sqrt{5}/3$
4825 7079 rt : (2,6,7,-5)
4825 8018 $e^x : 2e - 4\pi/3$
4825 8243 $J_2 : 1/\ln(\sqrt{2})$
4825 9626 $10^x : 4\sqrt{2}/3 - 2\sqrt{5}/3$
4825 9760 cr : $1/\ln(e/2)$
4826 1488 rt : (3,1,6,3)
4826 2011 as : $3 - 2\sqrt{3}$
4826 2053 $\Gamma : 4\sqrt[3]{2}$
4826 2275 $\theta_3 : 5/21$
4826 3708 at : $8\sqrt{2}$
4826 4165 $Ei : 7\sqrt[3]{3}$
4826 4490 $J_2 : 5\sqrt{3}/3$
4826 6483 $\exp(1/4) \div s\pi(1/3)$
4826 7764 id : $2\sqrt{3} - 4\sqrt{5}/3$
4826 8476 as : $1/\sqrt[3]{10}$
4826 8750 $\zeta : 1/\ln(\sqrt{2}\pi)$
4826 8945 $\ell_{10} : 1 + 3e/4$
4826 9595 rt : (1,6,4,-4)
4827 0553 rt : (9,5,-9,5)
4827 0960 $Ei : 7\sqrt{5}/4$
4827 4134 at : $\sqrt{13}\pi$
4827 4175 $\Gamma : \sqrt[3]{3}/10$
4827 4761 sr : $9\sqrt[3]{5}/7$
4827 7555 rt : (5,7,-4,-3)
4827 8022 $10^x : (2\pi)^{-1/3}$
4827 8311 id : $4e/3 - \pi$
4827 8869 rt : (1,9,-5,-4)
4828 3137 $3/5 \times \ln(\sqrt{5})$
4828 3494 at : $9\sqrt[3]{2}$
4828 5023 rt : (7,9,2,6)
4828 5923 rt : (1,1,-9,4)
4828 7071 cu : $2 + 3\sqrt{5}/2$
4828 7269 cu : $\sqrt{11}$
4828 7452 $J_1 : 23/9$
4828 8756 $J_1 : 8\sqrt{5}/7$
4828 9214 $\ell_2 : 5\sqrt{5}$
4829 0464 rt : (7,8,9,-7)
4829 0988 sq : $\sqrt[3]{13/2}$
4829 1909 $\ell_{10} : 4\sqrt{2}/3 + 2\sqrt{3}/3$
4829 2795 rt : (8,9,0,-3)
4829 3067 rt : (3,6,-4,-1)
4829 3592 rt : (5,3,-8,-4)
4829 4128 sr : $7\pi/10$
4829 5363 $\ell_2 : ((e + \pi)/3)^{1/2}$
4829 8397 cr : $2\sqrt{2} + \sqrt{3}/4$
4830 0875 rt : (8,5,4,-4)
4830 1129 rt : (9,7,-7,-4)
4830 2323 rt : (2,8,-5,-3)
4830 2351 sr : $4/3 + \sqrt{3}/2$
4830 4723 rt : (8,6,-6,4)
4830 5950 cr : $6e/5$
4830 7543 $J_1 : 8/7$
4830 8803 rt : (8,1,8,-5)
4830 9338 $\ell_2 : (\ln 3/3)^{1/3}$
4830 9530 at : $1/2 + 4e$
4831 1258 cr : $e/3 + 3\pi/4$
4831 1355 id : $5\pi^2/9$
4831 1404 $10^x : (3\pi/2)^{1/3}$
4831 2580 rt : (4,5,7,-5)
4831 3225 $erf : 11/24$
4831 6324 rt : (9,6,-5,0)
4831 6539 $e^x : 9\pi/4$
4831 7391 ln : $2\sqrt{2}/3 + 2\sqrt{3}$
4831 8344 cu : $1/3 - \sqrt{5}/2$
4831 9634 at : $4e\pi/3$
4831 9807 $2^x : (2e)^{-1/3}$
4831 9893 $e^x : 1/3 - 3\sqrt{2}/4$
4832 1469 $e^x : 21$
4832 2801 $\Psi : 8\ln 2/9$
4832 3969 sr : 11/5
4832 4499 $E : 2\ln 2/3$
4832 6600 rt : (9,7,9,-7)
4832 7614 rt : (4,9,-9,-6)
4832 8524 $\zeta : \exp(\sqrt[3]{4}/2)$
4832 9028 $2^x : 1/4 - 3\sqrt{3}/4$
4832 9046 rt : (5,4,1,9)
4832 9933 $J_2 : 8\zeta(3)/3$
4833 0357 at : $4\sqrt{3} + 2\sqrt{5}$
4833 0990 id : $3\sqrt{2}/4 - \sqrt{3}/3$
4833 1072 $J_1 : \sqrt{2} - \sqrt{3} - \sqrt{5}$
4833 1990 rt : (9,2,-1,-1)
4833 3607 rt : (1,8,4,1)
4833 6101 $Ei : \sqrt{20/3}$
4833 6463 rt : (1,8,-8,-9)
4833 7175 rt : (6,1,-7,7)
4833 7766 sr : $2/3 - \sqrt{3}/4$
4833 9354 rt : (5,9,1,-2)
4834 0705 sq : $\sqrt{2} + \sqrt{6} - \sqrt{7}$
4834 2880 $J_2 : 3e\pi/8$
4834 2898 $10^x : 13/14$
4834 6339 rt : (2,5,-8,-4)
4834 6678 $E : 6/13$
4834 7079 rt : (9,2,3,2)
4834 8443 cr : $2/3 + 3\sqrt{3}/2$
4834 8630 $Ei : (\sqrt[3]{3}/2)^{1/2}$
4834 8978 at : $2 + 3\pi$
4835 0403 rt : (6,6,-7,-4)
4835 0681 $\Psi : \sqrt{24}$
4835 2770 $J_2 : 16/5$
4835 4180 rt : (5,2,2,-2)
4835 5288 $t\pi : 1/2 + 4\pi/3$
4835 6674 sq : $(e/3)^{-2}$
4835 7686 rt : (6,8,3,-4)
4835 8828 $J_1 : \sqrt[3]{3/2}$
4835 8848 rt : (7,7,-3,3)
4835 9259 $J_1 : \ln(\pi)$
4835 9767 id : P_{32}
4835 9916 sinh : $\sqrt[3]{5/3}$
4836 1162 rt : (1,6,1,-2)
4836 1630 rt : (9,6,7,3)
4836 2637 rt : (9,5,-7,8)
4836 3560 sinh : $7\pi^2/10$
4836 4706 sr : $5\pi^2/8$
4836 4841 rt : (6,4,7,-5)
4836 5436 $J_0 : \ln(3\pi/2)$
4836 8240 $\ln(\sqrt{2}) \times \exp(1/3)$
4836 8257 rt : (4,8,-4,2)
4836 8486 rt : (5,2,6,3)
4836 9395 $2/3 \times \exp(4/5)$
4836 9429 id : $4\sqrt{2}/3 + 3\sqrt{3}/2$
4837 1562 $2^x : 21/16$
4837 1757 $e^x : 7(e + \pi)/3$
4837 5230 rt : (1,2,5,-3)
4837 5277 $J_2 : 10\sqrt{5}/7$
4837 6120 id : $2e + \pi/3$
4837 6343 rt : (2,8,-3,6)
4837 8312 at : $3e/4 + 3\pi$
4837 8500 $s\pi : 4e - 3\pi/2$
4837 8560 rt : (2,8,-9,-6)
4837 9033 sinh : 7/15
4838 1202 ln : $7\sqrt[3]{2}/2$
4838 1473 $E : (3\pi/2)^{-1/2}$
4838 1864 rt : (3,6,-4,-3)
4838 2342 ln : $4e/3 + \pi/4$
4838 4749 $J_2 : \sqrt{17/2}$
4838 8825 rt : (1,2,9,4)
4839 0061 $\ell_{10} : 2 + \pi/3$
4839 3192 $J : (2e/3)^{-3}$
4839 3194 sq : 16/23
4839 5036 $10^x : 4 + 2\sqrt{2}$
4839 5244 rt : (3,2,-8,-4)
4839 6322 $\ln(4/5) + s\pi(1/4)$
4839 6630 $\theta_3 : 3(1/\pi)/4$
4839 6860 rt : (9,9,8,-7)
4839 7588 at : $2\sqrt{2} - 3\sqrt{5}/2$
4839 8329 $\pi \div \exp(3/4)$
4839 8713 ln : $9\sqrt[3]{3}/8$
4839 9014 rt : (5,2,-4,6)
4840 1563 rt : (7,6,-7,-4)
4840 2541 $1/5 + \exp(1/4)$
4840 4612 ln : $1/3 + 3e/2$
4840 5798 at : 23/2
4840 6972 cr : P_{37}
4840 7589 rt : (8,7,3,-4)
4840 9296 $\theta_3 : \zeta(3)/2$
4841 0279 $\Psi : 1/\sqrt{11}$
4841 0765 rt : (4,1,-8,-1)
4841 4018 rt : (8,6,-4,7)
4841 4121 rt : (8,3,7,-5)
4841 6234 $J_2 : \sqrt[3]{25}$
4841 6695 at : $7\pi^2/6$
4841 7245 rt : (4,7,6,-5)
4841 7805 $J_2 : 10(1/\pi)$
4841 9709 id : P_{42}
4842 1061 sq : $3/2 - e$
4842 1825 $J_2 : 9\sqrt{2}/4$
4842 2057 $10^x : 2\sqrt{2}/3 + 4\sqrt{5}$
4842 4481 $e^x : 4\sqrt{2}/3 - 2\sqrt{5}/3$
4842 5031 $10^x : 25/9$
4842 6670 rt : (7,9,-6,0)
4842 9595 ln : $8\ln 2/9$
4843 0488 $10^x : 1/2 + 4\sqrt{3}$
4843 0875 rt : (7,9,4,9)
4843 1436 sr : $3e/2 + 2\pi/3$
4843 1564 $e^x : 23/5$
4843 1783 cu : $2/3 + 2\sqrt{3}$
4843 2318 sinh : $9\sqrt[3]{2}/4$
4843 2435 $e^x : 3 - \sqrt{5}/4$
4843 3256 $\ell_2 : \sqrt{2} + \sqrt{3} + \sqrt{6}$
4843 4220 at : $8\sqrt[3]{3}$
4843 5560 at : $2/3 + 4e$
4843 6746 $10^x : 25/18$
4843 6899 $J_2 : (e + \pi)/2$
4843 7033 rt : (3,6,-2,4)
4843 7464 at : $3\sqrt{3}/2 + 4\sqrt{5}$
4843 7941 $10^x : (\sqrt[3]{3})^{1/3}$
4843 9414 rt : (7,5,-2,-1)
4844 0403 sr : $\sqrt{2}/3 + \sqrt{3}$
4844 0423 $e^x : (1/(3e))^{-1/3}$
4844 0564 $\Psi : 1/17$
4844 1657 $10^x : 17/23$
4844 2778 $t\pi : 8\sqrt{3}$
4844 4261 $\Gamma : 7(e + \pi)/4$
4844 5439 rt : (3,9,-3,-1)
4844 6841 $J_1 : \sqrt{13/2}$
4844 7307 cu : $\pi/4$
4844 7495 $10^x : 3 - 2\sqrt{2}$
4844 7792 at : 10/19
4844 8362 cr : $-\sqrt{2} + \sqrt{5} + \sqrt{6}$
4844 9729 rt : (2,5,-6,3)
4845 1532 $2^x : \sqrt[3]{5}/3$
4845 1749 rt : (7,1,2,-2)

4845 1997 id : $10\sqrt{5}/9$
4845 2976 ℓ_{10} : $1/3+e$
4845 5218 10^x : $10\sqrt[3]{5}$
4845 5324 at : $1-4\pi$
4845 5575 sq : $8(1/\pi)$
4845 6582 $s\pi$: $4e/3-\pi/4$
4845 7586 rt : (6,8,0,2)
4845 8343 J_2 : $7e/6$
4845 9717 J_1 : $(\pi/3)^3$
4846 1497 id : $6\sqrt{3}/7$
4846 1999 rt : (8,4,-9,4)
4846 3337 rt : (1,9,0,-2)
4846 3458 rt : (5,0,-9,3)
4846 3518 rt : (6,6,6,-5)
4846 3699 rt : (7,3,6,3)
4846 4843 10^x : $2\sqrt{2}-\sqrt{3}$
4846 6360 rt : (2,4,-9,-2)
4846 8933 sr : $e+3\pi$
4846 9271 rt : (4,8,-2,7)
4846 9275 J_2 : $7\sqrt[3]{2}/3$
4846 9359 rt : (3,1,5,-3)
4846 9791 Γ : $(\ln 2/2)^3$
4847 2159 rt : (2,6,9,-6)
4847 3082 E : 11/24
4847 3203 Ei : $8(1/\pi)/5$
4847 3322 J_2 : 19/6
4847 4539 2^x : $4\sqrt{3}/3-3\sqrt{5}/2$
4847 4938 rt : (1,8,9,-1)
4847 7006 Li_2 : $\sqrt[3]{5}/4$
4847 8847 ℓ_2 : $3/2-\pi/4$
4848 0437 rt : (5,9,-3,-3)
4848 0779 rt : (3,3,9,4)
4848 1048 e^x : $\sqrt{2}+\sqrt{3}-\sqrt{5}$
4848 1647 sq : $1/\ln(\sqrt{2}\pi/3)$
4848 2968 ln : $3+\sqrt{2}$
4848 6173 2^x : $5\sqrt{3}/4$
4848 6195 rt : (1,7,8,-8)
4848 6864 J_2 : $\sqrt{10}$
4848 7771 sr : $7\sqrt[3]{2}/4$
4848 9965 $\exp(4)-\exp(\sqrt{2})$
4849 0664 ln : 12
4849 1535 rt : (5,5,-7,-4)
4849 1692 e^x : $1/\ln(3)$
4849 2907 rt : (7,7,-1,6)
4849 3112 ℓ_2 : $3+3\sqrt{3}/2$
4849 3131 tanh : 9/17
4849 4478 2^x : $4e/3+\pi/3$
4849 5395 ln : $3\sqrt{2}/4+3\sqrt{5}/2$
4849 6087 Γ : 1/15
4849 6803 rt : (9,9,-6,-4)
4849 7230 at : $3/4+4e$
4849 7667 ℓ_{10} : $4\sqrt{3}/3+\sqrt{5}/3$
4849 7815 Li_2 : $(\sqrt{5}/2)^{-1/3}$
4849 8407 cu : $5\sqrt{2}/9$
4849 8618 10^x : $\zeta(3)/7$
4850 1844 rt : (8,9,2,-4)
4850 2075 e^x : $4/3+e/4$
4850 2964 J : $7(1/\pi)/5$
4850 4126 J_1 : 23/20
4850 5822 Γ : $\exp(-\sqrt[3]{4})$
4850 5830 cu : 11/14
4850 7125 sr : 4/17
4850 7286 rt : (8,5,6,-5)
4850 9431 cu : $2+2\sqrt{2}/3$
4850 9771 J_2 : $\sqrt{3}-\sqrt{5}-\sqrt{6}$
4850 9897 rt : (4,9,5,-5)
4851 2073 ζ : $3\sqrt{5}/10$
4851 5223 rt : (4,5,9,-6)
4851 7040 10^x : $1/\ln(\sqrt{5})$
4851 7521 $4/5\times\exp(\sqrt{5})$
4851 7577 10^x : $3/2+3\sqrt{2}$
4851 7780 rt : (9,8,-6,0)
4851 9069 cr : $3\sqrt{2}/2+2\sqrt{3}/3$
4851 9565 Ψ : 25/2
4852 0302 id : $7\ln 2/10$
4852 2396 J_2 : $5\sqrt[3]{2}/2$
4852 2452 $4/5\div\exp(1/2)$
4852 2487 J_2 : $\sqrt{3}-\sqrt{5}-\sqrt{7}$
4852 3161 at : $4e+\pi/4$
4852 4307 cu : $4-\sqrt{2}/4$
4852 4440 at : $e/3-4\pi$
4852 4979 J_1 : $1/\ln(\sqrt{2}\pi/3)$
4852 5024 $t\pi$: $1/2+3\pi/4$
4852 5611 Ψ : $5\sqrt[3]{2}/3$
4852 5687 erf : $(3\pi/2)^{-1/2}$
4852 6319 J_1 : $8(1/\pi)$
4852 6465 rt : (5,2,-2,9)
4852 8137 id : $6\sqrt{2}$
4852 8446 rt : (9,4,-2,-1)
4852 9274 rt : (1,8,-4,1)
4852 9471 J_2 : $3\pi^2/10$
4853 2256 ℓ_{10} : $4-2\sqrt{2}/3$
4853 3512 rt : (5,8,-3,-1)
4853 3596 Ψ : 21/10
4853 3642 ℓ_2 : $10\sqrt[3]{2}/9$
4853 3973 $4\div\ln(5)$
4853 4949 rt : (1,3,-8,-6)
4853 6332 J_2 : $\sqrt{2}-\sqrt{3}-\sqrt{7}$
4853 6714 $t\pi$: $3\sqrt[3]{5}/10$
4853 8808 rt : (9,0,2,2)
4853 8814 $t\pi$: $\ln(\sqrt{3}/2)$
4853 8817 10^x : $5e$
4854 0282 J_2 : 22/7
4854 1011 $\ln(2)\times\exp(\sqrt{5})$
4854 1477 cu : $1/3-3e/2$
4854 2162 cr : $2-4\sqrt{2}/3$
4854 2682 ℓ_2 : 5/7
4854 3393 J_2 : π
4854 3732 rt : (5,4,1,-2)
4854 4700 ℓ_{10} : $9e/8$
4854 6170 rt : (6,6,-5,-1)
4854 6852 ζ : $8\sqrt{2}/9$
4854 8540 sq : $8\sqrt[3]{5}/5$
4854 8592 rt : (1,8,0,-2)
4854 8880 rt : (9,4,6,3)
4854 9177 sr : $\sqrt{2}/6$
4854 9420 e^x : $(e)^3$
4854 9488 sq : $\sqrt{2}+\sqrt{3}-\sqrt{6}$
4854 9727 J_1 : $9\sqrt{2}/5$
4854 9972 e^x : $\sqrt{2}+\sqrt{5}$
4855 0293 rt : (7,5,-6,3)
4855 0781 ln : 8/13
4855 1140 rt : (6,8,5,-5)
4855 1812 as : 7/15
4855 2336 rt : (8,4,-7,7)
4855 2736 $t\pi$: P_{75}
4855 3671 rt : (5,0,5,3)
4855 3692 $2/5+\exp(3)$
4855 4155 2^x : $\sqrt{3}+\sqrt{5}+\sqrt{6}$
4855 4796 ℓ_{10} : $4\sqrt{2}-3\sqrt{3}/2$
4855 5035 sq : $\sqrt[3]{19/2}$
4855 5599 ℓ_{10} : $\exp(\sqrt{5}/2)$
4855 6043 rt : (6,4,9,-6)
4855 8399 rt : (3,6,0,9)
4856 2066 10^x : $2\sqrt{2}+4\sqrt{3}$
4856 3067 rt : (3,8,-3,-3)
4856 3336 tanh : $3\sqrt{2}/8$
4856 3342 rt : (5,4,9,4)
4856 3346 sr : $3/2+\sqrt{2}/2$
4856 3641 ℓ_{10} : $3/4+4\sqrt{3}/3$
4856 6071 2^x : $4/3+\pi/2$
4856 6577 E : $(\sqrt{2}\pi/3)^{-2}$
4856 7644 at : $2(e+\pi)$
4856 7946 rt : (1,0,8,4)
4856 8765 J_2 : $7\sqrt{5}/5$
4856 9013 10^x : $(5/2)^3$
4857 1040 10^x : $3e/2+3\pi$
4857 1417 Ei : 17/20
4857 2493 rt : (3,4,-7,-4)
4857 6526 J_2 : $4\sqrt{5}/3$
4857 6698 J_1 : $1/\ln(\sqrt[3]{2}/3)$
4857 6983 rt : (7,8,-6,-4)
4857 7578 tanh : $5(1/\pi)/3$
4857 8596 e^x : $7\sqrt{5}/2$
4858 0049 J_2 : 25/8
4858 0781 $s\pi$: $3\sqrt{5}/8$
4858 1036 rt : (8,9,-9,-7)
4858 1253 rt : (5,0,-7,6)
4858 1511 ℓ_{10} : $2+3\sqrt{2}/4$
4858 2403 $\ln(3/5)\times s\pi(2/5)$
4858 2849 Ψ : 8/13
4858 4255 sq : $2/3-4\sqrt{2}/3$
4859 0257 rt : (8,7,5,-5)
4859 0282 rt : (5,8,-9,5)
4859 0553 Γ : $\ln(\sqrt{5}/2)$
4859 1109 J_2 : $9\ln 2/2$
4859 2462 rt : (4,1,-6,2)
4859 2877 at : $\sqrt{14}\pi$
4859 3745 J_2 : $9\sqrt{3}/5$
4859 4779 rt : (8,3,9,-6)
4859 6955 rt : (4,7,8,-6)
4859 7032 J_2 : $7\sqrt[3]{5}/4$
4859 8038 rt : (4,6,-9,2)
4859 8951 $\ln(2/5)-s\pi(\pi)$
4859 9428 2^x : 4/7
4860 2140 e^x : $1/3+\sqrt{3}/3$
4860 4634 sr : $9\sqrt{5}$
4860 5744 at : $2\sqrt{2}+4\sqrt{5}$
4860 5929 erf : 6/13
4860 6171 cu : $2-4\pi/3$
4860 6463 rt : (7,7,1,9)
4860 6797 id : $1/4+\sqrt{5}$
4860 7958 J_1 : $4\sqrt[3]{3}/5$
4860 9126 J_2 : 3
4860 9226 J_1 : 15/13
4861 0780 rt : (3,5,-8,-9)
4861 1336 10^x : $\ln(2\pi/3)$
4861 1658 J_2 : $8e/7$
4861 2425 rt : (7,7,-3,-1)
4861 3190 J_2 : $(\ln 2)^{-3}$
4861 3215 cu : $10(e+\pi)/3$
4861 3973 cu : $1/3-4\sqrt{5}$
4861 3998 E : $10(1/\pi)/7$
4861 6474 J_2 : $5\zeta(3)/2$
4861 6511 $t\pi$: $\sqrt{2}/2+4\sqrt{5}/3$
4861 7635 Li_2 : 3/7
4861 8762 ln : $\sqrt{2}/3+2\sqrt{3}/3$
4861 9998 ζ : $(\sqrt{2}\pi/2)^{-1/2}$
4862 0507 id : $\exp(-\sqrt[3]{3}/2)$
4862 1097 rt : (7,3,1,-2)
4862 1115 E : 5/11
4862 3346 $2\div\exp(\sqrt{2})$
4862 4331 as : $4/3-\sqrt{3}/2$
4862 5235 rt : (3,7,0,-2)
4862 5879 ℓ_2 : $1/3+4e$
4862 6061 $\ln(3/4)+s\pi(e)$
4862 6143 at : $1-\sqrt{2}/3$
4862 8333 rt : (1,6,-3,1)
4862 9440 rt : (6,8,2,5)
4862 9554 rt : (7,1,5,3)
4863 0214 J_0 : $9\zeta(3)/7$
4863 2501 J_1 : $2\sqrt{3}/3$
4863 2875 J_0 : 17/11
4863 3592 rt : (6,6,8,-6)
4863 3909 J_2 : $10e/9$
4863 5594 rt : (2,8,1,-9)
4863 5683 Γ : $9(1/\pi)/8$
4863 6914 at : $3/4-4\pi$
4863 7584 rt : (1,7,-3,-3)
4863 7806 rt : (7,5,9,4)
4863 7999 sr : $4+3e$
4863 8982 J_2 : $\sqrt{19/2}$
4863 9256 rt : (3,8,-8,2)
4863 9926 sq : $(2e/3)^{1/3}$
4864 0738 cr : $9e\pi/5$
4864 1676 rt : (1,6,3,1)
4864 1749 rt : (3,1,8,4)
4864 2041 J_2 : $9\sqrt[3]{5}/5$
4864 2432 J_2 : $7\sqrt{3}/4$
4864 5647 rt : (1,6,2,-7)
4864 6421 J_2 : $(3\pi)^{1/2}$
4864 6920 sr : $2\sqrt{2}+3\sqrt{5}/2$
4864 7328 J_1 : $5\ln 2/3$
4864 7720 ℓ_2 : $\sqrt[3]{22}$
4864 8490 ℓ_{10} : $4/3+\sqrt{3}$
4864 9501 rt : (5,7,-6,-4)
4864 9516 e^x : $9e\pi/8$
4864 9574 J_2 : $\exp(\sqrt{5}/2)$
4864 9599 J_2 : $1/\ln(\sqrt[3]{3}/2)$
4864 9664 J_2 : $9e/8$

4865 3293 cr : P_{78}
4865 3306 e^x : $3/4 - \sqrt{2}/4$
4865 3317 sr : $4\sqrt{3} - \sqrt{5}/3$
4865 4160 cr : $\sqrt{2} - 3\sqrt{3}/4$
4865 4402 10^x : $5e\pi/9$
4865 5673 rt : (7,5,-4,6)
4865 6180 at : $6\pi^2/5$
4865 6950 erf : $2\ln 2/3$
4865 8186 Li_2 : $(2e)^{-1/2}$
4865 8214 sq : $1/4 - 3e$
4865 9282 sπ : $1/3 + 2\sqrt{2}$
4865 9464 10^x : $4\sqrt{3} + 2\sqrt{5}$
4866 0649 Ψ : $(\sqrt{5}/2)^{1/2}$
4866 0931 10^x : $\sqrt[3]{11}$
4866 1013 rt : (9,7,-9,-5)
4866 1926 ln : $10\zeta(3)$
4866 3760 cu : $\sqrt{2}/4 + \sqrt{3}/4$
4866 4080 θ_3 : 6/25
4866 4600 rt : (8,9,4,-5)
4866 5409 cr : 23/7
4866 5633 e^x : P_{11}
4866 7666 Ei : 19/8
4866 8434 ln : $9\pi^2$
4866 8445 rt : (8,5,8,-6)
4867 0301 rt : (4,9,7,-6)
4867 1603 at : P_3
4867 3115 rt : (4,6,-7,7)
4867 5905 rt : (9,8,-8,0)
4867 6205 2^x : $1 - 3e/4$
4867 7082 at : $1 + 4e$
4867 8966 sq : $9\sqrt[3]{3}$
4867 9273 E : $e/6$
4868 2099 rt : (5,0,-5,9)
4868 3495 rt : (9,6,-3,-1)
4868 3741 Γ : $\zeta(3)/2$
4868 6702 at : $e/4 - 4\pi$
4868 7127 rt : (5,5,-9,-8)
4868 7930 tπ : $\sqrt[3]{3}/10$
4868 9923 at : 9/17
4869 0912 rt : (9,2,1,-2)
4869 3014 as : $3e/4 - \pi/2$
4869 3351 $1/4 \times \exp(2/3)$
4869 4463 rt : (5,9,3,1)
4869 4549 sinh : 19/16
4869 5760 at : $2/3 - 4\pi$
4869 6400 rt : (3,8,-6,9)
4869 8162 rt : (9,2,5,3)
4869 9022 at : $3\sqrt{3} + 3\sqrt{5}$
4869 9213 $\sqrt{2} \div s\pi(2/5)$
4869 9811 rt : (6,6,-3,2)
4870 0462 10^x : $4\sqrt[3]{5}/7$
4870 1574 rt : (7,3,-9,3)
4870 1635 rt : (5,2,4,-3)
4870 3264 cr : $\ln 2/6$
4870 3358 rt : (6,8,7,-6)
4870 4402 rt : (1,9,4,-3)
4870 5072 rt : (1,6,3,-3)
4870 5253 rt : (9,6,9,4)
4870 5704 λ : $(\ln 2/2)^3$
4870 6491 ℓ_{10} : $\sqrt{2}/3 + 3\sqrt{3}/2$
4870 7403 rt : (3,4,-7,4)
4870 8652 rt : (5,2,8,4)
4871 0196 at : $4e + \pi/3$
4871 0225 sinh : $5\sqrt[3]{3}/3$
4871 2015 rt : (1,2,7,-4)
4871 2824 ln : $\sqrt{2} - \sqrt{5} + \sqrt{6}$
4871 3264 E : $\exp(-\sqrt[3]{4}/2)$
4871 3304 sq : $5e\pi/8$
4871 3323 e^x : $1/\sqrt[3]{16}$
4871 3556 ℓ_{10} : $(3\pi)^{1/2}$
4871 4327 10^x : $\sqrt{3}/4 - \sqrt{5}/3$
4871 5344 rt : (3,6,-6,-4)
4871 8902 2^x : $4\sqrt{3} - 2\sqrt{5}$
4871 9286 J_1 : 22/19
4872 2022 id : $4\sqrt{3} + \sqrt{5}/4$
4872 4728 cu : $\sqrt{5} + \sqrt{6} - \sqrt{7}$
4872 5101 ℓ_{10} : $3/2 + \pi/2$
4872 5299 rt : (7,6,-9,-5)
4872 6622 $\ln(\sqrt{2}) + \exp(\pi)$
4872 6800 rt : (2,9,-1,-5)
4872 6898 at : $3 + 4\sqrt{5}$
4872 8031 cr : $\pi^2/3$
4872 8037 sr : $3/4 + 2e$
4872 8248 rt : (8,9,-7,-4)
4872 9145 tπ : $(1/(3e))^{-1/2}$
4873 0059 J_1 : $(\sqrt{5}/3)^{-1/2}$
4873 1761 10^x : $3/2 - 2e/3$
4873 4795 at : $7e\pi/5$
4873 4875 rt : (8,7,7,-6)
4873 5040 Ψ : $\sqrt{2}/6$
4873 5882 E : $\ln(\pi/2)$
4873 6470 rt : (4,1,-4,5)
4873 7243 sq : $\sqrt{2}/4 + \sqrt{3}/2$
4873 7581 λ : 1/24
4873 7581 tanh : $\exp(-\sqrt[3]{2}/2)$
4873 8787 sq : $2\pi/9$
4873 9163 cr : $9\sqrt[3]{5}$
4874 0792 ℓ_2 : $4 - 3\sqrt{3}/2$
4874 3229 θ_3 : $\sqrt[3]{3}/6$
4874 3551 cr : $\sqrt{24}\pi$
4874 4651 at : $7\sqrt[3]{5}$
4874 6613 rt : (7,5,-2,9)
4874 7367 ζ : $7\sqrt[3]{2}/4$
4875 0925 θ_3 : $\zeta(3)/5$
4875 0939 rt : (7,9,-4,-1)
4875 1806 Ψ : $7\zeta(3)/4$
4875 7024 2^x : $9(1/\pi)/5$
4875 7264 rt : (7,5,0,-2)
4875 8734 sπ : $\ln(2\pi)$
4875 8929 e^x : $2 - e$
4875 9557 $1/3 + \exp(e)$
4875 9967 Ei : 6/23
4876 0302 rt : (3,9,-1,-2)
4876 0520 cr : $3e/2 - \pi/4$
4876 1624 at : $3\sqrt{2}/8$
4876 3386 rt : (6,4,-8,-1)
4876 3461 rt : (7,1,4,-3)
4876 4407 rt : (9,8,-6,3)
4876 5509 at : 12
4876 6439 rt : (6,8,4,8)
4876 6631 J_0 : $(\sqrt[3]{2}/3)^{-1/2}$
4876 6838 $\ln(3) + \exp(2)$
4876 7157 E : $3\zeta(3)/8$
4876 8207 $1/5 - \ln(3/4)$
4876 9388 rt : (1,9,-2,-3)
4876 9536 rt : (7,3,8,4)
4877 0926 rt : (2,8,3,-2)
4877 2456 rt : (3,1,7,-4)
4877 2792 sr : $2 + 4\pi/3$
4877 3295 ln : $\sqrt{3} - \sqrt{5}/2$
4877 3583 e^x : 9/4
4877 3603 cr : $4 - \sqrt{2}/2$
4877 5348 rt : (1,8,7,-4)
4877 6170 at : $5(1/\pi)/3$
4877 6211 cu : $2 - \pi$
4877 8214 rt : (5,9,-5,-4)
4877 9670 at : $10\zeta(3)$
4878 4006 rt : (5,5,-9,-5)
4878 6612 rt : (7,3,-7,6)
4878 6792 rt : (9,9,-8,-5)
4878 7980 $\sqrt{2} \div \ln(2/3)$
4878 8191 sq : $e/3 - 3\pi/2$
4878 8826 rt : (2,9,1,4)
4878 9761 ℓ_2 : $7\zeta(3)/3$
4879 0007 sq : $2\sqrt{3} + 3\sqrt{5}/2$
4879 1369 rt : (3,4,-5,9)
4879 2351 ζ : $\zeta(9)$
4879 2498 tanh : 8/15
4879 5003 sr : 5/21
4879 5071 rt : (8,9,6,-6)
4879 5821 2^x : $5\pi^2/7$
4879 6451 rt : (4,1,-9,-2)
4879 6827 E : 9/20
4879 8950 id : $10\pi/7$
4879 9078 10^x : $((e+\pi)/2)^{1/2}$
4879 9127 10^x : $(\ln 2)^{-1/3}$
4879 9258 rt : (4,9,9,-7)
4880 0000 cu : $7\sqrt[3]{2}/5$
4880 1168 sπ : $\sqrt{10}$
4880 3387 id : $1/3 + 2\sqrt{3}/3$
4880 4738 J : $(\ln 3/2)^3$
4880 7656 id : $\sqrt{3}/2 - 3\sqrt{5}/2$
4880 7959 rt : (5,2,-8,-9)
4880 8671 $\ln(\sqrt{5}) \div \exp(1/2)$
4880 8980 rt : (9,8,-4,-1)
4880 9737 sr : $1/\ln(\pi/2)$
4881 0066 tπ : $e/2 + \pi/4$
4881 0488 E : $1/\sqrt[3]{11}$
4881 1032 at : $1/2 - 4\pi$
4881 1642 tπ : $1/3 - 4\pi/3$
4881 4095 sinh : $3e/2$
4881 4468 rt : (9,4,0,-2)
4881 5045 sinh : 8/17
4881 5383 sr : $3 - \pi/4$
4881 5569 tπ : $\exp(-\sqrt{5}/3)$
4881 6035 10^x : $3/4 - \sqrt{3}/3$
4881 6624 $\pi + \ln(\sqrt{2})$
4881 7087 sr : $\sqrt{15}\pi$
4881 7111 rt : (5,8,-1,-2)
4881 7881 id : $1/3 + 3e$
4881 9637 sr : $3\sqrt{3} - 4\sqrt{5}/3$
4881 9856 rt : (9,0,4,3)
4882 0854 sq : $\sqrt{2}/3 + 2\sqrt{3}$
4882 1536 sq : $3/2 + \pi/3$
4882 2242 rt : (1,9,-4,6)
4882 2451 rt : (5,4,3,-3)
4882 3740 rt : (6,6,-1,5)
4882 4651 sinh : 23/14
4882 5024 rt : (1,8,2,-3)
4882 5147 rt : (9,4,8,4)
4882 6250 ℓ_{10} : $9\sqrt[3]{5}/5$
4882 6363 rt : (6,8,9,-7)
4882 6375 J_0 : $\sqrt[3]{11/3}$
4882 7696 rt : (5,0,7,4)
4882 8125 cu : $5\sqrt[3]{2}/8$
4883 0223 rt : (1,4,6,-4)
4883 0735 sq : $e\pi/7$
4883 1335 cr : $3\sqrt{2}/4 + \sqrt{5}$
4883 2728 rt : (3,8,-5,-4)
4883 2887 ℓ_{10} : $1/4 + 2\sqrt{2}$
4883 7506 ln : $(\ln 2/3)^{1/3}$
4883 7793 rt : (3,4,-9,-5)
4883 7948 cr : P_{77}
4883 7983 2^x : $e/2 - \pi/4$
4883 8365 sq : $((e+\pi)/2)^{-1/3}$
4883 9719 cu : $4 - \sqrt{5}$
4884 0049 K : 15/16
4884 0235 rt : (7,8,-8,-5)
4884 0878 sr : $3\sqrt{2}/4 + 2\sqrt{3}/3$
4884 1648 tπ : $3\sqrt{2}/2 - \sqrt{3}/4$
4884 1690 rt : (9,8,-4,6)
4884 2347 J_1 : $(\pi/2)^{1/3}$
4884 2906 E : $\pi/7$
4884 4593 erf : $1/\sqrt[3]{10}$
4884 5895 $\exp(\pi) - \exp(\sqrt{3})$
4884 7387 rt : (8,9,-5,-1)
4884 7692 sπ : $(\pi/2)^{1/3}$
4884 7849 e^x : $3e/4 + 4\pi/3$
4885 0398 at : $7\sqrt{3}$
4885 2315 rt : (8,7,9,-7)
4885 3506 rt : (4,1,-2,8)
4885 6110 $2/5 \times \exp(1/5)$
4885 6241 ln : $3\sqrt{2}/2 + 4\sqrt{3}/3$
4885 6455 id : $2\sqrt[3]{5}/7$
4885 9358 id : $10e\pi/9$
4886 0251 sr : $3(1/\pi)/4$
4886 0276 ℓ_{10} : $2e - 3\pi/4$
4886 1093 rt : (7,3,-5,9)
4886 2121 tanh : $4\zeta(3)/9$
4886 3016 at : $e + 3\pi$
4886 6088 cr : $2 + 3\sqrt{3}/4$
4886 6302 10^x : $(\pi/3)^{-1/2}$
4886 7420 sq : $2e - 3\pi/4$
4886 7633 e^x : $5(1/\pi)/4$
4886 9083 rt : (7,7,-1,-2)
4887 0686 ζ : $3\pi/7$
4887 0948 at : $4 + 3e$
4887 1026 Γ : $\sqrt{2} + \sqrt{3} + \sqrt{7}$
4887 1633 10^x : $e/2 + \pi/4$
4887 1830 e^x : $4e - 2\pi$

4887 3303 $2^x : 2\sqrt{3}/3+\sqrt{5}$
4887 3756 rt : (7,3,3,-3)
4887 6012 rt : (9,6,-9,3)
4887 7902 ln : $\sqrt[3]{13/3}$
4887 8023 cr : $7\sqrt{2}/3$
4887 8224 $t\pi : \sqrt[3]{3/2}$
4887 8352 rt : (7,1,7,4)
4887 8546 $1/3-\exp(3/5)$
4887 9337 at : $\sqrt{15}\pi$
4888 0572 rt : (3,3,6,-4)
4888 1675 rt : (2,8,5,5)
4888 2774 rt : (1,7,-5,-4)
4888 2821 cu : $4\sqrt{2}/3+2\sqrt{5}/3$
4888 4313 $t\pi : \ln(\pi)$
4888 6180 $\ell_{10} : \sqrt{19/2}$
4888 7225 rt : (1,3,-9,-5)
4888 9376 rt : (5,7,-8,-5)
4888 9494 at : $\exp(5/2)$
4889 4787 id : $3e/2-4\pi$
4889 6103 sq : $3+3\sqrt{2}/4$
4889 9604 $2^x : 5\sqrt[3]{3}/4$
4890 0047 $10^x : 7\sqrt{2}/3$
4890 0941 at : $10e\pi/7$
4890 2083 rt : (8,9,8,-7)
4890 2315 sq : $4/3+\pi/4$
4890 2396 $e^x : (\pi/3)^{-2}$
4890 3127 rt : (4,1,-7,-5)
4890 3487 $10^x : 3/4-3\sqrt{2}/4$
4890 3505 $E : \sqrt{5}/5$
4890 4564 id : $\sqrt{2}/2-3\sqrt{3}$
4890 4958 $Ei : (\sqrt{2}\pi/3)^{1/2}$
4890 5504 at : $4/3+4e$
4890 5897 sinh : $\sqrt{2}/3$
4890 7942 ln : $4+\sqrt{3}/4$
4890 8598 sq : $10\sqrt[3]{5}/3$
4890 9780 rt : (9,8,-2,9)
4891 0429 $J_1 : 3e/7$
4891 1993 $2^x : (e+\pi)^{1/3}$
4891 2943 $t\pi : \sqrt{2}-\sqrt{5}/4$
4891 3513 $\Gamma : 1/14$
4891 3573 sinh : $2\zeta(3)$
4891 4857 cu : $2\sqrt{3}+4\sqrt{5}$
4891 6795 rt : (9,6,-1,-2)
4891 7437 ln : $3e/5$
4891 8813 rt : (5,5,-7,-5)
4891 9224 $\Gamma : 3/5$
4892 0718 sr : $1+3\sqrt{3}$
4892 0905 rt : (9,2,3,-3)
4892 2673 at : $9e/2$
4892 2893 rt : (5,9,5,4)
4892 3183 at : $1/3-4\pi$
4892 3682 cu : $1/3-3e$
4892 4954 rt : (9,2,7,4)
4892 5125 $10^x : 15/23$
4892 5892 rt : (6,6,1,8)
4892 6912 rt : (5,2,6,-4)
4892 7967 cu : $4/3-3\sqrt{2}/2$
4892 8560 $e^x : (e/2)^{-3}$
4892 8646 $J : \exp(-11\pi/5)$
4892 8857 rt : (2,6,-8,-8)
4893 0429 $\Gamma : \ln(1/(2e))$
4893 1924 $10^x : 11/5$
4893 2787 rt : (1,2,9,-5)
4893 3087 $J_1 : (e/2)^{1/2}$
4893 3934 $10^x : 4\ln 2/7$
4893 4690 rt : (3,6,-8,-5)
4893 6427 $\Psi : 1/18$
4893 6480 cu : $\sqrt{17/3}$
4893 8310 $\ell_{10} : 3e/4+\pi/3$
4893 8517 $2^x : 25/19$
4893 9620 at : $\exp(-\sqrt[3]{2}/2)$
4894 0240 rt : (9,6,-7,6)
4894 0314 $Ei : -\sqrt{2}+\sqrt{3}+\sqrt{6}$
4894 0379 $1/4 \div \ln(3/5)$
4894 3397 $\ln(3/4) \div s\pi(1/5)$
4894 5794 at : $1/3-\sqrt{3}/2$
4894 5871 rt : (8,9,-3,2)
4894 8681 rt : (4,9,-6,-9)
4894 8758 $2^x : 1/\ln(\pi/3)$
4895 2047 $J_1 : (\sqrt[3]{4})^{1/3}$
4895 4027 $e^x : 1/3+4\sqrt{2}$
4895 5098 sq : $3/4+\sqrt{5}/2$
4895 5735 $J_1 : 7/6$
4895 6375 rt : (7,1,-8,9)
4895 6665 $\Gamma : \sqrt{2}-\sqrt{5}-\sqrt{6}$
4895 6689 $e^x : 4+\sqrt{5}/4$
4895 8458 cr : $\sqrt{2}/2+3\sqrt{3}/2$
4895 8714 $E : 1/\ln(3\pi)$
4895 9348 cu : $3/2+2\pi$
4896 1451 sr : $1/3+4\sqrt{2}/3$
4896 2521 rt : (7,9,-2,-2)
4896 2749 $J_0 : 8\sqrt{3}/9$
4896 2821 $\zeta : 23/15$
4896 3639 $E : 7(1/\pi)/5$
4896 4899 $s\pi : (e+\pi)/7$
4896 5727 $2^x : 3\zeta(3)/2$
4896 6130 rt : (7,5,2,-3)
4896 6887 $\ell_{10} : 3\sqrt{2}-2\sqrt{3}/3$
4896 7879 rt : (3,9,1,-3)
4896 8073 $2^x : 3+3\sqrt{5}$
4896 8480 rt : (1,7,-3,5)
4896 9086 $\Psi : \exp(\sqrt{5}/3)$
4896 9652 rt : (6,4,-4,5)
4896 9688 rt : (7,1,6,-4)
4897 1413 rt : (3,5,5,-4)
4897 2698 $Ei : \ln(5/3)$
4897 3127 rt : (1,9,-4,-4)
4897 4804 $e^x : 1/3-\pi/3$
4897 4900 rt : (3,1,9,-5)
4897 6590 rt : (1,8,5,-7)
4897 7351 $\theta_3 : 4\sqrt{5}/9$
4897 8128 at : $1/4-4\pi$
4897 8270 rt : (5,9,-7,-5)
4897 9144 $2^x : \sqrt{2}+2\sqrt{5}/3$
4898 4243 $\ln(\pi) \times \exp(\pi)$
4898 4819 $2^x : 13/6$
4898 5386 cu : $3/2+3\sqrt{2}/2$
4898 5798 $t\pi : 2\sqrt{2}/3+\sqrt{5}/3$
4898 7620 $2^x : (\sqrt{3})^{1/2}$
4898 8027 $e^x : 2\sqrt{2}/3+\sqrt{5}/4$
4898 8156 rt : (8,7,-8,-1)
4898 9794 sr : 6/25
4899 1620 at : $5\pi^2/4$
4899 2298 rt : (4,1,-5,-8)
4899 3859 sq : $5\sqrt[3]{2}/9$
4899 4179 $t\pi : \sqrt[3]{5}/2$
4899 5426 cu : $7\sqrt{5}/8$
4899 5732 at : 8/15
4899 5945 at : $3e+4\pi/3$
4899 7460 rt : (9,6,-5,9)
4899 7795 $J_0 : 9\sqrt[3]{5}/10$
4899 9730 cr : 2/17
4900 0000 sq : 7/10
4900 0669 sinh : $(\sqrt{2})^{1/2}$
4900 1063 $10^x : 4e-\pi$
4900 2913 rt : (9,8,-2,-2)
4900 3382 ln : $1/3+3\sqrt{3}/4$
4900 3503 $Ei : 25/16$
4900 5991 rt : (2,7,-8,4)
4900 6119 rt : (9,4,2,-3)
4900 6454 $10^x : \sqrt{3}/10$
4900 6680 $t\pi : 10\pi/7$
4900 6743 $\Psi : (\sqrt{2}\pi)^{1/2}$
4900 7675 rt : (5,8,1,-3)
4900 8748 $E : 4/9$
4900 9283 rt : (9,0,6,4)
4900 9890 at : $4\sqrt{2}+3\sqrt{5}$
4901 0022 rt : (6,2,-9,2)
4901 0820 rt : (5,4,5,-4)
4901 2306 id : $9\sqrt[3]{3}/2$
4901 2347 rt : (1,8,4,-4)
4901 2473 sq : $3\sqrt{2}/4+3\sqrt{5}/2$
4901 2907 $\ln(2) \times s\pi(1/4)$
4901 3617 $\ell_{10} : \sqrt{2}+3\sqrt{5}/4$
4901 3895 rt : (2,6,-6,-9)
4901 3925 rt : (5,0,9,5)
4901 5129 at : $3/2+4e$
4901 5433 rt : (1,4,8,-5)
4901 6932 rt : (3,8,-7,-5)
4901 8063 $\ell_2 : 2e/3+3\pi$
4902 1798 cr : $1+4\sqrt{3}/3$
4902 3314 $\sqrt{5}+s\pi(\sqrt{3})$
4902 3813 $e^x : 4/3+e$
4902 4157 $t\pi : \sqrt{7}\pi$
4902 4576 sinh : $3\sqrt[3]{2}/8$
4902 4759 $e^x : (2\pi)^{-1/2}$
4902 6260 $\ell_2 : 1/2+4\sqrt{3}/3$
4902 6895 $J_0 : 20/13$
4902 8045 sr : $\sqrt[3]{3}/6$
4902 8429 $2^x : 4/3+4\sqrt{3}/3$
4902 8692 rt : (8,9,-1,5)
4902 9220 rt : (1,7,3,-2)
4902 9543 cu : $4/3+\pi/3$
4903 0889 rt : (4,9,-4,-4)
4903 1763 sr : $\zeta(3)/5$
4903 1830 tanh : $\ln(\sqrt[3]{5})$
4903 4295 $e^x : 5/4$
4903 4493 $10^x : \ln 2/4$
4903 6549 $s\pi : \sqrt{2}-\sqrt{3}/3$
4903 7938 at : $2\sqrt{3}+4\sqrt{5}$
4903 8030 $e^x : P_{59}$
4903 8186 ln : $1/4+4\pi/3$
4904 0448 $2^x : 3\sqrt{2}/2+3\sqrt{5}/2$
4904 1462 ln : $\sqrt{8/3}$
4904 3715 $e^x : 4\sqrt{2}/3-3\sqrt{3}/2$
4904 4361 $\exp(\sqrt{5})-s\pi(1/3)$
4904 4593 rt : (7,7,1,-3)
4904 5008 id : $(\sqrt{2}\pi/2)^{1/2}$
4904 5472 $10^x : 2-4\sqrt{3}/3$
4904 6692 $e^x : ((e+\pi)/3)^{1/3}$
4904 7107 $e^x : 4-3\pi/2$
4904 7403 rt : (7,3,5,-4)
4904 8510 at : $3+3\pi$
4904 8769 rt : (6,4,-2,8)
4905 0145 rt : (5,9,-9,-1)
4905 0180 rt : (7,1,9,5)
4905 0712 $\ell_{10} : 4-e/3$
4905 1531 rt : (3,3,8,-5)
4905 2873 rt : (1,7,-7,-5)
4905 3925 $\zeta : (2e/3)^{1/2}$
4905 4325 $\zeta : 11/5$
4905 7576 $\ell_{10} : 1+2\pi/3$
4906 0813 at : $4e+\pi/2$
4906 1732 $2^x : 2\sqrt{3}-\sqrt{5}/4$
4906 3616 cr : $3/2+2e/3$
4906 4702 rt : (8,7,-6,2)
4906 5850 id : $10\pi/9$
4906 6878 at : $4\zeta(3)/9$
4906 7135 $\theta_3 : 17/22$
4907 0812 $\Psi : 4\pi$
4907 1063 $2^x : 9e/2$
4907 1198 id : $2\sqrt{5}/3$
4907 1641 $e^x : \sqrt{2}/4+\sqrt{5}/4$
4907 2059 rt : (9,4,-8,9)
4907 2456 $erf : 7/15$
4907 3644 $e \times \ln(2/5)$
4907 5753 $J : \exp(-2\pi/5)$
4907 7343 $10^x : 1/3+4\pi$
4907 8928 rt : (9,6,1,-3)
4908 0168 rt : (5,5,-5,-2)
4908 0286 $s\pi : \exp(-2e/3)$
4908 1448 rt : (9,2,5,-4)
4908 2049 sr : $\sqrt{3}+2\sqrt{5}$
4908 2675 rt : (5,9,7,7)
4908 3286 rt : (6,2,-7,5)
4908 3940 rt : (9,2,9,5)
4908 4962 $2^x : 3e/4+\pi/3$
4908 5154 rt : (5,2,8,-5)
4908 6361 rt : (3,9,-6,-3)
4908 6603 $\ell_{10} : 2-3\sqrt{5}/4$
4908 8267 as : $\sqrt{2}/3$
4908 8866 $10^x : 1/4-\sqrt{5}/4$
4908 9672 $e^x : (5/2)^3$
4909 2922 sq : $(\ln 3/2)^{-1/3}$
4909 3196 $Ei : 3\sqrt{3}$
4909 5258 $10^x : 1/4+4\pi$
4909 5497 $t\pi : e/2-\pi/3$
4909 6229 rt : (5,9,-7,4)
4909 6634 at : 25/2
4909 7477 $J_1 : (e+\pi)/5$

4909 9341 rt : (8,9,1,8)
4910 0203 $\ln(2/3) - \exp(3)$
4910 1092 rt : (4,9,-2,1)
4910 1602 $s\pi$: $3\sqrt[3]{3}/2$
4910 2152 cr : $1/3 + 4\sqrt{5}/3$
4910 8566 J_1 : $7\sqrt[3]{3}/4$
4910 9209 ℓ_{10} : $1/2 + 3\sqrt{3}/2$
4910 9772 rt : (7,8,-9,-8)
4911 0500 ζ : $7\pi/10$
4911 1964 J_1 : $\exp(1/(2\pi))$
4911 2052 rt : (7,9,0,-3)
4911 3529 2^x : $1/3 - e/2$
4911 3785 e^x : $\sqrt{21}\pi$
4911 4308 rt : (7,5,4,-4)
4911 5408 rt : (3,9,3,-2)
4911 6542 rt : (7,1,8,-5)
4911 7549 sinh : $(\sqrt{2}\pi)^{1/3}$
4911 7630 rt : (3,5,7,-5)
4911 8714 rt : (1,9,-6,-5)
4912 1642 e^x : $e/4 + \pi/2$
4912 1807 sq : $9\pi/8$
4912 1971 rt : (5,9,-9,-6)
4912 3845 $3/5 \div \exp(1/5)$
4912 6247 sr : $\sqrt{3} - 2\sqrt{5}/3$
4912 6971 $t\pi$: $7e/8$
4912 8821 $t\pi$: $\sqrt{2}/4 - \sqrt{3}$
4912 9005 rt : (3,9,-4,4)
4912 9913 sinh : $(3\pi)^{-1/3}$
4913 0147 sr : $\sqrt[3]{11}$
4913 0347 id : $\ln(\sqrt{2}\pi)$
4913 0386 rt : (8,7,-4,5)
4913 0450 $s\pi$: $7(e+\pi)/6$
4913 1614 sr : $2e/3 - \pi/2$
4913 4757 sq : $2 - 3\sqrt{3}/4$
4913 8024 rt : (5,9,-5,9)
4913 8619 at : 4π
4914 1806 10^x : $3/4 - \sqrt{2}/4$
4914 1846 rt : (9,8,0,-3)
4914 2936 2^x : $\sqrt{2}/2 - \sqrt{3}$
4914 3751 cu : $7\sqrt{2}/6$
4914 3884 rt : (9,4,4,-4)
4914 4879 rt : (5,8,3,-4)
4914 4961 ℓ_{10} : $(\pi)^3$
4914 5148 Γ : $\sqrt{3} + \sqrt{5} + \sqrt{6}$
4914 5903 rt : (9,0,8,5)
4914 6379 rt : (6,2,-5,8)
4914 6889 rt : (5,4,7,-5)
4914 7000 ζ : $9\sqrt[3]{5}/7$
4914 7869 rt : (1,8,6,-5)
4914 8862 rt : (2,6,-2,5)
4915 0361 at : $4\sqrt{2} + 4\sqrt{3}$
4915 0818 rt : (3,8,-9,-6)
4915 4326 Li_2 : $3\sqrt[3]{3}/10$
4915 6351 at : $1/4 - \pi/4$
4915 8320 Ψ : $\pi^2/2$
4915 9005 10^x : $3\sqrt{2} + \sqrt{3}$
4915 9231 at : $10\sqrt[3]{2}$
4915 9380 rt : (8,5,-9,2)
4915 9795 sinh : 9/19
4916 1026 sq : $4\sqrt{2}/3 - 2\sqrt{3}$
4916 1590 2^x : $\sqrt{23/2}$
4916 1755 rt : (4,9,0,6)
4916 2072 cu : $2\sqrt{2}/3 - \sqrt{3}$
4916 4045 2^x : $2\sqrt[3]{3}/5$
4916 5486 ℓ_{10} : $\sqrt{3}/2 + \sqrt{5}$
4916 5552 id : $\exp(2\pi)$
4916 6965 $2 \times s\pi(\sqrt{3})$
4916 9298 ℓ_2 : $1 + 2e/3$
4917 1223 as : $4 - 2\sqrt{5}$
4917 2495 rt : (7,7,3,-4)
4917 3339 rt : (1,8,-5,-7)
4917 3352 10^x : $\pi^2/9$
4917 4318 rt : (7,3,7,-5)
4917 5208 rt : (3,7,6,-5)
4917 6050 E : 11/25
4917 6495 Ei : $\sqrt{7/2}$
4917 7305 J_0 : $\sqrt{3} + \sqrt{6} - \sqrt{7}$
4917 7814 Ei : 23/6
4917 7886 rt : (1,7,-9,-6)
4918 2252 tanh : 7/13
4918 2469 e^x : 2/5
4918 2900 sq : $3/4 + \sqrt{2}/3$
4918 3743 Ei : $\sqrt{2} + \sqrt{6} - \sqrt{7}$
4918 4578 sq : $5\sqrt[3]{5}/7$
4918 7151 cr : $4 - e/4$
4918 7384 rt : (8,7,-2,8)
4918 8671 rt : (4,7,-9,-4)
4918 8770 e^x : $3\pi^2/8$
4918 9503 e^x : 21/23
4919 0930 sinh : $5\pi/8$
4919 1362 $s\pi$: $9\zeta(3)/5$
4919 2701 $t\pi$: $1/3 + 2e/3$
4919 3338 sq : $2\sqrt{3} + \sqrt{5}$
4919 5221 at : $4 - 2\sqrt{3}$
4919 6704 2^x : γ
4919 7136 $\exp(1/2) - \exp(\pi)$
4919 8423 rt : (9,6,3,-4)
4919 8589 Li_2 : $\sqrt{3}/4$
4919 8919 Ψ : $(\ln 2/3)^{1/3}$
4919 9240 rt : (5,5,-3,1)
4919 9747 θ_3 : $1/\sqrt{17}$
4920 0081 rt : (9,2,7,-5)
4920 0891 rt : (5,6,6,-5)
4920 1938 Ei : $(e/3)^{-2}$
4920 3330 rt : (1,6,9,-6)
4920 3425 2^x : $3\sqrt{2}/2 + \sqrt{3}/3$
4920 4092 $s\pi$: $(\sqrt[3]{5})^{-1/3}$
4920 5423 cu : 15/19
4920 5672 Ψ : 19/9
4920 9165 as : $3\sqrt[3]{2}/8$
4921 0624 2^x : $\sqrt{3}/3$
4921 2628 J_1 : $2\sqrt[3]{2}$
4921 2758 rt : (8,5,-7,5)
4921 3176 ζ : $2\pi/5$
4921 3937 2^x : $4/3 - 3\pi/4$
4921 5349 at : $9\pi^2/7$
4921 6217 ζ : $3(e+\pi)/8$
4921 6727 J_1 : 20/17
4922 2037 rt : (7,8,-7,-5)
4922 2451 $e + s\pi(e)$
4922 2864 ℓ_{10} : $3/4 + 3\pi/4$
4922 3554 rt : (7,9,2,-4)
4922 5061 rt : (7,5,6,-5)
4922 5361 Li_2 : $5\ln 2/8$
4922 5798 rt : (6,0,-8,8)
4922 7287 rt : (3,5,9,-6)
4922 8012 Ei : $3\ln 2/8$
4922 8015 rt : (1,9,-8,-6)
4922 8521 sinh : 25/21
4922 8642 ℓ_{10} : $8e/7$
4922 8766 sr : $9\sqrt{3}/7$
4923 0895 sq : $1/3 - 4\sqrt{3}$
4923 1736 tanh : $7\ln 2/9$
4923 2404 ℓ_2 : $\sqrt{3}/3 + \sqrt{5}$
4923 2645 Ψ : $(\sqrt[3]{5}/2)^{-1/3}$
4923 3752 10^x : $8\sqrt[3]{3}/7$
4923 7251 cr : $3(1/\pi)/8$
4923 8386 rt : (4,7,-7,1)
4923 8611 $\ln(3/5) \times s\pi(\sqrt{2})$
4923 8798 2^x : $5\ln 2/6$
4923 8997 at : $9\sqrt{2}$
4924 0341 at : $\ln(\sqrt[3]{5})$
4924 2355 sinh : $(\sqrt{2}\pi)^{-1/2}$
4924 6355 J_1 : $(\sqrt[3]{3}/2)^{-1/2}$
4924 6465 rt : (9,8,2,-4)
4924 6644 sr : $3/2 + 3\pi/2$
4924 7145 rt : (5,3,-8,-2)
4924 7648 ln : 11/18
4924 7843 rt : (9,4,6,-5)
4924 7906 sr : $1/\sqrt{17}$
4924 8518 rt : (5,8,5,-5)
4924 9554 10^x : 4/23
4924 9880 rt : (5,4,9,-6)
4925 0548 rt : (1,8,8,-6)
4925 4328 $t\pi$: $9\sqrt[3]{3}/7$
4925 4631 $t\pi$: $4(e+\pi)/5$
4925 7594 sr : P_{15}
4925 7866 $\ln(\pi) \times s\pi(\pi)$
4925 7997 sinh : $\exp(-\sqrt{5}/3)$
4925 9695 J_1 : $3\pi/8$
4925 9718 rt : (8,5,-5,8)
4926 4068 id : $1/4 + 3\sqrt{2}$
4926 7179 $t\pi$: $2e/3 + 4\pi$
4926 9290 E : 7/16
4926 9971 rt : (7,7,5,-5)
4927 0533 sr : $7(1/\pi)$
4927 0588 rt : (3,7,-4,-7)
4927 0617 J_1 : $5\sqrt{2}/6$
4927 1137 cu : 8/7
4927 1221 rt : (7,3,9,-6)
4927 1834 rt : (3,7,8,-6)
4927 3885 rt : (1,8,3,-6)
4927 4089 $e - \exp(4/5)$
4927 7747 Ψ : $(3\pi)^{1/3}$
4927 7945 Γ : $1/\sqrt{24}$
4927 9548 sr : $10e\pi/7$
4928 2333 rt : (4,7,-5,6)
4928 4368 $s\pi$: $1/\ln(\sqrt[3]{4})$
4928 4789 e^x : $\zeta(3)/3$
4928 7138 $\pi - \exp(1/2)$
4928 8789 at : $3e\pi/2$
4929 0258 rt : (9,6,5,-5)
4929 0826 rt : (5,5,-1,4)
4929 1408 rt : (9,2,9,-6)
4929 1971 rt : (5,6,8,-6)
4929 2887 at : $1/4 + 4\pi$
4929 4360 J_1 : $\sqrt{19/3}$
4929 5648 E : $1/\sqrt[3]{12}$
4929 6235 tanh : $3\sqrt[3]{2}/7$
4930 1008 rt : (1,8,5,9)
4930 1242 sq : $2/3 + 3\sqrt{5}/4$
4930 6144 $5 \times \ln(3)$
4930 6833 cr : $1/2 + 2\sqrt{2}$
4930 6869 id : $\exp(-1/\sqrt{2})$
4930 8305 ln : $3\sqrt{3} - \sqrt{5}/3$
4930 8931 rt : (7,8,-5,-2)
4930 9993 rt : (7,9,4,-5)
4931 1049 rt : (7,5,8,-6)
4931 1568 rt : (3,9,7,8)
4931 2871 rt : (2,4,-9,5)
4931 4718 $1/5 - \ln(2)$
4931 5457 J_0 : 23/15
4931 6490 sr : $4\sqrt{2} + \sqrt{5}/4$
4931 6891 $e \times \ln(\sqrt{3})$
4932 0209 rt : (8,3,-8,8)
4932 0475 J_1 : $9\sqrt{5}/8$
4932 2638 θ_3 : $\exp(-\sqrt{2})$
4932 2689 e^x : $5\zeta(3)/4$
4932 3545 at : $3e + 3\pi/2$
4932 4241 cr : 3/25
4932 6069 $1/5 \div \ln(2/3)$
4932 6266 Γ : 5/14
4932 7081 at : $2 + 4e$
4932 7121 e^x : $\sqrt{2}/4 + 3\sqrt{5}$
4932 8172 rt : (9,8,4,-5)
4932 8654 rt : (5,3,-6,1)
4932 9147 rt : (9,4,8,-6)
4932 9626 rt : (5,8,7,-6)
4933 1105 $s\pi$: $3/4 + \sqrt{2}$
4933 3976 cu : $4/3 + \pi/2$
4933 5311 2^x : $\sqrt{2}/3 - 2\sqrt{5}/3$
4933 9471 rt : (1,9,-5,-3)
4933 9632 cu : $7\pi^2$
4933 9753 sq : $(\ln 2/2)^{1/3}$
4933 9814 $t\pi$: $1/2 + 3\sqrt{5}/2$
4934 1649 sq : $3\sqrt{2}/2 + \sqrt{5}/2$
4934 2448 Ψ : $10\sqrt[3]{2}$
4934 2874 2^x : $9(e+\pi)/8$
4934 2989 at : $1/3 + 4\pi$
4934 3257 ln : $10\pi^2/3$
4934 3364 $t\pi$: $1/2 + 2e/3$
4934 6780 rt : (7,7,7,-6)
4934 6858 $s\pi$: $4\sqrt{3} + \sqrt{5}$
4934 6872 $1/3 \div s\pi(\sqrt{5})$
4934 6939 as : 9/19
4934 7222 rt : (3,7,-2,-2)
4934 8235 ln : $3\sqrt{2}/4 + \sqrt{3}/3$
4934 9012 $\sqrt{3} - \exp(4/5)$
4935 2407 cr : $\zeta(3)/10$
4935 4896 10^x : $1/4 + 2\sqrt{5}$

4935 6494 $10^x : 4/3 + 2\pi/3$
4935 7623 $J_1 : 13/11$
4935 8937 $t\pi : e + \pi/4$
4935 9599 $Ei : \exp(-1/(2\pi))$
4936 1651 id : $1/\ln((e + \pi)/3)$
4936 1826 $10^x : \sqrt{2}/3 + 3\sqrt{3}/2$
4936 2626 tanh : $3\sqrt[3]{3}/8$
4936 3095 rt : (9,6,7,-6)
4936 3505 rt : (5,5,1,7)
4936 3832 $\ln(3) \div \exp(4/5)$
4936 4740 rt : (1,7,-8,-7)
4936 5040 $\exp(4) \times s\pi(\pi)$
4936 5670 $2^x : 4/3 + \sqrt{2}/3$
4936 7287 id : $3\sqrt{2}/4 + \sqrt{3}/4$
4936 7543 $e^x : \sqrt{7/2}$
4936 9283 $2^x : 3/2 + 2e$
4936 9522 at : $4 + 4\sqrt{5}$
4936 9931 $E : 10/23$
4937 3478 $10^x : 1/\sqrt[3]{16}$
4937 4766 at : $\sqrt{17}\pi$
4937 5896 $2^x : 11/19$
4937 7455 sr : $4/3 + 4e$
4937 8238 rt : (7,8,-3,1)
4937 8447 $J_1 : 4\pi/5$
4937 9010 rt : (7,9,6,-6)
4938 0158 id : $\sqrt[3]{10/3}$
4938 0670 $1/3 \div \ln(4/5)$
4938 1631 $10^x : 13/16$
4938 1975 $\Psi : 1/19$
4938 2151 sinh : $\pi^2/6$
4938 2716 sq : 11/9
4938 3291 at : $4e + 2\pi/3$
4938 3313 $\ell_{10} : 9\sqrt{3}/5$
4938 4837 sq : $4/3 + 4\pi/3$
4938 6829 $t\pi : 5\pi/2$
4938 8084 $\ell_{10} : 2 + \sqrt{5}/2$
4938 9531 cu : $3 + \pi/2$
4939 0038 $t\pi : e\pi/10$
4939 0806 at : $9\sqrt[3]{3}$
4939 1081 $\Gamma : 1/13$
4939 2181 $e^x : 4/3 - 3e/4$
4939 2885 $\Psi : 7e/9$
4939 3792 rt : (9,8,6,-6)
4939 4136 at : 7/13
4939 4146 rt : (5,3,-4,4)
4939 4860 rt : (5,8,9,-7)
4939 5592 $J : 4/23$
4939 5920 rt : (1,7,7,-7)
4940 0248 tanh : $(e/2)^{-2}$
4940 0495 sr : $1/2 + \sqrt{3}$
4940 1257 as : $e/2 - 3\pi/4$
4940 1497 $Ei : \sqrt{15}\pi$
4940 2443 at : 13
4940 3797 $\ell_{10} : 9\ln 2/2$
4940 4712 cr : $\sqrt{2}/4 + 4\sqrt{5}/3$
4940 7559 rt : (7,6,-8,-2)
4940 7984 sq : $10\zeta(3)$
4940 8894 rt : (7,7,9,-7)
4940 9221 rt : (3,7,0,3)
4941 0063 sq : $2(e + \pi)/5$
4941 0519 cu : $3 - e/4$
4941 2919 $Ei : 7\sqrt[3]{5}/3$
4941 3502 at : $3/2 - 3e/4$
4941 5788 cr : $\sqrt{3}/2 - \sqrt{5}/3$
4941 7209 $\ell_{10} : 2\sqrt[3]{3}/9$
4941 7779 $t\pi : 1/3 + \sqrt{5}/2$
4941 9387 cr : $e/3 - \pi/4$
4942 1989 rt : (5,1,-9,1)
4942 2303 rt : (9,6,9,-7)
4942 2471 $\ln(3) + \exp(1/3)$
4942 2805 $Ei : 14/3$
4942 3838 rt : (2,8,-6,-7)
4942 4853 tanh : 13/24
4942 4920 sr : $\sqrt[3]{5}/7$
4942 6145 $J_1 : 3(e + \pi)/7$
4942 6744 $\ell_{10} : 1 - e/4$
4942 7390 $10^x : \sqrt[3]{25}$
4942 7410 $4 \div \ln(\pi)$
4942 7582 $E : 5\ln 2/8$
4942 7709 $erf : 8/17$
4943 0681 $Li_2 : 10/23$
4943 1072 as : $(\sqrt{2}\pi)^{-1/2}$
4943 1249 at : $4e/3 + 3\pi$
4943 3834 $\ell_{10} : 1 + 3\sqrt{2}/2$
4943 4834 rt : (7,8,-1,4)
4943 5086 $E : \sqrt{3}/4$
4943 5414 rt : (7,9,8,-7)
4943 5700 rt : (3,5,-9,-7)
4943 8606 sr : $2e + \pi/4$
4943 9224 sinh : 10/21
4943 9901 cu : $1 - 3\pi/4$
4943 9999 sq : $\sqrt{3}/4 - 4\sqrt{5}/3$
4944 1287 at : $1/2 + 4\pi$
4944 3261 $J_1 : (e/2)^3$
4944 4454 tanh : $(2\pi)^{-1/3}$
4944 4740 at : $7\ln 2/9$
4944 6751 $10^x : 3\sqrt{3}/4 + 3\sqrt{5}/2$
4944 7014 as : $\exp(-\sqrt{5}/3)$
4944 7488 $E : 3\sqrt[3]{3}/10$
4944 7673 rt : (9,8,8,-7)
4944 7940 rt : (5,3,-2,7)
4944 9283 rt : (2,8,-4,2)
4945 2594 $t\pi : 1/\ln(2)$
4945 3829 rt : (1,9,5,6)
4945 7452 $J_1 : \sqrt[3]{5/3}$
4945 8529 $10^x : 4e/5$
4945 9162 rt : (7,6,-6,1)
4945 9696 cu : $\sqrt{2}/2 + \sqrt{5}$
4946 0426 rt : (3,7,2,8)
4946 2288 $\Psi : \ln(\ln 3/3)$
4946 2501 $10^x : \sqrt[3]{13/2}$
4946 9741 $e^x : 3 - 3\sqrt{3}/2$
4947 1155 rt : (5,1,-7,4)
4947 2642 $2^x : 1/4 + \pi$
4947 4352 sinh : $1/\ln(1/(3e))$
4947 9165 $10^x : 9(e + \pi)/4$
4947 9535 $\ell_{10} : 3\sqrt{2} - \sqrt{5}/2$
4948 0326 $\Psi : \sqrt{3}/9$
4948 1246 $\exp(\sqrt{2}) \div \exp(1/2)$
4948 1939 rt : (7,8,1,7)
4948 2605 rt : (3,5,-7,-2)
4948 3001 $t\pi : \sqrt{2} + \sqrt{3}$
4948 3084 $\exp(3/5) \times \exp(\sqrt{2})$
4948 5002 $\ell_{10} : 8/25$
4948 5289 $\zeta : 2\pi^2/9$
4948 7165 id : $2\sqrt{3}/7$
4949 3356 $\Psi : 7\sqrt{2}/2$
4949 3377 id : $9e/7$
4949 4740 $t\pi : 2\sqrt{2}/3 - 3\sqrt{5}/2$
4949 4802 $\Gamma : -\sqrt{2} + \sqrt{5} - \sqrt{7}$
4949 5120 at : $4\pi^2/3$
4950 1191 $t\pi : \sqrt{3}/4 - \sqrt{5}/3$
4950 1492 $erf : \sqrt{2}/3$
4950 2453 rt : (7,6,-4,4)
4950 3310 $t\pi : (\ln 2/2)^{1/2}$
4950 6191 $J_1 : 19/16$
4950 6236 sq : $3\sqrt{2}/4 + 3\sqrt{5}/4$
4950 6306 cu : $2/3 - 4\sqrt{3}$
4950 7625 $10^x : 5e/4$
4950 8875 $t\pi : 1/\ln(5)$
4950 9047 $\Psi : 7\zeta(3)/8$
4950 9379 id : $1/3 - 2\sqrt{2}$
4950 9587 at : $9(e + \pi)/4$
4951 0512 id : $5\ln 2/7$
4951 0706 at : $3\sqrt[3]{2}/7$
4951 0842 at : $3\sqrt{2} + 4\sqrt{5}$
4951 1649 rt : (9,9,-8,-2)
4951 2022 cr : $2e - 2\pi/3$
4951 2581 rt : (5,1,-5,7)
4951 3135 $Ei : 9\sqrt{3}$
4951 3827 $e + \ln(4/5)$
4951 4426 cu : $3/2 - 3\sqrt{3}$
4951 4736 $10^x : 3/2 + \sqrt{3}/3$
4951 4760 $\Gamma : 1/\sqrt[3]{22}$
4951 5992 cu : $4/3 + \sqrt{2}/2$
4951 9534 cr : $5e\pi$
4952 0637 $\Psi : 4/17$
4952 1412 rt : (7,4,-9,1)
4952 1629 ln : $7\sqrt{3}$
4952 1993 ln : $3 - e/2$
4952 2289 rt : (3,5,-5,3)
4952 2750 cu : $\sqrt{2}/3 - \sqrt{5}$
4952 3620 $\ln(2/3) \times \exp(1/5)$
4952 4528 $Ei : (2e/3)^{1/3}$
4953 0863 cr : $3\sqrt{3}/2 + \sqrt{5}/3$
4953 3611 $e^x : 4/3 + \sqrt{5}$
4953 4878 id : $(\sqrt{5})^{1/2}$
4953 5013 at : $4e + 3\pi/4$
4953 5599 id : $1/4 - \sqrt{5}/3$
4953 7124 at : $2/3 + 4\pi$
4953 8010 $e^x : 10e$
4953 8699 cu : $9\sqrt{2}/5$
4953 9300 rt : (7,6,-2,7)
4953 9736 $\Psi : 9(1/\pi)/7$
4953 9843 $s\pi : 3e/7$
4954 2000 $t\pi : 1/2 + 3\sqrt{2}/2$
4954 4444 at : $e/4 + 4\pi$
4954 4846 $J_1 : (2\pi)^{1/2}$
4954 7225 rt : (9,9,-6,1)
4954 7996 ln : P_{31}
4954 8245 $2^x : 2e/3 + 2\pi$
4954 9337 $2^x : (\ln 3/2)^{-3}$
4954 9437 $10^x : 22/7$
4954 9625 sr : $3e/4 + 4\pi/3$
4955 0599 $J_1 : (\sqrt{2})^{1/2}$
4955 4292 $t\pi : 1/2 + \sqrt{2}/4$
4955 4752 $\exp(5) \div s\pi(1/5)$
4955 5601 rt : (7,4,-7,4)
4955 6303 rt : (3,5,-3,8)
4956 1303 $\ell_{10} : \sqrt{5}/7$
4956 3196 $s\pi : 5\sqrt{3}/4$
4956 3662 rt : (5,1,-8,-7)
4956 3733 $\ell_2 : \pi^2/7$
4956 6247 at : $\sqrt{3}/3 - \sqrt{5}/2$
4956 7588 $10^x : 3\sqrt{3}/4 - \sqrt{5}/4$
4956 9268 $\theta_3 : \sqrt[3]{5}/7$
4957 0273 $J : 3\zeta(3)/10$
4957 5070 tanh : $e/5$
4957 7948 rt : (9,9,-4,4)
4957 8616 at : $3\sqrt[3]{3}/8$
4958 0752 $2^x : 9\ln 2$
4958 1423 sinh : $3(1/\pi)/2$
4958 1516 sr : $2/3 + \pi/2$
4958 3482 $1/4 + s\pi(\sqrt{3})$
4958 3544 $J_1 : 25/21$
4958 4147 at : $3/4 + 4\pi$
4958 5214 rt : (7,4,-5,7)
4958 5945 $E : (2e)^{-1/2}$
4958 5966 $2^x : 2 + e/3$
4959 0783 cr : $4\sqrt{2} - 4\sqrt{3}/3$
4959 1026 at : $\sqrt{18}\pi$
4959 1681 rt : (9,7,-9,1)
4959 4334 $s\pi : 9\sqrt[3]{2}/4$
4959 5247 $\Psi : (1/\pi)/7$
4959 6669 sq : $\sqrt{3}/2 + 2\sqrt{5}$
4959 7217 $E : 3/7$
4959 7749 $erf : 3\sqrt[3]{2}/8$
4959 9279 $Ei : (2e)^{-3}$
4960 0000 cu : $9\sqrt[3]{3}/5$
4960 0063 cr : $3/4 + 3\sqrt{3}/2$
4960 2702 $2^x : 9\zeta(3)/2$
4960 4751 rt : (9,9,-2,7)
4960 5736 rt : (1,8,9,-5)
4960 6370 cr : $4(e + \pi)/7$
4960 6798 ln : $1 + 2\sqrt{3}$
4960 8978 tanh : $1/\ln(2\pi)$
4960 9547 $\ell_{10} : 4 - \sqrt{3}/2$
4961 4413 $10^x : 3/2 + 2\sqrt{3}$
4961 4862 $\Psi : \sqrt[3]{19/2}$
4961 6608 cu : 19/24
4961 6820 rt : (9,7,-7,4)
4961 7103 at : $(e/2)^{-2}$
4961 8839 sq : $1/4 + 2\pi/3$
4962 2798 rt : (7,2,-8,7)
4962 2895 id : $7\sqrt[3]{5}/8$
4962 3228 rt : (3,3,-8,8)
4962 3727 $\Psi : \zeta(3)/4$
4962 5623 $t\pi : 3\sqrt{2}/2 + \sqrt{3}$
4962 9161 rt : (1,5,-2,4)
4963 1736 as : 10/21

4963 2602 tπ : $7\ln 2/10$
4963 5953 cu : $4\pi/3$
4963 6257 E : $\sqrt[3]{5}/4$
4963 9032 rt : (9,7,-5,7)
4963 9829 at : $6\sqrt{5}$
4964 1399 2^x : $2/3-3\sqrt{5}/4$
4964 1875 J_0 : $\sqrt{7/3}$
4964 2275 at : 13/24
4964 3688 ln : 14/23
4964 4451 at : $4+3\pi$
4964 6986 θ_3 : $2e/9$
4964 7160 sr : $4/3+e/3$
4965 3342 sq : $3/2-2e$
4965 4821 Ψ : 9/22
4966 0576 tπ : 5/16
4966 1477 2^x : $2e+\pi/3$
4966 1510 $\ln(4/5)\times\exp(4/5)$
4966 2329 at : $(2\pi)^{-1/3}$
4966 4732 Ei : $e\pi/10$
4966 6165 $\ln(2)\div\exp(1/3)$
4966 7548 as : $1/\ln(1/(3e))$
4966 7837 rt : (9,5,-8,7)
4966 8383 10^x : $e/5$
4967 0290 sinh : 11/23
4967 0672 at : $e/3+4\pi$
4967 5365 2^x : $3\sqrt{2}/4+\sqrt{5}/3$
4967 5779 ln : $e+3\pi$
4967 9045 Ei : $3\sqrt[3]{5}/10$
4968 1506 id : $(1/(3e))^{1/3}$
4968 1597 cr : $3+\sqrt{2}/4$
4968 3066 erf : $(3\pi)^{-1/3}$
4968 3890 sr : $4\sqrt{2}+\sqrt{3}/3$
4968 5728 at : $(\sqrt[3]{2}/3)^{-3}$
4968 6837 $\ln(2/5)+\exp(5)$
4968 6904 $s\pi(\pi)\div s\pi(1/3)$
4968 6918 cr : $3+4\pi$
4968 6929 at : $3e/2+3\pi$
4968 7907 $e-\exp(1/5)$
4968 9758 cr : $3\sqrt{5}/2$
4969 0079 sr : $3/4+2\sqrt{5}/3$
4969 0399 id : $2\sqrt{5}/9$
4969 3417 J : $\exp(-11\pi/15)$
4969 4344 Li_2 : $1/\sqrt[3]{12}$
4969 4506 $\exp(3/4)\times s\pi(1/4)$
4969 6044 as : $3\sqrt{2}/2-3\sqrt{3}/2$
4969 7501 sq : $3\sqrt{3}/4+\sqrt{5}$
4970 1398 ln : $4-3\pi/4$
4970 1488 Ψ : $(e/3)^{-1/2}$
4970 4018 $\ln(\sqrt{3})+\exp(2/3)$
4970 6398 10^x : $4\sqrt{5}/5$
4970 6794 $4/5\div\ln(5)$
4970 7248 erf : 9/19
4970 7479 ζ : $\ln(\pi)$
4970 9410 J_1 : 5/2
4970 9774 ζ : $\sqrt[3]{21/2}$
4971 0115 ln : $(\sqrt{2}\pi)^{1/3}$
4971 0574 tanh : 6/11
4971 0968 cu : $8\ln 2/7$
4971 4987 ℓ_{10} : π
4971 5745 E : $1/\sqrt[3]{13}$
4971 5819 E : $\exp(-\sqrt[3]{5}/2)$
4971 6164 $\exp(\pi)+\exp(\sqrt{5})$
4972 0434 rt : (1,6,-4,-5)
4972 0880 cr : $1+3\pi/4$
4972 1204 sr : $4+\sqrt{5}$
4972 1448 J_1 : $(\sqrt[3]{5})^{1/3}$
4972 1771 at : $1+4\pi$
4972 2321 $2/3\times s\pi(\sqrt{3})$
4972 2324 2^x : $2\sqrt{3}-2\sqrt{5}$
4972 3359 rt : (3,8,-9,-1)
4972 6293 rt : (5,8,-8,8)
4972 6449 tπ : $1/3+3\sqrt{5}/2$
4972 7327 10^x : $\ln(3\pi/2)$
4972 7489 $1/3\times\exp(2/5)$
4972 7991 sr : $2\sqrt{2}/3+3\sqrt{3}/4$
4972 8469 id : $3\sqrt{2}-\sqrt{5}/3$
4973 1112 sq : $2\sqrt{3}/3+\sqrt{5}$
4973 2256 Ei : $\sqrt{3}-\sqrt{5}+\sqrt{7}$
4973 2413 rt : (1,6,-2,8)
4973 2464 ℓ_{10} : 7/22
4973 5100 rt : (3,8,-7,6)
4973 5277 at : $5e$
4973 5454 Γ : $(3\pi/2)^{-1/3}$
4973 9200 e^x : $\sqrt{2}/4+2\sqrt{3}$
4974 0455 2^x : $2-e/4$
4974 1702 Ei : $e\pi/7$
4974 2922 sr : $\sqrt{3}/7$
4974 4464 e^x : $2\sqrt{2}/3+3\sqrt{3}/2$
4974 6597 sr : $9e/2$
4974 7254 E : $4(1/\pi)/3$
4974 7559 ln : $3\sqrt{2}-3\sqrt{3}/2$
4974 7728 tπ : $\exp(\sqrt[3]{2}/3)$
4974 9338 sπ : $(\ln 3/2)^3$
4974 9965 ℓ_2 : 17/24
4975 1394 cr : $\exp(-2\pi/3)$
4975 1541 J_1 : $7\sqrt[3]{5}/10$
4975 2346 ζ : $8(1/\pi)/5$
4975 2602 E : $3\sqrt{2}/10$
4975 5695 10^x : $e/2+2\pi$
4975 5817 sq : $4/3-3e/4$
4975 6844 sr : $7(1/\pi)/9$
4975 9426 at : $4\sqrt{3}+3\sqrt{5}$
4975 9526 cu : $1+\sqrt{3}/2$
4976 0909 $\exp(3/5)+s\pi(\sqrt{5})$
4976 4681 cr : $2+e/2$
4976 6383 sr : $9\ln 2$
4976 7426 2^x : $3\sqrt{2}/2+\sqrt{5}$
4976 8408 tπ : $4\sqrt{2}+3\sqrt{3}$
4976 9261 ℓ_{10} : $4/3+2e/3$
4976 9687 e^x : $4e\pi/9$
4977 0013 sπ : $(e/2)^{1/2}$
4977 0030 ln : $\pi^2/6$
4977 0256 ln : $4+\sqrt{2}/3$
4977 0938 e^x : $10e/9$
4977 2789 ln : $4+3e$
4977 3929 at : $8e\pi/5$
4977 4014 erf : $(\sqrt{2}\pi)^{-1/2}$
4977 4185 Ei : $\exp(\sqrt{3}/2)$
4977 4319 cr : $8\sqrt[3]{2}/3$
4977 4651 sinh : $3(e+\pi)/2$
4977 5167 $\exp(3)-s\pi(1/5)$
4977 6324 ℓ_2 : $2/3+\sqrt{5}/3$
4977 6714 as : $3(1/\pi)/2$
4977 7908 sr : $\ln(3\pi)$
4977 8714 id : $7\pi/4$
4977 8968 at : $7(e+\pi)/3$
4977 9228 sπ : $3\sqrt{3}/2+\sqrt{5}$
4977 9521 ℓ_{10} : $\sqrt{2}+\sqrt{3}$
4977 9604 ℓ_2 : $3e+\pi$
4978 2056 ζ : $\sqrt[3]{3/2}$
4978 2567 at : $8\sqrt[3]{5}$
4978 4499 Ψ : 1/20
4978 6613 ln : $2\sqrt{5}$
4978 6656 erf : $\exp(-\sqrt{5}/3)$
4978 7660 Li_2 : 7/16
4978 9242 cr : $4e+3\pi/2$
4978 9534 e^x : $2\sqrt{2}/7$
4979 0042 at : $\sqrt{19}\pi$
4979 0681 rt : (1,7,-4,-2)
4979 2283 $\sqrt{2}+\ln(2/5)$
4979 4581 Ψ : $5\sqrt[3]{2}/6$
4979 5310 e^x : $\exp(-e/3)$
4979 5826 e^x : $2\sqrt{2}-\sqrt{3}/3$
4979 5981 at : $e/5$
4979 9691 sr : $5\pi/7$
4980 4745 10^x : $8\sqrt{5}$
4980 4953 cr : $9\sqrt{3}$
4980 6167 $\pi+\exp(\sqrt{5})$
4981 5950 ζ : $4\zeta(3)/5$
4981 7179 Ei : $\exp(-2/3)$
4981 8956 e^x : $-\sqrt{2}-\sqrt{3}+\sqrt{6}$
4981 9339 Ψ : $3\sqrt{2}/2$
4982 3011 at : $2e/3-3\pi/4$
4982 4264 tπ : $\exp(-1/(2\pi))$
4982 4836 cr : $-\sqrt{3}+\sqrt{6}+\sqrt{7}$
4982 6989 sq : 12/17
4982 7631 cu : $9(e+\pi)/5$
4982 7863 ℓ_{10} : $-\sqrt{3}+\sqrt{5}+\sqrt{7}$
4982 7996 $\sqrt{3}\times\ln(3/4)$
4982 8334 ℓ_{10} : $5\sqrt[3]{2}/2$
4982 8905 J_1 : 6/5
4983 0166 10^x : $4/3+3\sqrt{5}/2$
4983 0683 at : $1/\ln(2\pi)$
4983 0707 2^x : 7/12
4983 0862 Ei : $\sqrt{23/3}$
4983 0949 cu : $3/4+3e/2$
4983 3224 sπ : P_1
4983 4578 J : $2(1/\pi)/9$
4983 4714 sinh : $\sqrt{23}$
4983 7073 rt : (1,8,-4,2)
4983 8001 E : $(4/3)^{-3}$
4984 2057 2^x : $\ln(\ln 3/3)$
4984 2830 Ei : $7\sqrt{5}/10$
4984 3240 2^x : $-\sqrt{2}-\sqrt{5}+\sqrt{7}$
4984 4959 sr : $\sqrt{5}/9$
4984 5119 sr : $3/2+\sqrt{5}/3$
4984 7583 cu : $3/2-\sqrt{2}/2$
4984 8981 ln : $4/3+\pi$
4984 9155 J_1 : $\sqrt{2}+\sqrt{5}-\sqrt{6}$
4985 2770 sr : $2+3\sqrt{2}$
4985 2865 J_1 : $(\sqrt[3]{3})^{1/2}$
4985 5001 at : $7\pi^2/5$
4985 5401 cr : $7\sqrt[3]{3}/3$
4985 7258 sπ : $4\sqrt{2}-2\sqrt{5}/3$
4985 7604 J_1 : $(\ln 2)^{-1/2}$
4985 8305 cu : $3\sqrt{2}-\sqrt{3}/2$
4986 0261 cu : $9(e+\pi)/8$
4986 1228 $2/5+\ln(3)$
4986 1939 ln : $2-\sqrt{2}/4$
4986 3476 $\exp(e)\div s\pi(1/3)$
4986 4550 sinh : 12/25
4986 5190 10^x : $1/2+\sqrt{5}/2$
4986 7267 E : 8/19
4986 7334 as : 11/23
4986 9416 rt : (1,9,-4,7)
4986 9602 $\exp(3)+\exp(5)$
4987 5244 at : $8\sqrt{3}$
4987 5498 ln : $\sqrt{15}\pi$
4987 6812 J_1 : $5\sqrt[3]{3}/6$
4987 8436 e^x : $(\ln 3/2)^{-2}$
4987 8565 J_1 : $\exp(1/(2e))$
4988 0740 tπ : $3\sqrt{3}/2-\sqrt{5}/3$
4988 1462 J_1 : $\zeta(3)$
4988 2111 ℓ_2 : $\exp(\sqrt{3})$
4988 2637 E : $\exp(-\sqrt{3}/2)$
4988 2901 Γ : $1/\ln(1/2)$
4988 3665 sinh : $\sqrt{17}\pi$
4988 3897 at : $3+4e$
4988 8118 $\sqrt{5}\div\exp(2/5)$
4989 4209 10^x : 7/8
4989 5813 ℓ_{10} : $2+2\sqrt{3}/3$
4989 6414 $\sqrt{5}\times\ln(4/5)$
4989 7609 at : $4/3+4\pi$
4990 0258 10^x : $3\sqrt{2}/4$
4990 5565 E : $\sqrt[3]{2}/3$
4990 7707 e^x : $3e-2\pi$
4991 0873 at : $e/2+4\pi$
4991 5176 sinh : $(\ln 2)^2$
4991 8728 e^x : $\beta(2)$
4992 0185 Ψ : $(\sqrt{3}/2)^{-1/3}$
4992 2040 2^x : $((e+\pi)/2)^{-1/2}$
4992 6797 cr : $3\sqrt{2}/4+4\sqrt{3}/3$
4992 7728 10^x : P_{14}
4992 8740 ℓ_{10} : $3\sqrt{3}/2+\sqrt{5}/4$
4992 9382 cu : $1/2-4e/3$
4993 2943 erf : 10/21
4993 4672 at : 6/11
4993 4948 as : $4/3-2e/3$
4993 8951 cu : $9\sqrt{3}/2$
4994 0058 sinh : $(\ln 2/3)^{1/2}$
4994 2818 Γ : 1/12
4994 3773 Ei : $\sqrt[3]{5}/2$
4994 8355 sinh : $\sqrt[3]{3}/3$
4994 8886 at : 14
4994 9303 ℓ_2 : $9\pi/5$
4995 0261 sinh : $3\sqrt[3]{5}$
4995 2456 10^x : $7\sqrt{3}/3$
4995 3437 Li_2 : $(e/3)^{1/3}$
4995 3774 $\ln(3)-\exp(4)$
4995 4473 $\ln(4)+\exp(\sqrt{2})$

4995 6205 cr : $4/3+3e/4$
4995 6350 at : $4e+\pi$
4995 6396 J_0 : $\exp(\sqrt[3]{2}/3)$
4995 6505 sinh : $2\zeta(3)/5$
4995 7402 e^x : $\sqrt[3]{13}$
4995 8005 e^x : $9(e+\pi)/8$
4995 9450 $\exp(\sqrt{5})\times s\pi(1/5)$
4996 1260 erf : $1/\ln(1/(3e))$
4996 1475 $t\pi$: $e-\pi/2$
4996 1577 10^x : $2\sqrt{3}/3+2\sqrt{5}$
4996 2551 $s\pi$: $(\sqrt[3]{4})^{1/3}$
4996 3556 cu : $19/14$
4996 5529 Ei : $\sqrt{19/2}$
4996 9387 e^x : $\exp(\sqrt[3]{5}/2)$
4996 9693 10^x : $5(1/\pi)/4$
4997 2946 e^x : $(\pi/2)^{-2}$
4997 3548 ln : $8\sqrt[3]{3}/7$
4997 3990 at : $\sqrt{20}\pi$
4997 5048 $\sqrt{2}+\exp(3)$
4998 2418 at : $3/2+4\pi$
4998 3474 2^x : $1/\sqrt[3]{5}$
4998 3880 J_1 : $(\sqrt[3]{5}/3)^{-1/3}$
4998 6235 e^x : $3+4\pi/3$
4999 2367 sinh : $5\sqrt{3}$
4999 2896 ℓ_{10} : $1/3+2\sqrt{2}$
4999 3173 $t\pi$: $4\sqrt{2}+2\sqrt{5}/3$
4999 4084 ln : $3/2+4\sqrt{5}/3$
4999 9006 at : $10\pi^2/7$
4999 9999 $s\pi$: $7/6$
5000 0000 id : $1/2$
5000 2253 e^x : $1/\sqrt[3]{15}$
5000 3086 cu : $3/4+\pi/2$
5000 4431 ln : $7\ln 2/8$
5000 5439 2^x : $2\sqrt{3}+\sqrt{5}/3$
5000 5840 J_1 : $(\ln 3)^2$
5000 6149 cu : $\ln(\pi)$
5000 7103 cr : $8(e+\pi)/3$
5000 7416 ζ : $\exp(-3\pi)$
5000 9306 e^x : $25/6$
5001 2225 sr : $e/4+\pi/2$
5001 2788 e^x : $9\sqrt{5}/2$
5001 3405 $\ln(\pi)\times\exp(4)$
5001 6909 e^x : $1/4-2\sqrt{2}/3$
5001 7253 sq : $2/3-3\sqrt{5}$
5001 7841 at : $9\pi/2$
5001 9410 sq : $3/4-3\sqrt{5}$
5001 9463 at : $3\sqrt{3}+4\sqrt{5}$
5001 9702 cr : $4\sqrt{2}/3+2\sqrt{5}/3$
5002 0314 at : $10\sqrt{2}$
5002 2242 E : $5/12$
5002 3926 cr : $3\sqrt{2}-\sqrt{3}/2$
5003 0616 10^x : $1/\ln(2\pi)$
5003 3657 e^x : $2/3-e/2$
5003 5616 J_1 : $4e/9$
5003 5890 sr : $2\sqrt{2}-\sqrt{3}/3$
5003 6266 $\ln(\sqrt{3})+s\pi(2/5)$
5003 9622 Ψ : $17/8$
5003 9940 Ei : $(\ln 3/2)^{-1/3}$
5004 3568 e^x : $1/\ln(\sqrt{2}\pi/2)$
5004 4732 ln : $4\sqrt{2}/3+3\sqrt{3}/2$
5004 7494 erf : $3(1/\pi)/2$
5004 8363 cu : $3/2+3e/2$
5004 9549 E : $3\ln 2/5$
5005 1305 id : $\sqrt{2}+\sqrt{3}-\sqrt{7}$
5005 2021 tanh : $11/20$
5005 5856 $\exp(4/5)\times\exp(\pi)$
5006 0235 ℓ_{10} : $6/19$
5006 5179 at : $5e\pi/3$
5006 5349 J_0 : $8\sqrt[3]{5}/9$
5006 5471 as : $12/25$
5006 8796 Ψ : $(\ln 3)^{1/2}$
5006 8907 at : $1/2-\pi/3$
5007 2427 ln : $3\sqrt{2}/7$
5007 3356 id : $e/2+\pi$
5007 6091 tanh : $1/\sqrt[3]{6}$
5007 6150 2^x : $\sqrt{3}+\sqrt{5}-\sqrt{7}$
5008 2122 Ψ : $11/18$
5008 3583 $\ln(\sqrt{2})+\exp(e)$
5008 3673 E : $1/\sqrt[3]{14}$
5008 5431 Ψ : $9\pi^2/7$
5008 5690 2^x : $2-\sqrt{2}$
5008 8119 10^x : $3\sqrt{2}/2-\sqrt{5}/4$
5008 8277 J_1 : $10\sqrt{5}/9$
5008 8337 cr : P_{53}
5009 2709 θ_3 : $(\ln 3/3)^{1/2}$
5009 3930 Ei : $1/\sqrt{15}$
5009 4001 e^x : $11/12$
5009 4501 10^x : $4-\sqrt{2}/3$
5009 6237 J_1 : $23/19$
5009 9033 10^x : $(\sqrt[3]{2}/3)^2$
5010 2029 as : P_{21}
5010 2219 2^x : $e/2-3\pi/4$
5010 3456 sq : $4/3-2\pi$
5010 6603 2^x : $(e+\pi)/10$
5010 9910 ζ : $(3\pi)^{-3}$
5011 0885 Ei : $\ln(\sqrt{5}/2)$
5011 2530 E : $(\sqrt{5}/3)^3$
5011 4217 e^x : $1/\ln(\sqrt{5}/2)$
5011 6771 Li_2 : $11/25$
5011 7117 as : $(\ln 2)^2$
5011 8191 cr : $2e/3+\pi/2$
5011 8980 erf : $11/23$
5011 9890 $\ln(\pi)+\exp(\sqrt{5})$
5012 0758 2^x : $(\sqrt{2}\pi/3)^3$
5012 1757 sq : $1/2+3\pi$
5012 3176 at : $2\sqrt{2}/3-2\sqrt{5}/3$
5012 4127 ℓ_{10} : $7e/6$
5012 5654 at : $4/3-\pi/4$
5012 7027 ℓ_{10} : $3\sqrt{2}/4-\sqrt{5}/3$
5013 0645 e^x : $4e/3$
5013 1072 10^x : $3/17$
5013 1932 cr : $\sqrt[3]{2}/10$
5013 4370 $t\pi$: P_{12}
5013 4882 2^x : $\sqrt{3}/3+\sqrt{5}/3$
5013 5913 ln : $\sqrt[3]{9/2}$
5013 6013 at : $2e/3+4\pi$
5013 6214 sr : $3\sqrt{2}/4+3\sqrt{3}$
5013 7942 J_1 : $6\sqrt{2}/7$
5014 0482 ln : $10\pi/7$
5014 1001 J_0 : $\sqrt{3}+\sqrt{5}-\sqrt{6}$
5014 2502 as : $(\ln 2/3)^{1/2}$
5014 4150 Ei : $5\sqrt[3]{5}/7$
5014 4290 J_1 : $7\sqrt{3}/10$
5014 4680 at : $\sqrt{21}\pi$
5014 5083 θ_3 : $5\sqrt[3]{3}/8$
5014 5109 at : $\sqrt{2}-\sqrt{3}/2$
5014 6452 cr : $7\sqrt{5}$
5014 6636 sr : $\sqrt{3}/3+3\sqrt{5}/4$
5014 6866 E : $(e+\pi)^{-1/2}$
5014 7240 cr : $2/3+e$
5014 8008 Γ : $2(1/\pi)/7$
5014 9944 Ψ : $1/21$
5015 0489 sq : $3e$
5015 0967 as : $\sqrt[3]{3}/3$
5015 1720 $t\pi$: $7\ln 2$
5015 4639 cu : $4\sqrt{5}/3$
5015 4720 sr : $3-\sqrt{5}/3$
5015 7099 at : $10\sqrt[3]{3}$
5015 7140 cr : $3/2+4\sqrt{2}/3$
5015 7721 2^x : $1/\ln(\ln 3/3)$
5015 8660 J_1 : $7\ln 2/4$
5015 9282 as : $2\zeta(3)/5$
5015 9751 J_0 : $\sqrt[3]{7/2}$
5016 1968 $s\pi$: P_{60}
5016 3882 10^x : $6\sqrt[3]{2}/5$
5016 4317 id : $3\sqrt{2}/4-\sqrt{5}/4$
5016 4445 cu : $1+\sqrt{5}/2$
5016 4523 at : $3e+2\pi$
5016 4652 sq : $4e/3+3\pi/2$
5016 8266 2^x : $4\sqrt{2}/3-3\sqrt{3}/4$
5016 9467 ℓ_2 : $2\sqrt{3}/3+3\sqrt{5}/4$
5016 9790 ζ : $(1/(3e))^3$
5017 1598 ℓ_2 : $\sqrt{13}\pi$
5017 1665 ℓ_{10} : $\sqrt[3]{2}/4$
5017 1957 ζ : $\exp(-2\pi)$
5017 3027 10^x : $(\pi/2)^3$
5017 3611 sq : $17/24$
5017 5227 at : P_{12}
5017 6783 $s\pi$: $(\sqrt[3]{3})^{-1/2}$
5017 6974 2^x : $3-3\sqrt{5}/4$
5017 8824 Ei : $15/7$
5018 0351 10^x : $7\sqrt{5}/5$
5018 2002 $s\pi$: $\sqrt{15}\pi$
5018 2603 id : $2\sqrt{2}/3+\sqrt{5}/4$
5018 2609 E : $2\sqrt[3]{3}/7$
5018 3982 sq : $1/4-3\sqrt{2}/2$
5018 4131 cr : $2\sqrt{2}+\sqrt{5}/4$
5018 4419 rt : (1,9,4,-2)
5018 6112 J : $\exp(-13\pi/9)$
5018 6562 Γ : $\ln(2e/3)$
5018 7523 Ei : $(\pi/2)^{-3}$
5018 9366 sinh : $(\sqrt[3]{5})^{1/3}$
5019 0727 J_1 : $17/14$
5019 2636 e^x : $3e/4-\pi/4$
5019 2962 e^x : $(\pi/2)^{1/2}$
5019 3235 E : $7/17$
5019 5649 e^x : $4+4e/3$
5019 6537 ln : $4/3+4e$
5019 8028 sr : $\sqrt[3]{2}/5$
5019 8930 $t\pi$: $\sqrt{2}/2-\sqrt{5}/4$
5020 1126 ln : $3+2\sqrt{5}/3$
5020 1568 10^x : $9\sqrt[3]{3}/5$
5020 2418 ℓ_{10} : $3/2+3\sqrt{5}/4$
5020 2856 $\exp(\sqrt{3})-\exp(e)$
5020 5097 2^x : $5(e+\pi)$
5020 5324 10^x : $(e/2)^{-3}$
5020 8495 cu : $1/\ln(5/3)$
5021 0441 at : $3/4-3\sqrt{3}/4$
5021 1717 $s\pi$: $(\ln 2)^{1/2}$
5021 2614 λ : $1/23$
5021 4764 Ψ : $\pi/3$
5021 5371 id : $4\sqrt{2}-2\sqrt{3}/3$
5021 8053 e^x : $3\sqrt{3}/2+\sqrt{5}/4$
5022 0070 id : $3e/2+3\pi$
5022 1887 $t\pi$: $(\sqrt[3]{4}/3)^3$
5022 4613 $s\pi$: P_{44}
5022 5061 cu : $5(e+\pi)/9$
5022 5259 at : $2+4\pi$
5022 5359 2^x : $1/4+2e$
5022 6550 ζ : $24/11$
5022 7372 ℓ_{10} : $2\sqrt{2}/3+\sqrt{5}$
5022 8223 ln : $(\ln 3/3)^{1/2}$
5022 9851 10^x : $1-3\sqrt{3}/4$
5022 9882 e^x : $10\sqrt{2}/3$
5023 0043 sq : $2/3+\sqrt{5}/4$
5023 0647 $\exp(2)+\exp(\sqrt{2})$
5023 1034 at : $\ln(\sqrt{3})$
5023 1418 cu : $\sqrt{3}+\sqrt{5}-\sqrt{6}$
5023 3244 cr : $2\sqrt{3}/3+\sqrt{5}$
5023 4465 ln : $\sqrt{2}/4+3\sqrt{3}/4$
5023 5306 2^x : $4-3\sqrt{3}/4$
5023 6812 rt : (1,8,4,2)
5023 6928 10^x : $\sqrt{2}/8$
5023 7348 10^x : $8\sqrt[3]{5}$
5023 7548 θ_3 : $\sqrt{3}/7$
5023 9100 sr : $\sqrt[3]{23/2}$
5024 1524 at : $4\sqrt{2}+4\sqrt{5}$
5024 2961 sinh : $\exp((e+\pi)/3)$
5024 3370 at : $3e/4+4\pi$
5024 4065 ln : $1/4+3\sqrt{2}$
5024 4228 10^x : $9\sqrt[3]{3}/7$
5024 5032 Ψ : $1/\sqrt{6}$
5024 5415 cr : $1/4+\pi$
5025 0034 ℓ_2 : $3/17$
5025 2091 Ei : $7e/8$
5025 2422 Ψ : $3/10$
5025 5406 $t\pi$: $3\sqrt{2}/4-2\sqrt{5}$
5025 7112 id : $5\zeta(3)/4$
5025 7313 J_1 : $(\sqrt{2}\pi/3)^{1/2}$
5025 8422 10^x : $5(1/\pi)/9$
5026 2862 ζ : $\exp(-(e+\pi))$
5026 3388 10^x : $e\pi/6$
5026 4120 at : $5(e+\pi)/2$
5026 6937 θ_3 : $7(1/\pi)/9$
5026 7424 e^x : $21/2$
5026 9751 ℓ_{10} : $2\sqrt{2}/9$
5027 2698 ℓ_2 : $\sqrt{2}/3+3\sqrt{3}$
5027 4003 $3/4\div\exp(2/5)$
5027 4183 2^x : $(3\pi/2)^{1/2}$
5027 4967 erf : $12/25$

5027 9861 $E : 9(1/\pi)/7$
5028 1264 $2^x : 3\sqrt{2} - 2\sqrt{3}/3$
5028 2081 rt : (3,9,-3,8)
5028 2376 $J_1 : \sqrt{2} + \sqrt{6} - \sqrt{7}$
5028 2441 $\Gamma : 6\ln 2/7$
5028 3583 $\text{t}\pi : 4\zeta(3)/7$
5028 3905 sq : $2/3 + \pi$
5028 3957 sinh : $8\sqrt[3]{3}/7$
5028 4010 $J_1 : (e/3)^{-2}$
5028 4321 at : 11/20
5028 5012 $\ell_{10} : \pi/10$
5028 5524 $E : 9/22$
5028 6588 $\text{t}\pi : 1/3 + \sqrt{2}/4$
5028 6927 $10^x : \exp(-\sqrt{3})$
5028 7101 $\ln(2/5) \div \exp(3/5)$
5029 0072 sq : $1/4 + \pi$
5029 1829 rt : (5,9,-6,7)
5029 1995 $10^x : 4\sqrt{2}/3 + 3\sqrt{3}/2$
5029 5393 rt : (3,9,-5,1)
5029 6628 $10^x : 1/4 + e/4$
5030 1736 $e^x : 9e\pi/7$
5030 2498 sq : $3\sqrt{3} - \sqrt{5}/4$
5030 3632 at : $\sqrt{22}\pi$
5030 6092 rt : (5,9,-8,2)
5030 6255 $\ell_{10} : 4\sqrt{2}/3 + 3\sqrt{3}/4$
5030 7134 sinh : $\sqrt[3]{17}$
5030 7959 cr : $2(1/\pi)/5$
5030 8978 at : $1/\sqrt[3]{6}$
5030 9436 $\text{t}\pi : 3\sqrt{3} + 2\sqrt{5}/3$
5031 0021 rt : (3,9,-7,-6)
5031 1182 sq : $3e - 2\pi$
5031 2957 $J_1 : (2e/3)^{1/3}$
5031 3374 $\text{t}\pi : (\pi/3)^3$
5031 3561 rt : (9,4,-9,8)
5031 3786 $2^x : 4/3 - \sqrt{5}/3$
5031 3820 $J_0 : 7\sqrt{3}/8$
5031 4154 rt : (1,7,4,5)
5031 4465 $E : 1/\sqrt{6}$
5031 5556 $erf : (\ln 2)^2$
5031 7455 cu : $2/3 - 2e/3$
5032 0465 $J_1 : 7\sqrt{2}/4$
5032 2936 $2^x : 1/2 - 2\sqrt{5}/3$
5032 5833 $\ell_2 : 9\sqrt[3]{2}$
5032 6112 $J_1 : \exp(e/3)$
5032 9437 $\ln(4) + \exp(3/4)$
5033 0763 rt : (1,7,2,-8)
5033 2412 $J_1 : e\pi/7$
5033 2412 $\ln(2/5) \times \ln(\sqrt{3})$
5033 2787 $J_1 : 10\sqrt{3}/7$
5033 3129 $\exp(4/5) \times s\pi(\sqrt{5})$
5033 5132 at : $3\pi^2/2$
5033 5495 $erf : (\ln 2/3)^{1/2}$
5033 7791 cr : $5e/4$
5033 8714 $2^x : 3\sqrt{2} + \sqrt{3}/2$
5033 9055 rt : (9,6,-6,8)
5034 0665 $2^x : 10/17$
5034 2143 $erf : \sqrt[3]{3}/3$
5034 2859 $\text{t}\pi : \sqrt{2}/3 + 3\sqrt{5}/4$
5034 8671 $erf : 2\zeta(3)/5$
5035 3527 rt : (7,1,-9,8)
5035 8435 rt : (9,6,-8,5)
5035 9089 $J_1 : (\ln 3/2)^{-1/3}$
5035 9118 $2^x : 2\sqrt{2}/3 + \sqrt{3}/2$
5036 0893 at : $2e + 3\pi$
5036 2004 ln : $1/2 + 2\sqrt{3}/3$
5036 3535 cu : $\sqrt{2}/3 + 4\sqrt{3}/3$
5036 6132 $s\pi : \exp(-1/(2e))$
5036 6201 at : $4 + 4e$
5036 7999 id : $e + \pi/4$
5036 8399 $J_1 : 5\sqrt[3]{5}/7$
5036 9057 rt : (9,8,-3,8)
5036 9342 $Ei : 11/9$
5036 9394 $10^x : 3\sqrt{2} + \sqrt{5}$
5036 9459 cr : 17/5
5037 0830 $s\pi : (2e/3)^{-3}$
5037 0952 as : $2\sqrt{3} - 4\sqrt{5}/3$
5037 2101 $\zeta : (2\pi)^{-3}$
5037 3401 cu : $4\sqrt{2} - \sqrt{5}/2$
5037 5363 $2^x : 4\sqrt{2} - 3\sqrt{3}/4$
5037 7839 $\zeta : \ln(5/3)$
5038 0418 $s\pi(\sqrt{3}) \times s\pi(\sqrt{5})$
5038 1246 $\ln(5) - \exp(\sqrt{2})$
5038 2995 as : $4e/3 - \pi$
5038 5927 $\ln(2/3) \div \ln(\sqrt{5})$
5038 6277 rt : (7,3,-6,8)
5038 6312 $Ei : 3\zeta(3)/7$
5038 8498 $J_1 : 11/9$
5038 9621 $2^x : (\ln 2/2)^{1/2}$
5038 9850 tanh : $4\ln 2/5$
5039 2125 rt : (9,8,-5,5)
5039 3022 cu : $5(1/\pi)/2$
5039 5456 rt : (1,9,-5,-2)
5039 6662 sinh : $\sqrt{2}\pi$
5039 6841 id : $2\sqrt[3]{2}/5$
5039 8183 at : $7e\pi/4$
5039 8198 $\text{t}\pi : 1/3 - 3\pi/2$
5039 8952 $10^x : \exp(1/(3e))$
5039 9138 $\ln(\pi) - \exp(1/2)$
5040 0037 $\Psi : 9\sqrt{2}$
5040 0576 sinh : $7\sqrt[3]{5}/10$
5040 1452 tanh : $(e + \pi)^{-1/3}$
5040 1505 sinh : $(e)^{1/2}$
5040 1716 id : $\sqrt{3} - \sqrt{5}$
5040 2054 $\zeta : \exp(-2e)$
5040 3330 sq : $2\sqrt{3} - \sqrt{5}/2$
5040 3847 sinh : 18/7
5040 5194 rt : (5,2,-9,-8)
5040 7598 $\ln(2/5) - s\pi(1/5)$
5040 7739 ln : 2/9
5040 7949 $e^x : \sqrt{3}/4 - \sqrt{5}/2$
5040 8180 $\Gamma : 1/\sqrt{15}$
5040 9060 $E : 1/\sqrt[3]{15}$
5040 9571 $E : \ln(3/2)$
5041 1625 rt : (7,3,-8,5)
5041 2183 rt : (3,4,-6,7)
5041 3414 $\ln(2) \times \exp(e)$
5041 5709 $E : (\pi/2)^{-2}$
5041 8058 $e^x : 1/\sqrt{6}$
5041 8254 rt : (9,8,-7,2)
5042 0016 $s\pi : 4e\pi/5$
5042 2816 at : 15
5042 2936 $s\pi : 6\ln 2/5$
5042 4039 ln : $e/2 + \pi$
5042 4123 sr : $8\sqrt{2}/5$
5042 5706 rt : (7,5,-3,8)
5042 9744 $s\pi : 2\sqrt{3}/3 + 3\sqrt{5}/4$
5042 9851 $\ell_{10} : 2/3 - \sqrt{2}/4$
5043 0253 $\text{t}\pi : 4\sqrt{2}/3 + 3\sqrt{3}/2$
5043 2377 $2^x : \exp(-\sqrt[3]{4}/3)$
5043 4094 rt : (1,6,4,7)
5043 4448 $\theta_3 : (\sqrt{5}/2)^{-1/2}$
5043 6678 $\Psi : \sqrt{2} - \sqrt{3} + \sqrt{6}$
5043 8696 $\ell_{10} : 10\sqrt{5}/7$
5044 1202 rt : (3,4,-8,2)
5044 1689 cr : 5π
5044 3158 as : $3\sqrt{2}/4 - \sqrt{3}/3$
5044 3179 $10^x : 3/4 - \pi/3$
5044 6178 ln : $3/4 + e/3$
5044 6629 sinh : $7\ln 2/10$
5044 8095 rt : (9,8,-9,-1)
5044 8815 rt : (5,0,-6,8)
5045 0101 at : $4e + 4\pi/3$
5045 0873 $J_1 : \sqrt{3/2}$
5045 2133 at : $\sqrt{23}\pi$
5045 3199 $\theta_3 : \sqrt{5}/9$
5045 3309 $2^x : (\sqrt[3]{5}/3)^{-1/2}$
5045 4582 cr : $2\sqrt{2} + \sqrt{3}/3$
5045 5463 $\zeta : (e/3)^{-1/3}$
5045 5588 ln : $4\sqrt{2} - 2\sqrt{3}/3$
5045 5961 $E : \exp(-e/3)$
5045 6071 cu : $\sqrt{10}\pi$
5045 6684 rt : (7,5,-5,5)
5045 7269 $E : 2\sqrt{2}/7$
5045 7548 $\Gamma : 8/3$
5045 9180 $\zeta : (e + \pi)^{-3}$
5045 9564 at : $4/3 - 4\sqrt{2}/3$
5046 1203 $\text{t}\pi : 3e - 4\pi$
5046 2263 sr : $1 - \sqrt{5}/3$
5046 2650 sr : $4(1/\pi)/5$
5046 6064 $s\pi : 2e/3 + 3\pi/4$
5046 6435 rt : (1,6,2,-4)
5046 7239 tanh : 5/9
5046 9924 $\Gamma : 8\sqrt{5}$
5047 0067 cu : $2\sqrt{2} + \sqrt{3}/3$
5047 1583 $e^x : 1/3 + 2e$
5047 4089 rt : (7,7,0,8)
5047 4333 $Ei : 7$
5047 4601 rt : (3,5,-9,-8)
5047 6032 as : P_{32}
5047 9064 $Li_2 : (\ln 3)^{-1/3}$
5048 1478 $\theta_3 : 3\sqrt{2}/7$
5048 3211 $\Psi : 1/22$
5048 3392 rt : (5,0,-8,5)
5048 3478 $J : \exp(-10\pi/13)$
5049 0382 at : $\exp(e)$
5049 2077 ln : $1/4 + \sqrt{2}/4$
5049 2500 rt : (7,5,-7,2)
5049 3456 rt : (3,6,-1,7)
5049 4484 rt : (1,8,-5,-6)
5049 4992 $e^x : \sqrt{2}/3 - 2\sqrt{3}/3$
5049 8259 $J_1 : \pi^2/4$
5049 8708 $s\pi : 5\pi^2/8$
5050 2748 rt : (9,9,9,-8)
5050 2952 rt : (5,2,-3,8)
5050 3519 cr : $\sqrt{3} + 3\sqrt{5}/4$
5050 6971 sr : $5e/6$
5050 7599 $2^x : 9e\pi/5$
5050 7866 id : P_{35}
5051 1249 $s\pi : P_{25}$
5051 2803 rt : (7,7,-2,5)
5051 4169 $\ell_2 : 4 + 3\sqrt{5}/4$
5051 4997 $\ell_{10} : 5/16$
5051 5779 $10^x : 25/21$
5051 7549 $J_0 : 6\sqrt[3]{2}/5$
5052 4222 $J_1 : 1/\ln(2/3)$
5052 6382 $J_1 : \sqrt[3]{15}$
5052 9401 ln : $2\sqrt{2} + 3\sqrt{5}/4$
5053 1429 $e^x : 3/2 + 2\sqrt{2}/3$
5053 1966 sr : $4e + 3\pi$
5053 3408 $e^x : 1/\ln(\ln 2/3)$
5053 4375 rt : (7,5,-9,-1)
5053 4669 $\text{t}\pi : 9\sqrt{3}$
5053 4861 rt : (7,9,3,8)
5053 5596 rt : (3,6,-3,2)
5053 6291 $Ei : 6/7$
5053 6457 rt : (2,8,-3,7)
5053 6702 cu : $2\sqrt{2}/3 + 4\sqrt{5}$
5053 7147 $\ell_2 : 4e/3 - \pi/4$
5053 8774 $\text{t}\pi : 2e + 3\pi/2$
5054 2817 $\exp(1/4) + \exp(1/5)$
5054 4528 as : P_{42}
5054 4857 $e^x : 9/22$
5054 6433 at : $e + 4\pi$
5054 6490 rt : (9,9,7,-7)
5054 6753 rt : (5,2,-5,5)
5054 7237 $e^x : 10\sqrt[3]{5}/7$
5054 7365 $\theta_3 : 3\zeta(3)/4$
5054 7562 $\ell_{10} : 3e\pi/8$
5054 7810 id : $2\sqrt{2} + 3\sqrt{5}/4$
5055 4097 $\exp(1/2) - \exp(e)$
5055 8359 rt : (7,7,-4,2)
5055 8887 sinh : $\exp(-\sqrt[3]{3}/2)$
5055 9007 cu : $4\sqrt{3}/3 + \sqrt{5}/3$
5055 9112 $10^x : 5\sqrt[3]{3}/4$
5056 0455 rt : (2,8,-5,-2)
5056 4092 $Ei : (\sqrt[3]{4})^{-1/3}$
5056 6577 at : $3\sqrt{3}/4 - \sqrt{5}/3$
5056 9647 $e^x : 9(1/\pi)/7$
5057 0626 $10^x : 4/3 - e/4$
5057 1160 $E : \zeta(3)/3$
5057 1625 rt : (9,7,8,-7)
5057 1633 cu : $3\sqrt{2} - 3\sqrt{3}/4$
5057 1840 $\text{t}\pi : 1/2 + 2\sqrt{2}/3$
5057 1926 rt : (5,4,0,8)
5057 2825 rt : (1,8,-9,-8)
5057 3355 $Ei : 1/9$
5057 4212 cu : $3\sqrt{2}/2 - 3\sqrt{5}$
5057 5800 $\zeta : (2e)^{-3}$
5057 7620 $10^x : (2\pi)^{-1/2}$
5057 7964 tanh : $7(1/\pi)/4$

5057 8706 cr : $2+\sqrt{2}$
5058 1770 10^x : $\sqrt{10}\pi$
5058 1986 cr : $\sqrt{3}/4+4\sqrt{5}/3$
5058 3045 e^x : $2\sqrt{3}/3-\sqrt{5}/3$
5058 3250 at : $9e\pi/5$
5058 3378 10^x : $7\sqrt[3]{2}/6$
5058 4609 rt : (7,9,1,5)
5058 5246 rt : (7,8,8,-7)
5058 5570 rt : (3,6,-5,-3)
5058 7485 Γ : 1/11
5058 8970 2^x : $1/\ln(2e)$
5058 9375 ℓ_{10} : $8\zeta(3)/3$
5059 0937 at : $9\sqrt[3]{5}$
5059 1279 at : $\sqrt{24}\pi$
5059 1517 ζ : $5\zeta(3)/7$
5059 1754 $t\pi$: $(\sqrt{2}/3)^{1/2}$
5059 2425 ℓ_{10} : $e/2-\pi/3$
5059 4086 sr : $9\sqrt[3]{2}/5$
5059 4161 E : 2/5
5059 7117 sr : $4-\sqrt{3}$
5059 7667 e^x : $1/2+4\sqrt{5}/3$
5059 8516 rt : (9,9,5,-6)
5059 8861 rt : (5,2,-7,2)
5059 8880 J_1 : 16/13
5059 9199 rt : (9,5,9,-7)
5059 9349 $\sqrt{3}+s\pi(e)$
5059 9546 rt : (5,9,8,-7)
5060 3159 ln : $2\sqrt{3}/3+3\sqrt{5}/2$
5060 3401 cr : $2e\pi/5$
5060 5611 $\exp(2)+\exp(3/4)$
5060 5706 cu : $3\sqrt{2}-\sqrt{5}/2$
5060 7503 rt : (4,5,-9,9)
5061 0197 2^x : $(\sqrt{2}/3)^{-3}$
5061 1766 Ψ : 5
5061 2733 rt : (7,7,-6,-1)
5061 4202 rt : (7,6,9,-7)
5061 4575 rt : (3,8,4,7)
5061 5380 J_0 : $5e/9$
5061 6314 rt : (1,7,-3,6)
5061 7824 $t\pi$: $1/4+2e$
5061 9555 2^x : 13/22
5062 3761 ζ : $\exp(-5)$
5062 5769 Ei : 15/8
5062 5957 2^x : $3/4-\sqrt{3}$
5062 6323 cr : $\sqrt{2}/2-\sqrt{3}/3$
5062 8796 rt : (9,7,6,-6)
5062 9000 10^x : P_{59}
5062 9196 rt : (5,4,-2,5)
5062 9556 E : $(2\pi)^{-1/2}$
5062 9991 rt : (5,7,9,-7)
5063 0511 at : $4\ln 2/5$
5063 1195 rt : (1,5,4,8)
5063 1945 Ψ : $e\pi/4$
5063 7226 rt : (1,7,-5,-9)
5063 8432 sq : $2/3+3e/2$
5064 2404 at : $(e+\pi)^{-1/3}$
5064 4486 rt : (7,9,-1,2)
5064 5338 rt : (7,8,6,-6)
5064 5614 sr : $2\sqrt{2}-\sqrt{5}/4$
5064 5771 rt : (3,6,-7,-8)
5064 6631 rt : (3,8,9,-7)
5064 6776 Γ : $e\pi/2$
5064 7958 2^x : $2-4\sqrt{5}/3$
5065 0666 e^x : $4\sqrt{2}-3\sqrt{5}/4$
5065 0808 $5\div s\pi(1/5)$
5065 1113 E : $(e/2)^{-3}$
5065 1721 θ_3 : $7\ln 2/8$
5065 2677 2^x : $3\sqrt{2}+4\sqrt{5}/3$
5065 3705 10^x : $3e-3\pi/4$
5065 5426 ζ : $(e/3)^{-3}$
5065 8983 e^x : $2-\sqrt{5}/3$
5065 9534 as : $7\ln 2/10$
5066 1403 rt : (9,9,3,-5)
5066 1772 $3/4\times s\pi(\sqrt{5})$
5066 1863 Γ : $\sqrt[3]{19}$
5066 1869 rt : (5,2,-9,-1)
5066 2324 rt : (9,5,7,-6)
5066 2641 id : $3e/2-\pi/2$
5066 2793 rt : (5,9,6,-6)
5066 2827 id : $(2\pi)^{1/2}$
5066 2828 θ_3 : $(e)^{-1/2}$
5066 3019 cr : $2\sqrt[3]{5}$
5066 3812 cu : $1/4-\pi/3$
5066 4194 rt : (1,9,9,-7)
5066 4343 at : $3+4\pi$
5066 4662 rt : (2,4,-8,9)
5066 4752 E : $5(1/\pi)/4$
5066 5041 2^x : $3\sqrt{2}/2+3\sqrt{3}$
5066 5395 J : $\exp(-16\pi/9)$
5066 8988 cr : $3\sqrt{2}/2+3\sqrt{3}/4$
5066 9707 2^x : $3\sqrt{2}/4+4\sqrt{5}$
5067 0411 J_1 : $\pi^2/8$
5067 2202 at : $4e+3\pi/2$
5067 3407 at : $9\sqrt{3}$
5067 5270 cr : $4\sqrt{2}-\sqrt{5}$
5067 5875 Γ : $(\pi/2)^{-3}$
5067 8735 rt : (7,7,-8,-4)
5068 0729 rt : (7,6,7,-6)
5068 1239 rt : (3,8,2,2)
5068 1369 10^x : $9\zeta(3)/8$
5068 2043 cu : $3/2-\sqrt{2}/4$
5068 3465 sr : $\exp(-e/2)$
5068 4614 2^x : π^2
5068 5215 J : $\exp(-16\pi/19)$
5068 8349 at : $(5/2)^3$
5068 8892 at : $8(e+\pi)/3$
5069 0071 rt : (8,3,-9,7)
5069 1386 rt : (4,7,-4,9)
5069 4444 sq : 19/12
5069 8571 rt : (9,7,4,-5)
5069 9121 rt : (5,4,-4,2)
5069 9250 E : $1/\sqrt[3]{16}$
5069 9538 at : $7\sqrt{5}$
5069 9657 rt : (9,3,8,-6)
5070 0210 rt : (5,7,7,-6)
5070 1141 ζ : $4e/5$
5070 1210 2^x : $(\ln 3)^3$
5070 1252 e^x : $2/3+3e$
5070 1864 rt : (1,5,2,-1)
5070 2625 cr : $4-\sqrt{3}/3$
5070 3500 Li_2 : 4/9
5070 7013 Ei : $(e/2)^{-1/2}$
5070 7030 $t\pi$: $10\sqrt[3]{5}/7$
5070 9164 J_1 : 21/17
5070 9850 at : 5/9
5071 0678 $1/5-s\pi(1/4)$
5071 1652 J : $\pi/10$
5071 2718 ℓ_{10} : $4-\pi/4$
5071 3553 e^x : $4\sqrt{3}/3+4\sqrt{5}/3$
5071 4296 ln : P_2
5071 4382 cr : 3/23
5071 4399 θ_3 : $(5/3)^{-1/2}$
5071 6896 $\ln(4/5)-\exp(1/4)$
5071 7900 rt : (7,9,-3,-1)
5071 9077 rt : (7,8,4,-5)
5072 0261 rt : (7,4,8,-6)
5072 0865 rt : (3,8,7,-6)
5072 2014 at : 5π
5072 3773 sinh : $7\sqrt{2}/6$
5072 4669 E : $4\ln 2/7$
5072 4771 tanh : $\sqrt{5}/4$
5072 8347 e^x : $8\sqrt[3]{2}/5$
5073 0124 sq : $e/3+4\pi$
5073 1489 J_1 : $6\sqrt[3]{3}/7$
5073 1668 rt : (4,7,-6,4)
5073 2056 ℓ_2 : P_{14}
5073 5026 J_1 : $(\sqrt[3]{4}/3)^{-1/3}$
5073 5276 10^x : $\sqrt{6}$
5073 6548 cu : $3+4\pi/3$
5073 7320 Γ : $-\sqrt{3}+\sqrt{6}-\sqrt{7}$
5073 8912 rt : (9,9,1,-4)
5074 0083 erf : $7\ln 2/10$
5074 0195 rt : (9,5,5,-5)
5074 0849 rt : (5,9,4,-5)
5074 1487 rt : (9,1,9,-6)
5074 1622 ζ : $(5)^{-3}$
5074 2146 rt : (5,5,8,-6)
5074 2807 rt : (1,9,7,-6)
5074 3451 $\exp(2/3)\times s\pi(e)$
5074 8717 cu : $3+3e/2$
5074 9805 sq : $4-3\pi/2$
5075 0378 $\ln(\pi)-\exp(\sqrt{3})$
5075 2249 rt : (8,5,-6,7)
5075 4883 J_1 : $5\sqrt{3}/7$
5075 5503 at : $8\pi^2/5$
5075 5672 sr : 25/11
5075 5710 ℓ_2 : $1/2+4e$
5075 5785 Ei : $\exp(-e/2)$
5075 5912 sr : $2\sqrt{3}/3+\sqrt{5}/2$
5075 6210 $\ln(5)-\exp(3/4)$
5075 9658 sq : $4\sqrt{2}/3-3\sqrt{3}/2$
5076 0853 $s\pi$: $(3\pi)^3$
5076 1123 J_1 : $7\sqrt{2}/8$
5076 2407 ℓ_{10} : $1/2+e$
5076 3295 rt : (7,6,5,-5)
5076 4013 rt : (3,8,0,-3)
5076 4713 rt : (7,2,9,-6)
5076 5436 rt : (3,6,8,-6)
5076 7850 2^x : $\sqrt[3]{7/3}$
5076 7968 2^x : $4\sqrt{2}+\sqrt{5}/2$
5077 1019 2^x : $2\sqrt{2}-\sqrt{5}$
5077 1442 ℓ_{10} : $4/3+4\sqrt{2}/3$
5077 2827 cr : $4\sqrt{3}/3+\sqrt{5}/2$
5077 4168 as : $\exp(-\sqrt[3]{3}/2)$
5077 4520 2^x : $\exp(\sqrt{5})$
5077 6905 rt : (4,7,-8,-1)
5077 7128 cr : $4/3+2\pi/3$
5078 2012 θ_3 : 1/4
5078 2638 Ψ : $5\sqrt[3]{5}/4$
5078 5398 ℓ_2 : $\sqrt[3]{23}$
5078 5590 rt : (9,7,2,-4)
5078 6371 rt : (5,4,-6,-1)
5078 7132 rt : (9,3,6,-5)
5078 7734 at : $4\sqrt{3}+4\sqrt{5}$
5078 7919 rt : (5,8,6,8)
5078 8314 rt : (6,0,-9,7)
5078 8370 Ψ : 1/23
5078 9479 rt : (5,3,9,-6)
5078 9488 cr : 24/7
5079 0277 rt : (1,7,8,-6)
5079 4908 sr : $(\pi/2)^{-3}$
5079 5139 tanh : $4\sqrt[3]{2}/9$
5079 6675 10^x : $7\sqrt[3]{3}/5$
5079 7743 tanh : 14/25
5080 0037 rt : (8,5,-8,4)
5080 2086 rt : (4,9,1,9)
5080 2777 2^x : $3/4+3\sqrt{2}/4$
5080 6304 sr : P_{28}
5080 6661 sq : $2\sqrt{3}-\sqrt{5}$
5080 7494 rt : (3,8,-8,3)
5080 9968 rt : (7,9,-5,-4)
5081 1081 2^x : $4\sqrt{2}/3-\sqrt{5}/4$
5081 1321 e^x : $1-3\sqrt{5}/4$
5081 1655 rt : (7,8,2,-4)
5081 2187 ℓ_2 : $3/2+4\pi/3$
5081 3274 sr : $1/\sqrt{15}$
5081 3357 rt : (7,4,6,-5)
5081 4227 rt : (3,8,5,-5)
5081 5950 rt : (3,4,9,-6)
5081 6390 rt : (2,6,-1,9)
5081 6831 rt : (1,8,-8,-6)
5081 7530 $2-\exp(2/5)$
5081 7697 rt : (1,6,-3,2)
5082 2653 cu : $3\pi^2$
5082 2834 ℓ_{10} : $\sqrt{3}+2\sqrt{5}/3$
5082 3388 at : $7(1/\pi)/4$
5082 3419 sinh : $2\sqrt[3]{5}/7$
5082 5392 id : $\sqrt{2}/4+2\sqrt{3}/3$
5082 6711 rt : (8,7,-3,7)
5082 8426 10^x : $\ln(\sqrt{5}/3)$
5082 9391 erf : $\exp(-\sqrt[3]{3}/2)$
5083 3403 rt : (5,8,-9,6)
5083 3678 ζ : $\exp(-3\pi/2)$
5083 4733 at : $3/2-2\sqrt{2}/3$
5083 5203 sq : $5e\pi/9$
5083 6309 E : $\pi/8$
5083 6346 2^x : $4/3-4\sqrt{3}/3$
5083 6568 E : $\ln(\sqrt{2}\pi/3)$
5083 6741 rt : (9,9,-1,-3)
5083 7751 at : 16

5083 8601 rt : (9,5,3,-4)
5083 9552 rt : (5,9,2,-4)
5084 0477 rt : (9,1,7,-5)
5084 1437 rt : (5,5,6,-5)
5084 2401 rt : (1,9,5,-5)
5084 4312 rt : (1,5,9,-6)
5084 5290 rt : (3,9,-8,-6)
5084 6161 e^x : $3\sqrt{3}+\sqrt{5}/3$
5084 6823 cu : $\ln(\sqrt{2}\pi/2)$
5084 9132 sr : $2\sqrt{2}+2\sqrt{3}$
5085 4569 cr : $3+\sqrt{3}/4$
5085 6745 rt : (4,9,-1,4)
5085 8050 ℓ_2 : $2-\sqrt{3}/3$
5086 0077 2^x : $6(e+\pi)/7$
5086 0876 Li_2 : $7(1/\pi)/5$
5086 0962 tπ : $\sqrt{3}-\sqrt{5}-\sqrt{7}$
5086 2181 cu : $8\ln 2$
5086 2480 tπ : $\sqrt[3]{18}$
5086 4886 10^x : $1+3\sqrt{5}$
5086 8428 rt : (7,6,3,-4)
5086 8895 sr : $4/3+2\sqrt{2}/3$
5086 9486 rt : (3,8,-2,-8)
5087 0513 rt : (7,2,7,-5)
5087 0531 rt : (6,2,-6,7)
5087 1581 rt : (3,6,6,-5)
5087 3175 rt : (2,6,-3,2)
5087 4464 tπ : $5\sqrt[3]{2}/2$
5087 4781 rt : (1,9,6,-8)
5087 4911 ζ : $(\ln 3)^{-3}$
5087 7454 J_1 : $5\sqrt{5}/9$
5087 8038 Li_2 : $1/\ln(3\pi)$
5088 0250 id : $2\sqrt{3}/3+3\sqrt{5}/2$
5088 0388 10^x : $4/3-2\sqrt{3}/3$
5088 1992 E : 9/23
5088 4160 cr : $8\pi^2/5$
5088 4730 rt : (8,7,-5,4)
5088 6280 $\exp(1/3)+\exp(\sqrt{2})$
5088 6727 tπ : $\sqrt[3]{19/3}$
5088 7109 tπ : $\sqrt{2}/4-\sqrt{3}/2$
5088 7300 J_1 : $1/\ln(\sqrt{5})$
5088 7312 e^x : $3\sqrt{2}/2-\sqrt{3}/2$
5088 7387 ζ : $(3\pi/2)^{-3}$
5088 7714 10^x : $10(1/\pi)/3$
5089 0725 $\sqrt{3}+\ln(4/5)$
5089 7046 rt : (9,7,0,-3)
5089 7648 tπ : $8\sqrt{2}$
5089 8169 ln : $1/3+4\pi/3$
5089 8209 rt : (5,4,-8,-4)
5089 9337 rt : (9,3,4,-4)
5089 9535 10^x : $3e/2$
5090 0351 tanh : $\exp(-\sqrt{3}/3)$
5090 0512 rt : (5,8,4,5)
5090 1652 rt : (9,1,8,5)
5090 2839 rt : (5,3,7,-5)
5090 3696 $s\pi(1/3)\times s\pi(1/5)$
5090 4033 rt : (1,7,6,-5)
5090 7114 rt : (1,9,5,7)
5090 7609 rt : (3,7,-9,-6)
5090 8483 $\ln(3/4)-\exp(1/5)$
5091 1101 at : $4e/3+4\pi$
5091 1175 cr : P_{49}
5091 1300 ln : $\zeta(3)/2$
5091 2142 rt : (1,9,-4,8)
5091 6989 J_1 : $\sqrt{6}$
5091 7485 rt : (8,9,0,7)
5091 7617 Γ : $10\ln 2$
5091 8968 sr : $\sqrt{2}+\sqrt{5}+\sqrt{7}$
5091 8993 tπ : $\sqrt{3}+\sqrt{5}/2$
5091 9340 rt : (4,9,-3,-1)
5092 1974 id : $e/4-4\pi/3$
5092 2866 ℓ_2 : $e\pi/3$
5092 8707 rt : (7,9,-7,-7)
5092 8801 id : $2-2\sqrt{5}/3$
5092 9038 e^x : 23/25
5092 9581 id : $8(1/\pi)/5$
5093 1245 rt : (7,8,0,-3)
5093 2768 2^x : $3/2-e/3$
5093 3810 rt : (7,4,4,-4)
5093 3885 sπ : $4\sqrt{2}/3+4\sqrt{5}$
5093 4556 cr : P_{64}
5093 5128 rt : (6,2,-8,4)
5093 5267 ln : $1/4+\sqrt{2}$
5093 6404 rt : (7,0,8,5)
5093 6946 Li_2 : $7\ln 2/5$
5093 7154 Ψ : 14/23
5093 7736 rt : (3,4,7,-5)
5093 8405 rt : (2,6,-5,-5)
5093 9076 rt : (1,8,-6,-5)
5093 9394 ℓ_{10} : $1/4+4\sqrt{5}/3$
5094 0152 Ei : $5\zeta(3)/7$
5094 1359 2^x : $3/4+\sqrt{3}/3$
5094 2420 sr : $\sqrt{2}-2\sqrt{3}/3$
5094 3092 rt : (5,8,-9,-6)
5094 5461 id : $\sqrt{2}+\sqrt{6}+\sqrt{7}$
5094 5769 $\exp(1/5)\times\exp(e)$
5094 6135 sinh : 6/5
5094 7817 ℓ_{10} : $3/2+\sqrt{3}$
5094 7922 e^x : 7/17
5094 9988 sq : $3\sqrt{2}/4+2\sqrt{5}/3$
5095 1415 rt : (8,7,-7,1)
5095 2544 tπ : 3/20
5095 5796 2^x : $6\ln 2/7$
5095 5830 tanh : $\ln(\sqrt[3]{5}/3)$
5095 5871 ζ : $(3\pi/2)^{1/2}$
5095 5974 at : $6e$
5095 6068 id : $7(e+\pi)/2$
5095 6493 ln : $3\sqrt{2}/4+2\sqrt{3}$
5095 6634 $\exp(1/4)+\exp(4/5)$
5095 8879 ℓ_2 : $(\ln 2/2)^{1/3}$
5095 9290 rt : (9,5,-7,9)
5095 9952 E : $1/\sqrt[3]{17}$
5096 0212 sq : $1/3-\pi/3$
5096 0673 E : 7/18
5096 2989 cu : $2\sqrt{3}/3-2\sqrt{5}$
5096 3477 cu : $4/3+e$
5096 3931 rt : (9,9,-3,-2)
5096 6766 rt : (9,5,1,-3)
5096 6799 2^x : 13/2
5096 8224 rt : (5,9,0,-3)
5096 8409 tπ : $5\sqrt[3]{5}/3$
5096 9486 cu : $4\sqrt{2}-4\sqrt{3}/3$
5096 9635 rt : (9,1,5,-4)
5097 1110 rt : (5,5,4,-4)
5097 1223 cr : $4-\sqrt{5}/4$
5097 1851 rt : (6,4,-3,7)
5097 2538 rt : (9,3,9,5)
5097 2594 rt : (1,9,3,-4)
5097 3967 at : $\sqrt{5}/4$
5097 4031 rt : (5,1,8,-5)
5097 5533 rt : (1,5,7,-5)
5097 7044 rt : (3,9,-6,-5)
5097 8930 E : $e/7$
5097 9633 id : $2\sqrt{2}/3-\sqrt{3}/4$
5098 1577 rt : (7,9,-9,-6)
5098 2997 tanh : 9/16
5098 3349 sr : $3\ln 2/8$
5098 3572 as : $2/3-2\sqrt{3}/3$
5098 3744 tπ : $3e/4-4\pi/3$
5098 6243 rt : (2,7,-8,5)
5098 6920 J_1 : $9e/10$
5098 8042 ζ : $\sqrt{7/3}$
5098 9396 rt : (8,9,-2,4)
5099 0144 sr : $5\sqrt[3]{2}$
5099 1722 rt : (4,9,-5,-6)
5099 2645 Γ : $(\ln 2/3)^3$
5099 2770 cr : $3e-3\pi/2$
5099 4210 e^x : $2\sqrt[3]{3}/7$
5099 5626 sr : $4\sqrt[3]{5}/3$
5099 8570 2^x : $\ln(2e/3)$
5099 8775 ℓ_2 : $1/2+3\sqrt{3}$
5099 8919 $\ln(4)\div\exp(1)$
5100 0386 10^x : $(\ln 3/2)^{1/2}$
5100 1763 ℓ_{10} : $1+\sqrt{5}$
5100 1944 id : $\sqrt{2}/3-4\sqrt{5}/3$
5100 4601 sr : $\sqrt{2}+\sqrt{3}/2$
5100 6699 rt : (7,6,1,-3)
5100 6744 sinh : $\sqrt[3]{9/2}$
5100 6787 J_1 : $9\ln 2/5$
5100 7834 at : $5\pi^2/3$
5100 8273 $\exp(1/4)-s\pi(e)$
5100 9941 rt : (7,2,5,-4)
5101 0518 rt : (9,5,-9,6)
5101 1611 rt : (3,6,4,-4)
5101 3225 rt : (7,2,9,5)
5101 4916 rt : (3,2,8,-5)
5101 5538 cu : $1/2-3\sqrt{3}/4$
5101 5657 id : $5e/9$
5101 6620 rt : (1,9,8,-5)
5102 0408 sq : 5/7
5102 3346 sinh : $\exp(\sqrt[3]{4}/2)$
5102 8840 rt : (8,7,-9,-2)
5103 0430 e^x : $4e+\pi$
5103 1606 e^x : $2\sqrt{2}/3+\sqrt{3}$
5103 2305 rt : (8,9,9,-8)
5103 2725 sπ : $3\sqrt{2}/2+3\sqrt{5}$
5103 3031 J_1 : 22/9
5103 3197 rt : (4,1,-6,-7)
5103 5489 Ψ : $\sqrt{3}-\sqrt{5}+\sqrt{7}$
5103 8810 rt : (9,7,-4,9)
5103 9268 J_0 : $5\zeta(3)/4$
5103 9337 erf : $2\sqrt[3]{5}/7$
5104 1751 e^x : 24/5
5104 2490 Γ : 13/22
5104 4221 $\sqrt{5}\times s\pi(\sqrt{5})$
5104 4383 as : $2\sqrt[3]{5}/7$
5104 4701 rt : (9,7,-2,-2)
5104 5845 ℓ_{10} : $3\sqrt{2}/2+\sqrt{5}/2$
5104 6160 at : $4\sqrt[3]{2}/9$
5104 7391 ζ : $(3\pi)^{-2}$
5104 8309 rt : (9,3,2,-3)
5104 8832 at : 14/25
5105 0170 rt : (5,8,2,2)
5105 0623 at : $4+4\pi$
5105 0704 10^x : $3\sqrt{3}-3\sqrt{5}/2$
5105 1967 rt : (9,1,6,4)
5105 2580 2^x : $\sqrt{2}/2-3\sqrt{5}/4$
5105 2948 rt : (6,4,-5,4)
5105 3855 rt : (5,3,5,-4)
5105 4331 tπ : $\zeta(3)/8$
5105 5554 J_1 : $7\pi/9$
5105 5756 rt : (1,7,4,-4)
5105 7591 rt : (5,1,9,5)
5105 7631 rt : (2,8,6,9)
5105 8987 id : $3\sqrt{2}-\sqrt{3}$
5105 9464 ℓ_{10} : $\sqrt{21/2}$
5105 9520 rt : (1,3,8,-5)
5106 1025 ζ : $(\sqrt{2}\pi)^{-3}$
5106 1463 rt : (3,7,-7,-5)
5106 2326 J_1 : 5/4
5106 3169 J_1 : $((e+\pi)/3)^{1/3}$
5106 4102 $\pi-\exp(\sqrt{3})$
5106 4804 $\sqrt{3}-\exp(1/5)$
5106 5421 Ψ : $3e\pi/2$
5106 5578 sq : $3/2-\pi/4$
5106 7449 e^x : $4\sqrt{3}/3-4\sqrt{5}/3$
5106 8835 Ψ : 1/24
5106 9211 id : $(e/2)^3$
5107 0139 Li_2 : $\sqrt{5}/5$
5107 0151 J_1 : $10\sqrt[3]{5}/7$
5107 0169 Ei : $\sqrt{3/2}$
5107 0801 $\ln(2/5)\times\exp(1/2)$
5107 3125 $\exp(\sqrt{5})+\exp(e)$
5107 3388 rt : (8,9,-4,1)
5107 5391 sr : 6/23
5107 8601 at : $3e/2+4\pi$
5107 8671 ℓ_2 : $\exp(\pi/3)$
5108 0779 Ψ : $(\sqrt{2}\pi)^{-1/3}$
5108 2190 10^x : $3+2e/3$
5108 2562 id : $\ln(5/3)$
5108 2703 $\ln(3)-s\pi(1/5)$
5108 3568 E : $5\ln 2/9$
5108 4729 rt : (7,2,-7,9)
5108 7295 cr : 2/15
5108 9396 E : $2\sqrt{3}/9$
5108 9550 sinh : $\sqrt{2}+\sqrt{5}-\sqrt{6}$
5108 9576 tπ : $\exp(\pi/3)$
5109 1440 rt : (7,8,-2,-2)
5109 2052 cu : $e-\pi/2$
5109 2168 θ_3 : 13/14
5109 4162 ℓ_2 : $5\sqrt[3]{5}/3$

5109 4896 $\zeta : \exp(-\sqrt{2}\pi)$
5109 5559 rt : (7,4,2,-3)
5109 6132 cu : $4/3 + \sqrt{3}/4$
5109 7164 id : $10(e+\pi)/9$
5109 7687 rt : (3,8,1,-3)
5109 8529 $E : 5/13$
5109 9074 at : $1/2 - 3\sqrt{2}/4$
5109 9085 rt : (9,7,-6,6)
5109 9740 rt : (7,0,6,4)
5109 9747 $e^x : 3\sqrt{2}/4 - \sqrt{3}$
5109 9786 rt : (6,6,0,7)
5109 9906 id : P_{13}
5110 0484 $\ell_2 : 2\sqrt{3} + \sqrt{5}$
5110 1539 $\exp(2) \times s\pi(\sqrt{3})$
5110 1675 $J : 1/10$
5110 1901 rt : (3,5,6,8)
5110 1975 $\Psi : 15/7$
5110 2988 rt : (2,6,-9,-7)
5110 3474 cu : $4/3 + \pi/4$
5110 3997 id : $3e + 3\pi/4$
5110 4080 rt : (1,8,-4,-4)
5110 6180 rt : (3,0,9,5)
5110 7931 tanh : $(\pi)^{-1/2}$
5110 8393 rt : (1,7,3,-8)
5111 0623 rt : (5,8,-7,-5)
5111 0848 $10^x : (3\pi)^{1/3}$
5111 2188 $\Gamma : 1/\ln(2e)$
5111 4008 $\Psi : 6(e+\pi)/7$
5111 5852 sinh : $(\sqrt[3]{3})^{1/2}$
5111 6468 cr : $\zeta(3)/9$
5111 8229 $\Psi : \sqrt[3]{2}/9$
5111 8841 $2^x : 2\sqrt{3}/3 - \sqrt{5}/4$
5111 9173 $10^x : 6/11$
5112 1324 $s\pi : (3\pi/2)^{1/2}$
5112 2293 $\ell_{10} : 9\sqrt[3]{3}/4$
5112 3348 cu : $\sqrt{22}\pi$
5112 4247 rt : (8,9,7,-7)
5112 5399 rt : (4,1,-8,-4)
5112 5920 id : $\sqrt{3}/4 - 4\sqrt{5}$
5112 8400 $t\pi : 6\pi$
5113 0720 $J_1 : 1/\ln(\sqrt{2}\pi/2)$
5113 2166 $s\pi : 3\sqrt{2} + 4\sqrt{3}$
5113 2658 rt : (9,9,-1,9)
5113 5673 rt : (9,9,-5,-1)
5113 5705 $\ell_{10} : 4\ln 2/9$
5113 6285 $J_1 : 2e\pi/7$
5113 6860 cu : $9(1/\pi)$
5113 7114 rt : (1,8,-4,3)
5113 7477 id : $3(e+\pi)/7$
5113 9060 $s\pi : (\sqrt[3]{5}/3)^{1/3}$
5114 0289 rt : (9,5,-1,-2)
5114 0466 rt : (1,5,-3,-1)
5114 0800 $J_1 : (\pi/2)^{1/2}$
5114 2680 rt : (5,9,-2,-2)
5114 4263 cr : $\sqrt{2}/3 + 4\sqrt{5}/3$
5114 4981 rt : (9,1,3,-3)
5114 7411 rt : (5,5,2,-3)
5114 7928 $e^x : 2/3 + 2\sqrt{5}$
5114 8634 rt : (6,4,-7,1)
5114 8893 $\zeta : (\ln 2/3)^3$
5114 9460 sinh : $(\ln 2)^{-1/2}$
5114 9709 $e^x : (e+\pi)^{-1/2}$
5114 9751 rt : (9,3,7,4)
5114 9862 rt : (1,9,1,-3)
5115 0511 $Ei : \sqrt{17/3}$
5115 0646 rt : (7,2,-9,6)
5115 2222 rt : (5,1,6,-4)
5115 3656 sq : $3/2 + 4\sqrt{3}/3$
5115 4122 at : $\exp(-\sqrt{3}/3)$
5115 4586 cr : $e/2 + 2\pi/3$
5115 4714 rt : (3,3,-9,6)
5115 7228 rt : (3,9,-4,-4)
5115 9648 rt : (1,1,9,-5)
5115 9847 $\Psi : \exp(-2\pi/3)$
5116 0280 cu : $4 + \sqrt{5}$
5116 0801 $e^x : 2\sqrt{2} + 3\sqrt{5}/4$
5116 1412 $\Psi : (2\pi)^3$
5116 2206 rt : (3,5,-8,-5)
5116 3084 $\ell_{10} : (\sqrt{2}\pi/3)^3$
5116 4786 rt : (7,9,-7,-5)
5116 5980 cu : $8\sqrt[3]{5}/9$
5116 6695 rt : (9,7,-8,3)
5116 7015 $\exp(\pi) \div \exp(2/5)$
5116 9065 cr : $1 - \sqrt{3}/2$
5117 0890 cr : $4/3 + 3\sqrt{2}/2$
5117 1601 $Ei : 7e\pi/10$
5117 2732 rt : (8,9,-6,-2)
5117 3650 $s\pi : \sqrt[3]{5}/10$
5117 4305 cu : $\sqrt{2}/2 + \sqrt{3}$
5117 4450 $2^x : 2e/3$
5117 5334 sr : $3/2 + \pi/4$
5117 7139 $\ln(3) + \exp(5)$
5117 7861 rt : (8,7,8,-7)
5117 9188 rt : (4,1,-3,7)
5118 2573 $\ln(\sqrt{5}) + s\pi(1/4)$
5118 2767 $J_0 : 3/2$
5118 2798 sinh : $7\sqrt{5}$
5118 2945 $E : 6(1/\pi)/5$
5118 3825 tanh : 13/23
5118 5212 sr : $4 + 4\sqrt{3}/3$
5118 5717 sq : $(\ln 3/3)^{1/3}$
5118 5789 sr : 16/7
5118 6430 $e^x : 3\sqrt{3}/2 - 3\sqrt{5}/4$
5118 7552 rt : (7,4,-4,9)
5118 8336 $\ell_{10} : 4/13$
5118 8643 $10^x : 2/5$
5119 0525 id : $6\sqrt[3]{2}/5$
5119 5635 sr : $((e+\pi)/3)^{-2}$
5119 5675 $E : 1/\sqrt[3]{18}$
5119 6355 rt : (7,6,-1,-2)
5119 7627 $\Psi : 8\pi/5$
5119 7843 $2^x : 5(e+\pi)/6$
5119 8579 rt : (2,9,1,5)
5120 0000 cu : 4/5
5120 0909 $2^x : (3\pi/2)^{-1/3}$
5120 1780 rt : (7,2,3,-3)
5120 2821 $2^x : 1/3 - 3\sqrt{3}/4$
5120 3666 rt : (1,8,3,-5)
5120 4010 $\theta_3 : \sqrt[3]{2}/5$
5120 4050 at : 17
5120 4596 rt : (9,9,-3,6)
5120 5150 id : $3\sqrt{3} - 3\sqrt{5}$
5120 6891 tanh : $\exp(-\sqrt[3]{5}/3)$
5120 7304 rt : (7,2,7,4)
5120 7372 rt : (6,8,8,-7)
5120 9833 tanh : $1/\ln(e+\pi)$
5120 9992 sq : $3\sqrt{3}/2 - 2\sqrt{5}$
5121 0173 rt : (3,2,6,-4)
5121 1061 at : $\ln(\sqrt[3]{5}/3)$
5121 3069 rt : (1,6,-5,-4)
5121 3183 $e^x : e\pi/2$
5121 5325 $E : 8/21$
5121 7472 $Ei : 8\sqrt{3}/5$
5121 7859 $\ell_{10} : 2\sqrt{2}/3 + 4\sqrt{3}/3$
5121 8359 tanh : $2\sqrt{2}/5$
5121 8805 rt : (1,2,-9,-5)
5121 9070 $J_1 : 2\pi/5$
5122 0208 $e^x : P_{56}$
5122 1785 rt : (5,6,-8,-5)
5122 5210 $e \div s\pi(e)$
5122 8647 $2^x : 1/4 + 3\sqrt{2}$
5122 9442 $J_1 : 8\sqrt{2}/9$
5123 0536 $s\pi : \sqrt{2}/2 + 2\sqrt{3}$
5123 1326 at : $2e\pi$
5123 1792 $4/5 + \ln(3/4)$
5123 3083 sinh : $3(e+\pi)/4$
5123 3659 $\zeta : 13/6$
5123 3937 rt : (8,9,5,-6)
5123 6919 rt : (8,5,9,-7)
5123 8250 at : $10\sqrt[3]{5}$
5123 8458 rt : (4,9,8,-7)
5123 8946 at : 9/16
5124 0736 sr : $5\pi^2/4$
5124 5655 $E : 2\sqrt[3]{5}/9$
5124 7201 id : $\sqrt{2}/4 - \sqrt{3}/2$
5124 7909 sr : $(\ln 3/3)^{-3}$
5124 8165 rt : (5,1,-7,-9)
5124 9067 rt : (9,7,-4,-1)
5125 0349 $\ell_{10} : 4 - \sqrt{5}/3$
5125 2130 $10^x : \sqrt[3]{25/3}$
5125 4421 $e^x : 3/2 + 3\pi$
5125 5205 rt : (9,3,0,-2)
5125 5410 $4/5 \times \exp(\pi)$
5125 7461 at : $4e + 2\pi$
5125 8398 rt : (5,8,0,-1)
5125 8594 rt : (2,9,-1,-4)
5125 9207 $\ell_2 : 2 - 3\sqrt{3}/4$
5126 0332 ln : $5\pi^2/4$
5126 1103 $2^x : \exp(\sqrt[3]{2})$
5126 1244 $\exp(1/3) + \exp(3/4)$
5126 1310 $\exp(3) - \exp(4)$
5126 1464 rt : (9,1,4,3)
5126 1580 $\ell_{10} : 5(e+\pi)/9$
5126 3017 $s\pi : 7e/6$
5126 3169 rt : (6,4,-9,-2)
5126 4721 rt : (5,3,3,-3)
5126 6363 rt : (6,8,3,7)
5126 6712 ln : $4\sqrt{2} - \sqrt{5}/2$
5126 6912 rt : (7,4,-6,6)
5126 7849 rt : (9,5,8,4)
5126 8013 rt : (1,7,2,-3)
5126 9589 rt : (6,6,9,-7)
5127 0783 $\zeta : 3\sqrt[3]{5}/10$
5127 1173 rt : (5,1,7,4)
5127 1256 rt : (2,8,2,-5)
5127 1637 $e^x : 3\sqrt{2}/2 + 2\sqrt{3}$
5127 3231 $t\pi : (\sqrt[3]{3}/2)^{1/2}$
5127 4532 rt : (1,3,6,-4)
5127 7928 rt : (3,7,-5,-4)
5127 8562 cu : $8(1/\pi)$
5128 0414 $Li_2 : \pi/7$
5128 0449 sr : $3\sqrt{3} + \sqrt{5}/2$
5128 1718 $e^x : 7\pi/9$
5128 2585 $\sqrt{2} + \ln(3)$
5128 4655 rt : (3,3,-9,-5)
5128 4876 sr : $1/4 + 3e/4$
5128 4972 $\ell_{10} : \sqrt{3}/2 - \sqrt{5}/4$
5128 5834 sinh : $5\sqrt[3]{3}/6$
5128 6156 rt : (9,9,-5,3)
5128 8161 rt : (7,7,-8,-5)
5129 0669 as : $1 - 2\sqrt{5}/3$
5129 1233 cu : $2\sqrt{2}/3 + \sqrt{3}/3$
5129 1978 rt : (8,9,-8,-5)
5129 3849 sq : $9(1/\pi)/4$
5129 4159 $\ln(4) \div \ln(2/5)$
5129 6017 $J_1 : \sqrt[3]{2}$
5129 6307 at : $7\pi^2/4$
5129 7241 sq : $3/2 - 4e/3$
5129 8297 sinh : $\exp(1/(2e))$
5129 8507 $\exp(\pi) \div \exp(1)$
5129 8810 $e^x : 1 + \sqrt{2}/2$
5129 8824 $e^x : (\sqrt{5}/3)^3$
5129 8827 rt : (8,7,6,-6)
5129 8896 $\Psi : 9\sqrt{5}/4$
5129 8917 sr : 5/19
5129 9278 id : $3\sqrt[3]{5}/10$
5129 9630 $s\pi : (\ln 3)^{-2}$
5129 9709 $t\pi : \sqrt{2}\pi$
5130 0607 rt : (4,1,-5,4)
5130 4106 rt : (4,7,9,-7)
5130 5009 rt : (4,6,-7,8)
5130 6507 $\sqrt{3} \div \ln(\pi)$
5130 6533 $Ei : 8\ln 2$
5130 6968 $\ln(\sqrt{2}) \div s\pi(\sqrt{5})$
5130 6988 $\ell_{10} : \ln(e/2)$
5130 8210 $10^x : (e+\pi)^{1/3}$
5130 8228 $J_1 : \sqrt{5} - \sqrt{6} + \sqrt{7}$
5130 8574 sr : $\sqrt[3]{12}$
5130 9292 $E : 3\sqrt[3]{2}/10$
5130 9665 $E : 1/\sqrt{7}$
5131 1843 rt : (7,6,-1,9)
5131 2532 at : $10\sqrt{3}$
5131 2555 $\Gamma : 1/\ln(\ln 3/3)$
5131 2959 cr : $4\sqrt{3} + 4\sqrt{5}$
5131 4060 ln : $3e + 4\pi/3$
5131 4554 cu : $2\sqrt{3}/3 + 2\sqrt{5}/3$
5131 6514 rt : (7,8,-4,-1)
5131 8025 $\exp(\sqrt{2}) \div \exp(1)$
5131 8889 sinh : $\zeta(3)$
5132 0867 $\ln(\sqrt{3}) - s\pi(\sqrt{2})$

5132 2300 $\ln(3/4) - \exp(4/5)$
5132 3708 rt : (7,4,0,-2)
5132 5037 $2/5 + \exp(\sqrt{2})$
5132 6538 sq : $3\sqrt{2} - \sqrt{3}/4$
5132 7412 id : $4\pi/5$
5132 7462 rt : (3,8,-1,-2)
5132 7487 Ψ : $1/25$
5132 8839 $s\pi$: $2\sqrt{2}$
5132 9183 sinh : $\exp(-1/\sqrt{2})$
5133 1061 rt : (7,0,4,3)
5133 1160 rt : (6,6,-4,1)
5133 2365 cr : $5\ln 2$
5133 2430 cu : $\sqrt{2}/4 + 3\sqrt{3}/4$
5133 4899 rt : (6,8,6,-6)
5133 5106 sr : $3e + 4\pi/3$
5133 6188 rt : (5,1,-9,-6)
5133 8578 rt : (7,4,8,4)
5133 8782 rt : (1,8,-2,-3)
5134 0638 rt : (2,8,9,-7)
5134 0937 ℓ_{10} : $2\sqrt{2} + \sqrt{3}/4$
5134 1711 id : $\exp(-2/3)$
5134 1877 ζ : $5\sqrt{3}/4$
5134 2504 rt : (3,0,7,4)
5134 3782 J_1 : $7\sqrt[3]{3}/8$
5134 4650 $t\pi$: $(\ln 3/2)^{-2}$
5134 6189 sq : $1/4 - 3\sqrt{3}/2$
5134 6477 rt : (1,7,5,-5)
5134 7268 ln : $e - \pi/3$
5134 7572 ℓ_{10} : $6e/5$
5134 8168 ln : $\sqrt[3]{14/3}$
5134 8275 cu : $\sqrt{2} + 2\sqrt{5}/3$
5134 9966 2^x : $4\sqrt{2}/3 + \sqrt{3}/3$
5135 0498 rt : (5,8,-5,-4)
5135 0769 Γ : $1/10$
5135 2194 10^x : $\sqrt[3]{2}/7$
5135 2235 ℓ_{10} : $e/3 + 3\pi/4$
5135 4401 sq : $9e$
5135 6128 10^x : $2/3 + 2\sqrt{3}$
5135 7042 cu : $3e/4 + 4\pi/3$
5135 7485 rt : (7,4,-8,3)
5135 8469 rt : (5,4,-9,-5)
5135 8562 e^x : $2\pi/5$
5135 9576 cr : $-\sqrt{2} + \sqrt{5} + \sqrt{7}$
5136 0083 Ψ : $\ln(\sqrt[3]{5}/2)$
5136 0289 $t\pi$: $1/3 + \sqrt{5}$
5136 1016 $2/5 \times \exp(1/4)$
5136 2637 rt : (9,8,-8,-5)
5136 2707 $t\pi$: $\pi/10$
5136 2736 sr : $\sqrt{3} + \sqrt{5}/4$
5136 3119 ln : $(2\pi)^3$
5136 3396 Ei : 14
5136 4167 rt : (3,5,-4,6)
5136 5692 2^x : $13/4$
5136 6974 rt : (8,9,3,-5)
5136 7201 at : $(\pi)^{-1/2}$
5136 9152 $s\pi$: $\zeta(3)/7$
5136 9413 cr : $3/4 + e$
5137 0880 Ψ : $3\ln 2/2$
5137 1021 rt : (8,5,7,-6)
5137 1462 J_1 : $24/19$
5137 3119 rt : (4,9,6,-6)
5137 6233 rt : (4,6,-9,3)
5137 8151 cu : $1/\ln(\sqrt{2}\pi/3)$
5137 8632 sinh : $19/7$
5137 9375 rt : (9,9,-7,0)
5138 2877 ℓ_2 : $(1/(3e))^{1/2}$
5138 4255 at : $4e/3 - 4\pi/3$
5138 4898 ℓ_{10} : $2/3 + 3\sqrt{3}/2$
5138 6371 rt : (5,1,-4,9)
5138 7588 rt : (9,5,-3,-1)
5139 0778 $t\pi$: $3\sqrt{3} + \sqrt{5}/2$
5139 1377 Ei : $(\pi/2)^{-1/3}$
5139 1884 rt : (5,9,-4,-1)
5139 2601 J_1 : $17/7$
5139 2635 θ_3 : $(\sqrt{2}\pi)^{-1/3}$
5139 2651 Li_2 : $1/\sqrt[3]{11}$
5139 3067 J_0 : $7\sqrt[3]{5}/8$
5139 4872 $s\pi$: $3e/2 + 2\pi/3$
5139 5997 rt : (9,1,1,-2)
5139 7352 at : $3(e + \pi)$
5139 8639 id : $7e/2$
5139 8966 id : $\sqrt{2} - 4\sqrt{3}$
5139 9414 2^x : $7\sqrt[3]{5}/9$
5140 0398 rt : (5,5,0,-2)
5140 2441 E : $3/8$
5140 4611 rt : (9,3,5,3)
5140 4856 rt : (1,9,-1,-2)
5140 4860 ζ : $(1/(3e))^2$
5140 6827 sq : $1/4 - 4\pi/3$
5140 6983 rt : (6,8,1,4)
5140 9122 rt : (5,1,4,-3)
5140 9205 rt : (7,6,-3,6)
5140 9422 ℓ_2 : $2 + 3\pi$
5141 0035 E : $1/\sqrt[3]{19}$
5141 0322 rt : (8,4,-7,8)
5141 0963 ζ : $1/\ln(\sqrt[3]{4})$
5141 1400 rt : (6,6,7,-6)
5141 3092 ln : $4\sqrt{3} - 3\sqrt{5}$
5141 3440 rt : (9,7,9,4)
5141 3692 rt : (2,4,-8,-4)
5141 4828 cr : $3 + \sqrt{2}/3$
5141 5744 sq : $2/3 + 4\sqrt{2}/3$
5141 7852 $t\pi$: $8(1/\pi)/3$
5141 8066 rt : (5,3,8,4)
5141 8323 rt : (3,9,-2,-3)
5141 8964 e^x : $\sqrt{3} - \sqrt{5} + \sqrt{7}$
5141 9001 sinh : $(\ln 3/3)^{-1/2}$
5142 0568 cu : $\sqrt{2}/4 - 2\sqrt{3}/3$
5142 2754 rt : (1,1,7,-4)
5142 3023 sr : $3 - \sqrt{2}/2$
5142 3036 sinh : $19/3$
5142 3939 e^x : $1/\sqrt[3]{14}$
5142 7506 rt : (3,5,-6,-4)
5142 8220 e^x : $4\zeta(3)$
5142 9597 $\exp(1/5) - s\pi(1/4)$
5143 2309 Ei : $1/23$
5143 2323 rt : (7,9,-5,-4)
5143 4008 ζ : $9\zeta(3)/5$
5143 4363 rt : (3,9,-4,5)
5143 6029 $\ln(\pi) \div \exp(4/5)$
5143 6445 $t\pi$: $7/15$
5143 7190 $s\pi$: $(e + \pi)/5$
5143 8770 erf : $\exp(-1/\sqrt{2})$
5143 9898 Li_2 : $9/20$
5144 0329 cu : $5\sqrt[3]{3}/9$
5144 1220 tanh : $(2e)^{-1/3}$
5144 1879 rt : (7,5,-9,-5)
5144 2123 J_0 : $(\sqrt{5})^{1/2}$
5144 2262 θ_3 : $6/7$
5144 2660 e^x : $e/3 - \pi/2$
5144 4076 $\ln(2/5) - \exp(4)$
5144 4264 λ : $(3\pi/2)^{-2}$
5144 5131 at : $13/23$
5144 5384 $t\pi$: $9e\pi/10$
5144 5914 cu : $(\pi/3)^3$
5144 6573 ζ : $\exp(-2/3)$
5144 6875 rt : (1,9,-5,-1)
5144 7042 rt : (8,7,4,-5)
5144 7112 Ψ : $3\sqrt{3}/5$
5144 9505 rt : (4,1,-7,1)
5144 9723 J_1 : $7\ln 2/2$
5145 0903 10^x : $\sqrt{3}/2 + 3\sqrt{5}$
5145 1845 rt : (8,3,8,-6)
5145 2802 J_1 : $19/15$
5145 3846 id : $7\pi^2/6$
5145 4156 10^x : $\sqrt[3]{3}/8$
5145 4341 rt : (4,7,7,-6)
5145 6249 ζ : $3\sqrt[3]{3}/2$
5145 6611 at : $9\pi^2/5$
5145 7317 ℓ_2 : $7/10$
5145 8908 rt : (1,7,-4,-1)
5146 0720 ζ : $(4)^{-3}$
5146 3543 $t\pi$: $7/17$
5146 3735 cu : $2\zeta(3)/3$
5146 5084 rt : (7,8,2,9)
5146 6010 cr : $1/3 + \pi$
5146 6119 sr : $2/3 + 4\sqrt{2}$
5146 6357 ℓ_2 : $5\sqrt[3]{2}/9$
5146 7813 e^x : $3/4 - \sqrt{2}$
5146 8541 2^x : $4e/5$
5146 8818 at : $\exp(-\sqrt[3]{5}/3)$
5147 0131 rt : (3,5,-6,1)
5147 0531 Γ : $\exp(-\sqrt[3]{4}/3)$
5147 1424 cr : $3/22$
5147 1570 rt : (7,6,-3,-1)
5147 1839 at : $1/\ln(e + \pi)$
5147 1862 sq : $3 - \sqrt{2}$
5147 2623 ℓ_{10} : $-\sqrt{2} + \sqrt{5} + \sqrt{6}$
5147 4539 E : $\sqrt{5}/6$
5147 4667 $\exp(1/2) + s\pi(1/3)$
5147 5215 J_1 : $7\sqrt{3}/5$
5147 5692 cu : $4 + \sqrt{2}/4$
5147 6928 $t\pi$: $2\sqrt{2}/9$
5147 8329 cr : $3(1/\pi)/7$
5147 8621 sinh : $\sqrt[3]{20}$
5147 9289 sq : $16/13$
5148 0595 at : $2\sqrt{2}/5$
5148 1615 rt : (7,2,1,-2)
5148 1974 rt : (5,9,-7,5)
5148 2219 $\sqrt{5} \div \ln(2/3)$
5148 2477 cr : $1/\ln(4/3)$
5148 6249 Ψ : $4\sqrt[3]{2}$
5148 6893 rt : (9,9,-9,-3)
5148 7460 ln : $4\sqrt{2} + 3\sqrt{5}$
5149 1617 sr : $2e - \pi$
5149 1936 rt : (7,2,5,3)
5149 2092 rt : (6,8,4,-5)
5149 3442 rt : (8,4,-9,5)
5149 5274 at : $8\sqrt{5}$
5149 5651 rt : (5,1,-6,6)
5149 6151 $\exp(\sqrt{2}) \div s\pi(\sqrt{3})$
5149 7364 rt : (6,4,8,-6)
5149 8317 $t\pi$: $2\sqrt{2} - 3\sqrt{5}/4$
5149 8760 sinh : $\exp(e/2)$
5150 0108 rt : (4,8,-2,8)
5150 2549 rt : (7,6,9,4)
5150 2808 rt : (3,9,-6,-2)
5150 2872 rt : (1,6,-3,-3)
5150 3226 sr : $5(1/\pi)/6$
5150 5262 ln : $2\sqrt{2} - 2\sqrt{3}/3$
5150 7486 10^x : $(\sqrt[3]{4}/3)^{-2}$
5150 8134 rt : (3,2,8,4)
5150 8415 Ei : $\exp(\sqrt[3]{2}/2)$
5151 3777 e^x : $8\sqrt{2}/9$
5151 3804 rt : (1,5,0,-8)
5151 3824 $t\pi$: $e/4 - 4\pi/3$
5151 5192 E : $7(1/\pi)/6$
5151 6724 id : $3\zeta(3)/7$
5151 7364 $\exp(\pi) \times \exp(\sqrt{5})$
5151 9562 rt : (5,6,-6,-4)
5152 1152 ℓ_2 : $3 - \pi/2$
5152 1933 rt : (7,6,-5,3)
5152 3131 10^x : $3\sqrt{3}/4 + \sqrt{5}$
5152 6244 $s\pi$: $2\sqrt{3} + 3\sqrt{5}$
5152 6598 $\ln(2) + \exp(3/5)$
5152 6960 rt : (2,5,-6,4)
5152 7604 sinh : $10(e + \pi)/7$
5152 8329 J_0 : $\sqrt[3]{10/3}$
5152 9185 $t\pi$: $\sqrt[3]{3}/3$
5152 9782 at : 18
5153 0684 at : $2e + 4\pi$
5153 1534 rt : (8,9,1,-4)
5153 1928 sinh : $2\sqrt{3}/7$
5153 4666 ℓ_{10} : $3\sqrt{2}/2 + 2\sqrt{3}/3$
5153 5338 tanh : $\sqrt[3]{5}/3$
5153 6370 $t\pi$: $3\sqrt{2}/5$
5153 6521 ln : $2(e + \pi)/7$
5153 7024 rt : (9,6,-9,-5)
5153 7211 rt : (8,5,5,-5)
5153 8638 J_0 : $1/\ln((e + \pi)/3)$
5154 0170 rt : (8,6,-4,8)
5154 0393 Ei : $\pi/2$
5154 1647 $t\pi$: $(1/(3e))^{-2}$
5154 2405 Li_2 : $3\zeta(3)/8$
5154 2972 rt : (8,1,9,-6)
5154 4673 sq : $3 - 4\pi$
5154 5976 rt : (4,5,8,-6)
5154 9003 rt : (9,7,-6,0)
5155 2696 ln : $3/2 + 4e$
5155 4445 id : $7\sqrt{3}/8$

5155 4944 rt : (5,9,-9,0)
5155 5881 cr : P_{24}
5155 6862 rt : (9,8,9,-8)
5155 7647 id : $9\sqrt{5}/8$
5155 8172 $\ell_2 : 3+e$
5155 8194 sinh : $5\ln 2/7$
5155 8914 rt : (5,3,-1,9)
5155 9795 $\ell_2 : -\sqrt{5}+\sqrt{6}+\sqrt{7}$
5156 0363 cr : $1/2+4\sqrt{5}/3$
5156 0609 rt : (9,3,-2,-1)
5156 0770 $\theta_3 : 14/23$
5156 1351 as : $\exp(-1/\sqrt{2})$
5156 2500 sq : 15/8
5156 6729 rt : (5,8,-2,-4)
5156 6784 $10^x : 2e\pi/5$
5156 7398 $J_1 : (e+\pi)^{1/2}$
5156 8807 $\Gamma : (\ln 2/2)^{1/2}$
5156 9083 rt : (1,4,-1,6)
5157 0510 $\ell_2 : 2e/3+\pi/3$
5157 1520 $e^x : 1/3+4\sqrt{5}/3$
5157 1656 $2^x : 3/5$
5157 2565 rt : (9,1,2,2)
5157 8051 rt : (3,9,-8,-9)
5157 8875 rt : (5,3,1,-2)
5157 9167 at : $1-\sqrt{3}/4$
5158 2068 rt : (6,8,-1,1)
5158 3814 $\exp(3) - s\pi(\pi)$
5158 3866 $s\pi : \exp(1/(2\pi))$
5158 3987 at : P_{80}
5158 4222 $K : 16/17$
5158 4891 rt : (9,5,6,3)
5158 5288 rt : (1,7,0,-2)
5158 5515 $10^x : \zeta(3)/3$
5158 7316 rt : (7,8,0,6)
5158 8333 rt : (6,6,5,-5)
5159 0318 $t\pi : 8\ln 2/3$
5159 1402 rt : (5,1,5,3)
5159 1537 rt : (7,9,9,-8)
5159 2205 $J_1 : 14/11$
5159 2468 id : $3(e+\pi)/5$
5159 2878 sq : $2-e$
5159 3691 $J_1 : 9\sqrt{2}/10$
5159 3721 sq : $7e/2$
5159 3732 rt : (3,5,-8,-4)
5159 4698 rt : (6,2,9,-6)
5159 4801 rt : (4,8,-4,3)
5159 4815 $e^x : 3/2-\sqrt{3}/3$
5159 6179 $2/3 \times s\pi(e)$
5159 6516 $s\pi : 9\pi/10$
5159 8022 rt : (2,6,8,-6)
5159 8164 $erf : 2\sqrt{3}/7$
5159 9381 $s\pi : 3/4+3e/2$
5160 3924 $J_1 : 4(1/\pi)$
5160 4333 rt : (5,5,9,4)
5160 4587 rt : (1,7,3,-1)
5160 4755 rt : (3,7,-3,-3)
5160 5416 $E : 7/19$
5160 5963 $E : 1/\sqrt[3]{20}$
5161 1174 rt : (1,1,8,4)
5161 5108 $\theta_3 : (\sqrt[3]{4})^{-1/3}$
5161 6972 rt : (6,1,-7,8)
5161 8135 rt : (3,3,-7,-4)
5161 8783 $erf : 5\ln 2/7$
5161 8883 $s\pi : 1/4+\sqrt{3}/3$
5162 1948 $E : \exp(-1)$
5162 3352 rt : (5,1,-8,3)
5162 4121 sq : $2/3-3\sqrt{3}$
5162 5219 rt : (7,7,-6,-4)
5162 5455 rt : (2,5,-8,-3)
5162 5792 sr : $1+3\sqrt{3}/4$
5162 6166 $10^x : 1/\ln((e+\pi)/2)$
5162 7811 sinh : $6e\pi/5$
5162 9073 $2/5+\ln(2/5)$
5163 1537 $\Psi : 4e\pi$
5163 2670 rt : (8,7,2,-4)
5163 4598 $s\pi : \sqrt{3}+\sqrt{6}+\sqrt{7}$
5163 6208 rt : (4,1,-9,-2)
5163 7140 rt : (2,8,-4,3)
5163 7399 $2^x : 3\sqrt{2}-3\sqrt{3}$
5163 9550 rt : (8,3,6,-5)
5163 9664 rt : (8,6,-6,5)
5163 9777 sr : 4/15
5163 9846 sr : $4\sqrt{2}+3\sqrt{5}$
5164 0644 $\ell_{10} : (2e/3)^2$
5164 0765 tanh : 4/7
5164 3148 rt : (4,7,5,-5)
5164 7243 $e^x : 2/3+e$
5165 0207 rt : (4,3,9,-6)
5165 0301 $Li_2 : \ln(\pi/2)$
5165 0716 $t\pi : 3\sqrt{2}-4\sqrt{3}$
5165 3848 rt : (1,9,4,-1)
5165 3901 rt : (7,8,-6,0)
5165 5676 $\Psi : (\sqrt{5}/2)^{1/3}$
5165 7508 sr : 23/10
5165 9682 $10^x : \exp(-\sqrt[3]{5})$
5166 1147 id : $\sqrt{19/3}$
5166 2979 $\ell_{10} : 7/23$
5166 3536 $E : \ln(\ln 2)$
5166 3860 $\exp(2/3) \div s\pi(e)$
5166 5765 $\Psi : 7\sqrt[3]{3}/2$
5166 5899 $J_1 : 8e/9$
5166 5995 rt : (3,7,1,6)
5166 7444 $J_0 : \ln(\sqrt{2}\pi)$
5166 8042 rt : (7,4,-2,-1)
5166 8958 $\zeta : \sqrt{14/3}$
5167 2911 $E : \ln(\sqrt[3]{3})$
5167 3166 ln : $(3\pi/2)^{1/3}$
5167 5539 rt : (3,8,-3,-1)
5167 8693 $J_1 : (\sqrt[3]{3}/3)^{-1/3}$
5167 9745 $2^x : \zeta(3)/2$
5168 1375 ln : $7(e+\pi)/9$
5168 1565 $\Gamma : 10/17$
5168 2671 rt : (7,0,2,2)
5168 2922 rt : (6,6,-8,-5)
5168 3413 $2/5 \div s\pi(e)$
5168 4016 sr : $2\zeta(3)/9$
5168 4704 cr : P_4
5168 8541 $10^x : 3+4\pi/3$
5168 9679 $e^x : 5/12$
5169 0435 rt : (6,8,2,-4)
5169 3282 id : $4e-3\pi/4$
5169 4250 cr : $10\pi/9$
5169 4374 $J_1 : (\sqrt{5}/3)^{-3}$
5169 5025 cr : $2+2\sqrt{5}/3$
5169 5518 rt : (9,8,7,-7)
5169 5658 $\ell_2 : ((e+\pi)/2)^{1/3}$
5169 6223 rt : (8,8,-1,8)
5169 7315 sr : $1/\sqrt{14}$
5169 7330 $\ln(2) \times s\pi(\sqrt{3})$
5169 7821 rt : (7,4,6,3)
5169 8083 rt : (6,4,6,-5)
5169 8170 rt : (5,3,-3,6)
5169 8345 rt : (1,8,0,-2)
5170 0001 $2/5+\exp(3/4)$
5170 0370 $J_0 : 2\sqrt{5}/3$
5170 2091 rt : (4,8,-6,-2)
5170 2317 $e^x : 12/13$
5170 2704 $\ln(\sqrt{2}) \times \exp(2/5)$
5170 3688 ln : $3\sqrt{5}/4$
5170 5870 rt : (3,0,5,3)
5170 6474 $10^x : 3\sqrt{3}/4-\sqrt{5}/2$
5170 6756 $J_1 : 4\sqrt{5}/7$
5170 7316 $J_1 : 23/18$
5170 8264 $\Psi : (e)^{-1/2}$
5170 9480 $Ei : 10\sqrt{2}/9$
5170 9529 at : $(2e)^{-1/3}$
5170 9953 rt : (2,4,9,-6)
5171 0702 ln : $4+\sqrt{5}/4$
5171 1180 $t\pi : e-4\pi$
5171 1323 rt : (2,8,-6,-6)
5171 3869 $\Gamma : 10\sqrt{2}$
5171 4076 rt : (2,8,-8,-6)
5171 4947 $J_0 : (\sqrt{2}\pi/2)^{1/2}$
5171 5257 cu : $2/3-3\sqrt{5}$
5171 6187 $t\pi : 1/4+3\sqrt{3}/2$
5171 7373 $\zeta : \exp(-4)$
5171 7849 $\ell_{10} : \pi^2/3$
5171 7865 $\Psi : 7\ln 2/8$
5171 9587 rt : (9,6,-7,7)
5172 1876 sr : P_{66}
5172 1881 rt : (3,4,9,4)
5172 2190 $s\pi : \sqrt{3}-\sqrt{5}/4$
5172 2444 rt : (5,8,-3,-3)
5172 2541 sr : P_{65}
5172 4076 $e^x : 3\sqrt{2}/2+\sqrt{5}/2$
5172 7066 rt : (6,1,-9,5)
5172 8237 $Li_2 : \exp(-\sqrt[3]{4}/2)$
5173 0404 rt : (1,0,-8,4)
5173 0855 ln : $8\sqrt[3]{5}/3$
5173 1383 rt : (7,8,-2,3)
5173 1509 $Ei : 15/4$
5173 4037 $Ei : 11/7$
5173 6851 rt : (7,9,7,-7)
5173 9101 rt : (5,4,-7,-4)
5173 9609 cu : $3/4-3\sqrt{5}$
5173 9999 rt : (8,9,-1,-3)
5174 1272 $10^x : 25/16$
5174 2742 ln : $2\sqrt{2}+\sqrt{3}$
5174 3282 $\exp(2/3) + s\pi(\pi)$
5174 4660 at : $\exp((e+\pi)/2)$
5174 4965 $J_1 : (2\pi/3)^{1/3}$
5174 5777 $\Gamma : 9(e+\pi)/4$
5174 6111 $E : 8(1/\pi)/7$
5174 6306 $\ell_{10} : 3e/2-\pi/4$
5174 6328 at : $(\ln 2/3)^{-2}$
5174 7980 rt : (9,8,-6,-4)
5174 8221 sr : $4\sqrt{2}-3\sqrt{5}/2$
5174 8293 rt : (8,5,3,-4)
5174 9153 rt : (9,9,-9,1)
5175 0519 $E : 4/11$
5175 0652 sinh : $(1/(3e))^{1/3}$
5175 1003 cu : $2\sqrt{3}-\sqrt{5}/4$
5175 2648 rt : (8,6,-8,2)
5175 2818 tanh : $9(1/\pi)/5$
5175 4563 sr : $3/2+4e$
5175 4941 cr : $\ln 2/5$
5175 6155 cr : $9e/7$
5175 6747 rt : (8,1,7,-5)
5175 7764 $\ell_{10} : 4-\sqrt{2}/2$
5175 7812 cu : 25/8
5175 8471 cr : $2/3+2\sqrt{2}$
5176 0666 sinh : $2\sqrt{5}/9$
5176 1187 rt : (4,5,6,-5)
5176 2130 $e^x : e\pi/5$
5176 3809 sr : $2-\sqrt{3}$
5176 5672 rt : (9,5,-5,0)
5176 8232 $s\pi : \sqrt{3}/10$
5176 8713 as : $2\sqrt{3}/7$
5176 9655 $erf : (1/(3e))^{1/3}$
5177 1471 $\theta_3 : 4(1/\pi)/5$
5177 4389 $\lambda : 1/22$
5177 4468 rt : (5,9,-6,0)
5177 7300 rt : (9,6,8,-7)
5177 7495 $erf : 2\sqrt{5}/9$
5177 9092 rt : (4,9,-8,-6)
5177 9204 $10^x : \sqrt{2}+\sqrt{6}-\sqrt{7}$
5177 9436 at : 6π
5178 0360 rt : (5,5,2,9)
5178 2818 rt : (9,1,-1,-1)
5178 4908 $\Psi : \sqrt[3]{10}$
5178 5867 $E : 1/\sqrt[3]{21}$
5178 6098 $t\pi : P_{71}$
5178 7655 $t\pi : P_{68}$
5178 8621 $\lambda : (1/\pi)/7$
5178 8798 $\Psi : (3)^{-3}$
5178 9454 rt : (1,7,-7,-9)
5179 0090 rt : (6,3,-4,8)
5179 0175 cr : $3\sqrt{2}-\sqrt{5}/3$
5179 0194 $s\pi : \ln 2/4$
5179 1960 rt : (5,5,-2,-1)
5179 4919 sq : $1/2-\sqrt{3}$
5179 5582 as : $5\ln 2/7$
5179 9494 $J_1 : \sqrt[3]{14}$
5180 0637 rt : (9,3,3,2)
5180 1292 rt : (1,9,-3,-1)
5180 2565 $2^x : 2/3+2\sqrt{3}$
5180 2587 $t\pi : ((e+\pi)/2)^3$
5180 2800 $J_1 : 8\sqrt[3]{3}/9$
5180 3917 $\Gamma : 10\sqrt[3]{5}/3$
5180 4406 $2^x : P_2$

5180 4571 ln : $2\sqrt{3}/3-\sqrt{5}/4$
5180 5700 rt : (6,8,-3,-2)
5180 5849 $\Psi : \zeta(5)$
5180 6247 at : $\sqrt[3]{5}/3$
5180 7625 rt : (1,8,-5,-5)
5180 8313 $\ell_{10} : 3\sqrt{2}/4+\sqrt{5}$
5181 0151 rt : (5,1,2,-2)
5181 0375 rt : (7,6,-9,-3)
5181 3666 $J_1 : 3\sqrt[3]{5}/4$
5181 4984 rt : (6,6,3,-4)
5181 7420 rt : (8,8,-3,5)
5181 7909 rt : (9,6,-9,4)
5181 9180 rt : (9,7,7,3)
5181 9869 rt : (1,5,1,-2)
5182 1326 at : 19
5182 2924 rt : (7,7,8,-7)
5182 4458 as : $1/4-\sqrt{5}/3$
5182 4459 rt : (6,2,7,-5)
5182 4634 rt : (4,8,-8,-7)
5182 5907 $\exp(\sqrt{3})+s\pi(1/3)$
5182 6233 rt : (3,7,-1,1)
5182 8182 $1/5-e$
5182 8584 $Ei : 3\sqrt{3}/10$
5182 9042 at : $7e$
5182 9094 rt : (5,3,6,3)
5182 9448 id : $\sqrt[3]{7/2}$
5182 9802 rt : (3,9,0,-2)
5183 2461 $10^x : 1/2-\pi/4$
5183 3503 sq : $4\sqrt[3]{2}/7$
5183 5064 rt : (3,8,-7,7)
5183 7152 ln : $2\sqrt{3}+4\sqrt{5}$
5183 8733 $\ell_{10} : 2+3\sqrt{3}/4$
5183 9230 rt : (1,2,6,8)
5184 0000 sq : 18/25
5184 0349 sq : $4+\sqrt{5}/3$
5184 0997 $E : 3\zeta(3)/10$
5184 2628 $E : \sqrt[3]{3}/4$
5184 2887 $\ell_2 : \pi/9$
5184 3288 $e^x : 2\sqrt{2}+2\sqrt{3}$
5184 3414 $\exp(4/5)-s\pi(1/4)$
5184 4385 rt : (2,6,-9,-6)
5184 5026 $t\pi : e+3\pi/2$
5184 5203 $Li_2 : e/6$
5184 6334 sr : $7e/3$
5184 8147 $10^x : 2/3+3\sqrt{2}$
5184 9178 $\ell_{10} : 7\sqrt{2}/3$
5184 9598 rt : (3,5,-4,-3)
5185 1851 cu : $7\sqrt[3]{5}/3$
5185 3808 ln : $1+e/4$
5185 7522 $\ln(4/5)\div s\pi(\pi)$
5185 9386 $E : 9/25$
5185 9439 rt : (1,3,9,4)
5185 9554 $e^x : 4\sqrt{2}-4\sqrt{5}/3$
5185 9984 cu : $3/4+\sqrt{5}/2$
5186 0058 $E : 2\sqrt[3]{2}/7$
5186 0209 rt : (7,9,-3,-3)
5186 0748 rt : (9,8,5,-6)
5186 1331 $\Gamma : (2e/3)^{-3}$
5186 1636 rt : (1,6,-2,9)
5186 2904 id : $\sqrt{3}+\sqrt{5}-\sqrt{6}$
5186 3417 $\Gamma : 10(1/\pi)/9$
5186 3702 $s\pi : 9\pi^2$
5186 4255 rt : (5,3,-5,3)
5186 5222 $\Psi : 3\sqrt{2}/7$
5186 5915 sq : $3e-\pi/3$
5186 6936 at : $1-\pi/2$
5186 7533 rt : (9,4,9,-7)
5186 7968 $J_1 : (\sqrt{2}/3)^{-1/3}$
5186 8401 id : $e/3-3\pi$
5187 0280 rt : (3,1,-8,-4)
5187 1092 rt : (5,8,8,-7)
5187 1483 rt : (8,7,0,-3)
5187 1548 $\ln(2/3)-\exp(\sqrt{2})$
5187 2464 $1/2\div s\pi(\sqrt{2})$
5187 3113 ln : $4\sqrt[3]{2}/3$
5187 3388 rt : (9,8,-4,7)
5187 4856 sq : $2/3-4\sqrt{5}$
5187 9894 at : $9e\pi/4$
5188 1383 rt : (7,5,-7,-4)
5188 1669 rt : (1,2,-9,-6)
5188 1806 rt : (8,3,4,-4)
5188 2642 cr : $e+\pi/4$
5188 3369 $e^x : \sqrt{2}/4+2\sqrt{3}/3$
5188 3544 $Ei : 4(1/\pi)/5$
5188 6256 $J_1 : 9/7$
5188 6741 $\ln(3)\times\exp(\sqrt{2})$
5188 7012 cu : $4/3+e/2$
5188 7252 rt : (4,7,3,-4)
5188 7413 $e^x : 1/4-e/3$
5188 9474 rt : (5,8,-8,9)
5188 9768 $2^x : 9\zeta(3)/4$
5189 2350 $\zeta : \exp(1/(3\pi))$
5189 2358 rt : (8,1,8,5)
5189 4769 $\ln(3)\div\exp(3/4)$
5189 5339 rt : (4,2,-7,-8)
5189 7434 $\ell_2 : 3e/4+3\pi$
5189 7515 $J : 3/8$
5189 7928 rt : (4,3,7,-5)
5190 0704 sinh : $7\pi/6$
5190 2604 id : $2\sqrt{2}-4\sqrt{3}/3$
5190 3565 rt : (7,8,-4,0)
5190 5139 $\ell_2 : 4+\sqrt{3}$
5190 5283 cu : $5\sqrt{3}$
5190 7844 sr : $8\sqrt[3]{3}/5$
5190 7899 $Ei : 3\ln 2/4$
5191 0143 id : $3\sqrt{3}-3\sqrt{5}/4$
5191 0812 rt : (7,9,5,-6)
5191 4611 at : 4/7
5191 4617 rt : (3,8,-9,0)
5191 4828 $2^x : \sqrt{19}$
5191 5827 $E : 9(1/\pi)/8$
5191 7866 $s\pi : 7\sqrt{5}/2$
5191 8169 rt : (7,5,9,-7)
5191 9550 $\ell_{10} : \sqrt{2}/2+3\sqrt{3}/2$
5192 0052 $2/3\div\exp(1/4)$
5192 0457 rt : (4,7,-9,-6)
5192 2033 rt : (3,9,8,-7)
5192 2238 sinh : $\pi^2/2$
5192 3856 cr : $\sqrt[3]{2}/9$
5192 5082 rt : (7,2,-1,-1)
5192 5506 $2^x : \sqrt{2}/4-3\sqrt{3}/4$
5192 5713 rt : (6,3,-6,5)
5193 0106 $J_1 : 2\zeta(3)$
5193 1585 cu : $3\sqrt{2}/4+3\sqrt{5}/4$
5193 6641 rt : (3,6,-2,-1)
5193 7902 $\ell_2 : 3\sqrt{2}+2\sqrt{5}/3$
5193 7964 $J_1 : 5\sqrt[3]{3}/3$
5194 0062 rt : (2,7,1,4)
5194 1224 $J : \exp(-7\pi)$
5194 3891 rt : (1,6,-4,-4)
5194 4073 $E : 5/14$
5194 5181 $\ln(\sqrt{2})-s\pi(1/3)$
5194 5354 sq : $5(e+\pi)/7$
5194 5441 $2^x : 1/4+\sqrt{2}/4$
5194 7586 rt : (7,2,3,2)
5194 8035 rt : (6,8,0,-3)
5195 0469 rt : (1,9,-4,9)
5195 1741 $E : 1/\sqrt[3]{22}$
5195 3063 $Ei : 13/25$
5195 5383 $\Gamma : \sqrt{2}/4$
5195 6871 rt : (8,8,-5,2)
5195 8395 $s\pi : 4/23$
5195 9698 rt : (9,6,6,-6)
5195 9763 $e^x : 1/2-2\sqrt{3}/3$
5195 9841 $Ei : (\sqrt[3]{3}/2)^2$
5196 1524 id : $3\sqrt{3}/10$
5196 3803 rt : (5,5,0,6)
5196 4534 at : $10(e+\pi)/3$
5196 4629 rt : (7,3,-7,7)
5196 5869 rt : (2,8,3,-4)
5196 6038 $t\pi : \sqrt{3}/2+4\sqrt{5}/3$
5196 7137 sr : $4\sqrt{3}/3$
5196 7432 $t\pi : 9\sqrt[3]{5}/4$
5196 8744 ln : $3+\pi/2$
5196 9270 ln : $3+3\pi$
5197 1166 rt : (7,6,7,3)
5197 1642 rt : (6,0,8,5)
5197 1646 $Ei : 4\ln 2$
5197 1801 rt : (5,6,9,-7)
5197 2120 rt : (1,6,-1,-2)
5197 4940 $\ell_{10} : 1+4\sqrt{3}/3$
5197 5052 sq : $3\sqrt{2}/2-3\sqrt{5}/2$
5197 6010 rt : (1,7,-9,-6)
5197 7751 $s\pi(\sqrt{3})-s\pi(e)$
5197 7967 rt : (2,4,7,-5)
5198 2794 $t\pi : 1/3+4\sqrt{5}/3$
5198 3882 rt : (3,2,6,3)
5198 4209 id : $2\sqrt[3]{2}$
5198 4374 rt : (2,8,-6,-5)
5198 6038 id : $3\ln 2/4$
5198 9823 $2^x : 1/\ln(\ln 2/2)$
5199 0417 rt : (9,8,-6,4)
5199 1108 $10^x : 2/11$
5199 1483 rt : (8,3,-8,9)
5199 2501 as : $(1/(3e))^{1/3}$
5199 4802 $\ell_{10} : e/9$
5199 6937 rt : (1,5,2,-5)
5199 7587 ln : $\ln(2e/3)$
5199 7861 id : $8\sqrt[3]{5}/9$
5199 8646 cu : $3e\pi/5$
5199 9083 $2^x : 2e/9$
5200 0000 id : 13/25
5200 0583 $s\pi : 4e-\pi/3$
5200 2095 id : $(\sqrt[3]{3}/2)^2$
5200 2749 as : $2\sqrt{5}/9$
5200 2980 sr : $10\ln 2/3$
5200 3597 rt : (6,8,-9,-6)
5200 3944 rt : (2,7,-9,1)
5200 3982 $J_1 : \sqrt{5/3}$
5200 4537 rt : (6,5,-1,8)
5200 4817 $10^x : \sqrt{22/3}$
5200 7187 $t\pi : 1/\ln(\sqrt[3]{2}/3)$
5200 9339 $\sqrt{5}+\exp(1/4)$
5201 0349 rt : (5,6,-4,-3)
5201 1496 $\ell_{10} : 3/2+2e/3$
5201 1951 rt : (8,9,-3,-2)
5201 3414 cu : $e+4\pi/3$
5201 4968 at : $1/3-e/3$
5201 5526 rt : (7,7,6,-6)
5201 5931 id : $2\sqrt{2}/3+\sqrt{3}/3$
5201 6722 $10^x : 4(1/\pi)/7$
5201 7868 sq : $3\zeta(3)/5$
5201 7900 at : $2\pi^2$
5201 8526 $J_1 : 12/5$
5202 0003 rt : (3,7,-3,-4)
5202 3051 rt : (1,2,-9,4)
5202 4704 rt : (8,5,1,-3)
5202 5531 $\ell_2 : 4\sqrt{3}/3+\sqrt{5}/4$
5202 5917 cu : $\sqrt{7}$
5202 6009 $\sqrt{2}\div e$
5202 6236 rt : (9,7,-8,1)
5202 6895 $\ln(4)-s\pi(1/3)$
5202 8552 $10^x : (e/3)^{-2}$
5202 8722 rt : (3,7,9,-7)
5202 9815 at : $9(1/\pi)/5$
5203 0933 cr : $3/2-e/2$
5203 1468 rt : (4,9,0,-3)
5203 4224 sq : $1/\ln(4)$
5203 5451 rt : (2,4,-9,6)
5203 5477 ln : $7\sqrt[3]{3}/6$
5203 7108 $\ell_{10} : (\ln 3/2)^2$
5203 7197 rt : (5,2,-8,-4)
5203 7783 rt : (8,1,5,-4)
5203 7992 $t\pi : 7(1/\pi)/4$
5203 9473 $J_0 : 6\sqrt{3}/7$
5204 0014 $t\pi : (e/2)^2$
5204 0851 tanh : $2\sqrt[3]{3}/5$
5204 2448 sq : $1/4+\sqrt{2}/3$
5204 4723 rt : (4,7,6,7)
5204 4902 $Li_2 : 5/11$
5204 5173 $\ell_{10} : 1/3+4\sqrt{5}/3$
5204 5781 $E : 10(1/\pi)/9$
5204 8229 rt : (4,2,-9,-5)
5204 9408 $E : \sqrt{2}/4$
5204 9987 $erf : 1/2$
5205 0300 sinh : $(\sqrt[3]{5}/3)^{-1/3}$
5205 0641 ln : $7\zeta(3)/5$
5205 0787 $e^x : 16$
5205 0997 $s\pi : \sqrt{10/3}$
5205 1202 rt : (8,3,9,5)

5205 1760 rt : (9,6,-7,-4)
5205 5091 tπ : $\sqrt{2}+\sqrt{3}/4$
5205 5308 cr : P_{70}
5205 5753 cr : P_{67}
5205 8328 rt : (4,1,8,-5)
5205 8800 sr : $1/2+2e/3$
5205 9370 cr : $3(e+\pi)/5$
5206 0791 rt : (9,8,3,-5)
5206 0948 Ei : $9\sqrt[3]{3}$
5206 1493 cu : $2/3+3\sqrt{3}$
5206 2750 $\ln(2/3)\times\exp(1/4)$
5206 3877 tanh : γ
5206 5082 sπ : $\sqrt{2}/4+2\sqrt{5}$
5206 5282 at : $7e\pi/3$
5206 5556 rt : (5,8,-4,-7)
5206 6304 ζ : $\sqrt[3]{4/3}$
5206 6360 ln : $6\ln 2/7$
5206 7256 E : $6/17$
5206 7573 ζ : $\sqrt[3]{10}$
5206 9168 Ei : $\exp(-\pi)$
5206 9301 Li_2 : $10(1/\pi)/7$
5206 9634 ℓ_{10} : $\sqrt{11}$
5206 9981 rt : (9,4,7,-6)
5207 0393 sq : $2+3\sqrt{5}/4$
5207 2888 rt : (4,9,-6,-5)
5207 3118 J_1 : $22/17$
5207 3441 sr : $3+3\sqrt{5}/2$
5207 3688 tanh : $\sqrt{3}/3$
5207 4832 rt : (5,8,6,-6)
5207 5529 $s\pi(2/5)+s\pi(\pi)$
5207 8076 10^x : $2\sqrt{2}/3+3\sqrt{3}/4$
5207 8583 rt : (9,1,0,1)
5208 1912 e^x : $15/8$
5208 1957 ln : $1/3+3\sqrt{2}$
5208 3111 rt : (6,3,-8,2)
5208 3793 at : 20
5208 5777 cu : $1/4-2e$
5208 7500 cu : $23/20$
5208 8891 at : $e/2-\pi/4$
5208 9262 rt : (1,8,9,-7)
5208 9748 $\exp(e)\times s\pi(\pi)$
5209 3112 rt : (5,3,-1,-1)
5209 3459 2^x : $2+2\pi$
5209 3537 tanh : $5\ln 2/6$
5209 3815 rt : (7,3,-9,4)
5209 3859 2^x : $e+2\pi$
5209 4917 rt : (8,9,-9,-6)
5209 5062 erf : $\sqrt{2}+\sqrt{3}-\sqrt{7}$
5209 8470 e^x : $4e+4\pi/3$
5209 9987 e^x : $(\sqrt[3]{5}/2)^{1/2}$
5210 0073 cr : $\sqrt{2}/10$
5210 0531 rt : (6,8,-5,-5)
5210 1276 rt : (2,7,-1,-3)
5210 3412 ln : $3/2-e/3$
5210 5033 at : $(e)^3$
5210 5153 cr : $3\sqrt{3}-3\sqrt{5}/4$
5210 5254 E : $1/\sqrt[3]{23}$
5210 6176 cr : $4(1/\pi)/9$
5210 6833 rt : (9,5,4,2)
5210 8057 rt : (1,7,-2,-1)
5210 8200 e^x : $15/4$
5210 8536 rt : (3,4,-7,5)
5210 9530 sinh : $1/2$
5211 1394 cu : $\ln(\sqrt{5})$
5211 2028 ln : $4+\sqrt{3}/3$
5211 2721 rt : (7,8,-6,-3)
5211 3398 rt : (4,7,-5,7)
5211 3478 2^x : $(\ln 3/3)^{1/2}$
5211 3611 ln : $1/2+3e/2$
5211 4120 sπ : $\sqrt{2}/3+3\sqrt{5}/2$
5211 4676 at : $9\sqrt{5}$
5211 4943 sr : $4+3\pi/4$
5211 5065 cu : $1/3+\sqrt{2}/3$
5211 5069 rt : (6,6,1,-3)
5211 8923 rt : (8,8,-7,-1)
5211 9426 ℓ_{10} : $4-e/4$
5212 0325 rt : (1,8,4,3)
5212 2171 rt : (5,1,3,2)
5212 2594 rt : (7,9,3,-5)
5212 2678 rt : (9,8,-8,1)
5212 3244 ln : $4e+\pi/2$
5212 3600 ℓ_2 : $\sqrt{2}+\sqrt{3}-\sqrt{6}$
5212 5944 E : $\exp(-\pi/3)$
5212 6049 Ψ : $\sqrt{14/3}$
5212 9257 id : $2\sqrt{2}/3-2\sqrt{3}$
5213 0011 rt : (6,2,5,-4)
5213 2651 rt : (7,5,7,-6)
5213 3976 rt : (8,8,9,-8)
5213 6657 rt : (9,9,8,3)
5213 7144 sr : $e/10$
5213 7970 rt : (4,0,-4,8)
5214 3216 10^x : $25/8$
5214 4208 tπ : $2\sqrt{2}-4\sqrt{5}/3$
5214 5381 rt : (6,2,9,5)
5214 5871 sr : $4\sqrt{2}/3+2\sqrt{5}$
5214 7169 J_1 : $7\sqrt[3]{5}/5$
5214 7312 E : $(1/(3e))^{1/2}$
5214 7332 cr : $\exp(-(e+\pi)/3)$
5214 9494 sπ : $4e/5$
5215 2016 10^x : $2\pi^2/7$
5215 2526 E : $7/20$
5215 2893 rt : (5,5,7,3)
5215 3574 rt : (2,2,8,-5)
5215 3801 rt : (1,9,-9,-7)
5215 4121 10^x : $3+2e$
5215 4257 rt : (3,7,-1,-2)
5215 5967 J : $(\sqrt{5})^{-3}$
5215 6996 at : $4e+3\pi$
5215 8470 10^x : $5\sqrt{2}/3$
5215 9082 J_1 : $9\sqrt[3]{3}/10$
5216 0493 sq : $13/18$
5216 1896 rt : (2,6,-7,-5)
5216 2217 id : $\sqrt{3}/3+4\sqrt{5}$
5216 2462 Ψ : $(e+\pi)^2$
5216 2987 rt : (8,5,-5,9)
5216 4543 sr : P_{30}
5216 7391 sinh : $\sqrt{2}+\sqrt{3}-\sqrt{7}$
5216 8044 rt : (7,5,-4,7)
5216 9642 rt : (1,1,6,3)
5217 1429 ℓ_2 : $3/2+3\sqrt{2}$
5217 2578 at : $(\ln 3/3)^{-3}$
5217 3913 id : $12/23$
5217 4913 ζ : $3\zeta(3)/7$
5217 5664 rt : (6,5,-3,5)
5217 8693 $4/5\times\exp(\sqrt{3})$
5217 9444 E : $\pi/9$
5218 1286 J_1 : $3\sqrt{3}/4$
5218 2779 rt : (9,6,4,-5)
5218 3792 rt : (6,9,9,-8)
5218 5679 $\exp(2/5)\times\exp(\pi)$
5218 5793 tπ : $3\sqrt{2}/2+4\sqrt{3}/3$
5218 6935 rt : (3,3,-5,-3)
5218 8455 rt : (5,5,-2,3)
5218 9184 rt : (8,7,-2,-2)
5218 9993 tanh : $11/19$
5219 0367 sπ : $3/4+3\pi$
5219 2150 id : $\exp(\sqrt[3]{2}/3)$
5219 3706 rt : (9,2,8,-6)
5219 5396 Γ : $1/\ln(1/\pi)$
5219 9233 Ψ : $1/\sqrt[3]{15}$
5219 9497 rt : (5,6,7,-6)
5220 1704 sq : $\pi^2/8$
5220 2324 J_1 : $13/10$
5220 3097 Ψ : $(\ln 3/3)^{1/2}$
5220 4807 rt : (7,7,-4,-3)
5220 5609 rt : (8,3,2,-3)
5220 7716 at : $7(e+\pi)/2$
5220 7759 rt : (7,8,-8,1)
5220 7793 sq : $3/2-3\sqrt{2}$
5220 8007 sinh : $(\ln 3)^2$
5220 9904 Ψ : $\ln(3/2)$
5221 2144 sπ : $\sqrt{2}/2+\sqrt{5}/2$
5221 2353 id : $1/3+4\pi/3$
5221 2994 2^x : $3\sqrt{2}/7$
5221 3266 tπ : $\sqrt{3}/4-\sqrt{5}/2$
5221 3496 ℓ_{10} : $\zeta(3)/4$
5221 3984 10^x : $6\sqrt{3}/7$
5221 4382 rt : (4,7,1,-3)
5221 4699 10^x : $2e/3-2\pi/3$
5221 4881 tπ : $\sqrt[3]{2}/4$
5221 5046 E : $8/23$
5221 6517 ln : $2\sqrt{3}+\sqrt{5}/2$
5221 7970 ζ : $(\sqrt[3]{5}/3)^{1/2}$
5221 8115 sr : $4/3-3\sqrt{2}/4$
5222 1698 rt : (3,1,-9,4)
5222 2246 J : $\exp(-16\pi/7)$
5222 2532 rt : (8,1,6,4)
5222 3296 sr : $3/11$
5222 3905 ℓ_{10} : $1/2+2\sqrt{2}$
5222 3943 Ψ : $\sqrt{17}\pi$
5222 4085 Γ : $(e+\pi)/10$
5222 6121 ln : $\sqrt{21}$
5222 8207 2^x : $1/2+3\sqrt{5}/4$
5222 8750 2^x : $2/3+3e/4$
5222 9434 sinh : $\sqrt[3]{23/3}$
5223 0706 2^x : $3\sqrt{3}/4-\sqrt{5}$
5223 1578 rt : (4,3,5,-4)
5223 1683 Li_2 : $(\sqrt{2}\pi/3)^{-2}$
5223 3339 rt : (1,6,3,2)
5223 3805 sr : $6(1/\pi)/7$
5223 5044 id : $2\sqrt{3}/3-3\sqrt{5}/4$
5223 7692 rt : (4,7,-7,2)
5223 9123 Ei : $(\sqrt{5}/3)^{1/2}$
5224 0774 rt : (7,4,-4,0)
5224 1647 cu : $4/3+2\sqrt{5}/3$
5224 1816 cr : $4-\sqrt{2}/3$
5224 2118 $\ln(\sqrt{3})\times s\pi(2/5)$
5224 2823 2^x : $\sqrt{2}+3\sqrt{5}/4$
5224 3172 $\sqrt{5}\times\exp(2)$
5224 3188 sq : $\sqrt{2}-\sqrt{3}-\sqrt{5}$
5224 6443 2^x : $4-\sqrt{2}/4$
5224 6512 $\ln(2/3)-\exp(3/4)$
5224 7816 E : $1/\sqrt[3]{24}$
5224 7816 sq : $3/2+3e/4$
5224 8131 θ_3 : $\exp(-e/2)$
5224 8362 sq : $\sqrt{19}\pi$
5224 9316 rt : (4,1,9,5)
5225 0185 J_1 : $2(e+\pi)/9$
5225 0434 rt : (1,7,-3,7)
5225 0872 E : $\ln 2/2$
5225 2556 rt : (7,7,4,-5)
5225 5213 sr : $2\sqrt{3}+4\sqrt{5}$
5225 5536 E : $\sqrt{3}/5$
5225 6496 2^x : $7\ln 2/8$
5225 6719 e^x : $5\pi^2/2$
5225 7402 at : $3e+4\pi$
5225 8812 rt : (3,8,-5,0)
5225 9007 2^x : $3+e/2$
5225 9332 2^x : $(e)^{-1/2}$
5226 4587 rt : (7,3,8,-6)
5226 4862 rt : (6,0,-8,9)
5226 8473 rt : (4,7,-7,-5)
5226 8813 Γ : $1/9$
5226 8924 sr : $9\sqrt{2}/2$
5226 9127 $\ln(4/5)-s\pi(\sqrt{3})$
5226 9179 rt : (9,5,-8,8)
5227 0508 sr : $4\sqrt{2}/3+\sqrt{3}/4$
5227 0977 rt : (3,7,7,-6)
5227 3247 sq : $\sqrt{2}/3-3\sqrt{3}/2$
5227 5795 id : $1/\sqrt[3]{7}$
5227 7021 rt : (6,7,2,8)
5227 9077 $s\pi(\sqrt{5})\times s\pi(e)$
5228 3387 sq : $3e/4+\pi/3$
5228 7135 sinh : $\sqrt[3]{14}$
5228 7560 λ : $\ln(\pi/3)$
5228 7874 ℓ_{10} : $3/10$
5228 7965 id : $4\sqrt{2}+\sqrt{3}/2$
5228 8465 ln : $4/3+\sqrt{2}/4$
5228 9532 e^x : $2+\sqrt{2}/4$
5229 0550 e^x : $\exp(-\sqrt{3}/2)$
5229 1283 rt : (2,5,-7,-8)
5229 2035 rt : (5,0,-7,7)
5229 2534 rt : (8,5,-7,6)
5229 4986 rt : (6,8,-2,-2)
5229 5563 e^x : $4(e+\pi)/9$
5229 6525 cu : $4\sqrt{2}+3\sqrt{3}/2$
5229 6745 tπ : $2e-2\pi$
5230 0002 $\ln(3)\times\ln(4)$
5230 0078 $e+\ln(\sqrt{5})$
5230 3069 $\exp(2)-s\pi(1/3)$

5230 3798 $10^x : 3 - 3\sqrt{3}/2$
5230 6786 $\Gamma : \exp(-e/2)$
5230 6845 $J_1 : (\sqrt{2}\pi/2)^{1/3}$
5230 7561 rt : (9,8,1,-4)
5230 9143 $\Psi : 3\sqrt[3]{3}/2$
5230 9289 $\ell_{10} : \sqrt{2}/4 + 4\sqrt{5}/3$
5230 9373 rt : (8,8,-9,-4)
5231 3061 rt : (7,4,4,2)
5231 3952 rt : (6,4,2,-3)
5231 4180 $t\pi : e\pi/3$
5231 4249 rt : (5,3,-9,-3)
5231 4845 rt : (1,8,2,-1)
5231 7929 $\ell_{10} : P_{79}$
5232 0412 rt : (9,4,5,-5)
5232 0508 $Ei : 19/22$
5232 1322 at : 21
5232 3694 cr : $9\pi/8$
5232 4125 rt : (2,8,1,-3)
5232 4277 $10^x : (\sqrt{3}/2)^{1/2}$
5232 5998 rt : (7,5,-6,4)
5232 6087 at : $2\sqrt[3]{3}/5$
5232 6128 rt : (1,7,-5,-8)
5232 7250 rt : (5,8,4,-5)
5232 8330 $\Psi : 9\zeta(3)/5$
5232 8819 rt : (8,8,7,-7)
5232 9561 sr : $3 - e/4$
5233 3550 rt : (9,0,9,6)
5233 4016 rt : (4,0,-6,5)
5233 5363 $2^x : (\sqrt[3]{2}/3)^{-1/3}$
5233 5392 cr : $3\sqrt{3}/4 + \sqrt{5}$
5233 8067 $\Psi : (\pi/2)^{-2}$
5233 9745 $E : 2\zeta(3)/7$
5234 0546 rt : (5,4,8,-6)
5234 1537 sr : $\sqrt[3]{25/2}$
5234 1600 sr : $3/4 + \pi/2$
5234 4075 rt : (2,4,5,-4)
5234 7627 rt : (1,8,7,-6)
5234 8220 $\Psi : 1/\ln(\sqrt[3]{4})$
5234 9781 at : γ
5235 0359 rt : (1,9,5,8)
5235 0396 sinh : $1/\ln(\pi/2)$
5235 0897 $e^x : 3 - 2\sqrt{5}/3$
5235 2050 $t\pi : \ln(\pi/2)$
5235 2854 rt : (7,8,8,3)
5235 3412 $\ell_2 : 9\sqrt{5}/7$
5235 3995 sr : $\sqrt{2}/2 - \sqrt{3}/4$
5235 4796 rt : (2,8,-4,-4)
5235 5013 $\Psi : (e/3)^{-1/3}$
5235 6195 $\ell_2 : 23$
5235 6446 $e^x : 8/19$
5235 7807 $t\pi : 1/\ln(1/(3e))$
5235 9877 id : $\pi/6$
5236 4216 ln : $2\sqrt{2} - \sqrt{5}$
5236 4723 rt : (2,0,9,5)
5236 6617 rt : (8,7,-2,9)
5236 8609 $J_1 : (\sqrt[3]{5})^{1/2}$
5236 9295 $J_1 : 17/13$
5237 1722 rt : (7,8,-8,-6)
5237 1901 $\ln(4) \div \ln(\sqrt{3})$
5237 2354 $t\pi : 4\sqrt{2}/3 - \sqrt{3}$
5237 2467 sr : $4\sqrt{3} - \sqrt{5}/4$
5237 2663 ln : $3\sqrt{2}/2 - \sqrt{3}/4$
5237 4815 rt : (3,4,7,3)
5237 5004 $\sqrt{5} - \ln(3/4)$
5237 5646 $e^x : 15/7$
5237 5832 rt : (2,4,-8,-5)
5237 6157 $e^x : 7\zeta(3)/3$
5237 6850 rt : (5,8,-1,-2)
5237 7203 $2^x : \sqrt[3]{6}$
5237 7252 rt : (4,7,-9,-3)
5237 7946 rt : (6,5,-5,2)
5237 9971 rt : (8,9,-5,-1)
5238 0303 at : $5\ln 2/6$
5238 0622 $E : \sqrt[3]{5}/5$
5238 0831 $e^x : \sqrt{2}/3 - \sqrt{5}/2$
5238 0952 id : 11/21
5238 3707 $2^x : 10\zeta(3)/9$
5238 5620 rt : (7,9,1,-4)
5238 6738 ln : $4e - 2\pi$
5238 7085 $t\pi : \exp(e/3)$
5238 7156 rt : (6,8,-7,-5)
5238 7161 cr : $3/2 + 3e/4$
5238 8227 cu : $4e/3 + 2\pi$
5238 8532 rt : (6,9,7,-7)
5238 9552 $\ln(2/3) \div s\pi(e)$
5239 0189 $e^x : 9e\pi/2$
5239 3674 id : $7\sqrt[3]{3}/4$
5239 3733 $\zeta : (2\pi)^{-2}$
5239 4268 $1/4 - s\pi(e)$
5239 5293 cu : $3e/4 + \pi/4$
5239 7639 rt : (1,6,6,-7)
5239 8032 $erf : 2\sqrt[3]{2}/5$
5239 9067 at : $5e\pi/2$
5239 9845 rt : (7,5,5,-5)
5239 9887 cr : P_{75}
5240 0899 rt : (8,5,-1,-2)
5240 2428 $\Psi : \ln 2/7$
5240 2834 $\ell_{10} : 2e - 2\pi/3$
5240 3573 rt : (1,6,1,-9)
5240 3883 rt : (9,5,-7,1)
5240 5030 rt : (2,3,-2,8)
5240 7435 rt : (3,9,4,1)
5240 7513 $e^x : (\sqrt[3]{2})^{-1/3}$
5240 7897 $\Psi : 5\sqrt{3}/4$
5240 9491 $\Gamma : 6/17$
5241 2160 rt : (4,9,-2,-2)
5241 2237 sr : $4 - 3\sqrt{5}/4$
5241 4111 $3/5 \div \ln(\pi)$
5241 4411 rt : (7,1,9,-6)
5241 4551 $e^x : 22/9$
5241 5982 $J : 1/14$
5241 8119 rt : (7,7,-1,7)
5241 8359 cr : $2\sqrt{2}/3 + 3\sqrt{3}/2$
5241 9091 $\ln(3/4) \times \exp(3/5)$
5241 9130 as : $\sqrt{2} + \sqrt{3} - \sqrt{7}$
5241 9252 $\ell_{10} : 3\sqrt{3}/2 + \sqrt{5}/3$
5241 9959 rt : (1,9,3,-9)
5241 9984 sq : $2\sqrt{3} - 3\sqrt{5}$
5242 1392 rt : (5,4,-5,-3)
5242 2188 rt : (3,5,8,-6)
5242 2338 sinh : $4e/9$
5242 2567 rt : (8,1,3,-3)
5242 5686 rt : (5,9,-8,1)
5242 6019 $\ln(\sqrt{3}) \times \exp(5)$
5242 7172 cu : $5\sqrt{2}/4$
5242 9519 rt : (1,8,-4,4)
5242 9806 $J : \exp(-11\pi)$
5243 0068 rt : (1,9,-7,-6)
5243 0184 cu : $2/3 - 2e$
5243 0281 $J_1 : \sqrt{17/3}$
5243 2610 $10^x : (\ln 2/2)^{-1/3}$
5243 4109 $\Gamma : 6\sqrt{5}/5$
5243 4242 rt : (4,7,4,4)
5243 4526 $t\pi : \sqrt{3}/4 - 4\sqrt{5}/3$
5243 8184 rt : (8,5,-9,3)
5243 9906 $e^x : 10\sqrt[3]{2}$
5244 1429 $\Psi : 9\sqrt[3]{3}$
5244 2107 cr : $\sqrt[3]{3}/10$
5244 2548 rt : (2,9,0,1)
5244 2574 $E : e/8$
5244 2896 sq : $2e - 3\pi/2$
5244 5027 rt : (8,3,7,4)
5244 5276 $2^x : (\sqrt{2\pi})^{-1/3}$
5244 5542 $\Gamma : \sqrt{21/2}$
5244 5586 rt : (8,6,8,-7)
5244 6147 rt : (9,8,-4,-3)
5245 1621 rt : (4,2,-1,8)
5245 7143 rt : (4,1,6,-4)
5245 7526 rt : (4,9,0,7)
5245 7686 id : $2e/3 + 3\pi/2$
5245 9294 rt : (9,7,-5,8)
5246 1369 rt : (9,6,2,-4)
5246 1730 sq : $3\sqrt{2}/2 + \sqrt{3}/4$
5246 5418 $t\pi : 4\sqrt[3]{3}/5$
5246 8459 $10^x : 3\sqrt{3}/7$
5246 9507 rt : (5,5,-4,0)
5247 1138 at : $1/\ln(\pi/3)$
5247 1298 $J_1 : 7e/8$
5247 1451 $\ell_{10} : 4\sqrt{2} - 4\sqrt{3}/3$
5247 2503 $J_1 : 21/16$
5247 6178 id : $3\sqrt{2}/4 + 2\sqrt{3}$
5247 6972 rt : (9,2,6,-5)
5247 8661 $2^x : 3\sqrt{3}/2 + \sqrt{5}$
5247 9330 $s\pi : 4\sqrt[3]{3}/7$
5247 9533 $\ell_{10} : 3/4 + 3\sqrt{3}/2$
5247 9577 at : 11/19
5248 1791 $e^x : (4/3)^{-3}$
5248 2128 rt : (4,9,-4,-4)
5248 4048 $t\pi : 2/13$
5248 4376 at : $8e$
5248 5026 $\ell_{10} : 4(e + \pi)/7$
5248 5318 rt : (5,9,8,9)
5248 5839 $\ell_2 : 3\sqrt{3} + \sqrt{5}/4$
5248 5902 $s\pi : P_{14}$
5248 7985 sr : $3 + 3\pi$
5248 7993 $2^x : 14/23$
5248 8176 $s\pi : 3e/4 + \pi/4$
5248 9535 rt : (6,2,-5,9)
5248 9790 $\theta_3 : (\pi/2)^{-3}$
5249 1435 rt : (9,3,1,1)
5249 3081 $J_1 : \exp(\sqrt{3}/2)$
5249 3254 $10^x : 3\sqrt{2} - \sqrt{5}$
5249 3781 rt : (1,9,-5,0)
5249 6388 rt : (9,4,-8,-4)
5249 9542 rt : (6,7,0,5)
5250 1535 $\Psi : 13/6$
5250 1538 rt : (5,2,9,-6)
5250 5077 rt : (6,8,-7,-8)
5250 5346 $s\pi : 1/3 + 2\sqrt{5}/3$
5250 6555 $\theta_3 : (e/2)^{-1/2}$
5250 6601 $\Psi : 1/\ln(1/(3\pi))$
5250 6914 rt : (4,5,-8,-5)
5250 8120 rt : (7,5,-8,1)
5250 8285 ln : $4 - 4\sqrt{3}/3$
5250 8394 tanh : 7/12
5251 0225 rt : (3,9,-3,9)
5251 1771 $\Gamma : 1/\sqrt[3]{5}$
5251 1845 rt : (6,7,8,-7)
5251 4316 id : $\exp(\sqrt[3]{2})$
5251 4704 $\pi \times \exp(4)$
5251 5164 $Ei : 12/23$
5251 5583 sq : $5e/3$
5251 5982 $10^x : 2\sqrt{2} + \sqrt{3}/3$
5251 6571 rt : (5,1,0,-1)
5251 7224 $e^x : 7\sqrt[3]{2}$
5251 8403 rt : (2,5,-9,-5)
5252 0337 rt : (8,9,-7,-5)
5252 2151 rt : (8,7,-4,6)
5252 9527 rt : (6,6,-1,-2)
5252 9686 $\theta_3 : 1/\sqrt{15}$
5253 2991 cu : $1/2 + 3\sqrt{5}$
5253 3447 rt : (3,6,-2,5)
5253 4291 cu : $7(e + \pi)/6$
5253 5478 at : 7π
5253 7304 at : 22
5253 8468 $\ell_2 : 7\pi^2/3$
5254 0176 rt : (9,7,5,2)
5254 2610 $J_1 : 25/19$
5254 2756 rt : (1,5,-1,-1)
5254 2775 cu : $2e\pi/7$
5254 3419 $J_1 : 19/8$
5254 5024 $s\pi : \sqrt{2} - \sqrt{5} + \sqrt{7}$
5254 6169 $t\pi : 8\sqrt[3]{3}/3$
5254 6512 $\exp(\sqrt{2}) - s\pi(1/5)$
5254 7965 $\theta_3 : 2e/7$
5254 8654 $J_1 : (\sqrt{3})^{1/2}$
5254 8713 $t\pi : 1/3 - \sqrt{5}/3$
5254 9719 as : $3\sqrt{2}/4 - \sqrt{5}/4$
5255 0522 $\ell_{10} : 3 + \sqrt{2}/4$
5255 0807 $Li_2 : 11/24$
5255 0854 rt : (7,7,2,-4)
5255 2941 $t\pi : 3\sqrt{2}/4 - \sqrt{5}/3$
5255 4941 rt : (6,2,3,-3)
5255 5145 sq : $2 + 3\pi$
5255 7109 $\exp(1/5) \times s\pi(\pi)$
5255 7431 sinh : $2\sqrt[3]{2}/5$
5255 7626 $\ell_{10} : 3\sqrt{5}/2$
5255 8437 cu : $\sqrt{3}/2 + 3\sqrt{5}$
5255 8777 $t\pi : 2\sqrt[3]{5}/9$
5255 8822 $t\pi : 2\ln 2/9$

5256 1104 ln : $e/3 + \pi/4$
5256 1789 rt : (8,8,5,-6)
5256 3800 ln : $2 + 3\sqrt{3}/2$
5256 6518 rt : (2,9,-2,-8)
5256 7353 rt : (5,3,4,2)
5256 7484 id : $2\sqrt{2} - 3\sqrt{5}/2$
5256 8258 rt : (7,3,6,-5)
5256 8440 rt : (9,9,8,-8)
5256 8712 rt : (4,0,-8,2)
5257 0036 rt : (1,5,-2,5)
5257 0074 rt : (3,9,2,-1)
5257 0246 θ_3 : 11/18
5257 2041 tanh : $((e+\pi)/2)^{-1/2}$
5257 2506 rt : (7,2,-8,8)
5257 2864 ln : 2/25
5257 3111 $1/2 \div s\pi(2/5)$
5257 4030 rt : (5,2,-4,7)
5257 5022 rt : (8,4,9,-7)
5257 5032 sr : $\sqrt{2} - \sqrt{3} + \sqrt{7}$
5257 7599 rt : (3,7,5,-5)
5257 9509 at : $9\pi^2/4$
5258 1392 rt : (6,2,7,4)
5258 1855 cu : $6\pi/7$
5258 2089 rt : (4,8,8,-7)
5258 4712 ℓ_{10} : $1 + 3\pi/4$
5258 6767 rt : (5,9,-6,8)
5259 1003 2^x : $\sqrt[3]{15}$
5259 2019 rt : (3,3,-8,9)
5259 2833 Ψ : $2e/9$
5259 2950 Ψ : $\ln(1/\sqrt{2})$
5259 3230 2^x : $3/4 - 3\sqrt{5}/4$
5259 5155 sq : 21/17
5259 5748 rt : (3,3,9,-6)
5259 7841 ℓ_2 : $8\sqrt[3]{2}/7$
5259 9301 $\ln(4) \times \exp(3/5)$
5259 9501 Ψ : 13
5259 9969 e^x : $1 - \sqrt{3}/3$
5260 0038 K : $3\pi/10$
5260 1651 Ψ : $(\ln 3)^{1/3}$
5260 2016 rt : (9,7,-7,5)
5260 5501 rt : (1,7,-8,-6)
5260 5630 ln : 5/23
5260 9309 ln : 13/22
5260 9852 2^x : $e + 3\pi/2$
5261 0434 rt : (2,6,-5,-4)
5261 0474 at : $10\sqrt{5}$
5261 0655 sq : $3/4 + 3\pi$
5261 2260 rt : (8,9,1,9)
5261 4465 10^x : $10\sqrt{5}/7$
5261 4797 tanh : $1/\sqrt[3]{5}$
5261 5544 rt : (7,7,-3,4)
5261 8455 $s\pi$: $(\sqrt[3]{2}/3)^2$
5261 8856 rt : (9,8,-1,-3)
5261 9151 ln : $1/3 + e/2$
5262 0459 rt : (6,5,-7,-1)
5262 0923 E : 1/3
5262 1342 10^x : P_{63}
5262 2441 rt : (5,7,8,3)
5262 2822 ℓ_{10} : $2 + e/2$
5262 3149 10^x : 23/10
5262 5466 rt : (3,5,-2,-2)
5262 5873 rt : (3,9,-5,2)
5262 6008 2^x : $1/\ln(3/2)$
5262 6571 rt : (4,9,-2,2)
5262 9362 E : $(\ln 2)^3$
5263 0123 rt : (8,7,-4,-1)
5263 1206 ℓ_{10} : $8\sqrt[3]{2}/3$
5263 1477 sq : $\sqrt{2}/3 + \sqrt{5}/2$
5263 1578 id : 10/19
5263 3403 sq : $3\sqrt{2}/2 - 2\sqrt{5}$
5263 4137 cu : $2\sqrt{2} - 3\sqrt{5}/4$
5263 4260 e^x : $4\sqrt{2}/3 + \sqrt{5}/4$
5263 4575 rt : (6,9,5,8)
5263 6501 2^x : 20/11
5263 7533 rt : (9,4,3,-4)
5263 8260 J_1 : $(4/3)^3$
5263 8930 ln : $3/4 + 2\sqrt{2}/3$
5263 9298 rt : (2,2,-9,-5)
5264 0948 $\exp(5) + \exp(\sqrt{2})$
5264 1929 sr : $6e/7$
5264 3216 $s\pi$: 3/17
5264 4325 $\exp(2/3) \div s\pi(\pi)$
5264 7579 rt : (5,8,2,-4)
5264 8996 rt : (6,5,9,-7)
5265 0888 $t\pi$: $\exp(e)$
5265 3419 rt : (1,3,7,3)
5265 5021 rt : (6,6,-8,-5)
5265 5376 sr : $2\ln 2/5$
5265 5903 Γ : $((e+\pi)/2)^{-1/2}$
5265 6629 rt : (7,9,-1,-2)
5265 6717 rt : (4,6,-8,6)
5265 6745 rt : (9,0,7,5)
5265 8265 10^x : $1/\ln(2\pi/3)$
5265 8329 rt : (8,3,0,-2)
5265 8903 ln : $\ln(2e)$
5265 9176 2^x : $9\sqrt{2}/7$
5265 9197 rt : (2,8,-3,8)
5266 0161 sq : $1/4 + 4\sqrt{3}$
5266 2420 rt : (6,2,-7,6)
5266 2790 rt : (7,6,-7,1)
5266 3726 $\exp(\sqrt{3}) \div s\pi(1/3)$
5266 4038 sr : $1/\sqrt{13}$
5266 7089 rt : (5,4,6,-5)
5267 1228 Γ : $5(e+\pi)/4$
5267 2529 10^x : $3 + e/3$
5267 3040 cu : $2e\pi/3$
5267 3671 rt : (4,7,-1,-2)
5267 4607 cr : $\sqrt{3}/3 + 4\sqrt{5}/3$
5267 5143 ℓ_{10} : $-\sqrt{3} + \sqrt{6} + \sqrt{7}$
5267 6628 cu : $3/2 + \pi/3$
5267 7598 rt : (1,8,5,-5)
5267 8079 cr : $3 + \sqrt{5}/4$
5268 1557 J_1 : $\sqrt{3} + \sqrt{5} - \sqrt{7}$
5268 1687 rt : (9,7,9,-8)
5268 3891 rt : (9,9,-2,8)
5268 5644 $3/4 + \ln(4/5)$
5268 5961 rt : (3,1,-6,-3)
5268 7728 rt : (2,6,-4,-1)
5268 7758 rt : (8,1,4,3)
5269 1214 at : $8e\pi/3$
5269 7216 sr : $(2e)^{1/2}$
5269 8012 rt : (1,4,9,-6)
5269 8699 $\ln(4) + \exp(\pi)$
5269 8955 rt : (8,7,-6,3)
5270 0360 tanh : $(e+\pi)/10$
5270 0746 2^x : $\sqrt{2}/3 + \sqrt{3}/2$
5270 0885 $\exp(\pi) \div \exp(e)$
5270 1720 ℓ_{10} : $7\sqrt[3]{3}/3$
5270 3794 rt : (8,6,6,-6)
5270 4627 sr : 5/18
5270 5332 ℓ_2 : $2 - \sqrt{5}/4$
5270 9025 rt : (3,8,-8,-6)
5270 9051 cu : $2e/3 + \pi/4$
5270 9160 rt : (5,9,-8,3)
5271 0011 cr : $1/2 - \sqrt{2}/4$
5271 0312 10^x : P_{47}
5271 1959 rt : (4,2,-3,5)
5271 5132 ℓ_{10} : $3\ln 2/7$
5271 8511 rt : (8,5,8,4)
5272 0102 $t\pi$: $\sqrt[3]{10}$
5272 0225 rt : (7,9,-1,-3)
5272 4954 $s\pi$: $\sqrt{2}/8$
5272 5461 Γ : $9\sqrt[3]{2}$
5272 7117 J_1 : $(\sqrt[3]{5}/3)^{-1/2}$
5272 7723 rt : (4,6,9,-7)
5272 8869 rt : (1,8,9,-4)
5273 0284 cr : $4\sqrt{2}/3 + 3\sqrt{5}/4$
5273 0461 10^x : $3\sqrt{3}/2 + \sqrt{5}$
5273 2824 rt : (7,9,2,7)
5273 4543 at : 23
5273 5302 rt : (4,1,7,4)
5273 8486 $s\pi(1/4) \times s\pi(\sqrt{3})$
5274 0022 at : $7\pi^2/3$
5274 1198 rt : (7,5,3,-4)
5274 1538 $s\pi$: $5(1/\pi)/9$
5274 2782 Ψ : $(3\pi/2)^{1/2}$
5274 3513 2^x : 11/18
5274 3572 2^x : $3\sqrt{2}/4 + \sqrt{5}/2$
5274 3960 e^x : $5e/9$
5274 5139 Ei : 16/13
5274 5656 rt : (7,8,9,-8)
5274 6394 $\sqrt{2} + \exp(\sqrt{2})$
5274 6416 10^x : $3\sqrt{2}/4 + 4\sqrt{5}/3$
5275 0029 cu : $4\sqrt{2} - 4\sqrt{5}$
5275 0709 2^x : $2\sqrt{2} - 2\sqrt{5}/3$
5275 2523 id : $\sqrt{7/3}$
5275 4057 $t\pi$: $2\sqrt[3]{3}/7$
5275 4192 Γ : $(\sqrt{2}\pi/2)^{-2}$
5275 6927 J_1 : $(\ln 3)^3$
5275 9459 sr : $4e + \pi/2$
5275 9776 rt : (9,4,-9,9)
5276 0927 at : $\exp(\pi)$
5276 1814 rt : (9,7,-9,2)
5276 2824 rt : (7,1,7,-5)
5276 3431 rt : (6,4,-2,9)
5276 3525 $s\pi$: $\exp(-\sqrt{3})$
5276 3771 ζ : $(e+\pi)^{-2}$
5276 3779 ℓ_{10} : $3\sqrt{2}/4 + 4\sqrt{3}/3$
5276 4922 J_1 : $\sqrt[3]{7/3}$
5276 5896 cu : $\ln(e+\pi)$
5276 5924 id : $3e/4 - 4\pi$
5276 7845 Ψ : $7\pi^2/2$
5276 8800 rt : (6,7,-2,2)
5277 0195 rt : (4,7,-5,-4)
5277 4512 rt : (3,5,6,-5)
5277 5103 sr : $7(1/\pi)/8$
5277 5316 cu : $4\sqrt{2}/7$
5277 5990 rt : (2,8,-5,-1)
5278 2144 sr : $3/4 - \sqrt{2}/3$
5278 2752 rt : (7,2,1,1)
5278 4575 rt : (6,8,-4,-1)
5278 5464 rt : (6,7,6,-6)
5278 6404 sq : $1 - \sqrt{5}$
5278 8701 $\pi + \ln(4)$
5278 9332 ℓ_{10} : $4/3 + 3e/4$
5278 9485 e^x : $1 + 3\sqrt{3}/2$
5279 0521 rt : (7,1,-9,-4)
5279 0941 ln : $\sqrt{3}/3 + \sqrt{5}/2$
5279 0983 Ei : $3\sqrt[3]{3}/5$
5279 1966 cr : $4 - \sqrt{3}/4$
5279 2020 rt : (9,9,6,-7)
5279 2216 $\exp(2/5) + s\pi(\sqrt{2})$
5279 3190 e^x : P_{34}
5279 3890 E : $\ln(\sqrt[3]{3}/2)$
5279 4458 E : $\exp(-\sqrt{5}/2)$
5279 8154 $s\pi$: $1/2 + 3\sqrt{5}/4$
5279 8462 Ei : $5\sqrt[3]{2}/4$
5279 9206 rt : (5,2,-6,4)
5280 2384 rt : (8,9,-1,6)
5280 4985 rt : (4,3,-9,-5)
5280 5384 rt : (1,7,4,6)
5280 6677 $t\pi$: $9\sqrt[3]{2}/7$
5280 7444 at : 7/12
5280 7888 J : 8/19
5280 9505 rt : (1,8,4,-9)
5280 9916 $\pi \times \ln(\sqrt{5})$
5281 1038 Ψ : $\ln(\sqrt[3]{5}/3)$
5281 1713 rt : (2,7,9,-7)
5281 3021 rt : (5,9,9,-8)
5281 4815 e^x : $3 + \sqrt{3}$
5281 5915 at : $4(e+\pi)$
5281 7536 J_0 : 25/17
5281 7930 Ψ : $-\sqrt{3} - \sqrt{5} + \sqrt{6}$
5281 8132 rt : (9,6,0,-3)
5281 8720 as : $2\sqrt[3]{2}/5$
5281 9760 rt : (7,4,-5,8)
5282 0037 rt : (4,9,-4,-3)
5282 2021 rt : (5,9,-8,-6)
5282 2484 sq : $6\sqrt[3]{3}/7$
5282 4088 rt : (8,7,-8,-5)
5282 4241 2^x : $2\sqrt{2}/3 + 4\sqrt{3}/3$
5282 4364 as : $\sqrt{3} - \sqrt{5}$
5282 5557 $\ln(2) - \exp(1/5)$
5282 6180 E : $(3\pi)^{-1/2}$
5282 6426 cr : $4/3 + \sqrt{5}$
5282 6881 $t\pi$: $2\sqrt{3}/3$
5282 7377 ℓ_{10} : $(3/2)^3$
5283 0335 rt : (5,5,-6,-3)
5283 2083 ℓ_2 : $\sqrt[3]{3}$
5283 4872 J_1 : $1/\ln(\sqrt{2}/3)$

5283 6516 rt : (2,8,-1,-2)
5283 7727 $10^x : e/3+2\pi$
5283 8661 $t\pi : (2\pi)^{1/3}$
5283 9418 $\Psi : 6\zeta(3)/7$
5284 0308 $J_1 : 7\sqrt[3]{5}/9$
5284 1276 $2^x : 2+3e/2$
5284 1407 rt : (9,2,4,-4)
5284 4040 $Ei : 1/\sqrt[3]{7}$
5284 4366 $2^x : 9\sqrt[3]{5}$
5284 4489 $\ell_{10} : \sqrt{2}-\sqrt{5}/2$
5284 4877 rt : (8,8,3,-5)
5284 5324 $Ei : 3e\pi/7$
5284 6333 cr : $2+\pi/2$
5284 6516 $e^x : 3\sqrt{2}/10$
5284 6718 sr : P_{72}
5284 6967 rt : (7,7,-5,1)
5284 7397 sr : P_{69}
5284 8158 $\ell_{10} : 3\sqrt{2}-\sqrt{3}/2$
5284 8665 $E : (\sqrt[3]{5}/3)^2$
5285 1278 rt : (7,6,5,2)
5285 3332 rt : (6,0,4,3)
5285 3959 $\ell_2 : 6\zeta(3)/5$
5285 4019 rt : (5,9,6,6)
5285 4215 rt : (9,9,-4,5)
5285 4556 $\ell_2 : 1/3+2e$
5285 5200 $J_0 : 7\sqrt[3]{2}/6$
5285 5354 cr : 25/7
5285 5395 rt : (1,6,1,-1)
5285 6562 $\zeta : 3\sqrt{5}/5$
5285 8547 sq : $(\sqrt[3]{4}/3)^{-1/3}$
5285 9547 id : $1-\sqrt{2}/3$
5286 0410 rt : (6,2,-9,3)
5286 0486 sinh : 23/19
5286 2190 $Li_2 : (3\pi/2)^{-1/2}$
5286 2522 tanh : 10/17
5286 2553 $K : 2\sqrt{2}/3$
5286 2929 rt : (8,4,7,-6)
5286 3079 $erf : 8(1/\pi)/5$
5286 3453 rt : (8,4,-8,7)
5286 4013 $\Gamma : 6\pi/5$
5286 4024 $\Psi : 5\sqrt[3]{3}/7$
5286 5457 rt : (9,2,8,5)
5286 8563 sr : $\sqrt{5}/8$
5286 9309 $e^x : 4(1/\pi)/3$
5286 9458 $Ei : (2\pi/3)^{-3}$
5287 0919 $\Gamma : 7/12$
5287 2659 rt : (4,8,6,-6)
5287 3015 at : $((e+\pi)/2)^{-1/2}$
5287 4631 $\zeta : 15/7$
5287 4824 sr : $7(e+\pi)/2$
5287 5528 $J : \zeta(3)/3$
5287 6637 $\ell_2 : \ln 2$
5287 7481 sq : $\sqrt[3]{13}$
5287 8503 rt : (5,2,7,-5)
5288 2848 $e^x : \sqrt[3]{5}$
5288 3081 cu : $7\ln 2/6$
5288 5139 $t\pi : 3e$
5288 6484 $\ell_2 : 3\sqrt{3}/2+4\sqrt{5}$
5288 8036 $\ell_2 : 1+4\sqrt{2}/3$
5288 8470 $10^x : 3-\sqrt{2}$
5288 8938 cr : $2\sqrt{2}+\sqrt{5}/3$
5288 9514 ln : $6\sqrt{2}/5$
5289 0330 rt : (6,4,8,4)
5289 0838 cu : $2\pi^2/5$
5289 1350 $\ln(3)-s\pi(\pi)$
5289 1795 rt : (2,6,-6,-8)
5289 2529 rt : (2,8,-2,-3)
5289 2561 sq : 8/11
5289 3474 cr : $4\sqrt{3}-3\sqrt{5}/2$
5289 6027 $t\pi : 2\sqrt{2}/3+3\sqrt{5}/4$
5289 6119 id : $\sqrt{2}/2-\sqrt{5}$
5289 6359 tanh : $(\ln 2/2)^{1/2}$
5290 0092 $\theta_3 : 3\ln 2/8$
5290 0432 sq : $1/3-3\sqrt{2}/4$
5290 1557 rt : (8,9,-7,0)
5290 3751 $e^x : \sqrt{2}+\sqrt{3}+\sqrt{5}$
5290 3956 $10^x : 4+2\sqrt{2}/3$
5290 6419 cr : $\sqrt{2}/2-\sqrt{5}/4$
5290 7532 $J_1 : 5\sqrt{2}/3$
5290 7780 sq : $\exp(-1/\pi)$
5290 9491 $J_1 : 4/3$
5290 9565 sq : P_{73}
5291 0694 rt : (2,0,7,4)
5291 0890 $10^x : \sqrt{2}/4+\sqrt{3}/3$
5291 3368 sr : $2\sqrt[3]{2}/9$
5291 5026 sr : 7/25
5291 5374 at : 24
5291 6025 rt : (6,5,-9,-4)
5291 7069 at : $1/\sqrt[3]{5}$
5291 7164 $10^x : 1/3+3e/4$
5291 7724 id : P_3
5291 9226 at : $(\ln 2/2)^{-3}$
5292 1476 $\exp(3/4)-s\pi(1/5)$
5292 2558 $\exp(3/5)+s\pi(1/4)$
5292 4033 $J_1 : 3\pi/4$
5292 5000 $1/4\times\exp(3/4)$
5292 5888 tanh : $\exp(-\sqrt[3]{4}/3)$
5292 6357 rt : (9,7,7,-7)
5292 8542 $t\pi : 4e-4\pi/3$
5292 9212 rt : (3,8,-8,4)
5292 9997 $\ell_{10} : 2e/3+\pi/2$
5293 1658 rt : (3,6,-9,-6)
5293 2691 id : $2\sqrt{2}/3-2\sqrt{5}$
5293 2798 $\ln(\sqrt{5})\times\exp(\sqrt{5})$
5293 4164 rt : (4,1,-5,-9)
5293 4655 rt : (5,4,-1,7)
5293 5385 rt : (6,9,3,5)
5293 6056 rt : (2,7,-8,6)
5293 6584 rt : (7,7,0,-3)
5293 8262 $s\pi : 3\sqrt{3}+4\sqrt{5}/3$
5293 8901 id : $2\sqrt{2}-3\sqrt{3}/4$
5293 8920 rt : (8,5,-3,-1)
5293 9789 cr : $3+\sqrt{3}/3$
5294 1176 id : 9/17
5294 2380 $s\pi : (\sqrt[3]{3}/2)^{-1/2}$
5294 2590 $10^x : 1/\ln(\ln 3)$
5294 3540 rt : (2,5,-5,8)
5294 3725 sq : $4-3\sqrt{2}/2$
5294 4684 $\ell_2 : \sqrt{3}/5$
5294 4898 cr : $8\sqrt{5}/5$
5294 5466 rt : (9,3,-6,1)
5294 7310 as : P_{35}
5294 7758 $\ln(\sqrt{3})\times s\pi(\sqrt{2})$
5294 8575 id : $1/3+3\sqrt{3}$
5294 8983 $10^x : 2/3-2\sqrt{2}/3$
5295 0749 $J_1 : (\sqrt[3]{2}/3)^{-1/3}$
5295 0849 cu : $1/4+\sqrt{5}/4$
5295 1224 rt : (1,7,2,-7)
5295 3255 rt : (6,8,-5,-4)
5295 3853 $\zeta : \sqrt{3}-\sqrt{5}+\sqrt{7}$
5295 5134 cr : $3/4+2\sqrt{2}$
5295 5206 $\ell_{10} : 2/3+e$
5295 5261 rt : (6,5,7,-6)
5295 6412 $J_1 : 10\zeta(3)/9$
5295 6603 $\exp(5)\times s\pi(1/3)$
5295 7817 $e^x : 2\sqrt{2}-2\sqrt{3}$
5295 8578 $e^x : \sqrt{3}/3+3\sqrt{5}/4$
5295 8702 $\exp(1/3)-s\pi(1/3)$
5295 9481 rt : (4,9,-4,-1)
5296 2986 rt : (7,3,4,-4)
5296 3796 $\ell_{10} : 3/2+4\sqrt{2}/3$
5296 4603 $E : 2\sqrt[3]{3}/9$
5296 4841 $e^x : 2+3\sqrt{5}/4$
5296 5134 $\Psi : 4e/5$
5296 6007 rt : (2,9,6,-3)
5297 0468 rt : (3,6,8,3)
5297 0878 rt : (5,1,-8,-8)
5297 1353 $e^x : 3\sqrt{2}-4\sqrt{5}/3$
5297 2314 $t\pi : (2e)^3$
5297 4710 cu : $2+\pi/2$
5297 4873 $\exp(2)+\exp(\pi)$
5297 5524 rt : (5,6,-2,-2)
5297 5935 $s\pi : 3\sqrt{2}/2-3\sqrt{3}/4$
5297 6478 $Ei : \sqrt{3}/2$
5297 7358 rt : (9,6,-6,9)
5297 7694 $E : 8/25$
5297 8223 rt : (8,1,1,-2)
5297 9216 rt : (1,6,-3,3)
5298 0122 $Li_2 : 6/13$
5298 1612 cr : $4e/3+4\pi$
5298 3005 ln : $(\ln 2/2)^{1/2}$
5298 5209 rt : (5,7,-7,1)
5298 5517 $\zeta : \exp(\sqrt[3]{2}/3)$
5298 6403 rt : (7,9,0,4)
5298 7213 $\ell_{10} : 2\sqrt{2}+\sqrt{5}/4$
5299 0278 at : $2-\sqrt{2}$
5299 0348 rt : (7,1,8,5)
5299 2377 $E : \sqrt{5}/7$
5299 3569 $\zeta : 1/\ln(5/2)$
5299 4357 at : $9e$
5299 5338 $e^x : 3e/2-3\pi/2$
5299 5568 $\ln(3/5)\div s\pi(\sqrt{2})$
5299 5870 rt : (6,4,-9,-5)
5299 6157 $erf : \ln(5/3)$
5299 6704 $10^x : 2+e$
5299 6795 sinh : $\pi^2/5$
5299 7372 $10^x : 7(1/\pi)/3$
5299 7947 rt : (2,9,-9,-7)
5299 8066 id : $4e/3-2\pi/3$
5299 8501 $Ei : 8\sqrt{5}/3$
5299 9637 ln : $5e/8$
5299 9907 rt : (4,7,2,1)
5300 2668 sq : $4/3-3e/2$
5300 3091 rt : (7,8,7,-7)
5300 3147 $e^x : \exp(-\sqrt[3]{5}/2)$
5300 3463 $e^x : 1/\sqrt[3]{13}$
5300 5196 $\theta_3 : (e+\pi)/6$
5300 5243 at : $(e+\pi)/10$
5300 5264 rt : (3,3,7,-5)
5300 5834 id : $\sqrt{2}-4\sqrt{5}$
5300 8559 rt : (7,4,-7,5)
5300 9653 $2^x : (\ln 2/3)^{1/3}$
5301 0767 $2^x : 2+\sqrt{2}/2$
5301 2089 rt : (3,6,-6,-5)
5301 2495 rt : (5,8,-9,7)
5301 3539 ln : $\sqrt{3}/8$
5301 5911 $\exp(3/4)+\exp(5)$
5301 6771 rt : (1,2,-7,3)
5301 8653 $2^x : 3/2+e/4$
5301 9664 rt : (8,3,5,3)
5302 0164 $J_1 : \exp(\sqrt[3]{5}/2)$
5302 0389 $J_1 : \sqrt[3]{13}$
5302 0487 rt : (1,7,-6,-5)
5302 0952 rt : (8,6,4,-5)
5302 1382 rt : (8,9,-3,3)
5302 1789 $E : (1/\pi)$
5302 2243 rt : (9,8,-3,-2)
5302 5120 $E : 7/22$
5302 5892 sr : $4\sqrt{2}+\sqrt{5}/3$
5302 6301 cu : $3/2-4\sqrt{3}/3$
5302 9002 at : $5\pi^2/2$
5302 9814 $\ell_{10} : 2\sqrt{3}/3+\sqrt{5}$
5303 0049 $\zeta : (1/\pi)/10$
5303 2351 rt : (4,2,-5,2)
5303 3008 id : $3\sqrt{2}/8$
5303 3857 tanh : $1/\ln(2e)$
5303 4536 $J_0 : 22/15$
5303 4892 $\ell_{10} : \sqrt{23/2}$
5303 8651 $e^x : 4e/3+2\pi/3$
5303 9805 rt : (3,5,-3,9)
5304 0368 $\ell_{10} : 1/4+\pi$
5304 2435 sq : $9\sqrt[3]{3}/4$
5304 2589 rt : (8,2,8,-6)
5304 3480 rt : (4,9,-6,-8)
5304 5982 $t\pi : 5\ln 2/3$
5304 6620 $2^x : \pi^2/4$
5304 6857 rt : (9,9,-6,2)
5304 6971 id : $5e/3$
5304 7321 rt : (1,9,-6,-9)
5304 9339 at : $4\sqrt{2}/3-3\sqrt{3}/4$
5305 0426 cu : 17/21
5305 0702 rt : (9,4,1,-3)
5305 1461 $2^x : \sqrt{3}-\sqrt{5}/2$
5305 1471 $\ell_2 : 9/13$
5305 1647 id : $5(1/\pi)/3$
5305 4321 rt : (4,6,7,-6)
5305 4912 tanh : 13/22
5305 5204 $Li_2 : 2\ln 2/3$
5305 6804 rt : (9,9,4,-6)
5306 0256 rt : (4,8,-3,6)

5306 0687 $e^x : 10(e+\pi)/3$
5306 1224 sq : $5\sqrt{3}/7$
5306 1394 tπ : $6/19$
5306 2612 cu : $1/2+2\sqrt{3}/3$
5306 2825 ln : $10/17$
5306 3304 $2^x : 1/2-\sqrt{2}$
5306 3469 rt : (8,7,9,4)
5306 4303 cu : $3/4+3\sqrt{3}/2$
5306 4875 $\pi+\exp(2)$
5306 5208 $e^x : P_{17}$
5306 6238 rt : (5,8,0,-3)
5306 6641 $\ln(5)\times s\pi(2/5)$
5306 8449 $10^x : \sqrt{19}\pi$
5307 1533 $\zeta : (\pi)^{-3}$
5307 1956 $e^x : \sqrt{2}/3+3\sqrt{3}/2$
5307 3784 $\ell_2 : \exp(1/e)$
5307 4679 rt : (9,5,8,-7)
5307 5711 $2^x : 4\sqrt[3]{3}$
5307 5763 $E : 1/\sqrt{10}$
5307 6361 $3\div\exp(\sqrt{3})$
5307 7012 sπ : $3\pi/8$
5307 7076 sr : $3e\pi/4$
5307 7129 sr : $3-e$
5307 9080 $J_1 : 3\sqrt{5}/5$
5308 0241 rt : (9,0,5,4)
5308 1763 at : 25
5308 2564 cu : $2e/3+2\pi$
5308 4237 tπ : $\sqrt[3]{11/3}$
5308 4281 sr : P_6
5308 4337 rt : (5,9,7,-7)
5308 6419 sq : $23/9$
5308 7076 $E : 6/19$
5308 7774 rt : (4,3,8,4)
5308 8272 $10^x : 3/4+3\pi$
5308 8340 $J_0 : 6\sqrt[3]{5}/7$
5308 8637 $Ei : 5\ln 2/4$
5308 9103 $e^x : 13/14$
5308 9337 $Ei : \sqrt[3]{20/3}$
5309 0655 rt : (4,9,-2,-3)
5309 2052 sr : $2/3+3\sqrt{5}/4$
5309 4645 $2^x : \sqrt{3}/3-2\sqrt{5}/3$
5309 5875 id : $2e+2\pi/3$
5309 6389 rt : (5,4,4,-4)
5309 8277 $2^x : 4+\pi/2$
5309 9633 sr : $2(e+\pi)/5$
5310 0604 rt : (6,7,-4,-1)
5310 1272 $\theta_3 : 6/23$
5310 2424 ln : 4π
5310 2856 at : 8π
5310 4594 rt : (6,6,1,9)
5310 5229 sπ : $1/4+4\sqrt{3}$
5310 5920 at : $((e+\pi)/2)^3$
5310 6686 ln : $4/3-\sqrt{5}/3$
5310 7281 rt : (9,5,2,1)
5310 7348 rt : (7,1,-9,9)
5310 7918 $E : \sqrt[3]{2}/4$
5311 0936 rt : (9,4,9,5)
5311 2887 rt : (1,9,4,0)
5311 3553 $e^x : 2\sqrt{2}-\sqrt{5}/2$
5311 4176 sr : $1/4+2\pi/3$
5311 4616 $10^x : 3e/2+2\pi/3$
5311 5225 $Ei : \pi/6$
5311 5662 tπ : $e/3+4\pi$
5311 7645 sr : $3/4+4\sqrt{2}$
5311 9184 rt : (7,6,-2,8)
5311 9311 rt : (9,6,-5,-3)
5311 9341 $10^x : 3/2+\pi/3$
5311 9391 ln : $3-3\sqrt{3}/4$
5311 9768 $J_0 : \ln(\ln 2/3)$
5312 0449 $\ell_{10} : 5e/4$
5312 1357 $\sqrt{2}+\exp(3/4)$
5312 1625 rt : (7,7,-7,-2)
5312 3480 sπ : $(\sqrt{2}\pi/3)^{-1/2}$
5312 3891 rt : (6,7,4,-5)
5312 4812 rt : (2,5,-7,1)
5312 5000 sq : $7\sqrt{2}/8$
5312 5413 cr : $8\pi/7$
5312 5472 $\ln(5)-\exp(\pi)$
5312 6172 $E : 2\sqrt{2}/9$
5312 7744 rt : (5,0,8,5)
5312 8424 $J_0 : (e+\pi)/4$
5312 8745 rt : (2,6,-8,-9)
5312 9005 $E : \pi/10$
5313 0194 rt : (8,6,-5,7)
5313 0451 $10^x : 4/3-\pi/4$
5313 1466 sq : $1/3-\pi/2$
5313 1811 $J_1 : (e/3)^{-3}$
5313 2928 cr : $3/20$
5313 3028 $e^x : 2/3-3\sqrt{3}/4$
5313 4139 ln : $1+4e/3$
5313 4382 rt : (6,6,-3,-1)
5313 8797 rt : (9,6,-8,6)
5313 8861 rt : (4,5,-6,-4)
5314 0687 $J_1 : \sqrt{11/2}$
5314 0715 sr : $\sqrt{11/2}$
5314 2802 cu : $1/\ln(\sqrt[3]{2}/3)$
5314 4928 rt : (1,4,7,-5)
5314 7713 $2^x : 3\sqrt{3}+4\sqrt{5}/3$
5314 7882 rt : (6,3,8,-6)
5314 7891 $\ell_{10} : 5/17$
5314 9249 $J_0 : (\pi)^{1/3}$
5314 9268 sπ : $\sqrt{3}/4+\sqrt{5}/3$
5315 2511 $Ei : 13/15$
5315 3054 $10^x : (2e/3)^{-1/2}$
5315 3240 at : $4/3-\sqrt{5}/3$
5315 6445 rt : (5,1,1,1)
5315 8457 rt : (7,9,-3,-2)
5316 0014 rt : (7,6,8,-7)
5316 0017 sinh : $8(1/\pi)/5$
5316 0937 rt : (2,7,7,-6)
5316 2505 rt : (3,8,-6,-5)
5316 3269 cr : $\zeta(3)/8$
5316 3323 sinh : $6\sqrt{2}/7$
5316 5226 $J_1 : 2(e+\pi)/5$
5316 5580 $\Psi : (\pi)^{-3}$
5316 5755 rt : (8,9,-5,-4)
5316 8121 rt : (8,9,8,-8)
5316 8783 sr : $2\sqrt{3}-\sqrt{5}/2$
5317 0522 $J_1 : (2e/3)^{1/2}$
5317 0537 rt : (3,8,3,5)
5317 1440 $E : 5/16$
5317 2406 at : $10/17$
5317 4046 $10^x : (\sqrt[3]{5}/3)^3$
5317 4969 $J_1 : 3\pi/7$
5317 6321 $2^x : 4e-\pi$
5317 6324 $2^x : \sqrt{2}/3+\sqrt{5}$
5317 6935 rt : (4,1,-7,-6)
5317 8291 at : $3e\pi$
5318 0972 sπ : $2/3+3e$
5318 2231 rt : (6,2,1,-2)
5318 2306 cr : $3/2+2\pi/3$
5318 2685 $\Psi : \exp(-e/3)$
5318 2958 sr : $\sqrt{2}/5$
5318 3164 $Ei : 11/21$
5318 4321 $erf : 3\sqrt[3]{5}/10$
5318 7212 sπ : $5\sqrt{2}/6$
5318 9942 $e^x : 3-\sqrt{5}/3$
5319 0086 $J_1 : 7\sqrt{3}/9$
5319 0828 tπ : $1/4-2\pi/3$
5319 1018 rt : (7,5,1,-3)
5319 1554 $e^x : 7\sqrt[3]{3}/5$
5319 2304 sr : $8(1/\pi)/9$
5319 5447 rt : (8,8,1,-4)
5319 6635 $2^x : 8/13$
5319 6895 $Ei : 7(e+\pi)/5$
5319 8313 $2^x : 1/\ln(\sqrt{3})$
5319 8920 $e^x : 1/2+3\pi/2$
5320 0116 $10^x : 4\sqrt[3]{5}/3$
5320 2387 rt : (9,9,6,2)
5320 5080 $1/5-\sqrt{3}$
5320 7296 at : $(\ln 2/2)^{1/2}$
5320 7700 $\Psi : 1/\ln(\sqrt[3]{2}/2)$
5320 8888 rt : (3,9,0,-3)
5320 9382 $2^x : 3+\pi/3$
5320 9517 sinh : $7\sqrt{3}/10$
5320 9685 $2^x : 1/\ln(1/3)$
5321 0242 $\Psi : 2\sqrt{2}/7$
5321 1265 $2^x : -\sqrt{2}-\sqrt{3}+\sqrt{5}$
5321 4705 $\exp(2)\div s\pi(1/3)$
5321 5468 rt : (7,8,-9,-6)
5321 5474 $\pi-\ln(5)$
5321 6772 $10^x : \sqrt{2}/3-\sqrt{5}/3$
5321 8478 rt : (9,7,5,-6)
5321 9002 rt : (8,5,-9,-5)
5321 9530 sπ : $2/3+2\sqrt{3}/3$
5322 0893 rt : (8,4,5,-5)
5322 1116 $erf : \exp(-2/3)$
5322 1625 $\ell_{10} : 2\sqrt{2}+\sqrt{3}/3$
5322 3577 rt : (7,4,-9,2)
5322 4755 id : $\sqrt{2}+\sqrt{5}/2$
5322 4933 rt : (7,1,5,-4)
5322 6051 rt : (6,4,-6,3)
5322 7609 $10^x : 4\sqrt{2}/5$
5322 7968 $\zeta : (2e)^{-2}$
5322 9527 rt : (5,4,-3,4)
5323 0750 rt : (9,2,-9,-4)
5323 1852 rt : (9,8,-3,9)
5323 2094 $J_1 : (\ln 3/2)^{-1/2}$
5323 3070 rt : (6,2,5,3)
5323 4581 cr : $1+3\sqrt{3}/2$
5323 4769 rt : (4,8,4,-5)
5323 5851 sπ : $3\sqrt{2}/4+\sqrt{5}/2$
5323 6014 $\ell_2 : \exp(\pi)$
5323 6766 $e^x : 7\sqrt[3]{3}/8$
5323 7745 at : $\exp(-\sqrt[3]{4}/3)$
5323 8232 $Ei : \sqrt[3]{2}/5$
5323 9359 rt : (9,3,9,-7)
5323 9374 $\ln(2)-\exp(4/5)$
5324 3125 tπ : $\sqrt{2}/4+2\sqrt{5}/3$
5324 3580 rt : (3,5,4,-4)
5324 7155 rt : (8,0,9,6)
5324 7687 $3\times\ln(3/5)$
5325 0698 rt : (5,7,8,-7)
5325 1019 ln : $4\sqrt{2}+4\sqrt{3}$
5325 1361 sq : P_{23}
5325 3019 $e^x : 16/5$
5325 4586 $10^x : \exp((e+\pi)/3)$
5325 5168 rt : (4,8,-5,1)
5325 6208 $\zeta : 5\sqrt[3]{5}/4$
5325 6610 $e^x : 2/3+4\sqrt{2}$
5325 7960 rt : (5,5,5,2)
5326 0312 rt : (7,3,9,5)
5326 1282 id : $\exp(-\sqrt[3]{2}/2)$
5326 1492 rt : (4,4,8,-6)
5326 1547 cu : $10\pi/9$
5326 1886 cr : $18/5$
5326 1999 rt : (3,5,-5,4)
5326 2675 rt : (1,9,-3,-4)
5326 3996 $\ell_{10} : \sqrt{3}+3\sqrt{5}/4$
5326 5047 rt : (3,7,1,-1)
5326 5984 $Li_2 : (e+\pi)/6$
5326 6327 rt : (9,9,-8,-1)
5326 6639 $Ei : 12$
5326 9207 id : $1/3-\sqrt{3}/2$
5327 1195 rt : (2,8,5,6)
5327 3803 sπ : $e/4+\pi$
5327 4147 cu : $1/4-3\sqrt{2}/4$
5327 5894 $2^x : 8\ln 2/9$
5327 6030 $Ei : \exp(-\sqrt{2}\pi/2)$
5327 6091 rt : (8,9,-5,0)
5327 6143 $2^x : 1/4+\pi/2$
5327 6249 sq : $\sqrt{3}/4-3\sqrt{5}/2$
5327 7647 sq : $2+2e/3$
5327 9612 id : $3\sqrt{2}/4+2\sqrt{5}$
5327 9802 rt : (3,1,8,-5)
5328 1915 at : $8\pi^2/3$
5328 3664 $E : 4\ln 2/9$
5328 4246 rt : (1,5,3,4)
5328 4472 sπ : $2\sqrt{2}/3+\sqrt{5}$
5328 5669 tanh : $6\ln 2/7$
5328 7267 rt : (6,6,9,4)
5328 8537 $E : (\sqrt{2}\pi/3)^{-3}$
5328 9150 rt : (9,6,-2,-2)
5328 9179 at : $9(e+\pi)/2$
5328 9855 rt : (7,9,-2,1)
5329 0962 rt : (2,6,-3,-3)
5329 1078 $Ei : 16/3$
5329 1494 id : $10(e+\pi)/3$
5329 2830 tπ : $2\sqrt[3]{5}/5$
5329 3027 $E : 4/13$

5329 3045 $J_1 : 9\zeta(3)/8$
5329 3501 rt : (6,1,-8,7)
5329 5137 sinh : 17/3
5329 8793 cu : $3e + \pi$
5329 9382 $\Psi : (\ln 2/2)^3$
5329 9777 rt : (1,8,6,-6)
5330 0607 $J_1 : 1/\ln(2\pi/3)$
5330 1960 $e^x : 4/3 - e/3$
5330 2014 sinh : $3\pi^2/5$
5330 2034 $10^x : 8\ln 2/3$
5330 2475 $e^x : 5\pi^2/7$
5330 2804 sq : $4/3 - 3e$
5330 2959 sq : $5(1/\pi)$
5330 5480 $J_1 : 23/17$
5330 6094 rt : (5,1,-5,8)
5330 8494 rt : (5,5,-8,-6)
5331 0709 rt : (6,9,1,2)
5331 2063 rt : (7,8,5,-6)
5331 2977 rt : (8,7,9,-8)
5331 4049 $E : \ln(e/2)$
5331 4194 sinh : $7\ln 2/4$
5331 4793 $10^x : 3 - \sqrt{3}$
5331 4930 tanh : $\ln(2e/3)$
5331 5505 $s\pi : (e/3)^2$
5331 5893 $\Psi : 3e\pi/5$
5331 7683 rt : (2,2,8,4)
5331 7949 $J_0 : 19/13$
5331 8530 id : $1/4 + 2\pi$
5331 8900 cu : $1/3 - 2\sqrt{2}$
5332 0075 cr : $1/4 + 3\sqrt{5}/2$
5332 0261 rt : (5,9,-6,-5)
5332 1500 $t\pi : 3\sqrt{2}/4 + \sqrt{5}/4$
5332 3091 rt : (4,1,-2,9)
5332 3696 $\ln(3) \times \exp(1/3)$
5332 5808 rt : (9,2,2,-3)
5332 6681 $\ell_2 : (2\pi/3)^{1/2}$
5332 9068 $\ell_{10} : 2 + \sqrt{2}$
5332 9456 rt : (2,5,-9,-6)
5332 9641 rt : (8,3,-2,-1)
5332 9992 sr : $\sqrt{3} + \sqrt{5} + \sqrt{6}$
5333 1006 $e^x : 2 + 3\pi/2$
5333 1906 $\ell_{10} : \sqrt{3}/4 + 4\sqrt{5}/3$
5333 1971 $Li_2 : 1/\sqrt[3]{10}$
5333 2378 rt : (8,6,-7,4)
5333 3333 id : 8/15
5333 4873 rt : (7,4,9,-7)
5333 6179 sq : $4e - \pi/2$
5333 6219 $10^x : \sqrt{2} - \sqrt{3}/2$
5333 9085 rt : (6,5,5,-5)
5333 9212 $10^x : 6\sqrt{5}/7$
5334 0319 rt : (7,4,-6,1)
5334 0623 sr : $\sqrt[3]{13}$
5334 0995 $e^x : \sqrt[3]{5}/4$
5334 1653 cr : $5\sqrt[3]{3}/2$
5334 2073 $e^x : 1/4 + e/4$
5334 4625 ln : $4\sqrt{2}/3 - 3\sqrt{3}/4$
5334 6029 rt : (5,9,4,3)
5334 6220 cu : $\sqrt{3}/2 - 3\sqrt{5}/4$
5334 7298 rt : (3,8,8,-7)
5334 9100 at : $1/\ln(2e)$
5334 9292 rt : (2,8,3,-7)
5334 9404 cr : $3\zeta(3)$
5335 0434 $\ell_{10} : 4e\pi$
5335 0905 rt : (7,6,-4,5)
5335 2047 $\sqrt{2} - \exp(2/3)$
5335 3524 cu : $4 + 2\sqrt{5}/3$
5335 4622 rt : (2,9,4,-5)
5335 9642 rt : (4,7,-3,-1)
5336 0131 $s\pi(\pi) - s\pi(\sqrt{2})$
5336 0617 $t\pi : (\sqrt{3})^{1/2}$
5336 1357 $t\pi : 3\sqrt[3]{2}/8$
5336 3415 ln : $10\sqrt[3]{2}$
5336 4116 rt : (9,2,6,4)
5336 4859 $\theta_3 : ((e+\pi)/3)^{-2}$
5336 7427 rt : (8,3,-9,8)
5336 8443 rt : (6,1,9,-6)
5337 0078 $J_1 : 7/3$
5337 0819 at : 13/22
5337 1283 $s\pi : 1/4 + \pi/2$
5337 2529 $E : (2e/3)^{-2}$
5337 2700 $erf : 3\zeta(3)/7$
5337 3741 rt : (5,9,9,3)
5337 4806 rt : (9,9,2,-5)
5337 5976 rt : (6,6,-1,6)
5337 6411 $E : 7/23$
5337 7621 at : $(3)^3$
5337 8598 $\ln(\pi) + \exp(2)$
5337 9653 cr : P_{71}
5338 0110 cr : P_{68}
5338 0155 $\zeta : (1/\pi)/9$
5338 1661 $s\pi : \sqrt{2}/2 + 2\sqrt{5}$
5338 2185 rt : (9,9,-9,-6)
5338 2312 $\Psi : 24/11$
5338 4540 rt : (2,5,8,-6)
5338 4809 $\Gamma : 1/\sqrt[3]{23}$
5338 5015 sr : $\sqrt[3]{5}/6$
5338 5291 rt : (5,2,5,-4)
5338 6198 rt : (7,3,-6,9)
5338 6745 rt : (8,1,2,2)
5338 8096 $t\pi : 1/4 + e/3$
5338 8396 $J_1 : 19/14$
5338 9692 $J_1 : \sqrt[3]{5/2}$
5339 2338 $\Psi : 7(e+\pi)/8$
5339 4159 $\Gamma : 1/8$
5339 6084 rt : (6,8,9,-8)
5339 6690 cr : $3e/4 + \pi/2$
5339 8347 rt : (3,8,-7,1)
5339 8740 rt : (9,5,6,-6)
5340 0957 rt : (2,9,-7,-6)
5340 1999 $\ell_{10} : 2\sqrt[3]{5}$
5340 2309 $J_1 : (2e)^{1/2}$
5340 2496 at : $10e$
5340 2541 $1/4 + \exp(1/4)$
5340 3872 $t\pi : \sqrt[3]{23}$
5340 6550 $\ell_2 : \sqrt{2} + \sqrt{3} + \sqrt{7}$
5340 7019 rt : (1,6,4,-4)
5340 7162 $\ell_{10} : 3\sqrt{2}/2 + 3\sqrt{3}/4$
5340 8001 $\ell_2 : 3 - 4\sqrt{3}/3$
5340 9343 sq : $9\pi^2/2$
5340 9681 ln : $3\sqrt{3} - \sqrt{5}/4$
5341 1801 rt : (5,9,5,-6)
5341 2433 id : $9(e+\pi)/7$
5341 2594 $\ell_{10} : 4\sqrt{2} - \sqrt{5}$
5341 2952 sr : $2 + \sqrt{2}/4$
5341 4356 $2^x : 1/4 - 2\sqrt{3}/3$
5341 5963 $2^x : 2/3 + 2\sqrt{3}/3$
5341 8862 rt : (8,6,2,-4)
5342 0990 $2^x : \sqrt{22/3}$
5342 2284 rt : (9,8,-5,6)
5342 4045 sq : $4\sqrt{2}/3 - 2\sqrt{3}/3$
5342 4751 id : $4\zeta(3)/9$
5342 6433 rt : (5,2,9,5)
5342 7584 $J_1 : e/2$
5342 9173 id : $9\pi/8$
5342 9972 $\lambda : 1/21$
5343 2446 cr : $3\sqrt{2}/2 + 2\sqrt{5}/3$
5343 3591 $E : e/9$
5343 4415 $J_1 : 6e/7$
5343 5021 $\exp(3/5) - \exp(\sqrt{5})$
5343 5391 rt : (4,2,-7,-1)
5343 5496 $2^x : 2/3 - \pi/2$
5343 5727 as : $2 - 2\sqrt{5}/3$
5343 6245 $\ell_{10} : 4 - \sqrt{3}/3$
5343 6633 as : $8(1/\pi)/5$
5343 6748 sq : $(\sqrt[3]{5}/2)^2$
5343 6802 rt : (5,5,9,-7)
5343 7212 $\zeta : e\pi/4$
5343 8962 $5 \div \exp(\sqrt{5})$
5343 9393 $2^x : 3\sqrt{5}/5$
5344 0817 $E : (\ln 3/2)^2$
5344 0963 $\ell_2 : 4e - 3\pi$
5344 2201 sr : $5e\pi$
5344 2876 rt : (1,5,8,3)
5344 3672 $\ln(3/4) + \exp(3/5)$
5344 4594 cu : $8\sqrt{3}/3$
5344 5690 ln : $(e+\pi)/10$
5344 6362 $E : 1/\sqrt{11}$
5344 6552 at : $\sqrt{2}/2 - 3\sqrt{3}/4$
5344 6953 $\Psi : 3\sqrt[3]{5}$
5344 7695 $10^x : 8\sqrt[3]{3}/3$
5344 7866 rt : (8,5,6,3)
5344 9254 rt : (1,2,8,-5)
5345 0053 rt : (8,2,6,-5)
5345 0467 rt : (1,9,8,-7)
5345 0784 rt : (8,9,6,-7)
5345 1149 rt : (8,8,-2,7)
5345 2248 sr : 2/7
5345 2396 $Li_2 : 4\sqrt[3]{5}/7$
5345 3066 tanh : $(3\pi/2)^{-1/3}$
5345 5707 $s\pi : 1/\ln(\sqrt{3})$
5345 7045 cu : $2 + 2\sqrt{5}/3$
5346 0969 sq : $4 + 3\sqrt{3}/2$
5346 2186 rt : (4,1,-9,-3)
5346 3595 sq : $1/4 + 4\sqrt{5}$
5346 7196 rt : (4,6,5,8)
5346 7400 $e^x : 3/4 + 4\pi$
5346 8693 $s\pi : 1/2 + e/4$
5347 0785 $E : \zeta(3)/4$
5347 2713 rt : (3,6,-7,-5)
5347 3112 $J_1 : \sqrt{2} - \sqrt{3} + \sqrt{7}$
5347 5064 $2^x : 3\sqrt{2} + 3\sqrt{5}/4$
5347 5899 rt : (4,8,-7,-4)
5347 6409 $t\pi : 4e + 2\pi$
5347 6605 sq : $10(e+\pi)/3$
5347 8228 at : $2\sqrt{2} - \sqrt{5}$
5347 8779 $s\pi : \sqrt{2}/3 + 3\sqrt{5}$
5347 9999 ln : $2 - \sqrt{2}$
5348 1757 $Li_2 : (\pi/3)^{-1/2}$
5348 2393 rt : (4,1,5,3)
5348 3028 $\ln(\pi) \div s\pi(\sqrt{3})$
5348 3346 $E : 3/10$
5348 3925 rt : (4,9,9,-8)
5348 4691 cu : $4 - 3\sqrt{5}/4$
5348 5988 $2^x : 3\sqrt[3]{3}/7$
5348 6428 cu : $3 - \sqrt{2}/4$
5348 7191 $2^x : 4/3 - \sqrt{5}$
5348 9051 rt : (7,8,1,8)
5349 1661 sr : $4/3 - \pi/3$
5349 4762 rt : (7,3,2,-3)
5349 4804 as : $2\sqrt{2}/3 - \sqrt{3}/4$
5349 6923 $\ell_{10} : 4\sqrt{3}/3 + \sqrt{5}/2$
5349 9006 sr : $3\pi/4$
5350 0195 rt : (4,2,9,-6)
5350 0640 $\ell_{10} : 4/3 + 2\pi/3$
5350 4000 rt : (7,6,6,-6)
5350 5755 ln : $3/2 + \pi$
5350 6300 $e^x : 3/7$
5350 6713 sq : $2e/3 + 2\pi$
5350 8609 $2^x : \exp(-\sqrt[3]{3}/3)$
5351 0608 id : $3\sqrt{3}/4 + \sqrt{5}$
5351 1320 $\ell_{10} : 7/24$
5351 2215 cr : $4\sqrt{2}/3 + \sqrt{3}$
5351 4034 $e^x : 1/\ln((e+\pi)/2)$
5351 5188 $J_1 : 15/11$
5351 6689 sr : $\sqrt{3}/2 + 2\sqrt{5}/3$
5351 8375 rt : (6,8,-6,0)
5351 9470 $10^x : 4 + \sqrt{2}$
5352 1586 cu : $4\sqrt{2} + \sqrt{3}/2$
5352 2646 $t\pi : 1/\sqrt{10}$
5352 3356 $s\pi : (2e/3)^{-1/3}$
5352 3365 $\exp(3/4) - \exp(\sqrt{3})$
5352 3452 $10^x : \exp(e/2)$
5352 3723 sr : $9(1/\pi)/10$
5352 5026 $10^x : 21/16$
5352 5978 sr : $5\sqrt{2}/3$
5352 6427 $\Psi : 3\sqrt[3]{5}/5$
5352 7203 $e^x : 4\sqrt{2} + \sqrt{3}/2$
5352 8208 cu : $2\sqrt{2}/3 - 3\sqrt{3}/2$
5352 9197 ln : $e\pi/5$
5353 6222 rt : (1,9,-5,1)
5353 6361 $e^x : 6\sqrt[3]{2}/5$
5353 7436 sq : $3\sqrt{2} - 2\sqrt{3}/3$
5353 9401 rt : (7,1,6,4)
5353 9816 id : $1/4 - \pi/4$
5354 1023 rt : (6,8,4,9)
5354 1915 $t\pi : 6\pi^2/5$
5354 2921 $e^x : \sqrt{3}/2 - 2\sqrt{5}/3$
5354 4329 at : 9π
5354 5580 $t\pi : \sqrt{7/3}$
5354 5880 rt : (3,6,9,-7)

5354 6366 $\zeta : (3)^{-3}$
5354 8277 sinh : 17/14
5354 8483 $10^x : 2\sqrt{2}/7$
5354 8545 $J_1 : 1/\ln(\sqrt[3]{3}/3)$
5355 0832 cr : $4\sqrt{2}/3 - \sqrt{3}$
5355 0857 rt : (4,5,9,4)
5355 1929 $\exp(2/3) + s\pi(1/5)$
5355 2166 rt : (6,7,2,-4)
5355 3390 id : $5\sqrt{2}/2$
5355 3974 $e^x : (2e)^{-1/2}$
5355 4645 $E : 3\ln 2/7$
5355 5934 rt : (4,7,-3,-3)
5355 6998 rt : (5,2,-9,-9)
5355 7623 $\Gamma : 10(e+\pi)/3$
5355 9269 sinh : $7\sqrt{2}/3$
5355 9292 rt : (2,9,1,6)
5355 9565 $e^x : 3 - 4e/3$
5356 0824 $Ei : \pi^2/8$
5356 1434 rt : (8,6,-9,1)
5356 2171 rt : (9,8,-5,-1)
5356 2523 $\theta_3 : 7/9$
5356 3846 cr : $3/2 + 3\sqrt{2}/2$
5356 4255 rt : (3,6,8,9)
5356 7541 $\ell_{10} : 3 + \sqrt{3}/4$
5356 8008 rt : (5,1,-7,5)
5356 8086 at : $10e\pi/3$
5356 8745 cu : 15/11
5357 0998 $10^x : \exp(-e/3)$
5357 1730 $\ell_2 : 3e - 3\pi/4$
5357 2591 ln : $4/3 - \sqrt{5}/2$
5357 2676 rt : (9,7,3,-5)
5357 3137 $E : (3/2)^{-3}$
5357 5577 rt : (8,9,-7,-3)
5357 5880 $K : 1/\ln(\ln 2/2)$
5357 5905 cu : $1 - 2e/3$
5357 6953 rt : (2,8,3,-1)
5357 7662 $\ell_{10} : (e+\pi)^2$
5357 7899 $\Gamma : 8\sqrt{2}/3$
5357 8250 $e^x : 5(e+\pi)/4$
5357 8923 id : $\sqrt{3} + \sqrt{6} - \sqrt{7}$
5357 9070 sinh : $3\sqrt[3]{5}/10$
5357 9358 $t\pi : (5/3)^{-1/3}$
5357 9687 $s\pi : \sqrt[3]{2}/7$
5357 9950 rt : (7,4,2,1)
5358 3228 cr : 2/13
5358 4874 rt : (6,4,-2,-1)
5358 5078 $s\pi : 3\sqrt{2} + \sqrt{3}/3$
5358 5974 $\Gamma : 6\pi/7$
5358 6096 $2^x : 13/21$
5358 6640 $e^x : 3/2 + 4\pi/3$
5358 7265 rt : (6,3,6,-5)
5358 7793 rt : (5,4,-5,1)
5358 8685 $e^x : 3 - \sqrt{2}/3$
5358 8800 rt : (4,7,-6,5)
5358 9838 id : $4 - 2\sqrt{3}$
5359 1959 $\theta_3 : 5/19$
5359 2661 at : $3/2 - e/3$
5359 4951 sr : $1 + e/2$
5359 6094 rt : (7,3,-8,6)
5359 7366 $J_0 : (\sqrt{2}/3)^{-1/2}$
5359 9099 $J_1 : 4\sqrt[3]{5}/5$
5359 9607 cr : $6e$
5360 0000 cu : $4\sqrt[3]{3}/5$
5360 1091 rt : (9,3,7,-6)
5360 1463 rt : (7,3,-6,-3)
5360 1899 cu : $1/4 + 3\sqrt{3}$
5360 2100 $t\pi : 1/\ln(\pi/3)$
5360 3385 $e^x : 2 - \pi/2$
5360 4266 $e^x : (\sqrt{3}/2)^{1/2}$
5360 4809 $10^x : ((e+\pi)/3)^2$
5360 4878 cr : $2\ln 2/9$
5360 6653 rt : (2,7,5,-5)
5360 6889 $J_1 : \sqrt[3]{25/2}$
5360 7023 cr : $4e/3$
5360 8148 rt : (9,4,-1,-2)
5360 8157 $10^x : (\pi/2)^{1/3}$
5360 8930 at : $6\ln 2/7$
5361 2592 rt : (3,1,9,5)
5361 2657 $s\pi : 2\pi^2/7$
5361 3094 sr : $3\sqrt{2}/4 + 3\sqrt{3}/4$
5361 4161 rt : (9,5,-9,7)
5361 4489 as : $\ln(5/3)$
5361 5676 $J : 2(1/\pi)/3$
5361 6696 rt : (5,7,6,-6)
5361 7997 sq : $1/3 + e/3$
5361 8414 rt : (7,6,-6,2)
5361 8570 cu : 15/13
5361 8803 $t\pi : ((e+\pi)/2)^{1/3}$
5362 0018 $t\pi : \ln(\sqrt[3]{5}/2)$
5362 2605 rt : (2,8,-3,-1)
5362 3077 $\ln(\sqrt{3}) - \exp(3)$
5362 3373 rt : (8,7,7,-7)
5362 4292 $e^x : 3/2 + 4\sqrt{2}/3$
5362 5397 $E : 5/17$
5362 5603 $4/5 \div \exp(2/5)$
5362 5636 ln : $(e+\pi)^2$
5362 7215 sinh : $\exp(-2/3)$
5362 8582 $\theta_3 : (\ln 2/3)^{1/3}$
5362 8805 at : $3/2 - 2\pi/3$
5363 0505 $\exp(1/3) + \exp(\pi)$
5363 0656 rt : (4,3,-7,-4)
5363 0754 $E : \ln(\sqrt{5}/3)$
5363 1177 $2/5 \div s\pi(\sqrt{3})$
5363 3778 rt : (5,8,-2,-2)
5363 4665 as : P_{13}
5363 5660 rt : (9,8,-7,3)
5363 6002 sr : $\ln(4/3)$
5363 6625 rt : (4,1,-4,6)
5363 7606 rt : (3,4,-6,8)
5363 7695 cu : 13/16
5363 8891 rt : (9,5,-9,2)
5363 9134 at : $\ln(2e/3)$
5363 9413 rt : (8,8,-1,-3)
5363 9621 rt : (1,6,1,-9)
5364 0837 tanh : $\ln(\ln 3/2)$
5364 2176 sq : $\ln(\sqrt[3]{3}/3)$
5364 3971 rt : (2,3,9,-6)
5364 5619 $\ln(2) \times s\pi(e)$
5364 5761 sr : $3e/2 + 3\pi/4$
5364 7930 id : $\ln(\sqrt[3]{5})$
5364 9687 $\Gamma : 9\sqrt[3]{3}/4$
5364 9716 $e^x : \sqrt{3} + \sqrt{5} + \sqrt{6}$
5365 1232 $\Psi : 3(1/\pi)/5$
5365 1712 rt : (6,3,-5,7)
5365 3326 sq : $(e+\pi)/8$
5365 5081 $10^x : P_{12}$
5365 6516 rt : (9,0,3,3)
5365 6546 $10^x : 8\sqrt[3]{2}/7$
5365 7199 $e^x : 24/19$
5365 7246 $s\pi : \sqrt[3]{3}/8$
5365 8366 $10^x : 1 + 2\sqrt{5}/3$
5365 8698 rt : (7,9,-4,-2)
5366 0797 rt : (8,5,-6,8)
5366 2962 rt : (1,7,9,-7)
5366 3105 id : $2\sqrt{2} + 3\sqrt{5}$
5366 3298 $s\pi : 3e/4 + \pi$
5366 4161 $10^x : \sqrt{2}/2 + 3\sqrt{5}/2$
5366 4641 rt : (2,7,-8,-6)
5366 7504 rt : (5,7,-7,-5)
5366 7914 at : $5(e+\pi)$
5366 8252 $\ell_{10} : 4 - \sqrt{5}/4$
5366 9419 $Li_2 : 7/15$
5367 2324 ln : $2\sqrt{2} - \sqrt{5}/2$
5367 2655 $\zeta : \sqrt{2} - \sqrt{3} + \sqrt{6}$
5367 2805 $s\pi : 5\sqrt{5}$
5367 2868 rt : (3,7,2,9)
5367 3399 $t\pi : 1/4 - 4\pi$
5367 3483 rt : (8,7,-6,-4)
5367 4258 $2^x : 2/3 + 4\sqrt{5}/3$
5367 6716 rt : (8,4,3,-4)
5367 6976 $\sqrt{3} + \ln(\sqrt{5})$
5367 9174 $t\pi : \sqrt{17/3}$
5367 9411 $t\pi : 3\sqrt{2} - \sqrt{5}/4$
5368 0054 $e^x : \sqrt{2}/4 + \sqrt{3}/3$
5368 3551 rt : (5,4,2,-3)
5368 3696 $E : 7/24$
5368 4779 rt : (8,9,-9,1)
5368 6189 rt : (3,9,-4,6)
5368 6246 $\theta_3 : \exp(-4/3)$
5368 6846 $\ell_{10} : 3e - 3\pi/2$
5368 8301 $\zeta : 8\sqrt[3]{5}/9$
5368 8480 $\Psi : \zeta(3)/2$
5368 9056 rt : (7,8,3,-5)
5369 2318 cu : $10e$
5369 5427 $t\pi : 1/2 + 4\sqrt{2}$
5369 7376 rt : (8,8,-4,4)
5369 7540 at : $(2e)^2$
5369 7952 sq : $2 + \sqrt{5}/3$
5369 9434 $10^x : \sqrt{2}/3 + 4\sqrt{3}$
5369 9721 $10^x : 4\sqrt{3}/3 - \sqrt{5}/2$
5370 0844 rt : (2,7,2,8)
5370 1442 cu : $3\sqrt{3}/2$
5370 3189 $10^x : (e)^{1/2}$
5370 3543 at : $3\pi^2$
5370 3763 $J_0 : 16/11$
5370 3836 sr : $1 + 2e$
5370 4956 tanh : 3/5
5370 6448 sq : $3\sqrt[3]{5}/7$
5370 7517 rt : (9,4,7,4)
5370 7533 sr : $\sqrt[3]{3}/5$
5370 9950 $J_1 : 8\zeta(3)/7$
5371 1384 $\ln(3/5) \div s\pi(2/5)$
5371 1415 rt : (1,8,1,-3)
5371 1783 ln : $2e - \pi/4$
5371 3783 rt : (9,8,8,-8)
5371 4403 rt : (1,8,-4,5)
5371 5562 rt : (8,0,7,5)
5371 9279 $2^x : e/2 + 4\pi$
5371 9577 rt : (7,5,-3,9)
5372 0426 rt : (7,4,7,-6)
5372 0507 $\pi + \exp(1/3)$
5372 0652 $t\pi : 1/4 - \sqrt{5}/3$
5372 3463 $e^x : 3/2 - 3\sqrt{2}/2$
5372 3683 rt : (6,8,7,-7)
5372 3959 at : $2\sqrt{3}/3 - \sqrt{5}/4$
5372 5008 $2^x : 24/11$
5372 6017 rt : (5,3,-2,8)
5372 7772 rt : (4,8,-9,-9)
5372 7992 $J_1 : (\sqrt[3]{4}/3)^{-1/2}$
5372 8081 rt : (7,8,6,2)
5372 8496 sr : $\sqrt{3}/6$
5372 9191 $s\pi : 7\sqrt[3]{2}$
5372 9369 $J_1 : \exp(1/\pi)$
5373 3119 $J_1 : 11/8$
5373 3679 rt : (2,9,-1,-3)
5373 4090 rt : (6,4,6,3)
5373 5175 at : $7e\pi/2$
5373 6106 rt : (5,0,6,4)
5373 6440 $10^x : 4\sqrt{2}/3 + \sqrt{5}/3$
5373 7116 rt : (4,4,6,-5)
5373 7723 rt : (3,8,6,-6)
5373 7992 $e^x : \exp(\sqrt[3]{2}/2)$
5373 9848 $3/4 \div \exp(1/3)$
5374 0157 rt : (2,8,0,-2)
5374 1026 $10^x : 8e\pi/5$
5374 1269 $\ell_2 : 2/3 + \sqrt{5}$
5374 1603 id : $3\sqrt{2}/4 - 3\sqrt{3}/2$
5374 9050 ln : $((e+\pi)/2)^{1/2}$
5375 3108 rt : (7,9,-5,-1)
5375 4146 $E : \sqrt{3}/6$
5375 4227 ln : $3\sqrt{3}/4 + 3\sqrt{5}/2$
5375 6727 $erf : 3\sqrt{3}/10$
5375 9190 rt : (8,5,-5,0)
5375 9419 $E : \sqrt[3]{3}/5$
5376 2928 rt : (9,9,0,-4)
5376 5617 rt : (1,4,5,-4)
5376 7730 $2^x : \sqrt{2} - 4\sqrt{3}/3$
5376 8207 $1/4 - \ln(3/4)$
5376 9882 $2^x : e/3 + \pi$
5377 6028 rt : (9,1,-5,1)
5377 7361 $E : \ln(4/3)$
5377 7777 sq : 11/15
5377 7840 $erf : 3\ln 2/4$
5377 8098 tanh : $\zeta(3)/2$
5377 8596 rt : (2,0,5,3)
5377 8667 $2^x : 1/3 + 4e/3$
5377 9975 rt : (7,8,-1,5)
5378 0807 $10^x : 1/4 + 3\sqrt{5}/4$
5378 1752 at : $(3\pi/2)^{-1/3}$
5378 3742 rt : (1,9,5,9)

5378 6058 rt : (2,8,-4,4)
5378 6605 rt : (8,9,4,-6)
5378 6837 sq : $2\sqrt{2}/3+3\sqrt{3}/2$
5378 9060 rt : (1,8,-9,-7)
5378 9863 erf : 13/25
5379 0489 rt : (6,9,-1,-1)
5379 1667 erf : $(\sqrt[3]{3}/2)^2$
5379 2180 $t\pi$: $3\sqrt{3}/2+\sqrt{5}/4$
5379 3334 rt : (5,9,-7,6)
5379 3818 ℓ_{10} : P_{82}
5379 5883 rt : (9,5,4,-5)
5379 6107 rt : (3,8,-4,-4)
5379 7297 J_1 : $10\ln 2/3$
5379 9656 id : $8\sqrt[3]{3}$
5379 9952 sq : $2\sqrt{2}/3+3\sqrt{5}$
5380 0637 cu : 23/13
5380 1683 rt : (4,9,-6,0)
5380 4031 rt : (7,9,8,-8)
5380 4376 $\exp(2/5)\div s\pi(1/5)$
5380 4672 2^x : $4-e/3$
5380 5374 E : $9(1/\pi)/10$
5380 6109 as : $\sqrt{2}/4-\sqrt{3}/2$
5380 7016 rt : (7.5,-1,-2)
5380 7099 $t\pi$: $1/3+\pi/3$
5380 9046 $t\pi$: $\sqrt{2}/9$
5380 9338 $\exp(1/4)-\exp(3/5)$
5381 0492 e^x : $e/2+\pi/4$
5381 0586 ℓ_2 : $2/3+\pi/4$
5381 3287 $s\pi$: $\exp(-\sqrt[3]{5})$
5381 4092 rt : (5,9,3,-5)
5381 4472 sq : $4e-3\pi/4$
5381 4821 rt : (8,5,8,-7)
5381 6598 2^x : $3/2+\sqrt{2}$
5381 7209 J_1 : $4\sqrt{3}/3$
5381 7229 rt : (3,5,-9,-6)
5381 7500 ℓ_{10} : $\sqrt{2}/3+4\sqrt{5}/3$
5381 8580 rt : (3,4,5,2)
5381 8642 ζ : $((e+\pi)/2)^{-3}$
5381 8900 sr : $3/2+\sqrt{3}/2$
5381 9058 $\exp(1/4)+s\pi(\sqrt{3})$
5382 1722 ζ : $(1/\pi)/8$
5382 3094 rt : (1,0,9,5)
5382 3111 E : 2/7
5382 5363 rt : (4,8,-8,-6)
5382 5887 sinh : $3\zeta(3)/7$
5382 6398 ℓ_{10} : $e/2+2\pi/3$
5382 9791 2^x : $1/\ln(5)$
5382 9996 rt : (9,1,8,-6)
5383 0356 rt : (4,9,7,-7)
5383 0809 sr : P_{82}
5383 2218 rt : (5,8,1,-1)
5383 2768 rt : (6,5,3,-4)
5383 6778 ℓ_{10} : $7\pi^2/2$
5383 7316 rt : (3,9,-2,-2)
5383 9045 Γ : $\exp(\pi)$
5383 9683 rt : (8,1,-1,-1)
5383 9727 E : $\sqrt[3]{5}/6$
5384 0162 e^x : $((e+\pi)/2)^{1/2}$
5384 0325 ζ : $3\sqrt[3]{2}/5$
5384 0451 ℓ_{10} : $4/3+3\sqrt{2}/2$
5384 0735 rt : (6,0,-9,8)
5384 0948 Ψ : $\sqrt[3]{21/2}$
5384 2934 ζ : 1/25
5384 6153 id : 7/13
5384 8757 $s\pi$: $3\sqrt{3}/4-\sqrt{5}/2$
5384 8873 rt : (5,5,7,-6)
5384 9865 cr : $1/2+\pi$
5385 0045 J_1 : $8\sqrt[3]{3}/5$
5385 0118 ℓ_2 : $\sqrt{2}+2\sqrt{5}/3$
5385 3004 rt : (4,6,-7,9)
5385 3866 $s\pi$: $(\ln 3/2)^{1/3}$
5385 5540 rt : (3,4,-8,-5)
5385 5596 at : $(\pi)^3$
5385 6318 rt : (2,9,2,-4)
5385 7798 rt : (3,9,-6,-1)
5385 8018 ℓ_2 : $2\sqrt{3}-\sqrt{5}/4$
5385 8217 $\ln(2/5)\times s\pi(1/5)$
5385 8339 rt : (5,5,-6,1)
5385 8470 rt : (1,6,4,8)
5385 9666 10^x : $3e/10$
5386 1593 10^x : $3+3\pi$
5386 4059 rt : (7,1,3,-3)
5386 4282 ℓ_2 : $4\sqrt{2}/3-\sqrt{3}/4$
5386 5279 2^x : $5e\pi/4$
5386 5942 cr : $4/3+4\sqrt{3}/3$
5386 6500 ln : $2\sqrt{3}/3+\sqrt{5}/4$
5386 6545 Ψ : $4\pi^2/3$
5386 6766 as : $3\sqrt[3]{5}/10$
5386 7384 rt : (2,4,9,4)
5386 8136 rt : (1,9,6,-6)
5386 9845 cu : $3+\sqrt{5}/3$
5387 0137 E : $\exp(-\sqrt[3]{2})$
5387 1134 rt : (1,7,-3,8)
5387 1137 id : $1/2+3e/4$
5387 3348 rt : (5,1,-9,2)
5387 4619 rt : (6,8,-3,-3)
5387 5059 ln : $2/3+\pi/3$
5387 6197 rt : (9,8,-9,0)
5387 6956 rt : (6,1,7,-5)
5387 8711 rt : (4,2,-8,-7)
5388 0191 rt : (8,5,-8,5)
5388 0396 sr : $2\sqrt{3}+4\sqrt{5}/3$
5388 1571 Ψ : $3\ln 2/7$
5388 2026 id : $4\sqrt{2}-\sqrt{5}/2$
5388 2943 rt : (3,4,-8,3)
5388 3856 Ei : $7\sqrt[3]{2}/3$
5388 4176 $s\pi(1/5)+s\pi(2/5)$
5388 5227 rt : (9,6,9,-8)
5388 5915 rt : (4,7,0,-2)
5388 6994 E : $8(1/\pi)/9$
5388 9160 rt : (7,8,-7,-5)
5388 9272 E : $\sqrt{2}/5$
5389 0164 10^x : $(\sqrt{2}\pi/3)^{-1/3}$
5389 2521 rt : (3,8,-7,8)
5389 2638 ℓ_2 : $3/4+4e$
5389 4994 rt : (1,9,3,-8)
5389 6227 rt : (6,8,2,6)
5389 7089 ln : $1-\pi/4$
5389 7608 at : 10π
5389 7835 id : $9\sqrt[3]{5}/10$
5389 8319 2^x : $(e/3)^{-3}$
5389 8329 $\exp(\sqrt{2})-\exp(\sqrt{3})$
5389 9650 ln : 7/12
5390 0308 e^x : $1/2-\sqrt{5}/2$
5390 0374 at : $2-3\sqrt{3}/2$
5390 1624 rt : (2,5,6,-5)
5390 7860 $t\pi$: $1/3-2\sqrt{5}/3$
5390 8134 ℓ_2 : $2+e/3$
5391 0094 e^x : $\sqrt{2}/4+4\sqrt{3}$
5391 0302 10^x : $3\sqrt{5}/4$
5391 1122 sq : $2\sqrt{2}/3-3\sqrt{5}/4$
5391 1447 id : $7\ln 2/9$
5391 3579 J_1 : 18/13
5391 5579 2^x : $e/4-\pi/2$
5391 6207 as : $\exp(-2/3)$
5391 6909 $t\pi$: $\sqrt{2}/4-3\sqrt{3}$
5391 8191 cr : $4-\sqrt{2}/4$
5391 8887 rt : (2,0,-8,4)
5391 9147 Ei : $9\pi^2/10$
5391 9946 rt : (9,7,-6,7)
5392 0824 Γ : $\exp(-\pi/3)$
5392 0876 e^x : $2+4\pi$
5392 1356 sq : $2+\sqrt{2}/4$
5392 1437 J : $(\pi/2)^{-3}$
5392 3874 sq : $3+e/4$
5392 4594 rt : (7,3,7,4)
5392 5381 rt : (3,9,-9,-7)
5392 6948 rt : (2,9,-5,-5)
5392 7048 rt : (6,6,8,-7)
5392 7242 rt : (8,3,3,2)
5392 7869 rt : (5,0,-6,9)
5392 8588 rt : (7,6,4,-5)
5392 8932 rt : (6,3,-7,4)
5392 9512 rt : (1,9,-1,-3)
5392 9947 $s\pi$: $4\sqrt{3}/3-2\sqrt{5}/3$
5393 0281 rt : (7,6,-8,-1)
5393 0844 rt : (8,6,0,-3)
5393 2049 rt : (8,9,-9,-6)
5393 2607 J_1 : $4\sqrt{3}/5$
5393 3904 10^x : 23/17
5393 4466 rt : (9,6,-4,-1)
5393 6622 sq : $1/4-2\sqrt{5}/3$
5393 9267 J : 4/15
5393 9473 erf : 12/23
5394 1320 cr : $2/3+4\sqrt{5}/3$
5394 1924 $\ln(\sqrt{5})\div\exp(2/5)$
5394 2154 $t\pi$: $\sqrt{11}$
5394 4079 rt : (2,8,1,-8)
5394 4718 J_1 : $2\ln 2$
5394 7976 Ei : 20/23
5394 8067 rt : (1,9,-7,1)
5394 8541 Ψ : $(\pi/3)^{1/2}$
5394 8881 rt : (3,8,-1,-5)
5394 8897 10^x : $2/3+\sqrt{3}/3$
5395 2642 J_1 : $\sqrt[3]{8/3}$
5395 3069 $t\pi$: $\sqrt[3]{2}/8$
5395 4068 E : 7/25
5395 4403 rt : (2,8,-6,-5)
5395 4466 E : $2\sqrt[3]{2}/9$
5395 4920 Γ : 11/19
5395 4955 sq : $((e+\pi)/3)^3$
5395 5943 rt : (4,2,-9,-4)
5395 8853 rt : (3,1,6,-4)
5395 9062 ℓ_{10} : $2\sqrt{3}$
5395 9590 cu : $\sqrt{15/2}$
5396 0071 id : $8\sqrt{3}/9$
5396 2346 cr : $\sqrt{2}/9$
5396 4928 e^x : $1+3\sqrt{5}$
5396 5201 E : $\sqrt{5}/8$
5396 6533 rt : (7,2,8,-6)
5397 1122 θ_3 : $5\zeta(3)/7$
5397 1230 $\exp(\sqrt{3})\div\exp(4/5)$
5397 2131 cr : $\sqrt{2}+\sqrt{5}$
5397 3302 rt : (7,5,-5,6)
5397 3389 rt : (6,4,-7,-4)
5397 3422 id : $e\pi$
5397 4028 rt : (5,9,-9,1)
5397 5699 at : $\ln(\ln 3/2)$
5397 5715 rt : (2,5,-9,-6)
5397 6268 rt : (2,7,-9,2)
5397 6755 $s\pi$: $9\zeta(3)$
5397 7864 rt : (8,2,4,-4)
5397 8877 2^x : 25/3
5397 9398 id : $1/3-4e$
5397 9546 ℓ_{10} : $5\ln 2$
5397 9781 rt : (8,8,-6,1)
5398 2356 $\ln(2)\div\exp(1/4)$
5398 2502 e^x : $2e+3\pi$
5398 4671 rt : (1,8,4,4)
5398 4797 rt : (7,7,9,-8)
5398 6584 cu : $\sqrt{2}+3\sqrt{5}/4$
5398 7253 J_1 : 23/10
5398 7502 E : $7(1/\pi)/8$
5398 7629 rt : (3,6,7,-6)
5398 8948 e^x : $3\sqrt{2}/4-3\sqrt{5}/4$
5399 0117 Ei : 10/19
5399 0135 rt : (2,7,0,1)
5399 0919 sr : $1/4+3\sqrt{2}/2$
5399 2621 J_1 : 25/18
5399 2652 10^x : 3/16
5399 2958 ℓ_{10} : $\sqrt[3]{3}/5$
5399 3413 tanh : $2e/9$
5399 4163 sq : $2e/3+\pi$
5399 4372 rt : (1,8,8,-3)
5399 5336 rt : (8,7,5,-6)
5399 6252 rt : (9,2,0,-2)
5399 6616 id : $3\sqrt[3]{2}/7$
5400 0035 rt : (3,7,0,4)
5400 0226 rt : (1,7,-5,-7)
5400 2971 ℓ_{10} : $-\sqrt{2}+\sqrt{5}+\sqrt{7}$
5400 2986 cr : $\sqrt[3]{2}/8$
5400 3079 rt : (6,9,-8,-6)
5400 4191 rt : (4,6,3,5)
5400 4237 E : 5/18
5400 4449 Ei : 21/17
5400 6172 sr : 7/24
5400 6737 ζ : $(\ln 2/2)^3$
5400 8633 10^x : $\exp(-\sqrt[3]{2}/3)$
5400 8680 e^x : $1/4-\sqrt{3}/2$
5400 9542 rt : (8,7,-3,8)

5400 9881 rt : (9,7,1,-4)
5400 9950 $\theta_3 : (\sqrt{2}/3)^{1/3}$
5401 0627 $\zeta : 1/24$
5401 1438 $\ell_{10} : 3/4 + e$
5401 3125 rt : (4,1,-6,3)
5401 3673 rt : (3,8,9,3)
5401 3843 $E : 1/\sqrt{13}$
5401 4403 $10^x : P_9$
5401 4437 sr : $1/3 + 3e/4$
5401 5890 $E : 2\ln 2/5$
5401 7347 sq : $4/3 - 2\sqrt{3}$
5401 9325 $2^x : 4 - \sqrt{2}/3$
5402 3106 rt : (8,7,7,3)
5402 3435 $2^x : 4/3 + 3e/2$
5402 4781 cu : $3\sqrt{3} + \sqrt{5}$
5402 5135 $3 \div \exp(2/3)$
5402 6024 tπ : $\sqrt{2}/3 - 2\sqrt{3}/3$
5402 6992 $erf : 1/\sqrt[3]{7}$
5402 7162 rt : (8,2,8,5)
5402 8519 rt : (8,3,9,-7)
5403 1242 rt : (5,9,2,0)
5403 3877 $2^x : 3e/2 + 4\pi$
5403 4231 $\ln(\pi) + \exp(1/3)$
5403 6035 $2^x : 5e/4$
5403 6203 $2^x : \sqrt{2} - \sqrt{5} + \sqrt{7}$
5403 6550 sinh : $(\sqrt{2}\pi/3)^{1/2}$
5403 9223 $4/5 \times s\pi(\sqrt{5})$
5403 9563 sπ : $9\sqrt{2}/7$
5404 0062 $e + \exp(3/5)$
5404 0932 at : $10\pi^2/3$
5404 1328 tπ : $6\sqrt{5}/5$
5404 1618 $\Psi : 2\pi^2/9$
5404 1852 $10^x : 23/12$
5404 1950 at : 3/5
5404 3108 $\theta_3 : 5(1/\pi)/6$
5404 4504 ln : 3/14
5404 4505 rt : (9,8,6,-7)
5404 5789 rt : (3,9,-8,-8)
5404 6911 rt : (4,7,8,-7)
5404 9193 cr : 3/19
5404 9770 rt : (9,3,5,-5)
5405 0352 $\zeta : 3\sqrt{3}/10$
5405 0522 $\ell_{10} : 3 + \sqrt{2}/3$
5405 4833 rt : (4,2,7,-5)
5405 5660 $\ln(4) + \exp(e)$
5405 6238 rt : (4,6,-9,4)
5405 6838 $\ell_2 : 11/16$
5405 7695 rt : (3,8,-9,1)
5405 7956 rt : (6,7,-8,-7)
5405 8329 sπ : $2\sqrt{3} + 3\sqrt{5}/2$
5405 8819 cu : $3/2 - 4\sqrt{5}$
5405 9872 sq : $e/2 - 2\pi/3$
5406 0323 rt : (5,3,-4,5)
5406 1921 $\Psi : 3/5$
5406 1968 rt : (9,2,4,3)
5406 2514 $e^x : 10\sqrt{2}/7$
5406 3705 rt : (1,1,-9,-5)
5406 4081 sπ : 2/11
5406 5660 $\Psi : 9(e + \pi)/4$
5406 8371 id : $\sqrt{3}/3 - \sqrt{5}/2$
5406 9263 $2/5 + \exp(\pi)$
5407 0263 tanh : $(\ln 3/3)^{1/2}$
5407 0911 ln : $10\zeta(3)/7$
5407 1998 rt : (5,8,5,7)
5407 2627 $2^x : 1/3 + 2\sqrt{5}/3$
5407 3210 $\ln(\pi) \div \exp(3/4)$
5407 6818 $\exp(2/5) - s\pi(2/5)$
5407 7419 ln : $9\pi^2/7$
5407 8953 rt : (4,3,6,3)
5408 2902 $\zeta : \zeta(3)$
5408 3299 rt : (6,6,-5,0)
5408 3421 sπ : $4(1/\pi)/7$
5408 4358 id : $3\sqrt[3]{3}/8$
5408 7539 tπ : $\sqrt{3}/2 - 3\sqrt{5}$
5408 7674 rt : (4,9,0,-2)
5408 8525 id : $2\sqrt{2}/3 + 3\sqrt{3}/2$
5409 0246 rt : (5,8,9,-8)
5409 0917 cr : $3\sqrt{2}/4 + 3\sqrt{3}/2$
5409 1258 rt : (9,1,9,6)
5409 3006 sinh : $\exp(\sqrt[3]{4})$
5409 3137 $10^x : 3 + 3\sqrt{3}$
5409 4144 $2^x : 1 + 2\pi/3$
5409 4555 $\ell_{10} : 1/3 + \pi$
5409 6167 $Ei : 3\pi/5$
5409 6325 $2^x : 9\ln 2/10$
5409 6465 rt : (6,5,-2,7)
5409 8323 rt : (1,9,-6,-8)
5409 9150 $erf : \pi/6$
5409 9329 rt : (5,2,3,-3)
5410 1064 $3/4 \div \ln(4)$
5410 2121 $2^x : e - 2\pi/3$
5410 3293 rt : (5,5,-8,-5)
5410 3362 $10^x : 2 - 2e/3$
5410 6624 $\exp(4/5) \div \exp(\sqrt{2})$
5410 6977 sπ : $9\sqrt{2}/4$
5410 7055 cr : $3e\pi/7$
5410 7716 rt : (7,2,-9,7)
5410 8050 $J_0 : (2\pi/3)^{1/2}$
5410 8720 $\ell_{10} : \ln(4/3)$
5410 9268 rt : (6,7,0,-3)
5411 0350 sr : 19/8
5411 2742 $e^x : e/4 + 2\pi$
5411 4257 $E : 6(1/\pi)/7$
5411 4419 rt : (5,3,8,-6)
5411 5154 rt : (7,9,-6,-5)
5411 5525 $J_1 : (e)^{1/3}$
5411 6678 $E : 3/11$
5411 7226 $erf : 11/21$
5411 7293 tπ : 3/19
5411 7537 at : $\zeta(3)/2$
5411 8541 rt : (6,8,5,-6)
5411 9610 sr : $1 - \sqrt{2}/2$
5412 0276 as : $3\zeta(3)/7$
5412 0519 cu : $1/3 + 4\pi$
5412 1117 rt : (9,7,3,1)
5412 1442 rt : (7,8,-3,2)
5412 5027 rt : (7,7,0,9)
5412 5558 $2^x : 1 - 4\sqrt{2}/3$
5412 7331 $J_1 : 4\pi/9$
5412 7581 $2^x : \exp(-\sqrt{2}/3)$
5412 8286 rt : (8,4,-7,9)
5413 1551 $e^x : 3\sqrt{2} + 3\sqrt{3}$
5413 2106 rt : (9,6,8,4)
5413 4113 id : $(e/2)^{-2}$
5413 5022 $e^x : 9\sqrt{2}/2$
5413 6465 $E : e/10$
5413 6550 cr : $9\pi^2/2$
5413 6997 tanh : $3\sqrt{2}/7$
5413 7499 $e^x : 3\sqrt[3]{3}/10$
5413 8126 rt : (1,7,7,-6)
5413 9556 tπ : $3\sqrt{3} - 3\sqrt{5}/2$
5414 2634 cr : $5(e + \pi)/8$
5414 4545 rt : (9,9,-7,-5)
5414 7212 rt : (9,7,-8,4)
5414 7450 $e^x : (3\pi)^{1/2}$
5414 9321 $J : \exp(-12\pi/19)$
5414 9591 ln : $4e/3 + \pi/3$
5415 0478 $J_1 : 5\sqrt{5}/8$
5415 1555 $J_1 : ((e + \pi)/3)^{1/2}$
5415 1569 $s\pi(1/3) + s\pi(\sqrt{5})$
5415 1680 $e^x : \exp((e + \pi)/2)$
5415 2468 rt : (7,9,6,-7)
5415 2976 at : $4e\pi$
5415 3654 cu : $e + 3\pi$
5415 4001 $J_1 : (\ln 3/3)^{-1/3}$
5415 4564 $\zeta : 17/8$
5415 4742 $\zeta : 3\ln 2/4$
5415 5007 rt : (6,4,9,-7)
5415 6032 $4 \div \ln(\sqrt{2})$
5415 7982 rt : (7,8,1,-4)
5415 8341 rt : (3,3,-9,7)
5415 8761 rt : (9,8,-2,-2)
5415 9265 $2/5 + \pi$
5416 0512 $\Gamma : 4\sqrt[3]{5}$
5416 2764 rt : (3,5,-9,-9)
5416 3202 rt : (6,3,4,-4)
5416 6134 tanh : $7\ln 2/8$
5416 6666 id : 13/24
5416 6871 $\zeta : \exp(-\pi)$
5416 7879 id : $3\sqrt{2} + 3\sqrt{3}/4$
5416 8033 tanh : $(e)^{-1/2}$
5416 8240 at : $(e + \pi)^2$
5416 9082 rt : (4,6,-9,-6)
5417 0378 rt : (7,7,-4,-1)
5417 2149 rt : (5,2,7,4)
5417 4487 $\zeta : (\sqrt[3]{5})^{-1/3}$
5417 4666 $J_1 : \sqrt[3]{12}$
5417 5279 cu : $4\sqrt{5}$
5417 5302 rt : (2,8,8,-7)
5417 5691 $\ell_{10} : 1/2 + 4\sqrt{5}/3$
5417 8079 cu : $5\ln 2/3$
5417 9207 $3/4 \times \exp(2)$
5417 9606 sq : $3 - \sqrt{5}/2$
5417 9627 $\zeta : \sqrt{3} + \sqrt{5} - \sqrt{6}$
5418 1483 cr : $7\pi/6$
5418 2389 cu : $2 + \sqrt{3}/2$
5418 2650 rt : (4,9,-1,5)
5418 2871 cr : $3\sqrt{2} - \sqrt{3}/3$
5418 2998 $K : 17/18$
5418 4972 $\ell_2 : \zeta(3)/7$
5418 5549 at : $7\pi^2/2$
5418 5589 $\ell_2 : 1/3 + \sqrt{2}/4$
5418 6853 rt : (6,2,-1,-1)
5418 9580 $e^x : \sqrt{3}/4$
5419 0208 sinh : 9
5419 1296 rt : (8,9,2,-5)
5419 2607 id : $(2\pi)^{-1/3}$
5419 3194 $J_1 : 10\sqrt[3]{2}/9$
5419 3617 rt : (2,7,3,-4)
5419 3620 $\zeta : 1/23$
5419 4771 $J_1 : 7/5$
5419 8535 $Li_2 : 8/17$
5419 8993 $e^x : P_{10}$
5420 1023 rt : (4,5,-4,-3)
5420 2166 id : $\sqrt[3]{11/3}$
5420 2601 rt : (7,4,5,-5)
5420 3237 sπ : $\sqrt{2}/4 + 2\sqrt{3}$
5420 4560 at : P_2
5420 4877 $\Gamma : \sqrt{2}/7$
5420 5083 $\theta_3 : (e/2)^{-1/3}$
5420 5268 $J : 3/20$
5420 8132 cu : $3/4 + e/3$
5420 9574 sq : $1/2 - 2\pi/3$
5421 2430 rt : (5,0,-8,6)
5421 3833 rt : (1,2,6,-4)
5421 4244 $\zeta : 13/25$
5421 4496 $e^x : 1 + 3\sqrt{5}/4$
5421 5641 $e^x : 1/2 + \sqrt{3}/4$
5421 6716 rt : (8,8,-3,-2)
5421 9523 sq : $\sqrt[3]{20/3}$
5421 9768 sq : $7\ln 2$
5422 0019 rt : (6,1,8,5)
5422 1082 $2^x : 5/8$
5422 1121 sinh : $\sqrt{2} + \sqrt{6} - \sqrt{7}$
5422 2274 ln : $7\pi^2/2$
5422 3178 $\zeta : (\sqrt[3]{3}/2)^2$
5422 3753 sr : $7e/8$
5422 3805 rt : (7,6,-9,2)
5422 6893 $\ln(2/3) + \exp(2/3)$
5422 7595 $e^x : 4 + e/2$
5422 7597 rt : (6,2,-6,8)
5422 9332 rt : (8,5,6,-6)
5423 0949 cu : $3e/10$
5423 2614 sr : 5/17
5423 3171 sinh : $(e/3)^{-2}$
5423 4812 id : $3\sqrt{3}/2 + 4\sqrt{5}$
5423 5908 $E : 1/\sqrt{14}$
5423 6197 at : $6(e + \pi)$
5423 6891 ln : $3\sqrt{2} + \sqrt{3}/4$
5423 7224 $2^x : (2e/3)^{1/2}$
5423 7750 rt : (5,3,2,1)
5423 7792 $J_1 : 7\zeta(3)/6$
5423 8874 $E : 2\zeta(3)/9$
5423 9554 $J_1 : 16/7$
5424 2559 sπ : $\exp(5/2)$
5424 4205 rt : (7,3,0,-2)
5424 5502 rt : (9,9,-2,-3)
5424 5975 sr : $3 + 2\sqrt{3}$
5424 6970 $10^x : \ln(\sqrt{3})$
5424 7402 $J_0 : \exp(1/e)$

5424 7453 rt : (2,6,-2,-1)
5424 8123 2^x : $2/3+e/4$
5424 8125 ℓ_2 : $(\sqrt{2}/3)^{1/2}$
5424 8593 cu : $5\sqrt{5}$
5424 8658 rt : (9,6,7,-7)
5424 8724 E : $4/15$
5424 9156 rt : (7,0,9,6)
5424 9753 rt : (6,3,-9,1)
5425 0542 rt : (4,9,5,-6)
5425 1831 sπ : $2\sqrt{2}+3\sqrt{5}/2$
5425 2154 rt : (2,3,7,-5)
5425 4326 Ei : $1/4$
5425 7289 rt : (3,9,4,0)
5425 8362 sπ : $7\pi^2/5$
5425 9680 J_0 : $13/9$
5426 0568 Ei : $6\sqrt[3]{3}/7$
5426 2605 rt : (7,5,-7,3)
5426 2728 sinh : $7/2$
5426 2741 $\ln(2)\times\exp(4/5)$
5426 3917 10^x : $(\pi/2)^{-2}$
5426 6777 ln : $3+3\sqrt{5}/4$
5426 7076 rt : (5,5,1,8)
5426 8117 rt : (8,9,-3,-3)
5427 2954 rt : (8,7,-5,5)
5427 4065 rt : (8,4,1,-3)
5427 5292 rt : (3,4,8,-6)
5427 6273 2^x : $3\pi/7$
5427 8833 tπ : $(\sqrt{5}/3)^{-1/2}$
5427 8960 E : $5(1/\pi)/6$
5428 1194 rt : (5,0,-7,3)
5428 1467 rt : (9,9,-3,7)
5428 3309 Ψ : $3(e+\pi)/8$
5428 4388 Ei : $(\pi/3)^{-3}$
5428 5320 rt : (2,7,-6,-5)
5428 6080 Γ : $5\ln 2/6$
5428 6470 rt : (3,7,-1,-2)
5428 7917 sr : $\sqrt{17/3}$
5428 8352 cr : $4/25$
5428 8510 rt : (8,7,-8,1)
5428 8610 rt : (2,6,-3,3)
5428 9691 sr : $4/3+\pi/3$
5429 0736 ℓ_{10} : $10\pi/9$
5429 0777 rt : (4,5,9,-7)
5429 1401 ℓ_{10} : $2+2\sqrt{5}/3$
5429 2079 $\sqrt{5}-\ln(2)$
5429 2328 tanh : $(\sqrt{2}\pi)^{-1/3}$
5429 2383 rt : (9,5,2,-4)
5429 3628 sq : $14/19$
5429 5503 10^x : P_{36}
5429 6223 ln : $\sqrt[3]{5}/8$
5429 7163 e^x : $14/15$
5429 8128 2^x : $4e-3\pi/2$
5429 9895 cu : $4e-2\pi/3$
5430 1184 Ei : $(\sqrt[3]{4}/3)^{-1/3}$
5430 1242 ζ : $(\pi/2)^{1/2}$
5430 1303 rt : (6,2,3,2)
5430 1829 ζ : $\sqrt[3]{7/2}$
5430 2100 rt : (1,8,-7,-6)
5430 2819 at : $1/4+\sqrt{2}/4$
5430 3813 ln : $3e/2-3\pi/4$
5430 5477 10^x : $4e/3-3\pi/4$
5430 6509 rt : (8,8,-8,-2)
5430 7348 cu : $2+e/3$
5430 7638 id : $\sqrt{3}/2+3\sqrt{5}/4$
5430 8184 id : $(\sqrt[3]{2}/3)^{-1/2}$
5430 8892 Li_2 : $\sqrt{2}/3$
5431 0660 ℓ_2 : $3+2\sqrt{2}$
5431 1453 cu : $4\sqrt{2}-2\sqrt{3}$
5431 2966 cr : P_{20}
5431 3271 rt : (9,8,-7,0)
5431 4405 E : $\exp(-4/3)$
5431 4475 10^x : $\sqrt{5/3}$
5431 6607 cr : $\sqrt[3]{3}/9$
5431 7989 e^x : $\sqrt{21/2}$
5431 8405 J_0 : $5\sqrt{3}/6$
5431 8710 rt : (5,9,1,-4)
5432 0632 ln : $4+e/4$
5432 0828 tanh : $14/23$
5432 0987 sq : $5\sqrt{5}/9$
5432 1241 rt : (1,7,-8,-8)
5432 1356 cr : $5\pi^2/3$
5432 3362 rt : (7,1,4,3)
5432 3738 E : $5/19$
5432 6601 tπ : $1/4+\sqrt{3}/4$
5432 6919 rt : (6,8,0,3)
5432 7214 rt : (2,7,-2,-6)
5432 7678 Ei : $4\sqrt{2}/3$
5432 8129 sπ : $10e$
5432 8181 rt : (4,1,-8,-4)
5433 1342 ℓ_2 : $4e+\pi/4$
5433 1563 sinh : $3\sqrt{3}/10$
5433 1834 $\exp(2)+\exp(e)$
5433 1857 rt : (3,6,-1,8)
5433 1886 erf : $10/19$
5433 4566 rt : (8,0,5,4)
5433 5765 sπ : $4\sqrt{3}-\sqrt{5}/3$
5433 8957 J_1 : $4\sqrt[3]{5}/3$
5434 0137 at : $2e/9$
5434 1140 $\sqrt{2}\div\ln(2/5)$
5434 1345 rt : (9,1,6,-5)
5434 1580 2^x : $\sqrt{2}/2+\sqrt{5}/2$
5434 2397 rt : (3,2,-9,-5)
5434 2782 ℓ_{10} : $4/3-\pi/3$
5434 3895 ℓ_{10} : $9e/7$
5434 4249 sπ : $\sqrt[3]{6}$
5434 5690 Ψ : $9\sqrt[3]{5}/7$
5434 5883 ℓ_{10} : $2/3+2\sqrt{2}$
5434 6055 θ_3 : $4/15$
5434 6163 E : $((e+\pi)/3)^{-2}$
5434 6317 10^x : $3+3\sqrt{2}$
5434 7601 cr : $2+3\sqrt{5}/4$
5435 1129 ζ : $(3\pi/2)^{-2}$
5435 1245 e^x : $2\sqrt{5}$
5435 2254 2^x : $4/3-\sqrt{2}/2$
5435 4249 2^x : $3/2+3\sqrt{2}$
5435 4389 Γ : $\sqrt{3}/3$
5435 5159 $\exp(\sqrt{2})-s\pi(\pi)$
5435 5793 J_0 : $1/\ln(2)$
5435 7984 $\ln(\sqrt{5})\times s\pi(\sqrt{5})$
5435 9180 rt : (8,4,-9,6)
5435 9463 sinh : $3\ln 2/4$
5435 9775 θ_3 : $\ln(\sqrt[3]{4}/2)$
5435 9925 rt : (9,6,-7,8)
5436 0342 ζ : $\exp(1/(2e))$
5436 1540 id : $7\pi^2/2$
5436 2554 $\sqrt{5}\div\exp(\sqrt{2})$
5436 3581 e^x : $1/3-2\sqrt{2}/3$
5436 3927 $\ln(2/3)\div s\pi(\sqrt{3})$
5436 5636 id : $e/5$
5436 7355 10^x : $2+4\sqrt{2}/3$
5436 8247 J_0 : $6\zeta(3)/5$
5436 8901 rt : (9,9,9,-9)
5436 9542 10^x : $\ln(3/2)$
5437 0262 ℓ_{10} : $1/2-\sqrt{2}/3$
5437 1556 J_1 : $\pi^2/7$
5437 2095 E : $6/23$
5437 3098 ℓ_{10} : $3\sqrt{2}-\sqrt{5}/3$
5437 3142 ℓ_2 : $\sqrt{17/2}$
5437 5355 sinh : $13/25$
5437 5942 e^x : $4+4\sqrt{2}$
5437 6404 θ_3 : $8/13$
5437 7740 sinh : $(\sqrt[3]{3}/2)^2$
5437 8342 10^x : $1/\sqrt[3]{15}$
5437 8618 Ψ : $7\pi/10$
5437 9397 sπ : $3/4+\sqrt{3}/4$
5437 9816 ln : $9\sqrt{2}$
5437 9930 e^x : $7\sqrt{3}/6$
5438 0259 J_0 : $\sqrt[3]{3}$
5438 1785 e^x : $3/4-e/2$
5438 2845 cr : $3+e/4$
5438 4522 rt : (5,2,-3,9)
5438 5677 rt : (6,7,-9,-6)
5438 6116 sq : $\sqrt{2}/4-\sqrt{5}$
5438 6491 rt : (4,0,-5,7)
5438 7875 tπ : $7\zeta(3)/10$
5438 8174 Γ : γ
5438 9027 rt : (3,7,-2,-1)
5439 1679 rt : (7,6,-8,-5)
5439 1817 E : $3\ln 2/8$
5439 2419 rt : (2,4,-9,-3)
5439 3077 rt : (9,4,-3,-1)
5439 4107 ζ : $1/22$
5439 4237 sq : $\sqrt{2}-\sqrt{3}+\sqrt{6}$
5439 5967 ζ : $(1/\pi)/7$
5439 7007 id : $3\sqrt{2}/2-\sqrt{3}/3$
5439 7185 rt : (1,9,4,-5)
5439 8557 rt : (8,4,9,5)
5440 0660 id : $2e/3-3\pi/4$
5440 2111 sπ : $10(1/\pi)$
5440 3220 cr : $-\sqrt{2}+\sqrt{6}+\sqrt{7}$
5440 3513 J_1 : $24/17$
5440 3929 sr : $2+2\sqrt{5}$
5440 4374 rt : (8,5,-7,-4)
5440 6804 ℓ_{10} : $2/7$
5440 7181 sr : $\sqrt{2}/2+3\sqrt{5}/4$
5440 8611 rt : (7,5,8,4)
5440 9162 2^x : $7\sqrt{3}/9$
5441 0394 rt : (1,8,-5,-3)
5441 0603 id : $1/\ln(2\pi)$
5441 3710 sinh : e
5441 4493 rt : (4,8,-2,9)
5441 5587 sq : $3-3\sqrt{2}$
5441 6464 rt : (1,7,-2,-3)
5441 9619 at : $(\ln 3/3)^{1/2}$
5442 1461 rt : (5,1,9,-6)
5442 1784 rt : (6,9,-3,-2)
5442 2382 sr : $\sqrt{2}-\sqrt{5}/2$
5442 2840 sq : $1/\ln(\sqrt{5})$
5442 3762 ζ : $3\sqrt{2}/2$
5442 7970 E : $1/\sqrt{15}$
5442 8981 rt : (6,6,7,3)
5442 9343 Ψ : $11/5$
5443 0737 rt : (8,9,0,8)
5443 1356 tπ : $3e\pi$
5443 1851 E : $(\pi/2)^{-3}$
5443 3054 2^x : $4-\sqrt{5}/3$
5443 3105 sr : $(3/2)^{-3}$
5443 4690 10^x : $4/3$
5443 4834 cu : $3-\sqrt{5}/4$
5443 4995 rt : (4,0,8,5)
5443 5451 rt : (7,2,-5,1)
5443 6564 Γ : $(\sqrt{2}\pi/3)^3$
5443 6755 rt : (9,8,4,-6)
5443 7011 rt : (2,6,9,-7)
5443 7691 rt : (7,7,-2,6)
5443 8096 e^x : $4\sqrt{2}+3\sqrt{3}/2$
5443 8236 sq : $3/2+\pi$
5443 8976 rt : (5,8,-4,-1)
5443 9511 cr : $3\sqrt{2}-\sqrt{5}/4$
5443 9695 cu : $4-3e/4$
5443 9892 sq : $8e/3$
5444 0020 cu : $3+3\sqrt{5}/2$
5444 1059 rt : (2,6,-1,-2)
5444 4446 θ_3 : $2\zeta(3)/9$
5444 6276 J_1 : $\sqrt{2}$
5444 6406 rt : (6,5,-4,4)
5444 6905 sinh : $(2e/3)^{1/3}$
5444 8089 10^x : $3/2+2\sqrt{3}/3$
5444 8141 rt : (8,7,3,-5)
5444 8464 e^x : $7\sqrt{3}$
5444 9169 rt : (9,4,-6,-3)
5445 1152 rt : (1,5,8,-6)
5445 2443 ℓ_{10} : $e+\pi/4$
5445 3106 Li_2 : $3\sqrt[3]{2}/8$
5445 4461 sr : $1/2+4\sqrt{2}/3$
5445 5311 E : $\exp(-e/2)$
5445 7649 rt : (5,3,-6,2)
5445 7922 rt : (3,1,7,4)
5446 0681 rt : (1,3,-9,-9)
5446 1573 ζ : $\ln(\pi/3)$
5446 2501 J_1 : $25/11$
5446 2723 e^x : $10/23$
5446 4116 rt : (7,6,2,-4)
5446 4168 cu : $3\sqrt{3}/4+\sqrt{5}/3$
5446 7517 Γ : $2\sqrt[3]{3}/5$
5446 8029 tπ : $7\zeta(3)/5$
5446 9764 10^x : $4\pi^2/9$
5447 2708 rt : (4,9,-3,0)
5447 2988 $2/5+\ln(\pi)$
5447 4048 θ_3 : $1/\sqrt{14}$

5447 5422 rt : (9,4,8,-7)
5447 6490 2^x : $\sqrt{17/2}$
5447 8005 at : $9e\pi/2$
5447 8122 Γ : $(1/(3e))^{1/2}$
5447 8843 rt : (9,0,1,2)
5448 0006 sq : $2\sqrt{3}+4\sqrt{5}/3$
5448 0577 rt : (6,9,8,-8)
5448 1188 10^x : $\exp(-5/3)$
5448 1684 rt : (3,9,-7,-6)
5448 2432 $\ln(5)-\exp(e)$
5448 4550 Γ : $-\sqrt{2}-\sqrt{3}+\sqrt{7}$
5448 5769 e^x : $4-3\sqrt{2}/2$
5448 8157 rt : (6,5,1,-3)
5448 8558 ln : $4/3+3\sqrt{5}/2$
5448 8640 cu : $4(e+\pi)/7$
5448 8652 at : $3\sqrt{2}/7$
5448 8723 e^x : $2+\pi/2$
5448 8875 J_1 : $17/12$
5448 9245 2^x : $\sqrt{10/3}$
5449 0770 sr : 4π
5449 0876 tanh : $11/18$
5449 1115 ℓ_{10} : $\exp(e+\pi)$
5449 2279 rt : (5,7,6,2)
5449 2761 cr : $1/3+3\sqrt{5}/2$
5449 3654 rt : (8,3,7,-6)
5449 5743 rt : (8,6,-4,9)
5449 6122 $\exp(\sqrt{5})\div s\pi(\sqrt{3})$
5449 7060 rt : (5,8,7,-7)
5449 8463 ln : $4(e+\pi)/5$
5449 8786 J : $4(1/\pi)/3$
5450 1324 E : $4(1/\pi)/5$
5450 1757 J_1 : $9\sqrt[3]{2}/8$
5450 2947 J : $3/11$
5450 3493 sr : $3\ln 2/7$
5450 4351 2^x : $3\sqrt{2}/4-\sqrt{3}/4$
5450 5225 rt : (2,2,6,3)
5450 5438 $\exp(3)\times s\pi(e)$
5450 7631 rt : (7,4,-6,7)
5450 8058 cr : $\sqrt{2}/2+4\sqrt{5}/3$
5450 8572 e^x : $2\sqrt{2}/3+4\sqrt{5}$
5450 8709 $s\pi$: $\sqrt{2/3}$
5450 8714 J_0 : $8\sqrt[3]{2}/7$
5450 8813 10^x : $4/3-\sqrt{2}/4$
5450 9224 10^x : $(e/3)^3$
5450 9590 rt : (1,6,3,-6)
5451 1571 e^x : $5\sqrt{5}/3$
5451 1783 rt : (4,8,-6,-5)
5451 4483 rt : (2,5,-6,5)
5451 6124 ℓ_{10} : $\sqrt[3]{5}/6$
5451 7459 ln : $1/2+4\pi/3$
5451 7744 id : $8\ln 2$
5451 8106 cu : $1/4+e/3$
5451 8398 rt : (3,5,0,-1)
5451 8797 at : $7\ln 2/8$
5451 9250 rt : (4,7,6,-6)
5451 9403 $t\pi$: $6\ln 2$
5451 9414 rt : (7,2,6,-5)
5451 9766 ζ : $6\zeta(3)/7$
5452 0762 at : $(e)^{-1/2}$
5452 1979 cr : $4/3+3\pi/4$
5452 1979 rt : (6,2,-8,5)
5452 6025 rt : (2,9,0,-3)
5452 7062 rt : (7,8,-5,-1)
5452 8163 sq : $2/3-2\pi$
5452 8426 rt : (5,4,0,-2)
5452 8438 ζ : $5\sqrt[3]{3}/6$
5452 9386 Ei : $5\sqrt{3}/7$
5453 0845 cu : $1/\ln(\sqrt[3]{3}/3)$
5453 2816 e^x : $4e-\pi/2$
5453 3032 rt : (8,1,0,1)
5453 3075 sr : $3\sqrt{2}+\sqrt{5}$
5453 3888 $t\pi$: $\sqrt{2}/2-\sqrt{3}/2$
5453 5031 e^x : $9(1/\pi)$
5453 7922 λ : $(\ln 3/3)^3$
5453 8165 $s\pi$: P_{63}
5454 0015 sq : $5e/4$
5454 0768 sq : $3\sqrt{2}/2+3\sqrt{3}$
5454 1334 $t\pi$: $4e\pi$
5454 1882 $s\pi$: $1/4+4\pi$
5454 4510 J_1 : $9\sqrt[3]{2}/5$
5454 5454 id : $6/11$
5454 5654 sq : $4\pi^2$
5454 6905 id : $4\sqrt{3}/3+\sqrt{5}$
5454 6941 $\exp(4)\div s\pi(e)$
5454 7144 at : $4\pi^2$
5454 9681 e^x : $2/3+3\sqrt{2}$
5455 0173 id : $9\zeta(3)/7$
5455 0736 rt : (3,6,5,-5)
5455 2122 ln : $\sqrt{22}$
5455 2571 rt : (9,9,-5,4)
5455 3519 10^x : $1+\sqrt{2}$
5455 4066 rt : (6,6,-7,-3)
5455 5467 Li_2 : $7\sqrt[3]{2}/9$
5455 5648 E : $\sqrt[3]{2}/5$
5455 8098 10^x : $3\zeta(3)/2$
5455 8441 id : $9\sqrt{2}/5$
5455 8580 rt : (6,1,5,-4)
5455 9140 $\ln(2/3)+s\pi(2/5)$
5455 9923 Ψ : $6(e+\pi)$
5456 0815 rt : (9,7,-1,-3)
5456 1726 rt : (1,9,4,1)
5456 3839 rt : (5,7,-5,-4)
5456 7833 $1/2\div\ln(2/5)$
5456 7914 rt : (7,9,4,-6)
5456 8746 rt : (3,6,-6,1)
5456 9772 $t\pi$: $(\pi)^{-1/3}$
5457 0347 Γ : $7\sqrt{5}$
5457 1110 as : $2\sqrt{2}-4\sqrt{3}/3$
5457 1444 rt : (9,4,5,3)
5457 1638 rt : (8,7,-7,2)
5457 3398 sinh : $12/23$
5457 4251 rt : (1,9,-5,2)
5457 4313 sq : $3+2\sqrt{2}/3$
5457 6234 2^x : $\pi/5$
5457 8563 rt : (8,6,9,-8)
5457 9514 2^x : $1/\ln(1/\pi)$
5458 0305 rt : (1,8,-1,-2)
5458 1162 Li_2 : $(3\pi)^{-1/3}$
5458 1947 Ψ : $3e/8$
5458 2050 $2/3\div\exp(1/5)$
5458 3433 2^x : $2+\sqrt{2}/3$
5458 3482 $1/5+s\pi(\sqrt{3})$
5458 7223 rt : (2,8,5,-4)
5458 8327 J_1 : $5e/6$
5458 8407 $\exp(2/3)+\exp(4)$
5458 8775 ln : $4\sqrt{3}-\sqrt{5}$
5459 0699 cr : $2\sqrt{2}+\sqrt{3}/2$
5459 0887 sinh : $e\pi/7$
5459 1246 $\exp(1/5)-s\pi(\sqrt{5})$
5459 2011 K : $3\sqrt[3]{2}/4$
5459 2178 rt : (5,4,-9,-5)
5459 4154 id : $(\sqrt{2}/3)^{-3}$
5459 5121 rt : (7,9,-7,0)
5459 5301 cu : $1+3\sqrt{5}/2$
5459 5725 E : $1/4$
5459 6220 10^x : $\sqrt{2}-3\sqrt{5}/4$
5459 6273 erf : $9/17$
5459 6273 rt : (9,6,-9,5)
5459 6337 sq : $4\pi/3$
5459 6648 rt : (2,6,-5,-4)
5459 8909 rt : (2,7,8,8)
5459 9897 2^x : $4\sqrt{2}/9$
5460 1170 Ei : $7\sqrt{2}/8$
5460 2596 rt : (6,8,3,-5)
5460 2687 J_1 : $e\pi/6$
5460 3956 ℓ_{10} : $6(e+\pi)$
5460 4794 $s\pi$: P_{47}
5460 8745 J_1 : $(\ln 2/2)^{-1/3}$
5460 9300 $t\pi$: $(\sqrt{2})^{-1/2}$
5460 9679 rt : (8,6,-2,-2)
5461 0245 rt : (7,5,8,-7)
5461 0484 rt : (3,2,9,-6)
5461 4353 $3/5\div\ln(3)$
5461 4697 Γ : $7/20$
5461 4722 ζ : $1/21$
5461 5774 $\ln(2/3)-\exp(\pi)$
5461 7497 Li_2 : $9/19$
5461 8913 rt : (9,3,3,-4)
5462 3382 10^x : $9e/10$
5462 5145 ln : $e/4+\pi/3$
5462 5293 rt : (5,0,4,3)
5462 6760 E : $\sqrt{5}/9$
5462 8412 rt : (7,9,3,9)
5462 9741 J_1 : $8\sqrt{2}/5$
5463 0248 $t\pi$: $(1/\pi)/2$
5463 0650 rt : (8,8,-9,-6)
5463 1379 sq : $17/23$
5463 1469 J_1 : $5\sqrt[3]{5}/6$
5463 2137 $t\pi$: $7\sqrt[3]{2}/6$
5463 2307 sr : $2e+\pi/3$
5463 3536 rt : (6,3,9,5)
5463 4033 rt : (3,9,7,9)
5463 9281 cu : $\sqrt{2}/2-3\sqrt{5}/2$
5464 0056 as : $3\sqrt{3}/10$
5464 0742 rt : (2,9,-3,-4)
5464 0773 $t\pi$: $10\sqrt{5}/9$
5464 0916 J_0 : $23/16$
5464 2228 at : $7(e+\pi)$
5464 3178 ℓ_{10} : $3\sqrt{3}-3\sqrt{5}/4$
5464 4033 2^x : $3e/4+2\pi/3$
5464 4266 E : $7(1/\pi)/9$
5464 4476 rt : (1,9,7,-7)
5464 6985 e^x : P_{29}
5464 7023 E : $\sqrt{3}/7$
5464 7908 id : $8(1/\pi)$
5464 9385 at : $(\sqrt{2}\pi)^{-1/3}$
5464 9467 e^x : $1/3+2e/3$
5465 1241 sq : $\ln(2\pi/3)$
5465 1838 Γ : $4(e+\pi)/3$
5465 1908 rt : (5,8,3,4)
5465 3028 id : $1/\ln(\sqrt{2}\pi/3)$
5465 3209 rt : (6,4,7,-6)
5465 4370 ln : $11/19$
5465 4375 $t\pi$: $\sqrt{2}+2\sqrt{3}/3$
5465 7359 $1/5+\ln(\sqrt{2})$
5465 7503 ln : $2/3+3\sqrt{2}/4$
5465 8654 cr : $\exp(-2e/3)$
5466 0234 rt : (1,3,5,2)
5466 2953 rt : (6,7,1,7)
5466 7216 tanh : $(\ln 2/3)^{1/3}$
5466 7762 $t\pi$: $6(e+\pi)$
5466 8750 as : $3\ln 2/4$
5467 0618 rt : (5,3,-9,-5)
5467 1964 rt : (9,8,9,4)
5467 4529 erf : $3\sqrt{2}/8$
5467 4827 ζ : $\sqrt[3]{19/2}$
5467 5863 rt : (6,6,-4,-3)
5467 6638 rt : (2,1,8,-5)
5467 7342 rt : (8,5,4,2)
5467 7754 sq : $\sqrt{2}/3+3\sqrt{5}$
5467 8884 at : $14/23$
5467 9996 rt : (9,1,7,5)
5468 1827 rt : (4,8,-4,4)
5468 2538 $t\pi$: $3\sqrt{2}+3\sqrt{3}/2$
5468 3386 cr : $3\pi^2/8$
5468 3401 rt : (9,6,5,-6)
5468 4546 rt : (8,2,2,-3)
5468 5095 as : $13/25$
5468 6962 rt : (8,9,0,-4)
5468 7500 cu : $(\sqrt[3]{4}/3)^{-3}$
5468 7548 as : $(\sqrt[3]{3}/2)^2$
5468 9493 sinh : $1/\sqrt[3]{7}$
5469 0403 erf : $5(1/\pi)/3$
5469 1444 rt : (2,4,-9,7)
5469 1816 rt : (7,9,1,-1)
5469 2224 J_1 : $10/7$
5469 4470 θ_3 : $8\ln 2/9$
5470 0252 rt : (6,4,-3,8)
5470 2418 rt : (6,7,9,-8)
5470 4317 e^x : $7\zeta(3)/9$
5470 5278 rt : (6,8,-8,1)
5470 5776 $\pi-\ln(2/3)$
5470 7515 Ψ : $7\sqrt[3]{2}/4$
5470 7661 e^x : $\sqrt{2}/2+\sqrt{5}/4$
5470 8097 rt : (5,5,-1,5)
5470 9352 E : $\sqrt[3]{5}/7$
5471 1366 at : $3/4-e/2$
5471 2134 10^x : $7\sqrt{5}/2$
5471 3369 $\exp(5)-s\pi(1/3)$
5471 3710 Ei : $\sqrt{5/2}$

5471 4786 rt : (5,3,6,-5)
5471 7675 ℓ_{10} : $\exp(\sqrt[3]{2})$
5471 7907 Li_2 : $(\sqrt{2}\pi)^{-1/2}$
5471 9755 id : $1/2+\pi/3$
5472 0510 rt : (9,9,7,-8)
5472 1328 J_1 : $\sqrt[3]{23/2}$
5472 1448 rt : (2,5,-7,-5)
5472 4475 sr : $7\sqrt[3]{5}/5$
5472 4579 rt : (2,5,-8,-7)
5472 5353 sr : $4/3+3\sqrt{2}/4$
5472 5592 rt : (6,1,-7,9)
5472 6012 $s\pi(1/4)\times s\pi(e)$
5472 7574 rt : (4,6,1,2)
5472 8934 rt : (9,2,9,-7)
5472 9108 $\exp(2)\div s\pi(e)$
5473 0123 $\exp(3/4)-s\pi(\pi)$
5473 0805 rt : (5,4,8,4)
5473 1903 J_1 : $((e+\pi)/2)^{1/3}$
5473 2177 E : $\exp(-\sqrt{2})$
5473 2312 e^x : $9e/10$
5473 3561 rt : (9,7,-8,-5)
5473 5781 at : $1/3-2\sqrt{2}/3$
5473 6299 rt : (9,8,-4,8)
5473 6935 Li_2 : $\exp(-\sqrt{5}/3)$
5473 8067 at : $5e\pi$
5473 9718 rt : (8,5,4,-5)
5473 9787 e^x : $1/4+3\pi/4$
5473 9799 $t\pi$: $7\sqrt[3]{3}/6$
5474 0238 rt : (5,2,-5,6)
5474 1312 rt : (3,8,-2,-3)
5474 1314 10^x : $\sqrt{3}/3+4\sqrt{5}/3$
5474 3514 E : $1/\sqrt{17}$
5474 3807 cr : $3\sqrt{3}-2\sqrt{5}/3$
5474 4564 sq : $\sqrt{2}+\sqrt{3}+\sqrt{7}$
5474 6206 rt : (2,8,-3,9)
5474 8177 rt : (3,9,-4,-1)
5474 8779 ℓ_2 : 13/19
5475 0661 rt : (1,7,5,-5)
5475 1975 rt : (9,5,-8,9)
5475 2264 2^x : $\sqrt[3]{2}/2$
5475 2620 rt : (8,9,-2,5)
5475 2933 rt : (1,7,-4,1)
5475 3306 sr : P_{79}
5475 4245 sr : $4\sqrt{2}+4\sqrt{3}$
5475 4511 rt : (7,8,-1,-3)
5475 4614 rt : (5,6,8,-7)
5475 4900 rt : (2,2,-9,4)
5475 6335 J_1 : $9(1/\pi)/2$
5475 7199 sr : $9\sqrt[3]{3}/2$
5475 7385 Γ : $9e\pi/8$
5475 7722 cu : $5\pi/3$
5475 7813 sq : $2/3+\sqrt{3}/3$
5475 9871 $t\pi$: $4\pi^2/3$
5476 0187 ℓ_{10} : $4-\sqrt{2}/3$
5476 1050 rt : (5,1,-9,-7)
5476 3124 sq : $3\sqrt{3}/4-\sqrt{5}/4$
5476 4321 $\ln(\pi)\times\exp(4/5)$
5476 4357 rt : (8,2,6,4)
5476 6808 rt : (2,8,-9,-7)
5476 7000 cr : $3+\sqrt{2}/2$
5476 7314 ζ : $(\ln 3/3)^3$
5476 8099 rt : (6,9,-6,-5)
5476 8296 rt : (8,8,6,-7)
5476 9247 $\exp(2/5)\div s\pi(\sqrt{2})$
5476 9614 rt : (8,6,-6,6)
5477 0848 cu : 9/11
5477 1264 sr : $2\sqrt{2}-\sqrt{3}/4$
5477 2255 sr : 3/10
5477 3137 e^x : $1/\sqrt[3]{12}$
5477 3404 $2/5-\exp(2/3)$
5477 4634 cr : $\sqrt{2}/4+3\sqrt{5}/2$
5477 4987 $s\pi$: $3e/10$
5477 6437 sinh : $9\sqrt[3]{5}/2$
5477 7406 id : $9(e+\pi)/5$
5477 9830 2^x : $(\ln 3/2)^{-1/2}$
5478 1490 Γ : $7e\pi/10$
5478 1984 10^x : 17/16
5478 4716 E : $\zeta(3)/5$
5478 5347 sinh : $\pi/6$
5478 5419 E : $\sqrt[3]{3}/6$
5478 6123 sinh : 11/3
5478 8537 $2\times s\pi(e)$
5478 8734 sinh : $(\ln 3/2)^{-1/3}$
5478 9297 2^x : $1/4-\sqrt{5}/2$
5479 0042 cu : $\sqrt{7/2}$
5479 0294 id : $2\sqrt{2}/3-2\sqrt{5}/3$
5479 0635 tanh : 8/13
5479 1199 rt : (4,0,-7,4)
5479 2651 E : 6/25
5479 2828 rt : (1,0,7,4)
5479 3111 id : $e/2+4\pi/3$
5479 3516 id : $4/3-\pi/4$
5479 4867 rt : (8,1,8,-6)
5479 5206 ℓ_2 : $\sqrt[3]{5}/5$
5479 5844 $t\pi$: 8/21
5479 5897 sq : $4\sqrt{2}+\sqrt{3}/2$
5479 6282 2^x : $\sqrt{11}\pi$
5479 6614 rt : (2,5,-8,-2)
5479 6784 rt : (7,1,1,-2)
5479 8356 Ei : $\sqrt[3]{10}$
5479 9098 rt : (7,7,-4,3)
5479 9497 $t\pi$: $\sqrt{2}+\sqrt{5}/3$
5480 3711 $t\pi$: $2\sqrt{3}/3-2\sqrt{5}$
5480 4339 rt : (4,9,-5,-5)
5480 6294 Γ : 1/7
5480 7178 2^x : $\ln(\sqrt[3]{2}/3)$
5480 9378 sinh : 11/21
5480 9571 rt : (7,4,-8,4)
5481 0425 rt : (4,2,5,-4)
5481 0961 $s\pi$: $((e+\pi)/3)^2$
5481 1973 $s\pi$: $4\sqrt{2}/3+3\sqrt{3}/4$
5481 2023 e^x : $1/4+4\sqrt{2}$
5481 3389 10^x : 11/20
5481 4899 rt : (5,9,-6,9)
5481 7012 E : $3(1/\pi)/4$
5481 8815 id : $\sqrt{2}-\sqrt{3}/2$
5481 9116 rt : (1,3,9,-6)
5481 9177 sr : $\zeta(3)/4$
5482 0577 rt : (7,4,3,-4)
5482 1507 $t\pi$: $(2\pi)^{-1/3}$
5482 1771 2^x : $3\sqrt{2}/2-2\sqrt{5}/3$
5482 6193 rt : (4,5,7,-6)
5482 8104 $s\pi$: $9(e+\pi)/4$
5482 8291 $s\pi$: $4\sqrt{2}-2\sqrt{5}$
5482 8431 at : $9\pi^2/2$
5482 9207 E : 5/21
5483 0239 ℓ_{10} : $9\pi/8$
5483 0916 e^x : $\sqrt{2}/2+2\sqrt{5}$
5483 1109 rt : (7,7,5,-6)
5483 1308 cu : $4\sqrt{2}+4\sqrt{3}/3$
5483 2876 sinh : $(\ln 3/2)^{-3}$
5483 3863 rt : (2,9,0,2)
5483 3995 cu : $4+4\sqrt{2}$
5483 4157 rt : (4,8,9,-8)
5483 4724 $t\pi$: $1/\ln(\ln 2/3)$
5483 6785 ζ : $(e)^{-3}$
5483 7370 cu : $1/2-3e$
5483 7835 J_1 : 9/4
5484 0006 Ψ : $(3/2)^{-3}$
5484 0245 ℓ_{10} : $3\sqrt{3}/4+\sqrt{5}$
5484 1046 J_1 : 23/16
5484 1126 id : $\sqrt{3}/4-4\sqrt{5}/3$
5484 1220 rt : (8,9,8,3)
5484 2834 tanh : $8\ln 2/9$
5484 3571 $t\pi$: $3\sqrt{2}/2+3\sqrt{3}$
5484 4476 rt : (4,7,-5,8)
5484 5500 ℓ_{10} : $\sqrt{2}/5$
5484 5958 $\ln(\sqrt{5})\times\exp(\sqrt{3})$
5484 7053 sinh : $3\sqrt{5}$
5485 2565 ln : $6\sqrt[3]{3}/5$
5485 4622 $t\pi$: $3e/4-3\pi/4$
5485 4940 at : 11/18
5485 6628 rt : (2,7,-8,7)
5485 6797 sq : $2+\sqrt{5}/4$
5485 7567 10^x : $7\sqrt[3]{2}/9$
5485 7898 sinh : $5\sqrt[3]{5}/7$
5485 7990 id : P_{12}
5485 8377 rt : (9,9,-4,-2)
5485 8654 ζ : 1/20
5486 1574 E : $\exp(-\sqrt[3]{3})$
5486 1820 Ψ : $3\sqrt{3}$
5486 4280 cr : $1/4+2\sqrt{3}$
5486 5424 sq : $3/4-2\sqrt{5}/3$
5486 8725 erf : $\exp(-\sqrt[3]{2}/2)$
5487 2786 cu : $4\sqrt{3}/3-2\sqrt{5}/3$
5487 3935 sinh : π
5487 4702 E : $\sqrt{2}/6$
5487 5461 2^x : $5\pi^2/2$
5487 7934 sr : $2/3+\sqrt{3}$
5488 0669 J_1 : $8\sqrt[3]{2}/7$
5488 0708 10^x : $\sqrt[3]{5}/9$
5488 1163 $1\div\exp(3/5)$
5488 1273 rt : (7,3,9,-7)
5488 1427 cr : $e/2+3\pi/4$
5488 2414 E : 4/17
5488 3029 e^x : 7/16
5488 3447 ln : $5\ln 2/6$
5488 4514 ℓ_{10} : $3/2+3e/4$
5488 8827 as : 12/23
5488 9570 sq : $4e-\pi/3$
5489 0366 rt : (7,0,7,5)
5489 0959 rt : (3,5,-4,7)
5489 2192 cr : $3\sqrt{3}/2+\sqrt{5}/2$
5489 9284 rt : (4,2,9,5)
5490 0281 e^x : 19/15
5490 0842 rt : (3,9,-5,3)
5490 1394 sq : $4/3+\sqrt{2}$
5490 2703 rt : (6,9,6,-7)
5490 3810 id : $1/4+3\sqrt{3}/4$
5490 4945 Ei : $\exp(-\sqrt{5})$
5490 7344 ln : $4+\sqrt{2}/2$
5490 8294 rt : (9,8,2,-5)
5490 9157 10^x : $\ln(3)$
5490 9704 rt : (3,7,8,-7)
5491 0048 sr : $1/\sqrt{11}$
5491 1185 ℓ_{10} : $2\sqrt{2}/3+3\sqrt{3}/2$
5491 1434 rt : (7,3,5,3)
5491 2748 rt : (8,7,-9,-1)
5491 7352 rt : (6,6,4,-5)
5491 8921 J_1 : $\sqrt[3]{3}$
5491 9333 sr : 12/5
5492 0654 $\exp(4)+s\pi(2/5)$
5492 0692 J_0 : $9(1/\pi)/2$
5492 2359 cr : $1+e$
5492 2485 J_1 : $6\zeta(3)/5$
5492 3060 cr : $4+4\pi$
5492 3311 rt : (1,9,1,-2)
5492 4273 e^x : $1+4\sqrt{5}$
5492 5696 Γ : 13/4
5492 5964 2^x : 12/19
5492 6179 J_1 : $1/\ln(2)$
5492 6568 ζ : $(\sqrt{2}\pi)^{-2}$
5492 7227 e^x : $4\sqrt{2}/3+2\sqrt{3}$
5492 7672 rt : (9,5,0,-3)
5492 7852 rt : (7,8,-5,-4)
5492 9921 erf : 8/15
5493 0230 $2/3\div s\pi(\pi)$
5493 0429 rt : (3,4,6,-5)
5493 0614 id : $\ln(\sqrt{3})$
5493 2874 e^x : $3e/2-\pi$
5493 3839 J_1 : $5\pi/7$
5493 4046 rt : (8,8,-1,9)
5493 6580 rt : (5,3,-8,-1)
5493 7253 J_1 : $5\sqrt{3}/6$
5493 7641 rt : (9,7,8,-8)
5493 9415 cu : $\sqrt{2}/3+3\sqrt{3}/4$
5494 1509 rt : (4,5,7,3)
5494 4191 J_1 : $\ln(3\pi)$
5494 6452 rt : (6,3,2,-3)
5494 6804 at : $8(e+\pi)$
5494 8677 ζ : $7e/9$
5495 0897 $s\pi$: $(\sqrt[3]{5}/3)^3$
5495 0975 id : $\sqrt{13/2}$
5495 3665 sr : $10\sqrt[3]{2}$
5495 3931 ln : γ
5495 4155 rt : (4,9,-9,-7)
5495 4605 J_1 : 13/9
5495 5382 e^x : $2\sqrt{3}/3+\sqrt{5}/4$
5495 7375 sr : $e/9$
5495 7464 Li_2 : 10/21

5495 8226 J_1 : $\exp(1/e)$
5495 8397 ζ : 12/23
5496 0264 rt : (9,4,6,-6)
5496 0325 rt : (5,4,0,9)
5496 0474 E : $(\sqrt[3]{3}/3)^2$
5496 0500 as : $2\sqrt{3}/3 - 3\sqrt{5}/4$
5496 0551 rt : (4,7,-1,-2)
5496 1796 E : $\ln 2/3$
5496 3525 10^x : $4e - \pi/2$
5496 4328 ζ : $1/\ln(\sqrt{2}\pi/2)$
5496 4607 cr : P_1
5496 6053 J : $\exp(-2e/3)$
5496 6976 E : 3/13
5496 7428 rt : (5,9,-1,-3)
5497 0581 id : $\sqrt{2}/4 + 3\sqrt{3}$
5497 1110 Ei : 25/9
5497 1600 rt : (9,2,-2,-1)
5497 1810 rt : (7,3,-7,8)
5497 1878 rt : (1,8,-5,-5)
5497 4652 rt : (2,9,9,8)
5497 4866 2^x : $e + \pi/2$
5497 5465 tπ : 4/25
5497 7443 $3/5 \times \ln(2/5)$
5497 7781 ln : $5\ln 2/2$
5497 9223 rt : (6,2,8,-6)
5497 9410 e^x : $2(e + \pi)$
5498 0860 tanh : $3\sqrt[3]{3}/7$
5498 2386 rt : (4,8,-6,-1)
5498 4434 Ei : 9/17
5498 6619 rt : (8,9,9,-9)
5498 6844 e^x : $2 - 3\sqrt{3}/2$
5498 7466 $\exp(1/3) + \exp(e)$
5498 7973 id : $5\sqrt[3]{5}$
5498 9774 rt : (5,8,5,-6)
5499 0130 2^x : $5(e + \pi)/9$
5499 0956 rt : (8,8,-5,-1)
5499 1879 rt : (1,8,-4,6)
5499 1938 E : $1/\sqrt{19}$
5499 2421 $\sqrt{5} \times \ln(2)$
5499 2634 rt : (7,6,-3,7)
5499 3879 sinh : $(\sqrt{5}/3)^{-3}$
5499 4573 tπ : $3\sqrt{3}/4 - 4\sqrt{5}/3$
5499 5618 tπ : $\sqrt[3]{4}$
5499 5693 tanh : $\exp(-\sqrt[3]{3}/3)$
5499 6339 sq : $\sqrt{2} - \sqrt{6} - \sqrt{7}$
5499 6387 cr : $\sqrt{2} + 4\sqrt{3}/3$
5499 6999 tπ : $2\sqrt[3]{2}/3$
5499 7622 rt : (2,7,1,-3)
5499 9173 J_1 : $(2\pi/3)^{1/2}$
5500 0000 id : 11/20
5500 0443 e^x : $2/3 + 3\pi/4$
5500 0674 J_0 : $((e + \pi)/2)^{1/3}$
5500 0942 rt : (9,1,4,-4)
5500 1568 ℓ_2 : $1/4 + \sqrt{3}/4$
5500 2114 rt : (1,8,3,-3)
5500 5015 rt : (4,6,-7,-5)
5500 5063 $\exp(4/5) - s\pi(\sqrt{5})$
5500 6811 ℓ_{10} : P_6
5500 6940 rt : (3,8,-8,5)
5500 7417 sinh : 11/9
5500 7500 erf : $4\zeta(3)/9$
5500 7712 rt : (4,1,-6,-8)
5500 8296 as : $1/\sqrt[3]{7}$
5500 8663 ln : $2\sqrt[3]{3}/5$
5500 8797 2^x : $4\sqrt{3}/3 - 3\sqrt{5}/4$
5500 9397 rt : (8,7,1,-4)
5501 0898 $\ln(3) - \exp(1/2)$
5501 3620 rt : (7,6,3,1)
5501 4594 rt : (2,6,7,-6)
5501 4833 tπ : $4\sqrt[3]{5}$
5501 5041 rt : (9,8,-6,5)
5501 5410 rt : (7,8,-7,-4)
5501 7781 sq : $2\sqrt{3} + 2\sqrt{5}/3$
5501 8383 E : $(2\pi/3)^{-2}$
5501 8513 ℓ_{10} : $3 - e$
5501 9172 rt : (8,6,7,-7)
5501 9287 rt : (5,7,-9,2)
5501 9499 id : $\ln(3\pi/2)$
5502 0126 2^x : $(5/2)^{-1/2}$
5502 0719 10^x : $3\sqrt{2} + 2\sqrt{3}/3$
5502 2434 sq : $(\sqrt{2}\pi/3)^3$
5502 2816 rt : (7,9,1,6)
5502 5438 id : $3\sqrt{3} + 3\sqrt{5}/2$
5502 6617 rt : (1,8,-9,-9)
5502 8483 E : $\exp(-\sqrt{2}\pi/3)$
5502 9482 E : $5(1/\pi)/7$
5502 9692 rt : (6,1,-9,6)
5502 9717 10^x : $7e$
5503 1148 E : 5/22
5503 1244 cu : $4e - \pi/4$
5503 2120 id : $1/\sqrt[3]{6}$
5503 2569 λ : $(e)^{-3}$
5503 3803 cu : $1/3 - 2\sqrt{5}/3$
5503 3971 $\ln(4/5) \div \ln(2/3)$
5503 5797 rt : (5,9,-8,4)
5503 6792 e^x : $2/3 + \pi/3$
5503 7507 10^x : $2\sqrt{2} - \sqrt{3}/4$
5503 7561 rt : (7,7,9,4)
5503 7600 at : $(\ln 2/3)^{1/3}$
5503 8376 rt : (5,9,0,-3)
5503 9986 Γ : $9\zeta(3)/4$
5504 0211 cu : $\sqrt{3}/4 - 4\sqrt{5}/3$
5504 0294 sr : $5\sqrt[3]{3}/3$
5504 0806 2^x : $7(e + \pi)/9$
5504 1068 rt : (6,1,6,4)
5504 2538 ln : $\sqrt{3} + 4\sqrt{5}/3$
5504 3125 rt : (5,5,3,-4)
5504 4282 sinh : $8\sqrt{3}/7$
5504 5282 rt : (5,4,9,-7)
5504 6032 rt : (4,2,-2,7)
5504 6060 Ei : $1/\ln(\pi)$
5504 6457 tanh : 13/21
5504 9138 Ei : $5e/3$
5504 9334 Ei : $\sqrt[3]{2/3}$
5504 9870 ℓ_2 : $(\pi)^{1/3}$
5505 1025 rt : (1,7,9,3)
5505 1250 Ei : $\sqrt{5}/9$
5505 1577 10^x : 4/21
5505 2049 sr : $2\zeta(3)$
5505 3387 sq : $1/4 + 4\sqrt{3}/3$
5505 3486 at : $5\pi^2$
5505 4639 ln : $10\sqrt{2}/3$
5505 5378 $\ln(\sqrt{5}) - s\pi(\sqrt{3})$
5505 8520 J_1 : $\sqrt{5}$
5506 4574 tπ : P_{20}
5506 4693 Γ : $9(e + \pi)/10$
5506 4770 rt : (6,4,-5,5)
5506 5672 rt : (7,4,-9,-5)
5506 6189 sπ : $5(e + \pi)/7$
5506 6228 at : $\sqrt{3} - \sqrt{5}/2$
5506 6370 rt : (7,8,8,-8)
5506 7886 sπ : $\sqrt[3]{5/3}$
5507 0482 rt : (7,9,2,-5)
5507 2330 ln : $1/4 + 4\pi$
5507 2407 e^x : $3\pi/4$
5507 3769 rt : (2,9,-2,-7)
5507 3850 rt : (8,3,5,-5)
5507 5829 sq : $4/3 + 3\pi/2$
5507 5909 10^x : $1/\sqrt[3]{6}$
5507 6516 tπ : $\sqrt{14/3}$
5507 6912 rt : (1,8,8,-7)
5507 7209 rt : (8,6,-8,3)
5507 7766 tπ : $\sqrt[3]{3}/9$
5507 8629 rt : (9,3,8,5)
5507 9913 $\ln(4/5) + s\pi(e)$
5508 4313 rt : (8,4,-1,-2)
5508 5243 cu : $2/3 + 4\sqrt{5}/3$
5508 6800 rt : (1,9,2,-4)
5508 6976 ℓ_2 : $(e + \pi)$
5508 9819 Ψ : $\exp(\sqrt[3]{4}/2)$
5509 0723 cu : $4\sqrt{2} - \sqrt{3}/4$
5509 1006 rt : (3,6,-5,-2)
5509 1154 sq : $1/2 + \sqrt{5}/3$
5509 1256 ζ : $(3\pi)^{1/3}$
5509 1298 2^x : $3\sqrt{2}/2 - 4\sqrt{5}/3$
5509 5354 sinh : 10/19
5509 5553 rt : (9,2,2,2)
5509 6333 rt : (2,3,5,8)
5509 6841 rt : (5,8,-8,-6)
5509 7335 E : $\sqrt{5}/10$
5509 7601 cr : P_{60}
5509 9253 rt : (7,5,-3,-2)
5510 2040 sq : $3\sqrt{3}/7$
5510 4667 cr : $3/4 + 4\sqrt{5}/3$
5510 4669 id : $\sqrt{3}/4 + \sqrt{5}/2$
5510 5651 $2/5 - s\pi(2/5)$
5510 5858 E : $\exp(-3/2)$
5510 6223 10^x : $4/3 + e/3$
5510 6958 as : $\pi/6$
5510 7294 10^x : 20/13
5511 0784 rt : (4,7,4,-5)
5511 1449 E : $7(1/\pi)/10$
5511 2447 rt : (8,5,-7,1)
5511 2799 rt : (3,3,8,4)
5511 3351 cr : $2 + \sqrt{3}$
5511 3797 rt : (1,6,-3,4)
5511 5890 sr : $3/2 + e/3$
5511 6350 J_1 : 16/11
5511 7765 Ψ : 4/21
5511 7840 10^x : $2\sqrt{3} + 4\sqrt{5}$
5511 7892 rt : (6,7,-1,4)
5511 8784 rt : (4,7,-7,3)
5511 9613 $5 \div \ln(3)$
5512 2042 E : 2/9
5512 2119 rt : (8,9,-4,2)
5512 2905 cr : P_{44}
5512 3528 rt : (5,1,7,-5)
5512 3772 sr : $e/2 + \pi/3$
5512 7318 rt : (5,9,-2,-3)
5512 7981 rt : (7,5,6,-6)
5512 8073 2^x : $1/2 - e/2$
5512 8219 at : $6e\pi$
5512 9298 Ψ : $8(e + \pi)/9$
5512 9808 ζ : 1/19
5512 9906 J_0 : 10/7
5513 0043 ℓ_{10} : $\sqrt{3}/3 + 4\sqrt{5}/3$
5513 0419 rt : (9,9,5,-7)
5513 0605 Li_2 : $3(1/\pi)/2$
5513 1307 rt : (3,9,-7,-4)
5513 1696 as : 11/21
5513 2122 rt : (5,8,-9,8)
5513 2889 $\sqrt{3} \div \pi$
5513 3006 ℓ_{10} : $3 + \sqrt{5}/4$
5513 4128 $\ln(5) \times s\pi(\sqrt{2})$
5513 6328 sr : $\sqrt{2} + \sqrt{6} + \sqrt{7}$
5513 6634 10^x : $4/3 + 2\sqrt{5}$
5513 7215 id : $3\sqrt{2}/4 + 2\sqrt{5}/3$
5513 7705 rt : (9,7,-5,9)
5513 8245 e^x : $1/4 + 4\sqrt{3}$
5513 8752 rt : (6,6,-9,-6)
5514 1414 cu : $4\sqrt{2} + 3\sqrt{5}$
5514 1658 rt : (8,1,9,6)
5514 4471 ln : $2 + e$
5514 4799 rt : (1,9,-6,-7)
5514 5356 sinh : $9\sqrt{5}/4$
5514 6790 J_1 : $(\sqrt{2}/3)^{-1/2}$
5515 4185 rt : (2,7,-4,-4)
5515 4203 cu : $4 + 4\pi$
5515 5425 Ei : $\sqrt{5} + \sqrt{6} + \sqrt{7}$
5515 5767 rt : (5,2,-7,3)
5515 6971 rt : (7,6,0,-3)
5515 8361 ℓ_2 : $6\sqrt[3]{5}/7$
5515 9453 ℓ_2 : $2/3 + 3\sqrt{3}$
5516 0810 rt : (3,9,5,4)
5516 3755 sq : $7(1/\pi)/3$
5516 5247 sq : $3\pi/4$
5516 5498 at : 8/13
5516 5831 rt : (9,1,-4,-1)
5516 6056 sr : $10(e + \pi)/9$
5516 7728 sr : 7/23
5516 8094 ln : $3\sqrt{2}/2 + 3\sqrt{3}/2$
5516 8250 e^x : $3/2 - 3\sqrt{2}/4$
5516 8778 $s\pi(1/5) - s\pi(\sqrt{2})$
5516 8986 cr : $3/2 + \sqrt{5}$
5516 9248 rt : (5,2,1,-2)
5516 9775 rt : (6,7,7,-7)
5517 0104 rt : (1,5,6,-5)
5517 0388 tπ : $\exp(3\pi/2)$
5517 1533 rt : (4,3,-5,-3)
5517 3241 ζ : $(1/\pi)/6$

5517 3796 rt : (2,8,-7,-9)
5517 5006 $\ln(\sqrt{5}) - \exp(\sqrt{5})$
5517 6741 rt : (9,5,9,-8)
5517 7548 ℓ_{10} : $4\sqrt{2}/3 + 3\sqrt{5}/4$
5518 0297 Ψ : $(4)^{-3}$
5518 0381 10^x : $(\pi/3)^{1/2}$
5518 0444 J_1 : $7(1/\pi)$
5518 0633 rt : (4,3,8,-6)
5518 0695 ζ : 19/9
5518 1916 sq : $(2e/3)^{-1/2}$
5518 3438 2^x : $3/2 - \sqrt{3}/2$
5518 3725 at : $9(e + \pi)$
5518 4745 rt : (8,0,3,3)
5518 7034 λ : 1/20
5518 9630 e^x : $3/2 - 2\pi/3$
5518 9767 10^x : $5\sqrt[3]{2}$
5518 9876 e^x : $7\sqrt{3}/8$
5519 2600 E : $1/\sqrt{21}$
5519 2943 rt : (1,7,9,-6)
5519 3376 Ψ : $\sqrt{18}\pi$
5519 6411 sinh : $7\sqrt{2}/5$
5519 6575 erf : $\ln(\sqrt[3]{5})$
5519 9477 J_1 : $9\sqrt{3}/7$
5520 1800 Ψ : $(\pi/3)^{1/3}$
5520 2978 rt : (8,5,-7,7)
5520 3025 rt : (5,9,8,-8)
5520 3195 rt : (2,1,9,5)
5520 4176 id : $3\sqrt{2} + 4\sqrt{3}/3$
5520 4922 cu : $\sqrt{2}/2 + 3\sqrt{5}/4$
5520 5546 rt : (9,9,-9,-2)
5520 5624 sr : $3e + 4\pi$
5520 5680 tanh : $1/\ln(5)$
5520 6999 E : 5/23
5520 7601 ζ : $(\ln 2/3)^2$
5520 7757 rt : (6,8,1,-4)
5520 8840 10^x : P_{19}
5520 9235 10^x : 25/17
5520 9332 ζ : $\exp(-(e + \pi)/2)$
5521 0425 10^x : P_{58}
5521 1238 rt : (9,6,3,-5)
5521 1400 $s\pi$: $4\sqrt{2}/3 + 4\sqrt{3}$
5521 1962 at : $1/4 - \sqrt{3}/2$
5521 2161 $\exp(1/3) - \exp(2/3)$
5521 4178 rt : (6,3,-4,9)
5521 5685 rt : (4,9,-8,1)
5521 5978 J : $\exp(-6\pi)$
5521 7684 cu : $10e/9$
5521 8482 rt : (3,9,-5,-5)
5521 9604 at : $8\ln 2/9$
5522 0856 rt : (7,7,-6,0)
5522 1146 cu : $2\sqrt{2}/3 - 4\sqrt{3}/3$
5522 2352 E : $\sqrt{3}/8$
5522 2532 rt : (3,5,9,-7)
5522 3492 cu : $3/2 - e/4$
5522 5942 J_1 : 19/13
5522 6123 ln : $1/4 + 2\sqrt{5}$
5522 6663 rt : (4,6,-8,7)
5522 8474 id : $1/3 - 4\sqrt{2}/3$
5523 0156 ℓ_{10} : $4 - \sqrt{3}/4$
5523 1108 E : $(5/3)^{-3}$
5523 2271 rt : (9,6,6,3)
5523 2626 10^x : $3/4 - \sqrt{5}/4$
5523 3309 $t\pi$: $\sqrt{2} - \sqrt{3}$
5523 3672 10^x : $3(1/\pi)/5$
5523 7997 sinh : $\exp(e)$
5523 8447 at : $3\sqrt{2}/4 - 3\sqrt{5}/4$
5523 8855 Li_2 : 11/23
5524 0776 2^x : $3/2 - 3\pi/4$
5524 1288 cu : 22/19
5524 1354 rt : (7,2,4,-4)
5524 2843 e^x : $9(e + \pi)/5$
5524 3639 rt : (8,8,4,-6)
5524 4165 J_1 : $\sqrt[3]{11}$
5524 5275 rt : (5,5,-3,2)
5524 6328 cr : $\sqrt{14}$
5524 7461 rt : (1,6,0,-2)
5524 8273 at : $\exp(4)$
5525 1769 $\exp(1/4) \times s\pi(\pi)$
5525 2605 rt : (3,5,-6,2)
5525 2980 Ψ : $1/\ln(\pi/2)$
5525 4102 ℓ_2 : 15/22
5525 5154 rt : (2,8,2,-1)
5525 7914 rt : (1,9,-8,-7)
5525 8587 rt : (1,7,4,7)
5525 9537 ℓ_{10} : $4/3 + \sqrt{5}$
5526 0332 K : $(\sqrt{5}/2)^{-1/2}$
5526 0592 E : 3/14
5526 3886 rt : (8,8,-3,6)
5526 6232 rt : (2,9,4,-8)
5526 6807 sq : $3\sqrt{2}/2 - 4\sqrt{5}$
5526 7252 10^x : $2 + e/3$
5526 9808 E : $\sqrt[3]{5}/8$
5527 0709 J_1 : 20/9
5527 0721 e^x : 11/25
5527 0787 e^x : 12/7
5527 1719 rt : (9,7,-3,-2)
5527 2453 $t\pi$: $8e\pi$
5527 3180 J_1 : $(\pi)^{1/3}$
5527 3492 rt : (9,2,7,-6)
5527 4130 rt : (4,0,-9,1)
5527 6305 J : $(\sqrt[3]{5}/3)^3$
5527 6507 ℓ_{10} : $2 + \pi/2$
5527 7909 $t\pi$: $4e - 3\pi/2$
5527 8640 rt : (5,5,-6,-4)
5527 8839 Ei : $3\sqrt{2}/8$
5527 8983 J_1 : $(e + \pi)/4$
5527 9128 E : $1/\sqrt{22}$
5528 0359 rt : (7,3,-9,5)
5528 0501 rt : (6,4,5,-5)
5528 0722 sinh : 5/3
5528 1298 rt : (6,9,-5,-1)
5528 1393 J_1 : $\ln(\ln 2/3)$
5528 4196 ℓ_{10} : 7/25
5528 6918 ℓ_{10} : $2\sqrt[3]{2}/9$
5528 9608 sq : $9e\pi/7$
5529 0134 J_1 : $6\sqrt[3]{5}/7$
5529 0384 rt : (6,6,0,8)
5529 0738 $\exp(\pi) - s\pi(1/5)$
5529 3312 rt : (8,4,7,4)
5529 6025 E : $2(1/\pi)/3$
5529 7465 cr : $3 + \sqrt{5}/3$
5529 7591 cu : $3/4 - \pi/2$
5529 9316 rt : (8,4,8,-7)
5530 2561 J : $5(1/\pi)/4$
5530 5066 J_1 : 22/15
5530 5361 rt : (8,9,-2,-3)
5530 5758 sq : $3\pi/5$
5530 6411 cu : $1/4 - 3\sqrt{2}/2$
5530 7801 rt : (5,1,-6,7)
5530 9196 rt : (5,6,6,-6)
5531 0970 rt : (3,5,-9,-6)
5531 2197 $s\pi$: $3/4 + 2e$
5531 2382 $\ln(3) \div s\pi(\pi)$
5531 2431 ζ : $7\sqrt{3}/8$
5531 2819 ℓ_{10} : $2\sqrt{2} + \sqrt{5}/3$
5531 3516 Ψ : $7\sqrt{5}/3$
5531 4185 $\ln(\sqrt{2}) \div \ln(4/5)$
5531 4318 rt : (5,2,5,3)
5531 6685 ℓ_{10} : $4\sqrt{3} - 3\sqrt{5}/2$
5531 6933 $s\pi$: $\sqrt{3}/3 + \sqrt{5}$
5531 7187 sinh : $3\pi^2/2$
5531 9570 cr : $3e/2 + 4\pi$
5531 9760 rt : (7,6,9,-8)
5532 1320 2^x : $9\zeta(3)/8$
5532 2493 rt : (4,8,-8,-6)
5532 3374 Γ : $\pi/9$
5532 4400 E : 4/19
5532 5773 e^x : $\sqrt{2}/2 - 3\sqrt{3}/4$
5532 6223 J_0 : $5\sqrt[3]{5}/6$
5532 6696 rt : (9,8,-8,2)
5533 0390 id : $4e/7$
5533 1136 rt : (4,8,7,-7)
5533 1371 rt : (7,2,8,5)
5533 2601 rt : (6,7,-7,-5)
5533 3460 E : $\sqrt[3]{2}/6$
5533 8565 Ei : $5(1/\pi)/3$
5533 9059 cu : $5\sqrt{2}$
5533 9663 rt : (3,3,-8,3)
5534 0026 sq : $(e/3)^3$
5534 0224 rt : (1,6,-5,-9)
5534 0452 ln : $7\sqrt{5}/9$
5534 7502 10^x : P_{52}
5534 8455 rt : (3,8,4,8)
5534 9305 rt : (6,5,-8,-2)
5535 0339 id : $2\sqrt{3}/3 - 3\sqrt{5}$
5535 0821 as : $2\sqrt{2} - 3\sqrt{5}/2$
5535 1947 10^x : $3 + \sqrt{2}$
5535 2913 Ψ : $(3\pi/2)^{-1/3}$
5535 4532 J_1 : $7\sqrt[3]{2}/6$
5535 5742 e^x : $3 - \sqrt{3}$
5535 6146 ℓ_{10} : $3 + \sqrt{3}/3$
5535 7378 rt : (6,0,9,6)
5535 7435 at : $1/2 - \sqrt{5}/2$
5535 8068 E : $1/\sqrt{23}$
5535 8945 e^x : 15/16
5535 9477 rt : (1,9,3,-7)
5536 0081 rt : (9,3,1,-3)
5536 0498 ℓ_{10} : $\sqrt{5}/8$
5536 1081 E : 5/24
5536 1625 cr : 15/4
5536 1767 rt : (7,3,-2,-1)
5536 2138 $\ln(3) - \exp(\sqrt{3})$
5536 2477 rt : (2,6,-9,-8)
5536 2715 at : $3\sqrt[3]{3}/7$
5536 3185 rt : (7,6,-5,4)
5536 4128 erf : 7/13
5536 4862 J_1 : 25/17
5536 5094 rt : (7,9,-8,-6)
5536 6032 10^x : $10\zeta(3)$
5536 7239 $\sqrt{2} \times \ln(3)$
5536 7550 E : $3\ln 2/10$
5536 8211 id : $3\sqrt{3}/4 - \sqrt{5}/3$
5536 8622 E : $\exp(-\pi/2)$
5536 9040 rt : (5,3,-5,1)
5536 9217 ℓ_{10} : $3/4 + 2\sqrt{2}$
5537 0677 rt : (9,9,-5,-4)
5537 1355 sinh : $9e/4$
5537 1478 rt : (3,4,-7,6)
5537 3277 at : $10(e + \pi)$
5537 4736 $t\pi$: $8\sqrt{3}/3$
5537 5505 cu : $4\sqrt{3}$
5537 7397 sr : $1 + \sqrt{2}$
5537 8097 at : $\exp(-\sqrt[3]{3}/3)$
5537 9077 tanh : $9\ln 2/10$
5537 9106 rt : (7,2,-8,9)
5538 0740 $\pi - s\pi(1/5)$
5538 0998 rt : (8,5,2,-4)
5538 2939 $s\pi$: $4(e + \pi)/3$
5538 3916 rt : (3,2,7,-5)
5538 4757 sq : $3 - \sqrt{3}/2$
5538 5126 10^x : $\sqrt{2/3}$
5538 5173 $\exp(5) + \exp(\pi)$
5538 6159 10^x : $2 + 3\pi$
5538 6929 J_1 : $1/\ln(\pi/2)$
5538 7771 rt : (9,7,6,-7)
5538 8418 e^x : $3\sqrt{2}/4 + 3\sqrt{3}$
5538 8582 Ei : 19/12
5538 8809 2^x : $1/\ln(2\pi/3)$
5539 0522 id : $\sqrt{2} - \sqrt{3} - \sqrt{5}$
5539 1106 at : $6\pi^2$
5539 2360 rt : (6,5,-1,-2)
5539 3586 cu : 25/7
5539 4297 sr : $\ln(e/2)$
5539 4475 $t\pi$: $4e/3 - \pi/4$
5539 4704 θ_3 : $9\ln 2/8$
5539 5279 ℓ_{10} : P_{69}
5539 5465 cu : $\sqrt{2}/3 + 2\sqrt{5}/3$
5539 6324 rt : (7,7,3,-5)
5539 6395 ℓ_{10} : P_{72}
5539 6907 ζ : $1/\sqrt[3]{7}$
5539 7963 rt : (8,9,7,-8)
5539 7966 sq : $3e/4 - 3\pi$
5539 8586 e^x : $1/\ln(1/(2e))$
5539 9088 E : $\sqrt[3]{3}/7$
5539 9173 J_0 : $(\ln 2/2)^{-1/3}$
5539 9288 sr : $4\sqrt{2} + \sqrt{3}/2$
5539 9360 tanh : $\exp(-\sqrt{2}/3)$
5540 1203 sr : $(\sqrt{5}/3)^{-3}$
5540 1913 rt : (2,4,8,-6)
5540 3491 $s\pi$: $3\sqrt{2} + 4\sqrt{5}$

5540 4769 rt : (9,8,-9,1)
5540 6936 at : $7e\pi$
5540 8037 rt : (6,9,4,7)
5540 8339 sr : $\sqrt{3}/2-\sqrt{5}/4$
5540 9137 10^x : $5\sqrt{5}/7$
5541 1573 rt : (5,8,1,1)
5541 2811 $5\times\ln(3/5)$
5541 3052 cu : $1/3-2\sqrt{3}/3$
5541 4921 rt : (1,6,9,-7)
5541 6960 $\sqrt{3}+\exp(3/5)$
5541 7282 rt : (5,4,-2,6)
5541 7527 sq : $4+4\sqrt{5}$
5541 7548 cu : $e/4+2\pi$
5541 8576 J_0 : $e\pi/6$
5541 9241 erf : $7\ln 2/9$
5542 0290 rt : (9,7,-7,6)
5542 1111 sq : $\sqrt{3}/2+2\sqrt{5}/3$
5542 1665 $t\pi$: $1/\ln(2e/3)$
5542 3428 e^x : $1-\sqrt{5}/4$
5542 4764 rt : (4,9,1,-4)
5542 4961 E : $\exp(-\sqrt[3]{4})$
5542 5974 10^x : $4e+3\pi/4$
5542 6183 as : 10/19
5542 6224 $2/5+\exp(e)$
5542 8297 ζ : $(\sqrt{2}\pi)^{1/2}$
5542 9162 J_1 : $\exp(\sqrt[3]{4}/2)$
5542 9421 ln : $1/4+2\sqrt{5}/3$
5543 0376 E : $1/\sqrt{24}$
5543 0749 at : 13/21
5543 2336 2^x : 23/17
5543 2512 sr : $2e/3+3\pi/2$
5543 3005 ζ : 1/18
5543 3304 id : $3\sqrt{2}/2+\sqrt{3}/4$
5543 4631 10^x : $2+3\sqrt{2}$
5543 5868 ln : $3+\sqrt{3}$
5543 8099 $s\pi$: $(2\pi/3)^3$
5543 8225 e^x : $4e/3-3\pi/4$
5543 8549 Ei : 7/8
5543 8669 rt : (1,9,9,-4)
5544 0628 2^x : 7/11
5544 2932 sr : $8e/9$
5544 6434 rt : (3,7,-3,-1)
5544 8021 rt : (2,8,-7,-6)
5544 9096 sinh : 9/17
5545 0664 ln : $1/2-\sqrt{2}/3$
5545 1774 id : $4\ln 2/5$
5545 3727 rt : (1,5,-2,6)
5545 4422 E : $(\sqrt{2}\pi/2)^{-2}$
5545 4428 rt : (9,1,5,4)
5545 7970 rt : (2,9,-2,-2)
5545 9220 θ_3 : $e/10$
5545 9972 tanh : 5/8
5546 0168 Γ : $\sqrt{22/3}$
5546 0985 rt : (8,1,6,-5)
5546 4299 E : $\sqrt{2}/7$
5546 5747 rt : (7,3,7,-6)
5546 6342 Γ : $9(1/\pi)/5$
5546 7325 id : $(e+\pi)^{-1/3}$
5546 7338 rt : (6,8,8,3)
5546 7721 rt : (7,5,-4,8)
5546 8227 2^x : $2(1/\pi)$
5546 8411 ln : $3e+3\pi/2$
5546 9356 sr : $3\sqrt{3}/4+\sqrt{5}/2$
5546 9364 rt : (6,5,8,-7)
5547 0019 sr : 4/13
5547 1737 Ei : 17/9
5547 1895 $1/4-\ln(\sqrt{5})$
5547 2204 rt : (5,7,9,-8)
5547 3251 sinh : $\sqrt{3/2}$
5547 3609 $1/3+\exp(1/5)$
5547 4003 10^x : $3/2+2\sqrt{5}$
5547 5175 $\pi+\exp(5)$
5547 5292 ζ : $\exp(\sqrt{5}/3)$
5547 5608 Li_2 : 12/25
5547 6732 rt : (5,6,9,4)
5547 8700 rt : (1,7,-3,9)
5548 3802 rt : (9,8,0,-4)
5548 6148 sr : $(\sqrt{2}\pi/3)^{-3}$
5548 6228 10^x : $1/4+4\sqrt{3}$
5548 6334 rt : (7,9,-1,3)
5548 7238 rt : (6,4,-7,2)
5548 9959 e^x : $3\sqrt{3}+\sqrt{5}$
5549 0619 $\sqrt{2}+\exp(\pi)$
5549 1072 erf : $3\sqrt[3]{2}/7$
5549 1381 E : $\zeta(3)/6$
5549 5813 rt : (4,8,-4,-4)
5549 6854 E : 1/5
5550 2256 $t\pi$: $\sqrt{2}-\sqrt{3}/2$
5550 2593 ℓ_{10} : $3/4-\sqrt{2}/3$
5550 3640 sr : $4\ln 2/9$
5550 4603 rt : (2,9,7,1)
5550 4725 Ei : $7(1/\pi)/9$
5550 4730 cr : $\sqrt[3]{5}/10$
5550 5823 rt : (3,7,6,-6)
5550 7436 rt : (4,7,7,9)
5550 8455 rt : (9,8,9,-9)
5550 9342 Ei : $3\pi/2$
5550 9669 rt : (5,3,4,-4)
5551 0797 $s\pi$: $e+2\pi/3$
5551 1719 cu : $\sqrt{2}-\sqrt{5}$
5551 2963 rt : (6,1,3,-3)
5551 3656 E : $5(1/\pi)/8$
5551 4181 ℓ_{10} : $8\pi/7$
5551 4198 ln : $2+4e$
5551 4614 rt : (8,5,-9,4)
5551 7259 at : $(4)^3$
5551 7359 ℓ_2 : $\sqrt{14}\pi$
5551 8312 rt : (7,1,2,2)
5552 6653 J_1 : $7\sqrt[3]{2}/4$
5552 6725 10^x : $1+e/3$
5552 7486 rt : (6,8,-4,-3)
5552 7521 rt : (1,7,2,-6)
5552 7925 E : $2\ln 2/7$
5552 8506 rt : (8,3,1,1)
5552 8584 sr : $4/3+3\sqrt{3}$
5553 1280 rt : (4,2,-7,-9)
5553 3369 rt : (7,8,-3,-2)
5553 7337 Li_2 : $(\ln 2)^2$
5553 8179 J : $3\ln 2/5$
5553 8615 10^x : $1+3\pi$
5553 9464 Γ : $4\sqrt{2}$
5553 9693 Ei : $5(e+\pi)/8$
5554 0696 θ_3 : $3\sqrt[3]{3}/7$
5554 0753 2^x : $1+\sqrt{2}/4$
5554 1935 rt : (8,6,-4,-1)
5554 2716 rt : (7,8,6,-7)
5554 2945 Γ : $(e)^{-3}$
5554 7400 10^x : $1/2+2\sqrt{5}$
5554 8363 rt : (8,6,5,-6)
5554 9216 ℓ_2 : $4-3\sqrt{2}/4$
5554 9857 rt : (8,9,-6,-1)
5555 0625 id : $8\sqrt{5}/7$
5555 4124 sinh : $3\sqrt{2}/8$
5555 5555 id : 5/9
5555 6238 e^x : P_{62}
5555 6911 2^x : $4\sqrt[3]{5}$
5555 7023 $s\pi$: 3/16
5555 7415 ln : $e/2-\pi/4$
5556 2180 rt : (1,5,4,9)
5556 2581 ℓ_{10} : $3/2+2\pi/3$
5556 5005 erf : $3\sqrt[3]{3}/8$
5556 6905 cr : $3-2\sqrt{2}$
5556 7275 rt : (1,7,3,-4)
5556 7679 Li_2 : $(\ln 2/3)^{1/2}$
5556 8108 Ei : $9\sqrt[3]{3}/5$
5557 1912 Γ : $4(1/\pi)/5$
5557 2575 ℓ_2 : $7\sqrt[3]{2}/3$
5557 3044 $s\pi$: P_9
5557 3693 J_1 : $6\sqrt{3}/7$
5557 5045 rt : (3,6,-7,-7)
5557 5100 cr : $e+\pi/3$
5557 5447 sinh : $5(1/\pi)/3$
5557 5842 $\ln(2/5)\div\exp(1/2)$
5557 6018 rt : (1,2,8,4)
5557 6020 at : $(1/(3e))^{-2}$
5557 6412 Ei : $\sqrt{3}/7$
5557 7673 10^x : $3/2+\sqrt{5}/4$
5557 7798 Li_2 : $\sqrt[3]{3}/3$
5557 9557 2^x : $\sqrt{2}+\sqrt{5}$
5558 0242 $t\pi$: $3\sqrt{5}/8$
5558 0621 $\pi+\sqrt{2}$
5558 1436 $\pi\div\exp(\sqrt{3})$
5558 3044 cr : $\zeta(3)/7$
5558 4711 rt : (8,4,-8,8)
5558 6559 sr : $(e+\pi)^{1/2}$
5558 7737 Li_2 : $2\zeta(3)/5$
5559 0484 rt : (9,9,-2,9)
5559 0528 rt : (8,3,-8,-4)
5559 1649 sq : $\sqrt{2}+3\sqrt{5}/4$
5559 2579 rt : (7,8,0,7)
5559 3762 λ : $\exp(-19\pi/20)$
5559 4889 at : $3/2-3\sqrt{2}/2$
5559 5264 rt : (4,1,9,-6)
5559 5942 at : $1/\ln(5)$
5559 6304 J_1 : 11/5
5559 7059 rt : (5,8,3,-5)
5559 8455 cu : $3\sqrt{2}/2-3\sqrt{3}/4$
5559 8589 $s\pi(\pi)\div s\pi(e)$
5560 0663 $t\pi$: $\sqrt{2}/4+2\sqrt{3}/3$
5560 0964 sr : $1/4+2\pi$
5560 1099 e^x : $4+2e$
5560 3037 $t\pi$: 7/22
5560 5274 rt : (5,6,-9,-6)
5560 6208 rt : (1,7,-6,1)
5560 6328 2^x : 17/5
5560 6897 erf : $(e/2)^{-2}$
5560 7035 ℓ_{10} : $1+3\sqrt{3}/2$
5560 7914 rt : (1,9,-5,3)
5560 8539 2^x : $2e-2\pi$
5560 8890 J_1 : $7\pi/10$
5560 9067 e^x : $\sqrt[3]{23/2}$
5561 0624 rt : (4,9,0,8)
5561 2239 rt : (7,8,-9,-7)
5561 3388 rt : (9,9,3,-6)
5561 4011 2^x : $4-2e/3$
5561 4432 rt : (8,2,9,-7)
5561 4932 E : $\sqrt{3}/9$
5561 5078 rt : (5,1,8,5)
5561 5991 at : $8e\pi$
5561 7040 J_1 : $9\sqrt[3]{5}/7$
5561 7808 rt : (2,4,7,3)
5561 9027 Γ : $9\pi^2/10$
5562 1241 rt : (2,8,4,3)
5562 2439 id : $(2e)^2$
5562 4767 rt : (1,7,-9,-7)
5563 0250 ℓ_{10} : 5/18
5563 1088 rt : (9,0,8,6)
5563 2081 θ_3 : $\exp(-\sqrt[3]{3}/3)$
5563 2288 at : $7\pi^2$
5563 2439 J_1 : $3(e+\pi)/8$
5563 4042 rt : (4,4,-8,-5)
5563 4294 erf : 13/24
5563 5165 rt : (3,8,9,-8)
5563 5985 rt : (7,4,1,-3)
5563 6111 cr : $6\pi/5$
5563 7289 E : $3(1/\pi)/5$
5563 8221 rt : (8,8,-5,3)
5563 8523 $1/4\times\exp(4/5)$
5563 8525 $s\pi$: $2e/3$
5563 9334 ℓ_2 : 17/25
5564 0387 Ψ : $\zeta(3)/3$
5564 1586 sq : $7\pi/8$
5564 1628 rt : (7,9,9,-9)
5564 4741 rt : (6,3,7,4)
5564 5032 E : 4/21
5564 5360 e^x : $\sqrt{3}/2+2\sqrt{5}/3$
5564 5572 e^x : $3+3\sqrt{2}/2$
5564 6599 rt : (5,8,-6,0)
5564 6901 $1/5+\exp(\sqrt{5})$
5564 7688 $1/3-\ln(4/5)$
5564 8634 sr : $3-\sqrt{3}/3$
5565 0564 rt : (6,8,-1,-2)
5565 2286 E : $\sqrt[3]{5}/9$
5565 2563 rt : (4,6,8,-7)
5565 3606 θ_3 : 3/11
5565 4343 cr : $8\sqrt{2}/3$
5565 5461 rt : (9,9,-6,-1)
5565 6046 rt : (8,7,8,-8)
5565 6119 erf : $(2\pi)^{-1/3}$
5565 8238 J_1 : $(\sqrt{2}\pi/2)^{1/2}$

5565 9408 ln : $1/2+3\sqrt{2}$
5565 9857 rt : (1,7,-5,-6)
5566 0450 rt : (2,9,-1,-3)
5566 0543 rt : (1,4,-1,7)
5566 1140 10^x : $3/4+e$
5566 2001 J_1 : $2\sqrt{5}/3$
5566 2956 λ : $(\sqrt{2}\pi)^{-2}$
5566 3489 cu : $3e-2\pi$
5566 4998 rt : (6,7,-3,1)
5566 6286 10^x : $9\zeta(3)/4$
5566 6776 sq : $9\ln 2/5$
5566 6790 $\exp(\sqrt{2})\div\exp(2)$
5566 9205 E : $\exp(-5/3)$
5566 9309 rt : (8,2,0,-2)
5566 9563 2^x : $3+2\sqrt{5}$
5566 9673 rt : (3,5,-8,-3)
5567 0488 J_1 : $\ln(\sqrt{2}\pi)$
5567 1052 Γ : $1/\ln(1/(3e))$
5567 1881 Γ : $7e\pi/5$
5567 3625 rt : (9,5,7,-7)
5567 4069 10^x : $4\sqrt{5}$
5567 4920 rt : (5,4,7,-6)
5567 7326 rt : (1,3,7,-5)
5567 7349 θ_3 : $6(1/\pi)/7$
5567 9706 ℓ_{10} : $1/4+3\sqrt{5}/2$
5568 0321 $t\pi$: $4\sqrt{2}/3+\sqrt{3}/3$
5568 2915 cr : $3/4-\sqrt{3}/3$
5568 3345 rt : (3,2,-7,-4)
5568 3434 10^x : $2+\sqrt{3}/2$
5568 7056 rt : (4,1,-3,8)
5568 7777 sq : $((e+\pi)/3)^2$
5568 8500 rt : (7,9,0,-4)
5568 9330 tanh : $\pi/5$
5568 9813 E : $3/16$
5568 9950 ln : $2/3+3e/2$
5569 1595 J_1 : $2\pi^2/9$
5569 3858 sq : $2e-4\pi/3$
5569 3901 e^x : $9\sqrt{3}/2$
5569 4012 rt : (7,5,6,3)
5569 4322 ln : $5\pi/9$
5569 4463 $\ln(\pi)-s\pi(1/5)$
5569 4514 as : $1-\sqrt{2}/3$
5569 4557 cu : $4\sqrt{2}+3\sqrt{3}/4$
5569 4674 Ψ : $20/9$
5569 5854 ln : $1+\sqrt{5}/3$
5569 6427 rt : (3,8,0,-3)
5569 7167 ℓ_{10} : $\sqrt{13}$
5569 7886 $t\pi$: $1/3+2\sqrt{2}$
5569 8042 ℓ_{10} : $5\sqrt[3]{3}/2$
5569 8051 sq : $3/2+3\sqrt{2}/4$
5569 8317 2^x : $7\pi^2/9$
5569 9953 rt : (4,9,-7,-6)
5570 0266 rt : (1,8,-5,-2)
5570 2373 rt : (2,7,1,5)
5570 3380 Ψ : $5\pi/3$
5570 3543 J_1 : $1/\ln((e+\pi)/3)$
5570 4006 rt : (8,7,-4,7)
5570 4230 id : $7(1/\pi)/4$
5570 4563 tanh : $4\sqrt{2}/9$
5570 4628 ℓ_{10} : $3\zeta(3)$
5570 4704 rt : (6,2,6,-5)
5570 6179 J_1 : $\sqrt[3]{10/3}$
5570 8295 rt : (5,9,6,-7)
5571 1454 ℓ_{10} : $2\ln 2/5$
5571 1533 $t\pi$: $1/4+2\pi$
5571 1615 2^x : $2\sqrt{5}/7$
5571 2042 2^x : $3\sqrt{2}+\sqrt{3}/4$
5571 2405 rt : (6,6,-7,1)
5571 4103 Ei : $(1/\pi)/3$
5571 5516 rt : (1,7,0,-2)
5571 6645 ln : $4+\sqrt{5}/3$
5571 7195 $\ln(5)+\exp(2/3)$
5571 7454 rt : (8,8,-7,-5)
5571 8247 rt : (7,7,-8,-3)
5571 9095 id : $3/2-2\sqrt{2}/3$
5571 9392 ℓ_2 : $1+\sqrt{2}/3$
5571 9563 rt : (8,7,-1,-3)
5571 9977 sr : $7\sqrt{3}/5$
5572 0051 2^x : $10\sqrt{3}/7$
5572 0191 rt : (1,8,6,-6)
5572 0436 ln : $1/3+4\pi$
5572 3329 rt : (9,0,-1,1)
5572 3768 rt : (2,1,6,-4)
5572 4142 E : $(\sqrt[3]{5}/3)^3$
5572 4928 cu : $2\sqrt{2}-\sqrt{3}/2$
5572 5805 rt : (7,1,-7,-3)
5572 5821 sinh : $8e/9$
5572 8158 J_1 : $(\sqrt{5})^{1/2}$
5572 9578 rt : (5,1,-8,4)
5573 0495 ℓ_2 : $e/4$
5573 0905 rt : (1,6,3,3)
5573 1367 10^x : $1+\pi$
5573 3868 rt : (9,7,-9,3)
5573 4830 rt : (6,7,5,-6)
5573 7320 ℓ_2 : $\sqrt{2}+2\sqrt{5}$
5573 7719 rt : (3,4,-9,1)
5573 9227 J_0 : $9\sqrt[3]{2}/8$
5574 0099 J_1 : $\sqrt[3]{21/2}$
5574 0618 J_1 : $7\sqrt[3]{5}/8$
5574 0772 $t\pi$: $(1/\pi)$
5574 2547 rt : (3,7,-6,-5)
5574 2555 cr : $\sqrt{3}/10$
5574 4797 ℓ_{10} : $3e/4+\pi/2$
5574 5377 rt : (4,6,-1,-1)
5574 6344 rt : (7,0,5,4)
5574 7224 ζ : $7\zeta(3)/4$
5574 9086 Ψ : $1/\ln(\sqrt[3]{2}/3)$
5574 9295 $\ln(3/5)\div\ln(2/5)$
5575 0666 rt : (3,6,6,2)
5575 1320 cr : $\ln 2/4$
5575 2872 cu : $e/4-4\pi$
5575 2951 rt : (6,6,-2,5)
5575 6705 sr : $7\ln 2/2$
5575 7044 10^x : $\sqrt{19/3}$
5575 7902 10^x : $\sqrt{3}/9$
5575 8214 rt : (2,6,5,7)
5575 8290 ℓ_2 : $3\sqrt{2}-3\sqrt{3}/4$
5576 0156 $2\div\exp(1/4)$
5576 0239 ζ : $\pi/6$
5576 2084 rt : (2,6,-2,7)
5576 2518 rt : (9,5,-2,-2)
5576 3063 as : P_3
5576 7916 10^x : $\sqrt{2}/4+\sqrt{5}/3$
5576 8015 id : $7(e+\pi)/9$
5576 8921 rt : (7,5,4,-5)
5577 1569 cr : $3\sqrt[3]{2}$
5577 2235 E : $4(1/\pi)/7$
5577 3294 E : $2/11$
5577 4283 ζ : $1/17$
5577 5162 ℓ_{10} : $3\sqrt{2}/2+2\sqrt{5}/3$
5577 5929 at : $9\ln 2/10$
5577 8591 rt : (9,6,-8,7)
5577 8598 at : $9e\pi$
5577 8867 $\ln(2)\times\ln(\sqrt{5})$
5577 9709 $s\pi$: P_{36}
5577 9799 J_0 : $17/12$
5577 9835 at : $e-2\pi/3$
5578 3739 rt : (1,8,9,-3)
5578 6801 cu : $3\sqrt{2}/2+\sqrt{5}/3$
5578 6967 E : $\exp(-\sqrt[3]{5})$
5578 7469 rt : (2,6,0,-7)
5578 8898 $\ln(\pi)+\exp(5)$
5578 9235 rt : (5,0,-7,8)
5578 9524 rt : (7,6,-7,1)
5578 9954 sr : $3/4+3\sqrt{5}/4$
5579 0703 as : $9/17$
5579 1826 rt : (8,7,5,2)
5579 3650 J_1 : $3/2$
5579 4102 Ψ : $\sqrt[3]{11}$
5579 4128 sq : $e\pi/4$
5579 5424 E : $\sqrt[3]{3}/8$
5579 6989 at : $\exp(-\sqrt{2}/3)$
5579 7946 $t\pi$: $\sqrt{2}/3-4\sqrt{3}/3$
5579 8075 2^x : $\sqrt{2}+3\sqrt{3}/4$
5579 9214 ln : $\sqrt[3]{16/3}$
5579 9613 E : $\sqrt[3]{2}/7$
5580 0061 rt : (7,6,-6,-4)
5580 1070 ζ : $\ln(\sqrt{5})$
5580 1538 erf : $e/5$
5580 2103 rt : (3,1,5,3)
5580 2503 tanh : $\sqrt[3]{2}/2$
5580 3680 $s\pi$: $4\sqrt{2}+2\sqrt{3}/3$
5580 6462 10^x : $\sqrt[3]{15}$
5580 7601 rt : (2,5,-5,-4)
5580 7874 rt : (2,5,-5,9)
5580 8139 rt : (6,3,9,-7)
5580 8646 rt : (8,2,4,3)
5580 9924 rt : (1,9,9,-8)
5581 0104 2^x : $10\sqrt[3]{3}$
5581 1695 sr : $1/3+2\pi/3$
5581 1721 rt : (6,8,8,-8)
5581 3185 at : $8\pi^2$
5581 4092 rt : (6,5,-1,9)
5581 4317 rt : (8,3,3,-4)
5581 4461 ln : $4/19$
5581 5030 at : $3-4e/3$
5581 5403 sinh : $\exp(-\sqrt[3]{2}/2)$
5581 5778 rt : (3,9,3,-1)
5581 7164 rt : (8,8,2,-5)
5581 8400 cr : $4/23$
5581 8723 $4/5\times\exp(2/3)$
5581 9306 Ei : $3\sqrt[3]{2}/2$
5582 1303 ln : $1/3+\sqrt{2}$
5582 3758 $t\pi$: $\sqrt{2}-4\sqrt{5}$
5582 4497 $2/5\times\exp(1/3)$
5582 5756 rt : (5,9,-3,-2)
5582 7631 2^x : $\exp(e/3)$
5582 9194 e^x : $2-3\sqrt{2}/4$
5582 9479 J_1 : $5\zeta(3)/4$
5583 0354 rt : (1,8,4,5)
5583 0648 10^x : $e/3+3\pi$
5583 2877 10^x : $7e\pi/9$
5583 2915 2^x : $16/25$
5583 7390 at : $\sqrt{3}/2-2\sqrt{5}/3$
5583 8744 sr : $17/7$
5583 9285 erf : $1/\ln(2\pi)$
5584 0768 rt : (7,5,-6,5)
5584 1668 cu : $10e/7$
5584 2881 ℓ_{10} : $4\sqrt{2}/3+\sqrt{3}$
5584 3131 E : $\exp(-\sqrt{3})$
5584 3742 $\ln(2/3)-s\pi(\sqrt{2})$
5584 4289 E : $5(1/\pi)/9$
5584 5162 E : $\sqrt{2}/8$
5584 6270 at : $(\ln 2/3)^{-3}$
5584 6576 rt : (7,9,-9,1)
5584 7182 $t\pi$: $\ln(2\pi)$
5584 7467 10^x : $1/3+4\sqrt{2}/3$
5584 7563 rt : (7,6,7,-7)
5584 8883 J_1 : $24/11$
5584 8941 cr : $3+\pi/4$
5584 9459 E : $3/17$
5584 9795 2^x : $\sqrt{3}+\sqrt{6}+\sqrt{7}$
5585 0718 rt : (2,7,8,-7)
5585 0758 E : $(\sqrt[3]{2}/3)^2$
5585 1512 ζ : $11/21$
5585 1821 cu : $14/17$
5585 1890 sr : $e/2-\pi/3$
5585 3942 rt : (7,1,8,-6)
5585 4942 rt : (3,8,-9,-7)
5585 8103 Γ : $4/7$
5585 9028 sr : $4-\pi/2$
5585 9448 rt : (2,2,9,-6)
5585 9931 at : $5/8$
5586 2955 rt : (9,6,1,-4)
5586 5599 rt : (7,4,9,5)
5586 9977 rt : (4,7,2,-4)
5587 2494 cr : $\sqrt{3}/4+3\sqrt{5}/2$
5587 3342 Ei : $1/24$
5587 3490 rt : (2,9,1,7)
5587 4791 Ψ : $6\sqrt{5}$
5587 7423 id : $\sqrt{3}/3+4\sqrt{5}/3$
5587 7726 rt : (9,1,2,-3)
5587 8151 rt : (7,4,-8,2)
5587 8968 rt : (8,9,5,-7)
5587 9395 rt : (7,4,-5,9)
5588 0272 rt : (1,8,-3,-4)
5588 2636 sq : $2/3-\sqrt{2}$
5588 4572 id : $9\sqrt{3}/10$
5588 5062 E : $4/23$
5588 5789 rt : (3,8,2,3)

5588 6694 ℓ_{10} : $3/2+3\sqrt{2}/2$
5588 6968 e^x : P_{46}
5588 7426 $1/3+\exp(4/5)$
5588 7615 rt : (3,9,-4,7)
5588 8690 $\ln(2/5) \div s\pi(1/5)$
5589 2548 rt : (8,4,6,-6)
5589 3585 $s\pi$: $4\pi/3$
5589 3699 E : $\ln 2/4$
5589 4072 λ : $\exp(-18\pi/19)$
5589 4344 $\sqrt{2}+\ln(\pi)$
5589 4824 E : $\sqrt{3}/10$
5589 6826 rt : (6,9,-4,-4)
5589 7802 rt : (2,8,-4,5)
5589 7932 sinh : 8/15
5589 8986 as : $3\sqrt{2}/8$
5590 0242 $\exp(\sqrt{2}) \div s\pi(\pi)$
5590 1699 id : $\sqrt{5}/4$
5590 3488 at : $\exp(\sqrt{2}\pi)$
5590 3664 rt : (3,5,7,-6)
5590 6048 rt : (6,7,-4,-1)
5590 6390 e^x : $3/2+\pi/2$
5590 8689 at : $10e\pi$
5590 9649 rt : (3,8,-7,9)
5591 0969 $t\pi$: $\sqrt{10}$
5591 1377 rt : (5,5,-5,-1)
5591 1944 rt : (9,4,3,2)
5591 3414 J_0 : $\sqrt{2}$
5591 3414 cu : $5\pi/4$
5591 3848 tanh : 12/19
5591 5135 E : $\zeta(3)/7$
5591 5436 rt : (8,1,7,5)
5591 5829 $s\pi$: $\exp(-5/3)$
5591 7258 $\ln(2)+s\pi(1/3)$
5591 8250 rt : (2,7,-9,3)
5591 8631 2^x : $7\sqrt{2}/4$
5591 9228 rt : (2,6,-8,-6)
5592 0973 as : $5(1/\pi)/3$
5592 2405 rt : (9,7,4,-6)
5592 3321 ℓ_{10} : $4e/3$
5592 3718 rt : (8,3,-9,9)
5592 4576 rt : (9,9,-4,6)
5592 5000 E : $\sqrt[3]{5}/10$
5592 5737 2^x : $2-e/2$
5593 1840 rt : (3,7,1,7)
5593 3629 J_1 : $5e/9$
5593 5045 e^x : $3\sqrt{2}/2-3\sqrt{5}/4$
5593 5384 sq : $3\sqrt{2}/4-4\sqrt{3}/3$
5593 5949 rt : (1,8,-3,-1)
5593 6411 rt : (4,8,5,-6)
5593 6705 rt : (3,0,8,5)
5593 7612 $3/4 \times s\pi(\sqrt{3})$
5593 9418 at : $(\sqrt{2}\pi)^3$
5594 0107 id : $1/4+4\sqrt{3}/3$
5594 0947 2^x : $4\sqrt[3]{3}/9$
5594 1543 rt : (8,5,9,-8)
5594 2596 rt : (7,7,-9,-6)
5594 4515 10^x : $3\sqrt{2}/4+4\sqrt{3}/3$
5594 5231 Ei : $5\sqrt{5}/9$
5594 5349 J_1 : $4e/5$
5594 5523 θ_3 : 13/21
5594 6179 rt : (5,5,1,-3)
5594 6489 e^x : $5\sqrt{2}/3$
5594 8084 at : $4/3-\sqrt{2}/2$
5594 8444 10^x : $7\sqrt[3]{2}$
5595 0671 rt : (9,2,5,-5)
5595 1333 rt : (4,2,-9,-6)
5595 2373 erf : 6/11
5595 2629 id : $6\sqrt[3]{2}$
5595 3889 at : $9\pi^2$
5595 5179 rt : (1,9,-6,-6)
5595 6525 sr : $2/3-\sqrt{2}/4$
5595 6853 cu : $2/3-2\sqrt{5}/3$
5595 7310 J_1 : $6\sqrt[3]{2}/5$
5595 8115 id : $9\ln 2/4$
5595 9274 sr : $\sqrt{5}-\sqrt{6}+\sqrt{7}$
5595 9568 cu : $4/3+4\sqrt{2}$
5596 0313 Ψ : $9\sqrt{3}/7$
5596 1578 ln : 4/7
5596 1631 rt : (5,4,-4,3)
5596 1733 rt : (9,8,7,-8)
5596 2197 rt : (8,6,8,4)
5596 2349 e^x : 4/9
5596 4925 E : $(2e/3)^{-3}$
5596 7018 2^x : $3\sqrt[3]{5}/8$
5596 8469 sinh : $1/\ln(\ln 3/2)$
5596 9384 $\sqrt{5} \times \ln(\pi)$
5596 9563 sr : $3\sqrt{2}+4\sqrt{3}/3$
5596 9929 Ei : $(\ln 2/2)^3$
5597 0166 2^x : $3/4+2\sqrt{3}$
5597 1258 rt : (2,1,-9,-5)
5597 2316 sinh : 20/9
5597 2734 $\ln(2/3)-\exp(e)$
5597 4072 tanh : $(5/2)^{-1/2}$
5597 4653 10^x : $\sqrt{3}/2-\sqrt{5}/2$
5597 5123 rt : (4,3,6,-5)
5597 6676 cu : $4\sqrt[3]{3}/7$
5597 7728 rt : (4,9,8,-8)
5597 8847 rt : (4,9,-2,3)
5598 0965 $t\pi$: $(\pi/2)^{1/3}$
5598 1175 sr : $2+\sqrt{3}/4$
5598 1649 rt : (6,8,-1,-3)
5598 3053 E : 1/6
5598 3523 ln : $4\sqrt{3}/3-\sqrt{5}/4$
5598 6595 cr : $4e\pi/9$
5598 8765 ln : $6(e+\pi)$
5599 0066 e^x : $2\sqrt{3}/3+4\sqrt{5}/3$
5599 3221 rt : (9,3,8,-7)
5599 3585 id : $8\sqrt[3]{5}/3$
5599 4698 rt : (5,3,-3,7)
5599 5197 E : $(\ln 3/2)^3$
5599 5804 J_1 : $(3\pi/2)^{1/2}$
5599 6491 id : $4\sqrt[3]{2}/9$
5599 7468 sr : $2\sqrt{2}/3+2\sqrt{5}/3$
5599 7733 Ψ : 21/4
5599 9815 rt : (4,8,-3,7)
5600 0000 id : 14/25
5600 0550 rt : (1,9,4,2)
5600 1482 rt : (4,9,2,-1)
5600 2074 rt : (5,9,7,8)
5600 2144 $s\pi$: $(\sqrt{2})^{1/2}$
5600 2304 rt : (9,3,6,4)
5600 2686 sinh : $4\zeta(3)/9$
5600 4907 10^x : $1/\sqrt{6}$
5600 6194 J_1 : $7\sqrt{3}/8$
5600 8711 rt : (6,3,-8,3)
5600 9628 Ei : $\exp(-\sqrt[3]{2}/2)$
5601 1537 rt : (1,7,4,-7)
5601 3955 cr : $1/3+2\sqrt{3}$
5601 6071 Ψ : $-\sqrt{3}+\sqrt{5}-\sqrt{7}$
5601 7999 rt : (1,9,0,-3)
5601 8184 rt : (4,4,9,-7)
5601 8513 rt : (3,5,9,4)
5602 0917 $s\pi$: $e/3+2\pi$
5602 1380 cu : $1/3+4\sqrt{5}$
5602 1653 ζ : 21/10
5602 2020 rt : (6,9,2,4)
5602 3200 e^x : $7\zeta(3)$
5602 6777 cr : $3\sqrt{2}/2+3\sqrt{5}/4$
5602 7176 E : $\exp(-2e/3)$
5602 8255 cr : P_{14}
5602 8470 rt : (6,2,-7,7)
5602 8629 rt : (5,0,2,2)
5602 9885 cu : $1/4+\sqrt{5}/2$
5603 0658 Ψ : $7(1/\pi)$
5603 1721 ζ : $5\sqrt[3]{2}/3$
5603 2816 $\ln(3/4) \times \exp(2/3)$
5603 4374 rt : (5,7,7,-7)
5603 6078 rt : (5,9,-7,7)
5603 6681 $s\pi$: $3/4+3\sqrt{2}/4$
5603 7375 rt : (7,9,-3,0)
5603 8683 J : $\sqrt{5}/5$
5603 9237 $t\pi$: $(2e)^{-1/3}$
5604 2780 J_1 : $\sqrt[3]{7/2}$
5604 3694 rt : (9,2,9,6)
5604 5889 sinh : $\sqrt{3}-\sqrt{5}-\sqrt{6}$
5604 6693 J_0 : 24/17
5604 6941 rt : (5,7,-3,-3)
5604 7087 sinh : $10\sqrt[3]{3}$
5604 7148 rt : (6,8,3,8)
5604 7210 J_1 : $\sqrt{3}+\sqrt{5}-\sqrt{6}$
5604 7793 id : $2\sqrt{2}+\sqrt{3}$
5604 9075 cr : 19/5
5604 9442 $\ln(5)+s\pi(2/5)$
5604 9560 rt : (8,9,-8,-4)
5604 9797 J_1 : 13/6
5604 9912 sr : $\pi/10$
5605 0084 at : $3\sqrt{2}/4-\sqrt{3}/4$
5605 0622 cr : $4\sqrt{3}/3+2\sqrt{5}/3$
5605 2275 e^x : $4+3\sqrt{2}$
5605 5548 rt : (4,2,7,4)
5605 7483 rt : (7,8,-2,4)
5605 7768 $\sqrt{5}-s\pi(\sqrt{5})$
5605 9511 Ei : $1/\ln(\sqrt{5})$
5605 9761 sr : $2\sqrt{2}/9$
5606 1068 rt : (6,3,0,-2)
5606 3817 Ψ : $5\sqrt{2}/7$
5606 4157 Ψ : $\ln(2e/3)$
5606 5034 J_1 : $8\sqrt[3]{5}/9$
5606 5336 ln : $4+4\sqrt{5}$
5606 6017 id : $1/2+3\sqrt{2}/4$
5606 6271 E : $\sqrt[3]{3}/9$
5606 6455 at : $10\pi^2$
5606 7953 10^x : $\sqrt{2}/4+2\sqrt{5}$
5606 9445 E : 4/25
5607 0544 J_1 : $5\sqrt{3}/4$
5607 1040 ln : $\sqrt[3]{2}/6$
5607 1221 Ei : $(\sqrt{2}\pi/3)^{-1/3}$
5607 1633 rt : (6,5,-8,-5)
5607 1729 rt : (7,7,-1,8)
5607 6651 cu : $5(e+\pi)/3$
5607 6814 rt : (5,1,5,-4)
5607 7593 rt : (9,9,4,1)
5607 8038 rt : (8,7,-6,4)
5608 0139 E : $(1/\pi)/2$
5608 0283 rt : (7,6,-2,-2)
5608 0866 sq : $(\sqrt[3]{2}/3)^{1/3}$
5608 3705 J_1 : $1/\ln(\sqrt[3]{4})$
5608 4234 $s\pi$: $\exp(\pi/2)$
5608 5170 $\ln(\sqrt{2}) \times \exp(2)$
5608 6459 sq : $1/4+4\sqrt{2}/3$
5608 6653 $t\pi$: $4\sqrt{2}/3+\sqrt{3}/4$
5608 7204 rt : (8,6,-5,8)
5608 8081 J_1 : $9\zeta(3)/5$
5608 9578 2^x : $3+4\sqrt{2}/3$
5609 0486 rt : (5,8,-6,-5)
5609 0565 J_1 : $\exp(\sqrt[3]{2}/3)$
5609 0610 rt : (6,4,3,-4)
5609 0689 cr : 3/17
5609 2298 J_1 : $3\sqrt[3]{3}/2$
5609 2989 rt : (8,9,-4,-2)
5609 3329 sq : $\sqrt{3}-4\sqrt{5}/3$
5609 5979 E : 3/19
5609 7842 rt : (9,5,-9,8)
5609 7988 10^x : $(\sqrt{2}\pi/3)^3$
5609 8211 at : $\pi/5$
5609 8433 rt : (5,2,8,-6)
5609 9662 $t\pi$: $4\sqrt{3}/3-2\sqrt{5}$
5610 0282 rt : (7,7,1,-4)
5610 1038 E : $\sqrt[3]{2}/8$
5610 5469 E : $\sqrt{2}/9$
5610 6043 rt : (6,5,6,-6)
5610 9065 rt : (5,9,-9,-7)
5611 1277 E : $\ln(\sqrt[3]{5}/2)$
5611 1857 rt : (7,8,4,-6)
5611 4042 at : $4\sqrt{2}/9$
5611 5528 rt : (6,6,9,-8)
5611 7045 10^x : $3+\sqrt{2}/2$
5611 8310 10^x : $\sqrt[3]{13}$
5611 9407 rt : (7,9,7,-8)
5611 9930 $\ln(3) \times \ln(3/5)$
5612 1009 2^x : $3\sqrt{5}$
5612 1046 sinh : 16/3
5612 3102 cr : $\sqrt{2}/8$
5612 4039 e^x : $4\sqrt{2}/3+3\sqrt{3}/2$
5612 4060 at : $(3\pi/2)^3$
5612 5618 cr : $7e/5$
5612 6698 rt : (9,8,7,3)
5612 9136 ℓ_{10} : $1/2+\pi$
5612 9676 cr : $5(1/\pi)/9$
5613 0110 Γ : $\sqrt{19}\pi$

5613 1011 tπ : $e/3 + 3\pi/2$
5613 1060 sπ : $\sqrt{3}/2 + 4\sqrt{5}$
5613 2288 J_1 : $\sqrt{14/3}$
5613 3655 ln : $\sqrt{17}\pi$
5613 8391 id : $\exp(-\sqrt{3}/3)$
5613 8454 Ψ : $2/5$
5614 1836 2^x : $9/14$
5614 1913 10^x : $3/2 + 3\sqrt{3}/4$
5614 2271 Ei : $10\sqrt[3]{5}$
5614 2750 ℓ_{10} : $4/3 + 4\sqrt{3}/3$
5614 3725 E : $2\ln 2/9$
5614 3967 $\ln(4) - \exp(2/3)$
5614 5744 J_0 : $\pi^2/7$
5614 5849 sπ : $\sqrt[3]{21/2}$
5614 6001 E : $2/13$
5614 7961 e^x : $7(1/\pi)/5$
5615 1168 ℓ_2 : $\sqrt{2}/2 + 3\sqrt{3}$
5615 1952 sr : $3\sqrt{2}/4 - \sqrt{5}/3$
5615 5281 rt : (3,9,-6,0)
5615 7517 Ψ : $7\sqrt[3]{3}/10$
5615 7582 tπ : $(e+\pi)/7$
5616 0081 10^x : $3\sqrt{3} - \sqrt{5}/3$
5616 1292 ζ : $1/16$
5616 2051 cr : $2/3 + \pi$
5616 3345 J_1 : $\sqrt{7/3}$
5616 4354 sπ : $4\sqrt{3} - \sqrt{5}/2$
5616 7454 cu : $7\pi^2/2$
5616 8204 e^x : $1/\ln(3\pi)$
5616 8471 as : $\exp(-\sqrt[3]{2}/2)$
5617 7337 2^x : $19/14$
5617 7655 cr : $3/2 + 4\sqrt{3}/3$
5617 7834 as : $1/3 - \sqrt{3}/2$
5617 8026 sr : $\sqrt{2}/2 + \sqrt{3}$
5618 0756 cr : $3\sqrt{2} - \sqrt{3}/4$
5618 1279 rt : (8,8,9,-9)
5618 1327 at : $\exp(3\pi/2)$
5618 6029 Li_2 : $7\ln 2/10$
5618 6278 rt : (8,7,6,-7)
5618 6798 rt : (1,9,-6,-6)
5618 6985 ℓ_{10} : $4 - \sqrt{2}/4$
5618 9049 2^x : $\sqrt[3]{5/2}$
5618 9130 rt : (9,9,1,-5)
5618 9239 E : $\zeta(3)/8$
5619 0033 rt : (7,4,8,-7)
5619 1689 cu : $3 + 2\sqrt{5}/3$
5619 2026 id : $3e\pi/10$
5619 2296 E : $3/20$
5619 3658 rt : (8,6,3,-5)
5619 3991 sinh : $\sqrt[3]{23}$
5619 5148 sr : $6/19$
5619 8125 rt : (9,8,-2,-3)
5619 9310 ln : $\sqrt{2} + 3\sqrt{5}/2$
5619 9584 rt : (7,3,5,-5)
5620 1112 tπ : $5\sqrt{3}/6$
5620 2562 sr : $2e\pi/7$
5620 3071 rt : (5,1,-8,-9)
5620 5256 $\exp(4) - s\pi(\sqrt{2})$
5620 5614 J_1 : $\sqrt[3]{10}$
5620 5646 rt : (8,5,0,-3)
5620 6561 ℓ_{10} : $2/3 + 4\sqrt{5}/3$
5620 6764 10^x : $1/\ln(\sqrt[3]{2})$
5620 7643 sπ : $\sqrt[3]{5}/9$
5620 7678 rt : (6,0,7,5)
5620 8483 ℓ_2 : $4 - \pi/3$
5620 9399 $\ln(2/3) \times \ln(4)$
5621 0034 ℓ_{10} : $\sqrt{2}/2 - \sqrt{3}/4$
5621 0211 rt : (5,0,-9,5)
5621 2549 e^x : $4e/3 + 3\pi/2$
5621 3298 id : $\ln(\sqrt[3]{5}/3)$
5621 3624 rt : (9,7,-5,-1)
5621 4161 E : $(\sqrt[3]{4}/3)^3$
5621 4684 cu : $3/2 + 4e/3$
5621 5123 Ψ : $6\ln 2/7$
5621 5247 rt : (4,0,-4,9)
5621 5731 cr : $2 + 2e/3$
5621 5848 at : $\sqrt[3]{2}/2$
5621 6371 rt : (7,2,2,-3)
5621 7932 $\exp(\sqrt{2}) \times s\pi(1/3)$
5621 8125 Ei : $8\sqrt{3}$
5621 8788 id : $1/4 - 2e/3$
5621 9392 rt : (5,3,9,5)
5622 3123 sr : $9\pi^2/7$
5622 5743 rt : (4,7,-8,-6)
5622 6932 rt : (4,1,-5,5)
5622 8321 λ : $\exp(-17\pi/18)$
5622 8343 Γ : $\sqrt[3]{5}/3$
5622 8513 rt : (8,4,-3,-1)
5622 9447 2^x : $\sqrt[3]{21/2}$
5623 0334 id : $3\sqrt{2}/2 - \sqrt{5}/4$
5623 0589 sq : $3/2 + 3\sqrt{5}/2$
5623 2636 ℓ_{10} : $\sqrt{2} + \sqrt{5}$
5623 2904 rt : (6,1,4,3)
5623 4132 sr : $1/\sqrt{10}$
5623 5749 rt : (3,8,7,-7)
5623 6455 sr : $3 - \sqrt{5}/4$
5623 7437 J_1 : $23/15$
5623 8121 rt : (3,9,-3,-4)
5623 9001 ℓ_2 : $1/4 + 4\sqrt{2}$
5623 9984 Ei : $8/15$
5624 1850 tanh : $7/11$
5624 2898 ℓ_2 : $-\sqrt{3} + \sqrt{5} + \sqrt{6}$
5624 4001 Γ : $19/7$
5624 4270 rt : (4,2,-6,1)
5624 4511 2^x : $3 - 3\pi/4$
5624 4795 ln : $4e + 2\pi/3$
5624 6652 rt : (9,6,-8,1)
5624 7022 sπ : $2\sqrt{2} + 4\sqrt{5}/3$
5624 7044 rt : (2,5,9,-7)
5624 8951 rt : (9,6,8,-8)
5624 9913 e^x : $3/2 - \sqrt{5}/4$
5625 0000 id : $9/16$
5625 0768 rt : (7,4,-7,6)
5625 2141 tanh : $\ln(\sqrt[3]{4}/3)$
5625 2457 10^x : $3(e+\pi)/8$
5625 3606 rt : (7,1,-1,-1)
5625 3624 as : $8/15$
5625 8290 rt : (3,7,4,-5)
5625 8619 sinh : $\ln(\sqrt[3]{5})$
5625 8879 cr : $((e+\pi)/3)^2$
5625 9360 tanh : $2(1/\pi)$
5625 9571 E : $\sqrt[3]{3}/10$
5626 0126 tπ : $1/\sqrt[3]{18}$
5626 0554 Γ : $\sqrt[3]{20}$
5626 1725 rt : (5,4,6,3)
5626 2059 rt : (1,8,-4,7)
5626 2124 sinh : $9e\pi/10$
5626 2213 at : $3\sqrt{2}/2 - 2\sqrt{5}/3$
5626 2279 rt : (4,6,-5,-4)
5626 3949 E : $\ln(\sqrt{3}/2)$
5626 5011 rt : (7,5,-8,2)
5626 5375 cu : $\sqrt{2}/4 - 3\sqrt{3}$
5626 5747 tπ : $\sqrt{2} - \sqrt{3}/3$
5626 6906 id : $4\sqrt{2}/3 + 3\sqrt{5}/4$
5626 8284 2^x : $19/7$
5626 8354 J_1 : $\sqrt{3} + \sqrt{6} - \sqrt{7}$
5626 9101 rt : (9,5,5,-6)
5627 0154 Γ : $8/23$
5627 2666 rt : (4,9,-1,-3)
5627 4458 erf : $\ln(\sqrt{3})$
5627 4847 10^x : $3 - 3\sqrt{2}/2$
5627 5063 rt : (6,5,-3,6)
5627 5112 E : $1/7$
5627 6190 2^x : $1/2 + 3\sqrt{3}/2$
5627 7053 $4 \times \exp(\pi)$
5627 7563 rt : (2,4,6,-5)
5627 9649 at : $(5)^3$
5628 1937 rt : (9,8,-5,7)
5628 4241 rt : (7,6,-9,-2)
5628 4292 ζ : $(1/\pi)/5$
5628 4775 rt : (8,2,7,-6)
5628 6858 10^x : $1 + 4\pi$
5628 6947 E : $\exp(-(e+\pi)/3)$
5628 9256 rt : (8,3,-6,1)
5629 0372 cr : $\sqrt{2}/4 + 2\sqrt{3}$
5629 0708 E : $4(1/\pi)/9$
5629 1264 E : $\sqrt{2}/10$
5629 3641 rt : (3,8,0,-2)
5629 4882 sr : $3/2 + 2\sqrt{2}/3$
5629 5322 sr : $10\sqrt[3]{5}/7$
5629 8353 rt : (9,9,-6,3)
5629 8475 10^x : $2/3 + 4e$
5629 9493 e^x : $16/17$
5630 0279 rt : (9,4,2,-4)
5630 1146 rt : (2,9,-8,-7)
5630 1625 sr : $3/4 - \sqrt{3}/4$
5630 1718 J_1 : $20/13$
5630 1947 rt : (4,7,9,-8)
5630 4128 rt : (1,6,-4,-2)
5630 4865 rt : (6,2,1,1)
5630 4961 sπ : $3\sqrt{2} - \sqrt{3}/4$
5630 5465 rt : (7,1,9,6)
5630 5730 rt : (1,1,8,-5)
5630 6415 rt : (5,3,-7,-4)
5630 7187 E : $\sqrt[3]{2}/9$
5630 7486 Ei : $6(e+\pi)/5$
5630 8137 J_1 : $9\sqrt[3]{5}/10$
5630 9615 rt : (8,9,-1,7)
5631 1113 Ei : $9\sqrt[3]{2}/2$
5631 1881 sπ : $3e\pi/2$
5631 4286 2^x : $2 - 2\sqrt{2}$
5631 4706 rt : (5,9,-9,2)
5631 5093 Ei : $\sqrt{14}\pi$
5631 5736 sr : $7\pi/9$
5631 5853 J_1 : $8\sqrt{3}/9$
5631 5939 rt : (5,9,4,-6)
5631 7079 $\ln(3/4) \div \ln(3/5)$
5631 8843 cr : $4/3 - 2\sqrt{3}/3$
5632 0135 ζ : $(5/2)^{-3}$
5632 2197 E : $\ln 2/5$
5632 3215 Li_2 : $\exp(-\sqrt[3]{3}/2)$
5632 4134 e^x : $3 + 2\sqrt{2}/3$
5632 4704 rt : (4,7,8,3)
5632 6714 rt : (8,1,4,-4)
5632 7607 rt : (4,8,-4,-1)
5632 8288 2^x : $\sqrt[3]{20}$
5633 1626 at : $12/19$
5633 1847 tπ : $\exp(-2e/3)$
5633 2005 sπ : $4/21$
5633 2336 erf : $11/20$
5633 2866 rt : (6,7,-5,-2)
5633 3111 ℓ_{10} : $3\sqrt{2}/4 + 3\sqrt{3}/2$
5633 5247 rt : (9,8,-9,-6)
5633 8230 cr : $e/4 + \pi$
5633 8466 rt : (6,9,9,-9)
5634 0649 sinh : $22/7$
5634 2517 rt : (4,7,-6,6)
5634 2867 ln : $9\sqrt[3]{3}$
5634 3545 sq : $3e\pi/10$
5634 5270 2^x : $\sqrt{2}/3 - 3\sqrt{3}/4$
5634 5716 J_1 : $\sqrt[3]{11/3}$
5634 6249 E : $3(1/\pi)/7$
5634 6756 ℓ_{10} : $3e\pi/7$
5634 6841 E : $3/22$
5634 7191 sr : $22/9$
5634 7817 J_1 : $15/7$
5635 0832 rt : (2,8,-5,-5)
5635 3224 rt : (9,1,9,-7)
5635 3288 sr : $4\sqrt{2}/3 + \sqrt{5}/4$
5635 3297 Γ : $5(e+\pi)/9$
5635 4914 rt : (9,3,-1,-2)
5635 6482 Ψ : $\zeta(7)$
5635 7779 10^x : $4\sqrt{2}/3 + 4\sqrt{3}$
5635 7890 E : $(e)^{-2}$
5635 8719 J_1 : $(\sqrt[3]{2}/3)^{-1/2}$
5635 9115 erf : $1/\sqrt[3]{6}$
5635 9487 2^x : $11/6$
5636 0120 10^x : $\sqrt{5}/3$
5636 0790 rt : (5,8,1,-4)
5636 1337 J_1 : $\sqrt{3} - \sqrt{5} + \sqrt{7}$
5636 1732 as : $4\zeta(3)/9$
5636 3859 sπ : $1/2 + 4\sqrt{3}/3$
5636 4084 tπ : $3\sqrt[3]{3}/2$
5636 4963 rt : (7,2,6,4)
5636 6033 rt : (4,8,-5,2)
5636 8363 rt : (3,3,8,-6)
5637 0165 E : $((e+\pi)/3)^{-3}$
5637 0373 cu : $\sqrt{3}/3 - \sqrt{5}$
5637 0992 E : $(\ln 3/3)^2$
5637 3749 ln : $\ln 2/9$
5637 3797 cu : $19/23$

5637 4910 rt : (6,8,6,-7)
5637 5832 $\zeta : (\ln 2)^{-1/2}$
5637 6744 $E : \zeta(3)/9$
5637 6833 $\ell_{10} : 5(e+\pi)/8$
5637 7625 rt : (4,5,-2,-2)
5637 9156 $E : 2/15$
5637 9349 rt : (1,7,-4,2)
5638 2614 $2^x : 1/\ln(3\pi/2)$
5638 6732 at : $4\sqrt{3}/3 - 3\sqrt{5}/4$
5638 7654 $J_1 : 17/11$
5638 8227 $J_1 : 9\zeta(3)/7$
5638 8257 at : $2/3 - 3\sqrt{3}/4$
5638 9257 rt : (6,8,-6,-6)
5639 0253 $2/5 + s\pi(\sqrt{2})$
5639 4264 at : $(5/2)^{-1/2}$
5639 4423 $\ln(\sqrt{3}) - \exp(\sqrt{2})$
5639 4831 $e^x : \sqrt{5}/5$
5639 6295 rt : (4,9,-4,-2)
5639 6639 $e^x : 1/3 - e/3$
5639 7324 rt : (3,5,-7,-5)
5639 7442 cr : $\sqrt{2}/3 + 3\sqrt{5}/2$
5639 9160 rt : (4,7,5,6)
5640 2120 $s\pi : 4e/3 + 4\pi$
5640 2657 rt : (5,5,8,-7)
5640 4508 id : $4\sqrt{3}/3 - \sqrt{5}/3$
5640 4635 $\ln(3) \div \exp(2/3)$
5640 5848 at : $\exp(5)$
5640 5856 rt : (7,0,-4,1)
5640 7607 sr : 7/22
5640 7731 $t\pi : 7(e+\pi)/6$
5640 7923 $\Gamma : 5\sqrt[3]{5}/2$
5640 8694 rt : (8,0,1,2)
5640 9216 rt : (9,0,6,5)
5640 9378 $E : 3/23$
5640 9666 $\ell_{10} : 7\pi/6$
5640 9973 rt : (1,3,8,9)
5641 0839 $\ell_{10} : 3\sqrt{2} - \sqrt{3}/3$
5641 1433 sr : $9e/10$
5641 1946 rt : (6,1,-8,8)
5641 2231 $J_1 : 5\sqrt[3]{5}/4$
5641 3257 $\Psi : (5)^{-3}$
5641 3347 rt : (6,0,-9,4)
5641 3413 tanh : $2\sqrt{5}/7$
5641 3870 $\sqrt{3} \div s\pi(\sqrt{5})$
5641 5537 ln : $3(e+\pi)/10$
5641 7535 rt : (3,6,-2,6)
5641 8958 id : $(\pi)^{-1/2}$
5642 2554 cu : $\sqrt{7/3}$
5642 3527 $\Psi : 4\sqrt[3]{2}/5$
5642 4078 rt : (8,5,-6,9)
5642 4290 rt : (7,7,8,-8)
5642 5264 rt : (6,5,8,4)
5642 6462 $t\pi : 8\pi^2/3$
5642 7143 $\ell_{10} : 3/11$
5642 9417 rt : (6,7,3,-5)
5643 3952 rt : (1,7,-7,-6)
5643 5760 $\ell_{10} : 4/3 - 3\sqrt{2}/4$
5643 7233 cr : $1 + 2\sqrt{2}$
5643 8239 ln : $(2e)^{1/3}$
5643 9180 rt : (9,9,8,-9)
5643 9262 sinh : $\exp(\sqrt{2})$
5644 0260 rt : (5,8,-1,-2)
5644 1064 $E : 2(1/\pi)/5$
5644 1443 id : $4e/3 - 4\pi/3$
5644 2171 $J_1 : e\pi/4$
5644 2999 sq : $5\zeta(3)/8$
5644 3384 $e^x : \sqrt[3]{7/2}$
5644 4778 $J_1 : \ln(3\pi/2)$
5644 7539 rt : (8,9,3,-6)
5644 7545 rt : (3,9,-8,-7)
5644 8728 rt : (6,2,-9,4)
5645 1822 rt : (2,7,-2,-3)
5645 2821 $\zeta : 2\pi/3$
5645 3720 rt : (1,9,7,-7)
5645 4393 $E : \sqrt[3]{2}/10$
5645 5217 rt : (3,2,5,-4)
5645 5414 $10^x : (2e/3)^{1/3}$
5645 6179 $t\pi : \sqrt{2} - \sqrt{6} - \sqrt{7}$
5645 7302 at : $(2e)^3$
5645 7430 rt : (8,3,8,5)
5645 7721 rt : (8,6,-7,5)
5646 0982 cr : $\sqrt[3]{2}/7$
5646 2599 rt : (7,9,-2,-3)
5646 3488 $s\pi : 1/4 + \sqrt{5}/4$
5646 4230 $E : 1/8$
5646 4247 $s\pi : 3(1/\pi)/5$
5646 4407 cu : $1/3 - 3\sqrt{5}/2$
5646 6088 rt : (3,7,-1,2)
5646 7920 rt : (5,5,3,1)
5647 1296 $2^x : 4 - 3\sqrt{5}/2$
5647 3064 rt : (4,2,-9,-5)
5647 3696 $2^x : 1/3 + 2e$
5647 4060 rt : (5,4,5,-5)
5647 4556 rt : (5,2,-4,8)
5647 5139 rt : (9,7,-6,-4)
5647 5211 $\Psi : \sqrt{5}$
5647 7239 $\zeta : \sqrt{2} + \sqrt{3} - \sqrt{5}$
5647 8396 rt : (4,7,-7,1)
5648 0534 $J_1 : \sqrt{2} - \sqrt{3} + \sqrt{6}$
5648 0676 sq : $3/4 + 2e/3$
5648 0717 rt : (7,6,-2,9)
5648 1745 $J_1 : 4e/7$
5648 2415 $E : \exp(-2\pi/3)$
5648 4389 rt : (7,6,5,-6)
5648 5586 $2^x : 2/3 - 2\sqrt{5}/3$
5648 6170 sinh : 7/13
5648 9868 rt : (2,4,-9,-6)
5648 9955 tanh : 16/25
5649 0347 id : $e - 2\pi$
5649 0911 rt : (9,8,5,-7)
5649 1548 cr : $\sqrt[3]{3}/8$
5649 4822 rt : (7,9,-6,-5)
5649 4935 ln : 13
5649 6392 id : $3\sqrt[3]{5}/2$
5649 7542 $\Psi : 9(e+\pi)/10$
5649 7781 rt : (6,9,-7,0)
5649 7909 as : $1/4 - \pi/4$
5649 8524 sinh : $\sqrt[3]{14/3}$
5649 9973 $t\pi : 9\zeta(3)/5$
5650 0962 rt : (8,7,-8,1)
5650 1200 $\ln(2/5) - \exp(1/2)$
5650 2090 $10^x : 9/22$
5650 2219 rt : (7,3,3,2)
5650 2347 $\Psi : \sqrt[3]{5}/9$
5650 2384 rt : (2,8,9,-8)
5650 2694 at : $3/2 - \sqrt{3}/2$
5650 4031 cr : 23/6
5650 7527 $e^x : 1 - \pi/2$
5650 8274 $J_1 : 14/9$
5650 8458 sr : $\sqrt{6}$
5650 9207 rt : (9,1,3,3)
5650 9966 sr : $2\pi^2/3$
5651 0659 $E : \zeta(3)/10$
5651 0812 $s\pi : 2\sqrt{2}/3 + \sqrt{3}/2$
5651 1536 $E : (\ln 2/2)^2$
5651 2609 $E : 3/25$
5651 2985 $s\pi : (\sqrt[3]{4}/3)^{1/3}$
5651 8599 $E : 3(1/\pi)/8$
5651 8871 sr : $\sqrt{5}/7$
5651 9264 $t\pi : (\sqrt[3]{5})^{-1/3}$
5651 9516 $t\pi : \exp(-\sqrt[3]{3}/3)$
5651 9565 $\Psi : (\sqrt[3]{2}/3)^{-3}$
5651 9771 rt : (8,8,0,-4)
5652 1739 id : 13/23
5652 4758 id : $7\sqrt{5}/10$
5652 4917 $\ell_{10} : P_{30}$
5652 9423 rt : (8,5,7,-7)
5653 0805 $2^x : 1 - \sqrt{2}/4$
5653 1340 rt : (7,7,-3,5)
5653 1507 $\zeta : \exp(-e)$
5653 1921 rt : (6,3,7,-6)
5653 1952 $Li_2 : (\pi/3)^{-1/3}$
5653 2051 $Ei : 4\zeta(3)/9$
5653 2375 $2^x : e/2$
5653 3501 rt : (5,3,-5,4)
5653 4685 $E : 2/17$
5653 6164 rt : (4,1,7,-5)
5653 6656 rt : (3,8,0,-2)
5653 7604 rt : (1,9,3,-1)
5654 1693 rt : (1,8,4,-5)
5654 3402 $10^x : 3\sqrt{2}/4 - \sqrt{3}/2$
5654 4626 $Ei : 7\sqrt{2}$
5654 4959 rt : (2,1,7,4)
5654 5392 $E : ((e+\pi)/2)^{-2}$
5654 5395 rt : (2,5,-9,-5)
5654 6667 $J_1 : 9\sqrt{3}/10$
5654 6761 $s\pi : P_{52}$
5654 8528 cu : $2e/3 + \pi$
5654 9965 $\ell_{10} : 2 + 3\sqrt{5}/4$
5655 1016 rt : (1,6,-4,-4)
5655 2203 $\ell_2 : 3\sqrt{2} + 3\sqrt{5}/4$
5655 2983 $s\pi : 7\ln 2/6$
5655 2997 id : $\exp(-\sqrt[3]{5}/3)$
5655 3773 cu : $1/3 - e$
5655 4220 $E : \ln 2/6$
5655 4585 sq : $1 + e/2$
5655 5188 $J_1 : 9\ln 2/4$
5655 5408 $J_0 : 7\zeta(3)/6$
5655 6220 sq : $\ln(\sqrt{2}/3)$
5655 6923 ln : $4 + \pi/4$
5655 6985 id : $1/\ln(e+\pi)$
5655 7040 $J_1 : 17/8$
5655 7146 as : $4 - 2\sqrt{3}$
5655 7992 tanh : $4\sqrt[3]{3}/9$
5655 8577 rt : (8,8,-9,-3)
5655 8970 id : $3\sqrt{2} - 3\sqrt{5}/4$
5656 0228 $\Gamma : (2e)^{-1/3}$
5656 1172 sinh : $7\ln 2/9$
5656 3035 $s\pi : 4\sqrt{3}/3 - \sqrt{5}/2$
5656 3833 $\ell_2 : 3\sqrt{3} - \sqrt{5}$
5656 4841 sq : $\sqrt{11}\pi$
5656 5606 rt : (9,7,2,-5)
5656 6957 cr : $3\sqrt{3}/4 - \sqrt{5}/2$
5656 6995 rt : (9,4,9,-8)
5656 7598 rt : (6,7,2,9)
5656 8429 sr : $4/3 + \sqrt{5}/2$
5656 8542 id : $2\sqrt{2}/5$
5657 0551 $\ell_{10} : e/10$
5657 1431 $2^x : \sqrt{2} - \sqrt{5}$
5657 2279 sq : $3e + 3\pi/2$
5657 3396 at : $3e/2 - 3\pi/2$
5657 4145 rt : (6,8,-8,-6)
5657 4392 tanh : $3\sqrt[3]{5}/8$
5657 7582 rt : (4,9,6,-7)
5657 7972 rt : (7,2,9,-7)
5657 8931 rt : (7,5,2,-4)
5657 9712 $\ell_{10} : 3 + e/4$
5658 0532 rt : (7,8,-5,-1)
5658 2661 at : $(e+\pi)^3$
5658 6280 sq : $6\pi^2/7$
5658 7214 $\ln(2/3) \times \exp(1/3)$
5658 8792 $J_1 : 25/16$
5658 9643 $E : \ln(\sqrt{5}/2)$
5658 9722 sinh : 16/13
5659 0539 $2^x : 1/3 - 2\sqrt{3}/3$
5659 2007 $e^x : 4 + 3e$
5659 2837 rt : (2,7,6,-6)
5659 3690 $E : 1/9$
5659 4020 sr : $3e - \pi/2$
5659 5599 rt : (5,0,9,6)
5659 6116 $e^x : \sqrt{3} + \sqrt{5} - \sqrt{6}$
5659 6906 $\ell_{10} : -\sqrt{2} + \sqrt{6} + \sqrt{7}$
5659 7244 $2^x : 11/17$
5659 7353 rt : (9,7,-6,8)
5659 7679 $Ei : 2/19$
5659 8117 $Ei : (4)^{-3}$
5659 8622 $J_1 : 3\sqrt{2}/2$
5659 9359 $10^x : 9(1/\pi)/7$
5660 0541 rt : (7,8,-4,1)
5660 2581 $\lambda : \exp(-16\pi/17)$
5660 2666 $\ell_2 : 3\pi^2/5$
5660 3181 rt : (5,3,2,-3)
5660 3882 $\zeta : 1/15$
5660 3944 rt : (2,6,-6,-7)
5660 6137 rt : (8,6,-8,-5)
5660 6496 sr : $1 - e/4$
5660 6865 rt : (7,3,-8,7)
5660 9215 $4/5 \div \ln(3/5)$
5661 0439 rt : (8,4,5,3)
5661 1097 $\Psi : 1/\ln(\ln 2)$

5661 1603 sinh : $7\sqrt{2}/2$
5661 1638 rt : (5,8,8,-8)
5661 2710 sr : $2\sqrt[3]{3}/9$
5661 2936 2^x : $3/4 - \pi/2$
5661 3320 E : $(2\pi/3)^{-3}$
5661 5859 Ei : $3\sqrt{3}/2$
5661 6717 E : $\exp(-\sqrt{2}\pi/2)$
5661 7497 Γ : 23/3
5661 8420 2^x : $1/3 + 4\sqrt{2}$
5661 9037 rt : (5,9,-2,-6)
5662 0107 J_1 : $7\sqrt{5}/10$
5662 3855 at : $2\sqrt{2} - 2\sqrt{3}$
5662 4684 rt : (8,4,4,-5)
5662 5189 $5 \times \exp(\sqrt{2})$
5662 5968 as : $\ln(\sqrt[3]{5})$
5662 6845 e^x : 21/4
5662 7525 ℓ_{10} : $3\sqrt{2} - \sqrt{5}/4$
5662 8560 rt : (3,2,9,5)
5663 0091 E : $\exp(-\sqrt{5})$
5663 1104 sq : $4\sqrt{3} - 2\sqrt{5}/3$
5663 1577 rt : (1,7,3,1)
5663 1600 Γ : 1/6
5663 2068 cu : $1/3 + 3e$
5663 2826 cu : $1/4 + \sqrt{3}/3$
5663 3239 e^x : $3\pi/10$
5663 3370 rt : (3,6,8,-7)
5663 3843 Ei : $4(e+\pi)/7$
5663 5586 θ_3 : $2\ln 2/5$
5663 6598 E : $(1/\pi)/3$
5663 6673 J_1 : $\sqrt[3]{19/2}$
5663 6785 tπ : $2\sqrt{2} + 2\sqrt{5}/3$
5663 7061 id : 4π
5663 7259 rt : (1,9,-5,4)
5663 9110 rt : (8,9,-1,-2)
5664 0180 e^x : $3 - \sqrt{5}/2$
5664 0966 Ei : $\sqrt[3]{4}$
5664 1026 tπ : $1/\ln(\sqrt[3]{4})$
5664 2969 e^x : $\pi/7$
5664 3600 E : 2/19
5664 4193 at : $\exp(2e)$
5664 6661 rt : (4,6,-9,5)
5664 6712 Li_2 : $2\sqrt[3]{5}/7$
5664 7698 $\exp(e) - s\pi(1/5)$
5664 7796 E : $(\sqrt{2}/3)^3$
5664 8526 rt : (3,9,1,-4)
5665 0746 rt : (9,8,-7,4)
5665 1101 K : 18/19
5665 1633 cr : 2/11
5665 1888 2^x : $3\sqrt{2}/2 + 4\sqrt{3}/3$
5665 1936 10^x : $2\sqrt{3}/3 - \sqrt{5}/3$
5665 2301 rt : (6,8,1,5)
5665 3330 sq : $4\sqrt{3} - \sqrt{5}/4$
5665 4068 10^x : $2e + 4\pi$
5665 5393 θ_3 : $1/\sqrt{13}$
5665 6769 $s\pi(1/5) \times s\pi(\sqrt{2})$
5665 9041 sinh : $3\sqrt[3]{2}/7$
5665 9233 cr : $4(1/\pi)/7$
5665 9784 sπ : $2/3 + \pi$
5666 1132 rt : (9,3,6,-6)
5666 1711 Γ : $10\sqrt{5}$
5666 6778 ζ : $6\sqrt[3]{2}/5$
5666 7992 sπ : $4\zeta(3)$
5666 8587 rt : (1,4,8,-6)
5666 9011 rt : (7,4,-9,3)
5667 0343 ln : $4\sqrt{2} - \sqrt{3}/2$
5667 2440 ℓ_{10} : $1/3 + 3\sqrt{5}/2$
5667 2921 at : 7/11
5667 3344 sr : $1/3 + 3\sqrt{2}/2$
5667 4596 rt : (2,9,6,-5)
5667 5709 E : $(\pi)^{-2}$
5667 6207 cr : $8\sqrt[3]{3}/3$
5667 6371 ln : $2\sqrt{2}/3 - \sqrt{3}/2$
5667 6490 at : $(2\pi)^3$
5667 6985 $\sqrt{3} \times \exp(4)$
5667 7386 J_1 : $7e/9$
5667 8883 rt : (9,5,7,4)
5667 9998 rt : (7,9,5,-7)
5668 1538 e^x : P_{41}
5668 2408 J_1 : $\pi/2$
5668 2788 rt : (9,6,-1,-3)
5668 3130 rt : (2,7,-3,-9)
5668 3635 at : $\ln(\sqrt[3]{4}/3)$
5668 4180 tanh : 9/14
5668 5304 J_1 : $1/\ln(\sqrt[3]{4}/3)$
5668 5340 ℓ_{10} : $\sqrt{2}/2 + 4\sqrt{5}/3$
5668 5512 J_0 : 7/5
5668 6194 E : 1/10
5668 6244 2^x : $3\sqrt{2}/2 + 2\sqrt{5}$
5668 7135 rt : (8,8,-2,8)
5668 8538 J_1 : $10\sqrt{2}/9$
5668 9428 J_1 : 11/7
5669 0266 J_0 : $10\sqrt[3]{2}/9$
5669 1026 cu : $\sqrt{2}/3 - 3\sqrt{3}/4$
5669 1150 at : $2(1/\pi)$
5669 3729 $\exp(2) - \exp(3/5)$
5669 3874 E : $\ln 2/7$
5669 5317 sπ : $4\sqrt{2}/7$
5669 5542 10^x : $\sqrt{3}/4 + \sqrt{5}/2$
5669 5854 cr : $9\sqrt[3]{5}/4$
5669 5901 cr : $\sqrt{3}/2 + 4\sqrt{5}/3$
5669 6213 2^x : $e/3 + \pi/2$
5669 7057 E : $\ln(e/3)$
5669 7078 ℓ_{10} : $4/3 + 3\pi/4$
5669 8219 rt : (6,5,-3,-1)
5669 8261 J_1 : $(3\pi)^{1/3}$
5669 8729 id : $1 - \sqrt{3}/4$
5670 0774 rt : (7,9,-5,-3)
5670 2956 rt : (1,7,1,-3)
5670 3421 rt : (8,8,7,-8)
5670 3753 rt : (7,1,6,-5)
5670 4411 rt : (4,7,-8,1)
5670 5100 id : P_{80}
5670 5277 2^x : $(\sqrt[3]{2}/3)^{1/2}$
5670 5729 e^x : $2 + \sqrt{3}/2$
5670 5734 rt : (6,6,5,2)
5670 6486 rt : (4,9,-5,-5)
5670 8105 erf : $4\ln 2/5$
5670 8495 10^x : $\exp(\sqrt[3]{5}/2)$
5670 8559 $5 \div \exp(2/3)$
5670 9818 e^x : $\sqrt[3]{20/3}$
5671 0053 $1/3 \div s\pi(1/5)$
5671 1244 J_1 : 19/9
5671 2004 tπ : $1/4 - \sqrt{2}$
5671 2374 rt : (1,4,9,4)
5671 2464 rt : (6,4,-4,7)
5671 3211 2^x : $4 + 3\sqrt{3}$
5671 4845 e^x : $\sqrt{11}$
5671 7765 rt : (5,8,6,9)
5671 8262 e^x : $2\sqrt{2}/3$
5671 8450 sr : $4\sqrt{3} - 2\sqrt{5}$
5671 8737 rt : (9,9,-8,0)
5672 0919 E : $3(1/\pi)/10$
5672 1007 erf : $(e+\pi)^{-1/3}$
5672 1040 id : $\sqrt{2} - 4\sqrt{5}/3$
5672 2834 E : 2/21
5672 3385 ζ : 23/11
5672 3488 rt : (9,7,9,-9)
5672 7692 J_1 : $5\sqrt[3]{2}/4$
5673 0603 ln : P_{80}
5673 0707 rt : (5,2,-4,-2)
5673 1712 E : $\ln(\ln 3)$
5673 4516 ℓ_2 : $-\sqrt{2} + \sqrt{3} + \sqrt{7}$
5673 4700 sq : $\sqrt{2}/2 - 4\sqrt{3}/3$
5673 5035 sq : $e/3 + 3\pi/2$
5673 5537 rt : (1,2,-8,-5)
5673 5941 tπ : $4\sqrt{3} + \sqrt{5}$
5673 6422 rt : (8,0,8,6)
5673 6740 e^x : 25/7
5673 7705 rt : (9,2,0,1)
5673 7850 $\ln(\sqrt{5}) \times \exp(2/3)$
5674 0459 10^x : $3/4 + 3\sqrt{2}/2$
5674 0695 10^x : $1 + 4\sqrt{5}$
5674 1272 ζ : $e/4$
5674 1837 ln : $1 - \sqrt{3}/4$
5674 1841 rt : (1,2,-9,-7)
5674 2134 $\exp(5) + \exp(e)$
5674 2660 rt : (6,6,7,-7)
5674 3630 rt : (3,4,-6,9)
5674 5074 rt : (3,8,-7,-6)
5674 6744 J_1 : $(\sqrt{2}\pi)^{1/2}$
5674 6825 Γ : e
5674 6837 rt : (6,0,-9,9)
5674 8273 θ_3 : 5/18
5674 9079 rt : (2,7,-8,8)
5674 9092 rt : (7,4,-1,-2)
5674 9446 ln : $4\sqrt{2}/3 - 3\sqrt{5}/4$
5675 0091 $\ln(2) \div \exp(1/5)$
5675 2129 ln : $7\sqrt[3]{2}/5$
5675 2344 E : $(\sqrt{2}\pi/2)^{-3}$
5675 3409 J_1 : $\exp(\sqrt{5}/3)$
5675 4251 rt : (5,5,-7,-4)
5675 4321 E : $2(1/\pi)/7$
5675 4542 ln : $4 - \sqrt{5}$
5675 4583 E : 1/11
5675 5009 ℓ_{10} : $2\sqrt{2} + \sqrt{3}/2$
5675 5891 sr : $9(e+\pi)/8$
5675 7352 cu : $3/4 - 2\sqrt{5}$
5675 7870 rt : (7,8,-3,-3)
5675 8524 $\exp(3) \times s\pi(\sqrt{5})$
5675 9910 sinh : $3\sqrt[3]{3}/8$
5676 2134 sr : $9\sqrt{2}$
5676 3294 rt : (7,5,9,-8)
5676 5000 E : $(\sqrt{5})^{-3}$
5676 6286 rt : (8,9,-3,4)
5676 7955 rt : (2,6,1,-1)
5676 9387 $\ln(\sqrt{3}) - \exp(3/4)$
5676 9706 10^x : $2/3 - \sqrt{2}/3$
5677 4245 rt : (2,7,-9,-7)
5677 4710 ln : $\sqrt{23}$
5677 5445 e^x : $1/\sqrt[3]{11}$
5677 6253 cr : $3\sqrt{2}/2 + \sqrt{3}$
5677 7263 sr : $3\sqrt{2}/2 + 2\sqrt{5}$
5677 7496 rt : (4,8,-7,-3)
5677 8858 rt : (3,5,5,-5)
5678 0542 rt : (6,9,0,1)
5678 2163 2^x : $7(e+\pi)/3$
5678 2273 E : 2/23
5678 3218 rt : (5,1,6,4)
5678 4414 E : $\ln 2/8$
5678 4562 rt : (8,3,1,-3)
5678 6163 cr : $1/2 + 3\sqrt{5}/2$
5678 7486 cu : $2e - \pi/3$
5678 9073 rt : (8,5,-8,6)
5678 9892 rt : (3,5,-4,-7)
5679 0123 sq : 17/9
5679 1522 J_1 : $7\zeta(3)/4$
5679 3387 rt : (7,7,7,3)
5679 3610 sinh : $\sqrt[3]{11}$
5679 4161 erf : 5/9
5679 4446 e^x : $2\sqrt[3]{5}$
5679 4473 at : $\exp(e+\pi)$
5679 5183 J_1 : $\sqrt{5/2}$
5679 5362 $\exp(3/5) - s\pi(\sqrt{3})$
5679 6160 rt : (2,4,-8,-5)
5679 7411 Ψ : $-\sqrt{2} + \sqrt{5} - \sqrt{6}$
5679 7634 $\ln(\sqrt{2}) + \exp(1/5)$
5679 7739 sq : $e/3 - 3\pi$
5680 0471 sq : $10\sqrt[3]{3}/9$
5680 1427 rt : (7,8,2,-5)
5680 2710 $\exp(1/2) \times s\pi(2/5)$
5680 3258 rt : (4,4,7,-6)
5680 5083 $2 \times \exp(1/4)$
5680 6568 E : 1/12
5680 7772 10^x : $3\sqrt{2} - \sqrt{5}/3$
5680 8084 J_0 : $(\ln 3/3)^{-1/3}$
5680 8138 ln : $4/3 + 2\sqrt{3}$
5680 8853 Ei : 22/25
5681 0117 rt : (6,5,-5,3)
5681 0712 rt : (9,2,3,-4)
5681 2459 rt : (3,9,8,-8)
5681 3606 rt : (1,9,3,-6)
5681 4533 cr : $3/2 + 3\pi/4$
5681 4698 E : $\exp(-5/2)$
5681 5129 rt : (6,2,8,5)
5681 5416 J_0 : $((e+\pi)/3)^{1/2}$
5681 5881 rt : (6,7,-5,-4)
5681 7128 sinh : $(e/2)^{-2}$
5681 7596 sq : $3\sqrt{2}/4 + 3\sqrt{3}/4$
5681 8420 rt : (8,7,4,-6)
5681 8550 J_1 : 19/12

5681 8642 $J_0 : 5\sqrt{5}/8$
5682 2468 rt : (5,3,9,-7)
5682 2661 rt : (7,9,2,8)
5682 2973 cu : $1/2 - 3\sqrt{3}$
5682 3070 rt : (2,9,-4,-1)
5682 4936 ln : $\sqrt[3]{11/2}$
5682 8002 $E : 2/25$
5682 8373 rt : (3,5,-5,5)
5682 8603 sr : $2 - 3\sqrt{5}/4$
5682 9213 $J_1 : 21/10$
5683 0581 $J_1 : 5\sqrt[3]{2}/3$
5683 0657 $E : (1/\pi)/4$
5683 1218 $e^x : 9/20$
5683 1832 sq : $7(e+\pi)$
5683 2516 rt : (9,6,6,-7)
5683 2799 sr : $(\pi)^3$
5683 3101 $\ell_{10} : 3\pi^2/8$
5683 3415 $\zeta : (\sqrt[3]{3})^{1/2}$
5683 4343 $10^x : 3 + \sqrt{3}/2$
5683 4863 tanh : $1/\ln(3\pi/2)$
5683 7288 rt : (1,5,-1,-2)
5683 7669 cr : P_{63}
5683 8683 $10^x : 1/2 - \sqrt{5}/3$
5683 9029 rt : (8,8,9,4)
5683 9592 $e^x : e/3 + 4\pi$
5684 1332 $e^x : 3e - 2\pi/3$
5684 2745 rt : (7,7,-8,1)
5684 2935 $\theta_3 : (\pi/2)^{-1/3}$
5684 3285 $s\pi : \sqrt{3}/9$
5684 3609 $J_1 : \ln(1/(3e))$
5684 6442 $E : \ln 2/9$
5684 7007 $E : 1/13$
5684 7580 $\sqrt{2} + \exp(e)$
5685 0228 rt : (5,6,-7,-5)
5685 1579 at : $2\sqrt{5}/7$
5685 2440 $\ln(2/5) - \exp(\sqrt{3})$
5685 2574 rt : (3,8,-2,-2)
5685 4249 cu : $2 + 2\sqrt{2}$
5685 4572 sinh : 13/24
5685 6555 $s\pi : 10\sqrt[3]{3}/3$
5685 7573 rt : (1,9,-4,-5)
5685 9731 $\exp(2/5) \div s\pi(2/5)$
5686 1030 as : 7/13
5686 1343 $J_1 : \sqrt[3]{4}$
5686 1591 ln : 5/24
5686 3772 cr : P_{47}
5686 3937 $E : (\sqrt[3]{2}/3)^3$
5686 4517 $2^x : 3\sqrt{3}/8$
5686 6524 sinh : 18
5686 6741 cu : $\sqrt{3}/4 - 4\sqrt{5}$
5686 7207 sr : $4 + 3\sqrt{3}/2$
5686 9305 $e^x : 4e/3 - 4\pi/3$
5687 1127 id : $(2e)^{-1/3}$
5687 1888 rt : (1,8,3,-2)
5687 2328 ln : $4e/3 + 3\pi$
5687 2930 rt : (8,6,-6,0)
5687 6555 sq : $(\ln 3)^{-3}$
5687 7771 sq : $5\sqrt[3]{5}/4$
5687 7945 rt : (9,6,-7,9)
5687 8364 sq : $4\zeta(3)/3$
5687 9083 $E : 1/14$
5688 0009 rt : (5,1,-5,9)
5688 2527 rt : (5,5,0,7)
5688 2911 $\Psi : \ln(3\pi)$
5688 2960 $E : 2(1/\pi)/9$
5688 3983 $\ell_{10} : 3\sqrt{3} - 2\sqrt{5}/3$
5688 4265 rt : (9,9,-1,-4)
5688 4413 sinh : $(2\pi)^{-1/3}$
5688 4960 rt : (1,7,8,-7)
5688 6896 $J_1 : 2\pi/3$
5688 7385 $\zeta : \ln 2/10$
5688 7823 rt : (5,9,5,5)
5688 7890 $J_0 : 4\pi/9$
5689 0597 rt : (6,9,7,-8)
5689 0681 as : $3/2 - 3e/4$
5689 0788 $E : \ln 2/10$
5689 1302 ln : $4/3 + \sqrt{3}/4$
5689 1410 id : $\sqrt{2} + 2\sqrt{3}/3$
5689 2472 rt : (9,5,-4,-1)
5689 2888 at : $\exp(2\pi)$
5689 4640 rt : (1,3,5,-4)
5689 5236 at : $(1/(3e))^{-3}$
5689 5563 $2^x : \sqrt{3} + \sqrt{5}/3$
5689 6120 ln : $3\sqrt{2} + \sqrt{5}/4$
5689 7477 rt : (7,8,9,-9)
5689 8540 $J_1 : 9\sqrt{2}/8$
5689 8791 $10^x : 3/4 + 4e$
5689 9194 rt : (6,5,4,-5)
5690 2223 $\ln(\sqrt{5}) \times s\pi(1/4)$
5690 3509 $\ell_{10} : 3 + \sqrt{2}/2$
5690 4288 $J_1 : 5(1/\pi)$
5690 4813 $e^x : 10\zeta(3)/7$
5690 4954 $E : 1/15$
5690 6162 rt : (5,2,-1,-1)
5690 8159 rt : (7,4,6,-6)
5690 8350 sq : $3\sqrt{2} - \sqrt{5}/4$
5690 8495 $E : \exp(-e)$
5690 9251 rt : (2,2,7,-5)
5690 9791 $\Psi : (2\pi)^{-1/2}$
5690 9793 $\theta_3 : 7(1/\pi)/8$
5690 9935 $\ell_{10} : \sqrt{2}/4 + 3\sqrt{5}/2$
5691 2978 rt : (3,9,1,-9)
5691 5411 rt : (8,3,8,-7)
5691 6819 $2^x : 13/20$
5691 7271 $erf : 7(1/\pi)/4$
5691 8659 $E : (5/2)^{-3}$
5691 9219 rt : (7,2,-9,8)
5691 9407 $\Psi : 5\pi/7$
5691 9757 rt : (3,1,9,-6)
5692 0356 $E : (1/\pi)/5$
5692 1458 $s\pi : 4\sqrt{2} - 2\sqrt{3}$
5692 1938 cu : $2\sqrt{3}$
5692 2153 $J_1 : 23/11$
5692 2609 $J : \exp(-13\pi/20)$
5692 3122 $J_0 : (e)^{1/3}$
5692 4081 rt : (6,1,1,-2)
5692 6122 $E : 1/16$
5692 6646 $\Psi : \exp(5/3)$
5693 1319 at : 16/25
5693 1321 rt : (7,6,-4,6)
5693 2344 $\ln(2/5) \div \ln(5)$
5693 2930 $10^x : 3 + e/4$
5693 5293 rt : (7,0,3,3)
5693 5392 $\theta_3 : 1/\ln(5)$
5693 6027 $s\pi : 1/4 + 2\sqrt{2}/3$
5693 8470 sr : $4\sqrt{2}/3 + \sqrt{3}/3$
5693 8535 as : $7\ln 2/9$
5694 0074 $e \div \sqrt{3}$
5694 0131 id : $1/3 + \sqrt{5}$
5694 1534 rt : (3,6,-4,1)
5694 2030 $10^x : 1/4 + e$
5694 2846 cu : $1/3 - 4\sqrt{5}/3$
5694 2971 rt : (8,1,5,4)
5694 3289 $\zeta : 10/19$
5694 3611 rt : (9,1,-7,2)
5694 3662 $E : 1/17$
5694 4187 cr : $2e - \pi/2$
5694 5173 $e^x : 2\sqrt{2} + 4\sqrt{5}/3$
5694 7680 cr : $3 + \sqrt{3}/2$
5694 8559 rt : (9,2,7,5)
5694 8757 sq : $3/2 - \sqrt{5}/3$
5694 8977 rt : (6,6,-6,-1)
5695 1751 $2^x : 1 - 2e/3$
5695 2235 $e^x : 3\zeta(3)/8$
5695 6689 rt : (9,9,6,-8)
5695 8359 $E : 1/18$
5695 9469 $2^x : 8\sqrt{3}/3$
5695 9688 rt : (9,7,-8,5)
5696 0182 at : $(3\pi)^3$
5696 1566 $J_1 : 5\sqrt{5}/7$
5696 1852 $t\pi : \sqrt{5}/7$
5696 2807 $\ell_2 : 1 + 4e$
5696 7590 $E : \exp(-(e+\pi)/2)$
5696 7660 $E : (\ln 2/3)^2$
5696 8665 tanh : 11/17
5696 9050 $E : (1/\pi)/6$
5696 9878 $1 + s\pi(\pi)$
5697 0497 rt : (2,9,3,-5)
5697 0795 $E : 1/19$
5697 1714 ln : $2\sqrt{2}/5$
5697 1954 rt : (3,8,5,-6)
5697 2694 sq : $1/\ln(\sqrt{2}\pi/2)$
5697 3128 rt : (4,7,7,-7)
5697 4094 rt : (5,6,9,-8)
5697 5470 ln : $1/3 + 2\sqrt{5}$
5697 5572 rt : (8,9,-8,-6)
5697 8798 $E : (\sqrt{2}\pi)^{-2}$
5697 8830 $\Psi : \sqrt{2} + \sqrt{5} - \sqrt{7}$
5698 0800 $s\pi(\sqrt{5}) \div s\pi(\pi)$
5698 1411 $E : 1/20$
5698 2247 $E : (e)^{-3}$
5698 2620 rt : (1,8,-5,-1)
5698 3330 $\exp(4/5) \div s\pi(1/3)$
5698 4029 rt : (4,4,8,4)
5698 4879 $E : (\ln 3/3)^3$
5698 5378 $\ell_{10} : 1/4 + 2\sqrt{3}$
5698 7727 $\ln(\sqrt{3}) \div s\pi(\sqrt{2})$
5698 9176 cu : $5\pi/8$
5698 9593 $J_1 : 8/5$
5699 0547 $E : 1/21$
5699 2146 ln : $\ln(e+\pi)$
5699 2240 at : $2 - e/2$
5699 2919 rt : (9,5,3,-5)
5699 3484 $10^x : 7\pi^2/9$
5699 3962 rt : (8,6,1,-4)
5699 5283 rt : (7,7,-1,-3)
5699 6078 $E : \ln(\pi/3)$
5699 7008 rt : (6,8,-3,-2)
5699 7136 $J_1 : 25/12$
5699 7653 $2^x : \sqrt{3}/2 - 3\sqrt{5}/4$
5699 8399 $E : (1/\pi)/7$
5699 8465 $E : 1/22$
5699 9198 id : $\sqrt[3]{5}/3$
5699 9358 rt : (5,9,-5,-1)
5699 9802 $\ell_{10} : e/2 + 3\pi/4$
5699 9968 $E : (3\pi/2)^{-2}$
5700 0336 $Ei : \exp(3/2)$
5700 1356 rt : (5,2,-6,5)
5700 2215 at : $4\sqrt[3]{3}/9$
5700 4180 ln : $1/2 + 4\pi$
5700 5372 $E : 1/23$
5700 5445 rt : (1,8,5,-8)
5700 6272 $E : \exp(-\pi)$
5700 6480 sinh : 13/2
5700 8310 $\zeta : 1/\ln(3)$
5700 8858 $\ell_{10} : 3\sqrt{3}/2 + \sqrt{5}/2$
5700 8882 $\ell_2 : 6\sqrt{3}/7$
5700 9053 rt : (2,6,-6,-5)
5701 0328 sq : $2e/3 + 3\pi/2$
5701 1433 $E : 1/24$
5701 1559 $E : (\ln 2/2)^3$
5701 1744 $J_1 : \exp(\sqrt{2}/3)$
5701 2091 $2^x : 1/4 - 3\sqrt{2}/4$
5701 3337 $J_1 : \sqrt{13/3}$
5701 4122 rt : (3,7,-4,-4)
5701 4203 $10^x : 1/3 - \sqrt{3}/3$
5701 4265 $J_1 : 10\sqrt[3]{3}/9$
5701 5083 rt : (8,7,-3,9)
5701 5580 $e^x : 2 + \sqrt{5}/3$
5701 6211 $J_1 : (\ln 2)^{-2}$
5701 6650 $J_1 : 4\zeta(3)/3$
5701 6781 $E : 1/25$
5701 7444 $E : (1/\pi)/8$
5701 7540 $E : ((e+\pi)/2)^{-3}$
5701 7690 ln : $10\sqrt[3]{3}/3$
5701 9307 at : $3\sqrt[3]{5}/8$
5702 1768 rt : (4,8,-2,-3)
5702 4452 $\lambda : \exp(-15\pi/16)$
5702 4823 rt : (2,3,-2,9)
5702 5074 $\Gamma : 1/\ln(1/(2\pi))$
5702 5517 $J_1 : (\ln 2/3)^{-1/2}$
5702 5750 $E : (3)^{-3}$
5702 6295 rt : (6,1,8,-6)
5702 6690 $t\pi : 3\sqrt[3]{3}/7$
5702 8496 rt : (1,7,-9,-7)
5702 8610 $J_1 : \sqrt[3]{9}$
5702 9802 $t\pi : 3e/7$
5703 0499 $E : (1/\pi)/9$
5703 2188 rt : (5,2,6,-5)
5703 2853 ln : $4\zeta(3)$

5703 4230 ℓ_{10} : $1+e$
5703 4622 rt : (5,9,0,-2)
5703 4631 $1/2 \times \exp(\pi)$
5703 4669 E : $(2e)^{-2}$
5703 4781 J_1 : $3\ln 2$
5703 5831 tanh : $(\sqrt[3]{2}/3)^{1/2}$
5703 6199 2^x : $(e+\pi)/9$
5703 7607 10^x : $3/4+e/2$
5703 8777 E : $(\pi)^{-3}$
5703 9690 as : $3\sqrt[3]{2}/7$
5703 9836 E : $(1/\pi)/10$
5703 9990 rt : (8,6,8,-8)
5704 0053 sq : $(2e)^2$
5704 0203 e^x : $3\sqrt{3}+3\sqrt{5}$
5704 0216 ζ : $2(1/\pi)/9$
5704 1780 cr : $\sqrt{15}$
5704 2990 rt : (3,8,-8,6)
5704 4173 J_1 : $6\sqrt{3}/5$
5704 4689 sr : $1/\ln(3/2)$
5704 4857 rt : (9,6,4,2)
5704 4922 rt : (7,7,6,-7)
5704 6322 E : $(e+\pi)^{-2}$
5704 8572 ln : $3\ln 2/10$
5704 8676 $s\pi$: $2\pi^2/9$
5705 0807 rt : (6,8,4,-6)
5705 3855 λ : $1/19$
5705 4433 E : $(2\pi)^{-2}$
5705 4485 ln : $13/23$
5705 5359 θ_3 : $(2\pi/3)^{-1/3}$
5705 7723 e^x : $14/11$
5705 7951 rt : (5,9,2,-5)
5705 7999 $t\pi$: $7\pi/9$
5705 9712 rt : (3,3,-9,8)
5706 0598 cr : $1/4+4e/3$
5706 0684 rt : (7,7,-5,2)
5706 1869 2^x : $3/2-4\sqrt{3}/3$
5706 2128 rt : (9,8,-9,1)
5706 2911 10^x : $1/2+\sqrt{3}/4$
5706 3390 $3/5 \times s\pi(2/5)$
5706 3417 2^x : $4+2\sqrt{3}$
5706 5439 $t\pi$: $5\sqrt{3}/4$
5706 6458 E : $\exp(-4)$
5706 6805 rt : (4,2,-8,-2)
5707 0044 E : $(4)^{-3}$
5707 0752 E : $(1/(3e))^2$
5707 1562 at : $\exp(3\pi)$
5707 2029 rt : (8,5,2,1)
5707 3199 sr : $(3\pi)^{-1/2}$
5707 3658 E : $(\ln 2/3)^3$
5707 4199 E : $\exp(-\sqrt{2}\pi)$
5707 4526 E : $(\sqrt{2}\pi)^{-3}$
5707 4655 E : $(3\pi)^{-2}$
5707 5056 rt : (1,5,-3,1)
5707 6046 E : $(3\pi/2)^{-3}$
5707 6463 E : $\exp(-3\pi/2)$
5707 7119 E : $(5)^{-3}$
5707 7849 E : $\exp(-5)$
5707 8111 E : $(2e)^{-3}$
5707 8662 E : $(e+\pi)^{-3}$
5707 8888 E : $\exp(-2e)$
5707 8994 E : $(2\pi)^{-3}$
5707 9313 E : $\exp(-(e+\pi))$
5707 9399 sq : $3e/4-\pi/4$
5707 9495 E : $\exp(-2\pi)$
5707 9499 E : $(1/(3e))^3$
5707 9576 E : $(3\pi)^{-3}$
5707 9632 id : $\pi/2$
5707 9688 K : $(3\pi)^{-3}$
5707 9766 K : $(1/(3e))^3$
5707 9769 K : $\exp(-2\pi)$
5707 9952 K : $\exp(-(e+\pi))$
5708 0270 K : $(2\pi)^{-3}$
5708 0377 K : $\exp(-2e)$
5708 0469 erf : $\sqrt{5}/4$
5708 0602 K : $(e+\pi)^{-3}$
5708 0909 e^x : $9\sqrt{2}/10$
5708 0946 rt : (8,4,-9,7)
5708 1153 K : $(2e)^{-3}$
5708 1366 rt : (4,8,-9,-7)
5708 1410 J_1 : $\ln(5)$
5708 1415 K : $\exp(-5)$
5708 2146 K : $(5)^{-3}$
5708 2553 rt : (5,9,9,-9)
5708 2801 K : $\exp(-3\pi/2)$
5708 2924 ℓ_2 : $3\sqrt{3}+\sqrt{5}/3$
5708 3209 e^x : $1/2-3\sqrt{2}/4$
5708 3218 K : $(3\pi/2)^{-3}$
5708 3564 sinh : $e/5$
5708 4252 rt : (9,1,0,-2)
5708 4559 rt : (6,3,-5,8)
5708 4610 K : $(3\pi)^{-2}$
5708 4738 K : $(\sqrt{2}\pi)^{-3}$
5708 5065 K : $\exp(-\sqrt{2}\pi)$
5708 5607 K : $(\ln 2/3)^3$
5708 7306 rt : (2,8,-5,1)
5708 8513 K : $(1/(3e))^2$
5708 9221 K : $(4)^{-3}$
5708 9866 rt : (3,4,9,-7)
5709 0651 rt : (4,2,-1,9)
5709 1278 J_1 : $\sqrt{2}-\sqrt{6}+\sqrt{7}$
5709 2808 K : $\exp(-4)$
5709 2874 rt : (1,6,-2,-1)
5709 3764 sq : $2+4\sqrt{3}/3$
5709 5786 rt : (5,9,-7,-6)
5709 6472 ℓ_{10} : $\sqrt{2}+4\sqrt{3}/3$
5709 8375 rt : (6,3,-9,-5)
5709 9384 $\exp(\pi)-s\pi(\pi)$
5710 0198 rt : (3,7,-3,-3)
5710 1056 rt : (9,8,-4,-2)
5710 1308 $\exp(2) \div s\pi(1/5)$
5710 3110 rt : (2,5,7,-6)
5710 4838 K : $(2\pi)^{-2}$
5710 5693 id : $1/\ln(\sqrt[3]{4}/3)$
5710 6320 sr : $\sqrt{2}+3\sqrt{3}$
5710 6938 rt : (8,2,-2,-1)
5710 8463 rt : (9,1,7,-6)
5710 9863 rt : (1,1,9,5)
5711 2953 K : $(e+\pi)^{-2}$
5711 2979 $t\pi$: $9\sqrt[3]{2}/4$
5711 3459 10^x : $1+\sqrt{2}/4$
5711 4947 ζ : $1/14$
5711 5028 rt : (9,8,3,-6)
5711 5184 ln : $3+2e/3$
5711 6639 rt : (8,5,9,5)
5711 8796 rt : (8,9,-6,-1)
5711 9444 K : $(1/\pi)/10$
5712 0142 rt : (1,6,5,-5)
5712 0503 K : $(\pi)^{-3}$
5712 2958 ln : $\sqrt{2}/3+3\sqrt{3}/4$
5712 3495 ζ : $10\ln 2/7$
5712 3574 rt : (8,2,5,-5)
5712 3636 rt : (3,9,-5,4)
5712 4485 θ_3 : $\sqrt{5}/8$
5712 4614 K : $(2e)^{-2}$
5712 4966 as : $\sqrt{3}/3-\sqrt{5}/2$
5712 5346 ln : $e+2\pi/3$
5712 7608 rt : (8,9,1,-5)
5712 8159 id : $\sqrt[3]{17}$
5712 8292 Ei : $\sqrt{14/3}$
5712 8789 K : $(1/\pi)/9$
5713 1695 rt : (8,9,8,-9)
5713 2659 10^x : $3/4+4\sqrt{3}/3$
5713 3177 rt : (8,8,-4,5)
5713 3375 ℓ_{10} : $5\sqrt{5}/3$
5713 3524 Ei : $(\sqrt{2}/3)^3$
5713 3542 K : $(3)^{-3}$
5713 3747 at : $9/14$
5713 4610 tanh : $3\sqrt{3}/8$
5713 4840 id : $10\sqrt{2}/9$
5713 5024 sinh : $\pi^2/8$
5713 5347 sinh : $1/\ln(2\pi)$
5713 7389 J_1 : $21/13$
5713 8443 e^x : $17/18$
5714 0325 $\ln(\sqrt{2}) \times \exp(1/2)$
5714 1761 K : $((e+\pi)/2)^{-3}$
5714 1858 K : $(1/\pi)/8$
5714 2521 K : $1/25$
5714 2764 Ei : $9\pi/4$
5714 2857 id : $4/7$
5714 3971 as : $3\sqrt[3]{3}/8$
5714 4587 10^x : $8\sqrt[3]{3}$
5714 5359 J_1 : $8\sqrt{2}/7$
5714 6437 sq : $3\sqrt[3]{2}/5$
5714 7749 K : $(\ln 2/2)^3$
5714 7876 K : $1/24$
5715 1039 sq : $4/3-\sqrt{3}/3$
5715 1312 Γ : $1/\sqrt[3]{24}$
5715 1438 $t\pi$: $(e+\pi)^{-1/2}$
5715 3044 K : $\exp(-\pi)$
5715 3446 2^x : $15/23$
5715 3945 K : $1/23$
5715 4187 sq : $4\sqrt{2}/3+\sqrt{3}/2$
5715 4988 J_1 : $\exp(\sqrt[3]{3}/3)$
5715 5510 J_1 : $7\ln 2/3$
5715 5842 rt : (6,7,0,6)
5715 6929 ζ : $17/25$
5715 7494 sq : $3\sqrt{3}/2-3\sqrt{5}/2$
5715 8657 rt : (6,4,8,-7)
5715 8682 erf : $4\sqrt[3]{2}/9$
5715 9272 cu : $\sqrt{3}/2-\sqrt{5}$
5715 9357 K : $(3\pi/2)^{-2}$
5716 0863 K : $1/22$
5716 0893 rt : (1,8,-8,-7)
5716 0928 K : $(1/\pi)/7$
5716 0935 e^x : $\sqrt{2}/3+3\sqrt{5}/4$
5716 1576 erf : $14/25$
5716 1723 rt : (6,7,-7,-5)
5716 3253 K : $\ln(\pi/3)$
5716 5236 id : $(\sqrt[3]{4}/3)^{-2}$
5716 6996 tanh : $13/20$
5716 8313 rt : (6,4,1,-3)
5716 8793 K : $1/21$
5717 0748 $\ln(3/4)-\exp(1/4)$
5717 1137 rt : (5,3,-7,1)
5717 1665 e^x : $\exp(-\sqrt[3]{4}/2)$
5717 2252 cu : $2-\sqrt{2}/3$
5717 2299 J : $\sqrt[3]{2}/8$
5717 3622 10^x : $3+\sqrt{5}/4$
5717 4456 sr : $\sqrt{3}+\sqrt{5}+\sqrt{7}$
5717 4472 K : $(\ln 3/3)^3$
5717 4666 rt : (1,8,-1,-3)
5717 6672 rt : (2,9,0,3)
5717 6781 rt : (4,1,-7,-7)
5717 6939 2^x : $4\sqrt{2}-2\sqrt{3}$
5717 7084 sr : $\exp(-\sqrt{5}/2)$
5717 7109 K : $(e)^{-3}$
5717 7945 K : $1/20$
5717 7947 rt : (5,5,6,-6)
5717 8791 2^x : $1+4\sqrt{3}$
5717 8912 J_1 : $9\sqrt[3]{2}/7$
5718 0541 rt : (3,4,-8,4)
5718 0553 cr : $10e/7$
5718 0564 K : $(\sqrt{2}\pi)^{-2}$
5718 1341 cu : $\sqrt{13/2}$
5718 1981 rt : (6,3,5,3)
5718 3550 rt : (2,9,-6,-6)
5718 4034 rt : (9,9,-3,8)
5718 5967 2^x : $3\sqrt{3}+\sqrt{5}/2$
5718 7096 cu : $2e+\pi/3$
5718 7459 ℓ_{10} : $3/4+4\sqrt{5}/3$
5718 8482 e^x : $2\sqrt{2}/3+3\sqrt{3}$
5718 8583 K : $1/19$
5718 9387 $\pi-s\pi(\pi)$
5719 0332 K : $(1/\pi)/6$
5719 1724 K : $(\ln 2/3)^2$
5719 1794 K : $\exp(-(e+\pi)/2)$
5719 2166 ζ : $\sqrt{2}+\sqrt{5}-\sqrt{6}$
5719 4754 ℓ_{10} : $2+\sqrt{3}$
5719 5164 10^x : $4-3\sqrt{2}$
5719 5644 ℓ_2 : $3/4+3\sqrt{3}$
5719 6343 rt : (5,8,-9,9)
5719 6640 $t\pi$: $1/4-3\sqrt{5}$
5719 7154 Γ : $5\pi^2/2$
5719 9326 rt : (1,4,0,-6)
5720 0643 rt : (1,6,4,9)
5720 0823 at : $3-3\pi/4$
5720 1046 K : $1/18$
5720 1160 $t\pi$: $1/2+\sqrt{5}/2$
5720 1324 rt : (3,7,9,-8)
5720 2021 rt : (4,6,4,7)

5720 2740 $J_1 : 9\sqrt[3]{3}/8$
5720 3135 as : $(e/2)^{-2}$
5720 3466 rt : (5,5,-1,-2)
5720 6282 tπ : $3\sqrt{2}/2+\sqrt{5}/4$
5720 7013 sr : $2+\sqrt{2}/3$
5720 7094 cu : $1/4+3\pi$
5720 8250 rt : (6,4,-6,1)
5721 0779 $J_1 : 10\sqrt[3]{3}/7$
5721 1451 $\ln(\sqrt{2})+\exp(4/5)$
5721 1851 $\pi\div\exp(1/5)$
5721 1880 $1/4-\exp(3/5)$
5721 2366 cr : $2+4\sqrt{2}/3$
5721 2743 $e^x : 8\sqrt[3]{5}/9$
5721 4308 rt : (9,4,7,-7)
5721 5162 tπ : $6(1/\pi)/5$
5721 5779 K : 1/17
5721 6445 rt : (7,8,1,9)
5721 6600 $Ei : \sqrt[3]{5}/7$
5721 9706 tπ : $3\sqrt{2}/2-3\sqrt{5}$
5721 9909 $\Psi : 5e$
5722 2305 rt : (1,6,-3,5)
5722 2395 rt : (2,9,1,-2)
5722 3344 rt : (2,8,7,-7)
5722 3990 rt : (4,1,8,5)
5722 4064 $2^x : 4+\sqrt{5}/4$
5722 4372 rt : (1,9,-6,-5)
5722 4835 J_1 : 13/8
5722 5469 sq : $4-e/3$
5722 5944 sr : $1/3+2\pi$
5722 7609 $\theta_3 : 2\sqrt[3]{2}/9$
5723 1427 θ_3 : 7/25
5723 2911 rt : (9,6,-9,6)
5723 3368 K : 1/16
5723 3976 $\Gamma : \ln 2/2$
5723 5712 cr : 3/16
5723 6494 ln : $(\pi)^{1/2}$
5723 7062 rt : (8,5,5,-6)
5723 8477 $2^x : 5\sqrt{3}$
5723 9048 sπ : $3\sqrt{2}/4+\sqrt{5}/3$
5723 9151 $K : (1/\pi)/5$
5724 0680 $e^x : 4(1/\pi)$
5724 0810 tanh : $(e+\pi)/9$
5724 0853 $K : (5/2)^{-3}$
5724 1262 rt : (7,8,-6,-2)
5724 1476 $\ell_{10} : 3/2+\sqrt{5}$
5724 1857 as : 13/24
5724 1952 cr : P_9
5724 2466 $J_1 : \exp(\sqrt[3]{3}/2)$
5724 2609 rt : (4,8,-9,-8)
5724 3197 sq : $3-3e$
5724 4102 $Ei : \ln(\sqrt[3]{5})$
5724 4487 rt : (8,5,-5,-3)
5724 4951 $\Gamma : 1/\ln(e/2)$
5724 5905 rt : (3,3,-8,-5)
5724 6627 rt : (5,2,3,2)
5724 7079 $2^x : 1/\ln(e/2)$
5724 8139 $J_1 : \sqrt{2}-\sqrt{5}+\sqrt{6}$
5724 9377 rt : (6,7,8,-8)
5725 0327 rt : (1,9,5,-6)
5725 1050 $K : \exp(-e)$
5725 1570 rt : (9,4,0,-3)
5725 2300 $\zeta : 4\sqrt[3]{2}/5$
5725 4603 K : 1/15
5725 4646 Ψ : 9/4
5725 5134 rt : (4,6,-3,-1)
5726 1342 $\zeta : 5\sqrt[3]{3}/8$
5726 2957 rt : (8,2,9,6)
5726 4028 $\ell_{10} : P_{65}$
5726 4797 rt : (7,6,3,-5)
5726 5144 $\ell_{10} : P_{66}$
5726 5405 rt : (2,7,2,9)
5726 5902 $J_1 : (\ln 2/3)^{-1/3}$
5726 6100 $Li_2 : \exp(-1/\sqrt{2})$
5726 6104 $e^x : 3\sqrt[3]{2}/4$
5726 7452 cr : $2-2e/3$
5726 8817 $K : \ln 2/10$
5727 0134 $\Gamma : \ln(\sqrt[3]{2}/2)$
5727 1346 $e^x : 3/2-\pi/3$
5727 1625 $J_1 : \sqrt[3]{13/3}$
5727 2719 as : $(2\pi)^{-1/3}$
5727 3222 rt : (3,0,6,4)
5727 3643 rt : (5,8,6,-7)
5727 3764 rt : (7,5,-5,7)
5727 3982 rt : (2,8,0,-3)
5727 4181 cu : $\sqrt{3}-2\sqrt{5}$
5727 5609 $erf : \exp(-\sqrt{3}/3)$
5727 6060 id : $1/3-e/3$
5727 6675 $K : 2(1/\pi)/9$
5727 7104 rt : (1,8,6,1)
5727 7205 ln : $3\sqrt{2}+\sqrt{3}/3$
5727 7212 $J_1 : 3e/5$
5728 0566 K : 1/14
5728 3568 $J_1 : 6\sqrt[3]{5}/5$
5728 4950 rt : (2,4,-9,8)
5728 5415 rt : (1,2,9,-6)
5728 6557 J_0 : 25/18
5728 7921 $2^x : 7e\pi/8$
5728 8509 rt : (8,9,-5,1)
5728 9551 rt : (9,7,7,-8)
5729 0709 sinh : 6/11
5729 0883 at : $1/\ln(3\pi/2)$
5729 2061 rt : (5,9,-8,5)
5729 2901 cr : $3/4+\pi$
5729 3925 rt : (8,5,-2,-2)
5729 4641 $J_1 : \sqrt{8/3}$
5729 5365 $e^x : 2\sqrt{2}/3+\sqrt{3}/3$
5729 5773 $K : (\sqrt[3]{2}/3)^3$
5729 5779 id : $9(1/\pi)/5$
5729 7152 tπ : $5\pi/6$
5729 7818 rt : (9,3,4,3)
5729 9342 rt : (6,7,1,-4)
5730 0635 $Ei : 7\sqrt[3]{5}/5$
5730 0931 sr : $10\sqrt{3}/7$
5730 2787 rt : (4,9,4,-6)
5730 3148 rt : (7,1,7,5)
5730 5172 rt : (2,5,-9,-6)
5730 5344 rt : (1,7,-5,-5)
5730 6401 $\ell_{10} : \sqrt{14}$
5730 9807 $e^x : e/6$
5731 1158 sq : $3\sqrt{2}-2\sqrt{5}/3$
5731 1611 $10^x : 4+3e/4$
5731 2775 K : 1/13
5731 3161 tanh : 15/23
5731 3218 id : $\sqrt{2}/2+\sqrt{3}/2$
5731 3234 sπ : $1/4+4\sqrt{5}$
5731 3343 $K : \ln 2/9$
5731 3670 $10^x : 2+\pi$
5731 5000 rt : (8,8,5,-7)
5731 7314 sr : $7\sqrt{2}/4$
5731 7806 ζ : 25/12
5731 8274 rt : (6,6,-9,-6)
5731 9575 $\Psi : -\sqrt{3}-\sqrt{5}+\sqrt{7}$
5731 9762 $\ell_2 : (2e/3)^3$
5732 2258 $Ei : 10\sqrt[3]{2}/3$
5732 2403 cr : P_{36}
5732 2504 $4/5\div\exp(1/3)$
5732 3289 J_1 : 18/11
5732 4555 $\zeta : 5e/9$
5732 4584 $\ln(2/5)\times\exp(\sqrt{5})$
5732 5795 tπ : $6e\pi/7$
5732 6628 $2^x : 4/3-e/4$
5732 8748 $\ell_{10} : 2\zeta(3)/9$
5732 9202 $K : (1/\pi)/4$
5733 1244 rt : (1,9,-2,-2)
5733 1868 K : 2/25
5733 2979 2^x : 15/11
5733 3126 $2^x : 2\pi^2/9$
5733 3561 cu : $4+\pi/3$
5733 4044 $\ln(5)-s\pi(\sqrt{2})$
5733 7258 $erf : \ln(\sqrt[3]{5}/3)$
5733 7876 cu : $2\sqrt{2}+2\sqrt{5}/3$
5733 8441 rt : (3,8,-2,-7)
5734 0736 $\ell_2 : 1/3+2\sqrt{3}/3$
5734 1286 rt : (5,4,-1,8)
5734 1425 $\sqrt{2}\times\ln(2/3)$
5734 1574 $Ei : 7\sqrt[3]{2}/10$
5734 1771 $\ell_2 : 3\sqrt{3}+3\sqrt{5}$
5734 1785 sπ : $10\sqrt{5}/7$
5734 3199 cu : $\sqrt{3}+3\sqrt{5}/2$
5734 3552 rt : (5,8,-1,-3)
5734 4480 10^x : 19/5
5734 4748 $\lambda : (1/\pi)/6$
5734 5064 rt : (7,9,3,-6)
5734 5238 $K : \exp(-5/2)$
5734 7605 sπ : $7e/5$
5734 7849 rt : (4,2,8,-6)
5734 8618 at : $4-3\sqrt{5}/2$
5734 9310 $\ell_{10} : 3+\sqrt{5}/3$
5735 0070 tπ : $(\ln 3/2)^3$
5735 0326 rt : (9,7,0,-4)
5735 1239 $\Gamma : 2\sqrt{2}/5$
5735 2129 rt : (7,6,-4,-1)
5735 3293 rt : (8,1,9,-7)
5735 3410 K : 1/12
5735 5194 $e-\ln(\pi)$
5735 6843 rt : (6,0,5,4)
5736 0284 $\Gamma : \sqrt{3}/5$
5736 0774 rt : (4,9,-1,6)
5736 3539 rt : (5,5,-8,2)
5736 7151 sinh : $2(e+\pi)/7$
5736 7259 cr : $9\sqrt{3}/4$
5736 7445 erf : 9/16
5736 8123 $\Gamma : 7e/3$
5736 9499 rt : (9,4,-7,1)
5736 9508 $\exp(4/5)\times s\pi(1/4)$
5737 1511 rt : (4,0,-8,3)
5737 3689 rt : (8,2,2,2)
5737 4275 id : $e/2-\pi/4$
5737 4288 $10^x : 1+\sqrt{3}$
5737 5342 cr : $\exp(-5/3)$
5737 5686 $K : \ln 2/8$
5737 5999 $J_1 : \sqrt[3]{17/2}$
5737 7168 J_1 : 23/14
5737 7839 K : 2/23
5737 8018 cu : $4-2\sqrt{2}/3$
5737 8311 id : $2\sqrt{2}+\sqrt{5}/3$
5737 8627 sr : $3+4e/3$
5737 9841 sπ : $1/3+2\sqrt{5}$
5738 1119 rt : (6,8,-1,2)
5738 1392 sr : $e/3+\pi/2$
5738 1644 $\Gamma : 1/\ln(e+\pi)$
5738 1750 tπ : $(e/2)^{1/2}$
5738 2474 $\Psi : (e/2)^{-3}$
5738 4109 rt : (1,5,9,-7)
5738 4223 $J_1 : \sqrt{5}+\sqrt{6}-\sqrt{7}$
5738 6056 $J_1 : (\sqrt{2}\pi)^{1/3}$
5738 7317 at : $1-\sqrt{2}/4$
5738 7432 rt : (3,6,-8,-6)
5738 9266 $2^x : \sqrt{2}/4-2\sqrt{3}/3$
5738 9429 sinh : $\sqrt{20/3}$
5739 1632 rt : (5,1,-7,6)
5739 1683 $\Psi : \zeta(9)$
5739 2140 $\Gamma : \exp(-\sqrt[3]{5}/3)$
5739 2996 $J_1 : 3e/4$
5739 3871 rt : (7,7,-7,-5)
5739 4036 $J_1 : \pi^2/6$
5739 4268 $1/5-s\pi(e)$
5739 4951 rt : (8,8,-2,-3)
5739 5148 rt : (7,2,7,-6)
5739 5213 $K : (\sqrt{5})^{-3}$
5739 5359 ln : $\sqrt{2}/4+2\sqrt{5}$
5739 5874 tπ : $3\sqrt{3}/2+\sqrt{5}$
5739 6861 $Ei : 10(e+\pi)/3$
5739 7795 rt : (1,6,-9,3)
5739 7801 sr : $\sqrt{3}+\sqrt{5}/3$
5740 0212 at : $\sqrt{2}/3-\sqrt{5}/2$
5740 0320 rt : (2,8,-7,-8)
5740 1320 $\zeta : (\sqrt[3]{2}/3)^3$
5740 1727 rt : (3,3,6,-5)
5740 1913 rt : (7,9,0,5)
5740 2118 rt : (4,9,-3,-2)
5740 3126 ℓ_{10} : 4/15
5740 3480 $J_0 : \sqrt[3]{8/3}$
5740 4350 tπ : $\exp(5)$
5740 5166 rt : (6,2,-6,9)
5740 5694 K : 1/11
5740 5958 $K : 2(1/\pi)/7$
5740 6612 sπ : $3\sqrt{2}/4-\sqrt{3}/2$
5740 7947 $K : (\sqrt{2}\pi/2)^{-3}$
5740 9337 Ψ : 13/22

5741 0126 id : $4\sqrt{3}-3\sqrt{5}/2$
5741 2379 Ei : $\sqrt{2}+\sqrt{5}+\sqrt{7}$
5741 3158 sq : $2-\sqrt{5}/3$
5741 4124 rt : (1,7,6,-4)
5741 4613 rt : (8,4,9,-8)
5741 4939 rt : (9,0,4,4)
5741 7098 2^x : $3\sqrt{2}+3\sqrt{5}/2$
5741 8833 rt : (2,5,-6,6)
5742 0878 J_1 : $8\sqrt[3]{3}/7$
5742 2933 id : $\sqrt{3}/2+3\sqrt{5}$
5742 3412 rt : (4,5,8,-7)
5742 4337 J_1 : $(e)^{1/2}$
5742 5145 rt : (5,4,-8,-3)
5742 5604 rt : (6,9,-9,-7)
5742 6580 J_0 : $2\ln 2$
5742 8674 cu : $\sqrt{2}/2-\sqrt{5}$
5742 8713 K : $\ln(\ln 3)$
5742 9539 rt : (1,9,4,3)
5743 0331 2^x : $4+4\sqrt{3}$
5743 0483 at : 11/17
5743 1273 ln : $3/4+3e/2$
5743 2036 sinh : 21/17
5743 2198 rt : (5,1,3,-3)
5743 3772 J_1 : $7\sqrt{2}/6$
5743 6170 e^x : $4/3+3\sqrt{5}/2$
5743 7033 rt : (6,7,-6,0)
5743 7415 sq : $4-4\sqrt{3}$
5743 7651 K : 2/21
5743 8524 $t\pi$: $\exp(e+\pi)$
5743 9579 K : $3(1/\pi)/10$
5743 9963 Ψ : $5\zeta(3)/6$
5744 1999 J_1 : $\sqrt[3]{9/2}$
5744 2034 Ei : $9\zeta(3)/2$
5744 3785 sr : $2/3+2e/3$
5744 4290 rt : (7,6,-6,3)
5744 4608 ζ : $(e+\pi)/7$
5744 6205 rt : (5,5,-4,-3)
5744 7310 rt : (1,8,9,-8)
5744 7678 rt : (6,3,5,-5)
5744 8974 sq : $\sqrt{2}/4+2\sqrt{3}$
5744 9182 rt : (8,7,-5,6)
5744 9829 ζ : $\sqrt{13/3}$
5745 0581 rt : (3,9,-7,-3)
5745 1888 rt : (7,9,-4,-2)
5745 2076 ln : $2+2\sqrt{2}$
5745 2703 rt : (9,8,-4,9)
5745 2932 Ei : $9\ln 2/5$
5745 3145 rt : (2,8,5,7)
5745 3376 Ei : 15/17
5745 3891 J_1 : $(\ln 3/3)^{-1/2}$
5745 4531 $\sqrt{2}\div\ln(\sqrt{3})$
5745 4691 2^x : $2/3+2e/3$
5745 4836 rt : (7,5,7,-7)
5745 5532 sq : $2\sqrt{2}+\sqrt{5}/2$
5745 6712 e^x : $4\sqrt{2}/3+4\sqrt{3}$
5745 8961 rt : (8,7,9,-9)
5745 9992 sq : $4\sqrt{3}-3\sqrt{5}/4$
5746 0947 rt : (4,7,-6,-5)
5746 1841 J_0 : $4\sqrt{3}/5$
5746 3611 K : $\ln(e/3)$
5746 4983 id : $\sqrt{2}/4-4\sqrt{3}$
5746 5544 rt : (3,9,-8,-7)
5746 6818 K : $\ln 2/7$
5747 1129 $\ln(3/5)+\exp(3)$
5747 1742 sinh : $4\pi^2/3$
5747 2210 e^x : $(\sqrt{5}/2)^{-1/2}$
5747 3224 2^x : $1/2-3\sqrt{3}/4$
5747 3393 ζ : $(\ln 2)^{-2}$
5747 4307 rt : (4,8,8,-8)
5747 4462 Γ : 13/23
5747 4556 K : 1/10
5747 5431 J_1 : $\exp(1/\sqrt{2})$
5747 6931 cu : $\sqrt{2}/2+2\sqrt{3}$
5747 8686 $t\pi$: P_1
5747 8739 as : $e/5$
5747 9102 cu : P_{25}
5748 0827 J_1 : $\sqrt[3]{25/3}$
5748 1432 rt : (6,7,9,4)
5748 2536 rt : (9,3,4,-5)
5748 3602 2^x : $\ln(2\pi)$
5748 4258 rt : (4,7,-5,9)
5748 5121 K : $(\pi)^{-2}$
5748 6465 rt : (8,1,2,-3)
5748 8033 cr : $3+e/3$
5748 8701 cr : $\sqrt[3]{5}/9$
5748 9090 Li_2 : $\pi^2/10$
5749 0131 id : $5\sqrt[3]{2}/4$
5749 2895 cu : $\sqrt{2}+\sqrt{5}+\sqrt{7}$
5749 4152 $\exp(e)\div\exp(1)$
5749 4606 cr : $2(e+\pi)/3$
5749 4770 rt : (2,8,-7,1)
5749 5892 10^x : 7π
5749 6397 rt : (9,7,-7,0)
5749 6691 rt : (7,8,7,-8)
5749 6771 cu : $4e+2\pi/3$
5749 7572 rt : (7,1,0,1)
5750 0582 at : $(\sqrt[3]{2}/3)^{1/2}$
5750 3110 $\ln(\pi)-s\pi(\pi)$
5750 3325 rt : (6,6,5,-6)
5750 3580 λ : $\exp(-14\pi/15)$
5750 4854 rt : (5,9,7,2)
5750 5229 Ψ : $\sqrt{2}-\sqrt{6}-\sqrt{7}$
5750 5449 $s\pi$: $4\sqrt{2}/3+4\sqrt{3}/3$
5750 6090 rt : (1,7,-5,-5)
5750 6251 erf : $(\pi)^{-1/2}$
5751 0357 rt : (7,9,-7,-6)
5751 1204 rt : (1,8,8,-9)
5751 3258 K : $(\sqrt{2}/3)^3$
5751 3968 rt : (8,8,-5,-4)
5751 4720 Li_2 : $2\sqrt{3}/7$
5751 5548 rt : (5,4,3,-4)
5751 5574 $t\pi$: $4\sqrt{2}-2\sqrt{5}/3$
5751 6831 sq : $5e\pi/3$
5751 7126 J_0 : 18/13
5751 7489 K : 2/19
5751 8629 Ψ : $1/\ln(2e)$
5752 0099 10^x : $5\sqrt[3]{2}/4$
5752 0475 as : $2e/3-3\pi/4$
5752 0937 rt : (3,9,6,7)
5752 1272 sq : $1/3+3\sqrt{3}$
5752 1326 rt : (8,5,-9,2)
5752 4551 K : $(1/\pi)/3$
5752 5052 rt : (1,8,-4,8)
5752 5234 rt : (2,9,-2,-6)
5752 6550 rt : (9,6,4,-6)
5752 7003 rt : (6,8,-8,-9)
5752 8104 sq : $1+2\pi/3$
5752 8194 cr : $2e\pi$
5753 0272 J_1 : $7\sqrt{3}/6$
5753 0941 rt : (2,5,-3,-3)
5753 1113 K : $\exp(-\sqrt{5})$
5753 2325 as : $1/\ln(2\pi)$
5753 2810 e^x : $(\sqrt[3]{4}/3)^{-2}$
5753 3005 $\ln(\sqrt{5})\div\ln(3/5)$
5753 3330 J_1 : $10\sqrt{2}/7$
5753 5164 ℓ_2 : $4e+\pi/3$
5753 6414 ln : 9/16
5753 6958 cr : 4/21
5753 7973 rt : (1,6,-5,-8)
5753 8182 rt : (9,7,8,4)
5753 8512 rt : (4,2,-8,-8)
5754 0545 cu : $3e+\pi/4$
5754 0915 rt : (6,9,5,9)
5754 1692 J_1 : $7\sqrt[3]{3}/5$
5754 4605 K : $\exp(-\sqrt{2}\pi/2)$
5754 4751 $s\pi$: $4e/3+\pi/2$
5754 5257 sr : $3/4+\sqrt{3}$
5754 5554 rt : (8,4,2,-4)
5754 5710 e^x : 5/11
5754 6143 sq : $2/3+4\pi/3$
5754 6594 2^x : $1/4-\pi/3$
5754 6659 cu : $6\ln 2/5$
5754 6943 Li_2 : $5\ln 2/7$
5754 8032 K : $(2\pi/3)^{-3}$
5754 9964 ζ : $(\ln 2/3)^{-1/2}$
5755 0765 cu : $\sqrt{24}$
5755 2063 $s\pi$: $7\zeta(3)/3$
5755 3396 $\ln(2/5)+\exp(2/5)$
5755 3421 rt : (3,6,-6,-4)
5755 7370 Γ : $5\sqrt{5}/2$
5755 7492 rt : (9,9,4,-7)
5755 9897 J_1 : 5/3
5756 2567 10^x : $2\sqrt{2}/3-\sqrt{5}/3$
5756 2912 rt : (5,8,4,6)
5756 3312 e^x : $4/3-4\sqrt{2}/3$
5756 5103 J_1 : $8\sqrt[3]{2}/5$
5756 6088 rt : (6,4,9,5)
5756 7073 10^x : $e/3+\pi$
5756 7314 rt : (5,8,-4,-4)
5756 7842 K : 1/9
5756 7884 $s\pi$: $1/3+\sqrt{2}/3$
5756 8010 rt : (6,6,-1,7)
5756 9993 cu : 9π
5757 0410 ln : $3\sqrt{3}/2+\sqrt{5}$
5757 0447 rt : (7,4,-6,8)
5757 1927 K : $\ln(\sqrt{5}/2)$
5757 2157 rt : (5,8,-8,1)
5757 2739 $s\pi$: $\ln(\sqrt{5})$
5757 3310 λ : $(\ln 2/3)^2$
5757 4536 e^x : $10(1/\pi)/7$
5757 4785 $t\pi$: 8/25
5757 4806 ℓ_2 : $(\sqrt{2}\pi/2)^{1/2}$
5757 5512 ζ : $\sqrt[3]{9}$
5757 5728 e^x : $9e\pi/4$
5757 6538 sq : $3\sqrt{2}/2-\sqrt{3}/2$
5757 7347 rt : (1,5,2,-3)
5757 7770 J_1 : $1/\ln(\ln 3/2)$
5758 1935 rt : (8,7,2,-5)
5758 2027 ℓ_{10} : $e+\pi/3$
5758 2390 cr : $7\sqrt{5}/4$
5758 4789 λ : $\exp(-(e+\pi)/2)$
5758 6130 e^x : $4+\sqrt{2}$
5758 6268 rt : (7,2,0,-2)
5758 6531 J_1 : $(1/(3e))^{-1/3}$
5758 7944 cr : $3/4-\sqrt{5}/4$
5758 8238 cr : $3(1/\pi)/5$
5758 8378 Ψ : 5/17
5758 8467 rt : (6,3,-7,5)
5758 8582 rt : (2,6,-3,4)
5758 9036 J_1 : $9\sqrt{5}/10$
5759 0119 cu : $\exp(-1/(2e))$
5759 0560 erf : 13/23
5759 1341 J_1 : $\sqrt[3]{14/3}$
5759 3337 rt : (9,4,8,5)
5759 3460 rt : (5,5,-2,4)
5759 7402 id : $1/3+3\sqrt{2}$
5759 7668 2^x : $3+\sqrt{3}$
5759 9979 rt : (2,8,-3,-4)
5760 0000 cu : 13/5
5760 0154 ℓ_2 : $2\sqrt{5}/3$
5760 3504 10^x : $3\sqrt{2}+4\sqrt{5}/3$
5760 3651 sinh : $6\sqrt[3]{3}/7$
5760 3704 at : $3\sqrt{3}/8$
5760 6591 rt : (4,7,3,3)
5760 7696 K : $\ln 2/6$
5760 8311 ln : P_{33}
5760 8379 rt : (6,1,9,6)
5761 1476 Ei : 13/5
5761 1834 rt : (8,8,-9,1)
5761 2182 rt : (9,9,-5,5)
5761 2295 e^x : $3\sqrt{3}$
5761 3198 J_1 : $2(e+\pi)/7$
5761 4620 rt : (2,6,8,3)
5761 4798 rt : (5,2,-8,2)
5761 5662 ln : P_{39}
5761 6181 erf : $\exp(-\sqrt[3]{5}/3)$
5761 6615 K : $((e+\pi)/2)^{-2}$
5761 9450 erf : $1/\ln(e+\pi)$
5761 9929 10^x : $2/3-e/3$
5762 0217 cr : P_{52}
5762 2016 rt : (1,8,2,-4)
5762 3601 sr : $10\sqrt{5}/9$
5762 4347 rt : (9,1,8,6)
5762 6645 ζ : $3\ln 2$
5762 6698 rt : (7,5,0,-3)
5762 7433 K : 2/17
5762 8825 J_1 : $(3\pi/2)^{1/3}$
5762 8920 erf : $2\sqrt{2}/5$
5763 0138 cr : $10\sqrt[3]{5}$
5763 0876 sinh : $(\sqrt[3]{4}/3)^{-1/3}$

5763 2261 rt : (8,9,6,2)
5763 2289 J_1 : $3\sqrt{5}/4$
5763 3111 ℓ_{10} : $6\pi/5$
5763 3738 rt : (6,2,9,-7)
5763 4577 J : $\exp(-\sqrt{2})$
5763 6874 10^x : $\sqrt{2}-\sqrt{3}/4$
5763 7522 at : 13/20
5763 7880 rt : (8,8,-6,2)
5763 8450 2^x : $2/3+4e/3$
5764 0305 Ψ : $\zeta(11)$
5764 0627 tanh : $\exp(-\sqrt[3]{2}/3)$
5764 3691 K : $3(1/\pi)/8$
5764 4074 rt : (4,8,-4,5)
5764 4220 rt : (9,2,8,-7)
5764 8373 ℓ_{10} : $8\sqrt{2}/3$
5764 9746 K : 3/25
5764 9842 rt : (7,8,0,-4)
5765 0143 J_1 : $5\zeta(3)/3$
5765 0637 rt : (6,5,9,-8)
5765 0831 K : $(\ln 2/2)^2$
5765 1339 J_1 : $4\sqrt[3]{2}/3$
5765 1718 K : $\zeta(3)/10$
5765 2945 $s\pi$: $3\pi^2/2$
5765 3473 rt : (1,3,-9,4)
5765 3956 Ψ : $\sqrt[3]{23/2}$
5765 6908 rt : (4,1,-9,-1)
5765 7387 ℓ_2 : $\ln(\sqrt{2}\pi)$
5765 7855 rt : (1,8,4,6)
5765 8043 rt : (9,5,8,-8)
5765 8058 $\sqrt{3} \div \ln(3)$
5765 8750 e^x : $7(e+\pi)/8$
5766 0632 $\ln(\sqrt{5}) \div \exp(1/3)$
5766 2000 rt : (5,6,7,3)
5766 2335 rt : (1,9,-5,5)
5766 2704 rt : (6,8,9,-9)
5766 3046 sinh : $7\zeta(3)/2$
5766 5143 J_1 : $1/\ln(2e/3)$
5766 7198 $t\pi$: $9\pi^2/2$
5766 8213 rt : (9,8,8,-9)
5766 8274 10^x : $\sqrt{24}\pi$
5766 9312 J_1 : $7\sqrt[3]{3}/6$
5767 0977 J_1 : $7\zeta(3)/5$
5767 2480 J_1 : 2
5767 2698 sr : $1/4+\sqrt{5}$
5767 5247 rt : (1,7,4,8)
5767 5341 rt : (7,7,-7,-1)
5767 7392 $t\pi$: $(\sqrt[3]{4})^{1/3}$
5767 8378 rt : (2,3,8,4)
5768 0279 K : $\exp(-2\pi/3)$
5768 1965 $\exp(1/2)-\exp(4/5)$
5768 2642 rt : (8,6,6,3)
5768 2708 Ψ : $5(1/\pi)/4$
5768 4189 10^x : $3\sqrt{2}/2+4\sqrt{5}/3$
5768 5644 $4/5+\ln(4/5)$
5768 6223 e^x : $7e\pi/10$
5768 6511 $\sqrt{2}\times\exp(3/5)$
5768 7042 2^x : $\exp(-\sqrt[3]{2}/3)$
5768 7065 sq : $2\sqrt{2}/3+4\sqrt{3}/3$
5768 7298 2^x : $\sqrt{2}+\sqrt{5}+\sqrt{7}$
5768 8507 2^x : $\sqrt{3}/2+3\sqrt{5}$
5768 8752 rt : (1,0,-9,5)
5768 9982 id : $2\sqrt[3]{3}/5$
5769 0030 $t\pi$: $4\sqrt[3]{2}/3$
5769 0528 cu : $3/2-2\sqrt{3}$
5769 1640 rt : (5,3,7,4)
5769 1863 $\exp(1/4)-s\pi(1/4)$
5769 3134 as : 6/11
5769 3902 $s\pi(\pi)\div s\pi(\sqrt{3})$
5769 4981 $\exp(1/4)\div\exp(4/5)$
5769 5898 10^x : 14/15
5769 6194 2^x : $\sqrt{2}+4\sqrt{3}$
5769 6674 cu : $1/2+4\sqrt{2}/3$
5769 7998 rt : (2,0,8,5)
5769 8677 K : 1/8
5769 9874 rt : (8,3,6,4)
5770 1351 cu : $3/2+2\sqrt{2}/3$
5770 2730 rt : (1,3,-9,-6)
5770 4167 rt : (5,0,7,5)
5770 4391 J_1 : $7\sqrt[3]{5}/6$
5770 4710 Ψ : $(\sqrt[3]{3}/3)^2$
5770 4836 ζ : $6\sqrt{3}/5$
5770 6652 sq : $4\sqrt{3}-4\sqrt{5}/3$
5770 7525 rt : (2,3,8,-6)
5770 7801 $2/5\div\ln(2)$
5770 8288 sr : $(\ln 2)^3$
5770 8574 rt : (8,0,6,5)
5770 8631 K : $\sqrt[3]{2}/10$
5770 8694 2^x : $4+\pi/4$
5770 8944 rt : (1,6,2,-2)
5770 9347 $s\pi$: $(\sqrt[3]{5})^{1/3}$
5771 0224 rt : (1,9,6,-9)
5771 1081 rt : (5,3,7,-6)
5771 1699 ζ : 1/13
5771 3112 $1/2\times\exp(e)$
5771 3153 rt : (2,6,8,-7)
5771 3404 10^x : $4/3+e/2$
5771 3822 rt : (8,3,6,-6)
5771 4184 ln : $4\pi^2/3$
5771 4408 cu : $\sqrt{3}/4+4\sqrt{5}$
5771 4612 at : $(e+\pi)/9$
5771 4895 rt : (1,9,-9,-8)
5771 5464 rt : (5,6,7,-7)
5771 6869 rt : (4,6,-8,8)
5771 7333 rt : (8,6,6,-7)
5771 7822 id : $4e/3-\pi/3$
5771 8490 rt : (5,9,7,-8)
5771 9409 rt : (1,8,-6,-9)
5771 9846 rt : (8,9,6,-8)
5772 0706 rt : (5,3,-2,9)
5772 1079 sr : $7e\pi/9$
5772 1566 id : γ
5772 1896 ζ : $\ln 2/9$
5772 2123 K : $2(1/\pi)/5$
5772 2398 rt : (5,0,-8,7)
5772 3204 rt : (8,9,0,9)
5772 3341 $s\pi(\sqrt{3})\times s\pi(e)$
5772 4382 rt : (8,6,-6,7)
5772 5947 cu : $10\sqrt[3]{5}/7$
5772 8797 cu : $2\sqrt{2}-3\sqrt{3}/4$
5773 0594 J_1 : 22/13
5773 0687 Ψ : 16/3
5773 1437 cr : $2\sqrt{2}/3+4\sqrt{5}/3$
5773 4909 e^x : $8\sqrt[3]{5}/3$
5773 5026 id : $\sqrt{3}/3$
5773 5595 sq : $3e/4+3\pi/2$
5773 5725 J_1 : $\ln(2e)$
5773 6162 $\exp(2/5)+\exp(3)$
5773 9080 cr : $4\sqrt{2}-\sqrt{3}$
5774 0064 Ei : 21/5
5774 0661 Γ : $\ln(\sqrt{3}/3)$
5774 2274 id : $1/2+3e/2$
5774 5887 Γ : $(\pi)^{-1/2}$
5774 6458 ℓ_{10} : $3\sqrt[3]{2}$
5774 7589 Ei : $9\sqrt{2}/8$
5774 9250 J_1 : $8\sqrt{5}/9$
5775 0504 e^x : $3/4-3\sqrt{3}/4$
5775 4211 K : 3/23
5775 7466 $\ln(\sqrt{3})\div s\pi(2/5)$
5775 8375 sq : $4\sqrt[3]{5}/9$
5775 9234 J_1 : $6\sqrt{2}/5$
5775 9319 10^x : $3\sqrt{2}/4-3\sqrt{3}/4$
5775 9464 2^x : $1/2+\sqrt{3}/2$
5776 0000 sq : 19/25
5776 1633 J_1 : $\exp(\sqrt[3]{4}/3)$
5776 2265 id : $5\ln 2/6$
5776 5311 sq : $1/\ln(\sqrt{5}/3)$
5776 6408 e^x : $(\sqrt{2}\pi/3)^{-2}$
5776 8378 cr : $5\pi/4$
5776 8601 J_1 : $(\ln 2/2)^{-1/2}$
5776 8777 e^x : $2e\pi$
5776 8918 10^x : $1/4+4\sqrt{3}/3$
5776 9129 Ei : $9(e+\pi)/8$
5776 9332 rt : (7,5,-7,4)
5776 9374 Ψ : $8e\pi/5$
5776 9959 10^x : $3+2\pi$
5777 0257 J_1 : $5e/8$
5777 0876 id : $8\sqrt{5}/5$
5777 2776 Ψ : $\exp(-\sqrt{2}\pi)$
5777 3899 rt : (7,8,-1,6)
5777 6403 10^x : $2\ln 2/7$
5777 6523 J_1 : 17/10
5777 7138 rt : (4,1,-9,-4)
5778 2849 ℓ_2 : P_{44}
5778 3147 Li_2 : $(1/(3e))^{1/3}$
5778 3887 sinh : $5\sqrt{3}/7$
5778 4235 rt : (4,2,-3,6)
5778 4833 K : 2/15
5778 5542 $t\pi$: $8\sqrt[3]{5}$
5778 7177 $\exp(1/5)+\exp(\sqrt{5})$
5778 7277 K : $\zeta(3)/9$
5778 8004 rt : (7,7,4,-6)
5779 0193 at : 15/23
5779 3108 K : $(\ln 3/3)^2$
5779 3150 cu : $2-2\pi$
5779 3946 K : $((e+\pi)/3)^{-3}$
5779 3976 J_1 : $7\sqrt{2}/5$
5779 4209 rt : (4,7,5,-6)
5779 5441 Li_2 : $2\sqrt{5}/9$
5779 6298 J_1 : $8\sqrt{3}/7$
5779 6354 $s\pi$: $3\sqrt{3}$
5779 6529 10^x : $-\sqrt{2}+\sqrt{3}+\sqrt{7}$
5779 6716 cu : $1/4-\sqrt{2}$
5779 7352 sq : $3/2+2\pi$
5779 7715 10^x : $4+4\sqrt{3}$
5779 8249 2^x : $4\sqrt{2}+\sqrt{5}/3$
5779 9745 rt : (7,4,4,-5)
5780 0116 $\exp(3/5)\times s\pi(1/3)$
5780 2063 rt : (1,7,6,-6)
5780 3989 10^x : $4(e+\pi)/5$
5780 6391 K : $(e)^{-2}$
5780 9254 rt : (4,4,5,-5)
5781 0216 cu : $3/2+4\sqrt{2}$
5781 0758 rt : (2,7,0,2)
5781 1156 ℓ_{10} : $3+\pi/4$
5781 2325 rt : (2,7,-7,-6)
5781 2413 2^x : P_{19}
5781 2578 2^x : P_{58}
5781 3133 Ψ : $5e\pi/8$
5781 5160 sinh : 11/20
5781 5631 id : $2e+\pi$
5781 5660 $\exp(1/5)\div s\pi(e)$
5781 6097 e^x : $2\sqrt{2}/3-2\sqrt{5}/3$
5781 7596 K : 3/22
5781 8170 rt : (7,1,4,-4)
5781 8196 K : $3(1/\pi)/7$
5781 9896 sr : $1+2\sqrt{5}/3$
5782 0035 ln : $\sqrt[3]{17/3}$
5782 1411 J_1 : $e\pi/5$
5782 2046 rt : (1,4,6,-5)
5782 2970 $t\pi$: $2\zeta(3)/5$
5782 3749 sr : P_7
5782 4700 ℓ_2 : $3+4\sqrt{5}$
5782 6305 rt : (5,7,-8,-6)
5782 6463 e^x : $9\sqrt{3}$
5782 6856 Ei : $\exp(-\sqrt{2})$
5782 6965 Ψ : $\ln(\sqrt[3]{3}/3)$
5782 6978 J_1 : $\pi^2/5$
5782 7530 cr : $4/3+3\sqrt{3}/2$
5782 8102 cr : $\sqrt{3}/3+3\sqrt{5}/2$
5783 0095 rt : (7,3,-7,9)
5783 0845 ℓ_{10} : $\sqrt{3}/4+3\sqrt{5}/2$
5783 1086 rt : (2,7,-9,4)
5783 2035 sinh : $7\sqrt{2}/8$
5783 2455 J_1 : $\sqrt[3]{5}$
5783 3961 rt : (6,4,-8,1)
5783 4151 $3/5\times s\pi(\sqrt{2})$
5783 4333 $\ln(3/4)+s\pi(1/3)$
5783 4420 10^x : $3\sqrt{3}/4-\sqrt{5}/3$
5783 4558 rt : (4,1,5,-4)
5783 5582 Ei : $9\sqrt[3]{5}/2$
5783 8196 J_1 : $\sqrt[3]{23/3}$
5783 9866 $\exp(\sqrt{3})\div s\pi(\sqrt{3})$
5784 0179 rt : (2,4,-7,-5)
5784 1222 Ψ : $7(e+\pi)/3$
5784 1738 J_1 : $((e+\pi)/2)^{1/2}$
5784 2595 K : $\ln 2/5$
5784 2712 id : $1/4-2\sqrt{2}$
5784 3077 cu : $3e/2+3\pi$
5784 4971 rt : (5,5,-9,-7)
5784 6483 rt : (8,7,-9,-6)

5784 6916 $2^x : 2e - \pi/2$
5784 7079 $\Psi : \ln 2/3$
5784 7226 rt : (4,9,-3,1)
5784 8353 id : $4\sqrt{2}/3 - 2\sqrt{3}$
5784 9598 $e^x : 3\sqrt{2}/2 + \sqrt{5}/3$
5785 0494 rt : (6,7,-2,3)
5785 0859 $\theta_3 : \sqrt{2}/5$
5785 1259 rt : (7,2,4,3)
5785 2266 sinh : $1/\sqrt[3]{6}$
5785 5351 $J_1 : 12/7$
5785 6895 $e^x : 1/2 - \pi/3$
5785 7827 $K : \sqrt[3]{2}/9$
5785 9013 rt : (1,1,6,-4)
5785 9241 rt : (9,8,1,-5)
5785 9407 rt : (8,7,-5,-1)
5786 2853 rt : (3,7,-5,-8)
5786 4813 $2^x : 2\sqrt{2}/3 - \sqrt{3}$
5786 6164 $\ln(\sqrt{5}) + s\pi(e)$
5786 7898 rt : (5,4,-8,-5)
5786 8038 rt : (9,8,-6,6)
5786 8717 rt : (2,5,-8,-1)
5786 8962 rt : (7,9,8,-9)
5787 0370 cu : 5/6
5787 0522 $J_1 : 10\zeta(3)/7$
5787 2113 cu : $\sqrt{24}\pi$
5787 2515 $Ei : 5\sqrt{2}/8$
5787 2549 $\theta_3 : 8(1/\pi)/9$
5787 3659 rt : (6,8,2,-5)
5787 3991 $K : \sqrt{2}/10$
5787 4556 $K : 4(1/\pi)/9$
5787 5381 $Ei : 7/13$
5787 6425 $erf : (2e)^{-1/3}$
5787 8374 $K : \exp(-(e+\pi)/3)$
5787 9462 cu : $3/4 - 3\sqrt{2}/2$
5788 0431 rt : (4,9,9,-9)
5788 0978 $\ell_2 : \ln((e+\pi)/3)$
5788 1090 $J_1 : 5\pi/8$
5788 1130 $10^x : 4/3 - \pi/2$
5788 2331 cr : $\sqrt{2}/3 + 2\sqrt{3}$
5788 2971 $e^x : 2 + e/4$
5788 3532 sq : $7\pi^2/8$
5788 4610 cu : $6\pi/5$
5788 6040 rt : (4,2,5,3)
5788 6422 $\ell_2 : 3\sqrt{2} + \sqrt{3}$
5788 6451 rt : (9,5,1,-4)
5788 6754 rt : (5,8,-3,-5)
5788 7956 sq : $4/3 + 4e/3$
5788 8727 rt : (1,9,-9,2)
5788 9958 rt : (8,9,-7,-2)
5789 0094 rt : (7,6,8,-8)
5789 0391 $K : 1/7$
5789 1410 $e^x : 18/19$
5789 1874 rt : (6,2,-8,6)
5789 1954 $4 \times \ln(\pi)$
5789 2555 $\Psi : 8\sqrt[3]{5}$
5789 3148 rt : (4,7,-7,4)
5789 3234 $\exp(3/4) \times s\pi(\sqrt{3})$
5789 4736 id : 11/19
5789 8853 $\ell_2 : \sqrt[3]{10/3}$
5789 9480 rt : (2,7,1,-8)
5790 1236 as : $1/2 - \pi/3$
5790 1592 at : $4/3 - e/4$
5790 1730 $K : \ln(\sqrt{3}/2)$
5790 3991 sq : $3\sqrt[3]{5}/2$
5790 5033 sr : $3e\pi/2$
5790 5687 rt : (4,6,9,-8)
5790 5840 ln : $9(e+\pi)/4$
5790 5930 $\ell_{10} : \exp(4/3)$
5790 6177 $K : \sqrt[3]{3}/10$
5790 6492 sq : $6e\pi/7$
5790 8446 rt : (6,5,2,-4)
5790 9677 $J_1 : 1/\ln(5/3)$
5791 0537 $J_1 : \sqrt[3]{15/2}$
5791 1958 rt : (6,5,-2,8)
5791 3670 sq : $2\pi/5$
5791 3841 cr : $4e + 2\pi$
5791 4662 $J_1 : 7\sqrt{5}/8$
5791 5161 $2^x : 4/3 - 3\sqrt{2}/2$
5791 5619 rt : (8,4,-5,0)
5791 9145 rt : (7,3,8,-7)
5791 9713 $J_1 : 19/11$
5792 0126 $Ei : 5(1/\pi)$
5792 1501 sq : $4/3 - 2\pi/3$
5792 2486 ln : $3\sqrt{2} + 4\sqrt{5}$
5792 5070 ln : $2/3 + \sqrt{5}/2$
5792 5372 $J_1 : 8\sqrt[3]{5}/7$
5792 6134 rt : (3,5,-9,-5)
5792 6183 $\ell_{10} : 4e\pi/9$
5792 7737 $\exp(4/5) \times \exp(\sqrt{3})$
5792 8050 rt : (7,5,4,2)
5792 8911 rt : (9,2,1,-3)
5792 9547 rt : (2,7,-3,-1)
5792 9807 $J_1 : (e+\pi)/3$
5793 0447 $10^x : 2\zeta(3)$
5793 3224 $10^x : 9/11$
5793 4784 rt : (5,3,-9,-2)
5793 5537 $J_1 : 6\sqrt[3]{3}/5$
5793 5997 rt : (8,7,-7,3)
5793 7942 rt : (3,5,3,-4)
5793 9722 ln : $7\ln 2$
5794 1566 rt : (4,3,9,-7)
5794 1644 $J_1 : \sqrt{3}$
5794 5300 $J_1 : 5\ln 2/2$
5794 6662 rt : (1,8,2,-9)
5794 7879 rt : (3,8,-5,-5)
5794 8520 sq : $9\sqrt[3]{2}$
5794 9034 $\ell_{10} : 1/3 + 2\sqrt{3}$
5795 0018 id : $3\sqrt{3}/2 + 4\sqrt{5}/3$
5795 0696 rt : (1,2,6,3)
5795 0982 $J_1 : 9\sqrt{3}/8$
5795 1297 rt : (9,7,5,-7)
5795 1458 rt : (8,9,-1,-4)
5795 2329 $K : (\sqrt[3]{4}/3)^3$
5795 2673 cr : $3\sqrt{2}/4 - \sqrt{3}/2$
5795 4571 $J_1 : \exp(2/3)$
5795 6819 rt : (3,8,3,6)
5795 8179 $\Gamma : 7(e+\pi)/5$
5795 9741 $\ell_{10} : 3\sqrt{2}/2 + 3\sqrt{5}/4$
5796 1256 $\ln(4/5) - \exp(\sqrt{5})$
5796 1592 rt : (7,0,8,6)
5796 1799 $\ell_2 : 7e\pi/5$
5796 2324 $\Psi : 8\sqrt{2}/5$
5796 2344 id : $3(e+\pi)$
5796 4145 ln : $1 + \pi/4$
5796 4290 $J_1 : \sqrt{3} - \sqrt{5} + \sqrt{6}$
5796 6320 rt : (8,5,-7,8)
5796 6522 rt : (6,2,2,-3)
5796 6682 $K : e\pi/9$
5796 7254 at : $1/2 - 2\sqrt{3}/3$
5797 0605 ln : $\sqrt[3]{3}/7$
5797 2548 $J_1 : 7\sqrt{5}/9$
5797 2871 rt : (6,7,6,-7)
5797 3626 id : $2\pi^2/3$
5797 4010 rt : (2,8,-4,6)
5797 4566 $K : 3/20$
5797 5404 $J_1 : \sqrt[3]{22/3}$
5797 7676 $K : \zeta(3)/8$
5797 8117 $10^x : \sqrt{2}/4 + \sqrt{3}/2$
5797 8359 $\ell_{10} : 5/19$
5797 9589 rt : (5,9,0,-4)
5797 9651 $\ell_{10} : 4\sqrt{3}/3 + 2\sqrt{5}/3$
5797 9929 cr : $3 + 2\sqrt{2}/3$
5798 0331 $J_1 : 5e/7$
5798 0926 $erf : \sqrt[3]{5}/3$
5798 1570 $\ell_2 : P_{60}$
5798 1849 ln : 14/25
5798 2411 ln : $3/2 + 3\sqrt{5}/2$
5798 3568 $e^x : 1/2 + 4\sqrt{5}$
5798 4140 cu : $1 + \pi/3$
5798 4799 $t\pi : P_{60}$
5798 5535 as : $2\sqrt{2}/3 - 2\sqrt{5}/3$
5798 6074 sq : $3/4 - 3\pi/4$
5798 8115 ln : $4\sqrt[3]{2}/9$
5798 9388 as : $4/3 - \pi/4$
5799 1410 rt : (9,4,5,-6)
5799 1601 $s\pi : 3\zeta(3)/2$
5799 2356 rt : (3,6,-1,9)
5799 2430 rt : (1,7,-4,3)
5799 4276 rt : (6,8,-6,-5)
5799 5399 $J : 2\ln 2/9$
5799 6313 $10^x : 2/3 + \sqrt{3}/4$
5799 6563 rt : (4,0,9,6)
5799 7648 $J_1 : 5\pi/9$
5799 9128 $e^x : 8(e+\pi)/3$
5799 9250 rt : (3,7,7,-7)
5799 9257 $\Psi : \sqrt{19}\pi$
5799 9589 sr : $1/4 + 4\pi$
5799 9940 sinh : $(3\pi/2)^{1/3}$
5800 0295 $\theta_3 : 18/23$
5800 2739 $\zeta : (1/\pi)/4$
5800 4392 rt : (9,1,1,2)
5800 4802 $J_1 : \sqrt[3]{16/3}$
5800 5016 rt : (8,6,-1,-3)
5800 5306 rt : (7,8,-8,-5)
5800 6104 rt : (1,7,2,-1)
5800 6454 rt : (5,1,-9,-8)
5800 7685 $t\pi : (\sqrt[3]{3})^{-1/2}$
5800 8803 sr : $1 + 4\sqrt{2}$
5800 9136 $s\pi : 7\sqrt[3]{5}/10$
5801 0320 ln : $2/3 + 4\pi/3$
5801 4873 cr : $2/3 - \sqrt{2}/3$
5801 5619 $J_1 : 7/4$
5801 5749 $t\pi : \sqrt{15}\pi$
5801 6165 rt : (2,9,1,-4)
5801 8523 rt : (5,4,-3,5)
5801 9634 as : $\sqrt{2} - \sqrt{3}/2$
5802 1679 $K : 2/13$
5802 2325 rt : (9,6,9,-9)
5802 2906 $\ell_2 : \sqrt{2} - \sqrt{5}/3$
5802 3996 $K : 2\ln 2/9$
5802 4122 $\zeta : 25/24$
5802 4691 sq : $8\sqrt{2}/9$
5802 5229 $\theta_3 : 9\ln 2/10$
5802 8305 $\ell_2 : 4e/3 + 3\pi/4$
5802 8690 cr : $2\sqrt{2} + \sqrt{5}/2$
5802 9481 rt : (7,3,8,5)
5803 1294 $J_1 : \sqrt{3} - \sqrt{6} + \sqrt{7}$
5803 2253 rt : (7,8,-7,0)
5803 2822 $\theta_3 : \exp(-\sqrt[3]{2})$
5803 2937 cr : $4\sqrt{3} - 4\sqrt{5}/3$
5803 3457 $e^x : 4\sqrt{2} - \sqrt{5}/2$
5803 4602 $Ei : \sqrt{22}\pi$
5803 4661 $J_0 : 11/8$
5803 6037 $e^x : 1/\ln(1/(2\pi))$
5803 7238 $\zeta : 24/23$
5803 7456 $e^x : 3\sqrt{2}/2 + 3\sqrt{3}/4$
5803 8817 rt : (8,8,3,-6)
5804 1485 rt : (3,9,-4,8)
5804 1808 $e^x : 2e/3 - 3\pi/4$
5804 2251 $\ell_{10} : 7e/5$
5804 2921 rt : (3,5,-5,-4)
5804 4183 $J_1 : 3(e+\pi)/10$
5804 4556 $J_1 : 10\sqrt{3}/9$
5804 5288 $J_0 : \exp(1/\pi)$
5804 5546 $J_1 : (2e)^{1/3}$
5804 5701 $3/4 \times s\pi(e)$
5804 5798 $t\pi : e/4$
5804 6230 $\Gamma : 6e/5$
5804 6559 $10^x : 3 + 3\sqrt{3}/2$
5804 7117 cr : $2\pi^2/5$
5804 7190 $J_1 : 9\sqrt[3]{5}/8$
5804 7238 $2^x : 1/3 - \sqrt{5}/2$
5804 7934 rt : (4,5,5,2)
5804 8202 $\ell_2 : (\sqrt{5})^{1/2}$
5804 8448 $s\pi : (e+\pi)^{1/3}$
5804 8480 rt : (7,4,-8,5)
5804 8642 $J_1 : 8\zeta(3)/5$
5804 8898 $\ell_2 : 3 + 4\sqrt{5}/3$
5804 9188 $J_0 : (\sqrt[3]{4}/3)^{-1/2}$
5804 9234 $\zeta : 2/25$
5804 9359 $J_1 : 25/13$
5804 9618 $J_1 : 4\sqrt[3]{3}/3$
5804 9780 rt : (9,1,5,-5)
5804 9886 sq : 16/21
5805 0277 rt : (5,9,3,2)
5805 0745 $\Psi : -\sqrt{2} - \sqrt{5} - \sqrt{6}$
5805 1543 $\zeta : 23/22$
5805 2384 $\lambda : \exp(-13\pi/14)$
5805 2679 $\Gamma : 8e\pi/3$
5805 3186 $\ell_2 : P_7$

5805 4622 sinh : $8\sqrt{5}/9$
5805 6656 ln : $2\sqrt{3}-3\sqrt{5}/4$
5805 7044 K : $\ln(\sqrt[3]{5}/2)$
5806 0012 $\exp(e)\div s\pi(e)$
5806 1551 $s\pi$: $5\sqrt[3]{3}/4$
5806 1563 rt : (1,7,2,-5)
5806 1679 $t\pi$: $(\ln 2)^{1/2}$
5806 1772 rt : (3,4,7,-6)
5806 2962 K : $\sqrt{2}/9$
5806 3726 rt : (8,1,-5,1)
5806 3750 J_1 : $7\sqrt[3]{2}/5$
5806 4726 at : $1/4-e/3$
5806 5089 J_1 : $\sqrt{2}-\sqrt{3}+\sqrt{5}$
5806 5689 rt : (1,9,-2,-4)
5806 6480 as : P_{12}
5806 6756 ζ : $10\zeta(3)/9$
5806 6851 rt : (9,8,-7,-5)
5806 7068 Ei : $\exp(-1/(3e))$
5806 7206 ζ : $22/21$
5806 7352 10^x : $2-\sqrt{5}$
5806 7477 K : $\sqrt[3]{2}/8$
5806 7800 J_1 : $\sqrt[3]{11/2}$
5806 8558 rt : (9,3,9,-8)
5806 9962 J_1 : $23/12$
5807 0076 J_1 : $6\sqrt{5}/7$
5807 2199 rt : (5,8,4,-6)
5807 2632 K : $3/19$
5807 2651 ℓ_{10} : $2/3+\pi$
5807 4756 rt : (3,7,-9,-7)
5807 5484 J_1 : $\sqrt{11/3}$
5807 5767 J_1 : $5\sqrt{2}/4$
5807 6855 J_1 : $\ln(e+\pi)$
5807 7205 $t\pi$: $4\sqrt{2}/3+\sqrt{3}$
5807 7229 J_1 : $\exp(\sqrt[3]{5}/3)$
5807 9150 rt : (7,9,-2,2)
5807 9953 Ψ : $\exp(-\sqrt[3]{4}/3)$
5808 0143 J_1 : $23/13$
5808 0185 $t\pi$: $\sqrt{2}+\sqrt{3}+\sqrt{5}$
5808 0842 sq : $5\pi^2/3$
5808 1088 e^x : $2+3\pi$
5808 1198 J_1 : $\sqrt[3]{7}$
5808 1595 cu : $3\sqrt{3}-3\sqrt{5}/4$
5808 1622 $t\pi$: P_{44}
5808 2977 rt : (6,1,2,2)
5808 3085 Ei : $7\ln 2/9$
5808 4430 ζ : $21/20$
5808 5670 ℓ_{10} : $3/2+4\sqrt{3}/3$
5808 6154 10^x : $7/17$
5808 6898 rt : (7,7,-2,7)
5808 7243 Ψ : $\sqrt[3]{3}/9$
5808 8256 ℓ_{10} : $3\sqrt{2}-\sqrt{3}/4$
5808 8599 2^x : e
5808 8778 K : $(1/\pi)/2$
5808 9466 J_1 : $(\pi)^{1/2}$
5808 9994 rt : (9,9,-7,2)
5809 0012 J_1 : $6(1/\pi)$
5809 1331 2^x : $\sqrt{2}/2-2\sqrt{5}/3$
5809 1544 rt : (4,3,9,5)
5809 2157 J_1 : $21/11$
5809 3703 rt : (8,3,-1,-2)
5809 3933 rt : (9,7,-7,7)
5809 7743 rt : (6,1,6,-5)
5809 7966 erf : $4/7$
5809 8836 $1/4\div s\pi(\pi)$
5809 8976 sq : $4/3+\sqrt{5}/4$
5809 9681 K : $4/25$
5810 0223 Ψ : $5e/6$
5810 0238 J_0 : $8\zeta(3)/7$
5810 1250 rt : (8,5,3,-5)
5810 1918 e^x : $\exp(\sqrt[3]{2}/3)$
5810 2917 K : $\sqrt[3]{3}/9$
5810 3081 $\exp(\pi)\div\exp(1/3)$
5810 3460 ζ : $20/19$
5810 3929 J_1 : $16/9$
5810 4298 10^x : $5(1/\pi)/8$
5810 4688 rt : (8,7,7,-8)
5810 4797 Ei : $5/4$
5810 6927 J_1 : $1/\ln(\sqrt[3]{5}/3)$
5810 7723 2^x : $3+\sqrt{3}/2$
5810 7725 cu : $3e/7$
5810 8819 J_1 : $7e/10$
5810 9055 2^x : $4\sqrt[3]{5}/5$
5811 2885 J_1 : $\exp(\sqrt{3}/3)$
5811 3883 id : $\sqrt{5/2}$
5811 4714 Ei : $((e+\pi)/3)^{1/3}$
5811 5707 J_1 : $19/10$
5811 5771 J_1 : $10\sqrt[3]{5}/9$
5811 6567 J_1 : $\sqrt[3]{17/3}$
5811 7429 ℓ_{10} : $2+2e/3$
5811 8579 2^x : $1/4+\sqrt{5}/2$
5812 1088 J_1 : $2e\pi/9$
5812 1293 as : $3/4-3\sqrt{3}/4$
5812 3144 $t\pi$: $2\sqrt[3]{3}/9$
5812 3322 J_1 : $25/14$
5812 4306 cu : $1+3\sqrt{3}/2$
5812 4597 ζ : $19/18$
5812 9286 $s\pi$: $3(e+\pi)/8$
5812 9317 $s\pi$: $2\sqrt{2}/3-\sqrt{5}/3$
5812 9337 2^x : $3\sqrt{2}+3\sqrt[3]{3}/4$
5813 0279 J_1 : $4\sqrt{5}/5$
5813 1940 Ei : $1/\sqrt{17}$
5813 2130 $t\pi$: $10\sqrt{3}$
5813 2518 at : $\exp(-\sqrt[3]{2}/3)$
5813 2526 rt : (9,0,9,7)
5813 3095 ℓ_2 : $7\sqrt[3]{5}$
5813 6593 rt : (4,9,-3,-4)
5813 7958 J_1 : $3\sqrt[3]{2}/2$
5813 9890 rt : (5,7,8,-8)
5813 9916 J_1 : $17/9$
5814 1068 cu : $3e/2+\pi/3$
5814 1587 sinh : $3\sqrt{5}/4$
5814 2519 rt : (9,2,5,4)
5814 2797 K : $\exp(-2e/3)$
5814 3090 J_1 : $4\pi/7$
5814 3494 rt : (2,9,1,8)
5814 3599 rt : (7,9,1,-5)
5814 3606 e^x : $11/24$
5814 4124 cu : $4e/3-2\pi/3$
5814 5365 $5\times\ln(2/5)$
5814 6083 J_1 : $4\sqrt{2}/3$
5814 7279 J_1 : $3\pi/5$
5814 8211 ζ : $18/17$
5815 0103 e^x : $4\sqrt{2}+\sqrt{5}/2$
5815 0885 10^x : $4e/3+\pi/4$
5815 0941 rt : (5,5,4,-5)
5815 1170 rt : (6,5,-6,-4)
5815 1504 J_1 : $10\sqrt[3]{2}/7$
5815 1695 J_1 : $9/5$
5815 2276 J_1 : $\sqrt[3]{20/3}$
5815 2982 e^x : $5\sqrt[3]{3}$
5815 3366 as : $\ln(\sqrt{3})$
5815 3411 ℓ_{10} : $((e+\pi)/3)^2$
5815 3528 θ_3 : $\exp(-\sqrt{2}/3)$
5815 3966 rt : (6,1,-9,7)
5815 5483 rt : (7,6,8,4)
5815 6290 J_1 : $5\sqrt[3]{3}/4$
5815 6371 J_1 : $(e+\pi)^{1/3}$
5815 6720 J_1 : $3\zeta(3)/2$
5815 7540 ln : $\sqrt{5}/4$
5815 8953 rt : (7,7,-3,-2)
5815 9448 J_1 : $\exp(\sqrt[3]{2}/2)$
5816 1870 ln : $3e/2+\pi/4$
5816 3092 J_1 : $15/8$
5816 3104 rt : (3,1,7,-5)
5816 3521 rt : (6,3,-9,2)
5816 3866 10^x : $\sqrt[3]{18}$
5816 7435 rt : (1,8,-6,-6)
5816 8513 J_1 : $\sqrt{7/2}$
5816 9263 J_1 : $2e/3$
5816 9691 rt : (8,4,7,-7)
5817 1478 sinh : $9\sqrt{3}/7$
5817 1511 rt : (4,1,-4,7)
5817 3639 J_1 : $\sqrt[3]{13/2}$
5817 4637 J_1 : $\sqrt[3]{6}$
5817 4764 ζ : $17/16$
5817 5251 e^x : $\sqrt{2}+\sqrt{5}/2$
5817 5442 K : $(\ln 3/2)^3$
5817 5571 rt : (9,4,1,1)
5817 5663 J_1 : $20/11$
5817 5750 J_1 : $9\sqrt{2}/7$
5817 5847 J_1 : $1/\ln(\sqrt[3]{5})$
5817 6415 J_1 : $5\sqrt{5}/6$
5817 7414 Ei : $(\ln 2)^{1/3}$
5817 7723 J_1 : $1/\ln(\sqrt{3})$
5817 8310 rt : (5,3,0,-2)
5817 8683 cr : $1/3+4e/3$
5817 9053 rt : (6,5,-9,-3)
5817 9667 ℓ_{10} : $\sqrt{2}/4+2\sqrt{3}$
5817 9969 2^x : $3/2+4\pi/3$
5818 0389 J_1 : $\sqrt{2}-\sqrt{5}+\sqrt{7}$
5818 0765 rt : (8,4,-8,9)
5818 1300 J_1 : $13/7$
5818 1629 J_1 : $\sqrt{10/3}$
5818 2570 rt : (2,7,9,-8)
5818 2983 J_1 : $9\sqrt[3]{3}/7$
5818 3182 rt : (4,7,1,-1)
5818 4867 J_1 : $\sqrt[3]{19/3}$
5818 5258 J_1 : $11/6$
5818 5456 J_1 : $8\ln 2/3$
5818 5682 e^x : $1+e/2$
5818 5764 J_1 : $(e/2)^2$
5818 6011 rt : (3,4,-9,-6)
5818 6015 J_1 : $24/13$
5818 6179 J_1 : $(2\pi)^{1/3}$
5818 6298 J_1 : $\ln(2\pi)$
5818 6994 $\ln(5)\times\exp(4/5)$
5818 7843 K : $1/6$
5818 8508 cu : $3\sqrt{3}/2-2\sqrt{5}$
5818 9058 2^x : $3+3\sqrt{5}/4$
5819 3454 cu : $2\sqrt{3}-4\sqrt{5}$
5819 3558 rt : (8,6,-8,4)
5819 4556 Γ : $9/16$
5819 5935 rt : (8,2,3,-4)
5819 8660 K : $(\sqrt[3]{5}/2)^{1/3}$
5819 8889 id : $\sqrt{20/3}$
5820 0797 e^x : $7\sqrt{5}/6$
5820 0994 $\exp(4/5)+\exp(\sqrt{5})$
5820 1150 rt : (7,8,5,-7)
5820 2357 Ei : $(\sqrt[3]{3})^{-1/3}$
5820 3537 rt : (6,4,-3,9)
5820 4842 ζ : $16/15$
5820 4980 Γ : $3\sqrt[3]{2}$
5820 6362 K : $(2e/3)^{-3}$
5820 7050 rt : (7,5,-7,1)
5820 8017 cu : $e/3+3\pi$
5820 8716 rt : (6,7,-9,-8)
5820 9827 rt : (3,8,-9,3)
5820 9956 rt : (1,6,-2,-3)
5821 1504 cu : $4/3-3\pi/2$
5821 2396 rt : (8,8,-8,-1)
5821 2559 at : P_{19}
5821 2665 at : P_{58}
5821 3560 id : $2\sqrt{3}+\sqrt{5}/2$
5821 5278 rt : (5,9,-7,8)
5821 6378 rt : (2,5,5,-5)
5821 8449 rt : (8,1,-9,-4)
5821 9263 rt : (5,4,8,-7)
5821 9557 ℓ_{10} : $e/4+\pi$
5822 2342 erf : $9(1/\pi)/5$
5822 2766 $s\pi$: $\sqrt{2}/2+2\sqrt{5}/3$
5822 3173 e^x : $2/3+e/2$
5822 4052 Li_2 : $1/2$
5822 5052 sq : $3\sqrt{2}-3\sqrt{5}/4$
5822 6852 rt : (9,3,9,6)
5822 7746 ln : $4+\sqrt{3}/2$
5822 8120 rt : (6,6,-3,4)
5822 8635 ζ : $(\sqrt{2}\pi/3)^{-1/3}$
5823 1049 cr : $2\sqrt{2}/3-\sqrt{5}/3$
5823 2370 e^x : $10e/7$
5823 3727 Ψ : $(\ln 2/2)^{1/2}$
5823 4995 e^x : $\sqrt{3}/3-\sqrt{5}/2$
5823 6423 as : $11/20$
5823 7209 10^x : $5\sqrt{2}/4$
5823 7701 rt : (7,6,1,-4)
5823 7901 $2/3\div\ln(\pi)$
5823 8843 rt : (4,8,-7,-6)
5823 9195 ζ : $15/14$
5824 2887 rt : (4,7,-2,-2)
5824 3519 e^x : $\sqrt{6}$
5824 3571 ln : $4e+3\pi/4$

5824 4970 4 × exp(1/3)
5824 5106 Ψ : $9\sqrt[3]{2}/5$
5824 5201 rt : (7,7,9,-9)
5824 6806 rt : (7,9,-4,-4)
5824 7170 K : $\sqrt[3]{5}/10$
5824 7431 rt : (8,9,-2,6)
5824 8637 ln : $4\sqrt{2}/3 + 4\sqrt{5}/3$
5824 9171 ℓ_2 : $5\zeta(3)/9$
5824 9826 rt : (4,8,6,-7)
5825 2633 $\exp(\sqrt{5}) - s\pi(e)$
5825 3432 Γ : $\zeta(3)/6$
5825 3812 sinh : $(e+\pi)^{1/2}$
5825 7257 K : $\zeta(3)/7$
5825 7569 id : $\sqrt{21}$
5825 9005 sπ : $\sqrt[3]{22}$
5826 0118 ℓ_2 : $1/3 + 4\sqrt{2}$
5826 1227 rt : (9,9,2,-6)
5826 2425 rt : (8,1,7,-6)
5826 2800 2^x : $\sqrt{2} + \sqrt{5} + \sqrt{6}$
5826 3801 cr : $1/2 + 2\sqrt{3}$
5826 3913 rt : (9,8,-6,-1)
5826 4156 $\sqrt{5} + \ln(\sqrt{2})$
5826 5655 5 ÷ exp(1/3)
5826 7022 2^x : $1 + 3\pi$
5826 8420 10^x : $2\sqrt[3]{3}/7$
5826 8652 rt : (8,0,-1,1)
5826 8894 ℓ_{10} : $\sqrt{2}/3 + 3\sqrt{5}/2$
5827 0134 rt : (3,5,-4,8)
5827 1251 rt : (6,8,-3,-1)
5827 1652 ln : $2/3 + 4\pi$
5827 3693 e^x : $4\sqrt{3} - \sqrt{5}/4$
5827 4889 as : $1/\sqrt[3]{6}$
5827 5607 rt : (3,6,-8,-9)
5827 5726 ln : $e/4 + 4\pi/3$
5827 6197 e^x : $e\pi/9$
5827 7647 rt : (5,2,4,-4)
5827 8034 K : $\sqrt{3}/10$
5827 8294 tanh : 2/3
5827 8806 ζ : 14/13
5827 9184 K : $\ln 2/4$
5827 9196 e^x : $1/3 + 2\sqrt{2}/3$
5827 9340 ζ : exp(−5/2)
5827 9357 2^x : $4/3 + e$
5827 9783 sπ : $2\ln 2/7$
5828 1743 rt : (2,4,9,-7)
5828 3055 rt : (1,5,-2,7)
5828 7742 rt : (6,9,-9,1)
5828 8022 K : 4/23
5828 8331 cu : $2/3 - 3e/4$
5828 8892 cr : $2\ln 2/7$
5828 9796 cr : $3e - 4\pi/3$
5829 0947 rt : (7,3,-9,6)
5829 1099 sr : $e/8$
5829 1472 rt : (7,5,5,-6)
5829 1866 rt : (1,3,-9,-8)
5829 2695 rt : (3,2,-5,-3)
5829 4107 2^x : P_{55}
5829 4917 rt : (9,6,-8,8)
5829 5193 Li_2 : $\sqrt{2} + \sqrt{3} - \sqrt{7}$
5829 5279 rt : (4,6,9,4)
5829 5739 $1/3 + \ln(2/5)$
5829 7298 tπ : $\sqrt{3}/3 - \sqrt{5}/3$
5829 7920 rt : (9,8,6,-8)
5830 0193 2^x : $2\sqrt{3} - \sqrt{5}/3$
5830 0801 tπ : $\exp(-1/(2e))$
5830 1380 rt : (7,8,4,1)
5830 2038 ℓ_{10} : $1 + 2\sqrt{2}$
5830 6570 rt : (1,8,7,-5)
5830 8088 tπ : $(2e/3)^{-3}$
5830 8302 ζ : 9/17
5830 9987 rt : (1,8,7,-7)
5831 0400 ζ : $(\sqrt[3]{2}/3)^{-1/3}$
5831 1534 rt : (3,7,0,5)
5831 2546 rt : (9,5,5,3)
5831 5637 rt : (6,9,3,6)
5831 5986 rt : (2,8,-6,-3)
5831 7244 cr : $\sqrt{3} + \sqrt{5}$
5831 8489 2^x : $2\sqrt{3}/3 + 4\sqrt{5}/3$
5831 9031 cu : $e/3 + 2\pi$
5832 0951 Ψ : $\ln 2/5$
5832 1105 θ_3 : $\sqrt[3]{5}/6$
5832 2024 rt : (9,7,-2,-3)
5832 3137 K : $(\sqrt[3]{2}/3)^2$
5832 3290 sr : $3e/2 - \pi/2$
5832 3348 sr : $(2\pi)^{1/2}$
5832 4468 K : 3/17
5832 4706 sq : $1/3 + 3\sqrt{5}$
5832 4981 ζ : 13/12
5832 6765 rt : (7,4,-3,-1)
5832 8279 rt : (9,8,-8,3)
5832 8868 K : $\sqrt{2}/8$
5832 9153 Γ : $1/\ln(\sqrt[3]{3})$
5832 9363 10^x : $5\pi^2/7$
5832 9762 K : $5(1/\pi)/9$
5833 0552 rt : (7,4,9,-8)
5833 0948 K : $\exp(-\sqrt{3})$
5833 1252 rt : (7,5,-9,1)
5833 3333 id : 7/12
5833 5014 sinh : $10\zeta(3)$
5833 7388 rt : (5,1,8,-6)
5833 7597 sinh : $4\ln 2/5$
5833 9013 rt : (7,8,-8,-6)
5834 1517 J : $2\sqrt[3]{3}/9$
5834 2623 rt : (2,9,-4,-5)
5834 2889 e^x : $(\sqrt[3]{5}/2)^{1/3}$
5834 6591 rt : (4,7,-9,-1)
5834 7665 10^x : $8\sqrt{3}/5$
5834 7665 rt : (1,5,3,5)
5834 8792 rt : (6,8,7,-8)
5834 9968 e^x : $3/2 - 3e/4$
5835 3072 J : $\exp(-5\pi/6)$
5835 3705 tanh : $5\zeta(3)/9$
5835 3840 Ei : $3\sqrt[3]{2}/7$
5835 4101 rt : (7,6,-3,8)
5835 5602 sinh : $(e+\pi)^{-1/3}$
5835 5793 rt : (3,2,7,4)
5835 6413 sπ : $3\sqrt{2} + \sqrt{5}/4$
5835 6474 rt : (8,1,3,3)
5835 6606 cu : $7e\pi/3$
5835 7658 ℓ_{10} : 6/23
5835 7769 10^x : 2π
5835 8789 rt : (5,1,-4,1)
5835 9213 sq : $3 - \sqrt{5}$
5836 1960 rt : (9,6,2,-5)
5836 2413 rt : (4,5,6,-6)
5836 4106 Ψ : $\sqrt{2} - \sqrt{3} - \sqrt{5}$
5836 4535 rt : (5,3,-4,6)
5836 6229 10^x : $4 + 2\sqrt{5}/3$
5836 6514 rt : (6,8,-5,-1)
5836 7212 10^x : $3\sqrt{3} + \sqrt{5}/2$
5836 8123 ℓ_2 : P_{26}
5836 8660 rt : (1,7,-1,-2)
5836 9117 ln : $e/4 + 4\pi$
5837 0824 Ψ : $(e)^{-3}$
5837 2459 rt : (4,6,-3,-3)
5837 5535 K : $\sqrt[3]{2}/7$
5837 7216 cr : $5(1/\pi)/8$
5837 9499 ζ : 12/11
5837 9829 K : $\sqrt[3]{3}/8$
5837 9869 rt : (9,6,9,5)
5838 1019 rt : (9,5,-7,-4)
5838 2740 rt : (3,9,4,2)
5838 4421 tπ : $4e\pi/5$
5838 5005 Ei : $3\sqrt[3]{3}/2$
5838 6307 rt : (2,8,-8,-7)
5838 7955 rt : (7,3,1,-3)
5838 8499 K : $\exp(-\sqrt[3]{5})$
5838 8954 tπ : $6\ln 2/5$
5839 1113 rt : (9,5,6,-7)
5839 1617 $\sqrt{3} \times \exp(2/5)$
5839 5039 rt : (2,4,7,9)
5839 5254 rt : (7,7,-9,-4)
5839 8088 rt : (4,9,-5,-4)
5839 8806 rt : (3,9,-6,1)
5839 9526 tπ : $2\sqrt{3}/3 + 3\sqrt{5}/4$
5840 0541 sinh : $9(e+\pi)/10$
5840 2519 K : 2/11
5840 3605 K : $4(1/\pi)/7$
5840 4768 rt : (8,9,4,-7)
5840 4915 id : $3e - \pi/2$
5840 5434 rt : (8,2,7,5)
5840 6128 Ψ : 3/13
5840 6679 rt : (3,1,-9,-5)
5840 6962 sπ : $9\sqrt[3]{5}/7$
5840 7323 sr : $4 - 2\sqrt{5}/3$
5840 9904 rt : (3,8,8,-8)
5841 0058 3/4 ÷ exp(1/4)
5841 0073 Ψ : 10/17
5841 1358 J_0 : $4\sqrt[3]{5}/5$
5841 1622 rt : (6,7,-1,-3)
5841 3802 e^x : $(\sqrt[3]{3}/3)^{-1/3}$
5841 5064 tanh : $(\sqrt{5})^{-1/2}$
5841 6570 rt : (7,8,-3,3)
5841 7068 2 ÷ sπ(e)
5841 7648 ζ : 1/12
5841 8433 rt : (8,5,-9,5)
5842 1249 id : $((e+\pi)/2)^{-1/2}$
5842 2544 rt : (1,6,3,-4)
5842 4039 rt : (8,8,8,-9)
5842 5176 rt : (7,2,5,-5)
5842 8452 e^x : 19/4
5842 9458 sπ : $2\zeta(3)/3$
5843 0723 $e + s\pi(1/3)$
5843 3546 rt : (6,4,2,1)
5843 3714 $\ln(\sqrt{5}) - \exp(2)$
5843 7578 rt : (2,1,9,-6)
5843 9179 rt : (4,5,-7,-5)
5844 0283 rt : (2,7,-2,-5)
5844 0880 rt : (6,6,3,-5)
5844 2314 10^x : $\sqrt{2} + \sqrt{5} + \sqrt{6}$
5844 2834 rt : (1,6,-4,-1)
5844 3058 sq : $\sqrt{2}/4 - \sqrt{5}/2$
5844 4846 ζ : 11/10
5844 6116 Ψ : $1/\sqrt[3]{16}$
5844 7061 10^x : $9\pi/4$
5844 7479 tπ : $\sqrt[3]{25/2}$
5844 7613 2^x : $5e\pi/7$
5844 8410 sr : $3\sqrt{2} - \sqrt{3}$
5844 9585 tπ : $1/4 - \pi/2$
5845 0222 rt : (7,1,9,-7)
5845 1111 rt : (4,7,-4,-8)
5845 1636 sr : $(e/2)^3$
5845 2950 K : $(\sqrt[3]{5}/3)^3$
5845 3223 id : $((e+\pi)/2)^2$
5845 3670 rt : (5,5,-4,1)
5845 4398 rt : (1,5,7,-6)
5845 5951 tπ : $2e/3 + 3\pi/4$
5845 6345 $\exp(e) - s\pi(\pi)$
5845 7778 sinh : 5/9
5845 9612 sq : $7\pi^2/6$
5846 0433 sπ : $5\sqrt[3]{3}/9$
5846 1363 rt : (5,9,-5,-5)
5846 1404 rt : (6,5,7,-7)
5846 5686 ℓ_{10} : $9e\pi/2$
5846 6134 tanh : $\ln((e+\pi)/3)$
5846 7637 rt : (9,5,-6,0)
5847 0222 rt : (7,9,8,3)
5847 0795 rt : (5,9,5,-7)
5847 2484 cr : $4\sqrt{2} - 3\sqrt{5}/4$
5847 3176 sr : $3(e+\pi)/7$
5847 4673 10^x : $4/3 + 4\sqrt{2}/3$
5847 4948 rt : (9,5,-9,9)
5847 5735 $\ln(3/5) \times \ln(\pi)$
5847 6023 sq : $1/3 + 4e$
5847 6568 sq : $2e + \pi$
5847 7508 sq : 13/17
5847 8173 θ_3 : 2/7
5847 9990 ln : $\sqrt{2} + 2\sqrt{3}$
5848 0354 id : $1/\sqrt[3]{5}$
5848 1248 tπ : $2e + \pi/3$
5848 3788 rt : (5,8,9,-9)
5848 4726 ln : $3/2 - 2\sqrt{2}/3$
5848 4748 rt : (8,9,-8,0)
5848 6452 rt : (5,8,-9,-7)
5848 8201 K : 3/16
5848 8769 rt : (3,9,-8,1)
5848 8859 Ei : $7\pi/6$
5848 9319 10^x : 1/5
5848 9611 e^x : $1/4 + 2\pi$
5849 0031 rt : (4,0,-5,8)

5849 1300 cr : $7\pi^2/4$
5849 2636 Γ : $\exp(-\sqrt{3}/3)$
5849 3069 rt : (8,8,-3,7)
5849 3993 cr : $1+4\sqrt{5}/3$
5849 4004 rt : (9,4,-2,-2)
5849 6250 ℓ_2 : 3
5849 8656 rt : (6,5,6,3)
5849 9175 $s\pi(1/3)\times s\pi(\sqrt{5})$
5850 0008 $5\times\exp(3/4)$
5850 0915 ℓ_{10} : $8\sqrt[3]{3}/3$
5850 2951 cr : P_{16}
5850 3073 cr : P_{40}
5850 4324 rt : (8,8,-4,-2)
5850 5747 id : $4\sqrt{2}+4\sqrt{3}$
5850 6702 tπ : $5\pi^2/8$
5850 9374 K : $\exp(-5/3)$
5850 9727 sπ : $5(1/\pi)/8$
5850 9996 rt : (3,4,-7,7)
5851 0238 rt : (9,3,2,-4)
5851 0239 as : $4/3-4\sqrt{2}/3$
5851 1409 ln : $4\pi/7$
5851 1802 e^x : $(3\pi/2)^{-1/2}$
5851 2400 rt : (7,9,-9,-9)
5851 3522 Ψ : 25/11
5851 3609 $4/5\div\ln(4/5)$
5851 3749 cr : $\zeta(3)/6$
5851 4327 ℓ_{10} : $3\ln 2/8$
5851 5278 rt : (5,1,4,3)
5851 6597 cr : $2\sqrt{2}+2\sqrt{3}/3$
5851 6957 Ei : $9\zeta(3)/5$
5851 7139 rt : (8,7,0,-4)
5851 7251 ℓ_{10} : $9\sqrt[3]{5}/4$
5851 7291 ℓ_{10} : $\sqrt{3}/2+4\sqrt{5}/3$
5851 8612 e^x : $4\sqrt{2}-3\sqrt{3}$
5851 8717 at : P_{55}
5851 8893 rt : (3,8,-4,-1)
5851 9218 2^x : $3\sqrt{3}-3\sqrt{5}/2$
5851 9766 ℓ_2 : $e/3+2\pi/3$
5852 1021 sq : $2e/3-\pi/3$
5852 1243 rt : (9,2,6,-6)
5852 3596 10^x : $4\ln 2/5$
5852 3750 cu : $8\sqrt{2}/5$
5852 3964 $e\times s\pi(2/5)$
5852 4604 ζ : 10/9
5852 5751 2^x : $8\pi/3$
5852 6181 rt : (8,6,4,-6)
5852 6207 tπ : P_{25}
5852 6337 rt : (2,9,-1,-1)
5852 6764 K : $\sqrt[3]{5}/9$
5852 9967 2^x : $3(e+\pi)/2$
5853 1882 rt : (5,2,8,5)
5853 2904 rt : (8,5,8,-8)
5853 3091 sr : $4\pi/5$
5853 3883 rt : (3,7,0,-3)
5853 4141 sinh : $e\pi/3$
5853 4221 K : 4/21
5853 4440 tanh : $1/\ln(\sqrt{2}\pi)$
5853 6096 rt : (7,9,6,-8)
5853 6711 tπ : $4\sqrt{2}/3-3\sqrt{3}/4$
5853 7337 rt : (9,7,-9,4)
5853 7968 θ_3 : 5/8
5853 8099 rt : (4,2,6,-5)
5853 8871 rt : (6,5,-4,5)
5854 0849 sr : $4e-4\pi/3$
5854 1953 erf : $2\sqrt[3]{3}/5$
5854 2182 K : $3(1/\pi)/5$
5854 2195 id : $3\sqrt{2}/2+2\sqrt{3}$
5854 2858 rt : (3,6,4,-5)
5854 3614 e^x : $1/4-\pi/4$
5854 5456 sq : $2\pi^2/5$
5854 8834 rt : (3,5,8,-7)
5855 1489 rt : (2,9,6,-2)
5855 1629 id : $4e+3\pi/2$
5855 1824 J_0 : $1/\ln(\sqrt[3]{3}/3)$
5855 1927 tanh : $3\sqrt{5}/10$
5855 3330 sπ : $7\pi/10$
5855 3692 $1/2+\exp(3)$
5855 9673 tanh : $(\sqrt{2}\pi/2)^{-1/2}$
5856 0685 cr : $4\sqrt{3}/3+3\sqrt{5}/4$
5856 3346 Γ : $7\sqrt[3]{2}$
5856 5176 K : $\sqrt{3}/9$
5856 6995 10^x : $\sqrt{3}+\sqrt{5}+\sqrt{7}$
5856 7498 erf : γ
5856 7510 e^x : $5\sqrt[3]{5}/9$
5856 8410 Γ : $\sqrt{19}$
5856 9042 Ei : $2e\pi/9$
5857 0965 e^x : 19/20
5857 1289 cu : $2-2e$
5857 3958 cr : $3/2-3\sqrt{3}/4$
5857 3981 10^x : P_{16}
5857 4439 10^x : P_{40}
5857 5666 tπ : $4e/3-2\pi/3$
5857 7575 10^x : 22/25
5857 8382 erf : $\sqrt{3}/3$
5857 8643 id : $2-\sqrt{2}$
5857 8821 cu : $4+\pi/4$
5858 0662 ln : $5(e+\pi)/6$
5858 3162 e^x : $4+\pi/3$
5858 3318 rt : (5,9,-9,3)
5858 3883 2^x : P_{81}
5858 4084 ℓ_{10} : $3\sqrt{2}/2+\sqrt{3}$
5858 4798 2^x : $9\pi/8$
5858 4992 rt : (7,5,-4,9)
5858 6446 rt : (9,9,7,-9)
5858 7814 rt : (8,9,-9,-5)
5858 9721 sr : $3\sqrt{3}+2\sqrt{5}/3$
5859 2320 ℓ_{10} : $1/2+3\sqrt{5}/2$
5859 8744 id : $(e+\pi)/10$
5860 0166 cu : $2\sqrt{2}/3+2\sqrt{3}$
5860 0401 erf : $5\ln 2/6$
5860 4166 sr : $2\zeta(3)/7$
5860 5640 2^x : $\pi^2/2$
5860 5689 sr : $9\sqrt{5}/8$
5860 5956 rt : (3,6,-3,4)
5860 6196 e^x : $3\pi/5$
5860 6603 cu : $3\pi^2/4$
5860 6959 10^x : $3/2-\sqrt{3}$
5860 6967 cr : $7\sqrt[3]{5}/3$
5860 8898 cu : $\sqrt{2}-\sqrt{3}/3$
5861 1609 rt : (4,9,7,-8)
5861 2156 cr : P_8
5861 4339 rt : (5,5,9,-8)
5861 4462 rt : (2,7,-8,9)
5861 4474 10^x : $\zeta(3)/6$
5861 5422 sπ : $1/3+\sqrt{3}/2$
5861 5892 ℓ_{10} : $3/2+3\pi/4$
5861 6556 rt : (4,2,-5,3)
5862 0212 rt : (5,6,5,-6)
5862 0559 2^x : $3\sqrt{3}/2+4\sqrt{5}$
5862 0653 Γ : 19/4
5862 2853 ln : $3/4+\pi/3$
5862 4129 ζ : 9/8
5862 4649 rt : (6,2,7,-6)
5862 5682 rt : (9,9,-9,-1)
5862 8397 rt : (5,8,2,3)
5862 9436 $1/5+\ln(4)$
5862 9580 ln : $3+4\sqrt{2}/3$
5862 9911 ℓ_2 : $(\ln 2)^3$
5863 0057 sinh : $7(1/\pi)/4$
5863 0147 rt : (7,2,9,6)
5863 0237 at : $3/4-\sqrt{2}$
5863 2571 Ei : $3\sqrt[3]{3}/8$
5863 4889 rt : (6,3,3,-4)
5863 6013 rt : (4,9,0,9)
5863 8314 sr : $\sqrt{19/3}$
5863 9453 rt : (8,9,-6,-5)
5864 0591 rt : (5,8,-3,-2)
5864 1265 rt : (4,8,-8,-5)
5864 1909 sq : $1-\pi$
5864 3331 rt : (7,1,5,4)
5864 4031 J_0 : 15/11
5864 5481 θ_3 : $9(1/\pi)/10$
5865 0986 rt : (6,4,-1,-2)
5865 1289 e^x : 6/13
5865 1787 id : $4\sqrt{2}/3-2\sqrt{5}$
5865 1903 rt : (5,2,-5,7)
5865 2001 10^x : $(e+\pi)^{-1/3}$
5865 2452 rt : (1,9,-7,-7)
5865 3786 Ei : $1/\ln(\sqrt[3]{4})$
5865 4635 rt : (7,5,-8,-5)
5865 4754 rt : (8,5,7,4)
5865 4757 K : $2\ln 2/7$
5865 6812 rt : (6,9,8,-9)
5865 7997 id : $4\sqrt{2}/3-3\sqrt{3}/4$
5866 0687 rt : (5,9,-7,0)
5866 3806 2^x : $\sqrt{3}/2+\sqrt{5}$
5866 3885 cu : $(e+\pi)/7$
5866 4148 at : $e/3-\pi/2$
5866 5571 rt : (7,0,1,2)
5866 6893 rt : (6,8,2,7)
5866 7058 e^x : $1/3+2\sqrt{3}$
5866 8509 ζ : $3\zeta(3)/4$
5866 8849 2^x : $3(e+\pi)/8$
5866 9463 K : $5(1/\pi)/8$
5867 1080 rt : (9,9,9,4)
5867 1770 ℓ_2 : $2\sqrt{2}/3+\sqrt{5}/4$
5867 2162 e^x : $3/4+3e$
5867 4116 rt : (7,6,6,-7)
5867 7544 10^x : $2\sqrt{2}+2\sqrt{5}/3$
5867 7596 cr : $\sqrt{2}/7$
5867 7951 as : $3\sqrt{3}/4-\sqrt{5}/3$
5867 9954 rt : (6,7,-4,0)
5868 1983 sπ : $10e\pi/7$
5868 2321 Γ : $7\pi^2/9$
5868 2626 rt : (8,2,8,-7)
5868 3187 rt : (1,9,-5,6)
5868 6784 K : 1/5
5868 7123 λ : $\exp(-12\pi/13)$
5868 7451 rt : (1,7,7,-5)
5868 7523 rt : (8,4,3,2)
5868 8358 id : $3\sqrt{2}/2-3\sqrt{5}$
5868 9588 rt : (7,7,2,-5)
5869 1088 2^x : $3/4+\sqrt{3}$
5869 2427 K : $\zeta(3)/6$
5869 3258 $\ln(4)\times\ln(\pi)$
5869 5604 Γ : $4e\pi/5$
5869 6146 rt : (1,2,7,-5)
5869 7000 rt : (7,7,-4,4)
5869 7025 Ψ : $10\pi^2$
5870 1247 rt : (8,3,4,-5)
5870 1853 at : P_{81}
5870 2166 10^x : $4\sqrt{3}/3-2\sqrt{5}/3$
5870 2253 e^x : $1/3-\sqrt{3}/2$
5870 2513 sq : $\sqrt{3}-\sqrt{5}+\sqrt{7}$
5870 3015 cu : $4\sqrt{2}+4\sqrt{5}/3$
5870 3147 rt : (1,9,8,-8)
5870 6210 $\sqrt{3}\times\ln(2/5)$
5870 6517 rt : (8,7,-4,8)
5870 7391 erf : 11/19
5870 9275 sr : $3e+3\pi/2$
5871 1031 rt : (7,8,-2,-3)
5871 1618 sπ : $10\sqrt[3]{2}/3$
5871 3824 rt : (9,6,7,-8)
5871 5645 rt : (9,1,6,5)
5871 6712 ζ : $3\sqrt{2}/8$
5871 6873 Γ : $5e\pi/2$
5871 8509 rt : (2,2,9,5)
5872 0363 K : $\sqrt{2}/7$
5872 2727 rt : (1,8,3,-1)
5872 2788 θ_3 : $(\sqrt[3]{3}/3)^{1/3}$
5872 3571 ℓ_{10} : $2e-\pi/2$
5872 4300 rt : (9,8,5,2)
5872 4926 cu : $3/2-4e/3$
5872 5213 ln : $1/2+3\sqrt{3}/4$
5872 5876 2^x : $8\sqrt[3]{5}/3$
5872 6470 ℓ_{10} : $3+\sqrt{3}/2$
5872 7970 rt : (8,4,0,-3)
5872 8624 sq : $1/3+\sqrt{3}/4$
5873 0369 rt : (6,9,-7,-6)
5873 0554 K : $(\sqrt{2}\pi/2)^{-2}$
5873 2092 $\sqrt{3}-\ln(\pi)$
5873 2437 rt : (9,7,3,-6)
5873 4023 cr : $10\sqrt{3}$
5873 4670 $\sqrt{5}-\exp(1/2)$
5873 5049 sinh : $5\sqrt{5}/9$
5873 5845 Ψ : $\sqrt{2}+\sqrt{3}+\sqrt{5}$
5873 6664 sq : $\sqrt{3}/3-3\sqrt{5}$
5873 6773 sq : $(\sqrt{2}\pi/2)^{-1/3}$
5873 9259 e^x : $e/2+2\pi$
5874 0105 id : $\sqrt[3]{4}$

5874 2299 $10^x : \ln(\sqrt[3]{4}/2)$
5874 2714 rt : (7,9,-6,-1)
5874 3328 $\ell_2 : 5\zeta(3)$
5874 3459 $Li_2 : 4\sqrt{3}/7$
5874 4745 $\Psi : 4\sqrt{5}/9$
5874 5595 cu : $2\sqrt{3} + 3\sqrt{5}$
5874 7342 rt : (6,6,-9,2)
5874 9785 $s\pi : 4\sqrt{3}/3 + 2\sqrt{5}/3$
5874 9879 rt : (9,0,2,3)
5875 1808 $\zeta : 8/7$
5875 5374 $K : 1/\sqrt{24}$
5875 6344 rt : (5,5,-9,-6)
5875 6500 sinh : $7(1/\pi)$
5875 6753 rt : (4,1,-6,-9)
5875 6976 rt : (9,8,-1,-4)
5876 0028 rt : (3,2,8,-6)
5876 0965 $K : \exp(-\sqrt[3]{4})$
5876 2186 sr : $3/2 - 2\sqrt{3}/3$
5876 2747 $10^x : 7\sqrt{2}/9$
5876 3090 rt : (3,8,1,1)
5876 9248 sr : $3/2 + 3\sqrt{3}$
5876 9526 rt : (8,5,-4,-1)
5877 2392 $\ln(\pi) \div \exp(2/3)$
5877 2400 ln : $10\sqrt[3]{2}/7$
5877 4145 rt : (1,8,9,-2)
5877 5001 $3/4 \times \exp(3/4)$
5877 5162 $Li_2 : 2\sqrt[3]{2}/5$
5877 5640 $e^x : 3\sqrt{2}/4 + 3\sqrt{3}/4$
5877 7219 rt : (2,7,2,-4)
5877 8333 as : $4\ln 2/5$
5877 8525 $s\pi : 1/5$
5877 8666 ln : 5/9
5877 9926 rt : (4,6,7,-7)
5878 0962 rt : (1,8,-4,9)
5878 2638 rt : (4,9,-8,-7)
5878 4302 $s\pi : ((e+\pi)/3)^{-1/3}$
5878 5172 $2^x : \sqrt{2}/2 + 2\sqrt{5}/3$
5878 5572 rt : (2,1,5,3)
5878 5614 sr : $8(e+\pi)/7$
5878 6353 $\zeta : \ln 2/8$
5878 7686 $K : \sqrt[3]{3}/7$
5878 7789 rt : (3,9,-8,-6)
5878 8723 $10^x : 1/2 + \sqrt{5}$
5879 0534 $Ei : (e/2)^{-2}$
5879 0793 rt : (9,9,-5,-2)
5879 1406 sr : $2 + 4e$
5879 1599 rt : (5,5,1,9)
5879 5382 cu : $3\sqrt{3}/2 - 4\sqrt{5}$
5879 6296 cu : 7/6
5879 7022 as : $(e+\pi)^{-1/3}$
5879 7734 id : $4/3 - \sqrt{5}/3$
5879 9274 rt : (6,4,-5,6)
5879 9805 $\zeta : 5(1/\pi)/3$
5880 0148 rt : (5,2,9,-7)
5880 0260 at : 2/3
5880 3981 $\ell_2 : P_{81}$
5880 4562 $\ell_{10} : \sqrt{15}$
5880 5293 $2^x : P_{26}$
5880 6799 rt : (9,1,-2,-1)
5880 7187 $s\pi : 10\sqrt[3]{2}/7$
5880 7223 $t\pi : 4\sqrt{2}/3 - 2\sqrt{5}$
5880 8228 rt : (6,6,8,-8)
5880 9792 $2^x : 4e$
5881 0865 sq : $3e + \pi/2$
5881 1959 sinh : $1/\ln(\sqrt{5})$
5881 3952 rt : (6,2,-6,-3)
5881 6478 $\ell_{10} : P_{28}$
5881 6555 rt : (2,9,-9,-8)
5881 6915 rt : (4,7,3,-5)
5881 8752 $\Gamma : \sqrt{10}\pi$
5881 9170 $K : \exp(-\pi/2)$
5881 9517 rt : (9,9,-4,7)
5882 0174 $\ell_{10} : 1/4 + 4e/3$
5882 0278 $K : 3\ln 2/10$
5882 1381 $\zeta : 2/23$
5882 1507 rt : (5,4,-5,2)
5882 3529 id : 10/17
5882 4912 cu : $e/4 - 3\pi/2$
5882 6965 $K : 5/24$
5882 9015 $10^x : (5/3)^3$
5882 9377 rt : (8,6,9,-9)
5882 9456 rt : (5,6,-5,-4)
5883 0081 $K : 1/\sqrt{23}$
5883 0318 rt : (4,8,-3,8)
5883 0751 cr : $10\zeta(3)/3$
5883 0976 $e^x : 3/4 + 3\sqrt{5}/2$
5883 3772 $2^x : 9\sqrt{5}$
5883 5963 $\ell_{10} : (\pi/2)^3$
5883 6399 rt : (2,8,-2,-2)
5883 7457 $s\pi : P_{16}$
5883 7662 $t\pi : 1/3 + 2\sqrt{3}/3$
5883 7775 $s\pi : P_{40}$
5883 9063 $Li_2 : 7\sqrt{2}/10$
5884 0737 $10^x : 3/2 - 3\sqrt{3}/4$
5884 1002 rt : (8,9,-4,3)
5884 1256 at : P_{26}
5884 2869 rt : (8,6,-8,1)
5884 3368 rt : (6,7,4,-6)
5884 5726 id : $9\sqrt{3}$
5884 5778 rt : (5,3,5,-5)
5884 8867 rt : (1,9,4,4)
5885 0736 rt : (4,1,6,4)
5885 4332 rt : (3,5,-6,3)
5885 4559 rt : (7,2,-7,2)
5885 5348 $J : \exp(-19\pi/5)$
5885 5528 $K : \sqrt[3]{2}/6$
5885 6619 sr : $\sqrt{3}/5$
5885 6761 $e^x : 4\sqrt{5}/7$
5885 9079 sinh : $\sqrt{5}/4$
5886 2257 rt : (8,7,5,-7)
5886 2546 $t\pi : 7\ln 2/3$
5886 4142 $\Gamma : 14/25$
5886 4901 $K : 4/19$
5886 5606 $e^x : 23/18$
5886 5621 $s\pi : \zeta(3)/6$
5886 5889 $2^x : 5\zeta(3)/9$
5886 6526 $2^x : \sqrt{2}/4 - \sqrt{5}/2$
5886 6816 rt : (3,5,7,3)
5886 6921 $10^x : 1/4 + 2\sqrt{2}/3$
5886 8710 cu : $7\pi/9$
5886 9026 sr : $7\sqrt[3]{3}/4$
5886 9560 rt : (7,3,6,-6)
5886 9662 $e^x : 2e/3 + 4\pi$
5887 0484 rt : (4,8,-1,-3)
5887 0501 id : $(\ln 2/2)^{1/2}$
5887 3589 $\Gamma : 4\sqrt[3]{2}/9$
5887 4696 rt : (3,6,-6,-5)
5887 6013 $Ei : \sqrt{18}\pi$
5887 8412 rt : (7,9,-4,-1)
5887 9330 at : $5\zeta(3)/9$
5887 9592 cr : $1/\sqrt{24}$
5888 1261 $10^x : (e+\pi)^{-1/2}$
5888 2612 rt : (6,1,7,5)
5888 3361 $10^x : 11/3$
5888 4410 $J_0 : e/2$
5888 5972 rt : (9,3,7,-7)
5888 7524 $10^x : \sqrt{3}/3 + \sqrt{5}/2$
5888 9429 $\zeta : (\pi/2)^{-1/3}$
5889 1148 rt : (6,8,0,-4)
5889 3786 rt : (8,6,-5,9)
5889 3849 $Ei : 13/24$
5889 4265 $K : 2(1/\pi)/3$
5889 4506 rt : (5,1,-6,8)
5889 5767 rt : (2,8,1,-7)
5889 8545 rt : (1,9,8,-6)
5889 9415 ln : $3/4 + 4\pi$
5890 1513 rt : (8,1,8,6)
5890 2512 cr : $\sqrt{2} + 3\sqrt{3}/2$
5890 2691 ln : $\sqrt{24}$
5890 3097 as : 5/9
5890 5014 rt : (8,8,1,-5)
5890 5609 $1/5 + \exp(2)$
5890 7297 $\zeta : 3\sqrt[3]{2}/4$
5890 8989 $\theta_3 : \ln(4/3)$
5890 9704 rt : (4,2,-7,-4)
5891 1284 $Ei : 1/\ln(\sqrt{2}\pi/2)$
5891 1351 $\Psi : 4\sqrt[3]{5}/3$
5891 1510 id : $\exp(-\sqrt[3]{4}/3)$
5891 1760 $K : 1/\sqrt{22}$
5891 2614 rt : (5,5,8,4)
5891 3268 rt : (2,6,-2,8)
5891 4963 rt : (5,4,1,-3)
5891 5472 $t\pi : (3\pi)^3$
5891 8534 sq : $4\sqrt{2}/3 - \sqrt{5}/2$
5891 9644 $\ell_{10} : 10e/7$
5891 9729 rt : (9,7,1,0)
5892 0660 $\exp(1/2) \times s\pi(\sqrt{2})$
5892 1412 $K : \sqrt[3]{5}/8$
5892 1553 $\zeta : 7/6$
5892 2906 sq : $\sqrt{2}/3 - \sqrt{3}$
5892 3261 rt : (3,6,9,-8)
5892 5411 $10^x : 11/10$
5892 8286 rt : (7,4,2,-4)
5892 8863 $\zeta : 15/22$
5892 9879 sinh : $4\sqrt[3]{2}/3$
5893 0957 $K : 3/14$
5893 0962 $\Psi : 7\pi^2/5$
5893 1254 rt : (7,5,9,5)
5893 1767 $e^x : 2e - \pi/3$
5893 3448 sq : $1/4 + e/2$
5893 3936 rt : (7,6,-5,5)
5893 4215 rt : (2,8,2,-9)
5893 4725 $10^x : 4\sqrt{2}/3 + 3\sqrt{3}$
5893 4764 ln : $5\sqrt[3]{3}/4$
5893 6800 rt : (9,4,3,-5)
5893 7000 rt : (1,7,-5,-4)
5893 7606 ln : $(e+\pi)^{1/3}$
5893 9519 $2^x : 2\sqrt{2} + \sqrt{3}/4$
5894 2378 sq : $3 - \sqrt{3}/4$
5894 3682 at : $(\sqrt{5})^{-1/2}$
5894 3850 id : $\sqrt{2}/3 + \sqrt{5}/2$
5894 5417 rt : (2,9,-6,0)
5894 6011 $\ell_{10} : 2 + 4\sqrt{2}/3$
5894 9801 rt : (4,1,-6,4)
5894 9928 ln : $3\zeta(3)/2$
5895 1337 rt : (9,9,-8,-6)
5895 1784 at : $\sqrt{2} - \sqrt{5}/3$
5895 2840 $10^x : 3\sqrt{2} - 2\sqrt{5}$
5895 3069 rt : (6,7,1,8)
5895 4892 rt : (4,9,-5,-1)
5895 6783 $2^x : \sqrt[3]{25}$
5895 7380 rt : (7,9,-9,-7)
5895 8823 cu : $3\sqrt{5}/8$
5895 9822 rt : (6,9,-4,-2)
5896 0564 $Li_2 : 10\ln 2/7$
5896 1508 $K : (5/3)^{-3}$
5896 1700 $e^x : \sqrt{2}/3 + 4\sqrt{5}/3$
5896 2136 id : $\sqrt{2}/4 + \sqrt{5}$
5896 2446 cu : $2\sqrt{3} + \sqrt{5}/3$
5896 2620 $\ln(\sqrt{2}) \div s\pi(1/5)$
5896 2845 rt : (8,9,-3,-3)
5896 5647 ln : $4\ln 2/5$
5896 7334 $e^x : 7e\pi/8$
5896 8432 $2^x : (\sqrt{5})^{-1/2}$
5896 9098 sinh : $4\sqrt[3]{2}/9$
5896 9407 $\theta_3 : 3\sqrt{5}/7$
5897 0584 $K : \sqrt{3}/8$
5897 3171 sinh : 14/25
5897 5451 rt : (8,0,4,4)
5897 6157 $Ei : (2\pi)^{-1/3}$
5897 6782 sr : 8/23
5897 9586 $s\pi : \sqrt{2} + \sqrt{5} - \sqrt{6}$
5898 0077 rt : (9,7,8,-9)
5898 0710 rt : (7,7,7,-8)
5898 1353 $2^x : \sqrt{2} - \sqrt{5}/3$
5898 1543 rt : (5,7,6,-7)
5898 2691 rt : (3,7,5,-6)
5898 4371 rt : (1,7,4,-5)
5898 4696 $10^x : P_8$
5898 5062 $\ln(3/5) \div s\pi(1/3)$
5898 6503 $K : 5/23$
5898 7068 rt : (1,7,-3,-4)
5898 7601 $J_0 : \sqrt[3]{5/2}$
5898 8859 $K : 5\sqrt[3]{5}/9$
5899 1122 $J_0 : 19/14$
5899 1576 ln : $\sqrt{18}\pi$
5899 2100 rt : (3,7,-2,-3)
5899 3888 rt : (9,8,-5,8)
5899 4200 id : $4e - 2\pi$
5899 4767 $\ell_2 : 4/3 + 3\sqrt{5}/4$

5899 7255 at : $\ln((e+\pi)/3)$
5899 9884 tπ : $\exp(\sqrt[3]{3}/3)$
5900 1116 Ei : $2\sqrt{3}$
5900 1123 K : 19/20
5900 1423 e^x : $3\sqrt{2}/2+\sqrt{5}/4$
5900 1436 K : $1/\sqrt{21}$
5900 2006 sr : $3\sqrt{5}$
5900 2050 rt : (4,7,-6,7)
5900 2454 rt : (6,6,-5,1)
5900 4408 Γ : $8\sqrt[3]{5}/5$
5900 4828 rt : (5,7,-1,-2)
5900 5940 ζ : $7\sqrt{2}/9$
5900 6190 2^x : $4/3-2\pi/3$
5900 6443 rt : (1,8,-6,-8)
5900 7615 rt : (9,5,-1,-3)
5901 2365 Ψ : $4\ln 2/7$
5901 2737 ℓ_{10} : $3/4+\pi$
5901 2767 rt : (8,3,9,-8)
5901 3300 e^x : $3\sqrt{3}/2+3\sqrt{5}/2$
5901 4365 cr : $2/3+3\sqrt{5}/2$
5901 4746 2^x : $9\sqrt[3]{5}/7$
5901 5104 tπ : $1/3-\pi/4$
5901 5580 sr : $3-\sqrt{2}/3$
5901 6406 sπ : $(\sqrt[3]{3})^{1/2}$
5901 6472 rt : (7,5,-2,-2)
5901 6574 rt : (1,7,-8,-9)
5901 6994 id : $5\sqrt{5}/2$
5901 9238 rt : (6,3,8,-7)
5902 0773 rt : (6,3,-6,7)
5902 2644 e^x : $2\sqrt{2}+\sqrt{5}/4$
5902 2731 sπ : $1/2+3\sqrt{3}/4$
5902 5979 rt : (4,9,-7,-9)
5902 6454 cu : $9\pi^2/2$
5902 6739 ℓ_{10} : $\exp(e/2)$
5902 7544 ℓ_2 : P_1
5902 8423 rt : (4,3,7,-6)
5902 8539 ln : $\sqrt{3}/4+2\sqrt{5}$
5902 9039 sq : $\ln(5)$
5902 9716 rt : (9,8,4,-7)
5903 0503 Ei : $(\pi/2)^{1/2}$
5903 2057 rt : (5,5,-3,-1)
5903 3244 2^x : $\sqrt{2}2\pi$
5903 6404 sq : $1/4-4\sqrt{5}$
5903 7339 rt : (9,4,6,4)
5903 8197 rt : (7,8,3,-6)
5903 9160 id : $8\pi/7$
5903 9297 λ : 1/18
5903 9481 rt : (3,8,-8,7)
5904 1205 sq : $2/3+2\sqrt{2}/3$
5904 1534 ln : $4/3+\sqrt{2}/3$
5904 2488 rt : (2,3,6,-5)
5904 2661 e^x : $4\sqrt{2}/3$
5904 3698 $\ln(3)+\exp(2/5)$
5904 5338 cr : $2\sqrt{3}+\sqrt{5}/4$
5904 5620 sq : $(1/(3e))^{-3}$
5904 6025 Ei : $7\zeta(3)$
5904 7809 ln : $4+e/3$
5904 9820 rt : (5,8,2,-5)
5905 3916 2^x : $\ln((e+\pi)/3)$
5905 4143 rt : (4,9,9,3)
5905 7411 rt : (8,8,-5,4)
5906 0041 e^x : $(e/3)^{1/2}$
5906 0451 erf : 7/12
5906 1175 2^x : $2+4\sqrt{5}/3$
5906 1610 id : $1/\ln(2e)$
5906 2498 rt : (7,4,5,3)
5906 2812 cr : $\sqrt[3]{3}/7$
5906 3430 sπ : $(\ln 2)^{-1/2}$
5906 4641 Ei : $5\sqrt{3}/4$
5906 6729 rt : (3,8,1,-4)
5906 6760 sq : $3\sqrt{2}-4\sqrt{5}/3$
5906 7568 e^x : $1/\sqrt[3]{10}$
5906 8923 at : $1/\ln(\sqrt{2}\pi)$
5906 9104 cr : $9\sqrt{5}/5$
5907 0063 2^x : $\sqrt{2}/4+2\sqrt{5}/3$
5907 1652 tπ : $e/2+4\pi/3$
5907 1685 10^x : $\sqrt{3}-\sqrt{5}+\sqrt{7}$
5907 1851 id : $\sqrt{2}/4-4\sqrt{5}$
5907 4314 ℓ_{10} : $9\sqrt{3}/4$
5907 4675 K : 2/9
5907 7307 θ_3 : $\sqrt[3]{3}/5$
5908 0534 rt : (8,4,5,-6)
5908 1795 sr : $\pi/9$
5908 2011 as : $7(1/\pi)/4$
5908 2785 sr : $2+3\pi/2$
5908 4371 Γ : 1/5
5908 5679 K : $7(1/\pi)/10$
5908 5873 rt : (5,1,9,6)
5908 6561 rt : (1,7,3,2)
5908 6882 rt : (2,6,-9,-7)
5908 7275 at : $3\sqrt{5}/10$
5908 7486 id : $8e\pi/9$
5908 7823 $\ln(2)-\exp(1/4)$
5908 9314 rt : (6,7,9,-9)
5908 9346 $\ln(\sqrt{5})-\exp(1/3)$
5909 0909 id : 13/22
5909 1488 K : $\exp(-3/2)$
5909 2873 rt : (8,1,0,-2)
5909 3249 rt : (9,9,0,-5)
5909 3591 rt : (1,8,0,-3)
5909 5315 Ψ : 4/25
5909 5403 at : $(\sqrt{2}\pi/2)^{-1/2}$
5909 5821 Ψ : $(\sqrt[3]{5}/3)^{-3}$
5909 8539 rt : (2,5,-7,3)
5909 8638 rt : (7,9,1,7)
5909 9025 id : $9\sqrt{2}/8$
5909 9912 as : $3/2-2\sqrt{2}/3$
5910 0345 K : $\sqrt{5}/10$
5910 0416 ζ : $(\sqrt{5})^{-3}$
5910 0904 rt : (7,7,0,-1)
5910 2129 rt : (5,0,-7,9)
5910 2168 rt : (6,4,4,-5)
5910 4030 $\exp(3/5)-\exp(5)$
5910 6511 rt : (4,7,8,-8)
5911 2340 ln : $3\sqrt{2}/4+\sqrt{5}/3$
5911 2913 rt : (9,6,-5,-1)
5911 3286 ln : $2/3+3\sqrt{2}$
5911 4700 rt : (7,9,-1,-4)
5911 5618 rt : (5,3,-6,3)
5911 5629 e^x : $2\sqrt{2}-3\sqrt{5}/2$
5911 6455 ln : $3\sqrt{3}/4-\sqrt{5}/3$
5911 8478 rt : (4,6,-8,-6)
5912 0133 rt : (3,1,8,5)
5912 0859 sπ : $3e-3\pi/4$
5912 2131 rt : (5,4,4,2)
5912 2460 sπ : P_8
5912 4007 tanh : $e/4$
5912 6591 at : $3\sqrt{2}/4-\sqrt{3}$
5912 6700 θ_3 : $\sqrt{3}/6$
5912 9309 Γ : $\sqrt{5}/4$
5912 9833 rt : (2,7,7,-7)
5913 0372 sr : $\sqrt{2}+\sqrt{5}/2$
5913 1005 erf : $((e+\pi)/2)^{-1/2}$
5913 1482 rt : (9,0,7,6)
5913 2734 cr : P_{39}
5913 3892 sinh : $\exp(-\sqrt{3}/3)$
5913 4182 cr : P_{33}
5913 5021 rt : (8,8,6,-8)
5913 6882 rt : (7,5,-6,6)
5913 7847 e^x : $3\sqrt{2}$
5913 8358 10^x : $(\ln 3/2)^{1/3}$
5913 8648 $\ln(\sqrt{3})-\exp(\pi)$
5913 9188 e^x : $3e/2-3\pi/4$
5914 0572 rt : (1,6,3,4)
5914 0914 id : $5e$
5914 2870 rt : (1,8,1,-2)
5914 5700 rt : (5,9,-2,-3)
5914 7494 2^x : $2/3+\sqrt{2}/2$
5914 8549 2^x : $8\zeta(3)/7$
5914 8718 e^x : $3e\pi/8$
5914 8778 rt : (9,7,-6,9)
5914 8937 rt : (6,1,-1,-1)
5915 1939 tanh : 17/25
5915 4943 rt : (4,6,-9,6)
5915 4943 id : $5(1/\pi)$
5915 6844 2^x : $2e/3+3\pi$
5915 7903 ln : $4\sqrt{2}-\sqrt{5}/3$
5915 8244 ζ : 6/5
5915 8791 rt : (6,8,5,-7)
5915 9372 10^x : P_{56}
5916 0797 sr : 7/20
5916 1578 sr : $1/3+4\pi$
5916 3263 rt : (7,8,-5,0)
5916 4044 θ_3 : 23/24
5916 7633 rt : (7,0,6,5)
5916 8442 2^x : $1/\ln(\sqrt{2}\pi)$
5916 9049 θ_3 : $1/\ln((e+\pi)/2)$
5916 9173 K : 5/22
5917 0159 at : $4\sqrt{3}/3-4\sqrt{5}/3$
5917 0906 K : $5(1/\pi)/7$
5917 1597 sq : 10/13
5917 1946 K : $\exp(-\sqrt{2}\pi/3)$
5917 2345 cu : $\sqrt{2}+3\sqrt{3}/2$
5917 3133 rt : (8,5,1,-4)
5917 3820 rt : (2,5,-8,-8)
5917 4268 ℓ_{10} : $3+e/3$
5917 5170 rt : (6,6,-7,-5)
5917 6062 sr : $(1/(3e))^{1/2}$
5917 6293 ζ : $10\sqrt[3]{3}/7$
5917 6893 rt : (9,4,8,-8)
5917 7161 tπ : P_{56}
5917 8102 $\ln(2/5)-s\pi(\sqrt{5})$
5917 8397 erf : $1/\sqrt[3]{5}$
5917 8540 rt : (4,9,-2,4)
5917 9687 cu : 15/8
5917 9705 ℓ_{10} : $2(e+\pi)/3$
5918 2457 K : $(2\pi/3)^{-2}$
5918 3673 sq : 15/7
5918 5282 rt : (1,1,7,4)
5918 6491 tπ : $4\sqrt{2}/3+4\sqrt{5}$
5918 6734 rt : (2,8,-5,2)
5918 7344 e^x : 20/21
5918 8436 rt : (3,3,9,-7)
5919 0887 rt : (6,6,0,9)
5919 1107 10^x : $7\pi/4$
5919 1697 sπ : $3\sqrt{2}/2+3\sqrt{5}/4$
5919 3132 id : $\sqrt{2}/2-3\sqrt{3}/4$
5919 4448 rt : (3,9,-3,-2)
5919 5199 cu : $\sqrt{8}\pi$
5919 6485 sr : $4+e$
5919 7522 2^x : $7\pi/10$
5919 7799 2^x : $3\sqrt{5}/10$
5919 8560 rt : (8,8,7,3)
5920 0000 cu : 24/5
5920 1514 rt : (9,3,2,2)
5920 4486 ln : $10\pi^2$
5920 7911 $\ln(5)\div\exp(1)$
5920 9536 rt : (7,4,7,-7)
5920 9618 rt : (8,9,2,-6)
5920 9989 K : $1/\sqrt{19}$
5921 0410 rt : (1,5,-5,1)
5921 0806 2^x : $(\sqrt{2}\pi/2)^{-1/2}$
5921 1557 rt : (6,2,-7,8)
5921 2523 2^x : $3\sqrt{3}/2-3\sqrt{5}/2$
5921 5210 rt : (2,7,-5,-5)
5921 5269 J_0 : 23/17
5921 6246 $2/5\div s\pi(\sqrt{5})$
5921 9358 rt : (4,0,-7,5)
5922 0933 sinh : $\ln(\sqrt[3]{5}/3)$
5922 1344 10^x : $9\sqrt[3]{5}/10$
5922 2949 Ψ : $8\sqrt{3}$
5922 3082 cu : $3+3\sqrt{3}$
5922 3151 rt : (6,5,0,-3)
5922 5601 rt : (5,0,5,4)
5922 5979 Ψ : 16/7
5922 8374 J_0 : $1/\ln(2\pi/3)$
5922 8650 $\pi-\ln(\sqrt{3})$
5923 2058 10^x : $\sqrt{2}/7$
5923 2068 rt : (7,8,8,-9)
5923 5879 id : $9(e+\pi)/8$
5923 5914 id : $2\sqrt{2}-\sqrt{5}$
5923 5991 K : 3/13
5923 6618 rt : (2,8,3,-5)
5923 8484 sr : $\exp(-\pi/3)$
5923 8960 2^x : $3\sqrt{3}/4+\sqrt{5}$
5924 0723 rt : (1,3,8,-6)
5924 1388 K : $\ln 2/3$
5924 2671 Ei : 8/9
5924 2765 K : $(\sqrt[3]{3}/3)^2$
5924 3088 sq : $8e\pi/3$

5924 3973 rt : (8,7,-6,5)
5924 4618 cr : $3\ln 2/10$
5924 5874 rt : (9,1,3,-4)
5924 6132 sπ : $\ln(\sqrt{2}\pi/2)$
5924 8696 J_0 : $9\zeta(3)/8$
5924 9937 rt : (6,9,1,3)
5925 0549 cr : $2+3e/4$
5925 1868 rt : (8,4,8,5)
5925 1914 Ei : $\zeta(3)/5$
5925 2305 ℓ_{10} : $7\sqrt{5}/4$
5925 3699 $\sqrt{5}+\exp(\sqrt{5})$
5925 3962 sπ : $5\sqrt[3]{3}/6$
5925 4017 Ei : $\sqrt{23}$
5925 4815 rt : (5,4,6,-6)
5925 6417 sq : $7\sqrt[3]{3}/8$
5925 6847 cr : $9\pi/7$
5925 7215 Ψ : $(e+\pi)/10$
5925 9259 sq : $4\sqrt{3}/9$
5926 1474 rt : (1,7,9,-8)
5926 2786 rt : (3,8,2,-1)
5926 3591 sinh : 9/16
5926 3755 rt : (6,5,-6,2)
5926 3971 rt : (9,5,4,-6)
5926 5762 ζ : 1/11
5926 5872 cu : $2/3+\sqrt{2}/2$
5926 8268 ln : $2\sqrt{2}/3+\sqrt{3}/2$
5926 8602 rt : (5,8,7,-8)
5926 8666 10^x : $3\sqrt{2}+3\sqrt{5}$
5926 9196 cu : $8\zeta(3)/7$
5926 9895 ζ : $2(1/\pi)/7$
5927 0013 tanh : 15/22
5927 0044 rt : (8,0,9,7)
5927 0400 cu : 21/25
5927 0761 sq : $4-\sqrt{5}/3$
5927 0834 2^x : $6e/5$
5927 1198 Ei : $\sqrt[3]{3}/6$
5927 1353 sπ : $\exp(1/(2e))$
5927 1534 rt : (5,9,-4,-9)
5927 3226 erf : $(e+\pi)/10$
5927 3802 rt : (9,9,5,-8)
5927 5153 2^x : $4/3+\sqrt{3}/2$
5928 1555 cr : 5/24
5928 2284 rt : (1,9,-4,-1)
5928 3897 Ψ : $9\zeta(3)/2$
5928 5287 rt : (9,7,-9,1)
5928 6031 tπ : $\ln(\sqrt{2}\pi)$
5928 6437 rt : (1,9,-6,-3)
5928 6561 cr : $7\sqrt{3}/3$
5928 7934 ℓ_2 : $\sqrt{2}/2+4\sqrt{3}/3$
5928 8645 tπ : $e/2-4\pi/3$
5929 3400 sπ : $\sqrt{2}/7$
5929 4867 cr : $3\sqrt{2}/4+4\sqrt{5}/3$
5929 5299 ℓ_2 : $(\ln 3/2)^3$
5929 6569 sπ : $3e+\pi/3$
5929 6614 e^x : $2-\pi/3$
5929 6971 e^x : $3\pi^2/4$
5929 7279 rt : (3,9,-5,5)
5929 8725 sr : $1/\sqrt[3]{23}$
5930 0077 sπ : $\zeta(3)$
5930 1019 ζ : $(\sqrt{2}\pi/2)^{-3}$
5930 3719 cr : $4\sqrt{2}/3-3\sqrt{5}/4$
5930 5558 rt : (1,6,-3,6)
5930 7033 rt : (8,6,-3,-2)
5930 9445 e^x : $1+3\pi/2$
5931 0377 e^x : $4+3\sqrt{5}/2$
5931 2480 e^x : $2\sqrt{3}/3-3\sqrt{5}/4$
5931 2733 rt : (7,4,-7,7)
5931 3910 rt : (7,8,0,8)
5931 4575 sq : $1/2-4\sqrt{2}$
5931 4870 rt : (3,6,-5,-1)
5931 4883 sπ : $(\pi/2)^{-1/2}$
5931 5186 sq : $10\ln 2/9$
5931 5701 rt : (7,9,4,-7)
5931 6191 sq : $1/3+3\sqrt{3}/2$
5931 6527 tanh : $1/\ln(\ln 2/3)$
5931 7284 rt : (3,8,6,-7)
5931 8752 rt : (7,5,3,-5)
5931 9309 2^x : $(\sqrt[3]{4}/3)^{-1/2}$
5931 9575 rt : (1,7,-8,-7)
5931 9977 as : $\sqrt{5}/4$
5932 0226 2^x : $(2\pi)^{1/3}$
5932 0616 rt : (2,7,1,6)
5932 1845 rt : (3,5,6,9)
5932 4169 K : 4/17
5932 5118 rt : (7,1,2,-3)
5932 6083 rt : (6,4,9,-8)
5932 7157 rt : (1,6,2,-8)
5932 7527 2^x : $3-2\sqrt{3}/3$
5932 9502 J : $\sqrt{3}/9$
5932 9516 e^x : $1+4\sqrt{5}/3$
5933 2218 K : $\sqrt{2}/6$
5933 2357 2^x : $\exp(1/\pi)$
5933 2634 tanh : $(\pi)^{-1/3}$
5933 3341 sr : $1/2+3e/4$
5933 3717 rt : (6,0,8,6)
5933 8415 rt : (5,2,-9,-5)
5933 9139 Ei : $9(e+\pi)/5$
5934 1738 tπ : $3\sqrt{2}/2+3\sqrt{5}$
5934 3025 rt : (9,9,-6,4)
5934 3483 Γ : $(3\pi)^{-3}$
5934 3537 Ψ : $10\ln 2/7$
5934 5629 id : $3\sqrt{2}/2+2\sqrt{5}$
5934 5923 K : $\exp(-\sqrt[3]{3})$
5934 6034 e^x : $4\sqrt{2}-\sqrt{5}$
5934 6615 rt : (7,3,1,1)
5934 7646 Γ : $\sqrt{15/2}$
5934 8249 rt : (4,8,-5,3)
5935 0165 Γ : $(1/\pi)/8$
5935 0250 10^x : $5\zeta(3)$
5935 1563 rt : (6,4,7,4)
5935 1873 tπ : $3\sqrt{2}+3\sqrt{3}/4$
5935 6359 rt : (5,2,-7,4)
5935 6390 10^x : 17/15
5935 7934 rt : (1,2,-5,-9)
5936 0126 10^x : $4+3\pi/4$
5936 0993 rt : (8,9,7,-9)
5936 1964 Ei : $10\sqrt[3]{5}/9$
5936 2010 cr : $3+\pi/3$
5936 3014 sq : $\sqrt{2}-\sqrt{6}+\sqrt{7}$
5936 5180 10^x : $1/2+\sqrt{5}/3$
5936 6732 2^x : 20/3
5936 7910 2^x : 11/8
5936 8429 cr : $e/3+\pi$
5936 9092 rt : (4,8,9,-9)
5936 9151 ln : $3/4+3\sqrt{2}/4$
5937 1222 $\exp(2/5)-\exp(3)$
5937 1300 rt : (5,9,3,-6)
5937 1368 Ei : 19/10
5937 2031 10^x : 25/7
5937 2320 sinh : $(\sqrt{2}\pi/2)^3$
5937 4598 rt : (6,8,-5,-4)
5937 5478 ℓ_{10} : $2\sqrt{2}/3+4\sqrt{5}/3$
5937 6196 rt : (8,5,6,-7)
5937 6390 2^x : $\ln(\sqrt{2}/3)$
5937 7345 ℓ_2 : P_{55}
5937 7651 rt : (6,1,-8,9)
5937 7684 sq : $3/2+3\pi/2$
5937 9727 K : 5/21
5938 0317 rt : (9,6,0,-4)
5938 0894 rt : (2,5,9,4)
5938 1366 10^x : 5/9
5938 1791 ℓ_{10} : $4\sqrt{2}-\sqrt{3}$
5938 2201 id : $4\sqrt{2}/3+3\sqrt{5}$
5938 5395 rt : (1,8,5,-6)
5938 6813 ℓ_2 : $3-2\sqrt{5}/3$
5938 6917 sπ : $3e\pi/8$
5938 7061 Ψ : $7\sqrt{2}/10$
5939 0095 rt : (4,4,8,-7)
5939 0605 id : $3/2-e/3$
5939 2469 K : $3(1/\pi)/4$
5939 4129 rt : (2,9,-1,-3)
5939 5643 e^x : $(\sqrt{3}/2)^{1/3}$
5939 6335 Γ : $2(e+\pi)$
5939 6820 rt : (7,7,-6,1)
5939 9222 rt : (5,5,2,-4)
5940 3148 rt : (2,6,-9,-9)
5940 3324 rt : (8,1,5,-5)
5940 5988 ℓ_{10} : $5\pi/4$
5940 6528 rt : (8,6,-7,6)
5940 6653 ℓ_{10} : $1-\sqrt{5}/3$
5940 7427 rt : (3,7,-7,-6)
5940 8852 sr : 6/17
5941 0743 tanh : $2\sqrt[3]{5}/5$
5941 1837 J_0 : $(\ln 3/2)^{-1/2}$
5941 2615 id : $6\ln 2/7$
5941 2880 rt : (6,9,6,-8)
5941 4543 rt : (9,5,9,-9)
5941 5572 rt : (6,6,-4,-1)
5941 5972 e^x : $9\sqrt{3}/5$
5941 6441 ζ : $(\sqrt[3]{4})^{1/3}$
5941 7933 K : 6/25
5941 8950 rt : (9,2,-1,-2)
5941 9972 cr : $4/3+e$
5942 0782 sq : $3e/2-3\pi$
5942 1486 Ψ : $4\sqrt{3}/7$
5942 2721 10^x : $7\sqrt[3]{2}/3$
5942 4071 ℓ_2 : $4+3e/4$
5942 4297 Ei : $10\sqrt{2}/3$
5942 4986 tanh : 13/19
5942 5494 K : $\sqrt[3]{3}/6$
5942 6230 K : $\zeta(3)/5$
5942 7673 e^x : $3e/4+3\pi$
5942 8543 sπ : $1/3+2\sqrt{3}$
5942 8827 Ψ : $\sqrt[3]{12}$
5942 9396 cu : $\sqrt{2}-2\sqrt{5}$
5942 9483 λ : $\exp(-11\pi/12)$
5942 9578 rt : (2,8,-1,-3)
5943 1036 2^x : $8\pi/5$
5943 1301 rt : (2,8,8,-8)
5943 2049 e^x : $\sqrt[3]{25/3}$
5943 2587 rt : (4,0,7,5)
5943 4344 as : $4\sqrt[3]{2}/9$
5943 7978 cr : $\sqrt[3]{2}/6$
5943 8580 as : 14/25
5943 9510 id : $1/2+2\pi/3$
5944 1447 rt : (6,8,6,2)
5944 1824 10^x : $e\pi/7$
5944 2020 rt : (4,1,-8,-6)
5944 6768 rt : (7,6,-9,-6)
5944 7550 rt : (9,1,8,-7)
5944 8053 rt : (6,5,5,-6)
5944 8076 sπ : $(\sqrt{2}\pi/2)^{-2}$
5944 8645 rt : (3,9,2,-3)
5945 0193 $\ln(5)\div\ln(3/4)$
5945 2914 erf : 10/17
5945 3489 id : $\ln(2e/3)$
5945 3545 cu : $7e\pi/2$
5945 4612 ζ : $((e+\pi)/3)^{1/3}$
5945 4828 ℓ_{10} : $4/3+3\sqrt{3}/2$
5945 5300 ℓ_{10} : $\sqrt{3}/3+3\sqrt{5}/2$
5945 6551 10^x : $(\sqrt{2}\pi/2)^{-2}$
5946 0076 sinh : $(\pi)^{-1/2}$
5946 0355 sr : $\sqrt{2}/4$
5946 2207 e^x : $(2\pi/3)^{1/3}$
5946 3874 cu : $3e-2\pi/3$
5946 4733 e^x : $5\pi^2/9$
5946 5226 tπ : $2\sqrt{2}/3-2\sqrt{5}/3$
5946 5436 rt : (8,3,4,3)
5946 6622 rt : (7,3,-8,8)
5946 6975 e^x : 7/15
5946 7637 rt : (1,8,4,7)
5946 8572 cu : $3e/2+4\pi$
5946 9330 K : $1/\sqrt{17}$
5946 9675 Ei : 6/25
5946 9707 10^x : $(\sqrt{5}/3)^3$
5946 9813 ℓ_2 : $5e/9$
5946 9997 rt : (5,1,1,-2)
5947 0260 sr : $\sqrt{3}/2+3\sqrt{5}/4$
5947 0803 sr : $10(1/\pi)/9$
5947 2519 10^x : $1/3-\sqrt{5}/4$
5947 2997 rt : (7,6,-1,-3)
5947 3470 rt : (2,9,0,4)
5947 6246 ζ : $\exp(\sqrt[3]{3}/2)$
5947 6701 rt : (2,7,2,-7)
5947 8302 rt : (7,5,8,-8)
5947 9341 2^x : 11/5
5947 9947 rt : (6,4,-7,3)
5948 0934 rt : (4,9,5,-7)
5948 1171 tπ : $(3\pi/2)^{1/2}$
5948 1196 K : $\exp(-\sqrt{2})$

5948 2786 rt : (2,4,7,-6)
5948 3261 rt : (5,9,-8,6)
5948 6989 id : $1/3 - 4\sqrt{3}$
5948 7516 rt : (1,9,5,0)
5948 8508 $s\pi : 3/4 + \pi/3$
5948 8508 $4 \times \exp(1/2)$
5948 8834 cr : 4/19
5949 0403 $erf : (\ln 2/2)^{1/2}$
5949 3698 rt : (9,8,-7,5)
5949 7128 rt : (7,7,-1,9)
5949 7246 rt : (4,7,-8,2)
5949 7801 rt : (3,3,-6,-4)
5949 8250 $t\pi : 3\sqrt{2} + 4\sqrt{3}$
5949 9527 $e^x : 8\pi$
5950 0059 $\ell_{10} : \sqrt{2}/3 + 2\sqrt{3}$
5950 1349 rt : (2,6,-4,1)
5950 2292 rt : (5,9,8,-9)
5950 3283 sinh : $(3/2)^3$
5950 3742 $2^x : e/3 + 3\pi/4$
5950 3917 rt : (8,6,2,-5)
5950 4161 $2^x : 4 + 4\sqrt{5}$
5950 5094 $K : \sqrt[3]{5}/7$
5950 5439 $10^x : 2e + 3\pi$
5950 9113 $t\pi : (\sqrt[3]{5}/3)^{1/3}$
5950 9410 rt : (5,5,-6,-2)
5950 9785 rt : (9,3,7,5)
5951 1182 $\zeta : 5/4$
5951 1214 sinh : $1/\ln(2e/3)$
5951 4005 rt : (6,7,-3,-3)
5951 5686 $2^x : 2\sqrt{2}/3 + \sqrt{3}/4$
5951 5733 at : $1 - 3\sqrt{5}/4$
5951 6947 rt : (8,9,-6,0)
5951 8071 $Li_2 : 8(1/\pi)/5$
5951 8208 rt : (6,1,4,-4)
5951 8284 as : $1/2 - 3\sqrt{2}/4$
5952 1531 rt : (3,5,-8,-2)
5952 3116 $erf : \exp(-\sqrt[3]{4}/3)$
5952 3249 rt : (9,6,5,-7)
5952 3603 $J_0 : 7\sqrt{3}/9$
5952 4147 $t\pi : 1/\ln(\sqrt[3]{5}/2)$
5952 4710 $Ei : e/5$
5952 5225 rt : (1,8,-5,1)
5952 8999 rt : (6,8,0,4)
5953 8516 cr : $3 + 3\sqrt{2}/4$
5953 8834 rt : (7,8,9,4)
5954 0349 rt : (2,3,-9,-6)
5954 0389 $2^x : 24/13$
5954 1336 rt : (7,1,7,-6)
5954 1340 $2^x : 3\sqrt{2}/2 - \sqrt{5}/3$
5954 2374 $\ln(\pi) - \ln(\sqrt{3})$
5954 3321 rt : (9,7,-4,-2)
5954 4603 $\Gamma : \sqrt[3]{2}/5$
5954 5700 cr : $\sqrt{2}/2 + 3\sqrt{5}/2$
5954 7390 cu : $2/3 + 3\sqrt{3}/4$
5954 8876 sr : $9\sqrt{2}/5$
5954 9473 rt : (8,5,-8,7)
5955 1360 rt : (4,7,6,8)
5955 5988 rt : (5,7,-6,-5)
5955 6786 sq : 24/19
5955 7166 rt : (5,1,-8,5)
5955 8496 $t\pi : 1/4 - 4\sqrt{3}$
5956 1242 $1/5 + \exp(1/3)$
5956 1545 $2^x : 2/3 - \sqrt{2}$
5956 3621 rt : (1,9,1,-4)
5956 3639 $t\pi : \sqrt[3]{5}/10$
5956 3691 $J_0 : 3\pi/7$
5956 3771 rt : (5,5,7,-7)
5956 4122 rt : (4,4,6,3)
5956 8354 id : $2\sqrt{3}/3 - \sqrt{5}/4$
5956 9268 $e^x : 4/3 - \sqrt{3}/2$
5957 0312 cu : 17/8
5957 0407 $K : \sqrt{3}/7$
5957 1891 rt : (4,7,-9,-7)
5957 3298 $K : 7(1/\pi)/9$
5957 4125 rt : (2,9,4,-6)
5957 4311 cu : $10\sqrt{5}/7$
5957 5411 id : $\sqrt{2} + \sqrt{3} + \sqrt{6}$
5957 5473 $J_0 : (2e/3)^{1/2}$
5957 5706 cu : $7\zeta(3)/10$
5957 6912 sr : $8(1/\pi)$
5957 8516 sr : $1/\ln(\sqrt{2}\pi/3)$
5957 8627 tanh : $(\sqrt{2}/3)^{1/2}$
5957 9685 sinh : 13/23
5958 0545 rt : (3,9,7,-8)
5958 0574 $\ell_{10} : 3 + 2\sqrt{2}/3$
5958 1872 rt : (6,2,-5,1)
5958 2428 rt : (5,4,9,5)
5958 2777 $2^x : 5\sqrt{2}/2$
5958 4125 $Ei : 4\sqrt{3}$
5958 5122 rt : (2,8,-7,-7)
5958 9516 $10^x : 5\sqrt{5}$
5959 1471 $\ell_2 : 2/3 + 3\pi/4$
5959 1657 $K : \sqrt{5}/9$
5959 1794 sq : $4\sqrt{2} + \sqrt{3}$
5959 1948 rt : (7,6,-7,2)
5959 2458 rt : (1,7,-4,4)
5959 2787 $\Gamma : 9\sqrt{5}/2$
5959 3918 rt : (8,9,-1,8)
5959 4987 rt : (7,2,-2,-1)
5959 8014 tanh : $4\zeta(3)/7$
5959 9144 rt : (8,2,1,-3)
5959 9422 sr : $3/2 + \pi/3$
5960 1185 rt : (9,2,4,-5)
5960 2453 rt : (7,2,-9,9)
5960 4922 id : $9\sqrt[3]{3}/5$
5960 5714 as : $\exp(-\sqrt{3}/3)$
5960 6148 $\exp(3/5) + s\pi(e)$
5960 7163 rt : (3,0,9,6)
5961 2304 $2^x : 2\sqrt{2} + \sqrt{3}$
5961 3377 rt : (8,6,7,-8)
5961 4735 rt : (7,6,4,-6)
5961 6073 sinh : $\exp(-\sqrt[3]{5}/3)$
5961 6850 rt : (6,6,1,-4)
5962 0607 rt : (5,9,1,-1)
5962 0717 sinh : $1/\ln(e + \pi)$
5962 0782 $\ell_{10} : 2\sqrt{2} + \sqrt{5}/2$
5962 1612 $\zeta : \ln(\ln 3)$
5962 3102 rt : (1,4,9,-7)
5962 4087 $e^x : (\ln 3)^{-1/2}$
5962 4222 $K : 1/4$
5962 4283 $\ell_{10} : 4\sqrt{3} - 4\sqrt{5}/3$
5962 4920 rt : (5,5,-1,6)
5962 5633 rt : (1,4,-1,8)
5962 6798 rt : (9,7,-8,6)
5962 7499 $\pi \times \ln(\pi)$
5962 9120 rt : (4,6,-5,0)
5963 1259 ln : $\pi^2/2$
5963 4172 sinh : $2\sqrt{2}/5$
5963 4229 rt : (1,8,-9,-8)
5963 5973 $\ell_{10} : 4\pi^2$
5963 6254 $\ln(3/5) - \exp(3)$
5963 6773 $\ell_2 : 6\sqrt[3]{2}/5$
5963 7355 tanh : 11/16
5964 2437 rt : (2,8,-6,-6)
5964 2715 $erf : 1/\ln(2e)$
5964 4256 sq : $2\sqrt{2} + 4\sqrt{5}$
5964 5531 $\exp(1/2) + \exp(2/3)$
5964 6681 id : $(3\pi/2)^{-1/3}$
5964 7842 ln : $6\sqrt{5}$
5964 8104 $2^x : 10\sqrt{5}/9$
5965 2415 $\ln(4) \times s\pi(\pi)$
5965 3251 $s\pi : \sqrt{2}/3 + \sqrt{3}$
5965 3395 $t\pi : \sqrt{2}/2 + 2\sqrt{3}$
5965 5000 sq : $\sqrt{3}/2 + \sqrt{5}/3$
5965 7167 rt : (8,3,9,6)
5965 7320 $t\pi : 2e/3 + \pi/2$
5965 7359 $1/4 + \ln(\sqrt{2})$
5965 7525 rt : (2,5,-9,-4)
5965 7793 rt : (3,8,-3,-4)
5965 8100 $\zeta : (\sqrt[3]{2})^{-1/2}$
5966 1886 rt : (6,3,-8,4)
5966 3013 $Ei : 5\sqrt{5}/7$
5966 3468 rt : (6,1,9,-7)
5966 3903 rt : (9,7,1,-5)
5966 5232 $\Gamma : 7(1/\pi)/4$
5966 5940 sq : $1/2 - 3e$
5966 6035 $erf : 13/22$
5966 6175 $e^x : 3e/4 - \pi/2$
5966 6303 $K : \sqrt[3]{2}/5$
5966 6786 $4/5 \times s\pi(\sqrt{3})$
5966 6953 rt : (3,6,-8,2)
5966 7137 $Ei : 1/\ln(2\pi)$
5966 8041 rt : (4,5,9,-8)
5966 9219 $10^x : 2\sqrt{5}$
5966 9848 $t\pi : 3e/4 + 2\pi$
5967 0773 rt : (9,7,6,3)
5967 1512 rt : (2,9,9,9)
5967 1843 sr : $\sqrt{13/2}$
5967 2987 $2/5 \times \exp(2/5)$
5967 4265 id : $3\sqrt{2} + 3\sqrt{5}/2$
5967 4284 $\Gamma : 2\zeta(3)/7$
5967 6184 rt : (8,4,-9,8)
5967 7716 rt : (7,3,6,4)
5967 7759 $J_0 : (e/3)^{-3}$
5967 9390 rt : (8,7,-2,-3)
5967 9779 $2^x : 9\sqrt[3]{3}$
5968 1367 $Ei : 9\sqrt{2}/2$
5968 1737 rt : (5,1,6,-5)
5968 2012 rt : (6,7,-6,-3)
5968 3708 $2^x : \exp(\sqrt{2}\pi/2)$
5968 4360 rt : (3,5,6,-6)
5968 6206 rt : (1,9,6,-7)
5968 6293 $2^x : 1/4 + 3\sqrt{3}$
5968 8287 at : $e/4$
5968 8544 sinh : $7\sqrt[3]{3}/6$
5969 1552 rt : (1,9,3,-4)
5969 3881 rt : (8,8,-7,1)
5969 4605 cu : $\sqrt{2}/3 - 3\sqrt{5}$
5969 6175 rt : (9,2,9,-8)
5969 6248 rt : (7,6,9,-9)
5969 6258 as : $\ln(\sqrt[3]{5}/3)$
5969 6828 rt : (4,8,0,-2)
5969 9859 rt : (1,9,-5,7)
5970 0736 $10^x : 3/2 + 4\sqrt{2}$
5970 4693 $t\pi : 7e/6$
5970 5550 $\Psi : 1/\sqrt[3]{5}$
5970 6490 rt : (7,7,-5,-1)
5970 7146 rt : (1,6,-5,-7)
5970 8042 $\ell_2 : \exp(3\pi)$
5970 9594 rt : (9,6,-8,-5)
5971 0210 ln : $4 + 3\pi$
5971 0248 $Ei : 13/6$
5971 0743 sq : 17/22
5971 2040 ln : $3/4 + 4\pi/3$
5971 2625 rt : (4,2,-2,8)
5971 4114 rt : (3,9,-7,-2)
5971 4753 rt : (6,6,6,-7)
5971 5895 rt : (2,7,-9,5)
5971 7159 cu : 16/19
5971 7543 rt : (8,2,6,-6)
5971 7665 at : 17/25
5971 7908 sq : $1/3 + 3\sqrt{5}/2$
5971 9141 id : $5\sqrt{5}/7$
5972 1205 rt : (7,2,-8,-4)
5972 3398 $K : 4(1/\pi)/5$
5972 5315 ln : $\sqrt[3]{6}$
5972 6903 rt : (6,3,3,2)
5972 7284 rt : (5,0,-9,6)
5972 9980 sq : $1/2 - 4\pi$
5973 0152 sr : $3\sqrt{2}/4 + 2\sqrt{5}/3$
5973 1458 $10^x : \sqrt{2}/4 - \sqrt{3}/3$
5973 2061 $Li_2 : \ln(5/3)$
5973 3980 rt : (9,7,6,-8)
5973 6138 $e^x : e + 4\pi$
5973 6255 rt : (3,8,-1,-4)
5973 9003 $\zeta : \exp(-\sqrt[3]{2}/2)$
5973 9342 rt : (8,8,-2,9)
5973 9679 sr : $1/\sqrt[3]{22}$
5974 0641 as : 9/16
5974 2954 rt : (5,9,6,7)
5974 4386 $\ell_{10} : 1/3 + 4e/3$
5974 5255 rt : (9,6,-9,7)
5974 6238 $s\pi : 7(e + \pi)/5$
5974 8963 $e^x : 21/22$
5975 0132 cu : $3 + e/3$
5975 1383 sinh : $9 \ln 2/5$
5975 1681 rt : (8,7,3,1)
5975 2396 rt : (7,2,3,-4)
5975 2498 rt : (6,7,-1,5)
5975 2633 rt : (1,9,0,-3)

5975 7251 ζ : $2/21$
5975 7499 rt : (7,5,-8,3)
5975 7742 cr : $3e/2$
5975 8287 rt : (6,7,-8,-6)
5975 8604 id : $2e/3+\pi/4$
5975 8716 sr : $2/3+4\sqrt{2}/3$
5975 9559 sinh : $7\zeta(3)/5$
5975 9863 rt : (8,7,3,-6)
5975 9925 $\ln(\pi)\times\exp(1/3)$
5976 1430 sr : $5/14$
5976 2557 tπ : $(\ln 3)^{-2}$
5976 6026 rt : (6,7,-8,1)
5976 6694 rt : (9,2,3,3)
5976 6827 cr : $3\sqrt{3}-\sqrt{5}/2$
5977 1806 K : $\exp(-e/2)$
5977 2096 Ψ : $\exp(-5/3)$
5977 3518 rt : (1,5,0,-2)
5977 6597 $e\times s\pi(1/5)$
5977 7068 rt : (3,8,-8,-7)
5977 9238 J : $\sqrt[3]{5}/8$
5978 0497 rt : (4,3,-8,-5)
5978 1487 sr : $4+4\sqrt{5}$
5978 3700 ln : $11/20$
5978 3750 θ_3 : $7/24$
5978 4446 rt : (5,4,-7,-1)
5978 6345 ζ : $3(1/\pi)/10$
5978 7016 rt : (9,8,-8,0)
5978 8591 rt : (4,9,-4,-1)
5978 8801 ln : $9\sqrt{2}/7$
5979 0658 cr : $\sqrt[3]{5}/8$
5979 1167 rt : (4,6,0,1)
5979 2588 Ei : $(\sqrt[3]{2})^{-1/2}$
5979 3176 rt : (3,8,4,9)
5979 3380 ln : $4+2\sqrt{2}/3$
5979 5142 rt : (1,7,3,-9)
5979 6504 K : $(\pi/2)^{-3}$
5979 8982 rt : (1,6,1,-7)
5979 9002 rt : (7,7,0,-4)
5980 0241 10^x : $4+\pi/4$
5980 0591 K : $1/\sqrt{15}$
5980 1771 rt : (2,5,8,-7)
5980 5550 rt : (5,6,8,-8)
5980 7621 id : $3\sqrt{3}/2$
5980 7750 2^x : $\sqrt{2}+\sqrt{3}/4$
5980 7991 rt : (8,7,8,-9)
5980 8177 ln : $\sqrt{2}/3+2\sqrt{5}$
5980 8752 tπ : $2\sqrt{2}$
5980 9424 sr : $-\sqrt{2}+\sqrt{3}+\sqrt{5}$
5980 9838 rt : (8,3,-8,2)
5981 4477 ℓ_{10} : $1/2+2\sqrt{3}$
5981 5003 id : $\exp(4)$
5981 5454 cu : $\sqrt{2}+4\sqrt{3}$
5981 6367 J_0 : $3\sqrt{5}/5$
5981 7168 2^x : $(e/2)^2$
5981 7178 rt : (6,2,-9,5)
5981 8575 sq : $4\sqrt{2}/3+4\sqrt{3}/3$
5981 8794 sq : $5(e+\pi)/9$
5981 8908 Ψ : $(2\pi/3)^{-3}$
5981 9349 rt : (9,3,0,-3)
5981 9676 rt : (2,4,-9,9)
5982 1074 rt : (1,8,2,-8)
5982 1567 rt : (7,2,8,-7)
5982 1755 sπ : $1/\sqrt{24}$
5982 2293 rt : (4,1,8,-6)
5982 2809 sr : $3\sqrt{2}/2+\sqrt{3}/4$
5982 3245 Γ : $4e\pi/3$
5982 3444 e^x : $18/5$
5982 3454 10^x : $1/2+3\pi/2$
5982 3727 rt : (1,0,8,5)
5982 5214 2^x : $\sqrt{2}/4+3\sqrt{5}$
5982 7879 rt : (2,1,-8,4)
5982 8796 sr : $3e/4+3\pi/2$
5983 0264 rt : (5,4,-2,7)
5983 0852 rt : (7,9,-6,-4)
5983 2786 rt : (2,4,-9,-4)
5983 2915 sπ : $\sqrt{23}$
5983 5237 cu : $9e/8$
5983 5876 ℓ_{10} : $3e-4\pi/3$
5983 6722 Ψ : $2e$
5983 6826 rt : (1,9,-5,-6)
5983 7518 sπ : $\sqrt{3}+2\sqrt{5}$
5983 8677 K : $3\ln 2/8$
5983 9838 sq : $1/4-4\sqrt{3}$
5984 0848 cr : $3/14$
5984 0850 10^x : P_{55}
5984 1342 sr : $9(1/\pi)/8$
5984 1889 at : $15/22$
5984 2068 2^x : $2\sqrt{2}+\sqrt{3}/3$
5984 3114 rt : (9,2,8,6)
5984 4338 rt : (4,8,-5,-5)
5984 4394 2^x : $3/4-2\sqrt{5}/3$
5984 5118 rt : (8,7,-8,2)
5984 6771 J : $\exp(-(e+\pi))$
5984 7109 rt : (7,7,5,-7)
5984 7214 sπ : $5(1/\pi)/2$
5984 8226 10^x : $3e\pi/5$
5984 8341 10^x : $4e+2\pi/3$
5984 8779 e^x : $3(1/\pi)$
5985 2619 rt : (9,8,-3,-3)
5985 2785 rt : (6,3,8,5)
5985 4552 rt : (4,6,5,9)
5985 7962 rt : (2,9,3,-1)
5985 8466 ℓ_{10} : $\sqrt{3}+\sqrt{5}$
5985 9466 K : $6/23$
5985 9508 sr : $8\sqrt{5}/7$
5986 1050 sr : $23/9$
5986 1228 $1/2+\ln(3)$
5986 1652 tanh : $(2\pi/3)^{-1/2}$
5986 2667 ℓ_{10} : $\sqrt[3]{2}/5$
5986 3192 rt : (7,7,-5,-4)
5986 4867 sπ : $5e/2$
5986 5047 rt : (6,7,-3,-2)
5986 6260 rt : (7,9,-1,4)
5986 7450 rt : (3,3,-9,9)
5986 8815 rt : (9,3,5,-6)
5987 0260 cr : $1-\pi/4$
5987 1466 rt : (8,2,0,1)
5987 2065 Ei : $9\ln 2/7$
5987 2554 tπ : $\zeta(3)/7$
5987 4482 id : $10(e+\pi)$
5987 5742 rt : (3,4,8,4)
5987 6169 ζ : $6\sqrt[3]{5}/5$
5987 6907 rt : (4,1,-8,1)
5987 8504 ζ : $(\pi)^{-1/3}$
5988 1257 $\sqrt{5}\times\ln(5)$
5988 2267 Γ : $25/7$
5988 4603 rt : (6,2,5,-5)
5988 4814 Γ : $(\sqrt[3]{5}/3)^{-3}$
5988 6812 K : $((e+\pi)/3)^{-2}$
5988 7155 Ei : $4(e+\pi)/9$
5988 7169 Ψ : $\pi^2/10$
5988 7285 rt : (9,8,2,-6)
5989 0842 at : $1/\ln(\ln 2/3)$
5989 2295 rt : (3,6,-3,-1)
5989 2793 Ei : $\exp(\sqrt{5})$
5989 4575 Γ : $(\ln 3/2)^3$
5989 5282 rt : (2,4,-5,-4)
5989 5397 10^x : $\sqrt{5}+\sqrt{6}-\sqrt{7}$
5989 6331 Γ : $\ln(\sqrt[3]{5}/3)$
5989 7524 $\ln(4/5)+\exp(3/5)$
5989 9781 rt : (7,4,-9,4)
5989 9824 Γ : $6e\pi/7$
5990 1297 e^x : $\sqrt{2}/4-\sqrt{3}/2$
5990 2049 rt : (4,1,-3,9)
5990 2827 rt : (8,7,8,4)
5990 4408 sr : $\sqrt{17}\pi$
5990 4785 sq : $e/2+\pi/4$
5990 5118 rt : (6,7,2,-5)
5990 5340 sπ : $\exp(-\sqrt[3]{4})$
5990 6041 cr : $3\sqrt{3}/2+2\sqrt{5}/3$
5990 7795 at : $(\pi)^{-1/3}$
5990 8738 rt : (9,8,7,-9)
5990 8898 Γ : $1/\ln(1/(3\pi))$
5990 9935 id : $\ln(\ln 3/2)$
5991 0320 Γ : $\exp(-(e+\pi)/3)$
5991 0473 K : $5/19$
5991 3278 rt : (8,3,-3,-1)
5991 3289 tπ : $3e/2+2\pi/3$
5991 5205 rt : (3,1,5,-4)
5991 9442 rt : (6,6,-7,-2)
5992 0323 K : $\exp(-4/3)$
5992 1049 id : $10\sqrt[3]{2}$
5992 1566 erf : $6\ln 2/7$
5992 3388 at : $1/4+\sqrt{3}/4$
5992 4073 rt : (2,9,-2,-5)
5992 4783 rt : (2,9,-2,-4)
5992 5684 rt : (9,9,-8,1)
5992 6160 rt : (4,8,-7,-2)
5992 6830 rt : (6,7,7,-8)
5992 7221 e^x : $5\zeta(3)$
5992 7394 ln : $1/4+\pi/2$
5992 8769 rt : (8,2,5,4)
5992 9940 Ψ : $((e+\pi)/2)^{-1/2}$
5993 1965 sq : $2\sqrt{2}+\sqrt{3}/3$
5993 2070 sπ : $4e\pi/9$
5993 3655 ln : $\sqrt{2}/7$
5993 4276 Ψ : $\sqrt{3}\pi$
5993 4562 rt : (8,8,-6,-1)
5993 5054 cr : $4/3-\sqrt{5}/2$
5993 5653 rt : (9,9,-3,9)
5993 9386 tanh : $\exp(-1/e)$
5994 2501 ℓ_2 : $3e-2\pi/3$
5994 2704 at : $\sqrt{2}/3-2\sqrt{3}/3$
5994 3781 rt : (8,3,2,-4)
5994 4098 10^x : $2/3+4\sqrt{3}$
5994 5136 as : $(\pi)^{-1/2}$
5994 6245 tanh : $9/13$
5994 7059 e^x : $1/4+4\sqrt{5}$
5994 7807 2^x : $e+\pi/3$
5994 8840 rt : (2,9,-7,-7)
5994 8892 $\ln(\sqrt{3})\div\ln(2/5)$
5995 3568 rt : (8,8,-1,-4)
5995 3962 erf : $\ln(2e/3)$
5995 4114 rt : (3,6,7,-7)
5995 4382 rt : (1,5,-3,2)
5995 6579 sq : $4/3-\sqrt{5}/4$
5995 7454 sq : $4/3-3\sqrt{3}/2$
5995 7748 K : $5(1/\pi)/6$
5995 8404 rt : (8,3,7,-7)
5995 8743 ln : $2/3+2\sqrt{3}/3$
5995 9286 Ei : $2\pi/5$
5996 0848 rt : (3,7,-4,-5)
5996 0937 cu : $11/8$
5996 4015 rt : (8,8,4,-7)
5996 4033 10^x : $1/\sqrt[3]{14}$
5996 8547 rt : (3,7,1,8)
5996 9010 2^x : $4\sqrt{5}$
5996 9278 sr : $2+\sqrt{5}/4$
5997 2372 as : $4e/3-4\pi/3$
5997 3008 Γ : $(\sqrt[3]{4}/3)^{-2}$
5997 4718 e^x : $\exp(1/\sqrt{2})$
5997 5926 rt : (8,6,-9,3)
5997 7778 $\exp(1/2)+s\pi(2/5)$
5997 9077 tanh : $2\sqrt{3}/5$
5997 9100 cr : $2+2\pi/3$
5998 0359 tπ : $(e+\pi)/5$
5998 0565 Γ : $7e/2$
5998 1282 sr : $1/4+4\sqrt{3}/3$
5998 2261 2^x : $3+\pi$
5998 3617 ℓ_2 : $7\sqrt{3}$
5998 4512 10^x : $2e+\pi/3$
5998 6160 ℓ_{10} : $4\sqrt{2}-3\sqrt{5}/4$
5998 6750 sinh : $(2e)^{-1/3}$
5998 9695 K : $4/15$
5999 0024 at : $2\sqrt[3]{5}/5$
5999 0489 2^x : $6(e+\pi)$
5999 2820 sπ : $4\pi/7$
5999 6771 $\exp(2/3)+\exp(\sqrt{3})$
5999 7313 e^x : $1/2+4\sqrt{3}/3$
5999 8120 sr : $2\sqrt[3]{2}/7$
6000 0000 id : $3/5$
6000 0106 K : $2\zeta(3)/9$
6000 1533 10^x : $1/\sqrt{24}$
6000 3241 K : $1/\sqrt{14}$
6000 3842 ℓ_{10} : $1+4\sqrt{5}/3$
6000 3988 Ψ : $23/10$
6000 5021 at : $13/19$
6000 6886 θ_3 : $\pi/5$
6000 7306 sπ : $7\sqrt[3]{2}/4$
6001 3700 tanh : $\sqrt[3]{1/3}$

6001 5103 ln : $2e/3 + \pi$
6001 6206 rt : (2,8,-4,7)
6001 7865 cr : $3(e + \pi)$
6001 8742 $\ln(\sqrt{5}) \times s\pi(\sqrt{3})$
6002 0629 sr : $3/2 + 3\sqrt{2}/4$
6002 0673 sπ : $5\sqrt{5}/4$
6002 2422 ℓ_{10} : $2\sqrt{2} + 2\sqrt{3}/3$
6002 5304 $\ln(2/3) \div s\pi(\sqrt{5})$
6002 7029 cr : $3/2 + 3\sqrt{3}/2$
6002 8306 $\ln(2) \times s\pi(1/3)$
6003 0343 rt : (2,7,-1,-1)
6003 0571 rt : (7,8,-2,5)
6003 4531 rt : (7,8,6,-8)
6003 5709 Li_2 : $3\sqrt[3]{5}/10$
6003 5954 ln : $2\sqrt{3} + 2\sqrt{5}/3$
6003 7342 2^x : $3/2 - \sqrt{5}$
6003 7789 rt : (7,8,-7,-3)
6003 9670 rt : (7,3,9,-8)
6004 2233 ln : P_{12}
6004 3563 rt : (2,6,9,-8)
6004 4030 rt : (7,8,1,-5)
6004 6328 rt : (9,8,-9,2)
6004 6847 sr : $\sqrt[3]{3}/4$
6004 8304 rt : (3,9,-9,-8)
6005 1400 sr : $3\zeta(3)/10$
6005 1588 cu : $2/3 + \sqrt{3}/2$
6005 2741 rt : (7,3,4,-5)
6005 3372 $\exp(1/3) \times s\pi(\pi)$
6005 6754 rt : (1,7,4,9)
6005 8654 ℓ_{10} : $4\sqrt{3}/3 + 3\sqrt{5}/4$
6005 9854 rt : (2,6,4,5)
6005 9997 sr : $3e\pi/10$
6006 0223 at : $\sqrt{3}/4 - \sqrt{5}/2$
6006 0732 rt : (7,8,-4,-2)
6006 1403 10^x : $1/4 - \sqrt{2}/3$
6006 2010 Ei : 11/3
6006 3785 ζ : 8/15
6006 4794 ln : $e/3 + 4\pi$
6006 5751 rt : (7,2,7,5)
6006 7027 Ei : 12/5
6006 7977 rt : (9,9,3,-7)
6006 8356 sr : $3/4 + 2e/3$
6006 9189 rt : (3,9,-4,-5)
6006 9675 as : 13/23
6007 1217 2^x : $e/2 - 2\pi/3$
6007 1604 $\exp(1/5) - \exp(3/5)$
6007 1611 Γ : 5/9
6007 3616 rt : (4,0,-9,2)
6007 5484 $\ln(4) \div s\pi(1/3)$
6007 6237 10^x : $4\sqrt{2} - 3\sqrt{5}/4$
6007 6943 cr : $7(e + \pi)/10$
6007 7193 rt : (2,1,7,-5)
6007 7912 rt : (9,4,6,-7)
6007 8645 rt : (7,3,-1,-2)
6007 9794 2^x : $4 + \sqrt{3}/4$
6008 2781 Ei : $8\sqrt{2}/9$
6008 3397 cu : $4\sqrt{3}/3 + \sqrt{5}/4$
6008 5414 rt : (4,7,6,-7)
6008 6291 rt : (9,9,-2,-4)
6008 8557 rt : (6,5,-3,7)
6008 8935 sq : $4\sqrt{2} - \sqrt{5}/2$
6009 1251 rt : (9,1,9,7)
6009 3472 rt : (3,4,-7,-5)
6009 3564 e^x : 8/17
6009 3775 ζ : 4/3
6009 3941 Ei : 6/11
6009 4375 ln : $4/3 + 4e/3$
6009 5234 Li_2 : $\exp(-2/3)$
6009 5588 rt : (2,6,-1,-2)
6009 6678 ℓ_{10} : $7\sqrt[3]{5}/3$
6009 7536 ln : $\sqrt{2} - \sqrt{5} + \sqrt{7}$
6009 7797 rt : (7,8,-9,1)
6009 8722 2^x : $8\ln 2/3$
6009 9590 rt : (6,8,8,-9)
6010 1089 Γ : $\exp((e + \pi)/3)$
6010 1720 rt : (4,2,9,-7)
6010 2051 sq : $\sqrt{2}/2 - 2\sqrt{3}$
6010 2845 id : $\zeta(3)/2$
6010 2954 rt : (9,4,1,-4)
6010 4462 sr : $4e + 2\pi/3$
6010 5247 θ_3 : $4\sqrt{2}/9$
6010 5424 cr : $3/4 + 3\sqrt{5}/2$
6010 5673 ln : $1/3 + 2\sqrt{5}/3$
6010 6872 erf : $(3\pi/2)^{-1/3}$
6010 7571 as : $\exp(-\sqrt[3]{5}/3)$
6010 8472 K : $e/10$
6010 8568 rt : (6,5,-8,-1)
6010 9258 10^x : $\sqrt{5} - \sqrt{6} + \sqrt{7}$
6011 1547 rt : (7,2,2,2)
6011 2407 as : $1/\ln(e + \pi)$
6011 2615 id : $4\sqrt{2} + 4\sqrt{5}$
6011 3380 2^x : $2\sqrt{2}/3 - 3\sqrt{5}/4$
6011 3669 ln : $\sqrt{2} - \sqrt{3}/2$
6011 4752 sr : $4e/3 + \pi$
6011 4912 rt : (1,5,-9,-7)
6011 5689 rt : (3,6,-2,7)
6011 6045 $\ln(\sqrt{2}) - \exp(2/3)$
6011 6477 rt : (4,7,1,-4)
6011 8083 rt : (9,9,-7,-1)
6011 8497 rt : (8,7,-7,-5)
6012 0481 10^x : $3 + \sqrt{5}/3$
6012 1695 tπ : $2\sqrt{3} + 3\sqrt{5}$
6012 1869 rt : (3,9,1,-2)
6012 3949 10^x : $\exp(-\sqrt[3]{4})$
6012 5630 rt : (6,8,3,-6)
6012 5657 Ei : 1/25
6012 5842 e^x : $3/4 + 4\pi/3$
6012 6421 as : $2\sqrt{2}/5$
6012 7597 rt : (9,1,4,4)
6012 8181 $e - \exp(3/4)$
6012 8548 cr : 5/23
6012 8751 sq : $e/4 + \pi$
6012 9437 K : 3/11
6013 0769 rt : (8,9,5,-8)
6013 2002 K : $6(1/\pi)/7$
6013 2053 rt : (6,8,-9,-7)
6013 2859 sq : $1/4 + 3\pi$
6013 3480 rt : (4,6,-8,9)
6013 3632 sq : $2e\pi/9$
6013 5641 J_0 : $10\zeta(3)/9$
6013 6145 sinh : $\sqrt[3]{5}/3$
6014 1189 rt : (3,6,-7,-6)
6014 2451 Ei : $3(1/\pi)/4$
6014 2697 rt : (5,2,-4,9)
6014 5863 10^x : $\sqrt{2}/4 + 3\sqrt{3}/2$
6014 7732 tπ : $\sqrt{2}/3 - 3\sqrt{3}/4$
6014 9062 rt : (8,4,8,-8)
6015 0215 e^x : $4 + \sqrt{3}$
6015 0276 J_0 : $(\sqrt[3]{2}/3)^{-1/3}$
6015 0391 rt : (3,9,-4,9)
6015 1649 rt : (9,4,-4,-1)
6015 4318 rt : (7,3,-6,1)
6015 5047 sr : $3\sqrt[3]{5}/2$
6015 9830 ln : $4/3 - \pi/4$
6016 0078 tanh : 16/23
6016 2765 rt : (3,7,8,-8)
6016 2946 Γ : $8\pi/3$
6016 2971 10^x : $1/2 + 3\pi$
6016 3817 sπ : $8\zeta(3)/3$
6016 4284 rt : (8,9,0,-5)
6016 5555 $\exp(3/4) \div s\pi(1/5)$
6016 5711 ln : $\sqrt{2}/2 + \sqrt{5}/2$
6016 6281 2^x : $e/4$
6016 6850 at : $(\sqrt{2}/3)^{1/2}$
6016 7120 sr : $3\sqrt{3}/2 - \sqrt{5}$
6016 7765 rt : (9,6,7,4)
6016 7913 rt : (3,9,-9,-9)
6016 8379 10^x : 15/2
6016 8984 rt : (7,7,-3,6)
6017 0068 rt : (6,8,-2,-3)
6017 3624 rt : (5,2,-9,1)
6017 3693 sr : $e - 3\pi/4$
6017 4583 sr : $3\sqrt{2} - 3\sqrt{5}/4$
6017 5785 $\exp(\pi) \times s\pi(1/5)$
6017 7026 sq : $1/2 - 4e$
6017 9256 rt : (1,8,9,-2)
6018 0944 10^x : $3\pi^2/8$
6018 2309 sq : $1/3 + \sqrt{5}$
6018 2940 rt : (4,7,-4,-1)
6018 3260 rt : (6,2,9,6)
6018 5417 10^x : $3\sqrt{2}/4 + 3\sqrt{3}/2$
6018 7078 at : $1/3 + \sqrt{2}/4$
6018 7277 at : $4\zeta(3)/7$
6018 8727 ln : $1/4 + 3\pi/2$
6018 8970 rt : (7,9,7,-9)
6018 9836 2^x : $\ln(\sqrt[3]{3}/3)$
6019 0034 e^x : $2 + \sqrt{5}/2$
6019 0257 sr : $4/3 + 2e$
6019 0893 ζ : $\ln 2/7$
6019 1811 rt : (7,7,5,2)
6019 1908 sinh : 5/4
6019 2164 J : $\exp(-(e + \pi)/3)$
6019 2254 rt : (4,9,-6,-6)
6019 4224 rt : (8,4,3,-5)
6019 6071 tπ : $\sqrt{3} + \sqrt{5} - \sqrt{7}$
6019 8481 Ei : $\exp((e + \pi)/2)$
6019 8615 sinh : $((e + \pi)/3)^{1/3}$
6019 8640 ln : $\sqrt{10/3}$
6019 9660 rt : (5,8,5,-7)
6019 9755 sq : $e/2 - 4\pi$
6020 4661 sr : $1/\sqrt[3]{21}$
6020 4748 2^x : $1 - \sqrt{3}$
6020 5000 rt : (7,7,-8,-2)
6020 5999 ℓ_{10} : 4
6020 6009 ℓ_2 : $e + 3\pi$
6020 7020 rt : (4,3,7,4)
6021 1857 rt : (9,1,-1,1)
6021 2653 rt : (1,1,9,-6)
6021 3289 tπ : $\exp(1/(2\pi))$
6021 3975 2^x : 17/25
6021 6024 rt : (6,3,1,-3)
6021 8175 rt : (6,4,-4,8)
6021 8204 sr : $3 - \sqrt{3}/4$
6022 0815 rt : (8,9,-5,-2)
6022 1271 $\sqrt{5} \div \exp(1/3)$
6022 1367 id : P_2
6022 3233 ln : $3/2 + 2\sqrt{3}$
6022 3577 rt : (3,8,-9,4)
6022 4299 id : $\exp(\sqrt{2}/3)$
6022 4919 rt : (6,8,-4,-4)
6022 8024 10^x : $\sqrt{2} + \sqrt{5}/2$
6022 8734 at : 11/16
6022 9429 id : $\sqrt{2}/2 - 4\sqrt{3}/3$
6023 1651 tanh : $\sqrt{2} + \sqrt{3} - \sqrt{6}$
6023 1763 ζ : $5\zeta(3)/4$
6023 3412 tπ : $9\pi/10$
6023 4904 rt : (5,3,8,-7)
6023 5169 rt : (7,9,2,-6)
6023 5635 sπ : $9\sqrt{3}/2$
6023 6347 K : $2\ln 2/5$
6023 6496 rt : (2,6,-6,1)
6023 7531 rt : (8,1,6,5)
6023 7970 tπ : $1/4 - 3e/2$
6023 8521 K : $1/\sqrt{13}$
6024 0928 rt : (3,5,-5,6)
6024 1854 tπ : $(2e)^2$
6024 2332 rt : (8,9,-3,5)
6024 2475 $\ln(3/5) + \exp(\sqrt{2})$
6024 5164 ℓ_2 : $\sqrt[3]{7/2}$
6024 5804 rt : (2,5,-6,7)
6024 5914 rt : (9,5,7,-8)
6024 8723 K : 5/18
6024 8925 2^x : $1/4 + \sqrt{5}$
6024 9952 id : $10\sqrt[3]{3}/9$
6025 0535 cu : $2 - 4e$
6025 3381 rt : (4,4,-9,-6)
6025 6616 J_0 : 4/3
6025 8060 rt : (1,6,1,-3)
6025 8705 rt : (1,9,4,5)
6025 9273 Ψ : $(\pi/3)^{-1/3}$
6026 1957 10^x : $2\sqrt{2} + 4\sqrt{5}$
6026 2809 e^x : 22/23
6026 2885 rt : (6,8,-7,0)
6026 4404 Ψ : 7/12
6026 5375 θ_3 : 19/21
6026 5864 rt : (5,8,0,-4)
6026 6500 K : $7(1/\pi)/8$
6026 7167 rt : (4,8,7,-8)
6026 8416 ln : P_8
6026 8968 ln : $(\sqrt[3]{2}/3)^3$

6026 9004 tπ : $1/4+\sqrt{3}/3$
6027 0493 rt : (1,0,-7,4)
6027 1325 cu : $1/2+3\pi$
6027 4253 id : $4\zeta(3)/3$
6027 4370 10^x : $9e/7$
6027 4434 rt : (3,2,6,-5)
6027 5769 rt : (7,4,5,-6)
6027 6950 ℓ_2 : $\sqrt{3}+\sqrt{5}-\sqrt{6}$
6027 8323 sr : $\sqrt{2}+2\sqrt{3}/3$
6027 8471 rt : (8,4,-2,-2)
6028 0377 ℓ_{10} : $10\zeta(3)/3$
6028 1086 sr : $9\sqrt[3]{3}$
6028 2468 e^x : $4-e$
6028 4382 as : $1-\sqrt{3}/4$
6028 5268 ln : $3e/2+3\pi$
6028 5966 rt : (1,8,-6,-7)
6028 6154 sr : $4\sqrt{2}+\sqrt{5}/2$
6028 6387 cr : $3+\sqrt{5}/2$
6028 6840 rt : (9,4,-9,2)
6028 7212 rt : (7,6,-4,7)
6029 0198 K : $\sqrt{5}/8$
6029 0714 10^x : $2(e+\pi)/3$
6029 1195 rt : (8,9,-8,-3)
6029 1809 Ψ : 14
6029 2116 as : P_{80}
6029 3120 $\ln(3)\div\exp(3/5)$
6029 3521 sr : $1/3+\sqrt{5}$
6029 4021 tπ : $\sqrt{3}+\sqrt{6}+\sqrt{7}$
6029 5404 sq : $2/3-4\pi$
6029 6073 rt : (6,2,4,3)
6029 8508 rt : (5,9,-8,-7)
6029 9914 rt : (6,9,4,8)
6030 0360 cu : $3/4+3\sqrt{3}/4$
6030 1609 K : $2\sqrt[3]{2}/9$
6030 1889 sπ : $\sqrt[3]{3}/7$
6030 2032 K : $7/25$
6030 2268 sr : $4/11$
6030 3751 ζ : $1/10$
6030 3810 rt : (8,5,9,-9)
6030 3841 sinh : $4/7$
6030 5759 rt : (8,6,9,5)
6030 6941 sq : $\sqrt{2}/2+\sqrt{5}/4$
6030 7837 rt : (2,7,5,-6)
6039 9024 λ : $\exp(-10\pi/11)$
6031 0392 2^x : $4\sqrt{2}+\sqrt{3}$
6031 0558 Ei : $\sqrt{3}+\sqrt{5}+\sqrt{6}$
6031 1024 rt : (1,9,-6,-2)
6031 1137 rt : (7,9,-3,-3)
6031 3677 sπ : $(\sqrt[4]{5}/3)^{-1/3}$
6031 3682 rt : (9,7,-9,-6)
6031 3954 Γ : $(e+\pi)^{-1/3}$
6031 4403 sr : $8(1/\pi)/7$
6031 4667 erf : $\ln(\ln 3/2)$
6031 4839 rt : (9,5,2,-5)
6031 8055 tanh : $2\pi/9$
6031 8825 sq : $2e/7$
6032 0610 rt : (4,9,-1,-3)
6032 3453 θ_3 : $5/17$
6032 5053 rt : (2,5,-6,-5)
6032 6130 sq : $1/3+2e/3$
6032 9005 rt : (3,8,9,-9)
6032 9014 cr : $2+3\sqrt{2}/2$
6032 9019 $\exp(\sqrt{3})+s\pi(2/5)$
6032 9478 rt : (6,3,-5,9)
6032 9706 J : $1/\sqrt[3]{19}$
6033 0543 rt : (3,7,-2,-2)
6033 2228 rt : (5,3,3,-4)
6033 3362 rt : (7,1,8,6)
6033 3757 cr : $4\sqrt{2}/3+\sqrt{5}$
6033 5424 rt : (5,9,-7,9)
6033 7030 sr : $1+\pi/2$
6033 7532 ℓ_2 : P_{58}
6033 7864 ℓ_2 : P_{19}
6033 9174 rt : (2,4,5,2)
6033 9229 ℓ_{10} : $\sqrt{2}+3\sqrt{3}/2$
6034 0979 Li_2 : $3\zeta(3)/7$
6034 3016 rt : (7,6,-9,-1)
6034 5964 ℓ_2 : $4+3e$
6034 7196 2^x : $8\sqrt[3]{3}$
6034 7448 $\ln(3)\times\ln(\sqrt{3})$
6034 8825 rt : (3,4,-8,5)
6034 9121 rt : (6,4,7,-7)
6035 0584 rt : (8,8,-4,6)
6035 1190 rt : (9,6,2,1)
6035 2162 sr : $\sqrt[3]{17}$
6035 2514 Γ : $(\ln 2/2)^{-2}$
6035 3921 $3/4\times\ln(\sqrt{5})$
6035 4753 10^x : $3\sqrt[3]{2}/2$
6035 5339 id : $1/4+\sqrt{2}/4$
6035 6354 Ei : $7e/10$
6035 6745 sr : $18/7$
6035 6762 Γ : $4\ln 2/5$
6035 7569 cu : $9\pi/10$
6035 9312 rt : (5,9,6,-8)
6036 1485 id : $e/4-2\pi$
6036 3383 tanh : $((e+\pi)/2)^{-1/3}$
6036 3847 10^x : $3e/2+\pi/3$
6036 3857 rt : (7,2,-3,-1)
6036 4162 2^x : $1/3+\pi/3$
6036 7985 sπ : $\sqrt[3]{1/2}$
6036 8043 cr : $4\sqrt{3}-3\sqrt{5}$
6036 8304 Ψ : $2(1/\pi)/9$
6036 8626 cr : $1/2+4e/3$
6036 9988 rt : (3,3,9,5)
6037 0965 K : $\sqrt{2}/5$
6037 1442 rt : (2,9,1,9)
6037 3391 K : $8(1/\pi)/9$
6037 6158 sπ : $\exp(4/3)$
6037 6886 rt : (8,1,1,2)
6037 8851 rt : (9,0,5,5)
6038 1515 rt : (2,2,8,-6)
6038 3882 e^x : $8\sqrt[3]{3}/9$
6038 3917 rt : (9,6,8,-9)
6038 4431 J : $\exp(-14\pi/9)$
6038 4723 rt : (8,5,4,-6)
6038 4951 ln : $1/2+2\sqrt{5}$
6038 5609 erf : $3/5$
6038 6341 rt : (7,4,0,-3)
6038 7248 cu : $1/4-2\pi$
6038 7547 rt : (1,7,7,-7)
6038 9395 rt : (7,5,-5,8)
6039 1256 rt : (6,9,-1,0)
6039 1339 K : $\exp(-\sqrt[3]{2})$
6039 5169 e^x : $3\sqrt[3]{2}/8$
6039 6092 10^x : $3e/2+2\pi$
6039 7313 rt : (5,8,-5,-1)
6039 8925 rt : (7,8,-6,-5)
6039 9282 cu : $2-2\sqrt{3}/3$
6040 2206 2^x : $1/3-3\sqrt{2}/4$
6040 5103 ℓ_2 : $8\sqrt[3]{5}/9$
6040 6262 id : $2e/9$
6040 6813 rt : (1,5,1,-1)
6040 7058 rt : (4,9,-1,7)
6040 7703 cu : $3e/4+2\pi/3$
6040 8345 sπ : $1/3+4e$
6040 8658 cr : $2\sqrt{2}+3\sqrt{3}/4$
6040 9901 e^x : $\sqrt{3}-\sqrt{5}$
6041 0196 id : $1/4+3\sqrt{5}/2$
6041 3844 sr : $4\sqrt{3}/3+2\sqrt{5}$
6041 4022 rt : (4,9,8,-9)
6041 5030 sq : $(\ln 3/3)^{-2}$
6041 5118 e^x : $1/4+\sqrt{2}/2$
6041 5331 rt : (8,8,-9,-2)
6041 5449 Ψ : $8\sqrt[3]{3}/5$
6041 5495 rt : (5,4,9,-8)
6041 5735 sinh : $\exp(\sqrt{2}\pi/3)$
6041 6015 2^x : $15/22$
6041 9514 e^x : $3+\sqrt{2}/4$
6042 0255 e^x : $4/3+\sqrt{5}/2$
6042 2252 ℓ_2 : $2\sqrt{2}/3+\sqrt{3}/3$
6042 3260 rt : (4,7,-9,2)
6042 3734 K : $\sqrt[3]{5}/6$
6042 6107 rt : (3,8,4,-6)
6042 7298 cu : $2\sqrt{2}/3+\sqrt{3}/4$
6042 8364 J : $\exp(-13\pi/17)$
6043 0590 e^x : $4/3+\sqrt{2}$
6043 0908 ℓ_{10} : $2/3+3\sqrt{5}/2$
6043 1960 rt : (4,3,5,-5)
6043 3616 tπ : $\sqrt{3}-\sqrt{5}/4$
6043 3855 J_0 : $7\sqrt[3]{5}/9$
6043 3993 tanh : $5\sqrt[3]{2}/9$
6043 6243 ζ : $(\sqrt{3})^{-1/2}$
6043 6777 tanh : $7/10$
6043 7061 rt : (9,5,-3,-2)
6043 8393 rt : (2,8,-6,-2)
6043 8757 $2-\exp(1/3)$
6043 8861 at : $3-4\sqrt{3}/3$
6043 8865 id : $4(e+\pi)/9$
6044 0625 e^x : $4\sqrt{2}+4\sqrt{5}/3$
6044 1029 e^x : $9\sqrt[3]{3}/2$
6044 1203 rt : (6,8,3,9)
6044 1443 K : $2/7$
6044 2913 rt : (5,2,6,4)
6044 4444 sq : $19/15$
6044 4533 10^x : $4\sqrt{3}-3\sqrt{5}/2$
6044 4707 rt : (8,7,-5,7)
6044 5489 sr : $1/2+2\pi$
6044 5706 sπ : $9\pi^2/4$
6044 6173 rt : (3,5,-8,-6)
6044 7725 J_0 : $1/\ln(\sqrt{2}/3)$
6044 8087 rt : (7,5,6,-7)
6044 8292 $\ln(\sqrt{2})-s\pi(2/5)$
6045 1437 cr : $2/3+2\sqrt{3}$
6045 2488 rt : (2,7,0,-3)
6045 3186 sr : $2\sqrt{2}/3-\sqrt{3}/3$
6045 6283 ℓ_{10} : $2\sqrt{3}+\sqrt{5}/4$
6045 6478 ζ : $(\pi)^{-2}$
6045 8418 rt : (8,0,7,6)
6045 8922 rt : (7,9,-8,0)
6045 9923 Γ : $\exp(-\pi)$
6046 0353 K : $9(1/\pi)/10$
6046 1084 tπ : $3\sqrt{3}/4-2\sqrt{5}$
6046 2871 rt : (5,9,1,-5)
6046 3628 rt : (8,7,-2,-2)
6046 3996 ln : $3/2-3\sqrt{3}/4$
6046 4336 tπ : $\exp(-\sqrt[3]{4}/2)$
6046 5220 at : $(2\pi/3)^{-1/2}$
6046 6523 erf : $\zeta(3)/2$
6046 8996 rt : (2,0,-9,5)
6046 9834 10^x : $2-\sqrt{5}/4$
6047 0770 rt : (6,7,7,3)
6047 1895 $1/5-\ln(\sqrt{5})$
6047 5645 rt : (9,5,8,5)
6047 5750 ℓ_{10} : $\sqrt{5}/9$
6047 6196 rt : (5,9,-3,-4)
6047 7339 ζ : $4\zeta(3)/9$
6047 8430 rt : (4,9,-6,-6)
6047 8607 rt : (7,4,-6,9)
6047 9361 rt : (9,6,3,-6)
6048 0602 $\pi\times\ln(3/5)$
6048 1663 Ei : $5/21$
6048 1705 cr : $3e/4+2\pi/3$
6048 2488 sinh : $9(1/\pi)/5$
6048 5331 sπ : P_{39}
6048 8286 $\ln(2/3)\times\exp(2/5)$
6048 9133 sπ : P_{33}
6049 0019 rt : (8,4,-7,1)
6049 0051 Γ : $(\sqrt{2}/3)^{-3}$
6049 0232 K : $\ln(4/3)$
6049 0245 sinh : $8\sqrt{2}$
6049 1493 Ei : $(1/(3e))^2$
6049 3093 cu : $4+3\sqrt{2}/4$
6049 3824 as : $(2e)^{-1/3}$
6049 3827 sq : $7/9$
6049 3900 rt : (4,8,-4,6)
6049 4142 ℓ_2 : $\sqrt{15}\pi$
6049 5609 $\ln(\sqrt{3})-\exp(e)$
6049 5916 rt : (9,9,-5,6)
6049 9121 sr : 10π
6049 9716 Ei : $7(e+\pi)/6$
6050 2478 J : $4\ln 2/9$
6050 2644 rt : (1,6,-5,-5)
6050 3703 cu : $23/15$
6050 5834 rt : (6,5,8,-8)
6050 6915 rt : (3,9,6,1)
6050 7096 tπ : $\sqrt{3}/10$
6050 7507 $\exp(2/5)+\exp(\sqrt{2})$
6050 8198 id : $3e/4+4\pi$
6050 8289 cu : $3\sqrt{2}/2-\sqrt{5}/3$
6050 9206 rt : (4,8,-3,-2)

6050 9217 $Ei : 5\ln 2$
6050 9380 $K : \sqrt[3]{3}/5$
6050 9592 rt : (7,1,3,3)
6051 1352 cu : $3/4 - 3\sqrt{2}$
6051 2762 $\Psi : 4\sqrt{3}/3$
6051 4799 id : $(\ln 3/3)^{1/2}$
6051 5007 $K : \sqrt{3}/6$
6051 8595 rt : (3,3,7,-6)
6052 0757 cr : $2\sqrt{3}/3 + 4\sqrt{5}/3$
6052 3449 $2^x : (\pi)^{-1/3}$
6052 3946 rt : (8,5,-1,-3)
6052 5421 id : $4\sqrt{3} + 3\sqrt{5}/4$
6052 5910 sq : $\sqrt{2}/2 + 4\sqrt{5}/3$
6052 7312 rt : (8,6,-6,8)
6052 9263 rt : (7,0,9,7)
6053 0528 $s\pi : (\ln 3)^2$
6053 1994 $Ei : 8/5$
6053 4624 rt : (8,6,5,-7)
6053 5921 sr : $4e/3 - \pi/3$
6053 7781 $t\pi : 2\sqrt{3}/3 - 4\sqrt{5}/3$
6053 7794 rt : (8,8,-8,-6)
6054 1115 $\Gamma : -\sqrt{2} + \sqrt{5} + \sqrt{6}$
6054 1280 sr : $2 + \sqrt{3}/3$
6054 1759 $2^x : \sqrt[3]{19/3}$
6054 2171 $t\pi : \ln 2/4$
6054 4197 $t\pi : 3\sqrt{3}/2 - 4\sqrt{5}/3$
6054 5495 sq : $9\pi^2/5$
6054 6510 $1/5 - \ln(2/3)$
6054 6916 $e^x : (3\pi)^{-1/3}$
6054 7228 at : $\exp(-1/e)$
6054 8338 $10^x : 3\ln 2/5$
6054 8891 $2^x : 1/4 + \sqrt{3}/4$
6055 1156 $\ln(2) \div \ln(\pi)$
6055 4466 at : $9/13$
6055 5127 id : $\sqrt{13}$
6055 6692 $10^x : 4 + 3\sqrt{2}/2$
6055 7752 $e^x : 3\sqrt[3]{5}/4$
6055 9244 rt : (6,8,-2,1)
6056 0497 rt : (5,4,4,-5)
6056 1825 rt : (6,9,-6,-1)
6056 2392 id : $5\sqrt[3]{3}/2$
6056 2848 rt : (4,2,8,5)
6056 4131 rt : (9,1,6,-6)
6056 5724 at : $2/3 - e/2$
6056 5925 rt : (5,3,-2,-1)
6056 9986 $s\pi : 1/2 + \sqrt{2}/2$
6057 0550 rt : (4,7,-7,5)
6057 0686 cr : $2/9$
6057 1562 $\Psi : 10\ln 2/3$
6057 1578 ln : $2 + 4\sqrt{5}/3$
6057 2612 rt : (9,8,-6,7)
6057 2637 cr : $8(e + \pi)$
6057 4162 rt : (4,8,-9,-7)
6057 6290 $t\pi : \sqrt{3}/3 + \sqrt{5}/3$
6057 8596 $10^x : \sqrt{2}/4 + 4\sqrt{5}/3$
6058 1137 $2^x : \sqrt{2}/3 + \sqrt{3}$
6058 2612 cu : $11/13$
6058 2958 rt : (8,6,4,2)
6058 5300 rt : (7,6,7,-8)
6058 5638 rt : (3,9,5,5)
6058 8349 at : $1/4 - 2\sqrt{2}/3$
6058 9111 at : $2\sqrt{3}/5$
6058 9395 $\ell_2 : \exp(\sqrt[3]{2}/3)$
6058 9978 $e^x : 9/19$
6059 0262 $K : 7/24$
6059 0325 tanh : $(\ln 2/2)^{1/3}$
6059 1465 cr : $1 + \pi$
6059 1580 rt : (3,9,-6,2)
6059 3076 $s\pi : 7\zeta(3)/2$
6059 4622 rt : (2,3,9,-7)
6059 7299 rt : (2,6,-7,-6)
6060 0033 rt : (7,5,1,-4)
6060 0401 rt : (8,5,-7,9)
6060 1571 rt : (6,9,-5,-5)
6060 1861 $10^x : (\ln 2)^{-2}$
6060 2259 $10^x : 5\sqrt[3]{5}/6$
6060 3745 $s\pi : 2e + 3\pi/4$
6060 3974 rt : (3,8,-1,-3)
6060 9152 id : $3\sqrt{2}/7$
6061 0904 rt : (9,7,4,-7)
6061 1055 id : $\sqrt{3}/2 - 2\sqrt{5}$
6061 1193 at : $\ln 2$
6061 2230 rt : (9,1,-6,-2)
6061 2308 sq : $2\sqrt{2} - 3\sqrt{3}$
6061 3580 ln : $6/11$
6061 3765 $10^x : 7(1/\pi)/4$
6061 4173 sinh : $7\sqrt{3}/5$
6061 4848 rt : (5,5,-3,3)
6061 5531 $s\pi : 3\sqrt{2}/4 + \sqrt{3}$
6061 7070 id : $3\zeta(3)$
6061 7439 $\ln(3/5) + \exp(3/4)$
6061 8902 rt : (6,1,5,4)
6061 9449 id : $1/4 + 3\pi/4$
6061 9635 rt : (7,9,2,9)
6062 1955 $e^x : -\sqrt{2} - \sqrt{3} + \sqrt{7}$
6062 4281 $\ell_{10} : 2 + 3e/4$
6062 4425 rt : (8,1,8,-7)
6062 4670 cr : $7(1/\pi)/10$
6062 5470 $J_0 : \sqrt[3]{7/3}$
6062 5653 at : $\sqrt[3]{1/3}$
6062 7136 $\theta_3 : \pi/4$
6062 8371 $\ln(\sqrt{5}) \div \ln(4/5)$
6062 9434 $\ell_{10} : 9\pi/7$
6062 9563 $2^x : \sqrt{2}/4 + 4\sqrt{3}$
6063 1001 cu : $22/9$
6063 1955 rt : (6,6,9,-9)
6063 7371 rt : (9,6,-2,-3)
6063 8704 rt : (2,9,7,2)
6064 0763 rt : (9,7,-7,8)
6064 1070 $\ell_2 : 4\sqrt{2} + \sqrt{3}/4$
6064 2828 rt : (7,4,-5,0)
6064 5701 $J_0 : (\ln 3)^3$
6064 6885 $K : \ln(\sqrt{5}/3)$
6064 6901 $1/4 + \exp(\sqrt{5})$
6064 9502 rt : (5,9,-4,-2)
6064 9609 as : $\sqrt[3]{5}/3$
6065 0378 id : $7\ln 2/8$
6065 2618 $K : 5/17$
6065 3065 id : $(e)^{-1/2}$
6065 3741 $\ell_{10} : \sqrt{3}/7$
6065 5718 rt : (8,7,6,-8)
6065 6706 $e \times \ln(4/5)$
6065 7727 $2^x : 2\sqrt[3]{5}/5$
6065 8593 $\Psi : \sqrt{20\pi}$
6065 9689 $t\pi : 9\pi^2$
6065 9909 rt : (2,5,-1,-2)
6066 0534 $\ell_{10} : 3\sqrt{2}/4 + 4\sqrt{5}/3$
6066 1541 $s\pi : 8\pi/9$
6066 5167 rt : (5,7,9,4)
6066 5536 rt : (4,2,-4,5)
6066 7278 rt : (6,5,3,-5)
6067 0583 rt : (8,0,2,3)
6067 2212 $\exp(4) \times s\pi(1/4)$
6067 3760 $\ell_2 : \exp(5/2)$
6067 6203 $e^x : \sqrt{7/3}$
6067 6771 $Ei : (1/\pi)/8$
6067 7373 rt : (3,7,-7,1)
6067 8659 $2^x : 1/2 + 2\sqrt{3}$
6067 8806 rt : (3,6,-9,-7)
6068 2245 $2^x : 13/19$
6068 2615 $2/3 \div \ln(3)$
6068 3149 cu : $2e - 2\pi$
6068 5011 sq : $3\sqrt{3}/2 + 3\sqrt{5}$
6068 5379 rt : (2,8,1,-4)
6068 5515 rt : (9,4,9,6)
6068 5683 sr : $\sqrt{20/3}$
6068 5722 sr : $5e/2$
6068 7210 rt : (1,9,9,-9)
6069 0128 rt : (1,7,8,-9)
6069 0325 $\Gamma : 7\pi/8$
6069 1277 rt : (6,7,-3,2)
6069 5414 rt : (3,4,8,-7)
6069 6223 sr : $1/\sqrt[3]{20}$
6069 7182 rt : (7,7,8,-9)
6069 7697 sr : $7/19$
6070 0716 cu : $\pi^2/3$
6070 0729 rt : (8,6,0,-4)
6070 1604 $\zeta : 4\sqrt[3]{5}/9$
6070 1729 rt : (9,6,-8,9)
6070 3436 ln : $3\sqrt{2} + \sqrt{5}/3$
6070 4462 $3/4 \div \ln(3/4)$
6070 4656 $erf : 2e/9$
6070 4843 $10^x : (2e)^{1/2}$
6070 4895 rt : (7,9,-7,-6)
6070 6528 rt : (9,2,7,-7)
6070 7317 $10^x : \sqrt[3]{3}/7$
6070 8121 $\Gamma : 7(e + \pi)/9$
6070 8577 $K : (3/2)^{-3}$
6070 8985 $e^x : (\sqrt{2}\pi)^{-1/2}$
6071 0662 rt : (5,1,7,5)
6071 1446 rt : (9,5,-8,1)
6071 1950 rt : (3,8,2,4)
6071 2398 rt : (1,9,-5,8)
6071 5132 $Li_2 : 4\sqrt{5}/9$
6071 5440 $\ell_{10} : 3 + \pi/3$
6071 7317 sq : $1/2 - 4\pi/3$
6071 8533 rt : (9,8,5,-8)
6071 8706 rt : (1,7,-6,-6)
6071 9091 $\zeta : 5\sqrt[3]{2}/6$
6071 9755 $10^x : (\sqrt{2}/3)^{-1/2}$
6072 0082 $\zeta : 19/25$
6072 0100 id : $9e\pi/8$
6072 0216 $\Gamma : \ln(\sqrt[3]{3}/3)$
6072 0687 $\ell_{10} : e/3 + \pi$
6072 0979 $J_0 : (\sqrt[3]{5}/3)^{-1/2}$
6072 3175 $\exp(e) \times s\pi(\sqrt{2})$
6072 5201 rt : (4,9,-2,-3)
6072 5627 $\Psi : 7/24$
6072 6307 cu : $1/3 + 2\sqrt{2}$
6072 7181 rt : (5,5,5,-6)
6072 8391 $K : 3\ln 2/7$
6072 9907 $e^x : 3\sqrt{5}/7$
6073 1539 $e^x : \exp(-\sqrt{5}/3)$
6073 2693 $\Gamma : 10(e + \pi)/9$
6073 4727 $e^x : 23/24$
6073 5321 rt : (1,4,7,3)
6073 5719 rt : (5,4,-4,4)
6073 7223 $10^x : 3/2 - 2\sqrt{2}/3$
6073 7293 $1/3 \times \exp(3/5)$
6073 7848 sinh : $1/\ln(\sqrt{2}\pi/2)$
6074 0104 $\theta_3 : \sqrt[3]{2}/2$
6074 1197 $10^x : \sqrt{2} + 2\sqrt{3}/3$
6074 4519 cr : P_{31}
6074 5931 rt : (5,4,-6,-4)
6074 6400 $t\pi : 7\sqrt{5}/2$
6074 7536 as : $1 - \pi/2$
6074 8303 $\ell_2 : 2 + \pi/3$
6074 8851 rt : (7,6,6,3)
6074 9226 $2^x : 5e\pi/2$
6075 0268 $\exp(2/5) \times \exp(e)$
6075 1063 $s\pi : 8\ln 2/7$
6075 2176 rt : (1,2,5,-4)
6075 3815 rt : (8,2,9,-8)
6075 4624 rt : (2,8,2,-3)
6075 5006 rt : (7,0,4,4)
6075 5616 rt : (8,8,7,-9)
6075 6134 rt : (1,8,3,-5)
6075 7060 $Ei : ((e + \pi)/2)^{-3}$
6075 7339 rt : (7,6,2,-5)
6075 7701 $Ei : 2\sqrt{5}/5$
6075 8567 rt : (7,9,-3,1)
6075 9398 ln : $1 + 4\pi$
6076 0707 cr : $3 + 2\sqrt{3}/3$
6076 2521 rt : (6,8,1,-1)
6076 2818 $\ell_{10} : 4/3 + e$
6076 2987 $s\pi : \exp(-\pi/2)$
6076 7405 $10^x : 3e/2 + \pi/2$
6076 8971 $10^x : 1 + \sqrt{2}/3$
6076 9515 sq : $3 - \sqrt{3}$
6076 9711 $\ln(4) + \exp(1/5)$
6077 2130 $\Psi : 5\pi^2/9$
6077 2529 ln : $\zeta(3)/6$
6077 6644 rt : (4,1,-5,6)
6077 9098 $s\pi : 3\ln 2/10$
6078 0199 at : $16/23$
6078 0912 rt : (4,5,7,-7)
6078 2089 $2^x : 2 - e$
6078 2773 rt : (2,6,-3,5)
6078 5052 rt : (2,7,-8,-7)
6078 5410 rt : (5,9,-9,4)

6078 5697 rt : (1,8,-5,2)
6078 5995 rt : (9,7,-1,-4)
6078 8879 rt : (4,1,9,6)
6078 9015 10^x : $\sqrt[3]{4/3}$
6078 9118 $s\pi$: $\sqrt{2}+\sqrt{3}+\sqrt{7}$
6078 9628 erf : $(\ln 3/3)^{1/2}$
6079 1288 rt : (3,9,0,-4)
6079 1598 rt : (8,9,-9,-7)
6079 5265 10^x : $2e+4\pi/3$
6079 5599 J : $\exp(-15\pi/7)$
6079 6184 rt : (9,1,1,-3)
6079 6496 ln : $3/4+3\sqrt{2}$
6079 7839 rt : (6,6,-4,3)
6079 8962 sinh : $7\sqrt[3]{5}/6$
6080 3807 sr : $4-\sqrt{2}$
6080 4249 θ_3 : $(3/2)^{-3}$
6080 4861 K : 3/10
6080 7334 sq : $9\ln 2/8$
6080 8101 cu : $4-2\sqrt{2}$
6080 8254 rt : (9,9,6,-9)
6080 8363 10^x : $\sqrt{21}\pi$
6080 8826 tanh : 12/17
6081 1347 $t\pi$: 4/23
6081 4464 rt : (3,7,-1,3)
6081 4635 cr : $6\ln 2$
6081 5675 cu : $3\sqrt{2}/4+2\sqrt{5}/3$
6081 8249 rt : (8,5,-6,0)
6081 8345 K : $\zeta(3)/4$
6081 8638 sinh : $(\pi/2)^{1/2}$
6081 9445 rt : (9,3,8,-8)
6082 1032 Γ : $\sqrt[3]{5}/5$
6082 1997 rt : (9,5,3,2)
6082 4269 $s\pi$: $4e/9$
6082 4347 rt : (5,5,-8,-5)
6082 4557 as : 4/7
6082 4883 ζ : $\sqrt[3]{17/2}$
6082 7266 ln : P_{40}
6082 7382 rt : (6,0,6,5)
6082 7893 ln : P_{16}
6082 8152 2^x : $4+3\sqrt{5}/2$
6082 8376 sq : $4e/3-3\pi/4$
6082 9144 id : $(\sqrt{2}\pi)^{-1/3}$
6082 9401 rt : (3,5,9,-8)
6082 9752 Ei : $(\pi)^{-2}$
6083 3846 rt : (3,5,-3,-3)
6083 4183 rt : (5,3,-5,5)
6083 4289 rt : (8,7,1,-5)
6083 4437 10^x : $3\sqrt{3}-\sqrt{5}/4$
6083 5569 J_0 : $\sqrt{3}+\sqrt{5}-\sqrt{7}$
6083 5942 Γ : 11/4
6084 2378 Ei : 17/19
6084 3886 $s\pi$: $1/2+3\sqrt{5}$
6084 4014 $\ln(\sqrt{5})-\exp(5)$
6084 4571 K : $1/\sqrt{11}$
6084 9528 rt : (2,9,2,-5)
6084 9532 $t\pi$: $3\sqrt{5}/4$
6085 0528 K : $(\ln 3/2)^2$
6085 1714 cr : $4/3+2\sqrt{2}$
6085 1892 $\pi\times\exp(e)$
6085 3195 rt : (7,8,-4,2)
6085 3512 $\exp(1/4)-s\pi(\sqrt{5})$
6085 3781 $\sqrt{2}\times s\pi(\pi)$
6085 4312 rt : (5,6,6,-7)
6085 5800 at : $\sqrt{2}+\sqrt{3}-\sqrt{6}$
6085 5865 ζ : $(\sqrt{2}/3)^3$
6085 6053 10^x : $7\sqrt[3]{2}/2$
6085 6353 10^x : $3/2+4e/3$
6085 7361 cu : $4-e/4$
6085 8290 K : $e/9$
6085 9101 rt : (5,4,-1,-2)
6085 9664 ℓ_{10} : $3+3\sqrt{2}/4$
6086 1113 ln : $\ln(2\pi)$
6086 1225 10^x : $7\zeta(3)/9$
6086 3295 rt : (8,1,3,-4)
6086 3406 erf : $3\sqrt{2}/7$
6086 5531 ℓ_{10} : $\sqrt{2}/2+3\sqrt{5}/2$
6086 7787 2^x : $\sqrt{10}\pi$
6086 7925 rt : (4,0,-6,7)
6086 9565 id : 14/23
6086 9949 10^x : $2+3e$
6087 1446 $t\pi$: $1/3+2\pi$
6087 3246 rt : (1,8,-7,-7)
6087 4028 rt : (6,6,8,4)
6087 4597 id : $7\sqrt{5}/6$
6087 5514 ℓ_2 : $\sqrt{2}+\sqrt{5}+\sqrt{6}$
6087 6142 $s\pi$: 5/24
6087 6434 ℓ_2 : $5e\pi/7$
6087 7805 Ei : $\sqrt[3]{2}$
6087 8080 rt : (7,7,3,-6)
6087 9013 $t\pi$: $4e-\pi/3$
6087 9035 rt : (4,2,-9,-3)
6088 1320 id : $3\pi^2$
6088 2121 10^x : $(\sqrt[3]{2}/3)^{1/3}$
6088 2471 $\exp(2/5)+\exp(3/4)$
6088 5681 rt : (6,5,-5,4)
6088 5936 tanh : $\sqrt{2}/2$
6088 9722 θ_3 : $5\sqrt{2}/9$
6089 0173 rt : (5,0,8,6)
6089 3469 $4\times\exp(\sqrt{3})$
6089 4757 rt : (6,4,-8,-5)
6089 5615 erf : $7\ln 2/8$
6089 6193 rt : (4,6,8,-8)
6089 7714 erf : $(e)^{-1/2}$
6089 8608 rt : (8,1,-4,-1)
6089 9338 rt : (3,6,-4,2)
6089 9375 sq : $1+\pi/2$
6090 0775 rt : (9,8,0,-5)
6090 1271 rt : (1,9,4,-6)
6090 2295 $\ln(3/4)\times\exp(3/4)$
6090 5357 rt : (7,5,-4,-1)
6090 6903 rt : (6,7,-8,-6)
6090 8146 cr : $4\sqrt{2}-2\sqrt{5}/3$
6090 9369 rt : (1,6,0,-2)
6091 1207 rt : (9,4,9,-9)
6091 2903 ζ : $\sqrt{5}+\sqrt{6}-\sqrt{7}$
6091 3974 e^x : $4\sqrt{3}/3+\sqrt{5}/4$
6091 4091 id : $1/4+e/2$
6091 4897 cr : 25/6
6091 5038 ζ : 2/19
6091 5963 rt : (5,2,-6,6)
6091 7984 rt : (6,7,5,-7)
6091 9449 θ_3 : $(\sqrt{3}/2)^{1/2}$
6091 9759 K : 7/23
6092 1265 $s\pi$: $1/\sqrt{23}$
6092 1716 cu : $1/4+\sqrt{2}$
6092 1794 rt : (7,1,5,-5)
6092 2307 rt : (2,8,-9,-8)
6092 3005 sr : $\sqrt{2}/4+\sqrt{5}$
6092 3934 K : $(2e/3)^{-2}$
6092 3972 rt : (4,8,-8,1)
6092 5171 rt : (9,7,-4,-3)
6092 7783 rt : (3,8,-3,-9)
6092 7893 θ_3 : 11/14
6093 0041 cr : $9\pi^2/5$
6093 0072 rt : (1,2,-9,-6)
6093 1644 rt : (8,5,5,3)
6093 2031 rt : (7,7,-5,3)
6093 4390 $s\pi$: $4\sqrt{2}/3-3\sqrt{5}/4$
6093 5394 e^x : $1/4-\sqrt{5}/3$
6093 6134 10^x : $10(e+\pi)/3$
6093 6981 cr : $2e/3+3\pi/4$
6093 8829 rt : (8,8,2,-6)
6093 9439 sr : $7(1/\pi)/6$
6093 9917 e^x : $8\sqrt{2}/5$
6094 1628 Ei : $10\sqrt{2}$
6094 3656 sinh : $2\sqrt[3]{3}/5$
6094 3791 id : $\ln(5)$
6094 4262 rt : (4,1,-7,-8)
6094 6745 sq : 21/13
6094 7097 at : $2\pi/9$
6094 7375 2^x : $(\sqrt{2}/3)^{1/2}$
6094 7570 id : $1/3-2\sqrt{2}/3$
6094 9460 $\sqrt{2}-\ln(\sqrt{5})$
6095 0769 id : $3e/4+\pi/2$
6095 0794 $s\pi$: $8\pi^2/5$
6095 1449 rt : (9,2,2,-4)
6095 1462 sq : $3e+\pi$
6095 2889 rt : (5,4,-9,-4)
6095 4509 rt : (5,7,7,-8)
6095 7303 ℓ_2 : $1/3+e$
6095 9355 rt : (6,4,-6,5)
6095 9953 $t\pi$: $\sqrt{10/3}$
6096 1179 rt : (4,9,4,0)
6096 3064 tanh : 17/24
6096 4159 2^x : $(e+\pi)^2$
6096 7312 Li_2 : $3\sqrt{3}/10$
6096 8492 2^x : $1/3-\pi/3$
6096 9549 rt : (1,3,6,-5)
6097 0773 rt : (3,5,-7,1)
6097 2208 10^x : $5\pi/4$
6097 2755 cu : $4\sqrt{2}/3+\sqrt{5}/4$
6097 3274 rt : (6,1,7,-6)
6097 3342 cr : $\sqrt{2}/2+2\sqrt{3}$
6097 3700 cu : $(e+\pi)/5$
6097 3707 θ_3 : $3\ln 2/7$
6097 3725 rt : (7,8,4,-7)
6097 3964 rt : (8,9,-5,2)
6097 4737 rt : (1,7,-3,-1)
6097 6691 rt : (7,9,-8,-7)
6097 7895 rt : (6,5,-2,-2)
6097 8642 10^x : P_{39}
6097 9739 $s\pi(1/5)\div s\pi(\sqrt{2})$
6098 0587 2^x : $1/3+\sqrt{2}/4$
6098 0646 sinh : $gamma$
6098 0914 2^x : $4\zeta(3)/7$
6098 2583 $t\pi$: $\sqrt{2}/4+2\sqrt{5}$
6098 2972 rt : (5,2,1,1)
6098 4276 10^x : P_{33}
6098 4826 10^x : $\sqrt{3}/2+\sqrt{5}/4$
6098 4978 rt : (5,1,-7,7)
6098 6741 $s\pi$: $8(e+\pi)/9$
6098 6880 K : $\ln(e/2)$
6098 6964 as : $1/3-e/3$
6098 8074 rt : (4,7,9,-9)
6098 9257 rt : (1,9,-8,-8)
6098 9555 sr : $4\sqrt{2}+2\sqrt{3}/3$
6098 9844 sq : $10\sqrt[3]{5}/9$
6099 2149 rt : (9,9,1,-6)
6099 2964 e^x : 10/21
6099 5004 at : $((e+\pi)/2)^{-1/3}$
6099 6413 sinh : $\sqrt{3}/3$
6099 8364 rt : (3,8,-8,8)
6099 8739 rt : (7,6,-6,4)
6099 9687 rt : (8,2,4,-5)
6100 0000 sq : 19/10
6100 1905 Li_2 : $3\ln 2/4$
6100 5850 rt : (6,8,6,-8)
6100 7497 ζ : $3e/4$
6100 9056 rt : (4,2,-8,-9)
6100 9525 K : 4/13
6101 1021 as : $9(1/\pi)/5$
6101 2871 2^x : $4/3+2\sqrt{3}/3$
6101 3330 ζ : $(1/\pi)/3$
6101 3791 rt : (8,6,-5,-1)
6101 4363 K : $(\sqrt{2}\pi/3)^{-3}$
6101 5721 10^x : 5/12
6101 5887 rt : (6,6,-9,-5)
6101 6159 cu : $4e/3$
6101 8736 rt : (7,5,7,4)
6101 8806 J : $\exp(-14\pi/3)$
6101 8951 rt : (5,1,9,-7)
6101 9215 rt : (2,6,-2,-3)
6101 9615 K : $4\ln 2/9$
6102 1482 sq : $2\sqrt{3}/3+\sqrt{5}/3$
6102 1610 Li_2 : 13/25
6102 2044 rt : (6,3,-7,6)
6102 2547 cu : $4/3+\pi$
6102 2918 rt : (8,9,3,-7)
6102 4568 Li_2 : $(\sqrt[3]{3}/2)^2$
6102 5331 Ψ : $10\pi^2/7$
6102 6122 cr : 5/22
6102 7030 $t\pi$: $8\zeta(3)$
6102 7943 2^x : $4\sqrt{2}/3-3\sqrt{3}/2$
6102 8137 sq : $4+3\sqrt{2}/4$
6102 8320 sinh : $5\ln 2/6$
6103 0385 2^x : $7\sqrt[3]{2}/4$
6103 0868 2^x : $4-3\pi/2$
6103 1729 rt : (3,8,-6,0)
6103 3693 10^x : $(2e/3)^{-1/3}$
6103 4308 cr : $5(1/\pi)/7$

6103 4640 rt : (8,8,-6,3)
6103 5096 erf : $(\sqrt{2}\pi)^{-1/3}$
6103 5268 cu : $2+\pi/4$
6103 5528 rt : (5,8,8,-9)
6103 6598 id : $\sqrt{2}+3\sqrt{3}$
6103 8574 ℓ_{10} : $3e/2$
6103 9214 cr : $\exp(-\sqrt{2}\pi/3)$
6103 9316 rt : (5,5,0,-3)
6104 2637 rt : (7,2,6,-6)
6104 3582 10^x : $10\sqrt[3]{3}/3$
6104 4007 rt : (5,0,-8,8)
6104 5885 $\exp(1/5)+\exp(2)$
6104 5983 ℓ_{10} : $3\sqrt{3}-\sqrt{5}/2$
6104 7358 sr : $\sqrt{5}/6$
6104 7513 id : $\sqrt{2}-\sqrt{6}+\sqrt{7}$
6104 9033 2^x : 11/16
6105 1126 rt : (2,3,6,3)
6105 1387 rt : (7,9,5,-8)
6105 5111 e^x : $2/3+4\sqrt{3}/3$
6105 5926 rt : (7,5,-7,5)
6105 8607 rt : (5,5,-7,-5)
6106 0619 rt : (1,5,-2,8)
6106 0724 rt : (9,4,4,3)
6106 1421 $t\pi$: $\sqrt{2}/3+3\sqrt{5}/2$
6106 1565 sq : 7π
6106 3033 rt : (2,4,5,-5)
6106 3418 $\exp(\sqrt{5})+s\pi(\sqrt{3})$
6106 3610 e^x : $\exp(2e/3)$
6106 4125 cu : $1/\ln(e/2)$
6106 4745 rt : (4,9,-3,2)
6106 5020 rt : (9,3,3,-5)
6106 5193 Ψ : $7\pi/4$
6106 6592 erf : 14/23
6106 7203 2^x : $4e-\pi/3$
6106 7429 e^x : $4e\pi/7$
6106 9652 at : $5\sqrt[3]{2}/9$
6107 0137 $1/2\times\exp(1/5)$
6107 0245 rt : (7,8,-1,-2)
6107 0517 rt : (3,9,-8,-5)
6107 0804 Ψ : $7\sqrt[3]{2}/9$
6107 1260 sr : $1/2+2\pi/3$
6107 2596 at : 7/10
6107 2751 ζ : $2\sqrt[3]{5}/5$
6107 6039 rt : (6,2,-8,7)
6107 7810 rt : (6,9,7,-9)
6108 0762 Γ : $(\ln 3/3)^{-2}$
6108 1139 rt : (6,2,8,-7)
6108 1767 rt : (5,9,-9,1)
6108 4178 $\ln(2)\div s\pi(\pi)$
6108 4749 $s\pi$: $4\sqrt{2}-\sqrt{3}/2$
6108 5145 e^x : $2e/3+3\pi/4$
6108 5153 cu : $\sqrt{2}/4-3\sqrt{5}$
6108 6048 $2/3\times\ln(2/5)$
6108 7014 rt : (8,7,-7,4)
6108 7932 ln : $3\sqrt{3}-3\sqrt{5}/2$
6108 8705 sq : $(2\pi/3)^{-1/3}$
6108 9699 rt : (6,5,9,5)
6109 1641 $t\pi$: $6\sqrt{2}$
6109 2044 rt : (4,7,4,5)
6109 2400 J : $\exp(-2\pi/3)$
6109 3857 id : $1/3-4\sqrt{5}$
6109 4025 cu : $3\sqrt{2}/5$
6109 5078 rt : (5,1,-9,-9)
6109 6971 e^x : $3\sqrt{3}/4+3\sqrt{5}/4$
6109 8898 rt : (9,7,-6,-1)
6109 9368 10^x : $3/2+3\sqrt{3}/2$
6110 1384 rt : (8,3,5,-6)
6110 2339 2^x : 18/13
6110 4133 ζ : $\exp(-\sqrt{5})$
6110 4658 rt : (3,6,-4,-4)
6110 5500 rt : (7,4,-8,6)
6110 5765 ℓ_2 : $4\sqrt{3}/3+\sqrt{5}/3$
6110 5935 cu : $3(e+\pi)/8$
6110 6828 as : $e/2-\pi/4$
6110 8178 2^x : $3/4+3\pi/4$
6110 8308 10^x : $4e+\pi$
6110 9314 Ψ : 11/2
6111 1111 id : 11/18
6111 2429 rt : (1,4,7,-6)
6111 3126 rt : (4,8,-6,1)
6111 5387 rt : (2,8,-4,-1)
6111 5638 rt : (9,9,-7,3)
6111 5735 rt : (1,9,3,-3)
6111 5883 sq : $2+2\pi$
6111 6900 sr : $2\sqrt{3}+3\sqrt{5}/2$
6111 8345 $t\pi$: $4e/5$
6111 8990 rt : (6,6,-1,-3)
6111 9621 ℓ_2 : $\sqrt{7/3}$
6112 2600 sr : $9\sqrt[3]{3}/5$
6112 3038 rt : (6,1,-9,8)
6112 4725 rt : (8,4,6,4)
6112 5978 Ψ : $\sqrt[3]{25/2}$
6112 8401 Ψ : $1/\sqrt{19}$
6113 2685 rt : (8,6,-8,5)
6113 3104 $t\pi$: $(e+\pi)/10$
6113 4266 rt : (7,3,7,-7)
6113 5643 rt : (6,5,-9,-6)
6113 7123 at : $2-3\sqrt{3}/4$
6113 7527 2^x : $2/3+3\sqrt{3}/2$
6113 7833 rt : (3,5,4,-5)
6113 8139 id : $\sqrt{3}/2+\sqrt{5}/3$
6114 0691 K : 5/16
6114 1439 rt : (9,0,-5,1)
6114 4071 rt : (8,7,-4,-2)
6114 5660 rt : (1,7,-1,-3)
6114 8174 $\exp(2/3)\div s\pi(\sqrt{3})$
6114 8891 rt : (7,3,-9,7)
6114 8901 sq : $\sqrt[3]{17}$
6115 1441 $\exp(1/3)\div s\pi(1/3)$
6115 1763 rt : (9,4,4,-6)
6115 2725 λ : 1/17
6115 3810 sq : $3/4+\sqrt{3}/2$
6115 5549 rt : (4,9,-7,0)
6115 5634 Γ : $\ln 2/7$
6115 5975 $s\pi$: $2\sqrt{3}+\sqrt{5}/3$
6115 7846 rt : (9,8,-8,4)
6115 8465 cu : $8(1/\pi)/3$
6115 8577 cr : $5(e+\pi)/7$
6115 9461 ℓ_{10} : $3\sqrt{3}/2+2\sqrt{5}/3$
6116 1532 rt : (5,9,4,4)
6116 4152 rt : (6,3,9,-8)
6116 4688 2^x : $4\sqrt{2}+3\sqrt{3}$
6116 6261 sq : $\sqrt[3]{21}$
6116 6701 J_0 : $(\sqrt{3})^{1/2}$
6116 9504 e^x : $2-\sqrt{2}/3$
6116 9647 e^x : 24/25
6117 0049 sr : P_{74}
6117 0284 sr : $2e/3+\pi/4$
6117 0451 10^x : $4+\sqrt{2}/2$
6117 0855 rt : (1,3,8,4)
6117 2501 e^x : $1/\ln(\pi/3)$
6117 2866 rt : (8,5,-9,6)
6117 3103 rt : (4,6,-6,-5)
6117 4112 Ei : $7e\pi/4$
6117 6710 rt : (5,9,2,-1)
6117 7989 rt : (2,5,6,-6)
6117 8386 rt : (7,4,8,5)
6117 9706 rt : (1,7,-4,5)
6118 0165 rt : (8,4,6,-7)
6118 0553 2^x : $3\sqrt{3}/2-\sqrt{5}/3$
6118 1652 J_0 : 25/19
6118 2354 rt : (1,8,-2,-2)
6118 3298 sq : $3\sqrt{2}/4+2\sqrt{5}$
6118 3528 $\ln(4)+\exp(4/5)$
6118 3566 sinh : 11/19
6118 3683 rt : (7,7,-2,-3)
6118 4169 $t\pi$: $1/4-3\pi$
6118 5230 $t\pi$: $3\sqrt{2}/4-3\sqrt{5}/4$
6118 5489 sr : $3\sqrt{3}/2$
6118 5831 rt : (6,1,0,1)
6118 5870 rt : (1,6,2,-1)
6118 6023 $s\pi(1/3)-s\pi(\sqrt{3})$
6118 6546 K : $\pi/10$
6118 9608 K : $2\sqrt{2}/9$
6119 1416 sq : $\sqrt{3}+\sqrt{6}+\sqrt{7}$
6119 5158 rt : (9,7,-9,5)
6119 6737 cu : $2+4\sqrt{5}/3$
6119 6768 sinh : $\exp(\pi/3)$
6119 7538 e^x : $e/2+2\pi/3$
6119 8256 e^x : $3(1/\pi)/2$
6119 8863 10^x : $1/3+3\sqrt{5}$
6119 9004 rt : (4,6,3,6)
6119 9195 cr : $4\pi/3$
6120 1791 e^x : 17/9
6120 3232 id : $3\sqrt{2}/2+2\sqrt{5}/3$
6120 4425 rt : (9,8,-5,-2)
6120 4948 Γ : $5(e+\pi)/7$
6120 5625 rt : (9,3,5,4)
6120 6152 rt : (7,4,8,-8)
6120 6800 rt : (2,7,-3,-4)
6120 8102 ln : $\sqrt{2}/4+2\sqrt{5}/3$
6120 8497 ln : $3e-\pi$
6120 9347 K : $\sqrt[3]{2}/4$
6121 0803 ℓ_{10} : $\sqrt[3]{5}/7$
6121 1309 sinh : $2\pi^2$
6121 2752 e^x : $1/3+2\sqrt{2}$
6121 3506 rt : (1,5,8,-7)
6121 5349 rt : (3,9,-5,-1)
6121 7332 sr : $1/\sqrt[3]{19}$
6121 8718 rt : (6,7,0,-4)
6121 8974 ℓ_{10} : $2+2\pi/3$
6121 9036 e^x : $1-2\sqrt{5}/3$
6121 9263 $t\pi$: $\sqrt{2}/2+\sqrt{5}/2$
6121 9965 $\ln(4/5)-\exp(2)$
6122 0208 rt : (9,5,5,-7)
6122 0666 ℓ_2 : $2-\sqrt{2}/3$
6122 1954 10^x : $9\sqrt[3]{2}/10$
6122 4059 rt : (3,2,5,3)
6122 4489 sq : 18/7
6122 6639 Ei : $\exp(\sqrt{2}/3)$
6122 9209 rt : (5,6,-8,-6)
6123 1893 K : 6/19
6123 2275 rt : (3,6,5,-6)
6123 5003 at : $(\ln 2/2)^{1/3}$
6123 5167 $\ln(4)-s\pi(e)$
6123 6124 sr : $4e/3+3\pi$
6123 7243 sr : 3/8
6123 7534 ζ : 3/2
6124 1306 J : 12/25
6124 3009 rt : (8,5,7,-8)
6124 4134 K : $1/\sqrt{10}$
6124 5154 sr : 13/5
6124 6343 2^x : $8e\pi/3$
6124 7591 rt : (8,3,7,5)
6124 7637 sq : 18/23
6124 9939 rt : (5,8,3,5)
6125 0887 rt : (2,8,-5,3)
6125 4482 erf : 11/18
6125 7411 rt : (3,7,-5,-5)
6125 8003 ℓ_{10} : $3/2+3\sqrt{3}/2$
6125 9007 id : $2e/3-3\pi$
6126 0143 rt : (1,8,4,8)
6126 1556 sq : $4/3+3\pi/4$
6126 2004 ℓ_2 : $9e$
6126 2023 rt : (2,6,7,-7)
6126 2380 $\exp(1/2)-s\pi(\sqrt{2})$
6126 2568 ln : $(2\pi)^{1/3}$
6126 4067 rt : (7,5,9,-9)
6126 4157 ln : $3-2\sqrt{3}/3$
6126 4860 rt : (2,9,-3,-2)
6126 6007 sinh : $7\ln 2/2$
6126 6406 Γ : $\exp(5/2)$
6126 6590 rt : (7,8,-1,-4)
6126 7281 Li_2 : 12/23
6126 9366 rt : (5,1,2,2)
6127 0590 e^x : $\sqrt{2}/2+\sqrt{3}/3$
6127 1315 $t\pi$: $9(e+\pi)/5$
6127 2849 sq : $3e-3\pi$
6127 4321 ln : $4\sqrt{3}+3\sqrt{5}$
6127 5606 rt : (9,6,6,-8)
6127 5628 cu : $7e\pi/10$
6127 7946 rt : (4,7,4,-6)
6127 9059 cr : $4\sqrt{2}/3+4\sqrt{3}/3$
6128 2308 rt : (9,9,-4,-3)
6128 3093 rt : (1,8,-2,-4)
6128 3970 rt : (7,3,9,6)
6128 6613 $\exp(3/5)\times\exp(e)$
6128 7285 Γ : $((e+\pi)/2)^{-3}$
6128 7438 $s\pi$: $\sqrt[3]{2}/6$
6128 7959 2^x : $4\sqrt{3}/5$

6128 8190 sr : $4 - 4e/3$
6128 8786 rt : (1,6,9,-8)
6128 9247 ζ : $\exp(-\sqrt{2}\pi/2)$
6129 0869 sr : $\sqrt{3}+\sqrt{6}+\sqrt{7}$
6129 1689 ζ : 13/19
6129 2997 rt : (6,8,1,-5)
6129 4316 rt : (8,6,8,-9)
6129 4744 $\pi \div \exp(2/3)$
6129 5542 ℓ_2 : $2\sqrt{2} - 3\sqrt{3}/4$
6129 7724 rt : (2,2,7,4)
6129 8209 ℓ_2 : $\exp(\sqrt{5}/2)$
6129 8635 ℓ_{10} : $7(e+\pi)$
6129 8970 K : 7/22
6130 0023 rt : (4,7,-7,-6)
6130 0400 cr : $2\sqrt{2} - 3\sqrt{3}/2$
6130 2276 Ψ : $9\pi/2$
6130 2579 K : $(1/\pi)$
6130 3070 rt : (9,2,6,5)
6130 3216 rt : (3,7,6,-7)
6130 3685 rt : (6,9,0,-2)
6130 5892 2^x : $3\sqrt{3}/4 + 2\sqrt{5}$
6130 6121 Ei : $10\sqrt[3]{3}/9$
6130 6563 rt : (1,9,-1,-3)
6130 7243 rt : (8,9,-2,-4)
6130 8345 rt : (7,0,-1,1)
6131 0344 Ei : $\ln(\sqrt{3})$
6131 0447 ln : 13/24
6131 2592 sr : $4 + 2\sqrt{2}$
6131 5910 $1/5 + \exp(5)$
6131 6063 ℓ_2 : $4/3 - e/4$
6131 6891 rt : (5,8,3,-6)
6132 1371 rt : (9,7,7,-9)
6132 1814 ℓ_{10} : $3/4 + 3\sqrt{5}/2$
6132 1989 rt : (2,8,-4,-5)
6132 4925 ℓ_2 : $3/4 + 4\sqrt{3}/3$
6132 5037 $1/2 + \exp(\sqrt{2})$
6132 6137 rt : (2,7,8,-8)
6132 6628 e^x : 11/23
6132 8176 2^x : $4/3 - 3e/4$
6132 9866 rt : (7,9,0,-5)
6133 0590 sq : $5(e+\pi)/2$
6133 1375 rt : (1,9,-6,-1)
6133 1733 tπ : $e/4 - 3\pi/4$
6133 2715 rt : (8,2,8,6)
6133 4454 K : $\sqrt{5}/7$
6133 5726 tanh : 5/7
6133 5939 ζ : $(2\pi/3)^{-3}$
6133 6409 rt : (5,7,-9,-7)
6133 7431 sπ : $3(e+\pi)/2$
6133 7485 cr : 3/13
6133 8619 rt : (4,8,5,-7)
6133 8688 Ψ : $10\sqrt{2}$
6133 9494 cr : $10\sqrt[3]{2}/3$
6134 2192 rt : (9,8,-5,-4)
6134 2864 cr : 21/5
6134 7026 rt : (1,5,2,1)
6134 7236 10^x : $3/2 - e/4$
6135 0374 K : 8/25
6135 0487 rt : (6,9,2,5)
6135 1342 ℓ_2 : $4e/3 - 2\pi/3$
6135 2722 sπ : $3\sqrt{3}/4 + 2\sqrt{5}/3$
6135 4375 J_0 : 21/16
6135 4612 rt : (1,2,9,5)
6135 4910 2^x : $8e/7$
6135 5419 rt : (3,8,-6,-6)
6135 5644 2^x : $3/2 + e$
6135 8465 rt : (3,8,7,-8)
6136 1589 ζ : $3(1/\pi)$
6136 2267 id : $(\ln 2/3)^{1/3}$
6136 3334 rt : (6,0,1,2)
6136 4333 rt : (1,6,-3,7)
6136 4571 K : $2\sqrt[3]{3}/9$
6136 5047 10^x : $2 - \sqrt{3}/2$
6136 7037 λ : $\exp(-9\pi/10)$
6136 8519 ln : $\sqrt{2} + \sqrt{3}/4$
6136 8584 sq : $(\sqrt[3]{3}/3)^{1/3}$
6136 9365 rt : (5,9,4,-7)
6137 0563 ln : $(e/2)^2$
6137 1370 cu : $\sqrt{2} - 2\sqrt{3}$
6137 1468 rt : (2,9,-3,-9)
6137 2997 rt : (1,9,-3,-5)
6137 3136 rt : (9,1,7,6)
6137 6668 rt : (2,8,9,-9)
6137 6880 sr : $4e\pi/5$
6137 7328 tπ : $2\sqrt{2}/3 - \sqrt{5}/2$
6137 8874 Ei : $(3\pi/2)^{1/2}$
6138 1184 sr : $4(e+\pi)/9$
6138 1424 Ei : $4\zeta(3)/3$
6138 2476 rt : (3,1,6,4)
6138 3208 e^x : $2/3 - 2\sqrt{3}/3$
6138 4286 ℓ_2 : $2 + 3\sqrt{2}/4$
6138 4467 rt : (4,8,-8,-7)
6138 6714 rt : (4,9,6,-8)
6138 7009 id : $\sqrt{3}+\sqrt{5}+\sqrt{7}$
6138 7631 rt : (7,9,-2,-3)
6138 8140 $3/5 \times \exp(\sqrt{5})$
6138 9199 2^x : $3/2 + \sqrt{2}/4$
6139 0012 Ei : $\exp(-\sqrt[3]{3})$
6139 0327 rt : (8,1,-2,-1)
6139 1098 10^x : $\exp(-\pi/2)$
6139 1536 10^x : $2\sqrt{2}/3 - 2\sqrt{3}/3$
6139 2116 Ei : $(\sqrt[3]{3}/2)^{1/3}$
6139 2600 J : $2\sqrt{2}/9$
6139 3424 rt : (1,2,-9,-8)
6139 5193 rt : (8,1,9,7)
6139 5339 2^x : $3 - 4\sqrt{3}/3$
6139 6314 rt : (6,8,1,6)
6139 7750 ln : $6(e+\pi)/7$
6139 8371 2^x : $10\pi^2/9$
6139 9906 rt : (2,9,-5,-6)
6140 1681 id : $\sqrt{3} - \sqrt{5}/2$
6140 2714 rt : (3,9,8,-9)
6140 3711 sq : $\sqrt{2}/2 - 2\sqrt{5}/3$
6140 4760 Ei : $3\sqrt{5}/2$
6140 4806 $3/4 \div \exp(1/5)$
6140 6381 2^x : $2\ln 2$
6140 7189 rt : (5,0,3,3)
6140 7799 tanh : $(\ln 3/3)^{1/3}$
6141 0429 cu : $\sqrt{3} - \sqrt{5}/4$
6141 1383 Li_2 : $1/\sqrt[3]{7}$
6141 1606 rt : (5,5,0,8)
6141 2117 tanh : $((e+\pi)/3)^{-1/2}$
6141 2500 cu : 17/20
6141 4016 tanh : $(\sqrt{5}/2)^{-3}$
6141 5098 10^x : $3\ln 2/10$
6141 8514 ℓ_{10} : $\exp(\sqrt{2})$
6142 0804 $\ln(2/5) \div \exp(2/5)$
6142 1271 sπ : 4/19
6142 1356 $1/5 + \sqrt{2}$
6142 2634 rt : (2,1,8,5)
6142 3619 rt : (3,9,-7,-7)
6142 4262 e^x : $2\sqrt{2}/3 + 4\sqrt{5}/3$
6142 4515 rt : (3,9,-5,6)
6142 4918 10^x : $e/2 - \pi/2$
6142 5958 rt : (9,0,8,7)
6142 7800 tπ : $(3\pi/2)^{1/3}$
6142 9240 rt : (7,1,0,-2)
6143 0034 e^x : $(\sqrt{2}/3)^{-1/3}$
6143 1633 ln : $8\ln 2/3$
6143 2627 at : $4/3 - 3e/4$
6143 2686 cu : $1/\ln(\sqrt[3]{3}/2)$
6143 3704 rt : (6,7,0,7)
6143 4997 cr : $7\zeta(3)/2$
6143 7123 sr : $1/4 + 3\pi/4$
6143 8905 2^x : $(2\pi/3)^{-1/2}$
6144 0070 $1/3 - \exp(2/3)$
6144 3012 rt : (4,0,5,4)
6144 3447 ℓ_2 : $\exp(2e/3)$
6144 4679 rt : (4,9,-9,-8)
6144 6345 rt : (5,4,-1,9)
6144 7328 rt : (8,9,-4,-4)
6144 8804 sinh : $2\pi/5$
6144 9026 rt : (9,2,-3,-1)
6144 9261 erf : $(\ln 2/3)^{1/3}$
6145 0749 Ei : $7\sqrt[3]{3}/8$
6145 1230 rt : (2,8,5,8)
6145 3452 rt : (7,9,0,6)
6145 4826 tanh : $9(1/\pi)/4$
6145 6444 tπ : $3\sqrt{2}/4 - \sqrt{5}$
6145 9702 $\exp(2) + \exp(4/5)$
6146 0307 tπ : $(\sqrt{5}/3)^3$
6146 1376 rt : (6,1,2,-3)
6146 2108 sinh : $5\sqrt[3]{5}/3$
6146 2515 ln : $3\sqrt[3]{3}/8$
6146 3878 cr : $2\sqrt{3} + \sqrt{5}/3$
6146 4795 rt : (6,6,-1,8)
6146 6295 at : 12/17
6146 7010 θ_3 : 12/19
6146 7213 rt : (9,1,9,-8)
6146 7401 sr : $7(e+\pi)/6$
6146 8992 ℓ_{10} : $3 + \sqrt{5}/2$
6146 9480 tπ : $\sqrt{2}/4 - 3\sqrt{5}/4$
6147 1291 2^x : $4 + 3\sqrt{2}/2$
6147 2841 rt : (3,0,7,5)
6147 3351 ln : $8\pi/5$
6147 4350 sr : $1/2 + 4\pi$
6147 5611 tanh : $(e)^{-1/3}$
6147 6653 2^x : $e/4 + 4\pi$
6147 8003 rt : (8,2,-1,-2)
6147 8815 sr : $1/\sqrt{7}$
6147 9523 rt : (2,7,0,3)
6147 9778 sr : $3\sqrt[3]{2}/10$
6148 1439 rt : (7,8,-1,7)
6148 3977 2^x : $\sqrt[3]{8/3}$
6148 7308 J : $\exp(-18\pi/7)$
6148 8379 rt : (5,1,4,-4)
6149 0478 K : $(\sqrt[3]{5}/3)^2$
6149 0669 cu : $4/3 + 4\sqrt{5}$
6149 1056 rt : (6,5,-2,9)
6149 1148 10^x : $4\sqrt{3}/3 + \sqrt{5}/4$
6149 1986 rt : (9,9,-6,-5)
6149 2814 as : $2\sqrt[3]{3}/5$
6149 3925 ζ : $\ln(\sqrt[3]{5})$
6149 5915 tπ : $1/3 - 4\sqrt{2}$
6149 8076 rt : (2,0,9,6)
6150 2555 rt : (7,2,1,-3)
6150 3607 rt : (2,6,-5,-2)
6150 3637 ℓ_{10} : $2 + 3\sqrt{2}/2$
6150 4600 rt : (1,7,3,3)
6150 5275 rt : (7,7,-2,8)
6150 5360 ℓ_{10} : P_{15}
6150 6414 e^x : $3 - 3e/4$
6150 7312 Ψ : $\sqrt{2} - \sqrt{3} + \sqrt{7}$
6150 7490 ℓ_{10} : $4\sqrt{2}/3 + \sqrt{5}$
6151 1341 $\exp(2) - s\pi(e)$
6151 1397 rt : (4,1,6,-5)
6151 4931 K : $(3\pi)^{-1/2}$
6151 5536 rt : (9,3,-2,-2)
6151 5759 10^x : $3e - \pi$
6151 6128 sr : $7\sqrt{5}/6$
6151 6720 rt : (3,5,-4,9)
6151 8260 rt : (8,9,-2,7)
6152 2371 Γ : $1/\sqrt[3]{6}$
6152 2446 ℓ_{10} : $\sqrt{17}$
6152 3233 cr : $3/4 + 2\sqrt{3}$
6152 3630 rt : (6,2,3,-4)
6152 4227 cu : $3 + 4\sqrt{3}$
6152 5822 rt : (7,6,-3,9)
6152 7556 ln : $\sqrt[3]{19/3}$
6152 9039 10^x : $2 + 2\sqrt{2}/3$
6152 9065 Ei : 11/20
6152 9478 e^x : $10\sqrt{2}$
6153 0409 Li_2 : $\pi/6$
6153 1258 rt : (3,1,8,-6)
6153 1486 as : γ
6153 2097 sr : $4\sqrt[3]{5}$
6153 2695 sinh : $8\sqrt{2}/9$
6153 4900 rt : (8,3,0,-3)
6153 4960 Ψ : $(\pi/3)^{-1/2}$
6153 5606 e^x : $2\sqrt{2} - 3\sqrt{3}/4$
6153 5755 rt : (3,4,-7,8)
6153 5822 ℓ_{10} : $1/2 + 4e/3$
6153 7114 rt : (8,8,-3,8)
6153 7615 Ψ : $\pi/8$
6153 8434 e^x : $4/3 + 4e$
6153 8461 id : 8/13
6153 9038 cu : $1/2 + 3\pi/2$
6153 9442 cu : $\sqrt{2} - \sqrt{3} + \sqrt{7}$
6154 1924 rt : (5,2,5,-5)
6154 2604 ln : $7(e+\pi)/3$

6154 3550 $\Psi : \ln(\sqrt{2}\pi/3)$
6154 6246 $t\pi : 8\sqrt{2}/7$
6154 7641 $sr : 3\sqrt{2} + 3\sqrt{3}/2$
6154 7877 $\Psi : 4\sqrt[3]{5}/7$
6154 7970 $at : \sqrt{2}/2$
6154 9028 $\sinh : \exp(5/3)$
6154 9448 $K : \exp(-\sqrt{5}/2)$
6155 0066 $K : \ln(\sqrt[3]{3}/2)$
6155 1802 rt : (7,3,2,-4)
6155 3626 rt : (8,7,-4,9)
6155 6387 $\Gamma : 4(1/\pi)/9$
6155 7957 rt : (4,2,7,-6)
6155 8092 rt : (1,8,-6,-6)
6155 9809 $10^x : 5/24$
6156 0257 $Li_2 : 11/21$
6156 0974 rt : (9,4,-1,-3)
6156 1620 $e^x : 2\sqrt[3]{3}/3$
6156 1682 rt : (4,8,-3,9)
6156 2810 rt : (9,9,-4,8)
6156 4917 $\ln : 9\sqrt{5}/4$
6156 5676 $cu : 4 - e/3$
6156 6688 rt : (6,3,4,-5)
6156 8342 $\ell_{10} : 2\sqrt{2} + 3\sqrt{3}/4$
6156 8984 $2^x : 2/3 + 3\pi/2$
6157 2126 rt : (3,2,9,-7)
6157 2797 $2^x : 1/4 + 2\pi$
6157 3731 rt : (2,7,-9,6)
6157 4623 $2^x : \exp(-1/e)$
6157 4689 $s\pi : 4\sqrt{5}/5$
6157 4829 rt : (8,4,1,-4)
6157 5357 rt : (4,7,-6,8)
6157 5511 $sq : 1/3 - \sqrt{5}/2$
6157 6370 rt : (9,8,-5,9)
6157 6624 $cr : 3/2 + e$
6157 9899 rt : (5,3,6,-6)
6157 9974 $sq : 7\ln 2/3$
6158 1334 $as : 5\ln 2/6$
6158 1502 $2^x : 9\sqrt[3]{3}/7$
6158 5547 $erf : 8/13$
6158 6105 $sq : \sqrt{2}/3 + 3\sqrt{5}/4$
6158 6614 $2^x : 9/13$
6158 7176 rt : (7,4,3,-5)
6158 7567 rt : (4,6,-9,7)
6159 0253 rt : (9,9,7,3)
6159 1706 rt : (4,3,8,-7)
6159 2011 $cr : 2/3 - \sqrt{3}/4$
6159 2084 $\ln : 8\sqrt[3]{5}$
6159 3134 $e^x : \sqrt[3]{25}$
6159 4001 rt : (9,5,0,-4)
6159 8249 rt : (6,4,5,-6)
6159 9761 $e^x : 4\zeta(3)/5$
6159 9993 $s\pi : 5\sqrt[3]{3}$
6160 0184 $cr : \sqrt{3}/2 + 3\sqrt{5}/2$
6160 0279 $cu : 1/3 - 3\sqrt{3}/2$
6160 2268 $\exp(\sqrt{5}) \times s\pi(1/4)$
6160 2540 $id : 1/4 - \sqrt{3}/2$
6160 3083 $\ell_{10} : 2/3 + 2\sqrt{3}$
6160 3346 $\zeta : 1/9$
6160 4406 rt : (8,5,2,-5)
6160 6398 $J_0 : 17/13$
6160 6787 $\ell_2 : 4/3 + \sqrt{3}$
6160 7440 $e^x : 12/25$
6160 8065 $J_0 : (\sqrt[3]{5})^{1/2}$
6160 8237 rt : (5,4,7,-7)
6161 0084 $s\pi : 3/4 + 3e/4$
6161 0301 $\ln(5) - \exp(4/5)$
6161 0747 $e^x : 1/2 + \pi/4$
6161 0809 $cu : 3 + 4\sqrt{2}/3$
6161 2085 $2^x : 1 + 2e$
6161 2426 $\Gamma : 11/20$
6161 3082 $id : 8\ln 2/9$
6161 3620 $\exp(\sqrt{3}) - s\pi(\sqrt{2})$
6161 3820 rt : (7,5,4,-6)
6161 3964 $sq : 3/4 - 2\pi$
6161 4084 rt : (5,9,-8,7)
6161 5428 $\exp(\sqrt{3}) \div s\pi(1/5)$
6161 6308 rt : (8,9,9,4)
6161 6410 $\Psi : 6e/7$
6161 7293 rt : (4,4,9,-8)
6161 9096 rt : (9,6,1,-5)
6162 2378 rt : (6,5,6,-7)
6162 3096 $J : \exp(-2\pi/7)$
6162 3937 $\theta_3 : 3/10$
6162 4407 $id : 8\sqrt{2}/7$
6162 4615 $Ei : 2\pi/7$
6162 4880 $\ln : 3\sqrt[3]{2}/7$
6162 7186 $10^x : 1/\sqrt{23}$
6162 7197 rt : (8,6,3,-6)
6162 7659 $\ell_{10} : 3e/4 + 2\pi/3$
6162 9693 $at : 17/24$
6163 0160 $\Psi : (e + \pi)/6$
6163 0194 rt : (5,5,8,-8)
6163 0276 $Ei : 1/\sqrt[3]{6}$
6163 2139 $t\pi : 7\pi^2/9$
6163 2823 $sq : 4\sqrt{2} - 4\sqrt{3}$
6163 4613 rt : (7,6,5,-7)
6163 5240 $t\pi : 2\sqrt{2}/3 - \sqrt{5}/4$
6163 5987 $cu : \sqrt{2}/2 + 2\sqrt{5}/3$
6163 8814 rt : (9,7,2,-6)
6163 9081 $id : 3\sqrt{2}/4 - 3\sqrt{5}/4$
6164 1426 rt : (6,6,7,-8)
6164 3177 $erf : 8\ln 2/9$
6164 3706 $sr : 2\sqrt[3]{5}/9$
6164 4040 $2^x : 2\sqrt{3}/5$
6164 5301 rt : (8,7,4,-7)
6164 5522 $id : 8\zeta(3)$
6164 6795 $10^x : 4\sqrt{2}/3 - 3\sqrt{5}/4$
6164 6988 rt : (9,8,8,4)
6164 7709 rt : (5,6,9,-9)
6165 1198 $t\pi : 4\sqrt[3]{3}/7$
6165 1293 rt : (7,7,6,-8)
6165 1864 $id : 1/3 + 2\pi$
6165 4717 rt : (9,8,3,-7)
6165 6845 rt : (6,7,8,-9)
6165 7586 $1/4 \div \ln(2/3)$
6165 7979 $\zeta : \ln(\sqrt{5}/2)$
6165 9215 rt : (1,9,4,6)
6165 9272 $e^x : P_{21}$
6165 9360 $\ell_{10} : 2\sqrt{3}/3 + 4\sqrt{5}/3$
6166 0028 rt : (8,8,5,-8)
6166 0276 $10^x : 3\sqrt[3]{5}/5$
6166 1853 $t\pi : P_{14}$
6166 3207 $\Gamma : 5(1/\pi)/8$
6166 4971 rt : (7,8,7,-9)
6166 5494 $\ln : 3\sqrt{3}/2 - \sqrt{5}/3$
6166 5540 $t\pi : 3e/4 + \pi/4$
6166 7136 $\ell_2 : 15/23$
6166 7816 rt : (9,9,4,-8)
6167 2244 rt : (8,9,6,-9)
6167 3299 $t\pi : \sqrt{3} + 4\sqrt{5}$
6167 6297 rt : (1,8,2,-7)
6167 8513 $s\pi : \exp(\sqrt[3]{4}/2)$
6167 9856 rt : (9,7,9,5)
6168 0667 $2^x : \ln 2$
6168 2349 $\Psi : \exp(\sqrt[3]{5})$
6168 4235 $e^x : 3 + \sqrt{2}$
6168 5027 $sq : \pi/4$
6168 8135 $\tanh : 4\sqrt[3]{2}/7$
6169 0930 $\tanh : 18/25$
6169 1625 $10^x : (e + \pi)$
6169 3389 $t\pi : 1/3 + 2\sqrt{5}/3$
6169 4937 $\ln : \sqrt{19}\pi$
6169 4995 $2^x : -\sqrt{2} - \sqrt{3} + \sqrt{6}$
6169 8333 $\sinh : 7/12$
6170 0001 $1/2 + \exp(3/4)$
6170 1518 $\exp(1/3) + \exp(1/5)$
6170 2413 $10^x : 9e\pi/8$
6170 3552 rt : (1,8,9,-1)
6170 4662 $2^x : \sqrt[3]{1/3}$
6170 4896 $4 \times \exp(e)$
6170 6523 $\Psi : (2e)^{1/2}$
6171 0454 $\ln : 3/2 + \sqrt{2}/4$
6171 3058 $sq : 4/3 + 4\sqrt{5}/3$
6171 5036 $\ln : 3 + 3e/4$
6171 6663 $e^x : (\ln 2/3)^{1/2}$
6171 6738 $\ell_{10} : 1 + \pi$
6172 0846 rt : (1,9,-5,9)
6172 1339 $sr : 8/21$
6172 3650 $e^x : 1 + 2\pi$
6172 3942 $cr : 3\pi^2/7$
6172 5078 $e^x : 9/7$
6172 6551 rt : (2,9,0,5)
6172 7792 $\ell_{10} : 2e/3 - \pi/2$
6172 8395 $sq : 5\sqrt{2}/9$
6172 8668 $id : \exp(\sqrt[3]{3}/3)$
6172 9149 $e^x : 4\sqrt{2}/3 + 3\sqrt{5}/2$
6172 9455 $K : (\ln 2)^3$
6173 0203 rt : (2,8,-7,-6)
6173 2894 $\ln(\sqrt{2}) + s\pi(\sqrt{2})$
6173 4342 $id : 7\ln 2/3$
6173 4693 $sq : 11/14$
6173 5791 $cr : 4/17$
6173 5875 $1/3 + \exp(1/4)$
6173 7281 $\ell_{10} : \sqrt{3} - 2\sqrt{5}/3$
6173 7979 $\theta_3 : \zeta(3)/4$
6173 8673 $K : 1/3$
6174 0459 $e^x : 2\zeta(3)/5$
6174 1501 $\exp(\sqrt{5}) \div \exp(e)$
6174 3710 $as : 11/19$
6174 7200 $Ei : 23/3$
6174 8342 $2^x : 3/2 + \sqrt{2}/2$
6175 1852 $\ln : 9\sqrt[3]{3}/7$
6175 2015 $t\pi : 4e + \pi/2$
6175 7617 $J_0 : (\sqrt{2}\pi/2)^{1/3}$
6175 7783 $t\pi : \sqrt{2} - \sqrt{5} + \sqrt{7}$
6175 8058 $e^x : 5\sqrt{3}/9$
6176 0553 $\tanh : \sqrt[3]{3}/2$
6176 1818 $Ei : \sqrt{2}/6$
6176 6889 $id : 4\sqrt{2}/3 + \sqrt{3}$
6176 7317 $\tanh : 3\zeta(3)/5$
6177 1467 $sr : 1/\sqrt[3]{18}$
6177 2318 $1/3 - s\pi(2/5)$
6177 4329 $\tanh : 1/\ln(4)$
6178 2734 $\ln : 7\ln 2/9$
6178 3748 $Ei : 24/19$
6178 9169 $s\pi : 2/3 + 3\sqrt{2}/2$
6179 0955 $Ei : 5\sqrt[3]{3}/3$
6179 5539 $erf : 3\sqrt[3]{3}/7$
6179 6621 $\Psi : 7/3$
6179 9387 $id : 5\pi/6$
6179 9461 $\ell_2 : 2\sqrt{2}/3 + 3\sqrt{3}$
6180 0935 $\Gamma : \sqrt{2}/10$
6180 1660 $\sinh : ((e + \pi)/2)^{-1/2}$
6180 2159 $sr : 5\pi/6$
6180 3398 $id : 1/2 + \sqrt{5}/2$
6180 3872 $sr : 6(1/\pi)/5$
6180 7370 $\Gamma : \ln(\sqrt{3})$
6180 8754 $e^x : 4e/3 - 2\pi/3$
6181 0695 $id : 3\sqrt[3]{3}/7$
6181 1909 $erf : \exp(-\sqrt[3]{3}/3)$
6182 2931 $\ell_2 : (3\pi)^{1/2}$
6182 5450 $t\pi : 8\ln 2/9$
6182 8391 $\tanh : 13/18$
6183 0381 $cu : 4\sqrt{2} - 3\sqrt{3}/2$
6183 1168 $s\pi : 6\sqrt{2}/7$
6183 1956 $id : \exp(-\sqrt[3]{3}/3)$
6183 6712 $cu : \sqrt{2}/3 - 3\sqrt{3}/2$
6183 6980 $s\pi : 2(1/\pi)/3$
6184 1606 $\Gamma : 3e/2$
6184 8292 $id : e/3 + 3\pi/2$
6184 8703 rt : (1,6,-5,-6)
6185 0313 $2^x : \ln(1/2)$
6185 3972 $\ell_{10} : 3 + 2\sqrt{3}/3$
6185 4081 $\ln : 1/3 + 3\pi/2$
6185 8482 $e^x : 3\sqrt[3]{2}/2$
6185 9816 $sr : 2\sqrt{2}/3 + 3\sqrt{5}/4$
6186 1282 $\ell_2 : 3 + \pi$
6186 2486 $\theta_3 : (5/2)^{-1/2}$
6186 4811 $2^x : 1/4 - 2\sqrt{2}/3$
6186 5569 $cu : 16/9$
6186 7934 $erf : 13/21$
6186 8621 $sq : \sqrt{2}/4 + \sqrt{3}/4$
6186 8950 $e^x : \sqrt{2}/4 + 3\sqrt{5}/4$
6187 0262 $s\pi : 6\zeta(3)$
6187 1153 $\sinh : 1/\sqrt[3]{5}$
6187 5148 $2^x : 25/2$
6187 5597 $\ln(4/5) - \exp(1/3)$
6187 6523 $Ei : 25/7$
6187 6915 $2^x : 25/18$
6187 7101 $t\pi : (\sqrt[3]{2}/3)^2$

6187 9170 2^x : $2/3 - e/2$
6188 0215 id : $8\sqrt{3}/3$
6188 2001 sπ : $1/2 + 3\pi/2$
6188 3314 ln : $4 + \pi/3$
6188 4064 Γ : $7e\pi/4$
6188 7040 sr : $\sqrt[3]{18}$
6188 9252 $\sqrt{2} \times \ln(\pi)$
6189 0238 rt : (9,5,8,-9)
6189 3496 sπ : $7\sqrt{3}/10$
6189 3997 J_0 : $2(e + \pi)/9$
6189 4549 $\ln(2/5) \times s\pi(\sqrt{5})$
6189 5432 rt : (8,4,9,-9)
6189 6706 ln : $7\sqrt[3]{3}/2$
6189 7671 ℓ_{10} : $6\ln 2$
6189 8024 ℓ_2 : $\sqrt{3} + \sqrt{6} - \sqrt{7}$
6189 8894 cu : $(\ln 3/2)^{-3}$
6189 8950 rt : (9,4,7,-8)
6189 9241 at : $4 - 3\pi/2$
6190 3828 at : $4\sqrt{2}/3 - 3\sqrt{3}/2$
6190 3920 ln : $7/13$
6190 4497 ℓ_{10} : $\zeta(3)/5$
6190 4761 id : $13/21$
6190 4920 sr : $1/2 + 3\sqrt{2}/2$
6190 5278 rt : (8,3,8,-8)
6190 5523 ℓ_2 : $(e + \pi)/9$
6190 9605 rt : (9,3,6,-7)
6191 1083 ℓ_{10} : $\sqrt[3]{3}/6$
6191 2504 rt : (7,2,9,-8)
6191 5414 $4/5 \times s\pi(e)$
6191 5662 Li_2 : $10/19$
6191 7378 tπ : $3/17$
6191 7481 rt : (8,2,7,-7)
6192 0266 id : $3e\pi$
6192 1568 sπ : $5\sqrt[3]{2}/8$
6192 2933 rt : (9,2,5,-6)
6192 5300 rt : (3,9,-7,-1)
6192 6621 rt : (7,1,8,-7)
6192 7588 ln : $2\ln 2/7$
6192 7707 ℓ_{10} : $4/3 + 2\sqrt{2}$
6192 8650 ln : $\sqrt{3}/3 + 2\sqrt{5}$
6192 8964 cu : $9e\pi/2$
6193 3005 rt : (8,1,6,-6)
6193 3119 Ψ : $(\sqrt{5})^{-3}$
6193 3853 K : $e/8$
6193 3951 ζ : $\exp(1/\sqrt{2})$
6193 7356 rt : (6,0,9,7)
6193 9676 sinh : $10\sqrt{3}$
6194 0082 rt : (9,1,4,-5)
6194 4932 rt : (7,0,7,6)
6194 5518 Ψ : $11/19$
6194 8251 tπ : $11/23$
6194 9258 e^x : $4/3 - 2e/3$
6194 9678 cu : $\sqrt{2}/4 - \sqrt{3}$
6195 0085 sq : $4\sqrt{2}/3 + 2\sqrt{3}$
6195 0952 sr : $2\sqrt{2}/3 - \sqrt{5}/4$
6195 3029 ln : $3\sqrt{3}/4 + \sqrt{5}/4$
6195 3413 rt : (8,0,5,5)
6195 8764 Ei : $2\zeta(3)$
6195 9281 θ_3 : $1/\sqrt{11}$
6195 9283 rt : (6,1,8,6)
6196 0880 rt : (2,5,-7,4)
6196 1641 2^x : $16/23$
6196 2059 cu : $3/4 + \pi/4$
6196 2969 rt : (9,0,3,4)
6196 2971 10^x : $3(e + \pi)/7$
6196 3373 10^x : $1/\ln(3/2)$
6196 5611 $\exp(e) \div \exp(\sqrt{5})$
6196 6953 $e - \ln(3)$
6196 7716 id : $3\sqrt{2}/4 + \sqrt{5}/4$
6196 9630 rt : (7,1,6,5)
6196 9680 sinh : $9e/8$
6197 1502 rt : (2,6,-2,9)
6197 1864 ζ : $(e/2)^{1/2}$
6197 3409 ℓ_{10} : $4\sqrt{2} - 2\sqrt{5}/3$
6197 6727 $5 \div s\pi(\pi)$
6197 6881 rt : (5,2,9,6)
6197 8875 ℓ_{10} : $6/25$
6197 9809 cr : $5/21$
6198 0561 λ : $\exp(-17\pi/19)$
6198 0590 cr : $17/4$
6198 1149 Ei : $4/17$
6198 1440 rt : (8,1,4,4)
6198 2039 10^x : $7\sqrt[3]{2}/10$
6198 3471 sq : $14/11$
6198 5059 sq : $4\sqrt{2} + 4\sqrt{5}/3$
6198 6002 id : $2\sqrt{2}/3 + 3\sqrt{5}/4$
6198 8905 cu : $2(e + \pi)/3$
6198 9101 e^x : $3\sqrt{5}/2$
6198 9719 sinh : $9\sqrt{3}/4$
6198 9788 rt : (6,2,7,5)
6198 9849 id : $9\sqrt[3]{2}/7$
6199 5044 rt : (9,1,2,3)
6199 6755 ℓ_{10} : $2e/3 + 3\pi/4$
6199 7033 at : $1/3 - \pi/3$
6199 7136 ζ : $\sqrt[3]{25/3}$
6200 0000 sq : $9\sqrt{2}/10$
6200 1444 Ψ : $5e\pi/3$
6200 1819 K : $\sqrt[3]{5}/5$
6200 2110 $\ln(5) \times \exp(\sqrt{2})$
6200 4416 sπ : $\sqrt{3}/4 + 3\sqrt{5}/2$
6200 4757 rt : (7,2,5,4)
6200 5874 Ψ : $8\ln 2$
6200 7585 rt : (1,3,-9,-7)
6200 7853 sq : $5\sqrt[3]{2}/8$
6200 8598 J_0 : $13/10$
6200 9447 θ_3 : $(\ln 3/2)^2$
6201 0415 sinh : $(e + \pi)/10$
6201 1070 2^x : $3 + 4\sqrt{2}$
6201 1450 ln : $1/2 + e/2$
6201 5153 2^x : $3\sqrt{2} + 3\sqrt{3}/2$
6201 5527 rt : (5,3,8,5)
6201 5858 10^x : $\sqrt{2}/4 + \sqrt{5}/2$
6201 5873 $e \times s\pi(\sqrt{2})$
6201 7367 sr : $5/13$
6201 8458 Γ : $\sqrt[3]{21}$
6201 8517 sr : $21/8$
6201 9370 10^x : $2 - \sqrt{5}/2$
6202 0198 sπ : $2\sqrt{3} - 3\sqrt{5}/4$
6202 1856 tπ : $\sqrt{2}$
6202 2321 rt : (8,2,3,3)
6202 4660 cu : $2e - \pi/4$
6202 4948 at : $5/7$
6202 5809 rt : (2,8,-4,8)
6202 6188 ℓ_{10} : $\sqrt{2}/2 + 2\sqrt{3}$
6202 6830 10^x : $4\sqrt{2} - 4\sqrt{5}/3$
6202 8962 ζ : $4\ln 2/3$
6203 4540 sπ : $7\ln 2/4$
6203 5049 cr : $3(1/\pi)/4$
6203 5122 rt : (6,3,6,4)
6203 5410 cu : $\exp(-1/(2\pi))$
6203 6602 rt : (5,9,7,-9)
6203 7285 sq : $1/\ln(e/2)$
6203 9089 rt : (1,8,-5,3)
6203 9912 sq : $1/3 - \sqrt{5}$
6204 0323 sr : $2\sqrt{3}/9$
6204 0747 e^x : $2\sqrt{3} - 4\sqrt{5}/3$
6204 3210 rt : (9,2,1,2)
6204 3619 erf : $1/\ln(5)$
6204 5183 rt : (6,9,5,-8)
6204 5877 at : $3/2 - \pi/4$
6204 6712 K : $2\zeta(3)/7$
6204 9546 rt : (4,4,9,5)
6204 9584 cu : $\sqrt{3} + 3\sqrt{5}/4$
6205 0492 tπ : $\sqrt{2}/8$
6205 0965 rt : (4,8,8,-9)
6205 3397 Ei : $8\sqrt{2}$
6205 3412 sr : $5e\pi/2$
6205 4413 rt : (7,9,3,-7)
6205 4956 sr : $5\ln 2/9$
6205 5261 2^x : $1 + 2\sqrt{5}/3$
6205 7839 e^x : $4e/3 - \pi$
6205 8672 rt : (7,3,4,3)
6205 8801 J_0 : $3\sqrt{3}/4$
6206 0029 sq : $e + 4\pi$
6206 0654 rt : (5,8,6,-8)
6206 3748 sq : $7e/10$
6206 4370 rt : (8,9,1,-6)
6206 5192 2^x : $(e)^2$
6206 7221 rt : (3,7,9,-9)
6207 0558 $3 \div \ln(\pi)$
6207 1127 rt : (6,8,4,-7)
6207 1639 10^x : $1/2 - \sqrt{2}/2$
6207 3337 sinh : $\sqrt[3]{2}$
6207 4139 id : $\sqrt[3]{18}$
6207 4770 θ_3 : $e/9$
6207 5142 rt : (9,9,-1,-5)
6207 6311 rt : (5,4,7,4)
6207 7246 2^x : $(e + \pi)/2$
6207 7529 tπ : $5(1/\pi)/9$
6207 8250 rt : (4,7,7,-8)
6207 9406 e^x : $3\sqrt{2}/2 - 3\sqrt{3}/2$
6208 1711 Ψ : $(2\pi)^{-3}$
6208 2121 sπ : $1/\sqrt{22}$
6208 2479 rt : (7,8,2,-6)
6208 3372 e^x : $(\sqrt{5}/2)^{-1/3}$
6208 4713 rt : (6,9,-8,-7)
6208 5840 tπ : $1/4 - \sqrt{3}/2$
6208 7493 rt : (8,3,2,2)
6208 7701 2^x : $\sqrt{2} + \sqrt{3} - \sqrt{6}$
6208 7843 2^x : $3e + 4\pi/3$
6208 8379 cr : $2e + 4\pi$
6209 0233 rt : (5,7,5,-7)
6209 2352 sq : $4/3 - 3\sqrt{2}/2$
6209 4828 rt : (8,8,0,-5)
6209 4960 e^x : $1/\ln(1/(3e))$
6209 5198 sinh : $7(e + \pi)/6$
6209 8440 rt : (3,6,8,-8)
6209 8538 Ei : $\exp(-1/(3\pi))$
6209 9415 Ei : $(\sqrt[3]{4}/3)^{-2}$
6210 1468 at : $(\ln 3/3)^{1/3}$
6210 1540 $1/3 - \ln(3/4)$
6210 2105 e^x : $3 + \pi/2$
6210 2859 10^x : $3e/7$
6210 3131 ζ : $9\ln 2/7$
6210 3298 rt : (6,7,3,-6)
6210 5672 rt : (5,9,-6,-6)
6210 6052 at : $((e + \pi)/3)^{-1/2}$
6210 8069 at : $(\sqrt{5}/2)^{-3}$
6210 8307 rt : (9,8,-2,-4)
6210 9544 rt : (6,4,5,3)
6211 1676 J_0 : $9\sqrt[3]{3}/10$
6211 2287 rt : (4,6,6,-7)
6211 2348 J : $(\sqrt{2}\pi/3)^{-2}$
6211 3389 tπ : $\exp(-\sqrt{3})$
6211 3893 sq : $4(1/\pi)$
6211 5335 $\exp(1/3) + \exp(4/5)$
6211 7597 rt : (7,7,1,-5)
6212 0794 rt : (7,8,-9,-7)
6212 1847 rt : (2,5,9,-8)
6212 3098 10^x : $9\sqrt[3]{5}/5$
6212 3550 rt : (9,3,0,1)
6212 7485 rt : (5,9,7,9)
6212 9262 ζ : $\ln 2/6$
6212 9981 rt : (4,9,-4,-5)
6213 2034 id : $1/2 + 3\sqrt{2}/2$
6213 2574 cr : $10\pi^2$
6213 3311 rt : (8,7,-1,-4)
6213 3493 id : $1/\ln(5)$
6213 5409 rt : (4,5,8,4)
6213 6517 J : $9/20$
6213 8033 rt : (3,5,7,-7)
6213 8120 sq : $8e\pi/9$
6213 9313 K : $\sqrt{3}/5$
6213 9399 tanh : $8/11$
6213 9516 2^x : $3/2 + 3e/4$
6213 9536 sπ : $\sqrt{2}/4 + \sqrt{3}/4$
6214 0275 $2/5 + \exp(1/5)$
6214 2568 $\exp(2/5) - \exp(\sqrt{2})$
6214 3231 e^x : $3\sqrt{2}/4 - \sqrt{3}/3$
6214 4238 rt : (6,6,2,-5)
6214 4447 K : $\ln 2/2$
6214 4650 cr : $6/25$
6214 5821 tanh : $\exp(-1/\pi)$
6214 7495 rt : (9,6,-9,8)
6214 7737 rt : (6,8,-7,-6)
6214 7811 K : $1/\sqrt[3]{24}$
6214 8179 tanh : $(\sqrt[3]{4}/3)^{1/2}$
6214 8837 ℓ_2 : $13/20$
6214 9694 ln : $4 + 3\sqrt{2}/4$
6215 0659 rt : (9,7,-3,-3)

6215 1408 at : $9(1/\pi)/4$
6215 1874 rt : (7,4,3,2)
6215 2675 Ei : $7\sqrt{5}/6$
6215 4769 ln : $\sqrt{2}/2 + 2\sqrt{3}/3$
6215 5935 rt : (4,5,5,-6)
6215 8502 rt : (3,9,-2,-4)
6215 9138 rt : (8,4,-9,9)
6216 0016 rt : (1,7,-5,-2)
6216 2795 rt : (7,6,0,-4)
6216 5407 sinh : $1/\ln(\ln 2)$
6216 6620 rt : (9,7,-8,7)
6216 6973 $\exp(1/3) - s\pi(e)$
6216 7950 sr : $3\sqrt{3} + 3\sqrt{5}/4$
6216 8003 Γ : $2\pi^2/5$
6216 8488 rt : (2,4,8,-7)
6216 8793 $t\pi$: $3\sqrt{2}/2 - 4\sqrt{5}$
6216 9903 $t\pi$: $1/2 + 3\sqrt{5}/4$
6217 0991 2^x : $3\pi^2/5$
6217 3485 at : $(e)^{-1/3}$
6217 5402 $\ln(\sqrt{5}) \times \exp(\pi)$
6217 5836 rt : (5,5,3,-5)
6217 6027 ℓ_{10} : $5(e+\pi)/7$
6217 6096 10^x : $\sqrt[3]{2}/6$
6217 6994 cr : $\sqrt[3]{3}/6$
6217 9619 rt : (8,5,-8,8)
6217 9639 rt : (5,8,-5,-5)
6218 0136 cr : $\zeta(3)/5$
6218 3461 rt : (8,6,-2,-3)
6218 3896 K : 8/23
6218 5862 rt : (5,5,6,3)
6218 5921 cu : $1/2 + \sqrt{2}/4$
6218 6043 $4 \times \ln(2/3)$
6218 7991 rt : (9,8,-7,6)
6218 9890 rt : (3,4,6,-6)
6218 9897 e^x : P_{32}
6219 0485 rt : (4,7,-8,3)
6219 2424 rt : (2,9,0,-3)
6219 3589 rt : (7,3,-8,9)
6219 6687 cu : $2/3 + 3\pi/4$
6219 7293 ℓ_2 : $9\sqrt[3]{5}/5$
6219 7501 sq : $\sqrt{3} + 3\sqrt{5}/4$
6219 8088 rt : (6,5,1,-4)
6219 8588 rt : (3,8,-9,5)
6219 9664 2^x : $\sqrt{3}/4 - \sqrt{5}/2$
6220 2598 rt : (8,6,-7,7)
6220 2776 sr : $4\sqrt{2}/3 + \sqrt{5}/3$
6220 3552 rt : (7,7,-8,-6)
6220 5082 rt : (1,3,9,-7)
6220 5115 id : $\sqrt{3} - 3\sqrt{5}/2$
6220 6608 rt : (9,6,-4,-2)
6220 7555 rt : (8,4,1,1)
6220 8860 ℓ_{10} : $4\pi/3$
6220 9690 2^x : $4 + 2\sqrt{3}/3$
6221 1880 $1/5 - \exp(3/5)$
6221 2028 rt : (9,9,-6,5)
6221 3924 rt : (4,8,8,9)
6221 4876 rt : (4,8,-5,4)
6221 6570 $s\pi$: $\sqrt[3]{5}/8$
6221 7996 rt : (4,8,-3,-4)
6221 8365 rt : (7,4,-7,8)
6221 9341 ℓ_2 : $1/2 + 4\sqrt{2}$
6222 3126 rt : (7,5,-1,-3)
6222 3128 K : $\pi/9$
6222 5444 ln : $2\sqrt{2} + \sqrt{5}$
6222 6008 10^x : $(e/3)^2$
6222 7083 rt : (9,8,6,-9)
6222 7471 rt : (3,6,9,4)
6222 8561 rt : (8,7,-6,6)
6223 1131 rt : (2,3,7,-6)
6223 2596 cr : $e\pi/2$
6223 3415 rt : (1,9,2,-2)
6223 4881 erf : $9\ln 2/10$
6223 5436 rt : (6,2,-7,9)
6223 6233 $\sqrt{5} + \ln(4)$
6223 6821 Ei : $-\sqrt{2} + \sqrt{5} + \sqrt{7}$
6223 8711 cu : $\sqrt{3} + \sqrt{6} - \sqrt{7}$
6223 9739 ln : $5\sqrt{5}/6$
6224 0241 2^x : $2\pi/9$
6224 0606 J : $\zeta(3)/4$
6224 1101 rt : (5,4,2,-4)
6224 1835 sr : P_{27}
6224 2544 rt : (4,9,-2,5)
6224 4889 ζ : $((e+\pi)/2)^{-2}$
6224 5279 e^x : $e/4 + \pi/3$
6224 5845 sr : $4/3 + 3\sqrt{3}/4$
6224 6495 rt : (7,5,-6,7)
6224 6957 $3/5 \div s\pi(\sqrt{2})$
6224 7313 rt : (6,7,-6,-5)
6225 0700 rt : (8,7,7,-9)
6225 1497 rt : (8,5,-3,-2)
6225 2810 K : 7/20
6225 3076 id : $9\sqrt[3]{3}/8$
6225 4113 rt : (6,5,4,2)
6225 5624 ℓ_2 : $3\sqrt{3}/8$
6225 7173 10^x : $\sqrt{5}/4$
6225 7249 erf : $\exp(-\sqrt{2}/3)$
6225 7766 cu : $3\sqrt{3} + \sqrt{5}/4$
6225 8127 rt : (8,8,-5,5)
6225 8561 K : $(1/(3e))^{1/2}$
6225 8888 rt : (9,2,9,7)
6225 9992 cr : $\sqrt{3} - 2\sqrt{5}/3$
6226 0744 rt : (3,3,5,-5)
6226 1731 rt : (3,4,-9,3)
6226 4526 cr : $2e/3 - \pi/2$
6226 4959 rt : (3,8,-1,-3)
6226 5759 Ei : $5\sqrt[3]{2}/7$
6226 6018 rt : (6,3,-6,8)
6226 6345 rt : (9,7,5,-8)
6226 8391 θ_3 : 18/19
6227 2069 rt : (6,4,0,-3)
6227 2729 ln : $\ln(\sqrt[3]{5})$
6227 3162 rt : (2,9,-2,-4)
6227 3394 ℓ_{10} : $4\sqrt{2}/3 + 4\sqrt{3}/3$
6227 4091 2^x : $\sqrt{2}/3 - 2\sqrt{3}/3$
6227 5067 sinh : 10/17
6227 7120 rt : (7,6,8,-9)
6227 7660 10^x : 3/2
6227 7771 cu : $e\pi/10$
6227 8707 rt : (7,6,-5,6)
6228 0634 $\ln(\sqrt{5}) \times s\pi(e)$
6228 1171 Ei : 9/10
6228 1281 Ei : 1/10
6228 1304 rt : (8,6,-9,-6)
6228 2138 K : $\exp(-\pi/3)$
6228 2305 rt : (1,2,8,-6)
6228 2658 as : 7/12
6228 3899 sq : $3e/4 - 4\pi/3$
6228 3903 rt : (9,5,-5,-1)
6228 7074 2^x : $\sqrt{3} + 3\sqrt{5}/4$
6228 7147 e^x : P_{42}
6228 7351 rt : (5,1,-6,9)
6228 8481 10^x : $(4/3)^3$
6228 9056 at : $2 - e$
6228 9465 2^x : 13/7
6229 0220 $\ln(4) \div \exp(4/5)$
6229 0256 sq : $2\sqrt{2}/3 - \sqrt{3}$
6229 2100 rt : (8,9,-4,4)
6229 4695 rt : (8,6,6,-8)
6229 6328 rt : (3,5,-6,4)
6229 6817 cr : $5\sqrt[3]{5}/2$
6229 8334 sq : $\sqrt{3}/2 + \sqrt{5}$
6229 9187 cr : $3\sqrt{3}/2 + 3\sqrt{5}/4$
6230 1249 rt : (6,4,-5,7)
6230 1412 e^x : $\sqrt[3]{10}$
6230 1722 rt : (5,7,-4,-4)
6230 2692 10^x : $2e + \pi$
6230 3751 $s\pi$: $3/4 + 2\sqrt{3}$
6230 4062 $4/5 \div \exp(1/4)$
6230 4975 K : $1/\sqrt[3]{23}$
6230 6881 rt : (6,5,9,-9)
6230 6927 2^x : $1/\ln(\ln 2/3)$
6230 6936 10^x : $\sqrt{2}/4 - \sqrt{5}/4$
6230 7663 rt : (7,4,-2,-2)
6231 0326 ℓ_2 : $2e - 3\pi/4$
6231 2479 $\exp(5) \times \exp(\sqrt{5})$
6231 2731 id : $1/4 - 4e$
6231 3308 rt : (4,6,7,3)
6231 3601 rt : (9,6,4,-7)
6231 4355 $2/5 - \ln(4/5)$
6231 5292 J_0 : 22/17
6231 5806 sr : $e/7$
6231 5956 rt : (7,7,-4,5)
6231 6017 sq : $1/4 + 3\sqrt{2}/2$
6231 9712 rt : (2,2,6,-5)
6232 0421 2^x : $((e+\pi)/2)^{-1/3}$
6232 0659 rt : (2,6,-7,-9)
6232 2207 ℓ_{10} : $10\sqrt[3]{2}/3$
6232 2433 $\exp(2/3) + s\pi(\sqrt{5})$
6232 2464 e^x : $5(e+\pi)$
6232 3745 rt : (2,8,1,-2)
6232 4088 erf : 5/8
6232 4248 10^x : $5\sqrt{2}$
6232 4929 ℓ_{10} : 5/21
6232 6051 rt : (5,2,-5,8)
6232 6767 rt : (7,5,7,-8)
6232 6869 sq : 15/19
6232 8128 rt : (9,3,8,6)
6232 8206 $t\pi$: $4\sqrt{2}/3 - 3\sqrt{5}$
6233 0409 sinh : $(\ln 2/2)^{1/2}$
6233 3993 rt : (5,3,1,-3)
6233 5412 sq : $3\sqrt{2}/4 + \sqrt{5}/4$
6233 6569 rt : (3,6,-3,5)
6233 8181 rt : (2,9,-8,-8)
6233 9751 $\ln(2/5) - s\pi(1/4)$
6234 1325 10^x : 3/4
6234 2276 rt : (6,5,-4,6)
6234 5183 rt : (7,6,-7,-5)
6234 5201 cu : $4 - 4\pi$
6234 5700 J : 7/23
6234 5867 sr : $3/4 + 4\sqrt{2}/3$
6234 6209 sq : $2/3 + 2\pi/3$
6234 6944 K : 6/17
6234 7538 rt : (1,9,-6,0)
6234 8261 rt : (8,5,5,-7)
6234 8980 $s\pi$: 3/14
6235 1324 rt : (7,5,2,1)
6235 2708 10^x : $4\sqrt{2} - \sqrt{5}/2$
6235 3473 rt : (4,0,-5,9)
6235 5831 Li_2 : 9/17
6235 7460 cu : $5e/6$
6235 8825 $s\pi$: $5\sqrt{2}/9$
6235 9521 rt : (7,8,-3,4)
6236 0956 sr : 7/18
6236 1019 Ψ : $2(e+\pi)/5$
6236 2738 sr : $1/\sqrt[3]{17}$
6236 3352 rt : (6,4,8,-8)
6236 5166 rt : (2,7,-2,-4)
6236 6295 rt : (1,9,-6,-7)
6236 6669 K : $\sqrt{2}/4$
6236 6670 cr : $3\sqrt{3}/4 + 4\sqrt{5}/3$
6236 6871 e^x : $4 - 2\sqrt{5}$
6236 7946 tanh : $(\sqrt[3]{5}/2)^2$
6236 8474 10^x : $10e\pi/9$
6236 9562 rt : (1,8,3,1)
6237 0678 K : $10(1/\pi)/9$
6237 0917 cr : P_{15}
6237 1147 rt : (4,7,-2,-3)
6237 1464 rt : (5,3,-4,7)
6237 1562 rt : (9,5,3,-6)
6237 3734 cu : $1/2 - 2\sqrt{2}$
6237 5769 θ_3 : $(\sqrt{5}/3)^{1/2}$
6237 7538 $s\pi$: $9e\pi/4$
6237 7673 10^x : 4/19
6237 8738 sinh : $\exp(-\sqrt[3]{4}/3)$
6237 9981 rt : (6,3,-1,-2)
6238 1071 ln : $4 - 2\sqrt{3}$
6238 3065 rt : (8,4,9,6)
6238 3246 id : $9\ln 2/10$
6238 3955 rt : (3,7,0,6)
6238 4193 ζ : 2/17
6238 5342 $s\pi$: $1/\ln(\pi/2)$
6238 6458 rt : (3,8,-9,-8)
6238 7170 Γ : $9\sqrt{5}$
6238 8008 rt : (7,4,6,-7)
6238 8145 10^x : $3(e+\pi)$
6238 8672 id : $e - 2\pi/3$
6239 0650 rt : (6,6,-3,5)
6239 0940 as : $((e+\pi)/2)^{-1/2}$
6239 1468 sr : $3\sqrt{2}/2 - \sqrt{3}$
6239 2382 $\ln(3/5) \times \exp(1/5)$

6239 2404 cu : $3\sqrt{2}/4-\sqrt{5}$
6239 3405 ln : $\sqrt[3]{13/2}$
6239 5372 sinh : 22/13
6239 6318 rt : (1,1,7,-5)
6239 6375 ℓ_2 : $\sqrt{19/2}$
6239 7478 rt : (9,4,-6,0)
6239 8999 tπ : $3\sqrt{3}+4\sqrt{5}/3$
6239 9307 ℓ_{10} : $7\zeta(3)/2$
6239 9334 at : $4\sqrt[3]{2}/7$
6239 9357 rt : (1,8,3,-1)
6240 2305 at : 18/25
6240 4022 rt : (4,1,-4,8)
6240 5223 ζ : 7/13
6240 5482 rt : (5,3,9,-8)
6240 5743 tπ : $(\sqrt[3]{3}/2)^{-1/2}$
6240 7112 sq : $9\sqrt[3]{2}/7$
6240 7600 $\ln(3/5)-\exp(\sqrt{2})$
6241 1151 rt : (7,9,-2,3)
6241 2505 id : $\exp(-\sqrt{2}/3)$
6241 4891 rt : (8,4,4,-6)
6241 8047 rt : (2,8,3,1)
6242 0024 rt : (2,8,-7,-7)
6242 2104 rt : (9,4,7,5)
6242 2333 ln : $e+3\pi/4$
6242 2614 ℓ_{10} : $2\sqrt{3}+\sqrt{5}/3$
6242 4741 10^x : $e+2\pi/3$
6242 5498 rt : (5,4,-3,6)
6242 6595 sπ : $\pi/4$
6242 7695 Ψ : $\sqrt{11/2}$
6242 8620 rt : (6,6,-5,-4)
6243 0056 rt : (2,9,-8,1)
6243 2251 rt : (1,9,2,-5)
6243 4060 rt : (6,3,7,-7)
6243 4292 e^x : $\sqrt{10}$
6243 4477 rt : (7,3,-3,-1)
6243 4883 10^x : $8\pi^2/5$
6243 5336 sq : $3e-3\pi/4$
6243 5792 tπ : $3\sqrt{2}+\sqrt{3}/4$
6243 7577 id : $4e/3$
6243 9595 Γ : $8\sqrt{5}/5$
6244 0207 e^x : P_{22}
6244 0573 rt : (3,8,3,7)
6244 1820 $\sqrt{3}-\exp(\sqrt{5})$
6244 1826 rt : (5,6,5,2)
6244 4308 rt : (9,4,2,-5)
6244 5540 2^x : $5\sqrt[3]{2}/9$
6244 8533 rt : (6,7,-2,4)
6245 0479 2^x : 7/10
6245 1462 Γ : $\sqrt{18}\pi$
6245 3644 tanh : $\ln(\sqrt[3]{3}/3)$
6245 4534 rt : (2,1,5,-4)
6245 5895 e^x : $3\sqrt{3}/4+3\sqrt{5}$
6245 7428 Ψ : $5\ln 2/6$
6245 7619 rt : (1,8,-5,-6)
6245 7665 sπ : $3\sqrt{3}-4\sqrt{5}/3$
6245 7899 cu : $4/3-3\sqrt{2}$
6245 8287 tanh : $(e+\pi)/8$
6246 0721 tπ : $3\sqrt{2}/2-3\sqrt{3}/4$
6246 1726 $e\div s\pi(1/5)$
6246 2682 rt : (3,7,0,-2)
6246 3571 sinh : $7\sqrt[3]{3}/8$
6246 3785 as : $1/\sqrt[3]{5}$
6246 4778 rt : (4,2,-3,7)
6246 5426 rt : (7,3,5,-6)
6246 7895 rt : (1,6,3,5)
6246 8257 cu : $2\sqrt{2}+4\sqrt{3}/3$
6246 8658 id : $\sqrt{3}/2-2\sqrt{5}/3$
6246 8933 rt : (3,9,-9,-8)
6247 0500 ℓ_{10} : $3/4+2\sqrt{3}$
6247 1256 rt : (2,9,0,-4)
6247 4734 K : $1/\sqrt[3]{22}$
6247 5539 cr : $e+\pi/2$
6247 6342 at : $\sqrt[3]{3}/2$
6247 6628 rt : (5,2,0,-2)
6247 7818 J_0 : $\sqrt{5/3}$
6247 8297 e^x : $4(e+\pi)/5$
6247 9554 sinh : $\sqrt{5}$
6248 0379 ln : $1+3e/2$
6248 0392 tanh : $3\sqrt[3]{5}/7$
6248 0589 sq : $3/4+3\sqrt{5}$
6248 1908 rt : (2,9,8,6)
6248 2303 ℓ_2 : $\sqrt[3]{11/3}$
6248 3227 K : 5/14
6248 3536 at : $3\zeta(3)/5$
6248 5241 rt : (3,7,-8,-7)
6248 6637 Li_2 : $3\sqrt{2}/8$
6248 8060 rt : (5,2,8,-7)
6248 8653 ln : $3/4+\sqrt{5}/2$
6249 0747 cu : $1/2-4\sqrt{3}$
6249 0855 rt : (5,5,-2,5)
6249 0994 at : $1/\ln(4)$
6249 1942 sr : $4/3-2\sqrt{2}/3$
6249 2035 ℓ_2 : $5\pi^2$
6249 2904 10^x : $3\sqrt{2}+2\sqrt{5}/3$
6249 4743 at : $1/4+\sqrt{2}/3$
6249 5958 J : $(e)^{-2}$
6249 9747 10^x : $3\sqrt{2}/2+\sqrt{3}$
6250 0000 id : 5/8
6250 0338 cr : $2/3+4e/3$
6250 5501 tπ : $3\sqrt{3}+\sqrt{5}$
6250 7202 sr : $4-e/2$
6250 9404 rt : (3,9,6,8)
6250 9690 10^x : $\sqrt{5}+\sqrt{6}+\sqrt{7}$
6251 0041 tanh : 11/15
6251 2085 cr : $\sqrt[3]{5}/7$
6251 2407 $\sqrt{5}+\exp(2)$
6251 3200 Li_2 : $5(1/\pi)/3$
6251 3559 ℓ_{10} : $3/2+e$
6251 4526 K : $9(1/\pi)/8$
6251 5297 rt : (3,9,-2,-3)
6251 5524 10^x : $(\ln 2/3)^{-1/3}$
6251 7531 $\exp(2/5)-\exp(3/4)$
6251 8620 2^x : $3\sqrt{3}/4+\sqrt{5}/4$
6251 9022 rt : (6,8,-1,3)
6251 9576 10^x : $9\sqrt{3}/7$
6252 1361 ln : $4\sqrt{2}-\sqrt{3}/3$
6252 4238 rt : (2,8,-6,-1)
6252 5145 rt : (9,3,9,-9)
6252 6229 Ψ : $(1/\pi)/2$
6252 9043 rt : (1,8,9,-9)
6252 9126 e^x : 19/11
6253 0290 rt : (1,9,3,-2)
6253 0439 sq : $1/4+3\pi/2$
6253 0684 e^x : $1+2\sqrt{3}/3$
6253 1483 rt : (2,7,-6,-6)
6253 2555 ℓ_{10} : $\sqrt{3}/2+3\sqrt{5}/2$
6253 3400 e^x : $7\pi/8$
6253 5386 id : $2e+4\pi/3$
6253 7622 sq : $3e/2$
6253 8285 rt : (9,3,1,-4)
6253 8306 sinh : $\exp(\sqrt{2}\pi/2)$
6253 8628 cu : $3/4-e$
6253 9172 rt : (4,9,4,-7)
6253 9331 tπ : $\sqrt{2}/2+3\sqrt{5}/4$
6254 2100 rt : (5,6,-3,-3)
6254 4243 2^x : $1-3\sqrt{5}/4$
6254 5756 cu : $\sqrt{2}-\sqrt{5}/4$
6254 8119 rt : (1,8,1,-4)
6254 8504 at : 13/18
6254 9330 e^x : $8(e+\pi)$
6255 1819 rt : (6,2,-2,-1)
6255 2285 rt : (4,1,9,-7)
6255 4092 $2/5+\exp(4/5)$
6255 4324 sr : 9/23
6255 5717 sinh : $1/\ln(2e)$
6255 6867 rt : (9,5,6,4)
6255 7959 cu : $3/4+\sqrt{5}$
6255 8826 2^x : $2-3\sqrt{3}/4$
6255 9566 e^x : $2/3+3\sqrt{2}/4$
6256 0559 rt : (2,8,7,-8)
6256 0990 Γ : 1/4
6256 2925 Ψ : $\sqrt{3}/3$
6256 4891 sr : $1/3+4\sqrt{3}/3$
6256 5385 rt : (4,9,-4,-2)
6256 6143 rt : (4,6,-9,-7)
6256 6373 rt : (7,2,4,-5)
6256 9548 ℓ_2 : $3e+4\pi/3$
6256 9660 Γ : $2(1/\pi)/9$
6257 0646 rt : (3,7,8,3)
6257 0976 sq : $1/3+\pi/2$
6257 1496 rt : (5,9,2,-6)
6257 4384 cr : $2\sqrt{2}/3+3\sqrt{5}/2$
6257 6378 K : $2\sqrt[3]{2}/7$
6257 6932 erf : $\pi/5$
6257 7092 sinh : $1/\ln(\sqrt[3]{3}/2)$
6257 7123 K : 9/25
6257 7271 Ei : 21/11
6257 9929 $\exp(2/5)-s\pi(1/3)$
6258 0124 rt : (9,3,-7,1)
6258 1458 ℓ_2 : $(\sqrt[3]{2}/3)^{1/2}$
6258 1681 rt : (1,0,6,4)
6258 4314 rt : (1,7,-4,-5)
6258 5007 as : $2-\sqrt{2}$
6258 8594 rt : (2,7,2,-1)
6258 8962 $\exp(\pi)\div\exp(\sqrt{2})$
6258 9235 Γ : $e/8$
6258 9384 Ψ : 9/23
6259 0021 θ_3 : 7/23
6259 0278 sinh : 13/22
6259 1205 θ_3 : $5\sqrt[3]{2}/8$
6259 1639 ζ : $3(1/\pi)/8$
6259 2103 ζ : $7\sqrt{3}/6$
6259 3200 ln : $7\pi^2/5$
6259 3720 erf : $4\sqrt{2}/9$
6259 4870 rt : (3,8,5,-7)
6259 5723 K : $\sqrt[3]{3}/4$
6259 6830 rt : (5,1,7,-6)
6259 7533 K : $3\zeta(3)/10$
6259 7631 sπ : $2/3+\sqrt{5}/2$
6259 8470 cu : $2/3+4\sqrt{2}/3$
6259 9668 rt : (7,9,6,-9)
6260 0569 rt : (7,6,9,5)
6260 1206 cr : $3+3\sqrt{3}/4$
6260 1985 rt : (9,2,8,-8)
6260 3195 rt : (2,8,-1,-3)
6260 4629 $\ln(3/4)\times\exp(\sqrt{3})$
6260 6711 rt : (6,9,0,2)
6260 9811 as : $(e+\pi)/10$
6261 0505 id : $\sqrt{2}/3+2\sqrt{3}/3$
6261 2457 rt : (8,2,2,-4)
6261 2664 sπ : $3\sqrt{2}/4+2\sqrt{3}/3$
6261 3343 e^x : $\exp(-\sqrt[3]{3}/2)$
6261 5090 Ψ : γ
6261 7470 10^x : $\sqrt{14/3}$
6261 9806 rt : (1,7,8,-8)
6262 2655 id : $4/3-\sqrt{2}/2$
6262 2816 rt : (5,9,-6,-1)
6262 4461 rt : (3,6,-7,-6)
6262 4888 θ_3 : $(2e/3)^{-2}$
6262 6574 tπ : $3\pi/8$
6262 6746 rt : (8,9,4,-8)
6262 9894 ζ : $10\sqrt{2}/7$
6263 1196 sinh : $\ln(2e)$
6263 2296 ℓ_{10} : $3\pi^2/7$
6263 2355 rt : (4,8,3,-6)
6263 3195 id : $8(e+\pi)/3$
6263 6102 ℓ_{10} : $\exp(\sqrt[3]{3})$
6263 7213 rt : (4,9,-7,-8)
6263 8148 ln : $\sqrt{7/2}$
6263 8888 10^x : $4\sqrt{3}+\sqrt{5}$
6263 9241 $\ln(3)\div s\pi(\sqrt{5})$
6263 9985 rt : (1,8,-7,1)
6264 0986 10^x : $3e/2-\pi$
6264 1849 rt : (1,6,-4,1)
6264 3175 10^x : P_{81}
6264 5211 rt : (7,9,-2,-4)
6264 6122 rt : (6,8,7,-9)
6264 6859 rt : (6,1,5,-5)
6264 7241 sr : $2\sqrt{3}/3+2\sqrt{5}/3$
6264 7719 tπ : $\sqrt{2}-\sqrt{5}$
6264 8552 rt : (1,1,-9,5)
6265 2170 ln : $\sqrt{3}+3\sqrt{5}/2$
6265 2222 10^x : $7\pi/9$
6265 3829 rt : (6,6,3,1)
6265 4681 $s\pi(2/5)+s\pi(\sqrt{5})$
6265 5678 rt : (9,9,2,-7)
6265 7656 sr : $\sqrt{7}$
6265 8772 K : $1/\sqrt[3]{21}$
6265 8986 tπ : $(\sqrt[3]{5}/3)^{-1/2}$
6265 9029 rt : (8,1,9,-8)

6265 9608 sq : $1/3+3e/4$
6265 9802 rt : (2,7,6,-7)
6265 9958 cu : $3\sqrt{2}/2-4\sqrt{5}$
6266 0497 2^x : $4+e/4$
6266 1449 sq : $3/4-4\pi$
6266 2870 λ : $\exp(-8\pi/9)$
6266 3006 $\sqrt{5}-\ln(5)$
6266 4279 rt : (9,2,0,-3)
6266 4479 ℓ_2 : $3\sqrt{2}-2\sqrt{3}/3$
6266 5076 sr : $\ln(\sqrt{2}\pi/3)$
6266 5381 Ei : $5\sqrt[3]{3}/8$
6266 5706 sr : $\pi/8$
6266 6321 at : $3\sqrt{3}-2\sqrt{5}$
6266 6801 rt : (3,8,-3,-2)
6266 8335 ζ : 3/25
6267 1575 $\ln(\sqrt{5})\div\exp(1/4)$
6267 2939 tπ : $1/4+4\sqrt{3}$
6267 3461 rt : (5,8,1,-5)
6267 3611 sq : 19/24
6267 4267 rt : (2,5,-9,-3)
6267 5742 rt : (8,6,7,4)
6267 6483 rt : (7,8,5,-8)
6267 6700 at : $2e-3\pi/2$
6267 9027 sr : $3-\sqrt{2}/4$
6268 2053 ζ : $(\ln 2/2)^2$
6268 2215 cu : $9\sqrt[3]{5}/7$
6268 2583 ln : $3e-2\pi$
6268 2953 as : $4\sqrt{2}/3-3\sqrt{3}/4$
6268 3649 id : $2\sqrt{3}/3+2\sqrt{5}$
6268 3775 $\ln(\sqrt{5})+\exp(3/5)$
6268 4527 rt : (4,0,8,6)
6268 6040 sinh : 2
6268 6786 $\ln(2/3)-\exp(1/5)$
6268 7462 rt : (8,9,-4,-3)
6268 8357 rt : (8,9,-1,9)
6268 9263 rt : (6,9,-8,0)
6268 9604 ln : $4\zeta(3)/9$
6269 0615 sinh : 24/19
6269 1115 tπ : $4\sqrt{2}-4\sqrt{5}/3$
6269 2225 rt : (2,6,-5,-5)
6269 3252 ζ : $\zeta(3)/10$
6269 6292 ℓ_{10} : $2+\sqrt{5}$
6269 7053 rt : (9,1,7,-7)
6269 7739 Ψ : $7\ln 2/5$
6269 8073 K : 4/11
6269 8401 rt : (5,7,8,-9)
6269 8585 10^x : 15/17
6270 1649 erf : $\sqrt[3]{2}/2$
6270 2732 Ei : $3\zeta(3)/4$
6270 2946 tπ : $(\sqrt{2}\pi/3)^{-1/2}$
6270 2975 K : $8(1/\pi)/7$
6270 3419 rt : (7,1,3,-4)
6270 3808 rt : (3,7,4,-6)
6270 5084 rt : (4,6,-1,-2)
6270 7145 ζ : $7\ln 2/9$
6270 8908 $\exp(1/3)\div\exp(4/5)$
6270 9047 rt : (8,8,3,-7)
6271 6117 rt : (1,7,0,-3)
6271 6482 2^x : $2/3+2\pi$
6271 8723 rt : (6,8,-1,-4)
6272 1172 tanh : $(5/2)^{-1/3}$
6272 2940 rt : (5,1,-1,-1)
6272 3344 tanh : 14/19
6272 3500 2^x : $(\ln 2/2)^{1/3}$
6272 5200 sinh : 17/7
6272 9510 $\exp(4)\times s\pi(\sqrt{2})$
6273 1541 Ei : $-\sqrt{3}+\sqrt{5}+\sqrt{6}$
6273 1753 cr : $2+4\sqrt{3}/3$
6273 1856 ζ : $7\sqrt[3]{5}/8$
6273 2688 rt : (6,7,6,-8)
6273 3842 ζ : $7\sqrt[3]{3}/5$
6273 4018 rt : (9,9,-6,-2)
6273 6508 rt : (3,9,-6,3)
6273 7568 Ψ : $2\sqrt[3]{3}/5$
6273 7642 sπ : $4\sqrt{2}+\sqrt{5}/4$
6273 7649 cu : $8e\pi/7$
6274 1005 rt : (4,8,-5,-1)
6274 1699 2^x : 9/2
6274 4054 rt : (9,8,1,-6)
6274 5255 rt : (4,5,-8,-6)
6274 5363 tπ : $\sqrt{3}/4+\sqrt{5}/3$
6274 6536 J : $(1/(3e))^3$
6274 6554 rt : (5,0,6,5)
6274 8021 rt : (8,9,-9,-4)
6274 8470 10^x : $4-3\sqrt{2}/2$
6274 9764 Ei : $7e/4$
6275 1786 Ψ : $\sqrt[3]{13}$
6275 1817 rt : (9,5,-2,-1)
6275 2098 J_0 : 9/7
6275 2389 Ψ : $\exp(\sqrt[3]{5}/2)$
6275 2437 rt : (4,7,2,-5)
6275 3046 sq : $8\ln 2/7$
6275 4439 rt : (1,7,-4,6)
6275 7688 rt : (4,6,9,-9)
6275 9503 ζ : $3\sqrt{5}/8$
6275 9872 sr : $4\pi^2/3$
6276 2603 2^x : $4\sqrt{3}/3-4\sqrt{5}/3$
6276 3532 id : $\sqrt{2}-\sqrt{5}+\sqrt{6}$
6276 3625 ℓ_{10} : $3\sqrt{2}$
6276 3939 rt : (8,2,-6,1)
6276 4746 id : $3\sqrt{2}/4-\sqrt{3}/4$
6276 4964 cu : $3/2-3\pi/4$
6276 5753 $\exp(\sqrt{3})\div\exp(1/5)$
6276 6031 rt : (9,6,5,3)
6276 6490 Ei : 19/15
6276 7828 rt : (8,1,1,-3)
6276 8779 rt : (7,8,-3,-3)
6276 9136 sπ : $(5/3)^{-3}$
6276 9157 rt : (8,0,8,7)
6276 9621 rt : (7,7,4,-7)
6277 1867 rt : (1,7,4,0)
6277 3576 10^x : $3/2+\sqrt{2}/3$
6277 4343 2^x : $7\pi^2/5$
6277 8978 rt : (1,5,-3,3)
6277 9935 cr : $\sqrt{3}/7$
6278 2564 $\ln(3/5)-\exp(3/4)$
6278 4461 K : $\ln(\sqrt[3]{3})$
6278 5858 rt : (9,4,-9,-5)
6278 7057 id : $\sqrt{2}/4-4\sqrt{5}/3$
6278 7367 Ψ : $\exp(-e)$
6278 8414 sq : $3/4+2\sqrt{3}/3$
6278 8690 sinh : $\exp(\sqrt{5}/2)$
6278 8925 rt : (2,6,5,8)
6278 8974 sπ : $e/2+3\pi$
6279 0115 2^x : $3\sqrt{2}/4-\sqrt{3}$
6279 1565 2^x : $1/2+e/2$
6279 1648 cr : $7(1/\pi)/9$
6279 3527 2^x : $3e-4\pi/3$
6279 4373 rt : (3,1,9,6)
6279 4429 J_0 : $(\sqrt{2}/3)^{-1/3}$
6279 4906 K : $\ln(\ln 2)$
6279 6725 rt : (5,6,7,-8)
6279 7546 tπ : $1/3-3e$
6279 7622 rt : (7,8,0,9)
6279 9146 cr : $4/3+4\sqrt{5}/3$
6279 9625 rt : (2,7,-2,-2)
6280 3122 ℓ_2 : 11/17
6280 3537 10^x : $4/3+\sqrt{2}/4$
6280 3671 e^x : P_{50}
6280 3820 rt : (7,7,-9,-3)
6280 5142 2^x : $1/3+3\sqrt{2}/4$
6280 6433 rt : (5,8,1,2)
6280 7527 ℓ_2 : $9\zeta(3)/7$
6280 7820 tπ : $5\sqrt{2}/6$
6280 8149 cu : $5\ln 2$
6280 8325 rt : (2,7,1,7)
6280 9510 rt : (8,7,2,-6)
6281 2276 Li_2 : $\exp(-\sqrt[3]{2}/2)$
6281 3090 sr : $e+4\pi/3$
6281 3852 $1/5\times\exp(\pi)$
6281 4093 ln : $3+2\pi/3$
6281 7666 rt : (9,0,6,6)
6281 9712 ℓ_2 : $4\sqrt{2}+3\sqrt{5}$
6281 9931 10^x : $1/2+\sqrt{3}$
6282 2934 rt : (1,8,-6,-5)
6282 3692 ln : $e/3+4\pi/3$
6282 4108 $\ln(\pi)\div\exp(3/5)$
6282 4323 erf : 12/19
6282 4401 rt : (8,8,-5,-2)
6282 7136 ℓ_2 : $4\sqrt{3}-\sqrt{5}/3$
6282 8568 rt : (5,8,-7,0)
6283 1853 id : $\pi/5$
6283 3299 cu : 20/17
6283 4286 rt : (9,6,7,-9)
6283 5502 rt : (3,5,-6,-5)
6283 8600 sq : $9e/2$
6283 8893 ℓ_{10} : 4/17
6283 8983 rt : (6,6,5,-7)
6283 9718 rt : (7,7,8,4)
6284 1267 K : $\exp(-1)$
6284 1554 sr : $4\sqrt{2}/3-2\sqrt{5}/3$
6284 1767 rt : (9,1,-1,-2)
6284 2348 tπ : $7\zeta(3)$
6284 4012 e^x : $3\sqrt{2}/2+3\sqrt{5}/4$
6284 4499 sr : $7\pi^2/10$
6284 6251 rt : (2,8,-9,2)
6284 7381 rt : (3,6,3,-5)
6284 8885 Ei : $6(1/\pi)$
6285 2709 rt : (9,7,0,-5)
6285 3936 id : $4\sqrt{2}/9$
6285 4089 rt : (7,1,9,7)
6285 4237 cr : $2\sqrt{2}+2\sqrt{5}/3$
6285 4670 rt : (3,6,-8,-8)
6285 5349 Γ : $5(e+\pi)/2$
6285 5360 2^x : $1/3+3\sqrt{3}/2$
6285 5598 as : $4/3-\sqrt{5}/3$
6285 9097 K : $1/\sqrt[3]{20}$
6285 9248 rt : (9,8,-9,3)
6285 9707 K : 7/19
6286 0865 ln : 8/15
6286 1058 tπ : $1/3-2\sqrt{3}/3$
6286 1950 tanh : 17/23
6286 5759 cr : $\sqrt{5}/9$
6286 6697 rt : (6,7,-2,-3)
6286 7965 sq : $3/2-\sqrt{2}/2$
6286 8581 10^x : $2+4\pi$
6286 9966 e^x : $3-2\sqrt{3}$
6287 0075 tanh : $\ln(2\pi/3)$
6287 0380 rt : (4,5,8,-8)
6287 0824 Γ : $8e/3$
6287 3381 rt : (4,1,7,5)
6287 4768 ln : $8\sqrt{3}$
6287 7599 10^x : $2-\sqrt{2}/2$
6287 9198 rt : (7,9,1,8)
6287 9628 at : 8/11
6288 0715 $\ln(\pi)\times\ln(\sqrt{3})$
6288 1638 ln : $4\sqrt{2}-\sqrt{5}/4$
6288 3167 at : $1/3-3\sqrt{2}/4$
6288 4865 rt : (7,6,3,-6)
6288 6471 at : $\exp(-1/\pi)$
6288 6526 rt : (9,8,-7,-1)
6288 7273 at : P_{73}
6288 7492 as : 10/17
6288 7957 tπ : $3\sqrt{2}/4+\sqrt{5}/2$
6288 8982 at : $(\sqrt[3]{4}/3)^{1/2}$
6289 0663 erf : $(5/2)^{-1/2}$
6289 0870 Ψ : $5\sqrt{5}/2$
6289 2110 rt : (1,5,6,-6)
6289 2890 sπ : $\sqrt{3}/8$
6289 3637 e^x : $3+3\sqrt{3}/4$
6289 5499 rt : (8,5,8,-9)
6289 8430 sq : $7e/4$
6289 9264 rt : (3,7,-4,-1)
6289 9812 Γ : 6/11
6289 9928 rt : (7,0,2,3)
6289 9945 Ei : $5\sqrt{5}/2$
6290 1697 rt : (8,6,-9,4)
6290 2793 rt : (4,7,6,2)
6290 4274 e^x : $3\sqrt{2}/2-2\sqrt{3}/3$
6290 5632 cu : $\sqrt{2}+4\sqrt{3}/3$
6290 7660 rt : (7,8,-8,-4)
6290 8978 rt : (5,4,-9,-6)
6291 1650 rt : (8,1,7,6)
6291 2944 rt : (4,6,1,3)
6291 3828 ℓ_2 : $3/2+4e$
6291 5203 Li_2 : 8/15
6291 5390 cu : $1/3-2\pi$
6291 6101 Ψ : $(\ln 3)^{-1/3}$
6291 6653 θ_3 : 19/22
6291 6953 sq : $4/3+4\sqrt{5}$

6291 7318 e^x : $2\sqrt{2}+\sqrt{3}$
6291 9114 rt : (9,2,-8,2)
6291 9224 rt : (5,5,6,-7)
6291 9694 J_0 : $3\sqrt[3]{5}/4$
6292 0391 sπ : $(\sqrt[3]{3}/3)^{1/3}$
6292 1434 rt : (3,8,-8,9)
6292 3602 Ei : $4e/5$
6292 5861 rt : (5,9,-9,5)
6292 8361 rt : (9,9,-8,2)
6293 2517 sr : $3/2+2\sqrt{3}/3$
6293 3238 rt : (6,8,-9,1)
6293 4338 rt : (7,7,-4,-2)
6293 4837 rt : (8,6,1,-5)
6293 5123 rt : (9,5,6,-8)
6293 5212 sr : $4\ln 2/7$
6293 6168 rt : (1,1,5,3)
6293 6790 e^x : $3+e/4$
6293 7738 rt : (6,7,1,9)
6293 7983 10^x : $3e/2-3\pi/4$
6293 8106 rt : (5,3,-9,-1)
6293 9283 rt : (4,8,-7,-1)
6293 9687 ℓ_2 : $1-\sqrt{2}/4$
6294 1212 rt : (1,4,-9,-7)
6294 2647 J : $\exp(-10\pi/17)$
6294 2939 ℓ_2 : $4-e/3$
6294 4354 rt : (2,5,-4,-4)
6294 4683 J_0 : $8\sqrt[3]{3}/9$
6294 5588 as : $(\ln 2/2)^{1/2}$
6294 7749 rt : (7,4,-9,5)
6294 9822 sq : $(\sqrt[3]{3}/3)^{-1/3}$
6295 0161 Ei : $4\ln 2/5$
6295 0548 tπ : $e/4+\pi$
6295 4948 sπ : $\sqrt{2}/3+\sqrt{5}/3$
6295 6036 rt : (3,4,9,-8)
6295 7861 rt : (3,6,1,-1)
6295 8283 cr : $1/4+3e/2$
6295 9423 2^x : $3\sqrt{2}/4+3\sqrt{3}/2$
6296 0442 cu : $9e\pi/7$
6296 0458 K : $7(1/\pi)/6$
6296 2962 id : $(5/3)^3$
6296 3764 rt : (2,7,8,9)
6296 4006 sr : $3/4-\sqrt{2}/4$
6296 4553 rt : (7,4,9,-9)
6296 4641 J : $\exp(-5\pi)$
6296 5585 rt : (5,1,5,4)
6296 5741 ℓ_2 : $1+2\pi/3$
6296 6597 id : $(\ln 2/3)^{-1/3}$
6296 6810 cu : $1/\ln(\sqrt[3]{5}/3)$
6296 8153 tπ : $2\sqrt{2}/3+\sqrt{5}$
6296 8172 sπ : $1/2+2\pi$
6296 9029 rt : (8,7,6,3)
6297 0128 sinh : $6\ln 2/7$
6297 0278 cu : $1+2\pi/3$
6297 0426 rt : (3,7,-5,-7)
6297 0889 cr : $3/2+2\sqrt{2}$
6297 2579 rt : (6,5,4,-6)
6297 3760 cu : 6/7
6297 5623 rt : (9,1,5,5)
6297 6695 rt : (8,7,-8,3)
6297 7849 ln : $3\sqrt{2}/2+4\sqrt{5}/3$
6298 1327 rt : (1,6,-1,-2)
6298 4061 tanh : $(\ln 3/2)^{1/2}$
6298 4253 2^x : $3/4+4\pi$
6298 4788 rt : (6,6,-8,-3)
6298 5442 rt : (1,9,-7,-9)
6298 6700 $\ln(3/5)+\exp(\pi)$
6298 6932 rt : (5,9,2,1)
6298 9446 rt : (9,6,-1,-4)
6299 2221 cr : $5\sqrt{3}/2$
6299 5368 ln : $3+4e$
6299 5649 J : $(\sqrt[3]{2}/3)^3$
6299 6052 id : $\sqrt[3]{2}/2$
6299 6330 as : $\exp(-\sqrt[3]{4}/3)$
6299 7480 e^x : $2\sqrt[3]{5}/7$
6299 7888 rt : (6,2,-9,6)
6299 8987 Ei : $(e+\pi)^{-1/3}$
6299 9916 J : 2/7
6300 0168 Ψ : $\sqrt{2}+\sqrt{3}+\sqrt{6}$
6300 0555 sπ : $(\sqrt{2}\pi/3)^{1/2}$
6300 0560 rt : (2,5,-6,8)
6300 4764 tπ : $5(e+\pi)/4$
6300 5917 K : $\sqrt{5}/6$
6300 6096 2^x : 8π
6300 7126 10^x : $2(1/\pi)/3$
6300 8125 Ψ : $3\pi/4$
6300 8708 sq : $(\ln 3/3)^{-3}$
6300 9384 rt : (8,4,7,-8)
6300 9872 rt : (1,7,2,-3)
6301 0722 rt : (8,7,-6,-1)
6301 0861 10^x : $\sqrt[3]{2}/3$
6301 3087 tπ : $\sqrt{3}/4+4\sqrt{5}/3$
6301 3163 rt : (4,4,7,-7)
6301 8159 sq : $((e+\pi)/2)^3$
6301 8436 sinh : $\ln(2e/3)$
6301 9384 tπ : $(e/3)^2$
6301 9932 rt : (4,9,-4,0)
6302 5119 rt : (7,2,8,6)
6302 9428 rt : (7,5,-8,4)
6302 9841 ζ : $8\sqrt[3]{2}/5$
6303 0121 $1/5-s\pi(\pi)$
6303 1074 rt : (7,5,2,-5)
6303 2441 id : $\sqrt[3]{13/3}$
6303 5301 rt : (7,9,-7,-5)
6303 5569 rt : (4,4,-7,-5)
6303 5700 cu : $3/2-3\sqrt{2}$
6303 5797 rt : (1,8,4,9)
6303 8854 Ei : $(e/2)^{-1/3}$
6303 9003 10^x : $3/2+3\sqrt{2}/4$
6304 0414 rt : (3,2,8,5)
6304 0506 at : P_{23}
6304 1435 ℓ_{10} : $5e\pi$
6304 1493 rt : (1,9,-9,-9)
6304 4013 sπ : $(e+\pi)^3$
6304 5121 rt : (6,8,2,8)
6304 5896 Li_2 : $4\zeta(3)/9$
6304 8720 10^x : $4\sqrt[3]{2}/9$
6305 0434 2^x : $10\pi^2/3$
6305 0559 rt : (1,9,4,7)
6305 0637 ζ : $\exp(-2\pi/3)$
6305 1742 Ψ : $5\sqrt{2}/3$
6305 2694 rt : (5,0,-9,7)
6305 3726 tanh : $3\sqrt{3}/7$
6305 5559 sπ : $\sqrt[3]{17/3}$
6305 6547 rt : (5,9,-9,-8)
6305 6919 $1/3+s\pi(\sqrt{2})$
6305 6921 ln : $4-3\sqrt{2}/2$
6305 7669 rt : (9,4,5,-7)
6305 9646 tπ : $(\sqrt[3]{5}/3)^2$
6305 9900 $\ln(\sqrt{2})+\exp(1/4)$
6306 0193 rt : (7,6,1,0)
6306 0835 id : $3\sqrt{2}/2-2\sqrt{5}/3$
6306 2166 ℓ_2 : $2-3\sqrt{5}/4$
6306 2602 e^x : $2/3+\sqrt{3}/2$
6306 2608 rt : (8,8,-7,2)
6306 3380 rt : (1,7,7,-4)
6306 8196 cu : $4-3\sqrt{2}/2$
6307 0993 rt : (6,6,-3,-2)
6307 3638 rt : (5,4,-8,-2)
6307 4412 cu : $1/2-3\sqrt{5}/4$
6307 4427 rt : (6,1,3,3)
6307 5283 rt : (9,7,-7,9)
6307 6013 rt : (5,4,5,-6)
6307 6623 rt : (4,9,7,-9)
6307 6702 $\pi+\ln(3/5)$
6307 7263 J_0 : $(2\pi/3)^{1/3}$
6307 7997 rt : (1,5,-2,-3)
6307 8054 10^x : 14/25
6307 8125 K : $1/\sqrt[3]{19}$
6307 8313 sr : $5(1/\pi)/4$
6307 8755 tanh : $7(1/\pi)/3$
6308 1062 rt : (6,8,9,4)
6308 1882 2^x : $e/3-\pi/2$
6308 3963 ζ : $(\sqrt{5})^{1/2}$
6308 6115 tanh : $(2e/3)^{-1/2}$
6308 6546 rt : (9,9,-9,-7)
6308 6633 K : 3/8
6308 7198 rt : (6,3,-8,5)
6308 7447 Ei : 4
6308 7733 $3/5\div s\pi(2/5)$
6308 9761 rt : (1,2,-8,4)
6309 2975 $\ln(2)\div\ln(3)$
6309 3001 ℓ_{10} : $5\sqrt[3]{5}/2$
6309 3833 ln : $3\sqrt{2}+\sqrt{3}/2$
6309 4241 rt : (7,3,8,-8)
6309 4904 ℓ_{10} : $3\sqrt{3}/2+3\sqrt{5}/4$
6309 5450 rt : (8,5,0,-4)
6309 6909 id : $3e/5$
6309 7407 id : $4\sqrt{2}/3+\sqrt{5}/3$
6309 7554 rt : (9,7,-8,0)
6310 0214 2^x : $(e)^{1/3}$
6310 0603 $\ln(2/3)-\exp(4/5)$
6310 1455 rt : (9,9,5,-9)
6310 2283 ζ : $3\sqrt[3]{2}/7$
6310 3109 rt : (8,2,6,5)
6310 3256 10^x : $8e/5$
6310 3259 2^x : $3/4-\sqrt{2}$
6310 4996 sq : $3\sqrt{3}-\sqrt{5}/2$
6310 5071 e^x : $(\ln 3/3)^{-2}$
6310 5593 10^x : $4\sqrt{2}-4\sqrt{3}/3$
6310 7467 Γ : $4e$
6310 7664 sr : $9(e+\pi)/4$
6310 8794 sπ : 5/23
6310 9639 rt : (9,0,-2,1)
6310 9929 rt : (1,4,5,-5)
6311 0621 cu : $1/3+\pi/3$
6311 0739 sr : $(e/2)^{-3}$
6311 0948 rt : (8,4,-7,-4)
6311 1556 tπ : $1/4+\pi/2$
6311 1912 10^x : $\ln(e+\pi)$
6311 2861 rt : (4,2,-9,-8)
6311 4196 2^x : 12/17
6311 4672 rt : (3,8,-2,-6)
6311 7606 at : $4\sqrt{2}/3-2\sqrt{3}/3$
6311 7960 ζ : $7\sqrt[3]{5}/9$
6311 8752 10^x : $7\sqrt[3]{5}/2$
6312 2329 rt : (2,6,-3,-1)
6312 3269 at : $(\sqrt[3]{5}/2)^2$
6312 3577 rt : (7,6,-7,3)
6312 3967 rt : (5,9,5,-8)
6312 4709 rt : (3,3,8,-7)
6312 7312 rt : (2,9,-1,1)
6312 8717 tπ : $\sqrt{2}/2+2\sqrt{5}$
6312 9230 sq : $3\sqrt{2}/2+4\sqrt{3}/3$
6313 1327 $\exp(\pi)\times s\pi(\sqrt{5})$
6313 2074 rt : (6,7,-7,-4)
6313 3002 sq : $4-2\pi/3$
6313 3204 rt : (4,9,-7,-7)
6313 3290 rt : (9,7,4,2)
6313 3317 sπ : $7\sqrt{5}/3$
6313 7263 ℓ_2 : $1+3\sqrt{3}$
6313 7888 sr : $3\sqrt{2}+4\sqrt{5}$
6313 8192 rt : (4,7,2,2)
6313 8786 sq : $1+3\pi/2$
6313 9381 10^x : $3\sqrt{2}/2+2\sqrt{3}$
6314 2592 e^x : $9(e+\pi)/2$
6314 2684 ℓ_{10} : $2/3-\sqrt{3}/4$
6314 5014 Γ : $-\sqrt{2}+\sqrt{3}+\sqrt{6}$
6314 5450 rt : (6,4,3,-5)
6314 7262 rt : (9,8,-6,8)
6314 8559 θ_3 : $\ln(e/2)$
6314 9065 ℓ_{10} : $3\sqrt{3}/4+4\sqrt{5}/3$
6314 9611 rt : (8,3,6,-7)
6314 9825 $\ln(\sqrt{2})\times\exp(3/5)$
6315 0094 tanh : $(e/3)^3$
6315 0774 rt : (5,1,-8,6)
6315 1813 cu : $2\sqrt{2}/3+4\sqrt{3}$
6315 7894 id : 12/19
6315 8342 e^x : $1/2+2e$
6315 8724 rt : (1,9,5,-7)
6315 9380 rt : (4,7,-7,6)
6315 9910 rt : (3,8,8,-9)
6316 0683 sinh : $3\pi^2/10$
6316 1877 sr : $(2\pi)^{-1/2}$
6316 1984 rt : (8,9,-6,1)
6316 2276 cr : $\sqrt[3]{2}/5$
6316 3120 10^x : $2+3\sqrt{5}/2$
6316 3185 J_0 : 23/18
6316 4459 J_0 : $4\sqrt{5}/7$
6316 4954 rt : (6,3,9,6)
6316 6070 sπ : $6\pi^2$

6316 6591 rt : (9,5,-2,-3)
6316 7083 rt : (1,7,-6,-9)
6316 7208 rt : (7,6,-5,-1)
6316 8040 $\sqrt{5}+\exp(1/3)$
6316 8572 rt : (5,7,-8,1)
6316 8926 sr : P_{59}
6317 0158 $\Psi : (e/3)^{1/3}$
6317 2839 $\Gamma : 9/2$
6317 3222 rt : (8,8,6,-9)
6317 4806 rt : (6,9,3,7)
6317 7142 rt : (4,2,-8,-1)
6317 7527 $2^x : \exp(\sqrt[3]{4}/2)$
6317 8298 $2^x : 2\sqrt{2}+3\sqrt{5}/2$
6317 8838 $e^x : (e/3)^{1/3}$
6317 9231 rt : (3,5,-8,-1)
6318 2212 as : $1/\ln(2e)$
6318 2668 rt : (1,3,-8,-6)
6318 5531 $erf : 7/11$
6318 5614 sr : $\sqrt{2}/4+4\sqrt{3}/3$
6318 5877 $\Psi : \sqrt{21}\pi$
6318 6754 ln : $4/3+4\pi$
6318 8968 $10^x : 2/3-\sqrt{3}/2$
6319 0679 $K : 1/\sqrt{7}$
6319 0848 rt : (6,4,-7,4)
6319 1097 $K : 3\sqrt[3]{2}/10$
6319 1222 rt : (9,2,4,4)
6319 1431 at : $1-\sqrt{3}$
6319 2179 rt : (6,2,9,-8)
6319 3993 rt : (3,4,-5,-4)
6319 6683 $10^x : 9\sqrt[3]{2}/4$
6319 6859 $erf : \ln(\sqrt[3]{4}/3)$
6319 7079 rt : (8,6,-6,9)
6319 7177 $e^x : 4\sqrt{2}+\sqrt{5}/4$
6319 8080 rt : (1,5,3,6)
6319 9891 rt : (4,3,6,-6)
6320 4581 rt : (7,1,1,2)
6320 4805 $erf : 2(1/\pi)$
6320 6172 rt : (1,8,-8,-8)
6320 7642 rt : (8,9,-7,-6)
6320 9711 rt : (9,3,4,-6)
6321 3699 rt : (4,8,6,-8)
6321 4696 at : $\ln(\sqrt[3]{3}/3)$
6321 4802 sr : $4\sqrt{3}$
6321 8524 as : 13/22
6321 8660 $\Psi : (\ln 2/3)^3$
6321 8811 rt : (9,8,4,-8)
6321 8958 $2^x : 4\pi/9$
6321 9495 rt : (3,9,-5,-6)
6321 9650 at : $(e+\pi)/8$
6322 1095 rt : (4,1,-8,-7)
6322 2510 rt : (7,4,1,-4)
6322 2693 $t\pi : 4e+3\pi/2$
6322 2956 $\Psi : \exp(-5/2)$
6322 5704 rt : (2,9,3,-6)
6322 7817 rt : (9,9,-5,7)
6322 8306 $J_0 : (\sqrt[3]{3}/3)^{-1/3}$
6322 9523 rt : (7,9,1,-6)
6323 1698 ln : $3-\sqrt{5}/2$
6323 1819 cr : $8e/5$
6323 1830 $\Psi : \exp(-\sqrt{2}\pi/2)$

6323 1934 cu : $e/4+3\pi$
6323 3264 rt : (7,7,-6,2)
6323 5009 id : $4\sqrt{3}/3-3\sqrt{5}/4$
6323 6395 $\ell_{10} : e+\pi/2$
6323 7016 $J : 5/19$
6323 7022 tanh : $\sqrt{5}/3$
6323 7143 id : $1/3+3\sqrt{3}/4$
6323 7332 ln : $\sqrt[3]{20/3}$
6324 1786 rt : (4,8,-4,7)
6324 3239 at : $3\sqrt[3]{5}/7$
6324 4969 $\ell_2 : \ln(3\pi/2)$
6324 5334 $s\pi : \sqrt{2}+\sqrt{6}-\sqrt{7}$
6324 5553 id : $(5/2)^{-1/2}$
6324 6908 sinh : $(3\pi/2)^{-1/3}$
6325 1283 $t\pi : 1/\ln(\sqrt{3})$
6325 1733 $\exp(2/5)+\exp(\pi)$
6325 2691 $2^x : \sqrt{2}/2$
6325 2876 cu : $4-\pi$
6325 3533 rt : (7,7,7,-9)
6325 5110 rt : (2,7,9,-9)
6325 6280 $\ell_{10} : 2/3+4e/3$
6325 6348 rt : (7,2,7,-7)
6325 8651 $10^x : 5\pi/9$
6325 9449 rt : (2,2,9,-7)
6326 0608 sq : $9\sqrt[3]{3}/8$
6326 1286 $s\pi : (e/3)^{-2}$
6326 2230 rt : (7,3,7,5)
6326 2583 $K : 2\sqrt[3]{5}/9$
6326 5306 sq : $4\sqrt{5}/7$
6326 5466 rt : (5,2,-7,5)
6326 6101 $\zeta : 9\ln 2/5$
6327 1005 rt : (2,3,9,5)
6327 1604 sq : 23/18
6327 1725 rt : (5,8,4,-7)
6327 2798 $t\pi : 1/2+e/4$
6327 4492 $t\pi : \sqrt{5}-\sqrt{6}+\sqrt{7}$
6327 4869 $10^x : \sqrt[3]{6}$
6327 4883 at : 11/15
6327 5882 $Ei : 5/9$
6327 6884 sr : $10\ln 2$
6327 7037 ln : $4+\sqrt{5}/2$
6327 7562 $\zeta : 1/8$
6327 8221 rt : (8,6,-7,0)
6328 1278 rt : (2,8,-5,4)
6328 2197 rt : (7,8,7,3)
6328 3719 rt : (5,3,4,-5)
6328 3942 rt : (8,7,-5,8)
6328 5507 rt : (1,8,-5,4)
6328 5627 cr : $4+\sqrt{2}/4$
6328 6468 $\Gamma : 1/\ln(2\pi)$
6328 8553 rt : (8,9,-1,-5)
6328 8781 id : $1/\ln(\ln 3)$
6328 9013 rt : (5,8,-8,-7)
6328 9515 $t\pi : \sqrt{2}/3+3\sqrt{5}$
6329 2485 cr : $1+3\sqrt{5}/2$
6329 6687 $K : 8/21$
6329 8143 cu : $5\zeta(3)/7$
6329 9316 id : $\sqrt{8/3}$
6329 9734 sr : $\zeta(3)/3$
6330 3396 $e^x : 6\sqrt{5}/5$

6330 4633 rt : (8,7,5,-8)
6330 7479 $\zeta : (1/(3e))^{-1/3}$
6330 8440 rt : (8,4,-1,-3)
6331 0017 $s\pi : 1/\sqrt{21}$
6331 1273 $e^x : 1/4-\sqrt{2}/2$
6331 2566 rt : (6,5,-6,3)
6331 5635 $\ell_{10} : 2\sqrt{2}/3+3\sqrt{5}/2$
6331 5842 $e^x : 7$
6331 6007 rt : (5,2,4,3)
6331 6787 rt : (3,7,7,-8)
6331 7273 rt : (2,9,-3,-5)
6331 8639 cr : $2+3\pi/4$
6331 8794 $K : 1/\sqrt[3]{18}$
6331 9411 sq : $1+e/3$
6332 0742 rt : (6,8,-6,-5)
6332 3828 $10^x : \sqrt[3]{17}$
6332 4211 $\ell_2 : \sqrt{3}+2\sqrt{5}$
6332 5565 $s\pi : 1/2+e$
6332 5701 sq : $2/3-3e/2$
6332 5739 sq : $5(1/\pi)/2$
6332 6426 rt : (8,2,5,-6)
6332 7803 $10^x : (\sqrt{2}\pi/3)^{-1/2}$
6333 1430 rt : (8,1,-7,2)
6333 2196 $e^x : e/4+3\pi/2$
6333 3121 $K : 6(1/\pi)/5$
6333 3557 cr : $3\sqrt{2}/2+\sqrt{5}$
6333 3943 at : $2\sqrt{2}/3-3\sqrt{5}/4$
6333 4044 sq : $2e-\pi/4$
6333 4481 rt : (6,8,2,-6)
6333 4822 rt : (4,9,-1,8)
6333 5249 rt : (2,7,-9,-8)
6333 6061 $\theta_3 : 4/13$
6333 6535 rt : (5,3,-8,-5)
6333 7129 $\ell_{10} : 3+3\sqrt{3}/4$
6333 7776 $e^x : 3/4+2\sqrt{3}$
6333 8680 $e^x : 5e/6$
6333 8902 cr : $4\sqrt{2}-3\sqrt{3}/4$
6334 0364 $\zeta : 9\sqrt{5}/10$
6334 1228 ln : $3+3\sqrt{2}/2$
6334 1735 $\pi+\exp(2/5)$
6334 2526 rt : (1,6,1,-6)
6334 3851 $\sqrt{3}-\ln(3)$
6334 4044 rt : (6,6,8,-9)
6334 4756 rt : (7,5,-5,9)
6334 5295 as : $\sqrt{2}/2-3\sqrt{3}/4$
6334 6562 $t\pi : 2\sqrt{2}/3+\sqrt{3}$
6334 8730 rt : (4,0,-7,6)
6335 1084 rt : (3,2,7,-6)
6335 2395 rt : (9,9,-3,-4)
6335 2430 sr : $\sqrt[3]{19}$
6335 3301 rt : (7,9,-5,-5)
6335 3903 $\Gamma : \sqrt{23/3}$
6335 5452 cr : $3+e/2$
6335 5597 $t\pi : \exp(\sqrt[3]{2})$
6335 6081 $s\pi : (2\pi/3)^{-1/3}$
6335 7720 rt : (1,8,4,-6)
6335 9272 rt : (9,7,3,-7)
6335 9556 rt : (1,9,-6,1)
6335 9643 $e^x : 23/15$
6336 0281 sq : $2+4e/3$

6336 1676 sinh : 19/15
6336 1750 $\exp(1/5)-s\pi(1/5)$
6336 2503 rt : (8,1,-1,1)
6336 2638 rt : (7,8,-5,1)
6336 3438 $t\pi : (2e/3)^{-1/3}$
6336 5430 $Li_2 : \ln(\sqrt[3]{5})$
6336 7365 $10^x : 4/3+2\sqrt{3}/3$
6336 8013 $Ei : \ln 2/7$
6336 9597 $s\pi : 4\sqrt{3}/3+2\sqrt{5}$
6337 0527 rt : (6,5,-4,-1)
6337 2069 $\Psi : 10\sqrt[3]{3}$
6337 2251 ln : $e/2+4\pi$
6337 3213 sr : $\sqrt{3}/4+\sqrt{5}$
6337 3568 sr : $3/2+2e$
6337 3681 rt : (8,3,5,4)
6337 4360 $erf : 2\sqrt{5}/7$
6337 6052 $\theta_3 : (\sqrt{2}\pi/3)^{-3}$
6337 7687 rt : (6,3,2,-4)
6338 0538 rt : (4,9,-1,-4)
6338 0686 $10^x : 1/\sqrt{22}$
6338 1839 rt : (6,1,8,-7)
6338 2266 rt : (8,8,-4,7)
6338 3674 rt : (4,7,5,-7)
6338 3733 cr : $1-\sqrt{5}/3$
6338 4057 cr : $4(1/\pi)/5$
6338 6878 rt : (2,5,-8,-9)
6338 7120 $Ei : 7\sqrt[3]{3}/2$
6339 0426 ln : $3\pi/5$
6339 1046 cu : $3\sqrt{2}/4-3\sqrt{3}/2$
6339 1545 $2^x : 17/24$
6339 3355 $10^x : 4e\pi/9$
6339 7459 id : $3/2-\sqrt{3}/2$
6339 7610 $J_0 : 4(1/\pi)$
6339 8386 as : $2\sqrt{2}-\sqrt{5}$
6339 9331 $\zeta : \sqrt[3]{2}/10$
6339 9368 rt : (1,6,-3,8)
6339 9680 at : $e/2-2\pi/3$
6340 0286 rt : (3,4,-8,6)
6340 0871 ln : $3/2+4e/3$
6340 1350 rt : (5,3,-6,4)
6340 1738 rt : (7,6,6,-8)
6340 2211 rt : (9,0,9,8)
6340 2535 rt : (7,8,0,-5)
6340 2580 $10^x : \exp(-\sqrt{3}/2)$
6340 3215 rt : (9,2,3,-5)
6340 3767 $2/3\times s\pi(2/5)$
6340 3834 $\lambda : 1/16$
6340 4758 rt : (9,4,-3,-2)
6340 5250 $10^x : 1+e/2$
6340 5580 rt : (2,7,-9,7)
6340 5643 ln : $3e/2+\pi/3$
6340 7429 rt : (9,6,-9,1)
6341 5176 cu : $1/2-e/2$
6341 5887 $\Gamma : e/5$
6341 8423 sq : $4/3+3\sqrt{3}$
6341 9447 $\theta_3 : 4\ln 2/9$
6342 0692 $J_0 : 9\sqrt{2}/10$
6342 1457 rt : (1,7,-7,-7)
6342 4043 $J_0 : 14/11$
6342 4228 $s\pi : \exp(\sqrt{3}/3)$

6342 4750 $\exp(4) + s\pi(\sqrt{2})$
6342 5566 ln : $3\sqrt{2}/8$
6342 5958 λ : $\exp(-15\pi/17)$
6342 6023 e^x : $2/3 + 3\pi$
6342 8221 K : 5/13
6342 8410 $\ln(\sqrt{3}) \div s\pi(1/3)$
6342 8872 rt : (1,9,-1,-4)
6343 6529 rt : (2,6,8,-8)
6343 8521 K : $2\sqrt{3}/9$
6344 0970 tπ : $1/4 + 3\pi$
6344 1691 ℓ_{10} : $2 + 4\sqrt{3}/3$
6344 2267 rt : (3,7,-2,1)
6344 5094 K : $5\ln 2/9$
6344 5819 rt : (2,8,2,-5)
6344 6872 rt : (5,5,9,-9)
6344 7528 tanh : $(\sqrt[3]{2}/3)^{1/3}$
6344 9122 rt : (3,3,7,4)
6344 9988 sq : $\sqrt{2}/3 + 3\sqrt{5}/2$
6345 1667 rt : (7,6,-4,8)
6345 2429 2^x : $5\sqrt{5}/8$
6345 2482 at : $3/2 - \sqrt{5}$
6345 4409 rt : (4,2,5,-5)
6345 6186 $\exp(1/5) + \exp(5)$
6345 6420 rt : (5,8,4,7)
6345 6959 tπ : $\sqrt[3]{2}/7$
6345 7256 $e - \ln(2/5)$
6345 7479 rt : (6,6,-5,2)
6345 8582 erf : 16/25
6345 8831 Ei : ln(5)
6345 9411 2^x : $1/4 - e/3$
6345 9679 ln : $7(e + \pi)/8$
6346 0898 rt : (1,5,-2,-1)
6346 2700 2^x : $\sqrt{2}/2 + 2\sqrt{3}/3$
6346 3307 2^x : $((e + \pi)/3))^{1/2}$
6346 3677 rt : (8,6,4,-7)
6346 4658 rt : (7,1,6,-6)
6346 5915 tπ : $3\sqrt{2} + \sqrt{3}/3$
6346 6711 rt : (6,4,8,5)
6347 0535 rt : (5,9,-3,-3)
6347 4028 rt : (1,4,-1,9)
6347 4115 2^x : $(\sqrt[3]{4}/3)^{-3}$
6347 5573 rt : (9,5,9,6)
6347 6542 rt : (8,8,-2,-4)
6347 8761 $\ln(3/5) \div \ln(\sqrt{5})$
6348 1179 rt : (9,8,-8,-6)
6348 3624 rt : (7,3,0,-3)
6348 4306 $\ln(\sqrt{3}) + \exp(3)$
6348 7083 id : $4\sqrt{3}/3 - 4\sqrt{5}$
6348 7888 rt : (6,7,-9,-7)
6348 8034 2^x : $(\ln 3/3)^{-1/3}$
6348 8229 sπ : $1/3 + 4\sqrt{2}/3$
6349 4458 rt : (8,9,-3,6)
6349 4466 e^x : $4/3 + \sqrt{5}/4$
6349 4719 2^x : 19/3
6349 5637 ℓ_{10} : $4/3 + 4\sqrt{5}/3$
6349 6623 id : $3e/2 - 3\pi/2$
6350 0352 at : $(5/2)^{-1/3}$
6350 1462 rt : (4,1,-6,5)
6350 2422 rt : (9,3,3,3)
6350 2673 at : 14/19
6350 3757 sπ : $3e/2 + \pi$
6350 7849 rt : (3,9,-5,7)
6350 9159 ln : $3\sqrt[3]{5}$
6350 9662 cu : $3\pi/8$
6351 0919 ζ : $3\sqrt[3]{3}/8$
6351 1135 rt : (9,4,8,-9)
6351 1750 tπ : $2\pi^2/7$
6351 2433 $\ln(2) \times \ln(2/5)$
6351 2555 rt : (6,5,7,-8)
6351 2841 rt : (1,8,2,-6)
6351 4529 rt : (3,6,6,-7)
6351 4816 ℓ_2 : $3 + 3\pi$
6351 4895 tanh : 3/4
6351 5290 cu : $7\pi/10$
6351 5835 rt : (3,5,-5,7)
6351 7457 rt : (7,9,-4,0)
6351 9482 J : 3/13
6352 0732 2^x : $1/2 - 2\sqrt{3}/3$
6352 2253 rt : (3,8,-4,-5)
6352 6701 rt : (4,6,-7,1)
6352 7581 rt : (6,4,-4,9)
6353 0316 rt : (9,6,2,-6)
6353 0317 ℓ_2 : $3 - 3\pi/4$
6353 1001 id : $\sqrt{2}/2 + 4\sqrt{3}$
6353 1546 rt : (6,9,-3,-4)
6353 1629 rt : (5,0,9,7)
6353 3435 erf : $4\sqrt[3]{3}/9$
6353 4011 ℓ_2 : $4e/7$
6353 4505 rt : (1,8,8,-7)
6353 5778 rt : (6,7,1,-5)
6353 5881 cu : $8\sqrt{2}/3$
6353 9323 θ_3 : $7\sqrt{2}/10$
6353 9719 ℓ_{10} : $2\sqrt{2} + 2\sqrt{5}/3$
6354 0023 rt : (2,1,8,-6)
6354 3634 sπ : $(2e/3)^{1/3}$
6354 3902 rt : (3,8,0,-4)
6354 4711 $\sqrt{2} \div \exp(4/5)$
6354 4901 sq : $1/4 - 3\sqrt{5}/2$
6354 5536 cr : $3/4 + 4e/3$
6354 7188 sπ : $\sqrt{2}/3 + 4\sqrt{3}/3$
6354 8152 rt : (2,8,1,-6)
6354 9987 sr : $2\sqrt{2}/3 + \sqrt{3}$
6355 0420 cu : $2/3 - e$
6355 1035 10^x : $(\ln 3/2)^{-1/3}$
6355 1218 Ei : $8\pi/9$
6355 1476 erf : $3\sqrt[3]{5}/8$
6355 2393 sq : $1/4 - \pi/3$
6355 3458 rt : (8,8,5,2)
6355 4310 2^x : $4e/3 + \pi/2$
6355 5194 rt : (1,5,0,-9)
6355 6109 rt : (8,1,4,-5)
6355 7258 rt : (9,8,-4,-3)
6355 8615 sinh : ln(ln 3/2)
6356 1632 $\sqrt{2} + \exp(1/5)$
6356 1808 id : $1/4 - 4\sqrt{2}/3$
6356 3249 K : $e/7$
6356 3264 ζ : $2(1/\pi)/5$
6356 4841 rt : (5,4,-5,3)
6356 5406 rt : (1,7,-8,2)
6356 5794 sr : $2\sqrt{2}/7$
6356 6894 e^x : $(\ln 3)^{-1/3}$
6356 7425 sr : $4\sqrt{2} - 4\sqrt{5}/3$
6356 7449 id : $2\sqrt{2} - 2\sqrt{3}$
6356 8124 tπ : $5e\pi/4$
6356 8827 sr : $\exp(-e/3)$
6357 0700 rt : (6,9,-5,-2)
6357 1664 rt : (5,2,3,-4)
6357 3978 Ψ : $9e\pi/2$
6357 4384 rt : (7,7,-3,7)
6357 7434 rt : (1,5,9,-8)
6357 9505 cu : $3/4 + 4\sqrt{3}/3$
6358 2210 Ei : 19/21
6358 3395 rt : (7,5,5,-7)
6358 3892 K : 7/18 .
6358 4364 e^x : $3/4 + 2\pi$
6358 4708 K : $1/\sqrt[3]{17}$
6358 5918 tπ : $\sqrt[3]{3}/8$
6358 6323 10^x : $\sqrt[3]{5}/8$
6359 0385 id : $7\pi^2/8$
6359 0524 as : $3/2 - e/3$
6359 0928 rt : (4,3,-6,-4)
6359 1527 tanh : $5\zeta(3)/8$
6359 5661 ℓ_2 : $4\sqrt{2} + \sqrt{5}/4$
6359 5990 tπ : $3e/4 + \pi$
6359 6538 rt : (2,6,-8,-7)
6359 8413 rt : (2,6,-6,-5)
6359 8876 ln : 9/17
6359 9916 rt : (6,9,6,-9)
6360 1204 10^x : $1 + \sqrt{5}/3$
6360 3786 rt : (8,3,-2,-2)
6360 6577 rt : (9,1,8,7)
6360 6797 $2/5 + \sqrt{5}$
6361 0103 10^x : $3e/2 + \pi$
6361 1801 Γ : $\sqrt{24}$
6361 1814 tπ : $5\sqrt{5}$
6361 2413 rt : (7,4,6,4)
6361 3503 rt : (6,0,-4,1)
6361 5267 rt : (3,8,1,2)
6361 6167 ℓ_{10} : $3\sqrt[3]{3}$
6361 6960 sr : $1 + 3\sqrt{5}/4$
6361 7884 as : $6\ln 2/7$
6361 8795 tπ : $10(e + \pi)/3$
6361 9693 rt : (8,3,9,-9)
6362 0159 10^x : $5e/2$
6362 2628 rt : (7,7,-1,-4)
6362 2712 cu : $\sqrt{3}/4 + \sqrt{5}/3$
6362 2932 ℓ_{10} : $1/4 + 3e/2$
6362 4032 10^x : $1/2 + 4\pi/3$
6362 4913 rt : (1,7,3,-5)
6362 5719 sr : $2/3 + 2\pi$
6362 7903 10^x : $2e/3 - \pi/4$
6362 8595 cu : $3\sqrt{2}/2 - 4\sqrt{5}/3$
6362 8724 rt : (5,7,-7,-6)
6362 9436 $1/4 + \ln(4)$
6362 9579 ℓ_{10} : $\ln 2/3$
6363 0265 $\ln(4/5) - \exp(5)$
6363 1105 rt : (6,0,7,6)
6363 1291 e^x : $1/3 - \pi/4$
6363 2834 rt : (6,7,-4,1)
6363 3010 ℓ_{10} : $3/2 + 2\sqrt{2}$
6363 6363 id : 7/11
6363 6413 sq : $2\pi^2$
6364 0097 e^x : $\sqrt{5/3}$
6364 0283 rt : (5,4,8,-8)
6364 0384 sπ : $\sqrt{2}/4 + \sqrt{3}/2$
6364 0716 id : $4\sqrt{3} + 3\sqrt{5}$
6364 0948 θ_3 : 7/11
6364 3183 ln : P_3
6364 3778 tπ : $3\sqrt{2}/2 + 2\sqrt{3}$
6364 5563 rt : (5,4,-3,-1)
6364 9192 10^x : $3 - \sqrt{3}/3$
6364 9353 rt : (3,6,-2,8)
6364 9798 Li_2 : 7/13
6365 0063 ℓ_{10} : $5\sqrt{3}/2$
6365 0821 at : 17/23
6365 1324 as : $3/2 - 2\pi/3$
6365 1416 id : $\ln(\sqrt[3]{4}/3)$
6365 3643 rt : (4,8,-2,-3)
6365 5458 $\ln(\pi) + \exp(2/5)$
6365 7511 rt : (9,1,2,-4)
6365 9508 at : $\ln(2\pi/3)$
6365 9980 rt : (8,5,3,-6)
6366 1444 rt : (7,9,4,-8)
6366 1977 $2 \div \pi$
6366 2413 rt : (6,5,-3,8)
6366 3046 ln : $2\sqrt{2} + 4\sqrt{3}/3$
6366 5089 10^x : 8/19
6366 5358 sinh : 3/5
6366 7176 rt : (2,8,-2,-4)
6366 8012 e^x : $4e/3 + 2\pi$
6366 8708 as : $\ln(2e/3)$
6366 9072 rt : (1,9,8,-9)
6366 9657 rt : (2,5,7,-7)
6367 0765 rt : (3,1,6,-5)
6367 1498 cr : P_{28}
6367 1872 2^x : $1/3 + 4\sqrt{5}$
6367 2233 erf : 9/14
6367 2944 K : 9/23
6367 3661 rt : (4,3,5,3)
6367 5690 e^x : $2/3 - \sqrt{5}/2$
6367 6142 sr : ln(3/2)
6367 6901 ζ : $(\sqrt{2}/3)^{1/2}$
6367 7321 sr : $1/\sqrt[3]{15}$
6367 8329 tπ : 5/13
6367 9608 Γ : $8\sqrt{3}/5$
6368 2009 ln : $2/3 + 2\sqrt{5}$
6368 2125 cu : $5\sqrt{2}/6$
6368 2209 ℓ_{10} : 3/13
6368 2708 rt : (7,9,-7,-1)
6368 7434 rt : (4,2,-5,4)
6368 8148 rt : (9,3,7,-8)
6368 8690 rt : (5,9,-8,8)
6369 2058 rt : (8,5,-8,1)
6369 3717 rt : (5,9,5,6)
6369 3935 sr : $2 + e/4$
6369 4527 rt : (8,8,-6,-5)
6369 6380 sq : $(2\pi/3)^{1/3}$
6369 8974 sq : $2\sqrt{2} + \sqrt{3}/4$
6370 0747 rt : (5,7,4,1)
6370 2057 $s\pi(\pi) \div s\pi(\sqrt{5})$

6370 5347 sq : $\ln(\sqrt{2}\pi/2)$
6370 5532 sq : $4\sqrt{2}+\sqrt{5}/4$
6370 5647 rt : (6,2,1,-3)
6370 5743 tπ : $7\sqrt[3]{2}$
6370 6003 ζ : $\sqrt[3]{10/3}$
6370 6238 rt : (1,9,-3,-2)
6370 6285 J : 1/11
6370 8397 at : $3\sqrt{3}/4-\sqrt{5}/4$
6371 0625 rt : (5,8,7,-9)
6371 3542 id : $3\sqrt{3}-\sqrt{5}/4$
6371 4527 10^x : $1/3+2\pi$
6371 4973 rt : (1,9,9,-5)
6371 6658 rt : (7,8,-2,6)
6371 7358 sr : $3\sqrt{2}/2+\sqrt{5}/4$
6371 7995 rt : (8,7,-3,-3)
6371 8588 $\sqrt{3}\div e$
6372 1346 sr : $4e+3\pi/4$
6372 2142 rt : (6,4,6,-7)
6372 2289 10^x : $4\sqrt{2}+\sqrt{3}/3$
6372 4423 K : $\ln(\sqrt{2}\pi/3)$
6372 4717 K : $\pi/8$
6372 5581 rt : (9,7,-9,6)
6372 7056 rt : (8,9,2,-7)
6372 7729 ln : $2\sqrt{3}+3\sqrt{5}/4$
6372 9591 rt : (9,1,-9,3)
6373 0433 e^x : $9(e+\pi)/4$
6373 1760 sinh : $6\sqrt{2}/5$
6373 2138 cr : $2e-\pi/3$
6373 3205 tπ : $9(1/\pi)/2$
6373 3207 sπ : $e\pi/7$
6373 3298 e^x : $\exp(-1/\sqrt{2})$
6373 4176 tπ : $(\ln 2/2)^{-2}$
6373 5711 $\sqrt{2}-\ln(4/5)$
6373 6288 ln : $2+\pi$
6373 6301 J_0 : 19/15
6373 6954 ℓ_2 : $2e+\pi/4$
6373 7040 rt : (7,2,0,1)
6373 8273 2^x : $\sqrt{3}/2+3\sqrt{5}/2$
6373 9809 rt : (5,5,9,5)
6374 0257 sr : $4\sqrt{2}+3\sqrt{3}/4$
6374 0972 rt : (9,3,-4,-1)
6374 1910 e^x : $3+2\pi$
6374 1961 rt : (7,0,5,5)
6374 2033 Ei : $7(1/\pi)/4$
6374 2228 sπ : $4\sqrt{3}-3\sqrt{5}$
6374 2984 rt : (9,5,1,-5)
6374 2992 ℓ_2 : 9/14
6374 3342 ζ : $(e/2)^{-2}$
6374 3583 Li_2 : $7\ln 2/9$
6374 3823 rt : (9,7,6,-9)
6374 5860 rt : (4,8,-9,-6)
6374 6150 $2\div\exp(1/5)$
6375 1947 rt : (1,7,-5,-1)
6375 2201 Ψ : $(4/3)^3$
6375 3023 at : $3/4-2\sqrt{5}/3$
6375 3182 ln : $1-\sqrt{2}/3$
6375 4133 rt : (5,9,-1,-3)
6375 7121 cu : $3\sqrt{2}/4+2\sqrt{3}$
6375 8391 θ_3 : $2(1/\pi)$
6375 9873 rt : (5,3,-3,9)
6376 1006 ℓ_{10} : $2\sqrt{2}-3\sqrt{3}/2$
6376 2743 tanh : $(\ln 3)^{-3}$
6376 3169 $\exp(1/5)\div s\pi(\sqrt{3})$
6376 5202 rt : (5,5,-4,2)
6376 6294 rt : (8,2,9,7)
6377 1395 rt : (3,5,5,-6)
6377 2418 sr : $2/3+4\pi$
6377 3225 sπ : $\sqrt{3}/2+3\sqrt{5}/2$
6377 5567 $\exp(4/5)-s\pi(1/5)$
6377 5814 tπ : $3e/2+3\pi$
6377 7071 rt : (5,8,-4,-2)
6377 8946 sq : $10\sqrt[3]{2}/3$
6377 9621 rt : (7,4,-5,-3)
6377 9954 rt : (6,8,5,-8)
6378 0672 ζ : $1/\ln((e+\pi)/3)$
6378 1124 Ei : $\sqrt{2}-\sqrt{6}+\sqrt{7}$
6378 1423 at : $(\ln 3/2)^{1/2}$
6378 1961 ln : $4/3+\sqrt{5}/4$
6378 3425 rt : (1,0,9,6)
6378 4077 sr : $1/4+3\sqrt{5}$
6378 4110 rt : (8,4,4,3)
6378 5170 cr : $\sqrt{2}-2\sqrt{3}/3$
6378 5759 2^x : $1/\ln(\sqrt[3]{3})$
6378 7311 sinh : $\zeta(3)/2$
6378 8820 rt : (1,5,-2,9)
6378 9225 rt : (4,3,9,-8)
6378 9370 10^x : 3/14
6379 3078 sπ : $9\ln 2/8$
6379 4123 rt : (4,7,-5,-5)
6379 6440 rt : (8,5,-9,7)
6379 7140 rt : (9,9,0,-6)
6379 9714 sπ : $3\sqrt[3]{2}$
6380 0991 cr : $3(e+\pi)/4$
6380 1044 id : $3\sqrt{2}/4+\sqrt{3}/3$
6380 5369 rt : (3,7,1,9)
6380 7251 sr : $6\sqrt{5}/5$
6380 8579 rt : (8,9,-9,0)
6380 8975 rt : (9,6,8,5)
6381 0080 cr : $\sqrt{2}+4\sqrt{5}/3$
6381 0945 tanh : $(\sqrt[3]{5}/3)^{1/2}$
6381 1133 rt : (7,4,4,-6)
6381 1638 as : $2\sqrt{3}/3-\sqrt{5}/4$
6381 2386 e^x : $4e-2\pi/3$
6381 5893 rt : (3,5,-9,-7)
6381 6701 sq : $3\sqrt{2}/4-2\sqrt{5}$
6381 7425 10^x : $-\sqrt{2}+\sqrt{6}+\sqrt{7}$
6381 8718 $e-\exp(\sqrt{5})$
6381 9330 cr : $3\ln 2/8$
6381 9920 rt : (6,6,-2,7)
6382 1448 rt : (8,2,8,-8)
6382 3080 rt : (9,7,-5,-2)
6382 6556 rt : (5,6,-9,2)
6382 7821 id : $4\sqrt{2}+4\sqrt{5}/3$
6382 8831 rt : (3,9,4,3)
6382 9105 cr : $\exp(\sqrt{2}\pi/3)$
6383 2042 rt : (6,9,8,3)
6383 3198 Ψ : 3/16
6383 5640 rt : (5,1,8,6)
6383 6274 rt : (4,7,8,-9)
6383 7179 rt : (1,8,0,-3)
6383 7412 rt : (2,8,-7,-5)
6383 7831 cu : $\sqrt{2}+\sqrt{5}$
6383 7922 erf : $1/\ln(3\pi/2)$
6383 9001 rt : (9,8,-8,5)
6384 1446 ℓ_{10} : $8e/5$
6384 2274 cr : $7\pi/5$
6384 3667 $\exp(3/4)\times s\pi(e)$
6384 5973 sinh : $\exp(\sqrt[3]{4}/3)$
6384 6055 tπ : $\exp(-\sqrt[3]{5})$
6384 6189 sq : $1/2-3\sqrt{3}/4$
6384 6395 sinh : $\sqrt{2}+\sqrt{3}+\sqrt{6}$
6384 9198 rt : (6,8,-3,0)
6385 1451 K : $4\ln 2/7$
6385 4160 rt : (7,8,3,-7)
6385 5796 cu : $4/3-3\sqrt{3}$
6385 5969 at : $3\sqrt{3}/7$
6385 6739 rt : (1,4,8,-7)
6385 6993 Γ : $4\ln 2$
6385 7610 Ψ : $4\pi^2/7$
6385 8039 rt : (8,6,7,-9)
6385 9664 sq : $2-4e/3$
6385 9881 rt : (7,2,-1,-2)
6386 0202 e^x : $7e\pi/9$
6386 1644 2^x : $5\sqrt{5}/6$
6386 2448 rt : (4,7,-9,1)
6386 2552 Γ : $\exp(-e)$
6386 3798 Ei : $8\zeta(3)$
6386 4254 cr : 22/5
6386 5348 ℓ_2 : $3e/4+4\pi/3$
6386 6004 Li_2 : $3\sqrt[3]{2}/7$
6386 6103 rt : (8,0,3,4)
6386 8262 cu : $e-3\pi$
6386 8678 tanh : $3\sqrt[3]{2}/5$
6386 9157 cu : $4/3+\sqrt{3}/2$
6387 0003 $\exp(4/5)+\exp(5)$
6387 0175 rt : (9,2,7,6)
6387 0342 sr : $e/4+2\pi$
6387 2856 rt : (7,3,-9,8)
6387 3124 cu : $2+2\pi/3$
6387 3855 rt : (3,9,-8,-8)
6387 6765 10^x : $7\sqrt{5}$
6387 7091 2^x : $\sqrt{2}/3-\sqrt{5}/2$
6387 7632 Ei : $7e\pi/3$
6387 7871 $\sqrt{5}\times\exp(\sqrt{3})$
6387 9604 rt : (2,6,-3,6)
6388 0349 K : $1/\sqrt[3]{16}$
6388 0744 rt : (2,9,-5,-1)
6388 2758 at : $7(1/\pi)/3$
6388 2800 rt : (3,7,-1,-3)
6388 3510 rt : (7,9,-1,5)
6388 4088 rt : (4,7,5,7)
6388 4387 ℓ_{10} : $4+\sqrt{2}/4$
6388 4839 10^x : $2\sqrt{2}$
6388 4970 rt : (5,3,7,-7)
6388 5005 cu : $10(e+\pi)/7$
6388 5536 2^x : $10\sqrt[3]{2}/9$
6388 6700 Γ : $2\ln 2/7$
6388 7114 rt : (1,7,3,4)
6388 7656 id : $2\sqrt{5}/7$
6388 8520 cu : $2/3+3\pi/2$
6388 9691 rt : (2,4,8,4)
6388 9859 ℓ_{10} : $1+3\sqrt{5}/2$
6389 0549 sπ : $\exp(\sqrt{2}\pi/2)$
6389 0636 at : $(2e/3)^{-1/2}$
6389 1740 rt : (4,9,-6,-5)
6389 1971 2^x : $2\sqrt{2}/3+3\sqrt{3}/2$
6389 2479 Ψ : $(\sqrt{5}/2)^{-1/3}$
6389 2529 2^x : $e/2+\pi$
6389 3694 $\ln(3)\times\exp(2/5)$
6389 4310 sr : $1/\sqrt{6}$
6389 5690 ζ : 13/24
6389 6117 cr : 6/23
6389 7370 rt : (1,8,7,-8)
6389 7515 sq : $1/3-3\sqrt{5}$
6389 9298 J : $3(1/\pi)/5$
6390 1582 2^x : 7/5
6390 5317 tπ : $3\sqrt{3}/4-\sqrt{5}/2$
6390 5609 $1/4+\exp(2)$
6390 5637 rt : (9,2,6,-7)
6390 5732 ln : 14
6390 8153 rt : (8,4,2,-5)
6390 8636 10^x : $1-\pi/4$
6390 9190 as : $(3\pi/2)^{-1/3}$
6391 0726 ℓ_{10} : $2+3\pi/4$
6391 1612 rt : (7,6,-2,-3)
6391 3855 tπ : $(\ln 3/2)^{1/3}$
6391 5030 rt : (5,7,6,-8)
6391 6165 Γ : $(2\pi)^{-1/3}$
6391 6428 e^x : $\sqrt{2}/2-2\sqrt{3}/3$
6391 6664 rt : (6,8,-6,-1)
6391 6694 J_0 : 24/19
6391 8426 10^x : $1/2+3\sqrt{3}$
6391 8704 rt : (8,6,-8,6)
6391 9593 K : $5(1/\pi)/4$
6391 9640 rt : (2,8,4,5)
6391 9755 e^x : $7e\pi/4$
6392 1641 $\exp(5)-s\pi(e)$
6392 2164 rt : (2,1,-8,-5)
6392 2626 ℓ_{10} : $3\sqrt{2}/2+\sqrt{5}$
6392 5790 10^x : $\sqrt{3}/3+3\sqrt{5}/4$
6392 6890 ℓ_{10} : $4\sqrt{2}-3\sqrt{3}/4$
6392 8776 rt : (9,6,5,-8)
6393 1109 cu : $2\sqrt{3}/3-3\sqrt{5}/2$
6393 3735 rt : (8,8,1,-6)
6393 4541 rt : (9,8,3,1)
6393 5117 K : $(e/2)^{-3}$
6393 5265 rt : (5,4,-2,8)
6393 5388 rt : (1,9,3,-1)
6393 5711 Ei : $\sqrt[3]{7}$
6393 6128 sinh : $1/\ln(\sqrt[3]{3})$
6393 7680 ℓ_{10} : $\sqrt{19}$
6393 8027 rt : (2,9,0,6)
6393 8962 rt : (6,5,7,4)
6393 9282 $s\pi(\sqrt{2})-s\pi(\sqrt{5})$
6394 0090 ℓ_{10} : $3+e/2$
6394 3010 rt : (6,6,-8,-6)
6394 6029 Ei : $e/3$
6394 8254 ζ : 3/23
6394 9736 sr : $e/4+4\pi$
6394 9997 cr : $2\sqrt{2}/3+2\sqrt{3}$

6395 0221 rt : (4,6,-9,8)
6395 4652 rt : (2,0,7,5)
6395 5526 rt : (6,1,-9,9)
6395 6045 $\Gamma : (2e/3)^2$
6395 9131 at : $(e/3)^3$
6395 9661 $K : (2\pi)^{-1/2}$
6396 0214 sr : 9/22
6396 3302 rt : (8,8,-9,-1)
6396 3550 rt : (1,6,-5,-5)
6396 3784 rt : (5,3,3,2)
6396 6075 rt : (2,5,-7,-6)
6396 6301 $2^x : 3e\pi/7$
6396 6481 rt : (9,9,-7,4)
6396 8227 rt : (7,8,-4,-4)
6397 1862 sq : $4 - 3\sqrt{2}/4$
6397 3002 rt : (6,1,6,5)
6397 3085 sr : $9(1/\pi)/7$
6397 5449 $\sqrt{2} + \exp(4/5)$
6397 5590 rt : (7,1,9,-8)
6397 7394 id : $4\pi^2/7$
6397 7785 $J_0 : 7\sqrt[3]{3}/8$
6397 7919 sinh : $5\zeta(3)/3$
6397 9028 cu : $1/3 - 4\pi$
6397 9934 $J : (\pi/2)^{-2}$
6398 0039 sr : $2\sqrt{3}/3 - \sqrt{5}/3$
6398 0200 rt : (3,6,9,-9)
6398 0957 $\zeta : 4\zeta(3)/7$
6398 2022 $Ei : \sqrt{21}$
6398 4874 cr : $7\sqrt[3]{2}/2$
6398 5003 $erf : 11/17$
6398 5497 cr : $4e/3 + \pi/4$
6398 5977 rt : (2,9,-6,-7)
6398 6921 rt : (7,5,8,-9)
6398 8712 rt : (9,4,2,2)
6398 8893 $s\pi(\sqrt{2}) \div s\pi(1/5)$
6398 9735 rt : (6,3,5,-6)
6399 0717 rt : (3,7,-3,-4)
6399 0902 ln : $4 + 2\sqrt{3}/3$
6399 1480 $\Gamma : 13/24$
6399 2118 cu : P_5
6399 2228 $Li_2 : 3\sqrt[3]{3}/8$
6399 2870 $s\pi : (\ln 3/2)^{-1/3}$
6399 3852 rt : (4,8,-6,2)
6399 4034 $\Psi : 19/8$
6399 6363 sq : $((e+\pi)/3)^{-1/3}$
6399 6604 rt : (5,1,2,-3)
6399 6808 $s\pi(1/3) + s\pi(e)$
6399 7670 cr : $1/3 + 3e/2$
6399 7799 sr : $1/3 + 3\pi/4$
6399 9685 rt : (6,7,4,-7)
6399 9986 $K : 2/5$
6400 0000 id : 16/25
6400 2346 rt : (1,6,2,-4)
6400 2627 $J : 2(1/\pi)/7$
6400 2966 sq : $\sqrt{2} + \sqrt{5} + \sqrt{7}$
6400 4146 rt : (2,8,-4,9)
6400 5067 rt : (7,4,-8,7)
6400 5145 rt : (1,9,-7,-8)
6400 5837 rt : (9,0,1,3)
6400 6281 rt : (6,7,-1,6)
6400 8390 $10^x : 5\zeta(3)/8$
6400 9449 rt : (5,5,1,-4)
6401 0820 ln : $4e + \pi$
6401 4237 rt : (9,4,0,-4)
6401 6301 rt : (5,9,0,-5)
6401 6583 $J : (5/3)^{-3}$
6401 6741 $2^x : 1/\ln(\sqrt[3]{5})$
6401 7247 $\zeta : (2\pi)^{-1/3}$
6401 9227 rt : (9,8,-1,-5)
6402 0548 $\ell_2 : 4\sqrt{2} + \sqrt{3}/3$
6402 4078 sq : $6e/5$
6402 4822 $\ln(3/4) \times \exp(4/5)$
6402 5250 sq : $8(e + \pi)$
6402 6207 $K : \zeta(3)/3$
6402 7087 $\sqrt{2} - s\pi(e)$
6402 8770 $e^x : 2\sqrt{3}/7$
6403 3521 $e^x : 1/\ln(1/(3\pi))$
6403 4177 $t\pi : (3\pi)^{-1/2}$
6403 4525 cu : $4/3 - \sqrt{2}/3$
6403 7372 cu : $\sqrt{2}/4 + 2\sqrt{3}$
6403 8820 rt : (8,6,-1,-1)
6403 9117 $\zeta : 16/21$
6404 0511 cr : $3 + \sqrt{2}$
6404 1143 $t\pi : 4\sqrt{3}/3 - 2\sqrt{5}/3$
6404 4768 cu : $2\sqrt{2}/3 - 3\sqrt{5}$
6404 5953 $\ln(4) + s\pi(\sqrt{3})$
6404 7306 cr : $3\sqrt{2}/4 + 3\sqrt{5}/2$
6404 7815 $\ell_2 : 9\sqrt{3}/5$
6404 8778 $s\pi : 4e - 2\pi/3$
6404 9443 $\exp(1/4) + \exp(\sqrt{5})$
6405 0022 sq : $4 + 4\sqrt{2}/3$
6405 1733 $t\pi : 2(e + \pi)/7$
6405 2231 at : $\sqrt{5}/3$
6405 6771 rt : (8,7,-7,5)
6405 7529 $e^x : 2 + 3\sqrt{2}/2$
6405 8817 $erf : (\sqrt[3]{2}/3)^{1/2}$
6406 2384 rt : (8,9,5,-9)
6406 2500 sq : 13/8
6406 3665 $\ell_2 : 4 + \sqrt{5}$
6406 3850 $Li_2 : (e/2)^{-2}$
6406 4248 $10^x : 4 - 3\sqrt{2}/4$
6406 4550 $\Psi : (2\pi/3)^{-2}$
6406 5006 rt : (4,2,-2,9)
6406 5248 ln : $2e\pi/9$
6406 6750 rt : (8,5,6,-8)
6406 7071 $2^x : 5/7$
6406 9263 $1/2 + \exp(\pi)$
6406 9359 rt : (4,7,-6,9)
6407 0513 rt : (4,6,7,-8)
6407 3103 rt : (8,1,7,-7)
6407 5448 rt : (7,8,-8,0)
6407 8951 rt : (4,2,8,-7)
6408 0598 rt : (1,8,-6,-4)
6408 1761 $s\pi : 3/4 + \sqrt{2}/3$
6408 2205 rt : (4,8,-9,-8)
6408 2404 cr : 5/19
6408 2854 $J_0 : \sqrt[3]{2}$
6408 3195 rt : (8,3,8,6)
6408 3417 $s\pi : 5\sqrt[3]{5}/7$
6408 5908 id : $2 - e/2$
6408 6111 $\Gamma : (e/2)^{-2}$
6408 7209 rt : (3,9,-7,0)
6408 7606 sr : $4/3 + e/2$
6408 8122 $\ln(2) + \exp(2/3)$
6408 8623 $\lambda : (1/\pi)/5$
6409 0867 rt : (7,7,2,-6)
6409 0926 $10^x : 7e/6$
6409 1608 $\ell_{10} : 3/4 + 4e/3$
6409 3516 rt : (4,2,9,6)
6409 5364 $s\pi : 3\sqrt{2} - 2\sqrt{3}$
6409 5690 $4 \div \ln(3)$
6409 6316 cu : $4e + \pi/4$
6409 7340 sr : $6\pi/7$
6409 7823 rt : (7,4,-7,1)
6409 8016 tanh : $(\sqrt{3})^{-1/2}$
6409 8982 $\Psi : \exp(\sqrt{3})$
6409 8984 rt : (6,2,-8,8)
6409 9385 $2^x : 1/\ln(\pi/2)$
6409 9480 $10^x : 1/2 + 3\sqrt{5}/2$
6409 9980 id : $4\sqrt[3]{3}/9$
6410 1711 rt : (8,7,9,5)
6410 3025 $2^x : 3/2 - \pi/4$
6410 4742 rt : (7,3,3,-5)
6410 6146 $\ell_2 : 3\sqrt[3]{5}/8$
6410 6394 $Ei : (\sqrt{5}/3)^{1/3}$
6410 7066 tanh : $4\sqrt[3]{5}/9$
6410 7696 tanh : 19/25
6410 8456 $\Psi : 5\sqrt{3}/9$
6410 9853 as : $2 - 3\sqrt{3}/2$
6411 0105 rt : (1,9,-4,-6)
6411 0729 $Li_2 : 13/24$
6411 3765 $Ei : 25/3$
6411 5862 $\ell_2 : 9\ln 2$
6411 6256 $2^x : 1/2 + 4\sqrt{5}$
6411 8038 cr : $\exp(-4/3)$
6411 8334 rt : (3,5,5,8)
6411 9566 $t\pi : 9\zeta(3)$
6412 1399 $\Psi : \exp(\sqrt{3}/2)$
6412 2567 rt : (6,9,-2,-1)
6412 3836 cu : $1/2 + e/4$
6412 4098 id : $3\sqrt[3]{5}/8$
6412 7476 rt : (2,3,-3,8)
6412 8704 rt : (7,1,4,4)
6413 1845 $2^x : 6\pi/5$
6413 2908 cu : 20/13
6413 3571 rt : (6,4,9,-9)
6413 3924 rt : (9,9,3,-8)
6413 5771 rt : (3,5,-6,1)
6413 6755 rt : (3,8,-9,6)
6413 9000 rt : (1,8,8,-4)
6414 0169 $Ei : 10e\pi/7$
6414 2763 sq : $1/4 + 3e$
6414 3068 rt : (4,9,-3,3)
6414 3488 rt : (8,9,-8,-2)
6414 4869 rt : (5,5,-1,7)
6414 7495 sinh : $2e/9$
6414 8095 $Li_2 : (2\pi)^{-1/3}$
6414 9157 $10^x : 3/4 - 2\sqrt{2}/3$
6414 9529 rt : (6,5,-1,-3)
6414 9917 $\Psi : 9\pi/5$
6415 2160 rt : (9,5,4,-7)
6415 3308 $\sqrt{5} - \ln(2/3)$
6415 3448 $2^x : 3 - \pi/4$
6415 3716 cu : $4 - 3\pi$
6415 5134 rt : (7,5,-7,6)
6415 5490 cu : $2\sqrt{3} + 2\sqrt{5}/3$
6415 6036 rt : (3,0,5,4)
6415 6231 $K : 2\sqrt{2}/7$
6415 7726 $K : \exp(-e/3)$
6415 8883 $10^x : 2/3$
6415 9265 id : $1/2 + \pi$
6415 9812 $s\pi : (\sqrt{2}/3)^{1/3}$
6416 0416 $\ell_2 : \sqrt[3]{3}/9$
6416 4832 $10^x : (4/3)^{-3}$
6416 5211 $e^x : 1/2 + \sqrt{2}/3$
6416 6487 rt : (1,8,3,2)
6416 7353 $erf : 3\sqrt{3}/8$
6416 8061 rt : (5,6,5,-7)
6416 8894 sr : 7/17
6417 1665 cu : $3\sqrt{2}/2 + 2\sqrt{5}$
6417 2096 $10^x : 4/3 - \sqrt{5}/2$
6417 3163 $\zeta : 5\zeta(3)/3$
6417 3399 rt : (9,9,-4,9)
6417 4243 rt : (5,6,-6,-5)
6417 4561 $10^x : 3\sqrt{2}/2 - 3\sqrt{3}/4$
6417 6303 $\Psi : 7e/8$
6417 7546 $e^x : 8e\pi/3$
6417 8302 $2^x : 2 + 3\sqrt{3}$
6417 8352 rt : (2,6,0,-3)
6417 9047 rt : (9,1,5,-6)
6417 9804 sinh : $(\ln 2/2)^{-1/2}$
6418 0023 rt : (7,5,5,3)
6418 0763 rt : (9,6,-6,-1)
6418 2136 $10^x : 8\sqrt{3}/9$
6418 3675 sq : $\sqrt{2}/4 - 2\sqrt{3}/3$
6418 3981 ln : $10\sqrt[3]{5}/9$
6418 4757 sq : $4e/3 - \pi/3$
6418 5388 ln : 10/19
6418 8345 $\Psi : 4\sqrt{2}$
6418 8363 ln : $2\sqrt{3}/3 + \sqrt{5}/3$
6418 9168 rt : (1,7,6,-7)
6418 9287 rt : (8,7,0,-5)
6418 9425 $\ell_2 : P_{20}$
6419 0014 rt : (7,0,-6,2)
6419 1653 $s\pi : 7e\pi$
6419 2092 $\ell_2 : 1 - e/4$
6419 2780 sr : $2\sqrt[3]{3}/7$
6419 2927 at : $2/3 - \sqrt{2}$
6419 3398 $\exp(2/5) \times s\pi(\pi)$
6419 3673 rt : (7,8,6,-9)
6419 4185 $2^x : 3\sqrt{3} - 4\sqrt{5}/3$
6419 8613 $2^x : (\ln 3/3)^{1/3}$
6420 0117 sq : $5\sqrt[3]{3}/9$
6420 0182 rt : (6,4,-9,1)
6420 1270 $1/2 \times \exp(1/4)$
6420 1521 rt : (5,0,-8,9)
6420 1550 id : $7e\pi/9$
6420 2071 rt : (6,1,0,-2)
6420 2932 $erf : 13/20$
6420 3757 $K : (\pi/2)^{-2}$

6420 4240 rt : (5,2,6,-6)
6420 6502 2^x : $((e+\pi)/3)^{-1/2}$
6420 7363 rt : (1,4,-1,-8)
6420 7958 sq : $7(e+\pi)/2$
6420 8883 $t\pi$: $7\sqrt{2}/4$
6420 9973 2^x : $(\sqrt{5}/2)^{-3}$
6421 0780 K : $\ln(2/3)$
6421 1365 K : $1/\sqrt[3]{15}$
6421 1722 ℓ_{10} : $4\pi^2/9$
6421 2108 rt : (3,9,7,-9)
6421 3795 Ψ : $4\zeta(3)/5$
6421 3876 rt : (8,8,-6,4)
6421 5642 ℓ_2 : $2+3\sqrt{2}$
6421 6071 $s\pi$: $2e+\pi/4$
6421 7747 rt : (5,8,-4,-6)
6421 8549 rt : (3,8,-7,-7)
6421 9591 sq : $2\zeta(3)/3$
6421 9753 tanh : 16/21
6422 2654 rt : (9,3,6,5)
6422 2983 e^x : $1/2-2\sqrt{2}/3$
6422 3843 sr : $4+4\sqrt{5}/3$
6422 4423 rt : (7,4,7,-8)
6422 4927 $t\pi$: $9\sqrt{2}/7$
6422 7549 rt : (2,7,-1,-3)
6422 8559 J_0 : $8\sqrt{2}/9$
6423 0110 rt : (6,8,0,5)
6423 0484 sq : $3/2+2\sqrt{3}$
6423 0628 2^x : $4+\sqrt{2}$
6423 0996 Γ : $3\sqrt[3]{3}/8$
6423 1415 rt : (8,3,1,-4)
6423 2622 id : $e/2+2\pi$
6423 5641 cu : $9e/10$
6423 5741 e^x : $3\sqrt{2}+\sqrt{5}/3$
6423 6877 10^x : $\exp(-\sqrt{3}/3)$
6423 7577 as : $\ln(\ln 3/2)$
6423 8587 rt : (7,9,-4,-3)
6423 9200 Ψ : $2\sqrt[3]{3}/3$
6424 0081 e^x : $1/3+3\sqrt{2}/2$
6424 0180 ℓ_{10} : $2e-\pi/3$
6424 0517 rt : (1,3,7,-6)
6424 0940 Ei : $(\sqrt[3]{3}/3)^2$
6424 2944 $s\pi(2/5)\times s\pi(\sqrt{5})$
6424 4744 cr : $3\sqrt{2}/2+4\sqrt{3}/3$
6424 8110 rt : (9,7,-7,-5)
6425 1183 J_0 : $2\pi/5$
6425 1569 rt : (3,5,8,-8)
6425 1683 sq : $7\sqrt[3]{2}/3$
6425 2440 cr : $5(1/\pi)/6$
6425 2582 sq : $e/3+3\pi/4$
6425 3721 2^x : $4-3\sqrt{3}/2$
6425 5491 10^x : $6\sqrt{3}$
6425 6437 $t\pi$: $2\sqrt{3}+3\sqrt{5}/2$
6425 7997 $s\pi$: $3/4+2\sqrt{5}$
6425 9602 ln : $\sqrt{20}\pi$
6425 9625 sinh : $5e/8$
6426 0168 $2/3\times s\pi(\sqrt{2})$
6426 1822 rt : (3,4,6,3)
6426 2297 rt : (7,9,-8,-7)
6426 2735 rt : (6,3,-7,7)
6426 3188 e^x : 11/4
6426 6097 $t\pi$: 2/11
6426 8566 rt : (5,2,7,5)
6427 0417 rt : (7,0,8,7)
6427 1282 rt : (9,2,9,-9)
6427 2903 sr : $(e+\pi)^{-1/2}$
6427 3051 cr : $4+\sqrt{3}/4$
6427 3441 id : $1/3+4\sqrt{3}/3$
6427 3667 rt : (8,8,4,-8)
6427 5666 Ψ : $9(1/\pi)/5$
6427 5778 rt : (7,7,-8,-1)
6427 6310 sr : $3\sqrt{2}/2+\sqrt{3}/3$
6427 6481 sinh : $(\ln 3/3)^{1/2}$
6427 7853 at : $(\sqrt[3]{2}/3)^{1/3}$
6427 8760 $s\pi$: 2/9
6427 9347 Ψ : $\sqrt{17/3}$
6427 9851 Ei : $\ln 2/3$
6428 0147 sq : $4-e$
6428 0871 rt : (1,8,-4,-1)
6428 1208 $\exp(4)\div s\pi(\sqrt{2})$
6428 3098 $\exp(3)\times s\pi(\pi)$
6428 4015 erf : $(e+\pi)/9$
6428 4599 2^x : $9(1/\pi)/4$
6428 4694 λ : $\exp(-7\pi/8)$
6428 5488 λ : $(5/2)^{-3}$
6428 5714 id : 9/14
6428 8069 rt : (1,6,-9,-8)
6428 8185 $\ln(4/5)+s\pi(1/3)$
6429 4957 ℓ_{10} : $3(e+\pi)/4$
6429 5747 rt : (9,7,-2,-4)
6429 7316 rt : (3,4,-8,-6)
6429 8588 $t\pi$: $4(1/\pi)/7$
6430 2187 ℓ_{10} : $\sqrt{2}+4\sqrt{5}/3$
6430 2638 rt : (4,9,5,-8)
6430 5606 rt : (9,7,7,4)
6430 5766 $t\pi$: $(\ln 3)^3$
6430 6273 rt : (8,1,2,3)
6430 6777 rt : (7,5,-3,-2)
6430 9168 sr : P_{56}
6430 9628 ζ : 2/15
6431 0333 cu : $1/2-3e/4$
6431 0413 rt : (1,6,5,1)
6431 0799 sq : $3/2-3e/2$
6431 2276 e^x : $4\sqrt{2}-\sqrt{3}$
6431 3959 rt : (2,7,4,-6)
6431 4333 rt : (3,1,9,-7)
6431 5095 10^x : $(\sqrt[3]{5}/3)^{-3}$
6431 6577 e^x : $4+\pi/2$
6431 6913 rt : (1,7,-4,7)
6431 7050 10^x : $3\sqrt{3}/4-2\sqrt{5}/3$
6431 7318 ℓ_{10} : $\exp(\sqrt{2}\pi/3)$
6431 7390 e^x : $3/4+\pi/4$
6431 9713 K : $1/\sqrt{6}$
6432 2033 rt : (8,4,5,-7)
6432 2126 $\exp(2)+s\pi(\sqrt{3})$
6432 2645 2^x : $(e)^{-1/3}$
6432 5857 sinh : 25/2
6432 6532 2^x : $\ln(\sqrt[3]{4}/3)$
6432 6891 rt : (7,6,-6,5)
6432 7666 $\sqrt{5}\times\ln(3/4)$
6432 7790 ℓ_{10} : $7\pi/5$
6432 8324 $s\pi(1/4)\div s\pi(\pi)$
6432 9057 $t\pi$: $5\pi^2/2$
6433 0842 rt : (6,8,-2,-3)
6433 1222 e^x : $2-\sqrt{2}/2$
6433 2505 ln : $7e/10$
6433 4055 rt : (4,9,-9,1)
6433 8183 $t\pi$: $9\sqrt{2}/4$
6433 8231 ζ : $\zeta(3)/9$
6433 9811 rt : (5,8,-1,-4)
6434 0910 2^x : $7\zeta(3)/6$
6434 1663 rt : (6,7,7,-9)
6434 1673 Γ : $\sqrt{3}\pi$
6434 2908 rt : (6,2,4,-5)
6434 3661 $\ln(3/5)+\exp(e)$
6434 4385 cr : $1/4+4\pi/3$
6434 5267 ℓ_{10} : 5/22
6434 5272 rt : (1,1,8,5)
6434 6034 sr : $4-3\sqrt{3}/4$
6434 6736 rt : (7,4,9,6)
6434 7854 e^x : $(1/(3e))^{1/3}$
6434 9272 rt : (6,3,1,1)
6434 9565 sr : $(\sqrt{5}/3)^3$
6435 0110 at : 3/4
6435 2302 sq : $8\sqrt[3]{3}/9$
6435 2905 K : 9/22
6435 8717 rt : (9,8,2,-7)
6435 9403 K : $9(1/\pi)/7$
6436 0322 Ei : $\sqrt{5}/4$
6436 2203 e^x : $\sqrt{23}\pi$
6436 2471 e^x : $2\sqrt{5}/9$
6436 3055 Γ : $4\pi/3$
6436 3476 erf : 15/23
6436 3982 2^x : $2\sqrt{2}-2\sqrt{3}$
6436 5312 rt : (4,5,6,-7)
6436 5606 rt : (8,9,-6,-2)
6436 5958 cr : 4/15
6436 7460 ℓ_2 : $3\sqrt{2}-\sqrt{5}/2$
6436 7473 rt : (1,9,-6,2)
6436 7718 rt : (7,9,6,2)
6437 0883 id : $2\sqrt{2}-2\sqrt{5}$
6437 1063 rt : (2,8,-5,-6)
6437 1415 rt : (9,3,-1,-3)
6437 5227 10^x : $9\sqrt{5}/5$
6437 7516 $2/5\times\ln(5)$
6437 7857 Ψ : 17/3
6437 8688 ln : $3/2+4\pi$
6437 9207 sq : $5e\pi/6$
6437 9335 id : $3e/2+4\pi$
6438 0433 rt : (2,8,8,-9)
6438 0550 id : $3-3\pi/4$
6438 0977 rt : (5,1,-7,8)
6438 2175 rt : (3,6,-2,-2)
6438 4046 tanh : 13/17
6438 5135 tanh : $(\sqrt[3]{5})^{-1/2}$
6438 5618 ℓ_2 : 25
6438 8265 rt : (7,6,1,-5)
6438 8427 Γ : $8\sqrt{5}/3$
6438 8673 sinh : $3\sqrt{2}/7$
6438 9628 sr : $4/3+4\sqrt{2}$
6439 0020 sinh : $\sqrt{3}-\sqrt{5}-\sqrt{7}$
6439 0675 2^x : $7\pi^2/6$
6439 1541 $t\pi$: $2\sqrt{3}/9$
6439 4186 rt : (8,9,-5,3)
6439 4881 id : $(\sqrt{2}\pi)^{1/3}$
6439 5487 rt : (8,0,6,6)
6439 5588 2^x : $3e/2-3\pi/2$
6439 7575 Li_2 : $e/5$
6439 7662 2^x : $3\sqrt{2}/4+2\sqrt{3}/3$
6439 9255 rt : (7,5,-9,-6)
6439 9417 2^x : $1+\sqrt{3}$
6439 9644 rt : (5,9,3,-7)
6440 2228 rt : (1,9,0,-4)
6440 2503 ln : $1/3+\pi/2$
6440 2714 cr : $2\zeta(3)/9$
6440 3355 10^x : $3\sqrt{2}/4+4\sqrt{5}$
6440 5504 rt : (8,4,-9,2)
6440 5628 2^x : $4e+2\pi/3$
6440 6325 ζ : $(\ln 3/3)^2$
6440 6477 rt : (6,3,8,-8)
6440 7163 Γ : 5π
6440 7846 rt : (3,6,-7,-5)
6440 8761 $s\pi$: $\sqrt{3}+2\sqrt{5}/3$
6440 9268 cu : $2\sqrt{2}+\sqrt{5}/3$
6440 9742 Ψ : $1/\sqrt[3]{17}$
6441 3034 θ_3 : 5/16
6441 3291 rt : (3,9,3,-1)
6441 3424 ℓ_{10} : $2\sqrt{2}/3+2\sqrt{3}$
6441 3761 sr : $1/\sqrt[3]{14}$
6441 4198 rt : (2,3,5,9)
6441 4877 rt : (5,6,8,4)
6441 5108 $\ln(3/4)-\exp(\sqrt{5})$
6441 5496 Γ : 23/7
6441 5852 cu : 19/22
6441 6100 ζ : $((e+\pi)/3)^{-3}$
6441 7529 $t\pi$: $3\sqrt{3}-3\sqrt{5}/4$
6442 0230 $2/3\div\ln(2/3)$
6442 0369 $\ln(2)+s\pi(2/5)$
6442 1097 rt : (8,8,-3,9)
6442 1250 J_0 : $(\pi/2)^{1/2}$
6442 1766 sq : $\sqrt{2}/3+2\sqrt{3}/3$
6442 1768 $s\pi$: $7(1/\pi)/10$
6442 3760 $2\times\exp(3/5)$
6442 4324 10^x : $(\sqrt[3]{2}/3)^{-1/3}$
6442 4332 rt : (6,7,-7,0)
6442 6744 Ψ : 7/18
6442 7103 rt : (9,4,3,-6)
6442 9127 θ_3 : $e/3$
6442 9924 sinh : $5\sqrt[3]{2}/2$
6443 0687 rt : (2,9,-9,-9)
6443 1753 rt : (7,7,5,-8)
6443 2336 at : $5\zeta(3)/8$
6443 2479 ln : $3/4+2\sqrt{3}/3$
6443 2514 Ei : 3/13
6443 2892 rt : (1,9,4,8)
6443 4161 cr : P_{66}
6443 4713 cr : P_{65}
6443 5302 Ψ : $9\sqrt[3]{2}/2$
6443 5919 rt : (9,2,-4,-1)
6443 7172 10^x : $(5/3)^{-3}$
6443 7711 sinh : $7\ln 2/8$

6443 7858 rt : (3,5,-7,2)
6443 8424 2^x : $2e+\pi/4$
6443 9224 10^x : $3/2+4\sqrt{3}/3$
6444 0908 sinh : $(e)^{-1/2}$
6444 1137 ℓ_{10} : $7\sqrt[3]{2}/2$
6444 1632 ℓ_{10} : $4e/3+\pi/4$
6444 3070 J_0 : $1/\ln(\sqrt{2}\pi/2)$
6444 5403 rt : (7,1,-2,-1)
6444 9263 rt : (5,4,0,-3)
6445 0930 2^x : $6\sqrt[3]{2}$
6445 1304 ℓ_{10} : $1/3+3e/2$
6445 1358 rt : (9,6,-9,9)
6445 6447 id : $3\sqrt{2}-3\sqrt{3}/2$
6445 7533 sr : $9\zeta(3)/4$
6445 8427 rt : (6,9,-6,-6)
6445 8735 rt : (3,4,-7,9)
6445 8862 rt : (9,5,7,-9)
6445 8888 K : 7/17
6446 0083 e^x : $6\sqrt[3]{3}/5$
6446 1637 $e\div s\pi(\sqrt{3})$
6446 2477 Li_2 : $1/\ln(2\pi)$
6446 2809 sq : 21/11
6446 5338 rt : (2,4,9,-8)
6446 5668 Ei : 14/11
6446 5885 ln : $\sqrt{2}/2+2\sqrt{5}$
6446 6555 sq : $4e/3-3\pi$
6446 7095 rt : (4,1,7,-6)
6446 8148 $t\pi$: $3\sqrt{2}/4+3\sqrt{5}/2$
6446 8982 cr : $2-\sqrt{3}$
6447 0049 rt : (6,2,5,4)
6447 1106 K : $2\sqrt[3]{3}/7$
6447 2988 $1/2+\ln(\pi)$
6447 5997 sq : $3\sqrt[3]{5}/4$
6447 6056 rt : (8,5,3,2)
6447 8271 rt : (6,8,-3,-3)
6447 8729 as : $\zeta(3)/2$
6447 9948 ln : $4-2\pi/3$
6448 0297 Γ : $(e+\pi)^{-3}$
6448 0334 sr : $2/3+3e/4$
6448 0459 at : $\ln(\sqrt{2}/3)$
6448 3076 tanh : $(\sqrt{2}\pi/2)^{-1/3}$
6448 3884 Ei : $9\sqrt{2}/10$
6448 4007 rt : (3,8,6,-8)
6448 4131 rt : (8,5,-5,-1)
6448 5334 ℓ_{10} : $3+\sqrt{2}$
6448 6374 ln : $3e/4+\pi$
6448 7243 Γ : $3\sqrt[3]{2}/7$
6448 9180 rt : (5,5,4,-6)
6448 9402 sr : $3\ln 2/5$
6449 0730 ℓ_{10} : $3\sqrt{2}/4+3\sqrt{5}/2$
6449 2427 cr : $3\sqrt{3}-\sqrt{5}/3$
6449 3406 id : $\pi^2/6$
6449 6415 rt : (7,9,-9,-8)
6449 6974 rt : (7,2,2,-4)
6449 7003 $s\pi$: $\exp(-3/2)$
6449 7254 rt : (4,0,-9,3)
6449 9009 e^x : $\sqrt{3}+\sqrt{6}-\sqrt{7}$
6450 0000 cu : $9\sqrt[3]{5}/10$
6450 0159 rt : (3,0,-9,5)
6450 0204 $t\pi$: $\sqrt{2}/4+2\sqrt{3}$
6450 0835 Ψ : 24/25
6450 3772 rt : (6,9,1,4)
6450 5051 rt : (9,9,-8,-1)
6450 5609 ln : $1+e/3$
6450 6633 rt : (6,1,9,7)
6450 8013 id : $1/\ln(3\pi/2)$
6450 9715 rt : (4,7,-8,-7)
6450 9868 rt : (5,6,8,-9)
6451 0054 rt : (9,1,0,2)
6451 1513 2^x : $2/3-3\sqrt{3}/4$
6451 1578 rt : (4,1,-7,-9)
6451 2218 K : $(e+\pi)^{-1/2}$
6451 2873 $t\pi$: $1/3-e$
6451 2918 rt : (8,6,-1,-4)
6451 4130 rt : (7,8,-7,-2)
6451 5008 cu : $1/3-4e/3$
6451 5569 sq : $4\sqrt{3}/3+4\sqrt{5}$
6451 7118 rt : (1,7,1,-2)
6451 9194 rt : (7,3,6,-7)
6451 9644 Ei : $8\sqrt[3]{3}/3$
6452 0210 sq : $5\pi^2/6$
6452 0385 rt : (8,4,7,5)
6452 3380 id : $3\sqrt{3}/4-4\sqrt{5}$
6452 5050 rt : (1,8,-5,5)
6452 5334 rt : (7,7,-5,4)
6452 5523 sinh : 14/11
6452 6930 2^x : $1+\sqrt{3}/2$
6452 7864 rt : (8,7,3,-7)
6452 8209 ℓ_2 : $4\sqrt{3}/3-\sqrt{5}/3$
6452 9689 rt : (2,4,-6,-5)
6453 0015 $2/5\times\exp(\sqrt{2})$
6453 2647 rt : (9,0,4,5)
6453 2877 sr : $2+\sqrt{2}/2$
6453 8025 sinh : $9\sqrt{2}/10$
6454 1252 id : $2\sqrt{3}/3+2\sqrt{5}/3$
6454 2536 rt : (1,8,-3,-5)
6454 2789 ℓ_2 : $3\sqrt{2}/4+3\sqrt{3}$
6454 3592 rt : (9,1,8,-8)
6454 3991 sr : $\sqrt{2}/3+\sqrt{5}$
6454 9299 $s\pi$: $2e/7$
6454 9722 sr : 5/12
6455 1742 K : $(\sqrt{5}/3)^3$
6455 2466 rt : (1,9,-7,-8)
6455 3537 $\ln(5)+s\pi(\sqrt{2})$
6455 3698 θ_3 : 15/19
6455 7393 e^x : $2\sqrt{2}+4\sqrt{5}$
6456 0408 sr : $\sqrt{22/3}$
6456 0728 ζ : $(e)^{-2}$
6456 1242 $1/4+\exp(1/3)$
6456 3193 sinh : 17/10
6456 4114 Γ : 25/9
6456 6493 $t\pi$: $\exp(5/2)$
6457 4810 $\ln(3)\times s\pi(1/5)$
6457 5131 id : $\sqrt{7}$
6457 9936 Ψ : $(3\pi/2)^3$
6458 2132 $t\pi$: $2\sqrt{2}+3\sqrt{5}/2$
6458 4982 K : $1/\sqrt[3]{14}$
6458 5094 2^x : $\sqrt[3]{13/2}$
6458 8794 J_0 : $((e+\pi)/3)^{1/3}$
6458 9168 rt : (5,2,-6,7)
6458 9803 id : $4-3\sqrt{5}/2$
6459 0608 J_0 : 5/4
6459 1190 $s\pi(1/3)\times s\pi(\sqrt{3})$
6459 2431 rt : (5,3,-8,1)
6459 2444 cu : $7e/4$
6459 3150 $t\pi$: $7\pi^2/5$
6459 3836 rt : (9,7,-8,8)
6459 5489 rt : (4,8,4,-7)
6459 8387 e^x : $4\sqrt{2}-\sqrt{3}/4$
6459 8715 rt : (8,9,-2,8)
6459 9859 rt : (3,4,7,-7)
6460 0105 ζ : 11/16
6460 2587 Ei : $e\pi$
6460 3155 rt : (4,7,0,-4)
6460 4267 $1/2\div s\pi(e)$
6460 8693 $\exp(2/5)+\exp(e)$
6460 9391 Ei : $4(1/\pi)$
6461 1363 $s\pi$: $\sqrt{5}/10$
6461 2165 rt : (6,6,6,-8)
6461 3471 ln : $10\pi^2/7$
6461 5741 rt : (7,9,-1,-5)
6461 5797 $t\pi$: $\sqrt{2}/2+\sqrt{5}/3$
6461 6168 at : $(\ln 3)^{-3}$
6461 6892 Ei : $\sqrt{11/3}$
6461 8379 id : $(3\pi/2)^3$
6461 8922 e^x : $3e+3\pi/4$
6461 9261 rt : (5,2,9,-8)
6461 9530 rt : (9,8,5,-9)
6461 9531 Ei : $(4/3)^{-1/3}$
6462 0079 rt : (4,6,-4,-1)
6462 1521 Ψ : $\sqrt{3}/6$
6462 4174 sinh : $4(1/\pi)$
6462 4316 K : $3\ln 2/5$
6462 4431 rt : (6,5,2,-5)
6462 6077 rt : (8,4,8,-9)
6462 7107 as : P_2
6462 9004 10^x : $\sqrt{3}/8$
6462 9721 rt : (2,1,6,4)
6462 9860 e^x : $1+2e/3$
6463 0549 rt : (9,7,1,-6)
6463 0972 rt : (7,8,-5,-2)
6463 4439 Ψ : $(1/\pi)/8$
6463 4502 2^x : $3+4\pi$
6463 6491 10^x : $1-\sqrt{3}/3$
6463 7857 rt : (5,1,5,-5)
6463 9087 ℓ_2 : $\sqrt{5}/7$
6464 1135 rt : (8,3,4,-6)
6464 3490 $3\div\exp(3/5)$
6464 4660 id : $1-\sqrt{2}/4$
6464 6665 at : $3/2-\sqrt{5}/3$
6464 7444 ℓ_{10} : $3\sqrt{2}/2+4\sqrt{3}/3$
6464 8193 tanh : 10/13
6464 8835 rt : (9,6,-3,-3)
6465 0056 ln : $1+4\pi/3$
6465 0480 sinh : $(\sqrt{2}\pi)^{-1/3}$
6465 5812 K : 5/12
6465 6175 cr : $1+2\sqrt{3}$
6465 6654 rt : (3,6,-4,3)
6465 6724 rt : (7,1,7,6)
6465 6779 Ei : $4\sqrt[3]{2}/9$
6465 7280 Li_2 : 6/11
6466 0182 $t\pi$: $2\sqrt{2}-2\sqrt{3}/3$
6466 1342 rt : (4,3,8,5)
6466 2716 ln : 11/21
6466 2946 id : $\sqrt{2}/3-\sqrt{5}/2$
6466 3125 10^x : $\sqrt{2}/3+2\sqrt{5}/3$
6466 4449 rt : (7,7,-9,1)
6466 7634 $s\pi$: $4\sqrt{2}-\sqrt{3}/4$
6466 7748 Ei : 14/25
6466 7950 at : $(\sqrt[3]{5}/3)^{1/2}$
6466 8816 rt : (8,2,0,-3)
6466 9897 ℓ_{10} : $4+\sqrt{3}/4$
6467 0906 rt : (6,5,-5,5)
6467 1167 rt : (9,3,9,7)
6467 6274 rt : (3,7,-4,-4)
6467 9003 rt : (1,5,9,4)
6468 1334 tanh : $4\sqrt{3}/9$
6468 3509 rt : (5,0,1,2)
6468 4612 rt : (5,9,6,-9)
6468 5094 rt : (9,5,-7,0)
6468 5824 10^x : $2/3+\sqrt{2}/2$
6468 6325 rt : (5,9,-4,-5)
6468 8398 rt : (4,5,-8,2)
6469 0194 ζ : 3/22
6469 3889 rt : (6,6,-7,-1)
6469 4360 rt : (8,5,-8,9)
6469 4921 rt : (2,7,7,-8)
6469 7114 ζ : $3(1/\pi)/7$
6469 8617 sinh : 14/23
6469 9518 id : $\sqrt{2}/2-3\sqrt{5}/2$
6470 0827 $s\pi$: $\sqrt[3]{11}$
6470 0949 sq : $1/3-3\sqrt{3}$
6470 2276 rt : (3,7,-6,-6)
6470 2453 tanh : $10\ln 2/9$
6470 2658 e^x : $3(e+\pi)/5$
6470 2716 sq : $3e/2+\pi/4$
6470 2902 rt : (4,5,9,-9)
6470 2958 ln : $\pi/6$
6470 3670 rt : (7,2,3,3)
6470 3796 rt : (1,6,-4,2)
6470 5392 Γ : $8\sqrt[3]{5}/3$
6470 5882 id : 11/17
6470 8792 2^x : $3+e$
6470 9351 $\exp(4)-s\pi(2/5)$
6470 9778 J_0 : $9\ln 2/5$
6471 0954 $t\pi$: $10e$
6471 3052 2^x : $4\sqrt[3]{2}/7$
6471 3583 rt : (1,8,5,-9)
6471 4185 sr : $2e+\pi/2$
6471 4464 rt : (3,9,-9,-7)
6471 5712 rt : (5,8,2,-6)
6471 6855 rt : (9,4,5,4)
6471 7301 $t\pi$: $3e/4-3\pi/2$
6471 7786 10^x : $8\zeta(3)/7$
6471 8203 2^x : 18/25
6471 8751 ζ : $\ln(\sqrt{2}\pi)$
6471 9509 sr : $\sqrt{2}+3\sqrt{3}/4$
6472 0111 sq : $1/3-4\pi$
6472 0901 rt : (6,3,-6,1)
6472 1082 $s\pi$: $3\sqrt{2}+4\sqrt{5}/3$

6472 1722 rt : (7,6,4,-7)
6472 2956 erf : $\exp(-\sqrt[3]{2}/3)$
6472 3859 $t\pi$: $4\sqrt{3}-\sqrt{5}/3$
6472 5293 sinh : $7\pi/5$
6472 6461 ℓ_{10} : $1/4+4\pi/3$
6472 6702 rt : (9,4,6,-8)
6472 6927 rt : (7,7,-2,9)
6472 7820 $\sqrt{3}\times s\pi(2/5)$
6472 9991 at : $3\sqrt[3]{2}/5$
6473 1359 rt : (6,7,-9,-7)
6473 1928 at : $4/3-\sqrt{3}/3$
6473 2658 e^x : $4/3+3\sqrt{2}/2$
6473 4645 at : $3\sqrt{3}/2-3\sqrt{5}/2$
6473 6354 rt : (8,9,-3,-4)
6473 6882 Γ : $7\ln 2/9$
6473 8200 $t\pi$: $\sqrt[3]{6}$
6473 9180 Ei : $\sqrt[3]{14}$
6473 9363 $s\pi$: P_{31}
6473 9571 10^x : $4\sqrt{2}/3+2\sqrt{5}/3$
6474 0249 cu : $1/4-3\sqrt{2}$
6474 2665 rt : (4,0,-6,8)
6474 3696 sq : $2e+4\pi/3$
6474 5530 10^x : $4-4\pi/3$
6474 5914 cr : $4+\sqrt{2}/3$
6474 6108 ln : $4/3+\sqrt{3}/3$
6474 7101 10^x : $5\sqrt{5}/3$
6474 7378 rt : (4,4,5,-6)
6474 7886 rt : (2,5,-7,5)
6474 8841 $t\pi$: $e/4-2\pi/3$
6475 0493 rt : (6,5,-7,-5)
6475 0894 sr : 19/7
6475 1032 rt : (6,2,7,-7)
6475 1225 ℓ_{10} : $9\pi^2/2$
6475 2292 rt : (9,8,-7,7)
6475 2316 $t\pi$: $\sqrt[3]{7/3}$
6475 3948 rt : (8,0,9,8)
6475 4376 id : $3\sqrt{2}/4-3\sqrt{5}$
6475 4897 sr : $\sqrt[3]{20}$
6475 6166 10^x : $4\sqrt{3}/3-\sqrt{5}/3$
6475 6261 sq : $6(1/\pi)$
6475 7093 rt : (7,8,-4,3)
6475 7259 sq : $\ln(\sqrt{5})$
6475 7702 rt : (4,3,-9,-6)
6476 0301 sq : $1/3+\sqrt{2}/3$
6476 2819 2^x : $1/4+2\sqrt{3}/3$
6476 4406 Ei : 10/11
6476 6487 ℓ_{10} : $\sqrt{2}\pi$
6477 0019 rt : (3,9,-4,-2)
6477 1442 rt : (5,8,-1,-1)
6477 2506 rt : (7,5,0,-4)
6477 2975 ln : $4e/3+\pi/2$
6477 3332 rt : (9,3,2,-5)
6477 6356 rt : (1,7,-6,-8)
6477 7328 rt : (2,6,-5,-1)
6477 7592 e^x : 22/17
6477 8590 cr : $e/10$
6477 9363 rt : (9,9,-9,1)
6477 9432 cu : $\sqrt{2}-4\sqrt{3}$
6477 9887 rt : (8,8,8,4)
6478 0169 2^x : $4\sqrt{3}-\sqrt{5}/3$
6478 0791 id : $\exp(3\pi)$
6478 3654 rt : (8,7,6,-9)
6478 3966 rt : (6,9,4,9)
6478 4335 rt : (4,1,-8,2)
6478 5998 θ_3 : $\pi/10$
6478 9152 cr : $4/3+\pi$
6478 9282 2^x : $4+2\sqrt{2}$
6479 0599 K : $\sqrt[3]{2}/3$
6479 0691 $2\div s\pi(\pi)$
6479 1539 $\ln(2/5)\times s\pi(1/4)$
6479 1843 ln : $3\sqrt{3}$
6479 2304 Ψ : 23/24
6479 2380 10^x : P_{26}
6479 2937 rt : (4,7,-8,4)
6479 5028 as : $1/4+\sqrt{2}/4$
6479 5508 10^x : $\exp(\sqrt[3]{5}/3)$
6479 5541 Ψ : $3\sqrt{5}/7$
6479 7509 rt : (3,0,8,6)
6479 7642 $t\pi$: $1/4-\sqrt{3}/4$
6479 7937 θ_3 : $2\sqrt{5}/7$
6479 9348 ℓ_2 : $2-\sqrt{3}/4$
6479 9893 rt : (8,8,-7,-1)
6480 0000 cu : 11/5
6480 0246 2^x : $2\sqrt{2}+2\sqrt{5}$
6480 0278 10^x : $2e+\pi/4$
6480 0956 rt : (1,2,6,-5)
6480 1283 cr : P_{30}
6480 2187 rt : (7,9,-6,-3)
6480 5376 id : $(\sqrt[3]{2}/3)^{1/2}$
6480 6079 rt : (2,8,-8,-8)
6480 9063 id : $1/3-4\sqrt{5}/3$
6481 0772 sq : $4e\pi/3$
6481 0840 θ_3 : $2\sqrt{2}/9$
6481 3044 Γ : $3\sqrt{5}$
6481 4497 rt : (3,6,-2,-3)
6481 6599 rt : (7,3,9,-9)
6481 7127 K : $\exp(-\sqrt{3}/2)$
6481 8795 sq : $2-4\pi$
6481 8854 rt : (2,7,-3,-7)
6482 0175 rt : (5,5,7,-8)
6482 1906 id : $3e/2+\pi/2$
6482 2184 10^x : $7\sqrt{3}/2$
6482 2645 rt : (8,1,5,5)
6482 3858 sq : $2e/3-4\pi/3$
6482 4524 rt : (3,7,5,-7)
6482 7070 rt : (4,8,-8,-3)
6482 8272 $\sqrt{5}-s\pi(1/5)$
6482 8522 id : $8\sqrt[3]{3}/7$
6482 8643 sq : $1/4-2\sqrt{2}$
6482 9593 rt : (6,1,3,-4)
6482 9925 rt : (1,9,3,-6)
6483 0891 rt : (5,8,0,-2)
6483 1663 ζ : $e/5$
6483 3502 rt : (5,3,-5,6)
6483 4153 $\sqrt{3}-\ln(2/5)$
6483 4916 K : 8/19
6483 6082 $t\pi$: $10(1/\pi)$
6483 6234 rt : (3,9,-6,4)
6483 9238 rt : (8,6,2,-6)
6484 1498 rt : (1,9,2,-9)
6484 3735 e^x : $2/3+2\sqrt{5}/3$
6484 3773 ℓ_{10} : $3\sqrt{3}-\sqrt{5}/3$
6484 4421 sq : $3/4+3\pi/4$
6484 5025 rt : (9,5,1,1)
6484 5615 rt : (6,8,0,-5)
6484 5642 cr : $4/3-3\sqrt{2}/4$
6484 6674 2^x : $\sqrt[3]{3}/2$
6484 8055 $t\pi$: $5\ln 2/9$
6484 9931 cr : 3/11
6485 1251 tanh : 17/22
6485 1386 10^x : $3\sqrt{2}/2-4\sqrt{3}/3$
6485 3527 rt : (2,7,3,-4)
6485 5225 sr : $\exp(-\sqrt{3}/2)$
6485 5486 $t\pi$: $4\sqrt{3}+\sqrt{5}/3$
6485 6065 2^x : $\sqrt{3}/2-2\sqrt{5}/3$
6485 7779 ζ : $4\sqrt{2}/7$
6485 7855 Ψ : $e/7$
6485 8631 cr : $6(1/\pi)/7$
6485 8912 as : $2e/9$
6485 9170 2^x : $3\zeta(3)/5$
6486 1231 rt : (9,2,-2,-2)
6486 1800 cu : $6e/7$
6486 3671 ln : $\sqrt[3]{7}$
6486 4350 Ei : 5π
6486 4916 rt : (8,6,-7,8)
6486 5872 $\ln(3)\times\exp(e)$
6486 6016 $s\pi$: $5\pi^2/6$
6486 6506 rt : (9,9,-5,-3)
6486 8806 K : $(4/3)^{-3}$
6486 8876 cr : $3/2+4\sqrt{5}/3$
6487 0039 2^x : $3-4e/3$
6487 2127 id : $(e)^{1/2}$
6487 3050 rt : (4,8,7,-9)
6487 6875 rt : (7,4,-4,-1)
6487 7073 Ψ : 4/7
6487 7098 sinh : $\sqrt{5}-\sqrt{6}+\sqrt{7}$
6487 8641 2^x : $1/4+\sqrt{2}/3$
6488 0728 ln : $9\pi/2$
6488 1017 rt : (9,7,4,-8)
6488 1814 rt : (2,4,-9,-5)
6488 2021 $\ln(4/5)\times\exp(2)$
6488 2279 10^x : $3/4+\sqrt{2}/2$
6488 2388 rt : (9,0,7,7)
6488 3836 $s\pi$: $\sqrt{3/2}$
6488 3898 rt : (2,3,8,-7)
6488 6191 sr : $2\sqrt{3}-\sqrt{5}/3$
6488 6530 10^x : $4+3\sqrt{5}$
6488 6545 rt : (5,6,-9,-7)
6488 6793 $\exp(2/5)-\exp(\pi)$
6488 7678 Ψ : $2e\pi/3$
6488 8568 sr : 8/19
6488 9677 rt : (7,9,2,-7)
6489 4415 rt : (7,2,5,-6)
6489 6712 cr : $4\sqrt{2}/3+3\sqrt{3}/2$
6489 7948 sq : $2\sqrt{2}+\sqrt{3}/2$
6489 8514 Ψ : $\exp(-\pi)$
6489 9959 rt : (5,9,-3,-7)
6490 3766 ln : $3\sqrt{3}+4\sqrt{5}$
6490 4720 rt : (7,5,8,5)
6490 9639 rt : (5,4,3,-5)
6491 0636 rt : (7,8,-7,-6)
6491 1064 sq : $2\sqrt{2}+\sqrt{5}$
6491 1174 rt : (8,7,-9,2)
6491 1745 Ψ : $1/\ln(\ln 3/3)$
6491 4800 rt : (3,7,-1,4)
6491 5868 ln : $10\sqrt{2}$
6491 6026 sr : $3/4+4\pi$
6491 6485 sq : $1/4-3e/2$
6491 7854 Ψ : $\sqrt[3]{3}/5$
6491 9675 sq : $\sqrt{2}-\sqrt{5}+\sqrt{6}$
6491 9691 cu : $5\pi^2$
6492 1894 rt : (2,9,5,0)
6492 2708 $\sqrt{5}+\exp(5)$
6492 4468 $1/3\times\exp(2/3)$
6492 8872 Γ : 7/13
6492 9530 rt : (9,9,-6,6)
6493 0686 ln : $1/2+\sqrt{2}$
6493 1041 rt : (8,3,7,-8)
6493 3297 rt : (5,3,6,4)
6493 5049 rt : (8,5,-2,-3)
6493 6223 $\exp(\sqrt{5})-s\pi(1/4)$
6493 7437 rt : (3,8,-8,1)
6494 2392 cu : $8\sqrt{3}/9$
6494 3591 rt : (1,6,8,8)
6494 6323 rt : (6,5,5,-7)
6494 7263 10^x : $(e/2)^{1/2}$
6494 7439 $t\pi$: $((e+\pi)/3)^3$
6494 7869 ℓ_{10} : P_{31}
6494 7982 10^x : $\sqrt[3]{17/3}$
6494 8958 Γ : $\sqrt{5}/9$
6494 9350 cr : $10\pi/7$
6495 0015 rt : (7,8,-1,8)
6495 1905 id : $3\sqrt{3}/8$
6495 3647 rt : (4,9,-2,-4)
6495 3750 $1/4\times\exp(4)$
6495 6626 rt : (1,8,-1,-3)
6495 6737 e^x : $\sqrt{2}+\sqrt{3}-\sqrt{7}$
6495 8085 cr : $\sqrt{2}/2-\sqrt{3}/4$
6495 8652 rt : (6,7,-4,-2)
6495 9438 tanh : $(5/3)^{-1/2}$
6496 0080 ζ : $2\sqrt{5}/3$
6496 1441 10^x : $1/2+2\sqrt{3}$
6496 2747 10^x : $3/4+\sqrt{2}/3$
6496 3729 rt : (6,9,-9,-8)
6496 4149 ln : $\sqrt{11/3}$
6496 4261 J_0 : $1/\ln(\sqrt{5})$
6496 4475 rt : (1,8,-6,-7)
6496 4807 10^x : 5/23
6496 5044 rt : (4,7,3,-6)
6496 5493 rt : (9,6,0,-5)
6496 7339 ζ : $7\sqrt[3]{5}/6$
6496 7820 K : $3\sqrt{2}/10$
6496 8620 id : $5(e+\pi)/2$
6496 8837 rt : (7,6,7,-9)
6496 9284 rt : (5,1,-3,-1)
6497 0784 θ_3 : $\sqrt[3]{2}/4$
6497 1499 $s\pi$: $4\sqrt{2}+\sqrt{5}/2$
6497 2118 2^x : 13/18
6497 3407 ℓ_{10} : $1+2\sqrt{3}$
6497 4028 K : $4(1/\pi)/3$

6497 6572 rt : (8,2,1,2)
6497 6615 ζ : $\ln 2/5$
6497 6628 at : $(\sqrt{3})^{-1/2}$
6497 6795 $\pi - \exp(2/5)$
6497 6942 rt : (4,8,-5,5)
6497 7759 $\exp(1/3) + s\pi(\sqrt{3})$
6497 8816 Ψ : $7\sqrt[3]{5}/5$
6497 8850 rt : (6,2,8,6)
6498 2697 cr : $3 + 2\sqrt{5}/3$
6498 2991 sq : $\sqrt{2}/2 + \sqrt{3}/3$
6498 3205 rt : (2,6,-1,-9)
6498 5117 J_0 : $5\sqrt{5}/9$
6498 5622 rt : (5,8,5,-8)
6498 6247 rt : (7,8,-2,-4)
6498 6367 at : $4\sqrt[3]{5}/9$
6498 6498 sinh : 11/18
6498 6813 rt : (7,4,-7,9)
6498 7044 at : 19/25
6498 7308 rt : (9,8,-9,0)
6498 8512 rt : (2,9,1,-5)
6498 9083 10^x : $5\sqrt[3]{5}/7$
6499 0037 Ψ : $5(e + \pi)/2$
6499 0713 rt : (4,9,-6,-1)
6499 1582 id : $7\sqrt{2}/6$
6499 4448 cu : $2 + 3\sqrt{3}$
6499 5175 as : $(\ln 3/3)^{1/2}$
6499 5565 2^x : $1 + \pi$
6500 0000 id : 13/20
6500 2289 rt : (8,9,0,-6)
6500 5680 rt : (3,8,-1,-3)
6500 6313 cr : $1/4 + 3\sqrt{2}$
6500 6675 rt : (9,1,3,4)
6500 7571 2^x : $3/2 - 3\sqrt{2}/2$
6500 8994 rt : (5,9,-9,6)
6501 0524 K : $\exp(-\sqrt[3]{5}/2)$
6501 0610 K : $1/\sqrt[3]{13}$
6501 1270 rt : (2,7,0,4)
6501 1516 sr : $1 - \sqrt{3}/3$
6501 2223 rt : (5,5,-9,-6)
6501 3548 rt : (4,0,6,5)
6501 6411 rt : (4,1,-5,7)
6501 6754 tπ : $\sqrt{2/3}$
6501 7439 Ψ : $10\sqrt[3]{5}/3$
6501 8659 2^x : $1/2 + e/3$
6502 0649 rt : (1,9,-7,-7)
6502 3526 rt : (1,1,-9,-6)
6502 3856 rt : (6,7,-6,-2)
6502 7972 rt : (5,1,8,-7)
6502 8153 id : $\sqrt{2} + \sqrt{5}$
6503 0409 2^x : $\sqrt{3} + \sqrt{5}$
6503 1147 rt : (7,9,-3,2)
6503 4044 rt : (6,0,-1,1)
6503 4514 ln : $8(e + \pi)/9$
6503 4795 2^x : $3/4 + \sqrt{5}/2$
6504 1103 sπ : $(5/3)^{-1/2}$
6504 1599 $\exp(2/3) - \exp(4)$
6504 2880 tπ : $4\sqrt{2}/3 - \sqrt{5}/4$
6504 3239 cu : $7\pi^2/6$
6504 3670 sr : $\sqrt{5}\pi$
6504 3735 cu : $e/2 - 4\pi$
6504 3877 cu : $5\ln 2/4$
6504 4331 rt : (4,9,-5,-2)
6504 4341 ζ : $1/\ln(2\pi)$
6504 4396 ℓ_{10} : $4 + \sqrt{2}/3$
6504 5843 rt : (7,1,1,-3)
6504 7539 e^x : $3\sqrt{2}/4 - 2\sqrt{5}/3$
6504 7687 Ei : $5\sqrt{5}/4$
6505 0797 rt : (8,2,3,-5)
6505 1228 ζ : $3e/7$
6505 1499 ℓ_{10} : $2\sqrt{5}$
6505 2315 e^x : $4 - \sqrt{3}/3$
6505 3522 rt : (9,3,5,-7)
6505 4068 cu : $1/\ln(\ln 3/2)$
6505 4313 at : $4/3 - 2\pi/3$
6505 5618 rt : (7,5,-9,3)
6505 6538 rt : (8,7,-6,7)
6505 6827 ln : $6\sqrt{5}/7$
6505 8498 ln : $(3\pi/2)^3$
6505 8756 ln : 12/23
6506 2919 rt : (4,4,8,-8)
6506 3861 cu : 13/11
6506 6216 Ei : $\sqrt{2} + \sqrt{3} - \sqrt{5}$
6506 6758 tπ : P_{63}
6506 6805 e^x : $8\pi^2/5$
6506 7122 ζ : $(\sqrt{2}\pi/2)^{1/2}$
6507 1182 sq : $4/3 + \sqrt{3}/3$
6507 3070 tπ : $1/4 + 4\pi$
6507 5169 $\pi \times \exp(3/4)$
6507 7910 Ei : $1/\ln(3)$
6507 8081 tanh : $2e/7$
6507 8407 rt : (8,6,5,-8)
6507 8585 ℓ_{10} : $4/3 + \pi$
6507 8857 e^x : $1/4 - e/4$
6507 9296 rt : (2,4,-9,-7)
6508 2255 rt : (7,5,3,-6)
6508 3315 $\pi \div s\pi(\sqrt{5})$
6508 4220 sr : $\sqrt{18}\pi$
6508 7377 rt : (9,1,-6,1)
6508 8646 rt : (6,4,1,-4)
6508 9055 rt : (3,7,8,-9)
6508 9852 rt : (3,3,-7,-5)
6509 3215 cu : $4\sqrt{3} + \sqrt{5}$
6509 4670 10^x : $\sqrt{2} + \sqrt{3}/2$
6509 6296 cu : 13/15
6509 6362 id : $\sqrt[3]{9/2}$
6509 8164 rt : (7,9,5,-9)
6510 0153 Ei : $\exp(-\sqrt{3}/3)$
6510 1204 sq : $8e/7$
6510 1297 id : $2\sqrt{2}/3 + 3\sqrt{5}$
6510 1346 rt : (5,3,-1,-2)
6510 3010 K : $\sqrt[3]{5}/4$
6510 3828 ln : $1/2 + 3\pi/2$
6510 5333 cr : $e/2 + \pi$
6510 5822 rt : (6,8,3,-7)
6510 7672 at : 16/21
6510 7973 sπ : $2/3 + \sqrt{5}/4$
6510 9340 rt : (2,8,2,-2)
6510 9716 id : $(e + \pi)/9$
6510 9968 Ψ : 22/23
6511 0680 $s\pi(\sqrt{2}) \times s\pi(\sqrt{5})$
6511 1051 cu : $2\sqrt{3}/3 - 4\sqrt{5}$
6511 3745 as : $3\sqrt{2}/7$
6511 3936 rt : (9,5,-4,-2)
6511 4231 rt : (6,6,-1,9)
6511 4519 rt : (3,7,-9,-8)
6511 6549 id : $2e - \pi/4$
6511 7729 rt : (5,8,2,4)
6511 9320 rt : (9,8,6,3)
6512 0517 rt : (1,9,6,-8)
6512 1106 e^x : $2\sqrt{2} + \sqrt{5}/3$
6512 2391 tπ : $2\sqrt{2} + 3\sqrt{5}$
6512 2398 rt : (9,9,-2,-5)
6512 2696 cr : $4\sqrt{2} - 2\sqrt{3}/3$
6512 4137 rt : (5,4,-4,5)
6512 8549 cu : $3e + 2\pi/3$
6512 9173 rt : (9,4,8,6)
6513 5556 sr : $3\sqrt{2}/10$
6513 7869 id : $\sqrt{2}/2 + 4\sqrt{5}$
6513 8781 rt : (8,8,-4,-3)
6514 0326 ln : $\sqrt{2} - \sqrt{3} + \sqrt{5}$
6514 1603 ℓ_{10} : $3/2 + 4\sqrt{5}/3$
6514 2935 tanh : 7/9
6514 3263 e^x : $3\sqrt{2}/4 - \sqrt{5}/4$
6514 4172 ℓ_{10} : $\exp(3/2)$
6514 7001 sr : $4(1/\pi)/3$
6514 7492 tπ : $1/\sqrt[3]{14}$
6514 8448 K : 3/7
6514 9533 ζ : $\sqrt[3]{2}/9$
6514 9612 ℓ_2 : π
6515 1226 rt : (8,1,8,7)
6515 1825 $\ln(3/5) - \exp(\pi)$
6515 3077 θ_3 : 6/19
6515 5822 2^x : $1/2 - \sqrt{5}/2$
6515 6985 2^x : $\sqrt{15}$
6516 0161 sq : $4 + \sqrt{3}/4$
6516 0504 rt : (8,5,6,4)
6516 1575 K : $(2e)^{-1/2}$
6516 2632 rt : (3,9,-1,-4)
6516 3329 cr : $2\sqrt{2} + 3\sqrt{5}/4$
6516 3599 ℓ_{10} : $4\sqrt{2}/3 + 3\sqrt{3}/2$
6516 5586 as : $7\ln 2/8$
6516 6164 sq : $1/2 - 3\pi$
6516 6408 id : $\exp(2e)$
6516 6852 rt : (5,9,-7,-7)
6516 7318 rt : (7,2,8,-8)
6516 7608 rt : (7,7,-6,-1)
6516 8966 as : $(e)^{-1/2}$
6517 0397 $\exp(1/5) - s\pi(\pi)$
6517 1252 sq : $3\sqrt{3}/4 + 3\sqrt{5}/2$
6517 1359 tanh : $(\sqrt{2}/3)^{1/3}$
6517 2944 Ei : $1/\sqrt{19}$
6517 4011 rt : (1,7,-2,-4)
6517 4127 10^x : 9/16
6517 6344 sr : $6(e + \pi)/5$
6517 7418 2^x : $3\sqrt{3} - 2\sqrt{5}$
6517 7433 tπ : $\sqrt{2} - 3\sqrt{3}/2$
6517 9579 rt : (6,5,8,-9)
6517 9992 tπ : P_{47}
6518 1033 rt : (4,7,9,4)
6518 4033 rt : (7,2,6,5)
6518 4906 rt : (4,9,-2,6)
6518 6065 rt : (6,8,-8,-8)
6519 1564 rt : (2,8,-8,-9)
6519 1577 sπ : $e/4 + 2\pi/3$
6519 4714 rt : (7,5,-6,8)
6519 4913 rt : (9,6,3,-7)
6519 5533 2^x : $2e - 3\pi/2$
6519 9258 rt : (6,1,6,-6)
6520 0309 ℓ_2 : $10\sqrt{2}/9$
6520 0355 Ψ : $\exp(-\sqrt{2}\pi/3)$
6520 1532 sr : $3/4 + 2\pi$
6520 1681 ln : $7\sqrt{5}/3$
6520 2893 rt : (8,9,3,-8)
6520 3941 cr : $2\sqrt{3}/3 + 3\sqrt{5}/2$
6520 5183 ℓ_{10} : $10\pi/7$
6520 6203 rt : (5,5,-5,-4)
6520 7139 cr : $2\ln 2/5$
6520 7669 ℓ_2 : 7/11
6520 8337 rt : (2,8,-3,-2)
6520 9345 rt : (7,9,0,7)
6520 9917 rt : (5,4,6,-7)
6521 0604 rt : (7,3,-8,2)
6521 1991 rt : (3,4,9,5)
6521 2376 rt : (2,7,-9,8)
6521 4132 sr : $\exp(-\sqrt[3]{5}/2)$
6521 4290 sr : $1/\sqrt[3]{13}$
6521 4671 rt : (6,3,-9,4)
6521 5060 Li_2 : $\ln(\sqrt{3})$
6521 5715 tπ : $3\sqrt[3]{3}$
6521 7391 id : 15/23
6521 7797 rt : (4,7,6,-8)
6522 1747 rt : (9,2,1,-4)
6522 3367 id : $\exp(\sqrt{3})$
6522 3859 rt : (3,8,2,5)
6522 4048 rt : (8,9,4,1)
6522 4843 sq : $1/2 + \pi/4$
6522 6878 10^x : $3\sqrt{2}/4 + 3\sqrt{5}/4$
6522 7015 rt : (2,1,9,6)
6522 7957 rt : (6,8,-5,-5)
6522 8978 rt : (8,5,1,-5)
6523 0074 2^x : $3\sqrt{2}/4 - 3\sqrt{5}/4$
6523 0225 J_0 : $7\sqrt{2}/8$
6523 1520 ℓ_{10} : $3 + 2\sqrt{5}/3$
6523 1666 rt : (6,6,-8,1)
6523 4337 rt : (7,8,1,-6)
6523 5813 at : $3 - \sqrt{5}$
6523 5888 rt : (1,2,9,-7)
6523 8466 rt : (6,7,-3,3)
6524 1746 rt : (5,9,-8,0)
6524 1827 $\pi - \ln(3/5)$
6524 3306 J_0 : $5\sqrt{3}/7$
6524 4616 sq : $2e/3 - 4\pi$
6524 5903 rt : (1,8,-9,-9)
6524 6361 Ei : $6\sqrt{5}/7$
6524 6598 2^x : $1/4 - \sqrt{3}/2$
6524 7584 id : $7\sqrt{5}$
6524 7794 cr : 5/18
6524 7961 rt : (7,6,4,2)
6524 8832 id : $(\ln 3/3)^{-1/2}$
6525 0168 ℓ_{10} : $1/4 + 3\sqrt{2}$

6525 1879 $\theta_3 : 1/\sqrt{10}$
6525 3839 $1/3 \div \ln(3/5)$
6525 5588 rt : (6,3,4,3)
6525 6545 sinh : $(\sqrt[3]{3}/3)^{-1/3}$
6525 7091 $\ln(2/3) \times \ln(5)$
6525 7703 $\lambda : \exp(-13\pi/15)$
6525 8648 tanh : $9\ln 2/8$
6525 9149 id : $\sqrt{2}/4 + 3\sqrt{3}/4$
6525 9340 rt : (5,0,4,4)
6525 9475 $Ei : 23/12$
6525 9922 $10^x : 2\sqrt{2}/3 + \sqrt{5}$
6526 0582 rt : (9,3,8,-9)
6526 0928 $\ell_2 : (e+\pi)^3$
6526 1573 rt : (4,3,4,-5)
6526 3053 rt : (3,6,4,-6)
6526 3221 $10^x : e/4 + \pi$
6526 3641 $4 \times \exp(5)$
6526 4107 rt : (2,9,4,-7)
6526 5790 $10^x : 1/4 + 4\sqrt{2}/3$
6526 9027 $10^x : 9\zeta(3)/2$
6526 9174 $\theta_3 : 4\sqrt[3]{5}/7$
6527 0364 rt : (3,9,0,-3)
6527 0446 at : $\sqrt{2}/4 - \sqrt{5}/2$
6527 3316 rt : (8,8,-5,6)
6527 9081 $10^x : 1/\sqrt{21}$
6527 9263 tπ : $4\sqrt{2} + 4\sqrt{3}$
6527 9317 rt : (2,8,-5,5)
6528 1468 $10^x : \exp(\pi)$
6528 2032 rt : (5,5,-6,-1)
6528 3979 rt : (6,7,-1,-4)
6528 4663 at : 13/17
6528 4886 $J_0 : (\sqrt[3]{4}/3)^{-1/3}$
6528 5046 rt : (2,8,5,9)
6528 5837 at : $(\sqrt[3]{5})^{-1/2}$
6528 6236 sinh : $(\ln 2/3)^{1/3}$
6528 6444 $2^x : 2\pi^2/3$
6528 6706 rt : (7,4,-1,-3)
6528 7921 sq : $\sqrt{18\pi}$
6528 9165 sr : $1 + \sqrt{3}$
6528 9869 $\Gamma : (e+\pi)$
6529 0116 $\Psi : 12/5$
6529 0422 tπ : $2\sqrt{2}/3 + 2\sqrt{5}$
6529 0650 ln : $3/4 + 2\sqrt{5}$
6529 0799 rt : (8,1,-1,-2)
6529 2285 $J_0 : 6\sqrt[3]{3}/7$
6529 7624 rt : (9,2,-1,1)
6529 7970 $\Gamma : \exp(-2\pi/3)$
6529 8530 cu : $\ln(\sqrt[3]{2}/3)$
6530 0238 rt : (6,8,6,-9)
6530 1774 sπ : $10\sqrt{3}/3$
6530 2609 at : $2e/3 - \pi/3$
6530 3907 rt : (7,5,6,-8)
6530 4525 rt : (9,8,-6,-2)
6530 5921 rt : (7,6,-8,2)
6530 5946 cr : $7(1/\pi)/8$
6530 6122 sq : 9/7
6530 7561 rt : (2,7,-7,-7)
6530 8363 $Ei : 21/13$
6530 8602 rt : (8,2,6,-7)
6530 9723 tπ : $3\sqrt{2}/2 - 3\sqrt{5}/4$
6531 1755 cr : $3/4 - \sqrt{2}/3$
6531 3137 $\Psi : 5(1/\pi)/7$
6531 3651 rt : (9,9,1,-7)
6531 4007 id : $3\sqrt{3}/4 + 3\sqrt{5}/2$
6531 4246 ln : $3/4 - e/4$
6531 4826 rt : (9,1,6,6)
6531 5763 $Li_2 : 11/20$
6531 6528 rt : (1,7,4,-8)
6531 8643 $\theta_3 : 16/25$
6532 0812 $Ei : 10\sqrt{3}$
6532 1251 $\ell_{10} : 2/9$
6532 2227 $\Gamma : 17/3$
6532 3080 $K : 3\sqrt[3]{3}/10$
6532 3305 ln : $4\sqrt{2} - \sqrt{3}/4$
6532 8330 $\ell_{10} : e/2 + \pi$
6532 8363 tπ : $(\sqrt{5}/3)^{-3}$
6532 9909 $\ell_2 : 2/3 + e/3$
6533 1193 rt : (1,9,3,0)
6533 1758 $\zeta : \sqrt{2}/10$
6533 2647 sq : $5(e+\pi)/4$
6533 6037 rt : (3,4,-9,4)
6533 6123 tπ : $1/2 + 4\sqrt{5}$
6533 7570 $K : \sqrt{3}/4$
6533 8103 $\zeta : 4(1/\pi)/9$
6533 8938 $J_0 : 21/17$
6533 9466 rt : (2,6,9,-9)
6534 2031 $\ell_{10} : 4\sqrt{2} - 2\sqrt{3}/3$
6534 2640 $1 - \ln(\sqrt{2})$
6534 4476 $\ln(4/5) + s\pi(\pi)$
6534 5659 rt : (6,3,-6,9)
6534 6340 $K : 5\ln 2/8$
6534 9742 id : $6\sqrt[3]{3}$
6535 0344 rt : (3,3,9,-8)
6535 1452 $\ln(5) \times \exp(1/2)$
6535 2576 sinh : $7\sqrt{3}/2$
6535 4313 rt : (4,9,-3,-3)
6535 5820 rt : (5,8,5,9)
6535 8950 sr : $1/3 + 3\sqrt{5}$
6535 9516 rt : (1,2,7,4)
6536 1945 rt : (2,9,-1,2)
6536 1989 tanh : $(2\pi/3)^{-1/3}$
6536 2404 $Li_2 : 1/\sqrt[3]{6}$
6536 3551 sr : $4/3 - e/3$
6536 3990 $\ell_2 : \sqrt{2} + \sqrt{3}$
6536 5012 cr : P_{72}
6536 5285 rt : (4,0,9,7)
6536 5572 cr : P_{69}
6536 6476 cr : $1/3 + 4\pi/3$
6536 9150 rt : (9,6,6,-9)
6537 1334 rt : (1,9,-6,3)
6537 1919 rt : (8,8,-1,-5)
6537 4001 rt : (5,9,0,-2)
6537 4088 $\ell_{10} : 2\sqrt{2} + 3\sqrt{5}/4$
6537 6287 id : $4/3 - e/4$
6538 0949 $\zeta : \exp(-(e+\pi)/3)$
6538 1795 $10^x : 5\zeta(3)/9$
6538 3024 sr : $\sqrt[3]{5}/4$
6538 5249 $2^x : 8\sqrt{2}/3$
6538 7091 rt : (5,3,9,6)
6538 7403 rt : (7,1,4,-5)
6538 7613 rt : (9,5,-1,-4)
6538 8075 at : $1/3 + \sqrt{3}/4$
6538 8616 ln : $4\sqrt[3]{3}/3$
6538 8684 rt : (2,4,-5,1)
6538 9005 ln : P_{52}
6538 9875 $\Psi : 3(1/\pi)$
6539 0620 as : $(\sqrt{2}\pi)^{-1/3}$
6539 0756 rt : (5,1,-9,5)
6539 1220 $10^x : 5\sqrt{2}/8$
6539 1424 at : $(\sqrt{2}\pi/2)^{-1/3}$
6539 2646 ln : 13/25
6539 4061 cu : $3\sqrt{2}/2 + \sqrt{3}/3$
6539 4993 sq : $7\ln 2/6$
6539 5369 rt : (4,9,-5,-6)
6539 8629 cr : $3\sqrt{2}/4 + 2\sqrt{3}$
6540 1557 rt : (8,9,-7,0)
6540 2872 sπ : $9\sqrt{3}/7$
6540 2937 rt : (6,7,0,8)
6540 3780 ln : $8\zeta(3)/5$
6540 4885 cu : $1/4 - \sqrt{5}/2$
6540 5733 rt : (6,3,-3,-1)
6540 6121 $\ell_{10} : 2\sqrt{3}/3 + 3\sqrt{5}/2$
6540 7421 sr : $8\sqrt[3]{5}/5$
6541 0639 sr : $1/2 + \sqrt{5}$
6541 1367 rt : (1,6,-3,9)
6541 3760 $K : 10/23$
6541 3920 rt : (8,2,4,4)
6541 7859 rt : (7,8,4,-8)
6541 8086 $2^x : 3 + 4\sqrt{3}/3$
6541 8817 tπ : $\exp(-\sqrt{5}/2)$
6541 9162 rt : (3,9,2,-2)
6541 9499 ln : $3\ln 2/4$
6541 9715 $J_0 : \pi^2/8$
6541 9959 sq : $(\sqrt[3]{4}/3)^{1/3}$
6542 0148 tanh : 18/23
6542 0557 cu : $2\sqrt{2} + 4\sqrt{5}$
6542 1326 cr : 7/25
6542 1813 rt : (5,4,9,-9)
6542 1897 rt : (6,6,9,5)
6542 2141 $erf : 2/3$
6542 2348 $\lambda : \exp(-e)$
6542 4915 rt : (1,5,7,-7)
6542 5434 $\Gamma : 1/\ln(\sqrt[3]{3}/2)$
6542 5647 sπ : $8\pi^2/9$
6542 6224 $1/2 + \exp(e)$
6542 6233 ln : $9\sqrt[3]{5}/8$
6542 6324 ln : $\sqrt{3}/4 + 2\sqrt{5}/3$
6542 7349 rt : (1,7,2,-2)
6542 8264 sq : $2/3 + 2\sqrt{5}/3$
6543 1204 rt : (7,6,-5,7)
6543 2036 rt : (3,6,-5,-5)
6543 5574 $e^x : (\pi)^3$
6543 6445 rt : (6,8,-5,-3)
6543 6541 sq : $1/3 - 3\pi$
6543 6563 rt : (8,5,4,-7)
6543 7260 rt : (2,5,7,3)
6543 9759 sinh : $3\pi/2$
6544 0955 rt : (9,4,-8,1)
6544 1556 as : 14/23
6544 3692 tπ : $\ln(\sqrt[3]{3}/2)$
6544 4005 rt : (6,1,9,-8)
6544 4306 $\ell_2 : \sqrt{2} + \sqrt{5} + \sqrt{7}$
6544 4489 $\Psi : \sqrt[3]{5}/3$
6544 5365 rt : (4,6,2,5)
6544 8534 $Ei : 9/16$
6545 0849 sq : $1/4 + \sqrt{5}/4$
6545 2890 cu : 18/13
6545 5944 rt : (9,5,4,3)
6545 6591 $e^x : 4\sqrt{2}/3 - 4\sqrt{3}/3$
6545 7526 $\Psi : 21/22$
6545 8391 rt : (2,2,7,-6)
6546 0302 sr : $3\sqrt{2}/4 + 3\sqrt{5}/4$
6546 0692 rt : (9,2,4,-6)
6546 1385 rt : (2,9,7,3)
6546 2258 rt : (9,9,4,-9)
6546 3016 tπ : $7(e+\pi)/3$
6546 4305 tanh : $(\sqrt[3]{3}/3)^{1/3}$
6546 5367 sr : 3/7
6546 6027 at : $4\sqrt{2}/3 - \sqrt{5}/2$
6546 6665 ln : $10\sqrt{3}/9$
6546 6714 $2^x : 2e + \pi/2$
6546 8149 cr : $5e/3$
6546 8166 sq : 8π
6547 0030 tπ : $3e/10$
6547 0053 id : $1/2 + 2\sqrt{3}/3$
6547 2461 $2^x : \sqrt{19}\pi$
6547 3810 rt : (7,2,9,7)
6547 5351 rt : (5,5,-3,4)
6547 7422 rt : (3,9,-8,-3)
6548 0783 rt : (3,5,-9,-3)
6548 1400 $3/5 \div \ln(2/5)$
6548 2756 rt : (5,3,2,-4)
6548 3079 $\Psi : 5\sqrt[3]{3}/3$
6548 3868 $10^x : (\pi/3)^{-1/3}$
6548 3908 $\ln(4) \div \exp(3/4)$
6548 4082 $\exp(2/3) + s\pi(1/4)$
6548 4548 id : $1/2 + 3e$
6548 6073 sπ : 5/22
6548 6257 rt : (7,9,9,4)
6548 6677 id : $9\pi/5$
6548 7504 $e^x : e - \pi$
6548 7545 sr : $\sqrt{15/2}$
6548 8332 rt : (3,6,7,-8)
6548 9078 sr : $(2e)^{-1/2}$
6549 1381 rt : (6,4,-8,3)
6549 2777 sπ : $2\sqrt{2} + 4\sqrt{5}$
6549 4793 rt : (8,2,9,-9)
6549 5244 sinh : $4\sqrt{5}/7$
6549 5885 $e^x : 3/2 + \sqrt{3}/2$
6549 6264 sq : $\sqrt{2}/3 + 4\sqrt{5}$
6549 6757 sinh : 8/13
6549 7493 rt : (3,8,-5,-1)
6549 7688 as : $3/4 - e/2$
6549 8169 sr : $10(e+\pi)$
6549 8460 $4/5 \div \exp(1/5)$
6549 9177 rt : (2,9,7,-4)
6550 0010 sinh : 23/18
6550 0756 $K : 1/\sqrt[3]{12}$
6550 1400 $\Psi : 5/22$
6550 1820 $\Psi : 2\zeta(3)$

6550 4005 rt : (6,7,2,-6)
6550 4753 $erf : 5\zeta(3)/9$
6550 6452 rt : (8,4,-3,-2)
6550 7789 $s\pi : 5(1/\pi)/7$
6550 8064 cu : $3\sqrt{3} - 4\sqrt{5}$
6551 0373 rt : (3,1,7,5)
6551 0448 $s\pi : 8e\pi/3$
6551 3010 sq : $3/2 - 4\sqrt{3}/3$
6551 3039 sr : $5\pi^2/7$
6551 3637 sr : $2 - \pi/2$
6551 4988 $\Gamma : \ln(\sqrt[3]{5})$
6551 5355 $\zeta : 1/7$
6551 5955 rt : (8,5,9,6)
6551 7639 $3/4 \div \ln(\pi)$
6551 8962 $s\pi : 4\ln 2$
6551 9391 rt : (9,8,-3,-4)
6552 0448 rt : (8,9,-4,5)
6552 0804 $s\pi : \exp(-\sqrt{2}\pi/3)$
6552 2014 rt : (1,7,-5,-6)
6552 3570 $\exp(3) + s\pi(\pi)$
6552 4124 rt : (4,3,7,-7)
6552 4485 rt : (9,8,-9,4)
6552 4578 $\ell_2 : -\sqrt{3} + \sqrt{5} + \sqrt{7}$
6552 5535 rt : (9,1,9,8)
6552 6142 $\ell_2 : 5\sqrt[3]{2}$
6552 6716 id : $2\sqrt{2}/3 - 3\sqrt{3}/2$
6552 7483 $2^x : 3/4 + 3\sqrt{3}$
6552 7708 $e^x : 2\sqrt[3]{2}/5$
6552 8706 $Ei : (\pi/3)^{-2}$
6552 9574 $Ei : (\sqrt[3]{3}/3)^{-1/3}$
6552 9933 rt : (5,7,7,-9)
6553 1486 $t\pi : ((e+\pi)/3)^2$
6553 1658 $K : 7/16$
6553 2625 rt : (5,4,2,1)
6553 2879 sq : 17/21
6553 3214 $t\pi : 4\sqrt{2}/3 + 3\sqrt{3}/4$
6553 4242 $\ell_{10} : 1/3 + 4\pi/3$
6553 5778 $\exp(\sqrt{5}) \div \exp(\sqrt{3})$
6553 9911 as : $1/3 - 2\sqrt{2}/3$
6554 0391 $s\pi : 3e/4 + 4\pi/3$
6554 0420 rt : (6,0,2,3)
6554 1953 rt : (7,4,2,-5)
6554 2255 $\Psi : (\ln 3)^{-1/2}$
6554 2608 rt : (8,8,2,-7)
6554 3134 $\exp(\sqrt{3}) \times \exp(e)$
6554 3147 $10^x : 3\sqrt{2}/2 + \sqrt{3}/3$
6554 3485 $2^x : 1/3 - 2\sqrt{2}/3$
6554 4238 rt : (2,8,0,-4)
6554 5441 $2^x : -\sqrt{2} + \sqrt{5} + \sqrt{6}$
6554 6510 $1/4 - \ln(2/3)$
6554 9574 rt : (3,9,-5,8)
6554 9760 $\Gamma : \pi^2/3$
6555 0655 $2^x : 8/11$
6555 0966 $s\pi : (\pi)^{1/2}$
6555 1489 rt : (1,5,-3,4)
6555 3395 $e^x : (e+\pi)/6$
6555 4868 cr : $3 - e$
6555 4935 $2^x : 1/3 + 4\sqrt{2}/3$
6555 5390 rt : (4,8,2,-1)
6555 5550 ln : $5\pi/3$
6555 7083 ln : $3 + \sqrt{5}$
6555 8696 $2^x : 3/4 - e/2$
6555 9572 $\ell_{10} : 3\sqrt{2}/4 + 2\sqrt{3}$
6556 0757 cr : P_6
6556 0789 $t\pi : 9(e+\pi)/4$
6556 1108 $t\pi : 4\sqrt{2} - 2\sqrt{5}$
6556 2661 $2^x : \exp(-1/\pi)$
6556 4069 $2^x : P_{73}$
6556 4246 rt : (6,4,7,-8)
6556 4881 cu : $3/4 + \sqrt{3}/4$
6556 5857 rt : (5,9,-5,-2)
6556 7069 $2^x : (\sqrt[3]{4}/3)^{1/2}$
6556 8142 $J_0 : 16/13$
6556 9373 rt : (9,1,-3,-1)
6556 9562 at : 10/13
6556 9750 cr : $4\sqrt{2} - \sqrt{5}/2$
6557 0278 rt : (1,8,5,-7)
6557 1587 rt : (8,7,-8,-6)
6557 1793 cu : $3 + \pi$
6557 1960 $erf : (\sqrt{5})^{-1/2}$
6557 3967 rt : (5,0,7,6)
6557 6016 $Ei : 8\sqrt{2}/7$
6557 8270 rt : (9,5,2,-6)
6557 9420 tanh : $\pi/4$
6558 0083 rt : (9,8,9,5)
6558 0120 ln : $2\sqrt{2} - 4\sqrt{3}/3$
6558 1118 sq : $(\ln 2/3)^{-1/3}$
6558 4214 $\exp(4/5) - s\pi(\pi)$
6558 5977 sinh : $8\ln 2/9$
6558 6884 rt : (4,9,0,-5)
6558 9202 $J : \sqrt[3]{2}/3$
6559 1278 rt : (8,5,7,-9)
6559 1308 $\ln(3/5) \times \exp(1/4)$
6559 3554 $\Psi : \sqrt{22}\pi$
6559 4579 rt : (8,1,2,-4)
6559 5149 tanh : $5\sqrt{2}/9$
6559 5896 rt : (3,5,-6,5)
6559 6177 $10^x : 3\sqrt{2}/4 + \sqrt{5}/4$
6559 7055 rt : (7,7,-7,1)
6559 7433 tanh : 11/14
6559 9084 ln : $1/4 + 3\sqrt{5}/4$
6560 0000 cu : 18/5
6560 0361 rt : (3,9,5,6)
6560 2058 rt : (4,4,7,4)
6560 2976 rt : (2,5,-9,-2)
6560 4799 rt : (6,4,-5,8)
6560 5337 at : $4\sqrt{3}/9$
6560 7009 rt : (9,8,-6,9)
6560 8997 rt : (7,1,7,-7)
6560 9394 id : $1/4 - e/3$
6561 1562 rt : (5,8,-3,-4)
6561 2607 $10^x : 1/4 - \sqrt{3}/4$
6561 3596 rt : (6,8,-2,2)
6561 4323 $\ell_{10} : 5e/3$
6561 6638 rt : (1,3,-9,-6)
6561 7024 $3 \div \ln(\sqrt{2})$
6561 7814 rt : (8,3,-3,-1)
6561 7916 $2^x : 2\sqrt{2}/3 + 3\sqrt{5}/2$
6562 0903 rt : (5,5,0,9)
6562 2016 $10^x : 3\sqrt{2}/10$
6562 5162 $t\pi : (e/2)^{-2}$
6562 6294 rt : (2,5,5,-6)
6562 6807 ln : $4\sqrt{2}/3 + 3\sqrt{5}/2$
6562 7889 $erf : \ln((e+\pi)/3)$
6562 8137 at : $10\ln 2/9$
6562 8381 cu : $4\sqrt{3} - 3\sqrt{5}/2$
6562 8975 rt : (6,8,-8,-7)
6563 1174 rt : (9,2,7,-8)
6563 6266 rt : (1,9,7,1)
6563 9355 rt : (6,6,-5,-1)
6564 0885 rt : (7,3,2,2)
6564 1118 $K : 11/25$
6564 1978 cr : $\sqrt{2}/5$
6564 4524 cu : $4/3 - e$
6564 5219 rt : (5,9,3,3)
6564 8135 rt : (6,3,7,5)
6564 8675 rt : (4,6,5,-7)
6564 9482 rt : (2,9,2,-2)
6564 9668 cr : $8(1/\pi)/9$
6565 0557 cr : $4\sqrt{3}/3 + \sqrt{5}$
6565 2097 $s\pi : (2\pi/3)^{-2}$
6565 3739 rt : (3,9,-3,-5)
6565 5099 $\ln(2/5) \div \exp(1/3)$
6565 5840 rt : (4,9,-8,-8)
6565 6658 sinh : 4π
6565 7566 sr : $1/\ln(\pi/3)$
6566 1958 $10^x : 1/4 + 4e$
6566 2043 rt : (1,6,7,-5)
6566 3100 rt : (6,7,5,-8)
6566 3315 rt : (7,7,0,-5)
6566 3639 rt : (8,7,-5,-2)
6566 4520 ln : $\sqrt{3} - \sqrt{6} + \sqrt{7}$
6566 6545 $s\pi : 2\sqrt{3} - \sqrt{5}$
6566 6905 rt : (4,7,-7,7)
6566 9790 rt : (8,2,7,6)
6566 9914 $\zeta : 8\sqrt{5}/9$
6567 2396 rt : (9,2,2,3)
6567 3171 rt : (8,8,5,-9)
6567 4517 rt : (8,6,-9,5)
6567 5033 rt : (9,8,0,-6)
6567 6165 cu : $e/2 - 4\pi/3$
6568 1785 $e^x : 4\sqrt[3]{5}/7$
6568 2531 ln : $1 + 3\sqrt{2}$
6568 4710 $\zeta : 6/11$
6568 5424 id : $4\sqrt{2}$
6568 7725 rt : (2,5,-6,9)
6569 0871 $\zeta : \sqrt[3]{3}/10$
6569 1158 sr : $2 + \sqrt{5}/3$
6569 2945 $\theta_3 : 7/22$
6569 4298 $\ell_{10} : 4\sqrt{2} - \sqrt{5}/2$
6569 6241 tanh : $5\sqrt[3]{2}/8$
6569 6814 rt : (1,2,6,9)
6569 6975 $10^x : 2/3 + 3\sqrt{5}/4$
6569 7525 $\Psi : (\sqrt{3}/2)^{1/3}$
6569 8478 rt : (5,2,-8,4)
6569 8659 $s\pi : 7(1/\pi)$
6570 0529 $\sqrt{3} - \exp(2)$
6570 1954 $e^x : (\pi/3)^{-1/2}$
6570 2495 rt : (7,7,-4,6)
6570 2682 $erf : 1/\ln(\sqrt{2}\pi)$
6570 4971 $\ln(2) - s\pi(\sqrt{2})$
6570 6411 id : $\exp(-\sqrt[3]{2}/3)$
6570 8006 rt : (3,2,5,-5)
6570 8099 $e^x : 8\sqrt{3}$
6571 0792 rt : (5,9,-8,9)
6571 1586 $e^x : P_{35}$
6571 3231 $10^x : 4(1/\pi)/3$
6571 3529 rt : (5,3,5,-6)
6571 4420 $t\pi : \sqrt{13}\pi$
6571 6241 $\ln(3/4) \times \exp(\pi)$
6571 6820 rt : (7,4,5,-7)
6571 6991 sq : $1/4 - 3\sqrt{2}/4$
6571 7882 rt : (1,6,3,6)
6571 9004 rt : (9,5,5,-8)
6572 1828 $erf : 3\sqrt{5}/10$
6572 1885 $\theta_3 : (1/\pi)$
6572 3676 $t\pi : 3\sqrt[3]{3}/10$
6572 3837 $t\pi : \sqrt{2} + \sqrt{6} + \sqrt{7}$
6572 4607 sq : $7e/8$
6572 6753 $2^x : \pi^2/7$
6572 8192 $e^x : e/4 + \pi$
6572 9810 rt : (8,4,0,-4)
6573 0309 $erf : (\sqrt{2}\pi/2)^{-1/2}$
6573 6417 rt : (1,8,8,-9)
6573 8040 rt : (1,7,-8,-8)
6573 9671 sr : $\sqrt{2}/4 + 3\sqrt{5}$
6574 0720 rt : (3,6,-8,-7)
6574 1949 $e^x : 20/13$
6574 2604 $2^x : \sqrt{7/2}$
6574 5776 rt : (6,8,1,7)
6574 5989 rt : (5,5,-8,-6)
6574 6355 as : 11/18
6574 8892 rt : (4,9,3,-7)
6575 0306 $\ell_2 : 1 + \sqrt{3}/3$
6575 0861 rt : (9,4,-5,-1)
6575 1623 cu : 20/23
6575 2280 $10^x : 1/4 + e/2$
6575 3520 rt : (2,8,3,-6)
6575 7100 rt : (9,5,7,5)
6575 7259 sr : $4/3 + \sqrt{2}$
6575 7745 rt : (9,9,-8,3)
6575 7817 rt : (1,8,-5,6)
6575 7870 $\ell_{10} : 4\sqrt{3} - 3\sqrt{5}$
6576 1111 rt : (7,4,-8,-5)
6576 7285 cu : $\sqrt{2} - \sqrt{3} - \sqrt{5}$
6576 8605 rt : (5,9,-2,-4)
6577 0121 rt : (7,6,7,4)
6577 1034 $t\pi : (\sqrt[3]{5}/3)^3$
6577 5856 rt : (2,6,-3,-4)
6577 6249 sq : $\sqrt{3}/2 - 3\sqrt{5}/4$
6577 8026 sr : $3\sqrt[3]{3}/10$
6578 0816 rt : (3,8,-2,-3)
6578 3868 $\theta_3 : 4\sqrt[3]{3}/9$
6578 5736 sinh : $\sqrt{2} - \sqrt{3} - \sqrt{7}$
6578 7509 rt : (8,1,5,-6)
6578 8605 rt : (2,6,-7,-8)
6578 8860 at : 17/22
6579 0799 rt : (5,2,-5,1)
6579 3159 rt : (9,8,3,-8)
6579 3621 rt : (4,6,8,-9)

6579 5769 sq : $\sqrt[3]{13/3}$
6579 6677 rt : (5,2,-5,9)
6579 7876 sr : $7\pi/8$
6579 8761 cr : $7(e+\pi)/9$
6579 9858 rt : (6,9,-7,-1)
6580 2153 10^x : $10\zeta(3)/9$
6580 2342 λ : $1/15$
6580 3700 sr : $\sqrt{3}/4$
6580 4249 rt : (1,7,-2,-2)
6580 4786 rt : (7,7,3,-7)
6580 5144 rt : (6,0,5,5)
6580 6365 rt : (1,9,4,9)
6580 6855 rt : (3,6,-6,-2)
6580 6921 rt : (5,7,7,3)
6580 8135 cr : $\sqrt[3]{5}/6$
6580 8146 rt : (2,5,8,-8)
6580 8339 sr : P_{10}
6580 9865 sq : $\sqrt{2}-\sqrt{3}-\sqrt{6}$
6581 1077 tanh : $15/19$
6581 1264 Ψ : $\sqrt[3]{14}$
6581 4444 Ei : $\sqrt{2}-\sqrt{3}+\sqrt{5}$
6581 4970 cr : $4+\sqrt{5}/4$
6581 5927 rt : (6,5,-7,2)
6581 6329 $s\pi(1/4)+s\pi(2/5)$
6581 6908 rt : (4,5,-3,-3)
6581 9221 sr : $5\ln 2/8$
6582 0477 Γ : $7e\pi/3$
6582 0630 Γ : $-\sqrt{2}+\sqrt{3}-\sqrt{5}$
6582 1068 id : P_{19}
6582 1220 id : P_{58}
6582 1435 sinh : $(2\pi/3)^{1/3}$
6582 1642 rt : (5,9,6,8)
6582 2429 sinh : $3\sqrt[3]{3}/7$
6582 2807 ln : $4/21$
6582 3236 rt : (9,1,0,-3)
6582 6109 cr : $8\sqrt[3]{5}/3$
6582 6813 Γ : $7\pi^2/5$
6582 7001 rt : (9,9,-5,8)
6582 8709 $s\pi$: $8\sqrt{3}/5$
6583 1239 sr : $11/4$
6583 2680 cr : $2\sqrt{2}+\sqrt{3}$
6583 3480 2^x : P_{23}
6583 4804 rt : (8,7,-2,-4)
6583 6622 10^x : $3+2\sqrt{5}/3$
6583 7066 cu : $4\sqrt{3}/3+3\sqrt{5}/2$
6583 7715 rt : (7,4,-9,6)
6583 7778 sq : $\sqrt{3}/4+4\sqrt{5}/3$
6583 8085 K : $4/9$
6583 8139 $\ln(5)-s\pi(2/5)$
6583 8309 rt : (4,1,5,4)
6583 9380 $s\pi$: $8\sqrt{2}/3$
6583 9407 Ψ : $20/21$
6583 9559 Ei : $24/11$
6584 1581 rt : (7,8,-9,-5)
6584 3355 Ei : $21/23$
6584 4290 e^x : $4\sqrt{3}$
6584 4320 rt : (7,4,8,-9)
6584 4662 10^x : $4\sqrt{3}-4\sqrt{5}/3$
6584 4752 ln : $4\sqrt{3}-3\sqrt{5}/4$
6584 6922 10^x : $7e\pi/10$
6584 7884 sinh : $\exp(-\sqrt[3]{3}/3)$
6584 8279 rt : (3,5,3,-5)
6585 2547 id : $4e+\pi/4$
6585 4064 $s\pi$: $3\sqrt{3}/4+2\sqrt{5}$
6585 4605 sq : $2/3+3\sqrt{3}/2$
6585 8677 rt : (1,7,6,-9)
6585 9132 rt : (8,6,2,1)
6585 9685 ℓ_2 : $3\sqrt{3}+\sqrt{5}/2$
6586 0568 cu : $4e-4\pi/3$
6586 3342 10^x : $3\sqrt{2}/2+\sqrt{5}/3$
6586 3375 cr : $2/7$
6586 4770 rt : (6,6,-2,-3)
6586 7374 rt : (1,7,-4,8)
6586 7531 rt : (1,6,2,-1)
6586 8077 sr : $9\pi/4$
6586 8381 rt : (4,9,6,-9)
6586 8593 rt : (5,3,8,-8)
6586 8866 e^x : $9\sqrt[3]{2}/5$
6586 9909 rt : (2,6,-4,3)
6587 1650 rt : (3,9,-6,-7)
6587 1770 id : $\sqrt{3}/3-\sqrt{5}$
6587 2522 J_0 : $\sqrt{3/2}$
6587 3638 id : $3\sqrt{2}/4+3\sqrt{3}/2$
6587 3678 rt : (2,9,4,-9)
6587 3682 Γ : $\exp(4/3)$
6587 4384 ℓ_{10} : $7(e+\pi)/9$
6587 6412 Ei : $4\sqrt{5}/7$
6587 9566 2^x : $3+4e/3$
6588 0453 rt : (8,4,3,-6)
6588 0729 rt : (9,7,-7,-1)
6588 0785 sr : $4\sqrt{2}/3+\sqrt{3}/2$
6588 0953 id : $4e/3-2\pi$
6588 3334 Ei : $23/18$
6588 4676 rt : (9,2,5,5)
6588 4815 $\ln(\sqrt{5})\div\exp(1/5)$
6588 7013 e^x : $4\sqrt{3}-2\sqrt{5}$
6588 7121 ℓ_{10} : $4+\sqrt{5}/4$
6588 8280 at : $4/3-\sqrt{5}/4$
6588 9381 sr : $3\sqrt{2}-2\sqrt{5}/3$
6589 1316 K : $7(1/\pi)/5$
6589 5747 cr : $4/3-\pi/3$
6589 5873 ℓ_{10} : $8\sqrt[3]{5}/3$
6589 6346 θ_3 : $3\sqrt[3]{5}/8$
6589 6708 rt : (4,8,-4,8)
6589 7131 K : $1/\ln(3\pi)$
6589 8705 sr : $3\sqrt{3}/2+2\sqrt{5}$
6590 0824 Ei : $\exp(\sqrt[3]{3}/3)$
6590 1035 ℓ_{10} : $2\sqrt{2}+\sqrt{3}$
6590 1301 ln : $e-\pi/4$
6590 2804 e^x : $1/4+\pi/3$
6590 4144 rt : (6,5,-4,7)
6590 5803 at : $(5/3)^{-1/2}$
6590 5943 rt : (3,2,8,-7)
6590 7042 $\exp(1/4)\div s\pi(e)$
6590 7818 rt : (5,9,1,-6)
6590 7977 ln : $3/2+\sqrt{3}/4$
6590 8464 rt : (2,8,-7,-4)
6591 1067 rt : (7,7,6,-9)
6591 3462 rt : (2,7,-7,1)
6591 4401 10^x : $16/15$
6591 4794 sr : $5\sqrt{2}$
6591 6737 $3/5\times\ln(3)$
6591 8505 Ei : $7\ln 2/3$
6591 8669 rt : (4,7,0,-1)
6591 9715 ln : $3\sqrt{2}-4\sqrt{3}/3$
6592 0373 rt : (6,3,3,-5)
6592 1219 rt : (8,1,8,-8)
6592 1993 cu : $\sqrt{2}-3\sqrt{3}/2$
6592 2076 cr : $9(1/\pi)/10$
6592 2214 $\ln(2)\times s\pi(2/5)$
6592 2767 $t\pi$: $1/4-\sqrt{3}/3$
6592 3734 rt : (2,2,5,3)
6592 4063 $\exp(1/3)\div\exp(3/4)$
6592 6262 Ψ : $(e/3)^{1/2}$
6592 6999 rt : (5,8,-6,-6)
6592 7824 rt : (8,7,-8,4)
6593 0571 sq : $\sqrt[3]{7}$
6593 1371 Ψ : $\ln(4/3)$
6593 1410 rt : (4,0,-8,5)
6593 1918 cu : $3/4-3\sqrt{5}/2$
6593 3733 sinh : $\ln(3\pi)$
6593 4357 rt : (9,4,-2,-3)
6593 5073 sinh : $13/21$
6593 5216 tanh : $19/24$
6593 8047 sr : $10/23$
6593 9615 rt : (7,8,-6,0)
6594 4291 rt : (7,3,5,4)
6594 5806 rt : (5,6,6,-8)
6594 6339 rt : (1,8,3,3)
6594 7309 rt : (2,9,-1,-4)
6594 8850 $2/5\times\exp(1/2)$
6595 2342 Ψ : $(2e)^{-1/3}$
6595 2455 rt : (1,7,9,8)
6595 2901 rt : (3,8,1,-5)
6595 3458 2^x : $3e-2\pi$
6595 5676 rt : (8,7,1,-6)
6595 7056 e^x : $4-\sqrt{3}$
6595 7655 cr : $3+\pi/2$
6595 8422 10^x : $4\sqrt{3}-3\sqrt{5}$
6596 2359 K : $\sqrt{5}/5$
6596 3556 tanh : $8\ln 2/7$
6596 3921 rt : (5,1,0,1)
6596 4323 10^x : $15/16$
6596 4916 sq : $1-2e/3$
6596 5787 rt : (6,9,-4,-3)
6596 6092 Ei : $(2\pi/3)^{-2}$
6596 8580 rt : (6,0,8,7)
6596 8990 $t\pi$: $5(e+\pi)/7$
6596 9431 2^x : $4\sqrt{2}/3-2\sqrt{3}/3$
6597 0219 10^x : $7\sqrt{2}/6$
6597 0863 rt : (1,1,8,-6)
6597 1216 $s\pi$: $4e+3\pi/4$
6597 1908 $t\pi$: $\sqrt[3]{5}/3$
6597 2168 cu : $9(e+\pi)/7$
6597 3037 Ψ : $23/4$
6597 3099 Li_2 : $4\ln 2/5$
6597 3891 e^x : $3/4+3\pi/2$
6597 5355 Ei : $(\pi)^{-1/2}$
6597 7050 θ_3 : $\sqrt{5}/7$
6597 7432 cu : $3\sqrt{2}/2+\sqrt{3}/2$
6597 7540 rt : (4,2,-8,-5)
6597 9427 2^x : $(\sqrt[3]{5}/2)^2$
6598 0022 ζ : $\zeta(5)$
6598 0942 rt : (9,1,3,-5)
6598 2713 e^x : $9\sqrt[3]{5}/10$
6598 6360 rt : (8,7,-5,9)
6598 8593 $\exp(4/5)\times s\pi(\sqrt{3})$
6598 8609 id : $3e\pi/7$
6598 9103 rt : (8,4,6,-8)
6599 1875 10^x : $7\sqrt[3]{2}/8$
6599 3390 $s\pi$: $1/\sqrt{19}$
6599 5778 Li_2 : $(e+\pi)^{-1/3}$
6599 7046 rt : (3,5,6,-7)
6599 9186 ℓ_{10} : $3+\pi/2$
6599 9714 J_0 : $11/9$
6600 0586 rt : (8,0,-5,1)
6600 3030 $\ln(3/5)\div s\pi(e)$
6600 5196 $\ln(3/4)+\exp(2/3)$
6600 6019 sq : $3e/5$
6600 6318 cu : $1/3+2\pi$
6600 6553 rt : (7,3,-2,-2)
6600 6581 cu : $3\pi^2/5$
6600 7748 rt : (8,9,7,3)
6600 8249 $t\pi$: $1/4+3e/2$
6600 8495 rt : (6,6,1,-5)
6600 9457 rt : (5,9,4,-8)
6601 2311 rt : (3,9,-9,-9)
6601 2505 sq : $2+2\sqrt{2}/3$
6601 4235 cr : $\ln(4/3)$
6601 5625 sq : $13/16$
6601 5969 rt : (6,2,-9,7)
6601 6595 2^x : $\ln(\ln 3/2)$
6601 6699 rt : (7,8,-3,5)
6601 6837 10^x : $13/2$
6601 9616 rt : (9,7,-4,-3)
6602 0295 cr : $1/3+3\sqrt{2}$
6602 1816 rt : (9,2,8,7)
6602 5403 id : $5\sqrt{3}$
6602 6815 $s\pi$: $\sqrt{2}/3+3\sqrt{3}/4$
6602 7667 id : $e/3-4\pi$
6602 8389 sr : P_{29}
6602 8607 rt : (1,7,1,-4)
6602 9897 $\ln(\pi)\div s\pi(\pi)$
6603 0008 cu : $1/2-4\sqrt{2}/3$
6603 0173 cu : $(\pi/3)^{-3}$
6603 2000 rt : (1,9,-7,-6)
6603 2050 e^x : $3/4+3\sqrt{2}/2$
6603 3168 cu : $4e/3-\pi/2$
6603 4048 K : $\pi/7$
6603 4130 at : $2e/7$
6603 4259 sr : $3+3e/2$
6603 6937 cr : $4+\sqrt{3}/3$
6603 7814 cr : $1/2+3e/2$
6603 9826 rt : (3,8,-9,7)
6604 0562 J_0 : $5\sqrt[3]{5}/7$
6604 1001 10^x : $4-\pi/2$
6604 1057 2^x : $5(e+\pi)/2$
6604 2892 $\exp(\sqrt{5})\div\exp(1/5)$
6604 3004 cu : $4\sqrt{3}/5$
6604 4224 rt : (4,8,-4,-2)

6604 4575 2^x : $3\sqrt{3}$
6604 5695 rt : (8,7,4,-8)
6604 6182 rt : (6,3,6,-7)
6604 7151 rt : (4,1,8,6)
6604 8355 ℓ_{10} : $1/3+3\sqrt{2}$
6605 0027 tanh : $\sqrt[3]{1/2}$
6605 2547 rt : (1,6,-5,-4)
6605 3316 rt : (4,9,-7,-7)
6605 3738 rt : (7,7,-6,-5)
6605 5538 rt : (2,5,-9,-8)
6605 6144 rt : (9,4,1,-5)
6605 7550 Ψ : $(\sqrt{5}/3)^{-3}$
6605 7621 sq : $1/4+3\sqrt{3}$
6605 7657 2^x : $2+\sqrt{2}$
6605 9197 rt : (7,9,-9,0)
6605 9462 J_0 : $(\ln 3/2)^{-1/3}$
6606 1414 ℓ_{10} : $4+\sqrt{3}/3$
6606 1927 rt : (7,0,3,4)
6606 2102 ℓ_{10} : $1/2+3e/2$
6606 2411 2^x : $24/17$
6606 2818 rt : (5,3,-7,3)
6606 3430 2^x : $2-3\sqrt{3}/2$
6606 4045 as : $(\ln 2/3)^{1/3}$
6606 8750 rt : (3,8,4,-7)
6607 1193 cu : $\sqrt{5}\pi$
6607 1402 rt : (2,9,-3,-7)
6607 2437 K : $1/\sqrt[3]{11}$
6607 2807 ℓ_2 : $1/3+2\sqrt{2}$
6607 2915 cr : $\sqrt[3]{3}/5$
6607 3556 rt : (5,5,-7,1)
6607 4527 ln : $\sqrt[3]{5}/9$
6607 7110 Ψ : $3\pi^2/2$
6607 7699 sq : $3/4-3e/4$
6607 7850 θ_3 : $3\sqrt[3]{3}/5$
6607 8048 rt : (6,9,-1,1)
6607 8611 rt : (5,4,5,3)
6607 9887 cu : $2\sqrt{3}/3+3\sqrt{5}/2$
6608 0243 rt : (1,4,6,-6)
6608 1858 rt : (4,8,-1,-3)
6608 4629 10^x : $14/17$
6608 5980 e^x : $1-\sqrt{2}$
6608 6994 10^x : $1+3e$
6608 8623 K : $9/20$
6608 8679 rt : (9,1,6,-7)
6609 0107 sr : $1/\sqrt[3]{12}$
6609 1223 $s\pi$: $3\pi^2/7$
6609 2640 $s\pi$: $10\ln 2/9$
6609 3587 rt : (4,7,3,4)
6609 4778 cr : $2\sqrt{3}+\sqrt{5}/2$
6609 6404 ℓ_2 : $\sqrt{10}$
6609 7357 rt : (5,8,-9,-8)
6609 9901 rt : (3,5,9,-9)
6610 0095 sr : $\sqrt[3]{21}$
6610 4191 θ_3 : $8/25$
6610 4316 at : $7/9$
6610 4980 rt : (8,3,0,1)
6610 6412 rt : (7,6,-4,-2)
6610 6793 ℓ_{10} : $2\sqrt{3}+\sqrt{5}/2$
6610 7179 Ei : $5\pi/4$
6610 7829 rt : (1,9,4,-1)
6610 8310 rt : (6,6,4,-7)
6610 8550 rt : (7,3,8,6)
6610 8967 rt : (4,9,-4,1)
6610 9849 rt : (2,9,0,7)
6610 9946 rt : (2,9,-4,-6)
6610 9987 2^x : $2/3+\sqrt{5}/3$
6611 0964 ℓ_{10} : $\sqrt{21}$
6611 2031 rt : (1,0,-8,5)
6611 3538 J_0 : $e\pi/7$
6611 3892 rt : (7,5,-8,5)
6611 3971 as : $\sqrt{3}-\sqrt{5}/2$
6611 5961 sr : $4\sqrt{2}/3+3\sqrt{3}$
6611 6504 rt : (9,7,-1,-5)
6611 6958 10^x : $3\sqrt{2}/2+3\sqrt{3}$
6612 0458 ℓ_2 : $4\sqrt{3}/3-3\sqrt{5}/4$
6612 1293 J : $\exp(-5\pi/9)$
6612 2130 2^x : $4\sqrt{2}-2\sqrt{3}/3$
6612 3040 rt : (2,7,-2,-3)
6612 3793 K : $3\zeta(3)/8$
6612 3848 Ψ : $8e/9$
6612 4509 Li_2 : $5/9$
6612 6635 rt : (3,7,-6,-9)
6612 8892 sq : $4\sqrt{3}/3+\sqrt{5}$
6612 8894 rt : (5,3,-4,8)
6612 8901 10^x : $\exp(2/3)$
6613 4007 cr : 6π
6613 4331 rt : (6,6,-9,-4)
6613 4723 10^x : $1/2-e/4$
6613 5085 at : $(\sqrt{2}/3)^{1/3}$
6613 5555 rt : (6,3,9,-9)
6613 6183 2^x : $4+3\pi$
6614 3011 rt : (9,4,4,-7)
6614 3397 rt : (9,7,-9,7)
6614 3782 sr : $7/16$
6614 6670 rt : (7,3,1,-4)
6614 9495 rt : (2,6,-9,-8)
6614 9771 2^x : $(e+\pi)/8$
6615 1733 at : $3\sqrt{2}-2\sqrt{3}$
6615 2113 $s\pi$: $6\pi/5$
6615 2137 rt : (3,8,7,-9)
6615 2903 10^x : $1+2\sqrt{2}/3$
6615 2904 J_0 : $(2e/3)^{1/3}$
6615 4273 sq : $3/2-3\sqrt{3}$
6615 5328 rt : (4,9,-1,9)
6615 5457 $s\pi$: $1/3+2e$
6615 7380 rt : (5,2,-2,-1)
6615 8625 2^x : $1+4\pi$
6615 9161 rt : (6,9,2,6)
6616 0042 rt : (1,7,4,-6)
6616 0890 K : $\ln(\pi/2)$
6616 2648 rt : (4,5,-6,-5)
6616 4429 sr : $2/3+2\pi/3$
6616 6797 tanh : $5(1/\pi)/2$
6616 7035 rt : (9,1,9,-9)
6616 9666 2^x : $3e+\pi/2$
6617 0213 rt : (4,8,-1,-4)
6617 3707 cu : $3/2+\sqrt{2}/3$
6617 3982 cr : P_{82}
6617 7081 rt : (2,7,1,8)
6617 7660 rt : (8,3,-7,1)
6617 8219 $s\pi$: $4\sqrt{3}/9$
6617 8600 $s\pi$: $\exp(\sqrt[3]{3})$
6618 0027 rt : (7,9,-6,-2)
6618 0503 rt : (8,6,5,3)
6618 0719 ℓ_{10} : $4e-2\pi$
6618 0811 Γ : $4\zeta(3)/9$
6618 1430 rt : (7,0,6,6)
6618 1737 rt : (6,6,7,-9)
6618 2349 rt : (1,6,-1,-3)
6618 7735 K : $\exp(-\sqrt[3]{4}/2)$
6618 8025 rt : (9,7,2,-7)
6618 9047 cr : $4e-2\pi$
6618 9604 2^x : $5(e+\pi)/8$
6619 1527 2^x : $3\sqrt[3]{5}/7$
6619 2994 rt : (1,4,9,-8)
6619 5590 rt : (3,7,-3,-1)
6619 7618 ζ : $(\sqrt[3]{4}/3)^3$
6620 0721 rt : (2,7,-4,-1)
6620 1542 rt : (6,6,-6,1)
6620 1951 rt : (4,7,6,9)
6620 2627 rt : (3,9,-7,1)
6620 2844 2^x : $8\sqrt[3]{5}/5$
6620 3169 rt : (2,2,8,5)
6620 4453 $t\pi$: $1/2+4\sqrt{2}/3$
6620 5554 e^x : $9\sqrt[3]{3}/10$
6620 8148 rt : (9,4,7,-9)
6620 9222 sinh : $1/\ln(5)$
6621 0609 rt : (7,6,-1,-4)
6621 1361 J_0 : $(e/3)^{-2}$
6621 1568 rt : (5,0,-9,8)
6621 2519 rt : (2,9,-7,-8)
6621 2556 $\ln(2/5)+s\pi(\sqrt{3})$
6621 3840 $s\pi$: $2\sqrt{2}-3\sqrt{3}/2$
6621 4658 J_0 : $\sqrt{2}+\sqrt{6}-\sqrt{7}$
6621 5763 rt : (8,8,-7,3)
6621 7407 θ_3 : $2\sqrt[3]{3}/9$
6621 7637 e^x : $3\sqrt{2}/4+3\sqrt{5}/2$
6621 7877 rt : (5,8,-9,1)
6621 9090 $t\pi$: $4\sqrt{2}/3+4\sqrt{3}$
6622 0073 rt : (5,1,3,3)
6622 0552 rt : (6,9,5,-9)
6622 2533 2^x : $\sqrt{3}/4+4\sqrt{5}/3$
6622 7112 sq : $10\pi/3$
6622 7766 sq : $\sqrt{2}/2+\sqrt{5}$
6622 8103 K : $e/6$
6622 9607 at : $9\ln 2/8$
6622 9816 rt : (4,2,9,-8)
6623 0926 e^x : $1/3-\sqrt{5}/3$
6623 2209 2^x : $3/2-2\pi/3$
6623 4762 ln : $3-3\sqrt{2}/4$
6623 5441 rt : (1,7,3,5)
6623 5897 rt : (8,0,-2,1)
6623 7513 rt : (7,3,4,-6)
6623 9104 rt : (1,6,3,-7)
6623 9187 Ei : $(5/3)^3$
6624 2155 id : $5(e+\pi)/8$
6624 3024 rt : (9,7,5,-9)
6624 3754 $\exp(2/5)-\exp(e)$
6624 4617 e^x : $10\sqrt{3}/3$
6624 4814 sinh : $5\pi/7$
6624 6259 J : $(\sqrt{2}\pi/2)^{-3}$
6624 6480 rt : (1,7,7,-8)
6624 7579 2^x : $11/15$
6624 7928 rt : (4,6,-9,9)
6624 8888 rt : (3,7,0,7)
6624 9196 10^x : $\exp(-\sqrt[3]{5}/2)$
6625 0462 10^x : $1/\sqrt[3]{13}$
6625 0886 rt : (1,9,2,-8)
6625 2008 2^x : $1/2+\sqrt{5}$
6625 3263 10^x : $4+4\sqrt{5}/3$
6625 3629 rt : (8,8,-4,8)
6625 5518 e^x : $4\sqrt{3}-3\sqrt{5}/2$
6625 5652 $\ln(\sqrt{3})+\exp(\sqrt{2})$
6625 6447 rt : (7,7,-9,-7)
6625 7616 ℓ_{10} : $2+3\sqrt{3}/2$
6626 0559 rt : (7,9,-3,-4)
6626 0755 id : P_{55}
6626 0961 Ψ : $19/20$
6626 1374 rt : (7,0,9,8)
6626 1783 $t\pi$: $4\sqrt{3}/5$
6626 3331 Ψ : $5\sqrt[3]{5}/9$
6626 5158 J_0 : $(\sqrt{2}\pi/3)^{1/2}$
6626 5954 cu : $1+2e$
6626 7281 rt : (5,5,-4,-1)
6626 7493 Ei : $8\sqrt[3]{2}/3$
6626 8827 rt : (5,4,8,5)
6626 9838 rt : (4,8,-4,-5)
6627 1406 Ei : $\exp(-\sqrt{2}\pi/3)$
6627 2815 e^x : $8\sqrt{3}/9$
6627 5758 rt : (4,5,7,-8)
6627 5783 ℓ_{10} : $5/23$
6627 6766 ln : $9(e+\pi)/10$
6627 7722 J : $\sqrt{2}/8$
6628 0129 rt : (8,6,-9,1)
6628 1534 rt : (8,3,3,3)
6628 1603 Ei : $9\sqrt[3]{5}/4$
6628 1996 rt : (7,6,2,-6)
6628 2017 cu : $4\sqrt{2}-2\sqrt{5}$
6628 2455 rt : (1,9,1,-3)
6628 2714 sr : $3/2-3\sqrt{2}/4$
6628 4150 sr : $6\sqrt{5}$
6628 5246 tanh : $(\pi/2)^{-1/2}$
6628 7162 cr : $2+3\sqrt{3}/2$
6628 7382 as : $8/13$
6628 9649 cu : $3+3\pi/4$
6629 2352 θ_3 : $(\sqrt{5}/3)^{1/3}$
6629 5446 id : $\sqrt{2}/4+4\sqrt{3}/3$
6629 5495 Ei : $13/23$
6629 5608 rt : (2,1,6,-5)
6629 6501 ℓ_2 : $3/19$
6629 7243 K : $5/11$
6630 0477 tanh : $\ln(\sqrt{2}\pi/2)$
6630 0656 rt : (7,9,-8,-6)
6630 1253 rt : (7,3,7,-8)
6630 1677 Ei : $5(1/\pi)/7$
6630 3519 e^x : $\sqrt{2}/2-\sqrt{5}/2$
6630 3551 Ψ : $4\pi^2$
6630 4209 rt : (8,5,-9,8)
6630 5709 K : $10(1/\pi)/7$
6630 6648 cu : $4/3-3e/2$

6630 6741 rt : (5,2,1,-3)
6630 7682 rt : (1,5,2,2)
6630 8614 rt : (4,5,-9,-7)
6630 8838 cu : $e/2-\pi$
6631 0349 cr : 23/5
6631 2243 e^x : $4\sqrt{2}/3+\sqrt{5}$
6631 2265 sπ : 3/13
6631 2980 rt : (4,8,5,-8)
6631 5910 $1/4+\exp(5)$
6631 6312 10^x : $1+3\sqrt{5}/2$
6631 6337 2^x : $3e-3\pi/4$
6631 6572 Γ : $7(1/\pi)/9$
6631 7273 rt : (8,3,-4,-1)
6631 7620 cr : 7/24
6631 7846 sinh : $8\sqrt[3]{3}/9$
6631 8136 rt : (7,9,0,-6)
6631 8309 rt : (8,6,8,5)
6631 8421 rt : (6,3,-8,6)
6631 8990 rt : (1,6,-4,-5)
6632 6368 ln : $3\zeta(3)/7$
6632 8796 sr : $8e$
6632 8825 rt : (2,7,-1,-3)
6632 8915 e^x : $9\sqrt[3]{3}/4$
6633 2183 $\ln(2)-\exp(\sqrt{5})$
6633 2495 sr : 11/25
6633 3181 Ψ : $4\sqrt[3]{3}$
6633 4001 rt : (7,6,5,-8)
6633 4210 e^x : $1/2+3e/4$
6633 7444 $\ln(3/4)+s\pi(2/5)$
6633 8156 rt : (5,8,-6,-1)
6633 9018 rt : (5,1,6,5)
6633 9302 rt : (1,7,-9,-9)
6634 1573 at : $(2\pi/3)^{-1/3}$
6634 1683 Li_2 : $7(1/\pi)/4$
6634 3744 rt : (4,1,-7,4)
6634 5116 2^x : $3\sqrt{3}+2\sqrt{5}$
6634 5415 2^x : $\sqrt{2}/2-3\sqrt{3}/4$
6634 5559 rt : (3,4,-8,7)
6634 6135 sq : $1/3+4e/3$
6634 7687 erf : $e/4$
6634 7814 rt : (8,0,1,3)
6634 8076 rt : (7,9,-5,-1)
6634 9119 rt : (1,3,-9,-7)
6634 9936 Ei : $(2\pi/3)^{1/3}$
6635 0304 id : $4\sqrt{3}/3+3\sqrt{5}/2$
6635 0952 ζ : $7\ln 2/6$
6635 1850 Ψ : $(e+\pi)^{1/2}$
6635 2190 Ei : 5/22
6635 2776 ln : $5e/7$
6635 2847 sr : $-\sqrt{2}+\sqrt{3}+\sqrt{6}$
6635 7387 rt : (2,4,4,-5)
6635 7475 id : $8e\pi/5$
6636 1372 rt : (7,9,3,-8)
6636 2159 K : $(\sqrt{2}\pi/3)^{-2}$
6636 4354 10^x : $3/4+3\sqrt{2}/4$
6636 5262 rt : (5,5,-1,-3)
6636 6365 rt : (4,8,-7,-7)
6636 6918 sπ : $4\sqrt[3]{3}$
6636 8580 λ : $\exp(-6\pi/7)$
6636 8703 as : $1/4-\sqrt{3}/2$
6637 0862 $\exp(1/4)-\exp(2/3)$
6637 1180 rt : (1,9,-6,4)
6637 1437 J : 11/24
6637 2502 rt : (1,7,-6,-7)
6637 2770 rt : (8,6,-6,-1)
6637 3281 e^x : $6\sqrt[3]{5}$
6637 3387 rt : (8,3,6,5)
6637 3794 rt : (1,9,-2,-5)
6637 4066 sq : $\sqrt{2}+\sqrt{5}/3$
6637 7972 10^x : $\sqrt[3]{15/2}$
6637 8042 sπ : $\ln 2/3$
6637 8220 erf : 17/25
6638 0415 10^x : $(\sqrt{5})^{-1/2}$
6638 1171 rt : (4,1,-4,9)
6638 1453 rt : (2,1,9,-7)
6638 2087 as : $8\ln 2/9$
6638 4587 rt : (7,9,-2,4)
6638 6997 e^x : $4/3-\sqrt{2}/4$
6638 8191 rt : (5,2,4,-5)
6639 0485 rt : (9,8,-8,6)
6639 2807 Ei : $\exp(-\sqrt[3]{5}/3)$
6639 3105 2^x : $2\sqrt{2}/3+2\sqrt{5}$
6639 3204 tπ : $1/4-2e$
6639 4072 2^x : $4\sqrt{3}-\sqrt{5}/4$
6639 4636 rt : (8,3,-1,-3)
6639 4809 sπ : $(\sqrt[3]{3}/3)^2$
6639 5448 rt : (1,9,8,-7)
6639 5973 sπ : $\sqrt{23/3}$
6639 8389 sr : $4+3\pi$
6639 8443 sπ : $3\sqrt{2}/4+3\sqrt{5}$
6639 8839 J_0 : 17/14
6639 8939 J : 1/6
6639 9357 sr : $\sqrt{23/3}$
6640 0000 cu : $9\sqrt[3]{2}/5$
6640 0732 J : $(\pi)^{-2}$
6640 1396 tπ : $\sqrt{3}/3+\sqrt{5}$
6640 1564 rt : (2,7,2,-5)
6640 2406 tanh : $((e+\pi)/3)^{-1/3}$
6640 3677 tanh : 4/5
6640 4616 at : 18/23
6640 5221 Ei : $1/\ln(e+\pi)$
6640 5925 2^x : $1/\ln(1/(2e))$
6640 6551 sr : $1-\sqrt{5}/4$
6640 7425 rt : (5,8,-3,-3)
6640 7848 rt : (5,1,9,7)
6641 0452 cr : $1-\sqrt{2}/2$
6641 0595 e^x : $2-2\sqrt{5}/3$
6641 1472 sinh : $3\sqrt[3]{5}/4$
6641 1893 e^x : $8(1/\pi)/5$
6641 3045 rt : (8,9,-8,-1)
6641 3442 rt : (8,0,4,5)
6641 3488 ln : $1+2\sqrt{2}/3$
6641 3574 rt : (7,9,1,9)
6641 4338 ln : $\sqrt[3]{22/3}$
6641 5099 as : $3\sqrt{2}/4-3\sqrt{5}/4$
6641 7599 Ψ : $(\sqrt[3]{5}/2)^{1/3}$
6641 8380 Ψ : $10\sqrt{3}/3$
6641 8990 ζ : $7\sqrt{2}/5$
6641 9427 rt : (2,4,7,-7)
6641 9721 cu : $2\ln 2$
6642 0827 $3\times\exp(1/5)$
6642 1356 id : $1/4+\sqrt{2}$
6642 4547 rt : (5,5,2,-5)
6642 6813 tπ : $1/\sqrt[3]{5}$
6642 7311 rt : (4,2,-9,-9)
6642 7370 sinh : $(1/(3e))^{-1/2}$
6642 9263 e^x : $7\sqrt[3]{2}/9$
6642 9470 rt : (8,6,-3,-3)
6642 9740 rt : (7,6,-7,4)
6642 9795 rt : (8,3,9,7)
6643 0009 rt : (1,9,-5,-7)
6643 2440 sr : $9\pi^2/2$
6643 6615 10^x : $9\sqrt{3}$
6643 8546 ζ : 3/20
6643 9530 rt : (5,2,7,-7)
6644 1193 Ei : $2\sqrt{2}/5$
6644 3879 rt : (8,3,2,-5)
6644 4448 cu : $4e+\pi$
6644 8123 cu : $4/3+2\sqrt{5}$
6644 8493 rt : (2,7,5,-7)
6645 0522 rt : (7,6,-4,9)
6645 2496 at : $(\sqrt[3]{3}/3)^{1/3}$
6645 2515 rt : (5,8,0,-5)
6645 2833 rt : (1,8,8,-6)
6645 2935 ℓ_{10} : $\sqrt{3}/8$
6645 5370 Γ : 8/15
6645 6397 rt : (8,9,-5,-3)
6645 6623 rt : (8,0,7,7)
6645 9368 ζ : $8\sqrt{3}/7$
6646 2475 ζ : $\sqrt[3]{7/3}$
6646 2970 J_0 : $7\ln 2/4$
6646 3469 Ψ : $e\pi/9$
6646 3651 sq : $1/3+3\sqrt{3}/4$
6646 4311 rt : (5,5,5,-7)
6646 5040 cu : $3/4-4\pi/3$
6646 5715 e^x : $4\pi^2/5$
6646 6088 sπ : $1/4+4\sqrt{5}/3$
6646 6386 at : $\sqrt{2}/2-2\sqrt{5}/3$
6646 7788 rt : (8,6,0,-5)
6646 8106 rt : (1,9,-8,-9)
6646 9891 2^x : $1+4e/3$
6647 0238 id : $e/3-\pi/2$
6647 1658 sr : $8\sqrt{3}/5$
6647 2085 ζ : $\zeta(3)/8$
6647 3487 tanh : $5\sqrt[3]{3}/9$
6647 3629 K : 11/24
6647 4202 J : $\exp(-11\pi/6)$
6647 4695 rt : (5,4,-9,-3)
6647 6501 rt : (4,7,-9,2)
6647 7999 rt : (8,3,5,-7)
6647 8053 rt : (7,3,-9,9)
6647 9615 sq : $3\sqrt{2}-3\sqrt{3}/4$
6648 0268 tanh : $2\zeta(3)/3$
6648 1292 rt : (2,7,8,-9)
6648 4222 rt : (5,8,3,-7)
6648 4841 10^x : $\sqrt{2}-\sqrt{5}+\sqrt{7}$
6648 6353 10^x : $\sqrt[3]{22/3}$
6648 7065 rt : (8,9,-2,-5)
6648 7408 sq : $\sqrt{3}/3+4\sqrt{5}/3$
6649 1656 J_0 : $7\sqrt{3}/10$
6649 2843 rt : (5,5,8,-9)
6649 3727 2^x : $4e/3+2\pi/3$
6649 5207 e^x : $2\sqrt{2}/3-\sqrt{3}/4$
6649 5428 rt : (8,6,3,-7)
6649 8135 rt : (2,8,0,-9)
6650 0569 rt : (5,4,-6,2)
6650 1504 sq : $3e/10$
6650 2171 cu : $2+3\pi/4$
6650 2865 cr : 5/17
6650 3043 rt : (8,3,8,-9)
6650 4318 J_0 : $6\sqrt{2}/7$
6650 5311 ℓ_{10} : $1+4e/3$
6650 6360 10^x : $\sqrt{2}-\sqrt{5}/3$
6650 7264 erf : 15/22
6650 7741 rt : (5,8,6,-9)
6650 8746 ℓ_2 : $7e/3$
6650 8786 $\ln(3/5)-\exp(e)$
6650 8962 sinh : $9\ln 2/10$
6650 9912 rt : (8,9,1,-7)
6651 0922 sr : $4\ln 2$
6651 4414 2^x : $\sqrt{2}$
6651 5231 cu : $4(e+\pi)/9$
6651 5653 tπ : $4(e+\pi)/3$
6651 6292 $4\times\ln(2/5)$
6651 6315 rt : (8,6,6,-9)
6651 7018 cu : $2+4\sqrt{2}/3$
6651 8380 ℓ_2 : $3\sqrt{2}/4-\sqrt{5}/3$
6651 8395 rt : (2,8,3,2)
6651 9142 id : $7\pi/6$
6651 9506 10^x : $3/2-3\sqrt{5}/4$
6651 9849 ℓ_2 : $3-\sqrt{2}$
6652 0290 rt : (5,4,-3,7)
6652 4616 id : P_{81}
6652 7211 sπ : $\sqrt{2}+3\sqrt{5}/2$
6652 7596 rt : (8,9,4,-9)
6652 9041 id : $3\sqrt{2}-\sqrt{3}/3$
6653 2596 rt : (8,9,-9,-3)
6653 3353 Γ : $\sqrt{3}/7$
6653 4220 at : $1/3-\sqrt{5}/2$
6653 6635 cr : $8\sqrt{3}/3$
6653 9000 Γ : $4\sqrt{3}$
6654 0881 cu : $4-\sqrt{3}$
6654 1906 sπ : $\exp(\sqrt[3]{5}/3)$
6654 2810 tπ : $\sqrt{2}-4\sqrt{5}/3$
6654 4105 sinh : $\exp(-\sqrt{2}/3)$
6654 4448 rt : (5,1,-8,7)
6654 5745 rt : (8,9,-6,2)
6655 0491 ln : $\sqrt{3}-\sqrt{5}+\sqrt{6}$
6655 1264 tπ : $3\sqrt{2}+4\sqrt{5}$
6655 2249 cu : $2e/3+4\pi$
6655 2260 sr : $e/4+2\pi/3$
6655 3981 cu : $3-\sqrt{5}/2$
6655 4624 ℓ_{10} : $(5/3)^3$
6655 6233 ℓ_2 : $9\pi^2/7$
6655 6544 rt : (8,9,-3,7)
6655 8088 erf : $1/\ln(\ln 2/3)$
6656 0278 cu : $4\sqrt{2}+3\sqrt{5}/2$
6656 3145 sq : $3+4\sqrt{5}$
6656 3375 Ψ : $7\sqrt{3}/5$
6656 3735 rt : (2,5,-8,2)

6656 5428 10^x : $\sqrt{3}/4+4\sqrt{5}$
6657 0800 rt : (8,6,-8,7)
6657 1142 $s\pi$: $\ln(e+\pi)$
6657 1772 2^x : $1/4+4e/3$
6657 3810 ln : $2e/3+4\pi$
6657 4820 rt : (1,8,-6,-2)
6657 5686 erf : $(\pi)^{-1/3}$
6657 6441 e^x : $6\pi^2$
6657 6888 e^x : $3\sqrt{3}/4$
6657 7101 $\exp(2) \div s\pi(\sqrt{2})$
6657 7375 at : $\pi/4$
6657 9156 Ψ : $((e+\pi)/2)^{-3}$
6658 3075 K : $(3\pi/2)^{-1/2}$
6658 7349 J_0 : 23/19
6658 9472 $t\pi$: $\sqrt{2}-\sqrt{3}+\sqrt{7}$
6659 2570 rt : (2,8,-6,1)
6659 3687 e^x : $4e/3+3\pi/4$
6659 4445 at : $5\sqrt{2}/9$
6659 6923 at : 11/14
6659 7419 ln : $4\sqrt{3}/3+4\sqrt{5}/3$
6659 7571 10^x : $(\pi)^{-1/2}$
6660 1274 Γ : $8\pi/9$
6660 1586 $e+\exp(2/3)$
6660 1724 sr : $3e-\pi/3$
6660 2540 $1/5-s\pi(1/3)$
6660 3596 cr : $1+4e/3$
6660 4559 cu : $3\sqrt{2}/2+\sqrt{3}/4$
6660 7919 10^x : $9\sqrt[3]{2}/2$
6661 1273 $t\pi$: $(2\pi/3)^3$
6661 1615 2^x : 20/9
6661 3092 $s\pi$: $1/2+\sqrt{3}$
6662 1560 Ψ : $7\ln 2/2$
6662 3943 as : $1/2-\sqrt{5}/2$
6662 4703 K : 6/13
6662 4757 e^x : $1/2-e/3$
6662 4977 ℓ_{10} : $3\sqrt{3}-\sqrt{5}/4$
6662 9073 $1/4+\ln(2/5)$
6663 0616 $\ln(\sqrt{3})+\exp(3/4)$
6663 0621 Li_2 : $\sqrt{5}/4$
6663 2159 sq : $4/3+3\sqrt{5}$
6663 3225 as : $3\sqrt[3]{3}/7$
6663 4203 $\ln(3) \div \exp(1/2)$
6663 6919 $\sqrt{5}-s\pi(\pi)$
6664 0755 Ei : $\beta(2)$
6664 0786 sq : $2+3\sqrt{5}/2$
6664 4481 10^x : $\exp(-1/(3e))$
6664 5588 $t\pi$: $4\sqrt{2}/3-\sqrt{3}/4$
6664 8265 2^x : $(5/2)^{-1/3}$
6664 9226 sinh : 5/8
6664 9568 at : $\sqrt{2}/4+\sqrt{3}/4$
6665 1255 K : $2\ln 2/3$
6665 2401 2^x : 14/19
6665 2698 θ_3 : 9/14
6665 3534 sr : $3\sqrt{2}/2-3\sqrt{5}/4$
6665 4071 cu : $e-\pi/3$
6665 5782 $s\pi$: $5\sqrt{2}/4$
6665 7869 J : $(2e/3)^{-2}$
6665 7911 cr : $\sqrt{2}-\sqrt{5}/2$
6665 8582 sinh : $9e/5$
6666 0196 sq : $3+\sqrt{5}/4$
6666 0274 as : $\exp(-\sqrt[3]{3}/3)$
6666 1014 erf : $2\sqrt[3]{5}/5$
6666 3933 cu : $1/\ln(\pi)$
6666 4020 $s\pi$: $9e/2$
6666 4265 $t\pi$: $4\sqrt{3}/3-4\sqrt{5}/3$
6666 6666 id : 2/3
6666 6702 ℓ_{10} : $3/2+\pi$
6667 2517 Γ : $\exp(-\sqrt[3]{2}/2)$
6667 5158 $\sqrt{3}\times\exp(3/4)$
6667 6571 erf : 13/19
6667 9216 e^x : $\sqrt{3}/3+\sqrt{5}$
6669 2729 sq : $4/3+\pi/3$
6669 3858 10^x : $8\pi/5$
6669 5575 e^x : P_{13}
6669 5729 ℓ_{10} : $4/3-\sqrt{5}/2$
6669 7300 cu : $3/4+4\sqrt{3}$
6669 8594 $s\pi$: $4\sqrt{2}/3-\sqrt{5}/2$
6669 9110 ln : $\sqrt{21}\pi$
6670 4192 at : $5\sqrt[3]{2}/8$
6670 7560 J_0 : $4e/9$
6670 8917 ln : $9\sqrt{3}/8$
6671 1654 θ_3 : 19/24
6671 5060 Ei : $9\sqrt[3]{2}/7$
6671 7657 e^x : $3/4-2\sqrt{3}/3$
6671 7863 rt : (1,9,3,1)
6672 3475 e^x : $1/4+3e$
6672 4125 cr : $3\ln 2/7$
6672 5900 id : P_{26}
6672 8712 Ψ : 18/19
6672 9466 Ei : $\ln(5/2)$
6673 2427 $\exp(5)+s\pi(\sqrt{3})$
6673 7290 at : $4/3-3\sqrt{2}/2$
6673 7492 $t\pi$: $e+2\pi/3$
6674 0601 e^x : $\sqrt{2}+\sqrt{5}/3$
6674 1725 rt : (1,6,-4,3)
6674 6268 e^x : $8\zeta(3)/3$
6674 9350 ln : $3\sqrt[3]{5}/10$
6674 9472 K : $1/\sqrt[3]{10}$
6675 1457 Ψ : 17/7
6675 2531 ln : $4+3\sqrt{3}/4$
6675 2944 as : 13/21
6675 4253 e^x : $3\sqrt{3}/2+2\sqrt{5}/3$
6675 5308 e^x : $4\sqrt{2}-\sqrt{5}/4$
6675 5694 id : $\sqrt{2}/3+3\sqrt{3}$
6675 5811 sr : $7(1/\pi)/5$
6675 6179 ℓ_{10} : $2e-\pi/4$
6675 6524 Γ : $\ln(\ln 3/3)$
6675 6687 cr : $3\sqrt{3}-\sqrt{5}/4$
6675 7476 sr : $\sqrt{2}/3+4\sqrt{3}/3$
6675 8754 $s\pi$: $-\sqrt{2}+\sqrt{3}+\sqrt{6}$
6676 3506 Γ : $1/\ln(1/(2e))$
6676 5520 sr : $1/\ln(3\pi)$
6676 5780 e^x : $\sqrt{2}-\sqrt{3}/4$
6676 6399 J_0 : $(\ln 3)^2$
6676 6661 sr : $5e\pi/6$
6676 6682 rt : (7,4,-8,8)
6676 7610 rt : (4,8,-6,3)
6676 7788 Ψ : $\sqrt{2}+\sqrt{3}+\sqrt{7}$
6676 8824 ℓ_2 : $3+3\sqrt{5}/2$
6676 9520 Li_2 : $4\sqrt[3]{2}/9$
6677 1789 ζ : $(\sqrt[3]{4}/3)^{1/3}$
6677 3737 $\sqrt{5}\times s\pi(\sqrt{3})$
6677 4612 ℓ_{10} : $3\sqrt{3}/4+3\sqrt{5}/2$
6677 4664 Li_2 : 14/25
6677 5491 10^x : $3/2-\sqrt{3}/4$
6677 6591 rt : (4,8,-9,-5)
6677 7950 $\ln(2/3)\times\exp(\sqrt{2})$
6678 0939 id : $5\zeta(3)/9$
6678 2290 rt : (7,7,-3,8)
6679 2386 10^x : $2e+3\pi/4$
6679 2530 $t\pi$: $1/\ln(\sqrt[3]{2})$
6679 2869 rt : (1,6,1,-5)
6679 4223 rt : (7,7,-6,3)
6679 4470 ζ : $7\sqrt[3]{5}/10$
6679 4867 cr : $5\pi^2$
6679 5507 e^x : $3\sqrt{3}-2\sqrt{5}/3$
6679 5743 rt : (4,1,-8,-8)
6679 9783 10^x : $2\sqrt{2}/3-\sqrt{5}/2$
6680 1617 2^x : 15/8
6680 2718 $\sqrt{2}\div\exp(3/4)$
6680 4159 sinh : $9\sqrt[3]{3}/5$
6680 7170 10^x : $4-\sqrt{3}/4$
6680 8753 $\ln(2/5)\div\ln(\sqrt{3})$
6680 8898 rt : (7,7,-9,-2)
6680 9579 $s\pi$: $5e\pi/3$
6680 9708 J_0 : $(\sqrt[3]{5}/3)^{-1/3}$
6681 0053 10^x : 2/9
6681 0099 cr : $3/2+\pi$
6681 2632 $\ln(2)\times s\pi(\sqrt{2})$
6681 4622 rt : (7,8,5,-9)
6681 4653 at : $2\sqrt{2}/3-\sqrt{3}$
6681 6326 ℓ_2 : $4+3\pi/4$
6681 7863 $t\pi$: 3/16
6681 9189 sinh : $e\pi/5$
6682 0251 ζ : $(\ln 3)^3$
6682 2975 rt : (4,2,-3,8)
6682 7483 rt : (7,5,7,-9)
6682 8837 id : $3\sqrt{3}+2\sqrt{5}$
6682 8941 at : 15/19
6683 1107 $\ln(3)+s\pi(\pi)$
6683 2034 Ei : 11/12
6683 2387 rt : (2,9,-2,-2)
6683 2969 sq : $2e\pi/5$
6683 4734 rt : (7,8,2,-7)
6683 6148 e^x : $4\sqrt{2}/3+\sqrt{3}/2$
6683 6656 ℓ_{10} : $1-\pi/4$
6683 7404 rt : (4,7,7,-9)
6683 7873 $\sqrt{2}+s\pi(\sqrt{3})$
6684 0164 id : $\sqrt[3]{19}$
6684 2796 rt : (7,2,9,-9)
6684 3117 10^x : $3\sqrt{3}+2\sqrt{5}/3$
6684 3913 $s\pi$: $2/3+4\pi$
6684 4352 erf : $(\sqrt{2}/3)^{1/2}$
6684 5599 rt : (4,2,-6,3)
6684 5741 $t\pi$: P_9
6684 8636 rt : (1,9,7,-9)
6684 9894 $\ln(2/3)\times\exp(1/2)$
6685 1526 rt : (7,5,4,-7)
6685 1720 ℓ_2 : $2\sqrt{2}/3+\sqrt{5}$
6685 4384 $t\pi$: $1/3+\pi$
6685 4774 rt : (4,4,9,-9)
6685 5297 tanh : $4\sqrt{2}/7$
6685 5974 $\ln(3/4)\div s\pi(\pi)$
6686 0728 Γ : $4e\pi/9$
6686 1160 rt : (7,8,-1,-5)
6686 4763 rt : (4,7,4,-7)
6686 5518 erf : $4\zeta(3)/7$
6686 8509 rt : (1,6,9,-9)
6686 9946 K : 7/15
6687 0163 rt : (8,7,-7,6)
6687 2052 rt : (7,2,6,-7)
6687 4030 id : $(\sqrt{5})^{-1/2}$
6687 5864 rt : (4,2,-9,-2)
6687 8370 e^x : $10\sqrt[3]{2}/3$
6687 8918 ln : $10\sqrt[3]{3}$
6688 0052 rt : (1,9,4,-7)
6688 1080 sinh : $(\sqrt{2}/3)^{-1/3}$
6688 4003 rt : (7,5,1,-5)
6688 4420 rt : (2,8,-9,-9)
6688 5685 tanh : $7\ln 2/6$
6688 5756 id : $\sqrt{2}-\sqrt{5}/3$
6688 6459 rt : (2,6,-3,7)
6688 8534 rt : (4,4,6,-7)
6689 3172 rt : (3,9,-6,5)
6689 4214 tanh : $(\sqrt[3]{4}/3)^{1/3}$
6689 5121 sr : $2+\pi/4$
6689 7415 rt : (7,8,-4,-3)
6689 7713 rt : (7,1,8,7)
6689 8014 rt : (1,7,-5,1)
6690 0678 ℓ_{10} : 3/14
6690 2540 rt : (4,7,1,-5)
6690 7908 rt : (3,9,-9,-6)
6690 8067 id : $\sqrt{3}/4+\sqrt{5}$
6690 8466 erf : 11/16
6690 9801 ln : $\ln 2/10$
6691 2932 rt : (7,2,3,-5)
6691 5569 $3/4\times\exp(4/5)$
6691 6946 2^x : 17/23
6691 7222 id : $1/\ln(\ln 3/2)$
6691 8801 rt : (4,1,8,-7)
6692 0558 rt : (9,7,-8,9)
6692 1923 10^x : $2+2\sqrt{5}$
6692 4554 rt : (1,9,1,-5)
6692 4697 cr : $2e-\pi/4$
6692 6549 ln : $3-\pi/3$
6692 6893 10^x : $4\sqrt{2}/3+2\sqrt{3}$
6692 7854 cr : P_{79}
6692 9666 e^x : 13/10
6693 0272 rt : (7,5,-2,-3)
6693 0644 rt : (5,2,-7,6)
6693 1017 rt : (2,8,-6,-7)
6693 2055 10^x : $3\sqrt{3}-\sqrt{5}/2$
6693 2490 2^x : $\ln(2\pi/3)$
6693 2748 tanh : 17/21
6693 3782 cu : $e+3\pi/4$
6693 3834 $e+s\pi(2/5)$
6693 7026 rt : (4,8,7,8)
6694 2123 Ψ : $\sqrt{5}-\sqrt{6}+\sqrt{7}$
6694 2148 sq : 9/11
6694 2520 ζ : 2/13

6694 3065 $3 \times \ln(4/5)$
6694 3295 cr : $3/10$
6694 3598 ln : P_{36}
6694 4155 rt : (1,3,8,-7)
6694 5366 tπ : $\sqrt{2}/2 + 3\sqrt{5}$
6694 6316 ℓ_{10} : $4e/3 + \pi/3$
6694 7069 J : $\exp(-11\pi/7)$
6694 7903 ln : $3 + 4\sqrt{3}/3$
6694 8315 cr : $3\sqrt{3}/4 + 3\sqrt{5}/2$
6695 0205 rt : (7,8,-7,-1)
6695 0758 rt : (7,1,5,5)
6695 1589 id : $\ln((e+\pi)/3)$
6695 2816 $\sqrt{3} \times s\pi(\sqrt{2})$
6695 8058 rt : (4,7,-2,-3)
6695 8761 Γ : $5\sqrt{5}/4$
6695 9056 $\ln(2/5) \times \exp(3/5)$
6695 9787 tπ : $2e/3$
6696 1886 sπ : $5(e+\pi)/3$
6696 3894 at : $19/24$
6696 4472 id : $9\sqrt[3]{2}/2$
6696 5203 Ei : $(\sqrt{5}/3)^{-3}$
6696 6384 rt : (1,6,3,-5)
6696 7088 ζ : $2\ln 2/9$
6696 7970 2^x : $17/12$
6697 0171 rt : (8,8,3,-8)
6697 1022 $\ln(5) \div s\pi(\sqrt{2})$
6697 1052 cr : $7e$
6697 2662 sr : $2/3 + 3\sqrt{2}/2$
6697 4047 rt : (7,2,0,-3)
6697 5229 rt : (2,5,-8,-7)
6697 5481 sπ : $(\sqrt{2}\pi/2)^{-1/3}$
6697 7689 Li_2 : $\exp(-\sqrt{3}/3)$
6698 0351 rt : (6,2,-8,9)
6698 1211 Ei : $5\pi/6$
6698 1522 cr : $\zeta(3)/4$
6698 3352 rt : (4,1,5,-5)
6698 3905 rt : (1,8,-5,7)
6698 4230 ℓ_{10} : $3\sqrt{2} + \sqrt{3}/4$
6698 5405 tπ : $3\sqrt{2}/2 - 3\sqrt{3}/2$
6698 6575 ln : $3e + 2\pi$
6698 7830 rt : (5,5,-2,6)
6698 7883 sπ : $1/3 + \sqrt{3}/4$
6699 1329 10^x : $4 + 4\sqrt{2}$
6699 2187 cu : $7/8$
6699 2433 rt : (1,9,-2,-3)
6699 2458 sr : $\pi/7$
6699 2500 ℓ_2 : $9\sqrt{2}$
6699 2613 $\exp(\sqrt{3}) \div \exp(3/4)$
6699 2853 rt : (2,9,8,7)
6699 4352 sr : $3/4 + 3e/4$
6699 4718 at : $8\ln 2/7$
6699 4995 rt : (2,7,-9,9)
6699 7210 ℓ_{10} : $3 + 3\sqrt{5}/4$
6699 7312 rt : (8,5,5,-8)
6699 8748 sπ : $\pi^2/8$
6700 1069 ln : $8\sqrt[3]{5}/7$
6700 1428 rt : (7,5,-5,-1)
6700 2340 rt : (7,4,7,5)
6700 2956 rt : (2,8,-3,-5)
6700 7918 ζ : $\pi^2/5$
6700 9998 ℓ_{10} : $\sqrt[3]{5}/8$
6701 0727 J_0 : $\zeta(3)$
6701 0830 sr : P_{41}
6701 1277 2^x : $3\sqrt{2}/4 + 3\sqrt{5}/4$
6701 2541 rt : (8,8,0,-6)
6701 4230 rt : (5,7,-8,-7)
6701 6390 J_0 : $\exp(1/(2e))$
6701 8397 rt : (5,7,5,-8)
6701 8834 Ψ : $(\sqrt{5}/2)^{-1/2}$
6701 9818 J_0 : $5\sqrt[3]{3}/6$
6702 0027 2^x : $3\sqrt{3}/4 - \sqrt{5}/4$
6702 0599 ℓ_{10} : $4 + e/4$
6702 4469 rt : (1,3,5,-5)
6702 5182 sq : $4\sqrt{3}/3 - 2\sqrt{5}/3$
6702 5402 $s\pi(1/3) \times s\pi(e)$
6702 5449 sr : $3\sqrt{3}/4 + 2\sqrt{5}/3$
6702 5768 10^x : $4\sqrt[3]{3}/7$
6702 6755 10^x : $\sqrt{19}$
6702 7590 rt : (1,7,1,-8)
6702 8251 e^x : $3\sqrt[3]{5}/10$
6702 9796 rt : (8,2,7,-8)
6703 1866 rt : (6,5,-3,9)
6703 1977 $\ln(4) + \exp(1/4)$
6703 2004 $1 \div \exp(2/5)$
6703 4984 10^x : $\exp(\sqrt{2}\pi/3)$
6703 5267 rt : (7,1,2,3)
6703 5674 rt : (5,5,-5,1)
6703 6217 rt : (3,6,-2,9)
6703 6462 Γ : $\exp((e+\pi)/2)$
6703 8630 10^x : $7(1/\pi)/10$
6703 8658 rt : (5,9,-9,7)
6703 8971 $1/2 \div s\pi(\sqrt{3})$
6703 9243 rt : (1,9,-7,-5)
6703 9249 at : $3/2 - \sqrt{2}/2$
6703 9886 sinh : $9/7$
6704 2380 rt : (2,9,5,-4)
6704 2664 as : $3/2 - 3\sqrt{2}/2$
6704 3196 ℓ_2 : $\pi/5$
6704 4526 as : $1/\ln(5)$
6704 4785 10^x : $1/4 + 4\pi/3$
6704 7554 rt : (4,7,-5,-1)
6704 8224 rt : (8,5,2,-6)
6704 8399 sinh : $\pi/5$
6704 8560 sr : $e/3 + 4\pi$
6704 8899 rt : (4,2,7,5)
6705 1566 Ψ : $7e\pi/4$
6705 5415 rt : (5,4,7,-8)
6705 5432 id : $1/\ln(\sqrt{2}\pi)$
6705 5522 sr : $1/\sqrt[3]{11}$
6705 7329 J_0 : $(\ln 2)^{-1/2}$
6706 0211 2^x : $4 - 3\sqrt{2}/4$
6706 0454 K : $8/17$
6706 2150 rt : (1,6,0,-3)
6706 3481 cu : $3 + 4\sqrt{3}/3$
6706 4059 tπ : P_{35}
6706 4685 rt : (8,9,-8,-7)
6706 5940 e^x : $2e\pi/9$
6706 6496 cu : $2/3 + 3\sqrt{2}/2$
6706 6575 J_0 : $(\sqrt[3]{3})^{1/2}$
6706 8589 rt : (8,8,-3,-4)
6706 9377 rt : (8,1,9,8)
6706 9590 Ei : $8\sqrt[3]{3}/9$
6706 9662 2^x : $\sqrt{2} + 3\sqrt{5}$
6707 1669 rt : (6,5,-6,4)
6707 2160 2^x : $3/4 + 4\pi/3$
6707 2264 sq : $(\ln 3/2)^{1/3}$
6707 3809 Γ : $5e$
6707 3812 J_0 : $\sqrt{2} + \sqrt{5} - \sqrt{6}$
6707 4436 $\ln(\sqrt{5}) + s\pi(1/3)$
6707 4988 sinh : $4\sqrt{2}/9$
6707 5216 rt : (7,2,-3,-1)
6707 5915 sπ : $4e/3 + \pi$
6707 6600 rt : (5,8,3,6)
6707 6875 rt : (7,7,9,5)
6707 7065 rt : (3,6,-5,1)
6707 7427 $e - \exp(2)$
6707 8002 rt : (2,5,-5,-5)
6707 8493 Ei : $7e/3$
6708 1651 rt : (9,8,4,-9)
6708 2039 id : $3\sqrt{5}/10$
6708 2082 tπ : $3\sqrt{2}/2 - 4\sqrt{3}/3$
6708 4962 rt : (2,6,7,-8)
6708 8797 at : $\sqrt[3]{1/2}$
6709 2133 rt : (8,2,4,-6)
6709 2403 $\ln(\sqrt{3}) \times \exp(1/5)$
6709 2446 rt : (7,5,-7,7)
6709 3528 ℓ_{10} : $4/3 + 3\sqrt{5}/2$
6709 3826 id : $(\sqrt{2}\pi/2)^{-1/2}$
6709 4047 cr : $e/9$
6709 6707 tanh : $13/16$
6709 6925 rt : (4,9,-3,4)
6709 7830 ℓ_{10} : $8(e+\pi)$
6709 9142 e^x : $\exp(-2/3)$
6710 0183 rt : (5,5,-8,-4)
6710 0436 K : $\sqrt{2}/3$
6710 0931 $s\pi(1/4) - s\pi(\sqrt{2})$
6710 1172 rt : (5,1,9,-8)
6710 3603 rt : (6,8,2,9)
6710 5777 2^x : $9\sqrt[3]{2}/8$
6710 6080 ℓ_{10} : $1/2 + 4\pi/3$
6710 8427 id : $e - \pi/3$
6710 8532 rt : (1,9,-5,-1)
6710 8551 sr : $8\pi/9$
6710 9579 rt : (2,9,2,-6)
6710 9931 id : $\sqrt[3]{14/3}$
6711 0636 rt : (1,0,7,5)
6711 0905 ℓ_2 : $4\sqrt{3} - \sqrt{5}/4$
6711 3274 J_0 : $6/5$
6711 3762 ℓ_2 : $4\sqrt{2}/3 + 3\sqrt{3}/4$
6711 4062 sr : $3\sqrt{2}/4 + \sqrt{3}$
6711 5237 rt : (9,5,6,-9)
6711 6201 sπ : $4\sqrt{2} + \sqrt{3}/3$
6711 6475 cu : $2 + 3\sqrt{2}/4$
6711 7562 rt : (8,5,-1,-4)
6711 8030 e^x : $4/3 - \sqrt{3}$
6711 8756 ln : $7\sqrt{5}/8$
6711 9428 rt : (1,8,-7,-8)
6712 1134 ℓ_{10} : $\sqrt{22}$
6712 1163 rt : (6,5,-9,-1)
6712 4639 rt : (7,5,-8,1)
6712 6972 rt : (7,4,4,3)
6712 7766 rt : (5,4,4,-6)
6712 8031 rt : (3,6,-8,-7)
6712 8181 tπ : $4/9$
6712 8565 rt : (2,8,0,-3)
6713 1818 rt : (1,8,2,-4)
6713 2336 $\sqrt{2} - \exp(3)$
6713 3749 rt : (9,8,1,-7)
6713 4745 tπ : $3\sqrt{2}/4 + 2\sqrt{5}$
6713 4805 tπ : $\sqrt{3} - \sqrt{5}/2$
6713 7034 cu : $\sqrt{2}/4 - \sqrt{5}$
6713 7052 ℓ_{10} : $4\sqrt{3} - \sqrt{5}$
6713 7848 rt : (4,9,-6,-4)
6713 8483 rt : (2,3,9,-8)
6713 8752 ln : P_{13}
6713 9063 id : $3\sqrt{2}/4 - \sqrt{3}$
6713 9506 sr : $3\zeta(3)/8$
6714 1022 rt : (6,7,6,-9)
6714 1624 Li_2 : $9/16$
6714 2057 10^x : $3\sqrt[3]{3}/2$
6714 6154 rt : (8,8,-6,-2)
6714 7490 rt : (8,1,6,6)
6714 8842 rt : (4,5,9,5)
6715 0682 rt : (5,7,-5,-5)
6715 0990 2^x : $(\ln 3/2)^{1/2}$
6715 2813 K : $3\sqrt[3]{2}/8$
6715 3233 erf : $(2\pi/3)^{-1/2}$
6715 4756 rt : (9,2,8,-9)
6715 7287 sq : $2 - \sqrt{2}/2$
6715 7332 id : $4e/3 + \pi/3$
6715 7756 rt : (5,8,0,1)
6715 9152 10^x : $\exp(-3/2)$
6715 9808 Ψ : $2\sqrt{2}/5$
6716 1761 rt : (6,8,-1,4)
6716 1922 Ψ : $3\sqrt[3]{2}/4$
6716 3434 ln : $\sqrt[3]{15/2}$
6716 3790 rt : (7,8,-2,7)
6716 7109 10^x : $\sqrt{20}\pi$
6716 8478 cr : $4e/3 + \pi/3$
6716 9562 sinh : $8(e+\pi)/5$
6716 9988 rt : (3,9,6,9)
6717 2699 ln : $\ln(5/3)$
6717 4761 J : $9/23$
6717 6720 rt : (9,5,3,-7)
6718 0143 rt : (8,2,1,-4)
6718 0510 2^x : $\sqrt[3]{11}$
6718 0705 2^x : $\exp(\sqrt[3]{4})$
6718 2106 cu : $1/2 - 2\sqrt{5}$
6718 2540 tπ : $e/3 - 2\pi/3$
6718 2905 rt : (1,5,-9,-8)
6718 5076 sr : $5\sqrt{5}/4$
6718 5477 rt : (6,4,8,-9)
6718 9785 sπ : $e + \pi/3$
6719 0653 rt : (7,1,-1,1)
6719 3580 rt : (5,1,6,-6)
6719 6573 rt : (2,8,3,-8)
6719 7362 ℓ_2 : $3\sqrt{2}/4 - \sqrt{3}/4$
6719 9074 tπ : $4e/3 + \pi/3$
6719 9444 K : $(3\pi)^{-1/3}$
6719 9607 tπ : $1/2 + 2\sqrt{2}$

6719 9903 sr : $\ln(\pi/2)$
6720 0344 λ : $\exp(-17\pi/20)$
6720 0706 rt : (9,8,-2,-5)
6720 1734 rt : (9,8,-7,8)
6720 2289 id : $4\sqrt{3}/3 - 4\sqrt{5}/3$
6720 5189 Ei : $3\sqrt[3]{5}/4$
6720 5903 rt : (2,9,-1,-4)
6720 6139 Ψ : $1/\ln(e+\pi)$
6720 6303 tπ : P_{36}
6720 8561 rt : (8,8,-6,5)
6720 9661 $\ln(\sqrt{3}) - \exp(1/5)$
6720 9739 θ_3 : $8\ln 2/7$
6721 0347 rt : (6,7,3,-7)
6721 2697 K : $9/19$
6721 4050 10^x : $\ln((e+\pi)/3)$
6721 4573 θ_3 : $(\sqrt[3]{5}/3)^2$
6721 4853 Γ : $6\sqrt[3]{3}$
6721 5436 rt : (4,7,-8,1)
6721 5923 at : $5(1/\pi)/2$
6721 5985 ζ : $\sqrt[3]{23/3}$
6721 7131 cr : $3\sqrt{2} + \sqrt{3}/4$
6721 7434 rt : (8,5,-4,-2)
6721 7860 Γ : $(3\pi/2)^{-2}$
6721 9127 rt : (7,8,-5,2)
6721 9461 rt : (8,4,8,6)
6722 0380 rt : (3,6,8,-9)
6722 0841 rt : (1,8,-4,-6)
6722 2131 Ψ : $\exp(-\sqrt[3]{5}/3)$
6722 3441 e^x : $4 - e/4$
6722 6868 rt : (4,0,-7,7)
6722 8421 rt : (9,2,5,-7)
6723 0422 θ_3 : $(\pi/3)^{-1/2}$
6723 0549 sr : $\sqrt{3}/4 + 3\sqrt{5}$
6723 1320 ln : $2 - 2e/3$
6723 2904 2^x : $1/3 - e/3$
6723 3006 rt : (5,4,1,-4)
6723 3790 cr : $3 + 3\sqrt{5}/4$
6723 4321 sπ : $9\pi^2/5$
6723 6722 rt : (4,7,-9,-8)
6723 7584 sr : $4 + \pi$
6723 7663 rt : (6,8,-4,-1)
6723 8025 erf : $\exp(-1/e)$
6723 8112 e^x : $\sqrt{2}/4 + 2\sqrt{5}$
6723 9348 10^x : $\sqrt[3]{4}$
6724 3470 sr : $\exp(-\sqrt[3]{4}/2)$
6724 4336 e^x : $1/4 + 4\pi/3$
6724 5506 erf : $9/13$
6724 6181 sinh : $\sqrt[3]{2}/2$
6724 6321 rt : (2,8,-5,6)
6724 6844 λ : $\ln 2/10$
6724 7201 rt : (1,6,-3,-1)
6724 8241 tπ : $4\sqrt{2} + 2\sqrt{3}/3$
6724 8783 rt : (3,9,3,1)
6724 9368 K : $(\sqrt{2}\pi)^{-1/2}$
6724 9581 rt : (2,3,6,-6)
6724 9851 $s\pi(1/4) \times s\pi(2/5)$
6725 0697 Ψ : $17/18$
6725 1867 sq : $2\sqrt{2}/3 - \sqrt{5}$
6725 1999 rt : (1,3,9,5)
6725 4462 cu : $3/4 + 3\sqrt{2}/2$
6725 5317 rt : (8,9,-5,-5)
6725 6325 K : $\exp(-\sqrt{5}/3)$
6725 7592 rt : (9,5,0,-5)
6726 0393 rt : (8,8,-9,0)
6726 0470 tanh : $3e/10$
6726 0771 sπ : $\sqrt[3]{11/2}$
6726 2929 rt : (8,1,3,4)
6726 3483 J_0 : $7\sqrt[3]{5}/10$
6726 3814 cr : $4 + e/4$
6726 5142 cr : $7/23$
6726 9548 rt : (6,4,5,-7)
6727 4085 rt : (7,2,-6,1)
6727 5405 ℓ_{10} : $4 + \sqrt{2}/2$
6727 6674 sq : $(2e/3)^{-1/3}$
6727 8635 rt : (5,8,-3,-4)
6727 9541 rt : (7,7,6,3)
6728 1312 erf : $2\sqrt{3}/5$
6728 1421 rt : (5,2,8,6)
6728 2370 rt : (2,8,-8,-8)
6728 3097 e^x : $3\sqrt{2} - 2\sqrt{5}/3$
6728 3324 rt : (2,5,-2,-3)
6728 4308 e^x : $\sqrt{2} + \sqrt{3} + \sqrt{7}$
6728 4930 2^x : $3\sqrt{3}/7$
6728 5534 $\ln(\pi) \times s\pi(1/5)$
6728 8230 rt : (7,8,-8,-3)
6728 9869 rt : (9,8,-5,-3)
6728 9966 cu : $8(e+\pi)/3$
6729 0597 sr : $3/2 - \pi/3$
6729 1824 rt : (9,1,7,7)
6729 2756 e^x : $1/2 + 2\sqrt{3}$
6729 8164 rt : (2,6,1,-2)
6729 9884 2^x : $2e\pi/5$
6730 1147 10^x : $8e/3$
6730 2979 at : $1/4 - \pi/3$
6730 3241 rt : (6,7,0,-5)
6730 3260 sr : $3/2 + 3\sqrt{3}/4$
6730 3340 rt : (7,9,-9,-8)
6730 3497 sπ : $2e/3 - \pi/3$
6730 4045 id : $7(e+\pi)/3$
6730 4128 erf : $\ln 2$
6730 6792 2^x : $1/2 + 4\pi$
6730 8121 Γ : $5(1/\pi)/3$
6730 8151 $\exp(1/4) + \exp(2)$
6730 8764 sr : $e/6$
6731 0463 sq : $3/2 - e/4$
6731 2207 rt : (4,7,-8,5)
6731 3311 2^x : $7\sqrt[3]{2}/3$
6731 3595 rt : (8,2,-2,-2)
6731 5851 tanh : $\sqrt{2/3}$
6731 6342 rt : (7,7,4,-8)
6731 7268 rt : (3,6,5,-7)
6731 8013 rt : (2,7,-1,1)
6731 8703 Γ : $5\sqrt[3]{2}$
6731 9068 erf : $\sqrt[3]{1/3}$
6731 9136 rt : (1,8,7,-6)
6732 0827 sq : $2/3 - 2\sqrt{2}$
6732 1674 J_0 : $(\sqrt[3]{5})^{1/3}$
6732 4113 ℓ_{10} : $3\pi/2$
6732 4507 2^x : $1 - \pi/2$
6732 4873 rt : (5,4,-7,-5)
6732 6342 Ψ : $2e\pi/7$
6732 7882 e^x : $\sqrt{14/3}$
6732 8068 rt : (9,2,2,-5)
6733 1878 sq : $4e\pi/5$
6733 2005 sr : $14/5$
6733 2217 $\exp(3) + s\pi(1/5)$
6733 2380 e^x : $3\sqrt{2} + 4\sqrt{3}/3$
6733 3125 2^x : $7(1/\pi)/3$
6733 4118 ℓ_{10} : $\sqrt{3} + 4\sqrt{5}/3$
6733 4198 10^x : $(\sqrt[3]{4})^{1/3}$
6733 5579 rt : (5,1,3,-4)
6733 6392 rt : (6,3,-7,8)
6733 6638 J : $e/6$
6733 7145 K : $10/21$
6733 9374 ℓ_{10} : $10\sqrt{2}/3$
6734 0624 θ_3 : $\sqrt{3}/2$
6734 0806 rt : (6,8,-7,-6)
6734 2709 10^x : $\sqrt{5}/10$
6734 3283 rt : (6,1,7,-7)
6734 4970 at : $(\pi/2)^{-1/2}$
6734 6938 sq : $6\sqrt{5}/7$
6734 7304 2^x : $(2e/3)^{-1/2}$
6734 7536 Ψ : $13/23$
6734 8148 tπ : $3e/4 - 3\pi$
6734 8199 tπ : $6\sqrt{3}/7$
6734 9326 sr : $4\sqrt{2} + 2\sqrt{5}/3$
6735 1433 rt : (1,9,-8,1)
6735 2858 K : $1/\ln(1/(3e))$
6735 5325 rt : (2,9,-4,-2)
6735 5922 id : $4\sqrt{3} + \sqrt{5}/3$
6735 6267 10^x : $(\ln 2)^{1/3}$
6735 6641 rt : (3,9,0,-5)
6735 7304 rt : (4,9,4,-8)
6735 7467 cr : $4/3 + 3\sqrt{5}/2$
6735 9305 rt : (2,0,8,6)
6736 0172 ln : $4 - 3e/4$
6736 1111 sq : $23/12$
6736 1572 at : $\ln(\sqrt{2}\pi/2)$
6736 2992 cr : $4(e+\pi)/5$
6736 3678 as : $9\ln 2/10$
6736 4321 Ψ : $5\ln 2/9$
6736 4910 Γ : $3\sqrt{2}/8$
6736 4934 ln : P_9
6736 5243 sπ : $(\sqrt[3]{5})^{-1/2}$
6736 6507 Ei : $3(1/\pi)/10$
6736 7054 rt : (1,9,-6,5)
6736 7760 id : $3e/4 - 3\pi/2$
6736 8632 rt : (9,5,-3,-3)
6736 9291 e^x : $2\sqrt{2} - \sqrt{5}/4$
6736 9564 sπ : $4/17$
6737 0621 as : $e - 2\pi/3$
6737 0661 sq : $3/4 - \pi/2$
6737 1427 rt : (9,4,9,7)
6737 2241 Ei : $8e/7$
6737 2658 id : $2\sqrt{2} - 2\sqrt{3}/3$
6737 3336 rt : (8,5,-7,0)
6737 3590 cr : $1/2 + 4\pi/3$
6737 4396 ln : $2\sqrt{2}/3 - \sqrt{3}/4$
6737 7326 ζ : $\sqrt{2}/9$
6737 7513 Γ : $\sqrt{8}\pi$
6737 7615 rt : (8,4,5,4)
6737 8387 ℓ_{10} : $2 + e$
6737 9079 rt : (7,4,6,-8)
6738 0556 rt : (1,8,-1,-4)
6738 1879 Ei : $(2e)^{-1/3}$
6738 3752 rt : (5,3,-6,5)
6738 5892 rt : (6,4,2,-5)
6738 8647 ℓ_{10} : $3\sqrt{2}/2 + 3\sqrt{3}/2$
6739 0142 Li_2 : $(\pi)^{-1/2}$
6739 1791 rt : (1,8,9,1)
6739 1843 e^x : $3\zeta(3)/7$
6739 2930 sr : $\sqrt[3]{22}$
6739 3549 e^x : $\sqrt{2}/3 - \sqrt{3}/2$
6739 6694 rt : (7,4,1,1)
6739 7643 ln : $3/16$
6739 7917 sinh : $\sqrt[3]{5}$
6739 8750 10^x : $3\sqrt{3}/4 + 2\sqrt{5}$
6739 9861 rt : (5,4,-2,-2)
6740 0840 K : $3(1/\pi)/2$
6740 1100 id : $5\pi^2/2$
6740 1120 as : $\exp(-\sqrt{2}/3)$
6740 2374 ln : $\sqrt{2}/3 + 2\sqrt{5}/3$
6740 2414 rt : (8,7,5,-9)
6740 2461 sinh : $9\sqrt{5}/10$
6740 3007 e^x : $\sqrt[3]{11/3}$
6740 3084 rt : (2,8,3,-1)
6740 4136 rt : (3,3,7,-7)
6740 5123 sq : $(e/3)^2$
6740 5753 tπ : $4\pi/3$
6740 5762 cu : $4/3 - 2\sqrt{3}$
6740 6061 rt : (1,7,-4,9)
6740 6843 θ_3 : $(3\pi)^{-1/2}$
6740 6955 10^x : $2 + 4\sqrt{5}$
6740 7905 tanh : $9/11$
6740 7973 rt : (4,7,-6,-6)
6740 8901 rt : (8,6,-7,9)
6741 2994 rt : (2,1,-9,5)
6741 3383 cr : $4\sqrt{3} - \sqrt{5}$
6741 3848 ℓ_{10} : $1/4 + 2\sqrt{5}$
6741 3984 rt : (7,7,1,-6)
6741 4307 rt : (9,8,-8,-1)
6741 5414 at : $1/2 - 3\sqrt{3}/4$
6741 5503 10^x : $2/3 + 2\sqrt{5}/3$
6741 6909 ln : $2\sqrt{2} - \sqrt{3}/2$
6741 7120 tπ : $3\sqrt{2}/4 - \sqrt{3}$
6741 7567 cu : $7(e+\pi)/9$
6741 7678 rt : (9,1,4,5)
6741 9986 sr : $5/11$
6742 3461 sr : $(\sqrt[3]{2}/3)^{-3}$
6742 4515 ζ : $\sqrt[3]{2}/8$
6742 4985 id : $2(e+\pi)/7$
6742 8266 2^x : $\sqrt{15/2}$
6742 8735 rt : (4,6,6,-8)
6742 8792 10^x : $(\sqrt{5}/3)^{-1/3}$
6743 3213 as : $3 - 4e/3$
6743 3553 sr : $10(1/\pi)/7$
6743 4035 rt : (6,7,-3,-3)
6743 4666 Ψ : 15
6743 5843 e^x : $2/3 - 3\sqrt{2}/4$
6743 7693 rt : (6,2,9,7)

6743 9812 $10^x : 2e/3 + 4\pi/3$
6744 0770 $K : 11/23$
6744 1301 $\sinh : 12/19$
6744 4274 rt : (5,9,5,-9)
6744 4710 rt : (4,8,-2,-7)
6744 4778 $t\pi : \exp(-5/3)$
6744 7073 $2^x : \exp(\sqrt[3]{2}/2)$
6744 7335 rt : (1,9,9,-9)
6744 8022 $10^x : 4/3 - e/3$
6744 9184 $\mathrm{cr} : \ln(e/2)$
6745 0404 rt : (8,1,0,2)
6745 1246 $\tanh : (\ln 3/2)^{1/3}$
6745 1288 $\mathrm{cu} : 1/2 + 3\sqrt{2}$
6745 3407 $\mathrm{sr} : 3e/2 + 3\pi$
6745 4980 rt : (6,6,-2,8)
6745 5301 rt : (7,1,8,-8)
6745 6054 cu : 19/16
6745 6267 rt : (4,5,6,3)
6745 8678 $\mathrm{cu} : 4e\pi/9$
6745 9418 $10^x : 1/\ln(4/3)$
6746 0322 rt : (1,2,-8,-6)
6746 0583 $\mathrm{cr} : \sqrt{3}/2 - \sqrt{5}/4$
6746 2803 rt : (7,6,-6,6)
6746 4261 $t\pi : \sqrt{18}\pi$
6746 4266 $s\pi : \sqrt{2}/6$
6746 4408 $\mathrm{sq} : 1/3 - 2\sqrt{3}/3$
6746 4463 rt : (5,7,-2,-3)
6746 4539 rt : (2,5,-7,6)
6746 6194 $10^x : 13/23$
6746 6242 rt : (8,4,7,-9)
6746 6777 $\mathrm{id} : 5e\pi/4$
6746 8770 rt : (4,9,1,-6)
6747 0192 rt : (9,2,-1,-3)
6747 0707 $2^x : (e/3)^3$
6747 2612 $\Psi : 10\sqrt[3]{5}/7$
6747 2625 $\ln : 5\pi/8$
6747 2708 $\mathrm{at} : ((e+\pi)/3)^{-1/3}$
6747 3010 $\mathrm{as} : \sqrt{3}/2 - 2\sqrt{5}/3$
6747 4048 sq : 22/17
6747 4094 at : 4/5
6747 4157 $\ln : 2 - 2\sqrt{5}/3$
6747 5354 $\mathrm{sr} : 7\zeta(3)/3$
6747 7270 rt : (5,8,-6,-9)
6747 7796 $\mathrm{id} : 1/4 + 3\pi$
6747 8307 $\zeta : 3/19$
6747 8650 $erf : 16/23$
6748 2541 $\zeta : 6\sqrt{3}/7$
6748 3747 rt : (5,2,5,4)
6748 5979 rt : (1,8,6,-8)
6748 5984 $\mathrm{id} : 2\sqrt{2}/3 + \sqrt{3}$
6748 8127 $\ln : \sqrt{3}/2 + 2\sqrt{5}$
6748 9954 $t\pi : \sqrt{3}/2 - 3\sqrt{5}/4$
6749 3311 rt : (6,1,4,-5)
6749 8173 rt : (7,4,3,-6)
6749 8532 $\exp(\sqrt{5}) \div \exp(1/2)$
6750 1116 rt : (8,7,2,-7)
6750 3317 $\ln(\sqrt{2}) \times \exp(2/3)$
6750 4786 $\Psi : 7\pi/9$
6750 4939 $\ell_{10} : 3 + \sqrt{3}$
6750 5646 $10^x : P_{62}$
6750 5727 $\Psi : 2\sqrt{3}/9$
6750 9062 $Ei : 4\sqrt[3]{3}/3$
6751 0295 $t\pi : \sqrt{3} + \sqrt{5} + \sqrt{7}$
6751 0638 cr : 4/13
6751 3087 rt : (3,9,-3,-3)
6751 3153 as : 5/8
6751 4714 rt : (9,9,-6,7)
6751 4919 rt : (8,9,-2,9)
6751 6026 rt : (5,6,7,-9)
6751 6767 rt : (4,3,8,-8)
6751 7262 $10^x : (\sqrt[3]{3})^{-1/3}$
6751 8008 rt : (3,0,9,7)
6751 9155 $\tanh : (2e/3)^{-1/3}$
6752 2975 $\mathrm{sr} : 3/2 + 4\sqrt{2}$
6752 3723 $\mathrm{sq} : (\sqrt{2}\pi/3)^{-1/2}$
6752 4341 $\ell_2 : 2/3 - \sqrt{2}/4$
6752 4651 $\mathrm{sq} : 1/4 - 4\sqrt{2}/3$
6752 5675 rt : (7,9,-6,-6)
6752 8393 $K : 12/25$
6752 9630 $\Psi : 9(1/\pi)/10$
6753 0271 rt : (9,5,-6,-1)
6753 0433 $Ei : 3\pi^2/10$
6753 0443 $s\pi : 5\pi/3$
6753 5176 $\zeta : \ln(\sqrt{3})$
6753 5470 rt : (9,4,6,5)
6753 6499 $Ei : 9\sqrt[3]{3}/8$
6753 6633 $Ei : 25/13$
6753 7914 $\mathrm{cr} : 4\ln 2/9$
6753 9052 rt : (8,6,-7,-5)
6753 9653 $\sinh : (1/(3e))^{-1/3}$
6754 1523 $Li_2 : 13/23$
6754 2268 rt : (8,1,9,-9)
6754 2898 rt : (7,6,-9,1)
6754 3027 $\mathrm{id} : 4\sqrt{2} - 4\sqrt{5}/3$
6754 3692 $\Psi : 2\sqrt{2}/3$
6754 4003 $\mathrm{at} : \sqrt{2}/4 - 2\sqrt{3}/3$
6754 4467 $\mathrm{sq} : \sqrt{2} - \sqrt{5}$
6754 5789 rt : (7,7,-2,-4)
6754 7058 $\sinh : (5/2)^{-1/2}$
6754 7117 rt : (8,7,7,4)
6754 8080 rt : (6,6,-5,3)
6754 8675 $10^x : P_{31}$
6754 9029 $s\pi : \sqrt{5}$
6755 0252 $\mathrm{at} : 5\sqrt[3]{3}/9$
6755 1025 $\mathrm{sq} : \sqrt{2}/4 - 2\sqrt{3}$
6755 1304 $K : (\ln 2)^2$
6755 1790 rt : (3,9,-5,9)
6755 2737 rt : (1,5,-3,-4)
6755 3722 $\ell_2 : 5\sqrt{5}/7$
6755 4350 $\Psi : 22/9$
6755 5444 rt : (5,9,2,-7)
6755 5509 rt : (2,9,-1,3)
6755 6654 $erf : \sqrt{2} + \sqrt{3} - \sqrt{6}$
6755 7243 rt : (6,4,-1,-3)
6755 7652 $\mathrm{at} : 2\zeta(3)/3$
6755 8886 $s\pi : 7\sqrt[3]{2}/5$
6755 9266 $e^x : 4\sqrt{3} - \sqrt{5}/2$
6756 1717 $\tanh : (e/3)^2$
6756 2033 rt : (3,7,-2,2)
6756 2576 $K : (\ln 2/3)^{1/2}$
6756 5338 $\mathrm{id} : 3\sqrt{2} + \sqrt{3}/4$
6756 6169 rt : (7,1,-4,-1)
6756 6336 $K : \sqrt[3]{3}/3$
6756 6418 $Ei : 8e/9$
6756 6734 rt : (4,6,3,7)
6756 6928 rt : (3,7,-7,-7)
6756 9497 $10^x : 2\pi^2$
6757 0030 $K : 2\zeta(3)/5$
6757 1470 rt : (9,7,3,-8)
6757 2562 rt : (2,8,7,-9)
6757 5413 $\ln : 4\pi^2$
6757 5569 $\sinh : 7(e+\pi)/9$
6758 1048 rt : (5,1,0,-2)
6758 2829 $s\pi : 6\sqrt[3]{3}/7$
6758 3750 $e^x : \sqrt{2}/4 - \sqrt{5}/3$
6758 4144 rt : (8,4,4,-7)
6758 4155 rt : (9,9,-9,2)
6758 5084 $\ln : 2/3 + 3\sqrt{3}/4$
6758 6231 rt : (3,6,7,9)
6758 6669 $J_0 : 25/21$
6758 7592 $Li_2 : \exp(-\sqrt[3]{5}/3)$
6758 8128 $\mathrm{cu} : 3 - 2e/3$
6758 8167 $2^x : 4/3 + 4\pi$
6758 8866 rt : (1,5,8,-8)
6759 1253 $\mathrm{cr} : 4 + \sqrt{2}/2$
6759 3289 $10^x : 3e/2 + \pi/4$
6759 3471 $Li_2 : 1/\ln(e+\pi)$
6759 3902 rt : (8,9,-5,4)
6759 6407 $t\pi : (\sqrt{2})^{1/2}$
6759 6960 rt : (4,4,-8,-6)
6759 9553 $\mathrm{sq} : 1 + 3\pi$
6760 0940 $\tanh : (\sqrt{2}\pi/3)^{-1/2}$
6760 2022 $\ell_{10} : 1/2 + 3\sqrt{2}$
6760 2476 rt : (5,3,9,-9)
6760 3129 $\Psi : 3\pi/10$
6760 3847 rt : (7,1,5,-6)
6760 4853 $10^x : \sqrt[3]{5}/4$
6760 5528 rt : (5,1,-7,9)
6760 6123 rt : (9,1,1,3)
6760 6748 rt : (7,9,-1,6)
6760 7765 rt : (3,9,-8,-2)
6761 0509 $Li_2 : 2\sqrt{2}/5$
6761 2680 $\mathrm{sr} : 1/2 + 4\sqrt{3}/3$
6761 2818 $Ei : 8\zeta(3)/5$
6761 3093 $\mathrm{sr} : (2e)^3$
6761 4807 $\mathrm{sq} : 3\sqrt{2}/2 - 3\sqrt{3}/4$
6761 5286 $\ell_{10} : 2/3 + 3e/2$
6761 6594 $s\pi : (\sqrt[3]{4}/3)^{-1/3}$
6761 7184 $\ell_{10} : 5e\pi/9$
6761 7497 rt : (2,9,-7,0)
6762 2046 $\ln(3/4) - s\pi(\sqrt{2})$
6762 2694 rt : (4,9,-2,-4)
6762 2983 $2^x : 4e/3 - 4\pi/3$
6762 3478 $\ln(\sqrt{3}) - \exp(4/5)$
6762 4451 rt : (6,9,3,8)
6762 4548 $\mathrm{cu} : 4\sqrt{3} + 2\sqrt{5}$
6762 4835 $s\pi : \exp(-\sqrt[3]{3})$
6762 6879 $\ell_{10} : 4 + \sqrt{5}/3$
6762 8005 rt : (2,0,5,4)
6762 9429 $t\pi : e/3 + 2\pi$
6763 0039 $e^x : 1 + 3\sqrt{2}/2$
6763 0052 rt : (8,7,-1,-5)
6763 1407 rt : (6,7,-6,-1)
6763 2271 $\mathrm{id} : \sqrt{3} + 4\sqrt{5}$
6763 5897 $\mathrm{id} : 7\pi^2/9$
6763 8673 rt : (6,2,6,5)
6763 8787 $2^x : \sqrt{5}/3$
6764 2224 $t\pi : 8\sqrt{5}/7$
6764 3856 rt : (4,8,-5,6)
6764 5361 $\Gamma : 9/17$
6764 6201 $\zeta : (1/\pi)/2$
6764 7544 $\lambda : \exp(-11\pi/13)$
6764 7895 rt : (1,8,3,-6)
6764 9078 $\Gamma : 14/5$
6764 9574 $J_0 : (\sqrt{2})^{1/2}$
6765 0409 rt : (5,6,4,-7)
6765 0799 $erf : 2\pi/9$
6765 0857 rt : (1,9,2,-7)
6765 2087 rt : (9,4,5,-8)
6765 3919 $\mathrm{id} : (3\pi/2)^{1/3}$
6765 5530 $\Psi : 9e/10$
6765 7168 $t\pi : 1/4 - 3\sqrt{2}/4$
6765 7903 $\sinh : 6\pi/5$
6766 0339 $t\pi : ((e+\pi)/2)^2$
6766 0498 rt : (8,4,2,2)
6766 0506 $Ei : 2/21$
6766 2278 $3 \times \exp(4/5)$
6766 2403 $\mathrm{id} : e/4 - 3\pi/4$
6766 4500 rt : (7,4,0,-4)
6766 5849 rt : (6,6,-8,-2)
6766 6794 $\mathrm{cr} : \sqrt{3} + 4\sqrt{5}/3$
6766 7641 $5 \div \exp(2)$
6766 8568 rt : (1,8,2,-2)
6766 9360 $\ell_{10} : 4/19$
6767 0376 $\mathrm{as} : 4/3 - \sqrt{2}/2$
6767 1797 rt : (2,5,9,-9)
6767 3558 $\mathrm{cr} : 10\sqrt{2}/3$
6767 3817 $\ln(3/4) - \exp(2)$
6767 3890 $\Gamma : 8\pi/7$
6767 4703 rt : (2,7,-8,-8)
6767 7492 $\theta_3 : \exp(-\sqrt{5}/2)$
6767 9622 rt : (6,9,-7,-7)
6768 0694 rt : (7,9,4,-9)
6768 4169 rt : (3,7,-5,-6)
6768 4367 $10^x : 3/4 + 4e/3$
6768 5598 $t\pi : P_{80}$
6768 5620 rt : (8,1,6,-7)
6768 8659 rt : (9,2,-4,-1)
6768 8927 rt : (8,9,-8,-1)
6769 2157 rt : (4,3,5,-6)
6769 5792 $\mathrm{sr} : 1 + 2e/3$
6769 5874 rt : (9,7,0,-6)
6769 7287 rt : (9,7,8,5)
6769 7842 $e^x : 3 + 2e$
6769 8629 $\tanh : 14/17$
6770 0178 $erf : ((e+\pi)/2)^{-1/3}$
6770 0320 sr : 11/24
6770 0517 $\theta_3 : 1/\ln(3\pi/2)$
6770 3296 rt : (5,9,-1,-5)
6770 4416 $\ln : 4\sqrt{2}/3 + 2\sqrt{3}$

6770 5098 id : $3\sqrt{5}/4$
6770 7221 cu : $3e/2 - \pi/4$
6770 8840 Ψ : $3/19$
6770 9537 rt : (1,8,3,4)
6771 0484 10^x : $3e\pi/8$
6771 2918 e^x : $(\pi/3)^{-1/3}$
6771 3772 sinh : $20/7$
6771 5407 rt : (6,6,5,-8)
6771 6906 rt : (5,5,7,4)
6771 9603 10^x : $9\sqrt[3]{2}/7$
6772 0221 rt : (7,9,-4,1)
6772 3774 cr : $2 + e$
6772 6540 rt : (4,7,-3,-4)
6772 7241 rt : (4,8,-8,-2)
6772 7702 rt : (6,1,1,-3)
6772 7743 Ψ : $5/13$
6773 0005 2^x : $\ln(\sqrt[3]{5}/3)$
6773 0771 10^x : $\exp(-\sqrt[3]{5}/3)$
6773 1831 tanh : $4\sqrt[3]{3}/7$
6773 2472 sr : $\sqrt{3}/3 + \sqrt{5}$
6773 2948 J : $\exp(-9\pi)$
6773 3138 rt : (7,7,-5,-2)
6773 3253 ℓ_{10} : $7e/4$
6773 4124 J_0 : $19/16$
6773 5715 e^x : $2(e + \pi)/9$
6773 6981 cr : $3\sqrt{2}/2 + 3\sqrt{3}/2$
6773 7809 rt : (4,1,-6,6)
6773 8354 2^x : $4 - 3\sqrt{2}/2$
6773 8891 $\ln(\pi) - \exp(3/5)$
6774 0489 rt : (7,2,7,6)
6774 0913 $t\pi$: $\exp(\pi/2)$
6774 1331 Ei : $23/25$
6774 2215 sq : $2\sqrt{3}/3 - 4\sqrt{5}$
6774 2320 rt : (8,4,1,-5)
6774 3762 rt : (6,4,-6,7)
6774 4219 rt : (3,4,-9,-7)
6774 4933 Ei : $9\pi/2$
6774 7581 rt : (8,7,-6,8)
6774 8569 rt : (9,1,7,-8)
6774 9594 rt : (1,6,-6,1)
6775 2530 rt : (1,7,-9,-9)
6775 6752 $s\pi$: $2e/3 + 3\pi$
6775 8093 rt : (3,5,-7,3)
6775 9085 ζ : $4/25$
6775 9604 sq : $3e + \pi/3$
6775 9814 rt : (3,9,-6,-1)
6776 0700 Ψ : $(\pi)^{-1/2}$
6776 0882 10^x : $(\ln 3)^{-3}$
6776 1173 at : $\ln(\sqrt{5})$
6776 2320 at : $1/3 + \sqrt{2}/3$
6776 4542 10^x : $1/\ln(e + \pi)$
6776 5069 rt : (6,9,0,3)
6776 5699 $t\pi$: $\sqrt[3]{14/3}$
6776 6511 Ei : $9\sqrt[3]{5}/8$
6776 7353 rt : (7,6,6,-9)
6776 8272 rt : (5,3,6,-7)
6776 8595 sq : $18/11$
6776 9429 cr : $1/4 + 2\sqrt{5}$
6777 0427 rt : (3,0,6,5)
6777 7087 erf : $5\sqrt[3]{2}/9$
6777 8756 rt : (5,9,-8,-8)
6777 9376 Ei : $\sqrt[3]{5}/3$
6778 0119 erf : $7/10$
6778 0791 ℓ_{10} : $\sqrt[3]{2}/6$
6778 1464 rt : (1,3,6,3)
6778 1897 rt : (5,4,-2,9)
6778 3718 $t\pi$: $e - \pi/3$
6778 5293 $\exp(3) \div s\pi(\pi)$
6778 6143 rt : (9,5,-9,1)
6778 6298 ln : $2 + 3\sqrt{5}/2$
6778 6654 cu : $3\pi^2/7$
6778 7643 rt : (3,8,5,-8)
6778 8481 2^x : $7\pi^2/2$
6779 2409 10^x : $9\sqrt{3}/4$
6779 2520 ζ : $\sqrt[3]{3}/9$
6779 3155 rt : (7,6,-8,-6)
6779 3717 K : $7\ln 2/10$
6779 5211 cu : $e/3 - 2\pi/3$
6779 7439 rt : (9,4,3,3)
6779 9184 2^x : $1/2 - 3\sqrt{2}/4$
6779 9267 10^x : $7/6$
6779 9348 rt : (2,8,4,6)
6780 1372 rt : (9,4,2,-6)
6780 5361 rt : (8,9,-2,-3)
6780 6021 $s\pi$: $5\sqrt{3}/7$
6780 7190 ℓ_2 : $5/8$
6780 8233 Ψ : $\sqrt{6}$
6780 9008 rt : (1,7,2,-1)
6781 1581 rt : (1,8,-6,-1)
6781 2510 rt : (8,2,8,7)
6781 3926 rt : (7,9,1,-7)
6782 0069 sq : $14/17$
6782 0124 cr : $e/2 - \pi/3$
6782 0203 cu : $2/3 - 3e/2$
6782 0323 id : $1/4 - 4\sqrt{3}$
6782 0409 rt : (7,1,2,-4)
6782 3450 $1/2 \times \exp(\sqrt{5})$
6782 3473 $t\pi$: $\sqrt{3}/2 + 4\sqrt{5}$
6782 5373 ln : $3 + 3\pi/4$
6782 5604 rt : (6,3,7,-8)
6782 6601 J_0 : $\sqrt[3]{5}/3$
6782 8834 rt : (7,2,-9,3)
6783 3692 rt : (6,4,-4,-1)
6783 4544 10^x : $\sqrt{2}/3 + 4\sqrt{3}/3$
6783 4715 rt : (5,9,4,5)
6783 5158 rt : (4,8,6,-9)
6783 6282 rt : (1,5,2,-6)
6783 6498 ℓ_{10} : $\sqrt{2} + 3\sqrt{5}/2$
6783 6928 tanh : $19/23$
6783 6933 Ψ : $16/17$
6784 0019 rt : (8,7,-9,3)
6784 0115 cr : $9e\pi/4$
6784 0917 cu : $3 - 3\sqrt{2}/2$
6784 2110 e^x : $9\pi/5$
6784 5369 K : $\exp(-\sqrt[3]{3}/2)$
6784 5772 $\ln(\sqrt{2}) \div \ln(3/5)$
6784 6595 rt : (6,5,8,5)
6784 6971 cu : $10e\pi/7$
6784 7851 rt : (4,9,-9,-9)
6784 8560 rt : (4,9,-5,-2)
6784 9567 $t\pi$: $\sqrt[3]{21/2}$
6785 2002 $4/5 \times \exp(4)$
6785 2763 as : $3\sqrt{2}/4 - \sqrt{3}/4$
6785 3536 rt : (8,9,2,-8)
6785 4677 10^x : $3e - \pi/3$
6785 5682 rt : (3,7,-4,-5)
6785 7005 ℓ_2 : $4\sqrt{2} + \sqrt{5}/3$
6785 7351 rt : (1,5,3,7)
6785 8109 rt : (6,4,-9,2)
6785 8597 rt : (4,0,7,6)
6785 9308 rt : (9,7,-3,-4)
6785 9851 10^x : $4\pi^2/7$
6786 0440 cr : $5/16$
6786 2421 10^x : $2\sqrt{2}/5$
6786 4404 rt : (7,9,-7,-4)
6786 5249 cu : $1/4 + 3\sqrt{2}$
6786 5508 $s\pi$: $7\sqrt{2}/8$
6786 5931 rt : (2,5,6,-7)
6786 6146 rt : (9,2,9,8)
6786 7200 rt : (7,7,3,1)
6787 0341 rt : (7,3,8,-9)
6787 0460 sq : $3 + 4\pi/3$
6787 1536 $s\pi$: $2/3 + \pi/2$
6787 1548 ln : $2 + 4\pi$
6787 1594 $s\pi(1/5) \div s\pi(1/3)$
6787 1854 sr : $(3\pi/2)^{-1/2}$
6787 1955 ζ : $11/20$
6787 4624 ln : $3/2 + \sqrt{2}/3$
6787 5019 sr : $4\sqrt{2} - 3\sqrt{3}$
6788 0368 ln : $4 + e/2$
6788 1410 rt : (5,2,2,2)
6788 2232 $t\pi$: $4\sqrt{3} - \sqrt{5}/2$
6788 2303 rt : (1,6,2,1)
6788 4334 rt : (8,1,3,-5)
6788 4699 Ei : $(\sqrt{2}/3)^{-1/3}$
6788 5265 cu : $7\sqrt{5}/5$
6788 6299 rt : (9,9,3,-9)
6788 6767 cr : $3 + \sqrt{3}$
6788 7286 rt : (7,7,-5,5)
6788 7710 rt : (4,1,-9,1)
6788 8773 rt : (1,8,0,-4)
6789 2206 sinh : $((e + \pi)/2)^{1/2}$
6789 2394 Ψ : $\sqrt{23}\pi$
6789 3853 Γ : $1/3$
6789 6064 ln : $\sqrt[3]{23/3}$
6789 6318 rt : (2,5,1,-1)
6789 8798 e^x : $(\sqrt[3]{2}/3)^{-1/2}$
6790 3041 ln : $3/2 - 2\sqrt{5}/3$
6790 4803 cr : $2/3 - \sqrt{2}/4$
6790 5068 sq : $2/3 - 2\sqrt{5}/3$
6790 8983 ln : $3\sqrt{2} + \sqrt{5}/2$
6790 9226 rt : (5,9,-4,-3)
6790 9416 rt : (3,8,-9,8)
6791 0550 Ei : $\exp(\sqrt{3})$
6791 1697 rt : (6,6,-9,-7)
6791 2447 rt : (7,4,-3,-2)
6791 2545 cu : $3\sqrt{3}/4 - 3\sqrt{5}/2$
6791 4795 rt : (3,5,7,-8)
6791 5367 ℓ_2 : $3e\pi$
6791 7809 rt : (9,1,-2,1)
6791 8381 cu : $25/18$
6791 8540 rt : (5,0,8,7)
6791 8757 $\ln(4) - s\pi(1/4)$
6791 9468 10^x : $\sqrt{3}/3 - \sqrt{5}/3$
6792 1104 sq : $4\sqrt[3]{3}/7$
6792 1690 sr : $1/4 + 4\sqrt{3}$
6792 2276 rt : (5,4,-5,4)
6792 2938 cu : $3/4 - \pi$
6792 4189 rt : (7,5,9,6)
6792 5190 sr : $2\pi^2/7$
6792 6043 rt : (7,6,3,-7)
6792 9772 Ei : $14/5$
6793 1088 rt : (9,1,4,-6)
6793 2273 $t\pi$: $2\ln 2$
6793 2422 rt : (2,7,-5,-6)
6793 6622 sr : $6/13$
6793 6850 Γ : $\sqrt[3]{22}$
6793 8992 as : $\pi/5$
6794 4947 rt : (2,8,1,-5)
6794 7304 erf : $(\ln 2/2)^{1/3}$
6794 7913 sr : $\sqrt{2}/3 + 3\sqrt{5}$
6795 0547 rt : (6,7,-1,7)
6795 2178 rt : (4,5,8,-9)
6795 3330 rt : (6,9,-3,-2)
6795 5799 rt : (1,7,-6,-6)
6795 7045 id : $e/4$
6795 7605 rt : (8,6,4,-8)
6795 8702 $t\pi$: $\sqrt[3]{5}/9$
6795 8855 $t\pi$: $(\sqrt{2}\pi/2)^{-1/2}$
6796 0914 sq : $\sqrt{2} - \sqrt{3} + \sqrt{5}$
6796 1610 rt : (6,7,-9,1)
6796 1617 ℓ_2 : $3/4 + 4\sqrt{2}$
6796 1936 rt : (6,0,9,8)
6796 3784 $\exp(1/3) + \exp(1/4)$
6796 5204 rt : (8,4,-2,-3)
6796 7207 ln : $\sqrt{2} + \sqrt{5}/4$
6796 7381 at : $4\sqrt{2}/7$
6796 7703 K : $2\sqrt[3]{5}/7$
6796 8882 tanh : $(\ln 3)^{-2}$
6797 0776 rt : (2,3,7,4)
6797 5268 ln : $2\sqrt{3} - 2\sqrt{5}/3$
6797 7799 sr : $2\ln 2/3$
6797 8365 $t\pi$: $4e\pi/3$
6797 8484 $\sqrt{3} + \exp(2/3)$
6797 9165 rt : (6,2,3,3)
6797 9614 10^x : $e/2 + 3\pi/2$
6798 0333 cr : $\pi/10$
6798 0757 id : $8\sqrt[3]{5}$
6798 3243 rt : (1,7,-6,-7)
6798 3276 rt : (9,6,5,-9)
6798 4152 e^x : $1 + 3\pi/4$
6798 5351 rt : (3,8,2,-6)
6798 6780 sinh : $8\sqrt[3]{5}/5$
6798 6845 rt : (7,9,-2,-5)
6798 8296 cr : $2\sqrt{2}/9$
6798 9473 id : $4\sqrt[3]{2}/3$
6798 9584 rt : (7,9,-3,-4)
6799 0692 $1/3 + \ln(\sqrt{2})$
6799 1586 ℓ_2 : $1/4 + 4\pi$
6799 1807 ℓ_{10} : $4 + \pi/4$

6799 5598 rt : (8,2,-8,2)
6799 7401 $\lambda : 2(1/\pi)/9$
6799 7840 $2^x : 9\sqrt{3}/5$
6799 9005 $J : \exp(-14\pi/17)$
6800 0000 id : 17/25
6800 0324 $t\pi : 3e/4 + 3\pi$
6800 0624 at : $7\ln 2/6$
6800 0887 tanh : $(\sqrt[3]{5}/3)^{1/3}$
6800 1579 rt : (7,5,-6,9)
6800 2185 ln : $\pi^2/5$
6800 2465 cu : $3 - 4\pi/3$
6800 2543 $2^x : 9\pi^2/7$
6800 3019 $e^x : 3/2 - 4\sqrt{2}/3$
6800 3025 rt : (9,4,-1,-4)
6800 4566 rt : (5,3,3,-5)
6800 4601 $2^x : (\pi/2)^3$
6800 6914 rt : (1,2,7,-6)
6800 7181 $\ln(2/3) + \exp(3)$
6800 9296 $\ell_2 : \exp(\sqrt{2}/3)$
6800 9956 at : $(\sqrt[4]{4}/3)^{1/3}$
6801 1912 cr : $1/2 + 3\sqrt{2}$
6801 2460 id : $\exp(e + \pi)$
6801 3177 rt : (8,9,-1,-6)
6801 4988 $J_0 : 13/11$
6801 5809 rt : (4,8,3,-7)
6801 7273 $s\pi : 5/21$
6801 7445 rt : (4,9,-2,7)
6801 7741 cu : $2 + 4\pi$
6801 8860 sq : $3\sqrt[3]{3}/2$
6801 9188 sinh : 7/11
6801 9368 rt : (7,7,-8,0)
6802 1497 at : $1/4 + \sqrt{5}/4$
6802 4086 rt : (8,7,4,2)
6802 7675 rt : (3,6,-4,-1)
6802 8218 $\zeta : 1/\sqrt[3]{6}$
6802 8337 $t\pi : 2\sqrt{2} + 4\sqrt{5}/3$
6802 9017 cr : $2/3 + 3e/2$
6803 0121 $1/4 - s\pi(\pi)$
6803 1466 cr : $5e\pi/9$
6803 2392 $\ell_2 : 10\sqrt[3]{3}/9$
6803 3733 id : $3\sqrt{2}/2 + \sqrt{5}/4$
6803 4344 rt : (7,2,4,4)
6803 4413 rt : (9,9,0,-7)
6803 6258 sq : $4e - 4\pi/3$
6803 6262 sinh : $\sqrt{3} + \sqrt{5} + \sqrt{7}$
6803 7394 sinh : $\ln(\sqrt[4]{4}/3)$
6803 7832 rt : (6,3,4,-6)
6803 8225 $\Psi : (e + \pi)$
6803 9023 $e^x : 2\sqrt{2} - 4\sqrt{3}/3$
6803 9500 cr : $\sqrt[3]{2}/4$
6803 9590 rt : (5,8,4,-8)
6804 1065 $\ell_{10} : 4\sqrt{2} - \sqrt{3}/2$
6804 2420 rt : (1,9,-7,-4)
6804 3432 $Ei : 10\sqrt{3}/9$
6804 3969 cr : $4 + \sqrt{5}/3$
6804 4707 at : $3/2 - 4\sqrt{3}/3$
6804 4919 rt : (1,5,0,-2)
6804 6363 rt : (1,7,8,-9)
6804 6942 rt : (2,2,8,-7)
6804 7793 $2^x : (\sqrt[3]{2}/3)^{1/3}$
6804 8298 rt : (3,3,8,5)
6804 8960 sr : $4/3 + 2\sqrt{5}/3$
6805 0167 sinh : $2(1/\pi)$
6805 0871 sr : $3e/4 + \pi/4$
6805 2064 $2/5 \div s\pi(1/5)$
6805 2122 at : 17/21
6805 3481 $\zeta : 5\pi/8$
6805 4269 $\ell_2 : 4\zeta(3)/3$
6805 5154 rt : (4,8,8,3)
6805 6161 rt : (5,2,-6,8)
6805 6812 rt : (8,7,-7,-1)
6805 7280 $Li_2 : (2e)^{-1/3}$
6805 8316 cu : $3\sqrt{2}/2 - \sqrt{3}/3$
6805 8674 rt : (6,8,5,-9)
6805 9578 $10^x : 3/2 - \sqrt{2}/3$
6806 0636 $10^x : \sqrt{2} + 2\sqrt{5}$
6806 0752 rt : (9,8,-9,5)
6806 2634 rt : (7,3,5,-7)
6806 2815 cr : $3\sqrt{2}/4 - \sqrt{5}/3$
6806 3072 rt : (6,9,-4,-5)
6806 4373 $10^x : e\pi/8$
6806 4398 $\ell_2 : e - 2\pi/3$
6806 4638 rt : (9,2,-7,1)
6806 9795 rt : (8,2,5,5)
6807 0082 $s\pi : 9\ln 2$
6807 0189 sinh : $\sqrt{5/3}$
6807 0524 $e^x : 3e/4$
6807 3438 rt : (4,6,-3,-2)
6807 4925 rt : (3,2,9,-8)
6807 5419 $\ell_{10} : 4\sqrt{2}/3 - 3\sqrt{5}/4$
6807 6946 $\ell_2 : 9\ln 2/10$
6807 7136 sinh : $1/\ln(\sqrt{3}/2)$
6807 7424 $2^x : 2 - \sqrt{3}/3$
6808 1836 rt : (8,3,6,-8)
6808 3003 $10^x : e - \pi/4$
6808 3065 rt : (9,7,-6,-2)
6808 5222 $\Gamma : \exp(-5/2)$
6808 6391 $\ell_{10} : \sqrt{23}$
6808 6674 rt : (9,7,5,3)
6808 8655 $s\pi : 6e\pi$
6809 0438 rt : (4,3,9,6)
6809 2535 cu : $2/3 + 2\pi$
6809 4497 rt : (9,2,6,6)
6809 5555 rt : (1,9,3,2)
6809 7144 rt : (9,3,7,-9)
6809 7716 cr : 6/19
6809 7872 $\exp(\sqrt{5}) - s\pi(\sqrt{5})$
6809 8770 cr : 19/4
6809 9467 rt : (4,7,-7,8)
6810 0439 $t\pi : 3\sqrt{5}/10$
6810 0537 $10^x : 11/9$
6810 0909 $\ell_{10} : 4/3 + 2\sqrt{3}$
6810 1797 $10^x : P_{54}$
6810 2496 rt : (5,9,1,0)
6810 2749 id : $\sqrt{2} - \sqrt{6} - \sqrt{7}$
6810 8739 $\theta_3 : 5\ln 2/4$
6810 8995 $2^x : 4\sqrt{2} - \sqrt{3}/2$
6810 9865 ln : $4\sqrt{2} + 4\sqrt{5}$
6811 0829 $\sqrt{3} - \exp(5)$
6811 1035 $\exp(e) \div \exp(\sqrt{3})$
6811 4184 $Ei : \sqrt{21/2}$
6811 4251 $Ei : 9/7$
6811 5863 rt : (6,7,-4,2)
6811 6510 rt : (1,6,-5,-3)
6811 7671 $\lambda : \exp(-16\pi/19)$
6811 9193 sr : $3 + 4\pi/3$
6812 0733 at : $1/4 - 3\sqrt{2}/4$
6812 1096 rt : (7,5,-9,4)
6812 2588 rt : (6,6,-1,-4)
6812 3838 $Ei : (3)^{-3}$
6812 4123 $\ell_{10} : 5/24$
6812 7069 rt : (4,9,-6,-7)
6812 8018 $4 \div \exp(2/5)$
6812 8313 sr : $e/3 + 2\pi$
6812 8537 rt : (8,8,-5,7)
6812 9206 sr : $1/\sqrt[3]{10}$
6813 0915 $t\pi : 3\sqrt{2} - \sqrt{3}/4$
6813 3064 rt : (1,5,2,-4)
6813 3301 sinh : $8\sqrt{3}/3$
6813 4297 $2^x : 9\sqrt{3}/7$
6813 6954 ln : $3e/4 + 4\pi$
6813 7309 rt : (7,6,0,-5)
6813 8060 $e^x : 3\sqrt{3}/10$
6813 9119 $\ell_{10} : 3\sqrt{2} + \sqrt{5}/4$
6813 9155 rt : (5,8,9,4)
6814 0297 rt : (4,9,-5,-1)
6814 2779 at : $\sqrt{3}/2 - 3\sqrt{5}/4$
6814 2881 tanh : $6\ln 2/5$
6814 2921 rt : (6,1,-2,-1)
6814 3175 $t\pi : 3e\pi/2$
6814 3534 rt : (3,8,1,3)
6814 3774 $t\pi : \sqrt{3}/4$
6814 3848 rt : (3,9,-7,-8)
6814 4820 rt : (9,8,-9,-7)
6814 5045 $Ei : 8\sqrt{5}/5$
6814 5583 $t\pi : (\ln 2/3)^{1/3}$
6814 7200 cu : 22/25
6814 8560 rt : (8,6,1,-6)
6814 9736 sr : $9\pi/10$
6815 0238 as : $\sqrt[3]{2}/2$
6815 2015 rt : (2,7,7,7)
6815 3107 $\Gamma : (\ln 2)^3$
6815 4092 tanh : $\exp(-1/(2e))$
6815 5886 $2 \div s\pi(\sqrt{3})$
6815 6032 rt : (2,9,-8,-9)
6815 7062 $2^x : (\ln 3/3)^{-2}$
6815 7438 rt : (9,6,2,-7)
6815 7583 $e^x : 3\sqrt{3}/2 - 4\sqrt{5}/3$
6815 8201 $s\pi(1/4) \times s\pi(\sqrt{2})$
6815 9485 $\zeta : (\sqrt[3]{5}/3)^{-1/2}$
6816 0818 sq : $9\zeta(3)/5$
6816 3876 $s\pi : 3(1/\pi)/4$
6816 4191 rt : (7,1,-1,-2)
6816 4913 rt : (3,5,4,-6)
6816 6411 rt : (4,4,-5,-4)
6816 8036 $e^x : 2 + 4\sqrt{5}/3$
6817 2247 sinh : $3\sqrt{3}/2$
6817 3580 $\ell_{10} : 1/3 + 2\sqrt{5}$
6817 4259 rt : (4,5,5,-7)
6817 7317 rt : (8,1,0,-3)
6817 8064 $J_0 : 5\sqrt{2}/6$
6817 8845 $t\pi : 4/21$
6817 9283 $2^x : 3/4$
6818 1224 cr : $7e/4$
6818 1343 rt : (5,5,6,-8)
6818 1818 id : 15/22
6818 1930 sq : $3\pi^2$
6818 3708 cr : $3/4 - \sqrt{3}/4$
6818 4532 tanh : $(\ln 2)^{1/2}$
6818 5077 $erf : 12/17$
6818 5930 $\ln(2/5) + \exp(4)$
6818 6034 $\zeta : (2\pi/3)^{-1/2}$
6818 6225 rt : (9,1,1,-4)
6818 6897 rt : (6,5,7,-9)
6818 8276 rt : (3,4,-6,-5)
6819 1411 tanh : $(\sqrt[3]{3})^{-1/2}$
6819 1916 $\ell_{10} : 10\sqrt[3]{3}/3$
6819 3930 $2^x : 4/3 - 4\sqrt{2}/3$
6819 4931 $\exp(2) - s\pi(1/4)$
6819 6244 $2^x : e\pi/6$
6819 8462 $J_0 : 3\pi/8$
6819 8501 $\ell_{10} : 4\zeta(3)$
6819 8706 id : $1/\ln(2e/3)$
6819 9308 rt : (2,4,-7,-6)
6820 1749 $\zeta : \exp(-2e/3)$
6820 2764 $e^x : 13/25$
6820 3220 rt : (3,9,-9,1)
6820 3407 rt : (1,8,-5,8)
6820 3578 $\ell_2 : \sqrt{3} + \sqrt{5} + \sqrt{6}$
6820 3951 $2^x : 2 + \sqrt{5}/2$
6820 4037 $K : \exp(-1/\sqrt{2})$
6820 4875 sq : $2/3 + 4\sqrt{3}$
6820 5328 $\ell_{10} : 3\ln 2/10$
6820 5957 rt : (1,4,-8,-7)
6820 6289 $e^x : (\sqrt[3]{3}/2)^2$
6820 6900 $2^x : 6\sqrt[3]{3}$
6821 0403 rt : (4,9,-8,0)
6821 1172 $s\pi : 2/3 + 2\pi/3$
6821 2410 rt : (5,2,-9,3)
6821 2852 at : $1 - 2e/3$
6821 4087 rt : (6,5,-5,6)
6821 4778 rt : (5,9,-7,-1)
6821 5509 rt : (7,8,-1,9)
6821 7721 rt : (6,9,-6,-2)
6821 8436 $e - s\pi(\sqrt{2})$
6821 8817 $\ell_{10} : \exp(\pi/2)$
6821 9837 rt : (7,9,-5,-3)
6822 1432 rt : (8,9,-4,-4)
6822 2677 rt : (9,9,-3,-5)
6822 3355 $J_0 : (\sqrt[3]{3}/2)^{-1/2}$
6822 4800 $Ei : 4/7$
6822 6179 tanh : 5/6
6823 1655 at : 13/16
6823 2780 rt : (9,0,9,9)
6823 3682 as : $3\sqrt{2}/2 - 2\sqrt{5}/3$
6823 4257 $\ell_{10} : 3 + 2e/3$
6823 5341 $t\pi : 1/2 + 4\sqrt{3}/3$
6823 5412 $e^x : \sqrt{3} - \sqrt{5}/3$
6823 8052 rt : (3,6,-4,4)
6823 8087 $e^x : 4 + 3\sqrt{3}/2$

6823 8586 id : $3\sqrt{3}/4 - 4\sqrt{5}/3$
6823 8670 ℓ_{10} : $e + 2\pi/3$
6824 0185 10^x : $7\pi/5$
6824 1965 sq : $19/23$
6824 3824 rt : (2,9,0,8)
6824 6785 Li_2 : $\sqrt[3]{5}/3$
6825 0371 10^x : $2/3 + 3\sqrt{3}/2$
6825 0603 ζ : $2\sqrt[3]{2}/3$
6825 0949 rt : (8,8,4,-9)
6825 1025 sq : $1/4 + 2\pi$
6825 1282 ln : $2/3 + 3\pi/2$
6825 2787 rt : (7,8,3,-8)
6825 3552 id : $1/\ln(\ln 2/3)$
6825 5038 rt : (6,8,2,-7)
6825 6452 sr : $2 + 3\sqrt{3}$
6825 7859 rt : (5,8,1,-6)
6826 0619 $\ln(3) \div \ln(5)$
6826 1500 rt : (8,8,-8,2)
6826 2307 2^x : $(\ln 2/2)^{-1/3}$
6826 2449 id : $7\sqrt[3]{3}/6$
6826 3182 $1/5 \times \exp(5)$
6826 6378 rt : (3,8,-1,-4)
6826 6781 cu : $4e - \pi/3$
6826 7941 $3/4 \div \ln(3)$
6826 8949 erf : $\sqrt{2}/2$
6826 9248 cr : $7/22$
6826 9579 10^x : $3/7$
6827 0169 10^x : $2\sqrt{3} - \sqrt{5}/4$
6827 2148 sq : $1/4 + \pi/3$
6827 3250 rt : (2,8,-2,-3)
6827 4091 rt : (3,9,-7,2)
6827 4358 rt : (1,5,-3,5)
6827 8088 sr : $2\sqrt{3}/3 + 3\sqrt{5}/4$
6827 8228 rt : (9,4,0,1)
6827 8406 id : $(\pi)^{-1/3}$
6827 8556 J_0 : $20/17$
6828 3658 rt : (1,8,-3,-2)
6828 3753 ln : $8\sqrt{3}/7$
6828 7425 rt : (4,9,-8,-9)
6828 7966 id : $7\zeta(3)/5$
6828 9118 rt : (3,7,7,-9)
6828 9585 rt : (9,4,-4,-2)
6828 9640 2^x : $3e/2 - \pi/2$
6829 0372 2^x : $(2\pi)^{1/2}$
6829 3578 J : $7/20$
6829 4916 $\exp(\sqrt{2}) + s\pi(\pi)$
6829 5395 e^x : $(2e/3)^2$
6829 6476 rt : (2,7,6,-8)
6829 6993 Γ : $7\sqrt{2}/3$
6829 9681 K : $2\sqrt{3}/7$
6830 0926 rt : (3,4,-9,5)
6830 1270 id : $1/4 + \sqrt{3}/4$
6830 3259 tπ : $4e/3 + 4\pi$
6830 4109 ln : P_{35}
6830 4582 ln : $7\sqrt{2}/5$
6830 4622 ℓ_{10} : $3\sqrt{2} + \sqrt{3}/3$
6830 4711 cu : $\sqrt{2} + \sqrt{5}/4$
6830 4953 rt : (9,3,4,-7)
6830 6045 rt : (1,7,5,-7)
6830 6672 e^x : $\pi^2/10$
6830 6704 sπ : $3\sqrt{2}/2 + \sqrt{5}/2$
6830 6807 rt : (9,5,8,6)
6830 7023 10^x : $3e/2 + 4\pi/3$
6830 7813 Ei : $13/8$
6830 7820 sq : $3\sqrt{2}/4 + \sqrt{3}/3$
6830 8020 sq : $1/\ln(\sqrt[3]{4})$
6830 8959 rt : (9,9,-5,9)
6830 9708 sq : $1 - 2e$
6831 0732 $\sqrt{3} + s\pi(2/5)$
6831 2111 K : $5\ln 2/7$
6831 2180 ln : $\sqrt{2} + \sqrt{3} + \sqrt{5}$
6831 3005 sr : $7/15$
6831 3669 rt : (8,6,-9,6)
6831 4413 rt : (8,3,3,-6)
6831 4550 cr : $\sqrt{2} + 3\sqrt{5}/2$
6831 4585 e^x : $3\sqrt{2}/2 - \sqrt{3}/3$
6831 8999 rt : (7,4,-6,0)
6832 0841 Ei : $7\sqrt[3]{2}$
6832 2422 rt : (8,5,7,5)
6832 2547 10^x : $3e\pi/7$
6832 3318 sinh : $2\sqrt{5}/7$
6832 3439 sπ : $1/3 + e/3$
6832 5082 sr : $17/6$
6832 5543 sr : $1 + 4\pi$
6832 6725 rt : (7,3,2,-5)
6832 7597 Γ : $7\zeta(3)/3$
6832 8157 id : $6\sqrt{5}/5$
6832 9119 rt : (6,7,-7,-3)
6832 9211 2^x : $5\zeta(3)/8$
6832 9449 $\ln(3/4) - \exp(1/3)$
6832 9601 id : $\sqrt{2}/3 - 2\sqrt{3}/3$
6833 2533 10^x : $1/\ln(\sqrt{2}\pi)$
6833 4870 2^x : $\ln(\sqrt{3}/3)$
6833 7520 rt : (1,8,5,0)
6833 7740 e^x : $3\sqrt{3}/2 - \sqrt{5}/4$
6834 0154 $\ln(2/5) \times s\pi(\sqrt{3})$
6834 0745 rt : (5,5,-1,8)
6834 1471 rt : (6,2,9,-9)
6834 3410 rt : (6,3,1,-4)
6834 3436 10^x : $4 + \sqrt{3}/2$
6834 4614 sq : $4\sqrt{2}/3 + 4\sqrt{3}$
6834 6385 rt : (7,5,6,4)
6834 7567 2^x : $3/4 - 3\sqrt{3}/4$
6834 9092 rt : (6,4,-7,1)
6834 9845 e^x : $2/3 - \pi/3$
6835 0261 10^x : $4 + \sqrt{3}/3$
6835 0658 rt : (2,9,-3,-6)
6835 2372 rt : (4,2,-5,5)
6835 2821 erf : $17/24$
6835 5936 ℓ_{10} : $\sqrt{2}/4 + 2\sqrt{5}$
6835 6167 rt : (5,2,8,-8)
6835 7453 λ : $1/14$
6835 8805 as : $12/19$
6835 8992 cr : $\sqrt{5}/7$
6835 8996 rt : (1,9,-6,6)
6835 9924 sr : $4/3 - \sqrt{3}/2$
6836 2369 id : $3\sqrt{2} - \sqrt{5}/4$
6836 6183 rt : (2,7,-2,-4)
6836 6341 tπ : $\sqrt{3}/2 - 3\sqrt{5}/2$
6836 6807 rt : (3,8,-2,-5)
6836 7299 rt : (5,3,0,-3)
6836 8695 $\exp(3) + \exp(4)$
6836 9307 sr : $9\sqrt[3]{2}/4$
6836 9692 id : $1/\ln(\pi/3)$
6837 1383 tπ : $2e/3 - \pi$
6837 1467 rt : (7,4,-2,-1)
6837 1533 ℓ_{10} : $3/4 + 3e/2$
6837 2088 rt : (2,3,-3,9)
6837 5462 rt : (4,2,7,-7)
6837 6183 rt : (3,6,-9,-8)
6837 8245 rt : (7,8,-4,4)
6837 9331 10^x : $21/2$
6838 0568 ℓ_{10} : $2 + 2\sqrt{2}$
6838 1825 tanh : $(\sqrt[3]{5})^{-1/3}$
6838 2034 sq : $3/4 + \sqrt{2}$
6838 3577 rt : (9,6,-1,-5)
6838 5921 rt : (8,5,6,-9)
6838 6831 10^x : $3/2 - \sqrt{5}/3$
6838 7827 rt : (6,5,5,3)
6838 9554 Ei : $\sqrt{5}/10$
6839 0572 sπ : $4\sqrt{2}/3 + 3\sqrt{5}/2$
6839 1569 rt : (6,5,-8,1)
6839 6913 rt : (1,1,-7,4)
6839 9037 id : $2\sqrt[3]{5}/5$
6840 1366 2^x : $2\sqrt{2}/3 - 2\sqrt{5}/3$
6840 1911 rt : (3,2,6,-6)
6840 2541 $2/5 + \exp(1/4)$
6840 2886 rt : (7,5,5,-8)
6840 3450 K : $(1/(3e))^{1/3}$
6840 4315 sr : $3e/4 - \pi/2$
6840 4335 rt : (8,6,-2,-4)
6840 5971 rt : (4,0,-6,9)
6840 6707 Ei : $25/4$
6840 6710 rt : (9,7,-9,0)
6840 8215 K : $2\sqrt{5}/9$
6840 8665 rt : (4,6,-8,-7)
6841 1168 at : $3e/10$
6841 1321 rt : (5,4,-8,2)
6841 2331 tπ : $1/4 + \sqrt{5}/4$
6841 3207 ln : $1/4 + \sqrt{3}$
6841 3680 tπ : $3(1/\pi)/5$
6841 6471 rt : (3,7,-1,-3)
6841 7657 rt : (7,6,-5,8)
6841 8611 rt : (2,7,0,5)
6841 9447 10^x : $22/17$
6842 0674 tπ : $1/\ln(\sqrt{2}\pi)$
6842 0824 sq : $\sqrt{2}/2 + 3\sqrt{5}/4$
6842 1052 id : $13/19$
6842 1072 Ei : $\sqrt[3]{18}$
6842 2190 rt : (3,6,-7,-4)
6842 4079 rt : (6,5,4,-7)
6842 4587 Ei : $5\zeta(3)$
6842 5474 $\exp(e) \div \exp(\sqrt{2})$
6842 7831 rt : (9,2,3,4)
6842 8322 tanh : $(e + \pi)/7$
6842 8854 rt : (8,3,-9,-5)
6842 9538 id : $(2e)^3$
6842 9628 cr : $1 - e/4$
6843 1397 rt : (1,9,-6,-8)
6843 1546 rt : (6,3,-9,5)
6843 1713 rt : (7,6,-3,-3)
6843 1960 ℓ_{10} : $3\sqrt{3}/2 + \sqrt{5}$
6843 3710 id : $4e - 4\pi/3$
6843 4637 cr : $2\sqrt[3]{3}/9$
6843 4819 2^x : $1/2 - \pi/3$
6843 5295 rt : (9,8,2,-8)
6843 7811 rt : (9,9,-8,4)
6843 7962 tπ : $4\pi^2/9$
6844 0405 rt : (2,2,5,-5)
6844 1891 ln : $5(e + \pi)/2$
6844 2619 Ψ : $9/16$
6844 8420 ℓ_{10} : P_{33}
6844 8945 Γ : $-\sqrt{3} + \sqrt{5} - \sqrt{6}$
6845 0126 rt : (2,5,-9,-1)
6845 0747 rt : (7,3,9,7)
6845 0846 sq : $1/4 + \sqrt{3}/3$
6845 1302 rt : (5,5,3,-6)
6845 1612 ℓ_{10} : P_{39}
6845 2664 rt : (1,7,-5,2)
6845 4710 sπ : $6/25$
6845 5076 cu : $2/3 + \sqrt{5}/2$
6845 5119 rt : (2,8,6,1)
6845 5744 ζ : $21/25$
6845 6257 rt : (8,8,1,-7)
6845 7158 sπ : $4\sqrt[3]{5}/9$
6845 8309 as : $4\sqrt{3}/3 - 3\sqrt{5}/4$
6845 9422 sinh : $16/25$
6845 9645 Li_2 : $4/7$
6846 0231 rt : (5,6,-7,-6)
6846 0793 rt : (9,7,-9,8)
6846 1064 as : $2/3 - 3\sqrt{3}/4$
6846 1456 10^x : $(2e)^{-1/2}$
6846 2478 ln : $3e/2 - 2\pi/3$
6846 4005 ℓ_2 : $4 - \pi/4$
6846 4261 rt : (2,9,-5,-7)
6846 6424 rt : (4,8,-4,9)
6846 7072 rt : (3,4,9,-9)
6846 7951 e^x : $2\sqrt{2} + 2\sqrt{3}/3$
6846 9473 rt : (9,9,-6,-3)
6847 1920 at : $\sqrt{2/3}$
6847 2398 rt : (8,2,2,3)
6847 2804 J_0 : $\exp(1/(2\pi))$
6847 6491 sinh : $8e\pi/7$
6847 6818 rt : (5,5,4,2)
6847 8590 Γ : $\ln(1/\pi)$
6848 0985 sr : $1/2 + 3\sqrt{5}$
6848 1873 $\exp(1/2) + s\pi(\sqrt{2})$
6848 2057 rt : (7,8,0,-6)
6848 5149 rt : (9,0,6,7)
6848 6789 sq : $9\sqrt[3]{3}/10$
6848 7557 rt : (4,7,4,6)
6849 0449 $\ln(3/5) \div s\pi(\sqrt{3})$
6849 0538 cu : $2\sqrt{3} - 2\sqrt{5}/3$
6849 0885 ln : $e/4 + 3\pi/2$
6849 1779 2^x : $\exp((e + \pi)/3)$
6849 2106 10^x : $2/3 + 4\sqrt{2}/3$
6849 2331 sπ : $(\sqrt{3})^{-1/2}$
6849 2509 Ψ : $\exp(e)$
6849 2616 rt : (5,7,6,-9)
6849 2658 sr : $4e/3 - \pi/4$

6849 3192 rt : (6,3,8,6)
6849 5546 e^x : 12/23
6849 6554 $t\pi$: $2\sqrt{2}/3+\sqrt{3}/2$
6849 7308 Γ : $(\sqrt{5})^{-3}$
6849 7735 sq : $\sqrt{2}/3-3\sqrt{3}/4$
6849 8218 10^x : $3\sqrt{3}+2\sqrt{5}$
6849 9552 J_0 : $(e+\pi)/5$
6850 0067 Ψ : 15/16
6850 0424 $t\pi$: $(\sqrt[3]{4}/3)^{1/3}$
6850 1591 e^x : $7\pi^2/4$
6850 1606 rt : (2,4,8,-8)
6850 1649 cu : $2/3+3\pi$
6850 2128 id : $\sqrt{3}/4-\sqrt{5}/2$
6850 2629 sq : $5\sqrt[3]{2}$
6850 2727 tanh : $3\sqrt{5}/8$
6850 2750 sq : $5\pi/2$
6850 3748 rt : (8,9,-7,-2)
6851 0178 rt : (3,9,-4,-6)
6851 0769 2^x : $5\sqrt[3]{5}/6$
6851 4494 ln : $\sqrt{3}/2+\sqrt{5}/2$
6851 4586 rt : (6,8,-1,-5)
6851 5308 cr : $4+\pi/4$
6851 8145 2^x : $3+3\sqrt{3}/4$
6852 2167 e^x : $3\sqrt{3}/4-3\sqrt{5}/4$
6852 2391 2^x : $\sqrt{3}/2+\sqrt{5}/4$
6852 2422 rt : (8,0,5,6)
6852 3050 rt : (9,1,-5,-1)
6852 3479 rt : (4,7,5,-8)
6852 4167 ln : $2\sqrt[3]{2}/5$
6852 4240 $s\pi$: $8e\pi/7$
6852 4890 rt : (5,9,-2,-5)
6852 5239 sinh : $((e+\pi)/3)^2$
6852 7822 rt : (4,7,0,-2)
6853 2144 10^x : $\sqrt[3]{3}$
6853 3104 ζ : $(\ln 3/2)^3$
6853 3651 10^x : ln(5)
6853 7666 sr : $5\sqrt[3]{3}$
6853 8233 rt : (6,8,0,6)
6853 9020 2^x : $7(1/\pi)$
6853 9478 $s\pi$: $\sqrt{21/2}$
6854 0526 $s\pi$: $\sqrt[3]{3}/6$
6854 3306 rt : (7,2,1,2)
6854 5810 10^x : $4\sqrt{2}+\sqrt{5}/3$
6854 6329 rt : (8,2,8,-9)
6854 7731 θ_3 : 13/15
6854 8096 rt : (1,4,7,-7)
6854 8864 $s\pi$: $\zeta(3)/5$
6854 9601 rt : (7,9,-8,-1)
6855 0230 10^x : $4\sqrt{3}/3+\sqrt{5}/2$
6855 0836 rt : (1,7,3,6)
6855 1798 Ψ : 2/7
6855 3692 $2/5-\exp(3)$
6855 4843 rt : (6,6,-6,-5)
6855 6173 rt : (5,3,7,5)
6855 6254 id : $3\sqrt{2}-4\sqrt{3}$
6855 6552 2^x : $4\sqrt{3}/3+3\sqrt{5}/2$
6855 6871 rt : (5,8,-2,-4)
6855 8027 sr : $6\zeta(3)$
6855 9922 rt : (8,9,-4,6)
6856 0601 $t\pi$: P_{52}
6856 2664 rt : (3,7,4,-7)
6856 3010 ℓ_2 : $3e+3\pi/2$
6856 6402 rt : (9,1,-2,-2)
6856 7300 10^x : 20/7
6857 0315 cu : $4-4e$
6857 1575 e^x : $10\sqrt[3]{5}/9$
6857 1692 $t\pi$: $7\ln 2/6$
6857 1964 rt : (5,5,-4,3)
6857 2605 rt : (7,0,4,5)
6857 2951 at : 9/11
6857 3613 sq : $e/3+2\pi$
6857 5035 K : 1/2
6857 6080 id : $9e\pi/10$
6857 6990 2^x : $3-\sqrt{5}/2$
6857 7952 rt : (2,8,-6,2)
6857 8118 tanh : $2\sqrt[3]{2}/3$
6857 8415 rt : (7,8,-7,-1)
6857 8882 rt : (4,9,-3,-5)
6857 9030 cr : $4\sqrt{2}-\sqrt{3}/2$
6858 0448 Ei : 12/13
6858 0622 sinh : $4\sqrt[3]{3}/9$
6858 0753 rt : (5,3,-5,7)
6858 0906 tanh : 21/25
6858 1623 2^x : $1/\ln(1/(2\pi))$
6858 1773 rt : (9,3,1,-5)
6858 3668 $t\pi$: P_{10}
6858 4400 rt : (6,8,-9,-8)
6858 4743 rt : (7,2,7,-8)
6858 4925 rt : (7,6,-8,3)
6858 6350 2^x : $2e/3-3\pi/4$
6858 9425 rt : (7,4,-9,7)
6858 9444 e^x : 19/10
6858 9614 $t\pi$: $4\sqrt{3}/3-\sqrt{5}/2$
6858 9646 rt : (9,5,4,-8)
6859 0646 ln : $3\sqrt{2}+2\sqrt{3}/3$
6859 2350 ℓ_{10} : $7\ln 2$
6859 9030 Γ : 10/19
6859 9434 sr : 8/17
6860 0000 cu : $7\sqrt[3]{2}/10$
6860 2854 K : $\sqrt{2}+\sqrt{3}-\sqrt{7}$
6860 3329 at : $4\sqrt{3}/3-2\sqrt{5}/3$
6860 4084 2^x : $\sqrt[3]{20/3}$
6860 4964 cu : $2-\sqrt{5}/2$
6860 5762 ℓ_{10} : $\sqrt[3]{3}/7$
6860 8510 cr : $2-3\sqrt{5}/4$
6860 9868 sinh : $3\sqrt[3]{5}/8$
6861 0610 cu : $\sqrt{2}-\sqrt{3}+\sqrt{6}$
6861 0889 ℓ_{10} : $3/2+3\sqrt{5}/2$
6861 4066 rt : (6,9,-9,0)
6861 4323 $t\pi$: $\sqrt{2}/3+4\sqrt{5}$
6861 7592 $s\pi$: $3/4+2\sqrt{5}/3$
6861 9539 10^x : $3\sqrt{5}/10$
6862 0539 at : $(\ln 3/2)^{1/3}$
6862 1332 cu : $4\sqrt{2}/3+3\sqrt{5}$
6862 2939 2^x : $7\pi/6$
6862 3010 ℓ_{10} : $2/3+4\pi/3$
6862 3590 sinh : 15
6862 7241 e^x : $2\sqrt{3}/3+\sqrt{5}/3$
6862 8357 tanh : $(\sqrt{2})^{-1/2}$
6862 9066 ℓ_2 : $1+2e$
6862 9150 sq : $4-\sqrt{2}$
6863 0193 rt : (8,5,3,-7)
6863 0778 sinh : 12/7
6863 3104 e^x : 21/5
6863 3885 rt : (6,2,6,-7)
6863 6200 rt : (8,3,0,-4)
6863 7687 cr : $\sqrt{23}$
6863 9754 $\ln(3)+s\pi(1/5)$
6863 9769 rt : (4,1,9,7)
6863 9895 ln : $(\sqrt[3]{5}/3)^3$
6864 0562 θ_3 : 11/17
6864 3802 rt : (6,0,3,4)
6864 5218 Ψ : $\sqrt[3]{15}$
6864 7099 sq : $(\ln 3)^{-2}$
6864 9335 rt : (8,1,-3,-1)
6864 9773 Ψ : $1/\ln(3/2)$
6865 3345 sr : $3/4+2\pi/3$
6865 3823 ζ : $1/\ln(5/3)$
6865 5692 ℓ_2 : ln(5)
6865 6346 Ei : $\exp(-3/2)$
6865 6365 id : $1/4+2e$
6865 6479 cr : $4/3+2\sqrt{3}$
6865 6778 rt : (1,8,-7,-9)
6865 7093 tanh : $7\zeta(3)/10$
6865 8158 ζ : 1/6
6865 8904 id : $(\sqrt{2}/3)^{1/2}$
6865 9080 ℓ_2 : $2/3+2\sqrt{2}/3$
6865 9363 rt : (4,3,6,4)
6866 0410 10^x : $2-\pi/2$
6866 2915 rt : (1,9,5,-8)
6866 5283 sr : $5e$
6866 5410 rt : (7,7,5,-9)
6866 5565 2^x : $(\ln 3)^{-3}$
6866 7301 e^x : $1/\sqrt[3]{7}$
6866 8188 as : $3/2-\sqrt{3}/2$
6866 9479 rt : (9,8,-1,-6)
6866 9879 $\exp(3/4)+s\pi(\pi)$
6867 0551 $\pi\times\exp(2/5)$
6867 2463 ζ : $\sqrt[3]{15/2}$
6867 3571 rt : (6,2,0,1)
6867 5259 rt : (5,4,8,-9)
6867 5913 rt : (1,9,7,-9)
6867 6183 $\sqrt{5}-\ln(\sqrt{3})$
6867 7394 rt : (7,8,-8,-7)
6867 8715 $\exp(2/3)\times s\pi(1/3)$
6868 0154 sr : $4-2\sqrt{3}/3$
6868 0487 2^x : $3\sqrt{2}/2+2\sqrt{3}/3$
6868 0942 2^x : $5\pi/3$
6868 1126 rt : (7,5,2,-6)
6868 1211 J : $3(1/\pi)/8$
6868 1814 sinh : 22/17
6868 2254 sr : $3e/2+\pi$
6868 2731 ζ : 17/21
6868 3793 $3/5\times\ln(\pi)$
6868 4456 rt : (1,7,2,-5)
6868 4786 sq : $2\sqrt{2}/3+3\sqrt{3}$
6868 6440 id : $3\sqrt{3}+2\sqrt{5}/3$
6868 6533 cr : 24/5
6868 6568 Li_2 : $9(1/\pi)/5$
6868 6957 sq : $8\pi^2/3$
6868 7255 rt : (8,8,9,5)
6868 8234 rt : (9,6,-4,-3)
6868 8288 cu : $1/3-2\sqrt{3}$
6868 8672 id : $1/3+\sqrt{2}/4$
6868 8827 ℓ_{10} : $3e/2+\pi/4$
6868 8965 id : $4\zeta(3)/7$
6869 2253 tanh : 16/19
6869 3035 rt : (5,9,-2,-4)
6869 3592 ln : $8\sqrt{5}/9$
6869 5133 at : $(2e/3)^{-1/3}$
6869 5298 cu : 15/17
6869 5728 rt : (3,8,-4,-2)
6869 6282 e^x : $(\sqrt{2}\pi/2)^{1/3}$
6869 8419 Ei : $9(1/\pi)/5$
6869 8699 10^x : $3e\pi/10$
6869 8978 rt : (5,2,5,-6)
6870 0692 rt : (8,1,-6,-2)
6870 0695 $s\pi$: $\sqrt[3]{21}$
6870 3975 sinh : $8\sqrt[3]{2}/5$
6870 4470 Ψ : $\pi^2/4$
6870 5184 Ei : $8(e+\pi)/7$
6870 5949 cr : $3\sqrt{2}+\sqrt{5}/4$
6870 7443 at : $3/2-e/4$
6870 8594 rt : (8,5,-9,9)
6870 8637 2^x : $3+3\sqrt{2}/4$
6870 9159 rt : (2,6,9,4)
6870 9987 2^x : $3\sqrt{2}-\sqrt{3}/3$
6871 0363 rt : (9,3,8,7)
6871 0574 $\ln(\sqrt{5})-\exp(2/5)$
6871 0694 rt : (7,3,-1,-3)
6871 2483 rt : (6,7,4,-8)
6871 5051 rt : (1,7,6,-5)
6871 5133 e^x : $-\sqrt{2}+\sqrt{6}+\sqrt{7}$
6871 7437 ℓ_{10} : $4+\sqrt{3}/2$
6871 8051 sr : $e\pi/3$
6871 8280 cu : 25/21
6871 9293 rt : (7,9,0,8)
6872 1188 rt : (8,8,-2,-5)
6872 1521 2^x : $3/2-\sqrt{5}/3$
6872 1804 $3/4\times\ln(2/5)$
6872 3397 rt : (8,7,-8,5)
6872 4822 sq : $4\ln 2$
6872 6141 rt : (9,4,-7,0)
6872 6509 ℓ_{10} : $4\sqrt{2}/3+4\sqrt{5}/3$
6872 8453 cu : $9(e+\pi)/10$
6872 9362 at : $3/4-\pi/2$
6872 9800 rt : (3,1,8,6)
6873 1432 $t\pi$: $3\sqrt{3}/4-2\sqrt{5}/3$
6873 2854 Γ : $7\zeta(3)$
6873 3876 rt : (4,4,7,-8)
6873 6481 sr : $3\sqrt[3]{2}/8$
6873 6583 rt : (6,4,-5,9)
6873 8274 ℓ_{10} : $e/4+4\pi/3$
6873 9956 rt : (1,5,-1,-2)
6874 1902 at : $(e/3)^2$
6874 2192 θ_3 : $\sqrt[3]{1/2}$
6874 3529 id : $1/3+3\sqrt{5}/2$
6874 3692 rt : (8,9,-7,1)
6874 3777 10^x : $9\ln 2/5$
6874 4503 2^x : $\sqrt{3}/3-\sqrt{5}/2$

6874 5746 sq : $(\sqrt[3]{5}/3)^{1/3}$
6874 6747 10^x : $(\sqrt{2}\pi/2)^{-1/2}$
6874 7019 rt : (6,5,1,-5)
6874 7522 2^x : $1/4+4\pi/3$
6874 8638 ℓ_2 : $\sqrt{2}-\sqrt{6}+\sqrt{7}$
6875 0000 id : 11/16
6875 0055 2^x : $4\sqrt{2}+\sqrt{5}$
6875 0356 $t\pi$: $1/\ln(\sqrt{2}/3)$
6875 0577 cr : $1/3+2\sqrt{5}$
6875 2685 rt : (5,0,2,3)
6875 3186 $t\pi$: $\ln(\sqrt[3]{5})$
6875 6045 cu : $3/4+3e/4$
6875 6357 rt : (1,6,-4,4)
6875 7778 erf : 5/7
6875 9797 J_0 : 7/6
6876 0180 Γ : $\exp(-2e/3)$
6876 0601 2^x : $(\sqrt[3]{5}/3)^{1/2}$
6876 0710 $s\pi$: $\sqrt{3}-2\sqrt{5}/3$
6876 1211 rt : (9,5,5,4)
6876 1247 10^x : 5/22
6876 1799 rt : (2,6,-7,-7)
6876 2197 e^x : $3+2\sqrt{3}$
6876 2295 tanh : $(5/3)^{-1/3}$
6876 2336 $t\pi$: $1/3-\pi$
6876 2431 ζ : $7\sqrt{5}/8$
6876 2442 sr : $1/4+3\sqrt{3}/2$
6876 2654 rt : (6,6,-4,5)
6876 3380 rt : (8,6,-5,-2)
6876 3461 at : $1/3-2\sqrt{3}/3$
6876 3536 10^x : $1/3+2\sqrt{5}/3$
6876 6534 J_0 : $(\sqrt[3]{4})^{1/3}$
6876 7475 rt : (3,5,-6,6)
6876 8207 $2/5-\ln(3/4)$
6877 1048 rt : (5,8,4,8)
6877 2739 $s\pi$: $2e/3-\pi/2$
6877 3827 rt : (9,2,6,-8)
6877 4328 cr : $10\sqrt[3]{3}/3$
6877 6201 sr : $3\sqrt{2}+4\sqrt{5}/3$
6877 6549 rt : (5,1,-9,6)
6877 7008 $t\pi$: $4\zeta(3)$
6878 1029 rt : (3,6,-7,1)
6878 2859 cr : $4\zeta(3)$
6878 5014 at : $(\sqrt{2}\pi/3)^{-1/2}$
6878 5453 10^x : $\exp(1/\sqrt{2})$
6878 6755 rt : (7,8,-3,-4)
6878 6903 θ_3 : $\ln(\sqrt[3]{5}/2)$
6878 8124 rt : (4,9,5,-9)
6878 8760 rt : (3,8,-9,-9)
6878 9295 rt : (4,2,4,-5)
6878 9958 id : $4(e+\pi)/5$
6879 1192 rt : (8,3,7,6)
6879 1510 K : $2\sqrt[3]{2}/5$
6879 2547 at : $\sqrt{2}-\sqrt{5}$
6879 2924 rt : (3,1,-8,-5)
6879 6481 as : $3e/2-3\pi/2$
6879 6675 cu : $2\sqrt{2}-\sqrt{5}/4$
6879 6798 10^x : $5(1/\pi)/7$
6879 7932 sq : $1/3-e$
6880 0886 rt : (3,7,-1,5)
6880 1139 J_0 : $(e/2)^{1/2}$
6880 1333 cu : $4\sqrt{2}/3+4\sqrt{3}$
6880 2800 rt : (6,8,-3,1)
6880 4294 rt : (4,5,3,1)
6880 5234 sr : $(3\pi)^{-1/3}$
6880 5925 rt : (9,9,-9,-1)
6880 5955 sinh : 9/14
6880 9179 e^x : $\pi/6$
6880 9787 rt : (7,1,9,8)
6880 9834 sq : $4\sqrt{2}/3+4\sqrt{5}/3$
6881 0415 rt : (3,4,6,-7)
6881 1157 ln : $9\zeta(3)/2$
6881 1204 rt : (1,9,-3,-6)
6881 3279 e^x : $1/3-\sqrt{2}/2$
6881 3802 rt : (9,7,4,-9)
6881 4420 $\exp(3/5)+s\pi(1/3)$
6881 4497 rt : (5,3,-8,2)
6881 6503 id : $2\sqrt{2}/3+\sqrt{5}/3$
6881 8077 at : $3\sqrt{2}/2-3\sqrt{3}/4$
6881 8111 10^x : $\exp(-\sqrt{2}\pi/3)$
6881 8242 sr : $5\sqrt[3]{5}/3$
6881 8814 rt : (6,3,-2,-2)
6882 1768 rt : (2,8,-5,-1)
6882 1941 sr : $\sqrt{3}+\sqrt{5}/2$
6882 4162 rt : (2,6,8,-9)
6882 4290 10^x : $(\sqrt[3]{5}/3)^{1/2}$
6882 4720 sr : 9/19
6882 5877 $t\pi$: $4\sqrt{2}/7$
6882 6985 ℓ_{10} : $\sqrt{2}+2\sqrt{3}$
6882 7466 rt : (9,8,-8,7)
6882 7742 Γ : $5\pi^2/4$
6882 9045 $s\pi$: $(2e)^{1/3}$
6882 9185 cr : $3+2e/3$
6882 9800 ℓ_2 : $\sqrt{3}+2\sqrt{5}/3$
6883 0764 id : $3\sqrt{2}/2-\sqrt{3}/4$
6883 1920 Ei : $7(1/\pi)/10$
6883 2014 rt : (9,0,3,5)
6883 3114 $\exp(\sqrt{3})+s\pi(\sqrt{2})$
6883 4904 cr : $e+2\pi/3$
6883 5583 rt : (5,5,0,-4)
6883 6042 erf : $(\ln 3/3)^{1/3}$
6883 8127 rt : (8,8,-7,-6)
6883 8550 2^x : $3/2-3e/4$
6883 8883 ln : $4/3+3e/2$
6884 0630 ℓ_{10} : $4e\pi/7$
6884 0730 erf : $((e+\pi)/3)^{-1/2}$
6884 1980 rt : (8,2,5,-7)
6884 2432 J_0 : $3e/7$
6884 2792 erf : $(\sqrt{5}/2)^{-3}$
6884 4087 ζ : $(2e/3)^{-3}$
6884 4759 e^x : 11/21
6884 5093 rt : (7,8,8,4)
6884 5899 rt : (4,7,2,-6)
6884 8987 rt : (7,4,7,-9)
6884 9068 Ei : 13
6884 9235 ln : $1/2+2\sqrt{5}/3$
6885 0564 $\exp(1/3)-s\pi(1/4)$
6885 2831 cu : $3\pi^2/2$
6885 2885 rt : (3,9,4,4)
6885 3052 10^x : $\sqrt{13/3}$
6885 3060 Ei : $5e\pi/8$
6885 3075 id : $\sqrt{2}/2+4\sqrt{5}/3$
6885 9407 rt : (3,3,5,3)
6886 0142 rt : (8,4,-8,1)
6886 5716 e^x : $\sqrt{3}+\sqrt{6}+\sqrt{7}$
6886 8068 rt : (7,6,-6,-1)
6886 9324 rt : (9,5,1,-6)
6887 0706 ℓ_{10} : $5(e+\pi)/6$
6887 1136 rt : (2,1,7,5)
6887 1185 cu : $2\sqrt{2}-2\sqrt{3}/3$
6887 2083 rt : (8,7,3,-8)
6887 2598 rt : (6,8,-4,-3)
6887 4509 10^x : 25/22
6887 4632 2^x : $3\sqrt[3]{2}/5$
6887 5529 rt : (5,5,-7,-2)
6887 5978 e^x : $2\sqrt{3}/3+\sqrt{5}$
6887 6242 rt : (1,1,9,-7)
6887 6574 $s\pi$: $2\sqrt{2}/3+3\sqrt{3}/4$
6887 8195 2^x : $4/3-\sqrt{3}/3$
6887 8516 sr : $(\sqrt{2}\pi)^{-1/2}$
6887 9020 id : $1/2+4\pi/3$
6888 1941 sq : $3+3\pi/4$
6888 2810 e^x : $10e\pi/7$
6888 3234 2^x : $9\ln 2/2$
6888 4637 10^x : $4\sqrt{3}+3\sqrt{5}/2$
6888 5207 $3/5\div\ln(4/5)$
6888 5647 Ei : $4\ln 2/3$
6888 5942 10^x : $\sqrt{3}/3+\sqrt{5}/4$
6888 6039 $s\pi$: $1/3+3\pi$
6888 7098 erf : $9(1/\pi)/4$
6888 7743 10^x : $2/3+2\sqrt{2}/3$
6888 8194 as : $2\sqrt{2}-2\sqrt{3}$
6888 8702 cr : $\exp(-\sqrt{5}/2)$
6888 9218 rt : (4,8,-8,-8)
6889 0506 2^x : $3+\sqrt{5}$
6889 1479 rt : (6,2,-9,8)
6889 1712 rt : (7,7,-4,7)
6889 1951 ℓ_{10} : $3+4\sqrt{2}/3$
6889 2017 ln : $\sqrt{2}+\sqrt{3}/3$
6889 2438 at : 14/17
6889 4948 Ψ : $\exp(-\sqrt{3}/3)$
6889 5061 rt : (1,6,3,7)
6889 6346 $e\div\ln(5)$
6890 1479 rt : (6,1,8,7)
6890 2047 cu : $8\sqrt{5}/7$
6890 2763 ln : $4+\sqrt{2}$
6890 4930 rt : (7,3,6,5)
6890 5234 Γ : $4\pi^2/7$
6890 5480 tanh : 11/13
6890 6961 2^x : $2+2\sqrt{2}/3$
6890 7436 e^x : π^2
6890 8576 cu : $9e/7$
6890 9532 rt : (3,9,-6,6)
6890 9662 erf : $(e)^{-1/3}$
6891 0751 $\ln(4)\times\exp(4)$
6891 1086 rt : (8,5,4,3)
6891 2051 e^x : $\sqrt{5}+\sqrt{6}-\sqrt{7}$
6891 2690 Γ : $\sqrt[3]{2}/9$
6891 4597 rt : (2,4,5,-6)
6891 5165 rt : (1,8,2,-3)
6891 6272 ln : $2\sqrt{2}/3+2\sqrt{5}$
6891 9952 $s\pi$: $3(e+\pi)/10$
6892 0389 cr : $3\sqrt{2}+\sqrt{3}/3$
6892 0748 rt : (6,9,-1,-3)
6892 3026 rt : (6,4,6,-8)
6892 3172 at : $2/3-2\sqrt{5}/3$
6892 5163 rt : (9,7,2,1)
6892 6605 ℓ_2 : $1/3+4\pi$
6892 8966 at : $4\sqrt[3]{3}/7$
6892 9758 rt : (7,2,4,-6)
6893 1817 rt : (2,9,-2,-5)
6893 3356 sq : $3\sqrt{2}/4-4\sqrt{5}/3$
6893 4760 rt : (2,9,3,-7)
6893 6122 rt : (6,9,4,-9)
6893 9835 rt : (8,0,2,4)
6893 9951 ℓ_{10} : $\exp(\sqrt[3]{4})$
6894 3347 rt : (7,7,2,-7)
6894 4924 rt : (6,4,-8,4)
6894 7209 sr : 7π
6894 9052 $\ln(2/3)-\exp(1/4)$
6894 9071 rt : (1,7,0,-3)
6894 9137 id : $9\pi^2/7$
6895 1920 rt : (7,9,-3,3)
6895 2782 id : $1/3+3\pi/4$
6895 3200 rt : (8,5,0,-5)
6895 6569 rt : (9,2,0,2)
6895 6698 ℓ_{10} : $9e/5$
6895 9866 sinh : $\sqrt{5}+\sqrt{6}+\sqrt{7}$
6896 0923 rt : (4,7,1,1)
6896 1894 Ei : $5e\pi/6$
6896 5349 Ψ : $7\zeta(3)/9$
6896 6117 J_0 : $(\pi/2)^{1/3}$
6896 6807 10^x : $1-\sqrt{3}/4$
6896 7433 rt : (9,3,-2,-3)
6897 0545 $s\pi$: $5\sqrt{5}/9$
6897 1729 10^x : $4+\sqrt{2}/3$
6897 2942 rt : (9,8,-4,-4)
6897 7500 as : 7/11
6898 0234 $1/3+\exp(\sqrt{5})$
6898 1844 $\ln(2/3)\div s\pi(1/5)$
6898 2903 10^x : $\sqrt[3]{13/3}$
6898 3481 rt : (3,5,-9,-2)
6898 4875 rt : (8,8,-4,9)
6898 6621 ln : $3\sqrt{2}/4-\sqrt{5}/4$
6898 6931 cr : $\sqrt{2}/4+2\sqrt{5}$
6898 7248 sr : $(1/(3e))^{-1/2}$
6898 9626 10^x : $3/4+\pi/3$
6898 9794 rt : (5,8,-5,-2)
6899 2449 $\exp(1/3)-\exp(3)$
6899 5200 2^x : $9(e+\pi)/10$
6899 5612 rt : (4,1,-5,8)
6899 5975 rt : (9,4,6,-9)
6899 6822 2^x : $1/4-\pi/4$
6899 6970 rt : (4,6,7,-9)
6899 7016 as : $\ln(\sqrt[3]{4}/3)$
6899 7530 rt : (9,1,8,8)
6899 8347 rt : (1,8,-8,-9)
6899 8628 cu : 23/9
6899 9154 rt : (4,1,9,-8)
6899 9877 $\ln(\sqrt{3})+\exp(\pi)$
6900 0000 sq : 13/10

6900 0058 $\zeta : 8\sqrt[3]{5}/7$
6900 2795 sr : $1/2 + 3\pi/4$
6900 3188 $10^x : 19/3$
6900 5635 rt : (4,7,-9,3)
6900 6555 sr : 10/21
6900 7162 cr : $3/4 + 3e/2$
6900 7470 sq : $4\sqrt{2}/3 - 2\sqrt{5}$
6901 0300 $e^x : 17/11$
6901 0562 $\ell_{10} : \sqrt{24}$
6901 0709 as : $2(1/\pi)$
6901 0889 $Ei : (\sqrt[3]{5}/2)^{1/2}$
6901 1323 rt : (5,9,3,-8)
6901 6847 $\Gamma : -\sqrt{2} - \sqrt{3} + \sqrt{5}$
6901 7295 rt : (5,4,5,-7)
6901 8095 $\zeta : 13/17$
6901 8297 rt : (5,9,-9,8)
6901 8882 cr : $2 + 2\sqrt{2}$
6902 0930 $10^x : P_{80}$
6902 3341 $\ln(2/5) - s\pi(e)$
6902 3807 rt : (6,6,-7,0)
6902 4957 sq : $\sqrt{2}/3 - 4\sqrt{3}$
6902 5111 ln : $\sqrt{22}\pi$
6902 7851 rt : (5,1,7,6)
6902 9975 tanh : $3\sqrt{2}/5$
6903 0850 sr : 20/7
6903 2432 $e^x : 9\zeta(3)/7$
6903 2470 rt : (6,7,1,-6)
6903 3221 $s\pi : 1/\sqrt{17}$
6903 3452 $10^x : (2\pi/3)^{-2}$
6903 4365 rt : (5,8,-7,-7)
6903 6454 $t\pi : \sqrt[3]{8/3}$
6903 6583 rt : (7,5,-8,6)
6903 7254 $t\pi : 6e/7$
6904 1345 $e^x : 2e + \pi/4$
6904 1575 id : $\sqrt{22}$
6904 3003 $Ei : \ln(\ln 3)$
6904 4645 at : 19/23
6904 5583 tanh : $8(1/\pi)/3$
6904 7054 rt : (6,2,3,-5)
6904 7145 rt : (2,9,-2,-1)
6905 0474 $\Psi : 10\sqrt{3}/7$
6905 1565 rt : (5,6,9,5)
6905 1571 $\zeta : (\sqrt[3]{5})^{-1/2}$
6905 2342 rt : (3,9,-9,-5)
6905 3396 cu : $5\sqrt{2}/8$
6905 4377 $e^x : 4\sqrt{3}/7$
6905 4909 $s\pi : P_{15}$
6905 7097 $s\pi : 3\sqrt{2}$
6905 7407 rt : (6,1,-5,1)
6905 7995 $Ei : 3\sqrt[3]{3}$
6905 9892 id : $3 - 4\sqrt{3}/3$
6906 2456 rt : (7,5,-1,-4)
6906 2998 ln : $7\sqrt[3]{5}/6$
6906 3389 rt : (8,9,3,-9)
6906 3728 $s\pi : 1/\ln(\sqrt{5})$
6906 4338 $\Psi : \exp(e/3)$
6906 4641 tanh : $(\sqrt[3]{3}/2)^{1/2}$
6906 4744 rt : (1,5,7,8)
6906 5217 $\ell_{10} : \sqrt{3}/4 + 2\sqrt{5}$
6906 8086 $t\pi : 7\sqrt[3]{5}/9$
6907 0317 rt : (7,1,9,-9)
6907 0829 rt : (5,3,-9,-6)
6907 1054 rt : (4,9,-4,2)
6907 2255 $Ei : \sqrt[3]{21/2}$
6907 3585 $\ell_{10} : 4 + e/3$
6907 3998 rt : (3,7,-4,-3)
6907 5014 $\theta_3 : (\ln 2)^3$
6907 5082 rt : (8,8,-5,-3)
6907 5963 sinh : $1/\ln(3\pi/2)$
6907 6062 $\Psi : 7\sqrt{2}/4$
6907 6782 rt : (6,3,5,4)
6907 9807 rt : (8,4,5,-8)
6908 2086 rt : (2,7,0,-4)
6908 5565 cr : $3\sqrt{3}/2 + \sqrt{5}$
6908 6866 $K : 8(1/\pi)/5$
6908 8026 rt : (9,7,1,-7)
6908 9879 $\pi + \ln(\sqrt{3})$
6908 9943 sr : $3/2 + e/2$
6909 0894 sr : $-\sqrt{5} + \sqrt{6} + \sqrt{7}$
6909 1034 $Ei : 8e\pi/5$
6909 1086 $t\pi : \sqrt{3}/9$
6909 2376 rt : (7,4,-9,2)
6909 4779 rt : (7,0,1,3)
6909 6212 rt : (3,6,6,-8)
6909 6812 $J : (2\pi/3)^{-3}$
6909 7174 sr : $2e/3 + \pi/3$
6909 7347 cu : $4\sqrt{3}/3 - \sqrt{5}/2$
6909 8300 rt : (4,6,-5,-5)
6909 8829 id : $(2\pi/3)^{-1/2}$
6910 0243 $\zeta : (e + \pi)/3$
6910 1901 $2^x : \sqrt{2}/2 + \sqrt{5}$
6910 2022 $\ell_{10} : 2/3 + 3\sqrt{2}$
6910 3462 $e^x : 7\sqrt[3]{3}/2$
6910 4529 rt : (8,6,-8,8)
6910 5040 $J : 2\ln 2/3$
6910 5656 rt : (8,1,7,7)
6910 6946 tanh : 17/20
6910 8265 rt : (8,3,-3,-2)
6910 9855 $e^x : 7\sqrt{2}/10$
6911 1354 cu : $\pi^2/5$
6911 3846 rt : (9,2,3,-6)
6911 4650 rt : (6,7,0,9)
6911 4918 $t\pi : 10\sqrt[3]{3}/3$
6911 5905 $\theta_3 : (\sqrt[3]{2}/3)^{1/2}$
6911 6290 rt : (3,9,-1,-4)
6911 6830 sinh : 9/4
6911 7230 $10^x : 4e/3 + 4\pi$
6911 7363 sq : $1/2 + 4\sqrt{2}/3$
6911 8745 rt : (9,6,-7,-1)
6911 9367 rt : (7,7,-7,2)
6911 9688 at : $1/4 + \sqrt{3}/3$
6911 9949 rt : (2,8,-7,-8)
6912 0167 $10^x : 7e\pi/3$
6912 0961 cu : 17/11
6912 1081 $t\pi : 4\sqrt{2}/3 + \sqrt{5}/4$
6912 1399 $\ell_{10} : 4\sqrt{2} - \sqrt{5}/3$
6912 1546 sr : $3 + 3\sqrt{2}$
6912 1838 cu : $1 + \pi/4$
6912 4017 $t\pi : ((e + \pi)/2)^{-1/2}$
6912 5478 rt : (1,1,6,4)
6912 6363 $2^x : 1/3 - \sqrt{3}/2$
6912 6588 $e^x : 2e/3 + 3\pi/2$
6912 6777 rt : (1,3,-9,-5)
6912 8274 $Ei : \sqrt[3]{22}$
6912 9497 ln : $2\sqrt{2} + 3\sqrt{3}/2$
6913 0699 rt : (2,6,-7,-7)
6913 1076 sq : P_{25}
6913 1818 $Ei : \sqrt{2} - \sqrt{5} + \sqrt{6}$
6913 3740 $\lambda : \exp(-5\pi/6)$
6913 4358 rt : (3,1,8,-7)
6913 6504 at : $\sqrt{2}/3 - 3\sqrt{3}/4$
6913 7488 $10^x : 3\sqrt{3}/2 + 2\sqrt{5}$
6913 8081 $10^x : 25/19$
6913 9918 $2^x : 3/2 + 3\sqrt{3}$
6914 0240 $s\pi : 7e/4$
6914 0280 $erf : 4\sqrt[3]{2}/7$
6914 1397 rt : (6,6,6,-9)
6914 3312 $erf : 18/25$
6914 5332 rt : (7,9,2,-8)
6914 5968 $\theta_3 : 1/3$
6914 7114 rt : (5,8,1,3)
6914 9210 id : $e/3 + \pi/4$
6915 1848 rt : (6,8,-6,-4)
6915 2337 rt : (8,6,9,6)
6915 4773 cu : $9\zeta(3)/7$
6915 6407 sr : 11/23
6915 8077 rt : (9,3,5,5)
6915 9233 rt : (4,8,-6,-1)
6916 2661 rt : (4,1,-9,-7)
6916 3286 $2/3 \div s\pi(\sqrt{2})$
6916 5188 $s\pi : \exp(-\sqrt{2})$
6916 5783 $Ei : 2/9$
6916 7773 rt : (1,8,7,-9)
6916 8832 $J_0 : (\sqrt{5}/3)^{-1/2}$
6916 8839 $\ln(3/4) \times \exp(\sqrt{5})$
6916 9190 cu : $2 + \sqrt{5}/3$
6917 0318 $\exp(5) \times s\pi(\sqrt{3})$
6917 0715 rt : (6,8,7,3)
6917 2699 $K : \ln(5/3)$
6917 8852 rt : (7,5,3,2)
6918 0038 $2^x : 10/7$
6918 2058 rt : (2,3,-9,-7)
6918 2469 $1/5 + \exp(2/5)$
6918 2738 rt : (6,1,8,-8)
6918 3013 $\Psi : 3\pi^2/5$
6918 3520 rt : (2,8,-5,7)
6918 3581 at : $2 - 2\sqrt{2}$
6918 3643 rt : (8,2,-1,1)
6918 4311 rt : (7,4,4,-7)
6918 4405 $Ei : -\sqrt{3} + \sqrt{6} + \sqrt{7}$
6918 4746 rt : (1,6,-5,-6)
6918 5234 sq : $6\ln 2/5$
6918 5649 rt : (8,7,0,-6)
6918 5829 $10^x : 2e - 4\pi/3$
6918 6096 rt : (8,8,-7,4)
6918 6800 rt : (9,8,-6,-5)
6918 7770 $\ell_2 : 13/21$
6918 7894 $10^x : \sqrt{7/3}$
6918 8198 $J_0 : 22/19$
6919 0004 at : $(\ln 3)^{-2}$
6919 0198 cu : $2\sqrt{2} + \sqrt{3}/4$
6919 0765 $J : 2\sqrt[3]{3}/7$
6919 3226 $e^x : 11/2$
6919 3785 rt : (3,4,-8,8)
6919 8530 rt : (5,4,-4,6)
6920 2087 $Ei : 7\zeta(3)/2$
6920 2147 rt : (2,6,-8,2)
6920 9029 $s\pi : 3\sqrt{3}/4 + 4\sqrt{5}$
6920 9365 $e^x : 3/4 - \sqrt{5}/2$
6920 9720 rt : (7,8,-6,-2)
6921 0609 rt : (6,5,-2,-3)
6921 1539 cu : $5e$
6921 1734 rt : (5,2,2,-4)
6921 2158 ln : $\sqrt{2} + \sqrt{3} - \sqrt{7}$
6921 2417 rt : (9,1,8,-9)
6921 3204 rt : (4,1,6,5)
6921 3525 id : $4\sqrt{3} - \sqrt{5}$
6921 6322 $s\pi : \ln(3\pi)$
6921 7004 $\ell_2 : 1/4 + 4\sqrt{5}/3$
6921 8827 $erf : \sqrt[3]{3}/2$
6921 9238 $10^x : 3 + 2\sqrt{5}$
6921 9381 cu : $2 + 4\sqrt{3}$
6922 0062 id : $\exp(-1/e)$
6922 2550 $s\pi : 2\sqrt{2} + 4\sqrt{3}$
6922 3086 rt : (3,9,1,-4)
6922 4067 rt : (5,0,-9,9)
6922 5143 rt : (2,6,5,9)
6922 5281 at : $(\sqrt[3]{5}/3)^{1/3}$
6922 6162 $erf : 3\zeta(3)/5$
6922 6316 $\ln(5) \div s\pi(2/5)$
6922 7649 rt : (1,3,9,-8)
6923 0769 id : 9/13
6923 1597 $t\pi : 4\sqrt{2} - 2\sqrt{3}$
6923 3766 $erf : 1/\ln(4)$
6923 5056 sr : $8e/3$
6923 5336 rt : (9,9,-7,6)
6923 7532 $e^x : e/4 - \pi/3$
6924 0959 rt : (9,5,-2,-4)
6924 1853 sq : $3 - e/2$
6924 2300 rt : (7,9,-6,-2)
6924 4753 rt : (8,2,2,-5)
6924 4866 $t\pi : 5\sqrt{3}/2$
6924 4987 $\ell_2 : 3 + 2\sqrt{3}$
6924 5191 rt : (6,9,1,5)
6924 6418 rt : (2,7,1,-2)
6924 7424 id : $1/3 + e/2$
6924 8531 rt : (5,6,5,-8)
6924 9368 rt : (7,1,6,6)
6925 0104 sinh : $2e\pi/7$
6925 0455 $\zeta : \sqrt[3]{5}/10$
6925 1019 rt : (2,8,2,-1)
6925 1304 rt : (9,8,7,4)
6925 2434 rt : (4,3,9,-9)
6925 4309 $\ln(5) \times s\pi(\pi)$
6925 6283 tanh : $\exp(-1/(2\pi))$
6925 6301 $\Psi : 14/15$
6925 6875 sr : $9(1/\pi)$
6925 7814 $t\pi : 1/4 + 2\sqrt{2}/3$
6925 8233 $e^x : 7\sqrt{5}/9$
6926 0067 rt : (9,9,1,-8)

6926 0484 rt : (6,5,-4,8)
6926 1087 sq : $4\sqrt{2}/3+3\sqrt{5}/4$
6926 3095 rt : (6,8,-6,-6)
6926 3781 rt : (8,6,5,-9)
6926 4507 ℓ_2 : $8\sqrt{2}/7$
6926 8174 10^x : $3/4+\sqrt{2}/4$
6926 8460 e^x : 10/19
6927 3147 Li_2 : $2\sqrt[3]{3}/5$
6927 4335 cu : $3/2-3\pi$
6927 5062 sr : $4\sqrt{3}+3\sqrt{5}$
6927 9287 rt : (5,6,-8,1)
6927 9370 id : $6\pi/7$
6928 0904 id : $1/4-2\sqrt{2}/3$
6928 1074 cu : $5\sqrt{5}/2$
6928 2032 id : $2\sqrt{3}/5$
6928 3311 sq : $4/3-4\sqrt{2}$
6928 4661 rt : (1,7,-1,-3)
6928 6950 2^x : $5\sqrt{5}$
6928 7824 rt : (3,8,-6,-7)
6929 2388 erf : 13/18
6929 3396 sr : $2+\sqrt{3}/2$
6929 3844 cr : $7\ln 2$
6929 5082 K : $3\sqrt[3]{5}/10$
6929 8029 2^x : $3-\pi/2$
6929 8053 Ei : $-\sqrt{2}+\sqrt{3}+\sqrt{7}$
6929 8511 sq : $1/4-4\pi$
6930 0046 rt : (8,6,-8,0)
6930 3706 as : $2\sqrt{5}/7$
6930 5171 sr : P_{21}
6930 5255 rt : (3,6,-3,7)
6930 7334 rt : (3,8,6,-9)
6930 8351 rt : (7,7,-1,-5)
6930 8503 rt : (2,5,-8,3)
6931 1080 cu : $2(e+\pi)/7$
6931 1988 cr : $10(e+\pi)/3$
6931 2172 rt : (3,5,4,7)
6931 2620 sr : $3\sqrt{2}/2+\sqrt{5}/3$
6931 3306 rt : (1,8,8,-5)
6931 3958 tanh : $e\pi/10$
6931 4718 id : $\ln 2$
6931 5275 rt : (7,3,-4,-1)
6931 5614 id : $2e\pi/3$
6931 6584 sinh : 11/17
6931 7181 J_0 : $5\ln 2/3$
6931 7935 cr : $3/2+3\sqrt{5}/2$
6931 8198 rt : (6,4,3,-6)
6931 9008 cu : $2/3-4\pi/3$
6931 9153 K : $\exp(-2/3)$
6932 0245 Li_2 : γ
6932 0500 cu : $3+4e/3$
6932 1802 rt : (9,4,3,-7)
6932 2375 $\ln(4)\times\exp(1/5)$
6932 2513 rt : (1,7,9,-7)
6932 4330 id : $4e-4\pi$
6932 6974 ℓ_{10} : $5\pi^2$
6932 7428 sinh : $7\sqrt[3]{5}/2$
6932 7512 sq : $8e\pi/5$
6932 7738 rt : (2,1,7,-6)
6932 9307 10^x : $4+e/4$
6932 9775 2^x : $(\sqrt{3})^{-1/2}$
6933 0007 2^x : $3\sqrt{2}-3\sqrt{3}/4$
6933 0065 rt : (5,1,7,-7)
6933 0774 sr : $(\ln 2/3)^{1/2}$
6933 1382 rt : (8,1,7,-8)
6933 1778 ln : $4\sqrt{3}-2\sqrt{5}/3$
6933 3687 cr : $2/3+4\pi/3$
6933 6127 id : $\sqrt[3]{1/3}$
6933 7482 rt : (2,8,-8,-7)
6933 9034 2^x : $4-2\sqrt{5}/3$
6934 0322 Li_2 : $\sqrt{3}/3$
6934 1384 sr : $2\zeta(3)/5$
6934 1555 2^x : $3\pi/5$
6934 3254 Ei : $(\sqrt[3]{2})^{-1/3}$
6934 3273 2^x : $\sqrt{2}/2+3\sqrt{5}/2$
6934 3678 J_0 : $2\sqrt{3}/3$
6934 7047 rt : (1,9,-6,7)
6934 7807 2^x : $4\sqrt[3]{5}/9$
6934 8518 e^x : $1/2-\sqrt{3}/2$
6934 8525 rt : (5,2,-8,5)
6934 9062 2^x : 19/25
6935 0223 ζ : $\zeta(3)/7$
6935 3517 ln : P_{47}
6935 4912 rt : (8,3,4,4)
6935 5575 10^x : $\sqrt{2}/2-\sqrt{3}/2$
6935 6887 $\exp(2/3)-s\pi(\sqrt{3})$
6935 7543 ℓ_2 : $\exp(\sqrt[3]{3}/3)$
6936 0001 $4/5\times\exp(3/4)$
6936 2057 ℓ_{10} : $3/4+4\pi/3$
6936 2211 2^x : $4\sqrt{3}+4\sqrt{5}$
6936 2604 ℓ_2 : $7\ln 2/3$
6936 2636 at : P_{25}
6936 3534 rt : (8,9,0,-7)
6936 4047 sr : $4\sqrt{3}/3+\sqrt{5}/4$
6936 4143 $s\pi$: $5\pi/7$
6936 6436 rt : (5,3,4,3)
6936 6503 2^x : $3/4+e/4$
6936 6632 tanh : $\sqrt[3]{5}/2$
6936 9162 $s\pi$: $2/3+\sqrt{3}/3$
6936 9554 rt : (5,9,0,-6)
6937 5025 rt : (7,6,4,-8)
6937 6049 $s\pi$: $3\sqrt[3]{2}/5$
6937 8966 rt : (2,9,0,-5)
6937 9102 Γ : $\ln(\sqrt{3}/2)$
6937 9849 rt : (7,6,8,5)
6938 0199 rt : (9,8,-7,-2)
6938 0394 Γ : 11/21
6938 0958 Li_2 : $5\ln 2/6$
6938 1865 rt : (6,3,-8,7)
6938 1885 at : $6\ln 2/5$
6938 4327 rt : (4,6,4,9)
6938 5219 J_0 : 15/13
6938 7479 J_0 : $4\sqrt[3]{3}/5$
6938 8285 rt : (6,3,8,-9)
6938 8489 id : $\sqrt{19}\pi$
6938 9859 Ψ : $\sqrt[3]{2}/8$
6939 1207 rt : (7,1,-7,-3)
6939 2195 Γ : $5e/2$
6939 4257 at : $\exp(-1/(2e))$
6939 5143 rt : (6,8,-7,-1)
6939 5757 rt : (1,9,0,-4)
6939 6974 Γ : $\ln(\ln 3/2)$
6939 7383 ℓ_{10} : $4+2\sqrt{2}/3$
6939 8560 rt : (8,5,-3,-3)
6939 9469 e^x : $4/3+\sqrt{2}/2$
6939 9481 rt : (1,6,4,-6)
6940 1118 $t\pi$: $3\sqrt{2}/2-4\sqrt{3}$
6940 1663 Ei : $\sqrt{3}-\sqrt{6}+\sqrt{7}$
6940 1974 rt : (3,3,8,-8)
6940 3603 ln : $\sqrt{3}\pi$
6940 3809 ℓ_{10} : $\sqrt{2}/3+2\sqrt{5}$
6940 5514 rt : (9,4,-3,-1)
6940 6200 rt : (7,4,-8,9)
6940 7142 cr : P_7
6940 7158 ℓ_2 : $3\sqrt[3]{3}/7$
6940 9910 rt : (6,7,-3,4)
6941 2685 cu : $3/2+4\sqrt{3}$
6941 3859 cu : P_{54}
6941 4572 rt : (4,6,8,4)
6941 6415 rt : (1,8,-5,9)
6941 7420 Γ : 19/5
6941 8811 K : $3\zeta(3)/7$
6941 9251 cr : $3e/2+\pi/4$
6941 9371 sq : $((e+\pi)/2)^2$
6942 0081 rt : (9,1,5,6)
6942 1162 rt : (7,2,1,-4)
6942 2771 rt : (5,5,-3,-2)
6942 4150 rt : (5,8,5,-9)
6942 4191 ℓ_2 : $1+\sqrt{5}$
6942 4198 cu : 25/14
6942 5417 rt : (3,8,-7,0)
6942 7190 id : $1/4-4\sqrt{5}$
6942 7852 at : $(\ln 2)^{1/2}$
6942 8550 rt : (2,6,-4,-5)
6942 9180 $s\pi$: $\sqrt[3]{5}/7$
6943 1377 $\ln(\pi)\div\exp(1/2)$
6943 2860 e^x : $4\sqrt{2}-\sqrt{3}/3$
6943 4227 rt : (4,9,0,-3)
6943 5445 at : $(\sqrt[3]{3})^{-1/2}$
6943 5568 J : $\exp(-12\pi/17)$
6943 6277 rt : (7,8,-3,6)
6943 6880 rt : (2,7,1,9)
6943 7678 sinh : $(\sqrt[3]{2}/3)^{1/2}$
6943 9384 2^x : $8\ln 2$
6943 9794 rt : (9,7,-2,-5)
6944 2543 rt : (1,7,1,-7)
6944 2852 cu : $3\sqrt{3}/2+4\sqrt{5}/3$
6944 3162 J_0 : $1/\ln(\sqrt[3]{2}/3)$
6944 3302 rt : (8,4,9,7)
6944 4218 rt : (4,8,-6,4)
6944 4444 sq : 5/6
6944 5252 id : $2\sqrt{2}+\sqrt{3}/2$
6944 6484 Γ : $\pi/6$
6944 9243 sinh : $9\sqrt[3]{3}/10$
6944 9675 rt : (6,1,5,5)
6944 9707 sr : $3/4+3\sqrt{2}/2$
6944 9826 as : 16/25
6945 0614 J : $\sqrt[3]{2}/9$
6945 1469 ζ : $\exp(-1/e)$
6945 2804 $1/2\times\exp(2)$
6945 4855 rt : (8,9,-3,8)
6945 5948 Ei : $10\pi/7$
6945 6002 cu : $4\sqrt{2}/3-4\sqrt{5}$
6945 6457 cr : $4+\sqrt{3}/2$
6945 6483 rt : (1,8,3,5)
6945 7909 Ψ : 14/25
6945 8304 ℓ_{10} : $\sqrt{2}/7$
6945 8748 e^x : $4+3\sqrt{2}/4$
6945 9679 rt : (8,4,2,-6)
6946 0743 cu : $1-4\sqrt{2}/3$
6946 0859 $t\pi$: $2\pi^2/9$
6946 3410 2^x : $2\sqrt{2}-3\sqrt{5}/2$
6946 4420 rt : (1,9,3,3)
6946 4881 Ei : $\sqrt{6}\pi$
6946 5173 $t\pi$: $\exp(1/e)$
6946 6502 rt : (1,5,9,-9)
6946 6518 $\ln(\pi)\div s\pi(\sqrt{5})$
6946 7174 rt : (6,7,-2,-4)
6946 8258 cr : $4\sqrt{2}/3+4\sqrt{5}/3$
6946 9651 rt : (3,8,2,6)
6947 0566 $t\pi$: $1/2+\pi/3$
6947 2213 Ψ : $4\sqrt[3]{2}/9$
6947 2383 sr : $1/3+4\sqrt{3}$
6947 3222 sinh : 13/5
6947 3827 at : 5/6
6947 4211 sinh : $10\zeta(3)/7$
6947 4846 rt : (3,3,-8,-6)
6947 5005 sr : $2\sqrt{3}-4\sqrt{5}/3$
6947 6157 rt : (1,8,-5,-7)
6947 7655 rt : (2,7,1,-9)
6947 7812 10^x : $5\pi^2/2$
6947 8267 tanh : 6/7
6948 1657 rt : (9,6,3,-8)
6948 2596 sr : $4e/3-\pi$
6948 2644 sq : $3/2-\pi$
6948 3220 rt : (7,1,6,-7)
6948 3497 tanh : $(\sqrt[3]{4})^{-1/3}$
6948 3561 cr : $e/4+4\pi/3$
6948 5979 ln : $5\zeta(3)/3$
6948 6785 rt : (5,4,9,6)
6948 7727 $e-\exp(5)$
6948 8492 sq : $e/4+2\pi/3$
6948 8870 sinh : $\sqrt{5}-\sqrt{6}-\sqrt{7}$
6948 9176 id : $9\sqrt[3]{5}/2$
6949 0108 rt : (7,9,-1,-6)
6949 0938 ln : $1/4+3\sqrt{3}$
6949 1263 ln : P_{63}
6949 2487 ln : $3\pi^2/2$
6949 3676 ℓ_{10} : $2e/3+\pi$
6949 4946 cu : $2/3+2\sqrt{2}$
6949 5867 rt : (5,4,2,-5)
6949 7483 $\ln(5)+\exp(3)$
6950 1344 rt : (8,3,7,-9)
6950 2732 ℓ_{10} : $2\sqrt{3}+2\sqrt{5}/3$
6950 5240 Γ : $6\sqrt[3]{2}$
6950 7103 rt : (5,6,-4,-4)
6950 9805 rt : (2,5,-9,-8)
6951 0377 tanh : $(e/2)^{-1/2}$
6951 1197 2^x : $4\sqrt{2}/3$
6951 1534 rt : (3,1,5,4)
6951 3154 rt : (6,6,3,-7)

6951 4221 rt : (3,7,-2,-3)
6951 5864 $10^x : 7\ln 2/2$
6951 6678 $\Psi : \sqrt[3]{5}/6$
6951 6705 $e^x : 3\sqrt{3}-\sqrt{5}/3$
6951 7897 $1/3 \times \exp(3)$
6951 9410 rt : (8,7,1,0)
6951 9640 $J : 8/17$
6952 0493 sr : $3\sqrt{2}/4-\sqrt{3}/3$
6952 1189 $1/5 \div \ln(3/4)$
6952 2313 $t\pi : 9\sqrt[3]{2}/2$
6952 2755 $\ell_2 : \sqrt{17}\pi$
6952 3283 rt : (7,8,4,-9)
6952 6516 rt : (1,7,-6,-5)
6952 8104 $\ell_{10} : 4/3+4e/3$
6952 8683 rt : (4,9,-1,-5)
6952 9159 $10^x : 2e/3+4\pi$
6952 9898 id : $\sqrt{6}\pi$
6953 0599 $10^x : \sqrt{2}/2+2\sqrt{5}/3$
6953 1160 $e^x : 3e/4-\pi/3$
6953 1582 rt : (4,1,6,-6)
6953 1744 $\Gamma : \ln(1/(3e))$
6953 1992 $2^x : e+3\pi/4$
6953 4963 rt : (4,8,-5,-6)
6953 6432 rt : (5,4,-7,1)
6953 7390 rt : (1,9,9,-8)
6953 7673 rt : (9,3,-5,-1)
6953 8425 id : $\sqrt{3}/3+\sqrt{5}/2$
6954 0110 $J : \sqrt{3}/6$
6954 0243 rt : (5,3,7,-8)
6954 1187 sr : P_{32}
6954 1262 rt : (6,5,-7,3)
6954 1635 $\Gamma : 1/\ln(1/3)$
6954 4561 rt : (7,6,-7,5)
6954 6958 rt : (8,7,-7,7)
6954 6969 rt : (9,5,2,2)
6954 8777 rt : (9,8,-7,9)
6954 9502 cu : $\sqrt{2}-\sqrt{5}-\sqrt{7}$
6955 2326 $e^x : 1/3+4\pi$
6955 3048 $s\pi : 3\sqrt{3}+\sqrt{5}/4$
6955 3603 $J : \sqrt[3]{5}/4$
6955 3745 $\Psi : 10\sqrt{5}/9$
6955 4188 tanh : $5\zeta(3)/7$
6955 4883 $\zeta : \sqrt{3}/10$
6955 5886 $t\pi : 7\pi/3$
6955 6614 sr : $9\sqrt{5}/7$
6955 8249 sr : 23/8
6956 1684 as : $2-e/2$
6956 2076 rt : (4,8,4,-8)
6956 3706 $\zeta : 9/13$
6956 5217 id : 16/23
6956 6185 $\zeta : \ln 2/4$
6956 9080 $\ell_{10} : 1/4+3\pi/2$
6956 9119 rt : (8,8,-8,-1)
6957 0628 $\ell_2 : 3\sqrt{2}+\sqrt{5}$
6957 1007 sq : $2(e+\pi)/9$
6957 1976 $J_0 : 23/20$
6957 2798 $2^x : 16/21$
6957 2993 rt : (7,6,-9,1)
6957 8749 $Li_2 : 11/19$
6957 8928 $e^x : 4e+2\pi/3$
6957 9527 $10^x : 24/7$
6957 9735 rt : (9,2,0,-4)
6958 0017 as : $4\sqrt[3]{3}/9$
6958 0518 $\ln(3/4)\times\exp(5)$
6958 4066 $\ell_{10} : 3/2+2\sqrt{3}$
6958 4272 sr : P_{42}
6958 7714 $t\pi : 3\ln 2/5$
6958 8438 rt : (8,0,-1,2)
6958 9854 rt : (2,4,9,5)
6959 0341 $\ell_2 : 9\sqrt[3]{2}/7$
6959 0891 $\ln(2)-\exp(2)$
6959 0994 rt : (9,9,-2,-6)
6959 2025 rt : (3,5,8,-9)
6959 3085 rt : (1,9,-1,-4)
6959 3991 $\zeta : 9\sqrt{3}/8$
6959 5334 $\sqrt{3}-s\pi(\sqrt{2})$
6959 6050 $10^x : 1/\sqrt{19}$
6959 6212 $s\pi : (\sqrt[3]{5}/3)^{1/2}$
6959 6407 rt : (5,5,-9,-7)
6959 8327 rt : (3,6,-6,-1)
6959 8999 cr : $\sqrt{2}+2\sqrt{3}$
6959 9236 rt : (8,7,-3,-4)
6960 0333 sinh : 13
6960 0657 $Ei : \sqrt{5/3}$
6960 0973 rt : (9,6,7,5)
6960 1940 $10^x : 4/3-2\sqrt{5}/3$
6960 3159 rt : (2,3,7,-7)
6960 3688 $\ell_{10} : P_8$
6960 4401 id : $10\pi^2$
6960 5160 $s\pi : 9\sqrt[3]{3}/4$
6960 6173 rt : (9,1,5,-7)
6960 7022 rt : (2,4,-9,-6)
6960 7095 sq : $2\sqrt[3]{5}$
6960 8414 rt : (9,8,3,-9)
6961 1442 as : $3\sqrt[3]{5}/8$
6961 2496 $t\pi : 2e\pi/5$
6961 4047 rt : (7,9,0,-2)
6961 5242 id : $1/2+3\sqrt{3}$
6961 5871 $\ell_2 : \sqrt{21/2}$
6961 6143 sinh : $3\sqrt{3}/8$
6961 6762 cr : $4e\pi/7$
6961 8907 $t\pi : 5\ln 2/8$
6961 9450 rt : (8,6,2,-7)
6962 1110 $2^x : ((e+\pi)/2)^{1/3}$
6962 1206 rt : (1,8,4,-7)
6962 1491 ln : $\sqrt{2}/2+3\sqrt{3}/4$
6962 2389 rt : (8,1,4,5)
6962 3659 $2^x : 2\sqrt{3}/3-3\sqrt{5}/4$
6962 4016 $\exp(1/4)-s\pi(1/5)$
6962 7043 $\Gamma : \sqrt{7}\pi$
6962 7452 $10^x : 8e\pi/9$
6962 7474 rt : (1,1,6,-5)
6962 9422 $erf : 8/11$
6963 1436 rt : (6,8,-9,-9)
6963 2339 rt : (8,8,6,3)
6963 6377 $erf : \exp(-1/\pi)$
6963 7385 rt : (7,4,1,-5)
6963 8852 tanh : $(\pi/2)^{-1/3}$
6963 8930 $erf : (\sqrt[3]{4}/3)^{1/2}$
6963 9277 rt : (1,5,1,-2)
6964 0526 $Ei : (e+\pi)^{1/2}$
6964 0642 $10^x : 2-3\sqrt{2}/4$
6964 2602 rt : (8,2,9,8)
6964 2729 sr : $8e\pi/5$
6964 2814 ln : $3\sqrt{2}-\sqrt{5}$
6964 4648 $s\pi : \sqrt{14}\pi$
6964 5772 at : $(\sqrt[3]{5})^{-1/3}$
6964 8082 $\exp(2)-\exp(3)$
6964 8472 rt : (6,9,-2,0)
6964 8700 cu : $3\sqrt{3}+3\sqrt{5}/4$
6964 8741 sinh : $3\sqrt{3}/4$
6965 0342 $10^x : 21/8$
6965 0487 $J_0 : (\pi/3)^3$
6965 2447 rt : (7,3,6,-8)
6965 2881 $\zeta : 4/23$
6965 4299 $\ell_{10} : 1/2+2\sqrt{5}$
6965 4779 $e^x : 1-\sqrt{2}/3$
6965 5921 cr : $5(e+\pi)/6$
6965 6516 sr : $7\ln 2/10$
6965 7458 rt : (7,3,3,3)
6966 0131 rt : (6,8,3,-8)
6966 6295 rt : (5,8,-8,0)
6967 0240 $e^x : \sqrt{3}/4+\sqrt{5}/4$
6967 1550 $s\pi : 1/2+\sqrt{5}/3$
6967 1624 rt : (6,2,0,-3)
6967 3276 rt : (5,3,-4,9)
6967 4752 sinh : 13/20
6967 4862 $K : 3\sqrt{3}/10$
6967 5098 $t\pi : e/3-3\pi/2$
6967 6027 rt : (7,4,8,6)
6967 6232 $\ln(3)+\exp(4)$
6967 7462 id : $\sqrt{2}+\sqrt{3}-\sqrt{6}$
6967 9204 $t\pi : \ln((e+\pi)/3)$
6967 9670 rt : (7,8,-5,-5)
6968 0366 $\zeta : \exp(2/3)$
6968 0719 rt : (5,8,-2,-2)
6968 1791 at : $\sqrt{2}-\sqrt{3}/3$
6968 1839 rt : (4,8,-9,-4)
6968 3095 $\ell_2 : 2e+\pi/3$
6968 3587 cr : $3+4\sqrt{2}/3$
6968 3732 rt : (6,1,5,-6)
6968 7408 $10^x : (\sqrt[3]{5})^{1/3}$
6968 8628 $\ell_{10} : 3/2-3\sqrt{3}/4$
6968 9090 $K : 3\ln 2/4$
6968 9134 $s\pi(2/5)-s\pi(\sqrt{3})$
6968 9244 rt : (5,9,5,7)
6969 0705 $\ln(\sqrt{5})\times s\pi(1/3)$
6969 3007 rt : (3,9,-8,-1)
6969 3845 sq : $3\sqrt{2}+\sqrt{3}$
6969 4875 rt : (5,5,7,-9)
6969 6258 $\lambda : (\sqrt[3]{2}/3)^3$
6969 7176 at : $(e+\pi)/7$
6969 7198 $K : 13/25$
6969 7620 $e^x : \sqrt[3]{17/2}$
6969 7789 rt : (4,2,-4,7)
6969 8416 $K : (\sqrt[3]{3}/2)^2$
6969 9940 id : $8(e+\pi)/7$
6970 3197 $e^x : 2+4\sqrt{2}/3$
6970 4945 rt : (3,5,-5,9)
6970 5627 id : $6\sqrt{2}/5$
6970 9924 rt : (2,9,-1,4)
6971 0758 $\Gamma : 1/\sqrt[3]{7}$
6971 2084 cu : $1/4+2\sqrt{2}/3$
6971 2261 rt : (8,9,-6,3)
6971 2467 $J : 6/23$
6971 5945 $10^x : \ln(\sqrt[3]{5}/2)$
6971 8451 $\exp(1/4)+\exp(5)$
6972 2436 rt : (2,9,1,-3)
6972 4267 $\Psi : (2e/3)^3$
6972 6562 cu : $(5/2)^3$
6972 8406 sr : $\exp(-\sqrt[3]{3}/2)$
6973 0384 cu : $3e/4+3\pi/2$
6973 1441 rt : (2,7,-2,-2)
6973 4014 cu : $\sqrt{6}\pi$
6973 5350 $\ell_{10} : 2+4\sqrt{5}/3$
6973 5576 cu : $3\pi/5$
6973 6166 rt : (3,8,3,-7)
6973 8320 rt : (4,3,6,-7)
6973 9481 rt : (3,9,-2,-4)
6974 1049 rt : (7,8,-6,1)
6974 1278 rt : (5,2,9,7)
6974 2933 rt : (8,8,2,-8)
6974 5243 rt : (9,3,5,-8)
6974 6113 id : $\exp(\sqrt[3]{4}/3)$
6974 6414 rt : (8,9,-3,-5)
6974 7024 rt : (3,8,-9,9)
6974 7591 $\exp(3/5)\div s\pi(\sqrt{5})$
6974 8138 rt : (5,1,4,4)
6974 9636 rt : (9,4,0,-5)
6975 1319 rt : (4,7,-2,-4)
6975 1320 $e^x : (\sqrt[3]{5})^{1/2}$
6975 3508 $e^x : P_3$
6975 4232 $\Psi : (\sqrt{3}/2)^{1/2}$
6975 4912 rt : (4,7,-8,6)
6975 6125 rt : (9,5,-5,-2)
6975 6887 rt : (6,6,7,4)
6975 7486 $2^x : 2\sqrt{2}+3\sqrt{5}$
6976 2850 rt : (5,0,-1,1)
6976 3086 $e^x : 17/13$
6976 3541 sr : $4-\sqrt{5}/2$
6976 3567 rt : (1,7,9,9)
6976 3846 $t\pi : 10\pi$
6976 4834 rt : (5,1,-8,8)
6976 6475 $\exp(1/2)-s\pi(2/5)$
6976 6767 $e^x : 3\sqrt{2}+3\sqrt{5}/2$
6976 7325 $Ei : (\ln 2/3)^{-1/3}$
6976 7933 cr : $9e/5$
6977 0745 sr : $7(e+\pi)/3$
6977 1301 rt : (4,6,-9,2)
6977 2334 sinh : 6π
6977 2888 $10^x : 23/7$
6977 3404 $1/4-\exp(2/3)$
6977 4924 rt : (6,8,1,8)
6977 7070 $e^x : 3\sqrt[3]{3}$
6977 7275 sinh : $4\pi^2/7$
6977 9315 $s\pi : (\ln 3)^{-3}$
6977 9468 at : $3\sqrt{5}/8$
6978 0615 rt : (7,5,6,-9)
6978 0621 cr : $e/8$
6978 3480 rt : (6,7,-6,-1)

6978 7239 rt : (2,5,7,-8)
6978 7689 rt : (8,0,9,9)
6978 8623 ln : $3/4 + 3\pi/2$
6978 8987 Ei : $24/5$
6978 9697 rt : (6,5,2,1)
6979 1314 rt : (7,6,1,-6)
6979 2615 ℓ_{10} : $3\sqrt{2} + \sqrt{5}/3$
6979 2679 sq : $4\sqrt[3]{3}/3$
6979 3323 e^x : $9/17$
6979 7453 tanh : $(\sqrt{5}/3)^{1/2}$
6979 8539 K : $12/23$
6980 1333 2^x : $1/2 + \sqrt{3}$
6980 1456 rt : (8,1,4,-6)
6980 1747 rt : (3,5,-8,-7)
6980 2623 Ψ : $6(1/\pi)/5$
6980 3286 sπ : $e/4 + 4\pi$
6980 4806 rt : (4,0,-8,6)
6980 5320 tπ : $3\sqrt{2}/4 + \sqrt{5}/3$
6980 6830 rt : (7,7,-4,-3)
6980 8138 10^x : $3/4 - e/3$
6980 8516 sinh : $(e + \pi)/9$
6980 9848 rt : (2,6,-3,8)
6981 1247 2^x : $3 - \sqrt{5}$
6981 2647 tanh : $19/22$
6981 3170 id : $2\pi/9$
6981 5099 rt : (9,4,7,6)
6981 5342 sq : $3\pi^2/8$
6981 6779 rt : (5,1,-6,2)
6981 9235 θ_3 : $3\sqrt{3}/8$
6981 9335 Ψ : $1/\ln((e + \pi)/2)$
6981 9443 ζ : $(\sqrt{5})^{1/3}$
6981 9730 rt : (7,7,-3,9)
6981 9971 ln : $2 + 2\sqrt{3}$
6982 2247 as : $9/14$
6982 2485 sq : $25/13$
6982 2572 rt : (8,2,-1,-3)
6982 2622 ℓ_{10} : $\zeta(3)/6$
6982 3897 $\sqrt{2} + \exp(1/4)$
6982 4027 rt : (3,1,5,-5)
6982 4583 ℓ_2 : $9\sqrt[3]{3}$
6982 5020 cu : $\sqrt{2}/3 + \sqrt{3}$
6982 5542 10^x : $3/2 + 4\sqrt{5}/3$
6982 5825 cu : $3\pi^2/8$
6982 6159 e^x : $7\sqrt[3]{3}$
6982 6845 10^x : $\sqrt{23}\pi$
6982 7217 J_0 : $\ln(\pi)$
6982 7594 e^x : $1 - e/2$
6982 7973 J_0 : $\sqrt[3]{3/2}$
6982 8215 10^x : $\sqrt{3} - \sqrt{5}/3$
6982 8501 e^x : $1/2 + \pi/3$
6982 9421 rt : (5,8,2,-7)
6983 1348 rt : (7,8,-9,0)
6983 2590 ln : $1/3 + 3\sqrt{5}/4$
6983 3031 ℓ_{10} : $3/4 + 3\sqrt{2}$
6983 3060 e^x : $4e - 3\pi/4$
6983 5417 rt : (2,7,-3,-2)
6983 8132 cr : $\sqrt{24}$
6983 8243 sinh : $13/10$
6984 2642 rt : (3,6,-3,-4)
6984 4407 rt : (6,3,5,-7)
6984 5531 rt : (9,3,2,3)
6984 5633 tπ : $\sqrt{2}/3 + 4\sqrt{5}/3$
6984 6394 ℓ_{10} : P_{40}
6984 6666 ℓ_{10} : P_{16}
6984 7301 sr : $\sqrt{2}/4 + 4\sqrt{3}$
6985 0904 2^x : $9\sqrt{5}/4$
6985 1583 sq : $5\pi^2/7$
6985 2256 rt : (5,9,-3,-4)
6985 3028 2^x : $3\sqrt{2} - \sqrt{3}$
6985 3145 rt : (9,5,5,-9)
6985 4960 rt : (4,4,8,5)
6985 7634 sq : $2/3 - 4\sqrt{3}/3$
6985 8194 K : $1/\sqrt[3]{7}$
6985 9042 rt : (8,5,-9,-6)
6985 9196 Ψ : $\sqrt{5}/4$
6986 1228 $2/5 - \ln(3)$
6986 2238 sr : $8\sqrt[3]{5}$
6986 2896 at : $2\sqrt[3]{2}/3$
6986 2997 $\pi \times \exp(\pi)$
6986 4277 rt : (7,2,8,7)
6986 4360 id : $(\ln 2/2)^{-1/2}$
6986 4409 rt : (3,8,-1,-2)
6986 5982 at : $21/25$
6986 7061 id : $3\sqrt{2}/2 + \sqrt{3}/3$
6986 7073 rt : (6,6,-3,7)
6986 7111 id : $5e\pi$
6986 9047 rt : (5,7,-9,-8)
6986 9137 ℓ_2 : $\ln 2/9$
6987 1071 sr : $1 + 4\sqrt{2}/3$
6987 3772 sr : $1 + 2\pi$
6987 4706 sq : $3 + e$
6987 4993 rt : (9,6,0,-6)
6987 5914 tπ : $3e - \pi/2$
6987 6331 ln : $2e + 3\pi$
6987 6863 erf : $(\sqrt[3]{5}/2)^2$
6987 9535 sinh : $8\pi^2/9$
6988 0294 tπ : $\sqrt{3}/3 + 4\sqrt{5}$
6988 3378 rt : (4,5,6,-8)
6988 4451 id : $((e + \pi)/2)^{-1/3}$
6988 5192 sq : $3\sqrt{2}/2 + 3\sqrt{3}/4$
6988 5482 $4/5 \div \ln(\pi)$
6988 8148 rt : (7,5,-7,8)
6988 8548 rt : (1,3,6,-6)
6989 1361 cu : $4e - 2\pi$
6989 2614 id : $5e/8$
6989 3412 2^x : $(e/2)^3$
6989 4296 2^x : $9(1/\pi)/2$
6989 7000 ℓ_{10} : 5
6989 7342 rt : (1,8,-7,-8)
6989 7393 sr : $2\sqrt[3]{5}/7$
6989 7904 10^x : $(\sqrt[3]{4}/3)^{-1/2}$
6989 7959 sq : $23/14$
6989 9020 2^x : $\sqrt{2}/2 + 3\sqrt{5}$
6989 9399 id : $(\sqrt{2}\pi)^3$
6990 0374 tanh : $3\sqrt[3]{3}/5$
6990 0656 Ei : $9\zeta(3)$
6990 2089 rt : (2,8,-8,1)
6990 2358 2^x : $13/17$
6990 3681 $\sqrt{2} - \exp(\sqrt{2})$
6990 4424 sr : $5\sqrt{3}/3$
6990 4550 2^x : $(\sqrt[3]{5})^{-1/2}$
6990 4844 sq : $8\zeta(3)/5$
6990 5584 rt : (9,7,-5,-3)
6990 6537 rt : (7,8,1,-7)
6990 7386 rt : (7,1,3,4)
6990 7588 K : $\pi/6$
6990 8202 rt : (5,0,9,8)
6990 9394 cr : $\sqrt{3}/4 + 2\sqrt{5}$
6991 0371 rt : (2,7,2,-7)
6991 6835 Ei : $2\sqrt[3]{3}/5$
6991 7657 sr : $\sqrt{3}/3 + 3\sqrt{5}$
6991 7733 J_0 : $8/7$
6991 8494 ζ : $\sqrt{3} - \sqrt{5} + \sqrt{6}$
6991 8516 at : $(\sqrt{2})^{-1/2}$
6991 8908 rt : (2,7,7,-9)
6991 9991 K : $11/21$
6992 0308 cr : $4 + e/3$
6992 0547 Γ : $8e\pi/9$
6992 0841 2^x : $11/3$
6992 1407 rt : (4,6,1,4)
6992 4256 $\ln(4) - \exp(3)$
6992 5973 rt : (6,5,-5,-1)
6992 6746 10^x : $6\zeta(3)/5$
6992 9810 rt : (8,3,4,-7)
6993 1578 sq : $(\sqrt[3]{5})^{-1/3}$
6993 1650 rt : (3,7,0,8)
6993 1693 cu : $7e\pi/4$
6993 4344 rt : (2,2,9,6)
6993 4910 tanh : $\sqrt{3}/2$
6993 5723 rt : (5,5,-3,5)
6993 5841 Ψ : $8\sqrt{5}/3$
6993 5844 ln : $2\sqrt{5}/9$
6993 5861 2^x : $2e/3 - \pi/3$
6993 6318 rt : (5,8,-4,-5)
6993 8659 rt : (4,9,-3,5)
6993 9872 sinh : $15/23$
6994 0005 tπ : $1/4 + 4\sqrt{5}$
6994 1725 rt : (7,9,-4,-4)
6994 2683 rt : (6,4,-7,6)
6994 4126 rt : (1,8,6,-8)
6994 6121 as : $3 - 3\pi/4$
6994 8340 rt : (2,8,-7,-2)
6994 9319 e^x : $3\sqrt{2}/8$
6995 0340 at : $7\zeta(3)/10$
6995 3742 ln : $(1/(3e))^{1/3}$
6995 5604 ln : $4 + 4e$
6995 5779 tanh : $5\ln 2/4$
6995 7243 rt : (8,6,6,4)
6995 7359 sinh : $7\sqrt[3]{3}/5$
6995 7398 cr : $2/3 + 3\sqrt{2}$
6995 7835 ln : $4/3 + e/4$
6995 9823 rt : (9,2,7,7)
6996 0088 ℓ_2 : $(\sqrt{2}\pi/3)^3$
6996 0770 rt : (5,1,4,-5)
6996 1562 rt : (6,5,5,-8)
6996 1640 10^x : $2\sqrt{2} - 3\sqrt{3}/2$
6996 2814 rt : (3,6,-9,-9)
6996 3633 Ψ : $9e\pi/5$
6996 6902 ln : $1 + 2\sqrt{5}$
6996 7658 tanh : $13/15$
6996 9595 erf : $\ln(\sqrt[3]{3}/3)$
6997 0499 rt : (3,7,2,-1)
6997 1337 rt : (9,8,0,-7)
6997 3497 Ei : $\sqrt[3]{13/3}$
6997 4048 ζ : $5\sqrt{3}/8$
6997 4618 erf : $(e + \pi)/8$
6997 5542 rt : (8,4,-1,-4)
6997 5948 e^x : $4\sqrt{2}/3 - \sqrt{3}/3$
6998 0999 e^x : $5(1/\pi)/3$
6998 2677 cr : $4\sqrt{2} - \sqrt{5}/3$
6998 2872 rt : (4,7,-4,-2)
6998 3101 rt : (9,9,-6,8)
6998 7934 rt : (7,0,-2,1)
6998 8138 $\exp(\pi) \div \exp(3/5)$
6998 9286 at : $16/19$
6999 0774 rt : (4,7,6,-9)
6999 1908 tπ : $10\sqrt{5}/7$
6999 1982 id : $10\sqrt[3]{5}/3$
6999 2531 rt : (3,7,-8,-8)
6999 4532 rt : (7,0,8,8)
6999 5389 rt : (1,7,-5,3)
6999 5466 ζ : $(\sqrt[3]{2}/3)^2$
6999 5613 id : $5\sqrt[3]{2}/9$
6999 8533 erf : $3\sqrt[3]{5}/7$
7000 0000 id : $7/10$
7000 2494 tπ : $7e/5$
7000 3266 rt : (3,5,-8,1)
7000 7347 rt : (3,3,5,-6)
7000 8392 ζ : $3/17$
7000 9974 2^x : $1 + \sqrt{3}/4$
7001 1612 rt : (6,6,0,-5)
7001 1960 ℓ_{10} : $3e - \pi$
7001 3271 $\ln(4) \times \exp(2/3)$
7001 3395 tanh : $\ln(\sqrt[3]{2}/3)$
7001 4302 Ei : γ
7001 4399 rt : (9,9,-5,-4)
7001 6959 id : $2\sqrt{3} + \sqrt{5}$
7001 7744 rt : (2,9,7,4)
7001 8750 10^x : $19/23$
7001 9117 rt : (5,3,-7,4)
7001 9566 rt : (6,4,7,5)
7002 0156 rt : (8,9,-9,-2)
7002 1685 rt : (9,1,2,4)
7002 6105 rt : (5,2,9,-9)
7002 6355 rt : (8,5,4,-8)
7002 6940 ln : $3\sqrt{2}/2 + 3\sqrt{5}/2$
7002 7681 rt : (3,6,-2,-6)
7003 0247 cr : $2\zeta(3)/7$
7003 0605 erf : $11/15$
7003 1218 e^x : $2\sqrt{2}/3 - 3\sqrt{3}/4$
7003 1887 rt : (2,9,-3,-3)
7003 2272 Γ : $12/23$
7003 2472 10^x : $1 - 2\sqrt{3}/3$
7003 3589 rt : (1,6,-2,-4)
7003 3639 e^x : $2 - 3\pi/4$
7003 4017 sq : $\sqrt{2} - \sqrt{3}/3$
7003 5089 rt : (5,9,-9,-9)
7003 6751 rt : (1,9,-7,-2)
7003 7608 rt : (4,3,-7,-5)
7004 2813 sinh : $\sqrt{15/2}$

7004 3971 $\ell_2 : 13$
7004 4230 rt : (8,8,-6,6)
7004 4658 $e^x : 3\sqrt[3]{3}/2$
7004 6984 rt : (8,5,-6,-1)
7004 8364 rt : (1,5,6,-7)
7005 0782 rt : (6,7,5,-9)
7005 0944 $\theta_3 : 13/20$
7005 1099 $\zeta : \sqrt{2}/8$
7005 2478 sq : $7\zeta(3)/2$
7005 2494 sr : $\sqrt{19}\pi$
7005 2643 sq : $3/4 - 3\pi/2$
7005 3299 $e^x : \sqrt{23/2}$
7005 4604 rt : (5,2,-1,-2)
7005 5833 $Ei : \sqrt{3}/3$
7005 6415 $J_0 : 2\sqrt[3]{5}/3$
7005 9771 $\zeta : 5(1/\pi)/9$
7006 1150 $t\pi : 1/3 + 2\sqrt{5}$
7006 1226 rt : (7,1,3,-5)
7006 1322 $10^x : \pi^2/7$
7006 2029 cu : $1 + 4\pi/3$
7006 2854 $J_0 : (\sqrt{2}\pi/3)^{1/3}$
7006 4222 rt : (9,0,7,8)
7006 6901 at : $(5/3)^{-1/3}$
7006 7596 rt : (7,9,-2,5)
7006 8199 $K : 10/19$
7006 8398 $\zeta : (\sqrt[3]{5}/2)^{1/2}$
7007 0994 sq : $9\sqrt[3]{5}/8$
7007 1272 $\zeta : \exp(-\sqrt{3})$
7007 1667 sq : $\sqrt{3}/4 + 2\sqrt{5}/3$
7007 2172 rt : (4,2,8,6)
7007 2253 $J : \exp(-4\pi/9)$
7007 6353 $Ei : 1/\ln(4/3)$
7007 7814 sq : $(e+\pi)/7$
7007 8044 rt : (9,1,-8,-3)
7007 8342 $Ei : 13/14$
7007 8692 $s\pi(\sqrt{5}) \div s\pi(\sqrt{2})$
7008 3424 rt : (8,6,-1,-5)
7008 4117 $\ln(3/4) - \exp(5)$
7008 4778 rt : (1,7,-7,-8)
7008 6309 $\Psi : 9\sqrt[3]{5}$
7008 7994 rt : (6,7,-5,-2)
7009 0025 rt : (7,2,8,-9)
7009 0817 $\zeta : 4\ln 2/5$
7009 1777 $\Psi : \sqrt{24}\pi$
7009 1915 rt : (3,8,-3,-5)
7009 2910 $10^x : P_{46}$
7009 3553 sq : $e/4 + 4\pi/3$
7009 4152 $\ell_{10} : 6(e+\pi)/7$
7009 4658 rt : (2,5,-9,-9)
7009 4981 sinh : $(\sqrt[3]{5}/3)^{-3}$
7009 5625 $2^x : 1/3 + \sqrt{3}/4$
7009 6189 id : $2 - 3\sqrt{3}/4$
7009 6382 rt : (5,3,4,-6)
7009 7481 $2/3 \div s\pi(2/5)$
7009 8789 $t\pi : 1/\ln(\ln 3/2)$
7009 8811 rt : (4,8,-9,1)
7010 1625 rt : (8,7,4,-9)
7010 1893 $2^x : (\sqrt{2}\pi/2)^{-1/3}$
7010 2023 $2^x : \sqrt{2}/4 - \sqrt{3}/2$
7010 2085 $10^x : 3\sqrt[3]{2}/5$
7010 2196 $\zeta : 2\sqrt{3}/5$
7010 2614 $\pi \times \ln(4/5)$
7010 3563 $\exp(\sqrt{2}) + s\pi(1/5)$
7010 3945 rt : (8,5,1,1)
7010 5268 ln : $8\sqrt[3]{2}/5$
7010 5651 $1/4 - s\pi(2/5)$
7010 7061 rt : (2,6,-6,-3)
7010 9766 rt : (6,8,0,-6)
7010 9898 $t\pi : 3\sqrt{2}/4 - \sqrt{3}/2$
7011 0165 id : $3\pi^2/8$
7011 2815 as : $1/\ln(3\pi/2)$
7011 5304 rt : (2,5,-7,7)
7011 5316 tanh : 20/23
7011 7204 rt : (2,0,9,7)
7011 7333 rt : (3,9,-8,-9)
7011 7696 rt : (5,9,1,-2)
7011 7715 $\exp(\sqrt{3}) - s\pi(2/5)$
7011 8257 rt : (7,8,-9,-4)
7011 8449 $e^x : 19/5$
7011 9169 $\Psi : 1/\sqrt[3]{18}$
7012 1321 $e^x : (\ln 2/2)^{-2}$
7012 4124 rt : (4,9,-4,-3)
7012 5427 $10^x : 3/13$
7012 5865 $2^x : 3/2 + 4\sqrt{2}$
7012 6106 rt : (7,7,-6,4)
7012 6735 $\Psi : 13/14$
7012 6780 rt : (9,0,-3,1)
7012 6985 $\ell_{10} : 8\pi/5$
7012 7858 rt : (3,5,5,-7)
7012 8018 $2^x : 10\sqrt{3}/3$
7012 9656 rt : (6,5,-8,-6)
7013 0161 $1 \div s\pi(1/5)$
7013 1397 rt : (8,4,6,5)
7013 3941 rt : (4,2,-7,2)
7013 5501 $e^x : (4/3)^3$
7013 8075 rt : (9,1,2,-5)
7013 8774 $s\pi : \sqrt{3}/7$
7013 9863 $Ei : 5\ln 2/6$
7014 0129 $e^x : 1/4 + 2\sqrt{5}/3$
7014 2038 $10^x : 4/3 - \sqrt{3}/3$
7014 2046 $\ell_2 : \sqrt{2}/3 + 2\sqrt{3}/3$
7014 3105 rt : (9,2,7,-9)
7015 0220 $e^x : 4\sqrt{5}/9$
7015 0226 cu : $3\sqrt{2} - 3\sqrt{5}/2$
7015 2777 $t\pi : e/3 + 3\pi$
7015 6077 cr : $3/2 - 2\sqrt{3}/3$
7015 6211 rt : (1,8,9,2)
7016 0822 $\Psi : \sqrt{2} - \sqrt{5} - \sqrt{7}$
7016 3122 id : $3\sqrt{2} - 4\sqrt{5}$
7016 3141 $2^x : 3(e+\pi)/7$
7016 3631 rt : (8,8,-1,-6)
7016 5028 $10^x : 1/\ln(5/3)$
7016 5531 rt : (7,3,3,-6)
7016 6669 rt : (8,7,-6,-2)
7016 6752 $\ell_{10} : 9\sqrt{5}/4$
7016 7310 ln : $5\pi^2/9$
7016 8012 $\zeta : (e+\pi)^{-1/3}$
7016 8738 rt : (6,2,7,6)
7016 9259 sq : $\sqrt{2} - \sqrt{5} - \sqrt{6}$
7016 9787 $s\pi : 7(1/\pi)/9$
7017 0885 rt : (7,2,-2,-2)
7017 3324 $Ei : 9e\pi/7$
7017 5411 $Ei : 3e/5$
7017 5882 at : $2 - 2\sqrt{3}/3$
7017 7601 tanh : $(\pi/3)^{-3}$
7017 7871 sq : $4\sqrt{2} - \sqrt{5}$
7017 9937 $\Psi : 7(e+\pi)$
7018 0340 $e^x : 3 - 3\sqrt{5}/2$
7018 0794 rt : (6,3,2,2)
7018 1892 rt : (1,9,-7,-9)
7018 3714 rt : (6,8,-2,3)
7018 3911 $2^x : 1 + 3\sqrt{2}/2$
7018 6333 cu : $2/3 + 2\sqrt{5}$
7018 7337 $2^x : 4/3 + 3\sqrt{2}$
7018 8026 sinh : $5\sqrt{2}$
7018 9395 $e^x : 3e/2 - \pi/3$
7018 9563 $\sqrt{2} - \ln(3/4)$
7019 0056 $s\pi : 9\ln 2/5$
7019 0429 rt : (5,5,4,-7)
7019 0679 rt : (7,1,-7,2)
7019 2379 cu : $1/4 + 3\pi/4$
7019 2555 rt : (4,1,-8,-9)
7019 3668 $s\pi : 6e\pi/5$
7019 4386 rt : (6,9,-5,-3)
7019 5541 sr : $2\sqrt{2} + 2\sqrt{5}$
7019 5885 rt : (1,8,-2,-5)
7019 6117 $4 \times s\pi(\sqrt{5})$
7019 8749 $\zeta : \sqrt[3]{22/3}$
7019 9418 rt : (8,7,-9,-7)
7020 1483 rt : (4,0,8,7)
7020 4102 sq : $\sqrt{2}/2 - 4\sqrt{3}$
7020 6663 $10^x : 3 - \pi/3$
7020 8673 $\Psi : 7e\pi/10$
7020 8676 rt : (5,4,-1,-3)
7020 9655 sq : $4 - 3\pi/4$
7021 0816 $t\pi : \sqrt{3}/4 + \sqrt{5}$
7021 3115 rt : (3,7,5,-8)
7021 4337 $s\pi : 2e - 4\pi/3$
7021 7580 $\ln(4) \times \exp(\sqrt{2})$
7021 8850 sr : $\exp(-1/\sqrt{2})$
7021 9902 as : $4 - 3\sqrt{5}/2$
7022 0466 rt : (2,9,6,-6)
7022 0757 rt : (9,3,2,-6)
7022 2509 $\zeta : \sqrt{3} + \sqrt{5} - \sqrt{7}$
7022 3721 rt : (4,8,-5,7)
7022 5693 at : 11/13
7022 8616 $\sqrt{3} \times \ln(2/3)$
7023 0441 rt : (2,2,9,-8)
7023 0834 $J_0 : 25/22$
7023 1219 cr : $\sqrt{3}/5$
7023 1929 rt : (5,8,6,2)
7023 1948 $\ell_{10} : 3 + 3e/4$
7023 3196 cu : 8/9
7023 3877 rt : (1,9,6,-9)
7023 4919 $t\pi : 1/3 - \sqrt{3}/2$
7023 5080 $10^x : \ln 2/3$
7023 5520 $Li_2 : 7/12$
7023 6826 rt : (4,9,-6,-3)
7023 8252 rt : (9,9,9,5)
7024 0070 rt : (8,2,6,6)
7024 0332 $\ell_{10} : 4\sqrt[3]{2}$
7024 0934 rt : (7,5,3,-7)
7024 1438 rt : (9,9,-9,3)
7024 1656 $2^x : 4\sqrt{2}/3 - \sqrt{5}/2$
7024 2261 id : $(\ln 2/2)^{1/3}$
7024 3358 rt : (1,7,3,-1)
7024 6524 $\ln(2) - \exp(1/3)$
7024 9250 $e^x : 2 + 4\sqrt{3}$
7024 9493 cr : $1/\sqrt[3]{24}$
7024 9668 rt : (8,9,-6,-3)
7024 9975 $\ell_2 : 4 - \sqrt{5}/3$
7025 1096 sq : $(\sqrt{2}\pi/2)^{1/3}$
7025 2240 rt : (9,2,-3,-2)
7025 2948 at : $2e - 2\pi$
7025 3092 $K : 9/17$
7025 3231 $e^x : 3/4 + \sqrt{5}/4$
7025 3667 $\ell_2 : \sqrt{2} + \sqrt{6} + \sqrt{7}$
7025 6769 cr : $2\pi^2$
7025 8899 $erf : (5/2)^{-1/3}$
7025 9489 rt : (5,7,4,-8)
7026 1247 $erf : 14/19$
7026 2467 $2^x : \sqrt{14}\pi$
7026 3056 $10^x : (\sqrt[3]{3}/3)^2$
7026 3664 rt : (6,3,-7,9)
7026 4346 $\lambda : \exp(-14\pi/17)$
7026 4899 rt : (1,8,-6,1)
7026 4927 rt : (6,0,7,7)
7026 7633 ln : $7\sqrt[3]{3}/5$
7026 8514 rt : (5,3,-6,1)
7026 9658 rt : (6,7,-9,-6)
7027 0126 rt : (1,8,1,-5)
7027 1136 sinh : $2(e+\pi)/9$
7027 1245 rt : (4,5,-7,-6)
7027 3718 cu : P_{43}
7027 6614 rt : (3,9,5,7)
7027 7749 $\ell_2 : \sqrt{2} - \sqrt{5} + \sqrt{6}$
7027 8645 $\sqrt{2} - \exp(3/4)$
7027 9723 rt : (7,4,-2,-3)
7028 0932 $s\pi : 3\sqrt{2} - 2\sqrt{5}/3$
7028 2541 rt : (9,5,2,-7)
7028 5848 sq : $1/4 + 4\pi/3$
7028 7254 rt : (4,2,8,-8)
7028 7285 $\ell_2 : 5(e+\pi)/9$
7028 9780 $s\pi : (\sqrt{2}\pi/3)^3$
7029 0169 $t\pi : 4\sqrt{2}/3 + 4\sqrt{3}/3$
7029 1781 as : $1 - \sqrt{2}/4$
7029 2334 $\ell_{10} : 1/3 + 3\pi/2$
7029 4056 $10^x : \exp(1/\pi)$
7029 6307 $10^x : 3\sqrt{2}/2 + 3\sqrt{5}/2$
7029 6945 cr : $3/4 + 4\pi/3$
7029 7715 $\ln(\sqrt{5}) \div \ln(\pi)$
7029 7995 rt : (7,7,3,-8)
7029 8515 ln : $5\ln 2/7$
7030 2262 rt : (8,3,1,2)
7030 3921 rt : (3,9,-7,3)
7030 4049 rt : (5,9,2,2)
7030 4260 $\ln(\sqrt{2}) + \exp(\sqrt{5})$
7030 5030 $\ell_{10} : 4 + \pi/3$
7030 5727 $\ell_2 : 4\sqrt{2}/3 - \sqrt{3}$
7030 5793 ln : $4 + 2\sqrt{5}/3$

7030 6353 $2^x : 2\sqrt[3]{5}$
7030 6538 rt : (3,9,2,-2)
7030 7425 rt : (2,4,9,-9)
7030 7742 $10^x : 2\sqrt{2} - 4\sqrt{5}/3$
7030 8324 $K : 3\sqrt{2}/8$
7031 0846 $\ell_{10} : 7\sqrt[3]{3}/2$
7031 2357 rt : (5,9,4,-9)
7031 2414 rt : (6,7,-8,-7)
7031 2500 cu : 21/4
7031 3837 $\Psi : 3(1/\pi)/7$
7031 4378 rt : (8,0,6,7)
7031 5664 $\Psi : 5/2$
7031 5750 as : $\sqrt{2}/3 - \sqrt{5}/2$
7031 7640 rt : (8,7,-6,9)
7031 8267 tanh : $1/\ln(\pi)$
7031 8870 tanh : $\sqrt[3]{2/3}$
7031 9298 cu : $6(e+\pi)/5$
7031 9556 $K : 5(1/\pi)/3$
7032 0801 $t\pi : \sqrt{2}/3 - 4\sqrt{5}$
7032 4258 $\ell_{10} : 2\ln 2/7$
7032 4719 $\ell_{10} : \sqrt{3}/3 + 2\sqrt{5}$
7032 4853 ln : $10\sqrt{2}/7$
7032 5225 ln : $1/2 - \sqrt{3}/4$
7032 5740 rt : (3,8,0,-5)
7032 5888 $\zeta : 5e/7$
7032 6777 cr : 8/23
7033 0140 $e^x : 9\zeta(3)/5$
7033 0470 rt : (9,7,2,-8)
7033 1247 rt : (1,9,-6,8)
7033 2109 rt : (6,2,7,-8)
7033 2494 cu : $3\sqrt{2}/4 - 2\sqrt{5}$
7033 3199 rt : (9,4,-3,-3)
7033 7712 $e^x : \exp(-\sqrt[3]{2}/2)$
7033 7882 $\Psi : 7\sqrt[3]{5}/2$
7033 8291 rt : (6,1,2,3)
7034 1626 rt : (2,9,0,9)
7034 2685 rt : (7,9,3,-9)
7034 3125 cr : $4 + 2\sqrt{2}/3$
7034 3363 rt : (9,1,-8,2)
7034 4646 $s\pi : 4\sqrt{2}/3 + \sqrt{3}/2$
7034 4796 rt : (9,7,9,6)
7034 5682 ln : $2\sqrt{3}/7$
7034 6316 $5 \times \exp(\pi)$
7034 6483 rt : (8,9,-9,-8)
7034 7115 sr : $2\sqrt{3}/7$
7034 7365 sq : $4/3 + 2\sqrt{5}$
7034 7926 $J_0 : 9\sqrt[3]{2}/10$
7034 8302 rt : (4,4,8,-9)
7034 8615 rt : (8,8,-9,1)
7034 9884 $2^x : 17/9$
7035 0322 $2^x : 4e + \pi/4$
7035 1527 cr : $\sqrt{2}/3 + 2\sqrt{5}$
7035 1966 rt : (7,6,-2,-4)
7035 2139 $10^x : \sqrt[3]{10}$
7035 4131 rt : (5,4,-3,8)
7035 6838 $\ln(\sqrt{3}) + \exp(e)$
7035 7752 rt : (2,5,-6,-6)
7035 9004 $\ln(\sqrt{5}) \times \exp(3/4)$
7036 1979 $t\pi : 4e/3 + \pi/2$
7036 3707 sr : $5\ln 2/7$
7036 3895 at : $3\sqrt{2}/5$
7036 4992 $\ell_2 : \sqrt{3} - \sqrt{5}/2$
7036 5310 rt : (3,6,7,3)
7036 5963 $s\pi : \sqrt{5}/9$
7036 7524 $Li_2 : ((e+\pi)/2)^{-1/2}$
7036 8406 rt : (8,2,6,-8)
7036 8739 rt : (9,9,2,-9)
7036 8934 rt : (5,8,-1,-5)
7036 9677 rt : (1,7,-4,-1)
7037 0370 cu : 7/3
7037 1284 cu : $1/2 + \pi/3$
7037 2050 as : 11/17
7037 4136 sr : $2/3 + \sqrt{5}$
7037 5348 $t\pi : 7\zeta(3)/3$
7037 6546 $J_0 : 17/15$
7037 6551 sinh : $8e/5$
7038 1231 at : $8(1/\pi)/3$
7038 1689 rt : (6,4,7,-9)
7038 3698 rt : (7,3,-7,1)
7038 4807 rt : (4,7,-7,-7)
7038 7972 rt : (5,2,-7,7)
7038 9716 $5 \div s\pi(\sqrt{3})$
7038 9807 rt : (9,6,-3,-4)
7039 0560 tanh : 7/8
7039 3968 rt : (2,3,4,8)
7040 0507 sinh : $10\sqrt{2}/7$
7040 0541 sinh : $7e$
7040 2401 at : $(\sqrt[3]{3}/2)^{1/2}$
7040 2669 rt : (8,1,1,3)
7040 3438 rt : (7,8,-2,-5)
7040 4283 $t\pi : 1/3 + \sqrt{2}/3$
7040 4412 cu : $4 - 3\sqrt{3}/4$
7040 4516 rt : (1,9,3,-2)
7040 5638 $\Gamma : \ln(2/3)$
7040 7982 rt : (6,9,-8,-8)
7040 8186 rt : (7,8,-2,8)
7040 9463 rt : (8,4,6,-9)
7041 0132 $Ei : \exp(-2\pi)$
7041 0234 cr : $\pi/9$
7041 0788 $e \div \ln(2/3)$
7041 1024 $erf : 17/23$
7041 2103 rt : (9,5,9,7)
7041 3163 $t\pi : \ln(\sqrt{5})$
7041 5071 sr : $4/3 + \pi/2$
7041 5945 rt : (5,5,-6,0)
7041 9802 $erf : \ln(2\pi/3)$
7042 0300 $\exp(\sqrt{5}) \div \exp(1/3)$
7042 0717 $\ell_{10} : 4 + 3\sqrt{2}/4$
7042 1567 $10^x : \pi^2/10$
7042 2793 cr : $7\sqrt{2}/2$
7042 3200 rt : (1,9,2,-5)
7042 3531 rt : (6,1,2,-4)
7042 3534 $\exp(\sqrt{3}) - \exp(\sqrt{5})$
7042 4521 $s\pi : 5\zeta(3)/8$
7043 1629 rt : (9,8,-3,-5)
7043 2797 $s\pi : 8e/3$
7043 2904 rt : (7,6,-6,7)
7043 4030 sq : $4e/3 + 2\pi/3$
7043 4372 $10^x : (2e)^{-1/3}$
7043 4567 ln : $7\pi/4$
7043 5047 rt : (2,7,-6,-7)
7043 6079 $2^x : 10/13$
7043 7112 $e^x : 4\sqrt{2}/3 - \sqrt{5}$
7043 8125 $\ell_2 : \ln(e/2)$
7043 8421 sr : $\sqrt{2} + 2\sqrt{5}/3$
7043 8422 rt : (2,8,3,3)
7043 9510 cu : $3/4 + 4e/3$
7044 0225 rt : (1,4,8,4)
7044 1721 rt : (4,9,9,9)
7044 1985 cu : $4\sqrt{2}/3$
7044 2339 rt : (3,7,-5,-1)
7044 2397 $e^x : 3 + e/3$
7044 3087 sr : $2\sqrt{3} - \sqrt{5}/4$
7044 6393 $K : \exp(-\sqrt[3]{2}/2)$
7044 6488 $\zeta : \ln 2$
7044 9406 at : 17/20
7044 9963 $\exp(2/3) - \exp(\sqrt{3})$
7045 1253 rt : (9,3,-8,1)
7045 1343 sq : $2/3 - 3\pi$
7045 1808 rt : (4,9,-7,-8)
7045 1834 $10^x : 1/2 + 2\pi$
7045 3615 $\ell_{10} : 2\sqrt{2} + \sqrt{5}$
7045 6337 $Li_2 : 1/\sqrt[3]{5}$
7045 7629 $\ell_2 : (\ln 2/3)^{1/3}$
7045 7726 rt : (2,5,4,-6)
7045 8485 rt : (9,3,9,8)
7045 9231 sq : $3\sqrt{2} - 3\sqrt{3}/2$
7046 0486 $e^x : 8/15$
7046 1048 rt : (7,8,5,2)
7046 2294 $e^x : 7e/10$
7046 2642 rt : (4,7,-7,9)
7046 2803 id : $9\zeta(3)/4$
7046 3396 rt : (8,1,1,-4)
7046 4355 rt : (1,9,-4,-2)
7046 5074 $2^x : 4\sqrt{3}/3 + 4\sqrt{5}$
7046 5819 $s\pi : 3e/4 + 3\pi/2$
7046 8808 $\exp(1/4) \div \exp(3/5)$
7046 9068 cr : $2e/3 + \pi$
7047 0775 $J_0 : 4\sqrt{2}/5$
7047 1919 rt : (7,9,-5,0)
7047 2694 sr : $2 + e/3$
7047 2987 cr : 7/20
7047 3237 cu : $3e + 2\pi$
7047 4809 ln : 2/11
7047 6993 rt : (6,3,2,-5)
7047 8343 $2^x : 3 + 4e$
7047 8526 $\exp(3/4) + s\pi(1/5)$
7048 0693 $Ei : 22/17$
7048 0917 cr : $2\sqrt{3} + 2\sqrt{5}/3$
7048 3179 rt : (9,8,-9,6)
7048 3532 cu : $e/2 - 3\pi/2$
7048 4303 $t\pi : \sqrt{2} - \sqrt{5}/3$
7048 4594 rt : (2,9,-6,-8)
7048 5109 sr : $(1/(3e))^{1/3}$
7048 8327 rt : (1,2,-9,5)
7048 9161 rt : (4,7,5,8)
7048 9442 $\Gamma : 2\pi^2/7$
7048 9507 rt : (5,8,-5,-7)
7048 9513 sinh : 20
7049 0203 $K : 8/15$
7049 1417 sr : $2\sqrt{5}/9$
7049 2390 rt : (9,1,9,9)
7049 4174 $10^x : (\sqrt{3})^{1/2}$
7049 5647 $Ei : 2\pi^2/9$
7049 5853 rt : (7,7,-9,-1)
7050 0117 rt : (2,7,4,-7)
7050 1268 $\zeta : \sqrt[3]{2}/7$
7050 2617 as : $(\sqrt[3]{2}/3)^{1/2}$
7050 2739 rt : (5,6,6,3)
7050 3382 $2^x : 4\sqrt{3}/9$
7050 4102 rt : (8,3,1,-5)
7050 4662 rt : (3,9,-5,-2)
7050 6817 tanh : $(\sqrt{2}\pi/3)^{-1/3}$
7050 8276 sr : $3\sqrt{2}/2 + 3\sqrt{3}$
7051 0033 rt : (9,5,-8,0)
7051 3663 rt : (6,5,2,-6)
7051 4058 $2^x : \sqrt{3} - \sqrt{5}$
7051 4119 cr : $4/3 + 4e/3$
7051 4372 $J_0 : \exp(1/(3e))$
7051 5907 $\ell_2 : \sqrt[3]{13/3}$
7051 6396 rt : (4,8,-8,-1)
7052 2372 rt : (4,7,3,-7)
7052 2585 rt : (7,7,-7,-1)
7053 0340 rt : (2,9,4,-7)
7053 1353 $10^x : \sqrt{22}\pi$
7053 2111 cu : $4 - \sqrt{2}/2$
7053 2305 $\ln(\sqrt{3}) \times \exp(1/4)$
7053 2432 rt : (3,4,7,4)
7053 2580 rt : (2,8,-6,3)
7053 3293 rt : (8,5,1,-6)
7053 3648 rt : (5,9,-6,-2)
7053 4668 cr : $\exp(-\pi/3)$
7053 4969 sinh : $8\sqrt{2}/3$
7053 5547 $Ei : 1/\ln((e+\pi)/2)$
7053 7227 $2^x : \sqrt{2} + 3\sqrt{3}$
7053 7467 sinh : $\exp(-\sqrt[3]{2}/3)$
7053 7803 id : $1/3 - 3e/4$
7053 8353 $10^x : 1/2 + \pi/2$
7053 9060 $J_0 : (\ln 2)^{-1/3}$
7053 9123 $\ell_{10} : e + 3\pi/4$
7054 0381 rt : (6,7,2,-7)
7054 1133 $e^x : 9\pi^2/2$
7054 1778 $s\pi : 3e + 2\pi/3$
7054 1931 rt : (9,8,4,2)
7054 2444 $\zeta : \sqrt[3]{3}/8$
7054 2914 $erf : (\ln 3/2)^{1/2}$
7054 4043 id : $3\sqrt{3} - 2\sqrt{5}/3$
7054 4637 $J_0 : (\sqrt[3]{3})^{1/3}$
7054 5951 $K : 4\zeta(3)/9$
7054 6309 $2^x : 10\ln 2/9$
7054 6922 rt : (4,9,3,-8)
7054 7061 rt : (9,7,-8,-1)
7054 8303 $Ei : 11/19$
7055 0775 sr : $6e\pi/7$
7055 0901 $\ell_2 : 2\sqrt{2} + \sqrt{3}/4$
7055 1157 sq : $2\sqrt[3]{2}/3$
7055 4673 rt : (1,2,8,5)
7055 5150 $2^x : 2 + \sqrt{5}/3$
7055 5252 rt : (8,7,1,-7)
7055 5490 rt : (7,9,-7,-2)

7055 8267 e^x : $3\pi^2/7$
7055 8775 10^x : $4\sqrt{3}+4\sqrt{5}$
7056 0000 sq : 21/25
7056 0046 tπ : $3\pi^2/2$
7056 0716 rt : (6,9,2,7)
7056 2005 sinh : $7\sqrt{3}/6$
7056 2031 rt : (7,6,5,3)
7056 4332 ℓ_{10} : $1+3e/2$
7056 5808 rt : (4,1,-7,5)
7056 6667 $1/3\times\exp(3/4)$
7056 7755 cr : $1/4+3\pi/2$
7056 8737 e^x : $1/4+\sqrt{5}/3$
7057 1963 rt : (1,0,-9,6)
7057 2370 rt : (8,9,1,-8)
7057 2539 rt : (9,9,-8,-2)
7057 2944 ℓ_2 : $3e/5$
7057 6712 rt : (6,7,-2,6)
7057 7703 Γ : $(\sqrt[3]{3}/2)^2$
7057 7832 rt : (5,4,6,4)
7058 0598 e^x : $9\pi/2$
7058 0808 sq : $\pi^2/6$
7058 1128 θ_3 : $(e+\pi)/9$
7058 2130 ℓ_{10} : $4\sqrt{2}-\sqrt{3}/3$
7058 2478 cr : $1/\sqrt[3]{23}$
7058 3521 e^x : 25/11
7058 4381 Γ : 13/25
7058 5982 rt : (2,9,-3,-5)
7058 7375 cr : $3/2+2\sqrt{3}$
7058 8235 id : 12/17
7058 8434 ℓ_2 : $4e/3+3\pi$
7058 9623 rt : (6,5,-6,5)
7059 0562 e^x : $2\sqrt{3}/3+\sqrt{5}/2$
7059 0585 rt : (3,2,7,5)
7059 1216 e^x : $e/2+4\pi/3$
7059 5032 rt : (9,6,4,3)
7059 6842 ln : $2/3+e/2$
7059 9303 cu : $e/2-3\pi$
7060 1097 rt : (1,0,8,6)
7060 4802 2^x : $3\sqrt[3]{2}/2$
7060 4819 rt : (7,4,5,4)
7060 5232 $\exp(5)-s\pi(1/4)$
7060 5566 10^x : $2+3\sqrt{2}/2$
7060 5939 sq : $10e\pi$
7060 6716 rt : (8,9,-5,5)
7060 7006 ζ : 5/9
7060 8011 2^x : $3\sqrt{2}/2+3\sqrt{3}/4$
7060 9670 Ψ : 8/21
7060 9912 rt : (1,2,-9,-7)
7061 0339 cu : $3-2\sqrt{2}/3$
7061 1072 10^x : $\sqrt{2}/4+\sqrt{5}$
7061 1766 Ψ : 6
7061 3074 rt : (5,2,6,5)
7061 3883 sq : $\sqrt{2}/4+\sqrt{5}$
7061 4663 rt : (8,7,-9,4)
7061 4961 cu : $5\pi^2/2$
7061 5453 at : $\exp(-1/(2\pi))$
7061 6389 e^x : $4\zeta(3)/9$
7061 6395 Ψ : 3/22
7061 8133 erf : $3\sqrt{3}/7$
7062 0131 rt : (3,0,7,6)
7062 2622 Ψ : $(\sqrt[3]{2})^{-1/3}$
7062 2700 rt : (9,4,4,4)
7062 3877 e^x : $4\sqrt{3}/3+2\sqrt{5}/3$
7062 5305 sπ : $4\sqrt{3}/3-\sqrt{5}/4$
7062 5454 ζ : $\exp(-\sqrt[3]{5})$
7062 6221 e^x : $1/\ln(\sqrt[3]{4})$
7062 6234 rt : (1,2,8,-7)
7062 8308 tπ : $(\sqrt{5})^{-1/2}$
7062 8475 rt : (7,2,5,5)
7062 8884 Γ : $3\ln 2/4$
7062 9115 10^x : $3/2+\sqrt{3}/4$
7063 1564 rt : (1,4,-9,-8)
7063 2526 Ei : $(\sqrt{3}/2)^{1/2}$
7063 3537 rt : (5,0,6,6)
7063 4395 Li_2 : $(e+\pi)/10$
7063 5051 θ_3 : $e/8$
7063 8010 rt : (3,2,7,-7)
7063 8940 ℓ_{10} : $\sqrt{3}+3\sqrt{5}/2$
7063 9681 rt : (9,2,4,5)
7064 0168 Ψ : $(2\pi)^{1/2}$
7064 1932 tanh : 22/25
7064 1992 rt : (1,4,8,-8)
7064 3488 rt : (7,0,5,6)
7064 4473 sr : $5(e+\pi)/4$
7064 5152 erf : $7(1/\pi)/3$
7064 5560 rt : (1,6,-9,-9)
7064 6907 rt : (5,2,6,-7)
7064 9613 id : $e-3\pi$
7064 9994 rt : (3,4,7,-8)
7065 1167 rt : (9,0,4,6)
7065 2795 rt : (1,6,8,9)
7065 3097 erf : $(2e/3)^{-1/2}$
7065 3864 rt : (7,2,5,-7)
7065 5337 at : $1/2+\sqrt{2}/4$
7065 6327 rt : (5,4,6,-8)
7065 7676 Ei : 21/8
7065 8584 rt : (3,6,7,-9)
7065 9453 rt : (9,2,4,-7)
7066 1181 e^x : $10\pi^2/3$
7066 1463 rt : (7,4,5,-8)
7066 1851 cu : $3\sqrt{2}+\sqrt{5}/4$
7066 3322 rt : (5,6,6,-9)
7066 3516 tπ : $(\sqrt[3]{5})^{1/3}$
7066 5713 rt : (9,4,4,-8)
7066 7269 rt : (7,6,5,-9)
7066 8882 Ψ : $7(1/\pi)/4$
7066 9840 cr : 6/17
7067 0079 rt : (3,8,3,9)
7067 0610 rt : (9,6,4,-9)
7067 2402 ζ : $\sqrt[3]{1/3}$
7067 4113 rt : (3,6,-5,2)
7067 4230 id : $3\sqrt{2}+2\sqrt{3}$
7067 5451 ln : $\sqrt[3]{25/3}$
7067 9357 cr : $1/2+2\sqrt{5}$
7067 9632 at : $e\pi/10$
7068 1662 rt : (1,8,2,-2)
7068 2806 K : $\ln(\sqrt[3]{5})$
7068 3831 10^x : $2\sqrt{2}-\sqrt{3}/2$
7068 5536 2^x : $-\sqrt{2}-\sqrt{3}+\sqrt{7}$
7068 6215 $\exp(\sqrt{5})\div s\pi(\sqrt{2})$
7068 7941 rt : (9,7,-9,9)
7068 8949 tπ : $\sqrt{5}+\sqrt{6}+\sqrt{7}$
7069 0895 $\ln(3/4)\times\exp(4)$
7069 4042 e^x : $1/4+3\sqrt{3}/4$
7069 4976 e^x : $2e-\pi/4$
7069 5172 as : $3\sqrt{3}/8$
7069 6937 ℓ_2 : $4/3+3\sqrt{3}$
7070 3547 cu : $2\sqrt{3}-3\sqrt{5}/4$
7070 7091 Γ : $3\sqrt{3}/10$
7070 9262 ℓ_{10} : $3+2\pi/3$
7071 0678 id : $\sqrt{2}/2$
7071 3431 ℓ_{10} : $e/3+4\pi/3$
7071 8567 sπ : $((e+\pi)/3)^{1/3}$
7071 8961 cr : $10(1/\pi)/9$
7072 2149 erf : $(e/3)^3$
7072 9633 cu : $2\sqrt{3}/3+3\sqrt{5}/4$
7073 4473 e^x : $1/3-e/4$
7073 8272 at : $\sqrt[3]{5}/2$
7073 8596 ℓ_{10} : $4\sqrt{2}-\sqrt{5}/4$
7073 9223 tanh : $7\sqrt[3]{2}/10$
7074 6797 sr : $7\pi/3$
7074 6947 sr : $\sqrt{2}+\sqrt{3}-\sqrt{7}$
7074 7249 id : $\sqrt{2}/3+\sqrt{5}$
7074 7648 sr : $\sqrt{17/2}$
7074 8378 rt : (1,6,-4,5)
7075 0220 rt : (4,9,-2,8)
7075 0321 at : $\sqrt{2}-\sqrt{5}/4$
7075 1874 ℓ_2 : $\sqrt{8/3}$
7075 2880 Ei : $7\zeta(3)/3$
7075 5345 J : 8/23
7075 8443 as : 13/20
7075 9342 ζ : 2/11
7075 9611 tanh : 15/17
7076 0463 rt : (2,7,-1,2)
7076 3901 sr : $\sqrt{5}+\sqrt{6}+\sqrt{7}$
7076 4407 sq : $e+4\pi/3$
7076 5535 id : $\sqrt{2}/4+3\sqrt{5}/2$
7076 8482 $4\div\exp(\sqrt{3})$
7076 9691 ζ : $4(1/\pi)/7$
7077 3631 $\ln(\pi)\times\exp(2/5)$
7077 4774 $\exp(1/2)-\exp(\sqrt{5})$
7077 5112 e^x : $3/4+\sqrt{2}$
7077 5926 J_0 : 9/8
7077 8527 $\ln(2/3)+\exp(\sqrt{2})$
7077 8656 ℓ_2 : $1/2+4\pi$
7077 9371 rt : (8,1,8,-9)
7077 9875 cu : $9\ln 2/7$
7078 0380 ℓ_{10} : $3\sqrt{2}/2+4\sqrt{5}/3$
7078 0780 cu : $6e/5$
7078 5568 cr : $2+4\sqrt{5}/3$
7078 6451 rt : (6,1,9,-9)
7078 6843 sinh : $(\sqrt{2}\pi/2)^{1/3}$
7078 7698 Γ : $(2\pi/3)^{-3}$
7078 8604 cu : $e/4-\pi/2$
7078 9060 Ψ : $\exp(3\pi/2)$
7078 9817 rt : (8,1,8,8)
7078 9848 rt : (7,6,-9,2)
7079 2110 sπ : $e/4+\pi/2$
7079 4684 id : $e\pi/5$
7079 6326 id : 5π
7079 7593 rt : (7,8,-5,3)
7079 7983 Ψ : $(\sqrt[3]{5}/3)^3$
7079 8075 Ψ : $5\zeta(3)$
7079 9364 rt : (6,1,9,8)
7080 1280 id : $\sqrt{22/3}$
7080 2099 sq : $7\zeta(3)/10$
7080 2379 cu : $(e+\pi)^3$
7080 4008 rt : (8,3,8,7)
7080 5020 ln : 15
7080 5266 K : 7/13
7080 7931 at : $3/2-3\pi/4$
7080 9675 Ei : $\sqrt{8/3}$
7081 0164 rt : (9,9,-1,-7)
7081 1839 sr : $4\sqrt{2}+3\sqrt{5}/4$
7081 1920 tπ : P_{27}
7081 5027 rt : (5,4,-6,3)
7081 5468 $\ln(3/5)\times\ln(4)$
7081 5945 erf : $\sqrt{5}/3$
7081 6344 10^x : $3\sqrt[3]{3}/10$
7081 7584 rt : (6,3,9,7)
7081 8730 rt : (7,9,0,-7)
7081 9566 e^x : $5\pi/6$
7082 0223 $\sqrt{3}\times e$
7082 0393 id : $3\sqrt{5}$
7082 2705 rt : (9,7,-1,-6)
7082 3433 tπ : $3\sqrt{3}$
7082 4237 sq : $2e\pi$
7082 4400 rt : (8,5,8,6)
7082 4494 rt : (3,7,-6,-8)
7082 4604 rt : (1,9,3,4)
7082 5826 tπ : $2e/3-3\pi$
7082 6773 sr : $3\sqrt{2}/4-\sqrt{5}/4$
7082 8913 rt : (5,9,1,-7)
7083 0751 ℓ_{10} : $3\sqrt{2}+\sqrt{3}/2$
7083 3190 10^x : $8\zeta(3)/9$
7083 3333 id : 17/24
7083 3346 ln : $\sqrt{2}/4+3\sqrt{5}/4$
7083 3687 rt : (7,7,0,-6)
7083 4481 rt : (1,7,3,7)
7083 4760 rt : (8,6,-9,7)
7083 5950 tanh : $5\sqrt{2}/8$
7083 6303 2^x : $9\pi^2/2$
7083 8614 Ψ : $(e/2)^3$
7083 8860 rt : (9,5,-1,-5)
7083 9803 sq : $1/4-3\sqrt{5}$
7084 1217 rt : (3,9,2,-1)
7084 2640 θ_3 : $5(1/\pi)/2$
7084 4075 rt : (8,8,-5,8)
7084 5225 rt : (6,5,9,6)
7084 5792 K : $7\ln 2/9$
7084 7059 rt : (5,8,2,5)
7084 7506 Ψ : $(\sqrt[3]{5}/2)^{1/2}$
7084 9647 2^x : 17/22
7085 1109 2^x : 23/16
7085 3445 rt : (7,5,0,-5)
7085 3681 sq : $e-2\pi$
7085 6063 $\ln(\pi)\times\exp(\sqrt{2})$
7085 6190 rt : (8,7,8,5)
7085 6387 rt : (1,9,3,-7)
7086 0455 rt : (9,3,-1,-4)

7086 0649 cr : $3\sqrt{2}+\sqrt{5}/3$
7086 2123 e^x : $1/4+3\sqrt{2}/4$
7086 2627 at : 6/7
7086 3700 rt : (6,4,-6,8)
7086 8456 at : $(\sqrt[3]{4})^{-1/3}$
7087 1333 tanh : $\exp(-1/(3e))$
7087 1782 rt : (5,5,1,-5)
7087 1912 Ψ : $3(e+\pi)/7$
7087 2922 Ψ : $\sqrt{2}/9$
7087 4552 e^x : $3/2+\sqrt{5}/2$
7087 5556 rt : (1,9,-4,-7)
7087 6248 ln : $4/3+4\pi/3$
7087 8119 rt : (6,6,-2,9)
7087 8395 λ : $\exp(-9\pi/11)$
7088 0760 rt : (7,3,0,-4)
7088 2935 cu : $3/2+2\sqrt{3}/3$
7088 3312 cu : $1/3+3\sqrt{2}/4$
7088 4979 rt : (1,7,3,-6)
7088 7343 rt : (3,9,-6,7)
7089 0104 rt : (4,0,-7,8)
7089 0789 rt : (9,1,-1,-3)
7089 1387 tanh : $(\ln 2)^{1/3}$
7089 2144 rt : (6,7,9,5)
7089 5537 rt : (3,5,2,-5)
7089 5918 tanh : $(\sqrt[3]{3})^{-1/3}$
7089 6441 Γ : $3\pi/2$
7089 7169 J_0 : $(\sqrt[3]{2})^{1/2}$
7089 8287 e^x : $4-2\sqrt{3}$
7089 8416 at : $(e/2)^{-1/2}$
7089 8533 10^x : $4+\sqrt{3}/4$
7089 8796 K : $3\sqrt[3]{2}/7$
7089 9958 10^x : $7\pi/3$
7090 0555 rt : (3,9,-5,-7)
7090 1307 rt : (8,8,3,-9)
7090 2909 as : $(e+\pi)/9$
7090 7446 rt : (5,3,1,-4)
7091 0258 rt : (4,2,-3,9)
7091 0316 ℓ_{10} : $4+\sqrt{5}/2$
7091 2629 rt : (8,9,8,4)
7091 2912 e^x : $4/3-3\sqrt{5}/4$
7091 3156 rt : (1,7,-4,-6)
7091 3658 cr : $3/4+3\sqrt{2}$
7091 4127 sq : 16/19
7091 4269 2^x : $4\pi/5$
7091 5386 tπ : $2\sqrt{2}+\sqrt{5}/4$
7091 5650 $\ln(2/5)\times s\pi(e)$
7091 6998 rt : (6,8,4,-9)
7091 7373 10^x : $2/3+3\sqrt{3}$
7092 0982 rt : (7,1,0,-3)
7092 3550 ln : $3\sqrt{3}/4-\sqrt{5}/2$
7092 4206 rt : (8,6,3,-8)
7092 5541 2^x : $4\sqrt{2}-\sqrt{5}$
7092 7535 rt : (2,8,0,-8)
7092 7860 sinh : $10\sqrt[3]{5}/7$
7093 1169 J_0 : $7\sqrt[3]{3}/9$
7093 1954 cr : $1/\sqrt[3]{22}$
7093 2355 Ψ : $4\ln 2/3$
7093 3107 $s\pi(2/5)\times s\pi(\sqrt{3})$
7093 4534 rt : (5,9,-6,-7)
7093 5475 at : $4-\pi$
7093 5513 rt : (4,8,5,-9)
7093 5769 rt : (8,9,-9,-1)
7093 6497 rt : (9,1,-1,2)
7093 7589 ℓ_2 : $(3\pi)^3$
7093 8194 ℓ_{10} : $3+3\sqrt{2}/2$
7093 8661 2^x : $1/4-\sqrt{5}/3$
7093 8925 e^x : $3/2+\pi$
7094 4095 rt : (6,6,4,-8)
7094 4676 $\exp(1/2)\times s\pi(\pi)$
7094 6749 rt : (3,5,-7,4)
7094 7261 at : $5\zeta(3)/7$
7094 9170 cr : 5/14
7094 9490 sπ : $2\sqrt{2}-\sqrt{3}/3$
7094 9846 rt : (1,5,-3,6)
7095 0584 10^x : $\sqrt{2}/2+4\sqrt{3}$
7095 1008 rt : (5,9,-9,9)
7095 1129 $\ln(2)\div\ln(2/3)$
7095 2253 rt : (3,7,-5,-6)
7095 3352 rt : (8,4,3,-7)
7095 3568 K : $3\sqrt[3]{3}/8$
7095 3624 Ei : $7\sqrt[3]{3}/3$
7095 4565 10^x : $1/2+3e/4$
7095 6036 sπ : $7\pi/8$
7095 7692 rt : (2,8,6,-9)
7095 9560 ℓ_{10} : $6e\pi$
7096 0901 sπ : $(\sqrt[3]{2}/3)^{1/3}$
7096 3366 rt : (5,1,1,-3)
7096 4096 ℓ_{10} : $3/2+4e/3$
7096 4498 Ψ : $4\pi/5$
7096 6169 ℓ_{10} : $3e/2+\pi/3$
7096 6179 sπ : $4\sqrt{3}-3\sqrt{5}/4$
7096 8085 rt : (4,6,5,-8)
7096 8469 rt : (6,9,-8,-1)
7096 9155 e^x : $2e/3+\pi$
7097 2994 rt : (1,8,9,-3)
7097 3069 Li_2 : 10/17
7097 3736 $s\pi(\sqrt{2})+s\pi(\sqrt{3})$
7097 5041 $\ln(\sqrt{3})\div s\pi(e)$
7097 7695 at : $1/2-e/2$
7097 9381 rt : (6,4,4,-7)
7097 9766 rt : (7,9,-8,-5)
7098 1629 rt : (1,8,-9,2)
7098 3436 rt : (7,9,-7,-7)
7098 4703 K : $(e/2)^{-2}$
7098 4748 rt : (8,9,-8,0)
7098 6065 rt : (7,1,0,2)
7098 7554 rt : (9,9,-8,5)
7098 8354 ℓ_2 : $-\sqrt{2}+\sqrt{5}+\sqrt{6}$
7098 9307 rt : (6,9,9,4)
7098 9636 ℓ_{10} : $7(e+\pi)/8$
7099 0733 sr : $2\sqrt[3]{2}/5$
7099 1704 rt : (8,2,3,-6)
7099 2036 10^x : $5\sqrt{3}$
7099 7594 id : $\sqrt[3]{5}$
7100 5103 K : 13/24
7100 5917 sq : 17/13
7100 9481 ln : $1/3+3\sqrt{3}$
7101 0205 rt : (5,7,-6,-6)
7101 0852 tπ : $\exp(-2/3)$
7101 1125 ℓ_{10} : $3\sqrt[3]{5}$
7101 1691 rt : (4,4,5,-7)
7101 2306 rt : (4,9,-7,-1)
7101 2404 cr : $9(1/\pi)/8$
7101 3181 rt : (9,3,-1,1)
7101 5054 rt : (2,6,-5,1)
7101 7957 J_0 : $8\sqrt[3]{2}/9$
7101 8498 rt : (2,8,-8,-9)
7102 0384 sq : $3\sqrt{2}+3\sqrt{3}/4$
7102 1376 K : $(2\pi)^{-1/3}$
7102 5252 $s\pi(\sqrt{5})\div s\pi(2/5)$
7102 5550 rt : (7,5,-9,5)
7102 7089 10^x : $\sqrt{3}/4$
7102 7227 rt : (9,3,6,-9)
7102 7987 rt : (1,9,-7,-1)
7102 8689 tπ : $2\sqrt{2}-4\sqrt{3}/3$
7103 1618 rt : (9,8,-6,-3)
7103 1629 2^x : $\sqrt{2}+\sqrt{3}+\sqrt{5}$
7103 3057 at : $3\sqrt{2}/2-4\sqrt{5}/3$
7103 4789 rt : (6,7,-8,0)
7103 4998 sq : $\sqrt{3}/3-3\sqrt{5}/2$
7103 7934 2^x : $4/3-\sqrt{5}/4$
7103 9141 $\exp(\pi)+s\pi(\pi)$
7103 9313 id : $2\sqrt{2}-\sqrt{5}/2$
7104 1700 at : $(\pi/2)^{-1/3}$
7104 2750 rt : (3,5,-5,-5)
7104 2953 erf : $(\sqrt[3]{2}/3)^{1/3}$
7104 3939 Li_2 : $(\ln 2/2)^{1/2}$
7104 4432 rt : (8,0,3,5)
7104 4860 as : 15/23
7104 5207 tπ : $\sqrt[3]{19}$
7104 6471 $\exp(1/2)\div s\pi(\sqrt{2})$
7104 8416 10^x : $3+\pi/2$
7104 9338 ℓ_2 : 11/18
7104 9894 sq : $3\sqrt{2}+3\sqrt{5}/2$
7105 2879 rt : (2,4,6,-7)
7105 3024 rt : (7,7,-5,6)
7105 3143 rt : (5,5,-9,-5)
7105 6382 rt : (7,3,7,-9)
7105 6884 cu : $1/3+\sqrt{5}/4$
7105 7713 $\ln(3/5)+\exp(1/5)$
7106 0015 rt : (9,9,-8,-7)
7106 0831 rt : (8,5,-9,1)
7106 1800 rt : (4,8,-9,-9)
7106 2396 rt : (1,5,2,3)
7106 2970 $\ln(\pi)\times\exp(\sqrt{5})$
7106 2998 2^x : $9e/5$
7106 3587 rt : (5,1,1,2)
7106 5190 10^x : P_{10}
7106 7811 cu : $4+\sqrt{2}$
7106 8886 sr : P_{35}
7106 9331 ln : $3\sqrt{2}/4+2\sqrt{5}$
7106 9702 rt : (7,8,-5,-3)
7106 9855 rt : (9,1,6,-8)
7107 1172 2^x : $(5/3)^{-1/2}$
7107 3081 rt : (4,2,5,-6)
7107 3526 rt : (6,7,-5,1)
7107 4156 rt : (2,9,-6,-1)
7107 4584 cu : $8e\pi/3$
7107 6617 Ψ : $9\sqrt{5}/8$
7107 6953 λ : 1/13
7107 7957 ℓ_{10} : $2\sqrt{2}+4\sqrt{3}/3$
7107 8643 sq : $2-\sqrt{2}/4$
7108 0406 $\ln(4)-s\pi(\sqrt{5})$
7108 2562 $1/5-\ln(3/5)$
7108 3145 rt : (2,6,-8,-8)
7108 4449 tanh : 8/9
7108 4912 rt : (1,7,-6,-4)
7108 5746 rt : (7,9,-1,7)
7108 6193 ℓ_{10} : $2/3+2\sqrt{5}$
7108 6511 rt : (8,9,-2,-6)
7108 7916 rt : (9,6,-6,-2)
7109 1484 rt : (5,3,8,-9)
7109 2067 rt : (2,8,-5,8)
7109 5781 rt : (6,0,4,5)
7110 1496 tπ : $(2e)^{1/2}$
7110 3619 θ_3 : 15/23
7110 5223 rt : (7,7,-7,-6)
7110 5842 Li_2 : $\exp(-\sqrt[3]{4}/3)$
7110 6049 ℓ_{10} : $2\sqrt{3}+3\sqrt{5}/4$
7110 6780 cr : $7e\pi/3$
7110 7219 rt : (4,9,-5,0)
7110 7883 rt : (7,1,7,-8)
7110 8270 J_0 : $\sqrt{5}/2$
7110 9210 rt : (7,3,0,1)
7110 9766 ℓ_{10} : $2+\pi$
7111 1313 2^x : $\sqrt{5}$
7111 2407 rt : (1,6,2,2)
7111 3095 rt : (5,3,-6,6)
7111 5563 erf : 3/4
7111 8077 rt : (5,8,-4,-3)
7112 1117 λ : $\ln 2/9$
7112 1466 rt : (8,2,3,4)
7112 2764 rt : (6,9,-1,2)
7112 5735 rt : (9,1,6,7)
7112 6066 $\ln(\pi)\div\ln(5)$
7112 6690 J_0 : 19/17
7112 6951 at : P_5
7112 6978 Ψ : $\sqrt{19/3}$
7112 9143 $\pi+s\pi(\pi)$
7113 0307 K : $e/5$
7113 1053 rt : (1,8,-7,-7)
7113 2464 $\ln(\sqrt{3})\times\exp(\pi)$
7113 4573 rt : (3,3,9,-9)
7113 4598 10^x : $2+\sqrt{2}/4$
7113 6380 cr : $2\sqrt[3]{2}/7$
7113 7866 cr : 9/25
7113 7871 at : $4/3-\sqrt{2}/3$
7113 8421 rt : (9,8,1,-8)
7113 9372 Ψ : 12/13
7113 9557 rt : (8,7,-2,-5)
7114 0041 $\exp(4)+\exp(\sqrt{2})$
7114 0477 Γ : 14/3
7114 1648 rt : (7,6,-5,-2)
7114 3170 cu : $4+3\sqrt{3}$
7114 4773 rt : (3,6,-8,-6)
7114 6757 rt : (6,5,-8,1)
7114 7018 $\exp(3/4)\times\exp(4/5)$
7114 7806 θ_3 : $\sqrt[3]{5}/5$
7114 8541 cr : $3e-\pi$
7114 8634 rt : (4,6,-9,-8)

7115 0082 $s\pi : \sqrt[3]{2}/5$
7115 0553 rt : (3,9,-9,-4)
7115 0747 $2^x : 3\sqrt{3} + \sqrt{5}$
7115 2584 $e^x : 1/4 + 3\sqrt{2}/2$
7115 3490 $\ln(2/3) + \exp(3/4)$
7115 4607 $10^x : 5\ln 2/8$
7115 4974 rt : (5,1,8,-8)
7115 6467 sq : $4\sqrt{2}/3 - \sqrt{3}/3$
7115 8725 $K : 1/\ln(2\pi)$
7115 9088 rt : (5,5,-2,7)
7116 2420 rt : (9,8,-8,8)
7116 3765 rt : (5,5,-6,-5)
7116 5917 rt : (4,0,5,5)
7116 6052 rt : (3,4,-9,6)
7116 6800 rt : (4,9,0,-6)
7116 7379 $2^x : 1 - 2\sqrt{5}/3$
7116 8053 rt : (9,5,-1,0)
7117 0594 id : $((e+\pi)/2)^{1/2}$
7117 3438 cu : $4\sqrt{3} + 3\sqrt{5}$
7117 4363 rt : (7,8,2,-8)
7117 4725 $\ln(4) \div \exp(2/3)$
7117 4890 cr : $\sqrt[3]{3}/4$
7117 5657 $\exp(5) \div s\pi(\sqrt{5})$
7117 6254 $\sqrt{5} \div \pi$
7117 7403 rt : (7,1,7,7)
7117 8488 cr : $3\zeta(3)/10$
7118 0356 cu : $3\sqrt{3}/4 + 2\sqrt{5}/3$
7118 1590 rt : (3,8,-3,-3)
7118 2281 rt : (2,4,-8,-7)
7118 3615 $t\pi : 3\zeta(3)/2$
7118 3734 tanh : $(\sqrt[3]{2})^{-1/2}$
7118 4475 sr : $4 + 3\sqrt{5}/2$
7118 4987 $t\pi : 3\sqrt{3} + 2\sqrt{5}$
7118 7436 rt : (6,7,-1,-5)
7118 7565 rt : (1,8,3,6)
7119 0696 rt : (9,6,1,-7)
7119 4455 $\ell_2 : 3\sqrt{2} + 4\sqrt{3}/3$
7119 4823 rt : (8,3,-9,2)
7119 7854 $t\pi : \sqrt{3}/4 - \sqrt{5}$
7119 8061 tanh : $9\ln 2/7$
7119 8136 $erf : 5\zeta(3)/8$
7119 8805 rt : (3,8,0,1)
7119 9934 $s\pi : 2\sqrt{2}/3 + 4\sqrt{3}/3$
7120 1013 $e^x : 3\sqrt{2} + \sqrt{5}/4$
7120 2172 rt : (9,3,6,6)
7120 3759 sinh : $\exp(5/2)$
7120 3842 rt : (8,8,-8,3)
7120 9433 $\Psi : (3\pi/2)^{-2}$
7120 9988 rt : (8,5,-2,-4)
7121 1730 sq : $3\sqrt{2}/4 + \sqrt{3}/2$
7121 1732 cu : $\sqrt{3}/3 + 3\sqrt{5}$
7121 3294 rt : (5,9,-1,-3)
7121 3595 sr : $1/3 + 3\sqrt{3}/2$
7121 4189 $\Psi : (\ln 3/2)^{-3}$
7121 4836 rt : (3,1,9,-8)
7121 6053 $t\pi : 7\sqrt[3]{5}/10$
7121 6943 rt : (5,8,3,-8)
7121 6953 ln : $4e + 4\pi/3$
7121 8783 at : $(\sqrt{5}/3)^{1/2}$
7121 8896 rt : (7,4,-9,8)
7122 0344 $\ell_{10} : 4 + 2\sqrt{3}/3$
7122 1445 rt : (2,9,1,-6)
7122 5431 $10^x : 6\sqrt[3]{5}/5$
7122 5779 $10^x : 4\sqrt{3}$
7122 9748 ln : $3\sqrt{2} + 3\sqrt{3}/4$
7123 1792 ln : $3e/4$
7123 5759 at : 19/22
7123 6452 rt : (7,6,2,-7)
7123 7518 $\zeta : (\sqrt[3]{5}/3)^3$
7123 8898 id : $3\pi/2$
7123 9527 $2^x : 4/3 + \sqrt{5}/4$
7124 2488 rt : (3,3,-5,-4)
7124 3874 rt : (5,1,8,7)
7124 3895 cu : $3\sqrt{3} - \sqrt{5}/4$
7124 4222 $K : 6/11$
7124 4562 rt : (8,4,3,3)
7124 5812 id : $4\sqrt{2}/3 - 3\sqrt{3}/2$
7124 5877 $J_0 : (\sqrt[3]{3}/2)^{-1/3}$
7124 6437 $erf : \ln(\sqrt{2}/3)$
7124 7422 rt : (4,7,0,-5)
7124 7699 ln : $\sqrt{23}\pi$
7124 8853 ln : $3\sqrt{3}/2 - \sqrt{5}/4$
7124 9467 rt : (7,4,-5,-1)
7124 9715 $2^x : 2e + 4\pi$
7125 3515 $s\pi : 1/3 + \sqrt{2}$
7125 4656 as : $4/3 - e/4$
7125 6543 cr : $6(e+\pi)/7$
7125 7493 rt : (9,4,1,-6)
7125 8936 rt : (7,6,-5,9)
7125 9221 $10^x : 2/3 - \sqrt{3}/4$
7126 2870 $t\pi : 8e/9$
7126 7573 rt : (2,0,6,5)
7126 8199 rt : (3,8,4,-8)
7126 8718 rt : (1,8,-2,-3)
7126 9900 cr : $3\sqrt{3}/2 - \sqrt{5}$
7127 2942 rt : (8,5,5,-9)
7127 3034 rt : (7,8,-8,-2)
7127 4226 rt : (1,8,8,2)
7127 5090 cr : $e - 3\pi/4$
7127 6122 rt : (6,5,-1,-4)
7127 6388 rt : (7,3,7,6)
7128 0889 sq : $2e + \pi/4$
7128 1292 sq : $2 + 4\sqrt{3}$
7128 1664 $\Psi : 5/9$
7128 2182 rt : (2,9,8,8)
7128 3498 rt : (6,4,-9,3)
7128 4015 $\Psi : 2\sqrt[3]{2}$
7128 5490 ln : $\sqrt{5} + \sqrt{6} - \sqrt{7}$
7128 8547 rt : (4,7,-5,-9)
7128 8830 $t\pi : (e+\pi)^{1/3}$
7129 1094 rt : (9,7,-8,-6)
7129 1458 $\ln(3/5) \times \exp(1/3)$
7129 1921 rt : (5,9,6,9)
7129 2862 ln : $8\ln 2$
7129 2937 rt : (4,4,-9,-7)
7129 3803 $2 \times \exp(\sqrt{5})$
7129 4744 rt : (5,3,1,1)
7129 6079 $J_0 : 7(1/\pi)/2$
7129 7873 sinh : $7\pi/9$
7129 7962 rt : (4,8,-7,2)
7129 9207 $\Gamma : 1/\ln(1/\sqrt{2})$
7129 9542 cr : $1/\sqrt[3]{21}$
7129 9611 $2^x : 2/3 - 2\sqrt{3}/3$
7129 9706 cr : $8\pi/5$
7130 1056 $\Psi : \exp(-\sqrt[3]{2})$
7130 3897 $2^x : 8\sqrt[3]{2}/7$
7130 5223 $10^x : 2e - 2\pi/3$
7130 7984 rt : (8,3,-2,-3)
7130 9395 $2^x : 8e$
7131 0441 cu : $e/4 - 3\pi/4$
7131 1635 rt : (1,9,-6,9)
7131 3027 rt : (9,5,6,5)
7131 3104 $t\pi : 5\sqrt[3]{3}/4$
7131 3665 rt : (3,6,-3,-2)
7131 5056 $2^x : 2e/7$
7131 5737 sinh : 6
7131 6552 ln : $4/3 + \sqrt{2}/2$
7131 7404 $\theta_3 : \ln(\sqrt{3}/2)$
7131 7702 rt : (7,4,2,-6)
7132 2662 rt : (4,2,5,4)
7132 2745 $\exp(1/5) + \exp(2/5)$
7132 3744 ln : $\sqrt[3]{3}/8$
7132 3750 $e^x : 1/4 + \pi$
7132 4018 rt : (6,5,6,-9)
7132 4815 rt : (2,7,1,-5)
7132 5037 $2/5 - \exp(\sqrt{2})$
7132 5166 id : $\sqrt{2} + 3\sqrt{3}/4$
7132 6870 cu : $5\pi^2/4$
7132 7885 rt : (2,9,4,-1)
7132 9238 $3/4 \times s\pi(2/5)$
7133 1115 rt : (1,8,5,-8)
7133 2639 $Li_2 : 1/\ln(2e)$
7133 2653 rt : (3,1,9,7)
7133 3815 at : $3\sqrt[3]{3}/5$
7133 5538 ln : $\sqrt[3]{17/2}$
7133 6888 $e^x : 7/13$
7133 8540 $s\pi : \sqrt[3]{16/3}$
7134 2508 ln : $e/2 + 4\pi/3$
7134 3153 rt : (6,6,-5,4)
7134 4484 rt : (5,2,-6,9)
7134 5816 rt : (9,2,1,-5)
7134 6262 $\Gamma : 2e\pi/3$
7134 7263 rt : (8,3,5,-8)
7134 7477 id : $\sqrt{3} + 4\sqrt{5}/3$
7134 9188 $s\pi : 1/\ln(\sqrt{2}\pi/2)$
7134 9968 $2^x : 2\sqrt{2} + 3\sqrt{5}/4$
7135 0767 $\zeta : 7(1/\pi)/4$
7135 1998 cr : $9\sqrt{5}/4$
7135 2549 sq : $3/4 + \sqrt{5}/4$
7135 3057 sinh : $(\sqrt[3]{5})^{1/2}$
7135 3793 rt : (2,2,-8,-6)
7135 6200 $e^x : 1/3 + \pi/2$
7135 6421 $1/3 \times \exp(\pi)$
7135 6580 $\exp(2) - s\pi(\sqrt{5})$
7135 7352 tanh : $2\sqrt{5}/5$
7135 8466 rt : (2,8,-8,-6)
7135 9371 sinh : 17/13
7136 0591 rt : (3,6,4,-7)
7136 0793 $\ln(2/5) \div \exp(1/4)$
7136 2592 rt : (4,7,2,3)
7136 4111 rt : (9,9,-4,-5)
7136 4417 sr : $2 - 2\sqrt{5}/3$
7136 4964 sr : $8(1/\pi)/5$
7136 6263 rt : (5,4,-4,-1)
7137 0023 sq : $3\sqrt{3} + 3\sqrt{5}$
7137 1351 $\Psi : 2\sqrt[3]{5}/9$
7137 1753 id : $2\sqrt{3}/3 + \sqrt{5}/4$
7137 2437 at : $\sqrt{3}/2$
7137 2547 tanh : 17/19
7137 3378 $10^x : 1/\ln(2)$
7137 3705 $10^x : 11/8$
7137 4491 ln : $\sqrt{2}/4 + 3\sqrt{3}$
7137 5183 rt : (6,3,-8,2)
7137 6003 $Ei : 14/15$
7137 6585 cr : 4/11
7137 6949 $Li_2 : 13/22$
7137 7258 rt : (5,3,8,6)
7137 8638 as : $1/2 - 2\sqrt{3}/3$
7138 2537 $erf : (\ln 3)^{-3}$
7138 2764 $e^x : 4 + 2\sqrt{5}$
7138 2995 rt : (2,9,-3,-6)
7138 3089 $\ln(4) \div \ln(3/5)$
7138 3153 rt : (8,7,-8,6)
7138 5399 rt : (4,5,7,-9)
7138 6160 cr : $8(1/\pi)/7$
7138 6421 id : $1/3 - \pi/3$
7138 9521 rt : (6,4,4,3)
7139 0253 $1/4 + s\pi(\sqrt{2})$
7139 2934 rt : (5,4,3,-6)
7139 5780 at : $5\ln 2/4$
7139 5798 rt : (1,8,1,-3)
7139 7290 $Ei : 1/\ln(\sqrt[3]{2})$
7139 8031 $J_0 : \exp(1/(3\pi))$
7139 8893 $e^x : 9\sqrt[3]{5}/5$
7139 9161 cu : $6(e+\pi)/7$
7140 0023 sr : $2\sqrt{2}/3 - \sqrt{3}/4$
7140 1275 rt : (5,4,-9,-2)
7140 3833 ln : $7(e+\pi)$
7140 4520 id : $10\sqrt{2}/3$
7140 5183 rt : (6,3,-1,-3)
7140 5450 sinh : $8e\pi/5$
7140 9069 at : 13/15
7141 0161 id : $1/4 + 2\sqrt{3}$
7141 0231 rt : (1,8,-6,-8)
7141 1291 $\Psi : 9\sqrt{3}$
7141 1878 rt : (6,8,-8,-7)
7141 3595 rt : (6,3,6,-8)
7141 5560 sq : $3\sqrt{3} + 2\sqrt{5}/3$
7141 5858 cu : $5\pi^2/8$
7141 6655 rt : (6,8,-1,5)
7141 9928 cr : $6e\pi$
7142 4175 $\Psi : \zeta(3)/10$
7142 6168 $e^x : 2/3 + 3\sqrt{3}$
7142 6626 $\ell_{10} : \sqrt{2}/2 + 2\sqrt{5}$
7142 8571 id : 5/7
7143 2113 $Ei : 1/\sqrt{21}$
7143 2663 $\exp(1/4) - s\pi(\pi)$
7143 4428 $erf : (\sqrt[3]{5}/3)^{1/2}$
7143 5524 $\ell_{10} : 3e/4 + \pi$
7143 5961 $t\pi : 3\sqrt{2}/4 - \sqrt{5}/4$

7143 6743 rt : (8,9,-4,7)
7143 7185 J_0 : 10/9
7143 7764 cr : $3+3e/4$
7143 8719 tπ : $3(e+\pi)/8$
7143 8775 tπ : $2\sqrt{2}/3-\sqrt{5}/3$
7143 9335 2^x : $(e)^3$
7144 0885 rt : (8,8,0,-7)
7144 1761 id : $\sqrt[3]{20}$
7144 2844 sπ : $3e/4-\pi/4$
7144 2866 $\ln(\pi)+s\pi(\pi)$
7144 3048 sπ : $(\pi/2)^{1/2}$
7144 4181 rt : (8,1,5,-7)
7144 5029 sr : $4-3\sqrt{2}/4$
7144 5134 ℓ_2 : $3-e/2$
7144 6052 rt : (8,8,-7,-2)
7144 7874 rt : (1,6,5,-5)
7144 8796 2^x : 7/9
7144 8814 10^x : $4/3+3\sqrt{3}$
7144 9684 10^x : $3e\pi/4$
7145 0343 rt : (9,7,3,-9)
7145 3117 sq : $4-\sqrt{3}/3$
7145 3536 rt : (8,1,-2,-2)
7145 5342 rt : (9,0,8,9)
7145 6729 rt : (4,7,-9,4)
7145 8045 rt : (9,7,-4,-4)
7145 8910 sr : $7\sqrt[3]{2}/3$
7146 0183 id : $3/2-\pi/4$
7146 0246 at : $\ln(\sqrt[3]{2}/3)$
7146 0596 rt : (2,5,8,-9)
7146 1492 e^x : $2\sqrt{3}/3-2\sqrt{5}/3$
7146 1812 rt : (3,7,-2,3)
7146 1842 2^x : $2\sqrt{3}/3+4\sqrt{5}$
7146 4542 rt : (1,6,-2,-2)
7146 7259 sr : $3e-\pi/4$
7146 8014 rt : (9,0,1,4)
7147 1044 tanh : $(\sqrt[3]{3}/2)^{1/3}$
7147 1530 $\ln(3/5)+\exp(4/5)$
7147 2066 sr : $\ln(5/3)$
7147 2226 rt : (8,6,3,2)
7147 7664 id : $3\sqrt{2}+2\sqrt{5}$
7147 8937 Γ : $2\pi^2$
7148 0237 10^x : $5\sqrt[3]{5}/8$
7148 0591 rt : (6,3,-9,6)
7148 1645 rt : (1,8,-6,2)
7148 2704 Ψ : $7\sqrt[3]{3}/4$
7148 4198 sr : P_{13}
7148 4915 θ_3 : $2\zeta(3)/7$
7148 6108 ln : $\sqrt[3]{2}/7$
7148 7100 at : $1/4-\sqrt{5}/2$
7148 8174 rt : (9,7,6,4)
7148 9681 rt : (3,5,5,9)
7149 0688 K : $\ln(\sqrt{3})$
7149 0804 rt : (8,1,-9,3)
7149 4333 cu : $\sqrt{2}/4+3\sqrt{3}/2$
7149 4587 10^x : $10\sqrt[3]{5}/3$
7149 5532 rt : (4,3,7,-8)
7149 5624 cr : $2\sqrt{2}/3-\sqrt{3}/3$
7149 6568 erf : $3\sqrt[3]{2}/5$
7149 6824 $\ln(4/5)-\exp(2/5)$
7149 7206 rt : (7,7,-8,1)
7150 0000 cu : $7\sqrt[3]{5}/10$
7150 1545 cu : $4\sqrt{2}+\sqrt{5}$
7150 3423 rt : (9,0,-6,1)
7150 5797 2^x : $2-\sqrt{5}/4$
7150 6611 ℓ_{10} : $1+4\pi/3$
7150 7258 rt : (5,9,-2,-5)
7150 7505 2^x : $(\sqrt{2}/3)^{1/3}$
7150 9447 rt : (6,8,1,-7)
7151 0046 rt : (4,9,-4,-6)
7151 0146 ln : $3\sqrt{3}/4+\sqrt{5}/3$
7151 0575 ζ : $(e/2)^{-1/3}$
7151 1087 rt : (7,7,4,-9)
7151 1273 cu : $10e\pi$
7151 2623 tanh : $2\pi/7$
7151 2844 rt : (3,8,-7,-8)
7151 3571 cu : $3+\sqrt{2}/4$
7151 3880 tπ : $1/4-\sqrt{3}$
7151 5004 $\exp(3/4)+\exp(4)$
7151 5539 e^x : $5\sqrt{3}/4$
7151 7240 cr : $1/3+3\pi/2$
7151 9613 Ei : $2e$
7152 1885 sπ : $4\sqrt{3}/3+4\sqrt{5}$
7152 2282 sinh : $7\pi^2/6$
7152 2523 rt : (3,3,9,6)
7152 4472 rt : (2,8,4,7)
7152 6425 rt : (1,7,-5,4)
7152 8370 10^x : $\sqrt[3]{5}/3$
7152 8487 λ : $\exp(-13\pi/16)$
7153 0139 rt : (4,9,-8,-8)
7153 2009 rt : (7,0,9,9)
7153 2705 rt : (8,6,-8,9)
7153 3540 id : $e/2+3\pi/4$
7153 3779 rt : (6,1,6,-7)
7153 3955 cr : $4+\pi/3$
7153 3983 sq : $2e+2\pi/3$
7153 5449 K : 11/20
7153 5532 J_0 : $8\ln 2/5$
7153 6360 rt : (5,2,3,-5)
7153 7322 Γ : $\sqrt[3]{5}/7$
7153 7776 sπ : $8e$
7153 8152 rt : (9,5,3,-8)
7153 9292 2^x : $3\sqrt{2}-2\sqrt{3}$
7154 0476 rt : (8,6,0,-6)
7154 1612 cr : $7\sqrt[3]{3}/2$
7154 2072 ζ : $1/\ln(\sqrt{5})$
7154 3614 rt : (7,7,-3,-4)
7154 4194 id : $(\ln 3/3)^{1/3}$
7154 5073 e^x : 21/16
7154 6175 sr : $2+2\sqrt{2}/3$
7154 6508 ℓ_2 : $(2e/3)^2$
7154 8080 rt : (6,8,-6,-2)
7155 1125 id : $((e+\pi)/3)^{-1/2}$
7155 1678 rt : (6,5,-5,7)
7155 4175 id : $(\sqrt{5}/2)^{-3}$
7155 4944 rt : (5,9,-9,0)
7155 6208 K : $1/\sqrt[3]{6}$
7155 6834 sr : $\sqrt{2}/2+\sqrt{5}$
7155 8801 e^x : $8\sqrt{5}/3$
7155 9878 cr : $\sqrt{3}/3+2\sqrt{5}$
7155 9993 ℓ_{10} : $4e/3+\pi/2$
7156 1274 e^x : $4+4\sqrt{3}/3$
7156 2656 rt : (1,9,9,-6)
7156 3125 as : $1/4-e/3$
7156 6845 rt : (2,5,-6,1)
7156 7075 2^x : $2/3+\pi/2$
7156 7092 rt : (9,2,8,8)
7156 7129 sr : $2/3+3\sqrt{5}$
7156 8159 $\exp(e)\times s\pi(1/4)$
7156 8188 ℓ_{10} : $3\sqrt{3}$
7156 8696 ζ : 3/16
7156 9303 sr : $3\sqrt{2}-3\sqrt{3}/4$
7157 1262 10^x : $\sqrt{2}+3\sqrt{3}/4$
7157 3786 Ei : $7\sqrt{3}/5$
7157 4241 2^x : $4/3+\sqrt{2}$
7157 4358 at : 20/23
7157 5183 rt : (2,7,-3,-5)
7157 5815 rt : (8,1,5,6)
7158 0315 rt : (7,9,-4,2)
7158 1355 rt : (4,2,-9,-6)
7158 2454 Ei : $9\sqrt[3]{3}/10$
7158 2825 rt : (5,7,5,-9)
7158 3319 rt : (9,5,-4,-3)
7158 4037 ζ : $5\sqrt[3]{3}/7$
7158 5217 10^x : $2/3+\pi$
7158 7147 e^x : $4\sqrt{5}/3$
7158 7371 Γ : $9\pi/10$
7158 8782 rt : (7,0,2,4)
7159 2376 rt : (4,8,2,-7)
7159 4880 e^x : $3\sqrt[3]{2}/7$
7159 6908 rt : (8,6,-7,-1)
7159 7233 tanh : $\exp(-1/(3\pi))$
7159 7458 $2-\exp(1/4)$
7159 7633 sq : 11/13
7159 8825 J_0 : $(e/2)^{1/3}$
7159 9080 rt : (3,8,0,-3)
7159 9405 rt : (2,3,8,-8)
7159 9666 rt : (5,5,8,5)
7160 1129 cu : $1/3-3\sqrt{2}/2$
7160 2188 rt : (1,9,9,-7)
7160 4938 sq : 25/9
7160 5519 rt : (3,9,-1,-5)
7161 0079 rt : (6,1,-1,-2)
7161 1020 id : $3\sqrt{3}/2+\sqrt{5}/2$
7161 2447 tπ : $\sqrt{2}/2+2\sqrt{5}/3$
7161 3240 cu : $2-4\pi$
7161 4994 rt : (7,5,4,-8)
7161 5156 J : $2\sqrt{2}/7$
7161 8865 rt : (1,4,5,-6)
7161 9724 id : $9(1/\pi)/4$
7162 0274 sq : $4/3-4\sqrt{5}/3$
7162 0703 ℓ_2 : 7/23
7162 3514 sr : $3\sqrt[3]{5}/10$
7162 3573 cu : $1/3-4\sqrt{3}/3$
7162 7041 tanh : $5\sqrt[3]{2}/7$
7162 8366 ln : $2\sqrt[3]{5}/7$
7162 8517 cu : 17/19
7162 9073 $1/5+\ln(2/5)$
7162 9787 tanh : 9/10
7163 0245 rt : (6,6,1,-6)
7163 1768 cu : $2+3\sqrt{5}/4$
7163 2825 rt : (1,6,-8,-9)
7163 6734 rt : (9,7,-1,-1)
7163 7207 rt : (6,7,-1,8)
7164 0421 e^x : $\sqrt{3}/4+\sqrt{5}/2$
7164 0735 rt : (6,2,-9,9)
7164 3812 rt : (4,4,5,3)
7164 4140 at : $(\pi/3)^{-3}$
7164 5277 10^x : $e/3+3\pi/2$
7164 6610 Ψ : $(e+\pi)^{-1/3}$
7164 7180 ln : $1+\pi/3$
7164 7406 10^x : $9\ln 2/2$
7164 8028 rt : (9,2,1,3)
7164 8649 rt : (4,1,7,-7)
7164 9538 rt : (9,3,3,-7)
7165 1513 rt : (5,8,-1,0)
7165 1525 cu : $3+2\sqrt{3}/3$
7165 3100 Ψ : $(5/2)^3$
7165 3131 id : $(e)^{-1/3}$
7165 6622 e^x : $3+3e$
7166 1902 Ψ : $8(e+\pi)/3$
7166 2701 rt : (4,9,3,-1)
7166 3546 rt : (3,6,-4,5)
7166 4694 rt : (8,9,2,-9)
7166 8885 rt : (3,7,6,-9)
7167 0251 rt : (7,2,9,8)
7167 0589 cu : $3\sqrt{5}/4$
7167 1383 rt : (8,4,0,-5)
7167 3579 ℓ_{10} : $8(e+\pi)/9$
7167 4329 rt : (7,6,-8,4)
7167 6821 sq : $2e-2\pi$
7167 7638 Ei : $(1/(3e))^3$
7167 9911 tπ : $\sqrt[3]{22}$
7168 1600 sπ : $\sqrt{3}/3+3\sqrt{5}/4$
7168 2742 rt : (3,2,4,-5)
7168 3231 rt : (4,8,-5,-2)
7168 3924 Ei : $e\pi/2$
7168 4083 cu : $2e/3+\pi/2$
7168 4417 sq : $8\sqrt[3]{3}/7$
7168 6333 cr : $4+3\sqrt{2}/4$
7168 6609 2^x : $3e/4+3\pi/2$
7168 7116 cr : $1/\sqrt[3]{20}$
7168 7739 ℓ_2 : P_{68}
7168 7962 rt : (9,9,-7,7)
7168 8141 rt : (1,6,-9,3)
7168 8163 2^x : $9\ln 2/8$
7168 8277 cr : 7/19
7168 9976 Ei : $8(e+\pi)/5$
7169 1445 ℓ_2 : P_{71}
7169 1751 as : $\exp(-\sqrt[3]{2}/3)$
7169 1850 rt : (8,1,-2,1)
7169 2398 rt : (7,7,7,4)
7169 4644 cu : $1/4+3\sqrt{3}/4$
7169 4784 rt : (2,8,3,-7)
7169 7802 rt : (6,9,-5,-6)
7169 8177 tanh : $5\sqrt[3]{3}/8$
7169 9572 cu : $4\sqrt{3}+2\sqrt{5}/3$
7170 0001 $2/5-\exp(3/4)$
7170 2287 rt : (7,5,-3,-3)
7170 2385 Γ : 18/5
7170 3459 rt : (6,1,6,6)

7170 3681 $\ell_{10} : 1/2 + 3\pi/2$
7170 3961 $\ell_2 : 4 - 3\pi/4$
7170 4819 tanh : $3\zeta(3)/4$
7170 5534 $10^x : 23/5$
7170 6705 rt : (6,7,-8,-4)
7170 7775 rt : (5,5,5,-8)
7171 0074 rt : (4,1,-6,7)
7171 0379 $\Psi : 23/25$
7171 1036 $\Psi : 4\ln 2/5$
7171 2369 rt : (2,7,0,6)
7171 3955 $J_0 : 21/19$
7171 5846 sinh : 2/3
7171 6537 $\ell_2 : (\sqrt{2\pi})^{1/3}$
7171 6576 rt : (9,8,-2,-6)
7171 7613 rt : (1,6,1,-2)
7171 8043 rt : (8,2,7,-9)
7171 8205 rt : (9,4,8,7)
7171 8255 sr : $7\pi^2/5$
7171 8621 tπ : $2\ln 2/7$
7172 0692 rt : (3,6,-7,-7)
7172 2414 id : $10\zeta(3)/7$
7172 4491 rt : (2,0,-8,5)
7172 9689 cr : $2\sqrt{2} + \sqrt{5}$
7173 1810 rt : (1,9,0,-5)
7173 2784 $\Gamma : 2\sqrt{2}$
7173 2902 $\Psi : 7\sqrt{3}/2$
7173 4755 sπ : $\sqrt{5}/3$
7173 5578 rt : (3,9,-8,0)
7173 5609 sπ : $4(1/\pi)/5$
7173 6227 $10^x : 4 + \sqrt{5}/4$
7173 6562 $\exp(2/5) + \exp(4/5)$
7173 6808 cu : $\sqrt{2} - 4\sqrt{3}/3$
7173 7046 ln : $3/4 + 3\sqrt{3}/4$
7173 9572 $e^x : 3/4 + 2\sqrt{3}/3$
7174 0607 sπ : $5\pi/9$
7174 0683 sq : $2 + 4e$
7174 2068 rt : (6,9,3,9)
7174 3268 $erf : (\sqrt{3})^{-1/2}$
7174 4261 $2^x : \sqrt[3]{3}$
7174 5507 $e^x : 3\sqrt[3]{3}/8$
7174 6178 $\ell_{10} : 7\sqrt{5}/3$
7174 6328 rt : (8,9,-5,-4)
7174 9339 rt : (6,6,-6,-1)
7174 9826 rt : (7,3,4,-7)
7175 2998 $erf : 4\sqrt[3]{5}/9$
7175 3675 $erf : 19/25$
7175 3801 ln : $4 + \pi/2$
7175 4710 $2^x : 4/3 - 2e/3$
7175 5899 rt : (5,0,3,4)
7175 8586 rt : (9,3,-4,-2)
7176 3795 $\exp(1/3) - \exp(\sqrt{2})$
7176 4454 rt : (7,8,-4,5)
7176 4535 tanh : $(e/2)^{-1/3}$
7176 4589 sπ : $7(e + \pi)/4$
7176 4793 rt : (8,3,5,5)
7176 5311 $\zeta : \sqrt{3} - \sqrt{6} + \sqrt{7}$
7176 6450 $\zeta : \exp(-5/3)$
7176 6854 rt : (4,9,-1,-8)
7177 0557 $\ell_2 : 3\sqrt{2} - 3\sqrt{3}/2$
7177 2965 rt : (5,5,-5,2)
7177 4111 rt : (1,7,7,-9)
7177 5151 sr : $3\zeta(3)/7$
7177 6092 rt : (8,7,2,-8)
7177 7959 tπ : $5\zeta(3)/9$
7177 9504 cu : $3 - \sqrt{3}/2$
7178 2422 rt : (7,0,-5,1)
7178 3008 sq : $1/4 + 3\sqrt{2}/4$
7178 4817 $\ell_{10} : 3/4 + 2\sqrt{5}$
7178 5461 $2^x : 6\zeta(3)/5$
7178 8201 $\exp(2/5) - s\pi(e)$
7179 0301 $10^x : 2/3 + 4\pi/3$
7179 1916 rt : (3,7,-1,-4)
7179 2252 $\exp(1/5) \times s\pi(1/5)$
7179 3924 rt : (6,4,1,-5)
7179 5422 rt : (9,1,3,-6)
7179 5763 cu : $2/3 + 4e$
7179 5886 rt : (1,9,2,-4)
7179 7424 sπ : $4\sqrt{2} + 3\sqrt{3}/2$
7179 8328 $\ln(3/4) + s\pi(\pi)$
7179 8943 rt : (4,7,-4,-5)
7179 8999 $\ell_{10} : 4\sqrt{2} - \sqrt{3}/4$
7180 0492 rt : (1,7,5,-7)
7180 1464 rt : (2,1,8,-7)
7180 1867 at : $1/\ln(\pi)$
7180 2544 at : $\sqrt[3]{2/3}$
7180 2975 $\ell_2 : \pi^2/3$
7180 3073 sr : $\sqrt{2}/4 + 3\sqrt{3}/2$
7180 5707 $10^x : \ln(\sqrt{3}/2)$
7180 5716 rt : (9,9,6,3)
7181 0150 $\Psi : 1/\ln(\sqrt{3}/2)$
7181 1865 $Ei : 7\zeta(3)/9$
7181 3061 rt : (7,8,-1,-6)
7181 7910 rt : (7,5,-8,7)
7181 8154 $Ei : 9\sqrt[3]{3}/4$
7181 8891 sq : $3\sqrt{3}/2 + 2\sqrt{5}/3$
7182 0216 rt : (6,2,8,-9)
7182 1465 rt : (1,7,1,-6)
7182 2114 rt : (3,5,6,-8)
7182 3262 rt : (3,5,-6,7)
7182 4405 $J_0 : (\sqrt{5}/3)^{-1/3}$
7182 4583 rt : (8,4,-7,0)
7182 5404 sr : $4\sqrt{2} + \sqrt{3}$
7182 6101 $2^x : 9\sqrt{5}/8$
7182 6968 rt : (2,9,-1,5)
7182 7532 $\ell_{10} : P_{52}$
7182 8182 id : e
7182 9588 rt : (5,7,-9,1)
7182 9617 $K : 4\ln 2/5$
7183 0888 tπ : $2\sqrt{2}/3 + 2\sqrt{5}/3$
7183 0979 $e^x : (e/2)^{-2}$
7183 3306 rt : (4,9,4,-9)
7183 4527 $\Psi : 7\sqrt{5}$
7183 7203 sr : $4 - \pi/3$
7183 8861 sq : $\sqrt{3} - \sqrt{6} + \sqrt{7}$
7183 9541 rt : (9,8,-9,-1)
7183 9828 $K : (e + \pi)^{-1/3}$
7184 0326 rt : (5,2,-9,4)
7184 0408 rt : (7,2,2,3)
7184 2431 cr : $e + 3\pi/4$
7184 3951 as : P_{19}
7184 4152 as : P_{58}
7184 6701 ln : $4/3 + 3\sqrt{2}$
7184 7643 $J_0 : 7\sqrt[3]{2}/8$
7184 9663 $2^x : 1/2 + 2\sqrt{2}/3$
7185 0024 rt : (7,8,-9,-8)
7185 1015 rt : (8,2,0,-4)
7185 3608 cu : $\sqrt{3} + 4\sqrt{5}/3$
7185 6156 rt : (9,6,-2,-5)
7185 6513 sinh : $5\zeta(3)/9$
7185 7700 sr : $-\sqrt{3} + \sqrt{5} + \sqrt{6}$
7185 8687 $Ei : \sqrt{2} + \sqrt{6} + \sqrt{7}$
7185 9167 $2^x : 3\sqrt{2}/2 - 3\sqrt{3}/2$
7185 9191 rt : (5,5,-2,-3)
7186 0862 tanh : 19/21
7186 1569 tπ : $3\sqrt{2} + \sqrt{5}/4$
7186 2383 rt : (2,6,-4,5)
7186 2614 rt : (6,9,-4,-3)
7186 4386 $Li_2 : 6\ln 2/7$
7186 4392 rt : (7,4,9,7)
7186 6189 rt : (8,0,7,8)
7186 6956 $Ei : 18/11$
7186 7215 rt : (9,4,5,-9)
7186 8420 $Ei : 3\sqrt{3}/4$
7186 8559 sinh : 22/9
7187 0591 $\exp(2) \times s\pi(e)$
7187 0969 rt : (2,8,-4,-2)
7187 1646 $2^x : 1/\ln(1/(3e))$
7187 2679 ln : $3/2 + 3e/2$
7187 2777 rt : (8,9,-7,2)
7187 3635 rt : (6,7,3,-8)
7187 3860 $e^x : \sqrt{3}/2 + 3\sqrt{5}/4$
7187 4128 $erf : 16/21$
7187 4463 rt : (5,3,5,-7)
7187 4562 id : $2\sqrt{3} - \sqrt{5}/3$
7187 5000 cu : $7\sqrt[3]{2}/4$
7187 5684 cr : $1 + 3e/2$
7187 6037 rt : (4,1,7,6)
7187 7080 sπ : $3\sqrt{2}/2 - \sqrt{3}/2$
7187 7087 id : $4e/3 + 2\pi/3$
7187 8494 cr : $7(1/\pi)/6$
7188 0086 ln : $6\sqrt[3]{5}/5$
7188 0194 rt : (3,5,9,5)
7188 1552 rt : (3,8,-3,-7)
7188 2911 $10^x : (\sqrt{2\pi}/2)^3$
7188 2999 at : 7/8
7188 5362 rt : (2,6,3,4)
7188 6925 $e^x : 13/24$
7189 1093 $10^x : \sqrt[3]{23/3}$
7189 1208 $s\pi(\sqrt{2}) \times s\pi(\sqrt{3})$
7189 3455 rt : (1,2,5,-5)
7189 7510 $\ell_2 : 3e - \pi/2$
7189 7798 $Ei : 7/12$
7189 7866 $K : 5/9$
7189 8888 cu : $e/3 + 3\pi/4$
7189 9165 cr : $4\sqrt{2} - \sqrt{3}/3$
7189 9862 $\ell_{10} : 5\pi/3$
7190 0370 rt : (1,9,-1,-5)
7190 0527 $\ell_{10} : 3 + \sqrt{5}$
7190 1215 $Ei : 3e/2$
7190 2752 $2^x : (2\pi/3)^{-1/3}$
7190 3972 $Ei : 5/23$
7190 4230 sq : $4 - \pi/3$
7190 5860 rt : (1,7,-8,-9)
7190 7220 $10^x : 4/17$
7190 7792 rt : (9,9,0,-8)
7190 9919 ln : $3\sqrt{3}/2 + 4\sqrt{5}/3$
7191 0505 $\ln(2) \div s\pi(\sqrt{2})$
7191 0830 rt : (7,7,-4,8)
7191 5525 rt : (8,7,-5,-3)
7191 6032 sq : $3 + \pi$
7191 6829 rt : (8,5,2,-7)
7191 7496 rt : (6,6,-8,-1)
7191 8760 sπ : $5(e + \pi)/9$
7191 9020 rt : (4,9,-4,3)
7192 0022 rt : (2,8,-7,-1)
7192 0173 $\pi \div \ln(\sqrt{3})$
7192 2359 rt : (2,7,4,0)
7192 5218 tanh : $e/3$
7192 6433 $Li_2 : \ln(2e/3)$
7192 8232 $\zeta : \sqrt[3]{5}/9$
7192 8766 $e^x : 1/3 + 3e/4$
7192 9566 sinh : 22/5
7193 0300 rt : (5,8,0,-6)
7193 0808 $\ell_{10} : 4\sqrt{2}/3 + 3\sqrt{5}/2$
7193 1519 $e^x : (2\pi)^{-1/3}$
7193 1745 rt : (7,1,4,-6)
7193 1944 $J_0 : \sqrt[3]{4/3}$
7193 3832 rt : (9,6,8,6)
7193 4922 rt : (7,3,-3,-2)
7193 5572 $\ell_2 : 2 - \sqrt{2}/4$
7193 8964 rt : (9,4,1,2)
7193 9655 id : $3\sqrt{2}/2 + 3\sqrt{3}/2$
7194 1062 $Ei : 5\pi^2/9$
7194 2331 rt : (1,8,1,-9)
7194 2571 rt : (2,9,5,-3)
7194 5885 rt : (4,2,9,-9)
7195 2393 rt : (9,2,-6,-2)
7195 2829 rt : (4,8,6,7)
7195 2889 rt : (3,4,-8,9)
7195 3080 rt : (5,6,-8,-7)
7195 3335 sq : $2\sqrt{3} + \sqrt{5}/3$
7195 3549 tanh : $(\sqrt{5}/3)^{1/3}$
7195 4671 cu : $\sqrt{2}/2 + 3\sqrt{3}$
7195 5009 $\ell_{10} : 1 + 3\sqrt{2}$
7195 6026 tπ : $9\sqrt[3]{5}/7$
7195 6455 $2^x : 5\sqrt{3}/6$
7195 8434 ln : $4e/3 - \pi/2$
7195 9221 $10^x : 5\sqrt[3]{3}/7$
7196 1044 rt : (6,3,6,5)
7196 2201 $J_0 : 11/10$
7196 2550 $e^x : \zeta(11)$
7196 2769 $1/3 + \ln(4)$
7196 3329 cr : $\sqrt{5}/6$
7196 4844 $J_0 : 7\sqrt{2}/9$
7196 5583 rt : (4,6,-5,-1)
7196 6195 sπ : $5e\pi/9$
7196 6261 rt : (5,4,-5,5)
7196 6685 rt : (1,5,7,-8)
7196 8095 sinh : $9\pi^2/5$
7196 9490 $e^x : 1/2 + 3\sqrt{3}$

7197 0431 $\zeta : (\sqrt{2})^{-1/2}$
7197 0595 $10^x : 1/4 + \sqrt{3}/3$
7197 1177 sinh : $(\sqrt{5})^{-1/2}$
7197 1542 sq : $3/4 - \pi$
7197 2015 rt : (7,4,6,-9)
7197 4134 cr : $\sqrt{3} + 3\sqrt{5}/2$
7197 4896 id : $2(e + \pi)$
7197 8235 $e^x : 7\sqrt{3}/4$
7197 9766 rt : (5,8,-1,-3)
7198 2395 rt : (1,7,0,-4)
7198 3069 rt : (7,6,-1,-5)
7198 4207 rt : (6,1,-1,1)
7198 4942 rt : (2,5,-8,4)
7198 7974 rt : (5,1,-9,7)
7198 8873 cu : $3/4 + e$
7199 1172 rt : (4,7,4,-8)
7199 4651 rt : (9,1,3,5)
7199 4749 rt : (2,8,-5,-7)
7199 5015 t$\pi : 6\sqrt{5}$
7199 5488 id : $4\sqrt[3]{2}/7$
7199 6082 $K : 7(1/\pi)/4$
7199 7441 $\zeta : 4/21$
7199 8103 t$\pi : 2\zeta(3)/3$
7199 8947 rt : (8,8,-7,5)
7200 0000 id : $18/25$
7200 0444 rt : (7,8,-8,-1)
7200 1192 $\ln(2) - \exp(5)$
7200 2954 rt : (6,0,8,8)
7200 3542 rt : (1,6,3,8)
7200 4342 cr : $9\sqrt{5}$
7200 5891 rt : (8,9,-6,-6)
7200 6696 rt : (7,9,1,-8)
7200 8700 rt : (4,9,-3,-4)
7201 1415 s$\pi : 2/3 + 3e/2$
7201 2213 $\exp(1/3) - s\pi(\sqrt{5})$
7201 3573 at : $(\sqrt{2}\pi/3)^{-1/3}$
7201 5325 rt : (1,9,-7,0)
7201 5930 $\ell_{10} : 21/4$
7201 5998 ln : $3\sqrt{2}/2 + 2\sqrt{3}$
7201 8134 rt : (9,2,5,-8)
7202 0697 rt : (9,1,-4,-1)
7202 0736 rt : (7,9,0,9)
7202 2200 sr : $\sqrt{2}/3 + 4\sqrt{3}$
7202 3861 $2^x : 18/23$
7202 5461 $\ell_{10} : 4\sqrt{3} - 3\sqrt{5}/4$
7202 7521 $J_0 : \ln(3)$
7202 8088 rt : (6,2,1,-4)
7203 0985 rt : (8,8,3,1)
7203 2106 rt : (4,8,-6,5)
7203 3731 rt : (3,3,6,-7)
7203 9133 sq : $4 + e/2$
7204 0346 rt : (9,4,-2,-4)
7204 3462 sr : $2\sqrt{2} - 4\sqrt{3}/3$
7204 3564 rt : (6,5,3,-7)
7204 4052 tanh : $(4/3)^{-1/3}$
7204 4930 rt : (5,7,8,4)
7204 6731 cu : $4/3 + 3\sqrt{2}/4$
7204 8959 sr : $3\sqrt{3} - \sqrt{5}$
7204 9680 rt : (9,7,0,-7)
7205 0619 sq : $8(1/\pi)/3$
7205 0631 $erf : 13/17$
7205 0779 s$\pi : (e/3)^3$
7205 1801 $erf : (\sqrt[3]{5})^{-1/2}$
7205 6079 t$\pi : 5\sqrt[3]{3}/9$
7205 7024 $\exp(\pi) \div s\pi(1/3)$
7205 8699 rt : (8,5,5,4)
7205 9717 cr : P_{74}
7206 5268 rt : (8,2,7,7)
7206 6756 sinh : $\ln((e + \pi)/3)$
7206 6982 cr : $3 + 2\pi/3$
7206 8851 $10^x : \sqrt{2}/6$
7206 8904 $J : \exp(-11\pi/8)$
7206 9563 tanh : $10/11$
7206 9904 sr : $3\pi^2/4$
7207 1208 $\zeta : 3(1/\pi)/5$
7207 2116 sr : $3\pi^2/10$
7207 2487 cr : $e/3 + 4\pi/3$
7207 5440 sq : $3\sqrt[3]{3}$
7207 5922 rt : (9,9,-7,-3)
7207 7305 $\Psi : (\ln 2/2)^2$
7207 7757 $e^x : \sqrt{2} + 3\sqrt{5}/2$
7207 9476 $\ell_2 : 9(e + \pi)$
7208 3121 $e^x : 1/4 - \sqrt{3}/3$
7208 4342 sr : $3\sqrt{3}/10$
7208 6114 $\zeta : (\sqrt{2}\pi/2)^{-1/3}$
7208 9649 $2^x : 4 - 2\sqrt{5}$
7209 1025 at : $3 - 3\sqrt{2}/2$
7209 1716 t$\pi : \sqrt{2}/3 + 3\sqrt{3}$
7209 4074 rt : (8,6,4,-9)
7209 6591 $\ell_2 : 8\sqrt[3]{3}/7$
7209 6846 cr : $1/\sqrt[3]{19}$
7209 8099 rt : (5,4,7,-9)
7210 0181 rt : (8,3,2,-6)
7210 0949 $10^x : 1/2 + 2\sqrt{2}/3$
7210 0968 ln : $5\sqrt{5}/2$
7210 1344 sr : $3\ln 2/4$
7210 3150 rt : (8,7,-7,8)
7210 3492 $\ell_2 : 3\sqrt{2} + 4\sqrt{5}$
7210 5065 t$\pi : \sqrt{2}/4 - 2\sqrt{3}/3$
7210 5727 cr : $4\sqrt{2} - \sqrt{5}/4$
7210 6638 $\sqrt{2} - \ln(2)$
7210 7111 $Ei : 7\ln 2/2$
7210 7950 rt : (5,1,5,-6)
7211 1025 sr : $13/25$
7211 1716 $Ei : 9\ln 2$
7211 2478 id : $\sqrt[3]{3}/2$
7211 3335 rt : (5,3,-5,8)
7211 5978 $2^x : (\sqrt[3]{3}/3)^{1/3}$
7211 8855 rt : (1,8,8,-4)
7212 0191 $\exp(4) \times s\pi(\sqrt{3})$
7212 0288 rt : (2,8,1,-4)
7212 0486 $4 \times s\pi(\pi)$
7212 1072 $J_0 : \pi^2/9$
7212 1609 id : $3e + 4\pi$
7212 2650 tanh : $\sqrt{2} + \sqrt{3} - \sqrt{5}$
7212 2825 id : $3e/2 - 3\pi/4$
7212 3045 rt : (2,1,8,6)
7212 3414 id : $3\zeta(3)/5$
7212 3723 rt : (5,9,2,-8)
7212 4705 tanh : $1/\ln(3)$
7212 5250 $\ell_{10} : \sqrt[3]{5}/9$
7212 6851 rt : (4,7,-8,7)
7212 7344 $K : \sqrt{5}/4$
7212 7732 rt : (8,8,-3,-5)
7212 8688 $2^x : \sqrt{21}\pi$
7213 3539 rt : (5,2,3,3)
7213 3876 $10^x : 10/23$
7213 4143 ln : $3 - 2\sqrt{2}/3$
7213 4752 id : $1/\ln(4)$
7213 5693 rt : (2,7,5,-8)
7213 7348 rt : (3,9,-2,-5)
7213 8759 $\exp(1/3) - \exp(3/4)$
7213 9888 $Ei : 13/10$
7214 0275 $1/2 + \exp(1/5)$
7214 0452 id : $1/4 + \sqrt{2}/3$
7214 0693 $\Gamma : 3\zeta(3)/7$
7214 1145 $\ell_2 : 7\ln 2/8$
7214 2093 rt : (5,9,3,4)
7214 8444 t$\pi : 5(1/\pi)/8$
7215 0023 rt : (8,5,-5,-2)
7215 0764 sr : $-\sqrt{2} + \sqrt{3} + \sqrt{7}$
7215 2469 cr : $4 - 4e/3$
7215 3914 rt : (7,2,6,-8)
7215 5671 rt : (1,5,-8,-8)
7215 6863 rt : (7,6,9,6)
7215 6937 $e^x : 1/2 + 4\pi/3$
7215 6963 $erf : (\sqrt{2}\pi/2)^{-1/3}$
7215 8000 $2^x : 13/9$
7215 9487 $\exp(1/4) \div s\pi(\sqrt{3})$
7216 0930 cr : $3\sqrt{2}/2 + 4\sqrt{5}/3$
7216 5485 at : $22/25$
7216 6010 $\Gamma : \ln(1/(2\pi))$
7216 7480 rt : (2,4,3,-5)
7216 7672 $\exp(2/3) + s\pi(e)$
7216 7792 $Ei : ((e + \pi)/2)^{-1/2}$
7216 8386 $2^x : 3\sqrt{2} - \sqrt{5}/2$
7216 9256 rt : (3,7,-8,1)
7216 9585 s$\pi : 2\pi/5$
7217 1463 sinh : $4e\pi/3$
7217 2388 rt : (1,6,-5,-1)
7217 3217 rt : (2,9,-4,-9)
7217 4795 rt : (7,7,1,-7)
7217 6250 rt : (1,9,3,5)
7217 6407 $\theta_3 : \sqrt{3}/5$
7217 7262 rt : (4,0,9,8)
7217 7862 $\exp(e) \div s\pi(\sqrt{2})$
7217 8415 $10^x : 3 - \pi$
7218 3375 rt : (5,3,-2,-2)
7218 3704 sq : $1/2 - 2e/3$
7218 5319 $\zeta : 10\sqrt{3}/9$
7218 6109 $J_0 : 23/21$
7218 6277 rt : (6,7,-5,-5)
7219 0480 rt : (7,1,4,5)
7219 0686 $K : 4\sqrt[3]{2}/9$
7219 1853 rt : (7,9,-7,-3)
7219 1977 $10^x : 4\sqrt{3} + 3\sqrt{5}$
7219 3035 $K : 14/25$
7219 4297 rt : (9,8,2,-9)
7219 4795 sinh : $1/\ln(\sqrt{2}\pi)$
7219 4961 $J : \exp(-3/2)$
7219 7945 rt : (2,9,-2,-4)
7219 9435 rt : (6,4,-8,5)
7219 9920 $\Psi : 5\pi$
7220 0149 $2^x : \exp(1/e)$
7220 0397 $2^x : 2 + 3\sqrt{2}$
7220 0639 rt : (4,5,4,-7)
7220 0811 ln : $\sqrt{2} + \sqrt{3} + \sqrt{6}$
7220 1411 $2^x : 7\pi/8$
7220 3847 tanh : $(\pi/3)^{-2}$
7220 4544 $\ell_2 : 2 + 3\sqrt{3}/4$
7220 5900 rt : (8,2,-7,1)
7220 7746 s$\pi : 3\sqrt{2}/4 + 3\sqrt{3}$
7220 8558 cu : $4\sqrt{3} + \sqrt{5}/4$
7220 8739 cu : $4\pi/9$
7220 9568 $2^x : 4/3 + e/3$
7221 0231 rt : (8,9,-3,9)
7221 1639 sq : $(3\pi)^3$
7221 1990 rt : (2,4,6,3)
7221 3082 $\ell_{10} : 9(e + \pi)$
7221 3595 id : $1/4 + 2\sqrt{5}$
7221 4134 rt : (9,0,5,7)
7221 4591 rt : (7,4,-1,-4)
7221 4592 $\theta_3 : \ln 2/2$
7221 5296 rt : (5,8,-7,-1)
7221 7316 $Ei : 3(e + \pi)/8$
7221 7759 rt : (6,6,5,-9)
7221 8741 rt : (4,2,-9,-1)
7221 9294 $Ei : (e + \pi)^{-3}$
7222 0069 $Li_2 : (3\pi/2)^{-1/3}$
7222 0318 rt : (3,6,-7,-3)
7222 2222 id : $13/18$
7222 2691 s$\pi : \exp(-e/2)$
7222 2868 ln : $3/2 + \sqrt{5}/4$
7222 4563 rt : (1,5,1,-8)
7222 7503 cr : $3\sqrt{2} + \sqrt{3}/2$
7222 7613 sinh : $3\sqrt{5}/10$
7222 8293 rt : (4,8,-6,-7)
7222 9269 $e^x : e/5$
7222 9464 rt : (9,5,0,-6)
7223 0246 t$\pi : 7\pi/10$
7223 0361 rt : (6,7,-4,3)
7223 1511 sr : $12/23$
7223 2726 rt : (7,9,-6,-3)
7223 5172 rt : (3,6,0,-2)
7223 6392 $\zeta : 5\sqrt{5}/9$
7223 6482 $2^x : \sqrt{19/3}$
7223 8943 $1/3 + \exp(2)$
7223 9242 $\ell_2 : 3\sqrt{2}/7$
7223 9609 $\theta_3 : 1/\sqrt[3]{24}$
7224 1652 $\sqrt{2} \times \ln(3/5)$
7224 1943 sr : $8\sqrt{3}$
7224 2154 sinh : $(\sqrt{2}\pi/2)^{-1/2}$
7224 2300 ln : $3 + 3\sqrt{3}/2$
7224 2548 rt : (2,9,-7,-9)
7224 4112 $\exp(4/5) \times s\pi(e)$
7224 4819 rt : (1,3,7,-7)
7224 9111 rt : (4,3,7,5)
7225 0000 sq : $17/20$
7225 0977 $J : \zeta(3)/8$
7225 2849 rt : (3,4,8,-9)

7225 4039 rt : (2,7,-3,-5)
7225 6343 rt : (1,7,-1,-4)
7225 7688 rt : (6,2,8,7)
7225 8982 tanh : 21/23
7225 9739 10^x : $e/3-\pi/3$
7226 0279 rt : (7,9,7,3)
7226 1078 exp(2/5) ÷ $s\pi(1/3)$
7226 1501 rt : (8,1,9,9)
7226 1902 rt : (1,8,2,-6)
7226 4465 rt : (3,9,3,2)
7226 5560 $s\pi$: $8\sqrt{2}/9$
7226 5625 sq : 21/16
7226 5941 rt : (5,5,-1,9)
7226 6176 rt : (9,8,8,5)
7226 6811 10^x : $8\sqrt[3]{2}/5$
7226 8097 2^x : $7e\pi/3$
7226 8829 ln(5) + $\exp(\sqrt{2})$
7227 0623 ln(4) ÷ $\ln(\sqrt{5})$
7227 1417 rt : (8,4,4,-8)
7227 1495 ζ : $9\sqrt[3]{5}/8$
7227 4000 e^x : $4\sqrt{3}+\sqrt{5}/3$
7227 4979 at : $7\sqrt[3]{2}/10$
7227 6175 at : $2-\sqrt{5}/2$
7227 9532 rt : (8,9,-1,-7)
7228 1919 rt : (6,8,-2,-5)
7228 2054 $s\pi$: $(2e/3)^{-1/2}$
7228 2692 rt : (9,3,3,4)
7228 3641 ζ : $\sqrt{3}/9$
7228 5910 K : $\exp(-\sqrt{3}/3)$
7228 6296 rt : (4,7,-3,-3)
7228 6885 e^x : $2\sqrt{3}+\sqrt{5}/2$
7228 7022 sr : $1/4+e$
7228 7271 $s\pi$: $\sqrt[3]{23/2}$
7228 7904 ln : $10\sqrt[3]{3}/7$
7229 1442 rt : (7,4,2,2)
7229 2689 3/4 × $s\pi(\sqrt{2})$
7229 3194 cu : $3/2-3e$
7229 3201 $t\pi$: $3\sqrt{2}/2+2\sqrt{5}/3$
7229 4241 rt : (3,9,-7,4)
7229 4297 rt : (9,2,-2,-3)
7229 4519 e^x : $5\zeta(3)/6$
7229 6122 cu : $1/4-3e/4$
7229 6926 rt : (2,6,-4,-1)
7229 7935 at : 15/17
7229 8480 rt : (6,3,-8,8)
7229 9688 10^x : $7\sqrt[3]{5}/5$
7230 0695 Ψ : $8(1/\pi)/9$
7230 2002 sr : $1/\sqrt[3]{7}$
7230 2640 ln : $1+3\sqrt[3]{2}/4$
7230 2757 cr : $3\sqrt[3]{2}/10$
7230 2897 rt : (9,7,-7,-2)
7230 5997 rt : (2,4,-1,8)
7230 6732 e^x : $1/\ln(2\pi)$
7230 7846 rt : (7,1,-3,-1)
7230 8588 $s\pi$: $7(1/\pi)/3$
7230 8694 λ : $(1/\pi)/4$
7230 9232 Γ : $(\ln 3/2)^{-2}$
7230 9697 2^x : $4-\sqrt{3}/3$
7231 0681 sr : $\sqrt{2}/2+3\sqrt{5}$
7231 1513 ln : $2\sqrt{2}/3+\sqrt{5}/2$
7231 4233 rt : (7,8,3,-9)
7231 4355 $1/2-\ln(4/5)$
7231 5219 sinh : 21/16
7231 6707 ln(5) ÷ exp(4/5)
7231 7846 cu : $2\pi/7$
7231 7967 rt : (7,8,-7,0)
7231 8786 ln : $7\ln 2/10$
7231 9402 ζ : $8\zeta(3)/5$
7232 0308 sq : $\sqrt{3}-\sqrt{5}-\sqrt{6}$
7232 0890 Ei : $7\sqrt{2}/2$
7232 1857 rt : (5,2,7,-8)
7232 6200 rt : (5,8,3,7)
7232 6463 $s\pi$: $1/2+3\sqrt{2}$
7233 0833 ℓ_2 : $\sqrt[3]{9/2}$
7233 1872 rt : (3,5,-9,-1)
7233 2192 Ψ : 11/12
7233 2221 rt : (7,6,-9,-7)
7233 2711 cr : $4+\sqrt{5}/2$
7233 4157 erf : 10/13
7233 6374 K : $\ln(\sqrt[3]{5}/3)$
7233 9838 sq : $4-\sqrt{3}/4$
7234 2618 rt : (2,6,-5,-6)
7234 3171 ζ : 25/13
7234 3683 10^x : $\exp(-\sqrt[3]{3})$
7234 3905 rt : (1,5,3,8)
7234 4894 10^x : $6\zeta(3)/7$
7234 5020 rt : (3,1,6,-6)
7234 6506 ζ : $\sqrt{5}/4$
7234 6859 rt : (8,4,7,6)
7234 6900 $t\pi$: $1/3+\sqrt{3}/2$
7234 7341 e^x : $4-2\pi/3$
7234 7452 rt : (9,9,-6,9)
7234 8789 rt : (8,1,2,-5)
7234 9212 Ei : $1/\sqrt[3]{5}$
7235 0245 rt : (8,6,-3,-4)
7235 0651 ln(3) − exp(3/5)
7235 1777 ζ : $4\sqrt[3]{3}/3$
7235 2340 ℓ_{10} : $4\sqrt{3}/3+4\sqrt{5}/3$
7235 6793 2^x : $\pi/4$
7235 8010 rt : (1,8,-7,-6)
7235 8371 rt : (2,6,-7,-6)
7236 0125 sr : $\pi/6$
7236 0679 rt : (5,5,1,0)
7236 1149 K : 9/16
7236 1463 id : $\sqrt{2}+4\sqrt{3}/3$
7236 4942 5 × $\ln(\pi)$
7236 5420 rt : (9,6,2,-8)
7236 5750 rt : (4,1,-9,2)
7236 5889 rt : (9,1,7,-9)
7236 6291 rt : (4,6,6,-9)
7236 7334 $t\pi$: $2\sqrt{3}/3-3\sqrt{5}/2$
7236 8056 10^x : $7\sqrt[3]{3}/3$
7236 8125 J_0 : $1/\ln(5/2)$
7236 9589 cr : $3+3\sqrt{2}/2$
7236 9708 erf : $4\sqrt{3}/9$
7237 3033 θ_3 : $(4/3)^{-1/3}$
7237 4669 e^x : $\zeta(9)$
7237 4686 sr : 11/21
7237 6134 sinh : 19/11
7237 6749 ln : $1/4+2e/3$
7237 9064 e^x : $4+e/4$
7238 2413 ℓ_{10} : exp(5/3)
7238 3925 at : $5\sqrt{2}/8$
7238 5002 rt : (6,5,-4,-2)
7238 9085 J_0 : 12/11
7238 9774 2^x : $5\sqrt{2}/9$
7239 0218 rt : (6,0,1,3)
7239 0888 Ei : $(\sqrt{2}\pi/2)^{-3}$
7239 2359 erf : $10\ln 2/9$
7239 2705 rt : (7,5,1,-6)
7239 4564 2^x : 11/14
7239 5648 rt : (1,5,0,-3)
7239 6102 exp(e) + $s\pi(\pi)$
7239 6123 sq : $1/4+4e$
7239 6919 rt : (7,0,6,7)
7239 7859 cr : $3e\pi/5$
7239 8019 rt : (4,9,-7,-5)
7239 8285 rt : (6,5,6,4)
7239 8325 tanh : $\beta(2)$
7239 8710 $s\pi$: $3\sqrt{3}/7$
7239 8883 rt : (3,8,1,4)
7240 0735 rt : (2,5,5,-7)
7240 1646 id : $3\sqrt{3}-2\sqrt{5}$
7240 2153 cu : $4-3e$
7240 2532 Ψ : ln(5/2)
7240 3861 cr : $3/2+4e/3$
7240 4722 Ψ : $\sqrt{2}+\sqrt{5}+\sqrt{6}$
7240 5414 Ψ : $5e\pi/7$
7240 6186 2^x : 25/8
7240 6603 cr : $3e/2+\pi/3$
7240 7297 J : 2/5
7240 7555 rt : (1,0,5,4)
7241 0710 Ei : $\sqrt{3}/8$
7241 3793 tanh : ln(5/2)
7241 4148 rt : (3,5,-9,-8)
7241 4260 π ÷ exp(3/5)
7241 6020 rt : (7,7,-7,3)
7241 7467 id : $2e-3\pi/2$
7241 9704 ℓ_{10} : $4+3\sqrt{3}/4$
7242 0604 e^x : $1/4+2\sqrt{2}$
7242 2113 10^x : $3/4+4\sqrt{3}$
7242 2907 rt : (4,5,-4,-4)
7242 3801 at : $\exp(-1/(3e))$
7242 4009 rt : (6,9,0,4)
7242 9495 as : P_{55}
7243 0809 rt : (5,9,-5,-3)
7243 1016 rt : (6,4,5,-8)
7243 1224 cr : $2\sqrt[3]{5}/9$
7243 1668 tanh : 11/12
7243 1721 rt : (1,8,2,-1)
7243 3402 cu : $4\sqrt{5}/5$
7243 4376 rt : (9,4,-9,1)
7243 5503 π × exp(3/5)
7243 6395 Ψ : $\sqrt{2}/5$
7243 7147 rt : (1,9,4,-8)
7243 7660 cr : $7(e+\pi)/8$
7243 9329 rt : (6,5,-4,9)
7244 2302 rt : (5,4,0,-4)
7244 4114 e^x : $9\sqrt{5}/7$
7244 4372 $t\pi$: $6\sqrt[3]{2}/5$
7244 6406 at : $(\ln 2)^{1/3}$
7244 7610 rt : (1,4,9,-9)
7244 7684 e^x : $3\sqrt{3}/4+\sqrt{5}/3$
7245 0365 sinh : $4(e+\pi)/9$
7245 1515 at : $(\sqrt[3]{3})^{-1/3}$
7245 3200 10^x : $3/4+3\sqrt{5}/2$
7245 3303 rt : (9,8,-5,-4)
7245 3574 Γ : 17/6
7245 5739 rt : (5,8,4,-9)
7245 7181 rt : (4,9,-8,-9)
7245 7672 rt : (2,8,-6,4)
7245 8306 rt : (5,1,5,5)
7245 9048 ln(4/5) + exp(2/3)
7246 1381 $t\pi$: $4e/3-3\pi$
7246 3123 $t\pi$: P_{26}
7246 3403 Ψ : $\beta(2)$
7246 4008 ℓ_2 : $(\ln 3/3)^{1/2}$
7246 4507 rt : (9,3,0,-5)
7246 6103 cr : $3\sqrt[3]{5}$
7246 6472 sr : $3+4e$
7246 7970 $s\pi$: $(\pi/2)^{-3}$
7247 0052 at : P_{54}
7247 0147 Ei : $10\zeta(3)/3$
7247 0238 rt : (1,7,2,1)
7247 2161 e^x : $1/3+3\pi/4$
7247 2165 $t\pi$: $10e\pi/7$
7247 3014 ℓ_2 : $\sqrt{2}+3\sqrt{3}$
7247 4487 rt : (4,8,-1,-5)
7247 4790 id : $\sqrt{2}/3-3\sqrt{3}$
7247 5627 K : $(\pi)^{-1/2}$
7247 8799 rt : (9,2,5,6)
7248 1223 at : $1-4\sqrt{2}/3$
7248 5313 rt : (9,6,1,1)
7248 6214 rt : (8,7,-1,-6)
7248 6724 Ei : 3π
7249 1897 rt : (4,3,4,-6)
7249 2024 cr : 8/21
7249 2033 Ei : $3\pi^2/2$
7249 2235 2^x : $3-2\sqrt{3}$
7249 3030 $s\pi$: P_{28}
7249 4475 rt : (7,6,-7,6)
7249 5820 rt : (8,3,-5,-1)
7249 6380 2^x : $\sqrt{2}/4+\sqrt{3}/4$
7249 7129 rt : (5,9,-3,-5)
7249 9604 rt : (3,7,-1,6)
7249 9920 $s\pi(2/5)+s\pi(e)$
7250 0049 λ : 2/25
7250 0640 rt : (8,2,4,-7)
7250 2683 ℓ_{10} : P_{36}
7250 4553 ℓ_{10} : $3+4\sqrt{3}/3$
7250 7274 ζ : ln(3)
7250 7506 θ_3 : 8/23
7250 8278 rt : (1,9,6,0)
7250 8356 $s\pi$: $1/\sqrt{15}$
7251 1093 Ei : 15/16
7251 2832 sr : $2/3+4\sqrt{3}/3$
7251 3368 rt : (3,7,3,-7)
7251 3451 sr : $3\sqrt{3}/4+3\sqrt{5}/4$
7251 5614 rt : (4,2,9,7)
7251 5940 2^x : $7\sqrt{3}$

7251 6283 rt : (7,6,3,-8)
7251 7087 rt : (3,8,-6,-1)
7251 8821 rt : (8,3,9,8)
7252 0260 rt : (7,7,-6,-2)
7252 3122 cu : $1/3+\sqrt{3}/2$
7252 4235 $\sqrt{5}+\ln(3/5)$
7252 6322 ln : P_{42}
7252 6619 Γ : $1/\ln(\sqrt[3]{2}/2)$
7252 6738 Ψ : $9\sqrt{2}/5$
7252 7008 rt : (8,2,0,2)
7252 8010 tπ : $10\sqrt[3]{2}/3$
7252 8027 sinh : $\exp(\sqrt{5})$
7252 8039 rt : (6,8,0,7)
7252 8059 cu : $\ln(3\pi/2)$
7253 1269 cr : $1/\sqrt[3]{18}$
7253 1623 ln : $1/3+\sqrt{3}$
7253 2994 10^x : $\exp(\sqrt{2}\pi/2)$
7253 3318 rt : (7,1,8,-9)
7253 4978 tπ : $3e/2+3\pi/4$
7253 5280 rt : (3,2,8,-8)
7253 9013 rt : (1,3,-8,-7)
7253 9247 e^x : 6/11
7253 9443 sπ : $\sqrt{14}$
7254 2412 e^x : 23/8
7254 4873 rt : (1,6,0,-9)
7254 5607 K : 13/23
7254 7302 sr : $1/2+4\sqrt{3}$
7254 7625 sr : 10/19
7254 7913 rt : (7,2,-1,-3)
7254 8220 rt : (2,6,7,-9)
7254 8892 rt : (5,4,-8,-6)
7254 9470 ℓ_2 : $\sqrt{3}+\sqrt{5}+\sqrt{7}$
7255 1898 erf : 17/22
7255 4092 $1/2+\exp(4/5)$
7255 4593 cr : $2\sqrt{2}+4\sqrt{3}/3$
7255 5589 rt : (2,7,-7,-8)
7255 6263 cr : $3/2-\sqrt{5}/2$
7255 6633 cr : $6(1/\pi)/5$
7255 7290 J_0 : $2e/5$
7255 8211 rt : (4,1,-9,-8)
7255 8691 rt : (7,5,-7,9)
7255 9799 10^x : $\sqrt{2}/3+\sqrt{5}/4$
7256 0479 rt : (9,9,-3,-6)
7256 4181 id : $3e+\pi/2$
7256 5501 cr : $2/3+2\sqrt{5}$
7256 6719 ℓ_2 : $4e+3\pi/4$
7256 6942 K : $\exp(-\sqrt[3]{5}/3)$
7256 8088 sq : $\sqrt[3]{9/2}$
7256 8867 rt : (5,5,-8,-3)
7256 9614 $\pi\times\ln(\sqrt{3})$
7256 9657 Ψ : $8(1/\pi)$
7256 9665 K : $1/\ln(e+\pi)$
7257 0811 rt : (6,8,-7,-7)
7257 2112 Ψ : $1/\ln(\sqrt{2}\pi/3)$
7257 2268 sinh : $\sqrt[3]{23/2}$
7257 3929 J_0 : 25/23
7257 4863 rt : (2,7,-2,-3)
7257 5197 10^x : $3\sqrt{2}/2+4\sqrt{3}$
7257 5558 Γ : $7e/5$
7257 6943 rt : (8,7,5,3)
7257 7560 K : $2\sqrt{2}/5$
7257 7911 rt : (3,6,-3,8)
7258 0015 sq : $4\sqrt{3}+3\sqrt{5}/2$
7258 0854 rt : (9,4,2,-7)
7258 4698 rt : (8,8,1,-8)
7258 7188 rt : (7,3,4,4)
7258 7671 rt : (7,9,-3,4)
7258 8136 $\sqrt{2}\times\exp(\pi)$
7258 9741 tanh : 23/25
7259 1740 rt : (8,1,-9,-4)
7259 1802 cr : $2\sqrt{3}+3\sqrt{5}/4$
7259 2567 sr : $e+3\pi/2$
7259 2688 rt : (6,1,3,-5)
7259 2976 2^x : $\ln(\sqrt[3]{2}/2)$
7259 6120 rt : (5,5,2,-6)
7259 6726 cr : $2+\pi$
7259 8118 tπ : $4\pi^2$
7259 8466 rt : (4,9,1,-7)
7259 9997 tπ : $4\sqrt{3}/3+2\sqrt{5}/3$
7260 2176 2^x : $5\sqrt[3]{2}/8$
7260 5871 rt : (9,1,7,8)
7260 6168 ln : $e/3+3\pi/2$
7260 6583 rt : (8,3,6,-9)
7260 7232 ℓ_2 : $1/3+2\pi$
7260 8145 $\sqrt{2}\div\exp(2/3)$
7260 9105 tπ : $\exp(-\sqrt[3]{2}/2)$
7261 0944 rt : (2,8,2,-2)
7261 7258 rt : (8,9,-8,-2)
7261 8663 rt : (9,5,-7,-1)
7262 0989 sr : $3\sqrt{3}+\sqrt{5}$
7262 3479 rt : (1,6,2,-5)
7262 3513 2^x : $2e\pi/9$
7262 4062 tπ : 4π
7262 5372 rt : (3,8,5,-9)
7262 7639 ℓ_{10} : $2-2e/3$
7262 7939 rt : (4,4,6,-8)
7263 0254 2^x : $3/4+2\sqrt{5}/3$
7263 1239 rt : (1,7,-6,-3)
7263 1618 rt : (5,0,7,7)
7263 3128 $\exp(1/5)-\exp(2/3)$
7263 3677 J : 7/25
7263 7328 rt : (6,4,8,6)
7264 0247 sq : $5e$
7264 1323 cu : $3\sqrt{2}/4-3\sqrt{3}$
7264 3085 $\exp(4/5)\times\exp(e)$
7264 3792 $\ln(5)+\exp(3/4)$
7264 4129 as : $3/4-\sqrt{2}$
7264 4669 at : $3\sqrt{2}-3\sqrt{5}/2$
7264 5520 id : $\exp((e+\pi)/2)$
7264 7026 rt : (7,8,-4,-4)
7264 7395 rt : (7,8,9,5)
7264 7907 $\sqrt{2}-\exp(\pi)$
7264 8661 sπ : $(\ln 3/2)^{1/2}$
7265 0197 ln : P_{32}
7265 0435 sinh : $8\pi^2/7$
7265 2343 rt : (9,5,-8,-5)
7265 4252 tπ : 1/5
7265 4414 rt : (1,9,-5,-8)
7265 4479 rt : (7,8,-3,7)
7265 4549 rt : (8,4,-3,-3)
7265 5550 rt : (2,6,-3,9)
7265 7014 ℓ_2 : $1+4\sqrt{3}/3$
7265 7999 sπ : $1/\ln(e/2)$
7265 8759 sq : $1/4+3e/2$
7266 1205 ln : $\sqrt{3}/3+2\sqrt{5}/3$
7266 2981 Ei : $9\sqrt[3]{5}/7$
7266 4234 at : 8/9
7266 4991 rt : (5,9,-4,-8)
7266 5163 tπ : $((e+\pi)/3)^{-1/3}$
7266 5493 rt : (9,5,4,-9)
7266 6035 rt : (9,0,-2,2)
7266 7832 erf : $(5/3)^{-1/2}$
7266 8004 sr : $4\sqrt{5}/3$
7266 9882 rt : (4,6,5,2)
7267 1233 10^x : $\sqrt{2}\pi$
7267 1700 cr : $2\sqrt{2}/3-\sqrt{5}/4$
7267 3278 Γ : $9\sqrt[3]{2}/4$
7267 3781 at : P_{43}
7267 3857 rt : (6,7,0,-6)
7267 6229 e^x : $1+e/3$
7267 6800 id : $e/4+\pi/3$
7267 7996 id : $5\sqrt{5}/3$
7267 8740 rt : (4,9,-3,6)
7267 8794 2^x : $(2\pi/3)^{1/2}$
7268 2372 rt : (1,1,7,-6)
7268 2966 e^x : 19/2
7268 4921 ln : $e+4\pi$
7268 5667 ℓ_{10} : P_9
7268 6201 rt : (7,3,1,-5)
7268 6380 rt : (4,2,-5,6)
7268 6777 rt : (3,7,-2,-4)
7268 9287 rt : (3,9,-4,-3)
7269 1428 sq : $e/3+3\pi$
7269 1907 rt : (1,8,-6,3)
7269 2188 rt : (6,2,-6,1)
7269 4608 rt : (8,9,-6,4)
7269 5601 $\ln(4)-\exp(\sqrt{2})$
7269 5736 Ψ : $9e/4$
7269 8174 rt : (5,6,4,-8)
7269 9872 ℓ_{10} : 3/16
7270 0635 sr : $2+2e$
7270 1385 Ei : $(5/3)^{-3}$
7270 2418 10^x : $4/3+\sqrt{5}/2$
7270 2874 10^x : $2\sqrt{3}/3+\sqrt{5}/4$
7270 3824 2^x : $7\pi/2$
7270 4201 rt : (2,5,-7,8)
7270 4567 rt : (8,1,2,4)
7270 4572 sr : $1-\sqrt{2}/3$
7270 6216 e^x : $4e/7$
7270 7953 sq : $2e+3\pi/4$
7270 8403 tπ : $10\sqrt[3]{2}/7$
7270 9539 as : $e/3-\pi/2$
7270 9721 ln : $3\sqrt{2}/4-\sqrt{3}/3$
7271 0996 ln : $2+4e/3$
7271 2373 Ei : $(e+\pi)/10$
7271 2439 Γ : $\exp(-2/3)$
7271 3120 rt : (4,7,-1,-2)
7271 4339 rt : (6,2,5,-7)
7271 4909 Ψ : $\sqrt{13/2}$
7271 6239 rt : (2,3,8,5)
7271 6585 10^x : $4e/3-\pi/3$
7271 7132 2^x : 11/4
7271 8440 rt : (1,6,-4,6)
7272 2121 Ei : $2(1/\pi)/7$
7272 2199 rt : (2,8,0,-5)
7272 2976 e^x : $3\sqrt{2}/2-\sqrt{5}/2$
7272 2996 ℓ_2 : $e/9$
7272 3573 rt : (4,8,-5,8)
7272 3630 cr : 5/13
7272 7272 id : 8/11
7272 7722 sq : $6\pi^2$
7273 1000 rt : (1,9,-8,-9)
7273 1444 rt : (9,6,-5,-3)
7273 2434 $\sqrt{2}\times\exp(1/5)$
7273 2437 ℓ_{10} : $5e\pi/8$
7273 2683 id : $1/3-3\sqrt{2}/4$
7273 4041 rt : (7,9,-2,-6)
7273 4778 rt : (9,5,3,3)
7273 4977 tanh : 12/13
7273 5813 rt : (5,3,-8,3)
7273 6298 10^x : $\sqrt{8}\pi$
7273 7734 id : $\exp(-1/\pi)$
7273 7854 rt : (7,2,6,6)
7273 7909 $\pi-\sqrt{2}$
7273 8961 id : P_{73}
7273 9169 ℓ_{10} : $\sqrt{3}/2+2\sqrt{5}$
7273 9529 rt : (5,1,9,-9)
7274 0061 10^x : $1/4+\pi$
7274 1123 rt : (8,8,-6,7)
7274 1575 id : $(\sqrt[3]{4}/3)^{1/2}$
7274 2963 J_0 : 13/12
7274 3274 cr : $4+2\sqrt{3}/3$
7274 3563 sπ : $1/4+2\sqrt{5}/3$
7274 4570 sr : P_3
7274 5204 rt : (6,9,-9,-9)
7274 6648 Ψ : $8\pi^2/5$
7274 7843 ℓ_2 : $e/4+4\pi$
7275 1092 rt : (1,7,4,-7)
7275 1994 rt : (4,1,-5,9)
7275 3013 cr : $5\ln 2/9$
7275 3222 rt : (6,8,2,-8)
7275 4738 ln : $2\sqrt{3}/3+2\sqrt{5}$
7275 4991 rt : (8,5,-1,-5)
7275 7111 $2/3\div\ln(2/5)$
7275 7764 rt : (5,1,-2,-1)
7275 8630 Li_2 : 3/5
7275 9372 10^x : 4/7
7275 9402 Ei : $2(e+\pi)/9$
7276 0687 sr : 9/17
7276 5636 tπ : P_{16}
7276 6052 10^x : $1/3-\sqrt{2}/3$
7276 6193 Ei : 1/11
7276 6239 tπ : P_{40}
7276 9273 e^x : 25/19
7276 9962 rt : (9,9,-9,4)
7277 2121 e^x : $\sqrt{2}-\sqrt{3}$
7277 4162 rt : (7,2,-5,-2)
7277 5203 rt : (8,6,7,5)
7277 6267 rt : (7,4,3,-7)
7277 6395 at : $(\sqrt[3]{2})^{-1/2}$

7277 6671 rt : (6,8,-9,0)
7277 8451 $\ell_2 : 3+4e/3$
7277 8705 $e^x : 5\pi/9$
7277 9535 rt : (2,9,-3,-4)
7278 0307 $J_0 : 5\sqrt{3}/8$
7278 2346 as : P_{81}
7278 2857 cr : $4e+3\pi$
7278 4525 rt : (9,1,0,-4)
7278 5448 $K : (2e)^{-1/3}$
7278 7178 rt : (2,3,5,-6)
7278 7650 tanh : $4\ln 2/3$
7278 9267 $\sqrt{5}+\exp(2/5)$
7279 0180 rt : (5,2,-8,6)
7279 1296 $\ln(4/5)+s\pi(2/5)$
7279 2049 $s\pi : \sqrt{2}-2\sqrt{3}/3$
7279 2206 id : $9\sqrt{2}$
7279 2589 at : $9\ln 2/7$
7279 3323 ln : $1/2+\pi/2$
7279 4021 $e^x : 1+\sqrt{5}/3$
7279 4631 at : $e/4-\pi/2$
7279 4906 $erf : 2e/7$
7279 5589 rt : (6,3,7,-9)
7279 7896 $e^x : 3e/4-3\pi/4$
7279 7905 $\theta_3 : \pi/9$
7280 0000 cu : 6/5
7280 0960 $3\div\ln(\sqrt{5})$
7280 2429 sr : $3/4+\sqrt{5}$
7280 3665 $10^x : \sqrt[3]{12}$
7280 4204 rt : (9,8,-9,7)
7280 5930 rt : (1,7,-7,1)
7280 6897 e^x : 23/7
7280 6979 $10^x : e/4+4\pi/3$
7280 7382 rt : (5,8,-4,-5)
7280 7704 rt : (9,7,-3,-5)
7280 8212 rt : (8,0,4,6)
7280 9032 cu : $2/3-3\sqrt{2}$
7280 9241 tanh : $(\sqrt[3]{5}/2)^{1/2}$
7281 1262 rt : (3,7,9,4)
7281 1996 $J_0 : 9\zeta(3)/10$
7281 3602 rt : (2,9,2,-7)
7281 5805 $2^x : 4+\sqrt{5}/2$
7281 6320 rt : (1,2,9,-8)
7281 7830 $t\pi : (\ln 2)^3$
7281 8143 rt : (2,9,0,-4)
7281 8775 ln : $4e/3-\pi$
7281 8931 $t\pi : \zeta(3)/6$
7281 9138 $4/5\div\ln(3)$
7281 9633 $J_0 : 3\sqrt[3]{3}/4$
7282 2356 $2^x : 5\pi^2/9$
7282 3067 sr : $4/3+4\pi$
7282 3765 sr : $3\sqrt{2}/8$
7282 4349 rt : (8,6,1,-7)
7282 5819 e^x : 19/6
7282 7717 $\zeta : 4\sqrt[3]{2}/9$
7282 8475 rt : (3,9,-6,8)
7282 9102 $J_0 : (\sqrt[3]{5}/2)^{-1/2}$
7282 9333 rt : (7,1,8,8)
7282 9510 rt : (3,2,-7,-5)
7283 0284 $10^x : 3\pi/5$
7283 2335 $\Psi : \exp(2e/3)$
7283 2740 rt : (1,8,6,-9)
7283 3102 $\ell_{10} : 4\sqrt{2}/3+2\sqrt{3}$
7283 3238 rt : (5,1,-8,9)
7283 3917 rt : (4,6,-1,1)
7283 6442 rt : (7,4,-8,1)
7283 6562 sr : $5(1/\pi)/3$
7283 9397 sr : $3\sqrt{2}/2+\sqrt{3}/2$
7283 9634 rt : (7,5,5,-9)
7284 0624 ln : $2\sqrt{3}-4\sqrt{5}/3$
7284 1677 id : $1/\ln(\ln 2)$
7284 2640 cu : $9e\pi/8$
7284 4378 2^x : 15/19
7284 4563 $2^x : 1/4-\sqrt{2}/2$
7284 4669 $\ell_2 : 1/4+\sqrt{2}/4$
7284 5571 ζ : 14/25
7284 7788 $\exp(\pi)+s\pi(1/5)$
7284 8289 rt : (3,1,6,5)
7285 1595 rt : (3,8,-4,-6)
7285 1726 $\Gamma : 3\sqrt[3]{5}/10$
7285 3041 $\exp(e)\times s\pi(e)$
7285 5339 sq : $1/2+\sqrt{2}/4$
7285 5437 rt : (9,4,5,5)
7285 7084 rt : (3,5,3,-6)
7285 7265 at : $1/3+\sqrt{5}/4$
7286 0187 rt : (9,2,2,-6)
7286 2754 rt : (9,8,-1,-7)
7286 3533 $\ell_2 : (\ln 3/2)^2$
7286 4342 erf : 7/9
7286 5818 $s\pi : 6\sqrt[3]{5}$
7286 6474 tanh : $(\sqrt[3]{2})^{-1/3}$
7286 8663 $\ell_{10} : 2+3\sqrt{5}/2$
7287 0804 sr : $10\sqrt{5}$
7287 0862 rt : (1,8,9,3)
7287 1355 rt : (6,9,2,-1)
7287 3763 $e^x : 2(e+\pi)/3$
7287 4126 $K : \sqrt[3]{5}/3$
7287 5145 rt : (8,7,3,-9)
7287 5270 $\ell_{10} : \exp(2\pi)$
7287 5356 $e^x : (\sqrt{3})^{1/2}$
7287 6222 rt : (8,1,6,-8)
7287 7495 rt : (2,4,7,-8)
7287 8156 rt : (8,5,9,7)
7287 8866 Ψ : 3/25
7287 9668 rt : (4,7,-8,-8)
7287 9881 $s\pi : \sqrt[3]{2}$
7288 1807 $\ln(2)\div s\pi(2/5)$
7288 1848 $s\pi : 3\ln 2/8$
7288 1956 $10^x : P_{29}$
7288 4250 $\zeta : \sqrt{2}-\sqrt{3}+\sqrt{5}$
7288 4752 $2/5\times\exp(3/5)$
7288 5582 $e^x : 4/3+3e$
7288 5633 $\ell_{10} : 3+3\pi/4$
7288 6297 cu : $5\sqrt[3]{2}/7$
7289 0197 rt : (5,8,-3,-4)
7289 0324 $\ell_2 : 1/3+4\sqrt{5}/3$
7289 2322 $s\pi : 3\sqrt{3}/4-\sqrt{5}/4$
7289 4575 $1/3+\exp(1/3)$
7289 4768 $erf : (\sqrt{2}/3)^{1/3}$
7289 5352 $J_0 : (\sqrt[3]{2})^{1/3}$
7289 5641 $2^x : 4e-3\pi$
7289 8447 $Ei : 7\pi/10$
7289 8854 $10^x : 4/3+4\sqrt{3}/3$
7289 9004 cr : P_{27}
7289 9590 sq : $4e/5$
7290 0000 cu : 9/10
7290 0142 rt : (5,2,0,-3)
7290 1276 $J_0 : 6\sqrt[3]{2}/7$
7290 3153 rt : (1,8,3,7)
7290 4378 rt : (9,9,1,-9)
7290 5583 $Ei : (1/\pi)/9$
7290 5884 sq : $3e-2\pi/3$
7290 6790 $\exp(3)-\exp(\sqrt{5})$
7290 7000 rt : (4,7,1,-6)
7290 9517 $\ell_{10} : 4+e/2$
7291 2728 rt : (9,3,4,-8)
7291 3836 e^x : 13/6
7291 5776 $Li_2 : \zeta(3)/2$
7291 7353 sq : $9\ln 2/2$
7291 7414 $5\times s\pi(\sqrt{3})$
7291 8151 rt : (1,5,-1,-3)
7292 1945 $\ell_{10} : 3\sqrt{2}+\sqrt{5}/2$
7292 2081 rt : (3,6,5,-8)
7292 4304 rt : (4,1,4,-5)
7292 5204 $\Psi : -\sqrt{2}+\sqrt{3}+\sqrt{5}$
7292 6730 rt : (9,3,7,7)
7292 7060 sq : $e\pi/10$
7293 0752 cu : $3/2+\sqrt{5}/4$
7293 1248 sinh : $\sqrt{2}+\sqrt{6}+\sqrt{7}$
7293 1676 rt : (6,8,-7,-5)
7293 2314 rt : (5,4,3,2)
7293 5020 $3\div\exp(\sqrt{2})$
7293 6751 sr : $3/2+2\sqrt{5}/3$
7293 6907 rt : (3,9,-6,-8)
7293 7208 $10^x : 3/2-\sqrt{5}/4$
7293 7790 $\pi+s\pi(1/5)$
7293 8942 rt : (7,5,-6,-1)
7294 1198 $e^x : 1/\ln(\ln 3)$
7294 2915 $\Psi : (3\pi)^3$
7294 3600 rt : (2,9,-9,1)
7294 4415 rt : (5,9,-1,-6)
7294 4881 rt : (3,0,8,7)
7294 9030 $\lambda : \exp(-4\pi/5)$
7295 1001 cu : $\sqrt{3}/3-4\sqrt{5}$
7295 3341 rt : (2,4,-5,-5)
7295 6007 rt : (4,8,3,-8)
7295 6750 cr : $e/7$
7295 7516 cu : $5\sqrt{5}/8$
7295 7526 rt : (6,4,-2,-3)
7295 9448 sq : $3e-4\pi/3$
7296 0388 $\Gamma : \sqrt{23}$
7296 3064 rt : (8,0,-7,2)
7296 7783 $e^x : 4/3-\pi/4$
7297 1514 sinh : 25/19
7297 1580 $\ell_2 : \sqrt{11}$
7297 2765 as : 2/3
7297 3046 rt : (2,8,-5,9)
7297 3211 rt : (1,8,-5,-1)
7297 3496 rt : (6,7,-7,-2)
7297 3530 id : P_{23}
7297 3660 rt : (5,3,2,-5)
7297 3841 rt : (9,2,9,9)
7297 4087 K : 4/7
7297 4269 rt : (4,5,7,4)
7297 7941 cu : $4/3-\sqrt{3}/4$
7298 0327 sr : $\exp(-\sqrt[3]{2}/2)$
7298 0791 rt : (3,3,-9,-7)
7298 2683 rt : (5,5,-4,4)
7298 3766 rt : (8,7,-8,-1)
7298 4398 ln : $4\pi^2/7$
7298 7222 sr : $7\sqrt[3]{5}/4$
7298 7801 rt : (4,2,6,-7)
7298 7826 $\Psi : 3\sqrt[3]{2}/10$
7298 8180 $erf : 9\ln 2/8$
7298 9965 at : 17/19
7299 1326 sq : $1/3+4\sqrt{3}$
7299 1985 cr : 7/18
7299 2419 sr : $((e+\pi)/3)^3$
7299 2715 tanh : 13/14
7299 3376 cr : $1/\sqrt[3]{17}$
7299 3455 $2^x : 3+3\pi/2$
7299 6746 rt : (7,8,-2,-3)
7299 7356 $\Psi : 1/\sqrt{7}$
7299 8256 $2^x : 3e\pi/4$
7299 8805 rt : (1,9,-7,1)
7300 1696 $\Psi : 8\sqrt{5}/7$
7300 2283 $2^x : 2\sqrt{2}/3+3\sqrt{3}/4$
7300 2652 ln : $(3\pi)^3$
7300 4050 Ψ : 23/9
7300 4103 sq : $1/2+4\pi$
7300 4204 rt : (6,6,-7,1)
7300 8275 rt : (6,5,0,-5)
7300 8540 $t\pi : e/6$
7301 0694 $\theta_3 : (\pi/2)^{-1/2}$
7301 1143 rt : (5,4,-4,7)
7301 1548 $e^x : \sqrt{2}-\sqrt{3}/2$
7301 1915 Ψ : 21/23
7301 2086 sr : $10\sqrt{5}/3$
7301 3495 $\Gamma : \sqrt{11}$
7301 4317 $e^x : 5\sqrt{5}$
7301 4989 at : $\sqrt{2}-4\sqrt{3}/3$
7301 5792 cr : $3\sqrt{2}/2-\sqrt{3}$
7301 5912 rt : (1,6,-3,-5)
7301 6991 cr : $\sqrt{2}/2+2\sqrt{5}$
7301 7023 θ_3 : 7/20
7301 8585 rt : (5,4,4,-7)
7301 9573 10^x : 5/21
7302 2124 rt : (6,2,1,2)
7302 5243 rt : (8,8,-6,-3)
7302 7511 $t\pi : \pi^2/4$
7302 7715 rt : (6,5,-7,4)
7302 7862 rt : (4,3,8,-9)
7302 8371 sinh : $(\sqrt{3})^{1/2}$
7302 8412 cu : $1/4+3e$
7302 8808 cr : $3e/4+\pi$
7302 8883 sq : $3\sqrt{3}-2\sqrt{5}/3$
7302 9674 sr : 8/15
7303 1829 rt : (2,5,-7,-7)
7303 3890 rt : (7,7,-2,-5)
7303 4282 sq : $2/3-3\sqrt{3}/2$
7303 4998 $t\pi : \sqrt{2}+\sqrt{5}-\sqrt{6}$

7303 5317 $2^x : \sqrt{3}+2\sqrt{5}$
7303 6148 sinh : $9e/10$
7303 6639 rt : (9,2,-9,2)
7304 0352 sinh : $9\zeta(3)/2$
7304 1066 J_0 : 14/13
7304 1580 rt : (1,9,-3,-3)
7304 1738 rt : (6,6,2,-7)
7304 1952 rt : (5,3,5,4)
7304 5460 sq : $7\pi^2/10$
7304 6002 rt : (1,7,-5,5)
7304 6294 rt : (6,4,-7,7)
7304 8893 rt : (5,5,6,-9)
7305 2263 as : P_{26}
7305 2728 $4/5 \times \exp(5)$
7305 3609 rt : (8,9,-4,-5)
7305 4150 cu : $4+3\sqrt{5}/2$
7305 4782 sπ : $\ln(2\pi/3)$
7305 7880 rt : (4,4,9,6)
7305 8845 $\sqrt{5} \times s\pi(e)$
7305 9425 $\theta_3 : (1/(3e))^{1/2}$
7305 9781 rt : (7,8,0,-7)
7306 0164 ζ : 23/12
7306 2420 $e^x : \sqrt{2}+\sqrt{5}-\sqrt{7}$
7306 3980 rt : (1,7,-5,-7)
7306 4330 $\zeta : 6\sqrt{5}/7$
7306 5467 rt : (6,7,4,-9)
7306 6789 sπ : $2\pi^2$
7306 7452 rt : (7,0,-1,2)
7306 9948 id : $6\sqrt[3]{3}/5$
7307 0565 $\ln(4) - \exp(3/4)$
7307 0603 $\ell_{10} : 2/3 + 3\pi/2$
7307 0996 tanh : $1/\ln((e+\pi)/2)$
7307 1767 $3 \div \ln(3)$
7307 3016 rt : (2,6,-9,-9)
7307 6079 cu : $4/3+\sqrt{5}/2$
7307 6402 sπ : $7\sqrt{5}/9$
7307 8864 rt : (7,9,2,-9)
7307 8963 rt : (6,1,3,4)
7307 9337 $e^x : P_{12}$
7308 1069 $K : 9(1/\pi)/5$
7308 3596 sπ : 6/23
7308 6080 rt : (9,3,-7,0)
7308 7492 rt : (1,7,3,8)
7308 7577 tanh : $(\sqrt{3}/2)^{1/2}$
7308 8594 sq : $1/2+2\pi/3$
7308 8689 rt : (5,2,7,6)
7309 0179 cu : $e\pi/4$
7309 1754 id : $4\sqrt{2}/3 - 2\sqrt{3}/3$
7309 2237 sr : $4\zeta(3)/9$
7309 2569 rt : (1,8,-7,-9)
7309 3521 $e^x : 3\sqrt{3}/2+\sqrt{5}$
7309 4160 cr : $4/3 - 2\sqrt{2}/3$
7309 4782 rt : (8,2,-3,-2)
7309 7051 $\ell_{10} : \sqrt{2}+\sqrt{3}+\sqrt{5}$
7309 7124 sr : $3/4+3\sqrt{5}$
7309 8723 $erf : (2\pi/3)^{-1/3}$
7309 8824 $\zeta : 2\sqrt[3]{5}/3$
7309 9587 $2^x : 1/3 - \pi/4$
7310 0443 id : $(\sqrt[3]{5}/2)^2$
7310 0572 $\Gamma : (1/\pi)/9$
7310 1519 at : $(\sqrt[3]{3}/2)^{1/3}$
7310 2292 rt : (7,1,1,-4)
7310 2516 $\zeta : 2\ln 2/7$
7310 4709 $\ell_{10} : 3 - 4\sqrt{5}/3$
7310 4949 tπ : $(\sqrt[3]{3})^{1/2}$
7310 5865 sq : $\sqrt{2}/4 + 3\sqrt{3}/4$
7310 6335 $e^x : \exp(\sqrt[3]{3})$
7310 7312 2^x : 19/24
7310 7476 sq : $7\sqrt[3]{5}/5$
7310 8841 rt : (6,0,5,6)
7310 9317 ln : $3/2+\sqrt{3}/3$
7310 9826 ζ : 16/23
7311 3057 rt : (9,4,-5,-2)
7311 3959 $\sqrt{5} \times \exp(1/5)$
7311 4603 rt : (5,1,9,8)
7311 6972 tπ : $1/2+3\sqrt{3}/4$
7311 8132 rt : (7,9,-4,-5)
7311 8340 rt : (8,3,-1,-4)
7312 0490 sq : $4/3+3\sqrt{2}/4$
7312 0956 sinh : $(\ln 3/2)^{-2}$
7312 3048 rt : (7,2,3,-6)
7312 3239 cr : $1+4\pi/3$
7312 3404 $3/5 \div \ln(\sqrt{2})$
7312 6183 as : $5\zeta(3)/9$
7312 7270 rt : (6,1,7,-8)
7312 9921 rt : (8,8,-8,-7)
7313 0055 rt : (9,5,-3,-4)
7313 0193 sq : 25/19
7313 1138 $J_0 : \ln((e+\pi)/2)$
7313 2157 sinh : $\sqrt[3]{25/3}$
7313 2605 $e^x : 4+\sqrt{2}/2$
7313 3602 rt : (8,4,1,-6)
7313 4028 ln : $3e/2+\pi/2$
7313 4937 $2^x : 2/3 - \sqrt{5}/2$
7313 6116 sq : $\sqrt{2} - \sqrt{5}/4$
7313 6701 $10^x : 3+\pi$
7313 6827 rt : (7,3,5,-8)
7313 6988 $\ell_2 : 4 - e/4$
7313 9344 sπ : $9(e+\pi)$
7314 1750 rt : (9,6,-1,-6)
7314 2282 cu : $\sqrt{2}+\sqrt{5}-\sqrt{6}$
7314 2396 id : $1/4 - 4\sqrt{5}/3$
7314 2411 cu : $\sqrt{3}/4+\sqrt{5}/2$
7314 2795 cr : 9/23
7314 4294 rt : (8,5,3,-8)
7314 8638 at : $2\pi/7$
7315 0188 $J_0 : (\sqrt{3}/2)^{-1/2}$
7315 0289 rt : (9,7,1,-8)
7315 3466 id : $4e - \pi$
7315 7992 sπ : $3\sqrt{2} - 4\sqrt{5}/3$
7315 9783 rt : (1,9,2,-3)
7316 0913 erf : 18/23
7316 2109 $\ell_2 : 7e\pi/9$
7316 2770 ln : $6\sqrt{3}/5$
7316 3239 sr : $7e\pi$
7316 5263 $\ell_2 : P_2$
7316 6198 $\Psi : \sqrt{5}/10$
7316 7501 $2^x : 8\ln 2/7$
7316 9017 sr : $e/2+4\pi$
7317 3028 $e^x : \sqrt{3}/4 - \sqrt{5}/3$
7317 3756 ln : $4/3+\sqrt{5}/3$
7317 4661 $\ell_{10} : e/4+3\pi/2$
7318 4401 $e^x : 3/2 - 2e/3$
7318 9179 cu : $3e/2 - 2\pi$
7319 4187 cr : $4e/3+\pi/2$
7319 4386 tπ : $(\ln 2)^{-1/2}$
7319 4499 sπ : $\sqrt{15/2}$
7320 0177 $Ei : (\ln 2/2)^{-2}$
7320 5080 id : $\sqrt{3}$
7320 5760 sπ : $2\sqrt{2}+\sqrt{3}/4$
7320 6282 $2^x : 10\sqrt[3]{5}/9$
7320 7215 $s\pi(\pi) \div s\pi(1/5)$
7320 8121 $erf : (\sqrt[3]{3}/3)^{1/3}$
7320 9936 ln : $3\ln 2$
7321 2462 $\exp(4) - s\pi(1/3)$
7321 3196 2^x : 19/10
7321 4422 tanh : 14/15
7321 7987 $\ell_{10} : 3\sqrt{2}+2\sqrt{3}/3$
7321 9197 sr : $e/3+2\pi/3$
7322 2838 $\ln(3/5) - \exp(1/5)$
7322 3533 $10^x : 3/4+3\sqrt{2}$
7322 5655 ln : $2\zeta(3)/5$
7322 6316 $e^x : 9\zeta(3)$
7322 6460 sπ : $1/3+4\sqrt{3}$
7322 7107 rt : (9,2,6,-9)
7322 7820 $2^x : 2\sqrt{3}/3+\sqrt{5}/3$
7322 9034 $\Psi : (\pi/3)^{-2}$
7322 9103 cr : $\ln(\sqrt{2}\pi/3)$
7322 9594 cr : $\pi/8$
7322 9785 rt : (8,0,8,9)
7323 0933 rt : (9,1,4,-7)
7323 1281 sq : $3\sqrt{2}/2+2\sqrt{3}/3$
7323 2082 id : $(\ln 2/3)^{-2}$
7323 3035 $\theta_3 : \exp(-\pi/3)$
7323 4701 rt : (8,1,6,7)
7323 5679 $\zeta : 5(1/\pi)/8$
7323 6368 rt : (9,0,2,5)
7323 8832 sq : $2 - 4e$
7323 9375 $\ell_{10} : (\sqrt[3]{5}/3)^3$
7323 9756 rt : (7,3,8,7)
7324 0819 $2/3 \times \ln(3)$
7324 2061 rt : (8,2,4,5)
7324 2187 cu : $5\sqrt[3]{3}/8$
7324 3053 $\ln(2/5)+\exp(1/2)$
7324 3272 tπ : $e/7$
7324 4577 at : $\exp(-1/(3\pi))$
7324 4694 rt : (9,1,0,3)
7324 4747 sr : $\ln(\sqrt[3]{5})$
7324 8431 id : $(e+\pi)/8$
7324 8676 $\ln(\sqrt{5}) \div \ln(3)$
7325 0287 rt : (7,4,6,5)
7325 0634 rt : (8,6,-9,8)
7325 1320 as : $(\sqrt{5})^{-1/2}$
7325 1513 $\ln(\pi)+s\pi(1/5)$
7325 1655 ln : $9\pi/5$
7325 4284 rt : (8,3,2,3)
7325 4547 $2^x : 3/2 - \sqrt{2}/2$
7325 4703 rt : (8,7,-9,5)
7325 6260 ln : $(\ln 2/3)^{1/2}$
7325 6464 cu : $3e/2 - 4\pi$
7325 7386 $10^x : \sqrt{2}+\sqrt{5}$
7325 8177 rt : (7,6,-6,8)
7325 9053 rt : (9,2,-2,1)
7325 9570 rt : (8,8,-9,2)
7326 1469 Ei : 11/5
7326 3223 rt : (6,6,8,5)
7326 3789 rt : (7,7,-6,5)
7326 4805 $\zeta : \sqrt{11/3}$
7326 5496 rt : (8,9,-9,-1)
7326 7093 as : $\sqrt{2} - \sqrt{5}/3$
7326 8716 rt : (6,6,-3,8)
7327 0001 rt : (7,5,4,3)
7327 0726 rt : (7,8,-6,2)
7327 3606 $10^x : 3(1/\pi)/4$
7327 5513 cu : $3\zeta(3)/4$
7327 6947 rt : (6,7,-3,5)
7327 8394 at : $5\sqrt[3]{2}/7$
7327 8561 rt : (8,4,0,1)
7327 9099 ln : $2/3+\sqrt{2}$
7327 9520 rt : (7,9,-6,-1)
7328 1510 at : 9/10
7328 4165 $3/5 \times \exp(1/5)$
7328 4473 rt : (5,9,3,-9)
7328 4683 id : $3\sqrt[3]{5}/7$
7328 5334 sr : $(\ln 2)^{-3}$
7328 6795 id : $5\ln 2/2$
7328 7613 $e^x : 3+2\sqrt{3}/3$
7328 7624 rt : (6,8,-3,2)
7328 7977 sq : $\sqrt{2}/3+4\sqrt{3}/3$
7328 8561 tanh : $7\zeta(3)/9$
7328 9708 rt : (9,3,-4,-1)
7329 2277 $\theta_3 : \ln(\sqrt{2}\pi/2)$
7329 3761 rt : (4,7,5,-9)
7329 3795 $\exp(3/5) \times s\pi(2/5)$
7329 5603 J_0 : 15/14
7329 6291 $e^x : 3/4 - 3\sqrt{2}/4$
7329 7709 rt : (5,8,1,-7)
7329 9221 at : $4/3 - \sqrt{3}/4$
7330 0135 Ei : 17/7
7330 0428 cu : $3\sqrt{2}/2+\sqrt{5}$
7330 0480 rt : (6,7,6,3)
7330 2023 rt : (6,9,-3,-1)
7330 2584 ln : $(\ln 2)^2$
7330 3258 $4/5 \times \ln(2/5)$
7330 3769 tπ : $3e - 3\pi/4$
7330 5512 rt : (3,5,7,-9)
7330 6819 tπ : P_8
7330 6900 sq : $3/2 - 3\pi/4$
7330 7557 $\Psi : 3e\pi/10$
7330 9769 sinh : $e/4$
7331 0474 sr : $4\sqrt{2}/3+\sqrt{5}/2$
7331 0603 rt : (4,6,3,8)
7331 2330 sπ : $6e/5$
7331 3754 $\ell_{10} : 9\zeta(3)/2$
7331 6236 rt : (5,8,0,2)
7331 6853 ln : $\sqrt{13/3}$
7331 8778 sπ : $7\sqrt[3]{3}/8$
7332 0141 rt : (7,6,2,1)
7332 0858 rt : (2,3,9,-9)
7332 1123 rt : (5,9,-7,-8)

7332 2500 rt : (6,8,-5,-3)
7332 5301 e^x : 11/20
7332 5795 ℓ_{10} : $4/3+3e/2$
7332 6162 10^x : 16/17
7332 6482 2^x : $\sqrt{2}/2-2\sqrt{3}/3$
7332 7673 rt : (3,4,5,-7)
7332 8542 $s\pi(\sqrt{3})\div s\pi(\pi)$
7332 9508 rt : (7,9,-9,-1)
7333 0131 ln : P_{21}
7333 0370 cu : $3\sqrt{2}+2\sqrt{3}$
7333 1142 erf : $\pi/4$
7333 3333 id : 11/15
7333 5267 id : $3\sqrt{2}+2\sqrt{5}/3$
7333 5337 rt : (4,7,3,5)
7334 3402 rt : (9,4,9,8)
7334 4015 rt : (5,9,0,-1)
7334 5243 cr : $8(e+\pi)/9$
7334 6745 $s\pi$: $((e+\pi)/3)^{-2}$
7334 6775 rt : (2,8,-8,-5)
7334 7947 erf : $5\sqrt{2}/9$
7334 9907 rt : (8,5,-2,-1)
7335 0387 erf : 11/14
7335 1343 rt : (2,2,7,-7)
7335 1524 2^x : $\sqrt[3]{1/2}$
7335 1879 rt : (6,7,-9,-8)
7335 2079 sr : $3+2\sqrt{5}$
7335 2623 rt : (8,9,0,-8)
7335 3461 sr : $5\zeta(3)/2$
7335 3538 ℓ_{10} : $4+\sqrt{2}$
7335 3924 rt : (6,7,-7,-1)
7335 4484 sr : $7e\pi/8$
7335 5686 as : $\ln((e+\pi)/3)$
7335 6813 $\exp(1/2)\div s\pi(2/5)$
7335 7171 rt : (2,9,-2,1)
7335 8742 $s\pi(1/4)\div s\pi(\sqrt{2})$
7335 9134 at : $5\sqrt[3]{3}/8$
7335 9404 ℓ_{10} : $2\sqrt{2}/3+2\sqrt{5}$
7335 9605 K : $2\sqrt[3]{3}/5$
7336 2375 rt : (3,6,6,8)
7336 3036 sinh : 17/25
7336 4712 cu : $3\sqrt{5}/2$
7336 6524 cr : $4\sqrt{2}/3-2\sqrt{5}/3$
7336 6675 at : $3\zeta(3)/4$
7336 7525 rt : (5,8,-5,-6)
7336 8535 rt : (7,7,2,-8)
7336 9629 $\ln(\sqrt{2})\times\exp(3/4)$
7337 0388 ln : $9\sqrt[3]{5}$
7337 1623 $\exp(2/5)-\exp(4/5)$
7337 2019 sinh : $6\sqrt[3]{3}/5$
7337 5192 rt : (8,8,-2,-6)
7337 5594 $\sqrt{5}\times\exp(3/4)$
7337 5680 ln : $\sqrt{24}\pi$
7337 7172 rt : (9,5,7,6)
7337 8766 cu : $2\sqrt{2}+3\sqrt{3}$
7337 9938 sr : 7/13
7338 0105 e^x : $2\sqrt{2}/3+3\sqrt{5}/4$
7338 0399 Li_2 : $2e/9$
7338 0984 e^x : $1/\sqrt[3]{6}$
7338 2095 K : γ
7338 2411 rt : (9,9,-6,-4)
7338 4331 ln : $2\sqrt{2}-\sqrt{5}/3$
7338 4994 rt : (5,9,8,3)
7338 5155 Γ : $\exp(-\sqrt{5}/2)$
7338 5299 cr : $1/2+3\pi/2$
7338 6370 rt : (1,0,9,7)
7338 6880 rt : (4,9,-1,-4)
7338 7117 $s\pi$: $e/3+3\pi/4$
7338 7205 $s\pi$: $3\sqrt{2}/4+3\sqrt{5}/4$
7338 8298 rt : (6,5,4,-8)
7338 8636 e^x : $2-4\sqrt{3}/3$
7339 0263 rt : (5,5,-5,-1)
7339 1688 K : $\sqrt{3}/3$
7339 1872 rt : (4,9,-2,9)
7339 2092 ζ : 1/5
7339 4059 e^x : $3/4+4\sqrt{3}$
7339 4788 10^x : $1/\sqrt[3]{12}$
7339 6698 rt : (7,6,0,-6)
7339 6917 ln : 12/25
7339 8703 10^x : $3e/4+\pi/3$
7340 0069 rt : (9,4,-6,-3)
7340 0912 θ_3 : $1/\sqrt[3]{23}$
7340 1066 Ei : 10/17
7340 3609 rt : (2,1,5,-5)
7340 4261 ln : $4\sqrt{3}/3+3\sqrt{5}/2$
7340 5752 rt : (9,8,-8,9)
7340 5897 rt : (8,7,-4,-4)
7340 7151 tanh : 15/16
7340 8193 sinh : $\exp(1/\sqrt{2})$
7340 8231 rt : (6,6,-9,1)
7340 9020 Ei : $8\sqrt{5}$
7341 0196 Γ : $(1/(3e))^3$
7341 1114 K : $5\ln 2/6$
7341 3509 rt : (5,3,6,-8)
7341 3557 e^x : $2e+\pi/2$
7341 4728 rt : (8,7,9,6)
7341 6014 rt : (9,8,-8,-2)
7341 6829 e^x : $1/4-\sqrt{5}/4$
7341 9344 cu : $1/4-4\sqrt{3}/3$
7341 9567 2^x : $1/\ln(1/(3\pi))$
7342 0055 J_0 : $5\sqrt[3]{5}/8$
7342 0814 Γ : $\exp(-\sqrt{2})$
7342 1316 e^x : $\sqrt{3}+3\sqrt{5}$
7342 2320 rt : (9,9,-8,6)
7342 4194 id : $2\sqrt{2}/3-3\sqrt{5}/4$
7342 4415 sr : $7\ln 2/9$
7342 4437 rt : (9,6,3,-9)
7342 4720 ℓ_{10} : $(1/(3e))^3$
7342 5819 $\exp(1/2)+\exp(3)$
7342 5846 λ : $\exp(-5/2)$
7342 5969 sq : $7e\pi/10$
7342 8818 rt : (3,8,-8,-9)
7342 9180 sinh : $5\pi^2/8$
7343 1130 J_0 : $8\zeta(3)/9$
7343 2390 at : $4/3-\sqrt{5}$
7343 2708 rt : (9,6,5,4)
7343 3612 cu : $4/3+4e$
7343 4496 at : $(e/2)^{-1/3}$
7343 4856 cu : $3e\pi/4$
7343 5125 rt : (6,6,-7,-6)
7343 6339 rt : (8,8,-5,9)
7343 6561 rt : (7,5,-2,-4)
7343 7500 cu : 15/4
7343 8106 rt : (1,6,1,-3)
7343 8927 $t\pi$: $3\sqrt{2}/2+3\sqrt{5}/4$
7343 9402 cr : $4\ln 2/7$
7344 1862 cr : $7\sqrt{5}/3$
7344 2948 ζ : $\zeta(3)/6$
7344 3556 rt : (2,9,-4,-7)
7344 4801 10^x : 24/5
7344 6784 rt : (4,1,8,-8)
7344 7877 rt : (8,4,5,-9)
7344 9480 ℓ_2 : $\zeta(3)/2$
7345 0087 rt : (8,6,-6,-2)
7345 1285 sq : $7\pi/3$
7345 2007 ℓ_{10} : $2\sqrt{2}+3\sqrt{3}/2$
7345 2377 Ψ : $3\sqrt[3]{5}/2$
7345 2896 rt : (3,7,0,9)
7345 4786 e^x : $5(e+\pi)/7$
7345 5921 erf : $5\sqrt[3]{2}/8$
7345 7630 rt : (9,5,1,-7)
7345 8618 rt : (8,9,-5,6)
7346 0105 ln : 3/17
7346 1011 rt : (4,0,-8,7)
7346 1056 Ψ : $1/\sqrt[3]{6}$
7346 1585 rt : (5,2,4,-6)
7346 1800 cr : $3/4-\sqrt{2}/4$
7346 7108 2^x : $1/3+\sqrt{5}/2$
7346 7617 rt : (1,4,-2,7)
7346 9387 sq : 6/7
7346 9821 rt : (6,8,4,1)
7347 1927 $\ln(\sqrt{3})-\exp(1/4)$
7347 2670 cu : $4e/3+\pi$
7347 4261 rt : (1,1,7,5)
7347 5280 $\exp(3)\div s\pi(\sqrt{5})$
7347 5815 ln : $\sqrt{2}/3+3\sqrt{3}$
7347 5940 rt : (5,7,-3,-4)
7347 6938 rt : (7,2,7,-9)
7347 6970 Γ : $8\pi^2/5$
7347 7928 rt : (7,8,-2,9)
7347 8293 rt : (6,3,0,-4)
7347 8573 J_0 : $e\pi/8$
7347 9192 2^x : $3\sqrt[3]{2}$
7348 2390 sr : $3\sqrt[3]{2}/7$
7348 2424 rt : (5,4,-7,1)
7348 2866 ζ : $\sqrt[3]{7}$
7348 3772 $s\pi$: $8\sqrt{2}/5$
7348 3913 2^x : $4\sqrt{2}/3+\sqrt{3}/2$
7348 4058 ℓ_2 : $1+4\sqrt{2}$
7348 6234 2^x : $3\sqrt{2}+\sqrt{5}/3$
7348 6517 rt : (3,8,2,7)
7348 6724 sq : $(\sqrt[3]{4})^{-1/3}$
7348 9211 rt : (8,3,3,-7)
7349 2697 2^x : $\ln(3\pi)$
7349 3306 cr : $3/4+2\sqrt{5}$
7349 5574 as : $1/\ln(\sqrt{2}\pi)$
7349 6316 Ei : $(\sqrt{2}\pi/2)^{1/3}$
7349 6684 rt : (3,7,-6,-7)
7349 7012 rt : (4,1,-8,4)
7349 7295 rt : (7,4,-4,-2)
7350 2628 rt : (9,4,-1,-5)
7350 3375 rt : (8,8,7,4)
7350 4135 θ_3 : $\exp(-\sqrt[3]{2}/3)$
7350 4591 sr : $4/3+3\sqrt{5}/4$
7350 4602 10^x : $2\sqrt{2}/3+3\sqrt{5}/4$
7350 5859 ln : $\sqrt{2}/4+\sqrt{3}$
7350 5868 K : 11/19
7350 9200 at : $2/3-\pi/2$
7350 9466 J : $\sqrt{3}/5$
7350 9478 rt : (7,9,-2,6)
7351 2192 cr : $4\sqrt{2}-\sqrt{3}/4$
7351 2380 Ei : 16/17
7351 2645 ln : $9\sqrt[3]{2}/2$
7351 2902 ℓ_2 : $3/4+4\pi$
7351 3925 rt : (6,0,9,9)
7351 5467 J_0 : 16/15
7351 9091 rt : (8,5,-8,0)
7351 9325 2^x : $2\sqrt[3]{2}$
7351 9497 rt : (1,9,3,6)
7352 2156 rt : (2,8,-2,-5)
7352 2752 $\ln(2/3)+\exp(\pi)$
7352 5020 at : $\sqrt{2}/3+\sqrt{3}/4$
7352 5418 id : $e/2-2\pi/3$
7352 8137 sq : $3/2+2\sqrt{2}$
7352 9839 rt : (7,1,5,-7)
7352 9967 10^x : e
7353 0678 rt : (3,9,4,5)
7353 1445 as : $3\sqrt{5}/10$
7353 2447 ℓ_{10} : $2e$
7353 4170 rt : (4,8,1,-2)
7353 4418 J : $(\sqrt{5}/3)^3$
7353 4469 $\exp(3/4)\div s\pi(e)$
7353 7273 rt : (1,9,1,-6)
7353 7569 id : $\sqrt{22}\pi$
7353 7756 rt : (6,8,1,9)
7353 7788 ζ : 25/17
7353 8418 rt : (5,1,2,-4)
7353 9857 ℓ_{10} : $4\sqrt{3}-2\sqrt{5}/3$
7354 0602 at : $1/4-2\sqrt{3}/3$
7354 0774 Ψ : $1/\ln(3)$
7354 0905 rt : (9,7,3,2)
7354 2068 sr : $3\sqrt[3]{3}/8$
7354 2967 $t\pi$: $\ln(\sqrt{2}\pi/2)$
7354 3976 at : 19/21
7354 4775 rt : (4,2,-8,1)
7354 4841 Ei : $(\ln 2/2)^{1/2}$
7354 6967 Li_2 : $(\ln 3/3)^{1/2}$
7354 7340 as : $(\sqrt{2}\pi/2)^{-1/2}$
7354 7455 rt : (8,2,1,-5)
7354 8871 Ψ : $\sqrt{2}+\sqrt{3}-\sqrt{5}$
7354 9298 ℓ_{10} : P_{47}
7355 0683 cr : $5(1/\pi)/4$
7355 2023 ζ : $\exp(-\sqrt{3}/3)$
7355 2433 rt : (1,9,2,-3)
7355 7158 cu : $1/2+3\sqrt{3}/2$
7355 7944 $t\pi$: $5\sqrt[3]{3}/6$
7356 2968 $\ln(\sqrt{2})+\exp(2)$
7356 6539 cu : $4/3-\sqrt{5}$
7356 7056 rt : (9,3,-3,-3)
7356 7152 Γ : $\ln(5/3)$
7357 0074 2^x : $1/2-2\sqrt{2}/3$

7357 1050 ℓ_{10} : $\sqrt{3}\pi$
7357 2391 sπ : 5/19
7357 3154 rt : (2,7,6,-9)
7357 3720 sr : $10\pi^2/3$
7357 5448 rt : (1,9,8,-8)
7357 5888 $2 \div e$
7357 6231 tanh : 16/17
7357 7834 e^x : $3+\sqrt{2}/2$
7357 8308 rt : (1,8,-7,-5)
7357 8515 erf : 15/19
7357 9660 rt : (9,0,6,8)
7358 0009 sπ : $(5/2)^{-1/3}$
7358 0052 rt : (1,5,-3,7)
7358 0509 rt : (4,5,-8,-7)
7358 4070 rt : (6,1,7,7)
7358 5857 rt : (6,9,1,6)
7358 8657 sinh : 15/22
7359 0311 at : $3-2\pi/3$
7359 1223 tπ : $\exp(1/(2e))$
7359 1245 rt : (3,8,2,-7)
7359 3225 rt : (3,2,8,6)
7359 4294 $\ln(2/5)+\exp(\sqrt{3})$
7359 5925 Ψ : 11/20
7359 6151 id : $8\sqrt[3]{5}/5$
7359 6705 2^x : $\exp(2e/3)$
7359 8007 sr : 13/24
7359 9355 2^x : $2/3+\pi/4$
7360 0934 2^x : $5(1/\pi)/2$
7360 0962 rt : (7,4,-9,-6)
7360 1391 θ_3 : $e\pi/9$
7360 3926 sq : $e-\pi/4$
7360 6323 $s\pi(2/5)\times s\pi(e)$
7360 6797 id : $1/2+\sqrt{5}$
7360 8222 rt : (7,0,3,5)
7360 8363 as : $3\sqrt{2}/4-\sqrt{3}$
7360 8979 ℓ_{10} : $1/4+3\sqrt{3}$
7360 9120 ℓ_{10} : P_{63}
7361 1111 cu : $5\sqrt[3]{3}/6$
7361 1171 rt : (4,9,-2,-5)
7361 2754 $\pi+\ln(2/3)$
7361 5628 sr : $(2\pi)^{-1/3}$
7361 7071 2^x : $3\sqrt{2}-2\sqrt{5}/3$
7361 7173 at : $e/3$
7361 8248 e^x : $6\sqrt{3}$
7361 8393 rt : (8,8,2,-9)
7362 1104 cr : P_{59}
7362 1480 Ψ : $5\pi^2/8$
7362 2418 2^x : $\sqrt{2}/4+3\sqrt{3}/2$
7362 3738 rt : (3,9,5,0)
7362 7853 sr : $4\sqrt{3}+\sqrt{5}/4$
7362 8443 rt : (1,5,8,-9)
7363 0181 rt : (1,8,-1,-4)
7363 1195 rt : (5,8,4,9)
7363 2587 rt : (8,2,8,8)
7363 3000 rt : (9,9,-2,-7)
7363 3433 tπ : $\sqrt{2}/7$
7363 5594 rt : (8,1,-1,-3)
7363 5859 tanh : $3\pi/10$
7363 6655 $\ln(3/5)-\exp(4/5)$
7363 9503 tπ : $3e+\pi/3$
7363 9651 10^x : $3\sqrt{2}-3\sqrt{5}/2$
7364 0895 cu : $\exp(1/(2e))$
7364 1695 rt : (4,2,-5,1)
7364 2183 $\ln(3)\div\exp(2/5)$
7364 2518 rt : (1,9,9,-6)
7364 3190 ln : $4+3\sqrt{5}/4$
7364 5863 ℓ_2 : $\sqrt{18}\pi$
7364 6222 tπ : $\zeta(3)$
7364 6565 cu : $1/3-\sqrt{3}$
7364 6569 cr : $5\pi/3$
7364 7456 cr : $3+\sqrt{5}$
7364 9410 at : $(\sqrt{5}/3)^{1/3}$
7364 9838 $\ln(\sqrt{3})\div s\pi(\sqrt{3})$
7365 0164 rt : (6,8,-1,-6)
7365 1019 tanh : $2\sqrt{2}/3$
7365 2509 rt : (2,6,4,7)
7365 3810 sq : $3/2+\sqrt{3}/4$
7365 5357 rt : (9,1,4,6)
7365 6648 rt : (2,4,-9,-8)
7366 0142 rt : (5,3,9,7)
7366 1955 rt : (7,6,4,-9)
7366 3957 rt : (2,9,-5,-2)
7366 5577 rt : (5,2,-7,8)
7366 5780 sπ : $\exp(-4/3)$
7366 6858 rt : (9,2,-5,-1)
7367 0328 Ei : $\exp(-\sqrt[3]{4}/3)$
7367 0863 rt : (7,9,-5,-4)
7367 4589 tπ : $(\pi/2)^{-1/2}$
7367 4657 e^x : $4+2\sqrt{2}$
7367 4833 2^x : $\sqrt{2}/2+\sqrt{5}/3$
7367 5995 ℓ_{10} : $\sqrt{3}/3-\sqrt{5}/4$
7367 7740 sinh : $1/\ln(\ln 2/3)$
7367 8013 rt : (2,7,0,-3)
7367 8594 e^x : $1/2+3\sqrt{2}$
7367 9415 rt : (3,7,0,-5)
7367 9838 rt : (8,7,0,-7)
7368 0629 id : $(5/2)^{-1/3}$
7368 0967 sr : $\sqrt{2}/2+4\sqrt{3}/3$
7368 2242 $\exp(2)-\exp(\sqrt{3})$
7368 3298 sq : $3\sqrt{2}+\sqrt{5}/2$
7368 3921 e^x : $3e/4+\pi$
7368 4210 id : 14/19
7368 6317 sq : $4-\pi$
7368 7818 cr : $4\sqrt{2}/3+3\sqrt{5}/2$
7368 8842 rt : (4,5,5,-8)
7368 9468 Ei : 3/14
7369 0106 cu : $\zeta(3)$
7369 1705 tanh : $1/\ln(\ln 2/2)$
7369 1926 Li_2 : $3\sqrt{2}/7$
7369 3709 as : $4\sqrt{3}/3-4\sqrt{5}/3$
7369 3985 ζ : $\sqrt{2}/7$
7369 5070 rt : (1,8,8,-5)
7369 5305 rt : (6,2,5,5)
7369 5861 sq : $4/3+3\pi$
7369 6559 ℓ_2 : 3/5
7369 7622 rt : (7,7,0,-1)
7369 9145 rt : (9,8,-4,-5)
7370 0454 cu : $2(e+\pi)$
7370 1887 2^x : $4\sqrt{2}/3-\sqrt{3}/4$
7370 3033 10^x : $\sqrt{3}/2+3\sqrt{5}/2$
7370 6921 2^x : $5\pi/7$
7370 7284 J_0 : 17/16
7370 8614 sinh : $(\pi)^{-1/3}$
7370 8654 θ_3 : 6/17
7370 9682 rt : (4,8,-4,-3)
7371 0966 erf : 19/24
7371 1422 rt : (1,8,1,-8)
7371 3190 sinh : 23/5
7371 3383 rt : (5,9,4,6)
7371 5827 rt : (6,4,6,-9)
7371 7792 ℓ_{10} : $\exp(4)$
7372 0084 cr : $1+3\sqrt{2}$
7372 0884 rt : (8,9,5,2)
7372 1393 rt : (1,9,-8,-8)
7372 1469 sq : $5\zeta(3)/7$
7372 2704 cr : $\zeta(3)/3$
7372 2967 rt : (5,3,-7,5)
7372 5758 tanh : 17/18
7372 7723 cr : $7(e+\pi)/2$
7372 7772 rt : (1,2,5,3)
7373 2036 sinh : $4e/3$
7373 3056 sr : $e/5$
7373 4082 rt : (8,3,6,6)
7373 5652 $\ln(2/5)\times\ln(\sqrt{5})$
7373 6205 rt : (7,1,1,3)
7373 6887 sπ : $1/2+\sqrt{5}$
7373 7023 rt : (1,4,6,-7)
7373 7698 rt : (3,9,-8,1)
7373 8231 rt : (7,5,2,-7)
7373 8262 ℓ_{10} : $3/4+3\pi/2$
7373 9326 rt : (7,4,-9,9)
7374 0481 rt : (6,7,-3,-4)
7374 1193 erf : $8\ln 2/7$
7374 1539 sq : $3\sqrt{2}-4\sqrt{3}/3$
7374 2880 10^x : $2+\sqrt{3}$
7374 3965 rt : (5,9,-8,-1)
7374 6532 Ei : 21/2
7374 8068 rt : (1,8,4,-1)
7374 8405 tanh : $3\sqrt[3]{2}/4$
7375 1876 ℓ_{10} : $2+2\sqrt{3}$
7375 2322 Ψ : $\sqrt[3]{17}$
7375 2446 at : $(4/3)^{-1/3}$
7375 3542 rt : (5,3,-9,3)
7375 5309 Li_2 : $7\ln 2/8$
7375 7195 rt : (9,4,3,-8)
7375 8041 Ψ : 10/11
7375 9290 Ψ : 18/7
7375 9442 Li_2 : $(e)^{-1/2}$
7375 9476 sπ : $8\sqrt[3]{5}/5$
7375 9894 ln : 11/23
7376 1444 J : $9(1/\pi)/10$
7376 2637 rt : (8,6,-2,-5)
7376 3543 sr : $1/\ln(2\pi)$
7376 4635 rt : (3,3,7,-8)
7376 5623 sinh : $\sqrt{3}$
7376 6880 $\exp(5)-s\pi(\sqrt{5})$
7376 7696 ℓ_2 : $\sqrt{2}/4+4\sqrt{5}/3$
7376 9177 rt : (9,2,2,4)
7377 0546 rt : (7,8,-7,-2)
7377 1115 id : $3\sqrt{2}/4+3\sqrt{5}/4$
7377 1372 Ei : $9(e+\pi)/7$
7377 1785 e^x : 14/9
7377 4696 J_0 : $10(1/\pi)/3$
7377 5370 rt : (4,2,-4,8)
7377 7059 $\exp(3)+\exp(\sqrt{3})$
7377 9914 ζ : $(\sqrt{2}\pi/3)^{1/3}$
7378 0082 10^x : 6/25
7378 1506 at : 10/11
7378 4181 rt : (5,2,8,-9)
7378 4325 rt : (8,0,-3,1)
7378 4521 $s\pi(\sqrt{2})-s\pi(e)$
7378 4910 tanh : $(\sqrt{5}/2)^{-1/2}$
7378 5273 ζ : $(\sqrt{2}\pi/2)^{-2}$
7378 7393 2^x : $e/4+2\pi$
7378 9317 rt : (9,7,-6,-3)
7378 9379 rt : (7,5,-9,6)
7379 0481 sr : $10e/9$
7379 1143 J : $\exp(-11\pi/9)$
7379 1815 J_0 : $3\sqrt{2}/4$
7379 4217 ln : P_{14}
7379 4921 rt : (1,7,-3,-2)
7379 6400 ℓ_2 : P_{79}
7379 6638 rt : (5,4,-7,2)
7379 7635 rt : (4,8,7,9)
7379 8562 cr : $3-3\sqrt{3}/2$
7379 9590 cu : $1+e/4$
7380 0884 sq : $7(e+\pi)/6$
7380 1025 sq : $2\sqrt{2}-\sqrt{3}/4$
7380 1332 cr : 21/4
7380 1393 rt : (3,3,6,4)
7380 1666 10^x : $3-e/2$
7380 2974 $1/2\div\ln(3/4)$
7380 3387 sq : $1/2+2\sqrt{3}/3$
7380 6893 10^x : $(\ln 3)^{-2}$
7381 0613 ln : $1/4+2e$
7381 2204 10^x : $3/2+e/3$
7381 2311 sq : $1/2-e/2$
7381 2764 tπ : $3e\pi/8$
7381 3076 rt : (6,3,4,-7)
7381 3511 2^x : $2e\pi/3$
7381 3932 $\ln(5)\div s\pi(1/5)$
7381 4046 cr : $4\sqrt{3}-3\sqrt{5}/4$
7381 4049 $\ln(2/3)\div\ln(\sqrt{3})$
7381 4087 e^x : $(2\pi/3)^3$
7381 5687 ℓ_{10} : $1+2\sqrt{5}$
7381 6911 rt : (7,2,-8,2)
7381 7422 10^x : 14/11
7381 7637 2^x : $\sqrt{15}\pi$
7381 9507 rt : (3,6,-2,-3)
7382 2601 rt : (4,4,-6,-5)
7382 2881 K : 7/12
7382 4413 rt : (8,2,5,-8)
7382 7020 Ei : $5\pi^2/4$
7382 8452 Ei : $\sqrt{13}\pi$
7382 8812 e^x : $\sqrt[3]{16/3}$
7383 0639 rt : (5,4,7,5)
7383 1406 rt : (1,9,5,-9)
7383 2776 rt : (6,4,-6,9)
7383 3210 ζ : $6(1/\pi)$
7383 3396 erf : $\sqrt[3]{1/2}$

7383 4651 rt : (5,5,-1,-4)
7383 5797 rt : (4,7,-9,5)
7383 7719 J_0 : ln(ln 2/2)
7383 9991 J : $9(1/\pi)/7$
7384 0820 ζ : $7\sqrt[3]{2}/6$
7384 0953 ln(2/5) − exp(3/5)
7384 1681 rt : (9,8,1,0)
7384 1762 ℓ_{10} : $3\sqrt{2}/2 + 3\sqrt{5}/2$
7384 1963 10^x : 7/16
7384 1999 at : $\sqrt{2} + \sqrt{3} - \sqrt{5}$
7384 4341 at : $1/\ln(3)$
7384 4916 rt : (7,4,0,-5)
7384 6334 rt : (7,5,8,6)
7384 8563 rt : (2,9,5,-8)
7384 9740 rt : (7,6,-9,3)
7384 9760 ln : $3/2 + 4\pi/3$
7385 0023 e^x : $3/4 + 4\sqrt{5}/3$
7385 0079 10^x : $4/3 + 3\sqrt{2}/4$
7385 2332 rt : (9,3,1,-6)
7385 2953 θ_3 : $\sqrt{2}/4$
7385 4894 sr : 6/11
7385 5000 2^x : $(\pi/2)^{-1/2}$
7385 6063 tπ : $\sqrt{2} - 4\sqrt{3}$
7385 6124 rt : (9,6,9,7)
7385 6412 cu : $3\sqrt{3}/2 + 4\sqrt{5}$
7385 8528 sinh : $2\sqrt[3]{5}/5$
7385 8574 Ψ : $(4/3)^{-1/3}$
7385 8935 tanh : 18/19
7386 1278 id : $\sqrt{15/2}$
7386 3773 sr : $2/3 + 3\pi/4$
7386 4002 2^x : $3e - 2\pi/3$
7386 6027 rt : (4,9,2,-8)
7386 6741 Ei : $3\pi/10$
7386 7020 5 × exp(2/3)
7386 8638 at : $1/3 + \sqrt{3}/3$
7386 8822 rt : (2,1,9,-8)
7386 9698 rt : (6,8,3,-9)
7387 0805 rt : (4,7,-6,-1)
7387 2326 $\sqrt{2} - s\pi(\sqrt{5})$
7387 6096 J_0 : 18/17
7387 6437 rt : (6,6,-5,-2)
7387 9612 rt : (7,9,-8,-8)
7387 9844 1/3 − ln(2/3)
7388 0158 rt : (8,5,-4,-3)
7388 0559 ln(2/3) × exp(3/5)
7388 2255 θ_3 : $10(1/\pi)/9$
7388 2596 at : P_{11}
7388 2830 e^x : $\sqrt{3}/3 + 4\sqrt{5}$
7388 3623 sπ : $\sqrt{22}\pi$
7388 3765 rt : (5,8,-9,-9)
7388 4266 exp(4) + exp(π)
7388 5898 sinh : 13/19
7388 6959 2^x : $1/3 + 3\sqrt{5}$
7388 7033 id : $9(e + \pi)$
7388 7041 K : $((e + \pi)/2)^{-1/2}$
7388 7710 Ψ : $\ln(\sqrt{3})$
7388 7749 2^x : $\ln(\sqrt{2}\pi/2)$
7388 9160 10^x : $4/3 + \pi$
7389 1658 rt : (4,8,-7,3)
7389 2736 tπ : $1/3 + 2\sqrt{3}$
7389 4625 rt : (5,5,-7,-1)
7389 5065 rt : (7,9,-1,-7)
7389 5412 rt : (9,8,0,-8)
7389 5774 rt : (1,8,-6,4)
7389 6333 $\ln(\sqrt{2})$ − exp(3)
7389 8362 rt : (8,4,4,4)
7390 2481 sinh : $7\pi^2/5$
7390 2724 ℓ_{10} : $5\pi^2/9$
7390 2984 rt : (6,5,-6,6)
7390 3124 ln : $3 - e/3$
7390 3829 e^x : $4/3 + 2\sqrt{2}/3$
7390 7814 Ei : 23/14
7390 8119 cu : $2/3 - \pi/2$
7390 8293 rt : (2,9,-1,6)
7390 8721 sπ : $2/3 + 3\sqrt{3}/2$
7390 8793 rt : (9,1,8,9)
7391 0373 rt : (7,0,7,8)
7391 2519 rt : (5,1,6,-7)
7391 3043 id : 17/23
7391 5597 rt : (3,2,5,-6)
7391 6398 id : $7\sqrt{5}/9$
7391 9356 rt : (9,6,-8,-1)
7392 0386 rt : (1,3,4,9)
7392 0880 id : $2\pi^2$
7392 1236 ζ : 21/11
7392 1843 10^x : $7\sqrt[3]{5}/10$
7392 2030 rt : (2,5,-9,1)
7392 3050 rt : (8,7,-8,7)
7392 3939 rt : (7,7,-9,0)
7392 5994 tπ : $4\sqrt{2} + 3\sqrt{5}/4$
7392 6477 id : ln(2π/3)
7392 6622 tanh : $e\pi/9$
7392 8861 rt : (2,8,-6,-8)
7392 9139 cr : $2\sqrt{2}/7$
7392 9376 cu : 25/11
7393 0172 10^x : $\sqrt[3]{3}/6$
7393 0292 K : $1/\sqrt[3]{5}$
7393 0295 tπ : $(\sqrt{2}\pi/2)^{-2}$
7393 1491 cr : $\exp(-e/3)$
7393 4581 at : $(\pi/3)^{-2}$
7393 6129 rt : (4,9,-5,-7)
7393 6899 cu : 17/9
7393 7531 2^x : $7e/10$
7393 8328 tanh : $(\sqrt[3]{5}/2)^{1/3}$
7393 9753 rt : (3,4,-9,7)
7394 0617 cu : $2\sqrt{3}/3 + \sqrt{5}/2$
7394 2665 J_0 : $(\sqrt{5}/2)^{1/2}$
7394 4772 10^x : $\zeta(3)/5$
7394 6673 rt : (8,1,3,-6)
7394 6932 rt : (5,6,5,-9)
7394 7158 sq : $9\sqrt[3]{3}/5$
7394 7948 rt : (3,5,-2,-3)
7395 3988 rt : (3,7,4,-8)
7395 5098 sq : $1/2 + 3\sqrt{5}/4$
7395 5813 rt : (6,2,2,-5)
7395 6937 Ei : $2\sqrt{2}/3$
7395 7847 erf : $5(1/\pi)/2$
7395 8345 rt : (4,9,-5,1)
7395 9219 rt : (9,3,0,2)
7396 0716 e^x : $4\sqrt{2}/3 + \sqrt{3}/3$
7396 2866 ℓ_{10} : $4 + 2\sqrt{5}/3$
7396 3444 rt : (1,8,3,-7)
7396 4681 e^x : $3\sqrt{3}/4 - \sqrt{5}/3$
7396 5736 λ : 1/12
7396 5878 rt : (5,4,-3,9)
7396 9219 rt : (8,6,2,-8)
7396 9793 10^x : $3 - \sqrt{5}/3$
7397 1681 at : $\sqrt{2}/4 + \sqrt{5}/4$
7397 3243 rt : (3,6,-9,-8)
7397 4268 π + exp(4)
7397 6776 rt : (1,9,-2,-6)
7397 7700 tanh : $5\sqrt[3]{5}/9$
7397 7824 sq : $3\sqrt{2}/2 - 4\sqrt{5}/3$
7397 8305 tanh : 19/20
7397 8466 rt : (1,9,-7,2)
7397 8578 rt : (6,7,1,-7)
7397 8661 cu : $\sqrt{2}/3 + \sqrt{3}/4$
7397 9093 ln : $1/2 + 3\sqrt{3}$
7398 1849 rt : (7,2,-1,1)
7398 6764 rt : (9,2,-1,-4)
7398 9063 e^x : $2e - \pi/2$
7399 0787 rt : (8,8,-8,4)
7399 0940 sq : $2\sqrt{2}/3 - 3\sqrt{3}/2$
7399 4224 e^x : $4\sqrt[3]{2}/5$
7399 5159 rt : (2,6,-5,2)
7399 7488 at : 21/23
7399 8257 cu : $3\sqrt{3} + \sqrt{5}/2$
7399 9066 rt : (9,7,-2,-6)
7400 1259 Ei : $\sqrt[3]{5}/8$
7400 2111 id : $3\sqrt{3}/4 - \sqrt{5}/4$
7400 3696 sq : $(\pi/2)^{-1/3}$
7400 3877 sinh : 5 ln 2/2
7400 4533 rt : (7,3,-2,-3)
7400 4699 e^x : $2 + 4\sqrt{5}$
7400 4867 $\sqrt{2}$ − exp(e)
7400 6966 ζ : $1/\sqrt{24}$
7400 7381 rt : (2,9,-1,-5)
7400 8102 tπ : $1/4 - \pi/3$
7400 8514 id : $\sqrt{3} - 2\sqrt{5}$
7401 0519 sq : $10\sqrt[3]{2}$
7401 1002 sq : 5π
7401 2745 sπ : $5e/6$
7401 2938 rt : (7,8,-3,-5)
7401 4673 cr : ln(3/2)
7401 4841 at : $\sqrt{3}/3 - 2\sqrt{5}/3$
7401 5587 cr : $1/\sqrt[3]{15}$
7401 6664 rt : (3,5,-7,5)
7401 7209 K : $(e + \pi)/10$
7401 7319 rt : (7,8,-9,-3)
7401 7320 Γ : $\sqrt[3]{23}$
7401 7538 cu : $3 + 3\pi/2$
7401 7685 sπ : $5(1/\pi)/6$
7401 8386 e^x : $2\sqrt{2}/3 + 2\sqrt{5}$
7401 8792 ℓ_{10} : $7\pi/4$
7402 0308 1/2 ÷ $s\pi(\sqrt{5})$
7402 2643 sr : $4/3 - \pi/4$
7402 3544 tπ : $\sqrt{3}/2 + \sqrt{5}/3$
7402 4252 rt : (6,2,9,8)
7402 5804 J_0 : 19/18
7402 5879 ln(5) ÷ $s\pi(\pi)$
7402 6490 cu : $1/4 + 4\sqrt{2}/3$
7402 6877 sq : $4 + 3e$
7402 8064 sq : $4 + 4\sqrt{5}/3$
7403 0481 Li_2 : $(\sqrt{2}\pi)^{-1/3}$
7403 1242 rt : (5,9,-4,-4)
7403 3482 exp(1/2) − exp(2)
7403 6268 ℓ_{10} : 2/11
7403 6661 cu : $1/3 - 4\sqrt{2}/3$
7403 7256 rt : (9,9,-7,8)
7403 9729 sr : $\sqrt{2} - \sqrt{3}/2$
7404 2022 rt : (9,3,5,-9)
7404 2174 cu : $9\sqrt{2}/2$
7404 4642 rt : (4,4,7,-9)
7404 5210 ln : $10\sqrt[3]{5}/3$
7404 5690 2^x : $3/4 + 4e/3$
7404 5891 rt : (9,2,6,7)
7404 8206 cu : $1/4 - 2\sqrt{3}/3$
7404 8947 rt : (1,5,-9,-9)
7404 9592 ln : $2\sqrt{3} + \sqrt{5}$
7404 9751 rt : (7,7,-5,7)
7405 0333 e^x : $1/3 + \sqrt{2}$
7405 0359 rt : (8,9,-7,-3)
7405 1194 rt : (4,1,8,7)
7405 3861 2^x : $4 + 3e/4$
7405 4092 rt : (9,7,7,5)
7405 4982 Γ : $\sqrt{13}$
7405 6779 ζ : $\exp(-\sqrt[3]{4})$
7405 8681 rt : (2,4,-9,-7)
7405 9715 rt : (8,4,-6,-1)
7406 0807 rt : (7,4,4,-8)
7406 1263 at : $1/2 - \sqrt{2}$
7406 2571 sq : $4/3 + \sqrt{5}$
7406 3276 cu : 19/21
7406 3734 tanh : $(e/3)^{1/2}$
7406 4527 cr : $9(e + \pi)/10$
7406 6180 sr : P_{12}
7406 9206 rt : (2,5,6,-8)
7406 9263 2/5 − exp(π)
7406 9550 10^x : $3 + 4\sqrt{5}$
7406 9648 2^x : $(2e/3)^2$
7407 0196 2^x : 16/11
7407 0280 rt : (4,6,9,5)
7407 1198 id : $1/4 + 2\sqrt{5}/3$
7407 1436 rt : (5,5,-3,6)
7407 2552 rt : (8,9,-8,1)
7407 3002 Ei : $9\sqrt[3]{3}/2$
7407 3398 rt : (7,1,5,6)
7407 4074 cu : $4\sqrt[3]{2}/3$
7407 4233 10^x : $9(1/\pi)/5$
7407 5072 ln : $2\sqrt{2}/3 + 2\sqrt{3}/3$
7407 5420 sr : $3e/2 - \pi/3$
7407 5898 ln(4/5) − $s\pi(\sqrt{2})$
7407 5900 rt : (3,5,-4,-1)
7407 6296 Γ : $8(1/\pi)/5$
7407 8449 ℓ_2 : $e - \pi/3$
7407 9747 ℓ_2 : $\sqrt[3]{14/3}$
7408 1822 exp(1/2) ÷ exp(4/5)
7408 4017 erf : $(\pi/2)^{-1/2}$
7408 4277 rt : (3,6,-8,-8)
7408 4869 rt : (2,7,-1,3)

7408 5249 rt : (5,5,3,-7)
7408 5893 Γ : $5\sqrt[3]{3}/2$
7408 5906 tanh : 20/21
7408 6985 rt : (8,1,-5,-1)
7409 2774 Li_2 : 14/23
7409 2830 rt : (2,0,7,6)
7409 7607 10^x : $9\sqrt{2}/10$
7409 8265 rt : (8,5,0,-6)
7410 0227 sr : $7\sqrt{3}/4$
7410 0236 erf : $\ln(\sqrt{2}\pi/2)$
7410 1332 Ψ : 16
7410 3699 rt : (6,7,-6,0)
7410 4663 rt : (7,6,6,4)
7410 7252 e^x : $\zeta(7)$
7410 7369 2^x : $((e+\pi)/3)^{-1/3}$
7411 0112 2^x : 4/5
7411 1481 θ_3 : 20/23
7411 1507 rt : (3,6,-5,3)
7411 1875 Γ : $\exp(-\sqrt{2}\pi/2)$
7411 3522 rt : (5,0,4,5)
7411 4763 J_0 : $(\sqrt[3]{5}/2)^{-1/3}$
7411 5190 id : $(\ln 3/2)^{1/2}$
7411 8356 rt : (3,6,2,-6)
7411 8546 $\ln(3/5) \times \exp(e)$
7412 0458 ζ : $7\zeta(3)/10$
7412 2125 tanh : $(\sqrt{3}/2)^{1/3}$
7412 3457 θ_3 : 10/11
7412 4178 e^x : $3e\pi/4$
7412 5063 ζ : 9/16
7412 6982 cu : $4/3 + \sqrt{2}$
7412 8539 rt : (8,0,1,4)
7412 9318 Ei : $1/\ln(2e)$
7412 9806 rt : (6,6,-1,-5)
7413 2986 ℓ_2 : $3\sqrt{3} + 2\sqrt{5}/3$
7413 3407 λ : $\exp(-15\pi/19)$
7413 4797 rt : (2,6,2,-1)
7413 7191 e^x : $(e+\pi)^{-1/3}$
7413 8348 rt : (7,9,-9,-6)
7413 8995 rt : (9,6,-4,-4)
7414 1353 rt : (5,7,-7,-7)
7414 1383 rt : (8,1,7,-9)
7414 4998 rt : (7,8,-5,4)
7414 5149 $\sqrt{5} \div \exp(1/4)$
7414 5615 rt : (3,1,-9,-6)
7414 6347 rt : (9,7,2,-9)
7414 9287 e^x : $7\pi/2$
7415 3115 e^x : $1 - 3\sqrt{3}/4$
7415 4093 2^x : $3/2 + \sqrt{5}/3$
7415 5690 rt : (2,7,-4,-6)
7415 6617 at : $\beta(2)$
7415 9265 $2/5 - \pi$
7415 9476 J_0 : 20/19
7416 1764 tanh : $(\ln 3)^{-1/2}$
7416 1984 sr : 11/20
7416 5738 id : $\sqrt{14}$
7416 6508 e^x : $(\sqrt{2}\pi/3)^3$
7416 7038 Ei : $5e/7$
7417 0694 rt : (4,8,4,-9)
7417 1685 rt : (8,9,-4,8)
7417 2321 rt : (3,2,9,-9)
7417 3169 rt : (4,7,-2,-4)
7417 4294 at : $\ln(5/2)$
7417 4859 rt : (7,8,1,-8)
7417 7541 rt : (6,2,6,-8)
7417 8818 sq : $1/\ln(\sqrt[3]{5}/2)$
7417 8888 rt : (7,7,-5,-3)
7418 0313 Ψ : $\exp(-3/2)$
7418 0581 rt : (4,6,-8,1)
7418 1783 rt : (9,2,3,-7)
7418 1840 sinh : $(\sqrt{2}/3)^{1/2}$
7418 2469 $1/4 + \exp(2/5)$
7418 3283 K : 10/17
7418 3396 tanh : 21/22
7418 3637 sr : $1/\sqrt[3]{6}$
7418 4697 $1/3 \times \exp(4/5)$
7418 5297 rt : (6,1,0,-3)
7418 5821 sinh : $7(e+\pi)/3$
7418 7468 ln : $5\sqrt[3]{2}/3$
7418 7615 rt : (8,4,8,7)
7419 0781 rt : (9,1,-3,-2)
7419 3338 sq : $4\sqrt{3} + \sqrt{5}/2$
7419 3734 ln : 10/21
7419 4412 rt : (2,5,8,4)
7419 4726 at : 11/12
7419 6091 J : $2\ln 2/5$
7419 7453 J_0 : $7\zeta(3)/8$
7419 7505 rt : (2,8,3,4)
7420 0251 $s\pi$: $\sqrt{2}/2 + \sqrt{5}/4$
7420 0668 tanh : $3(1/\pi)$
7420 3897 rt : (5,5,5,3)
7420 4749 $\exp(e) + s\pi(1/5)$
7420 8147 rt : (1,9,-6,-9)
7420 8622 rt : (8,5,2,2)
7420 8744 erf : $((e+\pi)/3)^{-1/3}$
7420 9597 rt : (2,9,3,-8)
7421 0096 erf : 4/5
7421 0193 $\ln(3) \times s\pi(\sqrt{5})$
7421 0611 ℓ_{10} : $4/3 + 4\pi/3$
7421 0785 rt : (5,9,0,-7)
7421 1776 rt : (8,9,-3,-6)
7421 5316 Ψ : $(\sqrt{5}/3)^{1/3}$
7421 6691 $2/3 \times \exp(\sqrt{2})$
7421 8160 K : $(\ln 2/2)^{1/2}$
7421 8601 $\ln(\sqrt{2}) + \exp(1/3)$
7421 8853 Ei : 13/22
7421 9029 $s\pi$: $3e/2 + 4\pi/3$
7421 9273 sinh : $4\zeta(3)/7$
7422 2719 rt : (9,0,-9,3)
7422 3512 2^x : $3\sqrt{2}/4 - 2\sqrt{5}/3$
7422 5949 Γ : $8\sqrt{3}/3$
7422 5965 rt : (7,3,2,-6)
7422 6360 rt : (4,3,5,-7)
7422 6866 rt : (1,3,8,-8)
7422 7409 cu : $3e + 4\pi/3$
7423 0748 id : $3\sqrt{3}/7$
7423 0752 rt : (6,7,-2,7)
7423 1154 ℓ_{10} : $3\sqrt{3}/4 - \sqrt{5}/2$
7423 1351 rt : (3,7,-3,1)
7423 2257 Ei : $(\sqrt{2}\pi/3)^3$
7423 3180 rt : (8,8,-9,-1)
7423 4466 sr : $e/4 + 3\pi/4$
7423 4629 rt : (5,8,-6,-2)
7423 4640 cr : 9/22
7423 6543 rt : (2,8,-3,-3)
7423 9192 rt : (1,8,0,-4)
7423 9682 10^x : $\sqrt{2}/2 + 4\sqrt{3}/3$
7424 1212 Ψ : $5e\pi$
7424 3097 rt : (4,8,-3,-5)
7424 4598 cr : $9(1/\pi)/7$
7424 4766 rt : (8,5,4,-9)
7424 7006 rt : (3,9,-7,5)
7424 8281 2^x : $1/4 - e/4$
7424 8660 K : $\exp(-\sqrt[3]{4}/3)$
7424 8716 rt : (9,3,4,5)
7424 9145 cu : $4 - \sqrt{5}/4$
7424 9432 rt : (7,8,-6,-6)
7424 9979 cr : $2\sqrt{3}/3 - \sqrt{5}/3$
7425 0674 cr : $4\sqrt{3}/3 + 4\sqrt{5}/3$
7425 1425 Ei : $(\sqrt{2}\pi)^{1/3}$
7425 1594 2^x : $4 - \pi/3$
7425 2394 10^x : $7e\pi/4$
7425 2802 rt : (6,3,7,6)
7425 4178 rt : (6,8,-6,-3)
7425 4845 J_0 : $(e/3)^{-1/2}$
7425 8285 $t\pi$: $2\sqrt{2}/3 + 3\sqrt{3}/2$
7425 8662 Ψ : $\sqrt{20/3}$
7425 8760 $2/5 \times \exp(\sqrt{5})$
7425 9324 sq : P_5
7426 1004 2^x : $5\sqrt[3]{3}/9$
7426 3311 ℓ_{10} : $\exp(\sqrt[3]{5})$
7426 3682 10^x : 8/9
7426 3732 ln : $1 + 3\pi/2$
7426 4068 id : $1/2 + 3\sqrt{2}$
7426 4272 rt : (7,9,-5,1)
7426 4714 sq : $3/4 + e/3$
7426 7219 10^x : $2/3 + \sqrt{2}/3$
7426 7763 rt : (7,1,9,9)
7426 8474 ℓ_{10} : $1/3 + 3\sqrt{3}$
7426 8702 rt : (7,2,-4,-1)
7426 9341 sinh : $7\sqrt{5}/2$
7426 9983 rt : (1,6,2,3)
7427 0489 cu : $3 - 2\pi/3$
7427 2136 tanh : 22/23
7427 2306 id : $7(1/\pi)/3$
7427 4985 10^x : 21/19
7427 5681 2^x : $2\zeta(3)/3$
7427 6337 $\ln(4) + \exp(\sqrt{5})$
7427 9199 rt : (8,4,-2,-4)
7427 9555 J_0 : 21/20
7428 2555 J_0 : $5\sqrt[3]{2}/6$
7428 3035 2^x : $1/3 + \pi/2$
7428 3853 rt : (9,6,0,-7)
7428 4381 erf : $5\sqrt[3]{3}/9$
7428 4531 id : $(2e/3)^{-1/2}$
7428 5506 cu : $1/4 + 4\sqrt{5}/3$
7428 5867 e^x : $1/4 + 3\sqrt{5}$
7428 6480 rt : (5,4,1,-5)
7428 7991 $\ln(\pi) + \exp(4)$
7428 8096 10^x : $4 + \sqrt{3}$
7428 9343 rt : (6,6,3,-8)
7428 9535 sinh : $\sqrt{3} + \sqrt{5} - \sqrt{7}$
7428 9723 e^x : $3/4 - \pi/3$
7429 0899 e^x : 5/9
7429 1595 erf : $2\zeta(3)/3$
7429 2127 sq : $4/3 - \sqrt{2}/3$
7429 4241 ζ : $\sqrt[3]{3}/7$
7429 4453 sq : $2e/3 + 4\pi$
7429 4466 ℓ_{10} : $3\sqrt{2}/4 + 2\sqrt{5}$
7429 5294 sinh : 11/16
7429 5920 rt : (3,6,6,-9)
7429 6509 $3/4 \div s\pi(\pi)$
7430 0695 rt : (8,9,9,5)
7430 1774 rt : (1,7,6,-7)
7430 4012 Ei : $(\sqrt[3]{5})^{1/2}$
7430 6387 ℓ_2 : $2\sqrt{2} - 2\sqrt{3}/3$
7430 8649 rt : (1,7,4,-9)
7431 0417 $s\pi$: $3\sqrt{2} + 2\sqrt{5}/3$
7431 0528 10^x : $\sqrt{2}/4 + 4\sqrt{3}$
7431 3010 Ei : 17/13
7431 4482 $s\pi$: 4/15
7431 6630 10^x : $\sqrt{3} - 2\sqrt{5}/3$
7431 7611 ζ : $\sqrt{2} + \sqrt{3} - \sqrt{6}$
7431 7996 rt : (3,9,-7,-1)
7431 8054 Ei : $1/\sqrt{22}$
7431 8644 Γ : $3\zeta(3)$
7431 8999 sr : $1 + 3e/4$
7431 9836 rt : (8,0,5,7)
7431 9887 J_0 : $(\sqrt{3}/2)^{-1/3}$
7432 1395 sinh : $9e\pi/8$
7432 5856 J : $\sqrt{5}/9$
7432 5985 rt : (8,9,1,-9)
7432 6035 $t\pi$: $\sqrt{2}/3 + \sqrt{3}$
7432 6761 rt : (2,8,5,-5)
7432 7018 Ψ : $e/3$
7432 7764 sq : $\sqrt{3} + \sqrt{5} + \sqrt{7}$
7433 1551 rt : (5,0,8,8)
7433 3236 ℓ_2 : $3/2 + 3\sqrt{3}$
7433 3470 rt : (7,2,3,4)
7433 5471 rt : (7,7,-1,-6)
7433 7799 10^x : $2e/3 - \pi/2$
7433 7952 rt : (9,5,-6,-2)
7434 0791 cr : $4 + 3\sqrt{3}/4$
7434 1250 $\ln(5) - s\pi(1/3)$
7434 3112 rt : (2,9,7,5)
7434 7112 rt : (4,7,2,-7)
7434 8422 cu : $10\sqrt[3]{2}/9$
7434 8987 rt : (2,7,-9,2)
7434 9606 $s\pi(1/4) \div s\pi(2/5)$
7435 1484 ℓ_2 : $2(e+\pi)/7$
7435 2425 tanh : $3\sqrt{5}/7$
7435 3252 tanh : 23/24
7435 3417 sq : $2 - e/4$
7435 3701 rt : (6,5,-3,-3)
7435 4371 rt : (2,8,-6,5)
7435 6007 rt : (4,2,6,5)
7435 7902 rt : (7,9,-1,8)
7436 0688 K : $1/\ln(2e)$
7436 1443 rt : (1,7,5,-8)
7436 2390 sr : $4e + \pi$
7436 3976 J_0 : $(\ln 3)^{1/2}$

7436 4134 $\ell_{10} : 3\sqrt{2}+3\sqrt{3}/4$
7436 4778 $10^x : 2/3+3\sqrt{2}/2$
7436 4996 $\ln : 7\zeta(3)/4$
7436 5094 $\mathrm{sr} : 4\sqrt{2}/3+2\sqrt{3}/3$
7436 6382 rt : (7,3,6,-9)
7436 6595 rt : (8,3,-8,1)
7436 6838 $\ln : 3+e$
7436 6875 $\Gamma : 1/\sqrt{17}$
7436 6922 rt : (4,0,-7,9)
7436 8531 rt : (3,9,1,-3)
7437 4455 rt : (6,8,-2,4)
7437 4804 $\mathrm{as} : 1-3\sqrt{5}/4$
7437 5390 $\ln : 4e/3+2\pi/3$
7437 5558 at : 23/25
7437 6282 $2^x : 3/2+3\sqrt{5}$
7437 6692 $10^x : 2\sqrt{2}+\sqrt{3}/4$
7437 6974 rt : (9,1,1,-5)
7437 9525 rt : (2,7,-8,-9)
7438 2628 $K : 13/22$
7438 2922 rt : (6,3,-9,7)
7438 3219 rt : (6,4,1,1)
7438 5315 rt : (2,9,-4,-8)
7438 5463 rt : (8,8,-5,-4)
7438 7522 rt : (3,8,-1,-1)
7438 8014 $J_0 : 22/21$
7438 8095 $\mathrm{t}\pi : 7e/2$
7439 0006 $\mathrm{sq} : 4+2\pi$
7439 0125 rt : (1,2,9,6)
7439 0877 $\mathrm{id} : (e/3)^3$
7439 1544 $\ell_{10} : 8\ln 2$
7439 3598 $e^x : 2/3+3\sqrt{5}/2$
7439 6020 cr : 7/17
7439 6630 rt : (6,1,4,-6)
7439 9849 $\exp(\sqrt{5}) \div s\pi(\pi)$
7440 0000 cu : 7/5
7440 2220 $Ei : 17/18$
7440 3065 rt : (5,8,-2,-5)
7440 4956 $\ell_{10} : \sqrt[3]{3}/8$
7440 5818 $5 \div \exp(3/5)$
7440 7197 $J_0 : \pi/3$
7440 8940 $\mathrm{id} : 1/3-3e/2$
7440 9818 $\mathrm{sr} : 3\sqrt{3}/4-\sqrt{5}/3$
7441 0492 $\mathrm{s}\pi : 2\zeta(3)/9$
7441 1528 rt : (9,3,8,8)
7441 2232 $\mathrm{s}\pi : 5\ln 2/2$
7441 3106 $\ell_{10} : e/2+4\pi/3$
7441 4481 $\mathrm{cr} : 2\sqrt[3]{3}/7$
7441 5111 rt : (8,4,2,-7)
7441 6177 $\sqrt{5} \times \exp(\pi)$
7441 6664 $\mathrm{s}\pi : 3\sqrt[3]{5}/7$
7441 7758 $\sinh : 9(1/\pi)$
7442 1385 rt : (6,9,-2,-4)
7442 2058 rt : (3,1,7,-7)
7442 2092 $\mathrm{sq} : 1/2-3\pi/2$
7442 2512 rt : (9,8,5,3)
7442 3396 $\Gamma : e\pi/3$
7442 4327 $\sqrt{5}-\exp(2/5)$
7442 4959 rt : (2,8,1,-6)
7442 5923 $\mathrm{sr} : 2e+2\pi/3$
7442 6753 rt : (8,1,-1,2)
7442 6804 $\Gamma : (3\pi)^{-1/2}$
7442 6996 $\ell_{10} : \sqrt{2}/4+3\sqrt{3}$
7442 7686 tanh : 24/25
7442 9150 $\mathrm{cu} : 3\sqrt{3}+\sqrt{5}/3$
7442 9269 rt : (7,6,-7,-1)
7442 9679 $\mathrm{id} : 5e\pi/9$
7443 1055 $\mathrm{at} : 3\sqrt{3}/2-3\sqrt{5}/4$
7443 1169 $2^x : 3/4+2\sqrt{3}/3$
7443 2557 $\ln(4) \times \exp(5)$
7443 3141 rt : (4,7,-8,8)
7443 4797 rt : (5,4,5,-8)
7443 7059 $2^x : (\sqrt{2}/3)^{-1/2}$
7443 9125 rt : (1,7,6,1)
7443 9340 $\mathrm{s}\pi : 1/\sqrt{14}$
7443 9646 $\pi \div \ln(\pi)$
7443 9789 rt : (9,9,-9,-2)
7444 1647 rt : (7,7,3,-9)
7444 3355 $e^x : 7\sqrt[3]{3}/10$
7444 5077 rt : (3,5,-6,-6)
7444 5171 $\mathrm{cu} : \sqrt{2}/2+3\sqrt{5}$
7444 9323 $J : \exp(-17\pi/6)$
7445 0275 $10^x : 2/3+\pi/3$
7445 0293 $\mathrm{sq} : 1/4-\pi/2$
7445 2945 rt : (1,8,-4,-7)
7445 4366 $\mathrm{cr} : 3+4\sqrt{3}/3$
7445 9001 rt : (2,4,8,-9)
7446 0770 $10^x : 7(e+\pi)/2$
7446 4428 rt : (6,9,-6,-3)
7446 5590 $Li_2 : 11/18$
7446 5948 $\mathrm{sr} : 4\ln 2/5$
7446 7849 rt : (9,5,-2,-5)
7446 9204 rt : (7,2,0,-4)
7447 0889 $\mathrm{cu} : 1/3-3\sqrt{2}$
7447 2287 $\Psi : (3\pi)^{-3}$
7447 2339 rt : (5,1,2,3)
7447 2988 $2/5-\ln(\pi)$
7447 4393 rt : (7,5,-8,8)
7447 4455 $Ei : 6\pi/5$
7447 5470 $\ell_{10} : \sqrt[3]{2}/7$
7447 5573 $\zeta : (\pi/2)^{1/3}$
7447 6203 rt : (8,5,6,5)
7447 6389 $\mathrm{sr} : (e+\pi)^{-1/3}$
7447 7134 $e^x : 4+3\pi/2$
7448 2607 rt : (1,5,1,-1)
7448 3593 $\mathrm{sr} : 9(e+\pi)/7$
7448 6462 $J_0 : 23/22$
7448 7007 rt : (8,8,-1,-7)
7448 7450 rt : (6,8,8,4)
7448 9404 $\sinh : 2\pi$
7448 9808 $erf : \ln(\sqrt{5})$
7449 0003 rt : (4,8,-7,-8)
7449 1252 $\mathrm{cu} : 2\sqrt{2}-\sqrt{3}/4$
7449 2603 $\mathrm{s}\pi : P_{66}$
7449 2694 $\mathrm{s}\pi : (e+\pi)/8$
7449 3165 rt : (9,1,5,-8)
7449 4043 $\mathrm{s}\pi : P_{65}$
7449 4106 rt : (6,5,1,-6)
7449 4507 $\tanh : 2\sqrt[3]{3}/3$
7449 5417 rt : (3,9,-3,-4)
7450 0701 $\mathrm{t}\pi : P_7$
7450 0995 $\tanh : 4\zeta(3)/5$
7450 2968 $e^x : 2-e/4$
7450 3793 rt : (7,2,7,7)
7450 4401 $\mathrm{cr} : P_{56}$
7450 5885 $2^x : 2e+3\pi/4$
7450 6130 $\mathrm{t}\pi : 7(e+\pi)/5$
7450 7716 rt : (4,2,3,-5)
7450 8020 $\exp(1/3)-\exp(\pi)$
7451 0481 rt : (5,8,2,-8)
7451 1285 $\ln : 3+4\pi$
7451 1578 rt : (6,4,-9,4)
7451 2955 rt : (6,5,-9,-7)
7451 4119 rt : (1,6,-1,-3)
7451 5038 rt : (1,9,2,-2)
7451 6600 $\mathrm{sq} : 4-4\sqrt{2}$
7451 8877 $\mathrm{at} : 3/2-\sqrt{3}/3$
7452 0598 rt : (6,1,8,-9)
7452 0750 $\mathrm{id} : e/4-3\pi$
7452 2070 rt : (4,1,-7,6)
7452 6714 rt : (3,5,4,-7)
7452 6899 rt : (5,5,9,6)
7452 6948 $\mathrm{s}\pi : (\ln 2/3)^{-2}$
7452 7900 $\tanh : 5\sqrt{3}/9$
7453 0357 $\sqrt{3} \times s\pi(\pi)$
7453 2925 $\mathrm{id} : 5\pi/9$
7453 5599 $\mathrm{id} : \sqrt{5}/3$
7453 7037 cu : 17/6
7453 7355 $Ei : 3\sqrt[3]{2}/4$
7453 8572 rt : (4,8,-6,6)
7454 1947 at : 12/13
7454 4400 rt : (2,1,5,4)
7454 5805 $Ei : (\sqrt{5})^{-3}$
7454 6731 $2^x : 4\sqrt{2}/3-4\sqrt{3}/3$
7454 6878 $J : e/7$
7454 8621 $\ell_2 : (3\pi/2)^{1/3}$
7455 0218 $e^x : 7(1/\pi)/4$
7455 0546 rt : (8,3,-4,-2)
7455 1899 rt : (8,7,-7,9)
7455 2169 $Ei : 6\sqrt{5}$
7455 2511 $\exp(2)+\exp(\sqrt{5})$
7455 3442 $\lambda : \exp(-11\pi/14)$
7455 4342 rt : (1,7,-5,6)
7455 4939 rt : (9,5,2,-8)
7455 6067 $\mathrm{cu} : 1/2-3\pi/2$
7455 6581 rt : (6,4,-9,2)
7455 7206 $2^x : 3/4+\sqrt{2}/2$
7455 7782 $10^x : 1/3+3e/2$
7455 9032 rt : (7,6,-3,-4)
7456 0192 $3/5 \div \ln(\sqrt{5})$
7456 0636 rt : (1,2,6,-6)
7456 0945 $\mathrm{cu} : 4-\pi/3$
7456 1669 $Ei : \pi^2/6$
7456 2239 $\mathrm{sr} : 2+\pi/3$
7456 4372 $\mathrm{s}\pi : 9\sqrt[3]{2}/5$
7456 4502 rt : (6,9,-2,1)
7456 5173 $\ln : (\sqrt{2\pi})^{1/2}$
7456 6224 $10^x : 4/3+3\sqrt{2}/2$
7456 8322 rt : (5,7,-8,0)
7456 8462 $10^x : P_{43}$
7456 9057 $\ell_2 : 3+\sqrt{2}/4$
7456 9871 $10^x : \sqrt{2}/2+2\sqrt{3}/3$
7457 1133 $2^x : e-\pi$
7457 2745 $\zeta : \exp(-\pi/2)$
7457 6168 $e^x : 3/2-2\sqrt{2}/3$
7457 6221 $J_0 : 24/23$
7457 8430 rt : (9,4,2,3)
7458 0812 $\Psi : 19/21$
7458 1350 rt : (8,1,3,5)
7458 2524 $\zeta : 3\ln 2/10$
7458 3065 $\tanh : (\sqrt{5}/2)^{-1/3}$
7458 3080 rt : (7,2,4,-7)
7458 3482 $\mathrm{s}\pi : \sqrt{3}$
7458 4223 rt : (6,5,5,-9)
7458 5162 $\mathrm{cr} : 1/\sqrt[3]{14}$
7458 6776 sq : 19/22
7458 9831 rt : (3,9,-3,-6)
7459 1094 $e^x : \sqrt[3]{18}$
7459 1234 $1/2 \times \exp(2/5)$
7459 1728 $\ell_{10} : 4+\pi/2$
7459 2654 $\ell_2 : 3\sqrt{5}$
7459 6392 $Ei : \sqrt[3]{22/3}$
7459 6669 $\mathrm{sq} : \sqrt{3}+\sqrt{5}$
7459 7341 rt : (2,5,-8,5)
7459 8022 $\pi-\exp(1/3)$
7459 8986 $\exp(2/5)+s\pi(\sqrt{3})$
7459 9347 rt : (3,9,1,-3)
7460 0380 rt : (7,6,-8,5)
7460 0531 $s\pi(\sqrt{2}) \times s\pi(e)$
7460 1665 rt : (4,6,0,3)
7460 1987 $e^x : 5\sqrt{2}/7$
7460 2349 $\mathrm{at} : 4\ln 2/3$
7460 3305 rt : (9,9,-1,-8)
7460 3607 rt : (1,8,3,8)
7460 5474 rt : (5,3,-1,-3)
7460 7337 $\ln : 4+\sqrt{3}$
7460 8022 $10^x : \sqrt{3}/4+\sqrt{5}$
7461 2899 rt : (6,0,-2,1)
7461 3890 rt : (9,8,9,6)
7461 6243 $e^x : 1+\pi/3$
7462 3035 rt : (4,9,-8,-7)
7462 5105 $K : 6\ln 2/7$
7462 5462 $\mathrm{id} : 8e$
7462 5948 rt : (4,2,7,-8)
7462 7118 $\mathrm{at} : (\sqrt[3]{5}/2)^{1/2}$
7462 7354 rt : (5,3,-6,7)
7462 8071 $\ln : 3/4+e/2$
7462 9021 $Ei : \exp(2\pi/3)$
7462 9751 $2^x : -\sqrt{3}+\sqrt{5}+\sqrt{6}$
7463 0046 $\ln : 3\sqrt{2}+2\sqrt{5}/3$
7463 0345 rt : (8,7,-7,-2)
7463 2032 $\mathrm{s}\pi : 4e/3-3\pi/4$
7463 2074 $\ell_{10} : 4/3+3\sqrt{2}$
7463 2964 rt : (6,9,2,8)
7463 3722 rt : (7,3,-3,-1)
7463 4204 $\ln : 4e+3\pi/2$
7463 5266 $\mathrm{sr} : 7(1/\pi)/4$
7463 5568 $\mathrm{cu} : 10\sqrt[3]{3}/7$
7463 6636 rt : (5,1,6,6)
7463 8059 rt : (2,2,-9,-7)
7463 9772 $\theta_3 : 1/\sqrt[3]{22}$

7464 0546 rt : (7,6,1,-7)
7464 1490 $\zeta : 5/24$
7464 2198 rt : (9,4,-8,0)
7464 3356 $\ell_{10} : 3/2 + 3e/2$
7464 3541 cr : $3\ln 2/5$
7464 4228 cu : $\sqrt{2}/3 + 4\sqrt{5}$
7464 5224 sr : $3/2 - 2\sqrt{2}/3$
7464 6266 rt : (2,7,-5,-1)
7464 8135 $\zeta : 7e/10$
7464 8888 $\Gamma : 10\sqrt{2}/3$
7464 9243 $1/3 + \exp(5)$
7465 1361 rt : (8,3,0,-5)
7465 2736 t$\pi : 1/\sqrt{24}$
7465 3072 ln : $9\sqrt{3}$
7465 6121 $K : \ln(2e/3)$
7465 6307 rt : (6,4,5,4)
7465 6824 rt : (1,6,3,-6)
7465 7359 $2/5 + \ln(\sqrt{2})$
7465 8394 $J_0 : 25/24$
7465 9529 $\ell_{10} : 3\sqrt{3}/2 + 4\sqrt{5}/3$
7466 3701 rt : (4,9,-4,4)
7466 5957 $2^x : 4 - 2\pi/3$
7466 6112 cr : $3e + 4\pi$
7466 6519 sinh : $10\sqrt{2}/3$
7466 7167 rt : (1,6,-4,7)
7466 7195 rt : (6,9,-6,-7)
7466 7697 rt : (3,9,5,8)
7466 8032 rt : (9,0,-1,3)
7466 8947 $\zeta : 1/\sqrt{23}$
7467 0080 rt : (8,1,7,8)
7467 0824 sq : $3\sqrt{2}/4 + \sqrt{5}/2$
7467 1490 rt : (6,5,-9,1)
7467 1812 ln : $\sqrt{3}/4 + 3\sqrt{5}/4$
7467 4389 rt : (8,8,-7,6)
7467 4427 t$\pi : \sqrt{23}$
7467 6032 rt : (1,1,-8,5)
7468 0547 $2^x : \ln(\sqrt{5})$
7468 2632 $\ell_2 : 4 + 3\pi$
7468 2780 rt : (4,6,4,-8)
7468 2835 $2^x : 1/3 + \sqrt{2}/3$
7468 3375 t$\pi : \sqrt{3} + 2\sqrt{5}$
7468 4981 rt : (5,8,-3,-3)
7468 8727 sr : $1/3 + e$
7468 9310 rt : (3,5,-5,-8)
7468 9360 $\sqrt{5} - \ln(3/5)$
7469 0079 cr : $5/12$
7469 0209 $erf : 4\sqrt{2}/7$
7469 0333 rt : (9,4,6,6)
7469 1399 s$\pi : 4e - \pi$
7469 2799 at : $(\sqrt[3]{2})^{-1/3}$
7469 5241 $e^x : 21/11$
7469 6221 rt : (3,9,-7,-9)
7470 0724 rt : (2,6,-8,-9)
7470 1319 rt : (8,7,-3,-5)
7470 1344 $\theta_3 : 5/14$
7470 2229 t$\pi : 5(1/\pi)/2$
7470 3110 $\ln(5) - \exp(\sqrt{5})$
7470 4447 rt : (7,7,4,2)
7470 5105 rt : (5,3,3,-6)
7470 5598 $\ell_{10} : 3\sqrt{2}/2 + 2\sqrt{3}$
7470 7881 rt : (1,9,-8,-7)
7471 1970 rt : (6,5,-5,8)
7471 2377 rt : (2,1,9,7)
7471 2447 $2^x : 4\sqrt{3} + \sqrt{5}$
7471 3705 $2 \div \ln(\pi)$
7471 4517 s$\pi : 3/4 + 4\sqrt{5}/3$
7471 5135 rt : (8,3,4,-8)
7471 6092 id : $\sqrt[3]{16/3}$
7471 7695 as : $e/4$
7471 7850 rt : (6,4,-5,-1)
7471 9839 rt : (5,6,-5,-5)
7472 1440 ln : $9/19$
7472 2023 $Ei : 13/2$
7472 2475 $erf : 7\ln 2/6$
7472 4641 rt : (4,2,-7,3)
7472 6068 $2^x : 2/3 + 2e$
7472 6216 rt : (9,4,-4,-3)
7472 6831 sinh : $(\sqrt[3]{5}/3)^{-1/2}$
7473 0314 sinh : $(2\pi/3)^{-1/2}$
7473 0817 $10^x : (\sqrt[3]{5}/3)^{1/3}$
7473 1089 rt : (2,6,-6,-7)
7473 1112 sq : $4\sqrt{3}/3 + \sqrt{5}/2$
7473 1530 $erf : (\sqrt[3]{4}/3)^{1/3}$
7473 3355 $\ln(3) + \exp(1/2)$
7473 4654 rt : (3,9,-6,9)
7473 6569 t$\pi : 5e/2$
7473 8199 $\ell_2 : 2\sqrt{3}/3 - \sqrt{5}/4$
7474 1525 rt : (9,0,3,6)
7474 2501 $\ell_{10} : 5\sqrt{5}/2$
7474 5328 sq : $2e/3 + \pi/4$
7474 6545 $J_0 : 3\ln 2/2$
7474 7741 t$\pi : 7/18$
7474 8965 $2^x : 3(e + \pi)$
7474 9372 rt : (8,7,1,-8)
7475 0957 $10^x : e/2 - \pi/4$
7475 1882 rt : (5,8,1,4)
7475 2291 rt : (1,7,2,2)
7475 2621 sinh : $(3\pi)^{1/2}$
7475 3854 rt : (7,7,-8,2)
7475 3905 cu : $2e/3 - 4\pi$
7475 4689 id : $1/3 + \sqrt{2}$
7475 5378 $Ei : (\sqrt{5}/2)^{-1/2}$
7475 6096 at : $3/4 - 3\sqrt{5}/4$
7475 6753 rt : (2,7,-1,-4)
7475 9767 cr : $5e\pi/8$
7476 1493 rt : (6,4,9,7)
7476 4494 rt : (5,3,7,-9)
7476 7042 rt : (9,8,-7,-3)
7476 7439 sr : $\sqrt{5}/4$
7476 7578 tanh : $(e/3)^{1/3}$
7476 8737 $J_0 : 3\sqrt{3}/5$
7476 8797 cr : $\sqrt{3}/2 + 2\sqrt{5}$
7477 0274 rt : (1,7,2,-2)
7477 2437 $erf : 17/21$
7477 3404 $1/5 - \exp(2/3)$
7477 3667 rt : (3,5,-6,8)
7477 3866 cu : $4e/7$
7477 4500 rt : (2,3,6,-7)
7477 4946 at : P_{34}
7477 6263 as : $17/25$
7477 8011 rt : (6,8,-8,-1)
7477 8072 ln : $(3\pi)^{1/3}$
7477 8198 rt : (9,4,0,-6)
7477 8570 rt : (3,4,8,5)
7477 8633 sr : $4\sqrt{3}/3 + \sqrt{5}/3$
7478 1453 rt : (7,7,-4,9)
7478 5269 rt : (7,1,-6,1)
7478 5862 $\ell_{10} : \sqrt{2} + \sqrt{3} + \sqrt{6}$
7478 7021 rt : (9,9,-9,-8)
7478 8323 rt : (9,0,7,9)
7479 1169 t$\pi : 4\sqrt{2} + \sqrt{3}$
7479 1915 ln : $3/2 + 3\sqrt{2}$
7479 2043 rt : (1,8,-7,-4)
7479 2892 rt : (6,4,-1,-4)
7479 2983 $2^x : 1 + e/3$
7479 5307 rt : (5,7,4,-9)
7479 6507 $\theta_3 : 14/15$
7479 7664 $10^x : 1/\sqrt{17}$
7479 9961 rt : (1,6,4,-6)
7480 2055 s$\pi : (\sqrt[3]{5}/2)^2$
7480 3355 $K : (3\pi/2)^{-1/3}$
7480 3880 $\ell_{10} : 3 + 3\sqrt{3}/2$
7480 4314 sq : $2/3 - 4e/3$
7480 4878 sinh : $\sqrt{6}$
7480 5509 rt : (3,8,-9,1)
7480 5761 rt : (9,8,-3,-6)
7480 5886 rt : (6,0,6,7)
7480 7454 t$\pi : 1/\sqrt[3]{17}$
7480 9211 sq : $\sqrt{5} + \sqrt{6} + \sqrt{7}$
7480 9231 $\ell_2 : 4 + e$
7481 1485 rt : (7,3,1,2)
7481 1948 id : $3\sqrt{3} - 4\sqrt{5}$
7481 2083 $\pi \div \ln(2/3)$
7481 3561 rt : (9,4,4,-9)
7481 3754 rt : (2,8,-9,-9)
7481 4580 rt : (2,7,3,-7)
7481 5363 t$\pi : \exp(-\sqrt[3]{4})$
7481 6210 $10^x : \sqrt{3}/4 - \sqrt{5}/4$
7481 8996 rt : (5,4,-6,4)
7481 9631 cu : $3/4 - 4e/3$
7482 0169 s$\pi : 4\sqrt{2}/3 - 2\sqrt{3}/3$
7482 1127 rt : (8,9,-7,3)
7482 7232 rt : (6,8,-4,-4)
7482 8353 sr : $\sqrt{20}\pi$
7482 9336 $J_0 : (\sqrt{5}/2)^{1/3}$
7483 0803 sr : $4\sqrt[3]{2}/9$
7483 1949 rt : (7,7,8,5)
7483 2129 cu : $\sqrt{3}/3 + 2\sqrt{5}$
7483 2481 tanh : $(\ln 3)^{-1/3}$
7483 2734 at : P_{17}
7483 3147 sr : $14/25$
7483 3424 rt : (9,8,1,-9)
7483 6078 $10^x : P_{15}$
7483 6356 rt : (6,4,3,-7)
7483 6747 rt : (5,6,-9,-8)
7483 7083 $\ell_2 : 4\sqrt[3]{2}/3$
7483 7804 at : $13/14$
7484 2604 $\Psi : 9\ln 2$
7484 4558 rt : (3,0,5,5)
7484 5119 cu : $3\sqrt{3}/4 + 3\sqrt{5}/2$
7484 8020 rt : (7,5,-9,1)
7484 8247 sr : $4 - 2\sqrt{2}/3$
7484 9018 $\Psi : 7(1/\pi)/10$
7485 3083 $10^x : \sqrt{2}/3 + \sqrt{3}/2$
7485 4279 $Li_2 : (\ln 2/3)^{1/3}$
7485 4481 rt : (1,9,3,7)
7485 5399 rt : (1,7,-2,-5)
7485 8562 rt : (6,8,0,-7)
7486 1594 rt : (7,1,-2,-2)
7486 1813 s$\pi : 1/\ln(\sqrt[3]{3})$
7486 5573 sq : $\sqrt{3} + \sqrt{5} - \sqrt{7}$
7486 5603 s$\pi : 6\sqrt[3]{3}/5$
7486 6749 $e^x : 1/4 + 3\sqrt{5}/2$
7486 7451 t$\pi : 4e\pi/9$
7486 7713 rt : (1,3,-9,-8)
7486 8557 ln : $7e/9$
7487 0260 rt : (3,8,-5,-2)
7487 1549 $10^x : 3e - \pi/4$
7487 2177 rt : (1,8,8,-3)
7487 2865 $J_0 : \zeta(5)$
7487 2929 rt : (7,3,5,5)
7487 3298 sr : $9e/8$
7487 5464 rt : (6,6,-9,-2)
7487 6253 ln : $4 - 4\sqrt{2}/3$
7487 9713 $\ln(2/3) + \exp(e)$
7488 0264 rt : (3,5,3,6)
7488 1060 $e^x : 1/4 + 3e/2$
7488 1714 sinh : $\exp(-1/e)$
7488 3017 sq : $3\sqrt[3]{3}/5$
7488 5906 $\Gamma : \exp(\pi/3)$
7488 7159 id : $(\sqrt[3]{2}/3)^{1/3}$
7488 7219 $3 \times \ln(2/5)$
7488 7822 rt : (2,7,-4,-8)
7488 8266 tanh : $7\ln 2/5$
7488 8428 s$\pi : 2\sqrt{2} - \sqrt{5}/4$
7488 8829 rt : (3,0,9,8)
7488 9357 id : $7\pi/8$
7489 0656 rt : (7,5,-5,-2)
7489 1424 at : $1/4 + e/4$
7489 1744 $Ei : 5\sqrt[3]{5}$
7489 2602 rt : (6,6,-5,5)
7489 2703 $\zeta : \sqrt[3]{2}/6$
7489 3625 sr : $4\sqrt{2} - 3\sqrt{3}/2$
7489 4479 rt : (1,7,-6,-8)
7489 4845 cr : $4\sqrt{2}/3 + 2\sqrt{3}$
7489 5090 sinh : $9/13$
7489 5242 $e^x : \sqrt{5}/4$
7489 5743 ln : $8(e + \pi)/3$
7489 6207 $Ei : 2(1/\pi)/3$
7489 7736 $10^x : 2\sqrt{3} + \sqrt{5}/4$
7489 7845 rt : (7,1,2,-5)
7489 9289 sq : $8\ln 2$
7490 1914 rt : (2,7,0,7)
7490 2636 rt : (3,8,-1,-5)
7490 3327 $e^x : 4 + \pi/4$
7490 3501 $\Gamma : \sqrt{2}\pi$
7490 4142 rt : (7,3,9,8)
7490 7409 sinh : $e\pi/2$
7490 8372 rt : (3,4,6,-8)
7490 8705 $2^x : \sqrt{2}\pi$

7490 9630 rt : (7,9,-8,-2)
7491 0462 sinh : $9(e+\pi)/8$
7491 1436 sr : $3/4+4\sqrt{3}/3$
7491 4593 rt : (7,5,-1,-5)
7491 7418 10^x : $7\sqrt{5}/10$
7491 9037 rt : (7,1,6,-8)
7491 9757 Ψ : $9\sqrt[3]{3}/5$
7491 9985 ln : $4/23$
7492 2073 rt : (3,8,3,-8)
7492 5557 sr : $\exp(-\sqrt{3}/3)$
7492 6623 rt : (7,9,-4,-5)
7492 7812 at : $1/\ln((e+\pi)/2)$
7492 7929 θ_3 : $9(1/\pi)/8$
7492 9926 rt : (7,5,3,-8)
7493 0614 $1/5+\ln(\sqrt{3})$
7493 2024 Γ : $5\sqrt[3]{5}/3$
7493 2222 sq : $4/3+2\pi/3$
7493 3314 rt : (2,9,-3,-3)
7493 4483 e^x : $8\zeta(3)$
7493 5649 10^x : $1/3+2\pi/3$
7493 6867 cu : $3/2+\sqrt{2}$
7493 7527 Γ : $\exp(-3\pi/2)$
7493 8234 rt : (7,9,0,-8)
7494 2240 rt : (3,6,-8,-5)
7494 2587 cr : $2+3\sqrt{5}/2$
7494 4781 rt : (7,8,-8,-1)
7494 5927 sr : $6\sqrt[3]{2}$
7494 6402 erf : $13/16$
7494 6886 at : $(\sqrt{3}/2)^{1/2}$
7494 7425 sr : $2+3\sqrt{2}/4$
7494 7585 tπ : $3\sqrt{2}-\sqrt{3}/3$
7494 8732 10^x : $1/4+2e$
7494 9682 rt : (3,6,-4,6)
7495 1235 cr : $8/19$
7495 1623 rt : (7,8,-4,6)
7495 4347 rt : (1,9,-7,3)
7495 5185 sq : $\sqrt{3}/3+\sqrt{5}/3$
7495 7242 $\exp(\sqrt{2})\div s\pi(1/3)$
7495 7490 cu : $3e/2-\pi/2$
7495 9146 sinh : $2\sqrt{3}/5$
7496 0824 Ψ : $(e/2)^{-1/3}$
7496 0994 sr : $(2\pi)^3$
7496 1906 ℓ_{10} : $e/3+3\pi/2$
7496 2878 10^x : $7\sqrt[3]{5}$
7496 2896 at : $\sqrt{2}/4+\sqrt{3}/3$
7496 3785 tπ : $e/3+\pi/2$
7496 5375 cr : $3+3\pi/4$
7497 4643 ζ : $19/10$
7497 4904 ζ : $4/19$
7497 6130 Ei : $5\sqrt{3}/2$
7497 7773 ζ : $10\sqrt[3]{5}/9$
7497 8019 ln : $3\sqrt[3]{2}/8$
7497 9858 2^x : $3/2+3e/2$
7498 1964 Ei : $7\pi^2/10$
7498 3039 ℓ_2 : $-\sqrt{3}+\sqrt{6}+\sqrt{7}$
7498 3491 sπ : $e\pi/2$
7498 4747 10^x : $4\sqrt{2}+\sqrt{5}/2$
7498 5978 tπ : $4\pi/7$
7499 7452 cr : $4+e/2$
7499 8227 ζ : $(\pi)^{-1/2}$
7499 9919 cu : $3/2+2\sqrt{5}/3$
7500 0000 id : $3/4$
7500 3249 Γ : $\exp(5/3)$
7500 4059 tπ : P_{81}
7500 4517 10^x : $3/2-3\sqrt{2}/4$
7500 5697 K : $\ln(\ln 3/2)$
7500 7432 ℓ_{10} : $2+4e/3$
7500 9450 10^x : $3-\sqrt{2}/3$
7500 9847 ln : $3\sqrt{3}+\sqrt{5}/4$
7501 0655 cr : $9(e+\pi)$
7501 3054 $\ln(5)+\exp(\pi)$
7501 4144 cr : $3\sqrt{2}+\sqrt{5}/2$
7501 4271 tπ : $7\sqrt[3]{2}/4$
7501 4349 sinh : $(\ln 3)^3$
7501 4709 Ψ : $3\sqrt{3}/2$
7501 6661 ℓ_2 : $\ln(2e/3)$
7501 9410 sq : $3-3\sqrt{5}/4$
7501 9540 $\ln(2/5)\div\exp(1/5)$
7502 2320 cu : $1+3\sqrt{2}/4$
7502 4524 as : $15/22$
7502 5619 ζ : $(\sqrt{5}/3)^{1/2}$
7502 6429 ℓ_{10} : $2\sqrt{3}/3+2\sqrt{5}$
7502 6763 sinh : $\sqrt[3]{1/3}$
7503 0990 sq : $e+3\pi/4$
7503 1301 J_0 : $(e/3)^{-1/3}$
7503 1709 10^x : $\exp(-\sqrt{2})$
7503 4189 rt : (9,8,-9,8)
7503 8408 id : $4\sqrt{3}/3-\sqrt{5}/4$
7504 0388 tπ : $5\sqrt{5}/4$
7504 1398 ℓ_2 : $3-4\sqrt{5}/3$
7504 1435 J : $1/\sqrt{14}$
7504 2845 2^x : $1-\sqrt{2}$
7504 3059 ln : $\sqrt[3]{19/2}$
7504 5352 Ψ : $25/4$
7504 5881 cr : $1-\sqrt{3}/3$
7504 6712 rt : (5,1,-9,8)
7504 7073 rt : (1,6,3,9)
7504 8829 ln : $1+\sqrt{5}/2$
7505 0704 sinh : $1/\ln(\sqrt{5}/2)$
7505 1604 sr : $3/2+4\pi$
7505 2737 sq : $3/2+e/4$
7505 6846 tπ : $3\sqrt[3]{3}/8$
7505 7310 Ei : $\sqrt{5}-\sqrt{6}+\sqrt{7}$
7505 8640 sq : $2+\pi/2$
7505 9073 rt : (5,5,-2,8)
7506 1107 e^x : $4\sqrt[3]{2}/9$
7506 2910 ln : $7\sqrt{5}$
7506 3246 rt : (9,7,-1,-7)
7506 6072 sπ : P_{23}
7506 7250 e^x : $14/25$
7506 7756 e^x : $9\sqrt{2}/5$
7506 9592 rt : (5,5,-6,1)
7507 0783 sq : $5\ln 2/4$
7507 1325 ℓ_2 : $7\sqrt[3]{3}/3$
7507 1939 e^x : $\sqrt{17}$
7507 1990 rt : (9,3,2,-7)
7507 3717 rt : (1,6,-9,-8)
7507 3934 tπ : $7\pi/6$
7507 5380 K : $3/5$
7507 5453 rt : (9,7,-5,-4)
7507 5767 at : $1/2+\sqrt{3}/4$
7507 7640 sq : $3-3\sqrt{5}$
7508 0328 rt : (5,9,5,8)
7508 1755 ln : $4/3+\pi/4$
7508 2384 sr : $4/3+\sqrt{3}$
7508 3540 rt : (9,1,5,7)
7508 4596 $\ln(3)+\exp(\sqrt{3})$
7508 5267 rt : (6,3,-8,9)
7508 6488 10^x : $3\sqrt{5}$
7508 8236 rt : (9,3,-2,-4)
7509 0855 rt : (1,9,2,-7)
7509 1652 sinh : $\sqrt[3]{7/3}$
7509 2906 at : $14/15$
7509 3076 2^x : $4\sqrt{2}/7$
7509 3201 ℓ_2 : $7\zeta(3)/5$
7509 3334 rt : (1,8,-6,5)
7509 3490 rt : (9,7,-9,-1)
7509 4534 e^x : $3\sqrt{2}-\sqrt{5}/2$
7509 4984 rt : (5,2,5,-7)
7509 9223 cu : $3\sqrt{3}+2\sqrt{5}$
7509 9505 rt : (9,5,8,7)
7510 1100 rt : (5,9,1,1)
7510 1318 e^x : $19/8$
7510 2079 J_0 : $(\ln 3)^{1/3}$
7510 3056 $\pi+\ln(5)$
7510 4745 Ψ : $13/5$
7510 5393 rt : (6,7,-1,9)
7510 5651 $1/5-s\pi(2/5)$
7510 6234 rt : (9,1,1,4)
7510 8233 10^x : $3\sqrt{2}-4\sqrt{3}/3$
7510 9388 sinh : $24/5$
7511 0035 id : $3e/4+3\pi/2$
7511 0070 rt : (2,8,4,8)
7511 1111 sq : $13/15$
7511 2554 sr : $(\pi)^{-1/2}$
7511 3934 rt : (9,3,-6,-1)
7511 5551 tπ : $3\sqrt{3}/4+\sqrt{5}/2$
7511 5879 ℓ_2 : $3\ln 2/7$
7511 6186 rt : (5,2,8,7)
7511 8363 rt : (1,9,-2,-4)
7512 0013 erf : $3e/10$
7512 1809 rt : (6,7,-5,2)
7512 2630 as : $1/\ln(\ln 2/3)$
7512 4169 10^x : $3e+\pi/2$
7512 5460 rt : (5,2,1,-4)
7512 6169 ℓ_{10} : $4\pi^2/7$
7512 6336 10^x : $4-2\sqrt{2}/3$
7512 7264 e^x : $3\sqrt{2}/2+3\sqrt{5}/2$
7512 7593 Li_2 : $8/13$
7512 8107 rt : (2,8,0,-7)
7512 8556 id : $5\zeta(3)/8$
7513 0872 rt : (3,8,-5,-7)
7513 1480 cu : $10/11$
7513 2335 sinh : $21/2$
7513 2703 J : $3\sqrt[3]{2}/8$
7513 3457 rt : (9,5,4,4)
7513 4443 sq : $\sqrt{3}/3-\sqrt{5}$
7513 6331 rt : (5,9,-3,-6)
7513 7949 rt : (6,7,2,-8)
7513 9173 2^x : $7e$
7513 9608 rt : (1,1,8,-7)
7514 1308 cr : $3\sqrt{2}/10$
7514 4272 rt : (6,7,-9,-5)
7514 5815 rt : (9,1,-3,1)
7514 9396 2^x : $7\sqrt[3]{3}/4$
7514 9566 rt : (4,8,-5,9)
7515 0110 cr : $4(1/\pi)/3$
7515 2416 cu : $3/2+\sqrt{2}/2$
7515 3207 rt : (1,7,9,-8)
7515 5239 K : $\zeta(3)/2$
7515 5554 cu : $2\sqrt{2}+4\sqrt{3}$
7515 6643 as : $(\pi)^{-1/3}$
7515 6893 2^x : $1/3-\sqrt{5}/3$
7515 9022 rt : (6,3,5,-8)
7515 9800 2^x : $7\ln 2/6$
7516 1020 tanh : $(e+\pi)/6$
7516 2435 rt : (7,9,-4,3)
7516 3653 $\exp(2)-\exp(\pi)$
7516 4348 id : $4\sqrt{2}/3+\sqrt{3}/2$
7516 4587 sq : $\sqrt{2}/4-3\sqrt{5}/4$
7516 5294 rt : (5,2,4,4)
7516 5381 rt : (4,8,-9,-2)
7516 7476 rt : (6,7,-2,-5)
7516 9334 ℓ_2 : $3/2-e/3$
7516 9715 rt : (1,9,-6,-1)
7516 9882 rt : (3,7,-2,4)
7517 0090 J_0 : $6\zeta(3)/7$
7517 1488 e^x : $1/2-\pi/4$
7517 2551 Γ : $(\sqrt[3]{5}/3)^2$
7517 4919 rt : (3,4,-8,-7)
7517 7116 J_0 : $5\sqrt[3]{3}/7$
7517 8338 at : $7\zeta(3)/9$
7517 8541 2^x : $(\sqrt[3]{4}/3)^{1/3}$
7517 8692 erf : $\sqrt{2/3}$
7517 9438 rt : (9,9,7,4)
7517 9660 rt : (2,6,5,-8)
7518 0941 sr : $13/23$
7518 1295 rt : (7,8,2,-9)
7518 2043 tanh : $4\sqrt[3]{5}/7$
7518 4649 rt : (5,2,-3,-1)
7518 4875 rt : (9,9,-9,5)
7518 5343 tanh : $(\pi/3)^{-1/2}$
7518 5472 ζ : $(\sqrt[3]{2})^{1/2}$
7518 7704 rt : (6,1,8,8)
7518 7941 as : $1/4+\sqrt{3}/4$
7518 8519 rt : (7,9,-8,-4)
7518 9829 e^x : $2\sqrt{2}+\sqrt{5}/2$
7519 1104 cu : $2e/3+3\pi/2$
7519 1153 ℓ_{10} : $3e/2+\pi/2$
7519 1729 sq : $2/3-2e$
7519 2870 id : $3\sqrt{2}-2\sqrt{5}/3$
7519 3337 θ_3 : $((e+\pi)/3)^{-1/3}$
7519 4188 ℓ_2 : $e/3+4\pi$
7519 4723 J : $10/23$
7519 6489 Ψ : $3\zeta(3)/4$
7519 6789 rt : (3,8,-1,-4)
7519 7455 tπ : $(e)^2$
7519 8247 Ei : $18/19$
7519 8641 rt : (3,7,-6,-7)
7519 9335 sr : $\sqrt{2}/3+3\sqrt{3}/2$

7519 9579 rt : (6,3,1,-5)
7520 0795 $Ei : 6\ln 2/7$
7520 1524 sinh : $\sqrt{21/2}$
7520 1726 $2^x : 1/4+\sqrt{5}/4$
7520 1726 cr : $\exp(-\sqrt[3]{5}/2)$
7520 1848 cr : $1/\sqrt[3]{13}$
7520 2438 rt : (9,5,0,1)
7520 3869 id : $\ln(\sqrt{2}/3)$
7520 4378 sr : $1/\ln(e+\pi)$
7520 5114 rt : (7,4,5,-9)
7520 6484 rt : (2,9,-5,-8)
7520 8144 $\exp(1/3)+\exp(\sqrt{5})$
7521 0923 $2^x : 23/7$
7521 1797 $t\pi : 1/3-4\sqrt{5}$
7521 2061 sr : $2\sqrt{2}/5$
7521 3038 rt : (6,7,-6,-2)
7521 3178 sr : $\sqrt{3}/2+3\sqrt{5}$
7521 3587 sr : $(3\pi)^{1/2}$
7521 3882 $e^x : 6(1/\pi)$
7521 3949 cr : $2/3+3\pi/2$
7521 3982 $2^x : \sqrt{2}/2-\sqrt{5}/2$
7521 4398 rt : (7,8,-2,-6)
7521 7177 rt : (2,2,8,-8)
7521 7224 $\theta_3 : 4/5$
7521 8241 rt : (3,9,-9,-2)
7522 0358 $1/3-\exp(3)$
7522 1263 $e^x : 3+\sqrt{3}/2$
7522 2010 $\ell_{10} : \exp(\sqrt{3})$
7522 2262 $10^x : (\sqrt{3})^{-1/2}$
7522 2705 $\Psi : 5\sqrt[3]{3}/8$
7522 6177 at : $3e/2-\pi$
7522 6737 as : $\sqrt{2}/3-2\sqrt{3}/3$
7522 7616 rt : (3,7,5,-9)
7522 9277 $2^x : 8(e+\pi)/7$
7522 9309 $10^x : 4e/7$
7522 9432 $e^x : \sqrt{3}+\sqrt{5}-\sqrt{7}$
7523 0117 $10^x : \sqrt{2}/3+3\sqrt{5}/4$
7523 1692 $\zeta : 2(1/\pi)/3$
7523 2092 rt : (9,1,-7,-2)
7523 2190 $10^x : 3/4+4\sqrt{2}$
7523 3217 rt : (2,6,1,-1)
7523 4364 rt : (7,7,-8,-7)
7523 4867 rt : (2,5,-7,9)
7523 6149 rt : (7,0,8,9)
7523 6877 sr : $3/2+\pi/2$
7524 1227 $\exp(4)+\exp(e)$
7524 2237 $\ell_{10} : 9\pi/5$
7524 2238 rt : (5,6,7,4)
7524 2505 $\zeta : 2e\pi/9$
7524 3505 $Li_2 : 8\ln 2/9$
7524 5299 $\ln(\sqrt{5})+\exp(2/3)$
7524 6920 rt : (6,1,4,5)
7524 8520 rt : (7,4,1,-6)
7524 9328 rt : (8,9,-2,-7)
7524 9519 cr : $\sqrt{2}+\sqrt{3}+\sqrt{5}$
7524 9845 ln : $4\sqrt[3]{3}$
7525 0767 cu : $\sqrt{2}/3-3\sqrt{5}/4$
7525 1782 rt : (1,5,-3,-1)
7525 5284 cu : $2\sqrt{2}/3+\sqrt{5}/2$
7525 7498 $\ell_{10} : 4\sqrt{2}$
7525 7704 rt : (2,5,-8,-8)
7525 8579 $Ei : 7\sqrt[3]{2}/4$
7526 0410 rt : (5,2,-9,5)
7526 2240 rt : (7,8,-6,-3)
7526 3285 $2^x : 17/21$
7526 5422 ln : $1/3+2e$
7526 6509 rt : (3,3,8,-9)
7526 6595 sq : $\ln(\sqrt[3]{2}/3)$
7526 6719 sinh : $8\sqrt{2}/5$
7526 7581 $e^x : 7\pi^2/9$
7526 7798 rt : (6,3,-3,-2)
7527 6188 $erf : 9/11$
7527 7481 $\ell_2 : 3\sqrt{2}/4+4\sqrt{3}/3$
7528 0580 rt : (2,9,-1,-5)
7528 0728 ln : $\ln 2/4$
7528 1772 $10^x : 4\sqrt[3]{3}/3$
7528 2356 rt : (6,4,-8,6)
7528 3251 rt : (8,5,1,-7)
7528 5963 rt : (5,2,0,1)
7528 7553 ln : $3\sqrt{3}/4+2\sqrt{5}$
7528 9693 $t\pi : \sqrt{2}/4+4\sqrt{5}/3$
7528 9695 $s\pi : -\sqrt{2}+\sqrt{5}+\sqrt{6}$
7528 9901 $10^x : 4\sqrt{2}+2\sqrt{3}$
7529 0100 at : $3\sqrt{3}/4-\sqrt{5}$
7529 3205 cu : $4e/3+\pi/4$
7529 5571 rt : (7,0,4,6)
7529 6487 rt : (8,9,-6,-4)
7529 6802 $\zeta : 22/15$
7529 8558 sr : $1-\sqrt{3}/4$
7530 0439 rt : (7,6,-7,7)
7530 0638 rt : (2,4,6,9)
7530 2788 sr : P_{80}
7530 3767 rt : (2,8,-8,-4)
7530 3932 tanh : $7\sqrt[3]{2}/9$
7530 5893 rt : (1,6,-8,-9)
7530 8513 $\ell_{10} : 4\sqrt{3}/3+3\sqrt{5}/2$
7530 9176 rt : (9,6,1,-8)
7530 9695 $e^x : \exp(-\sqrt{3}/3)$
7530 9842 $\Psi : 4(e+\pi)/9$
7531 0123 $t\pi : \sqrt{2}/4-\sqrt{5}/4$
7531 0919 rt : (1,7,3,9)
7531 3359 sinh : 16/23
7531 3762 rt : (1,7,-7,-9)
7531 5128 at : 15/16
7531 5130 rt : (7,4,-3,-3)
7531 5597 rt : (8,8,-6,8)
7531 6191 $e^x : 2+3e/4$
7531 6551 cr : $4/3-e/3$
7531 7017 rt : (6,8,-8,-8)
7532 0695 $t\pi : 8\zeta(3)/3$
7532 1879 as : $2\sqrt[3]{5}/5$
7532 2075 $erf : (\ln 3/2)^{1/3}$
7532 3747 rt : (6,5,7,5)
7532 4633 $10^x : 3e/5$
7532 5399 $Ei : \ln(2e/3)$
7532 6129 rt : (4,9,-3,7)
7532 7870 rt : (8,1,4,-7)
7532 7894 ln : $\sqrt{3}/10$
7533 0701 rt : (9,9,3,1)
7533 1509 cr : $\sqrt[3]{5}/4$
7533 2097 rt : (2,6,-3,-2)
7533 2766 $\ell_{10} : 3/17$
7533 3143 $e^x : 9\sqrt{3}/10$
7533 9580 rt : (5,9,1,-8)
7533 9589 $\ell_{10} : \sqrt{2}/3+3\sqrt{3}$
7534 4766 rt : (3,3,4,-6)
7534 5693 rt : (8,5,-3,-4)
7534 6833 $\lambda : \ln 2/8$
7534 8300 sq : $1/4-\sqrt{5}/2$
7534 8481 rt : (5,2,-7,2)
7535 0876 rt : (9,2,4,-8)
7535 2061 as : 13/19
7535 3944 cr : $e/4+3\pi/2$
7535 5584 $\ell_{10} : 9\sqrt[3]{2}/2$
7535 6216 at : P_{62}
7535 6577 $e^x : \sqrt{3}/3+\sqrt{5}/3$
7535 7529 $e^x : 17$
7535 7699 rt : (6,8,-1,6)
7535 8525 $e^x : 1/3+e/4$
7535 9075 rt : (5,9,-8,-9)
7536 0432 rt : (6,1,0,2)
7536 2366 $\ell_2 : 4/3+3e/4$
7536 4034 rt : (4,9,-7,-4)
7536 4631 rt : (7,4,7,6)
7536 5389 $J_0 : 3\sqrt[3]{5}/5$
7536 7060 rt : (9,6,-3,-5)
7536 7372 $e^x : \sqrt{20}\pi$
7537 1621 rt : (3,7,-3,-3)
7537 2058 rt : (4,4,4,-7)
7537 3382 rt : (1,3,7,4)
7537 4195 $\theta_3 : 2\sqrt[3]{2}/7$
7537 4507 $Ei : 9\sqrt{3}/5$
7537 6116 rt : (2,9,8,9)
7537 6409 $Ei : 7\pi^2/8$
7537 7180 ln : 8/17
7537 9559 $\theta_3 : 9/25$
7538 4164 $\zeta : 1/\sqrt{22}$
7538 4525 rt : (2,5,-4,-5)
7538 4552 sq : $2/3+\sqrt{3}$
7538 5458 $s\pi : 7\pi^2/4$
7538 7143 $e^x : 1/2+3\sqrt{2}/2$
7538 9201 rt : (8,3,7,7)
7538 9685 $s\pi : e/10$
7539 0604 rt : (5,5,4,-8)
7539 2641 $K : 2e/9$
7539 3952 $erf : (2e/3)^{-1/3}$
7539 4744 cr : 3/7
7539 5138 rt : (4,8,-3,-4)
7539 6331 rt : (3,8,3,-1)
7539 6565 cu : $7\sqrt{5}/6$
7539 9904 $\Psi : 3/8$
7540 0191 $t\pi : \sqrt{2}/3-3\sqrt{5}/4$
7540 1322 $\Gamma : 10\sqrt[3]{2}/3$
7540 1890 $2^x : 19/13$
7540 1951 $e^x : 5\pi/4$
7540 4027 rt : (6,6,4,-9)
7540 4654 $e^x : 1/3+3\sqrt{3}/2$
7540 5598 rt : (9,2,7,8)
7540 5900 cu : $\sqrt{2}+\sqrt{3}-\sqrt{5}$
7540 6259 rt : (6,3,-7,1)
7540 8474 $e^x : 3\sqrt{3}-3\sqrt{5}/4$
7541 0735 rt : (5,9,-3,-5)
7541 1740 $e^x : 2e/3-2\pi/3$
7541 2265 cr : $3\sqrt{2}+2\sqrt{3}/3$
7541 2279 $\ell_{10} : 4+3\sqrt{5}/4$
7541 2948 sr : $(2e)^{-1/3}$
7541 2958 at : $2-3\sqrt{2}/4$
7541 3797 rt : (1,6,-4,-6)
7541 4486 rt : (8,1,0,-4)
7541 4559 $e^x : 1+\sqrt{5}/4$
7541 6535 rt : (9,5,-4,-2)
7541 6546 id : $(\ln 3)^{-3}$
7541 6779 ln : 5π
7541 8272 id : $2e/3-4\pi$
7541 9057 sinh : $7\sqrt{5}/6$
7541 9486 sq : $\sqrt{2}/3+4\sqrt{3}$
7542 1836 rt : (6,8,-5,-1)
7542 2870 $10^x : 11/25$
7542 3119 rt : (2,2,7,5)
7542 4952 at : P_{46}
7542 5772 $10^x : 4\sqrt[3]{5}/9$
7542 6224 $2/5-\exp(e)$
7542 6389 $e^x : 4+\sqrt{2}/4$
7542 7315 rt : (9,2,0,-5)
7543 0141 rt : (7,4,-7,0)
7543 1800 cr : $2-\pi/2$
7543 4600 $\lambda : \exp(-7\pi/9)$
7543 5170 cu : $2\sqrt{2}-4\sqrt{5}$
7543 6309 $\ell_2 : 4/3+\sqrt{2}/4$
7543 6974 sq : $2\sqrt{2}+4\sqrt{5}/3$
7543 7390 rt : (2,9,3,-2)
7543 7929 $e+s\pi(\sqrt{2})$
7543 8988 $erf : (e/3)^2$
7543 9827 $\ln(4)-\exp(\pi)$
7543 9937 $10^x : 19/25$
7544 0072 cu : $3+4\sqrt{2}$
7544 0570 rt : (1,8,4,-8)
7544 1064 sr : $9\sqrt[3]{5}/5$
7544 1212 $s\pi : 9\sqrt{2}$
7544 2717 rt : (3,1,7,6)
7544 3196 rt : (8,5,-7,-1)
7544 7708 $e^x : 3-3\sqrt{5}/4$
7544 7804 rt : (2,9,4,-6)
7544 8629 $\exp(e)\div\exp(3)$
7544 8640 $s\pi : P_{30}$
7545 1189 $\sqrt{2}\times\exp(2/3)$
7545 1428 rt : (9,6,-7,-2)
7545 3115 rt : (3,8,0,2)
7545 3198 rt : (4,0,7,7)
7545 3925 sinh : $\sqrt{2}+\sqrt{3}-\sqrt{6}$
7545 4470 sr : $1/4+2\sqrt{2}$
7545 5659 $10^x : 2/3+2\sqrt{2}$
7545 6071 rt : (2,9,-2,2)
7545 7180 rt : (4,1,-6,8)
7545 7417 $t\pi : 3e/2-2\pi$
7545 9723 rt : (5,1,7,-8)
7546 0267 $e^x : 7/4$
7546 0391 rt : (5,3,-5,9)
7546 1757 $t\pi : 9\sqrt{3}/2$
7546 2962 cu : 19/6

7546 3291 as : $\sqrt{3}/4-\sqrt{5}/2$
7546 3906 rt : (1,8,1,-7)
7546 4176 rt : (6,2,7,-9)
7546 4400 id : $3/2-\sqrt{5}/3$
7546 4541 $Ei : -\sqrt{2}+\sqrt{6}+\sqrt{7}$
7546 6384 $2^x : 1/2-e/3$
7546 8121 $\zeta : \sqrt[3]{5}/8$
7546 8402 $2^x : (5/3)^3$
7547 3120 $\exp(1/4)\times s\pi(1/5)$
7547 3781 $\lambda : 2/23$
7547 6335 id : $\sqrt{14}\pi$
7547 7869 $\ell_{10} : P_{14}$
7547 8222 $K : (\ln 3/3)^{1/2}$
7548 0481 $erf : (\sqrt{2}\pi/3)^{-1/2}$
7548 0615 $e^x : \sqrt{2}+3\sqrt{3}$
7548 1443 rt : (2,6,-7,-5)
7548 4498 $J_0 : (\pi/3)^{1/2}$
7548 4990 $\ell_{10} : 1/4+2e$
7548 7766 rt : (9,9,0,-9)
7548 8750 $\ell_2 : (3)^3$
7549 2141 sr : $10\pi^2/7$
7549 2694 $\Psi : 9/10$
7549 4675 $e^x : 9\pi^2/7$
7549 7813 id : $(\sqrt[3]{5}/3)^{1/2}$
7549 9285 $2^x : \ln(2/3)$
7550 0180 at : $3/2-\sqrt{5}/4$
7550 1991 $\ell_{10} : 3/2+4\pi/3$
7550 2084 $10^x : \sqrt[3]{5}/7$
7550 3537 $\Psi : 5\sqrt[3]{2}/7$
7550 4774 $10^x : 3/2+\sqrt{2}$
7550 5465 $e^x : 9/16$
7550 9805 sr : $2e-3\pi/4$
7551 0204 sq : 24/7
7551 0440 at : 16/17
7551 1254 rt : (9,2,3,5)
7551 1265 tanh : $(\pi/3)^{-1/3}$
7551 2266 $\ell_2 : 3e/2+3\pi$
7551 3066 $\Psi : 7\sqrt{5}/6$
7551 3366 $\theta_3 : \sqrt[3]{3}/4$
7551 3813 rt : (6,8,-9,-8)
7551 5423 sr : $8e\pi/9$
7551 6941 id : $3\sqrt{3}+\sqrt{5}/4$
7551 7476 rt : (8,3,3,4)
7551 7915 rt : (5,9,-4,-6)
7551 8367 rt : (8,9,-6,5)
7551 8708 $\Psi : 6/11$
7551 9351 rt : (5,9,-7,-2)
7552 1342 rt : (7,7,-7,4)
7552 3020 rt : (2,5,7,-9)
7552 4958 rt : (6,5,-8,3)
7552 5124 $\Gamma : 7e/4$
7552 6381 $\theta_3 : 3\zeta(3)/10$
7552 7054 cu : $1/3+\sqrt{3}/3$
7552 7878 rt : (4,8,-7,-1)
7552 8183 rt : (7,4,3,3)
7552 8258 $5\times s\pi(2/5)$
7552 9446 rt : (5,3,-9,2)
7552 9829 $\pi-\ln(4)$
7553 0715 rt : (1,4,7,-8)
7553 1958 $10^x : \sqrt{2}/3+\sqrt{3}$
7553 2261 $J : (\ln 2)^2$
7553 2742 cu : $4-3\sqrt{3}/2$
7553 2761 $\zeta : 13/23$
7553 4606 rt : (6,8,-4,-5)
7553 5309 $\ell_{10} : 2e\pi/3$
7553 8030 rt : (1,8,9,4)
7553 9307 $2^x : 3/4-2\sqrt{3}/3$
7554 1247 cr : $9\zeta(3)/2$
7554 2709 rt : (3,8,-4,-9)
7554 4525 rt : (9,5,3,-9)
7554 5524 $s\pi : (\sqrt[3]{4}/3)^{1/2}$
7554 5594 $\ell_2 : 2\sqrt{2}-\sqrt{5}$
7554 5975 rt : (9,2,-4,-2)
7554 7784 $Ei : \sqrt{3}-\sqrt{5}+\sqrt{6}$
7555 0905 $s\pi : P_{73}$
7555 0925 $Li_2 : 3\sqrt[3]{3}/7$
7555 0950 rt : (6,5,3,2)
7555 1036 $\zeta : 3/14$
7555 2564 rt : (8,4,3,-8)
7555 2904 $K : 3\sqrt{2}/7$
7555 3106 rt : (2,9,0,-6)
7555 3429 $s\pi : \exp(-1/\pi)$
7555 7472 cr : $4/3+3e/2$
7555 7781 $\ell_2 : 3\sqrt{2}-\sqrt{3}/2$
7555 8160 $\ell_{10} : 1/2+3\sqrt{3}$
7556 2154 sr : $\sqrt{19/2}$
7556 2699 rt : (8,1,-4,-1)
7556 3256 rt : (7,3,3,-7)
7556 3824 $s\pi : 2/3+3\sqrt{2}/4$
7556 4007 rt : (1,2,-7,-6)
7556 5820 rt : (9,9,-4,-6)
7556 8355 rt : (1,8,0,-5)
7557 2001 rt : (8,6,-9,9)
7557 2364 $2^x : 21/11$
7557 4931 rt : (2,6,-7,1)
7557 4957 $s\pi : 3/11$
7557 6449 $s\pi : 2\sqrt{3}/3+\sqrt{5}/2$
7557 7357 rt : (8,8,-4,-5)
7557 7547 $\ln(\sqrt{5})+s\pi(2/5)$
7557 8174 rt : (6,2,3,-6)
7557 9401 at : $3\pi/10$
7558 0633 $e^x : \sqrt{2}+\sqrt{5}+\sqrt{6}$
7558 3787 $erf : 14/17$
7558 4040 $Li_2 : \exp(-\sqrt[3]{3}/3)$
7558 4535 $\zeta : 3\ln 2/2$
7558 5618 $K : 7\ln 2/8$
7558 6874 $\ell_{10} : 10\sqrt[3]{5}/3$
7558 7752 $K : (e)^{-1/2}$
7558 7915 sr : $2/3+4\sqrt{3}$
7558 8314 $s\pi : 9\sqrt{2}/10$
7558 8777 $\ell_{10} : 2\sqrt{3}+\sqrt{5}$
7559 0609 cu : P_{11}
7559 2147 $t\pi : \sqrt[3]{3}/7$
7559 2894 sr : 4/7
7559 4857 cr : $4+\sqrt{2}$
7559 5262 id : $3\sqrt[3]{2}/5$
7559 6941 at : $2\sqrt{2}/3$
7559 7067 rt : (5,8,3,-9)
7559 7536 $s\pi : 6(1/\pi)/7$
7559 8306 id : $4/3-\sqrt{3}/3$
7559 9023 $\Psi : 1/\sqrt[3]{19}$
7559 9198 rt : (2,8,-7,-9)
7560 0442 rt : (5,1,3,-5)
7560 1774 $10^x : 2\sqrt{2}/3+\sqrt{5}/4$
7560 2575 id : $3\sqrt{3}/2-3\sqrt{5}/2$
7560 2764 cr : $2\sqrt{2}/3+2\sqrt{5}$
7560 4091 $Ei : e\pi/9$
7560 6326 tanh : $\pi^2/10$
7560 9840 $10^x : \sqrt{3}/3+\sqrt{5}$
7561 2325 rt : (2,7,1,-9)
7561 3903 $\theta_3 : (\sqrt[3]{5}/2)^{1/3}$
7561 4297 rt : (4,7,3,-8)
7561 4366 sq : 20/23
7561 4949 rt : (6,6,-4,-3)
7561 5371 $t\pi : (\sqrt[3]{5}/3)^{-1/3}$
7561 6268 rt : (9,1,6,-9)
7561 7799 $Ei : 8\sqrt[3]{3}/7$
7561 8886 $erf : 4\sqrt[3]{3}/7$
7561 9402 $\Psi : 2\pi$
7562 1240 $\Psi : \exp(-1/(3\pi))$
7562 1890 sr : $3\sqrt{2}+3\sqrt{5}/2$
7562 2612 $\Gamma : 5e\pi/4$
7562 2940 $\ln(3/5)\div s\pi(\sqrt{5})$
7562 3999 sinh : $2\pi/9$
7562 5216 $2^x : 13/16$
7562 7484 $10^x : 2-3\sqrt{2}/2$
7562 7846 $2^x : 4+2\sqrt{2}/3$
7562 8067 rt : (3,7,-7,-8)
7562 8366 cu : $7e\pi$
7563 0649 $2^x : 10e/7$
7563 2135 rt : (1,5,2,4)
7563 2360 rt : (5,6,3,1)
7563 4492 $\exp(3/5)\times s\pi(\sqrt{2})$
7563 4607 cr : $3\sqrt[3]{3}/10$
7563 5580 cu : $8\sqrt{3}/7$
7563 6010 $10^x : 5\pi^2/3$
7563 6273 rt : (9,6,6,5)
7563 7266 rt : (8,0,6,8)
7563 7979 rt : (3,6,3,-7)
7563 8935 rt : (1,6,0,-3)
7564 2301 $\zeta : 8\ln 2/5$
7564 3581 rt : (9,5,-1,-6)
7564 4026 at : $1/\ln(\ln 2/2)$
7564 6901 $2/5+\exp(\sqrt{5})$
7564 7079 $\ln(2)\div\ln(2/5)$
7564 8036 ln : $\sqrt{2}+\sqrt{3}+\sqrt{7}$
7564 8710 rt : (5,5,-4,-2)
7564 9041 ln : $3/4-\sqrt{3}/3$
7564 9339 cu : $7\pi^2/10$
7565 4287 cr : $\sqrt{3}/4$
7565 6017 cu : $10\sqrt{2}/3$
7565 7517 $e^x : 3e-\pi/4$
7565 7842 cr : P_{10}
7565 9130 rt : (1,9,9,-5)
7566 1969 $J_0 : 3e/8$
7566 3035 id : $2\sqrt{2}+4\sqrt{3}$
7566 3731 rt : (6,1,-4,-1)
7566 6182 cr : $5\ln 2/8$
7566 6469 rt : (7,1,6,7)
7566 7553 sr : $3e/4+\pi/3$
7566 8953 $\Gamma : 21/5$
7567 0063 sq : $(\sqrt[3]{4}/3)^{-2}$
7567 0319 rt : (8,4,-1,-5)
7567 2574 rt : (2,7,7,8)
7567 3205 $Ei : 21/16$
7567 3334 rt : (9,9,-8,-3)
7567 4369 $Ei : (\sqrt[3]{5}/2)^{1/3}$
7567 6243 $s\pi : 5\sqrt{5}/3$
7567 7176 rt : (4,6,-7,-7)
7567 8698 $s\pi : e/4+\pi/3$
7567 8707 as : $(\sqrt{2}/3)^{1/2}$
7568 0249 $s\pi : 4(1/\pi)$
7568 1334 $e^x : 4\sqrt{3}/3-\sqrt{5}/4$
7568 1777 $\ell_{10} : 1+3\pi/2$
7568 2846 $2^x : 9/4$
7568 3450 at : 17/18
7568 4568 rt : (5,4,6,-9)
7568 5920 rt : (6,9,-1,3)
7568 8619 rt : (8,7,6,4)
7568 8645 rt : (1,7,-6,-1)
7568 9749 ln : $\sqrt{2}-\sqrt{3}+\sqrt{6}$
7569 0506 rt : (1,9,-8,-6)
7569 0519 rt : (9,2,-1,2)
7569 3975 sr : $9(1/\pi)/5$
7569 5835 $\zeta : \exp(-\sqrt[3]{5}/3)$
7569 7496 $Li_2 : 13/21$
7569 8117 rt : (1,5,-7,2)
7569 9319 id : $7e/4$
7569 9336 rt : (7,8,-3,8)
7569 9483 id : $\sqrt{2}/2-2\sqrt{3}$
7570 0171 $t\pi : e/3-\pi/2$
7570 1407 rt : (3,9,-8,2)
7570 1578 $\pi\times\exp(\sqrt{3})$
7570 6406 cu : $7\zeta(3)$
7570 6846 rt : (8,8,-8,-2)
7570 8129 rt : (7,3,-1,-4)
7570 9678 at : $3\sqrt[3]{2}/4$
7570 9854 rt : (9,3,9,9)
7571 3387 sinh : $((e+\pi)/2)^{-1/3}$
7571 6210 rt : (8,7,2,-9)
7571 6658 $\zeta : 1/\ln(e+\pi)$
7571 8352 rt : (5,8,-1,-6)
7571 9418 $\exp(\sqrt{2})\div\exp(2/5)$
7571 9658 as : $1/3+\sqrt{2}/4$
7572 0062 as : $4\zeta(3)/7$
7572 0117 $10^x : 5\sqrt{3}/6$
7572 1012 rt : (4,3,6,-8)
7572 1724 rt : (6,6,-4,7)
7572 2471 $t\pi : \sqrt[3]{1/2}$
7572 2673 rt : (2,9,-4,-3)
7572 5285 tanh : $4\sqrt{3}/7$
7572 5358 sr : $3\sqrt{2}-2\sqrt{3}/3$
7572 6555 $\ell_{10} : 3+e$
7572 7619 cr : $2\sqrt{2}+3\sqrt{3}/2$
7572 7892 rt : (3,6,-3,9)
7572 8065 $K : (\sqrt{2}\pi)^{-1/3}$
7572 9945 $erf : 19/23$
7573 0269 $\ell_{10} : 4e/3+2\pi/3$
7573 0379 $Ei : 13/4$
7573 4078 tanh : $7\sqrt{2}/10$

7573 5483 2^x : $7\sqrt{5}/5$
7573 5931 rt : (1,9,-3,-7)
7573 7441 10^x : $8\sqrt[3]{2}/3$
7573 7567 cu : $4\sqrt{2}-3\sqrt{3}/4$
7573 7892 ζ : $6\sqrt[3]{5}/7$
7573 8405 at : $\sqrt{2}/4-3\sqrt{3}/4$
7573 8601 tπ : $\exp(4/3)$
7574 0058 $\sqrt{2}\div\ln(\sqrt{5})$
7574 2896 sπ : $\exp((e+\pi)/2)$
7574 5196 tanh : $10\ln 2/7$
7574 5808 sr : $e/2-\pi/4$
7574 6157 rt : (9,1,2,-6)
7574 8076 rt : (7,6,2,-8)
7574 9413 rt : (5,4,-5,6)
7575 0854 cu : $4+\sqrt{5}/4$
7575 1967 at : $(\sqrt{5}/2)^{-1/2}$
7575 2747 rt : (8,4,9,8)
7575 3443 ℓ_2 : $4-4\sqrt{3}/3$
7575 4403 rt : (9,2,-8,1)
7575 4516 rt : (2,7,-3,-4)
7575 4714 rt : (9,9,-8,7)
7575 4833 J : 1/13
7575 5107 rt : (7,7,-8,-1)
7575 5324 Ei : $(e)^{1/2}$
7575 6902 rt : (2,5,0,-2)
7575 7224 cr : 10/23
7575 7695 tπ : $(\sqrt[3]{2}/3)^{-1/3}$
7576 0407 K : 14/23
7576 2530 ln : $3e-3\pi/4$
7576 3480 rt : (4,7,-1,-5)
7576 5679 rt : (6,2,-1,-3)
7577 0085 rt : (8,3,-1,1)
7577 0176 rt : (1,7,6,-9)
7577 0680 e^x : $4/3+\sqrt{3}/3$
7577 1434 sinh : $1/\ln(\sqrt{2}/3)$
7577 2454 2^x : $6(1/\pi)$
7577 3313 rt : (2,8,-7,1)
7577 3390 tπ : $3\sqrt{2}/2-\sqrt{3}$
7577 4039 rt : (3,2,6,-7)
7577 5971 rt : (8,7,-9,6)
7577 7023 ζ : $2\sqrt{2}/5$
7577 9499 rt : (5,5,-7,-6)
7578 1074 rt : (5,3,6,5)
7578 4541 rt : (4,2,-6,5)
7578 6204 rt : (9,5,-5,-3)
7578 9674 cu : $2+\sqrt{5}/4$
7578 9934 ζ : $2\pi/9$
7578 9970 10^x : $-\sqrt{3}+\sqrt{6}+\sqrt{7}$
7579 0275 rt : (6,5,2,-7)
7579 3387 rt : (7,8,-7,1)
7579 5538 rt : (4,1,-9,-9)
7579 5952 Γ : $(1/(3e))^{-1/2}$
7579 6234 id : $3(e+\pi)/10$
7579 7901 rt : (7,8,6,3)
7579 8624 ln : $3-\sqrt{3}/2$
7579 9011 rt : (8,0,2,5)
7579 9268 rt : (9,8,-2,-7)
7579 9680 rt : (1,8,-4,-2)
7580 2168 tπ : $1/3+4e$
7580 2247 e^x : $(\pi)^{-1/2}$
7580 2284 rt : (5,0,9,9)
7580 2293 rt : (4,8,-8,1)
7580 3796 2^x : $e/4+\pi/2$
7580 4076 as : 11/16
7580 4511 rt : (2,8,-3,-6)
7580 4655 10^x : $2+3\sqrt{3}/2$
7580 6661 sq : $4\sqrt{3}-\sqrt{5}/2$
7580 9220 rt : (8,3,5,-9)
7581 1129 id : $1/3+3\pi$
7581 3053 sπ : $9(e+\pi)/10$
7581 3113 e^x : $\sqrt{2}/4+3\sqrt{5}/2$
7581 3455 cu : $7\zeta(3)/6$
7581 5686 ζ : $(5/3)^{-3}$
7581 6188 rt : (7,5,9,7)
7581 9368 sq : $(\ln 3)^3$
7581 9923 sr : $\sqrt{2}+3\sqrt{5}/4$
7582 2924 Γ : $\ln 2/5$
7582 4522 sinh : $7\sqrt[3]{5}/9$
7582 6405 cr : P_{29}
7582 7771 sq : $(\pi/3)^{-3}$
7582 9644 ℓ_2 : $4e/3+\pi$
7583 0573 rt : (6,9,-5,-4)
7583 1003 ℓ_{10} : $4+\sqrt{3}$
7583 4191 J_0 : $(\pi/3)^{1/3}$
7583 6147 id : $(2e)^{1/3}$
7583 7771 at : 18/19
7584 0763 Ψ : $\sqrt{2}+\sqrt{5}+\sqrt{7}$
7584 0865 ℓ_{10} : $3\sqrt{2}+2\sqrt{5}/3$
7584 3552 ln : $e\pi/4$
7584 4153 rt : (8,7,-2,-6)
7584 4292 sq : $2+\pi/4$
7584 5756 sinh : $7\sqrt{5}/9$
7584 5926 10^x : 19/14
7584 6147 cr : $4\sqrt{3}-2\sqrt{5}/3$
7584 7681 cu : $1/\ln(2e/3)$
7584 8810 rt : (5,4,2,-6)
7585 1045 rt : (8,4,-5,-2)
7585 2864 sinh : $5\sqrt[3]{2}/9$
7585 3230 2^x : $4/3-\sqrt{3}$
7585 5453 sπ : $\sqrt{2}/2-\sqrt{3}/4$
7585 8331 rt : (2,1,6,-6)
7585 8370 sinh : 7/10
7586 0846 rt : (7,2,5,-8)
7586 1793 rt : (1,9,2,-1)
7586 3861 rt : (5,1,-1,-2)
7586 4910 10^x : $2\sqrt{2}/3+\sqrt{3}/4$
7586 4981 $\exp(1/3)-\exp(e)$
7586 5242 sq : $3/4+2\sqrt{3}$
7586 7743 Ψ : $-\sqrt{5}+\sqrt{6}-\sqrt{7}$
7586 9294 erf : $(\ln 3)^{-2}$
7587 0490 e^x : $2/3-2\sqrt{2}/3$
7587 0700 tπ : $\sqrt{2}/3-4\sqrt{5}/3$
7587 2135 rt : (8,1,-8,2)
7587 2391 2^x : $4\sqrt{3}/3+\sqrt{5}/2$
7587 2896 rt : (4,1,9,8)
7587 3649 cr : $1/\sqrt[3]{12}$
7587 5610 ln : $1/4+4\sqrt{2}/3$
7587 6027 tπ : $9\pi^2/4$
7587 6453 Γ : $2\sqrt[3]{2}/5$
7587 7048 rt : (1,6,0,-8)
7587 8064 rt : (5,4,-9,-1)
7587 8275 Ei : 4/19
7587 9029 rt : (7,1,2,4)
7587 9318 10^x : $\exp(\sqrt[3]{2})$
7588 0045 ln : $4/3+2\sqrt{5}$
7588 2401 cu : $4/3-3\pi$
7588 3377 rt : (7,9,1,-9)
7588 4094 rt : (1,8,2,1)
7588 5735 cu : $2\sqrt{3}/3+4\sqrt{5}/3$
7588 7709 rt : (9,3,5,6)
7588 8252 cr : $\sqrt{3}\pi$
7588 8347 cu : $1/2+\sqrt{2}/2$
7588 9001 $3/5\times\exp(4)$
7589 2417 id : $\sqrt[3]{21}$
7589 3528 rt : (2,8,2,-7)
7589 4089 ζ : $\sqrt{3}/8$
7589 5027 sr : $4-e/3$
7589 7774 rt : (4,6,5,-9)
7589 8138 tanh : $4\sqrt{5}/9$
7589 9190 ℓ_2 : 13/22
7589 9316 10^x : $4/3+\sqrt{2}/2$
7590 0789 rt : (5,8,-5,-3)
7590 2228 Ψ : $5\sqrt[3]{2}$
7590 3077 erf : $(\sqrt[3]{5}/3)^{1/3}$
7590 3435 rt : (8,9,-7,-7)
7590 4142 rt : (7,6,-2,-5)
7590 4522 sπ : 9π
7590 6481 rt : (9,4,1,-7)
7590 8928 sr : $1+2\pi/3$
7590 9493 sπ : $3e+\pi/2$
7590 9760 e^x : $4\sqrt{3}+\sqrt{5}$
7591 0924 Ei : $5\sqrt[3]{5}/9$
7591 0985 cu : $3/2+2e$
7591 1164 ℓ_{10} : $3/2+3\sqrt{2}$
7591 3388 ℓ_2 : $2/3+e$
7591 3878 Ei : $(3\pi/2)^{-1/3}$
7591 4561 Ei : 19/20
7591 4724 cr : 7/16
7591 6286 at : $e\pi/9$
7591 9615 rt : (6,6,9,6)
7592 1069 sq : $\sqrt[3]{7/3}$
7592 2975 rt : (1,3,9,-9)
7592 3945 rt : (1,8,7,-7)
7592 5700 rt : (3,7,-3,-5)
7592 6486 rt : (1,9,-7,4)
7592 9848 rt : (5,9,0,-3)
7592 9871 at : $(\sqrt[3]{5}/2)^{1/3}$
7593 0114 rt : (8,1,8,9)
7593 2240 rt : (6,1,5,-7)
7593 2715 tπ : $(e/2)^3$
7593 5446 rt : (4,9,8,8)
7593 7203 10^x : 18/7
7593 7748 rt : (6,8,1,-8)
7593 8769 rt : (9,6,2,2)
7593 9463 cr : $1/4+3\sqrt{3}$
7594 0834 rt : (2,7,3,-3)
7594 1923 ℓ_2 : $3/2+4\sqrt{2}/3$
7594 3044 Ψ : $5\pi/6$
7594 4488 rt : (9,1,-2,-3)
7594 6340 ln : $8\pi^2/5$
7594 6715 rt : (9,8,-6,-4)
7594 6852 ln : $3e/4-\pi/2$
7594 6870 $s\pi(1/5)\div s\pi(e)$
7594 6932 10^x : $3\pi/10$
7595 1782 rt : (7,4,-1,0)
7595 3922 sr : $2\sqrt[3]{3}/5$
7595 4449 tπ : P_{39}
7595 4709 K : 11/18
7595 5314 cu : $4+2\sqrt{2}/3$
7595 5397 cu : $3/2+3\pi/2$
7595 5433 rt : (2,9,-1,7)
7595 5434 Ψ : $2\pi/7$
7595 5494 ln : $2\sqrt{2}+4\sqrt{5}/3$
7595 5606 rt : (7,6,-6,9)
7596 0969 ln : $4\sqrt{3}-\sqrt{5}/2$
7596 1979 tπ : P_{33}
7596 2285 ln : $5\sqrt[3]{5}/4$
7596 5125 rt : (5,5,-8,1)
7596 5639 θ_3 : $1/\sqrt[3]{21}$
7596 6784 ℓ_{10} : 4/23
7596 7929 tanh : $1/\ln(\ln 3/3)$
7596 9625 id : $8e\pi/7$
7597 0073 10^x : $1/2+4\sqrt{2}$
7597 0738 ℓ_2 : $\ln(2e)$
7597 2307 rt : (3,5,5,-8)
7597 3269 $\ln(\sqrt{2})+\exp(5)$
7597 3550 2^x : $5e\pi/6$
7597 4710 sr : γ
7597 5573 at : $5\sqrt[3]{5}/9$
7597 6268 rt : (8,3,1,-6)
7597 6275 at : 19/20
7598 0409 sr : $4/3+2\pi$
7598 0421 rt : (4,9,-6,-8)
7598 3031 e^x : 13/23
7598 3142 Ψ : $\ln(\sqrt[3]{3}/2)$
7598 3568 id : $(\sqrt{3})^{-1/2}$
7598 4512 rt : (3,2,9,7)
7598 4858 rt : (1,7,2,-6)
7598 5388 2^x : $(\pi)^{1/3}$
7598 6060 rt : (2,9,7,1)
7598 6721 rt : (7,9,-3,5)
7598 7044 sq : $4\sqrt{2}/3-\sqrt{5}/4$
7598 7214 2^x : $4/3+\sqrt{3}/3$
7598 8396 rt : (6,5,-2,-4)
7598 8913 2^x : $3e/10$
7598 8983 rt : (8,4,5,5)
7598 9509 e^x : 25/8
7599 4241 sr : $9\pi/2$
7599 5707 ln : $4+2e/3$
7599 6493 rt : (7,5,4,-9)
7599 7466 cu : $\sqrt{2}/4+\sqrt{5}/4$
7599 8930 id : $4\sqrt[3]{5}/9$
7599 9261 rt : (6,4,-7,8)
7599 9306 rt : (1,8,-7,-3)
7599 9708 Ei : $8\sqrt{2}/3$
7600 0000 id : 19/25
7600 1490 sr : $5\ln 2/6$
7600 5242 10^x : $3/2+3e$
7600 5261 rt : (1,9,1,-4)
7600 5811 ℓ_{10} : $3\sqrt{3}+\sqrt{5}/4$

7600 7126 rt : (4,7,-5,-2)
7600 7901 $J : (3\pi)^{-1/3}$
7600 8564 rt : (9,5,-9,0)
7600 9455 rt : (5,8,2,6)
7601 3241 rt : (1,0,6,5)
7601 3528 sr : $1/2 + 3\sqrt{3}/2$
7601 3854 rt : (7,2,8,8)
7601 3890 cu : $4 + 3e$
7601 4361 rt : (6,2,2,3)
7601 7725 rt : (8,8,-9,3)
7601 9682 rt : (5,2,-4,-2)
7601 9714 $\ell_2 : 2\sqrt{2} + \sqrt{5}/4$
7601 9805 $Ei : 4\pi^2/9$
7602 0064 $\Gamma : 20/7$
7602 0990 cr : $3/2 - 3\sqrt{2}/4$
7602 2925 rt : (9,7,0,-8)
7602 3139 rt : (4,3,-2,9)
7602 4463 $\ln(5) \div \exp(3/4)$
7602 6970 rt : (8,7,-6,-3)
7602 7673 rt : (3,9,-2,-5)
7602 8326 $s\pi : 5\sqrt[3]{5}/2$
7602 8988 $10^x : 4(1/\pi)$
7603 1371 $\zeta : 5/23$
7603 2747 rt : (3,7,-1,7)
7603 3057 sq : 24/11
7603 4672 $t\pi : 2e - \pi/3$
7603 6897 $e^x : \sqrt{2}/3 - \sqrt{5}/3$
7603 7459 rt : (5,0,5,6)
7603 7555 sr : $3\sqrt{3} + 4\sqrt{5}$
7603 8037 $2^x : 2\sqrt{2} - \sqrt{3}/3$
7603 8048 $e^x : \exp(-\sqrt[3]{5}/3)$
7603 8423 rt : (7,9,-3,-6)
7603 9864 rt : (2,9,-8,0)
7604 3279 $e^x : 3/4 + 3\sqrt{3}/4$
7604 4006 $t\pi : (\ln 3)^2$
7604 5069 $e^x : 1/\ln(e+\pi)$
7604 6143 $e^x : (\sqrt[3]{5}/3)^{-1/2}$
7604 6239 rt : (6,7,-4,4)
7604 6729 $2^x : 1 + 2\pi$
7604 6983 $10^x : 1 - \sqrt{5}/4$
7604 7008 rt : (9,0,4,7)
7604 9486 rt : (4,2,8,-9)
7605 1324 rt : (6,9,-9,-1)
7605 1661 rt : (1,7,-5,7)
7605 1913 rt : (5,2,-8,7)
7605 2275 sinh : $5\sqrt{5}/3$
7605 2888 $erf : 6\ln 2/5$
7605 3524 $2^x : 6e\pi$
7605 4534 $Li_2 : 1/\ln(5)$
7605 6220 rt : (4,9,-6,-1)
7605 7466 $2^x : (e+\pi)/4$
7605 8366 rt : (9,2,-5,-1)
7605 9049 cr : 11/25
7606 0309 sr : $10\sqrt{2}$
7606 4712 $erf : \exp(-1/(2e))$
7606 5416 $e^x : 2\sqrt{2}/5$
7606 6541 $s\pi : 3\sqrt{3}/2 + 3\sqrt{5}/4$
7606 7798 $\zeta : (e+\pi)/4$
7606 8924 $2^x : \sqrt{2}/3 - \sqrt{3}/2$
7606 8985 rt : (8,0,-2,2)
7607 0439 $\Psi : \sqrt[3]{18}$
7607 1335 $J_0 : 5\sqrt{2}/7$
7607 1939 $e^x : (\pi/3)^{1/3}$
7607 3954 rt : (7,2,1,-5)
7607 5507 at : $(e/3)^{1/2}$
7607 6686 ln : $4/3 - \sqrt{3}/2$
7607 6836 rt : (6,4,4,-8)
7607 8715 $\sqrt{2} + \ln(\sqrt{2})$
7607 8969 $\Psi : 6e$
7607 9649 $10^x : 2/3 - \pi/4$
7608 0041 rt : (2,4,5,-7)
7608 1447 $\ln(4) \div \exp(3/5)$
7608 1783 rt : (1,3,-1,7)
7608 2562 $1/4 - \ln(3/5)$
7608 3171 $10^x : (\ln 3)^{1/3}$
7608 3281 cu : $5\sqrt{5}/3$
7608 4521 $4/5 \times s\pi(2/5)$
7608 7672 rt : (3,6,-9,-9)
7608 8591 sr : 11/19
7609 1190 $2^x : 4/3 + 2\pi/3$
7609 2542 $\exp(5) - \exp(\sqrt{3})$
7609 4320 $\Psi : 1/\ln(2\pi)$
7609 5695 cu : $1/3 + 2\sqrt{3}$
7609 5742 rt : (9,7,8,6)
7609 6810 $erf : (\ln 2)^{1/2}$
7609 6947 $J_0 : 7\sqrt[3]{3}/10$
7609 8199 rt : (8,6,0,-7)
7610 0620 rt : (9,4,-3,-4)
7610 1275 at : 20/21
7610 1431 rt : (2,8,1,-3)
7610 2011 $2^x : 2/3 - 3\sqrt{2}/4$
7610 2721 $\exp(3) + s\pi(\sqrt{5})$
7610 3896 sq : $3 - 3\sqrt{2}/4$
7610 4063 $erf : (\sqrt[3]{3})^{-1/2}$
7610 4202 rt : (3,8,4,-9)
7610 6176 id : $1/3 - 2\pi/3$
7610 8115 $2^x : 6e\pi/7$
7610 8367 $J : \exp(-17\pi/5)$
7610 8369 rt : (4,6,1,5)
7611 0040 $\ell_{10} : 4\sqrt[3]{3}$
7611 2410 $2^x : \sqrt{2/3}$
7611 3275 $t\pi : 10\zeta(3)/9$
7611 4132 cr : $3/4 + 3\pi/2$
7611 5344 rt : (5,4,-2,-3)
7611 5648 cr : $1 - \sqrt{5}/4$
7611 5722 cu : 21/23
7611 6669 rt : (4,0,-9,6)
7611 6805 $\ell_{10} : 1/3 + 2e$
7611 7766 rt : (1,8,-5,-8)
7611 8363 rt : (5,7,9,5)
7611 9713 $\Psi : (\sqrt[3]{3}/2)^{1/3}$
7612 0403 cu : $7\sqrt{2}/5$
7612 1891 rt : (5,5,-5,3)
7612 2287 $t\pi : 1/2 + \sqrt{2}/2$
7612 2722 sr : $4 + 4e/3$
7612 3373 at : $2 - \pi/3$
7612 3453 $\ell_{10} : \ln 2/4$
7612 3507 $\Psi : 2/9$
7612 6417 $\ell_{10} : 3\sqrt{3}/4 + 2\sqrt{5}$
7612 7606 sr : $\sqrt{3}/2 + \sqrt{5}$
7612 9406 rt : (6,8,-3,-5)
7613 2193 $Ei : 7\sqrt{2}/6$
7613 2536 cr : $2 + 2\sqrt{3}$
7613 2690 $Ei : \exp(2e/3)$
7613 6202 rt : (7,9,-7,-2)
7613 6765 rt : (6,3,8,7)
7613 6930 rt : (1,6,-5,1)
7614 0717 $erf : 5/6$
7614 1146 rt : (7,6,-6,-2)
7614 3021 rt : (7,1,7,-9)
7614 3383 at : $(\sqrt{3}/2)^{1/3}$
7614 3937 $\ell_{10} : \sqrt{3}/10$
7614 8130 rt : (4,7,-9,6)
7614 8351 rt : (5,8,-6,-7)
7615 0001 $\ln(2) \times \ln(3)$
7615 1228 $J_0 : \zeta(7)$
7615 5718 rt : (2,8,-2,-4)
7615 7018 rt : (9,3,3,-8)
7615 7560 rt : (8,1,4,6)
7615 8638 $K : (\ln 2/3)^{1/3}$
7615 8876 rt : (7,5,5,4)
7615 9415 tanh : 1
7615 9895 $\zeta : 1/\sqrt{21}$
7616 0537 at : $3\sqrt{2} - 3\sqrt{3}$
7616 1229 $\ell_2 : 2\sqrt{3}/3 + \sqrt{5}$
7616 1584 ln : $\sqrt{3} - \sqrt{5} + \sqrt{7}$
7616 2062 rt : (8,4,-9,1)
7616 2672 sinh : $(\ln 2/2)^{1/3}$
7616 4017 rt : (3,9,-7,6)
7616 5841 cu : $e/4 + 3\pi/2$
7616 6303 cu : $\sqrt{2}/2 - 3\sqrt{3}/2$
7616 6775 rt : (6,9,6,2)
7616 6928 rt : (1,5,-3,8)
7616 7889 $J : 5/14$
7616 8136 $t\pi : 7\zeta(3)/2$
7616 9486 $J_0 : 4\sqrt[3]{2}/5$
7617 0184 rt : (8,9,-5,7)
7617 2396 sq : $1/2 - 4e/3$
7617 2426 rt : (7,0,-8,3)
7617 2902 rt : (3,1,8,-8)
7617 3705 $\zeta : 3\sqrt[3]{2}/2$
7617 8097 $\ell_2 : \sqrt{23/2}$
7617 8385 rt : (6,7,3,-9)
7618 0162 tanh : $\zeta(11)$
7618 1334 rt : (1,9,3,8)
7618 2445 rt : (9,3,1,3)
7618 3298 $t\pi : e/2 - 4\pi$
7618 5873 sq : $3/4 + \sqrt{3}/3$
7618 6229 cu : $3\sqrt{3}/4 - 4\sqrt{5}/3$
7618 7017 rt : (3,6,-5,-1)
7618 7234 rt : (8,9,-1,-8)
7618 7351 rt : (5,3,4,-7)
7618 7597 ln : $\zeta(3)/7$
7618 7746 rt : (2,3,9,6)
7618 9332 $t\pi : 2e + 3\pi/4$
7618 9486 at : $(\ln 3)^{-1/2}$
7619 0476 id : 16/21
7619 1558 $10^x : \sqrt[3]{5/2}$
7619 3412 $\ln(2/3) - \exp(\sqrt{5})$
7619 4892 $Ei : \sqrt[3]{2}/6$
7619 5340 cu : $\sqrt{3}/3 - 2\sqrt{5}/3$
7619 6237 $2^x : 6\sqrt[3]{5}/7$
7619 6290 $\ell_2 : 1 + 4\pi$
7619 6392 $e^x : 1/2 + 3\sqrt{2}/4$
7619 7164 $\exp(1/2) + \exp(\sqrt{2})$
7619 7867 $J : \ln 2/9$
7619 8106 rt : (7,5,0,-6)
7619 8254 rt : (3,7,-5,-4)
7619 8819 rt : (4,6,-3,-4)
7620 1937 $10^x : 1 - \sqrt{5}/2$
7620 4733 $\sqrt{3} \div \ln(4/5)$
7620 4993 rt : (5,8,-9,0)
7620 6228 $\ln(3/5) \times \exp(2/5)$
7620 6423 rt : (9,7,-4,-5)
7620 8053 rt : (4,1,5,5)
7620 8731 $s\pi : 2e - 3\pi/2$
7620 9040 $e^x : 8(1/\pi)$
7620 9979 rt : (2,6,-6,-1)
7620 9992 sq : $3\sqrt{3} - \sqrt{5}$
7621 2754 $t\pi : 3\sqrt{2}/4 + \sqrt{3}$
7621 3672 $10^x : 2\sqrt{2}/3 + 2\sqrt{3}$
7621 4005 ln : 7/15
7621 4270 rt : (2,7,4,-8)
7621 4654 at : 21/22
7621 7121 rt : (8,7,2,1)
7621 7669 $2^x : \sqrt{2}/4 - \sqrt{5}/3$
7621 8821 cr : $1 + 2\sqrt{5}$
7622 0430 rt : (6,1,1,-4)
7622 1274 rt : (4,9,0,-7)
7622 2021 sinh : $\sqrt{2} + \sqrt{3} + \sqrt{5}$
7622 2693 rt : (2,8,1,-3)
7622 3749 rt : (2,8,-6,6)
7622 6748 rt : (6,7,-8,-3)
7622 7941 sq : $4e - \pi/4$
7622 8763 rt : (8,3,-3,-3)
7622 9672 cu : $9\sqrt{5}/7$
7623 1308 tanh : $5\zeta(3)/6$
7623 1683 as : $3 - 4\sqrt{3}/3$
7623 2905 $t\pi : e - 4\pi/3$
7623 4247 rt : (7,7,-6,6)
7623 4753 at : $3(1/\pi)$
7623 7565 rt : (9,6,2,-9)
7623 9302 rt : (9,4,7,7)
7623 9628 $\zeta : (\pi)^{1/3}$
7624 0903 $s\pi : 3\sqrt{3} - 2\sqrt{5}$
7624 1157 $e^x : 9\sqrt[3]{5}$
7624 3349 rt : (1,1,-9,-7)
7624 3634 tanh : $\zeta(9)$
7624 3992 sr : $3/4 + 3\pi/4$
7624 6555 $\theta_3 : 4/11$
7624 6759 $\exp(5) \div s\pi(e)$
7624 8495 $s\pi : 3\sqrt{2}/2 + 2\sqrt{3}/3$
7624 9537 cu : $3/2 + \sqrt{3}$
7624 9637 rt : (1,3,5,-6)
7625 3622 rt : (6,8,0,8)
7625 4091 cr : $3\sqrt{2}/2 + 3\sqrt{5}/2$
7625 5717 sr : $8e/7$
7625 6372 rt : (5,8,-2,-1)
7625 8385 rt : (7,9,-7,-3)
7626 4692 rt : (6,0,7,8)

7626 5541 rt : (8,2,3,-7)
7626 7585 Ψ : 21/8
7627 1378 rt : (4,4,-8,2)
7627 3185 $s\pi$: $1/3 + 2\sqrt{2}/3$
7627 4373 e^x : $1/\ln(\sqrt{2}\pi/3)$
7627 4717 ln : $3 + 2\sqrt{2}$
7627 6613 cu : $e/2 + 4\pi/3$
7628 1542 θ_3 : $8(1/\pi)/7$
7628 1848 10^x : $2\sqrt{2} - 2\sqrt{5}/3$
7628 2888 rt : (9,8,-7,-6)
7628 2972 ℓ_{10} : $\sqrt{2} + \sqrt{3} + \sqrt{7}$
7628 3409 ℓ_{10} : $3/4 - \sqrt{3}/3$
7628 4137 rt : (7,8,-1,-7)
7628 4395 10^x : $3\sqrt{3}/4 + \sqrt{5}/3$
7628 4673 rt : (1,8,-6,6)
7628 5288 rt : (2,7,-5,-7)
7628 5535 as : $(2\pi/3)^{-1/2}$
7628 5542 rt : (3,5,-8,3)
7628 5690 rt : (5,6,3,-8)
7628 6253 rt : (7,1,-2,1)
7628 6792 Ψ : $e/5$
7628 7674 rt : (3,4,7,-9)
7628 8955 rt : (4,5,-9,-8)
7628 9272 rt : (1,8,3,9)
7629 2399 rt : (5,3,2,2)
7629 2464 ln : $e/2 + \pi/4$
7629 2885 ζ : 17/9
7629 3810 rt : (9,9,-7,9)
7629 4780 e^x : $1 - \sqrt{3}/4$
7629 7874 rt : (8,8,8,5)
7630 2867 K : 8/13
7630 2948 sq : $9(e + \pi)/7$
7630 3440 ℓ_2 : $6\sqrt{2}/5$
7630 4260 cr : $3\sqrt{2}/2 - 3\sqrt{5}/4$
7630 4263 $t\pi$: $8\pi/9$
7630 4717 erf : $(\sqrt[3]{5})^{-1/3}$
7630 5617 Γ : $7\pi^2/8$
7630 6011 e^x : P_{80}
7630 7207 rt : (3,5,6,3)
7630 8579 rt : (8,9,-9,0)
7631 1600 rt : (6,5,-7,5)
7631 2197 sq : $1/\ln(\pi)$
7631 4282 cr : 4/9
7631 5325 $2/3 \times \ln(\pi)$
7631 5405 2^x : $3\pi^2/7$
7631 6391 rt : (6,5,-6,-1)
7631 6701 sq : $3\sqrt{2} - \sqrt{5}/2$
7631 7959 at : 22/23
7631 8250 2^x : 9/11
7631 8571 J_0 : $\ln(\ln 3/3)$
7632 0082 J_0 : $\sqrt{2} + \sqrt{5} - \sqrt{7}$
7632 0647 sr : $\sqrt{2}/2 + 4\sqrt{3}$
7632 2530 $s\pi$: $\sqrt{2} + 4\sqrt{3}/3$
7632 2565 rt : (7,2,4,5)
7632 7065 rt : (7,9,-9,-9)
7632 8358 rt : (8,6,-4,-4)
7632 9489 Ψ : $(e)^{-2}$
7632 9508 Γ : $-\sqrt{5} + \sqrt{6} + \sqrt{7}$
7632 9510 $t\pi$: $1/4 + \sqrt{2}$
7633 1756 rt : (9,1,6,8)
7633 2697 ℓ_{10} : $3e - 3\pi/4$
7633 2912 Ei : $\exp(\sqrt[3]{5})$
7633 3931 cu : $4e/9$
7633 4815 rt : (5,4,8,6)
7633 5575 $3 \times s\pi(1/5)$
7633 6579 cr : $5\pi^2/9$
7633 6736 10^x : $1/3 + 2\sqrt{3}/3$
7633 7097 rt : (6,4,0,-5)
7633 7854 ℓ_2 : $\exp(\sqrt[3]{4}/3)$
7633 8251 ln : $1/3 + 2e/3$
7634 0638 rt : (8,5,2,-8)
7634 1167 rt : (1,8,-8,1)
7634 5407 rt : (3,9,-6,-2)
7634 6670 rt : (5,5,-9,-4)
7634 8502 at : $1/4 + \sqrt{2}/2$
7634 9006 rt : (6,3,6,-9)
7634 9018 tanh : $\sqrt{2} + \sqrt{5} - \sqrt{7}$
7634 9449 $s\pi$: $(\sqrt[3]{3}/3)^{-1/3}$
7635 0445 tanh : $\ln(\ln 3/3)$
7635 3682 erf : $(e + \pi)/7$
7635 4746 Ψ : 7/25
7635 5169 rt : (5,9,2,-9)
7635 5289 ζ : $(\sqrt{3})^{1/2}$
7635 6201 rt : (8,7,-8,8)
7635 7648 rt : (3,8,0,-6)
7636 1574 Ei : $\exp(2/3)$
7636 3455 $1/3 - s\pi(\pi)$
7636 3463 rt : (1,7,-2,-3)
7636 4242 K : $8\ln 2/9$
7636 6197 rt : (2,0,8,7)
7636 7187 cu : 23/8
7637 0015 $\ln(5) + \exp(e)$
7637 2381 rt : (1,6,4,-7)
7637 4414 λ : $\exp(-10\pi/13)$
7637 4497 cu : $2/3 + 4\sqrt{5}$
7637 6261 sr : 7/12
7637 7374 $\pi \div \exp(\sqrt{2})$
7637 9125 cu : $4/3 + 3\sqrt{3}/2$
7637 9159 Ψ : $2\sqrt[3]{2}/9$
7638 0258 2^x : $4\sqrt{3}/3 - 2\sqrt{5}/3$
7638 2298 cr : $7(1/\pi)/5$
7638 2575 2^x : 22/15
7638 3733 ℓ_{10} : $4/3 + 2\sqrt{5}$
7638 8888 cu : $7\sqrt[3]{3}/6$
7638 8946 id : $7\sqrt[3]{2}/5$
7638 9703 cr : $1/\ln(3\pi)$
7639 0253 $1/5 + s\pi(\sqrt{2})$
7639 3202 id : $3 - \sqrt{5}$
7639 4813 cu : $\sqrt{2}/2 - 4\sqrt{5}/3$
7639 5894 rt : (8,8,1,-9)
7639 6280 rt : (9,2,5,-9)
7639 6423 rt : (8,9,-5,-5)
7639 6992 rt : (9,3,-1,-5)
7639 8132 rt : (9,4,-7,-1)
7640 2692 rt : (4,8,-7,4)
7640 4060 cu : $\sqrt{3} - \sqrt{5} - \sqrt{6}$
7640 6035 cu : 14/9
7640 6867 rt : (5,3,-8,4)
7640 7125 sq : $2 + 2\pi/3$
7640 8729 cu : $1/2 - \sqrt{2}$
7641 1514 at : $3\sqrt{5}/7$
7641 2202 rt : (6,7,-1,-6)
7641 2477 at : 23/24
7641 5409 2^x : $(\ln 3/2)^{1/3}$
7641 6069 rt : (7,1,3,-6)
7641 6500 ℓ_{10} : $2\sqrt{2} + 4\sqrt{5}/3$
7641 6694 rt : (4,8,-8,-9)
7641 7601 rt : (6,8,-7,-2)
7641 7997 cr : $4 + 2\sqrt{5}/3$
7641 8878 ℓ_{10} : $4\sqrt{3} - \sqrt{5}/2$
7642 0397 rt : (3,8,1,5)
7642 0638 2^x : $1 + 3e/2$
7642 1912 rt : (8,5,7,6)
7642 4369 rt : (7,2,-3,-2)
7642 8704 rt : (4,9,-2,-5)
7642 8866 Ei : $(e/3)^{1/2}$
7642 9078 rt : (7,5,-9,7)
7642 9655 Γ : $10\sqrt{5}/3$
7643 1323 J_0 : $\zeta(9)$
7643 2008 erf : $3\sqrt{5}/8$
7643 2551 rt : (4,7,5,9)
7643 3794 sr : $((e + \pi)/2)^{-1/2}$
7643 3964 ℓ_{10} : $4 + 2e/3$
7643 6329 rt : (8,4,1,2)
7643 6514 cu : $2e + 4\pi/3$
7643 8318 ℓ_2 : $(\ln 2/2)^{1/2}$
7644 0983 Ei : $9\ln 2/2$
7644 1999 rt : (5,1,7,7)
7644 2113 rt : (7,3,-9,2)
7644 3783 $\exp(1/2) - \exp(5)$
7644 3922 10^x : $1/2 + 4\sqrt{3}/3$
7644 4292 J_0 : $5\zeta(3)/6$
7644 5423 Li_2 : $9\ln 2/10$
7644 5897 rt : (4,6,-3,-2)
7644 6911 e^x : $2 + e/2$
7644 7504 rt : (1,9,3,-8)
7644 7575 sr : $e/2 + 2\pi$
7644 8059 id : $\sqrt{2}/4 - \sqrt{5}/2$
7644 8305 $t\pi$: $1/3 + 3\sqrt{3}$
7645 1479 rt : (3,3,-6,-5)
7645 3381 as : $\exp(-1/e)$
7645 3745 rt : (9,6,-2,-6)
7645 8648 ln : $4\sqrt{3} + 4\sqrt{5}$
7645 8831 e^x : $\sqrt{3}\pi$
7645 9766 λ : $(\sqrt{5})^{-3}$
7645 9801 rt : (2,3,7,-8)
7645 9813 ζ : $1/\ln((e + \pi)/2)$
7646 2313 ℓ_2 : $5e$
7646 2909 Ei : $\sqrt[3]{9/2}$
7646 4338 rt : (5,0,1,3)
7646 4606 rt : (6,8,-4,1)
7646 5181 $t\pi$: $\sqrt{2} - \sqrt{6} + \sqrt{7}$
7646 6345 id : $\sqrt{2}/3 - \sqrt{5}$
7646 7170 e^x : $2 + \sqrt{3}$
7646 7311 $t\pi$: $(e + \pi)^{-1/3}$
7646 8217 as : 9/13
7646 8387 rt : (9,7,-8,-2)
7647 0588 id : 13/17
7647 1862 sq : $3/2 - 2\sqrt{2}$
7647 2449 id : $(\sqrt[3]{5})^{-1/2}$
7647 2921 rt : (4,7,-9,1)
7647 4921 ln : $\sqrt{2}/3 + 3\sqrt{5}/4$
7647 5692 rt : (1,7,-7,-9)
7647 6575 rt : (7,4,2,-7)
7647 6660 rt : (1,7,1,-4)
7647 9057 sinh : $5e/6$
7647 9740 rt : (8,3,-5,-2)
7648 0687 Ψ : 19/3
7648 2665 $t\pi$: $8\ln 2/7$
7648 7824 rt : (6,6,-3,9)
7649 0732 tanh : $4\sqrt[3]{2}/5$
7649 1288 Li_2 : $\exp(-\sqrt{2}/3)$
7649 1301 as : $2/3 - e/2$
7649 2472 ζ : 22/25
7649 3364 rt : (8,2,6,7)
7649 3649 rt : (6,7,-6,-6)
7649 3740 cr : $7\pi/4$
7649 4664 $\exp(\sqrt{3}) \div \exp(2)$
7649 5110 rt : (9,9,-3,-7)
7649 8017 J_0 : $\zeta(11)$
7649 9033 id : $2e/3 - \pi/3$
7649 9283 at : 24/25
7649 9555 $s\pi$: $2\ln 2/5$
7649 9969 rt : (7,5,-4,-3)
7650 2632 rt : (6,6,5,3)
7650 2917 id : $\sqrt{2}/2 - 2\sqrt{5}$
7650 3537 sinh : 4/3
7650 4148 10^x : $\exp(5/2)$
7650 5966 rt : (5,2,6,-8)
7650 6460 $t\pi$: $\exp(-\pi/2)$
7650 7811 rt : (1,8,-1,-5)
7650 7840 tanh : $\zeta(7)$
7650 9016 Ψ : 17/19
7651 1338 erf : $2\sqrt[3]{2}/3$
7651 3018 rt : (4,9,-4,-4)
7651 4271 erf : 21/25
7651 7301 ℓ_{10} : $\zeta(3)/7$
7651 7416 id : $\sqrt[3]{11/2}$
7651 8010 $s\pi$: $1/\sqrt{13}$
7651 8934 rt : (7,7,1,-8)
7651 9768 J_0 : 1
7651 9824 θ_3 : $(\pi/3)^{-3}$
7652 0614 $\exp(2) - \exp(e)$
7652 4275 rt : (5,1,-5,1)
7652 7099 rt : (4,1,-9,3)
7652 7625 K : $3\sqrt[3]{3}/7$
7652 7895 $1/3 - \ln(3)$
7652 8792 id : $9\pi^2/5$
7652 9136 rt : (8,1,0,3)
7653 2646 e^x : $(3\pi/2)^{1/2}$
7653 2719 rt : (3,2,5,4)
7653 3320 $\ln(\sqrt{5}) \times s\pi(2/5)$
7653 6550 θ_3 : $5\sqrt[3]{3}/9$
7653 6686 sr : $2 - \sqrt{2}$
7653 7718 as : $1/4 - 2\sqrt{2}/3$
7653 8628 $t\pi$: $3\ln 2/10$
7653 9282 as : $2\sqrt{3}/5$
7653 9489 id : $2\sqrt{2}/3 - 3\sqrt{5}$
7653 9828 rt : (8,1,5,-8)
7654 3659 rt : (5,3,0,-4)

7654 3669 sinh : $\sqrt{11}$
7654 4344 cu : $1/4 + 4\sqrt{3}/3$
7654 5099 2^x : $3/2 - 4\sqrt{2}/3$
7654 5277 K : $\exp(-\sqrt[3]{3}/3)$
7654 7694 rt : (7,8,-5,-4)
7654 7937 id : $e + \pi/3$
7654 8405 $\exp(\sqrt{2}) + \exp(\sqrt{3})$
7654 8466 rt : (5,5,5,-9)
7654 9138 2^x : $3 + \pi/2$
7654 9816 sr : $(e + \pi)/10$
7655 2959 10^x : $4\sqrt{3} - \sqrt{5}/3$
7655 3474 ℓ_2 : 5/17
7655 5137 ℓ_{10} : $3 + 2\sqrt{2}$
7655 5182 10^x : $3/4 - \sqrt{3}/2$
7655 6443 rt : (2,9,5,-2)
7655 8605 tanh : $7\sqrt[3]{3}/10$
7655 8640 tπ : $\sqrt{2} + \sqrt{3} + \sqrt{7}$
7655 9635 J : $1/\sqrt{13}$
7656 0425 Γ : $\pi^2/2$
7656 2500 sq : 7/8
7656 2568 Ei : 20/21
7656 2706 cr : $\pi/7$
7656 4182 erf : $(\sqrt{2})^{-1/2}$
7656 6300 at : $3 - 3e/4$
7656 7962 2^x : $(2e/3)^{-1/3}$
7656 8636 ζ : $((e + \pi)/2)^{-1/3}$
7656 9053 rt : (9,5,9,8)
7656 9089 Ψ : $2\sqrt{5}/5$
7656 9591 sπ : $1/3 + 4\sqrt{5}$
7656 9857 sr : $9\sqrt{3}/5$
7657 2128 $\exp(1/2) + \exp(3/4)$
7657 2464 rt : (8,4,4,-9)
7657 2918 10^x : $8e/9$
7657 3422 2^x : $\sqrt[3]{7}$
7657 4340 rt : (7,8,-6,3)
7657 5641 rt : (3,6,-5,-6)
7657 6703 cr : P_{41}
7657 7076 cu : $\sqrt{3}/3 + 3\sqrt{5}/2$
7657 7269 at : $2\sqrt[3]{3}/3$
7657 7951 rt : (6,1,9,9)
7657 9556 sr : $2 + \sqrt{5}/2$
7658 0211 rt : (9,7,4,3)
7658 0468 rt : (8,2,-1,-4)
7658 0802 rt : (1,9,5,-1)
7658 2510 tanh : $5\sqrt{2}/7$
7658 3252 e^x : $(\ln 3)^3$
7658 4619 as : $\ln 2$
7658 4845 at : $4\zeta(3)/5$
7658 8509 sr : $4\sqrt{2}/3 - 3\sqrt{3}/4$
7658 8528 e^x : $\sqrt{21}$
7658 9338 rt : (5,9,2,3)
7658 9866 2^x : $2\sqrt{3}/3 + 3\sqrt{5}/2$
7659 2833 $\exp(1/3) \div \exp(3/5)$
7659 3169 2^x : $3/2 - e/4$
7659 4402 erf : $7\zeta(3)/10$
7659 5153 rt : (1,6,-4,8)
7659 5358 rt : (6,3,4,4)
7659 6842 id : $4e/3 + \pi$
7659 8021 sinh : 12/17
7659 8971 e^x : $(2e)^{-1/3}$
7659 9466 sπ : $4\sqrt{5}/7$
7660 1425 rt : (5,4,-4,8)
7660 1683 10^x : 11/2
7660 2002 rt : (9,2,8,9)
7660 4444 sπ : 5/18
7660 4478 sinh : $2\sqrt{5}$
7660 4645 sr : $2\sqrt{2}/3 + 3\sqrt{5}$
7660 4762 Ei : 25/19
7660 5651 cu : $7\zeta(3)/5$
7660 5832 K : 13/21
7660 8126 rt : (3,1,4,-5)
7661 0578 ln : $\sqrt[3]{5}/10$
7661 0747 cr : $1/\sqrt[3]{11}$
7661 1503 sr : $9\ln 2/2$
7661 3062 rt : (8,5,-2,-5)
7661 3629 ζ : 19/22
7661 4326 as : $\sqrt[3]{1/3}$
7661 5290 10^x : $2\sqrt{2}/3$
7661 5966 Γ : $\sqrt{20\pi}$
7661 6264 at : $5\sqrt{3}/9$
7661 6753 ℓ_2 : $4/3 - \sqrt{5}/3$
7661 9037 rt : (5,9,-2,-6)
7662 0347 tπ : $2\sqrt{3}/3 - 2\sqrt{5}/3$
7662 1862 sπ : $1/4 + 2\sqrt{5}$
7662 8472 rt : (3,9,-4,-7)
7662 8528 Li_2 : 5/8
7662 8893 tπ : $4e/9$
7662 9285 rt : (3,4,-9,8)
7663 0882 rt : (9,4,3,4)
7663 0943 cr : 9/20
7663 1100 rt : (2,8,-6,-1)
7663 1176 e^x : $\sqrt{3} + 3\sqrt{5}/2$
7663 1370 erf : 16/19
7663 4603 id : $1/3 + \sqrt{3}/4$
7663 5082 ℓ_2 : $3 - 3\sqrt{3}/4$
7663 5597 rt : (8,8,-3,-6)
7663 8379 ζ : 25/19
7663 8840 rt : (6,0,3,5)
7663 9919 id : $(\sqrt{2}\pi/2)^{-1/3}$
7664 0018 Ψ : $7e/3$
7664 2123 rt : (8,8,-8,5)
7664 3548 rt : (1,6,1,-2)
7664 3635 e^x : $2e - 3\pi/4$
7664 5192 10^x : $3\sqrt{2}/2 - \sqrt{5}/3$
7664 5747 id : $5(e + \pi)/3$
7664 5863 rt : (3,5,4,8)
7664 9075 rt : (3,6,-6,1)
7665 2109 rt : (8,3,-7,0)
7665 5744 cu : $4\sqrt{3} - 2\sqrt{5}/3$
7665 7995 Ei : $9\sqrt{3}/8$
7665 8705 10^x : $\exp(\sqrt[3]{4}/2)$
7665 8841 rt : (4,8,9,4)
7665 9184 rt : (9,1,2,5)
7665 9287 rt : (5,8,-6,-8)
7665 9814 10^x : $4\sqrt{3}/7$
7666 1848 $\ln(\sqrt{3}) \times \exp(1/3)$
7666 3773 2^x : $(e/3)^2$
7666 4079 rt : (6,3,2,-6)
7666 4870 e^x : $(\ln 3/3)^{-3}$
7666 5454 θ_3 : $2\zeta(3)/3$
7666 5650 2^x : $3\sqrt{3}/2 - 4\sqrt{5}/3$
7666 7678 $\ln(\sqrt{2}) - \exp(\sqrt{2})$
7666 8130 tπ : $1/2 + 3\sqrt{5}$
7666 8516 rt : (3,7,2,-7)
7666 8972 2^x : $4/3 + 3\sqrt{5}/2$
7666 9307 rt : (1,9,-8,-5)
7667 2588 sr : $1 + 3\sqrt{2}/2$
7667 3444 rt : (1,5,3,9)
7667 4702 cr : $3\zeta(3)/8$
7667 6956 Ψ : $-\sqrt{3} - \sqrt{5} - \sqrt{6}$
7667 7066 rt : (1,4,9,5)
7667 7254 rt : (9,2,1,-6)
7667 7311 rt : (8,6,-8,-1)
7667 8322 rt : (4,7,-8,9)
7667 9680 sr : $4/3 - \sqrt{5}/3$
7668 0567 rt : (6,6,1,-7)
7668 0715 at : $(\sqrt{5}/2)^{-1/3}$
7668 1819 sq : $3\pi^2/10$
7668 3621 rt : (3,8,-3,-6)
7668 4709 sr : $9e\pi$
7668 5376 Ei : $(\sqrt{3})^{1/2}$
7668 7554 ζ : $4\sqrt{2}/3$
7668 9678 rt : (2,0,-9,6)
7668 9792 rt : (9,5,0,-7)
7669 2182 rt : (6,9,0,5)
7669 2906 rt : (8,9,-9,-2)
7669 6498 sr : 10/17
7669 8927 rt : (5,4,-9,-7)
7669 9002 rt : (9,8,-1,-8)
7670 0807 rt : (6,6,-7,2)
7670 0999 ℓ_{10} : $\sqrt[3]{5}/10$
7670 1181 rt : (7,9,8,4)
7670 1366 Ei : $9\sqrt[3]{5}$
7670 4840 rt : (4,9,-5,2)
7670 4990 erf : $(5/3)^{-1/3}$
7670 6002 rt : (8,8,-9,-8)
7670 7772 rt : (4,5,8,5)
7671 0213 sr : $2 + 4\sqrt{2}$
7671 3015 rt : (1,1,8,6)
7671 7287 rt : (2,3,-8,-7)
7671 9379 J_0 : $1/\ln(\ln 3/3)$
7672 0154 2^x : 24/7
7672 0678 cr : $\ln(\pi/2)$
7672 0834 rt : (5,7,-8,-8)
7672 1772 rt : (7,6,7,5)
7672 2065 sq : $4e/3 + 3\pi/4$
7672 2531 $e - s\pi(2/5)$
7672 4627 rt : (4,2,7,6)
7672 7077 rt : (1,2,7,-7)
7672 7114 sr : $(\ln 2/2)^{1/2}$
7672 7483 tπ : $9\sqrt[3]{2}/8$
7672 7678 e^x : $\sqrt[3]{7/3}$
7672 9202 rt : (2,6,-7,-8)
7673 1573 rt : (7,3,6,6)
7673 1703 sq : $2e/3 - \pi$
7673 2698 id : $\sqrt{2} - \sqrt{3} - \sqrt{6}$
7673 3160 rt : (4,1,6,-7)
7673 3538 $\ln(\pi) \div \exp(2/5)$
7673 4574 rt : (1,5,6,-8)
7673 5842 rt : (2,9,-6,-9)
7673 7306 rt : (7,0,5,7)
7673 8313 rt : (4,4,5,-8)
7673 9233 rt : (1,8,5,-9)
7674 1067 rt : (7,3,4,-8)
7674 1764 rt : (4,7,4,-9)
7674 3725 rt : (7,6,3,-9)
7674 4235 rt : (4,2,-5,7)
7674 7232 rt : (7,9,-2,7)
7674 8450 rt : (7,6,-9,4)
7674 8986 10^x : $4 + 4e/3$
7675 1226 at : P_{22}
7675 2203 2^x : $(\sqrt{2}\pi/3)^{-1/2}$
7675 2314 sinh : $\sqrt{2}/2$
7675 2836 ln : $\sqrt[3]{10}$
7675 3778 cr : $4/3 + 4\pi/3$
7675 3834 cr : $\exp(-\sqrt[3]{4}/2)$
7675 4350 sπ : $7(1/\pi)/8$
7675 6177 sinh : $5(e + \pi)/4$
7675 8062 10^x : $\sqrt[3]{9/2}$
7675 8409 id : $4\sqrt{2}/3 - \sqrt{5}/2$
7675 9187 rt : (9,9,-1,-2)
7676 1833 tπ : $\sqrt{2} + 3\sqrt{3}$
7676 5175 ln : $1 + 2\sqrt{3}/3$
7676 5570 sr : $3\sqrt{2} - \sqrt{5}/2$
7676 7865 ζ : $3\pi/5$
7676 9316 sπ : $1/4 + \sqrt{2}/3$
7677 3429 ζ : 16/19
7677 6400 10^x : 25/13
7677 6695 id : $5\sqrt{2}/4$
7678 0752 10^x : $3\sqrt{2}/2 - \sqrt{5}$
7678 0790 sπ : $1/\ln(4)$
7678 1099 10^x : $\sqrt{3}/7$
7678 1190 Ei : $(\sqrt{3}/2)^{1/3}$
7678 1529 rt : (8,9,-4,9)
7678 5870 rt : (5,2,-7,9)
7678 6003 rt : (2,8,-9,-8)
7678 6589 ζ : 2/9
7678 7575 10^x : $2/3 + \pi/2$
7678 8831 ℓ_{10} : $(e + \pi)$
7678 9691 cr : $3/2 - \pi/3$
7679 0131 tπ : $\sqrt{21}$
7679 1623 J_0 : $4\sqrt{5}/9$
7679 2173 10^x : $9\zeta(3)$
7679 5314 at : $1/3 - 3\sqrt{3}/4$
7679 6788 $\ln(4) - \exp(e)$
7679 7181 K : $1/\ln(5)$
7679 7539 rt : (2,4,-1,9)
7679 7714 2^x : $\exp(\sqrt[3]{3})$
7679 7774 Ei : $8\pi^2/9$
7679 8414 ℓ_2 : $2\sqrt{2} + \sqrt{3}/3$
7680 0000 cu : 16/5
7680 0573 rt : (8,0,7,9)
7680 2373 tanh : $(\pi/3)^{1/3}$
7680 3511 cr : $e/6$
7680 3605 sπ : $3\zeta(3)/5$
7680 4790 sπ : $3e/2 - 3\pi/4$
7680 6809 rt : (5,1,8,-9)
7680 7235 sπ : $3e + 4\pi$
7680 9079 rt : (8,3,8,8)
7681 0649 ℓ_{10} : $2/3 + 3\sqrt{3}$

7681 2818 id : $\ln(e+\pi)$
7681 4826 $\ln(3)\times\ln(5)$
7681 4827 $\sqrt{3}+s\pi(\sqrt{2})$
7681 5210 sq : $1/\ln(\sqrt{2}/3)$
7681 5259 rt : (2,2,9,-9)
7681 5487 2^x : $2/3-\pi/3$
7681 5578 $2/3\times\exp(\sqrt{3})$
7681 6688 10^x : $3/4+3e/4$
7681 7086 e^x : $4/3+e/2$
7681 8148 ζ : $\pi/3$
7681 8432 rt : (5,2,9,8)
7682 0094 2^x : $3\sqrt{2}/2-3\sqrt{3}/4$
7682 1377 e^x : $4\sqrt{2}/3-\sqrt{5}/4$
7682 1536 rt : (7,7,-3,-5)
7682 1985 rt : (8,6,9,7)
7682 2850 at : P_{50}
7682 3121 tπ : $1/\sqrt{23}$
7682 5287 id : $\exp(\sqrt[3]{5}/3)$
7682 5602 sπ : $\sqrt[3]{3}/2$
7682 7136 sq : $3/4-3\sqrt{3}$
7683 1481 rt : (4,8,-2,-5)
7683 1552 id : $\sqrt{2}+3\sqrt{5}/2$
7683 2294 cr : $1/3+3\sqrt{3}$
7683 3877 $e\times\ln(4)$
7683 5026 10^x : $3+\sqrt{3}/3$
7683 5133 rt : (7,4,-2,-4)
7683 6561 e^x : $4\sqrt{3}-4\sqrt{5}/3$
7683 7495 10^x : $7(1/\pi)/9$
7684 2640 at : $3\sqrt{2}/2-2\sqrt{3}/3$
7684 3777 cu : $1+4e$
7684 5204 rt : (1,9,-1,-5)
7684 8595 $\sqrt{2}\div\ln(3/5)$
7684 8869 cu : $\beta(2)$
7684 9445 tπ : $4\sqrt{2}/3-3\sqrt{5}/4$
7685 0422 rt : (4,7,1,2)
7685 1284 rt : (1,7,-7,-8)
7685 1552 sr : $1/\ln(2e)$
7685 5397 erf : $11/13$
7685 6311 rt : (7,1,-1,-3)
7686 0425 tπ : $2\sqrt{2}/3+\sqrt{3}/3$
7686 0636 rt : (5,4,-8,1)
7686 1382 θ_3 : $\ln(\sqrt[3]{3})$
7686 2148 $\ln(\sqrt{5})-s\pi(\sqrt{2})$
7686 3056 ln : $2/3+3\sqrt{3}$
7686 4175 rt : (8,7,-1,-7)
7686 7575 cr : $3\sqrt{2}/4+2\sqrt{5}$
7686 7847 λ : $\exp(-13\pi/17)$
7686 8376 $\exp(\sqrt{5})-s\pi(1/5)$
7686 9523 Ψ : $\ln(1/(2\pi))$
7687 0611 sr : $13/22$
7687 3430 rt : (1,6,0,-4)
7687 7935 rt : (5,8,0,-7)
7688 0242 ζ : $7(1/\pi)/10$
7688 2359 tπ : $8\pi^2/5$
7688 2808 rt : (8,4,0,-6)
7688 4029 e^x : $17/2$
7688 5023 sq : $7\sqrt[3]{5}/9$
7688 6410 id : $3\sqrt{2}/4+3\sqrt{5}$
7688 6688 e^x : $\sqrt{2}-3\sqrt{5}/4$
7688 7462 id : $\sqrt{23/3}$
7688 8095 cr : $5/11$
7688 9389 ln : $2/3+2\sqrt{5}/3$
7689 1209 rt : (6,7,-3,6)
7689 1301 2^x : $\sqrt[3]{21}$
7689 1642 rt : (1,9,1,-9)
7689 1871 rt : (2,6,-5,3)
7689 3319 $\ln(2/5)\times\exp(\sqrt{2})$
7689 3852 rt : (7,2,0,2)
7689 4920 rt : (1,9,-7,5)
7689 5180 rt : (9,7,1,-9)
7689 5855 rt : (2,9,1,-7)
7689 6574 at : $(e/3)^{1/3}$
7689 7172 10^x : $3+\sqrt{2}/3$
7689 8410 cr : $10(1/\pi)/7$
7689 9025 rt : (7,7,-5,8)
7689 9828 id : $4\sqrt[3]{3}$
7690 2345 rt : (5,5,1,-6)
7690 6018 10^x : $4\sqrt{2}-3\sqrt{3}/4$
7690 6456 sπ : P_{72}
7690 6990 sinh : $17/24$
7690 7896 sπ : P_{69}
7690 8299 2^x : $1/2+\sqrt{2}$
7690 9479 rt : (8,1,1,-5)
7691 0309 rt : (6,8,2,-9)
7691 0728 rt : (4,9,-9,-9)
7691 1461 sinh : $(\sqrt[3]{2}/3)^{-1/3}$
7691 2000 rt : (2,7,-1,-4)
7691 3001 cu : $\sqrt{2}/2-4\sqrt{3}$
7691 5531 rt : (9,4,2,-8)
7692 3076 id : $10/13$
7692 3369 $\exp(3/4)+\exp(\sqrt{3})$
7692 4068 J : $\zeta(3)/6$
7692 4862 e^x : $4-4\sqrt{5}/3$
7692 5456 e^x : $9\sqrt[3]{3}$
7692 9235 rt : (4,0,-8,8)
7692 9631 ζ : $\exp(-3/2)$
7693 0734 cu : $\ln(5/2)$
7693 1336 $\exp(2)\div s\pi(2/5)$
7693 2053 sr : $7\sqrt{5}/5$
7693 2734 as : $16/23$
7693 5871 rt : (6,5,3,-8)
7693 8835 sπ : $(2\pi/3)^{1/3}$
7693 9167 ℓ_2 : $4\sqrt{3}+3\sqrt{5}$
7693 9385 rt : (5,2,2,-5)
7694 3041 rt : (9,1,3,-7)
7694 3321 rt : (7,8,-9,-1)
7694 4395 Ei : $(\ln 3/3)^{-1/2}$
7694 9003 J_0 : $10\ln 2/7$
7694 9674 2^x : $3\sqrt{3}/4-3\sqrt{5}/4$
7695 0812 rt : (8,2,2,4)
7695 2807 sπ : $\sqrt{5}/8$
7695 4543 tπ : $8(e+\pi)/9$
7695 7059 rt : (1,8,7,-9)
7695 8117 Ψ : $5\pi^2/3$
7696 0026 ℓ_2 : $4\sqrt{2}/3-3\sqrt{3}/4$
7696 0394 J_0 : $7\sqrt{2}/10$
7696 0397 tanh : $3e/8$
7696 0678 sq : $1/4+\sqrt{2}$
7696 0709 sπ : $8\pi^2/7$
7696 2175 cr : $3\sqrt{2}+3\sqrt{3}/4$
7696 2312 rt : (3,6,4,-8)
7696 2587 sinh : $25/7$
7696 2811 $\exp(5)+\exp(\sqrt{5})$
7696 4871 sr : $2\sqrt{2}-\sqrt{5}$
7696 6946 sq : $(\sqrt{2}\pi/3)^{-1/3}$
7696 7632 sinh : $10\zeta(3)/9$
7696 8607 rt : (1,7,-3,-6)
7696 9398 J_0 : $4\sqrt{3}/7$
7696 9912 cu : $1/4+2e/3$
7696 9981 rt : (4,8,-6,7)
7697 1838 rt : (6,2,4,-7)
7697 2039 rt : (8,8,-7,-3)
7697 2609 at : $(\ln 3)^{-1/3}$
7697 3017 2^x : $14/17$
7697 8585 rt : (7,5,1,1)
7697 8588 10^x : $\sqrt{2}/2+3\sqrt{3}$
7697 8752 rt : (3,5,-7,6)
7697 9207 ℓ_{10} : $4+4\sqrt{2}/3$
7697 9510 e^x : $3\sqrt{2}/2-\sqrt{5}/4$
7698 0035 id : $4\sqrt{3}/9$
7698 2292 rt : (9,2,4,6)
7698 2793 rt : (4,9,-8,-1)
7698 3682 rt : (8,9,-8,2)
7698 4198 rt : (2,3,3,7)
7698 4604 ℓ_{10} : $\sqrt{2}+2\sqrt{5}$
7698 5284 $\exp(2/3)+\exp(3/5)$
7698 6065 erf : $3\sqrt{2}/5$
7698 8128 Ei : $3/5$
7698 9699 id : $1/3+2e$
7699 0141 rt : (5,8,-2,-4)
7699 0892 ln : $\sqrt{2}+\sqrt{5}/3$
7699 1118 id : $6\pi/5$
7699 1438 rt : (9,8,-5,-5)
7699 2335 rt : (2,8,2,1)
7699 2658 sq : $5e/7$
7699 2888 rt : (7,2,6,-9)
7699 3664 $\exp(\sqrt{2})\times s\pi(\pi)$
7699 3860 ζ : $13/16$
7699 8098 rt : (7,9,-6,0)
7699 9408 cr : $8\ln 2$
7700 2269 rt : (5,1,3,4)
7700 2441 erf : $8(1/\pi)/3$
7700 3097 Ψ : $9\sqrt{2}/2$
7700 3504 rt : (1,7,2,3)
7700 4207 2^x : $7\sqrt[3]{2}/6$
7700 4865 ζ : $\sqrt{5}/10$
7700 7370 rt : (5,9,-6,-3)
7700 8045 K : $9\ln 2/10$
7700 9662 2^x : $3-\sqrt{2}/3$
7701 1372 rt : (3,3,5,-7)
7701 1899 sr : $4\sqrt{3}+\sqrt{5}/3$
7701 4216 at : $\sqrt{2}/2-3\sqrt{5}/4$
7701 6191 rt : (1,2,-8,-5)
7701 6353 id : $10\ln 2/9$
7701 9624 rt : (7,5,-8,9)
7702 0308 rt : (8,5,-6,-2)
7702 0744 Ei : $(\ln 3)^{-1/2}$
7702 2252 ln : $\sqrt{14/3}$
7702 2434 erf : $(\sqrt[3]{3}/2)^{1/2}$
7702 2894 rt : (6,9,-4,-2)
7702 3395 λ : $1/11$
7702 3455 rt : (8,5,3,3)
7702 5462 cu : $11/12$
7702 6189 rt : (6,1,5,6)
7702 7580 rt : (4,3,7,-9)
7702 8702 cr : $e/2+4\pi/3$
7702 9914 Ψ : $(2\pi)^{-1/3}$
7703 0353 sr : $4-\sqrt{3}/2$
7703 1852 rt : (4,8,-4,-6)
7703 2881 K : $\exp(-\sqrt{2}/3)$
7703 3590 rt : (7,9,-2,-7)
7703 3995 rt : (9,5,-4,-4)
7703 7293 λ : $2(1/\pi)/7$
7703 8008 at : $7\ln 2/5$
7703 9732 Ψ : $\sqrt{5}/8$
7704 0012 rt : (7,1,7,8)
7704 0063 rt : (6,5,-6,7)
7704 0961 Ψ : $\exp(-\sqrt{5})$
7704 1407 rt : (8,9,0,-9)
7704 1525 $3/5\times\exp(1/4)$
7704 2642 $\ln(\sqrt{2})-\exp(3/4)$
7704 2831 rt : (9,5,5,5)
7704 4262 id : $\sqrt{2}/3+3\sqrt{3}/4$
7704 4901 sπ : $4\sqrt[3]{5}/3$
7704 7576 cr : $\sqrt{2}/4+3\sqrt{3}$
7704 7810 sπ : $2\sqrt[3]{2}/9$
7704 8253 2^x : $4\sqrt[3]{3}/7$
7704 9161 rt : (2,9,-3,-2)
7704 9282 rt : (3,8,-6,-8)
7705 1324 sπ : $7/25$
7705 2362 $\ln(2/5)\times\exp(2)$
7705 3751 rt : (3,6,-9,2)
7705 4778 $e-\exp(2/3)$
7705 8850 rt : (1,7,7,-4)
7706 0357 sπ : $4\sqrt[3]{2}/7$
7706 2343 Ei : $1/\sqrt{23}$
7706 2429 sr : $7\pi^2/9$
7706 3410 cu : $2e-\pi/2$
7706 5300 sr : $3/2-e/3$
7706 6805 erf : $17/20$
7706 7033 Γ : $\sqrt{2}+\sqrt{3}-\sqrt{7}$
7706 8257 $\sqrt{2}+\exp(\sqrt{5})$
7706 9063 rt : (4,6,-7,0)
7707 0007 2^x : $3e+3\pi/2$
7707 0890 $\ln(\sqrt{3})+\exp(1/5)$
7707 2108 rt : (1,4,-2,8)
7707 2444 rt : (5,5,-4,5)
7707 3318 e^x : $25/16$
7707 4432 rt : (1,4,8,-9)
7707 5569 rt : (5,9,-2,-4)
7707 5708 2^x : $\sqrt{11/3}$
7707 8016 ℓ_2 : $\exp(4)$
7707 8150 J : $\exp(-20\pi/7)$
7707 9170 rt : (6,6,-3,-4)
7707 9495 e^x : $4/7$
7707 9579 sr : $6\ln 2/7$
7708 1457 rt : (7,6,-1,-6)
7708 3038 rt : (8,6,1,-8)
7708 4186 2^x : $7\sqrt[3]{2}$
7708 9100 as : $\sqrt{2}+\sqrt{3}-\sqrt{6}$

7708 9405 at : $1/2+\sqrt{2}/3$
7709 0799 J_0 : $\pi^2/10$
7709 1699 rt : (4,0,8,8)
7709 4731 e^x : $6e$
7709 5610 rt : (9,2,-3,-3)
7709 5709 sr : $3/4+4\sqrt{3}$
7709 6808 rt : (8,2,-5,-1)
7709 9065 rt : (7,2,-7,1)
7709 9156 $s\pi$: $\sqrt{2}+\sqrt{3}/2$
7710 1076 e^x : 23/2
7710 1083 cr : 11/24
7710 1570 $s\pi$: $2(e+\pi)$
7710 3790 e^x : $3/4+\sqrt{3}/3$
7710 4889 rt : (6,2,-9,3)
7710 5833 ℓ_2 : $(e+\pi)/10$
7710 6088 sr : $\ln(2e/3)$
7710 7317 K : 5/8
7710 8387 rt : (3,7,-4,-1)
7710 9185 ℓ_{10} : $\sqrt{2}/2+3\sqrt{3}$
7711 0937 rt : (7,4,8,7)
7711 2202 $4/5\times s\pi(\sqrt{2})$
7711 4071 10^x : $4/3+4\sqrt{3}$
7711 7406 id : $3\sqrt{3}/4+2\sqrt{5}$
7711 7846 rt : (6,4,6,5)
7711 8900 ζ : $\sqrt[3]{20/3}$
7712 1805 tanh : $(\pi/3)^{1/2}$
7712 3579 10^x : $7\sqrt{2}/10$
7712 3616 id : $8\sqrt{2}/3$
7712 4604 rt : (5,8,3,8)
7712 6277 e^x : $3/2+\pi/3$
7712 8129 id : $8\sqrt{3}/5$
7712 8644 rt : (6,9,-7,-8)
7713 1120 $5\times\exp(e)$
7713 1342 ℓ_2 : $\sqrt{3}+\sqrt{6}+\sqrt{7}$
7713 1370 $\ln(\sqrt{2})\times\exp(4/5)$
7713 1545 rt : (4,7,0,-6)
7713 1806 rt : (9,3,4,-9)
7713 2497 rt : (5,4,4,3)
7713 4867 2^x : 25/17
7713 5625 ℓ_{10} : $1/4+4\sqrt{2}$
7713 6234 2^x : $\sqrt{3}/3+3\sqrt{5}/4$
7713 6244 rt : (7,7,-9,1)
7713 7885 10^x : $2\sqrt{2}/3+\sqrt{5}/3$
7713 9064 rt : (8,3,2,-7)
7714 0853 e^x : $3e/8$
7714 1260 rt : (2,5,-9,-9)
7714 1650 λ : $(\sqrt{2}\pi/2)^{-3}$
7714 1665 Ψ : 13/24
7714 2198 rt : (3,7,-2,-4)
7714 3774 $s\pi$: $3\sqrt{3}/4+4\sqrt{5}/3$
7714 8372 rt : (9,8,6,4)
7714 9529 rt : (7,3,0,-5)
7714 9566 rt : (2,5,-8,6)
7715 0285 Li_2 : $\pi/5$
7715 0524 rt : (7,9,-5,-6)
7715 1599 Ei : 21/22
7715 1730 $t\pi$: $4\sqrt{2}-\sqrt{3}/2$
7715 4158 10^x : $9\sqrt{5}/4$
7715 5330 ℓ_2 : $2+\sqrt{2}$
7715 5639 rt : (6,3,-9,8)
7715 7104 $t\pi$: $8e\pi/5$
7716 0620 rt : (2,7,-4,-2)
7716 1300 rt : (3,5,-7,-7)
7716 4757 ℓ_2 : $\sqrt{3}/4+4\sqrt{5}/3$
7716 5928 rt : (6,3,-2,-3)
7716 6882 rt : (1,1,9,-8)
7716 6920 ln : $3\sqrt[3]{3}/2$
7716 8597 2^x : $5\sqrt[3]{2}$
7716 9096 $t\pi$: $9\sqrt[3]{5}$
7716 9358 Ei : 5/24
7717 2027 $s\pi$: $3\sqrt{2}/2+3\sqrt{3}/2$
7717 3206 cu : $1/4+2\sqrt{3}/3$
7717 6127 2^x : $1/3-\sqrt{2}/2$
7717 6774 rt : (9,2,-9,-4)
7717 7124 2^x : $1/2+3\sqrt{2}$
7717 8301 cu : $7\pi/8$
7717 8487 rt : (2,8,3,-8)
7717 9297 rt : (6,4,5,-9)
7717 9913 10^x : $\sqrt{2}-\sqrt{3}+\sqrt{7}$
7718 0537 sr : $2\sqrt{3}/3-\sqrt{5}/4$
7718 1879 rt : (9,8,-9,9)
7718 2084 ln : $9\zeta(3)/5$
7718 4450 rt : (8,8,4,2)
7718 4996 10^x : $2/3+3\pi$
7718 5074 Li_2 : $4\sqrt{2}/9$
7718 7262 J_0 : $(\pi/3)^{-1/3}$
7718 7559 2^x : $\sqrt{2}/3+2\sqrt{5}$
7718 9902 Γ : $9(1/\pi)$
7719 0365 2^x : $\sqrt{24}\pi$
7719 2159 rt : (3,9,-9,-1)
7719 2576 sq : $2\sqrt{2}+\sqrt{5}/3$
7719 2724 rt : (9,1,-2,2)
7719 3212 e^x : $\sqrt{14}\pi$
7719 5218 10^x : $\sqrt{5}/9$
7719 5295 rt : (5,3,-4,-1)
7719 5747 rt : (9,0,5,8)
7719 6123 rt : (4,3,9,7)
7719 7602 rt : (5,4,3,-7)
7719 7802 ln : $2\ln 2/3$
7719 8292 rt : (6,7,-7,-1)
7719 9140 Ψ : $9\ln 2/7$
7719 9422 e^x : 20/3
7720 0187 rt : (1,9,2,0)
7720 5704 ln : $3/4+\sqrt{2}$
7720 6541 e^x : $4\sqrt{2}/3-\sqrt{3}/2$
7720 6711 sr : $8e\pi/3$
7720 7793 sq : $3-3\sqrt{2}/2$
7720 8595 rt : (4,5,-5,-5)
7721 2533 rt : (5,3,-7,6)
7721 3450 rt : (9,7,-3,-6)
7721 6527 2^x : $3-\sqrt{5}/3$
7721 8505 rt : (8,0,3,6)
7722 2797 Ψ : $\sqrt{7}$
7722 3344 erf : $\exp(-1/(2\pi))$
7722 5576 rt : (4,8,5,6)
7722 6307 ℓ_2 : $e\pi/5$
7722 8634 rt : (8,8,-7,7)
7722 9327 tanh : $3\sqrt[3]{5}/5$
7722 9788 rt : (7,7,9,6)
7722 9908 ℓ_{10} : $3\sqrt{2}+3\sqrt{5}/4$
7722 9951 cu : $2/3+3e/2$
7723 0709 rt : (2,8,-8,-3)
7723 1264 sr : $(3\pi/2)^{-1/3}$
7723 2047 sr : $9e\pi/10$
7723 3665 cr : $4\sqrt{2}-3\sqrt{3}$
7723 6354 rt : (8,9,-3,-4)
7724 0762 rt : (3,5,6,-9)
7724 4837 sq : $7(e+\pi)/9$
7724 4969 ln : $5\sqrt{3}/4$
7724 5099 ℓ_{10} : $6\pi^2$
7724 5385 id : $(\pi)^{1/2}$
7724 5944 rt : (7,8,0,-8)
7724 7235 rt : (3,3,7,5)
7725 0378 $t\pi$: $4\sqrt{3}/3+3\sqrt{5}/2$
7725 1166 rt : (1,9,-4,-8)
7725 1176 ln : $4+4\sqrt{2}/3$
7725 3529 Ei : $4\pi^2/3$
7725 3994 rt : (7,0,1,4)
7725 5837 Ψ : $(\sqrt[3]{2})^{-1/2}$
7725 6147 Ei : $3(1/\pi)$
7725 8872 id : $4\ln 2$
7725 8987 θ_3 : 24/25
7726 2458 rt : (6,1,6,-8)
7726 2459 id : $8e\pi/3$
7726 3012 rt : (4,9,0,-6)
7726 3602 ln : $\sqrt{2}+2\sqrt{5}$
7726 3720 θ_3 : $\exp(-1)$
7726 5031 sr : $5e\pi/3$
7726 9856 rt : (7,7,-7,-2)
7726 9903 id : $2\sqrt{2}+4\sqrt{5}$
7727 0532 $\sqrt{5}\div\ln(3/4)$
7727 0548 2^x : $4\sqrt{2}+2\sqrt{5}$
7727 1571 cr : $4+\pi/2$
7727 2727 id : 17/22
7727 3616 rt : (6,8,-2,-6)
7727 5994 rt : (5,6,0,9)
7727 7634 rt : (2,5,4,-7)
7727 8469 as : $2\pi/9$
7728 0388 cr : 6/13
7728 1052 sr : 22/7
7728 2033 Ψ : $(e/2)^{-2}$
7728 3762 erf : $e\pi/10$
7728 6152 rt : (9,3,6,7)
7728 6555 rt : (1,4,0,-7)
7728 7025 2^x : 19/23
7728 7156 sq : $2/3+3\sqrt{2}/2$
7728 8939 e^x : $6\pi/7$
7729 1237 e^x : $\sqrt[3]{7}$
7729 1716 2^x : $1+\sqrt{2}/3$
7729 2962 rt : (8,5,3,-9)
7729 4461 rt : (9,4,-2,-5)
7729 5388 $t\pi$: $2\sqrt{3}+\sqrt{5}/3$
7729 6610 rt : (7,8,-5,5)
7729 6885 $s\pi$: $4e/3+2\pi/3$
7729 7702 sq : $9e\pi/2$
7729 7711 cu : $4e+2\pi$
7729 8168 id : $8\pi^2/9$
7730 0152 rt : (2,9,-2,-6)
7730 0348 Ei : $\zeta(3)/2$
7730 1269 rt : (2,7,-1,4)
7730 1917 $s\pi$: $2\sqrt{3}-\sqrt{5}/3$
7730 3875 rt : (3,8,2,8)
7730 5242 rt : (6,7,7,4)
7730 5640 Ψ : $\sqrt{5}/6$
7730 7039 rt : (5,1,4,-6)
7730 7603 rt : (4,9,1,-8)
7730 8684 rt : (1,8,3,-2)
7731 1612 cr : $2\ln 2/3$
7731 3375 rt : (5,8,-4,-4)
7731 3605 e^x : $5e\pi/7$
7731 4350 rt : (4,9,-4,5)
7731 7016 rt : (6,0,-1,2)
7731 7486 ζ : 10/13
7731 7531 $\exp(3/5)+s\pi(2/5)$
7731 8988 ln : 6/13
7731 9979 $\ln(3/5)+\exp(1/4)$
7732 3504 rt : (6,7,-9,0)
7732 5392 $\ln(\pi)\times s\pi(\sqrt{5})$
7732 6157 rt : (7,2,9,9)
7732 6218 ℓ_2 : $7(e+\pi)/3$
7732 6474 cr : $4/3+3\sqrt{2}$
7732 6865 2^x : $\sqrt{2}/4+\sqrt{5}/2$
7732 8770 at : $4/3-4\sqrt{3}/3$
7732 8779 rt : (7,5,1,-7)
7733 3224 rt : (6,6,-8,-7)
7733 8741 rt : (1,5,2,-5)
7733 8924 erf : $\sqrt[3]{5}/2$
7733 9884 rt : (8,3,4,5)
7734 0021 rt : (9,9,-2,-8)
7734 0913 rt : (8,4,-4,-3)
7734 1830 cr : $3/2+3e/2$
7734 5908 rt : (7,8,2,0)
7735 0269 id : $10\sqrt{3}/3$
7735 0496 e^x : $9(1/\pi)/5$
7735 2290 rt : (3,9,-1,-6)
7735 3457 rt : (3,2,7,-8)
7735 3512 ℓ_{10} : $1/2+2e$
7735 6173 sr : $4/3+2e/3$
7735 8362 at : $(e+\pi)/6$
7735 9147 rt : (1,6,2,4)
7736 0120 rt : (5,6,4,-9)
7736 0936 sinh : $(\sqrt[3]{4}/3)^{-2}$
7736 2816 rt : (9,3,-9,1)
7736 3846 cr : $3\sqrt{3}/2+4\sqrt{5}/3$
7736 8149 rt : (9,4,-1,1)
7736 9172 ζ : $(2e)^{-1/3}$
7737 5330 rt : (4,8,-6,-2)
7737 6581 $s\pi(\sqrt{3})\div s\pi(\sqrt{2})$
7737 6721 cu : $4e\pi$
7737 7122 sr : $\sqrt{2}+\sqrt{3}$
7737 8077 as : $((e+\pi)/2)^{-1/3}$
7737 8758 rt : (8,9,-4,-6)
7738 0085 rt : (6,5,-1,-5)
7738 0509 e^x : $2/3+\sqrt{2}/4$
7738 0679 10^x : $3e/2-4\pi/3$
7738 3090 rt : (8,2,4,-8)
7738 3095 at : $4\sqrt[3]{5}/7$
7738 3873 rt : (7,6,-8,6)
7738 5007 10^x : $1/3+3\sqrt{2}/4$
7738 5570 J : $8(1/\pi)/9$

7738 6674 rt : (6,2,7,7)
7738 6978 at : $(\pi/3)^{-1/2}$
7738 7374 cu : 23/19
7738 9671 ℓ_{10} : $3\sqrt{3}+\sqrt{5}/3$
7738 9684 θ_3 : $1/\sqrt[3]{20}$
7739 1941 K : $\pi/5$
7739 2147 Ψ : $3(1/\pi)/8$
7739 2628 rt : (3,9,0,-4)
7739 3825 rt : (1,6,-5,-7)
7739 3992 θ_3 : 7/19
7739 4268 sπ : e
7739 5985 J_0 : $7\sqrt[3]{2}/9$
7739 6414 rt : (2,3,5,3)
7739 6555 id : $e/4+2\pi/3$
7739 6679 sr : $9\sqrt[3]{5}/2$
7739 7603 ℓ_2 : $\sqrt[3]{5}$
7739 7982 tanh : $5\sqrt[3]{3}/7$
7739 9603 Ei : $3\ln 2/10$
7740 1168 rt : (9,8,-9,-2)
7740 1917 sπ : $\sqrt{2}/4+4\sqrt{3}$
7740 4019 sr : $\sqrt{6}\pi$
7740 4245 tanh : $6\zeta(3)/7$
7740 4476 rt : (6,8,-3,3)
7740 6566 rt : (3,6,-5,4)
7740 9169 Li_2 : $\sqrt[3]{2}/2$
7740 9370 sπ : P_6
7741 0258 $\ln(3)+s\pi(\sqrt{5})$
7741 0526 rt : (9,1,-1,-4)
7741 1011 K : $4\sqrt{2}/9$
7741 3838 cr : $5e\pi/2$
7741 4752 ℓ_2 : $3\sqrt{2}+3\sqrt{3}/2$
7741 4889 rt : (7,4,-6,-1)
7741 5177 rt : (4,6,2,7)
7741 5823 rt : (3,2,-8,-6)
7741 9984 sq : $4\sqrt{3}-3\sqrt{5}/2$
7742 0278 rt : (2,9,-3,-4)
7742 3603 ℓ_{10} : $3/4+3\sqrt{3}$
7742 3703 rt : (9,6,7,6)
7742 6075 rt : (9,6,-1,-7)
7742 6368 cr : $1/\sqrt[3]{10}$
7742 6572 cr : $3\sqrt{2}/2+2\sqrt{3}$
7742 8314 rt : (4,1,-8,5)
7743 1562 e^x : $3/2+4e$
7743 1633 id : $4/3-\sqrt{5}/4$
7743 1783 rt : (7,9,-6,-4)
7743 2796 ℓ_2 : $4\sqrt{2}-\sqrt{5}$
7743 4545 rt : (7,3,2,3)
7743 6574 rt : (9,1,7,9)
7743 7841 Ei : $\exp(-\pi/2)$
7743 8299 rt : (2,2,5,-6)
7744 0000 sq : 22/25
7744 1076 rt : (3,7,-8,-9)
7744 2336 2^x : $1/4+\sqrt{3}/3$
7744 5542 rt : (7,2,2,-6)
7744 5907 $3-\exp(4/5)$
7744 6812 rt : (2,7,5,-9)
7744 7717 10^x : $6e/7$
7744 9843 rt : (7,7,2,-9)
7745 0230 sπ : $8\sqrt[3]{3}/9$
7745 0697 sq : $1+2\sqrt{2}/3$
7745 1919 $\ln(3/5)\times\exp(2)$
7745 2136 rt : (7,9,-1,9)
7745 5768 erf : 6/7
7745 6707 e^x : $2\sqrt{2}/3+3\sqrt{5}$
7745 7116 sq : $\sqrt[3]{22/3}$
7745 9666 id : $(5/3)^{-1/2}$
7746 0966 ℓ_{10} : $(2e/3)^3$
7746 1241 erf : $(\sqrt[3]{4})^{-1/3}$
7746 4758 tanh : $(\ln 3)^{1/3}$
7746 5115 $1/3\div s\pi(\pi)$
7746 7592 ℓ_{10} : $3\sqrt{3}/2+3\sqrt{5}/2$
7746 9471 2^x : $5\sqrt[3]{5}$
7746 9875 rt : (1,8,-6,7)
7747 5871 sr : $-\sqrt{3}+\sqrt{5}+\sqrt{7}$
7747 6832 sr : $5\sqrt[3]{2}/2$
7747 6967 rt : (5,3,5,-8)
7748 3767 2^x : $3/4-\sqrt{5}/2$
7748 4707 $\ln(\sqrt{3})+\exp(4/5)$
7748 5112 10^x : $2\sqrt[3]{3}/5$
7748 5177 rt : (3,8,-8,0)
7748 7326 rt : (8,6,-3,-5)
7748 8823 id : $4\sqrt{2}+\sqrt{5}/2$
7748 9257 rt : (5,2,5,5)
7748 9362 erf : $(e/2)^{-1/2}$
7748 9763 e^x : $e/2-\pi/4$
7749 4885 10^x : $2(e+\pi)/5$
7749 5452 e^x : $2\pi^2/7$
7749 5652 Ei : $(2e)^{-2}$
7749 6210 $\sqrt{5}\times\ln(\sqrt{2})$
7749 6843 Ψ : $3\sqrt[3]{3}/8$
7749 7129 cu : $9\ln 2$
7749 8326 2^x : $4(e+\pi)/5$
7749 8618 rt : (9,9,-9,6)
7750 0184 rt : (1,9,3,9)
7750 0424 ln : $4\sqrt{2}-3\sqrt{3}$
7750 0945 10^x : $2-\sqrt{2}/3$
7750 0995 cu : $2\sqrt{2}/3+2\sqrt{5}$
7750 3856 rt : (8,1,-3,-2)
7750 5624 2^x : $e/4-\pi/3$
7750 8761 rt : (6,9,-8,-2)
7750 9749 ln : $(3\pi/2)^{1/2}$
7750 9794 rt : (8,1,5,7)
7751 0921 sπ : $4\sqrt{3}+3\sqrt{5}/2$
7751 1365 ℓ_2 : $4-\sqrt{3}/3$
7751 2549 rt : (6,7,0,-7)
7751 3254 10^x : $7\sqrt{3}/8$
7751 4351 J_0 : $(\pi/3)^{-1/2}$
7751 7634 J_0 : $4\sqrt[3]{5}/7$
7751 8523 at : $4/3-\sqrt{2}/4$
7751 9706 rt : (9,8,0,-9)
7751 9764 rt : (2,9,-2,3)
7752 0646 Ψ : $\ln(1/(3e))$
7752 5120 $3/5\div s\pi(e)$
7752 5443 sinh : $3e/4$
7752 5512 rt : (2,8,5,0)
7752 6024 sr : $\zeta(3)/2$
7752 6617 at : $7\sqrt[3]{2}/9$
7752 7515 tanh : $(e/3)^{-1/3}$
7752 8992 cu : $3\sqrt{2}-\sqrt{5}/3$
7753 1846 rt : (2,6,-4,7)
7753 2837 rt : (1,7,3,-7)
7753 3608 as : $5\sqrt[3]{2}/9$
7753 4132 K : $\sqrt[3]{2}/2$
7753 5046 $\ln(4)+\exp(2)$
7753 5182 rt : (3,1,8,7)
7753 5186 erf : $5\zeta(3)/7$
7753 5653 rt : (4,8,3,-9)
7753 5908 cr : $\sqrt{2}+\sqrt{3}+\sqrt{6}$
7753 6171 rt : (1,9,-5,-2)
7753 7699 rt : (6,0,8,9)
7753 8166 rt : (1,7,-5,8)
7753 8538 J_0 : $(e+\pi)/6$
7753 9749 as : 7/10
7754 0049 2^x : $6\sqrt{5}/7$
7754 1987 rt : (9,3,0,-6)
7754 3488 ℓ_2 : $((e+\pi)/2)^{1/2}$
7754 5373 ℓ_{10} : $8\sqrt{5}/3$
7754 5988 ℓ_2 : $\sqrt{19}\pi$
7754 5996 sπ : $3\sqrt[3]{5}/4$
7754 7034 rt : (9,5,-8,-1)
7754 8166 rt : (8,6,5,4)
7754 8953 10^x : $9\sqrt{5}/8$
7754 9725 2^x : 23/12
7755 0460 ln : $\sqrt{2}/2+3\sqrt{3}$
7755 0926 rt : (6,4,-8,1)
7755 6898 rt : (2,9,6,2)
7756 0463 cr : $3+3\sqrt{3}/2$
7756 1330 tπ : $\sqrt[3]{2}/6$
7756 1764 rt : (5,7,5,2)
7756 3152 rt : (1,8,-6,-9)
7756 4644 rt : (7,4,3,-8)
7756 5503 2^x : $\ln(\ln 2)$
7756 5559 cr : 7/15
7756 6026 $\ln(3/5)\div\ln(3/4)$
7756 7064 $\ln(\sqrt{5})\times s\pi(\sqrt{2})$
7756 9376 rt : (8,4,-3,-1)
7756 9619 rt : (6,4,-9,5)
7757 2885 rt : (9,4,8,8)
7757 5251 rt : (4,3,3,-6)
7757 5756 rt : (1,8,8,-2)
7757 5955 rt : (7,6,-5,-3)
7757 7176 2^x : $3/2+2\sqrt{5}$
7757 8443 rt : (4,7,-6,-7)
7758 1013 rt : (3,3,-8,3)
7758 3205 rt : (4,9,-8,-6)
7758 5011 $\exp(1/4)+\exp(2/5)$
7758 5052 rt : (8,8,-2,-7)
7758 6764 ζ : 5/22
7758 7154 rt : (1,8,2,2)
7758 8166 2^x : $(\ln 3)^{-2}$
7759 0533 rt : (9,2,0,3)
7759 0968 at : $\sqrt{2}-\sqrt{3}/4$
7759 1719 2^x : $1/2-\sqrt{3}/2$
7759 2314 rt : (6,5,8,6)
7759 3343 e^x : $3e-4\pi/3$
7759 9329 10^x : $9\sqrt{5}$
7759 9346 tπ : $4e/3+3\pi/2$
7759 9656 rt : (4,7,-3,-5)
7760 0370 ℓ_{10} : P_{44}
7760 1071 cr : $4/3-\sqrt{3}/2$
7760 1359 ζ : $5(1/\pi)/7$
7760 1816 rt : (5,8,-1,1)
7760 2334 at : $2-4\sqrt{5}/3$
7760 2427 sr : P_2
7760 2652 rt : (6,9,1,7)
7760 3123 $\exp(3/5)-\exp(4)$
7760 4259 sπ : $10\zeta(3)/7$
7760 4301 Ei : $9e\pi/4$
7760 4543 rt : (2,4,6,-8)
7760 9474 sq : $e/3-4\pi/3$
7761 0109 ζ : $\exp(-\sqrt{2}\pi/3)$
7761 0199 sr : $3\sqrt{2}+2\sqrt{3}$
7761 1341 ln : $1/4+4\sqrt{2}$
7761 2968 ℓ_{10} : $3/2+2\sqrt{5}$
7761 3685 $\sqrt{2}\div\exp(3/5)$
7761 3750 rt : (8,9,-7,4)
7761 4766 sr : $2+2\sqrt{3}/3$
7761 5973 rt : (8,1,6,-9)
7761 6276 rt : (7,8,-9,-2)
7761 7504 sπ : $\sqrt{2}/5$
7761 9012 rt : (4,2,-4,9)
7761 9925 θ_3 : $(e/3)^{1/3}$
7762 1970 rt : (6,8,-9,-9)
7762 2387 rt : (7,1,3,5)
7762 3704 erf : $(\pi/2)^{-1/3}$
7762 4735 Ei : 2/23
7762 5087 rt : (3,8,1,-7)
7762 7657 rt : (3,5,-6,9)
7762 8581 rt : (3,9,-8,3)
7762 9072 rt : (9,0,-8,2)
7763 0059 rt : (9,5,1,-8)
7763 1548 ℓ_{10} : $3\sqrt{2}+\sqrt{3}$
7763 3116 rt : (4,9,2,-2)
7763 3690 Ψ : $\ln(\sqrt[3]{2}/3)$
7763 4254 at : $3/4-\sqrt{3}$
7763 4662 cr : $3e/4-\pi/2$
7763 6523 sr : $1+3\sqrt{5}$
7763 6686 e^x : $4\sqrt[3]{5}/3$
7763 7192 sπ : $8(1/\pi)/9$
7763 9955 rt : (2,6,-2,-3)
7764 1848 rt : (8,3,-2,-4)
7764 2042 2^x : $(\pi)^3$
7764 2813 ζ : 19/13
7764 4323 rt : (1,9,-8,-4)
7764 5192 rt : (6,3,0,1)
7764 6145 rt : (1,9,4,-9)
7764 8715 cu : $4/3+\sqrt{5}/4$
7764 8840 Ψ : 8/9
7765 0251 rt : (5,0,6,7)
7765 2736 rt : (1,9,9,-4)
7765 3077 sq : $3\sqrt{2}/4-2\sqrt{3}$
7765 3792 rt : (1,6,3,-8)
7765 4239 ℓ_{10} : $7e\pi$
7765 5470 rt : (9,7,-7,-3)
7765 7255 rt : (2,8,-7,2)
7765 8137 rt : (9,9,8,5)
7765 9270 sinh : 5/7
7765 9977 sinh : $5\pi/9$
7766 0168 10^x : 9/2
7766 0190 ℓ_{10} : P_{60}

7766 1434 2^x : $(\sqrt[3]{5}/3)^{1/3}$
7766 2732 $t\pi$: $3(e+\pi)/2$
7766 2820 sq : $4e-\pi$
7766 4038 rt : (6,4,1,-6)
7766 4808 Li_2 : 12/19
7766 5195 id : $2e/7$
7766 7196 tanh : $\zeta(5)$
7766 7937 rt : (6,7,-2,8)
7766 9912 cu : $8\pi/9$
7767 1589 $t\pi$: $\sqrt{3}+\sqrt{5}+\sqrt{6}$
7767 3872 rt : (7,8,-4,-5)
7767 4259 ℓ_{10} : $4e/3+3\pi/4$
7767 4530 as : $2-3\sqrt{3}/4$
7767 4844 rt : (3,1,9,-9)
7767 5169 id : $\sqrt{3}/3-3\sqrt{5}/2$
7767 5173 K : 12/19
7767 5946 ζ : $\exp(\sqrt[3]{2}/2)$
7767 7653 rt : (3,6,8,4)
7768 0458 ℓ_{10} : $3+4\sqrt{5}/3$
7768 2109 sr : $3\sqrt{3}/2+\sqrt{5}/4$
7768 3619 $\ln(3)\times s\pi(1/4)$
7768 4038 rt : (4,5,4,-8)
7768 5321 $s\pi$: 2π
7768 5644 ln : $4e/5$
7768 7480 $\ln(\sqrt{2})-s\pi(\pi)$
7768 8698 sr : $1/4+\sqrt{2}/4$
7768 9357 Ei : 22/23
7769 0707 rt : (5,9,-1,-7)
7769 1111 sq : $3/4+3e/4$
7769 3624 rt : (3,4,-9,-8)
7769 3777 $t\pi$: $3\sqrt{3}/4+2\sqrt{5}/3$
7769 5481 Ψ : $3e\pi/4$
7769 8527 ζ : $(2\pi/3)^{-2}$
7769 9269 $t\pi$: $2\sqrt{2}/5$
7769 9594 10^x : $4+2\pi$
7770 0956 rt : (8,4,6,6)
7770 2276 2^x : $4\sqrt{2}+2\sqrt{3}$
7770 2832 rt : (2,8,-4,-7)
7770 5380 tanh : $(\sqrt{5}/2)^{1/3}$
7770 5804 ℓ_{10} : $7\sqrt[3]{5}/2$
7770 6170 rt : (3,8,-2,-3)
7770 6491 rt : (9,0,1,5)
7770 8763 sq : $2+4\sqrt{5}$
7771 0376 rt : (5,2,-2,-2)
7771 0452 10^x : $10\ln 2/7$
7771 1882 sr : $3+3\pi/2$
7771 2933 ℓ_2 : $2\sqrt{3}/3+\sqrt{5}/4$
7771 4692 Ei : $2e\pi/3$
7771 5471 rt : (6,6,-7,-1)
7771 7428 rt : (6,5,-5,9)
7772 1466 sr : $2e/9$
7772 3687 rt : (6,3,9,8)
7772 3806 e^x : $\sqrt{3}/2-\sqrt{5}/2$
7772 4329 2^x : $\sqrt{23}$
7772 5163 rt : (7,1,4,-7)
7772 5281 ℓ_2 : $2/3+\pi/3$
7772 6218 rt : (8,5,-1,-6)
7772 7010 rt : (9,9,-6,-5)
7772 7863 rt : (4,9,-7,-9)
7773 0651 e^x : $\exp(\sqrt{3}/2)$
7773 2568 rt : (1,4,-5,1)
7773 8978 rt : (4,7,-4,-3)
7774 0728 rt : (6,6,2,-8)
7774 1349 $s\pi$: $(e)^{-1/3}$
7774 2018 rt : (5,2,7,-9)
7774 3270 sq : $3+4\sqrt{5}/3$
7774 3677 2^x : $\sqrt{7\pi}$
7774 3793 rt : (2,1,6,5)
7774 4042 ℓ_{10} : $1/3+4\sqrt{2}$
7775 1578 rt : (5,4,-7,3)
7775 1937 K : $(5/2)^{-1/2}$
7775 7161 2^x : $4\sqrt{2}-3\sqrt{5}/4$
7775 7746 sinh : $9\sqrt[3]{2}/5$
7775 8401 tanh : $3\sqrt{3}/5$
7775 9574 rt : (2,8,-1,-5)
7775 9734 10^x : γ
7775 9735 10^x : $3\sqrt{2}+3\sqrt{3}/4$
7776 0757 ℓ_2 : 7/12
7776 2597 rt : (3,6,-9,-7)
7776 3760 rt : (4,8,-9,-1)
7776 6381 rt : (1,5,7,9)
7776 8015 e^x : $\sqrt{19\pi}$
7777 1219 at : $(\pi/3)^{-1/3}$
7777 1511 cu : $e\pi$
7777 3014 J : $\sqrt[3]{5}/10$
7777 4901 e^x : $\sqrt[3]{15}$
7777 5169 e^x : $4e-3\pi/2$
7777 5955 cr : $e/3+3\pi/2$
7777 7777 id : 7/9
7778 1090 rt : (8,8,-6,9)
7778 2223 cr : 8/17
7778 2381 $s\pi$: $\exp(-\sqrt[3]{2})$
7778 2651 sq : $7\sqrt[3]{2}/10$
7778 3176 rt : (8,7,0,-8)
7778 3854 $t\pi$: $2\sqrt{3}/3-4\sqrt{5}$
7778 5229 rt : (8,2,7,8)
7778 6137 rt : (8,7,-9,-1)
7778 6404 sq : $2-\sqrt{5}/2$
7778 7117 rt : (7,9,-5,2)
7778 8318 e^x : $2+2\pi$
7778 8359 rt : (1,8,6,-9)
7778 9259 $\exp(\pi)\div s\pi(\pi)$
7778 9397 erf : $(\sqrt{5}/3)^{1/2}$
7779 0040 rt : (7,3,5,-9)
7779 0448 ζ : $5\sqrt[3]{2}/9$
7779 1258 sr : $(\ln 3/3)^{1/2}$
7779 2355 rt : (8,2,-2,1)
7779 7120 ln : $1/2+3\sqrt{5}/4$
7779 8290 rt : (7,3,-4,-2)
7780 1061 sq : $5\sqrt[3]{3}/3$
7780 1512 rt : (6,8,-6,-3)
7780 1653 2^x : $4\sqrt{2}/3+2\sqrt{3}$
7780 3465 Li_2 : $(5/2)^{-1/2}$
7780 4331 cu : $10\pi^2/9$
7780 5262 erf : 19/22
7780 5717 sinh : $(\ln 3/3)^{1/3}$
7780 7177 cu : $2\sqrt{3}+4\sqrt{5}/3$
7780 7319 $s\pi$: $9(1/\pi)/4$
7780 7674 J_0 : $7\ln 2/5$
7780 8696 $3/4\div s\pi(\sqrt{2})$
7781 1006 e^x : $4\sqrt{3}/3-\sqrt{5}/3$
7781 1219 $2\times\exp(2)$
7781 1839 rt : (2,8,3,5)
7781 2383 rt : (1,5,2,-1)
7781 3398 sr : $1/3+2\sqrt{2}$
7781 3955 id : $7e\pi$
7781 4499 sinh : $((e+\pi)/3)^{-1/2}$
7781 5125 ℓ_{10} : 6
7781 5326 rt : (4,9,-3,-5)
7781 5848 Γ : 10/3
7781 6141 rt : (5,4,-1,-4)
7781 7582 rt : (6,1,1,3)
7781 8085 10^x : $\sqrt{3/2}$
7781 8363 sinh : $(\sqrt{5}/2)^{-3}$
7781 9140 θ_3 : $\sqrt{2}+\sqrt{3}-\sqrt{5}$
7782 0814 rt : (7,6,3,2)
7782 2204 ℓ_{10} : $2e/3+4\pi/3$
7782 2826 rt : (7,7,-8,3)
7782 4494 $s\pi$: $3\sqrt{3}/2+\sqrt{5}/2$
7782 4888 $\ln(3/4)\times\exp(3)$
7782 6059 $\sqrt{3}\div\exp(4/5)$
7782 6064 rt : (9,4,3,-9)
7782 7171 id : $(\sqrt{2}/3)^{1/3}$
7782 7941 10^x : 1/4
7782 8436 ln : $3\sqrt{2}+3\sqrt{5}/4$
7783 1823 rt : (3,5,2,-6)
7783 1892 sq : $1/3+2\pi$
7783 2872 sq : $4e/3+\pi$
7783 3124 $t\pi$: 4/19
7783 4641 J : $(3\pi/2)^{-3}$
7783 4781 2^x : $2+2\sqrt{5}$
7783 8086 cr : $2+4e/3$
7783 8663 rt : (4,0,4,5)
7783 8844 ζ : 7/10
7784 4571 rt : (8,9,-8,-3)
7784 5711 rt : (2,6,-8,-8)
7784 6070 $\exp(\sqrt{2})\times s\pi(\sqrt{5})$
7784 9829 10^x : $3\sqrt{3}/2+4\sqrt{5}$
7785 0092 2^x : $4-\sqrt{3}/2$
7785 1146 $s\pi$: $(2e/3)^2$
7785 1880 sr : $3\sqrt{2}/7$
7785 2265 rt : (9,4,-6,-2)
7785 2928 rt : (5,5,6,4)
7785 3907 id : $3\sqrt{2}-2\sqrt{3}$
7785 4472 rt : (6,8,-7,-4)
7785 4536 tanh : 25/24
7785 4671 sq : 15/17
7785 5126 10^x : $(\sqrt[3]{2})^{-1/2}$
7785 6551 rt : (2,1,7,-7)
7785 9687 rt : (1,9,-7,6)
7785 9724 $2-\exp(1/5)$
7786 1144 e^x : $3+\sqrt{3}/3$
7786 1967 rt : (3,6,-1,-3)
7786 2252 J_0 : $(\ln 3)^{-1/3}$
7786 3414 ln : $3\pi^2/5$
7786 4002 rt : (7,5,-3,-4)
7786 4017 cr : $2\sqrt{3}/3+2\sqrt{5}$
7786 5491 Γ : $\zeta(3)/5$
7786 7088 Ei : $7\pi^2/5$
7786 8410 $\ln(2)+\exp(3)$
7786 8800 cu : 23/25
7786 9430 $\sqrt{5}\div\ln(\sqrt{5})$
7787 0111 at : $\sqrt{3}-\sqrt{5}/3$
7787 2568 ln : $3\sqrt{2}/4+\sqrt{5}/2$
7787 3221 id : $4e-2\pi/3$
7787 4565 rt : (5,9,3,5)
7787 6474 Ψ : $3\sqrt[3]{2}/7$
7787 6835 10^x : $\sqrt{3}/3$
7787 7110 ζ : $(\pi/3)^{-2}$
7787 7841 rt : (3,4,9,6)
7787 8352 sr : $7\ln 2/8$
7787 9107 rt : (8,0,-1,3)
7787 9550 as : $(\ln 2/2)^{1/3}$
7788 0078 $1\div\exp(1/4)$
7788 2540 e^x : $1/\ln(3/2)$
7788 3569 at : $\pi^2/10$
7788 4124 rt : (3,7,3,-8)
7788 5784 cr : $3\sqrt[3]{2}/8$
7788 6119 Ei : $(2e/3)^3$
7788 8570 rt : (4,9,-3,8)
7788 9503 ℓ_{10} : $5\zeta(3)$
7789 1603 rt : (6,1,2,-5)
7789 2170 2^x : $4e/3+4\pi$
7789 2816 rt : (1,5,-7,-9)
7789 3848 id : $1/\ln(\sqrt[3]{5}/3)$
7789 4971 2^x : $2/3+2\pi/3$
7789 5842 rt : (5,7,-4,-5)
7789 6829 erf : $3\sqrt[3]{3}/5$
7789 7066 rt : (9,6,-5,-4)
7790 1437 sinh : $9(1/\pi)/4$
7790 1698 e^x : $\sqrt{2}+\sqrt{3}/2$
7790 2607 rt : (4,2,5,-7)
7790 4660 rt : (7,7,-2,-6)
7790 6698 rt : (9,5,1,2)
7790 9361 $2/5\times\exp(2/3)$
7790 9717 rt : (3,5,-7,1)
7791 1636 rt : (5,8,1,-8)
7791 1640 Ψ : $\sqrt{3}+\sqrt{5}+\sqrt{6}$
7791 2372 rt : (2,3,8,-9)
7791 2781 ln : $3/2+e/4$
7791 3963 $t\pi$: $5\pi^2/4$
7791 5164 cu : $1-4\sqrt{5}/3$
7791 8068 rt : (3,9,4,6)
7792 0479 rt : (7,8,-7,-7)
7792 1494 rt : (7,4,4,4)
7792 5359 rt : (8,2,0,-5)
7792 5551 J_0 : $(e/3)^{1/3}$
7792 5788 tanh : 24/23
7792 6086 Γ : $\sqrt[3]{3}/6$
7792 6993 rt : (9,8,-4,-6)
7792 9625 ζ : $1/\sqrt{19}$
7793 0576 sq : $3-2\pi$
7793 1025 rt : (1,6,-4,-1)
7793 2308 rt : (7,9,-1,-8)
7793 2802 rt : (6,3,3,-7)
7793 2863 erf : $\sqrt{3}/2$
7793 3917 rt : (5,3,7,6)
7793 6277 $s\pi$: $\sqrt{2}/2+\sqrt{3}/3$
7793 6513 $s\pi$: $(\sqrt{5}/2)^{-3}$
7793 7257 rt : (5,5,-3,7)

7793 7712 cr : $(3\pi)^{-1/3}$
7793 9547 rt : (4,4,6,-9)
7794 1879 rt : (3,7,-3,2)
7794 2516 $s\pi$: $((e+\pi)/3)^{-1/2}$
7794 3789 sinh : $(e)^{-1/3}$
7794 7418 2^x : P_{25}
7794 7506 rt : (1,7,1,-3)
7794 7519 ℓ_2 : $3+\sqrt{3}/4$
7795 0187 $t\pi$: $2\sqrt{2}/3-\sqrt{3}$
7795 0677 $\ln(3/4)-\exp(2/5)$
7795 1304 sr : $19/6$
7795 2412 $\ln(3/5)\times\exp(\sqrt{5})$
7795 2426 cr : $9/19$
7795 3331 sinh : $\sqrt{5}+\sqrt{6}-\sqrt{7}$
7795 3970 sq : $e/2+4\pi/3$
7795 4120 rt : (8,4,1,-7)
7795 4633 erf : $5\ln 2/4$
7795 6024 rt : (5,3,-6,8)
7795 6155 $s\pi$: $(\ln 3/3)^{1/3}$
7795 7035 sq : $\sqrt{3}/3-4\sqrt{5}/3$
7795 8017 rt : (9,3,2,4)
7795 9056 rt : (6,5,4,-9)
7795 9259 2^x : $\sqrt{2}-\sqrt{3}+\sqrt{5}$
7796 0639 rt : (3,8,-2,-5)
7796 1818 $\ln(5)-\exp(2)$
7796 2868 2^x : $1-e/2$
7796 3686 θ_3 : $1/\ln(3)$
7796 5436 rt : (7,2,5,6)
7796 6359 $t\pi$: $\sqrt{2}/4+4\sqrt{3}/3$
7796 6667 rt : (5,6,-9,1)
7796 7024 erf : $13/15$
7796 7503 rt : (7,8,-4,7)
7796 9288 10^x : $16/21$
7796 9818 rt : (2,9,-1,8)
7797 0364 rt : (5,1,-9,9)
7797 1355 rt : (5,9,-5,-7)
7797 1984 rt : (5,1,8,8)
7797 3739 rt : (8,6,2,-9)
7797 4725 $s\pi$: $e+4\pi$
7797 5060 ℓ_{10} : P_1
7797 6314 id : $3\sqrt[3]{2}$
7797 6319 sq : $2\zeta(3)$
7797 6367 rt : (3,7,-6,-1)
7797 7115 $s\pi$: $e/2+3\pi/4$
7797 8141 10^x : $10\sqrt[3]{3}$
7797 9057 id : $9\ln 2/8$
7797 9397 rt : (7,6,-7,8)
7798 0851 cu : $(2e/3)^3$
7798 4500 rt : (1,8,-3,-3)
7798 5624 rt : (9,1,3,6)
7798 7561 rt : (8,9,6,3)
7798 7585 2^x : $6\ln 2/5$
7799 0071 rt : (7,0,6,8)
7799 0105 ζ : $15/8$
7799 1419 rt : (1,9,0,-5)
7799 1704 $1/3-\exp(\sqrt{2})$
7799 2676 10^x : $13/6$
7799 3041 sr : $(\sqrt{2}\pi)^{-1/3}$
7799 5235 rt : (9,9,-8,8)
7799 5415 rt : (2,7,0,8)
7799 8500 rt : (6,8,9,5)
7799 8858 cu : $1/2+e/3$
7799 8900 $s\pi(\sqrt{5})\div s\pi(1/3)$
7800 0230 cr : $4\pi^2/7$
7800 0293 rt : (5,8,-8,-1)
7800 0517 sinh : $7\zeta(3)$
7800 0730 cr : $\exp(-\sqrt{5}/3)$
7800 2874 rt : (9,1,4,-8)
7800 3289 tanh : $23/22$
7800 3719 $\ln(4)\times\exp(1/4)$
7800 5125 rt : (3,9,-5,-3)
7800 7836 ℓ_2 : $10\zeta(3)/7$
7801 3413 2^x : $\exp(-1/(2e))$
7801 4722 erf : $\ln(\sqrt[3]{2}/3)$
7801 5847 10^x : $6\sqrt{2}/5$
7801 5855 ln : $11/24$
7801 6072 10^x : $23/13$
7801 8189 Ei : $\ln 2/8$
7801 8949 sr : $14/23$
7802 0180 e^x : $3+3\sqrt{2}$
7802 0878 Ei : $\exp(\sqrt[3]{4}/2)$
7802 2170 sq : $1+3e/2$
7802 2313 rt : (8,7,7,5)
7802 2824 2^x : $\pi^2/3$
7802 3058 $s\pi$: $(\sqrt{2}/3)^{-1/3}$
7802 4343 at : $4\sqrt{3}/7$
7803 3526 e^x : $4e-4\pi/3$
7803 4757 at : $7\sqrt{2}/10$
7803 6359 e^x : $9\pi/7$
7803 7712 rt : (8,5,8,7)
7804 2253 $s\pi$: $\sqrt[3]{5}/6$
7804 3274 $4/5\times\exp(4/5)$
7804 4789 2^x : $2+4\sqrt{2}/3$
7804 5364 e^x : $5\pi^2$
7804 5766 $\exp(e)\div\exp(2/3)$
7804 6405 rt : (8,3,9,9)
7804 6939 rt : (3,9,-7,7)
7804 7925 at : $10\ln 2/7$
7804 9904 ζ : $\sqrt[3]{5}/3$
7805 0997 e^x : $2\sqrt[3]{3}/5$
7805 5662 ℓ_{10} : $(\ln 3/2)^3$
7805 5698 2^x : $\sqrt[3]{23/2}$
7805 6373 sr : $4e-\pi$
7806 5839 $t\pi$: $4e-3\pi/4$
7806 6768 sinh : $7\pi/8$
7806 6812 rt : (2,8,-6,7)
7807 1443 tanh : $\pi/3$
7807 3252 at : $1/2-2\sqrt{5}/3$
7807 7640 rt : (8,8,-6,-4)
7808 0559 id : $\sqrt{2}/3+4\sqrt{3}/3$
7808 2250 sr : $7e/6$
7808 3593 2^x : $(\ln 2)^{1/2}$
7808 4759 $4/5\div\ln(3/4)$
7808 5951 10^x : $9e\pi/10$
7808 6598 rt : (1,6,-5,2)
7808 6894 $s\pi$: $3\sqrt{2}+2\sqrt{5}$
7808 7896 tanh : $22/21$
7808 9033 cr : $3e/2+\pi/2$
7808 9666 cr : $10/21$
7809 0735 cu : $3+\sqrt{3}/3$
7809 2918 sq : $1-4\pi$
7809 4427 ℓ_{10} : $4+3e/4$
7809 7265 e^x : $7\sqrt[3]{5}/9$
7809 7396 K : $7/11$
7809 8265 ℓ_2 : $3\sqrt{3}+3\sqrt{5}/4$
7809 9429 $\sqrt{3}-s\pi(2/5)$
7809 9464 2^x : $(\sqrt[3]{3})^{-1/2}$
7810 0813 cu : $3\sqrt{2}/4+\sqrt{3}$
7810 2998 θ_3 : $7(1/\pi)/6$
7810 4354 J_0 : $(\sqrt{5}/2)^{-1/3}$
7810 7241 e^x : γ
7810 8493 tanh : $(\ln 3)^{1/2}$
7810 9143 Γ : $7\sqrt{2}$
7811 0328 rt : (3,6,-4,7)
7811 0809 K : $\ln(\sqrt[3]{4}/3)$
7811 2357 e^x : $5\sqrt{5}/2$
7811 3045 ln : $1/2+2e$
7811 3325 rt : (9,9,-1,-9)
7811 3701 at : $3e/4-\pi/3$
7811 3906 10^x : $5\ln 2/6$
7811 6620 10^x : $3\sqrt{2}/4+2\sqrt{5}$
7811 6693 cu : $6\sqrt{2}/7$
7811 7354 10^x : $6e\pi/7$
7811 7694 rt : (7,8,1,-9)
7811 9116 rt : (9,7,-2,-7)
7812 0223 K : $2(1/\pi)$
7812 0383 2^x : $2\sqrt{2}/3-3\sqrt{3}/4$
7812 0960 erf : $20/23$
7812 1189 $s\pi$: $1/2+\pi/4$
7812 2255 2^x : $2-3\pi/4$
7812 3281 rt : (5,7,3,-9)
7812 5000 sq : $5\sqrt{2}/8$
7812 5137 rt : (7,6,0,-7)
7812 6156 cu : $4-\sqrt{3}/2$
7812 7149 rt : (9,5,-3,-5)
7812 7565 J : $5(1/\pi)/9$
7812 8379 rt : (1,5,0,-5)
7812 8494 e^x : $1/2+4\sqrt{3}$
7812 9416 cu : $3\sqrt{3}/2-3\sqrt{5}/4$
7813 0683 rt : (3,6,5,-9)
7813 1217 id : $\exp(\sqrt{3}/3)$
7813 3204 rt : (5,5,2,-7)
7813 4705 sq : $3/4-3\sqrt{5}/2$
7813 5116 10^x : $3\sqrt{2}-3\sqrt{5}/4$
7813 5977 rt : (7,4,-1,-5)
7813 6105 $t\pi$: $\sqrt{3}/3-\sqrt{5}/2$
7813 9040 rt : (9,3,-4,-3)
7813 9708 at : $\sqrt{3}/4+\sqrt{5}/4$
7814 0951 rt : (1,5,7,-9)
7814 2833 Γ : $((e+\pi)/3)^2$
7814 3785 10^x : $3\sqrt{2}/2-3\sqrt{5}/4$
7814 4457 $t\pi$: $4\ln 2/5$
7814 4575 rt : (3,4,4,-7)
7814 4819 ℓ_{10} : $4/3+3\pi/2$
7814 6022 $t\pi$: $4\sqrt{5}/5$
7814 6207 tanh : $(\sqrt{3}/2)^{-1/3}$
7814 7207 ζ : $3/13$
7814 8639 rt : (5,3,1,-5)
7815 0374 10^x : $9\zeta(3)/5$
7815 1781 $s\pi$: $\sqrt{3}/3+3\sqrt{5}$
7815 2801 sq : $4+2e/3$
7815 3225 rt : (7,2,-2,-3)
7815 3703 id : $4\sqrt{3}/3+2\sqrt{5}$
7815 4059 Γ : $\sqrt{3}/9$
7815 4555 cu : $\sqrt{17/2}$
7815 6155 rt : (1,4,-9,-9)
7815 6932 10^x : $e/4$
7815 7305 $s\pi$: $\sqrt[3]{20}$
7815 7483 J_0 : $5\sqrt{3}/9$
7815 8442 rt : (9,1,-5,-1)
7815 8874 cr : $9\pi/5$
7815 9264 id : $(2\pi/3)^{-1/3}$
7815 9833 rt : (1,8,9,5)
7816 0339 e^x : $1/2+\sqrt{2}$
7816 1804 rt : (1,3,6,-7)
7816 8326 rt : (3,7,8,9)
7816 9505 2^x : $3\sqrt{3}-\sqrt{5}$
7817 1979 rt : (8,3,3,-8)
7817 3595 sr : $11/18$
7817 5937 rt : (5,1,0,-3)
7817 6558 Ψ : $8/3$
7817 7229 Ei : $3\sqrt{5}/7$
7817 7588 sq : $3e/4+4\pi/3$
7817 8077 tanh : $5\sqrt[3]{2}/6$
7817 9743 2^x : $5/6$
7818 0635 tanh : $21/20$
7818 2259 Ei : $23/24$
7818 3148 $s\pi$: $2/7$
7818 3194 sq : $1-3\pi/2$
7818 3340 J_0 : $4\zeta(3)/5$
7818 4934 rt : (7,0,-3,1)
7818 5849 erf : $(\pi/3)^{-3}$
7818 6778 λ : $\ln(\ln 3)$
7818 7644 cu : $3/4+2e$
7818 7881 rt : (6,2,5,-8)
7818 9571 J_0 : $2\sqrt[3]{3}/3$
7819 0678 $\ln(4)+\exp(1/3)$
7819 1001 rt : (1,2,-8,-7)
7819 1476 rt : (8,7,-9,7)
7819 2291 ζ : $\ln 2/3$
7819 3321 rt : (8,1,2,-6)
7819 3886 id : $\sqrt{2}-3\sqrt{3}$
7819 4924 sinh : $3\sqrt{5}/5$
7819 5731 rt : (9,1,-6,-1)
7819 6110 e^x : $2\sqrt{3}/3+2\sqrt{5}$
7819 6304 ln : $3\sqrt{3}+\sqrt{5}/3$
7819 7085 $\exp(e)\div s\pi(1/5)$
7819 7766 $t\pi$: $5\sqrt[3]{3}$
7820 0094 $e\times\ln(3/4)$
7820 1006 sinh : $\sqrt[3]{16/3}$
7820 1695 tanh : $(e/3)^{-1/2}$
7820 2676 cr : $11/23$
7820 3146 rt : (1,1,5,-5)
7820 3791 rt : (8,9,-6,6)
7820 3794 ζ : $(\sqrt[3]{3}/3)^2$
7821 0242 rt : (4,1,7,-8)
7821 1991 rt : (4,9,-7,-3)
7821 4101 2^x : $3e/2+2\pi$
7821 5450 rt : (6,4,-8,7)
7821 8183 rt : (9,0,6,9)

7821 8412 $t\pi : 1/4 - 3e/4$
7821 9195 $s\pi : 1/4 + 2\sqrt{3}$
7822 0234 $Ei : 2e/9$
7822 0495 cu : $3 + \sqrt{3}/2$
7822 1892 $\Gamma : 9\zeta(3)$
7822 3701 $\Psi : 9\pi^2/2$
7822 4318 rt : (7,9,-9,-5)
7822 7128 rt : (8,1,1,4)
7822 8028 cu : $7\sqrt{3}/10$
7822 9255 at : $4\sqrt{5}/9$
7823 0235 $s\pi : 10\sqrt{2}/3$
7823 1613 $\ln(2/5) - s\pi(1/3)$
7823 3917 rt : (6,6,-5,6)
7823 4508 $5 \times \exp(\sqrt{5})$
7823 5648 $2^x : 3 - 3\sqrt{5}/2$
7823 7278 rt : (5,1,-1,1)
7824 1625 rt : (1,9,1,-8)
7824 1700 rt : (7,1,8,9)
7824 2331 $10^x : 1 + 2\pi/3$
7824 2595 $e^x : 1/2 - \sqrt{5}/3$
7824 2839 sr : $3/2 + 3\sqrt{5}/4$
7824 3573 $e^x : (\pi/3)^{1/2}$
7824 4008 rt : (2,0,9,8)
7824 5173 id : $e/2 - \pi$
7824 5963 $\Psi : 7\ln 2/9$
7824 7934 $e^x : 3\sqrt{2}/4 + 3\sqrt{5}$
7824 9524 cr : $4\sqrt{3}/3 + 3\sqrt{5}/2$
7824 9705 rt : (9,2,5,7)
7825 0490 $\ell_{10} : 3e - 2\pi/3$
7825 0509 tanh : $7\zeta(3)/8$
7825 0656 sq : $\exp(-1/(3e))$
7825 2222 rt : (4,1,-7,7)
7825 3513 $Ei : 7e\pi/8$
7825 3614 $J_0 : 24/25$
7825 5274 $\Psi : \sqrt[3]{19}$
7825 5940 $10^x : 4/9$
7825 6684 rt : (4,1,6,6)
7825 8096 rt : (6,8,-2,5)
7826 0869 id : 18/23
7826 2074 $10^x : \sqrt{2} - \sqrt{6} + \sqrt{7}$
7826 2303 rt : (7,2,-4,-1)
7826 2867 $\ell_{10} : 7\sqrt{3}/2$
7826 4375 rt : (8,9,-3,-7)
7826 5154 sq : $7\sqrt[3]{2}$
7826 5640 $s\pi : 2/3 + \pi/3$
7826 9586 $\exp(1/2) - s\pi(1/3)$
7827 0556 sq : $3\sqrt{3}/4 + 2\sqrt{5}/3$
7827 1430 rt : (6,2,3,4)
7827 4435 ln : $3/4 + 3\sqrt{3}$
7827 5395 rt : (2,4,-9,-8)
7827 9410 $10^x : 5/4$
7828 1567 rt : (1,0,-7,5)
7828 2076 $\ell_2 : 4 - \sqrt{5}/4$
7828 2708 id : $\sqrt[3]{17/3}$
7828 2735 tanh : 20/19
7828 2959 rt : (4,9,2,-9)
7828 3351 rt : (5,9,-5,-4)
7828 4028 sq : $4(e+\pi)/9$
7828 5424 $e^x : \sqrt[3]{21}$
7828 5713 sq : $\sqrt{2} - \sqrt{3} - \sqrt{7}$
7828 6334 rt : (7,3,7,7)
7828 8791 rt : (8,3,0,2)
7829 0200 ln : $4 - 2e/3$
7829 1117 rt : (6,8,-1,-7)
7829 2044 cr : $\sqrt{2}/3 + 3\sqrt{3}$
7829 3949 rt : (7,8,-8,-2)
7829 4055 sr : $2\sqrt{2}/3 + \sqrt{5}$
7829 4316 $s\pi : 2\sqrt{3}/3 + \sqrt{5}/4$
7829 5633 as : $4/3 - 3e/4$
7829 6473 rt : (9,3,-7,-3)
7829 7352 cr : 12/25
7829 9424 rt : (9,4,4,5)
7829 9942 rt : (8,7,-4,-5)
7830 3960 rt : (9,6,0,-8)
7830 4607 $2/5 \div \ln(3/5)$
7830 7075 at : $1/4 + \sqrt{5}/3$
7830 8827 rt : (1,7,6,-6)
7831 2113 at : $1/\ln(\ln 3/3)$
7831 3042 $e^x : 8\sqrt[3]{2}/3$
7831 3494 rt : (1,9,0,-6)
7831 3934 cr : $9\sqrt[3]{2}/2$
7831 4785 cr : P_{21}
7831 5788 rt : (4,8,-8,2)
7831 7071 $Ei : 3\pi^2/4$
7831 8530 id : $1/2 + 2\pi$
7832 0600 tanh : $(\sqrt[3]{5}/2)^{-1/3}$
7832 1976 sq : $(\ln 2)^{1/3}$
7832 2344 $K : 2\sqrt{5}/7$
7832 2779 $e^x : 6\sqrt[3]{5}/5$
7832 4342 rt : (2,2,8,6)
7832 4699 $J_0 : 23/24$
7832 5486 $J_0 : 3\sqrt{5}/7$
7832 6818 rt : (4,7,1,-7)
7832 7481 sq : $\sqrt{17}\pi$
7832 7521 rt : (3,8,-3,-4)
7832 8781 rt : (2,9,-7,-1)
7832 9813 $\ell_{10} : e/2 + 3\pi/2$
7833 0380 $\Gamma : 2\sqrt{5}/9$
7833 2296 $erf : 1/\ln(\pi)$
7833 2690 $s\pi : 9(1/\pi)/10$
7833 2924 $erf : \sqrt[3]{2/3}$
7833 4071 sr : $(\ln 2/3)^{1/3}$
7833 6165 rt : (7,5,2,-8)
7833 8103 id : $(\sqrt[3]{3}/3)^{1/3}$
7833 8905 rt : (6,6,-2,-5)
7834 0151 cu : $3\sqrt{3}/4 + 4\sqrt{5}$
7834 1736 $s\pi : \sqrt{3} + 4\sqrt{5}/3$
7834 2063 cr : $2\zeta(3)/5$
7834 2834 rt : (5,4,9,7)
7834 3296 rt : (5,8,-5,-6)
7834 3842 $\ell_2 : 3e - 3\pi/2$
7834 4318 sinh : $8\sqrt[3]{5}/3$
7834 4475 $t\pi : 5e\pi/8$
7834 5580 $10^x : P_{25}$
7834 6753 rt : (9,4,-1,-6)
7834 7436 $e^x : 1/3 - \sqrt{3}/3$
7834 7785 $10^x : 2\sqrt{2} + \sqrt{3}$
7835 2205 rt : (8,5,-5,-3)
7835 2751 $10^x : \sqrt{2}/3 - \sqrt{3}/3$
7835 2752 sinh : $\sqrt[3]{17/2}$
7835 3145 rt : (4,3,5,4)
7835 4098 rt : (3,9,-5,-8)
7835 4465 $\theta_3 : 2/3$
7835 8582 $t\pi : \exp(\sqrt[3]{4}/2)$
7835 9225 sr : $\sqrt{3} - \sqrt{5}/2$
7836 0462 $2^x : e/2 + 4\pi/3$
7836 0467 ln : $(2e/3)^3$
7836 0520 id : $\sqrt{2}/2 - 2\sqrt{5}/3$
7836 1162 rt : (7,8,-8,0)
7836 1361 $10^x : 7\sqrt[3]{5}/4$
7836 1770 $\Gamma : (1/(3e))^{1/3}$
7836 4278 rt : (7,5,6,5)
7836 5235 $Ei : (e+\pi)/3$
7836 6668 $1/3 - \exp(3/4)$
7836 6805 as : 12/17
7836 8983 cu : $1/2 - 3e/2$
7837 0202 $10^x : \sqrt{3} - \sqrt{6} + \sqrt{7}$
7837 1944 $\ell_{10} : 2 + 3e/2$
7837 3558 rt : (1,8,-2,-6)
7837 4050 $10^x : e/4 - \pi/4$
7837 5633 $10^x : 9\ln 2/7$
7837 5723 ln : $3\sqrt{3}/2 + 3\sqrt{5}/2$
7837 6906 rt : (2,6,3,5)
7837 7088 rt : (1,7,-7,-7)
7837 8315 sinh : $4\sqrt[3]{2}/7$
7837 9087 rt : (5,4,4,-8)
7837 9175 ln : $\sqrt[3]{21/2}$
7838 0621 rt : (8,2,5,-9)
7838 1067 sr : $9\sqrt{2}/4$
7838 4047 sinh : 18/25
7838 5279 $s\pi : \sqrt{2} + 3\sqrt{3}/4$
7838 5424 $\zeta : 4\sqrt{3}/9$
7838 5919 $e^x : 7\sqrt{5}/10$
7838 7752 sq : $10\zeta(3)/9$
7838 9429 rt : (9,6,3,3)
7839 1544 cr : $4 + 3\sqrt{5}/4$
7839 1887 id : $e/2 + 3\pi$
7839 2277 at : $e/2 - 3\pi/4$
7839 3551 rt : (7,3,1,-6)
7839 4471 rt : (4,7,2,4)
7839 4773 rt : (1,8,-7,-1)
7839 5577 $\Psi : 7(1/\pi)/6$
7839 5687 tanh : 19/18
7839 5702 $\Psi : (\sqrt[3]{3})^{-1/3}$
7839 5919 rt : (1,7,-1,-4)
7839 6641 sq : P_{54}
7839 8416 rt : (4,7,-9,7)
7839 8535 rt : (2,9,2,-3)
7840 1857 $J_0 : 22/23$
7840 2541 $1/2 + \exp(1/4)$
7840 4306 $t\pi : e/2 - \pi/2$
7840 5993 $\exp(3/5) \times s\pi(\pi)$
7840 7506 $erf : 7/8$
7840 9446 rt : (9,2,-2,-4)
7840 9869 $2^x : 4e + \pi/2$
7841 0378 rt : (2,7,-4,-7)
7841 2411 sr : $10(1/\pi)$
7841 3664 $\Psi : (\ln 2)^{1/3}$
7841 3729 $2^x : \sqrt{3}/3 + 4\sqrt{5}/3$
7841 4186 rt : (6,4,-3,-3)
7841 5937 $e^x : 11/19$
7841 6595 $t\pi : 3\sqrt{2} - 3\sqrt{5}/4$
7842 1386 $\theta_3 : \sqrt{5}/6$
7842 1414 $t\pi : P_{55}$
7842 1714 $s\pi(\sqrt{3}) \div s\pi(2/5)$
7842 1869 rt : (3,6,-4,-2)
7842 2361 $\exp(\pi) - \exp(\sqrt{5})$
7842 2542 $\Psi : 7(1/\pi)/8$
7842 3334 $Li_2 : 7/11$
7842 3632 $K : 16/25$
7842 4848 $10^x : ((e+\pi)/3)^{1/3}$
7842 8021 rt : (2,6,-8,-9)
7842 8381 sq : $4\sqrt[3]{5}$
7843 0566 rt : (6,1,7,-9)
7843 1938 sq : $1 - 4\sqrt{2}/3$
7843 2118 cu : $2/3 + e$
7843 2204 rt : (6,9,2,9)
7843 2863 $e^x : 10\pi^2/9$
7843 3946 $2^x : 4\sqrt{2}/3 - \sqrt{5}$
7843 6465 rt : (8,5,-1,0)
7843 8311 rt : (3,7,-7,-9)
7843 8865 cu : $9\zeta(3)/4$
7843 9169 rt : (3,3,6,-8)
7844 0721 rt : (5,8,-7,-8)
7844 2674 cr : $2\sqrt{3} - 4\sqrt{5}/3$
7844 3250 $2^x : 4\sqrt{2}/3 + 3\sqrt{5}/2$
7844 3986 ln : $4e/3 + 4\pi$
7844 4206 rt : (8,8,-1,-8)
7844 6454 sr : 8/13
7844 7320 rt : (8,0,4,7)
7844 8106 $\Gamma : 23/4$
7844 8388 cr : $4e/3 - \pi$
7845 0029 $\lambda : \exp(-3\pi/4)$
7845 2325 rt : (5,5,-7,0)
7845 4625 rt : (8,8,9,6)
7845 5336 $e^x : 4 - 3\sqrt{2}$
7845 6050 sr : $4\sqrt{2}/3 + 3\sqrt{3}/4$
7845 7079 $Ei : 4e\pi/7$
7845 8503 cu : 5π
7846 0117 rt : (5,2,3,-6)
7846 0780 $\ell_{10} : 4\sqrt{2} + \sqrt{3}/4$
7846 0969 sq : $2 + 3\sqrt{3}$
7846 1882 rt : (7,7,-7,5)
7846 3595 sq : $4 + \sqrt{5}/4$
7846 4057 $Li_2 : 2(1/\pi)$
7846 5567 tanh : $(\sqrt{5}/2)^{1/2}$
7846 5764 rt : (2,9,7,6)
7846 6852 $\exp(1/3) + \exp(2)$
7846 7449 rt : (7,9,-5,-5)
7846 9064 $\ln(2/5) \times \exp(2/3)$
7846 9575 $J_0 : 3(1/\pi)$
7846 9819 $Ei : \sqrt{3} + \sqrt{5} - \sqrt{7}$
7847 0065 id : $1/3 - \sqrt{5}/2$
7847 0676 $t\pi : \sqrt{2}/3 + \sqrt{3}/2$
7847 1791 $\exp(2/5) - s\pi(1/4)$
7847 2168 $2^x : \sqrt{2} + \sqrt{5}/2$
7847 3254 $10^x : -\sqrt{3} + \sqrt{5} + \sqrt{7}$
7847 4956 rt : (3,7,-4,-6)
7847 6910 cr : $3\sqrt{2}/4 - \sqrt{3}/3$
7848 0182 rt : (2,4,2,-5)

7848 0399 cu : $7\ln 2/4$
7848 1975 sq : $(2e/3)^2$
7848 2046 rt : (9,5,2,-9)
7848 2957 rt : (5,6,8,5)
7848 3585 rt : (7,1,0,-4)
7848 3817 rt : (9,8,-8,-3)
7848 4320 rt : (2,7,-8,1)
7848 5904 J_0 : 21/22
7848 6353 sq : $\sqrt{3}-\sqrt{5}+\sqrt{6}$
7849 1128 cr : $1/4+2e$
7849 2482 cr : P_{32}
7849 3060 ℓ_{10} : $4+2\pi/3$
7849 3157 Ψ : $\exp(-1/(3e))$
7849 4001 sr : $8\ln 2/9$
7849 5690 rt : (6,7,1,-8)
7849 9679 rt : (5,3,-2,-1)
7850 0239 rt : (9,3,7,8)
7850 0326 rt : (4,0,9,9)
7850 1238 rt : (5,2,-9,6)
7850 2967 rt : (1,6,-4,9)
7850 3512 cu : $7\sqrt{3}/2$
7850 6334 J_0 : $(\ln 3)^{-1/2}$
7850 8428 rt : (2,6,-7,-4)
7851 0046 rt : (9,0,-3,2)
7851 0920 ζ : $\sqrt{7/2}$
7851 2411 rt : (8,6,-2,-6)
7851 3460 rt : (7,9,-4,4)
7851 4051 rt : (3,9,0,-5)
7851 4153 K : $4\sqrt[3]{3}/9$
7851 4421 cr : $3/2+4\pi/3$
7851 5766 ln : $4\sqrt{2}-2\sqrt{3}$
7851 6176 rt : (1,6,-1,-4)
7851 7315 $1/4\times\exp(\pi)$
7852 1316 tanh : 18/17
7852 3104 rt : (6,9,-3,-5)
7852 4137 tπ : $2\sqrt{2}/3-2\sqrt{3}/3$
7852 4510 rt : (6,1,6,7)
7852 4900 cr : P_{42}
7852 7011 sinh : $\sqrt[3]{3}/2$
7852 8376 erf : $(\sqrt{2}\pi/3)^{-1/3}$
7852 9356 rt : (1,2,8,-8)
7852 9891 Ψ : 7/13
7853 0354 rt : (1,9,2,1)
7853 1355 ℓ_{10} : $\sqrt{2}+\sqrt{5}+\sqrt{6}$
7853 1631 ℓ_{10} : $5e\pi/7$
7853 3674 Ei : $\sqrt[3]{3}/7$
7853 4645 rt : (5,8,-3,-5)
7853 4780 rt : (7,7,5,3)
7853 6042 K : $3\sqrt[3]{5}/8$
7853 7412 $\sqrt{5}+\ln(\sqrt{3})$
7853 8237 ln : $2\pi^2/9$
7853 8502 rt : (7,4,4,-9)
7853 9676 10^x : $9e\pi/7$
7853 9816 id : $\pi/4$
7854 0457 2^x : $(\sqrt[3]{5})^{-1/3}$
7854 0916 sinh : $3\zeta(3)/5$
7854 2771 cu : $4\pi/7$
7854 3559 cu : $3/2-\sqrt{3}/3$
7854 3713 J_0 : $(\sqrt{3}/2)^{1/3}$
7854 8844 Ei : $(\ln 3/3)^{1/2}$
7854 9621 rt : (3,6,-8,-4)
7855 0099 rt : (6,2,-4,-1)
7855 0440 Γ : 6/25
7855 0441 2^x : $4\sqrt{2}+2\sqrt{5}/3$
7855 1555 rt : (8,2,3,5)
7855 3280 sπ : $3\pi/2$
7855 3373 tanh : $\ln(\ln 2/2)$
7855 4820 ln : $8\sqrt{5}/3$
7855 4931 rt : (2,4,7,4)
7855 5333 sinh : $1/\ln(4)$
7855 5975 ℓ_{10} : $2/3+2e$
7855 6772 rt : (9,3,1,-7)
7855 7225 rt : (7,7,-6,-3)
7855 7621 sq : $3+\sqrt{3}/4$
7856 0077 cr : $2e\pi/3$
7856 1305 e^x : $3e/2+\pi/2$
7856 2413 rt : (3,1,5,-6)
7856 4079 rt : (4,6,3,9)
7856 4519 at : $\zeta(11)$
7856 4813 rt : (5,4,-6,5)
7856 5603 tπ : $1/\ln(e+\pi)$
7856 6701 rt : (1,5,-2,-2)
7856 7092 sπ : $\ln(4/3)$
7856 7420 id : $5\sqrt{2}/9$
7856 8950 rt : (2,8,4,9)
7857 1428 id : 11/14
7857 3536 rt : (1,9,-5,-9)
7857 7802 J_0 : 20/21
7857 9035 rt : (2,4,-7,-7)
7857 9241 cr : $7\ln 2/10$
7858 2084 rt : (9,6,-9,-1)
7858 2500 rt : (1,9,9,2)
7858 5858 tπ : $1/3-3\sqrt{2}/2$
7858 6795 rt : (6,5,0,-6)
7858 7697 rt : (7,4,9,8)
7858 9395 2^x : $4\sqrt{3}$
7858 9838 sq : $1/2-2\sqrt{3}$
7859 1396 cr : $1/2+3\sqrt{3}$
7859 1639 tanh : $3\sqrt{2}/4$
7859 3047 rt : (7,8,-3,9)
7859 4709 rt : (8,1,-7,1)
7859 4943 e^x : $\sqrt{11/3}$
7859 6465 rt : (4,9,-7,-2)
7859 7445 e^x : $4\sqrt{3}-3\sqrt{5}/4$
7859 7874 10^x : $3\sqrt{3}+3\sqrt{5}$
7859 8246 rt : (9,9,-5,-6)
7859 8638 J_0 : $(e/3)^{1/2}$
7859 9447 2^x : $3\pi^2/10$
7859 9663 rt : (4,8,0,-3)
7860 0269 rt : (1,8,2,-7)
7860 0547 rt : (5,0,2,4)
7860 1295 rt : (9,8,2,1)
7860 2133 $\exp(3/5)-s\pi(\sqrt{2})$
7860 5888 tanh : $10(1/\pi)/3$
7860 7704 rt : (7,7,-9,-8)
7861 1672 rt : (8,4,-3,-4)
7861 2032 2^x : $3+\sqrt{5}/4$
7861 3061 λ : 2/21
7861 3378 rt : (5,3,6,-9)
7861 3434 tπ : $2\sqrt{3}+4\sqrt{5}/3$
7861 3591 sq : $1/\ln(\ln 3/2)$
7861 5592 rt : (1,9,-8,-3)
7861 6244 2^x : $\sqrt{2}-\sqrt{3}/3$
7861 6596 sq : $4\sqrt{2}+3\sqrt{5}/4$
7861 7613 rt : (8,7,1,-9)
7861 8154 rt : (9,5,6,6)
7861 9586 Ψ : $(e+\pi)^{-3}$
7861 9778 sr : $3\sqrt[3]{3}/7$
7862 0158 Ei : $2e\pi/7$
7862 0827 $\exp(\sqrt{3})-s\pi(1/3)$
7862 5447 at : $5\zeta(3)/6$
7862 9436 $2/5+\ln(4)$
7862 9964 rt : (3,8,-1,0)
7863 0092 10^x : 17/25
7863 0231 cu : $5e\pi/9$
7863 0761 cr : $10\sqrt[3]{5}/3$
7863 1637 $\exp(4/5)\div s\pi(1/5)$
7863 3298 sr : $\exp(-\sqrt[3]{3}/3)$
7863 3370 cr : $2\sqrt{3}+\sqrt{5}$
7863 3434 Ψ : $5\sqrt{2}/8$
7863 5694 Ei : 24/25
7863 8141 rt : (7,2,3,-7)
7864 0135 at : $\zeta(9)$
7864 2262 10^x : $\sqrt[3]{2}/5$
7864 2475 rt : (8,8,-9,4)
7864 2707 rt : (4,2,8,7)
7864 4076 10^x : $10\sqrt{2}/7$
7864 4970 2^x : $\ln(1/\sqrt{2})$
7864 4985 rt : (7,1,-1,2)
7864 7700 ℓ_{10} : $9e/4$
7864 8639 2^x : $(e+\pi)/7$
7864 9022 rt : (1,8,-6,8)
7865 0789 rt : (5,9,0,-8)
7865 2184 sq : $4-2e/3$
7865 2708 cu : 12/13
7865 3694 cu : $1/4+\sqrt{3}$
7865 5625 ζ : $(\sqrt{3}/2)^{1/3}$
7865 6609 id : $\sqrt{2}/4+\sqrt{3}/4$
7865 9403 rt : (2,5,5,-8)
7866 1881 tanh : 17/16
7866 3124 rt : (5,9,-1,-2)
7866 3314 2^x : $1/3-e/4$
7866 5227 rt : (9,1,0,-5)
7866 6120 tπ : $5(e+\pi)/8$
7866 6546 cu : $2/3-3\pi$
7866 6595 sinh : 13/18
7866 8731 erf : 22/25
7867 2317 tπ : $6\sqrt{2}/7$
7867 3195 rt : (7,8,-3,-6)
7867 3924 rt : (4,5,4,2)
7867 7023 $\exp(\sqrt{5})-s\pi(\pi)$
7867 8706 J_0 : 19/20
7867 9272 J_0 : $5\sqrt[3]{5}/9$
7867 9579 sr : 13/21
7868 1455 ln : P_{44}
7868 1901 cu : $9e/4$
7868 3267 K : 9/14
7868 3800 rt : (5,6,-6,-6)
7868 4288 tπ : $2(1/\pi)/3$
7868 4510 ℓ_{10} : $4+3\sqrt{2}/2$
7868 5887 sπ : $((e+\pi)/2)^{1/2}$
7868 6403 rt : (6,3,5,5)
7868 8430 rt : (3,8,-7,-9)
7868 8543 rt : (2,9,-4,-6)
7868 8707 rt : (8,1,6,8)
7869 1670 tπ : $\sqrt{3}-4\sqrt{5}$
7869 2054 rt : (6,6,3,-9)
7869 3534 rt : (8,7,-8,9)
7869 7361 rt : (9,7,-6,-4)
7869 9197 ℓ_2 : P_{82}
7870 0368 rt : (7,5,-7,-1)
7870 0583 rt : (1,7,-6,1)
7870 2319 ℓ_{10} : $\exp(2e/3)$
7870 3301 λ : $3(1/\pi)/10$
7870 3864 at : $3\sqrt{2}/2-\sqrt{5}/2$
7870 5063 id : $2\sqrt{3}-3\sqrt{5}/4$
7870 7062 rt : (2,7,3,-1)
7870 7851 e^x : $2/3-e/3$
7870 9037 rt : (3,7,-2,5)
7871 0464 ln : $3/2+2\sqrt{5}$
7871 1239 rt : (5,3,-8,2)
7871 1466 id : $\sqrt{3}/4+3\sqrt{5}/2$
7871 2289 rt : (1,5,-3,9)
7871 3427 as : 17/24
7871 4330 Ei : $8\sqrt[3]{5}/7$
7871 4465 rt : (6,6,-9,-1)
7871 5866 rt : (8,5,0,-7)
7871 6055 J_0 : $(\sqrt[3]{5}/2)^{1/3}$
7871 6095 sπ : $\sqrt[3]{3}/5$
7871 7530 rt : (3,2,8,-9)
7871 9010 $\ln(3)\div\exp(1/3)$
7872 1720 rt : (7,4,-5,-3)
7872 6821 sr : $3\sqrt{2}/4+3\sqrt{5}$
7872 6978 J_0 : $e\pi/9$
7872 7374 rt : (2,8,-9,-7)
7872 8364 sr : $10\sqrt{5}/7$
7872 8453 rt : (6,3,-1,-4)
7872 8862 sinh : $(e/3)^{-3}$
7872 8865 rt : (5,5,-2,9)
7872 9331 rt : (1,7,-4,-7)
7872 9893 ln : $3(e+\pi)/8$
7873 7153 rt : (8,4,2,3)
7873 8517 rt : (9,4,9,9)
7873 9635 10^x : $3\sqrt{2}/2-\sqrt{3}/4$
7874 0146 rt : (1,0,7,6)
7874 1430 e^x : $1/3+4\sqrt{5}$
7874 3215 rt : (3,8,2,-8)
7874 5065 id : $5\sqrt[3]{2}/8$
7874 5617 rt : (2,8,-5,-2)
7874 6539 ln : $\sqrt{2}/2+2\sqrt{5}/3$
7874 9935 rt : (7,9,0,-9)
7875 2893 tπ : $6\zeta(3)$
7875 3013 rt : (5,1,5,-7)
7875 3246 ln : $3\sqrt{2}+\sqrt{3}$
7875 4664 rt : (6,9,-7,-3)
7875 5450 rt : (1,7,1,-3)
7875 7992 rt : (9,2,3,-8)
7875 8027 rt : (8,9,-5,8)
7875 9019 sq : $2/3-3\sqrt{2}$
7875 9714 sπ : $\sqrt{3}/6$

7876 0292 cr : $2\sqrt[3]{5}/7$
7876 0924 cr : $1+3\pi/2$
7876 2443 Γ : $9\sqrt{5}/7$
7876 2823 $\sqrt{3}\times\ln(5)$
7876 5815 at : $\sqrt{2}+\sqrt{5}-\sqrt{7}$
7876 6728 sπ : $1/4+3e/4$
7876 7076 cu : $8\pi^2/3$
7876 7519 at : $\ln(\ln 3/3)$
7876 8207 $1/2-\ln(3/4)$
7876 8350 10^x : $1+\sqrt{3}/3$
7876 8634 10^x : $15/14$
7876 9256 rt : (8,2,-4,-2)
7876 9716 erf : $7\sqrt[3]{2}/10$
7877 0095 rt : (7,6,8,6)
7877 1089 Γ : $23/8$
7877 2362 rt : (9,8,-3,-7)
7877 3489 rt : (5,8,0,3)
7877 4429 rt : (6,5,-8,4)
7877 5382 sq : $3\sqrt{2}+4\sqrt{3}$
7877 5903 rt : (1,6,1,-5)
7877 7105 tπ : $1/2+3\pi/2$
7877 7867 ℓ_2 : $\sqrt{2}/3+4\sqrt{5}/3$
7877 8525 $1/5+s\pi(1/5)$
7877 9373 ln : $9\sqrt[3]{5}/7$
7878 1641 sinh : $19/4$
7878 3381 10^x : $4/3+\sqrt{2}/3$
7878 4723 rt : (3,5,-3,-4)
7878 5556 10^x : $1/4-\sqrt{2}/4$
7878 6771 e^x : $\sqrt{10}\pi$
7879 0007 J_0 : $18/19$
7879 0207 2^x : $(\ln 2/2)^{-2}$
7879 0479 rt : (4,5,5,-9)
7879 0485 e^x : $3\sqrt{2}/4-3\sqrt{3}/4$
7879 0869 erf : $15/17$
7879 1576 10^x : $\sqrt{2}/4+\sqrt{3}$
7879 2505 rt : (7,0,2,5)
7879 3375 J : $3\ln 2/10$
7879 8701 id : $1/3-3\sqrt{2}/2$
7879 9511 cu : $9\sqrt{3}/10$
7880 0809 2^x : $4/3-3\sqrt{5}/4$
7880 0825 tπ : $7\sqrt{3}/10$
7880 1377 2^x : $e/2+2\pi$
7880 1683 rt : (4,9,-6,0)
7880 2849 10^x : $6\sqrt{5}$
7880 5494 ln : $7\pi/10$
7880 6535 rt : (7,6,-4,-4)
7880 7426 ℓ_2 : $e+4\pi/3$
7880 7594 $5\div\ln(3/5)$
7880 9037 $s\pi(1/5)\div s\pi(\sqrt{3})$
7880 9490 ℓ_{10} : $2\sqrt{2}/3+3\sqrt{3}$
7881 0270 e^x : $4/3+\pi$
7881 2883 rt : (6,2,8,8)
7881 4500 rt : (3,9,5,9)
7881 6583 ln : $4/3+\sqrt{3}/2$
7881 8408 ζ : $4/7$
7881 9198 ln : P_{60}
7881 9204 rt : (6,4,-7,9)
7882 0207 tanh : $16/15$
7882 0822 rt : (1,9,-7,7)
7882 2155 2^x : $3\sqrt{5}/8$
7882 2372 cr : $3+e$
7882 2585 rt : (6,4,2,-7)
7882 3381 Li_2 : $2\sqrt{5}/7$
7882 4801 sr : $1/\ln(5)$
7882 6758 rt : (9,7,5,4)
7882 7469 cr : $4e/3+2\pi/3$
7882 8100 ℓ_{10} : $3+\pi$
7882 9361 rt : (6,8,-6,-2)
7883 4323 Ei : $3\sqrt{2}/7$
7883 4469 10^x : $\sqrt{15/2}$
7883 6692 rt : (9,1,-1,3)
7883 7623 rt : (4,8,-7,5)
7883 7668 sπ : $e+\pi/2$
7883 7988 rt : (4,3,-9,-7)
7884 1489 rt : (5,8,2,-9)
7884 1908 ℓ_2 : $7\pi^2/5$
7884 2732 rt : (9,5,-7,-2)
7884 5706 2^x : $8e\pi/9$
7884 5736 ln : $5/11$
7884 6281 tπ : $2\sqrt{2}-2\sqrt{5}/3$
7884 8492 rt : (4,9,-3,-6)
7884 8496 Ψ : $-\sqrt{3}-\sqrt{5}-\sqrt{7}$
7884 9589 ℓ_2 : $11/19$
7885 0443 sinh : $3\sqrt{2}$
7885 0491 tanh : $e\pi/8$
7885 1592 ln : $4e/3+3\pi/4$
7885 2660 rt : (2,3,4,9)
7885 4109 ℓ_2 : $2/3+3\sqrt{2}/4$
7885 4347 2^x : $3+\pi/4$
7885 4367 10^x : $6\ln 2/5$
7885 4576 cr : $1/\ln(\pi/3)$
7885 5330 rt : (8,3,5,6)
7885 5890 rt : (2,9,-2,-5)
7885 6556 $1/4\times\exp(e)$
7885 7591 rt : (8,3,-1,-5)
7885 8693 J_0 : $(\sqrt{5}/2)^{-1/2}$
7885 8782 tπ : $5\sqrt[3]{2}/8$
7885 9145 rt : (8,9,-7,-4)
7885 9666 $3/4\div s\pi(2/5)$
7886 2607 e^x : $4/3-\pi/2$
7886 5262 tπ : $\exp(-\sqrt[3]{5}/3)$
7886 5866 ln : $3+4\sqrt{5}/3$
7886 7097 e^x : $7e/8$
7886 7513 rt : (6,6,1,0)
7886 8711 cu : $3e/2$
7887 0045 erf : $5\sqrt{2}/8$
7887 0446 rt : (7,7,-1,-7)
7887 0963 rt : (7,1,5,-8)
7887 1137 id : $1/4-3e/4$
7887 1520 rt : (1,7,4,-8)
7887 1880 sq : $\sqrt{2}/3+\sqrt{3}/2$
7887 3839 rt : (5,7,-9,-9)
7887 4067 $\exp(\sqrt{2})+s\pi(\sqrt{5})$
7887 9151 rt : (6,7,-5,3)
7888 0483 ζ : $4/17$
7888 0553 $\exp(\sqrt{3})\div\exp(2/5)$
7888 5438 id : $4\sqrt{5}/5$
7888 7320 K : $1/\ln(3\pi/2)$
7888 9354 tanh : $8\zeta(3)/9$
7889 1272 rt : (5,4,-5,-1)
7889 2472 J_0 : $3\sqrt[3]{2}/4$
7889 7064 Γ : $(e+\pi)^{-2}$
7889 8414 tanh : $5\sqrt[3]{5}/8$
7889 9345 2^x : $5\pi^2/4$
7890 1031 rt : (4,8,-5,-3)
7890 1581 rt : (9,6,-4,-5)
7890 1836 10^x : $1/3+3\pi$
7890 2264 tπ : $2\sqrt{3}+\sqrt{5}/2$
7890 2738 rt : (5,2,1,2)
7890 5328 sπ : $\sqrt[3]{12}$
7890 5427 10^x : $\sqrt{3}/4+4\sqrt{5}/3$
7890 5609 $2/5+\exp(2)$
7890 6340 cu : $4\sqrt{2}+2\sqrt{3}$
7890 6729 erf : $\exp(-1/(3e))$
7890 7859 rt : (9,0,2,6)
7891 1023 rt : (8,4,2,-8)
7891 3397 J_0 : $17/18$
7891 4725 rt : (6,6,-4,8)
7891 6974 rt : (9,6,8,7)
7891 7044 $\sqrt{3}\times\exp(3)$
7891 8645 rt : (8,2,8,9)
7892 0256 tπ : $3e/2-4\pi$
7892 0652 rt : (1,8,1,-5)
7892 4176 id : $2\sqrt{2}/3-\sqrt{3}$
7892 4227 ln : $7\sqrt[3]{5}/2$
7892 4379 cu : $1+\sqrt{5}/4$
7892 4411 rt : (2,8,0,-4)
7892 5226 θ_3 : $1/\sqrt[3]{19}$
7892 6850 rt : (2,4,7,-9)
7892 7516 erf : $(\ln 2)^{1/3}$
7892 7710 rt : (3,0,7,7)
7892 7882 Ψ : $6\sqrt{5}/5$
7892 8806 sq : $3/2+e/3$
7892 9506 rt : (4,4,7,5)
7893 2213 erf : $(\sqrt[3]{3})^{-1/3}$
7893 5087 at : $4\sqrt[3]{2}/5$
7893 5201 sinh : $11/4$
7893 5601 rt : (5,8,7,3)
7893 5887 ℓ_{10} : $1/2+4\sqrt{2}$
7893 6176 Ψ : $15/17$
7893 9176 cu : $4\ln 2/3$
7893 9742 sπ : $2\sqrt{2}-\sqrt{5}/2$
7893 9923 at : $2\sqrt{3}-2\sqrt{5}$
7894 0317 rt : (9,7,-1,-8)
7894 1390 $\exp(1/2)+\exp(\pi)$
7894 4196 θ_3 : $5\zeta(3)/9$
7894 4412 rt : (3,6,-6,-7)
7894 4814 J_0 : $1/\ln(\ln 2/2)$
7894 7073 ζ : $\sqrt{2}/6$
7894 7368 id : $15/19$
7894 8179 sq : $\sqrt{2}/3-4\sqrt{5}$
7895 0105 sq : $3\sqrt{2}-3\sqrt{5}/2$
7895 1105 rt : (9,1,5,-9)
7895 1454 rt : (6,7,-8,-1)
7895 2516 sr : $3e\pi/8$
7895 4032 rt : (4,9,-2,-6)
7895 5542 at : $\zeta(7)$
7895 6897 2^x : $3\sqrt{3}/2-\sqrt{5}/2$
7895 7137 id : $2\sqrt{3}/3-4\sqrt{5}$
7895 8206 rt : (7,5,-9,8)
7895 9002 Ei : $7\ln 2/8$
7896 0823 rt : (2,8,4,-7)
7896 2184 rt : (3,7,4,-9)
7896 2668 rt : (5,5,-2,-4)
7896 3276 ℓ_{10} : $4e-3\pi/2$
7896 3685 rt : (7,3,-8,1)
7896 5785 cr : $4+\sqrt{3}$
7896 7129 Ei : $(e)^{-1/2}$
7896 7765 Γ : $5\ln 2/7$
7897 2273 sπ : P_{82}
7897 2537 rt : (4,3,4,-7)
7897 2698 e^x : $2-\sqrt{5}$
7897 3819 $\ln(2/3)\times\exp(2/3)$
7897 4486 10^x : $6\sqrt[3]{3}/5$
7897 5009 id : $3\sqrt{3}/4+2\sqrt{5}/3$
7897 6404 2^x : $7(e+\pi)/4$
7897 8409 rt : (6,1,-2,-2)
7897 9333 cr : $3\sqrt{2}+2\sqrt{5}/3$
7898 1172 id : $3(e+\pi)/2$
7898 2282 J_0 : $2\sqrt{2}/3$
7898 3065 sr : $9\ln 2/10$
7898 3607 sr : $3/2+2\pi$
7898 4368 e^x : $3\sqrt[3]{5}/5$
7898 4438 θ_3 : $3/8$
7898 6500 sr : $e-2\pi/3$
7898 7005 rt : (4,2,-7,4)
7899 0103 sπ : $\pi^2/3$
7899 0704 rt : (5,1,4,5)
7899 3899 rt : (2,9,-3,-7)
7899 4521 rt : (6,8,-5,-4)
7899 5589 $\sqrt{3}\times\exp(\sqrt{3})$
7899 6072 rt : (8,1,-8,-3)
7899 6223 J_0 : $3\pi/10$
7899 8483 2^x : $2\sqrt[3]{2}/3$
7899 9882 tanh : $15/14$
7900 0860 rt : (1,9,-9,1)
7900 1033 rt : (5,9,4,7)
7900 1585 sr : $\exp(-\sqrt{2}/3)$
7900 2602 Li_2 : $16/25$
7900 2673 ln : $\sqrt{2}/3+\sqrt{3}$
7900 5014 2^x : $21/25$
7900 8536 cu : $1/3+3\sqrt{2}/2$
7900 9504 rt : (4,1,-6,9)
7901 2272 ln : $1/3+4\sqrt{2}$
7901 2345 sq : $8/9$
7901 4061 rt : (1,7,-5,9)
7901 6126 rt : (7,4,-5,-2)
7901 6266 at : $7\sqrt[3]{3}/10$
7901 7054 Ψ : $7\sqrt[3]{2}/10$
7901 7976 ℓ_{10} : $5\pi^2/8$
7901 9040 10^x : $7(1/\pi)/5$
7901 9885 cu : $3/2+3\sqrt{2}/4$
7902 0132 sπ : $\sqrt[3]{5}$
7902 0742 rt : (6,9,-2,2)
7902 2124 e^x : $\exp(\pi/2)$
7902 3159 rt : (6,2,1,-5)
7902 8122 rt : (1,5,-3,-5)
7902 9361 rt : (5,0,7,8)
7903 0119 tπ : $\sqrt{3}/4+3\sqrt{5}/2$
7903 0886 rt : (6,5,4,3)

7903 1392 $\pi + \exp(1/2)$
7903 2103 $e^x : 3\sqrt{2} - \sqrt{5}/4$
7903 3991 $e^x : 13/4$
7903 5185 rt : (7,9,-8,-3)
7903 7799 rt : (2,3,-9,-8)
7903 8390 rt : (6,8,-1,7)
7903 8513 $\exp(\sqrt{5}) \times \exp(e)$
7903 8685 sr : $8\zeta(3)/3$
7903 8718 rt : (8,7,-8,-2)
7904 1309 $\ell_{10} : 3e/2 + 2\pi/3$
7904 1436 sinh : 7/4
7904 2734 sq : P_{43}
7904 3264 rt : (7,5,-2,-5)
7904 3566 $Ei : 5\sqrt[3]{3}$
7904 3681 $Ei : 2\sqrt[3]{3}/3$
7904 4875 at : $5\sqrt{2}/7$
7904 5189 cu : 17/14
7904 6481 sq : $e/2 + \pi/3$
7904 7372 rt : (6,3,4,-8)
7904 8726 rt : (7,8,-7,2)
7905 0869 $\Gamma : 2\sqrt{3}/7$
7905 0956 $J_0 : 16/17$
7905 6345 at : $2/3 - 3\sqrt{5}/4$
7905 6941 sr : 5/8
7905 8842 rt : (7,7,-6,7)
7905 9861 rt : (7,6,1,-8)
7906 0328 $\zeta : \exp(-\sqrt[3]{3})$
7906 2006 rt : (1,9,9,-7)
7906 2807 $t\pi : 2\sqrt{3} - 3\sqrt{5}/4$
7906 3229 rt : (7,2,1,3)
7906 6484 ln : $7\sqrt[3]{2}/4$
7906 8502 rt : (6,4,7,6)
7906 9028 rt : (8,9,-2,-8)
7906 9125 $\zeta : 10\ln 2/9$
7907 0473 $K : 11/17$
7907 2282 $2^x : 4 + 4\pi/3$
7907 5929 cr : $3/2 + 3\sqrt{2}$
7907 6044 rt : (9,3,-8,0)
7907 9509 rt : (8,1,-2,-3)
7908 0256 sq : $2 - 4\pi/3$
7908 2547 rt : (7,1,4,6)
7908 2884 id : $4\sqrt{2} - \sqrt{3}/2$
7908 3346 $Ei : 4\zeta(3)/5$
7908 6925 $Ei : (\sqrt[3]{5}/3)^{-1/2}$
7908 7941 $\zeta : \sqrt[3]{13/2}$
7908 9312 rt : (9,4,-5,-3)
7908 9984 $2^x : 1/2 + 4\pi/3$
7909 0244 $10^x : 1 + 2\sqrt{3}/3$
7909 1285 rt : (8,2,1,-6)
7909 2523 $t\pi : 7\ln 2/4$
7909 3069 rt : (7,0,7,9)
7909 3497 $\ln(\sqrt{5}) \times \exp(4/5)$
7909 3616 $4 \times \exp(2/3)$
7909 4529 $2^x : 2 + 3\sqrt{5}/4$
7909 7245 rt : (9,5,-2,-6)
7909 7761 cr : $2\sqrt{3}/7$
7909 8517 rt : (8,3,4,-9)
7910 0685 sinh : $\sqrt{10}$
7910 2337 $10^x : 1/\ln(3\pi)$
7910 2497 rt : (8,9,-8,-8)
7910 2523 rt : (9,6,1,-9)
7911 0154 $\exp(1/5) + s\pi(\pi)$
7911 0198 cr : $5\ln 2/7$
7911 4389 $t\pi : 3e/4 + \pi/2$
7911 6273 $2^x : (\sqrt{2})^{-1/2}$
7911 6616 $\ell_{10} : 2\sqrt{2} + 3\sqrt{5}/2$
7911 7674 tanh : $(\sqrt{3}/2)^{-1/2}$
7911 8635 $\Gamma : 2\sqrt[3]{3}/9$
7911 8851 $\ell_{10} : 4\sqrt{3} - \sqrt{5}/3$
7912 1062 $\ln(4) \div s\pi(e)$
7912 2485 $2 \times \exp(1/3)$
7912 3322 sinh : $(2e/3)^{1/2}$
7912 6760 $\Psi : 9\sqrt[3]{3}/2$
7912 7486 $erf : 8/9$
7912 8784 rt : (3,9,-9,0)
7913 3045 tanh : $\ln((e + \pi)/2)$
7913 4477 sr : $4/3 - \sqrt{2}/2$
7913 5523 $\ell_2 : 6\sqrt[3]{3}/5$
7913 6704 id : $8\pi^2/5$
7913 8603 rt : (9,1,4,7)
7914 0871 rt : (8,4,7,7)
7914 1655 rt : (9,2,1,4)
7914 1900 $e^x : 12$
7914 4019 $e^x : 8\sqrt{5}/5$
7914 4948 $\ell_{10} : 3/4 + 2e$
7914 5853 rt : (8,5,4,4)
7914 7458 rt : (9,3,-2,1)
7915 2393 cr : 23/4
7915 4117 rt : (7,8,7,4)
7915 5120 sr : $2e + 3\pi/4$
7915 6358 rt : (5,2,-8,8)
7915 7818 rt : (8,6,1,1)
7916 0576 $\ell_{10} : 2 + 4\pi/3$
7916 0808 rt : (5,3,-9,3)
7916 2297 $Li_2 : 4\sqrt[3]{3}/9$
7916 2783 rt : (9,4,-5,-2)
7916 3117 $K : (\sqrt[3]{2}/3)^{1/2}$
7916 5498 rt : (4,3,-3,8)
7916 6666 id : 19/24
7916 8250 ln : $3/2 + \sqrt{2}/2$
7916 8781 sinh : $3\pi/7$
7917 0718 sq : $8\sqrt[3]{3}/3$
7917 1865 $t\pi : 8/17$
7917 3832 rt : (8,8,-8,6)
7917 4098 $s\pi : 4\sqrt{3}/3 + 4\sqrt{5}/3$
7917 5946 ln : 6
7917 6138 $e^x : \pi^2/4$
7917 7151 $10^x : 1/4 + 3\pi$
7917 9543 $10^x : \sqrt[3]{23/2}$
7917 9600 rt : (8,9,-9,1)
7918 0077 $\ell_2 : 5\ln 2/6$
7918 0861 at : $1/3 + e/4$
7918 1457 sr : $9\sqrt{3}/2$
7918 1666 $10^x : \exp(-1/(2e))$
7918 3703 rt : (3,5,3,-7)
7918 3753 $2^x : 7\zeta(3)/10$
7918 5061 cu : $2 - \pi/4$
7918 5332 rt : (7,9,-3,6)
7918 7059 rt : (4,7,-3,-4)
7919 0697 sr : $2e/3 + 4\pi$
7919 0851 rt : (5,9,-9,-1)
7919 1210 $t\pi : 1/\sqrt{22}$
7919 2247 ln : $2e/3 + 4\pi/3$
7919 3977 rt : (7,9,4,1)
7919 4368 rt : (6,9,-8,-9)
7919 6320 $J : 2(1/\pi)/5$
7919 9117 rt : (2,2,6,-7)
7920 0182 $e^x : 7/12$
7920 0846 $Li_2 : 3\sqrt[3]{5}/8$
7920 0938 $10^x : \sqrt{22}$
7920 1167 cr : $(1/(3e))^{1/3}$
7920 1568 id : $\sqrt{2} + \sqrt{3} + \sqrt{7}$
7920 2757 $t\pi : 1/3 - 2\sqrt{2}/3$
7920 4505 cu : $3/4 - 4\sqrt{3}/3$
7920 4713 rt : (6,8,0,-8)
7920 5271 $J_0 : 15/16$
7920 5527 tanh : 14/13
7920 5893 cr : $2\sqrt{5}/9$
7920 6024 $Ei : 1/\ln(\pi/2)$
7920 6065 cr : $3\sqrt{3} + \sqrt{5}/4$
7920 6596 $s\pi : \sqrt{5/3}$
7920 8805 $e^x : 2e/3 - \pi/4$
7921 0635 $2^x : 4\sqrt[3]{3}/3$
7921 1272 rt : (4,6,-6,-1)
7921 2162 $Ei : 7e\pi/5$
7921 2209 $\ell_{10} : 1 + 3\sqrt{3}$
7921 2309 $t\pi : 1/\ln(\sqrt{2}\pi/3)$
7921 5799 $s\pi : 2/3 + 4e/3$
7921 6820 id : $8\ln 2/7$
7921 6916 $t\pi : \sqrt{2} - \sqrt{5} - \sqrt{7}$
7921 8269 rt : (5,5,3,-8)
7921 8898 $e^x : 2 - \sqrt{3}/4$
7922 0661 $s\pi : \sqrt{3} + \sqrt{5}/4$
7922 2665 rt : (6,7,-3,-5)
7922 2852 $2^x : 2\sqrt{3}/3 - 2\sqrt{5}/3$
7922 4205 sr : $3\sqrt{2}/4 - \sqrt{3}/4$
7922 4946 rt : (1,7,2,4)
7922 4972 sq : $1/4 + 3\pi/4$
7922 5555 rt : (8,7,-2,-2)
7922 6395 rt : (9,8,7,5)
7922 6574 rt : (1,1,9,7)
7922 7199 rt : (8,8,0,-9)
7922 7514 rt : (7,9,-9,-2)
7922 9329 cu : $4/3 + 2\sqrt{2}/3$
7922 9634 $2^x : 6e\pi/5$
7922 9934 ln : $3/2 - \pi/3$
7923 0215 $erf : (\sqrt[3]{2})^{-1/2}$
7923 1005 $2^x : 25/13$
7923 7723 rt : (4,2,6,-8)
7924 1011 rt : (3,4,-9,9)
7924 1490 rt : (6,8,-5,-6)
7924 1905 cu : $3/4 - 3\pi$
7924 2023 rt : (7,5,3,-9)
7924 4302 rt : (5,4,0,-5)
7924 5034 $erf : 9\ln 2/7$
7924 6674 rt : (8,7,-3,-6)
7924 7904 $Ei : 5\sqrt{3}/9$
7924 8125 $\ell_2 : \sqrt{3}$
7924 8946 $10^x : 5\zeta(3)/3$
7924 9031 $\exp(3/4) + s\pi(\sqrt{5})$
7925 1721 rt : (9,9,-9,-3)
7925 2266 sq : $e - \pi/3$
7925 2350 $\Gamma : \ln(1/(3\pi))$
7925 2680 id : $8\pi/9$
7925 6027 rt : (3,5,-8,4)
7925 6804 $2/5 \div \ln(4/5)$
7925 7290 sq : $\sqrt[3]{14/3}$
7926 1370 sq : $3\pi^2/4$
7926 2111 rt : (6,2,6,-9)
7926 4370 $\exp(3) + s\pi(1/4)$
7926 6419 $2^x : 16/19$
7926 6545 sr : $\pi/5$
7926 8001 rt : (3,1,9,8)
7926 8486 $\ell_{10} : \sqrt{3} + 2\sqrt{5}$
7926 8599 rt : (7,4,0,-6)
7926 9019 $10^x : 11/19$
7926 9807 rt : (4,5,-9,2)
7927 1097 id : $3\sqrt{2}/4 + \sqrt{3}$
7927 5270 rt : (1,8,2,3)
7927 5680 rt : (3,6,-7,-1)
7927 5744 rt : (8,6,-6,-3)
7927 5814 id : $2e + 3\pi/4$
7927 8911 rt : (4,1,3,-5)
7928 0474 sr : $4\sqrt{2}/9$
7928 1764 $\ell_2 : \gamma$
7928 5028 rt : (9,4,0,-7)
7928 6268 cu : $\sqrt{2}/4 + 4\sqrt{5}$
7928 7294 $2^x : 8\zeta(3)/5$
7928 9321 id : $3/2 - \sqrt{2}/2$
7928 9650 as : $4 - 3\pi/2$
7929 0583 $e^x : 3/2 - \sqrt{3}$
7929 0874 rt : (5,1,9,9)
7929 1133 $\Psi : 13/2$
7929 1659 rt : (5,3,-3,-2)
7929 2630 $\exp(\pi) + \exp(\sqrt{3})$
7929 3107 sr : $4 - \pi/4$
7929 3223 rt : (9,9,4,2)
7929 6177 rt : (2,3,-4,9)
7929 9504 as : $4\sqrt{2}/3 - 3\sqrt{3}/2$
7930 0229 $K : 3\sqrt{3}/8$
7930 0549 rt : (6,1,3,-6)
7930 1502 sq : $8\sqrt[3]{5}/3$
7930 2506 rt : (3,7,-6,-6)
7930 6745 rt : (6,5,-9,1)
7930 7228 $t\pi : \ln(5)$
7930 8412 at : $(\pi/3)^{1/3}$
7930 9738 $2^x : 2\sqrt{3}/3 + 3\sqrt{5}$
7931 0196 sinh : 8/11
7931 0481 $t\pi : \sqrt{2}/4 + \sqrt{3}/4$
7931 1228 rt : (2,8,-5,-8)
7931 1338 rt : (5,4,-5,7)
7931 1440 rt : (7,3,-3,-3)
7931 2735 $J_0 : 7\zeta(3)/9$
7931 3886 rt : (8,1,3,-7)
7931 4238 rt : (9,5,-8,-5)
7931 6172 $\ell_2 : 5\ln 2$
7931 7400 tanh : $6\sqrt[3]{2}/7$
7932 1650 rt : (1,2,6,4)
7932 2124 tanh : $(\sqrt[3]{2})^{1/3}$
7932 3250 rt : (9,3,-3,-4)

7932 3497 sinh : $7\sqrt{3}/9$
7932 3549 sinh : $\exp(-1/\pi)$
7932 3630 e^x : $3+\pi$
7932 3787 rt : (8,5,-9,0)
7932 5864 ℓ_{10} : $3/2+3\pi/2$
7932 8452 sinh : $(\sqrt[3]{4}/3)^{1/2}$
7933 1928 rt : (4,8,-6,8)
7933 1998 10^x : $(\sqrt[3]{2}/3)^{-3}$
7933 2246 10^x : $20/11$
7933 5334 $s\pi$: $7/24$
7933 5892 cu : $9\ln 2/4$
7933 8115 rt : (1,8,-8,-9)
7933 9003 ζ : $5/21$
7934 3107 $\ln(\sqrt{5})-\exp(4)$
7934 3183 rt : (3,2,6,5)
7934 5115 $\ln(\pi)+\exp(1/2)$
7934 5409 K : $13/20$
7934 5771 rt : (5,5,-6,2)
7934 6629 $\ln(2)\times\ln(\pi)$
7934 7208 ln : $5\zeta(3)$
7934 9286 $\exp(5)\div\exp(e)$
7934 9486 cr : $4\sqrt[3]{3}$
7935 0200 ℓ_{10} : $4\sqrt{2}+\sqrt{5}/4$
7935 1457 rt : (5,2,6,6)
7935 3105 $\ln(2/5)\times s\pi(1/3)$
7935 3716 ln : $9\pi^2/2$
7935 5530 Ψ : $6\pi/7$
7935 5835 rt : (7,2,6,7)
7935 6208 rt : (4,9,-5,3)
7935 7797 e^x : $((e+\pi)/2)^{-1/2}$
7935 8545 rt : (9,2,6,8)
7935 8798 cr : $1/3+2e$
7936 0075 $s\pi$: $3\sqrt{5}$
7936 0726 ℓ_2 : $\sqrt[3]{3}/5$
7936 6789 id : $\exp(4/3)$
7936 8172 Ei : $(3/2)^3$
7937 0052 id : $\sqrt[3]{1/2}$
7937 0129 e^x : $\sqrt{2}/2+2\sqrt{3}$
7937 2030 cr : $3\sqrt{3}/4+2\sqrt{5}$
7937 2172 $\ln(2/3)\times\exp(\sqrt{5})$
7937 4275 rt : (1,5,-8,-9)
7937 4285 ζ : $1/\ln(\sqrt[3]{5})$
7937 4393 $t\pi$: $4\sqrt{2}-2\sqrt{3}/3$
7937 4465 $\ln(2/3)\div\ln(3/5)$
7937 4867 tanh : $(\sqrt[3]{5}/2)^{-1/2}$
7937 5829 rt : (3,5,-8,-8)
7937 9196 rt : (5,5,-8,-7)
7937 9597 J_0 : $14/15$
7938 0373 e^x : $\exp(2\pi/3)$
7938 1810 rt : (6,7,2,-9)
7938 2392 tanh : $3\sqrt[3]{3}/4$
7938 2632 $\pi+\exp(\sqrt{3})$
7938 3960 rt : (4,7,2,-8)
7938 6137 rt : (9,2,-6,-1)
7938 7072 rt : (2,7,2,-7)
7938 8458 tanh : $9\zeta(3)/10$
7938 8943 rt : (9,9,-4,-7)
7938 9642 Γ : $\exp(\sqrt{5})$
7939 0123 cu : $e/4+\pi$
7939 0158 sq : $3/2+e$
7939 0977 cu : $2\sqrt{2}+3\sqrt{3}/2$
7939 1978 rt : (7,9,-4,-6)
7939 2734 ℓ_{10} : $2e+\pi/4$
7939 3375 $t\pi$: $(e+\pi)^2$
7939 3987 ℓ_2 : $-\sqrt{2}+\sqrt{5}+\sqrt{7}$
7939 5420 Ψ : $\ln(\sqrt[3]{5})$
7939 5703 sr : $1/2+e$
7939 6152 cr : $10\sqrt{3}/3$
7939 6177 rt : (5,9,-4,-5)
7939 6594 $s\pi$: $\sqrt{22/3}$
7939 7190 cr : $\sqrt{2}+\sqrt{3}-\sqrt{7}$
7940 0857 2^x : $9\sqrt[3]{5}/8$
7940 0861 rt : (2,7,-2,-5)
7940 1317 2^x : $\sqrt{3}/4+2\sqrt{5}/3$
7940 2364 rt : (3,9,-4,-4)
7940 3137 Ψ : $22/25$
7940 3733 $s\pi$: $3e/2-\pi/4$
7940 4478 rt : (5,2,-6,1)
7940 6054 $s\pi$: 5π
7940 6333 sq : $10\pi^2/7$
7940 6452 $t\pi$: $\sqrt{24}\pi$
7940 9190 $s\pi$: $e\pi/5$
7940 9678 erf : $2\sqrt{5}/5$
7941 0038 10^x : $17/4$
7941 0193 rt : (8,5,9,8)
7941 0853 sr : $3\sqrt{2}/2-2\sqrt{5}/3$
7941 1904 $\ln(\sqrt{2})-\exp(\pi)$
7941 2391 rt : (1,9,-4,-3)
7941 2871 sq : $5e\pi/2$
7941 3607 tanh : $5\sqrt{3}/8$
7941 4364 sr : $4/3+4\sqrt{2}/3$
7941 8263 rt : (3,7,-1,8)
7942 1016 rt : (6,5,-7,6)
7942 1824 sq : $9\ln 2/7$
7942 1850 rt : (4,7,-2,-4)
7942 2113 ℓ_2 : $3+4e$
7942 2424 rt : (6,5,9,7)
7942 2863 id : $9\sqrt{3}/2$
7942 5374 erf : $17/19$
7942 6066 rt : (5,4,5,-9)
7942 6646 θ_3 : $(\sqrt{5})^{-1/2}$
7942 7598 rt : (9,3,3,5)
7942 8334 sr : $\sqrt{21}\pi$
7942 8355 sq : $e/4-\pi/2$
7942 8963 rt : (8,6,-1,-7)
7942 9555 Ψ : $\sqrt{3}-\sqrt{6}-\sqrt{7}$
7943 1384 ℓ_{10} : $3e/4+4\pi/3$
7943 1435 2^x : $(5/3)^{-1/3}$
7943 1444 rt : (1,9,4,-2)
7943 6336 $t\pi$: $\sqrt{3}/2+2\sqrt{5}$
7943 8504 rt : (3,4,5,-8)
7944 0167 rt : (6,6,-1,-6)
7944 1412 rt : (9,8,-7,-4)
7944 3195 tanh : $13/12$
7944 3798 ζ : $3(1/\pi)/4$
7944 5009 $3\times\exp(4)$
7944 5382 rt : (4,5,9,6)
7944 6550 rt : (7,3,3,4)
7944 8225 Ψ : $\sqrt{2}+\sqrt{6}+\sqrt{7}$
7944 8804 K : $(e+\pi)/9$
7945 0076 rt : (3,7,-1,-7)
7945 2120 $\ln(2/3)-\exp(2)$
7945 2774 ζ : $5\sqrt{5}/6$
7945 5080 sq : $1/4+2\sqrt{3}$
7945 5645 rt : (7,8,-7,-3)
7945 6634 rt : (4,6,-1,2)
7945 6903 cr : $3\sqrt{2}/4-\sqrt{5}/4$
7945 8049 rt : (1,4,5,-7)
7945 9452 Li_2 : $9/14$
7946 0044 at : $4-4\sqrt{5}/3$
7946 3839 e^x : $1/\sqrt[3]{5}$
7946 4815 $s\pi$: $\sqrt{2}/4+3\sqrt{5}/2$
7946 5525 rt : (9,3,2,-8)
7946 7217 cu : $8\sqrt{5}/5$
7946 8155 rt : (6,6,-8,1)
7946 9690 Ψ : $5/18$
7947 0177 rt : (3,8,0,3)
7947 0834 $t\pi$: $\sqrt[3]{5}/8$
7947 1751 10^x : $4\sqrt{2}-3\sqrt{5}/2$
7947 1941 sr : $12/19$
7947 3408 Ψ : $10(e+\pi)/9$
7947 4679 10^x : $12/7$
7947 4900 rt : (8,1,-3,1)
7947 6971 rt : (5,8,-7,-2)
7947 8104 ℓ_{10} : $4\sqrt{2}+\sqrt{3}/3$
7947 8486 Ei : $\exp(-\sqrt[3]{4})$
7948 1707 rt : (3,9,4,-1)
7948 2921 Ψ : $(1/\pi)/9$
7948 3214 rt : (2,6,-1,-4)
7948 3561 rt : (7,3,2,-7)
7948 4117 rt : (5,3,3,3)
7948 5104 $\ln(3/5)-\exp(1/4)$
7948 5418 cu : $3e-3\pi/2$
7948 5801 2^x : $3e/2-\pi/4$
7948 9142 rt : (4,1,8,-9)
7949 0467 cu : $8e$
7949 1084 ℓ_{10} : $4+\sqrt{5}$
7949 1488 10^x : $1/3-\sqrt{3}/4$
7949 2212 Ei : $(\ln 3)^3$
7949 3236 rt : (1,7,6,-6)
7949 3427 J_0 : $(\sqrt{3}/2)^{1/2}$
7949 3473 e^x : $3\sqrt{2}-2\sqrt{5}$
7949 5930 at : $\sqrt{2}/3-2\sqrt{5}/3$
7949 8294 at : $3e/8$
7949 9261 Ei : $(\sqrt{2}\pi)^{-1/3}$
7949 9674 ln : $\ln(\pi/2)$
7949 9676 $s\pi$: $\sqrt{2}/3+\sqrt{5}$
7949 9912 $s\pi$: $2\sqrt{2}+2\sqrt{3}$
7950 0087 as : $1/3-\pi/3$
7950 1025 rt : (8,5,-4,-4)
7950 1906 $\pi-\ln(\sqrt{2})$
7950 3895 Γ : $9\sqrt[3]{5}$
7950 4238 rt : (2,5,9,5)
7950 4697 sq : $\sqrt[3]{21/2}$
7950 6480 rt : (7,6,-9,5)
7950 6797 ℓ_{10} : $9\ln 2$
7950 7262 ln : $3-\pi/4$
7950 8254 J_0 : $1/\ln((e+\pi)/2)$
7950 8497 id : $5\sqrt{5}/4$
7950 8658 rt : (9,1,-9,2)
7950 9916 at : $4\sqrt{2}/3-\sqrt{3}/2$
7951 0002 rt : (5,3,2,-6)
7951 2012 J : $3/23$
7951 2448 rt : (3,8,-7,-1)
7951 2980 ln : $3\sqrt{3}-4\sqrt{5}/3$
7951 4161 rt : (2,8,-7,3)
7951 8051 rt : (5,9,1,-9)
7951 9580 id : $4\pi/7$
7952 0209 ℓ_{10} : $\sqrt[3]{3}/9$
7952 0410 sr : $4(e+\pi)/3$
7952 0443 sr : $4\sqrt{3}/3-3\sqrt{5}/4$
7952 0550 sr : $\sqrt{3}+2\sqrt{5}/3$
7952 0952 rt : (3,9,-8,4)
7952 1244 rt : (8,6,6,5)
7952 2828 rt : (2,1,8,-8)
7952 3971 $\exp(4/5)+s\pi(\pi)$
7952 6376 rt : (8,0,5,8)
7952 7072 sr : $(5/2)^{-1/2}$
7952 7074 erf : $(\sqrt[3]{3}/2)^{1/3}$
7952 8941 ℓ_{10} : P_{20}
7953 0587 2^x : $1+4\sqrt{5}/3$
7953 3364 rt : (6,7,-2,-3)
7953 6833 ℓ_{10} : $2+3\sqrt{2}$
7954 0666 at : $2/3+\sqrt{2}/4$
7954 0768 sq : $3\sqrt{2}+3\sqrt{3}/2$
7954 1523 ln : $3\sqrt{2}/4+2\sqrt{3}/3$
7954 1950 rt : (9,8,-2,-8)
7954 3743 id : $4e\pi/9$
7954 4212 ln : P_1
7954 4728 rt : (3,9,1,-2)
7954 5569 2^x : $9e\pi/2$
7954 5759 rt : (9,2,-1,-5)
7954 6000 Ei : $7\sqrt{5}/8$
7954 6863 e^x : $4/3+4\sqrt{5}/3$
7954 9756 rt : (2,9,-2,4)
7955 0719 K : $15/23$
7955 1949 ℓ_2 : $3+\sqrt{2}/3$
7955 2516 rt : (3,6,5,7)
7955 5636 rt : (9,4,0,2)
7956 0295 as : $5/7$
7956 1242 $2/5+\exp(1/3)$
7956 2499 rt : (6,0,5,7)
7956 8626 rt : (9,3,8,9)
7956 9059 rt : (7,8,-2,-7)
7956 9320 $s\pi$: $\sqrt{2}/2$
7956 9982 erf : $2\pi/7$
7957 0286 rt : (4,0,-9,7)
7957 0457 id : $5e/2$
7957 5144 Γ : $8/25$
7957 5486 rt : (3,4,-5,-5)
7957 5569 e^x : $4e/3-\pi/2$
7957 6462 tanh : $25/23$
7957 7471 id : $5(1/\pi)/2$
7957 8091 J_0 : $13/14$
7957 9481 cr : $2\sqrt[3]{2}/5$
7958 3152 id : $\sqrt{23}$
7958 3756 cu : $4e/3-2\pi$
7958 5206 rt : (4,6,4,-9)
7958 5763 Ei : $(\sqrt{5}/2)^{-1/3}$
7958 6910 rt : (4,7,-2,-3)

7958 7694 cr : $\sqrt{2}+\sqrt{3}+\sqrt{7}$
7958 8001 ℓ_{10} : 4/25
7958 8763 rt : (7,2,-1,-4)
7958 9522 tanh : $2e/5$
7959 1836 sq : 23/7
7959 4148 10^x : $\sqrt[3]{21/2}$
7959 4218 rt : (9,9,9,6)
7959 5587 cu : $4+3\sqrt{3}/4$
7959 7891 rt : (7,1,-9,3)
7960 1117 Ei : $\sqrt[3]{7/3}$
7960 4358 rt : (8,5,1,-8)
7960 5475 as : $3/2-\pi/4$
7960 5500 2^x : $10\sqrt{3}/9$
7960 6325 rt : (5,8,-2,-6)
7960 9793 rt : (2,7,-7,-9)
7961 1279 tπ : $\sqrt{20/3}$
7961 1818 e^x : $e/2+3\pi$
7961 4086 rt : (6,7,-4,5)
7961 4152 rt : (7,3,8,8)
7961 7990 sinh : $3\pi^2/4$
7961 8265 rt : (7,7,-5,9)
7961 8309 rt : (8,4,-7,-1)
7961 9822 10^x : $3/4+\sqrt{5}/2$
7962 0361 rt : (4,0,5,6)
7962 1332 Ei : 14/23
7962 2521 sr : $3/2-\sqrt{3}/2$
7962 3488 rt : (9,7,-5,-5)
7962 4015 rt : (1,9,9,-3)
7962 8910 sq : $1/3+\sqrt{5}/4$
7963 2190 rt : (2,6,4,8)
7963 2637 10^x : $4e-\pi/4$
7963 5107 ζ : $7\zeta(3)/9$
7963 5314 ℓ_{10} : $3\sqrt{2}/4+3\sqrt{3}$
7963 7876 cr : P_{35}
7963 8623 sπ : $3\sqrt{2}+2\sqrt{3}$
7963 9270 rt : (9,2,4,-9)
7964 0318 e^x : $2-\sqrt{2}$
7964 2231 ζ : $9(1/\pi)/5$
7964 3202 rt : (8,1,2,5)
7964 3293 rt : (1,9,-1,-6)
7964 5111 e^x : $4/3+3\pi$
7964 5117 rt : (2,5,-8,7)
7964 7952 rt : (6,5,1,-7)
7965 2764 tπ : $1/4-2\sqrt{3}$
7965 2824 ζ : 6/25
7965 6247 cr : $3e-3\pi/4$
7965 6955 rt : (7,4,0,1)
7965 7256 erf : $\exp(-1/(3\pi))$
7965 9644 rt : (5,2,-1,-3)
7966 0213 $\exp(\pi)\times\exp(\sqrt{3})$
7966 0780 rt : (3,8,-2,-5)
7966 3578 rt : (2,6,1,-2)
7966 3763 2^x : $2-2\sqrt{3}/3$
7966 9907 at : $4/3-3\pi/4$
7967 2838 rt : (7,7,-5,-4)
7967 2942 cu : $3/4-3\sqrt{5}/4$
7967 4996 rt : (8,8,-7,8)
7967 5825 rt : (9,1,-4,-2)
7967 6432 e^x : $(e+\pi)/10$
7967 6886 rt : (1,6,-9,-9)
7967 7923 Ei : $1/\sqrt{24}$
7967 9507 ln : $3\zeta(3)/8$
7968 0501 rt : (3,3,7,-9)
7968 3366 2^x : 17/3
7968 4644 Ei : $2\pi^2/7$
7968 4739 10^x : $7\sqrt[3]{3}/2$
7968 6242 rt : (7,8,-6,4)
7968 6811 rt : (5,3,8,7)
7968 7042 10^x : $\ln(e/3)$
7968 7500 sq : $9\sqrt{3}/8$
7968 7990 erf : $5\sqrt[3]{2}/7$
7968 9042 rt : (7,2,4,-8)
7969 0047 rt : (9,7,0,-9)
7969 0151 J_0 : $(\sqrt[3]{2})^{-1/3}$
7969 0821 erf : 9/10
7969 1488 $\sqrt{3}\div s\pi(\sqrt{2})$
7969 2654 at : $(\pi/3)^{1/2}$
7969 2697 rt : (2,5,-4,-1)
7969 2968 e^x : $1+3\sqrt{5}/2$
7969 5599 Γ : $\exp(-1/\sqrt{2})$
7969 6044 rt : (4,7,-7,-8)
7969 6331 rt : (1,8,7,-8)
7969 6355 $\ln(\pi)+\exp(\sqrt{3})$
7969 6715 rt : (9,4,5,6)
7969 8224 ℓ_2 : $1/3+\pi$
7969 9895 2^x : $1/4-\sqrt{3}/3$
7970 0625 rt : (7,9,9,5)
7970 3474 ln : $1/3+4\sqrt{2}/3$
7970 5499 rt : (6,8,-5,0)
7970 6045 θ_3 : $1/\sqrt{7}$
7970 8934 θ_3 : $3\sqrt[3]{2}/10$
7970 9311 ℓ_2 : P_{75}
7971 1042 rt : (2,6,-5,4)
7971 1502 rt : (4,7,3,6)
7971 3444 e^x : $3/2-\sqrt{2}/3$
7971 4790 ζ : $\sqrt[3]{3}/6$
7971 5650 rt : (8,4,-2,-5)
7971 6292 10^x : $8\zeta(3)$
7971 9755 id : $1/4-\pi/3$
7972 0818 ζ : $\zeta(3)/5$
7972 0970 tanh : 12/11
7972 1556 rt : (9,9,-9,7)
7972 2392 2^x : $\ln(\sqrt[3]{3}/2)$
7972 2728 rt : (2,1,-8,-6)
7972 5096 $\ln(\sqrt{5})\div\ln(3/4)$
7972 5645 as : $(\ln 3/3)^{1/3}$
7972 6635 cr : $4/3+2\sqrt{5}$
7972 8105 rt : (2,0,5,5)
7972 9610 rt : (8,1,7,9)
7972 9805 ln : $(\ln 3/2)^3$
7973 0729 rt : (6,1,2,4)
7973 1585 $4\div\exp(4/5)$
7973 5566 as : $((e+\pi)/3)^{-1/2}$
7973 7277 tanh : $1/\ln(5/2)$
7973 8350 rt : (8,2,-6,-2)
7973 9605 rt : (7,7,0,-8)
7973 9697 10^x : $1-\sqrt{5}/3$
7973 9782 rt : (3,8,3,-9)
7973 9931 as : $(\sqrt{5}/2)^{-3}$
7974 0727 J_0 : $(\sqrt[3]{5}/2)^{1/2}$
7974 1233 cu : $2/3+3\sqrt{3}/2$
7974 1312 10^x : $4(1/\pi)/5$
7974 1434 cu : $3e/4-4\pi$
7974 2433 rt : (8,9,-8,3)
7974 3494 id : $1/3+2\sqrt{3}$
7974 4811 rt : (9,6,-8,-2)
7974 5277 ℓ_2 : $\ln(4/3)$
7974 6060 rt : (5,8,-4,-4)
7974 7338 tπ : 3/14
7974 7908 rt : (1,8,-2,-4)
7975 0921 rt : (3,9,-1,-5)
7975 2080 2^x : $3+4\sqrt{5}$
7975 4140 sr : $7\sqrt{5}/2$
7975 4271 10^x : $3\sqrt{3}/2+\sqrt{5}/4$
7975 4766 e^x : $4\sqrt{3}/3+\sqrt{5}/2$
7975 5159 sinh : $(\ln 3/2)^{-1/2}$
7975 5604 rt : (6,8,0,9)
7975 7885 rt : (9,1,1,-6)
7975 9771 J_0 : $4\ln 2/3$
7976 1307 erf : $5\sqrt[3]{3}/8$
7976 1430 rt : (5,2,4,-7)
7976 1619 sr : $1/4+4\sqrt{5}/3$
7976 3359 cu : P_{34}
7976 3647 rt : (8,9,-6,-5)
7976 6549 rt : (1,3,7,-8)
7976 7942 tπ : $5\sqrt{2}/9$
7976 8151 erf : $3\zeta(3)/4$
7976 9610 sr : $10\sqrt[3]{3}$
7977 0194 2^x : 11/13
7977 1591 at : $\sqrt{2}/2-\sqrt{3}$
7977 1841 cr : $2\sqrt{2}+4\sqrt{5}/3$
7977 2403 sr : 7/11
7977 3347 rt : (8,4,3,-9)
7977 5122 cr : $4\sqrt{3}-\sqrt{5}/2$
7977 5680 rt : (8,7,3,2)
7977 6032 rt : (7,9,-7,-1)
7977 7052 rt : (1,6,1,-1)
7977 8361 rt : (1,9,-7,8)
7977 9053 sr : $3/2+\sqrt{3}$
7977 9322 $\exp(\sqrt{5})-\exp(e)$
7978 1842 rt : (7,1,-4,-1)
7978 2927 e^x : $4\sqrt{2}/3-3\sqrt{3}/4$
7978 5485 $\ln(3/4)+\exp(3)$
7978 7037 sinh : $(\sqrt[3]{5}/2)^2$
7978 8371 tπ : $7(1/\pi)/5$
7978 8456 id : $(\pi/2)^{-1/2}$
7979 3196 rt : (1,8,7,1)
7979 3821 cu : $4+3\pi/4$
7979 5347 sq : $1/2-3e/2$
7979 5897 sq : $2\sqrt{2}+\sqrt{3}$
7979 5940 cr : $4+2e/3$
7979 6229 rt : (2,7,-2,1)
7979 7049 e^x : $1/3-\sqrt{5}/4$
7979 8526 rt : (9,6,-3,-6)
7980 0240 rt : (7,4,5,5)
7980 0845 e^x : $6\sqrt{5}/7$
7980 1722 sπ : 5/17
7980 3581 rt : (4,6,6,3)
7980 4912 rt : (8,9,-1,-9)
7980 5516 cu : $3e/2-2\pi/3$
7980 6143 J_0 : 12/13
7980 7122 tπ : $9e\pi/4$
7980 7353 rt : (5,8,1,5)
7980 8873 J : $\sqrt{2}/4$
7980 9439 rt : (7,9,-2,8)
7981 0590 rt : (6,1,7,8)
7981 2816 rt : (6,3,-9,9)
7981 3687 at : $1/3-e/2$
7981 5003 $1/5+\exp(4)$
7981 5629 id : $\ln(\sqrt{2}\pi/2)$
7981 5955 Li_2 : $1/\ln(3\pi/2)$
7981 7986 ℓ_{10} : 2π
7981 9064 ln : $4+3e/4$
7981 9221 rt : (1,8,3,-8)
7982 0594 sq : $8\pi/9$
7982 1279 rt : (3,3,8,6)
7982 2058 $\sqrt{3}\times\exp(2)$
7982 2195 rt : (1,8,-6,9)
7982 2359 at : $3\sqrt[3]{5}/5$
7982 3469 tπ : $1/\ln(\pi/2)$
7982 3560 ℓ_2 : $4\sqrt{2}+3\sqrt{3}/4$
7982 5979 e^x : 23/12
7982 9664 erf : $(e/2)^{-1/3}$
7983 0000 θ_3 : $\ln((e+\pi)/3)$
7983 3809 as : $9(1/\pi)/4$
7983 5129 λ : $\exp(-14\pi/19)$
7983 6032 rt : (3,9,-6,-9)
7983 7132 id : $3\sqrt{2}/2+3\sqrt{5}/4$
7983 9053 rt : (6,9,-6,-4)
7983 9396 ℓ_2 : $7\sqrt{5}/9$
7984 0037 rt : (7,6,-8,-1)
7984 0941 sinh : $5\zeta(3)$
7984 2234 cu : $4\sqrt{3}-3\sqrt{5}/4$
7984 2378 rt : (3,5,-7,7)
7984 2479 2^x : $6\sqrt{3}/7$
7984 5724 10^x : $9\sqrt{5}/7$
7984 6830 rt : (6,4,3,-8)
7985 0124 rt : (7,1,1,-5)
7985 0769 ln : 9/20
7985 0910 $\ln(4)-s\pi(1/5)$
7985 1419 10^x : $\sqrt{2}/3+4\sqrt{5}/3$
7985 2426 rt : (1,9,2,2)
7985 4021 ℓ_2 : $\exp(3\pi/2)$
7985 6312 rt : (8,2,-1,2)
7985 8500 cr : $2-2\sqrt{5}/3$
7985 8908 cr : $8(1/\pi)/5$
7986 0677 $s\pi(e)\div s\pi(\pi)$
7986 1455 rt : (4,9,-9,-8)
7986 1520 at : $2e/3-\pi/4$
7986 1668 e^x : $(3)^3$
7986 2230 Ei : $\sqrt[3]{15/2}$
7986 5099 id : $3e-3\pi/4$
7986 5371 rt : (2,5,6,-9)
7986 5626 rt : (5,3,-8,5)
7986 6007 rt : (3,7,-5,-2)
7986 6836 2^x : $2/3+3e/2$
7986 7751 $\ln(2)-\exp(2/5)$
7986 8247 rt : (8,7,8,6)
7987 1433 rt : (6,9,-1,4)
7987 1496 ℓ_2 : $1/4+3\sqrt{5}$

7987 2172 $s\pi : \exp(5/3)$
7987 2179 rt : (9,5,-3,-1)
7987 6434 rt : (3,2,4,-6)
7987 8200 tanh : 23/21
7987 8941 rt : (4,9,-4,6)
7987 9037 $t\pi : 2\sqrt{3}/3 + \sqrt{5}$
7988 0075 rt : (7,6,-3,-5)
7988 0397 rt : (6,7,-7,-7)
7988 1689 as : $(e)^{-1/3}$
7988 2520 $\ell_{10} : 2\sqrt{2} + 2\sqrt{3}$
7988 4509 rt : (2,8,-6,8)
7988 5060 cr : $2\sqrt{2}/3 - \sqrt{3}/4$
7988 5077 $\Psi : 9\zeta(3)/4$
7988 5310 $s\pi : 3\sqrt{3} - 2\sqrt{5}/3$
7988 5821 rt : (7,1,6,-9)
7988 6204 sr : $7\pi^2/3$
7988 6643 $e^x : 6\sqrt{5}$
7988 6759 rt : (4,8,-3,-8)
7988 8899 $e^x : 4e/5$
7989 0743 sr : $1 + \sqrt{5}$
7989 1078 $t\pi : 5\zeta(3)/4$
7989 1406 rt : (1,7,-7,-6)
7989 4295 rt : (8,3,-5,-2)
7989 7101 $s\pi : 2/3 + 3e/4$
7989 7312 rt : (3,9,-7,8)
7989 8987 cu : $2 + \sqrt{2}$
7990 0172 $e^x : 1/\ln(\sqrt[3]{2})$
7990 0885 rt : (8,2,4,6)
7990 2623 $2^x : 8\sqrt{2}/5$
7990 3810 id : $1/2 + 3\sqrt{3}/4$
7990 4069 rt : (7,6,2,-9)
7990 5585 rt : (3,7,0,-6)
7990 6698 $\ell_{10} : \sqrt{2} + \sqrt{5} + \sqrt{7}$
7990 8983 $10^x : (\sqrt{2}\pi/3)^{1/3}$
7990 9858 rt : (1,8,2,-3)
7990 9939 $t\pi : \pi/4$
7991 0537 rt : (2,8,1,-2)
7991 5741 $\lambda : \ln 2/7$
7991 6393 rt : (8,8,-9,-2)
7991 9024 rt : (1,3,-7,-7)
7992 0565 rt : (8,3,0,-6)
7992 2051 rt : (9,0,-7,1)
7992 2543 $10^x : 17/18$
7992 3461 sq : $3\sqrt{2}/4 + \sqrt{3}$
7992 6762 $10^x : 19/6$
7992 7844 $Ei : 1/\ln(5/3)$
7992 8261 tanh : $\pi^2/9$
7992 8830 $erf : 19/21$
7992 9003 rt : (9,5,2,3)
7992 9754 sr : $2\sqrt{5}/7$
7992 9842 ln : $\sqrt[3]{11}$
7993 0614 $1/4 + \ln(\sqrt{3})$
7993 1333 $\ell_{10} : 5\sqrt[3]{2}$
7993 3391 $J_0 : 23/25$
7993 3725 rt : (8,8,-4,-6)
7993 4759 $10^x : \sqrt{3}/2 + 4\sqrt{5}$
7993 5096 ln : $4/3 + 3\pi/2$
7993 8787 cr : $\ln(5/3)$
7993 8813 rt : (4,1,7,7)
7993 9685 $\Psi : (\sqrt{2}\pi/3)^{-1/3}$
7994 0628 $\sqrt{5} \times \ln(\sqrt{5})$
7994 2349 rt : (1,8,-3,-7)
7994 5346 $\ln(4) + \exp(5)$
7994 7605 rt : (9,0,-2,3)
7994 7750 $e^x : \sqrt{2}/4 - \sqrt{3}/3$
7994 7833 cr : P_{13}
7994 8303 rt : (3,6,-6,2)
7994 8791 $\zeta : 21/16$
7994 9342 at : $3/2 - \sqrt{2}/3$
7995 0220 rt : (4,9,-6,-3)
7995 0835 rt : (9,5,7,7)
7995 2784 rt : (2,9,-1,9)
7995 3144 rt : (4,4,3,-7)
7995 8346 $\zeta : 7\sqrt[3]{3}/9$
7995 8482 rt : (9,5,-6,-3)
7996 0535 rt : (9,0,3,7)
7996 2036 rt : (4,9,-1,-7)
7996 2995 $s\pi : 2e - \pi$
7996 3234 cr : $3 + 2\sqrt{2}$
7996 5841 $\ell_2 : e/2 + 4\pi$
7996 6698 sinh : $\ln(\sqrt[3]{3}/3)$
7996 7039 rt : (9,5,-1,-7)
7996 7751 $\ell_2 : 1/2 + 4\sqrt{5}/3$
7997 1145 $t\pi : 8(1/\pi)$
7997 4088 sr : $3e + 2\pi$
7997 5138 $t\pi : 3\sqrt{3} - 4\sqrt{5}/3$
7997 5404 cu : $7(e + \pi)/8$
7997 6444 sinh : $(e + \pi)/8$
7997 9808 cu : $4 - 2\sqrt{5}/3$
7998 2063 sr : $3\sqrt{2}/2 + \sqrt{5}/2$
7998 3137 $10^x : (e/3)^{-1/3}$
7998 4259 $Ei : 10\sqrt[3]{5}/7$
7998 8721 id : $10\sqrt[3]{2}/7$
7999 5042 $erf : e/3$
7999 7726 id : $((e + \pi)/3)^{-1/3}$
7999 8813 $\ell_{10} : 4 + 4\sqrt{3}/3$
8000 0000 id : 4/5
8001 0287 sr : $\sqrt{21/2}$
8001 1306 id : $4\sqrt{3}/3 + 2\sqrt{5}/3$
8001 1519 $\zeta : (\sqrt{2}/3)^{-1/2}$
8001 1862 $t\pi : 2\sqrt{2} - 3\sqrt{3}/4$
8001 5348 rt : (1,6,-5,3)
8001 7352 $\Gamma : 7\pi/3$
8001 8262 cu : $4/3 + \sqrt{2}/4$
8001 9200 $K : \exp(-\sqrt[3]{2}/3)$
8002 2870 sinh : $3\sqrt[3]{5}/7$
8002 3305 rt : (6,5,-6,8)
8002 4180 $erf : (\sqrt{5}/3)^{1/3}$
8002 6190 at : $5\sqrt[3]{3}/7$
8002 8539 ln : $5\pi^2/3$
8003 1586 rt : (6,8,1,-9)
8003 1739 $\ell_{10} : 3\sqrt{3} + \sqrt{5}/2$
8003 3768 at : $6\zeta(3)/7$
8003 4324 $e^x : 4/3 - \sqrt{5}/3$
8003 4597 rt : (1,5,2,5)
8003 5825 $10^x : \sqrt{5}/5$
8003 6028 $\Psi : \sqrt{22/3}$
8003 7511 rt : (6,3,5,-9)
8003 7742 $\ln(2) \div s\pi(1/3)$
8003 7989 at : $\sqrt{2}/3 + \sqrt{5}/4$
8003 8574 $s\pi : 9\zeta(3)/4$
8003 9353 rt : (6,8,-4,-5)
8003 9897 $2^x : 3 + \sqrt{3}/4$
8004 0218 id : $4e/3 - 3\pi$
8004 1176 cu : P_{17}
8004 1608 rt : (7,6,-8,7)
8004 3303 $\zeta : 9\zeta(3)/10$
8004 4140 rt : (1,7,5,-9)
8004 4274 $\ln(2/5) \div \ln(\pi)$
8004 4467 sq : $\sqrt{2}/4 - 4\sqrt{5}$
8004 6174 rt : (6,2,9,9)
8004 7886 tanh : $7\sqrt{2}/9$
8004 8994 rt : (6,3,0,-5)
8004 9902 tanh : 11/10
8005 0792 $\sqrt{2} + \ln(4)$
8005 1674 cr : $3\sqrt[3]{5}/10$
8005 2179 rt : (6,8,-9,-1)
8005 3674 sr : $2 - e/2$
8005 5401 sq : 17/19
8005 6647 rt : (1,2,9,-9)
8006 0570 $e^x : 3(e + \pi)/10$
8006 0948 rt : (1,7,0,-5)
8006 1870 $2^x : -\sqrt{2} + \sqrt{3} + \sqrt{7}$
8006 2057 ln : $9\sqrt{3}/7$
8006 2463 sr : $4\sqrt[3]{3}/9$
8006 3186 ln : P_{41}
8006 4892 rt : (6,2,4,5)
8006 5597 cu : 13/14
8006 6292 $2^x : 3\sqrt{2}/5$
8007 0594 rt : (6,3,-5,-1)
8007 0869 $J_0 : 11/12$
8007 1556 $10^x : (\ln 2)^{1/2}$
8007 1910 rt : (2,9,4,-5)
8007 2965 tanh : $\sqrt[3]{4/3}$
8007 3154 $\zeta : 1/\sqrt{17}$
8007 3277 rt : (4,6,-9,-9)
8007 3409 $e^x : 4/3 + 3e/2$
8007 3740 cr : $\exp(-2/3)$
8007 4503 $\Psi : 1/\sqrt{13}$
8007 5722 $2^x : 4 - \sqrt{2}/2$
8007 7523 sr : $3\sqrt[3]{5}/8$
8007 8212 rt : (7,5,-1,-6)
8008 0771 $e^x : 10/17$
8008 2672 $e^x : \sqrt{13/2}$
8008 5189 sinh : 11/15
8008 5212 cu : $2/3 + 3e/4$
8008 6349 $J_0 : \ln(5/2)$
8008 7682 rt : (1,2,4,-5)
8008 9900 $\Gamma : \sqrt{5}/7$
8009 0794 $\Gamma : 7\sqrt[3]{3}/2$
8009 0989 rt : (3,7,-5,-3)
8009 1927 $Ei : 5(e + \pi)/9$
8009 2738 rt : (2,9,-1,-6)
8009 5895 $\ell_{10} : 2/3 + 4\sqrt{2}$
8009 6071 rt : (6,7,8,5)
8009 7226 rt : (7,0,3,6)
8009 7618 rt : (8,7,-2,-7)
8009 8410 rt : (1,7,-5,-1)
8009 9734 $J_0 : \beta(2)$
8010 3475 rt : (7,5,-6,-2)
8010 3518 $2^x : 8(1/\pi)/3$
8010 7023 at : $(\ln 3)^{1/3}$
8010 7753 $\ln(2/3) - \exp(1/3)$
8010 8894 $\Gamma : \sqrt{2} - \sqrt{3} - \sqrt{7}$
8010 9087 rt : (6,2,-1,1)
8011 1443 rt : (9,9,-3,-8)
8011 4403 $10^x : 4 + 3\sqrt{3}/4$
8011 4469 $Ei : 5\sqrt{3}$
8011 4506 $10^x : 6\sqrt{3}/5$
8011 4714 id : $\sqrt{2}/4 - 2\sqrt{3}/3$
8011 6591 rt : (8,2,2,-7)
8011 7217 $erf : (4/3)^{-1/3}$
8011 8026 ln : $\pi/7$
8012 0748 rt : (2,4,3,-6)
8012 0909 $2^x : 7\pi^2/8$
8012 2498 rt : (8,7,-7,-3)
8012 3464 $10^x : 1/2 + 4\sqrt{5}$
8012 3782 cu : $1/2 + 3\sqrt{2}/4$
8012 4976 id : $5\sqrt[3]{3}/9$
8012 7084 $\exp(2) - s\pi(1/5)$
8012 8460 rt : (7,5,7,6)
8012 9485 rt : (9,4,1,-8)
8012 9731 $e^x : (\sqrt[3]{2}/3)^{-1/3}$
8013 0526 rt : (2,9,-6,-2)
8013 0573 $t\pi : 1/4 + e/2$
8013 2964 as : $2 - e$
8013 4059 $Li_2 : 11/17$
8013 4899 rt : (9,9,-8,-4)
8013 5246 rt : (4,6,-4,-5)
8013 5387 $\Psi : 1/\ln(1/(2e))$
8013 6068 sq : $4\sqrt{2} - \sqrt{3}/3$
8013 7051 tanh : $7\sqrt[3]{2}/8$
8013 7126 id : $2\zeta(3)/3$
8013 8103 rt : (3,6,2,-7)
8013 9243 $e^x : 1/4 - \sqrt{2}/3$
8013 9573 rt : (7,0,-2,2)
8014 0220 sinh : 25/11
8014 0543 sr : $9\sqrt[3]{3}/4$
8014 2141 cu : $\sqrt{2}/3 + \sqrt{5}/3$
8014 3430 $erf : 10/11$
8014 4722 rt : (8,3,6,7)
8014 7610 cu : $3\sqrt{2}/2 + 3\sqrt{5}/4$
8014 8694 $t\pi : 3\sqrt{2}/2 + \sqrt{3}/4$
8014 8794 rt : (4,8,1,-8)
8014 9002 $2^x : (\sqrt[3]{3}/2)^{1/2}$
8015 1150 $\exp(1/4) - \exp(3)$
8015 1745 $10^x : 2 + \pi/3$
8015 2407 $\ell_2 : e/2 - \pi/4$
8015 2734 rt : (9,9,-8,9)
8015 3806 sinh : $4e\pi/7$
8015 3918 rt : (8,2,-3,-3)
8015 4500 rt : (9,1,5,8)
8015 4671 tanh : $(\sqrt{5}/3)^{-1/3}$
8015 6040 rt : (5,5,-5,4)
8015 6278 rt : (1,3,8,5)
8015 6520 $e^x : 5\sqrt[3]{3}/7$
8015 9342 rt : (3,8,-3,-6)
8016 1276 rt : (6,6,-7,3)
8016 2258 rt : (9,4,-4,-4)
8016 2868 $s\pi : \sqrt{2} + \sqrt{5} + \sqrt{7}$

8016 3234 ℓ_{10} : $3/19$
8016 4621 cr : $3\zeta(3)/7$
8016 5236 rt : (7,7,-9,2)
8016 5378 e^x : $(\ln 2/2)^{1/2}$
8016 5768 id : $3\sqrt{2} + \sqrt{5}/4$
8016 6349 cr : 7π
8016 8553 Γ : $(\ln 2/2)^{-3}$
8016 9901 ζ : $\exp(-\sqrt{2})$
8017 1003 sq : $1/3 - 2\sqrt{3}$
8017 1073 Γ : $\sqrt[3]{3}/9$
8017 3036 rt : (2,1,7,6)
8017 8372 sr : $9/14$
8017 8411 ln : $3e - 2\pi/3$
8017 9442 rt : (3,1,6,-7)
8018 0773 2^x : $3\sqrt{2}/4 + \sqrt{3}/2$
8018 2825 rt : (4,3,5,-8)
8018 3058 at : $(e/3)^{-1/3}$
8018 4895 J : $\exp(-4\pi)$
8018 4915 rt : (5,5,4,-9)
8018 4979 cu : $2/3 + \sqrt{3}$
8019 0048 rt : (1,7,-6,2)
8019 0416 $s\pi$: $\sqrt{2} - \sqrt{5}/2$
8019 1503 rt : (9,6,9,8)
8019 3773 rt : (4,8,-4,-4)
8019 5387 $t\pi$: $e/2 + 2\pi/3$
8019 5909 2^x : $7\pi^2/4$
8019 6347 Ei : $5\sqrt[3]{5}/2$
8019 6628 e^x : $\sqrt{13}$
8019 7962 erf : $\sqrt{2} + \sqrt{3} - \sqrt{5}$
8019 9109 $e + \ln(2/5)$
8019 9206 λ : $\exp(-11\pi/15)$
8020 0073 erf : $1/\ln(3)$
8020 0291 e^x : $6\zeta(3)/7$
8020 2045 θ_3 : $2\sqrt[3]{5}/9$
8020 3536 rt : (9,1,0,4)
8020 3737 Ψ : $2\ln 2/5$
8020 3933 id : $\sqrt[3]{22}$
8020 4464 e^x : $8\sqrt[3]{3}$
8020 6911 ln : $7\sqrt{3}/2$
8020 7011 rt : (8,3,1,3)
8020 9912 sq : $7\zeta(3)$
8021 0572 sq : $3/2 - 3\pi$
8021 2319 $s\pi$: $(3/2)^{-3}$
8021 4132 e^x : $5e/3$
8021 4324 rt : (7,5,2,2)
8021 5130 $\exp(e) \div \exp(1/4)$
8021 7160 rt : (9,4,-9,0)
8021 9859 J_0 : $21/23$
8022 1520 $\exp(2) + \exp(5)$
8022 3928 rt : (1,7,-5,-8)
8022 4677 e^x : $\sqrt{2}/3 + \sqrt{5}/4$
8022 4917 rt : (7,9,-3,-7)
8022 5161 sr : $(\sqrt{2}\pi/3)^3$
8022 6244 rt : (8,2,-8,1)
8022 6921 rt : (5,0,8,9)
8022 7126 ℓ_{10} : $7e/3$
8022 7166 2^x : $\sqrt{2} - \sqrt{3}$
8023 1108 rt : (6,7,-2,-6)
8023 1891 rt : (1,8,8,-1)
8023 2781 rt : (4,0,-8,9)
8023 4804 e^x : $10\zeta(3)/9$
8023 6209 rt : (2,5,-6,-7)
8023 7071 rt : (4,2,9,8)
8023 7491 sr : $3 - 3\pi/4$
8023 8160 tanh : $21/19$
8023 8796 rt : (3,8,1,6)
8023 8921 $s\pi$: $3e + \pi$
8023 9278 e^x : $\exp(-\sqrt[3]{4}/3)$
8023 9680 rt : (6,7,3,1)
8024 0329 λ : $1/10$
8024 0371 rt : (5,5,-1,-5)
8024 6855 rt : (7,0,-7,2)
8024 6861 2^x : $3e/4 - 3\pi/4$
8024 6971 rt : (9,8,-1,-9)
8024 7004 rt : (2,4,-2,7)
8024 9560 sr : $5\pi/2$
8025 0092 2^x : $17/20$
8025 0575 cu : $4e - 3\pi/4$
8025 2083 2^x : $5e\pi/9$
8025 5749 rt : (8,6,0,-8)
8025 6976 ζ : $13/7$
8025 8875 rt : (2,6,-4,8)
8026 1714 Γ : $2\sqrt[3]{3}$
8026 1944 θ_3 : $\ln(\sqrt{5})$
8026 5391 rt : (3,3,-7,-6)
8026 7167 J_0 : $(\pi/3)^{-2}$
8026 7996 rt : (7,4,1,-7)
8026 9625 $t\pi$: $1/3 - \sqrt{5}/2$
8027 0463 rt : (8,9,-7,5)
8027 2911 10^x : $(\sqrt[3]{3})^{-1/2}$
8027 2975 rt : (9,6,4,4)
8027 4665 ℓ_{10} : $\sqrt[3]{2}/8$
8027 7078 2^x : $1/4 + 3\sqrt{5}/4$
8027 7563 sr : $13/4$
8027 8768 at : $1/4 + \pi/4$
8027 9692 rt : (7,8,-5,6)
8028 1196 id : $5\sqrt[3]{3}/4$
8028 1228 rt : (7,9,-8,-3)
8028 1324 erf : $(\pi/3)^{-2}$
8028 3979 sq : $9\sqrt[3]{5}/4$
8028 4889 Ei : $7\pi/9$
8028 6276 rt : (9,3,3,-9)
8028 6318 id : $(e + \pi)^{1/3}$
8028 6668 sq : $\sqrt{3}/2 + 4\sqrt{5}/3$
8028 8394 cu : $\pi^2/7$
8029 1447 rt : (6,7,-3,7)
8029 2080 ln : $1/2 + \sqrt{3}$
8029 2135 e^x : $(2e)^{1/3}$
8029 2629 $s\pi(e) \div s\pi(\sqrt{2})$
8029 3262 $s\pi$: $3\sqrt{2}/4 + \sqrt{5}$
8029 3436 rt : (3,8,-4,-8)
8029 4291 Li_2 : $(\sqrt[3]{2}/3)^{1/2}$
8029 6454 θ_3 : $\ln(e/3)$
8029 7911 rt : (9,8,-6,-5)
8029 8351 $\exp(1/2) + \exp(e)$
8030 0415 $\ln(\pi) - \exp(2/3)$
8030 1330 $t\pi$: $3\sqrt{2}/4 + 2\sqrt{3}/3$
8030 4083 rt : (6,7,-7,-2)
8030 4449 rt : (8,1,4,-8)
8030 5417 ℓ_{10} : $3 + 3\sqrt{5}/2$
8030 5527 id : $\sqrt{3}/4 - \sqrt{5}$
8030 5631 2^x : $1/4 + 4\sqrt{5}$
8030 6154 sq : $3\sqrt{2}/2 - 2\sqrt{3}$
8030 6930 rt : (5,6,-1,8)
8030 8535 id : $3\zeta(3)/2$
8031 1256 sq : $2(e + \pi)/7$
8031 1322 rt : (1,2,-9,-8)
8031 3417 10^x : $(e + \pi)/3$
8031 5189 Ψ : $19/7$
8031 5977 rt : (8,8,5,3)
8031 6505 rt : (5,0,3,5)
8031 6511 cr : $2/3 + 3\sqrt{3}$
8031 6732 rt : (2,8,1,-7)
8031 6881 sr : $1/\ln(3\pi/2)$
8031 7302 rt : (3,8,2,-2)
8031 9352 rt : (8,6,-5,-4)
8031 9717 ℓ_{10} : $4 + 3\pi/4$
8032 1051 Ψ : $\sqrt[3]{20}$
8032 4200 rt : (3,6,-8,1)
8032 4297 cu : $1/4 + e/4$
8032 4727 tanh : $(e/2)^{1/3}$
8032 5035 $\exp(1/3) \div s\pi(e)$
8032 6966 $\sqrt{2} + \exp(2)$
8032 7491 $s\pi$: $2\sqrt{2}/3 + 3\sqrt{5}/2$
8032 8249 rt : (4,7,3,-9)
8033 0114 cu : $\sqrt{2}/3 - 2\sqrt{3}$
8033 0372 ℓ_{10} : $4\sqrt{2}/3 + 2\sqrt{5}$
8033 1508 rt : (7,1,5,7)
8033 2385 10^x : $4\sqrt{3} - 2\sqrt{5}$
8033 3297 10^x : $2\sqrt[3]{5}/3$
8033 4855 J_0 : $1/\ln(3)$
8033 6610 J_0 : $\sqrt{2} + \sqrt{3} - \sqrt{5}$
8033 7900 erf : $21/23$
8033 8686 rt : (6,2,-6,-3)
8033 8851 sr : $2\sqrt{2}/3 + 4\sqrt{3}/3$
8034 0476 rt : (9,2,7,9)
8034 3100 rt : (5,5,-6,-1)
8034 4240 rt : (1,7,0,-4)
8034 6174 rt : (7,9,-6,-7)
8034 9085 rt : (6,7,-8,-2)
8034 9922 ℓ_2 : $5\pi/9$
8035 0146 Ei : $11/18$
8035 1332 Ψ : 17
8035 1907 rt : (7,4,-4,-3)
8035 2132 ℓ_2 : $1 + \sqrt{5}/3$
8035 2532 at : $\zeta(5)$
8035 4249 rt : (1,6,2,-6)
8035 5936 $s\pi$: $3\ln 2/7$
8035 7002 rt : (9,3,-2,-5)
8035 7601 rt : (5,8,-8,-9)
8035 9471 rt : (3,5,4,-8)
8036 1061 ln : $e/2 + 3\pi/2$
8036 2600 rt : (8,3,-4,-1)
8036 4278 cr : $2\sqrt{2} - 4\sqrt{3}/3$
8036 4419 ζ : $\sqrt[3]{5}/7$
8036 7781 sr : $4 - 3\sqrt{5}/2$
8037 1716 sq : $e/2 + 3\pi/4$
8037 2056 θ_3 : $1/\ln(\sqrt{2}\pi)$
8037 2118 tanh : $8\ln 2/5$
8037 2119 $\ln(5) - \exp(5)$
8037 2751 ℓ_{10} : $\sqrt{2}/9$
8037 2807 rt : (4,9,-3,9)
8037 3731 as : $4\sqrt[3]{2}/7$
8037 6088 rt : (8,4,8,8)
8037 6377 Ψ : $4\zeta(3)/9$
8037 7136 $\exp(\sqrt{5}) \div \exp(2/3)$
8037 8936 rt : (7,8,-1,-8)
8037 9510 rt : (5,9,-3,-6)
8038 0231 as : $18/25$
8038 1075 $s\pi$: $1/4 + \pi/3$
8038 1865 J_0 : $10/11$
8038 1875 $e + \exp(3)$
8038 3649 rt : (1,6,2,5)
8038 6047 sinh : $9\zeta(3)/8$
8039 1981 $\exp(\sqrt{5}) \div s\pi(1/3)$
8039 4676 cr : $3\sqrt{3}/10$
8039 5231 J : $\exp(-13\pi/16)$
8039 6701 2^x : $8\sqrt{3}$
8039 7187 e^x : $4 + \sqrt{3}/2$
8039 8917 at : $(\sqrt{5}/2)^{1/3}$
8039 8984 rt : (8,1,-1,-4)
8040 1636 rt : (4,2,7,-9)
8040 1707 Ψ : $7/8$
8040 1903 sr : $1 - \sqrt{2}/4$
8040 2504 Γ : $1/\ln(\sqrt{2})$
8040 3572 J_0 : $(4/3)^{-1/3}$
8040 6319 sr : $4 - \sqrt{5}/3$
8040 7232 rt : (6,2,-3,-2)
8040 7317 cr : $3\ln 2/4$
8040 8394 ℓ_{10} : $4\sqrt{3} - \sqrt{5}/4$
8040 8500 rt : (8,5,2,-9)
8040 8710 sr : $2\sqrt{3}/3 + 3\sqrt{5}$
8040 9596 rt : (2,3,5,-7)
8041 2710 cu : $4/3 + \sqrt{3}$
8041 4076 2^x : $3 + 3\sqrt{5}/2$
8041 4515 cr : $13/25$
8041 5001 rt : (6,6,0,-7)
8041 5595 sq : $(\sqrt[3]{3}/2)^{1/3}$
8041 7538 e^x : $3\sqrt{3} + 2\sqrt{5}/3$
8041 8382 rt : (2,7,-1,5)
8042 2606 $4 \times s\pi(2/5)$
8042 4764 rt : (4,9,1,-4)
8042 9648 sr : $5(e + \pi)/9$
8043 1387 cu : $1 + 3\pi/4$
8043 4086 10^x : $9(e + \pi)/10$
8043 4544 rt : (7,6,9,7)
8043 6548 θ_3 : $8/21$
8043 8403 at : $1 - 3e/4$
8043 9005 rt : (2,9,-4,-8)
8043 9966 sr : $11/17$
8044 0427 J : $11/23$
8044 0660 id : $3\pi^2/2$
8044 1066 rt : (2,8,-4,-3)
8044 5338 rt : (9,2,2,5)
8044 5480 tanh : $10/9$
8044 6766 $3/5 \div s\pi(\sqrt{3})$
8045 0085 rt : (7,3,3,-8)
8045 2004 $s\pi$: $(\ln 2/2)^{1/3}$
8045 2200 10^x : $1/3 + 2e/3$
8045 4435 rt : (6,6,-9,-8)

8045 4722 $10^x : 3\sqrt{3}/2 - 2\sqrt{5}/3$
8045 5093 rt : (9,7,-4,-6)
8045 6809 rt : (7,9,-6,1)
8045 7418 $e^x : 21/8$
8045 8071 $\ln : 2 + 3e/2$
8045 9421 $10^x : 1/3 + \sqrt{5}/4$
8046 3362 rt : (5,5,7,5)
8046 3367 at : $3\sqrt{3}/5$
8046 3756 $\Gamma : 10\pi^2/9$
8046 3987 rt : (3,0,8,8)
8046 4005 $e^x : 5\sqrt[3]{3}/2$
8046 4607 $10^x : 2 - 2\pi/3$
8046 4702 sinh : $1/\ln(2\pi/3)$
8046 5036 rt : (2,5,-1,-3)
8046 5595 $10^x : 4 + 3\sqrt{3}/2$
8046 7713 rt : (3,9,2,1)
8046 8519 rt : (5,4,1,-6)
8047 0018 rt : (7,1,0,3)
8047 1423 $10^x : 3e - \pi/2$
8047 1579 rt : (7,8,-6,-4)
8047 1895 id : $\ln(\sqrt{5})$
8047 3785 id : $1/3 + \sqrt{2}/3$
8047 4592 tanh : $\exp(1/(3\pi))$
8047 9229 rt : (6,4,-9,6)
8047 9944 id : $7\zeta(3)/3$
8048 0258 cu : $3/4 + \pi/3$
8048 0365 $J_0 : (\sqrt{5}/3)^{1/3}$
8048 0771 $erf : \beta(2)$
8048 4774 rt : (8,0,6,9)
8048 6933 at : $3\ln 2/2$
8048 7368 rt : (2,4,-7,2)
8048 9875 rt : (1,9,8,-9)
8049 2627 $\Psi : e$
8049 5871 $2^x : \sqrt{3}/4 + 3\sqrt{5}/2$
8049 6620 $erf : \ln(5/2)$
8049 6757 rt : (5,9,-8,-2)
8049 7216 $\exp(1/2) \div s\pi(1/5)$
8049 8238 $s\pi : \sqrt{2}/4 + 4\sqrt{5}$
8049 9182 rt : (6,8,-4,2)
8050 0324 cu : $1/\ln((e+\pi)/2)$
8050 1249 $\ell_2 : \sqrt[3]{16/3}$
8050 1786 sr : $(\sqrt[3]{2}/3)^{1/2}$
8050 3233 $10^x : 7(e+\pi)/10$
8050 3295 rt : (1,1,6,-6)
8050 4063 cr : 12/23
8050 4335 $J_0 : e/3$
8050 4958 $\Gamma : 7\sqrt[3]{3}$
8050 6435 $2^x : 1/3 + 2\sqrt{3}/3$
8050 7736 rt : (8,5,-3,-5)
8050 9945 $10^x : 9e$
8051 0021 $e^x : 1/\ln(2e)$
8051 1313 $\theta_3 : 3\sqrt{5}/10$
8051 1574 rt : (7,4,-9,1)
8051 1659 rt : (6,1,4,-7)
8051 2051 rt : (8,7,-9,8)
8051 3142 $s\pi : 4e + 3\pi$
8051 4068 sq : $3/4 + 2\sqrt{2}$
8051 4935 $erf : 11/12$
8051 5434 sinh : 23/17
8051 7392 rt : (5,3,-7,7)
8052 0646 $4 \div s\pi(1/5)$
8052 1487 at : $1/3 + \sqrt{2}/2$
8052 2003 rt : (5,0,-2,1)
8052 2817 $\ell_2 : 4/3 - \pi/3$
8052 3997 sr : $1/\ln(e/2)$
8052 6512 $\ell_2 : 9e/7$
8052 7241 $2^x : 10e\pi/3$
8052 7831 rt : (7,7,1,-9)
8052 9337 $J : \sqrt[3]{3}/8$
8053 0309 rt : (2,7,3,-8)
8053 0358 $10^x : 3\sqrt{2}/4 - 2\sqrt{3}/3$
8053 0612 $Li_2 : 3\sqrt{3}/8$
8053 0614 sinh : $(5/2)^{-1/3}$
8053 1516 rt : (8,4,3,4)
8053 3117 $\ell_2 : 1/3 + \sqrt{2}$
8053 3263 $2^x : \sqrt{3}/4 - \sqrt{5}/3$
8053 4241 $\zeta : (\ln 2/2)^{1/3}$
8053 4263 $\ln : 2/3 + \pi/2$
8053 5211 sinh : 14/19
8053 5372 $s\pi : 9\sqrt[3]{3}/10$
8053 5765 rt : (1,8,-8,-8)
8053 6458 rt : (3,7,-5,-7)
8054 1279 $e^x : 8\sqrt[3]{3}/3$
8054 1939 $2^x : 3/2 - 2e/3$
8054 2089 $\ln(\sqrt{2}) \div s\pi(\pi)$
8054 2179 rt : (6,6,-5,-3)
8054 2447 as : $\sqrt[3]{3}/2$
8054 3464 rt : (9,2,0,-6)
8054 4068 $e^x : 9(e+\pi)/7$
8054 5267 rt : (7,6,-7,9)
8054 6510 $2/5 - \ln(2/3)$
8054 6928 id : $1/3 + 2\sqrt{5}$
8054 7039 rt : (1,9,-8,-1)
8054 8241 rt : (3,4,5,3)
8054 8648 $e^x : 3(e+\pi)$
8054 9944 cr : $4 + 4\sqrt{2}/3$
8055 0149 tanh : $7(1/\pi)/2$
8055 0589 $t\pi : 10\sqrt{5}/3$
8055 3243 sr : $2\sqrt{2}/3 + 4\sqrt{3}$
8055 4200 rt : (7,2,7,8)
8055 5555 sq : $7\sqrt{5}/6$
8055 6214 rt : (4,2,2,-5)
8055 6431 cr : $1/\sqrt[3]{7}$
8055 7423 cr : $\sqrt{2} + 2\sqrt{5}$
8055 7712 sq : $1/3 - 3\sqrt{5}/4$
8055 7835 rt : (5,8,-1,-7)
8055 8233 as : $3\zeta(3)/5$
8055 8666 $J_0 : 19/21$
8055 9455 id : $7e/5$
8056 0283 rt : (1,6,-3,-2)
8056 2915 $e^x : 13/22$
8056 4099 $2^x : 2e + 4\pi/3$
8056 5483 $t\pi : 4\sqrt{2} + \sqrt{5}/4$
8056 6512 $e^x : 3\sqrt[3]{2}$
8056 7347 rt : (4,2,-5,8)
8056 8199 sq : $2\pi/7$
8056 8856 rt : (1,7,0,-9)
8057 0167 rt : (9,7,-9,-2)
8057 0675 rt : (3,4,6,-9)
8057 0813 $Ei : (\sqrt{2}\pi/2)^{-2}$
8057 1906 $\ln(3) + s\pi(1/4)$
8057 3055 $\theta_3 : (\sqrt{2}\pi/2)^{-1/2}$
8057 3740 rt : (3,6,-5,5)
8057 4602 as : $1/\ln(4)$
8057 6676 $\sqrt{5} + s\pi(\pi)$
8057 7774 rt : (8,9,-5,-6)
8057 8757 $Ei : (\ln 2/3)^3$
8057 9931 rt : (2,2,9,7)
8058 0349 at : 25/24
8058 2833 as : $1/4 + \sqrt{2}/3$
8058 3412 $10^x : (\ln 2/2)^{-2}$
8058 3723 rt : (1,4,-2,9)
8058 6924 rt : (7,3,-2,-4)
8058 7222 tanh : $(\sqrt[3]{3}/2)^{-1/3}$
8058 7326 $\zeta : 7\sqrt{2}/8$
8058 8299 $\theta_3 : 1/\sqrt[3]{18}$
8059 0825 rt : (4,7,-9,8)
8059 1634 rt : (6,5,2,-8)
8059 1778 rt : (9,7,6,5)
8059 2252 $J : 7/17$
8059 2744 cu : $(\sqrt{3}/2)^{1/2}$
8059 4568 sr : $2\sqrt{2} + \sqrt{3}/4$
8059 8238 $\exp(3) \times s\pi(1/5)$
8059 9597 cr : $\pi/6$
8060 1006 rt : (4,7,-7,-1)
8060 1616 id : $3\sqrt{2}/4 + \sqrt{5}/3$
8060 2628 rt : (3,9,-3,-5)
8060 6343 rt : (2,4,-8,-8)
8060 6839 rt : (9,6,-2,-7)
8060 7206 $Ei : 13/3$
8060 8268 $Li_2 : 13/20$
8060 8301 $2^x : \exp(-1/(2\pi))$
8060 8330 $\Psi : 2\pi^2/3$
8060 8366 sr : $6e/5$
8060 9712 $2^x : \sqrt{3} - \sqrt{6} + \sqrt{7}$
8061 0410 cr : 11/21
8061 2204 rt : (5,1,5,6)
8061 6137 rt : (6,3,1,2)
8061 7943 $\Gamma : 5\sqrt{3}/3$
8061 7997 $\ell_{10} : (4)^3$
8061 8062 sr : $e/3 + 3\pi/4$
8061 9215 rt : (6,8,-9,-7)
8062 0609 $e^x : 4e/3 + \pi$
8062 2015 $\ln : 4/3 + e/3$
8062 2577 sr : 13/20
8062 3524 $\ell_2 : 3\sqrt{2} - \sqrt{5}/3$
8062 4092 $e^x : (\ln 3)^{1/3}$
8062 4288 $\zeta : 9\sqrt[3]{3}/7$
8062 5063 rt : (7,2,5,-9)
8062 5675 rt : (1,8,1,-4)
8062 7272 $2^x : 3/4 - 3\sqrt{2}/4$
8062 9503 id : $e/3 - 3\pi/2$
8063 0071 rt : (2,8,2,2)
8063 1011 rt : (8,0,1,5)
8063 2210 $t\pi : (5/3)^{-3}$
8063 2302 $Ei : 7\sqrt[3]{5}/9$
8063 2993 $\ell_{10} : 4\sqrt{2} + \sqrt{5}/3$
8063 5572 rt : (3,1,-3,9)
8063 6096 rt : (7,5,-3,-2)
8063 8086 $10^x : 15/22$
8063 9111 rt : (2,8,-9,-6)
8063 9677 $J_0 : (e/2)^{-1/3}$
8063 9755 $\ln(2) + \exp(\sqrt{2})$
8064 2495 rt : (9,2,-3,1)
8064 4393 rt : (6,9,0,6)
8064 7625 $e^x : \sqrt{19/2}$
8064 7891 rt : (5,8,-3,-5)
8064 9520 $10^x : 3e/4 - \pi/3$
8064 9591 $\exp(\sqrt{3}) + \exp(e)$
8065 0561 $\ell_{10} : 3e\pi/4$
8065 3973 rt : (7,7,-4,-5)
8065 8407 $e^x : 4/3 + 2\pi/3$
8066 0831 rt : (5,4,-4,-2)
8066 2623 $\ln : 4\sqrt{2} + \sqrt{3}/4$
8066 4484 $\ell_{10} : 3/4 + 4\sqrt{2}$
8066 7154 at : 24/23
8066 7189 rt : (6,4,8,7)
8066 7504 rt : (2,9,1,-4)
8066 8053 $10^x : \exp(-e/2)$
8066 9913 $10^x : 2 + 4\sqrt{2}$
8066 9995 rt : (8,5,-8,-1)
8067 0400 cu : $\sqrt{2}/4 + \sqrt{3}/3$
8067 0467 $\lambda : (\pi)^{-2}$
8067 1858 cu : $\sqrt{2} + \sqrt{6} - \sqrt{7}$
8067 3521 $10^x : 3 - \sqrt{5}$
8067 4275 $t\pi : e/2 + 3\pi$
8067 4460 rt : (9,3,4,6)
8067 4893 tanh : 19/17
8067 6329 $\Gamma : 3(1/\pi)/4$
8067 6772 $erf : 23/25$
8067 9264 cu : $\sqrt{3}/4 + 4\sqrt{5}/3$
8067 9366 $\ln : 3/4 + 2\sqrt{5}/3$
8068 1506 rt : (2,2,7,-8)
8068 1896 sq : $5\pi^2/2$
8068 4243 $Ei : 25/8$
8068 5593 rt : (5,3,3,-7)
8068 5994 sr : $2/3 + 3\sqrt{3}/2$
8068 6540 $\theta_3 : 6(1/\pi)/5$
8068 6778 $\Psi : \sqrt[3]{2/3}$
8068 8288 id : $3\sqrt{2}/2 - 4\sqrt{3}$
8068 8334 rt : (8,4,-1,-6)
8068 8398 tanh : $\sqrt{5}/2$
8068 9178 $\Psi : 1/\ln(\pi)$
8068 9712 $J_0 : 3\zeta(3)/4$
8069 0592 sr : $(e+\pi)/9$
8069 4160 rt : (2,8,-6,-9)
8069 4599 $2^x : 1/2 + \sqrt{2}/4$
8069 5268 $J_0 : 5\sqrt[3]{3}/8$
8069 5676 rt : (1,9,2,-8)
8069 7669 $2^x : 2 - 4\sqrt{3}/3$
8069 7786 $s\pi : 3\pi^2/8$
8070 0983 as : 13/18
8070 1033 cu : $(e/3)^{-2}$
8070 2220 rt : (6,2,-8,2)
8070 5013 sq : $4\pi^2/7$
8070 7361 rt : (7,6,4,3)
8071 1317 rt : (4,7,8,4)
8071 1487 sq : $4\sqrt{2} - 4\sqrt{5}$
8071 5684 rt : (9,2,-5,-2)
8071 6287 $\ln(\sqrt{3}) - \exp(\sqrt{5})$

8071 9156 $2^x : 1/4 - \sqrt{5}/4$
8071 9476 rt : (6,1,-1,-3)
8072 0282 $Ei : (e/3)^{1/3}$
8072 2306 rt : (9,1,2,-7)
8072 2342 $10^x : \exp(5/3)$
8072 3708 $s\pi : 3\sqrt{3}/4$
8072 4187 rt : (2,8,-9,1)
8072 5280 rt : (5,8,-6,-3)
8072 5615 rt : (8,8,-3,-7)
8072 5650 at : $4\sqrt{3}/3 - 3\sqrt{5}/2$
8072 6245 rt : (6,0,6,8)
8072 9300 rt : (5,8,2,7)
8073 0014 ln : $2\sqrt{2}/3 + 3\sqrt{3}/4$
8073 0067 rt : (7,7,-8,4)
8073 0153 cr : $\sqrt{2}/2 + 3\sqrt{3}$
8073 1350 rt : (2,7,-2,-4)
8073 2339 rt : (1,9,-7,9)
8073 2672 $2^x : 5e/6$
8073 4191 cu : $3/2 + 4\sqrt{2}/3$
8073 4364 rt : (2,6,5,-9)
8073 5492 $\ell_2 : 7$
8073 5645 $2^x : 3 + 3\sqrt{2}/2$
8073 5929 $1/3 - \exp(\pi)$
8073 6951 ln : $4 + 2\pi/3$
8073 7322 $\ell_{10} : \sqrt{3} + \sqrt{5} + \sqrt{6}$
8073 7477 ln : $4 + 4\pi$
8073 7692 $10^x : 9\sqrt{2}/7$
8073 8242 rt : (1,8,9,6)
8073 8770 cr : 10/19
8073 9819 sq : $3/4 - 2\pi/3$
8074 1759 rt : (1,7,5,0)
8074 7215 $2^x : e\pi/10$
8074 8414 $Ei : 22/9$
8074 9856 id : $10\sqrt[3]{3}/3$
8075 2300 rt : (4,8,-8,3)
8075 2379 $J_0 : 9/10$
8075 3504 rt : (9,5,0,-8)
8075 4449 $e^x : 10\pi/9$
8075 4450 tanh : $8\sqrt[3]{2}/9$
8075 4669 $J_0 : 5\sqrt[3]{2}/7$
8075 6620 rt : (9,6,-7,-3)
8075 7285 sr : 15/23
8075 8302 rt : (5,4,-8,2)
8076 1543 rt : (6,4,4,-9)
8076 1672 at : 23/22
8076 3767 rt : (4,5,-6,-6)
8076 4088 $10^x : \sqrt{2}/2 + \sqrt{3}/4$
8076 4506 $e^x : 3/4 + \sqrt{5}$
8076 4703 sinh : $10\sqrt{2}$
8076 5406 rt : (1,8,-7,1)
8076 6362 rt : (7,3,9,9)
8076 6833 cr : $1/4 + 4\sqrt{2}$
8076 7152 $\ell_2 : 4\sqrt{3}/3 - \sqrt{5}/4$
8076 7233 rt : (8,9,-6,7)
8076 7877 rt : (6,5,-3,-4)
8076 8865 $\ln(\sqrt{2}) - \exp(e)$
8077 5457 rt : (9,9,-2,-9)
8077 6987 rt : (7,2,2,4)
8077 8176 $10^x : 7\pi^2/8$
8077 9499 $J_0 : \exp(-1/(3\pi))$
8078 0187 $\Psi : 8/15$
8078 2717 $\exp(1/3) - s\pi(1/5)$
8078 3411 rt : (2,9,-4,-5)
8078 5017 rt : (3,5,2,5)
8078 5588 $Li_2 : (e + \pi)/9$
8078 5652 $s\pi : 5(e + \pi)$
8079 5853 rt : (8,9,7,4)
8079 5958 rt : (6,9,-5,-3)
8079 6679 ln : $\ln(3\pi)$
8080 1757 rt : (6,6,-6,5)
8080 3738 rt : (7,1,-5,-1)
8080 5959 rt : (4,3,6,5)
8080 9874 rt : (7,2,0,-5)
8081 1678 $10^x : 3\pi^2/7$
8081 2203 id : $4\sqrt{2}/7$
8081 3234 $2^x : 4/3 + 2\sqrt{3}$
8081 4296 rt : (3,8,-1,-6)
8081 5269 $\Psi : 9(e + \pi)/8$
8081 6411 sq : $2\sqrt{2} - 4\sqrt{3}$
8081 7218 $\exp(1/5) \div s\pi(\sqrt{5})$
8081 7664 tanh : $7\sqrt[3]{3}/9$
8081 8682 cu : $3/2 - e$
8082 1293 rt : (4,9,-4,-7)
8082 2533 $\lambda : \exp(-8\pi/11)$
8082 2761 id : $4\zeta(3)$
8082 4931 $e^x : 2\sqrt{2} - \sqrt{5}$
8082 5127 ln : $\sqrt{2} + \sqrt{5} + \sqrt{6}$
8082 5281 $erf : 12/13$
8082 5764 ln : $5\pi/7$
8082 5932 id : $1/3 - \pi$
8082 8743 $s\pi : 5\sqrt[3]{2}$
8082 9035 rt : (7,6,-2,-6)
8082 9238 sinh : 17/23
8083 1657 $\Psi : 2e\pi$
8083 3406 rt : (8,5,5,5)
8083 4060 rt : (7,8,-4,8)
8083 6848 cu : $4e + \pi/3$
8083 7082 $t\pi : \sqrt{2}/2 - 2\sqrt{5}/3$
8083 7753 rt : (4,1,4,-6)
8083 9279 $Ei : 6\sqrt{2}$
8084 1579 rt : (7,3,-7,0)
8084 2359 tanh : $(\sqrt[3]{2})^{1/2}$
8084 4879 at : $\pi/3$
8084 5280 $2^x : -\sqrt{2} + \sqrt{3} + \sqrt{6}$
8084 5521 $\ell_{10} : 3e/2 + 3\pi/4$
8084 5719 rt : (3,9,-8,-1)
8084 6513 sinh : $\ln(2\pi/3)$
8084 7224 cu : $9\sqrt{5}/2$
8084 9789 $J_0 : 2\pi/7$
8085 1289 rt : (8,1,3,6)
8085 3696 rt : (4,7,4,8)
8085 4194 sq : $4 + 4\sqrt{3}/3$
8085 5173 cr : $1 - \sqrt{2}/3$
8085 5289 $\Psi : 7/19$
8085 5604 sr : $4/3 - e/4$
8085 8325 rt : (7,7,-9,-1)
8086 1540 $10^x : 24/17$
8086 1749 rt : (8,3,1,-7)
8086 3275 rt : (4,9,0,-8)
8086 3357 $s\pi : P_{79}$
8086 4978 at : 22/21
8086 5313 sinh : $3(e + \pi)/5$
8086 5406 $\ell_{10} : 1 + 2e$
8086 7171 id : $7\ln 2/6$
8086 8575 sr : $-\sqrt{2} + \sqrt{5} + \sqrt{6}$
8086 8614 rt : (8,7,-1,-8)
8087 0371 $s\pi : 2\sqrt{3} + \sqrt{5}$
8087 0372 $\Psi : 1/\sqrt[3]{20}$
8087 0600 $s\pi : 7\sqrt{2}/3$
8087 4367 $2^x : \sqrt[3]{5}/2$
8087 5789 rt : (3,2,-9,-7)
8087 7152 $\exp(1/3) + \exp(5)$
8087 8573 sq : $\sqrt[3]{14}$
8087 8865 rt : (7,8,-8,-8)
8087 9097 $erf : 4\ln 2/3$
8087 9759 sq : $\exp(-1/(3\pi))$
8088 1819 ln : $2/3 + 2e$
8088 2606 id : $(\sqrt[3]{4}/3)^{1/3}$
8088 3444 id : $2\sqrt{2}/3 + \sqrt{3}/2$
8088 3645 rt : (5,6,9,6)
8088 4229 $J_0 : (\sqrt[3]{3}/2)^{1/3}$
8088 4825 cr : P_3
8088 5194 rt : (1,0,8,7)
8088 6407 ln : $3/2 + \sqrt{5}/3$
8088 6498 rt : (3,6,-7,-8)
8088 7101 $\ell_2 : 4e + \pi$
8088 8777 rt : (9,8,8,6)
8088 9590 rt : (5,2,7,7)
8089 0103 $e^x : \sqrt{2} - \sqrt{3} + \sqrt{5}$
8089 0145 at : $(\ln 3)^{1/2}$
8089 0307 rt : (1,4,6,-8)
8089 1147 $Ei : 8\pi/7$
8089 1835 rt : (2,6,-8,-7)
8089 2137 rt : (9,4,6,7)
8089 2590 rt : (5,2,5,-8)
8089 3001 rt : (1,8,4,-9)
8089 3885 $\zeta : \sqrt{3}/7$
8089 4110 rt : (9,0,4,8)
8089 4398 rt : (5,6,3,-9)
8089 5045 $t\pi : \sqrt{3}/8$
8089 5409 rt : (9,4,2,-9)
8089 5608 rt : (5,5,-4,6)
8089 6772 cr : 9/17
8089 7163 $2^x : e/4 + 3\pi$
8089 7693 cr : $3\sqrt{2} + 3\sqrt{5}/4$
8090 0521 $2^x : \sqrt{2} - \sqrt{5}/4$
8090 1150 $erf : (\sqrt[3]{5}/2)^{1/2}$
8090 1699 id : $1/4 + \sqrt{5}/4$
8090 5247 $e^x : 2\sqrt{2}/3 - 2\sqrt{3}/3$
8090 7455 $10^x : 8\zeta(3)/5$
8090 7533 $\Gamma : \sqrt{2} + \sqrt{6} + \sqrt{7}$
8090 9579 $10^x : 3 + 4e$
8090 9797 $s\pi : 5\sqrt[3]{2}/9$
8091 5444 $e^x : 3e + 4\pi/3$
8091 6361 rt : (1,9,1,-6)
8091 6502 $s\pi : 10\sqrt[3]{5}/3$
8091 7250 $\zeta : 7(1/\pi)/9$
8091 8548 rt : (6,7,-2,9)
8091 8786 cr : $3\pi^2/5$
8091 8801 $\exp(1/5) + s\pi(1/5)$
8091 9095 at : $\sqrt{2}/3 + \sqrt{3}/3$
8092 4087 $e^x : \sqrt{2}/3 + \sqrt{3}/2$
8092 4350 $e^x : e/2 - \pi/2$
8092 4351 rt : (2,9,5,-1)
8092 4895 $\exp(e) \div \exp(4/5)$
8092 5833 $\ell_{10} : 2\sqrt{3} + 4\sqrt{5}/3$
8092 8456 $\zeta : 16/11$
8092 8757 $10^x : 3\sqrt[3]{2}/4$
8092 9295 rt : (9,9,-7,-5)
8092 9912 $e^x : 2 + 2\sqrt{5}/3$
8093 0107 tanh : 9/8
8093 0991 rt : (6,1,8,9)
8093 1859 $J : 7/15$
8093 2585 at : $1/4 - 3\sqrt{3}/4$
8093 5236 rt : (9,5,-5,-4)
8093 6248 at : $(\sqrt{3}/2)^{-1/3}$
8093 8280 rt : (2,1,9,-9)
8094 0107 id : $1/2 + 4\sqrt{3}/3$
8094 1025 $Ei : \sqrt{2}/7$
8094 1390 $t\pi : 9\sqrt[3]{2}$
8094 2313 rt : (5,8,-3,-2)
8094 3450 sr : $4\sqrt{2} + \sqrt{5}$
8094 3520 cr : $3\sqrt{2}/8$
8094 3791 $1/5 + \ln(5)$
8094 4436 rt : (9,1,-3,-3)
8094 4626 $\ln(\pi) \times s\pi(1/4)$
8094 5460 rt : (3,8,2,9)
8094 7501 sq : $3\sqrt{3}/2 + \sqrt{5}/2$
8094 8782 rt : (1,8,2,4)
8095 2380 id : 17/21
8095 3002 cr : $5(1/\pi)/3$
8095 3178 rt : (1,9,-3,-4)
8095 3317 $10^x : \sqrt{2} + 4\sqrt{5}/3$
8095 3543 rt : (4,9,-7,-2)
8095 3592 $t\pi : (\sqrt[3]{3}/3)^{1/3}$
8095 3932 $\Psi : 10\sqrt[3]{5}$
8095 6636 rt : (5,3,-2,-3)
8095 8886 sq : $3\sqrt{3} - \sqrt{5}/3$
8095 9588 $erf : (\sqrt[3]{2})^{-1/3}$
8095 9832 $Li_2 : 15/23$
8096 0133 id : $3e\pi/2$
8096 0590 rt : (9,3,-1,2)
8096 0746 as : $3\sqrt{3} - 2\sqrt{5}$
8096 2798 id : $3\sqrt{2} - \sqrt{3}/4$
8096 5543 $J_0 : 17/19$
8096 5784 rt : (6,8,-3,-6)
8096 6749 $K : 2/3$
8096 7903 cu : $e + 4\pi$
8096 9551 rt : (7,8,-9,-1)
8097 5227 at : $5\sqrt[3]{2}/6$
8097 6620 $2^x : (\sqrt{2}\pi/2)^{1/2}$
8097 8053 $J_0 : 2\sqrt{5}/5$
8097 8357 at : 21/20
8097 8592 rt : (1,5,-1,-3)
8098 1076 rt : (7,9,-2,-8)
8098 2379 rt : (6,4,-1,-5)
8098 2598 $\sqrt{2} + \exp(1/3)$
8098 3123 $t\pi : \sqrt{23/2}$
8098 3685 as : $2e - 3\pi/2$
8098 5109 id : $2\sqrt{2} + 4\sqrt{5}/3$

8098 6253 $\mathrm{t}\pi : 2\sqrt{5}/9$
8098 6557 rt : (5,1,0,2)
8098 9849 sq : $5\sqrt[3]{2}/7$
8099 1093 $J : (\ln 2)^3$
8099 2589 sr : $4\pi^2/5$
8099 6352 rt : (9,7,1,1)
8099 6549 $\mathrm{s}\pi : \zeta(3)/4$
8099 6976 sq : $2 - 2e$
8099 7316 rt : (7,5,0,-7)
8099 7377 $\mathrm{t}\pi : 7\sqrt{5}/10$
8099 7814 sr : $3\sqrt{2}/2 + 2\sqrt{3}/3$
8099 9105 rt : (2,8,-8,-1)
8100 0000 sq : $9/10$
8100 0053 rt : (2,7,0,9)
8100 2311 $\Psi : 4(1/\pi)/7$
8100 3257 rt : (1,7,1,-2)
8100 3792 rt : (2,6,0,-3)
8100 4125 at : $(e/3)^{-1/2}$
8100 4876 $e^x : \sqrt{17/3}$
8100 5372 $10^x : 20/9$
8100 5535 $2^x : 2 + 4\sqrt{2}$
8100 5550 $\mathrm{s}\pi : 2\sqrt{2} + 2\sqrt{5}$
8100 7072 rt : (8,6,1,-9)
8100 8316 $10^x : 5\pi^2/4$
8101 0568 rt : (6,0,1,4)
8101 4175 $e^x : \sqrt{15}\pi$
8101 4719 $\ln(2) + \exp(3/4)$
8101 6314 rt : (6,4,-8,8)
8101 6924 id : $4\sqrt{3} - \sqrt{5}/2$
8102 1911 rt : (7,1,2,-6)
8102 2886 $Ei : 4\sqrt{5}/3$
8102 3327 cu : $\sqrt{3}/2 + \sqrt{5}/2$
8102 3803 rt : (4,8,2,-9)
8102 4128 cu : $1/3 + 3e/2$
8102 7231 $\mathrm{t}\pi : \sqrt{2}/3 + \sqrt{5}/3$
8102 7630 $2^x : 2\sqrt{5}/3$
8102 8518 rt : (8,2,3,-8)
8102 9731 id : $\sqrt{3}/2 + 4\sqrt{5}$
8102 9861 rt : (3,7,0,-3)
8102 9915 rt : (3,9,-9,1)
8103 2548 sq : $2\pi^2/9$
8103 2748 $e^x : 2\sqrt{2} - 2\sqrt{5}/3$
8104 0719 rt : (2,8,-1,-5)
8104 2508 at : $4\sqrt{2} - 3\sqrt{5}$
8104 2836 $\Psi : 1/\ln(\sqrt[3]{3})$
8104 3147 $\zeta : 5\sqrt{3}/7$
8104 3360 $e^x : (\sqrt{2}/3)^{-3}$
8104 4069 $\mathrm{t}\pi : 3\pi^2$
8104 4756 rt : (1,9,-2,-7)
8104 6747 $10^x : 4\sqrt{2} + 3\sqrt{5}$
8104 7738 id : $\exp(\pi/2)$
8105 0817 $\ell_{10} : 3 + 2\sqrt{3}$
8105 3571 rt : (4,4,4,-8)
8105 3972 cu : $1/3 + \sqrt{3}$
8105 4640 $10^x : 5\zeta(3)/4$
8105 5430 $\mathrm{t}\pi : 1/2 + 2\pi$
8105 7723 sq : $4/3 - \sqrt{3}/4$
8105 9397 rt : (9,1,6,9)
8105 9491 sr : $\exp(-\sqrt[3]{2}/3)$
8105 9940 $10^x : \pi/7$
8106 0923 rt : (8,2,5,7)
8106 3533 rt : (7,3,4,5)
8106 3882 at : $7\zeta(3)/8$
8106 4054 $e^x : 4/3 + \pi/3$
8106 5337 $e^x : (e/3)^{-1/3}$
8106 5488 $\zeta : \sqrt{5}/9$
8106 6017 id : $1/4 - 3\sqrt{2}/4$
8106 9016 rt : (6,4,3,3)
8106 9391 cr : $1/2 + 2e$
8106 9701 sq : $1/4 + e$
8106 9758 $2^x : 1/2 + 3e/4$
8107 0949 rt : (9,8,-5,-6)
8107 1705 $10^x : 8/5$
8107 4726 rt : (8,7,-6,-4)
8107 8366 sq : $(3\pi/2)^{1/3}$
8108 0603 rt : (7,6,-7,-2)
8108 2023 $K : 5\zeta(3)/9$
8108 2102 rt : (1,2,-9,-8)
8108 5501 rt : (7,9,-5,3)
8108 7061 $\mathrm{s}\pi : (\sqrt{2}\pi)^3$
8108 7918 rt : (5,5,2,1)
8108 8386 $erf : 13/14$
8108 9328 sinh : $(\ln 3/2)^{1/2}$
8109 0373 rt : (5,1,7,-9)
8109 0998 rt : (6,7,-7,0)
8109 2880 $\ln(4/5) - s\pi(1/5)$
8109 3021 ln : $4/9$
8109 4756 $10^x : \sqrt{2}/3 + 4\sqrt{5}$
8109 6026 cr : $8/15$
8109 6140 tanh : $(\sqrt[3]{3})^{1/3}$
8109 7981 $10^x : 3 + \sqrt{5}$
8109 8646 $\ell_2 : \sqrt[3]{5}/3$
8109 8783 rt : (5,5,-9,-3)
8109 9271 $\Psi : \exp(-\sqrt[3]{2}/2)$
8109 9430 rt : (9,5,8,8)
8109 9535 $\mathrm{s}\pi : 5e/8$
8110 0121 tanh : $(\ln 2)^{-1/3}$
8110 2541 rt : (4,0,6,7)
8110 2557 id : $\sqrt{3}/2 - 3\sqrt{5}/4$
8110 3357 at : $20/19$
8110 4763 $\ell_{10} : 2 + 2\sqrt{5}$
8110 4845 $2^x : 3 + 2\sqrt{3}/3$
8110 4867 $e^x : 3/2 - e/3$
8110 6637 rt : (9,4,-3,-5)
8110 6772 $Ei : (\ln 2/3)^{1/3}$
8110 6814 sq : $e/4 - 3\pi/4$
8110 7958 rt : (1,8,9,-4)
8110 8665 $J_0 : 9\ln 2/7$
8110 9321 ln : $e/4 + \pi/2$
8111 0352 cu : $2\sqrt{2}/3 + \sqrt{5}/3$
8111 0656 rt : (4,9,-5,-4)
8111 4357 cu : $\sqrt{3} + 2\sqrt{5}$
8111 4373 rt : (5,4,-9,2)
8111 4535 $\mathrm{s}\pi : ((e+\pi)/2)^{-1/3}$
8111 5565 rt : (8,6,7,6)
8111 6235 rt : (7,8,0,-9)
8111 6988 cu : $2/3 - 4\sqrt{2}/3$
8111 7735 tanh : $\exp(1/(3e))$
8111 9586 sq : $1 + 3e$
8111 9650 cr : $3\sqrt{3} + \sqrt{5}/3$
8112 0365 $J_0 : (\sqrt[3]{2})^{-1/2}$
8112 1000 $Ei : (\ln 3)^{-1/3}$
8112 1272 rt : (8,3,-4,-3)
8112 2543 rt : (5,3,-6,9)
8112 4537 $\mathrm{t}\pi : (\sqrt{2}\pi/3)^{1/2}$
8112 6731 rt : (3,1,5,5)
8112 6916 $\Gamma : 4/25$
8112 9201 rt : (6,7,-1,-7)
8112 9751 $\Gamma : (1/\pi)$
8113 0184 sr : P_{19}
8113 0277 sr : P_{58}
8113 0544 $\ln(3) \times \exp(1/2)$
8113 0982 rt : (3,8,-6,-2)
8113 2423 ln : $3\sqrt{2}/2 - 3\sqrt{5}/4$
8113 2736 $\sqrt{2} - \exp(4/5)$
8113 3963 rt : (4,1,-8,6)
8113 6555 rt : (1,3,8,-9)
8113 9129 $10^x : (\pi/2)^{-3}$
8113 9655 $1/3 - \ln(\pi)$
8114 0870 ln : $2\sqrt{2} - \sqrt{3}/3$
8114 1951 rt : (9,2,-8,-3)
8114 2335 cr : $4\zeta(3)/9$
8114 2430 $\Gamma : \exp(\sqrt{2}\pi/2)$
8114 2873 $2^x : \ln(\sqrt{2}\pi)$
8114 4732 $2^x : 6/7$
8114 5070 $Ei : 23/4$
8114 6383 $\mathrm{s}\pi : 5e\pi$
8114 6475 $\mathrm{s}\pi : 3\sqrt{2}/2 + \sqrt{3}/3$
8114 7043 rt : (7,2,-5,-1)
8114 8791 tanh : $4\sqrt{2}/5$
8114 8845 $\ell_{10} : 3\sqrt{2} + \sqrt{5}$
8114 8932 rt : (5,9,1,2)
8114 9212 rt : (7,7,6,4)
8114 9491 rt : (8,8,-9,5)
8114 9761 at : $(\sqrt[3]{5}/2)^{-1/3}$
8114 9916 rt : (3,8,-3,-5)
8115 0729 rt : (5,3,9,8)
8115 1434 $\mathrm{s}\pi : (\ln 2/2)^{-1/2}$
8115 3047 rt : (4,8,-6,-8)
8115 5478 id : $4\sqrt{2} + 2\sqrt{3}/3$
8115 7233 rt : (3,6,-4,8)
8115 7335 $10^x : 7\pi^2/5$
8115 7430 $2^x : (\sqrt[3]{4})^{-1/3}$
8115 8002 $\zeta : 3\pi/10$
8115 9311 rt : (7,4,2,-8)
8116 1378 rt : (9,0,-1,4)
8116 6531 rt : (1,9,2,3)
8116 6827 cr : $3/4 + 3\sqrt{3}$
8116 6952 $\Psi : \sqrt{3} + \sqrt{5} + \sqrt{7}$
8116 8184 $erf : 1/\ln((e+\pi)/2)$
8116 9684 $\zeta : \sqrt[3]{19/3}$
8117 3769 rt : (2,9,0,-7)
8117 6335 $K : (\sqrt{5})^{-1/2}$
8117 7781 ln : $4 + 3\sqrt{2}/2$
8117 9868 $\mathrm{s}\pi : 1/\sqrt{11}$
8118 2583 rt : (4,7,-1,-1)
8118 2701 $\ell_{10} : 2e + \pi/3$
8118 5079 $erf : (\sqrt{3}/2)^{1/2}$
8118 6717 rt : (6,3,1,-6)
8118 7426 $10^x : P_{28}$
8118 9117 rt : (8,1,5,-9)
8119 8159 rt : (4,4,8,6)
8119 8324 $Ei : 9\pi/5$
8120 1227 $J_0 : 8/9$
8120 1435 $\mathrm{t}\pi : 1/\ln(3\pi)$
8120 2011 rt : (8,1,-2,2)
8120 2206 rt : (4,8,-7,6)
8120 3737 ln : $3e/2 + 4\pi$
8120 4431 rt : (6,1,-6,1)
8120 8046 rt : (9,7,-3,-7)
8120 9497 rt : (3,6,3,-8)
8121 0215 rt : (7,9,-7,-4)
8121 2891 rt : (3,5,-9,-9)
8121 3317 rt : (1,8,-1,-5)
8121 5700 tanh : $17/15$
8121 6149 sinh : $9\sqrt[3]{3}/4$
8121 6980 $10^x : 1/\sqrt{15}$
8121 7388 $\mathrm{t}\pi : (e+\pi)^3$
8121 7558 $10^x : 4\sqrt{2}/3 - \sqrt{5}/3$
8121 8788 id : $2e/3$
8121 9311 $10^x : P_{41}$
8121 9940 cu : $1/2 + \sqrt{3}/4$
8122 1288 $\mathrm{s}\pi : (\ln 3/2)^2$
8122 2725 $2^x : (e/2)^{-1/2}$
8122 2999 rt : (6,8,-3,4)
8122 3602 rt : (5,7,-5,-6)
8122 4025 rt : (7,0,4,7)
8122 5293 $\ell_{10} : 9\sqrt[3]{3}/2$
8122 5828 cr : $4 - 2\sqrt{3}$
8122 6419 $\mathrm{t}\pi : 10/23$
8122 6803 $e^x : 3\sqrt{2}/4 + 3\sqrt{3}/2$
8122 8004 cr : $3\sqrt{3}/2 + 3\sqrt{5}/2$
8123 2125 $\mathrm{t}\pi : 7\sqrt{5}/6$
8123 2288 cu : $3\sqrt{2}/2 - \sqrt{3}/4$
8123 3071 $\Psi : 2/11$
8123 3290 rt : (5,2,0,-4)
8123 5439 sq : $2/3 + e/4$
8123 5782 rt : (8,9,-5,9)
8123 5965 tanh : $9\sqrt[3]{2}/10$
8123 8159 sinh : $3\sqrt{3}/7$
8123 8705 $\ell_{10} : 2\ln 2/9$
8124 0954 rt : (8,6,-4,-5)
8124 1172 cu : $1/4 - 2e/3$
8124 1861 at : $19/18$
8124 2080 $\mathrm{t}\pi : \sqrt[3]{17/3}$
8124 4804 rt : (4,3,6,-9)
8124 5300 $\mathrm{s}\pi : 2\pi/9$
8124 6017 $\ell_2 : \sqrt{5}\pi$
8124 8261 $\Gamma : 7/22$
8124 8914 $\Psi : (\pi/3)^{-3}$
8125 0000 id : $13/16$
8125 2892 rt : (6,8,-8,-2)
8125 3274 sq : $5\sqrt[3]{3}/8$
8125 5167 cr : $\ln(\sqrt[3]{5})$
8125 5189 $K : \ln((e+\pi)/3)$
8125 5942 rt : (9,4,1,3)
8125 7585 rt : (5,9,-2,-7)
8125 8667 rt : (7,7,-7,6)
8126 0398 rt : (2,7,6,6)
8126 0868 rt : (8,3,7,8)

8126 2572 rt : (6,8,5,2)
8126 3610 rt : (1,3,-1,8)
8126 5393 sr : 23/7
8126 7693 id : $e+2\pi/3$
8126 7797 Ei : $\exp((e+\pi)/3)$
8127 1371 rt : (9,3,-1,-6)
8127 1627 rt : (1,4,-1,-9)
8127 2091 Γ : $\sqrt{3}-\sqrt{5}-\sqrt{6}$
8127 3951 Ψ : $8\sqrt[3]{5}/5$
8127 5148 $s\pi$: $e/9$
8127 7919 sq : $3\zeta(3)/4$
8127 8448 sq : $3\pi/7$
8127 9413 2^x : $1-3\sqrt{3}/4$
8128 3135 rt : (6,1,3,5)
8128 4130 rt : (1,8,-4,-8)
8128 6663 rt : (6,6,1,-8)
8128 8125 rt : (6,5,-9,3)
8128 8441 ln : $\sqrt{3}/3+3\sqrt{5}/4$
8128 9453 sq : $10(e+\pi)$
8129 1335 ℓ_{10} : 2/13
8129 1653 rt : (7,9,-1,-3)
8129 1709 sinh : $7(1/\pi)/3$
8129 1774 sq : $4e\pi/7$
8129 2069 10^x : 5/6
8129 3283 e^x : $1/2-\sqrt{2}/2$
8129 4991 2^x : $4\sqrt{2}-\sqrt{3}/3$
8129 5122 rt : (2,8,3,6)
8129 9209 ln : $3-\sqrt{5}/3$
8130 0521 e^x : $2\sqrt{3}/3+3\sqrt{5}/2$
8130 0744 rt : (3,9,3,4)
8130 3571 2^x : $4-\pi$
8130 3703 cu : 14/15
8130 4974 $s\pi$: $2(e+\pi)/9$
8130 6750 rt : (3,2,5,-7)
8130 7464 sinh : $(2e/3)^{-1/2}$
8130 9903 rt : (4,8,-3,-5)
8131 2193 Ψ : $\exp(-1)$
8131 3386 e^x : $10\sqrt{2}/9$
8131 4249 erf : 14/15
8131 5457 rt : (7,3,4,-9)
8131 5910 $2/5+\exp(5)$
8131 6363 rt : (8,9,-4,-7)
8131 6597 id : $4(e+\pi)/3$
8131 7087 cu : $\sqrt{2}/3+2\sqrt{5}$
8131 8599 tanh : 25/22
8131 9151 at : $2-2\sqrt{2}/3$
8131 9620 $\ln(\sqrt{3})\div s\pi(\sqrt{5})$
8132 1872 rt : (1,6,-8,2)
8132 4276 rt : (4,4,-8,-7)
8132 4457 $\ln(3/5)\times\exp(5)$
8132 5713 cu : $3\sqrt{2}/2-\sqrt{5}/4$
8132 7459 rt : (5,7,-9,0)
8132 7665 at : $(\sqrt{5}/2)^{1/2}$
8132 7840 ζ : 1/4
8132 9301 2^x : $5\zeta(3)/7$
8132 9471 Γ : $2\sqrt[3]{5}/7$
8132 9901 rt : (8,2,-2,-4)
8132 9962 $s\pi$: $4e-\pi/2$
8133 0248 2^x : $3+e/4$
8133 2146 rt : (7,4,6,6)
8133 2618 $\exp(4/5)+s\pi(1/5)$
8133 6136 rt : (3,5,7,4)
8133 6228 cr : $8\sqrt{5}/3$
8133 6439 λ : $\exp(-13\pi/18)$
8133 6556 θ_3 : 5/13
8134 0865 rt : (9,6,-1,-8)
8134 1824 id : $\sqrt{3}/3+\sqrt{5}$
8134 3149 at : $\sqrt{3}/4-2\sqrt{5}/3$
8134 3740 sq : $3-\pi/3$
8134 5026 rt : (2,8,-7,4)
8134 6065 rt : (5,5,0,-6)
8134 8394 rt : (8,3,-9,-5)
8135 1090 rt : (6,9,1,8)
8135 1974 e^x : 11/7
8135 2394 rt : (4,1,8,8)
8135 3745 J_0 : $(\sqrt[3]{3})^{-1/3}$
8135 4460 ℓ_{10} : $\sqrt{2}+\sqrt{6}+\sqrt{7}$
8135 5121 cr : 7/13
8135 6034 $t\pi$: 5/23
8135 7395 J_0 : $(\ln 2)^{1/3}$
8135 8515 rt : (9,1,-8,1)
8136 0650 rt : (3,6,-9,-6)
8136 1166 K : $1/\ln(\sqrt{2}\pi)$
8136 1859 cu : $\sqrt{2}/3-4\sqrt{5}/3$
8136 4580 ℓ_{10} : $10(e+\pi)/9$
8136 7648 rt : (9,1,1,5)
8136 7943 $s\pi$: $\exp(\sqrt[3]{4}/3)$
8137 0132 ℓ_{10} : $4\sqrt{2}/3-\sqrt{3}$
8137 1547 rt : (7,8,-5,-5)
8137 3033 $t\pi$: 9/23
8137 3536 J_0 : $\exp(-1/(3e))$
8137 5941 cu : 24/17
8137 5944 $\exp(2/3)+s\pi(1/3)$
8137 6218 rt : (8,5,0,1)
8137 7149 $s\pi(e)\div s\pi(2/5)$
8137 8836 2^x : $\sqrt{18}\pi$
8137 9872 rt : (6,4,-6,-1)
8137 9936 sr : $\pi^2/3$
8138 0119 rt : (3,9,-8,5)
8138 0352 rt : (8,7,9,7)
8138 1788 rt : (6,2,3,-7)
8138 2131 id : $4\sqrt{2}/3+4\sqrt{3}$
8138 3173 2^x : $3/4-\pi/3$
8138 4103 sinh : 19/14
8138 5933 rt : (2,9,6,0)
8138 7991 cr : $7\ln 2/9$
8138 8064 rt : (7,5,-9,9)
8138 8393 K : $3\sqrt{5}/10$
8138 8444 sq : $9(e+\pi)/10$
8138 9681 erf : $7\zeta(3)/9$
8139 0416 ℓ_2 : $3(e+\pi)/5$
8139 0603 Ψ : $\sqrt{15/2}$
8139 4463 rt : (1,7,-7,-5)
8139 4558 rt : (3,5,5,-9)
8139 6085 rt : (6,6,-5,7)
8139 6182 at : 18/17
8139 6788 cu : $\sqrt{2}/4+\sqrt{3}/2$
8139 6972 2^x : $2\sqrt{2}+2\sqrt{3}/3$
8139 7763 sinh : $\sqrt[3]{5/2}$
8139 8267 rt : (3,7,-4,-3)
8140 0465 K : $(\sqrt{2}\pi/2)^{-1/2}$
8140 0709 sr : P_{55}
8140 0883 rt : (8,5,-2,-6)
8140 1978 J_0 : $5\sqrt{2}/8$
8140 4257 ζ : $8\ln 2/3$
8140 4909 rt : (5,2,2,3)
8140 5302 at : $1/2+\sqrt{5}/4$
8140 6739 θ_3 : $2\sqrt{3}/9$
8140 6900 $s\pi$: $4\sqrt{2}-3\sqrt{5}/2$
8140 7663 cu : $1/\ln(e/3)$
8140 8593 $t\pi$: $7\sqrt{5}/3$
8141 0495 sinh : $3(e+\pi)/10$
8141 0739 rt : (5,8,0,-8)
8141 1567 ln : $\sqrt[3]{23/2}$
8141 7081 rt : (7,1,6,8)
8141 7627 rt : (1,7,2,5)
8141 7822 ℓ_2 : $3/2+4\pi$
8142 1356 $2/5+\sqrt{2}$
8142 2796 10^x : $2\sqrt{3}+2\sqrt{5}$
8142 3168 ℓ_2 : $(2e)^{1/3}$
8142 6850 $2/3\times\exp(1/5)$
8142 6973 e^x : $\sqrt{2}/4-\sqrt{5}/4$
8142 7065 e^x : $2\sqrt{3}/3-\sqrt{5}/4$
8142 8085 rt : (2,1,4,-5)
8142 9926 rt : (9,2,1,-7)
8143 0331 cr : $3/2+2\sqrt{5}$
8143 0827 cr : $3\sqrt[3]{2}/7$
8143 1335 rt : (7,1,-3,-2)
8143 3356 sq : $2\sqrt{2}/3-3\sqrt{5}/2$
8143 3994 as : 8/11
8143 5605 at : $\ln(\ln 2/2)$
8143 5726 Ei : 5/3
8143 6372 tanh : $(\sqrt{2}\pi/3)^{1/3}$
8143 8013 $t\pi$: $e/2-\pi$
8143 8575 rt : (6,5,3,-9)
8143 9372 sr : $3e/2-\pi/4$
8144 0116 10^x : $(e/2)^{1/3}$
8144 0869 tanh : $2\sqrt[3]{5}/3$
8144 1815 $s\pi$: $6\sqrt{2}/5$
8144 1878 as : $1/3-3\sqrt{2}/4$
8144 1924 rt : (5,0,-7,3)
8144 3936 ℓ_{10} : $4\sqrt{2}+\sqrt{3}/2$
8144 4573 sinh : $(e/3)^3$
8144 5313 rt : (9,4,-8,-1)
8144 6055 rt : (8,8,-2,-8)
8144 6761 rt : (2,6,-7,-3)
8144 7213 rt : (1,7,5,-8)
8144 8851 10^x : $2/3+\sqrt{5}/4$
8144 9238 as : $\exp(-1/\pi)$
8145 1025 as : P_{73}
8145 1186 rt : (4,9,-6,1)
8145 1504 θ_3 : $5\ln 2/9$
8145 1583 10^x : $3e+\pi/4$
8145 1954 rt : (8,4,9,9)
8145 2180 $s\pi$: $8(e+\pi)/7$
8145 2635 2^x : $10\sqrt{3}$
8145 2687 rt : (5,4,-7,4)
8145 4835 as : $(\sqrt[3]{4}/3)^{1/2}$
8145 5235 ℓ_{10} : $2e/3+3\pi/2$
8145 6009 rt : (6,7,-6,-3)
8145 6206 cr : $3\sqrt{2}+\sqrt{3}$
8145 8287 rt : (9,8,3,2)
8145 8476 sq : $2+4\sqrt{5}/3$
8145 8896 rt : (1,7,1,-6)
8146 3183 J_0 : 15/17
8146 3274 rt : (4,7,-8,-9)
8146 3308 sr : $4-\sqrt{2}/2$
8146 5558 ln : $2\sqrt{2}/3+3\sqrt{3}$
8146 6120 Ei : $7\ln 2/5$
8146 9657 rt : (9,5,1,-9)
8146 9726 cu : 25/16
8147 4911 cr : $3\sqrt[3]{3}/8$
8147 8018 rt : (6,5,5,4)
8147 8853 $t\pi$: $6\pi^2$
8147 9492 J_0 : $7\sqrt[3]{2}/10$
8148 0954 rt : (3,7,-3,3)
8148 1481 sq : $7\sqrt{3}/9$
8148 1882 rt : (8,2,0,3)
8148 2691 at : $3\sqrt{2}/4$
8148 4338 $\ln(4)\times s\pi(1/5)$
8148 7812 cr : $7e\pi/10$
8148 7897 ℓ_{10} : $4/3+3\sqrt{3}$
8149 0663 rt : (2,9,7,-7)
8149 2196 rt : (9,7,-8,-3)
8149 2536 rt : (7,4,-3,-4)
8149 2983 sq : $4/3-\sqrt{5}$
8149 3070 rt : (5,1,2,-5)
8149 3127 $s\pi$: $\sqrt{2}+\sqrt{3}-\sqrt{6}$
8149 4030 rt : (3,2,7,6)
8149 6921 Ψ : 20/23
8149 8605 rt : (6,8,-8,-5)
8149 9830 cu : $3e\pi/10$
8149 9887 cr : $(e/2)^{-2}$
8150 0231 at : $10(1/\pi)/3$
8150 0280 rt : (4,7,-1,-6)
8150 0906 $\exp(4)-\exp(5)$
8150 2848 rt : (2,4,4,-7)
8150 7290 rt : (1,9,-8,0)
8150 8409 ln : $3+\pi$
8151 0240 erf : 15/16
8151 1469 rt : (1,7,-6,-9)
8151 2497 ℓ_{10} : $1/4+2\pi$
8151 4224 10^x : $2\sqrt{3}/3+3\sqrt{5}/4$
8151 5701 cr : $4e/3+3\pi/4$
8151 6220 cr : 13/24
8151 6885 10^x : $1+4\pi/3$
8152 0566 Ei : $5e\pi/3$
8152 0710 ℓ_2 : $3\sqrt{3}-3\sqrt{5}/4$
8152 0746 rt : (1,6,3,-1)
8152 3661 sinh : $(\ln 3/3)^{-2}$
8152 4338 cr : $3+4\sqrt{5}/3$
8152 5405 $t\pi$: $e/3-4\pi$
8152 6922 rt : (7,7,-3,-6)
8152 8605 rt : (8,1,0,-5)
8152 9231 cr : $(2\pi)^{-1/3}$
8152 9717 sinh : $(2e)^{1/3}$
8153 0205 cu : $2/3+\sqrt{5}/3$
8153 1546 rt : (9,5,3,4)
8153 1690 rt : (4,2,-9,1)
8153 4523 rt : (5,4,2,-7)

8153 4766 id : $((e+\pi)/3)^2$
8153 5791 2^x : $(\pi/2)^{-1/3}$
8153 7394 tanh : $8/7$
8153 9338 rt : (6,2,5,6)
8153 9887 rt : (2,7,4,-9)
8154 1209 rt : (4,9,1,-3)
8154 6160 rt : (3,1,7,-8)
8154 7454 rt : (2,9,-2,5)
8154 8454 id : $3e/10$
8154 9324 rt : (7,8,8,5)
8155 4015 ζ : $(e/2)^2$
8155 4529 at : $3/4-2e/3$
8155 4964 rt : (8,4,0,-7)
8155 5950 tπ : $5\sqrt{3}$
8155 7109 J_0 : $22/25$
8155 8620 rt : (5,8,3,9)
8155 9654 cr : $7\sqrt[3]{5}/2$
8156 1164 rt : (8,0,-9,3)
8156 2623 sr : P_{81}
8156 4226 rt : (9,2,3,6)
8156 4280 rt : (7,8,-8,1)
8156 5224 cu : $4\sqrt{3}-2\sqrt{5}$
8156 6623 cu : $1/4+3\sqrt{5}/2$
8156 7846 cu : $e\pi/7$
8156 8370 rt : (6,1,5,-8)
8156 8944 sr : $3\sqrt{2}/4+\sqrt{5}$
8156 9192 at : $17/16$
8156 9228 e^x : $(3\pi/2)^{-1/3}$
8156 9329 10^x : $1+e/4$
8156 9780 2^x : $\ln(\sqrt{5}/3)$
8157 0652 sr : $1+4\sqrt{3}$
8157 0674 Γ : $5\sqrt{2}$
8157 1311 rt : (5,3,-7,1)
8157 1894 rt : (8,7,0,-9)
8157 3611 rt : (1,9,9,-2)
8157 4374 rt : (5,5,-3,8)
8157 6007 $\sqrt{3}+\ln(2/5)$
8157 6040 2^x : $\sqrt{23/3}$
8157 9362 rt : (2,6,-5,-1)
8158 2737 rt : (9,1,3,-8)
8158 4882 rt : (7,5,8,7)
8158 5480 rt : (5,2,-9,7)
8158 5644 Ei : $4/3$
8158 5907 rt : (1,9,3,-3)
8158 8615 $\sqrt{2}\times\exp(1/4)$
8158 9042 ℓ_2 : $1/3+3\sqrt{5}$
8159 3601 10^x : $2+3\pi/4$
8159 3765 $\ln(2/3)+\exp(1/5)$
8159 3985 2^x : $1/\ln((e+\pi)/3)$
8159 7676 rt : (8,3,-9,1)
8159 9040 rt : (8,8,-8,7)
8159 9377 2^x : $4\sqrt{2}/3+3\sqrt{5}/4$
8159 9543 tanh : $\sqrt[3]{3/2}$
8160 0066 tanh : $\ln(\pi)$
8160 1342 rt : (5,9,-4,-7)
8160 1829 rt : (7,2,8,9)
8160 4523 rt : (2,9,-5,-3)
8160 4985 2^x : $3\sqrt{2}/4+\sqrt{3}/4$
8160 5426 rt : (3,5,-9,2)
8160 6258 sπ : $1/2+3\sqrt{3}$
8160 7510 2^x : $9\sqrt[3]{2}/5$
8160 7562 10^x : $1/\sqrt[3]{11}$
8160 8960 tπ : $e/8$
8161 1521 Ei : $\sqrt{2}\pi$
8161 2852 rt : (8,6,-9,-1)
8161 2947 cr : $1/3+4\sqrt{2}$
8161 4726 rt : (5,9,-7,-3)
8161 5909 cr : $e/5$
8161 8292 $\exp(\pi)+s\pi(\sqrt{5})$
8162 1167 rt : (8,9,-9,-3)
8162 2733 e^x : $1/4+\pi/4$
8162 5239 Ψ : $7e\pi/9$
8162 5938 sπ : $10\pi^2$
8162 9358 rt : (4,8,-6,9)
8163 0109 2^x : $\sqrt[3]{10/3}$
8163 1305 sinh : $\sqrt{5}/3$
8163 2544 sr : $2+3\sqrt{3}/4$
8163 3440 $\exp(2/5)-s\pi(\sqrt{5})$
8163 6852 Ei : $8/13$
8163 7061 id : $1/4+4\pi$
8163 7665 ℓ_{10} : $3\sqrt{2}+4\sqrt{3}/3$
8163 7990 2^x : $4-\sqrt{3}$
8163 8405 cr : $1/\ln(2\pi)$
8163 8906 e^x : $\sqrt{3}+2\sqrt{5}$
8164 1062 Ei : $\sqrt{7}$
8164 2753 rt : (2,5,-9,4)
8164 5978 2^x : $10\zeta(3)$
8164 9188 tπ : $\sqrt{2}+\sqrt{6}-\sqrt{7}$
8164 9658 id : $\sqrt{2/3}$
8165 4387 sr : $7\sqrt{2}/3$
8165 7721 ln : $8\sqrt{2}/5$
8165 9149 sr : $2+4\pi$
8165 9557 $\ln(4)-s\pi(\pi)$
8166 2764 e^x : $2/3+2\pi/3$
8166 4321 J_0 : $(\sqrt{2}\pi/3)^{-1/3}$
8166 5784 ζ : $\sqrt[3]{2}/5$
8166 8611 rt : (7,9,-4,5)
8166 8788 rt : (1,7,-6,3)
8167 1963 rt : (7,9,-1,-9)
8167 4156 10^x : $1/4+3\sqrt{3}/2$
8167 5728 rt : (4,9,1,-9)
8167 6754 rt : (7,6,-1,-7)
8167 7734 rt : (2,8,-6,9)
8168 1023 rt : (1,9,3,-9)
8168 1186 cu : $\exp(\sqrt[3]{4}/2)$
8168 1906 erf : $16/17$
8168 2520 rt : (4,6,1,6)
8168 3518 tπ : $(e/3)^{-2}$
8168 4167 rt : (7,3,-1,-5)
8168 5922 sr : P_{26}
8168 9019 rt : (2,9,-5,-9)
8168 9411 $e+\ln(3)$
8168 9563 e^x : $e/4+3\pi/4$
8169 1405 rt : (1,6,3,-7)
8169 4097 rt : (4,9,7,7)
8169 6199 Ei : $9e/10$
8169 6989 sπ : $7/23$
8169 7160 rt : (7,0,-1,3)
8169 9684 rt : (2,9,6,3)
8170 1447 $\sqrt{2}\times\exp(\sqrt{2})$
8170 1755 ζ : $24/13$
8170 2493 rt : (5,9,-7,-9)
8170 3598 rt : (8,3,2,-8)
8170 4472 at : $2/3-\sqrt{3}$
8170 5794 cr : $6/11$
8170 5949 $\ln(2/3)\div\ln(4/5)$
8170 6249 cu : $3\sqrt{2}/2+3\sqrt{3}$
8170 6352 tπ : $\sqrt{2}-3\sqrt{3}$
8170 6880 rt : (2,6,-5,-7)
8170 6895 ζ : $(\sqrt{5}/2)^{-1/3}$
8170 8192 $s\pi(1/3)+s\pi(2/5)$
8170 8226 10^x : $(\pi)^{-1/3}$
8170 8294 $\pi+s\pi(\sqrt{5})$
8170 9132 10^x : $13/17$
8171 2059 id : $\sqrt[3]{6}$
8171 3286 sr : $2\sqrt{3}+2\sqrt{5}$
8171 5015 rt : (5,3,4,-8)
8171 6059 e^x : $3/4+2\sqrt{2}$
8171 6567 rt : (3,9,-7,9)
8171 8264 rt : (8,0,2,6)
8171 9605 sr : $5\zeta(3)/9$
8172 1690 tanh : $(\pi/3)^3$
8172 1932 cr : $2e/3+4\pi/3$
8172 2505 cu : $7\zeta(3)/9$
8172 2681 2^x : P_5
8172 5343 sπ : $(2e/3)^{-2}$
8172 5833 rt : (7,3,-1,1)
8172 6911 cr : $10\sqrt{5}$
8172 7138 rt : (1,8,-8,-7)
8173 0028 rt : (8,9,-9,-9)
8173 0194 rt : (9,0,5,9)
8173 1134 10^x : $\sqrt{2}/3-\sqrt{5}/4$
8173 2569 rt : (2,3,6,-8)
8173 4058 10^x : $(\sqrt[3]{5})^{-1/2}$
8173 7057 2^x : $23/4$
8173 7864 rt : (5,0,4,6)
8174 0632 rt : (5,6,-7,-7)
8174 2385 erf : $3\pi/10$
8174 2653 e^x : $2/3+3\pi/2$
8174 3877 rt : (8,3,2,4)
8174 4616 id : $7\pi^2/5$
8174 5044 sq : $2/3-\pi/2$
8174 5498 sπ : $\sqrt{3}/3+\sqrt{5}/2$
8174 6655 2^x : $4/3-\sqrt{2}/3$
8174 7150 rt : (6,0,7,9)
8174 7197 e^x : $9\pi^2/10$
8174 7614 λ : $(\sqrt{2}/3)^3$
8175 1959 rt : (9,3,5,7)
8175 3791 Γ : $5/21$
8175 4008 Li_2 : $\exp(-\sqrt[3]{2}/3)$
8175 4443 rt : (2,9,7,-5)
8175 4825 rt : (7,7,-6,8)
8175 6036 J_0 : $7/8$
8175 6543 ℓ_2 : $5\pi^2/7$
8175 6597 ln : $1/2+4\sqrt{2}$
8175 7756 erf : $2\sqrt{2}/3$
8175 8773 $\sqrt{3}+\exp(3)$
8176 0926 sπ : $\sqrt{6}\pi$
8176 2330 sinh : $5\pi/6$
8176 3042 rt : (1,3,-8,-8)
8176 4504 at : $16/15$
8176 5500 id : $\sqrt{2}/4+2\sqrt{3}$
8176 6156 10^x : $\sqrt{2}-2\sqrt{3}/3$
8176 7489 10^x : $2+4e/3$
8176 7844 ln : $5e/6$
8176 8185 ℓ_2 : $\exp(\sqrt[3]{2})$
8176 9206 cu : $e/2+3\pi/2$
8176 9947 rt : (6,7,-6,2)
8177 3167 rt : (3,0,9,9)
8177 3908 rt : (9,8,-4,-7)
8177 4076 rt : (8,8,-7,-4)
8177 5407 tanh : $23/20$
8177 6543 sr : $(\sqrt{5})^{-1/2}$
8177 9375 sπ : $(\sqrt{2}\pi/2)^{1/3}$
8177 9500 at : $3/2-\sqrt{3}/4$
8178 0977 rt : (6,3,7,7)
8178 3712 sr : $\sqrt{2}-\sqrt{5}/3$
8178 5262 rt : (5,3,4,4)
8178 5671 e^x : $2\sqrt{3}+2\sqrt{5}$
8178 6159 ℓ_{10} : P_{68}
8178 7121 rt : (4,1,-7,8)
8178 7274 ℓ_{10} : P_{71}
8178 8505 tπ : $1/\sqrt{21}$
8178 9833 rt : (9,6,5,5)
8179 1443 e^x : $4\sqrt{3}/3+4\sqrt{5}$
8179 3474 as : P_{23}
8179 3745 ζ : $2\sqrt[3]{3}/5$
8179 5449 rt : (6,8,-2,-7)
8179 7051 sq : $\sqrt{2}/3+\sqrt{3}/4$
8179 8310 sinh : $e/2$
8179 8997 erf : $1/\ln(\ln 2/2)$
8179 9958 rt : (8,6,2,2)
8180 0395 ℓ_2 : $\exp((e+\pi)/3)$
8180 1299 rt : (9,5,-4,-5)
8180 1604 rt : (5,8,-5,-4)
8180 1622 sr : $\sqrt{2}/2+3\sqrt{3}/2$
8180 1916 at : $e\pi/8$
8180 2898 $\exp(\sqrt{3})\times s\pi(\sqrt{5})$
8180 3162 rt : (2,7,-4,-6)
8180 9468 rt : (8,5,-7,-2)
8181 2347 J_0 : $\sqrt[3]{2/3}$
8181 2456 rt : (4,8,-8,-1)
8181 2820 J_0 : $1/\ln(\pi)$
8181 5067 rt : (4,4,-1,9)
8181 5075 10^x : $\sqrt{3}/4+\sqrt{5}/4$
8181 5823 cr : $5\zeta(3)$
8181 6286 $\sqrt{3}\div\exp(3/4)$
8181 7615 2^x : $e-\pi/4$
8181 8181 id : $9/11$
8181 9644 ζ : $(2\pi)^{1/3}$
8181 9662 ln : $4e-3\pi/2$
8182 0358 id : $2\sqrt{3}+3\sqrt{5}/2$
8182 0848 ℓ_{10} : $2\pi^2/3$
8182 1414 rt : (7,5,1,-8)
8182 2034 tπ : $1/2+e$
8182 2973 cu : $1/2+3\sqrt{3}$
8182 3950 sr : $\ln((e+\pi)/3)$
8182 6042 rt : (3,8,0,-7)
8182 7458 id : $9\sqrt{2}/7$
8182 9467 cr : $4/3-\pi/4$

8183 3035 rt : (3,3,9,7)
8183 3503 erf : 17/18
8183 4519 $s\pi : 9\sqrt[3]{5}/2$
8183 4662 rt : (6,5,-2,-5)
8183 8293 $10^x : 9/20$
8184 1354 rt : (7,6,-1,-1)
8184 1676 $\Psi : 7\pi/8$
8184 2059 cr : $\sqrt{2}-\sqrt{3}/2$
8184 4158 rt : (9,2,-4,-3)
8184 4803 rt : (2,8,-3,-4)
8184 4959 $\ell_2 : P_{80}$
8184 7146 rt : (8,2,4,-9)
8184 8014 $s\pi : \sqrt{2}/2+3\sqrt{3}/2$
8184 8081 rt : (1,5,-8,-8)
8184 8306 sq : $1/4-2\sqrt{3}/3$
8184 9306 $\ell_{10} : 3e-\pi/2$
8184 9952 at : $8\zeta(3)/9$
8185 0726 rt : (6,6,7,5)
8185 1212 id : $9\zeta(3)$
8185 1750 $3/4\div\ln(2/5)$
8185 1767 $2^x : 3/2+\sqrt{3}/4$
8185 1992 rt : (6,5,-8,5)
8185 2834 $e^x : 3\sqrt{2}+2\sqrt{3}/3$
8185 5805 rt : (1,0,-8,6)
8185 6447 $erf : 3\sqrt[3]{2}/4$
8185 7270 rt : (4,5,3,-8)
8185 7933 rt : (5,5,-5,-2)
8185 8291 cu : $3+e/4$
8185 9410 sq : 19/21
8185 9890 $2^x : 7\sqrt{3}/2$
8186 0643 $2^x : 3\sqrt{3}/2+4\sqrt{5}/3$
8186 1154 at : $5\sqrt[3]{5}/8$
8186 1167 $\ell_2 : 1-\sqrt{3}/4$
8186 1174 $Ei : 8\ln 2/9$
8186 1550 cr : P_{12}
8186 2421 $\ln(2/3)-\exp(5)$
8186 2426 rt : (9,9,-9,8)
8186 3129 tanh : $1/\ln(\sqrt[3]{2}/3)$
8186 4331 $2^x : 3/4+4\sqrt{3}$
8186 5122 sinh : $9(e+\pi)/4$
8186 7340 rt : (7,2,1,-6)
8186 8909 id : $4\sqrt{3}/3-2\sqrt{5}/3$
8186 9597 $\Psi : -\sqrt{3}+\sqrt{6}-\sqrt{7}$
8187 1796 $10^x : 8(e+\pi)/5$
8187 2141 rt : (9,9,5,3)
8187 2930 rt : (8,6,-9,-7)
8187 3075 $1\div\exp(1/5)$
8187 4894 ln : $1-\sqrt{5}/4$
8187 5029 $2^x : \sqrt{2}/2+4\sqrt{3}$
8187 6016 $\ell_2 : 7\sqrt[3]{2}/5$
8187 8296 cu : $1/3+3\sqrt{2}$
8187 9496 $\ell_2 : 4-\sqrt{5}$
8188 0158 $10^x : 5(e+\pi)/6$
8188 3572 ln : $9\sqrt[3]{2}/5$
8188 4167 $\ln(3)\div\ln(3/4)$
8188 4187 rt : (1,8,-6,-1)
8188 5494 cu : $4/3-4\sqrt{2}$
8188 6891 rt : (6,8,-2,6)
8188 7381 sr : $1/\ln(\sqrt{2}\pi)$
8188 7598 ln : $4-\sqrt{3}$
8188 7894 $t\pi : (2\pi/3)^{-1/3}$
8188 8165 rt : (5,2,6,-9)
8189 0102 Ψ : 11/4
8189 1712 rt : (2,3,6,4)
8189 3419 $erf : (\sqrt{5}/2)^{-1/2}$
8189 4537 rt : (9,7,-2,-8)
8189 7658 id : $(\ln 3/2)^{1/3}$
8189 9647 rt : (8,1,4,7)
8190 0891 tanh : $4\sqrt[3]{3}/5$
8190 1542 λ : 2/19
8190 2014 rt : (8,9,-9,2)
8190 2422 tanh : 15/13
8190 2684 rt : (6,2,-2,-3)
8190 3625 sr : $3\sqrt{5}/10$
8190 4083 $\ell_{10} : 9(e+\pi)/8$
8190 4813 $e^x : 3\sqrt{3}+\sqrt{5}/4$
8190 7664 sq : $2/3+3e$
8190 7992 cu : $\sqrt{19}$
8190 9404 $\ell_2 : 4-\sqrt{2}/3$
8190 9489 rt : (4,5,-8,1)
8191 0821 sr : $(\sqrt{2}\pi/2)^{-1/2}$
8191 1191 rt : (5,5,-8,-1)
8191 1313 $\ell_{10} : 3\sqrt{2}/2+2\sqrt{5}$
8191 1830 $2^x : 3\sqrt{2}-4\sqrt{3}/3$
8191 2114 rt : (1,5,5,-8)
8191 2520 sq : $8\sqrt[3]{5}/7$
8191 3747 $s\pi : 2\sqrt{2}+\sqrt{3}/2$
8191 7083 $t\pi : 4\sqrt{3}/3+2\sqrt{5}$
8191 7593 sr : $1+4\sqrt{3}/3$
8192 0623 rt : (4,9,-5,4)
8192 0691 rt : (9,1,-4,1)
8192 1475 rt : (8,7,-5,-5)
8192 1729 sq : $7\sqrt[3]{5}/2$
8192 2443 rt : (1,8,1,-4)
8192 2770 $J_0 : (\pi/3)^{-3}$
8192 3759 rt : (1,6,-5,4)
8192 4522 $2^x : (\sqrt{5}/3)^{1/2}$
8192 5594 $e^x : 2/3-\sqrt{3}/2$
8192 5920 rt : (3,8,-2,-2)
8192 8369 rt : (4,2,3,-6)
8193 0529 tanh : $2\sqrt{3}/3$
8193 1591 rt : (9,4,7,8)
8193 2127 cr : 11/20
8193 2301 $2^x : (\sqrt{5})^{1/2}$
8193 3711 $2^x : 3/4+\sqrt{5}/3$
8193 5731 $\exp(2)-s\pi(\pi)$
8193 6324 rt : (6,7,0,-8)
8193 8330 $\ln(3)\times s\pi(\sqrt{3})$
8194 0838 $10^x : 3\ln 2/8$
8194 1618 rt : (1,8,-7,2)
8194 1732 $\ell_{10} : 4+3\sqrt{3}/2$
8194 2485 rt : (5,6,4,2)
8194 4734 id : $7\sqrt[3]{2}$
8194 5614 ln : $5\pi^2/8$
8194 6282 $s\pi : 3/4+4\sqrt{5}$
8194 6733 rt : (7,1,1,4)
8194 6847 $\ln(\sqrt{3})\times\exp(2/5)$
8194 7526 rt : (9,4,-2,-6)
8194 8073 cr : $1/\sqrt[3]{6}$
8194 8429 tanh : $5\ln 2/3$
8194 8685 rt : (2,7,1,-8)
8194 9635 rt : (6,6,-4,9)
8195 0780 rt : (2,2,8,-9)
8195 1266 $5\times s\pi(\sqrt{2})$
8195 1994 ln : $2\sqrt{2}-\sqrt{5}/4$
8195 3957 rt : (3,8,-4,-7)
8195 5279 $\exp(1/5)+\exp(4)$
8195 7282 at : $1/2-\pi/2$
8195 7936 cu : $3e/2-\pi$
8195 9197 $3\div\exp(1/2)$
8195 9861 sinh : $7e\pi/4$
8196 1237 rt : (7,4,3,-9)
8196 1893 $10^x : 1/4+\sqrt{3}/4$
8196 1894 2^x : 19/22
8196 3755 rt : (7,7,-8,-2)
8196 6543 as : $4\sqrt{2}/3-2\sqrt{3}/3$
8196 6843 $Ei : \zeta(3)/6$
8196 7694 rt : (2,6,-6,1)
8196 7867 $\sqrt{2}-\ln(2/3)$
8196 7886 $\zeta : \gamma$
8196 8353 erf : 18/19
8196 8796 $2^x : 7\sqrt{5}$
8196 9257 rt : (5,1,6,7)
8196 9861 rt : (7,9,-9,-4)
8197 1228 J_0 : 20/23
8197 9276 as : $(\sqrt[3]{5}/2)^2$
8198 0390 rt : (5,8,-2,0)
8198 1053 $\ell_2 : \sqrt[3]{11/2}$
8198 4739 $\Psi : 7\pi^2/4$
8198 6726 at : 15/14
8198 8697 id : $2\pi^2/7$
8199 0100 rt : (8,4,4,5)
8199 0231 rt : (3,6,-8,-3)
8199 0880 rt : (8,9,-3,-8)
8199 3580 $e^x : 2/3+e/3$
8199 4172 sr : $3/2+2e/3$
8199 5371 rt : (3,7,2,-8)
8199 6926 rt : (8,4,-5,-3)
8199 7609 $2^x : \sqrt{22}$
8199 9095 id : $3\sqrt{2}+\sqrt{3}/3$
8199 9342 ln : $3e/2+2\pi/3$
8200 2724 $\ell_2 : \sqrt{2}/4+3\sqrt{5}$
8200 5679 rt : (3,6,9,5)
8200 6591 rt : (2,5,-7,-8)
8200 7582 $\ln(2/3)+\exp(4/5)$
8200 7955 $e\div s\pi(\sqrt{2})$
8200 8937 rt : (1,5,-1,-9)
8200 9570 rt : (6,4,9,8)
8201 1743 rt : (9,6,0,-9)
8201 2023 sq : $3-2\pi/3$
8201 5904 $s\pi : \sqrt{19}\pi$
8201 9410 rt : (8,9,2,0)
8202 2018 $\ln(\pi)+s\pi(\sqrt{5})$
8202 2359 id : $(2e/3)^{-1/3}$
8202 2550 $\ell_{10} : \sqrt{2}+3\sqrt{3}$
8202 3480 $\ln(\pi)\div\exp(1/3)$
8202 4560 rt : (8,8,-7,9)
8202 4915 Ψ : 20/3
8202 5101 rt : (4,4,5,-9)
8202 5678 $e^x : 1/2+3\sqrt{5}/4$
8202 6835 cu : $3/4+2e/3$
8202 9705 rt : (9,1,-2,-4)
8203 1132 $\Gamma : 3(1/\pi)/5$
8203 1810 $\Psi : 5(1/\pi)/3$
8203 2040 $\exp(1/3)\times s\pi(1/5)$
8203 2807 rt : (9,7,7,6)
8203 5196 $t\pi : \exp(\sqrt{3}/3)$
8203 5260 tanh : 22/19
8203 5715 $\exp(\sqrt{2})+s\pi(1/4)$
8203 6826 $erf : e\pi/9$
8203 7208 rt : (1,2,5,-6)
8203 8548 rt : (7,9,-6,-5)
8204 2182 $\zeta : \sqrt{3}/3$
8204 2753 rt : (7,1,3,-7)
8204 2954 id : $3/2-e/4$
8204 5558 $s\pi : 3\sqrt{3}/2+3\sqrt{5}$
8204 5566 $\ell_{10} : \sqrt{3}+\sqrt{5}+\sqrt{7}$
8204 7509 $e^x : \exp(\sqrt[3]{3}/2)$
8204 7845 id : $1/\ln(\sqrt{3})$
8204 8253 tanh : $(\sqrt{5}/3)^{-1/2}$
8204 8456 rt : (5,9,-1,-8)
8204 8664 $erf : (\sqrt[3]{5}/2)^{1/3}$
8204 9492 rt : (6,1,-2,1)
8205 0226 $J_0 : \ln(\sqrt[3]{2}/3)$
8205 1512 $2^x : 1/2-\pi/4$
8205 3609 $t\pi : 3e\pi/7$
8205 3830 $e^x : \zeta(5)$
8205 6997 rt : (4,7,-6,-2)
8205 7596 rt : (3,9,4,7)
8205 8302 $Ei : 5\pi/8$
8206 2954 $\ell_{10} : 1/3+2\pi$
8206 3738 rt : (8,6,-3,-6)
8206 4749 sr : $1/3+4\sqrt{5}/3$
8206 9500 rt : (2,8,4,-1)
8207 2051 rt : (2,7,-1,-5)
8207 2592 rt : (6,6,2,-9)
8207 3426 rt : (8,2,6,8)
8207 6800 $\ell_2 : 4/3+\sqrt{3}/4$
8207 8018 rt : (6,9,7,3)
8207 9632 id : $1/4+\pi/2$
8208 5238 rt : (7,4,-8,0)
8208 5273 Ψ : 13/15
8208 5556 J_0 : 13/15
8208 5633 sinh : $(\sqrt[3]{2}/3)^{1/3}$
8208 5874 $\ln(3/5)\times\exp(\pi)$
8208 7174 rt : (2,5,-8,8)
8208 8469 $erf : 5\sqrt[3]{5}/9$
8208 8703 rt : (9,3,0,-7)
8208 9080 erf : 19/20
8208 9091 $Ei : 7e/2$
8209 0262 $10^x : 2e/3-\pi/3$
8209 1809 rt : (5,4,-6,6)
8209 1906 $t\pi : 3e+3\pi/4$
8209 2890 $1/3-\exp(e)$
8209 2940 rt : (2,1,8,7)
8209 4721 $J_0 : 5\ln 2/4$
8209 5692 sq : $1+e/4$
8209 5741 rt : (9,4,-4,-1)
8209 7962 rt : (3,7,-2,6)
8209 8055 ln : 11/25

8209 8374 ln : $2\sqrt{3}/3+\sqrt{5}/2$
8209 9252 rt : (5,4,-3,-3)
8209 9704 θ_3 : $1/\ln(\pi)$
8210 0623 id : $(e/3)^2$
8210 1285 10^x : $1/4+4\sqrt{5}/3$
8210 2017 cr : $4+3e/4$
8210 3681 rt : (4,8,-1,-4)
8210 3770 rt : (9,5,9,9)
8210 4358 2^x : $7\sqrt[3]{5}/8$
8210 5547 rt : (8,8,-1,-9)
8210 9910 sπ : $\sqrt[3]{1/3}$
8211 0808 J_0 : $\sqrt{3}/2$
8211 1431 rt : (3,5,3,7)
8211 2867 rt : (6,9,-9,-2)
8211 4199 sr : $4\sqrt{2}+4\sqrt{5}$
8211 4496 ℓ_{10} : $3+4e/3$
8211 4558 cr : $3\sqrt{3}/4-\sqrt{5}/3$
8211 5025 Ψ : $3\sqrt{2}/8$
8211 6028 sr : $\sqrt{11}$
8211 6311 id : $e/4+\pi$
8211 6886 rt : (7,4,1,2)
8211 7171 rt : (7,8,-7,3)
8211 8599 $\sqrt{3}\div s\pi(2/5)$
8212 1622 rt : (4,9,-4,-5)
8212 1846 rt : (5,2,-8,9)
8212 2116 rt : (1,4,7,-9)
8212 2423 ζ : $4(1/\pi)/5$
8212 4076 θ_3 : $\sqrt[3]{2}/3$
8212 5965 2^x : $2\sqrt{2}-\sqrt{5}/4$
8212 6269 rt : (3,5,-8,5)
8212 8536 rt : (6,8,-6,-7)
8212 9358 $\ln(4/5)-\exp(4)$
8212 9534 rt : (2,9,1,-8)
8213 0079 rt : (8,1,-5,-1)
8213 2634 Ψ : $5\ln 2/4$
8213 2746 at : $(\sqrt{3}/2)^{-1/2}$
8213 3063 ζ : $(5/3)^{-1/3}$
8213 3419 λ : $\exp(-5\pi/7)$
8213 3793 rt : (5,4,6,5)
8213 6720 id : $1/3-2\sqrt{3}/3$
8213 7398 J_0 : $3\sqrt[3]{3}/5$
8214 0275 $2/5-\exp(1/5)$
8214 0392 rt : (7,6,-9,6)
8214 0874 rt : (6,1,0,-4)
8214 2113 ℓ_2 : 9π
8214 4626 rt : (7,6,-6,-3)
8214 6676 sπ : $2e\pi/3$
8214 8283 sπ : $\ln 2$
8215 0274 rt : (4,1,5,-7)
8215 1215 id : $1/3-3e$
8215 1737 rt : (5,9,2,4)
8215 1820 at : $\ln((e+\pi)/2)$
8215 4100 λ : $(1/\pi)/3$
8215 5847 cr : $4\ln 2/5$
8215 6825 rt : (1,5,-6,1)
8215 7419 Ei : $(\sqrt[3]{2}/3)^{-1/3}$
8215 8169 rt : (3,6,4,-9)
8216 3527 cr : $(e+\pi)^{-1/3}$
8216 5958 sr : $3e/4+4\pi$
8217 0129 cu : $e/4-3\pi$
8217 0331 10^x : $\sqrt[3]{19/3}$
8217 2065 Γ : $4e/3$
8217 2384 rt : (9,8,-9,-3)
8217 2463 cr : $4/3+3\pi/2$
8217 2710 e^x : $\sqrt{2}/2+\sqrt{3}/2$
8217 2742 ln : $2\sqrt{2}+3\sqrt{5}/2$
8217 2820 id : $(\sqrt{2}\pi/3)^{-1/2}$
8217 3828 tπ : $1/3+4\sqrt{2}/3$
8217 4191 rt : (9,2,-2,2)
8217 5350 ℓ_2 : $3\sqrt{3}+4\sqrt{5}$
8217 5400 erf : $(e/3)^{1/2}$
8217 6023 rt : (8,3,-3,-4)
8217 6058 Ei : $1/5$
8217 6146 sπ : $\sqrt{3}/2-\sqrt{5}/4$
8217 7040 rt : (7,8,-4,-6)
8217 7693 cu : $9\pi^2/5$
8217 7888 ln : $4\sqrt{3}-\sqrt{5}/3$
8217 8125 2^x : $3\sqrt[3]{3}/5$
8217 9676 sq : $4-\sqrt{3}/2$
8218 0591 rt : (1,5,1,-2)
8218 1087 sinh : $\sqrt{23}\pi$
8218 1331 rt : (5,8,1,-9)
8218 1533 rt : (6,3,2,-7)
8218 1884 rt : (7,2,3,5)
8218 2656 rt : (8,7,4,3)
8218 2751 tπ : $1/4+\pi$
8218 3585 tanh : $(\pi/2)^{1/3}$
8218 5221 as : $\ln(\sqrt[3]{3}/3)$
8218 5441 id : $\sqrt{2}-\sqrt{5}$
8218 8594 rt : (5,2,8,8)
8218 8982 rt : (1,7,6,-5)
8219 2672 ζ : $5\ln 2/6$
8219 2809 ℓ_2 : $5\sqrt{2}$
8219 2899 rt : (6,7,9,6)
8219 3845 sq : $3\sqrt{2}/4+4\sqrt{3}$
8219 6401 as : $(e+\pi)/8$
8219 7014 rt : (3,5,-4,-5)
8219 7791 erf : $20/21$
8219 8849 Γ : $\exp(-\sqrt[3]{3}/2)$
8219 9112 cu : $1/2+4\pi$
8220 1203 rt : (8,5,-1,-7)
8220 4631 sq : $4\sqrt[3]{2}/3$
8220 4752 J_0 : $19/22$
8220 6794 sπ : $2\sqrt{3}/5$
8220 7069 cr : $5/9$
8220 7506 tπ : $3e/2+\pi$
8220 8812 sπ : $3/4+2\sqrt{2}/3$
8220 9051 rt : (5,5,-9,-8)
8221 0115 $1/3\div\ln(2/3)$
8221 1016 rt : (7,0,5,8)
8221 1515 rt : (7,9,-3,7)
8221 1555 sπ : $6\pi/7$
8221 1880 e^x : $3/5$
8221 2668 cu : $3/4+\sqrt{2}/3$
8221 4212 $\ln(3/5)\times\ln(5)$
8221 5743 cu : $5\sqrt[3]{5}/7$
8221 5840 Ψ : $\sqrt{3}/2$
8221 6390 J_0 : $(\sqrt{5}/3)^{1/2}$
8222 0458 sr : $4-e/4$
8222 0601 2^x : $4+\sqrt{5}/3$
8222 2286 ℓ_2 : $\ln(e+\pi)$
8222 2300 rt : (8,5,6,6)
8222 2988 rt : (3,7,-9,1)
8222 4143 cu : $1-\pi$
8222 5158 rt : (9,2,2,-8)
8222 6277 cu : $3/4+2\sqrt{2}$
8222 8223 id : $3\sqrt{2}/2-3\sqrt{3}/4$
8222 9354 rt : (1,7,-4,-2)
8222 9988 ℓ_{10} : $7e\pi/9$
8223 1673 sinh : $3/4$
8223 2460 ℓ_2 : $\exp(\sqrt[3]{5}/3)$
8223 3194 2^x : $2e/3-2\pi/3$
8223 4359 erf : $(\sqrt{3}/2)^{1/3}$
8223 4574 rt : (1,7,-1,-5)
8223 6097 Ei : $10\zeta(3)/9$
8223 7978 ln : $3/4+2e$
8223 8932 rt : (3,7,-6,-8)
8224 0194 rt : (8,3,8,9)
8224 1387 rt : (1,9,1,-5)
8224 1827 at : $14/13$
8224 5554 sr : $4\sqrt{2}+4\sqrt{3}/3$
8224 7585 rt : (3,9,-7,-2)
8224 8206 ln : $4/3+2\sqrt{2}/3$
8224 9667 as : $3\sqrt[3]{5}/7$
8225 0117 rt : (1,9,-2,-5)
8225 2346 rt : (1,9,-3,-8)
8225 3359 θ_3 : $e/7$
8225 8584 id : $4\sqrt{2}/3-3\sqrt{5}$
8225 8888 Ei : $1/12$
8226 0343 $\sqrt{5}\div e$
8226 1603 at : $1/2+\sqrt{3}/3$
8226 2138 rt : (9,8,9,7)
8226 3465 2^x : $\sqrt{3}/2$
8226 4957 at : $3-3e/2$
8226 5007 $1/5\times\exp(\sqrt{2})$
8226 5538 cu : $1/2+3\sqrt{3}/4$
8226 5553 tanh : $3e/7$
8226 8651 sπ : $1/3+e/2$
8227 1894 Γ : $4(e+\pi)/7$
8227 2320 2^x : $4\sqrt{3}+4\sqrt{5}/3$
8227 3563 cu : $3\sqrt{3}/4-\sqrt{5}$
8227 3962 ln : $2+4\pi/3$
8227 4366 erf : $(\ln 3)^{-1/2}$
8227 4603 Ψ : $10\sqrt{3}$
8227 6438 cu : $2/3-3\pi/4$
8227 9849 Ψ : $\sqrt[3]{21}$
8228 0336 cr : $7(1/\pi)/4$
8228 1786 sπ : $8\sqrt[3]{3}/5$
8228 3147 ℓ_{10} : $(1/(3e))^2$
8228 7565 rt : (6,8,-7,-3)
8228 7655 cr : $3/2-2\sqrt{2}/3$
8229 2707 sπ : $(\sqrt[3]{5})^{1/2}$
8229 2818 tanh : $(e/2)^{1/2}$
8229 4071 tπ : $(2e/3)^{1/3}$
8229 4825 rt : (1,9,-9,-9)
8229 5156 id : $3\sqrt{2}/2-4\sqrt{5}$
8229 6192 erf : $21/22$
8229 8386 sπ : $4/13$
8230 0945 K : $e/4$
8230 1791 tπ : $\sqrt{2}/3+4\sqrt{3}/3$
8230 5015 rt : (1,6,0,-6)
8230 7340 rt : (9,9,-5,-7)
8230 8546 sq : $3-3\sqrt{3}$
8231 0967 rt : (7,7,-2,-7)
8231 1927 rt : (9,7,-7,-4)
8231 2040 tπ : $2\sqrt{2}/3-2\sqrt{5}$
8231 2223 ℓ_2 : $13/23$
8231 3616 erf : $3(1/\pi)$
8231 4355 $3/5-\ln(4/5)$
8231 5089 2^x : $5\ln 2/4$
8231 5616 rt : (1,8,1,-3)
8231 5627 tanh : $(\sqrt[3]{4})^{1/3}$
8231 6223 rt : (5,5,1,-7)
8231 6496 ℓ_{10} : $\zeta(3)/8$
8231 7489 sπ : $\exp(-1/e)$
8231 7676 rt : (7,5,-4,-4)
8231 9283 rt : (9,5,-9,-1)
8232 0064 tanh : $7/6$
8232 0274 cr : $3e-2\pi/3$
8232 0299 10^x : $4\sqrt{3}/3+4\sqrt{5}$
8232 0474 $\exp(4/5)\times\exp(\sqrt{5})$
8232 1197 as : $11/15$
8232 2300 cu : $3/4-4\sqrt{3}$
8232 2409 ℓ_2 : $3+3e/2$
8232 3233 sq : $\pi^2/3$
8232 3535 at : $1/3+\sqrt{5}/3$
8232 4524 rt : (3,3,4,-7)
8232 5128 $\ln(\pi)\div\ln(2/3)$
8232 5596 e^x : $(\sqrt{5}/2)^{1/3}$
8232 6904 ℓ_{10} : $1+4\sqrt{2}$
8232 6975 rt : (5,3,-1,-4)
8232 9148 sπ : $4\sqrt{3}-\sqrt{5}$
8232 9774 rt : (7,3,-6,-1)
8233 0310 sπ : $(\sqrt{2}\pi/3)^{-3}$
8233 4174 rt : (8,4,1,-8)
8233 4800 10^x : $6/23$
8233 7397 J_0 : $(\pi/2)^{-1/3}$
8233 7595 cr : $7\sqrt{3}/2$
8233 9142 cu : $3\sqrt{3}-\sqrt{5}/2$
8233 9303 10^x : $3\zeta(3)/8$
8233 9601 rt : (1,1,7,-7)
8233 9874 e^x : $2\pi^2/5$
8234 0050 tπ : $2\sqrt[3]{5}/7$
8234 4497 2^x : $13/15$
8234 4590 rt : (6,2,4,-8)
8234 6616 K : $17/25$
8234 7460 rt : (8,2,-1,-5)
8235 0633 rt : (5,1,-3,-1)
8235 2941 id : $14/17$
8235 3569 Ψ : $3\sqrt[3]{3}/5$
8235 7188 rt : (9,1,4,-9)
8235 7636 e^x : $2e/3+4\pi/3$
8235 8100 rt : (7,1,-8,2)
8235 9730 sq : $4e/3+\pi/3$
8236 1007 rt : (4,0,7,8)
8236 3910 $s\pi(1/3)\times s\pi(2/5)$
8236 4402 rt : (4,9,-4,7)
8236 4915 sπ : $4\ln 2/9$
8236 5483 sinh : $2e$
8236 5833 rt : (6,0,2,5)

8236 8080 rt : (8,9,-8,4)
8236 9096 $\exp(4)+\exp(4/5)$
8237 1332 rt : (8,0,-3,2)
8237 2673 rt : (7,1,7,9)
8237 5812 $e^x : 3+2\sqrt{2}$
8237 6926 rt : (9,1,2,6)
8237 7448 cr : $\sqrt{5}/4$
8237 8217 rt : (6,5,-7,7)
8237 8306 sq : $(3\pi/2)^3$
8238 0951 at : $6\sqrt[3]{2}/7$
8238 3996 $\lambda : \exp(-\sqrt{5})$
8238 5322 $\sqrt{5}+s\pi(1/5)$
8238 5682 $erf : 22/23$
8238 6042 rt : (1,8,7,-7)
8238 6360 sinh : 3π
8238 6832 at : $(\sqrt[3]{2})^{1/3}$
8238 8178 rt : (2,7,-3,-2)
8238 9689 id : $\sqrt{2}-\sqrt{5}+\sqrt{7}$
8239 0717 rt : (1,1,5,4)
8239 0874 $\ell_{10} : 3/20$
8239 2852 ln : $1+3\sqrt{3}$
8239 2898 $\theta_3 : 7/18$
8239 4570 rt : (6,7,-5,4)
8239 4641 $2^x : 3+2\sqrt{2}$
8239 4922 $10^x : 2/3+\sqrt{5}/3$
8239 5921 $3/4\times\ln(3)$
8239 6108 $e^x : 3-2\sqrt{2}/3$
8239 7460 cu : 15/16
8239 8181 sinh : $5\zeta(3)/8$
8239 8411 $\theta_3 : 1/\sqrt[3]{17}$
8239 9372 $e^x : \zeta(3)/2$
8240 0000 cu : 12/5
8240 0478 rt : (4,2,5,5)
8240 0657 $10^x : 2+2\sqrt{3}$
8240 0958 $s\pi : 4\sqrt{2}/3-\sqrt{3}/3$
8240 1637 $J_0 : 5\zeta(3)/7$
8240 2885 cu : $3/2+3\sqrt{3}/2$
8240 4152 rt : (7,3,5,6)
8240 4531 id : $1/3+2\sqrt{5}/3$
8240 8737 rt : (2,9,7,7)
8241 0679 $\exp(3/4)+s\pi(1/4)$
8241 0953 id : $3e/4+\pi/4$
8241 1009 $\ell_2 : 4\sqrt{2}/3+3\sqrt{3}$
8241 2262 rt : (9,3,0,3)
8241 3253 rt : (4,3,-4,7)
8241 4261 id : $4\sqrt[3]{3}/7$
8241 6137 ln : $4\sqrt[3]{5}/3$
8241 7200 rt : (9,8,-3,-8)
8241 8117 rt : (6,9,-3,1)
8241 8832 cu : $\sqrt{3}-\sqrt{5}+\sqrt{7}$
8242 0734 $\exp(4)-s\pi(e)$
8242 3310 $10^x : 2+2\pi/3$
8242 3368 rt : (8,9,-8,-4)
8242 3984 cr : $4\sqrt[3]{2}/9$
8242 4688 at : $1/3-\sqrt{2}$
8242 5705 cr : 14/25
8242 6763 rt : (8,2,-5,-1)
8242 8024 ln : $\sqrt{2}+\sqrt{3}/2$
8242 8584 $10^x : 1/4+\pi/3$
8243 0840 $\Psi : (5)^3$
8243 1311 cr : $e/2+3\pi/2$
8243 2984 cu : $3e/2-\pi/3$
8243 3274 $Ei : 20/9$
8243 4783 $J_0 : (e/2)^{-1/2}$
8243 6063 sr : $e/4$
8243 9774 sr : $1/2+2\sqrt{2}$
8243 9853 rt : (2,5,1,8)
8244 0252 cu : $3\sqrt{2}-\sqrt{3}$
8244 0612 rt : (7,6,0,-8)
8244 1949 rt : (8,6,8,7)
8244 3668 $s\pi : e/3+\pi/4$
8244 5030 rt : (5,1,-5,-2)
8244 6054 $Ei : (\pi)^{-3}$
8244 6734 rt : (9,6,-5,-5)
8244 7040 rt : (4,7,0,-7)
8244 7152 at : $2\sqrt{3}/3-\sqrt{5}$
8244 7482 rt : (1,8,-5,-9)
8245 2513 at : $(\sqrt[3]{5}/2)^{-1/2}$
8245 4631 $t\pi : 9e/7$
8245 4726 $Ei : 3\sqrt[3]{3}/7$
8245 4932 rt : (6,7,-5,-4)
8245 4938 as : $2\sqrt{2}/3-3\sqrt{5}/4$
8245 5087 $J_0 : (\sqrt[3]{4})^{-1/3}$
8245 5272 rt : (5,5,8,6)
8245 5532 sq : $\sqrt{2}/2+2\sqrt{5}$
8245 7239 rt : (1,8,0,-6)
8245 7883 rt : (2,6,-5,5)
8245 9035 $J_0 : 6/7$
8245 9384 sq : $4\sqrt{2}/3-4\sqrt{5}$
8246 0947 rt : (4,9,-5,-8)
8246 1889 at : $3\sqrt[3]{3}/4$
8246 2112 sr : 17/25
8246 2927 rt : (8,3,3,-9)
8246 3941 rt : (1,9,-8,1)
8246 3943 $e^x : 3/4-2\sqrt{2}/3$
8246 6585 $erf : 3\sqrt{5}/7$
8246 7418 $erf : 23/24$
8246 8249 rt : (3,8,-5,-3)
8246 8919 $e^x : 3\sqrt{2}/4+2\sqrt{5}/3$
8246 9316 $Ei : 10\pi^2/7$
8246 9447 at : $9\zeta(3)/10$
8247 2517 rt : (7,5,3,3)
8247 2641 cu : $4\sqrt{2}/3+4\sqrt{3}/3$
8247 2794 rt : (1,9,2,4)
8247 4359 rt : (5,4,3,-8)
8247 5906 rt : (2,9,0,-5)
8247 6973 $e^x : 3\zeta(3)$
8247 8404 $J : \exp(-\sqrt{5}/2)$
8248 3539 rt : (7,4,-2,-5)
8248 4515 rt : (2,9,-5,-9)
8249 0312 cr : $2+3e/2$
8249 0312 rt : (2,4,8,5)
8249 0424 tanh : $(e+\pi)/5$
8249 2459 rt : (2,7,7,9)
8249 2957 $e^x : 3/2+2\sqrt{2}$
8249 3548 cr : $\exp(-\sqrt{3}/3)$
8249 3654 rt : (9,5,-2,0)
8249 4359 sq : $2\sqrt{2}/3-4\sqrt{3}$
8249 5462 $\zeta : 3\sqrt[3]{3}/4$
8249 5791 sinh : $\ln(\sqrt{2}/3)$
8249 7782 $2^x : \pi$
8249 8844 rt : (6,8,-1,8)
8249 9000 $\zeta : (\sqrt[3]{4}/3)^{-1/3}$
8250 0792 at : $5\sqrt{3}/8$
8250 1355 $\theta_3 : 5\sqrt[3]{5}/9$
8250 1413 rt : (6,1,6,-9)
8250 4551 $t\pi : \sqrt{2}/4+\sqrt{3}/2$
8250 4570 $\exp(1/5)\times s\pi(\sqrt{5})$
8250 5635 rt : (4,7,0,1)
8250 5834 $\ell_{10} : 4e-4\pi/3$
8250 6767 rt : (3,6,-8,-9)
8250 7821 tanh : $\exp(1/(2\pi))$
8250 7948 $\zeta : \exp(-e/2)$
8250 9428 $t\pi : (\sqrt{2}\pi)^{-1/3}$
8251 0111 $e^x : 4+4\sqrt{2}/3$
8251 1655 cu : $\sqrt{2}+\sqrt{6}+\sqrt{7}$
8251 1849 rt : (8,1,1,-6)
8251 4076 id : $\sqrt{2}/2+\sqrt{5}/2$
8251 4154 rt : (3,7,-7,-8)
8251 4532 $\Psi : 8(e+\pi)/7$
8251 5803 $1/3+\exp(2/5)$
8251 8542 $Ei : \exp(-\sqrt[3]{3}/3)$
8252 0593 rt : (6,5,-7,-1)
8252 2251 $\ell_{10} : 3\sqrt{3}+2\sqrt{5}/3$
8252 2376 rt : (2,8,-9,-5)
8252 2434 ln : $\sqrt{3}+2\sqrt{5}$
8252 4139 rt : (9,7,-1,-9)
8252 5468 rt : (1,6,-3,-6)
8252 5733 $\Psi : 9/17$
8252 6671 sq : $2\sqrt{3}/3+3\sqrt{5}$
8252 7243 rt : (3,2,6,-8)
8252 7807 sq : $3/4-4\pi/3$
8253 1459 rt : (8,8,6,4)
8253 1488 $e^x : 3\sqrt{5}/5$
8253 3141 $s\pi : (2\pi/3)^{-1/2}$
8253 4017 sq : $\sqrt{2}/4-4\sqrt{3}/3$
8253 4080 $s\pi : 3/4+\sqrt{5}/4$
8253 4195 $\Gamma : \ln(e/3)$
8253 7385 $\exp(5)-s\pi(1/5)$
8253 7685 at : 13/12
8253 9871 rt : (9,2,4,7)
8254 0858 $K : 15/22$
8254 2364 $erf : 24/25$
8254 2519 rt : (5,2,1,-5)
8254 3076 $J_0 : \sqrt[3]{5}/2$
8254 4585 rt : (6,8,-1,-8)
8254 5994 $e^x : 7(e+\pi)$
8254 7601 rt : (1,6,-2,-3)
8254 8181 cr : 9/16
8254 9042 $10^x : \sqrt{2}/3+3\sqrt{5}$
8255 0648 id : $\sqrt{2}/3+3\sqrt{5}/2$
8255 4092 $2/5-\exp(4/5)$
8255 4653 $10^x : (\sqrt{5}/2)^{-1/2}$
8255 4670 rt : (8,8,-6,-5)
8255 4725 $10^x : 1/3+\sqrt{3}/2$
8255 6830 $\ln(3/4)+\exp(\sqrt{2})$
8255 7606 $e^x : 3\sqrt{3}/4-2\sqrt{5}/3$
8255 9402 $10^x : 13/9$
8255 9836 rt : (7,2,-4,-2)
8256 1136 $\ell_2 : P_{67}$
8256 1602 rt : (4,1,9,9)
8256 1975 sq : $1+e$
8256 4229 sq : $2/3-2\sqrt{3}$
8256 4696 rt : (6,7,-8,-8)
8256 4842 $\ell_2 : P_{70}$
8256 6233 rt : (7,5,2,-9)
8256 8318 sq : $7(e+\pi)/10$
8256 8934 id : $\sqrt{2}/4+2\sqrt{5}$
8257 0912 $\Gamma : 7\ln 2/10$
8257 2282 sr : 15/22
8257 4185 id : $\sqrt{10/3}$
8257 4806 $\ell_2 : (\pi)^{1/2}$
8257 7107 rt : (9,5,-3,-6)
8257 8875 cu : 11/9
8257 8915 rt : (6,1,4,6)
8257 9616 $2^x : 2/3-2\sqrt{2}/3$
8257 9625 rt : (9,2,-9,1)
8258 2533 $\ell_{10} : 3/2+3\sqrt{3}$
8258 2584 $J_0 : e\pi/10$
8258 2991 rt : (5,3,-9,4)
8258 7383 rt : (7,6,-8,8)
8258 8026 $\ell_{10} : 8(e+\pi)/7$
8258 8042 rt : (5,9,-6,-4)
8259 1364 $\exp(1/3)-s\pi(\pi)$
8259 2343 rt : (7,9,-8,-2)
8259 2976 id : $4e-\pi/3$
8259 3032 rt : (6,4,-2,-1)
8259 8514 rt : (8,1,-1,3)
8260 0517 rt : (4,6,2,8)
8260 1265 cu : P_{62}
8260 2151 $s\pi : 4\sqrt{3}/3$
8260 2499 cu : $4\sqrt{3}/3-\sqrt{5}/3$
8260 3644 $2^x : 1/4+e$
8260 4162 as : $e/2-2\pi/3$
8260 4828 rt : (4,7,-3,-5)
8260 8016 rt : (1,8,2,5)
8260 8642 $\theta_3 : 19/20$
8260 8695 id : 19/23
8260 9599 $erf : 2\sqrt[3]{3}/3$
8261 0381 rt : (8,9,-2,-9)
8261 1683 $1/2\times\exp(\sqrt{3})$
8261 1909 cu : $1/3-4\sqrt{3}$
8261 4784 cr : $4\sqrt{2}+\sqrt{3}/4$
8261 4836 rt : (7,4,7,7)
8261 5545 rt : (6,6,-3,-5)
8261 5652 rt : (3,5,-7,8)
8261 5683 $e^x : P_2$
8261 6125 $erf : 4\zeta(3)/5$
8261 7909 $K : 1/\ln(\ln 2/3)$
8261 9119 $s\pi : (1/(3e))^{-3}$
8261 9203 sr : $\sqrt{2}/4+4\sqrt{5}/3$
8261 9805 $10^x : 1/4+3\pi/4$
8262 2086 rt : (1,0,9,8)
8262 3635 $10^x : 3/4+3\sqrt{3}/2$
8262 3792 id : $7\sqrt{5}/2$
8262 4968 $2^x : -\sqrt{2}+\sqrt{6}+\sqrt{7}$
8262 5735 $J_0 : \exp(-1/(2\pi))$
8262 6602 rt : (5,3,5,-9)
8262 7901 $Ei : 6\sqrt{3}$
8263 0748 sr : $(\pi)^{-1/3}$

8263 1208 $\zeta : (\ln 3)^{-1/3}$
8263 1297 $\ln(5) \div \exp(2/3)$
8263 1820 $2 \times \exp(5)$
8263 2053 rt : (8,6,-8,-2)
8263 3010 rt : (7,8,-6,5)
8263 3351 rt : (3,8,-1,1)
8263 3544 tanh : $20/17$
8263 3579 cu : $(e/2)^3$
8263 4570 $s\pi : \sqrt{22}$
8263 6643 rt : (9,4,2,4)
8264 2773 rt : (7,3,0,-6)
8264 2831 $\sqrt{2} - s\pi(1/5)$
8264 3183 $erf : 5\sqrt{3}/9$
8264 3202 sq : $1/4 - 2\sqrt{5}$
8264 3960 id : $9\pi^2$
8264 4582 sr : $1/4 + \sqrt{3}/4$
8264 4628 sq : $10/11$
8264 4660 $K : (\pi)^{-1/3}$
8264 5462 $\Psi : -\sqrt{2} + \sqrt{3} + \sqrt{6}$
8264 5630 rt : (5,7,6,3)
8264 5774 sr : $3\sqrt{2}/4 + 4\sqrt{3}$
8264 6195 rt : (5,5,-7,1)
8265 2443 rt : (3,0,4,5)
8265 3448 $s\pi : 6e$
8265 4026 ln : $3/2 + \pi/4$
8265 4482 rt : (1,7,2,-7)
8265 4552 ln : $3/2 + 3\pi/2$
8265 8264 rt : (2,7,2,-2)
8265 8626 $10^x : \exp(\sqrt[3]{4}/3)$
8266 0034 cr : $4 + 2\pi/3$
8266 0524 rt : (9,3,-5,-3)
8266 0625 $\ell_{10} : 3\sqrt{5}$
8266 7209 rt : (6,7,1,-9)
8266 7857 ln : $7/16$
8266 9074 tanh : $(\sqrt[3]{3}/2)^{-1/2}$
8266 9733 rt : (8,0,3,7)
8267 0453 rt : (6,3,-9,2)
8267 1136 rt : (3,7,-1,9)
8267 2343 id : $2\sqrt{3}/3 - 4\sqrt{5}/3$
8267 5607 $e^x : 10\sqrt{5}$
8267 7387 rt : (3,7,-3,-4)
8268 0895 cr : $13/23$
8268 1751 rt : (8,7,-4,-6)
8268 2707 rt : (5,4,-5,8)
8268 4397 $2^x : 3\sqrt{2}/4 + \sqrt{5}$
8268 5068 tanh : $3\pi/8$
8268 6253 rt : (3,4,-6,-6)
8268 7692 rt : (2,4,5,-8)
8268 7711 $\ell_{10} : 2 + 3\pi/2$
8268 8696 rt : (5,0,-1,2)
8269 3413 rt : (4,9,-9,-7)
8269 4342 rt : (8,4,-7,-4)
8269 4786 $\Psi : 3\sqrt{5}$
8269 5795 $10^x : 10/7$
8269 6134 cr : $\exp(-\sqrt[3]{5}/3)$
8269 6428 rt : (9,3,6,8)
8269 8078 cr : $1/\ln(e + \pi)$
8269 8161 tanh : $5\sqrt{2}/6$
8269 8638 $erf : (\sqrt{5}/2)^{-1/3}$
8270 3256 rt : (5,8,-7,-9)
8270 3323 rt : (9,4,-1,-7)
8270 3414 rt : (4,3,7,6)
8270 3560 $\Psi : 19/22$
8270 3710 sr : $2\sqrt[3]{5}/5$
8270 4072 $e^x : 3\sqrt{3}/5$
8270 4074 at : $25/23$
8270 4086 $\zeta : (\pi/2)^{-3}$
8270 5118 $2^x : \sqrt{2}/3 - \sqrt{5}/3$
8270 7105 $t\pi : e\pi/7$
8270 9043 $2^x : 2 + 4\sqrt{3}/3$
8271 0589 ln : $4\sqrt{2} + \sqrt{5}/4$
8271 1218 $2^x : 20/23$
8271 2457 rt : (4,8,4,5)
8271 2658 $\Psi : \sqrt{23/3}$
8271 3674 rt : (4,8,-9,1)
8271 3730 cr : $\sqrt{2} + \sqrt{5} + \sqrt{6}$
8271 3978 $2^x : 8\sqrt{3}/5$
8271 4118 cr : $5e\pi/7$
8271 6675 rt : (4,8,-7,-9)
8271 7019 sr : $13/19$
8271 8704 rt : (7,9,-2,9)
8272 0400 at : $2e/5$
8272 0511 rt : (3,1,8,-9)
8272 3246 rt : (3,8,1,-8)
8272 4306 as : $3/2 - \sqrt{5}$
8272 5821 $\ell_{10} : 4 + e$
8272 6823 $t\pi : 4\sqrt{3} - 3\sqrt{5}$
8272 8461 $\sqrt{2} \div s\pi(e)$
8272 8672 $\ell_2 : P_6$
8272 9091 rt : (4,7,-9,9)
8272 9186 id : $\sqrt{3} + \sqrt{6} + \sqrt{7}$
8273 2683 rt : (7,7,1,0)
8273 2895 sinh : $15/11$
8273 3193 $\Psi : \ln(\sqrt[3]{3})$
8273 5026 id : $1/4 + \sqrt{3}/3$
8273 6512 $\zeta : 1/\sqrt{15}$
8273 6948 $J_0 : 17/20$
8273 7008 $Ei : 13/21$
8273 7266 $\sqrt{2} + \exp(5)$
8273 7319 sinh : $6\sqrt[3]{5}/5$
8274 0391 rt : (6,4,-5,-2)
8274 1120 cu : $9e\pi/4$
8274 2274 id : $1/4 - 3e/2$
8274 3338 id : $9\pi/10$
8274 3700 rt : (5,8,-4,-5)
8274 3822 $\exp(4) \div s\pi(\sqrt{5})$
8274 4166 rt : (6,3,2,3)
8274 6834 rt : (8,7,-9,9)
8274 7084 rt : (5,1,3,-6)
8274 8261 cr : $2/3 + 2e$
8274 8416 rt : (4,5,4,-9)
8274 9084 sr : $5(e + \pi)/2$
8275 1990 rt : (2,8,-2,-6)
8275 2487 at : $4e/3 - 3\pi/2$
8275 9202 $\zeta : 6\sqrt[3]{3}/7$
8276 3358 id : $\sqrt{2}/3 - 3\sqrt{3}/4$
8276 4200 $\Psi : (\sqrt{5}/3)^{1/2}$
8276 5511 rt : (2,1,-9,-7)
8276 5950 rt : (5,9,0,-9)
8276 6105 rt : (7,8,-9,-2)
8276 7107 cr : $1 - \sqrt{3}/4$
8276 7547 $\ell_2 : 3 - e$
8276 8319 $J_0 : (\sqrt[3]{3}/2)^{1/2}$
8276 9913 rt : (6,5,-1,-6)
8277 0206 cr : P_{80}
8277 1684 sinh : $(\ln 3)^{-3}$
8277 4904 $K : 2\sqrt[3]{5}/5$
8277 5009 $t\pi : \sqrt{2}/4 - \sqrt{5}/3$
8277 7017 rt : (7,1,-2,-3)
8278 1277 rt : (6,2,6,7)
8278 2431 $J_0 : 8(1/\pi)/3$
8278 3945 rt : (7,9,-5,-6)
8279 1345 $s\pi : 1/3 + 3\pi/4$
8279 3442 rt : (8,5,-6,-3)
8279 3980 $J_0 : 3\sqrt{2}/5$
8279 4100 $10^x : 9/4$
8279 4591 $s\pi : 10 \ln 2/3$
8279 4616 $t\pi : \sqrt{3}/2 + 3\sqrt{5}/2$
8279 5875 rt : (7,2,2,-7)
8279 6019 $e^x : 4 - 4\pi/3$
8279 7715 rt : (1,8,8,-6)
8279 7767 $s\pi : 9\pi^2/7$
8279 8511 rt : (7,6,5,4)
8279 8746 $K : 13/19$
8279 8893 ln : $1/4 + 3e/4$
8280 2406 tanh : $13/11$
8280 4868 rt : (8,6,-2,-7)
8280 7275 rt : (8,9,-7,6)
8280 7872 $Ei : 1/\ln(e/2)$
8280 8526 ln : $2e + \pi/4$
8281 0216 cu : $3/2 - 4\sqrt{2}$
8281 0965 rt : (9,8,-8,-7)
8281 2500 sq : $7\sqrt{5}/8$
8281 3147 rt : (8,2,1,4)
8281 5048 rt : (7,5,9,8)
8281 5988 sr : $2e - 2\pi/3$
8281 6098 $\zeta : \ln(2\pi)$
8281 7062 $\Psi : 8\sqrt{3}/5$
8281 8970 $2^x : \sqrt{2}/3 + 3\sqrt{3}$
8281 9052 $e^x : 6(e + \pi)/7$
8282 0501 rt : (9,2,-3,-4)
8282 1591 rt : (8,1,5,8)
8282 2523 $Ei : 5(1/\pi)/8$
8282 4455 $s\pi : 1/4 + 3\sqrt{2}/4$
8282 5631 rt : (9,3,1,-8)
8283 0221 ln : $\sqrt[3]{12}$
8283 0440 $\Psi : ((e + \pi)/3)^{-3}$
8283 0504 $2^x : \sqrt{3}/2 + 3\sqrt{5}/4$
8283 1373 ln : $(5)^3$
8283 3443 as : $(5/2)^{-1/3}$
8283 5325 sinh : $1/\ln(\sqrt[3]{5}/2)$
8283 8073 $t\pi : 9 \ln 2/8$
8283 8739 as : $14/19$
8284 2712 id : $2\sqrt{2}$
8284 4657 $2/3 \div \ln(\sqrt{5})$
8284 5092 $Ei : \sqrt[3]{14/3}$
8284 5747 sq : $\sqrt{2} + \sqrt{3} - \sqrt{5}$
8284 7066 rt : (9,4,8,9)
8284 8794 rt : (9,5,4,5)
8285 0545 sr : $3\sqrt{3}/2 + \sqrt{5}/3$
8285 0909 cr : $(2e)^{-1/3}$
8285 1776 rt : (8,9,8,5)
8285 1830 $10^x : ((e + \pi)/3)^{-2}$
8285 2604 $t\pi : 3\sqrt[3]{2}$
8285 2799 rt : (9,6,0,1)
8285 3544 id : $(\ln 3)^{-2}$
8285 4001 sinh : $8\sqrt{3}$
8285 4781 $e^x : 3\sqrt{2}/2 - 4\sqrt{3}/3$
8285 8941 rt : (4,0,-9,8)
8286 0501 $e^x : 1/4 + \sqrt{2}/4$
8286 0559 cu : $9\pi/5$
8286 0669 sr : $(\sqrt{2}/3)^{1/2}$
8286 5160 $2^x : 1/4 + 4\sqrt{3}$
8286 6578 $2^x : (\pi/3)^{-3}$
8286 7019 rt : (3,7,3,-9)
8286 7274 $10^x : \ln(\pi/2)$
8287 0849 rt : (6,5,-6,9)
8287 2222 rt : (9,7,-4,-3)
8287 3736 $\Psi : 4 \ln 2$
8287 4100 rt : (2,3,7,-9)
8287 5319 sq : $9\zeta(3)/8$
8287 6111 rt : (6,6,-8,2)
8287 6962 cr : $9e/4$
8287 7205 sinh : $(\sqrt[3]{5}/3)^{1/2}$
8287 7626 cu : $1/4 - 4e$
8287 8396 rt : (4,9,-9,-1)
8287 8629 sr : $1/3 + \sqrt{2}/4$
8287 8806 sr : $4\zeta(3)/7$
8287 9938 sq : $3(e + \pi)/8$
8288 2236 rt : (5,7,1,9)
8288 3524 cu : $2 - 3\sqrt{2}/4$
8288 3906 $erf : (e/3)^{1/3}$
8288 4905 at : $12/11$
8288 5810 $J_0 : 11/13$
8288 6477 rt : (1,7,-7,-4)
8289 1114 rt : (2,4,5,8)
8289 3466 rt : (5,8,-1,2)
8289 4255 $t\pi : 3\sqrt{2}/4 + 2\sqrt{3}$
8289 5559 cu : $4e - 3\pi/2$
8289 6059 rt : (3,5,-5,-1)
8289 7211 rt : (3,9,-9,2)
8289 7522 ln : $3e/4 + 4\pi/3$
8289 8012 $e^x : 2e\pi/3$
8289 8538 rt : (7,9,-7,-8)
8290 1799 ln : $\sqrt{3} + \sqrt{5}/4$
8290 4099 rt : (7,8,-3,-7)
8290 5337 at : $1/\ln(5/2)$
8290 8049 sq : $1/\ln(2e/3)$
8290 9104 rt : (4,9,8,9)
8291 1281 rt : (9,9,-4,-8)
8291 2319 rt : (1,8,-8,-6)
8291 3054 id : $(\sqrt[3]{5}/3)^{1/3}$
8291 3605 rt : (5,9,-3,-5)
8291 5619 sr : $11/16$
8291 8490 rt : (6,4,1,-7)
8292 0152 rt : (2,6,-4,9)
8292 0163 $e^x : 1/2 + 4e/3$
8292 1089 $s\pi : 1/2 + 4\pi/3$
8292 1897 tanh : $\sqrt[3]{5/3}$
8292 2335 rt : (7,7,-7,-3)

8292 2496 $\exp(1/4) - \exp(\sqrt{2})$
8292 4254 rt : (8,5,0,-8)
8292 6620 $2^x : \sqrt{3}/2 + 4\sqrt{5}$
8292 7393 $\zeta : 11/19$
8292 7559 rt : (9,8,-8,-4)
8292 8638 cr : $4 + 3\sqrt{2}/2$
8293 0377 $\ell_{10} : (\sqrt[3]{4}/3)^3$
8293 0616 rt : (1,2,7,5)
8293 1794 $\ln(3/4) + \exp(3/4)$
8293 3687 rt : (2,7,-2,2)
8293 4462 sq : $1/3 + \sqrt{3}/3$
8293 4495 t$\pi : 2\sqrt{3}$
8293 7222 $\ell_2 : 3e - \pi/3$
8293 7450 $\Gamma : 7(e + \pi)/10$
8293 7456 $\ell_{10} : 3e/4 + 3\pi/2$
8293 8677 rt : (3,1,6,6)
8294 1701 rt : (5,0,5,7)
8294 3285 rt : (7,1,4,-8)
8294 3310 cu : P_{46}
8294 4259 rt : (9,2,3,-9)
8294 7489 $\ell_{10} : \sqrt{2}/2 - \sqrt{5}/4$
8294 8354 rt : (8,4,-4,-4)
8294 8791 $10^x : 9\sqrt{3}/8$
8294 8993 $erf : (\ln 3)^{-1/3}$
8294 9141 $J : (4)^{-3}$
8295 2427 id : $3\sqrt{2}/2 + 3\sqrt{5}$
8295 2952 rt : (2,8,0,-5)
8295 3644 $e^x : 2e/9$
8295 6344 $\lambda : (2\pi/3)^{-3}$
8295 9923 $e^x : 3/2 + \pi/4$
8296 0464 sr : $4\sqrt{2} - 4\sqrt{3}/3$
8296 2552 $\lambda : \exp(-12\pi/17)$
8296 4929 id : $e/2 - 4\pi/3$
8296 6620 s$\pi : \sqrt{2}/2 + 4\sqrt{5}/3$
8296 6866 sinh : $10\sqrt{3}/3$
8296 7854 rt : (8,2,7,9)
8296 8985 $Ei : 7\pi^2/6$
8297 0705 sr : $3\sqrt{3}/4 + 3\sqrt{5}$
8297 1872 rt : (6,7,-4,6)
8297 3912 rt : (6,3,8,8)
8297 4119 rt : (9,1,-1,-5)
8297 4955 rt : (3,4,-9,3)
8297 5253 $\exp(e) + s\pi(\sqrt{5})$
8297 6227 rt : (4,2,-7,5)
8297 7490 sr : $3/4 + 3\sqrt{3}/2$
8297 8461 sq : $1/\ln(2\pi/3)$
8298 0190 tanh : 19/16
8298 0980 sq : P_{11}
8298 1443 ln : $3 - \sqrt{2}/2$
8298 2367 rt : (5,9,-2,-5)
8298 2563 rt : (2,9,2,-9)
8298 2653 cr : 4/7
8298 5431 rt : (4,4,9,7)
8298 8999 id : $4\sqrt{2}/3 + 4\sqrt{5}$
8298 9062 sr : $4(e + \pi)/7$
8299 0799 $J_0 : (5/3)^{-1/3}$
8299 2628 rt : (8,3,-8,0)
8299 2793 $s\pi(1/3) - s\pi(\sqrt{2})$
8299 2850 $\exp(1/3) - \exp(4/5)$
8299 3040 $\theta_3 : 9/23$
8299 5943 $4/5 \div s\pi(\sqrt{2})$
8299 9638 rt : (5,1,1,3)
8300 1111 rt : (6,8,-6,-1)
8300 3811 sinh : $3\sqrt[3]{2}/5$
8300 4904 $erf : 7\ln 2/5$
8300 5098 ln : $4\sqrt{2} + \sqrt{3}/3$
8300 5729 s$\pi : 3\sqrt{2}/2 - \sqrt{3}/4$
8300 7499 $\ell_2 : 9/16$
8300 7965 $10^x : 4/3 - \sqrt{2}$
8300 9092 rt : (9,8,-2,-9)
8301 1090 rt : (2,7,-4,-7)
8301 1487 $\exp(3/5) - \exp(\sqrt{3})$
8301 3617 rt : (4,9,-3,-6)
8301 6515 rt : (8,3,3,5)
8301 6943 rt : (6,2,-7,1)
8301 7078 ln : P_{29}
8302 0687 rt : (7,7,-1,-8)
8302 1458 cu : $3\sqrt{2}/4 + \sqrt{5}$
8302 1544 at : $2 - e/3$
8302 2683 rt : (1,5,6,-9)
8302 6490 $e^x : 5\sqrt[3]{2}/4$
8302 8787 rt : (9,0,-5,1)
8303 0121 $2/5 - s\pi(\pi)$
8303 0706 s$\pi : 2\sqrt{2}/3 + \sqrt{5}/3$
8303 1809 $\Gamma : 7\zeta(3)/2$
8303 2996 $\ell_{10} : 4e/3 + \pi$
8303 3280 tanh : $(\sqrt{2})^{1/2}$
8303 3895 rt : (2,8,-2,-5)
8303 4985 ln : $4 + \sqrt{5}$
8303 6090 rt : (7,7,-9,3)
8303 6655 $2^x : 10e\pi/7$
8303 7344 s$\pi : \sqrt{7}\pi$
8303 7757 rt : (5,9,3,6)
8303 8153 $\zeta : 3\ln 2/8$
8304 0121 rt : (8,4,2,-9)
8304 0732 sq : $\sqrt{2}/2 + 2\sqrt{5}/3$
8304 1912 $J_0 : 16/19$
8304 2219 sq : $3\sqrt{3}/4 - 4\sqrt{5}/3$
8304 3802 at : $1 - 2\pi/3$
8304 4982 sq : 23/17
8304 6075 $4 \div \ln(3/5)$
8304 6246 $e^x : 1/3 + \sqrt{2}/2$
8304 7427 rt : (6,9,-8,-3)
8304 7935 rt : (9,7,-6,-5)
8304 8100 $10^x : 2\sqrt[3]{5}/5$
8304 9680 s$\pi : e/2 - \pi/3$
8304 9950 rt : (9,6,6,6)
8305 1834 t$\pi : \exp(\sqrt{2}\pi/2)$
8305 3455 rt : (1,4,-7,-8)
8305 6015 rt : (4,6,-5,-6)
8305 6611 cr : $9(1/\pi)/5$
8305 7777 $K : (\sqrt{2}/3)^{1/2}$
8305 8206 $\ell_{10} : 4/3 + 2e$
8306 0286 $\sqrt{5} + \ln(2/3)$
8306 1683 rt : (6,3,3,-8)
8306 5378 rt : (3,5,1,-6)
8306 6309 $\sqrt{3} + \ln(3)$
8306 7511 $J_0 : 7\zeta(3)/10$
8306 9042 $\Gamma : 1/\sqrt{10}$
8307 1165 ln : $9\ln 2$
8307 2020 ln : $2e - \pi$
8307 2470 rt : (7,8,-3,-4)
8307 2649 tanh : 25/21
8307 3590 sq : $10e\pi/7$
8307 3761 $\sqrt{5} \div \exp(1/5)$
8307 7153 s$\pi : 4(e + \pi)/5$
8308 0021 rt : (4,8,-7,-2)
8308 2144 at : 23/21
8308 4067 $\ln(5) + \exp(1/5)$
8308 5397 rt : (7,0,6,9)
8308 6762 rt : (4,3,-3,9)
8308 8217 $\Gamma : 6\sqrt[3]{5}$
8308 8403 $J_0 : (\sqrt{2})^{-1/2}$
8309 0213 $\ell_{10} : 4\sqrt{2} + \sqrt{5}/2$
8309 0379 cu : $10\sqrt[3]{2}/7$
8309 0703 $K : 4\zeta(3)/7$
8309 1634 rt : (3,9,3,-2)
8309 1968 sinh : $1/\ln(\sqrt[3]{3}/3)$
8309 2445 s$\pi : 1/2 + 2e/3$
8309 4523 cr : $e/2 - \pi/4$
8309 6032 sq : $1/3 - 4e/3$
8309 6497 rt : (4,2,4,-7)
8309 7277 rt : (8,3,-2,-5)
8309 7713 sinh : $\sqrt{2} - \sqrt{6} - \sqrt{7}$
8309 7987 cu : $1/2 - 4\pi$
8309 8354 $\Psi : 25/9$
8309 9327 rt : (7,7,7,5)
8310 0360 rt : (4,1,-6,2)
8310 2048 ln : $\sqrt[3]{3}/9$
8310 2281 sr : $3 - 4\sqrt{3}/3$
8310 4198 cr : $2\sqrt{2}/3 + 3\sqrt{3}$
8310 7951 at : $4\sqrt{2}/3 - 4\sqrt{5}/3$
8311 1869 rt : (1,8,-7,3)
8311 3046 rt : (5,5,-4,-3)
8311 5020 rt : (6,9,-2,3)
8311 5676 $\ell_2 : 5e\pi/3$
8311 5821 rt : (7,8,-5,7)
8311 7720 rt : (4,8,-8,4)
8311 8684 $10^x : 7/12$
8312 1853 rt : (9,0,1,6)
8312 2155 ln : P_{20}
8312 2520 sq : $7\sqrt[3]{3}/6$
8312 2843 $J_0 : 21/25$
8312 4864 $J_0 : 2\sqrt[3]{2}/3$
8312 5387 $\sqrt{2} \times s\pi(1/5)$
8312 5705 sr : $(2\pi/3)^{-1/2}$
8312 6814 rt : (8,4,-1,1)
8312 7097 sr : $3 + \sqrt{2}/4$
8312 7262 rt : (5,1,7,8)
8312 7853 rt : (3,3,-8,-7)
8313 0354 cr : $3 + \pi$
8313 0532 $Ei : \sqrt{6}$
8313 2813 $\ell_{10} : 4\sqrt{3}/3 + 2\sqrt{5}$
8313 3862 at : $2\sqrt{2} - \sqrt{3}$
8313 3909 rt : (7,6,1,-9)
8313 7002 rt : (1,7,-6,4)
8313 7175 $\exp(3) - s\pi(\sqrt{3})$
8313 8041 $\ell_2 : \sqrt{3}/3 + 4\sqrt{5}/3$
8314 0327 ln : $2 + 3\sqrt{2}$
8314 1475 rt : (5,3,-8,6)
8314 2075 sr : $3\sqrt{5}/2$
8314 3368 $\ell_{10} : 1/2 + 2\pi$
8314 4574 rt : (2,7,-6,-1)
8314 5050 at : $\pi^2/9$
8314 5100 id : P_{25}
8314 5229 rt : (8,9,-7,-5)
8314 5664 rt : (6,2,-1,-4)
8314 6961 s$\pi : 5/16$
8314 7804 rt : (4,5,5,3)
8314 7882 $\ell_2 : 3 + \sqrt{5}/4$
8315 0525 sq : $4/3 + 2e$
8315 0786 rt : (2,8,-7,5)
8315 1719 $10^x : 2/3 + 3e$
8315 2324 $e^x : (\ln 3/3)^{1/2}$
8315 3220 $e + \exp(\sqrt{2})$
8315 3841 rt : (4,8,-1,-7)
8315 4456 sq : $(\pi/3)^{-2}$
8315 4716 sq : $\sqrt[3]{15/2}$
8315 5149 $\exp(1/2) \div s\pi(\pi)$
8315 7521 t$\pi : 3\sqrt[3]{5}/2$
8315 7687 $K : 11/16$
8315 8252 s$\pi : 1/3 + 3\sqrt{5}/2$
8315 9519 rt : (4,9,-8,-4)
8316 0720 rt : (9,1,-9,-3)
8316 0913 sq : $3e/4 - 4\pi$
8316 2231 rt : (2,1,5,-6)
8316 2853 $Ei : (e + \pi)/6$
8316 3585 $Ei : \sqrt[3]{11}$
8316 4655 rt : (9,6,-4,-6)
8316 6549 rt : (7,9,-7,0)
8317 3666 sq : $3/4 - 3e$
8317 4559 rt : (2,6,0,-3)
8317 4663 $\Gamma : (\ln 3/3)^3$
8317 5152 id : $2\sqrt{3}/3 + 3\sqrt{5}/4$
8317 6422 rt : (3,7,-6,-5)
8317 6946 rt : (5,5,2,-8)
8317 7661 id : $6\ln 2/5$
8317 7844 as : 17/23
8317 7883 rt : (7,1,2,5)
8317 8737 id : $4e\pi/5$
8317 9196 rt : (1,9,8,-9)
8317 9428 $J_0 : 3\sqrt{5}/8$
8318 0027 rt : (9,7,2,2)
8318 5585 sinh : $7\sqrt[3]{2}/5$
8318 5590 rt : (2,9,-3,1)
8319 3049 rt : (6,8,-6,-4)
8319 4257 cu : $4 - 4\sqrt{3}/3$
8319 7792 as : $\ln(2\pi/3)$
8319 8595 id : $\exp(-1/(2e))$
8319 9194 sr : $1 + 3\pi/4$
8319 9854 rt : (6,2,5,-9)
8320 0000 cu : 9/5
8320 0782 rt : (8,2,-6,-1)
8320 0936 $2^x : 2\sqrt{2}/3 + \sqrt{5}/4$
8320 2437 $\zeta : 6/23$
8320 5029 sr : 9/13
8320 6025 rt : (3,2,8,7)
8320 7569 rt : (3,9,-8,6)
8320 8396 sq : $7\zeta(3)/5$
8320 9520 cu : $e/4 - 2\pi/3$

8320 9713 rt : (8,4,5,6)
8321 0678 sq : $1+\sqrt{2}/4$
8321 0993 at : $3/2-3\sqrt{3}/2$
8321 4240 sinh : $4\pi^2/5$
8321 5112 rt : (3,9,-3,-7)
8321 6270 rt : (2,7,0,-6)
8321 6955 rt : (7,5,-3,-5)
8321 8881 $t\pi$: $e/4+3\pi/2$
8321 8924 2^x : $1/\ln(\pi)$
8321 9622 rt : (5,2,-3,-1)
8322 0440 2^x : $\sqrt[3]{2/3}$
8322 1165 rt : (1,7,1,-1)
8322 2118 rt : (8,9,-6,8)
8322 3449 ℓ_{10} : $5e/2$
8322 5703 rt : (6,8,0,-9)
8322 5722 sq : $1/\ln(5/3)$
8323 3100 J_0 : $(e+\pi)/7$
8323 3633 Γ : $-\sqrt{2}-\sqrt{5}+\sqrt{6}$
8323 3809 $t\pi$: $5\pi^2/9$
8323 4565 rt : (9,1,-3,2)
8323 5290 at : $\ln(3)$
8323 5777 2^x : 25/11
8323 5829 sr : $2\sqrt{3}/5$
8323 7307 tanh : $(\sqrt[3]{5})^{1/3}$
8323 8207 2^x : $2\sqrt{3}/3+\sqrt{5}/2$
8323 9053 rt : (1,9,8,1)
8324 0646 rt : (8,2,0,-6)
8324 1165 rt : (9,7,8,7)
8324 2185 rt : (5,8,8,4)
8324 6018 rt : (3,5,4,9)
8324 6635 J : $\exp(-11\pi/16)$
8324 6657 cr : $2\sqrt[3]{3}/5$
8324 8750 at : $\sqrt{2}/4+\sqrt{5}/3$
8324 9032 10^x : $\exp(-\sqrt[3]{4}/2)$
8324 9082 ln : $1+3\sqrt{3}/4$
8325 3335 $s\pi$: $4\zeta(3)/7$
8325 3846 $s\pi$: $1/3+\sqrt{2}/4$
8325 5080 rt : (2,6,-6,-8)
8325 5461 id : $(\ln 2)^{1/2}$
8325 6475 e^x : 24/7
8325 7545 sinh : $4\sqrt{5}/3$
8325 7706 rt : (1,9,-9,-8)
8325 7730 $s\pi$: $3\sqrt{3}+2\sqrt{5}/3$
8325 7873 rt : (6,4,-9,7)
8325 8146 ln : 4/25
8325 9197 rt : (8,8,-5,-6)
8325 9455 J : $(\ln 3/2)^2$
8326 0476 rt : (9,5,-8,-2)
8326 1625 cu : $3e/2+4\pi/3$
8326 1846 cr : γ
8326 4125 rt : (9,1,3,7)
8326 6551 J_0 : $(\sqrt[3]{5})^{-1/3}$
8326 7862 cu : $3+\sqrt{2}/3$
8326 8317 id : $(\sqrt[3]{3})^{-1/2}$
8327 0707 e^x : 16/7
8327 1052 2^x : $4e/3+\pi$
8327 1561 rt : (4,7,-5,-3)
8327 1610 2^x : 22/7
8327 3209 tanh : $7\sqrt[3]{5}/10$
8327 3639 rt : (7,8,-9,-9)
8327 4431 Ψ : $(\ln 3/3)^2$
8327 5088 rt : (1,8,9,7)
8327 5439 rt : (4,1,6,-8)
8327 5757 e^x : $1/4-\sqrt{3}/4$
8327 6515 $t\pi$: $1/4+\sqrt{5}$
8327 6888 sr : $2\sqrt{2}+3\sqrt{3}$
8327 6958 $t\pi$: $(\ln 3/2)^{-1/3}$
8327 7811 erf : $(e+\pi)/6$
8327 7928 rt : (9,5,-2,-7)
8327 8470 sq : $\sqrt{2}/4+\sqrt{5}/4$
8327 9592 sr : $2+e/2$
8327 9994 10^x : $\sqrt{2}/4-\sqrt{3}/4$
8328 1410 cr : $5\ln 2/6$
8328 1572 sq : $2+3\sqrt{5}$
8328 1716 rt : (3,8,0,4)
8328 1919 cr : $1/2+4\sqrt{2}$
8328 2317 rt : (5,4,-8,1)
8328 3616 at : $2/3+\sqrt{3}/4$
8328 4582 at : $\sqrt{3}/3-3\sqrt{5}/4$
8328 7302 cu : $3+e/2$
8328 7397 rt : (5,2,3,4)
8328 9225 rt : (4,7,1,-8)
8329 0465 $t\pi$: $1/\ln(\sqrt[3]{5}/3)$
8329 0912 ln : 10/23
8329 0913 rt : (1,9,2,-4)
8329 3023 10^x : 13/19
8329 4037 ℓ_2 : $\exp(\sqrt{3}/3)$
8329 4281 Ei : $4\sqrt[3]{5}/7$
8329 4482 rt : (1,3,-9,-9)
8329 5587 at : $7\sqrt{2}/9$
8329 5847 ℓ_2 : $4\sqrt{2}/3+3\sqrt{5}/4$
8329 6950 sq : $4e\pi$
8329 7085 e^x : $3\sqrt{2}/2+\sqrt{3}/2$
8329 7284 sr : $8\sqrt[3]{2}/3$
8329 7459 $\exp(1/4)-\exp(3/4)$
8329 8071 10^x : 5/19
8329 8126 at : 11/10
8329 8814 erf : $4\sqrt[3]{5}/7$
8330 2111 erf : $(\pi/3)^{-1/2}$
8330 2629 rt : (5,4,-2,-4)
8330 4542 rt : (5,2,9,9)
8330 5615 $s\pi$: $(\sqrt{2}/3)^{1/2}$
8330 5963 rt : (1,9,-4,-9)
8330 9055 $t\pi$: $\sqrt{2}/2-4\sqrt{3}$
8331 0028 $s\pi$: $1/4+2e$
8331 0174 as : $3\sqrt{3}/4-\sqrt{5}/4$
8331 1370 rt : (5,4,4,-9)
8331 2066 J : 3/25
8331 4921 Ei : $(\pi/3)^{-1/2}$
8331 5581 rt : (5,5,-6,3)
8331 9247 cu : $3/2-\sqrt{5}/4$
8332 0452 cr : $4e-3\pi/2$
8332 1334 ln : 17
8332 4625 ℓ_{10} : $4\sqrt{2}+2\sqrt{3}/3$
8332 5216 e^x : $3\sqrt{2}/7$
8332 7185 at : $\sqrt[3]{4/3}$
8332 7188 Li_2 : 2/3
8333 1008 θ_3 : $7\zeta(3)/9$
8333 2963 2^x : $4\ln 2$
8333 3156 $\ln(\sqrt{3})+\exp(1/4)$
8333 3333 id : 5/6
8333 7579 $\exp(3/4)\times s\pi(1/3)$
8333 8633 θ_3 : $\ln(\sqrt{2}\pi/3)$
8334 0602 θ_3 : $\pi/8$
8334 3178 sinh : $\sqrt{23/2}$
8334 4712 2^x : $\sqrt{2}-3\sqrt{5}/4$
8334 5027 cr : 11/19
8334 6915 rt : (1,6,2,6)
8334 7235 2^x : $5\zeta(3)/4$
8335 1065 10^x : $\sqrt[3]{7}$
8335 6190 rt : (7,4,-1,-6)
8335 7289 $s\pi$: $8\sqrt{2}$
8335 8034 sq : $9\sqrt[3]{5}/7$
8335 9213 sq : $2-3\sqrt{5}/2$
8336 1277 10^x : 20/3
8336 2112 rt : (1,9,-1,-6)
8336 3682 rt : (7,2,4,6)
8336 4341 Ψ : $(\sqrt[3]{4}/3)^{-3}$
8336 4839 sq : 21/23
8336 5460 tanh : 6/5
8336 5806 rt : (7,4,-7,-1)
8336 6996 ℓ_{10} : $2\sqrt{3}+3\sqrt{5}/2$
8336 7088 ln : $3\sqrt{2}/4+3\sqrt{3}$
8336 9868 rt : (8,7,-3,-7)
8337 0649 cu : 16/17
8337 5095 rt : (1,3,4,-6)
8337 6494 Ei : $2\ln 2/7$
8337 8030 J_0 : 5/6
8337 8447 10^x : $3\sqrt{2}/2+2\sqrt{5}$
8337 8481 rt : (7,8,9,6)
8337 8849 rt : (8,1,2,-7)
8337 9222 rt : (1,9,-7,-1)
8338 1194 rt : (8,7,-9,-2)
8338 3341 rt : (7,2,-2,1)
8338 3981 $\ln(2)+\exp(\pi)$
8338 7947 rt : (9,4,0,-8)
8338 9596 tanh : $\sqrt{2}+\sqrt{5}-\sqrt{6}$
8339 0028 sr : $-\sqrt{3}+\sqrt{6}+\sqrt{7}$
8339 0584 rt : (2,6,2,3)
8339 3630 e^x : 25/24
8339 3684 rt : (8,5,7,7)
8339 4017 tanh : $(\sqrt[3]{3})^{1/2}$
8339 5241 Γ : 4/21
8339 6783 sinh : $9\sqrt[3]{5}/5$
8339 7433 cr : $5\pi^2/8$
8339 7566 Ψ : $(\pi/2)^{-1/3}$
8339 7763 cu : $1/4-\sqrt{5}$
8339 7871 rt : (8,1,-4,-2)
8339 8659 rt : (3,9,0,-4)
8339 9664 tanh : $(\ln 2)^{-1/2}$
8340 0283 $t\pi$: $4e-2\pi/3$
8340 0296 rt : (5,8,-4,-6)
8340 0808 2^x : 7/8
8340 1093 sinh : $9\sqrt{5}/7$
8340 1314 rt : (9,2,5,8)
8340 2827 J_0 : $(\sqrt[3]{3})^{-1/2}$
8340 4376 $\ln(2/5)\div\ln(3)$
8340 4794 rt : (9,4,-6,-3)
8340 5737 e^x : $(e)^{-1/2}$
8340 5765 sr : 16/23
8340 7535 Ψ : $-\sqrt{2}-\sqrt{3}-\sqrt{6}$
8340 7729 J_0 : $(\ln 2)^{1/2}$
8340 7990 at : $7\sqrt[3]{2}/8$
8340 9441 rt : (1,3,9,6)
8341 0505 ln : $4\sqrt{2}-3\sqrt{5}/2$
8341 2952 as : $3/4-2\sqrt{5}/3$
8341 4418 id : $3\sqrt{3}/2+\sqrt{5}$
8341 4725 sq : $2e-4\pi$
8341 6221 rt : (2,0,7,7)
8341 6678 ℓ_2 : $\sqrt[3]{17/3}$
8341 7027 rt : (1,9,-8,2)
8341 8166 ζ : $((e+\pi)/3)^{-2}$
8341 8389 cu : $4\sqrt{2}/3+\sqrt{3}/2$
8341 8756 rt : (3,3,5,-8)
8341 8878 2^x : $3\sqrt{2}+4\sqrt{3}/3$
8342 0082 rt : (4,6,3,-9)
8342 0498 erf : $7\sqrt[3]{2}/9$
8342 0897 rt : (5,6,-2,7)
8342 1450 rt : (6,6,-7,4)
8342 1801 $\sqrt{5}+\exp(4)$
8342 2551 tanh : $5\sqrt[3]{3}/6$
8342 2727 Ei : $1/\ln(5)$
8342 2962 sq : $\sqrt{3}/3-2\sqrt{5}/3$
8342 4318 rt : (3,9,-6,-3)
8342 4641 tanh : $\exp(1/(2e))$
8342 4846 ℓ_{10} : $\sqrt{3}+\sqrt{6}+\sqrt{7}$
8342 7777 10^x : $2/3-\sqrt{5}/3$
8342 7962 10^x : $3e+4\pi/3$
8342 8092 tanh : $\zeta(3)$
8342 8275 sq : $1/\ln(e/3)$
8342 8314 at : $3e/4-\pi$
8342 9402 J_0 : $\exp(-1/(2e))$
8342 9540 rt : (9,2,-1,3)
8343 0222 at : $(\sqrt{5}/3)^{-1/3}$
8343 0281 cr : $3e/2+2\pi/3$
8343 2067 ℓ_{10} : $4+2\sqrt{2}$
8343 2334 cu : $2e\pi/9$
8343 2381 rt : (8,5,1,2)
8343 2409 $s\pi$: $(\ln 3/2)^{-2}$
8343 5434 $s\pi$: $\pi/10$
8343 7378 J_0 : $6\ln 2/5$
8343 7671 ζ : 11/6
8343 9592 Ψ : $1/\ln(\ln 2/3)$
8344 0135 $s\pi$: $3\sqrt{3}+\sqrt{5}/2$
8344 1813 10^x : $2/3+3\sqrt{5}/2$
8344 2001 rt : (7,8,3,1)
8344 3833 rt : (2,7,-1,6)
8344 4004 Γ : $9\sqrt[3]{3}/2$
8344 5778 Γ : $((e+\pi)/2)^2$
8344 6018 cu : $3+\sqrt{5}/2$
8344 6150 sr : $7\sqrt[3]{3}/3$
8344 6454 cr : $8e\pi/3$
8344 6576 rt : (5,9,-5,-5)
8344 6633 rt : (4,0,8,9)
8344 7328 rt : (1,7,8,7)
8344 8270 rt : (3,8,-5,-8)
8344 9274 $s\pi$: $9e\pi/10$
8345 0399 sinh : 23/8
8345 3434 ℓ_{10} : $8e\pi$
8345 3954 rt : (4,6,-3,-1)

8345 4550 sπ : $2\sqrt{2}/9$
8345 8629 at : $3/4 + \sqrt{2}/4$
8345 9841 rt : (7,9,-4,-7)
8346 1431 sq : $3/2 + 3\sqrt{3}/4$
8346 2553 2^x : $4e\pi/3$
8346 5733 rt : (4,5,-7,-7)
8346 8647 rt : (9,9,-3,-9)
8347 0607 ℓ_2 : $4 - \sqrt{3}/4$
8347 3027 sr : $\sqrt{2} + \sqrt{3} - \sqrt{6}$
8347 3143 tπ : $1/4 - \sqrt{2}/3$
8347 3983 $\pi + \ln(2)$
8347 4828 rt : (7,2,-8,-4)
8347 6194 e^x : $3\sqrt{2}/4 + \sqrt{5}/2$
8347 6408 $s\pi(1/3) \times s\pi(\sqrt{2})$
8347 6805 tπ : $5\sqrt[3]{5}/7$
8347 7423 rt : (8,6,-1,-8)
8347 8076 sq : $7e\pi/8$
8347 8457 as : $(\ln 3/2)^{1/2}$
8348 0492 rt : (9,8,4,3)
8348 0704 Γ : 6/19
8348 2236 id : $9\sqrt[3]{2}/4$
8348 3257 rt : (5,3,0,-5)
8348 3510 ℓ_{10} : $7(e+\pi)/6$
8348 3551 10^x : $\exp(-4/3)$
8348 5511 $\ln(\sqrt{5}) \div s\pi(\sqrt{2})$
8348 5540 rt : (2,5,-8,-9)
8348 5658 cu : $7\sqrt{5}/10$
8348 9563 sinh : $\sqrt{2} + \sqrt{3} + \sqrt{7}$
8349 0531 rt : (7,3,1,-7)
8349 4887 rt : (9,3,2,-9)
8349 6136 rt : (7,7,-8,5)
8349 7691 rt : (9,2,-7,-2)
8349 7884 $\ln(5) + \exp(4/5)$
8350 1324 rt : (9,9,-9,-4)
8350 1516 rt : (4,8,-7,7)
8350 2142 rt : (1,9,9,-1)
8350 3215 tπ : $3\sqrt{2} - 2\sqrt{3}$
8350 4999 ℓ_{10} : $4\sqrt[3]{5}$
8350 7605 rt : (6,7,-3,8)
8350 8871 e^x : $7\sqrt[3]{2}/5$
8350 9081 sinh : $(\sqrt{3})^{-1/2}$
8351 0161 ℓ_{10} : $3\sqrt{2} + 3\sqrt{3}/2$
8351 2234 rt : (8,0,4,8)
8351 4168 rt : (2,9,-2,6)
8351 5632 Li_2 : $5\zeta(3)/9$
8351 8552 2^x : $3/4 + 2e$
8352 1190 rt : (8,6,-7,-3)
8352 1817 tπ : $4e/3 - 2\pi$
8352 4375 rt : (6,6,-8,-1)
8352 6116 rt : (3,8,2,-9)
8352 8184 $e + \exp(3/4)$
8352 9096 sinh : $4\sqrt[3]{5}/9$
8353 0105 2^x : $3 - 3\sqrt{2}/4$
8353 0489 sinh : 19/25
8353 0490 rt : (1,8,1,-7)
8353 1075 rt : (4,6,-9,1)
8353 1222 e^x : $(e/3)^{-3}$
8353 3542 λ : $\exp(-7\pi/10)$
8353 3702 e^x : $4 - \sqrt{5}$
8353 3758 rt : (9,3,7,9)
8353 5655 at : 21/19
8353 6335 cr : $2\sqrt{2} + 3\sqrt{5}/2$
8353 8046 J_0 : $(\sqrt[3]{5}/3)^{1/3}$
8353 8851 sπ : $1/3 + 4\sqrt{5}/3$
8353 8888 rt : (1,8,0,-5)
8353 9483 cr : $4\sqrt{3} - \sqrt{5}/3$
8354 0894 rt : (7,3,6,7)
8354 3731 rt : (9,3,-4,-4)
8354 3930 K : $(2\pi/3)^{-1/2}$
8354 6962 10^x : $\sqrt[3]{11/3}$
8354 7923 rt : (6,7,-9,-3)
8354 9522 $\ln(5) - s\pi(e)$
8355 0056 tanh : $(\sqrt[3]{5}/3)^{-1/3}$
8355 1650 rt : (8,8,-9,6)
8355 3692 $1/4 - \exp(3)$
8355 4275 sr : $2\pi/9$
8355 4865 rt : (7,3,-5,-2)
8355 4965 cr : 7/12
8355 4979 sq : $3e/4 + \pi$
8355 5106 rt : (5,3,5,5)
8355 7495 rt : (2,5,4,-8)
8355 8359 rt : (1,9,1,-4)
8356 0650 J_0 : $(\ln 3)^{-2}$
8356 1601 rt : (3,8,1,-3)
8356 2907 e^x : $1/2 - e/4$
8356 5030 rt : (7,8,-2,-8)
8356 5909 rt : (2,8,1,-1)
8356 5967 $\ln(4) \times \exp(\sqrt{3})$
8356 8211 ℓ_2 : $4/3 + \sqrt{5}$
8356 9020 rt : (7,8,-4,9)
8356 9241 rt : (8,6,9,8)
8357 1644 sinh : $\sqrt[3]{11/2}$
8357 3206 Γ : $\exp(\sqrt[3]{5})$
8357 6043 sr : $4/3 + 3\sqrt{5}$
8357 6193 tanh : $(\ln 3)^2$
8357 6250 cr : $3/4 + 2e$
8357 6632 $\ln(2/3) + s\pi(\pi)$
8357 7265 sr : $3\sqrt{2}/4 + 4\sqrt{3}/3$
8357 7331 sπ : $\sqrt[3]{2}/4$
8357 7699 10^x : $1 + 2\pi$
8357 8166 rt : (5,8,-3,-6)
8357 8643 sq : $1/2 - \sqrt{2}$
8358 2025 rt : (8,5,1,-9)
8358 2504 rt : (1,7,2,6)
8358 4228 rt : (3,2,7,-9)
8358 4994 rt : (8,0,-2,3)
8359 1499 λ : 1/9
8359 4331 2^x : $\sqrt{24}$
8359 6920 sr : $((e+\pi)/2)^{-1/3}$
8359 7285 rt : (6,6,8,6)
8359 7591 rt : (1,5,1,1)
8359 8270 cr : $2 + 4\pi/3$
8359 8574 rt : (4,9,-7,-1)
8359 9320 rt : (6,5,0,-7)
8360 0005 rt : (3,8,-4,-4)
8360 3776 10^x : $6\sqrt[3]{2}$
8360 4472 ζ : 5/19
8360 6797 $2/5 - \sqrt{5}$
8360 7126 rt : (4,2,-6,7)
8360 7718 rt : (8,5,-5,-3)
8360 9261 rt : (9,8,-7,-5)
8361 0615 sq : $7\pi/10$
8361 1624 tanh : $4e/9$
8361 2438 Γ : $1/\ln(\sqrt[3]{3}/3)$
8361 4770 rt : (9,3,1,4)
8361 5211 θ_3 : $(\pi/3)^{-2}$
8361 6876 rt : (1,2,6,-7)
8361 6988 ℓ_2 : $\sqrt{3}/4 + 3\sqrt{5}$
8361 7299 $e \times s\pi(\sqrt{5})$
8362 0196 tπ : $e - 2\pi$
8362 0302 rt : (7,2,3,-8)
8362 0772 ln : $8\sqrt[3]{3}/5$
8362 2243 tπ : $3\sqrt{2}/4 + 3\sqrt{3}/2$
8362 4391 rt : (6,0,-3,1)
8362 4584 ℓ_2 : $4 + \pi$
8362 5103 id : $(\sqrt[3]{5})^{-1/3}$
8362 5231 rt : (3,6,-5,6)
8362 7124 erf : $(\pi/3)^{-1/3}$
8362 8009 rt : (4,7,1,3)
8363 0131 at : $3\sqrt{3}/2 - 2\sqrt{5}/3$
8363 1280 sr : $4/3 + 3e/4$
8363 3162 sπ : $3\sqrt{2}/4 - \sqrt{5}/3$
8363 3634 rt : (2,5,-2,-4)
8363 4357 rt : (7,8,-8,-3)
8363 9585 sinh : $4\sqrt[3]{5}/5$
8363 9826 e^x : $2/3 + 4\sqrt{2}/3$
8364 1117 rt : (8,1,6,9)
8364 3474 rt : (1,8,-5,-2)
8364 5135 at : $(e/2)^{1/3}$
8364 5874 tπ : $(\sqrt{2}/3)^{1/3}$
8364 7895 tπ : $\sqrt{3}/3 - \sqrt{5}$
8364 9908 sq : $4/3 - 2e$
8365 0126 ℓ_2 : 7/25
8365 0752 as : $3\sqrt{3}/7$
8365 0840 rt : (6,8,-5,1)
8365 1442 cu : $5e/2$
8365 1654 ln : $5\ln 2/8$
8365 2022 id : $7(e+\pi)/6$
8365 3512 J_0 : 19/23
8365 5033 rt : (2,1,9,8)
8365 5714 rt : (5,2,2,-6)
8365 6103 rt : (8,5,-5,-4)
8365 7531 rt : (3,6,-4,-9)
8365 8101 sinh : $\sqrt[3]{18}$
8365 8901 sr : $4\sqrt{3} + \sqrt{5}/2$
8365 9166 ℓ_2 : $\sqrt[3]{2}/9$
8366 0134 cu : $5\sqrt{5}/4$
8366 0998 rt : (5,3,-7,8)
8366 2841 10^x : $3/2 - \pi/3$
8366 3381 sr : $5\sqrt[3]{2}/9$
8366 6002 sr : 7/10
8366 7387 rt : (7,7,0,-9)
8366 8950 rt : (3,7,-7,-9)
8366 9344 Li_2 : $(\sqrt{5})^{-1/2}$
8366 9782 rt : (7,3,0,2)
8367 1045 cr : $1 + 3\sqrt{3}$
8367 1442 10^x : $4 - 3e/2$
8367 1926 cr : $2 - \sqrt{2}$
8367 6585 cu : $3/4 + 2\sqrt{3}$
8367 6951 rt : (5,8,-9,-1)
8367 7666 rt : (9,2,-2,-5)
8367 7778 sq : $9(e+\pi)/4$
8367 7932 10^x : $2\sqrt{2} - 3\sqrt{3}/4$
8367 9540 K : $\exp(-1/e)$
8367 9652 rt : (4,6,7,4)
8368 0423 cu : $1/3 - 4e$
8368 1496 cr : $(e+\pi)/10$
8368 1941 ζ : $\exp(-4/3)$
8368 3127 rt : (6,1,5,7)
8368 3729 tanh : 23/19
8368 4724 ln : P_{10}
8368 4889 rt : (2,7,2,-5)
8368 5406 10^x : $1/2 - \sqrt{3}/3$
8368 6329 id : $\sqrt{2} - \sqrt{3}/3$
8368 6605 Ψ : $8(e+\pi)$
8369 1551 K : 9/13
8369 1918 rt : (1,7,3,-8)
8369 4454 2^x : $(\sqrt{2}\pi/3)^{-1/3}$
8369 4890 sπ : $4e - 4\pi/3$
8369 8547 rt : (7,9,-6,2)
8369 8821 ln : $\sqrt{3}/4$
8369 9412 rt : (6,4,-8,9)
8370 2063 sπ : $(2e)^3$
8370 3083 tπ : $\sqrt{2} + \sqrt{5}/2$
8370 3217 e^x : $e/3 + 3\pi$
8370 3814 rt : (1,5,2,-7)
8370 3968 sinh : $4\sqrt[3]{5}/3$
8370 5139 at : $8\ln 2/5$
8370 5350 rt : (5,8,0,4)
8370 5990 rt : (6,5,-6,-2)
8370 9692 cr : $4\sqrt{2}/3 - 3\sqrt{3}/4$
8370 9867 at : $1/4 - e/2$
8371 0252 rt : (7,4,8,8)
8371 1730 sr : $(3/2)^3$
8371 2492 id : $(e+\pi)/7$
8371 2795 as : $7(1/\pi)/3$
8371 4161 rt : (9,7,-5,-6)
8371 4367 rt : (8,6,3,3)
8371 4993 $\pi \times \exp(1/5)$
8371 5919 ℓ_{10} : $3\sqrt{3} + 3\sqrt{5}/4$
8371 6466 tπ : $7e\pi$
8371 6647 sπ : 6/19
8371 6815 cu : $1/4 - 4\sqrt{3}$
8371 6947 cu : $3\pi/10$
8371 7581 rt : (2,4,6,-9)
8371 7885 sq : $4/3 + \sqrt{3}/2$
8371 8060 λ : $\ln(\sqrt{5}/2)$
8371 8084 2^x : $3/4 + 3\sqrt{2}$
8372 1708 erf : $\pi^2/10$
8372 3467 sr : $2 - 3\sqrt{3}/4$
8372 5466 sq : $2/3 + 3\pi$
8372 5650 rt : (6,4,2,-8)
8372 6300 Ei : $10\pi/9$
8372 7091 J_0 : $4\sqrt[3]{3}/7$
8372 8960 e^x : $(\sqrt{2}\pi)^{-1/3}$
8373 1054 as : $(2e/3)^{-1/2}$
8373 3040 rt : (9,8,-2,-2)
8373 3313 tanh : $6\sqrt{2}/7$
8373 4091 Ψ : $8\pi/9$
8373 4700 rt : (7,2,-3,-3)

8373 5959 $\Psi : 5\zeta(3)/7$
8373 6664 sq : $2\sqrt{3}/3 - 3\sqrt{5}/2$
8373 6745 $3/5 \times \exp(1/3)$
8374 0859 tanh : $7\sqrt{3}/10$
8374 2583 sq : $3e/2 - \pi/4$
8374 5208 $\ell_2 : 2\sqrt{2} + \sqrt{5}/3$
8374 5793 rt : (7,1,5,-9)
8374 5988 ln : $10\ln 2/3$
8374 6033 sinh : $9\pi/5$
8374 7927 sr : $4\sqrt{2}/3 + 2\sqrt{5}/3$
8374 9136 $K : 2\sqrt{3}/5$
8374 9947 rt : (2,9,3,-8)
8375 0267 $J_0 : 14/17$
8375 0397 cr : $\sqrt{3} + 2\sqrt{5}$
8375 1038 rt : (6,9,-1,5)
8375 4455 $s\pi : 2\sqrt[3]{5}/5$
8375 5687 sr : $3\sqrt{2} - \sqrt{3}/2$
8375 6543 rt : (8,0,-8,2)
8375 7939 tanh : $7\ln 2/4$
8375 8051 $\ell_2 : 2\sqrt{3} - 3\sqrt{5}/4$
8375 8177 rt : (9,9,6,4)
8376 2104 cu : $4e + 3\pi$
8376 3088 rt : (8,1,0,4)
8376 5456 rt : (7,7,-6,-4)
8376 5506 $s\pi : (\sqrt{3})^{1/2}$
8376 5780 rt : (4,1,4,5)
8376 7911 rt : (3,7,-2,-5)
8376 8906 $t\pi : \sqrt{5/2}$
8376 9337 sq : $3/4 + 3\pi/2$
8377 0651 $t\pi : 2e + \pi/4$
8377 1336 rt : (1,9,2,5)
8377 2233 rt : (1,8,6,0)
8377 2605 $\exp(3) \div s\pi(\sqrt{2})$
8377 3707 $e^x : 3/2 - 3\sqrt{5}/4$
8377 5677 $J : 2/15$
8377 6111 cr : $4/3 - \sqrt{5}/3$
8377 6870 ln : $3\sqrt[3]{3}/10$
8377 7784 $10^x : e/4 + 3\pi/4$
8377 8525 $1/4 + s\pi(1/5)$
8377 8826 sinh : 16/21
8378 1864 rt : (5,7,-6,-7)
8378 5135 rt : (3,7,-5,-2)
8378 5920 $K : \ln 2$
8378 6198 $\Gamma : 24/5$
8378 6281 rt : (8,4,-3,-5)
8378 7706 id : $\ln(2\pi)$
8378 8360 cr : 10/17
8379 1161 rt : (9,4,3,5)
8379 1879 $s\pi : 1/\sqrt{10}$
8379 3319 $10^x : 23/2$
8379 4083 $2^x : 1/3 + 4\pi$
8379 4433 rt : (4,8,5,7)
8379 6046 tanh : 17/14
8379 7372 $\exp(\sqrt{5}) \div s\pi(2/5)$
8379 7544 $Li_2 : \ln((e+\pi)/3)$
8379 8122 at : 10/9
8379 9445 $\Gamma : 6(e+\pi)/7$
8379 9655 id : $\sqrt{2}/3 - 4\sqrt{3}/3$
8379 9846 rt : (8,9,-6,-6)
8380 1321 $Ei : (1/\pi)/10$
8380 2889 $\Psi : 3(e+\pi)$
8380 3240 $e^x : 14/23$
8380 4564 rt : (9,3,-5,-1)
8380 4791 $s\pi : 1/\ln(\pi/3)$
8380 5248 cu : $2\sqrt{2}/3$
8380 5296 rt : (5,4,7,6)
8380 6937 rt : (9,1,0,-6)
8380 8748 rt : (4,9,-2,-7)
8380 9146 $\zeta : (\sqrt{3}/2)^{1/2}$
8381 0042 $K : \sqrt[3]{1/3}$
8381 0656 sr : $(\ln 2/2)^{1/3}$
8381 1232 $\exp(2/3) \times s\pi(\pi)$
8381 2385 rt : (1,6,-5,5)
8381 6185 rt : (9,6,-3,-7)
8381 6364 $s\pi : 3/4 + 4\pi$
8381 7340 $s\pi : 3\sqrt{2} - \sqrt{5}/4$
8381 8261 $J_0 : (\sqrt{2}\pi/3)^{-1/2}$
8381 9421 ln : $1/2 + 2e/3$
8382 0685 rt : (5,1,4,-7)
8382 1503 $10^x : \sqrt{2} - \sqrt{3} + \sqrt{5}$
8382 2263 rt : (9,2,-8,0)
8382 2598 $10^x : e/6$
8382 6096 $e^x : 3/2 + \sqrt{5}/4$
8382 6654 sq : $8e/9$
8382 8403 rt : (5,9,4,8)
8383 0113 cr : $\exp(-\sqrt[3]{4}/3)$
8383 1337 cr : $3/2 + 3\pi/2$
8383 3041 rt : (5,4,-9,1)
8383 5054 at : $\exp(1/(3\pi))$
8383 9828 $\ln(\sqrt{2}) + \exp(2/5)$
8383 9934 $erf : 4\sqrt{3}/7$
8384 0495 rt : (4,4,-9,-8)
8384 3008 $\exp(3/4) \div s\pi(\sqrt{3})$
8384 3953 $\Psi : 5\sqrt{5}/4$
8384 4595 $2^x : 9\sqrt{2}/5$
8384 4767 rt : (1,1,8,-8)
8384 5240 rt : (4,9,-6,-9)
8384 5480 $J_0 : (e/3)^2$
8384 5726 sq : $3/2 + 3\sqrt{3}$
8384 6673 rt : (6,2,7,8)
8384 6925 rt : (2,6,-8,-6)
8384 7908 rt : (6,3,4,-9)
8384 8315 $\exp(1/5) \div s\pi(\pi)$
8384 8337 $Ei : 2(e+\pi)/7$
8384 8668 $erf : 7\sqrt{2}/10$
8384 9712 $10^x : \sqrt{2} - 2\sqrt{5}/3$
8385 1648 rt : (5,8,2,-1)
8385 2549 id : $3\sqrt{5}/8$
8385 5789 $10^x : 9e\pi/5$
8385 9708 $erf : 10\ln 2/7$
8385 9890 $s\pi : \sqrt{11}$
8385 9901 $Ei : 19/3$
8386 3790 $t\pi : 1/4 - 2\sqrt{5}$
8386 5679 cr : $4\sqrt{2} + \sqrt{5}/4$
8386 6667 rt : (2,4,-3,5)
8386 6847 sr : $\sqrt{22}\pi$
8386 9177 $2^x : 3 - 3\sqrt{2}/2$
8387 1937 rt : (6,9,-7,-4)
8387 4964 $J_0 : (2e/3)^{-1/3}$
8387 5140 tanh : $(\sqrt{2}\pi/3)^{1/2}$
8387 5912 $s\pi : 6\sqrt{5}/5$
8387 8730 rt : (3,8,1,7)
8387 9046 rt : (6,4,-4,-3)
8388 3403 sq : $4\sqrt{2} - 3\sqrt{5}/4$
8388 3476 cu : $2 + \sqrt{2}/2$
8388 6070 $10^x : 4\sqrt{2}/3 + 4\sqrt{5}/3$
8388 9138 $\ell_2 : 3 + \sqrt{3}/3$
8389 0047 as : $(e/3)^3$
8389 0451 rt : (1,7,-3,-3)
8389 2066 rt : (7,6,-4,-5)
8389 3821 rt : (6,1,-1,2)
8389 5034 $10^x : ((e+\pi)/2)^{-1/2}$
8389 7760 id : $4e/3 - \pi/4$
8389 8364 $Ei : \exp(-5/2)$
8389 9296 sq : $\beta(2)$
8389 9674 rt : (8,8,-4,-7)
8390 1249 cr : $1/\ln(2e)$
8390 3255 rt : (2,9,-3,-5)
8390 3595 $\ell_2 : \sqrt{5}/4$
8390 4901 tanh : $\sqrt{2} + \sqrt{6} - \sqrt{7}$
8390 6150 rt : (7,1,-1,-4)
8390 6339 $e^x : 2 + \sqrt{2}/3$
8390 6439 ln : $10\sqrt[3]{5}$
8390 6841 tanh : $(e/3)^{-2}$
8390 7490 $e^x : 24/23$
8390 9963 $t\pi : 2/9$
8391 0164 $10^x : 1/3 + \sqrt{3}/4$
8391 1149 $\Psi : (e/2)^{-1/2}$
8391 2036 rt : (8,3,-1,-6)
8391 2258 $t\pi : 4e + \pi/4$
8391 4076 $e^x : 7e/6$
8391 5120 cr : 13/22
8391 5482 rt : (9,5,-1,-8)
8391 7305 $10^x : 1/2 + 4\pi$
8391 9273 $\Psi : 10/19$
8392 1875 $s\pi : 1/4 + \sqrt{3}/4$
8392 1894 $J_0 : (\ln 3/2)^{1/3}$
8392 4656 $\ln(4) - \exp(4/5)$
8392 4895 cu : $3(e+\pi)/7$
8392 5712 cr : $2e + \pi/4$
8392 6001 $2/5 \times \exp(4)$
8392 6349 sq : $1 - 3\pi/4$
8392 6820 $e^x : 2\sqrt{2}/3 - \sqrt{5}/2$
8392 8675 rt : (9,9,-9,9)
8392 8905 sr : $2e/3 + \pi/2$
8392 9398 $\ell_{10} : e + 4\pi/3$
8393 0993 at : $7(1/\pi)/2$
8393 1849 rt : (9,0,2,7)
8393 2422 $e^x : \pi^2/3$
8393 2559 $\ell_2 : 3/2 + 4\sqrt{2}$
8393 2965 rt : (8,2,2,5)
8393 5294 rt : (7,4,2,3)
8393 6301 ln : $2\sqrt{2} + 2\sqrt{3}$
8393 9720 $5 \div e$
8393 9778 $\ell_{10} : 7\pi^2$
8394 1205 tanh : $(2e/3)^{1/3}$
8394 1665 $\Gamma : 3\sqrt{5}/2$
8394 3149 rt : (6,6,2,1)
8394 3815 cu : $4\sqrt{3} + 4\sqrt{5}$
8394 4257 at : $3 - 4\sqrt{2}/3$
8394 4595 $s\pi : 7\zeta(3)/5$
8394 7576 rt : (5,5,3,-9)
8395 1773 $J_0 : 9/11$
8395 1794 $Ei : 3\sqrt{5}/5$
8395 3845 rt : (4,3,3,-7)
8395 7937 $\zeta : 17/22$
8395 7996 rt : (2,9,-1,-6)
8395 8870 sq : $\ln(5/2)$
8395 9498 rt : (9,5,5,6)
8396 0510 cu : $e/3 + \pi/4$
8396 0556 rt : (9,6,-9,-2)
8396 0913 $s\pi : (\pi)^{-1/3}$
8396 4297 tanh : $e\pi/7$
8396 5014 cu : 15/7
8396 6338 rt : (3,1,3,-5)
8396 9383 $Li_2 : 1/\ln(\sqrt{2}\pi)$
8397 2473 rt : (8,4,-9,0)
8397 2649 $2^x : \sqrt{3}/2 - \sqrt{5}/2$
8397 4995 $\ln(\sqrt{3}) - \exp(2)$
8397 5787 $\zeta : 5(1/\pi)/6$
8397 6089 rt : (8,7,5,4)
8397 8113 at : $(\sqrt[3]{3}/2)^{-1/3}$
8397 8771 rt : (5,0,6,8)
8398 0282 cr : $3e/4 + 4\pi/3$
8398 1646 $10^x : (\sqrt{2}\pi/2)^{-1/3}$
8398 1955 $\Psi : (1/\pi)/3$
8398 2295 sr : $2/3 + e$
8398 3025 rt : (3,7,-1,-4)
8398 3705 cr : $2\sqrt{2} - \sqrt{5}$
8398 5413 $2^x : e/4 + 2\pi/3$
8398 8132 $s\pi : 7\sqrt[3]{3}/6$
8399 0378 id : $4\sqrt[3]{5}$
8399 0819 rt : (1,8,1,-2)
8399 1218 $10^x : \exp(1/e)$
8399 1972 ln : $\sqrt{2} + \sqrt{5} + \sqrt{7}$
8399 2197 rt : (4,9,-8,-2)
8399 2510 rt : (1,6,-4,-7)
8399 4736 id : $2\sqrt[3]{2}/3$
8399 5156 rt : (6,8,-5,-5)
8399 5954 tanh : $(\ln 3/2)^{-1/3}$
8399 6883 rt : (8,7,-2,-8)
8399 8605 rt : (7,2,-9,2)
8400 0000 id : 21/25
8400 0434 rt : (9,1,-6,-1)
8400 0491 sr : $3/2 + 4\sqrt{2}/3$
8400 3457 rt : (6,3,9,9)
8400 4714 $s\pi : 3\sqrt{2}/2 + 3\sqrt{3}$
8400 7001 tanh : $5\sqrt[4]{5}/7$
8400 7181 $10^x : 6e/5$
8400 9633 $\theta_3 : 4\sqrt{2}/7$
8401 1449 $erf : 4\sqrt{5}/9$
8401 2190 rt : (9,4,1,-9)
8401 3067 rt : (4,9,-6,2)
8401 3447 $Li_2 : 3\sqrt{5}/10$
8401 3496 $10^x : 9\ln 2$
8401 4476 rt : (7,5,-2,-6)
8401 5050 $J_0 : \sqrt{2/3}$
8401 6805 sr : 12/17
8401 7507 at : $1/4 + \sqrt{3}/2$
8401 8382 rt : (5,9,-2,-3)

8401 8663 $\Psi : (\sqrt[3]{4})^{-1/3}$
8401 9135 $\ell_2 : P_{69}$
8402 1942 rt : (5,6,-1,9)
8402 2841 $\ell_2 : P_{72}$
8402 6576 rt : (3,7,-8,0)
8402 7777 sq : 11/12
8403 0846 tanh : 11/9
8403 1317 rt : (7,9,5,2)
8403 2973 $Li_2 : (\sqrt{2}\pi/2)^{-1/2}$
8403 3649 rt : (8,2,1,-7)
8403 6121 $t\pi : 4\sqrt{3} + 2\sqrt{5}/3$
8403 6404 sq : $4 - \sqrt{5}/4$
8403 6649 cu : $3/4 - 3e/2$
8403 7530 $2^x : 22/25$
8403 9587 $\Psi : 6/7$
8404 0046 rt : (5,5,9,7)
8404 2749 cu : $1/\ln(\ln 2/2)$
8404 2838 rt : (8,1,-6,-1)
8404 4295 rt : (6,3,-2,-4)
8404 4902 rt : (4,7,2,-9)
8404 5555 $\sqrt{5} - \exp(1/3)$
8404 6267 cr : $4\sqrt{2} + \sqrt{3}/3$
8404 6377 rt : (2,4,-4,-5)
8404 8356 $10^x : e/3 + \pi/2$
8404 8697 ln : $5\sqrt[3]{2}$
8405 0105 sr : $2\sqrt{2} + \sqrt{5}/4$
8405 2678 rt : (9,1,4,8)
8405 2998 $J_0 : 3e/10$
8405 4698 $\Psi : 14/5$
8405 6748 cr : $3/2 - e/3$
8405 6781 $s\pi : \exp(3\pi/2)$
8405 8156 $Ei : 7\sqrt[3]{2}/9$
8405 8700 rt : (2,8,-1,-6)
8406 2061 $\ell_{10} : 4\sqrt{3}$
8406 2352 $10^x : 3/2 + 2\sqrt{5}/3$
8406 4603 cr : $4 + \sqrt{5}$
8406 6312 cu : $3\sqrt{2} - 2\sqrt{5}/3$
8406 6532 rt : (4,8,-1,-4)
8406 7130 cr : $6\ln 2/7$
8406 9581 $K : 16/23$
8406 9943 rt : (7,0,1,5)
8407 1689 id : $3\sqrt{2} + 3\sqrt{3}/2$
8407 2994 $s\pi : 8e\pi$
8407 6121 rt : (3,4,6,4)
8407 6682 rt : (7,9,-3,-8)
8407 8549 rt : (6,9,-5,-7)
8408 0607 $erf : 1/\ln(\ln 3/3)$
8408 2546 $\ell_{10} : 10\ln 2$
8408 2859 rt : (3,5,2,-7)
8408 3003 $\ln(4) \div \exp(1/2)$
8408 6270 rt : (9,5,-7,-3)
8408 6404 cr : $\ln(2e/3)$
8408 6803 cr : $9\ln 2$
8408 9641 id : $(\sqrt{2})^{-1/2}$
8408 9666 at : 19/17
8409 0267 $\Gamma : 6\pi^2/5$
8409 0732 $\zeta : 25/21$
8409 1392 rt : (1,8,-8,-5)
8409 1424 $\Psi : 5e/2$
8409 1603 rt : (8,6,0,-9)
8409 5331 rt : (8,3,4,6)
8409 5958 $\ell_{10} : \sqrt[3]{3}/10$
8409 6656 $s\pi : 1/\ln(2e/3)$
8409 7683 ln : $4\sqrt{2}/3 + \sqrt{3}/4$
8409 8627 rt : (4,1,-9,5)
8410 0457 rt : (3,6,-4,9)
8410 1610 rt : (6,0,-9,4)
8410 1703 rt : (2,5,-9,5)
8410 4152 $\Psi : 2\ln 2/9$
8410 4825 tanh : $\sqrt{3/2}$
8410 5853 rt : (5,1,-2,-2)
8410 6792 rt : (4,2,5,-8)
8410 6867 at : $\sqrt{5}/2$
8410 8267 rt : (1,7,9,-8)
8411 4380 rt : (6,7,-3,-6)
8411 4437 $\ell_{10} : 3/2 + 2e$
8411 6977 $4/5 \div s\pi(2/5)$
8411 7063 rt : (7,8,-9,0)
8411 8969 $2^x : \sqrt{2}/4 + 3\sqrt{3}$
8412 0465 rt : (9,6,7,7)
8412 2207 rt : (1,6,-1,-4)
8412 5353 $s\pi : 7/22$
8412 9248 cr : $2 + 3\sqrt{2}$
8413 0047 cu : $10e\pi/3$
8413 0503 rt : (5,1,8,9)
8413 2491 rt : (5,6,-8,-8)
8413 2830 rt : (2,8,2,3)
8413 2947 rt : (7,4,0,-7)
8413 4509 cu : $2/3 + \sqrt{5}/4$
8413 4524 rt : (9,9,-8,-5)
8413 7856 at : $1/3 + \pi/4$
8413 9382 $2^x : 5e/7$
8413 9986 rt : (3,9,1,-1)
8413 9994 sq : $2\sqrt{3}/3 - 3\sqrt{5}$
8414 0395 sr : $2\sqrt{3}/3 + \sqrt{5}$
8414 0521 cr : $2\sqrt{3}/3 - \sqrt{5}/4$
8414 1383 $t\pi : 3\sqrt{5}/5$
8414 1829 rt : (2,5,-8,2)
8414 2012 $\Psi : \sqrt[3]{22}$
8414 3552 sr : $4(e + \pi)$
8414 3712 rt : (8,3,-7,-1)
8414 3983 id : $7\zeta(3)/10$
8414 4559 rt : (6,2,1,3)
8414 4579 sinh : 13/17
8414 4815 $e^x : 4\sqrt[3]{3}/3$
8414 6785 rt : (4,7,9,5)
8414 7011 sinh : $(\sqrt[3]{5})^{-1/2}$
8414 7098 $s\pi : (1/\pi)$
8415 1160 sr : $\sqrt{23/2}$
8415 1378 rt : (8,1,3,-8)
8415 3022 rt : (4,8,-6,-3)
8415 4937 rt : (6,7,-8,-1)
8415 7642 rt : (2,3,2,6)
8416 2541 sr : 17/24
8416 2771 sr : $1/4 + \pi$
8416 4682 $J_0 : 13/16$
8416 8760 rt : (2,8,-3,-7)
8416 9487 rt : (7,8,-1,-9)
8416 9674 rt : (9,2,6,9)
8417 0075 rt : (9,0,-4,2)
8417 0376 $Ei : 9\ln 2/10$
8417 5231 ln : $3 - e/4$
8417 7384 cr : $(3\pi/2)^{-1/3}$
8417 7829 rt : (5,5,-3,-4)
8417 8443 $2^x : 3/4 + \pi$
8417 9357 cu : $3e\pi/8$
8418 2669 rt : (7,5,4,4)
8418 3063 cu : $1 + 4\pi$
8418 3673 sq : 19/14
8418 4814 $Ei : 9\pi/10$
8418 6489 $2^x : 2(e + \pi)$
8418 6687 $10^x : 5(1/\pi)/6$
8418 7311 $\theta_3 : 4\ln 2/7$
8418 7896 rt : (4,2,-5,9)
8418 8800 rt : (8,7,-8,-3)
8418 9861 $2^x : 8(e + \pi)/3$
8419 0954 ln : $\sqrt[3]{25/2}$
8419 1037 ln : $3/4 + \pi/2$
8419 1046 at : $8\sqrt[3]{2}/9$
8419 2591 rt : (3,0,5,6)
8419 3346 rt : (7,9,-5,4)
8419 7555 $\ell_{10} : 2/3 + 2\pi$
8419 7700 $K : \sqrt{2} + \sqrt{3} - \sqrt{6}$
8419 7835 $e^x : 25/13$
8419 9773 $t\pi : \sqrt{3} + 2\sqrt{5}/3$
8420 0637 $10^x : 2\sqrt{5}/5$
8420 0828 $t\pi : 6\sqrt{3}$
8420 0892 $\ell_{10} : P_{75}$
8420 1038 rt : (1,8,7,-8)
8420 1525 $s\pi : 4\sqrt{2}/3 + \sqrt{3}/4$
8420 1574 sq : $\sqrt[3]{5/2}$
8420 2392 rt : (6,2,0,-5)
8420 4075 ln : $4 + 4\sqrt{3}/3$
8420 5045 id : $3\sqrt{3} - 3\sqrt{5}/2$
8420 6728 rt : (7,3,-6,-3)
8420 6775 $2^x : 8(1/\pi)$
8420 7733 rt : (9,4,-5,-4)
8420 7960 rt : (2,7,1,-7)
8420 8070 sr : $4 + 3e/2$
8421 0526 id : 16/19
8421 0851 $e^x : 1 + \sqrt{3}/3$
8421 1864 rt : (5,9,-4,-6)
8421 6947 rt : (1,9,-9,-7)
8421 7588 $\Gamma : 2\zeta(3)/5$
8421 7852 id : $\sqrt{3}/2 - 3\sqrt{5}$
8422 0202 rt : (8,8,7,5)
8422 0907 $\ln(3/5) \times \exp(1/2)$
8422 3341 $J : \exp(-8\pi)$
8422 3750 sq : $3\sqrt{3}/4 + \sqrt{5}/2$
8422 5866 rt : (3,7,0,-5)
8422 6026 $\zeta : 4/15$
8422 6715 rt : (7,1,3,6)
8422 7067 rt : (4,3,8,7)
8422 7505 $2^x : 1/\ln(\sqrt{2}\pi/3)$
8422 8013 sinh : $3\sqrt[3]{3}$
8422 8838 $t\pi : 7(1/\pi)/10$
8422 9143 rt : (9,8,-6,-6)
8422 9855 rt : (6,6,-1,-7)
8423 0582 sinh : $7\sqrt[3]{3}/2$
8423 0962 rt : (3,4,4,-8)
8423 2921 rt : (2,9,-9,0)
8423 5285 $\ell_{10} : 4\sqrt{2} + 3\sqrt{3}/4$
8423 6618 ln : $4e + 2\pi$
8423 7167 $\ln(5) \times \ln(\pi)$
8423 8741 $\Psi : \exp(-\sqrt[3]{5})$
8423 8844 rt : (1,7,4,-1)
8424 2112 cu : 17/18
8424 3923 $\Gamma : \sqrt[3]{2}/4$
8424 5277 $\Gamma : \sqrt[3]{3}/3$
8424 5898 rt : (6,8,-4,3)
8424 7718 rt : (7,3,2,-8)
8424 7745 $e^x : 11/18$
8424 9715 $\ell_{10} : 1/4 + 3\sqrt{5}$
8425 0064 rt : (3,1,-4,8)
8425 0720 $e^x : 3/2 + e/4$
8425 0803 rt : (8,4,6,7)
8425 2689 at : $1 - 3\sqrt{2}/2$
8425 3288 rt : (1,8,2,6)
8425 3759 rt : (4,1,7,-9)
8425 4747 $s\pi : 8\pi^2/3$
8425 6674 $\Gamma : \exp(-5)$
8425 7892 $Ei : \exp(-\sqrt{2}/3)$
8425 8985 $e^x : \sqrt[3]{11/2}$
8425 9072 $2^x : 2 + 3\sqrt{3}/4$
8425 9390 $s\pi : -\sqrt{2} + \sqrt{6} + \sqrt{7}$
8425 9903 id : $\sqrt{2}/4 - 3\sqrt{3}$
8426 0066 $\Psi : 7\zeta(3)/3$
8426 5454 rt : (8,0,5,9)
8426 6795 rt : (7,1,-7,1)
8426 8071 $\lambda : \exp(-9\pi/13)$
8426 8478 cr : $3\sqrt{2}/4 + 3\sqrt{3}$
8427 0013 $e^x : (2e/3)^{1/2}$
8427 0079 $erf : 1$
8427 0257 rt : (3,6,-6,-1)
8427 1703 at : $7\sqrt[3]{3}/9$
8427 3479 $\Gamma : (\ln 2/3)^{1/2}$
8427 4660 rt : (9,7,9,8)
8427 5713 $J_0 : 17/21$
8427 6236 rt : (1,8,-7,4)
8427 8116 $\ell_{10} : e/4 + 2\pi$
8427 8942 $e^x : 4\sqrt{2} - \sqrt{5}/3$
8427 9891 ln : $3\sqrt{3} + \sqrt{5}/2$
8428 0230 tanh : 16/13
8428 1857 rt : (7,5,-8,-1)
8428 3750 ln : $4 - 3\sqrt{5}/4$
8428 4873 rt : (1,5,2,6)
8428 5777 $2^x : 7\sqrt[3]{2}/10$
8428 7566 $s\pi : 2\sqrt{2} + 2\sqrt{5}/3$
8428 8495 $2^x : 2 - \sqrt{5}/2$
8428 8713 rt : (7,9,-9,-3)
8429 0583 $erf : \zeta(11)$
8429 3779 $e^x : 2/3 + e/4$
8429 7606 $e^x : 1 + 2\sqrt{3}$
8430 1695 $J_0 : (\sqrt[3]{4}/3)^{1/3}$
8430 2979 rt : (2,6,-7,-2)
8430 3237 at : $(\sqrt[3]{2})^{1/2}$
8430 4885 $10^x : 3/2 + 4\sqrt{3}$
8430 5717 sinh : 19/6
8430 7033 rt : (8,6,-3,-2)
8430 7440 $J_0 : 7\ln 2/6$

8430 7454 $\zeta : 2\zeta(3)/9$
8431 4575 sq : $4 - \sqrt{2}/2$
8431 7129 cu : 17/12
8432 0212 $3 \times \exp(2/3)$
8432 1347 rt : (9,7,-4,-7)
8432 4477 rt : (5,8,-2,-7)
8432 5216 rt : (9,3,-3,-5)
8432 7892 $J_0 : 4\sqrt{2}/7$
8432 9359 rt : (1,9,0,-7)
8432 9955 rt : (6,5,-9,4)
8433 0342 $\ell_2 : \exp(2e)$
8433 0583 rt : (5,4,-1,-5)
8433 1950 rt : (9,1,-2,3)
8433 1968 $\zeta : 1/\sqrt{14}$
8433 2641 sr : $5e/4$
8433 4661 rt : (6,9,0,7)
8433 6354 rt : (7,7,-6,9)
8433 7928 $2^x : 15/17$
8433 8104 $s\pi : \sqrt{5}/7$
8433 9829 $t\pi : 7\pi/4$
8434 0448 rt : (1,7,9,-7)
8434 1110 $erf : 5\zeta(3)/6$
8434 1800 rt : (6,5,1,-8)
8434 3266 id : $(5/3)^{-1/3}$
8434 3941 $e^x : 10\sqrt[3]{5}/3$
8434 4348 $t\pi : P_{58}$
8434 4353 $e^x : 8\zeta(3)/5$
8434 6031 $\exp(5) - s\pi(\pi)$
8434 6437 $t\pi : P_{19}$
8434 6601 rt : (9,5,-1,1)
8434 7079 rt : (3,7,-4,1)
8434 7229 $e^x : 24$
8435 1489 rt : (2,6,3,6)
8435 2216 sinh : $5\sqrt{2}/4$
8435 2528 id : $6\pi^2/5$
8435 3281 $erf : \zeta(9)$
8435 3447 $K : 2\pi/9$
8435 3913 rt : (6,1,2,-6)
8435 5166 $e^x : 3\pi/7$
8435 7304 $t\pi : 1/\sqrt[3]{10}$
8435 8115 $\Gamma : (\ln 2)^2$
8435 9004 rt : (7,2,4,-9)
8436 0760 $2^x : 1/2 - \sqrt{5}/3$
8436 2276 $\ell_2 : 1/4 + 4\sqrt{3}$
8436 4551 rt : (3,6,-3,-5)
8436 4936 tanh : $\pi^2/8$
8436 5294 sq : $3/4 + 3\sqrt{5}/2$
8436 6002 sinh : $(\sqrt{2}\pi/2)^{-1/3}$
8436 6583 rt : (1,9,-8,3)
8436 7658 rt : (1,7,-7,-3)
8436 8673 rt : (2,7,-5,-8)
8437 3427 rt : (3,3,6,-9)
8437 5000 cu : $3\sqrt[3]{2}/4$
8437 5624 $\ell_2 : 3/4 - \sqrt{2}/3$
8437 5959 $s\pi : 3\sqrt{2}/2 + \sqrt{5}/4$
8437 7025 rt : (7,2,5,7)
8437 8261 rt : (2,8,-9,-4)
8437 8516 $10^x : (3\pi)^{1/2}$
8437 8973 rt : (6,3,3,4)
8437 9584 $\theta_3 : 1/\sqrt[3]{16}$
8438 1033 $2^x : 4/3 + 2\sqrt{2}/3$
8438 1124 $\Gamma : (1/\pi)/2$
8438 4746 $10^x : \sqrt{3}/4 + 3\sqrt{5}/4$
8438 5524 rt : (5,4,1,1)
8438 6697 id : $\sqrt[3]{23}$
8438 7585 $Ei : 9\sqrt{3}/7$
8439 0516 $\ell_2 : \sqrt{2}/3 + 3\sqrt{5}$
8439 0889 sr : 17/5
8439 1429 cr : $\zeta(3)/2$
8439 4401 $\ell_{10} : 4 + 4\sqrt{5}/3$
8439 4422 $s\pi : 6e\pi/7$
8439 7199 $t\pi : \exp(-3/2)$
8439 7373 rt : (8,9,-5,-7)
8439 8950 rt : (1,4,-8,-9)
8439 9923 rt : (8,5,8,8)
8440 0231 $\ln(\sqrt{5}) - \exp(1/2)$
8440 2049 rt : (6,6,-5,8)
8440 2962 rt : (9,9,0,-1)
8440 3871 rt : (7,8,-7,-4)
8440 5171 rt : (4,9,-5,5)
8440 5981 rt : (3,1,7,7)
8440 7570 rt : (6,8,-7,-8)
8440 7639 $\zeta : 21/17$
8440 7949 rt : (5,4,-8,3)
8441 0810 tanh : 21/17
8441 1254 rt : (9,6,-2,-8)
8441 1810 $s\pi : \exp(e + \pi)$
8441 4120 $\ell_2 : 4\pi/7$
8441 4169 rt : (7,6,6,5)
8441 5206 rt : (6,7,-9,-1)
8441 5398 at : 9/8
8441 5895 $1/3 - \ln(3/5)$
8441 7851 $10^x : 1/\sqrt[3]{5}$
8442 0843 rt : (2,9,6,-7)
8442 1481 rt : (8,5,-4,-5)
8442 2950 $10^x : 7\zeta(3)/2$
8442 3102 $\sqrt{2} \times e$
8442 3703 $10^x : 3\sqrt{2} - 2\sqrt{5}/3$
8442 6537 id : $\sqrt{2}/4 + 2\sqrt{5}/3$
8442 7616 ln : $2/3 + 4\sqrt{2}$
8443 2159 rt : (3,9,-5,-4)
8443 2792 $s\pi : 8/25$
8443 3722 rt : (5,8,1,6)
8443 5629 $K : ((e + \pi)/2)^{-1/3}$
8443 6109 as : $2/3 - \sqrt{2}$
8443 7232 tanh : $6\sqrt[3]{3}/7$
8443 8121 rt : (2,2,5,4)
8443 8965 rt : (9,2,-1,-6)
8443 9097 $2^x : 1/3 - \sqrt{3}/3$
8443 9281 rt : (7,4,-6,-2)
8443 9510 id : $1/4 - 2\pi/3$
8444 1153 sinh : $10e\pi/7$
8444 1418 tanh : $(\sqrt[3]{4}/3)^{-1/3}$
8444 4711 $\ln(3) - s\pi(\sqrt{3})$
8444 4958 rt : (2,9,4,-4)
8444 5253 $2^x : 2 + \sqrt{5}$
8444 6657 $\ln(4) \times \exp(3)$
8444 6865 cr : P_2
8444 8868 rt : (8,9,9,6)
8444 8883 $\ell_{10} : 4/3 + 4\sqrt{2}$
8445 0419 rt : (6,4,3,-9)
8445 0508 $s\pi : 4\sqrt[3]{2}/3$
8445 1132 sinh : π^2
8445 1477 $\sqrt{2} - s\pi(\pi)$
8445 3480 $2^x : 1 + 2\sqrt{2}/3$
8445 4251 $J_0 : \ln(\sqrt{5})$
8445 4766 rt : (4,6,-2,1)
8445 5375 $\ln(2) \times \exp(4)$
8445 6473 sr : $4e\pi$
8445 7281 $erf : \sqrt{2} + \sqrt{5} - \sqrt{7}$
8445 7586 cu : $4/3 + 2\pi$
8445 7883 $2^x : \sqrt[3]{22/3}$
8445 8689 $erf : \ln(\ln 3/3)$
8446 0772 $t\pi : \sqrt{3}/3 - 3\sqrt{5}/2$
8446 1044 $\zeta : 12/17$
8446 1118 sinh : $\ln(e + \pi)$
8446 1513 rt : (6,7,4,2)
8446 3119 rt : (8,1,-3,-3)
8446 3866 rt : (2,8,-7,-1)
8446 4880 $10^x : 3\sqrt{3} + 3\sqrt{5}/2$
8446 4915 tanh : $5\sqrt{3}/7$
8446 5169 $s\pi : 8\sqrt[3]{5}$
8446 5546 $2^x : \sqrt{2}/4 + 2\sqrt{3}/3$
8446 8564 rt : (1,8,2,-8)
8446 9127 $e^x : 23/22$
8447 2299 tanh : $7\sqrt{2}/8$
8447 4956 $10^x : 4 - 2\sqrt{2}$
8447 5958 rt : (1,0,-9,7)
8447 7167 rt : (5,3,1,-6)
8447 8003 rt : (6,3,-8,1)
8447 8633 rt : (2,5,-8,9)
8447 9152 rt : (3,5,-9,3)
8448 5502 rt : (8,8,-3,-8)
8448 5583 $e^x : 4/3 + 2\sqrt{5}/3$
8448 5716 cu : $\sqrt{3}/3 + 2\sqrt{5}/3$
8448 6346 $\Psi : \sqrt[3]{5}/2$
8448 6725 rt : (9,2,0,4)
8448 9695 rt : (2,5,5,-9)
8449 0665 $\zeta : (2\pi/3)^{1/2}$
8449 2099 $\zeta : \sqrt{10/3}$
8449 2989 rt : (5,5,-4,7)
8449 4468 rt : (8,9,-9,3)
8449 7261 ln : $\sqrt{2} - \sqrt{3} + \sqrt{7}$
8449 7782 $2^x : 3/2 + \pi/3$
8449 8359 rt : (3,5,-4,-2)
8449 8717 sinh : $\exp(\sqrt[3]{5}/3)$
8449 8971 rt : (9,5,0,-9)
8449 9366 rt : (6,0,4,7)
8450 1198 rt : (3,5,8,5)
8450 5022 $s\pi : e/4$
8450 6489 cr : $7\pi^2/3$
8450 9441 cr : $1/4 + \sqrt{2}/4$
8450 9804 $\ell_{10} : 7$
8451 4476 $t\pi : 2e/7$
8451 5425 sr : 5/7
8451 6839 $s\pi : 2\sqrt[3]{3}/9$
8451 8211 $s\pi : 10\sqrt{3}$
8451 9498 $Ei : 5/8$
8451 9686 $2^x : 4 - 3\sqrt{2}$
8452 0787 cu : $\sqrt{2}/4 - 3\sqrt{3}/4$
8452 1368 rt : (7,3,7,8)
8452 2402 $10^x : 1/\ln(\pi/2)$
8452 3572 rt : (4,9,-1,-8)
8452 3693 sr : $2e/3 + 2\pi$
8452 7014 id : $(2\pi)^{1/3}$
8452 9946 id : $2 - 2\sqrt{3}/3$
8453 0118 $\sqrt{3} + \exp(\sqrt{2})$
8453 0617 $\Gamma : 12/25$
8453 1478 rt : (5,2,-9,8)
8453 2372 rt : (8,4,-2,-6)
8453 3202 cr : $2e/9$
8453 3593 $2^x : 5\sqrt{2}/8$
8453 4125 sr : $3/2 - \pi/4$
8453 4899 $e^x : \sqrt{3}/3 - \sqrt{5}/3$
8453 8291 $2^x : 2\sqrt{3} + 3\sqrt{5}/2$
8453 9786 rt : (5,6,2,-9)
8454 3673 $10^x : 4\sqrt{2}/3$
8454 3679 sq : $9\sqrt[3]{5}$
8454 6609 rt : (2,4,-2,8)
8454 7484 sr : $2\sqrt{2} + \sqrt{3}/3$
8454 9202 rt : (9,1,1,-7)
8455 0588 $\sqrt{5} + \ln(5)$
8455 1663 $10^x : 4\sqrt{2}/3 + \sqrt{5}$
8455 2264 rt : (8,3,-2,1)
8455 5442 $\ell_{10} : 2e + \pi/2$
8455 7400 rt : (2,9,-5,-8)
8455 8681 sq : $4/3 + \sqrt{2}/4$
8455 8743 rt : (2,6,-7,-9)
8456 4309 $K : 5\sqrt[3]{2}/9$
8456 4339 $\ln(3/5) + \exp(\sqrt{5})$
8456 5917 rt : (5,1,2,4)
8456 6033 $2^x : 4e - 4\pi/3$
8456 6257 $s\pi : \sqrt[3]{25/2}$
8456 6577 $s\pi : 3/4 + \pi/2$
8456 7117 rt : (6,6,-7,-2)
8456 9399 $K : 7/10$
8456 9774 rt : (1,7,-9,2)
8456 9959 $2^x : 1/2 + 3\sqrt{3}$
8457 1521 rt : (8,7,-1,-9)
8457 3905 $10^x : \ln(2\pi)$
8457 4900 $\ell_2 : 3 + 4\pi/3$
8457 5166 rt : (3,8,-3,-5)
8457 7409 rt : (2,1,6,-7)
8457 8120 $J_0 : 2\zeta(3)/3$
8457 8144 $Ei : (3\pi/2)^{1/3}$
8458 0378 rt : (2,7,-5,-9)
8458 2371 ln : $2 - \pi/2$
8458 2608 $J_0 : 5\sqrt[3]{3}/9$
8458 2669 ln : 3/19
8458 3801 sr : $(\ln 3/3)^{1/3}$
8458 4714 $\ell_2 : e/3 + 2\pi$
8458 4932 ln : $6e/7$
8458 5955 rt : (7,8,-8,2)
8458 7779 $e^x : 3 + 3\sqrt{5}/2$
8458 7898 sr : $((e + \pi)/3)^{-1/2}$
8458 8539 rt : (9,6,1,2)
8458 9701 sr : $(\sqrt{5}/2)^{-3}$
8458 9914 $e^x : 4 + \sqrt{5}$
8458 9951 rt : (7,3,-4,-3)
8459 0859 $\Psi : \sqrt{3} + \sqrt{6} + \sqrt{7}$

8459 2287 sq : $5(e+\pi)/6$
8459 2373 rt : (2,9,5,-1)
8459 3758 e^x : $3e/4+\pi/4$
8459 4886 rt : (1,7,-6,5)
8459 5705 e^x : $4e\pi/5$
8459 6697 rt : (4,7,2,5)
8459 6935 erf : $4\sqrt[3]{2}/5$
8459 9006 rt : (6,4,5,5)
8459 9885 id : $8\sqrt[3]{3}/3$
8460 2736 rt : (1,7,4,-9)
8460 7152 rt : (6,9,-6,-4)
8460 7835 rt : (3,2,9,8)
8460 9690 cu : $4+2\sqrt{3}$
8460 9892 tanh : $5\sqrt{5}/9$
8461 1001 cu : $4\sqrt{3}+\sqrt{5}/3$
8461 2571 rt : (8,0,-1,4)
8461 3158 $\ln(\sqrt{5}) \div s\pi(2/5)$
8461 3778 erf : $\zeta(7)$
8461 5384 id : 11/13
8461 8042 rt : (5,2,3,-7)
8461 8436 cr : $2\sqrt{2}+2\sqrt{3}$
8461 9274 Γ : $\ln(\sqrt[3]{4}/3)$
8462 1533 tanh : $1/\ln(\sqrt{5})$
8462 3589 rt : (7,6,-3,-6)
8462 3629 Ψ : $-\sqrt{2}-\sqrt{5}+\sqrt{6}$
8462 3917 θ_3 : $7\ln 2/6$
8462 4500 2^x : $\exp(-1/(3e))$
8462 7190 rt : (4,8,1,-9)
8462 7738 cr : $3\sqrt{2}/7$
8462 8116 at : $(\sqrt[3]{3})^{1/3}$
8462 8437 sr : $9(1/\pi)/4$
8462 8735 J_0 : 4/5
8462 9573 J_0 : $((e+\pi)/3)^{-1/3}$
8463 0281 sinh : $\exp(\sqrt[3]{3}/2)$
8463 0628 $\ln(\sqrt{3})-\exp(1/3)$
8463 1255 $t\pi$: $\sqrt[3]{5}/5$
8463 1653 rt : (7,7,8,6)
8463 3225 at : $(\ln 2)^{-1/3}$
8463 4981 rt : (9,3,2,5)
8463 5773 as : $(\sqrt[3]{2}/3)^{1/3}$
8463 7531 sr : $\sqrt{3}+3\sqrt{5}/4$
8463 9203 rt : (6,1,6,8)
8463 9915 rt : (8,3,0,-7)
8463 9981 e^x : $9\sqrt[3]{5}/8$
8464 0000 sq : 23/25
8464 0225 θ_3 : $5(1/\pi)/4$
8464 1177 e^x : $\sqrt{3}/4+2\sqrt{5}/3$
8464 2157 sinh : $\sqrt[3]{15}$
8464 2330 rt : (2,4,-6,1)
8464 2752 rt : (7,9,-2,-9)
8464 4988 rt : (4,1,-8,7)
8464 4993 e^x : $7\sqrt{3}/9$
8464 6922 cr : $7\ln 2/8$
8464 8172 cr : $(e)^{-1/2}$
8465 2699 cr : $\sqrt{2}+\sqrt{5}+\sqrt{7}$
8465 3269 rt : (1,6,2,-2)
8465 3922 $t\pi$: $\sqrt{5}/10$
8465 5131 rt : (7,9,-4,6)
8465 5831 at : $\exp(1/(3e))$
8465 6132 rt : (9,0,3,8)
8465 7359 $1/2+\ln(\sqrt{2})$
8465 7645 Γ : $\exp(-\sqrt[3]{3})$
8465 7807 id : $e\pi/3$
8465 8183 $\pi \times s\pi(1/5)$
8465 9055 rt : (2,8,3,7)
8466 0183 rt : (7,4,9,9)
8466 0587 rt : (2,8,-8,1)
8466 0901 cu : 22/13
8466 1187 $\ln(2)\times\exp(1/5)$
8466 1256 rt : (4,8,-2,-5)
8466 1737 Ψ : $4e\pi/5$
8466 2165 id : $2e-2\pi$
8466 3487 ℓ_{10} : $\sqrt{5}\pi$
8466 3738 erf : $7\sqrt[3]{3}/10$
8466 3787 rt : (7,6,-9,7)
8466 5308 sq : $4-3e/4$
8466 9517 2^x : $3-2\sqrt{5}/3$
8467 6083 2^x : $(\ln 2)^{1/3}$
8467 7783 rt : (5,5,-9,1)
8468 0000 $2/5\times\exp(3/4)$
8468 0700 rt : (2,7,-5,-2)
8468 2377 sr : $8(e+\pi)$
8468 3166 rt : (1,9,1,-5)
8468 3543 Γ : $\sqrt[3]{5}/9$
8468 7254 erf : $5\sqrt{2}/7$
8468 7617 cr : $5\sqrt[3]{2}$
8468 7624 cu : $\sqrt{2}/3+\sqrt{5}$
8468 7746 2^x : $(\sqrt[3]{3})^{-1/3}$
8469 0809 rt : (5,8,-8,-2)
8469 3954 Ei : $7\sqrt{5}/5$
8469 5707 at : $4\sqrt{2}/5$
8469 6228 Ei : $(e/3)^{-3}$
8469 6364 sinh : $1/\ln(2/3)$
8469 6673 J_0 : $\ln(\sqrt{2}\pi/2)$
8469 7279 Ψ : $e\pi/10$
8470 4435 e^x : 7π
8470 6675 J_0 : $(\pi/2)^{-1/2}$
8470 6955 ℓ_{10} : $6(e+\pi)/5$
8470 8152 2^x : $2/3-e/3$
8471 1078 e^x : $(\ln 2/3)^{1/3}$
8471 2708 rt : (9,9,-7,-6)
8471 5206 ℓ_{10} : $3/4+2\pi$
8471 9941 λ : $\exp(-11\pi/16)$
8472 1438 cu : $e\pi/2$
8472 2574 ℓ_2 : $2+3\sqrt{3}$
8472 2626 id : $\sqrt{2}+\sqrt{3}/4$
8472 5541 cu : $5\pi^2/9$
8472 6402 id : $(e/2)^2$
8472 8555 rt : (7,5,-1,-7)
8472 9786 ln : 3/7
8473 0005 cr : $(\sqrt{2}\pi)^{-1/3}$
8473 0077 2^x : P_{54}
8473 2085 2^x : $7(e+\pi)/5$
8473 2210 rt : (3,9,-9,3)
8473 3251 rt : (4,8,6,9)
8473 3904 $s\pi$: $3/4+4\sqrt{3}$
8473 4410 rt : (7,2,-2,-4)
8473 6302 Ψ : $7(e+\pi)/6$
8473 6808 sinh : 10/13
8473 8756 J : 3/7
8474 0265 e^x : $2\sqrt{2}/3+4\sqrt{3}/3$
8474 0942 $\ln(2)+\exp(e)$
8474 1387 Ei : $3\sqrt{5}/4$
8474 1903 rt : (1,3,5,-7)
8474 3181 θ_3 : $(e/2)^{-3}$
8474 4299 rt : (8,2,2,-8)
8474 4588 id : $9\sqrt[3]{5}/4$
8474 4937 id : $\sqrt{3}/2+4\sqrt{5}/3$
8474 7122 rt : (7,1,-3,1)
8474 8159 Ei : $\exp(-3\pi)$
8474 8769 cr : 14/23
8475 0569 $s\pi$: $3e/4+2\pi$
8475 1419 ζ : $\sqrt{2}-\sqrt{5}+\sqrt{7}$
8475 1471 $\ln(\sqrt{5})-\exp(\sqrt{3})$
8475 3622 rt : (5,1,5,-8)
8475 5916 rt : (8,1,1,5)
8475 6484 $t\pi$: $\sqrt{2}/4-\sqrt{3}/3$
8475 6972 ln : $8(e+\pi)$
8475 9941 rt : (9,1,5,9)
8475 9970 10^x : 17/19
8476 2755 tanh : $9\ln 2/5$
8476 3161 rt : (2,2,-8,-7)
8476 4949 sr : $3\pi^2/2$
8476 5484 ℓ_2 : $\sqrt{21}\pi$
8476 5625 cu : $9\sqrt[3]{2}/8$
8476 5769 cu : $2-\sqrt{3}/4$
8476 6663 λ : $\ln 2/6$
8476 6748 ℓ_{10} : $1/3+3\sqrt{5}$
8476 9786 2^x : $3\sqrt{2}/4-3\sqrt{3}/4$
8477 0759 rt : (2,0,8,8)
8477 1186 10^x : $2\sqrt{2}+3\sqrt{3}$
8477 3825 rt : (6,2,8,9)
8477 5482 rt : (5,2,4,5)
8477 5906 sr : $2+\sqrt{2}$
8477 6795 rt : (4,9,-4,8)
8477 7232 rt : (9,4,4,6)
8477 9047 Ei : $2\sqrt{2}$
8477 9554 rt : (6,5,-8,6)
8477 9941 $\exp(\pi)+s\pi(1/4)$
8478 0601 $t\pi$: $4\sqrt{2}-\sqrt{3}/4$
8478 1038 rt : (8,4,0,2)
8478 1697 at : 17/15
8478 1943 sr : $\sqrt{3}/4+4\sqrt{5}/3$
8478 3297 cr : $4+4\sqrt{3}/3$
8478 3894 e^x : $\sqrt{3}-\sqrt{5}/2$
8478 4237 J_0 : $5(1/\pi)/2$
8478 4979 10^x : 4/15
8478 9576 Ψ : $4\sqrt[3]{5}$
8479 0650 ℓ_2 : $10\sqrt[3]{2}/7$
8479 1118 rt : (3,6,-9,-5)
8479 3629 rt : (6,8,-3,5)
8479 3749 sinh : 23/13
8479 5706 rt : (7,4,-4,-2)
8479 6305 rt : (9,8,-5,-7)
8479 6865 θ_3 : $(\sqrt[3]{4}/3)^{1/3}$
8479 9690 ℓ_2 : 5/9
8480 0603 2^x : $7\sqrt{2}/3$
8480 1144 10^x : $2+3\sqrt{3}/4$
8480 3511 rt : (8,8,-9,-3)
8480 3586 10^x : 5/11
8480 4014 cu : $1/3+e/2$
8480 6207 as : 3/4
8480 6270 rt : (6,5,7,6)
8480 7621 id : $1/4+3\sqrt{3}/2$
8480 7721 Ψ : $8(1/\pi)/7$
8480 7762 at : $9\sqrt[3]{2}/10$
8480 8878 id : $e-4\pi$
8480 9759 at : $2-\sqrt{3}/2$
8481 1220 rt : (3,9,2,2)
8481 1480 sinh : $4\sqrt{3}/9$
8481 2939 cu : $7\sqrt{3}/4$
8481 4472 rt : (9,7,3,3)
8481 5003 $1/4+\exp(4)$
8481 5210 rt : (6,7,-2,-7)
8481 7170 ℓ_{10} : $5\pi^2/7$
8481 9534 rt : (9,5,-6,-4)
8482 0253 $\exp(\sqrt{2})\div\exp(4/5)$
8482 1364 sr : $2e\pi/5$
8482 3564 2^x : $4/3-\pi/2$
8482 5387 rt : (2,9,-2,-6)
8482 7486 rt : (2,3,7,5)
8482 8363 tanh : 5/4
8482 8747 sq : $3\sqrt{3}/2-3\sqrt{5}/4$
8482 8982 $s\pi$: $\sqrt{3}+\sqrt{5}-\sqrt{7}$
8482 9359 tanh : $((e+\pi)/3)^{1/3}$
8482 9371 $\exp(2/5)+\exp(\sqrt{5})$
8483 0001 cr : $3\sqrt{3}+\sqrt{5}/2$
8483 0371 ℓ_{10} : $\exp((e+\pi)/3)$
8483 0502 rt : (7,4,1,-8)
8483 1629 rt : (5,8,-5,-5)
8483 9248 id : $8\ln 2/3$
8483 9503 10^x : $3+2\sqrt{2}$
8484 0244 10^x : $7\sqrt{5}/9$
8484 0773 10^x : $\sqrt{3}+3\sqrt{5}/2$
8484 0950 2^x : $5e/9$
8484 3394 $t\pi$: $\sqrt{2}-\sqrt{3}-\sqrt{5}$
8484 5176 rt : (3,6,1,-7)
8484 5699 rt : (8,1,4,-9)
8484 5884 cu : $4\pi^2/3$
8484 6922 sq : $3\sqrt{2}/2+\sqrt{3}$
8484 7437 rt : (5,5,3,2)
8485 0155 sr : $4\sqrt[3]{2}/7$
8485 0186 10^x : $2/3+2\pi/3$
8485 2103 K : $(\ln 2/2)^{1/3}$
8485 2472 rt : (6,4,-3,-4)
8485 2813 id : $3\sqrt{2}/5$
8485 4812 rt : (1,9,-6,-2)
8485 5167 sinh : $8\zeta(3)/7$
8485 5437 $t\pi$: $\sqrt[3]{11}$
8485 6714 $t\pi$: $6\pi^2/7$
8485 9106 sinh : $10\ln 2/9$
8486 0320 J_0 : $\sqrt[3]{1/2}$
8486 0610 rt : (9,2,-7,-1)
8486 0722 cr : 11/18
8486 1228 $1/4-\ln(3)$
8486 3292 cu : $2\sqrt{2}+\sqrt{3}$
8486 4805 rt : (4,3,4,-8)
8486 7406 rt : (1,9,1,-3)
8486 7866 rt : (3,7,-3,4)
8486 7994 rt : (2,9,-4,-3)

8486 9469 e^x : $\exp(\sqrt[3]{5})$
8486 9943 $\pi + s\pi(1/4)$
8487 1623 $t\pi$: $2e - 2\pi/3$
8487 2127 $1/5 + \exp(1/2)$
8487 2238 rt : (1,8,-1,-6)
8487 2754 rt : (6,9,1,9)
8487 3134 rt : (7,1,0,-5)
8487 5552 10^x : $7\pi/6$
8487 5659 rt : (8,7,-1,-1)
8487 6213 rt : (8,9,-8,5)
8487 6762 e^x : $10\sqrt[3]{3}/7$
8487 6870 Ψ : $1/\sqrt{21}$
8487 7956 rt : (9,7,-3,-8)
8488 1597 at : $3/4 - 4\sqrt{2}/3$
8488 2636 id : $8(1/\pi)/3$
8488 3770 Ψ : $9\pi^2/5$
8488 4277 rt : (5,0,7,9)
8488 5004 rt : (2,6,-3,-3)
8488 5293 $s\pi$: $\sqrt{3}/3 + \sqrt{5}/3$
8489 1278 ℓ_{10} : $\sqrt{2}/4 + 3\sqrt{5}$
8489 2207 10^x : $7\sqrt[3]{5}/6$
8489 3192 10^x : $6/5$
8489 3619 rt : (8,2,3,6)
8489 9959 rt : (5,9,-3,-7)
8490 1145 $t\pi$: $3\sqrt{2} + 4\sqrt{5}/3$
8490 2963 Ψ : $2\pi^2/7$
8490 3224 erf : $(\pi/3)^{1/3}$
8490 4164 2^x : $4/3 + 2e/3$
8490 4668 rt : (3,5,-8,6)
8490 5082 $\sqrt{3} + \exp(3/4)$
8490 5623 2^x : $2 - \sqrt{5}$
8490 5791 θ_3 : $(2\pi)^{-1/2}$
8490 6814 Ei : $7(1/\pi)$
8490 7555 ln : $7\pi^2/4$
8490 7718 rt : (8,7,-7,-4)
8490 9134 tanh : $1/\ln(\sqrt{2}\pi/2)$
8491 0058 ln : $3 + 3\sqrt{5}/2$
8491 1017 sinh : $7\sqrt[3]{3}$
8491 2013 rt : (6,8,6,3)
8491 2175 $\ln(3) - \exp(2/3)$
8491 3930 rt : (9,5,6,7)
8491 4147 at : $25/22$
8491 4305 at : $\sqrt{3}/3 + \sqrt{5}/4$
8491 4822 rt : (6,6,0,-8)
8491 5270 ζ : $17/13$
8491 5806 Γ : $\zeta(3)/10$
8491 6364 Ψ : $2/13$
8491 6421 J_0 : $8\ln 2/7$
8491 6457 2^x : $\sqrt{2}/2 + 3\sqrt{3}$
8491 7534 Γ : $2\sqrt{2}/9$
8491 7876 2^x : $\sqrt{2} + 4\sqrt{5}$
8491 9066 id : $(\sqrt[3]{3}/2)^{1/2}$
8492 1036 tanh : $(\pi/2)^{1/2}$
8492 1036 cr : $2/3 + 4\sqrt{2}$
8492 1119 10^x : $1/4 + \pi/4$
8492 3160 rt : (9,4,-4,-5)
8492 3590 10^x : $10(1/\pi)/7$
8492 5505 sr : $3\zeta(3)/5$
8492 5584 $s\pi$: $3\sqrt{5}/4$
8492 7372 rt : (1,2,7,-8)
8492 8329 Ψ : $\exp(-1/(2\pi))$
8492 9593 rt : (7,3,3,-9)
8493 0143 cu : $\sqrt{2} - 4\sqrt{5}/3$
8493 1119 sr : $2\sqrt[3]{5}$
8493 2180 sr : $1/\ln(4)$
8493 2204 rt : (6,1,-4,-1)
8493 2325 rt : (2,8,-7,6)
8493 3069 e^x : $4 - \sqrt{5}/2$
8493 3239 ℓ_{10} : $9\pi/4$
8493 3915 Ψ : $4/11$
8493 4763 J_0 : $19/24$
8493 5193 2^x : $3\sqrt{2} - \sqrt{5}/4$
8493 5535 sr : $1/4 + \sqrt{2}/3$
8493 8607 rt : (5,4,-7,5)
8494 2111 sr : $3\sqrt{2}/2 + 3\sqrt{3}/4$
8494 2425 $t\pi$: P_{31}
8494 2984 ln : $4 + 3\pi/4$
8494 3244 ℓ_{10} : $3\sqrt{3}/2 + 2\sqrt{5}$
8494 4061 e^x : $4/3 + \sqrt{3}/4$
8494 4274 rt : (4,0,3,5)
8494 5849 rt : (8,9,-4,-8)
8494 6894 $\sqrt{5} - \exp(3)$
8494 6933 rt : (4,9,-7,-3)
8494 8500 ℓ_{10} : $5\sqrt{2}$
8494 8663 ζ : $(\sqrt[3]{5})^{1/2}$
8494 8828 at : $\sqrt{2}/4 - 2\sqrt{5}/3$
8495 2792 rt : (9,1,-8,-2)
8495 3677 sr : $4\sqrt{2} - \sqrt{5}$
8495 3985 rt : (2,8,0,-7)
8495 5419 rt : (7,9,-8,-9)
8495 5592 id : 6π
8495 7745 rt : (9,6,-1,-9)
8495 7881 10^x : $3/2 - \pi/2$
8496 3982 ℓ_2 : $1/2 + 3\sqrt{5}$
8496 4009 rt : (7,2,-1,2)
8496 5390 id : $\exp(\pi/3)$
8496 6972 rt : (5,9,-1,-1)
8496 7517 ln : $4\sqrt{2}/3 + 2\sqrt{5}$
8497 2063 rt : (2,8,3,-2)
8497 2759 $3/4 \div \ln(2/3)$
8497 2915 rt : (3,5,3,-8)
8497 3370 rt : (5,3,6,6)
8497 6818 cr : $(\ln 2/3)^{1/3}$
8497 7361 $\sqrt{5} - \ln(4)$
8497 9566 10^x : $2\zeta(3)/9$
8498 0107 sq : $8(e + \pi)/7$
8498 1407 rt : (8,4,-8,-1)
8498 1505 ln : $\sqrt[3]{5}/4$
8498 3658 sr : $13/18$
8498 6169 rt : (4,7,-4,-6)
8498 6844 rt : (6,3,-1,-5)
8498 7513 ℓ_{10} : $3 + 3e/2$
8498 8607 at : $2/3 + \sqrt{2}/3$
8499 0118 sq : $2\sqrt{2}/3 + \sqrt{5}/3$
8499 1698 rt : (4,2,6,-9)
8499 2062 e^x : $\sqrt{2}/2 + 3\sqrt{5}/4$
8499 2395 e^x : $3\sqrt{2}/4 + 2\sqrt{5}$
8499 3880 rt : (7,9,-8,-4)
8499 5008 cr : $\sqrt{3} - \sqrt{5}/2$
8499 5387 rt : (8,5,2,3)
8499 5991 id : $5\sqrt[3]{5}/3$
8499 7333 rt : (5,8,-1,-8)
8499 8552 sr : $\sqrt{2} + 3\sqrt{5}$
8500 0000 id : $17/20$
8500 0781 as : $5\zeta(3)/8$
8500 2117 rt : (6,6,9,7)
8500 2313 rt : (9,1,-5,-2)
8500 4046 sr : $4 - \sqrt{3}/3$
8500 4678 rt : (3,9,-8,7)
8500 6550 rt : (7,0,2,6)
8500 8479 id : $\sqrt{3} + \sqrt{5}/2$
8500 8872 rt : (8,6,-5,-5)
8501 0365 $s\pi$: $(3\pi/2)^{1/3}$
8501 1698 rt : (6,5,2,-9)
8501 2538 λ : $((e + \pi)/2)^{-2}$
8501 3432 tanh : $2\pi/5$
8501 3688 cu : $8\pi^2/5$
8501 4184 ℓ_{10} : $4\sqrt{2}/3 + 3\sqrt{3}$
8501 4661 rt : (5,5,-3,9)
8501 4850 J_0 : $15/19$
8501 6297 rt : (1,6,-7,1)
8501 6636 id : $\sqrt[3]{19/3}$
8502 0308 $s\pi$: $2/3 + 4\sqrt{2}$
8502 1985 ℓ_2 : $\sqrt{13}$
8502 2493 Γ : $\pi/10$
8502 3768 rt : (9,3,-2,-6)
8502 4182 rt : (8,8,-2,-9)
8502 4447 rt : (7,8,-7,4)
8502 4892 ℓ_2 : $5\sqrt[3]{3}$
8502 5658 rt : (1,5,0,-3)
8502 5673 tanh : $8\sqrt{2}/9$
8502 5839 λ : $\exp(-13\pi/19)$
8502 6095 rt : (8,3,5,7)
8502 6294 $\ln(3) \times s\pi(e)$
8502 6348 rt : (9,8,5,4)
8502 6971 cu : $18/19$
8502 8490 rt : (1,8,7,-6)
8502 8992 ℓ_2 : $(e + \pi)^{1/3}$
8502 9348 2^x : $3/4 + 3\sqrt{5}$
8503 2550 rt : (3,4,-7,-7)
8503 2686 rt : (7,6,-8,9)
8503 6814 e^x : $8/13$
8503 8134 10^x : $1/\sqrt{14}$
8503 8269 sq : $3\sqrt{2}/2 - \sqrt{3}/4$
8504 0166 $s\pi$: $7\pi^2/9$
8504 1078 ln : $4/3 - e/3$
8504 5181 2^x : $4\sqrt{3}/3 + 3\sqrt{5}/4$
8504 5480 rt : (9,6,8,8)
8504 6159 $s\pi$: $\sqrt{3} + 4\sqrt{5}$
8504 6357 θ_3 : $7/8$
8504 6769 ℓ_2 : $3\zeta(3)$
8504 8547 sinh : $(\ln 2/2)^{-2}$
8504 9903 rt : (1,8,-4,-3)
8505 0678 sq : $3e - 3\pi/2$
8505 2022 rt : (1,7,-3,-7)
8505 4979 sinh : $(\sqrt[3]{4}/3)^{-1/2}$
8505 7553 Ψ : $11/21$
8505 8074 cr : $8/13$
8505 8097 erf : $3e/8$
8505 9377 ℓ_{10} : P_{67}
8506 0493 ℓ_{10} : P_{70}
8506 2275 rt : (1,9,2,6)
8506 5080 $1/2 \div s\pi(1/5)$
8506 5098 ln : $\sqrt{2}/9$
8506 6059 at : $(\sqrt{2}\pi/3)^{1/3}$
8506 7605 2^x : $3/4 + 4\sqrt{2}$
8506 9446 ℓ_2 : $\ln 2/5$
8507 0249 sinh : $\exp(1/\pi)$
8507 1865 at : $2\sqrt[3]{5}/3$
8507 6828 2^x : $4\sqrt{3} - \sqrt{5}$
8507 7758 at : $\sqrt{2}/2 + \sqrt{3}/4$
8507 8105 rt : (4,8,-5,-4)
8508 3392 sq : $1/4 - 4e$
8508 3959 at : $4\sqrt{2}/3 - \sqrt{5}/3$
8508 5507 rt : (1,7,0,-7)
8508 5528 e^x : $22/21$
8508 7433 rt : (7,9,-3,8)
8508 7550 Ei : $\sqrt[3]{23/3}$
8508 8565 J_0 : $5\sqrt[3]{2}/8$
8508 8935 sq : $\sqrt{2}/2 - 4\sqrt{5}$
8508 9157 sr : $3\sqrt{3} - 2\sqrt{5}$
8509 1773 cu : $3/4 + 2\sqrt{2}/3$
8509 2083 rt : (5,8,2,8)
8509 2440 cr : $8\ln 2/9$
8509 2667 rt : (7,8,-6,-5)
8509 3668 2^x : $1/3 + 3\sqrt{2}$
8509 6239 rt : (3,4,5,-9)
8509 7235 rt : (5,5,-2,-5)
8509 8347 rt : (7,6,-9,-1)
8509 8453 sr : $2e - 3\pi/2$
8510 2051 sq : $2\sqrt{2} - \sqrt{3}/2$
8510 3914 rt : (1,1,9,-9)
8510 4234 tanh : $\sqrt[3]{2}$
8510 5979 rt : (3,7,7,8)
8510 7140 rt : (8,5,-3,-6)
8510 7389 cr : $7e/3$
8510 8402 rt : (5,3,-5,-1)
8510 8790 $\ln(2) \times \exp(\sqrt{2})$
8510 9804 2^x : $\sqrt{2}/3 + 4\sqrt{3}$
8511 1856 sinh : $11/8$
8511 4971 as : $\ln(\sqrt{2}/3)$
8511 5120 e^x : $1 + 3\sqrt{2}/4$
8511 6211 rt : (6,2,1,-6)
8511 8413 rt : (8,3,-6,-2)
8511 8743 e^x : $5(e + \pi)/2$
8512 1513 ℓ_{10} : $3/2 - e/2$
8512 1524 rt : (9,2,0,-7)
8512 3557 sq : $4e + \pi/2$
8512 7699 rt : (1,9,8,-8)
8512 8252 sq : $3/2 - \sqrt{3}/3$
8512 9453 rt : (1,1,6,5)
8513 0002 2^x : $3\sqrt{2} - 3\sqrt{5}/2$
8513 3332 sr : $4\sqrt{3}/3 + \sqrt{5}/2$
8513 3530 rt : (4,1,5,6)
8513 5036 rt : (7,1,4,7)
8513 7055 rt : (2,6,-5,6)
8513 7446 $\exp(\sqrt{5}) \div s\pi(\sqrt{5})$
8513 8307 rt : (9,0,-3,3)
8513 9294 cu : $\sqrt{3}/2 + \sqrt{5}$
8514 0878 sinh : $(\sqrt{2}\pi/3)^3$

8514 0932 $10^x : 4\sqrt[3]{2}/3$
8514 1255 sr : $4/3 + 2\pi/3$
8514 1404 $2^x : 4 - \sqrt{3}/4$
8514 1766 at : $2 - \pi$
8514 2372 $2^x : 3/2 - \sqrt{3}$
8514 2441 sq : $\sqrt[3]{22}$
8514 3980 rt : (8,1,-9,2)
8514 6392 $10^x : P_{66}$
8514 7171 ln : $4\sqrt{3} - \sqrt{5}/4$
8514 9322 $10^x : P_{65}$
8515 0825 rt : (5,1,-8,3)
8515 1305 rt : (4,1,-7,9)
8515 1708 $J_0 : 11/14$
8515 1742 $\zeta : e/10$
8515 3164 $J_0 : 5\sqrt{2}/9$
8515 3689 $\Psi : \pi/6$
8515 3712 rt : (8,4,7,8)
8515 5064 cu : $4e + 3\pi/2$
8515 6584 $s\pi : 3\sqrt{2} + \sqrt{3}/4$
8515 6767 $10^x : 1/4 + 3\sqrt{5}/2$
8516 0586 tanh : $7\sqrt[3]{3}/8$
8516 0767 rt : (5,4,8,7)
8516 2198 $10^x : 3e - 3\pi/2$
8516 3191 $J_0 : \pi/4$
8516 3921 $2^x : \sqrt{3} - \sqrt{5} + \sqrt{6}$
8516 4019 sr : 24/7
8516 5623 rt : (2,9,-3,2)
8516 7831 rt : (7,3,1,3)
8517 0813 rt : (4,9,0,-9)
8517 2514 $\theta_3 : 2/5$
8517 2576 cu : $4\sqrt{2} + \sqrt{3}/4$
8517 2585 rt : (1,9,-9,-6)
8517 2590 $\ell_{10} : 3e - \pi/3$
8517 2809 $e^x : 10\sqrt{3}/9$
8517 3682 sq : $1 + 4\sqrt{5}/3$
8517 3840 ln : $2/3 + 3\sqrt{5}/4$
8517 4626 rt : (6,8,-9,-6)
8517 4942 $2^x : 8/9$
8517 6204 rt : (9,2,-6,-1)
8517 7824 $\theta_3 : e/4$
8517 8929 $\ell_2 : 4/3 + \sqrt{2}/3$
8517 9746 rt : (2,4,9,6)
8518 0208 $\ell_2 : 3e + 2\pi$
8518 0519 rt : (1,9,1,-8)
8518 3316 cr : $3\sqrt[3]{3}/7$
8518 3666 $\ln(\pi) + s\pi(1/4)$
8518 3745 ln : $2(e+\pi)/5$
8518 5185 cu : $4\sqrt[3]{5}/3$
8518 6414 $2^x : 6\sqrt[3]{2}/5$
8518 8694 $s\pi : (\sqrt[3]{5}/3)^{-1/2}$
8518 9123 ln : $10\sqrt{3}$
8519 3082 cr : $\exp(-\sqrt[3]{3}/3)$
8519 3237 tanh : 24/19
8519 3309 $s\pi : 4\sqrt{2} - 4\sqrt{5}/3$
8519 5626 sinh : 17/22
8519 5653 cu : $2\sqrt{3} - \sqrt{5}$
8519 6227 $\Gamma : 11/23$
8519 6632 at : 8/7
8519 6882 $2^x : P_{43}$
8519 6984 rt : (8,6,4,4)
8519 8038 rt : (2,6,4,9)
8519 8493 rt : (2,9,-2,-7)
8520 1178 rt : (6,5,-7,8)
8520 1512 $2^x : \ln(\sqrt[3]{4}/2)$
8520 2678 rt : (8,4,-1,-7)
8520 2741 ln : $1/4 + 2\pi/3$
8520 3026 id : $7\ln 2$
8520 4485 rt : (4,7,-3,-5)
8520 7100 sq : 12/13
8520 7625 $3 \times \exp(1/4)$
8520 8189 sq : $4/3 - 2\sqrt{5}$
8521 5978 $erf : (\pi/3)^{1/2}$
8521 6578 rt : (9,1,2,-8)
8521 6645 $t\pi : \sqrt{2} + 4\sqrt{3}$
8521 8655 cr : $3 + 3\sqrt{5}/2$
8522 1169 rt : (6,2,-5,-2)
8522 2774 rt : (5,4,0,-6)
8522 3002 $\Psi : 9\pi/10$
8522 3347 rt : (1,6,-5,-8)
8522 3367 $1/5 + \exp(\sqrt{3})$
8522 5642 $e - s\pi(1/3)$
8522 5775 rt : (9,9,7,5)
8522 6310 $\ell_{10} : 5e\pi/6$
8522 6506 cr : 13/21
8522 7617 rt : (6,7,-6,3)
8522 9111 $t\pi : 5\pi^2/6$
8523 0937 rt : (7,5,-7,-2)
8523 1004 cu : $8\sqrt{5}/9$
8523 1520 $\exp(4) + s\pi(\sqrt{3})$
8523 4699 rt : (1,7,-2,-4)
8523 5115 $\zeta : 1/\ln(\sqrt{3})$
8523 6100 $e^x : (\ln 3)^{1/2}$
8523 6371 $J_0 : (\sqrt[3]{3}/3)^{1/3}$
8523 7333 rt : (8,9,-7,7)
8523 7404 ln : $\sqrt{11/2}$
8523 8251 $s\pi : 5(e+\pi)/4$
8523 8984 cr : $4 + 3\pi/4$
8524 0515 $\exp(2/3) \times s\pi(2/5)$
8524 0922 rt : (6,1,3,-7)
8524 3179 rt : (5,9,-3,-6)
8524 3198 $Ei : 6e/5$
8524 3635 rt : (9,9,-6,-7)
8524 5428 $Ei : (2e/3)^{1/2}$
8524 6096 $s\pi : (\sqrt[3]{5}/3)^2$
8524 7950 rt : (3,8,-9,0)
8524 9076 $Ei : 15$
8525 0058 rt : (8,2,-4,-3)
8525 4133 cr : $4\sqrt{2}/3 + 2\sqrt{5}$
8525 4213 rt : (5,6,5,3)
8525 7440 $\Gamma : \sqrt{17/2}$
8525 7664 cu : $2e + 4\pi$
8525 8698 $e^x : 2\sqrt{2}/3 + \sqrt{5}/2$
8525 8931 rt : (7,2,6,8)
8526 1128 rt : (4,4,-2,8)
8526 1167 $K : 12/17$
8526 4349 $J_0 : 18/23$
8526 4439 rt : (1,8,-8,-4)
8526 5070 $\Psi : 2\sqrt{2}$
8526 8503 rt : (6,9,-5,-2)
8526 9022 rt : (9,1,-1,4)
8526 9545 $t\pi : \sqrt{3/2}$
8527 2021 id : $3\sqrt{3}/2 - \sqrt{5}/3$
8527 2211 cu : $9\pi^2/10$
8527 4057 ln : $2\sqrt{3} - \sqrt{5}/2$
8527 4889 cu : $1/4 + 2\pi$
8527 5562 cu : $\sqrt{2} + 3\sqrt{3}$
8527 6488 rt : (4,7,-4,-6)
8527 6804 rt : (8,5,9,9)
8527 7087 at : $\sqrt[3]{3/2}$
8527 7764 at : $\ln(\pi)$
8528 0286 sr : 8/11
8528 1080 $\ell_2 : 3\sqrt{2} + 4\sqrt{5}/3$
8528 2105 rt : (7,6,-2,-7)
8528 3043 rt : (3,6,-8,-2)
8528 3909 sr : $3 + \sqrt{3}/4$
8528 5981 rt : (4,6,-6,-7)
8528 6420 id : $\exp(-1/(2\pi))$
8528 7013 $s\pi : 2\sqrt{2}/3 + \sqrt{3}$
8528 7017 $\ell_2 : 3\sqrt{3}/4 - \sqrt{5}/3$
8528 7139 sr : P_{73}
8528 7265 rt : (3,1,4,-6)
8528 8289 $10^x : 7\zeta(3)$
8528 8672 sr : $(\sqrt[3]{4}/3)^{1/2}$
8528 8847 $10^x : 2 - \sqrt{2}$
8528 9164 tanh : 19/15
8528 9294 rt : (7,5,-2,-1)
8529 0625 rt : (9,7,-9,-3)
8529 3324 $2^x : 3\sqrt{2} - 2\sqrt{5}$
8529 5628 rt : (8,3,1,-8)
8529 7225 rt : (1,8,3,-9)
8529 7288 cu : $9\pi^2/8$
8530 0023 rt : (6,8,-2,7)
8530 0442 $s\pi : 1/4 + 3\pi$
8530 0667 id : $4\sqrt{2} + 3\sqrt{3}$
8530 1056 $\ln(3/4) + \exp(\pi)$
8530 1121 $J_0 : (2\pi/3)^{-1/3}$
8530 3035 $\lambda : 2/17$
8530 4675 rt : (5,2,-3,-2)
8530 5649 rt : (3,5,-7,9)
8530 6345 $t\pi : \ln(\sqrt{2}\pi/3)$
8530 6577 $e^x : \sqrt{2}/2 - \sqrt{3}/2$
8530 8678 $Ei : 3\pi/7$
8530 8755 rt : (3,9,-4,-5)
8530 9073 rt : (9,0,4,9)
8531 0065 rt : (6,9,8,4)
8531 2002 at : $2/3 - 2e/3$
8531 2642 rt : (1,9,-8,4)
8531 3117 rt : (4,2,7,7)
8531 4398 $\zeta : 3/11$
8531 4401 cr : $9\sqrt{2}/2$
8531 5782 rt : (2,7,1,-3)
8531 6954 $3 \times s\pi(2/5)$
8531 7999 rt : (9,8,-4,-8)
8531 8503 $s\pi : 5e\pi/4$
8532 0975 $erf : 3\sqrt[3]{5}/5$
8532 6397 rt : (4,6,-1,3)
8532 8070 rt : (3,3,-9,-8)
8532 9014 $t\pi : \pi/8$
8533 1344 cr : $1/\ln(5)$
8533 1479 $10^x : 2 - \sqrt{3}$
8533 2522 rt : (6,1,0,3)
8533 4286 $\zeta : 6(1/\pi)/7$
8533 6394 rt : (3,8,-3,-4)
8533 7115 id : $3\sqrt{2}/2 + \sqrt{3}$
8534 0087 rt : (6,7,-8,-2)
8534 1305 rt : (9,4,-9,-5)
8534 2042 $s\pi : (\ln 2/2)^{-2}$
8534 3908 rt : (5,3,2,-7)
8534 5653 $\theta_3 : \zeta(3)/3$
8534 7820 sinh : $\pi^2/4$
8535 1777 $J : 4/13$
8535 2005 at : $3/2 - \sqrt{2}/4$
8535 2194 rt : (3,7,-2,7)
8535 5339 id : $1/2 + \sqrt{2}/4$
8535 8485 rt : (7,4,-5,-3)
8535 9872 rt : (6,8,-3,-7)
8536 0036 rt : (7,4,3,4)
8536 0107 $2^x : \sqrt{2} + \sqrt{3}$
8536 1288 rt : (6,0,5,8)
8536 3451 rt : (2,8,-4,-8)
8536 4886 $Ei : (\pi/3)^{-1/3}$
8536 5088 $\Gamma : \ln(1/3)$
8536 5105 cr : $4\sqrt{3} - \sqrt{5}/4$
8536 6239 $J_0 : 9\ln 2/8$
8537 2419 $\exp(1/4) + s\pi(\pi)$
8537 3034 rt : (7,5,0,-8)
8537 4983 $e^x : 4 + 2\pi$
8537 6736 rt : (8,1,-2,-4)
8537 7220 $\ell_{10} : \sqrt{3}/4 + 3\sqrt{5}$
8537 7683 sq : $4\sqrt{2}/3 + 3\sqrt{5}$
8537 7941 $\ln(\pi) \times s\pi(\sqrt{3})$
8537 8541 rt : (7,3,8,9)
8537 9507 $\ell_{10} : 4 + \pi$
8538 0826 $10^x : 2\sqrt{2} + 4\sqrt{3}/3$
8538 2523 cu : $\ln(2e)$
8538 3850 $10^x : 6e\pi/5$
8538 4438 $s\pi : (3\pi)^{-1/2}$
8538 5093 rt : (9,6,-7,-4)
8538 6122 rt : (8,2,3,-9)
8538 6914 $s\pi : 2(e+\pi)/7$
8538 6937 rt : (1,5,-7,-9)
8538 7061 $Ei : \exp(-\sqrt{2}\pi)$
8538 7225 rt : (8,7,6,5)
8538 9194 sq : $5\pi/6$
8538 9917 $\ell_{10} : \sqrt[3]{2}/9$
8539 0774 rt : (9,7,-2,-9)
8539 1058 $\pi + \ln(3/4)$
8539 1692 $10^x : 8\ln 2/5$
8539 3768 $\zeta : 7/12$
8539 4843 rt : (9,2,1,5)
8539 6871 at : $e - \pi/2$
8539 7342 id : $e\pi/10$
8539 8163 id : $5\pi/2$
8540 7467 $K : \sqrt{2}/2$
8540 9495 $e^x : \sqrt{2}/3 + \sqrt{3}/3$
8541 0196 id : $1/2 + 3\sqrt{5}/2$
8541 1350 $\theta_3 : 17/25$
8541 3869 sq : $4\ln 2/3$
8541 5818 $\ell_{10} : 4\sqrt{2} + 2\sqrt{5}/3$
8541 7240 rt : (4,8,-8,5)

8541 8505 rt : (4,9,-9,-6)
8541 8542 10^x : $2\sqrt{3}-\sqrt{5}/2$
8542 1025 J_0 : $(\sqrt{2}/3)^{1/3}$
8542 2648 rt : (3,9,3,5)
8542 2673 sπ : $(\ln 3)^3$
8542 4546 sr : P_{23}
8542 4814 10^x : $\sqrt{2}/3+\sqrt{5}/2$
8542 5946 sπ : $5\pi^2/2$
8542 7228 rt : (3,8,-2,-6)
8542 8461 e^x : $3e-3\pi/4$
8542 9041 cu : $e\pi/9$
8542 9606 sq : $3/4-2\sqrt{5}$
8543 0404 rt : (5,4,-6,7)
8543 1426 sinh : $2(e+\pi)/3$
8543 2087 id : $9\sqrt[3]{3}/7$
8543 3090 2^x : $(\sqrt[3]{2})^{-1/2}$
8543 4798 rt : (1,8,-7,5)
8543 5034 rt : (2,3,3,8)
8543 5307 2^x : $\sqrt{13/2}$
8543 5491 at : $(\pi/3)^3$
8543 5698 rt : (7,8,-6,6)
8543 6044 rt : (5,5,-9,-2)
8543 7139 rt : (3,7,-7,-1)
8543 8048 e^x : $4/3-2\sqrt{5}/3$
8543 8236 as : $(\ln 3)^{-3}$
8543 8822 J_0 : 7/9
8544 0375 10^x : $5\sqrt[3]{2}/3$
8544 1360 sinh : $(5/3)^{-1/2}$
8544 1432 rt : (7,9,-9,-3)
8544 4137 rt : (4,7,3,7)
8544 5525 cr : $9\ln 2/10$
8544 6049 rt : (5,9,-9,-2)
8544 8002 cr : $e-2\pi/3$
8544 8860 rt : (6,7,-1,-8)
8545 0426 sq : $2/3-3\pi/4$
8545 1125 rt : (2,9,-2,7)
8545 1751 rt : (8,9,-3,-9)
8545 3510 tanh : 14/11
8545 3767 e^x : $(\ln 3/2)^{-1/2}$
8545 5262 tanh : $9\sqrt{2}/10$
8545 5565 10^x : $3\sqrt{2}+2\sqrt{3}$
8545 6054 rt : (1,2,8,6)
8545 8881 cr : $\exp(-\sqrt{2}/3)$
8545 9612 rt : (3,0,6,7)
8546 0924 rt : (5,2,4,-8)
8546 1606 rt : (7,4,2,-9)
8546 3767 rt : (8,8,-8,-4)
8546 5193 cu : $4e-4\pi$
8546 7216 10^x : $(e+\pi)/10$
8546 7324 tanh : $4(1/\pi)$
8546 8809 tπ : $4\sqrt{2}+\sqrt{5}/2$
8547 0011 2^x : $\exp(e/2)$
8547 0435 2^x : $9\ln 2/7$
8547 0784 Li_2 : $e/4$
8547 2217 ℓ_{10} : $3/2+4\sqrt{2}$
8547 2361 sπ : $2\sqrt{2}-2\sqrt{3}/3$
8547 2513 Ψ : 17/6
8547 4996 $\sqrt{3}\times\exp(4/5)$
8547 5564 $\ln(\sqrt{2})\div\ln(2/3)$
8547 6418 Ψ : $-\sqrt{2}-\sqrt{6}-\sqrt{7}$
8547 6491 cu : $4\sqrt{2}/3+3\sqrt{5}/2$
8547 6966 rt : (9,5,-5,-5)
8547 7591 rt : (6,1,7,9)
8547 9351 J_0 : $2e/7$
8548 1349 rt : (7,3,-3,-4)
8548 1612 tπ : $1/4+3\sqrt{3}$
8548 3937 rt : (4,3,9,8)
8548 5240 sπ : $\sqrt[3]{7/3}$
8548 5382 erf : $5\sqrt[3]{3}/7$
8549 0090 rt : (5,1,-1,-3)
8549 1481 erf : $6\zeta(3)/7$
8549 2976 cu : $2\sqrt{3}+2\sqrt{5}$
8549 3500 cu : $4/3+4e/3$
8549 3715 sr : $4\sqrt{2}/3-2\sqrt{3}/3$
8549 8311 ln : $\sqrt[3]{13}$
8549 8797 id : $\sqrt[3]{5}/2$
8549 8868 sr : $4-\sqrt{5}/4$
8549 8882 Ei : $\exp(-2e)$
8549 9642 sπ : $4\sqrt{3}+\sqrt{5}/3$
8550 2688 Γ : $3(1/\pi)/2$
8550 4106 sr : $2e+3\pi$
8550 5273 at : 23/20
8550 6037 ℓ_2 : $4\sqrt{2}/3+\sqrt{3}$
8551 0489 Ei : $\pi/5$
8551 1138 as : $3/2-\sqrt{5}/3$
8551 2339 e^x : $(\sqrt{3}/2)^{-1/3}$
8551 3435 cu : $3e-4\pi$
8551 6112 rt : (9,3,3,6)
8551 8967 2^x : $1/3-\sqrt{5}/4$
8551 9653 rt : (1,6,7,7)
8551 9656 id : $\sqrt{2}-\sqrt{5}/4$
8552 1184 rt : (2,7,-6,-9)
8552 1538 sq : $\sqrt{2}/3+\sqrt{3}$
8552 3165 rt : (5,9,-2,-8)
8552 3879 Ei : $7\sqrt{3}/9$
8552 5045 rt : (5,8,-5,-7)
8552 5660 rt : (7,2,-8,1)
8552 5751 sπ : $4\sqrt{2}/3-\sqrt{5}/4$
8552 7011 $2-\ln(\pi)$
8552 7059 Ψ : 17/20
8553 1421 sq : $5\pi/8$
8553 5387 Ψ : $9\sqrt[3]{2}/4$
8553 5583 rt : (6,6,1,-9)
8553 7078 rt : (4,3,-9,3)
8553 7891 Ψ : $1/\sqrt[3]{7}$
8553 8581 sr : $3e-3\pi/2$
8554 1234 e^x : $3\sqrt[3]{3}/7$
8554 1744 rt : (3,7,0,-7)
8554 1842 rt : (7,5,5,5)
8554 2456 rt : (2,7,-3,-1)
8554 2740 Li_2 : 17/25
8554 5687 id : $1/3-4\pi/3$
8554 7882 e^x : $3/4-e/3$
8554 8561 Γ : $(\ln 2/2)^2$
8554 8812 10^x : $4e/3+2\pi$
8554 9187 ζ : $9\sqrt{2}/7$
8554 9775 sπ : $3\sqrt[3]{3}$
8555 0265 ζ : $(\sqrt[3]{5}/2)^{-1/2}$
8555 0379 erf : $(\ln 3)^{1/3}$
8555 1564 sq : $3/2-4e$
8555 2292 tπ : $7e/3$
8555 3212 J_0 : $(5/3)^{-1/2}$
8555 3609 rt : (8,9,-4,-5)
8555 4690 cr : $4/3-\sqrt{2}/2$
8555 4833 K : 17/24
8555 5442 tanh : $(\sqrt[3]{3}/3)^{-1/3}$
8555 7970 rt : (4,4,3,-8)
8556 0011 $\ln(3)\div\exp(1/4)$
8556 1013 $4\times s\pi(\sqrt{2})$
8556 2077 as : $(\sqrt[3]{5}/3)^{1/2}$
8556 2445 ζ : 20/11
8556 4063 rt : (4,5,-8,-8)
8556 4485 at : $2\sqrt{2}-3\sqrt{5}/4$
8556 6074 rt : (5,8,-7,-3)
8556 6379 rt : (9,4,-3,-6)
8556 6900 id : $(1/(3e))^{-1/2}$
8556 7286 rt : (8,8,8,6)
8557 4084 rt : (5,1,6,-9)
8557 6365 Ei : $4\sqrt{2}/9$
8557 7113 10^x : $4\sqrt{2}/3-\sqrt{5}/2$
8557 7250 rt : (1,5,2,-6)
8557 7935 sπ : $\exp(-\sqrt{5}/2)$
8557 9619 rt : (8,9,-6,9)
8558 0686 e^x : $\exp(-\sqrt[3]{3}/3)$
8558 1192 θ_3 : 17/21
8558 1210 rt : (9,4,-2,1)
8558 3026 rt : (6,5,-4,-4)
8558 4053 sπ : $7(e+\pi)/3$
8558 5297 sr : $(e+\pi)/8$
8558 5939 rt : (3,9,2,-3)
8558 8512 tanh : $4\sqrt{5}/7$
8558 9171 tanh : 23/18
8559 0951 rt : (2,9,-8,-1)
8559 2626 ln : $2+\sqrt{2}/4$
8559 4931 Ψ : $8\sqrt{5}$
8559 7794 rt : (5,0,-6,2)
8559 8299 ζ : $3e/10$
8559 9853 rt : (7,2,-1,-5)
8560 0654 rt : (2,2,5,-7)
8560 1574 ℓ_{10} : $1/4+4\sqrt{3}$
8560 1694 $2/3\times\exp(1/4)$
8560 1996 e^x : $\sqrt{2}/3+3\sqrt{5}/2$
8560 5735 cu : $2\pi^2/3$
8560 5861 10^x : $\sqrt{2}/2+\sqrt{5}/2$
8560 6473 sr : $3\sqrt[3]{5}/7$
8560 7911 cu : $\sqrt{2}+\sqrt{3}/2$
8560 8108 rt : (5,8,0,-9)
8561 0076 ℓ_{10} : $\sqrt{2}/3+3\sqrt{5}$
8561 1412 erf : $(e/3)^{-1/3}$
8561 6159 rt : (3,6,-5,-2)
8561 6193 e^x : $2\sqrt{3}+2\sqrt{5}/3$
8561 6217 tπ : $9e\pi/8$
8561 6556 rt : (8,1,2,6)
8561 7342 rt : (5,7,7,4)
8561 8784 sπ : $\sqrt{13}\pi$
8561 9350 cr : $3\sqrt{2}/4-\sqrt{3}/4$
8561 9370 10^x : $\sqrt{2}+3\sqrt{5}/4$
8561 9385 at : $1/\ln(\sqrt[3]{2}/3)$
8561 9449 id : $1/2+3\pi/4$
8561 9763 2^x : $1/3+\sqrt{5}/4$
8562 0249 J_0 : 17/22
8562 1202 cu : $4\sqrt{2}+4\sqrt{5}$
8562 3827 rt : (3,1,8,8)
8562 7445 tπ : $(5/3)^{-1/2}$
8562 8320 rt : (9,3,-8,-1)
8562 8659 rt : (7,9,-7,-5)
8562 9495 rt : (2,6,-1,-4)
8563 0883 2^x : $\sqrt{2}/4-\sqrt{3}/3$
8563 3138 rt : (9,4,5,7)
8563 3533 tanh : $(2\pi/3)^{1/3}$
8563 4883 sr : 11/15
8563 6853 rt : (2,9,-1,-7)
8563 8498 rt : (6,6,-9,1)
8564 0646 id : $8\sqrt{3}$
8564 1091 rt : (8,6,-4,-6)
8564 6901 $1/2+\exp(\sqrt{5})$
8564 7480 sπ : $3/4+\sqrt{3}/3$
8564 9059 Ei : $4\sqrt[3]{2}/3$
8564 9853 cr : $\pi/5$
8565 0092 tπ : $\sqrt{3}+\sqrt{6}-\sqrt{7}$
8565 0127 Γ : $8\sqrt[3]{2}/3$
8565 0977 rt : (3,9,-5,-9)
8565 1580 ℓ_2 : $3+3\sqrt{2}$
8565 1618 rt : (1,8,1,-1)
8565 2548 rt : (3,6,2,-8)
8565 3455 rt : (9,3,-1,-7)
8565 6832 rt : (6,3,-3,-1)
8565 6937 sr : $4+4e$
8565 7209 ℓ_2 : P_4
8565 7847 rt : (2,9,6,4)
8565 8141 cu : $7\sqrt{5}$
8565 9230 sπ : $1/4+3e/2$
8565 9886 cr : $4\sqrt{2}/9$
8566 4328 ln : $4\sqrt{2}+\sqrt{5}/3$
8566 5580 ℓ_{10} : $3+4\pi/3$
8566 5752 rt : (5,0,1,4)
8566 6721 $\exp(1/4)-\exp(\pi)$
8566 7157 2^x : $4\sqrt[3]{5}/3$
8566 7167 e^x : $1-2\sqrt{3}/3$
8566 8138 rt : (4,3,5,-9)
8566 8535 ℓ_{10} : $e/3+2\pi$
8566 8568 at : $4\sqrt[3]{3}/5$
8567 0562 at : 15/13
8567 2267 rt : (1,6,5,-5)
8567 9233 rt : (4,2,-9,2)
8567 9368 e^x : $1+3\sqrt{3}$
8568 1753 rt : (1,6,-5,6)
8568 1916 rt : (5,8,-4,-3)
8568 4923 cr : $4\sqrt{2}+\sqrt{5}/3$
8568 7964 rt : (6,4,-9,1)
8568 8424 10^x : $5\sqrt{5}/4$
8569 0148 rt : (6,3,4,5)
8569 1176 Γ : $9\pi^2/4$
8569 1663 e^x : $3e\pi/7$
8569 5343 10^x : $\sqrt[3]{17/2}$
8569 6798 Ψ : $(\sqrt[3]{3}/2)^{1/2}$
8569 7921 sq : $2/3+4\sqrt{3}/3$
8569 8159 rt : (6,4,-2,-5)
8570 1676 tanh : $8\sqrt[3]{3}/9$
8570 4781 ln : $3\pi/4$

8570 4904 rt : (6,7,-5,5)
8570 6798 cu : $3\sqrt{3}+4\sqrt{5}/3$
8570 7194 at : $2\sqrt{3}/3$
8570 7846 ζ : $(\sqrt{5}/2)^{-1/2}$
8570 9964 cr : $3e\pi/4$
8571 0035 ℓ_{10} : $2+3\sqrt{3}$
8571 0626 sq : $3\sqrt{3}/4+3\sqrt{5}/4$
8571 0812 as : $3\sqrt[3]{2}/5$
8571 1872 sinh : $2e/7$
8571 1962 J_0 : $10\ln 2/9$
8571 4285 id : 6/7
8571 4345 ζ : $\sqrt[3]{6}$
8571 4478 tanh : $3\sqrt[3]{5}/4$
8571 4691 J : $\exp(-20\pi/9)$
8571 5461 as : $4/3-\sqrt{3}/3$
8571 5847 e^x : 13/21
8571 8933 2^x : $6(e+\pi)/5$
8572 0480 rt : (1,7,2,7)
8572 1616 e^x : $-\sqrt{2}+\sqrt{3}+\sqrt{5}$
8572 1983 as : $3\sqrt{3}/2-3\sqrt{5}/2$
8572 3082 10^x : $(\sqrt{2}\pi/3)^{-2}$
8572 4337 λ : $3(1/\pi)/8$
8572 4398 id : $(\sqrt[3]{4})^{-1/3}$
8572 4933 J_0 : $4\sqrt{3}/9$
8572 5037 tπ : $3\sqrt{2}/2-2\sqrt{3}$
8572 6329 rt : (8,5,-2,-7)
8572 7820 ln : $\sqrt{3}/2+2\sqrt{5}/3$
8572 9810 cr : $3/4+4\sqrt{2}$
8573 0298 rt : (8,2,4,7)
8573 0535 at : $5\ln 2/3$
8573 1227 rt : (8,5,-9,-1)
8573 3547 rt : (9,9,-5,-8)
8573 3882 cu : $5\sqrt[3]{5}/9$
8573 6604 rt : (1,4,4,-7)
8573 6839 ln : $3/4+4\sqrt{2}$
8573 7278 e^x : $4e+3\pi$
8573 7500 cu : 19/20
8573 8309 rt : (9,2,1,-8)
8573 8774 sπ : $\sqrt{2}-\sqrt{3}+\sqrt{7}$
8573 9921 ln : $5\sqrt{2}/3$
8574 0056 rt : (4,8,-7,8)
8574 0295 rt : (8,2,-3,1)
8574 1457 rt : (3,6,-7,1)
8574 2553 $\exp(3/4)\times\exp(3/5)$
8574 4771 rt : (2,8,1,-8)
8574 5265 J_0 : 10/13
8574 6198 rt : (9,5,7,8)
8574 6311 e^x : $5\sqrt[3]{2}/6$
8574 7080 erf : $\zeta(5)$
8574 7959 ζ : $\exp(1/e)$
8574 8041 rt : (9,2,-6,-2)
8575 3773 cr : $3\sqrt{2}/2-2\sqrt{5}/3$
8575 6723 $\pi-\exp(1/4)$
8575 8407 2^x : $\sqrt{2}+\sqrt{3}/2$
8575 8711 rt : (2,1,7,-8)
8575 9864 rt : (3,5,4,-9)
8576 0464 rt : (4,3,-5,6)
8576 0800 id : $9e\pi$
8576 1802 2^x : 24/5
8576 2360 rt : (2,8,-6,-2)
8576 5111 e^x : 21/20
8576 6002 rt : (9,5,0,2)
8576 6419 sπ : $1/\ln(\sqrt[3]{2})$
8576 6870 at : $1/4+e/3$
8576 8086 sinh : $(\pi)^{1/2}$
8576 8151 2^x : $\exp(2/3)$
8576 8785 rt : (7,7,-9,4)
8576 9515 sq : $3/2-2\sqrt{3}$
8577 0027 rt : (6,8,-1,9)
8577 0238 rt : (8,9,3,1)
8577 2082 rt : (1,8,9,8)
8577 2999 2^x : $1/4-\sqrt{2}/3$
8577 3254 ℓ_2 : $2e/3$
8577 3281 Ψ : $8(1/\pi)/3$
8577 4280 Ei : $4\sqrt[3]{3}$
8577 5809 e^x : $5\sqrt{2}/4$
8577 6388 id : $(e/2)^{-1/2}$
8577 7376 cu : $2e\pi/5$
8577 8160 rt : (4,9,-6,-4)
8577 8452 tanh : $(\sqrt{2}/3)^{-1/3}$
8578 0201 tπ : $1/3-\sqrt{5}/4$
8578 2188 e^x : $8\sqrt{5}$
8578 2706 ℓ_{10} : $1/2+3\sqrt{5}$
8578 4096 cu : $4+\sqrt{5}/3$
8578 4124 erf : $(\sqrt{5}/2)^{1/3}$
8578 6019 rt : (6,8,-8,-4)
8578 7397 rt : (5,7,-7,-8)
8579 0854 10^x : $2\sqrt{2}+\sqrt{5}/3$
8579 2986 tπ : $1/\ln((e+\pi)/3)$
8579 3851 rt : (5,9,0,1)
8579 5186 2^x : 25/9
8579 7472 rt : (4,9,-8,-3)
8579 7746 cr : 12/19
8579 7756 rt : (7,7,-3,-7)
8579 9996 tanh : 9/7
8580 0096 rt : (9,8,-3,-9)
8580 1042 ℓ_{10} : $5\sqrt[3]{3}$
8580 5510 id : $3\sqrt{3}/4+\sqrt{5}/4$
8580 7627 ℓ_{10} : $6\zeta(3)$
8580 9432 rt : (8,4,0,-8)
8581 0461 cu : $2/3+4\sqrt{2}$
8581 2471 sq : $4-4\sqrt{3}/3$
8581 3242 e^x : $2\sqrt{2}-4\sqrt{5}/3$
8581 4454 ℓ_{10} : $\ln 2/5$
8581 7065 $e\div s\pi(2/5)$
8581 7881 sr : $\sqrt{2}/3+4\sqrt{5}/3$
8582 1049 rt : (9,1,3,-9)
8582 1546 sπ : $4e/3+\pi/3$
8582 1618 sπ : $1/2+2\sqrt{2}$
8582 1626 $\exp(3/5)+s\pi(\sqrt{2})$
8582 1819 at : $1/3-2\sqrt{5}/3$
8582 2414 rt : (7,8,-5,8)
8582 4908 rt : (7,0,3,7)
8582 4931 rt : (5,3,-9,5)
8582 9753 ln : $1+e/2$
8583 1811 Ei : $6e$
8583 2606 rt : (5,1,3,5)
8583 2651 cr : $4\sqrt{3}/3-3\sqrt{5}/4$
8583 3671 cr : $\sqrt{3}+\sqrt{5}+\sqrt{6}$
8583 5530 erf : $3\sqrt{3}/5$
8583 5930 Ψ : $3\sqrt{2}/5$
8583 6918 sr : $e/2+2\pi/3$
8583 6964 $\ln(2/3)\times\exp(3/4)$
8583 7421 sr : $(5/2)^{-1/3}$
8583 8211 rt : (1,7,-7,-2)
8583 9507 sr : 14/19
8584 0277 rt : (8,3,6,8)
8584 0734 id : $4-\pi$
8584 1882 $\ln(5)\div s\pi(1/3)$
8584 2913 Γ : $9(e+\pi)/5$
8584 3528 10^x : $4\sqrt{5}/9$
8584 3872 at : 22/19
8584 4255 rt : (1,6,1,1)
8584 4800 Ψ : $\exp(e+\pi)$
8584 6155 J_0 : $(\sqrt{2}\pi/2)^{-1/3}$
8584 6475 rt : (1,9,0,-6)
8584 7392 rt : (8,4,-7,-2)
8584 7755 Li_2 : 15/22
8584 7796 ℓ_{10} : $3e/2+\pi$
8584 9367 rt : (2,7,3,-9)
8584 9753 10^x : $(e+\pi)/5$
8585 0204 rt : (9,5,-7,-4)
8585 0338 $\exp(e)\div\exp(1/3)$
8585 0705 Ei : $\pi^2/5$
8585 1165 Ei : $3\sqrt[3]{2}$
8585 3376 ln : $3\sqrt{2}/4+3\sqrt{3}/4$
8585 4105 Γ : $\sqrt{2}/6$
8585 4308 erf : $3\ln 2/2$
8585 4741 $3/4\times\ln(\pi)$
8585 5543 rt : (9,6,9,9)
8585 7259 2^x : $9e$
8585 7346 sπ : $\sqrt{18}\pi$
8585 9480 ζ : 13/9
8585 9680 rt : (8,8,-9,7)
8586 0205 sinh : 7/9
8586 0849 at : $(\sqrt{5}/3)^{-1/2}$
8586 1207 id : $5\zeta(3)/7$
8586 3481 rt : (9,1,-4,-3)
8586 6790 10^x : 18/19
8586 6986 sr : $4/3+3\sqrt{2}/2$
8586 7026 Ψ : $7\pi^2/10$
8586 7789 rt : (6,9,-4,0)
8587 1056 cu : $10\sqrt[3]{5}/9$
8587 1496 $\ln(4)\div s\pi(\sqrt{3})$
8587 2010 2^x : $4\pi^2/7$
8587 2446 cu : $(2\pi)^3$
8587 3148 ζ : $\sqrt{2}/2$
8587 5612 10^x : $2\sqrt{3}+\sqrt{5}/3$
8587 6054 cu : $\sqrt{22/3}$
8587 6758 λ : 3/25
8587 8162 ℓ_{10} : $3\sqrt{2}+4\sqrt{5}/3$
8587 8251 rt : (7,7,9,7)
8587 8478 rt : (3,8,0,-4)
8587 9174 rt : (7,6,-1,-8)
8588 4566 10^x : $(\sqrt[3]{5})^{-1/3}$
8588 4900 rt : (1,8,2,7)
8588 7055 rt : (4,8,-4,-5)
8588 7169 2^x : $2\sqrt{5}/5$
8588 7830 rt : (5,4,-5,9)
8588 8907 rt : (1,3,6,-8)
8589 0502 rt : (8,3,2,-9)
8589 2357 rt : (7,0,-4,1)
8589 4498 ζ : $((e+\pi)/2)^{-1/2}$
8589 5659 cu : $1/\ln(\pi/2)$
8589 7446 $e+\exp(\pi)$
8589 7887 sπ : $\sqrt[3]{14/3}$
8590 0000 cu : 19/10
8590 0305 sπ : $e-\pi/3$
8590 0732 rt : (4,5,6,4)
8590 2122 rt : (5,5,-1,-6)
8590 3832 λ : $(\ln 2/2)^2$
8590 4554 ln : $\sqrt{3}+\sqrt{5}+\sqrt{6}$
8590 5677 J_0 : $(\sqrt[3]{5})^{-1/2}$
8590 6089 cr : $3/2-\sqrt{3}/2$
8590 6337 J_0 : 13/17
8590 6469 rt : (3,8,-2,-1)
8590 6705 2^x : $7\sqrt{3}/8$
8590 8520 $\exp(\sqrt{2})-s\pi(\sqrt{3})$
8590 9498 rt : (9,7,-8,-4)
8590 9833 rt : (2,0,9,9)
8591 1407 rt : (7,9,-8,-1)
8591 2696 cu : $1+2\sqrt{5}$
8591 4091 id : $1/2+e/2$
8591 6276 $4/5\div s\pi(\pi)$
8591 6463 Ψ : $\sqrt[3]{23}$
8591 6653 rt : (8,3,-1,2)
8591 7089 rt : (6,2,2,-7)
8591 7307 id : $\sqrt{5}-\sqrt{6}-\sqrt{7}$
8591 7430 cu : $1/4-3\pi/2$
8591 7794 rt : (2,5,-3,-5)
8591 9894 rt : (6,4,-9,8)
8592 0015 e^x : $(e/3)^{-1/2}$
8592 3781 sπ : $(\sqrt{2}\pi/2)^{-1/2}$
8592 5315 rt : (2,7,-4,-3)
8592 5318 sinh : $(\sqrt{2}/3)^{1/3}$
8592 5894 λ : $\zeta(3)/10$
8592 7071 2^x : 17/19
8592 8647 erf : 25/24
8592 9496 10^x : $-\sqrt{2}+\sqrt{5}+\sqrt{6}$
8593 1957 rt : (7,1,5,8)
8593 2348 ζ : $3\sqrt[3]{3}/5$
8593 5484 rt : (4,2,-7,2)
8593 6744 rt : (1,9,2,-9)
8593 7500 cu : $5\sqrt[3]{3}/4$
8593 8543 id : $2e/3+\pi/3$
8593 8676 tanh : $\sqrt{5/3}$
8594 0682 rt : (9,6,2,3)
8594 1358 tπ : $\exp(-\sqrt[3]{2}/3)$
8594 2352 sq : $3/4-3\sqrt{5}/4$
8594 2719 sπ : $3\sqrt{5}/10$
8594 3791 $1/4+\ln(5)$
8594 3985 rt : (7,6,-8,-2)
8594 5035 rt : (5,8,9,5)
8594 6727 rt : (8,4,8,9)
8594 7159 10^x : $(\sqrt{2}/3)^{1/2}$
8594 7467 sq : $2e+3\pi$
8594 7582 rt : (6,8,-2,-8)
8594 7960 $\exp(3)+s\pi(e)$
8594 8912 sq : $9\pi/2$
8595 0413 sq : 15/11

8595 1390 $\Psi : 1/\sqrt[3]{21}$
8595 2338 rt : (1,7,6,-4)
8595 6463 rt : (1,5,1,-9)
8595 8633 rt : (7,5,1,-9)
8595 9510 rt : (8,3,-5,-3)
8596 0086 $\pi \div \ln(3)$
8596 1224 cu : $2\sqrt{3}/3 + \sqrt{5}/3$
8596 2051 $J : (\ln 2/2)^2$
8596 4620 sr : $3\sqrt{3} + 4\sqrt{5}/3$
8596 6758 sinh : $5(e+\pi)/7$
8596 7160 $Ei : \pi^2/10$
8596 9630 rt : (4,1,2,-5)
8597 1671 tπ : $e/4 + 2\pi/3$
8597 2695 sr : 17/23
8597 4717 sinh : $\sqrt[3]{21}$
8597 4926 rt : (9,0,-2,4)
8597 6878 rt : (2,7,-2,3)
8597 9987 rt : (8,9,-9,-4)
8598 0508 sr : $\ln(2\pi/3)$
8598 2520 rt : (3,9,4,8)
8598 5424 sπ : $1/\ln(\sqrt{2}\pi)$
8598 7342 rt : (5,1,-4,-1)
8598 7448 id : $(e+\pi)$
8598 8063 cr : $3e/2 + 3\pi/4$
8598 8310 $2^x : 9\sqrt{3}/8$
8598 9694 $\ell_{10} : 3 + 3\sqrt{2}$
8599 1336 rt : (9,6,-6,-5)
8599 1388 $\ell_{10} : P_4$
8599 1464 rt : (5,2,5,6)
8599 2471 rt : (4,7,-2,-6)
8599 3487 $10^x : 4\sqrt{2}/3 - 3\sqrt{3}/4$
8599 4505 rt : (1,9,7,0)
8599 4633 $Ei : \exp(5/2)$
8599 5658 $\Gamma : 10/21$
8599 6394 $e^x : 3\sqrt{2}/2 + \sqrt{3}/3$
8599 7068 $e^x : 6e\pi/7$
8599 7585 $erf : 24/23$
8599 9216 rt : (6,9,-6,-8)
8599 9599 $\exp(3) - \exp(4/5)$
8600 0106 $Ei : \sqrt[3]{2}/2$
8600 1771 rt : (4,0,-9,9)
8600 4836 $\Psi : 12/23$
8600 5644 $J_0 : 16/21$
8600 6956 rt : (5,4,1,-7)
8600 7361 sq : P_{34}
8601 0077 $Li_2 : (\pi)^{-1/3}$
8601 0362 id : $3\sqrt{2}/2 - 4\sqrt{5}/3$
8601 3087 rt : (6,5,8,7)
8601 3862 cr : 7/11
8601 4705 rt : (5,5,-8,0)
8601 6451 cr : $1 + 2e$
8602 0114 tanh : 22/17
8602 1348 rt : (6,1,4,-8)
8602 3190 $2^x : 4 - \sqrt{5}/4$
8602 4479 sr : $3\sqrt{3}/4 - \sqrt{5}/4$
8602 5401 id : $(\pi/2)^{-1/3}$
8602 5490 rt : (6,7,0,-9)
8602 6321 $\ell_{10} : 8e/3$
8602 7485 $\ell_2 : 1/3 + 4\sqrt{3}$
8603 0411 $\Psi : e\pi/3$
8603 2588 rt : (5,6,-9,-9)
8603 4728 rt : (1,2,8,-9)
8603 5648 rt : (7,2,7,9)
8603 5751 rt : (1,5,-9,-7)
8603 5879 $10^x : 4(e+\pi)$
8603 7931 at : $(\pi/2)^{1/3}$
8603 7961 rt : (6,8,-9,-2)
8604 2628 rt : (1,7,-6,6)
8604 8793 $e^x : \sqrt{5/2}$
8605 2954 rt : (7,5,-6,-3)
8605 9375 rt : (1,9,-5,-3)
8606 0144 rt : (8,8,-7,-5)
8606 0243 $2 \times s\pi(\pi)$
8606 0519 $e^x : \exp(\sqrt[3]{5}/3)$
8606 0971 rt : (7,1,-2,2)
8606 7141 sπ : $6e/7$
8606 7846 rt : (8,2,-3,-4)
8606 7954 sinh : $10\sqrt[3]{3}/7$
8607 0797 $\exp(1/4) \div \exp(2/5)$
8607 1152 rt : (9,5,-4,-6)
8607 1210 sπ : $7\sqrt[3]{5}/9$
8607 2493 $erf : 23/22$
8607 2995 $J_0 : 19/25$
8607 3372 $J_0 : 4\sqrt[3]{5}/9$
8607 8798 $J_0 : (\sqrt{3})^{-1/2}$
8608 0585 rt : (3,9,-3,-6)
8608 0803 cu : $\sqrt{21}\pi$
8608 2638 rt : (9,1,0,5)
8608 3740 rt : (8,4,1,3)
8608 5717 cu : $3\sqrt{3}/4 - 4\sqrt{5}$
8608 7397 rt : (7,7,2,1)
8608 9013 rt : (6,6,-8,3)
8609 0179 sr : $(\ln 3/2)^{1/2}$
8609 1115 rt : (5,9,-1,-9)
8609 4508 sπ : $5\sqrt{3}/2$
8609 4614 rt : (4,6,0,5)
8609 6222 $\Gamma : 25/4$
8610 0198 $2^x : 1 + 3\sqrt{2}$
8610 1595 rt : (3,6,4,6)
8610 2744 cr : $2\sqrt{3} + 4\sqrt{5}/3$
8610 2852 $\ell_{10} : 1/3 + 4\sqrt{3}$
8610 3914 $10^x : 4/3 + 2\sqrt{2}/3$
8610 6274 rt : (9,7,4,4)
8610 8555 rt : (5,3,3,-8)
8611 2983 at : $1/4 - \sqrt{2}$
8611 4554 sq : $e/3 + \pi/4$
8611 4609 cu : $3/4 + 3\sqrt{5}$
8611 9629 rt : (2,8,-1,-8)
8612 0971 sr : $2\sqrt{3}$
8612 1150 ln : $1 - \sqrt{3}/3$
8612 1281 $\Psi : \sqrt[3]{3}/8$
8612 1376 tanh : $9\sqrt[3]{3}/10$
8612 1587 $s\pi(\sqrt{3}) \div s\pi(1/3)$
8612 2409 rt : (6,0,6,9)
8612 3770 $Ei : (\ln 3/2)^{-1/2}$
8612 4656 rt : (1,9,-9,-5)
8612 5672 sinh : $9\ln 2/8$
8612 6933 cr : $2\sqrt{5}/7$
8613 0964 sπ : $9\sqrt[3]{2}/2$
8613 1929 rt : (1,2,-9,-7)
8613 4161 id : $2e + 3\pi$
8613 5311 $\ln(4) \div \ln(5)$
8613 5367 sπ : $7\pi/3$
8613 5933 sq : $4e + 4\pi/3$
8613 5946 ln : $1 - 2\sqrt{2}/3$
8613 6368 rt : (1,7,-4,-8)
8613 6902 $\exp(e) + s\pi(1/4)$
8613 7645 rt : (7,7,-8,6)
8613 8302 $erf : \pi/3$
8613 8356 rt : (8,7,-5,-6)
8613 9022 $2/3 \div s\pi(e)$
8614 0660 $\zeta : 2\ln 2/5$
8614 0848 rt : (9,0,-9,2)
8614 1124 $e^x : 1/\ln(5)$
8614 1572 sq : $7\sqrt[3]{2}/4$
8614 2899 cu : $3 - \pi/4$
8614 3031 rt : (5,3,7,7)
8614 3942 rt : (6,7,-7,-3)
8614 5422 at : $3e/7$
8614 7533 tanh : $3\sqrt{3}/4$
8614 8567 rt : (3,7,-2,-5)
8614 8932 rt : (6,7,-4,7)
8614 9041 rt : (9,4,-2,-7)
8615 1521 sπ : $\ln((e+\pi)/3)$
8615 3693 ln : $3e/2 + 3\pi/4$
8615 4180 $erf : 22/21$
8615 5549 rt : (4,9,-7,0)
8615 7268 sr : $3\sqrt{3}/7$
8615 7405 $\zeta : 1/\sqrt{13}$
8615 8338 rt : (7,4,-4,-4)
8615 9534 $\Psi : \exp(\pi/3)$
8616 0622 sr : $4 + 4\pi/3$
8616 3464 rt : (3,3,5,4)
8616 4870 sr : $5\ln 2$
8616 5416 $\ell_2 : \sqrt[3]{6}$
8616 5786 rt : (5,7,0,8)
8616 7729 $\Psi : (\ln 3/3)^3$
8616 8652 rt : (1,9,1,-2)
8616 9913 rt : (3,6,-6,4)
8616 9919 $\Psi : 4\sqrt{3}$
8617 1952 rt : (8,8,-8,9)
8617 2315 tanh : 13/10
8617 2371 $\Psi : 5\sqrt[3]{5}/3$
8617 2640 rt : (8,1,-1,-5)
8617 3461 sq : $3\sqrt{2} + \sqrt{3}/4$
8617 3850 id : P_5
8617 4052 $erf : (\ln 3)^{1/2}$
8617 4771 cu : $\sqrt{3}/4 - \sqrt{5}$
8617 6496 $\ln(5) \times \exp(5)$
8617 7387 cr : 16/25
8617 7607 rt : (5,4,-6,-1)
8618 0161 rt : (3,8,-1,-7)
8618 0731 id : $\sqrt{2}/2 + 2\sqrt{3}/3$
8618 1216 at : $(e/2)^{1/2}$
8618 1382 sr : $7(1/\pi)/3$
8618 4294 rt : (2,4,-5,-6)
8618 6365 $2^x : (\sqrt[3]{3}/2)^{1/3}$
8618 6692 cr : $4(e+\pi)$
8618 6850 rt : (9,2,2,6)
8618 7548 rt : (9,9,-4,-9)
8618 8474 sr : $(2e/3)^{-1/2}$
8619 1087 rt : (2,9,7,8)
8619 1250 $\sqrt{5} \times \exp(5)$
8619 1543 $2^x : 3\sqrt{2}/4 + 2\sqrt{5}/3$
8619 2881 id : $4/3 - \sqrt{2}/3$
8619 3674 rt : (4,7,4,9)
8619 9441 rt : (6,1,-3,-2)
8619 9480 ln : $1 + 2e$
8620 0835 $\theta_3 : 2\sqrt{2}/7$
8620 3777 rt : (4,6,8,5)
8620 4108 cu : $3\zeta(3)/2$
8620 4285 rt : (7,9,-6,-6)
8620 6955 sq : P_{17}
8620 7105 rt : (5,2,5,-9)
8620 7480 $10^x : 9\sqrt[3]{3}/10$
8620 7790 rt : (2,8,-9,-3)
8620 8785 rt : (1,8,-3,-4)
8620 9058 $\exp(5) \div \exp(\sqrt{5})$
8621 0424 $erf : (\sqrt{3}/2)^{-1/3}$
8621 0632 $\theta_3 : \exp(-e/3)$
8621 1176 at : $(\sqrt[3]{4})^{1/3}$
8621 3089 $Li_2 : 2\sqrt[3]{5}/5$
8621 3314 sπ : $e/3 + 3\pi$
8621 4711 rt : (8,6,-3,-7)
8621 5083 sr : $-\sqrt{2} + \sqrt{5} + \sqrt{7}$
8621 5630 $J_0 : 3\sqrt[3]{2}/5$
8621 5929 cr : $2 - e/2$
8621 7005 at : 7/6
8621 9013 cu : $4e\pi/5$
8621 9479 rt : (1,1,-8,-7)
8622 0652 $\Psi : 10\ln 2$
8622 0861 sπ : $\sqrt{3}/4 + \sqrt{5}$
8622 1057 rt : (7,2,0,3)
8622 2074 $e^x : 1/4 + 3e/4$
8622 2239 cr : $4\sqrt[3]{3}/9$
8622 3591 rt : (8,3,-8,-4)
8622 3615 $\ell_{10} : \sqrt{2}/4 + 4\sqrt{3}$
8622 4489 sq : 13/14
8622 4879 tπ : $10\sqrt{3}/3$
8622 5088 rt : (9,3,0,-8)
8622 5428 rt : (1,3,-1,9)
8622 8698 tanh : $2(e+\pi)/9$
8622 8836 rt : (4,2,-8,4)
8623 1748 rt : (6,8,-8,-9)
8623 2135 $\ell_{10} : 1 + 2\pi$
8623 2342 $\zeta : 1/\sqrt[3]{5}$
8623 2591 rt : (5,9,-8,-3)
8623 3051 cr : $3\sqrt[3]{5}/8$
8623 3236 sr : $3/4 + e$
8623 3685 $\Psi : 18$
8623 5597 tπ : $4\sqrt{2}$
8623 5965 $\zeta : 5/18$
8623 6298 $\exp(5) \times s\pi(\pi)$
8623 7247 cu : $\sqrt{2}/4 - 2\sqrt{5}$
8623 9034 cu : $3(e+\pi)$
8623 9337 $\ln(4/5) + \exp(3)$
8624 0508 $e^x : 24/11$
8624 1145 $erf : 5\sqrt[3]{2}/6$
8624 1376 rt : (3,4,-8,2)
8624 2460 rt : (8,5,3,4)

8624 3610 $erf : 21/20$
8624 6259 $\ell_{10} : \sqrt{3}/3 + 3\sqrt{5}$
8624 6518 rt : (6,6,-5,-4)
8624 8707 rt : (3,2,3,-6)
8624 9647 $\ell_2 : 11/20$
8624 9876 $J_0 : (\sqrt[3]{5}/3)^{1/2}$
8625 0146 sr : $(e/3)^3$
8625 0171 $Li_2 : 13/19$
8625 2084 rt : (1,6,2,7)
8625 4139 rt : (2,9,-3,-8)
8625 5239 rt : (1,9,-8,5)
8625 6345 $s\pi : \sqrt{2} - \sqrt{5}/3$
8625 7007 $\ell_2 : 9\sqrt{2}/7$
8625 7697 rt : (5,3,-8,7)
8625 9750 $e^x : \sqrt{24\pi}$
8626 0358 rt : (7,3,-2,-5)
8626 3653 rt : (9,8,6,5)
8626 3902 $erf : (e/3)^{-1/2}$
8626 7568 $10^x : 4\sqrt{3}/3 + 3\sqrt{5}/4$
8626 8401 rt : (4,1,6,7)
8627 1794 $e^x : 3\sqrt{2}/2 + \sqrt{3}/4$
8627 2244 sq : $4/3 + 4\sqrt{2}$
8627 2534 rt : (1,9,9,-8)
8627 4110 rt : (8,0,1,6)
8627 4978 $s\pi : (\sqrt{5})^{-1/2}$
8627 5972 $2^x : 6\ln 2$
8627 6796 rt : (3,7,1,-8)
8627 8226 sq : $3/2 + 4\pi$
8627 8405 $J_0 : (\ln 3)^{-3}$
8627 8799 rt : (9,6,-5,-3)
8627 9162 rt : (7,8,-4,-7)
8627 9892 $e^x : 7\zeta(3)/8$
8628 0353 $10^x : 1/3 + \sqrt{2}/4$
8628 1356 cr : $3 + 2\sqrt{3}$
8628 1760 $K : 5/7$
8628 1908 id : $1/3 - 3\sqrt{3}$
8628 2090 id : $3e/2 + \pi/4$
8628 2814 $s\pi : \sqrt{5} + \sqrt{6} + \sqrt{7}$
8628 3484 $\exp(3/4) - s\pi(\sqrt{3})$
8628 3640 $10^x : 4\zeta(3)/7$
8628 3784 cu : $e/2 + \pi/4$
8628 7431 rt : (4,1,-5,1)
8628 7775 rt : (9,3,4,7)
8628 7922 rt : (5,4,9,8)
8628 9231 sr : $3 + 3\sqrt{3}$
8628 9294 rt : (8,5,-1,-8)
8628 9825 rt : (4,4,4,-9)
8629 0447 id : $2\sqrt{3}/3 + 3\sqrt{5}$
8629 3666 rt : (1,8,5,-1)
8629 4608 $\Gamma : \ln(\sqrt{2}/3)$
8629 5451 tanh : $(\sqrt{2}\pi/2)^{1/3}$
8629 6155 $2^x : 2\pi/7$
8629 6572 cu : $4 + 3\sqrt{5}$
8629 9378 rt : (9,2,2,-9)
8630 0515 rt : (4,3,-4,8)
8630 1171 $e + \ln(\pi)$
8630 4621 ln : $(4/3)^3$
8630 5437 cr : 9/14
8630 6033 as : $(\sqrt{3})^{-1/2}$
8631 0084 rt : (9,1,-7,-1)
8631 0915 $erf : 7\zeta(3)/8$
8631 1739 rt : (4,6,-7,-1)
8631 5468 $e^x : 1/4 + 2e/3$
8631 6535 $\ell_{10} : P_{24}$
8631 7055 sr : $3 + \sqrt{2}/3$
8631 7560 $Ei : 1/\ln(2e/3)$
8631 8332 rt : (9,4,-9,-1)
8632 5037 $1/4 - \exp(\sqrt{2})$
8632 5154 $\Gamma : 3/25$
8632 7888 $\exp(5) \times s\pi(e)$
8632 8724 $s\pi : \sqrt[3]{19}$
8632 9204 cu : $3\sqrt{3} - 4\sqrt{5}/3$
8632 9666 as : $4\sqrt[3]{5}/9$
8633 0131 rt : (2,6,-9,-9)
8633 1311 as : 19/25
8633 2495 rt : (5,8,-6,-4)
8633 4002 id : $(\sqrt{5}/3)^{1/2}$
8633 5635 $\ell_{10} : 2\sqrt{2} + 2\sqrt{5}$
8633 5895 $\Psi : 11/13$
8633 5959 $s\pi : (2e)^{1/2}$
8633 7740 sinh : 19/2
8633 8618 ln : $2\sqrt{3} + 4\sqrt{5}/3$
8633 8998 id : $5\sqrt{5}/6$
8634 0449 $2^x : 2\sqrt{2}/3 - 2\sqrt{3}/3$
8634 0525 $10^x : e/2$
8634 1935 $erf : 20/19$
8634 2435 rt : (6,8,-7,-2)
8634 4566 $e^x : 4\sqrt{3}/3 + 3\sqrt{5}/4$
8634 4690 ln : $1/4 + 3\sqrt{2}/2$
8634 5729 rt : (1,9,2,7)
8634 5896 rt : (6,5,-3,-5)
8634 6679 $s\pi : 3\sqrt{3} + 2\sqrt{5}$
8634 7526 sq : $e/2 + 3\pi/2$
8634 7583 $10^x : 3\pi^2/10$
8634 7856 cr : $3 - 3\pi/4$
8634 8267 rt : (7,9,-7,1)
8635 1656 rt : (1,7,-1,-5)
8635 2330 rt : (7,7,-2,-8)
8635 2944 $J_0 : \ln(\sqrt{2}/3)$
8635 3053 rt : (6,0,-1,3)
8635 4579 $2^x : e/2 - \pi/2$
8635 8007 rt : (2,9,0,-8)
8635 8501 cr : $2 + 2\sqrt{5}$
8635 9214 rt : (7,2,0,-6)
8636 0308 $\ln(5) + s\pi(\sqrt{3})$
8636 1175 rt : (3,7,-7,-7)
8636 2180 rt : (8,4,1,-9)
8636 3169 $\Gamma : 8\pi^2/9$
8636 3636 id : 19/22
8636 3641 sinh : $(2\pi/3)^{-1/3}$
8636 6109 rt : (4,9,-1,-5)
8636 6217 $\exp(1/5) \times s\pi(1/4)$
8636 6654 sq : $2\sqrt{2}/3 + 3\sqrt{5}/4$
8636 8225 tanh : $(\sqrt[3]{5})^{1/2}$
8636 9033 tanh : 17/13
8637 0652 rt : (3,6,3,-9)
8637 2452 rt : (8,1,3,7)
8637 2747 $\zeta : 7(1/\pi)/8$
8637 3416 rt : (7,3,2,4)
8637 5232 ln : $1/3 + 3e/4$
8637 6611 rt : (6,5,1,1)
8637 8363 $erf : (\sqrt[3]{5}/2)^{-1/3}$
8637 9297 $J_0 : 5\zeta(3)/8$
8638 3759 cu : 20/21
8638 3782 $\Psi : \zeta(3)/9$
8638 4337 rt : (2,7,-1,7)
8638 5295 $\ln(\sqrt{3}) - \exp(5)$
8638 5609 rt : (9,4,6,8)
8638 5897 rt : (2,8,-5,-9)
8638 9486 rt : (3,1,5,-7)
8639 0532 sq : 22/13
8639 0582 $\exp(\sqrt{3}) \div s\pi(\sqrt{2})$
8639 2465 $e^x : 2 + 2\sqrt{5}$
8639 3589 rt : (8,6,5,5)
8639 4545 $\zeta : 5\sqrt{3}/6$
8639 7113 rt : (9,8,-9,-4)
8639 9236 rt : (3,6,-4,-6)
8639 9525 sq : $2/3 + 3\sqrt{3}/4$
8640 0480 id : $1/\ln(\sqrt[3]{5})$
8640 2477 $\ln(2/5) - \exp(2/3)$
8640 3426 cu : $3\sqrt{2} + 2\sqrt{5}$
8640 4569 $\theta_3 : 15/22$
8640 4804 cr : $1/\ln(3\pi/2)$
8640 8160 rt : (4,2,8,8)
8641 0123 sq : $1/4 + e/4$
8641 1533 sr : $1/3 + \pi$
8641 1572 $\Psi : (1/(3e))^{-1/2}$
8641 2469 $J : (e + \pi)^{-3}$
8641 3416 $\exp(1/5) - \exp(3)$
8641 3574 rt : (9,9,8,6)
8641 4237 rt : (6,9,-3,2)
8641 5437 $10^x : 4/3 + \sqrt{5}/3$
8641 9817 rt : (7,8,4,2)
8642 0476 sq : $1/\ln(\sqrt[3]{3}/3)$
8642 0491 rt : (9,3,-7,-2)
8642 1565 cr : $3\sqrt{2} + \sqrt{5}$
8642 2484 $s\pi : 5\zeta(3)/9$
8642 3188 sr : $7(e + \pi)/5$
8642 3862 rt : (7,6,0,-9)
8642 4227 $J_0 : 3/4$
8642 4295 at : $4 - 2\sqrt{2}$
8642 5272 $K : (\ln 3/3)^{1/3}$
8642 5825 $\zeta : 2e/3$
8642 8652 $\ell_2 : \sqrt{2}/4 + 4\sqrt{3}$
8642 9354 rt : (5,8,-3,-1)
8643 0078 rt : (3,8,-1,2)
8643 1509 $10^x : 22/5$
8643 1540 rt : (1,8,-8,-3)
8643 1766 $\ell_2 : \ln(\sqrt{3})$
8643 1892 rt : (2,9,5,-9)
8643 3152 $e^x : 1/4 + 3\sqrt{3}$
8643 3899 $K : ((e + \pi)/3)^{-1/2}$
8643 6049 cu : 16/13
8643 6111 $\ell_{10} : 3\sqrt{2}/2 + 3\sqrt{3}$
8643 7696 $K : (\sqrt{5}/2)^{-3}$
8644 0755 rt : (3,8,-8,-1)
8644 1236 at : $(e + \pi)/5$
8644 1306 cr : $4 - 3\sqrt{5}/2$
8644 1934 sr : $1/\ln(4/3)$
8644 2261 rt : (6,4,-1,-6)
8644 5039 rt : (3,4,7,5)
8644 6907 sq : $1/3 + e/2$
8644 9110 rt : (8,1,-8,1)
8644 9665 rt : (2,8,2,-9)
8644 9756 $\ell_{10} : 6e\pi/7$
8645 0547 $erf : 19/18$
8645 2213 $2^x : \sqrt[3]{7/2}$
8645 3651 rt : (2,8,-8,2)
8645 4768 sq : $1/3 - 4\pi/3$
8645 5092 rt : (7,1,2,-7)
8645 6955 $\ell_2 : 1 + 2\pi$
8645 8019 $t\pi : 9\sqrt{3}/7$
8646 0519 rt : (4,7,-3,-4)
8646 2357 $s\pi : \sqrt{2}/3 + 3\sqrt{3}$
8646 3610 $J_0 : (\sqrt[3]{2}/3)^{1/3}$
8646 4179 at : $\exp(1/(2\pi))$
8646 4431 $\exp(2/5) - \exp(\sqrt{5})$
8646 5771 cr : $1 - \sqrt{2}/4$
8646 7847 rt : (8,2,5,8)
8647 0014 cr : $2e + \pi/3$
8647 0333 rt : (9,7,-7,-5)
8647 3232 $\Psi : 20/7$
8647 4838 cu : $3\sqrt{2}/4 + 4\sqrt{3}$
8647 8512 $e^x : 4\sqrt{3} - 2\sqrt{5}/3$
8647 8897 id : $9(1/\pi)$
8647 9832 $\ell_{10} : 5(e + \pi)/4$
8648 0102 rt : (5,2,-2,-3)
8648 0527 rt : (9,5,8,9)
8648 2233 $Ei : 12/19$
8648 4459 rt : (6,8,-1,-9)
8648 5826 at : $\sqrt{3} - \sqrt{5}/4$
8648 5843 rt : (7,7,-7,8)
8648 8515 rt : (1,6,1,-6)
8648 8653 $10^x : 3 + 3\sqrt{5}$
8649 0661 tanh : 21/16
8649 0741 $\sqrt{2} - \ln(\sqrt{3})$
8649 3058 cr : 11/17
8649 4542 rt : (4,9,-6,3)
8649 4836 as : $4/3 - 2\pi/3$
8649 6727 $2^x : 2 + 3\pi/2$
8649 7938 sinh : 18/23
8649 8503 cu : $2 - \pi/3$
8649 9507 rt : (6,1,1,4)
8649 9632 sr : $3/2 + 3\sqrt{5}$
8649 9743 ln : 8/19
8650 2181 $\ell_2 : 2/3 + 2\sqrt{3}/3$
8650 3872 $\ell_2 : \sqrt{3}/3 + 3\sqrt{5}$
8650 6222 rt : (8,5,-8,-2)
8650 6547 rt : (2,3,4,-7)
8650 8684 rt : (2,5,-9,6)
8650 9064 rt : (6,6,-7,5)
8650 9346 $s\pi : P_{26}$
8651 0650 $t\pi : 8\pi^2/9$
8651 0825 rt : (4,9,-5,-5)
8651 2043 $\theta_3 : (\pi/2)^{-2}$
8651 2665 $\ell_{10} : 7\pi/3$
8651 7655 $erf : (\sqrt{5}/2)^{1/2}$
8651 8115 $e^x : 20/19$
8651 8152 $\ell_{10} : \sqrt{5} + \sqrt{6} + \sqrt{7}$
8651 8648 $2^x : \sqrt{3} + \sqrt{5} - \sqrt{6}$

8651 8738 rt : (7,4,4,5)
8651 9383 rt : (9,2,-5,-3)
8651 9433 K : $9(1/\pi)/4$
8652 0182 2^x : $\exp(-1/(3\pi))$
8652 0790 rt : (2,6,-9,2)
8652 1331 Ei : $7\sqrt[3]{3}/6$
8652 1384 $4 \div \ln(2/3)$
8652 3342 rt : (5,9,1,3)
8652 4433 rt : (3,0,7,8)
8652 5403 cu : $3\sqrt{2}/2 + 4\sqrt{5}/3$
8652 5597 $e \div \pi$
8653 0142 ℓ_{10} : $3/22$
8653 0598 sq : $4\sqrt{2} + \sqrt{3}/3$
8653 0982 cr : $9\sqrt[3]{3}/2$
8653 2337 rt : (8,9,-8,-5)
8653 3529 ℓ_{10} : $4\sqrt{2} + 3\sqrt{5}/4$
8653 4974 id : $3\sqrt[3]{3}/5$
8653 5786 rt : (6,3,1,-7)
8653 6319 sq : $1/\ln((e+\pi)/2)$
8653 6336 rt : (3,9,-9,4)
8653 7367 sr : $(\sqrt[3]{2}/3)^{1/3}$
8653 7797 rt : (8,7,7,6)
8653 8806 rt : (2,5,0,6)
8653 8849 sq : $3e/2 - 2\pi$
8654 1877 rt : (9,6,-5,-6)
8654 4008 rt : (5,5,-7,2)
8654 6756 cu : $2/3 - \sqrt{5}$
8654 8158 rt : (7,0,4,8)
8655 0284 sinh : $16/7$
8655 4012 $s\pi$: $(\ln 2)^3$
8655 4538 rt : (7,1,-9,-4)
8655 4876 ζ : $\sqrt{5}/8$
8655 7968 θ_3 : $\ln(3/2)$
8655 8107 Ψ : $-\sqrt{5} + \sqrt{6} + \sqrt{7}$
8656 0245 sq : $3/4 + 2\sqrt{2}/3$
8656 0460 rt : (8,3,7,9)
8656 1185 K : $(e)^{-1/3}$
8656 1702 $3/5 \div \ln(2)$
8656 1794 θ_3 : $1/\sqrt[3]{15}$
8656 2056 Γ : $4/17$
8656 2068 rt : (3,6,-3,-7)
8656 2463 2^x : $2/3 + 4\sqrt{2}/3$
8656 3366 rt : (6,7,-3,9)
8656 4104 2^x : $4/3 + 3\pi$
8656 4230 Ψ : $6(1/\pi)/7$
8656 5860 cu : $\sqrt{2} + 3\sqrt{5}$
8656 7551 rt : (1,7,3,-2)
8657 1146 erf : $18/17$
8657 1450 rt : (3,6,-5,7)
8657 3300 tanh : $25/19$
8657 4451 e^x : $4e/3 + \pi/3$
8657 6733 id : $2e - \pi/2$
8657 7654 e^x : $1/2 + 4\sqrt{2}/3$
8658 0426 tanh : $(\sqrt{3})^{1/2}$
8658 1619 sr : $7e\pi/4$
8658 2403 rt : (7,3,-9,1)
8658 3857 rt : (3,7,-6,-2)
8658 4835 ℓ_2 : $4/3 - 3\sqrt{3}/4$
8658 5554 $\exp(5) \times \exp(\sqrt{3})$
8658 5624 at : $3\sqrt{2}/4 - \sqrt{5}$
8658 5743 sr : $1/2 + 4\sqrt{5}/3$
8658 5986 J_0 : $\sqrt{5}/3$
8658 7628 rt : (1,8,-7,6)
8658 9797 rt : (7,7,-9,-2)
8659 1892 10^x : $\sqrt{2}/4 + 2\sqrt{5}/3$
8659 5424 rt : (2,9,-5,-7)
8659 9304 2^x : $5\sqrt[3]{2}/7$
8660 0082 sinh : $(\sqrt[3]{3}/3)^{1/3}$
8660 1885 erf : $\ln(\ln 2/2)$
8660 2540 id : $\sqrt{3}/2$
8660 2917 Ei : $7\zeta(3)/5$
8660 4100 10^x : $3 - \pi/2$
8660 6598 2^x : $9/10$
8660 6721 sinh : $21/8$
8660 9174 Γ : $5/16$
8660 9922 $\sqrt{3} - \exp(4)$
8661 0587 λ : $\exp(-2\pi/3)$
8661 1499 rt : (5,6,-3,6)
8661 1820 rt : (9,5,-3,-7)
8661 2524 $4 \div \exp(1/3)$
8661 4616 rt : (5,5,0,-7)
8661 5206 rt : (9,1,-3,-4)
8661 6724 e^x : $e - 2\pi/3$
8661 9826 rt : (2,8,0,-4)
8662 1060 rt : (5,1,0,-4)
8662 2483 rt : (9,3,-3,1)
8662 2632 ℓ_2 : P_{12}
8662 3910 cr : $13/20$
8662 4890 as : $16/21$
8662 5557 id : $\sqrt[3]{13/2}$
8662 6404 ln : $3 + 2\sqrt{3}$
8662 6625 rt : (6,2,3,-8)
8662 7273 2^x : $1/2 - \sqrt{2}/2$
8662 8591 Γ : $\exp(-\sqrt{5}/3)$
8663 0226 at : $20/17$
8663 2943 rt : (3,5,-6,-7)
8663 3905 e^x : $23/13$
8663 6201 e^x : $9\zeta(3)/8$
8663 6219 J_0 : $(e/3)^3$
8663 6975 rt : (2,2,6,-8)
8663 7192 10^x : $(e)^{1/3}$
8663 8390 $t\pi$: $3\sqrt{3}/2 + \sqrt{5}/3$
8663 8559 erf : $3\sqrt{2}/4$
8663 8567 rt : (7,1,6,9)
8663 9420 rt : (6,2,3,5)
8664 2236 rt : (5,3,0,1)
8664 2489 ζ : $2\sqrt[3]{2}/9$
8664 3397 id : $5\ln 2/4$
8664 4651 10^x : $4\sqrt{2} - 2\sqrt{3}$
8664 5734 ζ : $7/25$
8664 6860 ln : $7e/8$
8664 7540 Γ : $\ln(\ln 2)$
8664 8075 2^x : $4/3 - \sqrt{3}/4$
8664 9127 ℓ_2 : $4 - \sqrt{2}/4$
8664 9385 rt : (9,7,-3,-2)
8665 0493 $t\pi$: $5/22$
8665 1492 Li_2 : $(\sqrt{2}/3)^{1/2}$
8665 2210 erf : $10(1/\pi)/3$
8665 2964 ℓ_{10} : $4 + 3\sqrt{5}/2$
8665 3991 rt : (3,1,9,9)
8665 4535 e^x : $5\pi^2/3$
8665 4563 at : $1/2 - 3\sqrt{5}/4$
8665 6174 $t\pi$: $2\zeta(3)/7$
8665 7359 e^x : $4\sqrt{5}$
8665 7623 rt : (7,5,6,6)
8665 8016 $\ln(3/4) + \exp(e)$
8665 8162 sq : $\sqrt{2}/4 + \sqrt{3}/3$
8666 1206 e^x : $\exp(-\sqrt{2}/3)$
8666 2317 rt : (5,3,-7,9)
8666 2944 rt : (2,2,-9,-8)
8666 6026 $t\pi$: $2\sqrt{2} + 4\sqrt{5}$
8666 6666 id : $13/15$
8666 7633 id : $3\sqrt{2}/2 + \sqrt{5}/3$
8667 0875 rt : (2,2,6,5)
8667 1196 rt : (8,7,-4,-7)
8667 1328 e^x : $3\sqrt{2}/4 + \sqrt{3}/2$
8667 2099 Γ : $\sqrt[3]{25}$
8667 2621 cr : $(e+\pi)/9$
8667 2666 Ei : $6(e+\pi)/7$
8667 3078 J_0 : $(2e/3)^{-1/2}$
8667 4047 ln : $3(e+\pi)$
8667 4737 sq : $\ln(2e)$
8667 5671 rt : (8,8,9,7)
8667 6730 sr : $5\zeta(3)/8$
8667 7231 at : $(\sqrt[3]{3}/2)^{-1/2}$
8667 7312 J_0 : $7(1/\pi)/3$
8667 8125 Γ : $-\sqrt{3} + \sqrt{6} + \sqrt{7}$
8668 0228 rt : (9,4,-1,-8)
8668 0447 cu : $4 + 4\sqrt{5}$
8668 1187 rt : (6,6,3,2)
8668 3470 Γ : $(\sqrt{2}\pi)^{-1/2}$
8668 3661 rt : (7,6,-7,-3)
8668 8359 Ei : $5(e+\pi)/2$
8668 8410 Ei : $2/25$
8668 9990 sq : $7\zeta(3)/3$
8669 0475 rt : (2,8,-7,7)
8669 1663 rt : (4,7,-1,-7)
8669 1702 J_0 : $3\sqrt{3}/7$
8669 1821 cu : $3\sqrt{2} - 3\sqrt{3}$
8669 2234 $s\pi$: $4\sqrt{2} + 3\sqrt{5}/4$
8669 5936 rt : (8,3,-4,-4)
8669 8401 at : $3\pi/8$
8669 8928 rt : (1,5,-8,-9)
8670 0001 $1/4 - \exp(3/4)$
8670 0821 $t\pi$: $5(1/\pi)/7$
8670 2301 Li_2 : $4\zeta(3)/7$
8670 2417 ℓ_2 : $\sqrt{2} - \sqrt{5} + \sqrt{7}$
8670 2822 rt : (2,9,-7,-2)
8670 3325 rt : (5,4,2,-8)
8670 4205 id : $4\sqrt{2}/3 + 4\sqrt{5}/3$
8670 4255 rt : (3,5,9,6)
8670 5826 erf : $17/16$
8670 6986 $t\pi$: $8e\pi/3$
8670 7287 sq : $4e - 4\pi$
8670 8139 rt : (9,0,-1,5)
8670 9767 at : $\sqrt{3}/4 + \sqrt{5}/3$
8671 4157 ℓ_2 : $1/3 + 2\sqrt{5}/3$
8671 4924 rt : (6,1,5,-9)
8671 5737 at : $5\sqrt{2}/6$
8671 5999 cr : $\sqrt{2} + \sqrt{6} + \sqrt{7}$
8671 7845 rt : (2,6,-8,-5)
8671 9420 rt : (7,9,6,3)
8672 0373 cr : $15/23$
8672 0671 rt : (3,6,-4,-3)
8672 1926 Ψ : $5/23$
8672 3446 id : $3e + 3\pi/2$
8672 3677 Ei : $4\sqrt{3}/7$
8672 4572 Ψ : $3/11$
8672 5693 ℓ_2 : $\sqrt{2} - \sqrt{3}/2$
8672 5995 2^x : $\sqrt{2}/4 - \sqrt{5}/4$
8672 6262 sinh : $\sqrt{17}$
8672 6730 $t\pi$: $4\ln 2$
8672 9464 $\ln(3/5) - \exp(\sqrt{5})$
8672 9540 rt : (7,9,-5,-7)
8673 0052 ln : $\sqrt{17/3}$
8673 0503 cr : $10(e+\pi)/9$
8673 1002 $t\pi$: $\exp(-\sqrt{2}\pi/3)$
8673 1678 J_0 : $(\ln 3/2)^{1/2}$
8673 2352 ln : $4/3 + \pi/3$
8673 3808 rt : (4,2,-7,6)
8673 4670 $\exp(1/4) \times s\pi(\sqrt{5})$
8673 4724 $\ln(2/5) - s\pi(2/5)$
8673 6658 ζ : $1/\ln(2)$
8673 7147 tanh : $\sqrt{3} + \sqrt{5} - \sqrt{7}$
8673 7325 ℓ_2 : P_{24}
8673 7985 J : $8/21$
8673 8371 rt : (8,6,-2,-8)
8674 3176 Ei : $(5/2)^{-1/2}$
8674 3491 ℓ_{10} : $3e - \pi/4$
8674 6002 rt : (1,4,5,-8)
8674 7165 rt : (9,3,1,-9)
8674 8506 rt : (8,1,-4,1)
8675 0618 ln : $2 + 2\sqrt{5}$
8675 4305 id : $3\sqrt{2}/4 - 4\sqrt{3}$
8675 5731 rt : (7,9,-6,3)
8675 5924 rt : (5,0,2,5)
8675 6322 id : $\ln(\sqrt[3]{2}/3)$
8675 7379 sq : $2\sqrt{2}/3 - 4\sqrt{3}/3$
8676 0051 at : $1/2 + e/4$
8676 2340 rt : (2,1,8,-9)
8676 2840 rt : (6,8,-8,-3)
8676 3139 $s\pi$: P_7
8676 6998 rt : (3,9,-2,-7)
8676 7299 rt : (9,4,-1,2)
8676 9420 J : $3\zeta(3)/8$
8677 1763 10^x : $\sqrt[3]{16/3}$
8677 2733 rt : (3,9,-8,8)
8677 3332 rt : (6,3,5,6)
8677 3709 10^x : $4e/3 + \pi$
8677 4742 rt : (7,5,-5,-4)
8677 5440 ℓ_{10} : $2/3 + 3\sqrt{5}$
8677 6447 $t\pi$: $3e/4 + 4\pi/3$
8677 8330 rt : (4,6,1,7)
8677 9724 Ei : $7\sqrt{2}/10$
8678 0830 2^x : $\exp(4/3)$
8678 1288 10^x : $\sqrt{3}/2 + \sqrt{5}/3$
8678 1597 rt : (1,5,-1,-8)
8678 3652 e^x : $1/\ln(2\pi/3)$
8678 6577 rt : (8,2,-2,-5)
8678 6799 2^x : $8\sqrt[3]{5}/9$

8678 6860 sr : $5\pi^2/6$
8678 7724 rt : (2,9,4,-2)
8678 8544 2^x : $5\sqrt[3]{3}/8$
8678 9603 rt : (5,3,4,-9)
8679 0598 rt : (7,6,8,7)
8679 0745 cr : $4/3 - e/4$
8679 0894 tanh : $(\sqrt[3]{5}/3)^{-1/2}$
8679 1484 10^x : $3/4 + 2\pi/3$
8679 1702 cu : $1/3 - 4\sqrt{2}$
8679 2288 ℓ_2 : $4/3 - \pi/4$
8679 2515 Ψ : $9(1/\pi)$
8679 5521 rt : (7,8,-3,-8)
8679 5806 Ψ : $(\sqrt[3]{3}/2)^2$
8679 6846 J_0 : $\ln(2\pi/3)$
8679 7475 e^x : $3 - \pi$
8679 8345 rt : (9,1,1,6)
8679 8667 e^x : $(\sqrt[3]{5}/2)^{-1/3}$
8680 0774 ℓ_2 : $\sqrt{2} + \sqrt{5}$
8680 0998 $t\pi$: $(\pi)^{1/2}$
8680 1479 J_0 : 17/23
8680 3398 id : $1/4 - \sqrt{5}/2$
8680 4124 rt : (8,5,0,-9)
8680 5480 Ψ : 13/25
8680 5521 Li_2 : 11/16
8680 6243 2^x : $3\zeta(3)/4$
8680 7590 sinh : $6\sqrt{3}$
8681 5090 rt : (2,5,-7,1)
8681 6043 rt : (4,9,-5,6)
8681 7910 $s\pi$: $3\sqrt{2} - \sqrt{3}/3$
8681 9954 rt : (1,6,1,-3)
8682 1326 2^x : $3e\pi/5$
8682 2720 2^x : $2\sqrt{2}/3 + \sqrt{3}/3$
8682 3675 rt : (4,7,-5,-7)
8682 4595 e^x : 5/8
8682 4809 $s\pi$: P_{81}
8682 6066 tanh : $(\ln 3)^3$
8682 8545 sq : $\sqrt[3]{18}$
8683 3040 sr : $10\pi/9$
8683 3339 $s\pi$: $7\pi/6$
8683 4471 sr : $2 + 2\sqrt{5}/3$
8683 4771 rt : (2,8,-5,-3)
8683 5500 tanh : $\sqrt[3]{7/3}$
8683 5890 rt : (1,9,-1,-7)
8683 6066 id : $e/4 + 4\pi/3$
8683 8605 2^x : $2/3 + 4\sqrt{3}/3$
8683 9182 J : $\ln 2/6$
8683 9301 10^x : $7\pi^2/4$
8683 9855 ℓ_2 : $3e/4 + 4\pi$
8683 9933 e^x : $e/3 - \pi/3$
8684 1649 sq : $3\sqrt{2}/4 + \sqrt{5}$
8684 1807 id : $4\sqrt{3}/3 + \sqrt{5}/4$
8684 1815 cu : $1/4 + 4\sqrt{3}$
8684 2700 sr : $(\ln 3)^{-3}$
8684 3405 Ψ : $(e + \pi)^{-2}$
8684 4272 cr : $4\sqrt{2} + \sqrt{3}/2$
8684 8279 ℓ_2 : $\sqrt{10/3}$
8685 0247 2^x : $3\sqrt{3}/4 + 3\sqrt{5}/4$
8685 0611 Ei : $10\ln 2/7$
8685 0872 ζ : $6\zeta(3)/5$
8685 1526 e^x : $3\sqrt{2} + \sqrt{5}/2$
8685 1709 rt : (6,8,-6,0)
8685 2121 ln : $3\sqrt{2} + \sqrt{5}$
8685 2199 10^x : $\sqrt{2} - \sqrt{3}/3$
8685 3939 at : 13/11
8685 7197 erf : 16/15
8685 8008 ℓ_{10} : $4\sqrt{2} + \sqrt{3}$
8685 8168 10^x : $4 + 4\pi$
8685 8269 Ei : $\sqrt{3}/9$
8685 8896 ℓ_{10} : $(e)^2$
8685 9627 $s\pi$: $\sqrt{2}/4 + 4\sqrt{5}/3$
8686 0127 rt : (7,7,-1,-9)
8686 0475 cr : $2e/3 + 3\pi/2$
8686 0696 rt : (2,9,-4,-2)
8686 2669 rt : (4,5,3,-9)
8686 3194 rt : (7,4,-3,-5)
8686 7096 sinh : $\pi/4$
8686 7412 rt : (1,3,7,-9)
8686 8511 rt : (4,2,1,-5)
8686 9776 rt : (7,1,-5,-1)
8686 9951 Ψ : $3\ln 2/4$
8687 0011 $t\pi$: $\sqrt{11}\pi$
8687 0248 sr : $3/2 - \sqrt{5}/3$
8687 1325 2^x : $3\sqrt{3} - \sqrt{5}/3$
8687 4616 rt : (8,1,0,-6)
8687 8760 e^x : 23/17
8688 0298 J_0 : 14/19
8688 1529 J_0 : $(5/2)^{-1/3}$
8688 3002 rt : (1,0,5,5)
8688 4592 ln : $\sqrt{2}/2 + 3\sqrt{5}/4$
8688 5181 rt : (5,1,4,6)
8688 5976 rt : (9,2,3,7)
8688 6111 erf : $e\pi/8$
8688 7685 rt : (8,2,-2,2)
8688 9477 sr : $(\sqrt[3]{5}/3)^{1/2}$
8689 2386 J : $\exp(-3\pi/5)$
8689 4009 rt : (7,2,-7,-3)
8689 5003 sq : $3\sqrt{3} + \sqrt{5}/2$
8689 9298 e^x : $3/2 + 4\sqrt{2}$
8690 1487 $1/4 \div \ln(3/4)$
8690 1719 rt : (6,4,7,7)
8690 1911 rt : (6,6,-6,7)
8690 1926 sinh : $(2\pi/3)^3$
8690 3663 sinh : $5\sqrt{2}/9$
8690 3750 at : $3/4 + \sqrt{3}/4$
8690 5439 rt : (9,5,1,3)
8690 6845 $\ln(3/5) \div s\pi(1/5)$
8690 7086 $\ln(3) \div s\pi(1/5)$
8690 7326 cr : $4/3 + 3\sqrt{3}$
8690 8974 sinh : 11/14
8691 1537 Ψ : $(5/3)^{-1/3}$
8691 1999 ζ : $(e + \pi)/10$
8691 3078 rt : (3,8,0,5)
8691 3846 rt : (2,3,8,6)
8691 5204 rt : (6,9,-2,4)
8691 6397 $s\pi$: $(\sqrt[3]{2}/3)^{-1/3}$
8691 7482 rt : (9,8,-8,-5)
8691 7696 cu : $3\sqrt{5}$
8691 8059 tanh : $1/\ln(\sqrt{2}/3)$
8691 8775 rt : (1,7,8,8)
8692 0869 ℓ_{10} : $\sqrt{2}/3 + 4\sqrt{3}$
8692 2288 e^x : $1/4 + 3\sqrt{5}/4$
8692 3041 rt : (5,4,2,2)
8692 3171 sq : $3 - \sqrt{3}/3$
8692 3197 erf : $8\zeta(3)/9$
8692 4476 tanh : $7\sqrt[3]{5}/9$
8692 5342 $t\pi$: $3e + 4\pi/3$
8692 6405 sq : $3 + 4\sqrt{2}/3$
8692 6915 rt : (5,9,-7,-4)
8692 9561 e^x : $\sqrt[3]{12}$
8693 0077 ln : $2e + \pi/3$
8693 0471 Γ : $(1/\pi)/10$
8693 1839 erf : $5\sqrt[3]{5}/8$
8693 4187 rt : (3,7,2,-9)
8693 5348 θ_3 : $(\pi)^{-1/3}$
8693 5402 rt : (8,5,-4,-2)
8693 6100 ℓ_{10} : $3\pi^2/4$
8693 6586 cr : $\exp(-\sqrt[3]{2}/3)$
8693 8067 rt : (6,6,-4,-5)
8693 8468 as : $3 - \sqrt{5}$
8693 9157 at : $\sqrt{2} - 3\sqrt{3}/2$
8694 2620 cr : $1/4 + 2\pi$
8694 5536 sr : $3\sqrt[3]{2}/5$
8694 5824 ln : $1/2 + 4\sqrt{2}/3$
8694 7286 sr : $4/3 - \sqrt{3}/3$
8694 7419 sr : $9e/7$
8694 9159 rt : (7,3,-1,-6)
8695 1699 sr : $2/3 + 2\sqrt{2}$
8695 4276 rt : (1,8,-3,-8)
8695 6521 id : 20/23
8695 7798 rt : (6,7,5,3)
8695 9302 $s\pi$: $10\zeta(3)/9$
8696 0193 rt : (8,0,2,7)
8696 0440 id : π^2
8696 1147 ζ : $\sqrt[3]{3}$
8696 1317 rt : (2,7,-3,-4)
8696 1889 cu : $2e + \pi/4$
8696 4064 10^x : $1 - 3\sqrt{2}/4$
8696 5603 2^x : $(e/2)^{-1/3}$
8696 7525 10^x : 11/16
8696 9384 rt : (5,6,-9,0)
8697 1168 rt : (9,3,5,8)
8697 1342 $5 \times s\pi(e)$
8697 2066 rt : (8,9,-9,4)
8697 3766 Γ : 9/19
8697 4079 cu : 21/22
8697 4771 at : $4\sqrt{2} - 2\sqrt{5}$
8697 5927 rt : (9,8,-1,-1)
8697 7420 rt : (3,5,1,4)
8698 1860 rt : (9,7,-6,-6)
8698 2554 10^x : $1/2 + 4\sqrt{5}/3$
8698 3224 Ψ : $3\sqrt{3}/10$
8698 5297 $s\pi$: $1/4 + \sqrt{2}$
8698 6621 rt : (6,3,-6,-1)
8698 7124 cr : P_{19}
8698 7191 cr : P_{58}
8698 9673 θ_3 : $10\ln 2/7$
8699 2806 K : $4\sqrt[3]{2}/7$
8699 3401 rt : (4,1,3,-6)
8699 4220 10^x : $e/10$
8699 4952 10^x : $4 + 3e$
8699 8540 K : 18/25
8699 9139 Ei : $9\zeta(3)/8$
8700 0744 J_0 : 11/15
8700 2807 10^x : $10\pi^2/9$
8700 3714 rt : (7,0,-3,2)
8700 4216 rt : (2,9,1,-9)
8700 4996 rt : (5,9,-4,-7)
8700 6166 tanh : 4/3
8700 9261 rt : (5,2,6,7)
8701 0285 sr : $3\sqrt{2} - \sqrt{5}/3$
8701 1269 rt : (5,8,-5,-5)
8701 2402 $\exp(1/2) + \exp(1/5)$
8701 2731 at : $\sqrt[3]{5/3}$
8701 2935 ℓ_{10} : $\sqrt{2}/2 + 3\sqrt{5}$
8701 2939 $\ln(\sqrt{5}) \div s\pi(\pi)$
8701 6091 $s\pi(1/5) \div s\pi(\sqrt{5})$
8701 6810 cu : $2\sqrt{2} + \sqrt{5}/4$
8701 7405 J_0 : $3\sqrt[3]{5}/7$
8701 9052 cu : $1/2 - \sqrt{3}$
8701 9745 ζ : $(e/2)^{1/3}$
8702 0665 rt : (8,3,0,3)
8702 2047 rt : (6,5,-2,-6)
8702 3522 as : $\sqrt{2}/4 - \sqrt{5}/2$
8702 3594 sq : $3/2 + 3\pi/4$
8702 3682 $\exp(1/4) - \exp(e)$
8702 4473 rt : (2,8,2,-3)
8702 5002 rt : (6,5,9,8)
8702 6501 λ : 1/8
8702 6930 2^x : $3 + 3e/4$
8702 8149 ln : $9\sqrt[3]{3}/2$
8702 8553 erf : 15/14
8702 9814 J_0 : $(e + \pi)/8$
8703 0265 rt : (9,4,-8,-2)
8703 0564 rt : (5,5,-6,4)
8703 1259 rt : (4,6,-7,-8)
8703 2419 J_0 : $\ln(\sqrt[3]{3}/3)$
8703 2769 rt : (7,2,1,-7)
8703 2907 e^x : $8(e + \pi)/9$
8703 3692 10^x : $9\pi^2/10$
8703 6221 $t\pi$: $(2\pi/3)^{-2}$
8703 6800 rt : (8,9,-7,-6)
8703 7476 rt : (9,6,3,4)
8704 1114 $t\pi$: $1/3 - 3\sqrt{5}/4$
8704 3436 rt : (8,1,4,8)
8704 4898 rt : (9,6,-4,-7)
8704 5458 10^x : $7(e + \pi)/4$
8704 9523 rt : (1,7,-8,1)
8705 0562 sq : $\sqrt{24}\pi$
8705 1270 sq : $1/2 + \sqrt{3}/4$
8705 3871 rt : (1,5,-1,-4)
8705 3886 e^x : $4/3 - \sqrt{2}/2$
8705 4049 rt : (9,4,7,9)
8705 4899 tanh : $(\sqrt[3]{2}/3)^{-1/3}$
8705 5018 rt : (1,9,8,-7)
8705 8477 as : 13/17
8705 9031 ln : $2\ln 2/9$
8706 0576 Ψ : $\sqrt[3]{2}/7$
8706 1365 as : $(\sqrt[3]{5})^{-1/2}$
8706 1590 tanh : $10\zeta(3)/9$
8706 7417 rt : (1,9,-4,-4)

8706 8420 rt : (1,7,-5,-9)
8706 9893 $t\pi : 2\sqrt{3}-\sqrt{5}$
8707 3193 rt : (1,9,-9,-4)
8707 3804 sinh : $7\pi/2$
8707 9108 $\Gamma : (3\pi)^{-1/3}$
8707 9142 id : $(\pi/3)^{-3}$
8708 0414 $J_0 : (\sqrt[3]{5}/2)^2$
8708 2869 id : $\sqrt{7/2}$
8708 4123 $s\pi : 8e\pi/5$
8708 6466 rt : (1,8,0,-8)
8708 7994 rt : (7,6,-9,8)
8708 8377 $\ell_{10} : 1/2+4\sqrt{3}$
8708 9104 rt : (7,8,-9,1)
8709 0345 at : 19/16
8709 0552 rt : (5,6,-2,8)
8709 3347 rt : (5,8,-2,1)
8709 3759 $2^x : 2/3-\sqrt{3}/2$
8709 5195 $s\pi : 4\sqrt{3}/3+3\sqrt{5}/2$
8709 8207 rt : (3,6,5,8)
8709 8752 rt : (8,8,-5,-7)
8709 9338 rt : (5,5,-7,-1)
8709 9994 sr : $4+3\sqrt{2}$
8710 1227 id : $2\sqrt{2}/3+4\sqrt{3}$
8710 2632 as : $2e/3-\pi/3$
8710 2802 $\ell_{10} : e+3\pi/2$
8710 3293 at : $3-2e/3$
8710 3668 $e^x : 1/3-\sqrt{2}/3$
8710 3772 rt : (6,4,0,-7)
8710 4798 rt : (8,6,-9,-2)
8710 6228 $\exp(3/5)-s\pi(2/5)$
8710 6648 rt : (9,5,-2,-8)
8710 8160 $Ei : 1/\ln(2\pi/3)$
8711 0713 rt : (1,7,-8,-9)
8711 1111 sq : 14/15
8711 1476 rt : (1,1,7,6)
8711 1858 $\ell_{10} : 3\sqrt{3}+\sqrt{5}$
8711 2823 $2^x : 4e+3\pi/2$
8711 3433 rt : (4,0,5,7)
8711 4145 rt : (7,1,3,-8)
8711 5154 rt : (2,9,-3,3)
8711 5675 rt : (9,3,-6,-3)
8711 5685 $e^x : 1+\sqrt{2}/4$
8711 6599 e^x : 19/12
8711 6811 $\sqrt{5}\times\exp(1/4)$
8711 7308 $10^x : P_{30}$
8712 0843 $\exp(1/3)\div s\pi(\sqrt{3})$
8712 1461 rt : (4,9,-4,9)
8712 2302 cr : $3\sqrt{2}+4\sqrt{3}/3$
8712 3573 at : $e/3-2\pi/3$
8712 4466 rt : (8,2,6,9)
8712 5797 $2^x : 3-\pi/3$
8712 6121 $2^x : 4/3+\sqrt{5}$
8712 8548 rt : (5,3,8,8)
8712 9380 $1/5\times\exp(\sqrt{5})$
8713 1882 rt : (7,1,-1,3)
8713 2034 id : $1/4-3\sqrt{2}/2$
8713 2969 $2^x : 3e-3\pi/2$
8713 3286 sq : $3\sqrt{2}+2\sqrt{5}/3$
8713 4547 $\ell_2 : 3\sqrt{2}/2+3\sqrt{3}$
8713 6503 $s\pi : 4e/3+3\pi/2$
8713 6944 rt : (7,9,-5,5)
8713 7135 sq : $4\sqrt[3]{5}/5$
8713 7229 $\ell_{10} : 2+2e$
8713 9154 sinh : $5\sqrt[3]{2}/8$
8714 0640 $erf : (\sqrt{3}/2)^{-1/2}$
8714 1880 $K : \sqrt[3]{3}/2$
8714 3481 sinh : 18/13
8714 3843 at : $3-4\pi/3$
8714 4427 $\ln(2/5)\times s\pi(2/5)$
8714 4798 $t\pi : 7(1/\pi)$
8714 7960 rt : (8,4,2,4)
8714 8579 $e^x : 4e/3+4\pi$
8715 1699 sq : $1/4+\sqrt{5}/2$
8715 5252 $erf : \ln((e+\pi)/2)$
8715 5856 $K : 3\zeta(3)/5$
8715 8998 rt : (6,0,-8,3)
8715 9459 rt : (8,7,-3,-8)
8715 9579 rt : (2,5,1,9)
8716 1116 at : $(\sqrt{2})^{1/2}$
8716 3122 rt : (4,9,-4,-6)
8716 3911 rt : (9,7,5,5)
8716 4669 rt : (2,9,-4,-9)
8716 6017 id : $3e-2\pi$
8716 7161 rt : (9,4,0,-9)
8716 8554 sr : $(\sqrt{3})^{-1/2}$
8717 0354 $K : 1/\ln(4)$
8717 3264 rt : (5,9,2,5)
8717 3276 $2^x : \exp(\sqrt[3]{2}/3)$
8717 3587 $s\pi : 5\pi^2/4$
8717 3786 $\zeta : \sqrt{2}/5$
8717 7365 sr : $4\sqrt[3]{5}/9$
8717 7978 sr : 19/25
8717 8464 Ei : 23/17
8717 8823 $2^x : \sqrt{2}/3+\sqrt{3}/4$
8717 9739 rt : (5,5,4,3)
8717 9748 $s\pi : \sqrt{2}/4+4\sqrt{3}/3$
8717 9874 $\ell_2 : 3e\pi/7$
8718 0217 ln : 2/13
8718 0387 cr : P_{55}
8718 1066 $\exp(1/4)+s\pi(1/5)$
8718 1195 sr : $e+\pi/4$
8718 1296 $10^x : 4\sqrt{3}/3+3\sqrt{5}$
8718 3356 rt : (6,3,2,-8)
8718 4993 $2^x : 10(e+\pi)/3$
8718 5055 rt : (1,8,-2,-5)
8718 6482 $\ln(4/5)-\exp(1/2)$
8718 8141 $2^x : 1/2+3\sqrt{5}$
8718 8186 rt : (8,5,-7,-3)
8719 2331 $\zeta : 8(1/\pi)/9$
8719 3364 Ψ : 16/19
8719 3399 rt : (7,0,5,9)
8719 4406 rt : (1,9,-8,6)
8719 5956 rt : (7,3,-5,-2)
8719 8693 rt : (9,2,-4,-4)
8719 9792 rt : (4,2,-6,8)
8720 0000 cu : 19/5
8720 2366 rt : (5,8,-6,-8)
8720 2753 $J_0 : (\sqrt[3]{4}/3)^{1/2}$
8720 3314 $2^x : 3\sqrt{3}+3\sqrt{5}/2$
8720 4059 $J_0 : \exp(-1/\pi)$
8720 6578 tanh : $3\sqrt{5}/5$
8720 7618 J_0 : 8/11
8720 8194 rt : (4,7,-8,-1)
8721 1029 rt : (6,8,7,4)
8721 1421 rt : (7,9,-4,-8)
8721 3650 at : 25/21
8721 4424 rt : (3,2,4,-7)
8721 5037 $\Psi : 9\sqrt{5}/7$
8721 6278 $\ln(2)\times\exp(5)$
8721 6658 cu : $\sqrt{13}$
8721 7340 Ψ : 23/8
8721 8966 rt : (8,6,-1,-9)
8722 1675 $2^x : -\sqrt{2}+\sqrt{3}+\sqrt{5}$
8722 3548 2^x : 19/21
8722 3829 rt : (5,4,-5,-2)
8722 4090 $s\pi : 5e\pi/8$
8722 4109 erf : 14/13
8722 4540 rt : (6,5,-9,5)
8722 6756 rt : (4,5,-9,-9)
8722 8954 rt : (4,1,7,8)
8723 0117 $\ell_{10} : ((e+\pi)/3)^3$
8723 2214 rt : (8,6,-2,-1)
8723 3086 $s\pi : P_{55}$
8723 3261 $\ln(5)\times\exp(4)$
8723 5993 $2^x : \sqrt{2}/3+4\sqrt{3}/3$
8723 6374 $\ell_{10} : 10\sqrt{5}/3$
8723 7440 $10^x : 4/3-\sqrt{5}/3$
8723 8841 $s\pi : \sqrt{2}/3+\sqrt{3}/2$
8723 9344 rt : (6,1,-2,-3)
8724 1303 rt : (4,8,-2,-7)
8724 3883 $\lambda : \sqrt[3]{2}/10$
8724 4748 rt : (2,7,0,-4)
8724 7514 id : $4\sqrt{3}+4\sqrt{5}$
8724 9491 $2^x : 7e\pi/9$
8725 0490 at : $4\sqrt{3}/3-\sqrt{5}/2$
8725 1127 rt : (5,5,1,-8)
8725 4406 $e+\exp(e)$
8725 4741 rt : (7,2,1,4)
8725 5360 $\ell_{10} : (\ln 3/3)^2$
8725 5497 $\ln(3)+s\pi(e)$
8725 7047 $2^x : (e+\pi)/3$
8725 9076 rt : (7,7,-8,-3)
8725 9799 rt : (3,5,-9,4)
8725 9841 cu : $3\sqrt{2}/4+2\sqrt{3}/3$
8726 0903 rt : (6,2,4,-9)
8726 1640 $s\pi : 5(e+\pi)/8$
8726 1954 $Ei : (1/\pi)/4$
8726 2561 rt : (6,7,-9,-2)
8726 3425 $\ell_{10} : 3/4+3\sqrt{5}$
8726 6105 $10^x : (e+\pi)/7$
8726 7799 rt : (9,9,1,0)
8726 8334 $\theta_3 : 1/\sqrt{6}$
8726 8947 tanh : $(e/3)^{-3}$
8726 9286 rt : (8,4,-5,-4)
8726 9953 rt : (7,8,-2,-9)
8727 0021 rt : (8,5,4,5)
8727 0333 rt : (6,6,-5,9)
8727 1712 rt : (1,7,8,-3)
8727 4344 $\sqrt{3}+\exp(\pi)$
8727 5565 sq : $3\sqrt{3}+4\sqrt{5}/3$
8727 8433 Ψ : 7
8727 9110 $s\pi(\sqrt{5})\div s\pi(e)$
8727 9444 rt : (9,1,-2,-5)
8727 9648 rt : (8,9,-8,6)
8727 9785 $\ell_2 : 5(e+\pi)$
8728 2463 K : 13/18
8728 2605 $s\pi : 2\sqrt{2}-2\sqrt{5}/3$
8728 3799 cu : $\sqrt{18}\pi$
8728 4556 $3\div\ln(3/5)$
8728 5187 rt : (9,8,7,6)
8728 7156 sr : 16/21
8729 3720 $10^x : 1/2-\sqrt{5}/4$
8729 4414 $\zeta : \pi^2/8$
8729 5154 ln : $7\sqrt[3]{5}/5$
8729 5956 cr : P_{81}
8729 6289 ln : $4/3+3\sqrt{2}/4$
8729 7754 $\ell_{10} : 4+2\sqrt{3}$
8729 8334 id : $\sqrt{15}$
8729 8483 10^x : 11/24
8729 9668 ζ : 17/24
8730 1571 cu : $2+4\sqrt{5}$
8730 2805 $\Gamma : 7\sqrt[3]{3}/3$
8730 5384 rt : (3,7,-5,-1)
8730 7671 $t\pi : 1/4-e/3$
8730 8533 $4/5\div\ln(2/5)$
8730 9007 cu : $\sqrt{3}/3+\sqrt{5}/2$
8731 0048 at : $1/4+2\sqrt{2}/3$
8731 2410 rt : (2,4,3,-7)
8731 2731 id : $4e$
8731 3395 as : $1/3+\sqrt{3}/4$
8731 3429 rt : (3,9,-9,-1)
8731 3947 sr : $4\sqrt{2}+3\sqrt{3}/2$
8731 4751 tanh : $(2e/3)^{1/2}$
8731 5931 $e^x : 7e\pi/6$
8731 7512 rt : (4,7,0,-8)
8731 9045 $e^x : 9\sqrt[3]{5}/4$
8731 9863 $e^x : 3\sqrt{2}/4-\sqrt{3}/4$
8732 0015 tanh : $3\pi/7$
8732 0147 rt : (6,8,-5,2)
8732 0340 id : $3\sqrt{3}+3\sqrt{5}/4$
8732 1079 $t\pi : 3\sqrt{3}/4-\sqrt{5}/3$
8732 1670 as : $(\sqrt{2}\pi/2)^{-1/3}$
8732 4624 rt : (1,2,9,7)
8732 5567 ln : $\sqrt{2}+\sqrt{6}+\sqrt{7}$
8732 5873 $10^x : (\sqrt{3}/2)^{-1/2}$
8732 6391 rt : (3,1,6,-8)
8732 7034 rt : (5,4,3,-9)
8732 8412 rt : (6,9,-9,-3)
8732 9482 $\zeta : \exp(-\sqrt[3]{2})$
8733 0237 $erf : 6\sqrt[3]{2}/7$
8733 2978 $2^x : 3-2\pi/3$
8733 4715 $erf : (\sqrt[3]{2})^{1/3}$
8733 4899 $\Psi : 7\zeta(3)/10$
8733 5422 $e^x : \sqrt{3}/2+4\sqrt{5}/3$
8733 5861 rt : (9,9,-9,-5)
8733 7907 tanh : $7\sqrt{3}/9$
8733 8339 rt : (7,6,-6,-4)
8733 9442 cu : $1/2-3\pi$
8733 9809 $\Gamma : 3(1/\pi)/7$
8734 0278 rt : (4,2,9,9)

8734 3284 rt : (5,3,-3,-3)
8734 4476 ℓ_{10} : $3+2\sqrt{5}$
8734 5240 ℓ_{10} : $7e\pi/8$
8734 5253 e^x : $\sqrt{2}/3+3\sqrt{3}/4$
8734 5976 $s\pi$: $(e+\pi)^2$
8734 8216 rt : (8,3,-3,-5)
8734 8869 ln : $10(e+\pi)/9$
8735 0934 $s\pi$: $\sqrt{3}/2+2\sqrt{5}$
8735 2052 cu : $3\sqrt{2}/2-3\sqrt{5}/2$
8735 4310 rt : (5,2,-9,9)
8735 5625 ln : $2\sqrt{2}-\sqrt{3}/4$
8735 6852 id : $1/\ln(\pi)$
8735 7113 e^x : $19/18$
8735 8046 id : $\sqrt[3]{2/3}$
8735 8396 10^x : $4/3-3\sqrt{2}/4$
8735 9103 10^x : $7e/2$
8735 9474 rt : (2,9,-2,8)
8736 0327 rt : (3,8,1,8)
8736 1125 sinh : $4\sqrt{3}/5$
8736 1653 ln : $4\sqrt{2}/3-\sqrt{3}$
8736 3643 $\exp(1/5)+\exp(\sqrt{3})$
8736 4346 $\pi+\sqrt{3}$
8736 9633 $t\pi$: $2(e+\pi)/5$
8736 9990 rt : (1,5,0,-3)
8736 9997 2^x : $4+\sqrt{3}/3$
8737 2701 rt : (7,3,3,5)
8737 3734 cu : $1/2+2\sqrt{2}$
8737 6145 rt : (4,8,-9,3)
8737 6647 rt : (6,9,-1,6)
8737 8376 rt : (8,2,-9,1)
8737 8879 J_0 : $13/18$
8738 0279 Γ : $9\pi^2/5$
8738 1742 10^x : $3/11$
8738 2563 rt : (3,9,-1,-6)
8738 3911 cr : P_{26}
8738 4013 sinh : $16/9$
8738 4682 erf : $(\sqrt[3]{5}/2)^{-1/2}$
8738 5577 cr : $2\pi^2/3$
8738 5792 $t\pi$: $1/3-2\sqrt{2}$
8738 7250 rt : (8,6,6,6)
8738 7640 tanh : $(\ln 3/2)^{-1/2}$
8738 8678 rt : (5,7,-8,-9)
8738 8855 ℓ_2 : $7\pi/3$
8738 8916 10^x : $1/3+\pi$
8739 1807 erf : $3\sqrt[3]{3}/4$
8739 1841 rt : (4,6,2,9)
8739 2338 cu : $(\sqrt{2}/3)^{-3}$
8739 2751 ℓ_2 : $3\sqrt{2}-\sqrt{3}/3$
8739 3999 rt : (9,8,-7,-6)
8739 5305 $t\pi$: $4e/3-4\pi/3$
8739 5838 2^x : $3\sqrt{2}/2+\sqrt{3}/4$
8739 6491 2^x : $e/3$
8739 7032 Li_2 : $(2\pi/3)^{-1/2}$
8739 7551 erf : $9\zeta(3)/10$
8740 1694 rt : (9,9,9,7)
8740 1845 rt : (5,2,-9,3)
8740 3204 sr : $3-\sqrt{5}$
8740 5913 rt : (6,8,-7,-4)
8740 5974 id : $3\sqrt{3}/2-2\sqrt{5}$
8740 7016 cu : $2/3+3\sqrt{5}/4$
8740 7081 ℓ_2 : $\sqrt{5}+\sqrt{6}+\sqrt{7}$
8740 7553 rt : (1,6,2,-7)
8740 7666 sinh : $15/19$
8740 7931 cr : $5\zeta(3)/9$
8740 8434 J_0 : $1/\ln(4)$
8740 8502 e^x : $10\sqrt{3}/7$
8740 9998 sq : $7\zeta(3)/9$
8741 2263 J_0 : $3\zeta(3)/5$
8741 3335 sq : $3/4-e$
8741 5523 rt : (7,5,-4,-5)
8741 5955 J_0 : $\sqrt[3]{3}/2$
8741 6035 rt : (5,6,6,4)
8741 6594 cu : $4/3-4\sqrt{5}$
8741 6686 cu : $2/3-4e/3$
8742 1017 rt : (2,3,5,-8)
8742 1357 erf : $5\sqrt{3}/8$
8742 5087 rt : (8,2,-1,-6)
8742 5184 rt : (5,4,-9,2)
8742 5439 rt : (3,9,1,-4)
8742 6219 $\exp(1/2)+\exp(4/5)$
8742 6511 cr : $3e-\pi/2$
8742 7240 rt : (2,6,-6,3)
8742 8887 sr : $4/3+4\sqrt{3}$
8742 9111 10^x : $6(1/\pi)/7$
8742 9639 rt : (3,8,-7,-2)
8743 1748 ℓ_{10} : $\zeta(3)/9$
8743 2060 ℓ_{10} : $4\sqrt{3}+\sqrt{5}/4$
8743 2778 sq : $\sqrt{3}/3+\sqrt{5}/2$
8743 3859 at : $(\sqrt[3]{5})^{1/3}$
8743 4394 rt : (3,0,8,9)
8743 4606 rt : (9,1,2,7)
8743 5130 rt : (6,1,-6,-2)
8743 7577 id : $1/4+4e/3$
8743 7745 2^x : $(2e/3)^3$
8744 3911 rt : (4,7,-2,-2)
8744 4414 rt : (6,9,9,5)
8744 5608 e^x : $\pi/5$
8744 6911 ℓ_2 : $3/11$
8744 7463 sr : $13/17$
8744 7818 at : $4-3\sqrt{3}$
8744 8527 sr : $(\sqrt[3]{5})^{-1/2}$
8744 8988 $t\pi$: $8\sqrt{3}/5$
8744 9352 erf : $13/12$
8745 0590 rt : (2,6,-2,-5)
8745 0609 Ψ : $(\sqrt{2})^{-1/2}$
8745 0916 rt : (7,8,-8,3)
8745 1008 rt : (9,7,-5,-7)
8745 1610 Γ : $3\sqrt[3]{2}/8$
8745 3638 cr : $\sqrt{2}-\sqrt{5}/3$
8745 3906 J_0 : $18/25$
8745 4265 cu : $4/3+4\sqrt{3}$
8745 5374 rt : (3,8,1,-9)
8745 5427 J_0 : $4\sqrt[3]{2}/7$
8745 6622 10^x : $1/3+3\sqrt{2}/2$
8745 8081 rt : (5,2,-1,-4)
8745 8162 ℓ_2 : $4\sqrt{2}+3\sqrt{5}/4$
8745 9836 tanh : $9\zeta(3)/8$
8746 2217 rt : (1,9,1,-1)
8746 3725 sr : $2e/3-\pi/3$
8746 3910 sinh : $8\zeta(3)$
8746 7512 10^x : $10/17$
8746 7841 rt : (7,4,-3,-1)
8746 8796 tanh : $1/\ln(2\pi/3)$
8747 2152 rt : (6,1,2,5)
8747 2780 2^x : $(\sqrt{5}/3)^{1/3}$
8747 4009 $t\pi$: $8\sqrt{2}/3$
8747 4571 tanh : $23/17$
8747 4736 10^x : $5\sqrt{3}/2$
8747 5441 rt : (1,6,4,-7)
8747 5538 ℓ_2 : $4/3-3\sqrt{2}/4$
8747 7858 $\exp(1/2)-s\pi(e)$
8747 8850 cu : $3e\pi/2$
8747 9781 rt : (5,5,-5,6)
8748 0410 rt : (1,7,-6,7)
8748 0589 sq : $3+3\sqrt{5}/4$
8748 1422 rt : (6,7,-5,-5)
8748 1980 at : $7\sqrt[3]{5}/10$
8748 2321 cr : $\ln((e+\pi)/3)$
8748 2916 $\ln(\sqrt{2})-\exp(1/5)$
8748 4050 θ_3 : $9/22$
8748 6124 rt : (7,4,5,6)
8748 7007 e^x : $4\sqrt{2}/9$
8748 9929 2^x : $1/2+3e/2$
8749 0056 2^x : $3/4-2\sqrt{2}/3$
8749 0732 rt : (7,4,-2,-6)
8749 0747 cu : $1/2+4\sqrt{3}$
8749 3420 ln : $2/3+\sqrt{3}$
8749 4454 id : $9\sqrt{5}/7$
8749 4586 rt : (7,9,-4,7)
8749 8629 2^x : $3/2+\pi/4$
8749 9387 cu : $2\sqrt{2}/3+3\sqrt{5}$
8750 0000 id : $7/8$
8750 3147 rt : (1,8,2,8)
8750 3826 e^x : $3+4\sqrt{5}$
8750 5327 cr : $9(e+\pi)/8$
8750 6126 ℓ_{10} : $2/15$
8750 6332 as : $4\sqrt{2}/3-\sqrt{5}/2$
8750 6932 rt : (9,6,-3,-8)
8750 7991 sr : $3(e+\pi)/5$
8750 8455 $t\pi$: $3\sqrt{3}/4+2\sqrt{5}$
8750 8697 e^x : $2\sqrt{3}+\sqrt{5}/4$
8750 9224 rt : (9,2,4,8)
8751 0224 sr : $3e/2+4\pi/3$
8751 2979 rt : (2,8,2,4)
8751 3141 2^x : $-\sqrt{3}+\sqrt{5}+\sqrt{7}$
8751 4824 rt : (9,2,-4,1)
8751 5410 cu : $22/23$
8751 5732 cr : $3\sqrt{2}/2+2\sqrt{5}$
8751 6630 2^x : $8\sqrt[3]{5}/7$
8751 7137 Ei : $10(e+\pi)/9$
8751 7331 $\exp(1/5)\div\exp(1/3)$
8752 1366 cu : $4\pi/5$
8752 2909 $s\pi$: $9\sqrt[3]{2}$
8752 5876 rt : (2,2,7,-9)
8752 6239 θ_3 : $9(1/\pi)/7$
8752 6459 rt : (3,2,-2,9)
8752 7526 cr : $1/\ln(\sqrt{2}\pi)$
8752 7707 rt : (1,5,-4,-1)
8753 0318 λ : $2(1/\pi)/5$
8753 1594 ln : $4\sqrt{2}+\sqrt{3}/2$
8753 2371 rt : (1,6,-5,7)
8753 4145 2^x : $5\sqrt[3]{2}/2$
8753 7722 $\ln(4/5)-\exp(\sqrt{3})$
8753 9101 cr : $3\sqrt{5}/10$
8754 1192 sr : $1/3+\sqrt{3}/4$
8754 1418 rt : (3,7,-5,-3)
8754 3568 rt : (6,9,-7,-9)
8754 4228 sr : $(\sqrt{2}\pi/2)^{-1/3}$
8754 5967 rt : (5,9,-6,-5)
8754 6084 10^x : $\exp(2e/3)$
8754 6732 $\ln(3/4)-s\pi(1/5)$
8754 6873 ln : $5/12$
8754 9029 $1/5+s\pi(\sqrt{5})$
8755 2372 10^x : $3\sqrt{2}/2+3\sqrt{3}/2$
8755 5035 rt : (8,8,-4,-8)
8755 6436 $t\pi$: $(e)^{-1/2}$
8755 7610 ln : $2e/3+3\pi/2$
8755 7677 Ψ : $3\zeta(3)/10$
8755 8881 $\exp(4/5)\div s\pi(e)$
8755 8923 2^x : $3\sqrt{3}/4-2\sqrt{5}/3$
8755 9518 cr : $4+3\sqrt{3}/2$
8756 1808 rt : (9,5,-1,-9)
8756 4061 rt : (7,3,0,-7)
8756 7849 Γ : $\exp(-\sqrt{5})$
8756 8559 rt : (5,1,1,-5)
8757 0615 J_0 : $(e)^{-1/3}$
8757 1317 rt : (1,7,1,1)
8757 1590 rt : (8,9,-7,8)
8757 2873 tanh : $19/14$
8757 3046 rt : (8,0,3,8)
8757 3471 J : $\exp(-1/\sqrt{2})$
8757 4410 tanh : $\sqrt[3]{5/2}$
8757 4806 rt : (6,5,-8,7)
8757 5288 erf : $25/23$
8757 6689 ζ : $\sqrt[3]{5}/6$
8757 7795 sq : $3e/2-\pi$
8757 8285 sinh : $13/4$
8757 8458 id : $(\pi/2)^3$
8757 9515 at : $1/3+\sqrt{3}/2$
8757 9571 rt : (4,1,-9,6)
8758 1704 e^x : $5\sqrt[3]{5}/2$
8758 1829 J_0 : $9(1/\pi)/4$
8758 1992 rt : (9,3,6,9)
8758 4913 10^x : $4\sqrt{2}/3+3\sqrt{3}/4$
8758 5971 rt : (9,5,-9,-2)
8758 6925 Ei : $5(e+\pi)/6$
8758 7615 erf : $2e/5$
8758 9618 cu : $5\pi^2/3$
8759 0014 2^x : $(2\pi/3)^3$
8759 1007 sinh : $\sqrt[3]{8/3}$
8759 2192 $s\pi$: $5\sqrt{3}$
8759 2682 sr : $3\sqrt{3}-3\sqrt{5}/4$
8759 2775 rt : (1,8,-8,-2)
8759 3093 ζ : $(5/3)^{-1/2}$
8759 5334 rt : (7,5,7,7)
8759 6782 e^x : $1+4e$
8759 8142 $s\pi$: $e/8$
8759 8534 rt : (3,5,-8,7)
8760 1351 θ_3 : $2\sqrt[3]{5}/5$
8760 2410 rt : (7,3,-8,0)

8760 3079 $\zeta : (\sqrt{2})^{1/2}$
8760 3238 $Li_2 : \exp(-1/e)$
8760 3820 $J_0 : (\sqrt{5}/2)^{-3}$
8760 4843 $J_0 : ((e+\pi)/3)^{-1/2}$
8760 5461 $2^x : 16/7$
8760 5557 $\Psi : \sqrt[3]{3}/4$
8760 5805 at : $6/5$
8760 7167 $J_0 : (\ln 3/3)^{1/3}$
8760 8588 rt : (4,4,5,4)
8760 9053 rt : (8,7,-2,-9)
8761 0954 $\Psi : 2\sqrt[3]{3}$
8761 1876 sr : $4\sqrt{2}/3 - \sqrt{5}/2$
8761 4856 rt : (1,4,6,-9)
8761 6608 rt : (8,0,-5,1)
8761 8044 rt : (5,8,-4,-6)
8761 8667 cu : $2 + 3e/4$
8761 9366 tanh : $e/2$
8762 0421 rt : (3,8,-1,-5)
8762 1464 $Li_2 : 9/13$
8762 1808 rt : (1,9,2,8)
8762 4627 $10^x : 3\sqrt{2}/4 - \sqrt{5}/2$
8762 5602 $3 \div s\pi(e)$
8762 5804 rt : (1,8,7,-5)
8762 6840 rt : (6,5,-1,-7)
8762 9849 cu : $3\sqrt{3}/2 + 3\sqrt{5}/2$
8763 0532 $\Gamma : 3/22$
8763 0703 rt : (9,3,-2,2)
8763 2818 ln : $4/3 + 3\sqrt{3}$
8763 4718 $t\pi : 7\ln 2/8$
8763 4835 rt : (5,7,8,5)
8763 5367 $10^x : 4/3 + 2\sqrt{5}/3$
8763 5597 rt : (7,2,2,-8)
8763 5981 rt : (2,9,-6,-3)
8763 8243 at : $\sqrt{2} + \sqrt{5} - \sqrt{6}$
8763 9479 cu : $1/3 + 2e/3$
8764 1733 $\Psi : 21/25$
8764 3684 rt : (2,5,-4,-6)
8764 4187 at : $(\sqrt[3]{3})^{1/2}$
8764 4315 rt : (8,1,5,9)
8764 5910 $J_0 : 5/7$
8764 7794 $\Psi : 1/\ln(\sqrt{2})$
8764 7901 $s\pi : 3e\pi/7$
8764 8274 $e^x : 1/2 + 2\pi$
8764 9079 rt : (3,6,-3,-4)
8764 9970 $10^x : 3\sqrt[3]{5}$
8765 1780 at : $(\ln 2)^{-1/2}$
8765 2964 $\Psi : 2\sqrt[3]{2}/3$
8765 3861 $\theta_3 : 21/23$
8765 5459 rt : (4,8,-8,6)
8765 5596 rt : (7,9,-3,-9)
8765 6089 rt : (5,1,-7,2)
8765 8729 $\Psi : \sqrt{5}\pi$
8765 9218 $2^x : 4 + 3\sqrt{3}/2$
8766 0276 rt : (9,4,-7,-3)
8766 0871 $J : 7/24$
8766 2067 $\Gamma : (e+\pi)/2$
8766 2091 rt : (4,9,6,6)
8766 6687 rt : (1,2,-7,-7)
8767 5018 rt : (5,0,3,6)
8767 5897 cr : $\sqrt{2} + 3\sqrt{3}$
8767 6091 cu : $1/4 + \sqrt{2}/2$
8767 6747 cu : $3\sqrt{3} - 2\sqrt{5}/3$
8767 8069 $\sqrt{3} + \ln(\pi)$
8767 8147 sinh : $\sqrt{2} + \sqrt{5} + \sqrt{6}$
8768 2563 at : $5\sqrt[3]{3}/6$
8768 3502 $10^x : 2 + \sqrt{3}/3$
8768 5026 $\ell_{10} : 2e + 2\pi/3$
8768 5375 at : $\exp(1/(2e))$
8768 5810 rt : (6,2,-4,-1)
8768 7255 $\exp(2/5) \times s\pi(1/5)$
8768 7862 rt : (6,3,6,7)
8768 8384 rt : (5,8,-1,3)
8768 9437 rt : (1,9,-2,-8)
8768 9462 ln : $1/4 + 2\pi$
8768 9695 $10^x : 5\pi^2/8$
8768 9798 cu : $4\sqrt{3}/3 + 2\sqrt{5}/3$
8769 0019 at : $\zeta(3)$
8769 6922 rt : (6,4,1,-8)
8769 8536 $\Psi : 2/15$
8769 9139 sinh : $19/24$
8770 0622 rt : (7,6,9,8)
8770 1455 rt : (6,7,-8,0)
8770 1665 sq : $\sqrt{3}/2 - \sqrt{5}$
8770 2971 ln : $5\sqrt[3]{3}/3$
8770 3278 $\ell_{10} : 9(e+\pi)/7$
8770 4054 $\Psi : 5\sqrt{3}/3$
8770 5423 rt : (7,1,4,-9)
8770 5801 sr : $10/13$
8770 8769 $Li_2 : 2\sqrt{3}/5$
8770 9053 cr : $\sqrt{3} + \sqrt{5} + \sqrt{7}$
8771 1407 $10^x : 1/3 + 4\sqrt{3}$
8771 1545 $erf : 12/11$
8771 1654 $\zeta : 2/7$
8771 4536 $10^x : \exp(1/(2\pi))$
8771 5070 rt : (7,5,-1,0)
8771 7128 $2^x : (4/3)^{-1/3}$
8771 8135 ln : $2\zeta(3)$
8771 8240 $\sqrt{5} - \exp(\sqrt{2})$
8771 8459 $Ei : 17/6$
8772 3268 $\theta_3 : 13/19$
8772 3378 tanh : $15/11$
8772 3902 rt : (8,6,-8,-3)
8772 4643 ln : $9\pi^2/5$
8772 6480 rt : (3,7,-4,2)
8772 6901 $erf : 1/\ln(5/2)$
8772 7534 $\Gamma : \exp(-5/3)$
8772 8590 rt : (8,1,-3,2)
8773 0807 id : $(\sqrt{2}\pi/3)^{-1/3}$
8773 2693 rt : (9,3,-5,-4)
8773 3854 ln : $3\ln 2/5$
8773 4106 cr : $1/3 + 2\pi$
8773 4112 $2^x : 4 - 4\pi/3$
8773 4802 rt : (1,8,-7,7)
8773 5006 $e^x : 3 + 3\pi/2$
8773 7393 sinh : $1/\ln(\sqrt[3]{5}/3)$
8773 8167 sr : $7\pi^2/2$
8773 8187 $J : \exp(-5\pi/8)$
8773 8267 sr : $4\sqrt{3}/9$
8774 0142 $e^x : \exp(e/3)$
8774 0374 rt : (2,8,-4,-4)
8774 2015 rt : (9,4,0,3)
8774 2577 $\exp(2/3) \times s\pi(\sqrt{2})$
8774 3883 rt : (8,8,2,1)
8774 7183 sinh : $9\pi^2/7$
8774 8737 rt : (4,9,-3,-7)
8775 0626 $Ei : 9e/7$
8775 2750 rt : (2,6,-5,7)
8775 2905 rt : (6,8,-4,4)
8775 3646 id : $\exp(\sqrt[3]{2}/2)$
8775 5102 sq : $22/7$
8775 6830 $2^x : 3e/2 + 4\pi/3$
8775 7065 rt : (5,9,3,7)
8775 8961 sr : $10\ln 2/9$
8775 9935 $10^x : 4 - \sqrt{2}/2$
8776 2650 $\zeta : 3\zeta(3)/2$
8776 3014 tanh : $1/\ln(\sqrt[3]{3}/3)$
8776 3641 as : $10/13$
8776 4430 $e^x : 2\sqrt{3} + 4\sqrt{5}/3$
8776 4466 $Li_2 : \ln 2$
8776 5357 rt : (6,3,3,-9)
8776 5859 sinh : $8\ln 2/7$
8776 6275 $\Psi : 6(e+\pi)/5$
8777 1326 cu : $\pi^2/8$
8777 1399 cu : $1/\ln(\sqrt[3]{4}/3)$
8777 1711 $\ell_2 : P_{30}$
8777 6472 cu : $4\sqrt{3}/3 + 2\sqrt{5}$
8777 7267 $2^x : 3\sqrt{2}/2 - 4\sqrt{3}/3$
8777 7460 $\exp(4/5) + \exp(\sqrt{3})$
8777 7731 rt : (5,1,5,7)
8777 8775 cu : $\sqrt{2} - \sqrt{6} - \sqrt{7}$
8778 0842 rt : (9,9,-8,-6)
8778 1753 $e^x : 8\sqrt{5}/7$
8778 2838 sinh : $\sqrt{21}$
8778 3369 $Ei : 3(1/\pi)/5$
8778 3904 $t\pi : 4e + 3\pi/4$
8778 5373 $10^x : \sqrt{2} + \sqrt{5} - \sqrt{6}$
8778 5790 rt : (1,8,-4,-9)
8778 6182 $2^x : 10/11$
8778 6223 rt : (7,8,-9,-3)
8778 6427 $t\pi : (\sqrt[3]{3}/2)^2$
8778 9314 $e^x : \sqrt{3} - \sqrt{6} + \sqrt{7}$
8778 9603 cu : $2(e+\pi)/5$
8779 0028 rt : (6,4,8,8)
8779 0962 rt : (7,8,-7,5)
8779 4787 rt : (8,5,-6,-4)
8779 5673 $\zeta : (e+\pi)^{1/3}$
8779 7009 $t\pi : (e/3)^{-3}$
8780 0466 ln : $3/2 + e/3$
8780 0961 $Li_2 : \sqrt[3]{1/3}$
8780 1607 $10^x : 10/13$
8780 2498 sq : $3\sqrt{3}/4 - \sqrt{5}$
8780 3290 $\zeta : 5\sqrt[3]{3}/4$
8780 3313 rt : (9,2,-3,-5)
8780 3460 rt : (6,9,0,8)
8780 3835 rt : (3,7,-3,-6)
8780 4026 $\ell_2 : \ln(2\pi)$
8780 5234 sr : $2 + 2\pi$
8780 8388 cr : $3 + 4e/3$
8781 0409 rt : (6,4,-7,-1)
8781 0628 ln : $e/2 + \pi/3$
8781 1905 $\ln(4) + \exp(2/5)$
8781 3340 sq : $4\sqrt{3} + 2\sqrt{5}/3$
8781 4393 rt : (4,5,7,5)
8781 6067 rt : (4,8,-1,-8)
8781 7326 cu : $4/3 + \sqrt{2}/3$
8782 1222 $s\pi : 3\sqrt{2}/4 + 3\sqrt{3}/2$
8782 3047 $\ln(3/5) + \exp(2)$
8782 3119 tanh : $4\sqrt[3]{5}/5$
8782 3391 rt : (7,1,-4,-2)
8782 3539 $\ln(\sqrt{5}) \div \ln(2/5)$
8782 4160 rt : (5,5,2,-9)
8782 5437 rt : (2,4,-6,-7)
8782 7443 $e^x : 2 - 2\sqrt{2}/3$
8782 8608 $Ei : 4\sqrt{5}/9$
8782 9071 $Ei : 9$
8782 9098 $\exp(\sqrt{3}) - s\pi(e)$
8783 0960 rt : (7,9,-3,9)
8783 1517 id : $\sqrt{2} + 2\sqrt{3}$
8783 2148 $e^x : 1 + 3e/4$
8783 2414 rt : (1,7,2,8)
8783 2566 rt : (9,8,-6,-7)
8783 6166 $t\pi : 1/\sqrt{19}$
8783 6287 rt : (8,2,-1,3)
8783 6576 at : $\sqrt{2}/3 - 3\sqrt{5}/4$
8784 1144 rt : (2,7,-2,-5)
8784 4500 $J_0 : 17/24$
8784 5236 $e^x : 23/9$
8784 5561 sr : $4 - \sqrt{2}/3$
8784 6812 rt : (1,1,-9,-8)
8784 7156 rt : (5,4,-8,4)
8784 9096 rt : (9,5,2,4)
8784 9458 $\ell_{10} : 6\sqrt[3]{2}$
8785 2831 as : $4\sqrt{3}/9$
8785 2892 $s\pi : 4e + \pi/4$
8785 3149 rt : (3,5,2,6)
8785 4351 at : $(\sqrt[3]{5}/3)^{-1/3}$
8785 4917 $\ell_2 : 2 + 3\sqrt{5}/4$
8785 5633 $\zeta : 9(1/\pi)/10$
8785 5636 rt : (7,7,-7,-4)
8785 6030 $\ln(2/3) + \exp(1/4)$
8785 7366 $e^x : 1/4 + 2e$
8785 9437 $erf : 23/21$
8786 1223 $\pi \times \ln(2/5)$
8786 1336 rt : (4,9,-2,-9)
8786 3946 rt : (8,4,-4,-5)
8786 5081 rt : (4,1,4,-7)
8786 6566 rt : (2,7,-5,-8)
8786 7965 id : $3 - 3\sqrt{2}/2$
8787 0029 $\sqrt{2} \div \ln(5)$
8787 2219 rt : (9,1,-1,-6)
8787 3462 $\Gamma : \sqrt{2}/3$
8787 3709 $2^x : 8\sqrt{5}/7$
8787 5318 $e^x : 3\sqrt{2}/2 - 2\sqrt{5}/3$
8787 6941 rt : (5,2,7,8)
8787 7831 $s\pi : 3\sqrt{5}/5$
8787 9344 $e^x : (\sqrt{5}/2)^{1/2}$
8788 0440 $t\pi : 3\sqrt{2} - 2\sqrt{5}$
8788 1870 rt : (4,7,1,-9)
8788 3317 $Ei : 8\sqrt{3}/7$
8788 3369 rt : (9,7,-4,-8)

8788 5241 $J_0 : \sqrt{2}/2$
8788 6168 sq : $3/2 - 2\pi$
8788 6809 $10^x : (\ln 2/2)^{1/2}$
8788 7233 $10^x : 3/2 + \sqrt{5}$
8788 8506 rt : (5,0,-5,1)
8788 8983 $4/5 \times \ln(3)$
8788 9627 at : $(\ln 3)^2$
8789 0625 sq : 15/16
8789 1769 rt : (1,9,6,-1)
8789 3796 $2^x : 23/9$
8789 5211 $t\pi : P_3$
8789 6051 at : $1/2 + \sqrt{2}/2$
8789 6131 rt : (5,5,-4,8)
8789 6240 $\ell_{10} : P_{64}$
8789 6733 $\ln(\sqrt{2}) - \exp(4/5)$
8789 9585 id : $8(e + \pi)$
8789 9863 $s\pi : P_{58}$
8790 0089 $s\pi : P_{19}$
8790 2806 rt : (5,9,-5,-8)
8790 4656 rt : (6,5,-7,9)
8790 4907 sr : 17/22
8790 4940 $Ei : 7/11$
8790 5763 rt : (1,5,1,2)
8790 6445 $erf : \pi^2/9$
8790 9439 $t\pi : 7\pi^2/6$
8790 9749 as : $10 \ln 2/9$
8791 3732 rt : (2,8,3,8)
8791 5025 $t\pi : \sqrt{2}/3 + 3\sqrt{3}/4$
8791 5286 rt : (6,9,1,-2)
8791 8073 cr : $e/4$
8792 1847 rt : (4,8,-7,9)
8792 3305 $\ell_2 : e/5$
8792 3388 rt : (7,6,-5,-5)
8792 5851 $J_0 : 12/17$
8792 6940 rt : (1,9,-3,-5)
8792 7301 rt : (8,9,-5,-8)
8792 7686 rt : (7,0,-2,3)
8793 0130 $2^x : \sqrt{2} + \sqrt{3} - \sqrt{5}$
8793 0911 $s\pi : \sqrt[3]{5}/5$
8793 1457 rt : (8,3,-2,-6)
8793 3281 rt : (9,6,-2,-9)
8793 3845 $\ell_{10} : \sqrt{3}/2 + 3\sqrt{5}$
8793 4667 cu : $2 - \sqrt{3}/3$
8793 5710 $2^x : 1/\ln(3)$
8793 6184 rt : (1,8,8,2)
8793 6593 cr : 17/25
8793 7480 at : $4e/9$
8793 8524 rt : (3,9,0,-3)
8793 8865 $K : 8/11$
8793 9491 rt : (9,0,1,7)
8794 0005 rt : (8,3,1,4)
8794 1688 rt : (7,6,1,1)
8795 0039 rt : (3,3,3,-7)
8795 2246 rt : (9,6,4,5)
8795 2629 $K : \exp(-1/\pi)$
8795 3735 $\ell_2 : 3 + e/4$
8795 5064 tanh : $8\zeta(3)/7$
8795 6062 $\ell_{10} : P_{49}$
8795 6574 rt : (4,9,-2,-6)
8795 6787 $\Psi : 3\sqrt{5}/8$
8795 6836 $s\pi : 2e - 2\pi/3$
8795 7684 $K : (\sqrt[3]{4}/3)^{1/2}$
8796 1487 rt : (4,0,6,8)
8796 6288 $10^x : 3e/4 - 2\pi/3$
8796 7723 rt : (8,9,4,2)
8796 8577 ln : $\sqrt[3]{14}$
8796 9841 sinh : $\sqrt[3]{1/2}$
8797 2396 $10^x : \sqrt{2}/2 - \sqrt{3}/4$
8797 2869 rt : (5,3,9,9)
8797 3741 $erf : \ln(3)$
8797 4942 cr : $7e\pi/9$
8797 5836 $\zeta : (\sqrt{2}\pi/2)^{1/3}$
8797 6479 rt : (8,8,-3,-9)
8797 6560 tanh : $(\sqrt[3]{4}/3)^{-1/2}$
8797 6905 rt : (3,6,-5,-7)
8797 7672 ln : $3\sqrt{2} + 4\sqrt{3}/3$
8797 7680 id : $2\sqrt{2} - 3\sqrt{5}$
8797 8201 tanh : $\exp(1/\pi)$
8798 0990 $Ei : 2(1/\pi)$
8798 1685 rt : (6,8,-6,-5)
8798 2451 cu : $3/4 - 4\pi$
8798 2669 tanh : 11/8
8798 4184 rt : (3,3,6,5)
8798 4812 id : $4e\pi/7$
8798 6009 rt : (4,1,-8,8)
8798 6820 $e - \exp(4)$
8798 7259 cu : $10\sqrt{2}/9$
8798 9555 rt : (7,5,-3,-6)
8799 0327 $J : 1/\sqrt[3]{10}$
8799 1043 $e^x : 5(e + \pi)/3$
8799 2429 rt : (4,8,-9,-1)
8799 3607 $2^x : 1/3 + \sqrt{3}/3$
8799 3913 $s\pi : \sqrt{2} + 4\sqrt{3}$
8799 5246 sr : $4/3 - \sqrt{5}/4$
8799 7034 rt : (2,3,-8,-8)
8799 7120 sr : $9\pi/8$
8799 7394 rt : (8,2,0,-7)
8799 8348 cu : $1/2 + \pi/2$
8800 0000 id : 22/25
8800 1665 rt : (3,1,-5,7)
8800 2335 $t\pi : 1/4 + 2\pi/3$
8800 4439 rt : (5,5,-6,-2)
8800 5200 rt : (9,1,3,8)
8800 6620 rt : (4,6,9,6)
8800 8506 cu : $3\sqrt{5}/7$
8801 0852 $\ell_2 : -\sqrt{2} + \sqrt{6} + \sqrt{7}$
8801 1087 sq : $3\sqrt{2} + \sqrt{5}/3$
8801 1173 sr : $(5/3)^{-1/2}$
8801 1926 rt : (2,8,-9,-2)
8801 3599 cu : 23/24
8801 4898 cr : 15/22
8801 6373 rt : (6,2,-3,-3)
8801 8232 rt : (1,9,-9,-3)
8801 8618 $erf : 7\sqrt{2}/9$
8801 8777 sr : $3\sqrt{3}/4 + \sqrt{5}$
8801 9326 $\ln(2/5) + s\pi(\sqrt{2})$
8802 0506 $erf : 11/10$
8802 0695 $e^x : 7\sqrt{2}/4$
8802 1602 rt : (1,8,-1,-6)
8802 3364 $2^x : 2\pi$
8802 6886 $2^x : P_{11}$
8802 8233 rt : (7,1,0,4)
8802 9183 $\ell_{10} : 8e\pi/9$
8802 9829 $s\pi : 7e/3$
8803 0154 sr : $5\sqrt{2}/2$
8803 0915 sinh : $\sqrt{19/2}$
8803 1346 $\Psi : \exp(-3\pi/2)$
8803 2725 rt : (2,7,6,7)
8803 4130 $Ei : 7\sqrt{2}/5$
8803 4986 at : 23/19
8803 5492 sq : P_{62}
8803 8927 $\Psi : 5\pi^2/7$
8803 9521 $\lambda : \exp(-13\pi/20)$
8804 0020 rt : (8,4,3,5)
8804 0265 $J_0 : (\ln 2/2)^{1/3}$
8804 0773 $t\pi : (\sqrt{2}/3)^{-3}$
8804 2107 $erf : \sqrt[3]{4/3}$
8804 2399 $\Psi : e/10$
8804 3784 rt : (3,2,5,-8)
8804 3831 $10^x : 5\sqrt[3]{5}/3$
8804 4125 $10^x : 5e\pi/2$
8804 6556 rt : (6,7,-4,-6)
8804 6647 cu : 11/7
8804 6795 rt : (4,9,-9,-5)
8805 1734 rt : (9,7,6,6)
8805 1948 sq : $3/4 - 3\sqrt{2}/2$
8805 1968 sinh : 25/18
8805 2033 $\ell_{10} : 2/3 + 4\sqrt{3}$
8805 2043 $\exp(4) \times s\pi(\sqrt{5})$
8805 2889 rt : (3,5,-9,2)
8805 4206 rt : (7,4,-1,-7)
8805 4781 rt : (4,1,8,9)
8805 6440 cr : $(\pi)^{-1/3}$
8805 7756 $e^x : 12/19$
8806 1826 rt : (8,1,2,-8)
8806 2741 $\ell_{10} : 3\sqrt{2} + 3\sqrt{5}/2$
8806 2780 $2^x : 3\sqrt{2}/2 + 4\sqrt{3}$
8806 2911 $2^x : 7 \ln 2$
8806 6268 cr : $1/4 + \sqrt{3}/4$
8806 7241 $t\pi : 3\pi^2/7$
8806 9416 rt : (9,2,5,9)
8806 9984 cu : $3/2 + 3\pi$
8807 0593 $t\pi : 10 \ln 2/9$
8807 1273 $s\pi : \exp(-\sqrt[3]{2}/3)$
8807 1580 $\Psi : \exp((e + \pi)/3)$
8807 3250 rt : (3,6,-9,-4)
8807 3970 rt : (9,4,-8,-4)
8807 5450 cu : $4 + 4\sqrt{2}/3$
8807 7859 rt : (4,2,-6,1)
8808 0812 rt : (1,8,4,-2)
8808 2627 rt : (8,8,-9,8)
8808 2774 $\exp(\pi) \div \exp(2/3)$
8808 2849 $\zeta : \ln(4/3)$
8808 6161 $2^x : 1/4 - \sqrt{3}/4$
8808 8079 $\exp(2) + \exp(2/5)$
8809 0218 rt : (9,1,-9,1)
8809 0998 rt : (4,7,-7,-2)
8809 6863 sr : $4e + 4\pi/3$
8809 8783 $\Psi : 9/25$
8810 2083 $erf : 7\sqrt[3]{2}/8$
8810 2132 at : $6\sqrt{2}/7$
8810 2482 $s\pi : 4\sqrt{2}$
8810 2496 rt : (5,9,-5,-6)
8810 6785 $Ei : 4/21$
8810 8044 rt : (6,7,-7,2)
8810 8268 cr : $2\sqrt[3]{5}/5$
8810 9966 rt : (6,6,-2,-7)
8811 1311 rt : (7,8,-6,7)
8811 2357 at : $7\sqrt{3}/10$
8811 2568 $\ell_2 : 3\sqrt{2} - \sqrt{5}/4$
8811 3248 rt : (1,7,1,-7)
8811 3925 rt : (6,1,-1,-4)
8811 4629 sr : $3/2 + 3e/4$
8811 4822 cr : $1 + 4\sqrt{2}$
8811 5172 $2^x : 3/2 + 4e/3$
8811 6818 rt : (2,9,3,-7)
8811 7407 rt : (7,3,1,-8)
8811 7719 cr : 13/19
8811 8214 $e^x : -\sqrt{3} + \sqrt{6} + \sqrt{7}$
8811 8563 $erf : (\sqrt{5}/3)^{-1/3}$
8811 8595 $\Psi : 2\sqrt[3]{2}/7$
8811 8601 rt : (3,7,-3,5)
8812 0088 $J_0 : 7/10$
8812 1193 $\ell_2 : \sqrt{22}\pi$
8812 1531 $J_0 : 5\sqrt[3]{2}/9$
8812 1674 rt : (4,8,-4,-7)
8812 2275 $2^x : 4/3 + 4\sqrt{3}$
8812 3940 rt : (2,4,4,-8)
8812 4818 rt : (8,0,4,9)
8812 5285 rt : (7,2,2,5)
8812 6811 rt : (6,4,0,1)
8812 7858 sr : $2e/7$
8812 9276 cu : $8e/3$
8813 0176 rt : (1,9,-8,7)
8813 1257 $\ell_2 : 3\sqrt{3} - 3\sqrt{5}/2$
8813 3029 $J : \zeta(3)/10$
8813 4573 rt : (3,1,7,-9)
8813 5254 $2^x : 7\sqrt{5}/8$
8813 5508 at : $7 \ln 2/4$
8813 6578 rt : (8,5,5,6)
8813 7358 ln : $1 + \sqrt{2}$
8813 7430 sq : $\exp(\sqrt[3]{4}/3)$
8814 0670 rt : (3,5,-7,-8)
8814 5018 $s\pi : 3\sqrt{3}/2 + \sqrt{5}/3$
8814 5148 rt : (2,1,5,5)
8814 6892 $s\pi : 2\zeta(3)/7$
8814 7803 rt : (9,8,8,7)
8814 7822 $\zeta : 8\sqrt[3]{2}/7$
8815 0810 rt : (7,7,3,2)
8815 0960 $2^x : (\pi/3)^{-2}$
8815 4249 rt : (6,8,-3,6)
8815 5642 $\ell_2 : 3e - \pi/4$
8815 6528 sr : $\sqrt{23}\pi$
8815 7428 cu : $4e/3 - \pi/4$
8815 8076 $J_0 : ((e + \pi)/2)^{-1/3}$
8815 8867 $10^x : e/2 + \pi/3$
8815 9360 $10^x : 1 + 3\pi/4$
8815 9744 rt : (9,5,-8,-3)
8816 1106 $e^x : \sqrt{3}/4 - \sqrt{5}/4$
8816 3986 $2^x : 3/4 + 3\pi$

8816 4709 rt : (5,8,-3,-7)
8816 6126 rt : (3,4,8,6)
8816 7999 ln : $(\sqrt{5}/3)^3$
8817 0586 θ_3 : 7/17
8817 1977 rt : (6,5,0,-8)
8817 2401 sr : $2\sqrt{2}/3+3\sqrt{3}/2$
8817 5027 sinh : $8\sqrt{5}/5$
8817 5103 rt : (3,9,-8,-2)
8817 5192 $s\pi$: $3e+4\pi/3$
8817 5650 ℓ_{10} : $4/3+2\pi$
8817 5949 λ : 3/23
8817 6237 rt : (2,4,-3,6)
8817 9218 rt : (7,2,3,-9)
8818 0354 2^x : $1/\ln(\ln 3)$
8818 1483 J_0 : $2\pi/9$
8818 6190 rt : (4,6,-5,-3)
8818 7193 at : 17/14
8818 7335 $s\pi$: $2/3+3\sqrt{5}/4$
8819 1710 sr : 7/9
8819 2119 Li_2 : 16/23
8819 4473 id : $7\sqrt[3]{2}/10$
8819 6578 erf : 21/19
8819 6601 id : $2-\sqrt{5}/2$
8819 7213 rt : (5,3,-2,-4)
8819 7962 Γ : 8/17
8819 8057 tanh : 18/13
8819 9967 at : $2-\pi/4$
8820 2054 rt : (1,6,3,-8)
8820 2832 e^x : $4\sqrt{3}/3-3\sqrt{5}/4$
8820 4731 Ei : $\sqrt{5}$
8820 7205 id : $\sqrt[3]{20/3}$
8820 8172 rt : (6,0,1,5)
8820 9432 10^x : $3\sqrt{3}+\sqrt{5}/4$
8821 0319 2^x : $3e/2$
8821 1460 $t\pi$: $6\pi/5$
8821 2294 sq : $2-3e$
8821 2343 rt : (2,3,6,-9)
8821 3350 ζ : 10/17
8821 5226 $s\pi$: $3/4+e/3$
8821 5893 rt : (8,7,-9,-3)
8821 7544 $\exp(1/4)+\exp(4)$
8821 9067 rt : (7,3,4,6)
8821 9390 $t\pi$: $1/3+2e$
8821 9709 sr : $(\sqrt{2}/3)^{1/3}$
8822 0429 ℓ_{10} : $4+4e/3$
8822 0809 tanh : $4\sqrt{3}/5$
8822 1696 ln : $8e/9$
8822 1697 $s\pi$: $2(e+\pi)/5$
8822 2677 e^x : $(5/2)^{-1/2}$
8822 2774 rt : (2,8,-8,3)
8822 2961 ζ : 9/5
8822 3308 rt : (9,4,-6,-4)
8822 5571 rt : (5,8,0,5)
8822 5988 J_0 : $\sqrt{2}+\sqrt{3}-\sqrt{6}$
8822 8272 ζ : $\sqrt[3]{3}/5$
8822 9900 rt : (8,6,7,7)
8823 0814 e^x : $1/2+3e$
8823 0824 rt : (1,8,9,9)
8823 2457 cr : $1/3+\sqrt{2}/4$
8823 2582 cr : $4\zeta(3)/7$
8823 2647 rt : (6,4,2,-9)
8823 4727 sq : $1/3+4\sqrt{2}$
8823 4861 sr : $3\sqrt{2}-2\sqrt{3}$
8823 5294 id : 15/17
8823 5435 sq : $3+4e/3$
8823 5596 Ei : $5\sqrt[3]{2}$
8823 5931 sq : $2-3\sqrt{2}/4$
8823 7995 rt : (9,8,-5,-8)
8823 9222 rt : (5,4,-7,6)
8823 9629 2^x : $\sqrt{2}/4+\sqrt{5}/4$
8823 9858 ζ : $10\sqrt[3]{2}/7$
8824 4772 tanh : $\sqrt[3]{8/3}$
8824 5444 rt : (2,7,-4,-4)
8824 6284 sinh : $5(1/\pi)/2$
8824 9509 θ_3 : $2\sqrt[3]{3}/7$
8825 0667 sq : $2/3-3e/4$
8825 1458 id : $\sqrt{2}/4-\sqrt{5}$
8825 3257 10^x : $\exp(-\sqrt[3]{4}/3)$
8825 5452 rt : (3,5,-6,-9)
8825 5692 ln : $3\sqrt{3}/4+\sqrt{5}/2$
8825 6825 Ψ : $(e+\pi)/7$
8825 8140 rt : (1,7,2,-3)
8825 8708 cr : 11/16
8826 1773 ℓ_2 : $2/3+3\sqrt{5}$
8826 1867 J : $8(1/\pi)/7$
8826 2741 J_0 : 16/23
8826 3947 $1/3+\ln(\sqrt{3})$
8826 4396 id : $9\pi^2/10$
8826 6189 $s\pi$: $(e/3)^{-3}$
8826 7581 rt : (3,8,-6,-3)
8826 9744 id : $\sqrt{2}/3-3\sqrt{5}/2$
8827 0989 ζ : $\sqrt{3}/6$
8827 3388 $t\pi$: $4\sqrt{3}/9$
8827 4291 $t\pi$: $\exp(\sqrt[3]{3})$
8827 4569 2^x : $2+\pi/2$
8827 7353 erf : $(e/2)^{1/3}$
8827 8221 rt : (8,6,-7,-4)
8827 8357 sq : P_{46}
8827 9618 id : $\sqrt{12}\pi$
8828 2147 cu : $4+\pi/2$
8828 2667 ℓ_{10} : $\sqrt{2}/2+4\sqrt{3}$
8828 3519 rt : (9,7,-3,-9)
8828 3593 λ : $\exp(-11\pi/17)$
8828 5138 rt : (5,9,4,9)
8828 5431 rt : (9,3,-4,-5)
8828 7085 at : $\sqrt{2}/3+\sqrt{5}/3$
8828 7480 $s\pi$: $1/4+2\pi/3$
8828 8184 rt : (1,5,5,-9)
8828 8802 rt : (5,2,0,-5)
8828 9501 rt : (9,1,-5,1)
8829 0740 10^x : $4e/3-2\pi/3$
8829 0886 2^x : $\sqrt{7/3}$
8829 2619 tanh : 25/18
8829 3593 ln : P_{56}
8829 3613 10^x : $5\sqrt[3]{5}/2$
8829 4618 at : $(\sqrt{2}\pi/3)^{1/2}$
8829 6022 rt : (3,4,-9,-9)
8829 6584 2^x : $1/2-e/4$
8829 7725 e^x : 18/17
8829 8095 rt : (1,2,5,9)
8829 9323 rt : (6,1,3,6)
8829 9998 $3-\exp(3/4)$
8830 0712 rt : (5,2,-1,1)
8830 1367 2^x : 21/23
8830 1762 sq : $6\sqrt[3]{3}$
8830 3189 sr : $\sqrt{7}\pi$
8830 4122 rt : (9,5,-6,-3)
8830 4626 ℓ_2 : $\sqrt{2}/2+4\sqrt{5}/3$
8830 5751 sr : $9\ln 2/8$
8830 9785 rt : (7,4,6,7)
8831 0917 rt : (3,9,-9,5)
8831 2667 as : 17/22
8831 3014 e^x : $3-\sqrt{2}$
8831 6162 10^x : $(\sqrt[3]{3})^{1/2}$
8831 7764 rt : (2,2,7,6)
8831 9590 ln : P_{68}
8832 0190 rt : (8,7,9,8)
8832 1495 $\ln(2/5)\times s\pi(\sqrt{2})$
8832 1522 erf : $8\ln 2/5$
8832 2159 ln : P_{71}
8832 2557 ℓ_{10} : $e/2+2\pi$
8832 2873 id : $5(e+\pi)/6$
8832 2931 rt : (6,5,2,2)
8832 3016 cu : $e\pi/6$
8832 3113 rt : (8,9,-4,-9)
8832 5975 id : $10e/7$
8832 6004 Ψ : $9\pi/4$
8832 7825 $\ln(5)\div\exp(3/5)$
8833 1816 cu : $2/3-3e$
8833 2120 rt : (7,8,-8,-4)
8833 4486 2^x : $1/3+3\sqrt{5}/2$
8833 5090 at : $\sqrt{2}+\sqrt{6}-\sqrt{7}$
8833 6127 10^x : 17/3
8833 6691 2^x : $\exp(\pi)$
8833 6722 rt : (4,8,0,-9)
8833 7592 J_0 : $\sqrt[3]{1/3}$
8833 7730 at : $(e/3)^{-2}$
8833 9161 rt : (8,5,-5,-5)
8834 3619 ℓ_2 : $4/3+3\pi/4$
8834 3749 $\exp(2/3)\div s\pi(\sqrt{5})$
8834 4576 J_0 : $\ln 2$
8834 4885 rt : (7,8,5,3)
8834 6175 rt : (9,2,-2,-6)
8834 6358 Ei : $7\sqrt[3]{5}/4$
8834 7175 rt : (4,9,-8,-2)
8834 8370 at : $3/2-e$
8835 1114 sq : $4\sqrt{3}+\sqrt{5}/3$
8835 3506 e^x : $1/2+\sqrt{5}/4$
8835 5235 J_0 : $2\sqrt{3}/5$
8835 5345 cu : $3\sqrt{2}/4-2\sqrt{3}$
8835 5941 $\exp(4)\div s\pi(\pi)$
8835 7063 rt : (3,7,-4,-4)
8835 7978 $t\pi$: $2\sqrt{2}-3\sqrt{3}/2$
8836 1173 id : $3\sqrt{2}/4-4\sqrt{5}$
8836 2182 rt : (7,4,-9,0)
8836 3746 Ψ : $5\sqrt{2}$
8836 5242 sq : $2+3\sqrt{3}/4$
8836 7287 rt : (4,8,3,4)
8836 8346 Ψ : $(1/(3e))^3$
8837 0553 2^x : $\sqrt[3]{15/2}$
8837 1893 ℓ_{10} : $2\sqrt{2}/3+3\sqrt{5}$
8837 1944 J_0 : 9/13
8837 2427 e^x : $\sqrt{3}/4+4\sqrt{5}$
8837 3046 rt : (8,1,-6,-1)
8837 3344 cr : $4e-4\pi/3$
8837 5315 at : $2/3-4\sqrt{2}/3$
8837 5432 J_0 : $\exp(-1/e)$
8837 7454 rt : (5,1,2,-6)
8837 8613 e^x : $1+4\pi$
8837 9482 rt : (7,7,-9,5)
8838 0611 Ei : 19/14
8838 0892 2^x : $3\sqrt{3}-\sqrt{5}/4$
8838 3204 ℓ_2 : $(2\pi)^{1/3}$
8838 3846 rt : (9,2,-3,2)
8838 3956 10^x : $10e/9$
8838 4204 Li_2 : $\sqrt{2}+\sqrt{3}-\sqrt{6}$
8838 4499 at : $(2e/3)^{1/3}$
8838 5496 ℓ_2 : $3-2\sqrt{3}/3$
8838 7426 rt : (3,8,-5,-9)
8838 7567 rt : (6,2,5,7)
8838 7956 cu : $\sqrt{2}/4+4\sqrt{3}/3$
8838 8347 id : $5\sqrt{2}/8$
8838 9825 erf : 10/9
8839 1119 cr : $3-4\sqrt{3}/3$
8839 1916 rt : (7,7,-6,-5)
8839 4019 rt : (4,5,0,9)
8839 4628 cu : $10\pi^2/7$
8839 5958 rt : (1,5,2,7)
8839 6139 $s\pi$: $9\pi/5$
8839 7082 cr : $3\sqrt{3}+2\sqrt{5}/3$
8839 7625 rt : (7,5,8,8)
8839 8563 sinh : $\sqrt[3]{12}$
8839 8769 rt : (8,4,-3,-6)
8839 9266 $s\pi$: $1/2+3e$
8839 9466 ln : $2\pi^2/3$
8839 9491 Ei : $\sqrt[3]{5/2}$
8840 0545 at : $\sqrt{2}/4+\sqrt{3}/2$
8840 1479 10^x : $(\sqrt[3]{3}/2)^{1/3}$
8840 2541 $2/5-\exp(1/4)$
8840 5038 ℓ_{10} : $2+4\sqrt{2}$
8840 5597 rt : (9,1,0,-7)
8840 6204 rt : (2,9,3,-3)
8840 6235 10^x : $1/2+2\sqrt{5}/3$
8840 6409 ln : $(e+\pi)^{1/2}$
8840 7103 $s\pi$: $\sqrt{11/2}$
8840 7413 $\ln(3)\times\ln(\sqrt{5})$
8840 7728 sr : $(2\pi/3)^{-1/3}$
8840 7789 $1/3\times\exp(\sqrt{3})$
8841 0136 rt : (5,9,-5,-8)
8841 1249 Ei : $\sqrt[3]{5}/9$
8841 1423 J : $\exp(-\sqrt[3]{3}/2)$
8841 3762 rt : (7,8,-5,9)
8841 4897 J_0 : $(2\pi/3)^{-1/2}$
8841 5099 cr : $10(e+\pi)$
8841 5949 at : $e\pi/7$
8841 6049 ln : $8\sqrt{5}$
8841 6906 erf : $\exp(1/(3\pi))$
8841 6938 $e\times\ln(2)$
8841 8084 10^x : $3\sqrt{2}+\sqrt{3}/2$
8841 9381 2^x : $1/\ln(5/3)$

8842 0546 $s\pi : 1/2 + 2\sqrt{3}/3$
8842 2602 rt : (2,6,-8,1)
8842 5060 rt : (1,6,0,-4)
8842 6024 rt : (2,8,-7,8)
8842 7005 rt : (4,6,-8,-9)
8843 3845 rt : (2,9,-5,-4)
8843 4064 $K : (\sqrt[3]{5}/2)^2$
8843 4418 rt : (1,9,-3,-9)
8843 4736 $10^x : 5(e+\pi)/9$
8843 9957 $\tanh : (e)^{1/3}$
8844 1557 $3/5 \times \exp(\pi)$
8844 3731 rt : (6,9,-7,-5)
8844 4425 $\Psi : (\sqrt[3]{5})^{-1/3}$
8844 8370 rt : (1,6,0,-4)
8844 9914 $\text{id} : 2\sqrt[3]{3}$
8845 0241 $10^x : \ln((e+\pi)/2)$
8845 0428 rt : (7,6,-4,-6)
8845 0917 $2^x : 3/2 - 3\sqrt{5}/4$
8845 2278 $\ell_2 : 13/24$
8845 3564 rt : (4,9,7,8)
8845 4129 $\tanh : 4\pi/9$
8845 4204 rt : (7,3,-7,-1)
8845 7102 rt : (8,3,-1,-7)
8845 7268 $\text{cu} : 1 + 3\sqrt{3}$
8845 9092 $\text{at} : (\ln 3/2)^{-1/3}$
8845 9401 $\text{id} : \exp(-1/(3e))$
8846 0278 $2^x : \sqrt{2}/2 + 3\sqrt{3}/2$
8846 0603 $5 \div \exp(\sqrt{3})$
8846 0658 $\ell_{10} : 3/23$
8846 1276 $\sinh : \exp(\sqrt{3}/3)$
8846 2220 rt : (1,0,6,6)
8846 2613 $\sinh : 9\sqrt[3]{3}$
8846 3346 rt : (5,0,4,7)
8846 3752 rt : (9,0,2,8)
8846 3961 cr : 9/13
8846 4617 rt : (8,0,-4,2)
8846 4992 $\ln : 3e - \pi/2$
8846 5173 sr : 18/23
8846 6644 $2^x : 8e\pi/7$
8847 3076 rt : (6,3,7,8)
8847 3600 cu : 24/25
8847 3879 $\text{at} : 3/4 + \sqrt{2}/3$
8847 4155 $\text{at} : 5\sqrt[3]{5}/7$
8847 4971 rt : (9,3,-1,3)
8847 6977 $\text{sr} : 3e/4 + 2\pi$
8847 7239 rt : (6,4,-9,9)
8847 7593 $\sqrt{3} \times \ln(3/5)$
8847 8680 $\ln(3/5) + \exp(1/3)$
8847 8924 $\sqrt{5} + \exp(1/2)$
8848 0530 rt : (2,3,9,7)
8848 1929 $\tanh : 5\sqrt{5}/8$
8848 2830 $2^x : 4e\pi/9$
8848 3222 $\tanh : ((e+\pi)/3)^{1/2}$
8848 3950 $10^x : 2 + 2\sqrt{2}$
8848 4269 $\text{cr} : 3/2 + 3\sqrt{3}$
8848 4408 $t\pi : 2\sqrt{2}/3 - 3\sqrt{3}/2$
8848 4815 rt : (3,7,-2,8)
8848 5267 rt : (5,3,1,2)
8848 5791 $\text{cr} : 2\sqrt{3}/5$
8848 6060 rt : (6,7,-6,4)
8848 6161 $\tanh : (\ln 3/3)^{-1/3}$
8848 6187 $\ln : 3 - \sqrt{3}/3$
8848 7105 $\zeta : (\ln 2/2)^{1/2}$
8848 7127 $erf : 7(1/\pi)/2$
8849 1396 rt : (8,3,-7,-3)
8849 2215 $\text{cr} : 8(e+\pi)/7$
8849 4924 $e^x : \sqrt{3} + \sqrt{5}$
8849 5135 $\text{cu} : 2e/3 + 3\pi$
8849 5559 $\text{id} : 3\pi/5$
8849 8916 rt : (3,1,-4,9)
8849 9684 $\ell_{10} : 4\sqrt{3} + \sqrt{5}/3$
8849 9704 $\text{id} : (\ln 2)^{1/3}$
8849 9898 cu : 21/17
8850 0826 $10^x : 4\sqrt{3} - \sqrt{5}/2$
8850 1183 rt : (6,8,-5,-6)
8850 4831 $2^x : 2 - \sqrt{2}/3$
8850 5686 rt : (6,6,4,3)
8850 6681 at : 11/9
8850 7708 rt : (7,5,-2,-7)
8850 7720 $e^x : 19/14$
8850 8815 $\text{id} : (\sqrt[3]{3})^{-1/3}$
8850 8817 $e^x : 3/2 - \sqrt{3}/2$
8851 2929 rt : (3,9,-8,9)
8851 3217 $e^x : 3e + 2\pi$
8851 4211 rt : (8,2,1,-8)
8851 4746 rt : (9,6,-4,-2)
8851 4766 $\theta_3 : (e+\pi)^{-1/2}$
8851 5527 $\ell_{10} : 7\pi^2/9$
8851 8230 rt : (2,8,-3,-5)
8852 0690 rt : (9,1,4,9)
8852 1547 $erf : (\sqrt[3]{3}/2)^{-1/3}$
8852 3715 $\text{cu} : 1/4 + 2\pi/3$
8852 5878 rt : (7,9,7,4)
8852 5953 $Ei : (\sqrt{2}\pi)^{-3}$
8852 5960 $\ell_{10} : 3/4 + 4\sqrt{3}$
8852 7725 $e^x : (\pi)^{1/2}$
8852 7747 rt : (6,8,-2,8)
8852 7781 $\Gamma : 13/2$
8852 7870 $\sinh : (\pi/2)^{-1/2}$
8852 8113 $J_0 : 11/16$
8853 0805 $10^x : \sqrt{2}/3 + \sqrt{5}$
8853 0882 $e^x : 7\sqrt{5}/5$
8853 3011 $s\pi : 2\sqrt{3} - \sqrt{5}/2$
8853 3101 rt : (6,5,-8,-1)
8853 3268 $\tanh : 10\sqrt[3]{2}/9$
8853 3344 $e^x : \sqrt[3]{5/2}$
8853 5164 $\tanh : 7/5$
8853 6059 $\ell_2 : 4\sqrt{2} + \sqrt{3}$
8853 6079 $\sinh : 20/3$
8853 7216 rt : (4,2,3,-7)
8853 8311 $t\pi : 1/4 + 3\pi/4$
8853 9008 $\text{id} : 1/\ln(\sqrt{2})$
8854 1878 $\text{id} : P_{54}$
8854 3150 rt : (7,2,-5,-2)
8854 4638 $\sinh : 5e\pi/7$
8854 4901 $\text{sq} : 3/2 - \sqrt{5}/4$
8854 6636 rt : (5,1,6,8)
8854 7870 $J_0 : 4\zeta(3)/7$
8854 8731 $s\pi : (2e/3)^{1/2}$
8854 8779 $e^x : \sqrt{3} + \sqrt{5}/4$
8854 9424 $2^x : ((e+\pi)/2)^2$
8855 0031 $\sinh : 23/2$
8855 2829 rt : (5,8,-7,-9)
8855 3144 rt : (8,1,-2,3)
8855 3692 $1/5 - \exp(3)$
8855 4942 $\text{sq} : 7\ln 2/2$
8855 6008 rt : (6,4,9,9)
8855 7444 rt : (6,7,-3,-7)
8855 7595 $J_0 : (\sqrt{2}/3)^{1/2}$
8855 7757 $s\pi : 2/3 + e/4$
8856 0319 $\Gamma : 19/13$
8856 0694 $\Gamma : (\pi)^{1/3}$
8856 0795 $\Gamma : (e+\pi)/4$
8856 1025 $\Gamma : 6\sqrt[3]{5}/7$
8856 1103 $4 \times \exp(1/5)$
8856 1403 $\Gamma : 22/15$
8856 1460 $\Gamma : (\sqrt{2}/3)^{-1/2}$
8856 1808 $\text{id} : 4\sqrt{2}/3$
8856 2476 $\Gamma : 16/11$
8856 2941 $2^x : 2\sqrt{2}/3 - \sqrt{5}/2$
8856 3084 rt : (9,4,1,4)
8856 3246 $\Gamma : 7\sqrt[3]{2}/6$
8856 3400 $\text{as} : 4/3 - \sqrt{5}/4$
8856 3747 $\Gamma : 25/17$
8856 3805 rt : (7,4,0,-8)
8856 4165 $\sinh : \ln(\sqrt{2}\pi/2)$
8856 5090 rt : (2,8,1,-4)
8856 7283 $\Psi : \ln(\ln 3/2)$
8856 8686 $\ell_{10} : 9e\pi$
8856 9281 $\Gamma : (2\pi/3)^{1/2}$
8857 0143 rt : (8,1,3,-9)
8857 1004 $\ln(2/5) \times \exp(e)$
8857 2627 rt : (9,9,-6,-8)
8857 2715 $\Gamma : \exp(1/e)$
8857 3031 $10^x : 4\sqrt{3}/9$
8857 3045 $\Gamma : 13/9$
8857 4682 $\Gamma : 5\sqrt{3}/6$
8857 5016 $e^x : 2 - 3\sqrt{2}/2$
8857 5776 $\Gamma : 1/\ln(2)$
8857 6149 $\Gamma : 6\zeta(3)/5$
8857 6514 $\Gamma : \sqrt[3]{3}$
8857 6587 $\text{id} : \sqrt{8}\pi$
8857 7838 $\ln : 7\sqrt{3}/5$
8857 9696 $\zeta : \sqrt{2/3}$
8858 0416 rt : (1,8,-9,1)
8858 0675 $\Gamma : 8\sqrt[3]{2}/7$
8858 1054 $s\pi : 3\pi/7$
8858 1314 sq : 16/17
8858 2673 rt : (1,5,-2,-5)
8858 2798 $\Gamma : 6\sqrt{3}/7$
8858 2995 $s\pi : \sqrt{3}/5$
8858 3200 rt : (1,7,-8,-8)
8858 3210 $\ln(3/4) - \exp(4)$
8858 5462 $\Gamma : 23/16$
8858 6925 $\tanh : 7\zeta(3)/6$
8859 1121 $\ln : 9(e+\pi)/8$
8859 2269 $t\pi : 3/13$
8859 5532 $\ln(\pi) \times s\pi(e)$
8859 5602 $\Gamma : (\sqrt{2}\pi/2)^{1/2}$
8859 6052 $\sqrt{5} \times \exp(e)$
8859 6243 $\Gamma : 2\sqrt{5}/3$
8859 6463 $s\pi : 6\sqrt[3]{3}$
8859 7277 $\text{cr} : 3\sqrt{5}$
8859 7286 $\Gamma : 9(1/\pi)/2$
8859 7713 $\Gamma : \ln(\sqrt{2}\pi)$
8860 0093 rt : (2,9,4,-3)
8860 1093 $\Gamma : ((e+\pi)/2)^{1/3}$
8860 2019 rt : (1,9,0,-9)
8860 2855 $erf : 19/17$
8860 3742 $\Gamma : 1/\ln((e+\pi)/3)$
8860 4244 $\Gamma : \sqrt[3]{10/3}$
8860 4684 rt : (5,4,-6,8)
8860 5560 $10^x : 3/4 + 3\sqrt{3}$
8860 6187 cr : 16/23
8860 6338 rt : (5,9,-4,-7)
8860 6806 $s\pi : \ln 2/2$
8860 7193 $\zeta : 7\zeta(3)/8$
8860 7645 $\Gamma : 10/7$
8860 7712 $\text{at} : \sqrt{3/2}$
8860 7768 $\ln : 3\sqrt{2}/2 + 2\sqrt{5}$
8860 7996 $t\pi : 3\zeta(3)$
8860 8549 $\Gamma : (\sqrt{5})^{1/2}$
8860 9264 rt : (3,5,3,8)
8860 9852 rt : (5,7,-9,-1)
8861 0391 $10^x : 1/3 + 2\sqrt{2}/3$
8861 1088 $\Gamma : 7\sqrt[3]{5}/8$
8861 2414 $Ei : 9\sqrt[3]{2}/4$
8861 2561 rt : (6,6,-1,-8)
8861 4259 rt : (9,8,-4,-9)
8861 5370 $erf : \sqrt{5}/2$
8861 6823 $Li_2 : 2\pi/9$
8861 7354 rt : (1,1,8,7)
8861 8390 rt : (4,1,5,-8)
8861 8459 $\Gamma : 23/6$
8861 8554 $\Gamma : 5\sqrt[3]{5}/6$
8861 8764 rt : (7,3,2,-9)
8861 9476 $2^x : 1/4 + 3e/4$
8861 9559 rt : (9,6,-9,-3)
8862 0251 $\ell_{10} : 9\sqrt[3]{5}/2$
8862 2391 $s\pi : 1/\sqrt[3]{24}$
8862 2549 $\ell_{10} : \sqrt{6}\pi$
8862 2643 $K : \ln(\sqrt[3]{3}/3)$
8862 2692 $\text{sr} : \pi/4$
8862 2903 $\Gamma : (\ln 2/2)^{-1/3}$
8862 4087 $\Gamma : e\pi/6$
8862 5004 $\ln : 7\ln 2/2$
8862 7112 $\ell_2 : 8\ln 2/3$
8862 7471 rt : (5,2,8,9)
8862 7949 $\lambda : \exp(-9\pi/14)$
8862 9213 rt : (7,1,-3,-3)
8862 9436 $1/2 + \ln(4)$
8863 1281 $\Gamma : 5\zeta(3)/4$
8863 1304 $\sqrt{3} + \exp(e)$
8863 1784 rt : (4,9,-7,1)
8863 2040 $\text{sq} : 1/3 - \pi$
8863 2905 $K : (e+\pi)/8$
8863 4413 $J_0 : 13/19$
8863 4951 $\text{id} : \sqrt{2} + 2\sqrt{5}$
8863 5003 $\text{sq} : 5e/8$
8863 6489 $\text{cr} : 2 + 3\pi/2$

8863 8264 sr : $5\sqrt{2}/9$
8863 8817 rt : (8,2,0,4)
8864 0526 sr : 11/14
8864 1512 J_0 : $2\sqrt[3]{5}/5$
8864 3150 10^x : 17/5
8864 3578 sq : $3/2+\sqrt{2}/3$
8864 4436 cu : $1/4+2e$
8864 5246 at : $2/3+\sqrt{5}/4$
8864 5303 Γ : $9\sqrt[3]{2}/8$
8864 7137 sr : $\sqrt{3}/3+4\sqrt{5}/3$
8864 8210 Γ : 17/12
8864 8369 rt : (9,5,3,5)
8864 8470 $s\pi$: $3\sqrt{3}/4+3\sqrt{5}/2$
8865 0587 Ei : $2\sqrt{5}/7$
8865 2300 rt : (2,4,-2,9)
8865 2746 $\exp(\pi)-s\pi(\sqrt{3})$
8865 3571 sr : $3+\sqrt{5}/4$
8865 3817 cr : $\sqrt{2}+\sqrt{3}-\sqrt{6}$
8865 5887 ln : $2\sqrt[3]{3}/7$
8865 7872 rt : (5,4,3,3)
8865 8142 Γ : $\sqrt{2}$
8865 9805 Γ : $5e/9$
8866 0506 rt : (5,8,-2,-8)
8866 3564 2^x : $2\sqrt{2}-3\sqrt{3}/4$
8866 6187 rt : (7,1,-8,-3)
8866 6577 rt : (6,5,1,-9)
8866 7053 Γ : $6\sqrt[3]{2}/5$
8866 7693 ln : $3/4+3\sqrt{5}/4$
8866 8239 rt : (7,7,-8,7)
8866 8597 Γ : 24/17
8867 0684 rt : (1,9,-2,-6)
8867 1666 ℓ_2 : $3\sqrt[3]{3}/8$
8867 4962 $5-\exp(\sqrt{2})$
8867 5134 id : $5\sqrt{3}/3$
8867 5668 rt : (9,5,-7,-4)
8867 6530 erf : $8\sqrt[3]{2}/9$
8867 6722 Γ : $\pi^2/7$
8867 6728 rt : (6,7,6,4)
8867 7812 ln : $4+3\sqrt{3}/2$
8867 9573 rt : (2,6,1,1)
8868 0015 id : $e/4-4\pi$
8868 0371 J_0 : $(\pi)^{-1/3}$
8868 1633 e^x : $4e$
8868 1831 K : $3\sqrt[3]{5}/7$
8868 2944 Γ : $7\sqrt{3}/8$
8868 3150 2^x : $\beta(2)$
8868 6553 rt : (7,9,-9,-2)
8868 7083 ℓ_{10} : $3\sqrt{2}+2\sqrt{3}$
8868 7665 rt : (8,4,-5,-2)
8868 8370 J_0 : $1/\ln(\ln 2/3)$
8868 8420 rt : (3,3,4,-8)
8868 8561 sr : $\sqrt{2}/4+\sqrt{3}/4$
8869 0695 $s\pi$: $7\sqrt{3}/9$
8869 1674 cr : $4+e$
8869 2028 e^x : $3/2+3\sqrt{2}$
8869 3949 rt : (5,9,-4,-6)
8869 5319 ℓ_{10} : $1+3\sqrt{5}$
8869 5602 ln : $1/3+2\pi/3$
8869 5673 Γ : $\sqrt[3]{7/2}$
8869 7223 rt : (4,0,7,9)
8869 7263 Γ : $\sqrt{3}+\sqrt{5}-\sqrt{6}$
8870 3283 rt : (6,3,-4,-3)
8870 3773 Γ : $8\sqrt[3]{5}/9$
8870 6423 2^x : $3e/2+\pi/3$
8870 6912 2^x : $3\sqrt{2}+\sqrt{3}$
8870 8095 id : $2\sqrt{2}/3+4\sqrt{5}$
8870 8878 rt : (9,7,-2,-1)
8871 1335 cr : $2\pi/9$
8871 1441 J_0 : 15/22
8871 2122 Ψ : $\sqrt{3}/8$
8871 2569 rt : (7,0,-1,4)
8871 3407 Γ : $\exp(\sqrt[3]{2}/3)$
8871 3574 Γ : $7\zeta(3)/6$
8871 3609 rt : (5,8,1,7)
8871 5466 rt : (2,7,-1,-5)
8871 8892 ℓ_{10} : $3+3\pi/2$
8871 9306 $\exp(\sqrt{2})+s\pi(e)$
8872 1807 rt : (8,3,2,5)
8872 2344 $t\pi$: $3e/4-\pi/2$
8872 2716 $t\pi$: $4\sqrt[3]{3}$
8872 4473 rt : (8,7,-8,-4)
8872 5348 sq : $2/3+\sqrt{2}/2$
8872 5678 2^x : $\ln(5/2)$
8872 6381 Γ : 7/5
8872 6632 ζ : $\exp(-\sqrt[3]{4}/3)$
8872 6859 Γ : $10\sqrt[3]{2}/9$
8872 6961 sq : $8\zeta(3)/7$
8872 8090 $s\pi$: $\sqrt{2}/4+3\sqrt{3}/4$
8872 9058 e^x : $4\sqrt{2}/3+\sqrt{5}/3$
8873 0319 ln : 7/17
8873 0579 $\ln(3/5)\times\exp(\sqrt{3})$
8873 0611 rt : (9,4,-5,-5)
8873 0998 rt : (9,6,5,6)
8873 1914 rt : (1,7,7,-5)
8873 4051 $\ln(\pi)\times\exp(1/2)$
8873 4562 sr : $4e/3+3\pi/2$
8873 4818 sq : $1-4e/3$
8873 4994 erf : $7\sqrt[3]{3}/9$
8873 5171 e^x : $3e/4+\pi/3$
8873 5418 cu : $3\sqrt{2}-3\sqrt{5}/4$
8873 7682 10^x : $e/3+4\pi$
8873 8416 sr : $5\sqrt[3]{2}/8$
8873 8938 Γ : $(\ln 3/3)^{-1/3}$
8873 9051 at : $2\sqrt{3}-\sqrt{5}$
8873 9177 Li_2 : $((e+\pi)/2)^{-1/3}$
8873 9704 Γ : $((e+\pi)/3)^{1/2}$
8874 0042 Γ : $5\sqrt{5}/8$
8874 1518 cr : $((e+\pi)/2)^{-1/3}$
8874 2748 $\pi-s\pi(\sqrt{3})$
8874 2917 Γ : $\sqrt{7/3}$
8874 3034 $s\pi$: $(\ln 3/3)^{-1/2}$
8874 3712 $\exp(1/3)+\exp(2/5)$
8874 4842 $s\pi$: $7\sqrt{5}$
8874 4879 ℓ_2 : $\sqrt{2}/3+4\sqrt{3}$
8874 7371 Γ : $4\pi/9$
8874 7624 K : 11/15
8874 8213 tanh : $\pi^2/7$
8874 8219 rt : (1,9,1,0)
8874 8964 10^x : $\zeta(5)$
8874 9125 λ : 2/15
8874 9300 $t\pi$: $\ln 2/3$
8875 0339 sr : $4\sqrt{2}/3+3\sqrt{5}/4$
8875 1158 Γ : $(e)^{1/3}$
8875 2207 cu : $6\sqrt{2}/5$
8875 4957 Ei : $(e)^2$
8875 6349 ln : $4-\pi/2$
8875 6799 rt : (2,5,3,-8)
8875 7815 erf : $(\sqrt[3]{2})^{1/2}$
8875 9882 rt : (4,9,-1,-9)
8876 5340 rt : (3,2,6,-9)
8876 5500 ℓ_2 : $\sqrt[3]{19/3}$
8876 6418 rt : (4,8,-8,-2)
8876 8207 $3/5-\ln(3/4)$
8876 9310 θ_3 : $(\sqrt{5}/3)^3$
8876 9823 J_0 : 17/25
8877 0945 $\exp(4/5)-\exp(\sqrt{2})$
8877 2444 rt : (7,9,-9,-4)
8877 4862 2^x : 11/12
8877 5449 rt : (5,5,-5,-3)
8877 6197 Γ : 23/15
8877 8449 rt : (8,6,-6,-5)
8877 9882 $s\pi$: $\exp(\sqrt{3})$
8878 1969 2^x : $4e/3-2\pi/3$
8878 3596 J_0 : $e/4$
8878 4412 $t\pi$: $\sqrt{3}/2-2\sqrt{5}$
8878 4434 rt : (9,3,-3,-6)
8878 5084 rt : (1,6,1,2)
8878 6828 tanh : 24/17
8878 8521 $s\pi$: 8/23
8878 8545 cr : $5\sqrt[3]{2}/9$
8878 9387 $t\pi$: $(\sqrt[3]{3}/3)^2$
8879 0243 $\sqrt{2}\times\exp(5)$
8879 0393 rt : (3,6,-7,2)
8879 0400 cr : 7/10
8879 1092 Γ : $\sqrt{3}+\sqrt{6}-\sqrt{7}$
8879 2092 $\exp(4)\div s\pi(1/5)$
8879 2170 $t\pi$: $\sqrt{23/3}$
8879 2558 Γ : 25/18
8879 3191 λ : $\zeta(3)/9$
8879 3375 rt : (7,1,1,5)
8879 5431 2^x : 25/7
8879 5475 ℓ_2 : $3\pi^2$
8879 8079 $t\pi$: $3\sqrt{2}/4+3\sqrt{5}$
8880 0337 rt : (6,6,-9,2)
8880 1588 2^x : $4+3\sqrt{2}$
8880 2253 J : $(1/\pi)$
8880 2271 rt : (8,4,4,6)
8880 3327 10^x : $7\sqrt{3}/6$
8880 4621 e^x : $2/3-\pi/4$
8880 6786 Γ : $\sqrt[3]{8/3}$
8880 6942 $1/3-\exp(1/5)$
8880 7558 sinh : $((e+\pi)/3)^{-1/3}$
8880 7854 Γ : 20/13
8880 9650 Γ : $2\ln 2$
8881 0263 sq : $\sqrt[3]{23/3}$
8881 0598 sinh : 4/5
8881 1104 rt : (4,4,-3,7)
8881 1124 rt : (9,7,7,7)
8881 1162 Γ : $9\sqrt[3]{5}/10$
8881 1804 cu : $2\sqrt{3}+3\sqrt{5}/4$
8881 4055 Γ : $4\sqrt{3}/5$
8881 5175 Γ : $8\sqrt{3}/9$
8881 6821 $s\pi$: $5\pi^2$
8881 9933 rt : (5,5,5,4)
8882 1043 Γ : 18/13
8882 2161 rt : (2,9,-4,-1)
8882 3600 rt : (1,7,2,-8)
8882 4370 $3/5\div s\pi(\sqrt{5})$
8882 4645 $s\pi$: $3/4+3\sqrt{3}/2$
8882 5483 rt : (7,8,-7,-5)
8882 6439 sq : $3\pi/10$
8882 6570 ℓ_{10} : $4e-\pi$
8882 7711 e^x : $3\sqrt{2}/4$
8882 7826 Ei : $10\sqrt[3]{3}/3$
8883 0015 rt : (6,1,-9,3)
8883 0165 $\sqrt{5}+\exp(\sqrt{3})$
8883 0358 cu : $3-3e/4$
8883 1051 cr : $2-3\sqrt{3}/4$
8883 1080 Γ : $\sqrt[3]{11/3}$
8883 1341 rt : (8,5,-4,-6)
8883 1890 rt : (2,4,5,-9)
8883 2421 sr : $\sqrt{2}+4\sqrt{3}$
8883 2754 e^x : $1/3+\sqrt{3}$
8883 3612 10^x : 14/3
8883 7180 rt : (9,2,-1,-7)
8883 7422 rt : (6,8,8,5)
8883 8194 Γ : $(\sqrt[3]{2}/3)^{-1/2}$
8883 8556 tanh : $\sqrt{2}$
8883 8648 rt : (6,7,-5,6)
8883 8722 $t\pi$: $9e/10$
8883 8823 erf : 9/8
8883 8842 $t\pi$: $4\sqrt{2}+\sqrt{3}/2$
8884 0242 $t\pi$: $3\sqrt{2}/7$
8884 1017 10^x : $(3\pi/2)^{-1/2}$
8884 1076 ζ : 7/24
8884 2590 rt : (3,9,0,-5)
8884 3945 e^x : $e+3\pi/4$
8884 5801 rt : (4,7,-6,-3)
8884 7484 cu : $3(e+\pi)/4$
8884 7977 at : 16/13
8884 9866 rt : (7,2,-6,-2)
8885 2331 sr : 15/19
8885 3872 id : $3\sqrt{2}-3\sqrt{5}/2$
8885 4381 id : $8\sqrt{5}$
8885 4443 Γ : 17/11
8885 4502 rt : (5,4,-3,-4)
8885 4770 Γ : $9\zeta(3)/7$
8885 5812 rt : (3,8,-4,-6)
8885 8152 rt : (2,6,-3,-6)
8886 2095 $s\pi$: $\exp(2e)$
8886 2412 2^x : $\sqrt[3]{12}$
8886 3161 rt : (6,1,0,-5)
8886 3902 ln : $\sqrt{2}+3\sqrt{3}$
8886 4694 sr : $4-\sqrt{3}/4$
8886 5884 cu : $1/3-\sqrt{5}$
8886 6478 rt : (2,5,-9,7)
8886 6584 e^x : $1-\sqrt{5}/2$
8886 7527 e^x : $3/4+3e/2$
8886 9459 sinh : $5\sqrt{5}/2$
8886 9489 rt : (8,5,-3,-1)

8886 9591 $10^x : 4\sqrt{2} - 3\sqrt{3}$
8887 1778 rt : (7,2,3,6)
8887 3134 $\Psi : 5\pi^2$
8887 3584 rt : (1,6,3,-9)
8887 4033 $10^x : 3/4 + e/4$
8887 6391 rt : (1,8,-7,8)
8887 7473 rt : (7,7,-5,-6)
8888 0354 rt : (8,5,6,7)
8888 3190 rt : (8,4,-2,-7)
8888 3951 sinh : $\sqrt{19}\pi$
8888 4012 $\Psi : \sqrt{17/2}$
8888 4945 ln : $\sqrt{5} - \sqrt{6} + \sqrt{7}$
8888 5683 $s\pi : 4(e + \pi)/7$
8888 6179 $2^x : 7\sqrt[3]{5}/3$
8888 7830 $10^x : 8\sqrt{5}/5$
8888 8261 $\Gamma : \ln(3\pi/2)$
8888 8888 id : 8/9
8889 0147 tanh : 17/12
8889 0624 rt : (1,9,2,9)
8889 1356 Γ : 11/8
8889 2713 cr : $(\ln 2/2)^{1/3}$
8889 2894 $\Gamma : \exp(1/\pi)$
8889 3459 $\Gamma : (\sqrt[3]{4}/3)^{-1/2}$
8889 6940 rt : (1,6,4,-7)
8889 7344 $\lambda : (\ln 3/3)^2$
8889 8906 at : $1/2 - \sqrt{3}$
8890 0903 $\Gamma : 8\zeta(3)/7$
8890 1906 rt : (4,9,-6,4)
8890 3180 $s\pi : \sqrt{2}/2 + 4\sqrt{5}$
8890 5275 rt : (3,9,2,3)
8890 5609 $1/2 + \exp(2)$
8890 5761 tanh : $9\sqrt[3]{2}/8$
8890 5908 $\ell_2 : 3\sqrt[3]{2}/7$
8890 5953 rt : (1,6,-5,-1)
8890 5981 id : P_{43}
8890 6062 cu : $1/2 + 2\sqrt{5}/3$
8890 6977 id : $7e\pi/2$
8890 8411 rt : (1,7,-6,8)
8890 9553 rt : (5,3,-9,6)
8891 1418 $\Gamma : 4e/7$
8891 2210 $\lambda : ((e + \pi)/3)^{-3}$
8891 3028 ln : $2 + \sqrt{3}/4$
8891 3205 $10^x : e\pi/9$
8891 4390 $10^x : \sqrt{2}/2 + \sqrt{3}$
8891 4424 rt : (3,9,-7,-3)
8891 6897 ln : $\sqrt{3} + \sqrt{5} + \sqrt{7}$
8891 8873 $10^x : \sqrt{3}/3 + 3\sqrt{5}/2$
8891 9065 $\sqrt{5} \div s\pi(e)$
8892 1282 cu : $6\sqrt[3]{3}/7$
8892 2855 rt : (6,9,-6,-3)
8892 3395 sq : $4/3 - 3\sqrt{5}$
8892 4112 sinh : $\sqrt[3]{17/3}$
8892 6453 $\sqrt{3} \div \exp(2/3)$
8892 7056 rt : (9,9,-5,-9)
8892 7920 at : $3\sqrt{2}/2 - 3\sqrt{5}/2$
8892 8264 $e^x : 3\sqrt{3}/2 + 3\sqrt{5}/4$
8892 8454 rt : (7,6,-3,-7)
8892 8592 sr : $4/3 + \sqrt{5}$
8892 8673 Γ : 14/9
8892 9334 cu : $4\zeta(3)/5$
8893 0218 $Li_2 : 5\sqrt[3]{2}/9$
8893 1247 rt : (5,3,-1,-5)
8893 1276 rt : (1,8,1,1)
8893 2876 rt : (2,7,-2,4)
8893 3330 as : $2e/7$
8893 3826 $s\pi : 2e - \pi/4$
8893 3916 cu : $7e/10$
8893 3919 ln : $2\sqrt{2}/3 + 2\sqrt{5}/3$
8893 4035 rt : (8,3,0,-8)
8893 5401 $e^x : 10(1/\pi)/3$
8893 5860 $Ei : 10(e + \pi)/7$
8893 7762 Li_2 : 7/10
8893 9599 rt : (9,0,3,9)
8894 1776 rt : (7,7,-7,9)
8894 3639 $s\pi : (e + \pi)/9$
8894 6621 $4 \div \exp(3/4)$
8894 7109 $\zeta : 4\pi/7$
8894 7912 rt : (7,3,5,7)
8894 8172 $\Gamma : 4\sqrt[3]{5}/5$
8894 9438 $\sqrt{5} - \ln(\sqrt{2})$
8895 2743 $Ei : e/2$
8895 4614 $\Gamma : 9\sqrt{3}/10$
8895 5725 $s\pi : 2\sqrt{2}/3 + 3\sqrt{5}$
8895 6186 rt : (8,6,8,8)
8895 6934 ln : $1/3 + 2\pi$
8895 9710 e^x : 7/11
8895 9806 rt : (1,9,-9,-2)
8896 0072 $t\pi : 1/4 + 4\sqrt{5}/3$
8896 0531 $\Gamma : 9 \ln 2/4$
8896 0549 rt : (8,3,-9,0)
8896 1750 rt : (1,9,8,-6)
8896 2807 $s\pi : \sqrt[3]{9/2}$
8896 4373 at : $\pi^2/8$
8896 4503 $\ell_2 : 3\sqrt{3} - 2\sqrt{5}/3$
8896 4530 $2^x : 3\sqrt{2} + \sqrt{5}/4$
8896 5508 sr : $2 + \pi/2$
8896 7035 $s\pi : \pi/9$
8896 9302 rt : (9,0,-6,1)
8897 2620 rt : (5,6,7,5)
8897 2994 rt : (6,8,-4,-7)
8897 3567 θ_3 : 13/16
8897 5652 sr : 19/24
8897 7815 sinh : $5\sqrt[3]{3}/9$
8897 8462 rt : (7,5,-1,-8)
8897 9223 rt : (3,8,-2,-6)
8897 9855 $2^x : (\sqrt[3]{4}/3)^{-2}$
8898 1021 $2/3 - \ln(4/5)$
8898 2236 sr : 25/7
8898 3026 $\theta_3 : 1/\sqrt[3]{14}$
8898 3594 Ei : 16/25
8898 3913 rt : (8,2,2,-9)
8898 4438 Γ : 25/16
8898 5676 Γ : 15/11
8898 8157 id : $3\sqrt[3]{2}/2$
8898 9570 rt : (3,8,-5,-4)
8899 1746 $erf : (\sqrt[3]{3})^{1/3}$
8899 3568 $s\pi : 8e/5$
8899 3589 $s\pi : (\ln 3/2)^{-1/2}$
8899 3856 rt : (2,5,-5,-7)
8899 4080 sinh : $2\zeta(3)/3$
8899 4554 $10^x : (\ln 2)^{-1/2}$
8899 5407 $erf : (\ln 2)^{-1/3}$
8899 7707 rt : (4,7,0,2)
8899 8426 cr : $3e/4 + 3\pi/2$
8900 0000 sq : 17/10
8900 1859 $\ln(2) \times \exp(1/4)$
8900 2562 rt : (6,0,-7,2)
8900 3831 sr : $8 \ln 2/7$
8900 5691 $s\pi : 5e\pi/2$
8900 5808 rt : (5,2,1,-6)
8900 7220 $2^x : \sqrt{3}/3 - \sqrt{5}/3$
8900 7562 $\Gamma : 7\sqrt{5}/10$
8900 8116 $e^x : 2(1/\pi)$
8900 8373 rt : (1,2,-8,-8)
8901 0136 $t\pi : 7\sqrt[3]{5}/5$
8901 0229 $10^x : 7\sqrt{5}/4$
8901 0682 $\ln(4/5) + \exp(\sqrt{2})$
8901 1601 $erf : \exp(1/(3e))$
8901 3404 $\ln(3/5) \times \exp(4)$
8901 3871 rt : (6,1,4,7)
8901 8367 $10^x : 4/3 + 3\pi$
8902 0788 rt : (7,3,-4,-1)
8902 1637 $2/5 \times \exp(4/5)$
8902 1898 rt : (7,4,7,8)
8902 2189 rt : (6,7,-2,-8)
8902 4598 $2^x : 3\sqrt{3} - \sqrt{5}/2$
8902 5345 $J : (1/\pi)/4$
8902 5587 $\ln(\sqrt{5}) + \exp(3)$
8902 6381 $\Gamma : e/2$
8902 7510 at : 21/17
8902 7531 rt : (7,4,1,-9)
8902 8305 tanh : $e\pi/6$
8902 8519 $\sqrt{3} \div \ln(2/5)$
8902 9367 $\ell_2 : 3 + \sqrt{2}/2$
8903 0351 rt : (7,9,-8,0)
8903 4279 cu : $2/3 + e/3$
8903 5383 rt : (2,9,-3,4)
8903 5585 $\exp(3/5) \div s\pi(\sqrt{2})$
8903 5672 tanh : $(\ln 2/2)^{-1/3}$
8903 5752 $\ell_{10} : 3\sqrt{2}/4 + 3\sqrt{5}$
8903 7175 ln : 18
8903 8419 cr : 12/17
8903 8820 rt : (8,6,-1,0)
8903 9873 $J : \exp(-\sqrt{5}/3)$
8904 0141 $erf : 4\sqrt{2}/5$
8904 3299 rt : (2,9,5,1)
8904 3668 $10^x : e - \pi/3$
8904 4471 $\Gamma : \sqrt[3]{5/2}$
8904 4521 sr : $2\sqrt{2} + \sqrt{5}/3$
8904 4551 sr : $3/2 - \sqrt{2}/2$
8904 5095 Γ : 19/14
8904 5260 $e^x : 3/4 - \sqrt{3}/2$
8904 7248 rt : (9,1,-4,2)
8904 7914 $\theta_3 : (\sqrt{2}/3)^{1/2}$
8905 0627 $\Psi : 5e\pi/6$
8905 0714 $\ell_2 : \sqrt{2}/2 + 3\sqrt{5}$
8905 2936 sr : $4\sqrt{3} - 3\sqrt{5}/2$
8905 3475 ln : $2e + 4\pi$
8905 4337 $\Psi : 3\zeta(3)/7$
8905 4718 rt : (1,5,-3,-2)
8905 6089 $\Gamma : \pi/2$
8905 6471 $\ln(\pi) - s\pi(\sqrt{3})$
8905 6662 rt : (9,9,2,1)
8905 6748 sq : $1/\ln(\ln 2/2)$
8905 7647 sq : $3/4 + 3\sqrt{5}/4$
8905 8134 at : $1 - \sqrt{5}$
8905 8317 $10^x : (\ln 3/3)^{-3}$
8905 8433 $t\pi : 9\pi/5$
8905 8743 $J_0 : (\sqrt{2}\pi/2)^{-1/2}$
8906 0462 $s\pi : \sqrt{2} + \sqrt{5}$
8906 0639 $s\pi : 4\sqrt{2}/3 + 2\sqrt{3}$
8906 1050 $\Gamma : 10\sqrt{2}/9$
8906 1773 Γ : 11/7
8906 2478 $J_0 : 3\sqrt{5}/10$
8906 2500 sq : 11/8
8906 2581 rt : (1,9,-8,8)
8906 2614 rt : (3,7,-3,-5)
8906 3322 tanh : $5\sqrt[3]{5}/6$
8906 3907 at : $6\sqrt[3]{3}/7$
8906 4959 $\Gamma : \sqrt{17}\pi$
8906 5238 rt : (5,9,-3,-8)
8906 5426 $10^x : 10 \ln 2/9$
8906 7278 cu : $3\sqrt{2}/4 + \sqrt{5}/3$
8906 9263 $1/4 - \exp(\pi)$
8906 9676 at : $(\sqrt[3]{4}/3)^{-1/3}$
8907 0475 rt : (6,6,0,-9)
8907 0908 $J_0 : 1/\ln(\sqrt{2}\pi)$
8907 2941 Ψ : 5/6
8907 5614 ln : $3 + 4e/3$
8907 8301 rt : (5,1,3,-7)
8908 2754 $e^x : \exp(4)$
8908 5702 Γ : 23/17
8908 6092 rt : (6,2,6,8)
8908 8131 $\Gamma : 1/\ln(2\pi/3)$
8908 9033 $t\pi : 1/2 + 3e$
8908 9091 $\ell_2 : 9\sqrt[3]{3}/7$
8908 9871 id : $(\sqrt[3]{2})^{-1/2}$
8909 0016 $t\pi : 1/3 + 3\sqrt{2}/4$
8909 1187 sq : $8\pi/7$
8909 1910 $\Gamma : 9\zeta(3)/8$
8909 2372 $e^x : 5\pi^2/6$
8909 2712 sq : $1/4 + 4\sqrt{2}$
8909 3559 $\Gamma : 5\sqrt[3]{2}/4$
8909 3849 rt : (7,5,9,9)
8909 4123 rt : (9,6,-8,-4)
8909 4270 $\exp(3/4) + s\pi(e)$
8909 6943 id : $\sqrt{2}/2 - 3\sqrt{3}/2$
8909 7263 cu : $5\sqrt{3}/9$
8910 0652 $s\pi$: 7/20
8910 1573 rt : (4,4,6,5)
8910 1574 erf : 17/15
8910 2041 rt : (1,6,2,8)
8910 2068 at : $5\sqrt{3}/7$
8910 2088 id : $\exp(\sqrt[3]{4})$
8910 3781 $J_0 : \ln((e + \pi)/3)$
8910 4325 $\exp(4) - s\pi(1/4)$
8910 6583 rt : (7,4,-8,-1)
8910 6784 $t\pi : \sqrt{2} + 3\sqrt{5}/2$
8910 8311 rt : (1,8,2,9)
8910 9399 rt : (3,7,-4,-7)

8911 2250 at : $7\sqrt{2}/8$
8911 2655 sπ : $7\sqrt{2}/6$
8911 2763 rt : (5,8,-1,-9)
8911 3282 at : $1/3-\pi/2$
8911 4744 rt : (8,1,-5,-2)
8911 5591 rt : (3,6,-6,5)
8911 5737 ℓ_{10} : $3/2+2\pi$
8911 6918 rt : (5,7,9,6)
8911 8923 id : $9\ln 2/7$
8912 0165 erf : $9\sqrt[3]{2}/10$
8912 2586 id : $e/4-\pi/2$
8912 2775 Γ : $(\ln 3/2)^{-1/2}$
8912 2866 rt : (9,2,-2,3)
8912 3193 rt : (2,4,-7,-8)
8912 6034 rt : (2,9,-4,-5)
8912 6401 sπ : $(1/(3e))^{1/2}$
8912 6890 Ei : $\exp(-5/3)$
8912 8305 J_0 : $(\sqrt{5})^{-1/2}$
8912 9655 λ : $(e)^{-2}$
8913 3645 ℓ_2 : $7\ln 2/9$
8913 3667 rt : (6,6,-8,4)
8913 7069 cr : $4e/3+\pi$
8913 7346 tanh : 10/7
8913 7697 rt : (8,8,-9,-4)
8913 8845 sr : $3+\sqrt{3}/3$
8914 1266 rt : (4,3,2,-7)
8914 1354 cr : 17/24
8914 1851 sinh : $3\sqrt[3]{2}$
8914 2102 tπ : $\exp(\sqrt[3]{5}/3)$
8914 2957 rt : (9,5,-6,-5)
8914 4467 Γ : $7\sqrt{3}/9$
8914 5364 sπ : $5(e+\pi)/2$
8914 6570 sq : $3\pi^2/7$
8914 6770 rt : (5,2,-8,2)
8914 7168 10^x : $3\sqrt{3}/4+4\sqrt{5}/3$
8914 8321 sr : $8\sqrt{5}/5$
8914 8833 rt : (5,0,5,8)
8915 0013 λ : $\exp(-7\pi/11)$
8915 0037 tπ : $(\pi)^{-1/2}$
8915 1082 10^x : $2e/3$
8915 1653 $\ln(\pi)\div\exp(1/4)$
8915 2018 10^x : $1/3+3\sqrt{3}/4$
8915 2357 Γ : $3\pi/7$
8915 3045 Γ : $\sqrt{5/2}$
8915 4687 Γ : $(2e/3)^{1/2}$
8915 6368 rt : (6,3,8,9)
8915 7705 J_0 : $5\zeta(3)/9$
8915 8683 rt : (4,9,-5,7)
8915 8810 cu : $1+4e/3$
8915 9112 e^x : $3\sqrt{2}/2-\sqrt{5}$
8915 9265 id : $1/4-\pi$
8915 9435 rt : (5,8,2,9)
8916 3760 rt : (6,1,-5,-1)
8916 5272 ln : $\sqrt{2}/2+\sqrt{3}$
8916 5766 tπ : $\sqrt{11/2}$
8916 7093 rt : (1,9,9,2)
8916 7119 e^x : $3/2+3\pi/2$
8916 7310 sr : $3/4+2\sqrt{2}$
8916 8478 rt : (6,7,-4,8)
8916 9119 ℓ_{10} : $2e+3\pi/4$
8916 9145 sπ : $3\sqrt{3}/8$
8917 0759 $3\times s\pi(\sqrt{2})$
8917 1747 cu : $(\ln 3/3)^{-3}$
8917 2523 cu : $2/3-2\sqrt{3}$
8917 2673 rt : (6,6,-9,-1)
8917 3111 cu : $3\sqrt{2}/4+3\sqrt{5}$
8917 3668 cr : $4/3+2e$
8917 4703 Γ : 19/12
8917 5123 Γ : $(e/3)^{-3}$
8917 7314 ℓ_{10} : $9\sqrt{3}/2$
8918 0002 sq : $4\sqrt{3}/3+3\sqrt{5}/4$
8918 0581 rt : (7,4,-2,0)
8918 2469 $2/5+\exp(2/5)$
8918 4762 $\exp(1/5)-\exp(\sqrt{2})$
8918 5735 rt : (8,7,-7,-5)
8918 5773 tanh : $((e+\pi)/2)^{1/3}$
8918 8453 rt : (8,0,-3,3)
8919 0807 at : $1/3+e/3$
8919 0873 rt : (9,4,-4,-6)
8919 0874 as : $(\sqrt{2}/3)^{1/3}$
8919 3746 J_0 : 2/3
8919 5444 rt : (2,8,-2,-6)
8919 5898 erf : 25/22
8919 6288 rt : (9,3,0,4)
8919 6691 ln : $2e\pi/7$
8919 7014 2^x : $8\pi^2/7$
8919 7235 rt : (8,7,1,1)
8919 7530 sq : 17/18
8920 0370 rt : (4,3,-6,5)
8920 2828 rt : (3,5,1,-7)
8920 3419 Γ : $3\sqrt{5}/5$
8920 5996 10^x : $\sqrt[3]{14/3}$
8920 6205 sr : $5(1/\pi)/2$
8920 7514 2^x : $2e/3+4\pi$
8921 0181 rt : (4,2,4,-8)
8921 1529 2^x : 23/25
8921 2427 tπ : $\ln(e+\pi)$
8921 2981 rt : (2,6,-9,-8)
8921 3452 Ψ : $(\sqrt[3]{3})^{-1/2}$
8921 3979 2^x : $9\sqrt[3]{3}/2$
8921 5568 sinh : $10\pi^2/9$
8921 5625 tanh : $9(1/\pi)/2$
8921 5850 Γ : $\sqrt[3]{4}$
8921 6306 θ_3 : $4\zeta(3)/7$
8921 7504 rt : (5,1,7,9)
8921 7906 sq : $3+\pi/2$
8921 8587 rt : (1,7,5,-9)
8922 0148 cr : $4\sqrt{2}+\sqrt{5}/2$
8922 0542 rt : (4,9,-3,-7)
8922 0759 $1/3-\exp(4/5)$
8922 1821 K : $(5/2)^{-1/3}$
8922 2702 $\ln(5)\times\exp(2)$
8922 4796 rt : (3,2,5,5)
8922 6752 K : 14/19
8922 7845 rt : (7,9,-8,-5)
8923 0064 tπ : $2\sqrt[3]{5}$
8923 1468 sπ : $\exp(-\pi/3)$
8923 2258 Ψ : $\sqrt[3]{25}$
8923 2882 rt : (8,6,-5,-6)
8923 3100 rt : (5,3,-8,8)
8923 3462 as : $3\sqrt{2}-2\sqrt{3}$
8923 3669 rt : (3,6,-6,-8)
8923 5032 id : $1/3+\sqrt{5}/4$
8923 5510 θ_3 : $3\ln 2/5$
8923 7904 rt : (9,3,-2,-7)
8923 9290 rt : (4,5,8,6)
8924 0082 ln : $3-\sqrt{5}/4$
8924 0296 rt : (2,9,-2,9)
8924 1257 Ψ : $(\ln 2)^{1/2}$
8924 1431 at : $1/4-2\sqrt{5}/3$
8924 1822 10^x : $4\sqrt{3}/3$
8924 4163 2^x : $4\sqrt[3]{2}$
8924 4834 rt : (6,5,-7,-2)
8924 5526 rt : (4,4,-2,9)
8924 5844 rt : (2,7,-1,8)
8924 6202 10^x : $9\sqrt[3]{5}/8$
8924 6701 rt : (2,3,-9,-9)
8925 2468 rt : (7,2,-4,-3)
8925 3237 Γ : $9\sqrt{2}/8$
8925 4117 10^x : 21/10
8925 7420 $4\times\ln(4/5)$
8925 9019 cu : $1/4+3\sqrt{5}$
8925 9152 Γ : $5(1/\pi)$
8926 0068 rt : (8,1,-1,4)
8926 2988 e^x : $2e+2\pi/3$
8926 3043 rt : (2,7,0,-7)
8926 7634 rt : (9,4,2,5)
8926 8168 2^x : $4\sqrt{2}+4\sqrt{5}$
8927 0192 rt : (3,4,3,-8)
8927 1264 Γ : $10\zeta(3)/9$
8927 3441 sq : $1/2+4\sqrt{3}/3$
8927 4245 rt : (7,8,-6,-6)
8927 4463 Γ : $(\sqrt[3]{2}/3)^{-1/3}$
8927 4762 cu : $3/4-3\sqrt{3}$
8927 7312 rt : (4,1,6,-9)
8927 9170 rt : (8,5,-3,-7)
8927 9780 Ei : $4\sqrt[3]{3}/9$
8927 9947 10^x : $\sqrt{3}/4+2\sqrt{5}/3$
8928 2027 cr : $4\sqrt{3}/3+2\sqrt{5}$
8928 2312 sπ : $(e)^{1/2}$
8928 2705 ℓ_{10} : $4(e+\pi)/3$
8928 3350 rt : (4,4,-9,2)
8928 4080 rt : (9,2,0,-8)
8928 4757 id : $\exp(e/2)$
8928 5189 rt : (3,8,-3,-3)
8928 8546 sq : $2\sqrt{2}/3+\sqrt{3}/4$
8929 0228 sπ : $\sqrt[3]{13}$
8929 0446 sinh : $1/\ln(\sqrt[3]{2})$
8929 0729 id : $9e/5$
8929 1309 sq : $3\sqrt[3]{2}/4$
8929 1843 sπ : $\exp(\sqrt[3]{5}/2)$
8929 2222 id : $4\sqrt{2}+\sqrt{5}$
8929 7361 cr : $1/2+2\pi$
8929 7563 tπ : $1/2+2\sqrt{3}/3$
8929 7867 $\ln(\pi)\times\exp(5)$
8929 7951 Γ : 4/3
8929 9241 rt : (5,1,-6,1)
8929 9856 Γ : 3/19
8930 0932 rt : (1,9,5,-2)
8930 1328 ℓ_2 : $1/2+4\sqrt{3}$
8930 2345 at : $5\sqrt{5}/9$
8930 3630 erf : $(\sqrt{2}\pi/3)^{1/3}$
8930 6016 2^x : $2+\sqrt{5}/4$
8930 7574 rt : (5,8,-7,-8)
8930 7739 erf : $2\sqrt[3]{5}/3$
8930 8479 ℓ_2 : 7/13
8931 0080 cu : $\sqrt{2}/2+\sqrt{3}/2$
8931 1907 ζ : 5/17
8931 3102 2^x : $\sqrt{2}/2+4\sqrt{5}/3$
8931 3464 tπ : $1/2+\sqrt{3}$
8931 4718 $1/5+\ln(2)$
8931 4862 ln : $3/2+2\sqrt{2}/3$
8931 4982 rt : (6,8,-1,-3)
8931 4988 sq : $3\sqrt{3}/2-3\sqrt{5}$
8931 5424 ln : $10\sqrt[3]{5}/7$
8931 7311 at : $3-3\sqrt{2}$
8931 8458 at : $1/\ln(\sqrt{5})$
8931 9246 10^x : 16/5
8931 9334 tanh : 23/16
8931 9806 ln : $2\sqrt{3}/3-\sqrt{5}/3$
8931 9812 rt : (7,7,-4,-7)
8932 0211 Γ : $5\sqrt{5}/7$
8932 0397 λ : 3/22
8932 0562 tπ : $2\sqrt{3}/3-3\sqrt{5}$
8932 1964 rt : (1,9,-1,-7)
8932 2357 rt : (7,1,-2,-4)
8932 4384 sr : $(\pi/2)^{-1/2}$
8932 4628 rt : (8,4,-1,-8)
8932 5216 2^x : $2/3+\sqrt{3}/2$
8932 7136 sq : $3-3\sqrt{3}/4$
8932 7789 sq : $3\sqrt{2}/2-\sqrt{5}/3$
8932 8919 rt : (3,9,3,6)
8932 9431 rt : (9,1,2,-9)
8932 9701 rt : (8,2,1,5)
8933 0489 λ : $3(1/\pi)/7$
8933 0576 rt : (7,5,0,1)
8933 1112 sr : $10(e+\pi)/7$
8933 1838 Γ : $7\sqrt[3]{2}/3$
8933 2511 sπ : $1/\sqrt[3]{23}$
8933 5846 rt : (3,3,5,-9)
8933 7014 rt : (9,5,4,6)
8933 7830 e^x : $4-\pi/4$
8933 8032 Γ : $7\sqrt[3]{5}/9$
8933 9593 sr : $\ln(\sqrt{2}\pi/2)$
8934 0090 rt : (4,8,-5,-8)
8934 0648 tπ : $\sqrt[3]{2}/3$
8934 1545 ln : $7\pi/9$
8934 2746 rt : (2,0,4,5)
8934 3928 sπ : $8\sqrt[3]{3}/7$
8934 4077 10^x : $2\sqrt{2}+4\sqrt{5}/3$
8934 6027 rt : (8,8,3,2)
8934 6037 2^x : $3\sqrt{3}/2-3\sqrt{5}/4$
8934 6332 $\ln(5)+\exp(1/4)$
8934 7193 10^x : $2\ln 2/5$
8934 7243 rt : (8,9,-9,5)
8934 9245 ℓ_2 : $2e+3\pi$
8935 1200 Ei : $3\sqrt[3]{5}/8$
8935 1534 Γ : 8/5
8935 1851 cu : $5\sqrt[3]{5}/6$
8935 1933 rt : (7,9,-7,2)

8935 2514 rt : (3,5,-8,-9)
8935 3262 $s\pi : 3e/2+\pi/2$
8935 5122 $Li_2 : (\ln 2/2)^{1/3}$
8935 5304 $\ell_{10} : 7\sqrt{5}/2$
8935 6463 rt : (3,3,7,6)
8935 9594 $e^x : 17/16$
8935 9848 rt : (6,9,-5,-1)
8936 3746 $10^x : 1/4+\sqrt{5}/3$
8936 3883 sq : $\sqrt{2}+\sqrt{5}/4$
8936 4155 cu : $5\sqrt{3}/7$
8936 4322 $\Psi : \exp(-1/(2e))$
8936 4573 rt : (7,6,-2,-8)
8936 4722 rt : (1,6,-5,8)
8936 7203 $s\pi(e) \div s\pi(1/3)$
8936 7956 tanh : $8\sqrt[3]{2}/7$
8936 8867 rt : (4,9,-9,-2)
8936 9285 rt : (8,3,1,-9)
8937 1370 at : $2/3+\sqrt{3}/3$
8937 1371 $s\pi : 2/3+4\sqrt{5}/3$
8937 2207 at : $\sqrt{3}/4-3\sqrt{5}/4$
8937 6059 rt : (5,9,-3,-4)
8937 6567 $s\pi : (\sqrt[3]{2}/3)^{1/2}$
8937 6993 $\Gamma : \exp(\sqrt{2}/3)$
8937 9328 $\ell_2 : 3\sqrt{3}+\sqrt{5}$
8937 9929 $\Gamma : 10\sqrt[3]{3}/9$
8938 1787 ln : 9/22
8938 2688 $\Gamma : \sqrt[3]{7/3}$
8938 2716 $\Gamma : 4\zeta(3)/3$
8938 3221 rt : (6,3,-3,-4)
8938 3426 $\zeta : 23/16$
8938 5295 $2/3 \div s\pi(\sqrt{3})$
8938 5646 $\ln(4/5)+\exp(3/4)$
8938 6412 rt : (1,8,1,-8)
8938 6535 sq : $1/3+2\pi/3$
8938 6970 $10^x : 1/\sqrt{13}$
8938 7484 $\Gamma : (\ln 3)^3$
8938 7691 sq : $3\sqrt{2}/2+4\sqrt{3}$
8938 9585 ln : $4\sqrt{2}/3+\sqrt{5}/4$
8938 9891 cu : $\sqrt{3}/2+\sqrt{5}/4$
8939 0353 cr : 5/7
8939 0962 sinh : $\sqrt{10}\pi$
8939 3029 $t\pi : 4\sqrt{3}+\sqrt{5}/4$
8939 3154 rt : (2,5,4,-9)
8939 4134 sq : $\sqrt{2}/4-3\sqrt{3}/4$
8939 5844 $erf : 8/7$
8939 7455 rt : (8,3,3,6)
8939 8171 cu : $\sqrt{3}/2-2\sqrt{5}$
8940 0391 $5 \div \exp(1/4)$
8940 3538 cr : $3/2-\pi/4$
8940 3931 rt : (6,8,-3,-8)
8940 4251 $Ei : 17$
8940 4528 rt : (9,6,6,7)
8940 5134 $e^x : 4-3\sqrt{3}/4$
8940 5464 $\Gamma : (\sqrt[3]{5}/3)^{-1/2}$
8940 8555 rt : (7,5,0,-9)
8940 9647 $2^x : 4-3e/4$
8940 9657 $\Psi : 6\ln 2/5$
8941 1183 $s\pi : \exp(3\pi)$
8941 3185 $e^x : 7e/2$
8941 3747 cr : $5e/2$
8941 4966 tanh : $\sqrt[3]{3}$
8941 6438 $t\pi : 5\sqrt{2}/4$
8941 8216 $2^x : \sqrt{3}+\sqrt{5}/4$
8941 9350 tanh : $6\zeta(3)/5$
8942 0933 $10^x : 23/8$
8942 1364 rt : (3,6,-5,8)
8942 2565 rt : (7,6,-9,9)
8942 3894 tanh : $1/\ln(2)$
8942 3897 at : $1/2+\sqrt{5}/3$
8942 6027 $t\pi : 10e\pi/9$
8942 6612 $10^x : 6/13$
8942 6663 sq : $2\sqrt{3}-2\sqrt{5}/3$
8942 7173 $2^x : 2+\pi/4$
8942 7945 rt : (4,9,-1,-7)
8942 8332 $s\pi : 9\zeta(3)/8$
8942 9513 rt : (1,8,3,-3)
8943 0373 sq : $4/3+2e/3$
8943 1994 $10^x : 5\sqrt{2}/2$
8943 3122 as : $9\ln 2/8$
8943 3249 $\Gamma : \sqrt{3}+\sqrt{5}-\sqrt{7}$
8943 5150 $e^x : 2\sqrt{5}/7$
8943 5760 rt : (4,8,-7,-3)
8943 6329 $t\pi : 9e/2$
8943 7367 $\theta_3 : 5/12$
8943 7521 tanh : $5\sqrt{3}/6$
8943 8560 cr : $(\ln 3/3)^{1/3}$
8944 0273 $2^x : 9e\pi/10$
8944 0501 sr : $8\pi/3$
8944 1448 sr : $((e+\pi)/3)^{-1/3}$
8944 1820 $10^x : (\sqrt[3]{5}/2)^{1/3}$
8944 2719 id : $2\sqrt{5}/5$
8944 3271 $e^x : 1/2+2\sqrt{2}$
8944 6093 rt : (6,6,-7,6)
8944 7258 rt : (6,7,-1,-9)
8944 9284 rt : (1,7,3,-9)
8944 9649 rt : (6,2,-1,-5)
8945 1310 cu : $\sqrt{3}/3-4\sqrt{5}/3$
8945 1467 rt : (2,9,6,5)
8945 2485 $erf : \sqrt[3]{3/2}$
8945 2961 $erf : \ln(\pi)$
8945 3847 $2^x : 23/15$
8945 4484 $t\pi : 12/25$
8945 4578 rt : (4,7,-7,-9)
8945 5190 cu : $7\pi/3$
8945 5824 rt : (1,2,3,-5)
8945 6551 rt : (7,1,2,6)
8945 7362 rt : (6,3,-1,1)
8945 8886 tanh : 13/9
8946 0267 $e^x : 2\sqrt[3]{3}$
8946 0995 $e^x : 3e/2-4\pi/3$
8946 1304 $\Gamma : \ln(5)$
8946 2925 $2^x : 1/4+4\sqrt{3}/3$
8946 3347 tanh : $\exp(1/e)$
8946 3425 rt : (8,4,5,7)
8946 3612 $\ell_2 : 1+e$
8946 3948 ln : $9e/10$
8946 6384 sq : $3\sqrt{2}+4\sqrt{5}$
8946 8831 rt : (2,1,6,6)
8947 0022 cr : $9(1/\pi)/4$
8947 0271 rt : (9,7,8,8)
8947 1878 rt : (7,6,2,2)
8947 2895 $\Psi : (e+\pi)/2$
8947 2988 $1/4-\ln(\pi)$
8947 3684 id : 17/19
8947 3791 $\Gamma : \sqrt{2}-\sqrt{6}+\sqrt{7}$
8947 5052 sq : $3\sqrt{2}/4-3\sqrt{5}$
8947 7040 $10^x : \sqrt{3}-\sqrt{5}/4$
8947 8568 id : $\sqrt{3}/3-2\sqrt{5}$
8948 1741 rt : (3,4,9,7)
8948 1873 $s\pi : 1/\ln(2\pi/3)$
8948 2225 rt : (1,7,0,-5)
8948 2381 rt : (9,7,-9,-4)
8948 2520 cu : $7\sqrt{2}/8$
8948 3286 sr : $8\pi/7$
8948 3931 cr : $(e)^{-1/3}$
8948 4525 $e^x : 4\sqrt{2}+\sqrt{3}/3$
8948 5407 rt : (5,9,-2,-9)
8948 6062 $10^x : 3\sqrt{5}/8$
8948 6268 rt : (8,9,5,3)
8948 8257 sinh : $(e)^{1/3}$
8949 3400 sinh : $10\sqrt{3}/7$
8949 4439 $J_0 : \exp(-\sqrt[3]{2}/3)$
8949 4522 cu : $1/3-\pi/2$
8949 5494 $10^x : 8/7$
8949 7794 $\exp(\sqrt{3}) \times s\pi(1/3)$
8950 0907 rt : (4,7,-5,-4)
8950 4279 rt : (9,2,-9,0)
8950 6959 $\Gamma : 5(e+\pi)/3$
8950 7644 rt : (5,4,-2,-5)
8950 8872 cu : $7e/6$
8950 8988 $\ell_{10} : 5\pi/2$
8950 9481 $3/5 \times \exp(2/5)$
8951 0651 rt : (2,6,-8,-4)
8951 1781 $Ei : 1$
8951 2555 sr : $5\sqrt[3]{3}/9$
8951 3832 tanh : $(2\pi/3)^{1/2}$
8951 4309 at : $9\ln 2/5$
8951 4355 rt : (6,1,1,-6)
8951 6329 $s\pi : 6/17$
8951 6369 $\Gamma : (\sqrt{3})^{1/2}$
8951 8554 at : $2e-4\pi/3$
8951 8751 id : $\sqrt{2}-4\sqrt{3}/3$
8951 9342 sr : $2\zeta(3)/3$
8951 9868 $t\pi : 4\sqrt{2}/3-\sqrt{5}/2$
8952 0222 $\Gamma : 25/19$
8952 0849 cu : $2\zeta(3)$
8952 1040 rt : (7,2,4,7)
8952 5984 rt : (9,6,-7,-5)
8952 6715 $\lambda : \exp(-12\pi/19)$
8952 7383 rt : (5,7,-1,7)
8952 7698 rt : (8,5,7,8)
8953 4030 $\Gamma : 21/13$
8953 6141 $2^x : 3+4\pi/3$
8953 7054 $10^x : 4e\pi/5$
8954 0736 ln : $\zeta(3)/8$
8954 3665 $K : 17/23$
8954 4768 $\Gamma : 8\sqrt{2}/7$
8954 6808 $2 \times \exp(2/3)$
8955 0489 rt : (3,5,-8,1)
8955 1947 rt : (1,7,1,-4)
8955 3128 rt : (5,9,-6,-9)
8955 6372 at : $3\sqrt{2}/4-4\sqrt{3}/3$
8955 7838 rt : (3,9,-6,-4)
8955 7871 $\Gamma : \exp(\sqrt[3]{3}/3)$
8955 8299 $\ell_{10} : 2\sqrt{3}/3+3\sqrt{5}$
8955 8522 $\Psi : \ln 2/10$
8955 8586 $\Gamma : 7\ln 2/3$
8955 8878 $\theta_3 : 11/16$
8955 9364 $2^x : 3/2-\sqrt{3}/3$
8955 9725 $\exp(1/5)-\exp(3/4)$
8956 0531 $\ln(2) \div s\pi(e)$
8956 0895 cr : $4\sqrt{2}+2\sqrt{3}/3$
8956 1242 $1/2+\exp(1/3)$
8956 2380 $K : \ln(2\pi/3)$
8956 3602 $erf : (\pi/3)^3$
8956 4392 rt : (8,8,-8,-5)
8956 4703 ln : $1+4\sqrt{2}$
8956 5366 $\Gamma : 21/16$
8956 8352 id : $4\pi^2/5$
8956 8828 rt : (9,5,-5,-6)
8956 9619 $2^x : \sqrt{2}/2-\sqrt{3}/2$
8957 0920 rt : (5,3,0,-6)
8957 1300 rt : (9,1,-7,-1)
8957 2326 $10^x : 1/\ln(\sqrt[3]{4})$
8957 2982 $10^x : 5\sqrt[3]{2}/2$
8957 3565 $10^x : 5/18$
8957 6747 rt : (1,1,5,-6)
8957 7424 rt : (6,0,3,7)
8957 8179 rt : (5,1,-2,1)
8957 9985 rt : (9,3,-9,-4)
8958 1069 at : $\sqrt{3}-4\sqrt{5}/3$
8958 3902 rt : (7,3,6,8)
8958 7973 ln : $\sqrt{6}$
8958 8900 sr : $3/2+2\pi/3$
8958 9014 rt : (2,2,8,7)
8958 9069 rt : (8,9,-8,7)
8959 0355 rt : (8,6,9,9)
8959 1066 $\Gamma : 9\sqrt[3]{2}/7$
8959 1851 rt : (6,4,1,2)
8959 7453 $10^x : 1/\ln(2e)$
8960 1893 $s\pi : \sqrt{2}/4$
8960 3058 $\ell_{10} : 2\sqrt{2}/3+4\sqrt{3}$
8960 3636 cu : $4(e+\pi)$
8960 5094 rt : (4,3,-5,7)
8960 5411 rt : (7,7,4,3)
8960 5538 at : 5/4
8960 6596 rt : (8,7,-6,-6)
8960 6924 at : $((e+\pi)/3)^{1/3}$
8960 7475 $\zeta : 1/\ln(2e)$
8960 7744 rt : (1,7,6,-3)
8961 0940 rt : (9,4,-3,-7)
8961 2384 $erf : 23/20$
8961 2473 rt : (4,8,-9,4)
8961 4845 $\sqrt{2} \div s\pi(\sqrt{3})$
8961 5502 $2^x : 12/13$
8961 6302 $e-\exp(3/5)$
8961 9220 $s\pi : 10(1/\pi)/9$
8961 9941 rt : (3,8,-4,-5)
8962 2552 cr : $2\sqrt{3}+3\sqrt{5}/2$
8962 3330 $Ei : 22/13$

8962 5055 $\Gamma : 9\sqrt[3]{3}/8$
8962 6222 cr : $4\sqrt[3]{2}/7$
8962 6295 rt : (4,7,1,4)
8962 7773 sinh : $4\pi/9$
8962 8094 cr : 18/25
8962 8402 $1/5 \div \ln(4/5)$
8962 9829 rt : (8,3,-8,-1)
8963 2625 rt : (5,2,2,-7)
8963 2865 cu : $3\zeta(3)$
8963 3171 $2^x : \sqrt{2}/3 + 2\sqrt{5}/3$
8963 3342 $\Gamma : 17/13$
8963 3800 $\Gamma : (\sqrt[3]{5})^{1/2}$
8963 4357 cu : $\sqrt{3}/4 - 2\sqrt{5}$
8963 6364 sq : $\pi^2/5$
8963 6510 rt : (9,0,-5,2)
8963 8523 $s\pi : (3\pi/2)^3$
8963 8932 rt : (6,1,5,8)
8963 9404 cu : $1/3 - 2\sqrt{5}$
8964 0992 $\Gamma : 7\pi^2/3$
8964 1026 rt : (3,7,-8,-9)
8964 3824 rt : (7,9,-7,-6)
8964 5214 rt : (7,4,8,9)
8964 5700 $J : (\sqrt{2}\pi)^{-3}$
8964 6078 $J_0 : 15/23$
8964 6116 $Ei : \zeta(11)$
8964 8087 $e^x : 16/25$
8964 8088 rt : (8,6,-4,-7)
8965 2342 rt : (9,3,-1,-8)
8965 3550 rt : (7,9,-6,4)
8965 4376 $\Gamma : 3e\pi/4$
8965 6933 rt : (2,9,2,-4)
8965 7428 $\Gamma : 13/8$
8965 8801 tanh : 16/11
8966 0396 sinh : $\exp(e/3)$
8966 2532 $10^x : 2 + 4\sqrt{5}/3$
8966 4595 ln : $4/3 + \sqrt{5}/2$
8966 5279 $2^x : 4/3 - 2\sqrt{5}/3$
8966 5475 $t\pi : \sqrt{2} - \sqrt{3} - \sqrt{6}$
8966 7410 rt : (2,7,-9,1)
8966 8071 $10^x : 2 - \sqrt{3}/4$
8966 8770 rt : (1,6,-1,-5)
8967 0377 $\ell_2 : \sqrt{2} + 4\sqrt{3}/3$
8967 2753 $\zeta : \sqrt{3}/2$
8967 4743 id : $(\sqrt[3]{3}/2)^{1/3}$
8967 5339 $\Gamma : (\sqrt{2}\pi/2)^{1/3}$
8967 8282 $s\pi : 3\sqrt{5}/2$
8967 9275 cr : $3\zeta(3)/5$
8967 9331 $J_0 : (e+\pi)/9$
8968 0525 rt : (3,8,-2,0)
8968 3974 cr : $1/\ln(4)$
8968 4713 rt : (7,8,-5,-7)
8968 5956 sr : $1 + 3\sqrt[3]{3}/2$
8968 6336 cr : $1/4 + \sqrt{2}/3$
8968 6689 rt : (4,8,5,8)
8968 7043 rt : (7,5,-9,-1)
8968 8891 rt : (8,5,-2,-8)
8968 9305 $\exp(1/5) + s\pi(\sqrt{5})$
8969 0603 $Ei : 3\pi^2/5$
8969 1933 $erf : 1/\ln(\sqrt[3]{2}/3)$
8969 2199 rt : (1,0,7,7)
8969 2498 $\Gamma : \sqrt{2} - \sqrt{5} + \sqrt{6}$
8969 2831 rt : (5,1,4,-8)
8969 3058 rt : (9,2,1,-9)
8969 3536 rt : (8,2,-6,-2)
8969 6586 tanh : $(\sqrt{2}/3)^{-1/2}$
8969 8668 $s\pi : \sqrt{7}$
8969 8951 rt : (6,2,7,9)
8970 0003 rt : (9,1,-3,3)
8970 2066 $10^x : 1 - \pi/3$
8970 5443 rt : (1,7,1,2)
8970 6128 sr : $\ln(\sqrt{5})$
8970 6444 rt : (5,2,0,2)
8970 6767 cr : $\sqrt{3} + \sqrt{6} + \sqrt{7}$
8970 7182 sr : $1/3 + \sqrt{2}/3$
8970 8401 cu : $1 + 3e/2$
8970 9465 $2^x : \ln(\sqrt[3]{5}/2)$
8970 9665 $Ei : (\ln 2/3)^{-2}$
8971 0207 $2^x : 2\sqrt{2} - \sqrt{3}/2$
8971 0736 $Ei : \sqrt{2} + \sqrt{3} + \sqrt{6}$
8971 1431 id : $9\sqrt{3}/4$
8971 1998 ln : 3/20
8971 3164 $J_0 : 13/20$
8971 4001 $\Gamma : 2(e+\pi)/9$
8971 7281 cr : $4 + 2\sqrt{2}$
8971 8060 at : $1/\ln(\sqrt{2}\pi/2)$
8971 9252 rt : (6,5,3,3)
8971 9408 rt : (3,9,4,9)
8971 9887 $\Gamma : (\ln 2/3)^{-1/3}$
8972 0210 cr : 13/18
8972 0827 rt : (6,8,-8,-3)
8972 1492 as : $(2\pi/3)^{-1/3}$
8972 3140 $\exp(2) - \exp(2/5)$
8972 3782 $\ell_{10} : 4\sqrt{2} + \sqrt{5}$
8972 4195 rt : (9,4,-7,-3)
8972 4931 rt : (7,7,-3,-8)
8972 6133 $erf : 4\sqrt[3]{3}/5$
8972 6421 $e^x : 3e/2 - \pi/4$
8972 7048 $t\pi : 3\sqrt{2}/2 - 3\sqrt{5}/2$
8972 7518 $erf : 15/13$
8972 7978 $J_0 : 3\sqrt{3}/8$
8972 8836 $\Gamma : \sqrt[3]{13/3}$
8972 8980 $\ln(2/3) - \exp(2/5)$
8972 9024 rt : (8,4,0,-9)
8972 9645 $\lambda : \ln 2/5$
8973 1960 rt : (7,8,6,4)
8973 3294 $\zeta : (3/2)^{-3}$
8973 3653 rt : (2,9,-3,-6)
8973 4625 at : $3e/4 - \pi/4$
8973 4661 at : $(\pi/2)^{1/2}$
8973 6659 sr : 18/5
8973 7631 $\Gamma : 3e/5$
8973 8973 $\ell_{10} : 8\pi^2$
8973 9659 rt : (6,6,-6,8)
8974 5161 $2^x : 3/4 - e/3$
8974 5644 rt : (4,3,3,-8)
8974 5666 $s\pi : 2\sqrt{3}/3 + 2\sqrt{5}/3$
8974 7069 $\Gamma : 13/10$
8974 8396 cr : $4e\pi/5$
8974 9309 rt : (7,4,-7,-2)
8975 0692 sq : 18/19
8975 1607 rt : (5,0,6,9)
8975 2110 cu : $8e\pi/5$
8975 2956 $erf : 2\sqrt{3}/3$
8975 5598 rt : (8,1,-4,-3)
8975 9790 id : $2\pi/7$
8976 0476 rt : (6,9,-4,1)
8976 1723 $\Gamma : 3\sqrt{3}/4$
8976 1862 rt : (9,2,-1,4)
8976 2673 $2^x : 4\ln 2/3$
8976 2779 $\Gamma : \sqrt{22}\pi$
8976 4498 rt : (7,6,-1,-9)
8976 5293 $2^x : 4\sqrt{5}/3$
8976 5452 $\Gamma : \sqrt{8/3}$
8976 6690 sq : $\sqrt{2}/3 - 3\sqrt{5}$
8976 9149 $erf : 5\ln 2/3$
8976 9246 rt : (2,8,0,-5)
8977 1871 id : $2e\pi/9$
8977 2178 $\ell_2 : ((e+\pi)/3)^3$
8977 3054 $J_0 : (\sqrt[3]{2}/3)^{1/2}$
8977 4436 $\Gamma : 7/15$
8977 6647 $\Psi : 9(1/\pi)/8$
8977 7268 $\Gamma : 9\sqrt[3]{3}/10$
8977 7708 cr : $6\pi^2$
8978 0172 $\zeta : 13/22$
8978 0448 rt : (1,5,-3,-6)
8978 0630 $Ei : 5\pi/3$
8978 0714 rt : (1,8,-9,-9)
8978 6358 at : $2 - \sqrt{5}/3$
8978 8737 $t\pi : 5e\pi/3$
8979 1576 rt : (2,8,-1,-7)
8979 1678 $s\pi : 1/\ln(3\pi/2)$
8979 2203 cr : $7(e+\pi)/6$
8979 2964 $\ell_2 : 5\sqrt{5}/3$
8979 4447 cr : $3\sqrt{3} - 2\sqrt{5}$
8979 5098 tanh : 19/13
8979 5918 sq : 17/7
8979 7409 rt : (3,5,2,-8)
8979 8568 ln : $1/3 + 3\sqrt{2}/2$
8979 9491 rt : (6,8,-2,-9)
8979 9825 $10^x : 2\ln 2/3$
8980 0987 cr : $2e - 3\pi/2$
8980 1666 sinh : $7\sqrt{2}/4$
8980 2631 rt : (1,1,9,8)
8980 3223 rt : (4,2,5,-9)
8980 3609 $J_0 : 11/17$
8980 3881 rt : (6,6,-8,-2)
8980 8053 sinh : 25/14
8980 9999 rt : (7,3,-5,-3)
8981 1081 $e^x : 2 - e/2$
8981 1637 at : $3\sqrt{2}/2 - \sqrt{3}/2$
8981 1868 $s\pi : \pi^2/6$
8981 2482 $\Gamma : 18/11$
8981 6095 rt : (8,0,-2,4)
8981 7901 $Ei : (3\pi)^{-2}$
8981 9279 rt : (8,9,-7,9)
8982 2167 rt : (9,3,1,5)
8982 2176 cu : P_{22}
8982 3103 $2^x : (\sqrt[3]{5}/2)^{1/2}$
8982 3509 cr : $4\sqrt[3]{5}$
8982 6593 $K : (\ln 3/2)^{1/2}$
8982 8217 rt : (5,3,2,3)
8982 8787 rt : (2,9,7,9)
8982 9560 $Ei : 9/14$
8983 1030 cr : $3\sqrt{2} + 3\sqrt{3}/2$
8983 1881 $2^x : 1 - 2\sqrt{3}/3$
8983 4015 rt : (5,5,-8,1)
8983 6475 rt : (4,8,-8,7)
8983 7794 $e^x : 4\sqrt[3]{3}/9$
8983 7966 sq : $4 + e/4$
8983 8200 $\Gamma : 22/17$
8984 0250 rt : (6,6,5,4)
8984 0557 $\ell_2 : \ln(\sqrt[3]{5})$
8984 2155 $2^x : 4/3 + 2\sqrt{2}$
8984 4599 $e^x : \sqrt{22}$
8984 4725 sr : $1/4 + 3\sqrt{5}/2$
8984 5744 $s\pi(1/3) \div s\pi(\sqrt{2})$
8984 6348 rt : (8,2,-8,-3)
8984 7614 $erf : 22/19$
8984 8165 rt : (2,7,1,-8)
8985 2195 rt : (7,9,8,5)
8985 3082 $\exp(\sqrt{5}) \times s\pi(2/5)$
8985 3835 rt : (3,4,4,-9)
8985 4076 tanh : $(\pi)^{1/3}$
8985 6033 $\ell_2 : (\ln 3/3)^2$
8985 6140 ln : $4\sqrt{3} - 2\sqrt{5}$
8985 7300 rt : (5,8,-9,-2)
8985 7849 $\Psi : (5/3)^{-3}$
8985 9343 $erf : (\sqrt{5}/3)^{-1/2}$
8985 9647 sinh : $7e/6$
8985 9746 rt : (9,5,-5,-2)
8986 0370 $10^x : 13/22$
8986 1228 $1/5 - \ln(3)$
8986 1330 tanh : $(e+\pi)/4$
8986 2867 $s\pi : 3\sqrt{2} - 3\sqrt{3}/2$
8986 3255 rt : (6,5,-6,-3)
8986 3384 $\ln(4/5) - s\pi(\sqrt{5})$
8986 3709 at : $2\pi/5$
8986 4250 $J_0 : 1/\ln(3\pi/2)$
8986 4333 cu : $\sqrt{11/2}$
8986 4344 tanh : $\ln(\ln 2/3)$
8986 5792 $2 \div \exp(4/5)$
8986 8133 id : $10\pi^2/3$
8986 8419 $2^x : 3/4 + \pi/4$
8986 9189 rt : (7,2,-3,-4)
8986 9762 $2^x : 9\sqrt{3}/4$
8987 1894 rt : (5,6,-4,5)
8987 2127 $1/4 + \exp(1/2)$
8987 2133 $t\pi : 1/3 - 4\pi$
8987 3202 $\Psi : 7\sqrt[3]{2}/3$
8987 4720 $\Psi : 2/17$
8987 5101 rt : (8,1,0,5)
8987 5278 tanh : $6\sqrt[3]{5}/7$
8987 6771 rt : (2,7,-2,-6)
8987 7575 $\Psi : \exp(-2/3)$
8987 9055 $\exp(2/3) + s\pi(2/5)$
8988 0829 at : $8\sqrt{2}/9$
8988 0991 rt : (9,4,3,6)
8988 2249 $\zeta : 3\ln 2/7$
8988 2723 rt : (9,7,-8,-5)
8988 2824 $\ell_2 : 3/4 + 3\sqrt{5}$

8988 2892 sr : $\sqrt{13}$

8988 3583 e^x : $3\sqrt[3]{5}/8$

8988 4543 as : 18/23

8988 4805 sr : $5\sqrt[3]{3}/2$

8988 7393 rt : (1,7,6,-7)

8988 8065 Γ : $\sqrt{5/3}$

8988 9768 ln : $5\pi^2$

8989 2837 Ei : $\ln(2e)$

8989 3967 tanh : 22/15

8989 5608 sr : $4\sqrt{2}/7$

8989 7948 id : $\sqrt{24}$

8989 7978 cu : $\sqrt{2}/2 - 4\sqrt{5}$

8989 8332 10^x : $7(1/\pi)/8$

8989 9202 sr : $3\zeta(3)$

8989 9808 sinh : $4\sqrt{2}/7$

8990 1317 10^x : $3/2 + 3e/2$

8990 2138 sinh : $5\sqrt{5}/8$

8990 2872 θ_3 : $(\sqrt{2}\pi/3)^{-1/3}$

8990 3480 rt : (2,6,3,7)

8990 5555 Γ : 23/14

8990 7968 10^x : $9\pi^2/5$

8990 9191 rt : (5,7,0,9)

8990 9293 rt : (7,8,-9,2)

8990 9610 rt : (2,5,-5,-1)

8991 1064 sq : $4\sqrt{2} + \sqrt{5}/2$

8991 1115 sr : $1/4 + 3e$

8991 2274 cu : $6(e + \pi)$

8991 4922 sinh : $((e + \pi)/3)^{1/2}$

8991 5410 rt : (5,8,-6,-6)

8991 7013 ζ : $4\sqrt{5}/5$

8991 7338 rt : (8,9,-9,-5)

8991 7477 ℓ_{10} : $1 + 4\sqrt{3}$

8991 9118 rt : (1,7,2,9)

8991 9239 cu : $\sqrt{2} + \sqrt{3}/3$

8992 1189 rt : (9,6,-6,-6)

8992 1520 Γ : $(\sqrt{2}\pi)^{1/3}$

8992 3573 e^x : $e + \pi/2$

8992 6175 sr : $7\ln 2/6$

8992 6947 rt : (7,1,-1,-5)

8992 8862 cr : 8/11

8993 0835 10^x : $3/4 - \sqrt{2}/3$

8993 2180 J_0 : 9/14

8993 2685 rt : (8,2,2,6)

8993 3174 id : $\exp(-1/(3\pi))$

8993 3680 cr : P_{73}

8993 4378 Ψ : $-\sqrt{5} - \sqrt{6} - \sqrt{7}$

8993 4757 sr : $(\sqrt[3]{4}/3)^{1/3}$

8993 6009 Γ : $\pi^2/6$

8993 7142 rt : (7,9,-5,6)

8993 8027 2^x : $2\sqrt{2} - 4\sqrt{5}/3$

8993 8403 rt : (9,5,5,7)

8994 2254 rt : (4,2,-8,5)

8994 2916 J : $\exp(-17\pi/3)$

8994 3247 sinh : $6\sqrt{5}$

8994 3985 sinh : $(\ln 3/3)^{-1/3}$

8994 4100 rt : (5,4,4,4)

8994 5371 sr : $1/4 + \sqrt{5}/4$

8994 6766 $\ln(3) \div \exp(1/5)$

8994 6903 10^x : $2\pi/7$

8994 7005 2^x : $\sqrt{3} + \sqrt{6} - \sqrt{7}$

8994 7344 e^x : $\sqrt{2}/3 - \sqrt{3}/3$

8994 7553 sπ : $(\sqrt{2}\pi)^{1/3}$

8994 9493 id : $7\sqrt{2}$

8995 2190 rt : (1,8,2,-9)

8995 3523 Li_2 : 12/17

8995 3564 10^x : $2 + 3e/2$

8995 4883 rt : (3,5,-9,5)

8995 5268 rt : (8,8,-7,-6)

8995 5436 rt : (6,7,7,5)

8995 5988 tanh : $7\sqrt[3]{2}/6$

8995 7962 e^x : $e/4 - \pi/4$

8995 8504 e^x : $3 + 3\pi$

8995 9033 sq : $8e$

8995 9043 rt : (9,5,-4,-7)

8996 1465 ℓ_{10} : $2\sqrt{3} + 2\sqrt{5}$

8996 2513 rt : (7,0,-9,3)

8996 3760 cu : $9\pi/7$

8996 5666 ℓ_{10} : $\sqrt[3]{2}/10$

8996 6505 rt : (4,9,-8,-3)

8996 7216 sπ : $3\pi/4$

8996 8493 λ : $\sqrt[3]{2}/9$

8996 8778 rt : (2,8,-8,4)

8996 8883 sπ : $3e/2 + 4\pi$

8996 8960 tanh : 25/17

8996 9726 rt : (3,9,-1,-6)

8997 0394 id : $1/3 + 4\pi$

8997 2136 rt : (5,9,-2,-2)

8997 2629 ℓ_2 : $3/4 + 4\sqrt{5}/3$

8997 3541 sr : 17/21

8997 3736 sinh : $7\ln 2/6$

8997 4717 Γ : 9/7

8997 4977 $\exp(\pi) \div s\pi(e)$

8997 5110 rt : (8,3,-6,-2)

8997 6704 ln : $4e - 4\pi/3$

8997 7596 2^x : $3/2 + 3\sqrt{2}/4$

8997 7718 Ei : $5\zeta(3)/6$

8997 7747 rt : (6,3,-2,-5)

8997 8506 rt : (4,3,-4,9)

8997 9521 cu : $2\sqrt{2} + \sqrt{5}$

8997 9879 rt : (2,6,-4,-7)

8998 1322 erf : $(\pi/2)^{1/3}$

8998 1435 J_0 : $3\sqrt[3]{5}/8$

8998 1477 e^x : $5e/4$

8998 3339 rt : (7,0,1,6)

8998 3577 2^x : $(\sqrt[3]{2})^{-1/3}$

8998 4213 Ψ : $(\sqrt[3]{5}/3)^{1/3}$

8998 4364 sq : $(\ln 2/3)^{-2}$

8998 4584 e^x : $3/2 + 2\sqrt{5}/3$

8998 5846 Γ : $8\sqrt[3]{3}/7$

8998 7044 sr : $3e/4 + \pi/2$

8998 7604 rt : (9,6,-3,-1)

8998 8172 $\ln(3/5) - \exp(2)$

8998 8373 Γ : $(\sqrt{2}/3)^{-1/3}$

8998 8775 J_0 : $4\sqrt[3]{3}/9$

8998 8912 rt : (8,3,4,7)

8998 9182 rt : (6,5,-9,6)

8998 9603 K : $3\sqrt{3}/7$

8998 9969 rt : (1,3,-7,-8)

8999 0834 at : $\sqrt[3]{2}$

8999 1653 rt : (1,9,-8,9)

8999 2394 Γ : $(e)^{1/2}$

8999 2602 rt : (8,7,-5,-7)

8999 4360 id : $5\sqrt[3]{2}/7$

8999 4465 rt : (9,6,7,8)

8999 4501 sinh : $(\sqrt[3]{4}/3)^{1/3}$

8999 6305 rt : (9,4,-2,-8)

8999 6862 ℓ_2 : $2 + \sqrt{3}$

8999 7327 id : $10\sqrt[3]{5}/9$

9000 0000 id : 9/10

9000 3558 sq : $4 + \pi/4$

9000 4834 sπ : $\exp(\sqrt{5})$

9000 5254 rt : (1,4,-7,-8)

9000 5653 id : $2\sqrt{3}/3 + \sqrt{5}/3$

9000 5753 2^x : $5\pi/8$

9000 8708 as : $(\sqrt[3]{3}/3)^{1/3}$

9000 9222 Ei : 3/16

9000 9707 $\exp(e) - s\pi(\sqrt{3})$

9001 0408 Γ : $7\sqrt{2}/6$

9001 0679 rt : (3,7,-7,-6)

9001 1072 rt : (1,7,-6,-1)

9001 2467 rt : (1,8,-7,9)

9001 4506 ln : $3\sqrt{3} + 2\sqrt{5}/3$

9001 4657 ℓ_2 : $\sqrt[3]{13/2}$

9001 8854 at : $\sqrt{2}/3 - \sqrt{3}$

9001 9180 J_0 : 16/25

9001 9316 cu : $1/3 + 4\sqrt{3}$

9002 0109 id : $2e/3 - 3\pi/2$

9002 1141 e^x : $7e\pi/5$

9002 1600 10^x : $2\pi^2/3$

9002 2070 rt : (4,8,-6,-4)

9002 5512 sq : $\sqrt{2}/4 - \sqrt{3}$

9002 5719 rt : (7,9,-6,-7)

9002 5896 10^x : $2 + 3\sqrt{2}/4$

9002 6298 Γ : $\sqrt[3]{9/2}$

9002 6773 rt : (1,9,1,1)

9002 7181 J : $\exp(-19\pi/7)$

9002 7540 rt : (5,5,-3,-5)

9002 9229 Γ : $3\sqrt[3]{5}/4$

9002 9359 rt : (8,6,-3,-8)

9003 0249 cr : P_{23}

9003 1883 10^x : 22/3

9003 2063 id : $4/3 - \sqrt{3}/4$

9003 2992 rt : (9,3,0,-9)

9003 3408 sq : $e\pi/9$

9003 3465 λ : $\exp(-5\pi/8)$

9003 7460 Γ : $8\sqrt[3]{3}/9$

9003 7861 2^x : $3 - \sqrt{2}/2$

9003 8425 rt : (7,1,3,7)

9004 0879 at : $3\sqrt{2} - 4\sqrt{5}/3$

9004 1544 sπ : $\sqrt{3}/2 + 2\sqrt{5}/3$

9004 2528 ln : $\sqrt{3} - 3\sqrt{5}/4$

9004 3841 rt : (8,4,6,8)

9004 4256 sq : $1/4 - 2e$

9004 4782 as : $\sqrt{2}/2 - 2\sqrt{5}/3$

9004 5164 rt : (1,7,-8,-7)

9004 5939 rt : (3,8,-1,3)

9004 8453 K : $7(1/\pi)/3$

9004 9238 rt : (9,7,9,9)

9004 9572 Γ : $(\ln 3/3)^{-1/2}$

9004 9890 10^x : $2e/3 + \pi/4$

9005 3293 J_0 : $2\sqrt{5}/7$

9005 3474 sr : $3\sqrt{2}/2 + 2\sqrt{5}/3$

9005 4617 rt : (5,5,6,5)

9005 5034 erf : $3e/7$

9005 7187 rt : (3,9,-9,6)

9005 7719 Ei : $\zeta(9)$

9006 0246 cu : $1/3 - 3\sqrt{3}/4$

9006 1426 sπ : $1/\sqrt[3]{22}$

9006 1983 rt : (7,8,-4,-8)

9006 5321 rt : (6,8,9,6)

9006 5555 rt : (8,5,-1,-9)

9006 5788 K : $(2e/3)^{-1/2}$

9006 9238 rt : (3,8,-3,-7)

9006 9882 at : $7\sqrt[3]{3}/8$

9007 5823 sr : $7\zeta(3)$

9007 6362 rt : (4,7,-4,-5)

9007 7894 Ψ : $3\sqrt[3]{5}/10$

9007 8842 cr : $4\sqrt{2}/3 - 2\sqrt{3}/3$

9007 8897 rt : (2,5,-6,-8)

9007 9013 ℓ_{10} : P_{53}

9007 9524 erf : $(e/2)^{1/2}$

9008 0488 sπ : $5\sqrt{2}/3$

9008 1574 Γ : $(2\pi/3)^{1/3}$

9008 1680 rt : (5,4,-1,-6)

9008 2411 sq : $(\sqrt[3]{5}/2)^{1/3}$

9008 3836 2^x : $4/3 + 4\sqrt{5}/3$

9008 5103 rt : (7,1,-7,-2)

9008 6980 rt : (6,1,2,-7)

9008 7915 $\exp(1/4) \div s\pi(\sqrt{5})$

9008 8392 sinh : 17/21

9008 8500 rt : (1,2,-9,-9)

9009 2262 rt : (7,2,5,8)

9009 4186 rt : (6,8,-7,-1)

9009 6065 e^x : $7(e + \pi)/5$

9009 6886 sπ : 5/14

9009 7528 rt : (8,5,8,9)

9009 7701 rt : (7,7,-2,-9)

9009 9629 e^x : $\sqrt{2}/3 + \sqrt{5}/2$

9010 0002 erf : $(\sqrt[3]{4})^{1/3}$

9010 3048 sq : $4 - \pi/2$

9010 3122 $\exp(4/5) + s\pi(\sqrt{5})$

9010 3984 erf : 7/6

9010 5874 sπ : $\sqrt[3]{5/2}$

9010 6141 cu : $10\zeta(3)$

9010 8580 rt : (9,7,-1,0)

9011 0568 Γ : 23/18

9011 1000 Γ : $4\sqrt{5}/7$

9011 3609 sπ : $1/3 + 4\sqrt{3}/3$

9011 3816 Ψ : $(\ln 3)^{-2}$

9011 5738 at : 24/19

9011 9043 rt : (1,6,-4,-2)

9012 1660 J_0 : $2(1/\pi)$

9012 4248 rt : (3,9,-5,-5)

9012 4853 J_0 : $\ln(\sqrt[3]{4}/3)$

9012 4939 sr : $(\ln 2/2)^{-3}$

9012 5422 ℓ_{10} : $4\sqrt{2} + 4\sqrt{3}/3$

9012 9405 J_0 : 7/11

9012 9437 rt : (6,9,-3,3)

9013 0318 sπ : $3\sqrt{2}/2 + \sqrt{5}$

9013 1058 cr : $3\sqrt{3} + 3\sqrt{5}/4$

9013 2755 $\Gamma : (\sqrt[3]{3}/3)^{-1/3}$
9013 4609 rt : (5,3,1,-7)
9013 6373 sq : $9\pi^2/10$
9013 6726 ln : $4\sqrt{2}/3 + \sqrt{3}/3$
9013 8771 $2 - \ln(3)$
9013 8781 sr : 13/16
9013 9717 rt : (2,8,-7,9)
9013 9765 rt : (6,0,4,8)
9014 0325 cu : $3/4 + 2\pi$
9014 0598 id : $5\sqrt[3]{3}/8$
9014 1837 $10^x : 1/\ln(\sqrt{2}\pi/2)$
9014 3159 cr : $(e + \pi)/8$
9014 4903 rt : (7,3,7,9)
9014 5417 rt : (2,8,0,-3)
9015 0026 rt : (4,3,5,5)
9015 2070 $\ell_2 : 3 + 2\sqrt{5}$
9015 3309 ln : $3/2 + 3\sqrt{3}$
9015 3669 sr : $4\sqrt{3} + 2\sqrt{5}/3$
9015 4267 id : $3\zeta(3)/4$
9015 4609 $\ell_2 : 7e\pi$
9015 8028 cr : $3\sqrt[3]{5}/7$
9015 8366 rt : (8,4,-9,-1)
9015 8697 cu : $2 + 4\sqrt{2}$
9015 9477 tπ : $5(e + \pi)/3$
9016 0221 rt : (5,6,8,6)
9016 2788 $e^x : 1/4 - \sqrt{2}/4$
9016 3735 rt : (9,1,-6,-2)
9016 4917 rt : (3,7,-5,-8)
9016 5957 ln : $8(e + \pi)/7$
9016 5977 $Li_2 : \sqrt{2}/2$
9016 6292 J : 8/25
9016 9130 sπ : $e/2 + 2\pi$
9017 0977 $10^x : 3\sqrt{2}/4 + \sqrt{5}/2$
9017 4122 rt : (2,4,-8,-9)
9017 6297 rt : (3,8,-3,-6)
9017 6756 at : $4/3 - 3\sqrt{3}/2$
9017 7974 cr : 11/15
9017 8823 rt : (7,8,-8,4)
9017 9843 $2^x : 9\sqrt[3]{2}/2$
9017 9863 cu : $5e\pi/2$
9017 9961 rt : (1,8,7,-4)
9018 0029 sπ : $4\sqrt{2}/3 + 2\sqrt{5}$
9018 1346 rt : (4,7,2,6)
9018 4746 $2^x : P_{34}$
9018 6379 rt : (5,2,3,-8)
9018 8052 rt : (1,7,-7,1)
9018 8459 sπ : $4\sqrt{2} - 3\sqrt{3}/4$
9019 0187 rt : (6,1,-8,2)
9019 0714 e^x : 9/14
9019 1303 $\Gamma : 4(1/\pi)$
9019 1396 rt : (6,1,6,9)
9019 2670 tπ : $(\sqrt{2}\pi/2)^{-1/3}$
9019 5064 rt : (5,5,-7,3)
9019 5352 $\exp(\sqrt{3}) \div \exp(2/3)$
9019 7211 sq : $2/3 - 4\sqrt{2}$
9019 9383 $\Gamma : 9\sqrt{2}/10$
9020 0557 Γ : 14/11
9020 1325 rt : (7,2,-5,-1)
9020 1705 sr : $4\sqrt{2}/3 + \sqrt{3}$
9020 2417 $e^x : 9\pi/10$
9020 4389 rt : (7,9,-4,8)
9020 4477 $2^x : 4/3 + 3\pi/4$
9020 5026 cu : $7e\pi/5$
9020 5121 sinh : $2\pi^2/5$
9020 5665 $\zeta : 7\sqrt[3]{2}/10$
9020 7038 $2^x : 2 + 3\sqrt{5}/2$
9020 9252 cu : P_{50}
9020 9693 cu : $4 - 2\pi$
9021 0569 $J : \zeta(3)/7$
9021 1293 sπ : $7e\pi/9$
9021 1303 $2 \times s\pi(2/5)$
9021 2382 rt : (8,5,-2,0)
9021 3720 $\lambda : \sqrt{2}/10$
9021 4275 rt : (3,5,-8,8)
9021 5803 rt : (9,8,-9,-5)
9021 6371 $4 \times \exp(4/5)$
9021 7031 $K : (e/3)^3$
9021 7598 rt : (9,0,-4,3)
9021 8570 tπ : $5\sqrt[3]{3}/2$
9022 0935 $\Psi : 3(1/\pi)/10$
9022 1306 as : $1/3 - \sqrt{5}/2$
9022 2141 $\lambda : 4(1/\pi)/9$
9022 2289 rt : (1,5,-2,-3)
9022 2963 tπ : $1/3 + \sqrt{3}/4$
9022 3359 rt : (9,8,1,1)
9022 3367 $1/4 + \exp(\sqrt{3})$
9022 6759 sπ : $9(1/\pi)/8$
9022 6986 $\ell_2 : P_{65}$
9022 7220 rt : (3,7,-1,-7)
9022 8205 rt : (8,8,-9,9)
9022 8860 rt : (5,6,-3,7)
9022 9367 $10^x : P_{72}$
9022 9840 at : $\sqrt{2}/2 + \sqrt{5}/4$
9023 0693 $\ell_2 : P_{66}$
9023 2136 rt : (4,4,2,-8)
9023 2510 $10^x : P_{69}$
9023 2904 tanh : $6\sqrt{3}/7$
9023 3113 $\ln(5) - s\pi(1/4)$
9023 3347 $\ln(3/5) + \exp(5)$
9023 3459 $10^x : e/4 + 3\pi/2$
9023 7036 rt : (5,1,5,-9)
9023 7865 sq : $3e/2 + \pi/2$
9023 7919 $\ln(2/3) \times \exp(4/5)$
9024 1096 Ei : 15/11
9024 1922 rt : (3,1,4,5)
9024 4604 $10^x : (\sqrt[3]{3}/3)^{-1/3}$
9024 4850 $2^x : 2\sqrt{3} + 3\sqrt{5}$
9024 4937 $10^x : 4 + 2\sqrt{3}/3$
9024 7239 $J_0 : (5/2)^{-1/2}$
9024 7461 sq : $5\sqrt[3]{5}/9$
9024 8197 rt : (4,9,-4,-8)
9024 8499 $\ell_{10} : 3\sqrt{2}/4 + 4\sqrt{3}$
9024 9515 tπ : $\pi^2/8$
9025 0000 sq : 19/20
9025 0467 rt : (9,7,-7,-6)
9025 0690 at : 19/15
9025 1647 rt : (4,4,7,6)
9025 2275 rt : (6,5,-8,8)
9025 6597 $erf : (e + \pi)/5$
9025 7030 rt : (3,6,-7,-9)
9026 0067 rt : (7,5,-8,-2)
9026 1684 $10^x : 4\sqrt{2} + 2\sqrt{5}/3$
9026 5154 rt : (8,2,-5,-3)
9026 5391 sq : $7\pi/2$
9026 6125 $\zeta : 2\sqrt{5}/5$
9026 8340 ln : $\sqrt[3]{15}$
9026 8536 sπ : $1/2 + \pi$
9027 0225 rt : (9,1,-2,4)
9027 0276 cu : $e + \pi/2$
9027 2045 ln : $\ln(3/2)$
9027 2150 $erf : \exp(1/(2\pi))$
9027 2262 rt : (2,9,-2,-7)
9027 3464 id : $1/3 - \sqrt{5}$
9027 3580 J_0 : 12/19
9027 4323 sinh : $3(e + \pi)$
9027 4529 Γ : 5/3
9027 7064 rt : (3,6,1,-8)
9027 7288 id : $(e/2)^{-1/3}$
9027 8804 $\lambda : \exp(-(e + \pi)/3)$
9027 9094 sπ : $\sqrt{2} + 4\sqrt{5}$
9027 9727 id : $7e/10$
9028 1322 rt : (8,9,-8,-6)
9028 1852 rt : (4,3,4,-9)
9028 4127 tπ : $4\sqrt{3}/3 - \sqrt{5}/3$
9028 4619 rt : (9,6,-5,-7)
9028 5230 $\sqrt{3} \times \ln(3)$
9028 5837 $\Gamma : (3/2)^3$
9028 8644 $\exp(1/2) + s\pi(\sqrt{3})$
9029 0764 rt : (5,3,-9,2)
9029 1861 rt : (4,2,-7,7)
9029 2428 $e^x : 1 + \pi$
9029 7670 sr : $3/2 + 3\sqrt{2}/2$
9029 8044 $\theta_3 : \sqrt[3]{2}/3$
9029 9082 cr : $9e$
9029 9903 at : $3 - \sqrt{3}$
9029 9948 $\Psi : \exp((e + \pi)/2)$
9030 1272 $J : 5\ln 2/9$
9030 1314 rt : (6,0,-6,1)
9030 4183 sr : $3e/10$
9030 4728 rt : (1,7,8,9)
9030 6798 $2^x : 4\sqrt{2} + 3\sqrt{5}/2$
9030 6828 rt : (6,7,-9,-2)
9030 8753 rt : (2,6,-5,8)
9030 8802 at : $4e/3 - 3\pi/4$
9030 8998 ℓ_{10} : 8
9031 1791 rt : (7,4,-6,-3)
9031 2216 Γ : 19/15
9031 2813 $e^x : 2/3 + 3e/2$
9031 3679 sr : $3/2 + 4\sqrt{3}$
9031 5035 rt : (8,8,-6,-7)
9031 6021 sπ : $3\sqrt[3]{5}/8$
9031 6493 cu : $3\sqrt{2}/2 - 2\sqrt{3}/3$
9031 6541 ln : $\pi^2/4$
9031 6738 rt : (8,1,-3,-4)
9031 8273 rt : (9,5,-3,-8)
9032 0107 cr : $(5/2)^{-1/3}$
9032 1192 rt : (2,8,0,-8)
9032 1570 cr : 14/19
9032 1669 rt : (9,2,0,5)
9032 2125 $J_0 : \sqrt[3]{2}/2$
9032 2183 cu : $1/3 - 2e$
9032 2196 rt : (8,6,0,1)
9032 2976 $\Gamma : \sqrt{2} - \sqrt{3} - \sqrt{5}$
9032 4540 $10^x : 3\sqrt{2} - \sqrt{3}/3$
9032 5872 rt : (3,5,3,-9)
9032 5920 id : $\sqrt{2}/2 + 3\sqrt{3}$
9032 6574 $2^x : P_{17}$
9032 6803 rt : (1,7,-6,9)
9033 0538 rt : (2,5,7,9)
9033 0547 $10^x : \sqrt{5}/8$
9033 2116 $\Psi : (\ln 2/3)^{-2}$
9033 2531 rt : (9,9,3,2)
9033 3124 ln : $3\sqrt{5}$
9033 3911 as : $\pi/4$
9033 4884 sπ : $\sqrt{19}$
9033 6272 $e^x : 4 - 3\sqrt{2}/4$
9033 7378 rt : (4,8,-6,-9)
9033 8643 cu : $3e/2 - 3\pi$
9033 8793 $\sqrt{2} + \ln(3/5)$
9033 9030 2^x : 13/14
9033 9829 rt : (3,2,6,6)
9034 0457 $10^x : \sqrt{3}/2 + 2\sqrt{5}$
9034 0752 tanh : $(\sqrt{2}\pi/2)^{1/2}$
9034 5566 tanh : $2\sqrt{5}/3$
9034 8014 $\Gamma : \sqrt[3]{14/3}$
9034 8262 rt : (8,7,-4,-8)
9034 8295 $\ell_{10} : 3\sqrt{3}/4 + 3\sqrt{5}$
9034 8522 sπ : $4\sqrt[3]{3}/9$
9034 9065 rt : (4,5,9,7)
9035 1443 rt : (9,4,-1,-9)
9035 3253 rt : (3,7,-6,-3)
9035 4940 $e^x : 3 - \sqrt{2}/2$
9035 6430 tanh : $\ln(\sqrt{2}\pi)$
9035 7540 rt : (6,6,-7,-3)
9036 0200 sr : $\sqrt{2/3}$
9036 2368 rt : (7,3,-4,-4)
9036 3321 $J : \exp(-17\pi/7)$
9036 4662 $J_0 : 4\sqrt{2}/9$
9036 5379 $\exp(1/5) \times \exp(\sqrt{3})$
9036 7182 rt : (8,0,-1,5)
9036 7279 $10^x : 4\pi/9$
9036 7462 sπ : $e/2$
9036 7746 $\ell_2 : \sqrt{14}$
9036 9121 rt : (2,7,2,-9)
9036 9553 cu : $5e/8$
9037 1170 $e^x : 3 - 3\pi/4$
9037 1264 $J_0 : \pi/5$
9037 1699 sq : $1/\ln(\pi/2)$
9037 1980 rt : (9,3,2,6)
9037 5885 at : $3e - 3\pi$
9037 7884 rt : (7,9,-5,-8)
9037 7933 sr : $4e/3$
9037 7988 $\pi \times \ln(3/4)$
9037 8518 as : $5\sqrt{2}/9$
9037 8705 Γ : 24/19
9037 9157 Li_2 : 17/24
9038 1015 rt : (8,6,-2,-9)
9038 2149 cu : $1/3 + \pi/2$
9038 4373 erf : 20/17
9038 4881 rt : (3,8,0,6)

9038 4998 as : $11/14$
9039 5492 ln : $2+3\pi/2$
9039 5781 rt : (1,8,0,-6)
9039 6253 $\pi \div \ln(\sqrt{5})$
9039 6723 cu : $2+e/2$
9039 7201 rt : (5,2,-7,1)
9039 8159 rt : (1,9,4,-3)
9039 8335 cu : $1/3+e/3$
9039 8802 tanh : $1/\ln((e+\pi)/3)$
9039 9380 10^x : $6(e+\pi)/7$
9040 1157 Γ : $2(e+\pi)/7$
9040 1555 Γ : $7\sqrt[3]{3}/8$
9040 2185 tanh : $\sqrt[3]{10/3}$
9040 2433 rt : (5,8,-8,-3)
9040 3307 ζ : $25/14$
9040 3944 $\exp(2/5)-s\pi(1/5)$
9040 6617 2^x : $1/3+2\sqrt{3}$
9040 7148 rt : (6,7,-3,-4)
9040 7265 2^x : $4(e+\pi)/3$
9040 8725 tπ : $2\sqrt{3}-\sqrt{5}/2$
9041 0231 rt : (7,8,-3,-9)
9041 1282 sinh : $10\sqrt[3]{2}/9$
9041 1623 rt : (1,9,1,-9)
9041 1848 rt : (7,2,-2,-5)
9041 2966 id : $1/3+\pi/2$
9041 4974 cr : $17/23$
9041 6031 erf : $(\sqrt[3]{3}/2)^{-1/2}$
9041 6085 rt : (1,3,2,5)
9041 6534 rt : (8,1,1,6)
9041 7030 rt : (7,4,-1,1)
9041 8798 e^x : $e/3+\pi/2$
9042 0451 cr : $\ln(2\pi/3)$
9042 1206 rt : (9,4,4,7)
9042 2392 rt : (1,8,8,3)
9042 3798 $4 \div \ln(3/4)$
9042 4141 K : $\sqrt{5}/3$
9042 4972 rt : (2,0,5,6)
9042 5404 Ψ : $-\sqrt{3}+\sqrt{5}+\sqrt{6}$
9042 6224 $1/4-\exp(e)$
9042 6848 rt : (8,7,2,2)
9043 0063 at : $4\sqrt{2}-4\sqrt{3}$
9043 0150 sinh : $7/5$
9043 0273 erf : $3\pi/8$
9043 0420 tanh : $(\sqrt{5})^{1/2}$
9043 0821 e^x : $2+3e/2$
9043 0921 10^x : $\exp(\sqrt{3})$
9043 1852 ℓ_2 : $3e-2\pi$
9043 3222 rt : (7,8,-7,6)
9043 3774 tπ : $\sqrt{13}$
9043 3809 rt : (3,3,8,7)
9043 5635 id : $3\sqrt{3}+3\sqrt{5}$
9043 8390 tπ : $4e/3+\pi$
9043 9098 rt : (6,8,-6,1)
9044 0248 Γ : $(3\pi/2)^{1/3}$
9044 0739 tπ : P_{29}
9044 1251 Γ : $\sqrt[3]{2}$
9044 1722 id : $\sqrt{2}/3+\sqrt{3}/4$
9044 1929 erf : $5\sqrt{2}/6$
9044 1980 ℓ_2 : $\zeta(3)/9$
9044 2228 ℓ_{10} : $2\sqrt{2}+3\sqrt{3}$
9044 2305 sπ : $3\sqrt{2}/4+3\sqrt{3}/4$
9044 2845 cr : $e+4\pi/3$
9044 3018 ℓ_2 : $4\sqrt{3}+\sqrt{5}/4$
9044 6129 sq : $3-\pi/4$
9044 6446 tanh : $7\sqrt[3]{5}/8$
9044 9042 Γ : $3\sqrt{5}/4$
9045 0849 $s\pi(2/5) \times s\pi(3/5)$
9045 1105 rt : (5,8,-5,-4)
9045 1277 cr : $3\sqrt{3}/4-\sqrt{5}/4$
9045 3403 sr : $9/11$
9045 4265 λ : $1/7$
9045 4523 sπ : $8\sqrt[3]{2}/3$
9045 4725 10^x : $3-4\sqrt{3}/3$
9045 5246 2^x : $7\sqrt{2}/2$
9045 5697 rt : (6,4,-3,-5)
9045 5925 ζ : $3/10$
9045 7633 sr : $3+2e$
9045 8018 cr : $7\pi^2/10$
9046 0275 rt : (7,1,0,-6)
9046 2465 $\sqrt{5}-\exp(\pi)$
9046 4839 rt : (8,2,3,7)
9046 6764 Ei : $5\pi^2/6$
9046 6841 θ_3 : $\exp(-\sqrt{3}/2)$
9046 9389 rt : (9,5,6,8)
9047 0053 id : $1/4-2\sqrt{3}/3$
9047 0222 J_0 : $5/8$
9047 0608 rt : (6,9,-2,5)
9047 0643 Ψ : $5\sqrt[3]{3}$
9047 0881 2^x : $1/4+e/4$
9047 4391 rt : (2,8,1,1)
9047 6190 id : $19/21$
9047 8662 ln : $2+\sqrt{2}/3$
9047 9687 sπ : $2\sqrt[3]{2}/7$
9048 1439 sr : $4\sqrt{3}/3-2\sqrt{5}/3$
9048 2705 sπ : $9/25$
9048 2708 at : $14/11$
9048 3244 ln : $4+e$
9048 3741 $\exp(1/2) \div \exp(3/5)$
9048 4372 sq : $1/3-4\sqrt{3}/3$
9048 4548 id : $1/4-3e$
9048 4571 2^x : $20/13$
9048 5187 at : $9\sqrt{2}/10$
9048 6834 Ei : $1/\ln(3\pi/2)$
9048 6906 Ψ : $6\zeta(3)$
9048 9296 rt : (4,8,-4,-9)
9048 9373 sinh : $13/16$
9049 0198 rt : (1,8,2,-4)
9049 2554 id : $\sqrt{2}+2\sqrt{5}/3$
9049 3126 2^x : $3e\pi/10$
9049 4269 rt : (4,9,-7,-4)
9049 6234 J_0 : $\exp(-\sqrt{2}/3)$
9049 7134 Γ : $8\sqrt{2}/9$
9049 7325 sr : $(\ln 3/2)^{1/3}$
9049 7666 10^x : $1/3+3\sqrt{3}/2$
9049 8268 Γ : $4\sqrt[3]{2}/3$
9049 8380 $\exp(2/5)+\exp(5)$
9049 8756 rt : (5,9,-1,0)
9050 0285 $\ln(2)-\exp(4)$
9050 2211 sq : $3\sqrt{3}-4\sqrt{5}/3$
9050 2257 at : $4(1/\pi)$
9050 2767 2^x : $1+4\sqrt{2}$
9050 3229 rt : (6,3,-1,-6)
9050 4925 J_0 : $9\ln 2/10$
9050 5899 Γ : $2\pi/5$
9050 7267 rt : (1,2,4,-6)
9050 7690 rt : (7,0,2,7)
9050 8157 rt : (6,2,-2,1)
9050 8462 id : $2\sqrt{3}-\sqrt{5}/4$
9051 0285 ℓ_2 : $3+\sqrt{5}/3$
9051 0618 2^x : $\ln(\sqrt{3}/2)$
9051 2137 rt : (8,3,5,8)
9051 2758 e^x : $1/3-\sqrt{3}/4$
9051 2920 sπ : $4\pi^2/7$
9051 4825 tanh : $3/2$
9051 4865 id : $\sqrt{3}/4+2\sqrt{5}$
9051 5732 rt : (2,1,7,7)
9051 6572 rt : (9,6,8,9)
9051 7500 rt : (7,5,1,2)
9051 8105 sinh : $(\sqrt{2}/3)^{-3}$
9052 1165 sr : $\sqrt{3}+3\sqrt{5}$
9052 2820 as : $\sqrt{2}/4+\sqrt{3}/4$
9052 6301 cu : $4+\sqrt{3}/3$
9052 6785 rt : (8,8,4,3)
9052 7388 $\exp(5) \div s\pi(\pi)$
9052 9469 rt : (5,5,-6,5)
9053 0106 10^x : $3e+3\pi/2$
9053 3907 ℓ_{10} : $4/3+3\sqrt{5}$
9053 4611 erf : $13/11$
9053 4881 Γ : $1/\ln(2e/3)$
9053 7209 tπ : $4\sqrt{2}+\sqrt{3}/3$
9053 8374 10^x : $2\sqrt[3]{2}/9$
9054 1051 rt : (4,8,-5,-5)
9054 4334 cr : $3\sqrt{3}/7$
9054 5424 rt : (5,5,-2,-6)
9054 6071 10^x : $7/25$
9054 6102 Γ : $7\sqrt[3]{3}/6$
9054 6510 $1/2-\ln(2/3)$
9054 7226 $\pi \div \exp(1/2)$
9054 7400 rt : (1,6,7,8)
9054 9785 rt : (6,2,1,-7)
9055 0602 Γ : $7\zeta(3)/5$
9055 0741 rt : (1,8,1,2)
9055 2422 2^x : $2+2\sqrt{3}/3$
9055 2467 $\pi-\sqrt{5}$
9055 4134 rt : (7,1,4,8)
9055 6886 ζ : $\zeta(3)/4$
9055 7791 sπ : $\sqrt[3]{3}/4$
9055 8407 cu : $2+3\sqrt{3}/4$
9055 8470 rt : (8,4,7,9)
9055 8859 rt : (9,8,-8,-6)
9055 9283 ℓ_{10} : $4\sqrt{3}+\sqrt{5}/2$
9055 9775 rt : (5,6,-2,9)
9056 0485 2^x : $1/\ln((e+\pi)/2)$
9056 0489 id : $3-2\pi/3$
9056 0989 sq : $10e$
9056 1010 sq : $8(e+\pi)/5$
9056 1179 tanh : $5\zeta(3)/4$
9056 1228 cr : $7(1/\pi)/3$
9056 3874 sπ : $3e/2+2\pi$
9056 5076 sπ : $3\zeta(3)/10$
9056 5264 tπ : $(2e/3)^{1/2}$
9056 5272 $\ln(\sqrt{3}) \times \exp(1/2)$
9056 6188 Ψ : $\ln(2/3)$
9056 6196 sr : $(2e/3)^{-1/3}$
9056 7774 e^x : $16/15$
9056 8349 $s\pi(\sqrt{5}) \div s\pi(\sqrt{3})$
9057 0378 sq : $\sqrt{2}/4-4\sqrt{5}/3$
9057 1867 10^x : $2\sqrt{3}-\sqrt{5}$
9057 2174 sq : $4+3\pi/2$
9057 2542 Γ : $(\pi/2)^{1/2}$
9057 2748 sπ : $3\sqrt{2}+\sqrt{5}/2$
9057 2808 rt : (2,9,-5,-5)
9057 3427 sπ : $10\sqrt{5}$
9057 7516 $\ln(\sqrt{3})+\exp(\sqrt{5})$
9057 7565 sr : $3/2-e/4$
9057 7872 rt : (3,6,-9,1)
9057 8723 rt : (1,7,0,-5)
9057 8965 J_0 : $1/\ln(5)$
9057 9925 θ_3 : $8/19$
9058 1190 Γ : $1/\ln(\sqrt{2}\pi/2)$
9058 2806 $\ln(2/5)+\exp(3/5)$
9058 5333 sq : $2e-3\pi$
9058 6552 sq : $1/3+\pi/3$
9058 6884 rt : (4,9,0,-5)
9058 8649 2^x : $9\sqrt[3]{5}/10$
9058 9776 rt : (9,7,-6,-7)
9059 1149 rt : (5,4,0,-7)
9059 1370 rt : (9,3,-9,-1)
9059 4977 tπ : $3\sqrt{3}/2-4\sqrt{5}$
9059 5002 rt : (1,1,6,-7)
9059 5402 rt : (6,1,3,-8)
9059 5843 rt : (5,0,-3,1)
9059 8108 ln : $10\sqrt{3}/7$
9059 9643 rt : (7,2,6,9)
9060 2672 2^x : $3/4+2e/3$
9060 3078 rt : (2,2,9,8)
9060 3471 2^x : $2/3+3\sqrt{3}/4$
9060 3826 $\sqrt{2}+\exp(2/5)$
9060 4748 rt : (6,3,0,2)
9060 5040 ζ : $(\sqrt{5}/3)^{-1/2}$
9060 7496 2^x : $(\sqrt{3}/2)^{1/2}$
9060 9394 id : $e/3$
9061 2848 at : $1/3+2\sqrt{2}/3$
9061 3601 rt : (7,6,3,3)
9061 3977 e^x : $1/\ln(3\pi/2)$
9061 5294 cu : $1/4-3\sqrt{5}/4$
9061 6449 rt : (4,2,-6,9)
9061 7428 rt : (8,9,-7,-7)
9061 8937 ln : $2\sqrt{2}/7$
9062 0272 rt : (9,6,-4,-8)
9062 2402 rt : (8,9,6,4)
9062 5000 cu : $5\sqrt[3]{2}/4$
9062 5409 Ei : $18/5$
9062 6136 Ψ : $5/14$
9062 7136 at : $(\sqrt[3]{3}/3)^{-1/3}$
9062 7632 rt : (3,9,-4,-6)
9063 1805 rt : (4,6,-1,4)
9063 2414 sπ : $2\sqrt{5}/7$
9063 5037 sq : $1/2+3e$
9063 5966 rt : (5,3,2,-8)

9063 6858 cr : $4\sqrt{3}$
9063 8470 rt : (9,2,-7,-2)
9063 9515 $\Gamma : ((e+\pi)/3)^{1/3}$
9063 9884 $\exp(\sqrt{3}) + s\pi(\sqrt{3})$
9064 0115 rt : (6,0,5,9)
9064 0247 $\Gamma : 5/4$
9064 0580 $erf : \sqrt[3]{5/3}$
9064 2907 $J : \ln(\ln 3)$
9064 3804 $\ln(3/5) - \exp(1/3)$
9064 6537 $J_0 : 13/21$
9064 6977 $2^x : \sqrt{2}/4 + \sqrt{3}/3$
9064 7561 rt : (8,8,-5,-8)
9064 8561 $\Psi : 19/23$
9064 9225 sr : $(\sqrt{2}\pi/3)^{-1/2}$
9065 0357 rt : (9,5,-2,-9)
9065 1785 $2^x : 3 - \pi$
9065 2808 rt : (1,5,-4,8)
9065 5282 $t\pi : 1/3 - e/4$
9065 6146 rt : (2,9,1,-5)
9065 8298 id : $2(e+\pi)/3$
9066 0954 $e^x : 3/2 - \sqrt{3}/4$
9066 3980 rt : (1,6,-2,-6)
9066 4754 $\sqrt{5} \times \ln(2/3)$
9066 5428 $2^x : 2/3 + 3\sqrt{2}/2$
9066 6199 as : $5\sqrt[3]{2}/8$
9066 6427 $e^x : 3 + 3\sqrt{3}/2$
9066 6833 cr : $10\ln 2$
9066 7999 $J_0 : \exp(-\sqrt[3]{3}/3)$
9066 8114 id : $(\sqrt{5}/3)^{1/3}$
9067 0974 $e^x : 4/3 + 3\sqrt{3}/4$
9067 1753 cu : $\sqrt{12}\pi$
9067 1775 rt : (3,8,-2,-7)
9067 3812 rt : (7,8,-6,8)
9067 4080 at : $4\sqrt{5}/7$
9067 4261 $J_0 : 3\sqrt[3]{3}/7$
9067 5016 at : $23/18$
9067 5847 rt : (4,7,3,8)
9067 7291 rt : (8,7,-3,-9)
9067 8268 $\lambda : \sqrt[3]{3}/10$
9067 9530 rt : (1,0,8,8)
9067 9779 sr : $3\sqrt{2}/2 - 3\sqrt{3}/4$
9067 9908 rt : (5,2,4,-9)
9068 0325 rt : (8,4,-8,-2)
9068 2520 $2^x : e/3 - \pi/3$
9068 4581 rt : (9,1,-5,-3)
9068 4906 $Ei : 4\zeta(3)$
9068 5424 id : $1/4 + 4\sqrt{2}$
9068 8697 $\Gamma : 9\ln 2/5$
9068 8826 rt : (5,1,-1,2)
9068 9059 $\ell_2 : 15$
9069 2171 $erf : 19/16$
9069 6683 tanh : $5e/9$
9069 7279 rt : (6,4,2,3)
9069 8640 $\Gamma : 3(1/\pi)/8$
9070 0268 rt : (3,8,1,9)
9070 0384 ln : $e/3 + \pi/2$
9070 2947 sq : $20/21$
9070 3920 rt : (7,9,-4,-9)
9070 5684 rt : (7,7,5,4)
9070 7203 id : $e + 4\pi/3$
9070 9445 $\exp(2) \div s\pi(\sqrt{3})$
9070 9982 $e^x : 3/2 + 2\pi$
9071 0678 $1/5 + s\pi(1/4)$
9071 1690 $s\pi : 4\sqrt{2} + 4\sqrt{5}/3$
9071 3509 cr : $3/2 + 2e$
9071 4033 $2^x : 8\sqrt{3}/9$
9071 5066 rt : (3,7,0,-8)
9071 8079 $t\pi : e + \pi/3$
9071 9042 rt : (4,4,3,-9)
9072 0358 $\Gamma : 22/13$
9072 1236 ln : $\sqrt{3} + \sqrt{5}/3$
9072 1432 rt : (7,6,-9,-2)
9072 1591 $\Psi : 3\pi^2/10$
9072 3544 $10^x : 3/4 + 4\sqrt{5}$
9072 5593 rt : (8,3,-6,-3)
9072 7281 $\ell_{10} : 4 + 3e/2$
9072 7657 tanh : $6\sqrt[3]{2}/5$
9072 9742 rt : (9,0,-3,4)
9073 0249 rt : (6,7,-9,-1)
9073 2380 $J_0 : 8\ln 2/9$
9073 5801 $\Gamma : \ln(2e)$
9073 7185 $2^x : 2\sqrt{2} + \sqrt{5}/3$
9073 8059 at : $(2\pi/3)^{1/3}$
9073 8699 rt : (2,8,-1,-6)
9073 9098 $erf : (\sqrt{2})^{1/2}$
9074 3229 $Ei : \sqrt{2} + \sqrt{5} - \sqrt{7}$
9074 4127 $2^x : 2e - \pi$
9074 5185 $\exp(e) \times s\pi(1/5)$
9074 6190 rt : (1,8,9,-5)
9074 7064 $s\pi : 3\sqrt{2}/4 + \sqrt{3}/3$
9074 7694 $10^x : (e)^3$
9074 8521 sr : $14/17$
9074 9529 $s\pi : 3\sqrt{3}/2 - \sqrt{5}$
9075 0369 $\lambda : \exp(-8\pi/13)$
9075 3098 $\zeta : 1/\sqrt{11}$
9075 3264 as : $4/3 - 3\sqrt{2}/2$
9075 3645 rt : (2,9,-1,-8)
9075 4282 $J_0 : 8/13$
9075 6107 id : $4e/3 + 2\pi$
9075 7538 rt : (3,6,2,-9)
9075 8919 rt : (6,8,-5,3)
9075 9965 $s\pi : e - 3\pi/4$
9076 0519 $e^x : 10\pi/3$
9076 1816 rt : (3,5,-4,-5)
9076 5416 $J : \exp(-e/2)$
9076 5572 $\exp(3) + \exp(3/5)$
9076 5885 rt : (7,5,-7,-3)
9076 6478 $t\pi : 2\sqrt{2}/3 - 3\sqrt{5}$
9076 9943 $e^x : 4 - 3\sqrt{5}/2$
9076 9943 rt : (8,2,-4,-4)
9077 2160 sinh : $4\sqrt{5}/5$
9077 2280 cu : $((e+\pi)/3)^3$
9077 3434 $10^x : 4/3 + 4\pi$
9077 3859 $erf : 25/21$
9077 3990 rt : (9,1,-1,5)
9077 8026 rt : (5,2,1,3)
9077 9657 ln : $2/3 + 2e/3$
9078 1138 rt : (2,8,2,5)
9078 2300 sr : $4\sqrt[3]{3}/7$
9078 2307 sq : $3\sqrt{2}/4 + 2\sqrt{3}/3$
9078 3250 sq : $2 - \pi/3$
9078 6065 rt : (6,5,4,4)
9078 7224 rt : (6,9,-1,7)
9078 9801 $\zeta : 2(e+\pi)/9$
9079 1804 tanh : $7\sqrt{3}/8$
9079 4060 rt : (7,8,7,5)
9079 4307 $\exp(1/4) \times s\pi(1/4)$
9079 4423 $\Gamma : 1/\ln(\sqrt{5})$
9079 5132 $\theta_3 : (4/3)^{-3}$
9079 5415 rt : (2,8,1,-9)
9079 7631 $\zeta : (\ln 3/2)^2$
9079 8013 $\Gamma : 1/\sqrt[3]{10}$
9079 8549 rt : (9,3,-8,-3)
9079 9684 $Ei : \ln 2/9$
9080 0354 $e^x : e\pi/8$
9080 3225 $\Gamma : 5\sqrt{5}/9$
9080 5483 rt : (6,7,-8,-3)
9080 5903 $J_0 : (\ln 2/3)^{1/3}$
9080 8420 $\Gamma : 6\sqrt{2}/5$
9080 9071 $s\pi : 1/\sqrt[3]{21}$
9080 9452 rt : (7,4,-5,-4)
9081 0056 $s\pi : (\ln 3/3)^{-3}$
9081 0441 cr : $(\sqrt[3]{2}/3)^{1/3}$
9081 2619 rt : (5,4,-9,3)
9081 3411 rt : (8,1,-2,-5)
9081 6008 $\Gamma : \exp(\sqrt[3]{4}/3)$
9081 6011 rt : (4,9,-8,-1)
9081 7360 rt : (9,2,1,6)
9081 7674 $10^x : 4e/3$
9081 8291 $t\pi : 2\pi^2/3$
9081 8379 rt : (2,7,-3,-7)
9082 3687 $\ell_{10} : 2e/3 + 2\pi$
9082 4400 at : $4 - e$
9082 5552 rt : (1,7,5,-9)
9082 7788 $t\pi : 9\pi^2/5$
9082 9574 sr : $1/2 + \pi$
9083 0108 $\Psi : -\sqrt{2} + \sqrt{3} + \sqrt{7}$
9083 1425 sq : $1/3 - 3e/4$
9083 1815 sq : $4e + 3\pi/2$
9083 2691 rt : (1,9,-9,0)
9083 2761 $2^x : 2\sqrt{2}/3 + 4\sqrt{5}$
9083 5048 at : $8\sqrt[3]{3}/9$
9083 5214 cr : $2/3 + 2\pi$
9083 6419 rt : (2,5,-1,4)
9083 7616 cu : $4e/3 + 3\pi/4$
9083 8240 $\Gamma : (\ln 2/2)^{-1/2}$
9083 8881 sq : $(2\pi)^3$
9083 9997 tanh : $\sqrt[3]{7/2}$
9084 0070 sq : $1 + 2e/3$
9084 0258 rt : (5,5,-5,7)
9084 0511 rt : (2,6,-6,-1)
9084 3568 $\Gamma : 5e/8$
9084 4406 rt : (5,9,-9,-3)
9084 5844 tanh : $\sqrt{3} + \sqrt{5} - \sqrt{6}$
9084 8011 $10^x : 3\sqrt{3}/4$
9084 8291 rt : (9,9,-9,-6)
9084 8465 rt : (1,9,8,-5)
9085 0723 $\exp(\sqrt{2}) \times s\pi(1/4)$
9085 2165 rt : (7,3,-3,-5)
9085 3290 at : $3\sqrt[3]{5}/4$
9085 4015 $\zeta : \sqrt[3]{17/3}$
9085 4962 sr : $6\pi^2/7$
9085 5499 $J : e/10$
9085 5649 $\zeta : e/9$
9085 6029 id : $(4/3)^{-1/3}$
9085 7872 $\Psi : 1/\sqrt[3]{22}$
9085 9487 sr : $4/3 + 4\sqrt{3}/3$
9085 9883 rt : (9,3,3,7)
9086 0293 $J : \sqrt[3]{3}/10$
9086 0730 $e^x : 9\sqrt{2}/8$
9086 2208 $s\pi : 3\sqrt{3} - \sqrt{5}/4$
9086 2231 $Ei : 8\sqrt{5}/9$
9086 3613 $10^x : 3 - \pi/4$
9086 3727 rt : (5,3,3,4)
9086 3873 $\Gamma : 17/10$
9086 6659 sinh : $9\pi^2/10$
9086 8528 $2^x : 19/4$
9086 9393 tanh : $8\sqrt[3]{5}/9$
9087 1385 rt : (6,6,6,5)
9087 2308 id : $7\pi^2/10$
9087 3327 $2^x : 1/3 - \sqrt{2}/3$
9087 4624 $e^x : 1 - \sqrt{2}/4$
9087 6369 rt : (9,8,-7,-7)
9087 7493 cu : $10e/3$
9087 9004 rt : (7,9,9,6)
9087 9254 $J_0 : 11/18$
9087 9727 $J : (\sqrt[3]{2}/3)^2$
9088 1469 rt : (7,7,-9,6)
9088 4078 rt : (8,1,-9,-3)
9088 4092 $\ell_2 : \exp(\sqrt[3]{2}/2)$
9088 6480 rt : (5,8,-7,-4)
9088 8086 $t\pi : 3\pi/7$
9088 8460 rt : (3,9,-2,-7)
9088 8738 $\ell_{10} : (\ln 2/3)^3$
9088 9325 sr : $19/23$
9089 0273 rt : (6,5,-4,-5)
9089 0486 cr : $4\sqrt{2} + 3\sqrt{3}/4$
9089 0800 $e^x : 7\sqrt{3}/3$
9089 2078 rt : (4,1,-9,7)
9089 2285 sinh : $3e/10$
9089 2465 rt : (9,4,-6,-2)
9089 3028 $t\pi : \sqrt[3]{11/2}$
9089 4056 rt : (7,2,-1,-6)
9089 4649 $10^x : (2\pi/3)^{-1/2}$
9089 5368 rt : (2,6,-5,-8)
9089 6062 $\Gamma : \sqrt[3]{2}/8$
9089 7829 rt : (8,1,2,7)
9090 1592 rt : (9,4,5,8)
9090 2245 rt : (1,3,-8,-9)
9090 2449 $10^x : 2 + 2\sqrt{3}/3$
9090 2912 $e^x : 2 + 2e$
9090 3193 tanh : $\exp(\sqrt[3]{2}/3)$
9090 4088 rt : (9,7,-5,-8)
9090 4889 $s\pi : -\sqrt{3} + \sqrt{6} + \sqrt{7}$
9090 7505 $t\pi : \sqrt{3}/5$
9090 7911 cr : $5\zeta(3)/8$
9090 8254 $\Gamma : 7\sqrt{2}/8$
9090 8515 ln : $3/4 + \sqrt{3}$
9090 9090 id : $10/11$
9091 0340 sq : $10\pi^2$

9091 1629 cr : $1/4+3\sqrt{5}$
9091 2664 rt : (1,5,-9,-8)
9091 2917 10^x : $3e+2\pi/3$
9091 3942 Γ : $5\sqrt{3}/7$
9091 6584 rt : (1,8,-7,-1)
9091 8462 sq : $3\sqrt{2}-3\sqrt{3}$
9091 8901 erf : $(\sqrt[3]{5})^{1/3}$
9092 4040 rt : (2,5,-4,-2)
9092 5883 2^x : $1/2+\sqrt{3}/4$
9092 7437 rt : (2,9,-3,5)
9092 7754 rt : (5,8,-4,-2)
9092 7898 at : $\sqrt{2}/2+\sqrt{3}/3$
9092 8990 rt : (8,9,-6,-8)
9092 9742 $s\pi$: $2(1/\pi)$
9093 0566 Ei : $1/13$
9093 0735 id : $1/3-3\sqrt{2}$
9093 1458 rt : (9,6,-3,-9)
9093 2076 Γ : $(\sqrt[4]{4}/3)^{-1/3}$
9093 3551 K : $(\sqrt[3]{2}/3)^{1/3}$
9093 5153 rt : (7,1,1,-7)
9093 5312 Γ : $6\sqrt[3]{3}/7$
9093 8151 Γ : $4\ln 2/9$
9093 8596 rt : (1,8,-9,-8)
9093 8839 rt : (8,2,4,8)
9094 0595 e^x : $e-\pi/4$
9094 1623 cu : $1/4-3\sqrt{5}/2$
9094 2083 $\ln(\sqrt{3})\div\ln(3/4)$
9094 2516 rt : (9,5,7,9)
9094 3504 J : $4(1/\pi)/5$
9094 4546 at : $(\sqrt{2}/3)^{-1/3}$
9094 4880 $t\pi$: $\sqrt{2}/3-\sqrt{3}/2$
9094 6183 rt : (5,4,5,5)
9094 6854 sinh : $7\zeta(3)/6$
9094 8169 e^x : $\sqrt{3}/3+2\sqrt{5}/3$
9094 9542 J_0 : $14/23$
9095 0451 erf : $7\sqrt[3]{5}/10$
9095 1596 rt : (1,2,-9,-6)
9095 3249 cr : $e/4+2\pi$
9095 3490 rt : (6,7,8,6)
9095 4037 2^x : $5\pi^2/3$
9095 4250 ln : $(\sqrt[3]{4}/3)^3$
9095 5778 Γ : $21/17$
9095 5911 sr : $e+4\pi$
9095 6057 rt : (8,8,-4,-9)
9095 6712 sr : $4-\sqrt{2}/4$
9095 6979 $\exp(\pi)\times s\pi(e)$
9095 7327 as : $2\sqrt{2}/3-\sqrt{3}$
9095 7518 $s\pi$: $4\sqrt{3}+3\sqrt{5}$
9095 8423 ℓ_{10} : $\exp(2\pi/3)$
9095 8796 sr : $1/4+\sqrt{3}/3$
9096 1279 J_0 : $(\sqrt{2}\pi)^{-1/3}$
9096 2720 id : $4\sqrt{3}+4\sqrt{5}/3$
9096 3199 $s\pi$: $4/11$
9096 3203 rt : (3,8,-4,-8)
9096 3398 at : $1/2+\pi/4$
9096 4628 rt : (4,9,-6,-5)
9096 5718 10^x : $10\sqrt[3]{3}/7$
9096 8257 rt : (5,9,0,2)
9096 8320 2^x : $14/15$
9096 8530 ℓ_{10} : $\sqrt{2}+3\sqrt{5}$
9096 8814 rt : (3,7,-4,3)
9096 9833 rt : (2,5,-7,-9)
9097 0550 ln : $3e/4+3\pi/2$
9097 0961 cu : $2\sqrt{2}-4\sqrt{3}$
9097 1876 rt : (6,3,0,-7)
9097 1908 ℓ_2 : $4-3\sqrt{2}/2$
9097 2905 Ψ : 6π
9097 4342 rt : (8,2,-7,-2)
9097 5315 at : $9/7$
9097 5487 rt : (7,0,3,8)
9097 6246 $t\pi$: $\sqrt{2}/2-2\sqrt{5}$
9097 7426 cu : $3\sqrt{2}+3\sqrt{3}$
9097 9089 rt : (8,3,6,9)
9098 1850 cu : $1/2+3e/2$
9098 2291 $s\pi$: $8(1/\pi)/7$
9098 5931 id : $6(1/\pi)$
9099 1013 cu : $1/4-2\sqrt{5}/3$
9099 1466 Γ : $\pi^2/8$
9099 1516 e^x : $11/17$
9099 2317 e^x : $2-2\pi/3$
9099 3652 ln : $\sqrt{2}/2-\sqrt{5}/4$
9099 5103 as : $15/19$
9099 7349 rt : (1,7,-5,-2)
9099 8549 $t\pi$: $2e/3-\pi/3$
9099 9754 sr : $2/3+4\sqrt{5}/3$
9099 9798 2^x : $4\sqrt{3}-3\sqrt{5}/2$
9100 0013 tanh : $\sqrt{7/3}$
9100 1476 10^x : $3\sqrt{2}+3\sqrt{3}$
9100 3287 cu : $3/4+2\sqrt{3}/3$
9100 3637 cr : $7e\pi$
9100 4468 rt : (4,8,-4,-6)
9100 5523 $s\pi$: $9\sqrt{2}/2$
9100 7947 ln : $10\sqrt{5}/9$
9100 8015 rt : (5,5,-1,-7)
9101 1553 rt : (6,2,2,-8)
9101 2290 2^x : $3/4+3e/4$
9101 2326 J_0 : $(e)^{-1/2}$
9101 3104 J_0 : $7\ln 2/8$
9101 5082 rt : (7,1,5,9)
9101 6861 Γ : $e\pi/5$
9101 6987 rt : (1,9,9,3)
9101 8604 rt : (4,2,4,5)
9101 9639 id : $\sqrt{2}+\sqrt{3}-\sqrt{5}$
9102 3109 $s\pi$: $7\pi^2/8$
9102 3922 id : $1/\ln(3)$
9102 4601 e^x : $3\sqrt{2}/4-2\sqrt{3}/3$
9102 5036 J_0 : $3\sqrt{2}/7$
9102 5620 rt : (5,5,7,6)
9102 6888 Ψ : $8e/3$
9102 9093 sinh : $\sqrt{2/3}$
9102 9375 cu : $2/3-2\pi/3$
9102 9757 e^x : $3/2+\sqrt{3}/4$
9103 0238 10^x : $2+3\sqrt{5}/4$
9103 0626 rt : (6,7,-8,1)
9103 1397 erf : $6/5$
9103 3650 rt : (4,9,-7,2)
9103 3844 e^x : $\sqrt{3}+\sqrt{5}/3$
9103 5555 rt : (3,7,-6,-9)
9103 8706 e^x : $15/11$
9103 8898 e^x : $4-\sqrt{5}/3$
9104 2036 rt : (1,8,-8,1)
9104 2428 $t\pi$: $6\sqrt[3]{3}$
9104 2999 rt : (2,7,-5,-7)
9104 3171 cr : $3/2-\sqrt{5}/3$
9104 3587 rt : (4,7,-2,-7)
9104 7055 rt : (5,4,1,-8)
9105 0258 10^x : $7\pi/8$
9105 0514 rt : (6,1,4,-9)
9105 2317 J_0 : $(\ln 3/3)^{1/2}$
9105 2545 erf : $\sqrt{2}+\sqrt{5}-\sqrt{6}$
9105 3221 rt : (7,0,-8,2)
9105 5281 e^x : $1/\ln(\sqrt{2})$
9105 6418 erf : $(\sqrt[3]{3})^{1/2}$
9105 6470 rt : (6,8,-4,5)
9105 6605 sr : $(\sqrt[3]{5}/3)^{1/3}$
9105 6693 Γ : $\sqrt[3]{5}$
9105 6740 cu : $9(e+\pi)/2$
9105 7099 sr : $\sqrt{2}+\sqrt{5}$
9105 7888 Γ : $16/13$
9105 9299 $s\pi(1/3)\div s\pi(2/5)$
9105 9823 λ : $\exp(-11\pi/18)$
9106 0249 $s\pi$: $3/4+4\sqrt{2}/3$
9106 0811 rt : (8,3,-5,-1)
9106 1364 erf : $(\ln 2)^{-1/2}$
9106 6034 2^x : $3\sqrt{2}+2\sqrt{3}$
9106 7926 rt : (2,8,3,9)
9106 8360 id : $1/3+\sqrt{3}/3$
9106 9338 sr : $4+2\sqrt{5}$
9107 0233 ln : $1/4+\sqrt{5}$
9107 1196 Ei : $11/17$
9107 4931 Ψ : $4\sqrt[3]{3}/7$
9107 5200 rt : (1,6,-3,-3)
9107 5996 e^x : $3+3e/2$
9107 7016 Ei : $\sqrt[3]{15}$
9107 8608 rt : (3,9,-3,-7)
9108 0206 sinh : $6(e+\pi)/7$
9108 1402 erf : $5\sqrt[3]{3}/6$
9108 2008 rt : (6,9,0,9)
9108 2562 $2/5-\ln(3/5)$
9108 3231 erf : $\exp(1/(2e))$
9108 3652 J_0 : $2e/9$
9108 5342 10^x : $9\sqrt{5}/10$
9108 5400 rt : (5,3,3,-9)
9108 6242 10^x : $5\zeta(3)/2$
9108 6252 erf : $\zeta(3)$
9108 8118 at : $3/4-3e/4$
9108 8784 rt : (3,0,3,5)
9109 0901 Γ : $((e+\pi)/2)^{1/2}$
9109 1883 ζ : $\exp(\sqrt{3}/3)$
9109 3897 id : P_{11}
9109 4768 $\exp(3/4)\times s\pi(\pi)$
9109 5529 rt : (4,3,6,6)
9109 5766 cr : $3\sqrt[3]{2}/5$
9109 6471 $\exp(1/5)\times s\pi(\sqrt{3})$
9109 6988 cr : $4/3-\sqrt{3}/3$
9109 8978 K : $3/4$
9109 9284 rt : (7,7,-8,8)
9109 9328 tanh : $23/15$
9109 9502 rt : (4,9,-5,-9)
9109 9522 e^x : $8\zeta(3)/9$
9110 0205 $s\pi$: $\sqrt{2}/2+4\sqrt{3}$
9110 2242 rt : (5,6,9,7)
9110 4646 rt : (8,5,-9,-2)
9110 5013 rt : (5,4,-8,5)
9110 5366 Ψ : $\ln(5/3)$
9110 5940 $1/2\times\exp(3/5)$
9110 8156 rt : (9,2,-6,-3)
9110 8707 10^x : $2/3-\sqrt{2}/2$
9111 5702 sq : $21/22$
9111 6428 rt : (3,8,-1,-8)
9111 9755 rt : (4,5,2,-9)
9112 0661 Ei : $1/\ln(3/2)$
9112 3755 cr : $4+4\sqrt{5}/3$
9112 4487 $4/5\times\exp(2)$
9112 5279 cr : $5\pi^2/2$
9112 5924 10^x : $3\sqrt{2}/2+2\sqrt{5}/3$
9112 9305 rt : (6,2,-9,2)
9112 9990 rt : (5,5,-4,9)
9113 0663 cu : $\sqrt{19}\pi$
9113 0678 Γ : $(\sqrt{2}\pi/3)^{-3}$
9113 5283 e^x : $5(1/\pi)$
9113 6562 rt : (7,1,-6,-1)
9114 1008 tanh : $\sqrt{3}+\sqrt{6}-\sqrt{7}$
9114 1573 ℓ_{10} : $3e$
9114 2333 Γ : $12/7$
9114 3782 rt : (8,4,-3,0)
9114 6035 rt : (9,9,-8,-7)
9114 6204 $t\pi$: $\ln 2/2$
9114 7212 rt : (9,1,-4,-4)
9114 7788 Γ : $6e$
9114 7847 $s\pi$: $4\sqrt{2}+3\sqrt{5}$
9114 9023 Ei : $6\sqrt{2}/5$
9114 9278 ln : $3-3\sqrt{3}/2$
9114 9825 id : $4\sqrt{2}-\sqrt{5}/3$
9115 0331 rt : (1,5,-1,-4)
9115 0965 rt : (9,7,0,1)
9115 1341 $t\pi$: $(\sqrt[3]{5})^{-1/2}$
9115 3594 rt : (3,7,1,-9)
9115 3728 10^x : $5(e+\pi)/7$
9115 5827 $\sqrt{5}+s\pi(\sqrt{5})$
9115 6850 rt : (2,4,4,7)
9115 8214 10^x : $\sqrt{3}/3+3\sqrt{5}$
9115 9970 $t\pi$: $8\sqrt[3]{5}/9$
9116 2045 $t\pi$: $4/17$
9116 2784 rt : (1,6,-1,-9)
9116 3340 rt : (3,1,5,6)
9116 3657 $\exp(2/3)-s\pi(\sqrt{2})$
9116 4242 10^x : $2\sqrt{2}-\sqrt{5}$
9116 5975 10^x : $(\sqrt{5}/2)^{1/3}$
9116 9087 10^x : $2/3+\sqrt{5}/2$
9116 9279 e^x : $2e+4\pi/3$
9116 9377 e^x : $5\sqrt[3]{5}/8$
9116 9800 rt : (4,4,8,7)
9117 0977 J_0 : $\zeta(3)/2$
9117 1305 rt : (9,8,-6,-8)
9117 2490 $s\pi$: $7\sqrt[3]{3}/3$
9117 3829 at : $\sqrt{5/3}$
9117 6947 rt : (4,1,-8,9)
9117 7635 rt : (2,5,-9,8)
9117 8189 10^x : $3/4+2e/3$

9117 8217 $10^x : 1/\sqrt[3]{10}$
9117 8840 rt : (7,6,-8,-3)
9117 9273 rt : (1,6,-5,9)
9118 0204 $2^x : 7\zeta(3)/9$
9118 1636 $e^x : (\sqrt[3]{2}/3)^{1/2}$
9118 2201 rt : (8,3,-5,-4)
9118 3934 sr : P_{25}
9118 5553 rt : (9,0,-2,5)
9118 6148 tanh : 20/13
9118 6242 rt : (1,9,0,-7)
9118 6453 $t\pi : \sqrt{2}/3 - \sqrt{5}$
9118 6576 $e^x : 3\sqrt{2} - 4\sqrt{3}/3$
9118 6932 rt : (2,9,0,-9)
9118 9065 id : $(\pi/3)^{-2}$
9119 0515 $e^x : 4\sqrt{2}/3 + 2\sqrt{3}/3$
9119 0540 ln : $4e/3 + \pi$
9119 2831 $erf : (\sqrt[3]{5}/3)^{-1/3}$
9119 3357 $\exp(\sqrt{2}) \times s\pi(2/5)$
9119 4853 tanh : $9\sqrt[3]{5}/10$
9119 6276 rt : (9,7,-4,-9)
9119 7562 $\Gamma : \sqrt{3/2}$
9119 9543 $s\pi : 2\sqrt{2}/3 - \sqrt{3}/3$
9120 0486 $J_0 : 3/5$
9120 1532 $\Gamma : 10\zeta(3)/7$
9120 1788 sr : $6\ln 2/5$
9120 2283 $2^x : \sqrt[3]{11/3}$
9120 3693 cr : $4/3 + 4\sqrt{2}$
9120 4698 sq : $1 - 2\pi$
9120 5325 tanh : $8\sqrt{3}/9$
9120 9717 $\Psi : 14/17$
9120 9725 rt : (6,1,-7,1)
9121 3264 sr : $\exp(-1/(2e))$
9121 3346 rt : (6,8,-9,-3)
9121 5630 $erf : (\ln 3)^2$
9121 6640 rt : (7,9,-2,-4)
9121 8798 rt : (8,9,-5,-9)
9121 9926 rt : (8,2,-3,-5)
9122 2917 rt : (3,6,-9,-3)
9122 3205 rt : (9,1,0,6)
9122 3515 $10^x : 5\sqrt[3]{5}/9$
9122 3520 rt : (8,5,-1,1)
9122 4057 $\ell_2 : 3 - \sqrt{5}/2$
9122 4589 $\zeta : 2\sqrt{2}/3$
9122 6290 $J_0 : \ln(\ln 3/2)$
9122 9168 rt : (2,1,4,-6)
9123 0366 rt : (9,8,2,2)
9123 1306 $2^x : 9\zeta(3)$
9123 2186 $\ell_2 : \sqrt[3]{20/3}$
9123 5391 rt : (3,2,7,7)
9123 6201 $e^x : 7\sqrt[3]{2}/3$
9123 8582 $2^x : 9(e + \pi)/5$
9124 0685 rt : (4,9,-6,5)
9124 4430 sr : $(\ln 2)^{1/2}$
9124 4967 at : $2 - \sqrt{2}/2$
9124 5945 tanh : $\sqrt[3]{11/3}$
9124 6512 $erf : 4e/9$
9124 6575 rt : (3,7,-3,6)
9124 7064 $Ei : 4\sqrt{2}$
9124 8587 ln : $4/3 + 2e$
9125 0545 rt : (6,7,-7,-4)
9125 0938 $10^x : 19/20$
9125 1475 sr : $(\sqrt[3]{3})^{-1/2}$
9125 1550 cu : $\sqrt{2}/2 - 3\sqrt{5}/4$
9125 3767 rt : (7,4,-4,-5)
9125 4636 $Ei : \ln(3\pi)$
9125 6860 ln : $1 + 2\sqrt{5}/3$
9125 6982 rt : (8,1,-1,-6)
9125 7038 id : $\sqrt{2}/4 + \sqrt{5}/4$
9125 7111 sinh : 9/11
9125 7323 $\Gamma : 11/9$
9125 7624 cr : $4\sqrt[3]{5}/9$
9125 8052 cr : 19/25
9125 8306 rt : (2,5,0,7)
9125 8654 at : $2\sqrt{2}/3 - \sqrt{5}$
9126 0190 rt : (9,2,2,7)
9126 2460 $\ell_{10} : 3\sqrt{3} + 4\sqrt{5}/3$
9126 2592 $\sqrt{5} \times \exp(3)$
9126 3677 tanh : $(\sqrt[3]{2}/3)^{-1/2}$
9126 4845 $e^x : 2\sqrt{3} + 3\sqrt{5}/4$
9126 4952 cu : $4/3 - \pi$
9126 8561 $J : 2/25$
9127 1673 $2^x : 8\sqrt[3]{2}$
9127 2419 $s\pi : 1/2 + \sqrt{3}/2$
9127 6692 $\Gamma : 5\sqrt[3]{5}/7$
9127 8236 sr : $3\sqrt{2}/4 + 3\sqrt{3}/2$
9127 9239 $Ei : \exp(\sqrt[3]{4}/3)$
9127 9804 $t\pi : e/3 - \pi$
9128 0462 rt : (5,3,-8,1)
9128 1042 rt : (1,9,3,-4)
9128 3354 $\ell_2 : e + \pi/3$
9128 3435 cu : $2\sqrt{3} - \sqrt{5}/2$
9128 3932 rt : (5,9,-8,-4)
9128 5683 $\Gamma : (\ln 3/2)^{-1/3}$
9128 7092 sr : 5/6
9128 8275 $2^x : (2e)^2$
9128 8601 $K : 5\zeta(3)/8$
9129 0245 rt : (7,3,-2,-6)
9129 0433 $2^x : 4e\pi/5$
9129 0772 at : 22/17
9129 1541 sq : $1/4 - 3\pi/2$
9129 1721 $t\pi : \sqrt{2}/4 - \sqrt{5}/2$
9129 3118 id : $\sqrt[3]{7}$
9129 3391 rt : (8,0,1,7)
9129 3691 rt : (7,3,-2,1)
9129 3811 $\lambda : (\sqrt[3]{4}/3)^3$
9129 5063 sr : $6\sqrt{2}$
9129 5342 $s\pi : \ln(\sqrt[3]{3})$
9129 6530 rt : (9,3,4,8)
9129 7992 rt : (1,9,1,2)
9129 9100 $2^x : 3e/2 - \pi$
9129 9118 rt : (2,0,6,7)
9130 0000 cu : 17/10
9130 0258 rt : (8,6,1,2)
9130 1410 $10^x : 3 - e$
9130 1511 $J_0 : (3\pi/2)^{-1/3}$
9130 2807 $t\pi : 1/\sqrt[3]{24}$
9130 3103 $t\pi : 3\sqrt{2}/2 + 3\sqrt{3}/4$
9130 3238 tanh : 17/11
9130 4023 tanh : $9\zeta(3)/7$
9130 4347 id : 21/23
9130 5093 rt : (3,3,9,8)
9130 6795 rt : (9,9,4,3)
9130 8287 sr : $3e\pi/7$
9130 8362 $\Gamma : 4/13$
9130 9277 $erf : 23/19$
9130 9350 $2^x : 3\sqrt{2}/2 + 3\sqrt{5}/4$
9131 0749 rt : (6,7,-7,3)
9131 1154 sq : $e/3 - 2\pi$
9131 1510 $\Gamma : e\pi/7$
9131 1896 id : $7\sqrt{5}/4$
9131 4034 rt : (2,7,-4,-3)
9131 4461 $\zeta : 7/23$
9131 5910 $1/2 + \exp(5)$
9131 6703 $t\pi : 4\sqrt{3} + 3\sqrt{5}/4$
9131 8627 $\Gamma : 3\pi^2/4$
9131 9916 rt : (5,8,-6,-5)
9132 1974 $\ell_{10} : 4 + 4\pi/3$
9132 2287 ln : $4\sqrt{2} + \sqrt{5}/2$
9132 3009 rt : (6,5,-3,-6)
9132 5037 $1/5 - \exp(\sqrt{2})$
9132 6095 rt : (7,2,0,-7)
9132 7710 $\ell_2 : \sqrt{23}\pi$
9132 9175 rt : (8,1,3,8)
9133 0406 $\Gamma : (2e/3)^{1/3}$
9133 0616 $10^x : 2\sqrt{2} + \sqrt{5}$
9133 2248 rt : (9,4,6,9)
9133 3112 $2^x : 3\sqrt{3}/2 + 3\sqrt{5}/2$
9133 4142 rt : (6,8,-3,7)
9133 4228 cr : 16/21
9133 4855 $10^x : P_6$
9133 6171 id : $\sqrt{3}/3 - 2\sqrt{5}/3$
9134 3984 $\ell_2 : 9(e + \pi)/7$
9134 4792 sr : $1/3 + 3e$
9134 5585 $\zeta : (2e/3)^{-2}$
9134 8922 rt : (4,5,-9,1)
9134 9473 rt : (1,8,1,-5)
9135 2373 $erf : 6\sqrt{2}/7$
9135 3217 as : 19/24
9135 4521 $2^x : 1 + 4\sqrt{3}/3$
9135 5287 rt : (5,8,-3,0)
9135 6287 $e - \ln(\sqrt{5})$
9135 6522 $J_0 : \ln(2e/3)$
9135 8185 rt : (1,5,-1,-7)
9135 8315 rt : (6,4,-1,-7)
9135 8615 $\Gamma : (e/3)^{-2}$
9135 8926 $erf : 7\sqrt{3}/10$
9136 0138 cr : $2e + \pi/2$
9136 0211 $\Gamma : \sqrt{2} + \sqrt{6} - \sqrt{7}$
9136 0992 $J : \exp(-10\pi/3)$
9136 1002 $\ell_{10} : 3 + 3\sqrt{3}$
9136 1076 rt : (1,2,5,-7)
9136 1337 rt : (7,1,2,-8)
9136 1623 rt : (6,1,-3,1)
9136 3590 $10^x : 2 - e/4$
9136 4230 $s\pi(1/4) \div s\pi(e)$
9136 4353 rt : (8,2,5,9)
9136 4736 sinh : $(\ln 3/2)^{1/3}$
9136 4792 $Ei : (\sqrt[3]{2}/3)^{1/2}$
9136 6843 rt : (2,1,8,8)
9136 7041 sq : 4π
9136 7849 $10^x : 3\sqrt{2}/4 + 3\sqrt{5}$
9136 7930 rt : (7,4,0,2)
9136 8140 $J_0 : 6\ln 2/7$
9137 3756 $erf : 7\ln 2/4$
9137 4208 rt : (8,7,3,3)
9137 4542 sr : $5(e + \pi)/8$
9137 7335 $\Gamma : 4\pi$
9137 8195 rt : (5,4,-7,7)
9138 1765 tanh : $\ln(3\pi/2)$
9138 1964 cu : $7\ln 2/5$
9138 4545 $2^x : 9(e + \pi)/2$
9138 4733 $\Gamma : (\sqrt{2}\pi/3)^{1/2}$
9138 7092 rt : (4,9,-5,-6)
9139 0065 rt : (5,9,1,4)
9139 0610 $e^x : 3\sqrt{3}/4 + 3\sqrt{5}/2$
9139 3031 rt : (6,3,1,-8)
9139 5901 $t\pi : 3(e + \pi)$
9139 5991 rt : (7,0,4,9)
9139 7100 $t\pi : \sqrt{2}/6$
9139 7960 $s\pi : \sqrt{8/3}$
9140 0266 $K : \ln(\sqrt{2}/3)$
9140 1634 $\ell_{10} : 7(e + \pi)/5$
9140 5748 at : $1/4 + \pi/3$
9140 6819 $erf : 17/14$
9140 8926 $\Gamma : 19/11$
9141 1385 $s\pi : 1/\ln(\ln 3)$
9141 5163 cr : $3 - \sqrt{5}$
9141 5764 rt : (1,7,-1,-6)
9141 6354 $2^x : (\sqrt[3]{2}/3)^{-1/2}$
9141 8831 $Li_2 : 5/7$
9142 0377 ln : $4\sqrt{3}/3 + 2\sqrt{5}$
9142 1356 id : $1/2 + \sqrt{2}$
9142 2109 $\theta_3 : 3\sqrt{2}/10$
9142 3215 rt : (1,6,0,-3)
9142 3327 rt : (9,9,-7,-8)
9142 3662 $\Gamma : \ln 2/10$
9142 4268 rt : (5,5,0,-8)
9142 4813 $\ell_{10} : 3/2 + 3\sqrt{5}$
9142 6925 rt : (1,1,7,-8)
9142 7174 rt : (6,2,3,-9)
9142 7446 rt : (5,1,-4,-1)
9142 8439 $e^x : 3\sqrt{3}/4 + 2\sqrt{5}$
9143 2897 tanh : $4e/7$
9143 3509 rt : (6,2,-1,2)
9143 5352 as : $8\ln 2/7$
9143 5504 $10^x : 3/4 + 2\pi$
9143 6565 at : $9\sqrt[3]{3}/10$
9143 7908 rt : (4,9,-5,8)
9143 9546 rt : (7,5,2,3)
9144 4680 ln : $1/2 + 2\pi$
9144 5557 rt : (8,8,5,4)
9144 6020 cr : 13/17
9144 6170 rt : (9,8,-5,-9)
9144 6762 sr : $(\sqrt[3]{5})^{-1/3}$
9144 6896 sr : $7\pi/6$
9144 9482 sr : $3\sqrt{2} - \sqrt{3}/3$
9145 0291 $\Gamma : 17/14$
9145 3053 $\ell_2 : 3\pi/5$
9145 4993 $\ln(2) + \exp(1/5)$
9145 5061 rt : (4,7,-1,-8)

9145 7357 cr : $2e/3 - \pi/3$
9145 7811 ln : $\zeta(3)/3$
9145 7915 rt : (5,4,2,-9)
9145 9326 J_0 : $13/22$
9145 9605 rt : (9,4,-9,-2)
9146 1330 θ_3 : $4(1/\pi)/3$
9146 1948 J : $\exp(-13\pi/2)$
9146 1977 e^x : $3\sqrt{3}/8$
9146 2850 e^x : $5\pi/3$
9146 3531 $\exp(\pi) + s\pi(e)$
9146 3690 e^x : $1 + 4\sqrt{2}/3$
9146 6363 sπ : $(5/2)^{-1/2}$
9146 7608 J_0 : $1/\ln(2e)$
9146 9749 tanh : $14/9$
9146 9906 tπ : $1/3 + 3e$
9147 2095 10^x : $2 - 3e/4$
9147 4295 at : $3\sqrt{3}/4$
9147 4634 rt : (2,9,0,-6)
9147 5344 erf : $(\sqrt{2}\pi/3)^{1/2}$
9147 6225 J : $\exp(-\sqrt[3]{5}/2)$
9147 7039 sπ : $1/3 + 3\sqrt{3}/4$
9147 9748 sπ : $4\sqrt{3}/3 - 3\sqrt{5}/4$
9148 0038 rt : (1,6,-3,-7)
9148 0232 sr : $\sqrt{2} - \sqrt{3}/3$
9148 1403 Γ : $6\sqrt[3]{3}/5$
9148 2076 Γ : $7\ln 2/4$
9148 5421 id : $\sqrt{11/3}$
9148 5815 Ei : $3(e + \pi)/4$
9148 7626 Ei : $4\sqrt[3]{5}/5$
9148 8224 rt : (4,6,1,8)
9149 0784 rt : (1,0,9,9)
9149 1281 rt : (4,9,1,-3)
9149 3383 sq : $22/23$
9149 4202 rt : (9,3,-7,-3)
9149 4531 sr : $(e + \pi)/7$
9149 5904 $s\pi(2/5) - s\pi(\sqrt{2})$
9149 6365 Γ : $7\sqrt{3}/10$
9149 6716 cu : $3 - \sqrt{3}/4$
9149 6801 rt : (1,7,-8,-6)
9149 7117 rt : (5,0,-2,2)
9150 1092 erf : $\sqrt{2} + \sqrt{6} - \sqrt{7}$
9150 2685 Γ : $6\sqrt{2}/7$
9150 2770 erf : $(e/3)^{-2}$
9150 2927 rt : (6,3,1,3)
9150 3749 ℓ_2 : $3\sqrt{2}/8$
9150 5927 Ei : $(\sqrt[3]{5}/3)^3$
9150 6957 rt : (3,7,-2,9)
9150 8713 rt : (7,6,4,4)
9150 8850 sπ : $\exp(-1)$
9150 9980 J_0 : $\exp(-\sqrt[3]{4}/3)$
9150 9988 2^x : $3\sqrt{3}/4 + 2\sqrt{5}/3$
9151 0070 at : $13/10$
9151 0229 Γ : $\sqrt{3}$
9151 0821 10^x : $1 + 2e/3$
9151 1347 Γ : $3/16$
9151 1351 cr : $1/3 + \sqrt{3}/4$
9151 1849 ℓ_{10} : $5\pi^2/6$
9151 3468 cr : $(\sqrt{2}\pi/2)^{-1/3}$
9151 4474 rt : (8,9,7,5)
9151 5170 $\exp(4/5) - \exp(\pi)$
9151 8114 rt : (3,8,0,-9)
9151 8893 cr : $\sqrt{5}\pi$
9152 0184 10^x : $4/3 + 3\sqrt{3}/4$
9152 0656 2^x : $15/16$
9152 1540 J_0 : $(\ln 2/2)^{1/2}$
9152 1681 sπ : $4\sqrt[3]{5}/5$
9152 3037 $2 \div \ln(3/5)$
9152 3326 tanh : $9\sqrt{3}/10$
9152 4942 cu : $4\sqrt{3}/3 + \sqrt{5}$
9152 5355 rt : (8,5,-8,-3)
9152 7727 Γ : $5\ln 2/2$
9152 8215 rt : (9,2,-5,-4)
9152 8418 sπ : $1/4 + \sqrt{5}/2$
9152 8848 Ei : $5\pi/7$
9153 2223 sq : $\sqrt{3}/3 + \sqrt{5}$
9153 2474 erf : $(2e/3)^{1/3}$
9153 3719 sinh : $(2e/3)^{-1/3}$
9153 4480 θ_3 : $(2\pi/3)^{-1/2}$
9153 4771 J_0 : $10/17$
9153 5257 tanh : $9\ln 2/4$
9153 5883 2^x : $4\pi^2/9$
9153 7200 rt : (2,8,-2,-7)
9154 2131 Ψ : $4\sqrt{5}/3$
9154 2407 rt : (1,8,8,-7)
9154 4345 Γ : $23/19$
9154 5189 cu : $10/7$
9154 7594 id : $\sqrt{17/2}$
9154 9665 10^x : $10/9$
9155 0295 rt : (3,2,-3,8)
9155 2420 erf : $e\pi/7$
9155 3239 rt : (3,5,-6,-1)
9155 3533 $\pi + s\pi(e)$
9155 4082 e^x : $13/20$
9155 4239 as : $3/2 - \sqrt{2}/2$
9155 6052 rt : (7,7,-9,-3)
9155 8860 rt : (8,4,-6,-4)
9156 0605 cr : $4\sqrt{2}/3 - \sqrt{5}/2$
9156 1662 rt : (9,1,-3,-5)
9156 1849 tπ : $5\pi/3$
9156 2397 J : $\pi/7$
9156 4458 rt : (5,1,0,3)
9156 5550 tπ : $3\sqrt{3}/4 + 3\sqrt{5}/2$
9156 6414 cu : $1/2 - e$
9156 8137 tπ : $4e/7$
9156 8501 rt : (2,7,-3,1)
9156 8526 e^x : $4 - 2e/3$
9156 9488 10^x : $\sqrt{2}/2 - \sqrt{5}/3$
9157 0033 rt : (6,4,3,4)
9157 0984 λ : $3/20$
9157 1037 sr : $3\sqrt{5}/8$
9157 2726 rt : (6,7,-6,5)
9157 5073 sπ : $1/\sqrt[3]{20}$
9157 5585 rt : (7,7,6,5)
9157 7332 sπ : $7/19$
9157 8390 $4/5 \times \ln(\pi)$
9157 9746 erf : $(\ln 3/2)^{-1/3}$
9158 2454 tanh : $25/16$
9158 2801 cr : $6(e + \pi)/5$
9158 8908 $\exp(3/5) \div s\pi(2/5)$
9158 9048 $\sqrt{2} \div \ln(3/4)$
9158 9064 rt : (7,6,-7,-4)
9158 9276 erf : $5\sqrt[3]{5}/7$
9159 1557 at : $2(e + \pi)/9$
9159 1815 rt : (8,3,-4,-5)
9159 1873 J : $3/16$
9159 2104 ζ : $5\sqrt{3}/9$
9159 3980 rt : (6,8,-2,9)
9159 4561 rt : (9,0,-1,6)
9159 4933 cr : $3/4 + 2\pi$
9159 5827 2^x : $3\sqrt{2}/2 - \sqrt{3}/3$
9159 6559 id : $\beta(2)$
9159 7952 rt : (2,7,-4,-8)
9159 7955 J_0 : $(e + \pi)/10$
9159 8361 sq : $8e/5$
9160 2974 rt : (1,4,-7,-9)
9160 3335 e^x : $-\sqrt{2} + \sqrt{3} + \sqrt{6}$
9160 5172 $\pi - \exp(4/5)$
9160 5206 $3/4 \times \exp(1/5)$
9160 5313 Γ : $4e/9$
9160 5339 sq : $1/4 + \sqrt{2}/2$
9160 6528 Ψ : $(\sqrt{2}\pi/3)^{-1/2}$
9160 6636 ℓ_{10} : $4 + 3\sqrt{2}$
9160 8191 tπ : $\sqrt{5}$
9160 8809 λ : $\zeta(3)/8$
9160 9837 erf : $11/9$
9161 0582 rt : (2,4,-5,1)
9161 1582 e^x : $\sqrt{2}/3 - \sqrt{5}/4$
9161 3419 rt : (2,7,-7,-1)
9161 8014 rt : (1,7,-7,2)
9161 8229 2^x : $3e - \pi/3$
9161 8839 rt : (6,8,-8,-4)
9162 1541 rt : (7,5,-5,-5)
9162 1782 Li_2 : $(\ln 3/3)^{1/3}$
9162 2701 sq : $\sqrt{5} - \sqrt{6} + \sqrt{7}$
9162 3201 tπ : $\exp(-1/\sqrt{2})$
9162 3209 2^x : P_{62}
9162 3693 Ψ : $2/19$
9162 4238 rt : (8,2,-2,-6)
9162 6032 cr : $10/13$
9162 6652 tanh : $7\sqrt{5}/10$
9162 6929 rt : (9,1,1,7)
9162 8370 Γ : $\exp(-2\pi)$
9162 9073 id : $\ln(5/2)$
9162 9614 rt : (5,2,2,4)
9163 0378 rt : (8,9,-9,6)
9163 1144 J_0 : $1/\sqrt[3]{5}$
9163 2781 tπ : $7\sqrt[3]{2}/5$
9163 3571 tπ : $(\ln 3/3)^{1/2}$
9163 3959 Li_2 : $((e + \pi)/3)^{-1/2}$
9163 4100 rt : (5,4,-6,9)
9163 4970 rt : (6,5,5,5)
9163 5439 Γ : $(\ln 3)^2$
9163 6591 2^x : $\sqrt{3}/4 - \sqrt{5}/4$
9163 8155 rt : (1,7,0,-4)
9163 9317 Li_2 : $(\sqrt{5}/2)^{-3}$
9163 9846 sinh : $(e/3)^2$
9164 0304 rt : (7,8,8,6)
9164 5053 Ei : $\pi^2/4$
9164 7691 J_0 : $((e + \pi)/2)^{-1/2}$
9164 8320 Γ : $2\ln 2/3$
9164 8642 sr : $2\sqrt[3]{2}/3$
9164 9334 rt : (9,2,-9,-3)
9165 0592 ζ : $16/9$
9165 0851 rt : (6,7,-6,-5)
9165 1513 sr : $21/25$
9165 2123 rt : (3,9,-3,-8)
9165 3499 sπ : $4\sqrt{2}/3 + \sqrt{5}/3$
9165 3500 rt : (7,4,-3,-6)
9165 4125 sπ : $3e/5$
9165 6143 rt : (8,1,0,-7)
9165 6987 rt : (2,6,-6,-9)
9165 7733 Γ : $(\sqrt[3]{5}/3)^{-1/3}$
9165 8782 rt : (9,2,3,8)
9165 9685 Ei : $(\ln 2/2)^{-1/2}$
9166 0196 sq : $3/4 + \sqrt{5}$
9166 0217 ζ : $2e/7$
9166 2429 cu : $4 + e/2$
9166 2969 id : $6\sqrt{5}/7$
9166 3052 cr : $10\ln 2/9$
9166 4248 Γ : $7\sqrt{5}/9$
9166 4328 cu : $1/2 + \sqrt{2}/3$
9166 6666 id : $11/12$
9166 8319 rt : (1,6,1,3)
9166 9145 e^x : $3/2 + e$
9166 9260 Ei : $4\sqrt[3]{2}/5$
9166 9359 e^x : $16/9$
9166 9412 sq : $9\ln 2$
9167 0743 cr : $1/3 + 3\sqrt{5}$
9167 1341 ℓ_{10} : $4\sqrt{2} + 3\sqrt{3}/2$
9167 1917 rt : (1,9,-8,-1)
9167 2274 J_0 : $7/12$
9167 2578 $s\pi(2/5) \times s\pi(\sqrt{2})$
9167 3548 erf : $\sqrt{3/2}$
9167 3631 sπ : $4\sqrt{3} - \sqrt{5}/4$
9167 4788 cu : $3 + 4\pi$
9167 7147 rt : (2,6,-5,-2)
9167 9755 rt : (5,9,-7,-5)
9168 0724 Γ : $9\sqrt{2}$
9168 2357 rt : (9,9,-6,-9)
9168 3145 Γ : $-\sqrt{3} + \sqrt{5} + \sqrt{6}$
9168 4955 rt : (7,3,-1,-7)
9168 6838 as : $\sqrt[3]{1/2}$
9168 7547 rt : (8,0,2,8)
9168 8207 at : $(\sqrt{2}\pi/2)^{1/3}$
9169 0135 rt : (9,3,5,9)
9169 1721 θ_3 : $\exp(-\sqrt[3]{5}/2)$
9169 2265 θ_3 : $1/\sqrt[3]{13}$
9169 2437 2^x : $4e/3 + 3\pi$
9169 2439 rt : (2,8,-8,5)
9169 2551 tπ : $6\sqrt[3]{3}/7$
9169 2718 rt : (5,3,4,5)
9169 3071 ζ : $6\ln 2/7$
9169 5795 e^x : $3 + 3\pi/4$
9169 7208 cu : $\sqrt{2}/3 + 3\sqrt{5}/4$
9169 7868 rt : (6,6,7,6)
9169 9391 sπ : $3\sqrt{2}/2 - 2\sqrt{5}/3$
9170 0001 $1/5 - \exp(3/4)$
9170 0404 sr : $(\sqrt{2})^{-1/2}$
9170 4845 sπ : $9(e + \pi)/2$
9170 6082 ℓ_{10} : $4/3 + 4\sqrt{3}$

9170 6367 $s\pi : 3e - \pi/4$
9170 6788 $cu : 5\sqrt{5}/9$
9170 8242 $sq : e\pi/5$
9170 9696 rt : (3,8,-5,-9)
9171 0821 rt : (5,8,-5,-6)
9171 3374 rt : (6,5,-2,-7)
9171 5233 $\tanh : \pi/2$
9171 5922 rt : (7,2,1,-8)
9171 5976 sq : 18/13
9171 7735 rt : (9,3,-7,-2)
9171 7926 $K : (\ln 3)^{-3}$
9171 8465 rt : (8,1,4,9)
9171 9371 $\tanh : 1/\ln(\sqrt[3]{4}/3)$
9172 3997 $\tanh : 10\sqrt{2}/9$
9172 5269 $\tanh : 11/7$
9172 8599 rt : (1,5,0,-2)
9173 0029 $sr : 7\zeta(3)/10$
9173 0658 $\ell_{10} : 3e/2 + 4\pi/3$
9173 1942 rt : (3,6,-7,3)
9173 3869 rt : (1,8,-6,-2)
9173 4938 $s\pi : \sqrt[3]{13/3}$
9173 7797 $\sinh : (\sqrt{2}\pi/3)^{-1/2}$
9173 8901 rt : (2,7,1,-7)
9173 9046 $10^x : 1/3 + \sqrt{2}$
9174 0204 $2^x : 3\sqrt[3]{5}/2$
9174 0575 rt : (7,9,-9,-1)
9174 1411 rt : (5,8,-2,2)
9174 3915 rt : (6,4,0,-8)
9174 4289 $10^x : 4/3 + \pi/2$
9174 4936 $cr : 5\pi^2/7$
9174 6415 rt : (7,1,3,-9)
9174 7131 $10^x : 2\sqrt[3]{2}/3$
9174 8911 rt : (4,1,3,5)
9174 9206 $10^x : 5\sqrt[3]{3}/6$
9175 0617 $Ei : 5e/8$
9175 3783 $\ell_2 : 9/17$
9175 3887 rt : (5,4,6,6)
9175 4536 $Li_2 : 9(1/\pi)/4$
9175 4908 $\ln(4/5) + \exp(\pi)$
9175 6381 $sr : 2 + 3\sqrt{5}/4$
9175 8846 rt : (6,7,9,7)
9176 2548 $\Gamma : \zeta(3)$
9176 4067 rt : (1,9,-9,1)
9176 4364 $e^x : (e + \pi)/9$
9176 4650 cr : 17/22
9176 5052 $2^x : 2 - 3\sqrt{2}/4$
9176 5533 $\Gamma : \exp(1/(2e))$
9176 5954 $\Psi : (e/3)^2$
9176 6293 sr : 16/19
9176 7341 $\Gamma : 5\sqrt[3]{3}/6$
9176 9074 rt : (4,9,-4,-7)
9177 1537 rt : (5,9,2,6)
9177 3995 rt : (6,3,2,-9)
9177 6300 rt : (3,9,-9,7)
9177 6929 $t\pi : (\sqrt[3]{4}/3)^{-1/3}$
9177 8419 rt : (8,1,-8,-2)
9178 0188 $\tanh : 5\sqrt[3]{2}/4$
9178 0387 $s\pi : \sqrt[3]{2}/2$
9178 1376 $Ei : \zeta(7)$
9178 2983 $\ln(2) \times \exp(\sqrt{3})$
9178 3102 $s\pi : 3\sqrt{2}/4 + 4\sqrt{3}/3$
9178 3762 rt : (9,4,-5,-1)
9178 4529 $\ln(4/5) \times \exp(\sqrt{2})$
9178 7165 $\Gamma : (\ln 2)^{-1/2}$
9178 9241 $\Psi : 19$
9179 1252 rt : (4,8,-9,5)
9179 2063 $\Gamma : (\sqrt[3]{3})^{1/2}$
9179 3795 $at : (\sqrt[3]{5})^{1/2}$
9179 3948 rt : (1,7,-4,-3)
9179 4412 $J_0 : 11/19$
9179 4472 $t\pi : 3(e + \pi)/4$
9179 4969 at : 17/13
9179 5069 $2^x : P_{46}$
9179 5179 $t\pi : 4\sqrt{2}/3 - 2\sqrt{5}/3$
9179 5899 $\Gamma : \sqrt{2} + \sqrt{5} - \sqrt{6}$
9179 6894 $Ei : 3\sqrt{3}/8$
9179 6933 $\ln(\sqrt{5}) + \exp(\sqrt{2})$
9179 7398 $10^x : \sqrt{2}/5$
9179 7539 $t\pi : \exp(-\sqrt[3]{3})$
9179 8795 rt : (6,6,-9,3)
9180 0847 $\Gamma : 5\pi/9$
9180 1213 rt : (5,5,1,-9)
9180 3625 rt : (3,8,9,9)
9180 7927 rt : (2,7,-2,5)
9180 8233 $sq : 4/3 - e$
9180 8438 rt : (4,2,5,6)
9180 9097 $e^x : 2\sqrt{3} + \sqrt{5}$
9180 9390 $10^x : 1/4 + 2\sqrt{2}$
9181 3233 rt : (5,5,8,7)
9181 3297 $Li_2 : (e)^{-1/3}$
9181 3442 rt : (1,7,1,3)
9181 3491 rt : (8,9,-8,8)
9181 4198 $\zeta : \ln(e/2)$
9181 6199 $at : 4\sqrt{2}/3 - \sqrt{3}/3$
9181 6874 $\Gamma : 6/5$
9181 7087 $s\pi : (\ln 2/3)^{-1/3}$
9181 7586 $\exp(\sqrt{5}) \div s\pi(1/5)$
9181 7705 $\ell_2 : P_3$
9181 8365 rt : (6,7,-5,7)
9181 9426 $\sqrt{5} - \exp(e)$
9181 9737 $\ell_{10} : 2 + 2\pi$
9182 1610 $s\pi : (4/3)^3$
9182 2064 $sr : 3 + e/4$
9182 3073 $id : \sqrt{2} - \sqrt{3} + \sqrt{5}$
9182 4111 $erf : 16/13$
9182 7509 $cr : 4/3 - \sqrt{5}/4$
9182 8078 rt : (4,7,0,-9)
9182 8182 $1/5 + e$
9182 9295 $cu : 2\sqrt{2}/3 + \sqrt{3}/2$
9182 9583 $\ell_2 : 3\sqrt[3]{2}$
9183 0791 $\ell_{10} : e/3 - \pi/4$
9183 0801 $t\pi : 3e/4 + 4\pi$
9183 0970 $10^x : 21/25$
9183 1137 $J_0 : 5\ln 2/6$
9183 2074 $cu : (3\pi/2)^3$
9183 5130 $\Psi : 8(1/\pi)/5$
9183 6734 $sq : 3\sqrt{5}/7$
9183 7298 rt : (7,1,-9,2)
9183 7611 $\lambda : \exp(-3\pi/5)$
9183 7845 $sr : 4e - 3\pi/4$
9183 8590 $sr : (5/3)^{-1/3}$
9183 8678 $J_0 : \sqrt{3}/3$
9184 0226 $K : (\sqrt[3]{5}/3)^{1/2}$
9184 0277 sq : 23/24
9184 0280 $\ell_{10} : \sqrt{3}/2 - \sqrt{5}/3$
9184 1305 $10^x : 8(1/\pi)/9$
9184 1985 $\Gamma : \sqrt[3]{16/3}$
9184 2404 $J_0 : \gamma$
9184 2444 rt : (8,2,-6,-1)
9184 3818 $at : 3/4 + \sqrt{5}/4$
9184 4910 $\pi + \ln(4/5)$
9184 7569 rt : (9,5,-3,0)
9185 1142 $J_0 : 2\sqrt[3]{3}/5$
9185 2259 rt : (1,6,-2,-4)
9185 4031 $cr : \sqrt{2}/4 + 3\sqrt{5}$
9185 4598 rt : (3,9,-1,-5)
9185 5377 rt : (5,3,-9,7)
9185 6933 rt : (2,3,1,5)
9185 7970 $at : 1 - 4\sqrt{3}/3$
9186 0039 $sr : -\sqrt{2} + \sqrt{6} + \sqrt{7}$
9186 1590 rt : (3,0,4,6)
9186 2040 $\ln : 4\sqrt{2} + 2\sqrt{3}/3$
9186 6231 rt : (4,3,7,7)
9186 7257 $\ln(\pi) + s\pi(e)$
9186 8810 $\ln(2) + \exp(4/5)$
9187 1536 $\ln : P_{59}$
9187 3948 $e^x : 8\sqrt[3]{5}$
9187 4973 rt : (8,6,-9,-3)
9187 7378 rt : (9,3,-6,-4)
9187 7955 $\tanh : \sqrt{5/2}$
9188 0692 $\Gamma : 6/13$
9188 2867 $e^x : 2\sqrt{2}$
9188 3080 rt : (3,1,-6,6)
9188 3092 $cu : 3 + 3\sqrt{2}$
9188 8107 $\zeta : (\sqrt[3]{2})^{-1/3}$
9188 8237 $cu : 9\sqrt{5}/8$
9189 3779 $\ln : 3e/2 - \pi/2$
9189 3853 $\ln : (2\pi)^{1/2}$
9189 6021 $2^x : 17/11$
9189 6569 $erf : \pi^2/8$
9189 6610 $cu : 1/\ln(\sqrt{5})$
9189 7502 $\Gamma : 7\sqrt[3]{5}/10$
9189 7683 $10^x : 3e/4 - \pi/4$
9189 9432 rt : (7,0,-7,1)
9189 9444 rt : (1,6,2,9)
9190 1517 $10^x : (\pi/2)^{1/2}$
9190 4325 $at : 1/4 + 3\sqrt{2}/4$
9190 4374 rt : (8,5,-7,-4)
9190 5568 $2^x : 9\zeta(3)/7$
9190 6252 $\Gamma : 7/4$
9190 6734 rt : (9,2,-4,-5)
9190 8899 rt : (1,5,0,-5)
9190 9297 rt : (9,6,-1,1)
9191 2087 $\tanh : 19/12$
9191 3409 rt : (2,4,5,9)
9191 5829 $cr : 9\pi/4$
9191 7904 rt : (3,1,6,7)
9191 9745 $cr : 2e/7$
9192 2383 rt : (4,4,9,8)
9192 6923 $cu : 2 + 4e/3$
9192 7686 $sr : 3\sqrt{2} - \sqrt{5}/4$
9192 8987 $\exp(1/5) - \exp(\pi)$
9192 9077 $\Gamma : (\sqrt[3]{5})^{1/3}$
9193 0568 $cr : 3\sqrt{3}/2 + 2\sqrt{5}$
9193 0622 $s\pi : 10(e + \pi)/7$
9193 1013 rt : (7,7,-8,-4)
9193 2449 $cu : 3 - \pi/2$
9193 3333 rt : (8,4,-5,-5)
9193 4588 $2^x : 2 - 3\sqrt{2}/2$
9193 5650 rt : (9,1,-2,-6)
9193 5740 $erf : 21/17$
9193 5945 $\zeta : \ln((e + \pi)/2)$
9193 5994 $\sinh : 2\sqrt[3]{3}$
9193 8310 $cr : 5\sqrt{2}$
9193 8313 $\zeta : \ln(2e/3)$
9193 8646 $Ei : 13/20$
9193 9019 $\Psi : (2e/3)^{-1/3}$
9193 9408 $s\pi : 1/4 + 3\sqrt{2}/2$
9194 0168 $sr : 2 - 2\sqrt{3}/3$
9194 0294 $\Psi : 7e$
9194 1996 $t\pi : e/2 + \pi$
9194 3165 $e^x : 7\pi^2/3$
9194 4497 $s\pi : 7(1/\pi)/6$
9195 2958 $\zeta : 5\ln 2/4$
9195 4730 $e^x : 15/14$
9195 4817 rt : (6,2,-8,1)
9195 5080 $sq : 2 + \sqrt{3}/4$
9195 6736 $s\pi : 4\sqrt{2}/9$
9195 7303 rt : (6,9,-9,-4)
9195 8279 $erf : 6\sqrt[3]{3}/7$
9195 9586 rt : (7,8,-4,-5)
9195 9603 $\ln : 2\sqrt{3} + 3\sqrt{5}/2$
9195 9844 $J_0 : 9(1/\pi)/5$
9196 0495 $at : 1/2 - 2e/3$
9196 1849 $erf : (\sqrt[3]{4}/3)^{-1/3}$
9196 1864 rt : (8,3,-3,-6)
9196 3000 $1/5 \times \exp(4)$
9196 3408 $cu : 7\sqrt{5}/4$
9196 3956 rt : (4,8,-8,8)
9196 4139 cr : 7/9
9196 4259 $s\pi : 9e\pi/5$
9196 4337 rt : (8,4,-2,1)
9196 5213 $sq : 7\sqrt[3]{5}/3$
9196 7615 $sq : 3/2 + 2\pi/3$
9196 8186 $e^x : 8\sqrt[3]{2}$
9196 8312 rt : (2,1,5,-7)
9196 9071 rt : (9,7,1,2)
9196 9167 $id : 3\sqrt{2} + 3\sqrt{5}/4$
9196 9889 $\ell_{10} : \sqrt{7}\pi$
9197 0957 $e^x : 15/23$
9197 1960 at : 21/16
9197 2653 rt : (3,2,8,8)
9197 4030 $e^x : 1/2 + \sqrt{3}/2$
9197 4989 $\tanh : \sqrt[3]{4}$
9197 6334 $cu : \sqrt{2}/3 + 3\sqrt{3}/2$
9197 6400 $\ell_2 : 1 - \sqrt{2}/3$
9197 8144 rt : (7,9,-8,1)
9198 0131 $\Psi : 7\sqrt[3]{5}/4$
9198 0950 $\exp(3/4) \div s\pi(\pi)$
9198 1607 $Ei : \sqrt{3}\pi$

9198 1877 $erf : 5\sqrt{3}/7$
9198 2379 sinh : 14/17
9198 2514 $\zeta : 4/13$
9198 3590 $2^x : 3/2 - \sqrt{5}/4$
9198 3602 cr : $(\sqrt{2}/3)^{1/3}$
9198 3974 $s\pi : \pi/5$
9198 4990 $\ell_2 : P_{64}$
9198 5499 rt : (6,8,-7,-5)
9198 6621 sr : 11/13
9198 7133 sq : $e/2 + 4\pi$
9198 7555 $K : 3\sqrt[3]{2}/5$
9198 7741 rt : (7,5,-4,-6)
9198 8028 cu : $4 - 2\pi/3$
9198 8168 $erf : 7\sqrt{2}/8$
9198 9978 rt : (8,2,-1,-7)
9199 2212 rt : (9,1,2,8)
9199 2767 $t\pi : 7\sqrt{3}/9$
9199 3747 sq : $1/2 - 4\sqrt{2}/3$
9199 4134 cr : $3\sqrt{2} - 2\sqrt{3}$
9199 5792 cr : $3 + 3e/2$
9199 6933 $2^x : 3\sqrt{2} - 3\sqrt{5}/4$
9199 8528 $10^x : 1/\ln(\sqrt[3]{3})$
9199 9905 ln : $4 - 2\sqrt{5}/3$
9200 0000 id : 23/25
9200 1118 $10^x : \exp(1/(2e))$
9200 1828 $J_0 : 4/7$
9200 4441 $\exp(1/4) \div \exp(1/3)$
9200 7497 $\ell_{10} : \zeta(3)/10$
9200 8681 rt : (5,4,-9,1)
9200 9050 rt : (1,9,2,-5)
9200 9337 $2^x : 16/17$
9201 3288 rt : (6,7,-5,-6)
9201 5489 rt : (7,4,-2,-7)
9201 6105 rt : (2,6,-9,-7)
9201 7512 rt : (1,3,4,-7)
9201 7687 rt : (8,1,1,-8)
9201 7877 rt : (7,2,-3,1)
9201 7919 $\ell_2 : 4/3 + \sqrt{5}/4$
9201 8286 $\sqrt{3} - \exp(\sqrt{3})$
9201 8450 $\zeta : (\sqrt{2}\pi/3)^{-3}$
9201 8989 $\pi - \exp(1/5)$
9201 9881 rt : (9,2,4,9)
9202 1722 rt : (2,0,7,8)
9202 2231 $\ell_{10} : 3e/4 + 2\pi$
9202 2451 rt : (8,5,0,2)
9202 6959 sr : $1/3 + 3\sqrt{5}/2$
9202 7008 rt : (9,8,3,3)
9202 8577 as : $5(1/\pi)/2$
9202 8680 $s\pi : 1/3 + 3e/4$
9202 8812 $t\pi : 3\sqrt{3}/2 + 4\sqrt{5}/3$
9202 9467 rt : (1,8,0,-5)
9203 0102 tanh : $9\sqrt{2}/8$
9203 0603 $2^x : 3\sqrt{3}/2 + \sqrt{5}/4$
9203 1343 rt : (6,6,-8,5)
9203 1709 $t\pi : \sqrt{2}/3 - 3\sqrt{5}$
9203 2486 id : $4e + \pi/3$
9203 3652 rt : (2,7,-1,9)
9203 5100 cr : $4\sqrt{2}/3 + 3\sqrt{3}$
9203 6297 $\pi \div \ln(3/4)$
9203 8513 rt : (5,9,-6,-6)
9203 8656 tanh : $5(1/\pi)$
9203 8750 $\zeta : 25/22$
9203 8762 $Ei : 8\pi^2/5$
9204 0679 rt : (6,6,-3,-7)
9204 0906 $\ell_{10} : (\ln 2/2)^2$
9204 1175 $J_0 : \sqrt[3]{5}/3$
9204 2842 rt : (7,3,0,-8)
9204 3401 cr : $9\ln 2/8$
9204 4505 $\ell_2 : 3 + \pi/4$
9204 5001 rt : (8,0,3,9)
9204 7350 $\zeta : 9(1/\pi)/2$
9204 9232 rt : (6,7,-4,9)
9205 1773 ln : $3\sqrt{2} - \sqrt{3}$
9205 5480 sr : $\sqrt{2}/2 + 4\sqrt{5}/3$
9205 5845 ln : $(e/2)^3$
9205 7458 $\zeta : 4\ln 2/9$
9206 1183 $10^x : (\sqrt[3]{2}/3)^{-1/2}$
9206 1457 rt : (1,8,0,-6)
9206 3838 $e^x : 4 - \sqrt{2}/2$
9206 5321 sr : $9e\pi/5$
9206 5556 rt : (5,8,-4,-7)
9206 5712 sinh : $4\sqrt[3]{3}/7$
9206 6473 $s\pi : 3\sqrt{2}/4 - \sqrt{3}/4$
9206 7600 rt : (1,5,1,3)
9206 7685 rt : (6,5,-1,-8)
9206 7801 $t\pi : 5\sqrt{2}/2$
9206 7962 $s\pi : \sqrt{2} - \sqrt{5} + \sqrt{6}$
9206 9057 $10^x : 23/6$
9206 9641 rt : (1,2,6,-8)
9206 9810 rt : (7,2,2,-9)
9206 9993 rt : (6,0,-4,1)
9207 1861 $\Psi : 6e\pi/7$
9207 2846 $J : \exp(-19\pi/2)$
9207 3108 $\ln(3) + \exp(3/5)$
9207 3715 rt : (2,1,9,9)
9207 4414 rt : (7,3,-1,2)
9207 5284 $\Gamma : 25/21$
9207 6176 $J_0 : (2e)^{-1/3}$
9207 6379 id : $3\sqrt{2}/4 - 4\sqrt{5}/3$
9207 7427 sq : $1/2 - e$
9207 8821 rt : (8,6,2,3)
9208 1436 sr : $4/3 + 3\pi/4$
9208 1875 $\ell_{10} : 3/25$
9208 3032 ln : $3(e + \pi)/7$
9208 3134 rt : (5,3,-8,9)
9208 3212 rt : (9,9,5,4)
9208 9755 $\Gamma : 3(e + \pi)/10$
9209 0707 rt : (1,8,-9,-7)
9209 2222 rt : (5,8,-1,4)
9209 2587 at : 25/19
9209 2808 ln : $\sqrt{3} + \sqrt{6} + \sqrt{7}$
9209 4314 rt : (6,4,1,-9)
9209 5178 cu : $(\sqrt{2}\pi)^3$
9209 5430 $\zeta : 4\sqrt{3}/7$
9209 6320 $Ei : 17/10$
9209 9080 $\Gamma : (2e)^{1/3}$
9209 9134 $\zeta : 7\sqrt{2}/10$
9209 9709 $2^x : 2/3 - \pi/4$
9209 9754 $\ell_{10} : 4e/3 + 3\pi/2$
9209 9788 sq : $3\sqrt{2} + 3\sqrt{5}$
9210 2522 id : $3\sqrt{3}/2 - 3\sqrt{5}/4$
9210 3004 at : $(\sqrt{3})^{1/2}$
9210 4148 sq : $\sqrt{3} - \sqrt{5} - \sqrt{7}$
9210 5166 $erf : 5\sqrt{5}/9$
9210 6338 $s\pi : \sqrt{5}/6$
9210 8926 id : $\sqrt{3}/4 - 3\sqrt{5}/2$
9210 9435 ln : $4 + 2\sqrt{2}$
9210 9910 $\ell_2 : \sqrt{3}/2 + 3\sqrt{5}$
9211 0578 $\Gamma : (\sqrt{2})^{1/2}$
9211 1858 $1/3 + s\pi(1/5)$
9211 2493 rt : (1,7,-2,-7)
9211 4250 cr : $(2\pi/3)^{-1/3}$
9211 4494 $Ei : 7\sqrt[3]{3}/10$
9211 5044 $erf : 1/\ln(\sqrt{5})$
9211 5587 sr : $3\sqrt{2}/5$
9211 6460 rt : (4,9,-3,-8)
9211 8521 rt : (5,9,3,8)
9211 8583 $\ln(\pi) \times \ln(\sqrt{5})$
9212 0416 rt : (1,1,8,-9)
9212 0559 $\lambda : 2/13$
9212 0753 rt : (5,2,-5,-1)
9212 1630 sq : $4e + \pi/4$
9212 4468 tanh : $5\sqrt{5}/7$
9212 5032 rt : (6,1,-2,2)
9212 5657 sq : $5\sqrt[3]{2}/2$
9212 8328 $t\pi : 2e/3 + 3\pi$
9212 9188 $\ell_{10} : \sqrt{2} + 4\sqrt{3}$
9212 9296 rt : (7,4,1,3)
9212 9536 $2^x : 1 + 3\sqrt{3}/4$
9213 1773 sr : $8(1/\pi)/3$
9213 3440 $\exp(2/5) - \exp(5)$
9213 3545 rt : (8,7,4,4)
9213 4434 cu : $3 - 3\sqrt{3}/4$
9213 7131 sq : $4/3 - 3\sqrt{3}$
9213 8716 $e^x : 4 + 3e/2$
9214 2436 rt : (4,8,-1,-9)
9214 4247 $2^x : 1 - \sqrt{5}/2$
9214 4800 $\Gamma : \ln(e/2)$
9214 6057 $\Psi : 5(e + \pi)/4$
9214 6347 $\lambda : 2\ln 2/9$
9214 9706 $2^x : 3/2 + \sqrt{2}/3$
9215 1541 sr : $(\sqrt[3]{3}/2)^{1/2}$
9215 4148 cr : 18/23
9215 7019 rt : (1,8,1,3)
9215 7267 cu : $3/4 + e/4$
9215 8371 rt : (2,9,-1,-7)
9215 8377 $\Gamma : 19/16$
9215 8584 $J_0 : 2\sqrt{2}/5$
9215 8633 ln : $4\pi/5$
9216 0000 sq : 24/25
9216 1214 $s\pi : 1/2 + 4e$
9216 1724 $J_0 : 1/\ln(e + \pi)$
9216 2225 rt : (1,6,-4,-8)
9216 2468 $4 \div s\pi(\sqrt{5})$
9216 2807 $J_0 : \exp(-\sqrt[3]{5}/3)$
9216 5171 $\ell_2 : \exp(e)$
9216 5627 $s\pi : 2\sqrt{3}/3 + 2\sqrt{5}$
9216 5943 $10^x : \exp(-\sqrt[3]{2})$
9216 6068 rt : (3,7,-9,0)
9216 6855 tanh : 8/5
9216 8063 rt : (4,5,-1,8)
9216 8258 $\zeta : 4\sqrt{5}/9$
9217 0075 $\sqrt{5} \times \exp(\sqrt{5})$
9217 0224 rt : (4,8,-2,-5)
9217 1295 $J_0 : 13/23$
9217 1897 $\ln(\sqrt{5}) + \exp(3/4)$
9217 2297 rt : (9,5,-9,-3)
9217 2599 $10^x : 3/4 - \pi/4$
9217 4366 rt : (5,1,-3,-2)
9217 6264 id : $3\pi^2/5$
9217 8496 rt : (6,2,0,3)
9218 0554 rt : (1,8,-8,2)
9218 1205 sq : $2\ln 2$
9218 1842 sq : $7\pi^2/5$
9218 2106 cu : $1/2 + 2\sqrt{5}$
9218 2609 $2^x : 3\pi/10$
9218 2611 rt : (7,5,3,4)
9218 2759 rt : (3,6,-5,9)
9218 3714 $\ell_2 : P_{49}$
9218 4453 cr : $(\sqrt[3]{3}/3)^{1/3}$
9218 6713 rt : (8,8,6,5)
9219 1386 rt : (3,9,0,-2)
9219 2384 $10^x : 4 + 2\sqrt{5}$
9219 2415 $\zeta : 22/19$
9219 5444 sr : 17/20
9219 8572 rt : (9,4,-7,-4)
9219 9174 $J_0 : (\pi)^{-1/2}$
9220 0548 tanh : $\exp(\sqrt{2}/3)$
9220 1528 rt : (7,9,-7,3)
9220 1621 $J : \sqrt[3]{5}/5$
9220 2458 sq : $4\sqrt{2}/3 + \sqrt{5}/3$
9220 4393 tanh : $10\sqrt[3]{3}/9$
9220 6978 rt : (2,8,-3,-8)
9220 8033 tanh : $4\zeta(3)/3$
9220 8531 sq : $\exp(\sqrt[3]{4})$
9220 9586 sr : $2\sqrt{2} + \sqrt{3}/2$
9221 0713 rt : (1,8,7,-9)
9221 1131 $\Gamma : \sqrt[3]{5/3}$
9221 1551 $e \div \sqrt{2}$
9221 2591 $\exp(2/5) - s\pi(\pi)$
9221 3133 $2^x : 3/2 + 3\sqrt{5}/2$
9221 5295 $\Psi : (\ln 3/2)^{1/3}$
9221 5717 $\pi \times \exp(\sqrt{2})$
9221 6371 rt : (3,1,-5,8)
9221 8465 rt : (3,6,-7,-1)
9222 2479 rt : (8,6,-8,-4)
9222 3328 $\ln(2) \times \exp(3)$
9222 4481 rt : (9,3,-5,-5)
9222 5848 $10^x : 3/4 + 2\sqrt{5}$
9222 6480 rt : (5,0,-1,3)
9222 6740 $2^x : 2\sqrt{2}/3$
9222 7012 $s\pi : 8(e + \pi)/3$
9222 7051 $\Psi : 7\pi/3$
9222 7687 $\ln(\sqrt{2}) \times \exp(4)$
9222 7886 ln : $7(e + \pi)/6$
9222 8236 $\theta_3 : \exp(-1/e)$
9222 8236 sr : $e\pi$
9222 9476 $\Gamma : 7\sqrt[3]{2}/5$
9223 0415 $e^x : 4/3 - \sqrt{2}$
9223 0468 rt : (6,3,2,4)

9223 0716 rt : (2,6,-8,-3)
9223 4444 rt : (7,6,5,5)
9223 4621 erf : $9\ln 2/5$
9223 6203 sinh : $\exp(\sqrt[3]{5})$
9223 8408 rt : (8,9,8,6)
9223 9807 sπ : $2/3+\sqrt{2}/2$
9224 0519 sπ : $8\zeta(3)/7$
9224 0585 Ψ : $\sqrt{5}+\sqrt{6}+\sqrt{7}$
9224 4140 tπ : $\sqrt{2}/2-4\sqrt{5}$
9224 4902 J_0 : $9/16$
9224 6094 rt : (7,8,-9,-4)
9224 7162 Γ : $(3\pi/2)^{-1/2}$
9224 8065 rt : (8,5,-6,-5)
9224 8891 2^x : $1/2+\pi/3$
9225 0034 rt : (9,2,-3,-6)
9225 0199 ln : $9\sqrt{5}/8$
9225 0289 rt : (6,6,-7,7)
9225 2292 tπ : $5\sqrt{3}/7$
9225 2812 J : $11/25$
9225 4393 rt : (2,7,-5,-9)
9225 4542 sπ : $\sqrt{2}/3+2\sqrt{3}/3$
9225 4818 J_0 : $\ln(\sqrt[3]{5}/3)$
9225 8774 tπ : $3\sqrt[3]{2}/7$
9226 0112 Γ : $\sqrt[3]{11/2}$
9226 1884 Ei : $(e+\pi)/9$
9226 2093 at : $2-e/4$
9226 3503 rt : (2,9,3,-6)
9226 3507 cr : $\pi/4$
9226 3890 as : $1/4-\pi/3$
9226 4623 2^x : $\sqrt[3]{23/3}$
9226 4973 id : $3/2-\sqrt{3}/3$
9226 5532 rt : (2,8,-8,-1)
9226 5973 $\exp(4)-s\pi(\sqrt{5})$
9226 6891 10^x : $\exp(-1/e)$
9226 8720 cr : $3e-\pi/3$
9226 9424 rt : (3,5,-5,-2)
9227 0161 Ψ : $22/3$
9227 1366 rt : (7,7,-7,-5)
9227 1576 Ei : $5\sqrt{2}/7$
9227 2624 2^x : $3/4-\sqrt{3}/2$
9227 3304 rt : (8,4,-4,-6)
9227 3448 Γ : $10\sqrt{3}/3$
9227 4007 θ_3 : $\sqrt[3]{5}/4$
9227 4315 cr : $5\sqrt{2}/9$
9227 5039 J_0 : $\exp(-\sqrt{3}/3)$
9227 5239 rt : (9,1,-1,-7)
9227 5460 at : $1/4-\pi/2$
9227 5884 cr : $11/14$
9227 6235 e^x : $4/3-e/4$
9227 7172 rt : (5,1,1,4)
9227 7366 ln : $4\sqrt[3]{5}$
9227 8433 Ψ : 3
9227 9027 ℓ_{10} : $10(e+\pi)/7$
9228 1028 rt : (6,4,4,5)
9228 1338 ℓ_{10} : $(3\pi)^3$
9228 4872 rt : (7,7,7,6)
9228 7567 rt : (2,4,-3,7)
9228 7779 2^x : $4+3\pi/4$
9228 8759 sπ : P_{74}
9228 9253 ln : $3\sqrt{2}+3\sqrt{3}/2$
9228 9684 θ_3 : $9/13$
9229 0012 erf : $5/4$
9229 0852 erf : $((e+\pi)/3)^{1/3}$
9229 1334 ln : $\sqrt{19/3}$
9229 4334 tπ : $3e/2-3\pi$
9229 4388 rt : (6,9,-8,-5)
9229 6297 rt : (7,6,-5,-6)
9229 7145 rt : (3,9,-4,-9)
9229 8145 sr : $9\sqrt[3]{5}$
9229 8204 rt : (8,3,-2,-7)
9229 9942 id : $4\sqrt[3]{3}/3$
9230 0107 rt : (9,0,1,8)
9230 1492 cu : $2+\pi$
9230 3726 10^x : $1/2+3\sqrt{5}$
9230 4611 Γ : $\sqrt{2}/9$
9230 7692 id : $12/13$
9230 8305 J : $\exp(-13\pi/18)$
9230 8526 sr : $\sqrt{24}\pi$
9230 9218 cr : $\sqrt{2}/4+\sqrt{3}/4$
9231 0758 e^x : $2+\sqrt{5}/4$
9231 1629 2^x : $3/4+\sqrt{5}$
9231 1860 ℓ_{10} : $8\pi/3$
9231 2307 cu : $1+3\pi$
9231 2331 J_0 : $14/25$
9231 2728 sπ : $3/4+4e/3$
9231 3276 J_0 : $4\sqrt[3]{2}/9$
9231 5255 rt : (2,7,-6,-2)
9231 9019 rt : (6,8,-6,-6)
9232 0142 Γ : $13/11$
9232 0897 rt : (7,5,-3,-7)
9232 2323 Γ : $5\sqrt{2}/4$
9232 2772 rt : (8,2,0,-8)
9232 3021 tanh : $\sqrt{2}-\sqrt{6}+\sqrt{7}$
9232 4187 2^x : $3e/2+3\pi$
9232 4645 rt : (9,1,3,9)
9232 6514 rt : (5,2,3,5)
9232 9104 id : $8\zeta(3)/5$
9233 0175 sinh : $19/23$
9233 0245 rt : (6,5,6,6)
9233 1031 Γ : $\ln(e+\pi)$
9233 2664 at : $\sqrt{3}+\sqrt{5}-\sqrt{7}$
9233 3965 rt : (7,8,9,7)
9233 4039 Γ : $\exp(\sqrt[3]{5}/3)$
9233 4321 $\ln(3)\div\ln(4/5)$
9233 7393 cu : $9\sqrt{2}$
9233 8769 J_0 : $\sqrt{5}/4$
9234 3327 rt : (6,7,-4,-7)
9234 3808 cr : $5\sqrt[3]{2}/8$
9234 4987 at : $\sqrt{3}/3+\sqrt{5}/3$
9234 5174 rt : (7,4,-1,-8)
9234 5353 sπ : $2e+4\pi/3$
9234 7019 rt : (8,1,2,-9)
9234 8012 cr : $5e\pi/6$
9234 8500 cu : $\sqrt{2}+\sqrt{3}+\sqrt{5}$
9235 0647 sr : $\exp(-1/(2\pi))$
9235 2484 sq : $3/4+4\sqrt{5}/3$
9235 3815 at : $3-3\sqrt{5}/4$
9235 4378 sq : $1+4\pi/3$
9235 4384 2^x : $3\sqrt{2}/2-\sqrt{5}$
9235 5415 sπ : $(\sqrt[3]{4}/3)^{-1/2}$
9235 7672 Γ : $23/13$
9235 8060 erf : $1/\ln(\sqrt{2}\pi/2)$
9235 8612 sπ : $1/\sqrt[3]{19}$
9235 9338 ℓ_2 : $\exp(4/3)$
9236 0019 rt : (1,9,-7,-2)
9236 1559 e^x : $\sqrt{2}/4-\sqrt{3}/4$
9236 3674 rt : (2,6,-4,-3)
9236 4158 sπ : $\exp(1/\pi)$
9236 5497 rt : (5,9,-5,-7)
9236 7318 rt : (6,6,-2,-8)
9236 7703 rt : (2,9,-6,-9)
9236 7875 cu : $e/4-\pi$
9236 8075 erf : $(\pi/2)^{1/2}$
9236 9136 rt : (7,3,1,-9)
9237 0951 rt : (4,0,2,5)
9237 2294 id : $9\sqrt[3]{5}/8$
9237 2388 sπ : $2/3+3\sqrt{5}$
9237 2468 id : $\sqrt{3}/4+2\sqrt{5}/3$
9237 3000 tπ : $\sqrt{2}/4+3\sqrt{3}/4$
9237 3757 at : $\sqrt{2}/4-3\sqrt{5}/4$
9237 4048 cu : $4/3-2\sqrt{5}$
9237 4538 sq : $1/3+4\sqrt{2}/3$
9237 4574 rt : (5,3,5,6)
9237 6043 $4/5\div s\pi(1/3)$
9237 7772 as : $(\pi/2)^{-1/2}$
9237 8186 rt : (6,6,8,7)
9237 9129 rt : (1,5,2,8)
9238 2474 sr : $3\pi^2/8$
9238 5655 cu : $1/2-4\sqrt{3}/3$
9238 7953 sπ : $3/8$
9238 8193 Ψ : $(\ln 2)^{-3}$
9238 8263 10^x : $9/13$
9238 9209 rt : (5,8,-3,-8)
9239 1001 rt : (6,5,0,-9)
9239 1713 Ψ : $9/11$
9239 1749 J_0 : $7(1/\pi)/4$
9239 5180 tanh : $21/13$
9239 6043 rt : (9,2,-8,-2)
9239 8472 $\ln(3/5)-\exp(5)$
9240 1773 id : $\sqrt[3]{25}$
9240 2274 tπ : $7\sqrt{2}/8$
9240 2949 ζ : $19/16$
9240 4439 id : $e/2-2\pi$
9240 7313 rt : (1,8,-5,-3)
9240 7582 sq : $3-3e/4$
9240 7744 tanh : $8\sqrt{2}/7$
9240 8180 sr : $3\sqrt{3}+3\sqrt{5}/2$
9240 8784 ζ : $8\sqrt[3]{2}/9$
9241 0682 sr : $e\pi/10$
9241 0732 sq : $7\sqrt[3]{3}$
9241 0853 rt : (2,7,2,-4)
9241 1681 at : $(\sqrt[3]{5}/3)^{-1/2}$
9241 2031 rt : (7,9,-6,5)
9241 2620 rt : (5,8,0,6)
9241 4383 rt : (3,7,6,7)
9241 4824 e^x : $2e-\pi$
9241 6198 Γ : $5\sqrt{2}/6$
9241 7089 Li_2 : $4\sqrt[3]{2}/7$
9241 7358 10^x : $\zeta(3)$
9241 7491 tπ : $1/3-\pi/2$
9241 7903 rt : (4,1,4,6)
9241 8223 10^x : $1+4\sqrt{3}/3$
9241 9624 id : $4\ln 2/3$
9242 1413 rt : (5,4,7,7)
9242 1888 2^x : $3\sqrt{3}+4\sqrt{5}$
9242 2820 cr : $15/19$
9242 2864 as : $\ln(\sqrt{2}\pi/2)$
9242 2960 tanh : $\exp(\sqrt[3]{3}/3)$
9242 3301 id : $2\sqrt{2}/3+4\sqrt{5}/3$
9242 3461 sq : $3\sqrt{2}/2+\sqrt{3}/2$
9242 3788 tanh : $7\ln 2/3$
9242 5065 Li_2 : $18/25$
9242 6616 ℓ_2 : $2e\pi/9$
9242 8325 Γ : $3\pi/8$
9243 1524 J_0 : $5/9$
9243 2704 e^x : $2/3-\sqrt{5}/3$
9243 3703 rt : (2,6,-7,1)
9243 3997 rt : (4,9,-2,-9)
9243 6076 Γ : $(\pi)^{1/2}$
9243 8984 tπ : $4\sqrt{2}-4\sqrt{3}/3$
9244 2502 rt : (8,0,-9,2)
9244 3157 Γ : $(\sqrt[3]{3}/2)^{-1/2}$
9244 4767 2^x : $17/18$
9244 5695 erf : $2\pi/5$
9244 6247 rt : (9,3,-6,-1)
9244 8169 sq : $2\sqrt[3]{3}/3$
9245 0089 id : $10\sqrt{3}/9$
9245 2973 ℓ_{10} : $1/4+3e$
9245 3420 rt : (1,7,-3,-4)
9245 4038 θ_3 : $3e/10$
9245 5081 J_0 : $(e+\pi)^{-1/3}$
9245 5964 erf : $8\sqrt{2}/9$
9245 6852 rt : (6,6,-6,9)
9245 7765 2^x : $23/10$
9245 9230 J_0 : $4\ln 2/5$
9246 0274 rt : (3,1,3,-6)
9246 0949 tanh : $9\sqrt[3]{2}/7$
9246 2822 sπ : $4e/3$
9246 3464 at : $(\ln 3)^3$
9246 3686 rt : (4,2,6,7)
9246 5559 id : $(\sqrt[3]{5}/2)^{1/2}$
9246 7018 rt : (1,9,0,-6)
9246 7089 rt : (5,5,9,8)
9247 4576 J : $\zeta(3)/9$
9247 6181 Γ : $20/17$
9247 6211 sq : $4\zeta(3)/5$
9247 6838 sr : $\sqrt{2}-\sqrt{5}/4$
9247 7172 Γ : $7\sqrt[3]{5}/3$
9247 7363 at : $\sqrt[3]{7/3}$
9247 7737 rt : (3,8,-5,-8)
9247 7796 id : $1/2+3\pi$
9248 0069 2^x : $4e+\pi/3$
9248 0344 id : $4\sqrt{2}-\sqrt{3}$
9248 1315 Ψ : $5\zeta(3)/2$
9248 3741 sπ : $8(e+\pi)/5$
9248 6375 at : $4\sqrt{2}/3-\sqrt{5}/4$
9248 6438 sπ : $2\sqrt{2}/3+\sqrt{3}/4$
9249 1440 rt : (8,1,-7,-1)
9249 2792 sπ : $\exp(-\sqrt{2}/3)$
9249 5059 rt : (9,5,-8,-4)

9249 5206 sr : $3\sqrt{3}-2\sqrt{5}/3$
9249 6090 cu : $\sqrt{3}/4-3\sqrt{5}/2$
9249 8396 rt : (1,6,-1,-5)
9249 9049 tanh : $9\sqrt[3]{3}/8$
9250 0061 rt : (3,9,1,1)
9250 1724 rt : (2,3,2,7)
9250 2206 ζ : $(\pi)^{1/2}$
9250 2306 ℓ_{10} : $7\zeta(3)$
9250 2523 10^x : $(\ln 3/3)^{-1/2}$
9250 2525 ℓ_2 : $1/3+2\sqrt{3}$
9250 3466 $s\pi$: $3\sqrt{2}/2-\sqrt{5}/3$
9250 3918 $\sqrt{2}-\ln(3/5)$
9250 5043 rt : (3,0,5,7)
9250 5879 rt : (1,5,-7,-9)
9250 8318 cr : 19/24
9250 8354 rt : (4,3,8,8)
9251 0507 10^x : $4/3+3e$
9251 0987 2^x : $3\sqrt[3]{2}/4$
9251 3511 at : $3/4+\sqrt{3}/3$
9251 4876 rt : (2,9,-5,-4)
9251 5799 e^x : $4/3+4\sqrt{2}$
9251 6333 rt : (8,7,-9,-4)
9251 8043 rt : (9,4,-6,-5)
9252 0930 cu : $1+e/3$
9252 1229 $s\pi$: $e-2\pi/3$
9252 1341 cu : $2/3+\sqrt{3}/3$
9252 1390 ln : $3/4-\sqrt{2}/4$
9252 1772 erf : $\sqrt[3]{2}$
9252 5449 $t\pi$: $(\ln 3/3)^{-1/2}$
9252 5613 ℓ_{10} : $4\sqrt{3}+2\sqrt{5}/3$
9252 7696 $s\pi$: $9\ln 2/10$
9252 7849 cr : $8\ln 2/7$
9253 1251 cu : $\sqrt{3}/4-3\sqrt{5}/4$
9253 4622 tanh : 13/8
9253 5399 Ei : 7/2
9253 5543 rt : (7,1,-8,1)
9253 8094 ℓ_2 : $3\sqrt{2}+3\sqrt{5}/2$
9253 8483 sr : $3+\sqrt{2}/2$
9253 9052 rt : (8,6,-7,-5)
9254 0007 sq : $3\sqrt{3}/4+4\sqrt{5}$
9254 0736 rt : (9,3,-4,-6)
9254 2291 rt : (1,7,9,-6)
9254 2378 $t\pi$: $3\sqrt{2}/2+3\sqrt{5}/2$
9254 2552 rt : (9,5,-2,1)
9254 3914 $t\pi$: $7\sqrt{5}$
9254 4467 sq : $2\sqrt{2}-\sqrt{5}/2$
9254 5522 rt : (2,2,4,-7)
9254 7051 $s\pi$: $4\sqrt{2}/3+2\sqrt{5}/3$
9254 8743 rt : (3,1,7,8)
9254 9029 $1/4+s\pi(\sqrt{5})$
9254 9852 e^x : $4-3e/2$
9255 1152 id : $e/2+4\pi$
9255 2481 at : $3/2-2\sqrt{2}$
9255 2729 sr : $\sqrt{2}/4+3\sqrt{5}/2$
9255 3097 10^x : 17/22
9255 5180 cu : $1/3+4\sqrt{2}/3$
9255 6069 cr : $3/2-\sqrt{2}/2$
9255 6559 e^x : $1/2-\sqrt{3}/3$
9255 9286 θ_3 : 3/7
9255 9828 rt : (7,8,-8,-5)
9256 0011 10^x : $3/2-e/3$
9256 1487 rt : (8,5,-5,-6)
9256 1985 rt : (1,9,1,3)
9256 3144 rt : (9,2,-2,-7)
9256 3813 λ : $\sqrt{2}/9$
9256 3849 rt : (2,9,4,-2)
9256 7314 Γ : 16/9
9256 8047 $4\div\ln(4/5)$
9256 8832 Γ : $(2\pi)^{-2}$
9256 8883 erf : $7\sqrt[3]{3}/8$
9256 9376 as : $1/2-3\sqrt{3}/4$
9257 0702 2^x : $3-\sqrt{3}/4$
9257 0777 J_0 : $1/\sqrt[3]{6}$
9257 0933 cr : $\sqrt{3}/4+3\sqrt{5}$
9257 0979 rt : (3,5,-9,6)
9257 1020 2^x : $3e/2-4\pi/3$
9257 2409 tanh : $\sqrt{2}-\sqrt{5}+\sqrt{6}$
9257 3499 ℓ_{10} : $3/2+4\sqrt{3}$
9257 4313 cr : $4+\pi$
9257 8342 sinh : $\pi^2/7$
9257 8603 rt : (6,3,-9,1)
9257 8922 Ei : 15/23
9257 9282 J_0 : 11/20
9258 0282 Γ : $9\sqrt{2}/2$
9258 0957 $s\pi$: $3\sqrt{2}-\sqrt{3}/2$
9258 1988 ln : $7\sqrt[3]{3}/4$
9258 2009 sr : 6/7
9258 2053 K : $(\sqrt{3})^{-1/2}$
9258 3644 rt : (8,4,-3,-7)
9258 4303 θ_3 : $2\sqrt{3}/5$
9258 5155 rt : (1,5,6,7)
9258 5276 rt : (9,1,0,-8)
9258 5406 rt : (8,3,-3,1)
9258 5896 $t\pi$: $2e/3+3\pi/2$
9258 7471 id : $(\sqrt[3]{2})^{-1/3}$
9258 7826 at : $2e/3-\pi$
9258 8292 rt : (2,1,6,-8)
9258 8793 rt : (9,6,0,2)
9258 8941 cu : $3+3e/4$
9259 1421 rt : (3,2,9,9)
9259 2592 sq : $5\sqrt{3}/9$
9259 2636 $\exp(4)\times s\pi(2/5)$
9259 3853 Γ : $\exp(1/(2\pi))$
9259 6144 erf : 24/19
9259 6461 rt : (7,8,-9,3)
9259 7640 J_0 : $\ln(\sqrt{3})$
9259 7912 ℓ_2 : $10\sqrt[3]{5}/9$
9259 9172 at : $1/\ln(\sqrt{2}/3)$
9259 9941 ℓ_2 : 5/19
9260 1399 tanh : $(\ln 2/3)^{-1/3}$
9260 2307 rt : (6,9,-7,-4)
9260 3739 $2/3\times\exp(2)$
9260 3919 rt : (7,6,-4,-7)
9260 4094 sq : $1-4\sqrt{5}/3$
9260 4234 ℓ_2 : $2\sqrt{3}/3+\sqrt{5}/3$
9260 4651 id : $7e\pi/3$
9260 5528 rt : (8,3,-1,-8)
9260 5685 Ei : $4\sqrt{5}$
9260 5821 K : $4\sqrt[3]{5}/9$
9260 7136 rt : (9,0,2,9)
9260 7475 K : 19/25
9260 8653 at : $7\sqrt[3]{5}/9$
9261 0236 Γ : $(e+\pi)/5$
9261 0284 λ : $\sqrt[3]{2}/8$
9261 0776 tanh : $\sqrt[3]{13/3}$
9261 0798 rt : (7,9,-5,7)
9261 1494 $s\pi$: $3/4+4e$
9261 2870 ln : $4\ln 2/7$
9261 5543 sr : $(e/2)^{-1/2}$
9261 6558 ℓ_{10} : $3+2e$
9261 6779 $\exp(\sqrt{5})+s\pi(\pi)$
9261 7877 2^x : $(\sqrt{5}/2)^{-1/2}$
9261 8432 λ : $\exp(-10\pi/17)$
9261 9946 tanh : $3e/5$
9262 0911 rt : (1,9,1,-6)
9262 1973 2^x : $1/4+3\sqrt{3}/4$
9262 3975 rt : (6,8,-5,-7)
9262 4086 Li_2 : $\sqrt[3]{3}/2$
9262 4735 10^x : $3e-2\pi/3$
9262 5562 rt : (7,5,-2,-8)
9262 6013 rt : (2,6,-6,5)
9262 7031 rt : (1,3,5,-8)
9262 7148 rt : (8,2,1,-9)
9262 7274 rt : (7,1,-4,1)
9262 7991 cr : $4\sqrt{2}+2\sqrt{5}/3$
9263 0079 rt : (2,0,8,9)
9263 0564 rt : (8,4,-1,2)
9263 3844 rt : (9,7,2,3)
9263 5437 e^x : $\sqrt{2}-2\sqrt{5}/3$
9263 5555 ℓ_{10} : $\sqrt{3}+3\sqrt{5}$
9263 8096 rt : (6,5,-9,7)
9263 8601 sq : $4/3-4\sqrt{5}$
9264 1084 e^x : $7\pi^2/2$
9264 1605 θ_3 : $(2e)^{-1/2}$
9264 3453 Li_2 : $3\zeta(3)/5$
9264 5382 rt : (6,7,-3,-8)
9264 6103 sq : $3/4+3\sqrt{2}$
9264 6369 2^x : $\sqrt{2}+\sqrt{5}/4$
9264 6946 rt : (7,4,0,-9)
9264 7879 $t\pi$: $\exp(-\sqrt{3}/2)$
9264 8665 tanh : $\sqrt{8/3}$
9264 8716 sq : $7\pi^2/9$
9265 0275 sr : $4-\pi$
9265 5628 Γ : $\exp(\sqrt{3}/3)$
9265 6321 rt : (2,9,-4,1)
9266 0181 sinh : $\sqrt{2}-\sqrt{3}-\sqrt{6}$
9266 0235 $s\pi$: $\sqrt{3}/4+4\sqrt{5}$
9266 1322 sr : $5\zeta(3)/7$
9266 2013 rt : (1,8,-1,-7)
9266 2877 λ : 3/19
9266 3535 Li_2 : $1/\ln(4)$
9266 3707 sinh : $(\ln 3)^{-2}$
9266 4991 rt : (5,9,-4,-8)
9266 6534 rt : (6,6,-1,-9)
9266 7962 rt : (1,2,7,-9)
9266 8054 cr : $5(1/\pi)/2$
9266 8197 rt : (6,1,-5,-1)
9266 8557 id : $3\sqrt{2}/4+\sqrt{3}/2$
9266 9138 cu : $\sqrt{2}/4+3\sqrt{3}$
9266 9930 Ψ : $6e\pi$
9267 1394 rt : (7,2,-2,2)
9267 4583 rt : (8,5,1,3)
9267 6079 erf : 19/15
9267 7763 rt : (9,8,4,4)
9267 8897 $s\pi$: $\exp(\sqrt{3}/2)$
9268 2030 $s\pi$: $9\sqrt[3]{3}/8$
9268 5459 2^x : $4/3+2\sqrt{5}$
9268 5917 rt : (5,8,-2,-9)
9268 9665 2^x : $2\sqrt{3}-2\sqrt{5}/3$
9269 1101 sq : $e/2+2\pi/3$
9269 2107 rt : (1,9,-9,2)
9269 2801 rt : (1,8,7,-3)
9269 3772 Γ : $\sqrt[3]{17/3}$
9269 5140 $s\pi$: $8\pi/3$
9269 5933 e^x : $5e\pi/9$
9269 6250 tanh : 18/11
9269 9081 id : $5\pi/4$
9269 9155 J_0 : 6/11
9270 2201 rt : (1,7,-3,-8)
9270 3415 $t\pi$: $2\sqrt{2}/3-2\sqrt{3}$
9270 5098 id : $1/4+3\sqrt{5}/4$
9270 6059 sq : $e\pi$
9270 6598 rt : (5,8,1,8)
9270 8215 rt : (5,3,-6,-1)
9270 9069 ζ : 16/13
9271 1324 rt : (6,0,-3,2)
9271 1393 cr : $3/2+4\sqrt{2}$
9271 1847 10^x : $\sqrt[3]{22}$
9271 4424 rt : (7,3,0,3)
9271 5705 rt : (1,9,8,-4)
9271 7517 rt : (8,6,3,4)
9271 9002 e^x : $\exp(\sqrt{3})$
9272 0046 sr : $1/4+2\sqrt{3}$
9272 0601 rt : (9,9,6,5)
9272 5562 rt : (4,3,-7,4)
9272 5733 as : $((e+\pi)/3)^{-1/3}$
9272 7926 sinh : $4\pi/7$
9272 9521 at : 4/3
9273 3185 $\sqrt{3}-\ln(\sqrt{5})$
9273 4542 J_0 : $1/\ln(2\pi)$
9273 5295 ℓ_{10} : $6\pi^2/7$
9273 5990 sinh : $1/\ln(\sqrt{2})$
9273 7498 $\exp(4/5)\times s\pi(1/3)$
9273 8687 rt : (2,9,-2,-8)
9274 0154 id : P_{34}
9274 0340 $s\pi$: $1/\sqrt{7}$
9274 1228 e^x : $4\sqrt[3]{5}/5$
9274 1513 rt : (1,6,-5,-9)
9274 1732 $s\pi$: $3\sqrt[3]{2}/10$
9274 4854 sinh : $(\sqrt[3]{5}/3)^{1/3}$
9274 5792 rt : (4,4,-3,8)
9274 6323 J_0 : $e/5$
9274 7365 rt : (4,7,-5,-7)
9274 9879 sr : $(\pi/2)^{-1/3}$
9275 0389 rt : (5,2,-4,-2)
9275 0712 10^x : $\sqrt[3]{5}/6$
9275 2053 sr : $e/2+3\pi/4$
9275 3406 rt : (6,1,-1,3)
9275 6415 rt : (7,4,2,4)
9275 6482 $\exp(2/5)\div s\pi(e)$

9275 7015 $10^x : 3\sqrt{3}+\sqrt{5}/3$
9275 7540 rt : (3,5,-8,9)
9275 9008 $10^x : 6\ln 2/7$
9275 9415 $e^x : 2/3+2e/3$
9275 9416 rt : (8,7,5,5)
9276 0407 cr : $\ln(\sqrt{2}\pi/2)$
9276 0847 $10^x : \exp((e+\pi)/2)$
9276 2135 $e^x : 1/4+\sqrt{5}/2$
9276 3028 ln : $3\sqrt{3}+3\sqrt{5}/4$
9276 3531 $10^x : 4\sqrt{2}-3\sqrt{3}/2$
9276 4354 rt : (3,8,-4,-5)
9276 6400 ln : $3-\sqrt{2}/3$
9276 6439 $10^x : 3\sqrt{2}/4+3\sqrt{3}/4$
9276 6663 $\Psi : \sqrt{2/3}$
9276 6952 $\Gamma : 25/14$
9276 9337 $e^x : 3\sqrt{3}/2+3\sqrt{5}$
9276 9609 rt : (9,6,-9,-4)
9277 1933 $\Gamma : 7/6$
9277 2150 sr : $3\sqrt{3}/2+\sqrt{5}/2$
9277 2507 $\theta_3 : \ln 2$
9277 6175 $\Gamma : (\sqrt[3]{4})^{1/3}$
9277 6412 $10^x : 10e/7$
9277 7233 rt : (2,8,-4,-9)
9278 0506 $2\times s\pi(\sqrt{2})$
9278 2305 $10^x : 3\sqrt{2}+\sqrt{5}/2$
9278 4151 rt : (3,9,2,4)
9278 5606 rt : (7,8,-8,5)
9278 5681 rt : (3,7,-8,-1)
9278 6440 $t\pi : 5/21$
9278 7092 tanh : 23/14
9278 8625 rt : (4,8,-1,-3)
9279 0094 rt : (9,5,-7,-5)
9279 0481 $t\pi : 3\sqrt{2}+\sqrt{5}$
9279 1561 rt : (5,1,-2,-3)
9279 1574 $J_0 : (2\pi)^{-1/3}$
9279 2375 rt : (2,9,-3,6)
9279 3895 $Ei : -\sqrt{2}+\sqrt{5}+\sqrt{6}$
9279 4491 rt : (6,2,1,4)
9279 5649 sq : $3e+\pi/4$
9279 7412 rt : (7,5,4,5)
9279 8006 $\Gamma : (e/2)^{1/2}$
9279 8347 $J_0 : 13/24$
9279 8833 rt : (7,9,-4,9)
9279 9291 $\ell_{10} : 4+2\sqrt{5}$
9280 0327 rt : (8,8,7,6)
9280 1778 at : $(\sqrt[3]{2}/3)^{-1/3}$
9280 2234 cu : $2/3-3\sqrt{3}$
9280 2257 tanh : $(\sqrt{2}\pi)^{1/3}$
9280 2744 $s\pi : 7e/8$
9280 6842 $J_0 : (e/2)^{-2}$
9280 7207 $e^x : 1/4+4\sqrt{3}/3$
9280 8492 rt : (2,6,-5,9)
9280 8892 rt : (8,7,-8,-5)
9280 9986 $s\pi : 2e/3+4\pi$
9281 0340 rt : (9,4,-5,-6)
9281 1707 at : $10\zeta(3)/9$
9281 2186 $\ell_2 : 7e/5$
9281 2486 $erf : 14/11$
9281 3936 $erf : 9\sqrt{2}/10$
9281 5917 tanh : $\pi^2/6$
9281 8576 $Li_2 : 13/18$
9281 9816 $J_0 : 3\sqrt[3]{3}/8$
9282 0323 id : $4\sqrt{3}$
9282 0492 $\exp(e)+s\pi(e)$
9282 1439 $\zeta : ((e+\pi)/2)^{1/3}$
9282 1630 $t\pi : P_{13}$
9282 2446 $s\pi : 1/\ln(5)$
9282 3197 rt : (2,9,-9,-1)
9282 3919 $erf : 4(1/\pi)$
9282 4057 rt : (6,5,-8,9)
9282 4151 $s\pi : 1/2+3\sqrt{2}/2$
9282 4156 $\Gamma : 3e/7$
9282 4463 $\lambda : (1/\pi)/2$
9282 5985 $t\pi : 3\sqrt{3}/2-\sqrt{5}/2$
9282 6063 rt : (3,6,-6,-2)
9282 7494 rt : (7,9,-9,-5)
9282 8468 sr : $1+e$
9282 8923 rt : (8,6,-6,-6)
9282 9871 sr : P_5
9283 0351 rt : (9,3,-3,-7)
9283 0897 cr : $((e+\pi)/3)^{-1/3}$
9283 1237 id : $\sqrt{3}-\sqrt{6}+\sqrt{7}$
9283 1776 sq : $(\sqrt{5}/2)^{-1/3}$
9283 4293 $2^x : \pi^2/5$
9283 4622 rt : (6,3,3,5)
9283 5199 $2^x : 18/19$
9283 7438 $Ei : \pi$
9283 7461 rt : (7,6,6,6)
9284 0121 sr : $4/3-\sqrt{2}/3$
9284 0293 rt : (8,9,9,7)
9284 0876 rt : (2,9,5,2)
9284 2199 $2^x : 8\pi/9$
9284 2669 $J_0 : 3\sqrt[3]{2}/7$
9284 6974 $e^x : (4)^3$
9284 7262 $\Gamma : 4\sqrt{5}/5$
9284 7316 rt : (7,8,-7,-6)
9284 7701 id : P_{17}
9284 8508 rt : (1,9,9,4)
9284 8725 rt : (8,5,-4,-7)
9285 0132 rt : (9,2,-1,-8)
9285 0953 $\exp(1/5)+s\pi(1/4)$
9285 2540 sq : $1/4+\sqrt{3}$
9285 4622 cu : $\sqrt{3}+4\sqrt{5}$
9285 5670 $Ei : 8/3$
9285 6718 $2^x : \ln(3\pi/2)$
9285 7142 id : 13/14
9286 1681 $10^x : 4\sqrt{2}+\sqrt{3}/4$
9286 2193 tanh : $8\sqrt[3]{3}/7$
9286 2735 rt : (2,8,-7,-2)
9286 4043 $\zeta : 23/25$
9286 4456 $10^x : 7/15$
9286 4819 $J_0 : 7\ln 2/9$
9286 5522 rt : (3,5,-4,-3)
9286 6624 $\ell_{10} : 6\sqrt{2}$
9286 6913 rt : (7,7,-5,-7)
9286 8025 $s\pi : 2/3+3\pi/2$
9286 8194 tanh : $(e)^{1/2}$
9286 8302 rt : (8,4,-2,-8)
9286 9689 rt : (9,1,1,-9)
9287 0423 rt : (1,8,8,4)
9287 1075 rt : (5,1,2,5)
9287 2131 $\exp(4/5)-\exp(e)$
9287 3307 $e^x : (\sqrt{3}/2)^{-1/2}$
9287 3842 rt : (6,4,5,6)
9287 6602 rt : (7,7,8,7)
9287 6709 at : $\sqrt{2}/3+\sqrt{3}/2$
9288 0092 cr : $5\sqrt[3]{3}/9$
9288 0901 $2^x : 19/5$
9288 1452 $\ell_{10} : 1/3+3e$
9288 1778 $J_0 : 7/13$
9288 4606 tanh : $7\sqrt{2}/6$
9288 4786 cr : $2\zeta(3)/3$
9288 4862 $J : \exp(-4\pi/7)$
9288 4897 sr : $2e+\pi$
9288 4917 rt : (6,9,-6,-2)
9288 6289 rt : (7,6,-3,-8)
9288 6935 at : $2\sqrt{2}-2\sqrt{5}/3$
9288 7247 $\Psi : -\sqrt{5}-\sqrt{6}+\sqrt{7}$
9288 7658 rt : (8,3,0,-9)
9289 1654 $s\pi : \sqrt[3]{18}$
9289 5403 cu : $4\sqrt{3}+\sqrt{5}/2$
9289 5930 $\theta_3 : \sqrt[3]{1/3}$
9289 6682 $2^x : 4+2e$
9289 6741 $erf : (\sqrt[3]{3}/3)^{-1/3}$
9289 8973 tanh : $\sqrt[3]{9/2}$
9290 1234 sq : 25/18
9290 1386 rt : (2,7,-5,-3)
9290 2570 $t\pi : \exp(\sqrt{3})$
9290 2823 cr : $1/4+4\sqrt{3}$
9290 3122 $\Gamma : (\pi/2)^{1/3}$
9290 3781 $K : 16/21$
9290 4096 rt : (6,8,-4,-8)
9290 5449 rt : (7,5,-1,-9)
9290 5527 $2^x : 5\pi^2/8$
9290 6800 rt : (4,1,1,-5)
9290 9497 rt : (5,2,4,6)
9291 0726 ln : $\sqrt{2}+\sqrt{5}/2$
9291 2032 $e^x : \exp(-\sqrt[3]{2}/3)$
9291 2188 rt : (6,5,7,7)
9291 3172 $\ell_2 : 2/3+\pi$
9291 4230 rt : (1,7,5,-8)
9291 5255 $\zeta : 21/23$
9291 5409 cr : $\sqrt{2}/3+3\sqrt{5}$
9291 6092 sr : $(\sqrt{5}/3)^{1/2}$
9291 8377 cr : 8π
9291 9561 $2^x : \sqrt{2}/3-\sqrt{3}/3$
9291 9829 tanh : $(\ln 3/3)^{-1/2}$
9292 0636 $t\pi : 9\ln 2$
9292 0941 $\ln(\sqrt{5})\div s\pi(1/3)$
9292 0957 as : $\sqrt{2}/4-2\sqrt{3}/3$
9292 1515 $\sqrt{5}+\ln(2)$
9292 3063 rt : (6,7,-2,-9)
9292 4018 $erf : 4\sqrt{5}/7$
9292 4561 $erf : 23/18$
9292 5127 sq : $3\sqrt{2}+4\sqrt{3}/3$
9292 6089 $\ln(4)\div\exp(2/5)$
9292 7164 $2^x : e/4-\pi/4$
9292 8881 $10^x : 3/2+\pi/4$
9293 0653 $2^x : 3\sqrt{2}/4+\sqrt{3}$
9293 0673 $\lambda : \exp(-7\pi/12)$
9293 0958 $\lambda : 4/25$
9293 2037 sr : 19/22
9293 3147 $J_0 : \ln(\sqrt[3]{5})$
9293 5748 $\ln(2)\div s\pi(\sqrt{3})$
9293 6543 rt : (1,9,-6,-3)
9293 7939 sq : $4/3+3\sqrt{3}/4$
9293 8105 as : $5\sqrt[3]{3}/9$
9293 8301 rt : (1,7,-8,-5)
9293 8401 cu : $3/2+3\sqrt{3}/4$
9293 9186 rt : (2,6,-3,-4)
9294 0505 rt : (5,9,-3,-9)
9294 1229 $\Psi : 9e\pi/4$
9294 1822 rt : (3,6,3,5)
9294 1892 $\ell_{10} : 2/17$
9294 4453 rt : (4,0,3,6)
9294 7078 rt : (5,3,6,7)
9294 9397 cu : $e-3\pi/2$
9294 9697 rt : (6,6,9,8)
9295 1848 $t\pi : 3\sqrt{2}/4-3\sqrt{3}/4$
9295 1954 $\ell_{10} : \exp(\sqrt{2}\pi)$
9295 2332 sr : $6(e+\pi)$
9295 4693 $\zeta : 5/16$
9295 6419 $\ell_2 : 3/2+4\sqrt{3}/3$
9295 7045 id : $1/4+e/4$
9295 7093 cu : $3/2+e/3$
9295 8138 $2/5\div s\pi(\pi)$
9295 8416 as : $2\zeta(3)/3$
9295 9079 rt : (5,7,-2,6)
9296 1108 $erf : (2\pi/3)^{1/3}$
9296 2175 $\lambda : \sqrt[3]{3}/9$
9296 4794 rt : (7,8,-7,7)
9296 5012 $\ell_2 : 3\sqrt{2}-\sqrt{3}/4$
9296 5418 sinh : $6e\pi/7$
9296 5451 rt : (9,1,-9,-2)
9296 6697 sr : $\sqrt{2}+4\sqrt{3}/3$
9296 7733 $\Gamma : 3\pi^2/10$
9296 7896 $t\pi : 6e\pi$
9296 9809 $10^x : 2\sqrt{3}/5$
9296 9848 $\Gamma : \exp(\sqrt{2})$
9297 0390 $e^x : 3+\sqrt{5}$
9297 0515 $Ei : (e+\pi)^{-2}$
9297 3596 rt : (1,8,-4,-4)
9297 3856 sinh : 24/17
9297 6169 rt : (2,7,3,-1)
9297 6282 $\zeta : 19/21$
9297 7030 $s\pi : 2\sqrt[3]{5}/9$
9297 8736 rt : (3,7,7,9)
9298 1297 rt : (4,1,5,7)
9298 3852 rt : (5,4,8,8)
9298 9383 $s\pi : 9\sqrt[3]{2}/7$
9299 0063 $2^x : 4\sqrt{3}/3+2\sqrt{5}/3$
9299 0773 $J_0 : 4\zeta(3)/9$
9299 0778 cu : $4/3-4\sqrt{3}/3$
9299 1239 $t\pi : 8/23$
9299 2401 $\Psi : 3e/10$
9299 3724 id : $(e+\pi)/2$
9299 3827 $s\pi : 2\sqrt{2}/3+3\sqrt{5}/4$
9299 4365 rt : (4,3,-6,6)
9299 4704 $\ln(5)\div\ln(\sqrt{3})$
9299 7612 cr : $3+4\pi/3$

9300 1988 cr : $e/3 + 2\pi$
9300 4646 rt : (9,2,-7,-1)
9300 6450 cu : $3e + \pi/2$
9300 8930 2^x : $9e\pi/8$
9300 9863 rt : (1,7,-2,-5)
9301 1735 Γ : $4\pi/7$
9301 2369 rt : (2,8,9,8)
9301 3676 2^x : $3\sqrt{2}/2 + \sqrt{3}/2$
9301 3947 cr : $\ln(\sqrt{5})$
9301 4314 J_0 : $8/15$
9301 4675 cr : $1/3 + \sqrt{2}/3$
9301 4869 rt : (3,1,4,-7)
9301 4936 $s\pi$: $3\sqrt{2}/4 + \sqrt{5}/4$
9301 7145 erf : $8\sqrt[3]{3}/9$
9301 7364 rt : (4,2,7,8)
9302 3716 ζ : $23/13$
9302 4176 sr : $3\sqrt[3]{3}/5$
9302 4899 id : $1/\ln((e + \pi)/2)$
9302 4902 ℓ_2 : $4 - 2\pi/3$
9302 5033 $\ln(4/5) - s\pi(1/4)$
9302 5041 ln : $(\ln 2/3)^2$
9302 7401 at : $3\sqrt{5}/5$
9302 7481 $\exp(3) \div s\pi(\sqrt{3})$
9302 7659 erf : $3\sqrt[3]{5}/4$
9302 8322 ℓ_{10} : $4e - 3\pi/4$
9302 8949 rt : (3,8,-3,-2)
9302 9658 2^x : $\sqrt{3}/4 + \sqrt{5}/2$
9303 0121 $1/2 - s\pi(\pi)$
9303 2610 $s\pi$: $\sqrt{17/3}$
9303 2841 J_0 : $\exp(-\sqrt[3]{2}/2)$
9303 4585 2^x : $e\pi/9$
9303 4643 Γ : $(\sqrt{5}/3)^{-1/2}$
9303 8282 rt : (1,7,-7,3)
9303 8915 $s\pi$: $1/3 + \pi/3$
9303 9421 10^x : $5\pi^2/6$
9303 9438 rt : (1,5,-4,9)
9304 0292 rt : (8,0,-8,1)
9304 0828 $2/3 \times \exp(1/3)$
9304 2928 rt : (9,6,-8,-5)
9304 5374 rt : (1,6,0,-6)
9304 5919 2^x : $1/4 + 3e$
9304 6189 rt : (4,9,-9,-3)
9304 6596 rt : (3,9,3,7)
9304 7128 Ψ : $-\sqrt{2} + \sqrt{3} - \sqrt{7}$
9304 7345 Γ : $22/19$
9304 7816 rt : (2,3,3,9)
9304 8697 sr : $5\sqrt{5}/3$
9305 0099 2^x : $3\sqrt{2}$
9305 0243 ζ : $17/19$
9305 0252 rt : (3,0,6,8)
9305 2683 rt : (4,3,9,9)
9305 7003 10^x : $9\sqrt[3]{3}/8$
9305 9814 rt : (8,8,-9,-5)
9306 0136 Ψ : $2/21$
9306 0485 id : $(\sqrt{3}/2)^{1/2}$
9306 1067 rt : (9,5,-6,-6)
9306 1922 ℓ_2 : $1 + e/3$
9306 3474 cr : $2 + 3\sqrt{3}$
9306 8110 at : $3\sqrt{2}/2 - 2\sqrt{3}$
9306 9134 2^x : $(\sqrt[3]{5}/2)^{1/3}$
9306 9568 $s\pi$: $3e\pi$
9306 9772 10^x : $2/7$
9307 0016 J : $\exp(-\sqrt[3]{4}/2)$
9307 3773 2^x : $1/4 - \sqrt{2}/4$
9307 5206 rt : (7,2,-9,1)
9307 7505 Ψ : $10e/9$
9307 7774 rt : (8,9,-5,-6)
9307 8497 $\ln(3/5) \times \exp(3/5)$
9307 9010 rt : (9,4,-4,-7)
9308 0134 erf : $(\sqrt{2}/3)^{-1/3}$
9308 0160 rt : (1,5,2,-7)
9308 0337 rt : (9,4,-3,1)
9308 0882 Ψ : $(e)^2$
9308 2435 sr : $5\ln 2/4$
9308 2540 rt : (2,2,5,-8)
9308 4915 rt : (3,1,8,9)
9308 6613 J_0 : $5(1/\pi)/3$
9308 7374 $s\pi$: $8/21$
9308 7900 cu : $\sqrt{2} + 4\sqrt{5}/3$
9308 8329 cu : $\sqrt{2}/2 + 2\sqrt{5}$
9308 8920 $\exp(\pi) \div \exp(3/4)$
9309 0365 id : $\sqrt{2}/4 + \sqrt{3}/3$
9309 1384 J_0 : $3\sqrt{2}/8$
9309 4148 sq : P_{22}
9309 4322 rt : (7,9,-8,-6)
9309 4933 sr : $13/15$
9309 5208 sinh : $5\sqrt[3]{5}/2$
9309 5543 rt : (8,6,-5,-7)
9309 5611 J : $\ln(\pi/2)$
9309 6763 rt : (9,3,-2,-8)
9309 7288 rt : (2,9,6,6)
9309 7782 erf : $9/7$
9310 1501 at : $1/3 - 3\sqrt{5}/4$
9310 4276 ζ : $(3\pi/2)^{-1/3}$
9310 6071 sinh : $6\ln 2/5$
9310 6933 10^x : $\exp(-1/(3\pi))$
9311 0960 tanh : $5/3$
9311 1869 $\ln(4/5) + \exp(e)$
9311 1920 rt : (7,8,-6,-7)
9311 3125 rt : (8,5,-3,-8)
9311 3638 e^x : $1/2 + \pi/2$
9311 3866 cu : $3\sqrt{2} - 4\sqrt{5}$
9311 4246 rt : (1,4,4,-8)
9311 4329 rt : (9,2,0,-9)
9311 4418 rt : (8,2,-4,1)
9311 4870 J_0 : $9/17$
9311 5520 $s\pi$: $8\sqrt{3}/3$
9311 6567 rt : (2,1,7,-9)
9311 6910 rt : (9,5,-1,2)
9312 0501 at : $(e/3)^{-3}$
9312 1728 Γ : $(e)^{-2}$
9312 5640 at : $3/4 - 2\pi/3$
9312 7468 $\exp(3) - \exp(e)$
9312 8142 rt : (6,8,-9,-4)
9312 8828 10^x : $\ln(2e/3)$
9312 9333 rt : (7,7,-4,-8)
9312 9864 sinh : $\sqrt{24}\pi$
9313 0523 rt : (8,4,-1,-9)
9313 2564 Γ : $5\ln 2/3$
9313 3348 e^x : P_{19}
9313 3640 e^x : P_{58}
9313 4675 sinh : $\exp(-1/(2e))$
9313 4828 rt : (7,8,-6,9)
9313 5383 Γ : $10\sqrt[3]{2}/7$
9313 8377 Γ : $9/5$
9314 0954 id : $1/3 + 3\sqrt{3}/2$
9314 0960 2^x : $3e/4 + 3\pi$
9314 1739 ζ : $15/17$
9314 3116 rt : (1,9,0,-7)
9314 3699 cu : $1/2 + \sqrt{5}/3$
9314 4215 tanh : $1/\ln(\ln 3/2)$
9314 4435 ℓ_{10} : $e\pi$
9314 4879 cr : $4\sqrt{2}/7$
9314 5223 id : $\sqrt{3}/3 + 3\sqrt{5}/2$
9314 5390 rt : (6,9,-5,0)
9314 6566 rt : (7,6,-2,-9)
9314 7180 id : $10\ln 2$
9314 7660 rt : (1,3,6,-9)
9314 7826 rt : (7,0,-5,1)
9314 8336 $1/3 + \exp(4)$
9315 0014 cu : $7\sqrt{2}/3$
9315 0204 Γ : $2\sqrt{3}/3$
9315 0257 rt : (8,3,-2,2)
9315 1053 e^x : $9\sqrt[3]{2}/2$
9315 2041 $s\pi$: $e/3 + 3\pi/2$
9315 2213 sr : $4\sqrt{2}/3 + 3\sqrt{5}$
9315 2572 $t\pi$: $(e + \pi)^{1/2}$
9315 2683 rt : (9,6,1,3)
9315 3671 Ei : $8\zeta(3)/7$
9315 5968 2^x : $3e/4 + 4\pi/3$
9315 6076 cu : $(e + \pi)/6$
9315 8248 $s\pi$: $1/\sqrt[3]{18}$
9315 9645 $t\pi$: $3(1/\pi)/4$
9316 2461 rt : (6,8,-3,-9)
9316 3001 rt : (1,7,6,-2)
9316 5162 e^x : $3/2 - \pi/2$
9316 5661 ln : $1/2 + 3e/4$
9316 5992 cr : $7\ln 2/6$
9316 8940 sr : $3/4 + 4\sqrt{5}/3$
9316 9687 tanh : $\sqrt[3]{14/3}$
9317 0693 $s\pi$: $\exp(-\sqrt[3]{3}/3)$
9317 1189 cr : $1/2 + 3\sqrt{5}$
9317 1920 cr : $(\sqrt[3]{4}/3)^{1/3}$
9317 5991 rt : (1,8,-2,-8)
9317 7951 Γ : $15/13$
9317 8210 rt : (5,5,-9,0)
9317 9250 cr : $1/4 + \sqrt{5}/4$
9317 9464 Γ : $4\sqrt[3]{3}/5$
9318 0587 rt : (6,2,-6,-1)
9318 1452 ℓ_2 : $((e + \pi)/3)^2$
9318 2842 ζ : $\exp(\sqrt[3]{5}/3)$
9318 2959 rt : (7,1,-3,2)
9318 5165 sr : $2 + \sqrt{3}$
9318 5326 rt : (8,4,0,3)
9318 5476 2^x : $5\sqrt[3]{5}/9$
9318 5505 e^x : $3\sqrt{2} - 2\sqrt{3}/3$
9318 7265 2^x : $19/20$
9318 7688 rt : (9,7,3,4)
9318 8996 at : $(2e/3)^{1/2}$
9319 1195 at : $2/3 + e/4$
9319 1550 Ei : $8(e + \pi)/9$
9319 2461 10^x : $21/5$
9319 3769 J_0 : $10/19$
9319 4921 ζ : $13/10$
9319 4932 $s\pi$: $3\sqrt[3]{3}/7$
9319 4950 rt : (5,6,-5,4)
9319 6000 ℓ_{10} : $5\sqrt[3]{5}$
9319 6874 at : $3\pi/7$
9319 7903 ℓ_{10} : $3\sqrt{3} + 3\sqrt{5}/2$
9319 8377 cr : $5\sqrt[3]{3}$
9319 8704 cr : $17/21$
9320 0240 $3/4 \div \ln(\sqrt{5})$
9320 1836 2^x : $2\sqrt{3} + 2\sqrt{5}$
9320 3162 ζ : $\ln(e + \pi)$
9320 3242 $s\pi$: $1/2 + \sqrt{5}/2$
9320 3476 λ : $\exp(-11\pi/19)$
9320 3908 $s\pi$: $6(1/\pi)/5$
9320 5080 $1/5 + \sqrt{3}$
9320 7807 $s\pi$: $5\pi/6$
9320 8142 cr : $6\zeta(3)$
9320 8236 rt : (1,7,-4,-9)
9320 9522 e^x : $\sqrt{7}\pi$
9321 0403 rt : (4,9,-8,0)
9321 1084 erf : $\sqrt{5/3}$
9321 1134 tanh : $2(e + \pi)/7$
9321 1523 rt : (5,7,-1,8)
9321 2399 sinh : $(\ln 2)^{1/2}$
9321 2723 rt : (5,4,-7,-1)
9321 3318 Γ : $5\sqrt[3]{3}/4$
9321 4688 Γ : $(e + \pi)^{1/3}$
9321 5039 rt : (6,1,-4,-2)
9321 7349 rt : (7,2,-1,3)
9321 9655 rt : (8,5,2,4)
9322 0636 Γ : $3\zeta(3)/2$
9322 0994 e^x : $5(e + \pi)/9$
9322 1956 rt : (9,8,5,5)
9322 3533 Γ : $11/24$
9322 3663 at : $7\sqrt{3}/9$
9322 6825 rt : (4,2,-9,4)
9322 7602 sq : $3 - 3\pi/2$
9322 9901 2^x : $2\sqrt{2} + 4\sqrt{5}$
9322 9976 sinh : $(\sqrt[3]{3})^{-1/2}$
9323 7122 rt : (1,8,-9,-6)
9323 7754 rt : (2,9,-3,-9)
9323 8002 Ei : $(2\pi)^{-3}$
9323 9862 sq : $3e/2 - 2\pi/3$
9323 9875 rt : (1,9,9,-9)
9324 1099 tanh : $(3\pi/2)^{1/3}$
9324 3085 rt : (4,3,-5,8)
9324 4257 rt : (4,6,-8,-1)
9324 4745 $s\pi$: $4\sqrt{2}/3 + \sqrt{3}$
9324 4885 $s\pi$: $\sqrt{2} + \sqrt{3} + \sqrt{5}$
9324 6518 rt : (5,3,-5,-2)
9324 7780 tanh : $3\sqrt{5}/4$
9324 8775 rt : (6,0,-2,3)
9325 0480 sr : $20/23$
9325 1026 rt : (7,3,1,4)
9325 3273 rt : (8,6,4,5)
9325 4581 ln : $e + 4\pi/3$
9325 5516 rt : (9,9,7,6)

9325 7244 $10^x : 4/3-\pi/3$
9325 7325 $J_0 : 11/21$
9325 7840 $\zeta : 13/15$
9325 8106 rt : (3,7,-8,-8)
9325 8570 sq : $1/3-3\sqrt{3}/4$
9325 8707 $\ln(5)\times\exp(3/5)$
9326 0434 $10^x : (\sqrt{2})^{-1/2}$
9326 0692 $2\div\ln(2/3)$
9326 2069 $\zeta : 5\sqrt{2}/4$
9326 2657 $J_0 : \pi/6$
9326 4770 $\exp(4/5)+s\pi(1/4)$
9326 6408 $Ei : 2\pi^2/5$
9326 7719 cr : $3e/2+\pi$
9326 8673 $\ell_2 : \sqrt{2}/2+4\sqrt{3}$
9327 1210 $\Psi : 3\pi^2/4$
9327 1340 $2^x : 3/2+2e/3$
9327 4065 rt : (3,8,-2,1)
9327 4668 $\exp(1/2)+\exp(1/4)$
9327 5209 rt : (3,8,-9,-1)
9327 7379 $erf : 22/17$
9327 7418 rt : (4,7,-4,-5)
9327 8482 ln : $7\pi^2/10$
9327 8492 $e^x : 3/2+\sqrt{5}$
9327 8521 rt : (9,7,-9,-5)
9327 9623 rt : (5,2,-3,-3)
9328 0493 $t\pi : 1/3-2\pi/3$
9328 0789 rt : (2,8,-9,1)
9328 1639 $s\pi : 7\ln 2/3$
9328 1822 rt : (6,1,0,4)
9328 2052 $2^x : 1/3+3\pi$
9328 2380 $t\pi : 5\pi^2$
9328 3907 $J_0 : 1/\sqrt[3]{7}$
9328 4018 rt : (7,4,3,5)
9328 4789 tanh : $4\sqrt[3]{2}/3$
9328 5826 rt : (7,7,-9,7)
9328 5948 $\exp(1/5)-\exp(e)$
9328 6209 rt : (8,7,6,6)
9328 8061 $s\pi : \exp(\sqrt[3]{3}/3)$
9328 8366 id : $e-\pi/4$
9328 8580 $\ell_2 : 11/21$
9328 9109 sr : $3/2+\sqrt{5}$
9329 3510 $\zeta : \pi/10$
9329 4460 cu : $4\sqrt[3]{5}/7$
9329 4836 rt : (9,6,-7,-6)
9329 6576 cu : $4\sqrt{2}/3-2\sqrt{3}$
9329 7208 $10^x : 4/3-\sqrt{3}/2$
9329 8209 at : $(\ln 3/2)^{-1/2}$
9330 1270 id : $1/2+\sqrt{3}/4$
9330 4093 $\Gamma : 23/20$
9330 4884 $10^x : 2\sqrt{2}/3+3\sqrt{5}$
9330 6507 $e^x : \sqrt[3]{21/2}$
9330 6584 $\Psi : \ln(\sqrt{2}/3)$
9330 7758 rt : (3,7,-7,-2)
9330 9615 $J_0 : 12/23$
9330 9913 rt : (8,8,-8,-6)
9331 0988 rt : (9,5,-5,-7)
9331 1893 tanh : $1/\ln(2e/3)$
9331 2063 rt : (5,1,-1,-4)
9331 2768 cr : $3\sqrt{2}+4\sqrt{5}/3$
9331 2778 cr : $13/16$
9331 3574 rt : (1,8,-8,3)
9331 4208 rt : (6,2,2,5)
9331 6117 $\zeta : 2\sqrt{2}/9$
9331 6205 sr : $(\pi/3)^{-3}$
9331 6350 rt : (7,5,5,6)
9331 8487 rt : (8,8,8,7)
9331 8680 $s\pi : 2e/3+\pi/2$
9331 8884 sinh : $5/6$
9332 0001 rt : (2,8,-1,-6)
9332 0130 tanh : $7\sqrt[3]{3}/6$
9332 1702 $t\pi : 2\sqrt{2}-\sqrt{3}/4$
9332 2155 sinh : $4\sqrt{5}$
9332 3424 tanh : $7\zeta(3)/5$
9332 3961 id : $3\sqrt{2}-4\sqrt{3}/3$
9332 4037 $2^x : 1/3-\sqrt{3}/4$
9332 5918 rt : (8,7,-6,-7)
9332 6980 rt : (9,4,-3,-8)
9332 7264 sinh : $5\pi^2/3$
9333 2482 cu : $6\ln 2$
9333 2504 $\lambda : \exp(-2e/3)$
9333 3333 id : $14/15$
9333 7452 ln : $\sqrt{3}/2+3\sqrt{5}/4$
9333 7557 rt : (2,9,-8,-2)
9333 9395 $\ell_{10} : 2e+\pi$
9333 9664 rt : (3,6,-5,-3)
9334 0715 rt : (7,9,-7,-7)
9334 0966 $10^x : \ln 2$
9334 1766 rt : (8,6,-4,-8)
9334 2815 rt : (9,3,-1,-9)
9334 3864 rt : (5,0,1,5)
9334 4960 $K : 13/17$
9334 5958 rt : (6,3,4,6)
9334 6637 $\ell_2 : \pi/6$
9334 6669 sq : $\sqrt{3}/4+4\sqrt{5}$
9334 7135 sr : $4\sqrt{3}+3\sqrt{5}/4$
9334 7913 $K : (\sqrt[3]{5})^{-1/2}$
9334 8047 rt : (7,6,7,7)
9335 1831 sinh : $6\sqrt{3}/5$
9335 2862 $J_0 : (\sqrt[3]{3}/2)^2$
9335 3389 $J_0 : 13/25$
9335 6421 rt : (7,8,-5,-8)
9335 6897 $J_0 : 3\ln 2/4$
9335 7459 rt : (8,5,-2,-9)
9335 7810 $\Gamma : (\pi/3)^3$
9335 9559 $erf : 9\sqrt[3]{3}/10$
9335 9774 $e^x : 4\sqrt{2}+\sqrt{3}$
9336 1407 sq : P_{50}
9336 2643 $2^x : \sqrt{2}+2\sqrt{3}/3$
9336 3054 $t\pi : 1/4-3\sqrt{3}/2$
9336 3055 $J_0 : 3\sqrt{3}/10$
9336 6612 cu : $\exp(\sqrt{2}\pi/2)$
9336 6738 rt : (4,9,-7,3)
9336 8893 rt : (2,8,-6,-3)
9336 9669 $Ei : 9/2$
9337 0949 rt : (6,8,-8,-2)
9337 1663 $\ell_{10} : ((e+\pi)/2)^2$
9337 1975 rt : (7,7,-3,-9)
9337 2068 $\ln(3/4)+\exp(1/5)$
9337 2775 $\ell_{10} : P_{77}$
9337 3001 rt : (4,2,0,-5)
9337 3943 $\zeta : \pi^2/9$
9337 4698 $s\pi : 1/3+2\pi$
9337 5049 rt : (5,1,3,6)
9337 7093 rt : (6,4,6,7)
9337 9133 rt : (7,7,9,8)
9338 0742 $erf : 3\sqrt{3}/4$
9338 1820 cu : $e/2+\pi/3$
9338 1827 $s\pi : 8\zeta(3)$
9338 3257 $Ei : 3$
9338 6366 rt : (6,9,-4,2)
9338 8101 sr : $(3\pi)^3$
9339 3089 $2^x : \ln(e/3)$
9339 4490 rt : (2,8,-8,6)
9339 4761 $\ln(5)-s\pi(\sqrt{5})$
9339 5184 $e-\exp(\sqrt{3})$
9339 8142 $2^x : 4\sqrt{2}-3\sqrt{5}/2$
9339 9629 rt : (2,7,-4,-4)
9340 0794 $erf : 13/10$
9340 1184 $\ell_2 : e+4\pi$
9340 1637 rt : (3,5,0,3)
9340 3640 rt : (4,1,2,-6)
9340 5538 $s\pi : 8\sqrt{2}/7$
9340 5640 rt : (5,2,5,7)
9340 6646 at : $9\zeta(3)/8$
9340 7637 rt : (6,5,8,8)
9340 8889 $\ell_2 : 4/3+\sqrt{3}/3$
9340 9589 $s\pi : 2\sqrt{2}/3-\sqrt{5}/4$
9340 9987 $10^x : 9(1/\pi)/10$
9341 0015 $\zeta : 11/13$
9341 3428 $\exp(e)\div s\pi(2/5)$
9341 5772 rt : (5,5,-8,2)
9341 8237 $s\pi : 8\ln 2/9$
9341 8635 $\ell_{10} : 4\sqrt{2}/3+3\sqrt{5}$
9342 0123 at : $1/\ln(2\pi/3)$
9342 3965 sq : $2/3+3\pi/2$
9342 6894 cr : $3e/10$
9342 6967 $Ei : (\sqrt[3]{4}/3)^{-1/2}$
9342 7822 rt : (1,9,-5,-4)
9342 8811 at : $23/17$
9342 9787 rt : (2,6,-2,-5)
9343 0049 $s\pi : 3/4+\sqrt{3}/2$
9343 0768 rt : (5,6,-4,6)
9343 1748 rt : (3,6,4,7)
9343 3642 sr : $\sqrt{14}$
9343 3706 rt : (4,0,4,7)
9343 5384 sq : $3\sqrt{2}/2-2\sqrt{3}/3$
9343 5660 rt : (5,3,7,8)
9343 5884 $\ln(\sqrt{2})+s\pi(1/5)$
9343 6025 ln : $9\sqrt{2}/5$
9344 0364 rt : (7,7,-8,9)
9344 0471 $2^x : (e/3)^{1/2}$
9344 2537 $\Gamma : -\sqrt{2}+\sqrt{3}+\sqrt{7}$
9344 4050 tanh : $22/13$
9344 4492 rt : (2,5,-9,9)
9344 5483 $Ei : \sqrt{23/2}$
9344 6352 $erf : 2(e+\pi)/9$
9344 7846 $Ei : \exp(1/\pi)$
9344 9310 cu : $4-\sqrt{2}/3$
9345 0434 at : $1+\sqrt{2}/4$
9345 0547 $s\pi : \sqrt{2}/2+3\sqrt{5}/4$
9345 0674 $10^x : \sqrt{20/3}$
9345 3403 $2^x : 1/4+4\pi$
9345 3543 $\ln(\sqrt{3})\div s\pi(1/5)$
9345 4688 tanh : $\ln(2e)$
9345 7467 rt : (1,8,-3,-5)
9345 8826 id : $\sqrt{10}\pi$
9345 9387 rt : (4,2,-8,6)
9346 1304 rt : (3,2,3,-7)
9346 1786 $\zeta : \sqrt[3]{2}/4$
9346 1976 $e^x : 7\sqrt[3]{5}/4$
9346 3202 sq : $4/3+3\sqrt{2}/2$
9346 3217 rt : (4,1,6,8)
9346 4887 sr : $1/\ln(\pi)$
9346 5127 rt : (5,4,9,9)
9346 5526 sr : $\sqrt[3]{2/3}$
9346 6832 $Ei : 6\ln 2$
9346 7551 $\Gamma : 2e/3$
9346 9336 rt : (3,9,-9,8)
9346 9799 at : $2-3\sqrt{5}/2$
9347 1165 ln : $\pi/8$
9347 2963 $\ln(4)\times\exp(1/3)$
9347 3175 ln : $\ln(\sqrt{2}\pi/3)$
9347 4315 $J_0 : 3\zeta(3)/7$
9347 6370 $\ell_{10} : 4\sqrt{3}+3\sqrt{5}/4$
9347 8322 cr : $3+3\sqrt{2}$
9347 8518 $\ln(4)\times\exp(3/4)$
9348 0220 id : $\pi^2/2$
9348 0237 $\Gamma : \ln(\pi)$
9348 0765 $\Gamma : \sqrt[3]{3/2}$
9348 2677 rt : (9,1,-8,-1)
9348 3219 $\Gamma : 10e\pi/7$
9348 3605 $J : \exp(-\sqrt[3]{4})$
9348 5691 $e^x : 1/4+3\pi/2$
9348 6572 rt : (1,7,-1,-6)
9348 7511 rt : (3,7,-7,-5)
9348 8449 rt : (2,4,2,-7)
9348 8485 $2^x : 4e/7$
9349 0232 $Ei : 9(e+\pi)/4$
9349 0322 rt : (3,1,5,-8)
9349 2193 rt : (4,2,8,9)
9349 3314 id : $7\zeta(3)/9$
9349 5040 $e^x : 5\sqrt{3}/3$
9349 8851 $\Psi : 7\sqrt{3}/4$
9349 9375 ln : $3/2+\pi/3$
9349 9779 $10^x : (\sqrt{2}\pi/2)^{1/2}$
9350 0174 $erf : (\sqrt{2}\pi/2)^{1/3}$
9350 1624 $s\pi : 5/13$
9350 1856 rt : (3,8,-1,4)
9350 4006 tanh : $6\sqrt{2}/5$
9350 4732 $Ei : 11/8$
9350 6355 $2^x : 20/21$
9350 6567 $t\pi : 9\sqrt{2}/5$
9350 6682 sinh : $\sqrt{2}$
9350 6719 sq : $2+3e/2$
9350 7076 $t\pi : 4\sqrt{2}-\sqrt{5}$
9350 9093 tanh : $\exp(\sqrt[3]{4}/3)$
9351 0150 $J : \sqrt[3]{5}/7$
9351 1346 rt : (8,1,-9,1)
9351 1666 cu : $3/4-2e$
9351 3316 rt : (9,7,-8,-6)

9351 5155 rt : (1,6,1,-7)
9351 5334 cu : $3\sqrt{2}+\sqrt{5}$
9351 5767 rt : (4,9,-6,6)
9351 6990 rt : (2,3,4,-8)
9351 7847 J_0 : $\exp(-2/3)$
9351 8822 rt : (3,0,7,9)
9352 0182 Γ : $8\sqrt{3}$
9352 0190 as : $\ln(\sqrt{5})$
9352 1541 10^x : $1+2e$
9352 3373 as : $1/3+\sqrt{2}/3$
9352 3929 tanh : $(\ln 2/2)^{-1/2}$
9352 5258 tπ : $3\sqrt{2}/2+\sqrt{5}/2$
9352 6967 rt : (8,9,-9,-6)
9352 7469 tanh : $5e/8$
9352 7907 rt : (9,6,-6,-7)
9352 8381 J_0 : $3\sqrt[3]{5}/10$
9352 9222 sr : $3+\sqrt{5}/3$
9352 9786 cr : $9/11$
9353 2721 cr : $8e/3$
9353 3313 sπ : $2\sqrt{3}/9$
9353 6917 $\exp(3/5)+\exp(\sqrt{2})$
9353 8682 sπ : $2/3+e$
9354 0907 tanh : $17/10$
9354 1434 sr : $7/8$
9354 2364 rt : (9,5,-4,-8)
9354 3233 rt : (2,8,0,-2)
9354 3357 rt : (9,3,-4,1)
9354 3572 at : $1-3\pi/4$
9354 3756 Γ : $8/7$
9354 5028 rt : (2,2,6,-9)
9354 6707 sq : $4\sqrt{3}+4\sqrt{5}$
9354 9111 cr : $4\sqrt{3}/3-2\sqrt{5}/3$
9355 0613 id : $\sqrt{2}/3+2\sqrt{3}$
9355 0698 $\exp(1/3)\div\exp(2/5)$
9355 3480 sπ : $5\ln 2/9$
9355 5769 rt : (8,7,-5,-8)
9355 6688 rt : (9,4,-2,-9)
9355 8710 erf : $(\sqrt[3]{5})^{1/2}$
9355 9360 erf : $17/13$
9356 0050 ln : $4\sqrt{3}$
9356 0060 cr : $(\ln 3/2)^{1/3}$
9356 2898 2^x : $2-\pi/3$
9356 3060 2^x : $1/3+\sqrt{5}$
9356 3292 e^x : $14/13$
9356 5076 ℓ_2 : $2\sqrt{2}/3+3\sqrt{5}$
9356 8135 Ei : $7\sqrt[3]{5}/6$
9356 8228 tπ : $1/3+e/3$
9356 9063 rt : (7,9,-6,-8)
9356 9843 10^x : $\sqrt{2}/4+3\sqrt{3}/4$
9356 9972 rt : (8,6,-3,-9)
9357 0823 rt : (1,4,5,-9)
9357 0944 rt : (8,1,-5,1)
9357 2821 rt : (9,4,-2,2)
9357 6959 at : $19/14$
9357 8497 ℓ_2 : $\sqrt[3]{7}$
9357 9279 at : $\sqrt[3]{5/2}$
9358 2053 J_0 : $\ln(5/3)$
9358 3008 id : $3e/2-\pi$
9358 3149 rt : (7,8,-4,-9)
9358 4228 10^x : $\sqrt[3]{1/3}$
9358 5323 10^x : $4\sqrt{3}+3\sqrt{5}/4$
9359 0108 ln : $\sqrt{13/2}$
9359 4492 rt : (1,9,-1,-8)
9359 6218 rt : (6,8,-7,0)
9359 8058 rt : (7,1,-6,-1)
9359 9896 rt : (8,2,-3,2)
9360 0410 e^x : $4\sqrt{2}/3+4\sqrt{5}/3$
9360 1729 rt : (9,5,0,3)
9360 3999 Γ : $\sqrt[3]{6}$
9360 7217 ln : $10\ln 2$
9360 7522 cr : $(2e/3)^{-1/3}$
9361 0063 rt : (6,9,-3,4)
9361 0475 rt : (1,4,-3,8)
9361 1572 Γ : 2π
9361 2863 sπ : $1/2+4\sqrt{2}/3$
9361 4136 2^x : $(\sqrt{3}/2)^{1/3}$
9361 4855 K : $(\sqrt{2}\pi/2)^{-1/3}$
9361 5355 sπ : $4\sqrt{3}/5$
9361 5356 cr : $3/2-e/4$
9361 6630 $\ln(\sqrt{5})\times\exp(4)$
9361 6843 rt : (1,9,-9,3)
9361 7067 λ : $(\ln 3/2)^3$
9361 8158 ζ : $9/11$
9361 9771 10^x : $3\sqrt{2}/4+2\sqrt{5}/3$
9361 9813 J_0 : $8(1/\pi)/5$
9361 9918 cu : $3/4+\pi$
9362 1225 rt : (1,8,-3,-9)
9362 3173 cu : $3/4+3\sqrt{2}/4$
9362 3757 id : $2\sqrt{3}+2\sqrt{5}$
9362 4716 rt : (6,3,-7,-1)
9362 5909 $\exp(1/4)+\exp(\sqrt{3})$
9362 6514 rt : (7,0,-4,2)
9362 8057 ζ : $6/19$
9362 8310 rt : (8,3,-1,3)
9362 9670 e^x : $6\sqrt[3]{2}$
9363 0102 rt : (9,6,2,4)
9363 0779 ℓ_{10} : $7\pi^2/8$
9363 3597 Γ : $20/11$
9363 6188 Γ : $9\sqrt{2}/7$
9363 6527 rt : (5,5,-7,4)
9363 7286 sq : $(e/3)^{1/3}$
9363 8099 ln : $\sqrt[3]{3}/10$
9363 9515 tanh : $e\pi/5$
9364 0334 $\ln(3/4)-\exp(1/2)$
9364 2158 Γ : $2\sqrt[3]{5}/3$
9364 2718 ℓ_{10} : $4\sqrt{2}+4\sqrt{5}/3$
9364 2838 $\ln(4)\times s\pi(\sqrt{5})$
9364 6436 cr : $1/3+4\sqrt{3}$
9364 6760 Γ : $(\sqrt{2}\pi/3)^{1/3}$
9364 7200 at : $e/2$
9364 8935 ln : 6π
9364 9167 sr : $15/4$
9364 9580 rt : (1,8,0,-4)
9365 0020 rt : (5,6,-3,8)
9365 0897 θ_3 : $3\sqrt[3]{3}/10$
9365 0931 rt : (5,5,-8,-1)
9365 1800 10^x : $3/2+\sqrt{5}/3$
9365 2660 θ_3 : $\sqrt{2/3}$
9365 2692 rt : (6,2,-5,-2)
9365 3136 sπ : $\sqrt{3}-\sqrt{5}/2$
9365 4450 rt : (7,1,-2,3)
9365 5568 cu : $3\sqrt{3}-\sqrt{5}$
9365 6205 rt : (8,4,1,4)
9365 6365 id : $1/2+2e$
9365 6658 λ : $\exp(-4\pi/7)$
9365 6857 erf : $21/16$
9365 7957 rt : (9,7,4,5)
9365 8008 rt : (4,9,-5,9)
9366 0631 Ψ : $13/16$
9366 2146 Ei : $4(1/\pi)/7$
9366 3131 ln : $3\sqrt{2}/4+2\sqrt{5}/3$
9366 4459 tanh : $\sqrt[3]{5}$
9366 4725 sr : $(\sqrt{2}\pi/3)^{-1/3}$
9366 6489 2^x : $2-2\pi/3$
9366 8098 10^x : $\sqrt{2}+4\sqrt{3}/3$
9366 9286 sπ : $\sqrt{3}+\sqrt{5}+\sqrt{7}$
9367 3691 cu : $3/2+\pi/4$
9367 5179 ℓ_2 : $1+2\sqrt{2}$
9367 5825 rt : (4,2,-7,8)
9367 6717 rt : (4,7,-9,-1)
9367 8441 rt : (5,4,-6,-2)
9367 9049 cu : $8\zeta(3)/3$
9368 0163 rt : (6,1,-3,-3)
9368 0650 ln : $3/2+2e$
9368 1883 rt : (7,2,0,4)
9368 2605 cu : $4-\sqrt{5}/2$
9368 2778 sq : $2\sqrt{3}/3+\sqrt{5}/4$
9368 3599 rt : (8,5,3,5)
9368 4245 sq : $3-2e$
9368 4998 2^x : $3+\sqrt{3}/3$
9368 5312 rt : (9,8,6,6)
9368 5651 tanh : $((e+\pi)/2)^{1/2}$
9368 5771 cr : $3\sqrt{2}/2-3\sqrt{3}/4$
9368 6573 ζ : $\sqrt[3]{11/2}$
9368 7364 sπ : $2\ln 2$
9368 8727 e^x : $1/2+\sqrt{3}/3$
9368 9334 sπ : $4e\pi/3$
9368 9523 2^x : $3\sqrt{2}/4-2\sqrt{3}/3$
9369 0378 2^x : $\sqrt{23}\pi$
9369 3744 J : $9/19$
9369 6180 sinh : $23/10$
9369 6474 sπ : $(\ln 2/3)^{1/3}$
9369 7947 Γ : $1/\ln(\sqrt{3})$
9369 8893 ln : $2/3+4\sqrt{2}/3$
9370 1212 rt : (3,7,-6,-2)
9370 1897 e^x : $4+4\sqrt{5}$
9370 2987 id : $3\sqrt{3}/4-\sqrt{5}$
9370 3776 rt : (4,6,-7,-2)
9370 3876 sq : $7e\pi/5$
9370 5463 rt : (5,3,-4,-3)
9370 7148 rt : (6,0,-1,4)
9370 7505 10^x : $3e/4-\pi/2$
9370 8830 rt : (7,3,2,5)
9370 8930 sπ : $4\pi^2/9$
9371 0414 Ei : $2/11$
9371 0508 rt : (8,6,5,6)
9371 2185 rt : (9,9,8,7)
9371 4159 rt : (3,8,0,7)
9371 7125 tanh : $12/7$
9371 7583 Li_2 : $8/11$
9371 7962 rt : (9,8,-9,-6)
9371 8286 ζ : $1/\sqrt{10}$
9371 8360 sinh : $(\sqrt[3]{5})^{-1/3}$
9371 9166 ℓ_{10} : $6\sqrt[3]{3}$
9372 1059 λ : $1/6$
9372 2556 $\ln(\pi)\div\exp(1/5)$
9372 2860 erf : $25/19$
9372 2920 Ei : $(\pi/3)^{1/3}$
9372 3455 ℓ_2 : $\sqrt{11/3}$
9372 5931 ℓ_{10} : $1/2+3e$
9372 8543 erf : $(\sqrt{3})^{1/2}$
9372 8708 rt : (3,8,-8,-2)
9373 0362 rt : (4,7,-3,-3)
9373 1188 rt : (9,7,-7,-7)
9373 2013 rt : (5,2,-2,-4)
9373 2310 2^x : $(\ln 3)^{-1/2}$
9373 2578 ℓ_{10} : $\ln 2/6$
9373 3055 sq : $2/3+\pi/3$
9373 3112 cr : $14/17$
9373 3661 rt : (6,1,1,5)
9373 4316 sπ : $\sqrt[3]{8/3}$
9373 5307 rt : (7,4,4,6)
9373 6010 ℓ_{10} : $3+4\sqrt{2}$
9373 6275 Li_2 : $\exp(-1/\pi)$
9373 6950 rt : (8,7,7,7)
9373 7914 sr : $3-3\sqrt{2}/2$
9374 0390 rt : (1,9,0,-5)
9374 0718 Ψ : $10(1/\pi)/9$
9374 0881 tπ : $4\sqrt{2}/3+3\sqrt{5}/2$
9374 1133 θ_3 : $\sqrt{3}/4$
9374 1390 2^x : $3\sqrt{2}/2+3\sqrt{5}$
9374 3138 Li_2 : $(\sqrt[3]{4}/3)^{1/2}$
9374 3482 rt : (8,9,-8,-7)
9374 4299 rt : (9,6,-5,-8)
9374 5182 $\ln(\sqrt{2})-\exp(1/4)$
9375 0000 id : $15/16$
9375 0308 tπ : $\exp(2e)$
9375 0387 rt : (1,8,1,4)
9375 0481 J_0 : $2\sqrt[3]{2}/5$
9375 1573 rt : (2,8,1,2)
9375 2791 tanh : $10\zeta(3)/7$
9375 3063 ℓ_{10} : $5\sqrt{3}$
9375 4871 rt : (3,7,-6,-3)
9375 6371 cr : $4\sqrt[3]{3}/7$
9375 6489 rt : (8,8,-6,-8)
9375 7298 rt : (9,5,-3,-9)
9375 8105 rt : (5,1,0,-5)
9375 9719 rt : (6,2,3,6)
9376 1330 rt : (7,5,6,7)
9376 2364 ln : $-\sqrt{2}+\sqrt{3}+\sqrt{5}$
9376 2938 rt : (8,8,9,8)
9376 4252 e^x : $\exp(\sqrt{3}/3)$
9376 4801 e^x : $1/3+3\pi$
9376 7046 rt : (1,7,0,-3)
9376 7799 Γ : $25/22$
9376 9252 Ei : $22/7$
9376 9386 rt : (8,7,-4,-9)
9377 5043 Ψ : $3/14$
9377 6641 10^x : $14/13$
9377 9001 rt : (2,9,-7,-3)

9377 9115 ln : $3\sqrt{2}/2 + \sqrt{3}/4$
9378 0588 rt : (3,6,-4,-4)
9378 1381 rt : (7,9,-5,-9)
9378 2173 rt : (4,9,5,5)
9378 3755 rt : (5,0,2,6)
9378 5335 rt : (6,3,5,7)
9378 5885 $\ln(3/4) + \exp(4/5)$
9378 6912 rt : (7,6,8,8)
9379 2542 Ei : $\sqrt[3]{19}$
9379 4476 Γ : $\sqrt{2} - \sqrt{5} + \sqrt{7}$
9379 5726 θ_3 : $5\ln 2/8$
9379 6894 2^x : 21/22
9380 4321 rt : (2,8,-5,-4)
9380 4749 at : 15/11
9380 5875 rt : (6,8,-6,2)
9380 7428 rt : (4,2,1,-6)
9380 7547 10^x : $5\pi/8$
9380 8315 sr : 22/25
9380 8757 $s\pi$: P_{27}
9380 8977 rt : (5,1,4,7)
9381 0525 rt : (6,4,7,8)
9381 3055 $s\pi$: $2\sqrt{2} + \sqrt{5}/4$
9381 3663 10^x : 14/9
9381 8362 rt : (6,9,-2,6)
9381 8859 rt : (1,9,1,4)
9382 5034 ln : $8\sqrt{5}/7$
9382 6008 cr : $\sqrt{2}/4 + 4\sqrt{3}$
9382 6963 ln : 9/23
9382 7160 sq : 20/9
9382 7231 id : P_{62}
9382 9223 rt : (2,7,-3,-5)
9383 0044 cr : 19/23
9383 0238 rt : (8,9,-9,7)
9383 0745 rt : (3,5,1,5)
9383 2266 rt : (4,1,3,-7)
9383 3784 rt : (5,2,6,8)
9383 4544 J_0 : $\sqrt{2} + \sqrt{3} - \sqrt{7}$
9383 5300 rt : (6,5,9,9)
9383 6224 $\ln(\sqrt{3}) + \exp(2)$
9383 8165 sinh : $(e+\pi)/7$
9383 8684 cr : $1 + 2\pi$
9384 2274 rt : (5,5,-6,6)
9384 6946 Γ : $\sqrt{10/3}$
9384 6980 J_0 : 1/2
9384 8511 2^x : $3(1/\pi)$
9385 2224 rt : (1,9,-4,-5)
9385 2500 $t\pi$: $9e$
9385 3145 erf : $\sqrt{3} + \sqrt{5} - \sqrt{7}$
9385 3335 Γ : $9\sqrt[3]{2}/10$
9385 3429 Ψ : $\sqrt{2}/4$
9385 3719 rt : (2,6,-1,-4)
9385 3936 cu : $\sqrt{19/3}$
9385 5211 rt : (3,6,5,9)
9385 6701 rt : (4,0,5,8)
9385 7162 ℓ_2 : $6\sqrt{5}/7$
9385 8189 rt : (5,3,8,9)
9385 9697 cr : $\sqrt{3}/3 + 3\sqrt{5}$
9385 9945 ℓ_2 : 3/23
9386 4005 ℓ_{10} : $\sqrt{2} - 3\sqrt{3}/4$
9386 4936 at : $1/\ln(\sqrt[3]{3}/3)$
9386 6327 ℓ_{10} : P_{78}
9386 9083 sr : $7\pi^2/8$
9386 9884 $s\pi$: $3\sqrt{2}/2 + 2\sqrt{5}/3$
9387 1396 Ψ : $\ln(1/\pi)$
9387 2036 ln : $2/3 + 2\pi$
9387 2336 λ : $(2e/3)^{-3}$
9387 3281 tanh : 19/11
9387 4388 Γ : 17/15
9387 6350 e^x : $4\sqrt{2} + 2\sqrt{3}$
9387 6357 rt : (1,8,-2,-6)
9387 7551 sq : 12/7
9387 7822 rt : (2,7,5,5)
9387 7850 cr : $1/4 + \sqrt{3}/3$
9387 9020 id : $1/4 - 4\pi/3$
9387 9284 rt : (3,2,4,-8)
9387 9720 ln : P_{75}
9388 0424 $\exp(2) \div s\pi(\sqrt{5})$
9388 0737 e^x : $e + 3\pi/2$
9388 0745 rt : (4,1,7,9)
9388 2930 J : $\exp(-\sqrt{3})$
9388 8201 at : $1/2 + \sqrt{3}/2$
9388 8321 Γ : $\sqrt{15}\pi$
9389 2107 cu : $9(1/\pi)/2$
9389 2626 $5 \times s\pi(1/5)$
9389 3904 $t\pi$: $(1/(3e))^{1/3}$
9389 5712 erf : $(\sqrt[3]{5}/3)^{-1/2}$
9389 6000 Γ : $(1/(3e))^2$
9389 6134 rt : (1,7,1,4)
9389 7135 rt : (9,0,-9,1)
9389 7601 sinh : $\sqrt{23/3}$
9389 8017 ζ : $7\sqrt[3]{2}/5$
9390 0107 rt : (1,7,0,-7)
9390 0825 rt : (3,7,-5,1)
9390 1543 rt : (2,4,3,-8)
9390 2978 rt : (3,1,6,-9)
9390 6250 $t\pi$: 6/25
9390 7966 at : $2\sqrt{2}/3 - 4\sqrt{3}/3$
9390 8630 $s\pi$: $e/7$
9390 9479 sr : $4\sqrt{2} + 4\sqrt{5}/3$
9391 1881 $\exp(3/4) + \exp(3/5)$
9391 1912 sr : $7\sqrt[3]{2}/10$
9391 1989 e^x : $1 + 4e/3$
9391 2570 $t\pi$: $4\sqrt[3]{5}/9$
9391 3045 sr : $2 - \sqrt{5}/2$
9391 3854 tanh : $6\sqrt[3]{3}/5$
9391 4091 e^x : $5\sqrt{5}/7$
9391 5115 e^x : $3\sqrt{3}/2 + 4\sqrt{5}/3$
9391 6207 rt : (4,8,-9,6)
9391 6907 $t\pi$: $4(e+\pi)/9$
9392 0090 ζ : 7/9
9392 0189 $t\pi$: $(e)^{1/3}$
9392 1769 J_0 : $2\sqrt{5}/9$
9392 2076 rt : (9,8,-8,-7)
9392 2655 sq : $(\ln 3)^{-1/3}$
9392 3486 rt : (1,6,2,-8)
9392 3523 erf : $(\ln 3)^3$
9392 3911 J_0 : $(1/(3e))^{1/3}$
9392 4350 2^x : $3e - \pi/2$
9392 4895 rt : (2,3,5,-9)
9392 9781 tanh : $\sqrt{3}$
9393 0977 erf : $\sqrt[3]{7/3}$
9393 2651 cu : $3e/4 - 3\pi$
9393 3643 sr : 15/17
9393 3982 id : $2 - 3\sqrt{2}/4$
9393 4004 rt : (9,7,-6,-8)
9393 9393 tanh : $5\ln 2/2$
9394 0186 $s\pi$: $\sqrt{3}/2 + \sqrt{5}/3$
9394 0489 Γ : $5\pi^2/9$
9394 4113 Γ : $4\sqrt{2}/5$
9394 4209 2^x : $7e\pi/5$
9394 5122 rt : (8,9,-7,-8)
9394 5137 cr : $(\sqrt[3]{5}/3)^{1/3}$
9394 5835 rt : (9,6,-4,-9)
9394 6505 rt : (2,8,2,6)
9394 6556 10^x : $\ln(4/3)$
9394 6594 rt : (9,2,-5,1)
9394 6826 cu : $3e/4 - 4\pi/3$
9394 6898 2^x : 14/9
9394 9942 Γ : $(\sqrt[3]{3}/3)^2$
9395 3823 sq : $1/3 + 3\sqrt{2}$
9395 6363 at : $4\sqrt[3]{5}/5$
9395 6563 id : P_{46}
9395 6866 rt : (8,8,-5,-9)
9395 8216 at : $1/4 + \sqrt{5}/2$
9395 8912 ln : $4\sqrt{2} + 3\sqrt{3}/4$
9395 9403 10^x : $6\pi/7$
9395 9667 sinh : $5\sqrt{3}/3$
9396 0600 rt : (8,9,-8,9)
9396 2319 ln : $2 + \sqrt{5}/4$
9396 5023 J_0 : $5\ln 2/7$
9396 7817 rt : (7,3,-9,0)
9396 8143 Γ : $(\sqrt{2}\pi/2)^3$
9396 8458 $e + \exp(1/5)$
9396 9262 $s\pi$: 7/18
9397 0560 10^x : 23/14
9397 0626 J_0 : $2\sqrt{3}/7$
9397 0704 rt : (9,3,-3,2)
9397 0999 $s\pi$: $4\sqrt{2} + \sqrt{3}$
9397 1650 $s\pi$: $1/\sqrt[3]{17}$
9397 6587 Γ : $\exp(1/(3e))$
9397 7327 ln : $1/4 + 4\sqrt{3}/3$
9397 7626 ℓ_2 : $\sqrt{2} - \sqrt{3} + \sqrt{5}$
9397 8749 $\ln(2/5) \times \exp(3/4)$
9398 1481 cu : $7\sqrt[3]{5}/6$
9398 1578 id : $7\sqrt[3]{2}/3$
9398 1653 2^x : $8\sqrt{5}/5$
9398 4398 e^x : P_{55}
9398 4644 cu : $3\sqrt{2}/4 + 3\sqrt{3}$
9398 7215 $s\pi$: $(e)^2$
9398 7786 $s\pi$: $2/3 + 4\sqrt{5}$
9398 8837 rt : (1,9,-2,-9)
9398 8840 id : $\sqrt{2} - 3\sqrt{5}/2$
9398 9589 ℓ_2 : $4\sqrt{3} + \sqrt{5}/3$
9399 0171 rt : (6,7,-9,0)
9399 1574 $\ln(3/4) - \exp(\sqrt{3})$
9399 1589 rt : (7,2,-7,-1)
9399 2138 ln : $1/4 + 3\sqrt{5}$
9399 2729 cr : $2\sqrt{2} + 2\sqrt{5}$
9399 2859 ℓ_{10} : $2 + 3\sqrt{5}$
9399 3004 rt : (8,1,-4,2)
9399 4418 rt : (9,4,-1,3)
9399 5036 Γ : $(\ln 2)^{-1/3}$
9399 5167 $t\pi$: $4(e+\pi)/7$
9399 5822 Ψ : $((e+\pi)/3)^3$
9399 6090 erf : $1/\ln(\sqrt{2}/3)$
9399 6235 $t\pi$: $\sqrt{2} + 4\sqrt{5}/3$
9399 9209 Γ : $(\sqrt[3]{3})^{1/3}$
9400 1142 erf : $7\sqrt[3]{5}/9$
9400 1564 rt : (6,8,-5,4)
9400 3439 $t\pi$: $(\sqrt{3})^{-1/2}$
9400 7306 10^x : $\exp(1/(3\pi))$
9401 0094 $s\pi$: $3\sqrt{2}/2 - \sqrt{3}$
9401 1008 sinh : $\sqrt[3]{9}$
9401 1239 Ψ : $10\sqrt{5}/3$
9401 2869 rt : (6,9,-1,8)
9401 2960 tanh : $7\sqrt{5}/9$
9401 3586 rt : (6,4,-8,-1)
9401 3725 ℓ_{10} : $4 + 3\pi/2$
9401 3813 J_0 : $\exp(-1/\sqrt{2})$
9401 4975 rt : (7,1,-5,-2)
9401 5077 sr : $5\sqrt{2}/8$
9401 6362 rt : (8,2,-2,3)
9401 6516 Ei : $\exp(-\sqrt[3]{2}/3)$
9401 7748 rt : (9,5,1,4)
9402 0432 $s\pi$: $2e - \pi/3$
9402 3418 rt : (5,4,-9,4)
9402 4364 id : $3e + \pi/4$
9402 5626 ℓ_{10} : $3\sqrt{2} + 2\sqrt{5}$
9402 5894 cu : $6\sqrt{2}$
9402 6510 ln : $3/2 + 3\sqrt{2}/4$
9402 8098 at : $\sqrt{3}/2 - \sqrt{5}$
9403 0321 sinh : $3\sqrt{5}/8$
9403 1758 10^x : $7\zeta(3)/4$
9403 2695 cr : P_{25}
9403 4557 rt : (5,5,-5,8)
9403 5261 rt : (5,9,-6,-9)
9403 6619 Ψ : $1/\ln(\sqrt[3]{5}/3)$
9403 6625 rt : (6,3,-6,-2)
9403 7409 $s\pi$: $\sqrt{2} - \sqrt{6} + \sqrt{7}$
9403 7987 rt : (7,0,-3,3)
9403 9347 rt : (8,3,0,4)
9404 0705 rt : (9,6,3,5)
9404 1203 rt : (4,8,-8,9)
9404 1596 sinh : 17/12
9404 1618 2^x : $7\pi/3$
9404 2217 ℓ_2 : $7\pi^2/9$
9404 4969 cr : $6\ln 2/5$
9404 8431 sr : $e + \pi/3$
9404 9067 $s\pi$: $\sqrt{2} + 3\sqrt{3}$
9405 2858 sr : $\exp(-1/(3e))$
9405 4176 sq : $9e/5$
9405 5811 cu : $4/3 - \sqrt{2}/4$
9405 5933 rt : (4,1,-9,8)
9405 6907 Ei : 9/4
9405 7533 ln : $e/4 + 2\pi$
9405 7963 rt : (5,5,-7,-2)
9405 8018 Ψ : $(\ln 3/3)^{-2}$
9405 9300 rt : (6,2,-4,-3)
9406 0636 rt : (7,1,-1,4)
9406 1010 Ei : $\sqrt[3]{23}$

9406 1970 rt : (8,4,2,5)
9406 2550 2^x : 22/23
9406 3301 rt : (9,7,5,6)
9406 5017 $s\pi$: $9\sqrt[3]{5}$
9406 5356 erf : 4/3
9406 5585 Γ : 11/6
9406 8459 e^x : $\sqrt{23/3}$
9406 9263 $1/5 - \exp(\pi)$
9407 0256 e^x : $7\sqrt[3]{3}/3$
9407 1671 K : 10/13
9407 1827 2^x : $5\sqrt{5}/4$
9407 2484 at : $3/4 - 3\sqrt{2}/2$
9407 4281 sr : $(\ln 2)^{1/3}$
9407 5708 ln : $3e\pi/10$
9407 6874 ℓ_2 : $3/4 + 4\sqrt{3}$
9407 8662 Γ : $\ln 2/3$
9407 8997 rt : (4,7,-8,-2)
9407 9123 sr : $(\sqrt[3]{3})^{-1/3}$
9407 9175 sq : $\sqrt{2}/2 - 3\sqrt{5}/4$
9408 0311 rt : (5,4,-5,-3)
9408 0703 $\ln(\pi) - \exp(3)$
9408 1623 rt : (6,1,-2,-4)
9408 2256 e^x : $1/3 + \sqrt{5}/3$
9408 2933 rt : (7,2,1,5)
9408 4160 tanh : $5\pi/9$
9408 4208 ζ : 10/7
9408 4241 rt : (8,5,4,6)
9408 5547 rt : (9,8,7,7)
9408 5985 $t\pi$: $8e\pi/7$
9408 6153 ln : $3/4 + 2e/3$
9408 7744 rt : (3,7,-4,4)
9408 9089 cu : $4\sqrt{2}/3 + \sqrt{3}/3$
9409 5679 as : $4\sqrt{2}/7$
9409 6693 sr : P_{54}
9409 7622 at : $2/3 - 3e/4$
9409 8300 id : $3/2 - \sqrt{5}/4$
9409 9736 rt : (3,9,-9,-2)
9410 0389 10^x : 21/4
9410 1027 rt : (4,6,-6,-3)
9410 1508 cu : $7\sqrt[3]{2}/9$
9410 1671 rt : (9,9,-9,-7)
9410 2315 rt : (5,3,-3,-4)
9410 3568 erf : $(\sqrt[3]{2}/3)^{-1/3}$
9410 3602 cr : 5/6
9410 4888 rt : (7,3,3,6)
9410 5155 tanh : $\sqrt[3]{16/3}$
9410 6171 rt : (8,6,6,7)
9410 7453 rt : (9,9,9,8)
9410 7885 2^x : $\sqrt{2}/3 - \sqrt{5}/4$
9410 8809 erf : $10\zeta(3)/9$
9411 2549 sq : $3 + 4\sqrt{2}$
9411 2554 rt : (9,8,-7,-8)
9411 2684 $\ln(3/5) + s\pi(\pi)$
9411 2823 e^x : $1/2 + 4\sqrt{2}$
9411 3270 ζ : $(\sqrt{3}/2)^{-1/3}$
9411 4301 e^x : $1 - 3\sqrt{2}/4$
9411 7647 id : 16/17
9411 7965 sq : $1/4 - 3\sqrt{2}$
9411 8879 rt : (2,7,-5,-6)
9412 1052 J_0 : $2\sqrt[3]{5}/7$
9412 1455 rt : (3,8,-7,-3)
9412 2020 ζ : 7/22
9412 2722 rt : (4,7,-2,-1)
9412 3354 rt : (9,7,-5,-9)
9412 3986 rt : (5,2,-1,-5)
9412 5249 rt : (6,1,2,6)
9412 5439 $t\pi$: $\sqrt{21/2}$
9412 6510 rt : (7,4,5,7)
9412 7770 rt : (8,7,8,8)
9412 8152 $t\pi$: $\sqrt[3]{3}/6$
9412 8429 10^x : $7\zeta(3)/10$
9413 3447 rt : (8,9,-6,-9)
9413 7284 2^x : $1 + \pi/2$
9413 7553 tanh : 7/4
9414 0353 $s\pi$: $3e/4 + \pi/2$
9414 1262 2^x : $1/4 + \sqrt{2}/2$
9414 1603 sinh : $(\ln 2/3)^{-1/2}$
9414 2389 cr : $3\sqrt{2}/2 + 3\sqrt{3}$
9414 2537 $e - \ln(4/5)$
9414 2849 rt : (3,7,-5,-4)
9414 3743 $s\pi$: $2/3 + 2\sqrt{2}/3$
9414 4092 rt : (4,8,2,3)
9414 5333 rt : (5,1,1,-6)
9414 6573 rt : (6,2,4,7)
9414 7746 $s\pi$: $\ln(5)$
9414 7811 rt : (7,5,7,8)
9414 8565 ζ : $(1/\pi)$
9414 9753 $t\pi$: $\zeta(3)/5$
9415 0841 id : $3\sqrt{3} + \sqrt{5}/3$
9415 1544 $s\pi$: $\sqrt{24}\pi$
9415 1551 $\exp(1/4) - \exp(4/5)$
9415 7547 at : $2/3 + \sqrt{2}/2$
9415 7750 at : $8\zeta(3)/7$
9415 9265 $1/5 - \pi$
9416 1448 $\exp(1/2) - s\pi(1/4)$
9416 2591 sr : $6\pi/5$
9416 2704 rt : (2,9,-6,-4)
9416 2723 cr : $6e\pi/7$
9416 2987 id : $5e/7$
9416 3926 rt : (3,6,-3,-5)
9416 4262 ℓ_{10} : $2 - 4\sqrt{2}/3$
9416 4406 K : $4\sqrt{3}/9$
9416 5146 rt : (4,9,6,7)
9416 6364 rt : (5,0,3,7)
9416 7582 rt : (6,3,6,8)
9416 8275 sr : $6\sqrt[3]{3}$
9416 8790 sq : $7\ln 2/5$
9416 8797 rt : (7,6,9,9)
9416 9096 cu : 16/7
9416 9586 $s\pi$: $2\sqrt{3}/3 + \sqrt{5}$
9416 9736 rt : (1,6,-1,-8)
9416 9978 10^x : $2\sqrt{3}/3 - \sqrt{5}/4$
9417 1000 λ : $\exp(-9\pi/16)$
9417 4269 Γ : 9/8
9417 4348 rt : (6,7,-8,2)
9417 6861 J_0 : $\exp(-\sqrt[3]{3}/2)$
9417 7280 $t\pi$: $\sqrt{2}/2 + 4\sqrt{5}$
9417 9156 $s\pi$: $1/4 + e/2$
9418 2382 sinh : $10\sqrt[3]{2}/7$
9418 3495 rt : (2,8,-4,-5)
9418 4695 rt : (6,8,-4,6)
9418 5893 rt : (4,2,2,-7)
9418 7089 rt : (5,1,5,8)
9418 8285 rt : (6,4,8,9)
9418 9060 as : $7\ln 2/6$
9418 9574 λ : $\sqrt[3]{5}/10$
9419 0649 at : $(\sqrt[3]{4}/3)^{-1/2}$
9419 1187 sr : $1/2 + 3e$
9419 3161 at : $\exp(1/\pi)$
9419 4441 ln : $3\sqrt[3]{5}/2$
9419 6638 Ψ : $(1/\pi)/10$
9419 6708 sr : $8\sqrt{2}/3$
9419 8563 Γ : $\ln(2\pi)$
9419 9110 cu : $1/3 + 4\sqrt{2}$
9419 9484 id : $4e/3 - 4\pi$
9420 0004 at : 11/8
9420 0485 J_0 : $7\ln 2/10$
9420 3986 rt : (2,7,-2,-6)
9420 4172 sinh : $9\sqrt[3]{2}/8$
9420 4575 rt : (5,4,-8,6)
9420 5164 rt : (3,5,2,7)
9420 6341 rt : (4,1,4,-8)
9420 7516 rt : (5,2,7,9)
9420 7549 cr : $5(e + \pi)/4$
9420 8469 $e \times \ln(\sqrt{2})$
9421 0606 $2/3 \times \exp(5)$
9421 1422 $s\pi$: $\sqrt{23/2}$
9421 3301 cr : $(\sqrt[3]{5})^{-1/3}$
9421 3718 $s\pi$: $3\pi^2$
9421 4803 θ_3 : 10/23
9421 5303 as : $(\sqrt[3]{4}/3)^{1/3}$
9421 7428 sinh : 9/5
9421 8807 ℓ_2 : $9e\pi/5$
9421 8835 ln : $3\sqrt{2} - 3\sqrt{5}/4$
9421 9977 cu : $9\ln 2/5$
9422 0796 $s\pi$: $7\sqrt{5}/6$
9422 2042 erf : $3\sqrt{5}/5$
9422 2788 sinh : $7\sqrt[3]{5}/4$
9422 3026 rt : (1,9,-3,-6)
9422 3683 K : $10\ln 2/9$
9422 4047 θ_3 : 16/23
9422 4185 rt : (2,6,0,-1)
9422 5098 10^x : $5\sqrt[3]{2}/7$
9422 5326 sr : $3 + 4\sqrt{2}$
9422 5342 rt : (3,3,3,-8)
9422 5545 $\sqrt{2} - \exp(\sqrt{5})$
9422 5590 sinh : $2\sqrt[3]{2}/3$
9422 6092 $s\pi$: 9/23
9422 6499 rt : (4,0,6,9)
9422 7498 tanh : $3(e + \pi)/10$
9422 7551 e^x : $1/3 + 3\sqrt{5}/2$
9422 8203 $\ln(2/3) \div s\pi(\pi)$
9422 8420 at : $2\sqrt{2}/3 + \sqrt{3}/4$
9423 0700 10^x : $4/3 - e/2$
9423 1972 tanh : $(2e)^{1/3}$
9423 2450 e^x : $\sqrt{3}/3 + 2\sqrt{5}$
9423 2822 sinh : 21/25
9423 3350 at : $3\sqrt{2}/2 - \sqrt{5}/3$
9423 6288 cr : $\sqrt{2} - \sqrt{3}/3$
9423 7559 Γ : $(\sqrt{2}\pi/3)^{-2}$
9424 2960 rt : (1,8,-1,-7)
9424 4099 rt : (2,7,6,8)
9424 4766 e^x : $10\pi/7$
9424 5237 rt : (3,2,5,-9)
9424 6107 cr : $(e + \pi)/7$
9424 7779 id : $3\pi/10$
9425 1855 $\ln(\sqrt{3}) - \exp(2/5)$
9425 3388 rt : (3,6,-9,-2)
9425 6384 $s\pi$: $1/4 + \pi$
9425 6496 cr : $7\pi/3$
9426 0413 ζ : $3\sqrt{3}/4$
9426 1018 Ψ : 2/23
9426 2331 sr : $3\sqrt{2} - 3\sqrt{5}/2$
9426 2618 rt : (1,7,1,-8)
9426 3178 rt : (3,7,-3,7)
9426 3737 rt : (2,4,4,-9)
9426 4670 2^x : $4\pi^2$
9426 4677 cr : $\sqrt{5} + \sqrt{6} + \sqrt{7}$
9426 4745 λ : $\zeta(3)/7$
9426 6124 10^x : 19/13
9426 7591 Γ : $(\sqrt[3]{2})^{1/2}$
9426 8540 $s\pi$: $(\sqrt{2}\pi)^{-1/3}$
9426 9074 e^x : $1/2 - \sqrt{5}/4$
9427 0555 erf : $(e/3)^{-3}$
9427 0663 cu : $2e - 4\pi/3$
9427 2264 $\exp(1/3) \times s\pi(\sqrt{5})$
9427 2798 cu : $1 + \sqrt{3}/4$
9427 3295 ln : $3 - \sqrt{3}/4$
9427 7124 $1/3 + \ln(5)$
9427 9455 Ψ : $7e\pi/8$
9428 0904 id : $2\sqrt{2}/3$
9428 2005 rt : (1,6,3,-9)
9428 2556 id : $\sqrt[3]{22/3}$
9428 3095 sr : $5\sqrt{3}$
9428 7606 cr : $4\sqrt{2} + 3\sqrt{5}/4$
9428 9761 10^x : $\sqrt[3]{3}/5$
9428 9968 sr : P_{43}
9429 0789 rt : (9,8,-6,-9)
9429 1600 rt : (2,7,-4,-2)
9429 3582 tanh : $7\sqrt[3]{2}/5$
9429 3954 Γ : $7\sqrt[3]{3}/9$
9429 4811 $s\pi$: $e/4 + 3\pi/2$
9429 6190 $\exp(\sqrt{2}) \div \exp(3/4)$
9429 6520 e^x : $4\pi/3$
9429 7384 2^x : $\sqrt{3} + \sqrt{5} + \sqrt{7}$
9429 8638 cr : $3\sqrt{5}/8$
9429 9461 ℓ_{10} : $(\sqrt{2}\pi)^3$
9430 0046 rt : (8,2,-9,0)
9430 1129 rt : (1,4,2,7)
9430 1195 rt : (9,1,-6,1)
9430 3216 J_0 : $2\zeta(3)/5$
9430 3898 2^x : $3\sqrt{5}/7$
9430 4918 J_0 : $\sqrt[3]{3}/3$
9430 6107 erf : $(2e/3)^{1/2}$
9430 6388 2^x : 23/24
9430 6652 J_0 : $(\ln 2/3)^{1/2}$
9430 7809 tanh : $\sqrt[3]{11/2}$
9431 0188 erf : $3\pi/7$
9431 1019 $\exp(\sqrt{3}) \div s\pi(2/5)$
9431 1848 J_0 : $(\ln 2)^2$

9431 3153 as : $3/2 - 4\sqrt{3}/3$
9431 4718 $1/4 + \ln(2)$
9431 4722 ℓ_{10} : $8\pi^2/9$
9431 7475 id : $\sqrt{2}/2 + \sqrt{5}$
9431 9642 rt : (2,9,-6,-8)
9432 0061 rt : (8,1,-7,-1)
9432 0794 at : $\sqrt{2}/4 - \sqrt{3}$
9432 1189 rt : (9,2,-4,2)
9432 1789 sq : $1/3 + 3\sqrt{2}/4$
9432 2415 J_0 : 12/25
9432 4055 erf : $7\sqrt{3}/9$
9432 5290 ln : $4 + 4\sqrt{5}/3$
9432 7982 e^x : $3/2 + 3\sqrt{5}$
9432 8060 tπ : $1/4 - 2\sqrt{5}/3$
9432 8234 10^x : 9/10
9433 0753 Ψ : 17/21
9433 0884 sπ : $6\sqrt{3}$
9433 4055 as : 17/21
9433 4836 Γ : $(1/\pi)/3$
9433 5833 ℓ_2 : $4\sqrt[3]{3}/3$
9433 6416 tanh : $5\sqrt{2}/4$
9433 7143 Ei : $\exp(-\sqrt[3]{5})$
9433 7161 2^x : $\sqrt[3]{17}$
9433 8091 rt : (1,6,0,-2)
9433 8677 rt : (7,3,-8,-1)
9433 9786 rt : (8,0,-5,2)
9433 9868 sinh : $4e$
9434 0390 tanh : $\ln(e + \pi)$
9434 0894 rt : (9,3,-2,3)
9434 1647 ℓ_2 : 13/25
9434 1762 tanh : $\exp(\sqrt[3]{5}/3)$
9434 3180 ℓ_{10} : $4e - 2\pi/3$
9434 4715 cu : $5\pi/6$
9434 7060 rt : (6,7,-7,4)
9434 8328 ln : $\sqrt{2} + 2\sqrt{3}/3$
9435 1908 cr : $2\sqrt[3]{2}/3$
9435 2505 tanh : 23/13
9435 2731 2^x : $8\sqrt{3}/7$
9435 3879 cr : 21/25
9435 4047 id : $\sqrt{2}/3 + 2\sqrt{5}$
9435 6031 sinh : $(\sqrt{2})^{-1/2}$
9435 6488 rt : (6,8,-3,8)
9435 7049 rt : (6,5,-9,-1)
9435 7709 ℓ_2 : $8\zeta(3)/5$
9435 8140 rt : (7,2,-6,-2)
9435 9230 rt : (8,1,-3,3)
9436 0258 id : $3\sqrt{2} - 3\sqrt{3}/4$
9436 0318 rt : (9,4,0,4)
9436 1637 Γ : $8\sqrt[3]{2}/9$
9436 2547 erf : $(\ln 3/2)^{-1/2}$
9436 2674 sq : $1/2 + \sqrt{2}/3$
9436 2894 J_0 : 11/23
9436 7291 ln : $1/3 + \sqrt{5}$
9436 9848 rt : (1,7,-8,-4)
9436 9883 id : $1/\ln(\ln 2/2)$
9437 1100 sπ : $\ln(\sqrt{2}\pi/3)$
9437 1920 sπ : $\pi/8$
9437 4631 rt : (5,4,-7,8)
9437 6257 rt : (6,4,-7,-2)
9437 7329 rt : (7,1,-4,-3)
9437 7914 rt : (1,8,-9,-5)
9437 8399 rt : (8,2,-1,4)
9437 9469 rt : (9,5,2,5)
9438 0387 ℓ_2 : $3\ln 2/4$
9438 1376 J_0 : $3(1/\pi)/2$
9438 2275 $e + \exp(4/5)$
9438 2298 sπ : $9e\pi/8$
9438 2896 ζ : $\sqrt{5}/7$
9438 6068 Li_2 : $(\sqrt[3]{5}/2)^2$
9438 7431 sr : $(\sqrt[3]{2})^{-1/2}$
9438 7774 tanh : $(\pi)^{1/2}$
9439 0103 ℓ_2 : $9\sqrt[3]{5}$
9439 0234 ℓ_2 : $\sqrt{3}/2 + 4\sqrt{5}/3$
9439 0543 10^x : $\sqrt{3}/6$
9439 0840 at : $1/3 + \pi/3$
9439 4142 rt : (5,9,-5,-7)
9439 4426 $\pi - \exp(3)$
9439 5121 $\exp(5) \times s\pi(1/4)$
9439 5196 rt : (6,3,-5,-3)
9439 6250 rt : (7,0,-2,4)
9439 7302 rt : (8,3,1,5)
9439 7417 e^x : $3 + 4e$
9439 7574 ζ : 5/7
9439 7715 2^x : 18/7
9439 7737 ℓ_2 : $\sqrt{6}\pi$
9439 7957 ℓ_{10} : $3(e + \pi)/2$
9439 8353 rt : (9,6,4,6)
9440 2819 sr : $9\ln 2/7$
9440 3614 J_0 : $1/\ln(1/(3e))$
9440 5816 Ψ : $2\sqrt[3]{2}/5$
9440 7759 cr : $7\zeta(3)/10$
9440 9781 Ψ : 6/17
9441 0903 J_0 : 10/21
9441 1800 rt : (4,8,-9,-2)
9441 2685 cu : $3/2 - 4\sqrt{3}$
9441 2838 rt : (5,5,-6,-3)
9441 3875 rt : (6,2,-3,-4)
9441 4910 rt : (7,1,0,5)
9441 5705 sq : $2e - \pi/2$
9441 5718 λ : $\sqrt{3}/10$
9441 5945 rt : (8,4,3,6)
9441 6129 sr : $3\sqrt[3]{2}$
9441 6978 rt : (9,7,6,7)
9441 8288 erf : $9\zeta(3)/8$
9441 8319 Γ : $(2\pi)^{1/3}$
9441 9235 rt : (3,6,-8,1)
9442 0827 2^x : $9e\pi/4$
9442 1570 ln : $1 + \pi/2$
9442 2651 10^x : $4e + \pi/4$
9442 3930 λ : $\ln 2/4$
9442 4103 e^x : $3\sqrt{2}/4 - \sqrt{5}/2$
9442 5195 erf : $1/\ln(2\pi/3)$
9442 5426 $\exp(\sqrt{5}) + s\pi(1/5)$
9442 7190 id : $4\sqrt{5}$
9442 7519 rt : (2,9,-5,-3)
9442 9645 erf : 23/17
9443 0259 rt : (4,7,-7,-3)
9443 0758 sinh : $7\zeta(3)/10$
9443 1280 rt : (5,4,-4,-4)
9443 2299 rt : (6,1,-1,-5)
9443 2639 cr : 16/19
9443 2663 Γ : $\sqrt{5}/2$
9443 3317 rt : (7,2,2,6)
9443 3914 $e - s\pi(e)$
9443 4335 rt : (8,5,5,7)
9443 5351 rt : (9,8,8,8)
9443 5817 sq : $3/2 + 2\sqrt{5}/3$
9443 5999 e^x : $3/2 + 3\sqrt{2}/4$
9444 0444 ln : $\sqrt[3]{17}$
9444 1019 10^x : $\sqrt{2}/3 + 3\sqrt{3}/4$
9444 1165 rt : (1,8,-8,4)
9444 2702 e^x : $3\sqrt{2}/4 + 4\sqrt{3}$
9444 3897 ln : 19
9444 4444 id : 17/18
9444 4867 Γ : 24/13
9444 5565 tanh : 16/9
9444 6160 ln : 7/18
9444 6601 sq : $1/4 - \sqrt{5}$
9444 7224 Γ : 19/17
9444 7465 rt : (3,9,-8,-3)
9444 8028 e^x : $6\sqrt[3]{2}/7$
9444 8434 ℓ_2 : $10\sqrt{3}/9$
9444 8450 J_0 : $\exp(-\sqrt{5}/3)$
9444 8470 rt : (4,6,-5,-3)
9444 8971 rt : (9,9,-7,-9)
9444 9025 rt : (1,7,-7,4)
9444 9473 rt : (5,3,-2,-5)
9445 0476 rt : (6,0,1,6)
9445 0739 ln : $4/3 + 4\sqrt{2}$
9445 1477 rt : (7,3,4,7)
9445 1680 sπ : $(e)^{-1/2}$
9445 1687 J_0 : $(\sqrt{2}\pi)^{-1/2}$
9445 2477 rt : (8,6,7,8)
9445 3488 id : $7e\pi/4$
9445 4205 rt : (2,7,-3,2)
9445 4453 sπ : $7\ln 2/8$
9445 8087 tanh : $1/\ln(\sqrt[3]{5}/3)$
9446 4296 sr : $1/3 + \sqrt{5}/4$
9446 5449 rt : (3,8,-6,-4)
9446 5461 2^x : $7\sqrt{2}/5$
9446 5616 cu : $\sqrt{2} - \sqrt{3}/4$
9446 5791 cr : $4 + 3\sqrt{5}/2$
9446 6438 rt : (4,7,-1,1)
9446 7425 rt : (5,2,0,-6)
9446 8411 rt : (6,1,3,7)
9446 8757 J_0 : 9/19
9446 9396 rt : (7,4,6,8)
9447 0380 rt : (8,7,9,9)
9447 2362 $\ln(\sqrt{2}) + \exp(4)$
9447 2988 $1/5 - \ln(\pi)$
9447 3310 $2 \div \exp(3/4)$
9447 4930 J_0 : $(3\pi)^{-1/3}$
9447 8306 Γ : $(e/2)^2$
9448 2230 cr : $(5/3)^{-1/3}$
9448 2544 2^x : $e/3 + 2\pi$
9448 3197 rt : (3,7,-4,-5)
9448 3610 tanh : $\exp(\sqrt{3}/3)$
9448 3888 tπ : $2e/3 - 2\pi$
9448 4169 rt : (4,8,3,5)
9448 5140 rt : (5,1,2,-7)
9448 5546 e^x : $(\sqrt[3]{2})^{1/3}$
9448 6110 rt : (6,2,5,8)
9448 6316 sπ : $1/4 + 3\pi/4$
9448 6501 λ : 4/23
9448 7079 rt : (7,5,8,9)
9448 8407 Ψ : $(\sqrt[3]{4}/3)^{1/3}$
9448 8762 sπ : $3\zeta(3)$
9449 0046 e^x : $3\sqrt{2}/4 + 4\sqrt{5}/3$
9449 1856 $\ln(\sqrt{3}) + \exp(1/3)$
9449 4078 id : $3\sqrt[3]{2}/4$
9449 6672 J_0 : $3\sqrt[3]{2}/8$
9449 6823 rt : (1,6,1,4)
9449 6904 sπ : $3\sqrt{2}/7$
9449 6923 e^x : P_{81}
9449 7213 tπ : $2e - \pi/4$
9449 9756 rt : (2,9,-5,-5)
9449 9840 tanh : $\sqrt[3]{17/3}$
9449 9848 ζ : 8/25
9450 0714 rt : (3,6,-2,-6)
9450 1670 rt : (4,9,7,9)
9450 2626 rt : (5,0,4,8)
9450 2848 sπ : $7\sqrt[3]{5}/5$
9450 3580 rt : (6,3,7,9)
9450 5233 erf : 19/14
9450 5640 sπ : $1/3 + 3\sqrt{2}/4$
9450 6413 erf : $\sqrt[3]{5/2}$
9450 9387 rt : (6,7,-6,6)
9451 0468 Γ : $(2e/3)^{-2}$
9451 2391 Γ : $8\ln 2/3$
9451 2689 $\exp(\sqrt{3}) - s\pi(1/4)$
9451 6424 ℓ_{10} : $4\sqrt{2}/3 + 4\sqrt{3}$
9451 7063 rt : (2,8,-3,-6)
9451 8005 rt : (3,5,0,-7)
9451 8517 10^x : $9\sqrt[3]{5}/4$
9451 8946 rt : (4,2,3,-8)
9451 9887 rt : (5,1,6,9)
9452 1130 J_0 : $\sqrt{2}/3$
9452 2303 sinh : 16/19
9452 3310 Ψ : $7\ln 2/6$
9452 7622 ℓ_{10} : P_{37}
9452 7873 as : $1/4 - 3\sqrt{2}/4$
9452 8049 $5 \times \exp(2)$
9452 8520 θ_3 : $2\sqrt[3]{3}/3$
9452 9573 $\exp(1/5) \times s\pi(e)$
9453 0070 cu : $2 - 4\sqrt{5}/3$
9453 0642 tanh : 25/14
9453 0989 2^x : 24/25
9453 1128 at : 18/13
9453 1290 rt : (2,9,-4,2)
9453 2787 sinh : $5\pi^2/6$
9453 3754 sinh : $(\ln 2)^{-2}$
9453 4149 rt : (2,7,-1,-7)
9453 5077 rt : (3,5,3,9)
9453 6003 rt : (4,1,5,-9)
9453 7094 10^x : $3\sqrt{3}/5$
9453 8306 rt : (1,9,-9,4)
9453 9825 J_0 : 8/17
9454 0877 erf : $e/2$
9454 1158 $\ln(\sqrt{5}) + \exp(\pi)$
9454 2060 Γ : $(\sqrt[3]{3}/2)^{-1/3}$

9454 2400 sr : $3+4\pi$
9454 2545 at : $4/3-e$
9454 3648 cu : $3+\sqrt{2}/2$
9454 4137 ℓ_{10} : $7\sqrt[3]{2}$
9454 4301 tπ : $\sqrt[3]{21}$
9454 4862 sπ : $5\sqrt[3]{3}/2$
9454 6099 Ψ : $9e/8$
9454 7257 id : $\sqrt{3}-\sqrt{5}+\sqrt{6}$
9454 8060 2^x : $4/3-\sqrt{2}$
9454 8471 id : $\sqrt{2}/4-3\sqrt{3}/4$
9455 0106 rt : (1,9,-2,-7)
9455 1020 rt : (2,6,1,2)
9455 1886 cr : $2-2\sqrt{3}/3$
9455 1933 rt : (3,3,4,-9)
9455 2295 sπ : $\sqrt{13}$
9455 3162 λ : $\exp(-5\pi/9)$
9455 4303 ℓ_{10} : $2/3+3e$
9456 0997 sr : $3+\pi/4$
9456 1170 10^x : $1+\sqrt{2}/2$
9456 3952 tanh : $4\sqrt{5}/5$
9456 3991 rt : (1,9,8,-3)
9456 5484 at : $1/2-4\sqrt{2}/3$
9456 6171 Γ : $\sqrt[3]{19/3}$
9456 6257 at : $4\sqrt{3}/5$
9456 6781 rt : (1,8,0,-8)
9456 7681 rt : (2,5,3,-9)
9457 4160 id : $(\sqrt{5}/2)^{-1/2}$
9457 5488 rt : (3,6,-7,4)
9457 5735 Ψ : $\exp(\sqrt{5}/2)$
9457 6320 2^x : $3/4+2\sqrt{2}$
9458 0402 2^x : $2e+2\pi/3$
9458 2326 Γ : $7(1/\pi)/2$
9458 2636 sπ : $4\sqrt{3}+3\sqrt{5}/4$
9458 3251 rt : (1,7,2,-9)
9458 3482 $1/5-s\pi(\sqrt{3})$
9458 3731 cr : 11/13
9458 3822 e^x : $3e/4-2\pi/3$
9458 4185 Γ : 3/13
9458 5300 Ei : $5e$
9458 8637 at : $2\ln 2$
9459 0040 tπ : $(3\pi)^{-1/3}$
9459 0308 as : $\sqrt{3}/2-3\sqrt{5}/4$
9459 0530 sr : 17/19
9459 1014 ln : 7
9459 1635 $\exp(4)-\exp(\sqrt{3})$
9459 3465 sπ : $(\ln 3/3)^{1/2}$
9459 3730 ℓ_{10} : $3\sqrt{2}/2+3\sqrt{5}$
9459 8646 rt : (9,1,-9,0)
9459 8952 sπ : $3(e+\pi)/4$
9459 8976 sπ : $4\sqrt{2}/3-2\sqrt{5}/3$
9459 9520 rt : (1,6,4,-6)
9459 9670 $\exp(\pi)\div\exp(1/5)$
9459 9927 tπ : $(e+\pi)/9$
9460 0189 sπ : $3e/4+4\pi$
9460 0386 $\ln(5)\times s\pi(1/5)$
9460 0957 cr : $3e-\pi/4$
9460 2871 cu : $1/4-3\sqrt{3}/2$
9460 3289 at : $\sqrt[3]{8/3}$
9460 4099 ζ : $2\sqrt[3]{3}/9$
9460 5104 sr : $\sqrt{3}/4+3\sqrt{5}/2$
9460 7580 rt : (2,7,-2,6)
9460 8744 sq : $\sqrt[3]{11}$
9460 9953 10^x : $7e/10$
9461 1351 $\ln(\sqrt{5})\times\exp(2)$
9461 2115 ℓ_2 : $3\sqrt{2}+2\sqrt{3}$
9461 5242 id : $1/4-3\sqrt{3}$
9461 5641 rt : (9,0,-7,1)
9461 6174 2^x : $2\sqrt{2}+\sqrt{3}/2$
9461 6381 sr : $9e$
9461 7488 2^x : $(\ln 3/2)^{-2}$
9461 8030 2^x : $9\sqrt{3}/10$
9462 0368 erf : 15/11
9462 8481 Ψ : $10(e+\pi)/3$
9462 9213 J_0 : 7/15
9463 0622 tanh : $4\pi/7$
9463 0625 rt : (7,9,-9,0)
9463 1160 $\pi+\ln(\sqrt{5})$
9463 1191 rt : (2,9,-3,7)
9463 1521 rt : (8,2,-8,-1)
9463 2416 rt : (9,1,-5,2)
9463 9474 ℓ_2 : $1+3\sqrt{5}$
9464 1227 2^x : $\sqrt{2}/4-\sqrt{3}/4$
9464 5477 K : 17/22
9464 6111 id : $2\sqrt{2}+\sqrt{5}/2$
9464 6628 cu : $(1/(3e))^{-3}$
9464 7213 rt : (7,4,-9,-1)
9464 7691 Ψ : $4\sqrt{2}/7$
9464 8095 rt : (8,1,-6,-2)
9464 8268 sq : $3/4+4\sqrt{2}/3$
9464 8682 cr : $2/3+3\sqrt{5}$
9464 8976 rt : (9,2,-3,3)
9465 0569 erf : $1/\ln(\sqrt[3]{3}/3)$
9465 0628 sπ : $2\sqrt{2}-\sqrt{3}/4$
9465 2177 Li_2 : $(e+\pi)/8$
9465 3008 2^x : $1+\sqrt{5}/4$
9465 4328 rt : (6,6,-9,4)
9465 4660 sinh : $\sqrt{13/3}$
9465 7359 $3/5+\ln(\sqrt{2})$
9465 9379 Ei : $e\pi/5$
9466 2127 rt : (1,9,9,5)
9466 2276 rt : (6,7,-5,8)
9466 3591 rt : (7,3,-7,-2)
9466 4431 e^x : $\sqrt[3]{17/3}$
9466 4459 rt : (8,0,-4,3)
9466 4691 Γ : $\exp(1/(3\pi))$
9466 5326 rt : (9,3,-1,4)
9466 9508 ln : $4e/3-\pi/3$
9466 9542 Γ : 7/23
9467 0572 sπ : $4(e+\pi)/9$
9467 0682 sπ : $(e)^{1/3}$
9467 2115 cr : $3\sqrt{2}/5$
9467 3222 sπ : $\sqrt{2}+4\sqrt{5}/3$
9467 4083 $\sqrt{3}\div s\pi(1/5)$
9467 5748 Ψ : 15/2
9467 6184 ln : $2+\sqrt{3}/3$
9467 7327 at : 25/18
9467 7612 rt : (5,3,-9,8)
9467 7678 cu : $1/4-3e$
9467 7926 id : $4\sqrt{3}-4\sqrt{5}/3$
9467 8908 rt : (6,5,-8,-2)
9467 9433 tanh : $10\sqrt[3]{2}/7$
9467 9764 rt : (7,2,-5,-3)
9468 0601 tanh : 9/5
9468 0618 rt : (8,1,-2,4)
9468 1472 rt : (9,4,1,5)
9468 2393 sinh : $5(e+\pi)/9$
9468 3205 cr : $8(1/\pi)/3$
9468 5259 J : $\exp(-2\pi/9)$
9468 6005 J_0 : $1/\sqrt[3]{10}$
9469 1752 2^x : $2/3-\sqrt{5}/3$
9469 3075 Γ : $9\sqrt[3]{3}/7$
9469 3702 cu : $2\sqrt{2}/3-3\sqrt{3}$
9469 4050 rt : (5,8,-8,-9)
9469 4893 rt : (6,4,-6,-3)
9469 5735 rt : (7,1,-3,-4)
9469 6102 ln : $2e+\pi/2$
9469 6268 erf : $4\sqrt[3]{5}/5$
9469 6534 Γ : 10/9
9469 6577 rt : (8,2,0,5)
9469 6749 sr : $(\sqrt[3]{3}/2)^{1/3}$
9469 6805 Γ : $4(e+\pi)$
9469 7418 rt : (9,5,3,6)
9469 9544 sπ : $1/4+3\sqrt{5}/2$
9470 0901 tπ : $\sqrt{3}-2\sqrt{5}/3$
9470 2902 sinh : $7e/2$
9470 3493 2^x : $\exp(e)$
9470 3513 sπ : $2e/9$
9470 4823 2^x : $3-3e/4$
9470 5040 sinh : $(5/3)^{-1/3}$
9470 9644 tanh : $5\sqrt[3]{3}/4$
9470 9851 rt : (5,9,-4,-5)
9471 0171 tanh : $(e+\pi)^{1/3}$
9471 0682 rt : (6,3,-4,-4)
9471 1316 cu : $3/4-\sqrt{3}$
9471 1511 rt : (7,0,-1,5)
9471 2340 rt : (8,3,2,6)
9471 2459 tanh : $3\zeta(3)/2$
9471 3168 rt : (9,6,5,7)
9471 7454 Li_2 : $3\sqrt[3]{5}/7$
9471 7781 ℓ_2 : $3+3\pi/2$
9471 8307 sπ : $4\ln 2/7$
9472 2893 cu : $3\sqrt{2}/4-4\sqrt{3}/3$
9472 2980 rt : (3,6,-6,7)
9472 4642 rt : (4,8,-8,-3)
9472 5197 10^x : $4/3-\sqrt{5}/4$
9472 5460 rt : (5,5,-5,-4)
9472 6278 rt : (6,2,-2,-5)
9472 6521 λ : $(\sqrt[3]{2}/3)^2$
9472 6602 tπ : $2\sqrt{2}/3+3\sqrt{5}$
9472 6823 cr : 17/20
9472 7096 rt : (7,1,1,6)
9472 7268 $\exp(\sqrt{2})\div\exp(1/3)$
9472 7840 Ei : $\sqrt[3]{3}/8$
9472 7912 rt : (8,4,4,7)
9472 8728 rt : (9,7,7,8)
9473 2332 tπ : $2e/3-\pi/2$
9473 2457 J_0 : $2\ln 2/3$
9473 3192 sq : $2\sqrt{2}+3\sqrt{5}$
9473 3313 2^x : $2\sqrt[3]{3}/3$
9473 3877 ℓ_2 : $\sqrt{3}-\sqrt{6}+\sqrt{7}$
9473 4678 e^x : $4(e+\pi)/3$
9473 5356 λ : 3/17
9473 6356 sπ : $4\pi/9$
9473 6842 id : 18/19
9473 6966 e^x : $\sqrt{13}\pi$
9474 0075 rt : (4,7,-6,-4)
9474 0882 rt : (5,4,-3,-5)
9474 1643 sr : $2\pi/7$
9474 1688 rt : (6,1,0,-6)
9474 2493 rt : (7,2,3,7)
9474 3297 rt : (8,5,6,8)
9474 4101 rt : (9,8,9,9)
9474 5038 J_0 : 6/13
9474 5089 rt : (2,6,-9,-6)
9474 6792 10^x : 19/7
9475 2996 2^x : $4\zeta(3)/5$
9475 3080 $\ln(\pi)\div s\pi(1/5)$
9475 3975 θ_3 : $1/\sqrt[3]{12}$
9475 4530 rt : (3,9,-7,-4)
9475 4767 sπ : $1/4+\sqrt{2}/4$
9475 5325 rt : (4,6,-4,-1)
9475 6120 rt : (5,3,-1,-6)
9475 6914 rt : (6,0,2,7)
9475 7707 rt : (7,3,5,8)
9475 8500 rt : (8,6,8,9)
9475 8814 Γ : $10(1/\pi)/7$
9476 3697 J : 3/19
9476 4435 λ : $\sqrt{2}/8$
9476 4783 J_0 : $(3\pi/2)^{-1/2}$
9476 8070 2^x : $2+4\pi/3$
9476 8253 2^x : $9\ln 2/4$
9476 8768 sπ : $\sqrt{21}\pi$
9476 9612 rt : (3,8,-5,-5)
9477 0319 λ : $5(1/\pi)/9$
9477 0396 rt : (4,7,0,3)
9477 1179 rt : (5,2,1,-7)
9477 1961 rt : (6,1,4,8)
9477 2076 cr : $4\sqrt{2}+\sqrt{3}$
9477 2743 rt : (7,4,7,9)
9477 3404 id : $\exp(2/3)$
9477 3527 Ei : $3e/8$
9477 4550 e^x : $2\sqrt{3}$
9477 4921 2^x : $4-3e/2$
9477 7040 Γ : $8\ln 2/5$
9477 7476 rt : (7,9,-8,2)
9477 8111 λ : $\exp(-\sqrt{3})$
9477 9682 2^x : $1/2-\sqrt{3}/3$
9478 0034 Γ : 13/7
9478 4094 tπ : $9\zeta(3)/7$
9478 4176 id : $2\pi^2/5$
9478 4517 rt : (3,7,-3,-6)
9478 4957 sr : $4e+3\pi/2$
9478 5290 rt : (4,8,4,7)
9478 5883 e^x : $3e/4+\pi/2$
9478 6062 rt : (5,1,3,-8)
9478 6833 rt : (6,2,6,9)
9478 7342 tπ : $1/4+3\sqrt{5}/2$
9478 8375 10^x : $\sqrt{3}\pi$
9478 9264 as : $1-2e/3$
9479 4575 cu : $7e\pi/6$

9479 5222 sπ : $1/\sqrt[3]{16}$
9479 6176 erf : $8\zeta(3)/7$
9479 6192 cr : $3e\pi$
9479 7570 $\sqrt{2} \div \exp(2/5)$
9479 8489 rt : (2,9,-4,-6)
9479 9251 rt : (6,6,-8,6)
9480 0013 rt : (4,3,2,-8)
9480 0773 rt : (5,0,5,9)
9480 0916 tπ : $\sqrt[3]{9/2}$
9480 5109 Li_2 : 11/15
9480 5371 tanh : $2e/3$
9480 7422 $s\pi(1/4) \div s\pi(\sqrt{3})$
9481 0603 e^x : $9e/7$
9481 2399 erf : $(\sqrt[3]{4}/3)^{-1/2}$
9481 3065 rt : (2,8,-2,-7)
9481 3637 erf : $\exp(1/\pi)$
9481 3816 rt : (3,5,1,-8)
9481 4567 rt : (4,2,4,-9)
9481 6330 ζ : $(2e)^{1/3}$
9481 6814 J_0 : 11/24
9481 7007 erf : 11/8
9481 8824 sr : $4e\pi/9$
9481 9329 10^x : $5\pi/6$
9482 2203 sr : $9\sqrt{3}$
9482 7082 sπ : $\exp(\sqrt{2}\pi/3)$
9482 7476 rt : (2,7,0,-8)
9482 8217 rt : (3,4,3,-9)
9482 8561 ln : P_{27}
9482 9247 Γ : $(e/2)^{1/3}$
9483 3103 sr : $\exp(-1/(3\pi))$
9483 4673 2^x : $5\sqrt{3}/9$
9483 5662 2^x : $\sqrt{2} - 2\sqrt{5}/3$
9483 5877 sπ : $4\zeta(3)/3$
9483 7447 Γ : 5/11
9483 8590 $\sqrt{5} + \ln(3/4)$
9484 0994 rt : (1,9,-1,-8)
9484 1725 rt : (2,6,2,5)
9484 2503 10^x : 16/9
9484 2783 as : 13/16
9484 4221 sπ : $3\sqrt{2} + 2\sqrt{3}/3$
9484 4321 sπ : $10e\pi$
9484 5322 tπ : $\pi/9$
9484 7640 10^x : $8(1/\pi)$
9484 7643 λ : $\exp(-11\pi/20)$
9484 9422 cu : $3\sqrt{3}/4 - 2\sqrt{5}$
9485 1184 at : $1/3 + 3\sqrt{2}/4$
9485 3217 sq : $5e\pi/4$
9485 4225 ℓ_{10} : $9\pi^2$
9485 5054 tanh : $\sqrt[3]{6}$
9485 5095 rt : (1,8,1,-9)
9485 5715 id : $9\sqrt{3}/8$
9485 5999 ln : $\sqrt{20/3}$
9485 7349 $e \div \ln(\sqrt{3})$
9485 8640 cr : $1/2 + \sqrt{2}/4$
9486 0066 sπ : $10\sqrt[3]{3}/9$
9486 1252 $3/4 \times \exp(4)$
9486 3897 Ψ : $1/\sqrt{14}$
9486 4242 sπ : $5\sqrt{5}/8$
9486 5357 sr : $5\sqrt[3]{2}/7$
9486 5529 10^x : $\sqrt{10/3}$
9486 5682 tanh : 20/11
9486 6072 cr : $\sqrt{2}/3 + 4\sqrt{3}$
9486 6174 2^x : $3/2 + 2\sqrt{5}/3$
9486 6610 tanh : $9\sqrt{2}/7$
9486 8329 sr : 9/10
9486 9041 rt : (1,4,-8,-7)
9486 9486 ℓ_{10} : $\sqrt{8}\pi$
9486 9983 J_0 : $(\sqrt{2}\pi/3)^{-2}$
9487 0083 sr : $1/3 + 2\sqrt{3}$
9487 0160 sπ : $((e+\pi)/3)^{1/2}$
9487 4198 cr : $e\pi/10$
9487 5519 e^x : $9\zeta(3)/4$
9487 9657 tπ : $(2e)^{1/3}$
9487 9785 θ_3 : $\sqrt{2} + \sqrt{3} - \sqrt{6}$
9488 1528 10^x : $(3\pi/2)^{-1/3}$
9488 2835 rt : (2,6,-8,-2)
9488 3171 ζ : $3(e+\pi)/10$
9488 3601 sπ : $(\ln 3/3)^{-1/3}$
9488 3737 10^x : P_{82}
9488 5227 sr : $4/3 - \sqrt{3}/4$
9488 5540 sπ : $\exp(\sqrt{2}/3)$
9488 5877 sq : $10\zeta(3)/7$
9488 5935 id : $e\pi/9$
9488 8449 sπ : P_2
9488 8610 tanh : $1/\ln(\sqrt{3})$
9488 8809 e^x : P_{26}
9488 8854 cr : $3\pi^2/4$
9489 1069 tπ : 8/19
9489 4107 sr : $3\sqrt{2}/2 + 3\sqrt{5}/4$
9489 4987 sπ : $5e/4$
9489 6518 rt : (9,1,-8,-1)
9489 6977 J_0 : $10(1/\pi)/7$
9489 8461 sπ : $5(1/\pi)/4$
9490 1029 J_0 : 5/11
9490 5195 e^x : $(\sqrt[3]{5}/2)^{-1/2}$
9490 6147 at : $(e)^{1/3}$
9490 6190 2^x : $8\zeta(3)$
9490 7137 tπ : $2e/9$
9490 9449 e^x : $4\sqrt{2}/3 + 2\sqrt{5}$
9491 0016 rt : (8,3,-9,-1)
9491 0736 rt : (9,0,-6,2)
9491 1754 id : $(\sqrt[3]{5}/2)^{1/3}$
9491 4888 10^x : $4/3 - \sqrt{3}/4$
9491 5169 10^x : $3(e+\pi)/4$
9491 6171 rt : (7,9,-7,4)
9491 8843 ζ : $(\sqrt{2}/3)^{1/3}$
9491 9472 cr : $\sqrt{2} - \sqrt{5}/4$
9492 1077 2^x : $1/3 + 2\sqrt{2}$
9492 2252 Ei : $\sqrt[3]{2}/7$
9492 2553 tanh : $\sqrt{2} - \sqrt{5} + \sqrt{7}$
9492 4082 rt : (8,2,-7,-2)
9492 4792 rt : (9,1,-4,3)
9492 5014 Γ : 21/19
9492 8224 at : $4\pi/9$
9493 0614 $2/5 + \ln(\sqrt{3})$
9493 0627 2^x : $2/3 + 4\pi/3$
9493 2321 sπ : $7\pi/5$
9493 4162 J_0 : $e/6$
9493 5886 sr : 19/5
9493 6228 rt : (6,6,-7,8)
9493 7289 rt : (7,4,-8,-2)
9493 7989 rt : (8,1,-5,-3)
9493 8689 rt : (9,2,-2,4)
9493 8786 sr : $4\sqrt{3}/3 + 2\sqrt{5}/3$
9493 8918 sπ : $(e/2)^{-3}$
9494 0781 tanh : $\sqrt{10/3}$
9494 2402 sr : $5\sqrt[3]{3}/8$
9494 2876 $\exp(1/2) - \exp(4)$
9494 4884 ln : $\sqrt{5}\pi$
9494 5075 θ_3 : 7/16
9494 8974 sq : $\sqrt{2}/2 + \sqrt{3}$
9494 9601 sr : $3\zeta(3)/4$
9495 0358 rt : (6,6,-9,-2)
9495 1049 rt : (7,3,-6,-3)
9495 1473 10^x : $1/2 + \sqrt{3}/3$
9495 1740 rt : (8,0,-3,4)
9495 2430 rt : (9,3,0,5)
9495 3539 J_0 : $\exp(-\sqrt[3]{4}/2)$
9495 4943 10^x : 15/7
9495 5148 sq : $4\pi/9$
9495 6774 K : $(5/3)^{-1/2}$
9496 2638 ℓ_{10} : $3/4 + 3e$
9496 3975 rt : (6,5,-7,-3)
9496 4656 rt : (7,2,-4,-4)
9496 5194 e^x : $3\sqrt[3]{3}/4$
9496 5337 rt : (8,1,-1,5)
9496 6018 rt : (9,4,2,6)
9496 6438 J_0 : $\ln(\pi/2)$
9497 1563 at : $5\sqrt{5}/8$
9497 3322 e^x : 23/3
9497 3580 at : $((e+\pi)/3)^{1/2}$
9497 4728 Γ : $5\sqrt{5}/6$
9497 4746 id : $7\sqrt{2}/2$
9497 6768 rt : (5,8,-7,-7)
9497 6804 Ei : $10\sqrt{3}/7$
9497 6870 2^x : $\sqrt{2}/3 + 4\sqrt{5}/3$
9497 7441 rt : (6,4,-5,-4)
9497 8114 rt : (7,1,-2,-5)
9497 8164 at : $(\ln 3/3)^{-1/3}$
9497 8663 erf : 18/13
9497 8786 rt : (8,2,1,6)
9497 9457 rt : (9,5,4,7)
9498 0371 sπ : $2/3 + \sqrt{3}$
9498 3016 10^x : $1/3 + 3\sqrt{5}/2$
9498 4279 J_0 : $3\zeta(3)/8$
9498 5197 id : $1/3 - 2\pi$
9498 8798 2^x : $1/2 + 3\sqrt{2}/4$
9498 9433 rt : (4,9,-9,-3)
9499 0097 rt : (5,9,-3,-3)
9499 0761 rt : (6,3,-3,-5)
9499 0872 10^x : $2\sqrt{2} + 2\sqrt{3}/3$
9499 1425 cr : 6/7
9499 1923 rt : (2,8,-9,2)
9499 2087 rt : (8,3,3,7)
9499 2750 rt : (9,6,6,8)
9499 4052 Γ : $1/\ln(\sqrt[3]{5})$
9499 5160 cr : $(\sqrt[3]{4})^{-1/3}$
9499 5647 erf : $4\sqrt{3}/5$
9499 5728 sπ : $4\sqrt{2} + 4\sqrt{5}$
9499 6103 e^x : $5\zeta(3)/9$
9499 8663 id : $5\sqrt[3]{5}/9$
9500 0000 id : 19/20
9500 0530 Ei : $(\sqrt[3]{2}/3)^3$
9500 1212 J_0 : 9/20
9500 2442 sπ : $(2\pi)^{-1/2}$
9500 2481 2^x : $(\sqrt{5}/2)^{-1/3}$
9500 2627 rt : (4,8,-7,-4)
9500 2969 ln : $4 - \sqrt{2}$
9500 3282 rt : (5,5,-4,-5)
9500 3819 cr : $\sqrt{2}/2 + 3\sqrt{5}$
9500 3937 rt : (6,2,-1,-6)
9500 4274 tπ : $2\sqrt{2}/3 + 3\sqrt{3}/4$
9500 4592 rt : (7,1,2,7)
9500 5116 Ψ : $(3\pi)^{1/2}$
9500 5246 rt : (8,4,5,8)
9500 5312 sπ : $\zeta(3)/2$
9500 5899 rt : (9,7,8,9)
9500 6452 erf : $2\ln 2$
9500 6470 rt : (1,9,0,-4)
9500 9012 J_0 : $1/\sqrt[3]{11}$
9501 1171 sπ : P_{59}
9501 1329 at : $1/3 - \sqrt{3}$
9501 3386 rt : (2,6,-7,2)
9501 3518 erf : $\sqrt[3]{8/3}$
9501 3582 e^x : $9\zeta(3)/10$
9501 4361 sr : $(e/2)^{-1/3}$
9501 5114 tanh : 11/6
9501 5679 rt : (4,7,-5,-5)
9501 6326 rt : (5,4,-2,-6)
9501 6972 rt : (6,1,1,-7)
9501 7618 rt : (7,2,4,8)
9501 7870 Γ : $(\sqrt{5}/3)^{-1/3}$
9501 8264 rt : (8,5,7,9)
9501 8392 Ψ : $\sqrt[3]{5}/8$
9501 8817 cu : $\sqrt{3} - 4\sqrt{5}/3$
9502 2858 e^x : $2/3 + \sqrt{2}/2$
9502 5177 e^x : $8\zeta(3)/7$
9502 7524 J_0 : $\pi/7$
9502 7954 rt : (3,9,-6,-5)
9502 8593 rt : (4,6,-3,1)
9502 8840 Ei : $8e/3$
9502 9117 tπ : $1/3 + 3\pi$
9502 9232 rt : (5,3,0,-7)
9502 9869 rt : (6,0,3,8)
9503 0507 rt : (7,3,6,9)
9503 1271 10^x : $\sqrt{3} + 3\sqrt{5}$
9503 7529 Γ : $7\sqrt[3]{2}/8$
9503 8113 cr : $4 - \pi$
9504 0741 rt : (3,8,-4,-6)
9504 1371 rt : (4,7,1,5)
9504 1918 2^x : $1/3 + 4\sqrt{5}/3$
9504 2001 rt : (5,2,2,-8)
9504 2551 2^x : $4e + 4\pi/3$
9504 2631 rt : (6,1,5,9)
9504 4380 rt : (1,8,6,-9)
9504 4974 ln : $6(e+\pi)/5$
9504 5668 cr : $5\zeta(3)/7$
9504 7397 rt : (7,9,-6,6)
9504 9141 erf : 25/18
9505 1720 at : $10\sqrt[3]{2}/9$

9505 3394 rt : (3,7,-2,-7)
9505 3498 cu : $4e/3 + \pi/3$
9505 4017 rt : (4,8,5,9)
9505 4257 2^x : $1/4 + \sqrt{3}$
9505 4638 rt : (5,1,4,-9)
9505 4684 at : 7/5
9505 6963 $\sqrt{3} \div \exp(3/5)$
9505 8601 2^x : $8\sqrt[3]{3}/5$
9505 9095 tanh : $\ln(2\pi)$
9506 0517 $\ln(2/5) \div s\pi(\sqrt{2})$
9506 0744 λ : $\sqrt[3]{2}/7$
9506 1728 sq : $8\sqrt{5}/9$
9506 2153 J_0 : $\sqrt{5}/5$
9506 3970 ln : $3/4 + 2\pi$
9506 5044 Γ : $\sqrt[3]{13/2}$
9506 5302 rt : (2,9,-3,-7)
9506 5917 rt : (3,6,0,-8)
9506 6531 rt : (4,3,3,-9)
9506 7499 $s\pi$: $\sqrt{2}/3 + 4\sqrt{3}$
9506 8717 rt : (1,9,1,5)
9506 9208 $1/3 - \exp(1/4)$
9507 2630 Ψ : $2\zeta(3)/9$
9507 5480 ℓ_2 : $4e - \pi$
9507 5629 rt : (2,8,-8,7)
9507 6406 ℓ_{10} : $2 + 4\sqrt{3}$
9507 7704 rt : (2,8,-1,-8)
9507 8311 rt : (3,5,2,-9)
9507 9331 sr : $7e/5$
9508 0174 sinh : 11/13
9508 1163 λ : $\exp(-6\pi/11)$
9508 4461 id : $3\sqrt{2} + 3\sqrt{5}$
9508 5111 ℓ_2 : $3 + \sqrt{3}/2$
9508 6941 λ : $\sqrt[3]{3}/8$
9508 9980 rt : (2,7,1,-9)
9509 2403 sinh : $5\sqrt[3]{3}/4$
9509 3726 J_0 : $1/\ln(3\pi)$
9509 6544 J_0 : $7(1/\pi)/5$
9509 6660 sr : $2 + 3\sqrt{5}$
9509 7131 $s\pi$: $10\sqrt[3]{2}/9$
9510 0390 $\ln(2/3) + \exp(\sqrt{5})$
9510 0852 sr : $\sqrt{2}/3 + \sqrt{3}/4$
9510 1541 rt : (1,9,0,-9)
9510 2045 ℓ_2 : $3\sqrt{2} - 4\sqrt{3}/3$
9510 2133 rt : (2,6,3,8)
9510 5651 $s\pi$: 2/5
9510 6215 cr : $(\pi/2)^{-1/3}$
9510 8364 sinh : $(e + \pi)^{1/3}$
9510 9206 10^x : $(5/3)^{-1/2}$
9510 9210 Γ : $\sqrt[3]{4/3}$
9511 0827 Ei : $\exp(e/3)$
9511 3872 rt : (3,5,-9,7)
9511 3942 2^x : $7\sqrt{5}/2$
9511 6767 cr : $1/2 + 4\sqrt{3}$
9511 7735 $t\pi$: $2e$
9511 8208 $t\pi$: $3(e + \pi)/10$
9511 8973 sr : 19/21
9511 9617 at : $4 - 3\sqrt{3}/2$
9512 2361 J_0 : 4/9
9512 2942 $\exp(1/5) \div \exp(1/4)$
9512 4520 $t\pi$: $8e/5$
9512 4740 $t\pi$: $(\ln 3/2)^{-1/2}$
9512 5500 rt : (1,4,-7,-9)
9512 7019 id : $(2e/3)^3$
9512 8771 10^x : $1/4 + \sqrt{3}$
9512 9849 tanh : $(2\pi)^{1/3}$
9513 1348 Γ : $4\pi^2/9$
9513 4935 ℓ_{10} : $3e + \pi/4$
9513 5076 Γ : 11/10
9513 5663 at : $7\zeta(3)/6$
9513 6196 10^x : $4\sqrt{2} - \sqrt{5}/3$
9513 7303 rt : (3,9,-9,9)
9513 7331 10^x : $(e/3)^{1/2}$
9513 7339 Γ : $7\sqrt{2}/9$
9513 8239 tanh : 24/13
9513 8370 cr : $e + 3\pi/2$
9513 8525 $s\pi$: $4\sqrt{3} + 2\sqrt{5}$
9513 9304 λ : $\exp(-\sqrt[3]{5})$
9514 2615 $\ln(3) \times s\pi(1/3)$
9514 7619 sr : $2/3 + \pi$
9514 8761 tanh : $(e/2)^2$
9514 9020 rt : (9,2,-9,-1)
9515 1167 ln : $\sqrt{2}/4 + \sqrt{5}$
9515 1934 cr : $3\sqrt{3} + \sqrt{5}$
9515 2765 sq : $2e/3 - 3\pi$
9515 4499 ℓ_{10} : $4\sqrt{5}$
9515 6232 $\exp(3) + s\pi(1/3)$
9515 8345 erf : $(e)^{1/3}$
9515 9432 tanh : $8\ln 2/3$
9516 0030 sq : $3\sqrt{3} + 4\sqrt{5}$
9516 0890 cr : P_5
9516 1181 rt : (9,1,-7,-2)
9516 1237 Ψ : $9(e + \pi)/7$
9516 2960 id : $\sqrt{2}/4 + 3\sqrt{3}/2$
9516 3275 sr : $3 - 2\pi/3$
9516 5901 rt : (7,8,-9,4)
9516 6037 rt : (1,8,7,-2)
9516 7562 sr : $4 + 3\pi/2$
9516 7895 cr : $4/3 - \sqrt{2}/3$
9516 8810 erf : $4\pi/9$
9517 1764 rt : (7,9,-5,8)
9517 1992 $s\pi$: $\zeta(3)/3$
9517 2639 rt : (8,3,-8,-2)
9517 3217 rt : (9,0,-5,3)
9517 6162 tanh : $\sqrt[3]{19/3}$
9517 6253 sq : $2\pi^2/7$
9517 6870 sr : $3/2 + 4\sqrt{3}/3$
9517 7596 sinh : $3\zeta(3)/2$
9517 7712 e^x : $(\sqrt{5})^{-1/2}$
9518 2004 $s\pi$: $10\sqrt[3]{2}$
9518 2650 ln : $1/3 + 3\sqrt{5}$
9518 2683 sr : $3\sqrt{2} - \sqrt{3}/4$
9518 3124 rt : (6,5,-9,8)
9518 3860 e^x : $5e\pi/3$
9518 3989 rt : (7,5,-9,-2)
9518 4560 rt : (8,2,-6,-3)
9518 5131 rt : (9,1,-3,4)
9518 6475 2^x : P_{22}
9518 8966 id : $(e/3)^{1/2}$
9518 9317 erf : $5\sqrt{5}/8$
9518 9941 cr : $2 + 2e$
9519 0270 erf : $((e + \pi)/3)^{1/2}$
9519 1169 Γ : $\ln(3)$
9519 2436 erf : $(\ln 3/3)^{-1/3}$
9519 2796 10^x : 16/19
9519 3325 e^x : $3/4 + 4\sqrt{2}/3$
9519 5796 rt : (7,4,-7,-3)
9519 6360 rt : (8,1,-4,-4)
9519 6880 e^x : $7e/5$
9519 6924 rt : (9,2,-1,5)
9519 8126 $\pi \div \ln(5)$
9520 0000 cu : 14/5
9520 0601 e^x : $\sqrt{2} - \sqrt{5}/3$
9520 3536 sq : $7(e + \pi)/3$
9520 4102 sq : $4\sqrt{2} - \sqrt{3}/2$
9520 4256 $\sqrt{5} - \exp(1/4)$
9520 6928 rt : (6,6,-8,-3)
9520 7485 rt : (7,3,-5,-4)
9520 8005 sr : $3\sqrt{2} + 2\sqrt{5}$
9520 8042 rt : (8,0,-2,5)
9520 8599 rt : (9,3,1,6)
9521 0495 Γ : $\sqrt{7/2}$
9521 1231 2^x : $3/2 - \pi/2$
9521 1899 id : $1/\ln(\sqrt{3}/2)$
9521 3133 at : $1/4 + 2\sqrt{3}/3$
9521 3548 sq : $4 + \sqrt{3}/3$
9521 4420 e^x : $5\sqrt{3}/8$
9521 5122 tanh : $9\sqrt[3]{3}/7$
9521 7817 id : $3\sqrt{3}/2 + 3\sqrt{5}/2$
9521 7956 rt : (5,8,-9,-3)
9521 8250 J_0 : 11/25
9521 8508 rt : (6,5,-6,-4)
9521 9058 rt : (7,2,-3,-5)
9521 9609 rt : (8,1,0,6)
9521 9805 sr : $(\sqrt{5}/3)^{1/3}$
9522 0159 rt : (9,4,3,7)
9522 2512 λ : 2/11
9522 2786 $\exp(3) \div s\pi(e)$
9522 4193 Ei : $7\sqrt{2}/4$
9522 4758 cu : $3\sqrt{2}/2 + 4\sqrt{5}$
9522 6765 $s\pi$: $10(e + \pi)$
9522 7117 erf : $10\sqrt[3]{2}/9$
9522 8511 erf : 7/5
9522 8881 λ : $4(1/\pi)/7$
9522 9429 rt : (5,8,-6,-5)
9522 9973 rt : (6,4,-4,-5)
9523 0518 rt : (7,1,-1,-6)
9523 0699 cr : 19/22
9523 1061 rt : (8,2,2,7)
9523 1604 rt : (9,5,5,8)
9523 8095 id : 20/21
9524 0251 rt : (4,9,-8,-4)
9524 0790 rt : (5,9,-2,-1)
9524 1327 rt : (6,3,-2,-6)
9524 1411 tanh : 13/7
9524 1865 rt : (7,0,1,7)
9524 1961 2^x : $\sqrt{10}$
9524 2402 rt : (8,3,4,8)
9524 2939 rt : (9,6,7,9)
9524 4565 ℓ_{10} : $\ln(\sqrt{5}/2)$
9524 6717 Ei : $(\ln 2)^{-3}$
9524 8249 sr : $2 + 2e/3$
9524 9244 sq : $1 - e$
9525 1339 $t\pi$: $5\sqrt{5}/9$
9525 1508 rt : (4,8,-6,-5)
9525 2040 rt : (5,5,-3,-6)
9525 2403 $t\pi$: $5e\pi/2$
9525 2572 rt : (6,2,0,-7)
9525 3103 rt : (7,1,3,8)
9525 3633 rt : (8,4,6,9)
9525 6412 rt : (1,8,0,-3)
9525 9968 at : $1/2 + e/3$
9526 2314 10^x : 13/10
9526 2657 rt : (4,7,-4,-6)
9526 3183 rt : (5,4,-1,-7)
9526 3464 $t\pi$: $(\pi)^{1/3}$
9526 3708 rt : (6,1,2,-8)
9526 4232 rt : (7,2,5,9)
9526 6144 ζ : 3/5
9526 6525 erf : $7\zeta(3)/6$
9527 0823 sq : $4/3 - 4\sqrt{3}/3$
9527 1784 J_0 : 7/16
9527 2106 Γ : $\pi^2/9$
9527 3180 rt : (3,9,-5,-6)
9527 3700 rt : (4,6,-2,3)
9527 4219 rt : (5,3,1,-8)
9527 4738 rt : (6,0,4,9)
9527 9896 sq : $1/2 + 3e/2$
9527 9933 Ψ : 2/25
9528 0244 id : $2 - \pi/3$
9528 1778 rt : (1,8,8,5)
9528 3418 sq : $2\sqrt{2}/3 + 4\sqrt{3}$
9528 3640 $s\pi$: $\exp(4)$
9528 4125 rt : (3,8,-3,-7)
9528 4381 rt : (7,8,-8,6)
9528 4638 rt : (4,7,2,7)
9528 4707 sr : $(5/2)^3$
9528 5151 rt : (5,2,3,-9)
9528 6145 sq : $3 + 2\pi/3$
9528 6593 2^x : $4\sqrt{2} - 2\sqrt{5}/3$
9528 6930 J_0 : $1/\sqrt[3]{12}$
9529 0297 $t\pi$: $4\sqrt{3}/3 + \sqrt{5}$
9529 0675 $s\pi$: $3\sqrt{3}/2$
9529 3634 cr : $3\sqrt[3]{3}/5$
9529 4966 rt : (3,7,-1,-8)
9529 5474 rt : (4,4,2,-9)
9529 6410 Γ : $9(e + \pi)/7$
9529 6635 10^x : $\sqrt{8/3}$
9529 8753 ln : $5\pi^2/7$
9529 9107 tanh : $5\sqrt{5}/6$
9530 1555 sr : $8(e + \pi)/3$
9530 1910 2^x : $2\sqrt{3} + \sqrt{5}/2$
9530 2737 K : $2e/7$
9530 3242 e^x : 8/5
9530 4749 tanh : $1/\ln(\sqrt[3]{5})$
9530 5206 rt : (2,9,-2,-8)
9530 5708 rt : (3,6,1,-9)
9530 8532 cu : $\sqrt{2} + 2\sqrt{3}/3$
9531 1298 Ψ : $9\sqrt[3]{5}/5$
9531 1834 id : $\sqrt{17}\pi$
9531 2500 sq : $5\sqrt{5}/8$

9531 5169 $Ei : \sqrt[3]{5}$
9531 5853 rt : (2,8,0,-9)
9531 7269 $s\pi : 3\pi^2/4$
9531 7928 $s\pi : 4\sqrt{2} + \sqrt{5}/3$
9531 8020 $10^x : 3/4 + \sqrt{2}$
9531 8429 id : $(\sqrt{3}/2)^{1/3}$
9532 0718 $\Gamma : \sqrt{23/2}$
9532 1379 $J : \sqrt{3}/7$
9532 1834 $2^x : 3e/2 - 2\pi/3$
9532 1966 $J : \exp(-11\pi/4)$
9532 2943 sq : $2e\pi/7$
9532 4971 $2^x : 3\sqrt{2}/2 - \sqrt{5}/4$
9532 5344 tanh : $\sqrt[3]{13/2}$
9532 6404 rt : (2,5,-2,2)
9532 6626 $2^x : 2/3 + 4\sqrt{5}$
9532 8015 cu : $\sqrt{3}/2 + 4\sqrt{5}/3$
9532 8794 $\Gamma : 23/21$
9532 9149 id : $(e + \pi)/3$
9532 9645 $J_0 : 10/23$
9533 1114 rt : (1,8,1,5)
9533 3416 cr : $5\ln 2/4$
9533 4026 sinh : 25/12
9533 5338 ln : $1/2 + 2\pi/3$
9533 5860 $\sqrt{5} \div \ln(\pi)$
9533 5912 $s\pi : 7\zeta(3)/6$
9533 7262 $s\pi : 2e/3 + \pi/4$
9533 7855 $e^x : 2/3 + 2\sqrt{2}$
9533 8530 cr : $10\sqrt{5}/3$
9534 1569 cu : $6\sqrt{5}$
9534 1950 cr : 13/15
9534 2712 sq : $4 + \sqrt{2}/4$
9534 4529 $\ell_2 : \sqrt{15}$
9534 4581 $\Gamma : 15/8$
9534 5356 $\sqrt{3} + \exp(1/5)$
9534 5484 $2^x : 3/2 + 3\pi$
9534 6258 sr : 10/11
9534 8823 $J : \exp(-2e)$
9535 0691 id : $\sqrt{3} - \sqrt{5} - \sqrt{6}$
9535 1173 id : $3\sqrt{2} - 3\sqrt{3}$
9535 4751 $t\pi : 4\ln 2/7$
9535 6485 $10^x : 4 + 3e/2$
9535 6569 $e^x : 3e/4 + 2\pi$
9535 6589 as : $3e/10$
9535 6698 $\Gamma : 7\sqrt[3]{5}/2$
9536 1090 $10^x : 3/4 + 3\sqrt{3}/4$
9536 2825 $J_0 : 5\ln 2/8$
9536 5229 $2^x : 25/16$
9536 5836 $10^x : 2\sqrt{5}/3$
9536 6919 tanh : $\sqrt{7/2}$
9536 6997 sq : $(\ln 3/3)^{-1/3}$
9536 7146 $J_0 : \sqrt{3}/4$
9536 7194 rt : (1,6,7,9)
9537 0491 $J : 5\ln 2/7$
9537 3807 $2^x : P_{50}$
9537 4287 $J_0 : 3\sqrt[3]{3}/10$
9537 4605 $s\pi : 5\sqrt{5}/7$
9537 8053 id : $2e/3 + \pi$
9537 8843 sinh : $10\pi/3$
9537 9091 cr : $3/4 + 3\sqrt{5}$
9538 3691 sq : $(e + \pi)/6$
9538 4113 $\ell_2 : P_{28}$
9538 4370 $erf : \pi^2/7$
9538 7795 rt : (9,2,-8,-2)
9538 8203 sr : $\sqrt{2}/4 + 2\sqrt{3}$
9538 9038 at : $\pi^2/7$
9538 9893 $e^x : 1 - \pi/3$
9539 0548 $\zeta : 9\sqrt[3]{3}/10$
9539 0813 $\Gamma : \sqrt{3} - \sqrt{5} - \sqrt{7}$
9539 3354 $e^x : 5\sqrt{3}/2$
9539 3976 cu : $\sqrt{2}/3 + 2\sqrt{3}$
9539 6388 $\ell_2 : 1/4 + 4e/3$
9539 6987 rt : (7,8,-7,8)
9539 7707 rt : (8,4,-9,-2)
9539 8183 rt : (9,1,-6,-3)
9539 9076 ln : $9\sqrt[3]{3}/5$
9540 0756 $e^x : (\sqrt[3]{4}/3)^{-1/2}$
9540 4213 sr : $\sqrt{2} + \sqrt{3} - \sqrt{5}$
9540 4526 tanh : 15/8
9540 6458 id : $(\ln 3)^{-1/2}$
9540 8005 rt : (8,3,-7,-3)
9540 8062 sinh : $3\sqrt{2}/5$
9540 8476 rt : (9,0,-4,4)
9541 0940 $2^x : 7(e + \pi)/8$
9541 2447 $erf : 24/17$
9541 5851 $e^x : 3 - 3\sqrt{2}/4$
9541 5923 $2^x : 4/3 + 4\pi/3$
9541 6634 $t\pi : 1/\sqrt{17}$
9541 6891 $s\pi : 3\sqrt{2} + 3\sqrt{5}/2$
9541 7743 rt : (7,5,-8,-3)
9541 8209 rt : (8,2,-5,-4)
9541 8674 rt : (9,1,-2,5)
9542 0650 $\Psi : \ln(\sqrt{5})$
9542 3435 $Ei : 2$
9542 4250 $\ell_{10} : 9$
9542 5645 $2^x : 3\sqrt{2}/2 - 2\sqrt{3}/3$
9542 5822 id : $8\sqrt[3]{5}/7$
9542 6224 $1/5 - \exp(e)$
9542 6775 $\Gamma : \exp(\sqrt[3]{2}/2)$
9542 7247 tanh : $\exp(\sqrt[3]{2}/2)$
9542 7399 rt : (6,7,-9,-3)
9542 7439 $\lambda : \exp(-7\pi/13)$
9542 7860 rt : (7,4,-6,-4)
9542 8320 rt : (8,1,-3,-5)
9542 8780 rt : (9,2,0,6)
9542 9459 $e^x : \exp(1/\pi)$
9542 9743 sr : $1/3 + \sqrt{3}/3$
9542 9958 ln : $5\ln 2/9$
9543 0577 cr : $4 + 2\sqrt{3}$
9543 1157 rt : (4,9,-9,-2)
9543 1330 $\ell_{10} : e + 2\pi$
9543 1718 cu : $9e\pi$
9543 2685 $Li_2 : (5/2)^{-1/3}$
9543 4873 $2/3 - \ln(3/4)$
9543 4980 $\sqrt{5} + e$
9543 7430 rt : (6,6,-7,-4)
9543 7885 rt : (7,3,-4,-5)
9543 8340 rt : (8,0,-1,6)
9543 8795 rt : (9,3,2,7)
9543 9172 $Li_2 : 14/19$
9544 2788 $s\pi : (3\pi/2)^{-1/3}$
9544 3122 sr : P_{11}
9544 6920 rt : (5,8,-8,-4)
9544 7371 rt : (6,5,-5,-5)
9544 7821 rt : (7,2,-2,-6)
9544 8121 cr : 20/23
9544 8271 rt : (8,1,1,7)
9544 8721 rt : (9,4,4,8)
9544 8838 $\ell_2 : (\pi)^3$
9544 9285 sinh : $8(1/\pi)/3$
9544 9879 $10^x : \sqrt[3]{3/2}$
9544 9938 at : 24/17
9544 9973 $erf : \sqrt{2}$
9545 0881 $10^x : 1/3 - \sqrt{2}/4$
9545 1152 $e^x : 13/12$
9545 4102 $J_0 : (2e)^{-1/2}$
9545 4545 id : 21/22
9545 5351 $10^x : \sqrt{2}$
9545 6525 rt : (1,9,-9,5)
9545 6778 rt : (5,8,-5,-3)
9545 7223 rt : (6,4,-3,-6)
9545 7434 cu : $3/4 + 4\sqrt{5}/3$
9545 7669 rt : (7,1,0,-7)
9545 8114 rt : (8,2,3,8)
9545 8257 ln : $2e/3 + \pi/4$
9545 8556 at : $2/3 + \sqrt{5}/3$
9545 8559 rt : (9,5,6,9)
9546 0607 $J_0 : 3/7$
9546 1904 $\exp(4) + \exp(\sqrt{5})$
9546 3028 $s\pi : 5\sqrt[3]{3}/3$
9546 5346 sinh : $5\pi/3$
9546 6107 rt : (4,9,-7,-5)
9546 6549 rt : (5,9,-1,1)
9546 6989 rt : (6,3,-1,-7)
9546 7430 rt : (7,0,2,8)
9546 7604 tanh : $\sqrt[3]{20/3}$
9546 7870 rt : (8,3,5,9)
9546 9393 ln : $\sqrt{2}/4 + 3\sqrt{5}$
9546 9681 sq : $4e - 3\pi/2$
9547 3928 $t\pi : P_{15}$
9547 4069 $\Psi : \sqrt{19/2}$
9547 5798 rt : (4,8,-5,-6)
9547 6234 rt : (5,5,-2,-7)
9547 6670 rt : (6,2,1,-8)
9547 7087 $\ell_{10} : 4\sqrt{2} + 3\sqrt{5}/2$
9547 7106 rt : (7,1,4,9)
9547 7125 ln : $2\sqrt{3}/9$
9547 7955 sr : $e/4 + \pi$
9547 8130 sq : $4\sqrt[3]{5}/7$
9547 9710 $t\pi : 3\sqrt{2}$
9548 0432 sq : $7\sqrt[3]{5}/4$
9548 0484 sq : $3/4 + 4\sqrt{3}$
9548 1360 id : $2\sqrt{3} + 2\sqrt{5}/3$
9548 1859 $s\pi : 9\sqrt[3]{3}/5$
9548 3142 $J_0 : \sqrt[3]{5}/4$
9548 4114 $\Gamma : e/6$
9548 5404 rt : (4,7,-3,-7)
9548 5836 rt : (5,4,0,-8)
9548 6267 rt : (6,1,3,-9)
9548 7305 $erf : 17/12$
9548 9294 $\Gamma : 1/\ln(5/2)$
9549 1483 sinh : $e\pi/6$
9549 2146 $s\pi : 2\sqrt{2}/7$
9549 2965 id : $3/\pi$
9549 3080 tanh : $3\pi/5$
9549 3919 $K : 7/9$
9549 4501 rt : (3,9,-4,-7)
9549 4928 rt : (4,6,-1,5)
9549 5355 rt : (5,3,2,-9)
9549 5742 $s\pi : \exp(-e/3)$
9549 6465 $10^x : 4\sqrt{2} + 4\sqrt{5}$
9549 7068 $s\pi : 2\zeta(3)$
9549 7237 $t\pi : 1/\ln(\sqrt{5})$
9549 8583 $erf : 9\sqrt[3]{2}/8$
9549 8914 tanh : $4\sqrt{2}/3$
9549 9653 sinh : $(\sqrt[3]{3}/2)^{1/2}$
9550 0144 $10^x : \ln(\pi)$
9550 0672 cr : $3 + 2\sqrt{5}$
9550 1818 cr : $7e\pi/8$
9550 3947 rt : (3,8,-2,-8)
9550 4370 rt : (4,7,3,9)
9550 5596 cu : $\sqrt{2}/3 - 3\sqrt{5}/2$
9550 7393 $\Gamma : (5/3)^3$
9550 7531 $\lambda : (\sqrt[3]{5}/3)^3$
9550 7672 $e^x : 11/8$
9550 7953 $\Gamma : 12/11$
9550 9120 $\ell_{10} : 4\sqrt{3}/3 + 3\sqrt{5}$
9550 9371 $s\pi : \sqrt{2} + \sqrt{3} + \sqrt{6}$
9551 2586 $10^x : 3/4 + 4\sqrt{5}/3$
9551 3155 rt : (1,8,-9,-4)
9551 3314 rt : (3,7,0,-9)
9551 4000 $2^x : e/2 + 2\pi/3$
9551 5277 $t\pi : 5/11$
9551 5938 $s\pi : 2\sqrt{3}/3 - \sqrt{5}/4$
9551 7175 $t\pi : 1/4 - \pi/4$
9551 9214 cu : $2 - 3\pi/2$
9551 9435 $\sqrt{3} - \ln(4/5)$
9552 0923 $10^x : 8/17$
9552 1110 $\Psi : 6\sqrt[3]{2}$
9552 2188 rt : (2,9,-1,-9)
9552 2602 rt : (3,3,-1,9)
9552 6691 $\zeta : (\sqrt[3]{5}/3)^2$
9552 7606 tanh : 17/9
9552 8549 sr : $\sqrt{2}/4 + \sqrt{5}/4$
9552 8753 $\sqrt{2} \times s\pi(\sqrt{5})$
9552 9063 $J_0 : 1/\sqrt[3]{13}$
9552 9105 $J_0 : \exp(-\sqrt[3]{5}/2)$
9553 1403 rt : (2,4,-7,-4)
9553 1661 at : $\sqrt{2}$
9553 2090 $e^x : 1/\ln(\sqrt{2}\pi)$
9553 5981 rt : (4,9,-8,1)
9553 6278 tanh : $3\sqrt[3]{2}/2$
9554 0541 rt : (2,5,-1,5)
9554 3463 $10^x : 1 + 3\sqrt{2}$
9554 6924 cu : $\sqrt{2} + \sqrt{5} + \sqrt{6}$
9554 7279 $J_0 : 4(1/\pi)/3$
9555 0372 $J_0 : 3\sqrt{2}/10$
9555 1144 ln : 5/13
9555 1596 $s\pi : 1/4 + 2\sqrt{3}/3$
9555 2025 at : $e/4 - 2\pi/3$
9555 3308 sr : 21/23

9555 3861 $10^x : 3\pi$
9555 3923 $10^x : 3e/2 - \pi/4$
9555 6163 $e^x : 5(e+\pi)/8$
9555 6562 $\ell_2 : 3 - 3\sqrt{2}/4$
9555 7409 $\ln(2) - \exp(1/2)$
9555 8195 rt : (1,5,-9,-9)
9555 9005 $J : (2\pi/3)^{-2}$
9556 0869 $s\pi : 3e\pi/4$
9556 2243 $2/5 \times \exp(2)$
9556 2791 id : $7e\pi/5$
9556 3394 rt : (1,8,-8,5)
9556 5016 $s\pi : 1/4 + 3e$
9556 6010 $\ln : 9\pi/4$
9556 7089 $2^x : (e/3)^{1/3}$
9556 7115 rt : (1,2,-9,-5)
9556 9518 $\sinh : (\ln 2/2)^{-1/3}$
9557 4998 $\Gamma : \sqrt[3]{20/3}$
9557 8175 $K : (\sqrt{2}/3)^{1/3}$
9558 4122 $e^x : 3\sqrt{5}/10$
9558 4212 $\Psi : 9(e+\pi)$
9558 4768 id : $\sqrt{2}/4 - 4\sqrt{3}/3$
9558 5434 $10^x : 3\sqrt{2}/2 + 3\sqrt{5}/4$
9558 6797 $erf : e\pi/6$
9558 9020 sr : $\sqrt{2}/3 + 3\sqrt{5}/2$
9558 9048 $\ln : 3\sqrt{3}/2 + 2\sqrt{5}$
9558 9235 id : $4\sqrt{2} + 3\sqrt{3}/4$
9559 1311 $s\pi : 2/3 + 4\sqrt{3}$
9559 1861 $10^x : 2\sqrt{3} + 3\sqrt{5}/2$
9559 2082 $erf : (\ln 2/2)^{-1/3}$
9559 3865 rt : (9,3,-9,-2)
9559 4372 cr : $1/\ln(\pi)$
9559 4807 cr : $\sqrt[3]{2/3}$
9559 7201 $10^x : 2\sqrt{2} - \sqrt{5}/4$
9559 7429 cu : $\sqrt{2}/2 - 2\sqrt{3}$
9559 9314 $\ln(4) + s\pi(\pi)$
9559 9761 $\sinh : 10\sqrt{5}/9$
9559 9787 $J_0 : (4/3)^{-3}$
9560 1150 $\ln : 5\sqrt{2}$
9560 1902 rt : (7,7,-9,8)
9560 2895 rt : (9,2,-7,-3)
9560 3642 $\Psi : 1/\sqrt[3]{23}$
9560 4173 $\tanh : 2e\pi/9$
9560 4652 $2^x : \sqrt{3}/2 + \sqrt{5}/2$
9560 5565 $s\pi : (\pi/2)^{-2}$
9560 6733 cu : $3 + 3\sqrt{2}/4$
9560 7177 $e^x : (\sqrt{2}\pi/2)^{-1/2}$
9560 9339 $\exp(3/5) - s\pi(1/3)$
9561 1458 rt : (8,4,-8,-3)
9561 1595 $\sinh : 17/20$
9561 1850 rt : (9,1,-5,-4)
9561 1902 $erf : 5\sqrt[3]{5}/6$
9561 3337 at : 17/12
9561 4191 $10^x : 1/2 + 3\sqrt{3}/4$
9561 6735 $J_0 : 8/19$
9561 9955 rt : (7,6,-9,-3)
9562 0211 $\theta_3 : 11/25$
9562 0342 rt : (8,3,-6,-4)
9562 0730 rt : (9,0,-3,5)
9562 2164 $s\pi : \ln(3/2)$
9562 3517 $\tanh : 10\sqrt[3]{5}/9$
9562 3545 $s\pi : 1/\sqrt[3]{15}$
9562 3589 sq : $\sqrt{2}/3 - 2\sqrt{3}$
9562 3745 $\tanh : 19/10$
9562 5638 $J_0 : \exp(-\sqrt{3}/2)$
9562 8065 $10^x : 4\sqrt{5}/7$
9562 8685 $10^x : 4\pi$
9562 8769 rt : (7,5,-7,-4)
9562 9153 rt : (8,2,-4,-5)
9562 9536 rt : (9,1,-1,6)
9563 0466 cu : $e/4 - 2\pi$
9563 2099 sr : $7\sqrt{5}$
9563 2138 cr : $4\sqrt{3} + \sqrt{5}/4$
9563 5006 $s\pi : 1/2 + 2\pi/3$
9563 5982 rt : (4,9,-7,4)
9563 7132 rt : (6,7,-8,-4)
9563 7512 rt : (7,4,-5,-5)
9563 7891 rt : (8,1,-2,-6)
9563 8088 at : $9\sqrt[3]{2}/8$
9563 8271 rt : (9,2,1,7)
9563 8923 $J_0 : \sqrt[3]{2}/3$
9564 1057 sq : $1/3 - \sqrt{3}$
9564 2974 $10^x : 3\sqrt{3} - 4\sqrt{5}/3$
9564 5430 rt : (5,9,-9,-4)
9564 5807 rt : (6,6,-6,-5)
9564 6183 rt : (7,3,-3,-6)
9564 6559 cr : 7/8
9564 6901 $2/5 - \exp(\sqrt{5})$
9564 6934 rt : (9,3,3,8)
9564 7629 $\tanh : 7e/10$
9565 0015 $2^x : 4\sqrt{2} + \sqrt{3}/2$
9565 0828 $s\pi : 2\sqrt{2} + \sqrt{3}/3$
9565 2173 id : 22/23
9565 4039 rt : (5,8,-7,-5)
9565 4411 rt : (6,5,-4,-6)
9565 4784 rt : (7,2,-1,-7)
9565 5156 rt : (8,1,2,8)
9565 5528 rt : (9,4,5,9)
9565 5948 id : $7\sqrt{5}/8$
9565 9023 $\Gamma : 2e/5$
9565 9663 $s\pi : 6\ln 2/7$
9566 2571 $\ell_{10} : 3\sqrt{2}/2 + 4\sqrt{3}$
9566 2578 rt : (5,8,-4,-1)
9566 2875 $Ei : \exp(\sqrt{2}\pi/3)$
9566 2947 rt : (6,4,-2,-7)
9566 3316 rt : (7,1,1,-8)
9566 3668 sr : $1 + 2\sqrt{2}$
9566 3684 rt : (8,2,4,9)
9566 4824 $erf : 10/7$
9566 9850 $\theta_3 : 9/11$
9567 0064 $\Gamma : 3\pi/5$
9567 0102 $3 \times \exp(\sqrt{3})$
9567 0684 rt : (4,9,-6,-6)
9567 1050 rt : (5,9,0,3)
9567 1415 rt : (6,3,0,-8)
9567 1421 $\lambda : \exp(-8\pi/15)$
9567 1780 rt : (7,0,3,9)
9567 4097 $\Gamma : 25/23$
9567 7049 $\theta_3 : 2\pi/9$
9567 7260 $2^x : 4\sqrt{3}/3$
9567 7569 cu : $\sqrt{2} + 2\sqrt{5}$
9567 9092 rt : (4,8,-4,-7)
9567 9454 rt : (5,5,-1,-8)
9567 9790 $s\pi : 1/2 + e/3$
9567 9815 rt : (6,2,2,-9)
9568 1071 $J : 3\sqrt[3]{3}/10$
9568 1726 $2^x : 4\sqrt{3}/3 - \sqrt{5}/3$
9568 3520 id : $8\pi^2$
9568 6636 $e^x : 7\sqrt[3]{5}/5$
9568 7140 $2^x : 5(e+\pi)/3$
9568 7434 rt : (4,7,-2,-8)
9568 7463 $s\pi : 4\sqrt{2}/3 + 3\sqrt{5}$
9568 7792 rt : (5,4,1,-9)
9568 8722 $\ell_2 : 3\zeta(3)/7$
9569 0982 $\ln : 3 + 3e/2$
9569 1997 $\Gamma : 4\sqrt{2}/3$
9569 4299 $\lambda : 3/16$
9569 5113 cu : $4e + \pi/2$
9569 5355 rt : (3,9,-3,-8)
9569 5710 rt : (4,6,0,7)
9569 6474 cu : $3/2 + \pi/2$
9569 8841 $t\pi : \sqrt{2}/2 - 2\sqrt{3}$
9569 9336 $erf : ((e+\pi)/2)^{1/3}$
9569 9825 $t\pi : 7e/4$
9570 0440 $Ei : e\pi/3$
9570 0900 $\tanh : 21/11$
9570 2097 $s\pi : e/2 + \pi/3$
9570 3571 rt : (3,8,-1,-9)
9570 5487 $\zeta : (3\pi)^{-1/2}$
9570 6091 sr : $\beta(2)$
9570 6283 $\Psi : 2\pi^2$
9570 6590 $J_0 : 5/12$
9570 7360 $\tanh : 6(1/\pi)$
9570 8427 $2^x : 2e - \pi/3$
9570 8454 $10^x : 3/2 + \pi$
9571 0678 id : $1/4 + \sqrt{2}/2$
9571 1723 rt : (3,2,-4,7)
9571 6417 $e^x : 4/3 + 3\sqrt{2}/4$
9571 7322 $\ell_{10} : 10e/3$
9571 8892 $10^x : 3\sqrt{2}/2 + 4\sqrt{5}$
9571 9796 $\ln : 4(e+\pi)/9$
9572 0566 $erf : 9(1/\pi)/2$
9572 0777 $s\pi : 3\sqrt{2}/2 + 2\sqrt{5}$
9572 2443 $J_0 : 3\ln 2/5$
9572 2997 $t\pi : 1/3 - 4e$
9572 3076 sr : $\ln(5/2)$
9572 6821 $\ell_2 : 5e/7$
9572 7500 rt : (2,4,-6,-1)
9573 0584 cr : $(\sqrt{2}\pi/3)^{-1/3}$
9573 1161 $10^x : 3 + 3e/4$
9573 1493 rt : (4,9,-6,7)
9573 3091 $\tanh : \sqrt[3]{7}$
9573 4178 $10^x : 7/24$
9573 4448 $10^x : 14/5$
9573 5471 rt : (2,5,0,8)
9573 5652 $10^x : 23/18$
9573 7428 $10^x : \sqrt{3}$
9573 8689 $\theta_3 : 22/25$
9574 2260 $J_0 : 1/\sqrt[3]{14}$
9574 2710 sr : 11/12
9574 3382 id : $\sqrt[3]{15/2}$
9574 5624 $e^x : e/4 + 4\pi$
9574 6820 $\exp(\pi) \times s\pi(\pi)$
9574 8966 $s\pi : 3/4 + 4\sqrt{2}$
9574 9122 $\tanh : \sqrt{11/3}$
9575 0822 cu : $3\pi^2/10$
9575 2394 $\ln : 4\sqrt{2}/3 + 3\sqrt{3}$
9575 3270 $e^x : 5e\pi/8$
9575 4076 $s\pi : 2\sqrt{2}/3 + 2\sqrt{3}$
9575 8698 rt : (1,5,-1,-6)
9575 9027 $J_0 : (\sqrt{5}/3)^3$
9575 9141 $10^x : 9\sqrt[3]{5}/7$
9575 9173 $\sqrt{3} + \exp(4/5)$
9576 1518 id : $1/\ln(5/3)$
9576 2241 $10^x : e/4 + 2\pi$
9576 3871 $\tanh : 6\sqrt{5}/7$
9576 4177 $\tanh : 23/12$
9576 5297 $\exp(4/5) \times s\pi(\pi)$
9576 5441 $\ln : 2\sqrt{2}/3 - \sqrt{5}/4$
9576 5721 $10^x : 2\sqrt{2} + 2\sqrt{3}$
9576 6014 $t\pi : \exp(-\sqrt{2})$
9576 6921 $2^x : (\ln 3)^{-1/3}$
9576 7087 $\exp(1/4) \times s\pi(\sqrt{3})$
9576 8033 $\Psi : 4/15$
9577 0367 $\ell_2 : 3\sqrt{2}/4 + 3\sqrt{5}$
9577 0910 id : $1/3 + 4e/3$
9577 3443 $10^x : 3\ln 2/2$
9577 7128 $\tanh : \sqrt{2} - \sqrt{3} + \sqrt{5}$
9577 8985 $J_0 : (e+\pi)^{-1/2}$
9577 9624 $e^x : 2/3 + \sqrt{5}/2$
9578 0446 cr : $3 - 3\sqrt{2}/2$
9578 1587 rt : (1,7,-1,-9)
9578 4447 sq : $e/3 + 4\pi/3$
9578 9002 sr : 23/6
9578 9110 $\ln : 1/4 + 3\pi/4$
9578 9996 rt : (9,3,-8,-3)
9579 1620 rt : (1,7,-8,-3)
9579 2637 cu : 21/11
9579 4060 $erf : 23/16$
9579 7508 rt : (8,5,-9,-3)
9579 7836 rt : (9,2,-6,-4)
9579 8704 $e^x : 8$
9579 9770 $J_0 : 2\sqrt[3]{3}/7$
9580 0785 $\Gamma : 17/9$
9580 1872 $\lambda : \exp(-5/3)$
9580 5291 rt : (8,4,-7,-4)
9580 5616 rt : (9,1,-4,-5)
9580 5952 $J_0 : 7/17$
9581 0812 $\sqrt{3} - s\pi(e)$
9581 0986 $\zeta : (\ln 3/3)^{1/3}$
9581 1752 at : $2 - \sqrt{3}/3$
9581 2693 rt : (7,6,-8,-4)
9581 2806 $10^x : \exp(\sqrt[3]{3}/2)$
9581 3015 rt : (8,3,-5,-5)
9581 3337 rt : (9,0,-2,6)
9581 4410 $\ell_2 : 1 + 2\sqrt{2}/3$
9581 5637 $\ell_2 : \sqrt[3]{22/3}$
9581 6373 $\tanh : 4\sqrt[3]{3}/3$
9581 7008 $\tanh : 25/13$
9581 8761 $\tanh : 8\zeta(3)/5$
9581 9959 $s\pi : 2\sqrt{2} - \sqrt{5}$

9581 9991 $s\pi : 9(e+\pi)/8$
9582 0043 rt : (6,8,-9,-4)
9582 0363 rt : (7,5,-6,-5)
9582 0393 id : $1/4 + 3\sqrt{5}$
9582 0682 rt : (8,2,-3,-6)
9582 1001 rt : (9,1,0,7)
9582 1156 $\exp(2/5) \times \exp(\sqrt{5})$
9582 2219 sinh : $2\sqrt{3}$
9582 2296 tanh : $9\sqrt[3]{5}/8$
9582 4158 as : 9/11
9582 5650 sinh : $8\sqrt{3}/5$
9582 7658 rt : (6,7,-7,-5)
9582 7975 rt : (7,4,-4,-6)
9582 8291 rt : (8,1,-1,-7)
9582 8376 $erf : 8\sqrt[3]{2}/7$
9582 8397 cr : 22/25
9582 8568 $\Gamma : 13/12$
9582 8607 rt : (9,2,2,8)
9582 8656 tanh : $10\sqrt{3}/9$
9582 8688 $2^x : 7\sqrt[3]{5}/4$
9583 1484 id : $3\sqrt{5}/7$
9583 2656 $t\pi : \sqrt{2}+\sqrt{5}$
9583 2811 $e^x : 2\sqrt{2}/3+\sqrt{3}/4$
9583 2888 at : $e\pi/6$
9583 3333 id : 23/24
9583 3967 $\Gamma : 3\sqrt[3]{2}/2$
9583 4537 $t\pi : 4\sqrt{2}/3+2\sqrt{3}$
9583 4904 rt : (5,9,-8,-5)
9583 5217 rt : (6,6,-5,-6)
9583 5531 rt : (7,3,-2,-7)
9583 5844 rt : (8,0,1,8)
9583 6158 rt : (9,3,4,9)
9583 8720 $K : 9\ln 2/8$
9583 9803 sq : $3-\sqrt{5}/4$
9584 2410 rt : (5,8,-6,-6)
9584 2721 rt : (6,5,-3,-7)
9584 3031 rt : (7,2,0,-8)
9584 3342 rt : (8,1,3,9)
9584 4629 at : $(\ln 2/2)^{-1/3}$
9584 9861 rt : (5,8,-3,1)
9585 0169 rt : (6,4,-1,-8)
9585 0409 rt : (1,7,-7,5)
9585 0477 rt : (7,1,2,-9)
9585 2345 $\lambda : \exp(-9\pi/17)$
9585 3206 $Ei : (\pi/3)^{1/2}$
9585 4477 $Li_2 : 17/23$
9585 6376 $J_0 : 9(1/\pi)/7$
9585 6455 ln : P_{67}
9585 6953 rt : (4,9,-5,-7)
9585 7258 rt : (5,9,1,5)
9585 7563 rt : (6,3,1,-9)
9585 9023 ln : P_{70}
9585 9675 $J_0 : 9/22$
9585 9677 tanh : $\sqrt{3}-\sqrt{6}+\sqrt{7}$
9586 0870 $\ell_{10} : 2\sqrt{3}-3\sqrt{5}/2$
9586 1469 $erf : \sqrt[3]{3}$
9586 3024 $\Gamma : 5\sqrt{3}/8$
9586 3060 sinh : $5\sqrt[3]{5}/6$
9586 3498 $e^x : \exp(\sqrt[3]{2})$
9586 4299 rt : (4,8,-3,-8)
9586 4551 $erf : 6\zeta(3)/5$
9586 4602 rt : (5,5,0,-9)
9586 5437 $\Gamma : \exp(-\sqrt[3]{4}/2)$
9586 7744 $erf : 1/\ln(2)$
9586 9771 rt : (1,7,0,-2)
9587 1593 rt : (4,7,-1,-9)
9587 4382 $s\pi : 1/\sqrt{6}$
9587 4726 $Ei : ((e+\pi)/2)^{1/2}$
9587 5488 $\exp(2) + s\pi(\pi)$
9587 6535 $J_0 : 1/\sqrt{6}$
9587 7316 $erf : 5\sqrt{3}/6$
9587 8537 rt : (3,9,-2,-7)
9587 8755 $10^x : 7\pi^2/3$
9587 8835 rt : (4,6,1,9)
9587 8903 $Li_2 : \ln(2\pi/3)$
9588 2491 $t\pi : 3\sqrt{3}/4+4\sqrt{5}$
9588 2526 $2^x : 1-3\sqrt{2}/4$
9588 5558 $s\pi : 2\sqrt{3}+4\sqrt{5}$
9588 5730 rt : (3,1,-7,5)
9588 6963 ln : $7\sqrt{5}/6$
9588 7755 $\lambda : \sqrt[3]{5}/9$
9588 8727 at : $5\sqrt[3]{5}/6$
9588 9264 $e^x : 3\sqrt{2}/2-\sqrt{5}/3$
9589 0788 at : $\sqrt{3}/2+\sqrt{5}/4$
9589 1491 $\ln(\sqrt{2}) \times \exp(\sqrt{3})$
9589 2309 $erf : 13/9$
9589 2359 $\Gamma : 9\zeta(3)/10$
9589 2427 $s\pi : 5(1/\pi)$
9589 2873 rt : (3,2,-3,9)
9589 5438 $erf : \exp(1/e)$
9589 6080 $\zeta : ((e+\pi)/3)^{-1/2}$
9589 8119 $\ln(\sqrt{5}) + \exp(e)$
9589 8936 cr : $7\sqrt[3]{2}/10$
9589 9441 $\Gamma : 3\sqrt[3]{3}/4$
9589 9707 cr : $2-\sqrt{5}/2$
9590 1884 $t\pi : \ln(3\pi)$
9590 2716 $\zeta : \zeta(3)/2$
9590 3368 $t\pi : 4\pi/9$
9590 4919 $s\pi : 5e$
9590 6721 rt : (2,4,-5,2)
9590 8229 $\Gamma : (\sqrt[3]{5}/2)^{-1/2}$
9590 8327 $Ei : (\sqrt[3]{2}/3)^{-3}$
9590 8649 $\ln(2) - \exp(\sqrt{3})$
9591 1245 cu : $4(e+\pi)/3$
9591 2442 as : $4\sqrt{3}/3-2\sqrt{5}/3$
9591 3729 cr : 15/17
9591 6630 sr : 23/25
9591 8367 sq : $9\sqrt{3}/7$
9591 8452 $t\pi : 2\sqrt{2}+4\sqrt{3}$
9592 3924 $\lambda : 4/21$
9592 7177 sq : $4e/3-4\pi$
9592 8282 $2^x : 7\sqrt{5}/10$
9593 0022 cu : $4\sqrt{2}-\sqrt{3}/2$
9593 0787 $erf : (2\pi/3)^{1/2}$
9593 1691 $J_0 : 1/\sqrt[3]{15}$
9593 1990 $J_0 : \ln(3/2)$
9593 3534 $\zeta : (\sqrt{5}/2)^{-3}$
9593 4144 rt : (1,5,0,-1)
9593 5571 $J_0 : (\pi/2)^{-2}$
9593 6916 sinh : $8\sqrt[3]{5}$
9593 9199 $2^x : 7\ln 2/5$
9593 9277 $2^x : 25/6$
9594 2110 $s\pi : 9\sqrt{2}/8$
9594 9297 $s\pi : 9/22$
9595 0260 $s\pi : \sqrt{3}+3\sqrt{5}/4$
9595 1571 $\exp(1/4) + s\pi(\sqrt{5})$
9595 2326 $s\pi : 8e\pi/9$
9595 4302 rt : (1,7,1,5)
9595 4305 rt : (8,9,-9,8)
9595 4689 $J : (\sqrt{2}\pi/3)^{-3}$
9595 6999 at : $1/4-3\sqrt{5}/4$
9595 7665 $\zeta : \exp(-\sqrt{5}/2)$
9595 9064 $J_0 : \exp(-e/3)$
9595 9828 $J_0 : 2\sqrt{2}/7$
9596 1684 rt : (9,4,-9,-3)
9596 2103 $\lambda : 3(1/\pi)/5$
9596 2525 as : $(\ln 3/2)^{1/3}$
9596 3455 $\Psi : 8e\pi/9$
9596 3761 sq : $4\sqrt{2}/3-3\sqrt{3}$
9596 3858 $s\pi : 9(1/\pi)/7$
9596 3902 $s\pi : 9\zeta(3)/2$
9596 4566 $\ell_{10} : P_{76}$
9596 6304 tanh : $5e/7$
9596 8587 rt : (9,3,-7,-4)
9596 9155 cr : $5\sqrt{2}/8$
9596 9935 $\Gamma : (\sqrt[3]{2})^{1/3}$
9597 0059 sr : $3\sqrt{3}/2-3\sqrt{5}/4$
9597 1311 $\ln(3) \div \ln(\pi)$
9597 1716 $s\pi : 2\sqrt{3}/3-\sqrt{5}/3$
9597 5168 rt : (8,5,-8,-4)
9597 5188 $s\pi : 1/\ln(2e)$
9597 5438 sq : $10\sqrt[3]{2}/9$
9597 5443 rt : (9,2,-5,-5)
9597 5472 $\Gamma : 6\sqrt[3]{2}/7$
9597 5745 tanh : $\sqrt[3]{22/3}$
9597 6264 $2^x : 3/2+3\sqrt{3}/4$
9597 6523 $10^x : 4e+4\pi/3$
9597 7525 cu : $1/3-3e/4$
9597 9302 at : $2/3-2\pi/3$
9598 0024 $10^x : 3+\pi/4$
9598 0871 cu : $5e\pi/7$
9598 1705 rt : (7,7,-9,-4)
9598 1978 rt : (8,4,-6,-5)
9598 1996 sinh : $8\sqrt[3]{2}$
9598 2250 rt : (9,1,-3,-6)
9598 4237 $e^x : 4\sqrt{2}-2\sqrt{3}$
9598 4855 $2^x : e/3+3\pi$
9598 8471 rt : (7,6,-7,-5)
9598 8741 rt : (8,3,-4,-6)
9598 9011 rt : (9,0,-1,7)
9598 9206 $e^x : 2+3\pi/4$
9599 0684 rt : (4,8,-9,7)
9599 1739 $\lambda : \exp(-10\pi/19)$
9599 1795 $2^x : 1/2-\sqrt{5}/4$
9599 4864 cr : $\exp(-1/(3e))$
9599 4922 rt : (6,8,-8,-5)
9599 4972 $s\pi : 8\pi/7$
9599 5190 rt : (7,5,-5,-6)
9599 5458 rt : (8,2,-2,-7)
9599 5726 rt : (9,1,1,8)
9599 5744 $e^x : 6\sqrt{2}$
9599 6567 tanh : $\sqrt{3}-\sqrt{5}+\sqrt{6}$
9599 6873 sq : $4/3-\sqrt{2}/4$
9599 7105 $e^x : 2/3+3e/4$
9599 9528 ln : $3/2-e/2$
9600 0000 id : 24/25
9600 1598 rt : (6,7,-6,-6)
9600 1864 rt : (7,4,-3,-7)
9600 2129 rt : (8,1,0,-8)
9600 2395 rt : (9,2,3,9)
9600 4035 $\ell_{10} : 4\sqrt{2}+2\sqrt{3}$
9600 5113 $s\pi : 7\sqrt[3]{2}/2$
9600 7036 at : 10/7
9600 7964 rt : (5,9,-7,-6)
9600 8228 rt : (6,6,-4,-7)
9600 8260 sinh : $\exp(-1/(2\pi))$
9600 8444 id : $3\sqrt{3}-\sqrt{5}$
9600 8492 rt : (7,3,-1,-8)
9600 8756 rt : (8,0,2,9)
9600 9440 cr : $(\ln 2)^{1/3}$
9600 9530 $s\pi : 4e/3+\pi/4$
9601 0392 $10^x : 5\sqrt[3]{3}/2$
9601 1566 cu : $1/3+\pi$
9601 2064 $\ell_2 : \sqrt{3}-\sqrt{5}+\sqrt{6}$
9601 2344 cr : $2e+2\pi/3$
9601 2735 cr : $(\sqrt[3]{3})^{-1/3}$
9601 4274 tanh : $\exp(2/3)$
9601 4459 $s\pi : 5\sqrt{5}/2$
9601 4553 rt : (5,8,-5,-7)
9601 4814 rt : (6,5,-2,-8)
9601 5076 rt : (7,2,1,-9)
9602 0699 tanh : $9\sqrt{3}/8$
9602 1097 rt : (5,8,-2,3)
9602 1356 rt : (6,4,0,-9)
9602 4413 $s\pi : \pi^2/7$
9602 4689 cr : P_{54}
9602 6371 $J_0 : \zeta(3)/3$
9602 6961 $\zeta : (\ln 3/2)^{1/3}$
9602 7340 rt : (4,9,-4,-8)
9602 7597 rt : (5,9,2,7)
9602 7822 at : $3-\pi/2$
9602 7964 sq : $7\sqrt[3]{2}/9$
9603 1480 $\zeta : 5\sqrt[3]{5}/6$
9603 1742 $erf : 16/11$
9603 2704 $2^x : \sqrt{21}$
9603 3800 rt : (4,8,-2,-9)
9603 4449 $s\pi : 4e-2\pi$
9603 6070 $\ell_2 : 3+4\pi$
9603 6667 $e^x : 2/3-\sqrt{2}/2$
9603 9804 cr : $9(e+\pi)/7$
9603 9822 $J_0 : 2/5$
9603 9874 at : $3/4+e/4$
9604 0217 rt : (4,5,-2,7)
9604 1828 $s\pi : \sqrt[3]{14}$
9604 2731 $10^x : 3\sqrt{5}/5$
9604 4010 sq : 10π
9604 6341 rt : (3,9,-1,-4)
9604 7371 $J : \exp(-10\pi/19)$
9604 9894 sq : $3\sqrt{2}/2+3\sqrt{5}$
9605 1628 $\sqrt{2} \times \ln(4)$

9605 1738 2^x : $10\ln 2/3$
9605 2677 rt : (3,1,-6,7)
9605 4657 sr : $3/2-\sqrt{3}/3$
9605 6460 tπ : $\sqrt{2}/2-\sqrt{5}$
9605 7461 tanh : $(e+\pi)/3$
9605 7919 erf : $(\sqrt{2}/3)^{-1/2}$
9606 0530 J_0 : $(2\pi)^{-1/2}$
9606 1317 cu : $\sqrt{3}-\sqrt{5}/3$
9606 2486 sπ : $\sqrt{2}/4+\sqrt{5}$
9606 4927 tanh : $8\sqrt[3]{5}/7$
9606 4981 rt : (2,9,2,-9)
9606 8140 e^x : $3e\pi/10$
9606 9945 λ : $\sqrt{3}/9$
9607 1198 rt : (2,4,-4,5)
9607 3146 J_0 : $(e/2)^{-3}$
9607 4853 2^x : $1/2+\sqrt{2}/3$
9607 6892 sr : 12/13
9607 6897 10^x : $\sqrt{2}/3$
9607 8432 sπ : $\sqrt{2}/3+\sqrt{5}/2$
9608 0666 2^x : $3/2+\pi$
9608 1131 J_0 : $5(1/\pi)/4$
9608 1234 $2\div s\pi(\sqrt{5})$
9608 2585 ℓ_2 : $\exp(e/2)$
9608 2642 tanh : $7\sqrt{5}/8$
9608 4627 at : $((e+\pi)/2)^{1/3}$
9608 5659 $\exp(1/3)-\exp(\sqrt{5})$
9608 8132 id : $3\pi^2/10$
9608 9353 tanh : $\sqrt[3]{15/2}$
9609 0602 $\ln(2)\times\ln(4)$
9609 0743 tanh : $1/\ln(5/3)$
9609 1519 Γ : $9\pi/2$
9609 5365 sπ : $1/3+3e/2$
9609 5438 rt : (1,5,1,4)
9609 7792 θ_3 : $((e+\pi)/2)^{-1/3}$
9609 8039 Ψ : $\sqrt{2}+\sqrt{3}-\sqrt{7}$
9609 8638 Γ : $2e\pi/9$
9610 1188 2^x : $3\sqrt{2}/4-\sqrt{5}/2$
9610 1332 J_0 : $1/\sqrt[3]{16}$
9610 5897 10^x : $8\pi/7$
9610 6556 sπ : $\exp(-\sqrt[3]{4}/3)$
9610 7006 Γ : 14/13
9611 1664 $\ln(\sqrt{2})\times\exp(3)$
9611 2183 sr : $8\sqrt[3]{3}/3$
9611 6223 J_0 : $4\ln 2/7$
9611 7137 ln : $3e-\pi/3$
9611 8084 J : $7(1/\pi)/6$
9612 0575 Γ : $\ln(\pi/2)$
9612 5901 erf : 19/13
9612 5914 rt : (9,4,-8,-4)
9612 8862 id : $3-3e/4$
9613 1751 rt : (8,6,-9,-4)
9613 1986 rt : (9,3,-6,-5)
9613 2535 at : $9(1/\pi)/2$
9613 4482 cu : $4\sqrt{2}+\sqrt{5}/2$
9613 5125 sr : $4\ln 2/3$
9613 5565 tanh : $5\pi/8$
9613 7343 cr : $3\sqrt{2}-3\sqrt{5}/2$
9613 7380 2^x : $2\sqrt{2}+2\sqrt{5}/3$
9613 7787 rt : (8,5,-7,-5)
9613 8019 rt : (9,2,-4,-6)
9613 8919 cu : $\pi^2/10$
9614 2076 sπ : $(\ln 2/2)^{1/2}$
9614 3553 rt : (7,7,-8,-5)
9614 3783 rt : (8,4,-5,-6)
9614 3889 2^x : $3+3\pi/4$
9614 4014 rt : (9,1,-2,-7)
9614 9072 sr : $9\sqrt[3]{5}/4$
9614 9161 sr : $\sqrt{3}/2+4\sqrt{5}/3$
9614 9285 rt : (6,9,-9,-5)
9614 9513 rt : (7,6,-6,-6)
9614 9742 rt : (8,3,-3,-7)
9614 9971 id : $2\sqrt[3]{3}/3$
9615 0501 10^x : 20/21
9615 0733 K : $(2\pi/3)^{-1/3}$
9615 2422 cu : $3+\sqrt{3}$
9615 2787 at : $1+\sqrt{3}/4$
9615 5210 rt : (6,8,-7,-6)
9615 5354 Ψ : $(\sqrt[3]{2}/3)^3$
9615 5437 rt : (7,5,-4,-7)
9615 5664 rt : (8,2,-1,-8)
9615 5891 rt : (9,1,2,9)
9615 6133 cr : P_{43}
9615 9013 sr : $(\sqrt[3]{5}/2)^{1/2}$
9616 0811 Ei : $4e\pi/5$
9616 1098 rt : (6,7,-5,-7)
9616 1323 rt : (7,4,-2,-8)
9616 1520 Γ : $8\sqrt[3]{3}/3$
9616 1548 rt : (8,1,1,-9)
9616 2087 sinh : $e\pi/10$
9616 3456 sπ : $9\sqrt{3}$
9616 4552 id : $4\zeta(3)/5$
9616 5101 ℓ_{10} : $1+3e$
9616 6416 erf : $(\pi)^{1/3}$
9616 6726 rt : (5,9,-6,-7)
9616 6949 rt : (6,6,-3,-8)
9616 7173 rt : (7,3,0,-9)
9617 1389 erf : $(e+\pi)/4$
9617 2543 rt : (5,8,-4,-8)
9617 2765 rt : (6,5,-1,-9)
9617 3455 erf : $\ln(\ln 2/3)$
9617 5667 Γ : $10\sqrt[3]{5}/9$
9617 6583 Γ : 19/10
9617 8324 rt : (5,8,-1,5)
9617 9669 ℓ_2 : $\exp(2/3)$
9618 0178 as : $(2e/3)^{-1/3}$
9618 0946 erf : $6\sqrt[3]{5}/7$
9618 1685 J_0 : $\pi/8$
9618 1837 J_0 : $\ln(\sqrt{2}\pi/3)$
9618 2000 cu : $2\sqrt{2}/3-2\sqrt{5}$
9618 2564 sπ : 7/17
9618 3851 rt : (4,9,-3,-9)
9618 4070 rt : (5,9,3,9)
9618 5952 Ψ : $2\zeta(3)/3$
9618 9564 rt : (4,4,-5,5)
9619 2195 sr : $8\pi^2/9$
9619 2672 Γ : $\ln((e+\pi)/2)$
9619 3739 erf : 22/15
9619 4760 10^x : 16/23
9619 5242 rt : (4,5,-1,9)
9619 5330 e^x : $1+\pi/4$
9619 7281 cu : $4e-\pi/2$
9619 8213 tanh : $\sqrt[3]{23/3}$
9620 0673 rt : (3,9,0,-1)
9620 2833 e^x : $2-3e/4$
9620 4709 sπ : $2/3+\sqrt{5}/3$
9620 5205 sq : $e/3-4\pi$
9620 6284 rt : (3,1,-5,9)
9620 8500 J_0 : 9/23
9620 8879 sπ : $2\sqrt[3]{3}/7$
9620 9793 ℓ_{10} : $4\sqrt{3}+\sqrt{5}$
9621 0884 Γ : $(\sqrt{3}/2)^{-1/2}$
9621 1650 id : $\sqrt{2}/3+2\sqrt{5}/3$
9621 3404 ℓ_2 : $2e+3\pi/4$
9621 3799 tanh : $\pi^2/5$
9621 3822 Ψ : $5\sqrt[3]{3}/9$
9621 3835 2^x : $3e/4-2\pi/3$
9621 6189 as : $3/2-e/4$
9621 7199 rt : (2,9,3,-5)
9621 8277 $\sqrt{5}\times s\pi(\pi)$
9622 0126 J : $(e+\pi)^{-1/2}$
9622 2383 sr : $(\sqrt[3]{2})^{-1/3}$
9622 2538 Li_2 : $(\ln 3/2)^{1/2}$
9622 2713 rt : (2,4,-3,8)
9622 5044 id : $5\sqrt{3}/9$
9623 2009 ζ : 7/4
9623 2841 cr : $9\ln 2/7$
9623 6092 erf : $7\sqrt[3]{2}/6$
9623 7424 id : $(\sqrt{2}\pi/2)^3$
9623 8898 id : $1/4+3\pi/2$
9624 0172 id : $2\sqrt{2}-\sqrt{3}/2$
9624 0625 ℓ_2 : $9\sqrt{3}$
9624 0832 ln : $5\pi/6$
9624 1558 rt : (2,9,-6,-7)
9624 2365 ln : $3/2+\sqrt{5}/2$
9624 3549 10^x : $(\ln 2/2)^{-1/2}$
9624 4249 rt : (1,5,2,9)
9624 4931 erf : 25/17
9624 6524 $\ln(\sqrt{3})+\exp(5)$
9624 7305 e^x : $\sqrt{2}/2-\sqrt{5}/3$
9625 3956 sπ : $\sqrt[3]{4}$
9625 4299 J_0 : $1/\sqrt[3]{17}$
9625 4723 J_0 : 7/18
9625 4926 tanh : $8\sqrt{3}/7$
9625 7956 tanh : $7\sqrt{2}/5$
9625 9156 Ψ : $\exp(-\pi/3)$
9625 9879 cr : $6\sqrt[3]{2}$
9626 0988 θ_3 : $4\zeta(3)/5$
9626 1050 tπ : 7/20
9626 2669 sq : $3e/2-3\pi/4$
9626 5366 rt : (1,9,0,-3)
9626 5455 J_0 : $e/7$
9626 8884 at : $4-2e$
9627 1348 rt : (9,5,-9,-4)
9627 2233 sq : $\sqrt{2}/4-2\sqrt{5}$
9627 2715 Γ : $7e/10$
9627 3291 cu : $1/3+\sqrt{5}$
9627 4172 Ei : 18/13
9627 4616 cr : $1/3+\sqrt{5}/4$
9627 5512 sq : $\sqrt{2}-\sqrt{3}/4$
9627 5576 id : $e/4+2\pi$
9627 6751 rt : (9,4,-7,-5)
9627 6938 tπ : $e/4-2\pi$
9628 0264 Ei : $\sqrt{13}$
9628 0370 as : $3/4-\pi/2$
9628 1143 $\exp(3/5)+\exp(\pi)$
9628 1924 rt : (8,6,-8,-5)
9628 2123 rt : (9,3,-5,-6)
9628 3811 tπ : $3\sqrt{3}/2-3\sqrt{5}/2$
9628 4023 id : $1/\ln(\sqrt{5}/2)$
9628 4793 id : $8\sqrt{5}/3$
9628 5291 Ψ : $1/\sqrt{22}$
9628 5351 2^x : $2-\sqrt{3}/4$
9628 7068 rt : (7,8,-9,-5)
9628 7266 rt : (8,5,-6,-6)
9628 7464 rt : (9,2,-3,-7)
9628 7759 10^x : $1-\sqrt{2}/2$
9628 9254 sr : $4e-2\pi/3$
9628 9988 Ψ : $(\sqrt{2}/3)^3$
9629 1019 tπ : $1/\sqrt[3]{7}$
9629 2381 rt : (7,7,-7,-6)
9629 2577 rt : (8,4,-4,-7)
9629 2773 rt : (9,1,-1,-8)
9629 3420 sq : $1/2+2\sqrt{5}/3$
9629 6227 tπ : $5\pi/7$
9629 6296 cu : 8/3
9629 6570 sπ : $(e+\pi)^{-1/2}$
9629 7467 rt : (6,9,-8,-5)
9629 7662 rt : (7,6,-5,-7)
9629 7857 rt : (8,3,-2,-8)
9629 8052 rt : (9,0,1,9)
9629 8956 ℓ_2 : $3\sqrt[3]{5}/10$
9629 9433 at : 23/16
9630 0054 sr : $3\sqrt{2}/2+\sqrt{3}$
9630 1522 sπ : $\exp(5)$
9630 1689 sr : P_{34}
9630 1906 $3/4\times\exp(1/4)$
9630 2325 id : $7e\pi/6$
9630 2719 rt : (6,8,-6,-7)
9630 2889 sinh : $\sqrt[3]{5}/2$
9630 2913 rt : (7,5,-3,-8)
9630 3106 rt : (8,2,0,-9)
9630 6659 sπ : $9\pi^2/2$
9630 7941 rt : (6,7,-4,-8)
9630 8134 rt : (7,4,-1,-9)
9630 9660 tπ : $1/3-\sqrt{3}/3$
9631 1662 rt : (1,9,1,6)
9631 1883 rt : (2,9,-5,-2)
9631 2089 ln : $2\sqrt{2}/3+3\sqrt{5}/4$
9631 2943 rt : (5,9,-5,-8)
9631 3134 rt : (6,6,-2,-9)
9631 4211 tanh : $8\sqrt{5}/9$
9631 7127 as : $(e/3)^2$
9631 7236 $\ln(3/4)-s\pi(\sqrt{5})$
9631 7584 ℓ_{10} : $(2\pi/3)^3$
9631 8107 rt : (5,8,-3,-9)
9631 8668 sr : $1/2+3\sqrt{5}/2$
9631 8726 e^x : 19
9631 9300 sq : $2-4\sqrt{5}/3$
9632 2586 tπ : $\sqrt{3}/4-3\sqrt{5}/4$
9632 3241 rt : (5,8,0,7)

9632 3576 $s\pi : 4\sqrt{2}/3 - 3\sqrt{3}/4$
9632 6931 rt : (1,6,-8,-9)
9632 7017 $J_0 : 5\ln 2/9$
9632 8056 $K : 18/23$
9632 8091 $t\pi : 3\sqrt[3]{2}/5$
9632 8160 rt : (4,3,-8,3)
9633 0449 $J_0 : 2\sqrt{3}/9$
9633 0777 $2^x : \sqrt{11}$
9633 2729 sr : 5π
9633 3239 rt : (4,4,-4,7)
9633 5827 $J_0 : 5/13$
9633 5964 $s\pi : P_{56}$
9633 8104 rt : (3,8,-5,-7)
9634 2729 $2^x : 1/3 + 2\sqrt{5}$
9634 3127 rt : (3,9,1,2)
9634 4509 $\ln(2/5) \div s\pi(2/5)$
9634 5725 ln : $\sqrt[3]{18}$
9634 9248 id : $(\sqrt{5}/2)^{-1/3}$
9634 9540 id : $5\pi/8$
9635 0965 $\Gamma : 15/14$
9635 1002 $\Psi : 1/2$
9635 1734 $\ell_{10} : 1/4 + 4\sqrt{5}$
9635 1970 $\sqrt{2} + \ln(\sqrt{3})$
9635 4421 $Ei : 5\sqrt[3]{3}/2$
9635 7511 sr : P_{17}
9635 7852 rt : (2,9,4,-1)
9635 8855 id : $\sqrt{2} - \sqrt{3} - \sqrt{7}$
9636 0365 cr : 17/19
9636 0616 $\ell_{10} : 4 + 3\sqrt{3}$
9636 1377 $t\pi : 2\sqrt{3} - 3\sqrt{5}$
9636 2411 sr : 13/14
9636 3027 $2^x : 4/3 + 3\sqrt{2}/2$
9636 5439 $e^x : 3\sqrt{3}/4 + 4\sqrt{5}$
9636 7061 tanh : $7\sqrt[3]{5}/6$
9636 7813 ln : $1/2 + 3\sqrt{2}/2$
9636 8225 $s\pi(\sqrt{3}) \div s\pi(e)$
9637 1145 $\pi + \exp(3/5)$
9637 1531 rt : (1,9,-9,6)
9637 1955 sr : $3/2 + 3\pi/4$
9637 3420 $s\pi : (e+\pi)/10$
9637 6941 $e^x : 10\sqrt[3]{5}$
9637 7930 at : $8\sqrt[3]{2}/7$
9637 9588 rt : (2,9,-4,3)
9637 9630 $s\pi : (\sqrt{5}/3)^3$
9637 9926 $2^x : \sqrt{3}/4 + 2\sqrt{5}$
9638 0382 as : $1/3 - 2\sqrt{3}/3$
9638 3831 $e^x : 25/14$
9638 5564 $J_0 : 6(1/\pi)/5$
9638 7036 cr : $\sqrt{3}/2 + 3\sqrt{5}$
9638 7891 sq : $3/4 - 2e$
9638 8426 $\ell_{10} : 3e + \pi/3$
9638 9424 $t\pi : 7\sqrt{2}/6$
9639 0029 sinh : $4\pi/3$
9639 0253 $s\pi : \sqrt{2}$
9639 3070 $J_0 : 1/\sqrt[3]{18}$
9639 3821 rt : (1,9,8,-2)
9639 6679 cu : $2\sqrt{3} + 3\sqrt{5}/2$
9639 6968 $e^x : 3\sqrt{2} + \sqrt{3}/3$
9640 2022 $e^x : \sqrt{3}/4 + 3\sqrt{5}$
9640 2758 tanh : 2
9640 3072 sinh : $9\sqrt[3]{2}/2$
9640 3492 $\Psi : 8e/7$
9640 4657 $J_0 : 8/21$
9640 5691 $s\pi : 7\zeta(3)$
9640 6424 $10^x : 3 - 3\sqrt{2}/4$
9640 8891 $s\pi : \sqrt{3}/4 + 4\sqrt{5}/3$
9641 0004 $e^x : 17/5$
9641 0161 id : $1/2 + 2\sqrt{3}$
9641 1026 rt : (9,5,-8,-5)
9641 2815 $s\pi : 4e + 3\pi/2$
9641 2833 at : $2 - \sqrt{5}/4$
9641 4234 sr : $1/4 + e/4$
9641 5038 $e^x : 3/4 + 2e/3$
9641 5453 $e^x : \exp(\sqrt{2}/3)$
9641 5659 rt : (8,7,-9,-5)
9641 5830 rt : (9,4,-6,-6)
9642 0438 rt : (8,6,-7,-6)
9642 0609 rt : (9,3,-4,-7)
9642 0677 $s\pi : 3\sqrt{2}/2 + 2\sqrt{3}$
9642 2457 $Ei : \exp(\pi/2)$
9642 2549 $J_0 : 2\sqrt[3]{5}/9$
9642 3266 $erf : 6\sqrt{3}/7$
9642 5022 rt : (7,8,-8,-6)
9642 5103 $2^x : 3/2 + 4e$
9642 5191 rt : (8,5,-5,-7)
9642 5361 rt : (9,2,-2,-8)
9642 5876 $e^x : 2\pi^2/9$
9642 6898 tanh : $5\zeta(3)/3$
9642 9751 rt : (7,7,-6,-7)
9642 9919 rt : (8,4,-3,-8)
9643 0087 rt : (9,1,0,-9)
9643 2489 cr : $(\sqrt[3]{3}/2)^{1/3}$
9643 3447 $Li_2 : 3\sqrt{3}/7$
9643 4287 rt : (6,9,-7,-3)
9643 4454 rt : (7,6,-4,-8)
9643 4622 rt : (8,3,-1,-9)
9643 5996 $s\pi : 3\sqrt{2}/4 + 3\sqrt{5}/2$
9643 8967 rt : (6,8,-5,-8)
9643 9133 rt : (7,5,-2,-9)
9644 0450 $2^x : 14/5$
9644 2095 $2^x : 4 + 2\sqrt{5}/3$
9644 2378 sq : $3/4 - \sqrt{3}$
9644 3621 rt : (6,7,-3,-9)
9644 3699 as : $(\sqrt{2}\pi/3)^{-1/2}$
9644 4819 rt : (2,9,-3,8)
9644 8087 rt : (5,9,-4,-9)
9644 8542 cu : $1 - 3\pi$
9644 8548 $s\pi : 1/\sqrt[3]{14}$
9644 9416 sr : $1/\ln((e+\pi)/2)$
9645 0975 $s\pi : 4\sqrt{2} + 4\sqrt{3}$
9645 1180 $s\pi : 2\sqrt{2}/3 + 2\sqrt{5}$
9645 1879 $s\pi : (\sqrt{5}/3)^{-3}$
9645 2694 rt : (5,7,-3,5)
9645 3979 at : $\sqrt[3]{3}$
9645 6440 cu : $3/2 + \sqrt{3}/3$
9645 7277 rt : (5,8,1,9)
9645 7915 $\lambda : 2\ln 2/7$
9645 8296 rt : (1,9,9,6)
9645 9295 $e^x : 1 + 3\sqrt{3}/4$
9645 9369 sinh : $\exp(\sqrt[3]{2})$
9646 0113 $J_0 : 3\sqrt[3]{2}/10$
9646 0332 $J_0 : 1/\sqrt{7}$
9646 1079 at : $6\zeta(3)/5$
9646 1674 rt : (4,3,-7,5)
9646 2965 cr : $2\pi/7$
9646 3528 $K : (\sqrt[3]{3}/3)^{1/3}$
9646 3769 $\ln(3) + s\pi(1/3)$
9646 5816 cu : $4 - 2e$
9646 5840 $t\pi : 1/4 + \sqrt{2}/4$
9646 5850 as : $\sqrt{2} - \sqrt{5}$
9646 6210 rt : (4,4,-3,9)
9646 7862 sr : $(\sqrt{3}/2)^{1/2}$
9646 8439 at : $1/\ln(2)$
9647 0469 $t\pi : \sqrt[3]{5}/7$
9647 0564 rt : (3,8,-4,-4)
9647 2009 $s\pi : 1/\sqrt[3]{5}$
9647 2056 $e^x : \sqrt{3} + 4\sqrt{5}$
9647 2138 at : $1/2 + 2\sqrt{2}/3$
9647 2370 $\Gamma : 5\sqrt[3]{5}/8$
9647 3799 sq : $7(1/\pi)$
9647 5055 rt : (3,9,2,5)
9647 5203 $\Gamma : 3\zeta(3)/8$
9647 5977 $\ell_{10} : \exp(\sqrt{2}\pi/2)$
9647 6166 sr : $3(e+\pi)/2$
9647 7245 $\exp(\sqrt{2}) \times s\pi(\sqrt{2})$
9648 1453 $s\pi : \sqrt{2}/2 + 3\sqrt{5}$
9648 3242 $\Gamma : 8\zeta(3)/9$
9648 3348 sr : $\sqrt{2}/4 + \sqrt{3}/3$
9648 5309 id : P_{22}
9648 7228 sq : $9\pi/4$
9648 8234 rt : (2,9,5,3)
9648 9557 $Ei : 9\pi^2/8$
9648 9747 tanh : $9\sqrt{5}/10$
9649 0520 at : $5\sqrt{3}/6$
9649 1248 $\Gamma : 21/11$
9649 2231 tanh : $(1/(3e))^{-1/3}$
9649 4414 $s\pi : ((e+\pi)/2)^2$
9649 5298 $erf : (\sqrt{2}\pi/2)^{1/2}$
9649 8501 $erf : 2\sqrt{5}/3$
9650 0856 $\Psi : 4/5$
9650 5553 rt : (1,7,5,-7)
9650 5728 $erf : \ln(\sqrt{2}\pi)$
9650 6083 $\Psi : ((e+\pi)/3)^{-1/3}$
9650 7528 $e^x : 4 - \sqrt{3}/2$
9650 8089 ln : 8/21
9650 9385 $Li_2 : 7(1/\pi)/3$
9651 1598 $s\pi : \sqrt{2}/3 + 4\sqrt{5}$
9651 3207 tanh : $8\sqrt[3]{2}/5$
9651 5153 $J_0 : 3/8$
9651 5308 cu : $4 + 3\sqrt{5}/4$
9651 7069 $\lambda : 5(1/\pi)/8$
9651 8143 $\Gamma : 6(1/\pi)$
9651 9643 $J_0 : 1/\sqrt[3]{19}$
9652 0729 $s\pi : ((e+\pi)/2)^{-1/2}$
9652 0773 sq : $8\pi^2/9$
9652 2102 $e^x : 3/4 - \pi/4$
9652 2656 cu : $4 + 2\sqrt{3}/3$
9652 3569 $e^x : 25/23$
9652 5036 cr : $\exp(-1/(3\pi))$
9652 5166 at : 13/9
9652 9005 $s\pi : 3\ln 2/5$
9652 9446 $s\pi : 2e\pi/5$
9652 9944 $\Gamma : e\pi/8$
9653 0794 cr : $8e\pi/9$
9653 1120 $\exp(1/3) + s\pi(\pi)$
9653 1733 $Li_2 : (2e/3)^{-1/2}$
9653 2141 $s\pi : 10\pi$
9653 2404 at : $\exp(1/e)$
9653 3865 $erf : 1/\ln((e+\pi)/3)$
9653 4134 $s\pi : 3e - \pi/2$
9653 5276 ln : $9(e+\pi)$
9653 5581 tanh : $7\sqrt[3]{3}/5$
9653 5914 rt : (9,6,-9,-5)
9653 6107 $erf : \sqrt[3]{10/3}$
9653 6338 sinh : 17/5
9653 6697 $t\pi : (1/(3e))^{1/2}$
9653 9030 sq : $4 - 3\sqrt{3}/2$
9654 0228 rt : (9,5,-7,-6)
9654 0987 as : $3\sqrt{2}/2 - 3\sqrt{3}/4$
9654 2812 $e^x : 10\sqrt[3]{3}/9$
9654 3441 tanh : $10\sqrt{2}/7$
9654 4373 rt : (8,7,-8,-6)
9654 4520 rt : (9,4,-5,-7)
9654 6299 tanh : $7\sqrt{3}/6$
9654 6921 cr : $5\sqrt[3]{2}/7$
9654 8497 rt : (7,9,-9,-6)
9654 8644 rt : (8,6,-6,-7)
9654 8791 rt : (9,3,-3,-8)
9654 8938 cr : 9/10
9655 0269 cu : $2 - 3e/2$
9655 2748 rt : (7,8,-7,-7)
9655 2894 rt : (8,5,-4,-8)
9655 3040 rt : (9,2,-1,-9)
9655 4600 $\ell_2 : 3 - \pi/3$
9655 4810 $erf : (\sqrt{5})^{1/2}$
9655 6977 rt : (7,7,-5,-8)
9655 7123 rt : (8,4,-2,-9)
9655 7802 $J_0 : \sqrt{5}/6$
9655 8662 $s\pi : 8e/9$
9656 0402 cr : $4/3 - \sqrt{3}/4$
9656 1042 rt : (6,9,-6,-1)
9656 1186 rt : (7,6,-3,-9)
9656 5230 rt : (6,8,-4,-9)
9656 5266 cr : $2/3 + 4\sqrt{3}$
9656 5411 $erf : 7\sqrt[3]{5}/8$
9656 6274 $3/5 \times \ln(5)$
9656 6405 $\Gamma : 16/15$
9656 9904 $Ei : 4\sqrt{3}/5$
9657 0477 id : $1/3 - 3\sqrt{3}/4$
9657 1507 $s\pi : 6\sqrt{5}$
9657 2664 $\ell_2 : 3 + e/3$
9657 3404 rt : (5,6,-6,3)
9657 6176 $Ei : 3\sqrt[3]{5}/5$
9657 7190 $\ln(5) \times \exp(1/5)$
9657 7532 rt : (5,7,-2,7)
9657 7878 $e^x : 3/4 + \sqrt{3}$
9657 8428 $\ell_2 : (5)^3$
9658 1115 $2^x : 8\sqrt{5}/9$
9658 1421 cr : $3\sqrt{2} + 3\sqrt{5}/2$
9658 1868 $J_0 : 7(1/\pi)/6$

9658 2804 $\mathrm{t}\pi : 4\pi/5$
9658 5219 $\lambda : 1/5$
9658 5590 rt : (4,3,-6,7)
9658 6784 $\mathrm{t}\pi : 3e/4 - 2\pi$
9658 8315 $\ln : \sqrt{3}/4 + 3\sqrt{5}$
9658 8898 $e^x : 7\pi/3$
9659 0692 $\ln(5) + \exp(\sqrt{5})$
9659 0726 $\ell_2 : (e+\pi)/3$
9659 1317 $\tanh : \sqrt[3]{25/3}$
9659 2482 $\sinh : \sqrt{7}\pi$
9659 2582 $\mathrm{sr} : 1/2 + \sqrt{3}/4$
9659 3571 rt : (3,8,-3,-1)
9659 3581 $\ln : 4 + \pi$
9659 6101 $\tanh : \exp(1/\sqrt{2})$
9659 7602 rt : (3,9,3,8)
9659 9188 $\mathrm{cr} : 5\sqrt[3]{3}/8$
9660 2279 $\sinh : 6/7$
9660 4071 $\mathrm{cr} : 3\zeta(3)/4$
9660 5472 rt : (2,8,-2,-9)
9660 5528 $\mathrm{id} : 3e - 4\pi/3$
9660 7072 $\lambda : \zeta(3)/6$
9660 9178 $\mathrm{sr} : 14/15$
9660 9447 rt : (2,9,6,7)
9661 0514 $erf : 3/2$
9661 4311 $\Gamma : 4\sqrt{5}/3$
9661 4412 $\mathrm{at} : (2\pi/3)^{1/2}$
9661 5546 $\mathrm{sr} : 2e - \pi/2$
9661 6340 $\sinh : (\sqrt[3]{4})^{-1/3}$
9661 6473 $\mathrm{t}\pi : \sqrt{2}/4 - 3\sqrt{3}/2$
9661 7551 $\ln : \sqrt[3]{2}/9$
9662 2109 $\mathrm{sr} : 3 + \sqrt{3}/2$
9662 2711 $\mathrm{id} : 10\pi^2/9$
9662 3290 $\zeta : 10(1/\pi)/3$
9662 3707 $\mathrm{id} : P_{50}$
9662 5029 rt : (1,7,6,-1)
9662 5326 $10^x : 3 + 4\sqrt{5}/3$
9662 5469 $\mathrm{s}\pi : 3\sqrt{3}/4 + \sqrt{5}/2$
9662 5532 $\mathrm{id} : 4\sqrt{2} + 4\sqrt{3}/3$
9662 6124 $\Gamma : \sqrt[3]{7}$
9662 9212 $e^x : 2e/5$
9663 0528 $\theta_3 : (\ln 3/2)^{1/3}$
9663 1631 $10^x : 3\sqrt{5}/2$
9663 2649 $\mathrm{sq} : \sqrt{2} + \sqrt{3}/3$
9663 3143 $\mathrm{cu} : 6(1/\pi)$
9663 5326 $J_0 : 7/19$
9663 5651 $J_0 : 1/\sqrt[3]{20}$
9663 5670 $2^x : 1/2 + 2e/3$
9664 0975 $erf : 5\zeta(3)/4$
9664 2916 rt : (1,8,-9,-3)
9664 5128 $J_0 : \exp(-1)$
9664 7991 $\mathrm{cr} : (e/2)^{-1/3}$
9665 1454 $\mathrm{at} : 4e - 3\pi$
9665 2855 $\mathrm{s}\pi : 9\sqrt[3]{2}/8$
9665 3912 $\mathrm{s}\pi : \sqrt{21}$
9665 4142 $\sinh : 10/7$
9665 5479 $10^x : 2e\pi$
9665 6259 rt : (9,6,-8,-6)
9666 0000 rt : (8,8,-9,-6)
9666 0129 rt : (9,5,-6,-7)
9666 1499 $e \div \ln(2/5)$
9666 1980 $\mathrm{id} : 3\sqrt{2}/2 - 2\sqrt{3}/3$
9666 2110 $\ell_2 : 8\sqrt[3]{5}/7$
9666 3494 $e^x : 4\zeta(3)/3$
9666 3852 rt : (8,7,-7,-7)
9666 3981 rt : (9,4,-4,-8)
9666 6290 $\tanh : 3e/4$
9666 7559 rt : (7,9,-8,-7)
9666 7687 rt : (8,6,-5,-8)
9666 7814 rt : (9,3,-2,-9)
9666 8742 $\mathrm{s}\pi : \sqrt{3} + \sqrt{5} + \sqrt{6}$
9666 9117 $\mathrm{cu} : 1/\ln(\sqrt{2}\pi/2)$
9666 9797 $J_0 : \ln(\ln 2)$
9667 0455 $10^x : \sqrt{3}/3 + 2\sqrt{5}/3$
9667 1377 rt : (7,8,-6,-8)
9667 1503 rt : (8,5,-3,-9)
9667 2497 $\mathrm{sq} : 7\zeta(3)/6$
9667 2513 $\Psi : 7e\pi/3$
9667 3462 $\tanh : \sqrt{5} + \sqrt{6} - \sqrt{7}$
9667 5050 rt : (6,8,-9,-3)
9667 5176 rt : (7,7,-4,-9)
9667 5359 $J_0 : \ln(\sqrt[3]{3})$
9667 7191 $\ln : 4\sqrt{2} + 2\sqrt{5}/3$
9667 8833 rt : (6,9,-5,1)
9668 0136 $\tanh : \sqrt[3]{17/2}$
9668 0327 rt : (1,8,-8,6)
9668 0717 $10^x : 3 - \sqrt{3}/4$
9668 1719 rt : (2,8,-9,3)
9668 1740 $10^x : 8\sqrt[3]{2}$
9668 4868 $\ln(\pi) + \exp(3/5)$
9668 6223 rt : (5,5,-9,1)
9668 8645 $\sinh : (e/2)^{-1/2}$
9668 9286 $\mathrm{s}\pi : 2\sqrt{3} + \sqrt{5}/2$
9668 9954 rt : (5,6,-5,5)
9669 1070 $Ei : 5\zeta(3)/3$
9669 1941 $\mathrm{sr} : 7\zeta(3)/9$
9669 2192 $10^x : 1 + 2\sqrt{5}$
9669 3669 rt : (5,7,-1,9)
9669 4100 $\Gamma : \sqrt{11/3}$
9669 7245 rt : (4,2,-9,5)
9670 0927 rt : (4,3,-5,9)
9670 1037 $\mathrm{s}\pi : \sqrt{20/3}$
9670 1841 $2^x : 3e + \pi/3$
9670 4472 rt : (3,7,-8,-7)
9670 6635 $\mathrm{cr} : \sqrt{2}/3 + \sqrt{3}/4$
9670 7940 $\mathrm{at} : e/3 - 3\pi/4$
9670 8121 rt : (3,8,-2,2)
9670 8250 $\mathrm{cu} : 4e + 4\pi/3$
9670 9057 $\Gamma : 1/\ln(\ln 2)$
9670 9885 $Ei : 12/7$
9671 2795 $\lambda : \sqrt{2}/7$
9671 8817 $J_0 : 8(1/\pi)/7$
9671 8855 rt : (2,8,-1,-5)
9671 8919 $\mathrm{cr} : 19/21$
9672 0158 $\zeta : \sqrt[3]{16/3}$
9672 1435 $J_0 : 4/11$
9672 3750 $\mathrm{sq} : 3\sqrt{3}/2 + \sqrt{5}/4$
9672 3758 $\exp(1/5) - s\pi(\sqrt{3})$
9672 6313 $Li_2 : (e/3)^3$
9672 9413 $\mathrm{sr} : 5\pi^2/2$
9672 9481 $erf : 5e/9$
9672 9879 $4 \times \exp(2/5)$
9673 1601 $\mathrm{sq} : 3/2 + 2\sqrt{2}/3$
9673 5414 $\ln : 4\sqrt{2}/3 + \sqrt{5}/3$
9673 6481 $\ln(2) \times \exp(1/3)$
9673 6514 rt : (2,8,-8,8)
9673 8311 $\mathrm{sr} : 3e/2 - \pi$
9673 8313 $\mathrm{sq} : 10\sqrt[3]{5}/7$
9674 0279 $\mathrm{t}\pi : 5(e+\pi)/2$
9674 0541 $\zeta : 9(1/\pi)/4$
9674 1291 $\exp(2) - \exp(\sqrt{5})$
9674 1416 $J : \exp(-4\pi/5)$
9674 2436 $J_0 : 1/\sqrt[3]{21}$
9674 3588 $\ell_{10} : 1/3 + 4\sqrt{5}$
9674 8742 $\mathrm{at} : 1/3 + \sqrt{5}/2$
9674 8948 $\mathrm{cr} : 3 - 2\pi/3$
9674 9599 $erf : 6\sqrt[3]{2}/5$
9675 0372 $\lambda : (\sqrt{2}\pi/2)^{-2}$
9675 1304 $Ei : 7\sqrt[3]{5}$
9675 1855 $\mathrm{cr} : 4/3 + 2\pi$
9675 2120 $\tanh : 6\sqrt[3]{5}/5$
9675 2241 $\mathrm{id} : 4e + 2\pi/3$
9675 6670 $\theta_3 : 5\sqrt[3]{2}/9$
9675 7120 $\Gamma : 6\sqrt{5}/7$
9675 8006 $\Gamma : 17/16$
9675 8435 $\Gamma : 23/12$
9675 8509 $Ei : 2\ln 2$
9675 9809 $\ln : 2\sqrt[3]{5}/9$
9676 0492 $\mathrm{as} : 14/17$
9676 4157 $\zeta : (\ln 2/2)^{-1/3}$
9676 4741 rt : (9,7,-9,-6)
9676 6360 $\mathrm{id} : (e/3)^{1/3}$
9676 8242 rt : (9,6,-7,-7)
9676 8721 $\mathrm{s}\pi : \sqrt{5/2}$
9676 9701 $\ell_{10} : 3 + 2\pi$
9677 0216 $\mathrm{cu} : \sqrt{2} - 4\sqrt{5}$
9677 1188 $\mathrm{at} : 2/3 + \pi/4$
9677 1326 $2^x : 4e/3 - \pi/3$
9677 1616 rt : (8,8,-8,-7)
9677 1728 rt : (9,5,-5,-8)
9677 3001 $\mathrm{s}\pi : 4\sqrt{3} + 2\sqrt{5}/3$
9677 3353 $\mathrm{sq} : 2\sqrt{3} + 4\sqrt{5}$
9677 5087 rt : (8,7,-6,-8)
9677 5199 rt : (9,4,-3,-9)
9677 5212 $J_0 : 3\zeta(3)/10$
9677 6182 $J_0 : \sqrt[3]{3}/4$
9677 8433 rt : (7,9,-7,-8)
9677 8544 rt : (8,6,-4,-9)
9678 1449 $2^x : 1 - \pi/3$
9678 1874 rt : (7,8,-5,-9)
9678 2488 $\tanh : \exp(\sqrt[3]{3}/2)$
9678 2736 $\theta_3 : 7/10$
9678 3988 $\mathrm{at} : \sqrt{2}/2 + \sqrt{5}/3$
9678 5191 rt : (6,8,-8,-1)
9678 6149 $J_0 : 9/25$
9678 6549 $J_0 : 2\sqrt[3]{2}/7$
9678 7259 $\mathrm{cr} : (\sqrt{5}/3)^{1/3}$
9678 8477 $2^x : (e+\pi)/6$
9678 8512 $\ln : 4/3 + 3\sqrt{3}/4$
9678 8574 $\mathrm{at} : 4\sqrt{2}/3 - \sqrt{3}/4$
9678 8604 rt : (6,9,-4,3)
9679 1126 $erf : 7\sqrt{3}/8$
9679 3911 $e^x : 1 + 2\sqrt{5}$
9679 5277 rt : (5,5,-8,3)
9679 8647 rt : (5,6,-4,7)
9679 8728 $\ell_{10} : 2/3 - \sqrt{5}/4$
9679 8967 $\mathrm{sr} : \sqrt{15}$
9680 3588 $\mathrm{t}\pi : 3\sqrt{3} + \sqrt{5}/4$
9680 4041 $\Gamma : 7\sqrt{5}/2$
9680 4440 $10^x : 3\sqrt[3]{2}/8$
9680 5237 rt : (4,2,-8,7)
9680 5272 $\tanh : 10\sqrt[3]{3}/7$
9680 5646 $\mathrm{cu} : \sqrt{2}/4 + 3\sqrt{5}/2$
9680 6663 $\sinh : 5\zeta(3)/7$
9680 7053 $\ln : 3/2 + 4\sqrt{2}$
9681 1773 rt : (3,7,-7,-4)
9681 1878 $\mathrm{id} : \sqrt{3} + \sqrt{5}$
9681 3640 $\Gamma : 9/20$
9681 4161 $\Gamma : \sqrt{2} - \sqrt{3} + \sqrt{5}$
9681 5074 rt : (3,8,-1,5)
9681 7880 $\mathrm{s}\pi : \sqrt{11}\pi$
9681 9228 $Ei : 1/\ln(\sqrt{5}/2)$
9681 9487 $\mathrm{cr} : 4 + 4e/3$
9681 9737 $J_0 : 9(1/\pi)/8$
9682 0159 $K : \pi/4$
9682 2205 $erf : \sqrt[3]{7/2}$
9682 2862 $\Psi : 9\sqrt{3}/5$
9682 4583 $\mathrm{sr} : 15/16$
9682 4790 rt : (2,8,0,-1)
9682 5968 $erf : \sqrt{3} + \sqrt{5} - \sqrt{6}$
9682 6160 $\Gamma : 10(1/\pi)/3$
9682 7025 $e^x : 7\pi^2/10$
9682 8182 $\mathrm{id} : 1/4 + e$
9682 8369 $\theta_3 : 4/9$
9682 8844 $Ei : 2/3$
9683 1896 $\ell_2 : 7\sqrt{5}$
9683 2000 $Ei : 5\zeta(3)/2$
9683 4340 $\mathrm{sr} : 1/4 + 4e/3$
9683 6555 $J_0 : 5/14$
9683 8138 $\ell_{10} : \sqrt{2}/4 + 4\sqrt{5}$
9683 8511 $e^x : 3 + 3\sqrt{3}$
9683 9749 $\lambda : 1/\sqrt{24}$
9684 0790 rt : (1,4,-3,9)
9684 1110 $erf : 8\sqrt[3]{5}/9$
9684 1121 $J_0 : 1/\sqrt[3]{22}$
9684 1944 $10^x : 5/17$
9684 2499 $2^x : 2 + \sqrt{3}/3$
9684 3536 $\Gamma : 3\sqrt{2}/4$
9684 9769 $e^x : 9(e+\pi)$
9685 0246 rt : (1,8,0,-2)
9685 0898 $\mathrm{at} : 16/11$
9685 1497 $\mathrm{as} : 2/3 - 2\sqrt{5}/3$
9685 2049 $\ln(\pi) - \exp(\sqrt{2})$
9685 3690 $\exp(3) - \exp(3/4)$
9685 4079 $\mathrm{cr} : (4/3)^{-1/3}$
9685 4371 $\mathrm{at} : 2/3 - 3\sqrt{2}/2$
9685 4420 $2^x : 4\sqrt[3]{5}/7$
9685 4556 $\mathrm{s}\pi : 2\sqrt[3]{5}$
9685 6259 $\mathrm{s}\pi : \sqrt[3]{2}/3$
9685 6730 $10^x : 13/7$
9685 9178 $\ell_{10} : 4e - \pi/2$

9685 9471 $\lambda : \exp(-\sqrt[3]{4})$
9686 0745 $\ell_2 : P_{13}$
9686 2800 $erf : \exp(\sqrt[3]{2}/3)$
9686 4457 sr : P_{62}
9686 4779 $2^x : (\pi/3)^{-1/2}$
9686 8673 as : $4\sqrt[3]{3}/7$
9686 9282 $K : 5\sqrt{2}/9$
9686 9567 rt : (9,7,-8,-7)
9686 9686 cu : $3e/4 - \pi/4$
9687 0124 sr : $(\pi/2)^3$
9687 2634 rt : (8,9,-9,-7)
9687 2733 rt : (9,6,-6,-8)
9687 2930 cr : 10/11
9687 5000 sinh : $4\ln 2$
9687 5786 rt : (8,8,-7,-8)
9687 5885 rt : (9,5,-4,-9)
9687 6422 $K : 11/14$
9687 7612 $\ell_{10} : 3\sqrt{3}/2 + 3\sqrt{5}$
9687 8384 $\Psi : 9\ln 2/2$
9687 8887 s$\pi : 2\pi^2/3$
9687 8926 rt : (8,7,-5,-9)
9688 0107 $10^x : 7\zeta(3)/3$
9688 0806 sr : $4\sqrt{2}/3 + 4\sqrt{3}$
9688 1956 rt : (7,9,-6,-9)
9688 2065 $Ei : \sqrt[3]{8/3}$
9688 5114 $10^x : 7e/4$
9688 6259 s$\pi : 3\sqrt{2}/2 + 3\sqrt{3}/4$
9688 7402 cu : $3\sqrt{3}/2 + \sqrt{5}$
9688 7667 s$\pi : 3(e + \pi)$
9688 8077 rt : (6,8,-7,1)
9689 1167 rt : (6,9,-3,5)
9689 5499 t$\pi : \sqrt{21}\pi$
9689 6353 $\ell_2 : \sqrt[3]{15/2}$
9689 7166 $J_0 : 10(1/\pi)/9$
9689 7215 rt : (5,5,-7,5)
9689 7246 s$\pi : 3\sqrt{3}/2 + 4\sqrt{5}/3$
9689 7838 $\ln(4/5) + s\pi(\sqrt{3})$
9689 9329 $J_0 : \sqrt{2}/4$
9689 9459 rt : (1,8,1,6)
9689 9575 $10^x : 3\sqrt{2} + \sqrt{3}/3$
9690 0269 rt : (5,6,-3,9)
9690 3953 $10^x : 5\sqrt[3]{3}/8$
9690 5641 $e^x : 2 + 2\sqrt{2}/3$
9690 6247 rt : (4,2,-7,9)
9690 6400 $3/4 \div s\pi(e)$
9690 6573 s$\pi : \exp(-\sqrt{3}/2)$
9690 6865 $2^x : \sqrt{2}/3 + 4\sqrt{5}$
9690 8086 $10^x : 2(e + \pi)$
9690 9404 $\exp(1/2) \times s\pi(1/5)$
9690 9720 $\ell_2 : \ln(5/3)$
9690 9974 $J_0 : 6/17$
9691 1772 ln : $3/4 + 4\sqrt{2}/3$
9691 2178 rt : (3,7,-6,-1)
9691 2181 cr : $\sqrt{2} + \sqrt{3} - \sqrt{5}$
9691 2664 $Ei : 3\zeta(3)$
9691 2782 at : $(\sqrt{2}/3)^{-1/2}$
9691 3530 cr : $\sqrt{2}/2 + 4\sqrt{3}$
9691 3701 id : $(\ln 3)^{-1/3}$
9691 3964 $2^x : \sqrt{3} + 3\sqrt{5}/2$
9691 4147 s$\pi : (e + \pi)^{1/2}$
9691 5173 rt : (3,8,0,8)
9691 7129 tanh : $6\sqrt{3}/5$
9691 9450 s$\pi : 4\sqrt{2} - \sqrt{5}$
9691 9545 sr : $2 - 3\sqrt{2}/4$
9692 0063 t$\pi : (\sqrt[3]{5}/3)^{1/2}$
9692 3076 tanh : $3\ln 2$
9692 3998 rt : (2,8,1,3)
9692 4643 $erf : \sqrt{7/3}$
9692 5566 $\Psi : \ln(\sqrt{2\pi}/2)$
9692 6966 tanh : $\sqrt[3]{9}$
9692 8909 tanh : $(\ln 2/3)^{-1/2}$
9692 9470 cr : $1/3 + \sqrt{3}/3$
9692 9486 $\Gamma : 18/17$
9693 1193 sr : P_{46}
9693 2267 t$\pi : e\pi$
9693 2646 $J_0 : 1/\sqrt[3]{23}$
9693 2870 cu : 19/12
9693 3006 at : $3/4 + \sqrt{2}/2$
9693 4733 tanh : $(\ln 2)^{-2}$
9693 6229 $\ln(5) \div \ln(2/3)$
9693 6526 tanh : $\sqrt{13/3}$
9693 8529 cr : P_{11}
9694 0026 s$\pi : 8/19$
9694 0245 $\Psi : (1/(3e))^{1/2}$
9694 0276 t$\pi : 1/\sqrt[3]{12}$
9694 4233 t$\pi : 9\sqrt[3]{3}/4$
9694 4996 $J_0 : \exp(-\pi/3)$
9694 6569 tanh : 25/12
9694 8710 t$\pi : 4\sqrt{2}/3 - \sqrt{5}$
9695 0149 sq : $\sqrt{2} + \sqrt{3} + \sqrt{5}$
9695 1039 $e^x : 3e\pi/2$
9695 1718 $\lambda : \sqrt[3]{3}/7$
9695 4447 cu : $4\sqrt{3}/7$
9695 7753 $J_0 : (1/(3e))^{1/2}$
9695 9866 $\zeta : e\pi/6$
9696 0867 $J_0 : 7/20$
9696 4725 rt : (9,8,-9,-7)
9696 7607 rt : (9,7,-7,-8)
9697 0132 $\Gamma : 1/\sqrt[3]{11}$
9697 0392 rt : (8,9,-8,-8)
9697 0478 rt : (9,6,-5,-9)
9697 2275 sq : $(\pi/3)^{-1/3}$
9697 3252 rt : (8,8,-6,-9)
9697 3827 cr : $e/2 + 2\pi$
9697 5543 sr : $7\sqrt[3]{2}$
9697 6945 $J_0 : \pi/9$
9697 7049 $Ei : \exp(-\sqrt{3})$
9697 8725 $2^x : \sqrt{2}/3 + 3\sqrt{5}$
9698 0019 s$\pi : 3/4 + 2\sqrt{2}$
9698 0513 $e^x : 3 + \sqrt{3}/4$
9698 0746 $\Gamma : 3e\pi/5$
9698 3162 id : $7\sqrt[3]{5}$
9698 4417 rt : (6,8,-6,3)
9698 5257 $\Gamma : 4\sqrt[3]{3}/3$
9698 7225 rt : (6,9,-2,7)
9698 7635 $erf : 23/15$
9698 8052 $\Gamma : 25/13$
9698 8290 $\Psi : (\pi/2)^{-1/2}$
9699 1615 $Li_2 : \sqrt{5}/3$
9699 1793 tanh : 23/11
9699 2725 rt : (5,5,-6,7)
9699 4420 id : $\sqrt{2}/2 - 3\sqrt{5}/4$
9699 4424 cu : $3/4 + 3\pi/4$
9699 5778 $\Gamma : 8\zeta(3)/5$
9699 6315 t$\pi : 3\sqrt{3}/8$
9699 6364 cr : $\sqrt{2}/4 + \sqrt{5}/4$
9699 7854 $\Gamma : (\sqrt{5}/2)^{1/2}$
9699 8218 $J_0 : 8/23$
9700 0734 s$\pi : 2e + \pi$
9700 3125 s$\pi : (4/3)^{-3}$
9700 4278 sr : $3/2 - \sqrt{5}/4$
9700 6342 rt : (3,7,-5,2)
9700 8602 $\Gamma : 6\sqrt{3}$
9701 0213 $e^x : 5e/7$
9701 0305 sr : $2/3 + 3e$
9701 1373 $\Gamma : 9\sqrt[3]{5}/8$
9701 2382 tanh : $2\pi/3$
9701 3123 cr : 21/23
9701 3935 $erf : \sqrt{3} + \sqrt{6} - \sqrt{7}$
9701 4250 sr : 16/17
9701 4411 rt : (2,7,-6,-9)
9701 5050 cu : $7\sqrt{2}/10$
9701 7108 rt : (2,8,2,7)
9701 7465 $\ln(4) - \exp(\sqrt{5})$
9701 7808 $J_0 : 1/\sqrt[3]{24}$
9701 9636 $J_0 : \ln 2/2$
9702 0666 $10^x : 2 - \pi/3$
9702 2425 $J_0 : \sqrt{3}/5$
9702 4915 $\Psi : 23/3$
9703 2194 $\Gamma : 9\sqrt[3]{5}/4$
9703 2635 $Ei : 5(1/\pi)/9$
9703 3802 $10^x : 4\sqrt{2}/3 + 2\sqrt{5}$
9703 4815 s$\pi : 8\sqrt{5}/5$
9703 5321 sinh : $(\pi/2)^{-1/3}$
9703 7103 tanh : $\ln(1/(3e))$
9703 7139 $\zeta : 5\pi/9$
9703 8383 sinh : $5\sqrt{3}/2$
9703 9500 $\Gamma : 10\sqrt{3}/9$
9704 0605 id : $7\ln 2/5$
9704 1351 s$\pi : 5\ln 2/6$
9704 2328 $erf : 20/13$
9704 3505 $\Gamma : e/9$
9704 4427 tanh : $5\sqrt[3]{2}/3$
9704 5193 tanh : 21/10
9704 5898 cu : 23/16
9704 6812 cu : $1/3 + 2\sqrt{5}$
9704 7299 $\Gamma : 1/\sqrt{19}$
9704 7792 $erf : 9\sqrt[3]{5}/10$
9704 8430 cr : $2\sqrt{2}/3 + 3\sqrt{5}$
9704 8525 $\Gamma : 9(e + \pi)/8$
9704 9659 $e^x : 5\pi/2$
9705 0142 sq : $7\pi/9$
9705 0321 s$\pi : 10\sqrt[3]{3}$
9705 1003 t$\pi : \sqrt{14}\pi$
9705 2399 $\Psi : 20$
9705 4361 $erf : 8\sqrt{3}/9$
9705 6274 sq : $2 + 3\sqrt{2}$
9705 6495 s$\pi : 1/2 + 3e/2$
9705 6893 rt : (9,8,-8,-8)
9705 8858 sq : $3/2 + 2e/3$
9705 9516 rt : (9,7,-6,-9)
9705 9876 $\lambda : 3\ln 2/10$
9705 9883 sr : $10e/7$
9706 1976 s$\pi : \sqrt{3}/3$
9706 2053 rt : (8,9,-7,-9)
9706 6078 tanh : $7\zeta(3)/4$
9706 7947 $4 \div \ln(\sqrt{5})$
9706 8642 $2^x : \pi/2$
9707 2143 s$\pi : \gamma$
9707 2265 rt : (6,7,-9,1)
9707 2498 $10^x : 2/3 - e/4$
9707 2809 $J_0 : 2\zeta(3)/7$
9707 4571 $Ei : \sqrt{2}/8$
9707 4611 at : 19/13
9707 4833 rt : (6,8,-5,5)
9707 4968 s$\pi : 4e/3 - \pi/3$
9707 7392 rt : (6,9,-1,9)
9707 8034 t$\pi : 3\zeta(3)/7$
9707 9793 $erf : \sqrt[3]{11/3}$
9707 9868 rt : (5,4,-9,5)
9708 1295 sr : $3\pi/10$
9708 1495 $\lambda : 5/24$
9708 2023 $\ln(4) \times \exp(\sqrt{5})$
9708 2409 rt : (5,5,-5,9)
9708 3836 $\Gamma : 19/18$
9708 6773 tanh : $\exp(\sqrt{5}/3)$
9708 7392 rt : (4,1,-9,9)
9708 7630 cu : $8\pi/3$
9709 0058 $Ei : \sqrt[3]{23/2}$
9709 0350 tanh : $(\sqrt{2}\pi)^{1/2}$
9709 0871 $erf : (\sqrt[3]{2}/3)^{-1/2}$
9709 1505 $\lambda : 1/\sqrt{23}$
9709 1735 cu : $10\ln 2/7$
9709 1859 cu : $(2e)^3$
9709 2350 $\Psi : \exp(-\sqrt{3})$
9709 4838 rt : (3,7,-4,5)
9709 5929 s$\pi : 2\sqrt[3]{3}/5$
9709 7287 $J_0 : \sqrt[3]{5}/5$
9709 8354 sr : $2\sqrt{2}/3$
9709 8446 $\Psi : 25/8$
9709 8563 cr : $2 + 4\sqrt{2}$
9710 2209 rt : (2,7,-5,-5)
9710 4910 ln : $1/4 + 4\sqrt{3}$
9710 7086 $\Psi : 7/20$
9710 9193 tanh : 19/9
9711 0107 s$\pi : e\pi/6$
9711 0427 ln : $4 - e/2$
9711 1198 $\ell_{10} : \exp(\sqrt{5})$
9711 2679 $\exp(5) \div s\pi(\sqrt{2})$
9711 4376 rt : (1,6,-1,-7)
9711 5221 sq : $1 + 4e$
9711 5532 $erf : 17/11$
9711 5999 tanh : $(3\pi)^{1/3}$
9711 6020 $erf : 9\zeta(3)/7$
9711 6506 cr : $\beta(2)$
9711 9081 $2^x : 9\sqrt{3}/2$
9711 9354 $e^x : 2/3 + 4\sqrt{3}$
9711 9711 sr : $2 + 4\sqrt{2}/3$
9712 3841 t$\pi : 1/2 + \sqrt{5}/3$
9712 4484 ln : $\sqrt{2}/3 + 3\sqrt{5}$

9712 6849 tanh : $7e/9$
9712 7996 cr : $\ln(5/2)$
9713 4410 $J_0 : e/8$
9713 6689 $s\pi : (\ln 2/2)^{-1/3}$
9713 7575 $\exp(4/5) - s\pi(\sqrt{3})$
9714 0275 $1/4 - \exp(1/5)$
9714 0452 id : $1/2 + \sqrt{2}/3$
9714 1071 rt : (9,9,-9,-8)
9714 1278 cr : $11/12$
9714 3474 rt : (9,8,-7,-9)
9714 5154 sr : $3\sqrt{2}/2 + 3\sqrt{5}$
9714 7688 tanh : $\sqrt[3]{19/2}$
9715 3271 $\zeta : (e)^{-1/3}$
9715 3434 $\theta_3 : 7(1/\pi)/5$
9715 4821 $10^x : 3\zeta(3)/4$
9715 6244 at : $2 - 2\sqrt{3}$
9715 7518 rt : (6,7,-8,3)
9715 8759 $\zeta : (\sqrt{5}/2)^{1/3}$
9715 9848 $\Psi : 7\pi^2/9$
9715 9865 rt : (6,8,-4,7)
9716 2607 $Ei : 5\zeta(3)/9$
9716 4263 $erf : \ln(3\pi/2)$
9716 4470 rt : (5,4,-8,7)
9716 5103 $s\pi : 1/3 + 3\sqrt{2}$
9716 5772 $\pi - \exp(\sqrt{2})$
9716 6792 tanh : $3\sqrt{2}/2$
9716 7807 $2^x : \exp(4)$
9717 1737 $\lambda : \sqrt[3]{2}/6$
9717 1836 at : $(\pi)^{1/3}$
9717 5801 sinh : $((e+\pi)/2)^{1/3}$
9717 5885 rt : (3,6,-9,-1)
9717 6578 $\Gamma : (\sqrt[3]{5}/2)^{-1/3}$
9717 7291 $t\pi : 3(e+\pi)/5$
9717 7999 $\Gamma : \sqrt{3} - \sqrt{6} + \sqrt{7}$
9717 8040 $e^x : 8(e+\pi)/7$
9717 8172 rt : (3,7,-3,8)
9718 0186 $\ell_2 : 4 - 3e/4$
9718 1339 $e^x : 2\sqrt{3} - 3\sqrt{5}/4$
9718 1414 ln : $1/3 + 4\sqrt{3}/3$
9718 2343 $2^x : 10\sqrt{2}/9$
9718 2531 sr : $17/18$
9718 2723 id : $\sqrt[3]{23/3}$
9718 2759 $s\pi : 3\sqrt{2}/10$
9718 3404 $\Gamma : 7\pi^2/4$
9718 3813 at : $(e+\pi)/4$
9718 4924 rt : (2,7,-4,-1)
9718 6036 $3 \div s\pi(\pi)$
9718 7274 tanh : $17/8$
9718 7287 $K : 5\sqrt[3]{2}/8$
9718 8790 at : $\ln(\ln 2/3)$
9718 8908 $\theta_3 : 1/\ln(3\pi)$
9719 2013 $t\pi : 3(e+\pi)/7$
9719 3790 $s\pi : 4(1/\pi)/3$
9719 5836 $erf : 4e/7$
9719 6076 rt : (1,6,0,-1)
9719 7930 $t\pi : 1/3 + 3\sqrt{2}/2$
9719 8857 $2^x : 11/7$
9720 0622 $\lambda : 4/19$
9720 0707 $\ell_2 : 2\sqrt{2}/3 - \sqrt{3}/4$
9720 0829 $\ell_{10} : 8(e+\pi)/5$
9720 3791 rt : (1,7,-8,-2)
9720 4968 sq : $4/3 + 3\sqrt{5}/2$
9720 6851 at : $6\sqrt[3]{5}/7$
9720 7709 $\ell_{10} : \sqrt{3}/4 + 4\sqrt{5}$
9720 8064 sr : $3\sqrt[3]{2}/4$
9721 0098 ln : $10(e+\pi)/3$
9721 2827 as : $19/23$
9721 3595 id : $1/2 + 2\sqrt{5}$
9721 6456 $2^x : 4/3 - \sqrt{2}/4$
9721 8514 $erf : 14/9$
9722 0684 $s\pi : 3\pi$
9722 2773 rt : (9,9,-8,-9)
9722 3484 $\Gamma : 20/19$
9722 3934 tanh : $\sqrt{2} - \sqrt{3} + \sqrt{6}$
9722 7013 $\ln(\pi) - \exp(3/4)$
9722 7532 $s\pi : 7\sqrt{3}/5$
9722 8654 $\exp(3) - \exp(\sqrt{2})$
9723 5521 $s\pi : 5\sqrt[3]{5}/6$
9723 5827 $2^x : 2/3 - \sqrt{2}/2$
9723 7743 at : $22/15$
9723 7840 rt : (6,7,-7,5)
9723 8145 $2^x : 7\sqrt[3]{2}/9$
9723 9395 cu : $1/2 - 2\sqrt{5}/3$
9723 9991 rt : (6,8,-3,9)
9724 0100 $s\pi : \sqrt{3}/2 + \sqrt{5}/4$
9724 1070 $\ell_2 : \sqrt{2}/3 + 2\sqrt{5}/3$
9724 1452 $J_0 : 1/3$
9724 1796 cr : $4\sqrt{3} + \sqrt{5}/3$
9724 1840 tanh : $e\pi/4$
9724 2596 rt : (1,7,-7,6)
9724 4214 rt : (5,4,-7,9)
9724 4225 $s\pi : 5\sqrt[3]{2}/4$
9724 4699 $e + s\pi(\sqrt{3})$
9724 6524 $J_0 : (\ln 2)^3$
9724 6693 $4 \div \exp(\sqrt{2})$
9724 9247 sr : $(\sqrt{5}/2)^{-1/2}$
9725 1366 $erf : 9\sqrt{3}/10$
9725 2289 ln : $3 + 4\pi/3$
9725 4689 rt : (3,6,-8,2)
9725 5605 tanh : $5\sqrt[3]{5}/4$
9725 8096 $s\pi : \exp(-\sqrt[3]{5}/2)$
9725 8247 $s\pi : 1/\sqrt[3]{13}$
9725 8663 $erf : 9\ln 2/4$
9725 8882 cr : $23/25$
9725 9092 ln : $e/3 + 2\pi$
9726 2040 $\ell_2 : 4\sqrt{2} - \sqrt{3}$
9726 2988 rt : (2,7,-3,3)
9726 3479 $\Gamma : 7\zeta(3)/8$
9726 5782 cr : $7\pi^2/9$
9726 6680 $Ei : 3\pi^2/8$
9726 9032 $\zeta : (\ln 2)^3$
9727 1200 sr : $3/4 + \pi$
9727 3237 rt : (1,6,1,5)
9727 4447 $e^x : 2\sqrt{2} - \sqrt{3}/4$
9727 8594 tanh : $\sqrt{3} - \sqrt{5} + \sqrt{7}$
9728 0748 $e^x : 7(e+\pi)/2$
9728 1362 $Ei : 3/17$
9728 1578 cr : $3/4 + 4\sqrt{3}$
9728 2701 ln : $2\sqrt{3}/3 + 2\sqrt{5}/3$
9728 3352 rt : (1,9,-9,7)
9728 4616 tanh : $15/7$
9728 7461 $erf : 25/16$
9728 8869 $\lambda : 2(1/\pi)/3$
9729 0563 $\Psi : 9e\pi/10$
9729 2374 ln : $3\sqrt[3]{2}/10$
9729 3226 $s\pi : \sqrt{3}/2 + 3\sqrt{5}$
9729 4996 cr : $3\sqrt{3}/2 - 3\sqrt{5}/4$
9729 5363 at : $4\sqrt{2}/3 - 3\sqrt{5}/2$
9729 5507 ln : $\sqrt{7}$
9730 2516 $s\pi : 4\sqrt{3} - 3\sqrt{5}/2$
9730 3004 $e^x : e/4$
9730 4276 $4/5 \div \ln(2/3)$
9730 5158 $\Psi : 7\sqrt{5}/5$
9731 0946 $s\pi : 7\ln 2/2$
9731 3654 rt : (6,7,-6,7)
9731 4330 $erf : 7\sqrt{5}/10$
9731 4355 $3/4 - \ln(4/5)$
9731 8360 sq : $1/4 + 2\sqrt{3}/3$
9732 0873 $10^x : (5/3)^{-1/3}$
9732 1783 ln : $3 - \sqrt{2}/4$
9732 3055 id : $\sqrt{2} + \sqrt{5}/4$
9732 3097 $\exp(1/2) - s\pi(\sqrt{5})$
9732 3103 $t\pi : 4\sqrt{2}/3 - 2\sqrt{3}$
9732 4187 $\Gamma : (e/3)^{-1/2}$
9732 5526 $s\pi : 2\sqrt{2} + \sqrt{5}/3$
9732 6137 $10^x : ((e+\pi)/2)^{1/3}$
9732 6285 $10^x : 2e/3 + 3\pi$
9732 8438 $s\pi : e/2 - \pi/4$
9732 9149 rt : (3,6,-7,5)
9733 2852 sr : $18/19$
9733 3318 cu : $1/3 - 3\pi/2$
9733 6024 $\zeta : 1/3$
9733 6790 rt : (2,7,-2,7)
9733 8963 id : $2\sqrt{3} - 2\sqrt{5}/3$
9733 9874 $\lambda : 1/\sqrt{22}$
9734 0448 at : $7\sqrt[3]{2}/6$
9734 2422 $\ell_2 : 5\pi$
9734 3162 $\Gamma : \pi/7$
9734 4040 $Ei : (\sqrt[3]{2}/3)^2$
9734 4632 $\ell_2 : 1 - \sqrt{5}/3$
9734 5505 $J_0 : \ln(\sqrt[3]{3}/2)$
9734 5847 $J_0 : \exp(-\sqrt{5}/2)$
9734 5944 tanh : $\sqrt[3]{10}$
9734 6160 $s\pi : 2\sqrt{2} + 3\sqrt{3}/2$
9734 6284 cr : $9e\pi/10$
9734 7612 $\Gamma : (\ln 3/2)^2$
9734 9174 $t\pi : (2\pi/3)^{1/2}$
9735 0426 $\Gamma : 21/20$
9735 2166 cr : $3/2 - \sqrt{3}/3$
9735 2411 $2^x : 2 - 3e/4$
9735 3615 $\Gamma : 5\sqrt[3]{2}/6$
9735 4649 ln : $2 + 3\sqrt{3}$
9735 6665 sq : $\sqrt{3} - \sqrt{5}/3$
9735 9441 at : $e - 4\pi/3$
9736 1966 at : $25/17$
9736 4956 $J_0 : (3\pi)^{-1/2}$
9736 5926 $Ei : 22/3$
9736 6390 $\Psi : 5(1/\pi)/9$
9736 6596 sq : $3\sqrt{2} + \sqrt{5}$
9736 7188 cr : $12/13$
9736 7527 $\lambda : \sqrt[3]{5}/8$
9736 7881 sinh : $\sqrt{13}\pi$
9736 7892 $erf : \pi/2$
9736 9401 $\sqrt{2} \times \exp(1/3)$
9737 0385 $erf : 1/\ln(\sqrt[3]{4}/3)$
9737 2300 $s\pi : \sqrt{2}/2 + \sqrt{3}/2$
9737 3170 $erf : 10\sqrt{2}/9$
9737 3936 $erf : 11/7$
9737 6215 tanh : $\sqrt{14/3}$
9737 7997 $e^x : 10\zeta(3)/3$
9737 8051 rt : (7,9,-9,1)
9737 8506 $J_0 : (\sqrt[3]{5}/3)^2$
9738 3511 rt : (6,6,-9,5)
9738 3530 sr : $8\pi^2/5$
9738 3603 $2^x : \sqrt{2}/2 - \sqrt{5}/3$
9738 4763 $s\pi : 9(1/\pi)/5$
9738 5152 $\ell_{10} : \sqrt{2}/3 + 4\sqrt{5}$
9738 5337 rt : (6,7,-5,9)
9738 5389 $s\pi : 3/4 + 3\sqrt{5}/4$
9738 6702 $1/2 \times \exp(2/3)$
9738 7767 at : $1 + \sqrt{2}/3$
9738 7773 $e^x : 17/25$
9738 8926 rt : (5,3,-9,9)
9738 9957 $J : \exp(-3\pi/4)$
9739 2088 id : $\pi^2/5$
9739 2362 tanh : $3\sqrt[3]{3}/2$
9739 3383 $\Gamma : (\sqrt{3}/2)^{-1/3}$
9739 3544 at : $\sqrt{2}/4 + \sqrt{5}/2$
9739 3689 cu : $20/9$
9739 4050 tanh : $9\zeta(3)/5$
9739 4268 $1/5 + s\pi(e)$
9739 4540 $\lambda : 3/14$
9739 5799 tanh : $1/\ln(\sqrt[3]{4})$
9739 6979 $10^x : 2 - \sqrt{5}/3$
9739 8819 $s\pi : 2/3 + e/3$
9739 9622 rt : (3,6,-6,8)
9740 1042 tanh : $5\sqrt{3}/4$
9740 4904 rt : (2,6,-9,-5)
9740 5852 $t\pi : 2e/3 - 4\pi$
9740 6528 cr : $4\ln 2/3$
9740 6921 $erf : 5\sqrt[3]{2}/4$
9740 7961 $\ln(4) + s\pi(1/5)$
9740 8303 $J : (1/(3e))^{1/3}$
9740 9091 sq : $\pi^2/10$
9740 9254 tanh : $13/6$
9740 9412 sr : $e\pi/9$
9741 0791 $2^x : \sqrt{2} - \sqrt{3}/4$
9741 0869 at : $3 - 2\sqrt{5}$
9741 1101 sr : $9\sqrt{3}/4$
9741 2730 $s\pi : 4\sqrt{3}/3 + \sqrt{5}/2$
9741 3671 sq : $4 - e/2$
9741 6421 $t\pi : (\ln 3)^{-3}$
9741 6912 $s\pi : \sqrt[3]{5}/4$
9741 7349 $\Psi : 9\sqrt[3]{5}/2$
9741 8245 $e^x : 23/10$
9742 2663 sr : $(\sqrt[3]{5}/2)^{1/3}$
9742 3000 $\Psi : \sqrt{6}\pi$
9742 3872 cu : $\sqrt{2} + 3\sqrt{3}/4$
9742 4405 cr : $9\sqrt[3]{5}/2$
9742 5454 $Ei : 9\sqrt{5}/5$

9742 5604 $2^x : \sqrt[3]{22}$
9742 7112 $\ell_{10} : 3\pi$
9742 7798 $J : 1/\sqrt[3]{15}$
9742 7887 cr : $\sqrt{6}\pi$
9743 0328 tanh : $(3\pi/2)^{1/2}$
9743 3087 $2^x : 1/2 + 2\sqrt{5}/3$
9743 3518 sπ : $1/3 + 2\pi/3$
9743 3604 $e^x : 1/4 + 4\pi$
9743 4365 $Ei : (\sqrt{5})^{-1/2}$
9743 6842 $\Gamma : 5e/4$
9743 7370 as : $1/4 + \sqrt{3}/3$
9744 0527 $\Gamma : (\ln 3)^{1/2}$
9744 6498 rt : (7,9,-8,3)
9744 8434 $J_0 : 2\sqrt[3]{3}/9$
9744 9644 tanh : $4e/5$
9745 1407 $10^x : (3\pi)^{-1/3}$
9745 1538 rt : (6,6,-8,7)
9745 2258 $e^x : 4/3 - e/2$
9745 6063 cr : $8\pi^2/3$
9745 6337 $J_0 : 8/25$
9746 3252 sinh : 23/3
9746 3891 $\Gamma : 9\pi/4$
9746 5202 $J_0 : \sqrt{5}/7$
9746 5268 $erf : \sqrt{5/2}$
9746 5460 cr : $(\sqrt[3]{2})^{-1/3}$
9746 5791 sinh : $(\sqrt{5}/3)^{1/2}$
9746 6312 $\Gamma : 22/21$
9746 6986 sπ : $1/2 + 4\sqrt{3}$
9746 7257 sr : $5\sqrt[3]{5}/9$
9746 7943 sr : 19/20
9746 9149 id : $3\sqrt{2} + \sqrt{3}$
9747 1306 rt : (2,6,-8,-1)
9747 3387 $2^x : 2/3 + e/3$
9747 5689 cu : $3e/4 - \pi/3$
9747 6423 $\Psi : 5(1/\pi)/2$
9747 6963 tπ : $\sqrt{5}/5$
9747 7122 sπ : $(\sqrt[3]{4}/3)^{-2}$
9747 8822 $\lambda : (5/3)^{-3}$
9747 8856 $10^x : \sqrt{2} + \sqrt{3} - \sqrt{6}$
9748 1387 $\ell_{10} : 4 + 2e$
9748 1675 tπ : $e/4 + 4\pi$
9748 2965 $J_0 : (1/\pi)$
9748 4253 $\sqrt{2} - \exp(2)$
9748 4978 $J_0 : 7/22$
9748 5523 $erf : 19/12$
9748 5612 tanh : 24/11
9748 6930 $\Gamma : \pi/3$
9748 7373 cu : $1 + \sqrt{2}/2$
9748 7829 as : $\sqrt{2}/3 - 3\sqrt{3}/4$
9748 9995 $\Psi : (e)^3$
9749 1646 $\ell_{10} : 3\sqrt{2} + 3\sqrt{3}$
9749 2791 sπ : 3/7
9749 6780 $10^x : 3e/4 + \pi/4$
9749 7476 cu : $2 - \sqrt{5}/3$
9749 8392 sπ : $10\sqrt{2}/9$
9749 9021 sinh : $9(1/\pi)/2$
9750 3055 sπ : $\sqrt[3]{17}$
9750 3235 $\lambda : \sqrt{3}/8$
9750 4666 $\ell_2 : 2/3 + 3\sqrt{3}/4$
9750 6232 $\ell_2 : 2\sqrt{3}/3 + 3\sqrt{5}$
9750 7177 sinh : 19/22
9750 7403 $Ei : 25/18$
9750 9023 sπ : $9e\pi/2$
9751 0532 tπ : $3\sqrt{2}/2 - 2\sqrt{5}$
9751 1391 rt : (7,9,-7,5)
9751 4451 sπ : $(2e)^{-1/2}$
9751 5581 $J_0 : 1/\sqrt{10}$
9751 6053 rt : (6,6,-7,9)
9751 6717 $e^x : 2\sqrt{3}/3 + 3\sqrt{5}/4$
9751 6848 $\ell_{10} : 1/2 + 4\sqrt{5}$
9751 7184 rt : (1,9,0,-2)
9751 9007 cr : P_{34}
9752 1981 ln : $1/2 + 3\sqrt{5}$
9752 2420 $J_0 : 6/19$
9752 2699 $erf : \sqrt[3]{4}$
9752 3195 sq : $4/3 + 2\sqrt{5}/3$
9752 4580 $J : \exp(-1)$
9752 4741 tanh : $\sqrt[3]{21/2}$
9752 5701 cr : $3\sqrt{2} + 2\sqrt{3}$
9753 0864 sq : 22/9
9753 4358 rt : (2,6,-7,3)
9753 5023 $J_0 : \sqrt[3]{2}/4$
9753 5522 cu : $4/3 + \sqrt{3}/3$
9753 6797 sπ : $\pi/2$
9753 8187 cr : $1 + 3\sqrt{5}$
9753 8414 sinh : $8\sqrt[3]{3}/5$
9754 1727 tanh : $2\pi^2/9$
9754 1852 rt : (1,8,6,-8)
9754 5385 $\lambda : 5/23$
9754 6064 $J_0 : 2\sqrt{2}/9$
9754 7777 $J_0 : \pi/10$
9754 7793 rt : (1,9,1,7)
9755 0013 $2^x : \sqrt{2}/2 + \sqrt{3}/2$
9755 1280 $J : \exp(-5/3)$
9755 3417 $K : 15/19$
9755 4092 $1/4 - \exp(4/5)$
9755 5105 $erf : 9\sqrt{2}/8$
9755 6688 cr : P_{17}
9755 9466 sq : $3e/4 + \pi/4$
9755 9995 cr : 13/14
9756 0120 $erf : 5(1/\pi)$
9756 2075 tanh : $3(e + \pi)/8$
9756 2150 sπ : $3/4 + e/4$
9756 3927 $Ei : \exp(\pi/3)$
9756 4200 ln : $5\sqrt[3]{3}$
9756 4833 sr : $(e/3)^{1/2}$
9756 7308 tanh : $9\sqrt[3]{5}/7$
9756 9384 $e^x : 2 + e$
9757 0067 tanh : $7\pi/10$
9757 2521 $\Gamma : 23/22$
9757 3004 rt : (7,9,-6,7)
9757 3230 $\Psi : \sqrt{2}/8$
9757 3454 $J_0 : 5/16$
9757 3500 cu : $4/3 + 3\pi/2$
9757 3931 cr : $3 + 3\pi/2$
9757 4313 tanh : 11/5
9757 6242 $2^x : 3/4 - \pi/4$
9757 9364 ln : $6\zeta(3)$
9758 1638 $\Gamma : 1/\sqrt{11}$
9758 4167 $\lambda : 1/\sqrt{21}$
9758 6339 $\theta_3 : \sqrt{5}/5$
9758 6466 $10^x : 3\sqrt{2}/4 + \sqrt{5}/3$
9758 8691 rt : (3,5,-9,8)
9758 9009 cu : $3e - 3\pi/4$
9758 9931 $10^x : 1/3 + \sqrt{2}/2$
9759 0007 sr : 20/21
9759 2225 sπ : $\sqrt[3]{5}/3$
9759 4311 rt : (2,6,-6,7)
9759 4373 $e^x : 3/2 + 3\sqrt{5}/4$
9759 4970 cr : $1/4 + e/4$
9759 5083 ln : $\ln 2/5$
9759 6261 $e^x : 9\sqrt{5}/5$
9759 7503 tanh : $7\sqrt[3]{2}/4$
9759 9498 $e^x : \sqrt{2}/2 + \sqrt{5}$
9760 1270 rt : (1,8,7,-1)
9760 6774 id : $1/3 - 4\sqrt{3}/3$
9760 8908 id : $3\sqrt{3}/4 + 3\sqrt{5}/4$
9760 9601 $\ell_{10} : P_{38}$
9761 0228 $erf : 5\sqrt{5}/7$
9761 1600 sr : $2 - \pi/3$
9761 5312 id : $\sqrt{3} - 3\sqrt{5}$
9761 7343 $\ln(5) \div \exp(1/2)$
9761 8711 cr : $1/\ln((e + \pi)/2)$
9762 0785 tπ : $1/4 - 2\sqrt{2}$
9762 2340 at : $1/3 - 2e/3$
9762 2493 $Ei : 7\pi/5$
9762 4066 sq : $4\sqrt{2}/3 + \sqrt{5}/4$
9762 6806 $\zeta : (\pi/3)^{-1/3}$
9762 7915 cu : $\sqrt{3}/4 + \sqrt{5}/4$
9762 9118 tanh : $\exp(\sqrt[3]{4}/2)$
9762 9360 as : $2 - 2\sqrt{2}$
9763 0231 rt : (7,8,-9,5)
9763 1157 sr : $(\sqrt{3}/2)^{1/3}$
9763 1586 rt : (7,9,-5,9)
9763 2532 sπ : $1/3 + \sqrt{5}$
9763 3185 ln : $3/2 + 2\sqrt{3}/3$
9763 4254 rt : (6,5,-9,9)
9763 4838 $erf : 8/5$
9763 5144 $10^x : 9/19$
9763 6201 sinh : $5e\pi/8$
9763 7431 sπ : $e + 3\pi/2$
9763 8104 $\exp(3/5) + \exp(e)$
9763 8405 sr : $3 + e/3$
9763 8548 $Li_2 : (\sqrt[3]{2}/3)^{1/3}$
9764 0252 sq : $4(e + \pi)/5$
9764 0865 sπ : $3\sqrt{2}/2 + 4\sqrt{3}/3$
9764 1428 $J_0 : 4\ln 2/9$
9764 1605 cr : $\sqrt{2}/4 + \sqrt{3}/3$
9764 2522 tanh : $1/\ln(\pi/2)$
9764 3043 sq : $2 + 3\pi/4$
9764 4382 $J_0 : (\sqrt{2}\pi/3)^{-3}$
9764 6080 $\sqrt{5} \times \exp(4/5)$
9764 6154 cr : $9(e + \pi)/2$
9764 6333 sq : $e - 3\pi/2$
9764 7104 $J_0 : 4/13$
9764 8703 as : $(\ln 3)^{-2}$
9765 0777 sr : $2(e + \pi)/3$
9765 4333 $erf : \exp(\sqrt{2}/3)$
9765 4918 $\ell_2 : \sqrt{2}/3 + 2\sqrt{3}$
9765 5228 sπ : $((e + \pi)/2)^{1/3}$
9765 6554 $erf : 10\sqrt[3]{3}/9$
9765 7864 rt : (1,8,8,6)
9765 8656 $erf : 4\zeta(3)/3$
9765 9585 at : $3\sqrt{3}/2 - \sqrt{5}/2$
9765 9850 $J_0 : \ln(e/2)$
9766 0685 $Ei : \ln((e + \pi)/3)$
9766 2251 $Ei : 10\zeta(3)/7$
9766 4574 id : $(e + \pi)/6$
9766 5417 $10^x : 2\sqrt[3]{5}$
9766 5526 sπ : $\sqrt{2} + 2\sqrt{3}/3$
9766 7822 $2^x : \sqrt{2} + \sqrt{3}/3$
9767 0211 $\Gamma : 24/23$
9767 0983 $\Gamma : 5e/7$
9767 1067 tπ : $\exp(-\pi/3)$
9767 1857 ln : $3e/2 + \pi$
9767 2742 $2^x : 1/2 + 3\pi$
9767 4337 $\exp(\sqrt{3}) - s\pi(\sqrt{5})$
9767 6229 sr : $(\ln 3)^{-1/2}$
9767 8101 cu : $3 - 3e$
9767 8536 tanh : 20/9
9767 9194 sπ : $(2e)^{-1/3}$
9768 4135 $\exp(2) + s\pi(1/5)$
9768 6096 rt : (7,8,-8,7)
9768 6589 tanh : $\sqrt[3]{11}$
9768 8368 sq : $1 - 3\pi$
9768 9262 $\Psi : 9\sqrt{5}$
9768 9838 $Ei : \sqrt{15}$
9769 2349 $\ell_{10} : P_{57}$
9769 4712 $\theta_3 : 15/16$
9769 5326 $J_0 : (2e/3)^{-2}$
9769 7681 $J_0 : 7/23$
9769 7918 $e^x : 12/11$
9769 8657 cu : $3\sqrt{2}/4 - 4\sqrt{5}$
9770 0007 tanh : $9\sqrt{3}/7$
9770 0842 sr : 21/22
9770 1240 $e^x : 1/3 + \pi/3$
9770 2865 at : $3/2 - 4\sqrt{5}/3$
9770 3112 $\Gamma : \ln(1/\sqrt{2})$
9770 5669 tanh : $7(1/\pi)$
9770 7961 $2^x : 1/3 + 4\pi/3$
9770 9037 sq : $e - 3\pi$
9771 0017 sq : $1/2 + e/3$
9771 1908 tπ : $1/\sqrt[3]{16}$
9771 2220 $4/5 \times \exp(1/5)$
9771 2722 $10^x : 5\sqrt{5}/8$
9771 2911 id : $4\sqrt[3]{5}/7$
9771 3281 cu : $3\sqrt{2}/4 + 3\sqrt{3}/2$
9771 5288 cr : $1/2 + \sqrt{3}/4$
9771 5931 $erf : \ln(5)$
9771 5936 $\Gamma : \sqrt[3]{22/3}$
9771 6311 $Ei : 5\sqrt[3]{3}/7$
9771 7259 $\ln(2) + \exp(1/4)$
9771 8286 $\ell_{10} : 4/3 + 3e$
9772 0184 $\ell_{10} : 10e\pi/9$
9772 0502 id : $(\pi/3)^{-1/2}$
9772 1150 cu : $e/4 + 3\pi/4$
9772 1330 $\Psi : \pi$
9772 2481 at : $1/4 - \sqrt{3}$
9772 4694 $erf : \sqrt{2} - \sqrt{6} + \sqrt{7}$
9772 6480 cr : 14/15

9772 7991 $J : \exp(-19\pi/9)$
9773 2396 $J_0 : e/9$
9773 6785 $J_0 : (\ln 3/2)^2$
9773 7286 cr : $4e - \pi$
9773 9344 rt : (7,8,-7,9)
9774 0153 $J_0 : 1/\sqrt{11}$
9774 1223 tanh : $\sqrt{5}$
9774 1468 $s\pi : 3\sqrt{3} + \sqrt{5}$
9774 1776 ln : $3\sqrt{2} + 4\sqrt{5}/3$
9774 6627 sinh : $3\sqrt[3]{3}/5$
9774 6986 $e^x : 15/22$
9774 7435 rt : (4,9,-9,-1)
9774 8709 $s\pi : \sqrt{5} - \sqrt{6} + \sqrt{7}$
9774 9522 $Ei : 5\sqrt[3]{5}/3$
9775 1825 sq : $9\pi/5$
9775 3011 $s\pi : 9(1/\pi)/2$
9775 4991 $J_0 : \zeta(3)/4$
9775 5066 as : $(\sqrt[3]{5}/3)^{1/3}$
9775 7353 $10^x : \sqrt{2}/3 + 2\sqrt{3}$
9775 8796 $Ei : 6\zeta(3)/7$
9775 9183 $\zeta : \sqrt[3]{5/3}$
9776 0365 $\Gamma : 25/24$
9776 2624 $J_0 : 3/10$
9776 4237 $\Psi : \ln(1/2)$
9776 4247 $\lambda : 2/9$
9776 5776 $erf : 21/13$
9776 7266 rt : (1,8,-9,-2)
9776 8642 $\Psi : 22/7$
9777 1541 $s\pi : 3\sqrt[3]{3}/10$
9777 2360 $\ell_{10} : 2/19$
9777 2902 $erf : 8\sqrt{2}/7$
9777 3484 tanh : $\ln(3\pi)$
9777 6355 tanh : $5\pi/7$
9777 8725 $10^x : \sqrt{2} - \sqrt{5}/2$
9778 1395 id : $7e\pi/10$
9778 1519 $erf : \exp(\sqrt[3]{3}/3)$
9778 1987 $erf : 7\ln 2/3$
9778 2285 cr : $7\zeta(3)/9$
9778 9585 $s\pi : P_{80}$
9778 9921 $\lambda : 7(1/\pi)/10$
9779 2030 rt : (1,8,-8,7)
9779 2206 sq : $3/2 + 3\sqrt{2}$
9779 3103 $e^x : 3\sqrt{3}/2 + 4\sqrt{5}$
9779 3767 $s\pi : \sqrt{3}/4$
9779 7712 rt : (4,9,-8,2)
9779 7772 $s\pi : P_{10}$
9780 0268 $\zeta : 2e/9$
9780 1174 sq : $1/2 - \pi$
9780 1929 sr : $22/23$
9780 2605 at : $6\sqrt{3}/7$
9780 2611 tanh : $9/4$
9780 2978 $erf : 9\sqrt[3]{2}/7$
9780 3335 $\lambda : \exp(-3/2)$
9780 5031 cu : $3\sqrt{2}/2 - \sqrt{3}/2$
9780 5976 $J_0 : 3\ln 2/7$
9780 6212 $5 \times \exp(1/3)$
9780 7154 $s\pi : 5\ln 2/8$
9780 9699 $10^x : (\sqrt{3}/2)^{1/3}$
9781 3545 cr : $3e/2 - \pi$
9781 5858 $\Gamma : \sqrt{3} - \sqrt{5} + \sqrt{6}$
9781 6049 sr : $7\sqrt{5}/4$
9781 7226 $J_0 : (3/2)^{-3}$
9781 8923 $10^x : e/4 + \pi/2$
9782 3605 $\lambda : \sqrt{5}/10$
9782 4422 $erf : 9\sqrt[3]{3}/8$
9782 7002 $s\pi : 2\sqrt{2}/3 + 2\sqrt{5}/3$
9783 1191 $e^x : 1/\ln(5/2)$
9783 1834 sr : $1/4 + \sqrt{2}/2$
9783 1888 $10^x : (3/2)^{-3}$
9783 2587 $e^x : 1 + 2\sqrt{2}/3$
9783 3264 $s\pi : 3e/2 + 3\pi/4$
9783 3595 tanh : $\sqrt[3]{23/2}$
9783 4055 $s\pi : 4\pi$
9783 6252 $\ell_{10} : 7e/2$
9783 7672 rt : (7,7,-9,9)
9783 8047 $\exp(\sqrt{5}) \times s\pi(\sqrt{3})$
9784 1132 sinh : $\sqrt{3}/2$
9784 1554 $2^x : 8\sqrt[3]{5}$
9784 1858 $2^x : 3/4 + 2\pi$
9784 4115 $e^x : \sqrt[3]{22/3}$
9784 4373 $erf : 13/8$
9784 5766 rt : (4,9,-7,5)
9784 6939 $Li_2 : 3/4$
9784 7923 at : $3/4 - \sqrt{5}$
9784 9034 $J_0 : 5/17$
9785 1205 $2^x : 10\sqrt[3]{2}$
9785 1277 cu : $3/4 - 2\sqrt{2}$
9785 2296 $J_0 : \ln(\sqrt{5}/3)$
9785 7308 tanh : $8\sqrt{2}/5$
9785 7850 $\Gamma : 3\ln 2/2$
9786 2027 $10^x : 3/4 + 3\pi/2$
9786 5150 $\zeta : 5\zeta(3)/6$
9786 5489 $erf : \sqrt{2} - \sqrt{5} + \sqrt{6}$
9786 7060 $10^x : 3\sqrt{2} + \sqrt{5}/3$
9786 7852 tanh : $5e/6$
9787 1094 $\ell_{10} : \sqrt{3}/3 + 4\sqrt{5}$
9787 1691 cr : $15/16$
9787 4846 $\Gamma : 3(e + \pi)$
9787 8387 $s\pi : 2\sqrt{2}/5$
9787 8392 $e^x : 4/3 + 3\sqrt{3}/2$
9787 8890 tanh : $9\sqrt[3]{2}/5$
9788 0759 $1/2 \div \ln(3/5)$
9788 1637 $erf : (\ln 2/3)^{-1/3}$
9788 2518 $\Gamma : 3\sqrt{3}/5$
9788 4544 $J_0 : 7/24$
9788 4544 $s\pi : 3\sqrt{2} - 3\sqrt{5}/4$
9788 5820 $s\pi : 1/\ln(e + \pi)$
9788 6851 $erf : \sqrt[3]{13/3}$
9788 6933 $\Psi : 2\sqrt{5}/9$
9788 8382 $s\pi : \exp(-\sqrt[3]{5}/3)$
9789 0786 sq : $3/4 + 4\sqrt{5}$
9789 1743 rt : (4,9,-6,8)
9789 1944 $erf : 3e/5$
9789 3556 sr : $3\sqrt{5}/7$
9789 4501 sr : $23/24$
9789 5138 cu : $6\sqrt{3}/5$
9789 6352 $2^x : (\pi/3)^{-1/3}$
9789 8301 sinh : $5\ln 2/4$
9789 8559 cr : P_{62}
9789 9231 tanh : $25/11$
9790 1671 $\Gamma : \exp(2/3)$
9790 3870 rt : (1,7,-1,-8)
9790 4890 sinh : $9\sqrt{5}/5$
9790 6478 $s\pi : 7\sqrt{5}/10$
9790 7516 $J : \exp(-11\pi/17)$
9790 7866 $erf : \sqrt{8/3}$
9790 8408 $s\pi : 10/23$
9790 9140 at : $1/3 + 2\sqrt{3}/3$
9790 9478 $e^x : 8\sqrt{3}/5$
9791 4536 $\lambda : \exp(-9\pi/19)$
9791 4919 $\ln(\pi) \div \ln(3/4)$
9791 5113 $2^x : 5\sqrt[3]{2}/4$
9791 6602 ln : $4 - 4e/3$
9791 9538 $J : 4/21$
9792 2383 $\ell_2 : 3 + 2\sqrt{2}/3$
9792 4578 $s\pi : 3\sqrt[3]{5}/2$
9792 4851 $\Psi : 5(1/\pi)/6$
9792 5293 sq : $3/4 - 4\sqrt{5}/3$
9792 6238 $e^x : 3e\pi/5$
9792 7492 $J_0 : \sqrt{3}/6$
9792 7578 $\exp(\sqrt{2}) + s\pi(1/3)$
9792 9121 tanh : $4\sqrt[3]{5}/3$
9793 0708 $J_0 : \sqrt[3]{3}/5$
9793 0867 sinh : $13/15$
9793 1294 $\Psi : (1/(3e))^{1/3}$
9793 2212 $10^x : 2e/7$
9793 3006 $\Gamma : 9\sqrt{3}/8$
9793 4147 $erf : 18/11$
9793 4553 $\Psi : \exp(-5)$
9793 5673 cr : $2 - 3\sqrt{2}/4$
9793 6147 $2^x : 4 + 3\sqrt{5}$
9793 8078 $e^x : (\pi)^{-1/3}$
9793 9498 $\ell_{10} : 2\sqrt{2} + 3\sqrt{5}$
9794 1653 $J_0 : \ln(4/3)$
9794 3519 cr : P_{46}
9794 3620 ln : $\sqrt{2}/4 + 4\sqrt{3}/3$
9794 7160 rt : (1,7,0,-1)
9794 8663 id : $8\sqrt{3}/7$
9795 0141 $\Gamma : (\sqrt{5}/2)^{1/3}$
9795 2542 tanh : $16/7$
9795 3315 $\ell_2 : \sqrt[3]{23/3}$
9795 3768 cu : $3\sqrt{2} + \sqrt{3}/3$
9795 5194 $K : 19/24$
9795 8259 $\Psi : \sqrt[3]{1/2}$
9795 8746 $J_0 : 9(1/\pi)/10$
9795 9183 sq : $4\sqrt{3}/7$
9796 0736 $\ln(\sqrt{3}) - s\pi(\pi)$
9796 3566 $Ei : 1/\ln(\sqrt{2}\pi)$
9796 5828 id : $9e\pi/7$
9796 7541 tanh : $\sqrt[3]{12}$
9796 9572 $J_0 : 2/7$
9796 9638 $2^x : 19/6$
9797 2372 $\Psi : \pi/9$
9797 3593 $s\pi : (\pi)^{-1/2}$
9797 3896 $\lambda : 5/22$
9797 7523 $\lambda : 5(1/\pi)/7$
9797 7994 id : $4/3 - \sqrt{2}/4$
9797 9589 sr : $24/25$
9797 9693 $\lambda : \exp(-\sqrt{2}\pi/3)$
9797 9716 $J_0 : \sqrt[3]{5}/6$
9798 0326 id : $4\sqrt{2} - 3\sqrt{5}/4$
9798 1877 $\ell_{10} : (\sqrt{2}/3)^3$
9798 2675 $s\pi : 4\sqrt{3}/3 - \sqrt{5}/3$
9798 3340 $e^x : 1/4 + \sqrt{3}/4$
9798 3925 $s\pi : P_{29}$
9798 3968 $erf : 23/14$
9798 4228 at : $(\sqrt{2}\pi/2)^{1/2}$
9798 8689 rt : (1,7,1,6)
9798 9898 id : $7\sqrt{2}/5$
9799 1303 sq : $7\sqrt[3]{5}/6$
9799 2240 $erf : (\sqrt{2}\pi)^{1/3}$
9799 2357 at : $2\sqrt{5}/3$
9799 2745 cr : $3/2 - \sqrt{5}/4$
9799 3602 $\Gamma : 8\sqrt[3]{2}$
9799 3859 id : $7\sqrt[3]{2}/9$
9799 5288 $Ei : \sqrt{3} + \sqrt{5} + \sqrt{7}$
9799 8288 $J_0 : \exp(-\sqrt[3]{2})$
9799 8299 $e^x : 1/3 - \sqrt{2}/4$
9799 8587 ln : $3 + 3\sqrt{2}$
9799 8951 $\Gamma : \zeta(5)$
9799 9461 cr : $16/17$
9799 9681 $erf : \pi^2/6$
9800 0000 sq : $7\sqrt{2}/10$
9800 1492 $\lambda : (2\pi/3)^{-2}$
9800 2489 ln : P_4
9800 2680 $\lambda : \exp(-8\pi/17)$
9800 4307 $\Gamma : 1/\ln(\sqrt[3]{5}/3)$
9800 8433 $\Gamma : (\sqrt[3]{5}/3)^3$
9800 8585 $J_0 : 8(1/\pi)/9$
9800 9026 rt : (8,9,-9,9)
9800 9639 tanh : $23/10$
9800 9977 $J_0 : \sqrt{2}/5$
9801 0709 at : $\ln(\sqrt{2}\pi)$
9801 7703 rt : (4,8,-9,8)
9801 8189 $\zeta : 9\ln 2/8$
9802 0715 $s\pi : 2e$
9802 2115 $\theta_3 : \pi/7$
9802 4613 id : $9\sqrt[3]{3}$
9802 4808 $erf : 8\sqrt[3]{3}/7$
9802 4947 sq : $3\sqrt{2}/2 + 3\sqrt{5}/2$
9802 5814 $\sqrt{2} \times \ln(2)$
9802 6727 sinh : $2e/3$
9802 6843 $\Psi : -\sqrt{3} + \sqrt{5} + \sqrt{7}$
9802 8057 $erf : (e)^{1/2}$
9802 8117 $\Psi : 5\sqrt[3]{2}/2$
9803 4782 $s\pi : 1/\sqrt[3]{12}$
9803 6934 $erf : 7\sqrt{2}/6$
9803 7647 sr : $9\pi^2/10$
9803 9367 tanh : $8\sqrt[3]{3}/5$
9804 1145 $Ei : 3\sqrt{5}/10$
9804 4039 $\zeta : e\pi/9$
9804 4606 cr : $3\pi/10$
9804 4692 $erf : \sqrt[3]{9/2}$
9804 5327 sr : $3 - 3e/4$
9804 6350 tanh : $4\sqrt{3}/3$
9804 6902 $\Gamma : \sqrt{5}/5$
9804 7807 $K : 8\ln 2/7$
9804 9300 $\exp(3) \times s\pi(\sqrt{3})$
9804 9583 $J_0 : 7/25$
9804 9826 $J_0 : 2\sqrt[3]{2}/9$

9805 0561 tanh : $10\ln 2/3$
9805 1635 sq : $10\ln 2/7$
9805 5013 cr : $3\sqrt{2}/4+3\sqrt{5}$
9805 5318 10^x : $((e+\pi)/3)^{1/2}$
9805 5932 erf : $(\ln 3/3)^{-1/2}$
9805 5953 ℓ_2 : $4\sqrt{2}+\sqrt{5}$
9805 6091 sr : $2\sqrt[3]{3}/3$
9805 6391 J_0 : $\sqrt{5}/8$
9805 6394 sinh : $\ln(\sqrt[3]{2}/3)$
9805 7026 id : $4e/3+3\pi/4$
9805 7310 λ : $1/\sqrt{19}$
9805 7489 2^x : $2/3+3\sqrt{5}$
9806 3526 sr : $4\zeta(3)/5$
9806 5118 cu : $3+e$
9806 6991 10^x : $1+4\sqrt{2}$
9806 7583 ℓ_2 : $2\sqrt{3}-2\sqrt{5}/3$
9806 8152 $s\pi$: $4\sqrt{2}/3+3\sqrt{5}/4$
9807 0031 J_0 : $7(1/\pi)/8$
9807 1065 2^x : $1/\ln(\sqrt{5}/2)$
9807 5512 Ei : $(\sqrt{2\pi}/2)^{-1/2}$
9807 6211 cu : $2+\sqrt{3}$
9807 6348 2^x : $2e/3+3\pi/4$
9807 7991 $s\pi$: $4\sqrt{3}-2\sqrt{5}/3$
9807 8528 $s\pi$: $7/16$
9807 8732 cu : $3\sqrt{2}/2+4\sqrt{3}/3$
9808 0270 J_0 : $5/18$
9808 0545 ζ : $(\ln 3)^{1/3}$
9808 2377 at : $1/\ln((e+\pi)/3)$
9808 2925 ln : $3/8$
9808 4121 at : $3\sqrt{2}/4+\sqrt{3}/4$
9808 4799 Li_2 : $5\zeta(3)/8$
9808 6148 J_0 : $1/\sqrt{13}$
9808 7401 J_0 : $2\ln 2/5$
9808 8104 at : $\sqrt[3]{10/3}$
9808 8325 Γ : $3e\pi/7$
9808 9944 tanh : $\sqrt[3]{25/2}$
9809 0017 sr : $\sqrt{8\pi}$
9809 0561 $s\pi$: $3\sqrt{2}/2-\sqrt{5}/4$
9809 4365 sr : $5\sqrt{3}/9$
9809 6029 e^x : $\sqrt{3}/3+3\sqrt{5}/2$
9809 6769 sr : $2\sqrt{2}/3+4\sqrt{5}/3$
9809 7594 e^x : $e+2\pi$
9809 7610 $s\pi$: $3/4+2e/3$
9809 9907 $\ln(3/5)+\exp(2/5)$
9810 2298 Γ : $8\pi/5$
9810 6416 ℓ_2 : $\pi^2/5$
9810 7170 10^x : $3/5$
9810 8369 λ : $3/13$
9811 0609 λ : $\exp(-7\pi/15)$
9811 0923 $s\pi$: $3e+2\pi$
9811 1166 sr : $4\sqrt{2}-\sqrt{3}$
9811 2755 cr : $17/18$
9811 3896 $s\pi$: $3e\pi/10$
9811 4292 Γ : $(e+\pi)/3$
9811 6023 ζ : $7\sqrt{5}/9$
9811 6695 tanh : $\sqrt{2}-\sqrt{3}+\sqrt{7}$
9811 8772 λ : $\ln 2/3$
9812 0086 id : $\sqrt{2}-\sqrt{3}/4$
9812 0642 Γ : $17/5$
9812 1417 λ : $(\sqrt[3]{3}/3)^2$
9812 4298 tanh : $6e/7$
9812 6442 2^x : $10\sqrt[3]{5}/3$
9812 8069 at : $4/3-2\sqrt{2}$
9812 9939 cr : $3\sqrt[3]{2}/4$
9813 0561 tanh : $(2e)^{1/2}$
9813 4640 rt : (2,9,-6,-6)
9813 5950 at : $(\sqrt{5})^{1/2}$
9813 6172 at : $3/4+\sqrt{5}/3$
9813 6808 tanh : $7/3$
9813 7796 Ei : $8\sqrt[3]{5}$
9814 2396 id : $4\sqrt{5}/3$
9814 3261 10^x : $(\sqrt{2\pi})^{-1/2}$
9814 6329 $s\pi$: $\exp(-\sqrt{3}/3)$
9814 7639 J_0 : $6(1/\pi)/7$
9814 7965 ln : $\sqrt[3]{19}$
9814 9122 J_0 : $3/11$
9815 0155 θ_3 : $(2e/3)^{-1/3}$
9815 1023 sq : $1/2-2\sqrt{5}/3$
9815 1530 Γ : $8\sqrt[3]{5}/7$
9815 3875 cu : $4\sqrt{5}/9$
9815 5230 Ei : $9e\pi/10$
9815 6798 $s\pi$: $1/4+4\pi/3$
9815 6972 $s\pi$: $3\sqrt{2}+3\sqrt{3}$
9815 7652 sr : $(\sqrt{5}/2)^{-1/3}$
9815 7787 erf : $5/3$
9816 1249 J_0 : $e/10$
9816 3137 at : $7\sqrt[3]{5}/8$
9816 3988 2^x : $\sqrt{3}-\sqrt{5}/3$
9816 6364 sr : $5\pi/4$
9816 9040 rt : (2,9,-5,-1)
9816 9818 Ei : $(\ln 3)^{1/3}$
9817 1294 Γ : $3(1/\pi)/10$
9817 1420 as : P_{25}
9817 2723 rt : (1,9,7,-9)
9817 2775 sq : $e/4+\pi/3$
9817 2870 Ei : $8\pi/5$
9817 3409 ln : $\sqrt{3}/4+\sqrt{5}$
9817 5292 erf : $1/\ln(\ln 3/2)$
9817 5595 tanh : $2(e+\pi)/5$
9817 6638 cr : $3/2+2\pi$
9817 6998 e^x : $2\sqrt[3]{5}/5$
9817 8299 Γ : $(e/3)^{-1/3}$
9817 8790 $s\pi$: $\sqrt{2}/2+\sqrt{3}$
9817 9108 2^x : $9(e+\pi)/4$
9818 0138 tanh : $\sqrt{11/2}$
9818 1484 10^x : $3\ln 2/7$
9818 4561 cu : $2\sqrt{2}/3+3\sqrt{5}/4$
9818 5501 Γ : 3π
9818 7645 Ei : $7\sqrt[3]{5}/2$
9818 8657 erf : $\sqrt[3]{14/3}$
9818 9651 $s\pi$: $1/2+3\sqrt{2}/4$
9819 0280 $\sqrt{5}-s\pi(\sqrt{3})$
9819 0340 Ei : $(\sqrt{2}/3)^{-3}$
9819 0934 cu : $8e/7$
9819 1121 cu : $4/3+\sqrt{5}/3$
9819 2019 rt : (1,9,-9,8)
9819 7786 e^x : $4e/3+3\pi$
9819 9047 $s\pi$: $4(e+\pi)$
9820 0477 2^x : $\pi^2/10$
9820 0479 $s\pi$: $2\sqrt{2}+\sqrt{3}$
9820 2102 tanh : $\sqrt[3]{13}$
9820 2143 tanh : $\exp(\sqrt[3]{5}/2)$
9820 2185 rt : (2,9,-4,4)
9820 5080 id : $1/4+\sqrt{3}$
9820 5460 $1/3+\exp(1/2)$
9820 5672 rt : (1,9,8,-1)
9821 0322 erf : $2(e+\pi)/7$
9821 3804 Ei : $4(e+\pi)/5$
9821 3902 cr : $18/19$
9821 8412 ℓ_{10} : P_{51}
9821 9338 tanh : $3\pi/4$
9821 9856 $\ln(3)\times\exp(4)$
9822 0631 e^x : $13/19$
9822 2172 cu : $e\pi/5$
9822 2241 J_0 : $1/\sqrt{14}$
9822 2258 tanh : $5\sqrt{2}/3$
9822 3326 θ_3 : $(e/3)^{1/2}$
9822 3409 $s\pi$: $8\sqrt[3]{2}/7$
9822 4062 J_0 : $2\zeta(3)/9$
9822 4252 $s\pi$: $2e\pi/7$
9822 5654 $t\pi$: $(e)^{1/2}$
9822 5922 erf : $(3\pi/2)^{1/3}$
9822 6935 sr : P_{22}
9822 7059 2^x : $4/3-e/2$
9822 8725 $s\pi$: $11/25$
9822 9393 erf : $3\sqrt{5}/4$
9823 0052 as : $6\ln 2/5$
9823 0107 J_0 : $4/15$
9823 0648 θ_3 : $(\ln 2/2)^{1/3}$
9823 0790 $s\pi$: $4\sqrt[3]{2}/9$
9823 2498 $s\pi$: $8\sqrt[3]{5}/3$
9823 3712 e^x : $2+4e$
9823 4145 rt : (2,9,-3,9)
9823 4546 10^x : $3/2+3\pi/2$
9823 7448 rt : (1,9,9,7)
9823 9613 10^x : $\exp(-\sqrt{5}/3)$
9824 0065 10^x : $1+2\sqrt{3}$
9824 0478 Γ : $7\sqrt{5}/8$
9824 3689 $s\pi$: $\sqrt{3}+3\sqrt{5}$
9824 5577 λ : $\exp(-6\pi/13)$
9824 8572 erf : $4\sqrt[3]{2}/3$
9824 8672 J_0 : $5(1/\pi)/6$
9825 3296 $s\pi$: $9\ln 2/4$
9825 4351 θ_3 : $\beta(2)$
9825 5035 θ_3 : $1/\sqrt[3]{11}$
9825 6501 $s\pi$: $6\sqrt[3]{2}$
9825 7853 cr : $2e+3\pi/4$
9825 9145 ln : $1/3+4\sqrt{3}$
9825 9281 Γ : $(\ln 3)^{1/3}$
9825 9477 e^x : $4\sqrt{5}/5$
9825 9687 ℓ_{10} : $9e\pi/8$
9826 0695 ln : $2\pi^2$
9826 0798 Γ : $7e\pi/9$
9826 1683 sinh : $7\pi/3$
9826 2566 erf : $1/\ln(2e/3)$
9826 3808 $s\pi$: $1/4+4\sqrt{3}/3$
9826 5397 cr : $e\pi/9$
9826 6582 2^x : $8(e+\pi)/9$
9826 6810 erf : $7\sqrt[3]{3}/6$
9826 7774 as : $\exp(-1/(2e))$
9826 8507 erf : $7\zeta(3)/5$
9826 8683 tanh : $(4/3)^3$
9827 0322 cr : $9\sqrt{3}/2$
9827 0328 λ : $4/17$
9827 0442 J_0 : $\exp(-4/3)$
9827 3032 $\sqrt{3}\times\ln(\pi)$
9827 4309 cr : $(\sqrt[3]{5}/2)^{1/3}$
9827 4384 Γ : $\sqrt[3]{15/2}$
9827 6177 J_0 : $5/19$
9827 6580 ℓ_{10} : $2/3+4\sqrt{5}$
9827 7823 sr : $4/3+3\sqrt{3}/2$
9827 8900 sr : $\sqrt{3}/3+3\sqrt{5}/2$
9827 9372 at : $3/2$
9828 1425 Γ : $1/\ln(5/3)$
9828 4297 λ : $\sqrt{2}/6$
9828 4502 tanh : $19/8$
9828 5263 10^x : $8\pi^2/9$
9828 5788 $\exp(5)+s\pi(\pi)$
9828 6123 $s\pi$: $\sqrt{5}/4$
9828 8595 $\ln(\sqrt{5})\times\exp(1/5)$
9828 9958 J_0 : $((e+\pi)/3)^{-2}$
9829 2694 10^x : $\sqrt{2}/4+4\sqrt{5}$
9829 2775 2^x : $7e/5$
9829 2791 tanh : $\exp(\sqrt{3}/2)$
9829 6028 $s\pi$: $9\sqrt{3}/10$
9829 6356 tanh : $7e/8$
9829 7358 sr : P_{50}
9829 7466 e^x : $\sqrt{3}/4+2\sqrt{5}$
9830 0154 $s\pi$: $\sqrt{3}/3+4\sqrt{5}/3$
9830 1501 ℓ_{10} : $8\zeta(3)$
9830 2502 ln : P_{74}
9830 2542 $\exp(3/4)+s\pi(1/3)$
9830 2764 id : $3e/2-2\pi/3$
9830 3030 tanh : $\sqrt{17/3}$
9830 4296 cr : $5\sqrt[3]{5}/9$
9830 4757 cr : $19/20$
9830 5899 J_0 : $6/23$
9830 7786 λ : $\exp(-\sqrt[3]{3})$
9830 9965 sq : $3e/4-\pi/3$
9831 0079 $s\pi$: $\sqrt{3\pi}$
9831 2341 $t\pi$: $\sqrt[3]{13}$
9831 2766 id : $2\sqrt{2}+2\sqrt{3}/3$
9831 5468 Ψ : $8\ln 2/7$
9831 6824 sr : $3\sqrt{2}/2-2\sqrt{3}/3$
9831 8026 J_0 : $3\ln 2/8$
9832 0732 cu : $\sqrt{2}/2+3\sqrt{5}/2$
9832 8730 e^x : $3/2+\sqrt{3}/3$
9833 0030 $t\pi$: $\exp(\sqrt[3]{5}/2)$
9833 0171 erf : $22/13$
9833 2457 K : $\sqrt[3]{1/2}$
9833 2893 10^x : $\sqrt[3]{7/2}$
9833 3931 $4\times s\pi(\sqrt{3})$
9833 5510 at : $2\sqrt{2}/3+\sqrt{5}/4$
9833 5567 erf : $\ln(2e)$
9833 6977 sinh : $20/23$
9833 7604 Γ : $6\zeta(3)/7$
9833 9750 cu : $3\sqrt{2}-\sqrt{5}/4$
9833 9767 $\exp(1/3)+s\pi(1/5)$
9834 0264 J_0 : $1/\sqrt{15}$
9834 1670 ℓ_{10} : $2e+4\pi/3$

9834 2653 $J_0 : (\pi/2)^{-3}$
9834 3864 $2^x : 4 + e/3$
9834 4789 $10^x : 3\sqrt{3}/4 + 3\sqrt{5}$
9834 5724 $\Gamma : 5\sqrt[3]{3}/7$
9834 7107 sq : 19/11
9834 7830 tanh : $7\sqrt[3]{5}/5$
9835 0861 rt : (2,8,-9,4)
9835 3152 $\theta_3 : 9/20$
9835 3586 sq : $(\ln 3/2)^{-2}$
9835 7090 $J_0 : \exp(-e/2)$
9835 8390 at : $5\zeta(3)/4$
9835 8697 $s\pi : \sqrt[3]{3}$
9835 9099 $\ln(2/3) + \exp(2)$
9836 0492 $erf : 6\sqrt{2}/5$
9836 2678 $s\pi : 7(e + \pi)/9$
9836 3055 $erf : \exp(\sqrt[3]{4}/3)$
9836 4159 $\lambda : 5/21$
9836 4535 $t\pi : 10(1/\pi)/7$
9836 4939 $2 \times \exp(2/5)$
9836 5800 sq : $2/3 + 3\sqrt{2}/4$
9836 7485 tanh : 12/5
9836 9894 sr : $(e/3)^{1/3}$
9837 0351 as : $(\ln 2)^{1/2}$
9837 0407 $s\pi : 3e - 3\pi/2$
9837 0520 $erf : (\ln 2/2)^{-1/2}$
9837 1072 $s\pi : 6\zeta(3)/5$
9837 2299 $erf : 5e/8$
9837 7772 rt : (2,8,-8,9)
9837 8409 $J : \exp(-3\pi/7)$
9837 9045 $erf : 17/10$
9837 9582 tanh : $5\sqrt[3]{3}/3$
9838 0754 tanh : $2\zeta(3)$
9838 1101 sr : $\sqrt{2}/3 + 2\sqrt{3}$
9838 3150 $\exp(2/3) + s\pi(\sqrt{2})$
9838 3852 $s\pi : 1/\ln(2)$
9838 4846 $\lambda : 3(1/\pi)/4$
9838 5419 $J_0 : 4(1/\pi)/5$
9838 6676 sq : $4\sqrt{3} + \sqrt{5}$
9838 6814 cr : 20/21
9838 9698 ln : $2\sqrt{2}/3 + \sqrt{3}$
9839 0259 $s\pi : 1/2 + 2\sqrt{2}/3$
9839 1030 $s\pi : 10\sqrt[3]{5}/7$
9839 1234 $\sqrt{2} + s\pi(\pi)$
9839 1990 $10^x : -\sqrt{2} + \sqrt{5} + \sqrt{7}$
9839 3564 as : $(\sqrt[3]{3})^{-1/2}$
9839 4405 $s\pi : \sqrt{2}\pi$
9839 5950 $\exp(4/5) \div s\pi(\sqrt{3})$
9839 8594 $s\pi : 7(1/\pi)/4$
9840 0004 tanh : $\sqrt[3]{14}$
9840 1326 cr : $2 - \pi/3$
9840 1530 $\exp(1/4) \div s\pi(\pi)$
9840 1724 $t\pi : \sqrt{3}/7$
9840 2750 sr : $4\sqrt{3} + 4\sqrt{5}$
9840 4516 sq : $1/3 + 4\sqrt{3}/3$
9840 5939 id : $\sqrt{3}/2 + \sqrt{5}/2$
9840 6253 rt : (1,8,-1,-9)
9840 9877 sr : $3/4 + 3e$
9841 1021 ln : $4\sqrt{2} - 4\sqrt{5}/3$
9841 1508 $e^x : 3\sqrt{2} - 3\sqrt{3}/4$
9841 2291 sq : $\sqrt{3}/4 + \sqrt{5}/4$
9841 4470 cr : $(\sqrt{3}/2)^{1/3}$
9841 5205 tanh : $(\sqrt{5}/3)^{-3}$
9841 8621 $\lambda : \exp(-5\pi/11)$
9841 8633 $2^x : 3e/2 + \pi$
9841 8887 $J_0 : \sqrt[3]{2}/5$
9841 9278 tanh : $8e/9$
9842 1249 $2^x : 1 + \sqrt{3}/3$
9842 1915 $s\pi : 5\sqrt{3}/6$
9842 5295 $\lambda : 6/25$
9842 6652 $s\pi : 7\pi/9$
9842 8213 $erf : e\pi/5$
9843 0769 cr : $4(e + \pi)/3$
9843 1354 rt : (1,8,0,-1)
9843 2617 $\Psi : 19/24$
9843 3228 tanh : $(e + \pi)^{1/2}$
9843 4684 sinh : 23/11
9843 7082 $\lambda : \sqrt[3]{3}/6$
9843 7423 cu : $\sqrt{2}/3 + 3\sqrt{5}/2$
9843 7500 sinh : $5 \ln 2$
9843 8224 $\lambda : \zeta(3)/5$
9844 0170 cu : $2\pi/5$
9844 0383 $e^x : 3/4 + 4\sqrt{2}$
9844 0554 $erf : \sqrt[3]{5}$
9844 3592 $J_0 : 1/4$
9844 4070 $s\pi : (2e)^2$
9844 4439 $\ln(4) + \exp(4)$
9844 4756 sr : $(\ln 3)^{-1/3}$
9844 6088 tanh : $7\sqrt{3}/5$
9844 9612 tanh : $7 \ln 2/2$
9845 1010 $erf : ((e + \pi)/2)^{1/2}$
9845 2080 cu : $3\sqrt{2} + 4\sqrt{3}$
9845 2231 $s\pi : 4e + \pi/2$
9845 3596 sinh : $4\sqrt{3}/3$
9845 5673 rt : (1,8,1,7)
9845 6008 $e^x : 4\sqrt{3} + 2\sqrt{5}/3$
9845 7457 tanh : 17/7
9845 8935 $\ell_{10} : \sqrt{2}/2 + 4\sqrt{5}$
9846 1293 cr : 21/22
9846 2712 $\Psi : \ln 2/8$
9846 2733 $J_0 : \sqrt{5}/9$
9846 5484 $Ei : 5\sqrt{2}$
9846 6486 $erf : 12/7$
9846 6557 $\ell_2 : 1/3 + 4e/3$
9846 8114 $\ln(\sqrt{5}) \div \ln(2/3)$
9846 8256 $10^x : 10\pi/9$
9846 8919 tanh : $\sqrt{5} - \sqrt{6} + \sqrt{7}$
9847 1209 $s\pi : 3\sqrt{2}/2 - 3\sqrt{5}/4$
9847 1348 $s\pi : 1/2 + 4\sqrt{5}$
9847 1580 ln : $1 + 3\sqrt{5}/4$
9847 2059 cu : $2\sqrt{3}/3 + \sqrt{5}$
9847 3533 $J_0 : 7(1/\pi)/9$
9847 4502 id : $(\pi/3)^{-1/3}$
9847 4532 $\ell_{10} : 3/2 + 3e$
9847 5234 $J_0 : \sqrt{3}/7$
9847 5358 sq : $1/2 + 4\pi/3$
9847 5382 sinh : $3e\pi/2$
9848 0775 $s\pi : 4/9$
9848 3463 $s\pi : 8\sqrt{5}/7$
9848 3567 $\ell_{10} : 4 + 4\sqrt{2}$
9848 3948 $erf : 10\zeta(3)/7$
9848 5273 sq : $e - 4\pi$
9848 6219 $\zeta : (\ln 3/3)^{1/2}$
9848 7420 $t\pi : 7(1/\pi)/9$
9849 0382 $\Psi : ((e + \pi)/2)^{-2}$
9849 1157 $s\pi : 4\sqrt{2}/3 + \sqrt{5}/4$
9849 1581 id : $7\sqrt[3]{5}/2$
9849 1825 tanh : $2e\pi/7$
9849 2505 $\Psi : \sqrt{10}$
9849 2939 $s\pi : \exp(1/e)$
9849 9975 $e^x : e\pi$
9850 0478 tanh : $10\sqrt[3]{5}/7$
9850 2377 tanh : $7\pi/9$
9850 3460 $\lambda : 1/\sqrt{17}$
9850 5297 tanh : 22/9
9850 5419 sinh : $5\pi/2$
9850 8748 cu : $3/4 + 3\pi/2$
9850 9028 sinh : $(\pi/3)^{-3}$
9850 9190 sr : $7 \ln 2/5$
9851 1078 as : 5/6
9851 1140 $\Gamma : 5\pi/8$
9851 1247 tanh : $9e/10$
9851 2632 $\ell_2 : 8\sqrt{3}/7$
9851 3544 sq : $9e/10$
9851 3709 $J_0 : \sqrt[3]{5}/7$
9851 5449 $e^x : 1/2 + e$
9852 0193 tanh : $\sqrt{6}$
9852 0869 $\lambda : \exp(-\sqrt{2})$
9852 4466 $\lambda : \exp(-9\pi/20)$
9852 7289 sinh : $8(e + \pi)/7$
9852 7805 $J_0 : \exp(-\sqrt{2})$
9852 8137 sq : $2 + 3\sqrt{2}/2$
9852 8529 $s\pi : (e + \pi)^{-1/3}$
9852 9198 cr : 22/23
9852 9430 $Ei : \sqrt{2} + \sqrt{5} + \sqrt{6}$
9853 2375 at : $\sqrt{2}/4 + 2\sqrt{3}/3$
9853 4809 $J_0 : 1/\sqrt{17}$
9853 4959 $\ell_{10} : 3\sqrt{3} + 2\sqrt{5}$
9853 6867 $s\pi : 4 \ln 2/5$
9853 7212 ln : $\sqrt{2}/4 + 4\sqrt{3}$
9853 9186 $s\pi : 2\sqrt{3} + 4\sqrt{5}/3$
9853 9418 id : $2\sqrt{2}/3 - 4\sqrt{3}$
9854 1370 cr : $7\sqrt{5}/2$
9854 2000 $\ell_2 : P_{35}$
9854 2274 cu : $8\sqrt[3]{2}/7$
9854 2346 $erf : 19/11$
9854 2682 $\ell_2 : 7\sqrt{2}/5$
9854 3487 $t\pi : 9 \ln 2/5$
9854 4972 $s\pi : 7(1/\pi)/5$
9854 6740 $s\pi : 3\sqrt{2}/2 + \sqrt{3}/4$
9854 9281 cr : $1/4 + \sqrt{2}/2$
9855 1886 $s\pi : 1/\ln(3\pi)$
9855 3484 $t\pi : 6e\pi/5$
9855 5235 $\lambda : \sqrt[3]{5}/7$
9855 6700 $1/3 + \exp(\sqrt{3})$
9855 6830 ln : $1 + 2\pi$
9855 9855 sr : $1/2 + \sqrt{2}/3$
9856 0270 $J_0 : \zeta(3)/5$
9856 0705 $J_0 : \sqrt[3]{3}/6$
9856 1802 $erf : 6\sqrt[3]{3}/5$
9856 3936 at : $3 - 2\sqrt{5}/3$
9856 4100 $\ell_{10} : 1/4 + 3\pi$
9856 5078 sr : $3 + 2\sqrt{2}/3$
9856 5175 $J_0 : 6/25$
9856 5309 $\Gamma : 3\sqrt[3]{5}/5$
9856 5650 ln : $2 + e/4$
9856 6161 $\theta_3 : 3\zeta(3)/8$
9856 8518 tanh : $\sqrt[3]{15}$
9856 8778 tanh : $1/\ln(2/3)$
9856 9412 $erf : \sqrt{3}$
9856 9482 $s\pi : -\sqrt{2} + \sqrt{3} + \sqrt{5}$
9857 1894 tanh : $\pi^2/4$
9857 2532 $s\pi : 1/4 + 3\sqrt{3}$
9857 3996 $erf : 5 \ln 2/2$
9857 6022 $J : \exp(-\sqrt{3}/2)$
9858 0238 $J_0 : 3(1/\pi)/4$
9858 1271 $s\pi : 3\sqrt{3}/4 - \sqrt{5}/3$
9858 3164 $2^x : 4\sqrt{3}/7$
9858 5537 $e^x : 2 + \sqrt{2}/2$
9858 6771 rt : (1,7,-9,-9)
9858 7779 $J_0 : 5/21$
9858 8323 ln : $3 - 4\sqrt{5}/3$
9858 8421 $s\pi : 9e/10$
9858 9351 ln : $\sqrt{3}/3 + 3\sqrt{5}$
9859 0423 at : $5e/9$
9859 0727 cr : $3\sqrt{5}/7$
9859 1361 cr : 23/24
9859 1447 $e^x : 2 - e/3$
9859 1488 tanh : $10\sqrt{3}/7$
9859 1900 $2^x : e + 3\pi$
9859 2269 tanh : $\exp(e/3)$
9859 2929 tanh : $7\sqrt{2}/4$
9859 4265 ln : $3\sqrt{2}/2 + \sqrt{5}/4$
9859 4352 $\Psi : 3/17$
9859 7120 $Ei : (e/3)^{-1/3}$
9860 1145 $s\pi : 4e/7$
9860 5431 $e^x : \sqrt{2}/2 + 4\sqrt{5}/3$
9860 6530 rt : (1,7,-8,-1)
9860 6766 $\Psi : 2(1/\pi)/3$
9860 6797 id : $1/4 - \sqrt{5}$
9860 7802 $J_0 : \exp(-\sqrt[3]{3})$
9860 8228 $2^x : 1/3 - \sqrt{2}/4$
9860 8351 $2^x : 7\sqrt[3]{5}/6$
9860 8884 $erf : 7\sqrt{5}/9$
9861 0574 $10^x : 2\sqrt{3} + 4\sqrt{5}/3$
9861 0713 $t\pi : 2e - 4\pi/3$
9861 1545 $2^x : 7\sqrt{2}/10$
9861 3257 cu : $1/4 + \sqrt{5}/3$
9861 3369 sq : $3 - 4e$
9861 5120 $\Gamma : 7\sqrt[3]{5}/4$
9861 5926 $J_0 : \sqrt{2}/6$
9861 9058 $\Gamma : \zeta(3)/4$
9861 9386 $Li_2 : (\ln 3)^{-3}$
9861 9629 tanh : $10\sqrt{5}/9$
9862 0165 $10^x : 2\sqrt{3} + \sqrt{5}$
9862 0698 $J_0 : 4/17$
9862 1597 $\sqrt{3} + s\pi(\sqrt{3})$
9862 3102 $\sqrt{3} - e$
9862 5742 rt : (1,7,-7,7)
9862 7537 $s\pi : (2\pi/3)^{1/2}$
9862 8112 $s\pi : \sqrt{5}/5$

9862 9436 $2/5 - \ln(4)$
9863 1821 sinh : $23/16$
9863 3782 $e \times \ln(3)$
9864 1543 $10^x : 1/3 + \sqrt{5}/3$
9864 2314 $erf : 5\pi/9$
9864 3039 $\exp(\pi) - \exp(e)$
9864 3067 cu : $1/\ln(\ln 3/3)$
9864 3684 at : $6\sqrt[3]{2}/5$
9864 4616 $\lambda : \sqrt{3}/7$
9864 4702 $\Gamma : 1/\ln(2/3)$
9864 5205 id : $4\sqrt{3}/3 + 3\sqrt{5}/4$
9864 6435 cu : $6e\pi$
9864 6934 $\lambda : \exp(-4\pi/9)$
9864 7447 $2^x : 10\ln 2/7$
9864 7595 cu : $3/2 + 3e$
9864 8135 at : $3\sqrt{3} - 3\sqrt{5}$
9864 8424 $\lambda : 7(1/\pi)/9$
9864 8482 cr : $24/25$
9864 9477 cu : $8\sqrt{2}/9$
9865 1519 $\ell_{10} : 3/4 + 4\sqrt{5}$
9865 1903 $10^x : 3/2 + \sqrt{5}/2$
9865 2109 $erf : \sqrt[3]{16/3}$
9865 2138 $2/5 \div \ln(2/3)$
9865 4005 $s\pi : 2/3 + 4\sqrt{2}/3$
9865 5381 $\Psi : 19/6$
9865 7018 sr : $2\sqrt{2} + \sqrt{5}/2$
9865 7036 cu : $2\sqrt{3}/3 + 4\sqrt{5}$
9866 1429 tanh : $5/2$
9866 5026 sr : $4\sqrt{3} - 4\sqrt{5}/3$
9866 5463 sinh : $3e\pi/5$
9866 6459 $s\pi : 3\sqrt{2} + 4\sqrt{3}/3$
9866 7167 $erf : 7/4$
9866 7290 $s\pi(2/5) \div s\pi(\sqrt{2})$
9866 9035 $J_0 : (\sqrt[3]{3}/3)^2$
9866 9481 id : $\sqrt{3} - \sqrt{5}/3$
9866 9854 $J_0 : \ln 2/3$
9866 9963 $10^x : 4\sqrt{2} - 2\sqrt{3}/3$
9867 2329 $\lambda : \sqrt{5}/9$
9867 3063 $J_0 : 3/13$
9867 8941 tanh : $(2\pi)^{1/2}$
9868 3298 sq : $3\sqrt{2}/2 + \sqrt{5}$
9868 4700 $\exp(1/3) \times s\pi(1/4)$
9868 6384 $s\pi : 4e - 3\pi$
9868 8532 $J_0 : 1/\sqrt{19}$
9868 9564 tanh : $(e/2)^3$
9869 1340 tanh : $3(e+\pi)/7$
9869 1677 sq : $2/3 + 4\sqrt{2}$
9869 1765 sr : $2\pi^2/5$
9869 2463 $\ln(3) - \exp(3)$
9869 2602 cr : $3 - 3e/4$
9869 2665 $e^x : (\sqrt{2}/3)^{1/2}$
9869 3744 $2/3 \div s\pi(\sqrt{5})$
9869 5424 $\Gamma : 1/\ln(3\pi)$
9869 6044 id : $\pi^2/10$
9869 6270 tanh : $4\pi/5$
9869 9394 $\ell_2 : 1 + 4\sqrt{3}$
9869 9825 cr : $2\sqrt[3]{3}/3$
9870 0481 $s\pi : 3\sqrt{2}/4 + 2\sqrt{5}/3$
9870 0746 cu : $4\sqrt{2}/3 + 2\sqrt{5}$
9870 2220 tanh : $9\sqrt{5}/8$
9870 2341 $10^x : 4\sqrt{2} + 3\sqrt{3}$
9870 3258 $t\pi : \exp(\sqrt{2}\pi/3)$
9870 4051 ln : $\sqrt{5}/6$
9870 4814 cr : $4\zeta(3)/5$
9870 4887 tanh : $\sqrt{19/3}$
9870 4925 $J_0 : (2\pi/3)^{-2}$
9870 6240 $\Gamma : (\pi/3)^{1/2}$
9870 8607 $erf : 3(e+\pi)/10$
9870 9105 $s\pi : \pi/7$
9871 0654 $erf : (2e)^{1/3}$
9871 1188 $J_0 : \exp(-\sqrt{2}\pi/3)$
9871 1807 $J_0 : 5(1/\pi)/7$
9871 2840 $J_0 : 5/22$
9871 3174 tanh : $2\sqrt[3]{2}$
9871 3560 $\lambda : 1/4$
9871 3889 $\ell_{10} : 3 + 3\sqrt{5}$
9871 5252 cu : $3/4 + 4\sqrt{2}$
9871 6860 $s\pi : \sqrt{3}/4 + \sqrt{5}/2$
9871 7910 $e^x : 2/3 - e/4$
9872 0613 $\exp(2) + \exp(4)$
9872 1462 $s\pi : P_{41}$
9872 1893 $K : 5(1/\pi)/2$
9872 3602 tanh : $7\sqrt[3]{3}/4$
9872 5507 cr : $5\sqrt{3}/9$
9872 7182 $t\pi : \ln(\sqrt{2}/3)$
9873 4482 $10^x : 3e/4 + 4\pi/3$
9873 4574 id : $3\sqrt{2}/2 + \sqrt{3}/2$
9873 6219 $s\pi : 5\pi^2/3$
9873 8711 $erf : 7\sqrt[3]{2}/5$
9874 3630 $s\pi : \sqrt{6}$
9874 5153 $erf : \sqrt[3]{11/2}$
9875 1128 $\zeta : e/8$
9875 1169 ln : P_{24}
9875 1292 $s\pi : 1/\sqrt[3]{11}$
9875 1819 $e^x : 1/3 + \sqrt{2}/4$
9875 2403 $e^x : 4\zeta(3)/7$
9875 2997 $s\pi : 1/\sqrt[3]{6}$
9875 3624 $\Gamma : 7(1/\pi)/5$
9875 3900 $J_0 : \sqrt{5}/10$
9875 4252 at : $7\sqrt{3}/8$
9875 6300 $s\pi : 3\sqrt{3} + 3\sqrt{5}/2$
9875 8066 $erf : 5\sqrt{2}/4$
9875 9190 $J_0 : \exp(-3/2)$
9875 9232 $s\pi : \ln(3\pi/2)$
9875 9856 $erf : \ln(e+\pi)$
9876 0473 $erf : \exp(\sqrt[3]{5}/3)$
9876 0955 cu : $3\sqrt{3}/2 + 3\sqrt{5}$
9876 1170 $2^x : 7\zeta(3)/3$
9876 1332 $2^x : 4\sqrt{3} + 2\sqrt{5}$
9876 1341 $2/3 \div \ln(4/5)$
9876 1597 id : $8\sqrt{5}/9$
9876 2028 rt : (1,9,0,-1)
9876 2661 $J_0 : 7(1/\pi)/10$
9876 4586 $2^x : \sqrt{20/3}$
9876 4711 $\lambda : \sqrt[3]{2}/5$
9876 5305 $erf : 23/13$
9876 5432 sq : $4\sqrt{5}/9$
9876 7966 cr : $(\sqrt{5}/2)^{-1/3}$
9876 8834 $s\pi : 9/20$
9876 9237 $J_0 : 2/9$
9877 0478 $\ell_2 : 3e - 4\pi/3$
9877 0676 $Li_2 : (\sqrt[3]{5}/3)^{1/2}$
9877 4737 $s\pi : 5\sqrt[3]{5}$
9877 5701 cr : $5\pi/2$
9877 6870 $t\pi : 1/\sqrt[3]{23}$
9877 7207 rt : (1,9,1,8)
9877 7354 tanh : $9\sqrt{2}/5$
9877 8525 $2/5 + s\pi(1/5)$
9877 9527 tanh : $8(1/\pi)$
9877 9651 tanh : $1/\ln(\sqrt{2}\pi/3)$
9878 0669 cu : $3 - \sqrt{2}$
9878 1111 $erf : (\pi)^{1/2}$
9878 3216 $\pi \times s\pi(2/5)$
9878 3249 $s\pi : \sqrt{2}/4 + 3\sqrt{3}$
9878 6400 $\Psi : 9\sqrt{3}/2$
9878 6858 tanh : $\sqrt{13/2}$
9878 6877 $\Psi : 5\ln 2/7$
9878 7603 $e^x : 2 + \pi$
9878 9668 $\lambda : \exp(-7\pi/16)$
9879 0007 sq : $3\sqrt{3}/2 + 2\sqrt{5}$
9879 0573 $\theta_3 : \ln(\pi/2)$
9879 1827 $\ell_{10} : 3e + \pi/2$
9879 2810 $s\pi : \sqrt{13/2}$
9879 4063 sq : $\pi^2/7$
9879 4467 sq : $4\sqrt{2} - \sqrt{5}/4$
9879 5147 ln : $2\sqrt{2} + 2\sqrt{5}$
9879 5444 $t\pi : 3\sqrt{2} - 2\sqrt{5}/3$
9879 7412 tanh : $\sqrt{2} - \sqrt{3} - \sqrt{5}$
9879 9667 id : $3\sqrt{2} + \sqrt{5}/3$
9880 0991 sr : $2 + 4\sqrt{3}$
9880 1233 tanh : $8\sqrt{5}/7$
9880 1351 tanh : $23/9$
9880 2699 $s\pi : \ln(\sqrt{3})$
9880 6451 $s\pi : 3\zeta(3)/8$
9880 6737 $10^x : (\sqrt{5}/3)^{-3}$
9880 6828 $erf : 16/9$
9880 7665 $s\pi : 3\sqrt{3} - \sqrt{5}/3$
9881 2370 $erf : 1/\ln(\sqrt[3]{5}/3)$
9881 3062 $J_0 : 1/\sqrt{21}$
9881 3534 $Ei : 5e\pi/7$
9881 4436 cr : P_{22}
9881 5655 $s\pi : 1/4 + 3\sqrt{3}/4$
9881 6418 cu : $3\sqrt{2} - 3\sqrt{5}$
9881 6422 tanh : $3e\pi/10$
9882 0025 $t\pi : (\sqrt{2}\pi/3)^3$
9882 1430 id : $2e - 3\pi$
9882 2010 $J_0 : 5/23$
9882 3563 tanh : $3\sqrt[3]{5}/2$
9882 3634 $erf : \exp(\sqrt{3}/3)$
9882 5388 sr : $(e+\pi)/6$
9882 7018 $2^x : 3e/4 - \pi/3$
9882 8106 $\Psi : 7e/6$
9882 8310 $2^x : 6\sqrt{3}$
9882 9598 sq : $4\sqrt{2} + \sqrt{5}/3$
9883 0480 $\lambda : 4(1/\pi)/5$
9883 0772 $erf : \sqrt[3]{17/3}$
9883 1553 $J_0 : \sqrt{3}/8$
9883 4235 $10^x : (\ln 3/3)^{-1/3}$
9883 5120 $s\pi : 1/3 + \sqrt{5}/2$
9883 6996 $J_0 : (5/3)^{-3}$
9883 7560 at : $\sqrt[3]{7/2}$
9883 7642 $s\pi : P_{12}$
9883 8248 tanh : $\sqrt[3]{17}$
9883 8588 tanh : $18/7$
9884 1409 $\Gamma : \sqrt[3]{23/3}$
9884 3269 $\Gamma : 1/\ln(\sqrt{5}/2)$
9884 4272 $erf : 25/14$
9884 5395 $s\pi : \ln(\pi/2)$
9884 5521 $\ell_2 : \sqrt{3} + \sqrt{5}$
9884 7681 at : $\sqrt{3} + \sqrt{5} - \sqrt{6}$
9884 9841 sr : $4\sqrt[3]{5}/7$
9885 0948 cr : $2\sqrt{3}/3 + 3\sqrt{5}$
9885 3680 sr : $(\pi/3)^{-1/2}$
9885 5331 $J_0 : 3/14$
9885 6277 $s\pi : \sqrt{2} - \sqrt{3}/2$
9885 8544 $2^x : 4\sqrt{2}/3 + \sqrt{3}/4$
9885 8796 $erf : 4\sqrt{5}/5$
9885 9476 $\ell_2 : \sqrt[3]{2}/5$
9885 9686 cu : $5e/3$
9886 1062 $J_0 : \sqrt[3]{5}/8$
9886 1659 cr : P_{50}
9886 2722 tanh : $\sqrt{20/3}$
9886 6860 $J_0 : 1/\sqrt{22}$
9886 8232 $s\pi : 2/3 + \pi/4$
9886 8422 $s\pi : e/2 + 4\pi/3$
9887 1212 $\ln(5) - \exp(4)$
9887 2731 cu : $1/4 + 3e/4$
9887 3109 $s\pi : \exp(-\sqrt[3]{4}/2)$
9887 3746 $e^x : 11/16$
9887 4603 $J : 1/\sqrt[3]{24}$
9887 4711 cr : $3\sqrt{2}/2 - 2\sqrt{3}/3$
9887 5812 $s\pi : 9(e+\pi)/5$
9887 6500 $Ei : -\sqrt{3} + \sqrt{5} + \sqrt{7}$
9887 7373 $J_0 : 2(1/\pi)/3$
9888 2092 cu : $9(e+\pi)/4$
9888 3163 $\lambda : \exp(-e/2)$
9888 4668 $e^x : 3/4 + \pi$
9888 6271 rt : (1,8,-9,-1)
9888 6340 id : $3\sqrt{2}/4 + 4\sqrt{3}$
9888 6910 $s\pi : \sqrt{2}/2 + \sqrt{5}/3$
9888 7635 $erf : 4\pi/7$
9888 8002 $s\pi : ((e+\pi)/3)^3$
9888 8474 at : $8\sqrt[3]{5}/9$
9889 3566 $s\pi : 4\sqrt{2}/3 - \sqrt{3}/4$
9889 3746 $\Psi : (\ln 3/3)^{-3}$
9889 3932 at : $2\sqrt{2}/3 + \sqrt{3}/3$
9889 4083 tanh : $9\sqrt[3]{3}/5$
9889 5032 $J_0 : 4/19$
9889 8127 $2^x : \sqrt{3}/4 + \sqrt{5}/4$
9889 8533 tanh : $3\sqrt{3}/2$
9889 8567 rt : (1,8,-8,8)
9889 9232 sinh : $1/\ln(\pi)$
9890 0672 $J_0 : \sqrt[3]{2}/6$
9890 0911 sinh : $\sqrt[3]{2/3}$
9890 1791 $Ei : 5\sqrt[3]{2}/2$
9890 2731 $s\pi : 1/2 + \pi/3$
9890 2740 tanh : $13/5$
9890 2755 $t\pi : 8\sqrt[3]{3}/7$
9890 3939 $s\pi : \sqrt{2}/3 + 4\sqrt{5}/3$
9890 4088 $\Psi : 2\sqrt{3}/7$

9890 4237 $\Psi : (\sqrt[3]{2}/3)^2$
9890 5609 $2/5 - \exp(2)$
9890 8551 $erf : 10\sqrt[3]{2}/7$
9890 9011 $\lambda : (\pi/2)^{-3}$
9890 9050 $erf : 9/5$
9891 0288 cr : $(e/3)^{1/3}$
9891 1706 $Ei : 1/14$
9891 2276 tanh : $4(e+\pi)/9$
9891 3223 $\lambda : 1/\sqrt{15}$
9891 4051 $s\pi : e/6$
9891 5993 $J_0 : 1/\sqrt{23}$
9891 7870 $J_0 : 5/24$
9891 9169 $\Gamma : 3e/8$
9891 9272 cr : $2\sqrt{2}/3 + 4\sqrt{3}$
9892 1413 $erf : 5\sqrt[3]{3}/4$
9892 1638 $erf : (e+\pi)^{1/3}$
9892 1663 tanh : $7\sqrt{5}/6$
9892 1898 $J_0 : 3\ln 2/10$
9892 2566 $J_0 : \exp(-\pi/2)$
9892 2609 $erf : 3\zeta(3)/2$
9892 3276 $3/5 \times \exp(1/2)$
9892 4451 sq : $2\sqrt{3}/3 + 4\sqrt{5}$
9892 5284 $\Gamma : \pi^2/5$
9892 9985 $\ell_{10} : 2\sqrt{2} + 4\sqrt{3}$
9893 1139 $\exp(5) \div s\pi(\sqrt{3})$
9893 3483 $s\pi : 1/\ln(\sqrt{2}\pi/3)$
9893 5824 $s\pi : 8(1/\pi)$
9893 6515 $s\pi : e/2 + 2\pi/3$
9893 6564 ln : $1/3 + 3\pi/4$
9893 6576 $\ell_{10} : 1/3 + 3\pi$
9893 6959 $e^x : 1/2 + 4\pi$
9893 7607 $s\pi : 10\sqrt{5}/3$
9893 9918 sr : $1/3 + 4e/3$
9894 0470 $10^x : 9e\pi/4$
9894 1320 tanh : $5\pi/6$
9894 1545 $J_0 : \sqrt[3]{3}/7$
9894 2093 $10^x : 15/8$
9894 3630 $\ell_{10} : 8e\pi/7$
9894 6202 $\Psi : 15/19$
9894 7091 tanh : $\sqrt[3]{18}$
9894 7114 at : $\exp(\sqrt[3]{2}/3)$
9895 1610 $\lambda : 3\ln 2/8$
9895 2350 $Li_2 : 3\sqrt[3]{2}/5$
9895 2800 $\theta_3 : \exp(-\sqrt[3]{4}/2)$
9895 5098 sq : $1/4 + 3\sqrt{5}/2$
9895 5974 tanh : $21/8$
9895 6356 $10^x : 9(e+\pi)/5$
9895 6960 $\lambda : \exp(-3\pi/7)$
9895 7667 $J_0 : \exp(-\sqrt[3]{4})$
9895 7956 $2^x : e/4 + 3\pi/2$
9896 0217 $\ln(2/5) \times \exp(5)$
9896 0254 $s\pi : (\sqrt{2}/3)^{-3}$
9896 0465 cr : $(\ln 3)^{-1/3}$
9896 1042 $J_0 : 1/\sqrt{24}$
9896 1743 $erf : 2e/3$
9896 1822 $2^x : 4 - e/4$
9897 1921 $\lambda : 6/23$
9897 2661 $t\pi : 4\sqrt{2}/3 + \sqrt{3}/2$
9897 3706 $\ell_{10} : 5(e+\pi)/3$
9897 4331 id : $4\sqrt{3}/7$
9897 6033 $J_0 : (\sqrt{2}\pi/2)^{-2}$
9897 6214 $Ei : 6$
9897 6329 $s\pi : 9\sqrt{2}/5$
9897 8492 $t\pi : 4\zeta(3)/3$
9898 0033 $s\pi : 9\zeta(3)/7$
9898 1312 sq : $4e/3 + \pi/2$
9898 1495 $s\pi : 4\sqrt{3}/3 + \sqrt{5}$
9898 1794 sq : $2e/3 - 2\pi$
9898 2144 $s\pi : 5/11$
9898 2191 $J_0 : \sqrt{2}/7$
9898 2417 $erf : \sqrt[3]{6}$
9898 3834 sr : $4/3 - \sqrt{2}/4$
9898 6817 $erf : 20/11$
9898 6977 $s\pi : 1/3 + 3\sqrt{2}/2$
9898 7200 $erf : 9\sqrt{2}/7$
9898 9447 $e^x : 23/21$
9898 9873 cu : $1 + 4\sqrt{2}$
9899 0131 $10^x : 1/2 + 3e/2$
9899 0307 $s\pi : 10(1/\pi)/7$
9899 1847 sr : $7\sqrt[3]{2}/9$
9899 4387 id : $7\sqrt[3]{5}/3$
9899 4495 $s\pi : 8\ln 2$
9899 4949 id : $7\sqrt{2}/10$
9899 6280 $erf : 1/\ln(\sqrt{3})$
9899 6412 $10^x : 1/\ln(\sqrt{2}\pi/3)$
9899 7971 $\lambda : ((e+\pi)/3)^{-2}$
9899 8205 tanh : $\sqrt{7}$
9899 9083 $J_0 : \zeta(3)/6$
9900 2402 sr : $3e + \pi/4$
9900 2497 $J_0 : 1/5$
9900 2589 $10^x : 1 + 3\sqrt{2}/4$
9900 3640 cr : $7\ln 2/5$
9900 5802 $t\pi : 3e/2 + \pi/2$
9900 7791 sinh : $10\ln 2/3$
9900 8122 $10^x : 2\sqrt{3}/3 + 2\sqrt{5}/3$
9901 0220 $erf : \sqrt{2} - \sqrt{5} + \sqrt{7}$
9901 1969 $10^x : 3/4 + \pi$
9901 2980 $J_0 : 5(1/\pi)/8$
9901 4455 $e^x : 1 + 2\sqrt{2}$
9901 4731 $\ln(2/3) + \exp(1/3)$
9901 7672 $erf : \sqrt{10/3}$
9901 8113 $\Gamma : 9\sqrt[3]{2}/2$
9901 8758 id : $1/3 + 4\sqrt{2}$
9901 9916 $\lambda : 5/19$
9902 0767 $Ei : 4/23$
9902 1025 id : $10\ln 2/7$
9902 1884 $J_0 : 2\ln 2/7$
9902 6460 $2^x : 2\sqrt{3} + \sqrt{5}$
9902 6501 ln : $3\sqrt{2}/2 + 3\sqrt{3}$
9902 8893 $\lambda : \exp(-4/3)$
9903 2060 $t\pi : \sqrt{5}/9$
9903 5798 $10^x : 2\pi/9$
9903 7584 cr : $1/2 + \sqrt{2}/3$
9903 7686 cu : $1 + \pi/2$
9903 9049 tanh : $8/3$
9904 1029 as : $(\sqrt[3]{5})^{-1/3}$
9904 1547 $s\pi : 1/\ln(2\pi)$
9904 2362 tanh : $\sqrt[3]{19}$
9904 3774 $s\pi : (\sqrt{2}\pi/3)^{-2}$
9904 4475 $\Psi : -\sqrt{2} - \sqrt{3} - \sqrt{5}$
9904 5283 $\Psi : 4(e+\pi)/3$
9904 6056 ln : $4/3 + e/2$
9904 7437 $s\pi : 3\sqrt{2}/2 - \sqrt{3}/3$
9904 7810 $erf : 11/6$
9904 9052 $s\pi : 4\sqrt{3} - 2\sqrt{5}$
9905 1044 $10^x : \zeta(3)/2$
9905 5583 sr : $\sqrt{2} - \sqrt{3}/4$
9905 5834 $t\pi : 2e + \pi$
9905 6154 sq : $4\sqrt{2} - 3\sqrt{3}/4$
9905 7920 ln : $6\pi/7$
9906 0960 $s\pi : e/5$
9906 2174 $\lambda : 5(1/\pi)/6$
9906 2718 $s\pi : 7\pi^2/2$
9906 3314 $10^x : 4/3 + e$
9906 5452 $erf : \ln(2\pi)$
9906 6607 $s\pi : (\sqrt{2}/3)^{-1/2}$
9906 8313 cu : $3 + 3\sqrt{5}$
9906 9756 sr : $4\sqrt{5}$
9906 9766 $10^x : 2/3 + 3e/2$
9907 0310 tanh : $6\sqrt{5}/5$
9907 1198 id : $1/2 + 2\sqrt{5}/3$
9907 3355 sq : $1/4 + \sqrt{5}/3$
9907 4661 $J : \sqrt[3]{3}/3$
9907 5016 $5 \times \exp(4)$
9907 6215 $J_0 : \sqrt{3}/9$
9908 1269 $s\pi : (\ln 3/3)^{-2}$
9908 1422 $\Psi : \exp(4)$
9908 3159 $J : \exp(-17\pi/9)$
9908 4348 $\lambda : \exp(-8\pi/19)$
9908 5475 $s\pi : (\sqrt[3]{2}/3)^{-1/2}$
9908 5639 $\zeta : 3\sqrt{2}/7$
9908 5707 $s\pi : \sqrt{3}/2 + 3\sqrt{5}/4$
9908 6954 $e^x : \sqrt{6}\pi$
9908 7750 tanh : $6\pi/7$
9908 9581 $\lambda : 4/15$
9909 0186 $J_0 : 3(1/\pi)/5$
9909 3320 sq : $1/\ln(\ln 3/3)$
9909 3454 $s\pi : 3/4 + \sqrt{2}/2$
9909 3534 $erf : (2\pi)^{1/3}$
9909 5025 $J_0 : 4/21$
9909 6840 $erf : 24/13$
9909 7559 rt : (1,9,-9,9)
9909 8318 $\lambda : 2\zeta(3)/9$
9909 9559 $J_0 : \sqrt[3]{5}/9$
9910 0517 sr : $1/2 + 2\sqrt{3}$
9910 0663 sinh : $7/8$
9910 0932 $\lambda : 1/\sqrt{14}$
9910 0978 $erf : (e/2)^2$
9910 3674 cr : $4\sqrt{2} + \sqrt{5}$
9910 3904 $\ell_2 : 8\sqrt{5}/9$
9910 5166 $erf : 8\ln 2/3$
9910 6092 $e^x : 1 + 3\pi$
9910 6737 $2^x : 2e/3 + \pi$
9910 8993 tanh : $9\zeta(3)/4$
9910 9565 $2^x : 2/3 - e/4$
9910 9928 $\ln(\sqrt{3}) \times \exp(4)$
9911 0136 $J_0 : \exp(-5/3)$
9911 0829 $10^x : 24/11$
9911 1663 $\theta_3 : (e/3)^2$
9911 1715 $erf : \sqrt[3]{19/3}$
9911 2121 sinh : $1/\ln(e/2)$
9911 2918 cu : $1/3 - 3\sqrt{3}$
9911 3219 $\exp(1/4) + s\pi(1/4)$
9911 4857 id : 7π
9911 4978 tanh : $\sqrt{22/3}$
9911 5658 at : $\sqrt{7/3}$
9911 6315 $s\pi : 3\sqrt{3}/2 + 4\sqrt{5}$
9911 8674 cu : $e/2 - 3\pi/4$
9912 2987 $K : (\pi/2)^{-1/2}$
9912 3023 $J_0 : 3/16$
9912 3567 $\exp(2) \times s\pi(\sqrt{5})$
9912 5806 cu : $3e/2 + \pi/4$
9912 5964 tanh : $19/7$
9912 6193 tanh : $\sqrt[3]{20}$
9912 6885 $erf : 9\sqrt[3]{3}/7$
9912 6889 cr : $4\pi^2/5$
9912 7173 ln : $5(e+\pi)/4$
9912 7230 $\Psi : 8/23$
9912 9221 $\Gamma : (\pi/3)^{1/3}$
9912 9867 $s\pi : \sqrt[3]{11/3}$
9913 2891 tanh : e
9913 3696 $e^x : \sqrt{2}/3 + \sqrt{5}$
9913 3816 $s\pi : (2\pi)^{-1/3}$
9913 4092 $e^x : 2e + 3\pi/4$
9913 6516 $\ln(\pi) \times s\pi(1/3)$
9913 7057 $erf : 13/7$
9913 9171 $s\pi : 3/4 + 3\sqrt{5}$
9913 9233 cu : $3\sqrt{2}/2 - 2\sqrt{5}$
9914 2362 cu : $1/3 + 4e/3$
9914 3444 $2^x : 4\sqrt{5}/9$
9914 3988 $s\pi : 3\sqrt{2} + 3\sqrt{3}/4$
9914 3994 $\Gamma : (\sqrt[3]{2}/3)^3$
9914 4486 $s\pi : 11/24$
9914 4497 $J_0 : (\sqrt[3]{5}/3)^3$
9914 7750 at : $2 - \sqrt{2}/3$
9914 9573 sr : $3e - 4\pi/3$
9914 9988 $\Gamma : 8\sqrt{3}/7$
9915 0216 tanh : $1/\ln(\ln 2)$
9915 1382 id : $3e/4 - \pi/3$
9915 2780 as : $\sqrt{2} - \sqrt{3}/3$
9915 3218 $\lambda : \exp(-5\pi/12)$
9915 4101 tanh : $1/\ln(\sqrt[3]{3})$
9915 6383 id : $\sqrt{2} + \sqrt{3}/3$
9915 6598 $10^x : 4\sqrt{3} + \sqrt{5}/2$
9915 6898 $\Gamma : 3/10$
9915 7783 $s\pi : (e/2)^{-2}$
9915 8709 at : $\sqrt{2}/2 - \sqrt{5}$
9915 9200 $erf : 5\sqrt{5}/6$
9916 1352 $erf : 1/\ln(\sqrt[3]{5})$
9916 2890 tanh : $8\sqrt[3]{5}/5$
9916 6505 $10^x : \sqrt{2}/2 + 3\sqrt{5}$
9916 6560 cu : $3\ln 2$
9916 6736 $\Gamma : 7\sqrt{2}/5$
9916 7298 tanh : $\sqrt{15/2}$
9916 8216 $\ell_{10} : \sqrt{3}/2 + 4\sqrt{5}$
9916 8293 sinh : $8\sqrt[3]{2}/7$
9916 9186 $erf : \sqrt[3]{13/2}$
9917 1281 $2^x : 6\pi^2/5$
9917 1524 at : $2\sqrt{2} - 3\sqrt{3}/4$
9917 1932 $2^x : 3 + e/3$

9917 4303 $\pi \times \exp(4/5)$
9917 4597 $J_0 : 4(1/\pi)/7$
9917 5015 $K : \ln(\sqrt{2}\pi/2)$
9917 5259 $J_0 : 2/11$
9917 6230 $s\pi : 2\sqrt{2}/3 + 3\sqrt{3}/2$
9917 7905 $s\pi : 3\sqrt[3]{3}/8$
9918 2469 $1/2 + \exp(2/5)$
9918 3820 $J_0 : \exp(-\sqrt[3]{5})$
9918 3943 $\lambda : e/10$
9918 4176 $\tanh : 7\pi/8$
9918 4902 $erf : \sqrt{7/2}$
9918 5972 $\tanh : 11/4$
9918 9116 $J_0 : \sqrt[3]{3}/8$
9918 9238 $\mathrm{at} : 4e/3 - 2\pi/3$
9918 9416 $\mathrm{cu} : 3\sqrt{2}/2 + \sqrt{5}/2$
9919 1739 $J_0 : \sqrt[3]{2}/7$
9919 5500 $t\pi : 5\zeta(3)/8$
9919 6477 $\theta_3 : e/6$
9919 8062 $s\pi : 6\pi^2/7$
9919 8858 $e^x : 3\sqrt{2}/2 + 4\sqrt{3}/3$
9919 9005 $erf : 15/8$
9919 9444 $\lambda : 3/11$
9920 0315 $\tanh : \sqrt[3]{21}$
9920 0592 $\mathrm{as} : (e+\pi)/7$
9920 1319 $\lambda : 6(1/\pi)/7$
9920 1375 $\mathrm{sr} : \sqrt{3} + \sqrt{5}$
9920 2774 $\ln : 7\pi/3$
9920 2969 $\mathrm{id} : \sqrt{3}/4 + \sqrt{5}/4$
9920 5942 $2^x : \sqrt{5/2}$
9920 6104 $t\pi : 1/3 - 4\sqrt{5}/3$
9920 7474 $erf : \exp(\sqrt[3]{2}/2)$
9921 0325 $\mathrm{cu} : 2 - \sqrt{5}/4$
9921 0544 $10^x : 3\sqrt{2}/2 - \sqrt{3}/3$
9921 2000 $t\pi : 3\sqrt{2}/4 - 4\sqrt{3}/3$
9921 2511 $\Gamma : \sqrt{21}\pi$
9921 2801 $s\pi : 3\sqrt[3]{2}/7$
9921 3589 $\tanh : \sqrt{2} - \sqrt{3} - \sqrt{6}$
9921 5310 $10^x : 4e + \pi/3$
9921 5384 $\mathrm{cr} : (e+\pi)/6$
9921 5408 $\ln : \sqrt{5} + \sqrt{6} + \sqrt{7}$
9921 5926 $e^x : 6\sqrt{3}/5$
9921 6010 $\tanh : \sqrt{23/3}$
9921 8631 $t\pi : 8e/3$
9921 8750 $\sinh : 6\ln 2$
9921 9001 $J_0 : \exp(-\sqrt{3})$
9921 9561 $s\pi : 2/3 + 4e$
9921 9727 $J_0 : 5(1/\pi)/9$
9921 9760 $\tanh : 8\sqrt{3}/5$
9922 0274 $J_0 : \sqrt{2}/8$
9922 1631 $\Psi : 9\sqrt{2}/4$
9922 1789 $\tanh : 4\ln 2$
9922 1900 $s\pi : e\pi$
9922 2378 $t\pi : (4/3)^{-3}$
9922 2417 $erf : \sqrt[3]{20/3}$
9922 2967 $J_0 : 3/17$
9922 3592 $\Psi : 7\sqrt{5}/2$
9922 3781 $J_0 : (\sqrt[3]{2}/3)^2$
9922 5631 $\lambda : \exp(-7\pi/17)$
9922 7113 $s\pi : 8\sqrt{3}/9$
9922 9793 $\tanh : 25/9$
9923 1750 $\mathrm{cr} : 4\sqrt[3]{5}/7$
9923 1785 $erf : 3\pi/5$
9923 3923 $erf : 4\sqrt{2}/3$
9923 4319 $\mathrm{sr} : (\pi/3)^{-1/3}$
9923 6006 $\ell_2 : P_{53}$
9923 7180 $\zeta : 5\ln 2/2$
9923 7200 $s\pi : (3\pi/2)^{-1/2}$
9923 7365 $\ell_{10} : 4e - \pi/3$
9923 8863 $s\pi : 4\sqrt{2} - 3\sqrt{3}$
9924 0210 $\sinh : e\pi$
9924 0238 $t\pi : 3\sqrt{2} + 2\sqrt{3}/3$
9924 2459 $\zeta : \sqrt[3]{5}/5$
9924 3016 $\ln : 3/22$
9924 3369 $t\pi : 10e\pi$
9924 4401 $erf : 17/9$
9924 5284 $J_0 : 4/23$
9924 5790 $\mathrm{id} : 7\sqrt[3]{5}/4$
9924 5953 $s\pi : 7\ln 2/9$
9924 7556 $erf : 3\sqrt[3]{2}/2$
9925 0699 $J_0 : \ln 2/4$
9925 0814 $\ln : 4\sqrt{2} + 3\sqrt{5}/4$
9925 1186 $s\pi : 9\sqrt[3]{5}/10$
9925 1405 $J_0 : \sqrt{3}/10$
9925 1438 $\ln(\pi) \times \exp(3)$
9925 2097 $\tanh : 8\pi/9$
9925 5854 $\Gamma : 5(e+\pi)/8$
9925 5900 $\tanh : 5\sqrt{5}/4$
9925 6131 $\mathrm{cu} : 4/3 + 3\sqrt{5}/2$
9925 7240 $s\pi : 4\sqrt{2} - \sqrt{5}/2$
9926 1396 $s\pi : 1/2 + 3e/4$
9926 2860 $\Psi : 10(1/\pi)$
9926 3152 $\tanh : 14/5$
9926 3667 $t\pi : (\sqrt[3]{2}/3)^{1/2}$
9926 4068 $\mathrm{id} : 1/4 - 3\sqrt{2}$
9926 4142 $J_0 : \zeta(3)/7$
9926 6140 $\tanh : \sqrt[3]{22}$
9926 7622 $10^x : 25/6$
9926 9709 $\mathrm{id} : \sqrt{2}/3 - 2\sqrt{3}$
9926 9711 $\ell_2 : 4\sqrt{2} - 3\sqrt{5}/4$
9927 0165 $\tanh : 7\zeta(3)/3$
9927 0297 $\mathrm{at} : 2/3 + \sqrt{3}/2$
9927 0330 $J_0 : \sqrt[3]{5}/10$
9927 0887 $s\pi : 6/13$
9927 2049 $erf : 2e\pi/9$
9927 3604 $\lambda : 2\ln 2/5$
9927 5031 $\lambda : 1/\sqrt{13}$
9927 5928 $\ln : 3\sqrt{2}/2 + \sqrt{3}/3$
9927 8961 $erf : 10\sqrt[3]{5}/9$
9927 9042 $erf : 19/10$
9928 1687 $\lambda : 5/18$
9928 2967 $\mathrm{sq} : 4\sqrt{3}/3 + 4\sqrt{5}/3$
9928 4930 $\zeta : (\sqrt{3}/2)^{-1/2}$
9928 7536 $erf : 7e/10$
9928 8389 $s\pi : 8\sqrt[3]{3}$
9928 8561 $e^x : \sqrt{18}\pi$
9928 9352 $\Gamma : 4/9$
9928 9438 $\mathrm{at} : 23/15$
9929 1782 $\tanh : 2\pi^2/7$
9929 1926 $s\pi : 2\ln 2/3$
9929 3125 $\lambda : 7(1/\pi)/8$
9929 5381 $J_0 : (2e/3)^{-3}$
9930 1585 $\mathrm{cu} : 3/4 - 2\sqrt{3}$
9930 2354 $\tanh : 9\pi/10$
9930 2740 $s\pi : 3/4 + 3\pi/2$
9930 3170 $2^x : 3e/4 + 2\pi$
9930 3734 $\tanh : 2\sqrt{2}$
9930 6318 $erf : 21/11$
9930 6760 $J_0 : 1/6$
9930 7958 $\mathrm{sq} : 24/17$
9930 8063 $\lambda : \sqrt{5}/8$
9930 8360 $2^x : 9\sqrt[3]{2}$
9930 8580 $erf : 6(1/\pi)$
9931 0509 $\tanh : 17/6$
9931 1007 $t\pi : 3e/4 + 3\pi/2$
9931 2552 $\tanh : 9\sqrt[3]{2}/4$
9931 4383 $J_0 : (\ln 3/2)^3$
9931 5132 $\lambda : 2\sqrt[3]{2}/9$
9931 5393 $\lambda : 7/25$
9931 6189 $Ei : 5e/4$
9931 7558 $erf : \sqrt[3]{7}$
9932 1403 $\mathrm{cr} : 4/3 - \sqrt{2}/4$
9932 3125 $erf : \sqrt{11/3}$
9932 3301 $e^x : 3e/2$
9932 4032 $s\pi : 4\sqrt{2}/3 + \sqrt{3}/3$
9932 4834 $\tanh : \sqrt[3]{23}$
9932 5036 $\ln(2/3) - s\pi(1/5)$
9932 6764 $\mathrm{cr} : 7\sqrt[3]{2}/9$
9932 8229 $erf : 6\sqrt{5}/7$
9932 8335 $erf : 23/12$
9932 8450 $\ell_2 : 1 + 4\sqrt{5}/3$
9932 8473 $\tanh : e\pi/3$
9932 8972 $e^x : 18/13$
9932 9957 $e^x : 2\sqrt{2} - \sqrt{3}$
9933 2331 $\ln(4) \div \exp(1/3)$
9933 2393 $10^x : 9\sqrt{2}/8$
9933 2513 $\mathrm{sr} : \sqrt{3} - \sqrt{5}/3$
9933 2578 $\tanh : \exp(\pi/3)$
9933 2802 $erf : \sqrt{2} - \sqrt{3} + \sqrt{5}$
9933 2985 $\tanh : 5\sqrt[3]{5}/3$
9933 4466 $J_0 : \exp(-2e/3)$
9933 8561 $s\pi : 2\sqrt{2} + 3\sqrt{5}$
9934 0533 $\tanh : (1/(3e))^{-1/2}$
9934 2467 $\tanh : 20/7$
9934 2897 $s\pi : 3e/4 + 3\pi$
9934 4024 $s\pi : \ln(\sqrt[3]{5})$
9934 4367 $\exp(1/3) - \exp(2)$
9934 5123 $\tanh : \sqrt{5} - \sqrt{6} - \sqrt{7}$
9934 5838 $\mathrm{sq} : 4 + \sqrt{2}/3$
9934 5882 $\mathrm{sr} : \pi^2/10$
9934 6256 $erf : 4\sqrt[3]{3}/3$
9934 6472 $erf : 25/13$
9934 6551 $\sqrt{2} \times \exp(\sqrt{3})$
9934 7070 $erf : 8\zeta(3)/5$
9934 8275 $erf : 9\sqrt[3]{5}/8$
9934 8457 $\zeta : 7\ln 2/8$
9934 8563 $10^x : \exp(2\pi/3)$
9935 0353 $e^x : 4e + 3\pi/4$
9935 0441 $erf : 10\sqrt{3}/9$
9935 0997 $\mathrm{at} : 3/4 + \pi/4$
9935 2414 $\tanh : 9(1/\pi)$
9935 6427 $\lambda : \sqrt{2}/5$
9935 6526 $10^x : 7\sqrt{3}/5$
9935 7238 $2^x : 1/4 + \sqrt{5}/3$
9935 7729 $10^x : 10/21$
9935 7821 $\lambda : 8(1/\pi)/9$
9935 9028 $J_0 : \sqrt[3]{3}/9$
9936 0796 $\ln : 4 - 3\sqrt{3}/4$
9936 0957 $erf : \sqrt{3} - \sqrt{6} + \sqrt{7}$
9936 1023 $J_0 : 4/25$
9936 2643 $\mathrm{at} : \sqrt{3} + \sqrt{6} - \sqrt{7}$
9936 4728 $s\pi : 2\sqrt{3}$
9936 5393 $\tanh : 9\sqrt{5}/7$
9936 5463 $\tanh : 23/8$
9936 5610 $\zeta : (e)^{-1/2}$
9936 6751 $s\pi : 1/\sqrt[3]{10}$
9936 7744 $J_0 : (1/\pi)/2$
9936 8033 $\lambda : \exp(-\sqrt[3]{2})$
9936 8581 $s\pi : \sqrt{3} + \sqrt{6} - \sqrt{7}$
9936 9393 $\mathrm{cr} : \sqrt{2} - \sqrt{3}/4$
9937 3762 $\Gamma : ((e+\pi)/3)^{-3}$
9937 6207 $\exp(1/4) \times s\pi(e)$
9937 6245 $\exp(1/3) + \exp(4)$
9937 7367 $\tanh : 2\sqrt[3]{3}$
9937 7549 $s\pi : 5\sqrt{2}/2$
9937 7701 $J_0 : 3/19$
9937 8472 $\tanh : 1/\ln(\sqrt{2})$
9937 8875 $\mathrm{cu} : 1 + 3\sqrt{5}$
9938 0009 $s\pi : 9e$
9938 0029 $\mathrm{sr} : 1/\ln(\sqrt{5}/2)$
9938 0157 $\tanh : 5\sqrt{3}/3$
9938 0798 $\mathrm{id} : 4\sqrt{5}/9$
9938 0799 $\lambda : \exp(-2\pi/5)$
9938 0882 $J_0 : \sqrt[3]{2}/8$
9938 1859 $2^x : 7(e+\pi)$
9938 1944 $s\pi : (\pi)^{1/3}$
9938 2291 $s\pi : 3/4 + \pi/4$
9938 3667 $J_0 : \sqrt{2}/9$
9938 4154 $\zeta : \sqrt{3}$
9938 6018 $\lambda : \sqrt[3]{5}/6$
9938 7320 $J_0 : \ln(\sqrt[3]{5}/2)$
9938 9013 $\sqrt{2} \times \exp(3/4)$
9939 0172 $\ell_2 : \sqrt{2} + \sqrt{3}/3$
9939 2432 $s\pi : 3\sqrt{3}/4 + \sqrt{5}$
9939 5013 $s\pi : (e+\pi)/4$
9939 5612 $\lambda : 2/7$
9939 6510 $erf : 5e/7$
9939 8453 $e^x : 4/3 + \sqrt{5}/3$
9939 9613 $erf : \sqrt[3]{22/3}$
9939 9894 $\mathrm{cr} : 1 + 4\sqrt{3}$
9940 3720 $e^x : \pi^2/9$
9940 5677 $\lambda : 9(1/\pi)/10$
9940 6432 $erf : \sqrt{3} - \sqrt{5} + \sqrt{6}$
9940 7727 $J_0 : 2\ln 2/9$
9940 9158 $J_0 : 2/13$
9941 0715 $\mathrm{id} : e - 3\pi/2$
9941 1043 $\mathrm{at} : 3\sqrt{2}/4 - 3\sqrt{3}/2$
9941 1583 $\mathrm{sq} : e/2 - 3\pi/4$
9941 2202 $erf : \exp(2/3)$
9941 3589 $4/5 \div \ln(\sqrt{5})$
9941 4290 $erf : 9\sqrt{3}/8$

9941 4661 tanh : $\sqrt{17/2}$
9941 6082 10^x : $(e/2)^{-1/3}$
9941 9773 $s\pi$: $6\sqrt[3]{5}/7$
9942 0266 $s\pi$: $9\pi/8$
9942 1199 $s\pi$: $5\ln 2$
9942 1208 λ : $\ln(4/3)$
9942 1759 $s\pi$: $4\zeta(3)/9$
9942 1944 Ψ : $5\sqrt[3]{2}/8$
9942 2711 $s\pi$: $10e\pi/3$
9942 4079 Γ : $5\sqrt{2}/7$
9942 4548 tanh : $\sqrt[3]{25}$
9942 5874 sinh : $(\sqrt{2}\pi/3)^{-1/3}$
9942 5907 $s\pi$: $9(e+\pi)/7$
9942 6168 erf : $(e+\pi)/3$
9942 8566 erf : $8\sqrt[3]{5}/7$
9942 9974 ℓ_{10} : π^2
9943 0897 10^x : $e/4+4\pi$
9943 0904 10^x : P_{79}
9943 0927 λ : $\sqrt[3]{3}/5$
9943 1301 tanh : $(e+\pi)/2$
9943 3750 λ : $\sqrt{3}/6$
9943 4240 erf : $7\sqrt{5}/8$
9943 6365 J_0 : $\zeta(3)/8$
9943 6382 erf : $\sqrt[3]{15/2}$
9943 6825 erf : $1/\ln(5/3)$
9943 7159 $s\pi$: $\sqrt[3]{15}$
9943 7956 sq : $9\pi/10$
9943 8290 J_0 : $3/20$
9944 0197 $s\pi$: $1/\ln(3/2)$
9944 2110 at : $20/13$
9944 2395 tanh : $7\sqrt[3]{2}/3$
9944 8881 Ei : $8\sqrt{2}/5$
9944 9530 at : $1/2-3e/4$
9944 9899 Ei : $\ln 2/4$
9945 0648 Ei : $(e)^{1/3}$
9945 1033 erf : $5\pi/8$
9945 2055 J_0 : $(\sqrt[3]{4}/3)^3$
9945 2189 $s\pi$: $7/15$
9945 4057 10^x : $\exp(\sqrt{5}/3)$
9945 4979 $2/3\times\exp(2/5)$
9945 6321 Γ : $7\sqrt[3]{3}/10$
9945 7039 $s\pi$: $1/4+2\pi$
9945 7135 as : $3\sqrt{5}/8$
9945 7416 tanh : $\sqrt{3}-\sqrt{5}-\sqrt{6}$
9945 7456 at : $9\sqrt[3]{5}/10$
9946 0590 2^x : $2\sqrt{2}/3+3\sqrt{5}$
9946 5338 tanh : $3\pi^2/10$
9946 5843 $s\pi$: $10(e+\pi)/3$
9946 7228 cr : $2\sqrt{3}+2\sqrt{5}$
9946 7483 θ_3 : $\ln(5/2)$
9946 8218 tanh : $\sqrt{2}-\sqrt{3}-\sqrt{7}$
9946 9688 $s\pi$: $3\sqrt{2}/4+2\sqrt{5}$
9947 0067 λ : $7/24$
9947 0611 erf : $\sqrt[3]{23/3}$
9947 3045 $s\pi$: $2/3+\sqrt{3}/2$
9947 5430 erf : $\pi^2/5$
9947 5595 $s\pi$: $\exp(-\sqrt[3]{2}/2)$
9947 5928 at : $8\sqrt{3}/9$
9947 6042 $s\pi$: $\pi^2/4$
9947 8885 $\ln(3/4)-s\pi(1/4)$
9948 0654 J_0 : $\sqrt[3]{3}/10$
9948 0911 2^x : $3-e/4$
9948 2594 $s\pi$: $-\sqrt{2}+\sqrt{5}+\sqrt{7}$
9948 2723 Γ : $8\sqrt{5}/9$
9948 3412 J_0 : $\ln(\sqrt{3}/2)$
9948 4250 10^x : $7/9$
9948 5844 sr : $4\sqrt{3}/7$
9948 6805 tanh : $4\sqrt{5}/3$
9948 7266 $s\pi$: $\sqrt{2}+\sqrt{5}/2$
9948 8049 erf : $8\sqrt{3}/7$
9948 8892 cr : $(\pi/3)^{-1/3}$
9948 8973 erf : $7\sqrt{2}/5$
9949 0446 J_0 : $1/7$
9949 0997 e^x : $3-4\sqrt{3}/3$
9949 2419 $s\pi$: $3e/4-\pi/2$
9949 4442 sr : $4\sqrt{2}-3\sqrt{5}/4$
9949 4936 cu : $3+2\sqrt{2}$
9949 5050 10^x : $5e/8$
9949 5245 $\ln(3/5)\times\exp(2/3)$
9949 6205 sr : $7\sqrt{2}/10$
9949 6439 ln : $9\zeta(3)/4$
9949 7193 id : $7\sqrt[3]{5}/6$
9949 7904 J_0 : $\exp(-(e+\pi)/3)$
9949 7978 tanh : $7\sqrt[3]{5}/4$
9949 8241 λ : $5/17$
9950 0274 J_0 : $4(1/\pi)/9$
9950 0624 J_0 : $\sqrt{2}/10$
9950 3948 $s\pi$: $3/4+e$
9950 4952 λ : $\exp(-7\pi/18)$
9950 5475 tanh : 3
9950 5988 Ei : $\sqrt{3}/10$
9950 5989 erf : $8\sqrt{5}/9$
9950 6809 ℓ_{10} : $2\sqrt{2}/3+4\sqrt{5}$
9950 8211 tanh : $(\ln 2)^{-3}$
9950 9309 sr : $10\ln 2/7$
9950 9389 2^x : $8\pi^2/3$
9951 0522 tanh : $5\zeta(3)/2$
9951 0531 ℓ_2 : $4\sqrt{3}/3+3\sqrt{5}/4$
9951 0661 J_0 : $\sqrt[3]{2}/9$
9951 1615 cu : $9\sqrt{3}$
9952 0123 J_0 : $\ln 2/5$
9952 0729 sinh : $\sqrt{6}\pi$
9952 1729 erf : $7\sqrt[3]{5}/6$
9952 2145 λ : $(3/2)^{-3}$
9952 3981 $t\pi$: $3e+2\pi/3$
9952 4165 ln : $2/3+3e/4$
9952 4909 Γ : $\zeta(7)$
9952 5116 tanh : $10e/9$
9952 5824 ln : $4+3\sqrt{5}/2$
9952 5873 K : $((e+\pi)/3)^{-1/3}$
9952 6231 10^x : $3/10$
9952 6656 ℓ_2 : $3\sqrt{2}/4-\sqrt{5}/4$
9952 7123 J : $3/14$
9952 7399 $s\pi$: $2e+2\pi/3$
9952 9486 $\ln(\sqrt{2})+\exp(1/2)$
9953 0277 K : $4/5$
9953 0310 λ : $3\ln 2/7$
9953 2070 sq : $6\sqrt[3]{3}/5$
9953 2226 erf : 2
9953 2616 e^x : $2/3+2\pi$
9953 4544 $\exp(1/5)+s\pi(e)$
9953 5058 sr : $1+4\sqrt{5}/3$
9953 5218 tanh : $7\sqrt{3}/4$
9953 5290 J_0 : $3(1/\pi)/7$
9953 5599 id : $1/4+\sqrt{5}/3$
9953 5648 10^x : $3+3e/2$
9953 5663 J_0 : $3/22$
9953 6079 e^x : $10\sqrt{5}/9$
9953 9263 erf : $5\zeta(3)/3$
9953 9585 cu : $4e/3-4\pi$
9954 0795 $s\pi$: $5(1/\pi)/3$
9954 1772 J : $\exp(-12\pi/7)$
9954 2200 $s\pi$: $5e/3$
9954 2632 J_0 : $(e)^{-2}$
9954 4475 10^x : $3/4+2\sqrt{3}$
9954 5627 id : $1/\ln(\ln 3/3)$
9954 6383 $s\pi$: $3\sqrt{2}/8$
9954 6438 J : $\sqrt{3}/4$
9954 7356 λ : $\exp(-5\pi/13)$
9954 7679 at : $\sqrt[3]{11/3}$
9954 8058 Γ : $4\sqrt[3]{2}/5$
9955 0376 J_0 : $((e+\pi)/3)^{-3}$
9955 0898 J_0 : $(\ln 3/3)^2$
9955 2026 Ei : $\zeta(5)$
9955 3469 $s\pi$: $7\sqrt[3]{2}/6$
9955 4346 10^x : $3\sqrt{2}$
9955 4512 cr : $\sqrt{3}-\sqrt{5}/3$
9955 4527 J_0 : $\zeta(3)/9$
9955 6049 J_0 : $2/15$
9955 6767 $s\pi$: $4e/3-2\pi/3$
9955 7347 erf : $9\sqrt{5}/10$
9955 7428 id : $7\pi/2$
9955 8055 erf : $(1/(3e))^{-1/3}$
9955 9577 tanh : $9e/8$
9956 0077 tanh : $1/\ln(\sqrt[3]{3}/2)$
9956 0251 tanh : $\exp(\sqrt{5}/2)$
9956 0436 λ : $3/10$
9956 1242 $2/5-\exp(1/3)$
9956 1303 ℓ_{10} : $7\sqrt{2}$
9956 3444 cr : $\pi^2/10$
9956 4010 erf : $8\sqrt[3]{2}/5$
9956 4530 e^x : $3(e+\pi)/2$
9956 5528 λ : $\zeta(3)/4$
9956 6613 ζ : $2\zeta(3)/7$
9956 8689 e^x : $(2\pi/3)^{-1/2}$
9956 9924 tanh : $(3\pi)^{1/2}$
9957 0320 erf : $7\sqrt[3]{3}/5$
9957 1270 $s\pi$: $1/3+3\sqrt{3}$
9957 2526 erf : $10\sqrt{2}/7$
9957 3227 ln : 20
9957 3327 erf : $7\sqrt{3}/6$
9957 3417 $s\pi$: $8/17$
9957 4076 $s\pi$: $2\sqrt{2}-3\sqrt{3}/4$
9957 4786 sr : $3e/4-\pi/3$
9957 5121 J_0 : $3/23$
9957 5251 λ : $1/\sqrt{11}$
9957 6717 tanh : $9\sqrt[3]{5}/5$
9957 7426 λ : $(\ln 3/2)^2$
9957 7745 sr : $2\sqrt{2}+2\sqrt{3}/3$
9957 9050 at : $(\sqrt[3]{2}/3)^{-1/2}$
9958 0188 $s\pi$: P_3
9958 0243 λ : $e/9$
9958 0292 tanh : $\sqrt{19/2}$
9958 0987 sinh : $\sqrt[3]{6}$
9958 3482 $1/4-s\pi(\sqrt{3})$
9958 4469 10^x : $\ln(\sqrt{2}\pi)$
9958 5849 erf : $\sqrt[3]{25/3}$
9958 7169 erf : $\exp(1/\sqrt{2})$
9958 8045 ln : $2+\sqrt{2}/2$
9959 0157 $s\pi$: $\exp(\sqrt[3]{5})$
9959 5125 J_0 : $2(1/\pi)/5$
9959 5128 Ei : $6\zeta(3)$
9959 6274 $\ln(\pi)-\exp(\pi)$
9959 6669 sq : $2\sqrt{3}+\sqrt{5}/2$
9959 6675 ℓ_{10} : $4e/3+2\pi$
9959 6751 $s\pi$: $\sqrt{2}/3$
9960 0243 tanh : $8e/7$
9960 0442 $\ln(2/3)\times\exp(2)$
9960 0687 sr : $\sqrt{3}/4+\sqrt{5}/4$
9960 0986 Ψ : $5\pi/2$
9960 1554 ln : $\sqrt{2}/3+\sqrt{5}$
9960 1847 λ : $7/23$
9960 1889 $s\pi$: $\sqrt{2}/4+\sqrt{5}/2$
9960 3270 λ : $(2e/3)^{-2}$
9960 3543 J_0 : $\sqrt[3]{2}/10$
9960 4032 J : $\ln(3/2)$
9960 5309 at : $3\sqrt{2}/2-\sqrt{3}/3$
9960 5731 ℓ_{10} : $4\sqrt{3}+4\sqrt{5}/3$
9960 6309 erf : $3e/4$
9960 6919 $t\pi$: $\sqrt{3}-4\sqrt{5}/3$
9960 7251 2^x : $\sqrt[3]{25/2}$
9960 7912 2^x : $3/4+\pi/2$
9960 8240 erf : $\sqrt{5}+\sqrt{6}-\sqrt{7}$
9960 8990 tanh : $9\sqrt{3}/5$
9960 9375 sinh : $7\ln 2$
9960 9756 J_0 : $1/8$
9961 0033 erf : $\sqrt[3]{17/2}$
9961 0136 tanh : $9\ln 2/2$
9961 2684 $s\pi$: $10\pi/3$
9961 3086 θ_3 : $5/11$
9961 4653 tanh : $25/8$
9961 7104 $s\pi$: $2\sqrt{5}$
9961 8856 tanh : $7\sqrt{5}/5$
9962 0707 $s\pi$: $7e\pi/8$
9962 1243 J_0 : $\exp(-2\pi/3)$
9962 1508 ln : $\sqrt{22/3}$
9962 4078 λ : $\ln(e/2)$
9962 5036 $\exp(3/4)-\exp(\sqrt{2})$
9962 5181 10^x : $(1/(3e))^{-1/3}$
9962 5308 ℓ_{10} : P_{45}
9962 6076 $s\pi$: $e/3+4\pi$
9962 6234 $s\pi$: $3\sqrt[3]{2}/8$
9962 6353 $s\pi$: $\sqrt{7/3}$
9962 7207 tanh : π
9962 7954 ζ : $6\sqrt[3]{3}/5$
9962 8147 tanh : $22/7$
9962 9123 erf : $6\sqrt[3]{5}/5$
9963 1274 λ : $4/13$
9963 1351 10^x : $(\ln 3)^{-1/2}$
9963 1421 Ψ : $7(e+\pi)/2$

9963 2730 $\lambda : \exp(-3\pi/8)$
9963 2792 $\lambda : (\sqrt{2}\pi/3)^{-3}$
9963 3242 tanh : $\sqrt{3}-\sqrt{5}-\sqrt{7}$
9963 3267 tanh : $5\sqrt[3]{2}/2$
9963 4334 $\ln(3/4)+\exp(1/4)$
9963 4433 $\lambda : 4\ln 2/9$
9963 6845 $\ell_2 : 7\sqrt[3]{5}/3$
9963 7038 $erf : \exp(\sqrt[3]{3}/2)$
9963 8990 $Ei : 4\pi/9$
9963 9090 $J_0 : \zeta(3)/10$
9963 9645 $J_0 : (\ln 2/2)^2$
9964 0323 $J_0 : 3/25$
9964 1014 $10^x : \sqrt{3}+4\sqrt{5}$
9964 2288 tanh : $\sqrt{10}$
9964 2923 $erf : 10\sqrt[3]{3}/7$
9964 4109 $J_0 : 3(1/\pi)/8$
9964 4199 sinh : $\sqrt{18}\pi$
9964 5409 tanh : 19/6
9964 7010 $t\pi : 8\sqrt{3}/9$
9964 8167 $t\pi : \exp(3\pi)$
9964 8694 tanh : $7e/6$
9964 9149 at : 17/11
9965 0542 at : $9\zeta(3)/7$
9965 1456 $s\pi : (3\pi)^{-1/3}$
9965 4278 $J_0 : 2/17$
9965 6086 tanh : $9\sqrt{2}/4$
9965 6638 $\Gamma : 2/25$
9965 6853 tanh : $10(1/\pi)$
9965 6935 cr : $4\sqrt{3}/7$
9965 8449 $s\pi : 9/19$
9965 9927 cu : $3e-\pi$
9966 1014 sr : $4\sqrt{3}/3+3\sqrt{5}/4$
9966 1047 $J_0 : ((e+\pi)/2)^{-2}$
9966 1415 $2^x : 19/12$
9966 2724 $\Psi : \exp(5)$
9966 3854 cr : $7\sqrt{2}/10$
9966 4009 $e^x : 10\sqrt{5}/3$
9966 4037 $\theta_3 : 10(1/\pi)/7$
9966 4497 tanh : $10\sqrt{5}/7$
9966 6630 $J_0 : \ln 2/6$
9966 7752 $\exp(2/3)-s\pi(2/5)$
9966 8239 tanh : 16/5
9966 9826 tanh : $3e\pi/8$
9967 0152 $\lambda : 5/16$
9967 0436 $10^x : 4+e$
9967 1138 $erf : 6\sqrt{3}/5$
9967 1853 tanh : $8\zeta(3)/3$
9967 2079 $\ell_{10} : 1/2+3\pi$
9967 2604 cr : $10\ln 2/7$
9967 2607 $erf : 3\ln 2$
9967 2919 sq : $1/3+2\sqrt{2}$
9967 3275 $\exp(\sqrt{3})\times s\pi(1/4)$
9967 3565 $erf : \sqrt[3]{9}$
9967 4044 $erf : (\ln 2/3)^{-1/2}$
9967 5476 $erf : (\ln 2)^{-2}$
9967 5713 $s\pi : 10\sqrt{3}/7$
9967 5916 $erf : \sqrt{13/3}$
9967 7397 $s\pi : (\sqrt{2}\pi)^{-1/2}$
9967 7910 $\Psi : 10\sqrt{5}/7$
9967 8377 $erf : 25/12$
9967 9058 $Li_2 : (\sqrt{3})^{-1/2}$
9968 0208 $2^x : 2(e+\pi)/3$
9968 0926 $s\pi : \exp(-\sqrt{5}/3)$
9968 2688 $\lambda : \pi/10$
9968 2734 $s\pi : \exp(e/3)$
9968 3506 $\lambda : 2\sqrt{2}/9$
9968 3681 $\lambda : \exp(-7\pi/19)$
9968 5131 $\ell_{10} : P_{61}$
9968 6276 $\pi-\ln(\pi)$
9968 7065 $\ell_{10} : 3+4\sqrt{3}$
9968 8194 $s\pi : \exp(\sqrt[3]{2})$
9968 8613 $s\pi : 7\sqrt{2}/4$
9968 8730 $\lambda : \sqrt[3]{2}/4$
9968 9035 $J_0 : \ln(\sqrt{5}/2)$
9968 9343 $erf : 23/11$
9968 9906 $s\pi : 1/3+\pi$
9968 9918 sr : $4\sqrt{5}/9$
9969 0294 sinh : $\sqrt[3]{3}$
9969 1596 $J_0 : 1/9$
9969 3763 $e^x : \sqrt{3}-\sqrt{5}+\sqrt{6}$
9969 3934 tanh : $\sqrt{21/2}$
9969 4274 $erf : 2\pi/3$
9969 4585 $\lambda : 6/19$
9969 6787 tanh : $9\sqrt[3]{3}/4$
9969 7501 sq : $\sqrt{3}/4+3\sqrt{5}$
9969 7577 $s\pi : 3\sqrt{2}/4+2\sqrt{3}$
9969 7714 $\lambda : 1/\sqrt{10}$
9969 8628 tanh : $(\sqrt{2}\pi/3)^3$
9969 9763 tanh : 13/4
9970 0144 $erf : \ln(1/(3e))$
9970 0548 at : $1/2+\pi/3$
9970 1872 $erf : 5\sqrt[3]{2}/3$
9970 2053 $erf : 21/10$
9970 2054 $s\pi : 3\sqrt{2}/2+3\sqrt{5}/2$
9970 2074 $s\pi : 2e/3+3\pi/2$
9970 3034 tanh : $5(e+\pi)/9$
9970 3472 $10^x : e+3\pi/4$
9970 4017 $J_0 : (2\pi/3)^{-3}$
9970 5047 tanh : $1/\ln(e/2)$
9970 5357 id : $e/2-3\pi/4$
9970 6166 $J_0 : \exp(-\sqrt{2}\pi/2)$
9970 6675 $\Gamma : (2\pi/3)^{-2}$
9970 6836 tanh : $6e/5$
9970 6953 $erf : 7\zeta(3)/4$
9970 7901 $Li_2 : 4\sqrt[3]{5}/9$
9970 8454 cu : 19/7
9970 8623 $\ln(4/5)-s\pi(e)$
9970 9908 $Li_2 : 19/25$
9971 1318 $\lambda : 7/22$
9971 1769 $erf : \exp(\sqrt{5}/3)$
9971 2190 $\lambda : (1/\pi)$
9971 2292 tanh : $\sqrt{2}-\sqrt{5}-\sqrt{6}$
9971 2597 $erf : (\sqrt{2}\pi)^{1/2}$
9971 3775 $\Gamma : 5\sqrt{3}$
9971 4631 $J_0 : \exp(-\sqrt{5})$
9971 4987 $\ell_{10} : \sqrt{10}\pi$
9971 6323 cr : $3e/4-\pi/3$
9971 6942 $erf : 19/9$
9971 7205 $\lambda : \exp(-4\pi/11)$
9971 7296 $s\pi : 1/\ln(4/3)$
9971 7384 $s\pi : 7\sqrt[3]{3}/4$
9971 8398 cr : $4\sqrt{2}+4\sqrt{3}/3$
9971 8503 $erf : (3\pi)^{1/3}$
9971 8622 as : $2\sqrt[3]{2}/3$
9971 8750 $J_0 : (1/\pi)/3$
9971 9487 tanh : $(2e/3)^2$
9971 9772 $\lambda : \sqrt{5}/7$
9972 0379 $s\pi : 10/21$
9972 0432 tanh : 23/7
9972 0983 $erf : 7e/9$
9972 2741 tanh : $\pi^2/3$
9972 3183 $J_0 : 2/19$
9972 3479 $\lambda : 8/25$
9972 5305 $s\pi : \pi/6$
9972 5716 $erf : \sqrt[3]{19/2}$
9972 5839 $J_0 : (\sqrt{2}/3)^3$
9972 6741 $\lambda : 2\sqrt[3]{3}/9$
9972 8204 tanh : $7\sqrt{2}/3$
9972 8322 as : 21/25
9973 0020 $erf : 3\sqrt{2}/2$
9973 3553 $10^x : 4\sqrt{3}/3+3\sqrt{5}/2$
9973 3614 cr : $\sqrt{3}/4+\sqrt{5}/4$
9973 4271 ln : $3e-\pi/4$
9973 4597 $erf : 17/8$
9973 5863 tanh : $(\ln 3/2)^{-2}$
9973 6567 $s\pi : e/3+\pi/2$
9973 7169 tanh : $\sqrt{11}$
9973 8599 $e^x : 4\sqrt{3}/5$
9973 9011 $t\pi : 10\sqrt[3]{3}/9$
9973 9144 sinh : $6\zeta(3)/5$
9974 0529 $\Gamma : \sqrt{2}+\sqrt{5}-\sqrt{7}$
9974 1784 $s\pi : 4\sqrt{2}+\sqrt{3}/2$
9974 2691 $erf : \sqrt{2}-\sqrt{3}+\sqrt{6}$
9974 3515 $J_0 : (\pi)^{-2}$
9974 4523 $s\pi : 1/\sqrt[3]{7}$
9974 5448 sq : $e/3-\pi$
9974 5796 tanh : 10/3
9974 6600 $erf : e\pi/4$
9974 8207 $s\pi : \sqrt{3}+\sqrt{5}/3$
9974 8438 sr : $7\sqrt[3]{5}/3$
9974 9498 $s\pi : 3(1/\pi)/2$
9974 9564 cu : $3\sqrt{3}+2\sqrt{5}/3$
9974 9584 $erf : 5\sqrt[3]{5}/4$
9975 0156 $J_0 : 1/10$
9975 3382 tanh : $4(e+\pi)/7$
9975 3940 $\lambda : (\sqrt[3]{5}/3)^2$
9975 4529 $erf : \sqrt{3}-\sqrt{5}+\sqrt{7}$
9975 4735 at : $1/4+3\sqrt{3}/4$
9975 5021 $J_0 : \ln 2/7$
9975 5816 $erf : 15/7$
9975 6126 tanh : $3\sqrt{5}/2$
9975 7038 $J_0 : \ln(e/3)$
9975 7299 $\ell_{10} : 1+4\sqrt{5}$
9975 8264 $\lambda : \exp(-5\pi/14)$
9975 8562 $s\pi : 1/3+4\pi/3$
9975 8881 $\lambda : (3\pi)^{-1/2}$
9975 8881 tanh : $8\sqrt[3]{2}/3$
9975 8964 $t\pi : 4\sqrt{3}/3-\sqrt{5}/4$
9976 0514 tanh : $\sqrt{3}-\sqrt{6}-\sqrt{7}$
9976 1497 tanh : $7\sqrt[3]{3}/3$
9976 2620 $10^x : \zeta(3)/4$
9976 2950 $s\pi : \exp(\sqrt[3]{2}/3)$
9976 4222 $\sqrt{3}\div\ln(\sqrt{2})$
9976 5677 $\lambda : \exp(-\sqrt{5}/2)$
9976 6097 tanh : $(3/2)^3$
9976 6876 $s\pi : 11/23$
9976 7529 sr : $1/4+\sqrt{5}/3$
9976 7568 $e^x : \sqrt{17}\pi$
9976 7594 $10^x : 3\sqrt{3}/4+4\sqrt{5}$
9976 8153 sq : $4/3+4e$
9976 8732 $erf : \sqrt[3]{10}$
9976 9377 $s\pi : \sqrt{3}/3+4\sqrt{5}$
9977 0225 $s\pi : 4\pi^2$
9977 1979 $10^x : \sqrt{2}+3\sqrt{5}$
9977 2157 $J_0 : 3(1/\pi)/10$
9977 2626 $J : 4/25$
9977 3356 $K : 5\sqrt[3]{3}/9$
9977 3371 $J_0 : 2/21$
9977 3530 tanh : $\sqrt{23/2}$
9977 4977 $erf : \sqrt{14/3}$
9977 6378 $s\pi : 3\sqrt{2}+\sqrt{5}$
9977 6536 tanh : $5e/4$
9977 7492 tanh : 17/5
9977 8273 $erf : 3\sqrt[3]{3}/2$
9977 8616 $erf : 9\zeta(3)/5$
9977 8643 tanh : $1/\ln(\sqrt{5}/3)$
9977 8971 $erf : 1/\ln(\sqrt[3]{4})$
9977 8997 $J_0 : \ln(\ln 3)$
9977 9431 $s\pi : 2/3+2e/3$
9978 0035 $erf : 5\sqrt{3}/4$
9978 1695 $erf : 13/6$
9978 2272 $\lambda : \exp(-6\pi/17)$
9978 4446 tanh : $2e\pi/5$
9978 5926 $erf : (3\pi/2)^{1/2}$
9978 6187 tanh : $2\sqrt[3]{5}$
9978 7951 $\exp(\sqrt{2})\div s\pi(1/5)$
9978 8461 $\Gamma : 7\sqrt[3]{5}/6$
9978 8748 at : $\ln(3\pi/2)$
9978 9768 $erf : 4e/5$
9978 9800 sinh : $1/\ln(2)$
9978 9838 tanh : 24/7
9979 2075 $J_0 : (\sqrt{2}\pi/2)^{-3}$
9979 3172 cr : $4\sqrt{5}/9$
9979 3328 $J_0 : 2(1/\pi)/7$
9979 3495 $J_0 : 1/11$
9979 6828 $erf : 24/11$
9979 7088 $K : 2\zeta(3)/3$
9979 7545 sinh : 10/3
9979 7948 $\lambda : \exp(-7\pi/20)$
9979 7978 $\lambda : (\ln 2)^3$
9979 9026 $\ell_2 : 3\sqrt{2}/4+4\sqrt{3}$
9979 9500 $\lambda : 1/3$
9979 9517 $s\pi : 2\sqrt{2}/3+\sqrt{3}/3$
9980 0099 $J_0 : (\sqrt{5})^{-3}$
9980 0890 $\Gamma : (\ln 3/3)^2$
9980 2259 $s\pi : (\sqrt[3]{3}/2)^2$
9980 2672 $s\pi : 12/25$
9980 3036 sinh : $10\sqrt{5}/3$
9980 3094 $s\pi : 8\sqrt[3]{5}/9$
9980 3504 $s\pi : 3\sqrt{3}/2-\sqrt{5}/2$

9980 4239 tanh : $2\sqrt{3}$
9980 4370 erf : $\sqrt[3]{21/2}$
9980 4687 sinh : $8\ln 2$
9980 4878 tanh : $5\ln 2$
9980 5417 $s\pi$: $3\ln 2/4$
9980 5605 tanh : $\sqrt{2}-\sqrt{5}-\sqrt{7}$
9980 5775 $s\pi$: $2\sqrt[3]{2}$
9980 5839 sinh : 22/25
9980 7395 $\sqrt{5}\div s\pi(\sqrt{3})$
9980 7598 erf : $2\pi^2/9$
9980 7835 ln : $2/3+3\sqrt{5}$
9980 8861 tanh : $1/\ln(4/3)$
9980 8947 $s\pi$: P_{21}
9981 0189 $s\pi$: $3\sqrt{3}/10$
9981 0778 e^x : $\exp(-1/e)$
9981 1053 J_0 : 2/23
9981 1430 erf : $3(e+\pi)/8$
9981 1508 $s\pi$: $(\ln 2)^2$
9981 2408 erf : $9\sqrt[3]{5}/7$
9981 2411 J_0 : $\ln 2/8$
9981 2439 Ψ : $\exp(-1/\sqrt{2})$
9981 2924 erf : $7\pi/10$
9981 3715 erf : 11/5
9981 3766 at : $\sqrt{3}/4+\sqrt{5}/2$
9981 3992 Ei : $(\sqrt{5}/2)^{1/3}$
9981 4356 tanh : $10\pi/9$
9981 4779 ln : $\sqrt{2}+3\sqrt{3}/4$
9981 5003 $2/5+\exp(4)$
9981 5775 $s\pi$: $(\ln 2/3)^{1/2}$
9981 5935 tanh : $9e/7$
9981 7187 $s\pi$: $\sqrt[3]{3}/3$
9981 7789 tanh : 7/2
9981 8007 erf : $7\sqrt[3]{2}/4$
9981 8569 $s\pi$: $2\zeta(3)/5$
9982 0000 $s\pi$: $3\sqrt{3}-3\sqrt{5}/4$
9982 0495 ζ : 22/17
9982 0741 sq : $2\sqrt{3}/3+2\sqrt{5}/3$
9982 1418 $s\pi$: $2\sqrt{2}-4\sqrt{3}/3$
9982 3496 tanh : $3(e+\pi)/5$
9982 3775 erf : $\exp(\sqrt[3]{4}/2)$
9982 5742 e^x : $\sqrt{22}\pi$
9982 5920 $\ln(4/5)+\exp(1/5)$
9982 6192 erf : $1/\ln(\pi/2)$
9982 6464 J_0 : 1/12
9982 6718 tanh : $\exp(\sqrt[3]{2})$
9982 8790 $s\pi$: $\sqrt{3}+\sqrt{5}-\sqrt{6}$
9982 9043 λ : $e/8$
9982 9763 $s\pi$: $1/2+4\sqrt{5}/3$
9982 9857 tanh : $9\pi/8$
9983 0279 tanh : $5\sqrt{2}/2$
9983 0804 cu : $3e\pi$
9983 1622 J_0 : $\exp(-5/2)$
9983 1785 Ψ : 11/14
9983 2172 e^x : 9/13
9983 2600 erf : 20/9
9983 4016 erf : $\sqrt[3]{11}$
9983 4586 $s\pi$: $\exp(3/2)$
9983 4883 $s\pi$: $\sqrt[3]{7/2}$
9983 6362 erf : $9\sqrt{3}/7$
9983 7347 erf : $7(1/\pi)$
9983 8224 λ : $\sqrt[3]{5}/5$
9983 9114 ℓ_{10} : $7e\pi/6$
9983 9306 $t\pi$: $9\zeta(3)/8$
9983 9670 $t\pi$: $8\ln 2$
9983 9740 cu : $\sqrt{23/2}$
9984 0063 J_0 : 2/25
9984 1055 $s\pi$: $3/4+\sqrt{3}$
9984 1262 Ψ : $5\sqrt{2}/9$
9984 1748 J_0 : $(1/\pi)/4$
9984 2026 tanh : 25/7
9984 2097 tanh : $(\sqrt[3]{4}/3)^{-2}$
9984 3459 erf : $\sqrt{5}$
9984 3997 tanh : $8\sqrt{5}/5$
9984 4003 λ : $2\zeta(3)/7$
9984 4959 cr : $1/4+\sqrt{5}/3$
9984 7901 tanh : $8\pi/7$
9984 8842 2^x : $1/4+4\sqrt{2}$
9984 8901 erf : $\ln(3\pi)$
9984 9380 erf : $5\pi/7$
9984 9401 $\ln(5)+\exp(2)$
9984 9467 ζ : $(2e/3)^{-1/3}$
9985 0097 at : $1/3-4\sqrt{2}/3$
9985 0794 tanh : 18/5
9985 1522 e^x : $1/4+e/2$
9985 1766 J_0 : $\ln 2/9$
9985 1960 $s\pi$: $2\sqrt{3}-4\sqrt{5}/3$
9985 2038 ℓ_2 : $\sqrt{2}+\sqrt{3}-\sqrt{7}$
9985 2125 J_0 : 1/13
9985 2440 tanh : $\sqrt{13}$
9985 2461 tanh : $5\sqrt[3]{3}/2$
9985 2622 tanh : $3\zeta(3)$
9985 2883 ln : 7/19
9985 3728 erf : 9/4
9985 3757 $s\pi$: $4e/3-\pi$
9985 5251 λ : $\sqrt{3}/5$
9985 5545 10^x : $((e+\pi)/2)^{-1/3}$
9985 5849 λ : $\ln 2/2$
9985 6239 λ : $1/\sqrt[3]{24}$
9985 7742 ln : $\sqrt[3]{20}$
9985 7888 tanh : $4e/3$
9985 8542 $s\pi$: $4e-3\pi/4$
9985 8773 erf : $\sqrt[3]{23/2}$
9985 9316 $s\pi$: $5\pi^2/9$
9985 9330 10^x : $2\sqrt{2}/3+\sqrt{3}$
9986 0360 λ : 8/23
9986 2567 $s\pi$: $3\sqrt{2}/4-\sqrt{3}/3$
9986 2572 erf : $8\sqrt{2}/5$
9986 2872 J_0 : $(\sqrt[3]{2}/3)^3$
9986 3859 $s\pi$: $\sqrt{19/3}$
9986 4244 erf : $5e/6$
9986 4700 λ : $\pi/9$
9986 5982 erf : $9\sqrt[3]{2}/5$
9986 7083 Ei : 16
9986 7265 $s\pi$: P_{32}
9986 7625 tanh : $3e\pi/7$
9986 7889 λ : 7/20
9986 8052 Ei : $10\sqrt{5}/9$
9986 8294 tanh : $5(e+\pi)/8$
9986 8498 λ : $(1/(3e))^{1/2}$
9986 8824 $s\pi$: $4\sqrt{2}/3+3\sqrt{3}/2$
9986 9021 tanh : $7\pi/6$
9986 9134 sinh : $2\pi/3$
9986 9155 erf : 25/11
9986 9406 tanh : 11/3
9986 9898 $s\pi$: $2e+\pi/3$
9987 0830 $t\pi$: $5\sqrt{5}/8$
9987 0964 λ : $\exp(-\pi/3)$
9987 2489 J_0 : 1/14
9987 3102 tanh : $\sqrt{2}-\sqrt{6}-\sqrt{7}$
9987 3306 λ : $1/\sqrt[3]{23}$
9987 3746 erf : $4\sqrt[3]{5}/3$
9987 4881 $s\pi$: $3(e+\pi)/5$
9987 4951 J_0 : $2(1/\pi)/9$
9987 6787 $s\pi$: P_{42}
9987 7282 erf : 16/7
9987 7495 λ : 6/17
9987 8092 tanh : $3\pi^2/8$
9987 9414 λ : $\sqrt{2}/4$
9987 9519 erf : $\sqrt[3]{12}$
9987 9800 λ : $10(1/\pi)/9$
9987 9922 J_0 : $\ln 2/10$
9987 9974 at : $4e/7$
9988 0292 $s\pi$: $9\sqrt{5}/8$
9988 0726 $\ln(\sqrt{2})+\exp(\sqrt{3})$
9988 0784 $s\pi$: $7\sqrt{3}/8$
9988 1770 $s\pi$: $10\sqrt{5}/9$
9988 3216 $s\pi$: $6\sqrt{3}/7$
9988 3889 Ψ : 16/5
9988 4192 tanh : $5\sqrt{5}/3$
9988 4470 Γ : $\zeta(9)$
9988 4684 cu : $\sqrt{2}/4-4\sqrt{5}$
9988 5682 erf : 23/10
9988 6498 $s\pi$: $3\zeta(3)/7$
9988 7585 tanh : $\sqrt{14}$
9988 8919 J_0 : 1/15
9988 9385 λ : $1/\sqrt[3]{22}$
9988 9444 tanh : 15/4
9988 9930 erf : $8\sqrt[3]{3}/5$
9989 0131 λ : 5/14
9989 0251 sinh : $5e/2$
9989 0916 erf : $4\sqrt{3}/3$
9989 1169 J_0 : $\exp(-e)$
9989 1508 erf : $10\ln 2/3$
9989 1971 $s\pi$: $7\ln 2/10$
9989 2360 e^x : $5e/2$
9989 2837 λ : $9(1/\pi)/8$
9989 3112 $s\pi$: $6\sqrt{2}$
9989 3745 as : $(\sqrt{2})^{-1/2}$
9989 3758 tanh : $6\pi/5$
9989 4039 tanh : $8\sqrt{2}/3$
9989 4554 $\sqrt{2}-\exp(5)$
9989 5712 $s\pi$: $7\pi^2/6$
9989 5829 tanh : $3\sqrt[3]{2}$
9989 6962 erf : $\sqrt[3]{25/2}$
9989 7626 J_0 : $(5/2)^{-3}$
9989 7981 λ : $2\sqrt[3]{2}/7$
9989 8042 λ : 9/25
9989 8685 tanh : $\exp(4/3)$
9989 8704 J_0 : $(1/\pi)/5$
9989 9043 tanh : $4e\pi/9$
9989 9537 λ : $\sqrt[3]{3}/4$
9989 9681 λ : $3\zeta(3)/10$
9989 9959 tanh : 19/5
9990 0581 erf : $\sqrt{2}-\sqrt{3}+\sqrt{7}$
9990 1072 tanh : $7e/5$
9990 1350 Γ : $5\zeta(3)/6$
9990 1597 erf : $6e/7$
9990 2343 sinh : $9\ln 2$
9990 2367 J_0 : 1/16
9990 2430 erf : $(2e)^{1/2}$
9990 2982 tanh : $((e+\pi)/3)^2$
9990 3257 erf : 7/3
9990 3481 $s\pi$: $7e/2$
9990 3757 ln : $\sqrt{3}/3-\sqrt{5}/4$
9990 4230 $s\pi$: $1/4+\sqrt{5}$
9990 4439 λ : $1/\sqrt[3]{21}$
9990 4478 e^x : $3e/4+3\pi/2$
9990 6105 $s\pi$: $\exp(-\sqrt[3]{3}/2)$
9990 6408 tanh : 23/6
9990 6551 Ψ : $\pi/4$
9990 7151 cr : $3\sqrt{2}/4+4\sqrt{3}$
9990 7368 λ : 4/11
9990 7727 λ : $8(1/\pi)/7$
9990 8306 erf : $2(e+\pi)/5$
9990 8748 tanh : $8\sqrt[3]{3}/3$
9990 8888 erf : $\sqrt{11/2}$
9990 9012 tanh : $9\sqrt[3]{5}/4$
9990 9997 $\pi-\exp(\pi)$
9991 1177 $s\pi$: $\exp(-2/3)$
9991 1673 erf : $\sqrt[3]{13}$
9991 1678 erf : $\exp(\sqrt[3]{5}/2)$
9991 2851 sinh : $\sqrt{22}\pi$
9991 3060 $s\pi$: $4\pi/5$
9991 3484 λ : $\ln(\sqrt[3]{3})$
9991 3513 J_0 : 1/17
9991 3540 tanh : $\sqrt{15}$
9991 3825 erf : $3\pi/4$
9991 4023 tanh : $(\pi/2)^3$
9991 4187 erf : $5\sqrt{2}/3$
9991 5298 tanh : $10e/7$
9991 6323 sinh : 20/11
9991 6705 $s\pi$: $3\sqrt[3]{5}/10$
9991 6906 tanh : $\exp(e/2)$
9991 7277 λ : $\exp(-1)$
9991 7612 tanh : $9\sqrt{3}/4$
9991 8431 λ : $1/\sqrt[3]{20}$
9991 8470 λ : 7/19
9991 9157 tanh : $2(e+\pi)/3$
9991 9414 $s\pi$: $4\sqrt{3}+\sqrt{5}/4$
9991 9830 erf : $(4/3)^3$
9992 0207 tanh : $7\sqrt{5}/4$
9992 1706 erf : 19/8
9992 2389 tanh : $5\pi/4$
9992 2679 erf : $\exp(\sqrt{3}/2)$
9992 2854 J_0 : 1/18
9992 3095 erf : $7e/8$
9992 3540 λ : $\exp(-6\pi/19)$
9992 3871 erf : $\sqrt{17/3}$
9992 4685 λ : $7(1/\pi)/6$
9992 5558 tanh : $2\pi^2/5$

9992 7327 $\lambda : \sqrt{5}/6$
9992 8722 $J_0 : \exp(-(e+\pi)/2)$
9992 8767 $J_0 : (\ln 2/3)^2$
9992 8822 $s\pi : 10\pi/7$
9992 8971 $erf : 7\sqrt[3]{5}/5$
9992 9347 $s\pi : 1/3+2\sqrt{3}/3$
9992 9650 $J_0 : (1/\pi)/6$
9993 0064 $s\pi : 6\sqrt[3]{2}/5$
9993 0759 $J_0 : 1/19$
9993 1048 $s\pi : 1/3+3e$
9993 1140 $\lambda : \exp(-5\pi/16)$
9993 1148 $erf : 12/5$
9993 1328 $\lambda : 1/\sqrt[3]{19}$
9993 1567 $\tanh : 7\sqrt[3]{5}/3$
9993 1785 $\lambda : 3/8$
9993 2469 $erf : 5\sqrt[3]{3}/3$
9993 2596 $erf : 2\zeta(3)$
9993 2929 $\tanh : 4$
9993 3843 $\tanh : 10\zeta(3)/3$
9993 4113 $e^x : 5\pi$
9993 4639 $e^x : 2\sqrt{3}/5$
9993 4668 $erf : \sqrt[3]{14}$
9993 5474 $s\pi : 2\sqrt[3]{5}/7$
9993 5801 $s\pi : 10e\pi/9$
9993 5847 $J_0 : (\sqrt{2}\pi)^{-2}$
9993 6157 $s\pi : 3(e+\pi)/7$
9993 6190 $\tanh : 9\sqrt{5}/5$
9993 6278 $erf : (\sqrt{5}/3)^{-3}$
9993 6706 $erf : 8e/9$
9993 7118 $\lambda : 1/\sqrt{7}$
9993 7138 $\lambda : 3\sqrt[3]{2}/10$
9993 7509 $J_0 : 1/20$
9993 7984 $\tanh : 9\pi/7$
9993 8040 $J_0 : (e)^{-3}$
9993 8159 $erf : (e+\pi)^{1/2}$
9993 8264 $\tanh : 7\sqrt{3}/3$
9993 9482 $erf : 7\sqrt{3}/5$
9993 9714 $J_0 : (\ln 3/3)^3$
9993 9842 $erf : 7\ln 2/2$
9993 9860 $s\pi : 3e+3\pi/4$
9994 0177 $\ell_{10} : P_{48}$
9994 0305 $s\pi : P_{13}$
9994 0458 $\Gamma : 7(e+\pi)/3$
9994 0551 $\lambda : 2\sqrt[3]{5}/9$
9994 0602 $s\pi : 10(e+\pi)/9$
9994 0639 $erf : 17/7$
9994 1146 $\lambda : \exp(-4\pi/13)$
9994 1792 $erf : \sqrt{5}-\sqrt{6}+\sqrt{7}$
9994 1913 $\sinh : 5\sqrt{3}/6$
9994 2111 $\lambda : 8/21$
9994 2172 $s\pi : \ln(5/3)$
9994 2548 $\tanh : 3e/2$
9994 3100 $\lambda : 1/\sqrt[3]{18}$
9994 3318 $J_0 : 1/21$
9994 3589 $s\pi : (e/2)^3$
9994 3732 $\lambda : 6(1/\pi)/5$
9994 3901 $e^x : \sqrt{22/3}$
9994 4060 $erf : 2e\pi/7$
9994 4663 $s\pi : 3\sqrt{2}-\sqrt{3}$
9994 4903 $erf : 10\sqrt[3]{5}/7$
9994 5088 $erf : 7\pi/9$
9994 5293 $\tanh : 7(e+\pi)/10$
9994 5370 $erf : 22/9$
9994 5651 $\sinh : 9\sqrt{2}/7$
9994 5884 at : 14/9
9994 5943 $erf : 9e/10$
9994 6520 $\tanh : \exp(\sqrt{2})$
9994 6799 $erf : \sqrt{6}$
9994 6836 $J_0 : \ln(\pi/3)$
9994 7563 $\tanh : \sqrt{17}$
9994 7745 $\lambda : 5/13$
9994 8162 $\lambda : 2\sqrt{3}/9$
9994 8312 $J_0 : (1/\pi)/7$
9994 8353 $J_0 : 1/22$
9994 8361 $\exp(4/5)+s\pi(e)$
9994 8426 $\lambda : 5\ln 2/9$
9994 9098 $s\pi : 5e/9$
9994 9310 $J_0 : (3\pi/2)^{-2}$
9994 9666 $\ell_{10} : P_{18}$
9995 1171 $\sinh : 10\ln 2$
9995 1183 $\tanh : 6\ln 2$
9995 1289 $erf : \sqrt[3]{15}$
9995 1313 $erf : 1/\ln(2/3)$
9995 1595 $erf : \pi^2/4$
9995 1862 $s\pi : 9\sqrt[3]{3}/2$
9995 1937 $\tanh : 25/6$
9995 2645 $s\pi : 2\sqrt{2}/3-\sqrt{3}/4$
9995 2746 $J_0 : 1/23$
9995 2945 $\lambda : e/7$
9995 3319 $J_0 : \exp(-\pi)$
9995 3346 $erf : 10\sqrt{3}/7$
9995 3415 $erf : \exp(e/3)$
9995 3474 $erf : 7\sqrt{2}/4$
9995 3692 $\lambda : 7/18$
9995 3721 $\lambda : 1/\sqrt[3]{17}$
9995 3725 $\tanh : 5(e+\pi)/7$
9995 4017 $\tanh : 4\pi/3$
9995 4699 $\lambda : \exp(-3\pi/10)$
9995 4895 $s\pi : 7(e+\pi)/2$
9995 4997 $s\pi : (\sqrt{2}\pi/2)^{1/2}$
9995 5012 $\tanh : 10\sqrt[3]{2}/3$
9995 5036 $\tanh : 21/5$
9995 5679 $\tanh : 7\zeta(3)/2$
9995 5799 $erf : 10\sqrt{5}/9$
9995 5891 $s\pi : \sqrt{2}+\sqrt{6}+\sqrt{7}$
9995 6601 $J_0 : 1/24$
9995 6682 $J_0 : (\ln 2/2)^3$
9995 6777 $\lambda : 9/23$
9995 6940 $s\pi : 10\pi/9$
9995 7360 $s\pi : 8(1/\pi)/5$
9995 7431 $s\pi : 2\sqrt{5}/3$
9995 7640 $\tanh : 3\pi^2/7$
9995 7671 $\tanh : \exp(\sqrt[3]{3})$
9995 8464 $\lambda : \ln(\sqrt{2}\pi/3)$
9995 8473 $\lambda : \pi/8$
9995 8528 $Ei : 2(1/\pi)/9$
9995 8711 $\tanh : 3\sqrt{2}$
9995 9304 $erf : 5/2$
9995 9314 $\tanh : 17/4$
9996 0003 $J_0 : 1/25$
9996 0425 $J_0 : (1/\pi)/8$
9996 0486 $J_0 : ((e+\pi)/2)^{-3}$
9996 0724 $erf : (2\pi)^{1/2}$
9996 0899 $\tanh : e\pi/2$
9996 1293 $\tanh : 5\sqrt[3]{5}/2$
9996 1572 $erf : (e/2)^3$
9996 1713 $erf : 3(e+\pi)/7$
9996 1765 $s\pi : 2\sqrt{3}/3+3\sqrt{5}/2$
9996 2102 $erf : 4\pi/5$
9996 2344 $\lambda : 4\ln 2/7$
9996 2568 $erf : 9\sqrt{5}/8$
9996 2680 $s\pi : \ln(\sqrt{2}\pi)$
9996 2776 $erf : \sqrt{19/3}$
9996 3173 $\lambda : 1/\sqrt[3]{16}$
9996 3258 $\lambda : \exp(-5\pi/17)$
9996 3418 $erf : 2\sqrt[3]{2}$
9996 4218 $erf : 7\sqrt[3]{3}/4$
9996 4270 $\lambda : 5(1/\pi)/4$
9996 4695 $\lambda : (e/2)^{-3}$
9996 5103 $\tanh : 3\sqrt[3]{3}$
9996 5196 $\tanh : 1/\ln(\sqrt[3]{2})$
9996 5337 $\tanh : 5\sqrt{3}/2$
9996 5356 $\lambda : (2\pi)^{-1/2}$
9996 5559 $\tanh : 13/3$
9996 5640 $s\pi : \exp(2\pi)$
9996 5709 $J_0 : (3)^{-3}$
9996 6382 $s\pi : \sqrt{2}/4+2\sqrt{3}/3$
9996 6414 $\lambda : 2/5$
9996 6638 $\tanh : 8e/5$
9996 7049 $\sinh : 5\pi$
9996 7085 $\lambda : \zeta(3)/3$
9996 7275 $\tanh : \sqrt{19}$
9996 8178 $erf : 9\sqrt{2}/5$
9996 8332 $erf : 8(1/\pi)$
9996 8341 $erf : 1/\ln(\sqrt{2}\pi/3)$
9996 8730 $J_0 : (1/\pi)/9$
9996 8850 $erf : \sqrt{13/2}$
9996 9032 $\tanh : 4\pi^2/9$
9996 9549 $\tanh : 3(e+\pi)/4$
9996 9588 $erf : \sqrt{2}-\sqrt{3}-\sqrt{5}$
9996 9686 $\tanh : \exp(\sqrt{2}\pi/3)$
9996 9751 $\tanh : 7\pi/5$
9996 9852 $erf : 8\sqrt{5}/7$
9996 9857 $\tanh : 22/5$
9996 9860 $erf : 23/9$
9997 0213 $\lambda : 2\sqrt{2}/7$
9997 0248 $\lambda : \exp(-e/3)$
9997 0438 $\tanh : 7\sqrt[3]{2}/2$
9997 0890 $erf : 3e\pi/10$
9997 1280 $\lambda : (\pi/2)^{-2}$
9997 1371 $erf : 3\sqrt[3]{5}/2$
9997 1383 $J_0 : (2e)^{-2}$
9997 1434 $\lambda : \ln(3/2)$
9997 1447 $\lambda : 1/\sqrt[3]{15}$
9997 1498 $\Gamma : \zeta(11)$
9997 1780 $\Psi : 3e\pi/8$
9997 2334 $\tanh : \sqrt{2}\pi$
9997 2346 $erf : \sqrt[3]{17}$
9997 2368 $erf : 18/7$
9997 3162 $\lambda : \exp(-2\pi/7)$
9997 3274 $s\pi : 1/4+3\sqrt{2}$
9997 3722 $\lambda : 1/\sqrt{6}$
9997 3906 $\tanh : 2\sqrt{5}$
9997 3927 $erf : \sqrt{20/3}$
9997 3997 $J_0 : (\pi)^{-3}$
9997 4381 $\lambda : 9/22$
9997 4400 $\tanh : \exp(3/2)$
9997 4508 $\lambda : 9(1/\pi)/7$
9997 4671 $J_0 : (1/\pi)/10$
9997 4721 $\tanh : 10\pi/7$
9997 5321 $\tanh : 9/2$
9997 5874 $erf : 9\sqrt[3]{3}/5$
9997 6143 $erf : 3\sqrt{3}/2$
9997 6292 $s\pi : \exp(-1/\sqrt{2})$
9997 6376 $\lambda : 7/17$
9997 6396 $erf : 13/5$
9997 6595 $\lambda : 2\sqrt[3]{3}/7$
9997 6779 $\tanh : 5e/3$
9997 6964 $erf : 4(e+\pi)/9$
9997 7319 $\lambda : (e+\pi)^{-1/2}$
9997 7514 $erf : 7\sqrt{5}/6$
9997 7994 $\lambda : (\sqrt{5}/3)^3$
9997 8009 $\tanh : 7(e+\pi)/9$
9997 8108 $\tanh : 8\sqrt[3]{5}/3$
9997 8320 $s\pi : (2\pi)^{1/2}$
9997 8332 $s\pi : 3e/2-\pi/2$
9997 8545 $\lambda : 1/\sqrt[3]{14}$
9997 8643 $erf : 5\pi/6$
9997 8798 $J_0 : (e+\pi)^{-2}$
9997 8968 $erf : \sqrt[3]{18}$
9997 9077 $\tanh : \sqrt{21}$
9997 9179 $\lambda : 3\ln 2/5$
9997 9462 $erf : 21/8$
9997 9673 $\lambda : 5/12$
9997 9794 $\tanh : 23/5$
9997 9891 $s\pi : 1/\ln((e+\pi)/3)$
9998 0245 $s\pi : 3\sqrt{2}/4+\sqrt{3}/4$
9998 0397 $\lambda : \exp(-5\pi/18)$
9998 0539 $\tanh : 8\sqrt{3}/3$
9998 0956 $\tanh : (5/3)^3$
9998 1040 $s\pi : \sqrt[3]{10/3}$
9998 1658 $\lambda : \sqrt[3]{2}/3$
9998 1718 $erf : \sqrt{7}$
9998 2024 $\lambda : \exp(-\sqrt{3}/2)$
9998 2266 $\lambda : 8/19$
9998 2316 $\tanh : 14/3$
9998 2718 $\lambda : (4/3)^{-3}$
9998 3051 $\tanh : 4(e+\pi)/5$
9998 3136 $\tanh : \sqrt{22}$
9998 3755 $erf : 8/3$
9998 3861 $\tanh : 3\pi/2$
9998 3914 $\tanh : 10\sqrt{2}/3$
9998 3914 $erf : \sqrt[3]{19}$
9998 3960 $J_0 : (2\pi)^{-2}$
9998 3971 $\lambda : 3\sqrt{2}/10$
9998 4047 $\lambda : 4(1/\pi)/3$
9998 4100 $\lambda : \exp(-3\pi/11)$
9998 4483 $\lambda : \exp(-\sqrt[3]{5}/2)$
9998 4484 $\lambda : 1/\sqrt[3]{13}$
9998 4859 $\tanh : 5e\pi/9$

9998 5030 tanh : $19/4$
9998 5191 $s\pi$: $2\sqrt{2}+3\sqrt{5}/4$
9998 5219 erf : $6\sqrt{5}/5$
9998 5238 tanh : $7e/4$
9998 5536 λ : $\sqrt[3]{5}/4$
9998 6000 erf : $6\pi/7$
9998 6027 λ : $3/7$
9998 6165 λ : $(2e)^{-1/2}$
9998 6341 tanh : $\sqrt{23}$
9998 6455 tanh : $24/5$
9998 6656 tanh : $10\sqrt[3]{3}/3$
9998 6676 tanh : $4\zeta(3)$
9998 6736 tanh : $\exp(\pi/2)$
9998 6918 erf : $9\zeta(3)/4$
9998 7021 $s\pi$: $2\sqrt{3}/7$
9998 7170 erf : $\sqrt{22/3}$
9998 7272 $s\pi$: P_{35}
9998 7334 $s\pi$: $9e/7$
9998 7625 erf : $19/7$
9998 7634 erf : $\sqrt[3]{20}$
9998 7760 λ : $3\sqrt[3]{3}/10$
9998 7762 λ : $\exp(-4\pi/15)$
9998 7793 tanh : $7\ln 2$
9998 7894 λ : $\sqrt{3}/4$
9998 7906 erf : e
9998 7974 λ : $5\ln 2/8$
9998 8121 $s\pi$: $2/3+2\sqrt{2}$
9998 8176 $s\pi$: $5\ln 2/7$
9998 8454 tanh : $4e\pi/7$
9998 8532 tanh : $5(e+\pi)/6$
9998 8573 λ : $10/23$
9998 8594 erf : $1/\ln(\ln 2)$
9998 8709 tanh : $\exp(\sqrt[3]{4})$
9998 8745 erf : $1/\ln(\sqrt[3]{3})$
9998 8751 tanh : $9e/5$
9998 8887 tanh : $\sqrt{24}$
9998 9082 erf : $8\sqrt[3]{5}/5$
9998 9248 erf : $\sqrt{15/2}$
9998 9302 λ : $1/\sqrt[3]{12}$
9998 9324 $s\pi$: $(\sqrt{5})^{1/2}$
9998 9357 $s\pi$: $3/4+\sqrt{5}/3$
9998 9541 λ : $\exp(-5\pi/19)$
9998 9550 λ : $7/16$
9998 9655 tanh : $\pi^2/2$
9998 9872 erf : $7\pi/8$
9998 9937 erf : $11/4$
9998 9960 tanh : $7\sqrt{2}/2$
9999 0381 λ : $11/25$
9999 0448 erf : $\sqrt[3]{21}$
9999 0906 erf : $\sqrt{2}-\sqrt{3}-\sqrt{6}$
9999 0920 tanh : 5
9999 0988 erf : $\sqrt{23/3}$
9999 1115 erf : $8\sqrt{3}/5$
9999 1182 erf : $4\ln 2$
9999 1324 tanh : $6(e+\pi)/7$
9999 1389 tanh : $8\pi/5$
9999 1447 erf : $25/9$
9999 1468 tanh : $9\sqrt{5}/4$
9999 1613 J_0 : $\exp(-4)$
9999 1613 tanh : $4\sqrt[3]{2}$
9999 1712 λ : $4/9$
9999 1749 tanh : $7\sqrt[3]{3}/2$
9999 2039 λ : $7(1/\pi)/5$
9999 2074 λ : $1/\ln(3\pi)$
9999 2159 erf : $8\pi/9$
9999 2228 $s\pi$: $2\sqrt[3]{2}/5$
9999 2277 erf : $5\sqrt{5}/4$
9999 2455 λ : $\sqrt{5}/5$
9999 2498 erf : $14/5$
9999 2588 erf : $\sqrt[3]{22}$
9999 2709 erf : $7\zeta(3)/3$
9999 2853 λ : $\pi/7$
9999 2912 tanh : $3e\pi/5$
9999 2962 tanh : $7(e+\pi)/8$
9999 2982 $s\pi$: $7\sqrt[3]{5}/8$
9999 2998 tanh : $3\sqrt[3]{5}$
9999 3057 λ : $1/\sqrt[3]{11}$
9999 3141 λ : $9/20$
9999 3317 $s\pi$: $e+\pi/4$
9999 3321 λ : $3\zeta(3)/8$
9999 3334 erf : $2\pi^2/7$
9999 3506 λ : $\ln(\pi/2)$
9999 3628 erf : $9\pi/10$
9999 3636 λ : $\exp(-\sqrt[3]{4}/2)$
9999 3665 erf : $2\sqrt{2}$
9999 3827 λ : $e/6$
9999 3848 erf : $17/6$
9999 3866 tanh : $3\sqrt{3}$
9999 3896 J_0 : $(4)^{-3}$
9999 3903 erf : $9\sqrt[3]{2}/4$
9999 4019 tanh : $8(e+\pi)/9$
9999 4122 tanh : $7\sqrt{5}/3$
9999 4141 λ : $5/11$
9999 4178 λ : $10(1/\pi)/7$
9999 4225 erf : $\sqrt[3]{23}$
9999 4234 as : $7\zeta(3)/10$
9999 4318 erf : $e\pi/3$
9999 4336 tanh : $5\pi/3$
9999 4347 J_0 : $(1/(3e))^2$
9999 4420 λ : $\exp(-\pi/4)$
9999 4421 λ : $(\sqrt{2}\pi/3)^{-2}$
9999 4422 erf : $\exp(\pi/3)$
9999 4433 erf : $5\sqrt[3]{5}/3$
9999 4492 tanh : $21/4$
9999 4621 erf : $(1/(3e))^{-1/2}$
9999 4668 erf : $20/7$
9999 4733 erf : $\sqrt{5}-\sqrt{6}-\sqrt{7}$
9999 4749 tanh : $9(e+\pi)/10$
9999 4871 λ : $11/24$
9999 4909 erf : $9(1/\pi)$
9999 4961 tanh : $\exp(5/3)$
9999 4994 $s\pi$: $(1/(3e))^{1/3}$
9999 5212 erf : $9\sqrt{5}/7$
9999 5214 erf : $23/8$
9999 5269 $s\pi$: $2\sqrt{5}/9$
9999 5278 λ : $(3\pi/2)^{-1/2}$
9999 5338 tanh : $16/3$
9999 5375 tanh : $5e\pi/8$
9999 5424 λ : $6/13$
9999 5482 erf : $2\sqrt[3]{3}$
9999 5507 erf : $1/\ln(\sqrt{2})$
9999 5515 λ : $2\ln 2/3$
9999 5544 erf : $5\sqrt{3}/3$
9999 5773 tanh : $\sqrt{2}+\sqrt{3}+\sqrt{5}$
9999 5835 λ : $1/\sqrt[3]{10}$
9999 5920 tanh : $(\sqrt[3]{5}/3)^{-3}$
9999 5995 tanh : $9\zeta(3)/2$
9999 6196 J_0 : $(\ln 2/3)^3$
9999 6197 λ : $7/15$
9999 6207 tanh : $2e$
9999 6244 tanh : $\sqrt{3}\pi$
9999 6262 erf : $\sqrt{17/2}$
9999 6361 $s\pi$: $3\sqrt{2}-\sqrt{5}/3$
9999 6453 erf : $\sqrt[3]{25}$
9999 6541 J_0 : $\exp(-\sqrt{2}\pi)$
9999 6544 tanh : $5\pi^2/9$
9999 6580 erf : $(e+\pi)/2$
9999 6644 tanh : $7\pi/4$
9999 6659 tanh : $11/2$
9999 6706 λ : $8/17$
9999 6737 $s\pi$: $5\zeta(3)/4$
9999 6749 J_0 : $(\sqrt{2}\pi)^{-3}$
9999 6782 erf : $7\sqrt[3]{2}/3$
9999 6804 λ : $\sqrt{2}/3$
9999 6831 J_0 : $(3\pi)^{-2}$
9999 6846 tanh : $\exp(\sqrt[3]{5})$
9999 6928 λ : $3\sqrt[3]{2}/8$
9999 6948 tanh : $8\ln 2$
9999 7034 λ : $(3\pi)^{-1/3}$
9999 7044 erf : $\sqrt{3}-\sqrt{5}-\sqrt{6}$
9999 7064 λ : $9/19$
9999 7144 λ : $(\sqrt{2}\pi)^{-1/2}$
9999 7159 λ : $\exp(-\sqrt{5}/3)$
9999 7177 erf : $3\pi^2/10$
9999 7210 tanh : $5\sqrt{5}/2$
9999 7224 erf : $\sqrt{2}-\sqrt{3}-\sqrt{7}$
9999 7241 tanh : $\sqrt{2}+\sqrt{3}+\sqrt{6}$
9999 7327 λ : $10/21$
9999 7452 λ : $3(1/\pi)/2$
9999 7456 λ : $\exp(-4\pi/17)$
9999 7474 tanh : $4\pi^2/7$
9999 7517 erf : $4\sqrt{5}/3$
9999 7528 λ : $11/23$
9999 7536 tanh : $\exp(\sqrt{3})$
9999 7549 tanh : $9\pi/5$
9999 7559 tanh : $4\sqrt{2}$
9999 7583 $s\pi$: $7\pi/4$
9999 7606 tanh : $17/3$
9999 7610 $s\pi$: $3e/2+3\pi$
9999 7620 tanh : $9\sqrt[3]{2}/2$
9999 7683 erf : $7\sqrt[3]{5}/4$
9999 7686 λ : $12/25$
9999 7711 $s\pi$: $4\sqrt{2}-2\sqrt{3}/3$
9999 7717 J_0 : $(3\pi/2)^{-3}$
9999 7726 λ : $(\ln 2)^2$
9999 7730 tanh : $2e\pi/3$
9999 7745 λ : $(\ln 2/3)^{1/2}$
9999 7752 λ : $\sqrt[3]{3}/3$
9999 7758 λ : $2\zeta(3)/5$
9999 7760 tanh : $10\sqrt[3]{5}/3$
9999 7790 erf : 3
9999 7829 erf : $(\ln 2)^{-3}$
9999 7861 erf : $5\zeta(3)/2$
9999 7955 ln : $4\sqrt{2}+\sqrt{3}$
9999 7974 tanh : $23/4$
9999 7982 J_0 : $\exp(-3\pi/2)$
9999 8041 λ : $\exp(-3\pi/13)$
9999 8049 tanh : $4\sqrt[3]{3}$
9999 8057 erf : $10e/9$
9999 8067 tanh : $10\sqrt{3}/3$
9999 8107 λ : $7\ln 2/10$
9999 8137 tanh : $\sqrt{2}+\sqrt{3}+\sqrt{7}$
9999 8179 λ : $\exp(-\sqrt[3]{3}/2)$
9999 8185 erf : $7\sqrt{3}/4$
9999 8340 λ : $2\sqrt[3]{5}/7$
9999 8354 $s\pi$: $2\sqrt{2}/3+\sqrt{5}/4$
9999 8373 tanh : $(e+\pi)$
9999 8400 J_0 : $(5)^{-3}$
9999 8473 erf : $9e/8$
9999 8478 erf : $1/\ln(\sqrt[3]{3}/2)$
9999 8480 erf : $\exp(\sqrt{5}/2)$
9999 8563 tanh : $3\pi^2/5$
9999 8585 erf : $(3\pi)^{1/2}$
9999 8612 λ : $\exp(-1/\sqrt{2})$
9999 8645 tanh : $(2e/3)^3$
9999 8656 erf : $9\sqrt[3]{5}/5$
9999 8667 $s\pi$: $3\sqrt{2}/4-\sqrt{5}/4$
9999 8676 tanh : $8\sqrt{5}/3$
9999 8692 erf : $\sqrt{19/2}$
9999 8709 λ : $2\sqrt{3}/7$
9999 8715 tanh : $7e\pi/10$
9999 8721 λ : $5\ln 2/7$
9999 8733 tanh : $7\sqrt[3]{5}/2$
9999 8771 tanh : 6
9999 8796 tanh : $5\zeta(3)$
9999 8806 λ : $(1/(3e))^{1/3}$
9999 8810 λ : $2\sqrt{5}/9$
9999 8840 λ : $\exp(-2\pi/9)$
9999 8850 tanh : $(\ln 3/2)^{-3}$
9999 8865 J_0 : $\exp(-5)$
9999 8882 $s\pi$: $(1/(3e))^{-2}$
9999 8884 erf : $8e/7$
9999 8914 tanh : $7\sqrt{3}/2$
9999 8952 λ : $1/2$
9999 8961 erf : $9\sqrt{3}/5$
9999 8971 erf : $9\ln 2/2$
9999 8993 tanh : $5e\pi/7$
9999 9010 erf : $25/8$
9999 9025 tanh : $9e/4$
9999 9031 J_0 : $(2e)^{-3}$
9999 9040 tanh : $\exp(2e/3)$
9999 9045 erf : $7\sqrt{5}/5$
9999 9112 erf : π
9999 9119 erf : $22/7$
9999 9122 tanh : $5\pi^2/8$
9999 9158 erf : $\sqrt{3}-\sqrt{5}-\sqrt{7}$
9999 9159 erf : $5\sqrt[3]{2}/2$
9999 9225 erf : $\sqrt{10}$
9999 9237 tanh : $9\ln 2$
9999 9247 erf : $19/6$

9999 9254 tanh : $25/4$
9999 9270 erf : $7e/6$
9999 9302 tanh : 2π
9999 9312 λ : $\exp(-3\pi/14)$
9999 9320 tanh : $\sqrt{2}+\sqrt{5}+\sqrt{7}$
9999 9325 tanh : $5\sqrt[3]{2}$
9999 9358 cu : $3+4\sqrt{5}/3$
9999 9369 tanh : $19/3$
9999 9374 erf : $10\sqrt{5}/7$
9999 9380 tanh : $7e/3$
9999 9382 J_0 : $(e+\pi)^{-3}$
9999 9397 erf : $16/5$
9999 9406 tanh : $9\sqrt{2}/2$
9999 9407 erf : $3e\pi/8$
9999 9419 erf : $8\zeta(3)/3$
9999 9428 tanh : $1/\ln(\sqrt[3]{5}/2)$
9999 9453 tanh : $3e\pi/4$
9999 9466 tanh : $\sqrt{3}+\sqrt{5}+\sqrt{6}$
9999 9471 λ : $\exp(-4\pi/19)$
9999 9525 J_0 : $\exp(-2e)$
9999 9538 tanh : $9\sqrt[3]{3}/2$
9999 9540 erf : $\sqrt{21/2}$
9999 9547 tanh : $13/2$
9999 9555 erf : $9\sqrt[3]{3}/4$
9999 9556 tanh : $\sqrt{2}+\sqrt{6}+\sqrt{7}$
9999 9557 tanh : $10(e+\pi)/9$
9999 9564 erf : $(\sqrt{2}\pi/3)^3$
9999 9569 erf : $13/4$
9999 9585 erf : $5(e+\pi)/9$
9999 9593 J_0 : $(2\pi)^{-3}$
9999 9594 erf : $1/\ln(e/2)$
9999 9603 erf : $6e/5$
9999 9614 tanh : $2\pi^2/3$
9999 9624 tanh : $9(e+\pi)/8$
9999 9627 erf : $\sqrt{2}-\sqrt{5}-\sqrt{6}$
9999 9640 tanh : $\sqrt{3}+\sqrt{5}+\sqrt{7}$
9999 9658 erf : $(2e/3)^2$
9999 9659 tanh : $7e\pi/9$
9999 9662 erf : $23/7$
9999 9672 erf : $\pi^2/3$
9999 9676 tanh : $20/3$
9999 9693 erf : $7\sqrt{2}/3$
9999 9695 tanh : $8(e+\pi)/7$
9999 9701 tanh : $3\sqrt{5}$
9999 9722 erf : $(\ln 3/2)^{-2}$
9999 9725 tanh : $(\sqrt[3]{4}/3)^{-3}$
9999 9727 erf : $\sqrt{11}$
9999 9734 sπ : $e/2+\pi$
9999 9749 tanh : $5e/2$
9999 9757 erf : $10/3$
9999 9758 λ : $\exp(-\pi/5)$
9999 9765 tanh : $\sqrt{3}+\sqrt{6}+\sqrt{7}$
9999 9767 tanh : $4e\pi/5$
9999 9769 tanh : $7(e+\pi)/6$
9999 9770 tanh : $4\sqrt[3]{5}$
9999 9781 erf : $4(e+\pi)/7$
9999 9789 erf : $3\sqrt{5}/2$
9999 9796 J_0 : $\exp(-(e+\pi))$
9999 9798 erf : $8\sqrt[3]{2}/3$
9999 9800 tanh : $7\pi^2/10$
9999 9802 erf : $\sqrt{3}-\sqrt{6}-\sqrt{7}$
9999 9805 erf : $7\sqrt[3]{3}/3$
9999 9808 tanh : $4\sqrt{3}$
9999 9809 tanh : $10\ln 2$
9999 9816 tanh : $1/\ln(\sqrt{3}/2)$
9999 9818 erf : $(3/2)^3$
9999 9833 tanh : 7
9999 9837 erf : $\sqrt{23/2}$
9999 9841 tanh : $\sqrt{5}\pi$
9999 9843 tanh : $6(e+\pi)/5$
9999 9845 erf : $5e/4$
9999 9847 erf : $17/5$
9999 9849 tanh : $5\pi^2/7$
9999 9850 tanh : $\exp((e+\pi)/3)$
9999 9855 tanh : $9\pi/4$
9999 9864 erf : $2e\pi/5$
9999 9867 erf : $2\sqrt[3]{5}$
9999 9868 tanh : $5e\pi/6$
9999 9870 sπ : $\sqrt{2}+\sqrt{3}-\sqrt{7}$
9999 9875 erf : $24/7$
9999 9891 tanh : $6\zeta(3)$
9999 9898 tanh : $8e/3$
9999 9903 erf : $2\sqrt{3}$
9999 9904 erf : $5\ln 2$
9999 9906 erf : $\sqrt{2}-\sqrt{5}-\sqrt{7}$
9999 9911 erf : $1/\ln(4/3)$
9999 9912 J_0 : $\exp(-2\pi)$
9999 9913 tanh : $5(e+\pi)/4$
9999 9914 J_0 : $(1/(3e))^3$
9999 9915 λ : $\exp(-3\pi/16)$
9999 9920 erf : $10\pi/9$
9999 9922 erf : $9e/7$
9999 9923 tanh : $(e)^2$
9999 9925 tanh : $3\pi^2/4$
9999 9932 tanh : $10\sqrt{5}/3$
9999 9933 tanh : $(\ln 3/3)^{-2}$
9999 9935 tanh : $7e\pi/8$
9999 9938 tanh : $15/2$
9999 9942 tanh : $9(e+\pi)/7$
9999 9945 tanh : $6\sqrt[3]{2}$
9999 9948 tanh : $8e\pi/9$
9999 9949 λ : $\exp(-2\pi/11)$
9999 9955 erf : $25/7$
9999 9956 tanh : $23/3$
9999 9957 tanh : $7\pi^2/9$
9999 9958 tanh : $\sqrt{6}\pi$
9999 9961 erf : $8\pi/7$
9999 9964 J_0 : $(3\pi)^{-3}$
9999 9965 erf : $\sqrt{13}$
9999 9966 tanh : $9\sqrt{3}/2$
9999 9967 tanh : $4(e+\pi)/3$
9999 9968 tanh : $7\sqrt{5}/2$
9999 9969 tanh : $5\pi/2$
9999 9970 λ : $\exp(-3\pi/17)$
9999 9972 tanh : $4\pi^2/5$
9999 9977 tanh : 8
9999 9978 erf : $11/3$
9999 9980 erf : $\sqrt{2}-\sqrt{6}-\surd$
9999 9982 tanh : $\exp(2\pi/3)$
9999 9983 tanh : $3e$
9999 9985 tanh : $5\pi^2/6$
9999 9986 erf : $5\sqrt{5}/3$
9999 9987 tanh : $\sqrt{7}\pi$
9999 9988 tanh : $25/3$
9999 9989 λ : $\exp(-\pi/6)$
9999 9989 tanh : $8\pi/3$
9999 9990 tanh : $7\zeta(3)$
9999 9991 tanh : $17/2$
9999 9992 tanh : $e\pi$
9999 9993 tanh : $7\pi^2/8$
9999 9994 erf : $23/6$
9999 9995 tanh : $8\pi^2/9$
9999 9996 tanh : 9
9999 9997 erf : $5\pi/4$
9999 9998 erf : $7\sqrt[3]{5}/3$
9999 9999 J_0 : $\exp(-3\pi)$